Atlas of Musculoskeletal and Small Parts Ultrasound with Color Flow Imaging

Atlas of Musculoskeletal and Small Parts Ultrasound with Color Flow Imaging

Third Edition

PK Srivastava
MD FICRI FICMU
Professor
Whole Body CT Scan and Ultrasound Unit
Department of Radiotherapy
King George's Medical University
Lucknow (UP)
India

© 2007 PK Srivastava

First published in India by

Jaypee Brothers Medical Publishers (P) Ltd
EMCA House, 23/23B Ansari Road, Daryaganj, New Delhi 110 002, India
Phones: +91-11-23272143, +91-11-23272703, +91-11-23282021, +91-11-23245672
Fax: +91-11-23276490, +91-11-23245683 e-mail: jaypee@jaypeebrothers.com
Visit our website: www.jaypeebrothers.com

First published in USA by The McGraw-Hill Companies, 2 Penn Plaza, New York, NY 10121-2298. Exclusively worldwide distributor except South Asia (India, Nepal, Sri Lanka, Bhutan, Pakistan, Bangladesh).

All rights reserved. No part of this publication may be reproduced, stored in a retrieval system, or transmitted, in any form or by any means, electronic, mechanical, photocopying, recording, or otherwise, without the prior permission of the publisher or in accordance with the provisions of the Copyright, Designs and Patents Act 1988 or under the terms of any licence permitting limited copying issued by the Copyright Licensing Agency, 90 Tottenham Court Road, London W1P 0LP.

NOTICE

Medicine is an ever-changing science. As new research and clinical experience broaden our knowledge, changes in treatment and drug therapy are required. The authors and the publisher of this work have checked with sources believed to be reliable in their efforts to provide information that is complete and generally in accord with the standards accepted at the time of publication. However, in view of the possibility of human error changes in medical science, neither the editors nor the publisher nor any other party who has been involved in the preparation or publication of this work warrants that the information contained herein is in every respect accurate or complete, and they disclaim all responsibility for any errors or omissions or for the results obtained from use of the information contained in this work. Readers are encouraged to confirm the information contained herein with other sources. For example and in particular, readers are advised to check the product information sheet included in the package of each drug they plan to administer to be certain that the information contained in this work is accurate and that changes have not been made in the recommended dose or in the contraindications for administration. This recommendation is of particular importance in connection with new or infrequently used drugs.

ISBN 0-07-148583-X
ISBN 13 9780071485838

This Atlas is dedicated
to
My wife Dr Yashodhara Pradeep and
to
My Children, Divya and Ayush

Preface to the Third Edition

It is a movement of pleasure and pride for me to bring out the Third Edition of the book titled **"An Atlas of Small Parts and Musculoskeletal Ultrasound with Color Flow Imaging".** The second edition of the book was a huge success. The tremendous response received by the book, was the main driving force to bring out the third edition in less than four years time. Rapid growth on technology front with evolvement of 3D imaging has given a new vision to look into the pathological conditions in much better way. Introduction of 3D imaging has definitely given a new and exciting way to peep deeper into many pathological conditions from different angles. It has definitely enhanced the confidence level to diagnose many tricky situations.

This is the first Atlas where **3D ultrasound imaging** has been used widely for its non-obstetrical applications for better subject understanding. The size and format of the Atlas have been totally changed for better image quality. The Atlas has become more comprehensive with **more than 3600 images**. Like the previous edition, the focus of the atlas is on images and not on text. The images are self-explanatory. Incorporation of clinical images makes it easier to understand. The aim of the book is to provide most illustrative and comprehensive Atlas on the subject. I wish this Atlas would serve its purpose and work as a reference book for all the practicing sonologists having interest in small parts and musculoskeletal ultrasound.

PK Srivastava
Yashdeep
C-212 Nirala Nagar
Lucknow-226020
India
yashdeep@sancharnet.in
pks_yash@yahoo.co.in

Preface to the First Edition

The development of ultrasonographic imaging has been one of the most important breakthrough in the field of diagnostic imaging in last two decades. The ultrasonographic imaging has crossed its infancy and attained adulthood after improvement in technology and invention of high resolution transducers. The application of high resolution sonography in small parts imaging has opened a new horizon in the diagnosis, treatment and management of small parts parthology. It was a long-felt need to have a book on small parts imaging. This book has been written to give a detailed account of ultrasonographic applications in small parts, and the main focus of the book is on the commonly encountered problems involving small parts. This book is intended for practising sonologists, physicians and medical students. The book contains ten chapters covering almost all the small parts of body which are superficially placed and need high resolution sonography for their imaging.

The authors who have contributed chapters in the book are recognised experts in their fields. They have provided most up-to-date information available in the literature to make this book comprehensive.

I hope this book will become a reference work for all who use ultrasonographic imaging in their clinical practice and further gain expertise in small parts imaging.

PK Srivastava

Acknowledgements

This would have not been possible without the help of many individuals.

I am extremely thankful to Professor Mukund Joshi MD, FICRI, FAMS for his encouragement and valuable guidance in completing the book. He is generous enough to contribute a chapter in the book.

I extend my sincere thanks to Mr. Sunil Kumar Gupta, Mr. Vivek Tripathi and Mr. Prem Chand Joshi who have done an excellent job in typing manuscript and formatting the Atlas and made the work simple and easy. I am also thankful to Mr. OP Verma, Rizwan Ahamad and Mr. Ashutosh Mishra, the clinic Staff members of Yashdeep Ultrasound Centers for their contribution and support in writing the book.

Lastly, I am extremely thankful to Shri JP Vij, Chairman and Managing Director, Jaypee Brothers Medical Publishers (P) Ltd and his staff for his constant support and encouragement in giving the final shape to the Atlas and publishing it.

Contents

Eye and Orbit

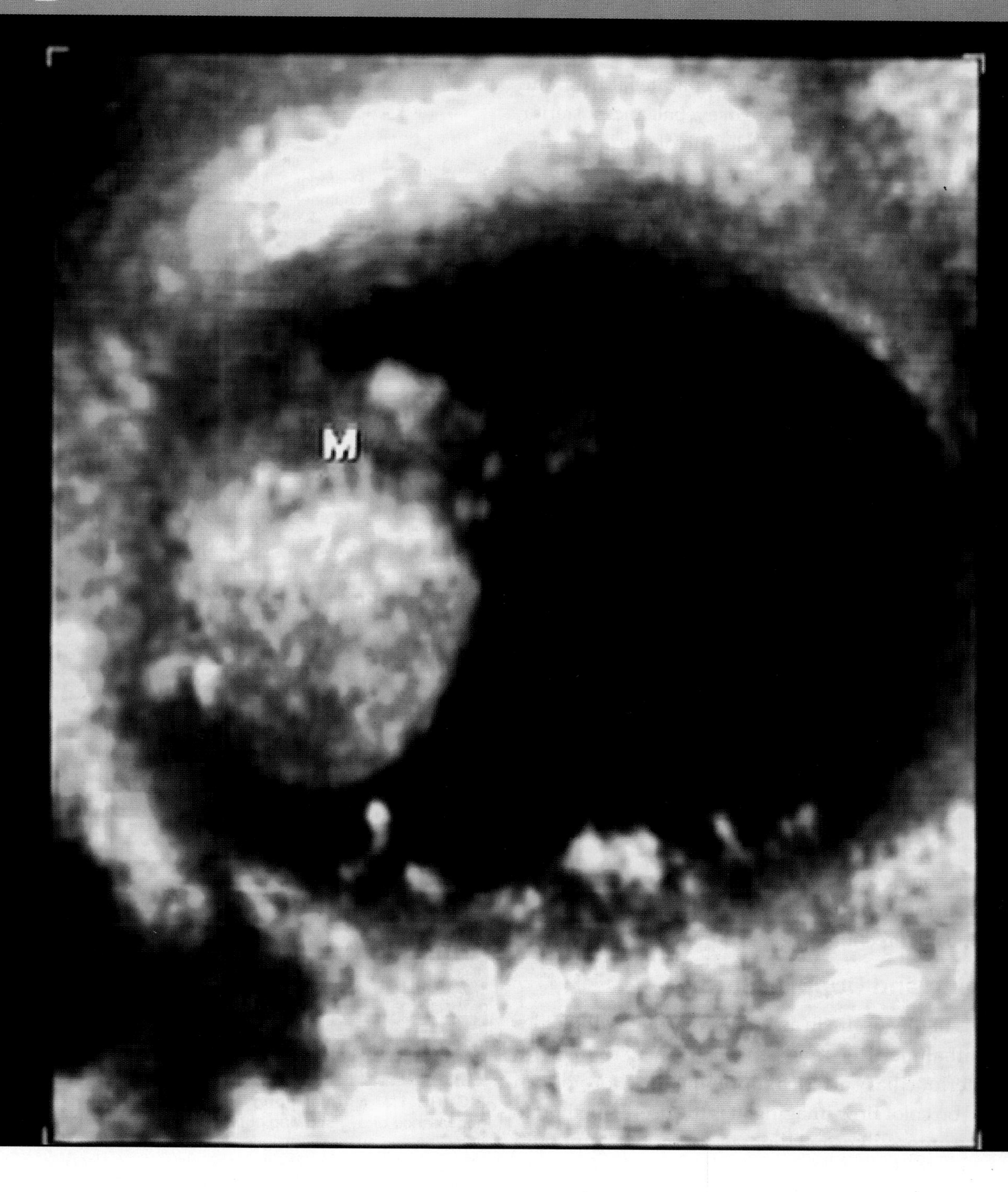

Introduction

Ultrasonic appearance of eye and orbit closely resemble the normal anatomical structures. Eye is the easiest structure to visualize within the orbit as its fluid content and superficial position makes it ideal for USG examination. Ultrasound is only practical method of obtaining images of the posterior segment of eye when the light conducting media are opaque and is the most useful investigation prior to vitrectomy. Ultrasound contributes more to tissue diagnosis than Computed tomography. The characteristics of motion and topography of pathological intraocular structures enables differentiation between retinal detachment and vitreous membrane or tumor and hemorrhage. The 2D ultrasound image displays ocular and orbital structures much more than a CT scan.

Normal Anatomy

Eye is fluid filled structure. It is situated in the anterior part of the orbit and embedded in the fat. The Tenon's capsule separates it from the orbital wall. The anterior segment forms 1/6th of eyeball and posterior segment forms 5/6th of eyeball. Normal axial length of eye is 22 mm.

Refractive Media

Cornea, lens, aqueous humor and vitreous gel form the refractive media. Aqueous humor is a saline solution, which fills the anterior segment. Lens is a transparent biconvex body, which rests over the vitreous gel. It is 3 to 4 mm in thickness. The vitreous is transparent gel, which fills the posterior segment. All of them are very good sound conducting media and stand out clearly on HRSG. The retrobulbar area is predominantly filled with retrobulbar fat and optic nerve is loosely embedded in it. It comes out as an echo poor strap in longitudinal axis. The retina is closely applied to the ocular coat and cannot be seen separately in normal eye. It is firmly adherent at ora serrata and optic nerve head.

Examination Technique

2-D Imaging

Short focus 7.5 mhz and 12 mhz real time small part sector transducers are pre-requisite for eye and orbital imaging. The quality of images produced by general purpose Ultrasound with these probes are good and made eye scanning a practical proposition. Sector imaging is done with patient lying supine. The main method of examination is contact method. The probe is directly placed on closed eyelid with intervening coupling gel.

Complete visualization of the ocular structures is achieved with careful movement and orientation of transducers with eye fixed in primary position and in all directions of gaze (Straight head position).

Color Flow Imaging and Doppler Study

Color Doppler study is highly useful in assessing vascular pathologies of eye and orbit. Ophthalmic artery and central artery of retina with short ciliary and posterior ciliary arteries supplying the choroids can be easily documented on color flow imaging. It is the useful method to examine intraocular tumors like melanomas, retinoblastoma and

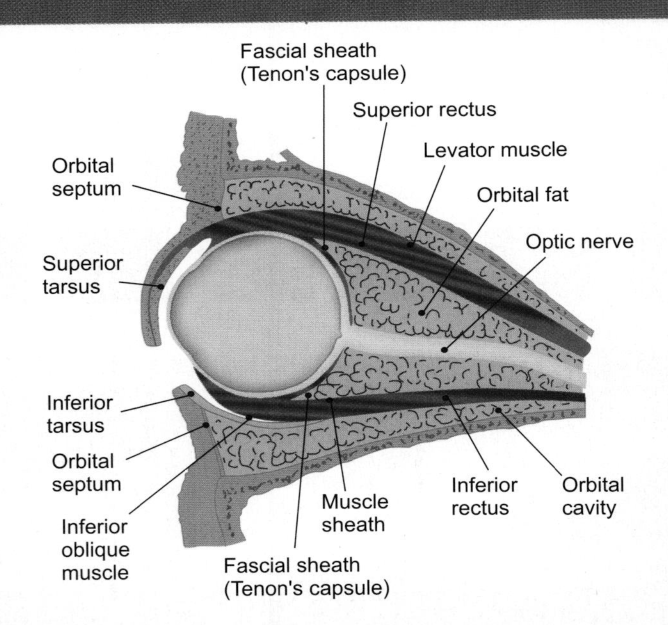

FIGURE 1.1: Sagittal section of orbit. Sagittal section of the orbit shows eyeball, retrobulbar fat, optic nerve and ocular muscles.

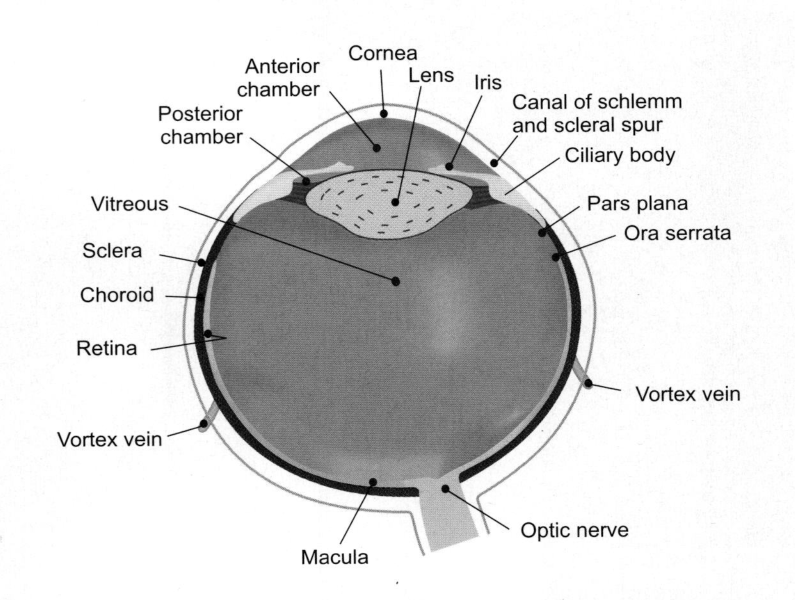

FIGURE 1.2: Cross section of eye. Cross section of the eye shows lens, A.C. chamber, refractive media and posterior segment.

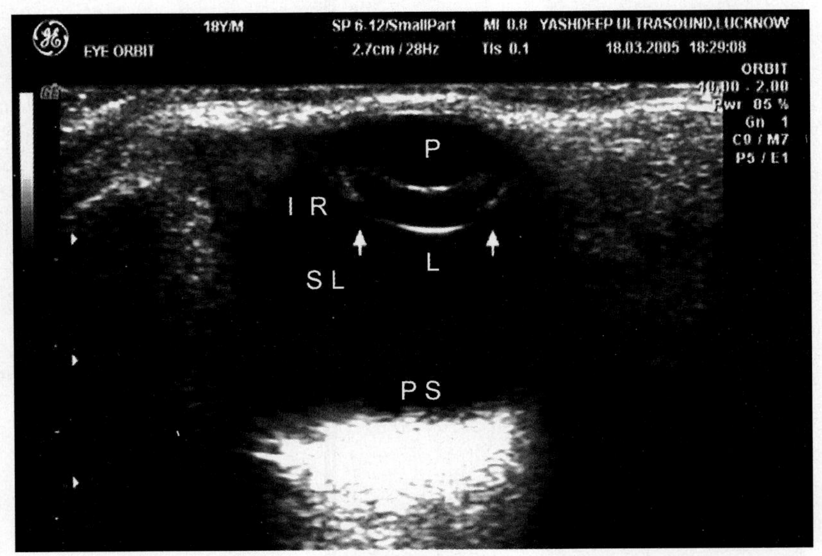

FIGURE 1.3: Normal sono appearance of eye. HRSG shows normal sono appearance of eye showing pupil, lens- echogenic convex body sitting over vitreous body, iris and suspensory ligaments holding the lens.

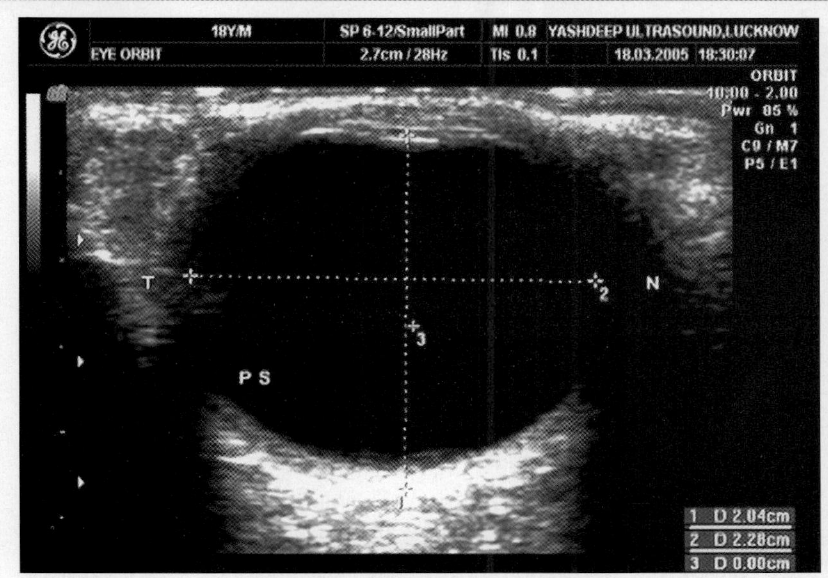

FIGURE 1.4: Normal image of eye HRSG. shows normal image of eye showing lens, posterior chamber filled with anechoic vitreous gel.

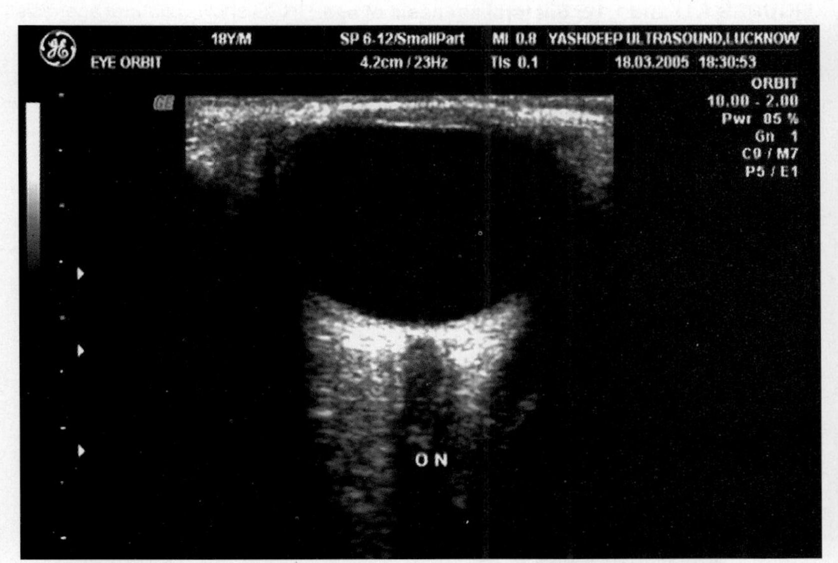

FIGURE 1.5: Normal eye with retrobulbar space. HRSG shows normal eyeball with optic nerve in long axis as a hypoechoic strap traversing through echogenic retrobulbar fat.

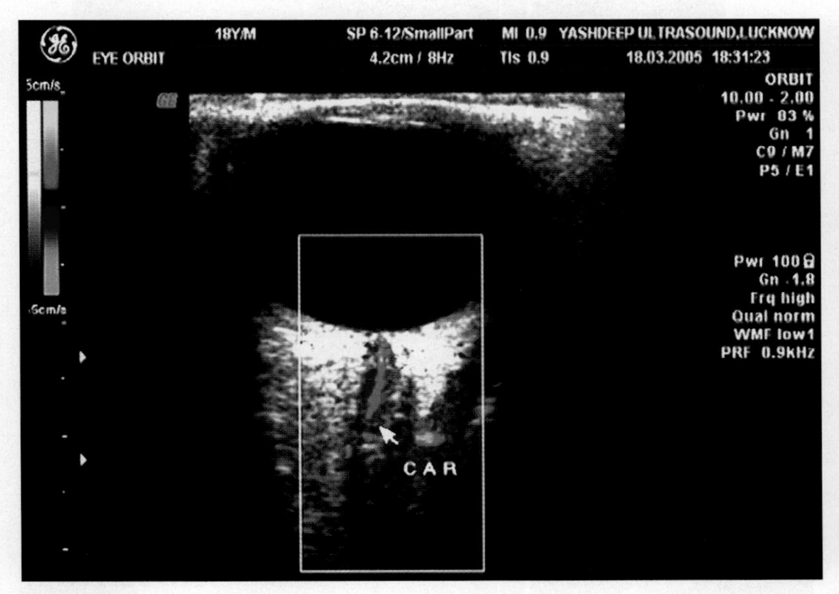

FIGURE 1.6: Color Doppler study shows central artery of retina and vein traversing through optic nerve.

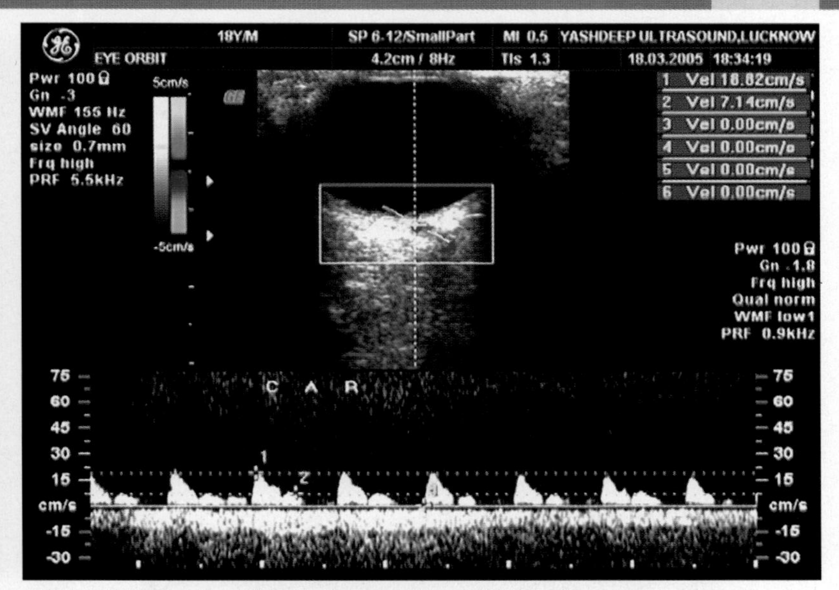

FIGURE 1.7: Spectral Doppler tracing shows flow in the central artery of retina. Short systolic and prolonged diastolic flow is seen in the central artery of retina typical of small arterial flow pattern.

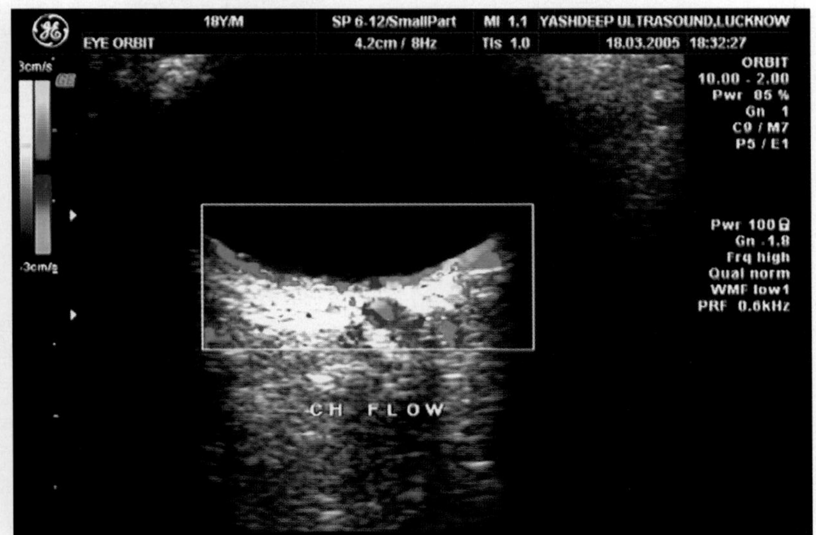

FIGURE 1.8: Normal choroidal flow. Color Doppler flow study shows central artery of retina and its branches short and long posterior ciliary arteries supplying the choroid.

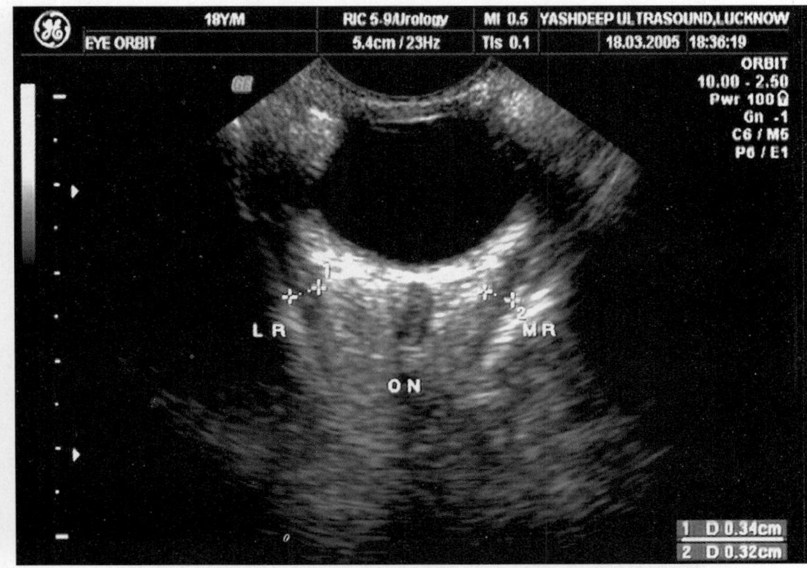

FIGURE 1.9: Extra ocular muscle. HRSG shows normal extra ocular muscles as a hypoechoic bands. The medial and the lateral rectus muscles are seen on nasal and temporal side. Maximum normal thickness of the muscles is 3 mm.

metastasis. It is an important investigation in retinopathies. Neovascularisation of tumor and infiltration in adjoining tissues can be better evaluated on color flow imaging.

Indications for Ocular Ultrasound

All the pathological conditions where direct ophthalmoscopy is not possible are the indication for high-resolution sonography (HRSG) of the globe. They are grouped as under:

Congenital Anomalies

High-resolution sonography is an excellent tool for evaluation of congenital developmental anomalies of eye and orbit. The following conditions can be very well evaluated on HRSG:

1. Anophthalmia.
2. Agenesis of eye.

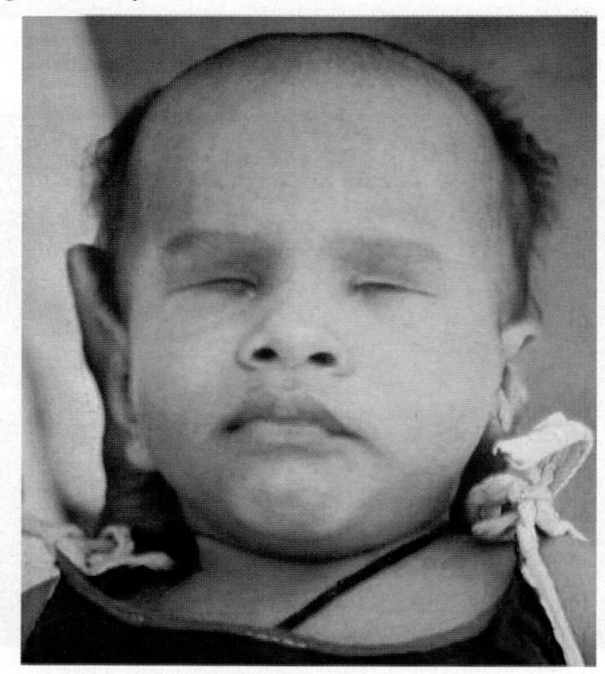

FIGURE 1.10: A young child was born with small orbital sockets and thin palpebral fissure.

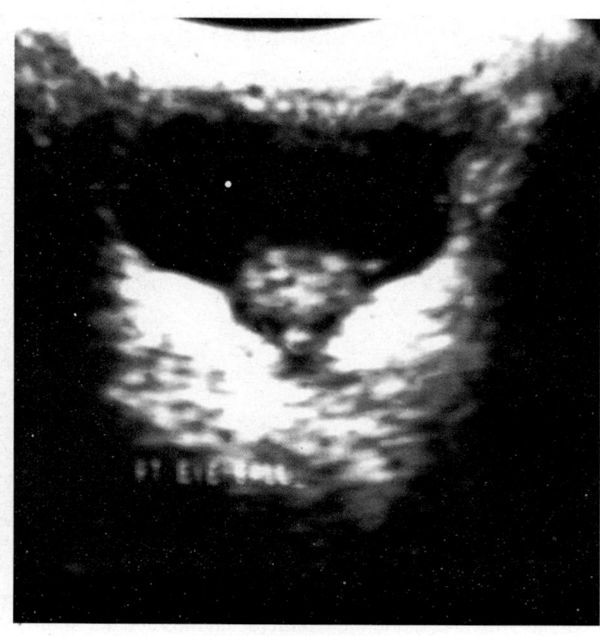

FIGURE 1.11

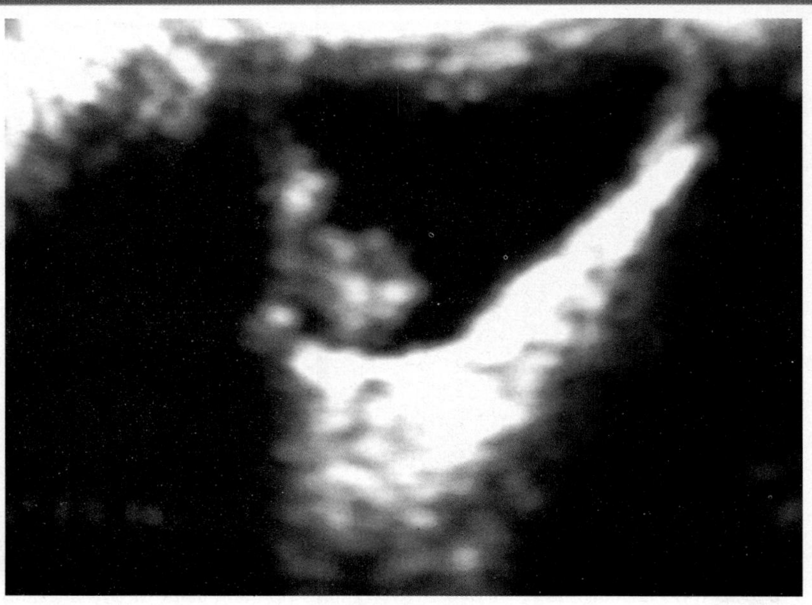

FIGURE 1.12

FIGURES 1.11 and 1.12: Bilateral agenesis of eye. HRSG shows bilateral agenesis of eyeballs with small palpebral fissure in the orbital socket. Irregular cystic spaces are seen in both the orbits with small echogenic retrobulbar fat.

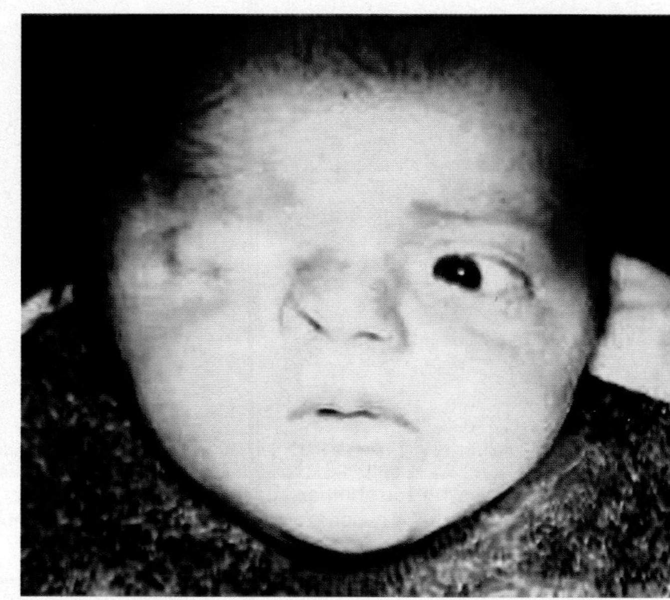

FIGURE 1.13: Cryptophthalmus. Rt eye covered with a skin fold in an infant.

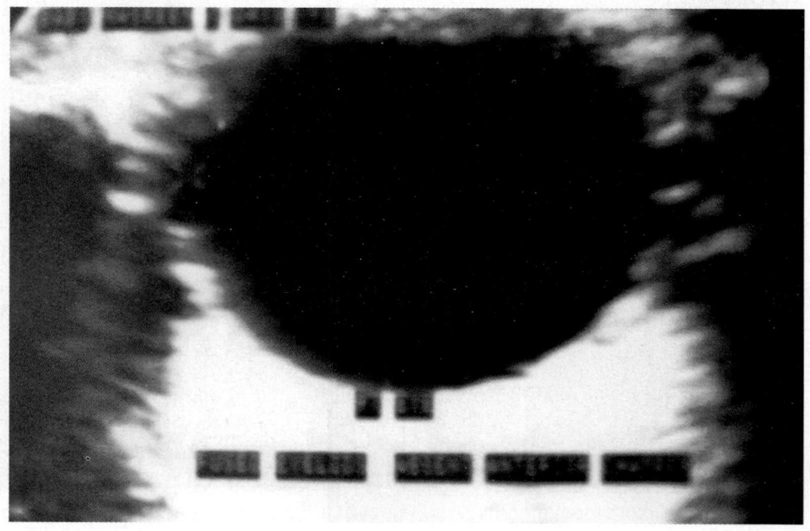

FIGURE 1.14: Cryptophthalmus. HRSG reveals a small cystic mass with absent lens and iris in Rt orbit.

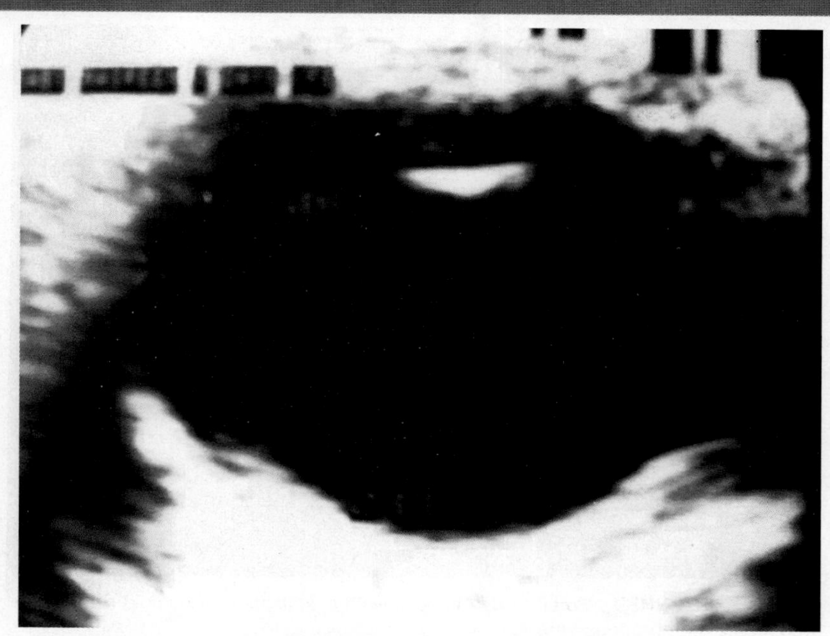

FIGURE 1.15: Normal Lt eye of the same patient.

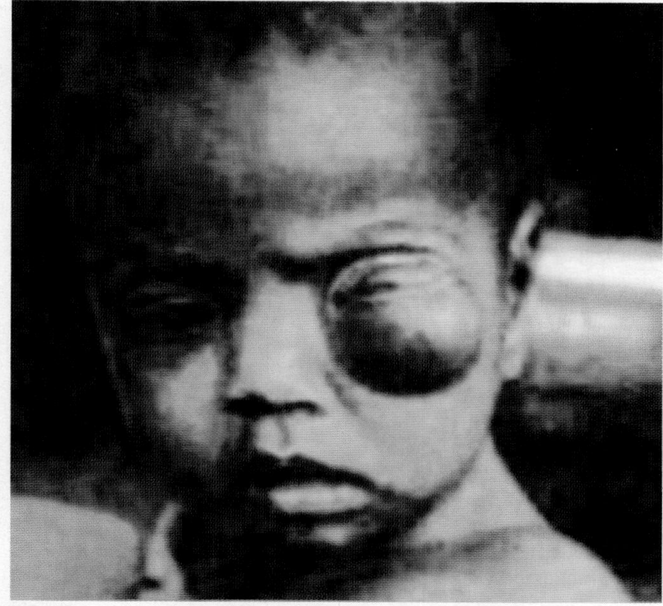

FIGURE 1.16: Congenital cystic eye. Cystic mass is seen over the Lt orbit.

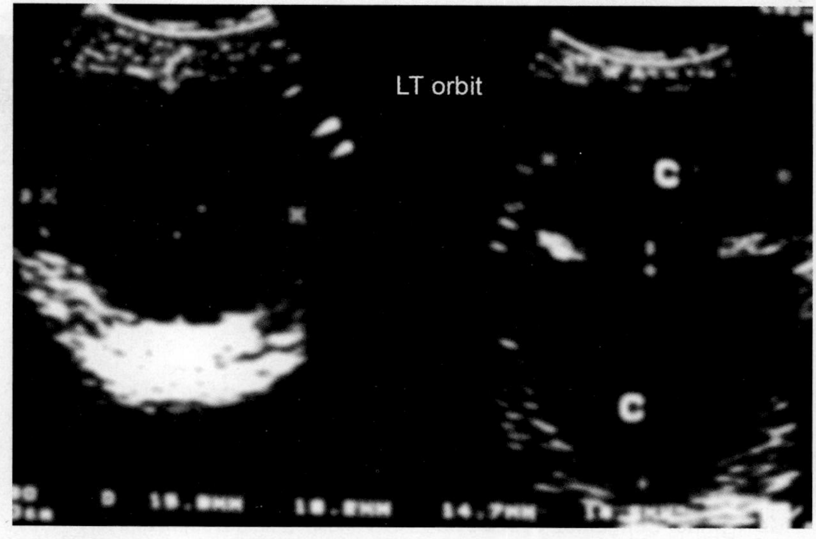

FIGURE 1.17: Lt orbit shows two big cysts one in anterior part another in retrobulbar part communicating with each other. Normal eye is absent.

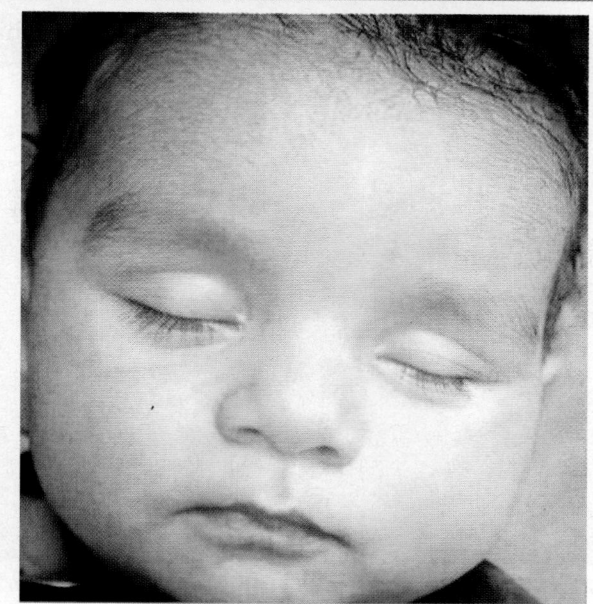

FIGURE 1.18: Persistent hyperplasia primary vitreous (PHPV). A young child was born with white eye reflex on ophthalmic examination in Lt eye.

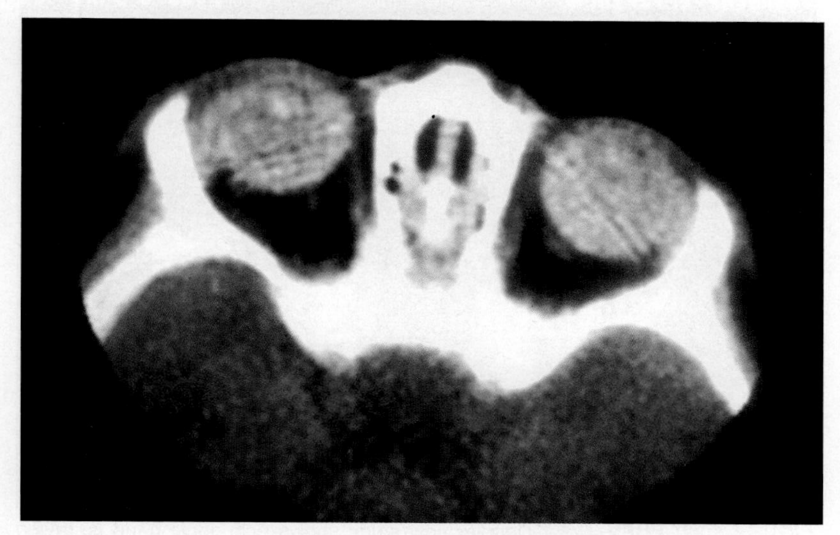

FIGURE 1.19: PHPV. CT scan of the same patient shows thick lens in Lt eye with small eye.

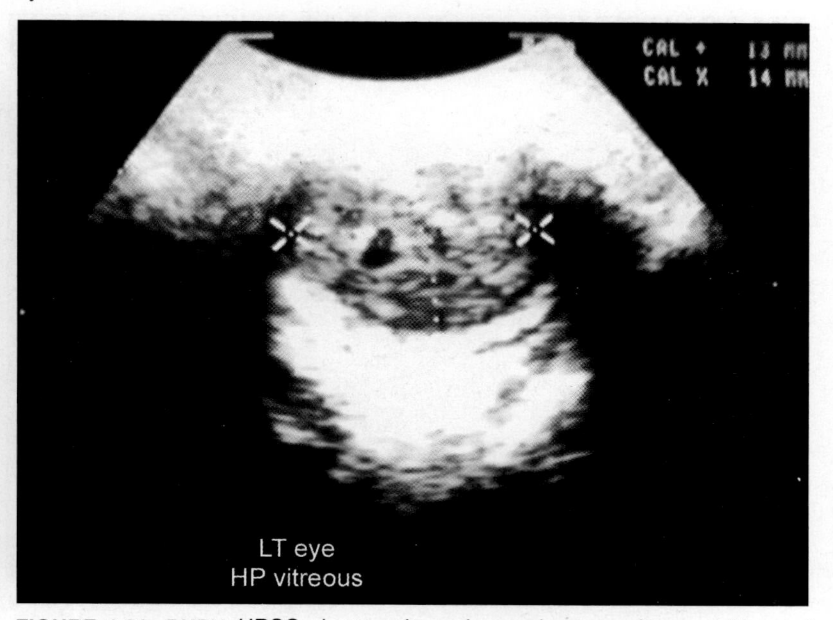

FIGURE 1.20: PHPV. HRSG shows echogenic membranous vitreous with small Lt eye. The embryonic vitreous did not develop into normal transparent vitreous.

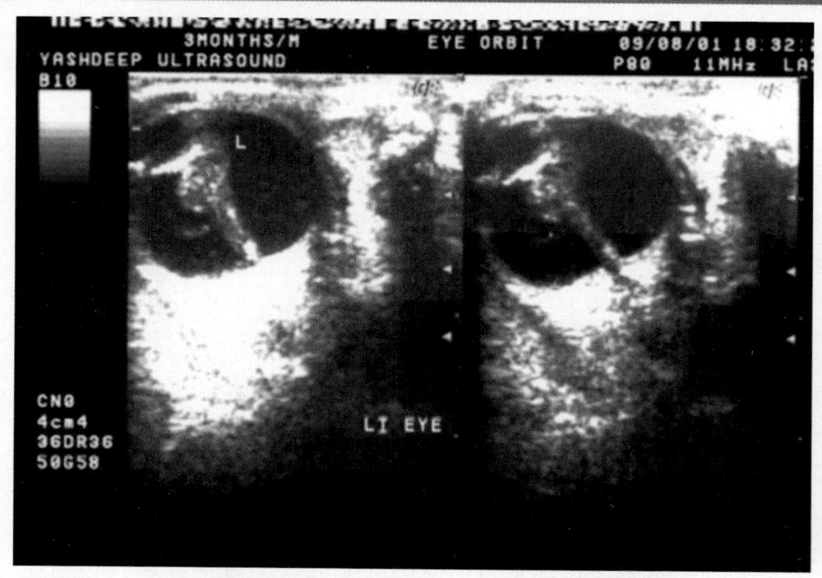

FIGURE 1.21: PHPV. HRSG shows small Lt eye with thick lens and an echogenic band going from the lens to the posterior pole.

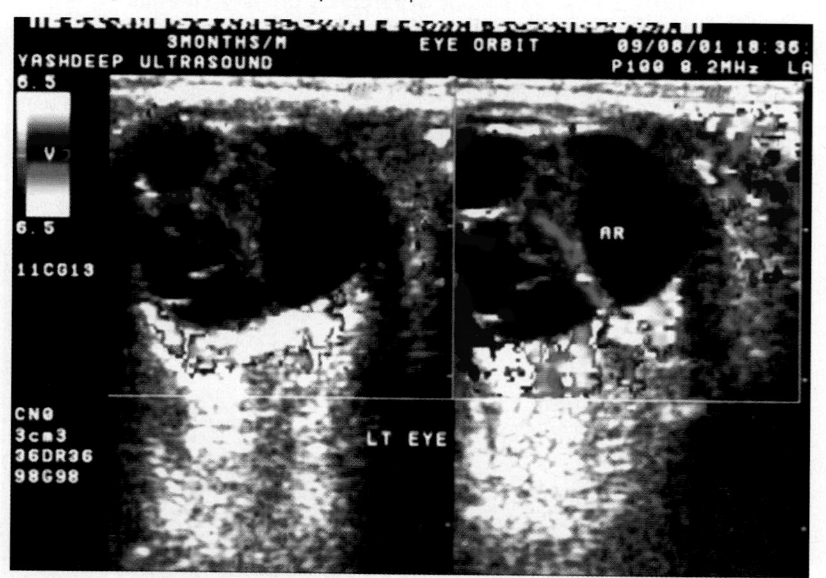

FIGURE 1.22: PHPV. Color Doppler flow study shows echogenic band is having hyaloid artery.

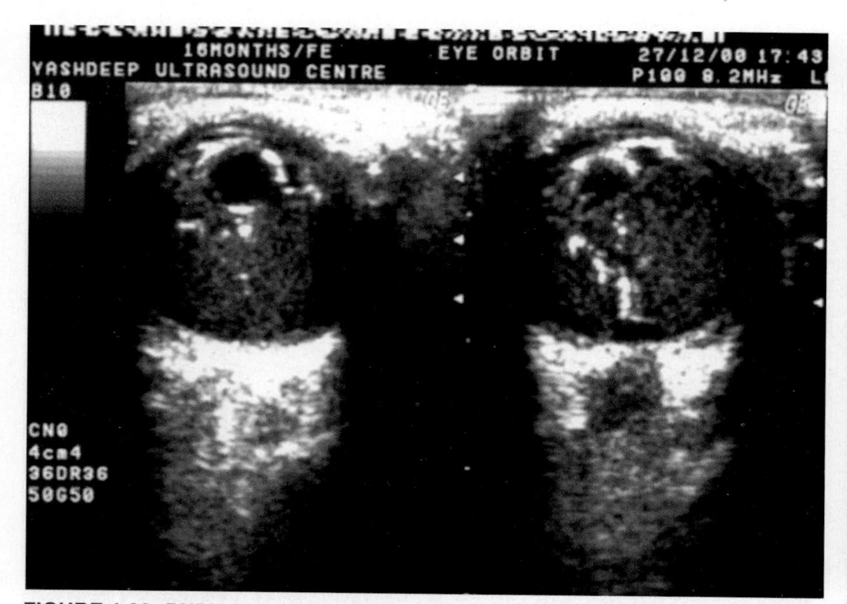

FIGURE 1.23: PHPV. Small eye is seen with echogenic membranous vitreous filling the posterior segment.

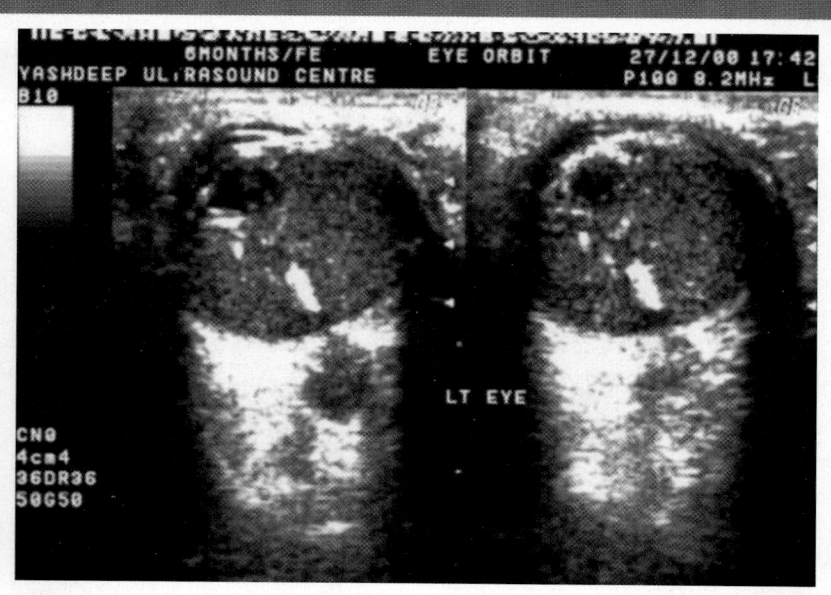

FIGURE 1.24: PHPV. Small Lt eye is seen with calcification of hyaloid artery and echogenic membranous vitreous.

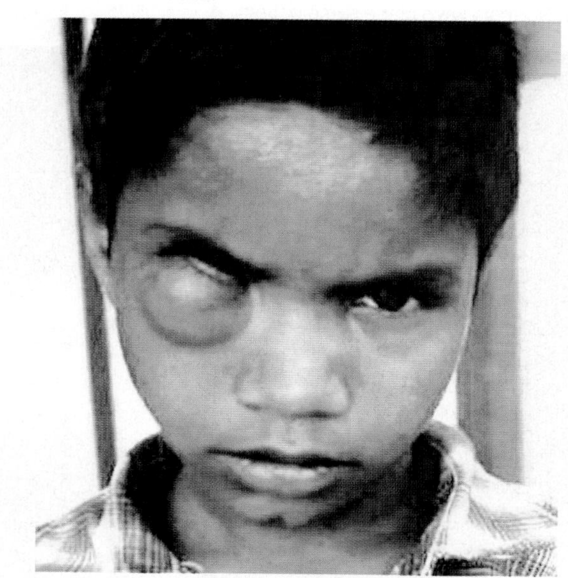

FIGURE 1.25: Congenital cystic eye. A young boy was presented with a cystic mass over the Rt orbit.

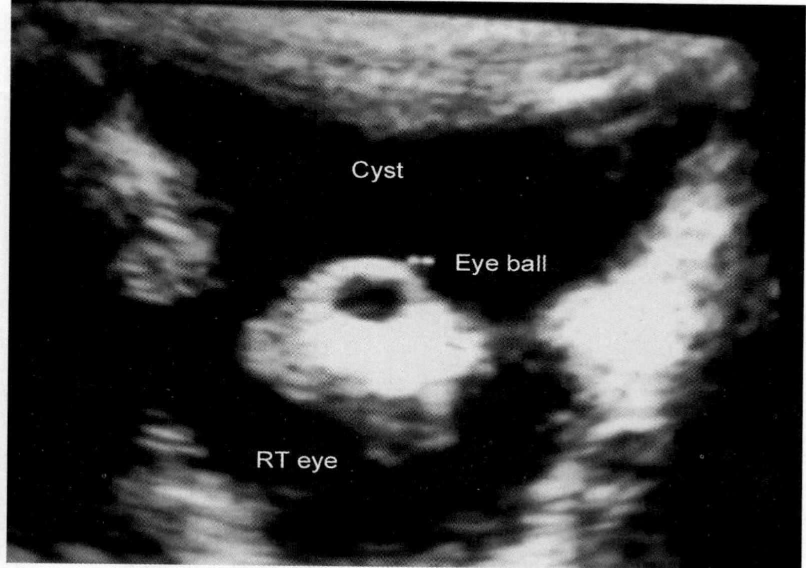

FIGURE 1.26: Congenital cystic eye. HRSG shows a small rudimentary cystic eye which is engulfed by a big orbital cyst. Small amount of retrobulbar fat is seen posterior to the rudimentary eye.

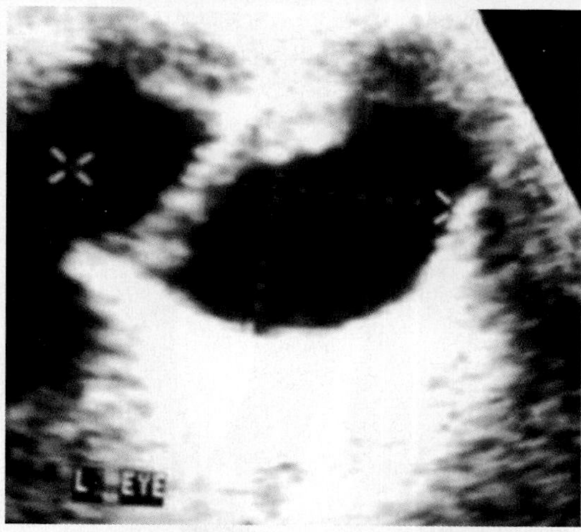

FIGURE 1.27: Retrolental fibroplasia. HRSG shows a thick opaque lens with a posterior membrane and an echogenic band going from the membrane to the posterior pole in a newborn child.

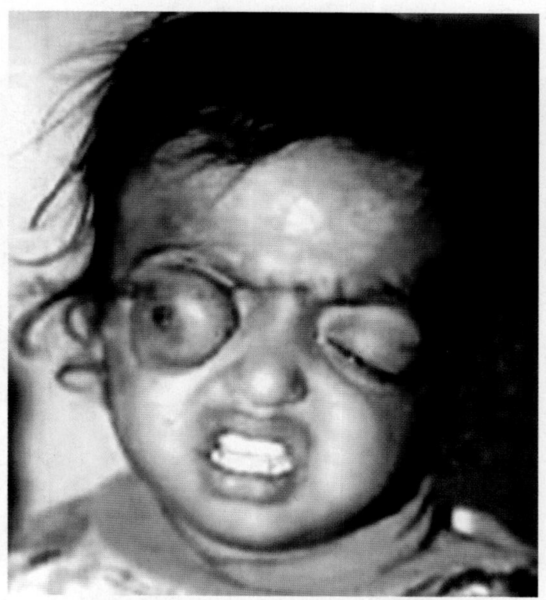

FIGURE 1.28: Congenital exophthalmoses. Patient shows bilateral exophthalmos with exposure keratitis of Rt eye in a case of craniostenosis—Crouzon's disease.

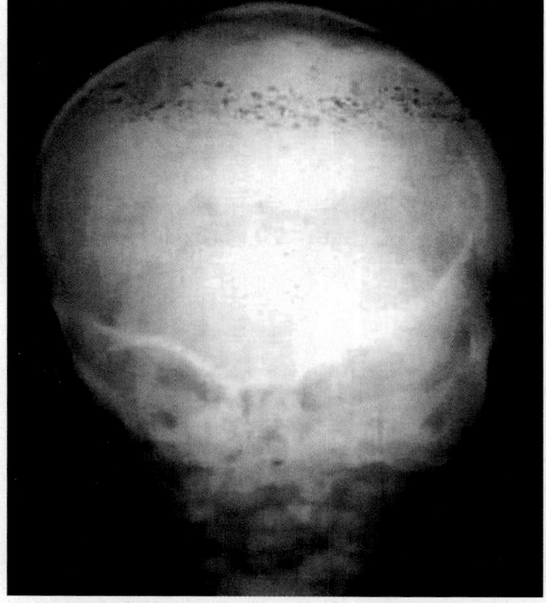

FIGURE 1.29: Plain X-ray skull of same patient (Figure 1.19) shows "Tower Skull" of Crouzon's disease.

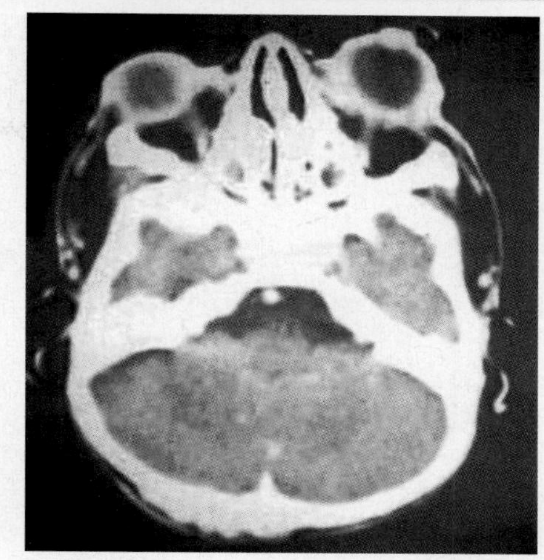

FIGURE 1.30: CT scan of the same patient shows bilateral proptosis of eyes more marked on Rt side with normal retrobulbar fat.

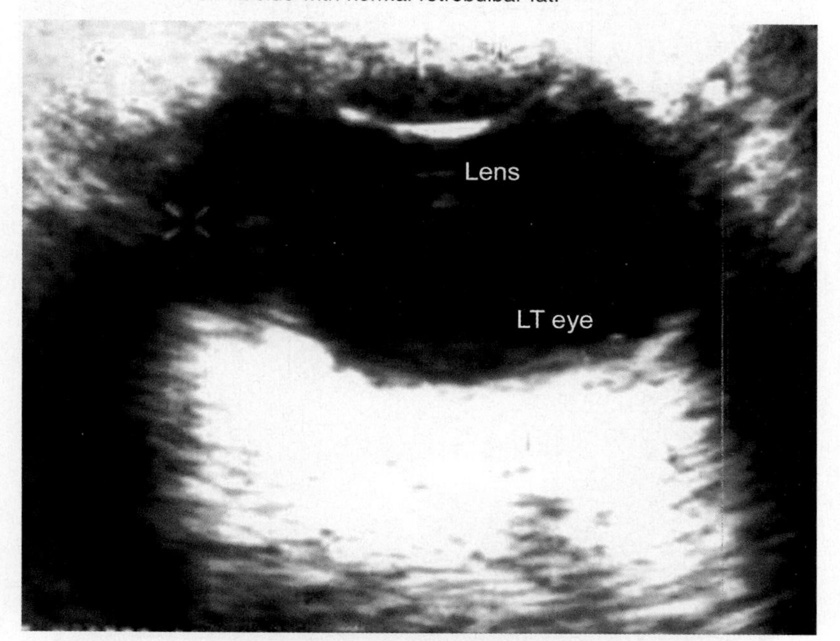

FIGURE 1.31: HRSG shows normal appearance of the eye in the same patient.

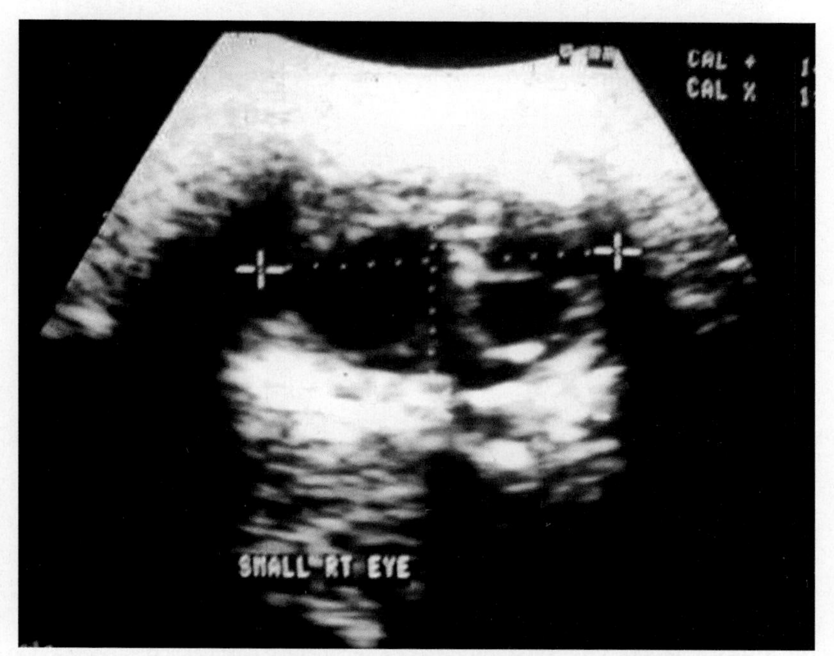

FIGURE 1.32: Congenital microphthalmus. Rt eye is small and deshaped with multiple bands are seen in the vitreous cavity.

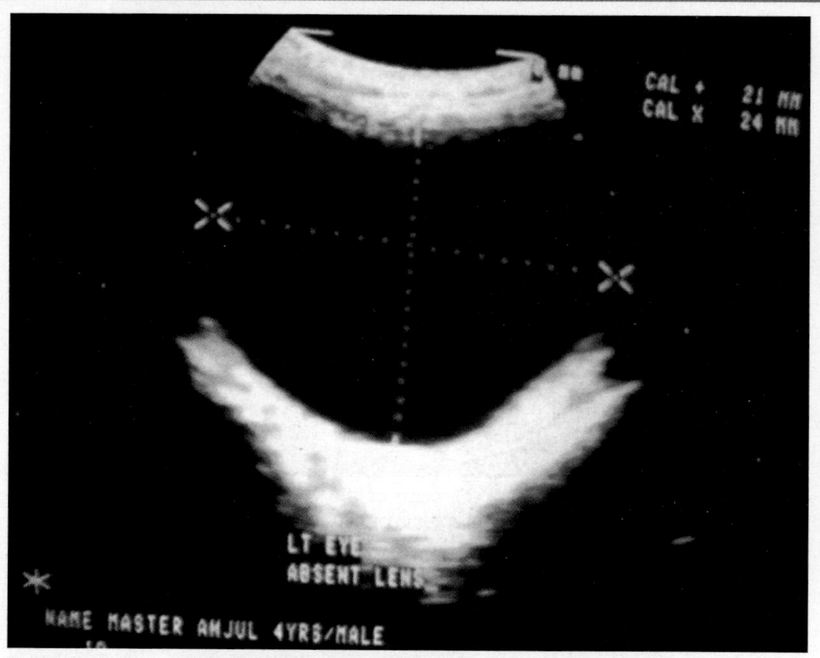

FIGURE 1.33: Congenital macrophthalmus. Lt eye is enlarged and cystic. No lens is seen in the Lt eye in the same patient (Figure 1.23).

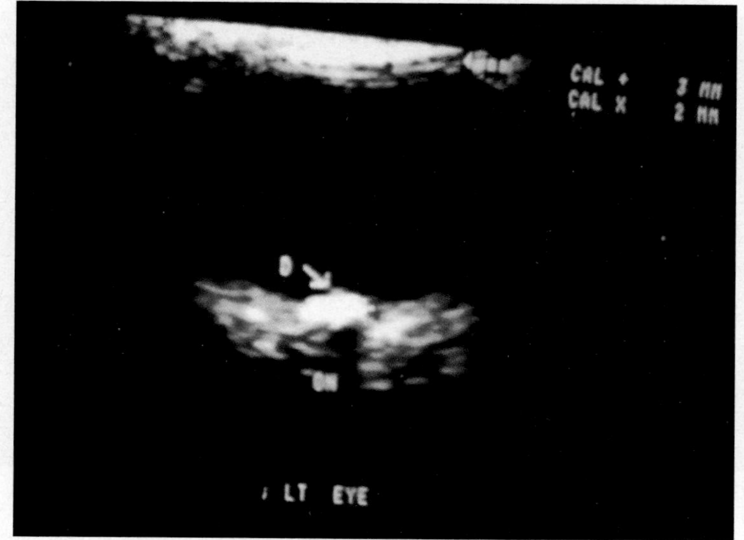

FIGURE 1.34

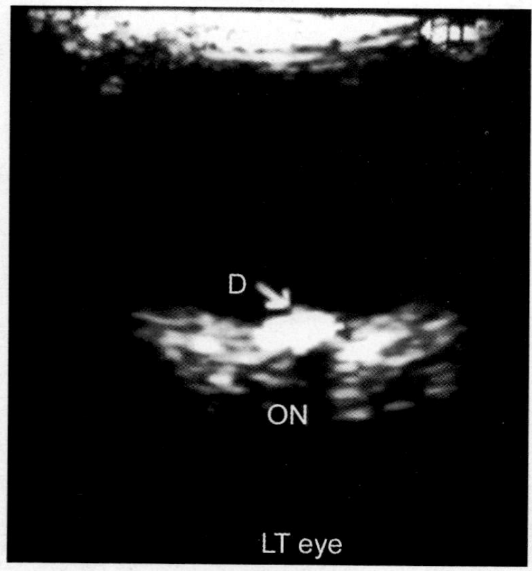

FIGURE 1.35

FIGURES 1.34 and 1.35: Bilateral drusens. Echogenic calcification is seen at the optic disc in both eyes. It is accompanied with acoustic shadowing. Drusens is a calcification of optic disc due to deposition of calcium specks over optic disc.

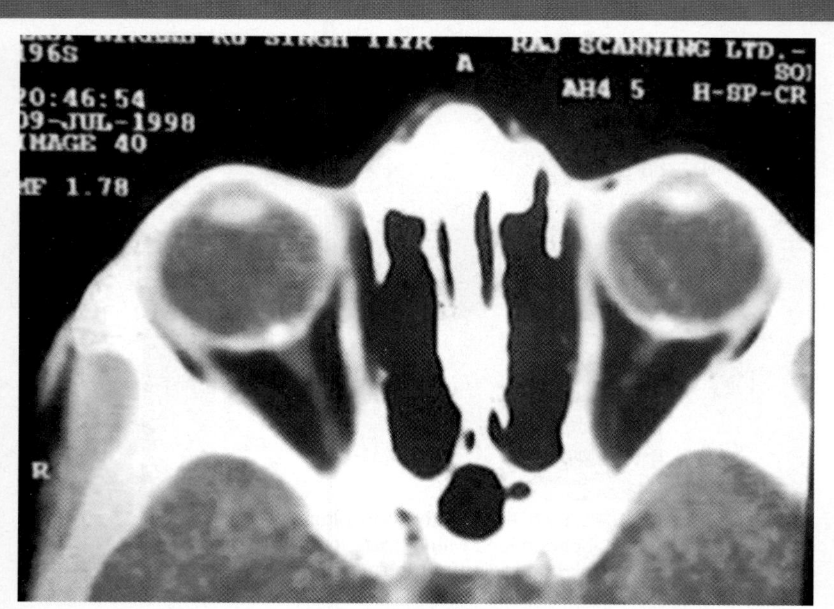

FIGURE 1.36: CT scan of the same patient (Figures 1.34 and 1.35) shows bilateral optic disc calcification.

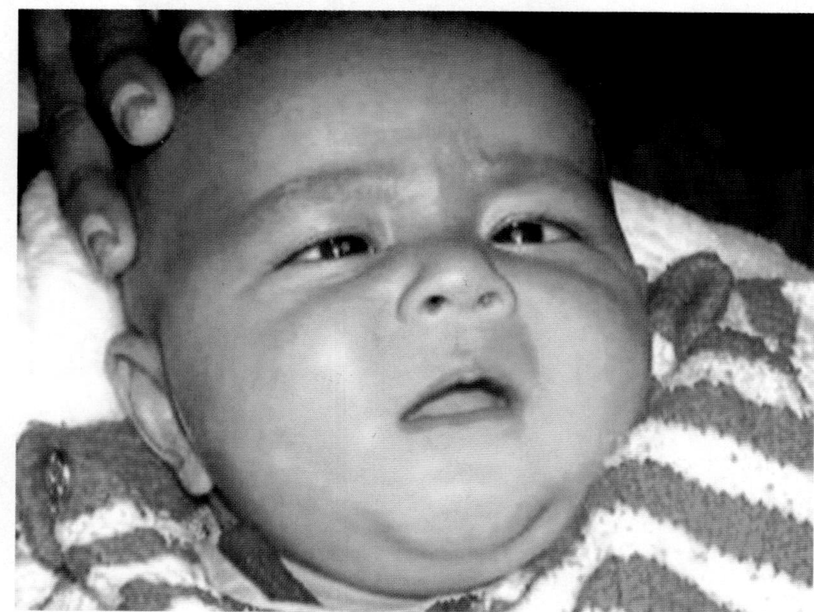

FIGURE 1.37: Bilateral congenital cataract. An infant was born with white eye reflex in both eyes.

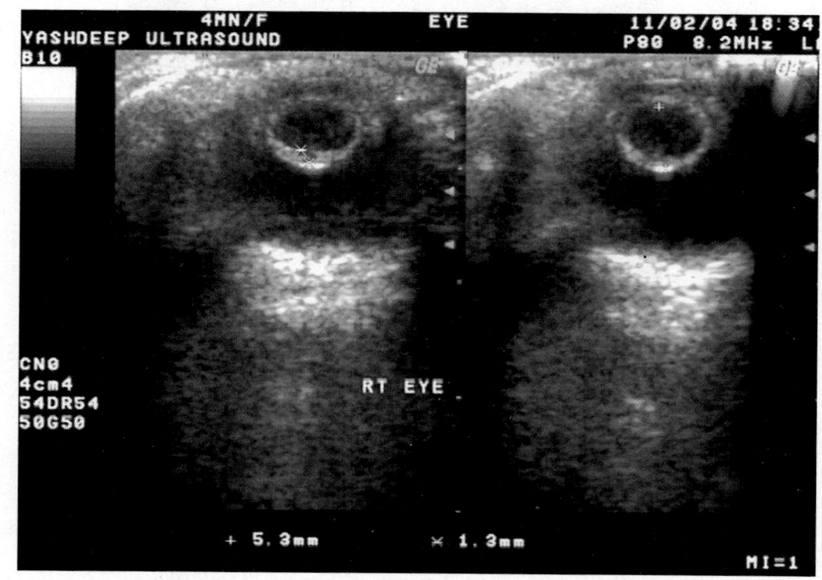

FIGURE 1.38: HRSG shows echogenic thickened opaque lens in both eyes. Posterior segment of the eye is free.

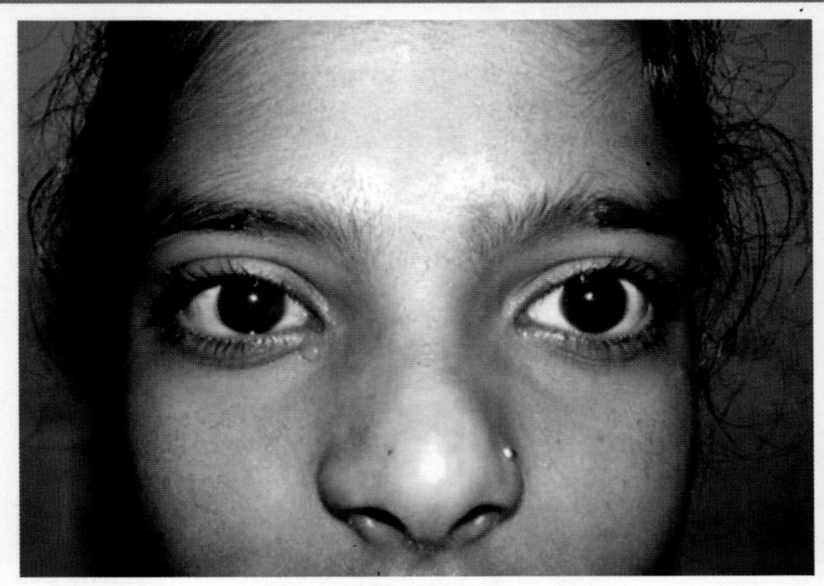

FIGURE 1.39

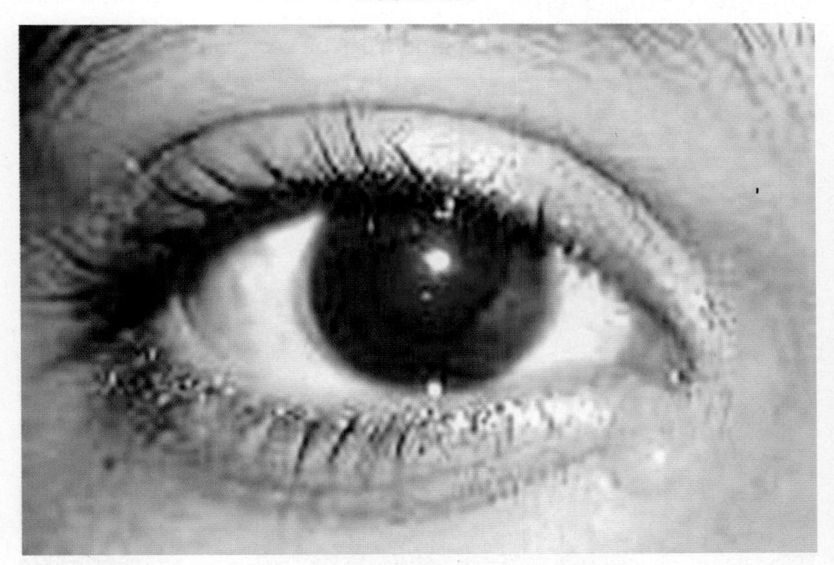

FIGURE 1.40

FIGURES 1.39 and 1.40: Opaque corneal membrane. Young girl presented with opaque membrane covering the cornea.

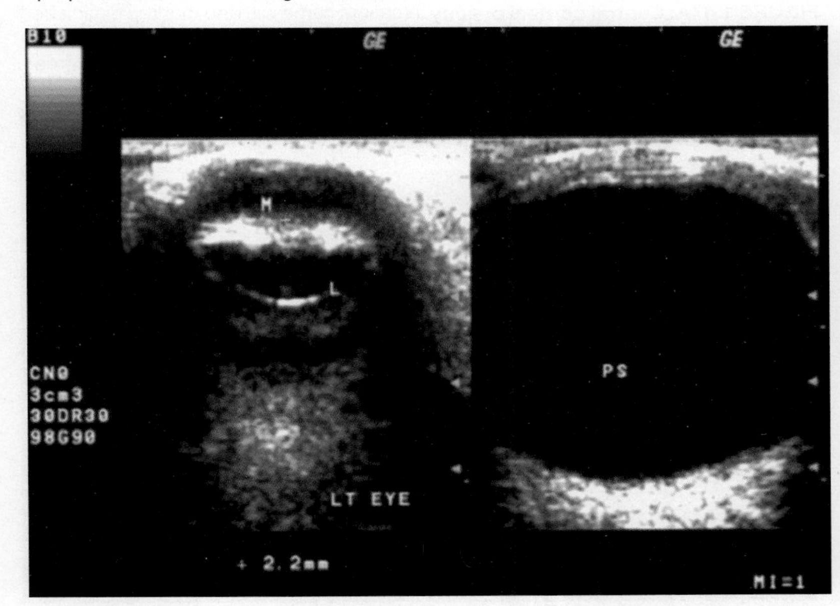

FIGURE 1.41: HRSG shows thick echogenic membrane covering the cornea. However, posterior segment of the eye is free. No intragel pathology is seen.

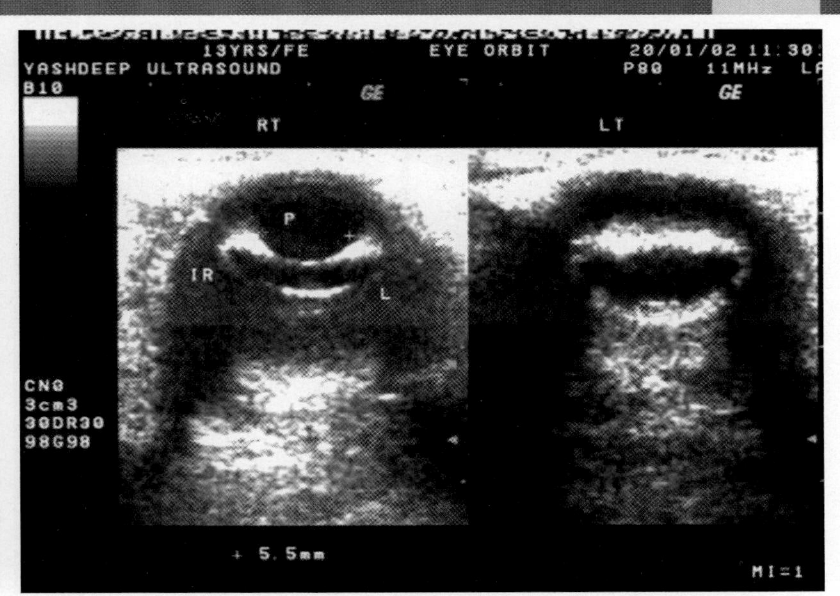

FIGURE 1.42: HRSG shows normal pupil and cornea in the Rt eye and thickened echogenic membrane seen in the Lt eye covering the cornea.

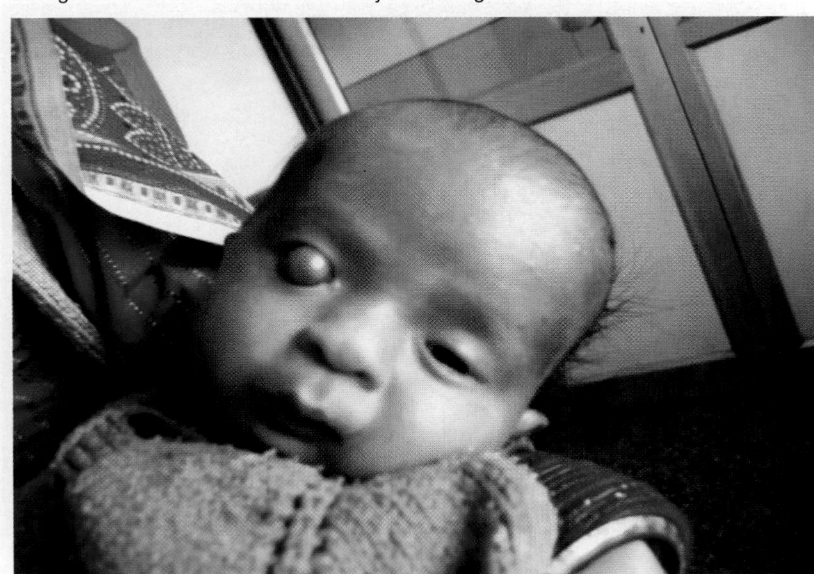

FIGURE 1.43: Anterior staphyloma. An infant is seen born with Rt eye covered with membrane.

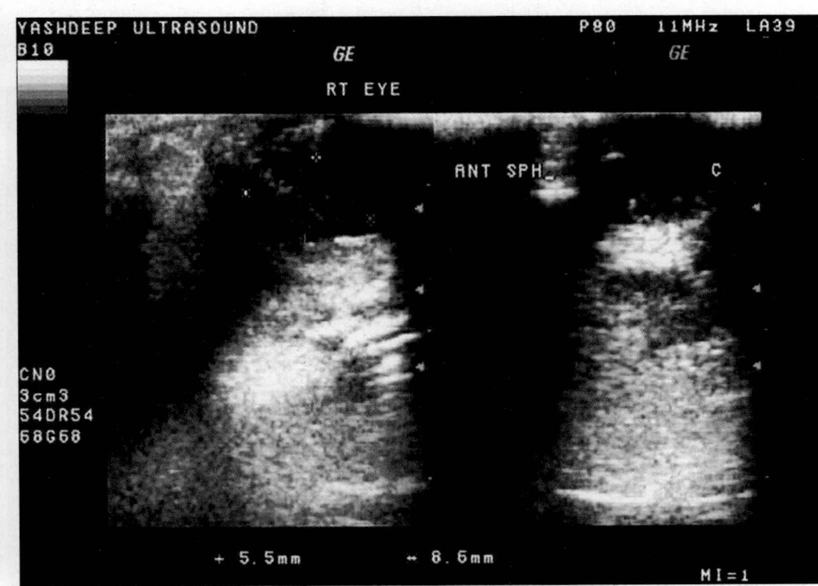

FIGURE 1.44: HRSG shows a cystic mass seen coming out from the Iris and covering the cornea suggestive of anterior staphyloma.

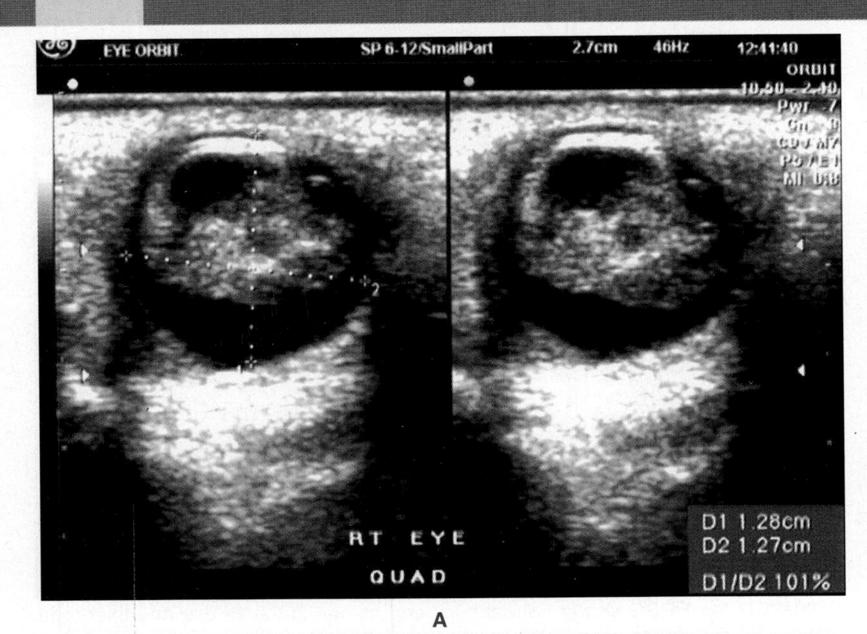

A

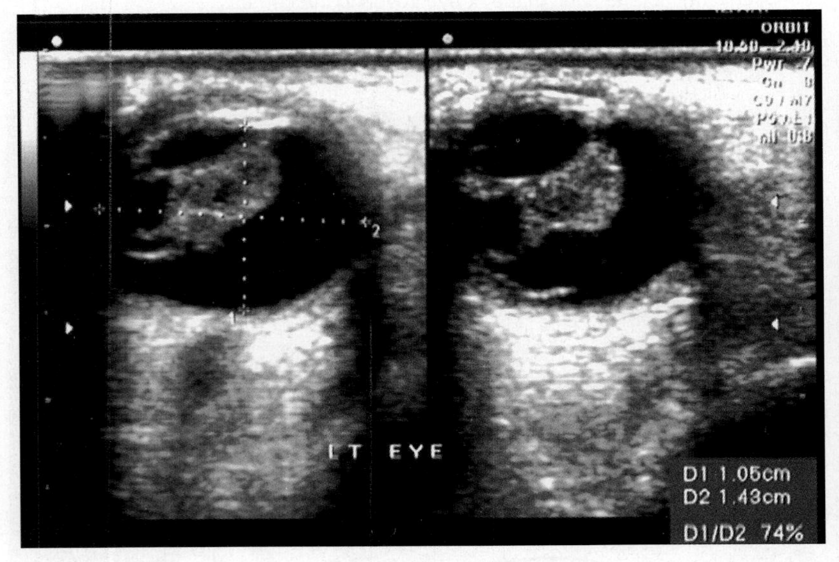

B

FIGURES 1.45A and B: Retinopathy of prematurity. A baby was born with white eye reflex in both eyes. HRSG shows echogenic membranous shadows filling the posterior segment of the eye. The eye is small in size. Multiple cysts are seen in it. No calcification is seen. Findings are suggestive of retinopathy of prematurity ROP.

3. Cryptophthalmos.
4. Persistent hyperplastic primary vitreous.
5. Retrolental hyperplasia.
6. Congenital cataract.
7. Micro-ophthalmos.
8. Macro-ophthalmos
9. Retinopathy of prematurity.
10. Congenital orbital cyst.
11. Congenital ocular tumors.
12. Congenital exophthalmoses.
13. Anterior staphyloma

Acquired Pathologies

All the pathological conditions where direct ophthalmoscopy is not possible are the indications for HRSG. They are grouped as under:

Opaque Ocular Media

Anterior segment: The anterior segment can be very well evaluated with

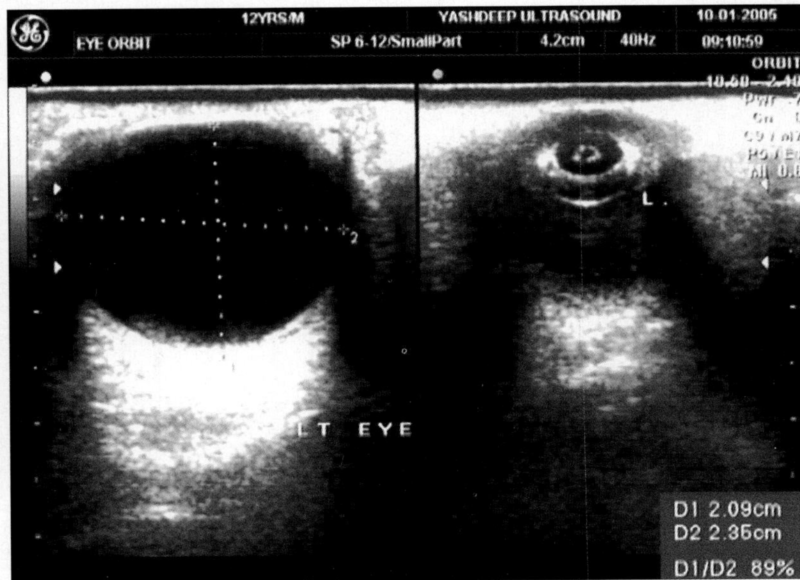

FIGURE 1.47A: Central corneal opacity. A patient came with the marked diminished vision in one eye with central corneal opacity. HRSG shows an echogenic shadow in the central part of the cornea. The posterior segment of the eye and lens are free.

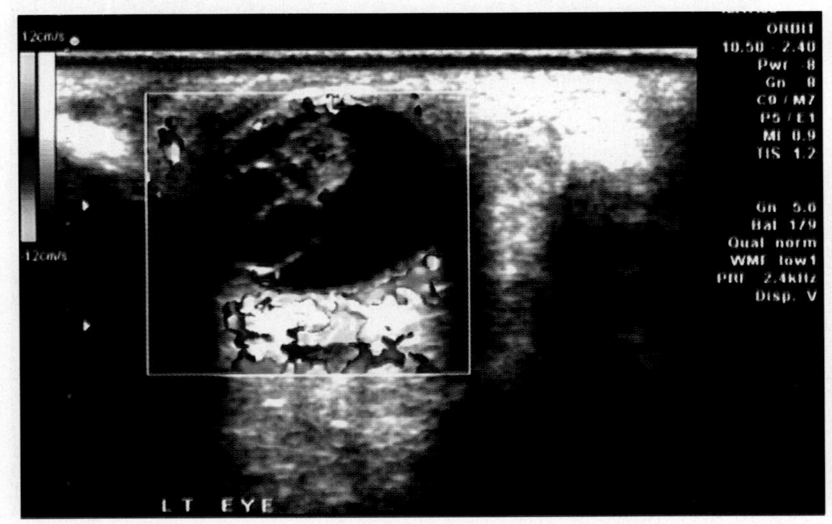

FIGURE 1.46: Color Doppler of same patient doesn't show any flow in the retinal membrane.

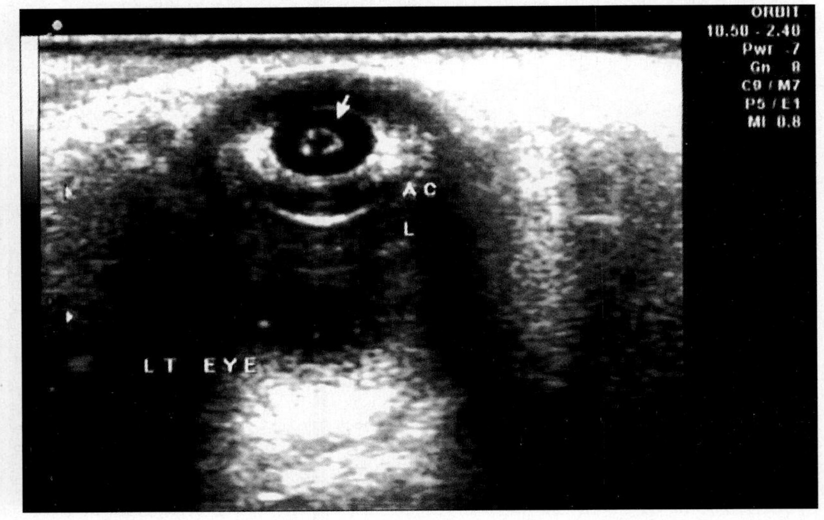

FIGURE 1.47B: 3D imaging of the eye shows fine details of the corneal opacity.

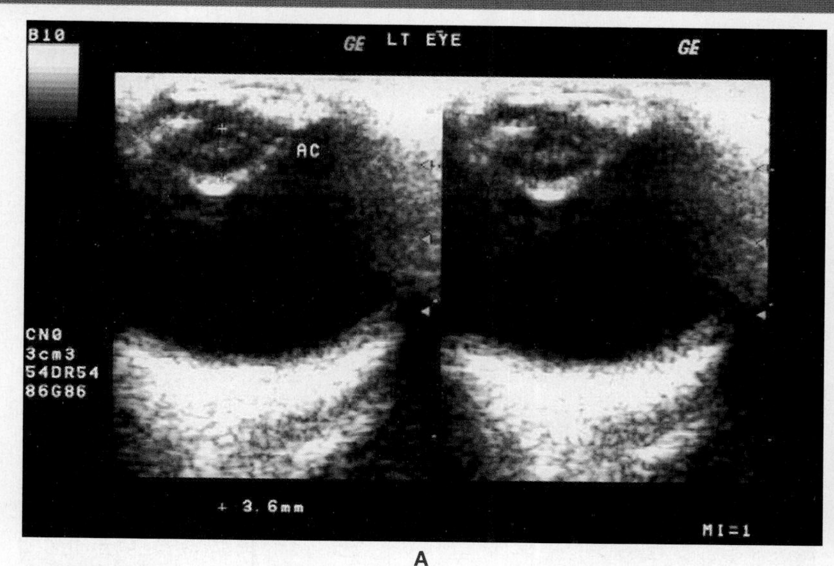

A

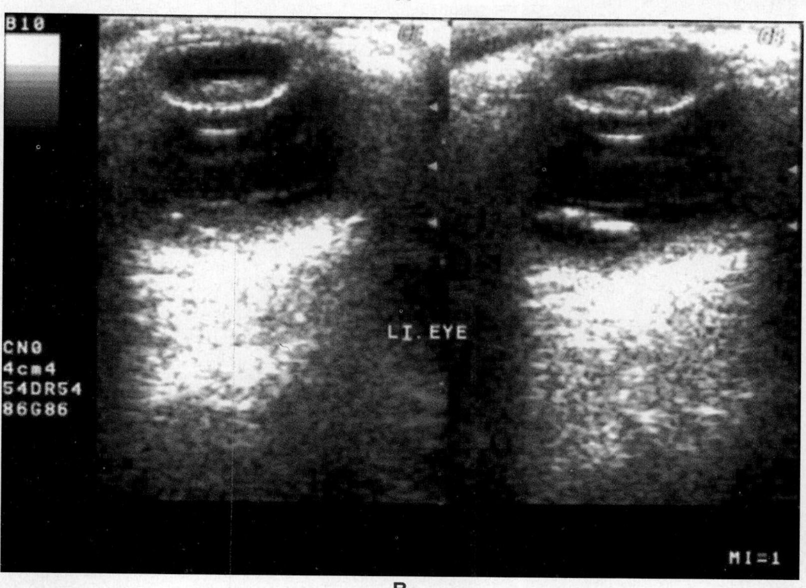

B

FIGURES 1.48 A and B: Anterior chamber hyphema. HRSG shows echogenic collection in anterior chamber in case of trauma to eye. Echogenic collection is a blood in anterior chamber. Associated cataract lens is also seen.

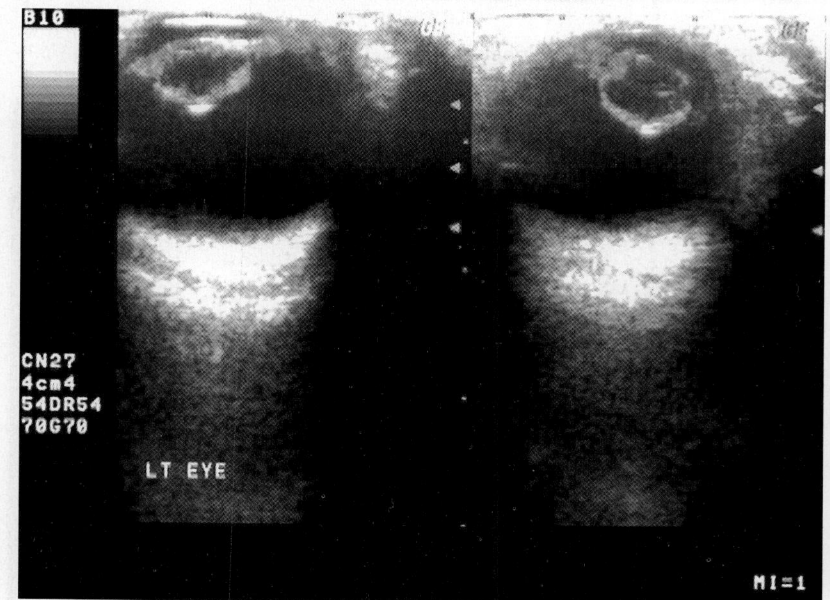

FIGURE 1.49: Bilateral congenital cataract. A young child presented with bilateral cataract. HRSG shows thickened opaque lens in both eyes. Rt lens is 2.5 mm in thickness and shows nucleus calcification. Lt lens is 2.9 mm in thickness. Posterior segment of the eye is free. No intragel pathology is seen.

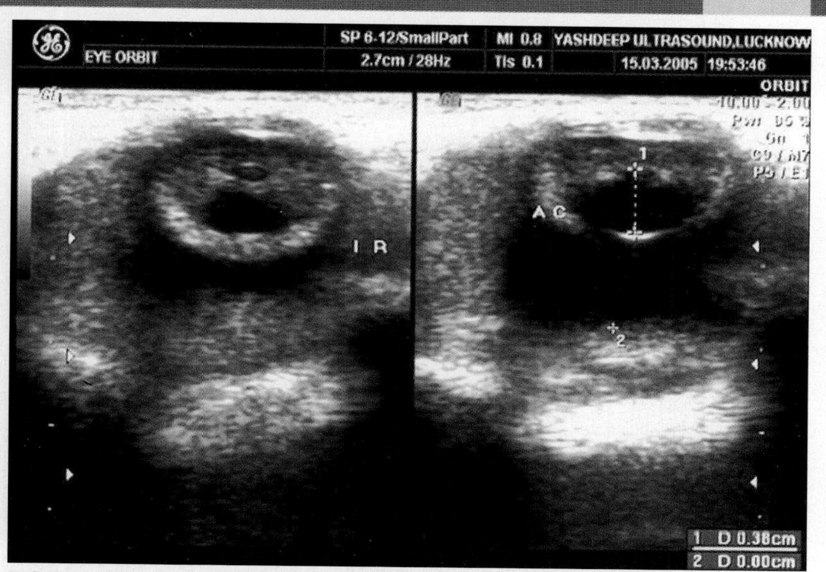

FIGURE 1.50: Congenital cataract. HRSG shows thickened opaque lens in the eye. The anterior chamber is also seen increased in depth. The iris is thickened and irregular in outline suggestive of iritis.

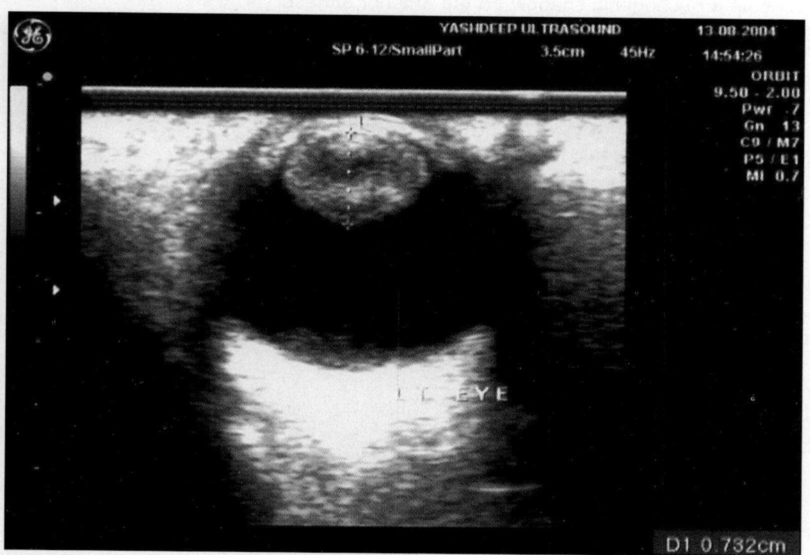

FIGURE 1.51: Post traumatic cataract. HRSG shows thickened opaque lens in Lt eye in a case of trauma. Lens is 7.3 mm in thickness. Multiple intralenticular opacities are present. Posterior segment of the eye is free. Lt eye shows normal lens.

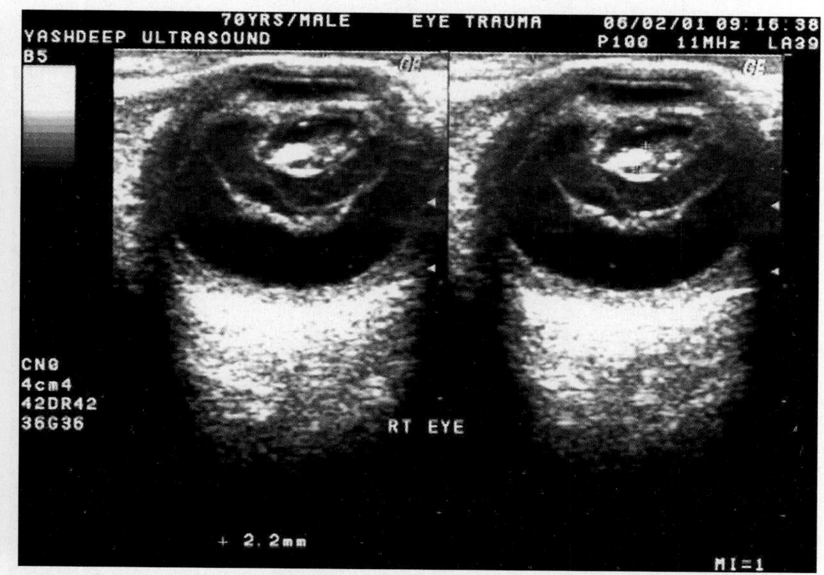

FIGURE 1.52: Hypermature cataract. HRSG shows thickened opaque lens in Rt eye with echogenic-calcified nucleus. Associated intragel hemorrhage is also seen in the posterior segment. But no retinal detachment is seen.

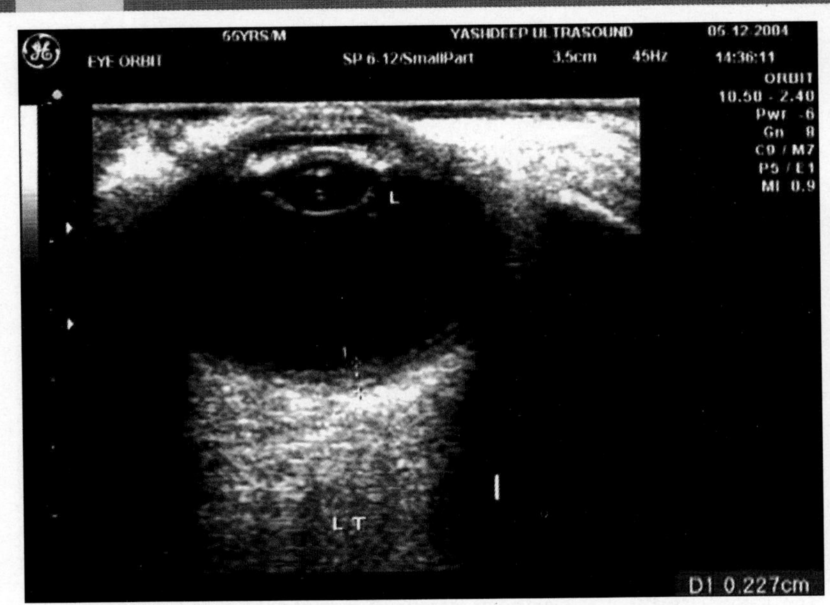

FIGURE 1.53: Early cataract. HRSG shows early calcification of the nucleus in the lens as an echogenic shadow on the anterior surface.

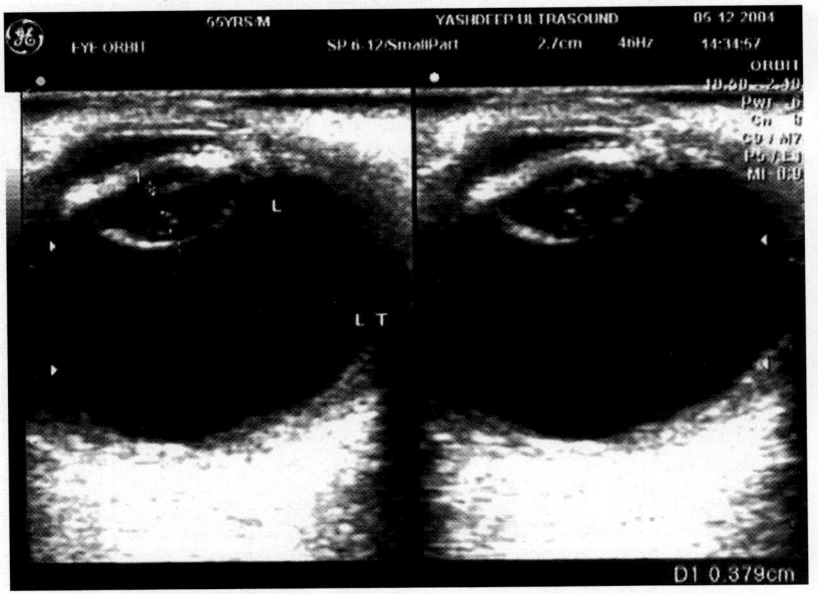

FIGURE 1.54: Early cataract. HRSG shows multiple echogenic specks in the lens.

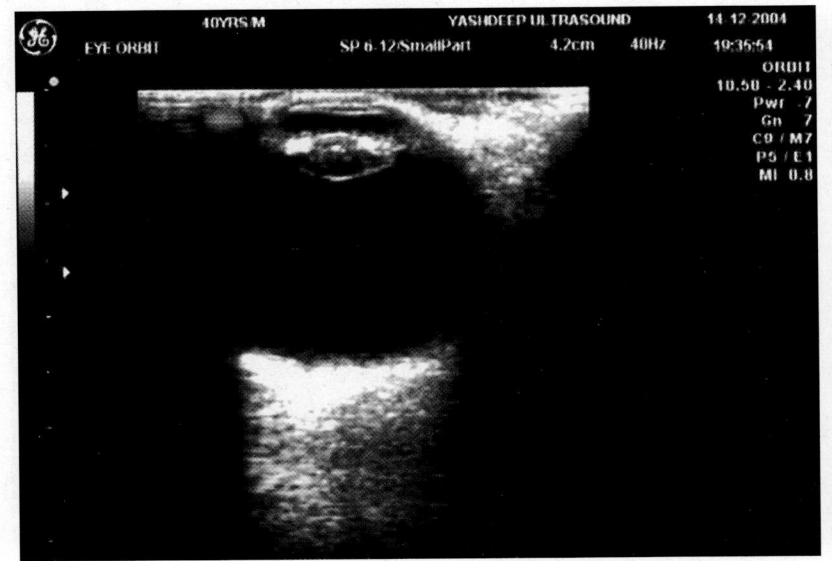

FIGURE 1.55: Mature cataract. Thickening of the lens with evidence of calcification of the nucleus is seen on HRSG.

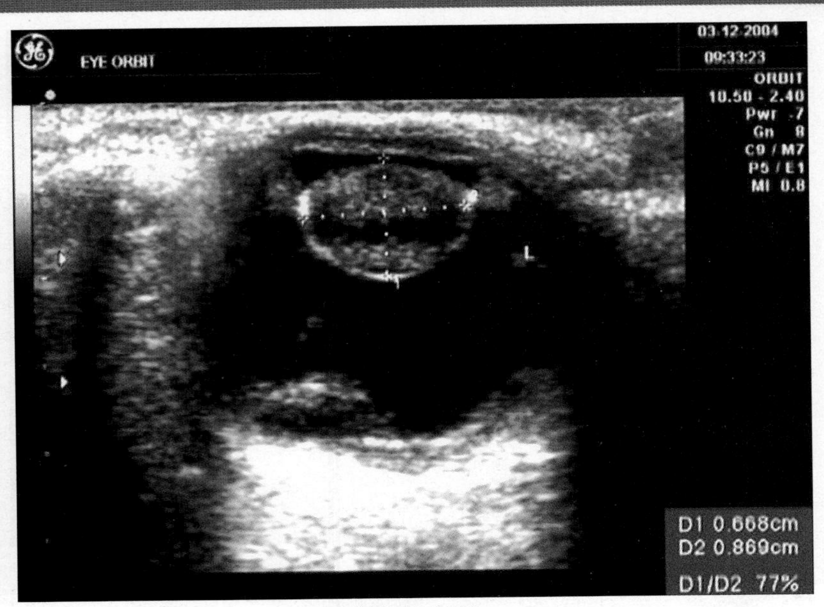

FIGURE 1.56A

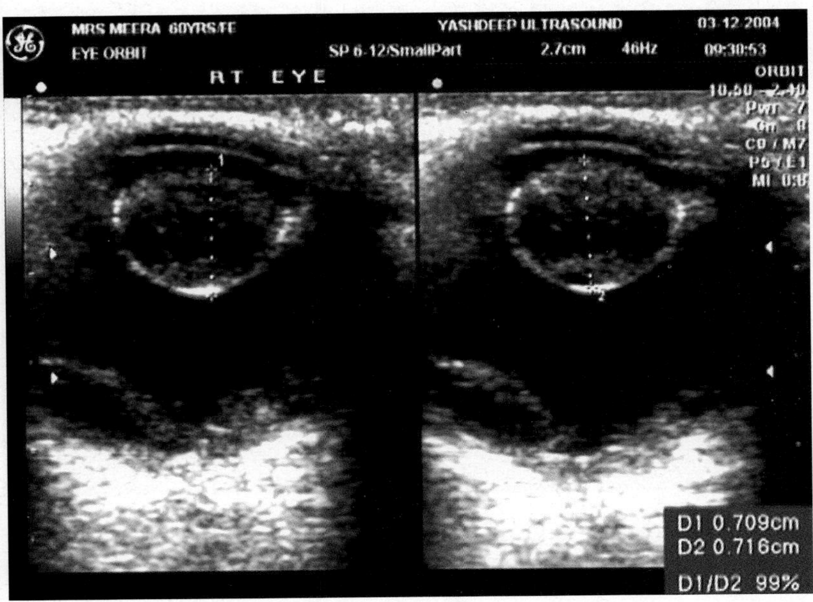

FIGURE 1.56B

FIGURES 1.56A and B: Hypermature cataract. HRSG shows markedly swollen lens and with echogenic shadows opacifying the lens from anterior and posterior surface.

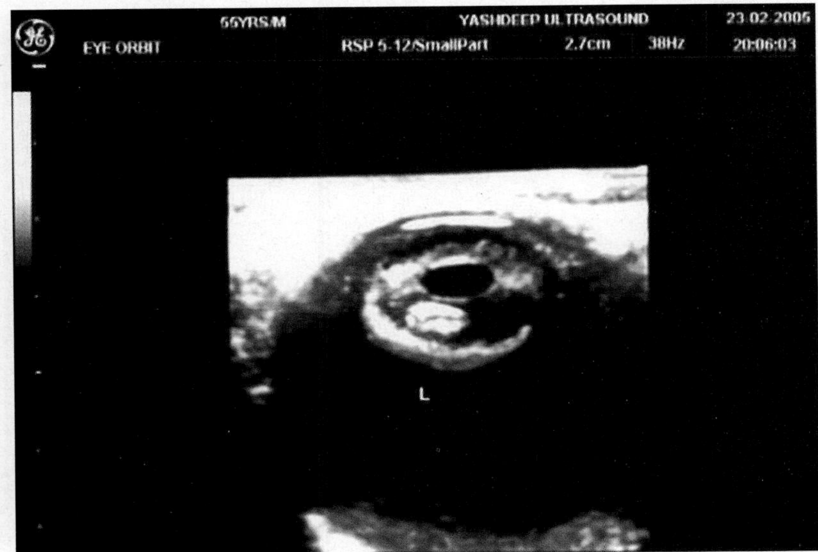

FIGURE 1.57A

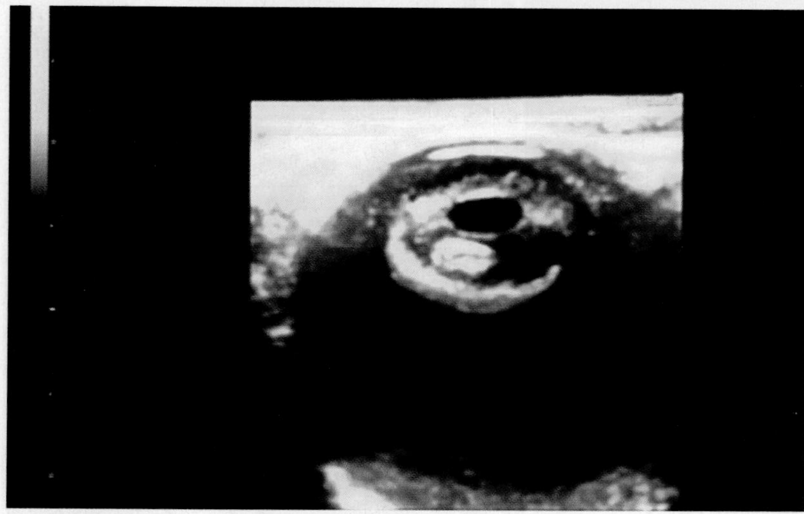

FIGURE 1.57B

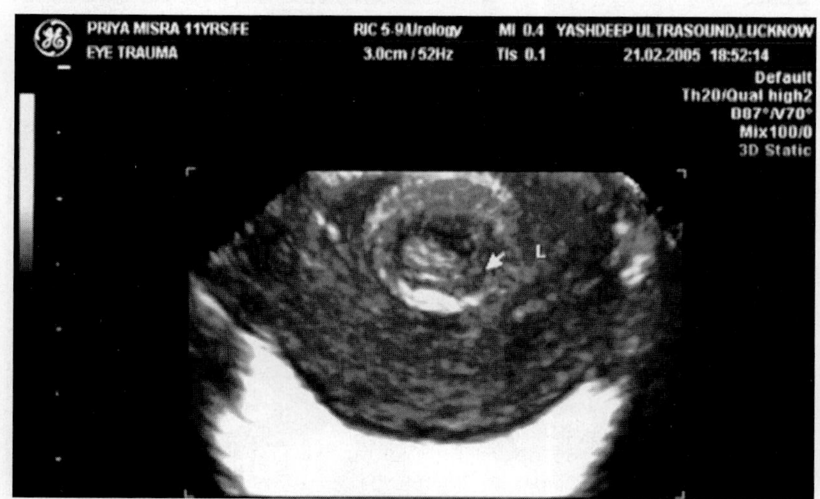

FIGURE 1.57C

FIGURE 1.57A to C: 3D of the mature cataract. The 3D imaging of the eye shows the calcified nucleus in the hypermature cataract. Finer details of anterior segments are seen on 3D imaging.

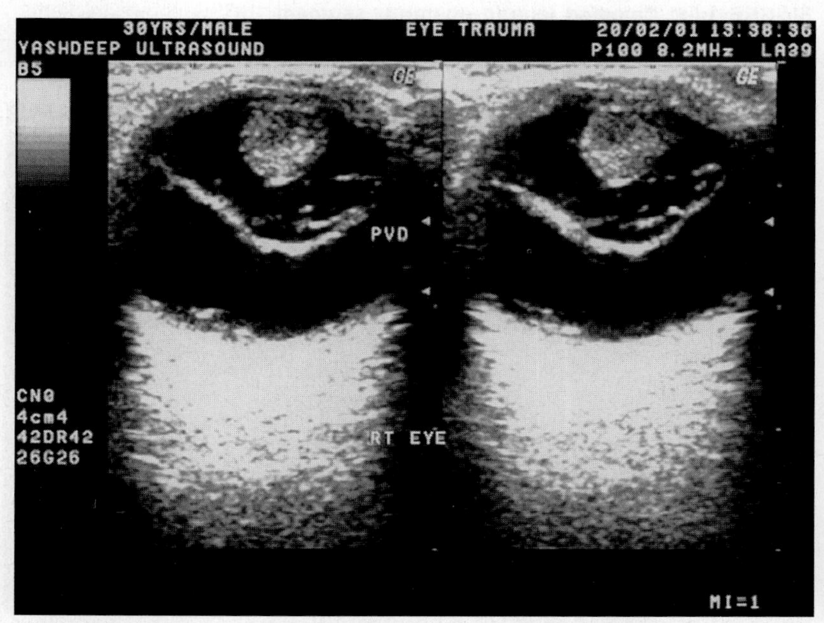

FIGURE 1.58: Post-traumatic cataract. HRSG shows thickened opaque lens. It is 5 mm in thickness. Lens capsule is also thickened. Posterior segment of eye shows intragel hemorrhage and associated posterior vitreous detachment seen as an echogenic membrane. It shows flappy movement on dynamic scanning.

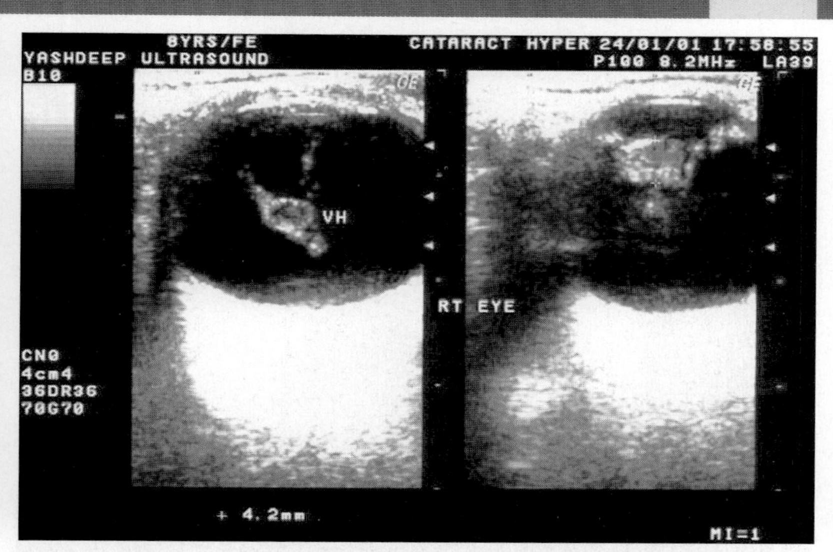

FIGURE 1.59: Hypermature cataract. A young girl presented with cataract in Rt eye. HRSG shows hypermature thickened opaque lens. Nucleus is also calcified and thickened. Posterior segment eye shows intragel hemorrhage.

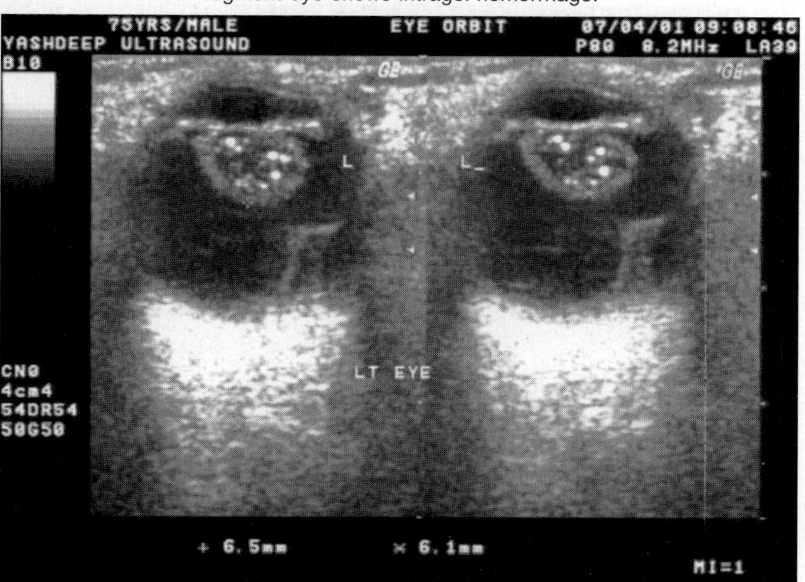

FIGURE 1.60: Post-traumatic cataract. HRSG shows markedly thickened opaque lens. The nucleus is seen broken into multiple small echogenic calcified dots. Associated intragel hemorrhage is also seen.

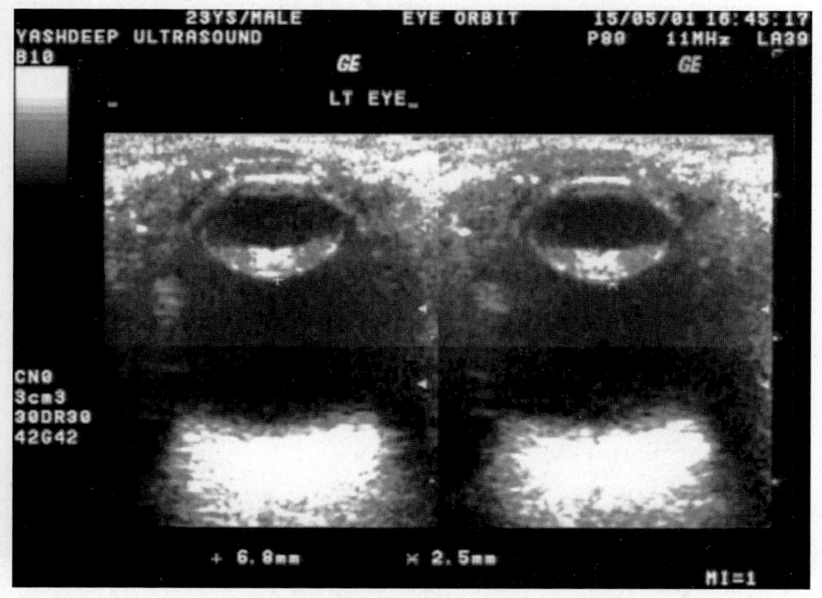

FIGURE 1.61: Post-traumatic cataract. HRSG shows opaque calcified lens in Lt eye. However, posterior segment of the eye is free. No intragel hemorrhage or retinal detachment is seen.

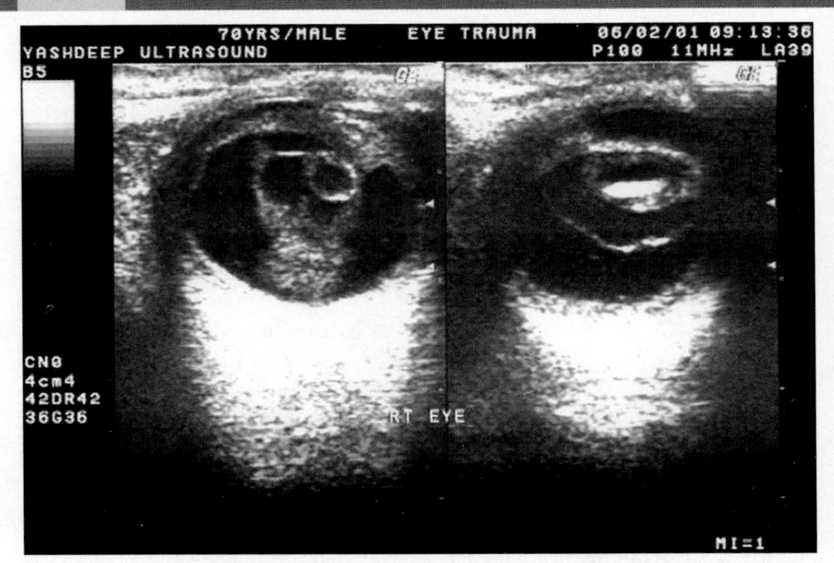

FIGURE 1.62: Post-traumatic complicated cataract. HRSG shows complicated posttraumatic cataract. Nucleus is calcified. Lens capsule is also thickened. Iris is seen prolapsed on nasal side. Posterior segment of the eye shows intragel hemorrhage with retraction of vitreous. Eye coat is also thickened and edematous.

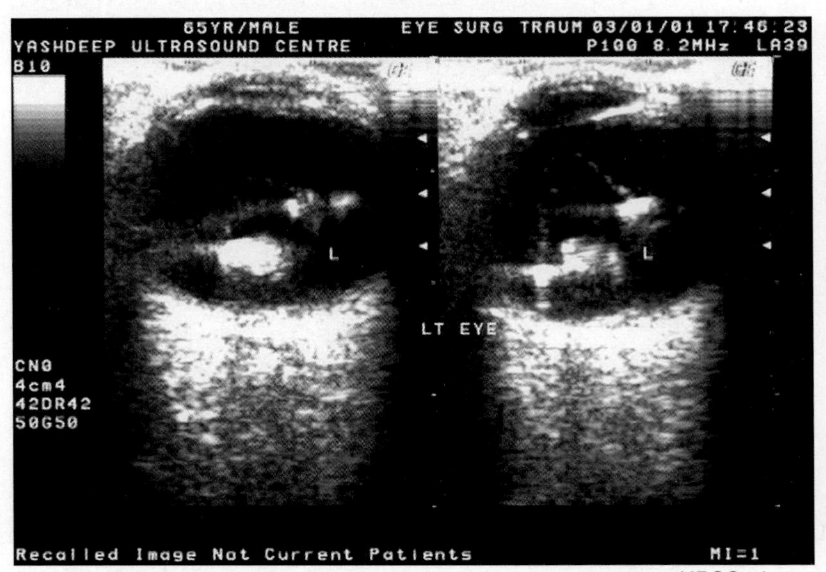

FIGURE 1.63: Post-surgical trauma lens in posterior segment. HRSG shows a hyper-reflective echogenic shadows in posterior segment of eye suggestive of displaced intraocular lens in posterior segment. Multiple echogenic shadows seen in the posterior segment are air bubbles. Intragel hemorrhage is also seen.

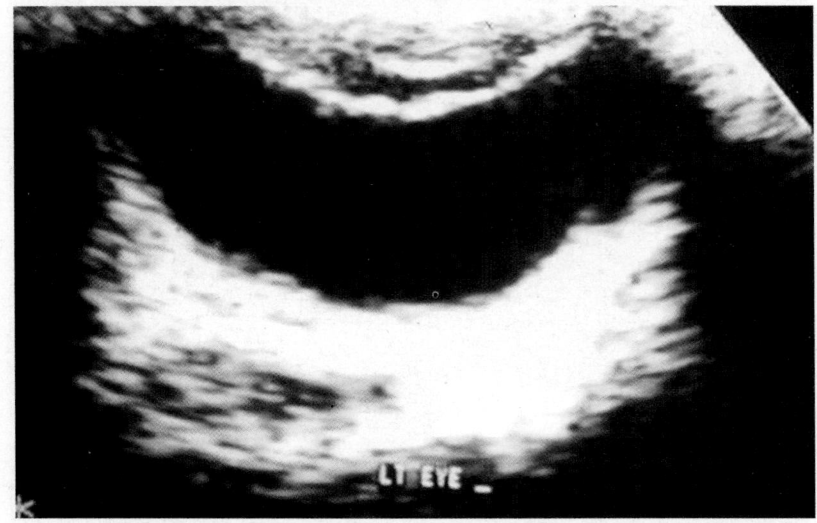

FIGURE 1.64: Retrolental membrane. A thick echogenic membrane is seen in the postoperated case of congenital cataract. The membrane is seen posterior to the lens capsule and was the case of diminished vision. However, posterior segment of the eye is free.

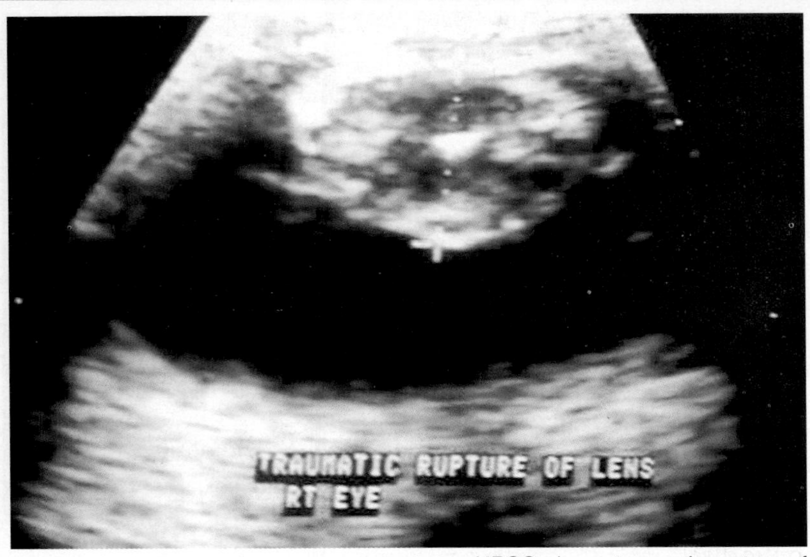

FIGURE 1.65: Traumatic rupture of the lens. HRSG shows traumatic rupture of the lens and avulsion of lens capsule and prolapse of lens material nucleus calcification is also seen in the lens.

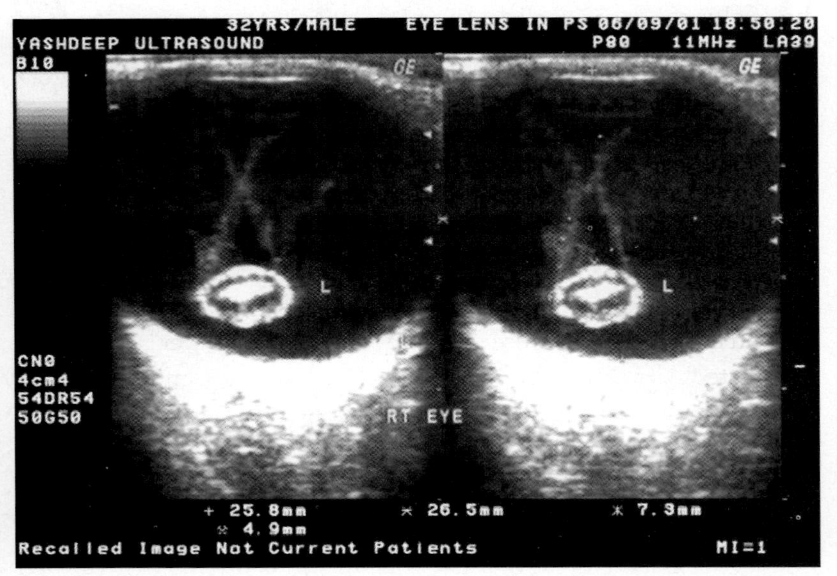

FIGURE 1.66: Calcified lens in posterior segment. HRSG shows thickened opaque lens with calcified nucleus displaced in posterior segment. The membranous ligaments holding the lens are also seen as twin echogenic bands. No intragel hemorrhage is seen.

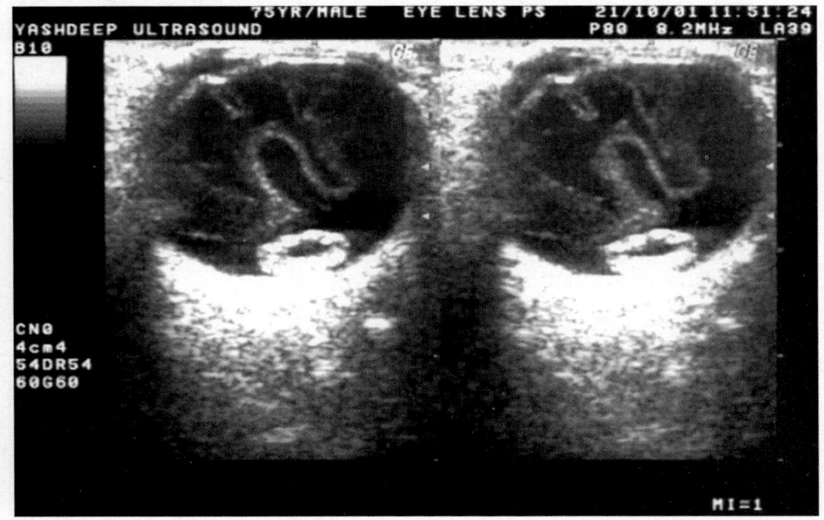

FIGURE 1.67: Lens in posterior segment. HRSG shows thickened opaque lens displaced in posterior segment in case of blunt trauma to eye. The calcified lens is seen at the posterior pole. Associated vitreous hemorrhage is also seen. But no retinal detachment is seen.

high-resolution sonography. Following are the main conditions, which can be easily evaluated on HRSG:

i. Corneal opacification
ii. Hyphema or hypopyon
iii. Miosis
iv. Cataract
v. Pupillary or retrolental membrane.

Examination of Vitreous

The vitreous is a transparent gel. It does not produce any echo on B-scanning in normal conditions. However, scattered opacities of low density may be noted in aging eye due to degenerative changes. The vitreous evaluation should include assessment of vitreous body posterior hyaloid, subvitreal space, retina, choroid, sclera and optic disc. The anterior segment should also be examined in cases of hyphema lens, cataract and for lesions of pupils and ciliary body. As stated earlier, normal vitreous is an echo free cavity on HRSG. It should be evaluated for the presence of opacities, bands and membrane. Vitreous opacities produce dot and short lines on 'B' scanning. These opacities may be originated from liquefied vitreous, hemorrhage, inflammation, infections or calcium soaps (asteroid hyalosis). In vitreous hemorrhage, the bright echoes of dots are produced due to clump or cells having blood. Small, subtle hemorrhage is difficult to see on HRSG. However, low intensity echoes are seen on high gain setting. Medium to dense level echoes are seen in vitreous hemorrhage depending upon the severity of hemorrhage. Vitreous hemorrhage is second most common cause of opaque ocular media after cataract. It is associated with a number of conditions. The most common causes are trauma, diabetic retinopathy, macular degeneration, vein occlusion, retinal tear and intraocular tumors. Ultrasonography is a useful tool to examine it and helps in establishing the density and location of hemorrhage, and cause of hemorrhage, which is not clinically explained. In fresh hemorrhage, dense echogenic collection is seen in posterior segment. Layering of collection is noted in organization of the hemorrhage. Dense interfaces are formed due to layering. They stand out clearly on HRSG. Pseudomembranes are also formed. Posterior vitreous detachment commonly takes place with vitreous hemorrhage.

The asteroid hyalosis is senile degenerative condition of unknown etiology. It is usually unilateral in 75% of the cases. It is possible to diagnose it on HRSG. The findings are quite typical. Calcium soaps from asteroid bodies which are scattered throughout the vitreous. These bodies show multiple bright echogenic echoes and demonstrate considerable movements.

Posterior Vitreous Detachment

Posterior vitreous detachment (PVD) may be focal or diffuse. It is most commonly associated with synchysis senilis a degenerative disorder. The gel loses it volume and becomes hypermobile. It remains suspended from the vitreous base. It is also seen associated with cataract eye. The posterior hyaloid may separate completely from the posterior pole. It may remain attached at optic disc giving a funnel-shaped appearance. Vitreoretinal adhesions may also be noted in areas of retinal tear. On "B" scan imaging, PVD is usually smooth and may be thick when blood is layered posteroinferiorly. Real-time HRSG shows an undulating movement in PVD. This differentiates it from retinal detachment.

Subvitreal Hemorrhage

At times vitreous hemorrhage is associated with subvitreal hemorrhage. It may also present alone. In mild subvitreal hemorrhage, very high gain setting is required to detect it. Subvitreal blood usually does not clot. Therefore, it is of low reflectivity and mobile even in chronic cases.

Posterior Hyphema

Subvitreal hemorrhage has the tendency to layer out similar to anterior chamber hyphema. The surface of this layered blood stands out as smooth, dense echogenic membrane on "B" scans. With the kinetic movements of eye, this hyphema normally slides across the fundus surface of eye. In sitting upright position, this hyphema changes its position in the eye. The probe position should not be changed while changing the position of patient from lying to sitting position. This

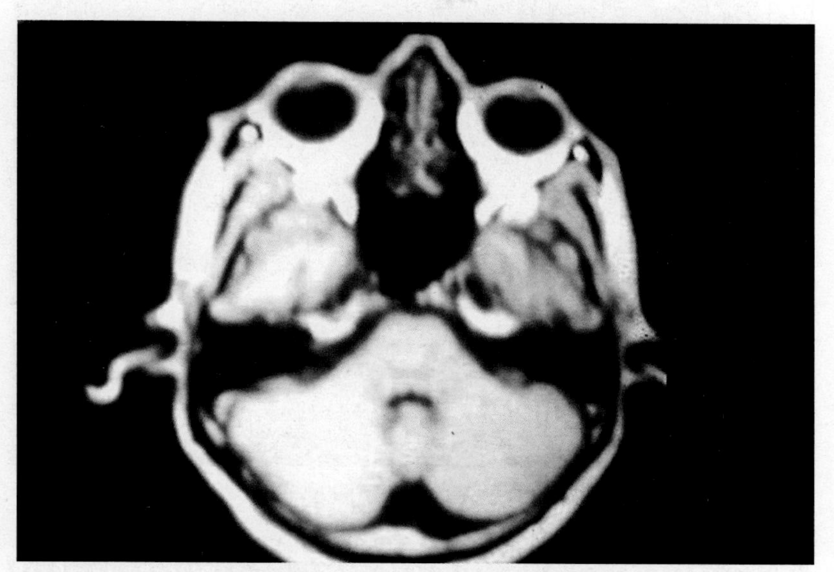

FIGURE 1.68: Vitreous hemorrhage. MRI of a patient with painful Rt eye and blurred vision. No hemorrhage is noted on MRI.

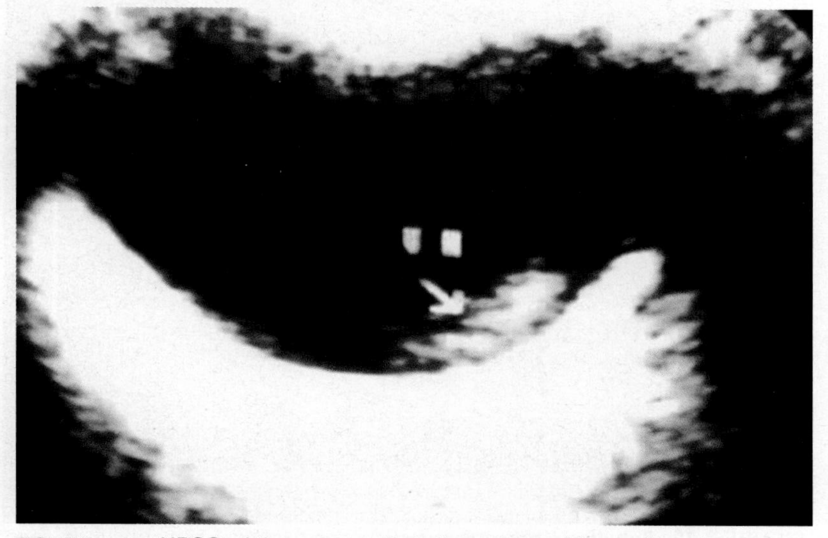

FIGURE 1.69: HRSG of the same patient shows intragel hemorrhage in posterior segment on nasal side partially covering the macula.

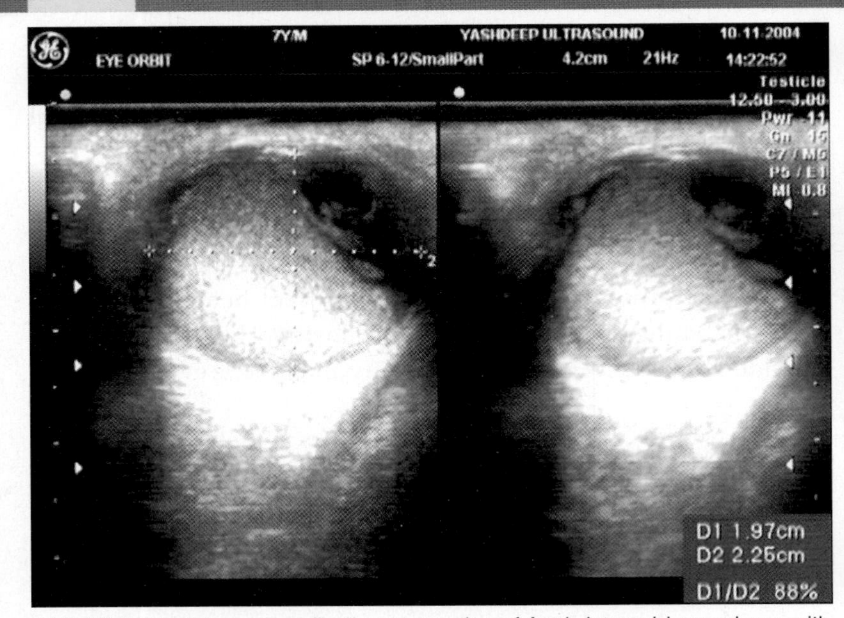

FIGURE 1.70: Echogenic collection suggestive of fresh intragel hemorrhage with layering of the blood is seen in vitreous cavity.

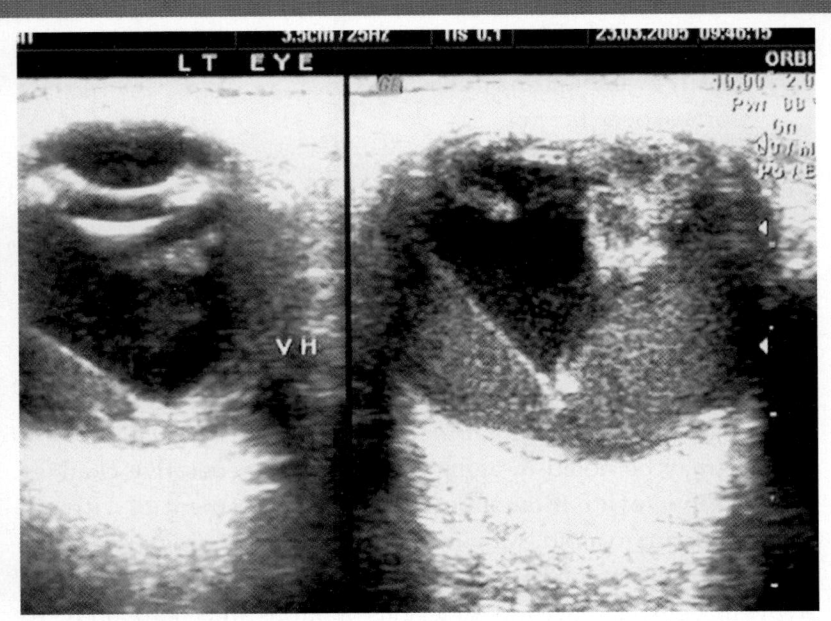

FIGURE 1.73: Partial subvitreal hemorrhage confined to the nasal side with intragel hemorrhage. No organization of blood is seen.

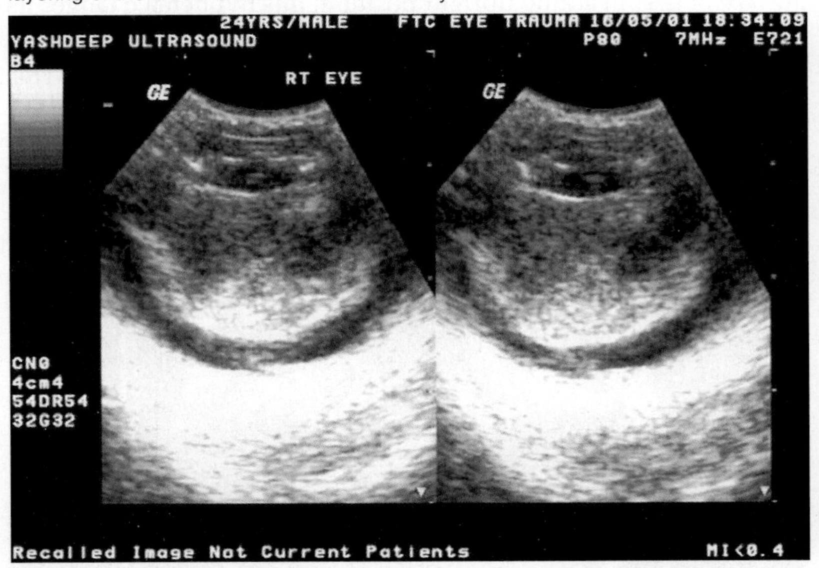

FIGURE 1.71: Fresh intragel hemorrhage echogenic collection in the vitreous cavity suggestive of intragel hemorrhage.

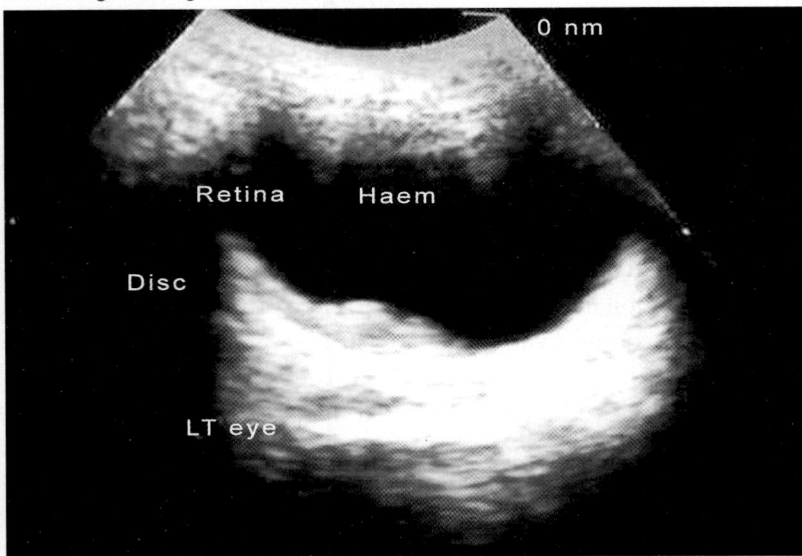

FIGURE 1.74: Macular hemorrhage. Echogenic intragel hemorrhage covering part of macula.

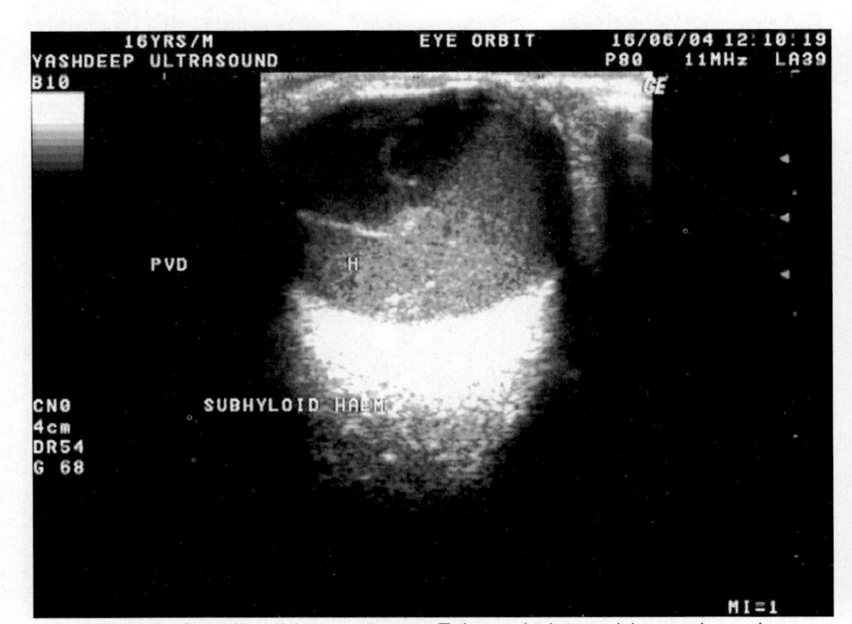

FIGURE 1.72: Subvitreal hemorrhage. Echogenic intragel hemorrhage is seen confined to the subvitreal space with layering of the blood.

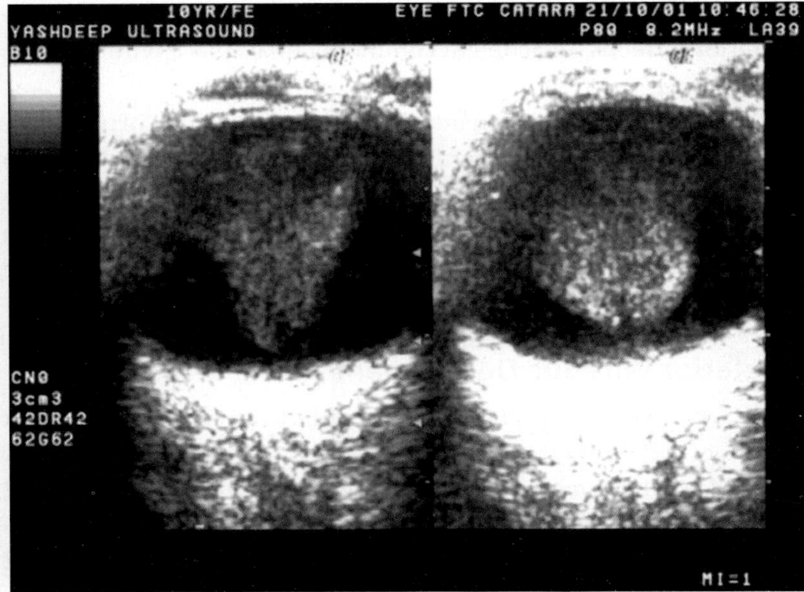

FIGURE 1.75: Fresh intragel hemorrhage. Echogenic fresh intragel hemorrhage filling the vitreous cavity.

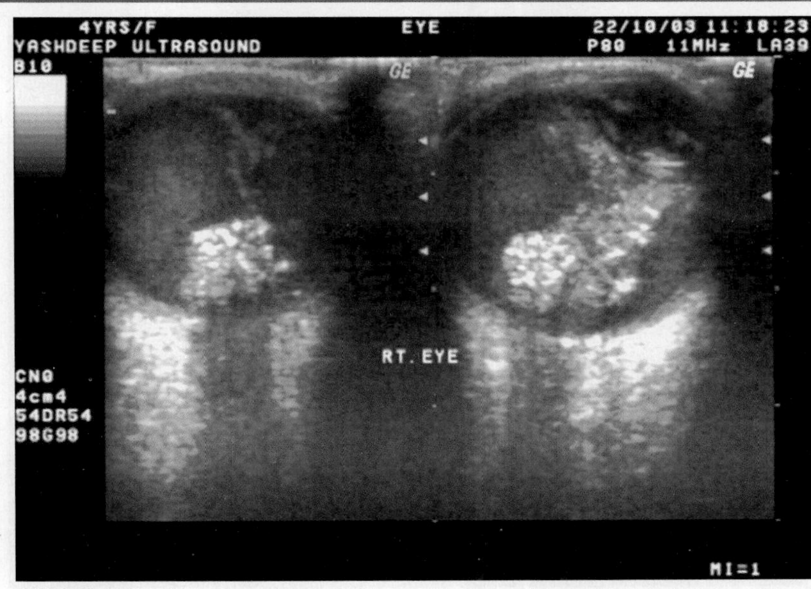

FIGURE 1.76: Old hemorrhage. Old hemorrhage confined to the posterior segment of the eye with organization of the blood. HRSG shows membranous vitreous retracted from its surface and hanging from the base.

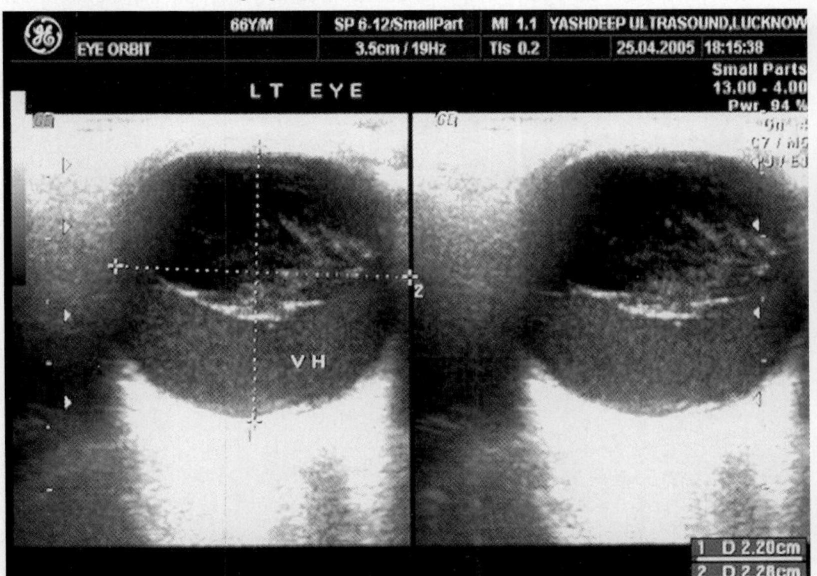

FIGURE 1.77: Vitreous hemorrhage with posterior vitreous detachment. Echogenic intragel hemorrhage with a thin membrane seen detached from its surface and folded forming a cyst suggestive of posterior vitreous detachment.

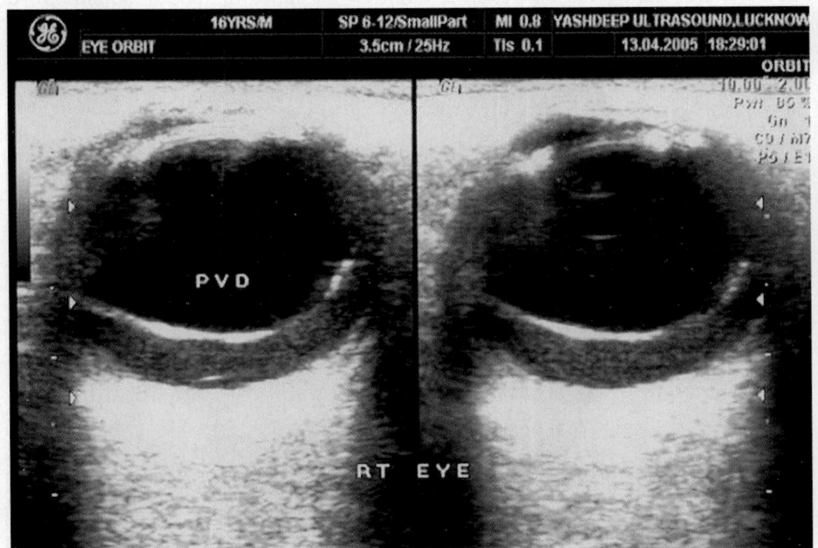

FIGURE 1.78: Complete posterior vitreous detachment. HRSG shows an echogenic membrane lifted from its surface suggestive of complete PVD with vitreoretinal adhesions.

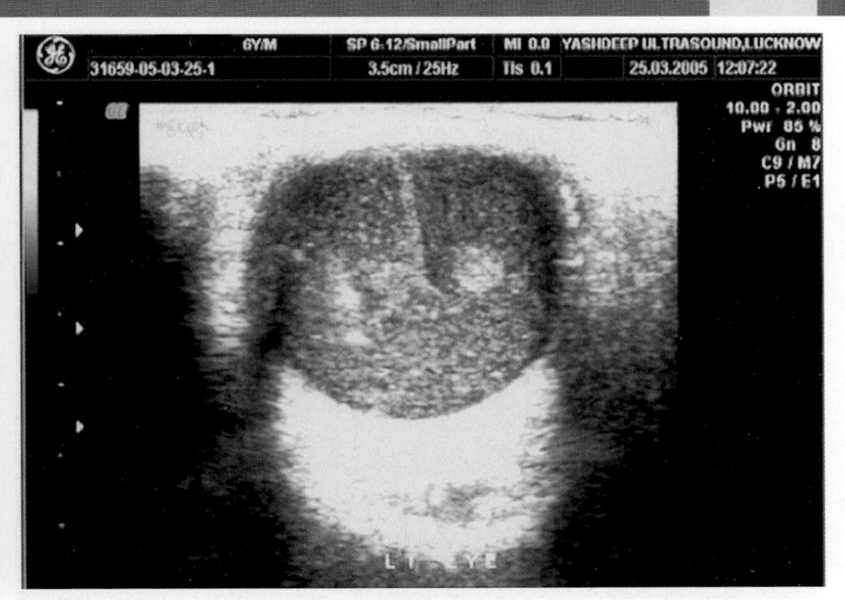

FIGURE 1.79: Hemorrhagic cystic eye. Fresh intragel hemorrhage filling whole of the vitreous cavity. HRSG shows homogeneous echogenic collection in the posterior segment.

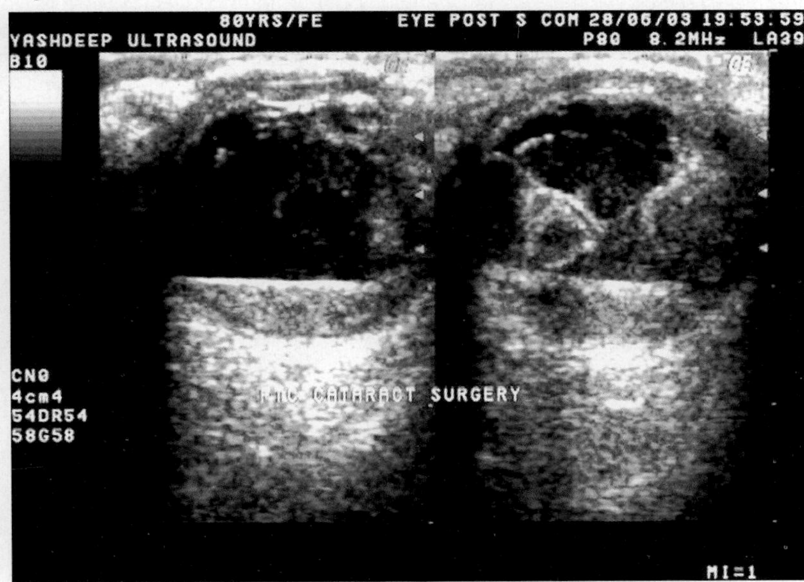

FIGURE 1.80: Posterior hyphema. HRSG shows homogeneous echogenic collection confined to subvitreal space and limited with vitreous membrane.

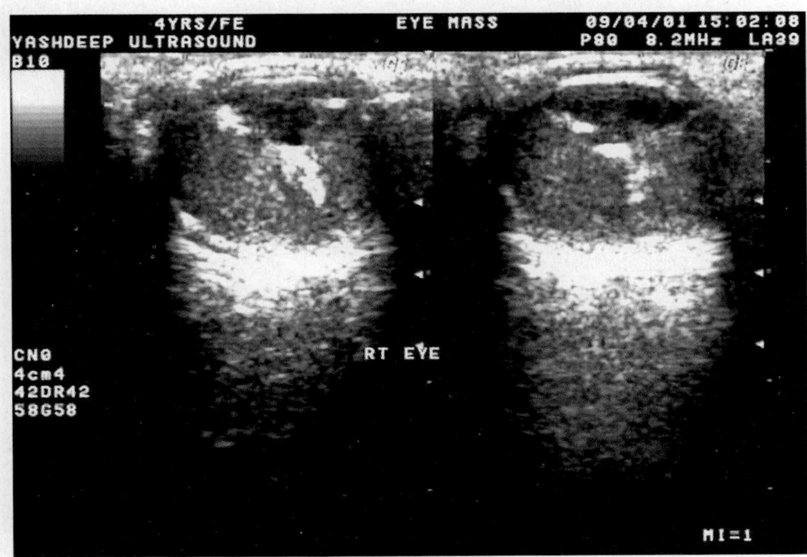

FIGURE 1.81: Fresh intragel hemorrhage. HRSG shows homogeneous echogenic collection filling whole of the vitreous cavity. There is also suggestion of organization of blood seen in the hemorrhage shown as echogenic sedimentation.

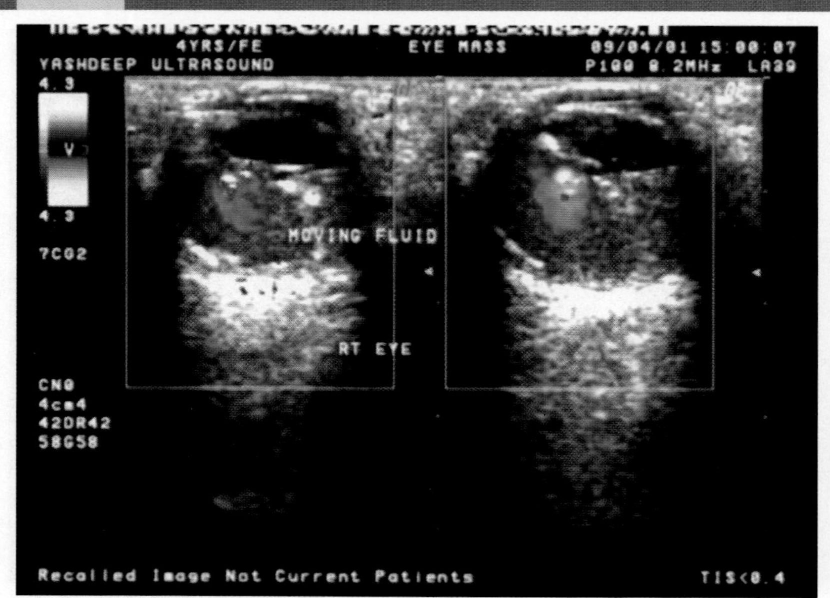

FIGURE 1.82: Color Doppler study of Figure 1.81 shows movement of the hemorrhage on dynamic scanning.

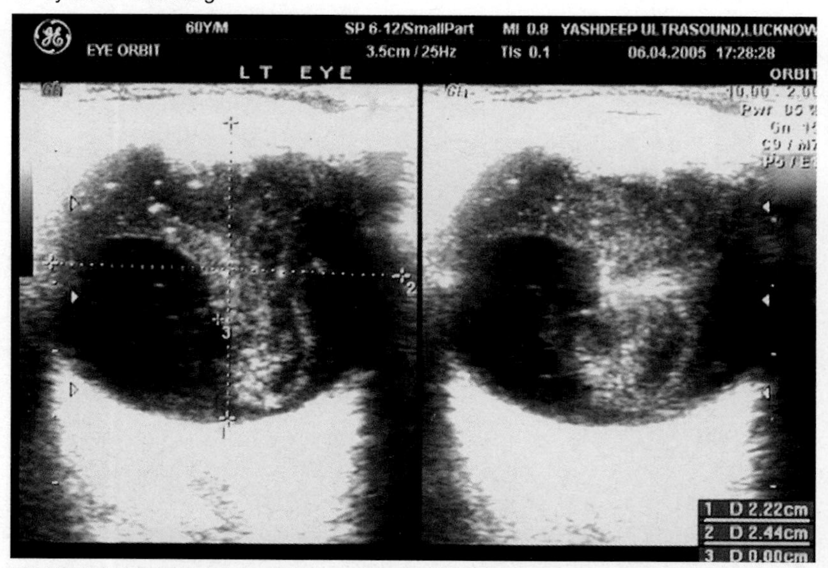

FIGURE 1.83: **Old intragel hemorrhage.** HRSG shows old intragel hemorrhage with echogenic membranous vitreous retracted from its surface and forming a dense membrane.

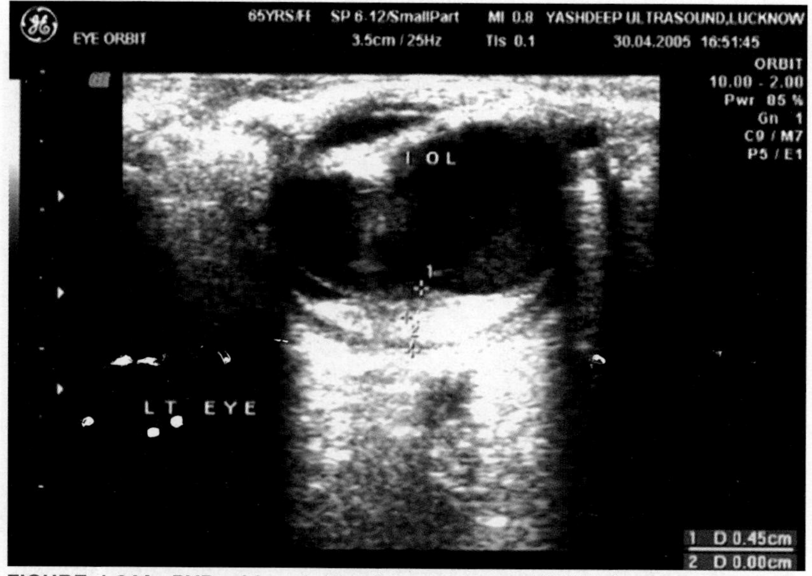

FIGURE 1.84A: **PVD with subvitreal hemorrhage.** HRSG shows an echogenic membrane detached from its surface suggestive of PVD. Echogenic hemorrhage is seen in the subvitreal space with flat membrane limiting the hemorrhage also known as posterior hyphema.

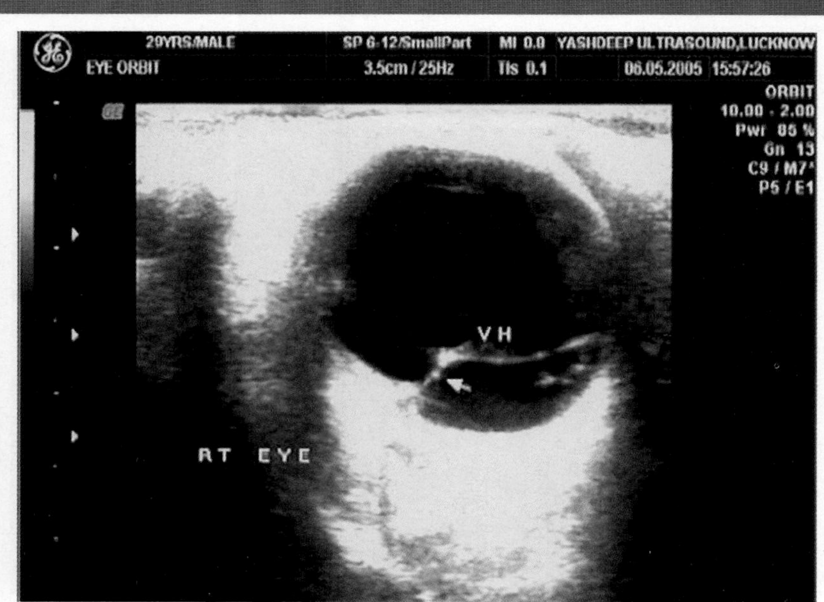

FIGURE 1.84B: **PVD with vitreoretinal band.** HRSG shows an echogenic membrane detached from its surface on posterior side. It shows limited movements suggestive of PVD. There is thin echogenic membranous band seen going from this membrane to the posterior pole and attached with retina typical vitreoretinal band. The limited movement of PVD is due to adherence of vitreous membrane with the retina by the band.

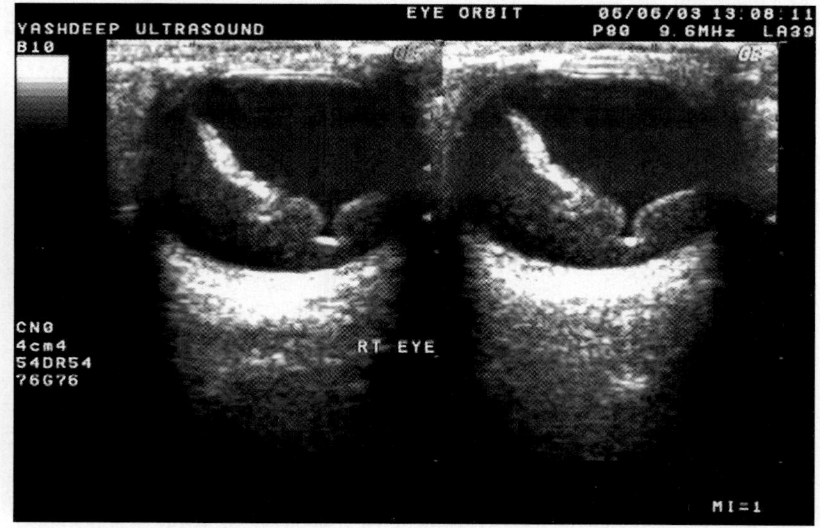

FIGURE 1.85: **Complete PVD.** HRSG shows an echogenic membrane detached from all side in the posterior segment suggestive of complete posterior vitreous detachment. Echogenic collection is seen in the subvitreal space suggestive of subvitreal hemorrhage.

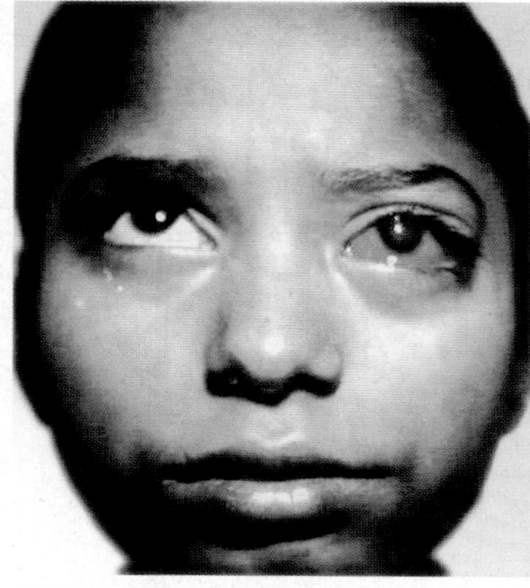

FIGURE 1.86: **Pyophthalmos.** Patient presented with painful red eye with diminished vision.

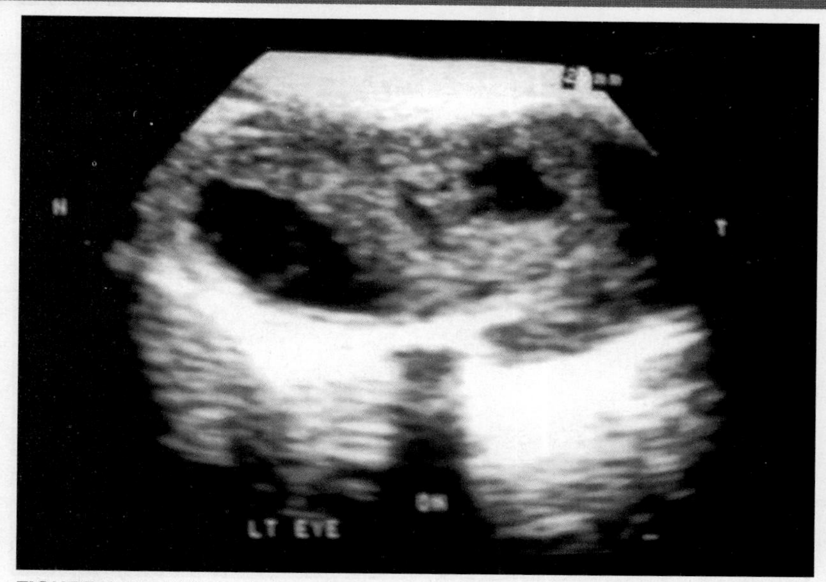

FIGURE 1.87: HRSG shows thick echo collection in the posterior segment with edematous eye coat suggestive of pyogenic collection. The collection is also seen in the episcleral space due to perforation of the sclera and seeping of the pus in the episcleral space.

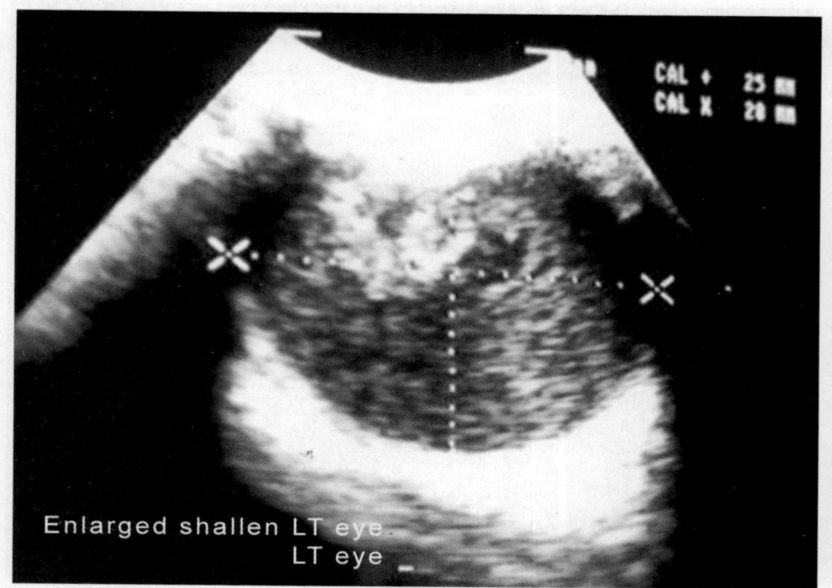

FIGURE 1.88: Pyophthalmos. Dense echogenic collection filling whole of the eyeball and increased length of the eye in a young boy.

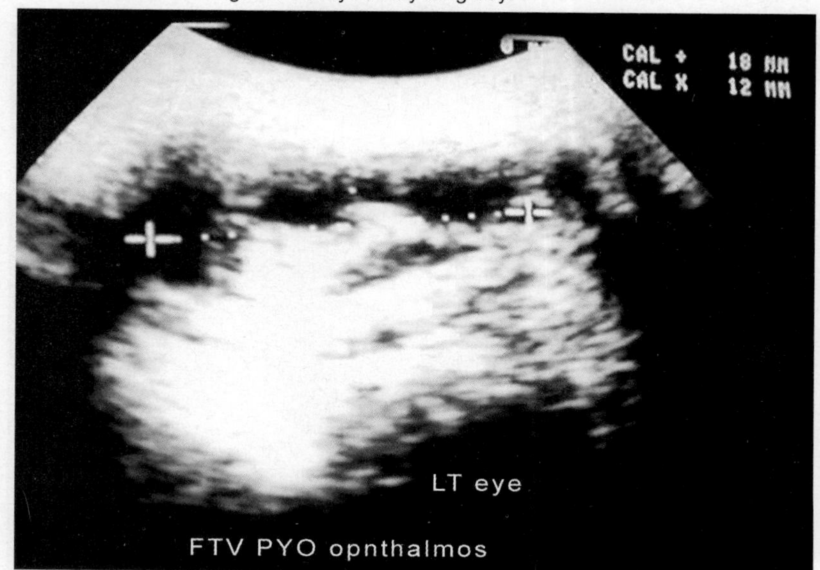

FIGURE 1.89: Small collapsed same eye after 3 months resulting into endophthalmitis due to pyophthalmos.

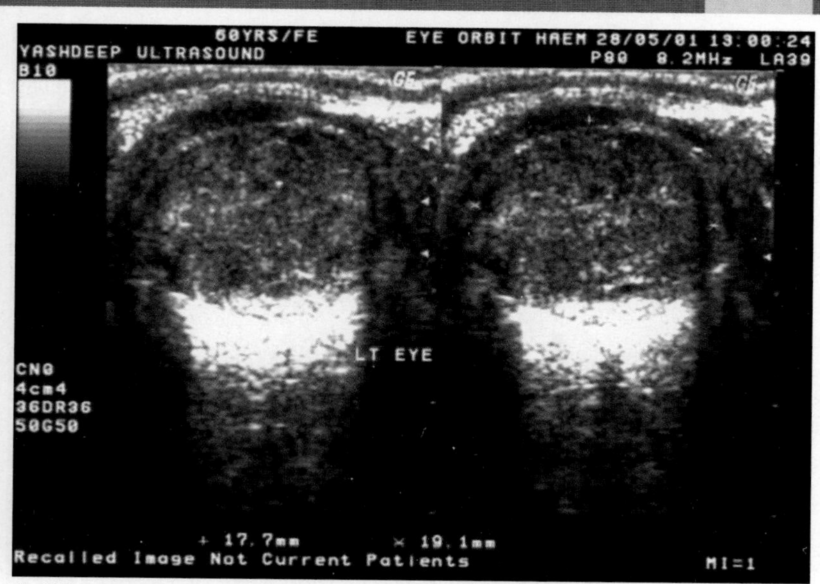

FIGURE 1.90: Pyophthalmos. Dense homogeneous low level collection filling whole of the eyeball with increased length of the eyeball in a case of post-traumatic pyophthalmos.

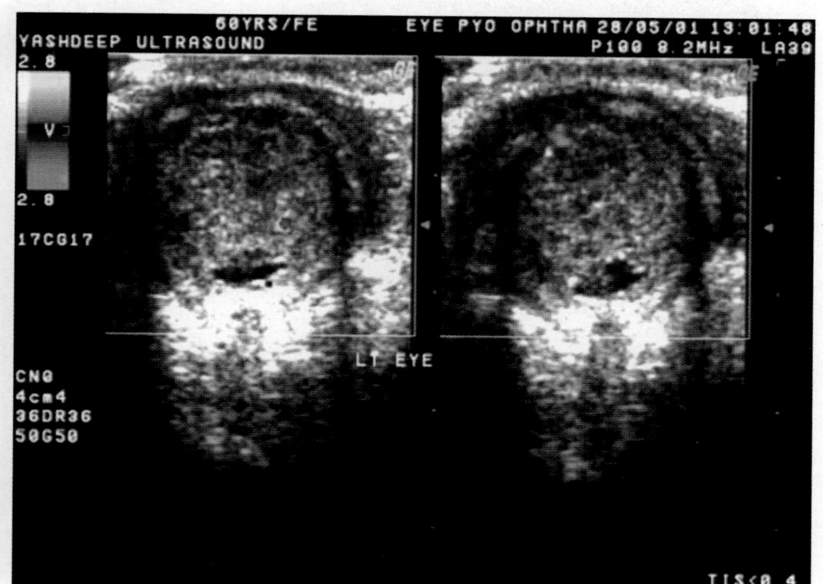

FIGURE 1.91: Color Doppler imaging of the same patient does not show any vascularity with no movement of the collection suggestive of thick pyogenic collection.

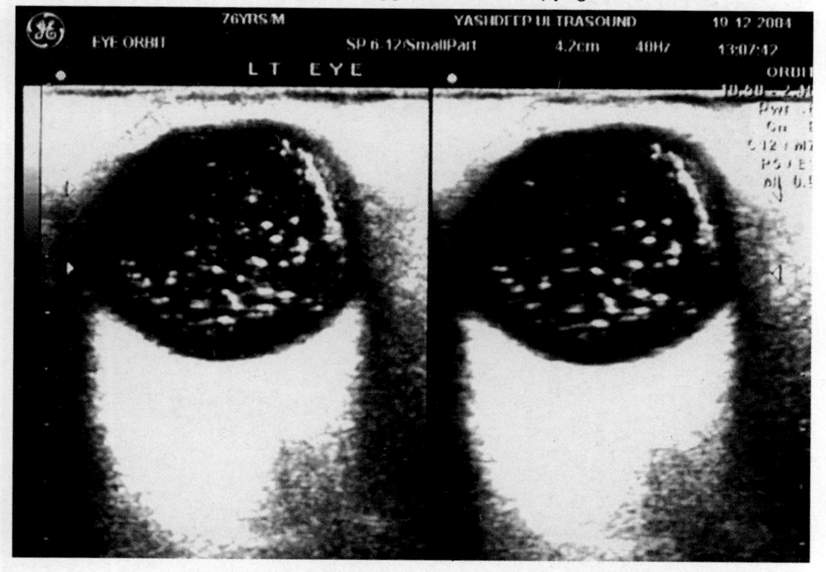

FIGURE 1.92: Asteroid hyalosis (Starry eye). HRSG shows multiple bright echogenic specks in the posterior segment. Calcium specks are seen as bright dots. Dynamic study shows movements of the dots.

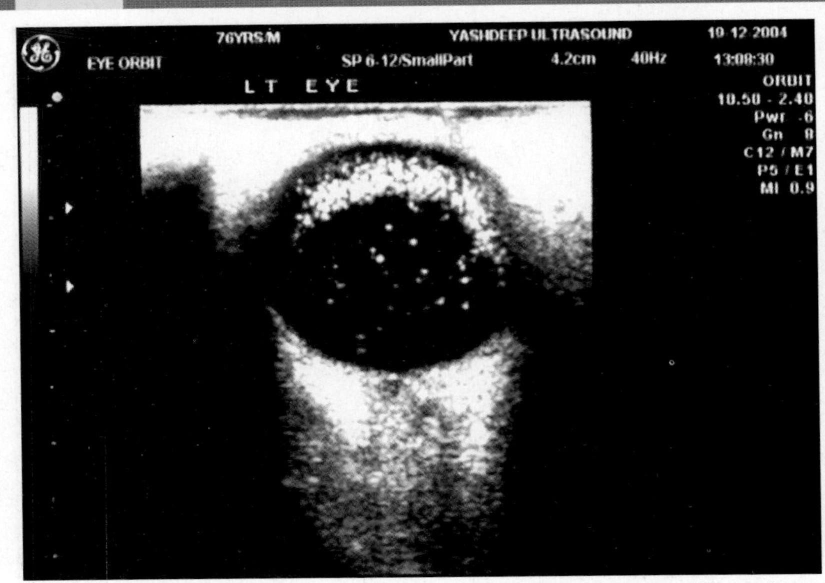

FIGURE 1.93: Bright echoes are seen in the posterior segment arranged in semilunar session in a form of galaxy. It gives typical starry eye appearance.

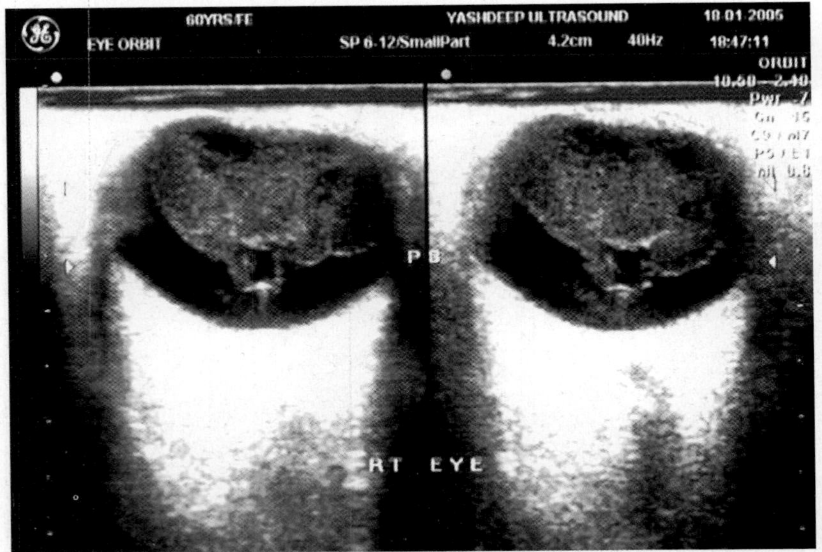

FIGURE 1.94: Degenerative vitreous. Thick membranous vitreous seen with multiple fine interlacing netting pattern forming dense net suggestive of degenerative vitreous in a patient of glaucoma involving both eyes.

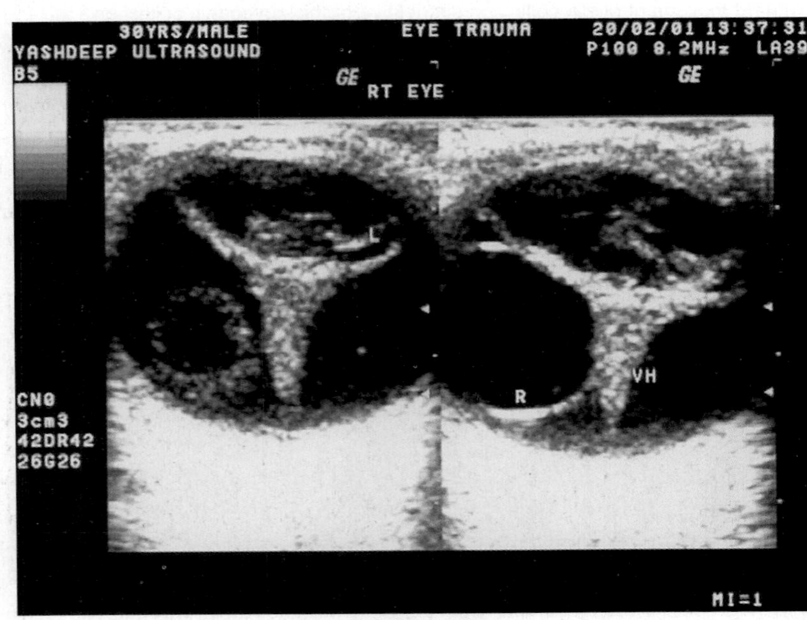

FIGURE 1.95

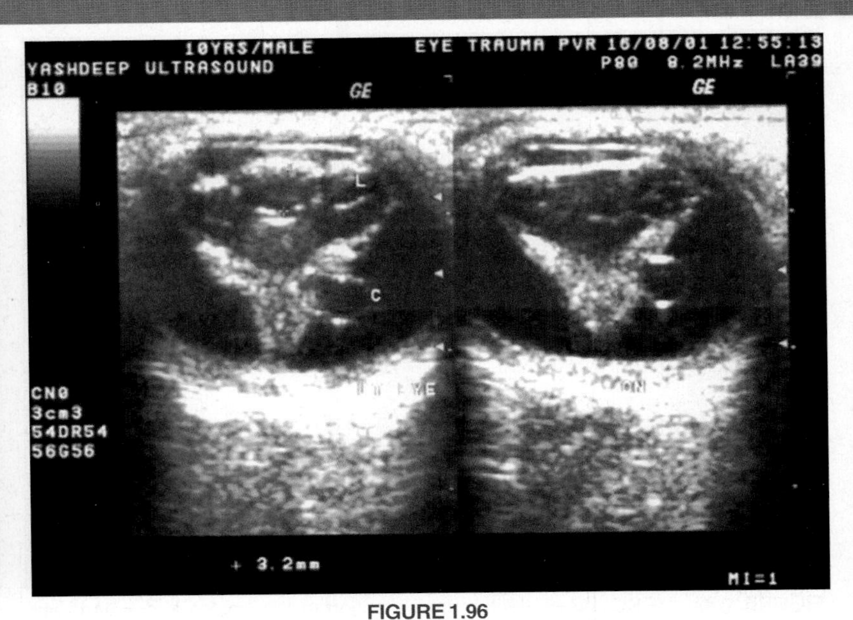

FIGURE 1.96

FIGURES 1.95 and 1.96: Proliferative vitreoretinopathy. HRSG shows typical triangle sign of PVR with multiple vitreoretinal bands and retinal cyst formation.

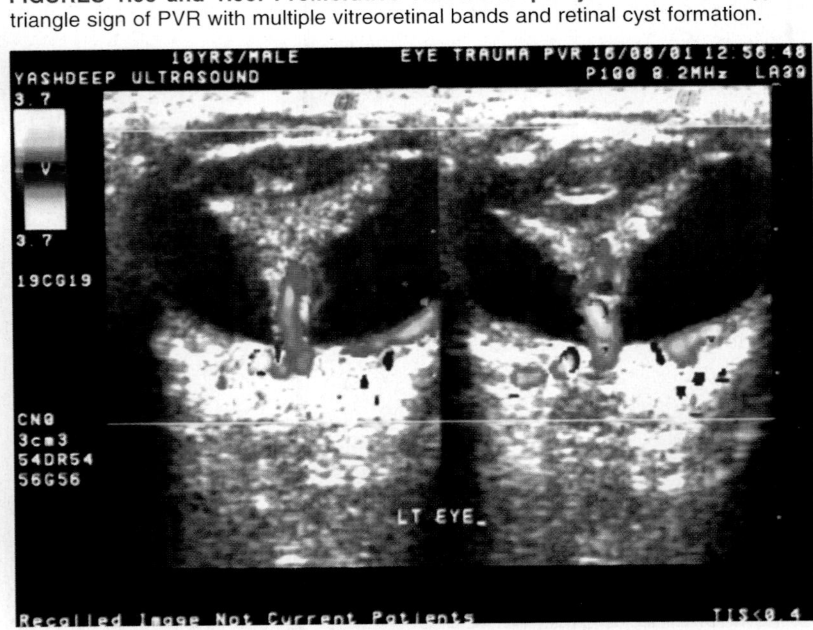

FIGURE 1.97: Proliferative vitreoretinopathy. Color Doppler study shows dragging of the blood vessels in the triangle with typical triangle sign of PVR.

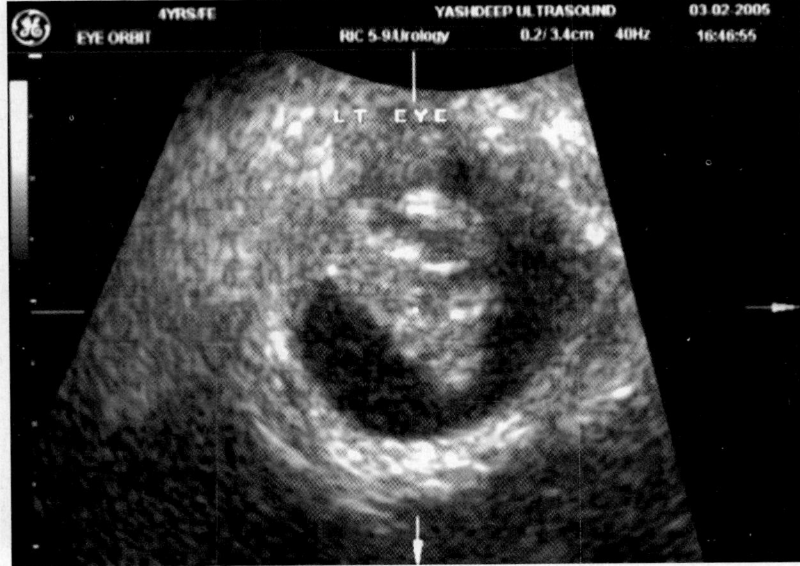

FIGURE 1.98

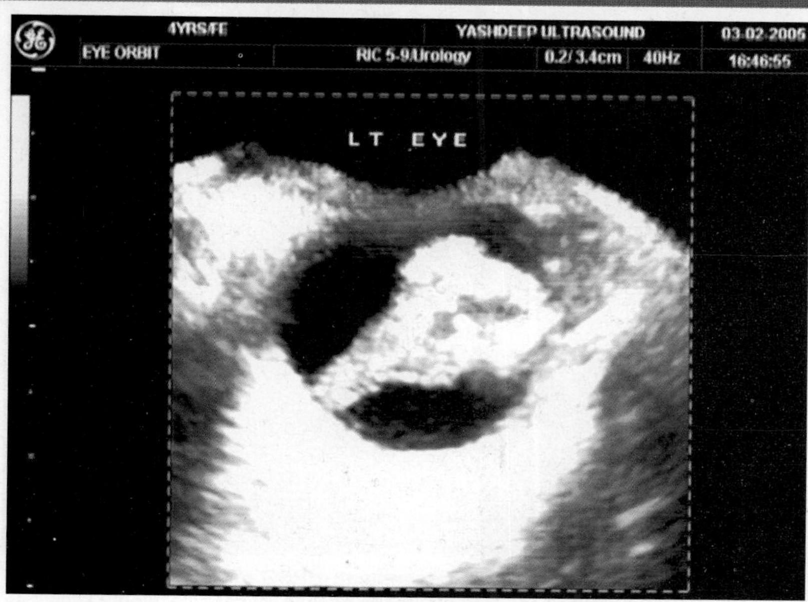

FIGURE 1.99

FIGURES 1.98 and 1.99: 3D imaging of vitreous hemorrhage. 3D imaging of the eye shows details of the retracted clot in old vitreous hemorrhage. No suggestion of any retinal detachment is seen.

change of position also excludes underlying retinal detachment or retinochoroid layer thickening.

The characteristic motion of vitreous and retina normally differentiates their pathologies. However, if posterior hyaloid surface is thickened and attached to the optic nerve head, very careful monitoring and observation are required dynamic scanning to differentiate it from detached retina.

Retina

One of the most important role of HRSG is in evaluating the pathology of retina in opaque ocular media, retinal tears, retinal detachments, retinoschisis and other disorders.

Retinal tears and retinal detachments can be easily evaluated on HRSG. Small retinal tears show flappy movements on dynamic scanning. They are usually present in superior peripheral fundus. They may be associated with posterior vitreous detachment but giant retinal tears show varied presentation and usually associated with severe injury to eye. When an unusual insertion of membrane is seen on HRSG possibility of giant retinal tear is to be kept.

Retinal Detachment

Retina is closely applied to choroid. It is firmly attached at optic nerve head and ora serrata. Retinal detachment (RD) typically produces a bright continuous normally folded membrane on 'B' scan. It is usually attached at optic nerve head and ora serrata giving a funnel shaped appearance. The RD shows little or more restricted movement than posterior vitreous detachment. Extensive detachment may present as funnel shaped membrane, which may be open or close type. Long-standing retinal detachment may develop retinal cyst. It may be partially calcified and cholesterol debris may accumulate in subretinal. Three main type of retinal detachments are seen.

1. *Rhegmatogeneous retinal detachment:* It is caused due to the brake in the continuity of retina due to the weakness in the peripheral retina due to degeneration as in myopia, diabetic retinopathy or vitreoretinal traction in detached vitreous.
2. *Nonrhegmatogeneous retinal detachment:* It is further devided into two types:
   a. Tractional RD—It is caused due to fraction of detached vitreous pulling the retina from the pigmented epithelium through vitreo-retinal bands resulting into angular or fix retinal detachment.
   b. Exudative retinal detachment—It is caused due to collection of fluid in the subvitreal space due to inflammation-uveal effusion or tumors.
3. Exudative retinal detachment (Coats' disease)—Coats disease is a unilateral condition mainly found in children. Usually present in first decade of life. It is more common in males than females. It is to be differentiated from retinoblastoma, retinopathy of prematurity, toxocara and PHPV.

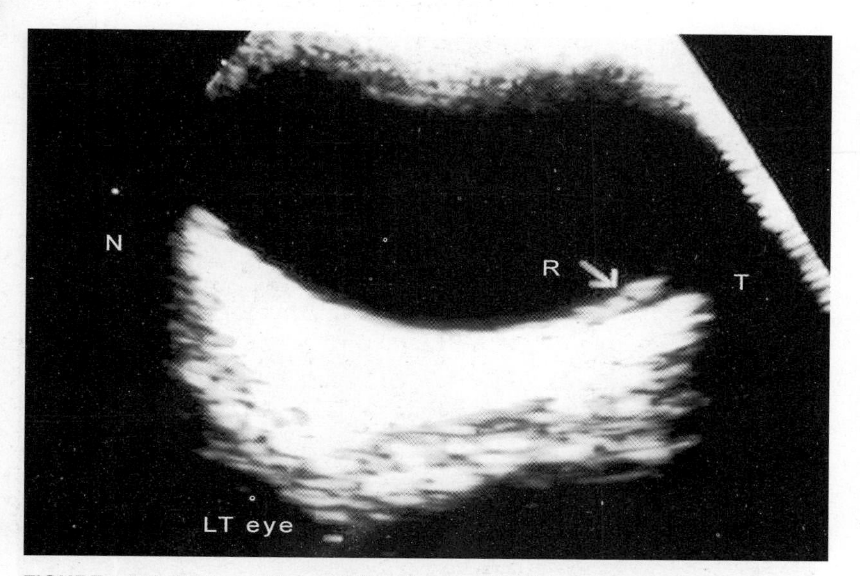

FIGURE 1.100: Retinoschisis. HRSG shows a thin dome shaped membrane in the temporal inferior quadrant. It is more focal, smooth and thin in outline and not seen inserting at the optic nerve head suggestive of retinoschisis. In this condition retina splits into two thin layers. One layer is lifted up and another layer is seen down in relation to the choroid.

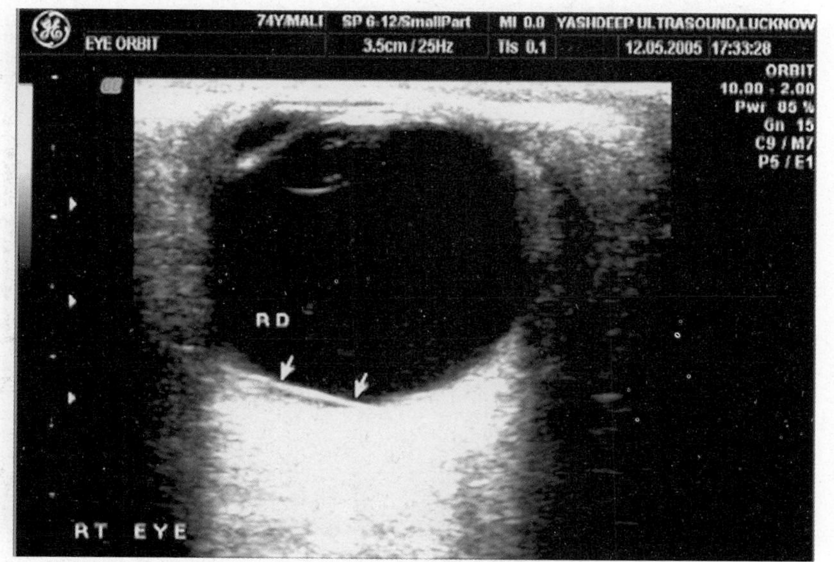

FIGURE 1.101: Retinoschisis. HRSG shows a thin echogenic membrane lifted up in the inferiotemporal quadrant of eye. It is smooth in outline suggestive of retino schisis.

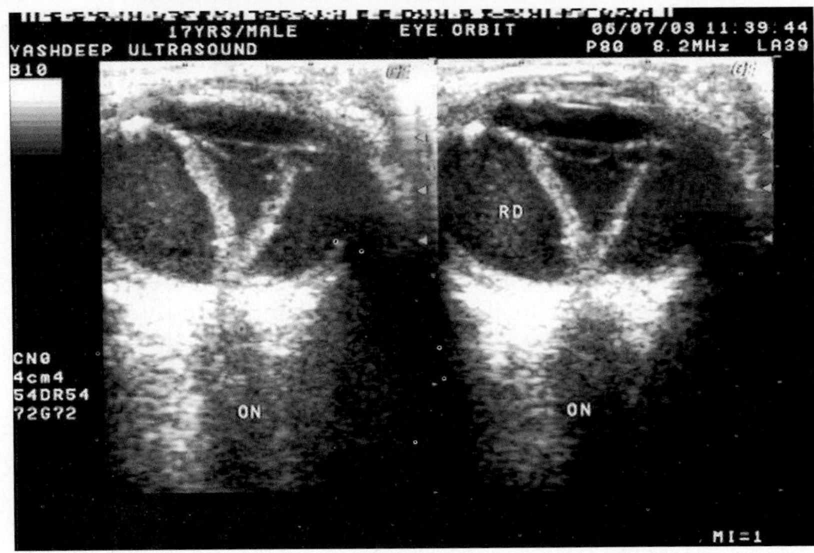

FIGURE 1.102

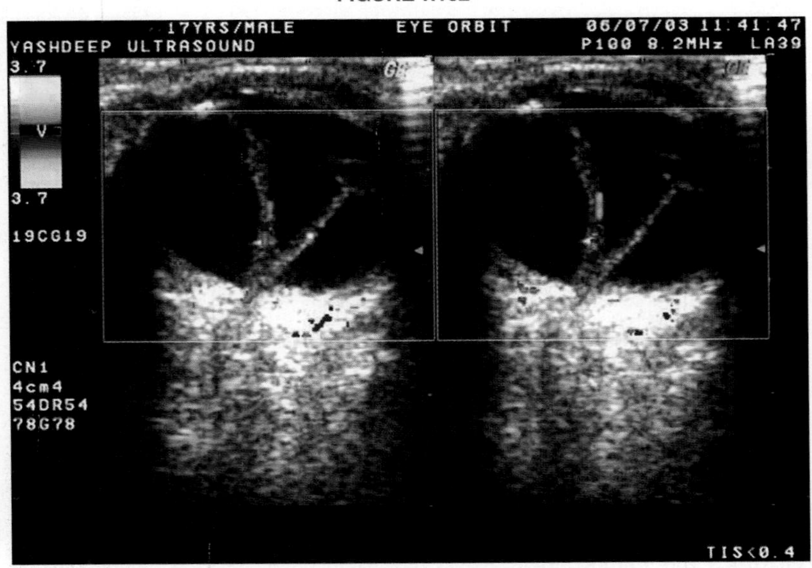

FIGURE 1.103

FIGURES 1.102 and 1.103: Retinal detachment. HRSG shows a typical 'V' shaped echogenic membrane in the posterior segment of the eye. This 'V' shape is due to insertion of the retina at optic nerve and ora serrata. On dynamic scanning little movement is seen due to the insertion of retina at ora serrata at optic nerve head.

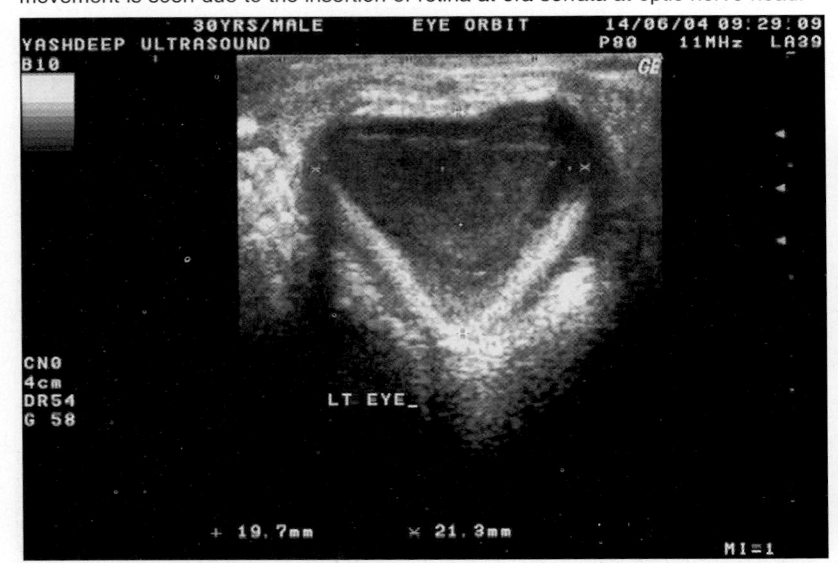

FIGURE 1.104: Tent like retinal detachment. HRSG shows a typical tent like retinal detachment caused due to the traction over the retina. The apex of the tent is seen as the point like attachment at the optic nerve head and wide base of the tent is seen fixed at ora serrata.

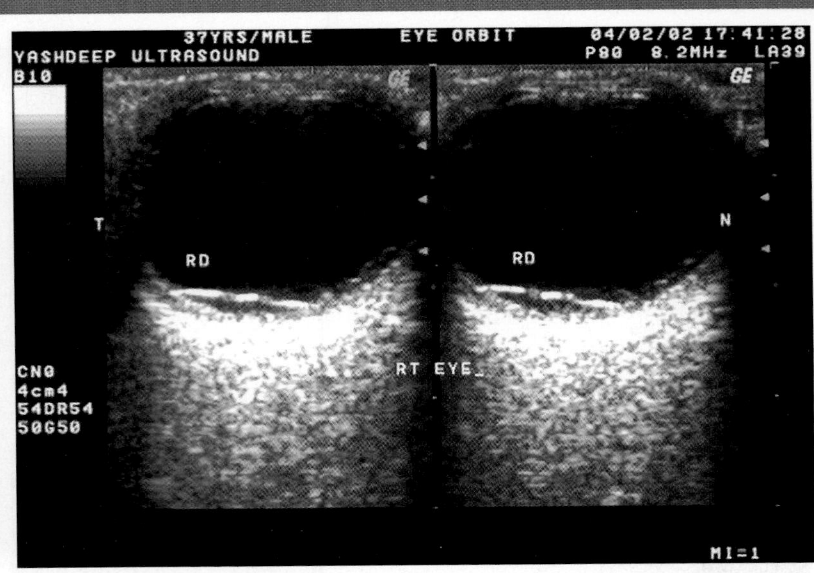

FIGURE 1.105: Table top retinal detachment. HRSG shows broad wide retinal detachment with the wide base also due to the traction retinal detachment known as table top detachment. Multiple vitreoretinal bands are also seen.

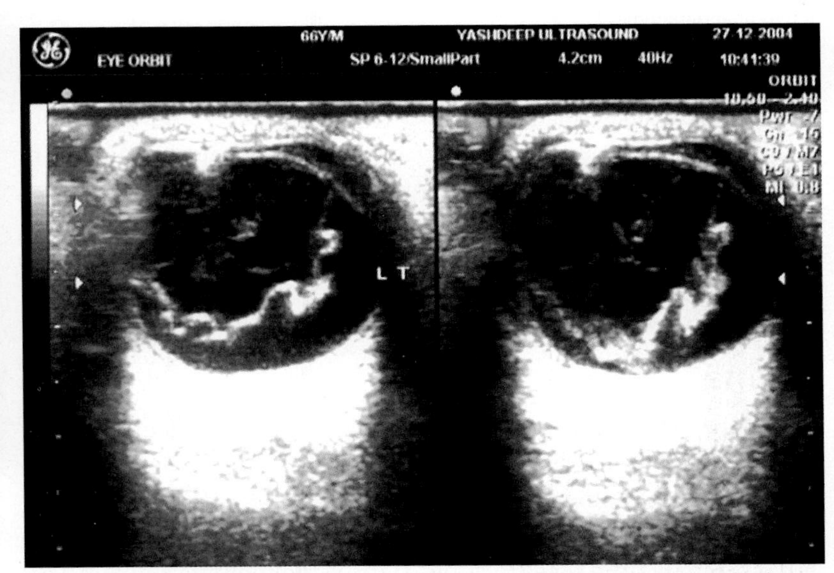

FIGURE 1.106: Giant retinal tear. HRSG shows a thick echogenic membrane detached from its surface from all side with folding of the membrane suggestive of giant retinal tear. There is also evidence of echogenic collection seen in subretinal space suggestive of subretinal hemorrhage.

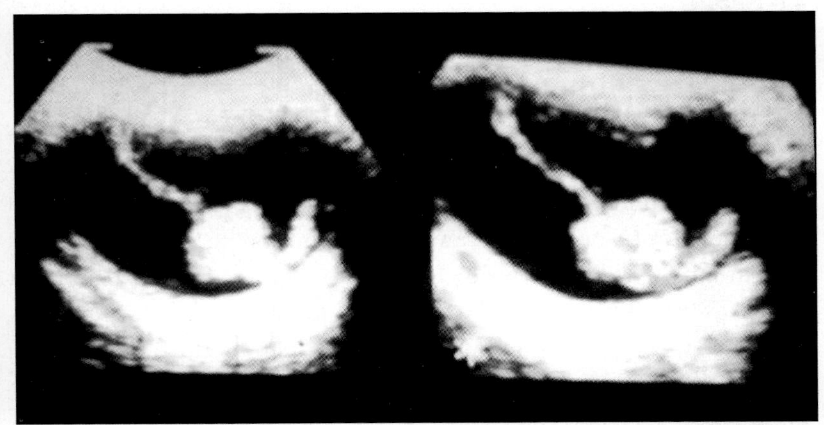

FIGURE 1.107: Exudative retinal detachment (Coats disease). HRSG shows thick echogenic membrane detached from its surface. There is evidence of large intra-retinal and subretinal exudative collection seen responsible for the detachment. Typical finding of Coats disease is found in children. It is unilateral condition.

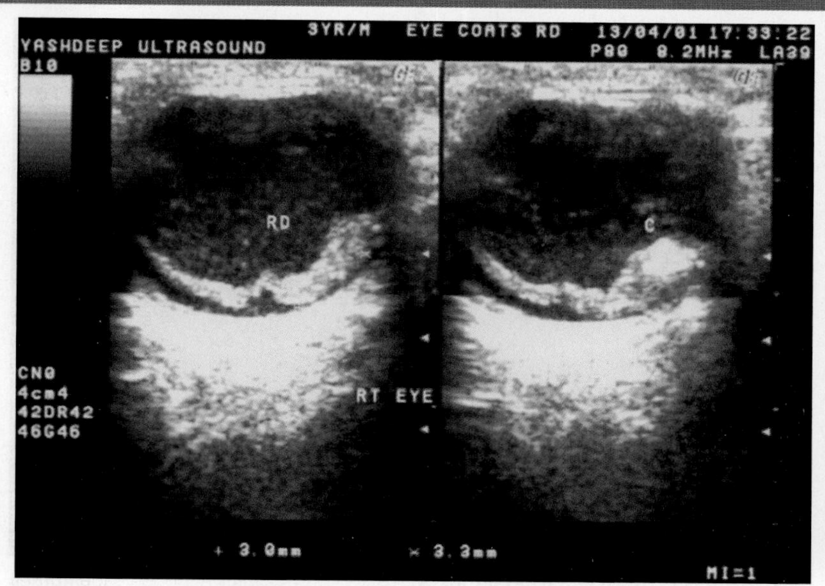

FIGURE 1.108: Exudative Coats disease. Thick echogenic membrane is seen detached from its surface. Echogenic exudative collection is also seen in subretinal space and also in intraretinal space between the two layers of retina suggestive of Coats disease.

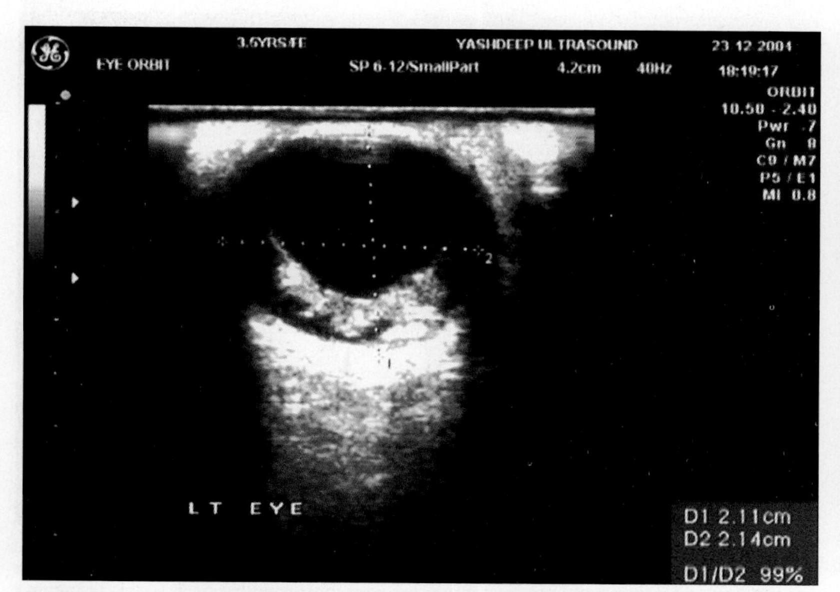

FIGURE 1.109: Exudative RD. HRSG shows thick echogenic membranous shadow detached from the posterior surface suggestive or retinal detachment. Dense exudation is seen attached with the retina typical of Coats disease.

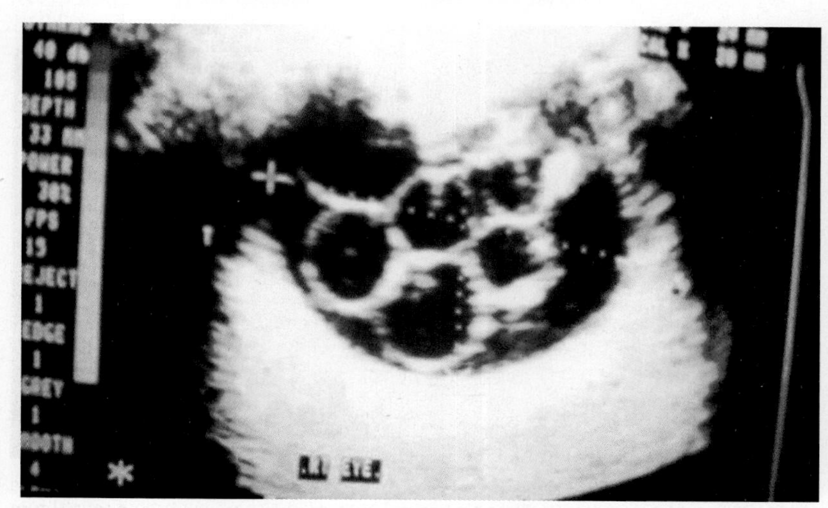

FIGURE 1.110: Chronic retinal tear or retinal cysts. HRSG shows multiple retinal cysts formation in the posterior segment due to the folding of the retina in chronic long standing tear.

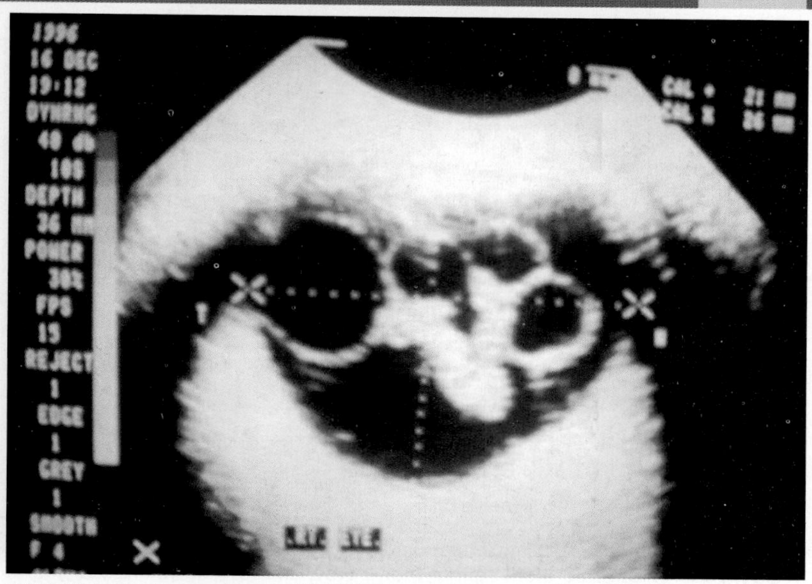

FIGURE 1.111: Retinal cysts. HRSG shows multiple retinal cysts formed in chronic retinal tear with calcification of retina are also seen in the posterior segment.

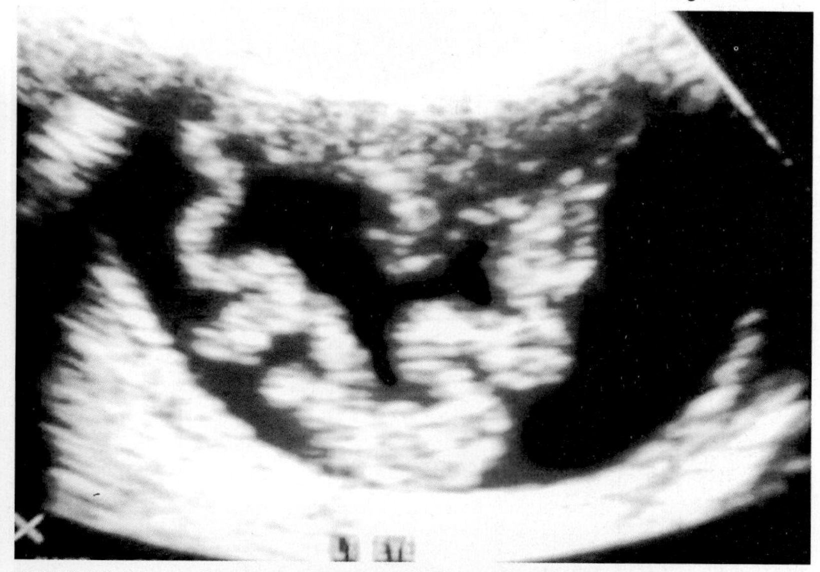

FIGURE 1.112: Coiled retinal tear. HRSG shows a thick echogenic membrane detached from its surface in a serpentine fashion suggestive of chronic retinal detachment. It shows limited movement on dynamic scanning. Associated intragel hemorrhage is also seen.

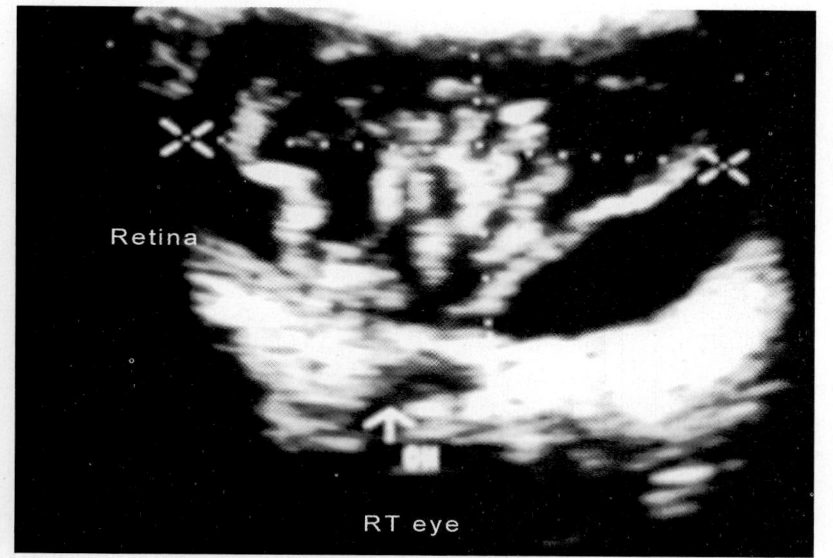

FIGURE 1.113: Chronic multilayered retinal tear. HRSG shows a chronic retinal detachment where retina has detached from its surface from all side and arranged itself into multiple layers in a cone fashion. It is a feature of chronic retinal detachment and also one of the type of the serpentine tear.

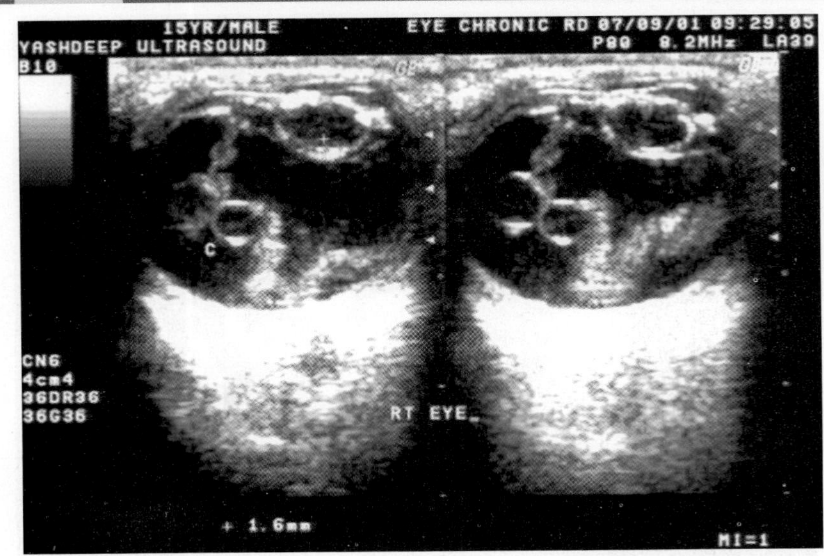

FIGURE 1.114: Chronic giant retinal tear. HRSG shows giant retinal tear due to long standing cataract. The retina is detached completely and seen as a thick echogenic membrane. Part of the retina shows cyst formation. Lens is thickened and opaque.

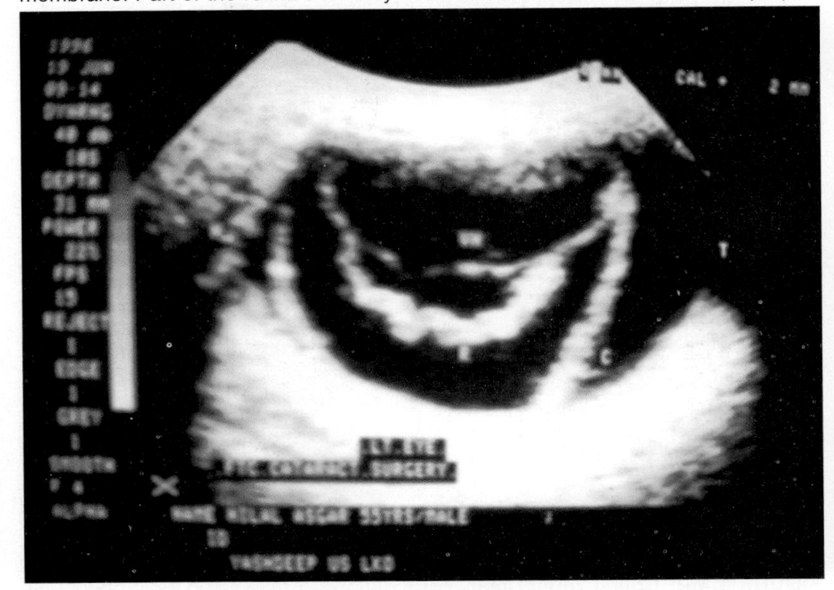

FIGURE 1.115: Chronic retinal tear. HRSG shows chronic retinal tear where retina is seen detached from its surface. In the center calcification is seen in the retina. There is also thick echogenic membrane seen at the periphery suggestive of an associated choroidal tear.

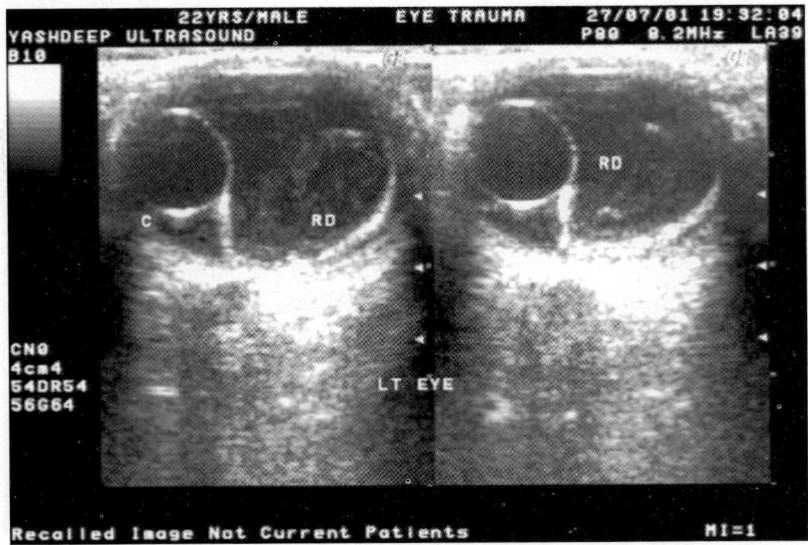

FIGURE 1.116: Giant retinal tear with retinal cyst formation. HRSG shows total retinal detachment with folding of the retina forming the retinal cyst. However, part of the retina is seen attached at the optic nerve head.

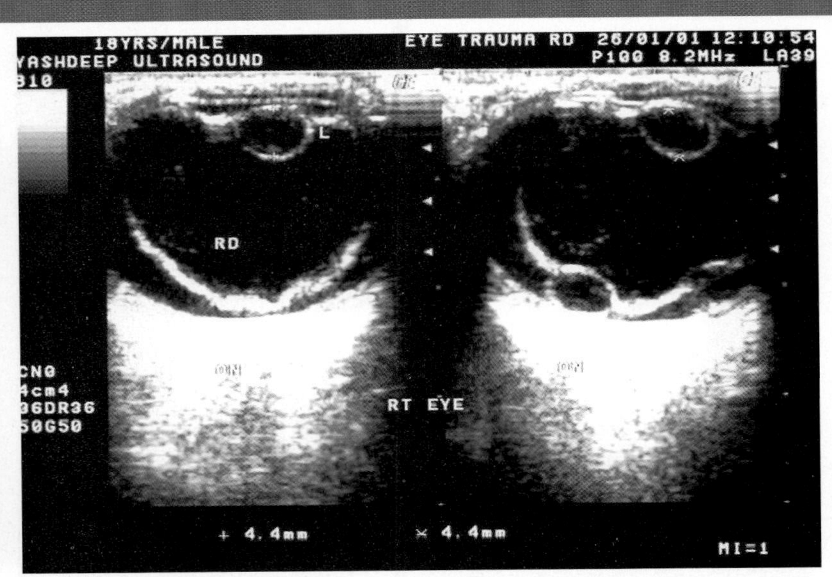

FIGURE 1.117: Total retinal detachment. HRSG shows a thick echogenic membrane detached from all side of the globe with irregular wavy outline and retinal cyst formation. Lens is also thickened and opaque due to the following blunt trauma to the eye.

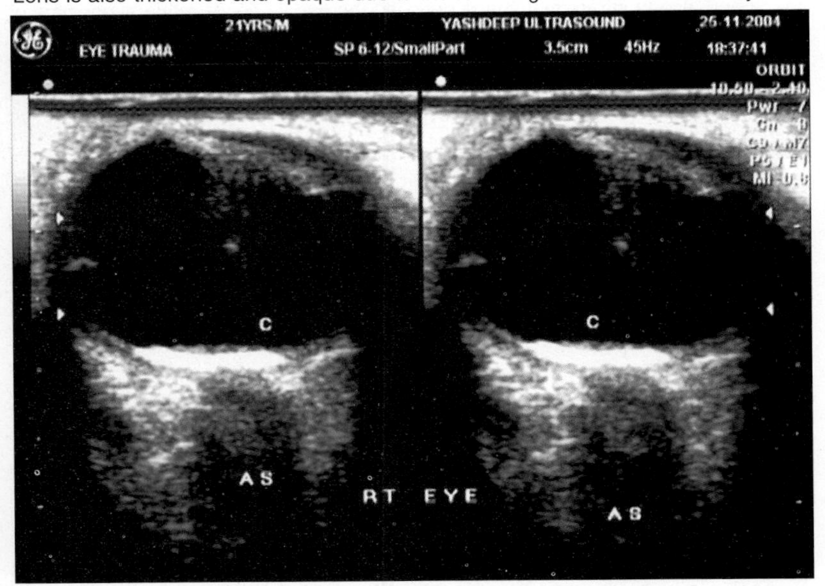

FIGURE 1.118

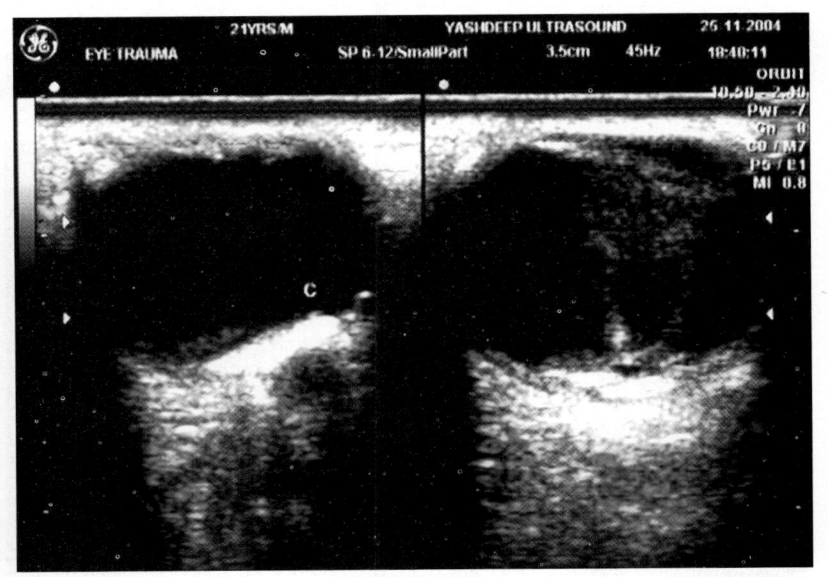

FIGURE 1.119

FIGURES 1.118 and 1.119: Chronic retinal tear. HRSG shows chronic retinal tear in an old case of retinal detachment. Dense linear calcification is seen in the retinal membrane. It is accompanied with dense acoustic shadowing.

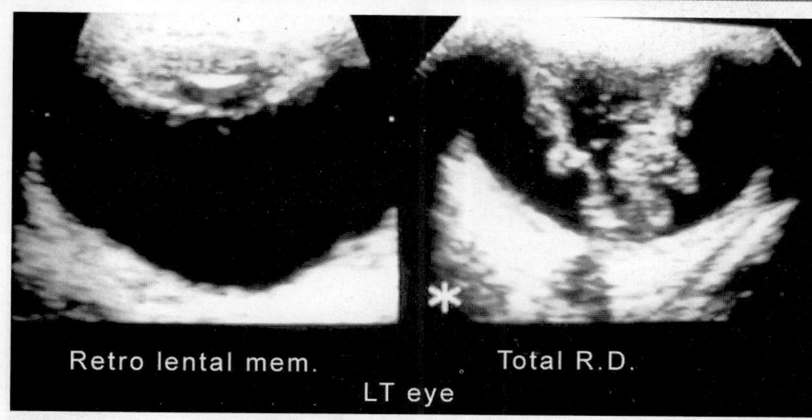

FIGURE 1.120: Giant retinal tear in operated case of congenital cataract. HRSG shows a thick retrolental membrane in a case of operated congenital cataract. Posterior segment of the eye also shows an echogenic-thickened membrane, which is showing complete detachment and partial folding suggestive of chronic retinal detachment.

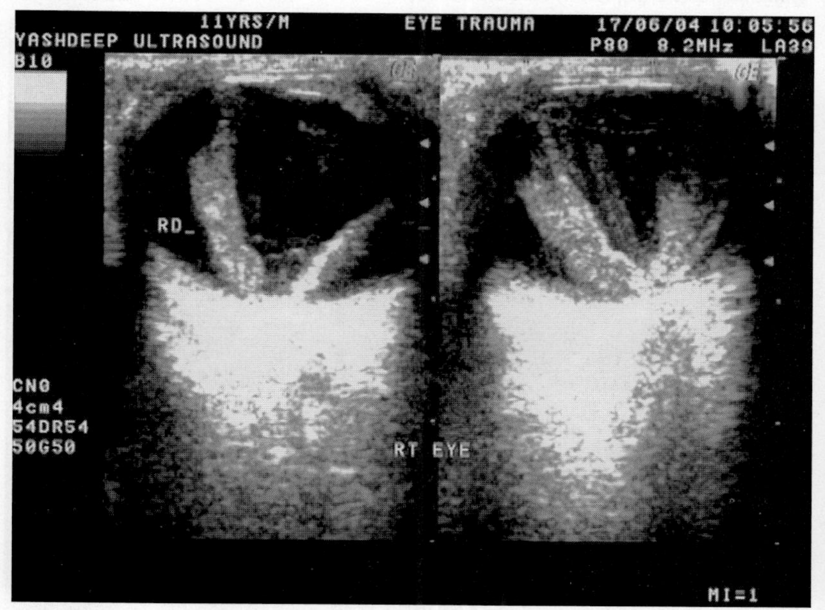

FIGURE 1.121: Chronic retinal detachment. HRSG shows a thick membranous shadow with open funnel shaped pattern. Which is seen attached at the optic nerve suggestive of tractional retinal detachment.

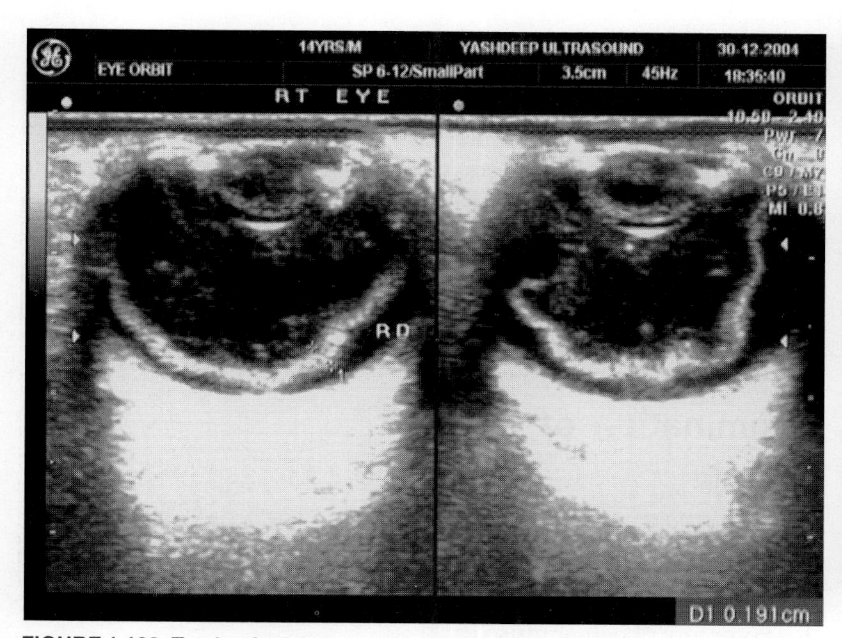

FIGURE 1.122: Total retinal detachment. HRSG shows complete retinal detachment from all side. It shows undulating movements on dynamic scanning.

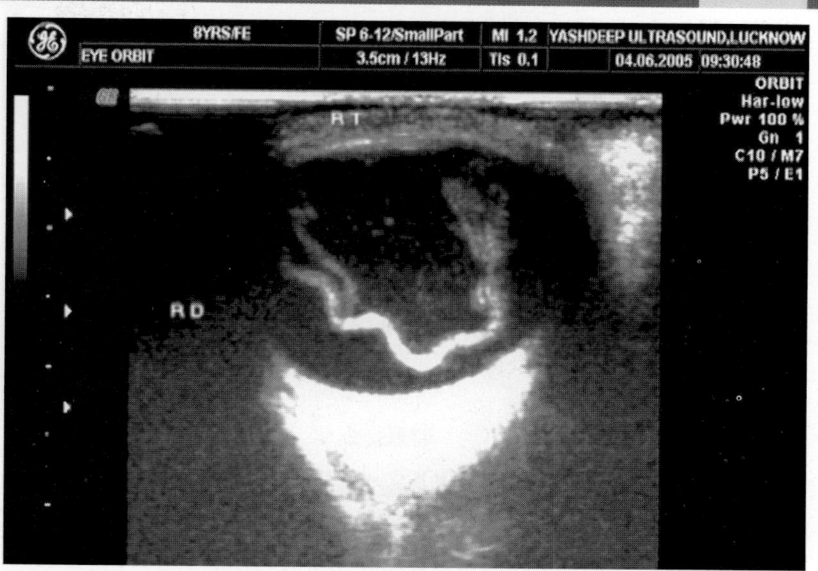

FIGURE 1.123: Total RD. Same case after dynamic scanning shows total RD.

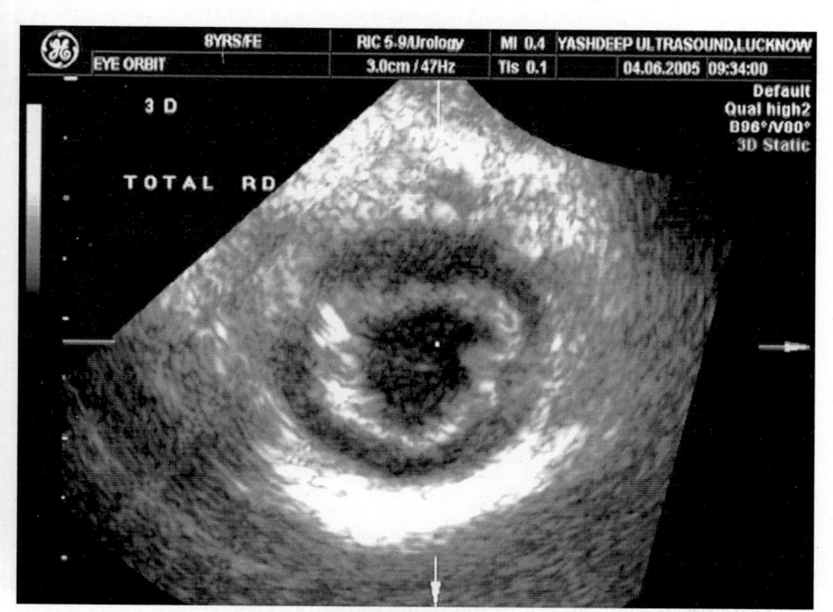

FIGURE 1.124

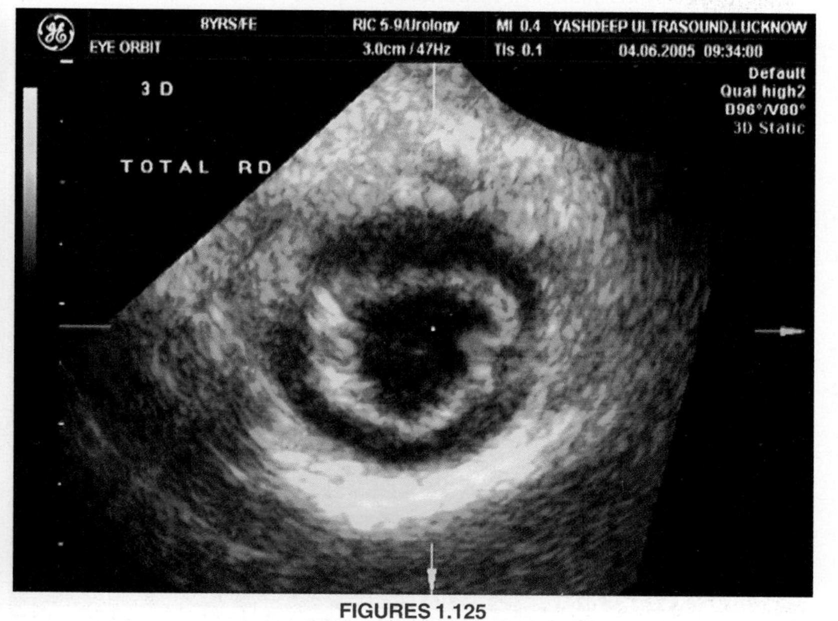

FIGURES 1.125

FIGURES 1.124 and 1.125: 3D imaging of total RD. 3D imaging shows finer details of the total RD with depth of the globe.

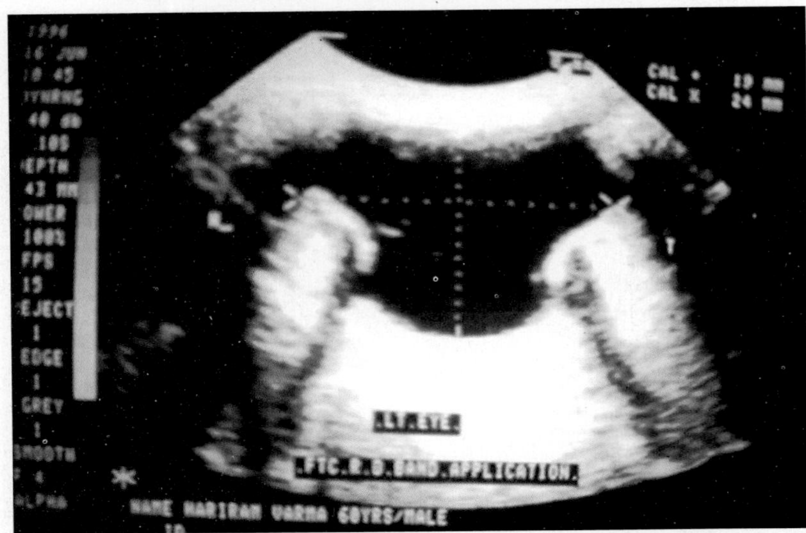

FIGURE 1.126: Retinal buckle applications in a case of retinal detachment. HRSG shows buckle applications at the equator in a case of retinal detachment. The buckles are seen in position and also keeping the retina close to the choroid. HRSG is a valuable investigation for the evaluation of success of post-surgical treatment of retinal detachment.

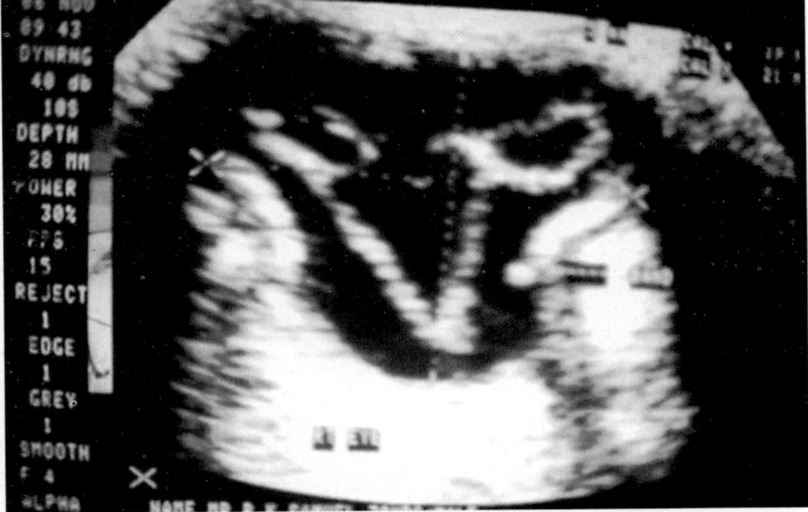

FIGURE 1.127: Slipping out of the retina under buckles. HRSG shows slipping out the retina from the buckle in a case of failed surgical treatment of retinal detachment. The slipped retina is seen as a 'V' shaped membrane and folding of the retina is also seen at the base. The buckles are seen in position at the equator as echogenic bands applied at the equator.

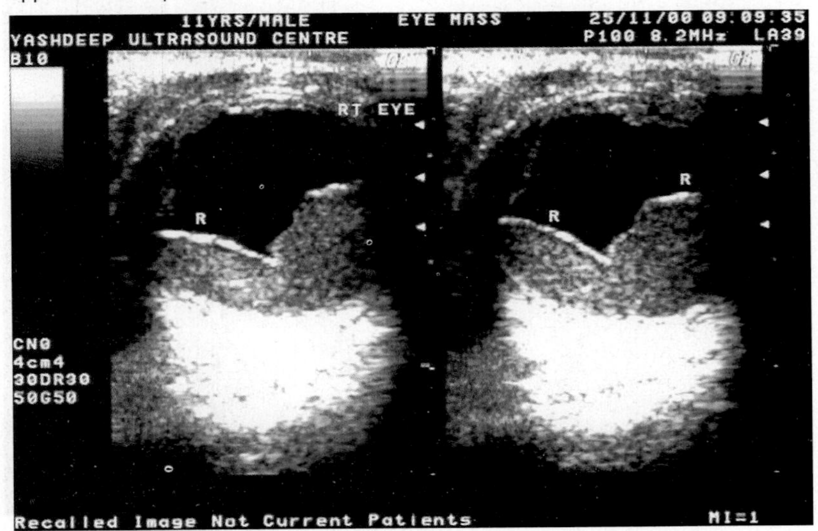

FIGURE 1.128: Giant retinal tear with subretinal hemorrhage. HRSG shows a 'V' shaped giant retinal tear fixed at ora serrata and dense low level echo collection seen in the subretinal space suggestive of subretinal hemorrhage. The vitreous cavity does not show any intragel hemorrhage.

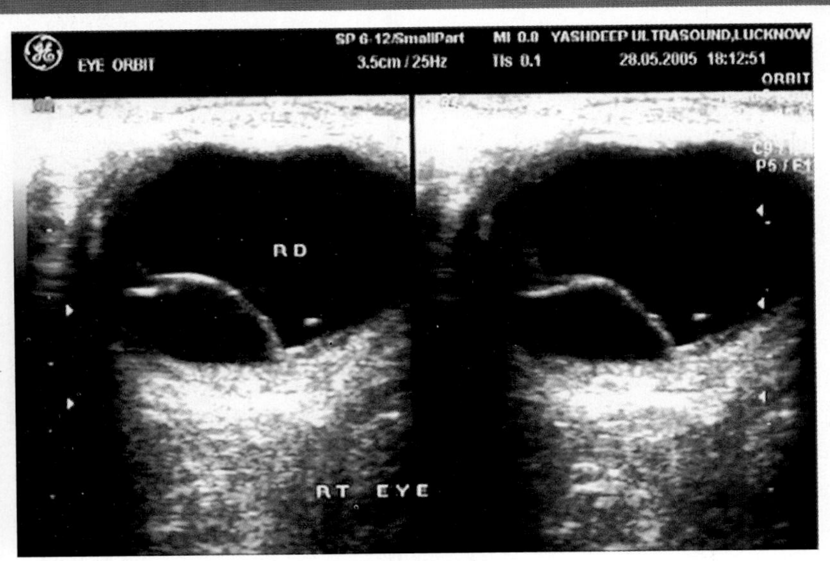

FIGURE 1.129A

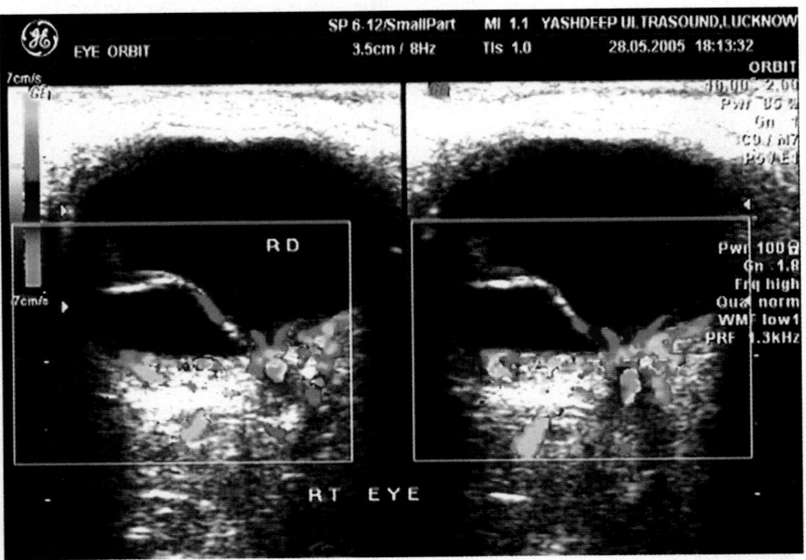

FIGURE 1.129B

FIGURES 1.129A and B: Giant retinal tear. HRSG shows a big retinal tear in the eye. On color flow imaging dragging of the vessels along with the tear is seen.

Proliferative Vitreoretinopathy (PVR)

In proliferative vitreoretinopathy thin membranes are formed on inner retinal surface and posterior hyaloid interface. Contraction of the membrane leads to gel retraction, retinal traction and hemolization. In moderate PVR, HRSG shows flappy movements of retina. But in a severe PVR funnel like retina is seen at postequatorial retina. A fibrous membrane is seen starching across the anterior retina giving a classical triangle appearance. It is an important indication with poor surgical prognosis.

Examination of Choroid

Choroid

Choroidal thickening may be focal or diffuse. It is associated with many conditions. It can be due to swelling or edema. Following are the main causes of choroidal swelling or edema:

1. Hypotoni of ocular muscles.
2. Vascular congestion
3. Endophthalmitis.

4. Uveitis.
5. Scleritis.

HRSG shows a highly reflective or low to medial reflective band in choroidal thickening. However, small choroidal tumors are difficult to be differentiated from it like choroidal melanomas, metastatic deposit or lymphomas.

Choroidal Detachment

The choroidal detachment may take place due to trauma, post-surgery or spontaneously. HRSG findings are characteristic in choroidal detachment (CD) on B scanning. It presents as a smooth dome shaped thick echogenic membrane in the periphery. It shows little or no movement on dynamic scanning. In shallow CD it is more flat than dome shape. Extensive choroidal detachment shows classical kissing bullae sign on 'B' scanning. Anterior choroidal detachment extends to ciliary body and known as cilio choroidal detachment.

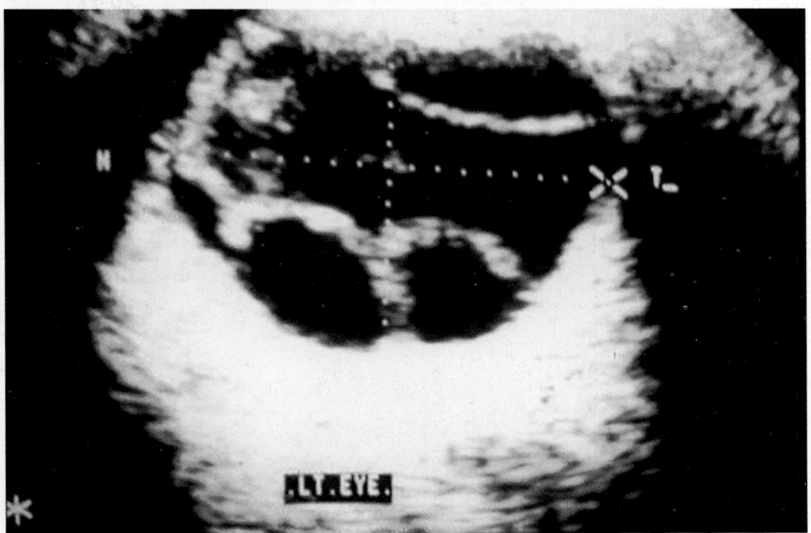

FIGURE 1.132: Multiple choroidal detachment. HRSG shows multiple thick echogenic dome shaped membranous shadows seen at the periphery in the posterior pole, upper pole and nasal side of the Lt eye with smooth outline suggestive of multifocal choroidal detachment following trauma. Associated intragel hemorrhage is also seen.

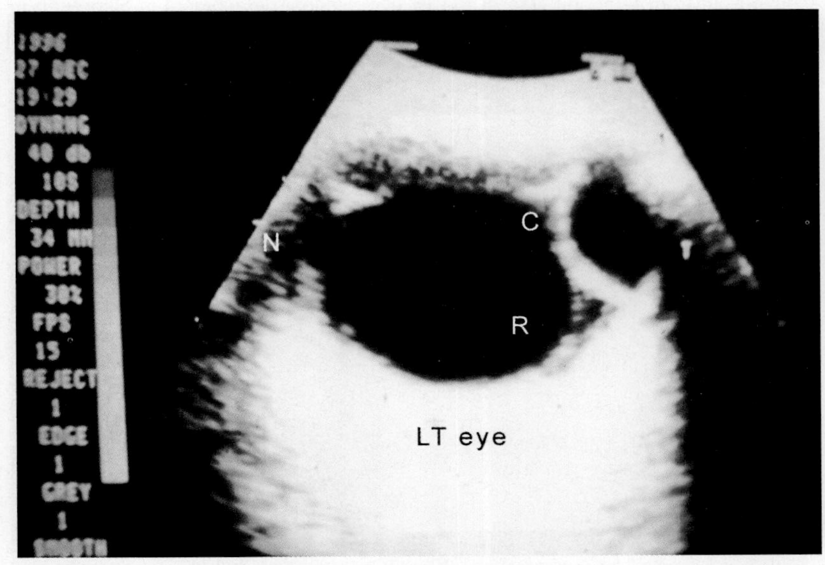

FIGURE 1.130: Choroidal detachment. HRSG shows a thick echogenic membrane due to trauma in the upper temporal quadrant of the Lt eye. It is smooth in outline and dome shaped seen at the periphery. No movement is seen on dynamic imaging.

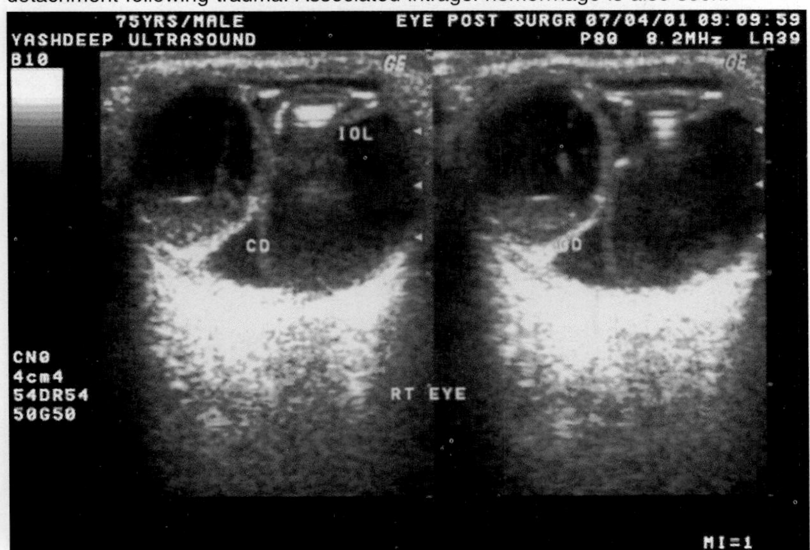

FIGURE 1.133

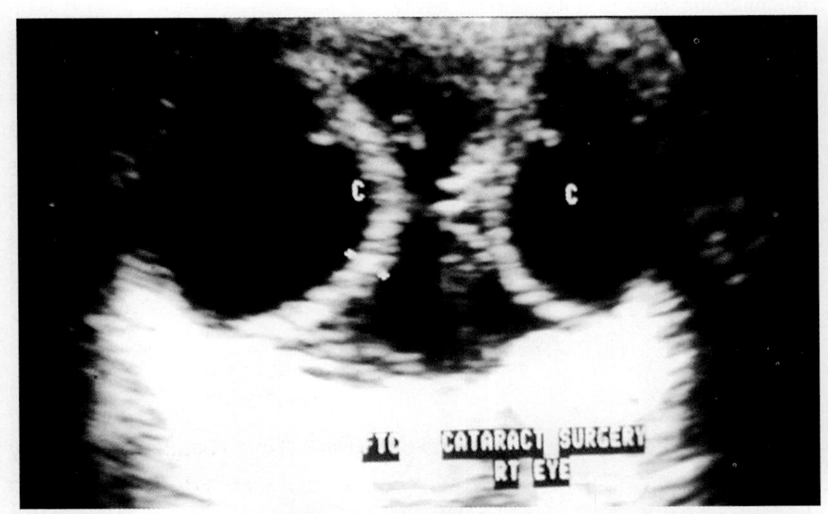

FIGURE 1.131: Postsurgical choroidal detachment. Patient complaints of sudden loss of vision after the surgery for cataract. HRSG shows two well-defined thick echogenic dome shaped membranes seen detached on both temporal and nasal side at the periphery. They are seen coming into the center and give a kissing bullae sign. Associated intragel hemorrhage is also seen.

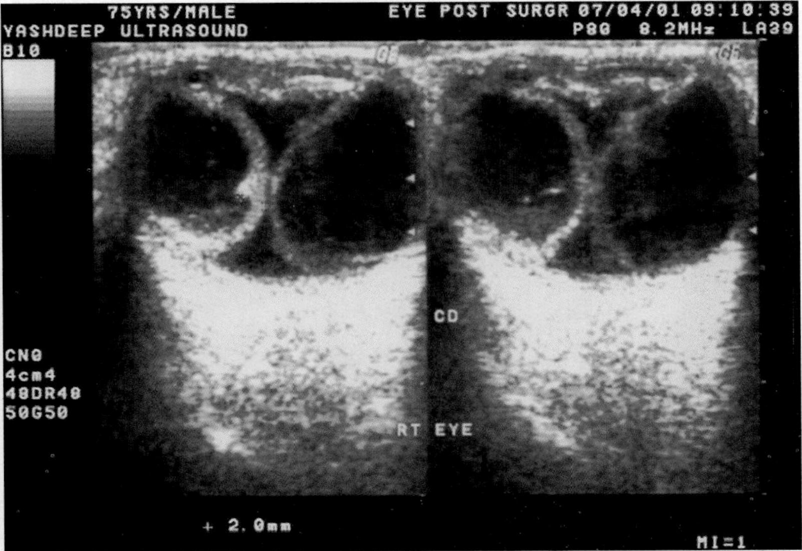

FIGURE 1.134

FIGURES 1.133 and 1.134: Post-surgical choroidal detachment following surgery. HRSG shows well defined dome shaped echogenic membranous shadows coming into the center from the periphery suggestive of choroidal detachment. An echogenic hyper-reflective shadow is seen at the junction of anterior and posterior segment suggestive of intraocular lens in position.

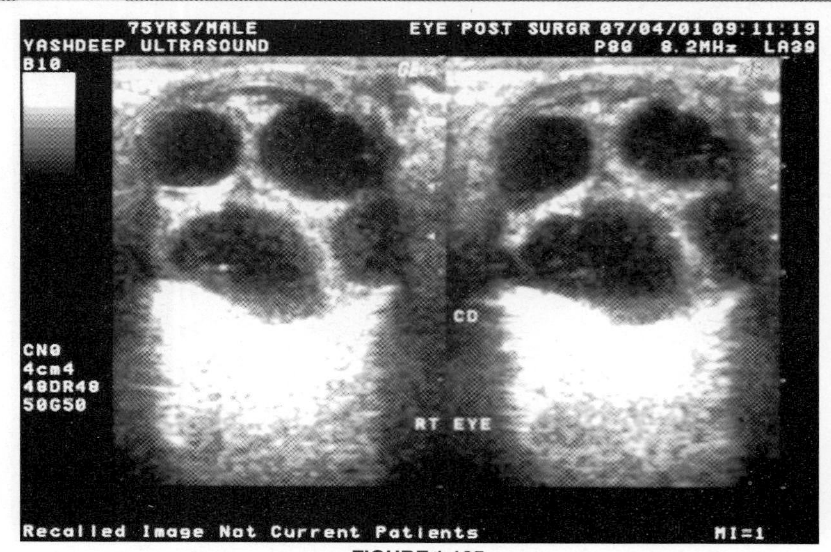

FIGURE 1.135

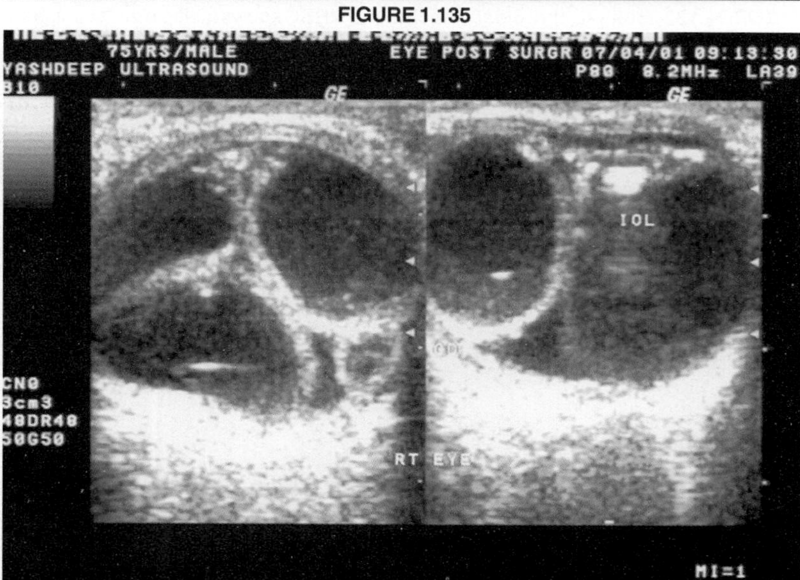

FIGURE 1.136

FIGURES 1.135 and1.136: Multifocal choroidal detachment. HRSG shows multiple dome shaped thick echogenic membranous shadows coming out from all side of the periphery to the center in the globe. The membranes are smooth in outline and show limited movement suggestive of total choroidal detachment. IOL is seen in position.

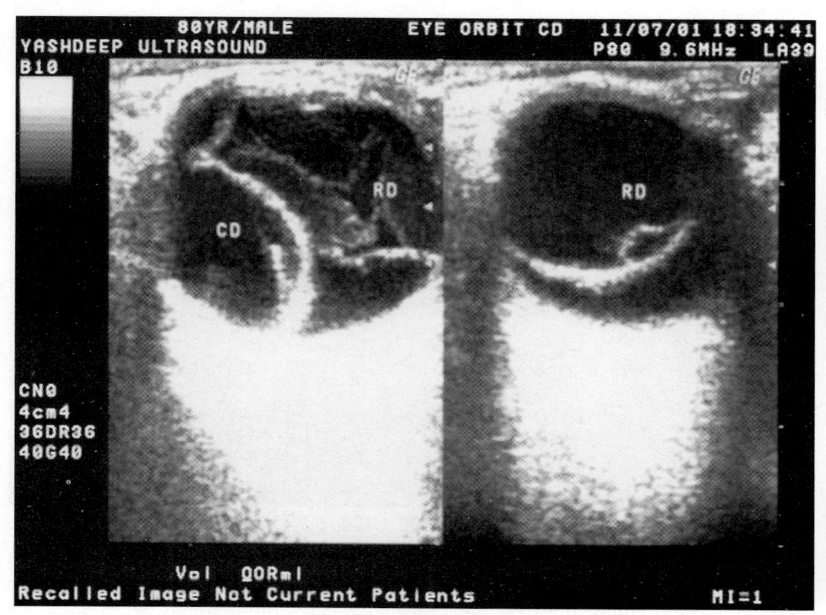

FIGURE 1.137

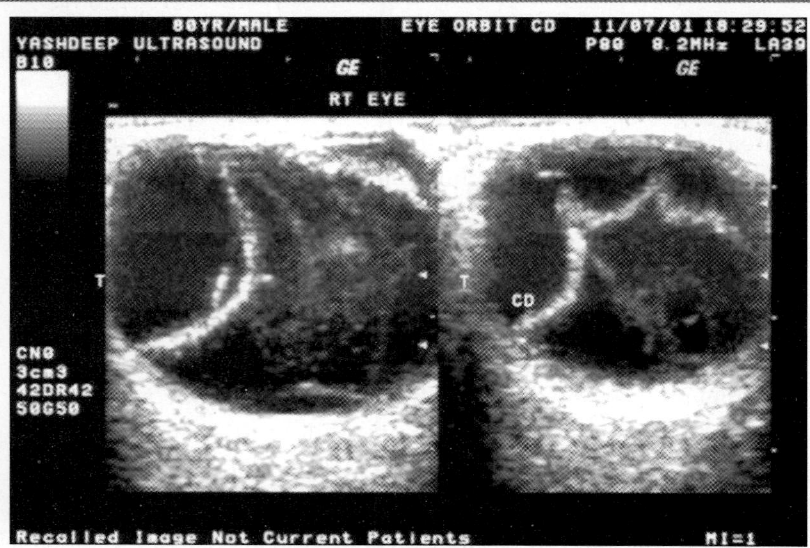

FIGURE 1.138

FIGURES 1.137 and 1.138: Choroidal detachment with associated retinal detachment. HRSG shows a typical dome shaped echogenic membranous shadow coming out from temporal side of the Rt eye. Retinal detachment is also seen in association to the choroidal detachment. It shows limited movement on dynamic scanning in the detached retina.

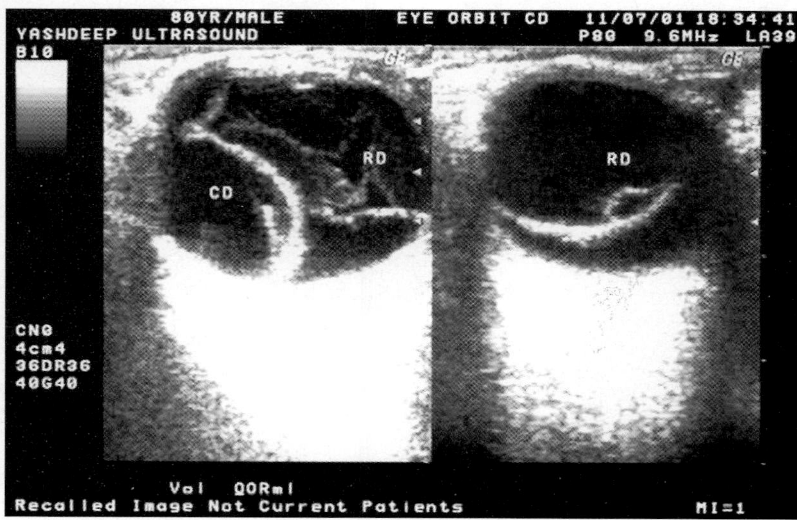

FIGURE 1.139: Choroidal detachment, RD with vitreous hemorrhage. HRSG shows dome shaped choroidal detachment, wavy retinal detachment in the same patient of Figure 1.137. However, associated intragel hemorrhage is also seen as a membranous shadow in the vitreous cavity.

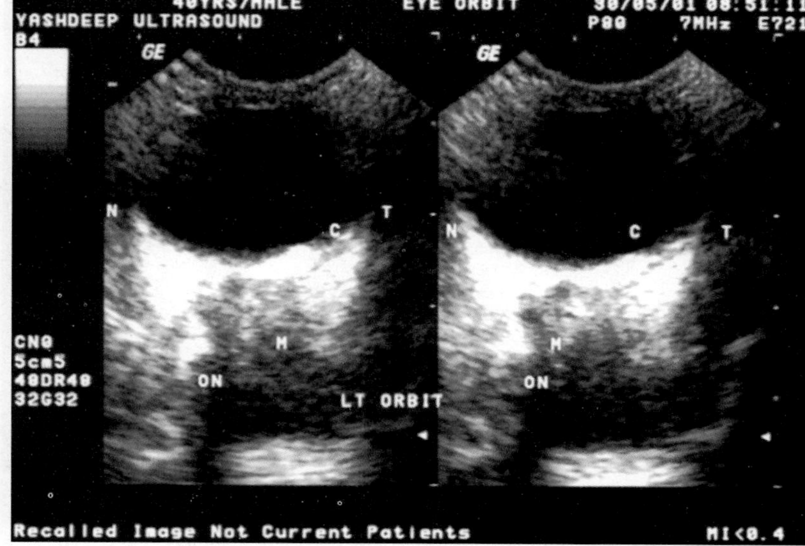

FIGURE 1.140

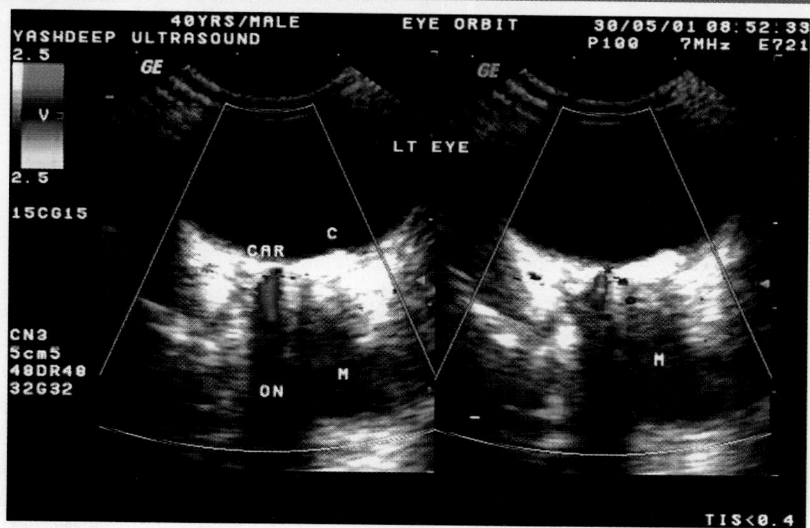

FIGURE 1.141

FIGURES 1.140 and 1.141: Choroid osteoma. HRSG shows an echogenic calcified raised shadow at the posterior pole of the eyeball. It is accompanied with dense acoustic shadowing. The shadow masks the details in the retrobulbar space suggestive of choroid osteoma.

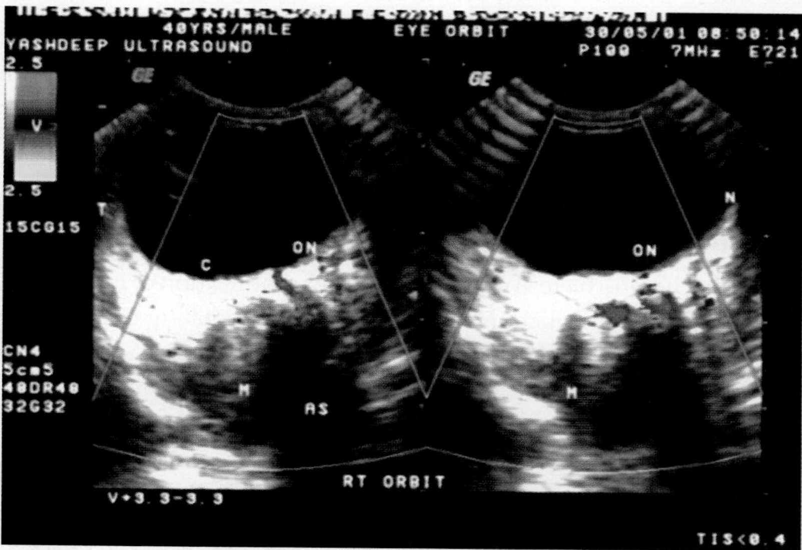

FIGURE 1.142: On color Doppler imaging in the same patient in Figure 1.141, little flow is seen in the calcified osteoma. However, central artery of retina shows normal flow. But choroidal flow is decreased causing arterial insufficiency.

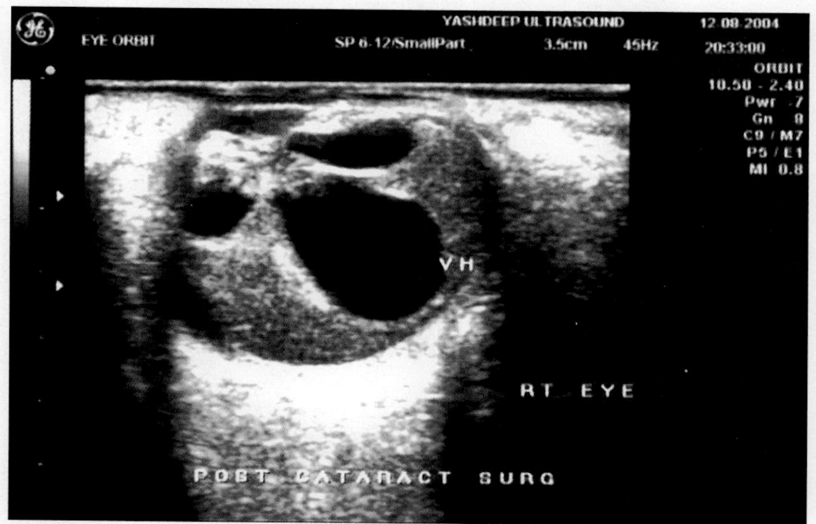

FIGURE 1.143: Choroidal detachment with expulsive hemorrhage. A patient following cataract surgery complained of sudden loss of vision and severe pain in the eye. HRSG shows dome shaped echogenic membranes coming out from the periphery on temporal and nasal side suggestive of choroidal detachment. Echogenic collection is seen under the membrane suggestive of expulsive hemorrhage on both sides.

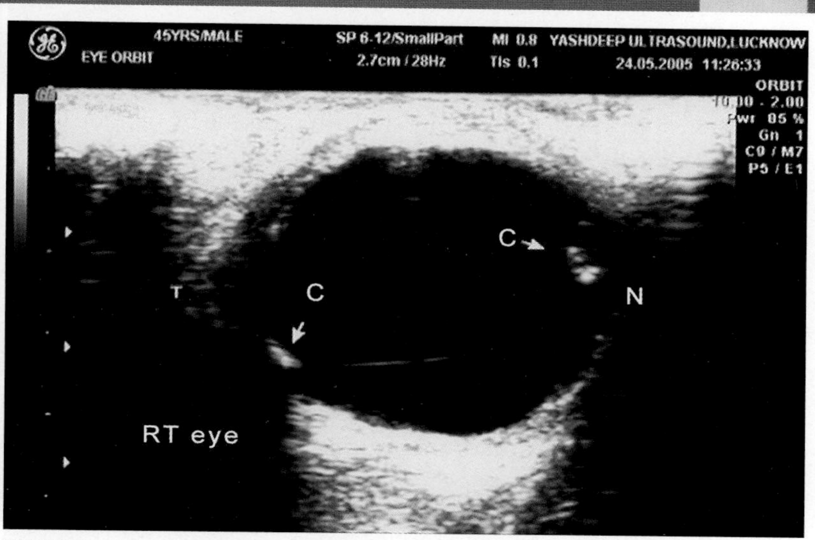

FIGURE 1.144: Choroidal calcification. A patient presented with gradual loss of vision. HRSG shows multiple fine calcified specks fixed with the choroid. They are seen as bright echogenic spots in choroid.

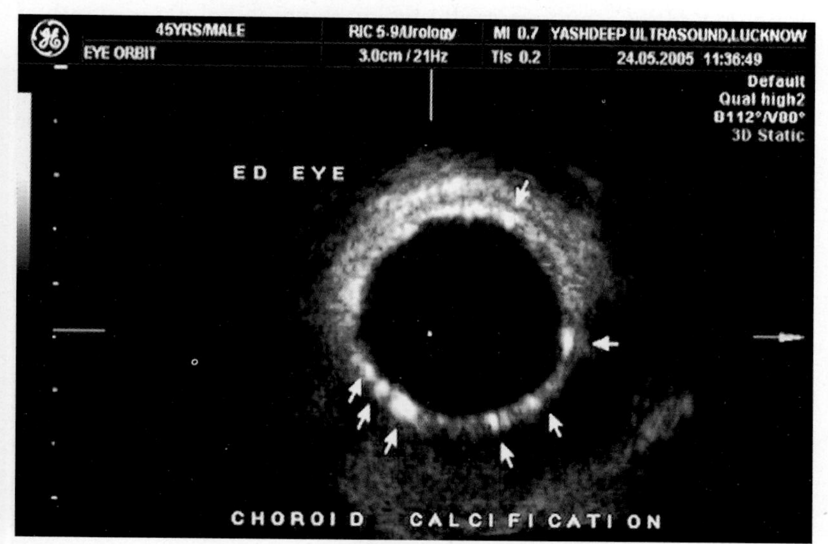

FIGURE 1.145

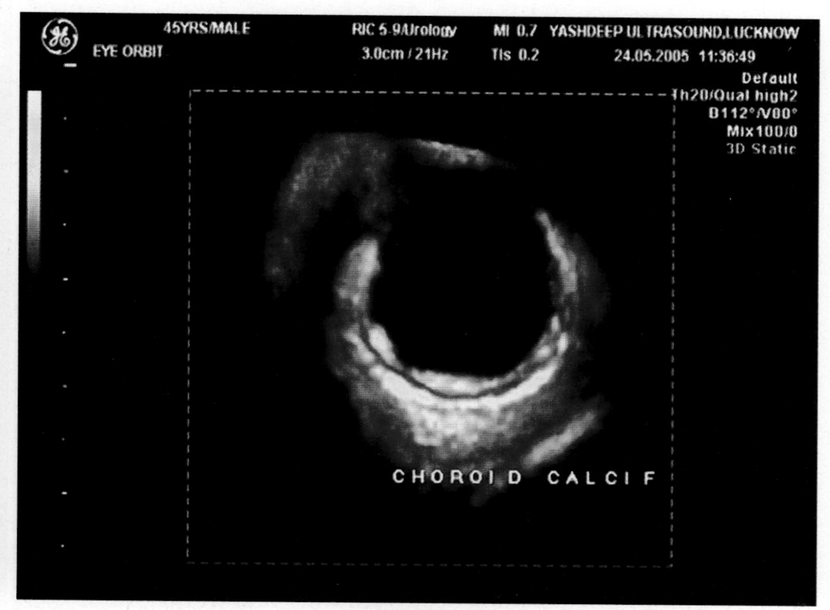

FIGURE 1.146

FIGURES 1.145 and 1.146: 3D imaging of choroid calcification. 3d imaging of the same patient shows finer details of choroid calcification, which are seen embedded in the choroidal layer.

Eye Trauma

HRSG echography is the best practical method to evaluate eye traumas, as direct vision is frequently hampered after trauma by opaque light-conduction media. HRSG can very well assess the extent and degree of damage. Early assessment of globe can result into early surgical repair and other microsurgical procedures to save the eye.

Severe blunt trauma to eye leads to sudden compression of the eyeball anteroposteriorly and corresponding expansion of the equatorial plane. It may lead to rupture of globe at equator with collapse of the eyeball. Retraction of gel with retinal tearing or dialysis can take place. This traumatic dialysis can take place anywhere, but it is more common in upper nasal quadrant of globe.

Anterior Segment Trauma

Anterior segment trauma can be better evaluated with immersion technique. This is important in corneal opacification or anterior chamber hyphema. HRSG can very well display the blood clot and depth of anterior chamber. Even status of lens can be assessed. The trauma can also injure lens. Although mild cataractous changes are difficult to appreciate, but moderately dense post-traumatic cataract can easily be picked up on HRSG. Location of lens or subluxation of lens can be detected on HRSG. This helps in the proper management in restoring the vision.

Posterior Segment Trauma

Eye trauma can lead to many posterior segment abnormalities including vitreous hemorrhage, PVD, retinal detachment and retinal tears. The findings are already described in various heads. Penetrating ocular trauma may lead to retinal detachment in almost one-fifth cases. Small or subtle hemorrhage in vitreous is difficult to detect. However, with high gain setting, it can be picked up on HRSG. Hemorrhage may be confined to vitreous gel, retrohyaloid space or including both. Dynamic scanning reveals vitreoretinal adhesions. Long standing posttraumatic retinal detachment may lead to proliferative vitreoretinopathy (PVR). Thickening of retinochoroid layer secondary to post-traumatic edema can lead to severe visual loss when macula is involved. It can be demonstrated by HRSG.

Posterior Scleral Rupture

Severe blunt trauma can lead to posterior scleral rupture which may be difficult to detect clinically. The patient usually presents with normal intraocular pressure but marked hemorrhage chemosis and vitreous hemorrhage. On HRSG, the sclera in the area of rupture shows moderately irregular contour and low echogenicity. There may be indirect signs of rupture seen on HRSG. They are as follows:

Indirect Signs of Scleral Rupture

1. Vitreous incarceration into fundus with vitreous hemorrhage and PVD.
2. Thickening or detachment of surrounding retina or choroid.
3. Hemorrhage in the episcleral space.
4. Vitreous traction towards the site of incarceration.

Rupture of the Globe

Severe blunt trauma may lead to rupture of globe. It usually takes place at equator. The eye looses its normal shape and loss of ocular volume. Associated intravitreal hemorrhage and intraocular air are also seen. The eye gets separated from the orbital wall and shrinks. Air may be seen in globe if associated blowout fracture of ethmoid is present with eye injury.

Pyophthalmos

Acute infection of the eyeball results into enlarged, swollen painful eye. The eye size increases in acute stage. The eyeball is filled with low-level homogeneous echoes suggestive of pus collection. It can be very well appreciated on HRSG. Ultimately, this eye goes into endophthalmitis due to pyogenic process.

Endophthalmitis

Endophthalmitis may result due to an endogenous infection presenting elsewhere in body. However, it most commonly occurs following surgical trauma or penetrating injury to eye. HRSG is highly useful in evaluating endophthalmitis. On "B" scan imaging dense opacities are seen in the posterior segment. Thick echogenic exudative collection is seen in endophthalmitis. Diffuse thickening of retinochoroid layer as well as tractional or exudative retinal detachment can occur.

Expulsive Hemorrhage

Expulsive hemorrhage is most devastating complication of intraocular surgery. Massive subchoroidal haemorrhagic detachment fills the vitreous cavity. This may include only one or two quadrant of eye in few cases. HRSG is also very helpful in detecting delayed expulsive hemorrhage. In most of the cases, delayed hyperechoic dots can present in subchoroidal space. On following scan these clots diminish in size and also become less reflective.

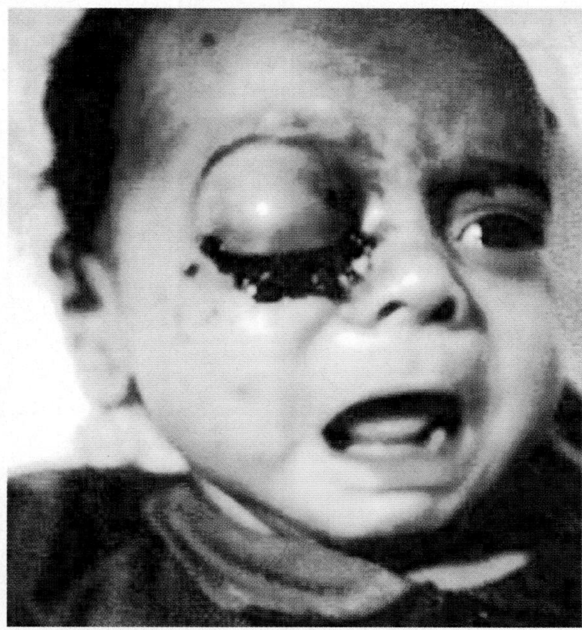

FIGURE 1.147: A young child came with markedly swollen eyelid with chymosis. Hemorrhagic specks are also seen over the eyelid. Clinical examination was not possible due to marked lid edema.

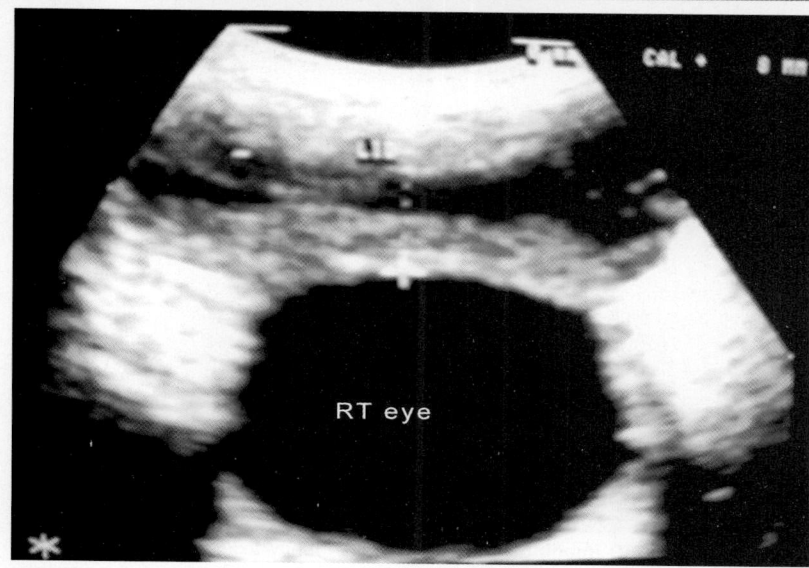

FIGURE 1.148: HRSG shows low-level collection over the upper eyelid. Echogenic shadows are also seen in the collection suggestive of hematoma formation. However, the eyeball shows normal anechoic posterior segment. Although it has been pushed down by the edematous eyelid.

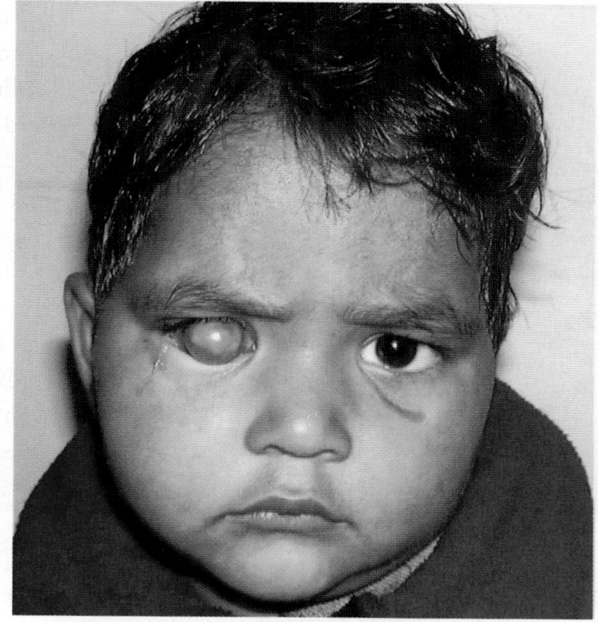

FIGURE 1.149: Traumatic anterior staphyloma. A young child presented with white glistening shadow in the Rt eye.

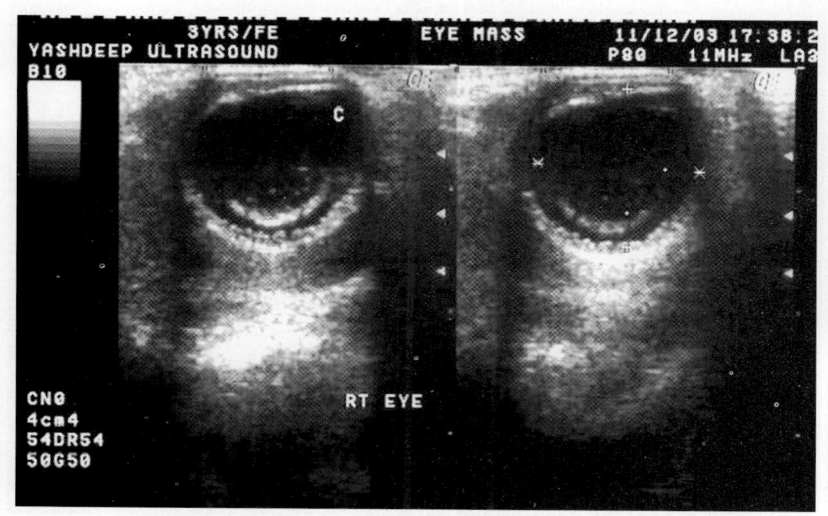

FIGURE 1.150: HRSG shows well-defined cystic mass coming out from the iris and bulging out side. Few internal echoes are seen in it suggestive of anterior staphyloma.

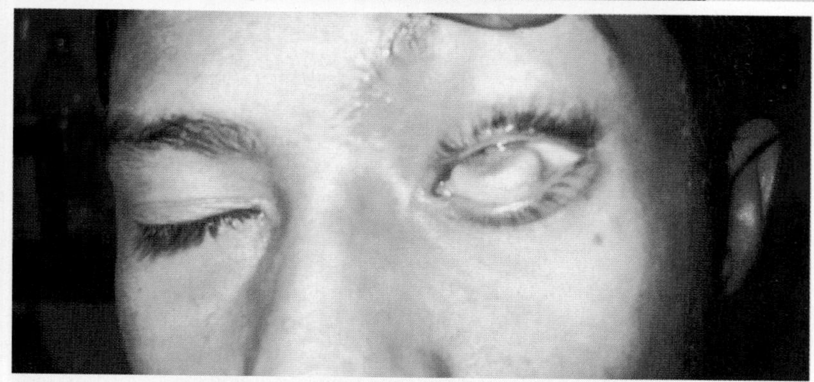

FIGURE 1.151: Complicated traumatic anterior staphyloma. Patient presented with bulging shining membranous shadow in the Lt eye.

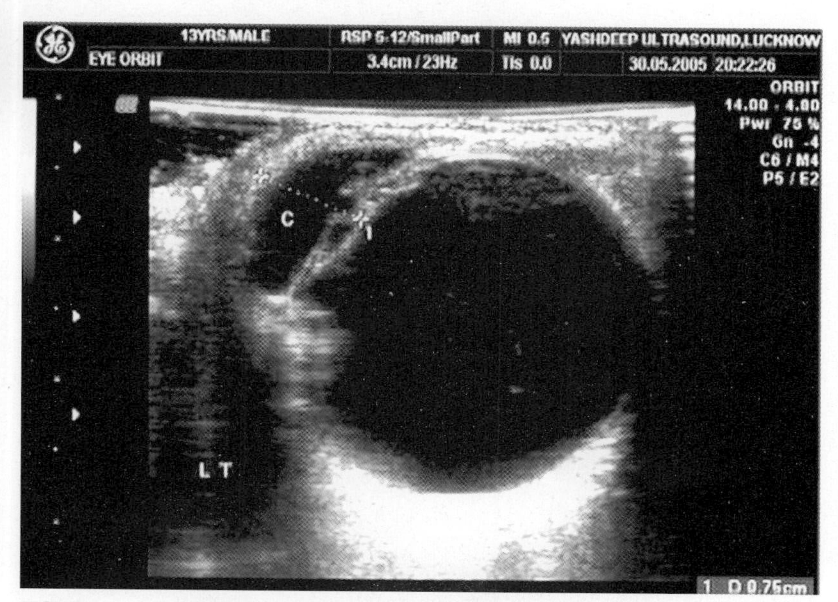

FIGURE 1.152: HRSG shows a thick wall cystic mass coming out from the anterior chamber. Multiple internal echoes are also seen in it.

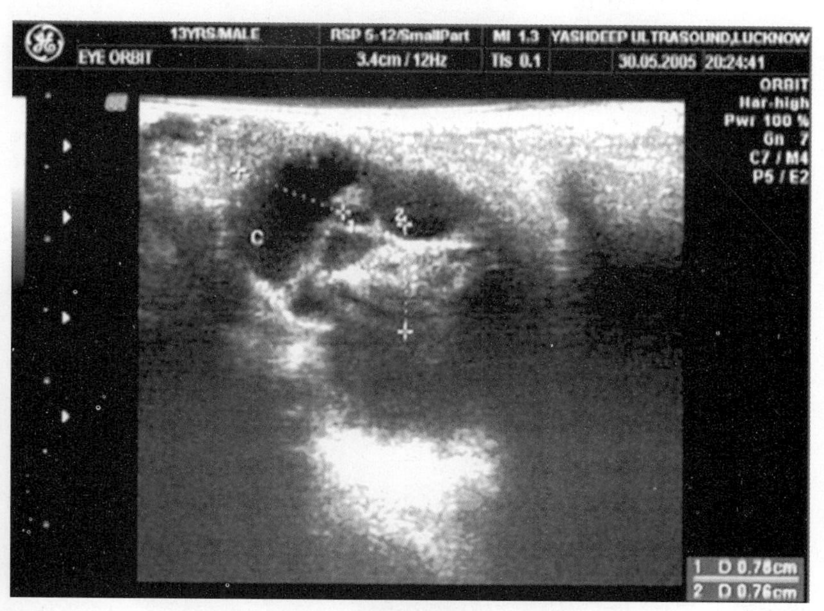

FIGURE 1.153

Lens Material

Normal lens appears as a highly reflective concave shadow on HRSG. It is seen sitting over the vitreous gel. However, cataract lens is seen as thickened biconvex reflective body. HRSG can very well delineate the

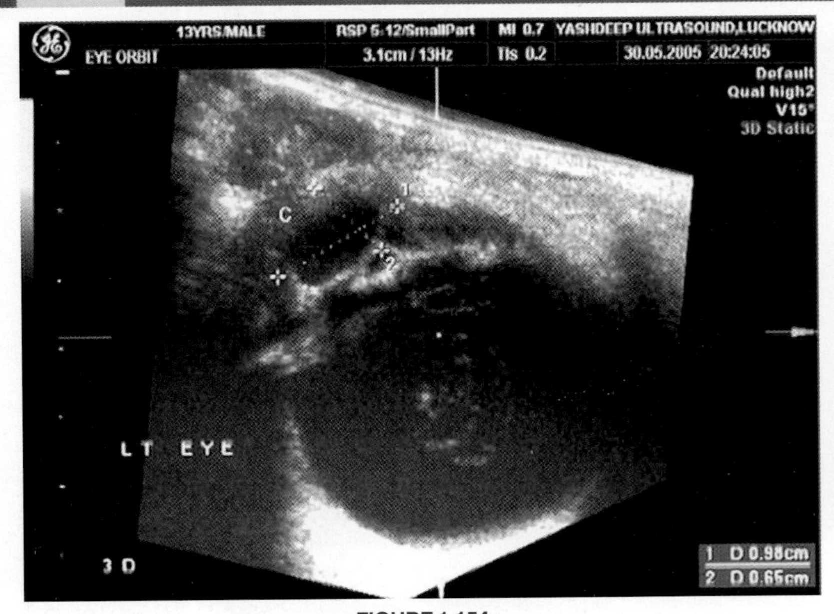

FIGURE 1.154

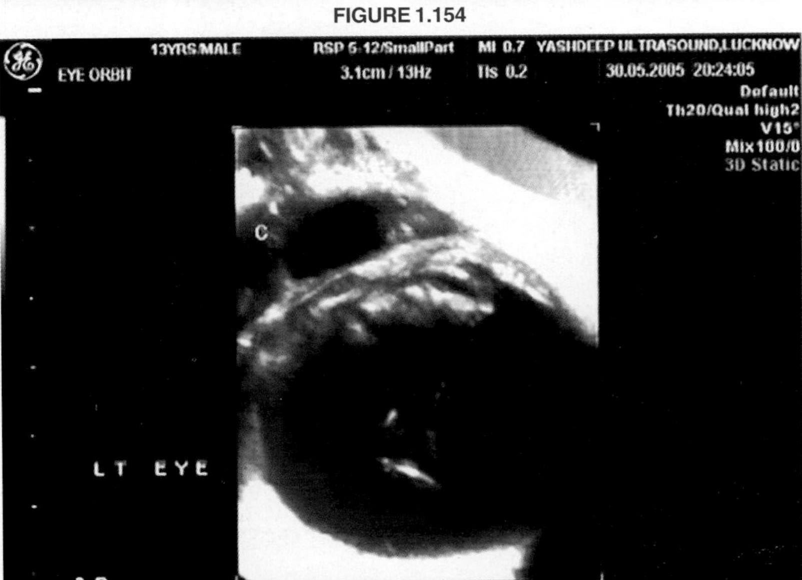

FIGURE 1.155

FIGURES 1.153 to 1.155: 3D imaging of the same patient shows details of the complicated staphyloma. Multiple echogenic bands are seen going from the lens to the cyst. The cyst is adhered with the lens. Lens is thickened and opaque with calcification of the nucleus.

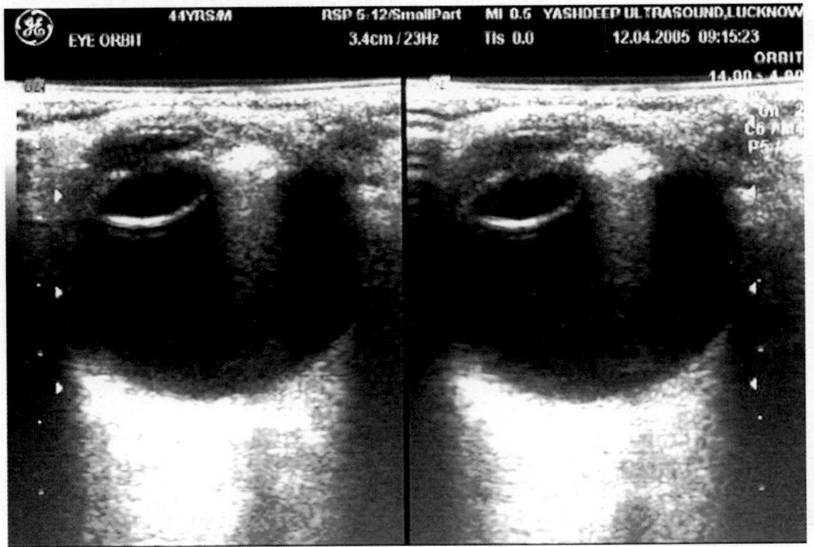

FIGURE 1.156A: A patient presented with trauma to the eye with an open wound in the sclera. HRSG shows anterior chamber hyphema with evidence of air pocket in the open wound of the sclera.

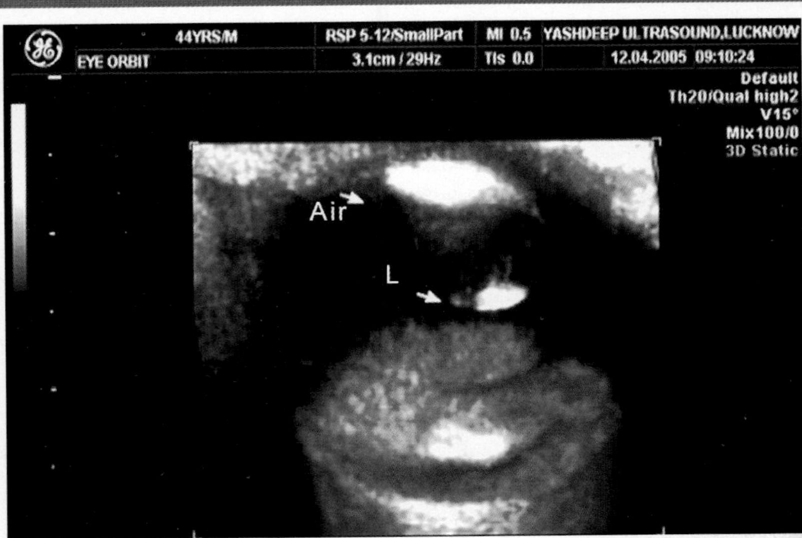

FIGURE 1.156B: 3D imaging of the same patient shows details of the air pocket, which was mistaken as foreign body. The lens was seen subluxated in the posterior segment associated intragel hemorrhage is also seen.

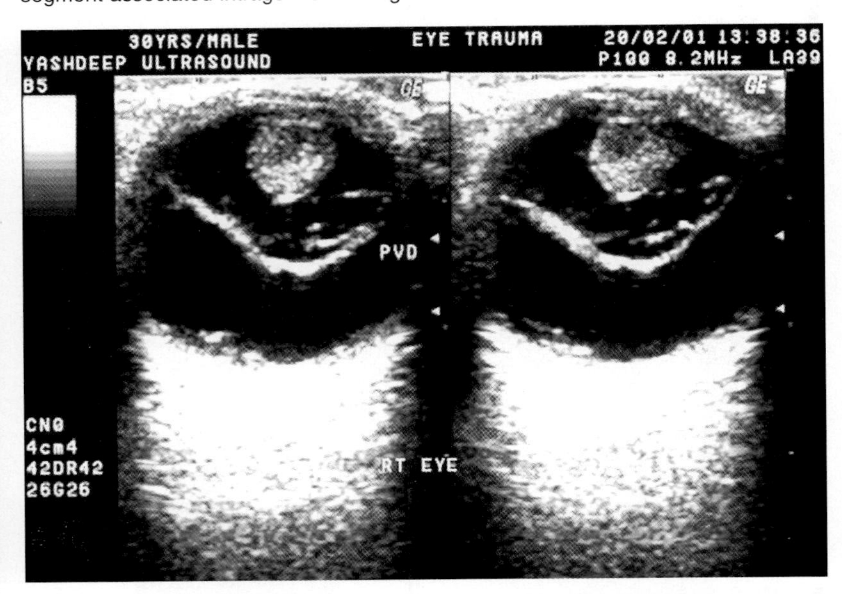

FIGURE 1.157: Traumatic cataract with vitreous hemorrhage. HRSG shows markedly thickened opaque lens. The lens capsule is broken and prolapsed of the lens material is seen. Associated intragel hemorrhage is also seen with retraction of the vitreous. But no retinal detachment is seen.

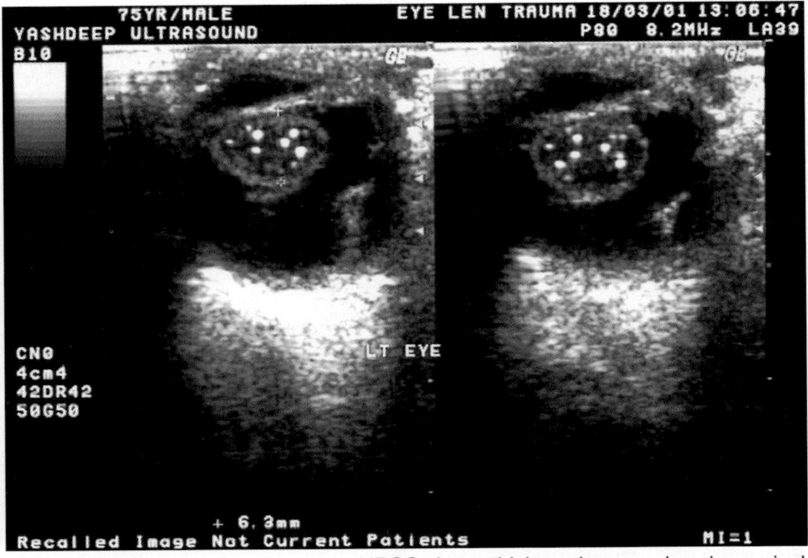

FIGURE 1.158: Traumatic cataract. HRSG shows thickened opaque lens in surgical trauma to the eye. Multiple small-calcified specks are seen in the lens suggestive of breaking down of the nucleus into small pieces. Posterior segment of the eye is free.

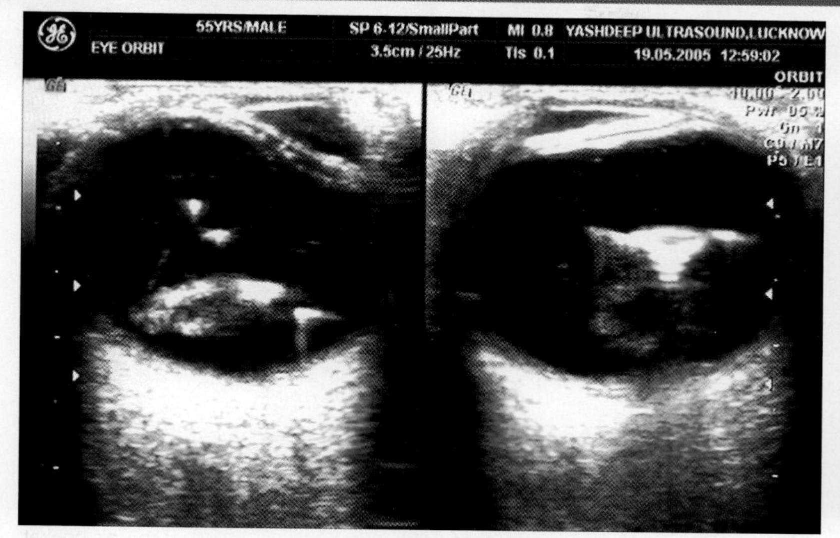

FIGURE 1.159

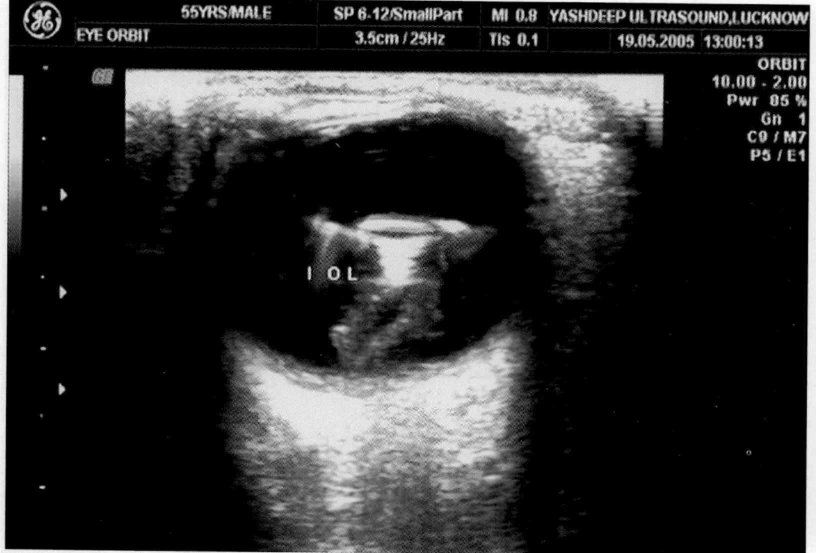

FIGURE 1.160

FIGURES 1.159 and 1.160: Surgical trauma to the eye with IOL in posterior segment. HRSG shows an echogenic hyper-reflective shadow displaced in the posterior segment after post-cataract surgery, suggestive of displaced intraocular lens in the posterior segment. Associated intragel hemorrhage is also seen.

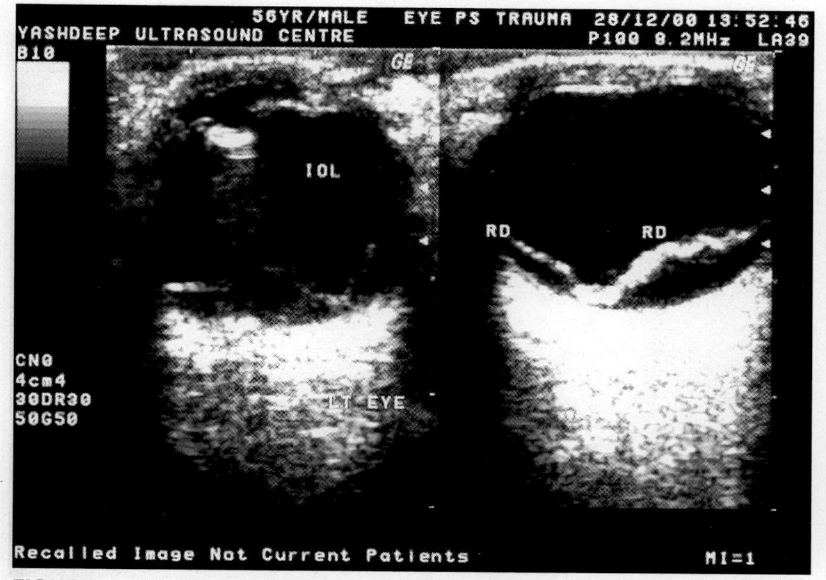

FIGURE 1.161: Post-surgical trauma after cataract surgery. HRSG shows displaced subluxated intraocular lens from its position. Posterior segment of the eye shows 'V' shaped echogenic membrane suggestive of retinal detachment.

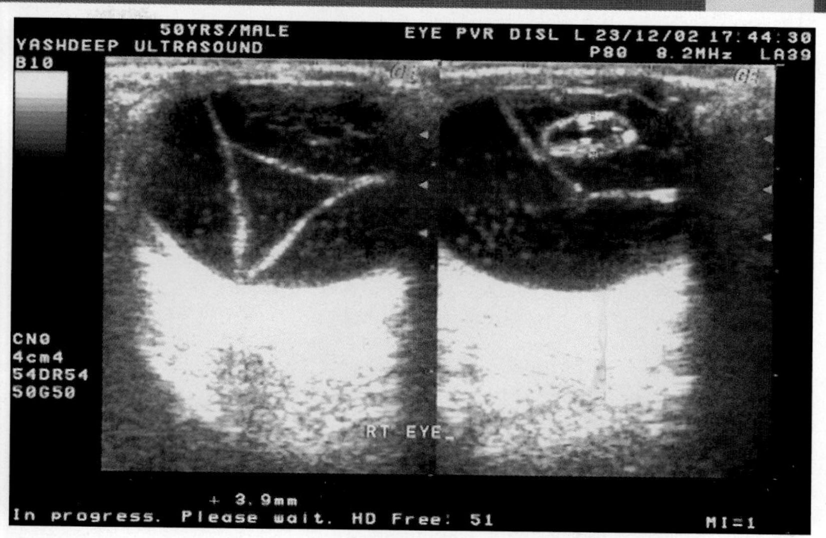

FIGURE 1.162

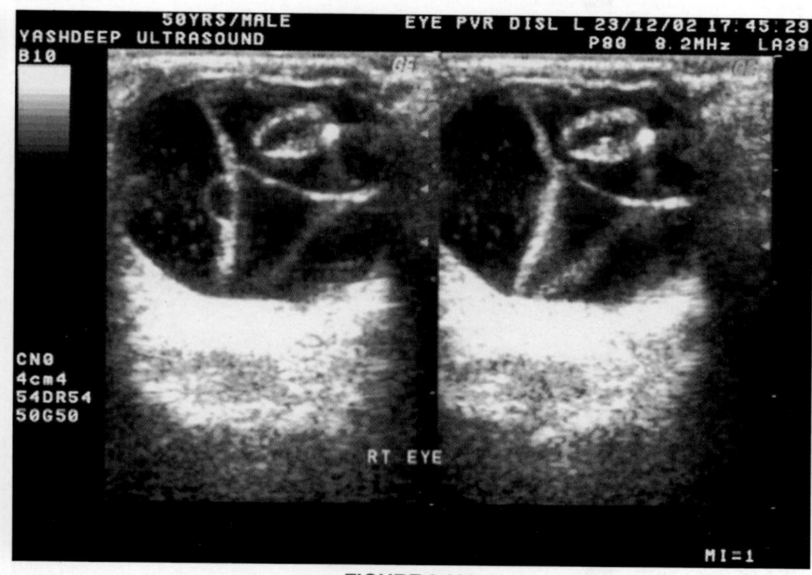

FIGURE 1.163

FIGURES 1.162 and 1.163: Blunt trauma to the eye. HRSG shows subluxated thickened opaque lens in a patient sustained blunt trauma to the eye associated retinal detachment is also seen. Intragel hemorrhage is also seen.

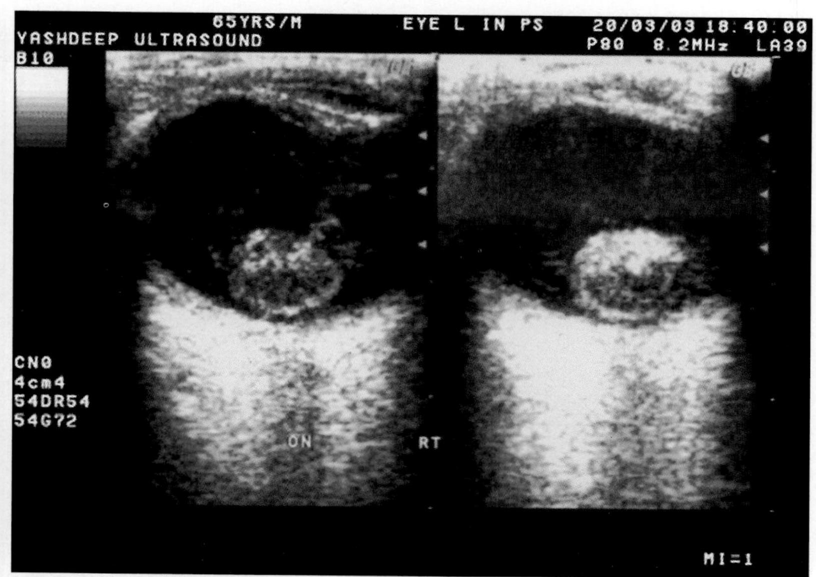

FIGURE 1.164: Blunt trauma to the eye dislocated lens in the posterior segment. HRSG shows thickened opaque cataractous lens displaced in the posterior segment of the eye due to blunt trauma.

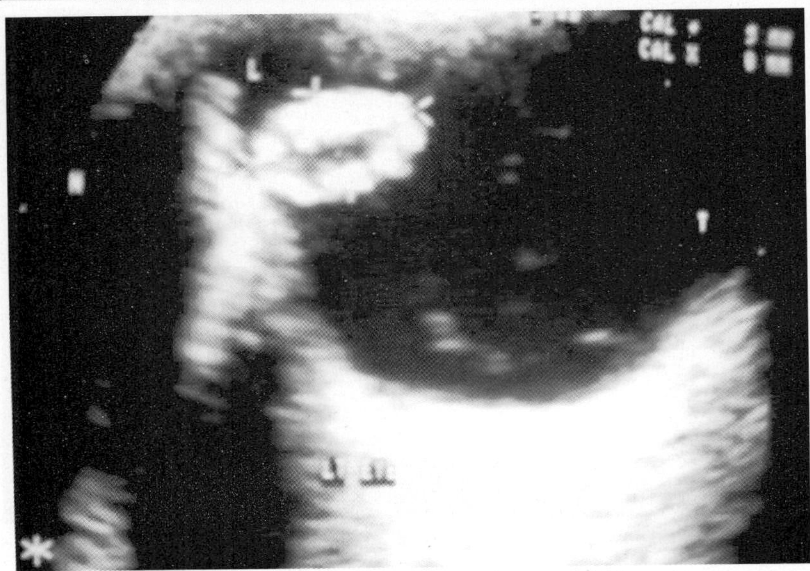

FIGURE 1.165: Blunt trauma to the eye. HRSG shows thickened opaque lens suggestive of traumatic cataract with subluxation.

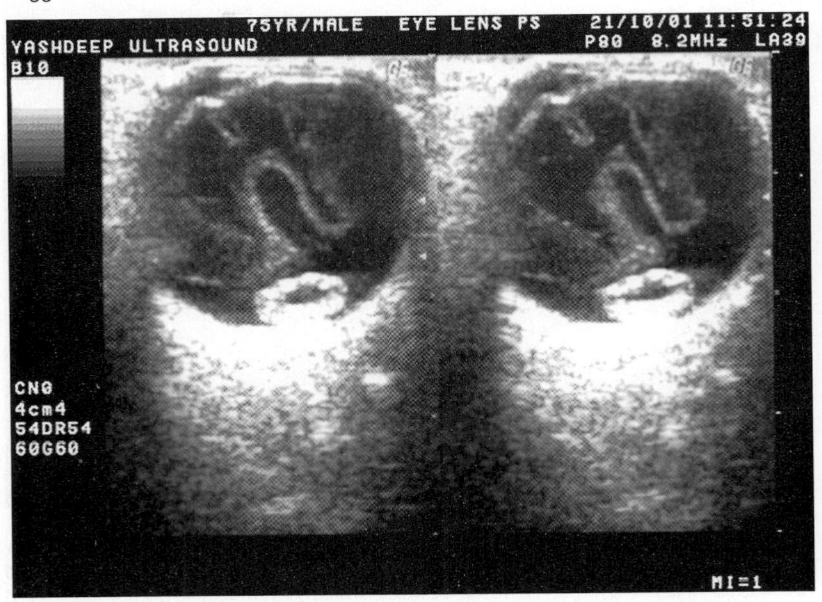

FIGURE 1.166: Blunt trauma to the eye. HRSG shows echogenic thickened opaque lens displaced in the posterior segment and lying at the posterior pole. Traumatic cataract of the lens is seen.

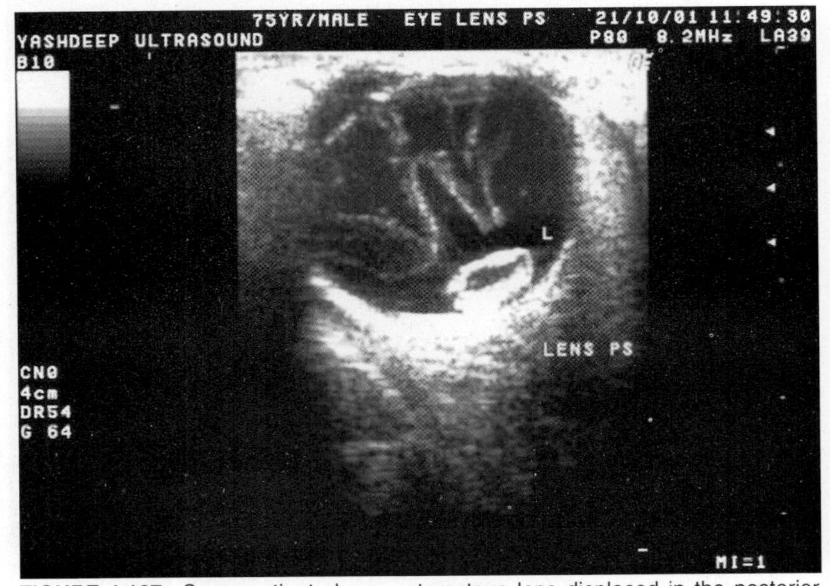

FIGURE 1.167: Same patient shows cataractous lens displaced in the posterior segment with multiple thin bands holding the lens.

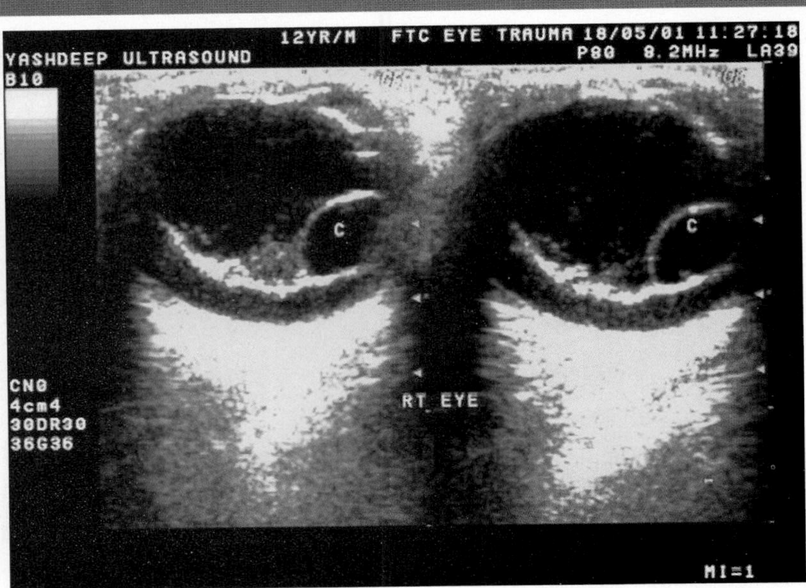

FIGURE 1.168: Old trauma to the eye shows complete retinal detachment with retinal cyst formation.

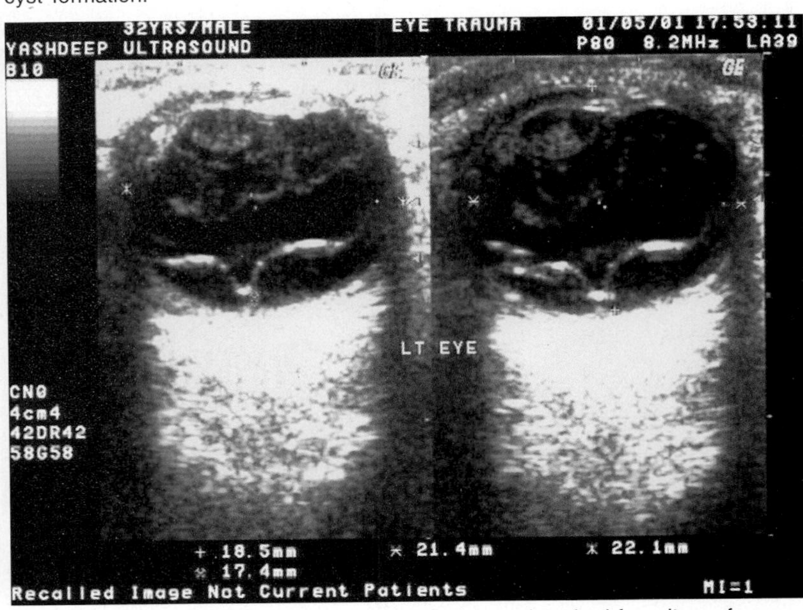

FIGURE 1.169: HRSG shows echogenic membrane detached from its surface and attached at the optic nerve head in blunt trauma to eye suggestive of retinal detachment. Associated intragel hemorrhage is also seen. Vitreoretinal bands are present.

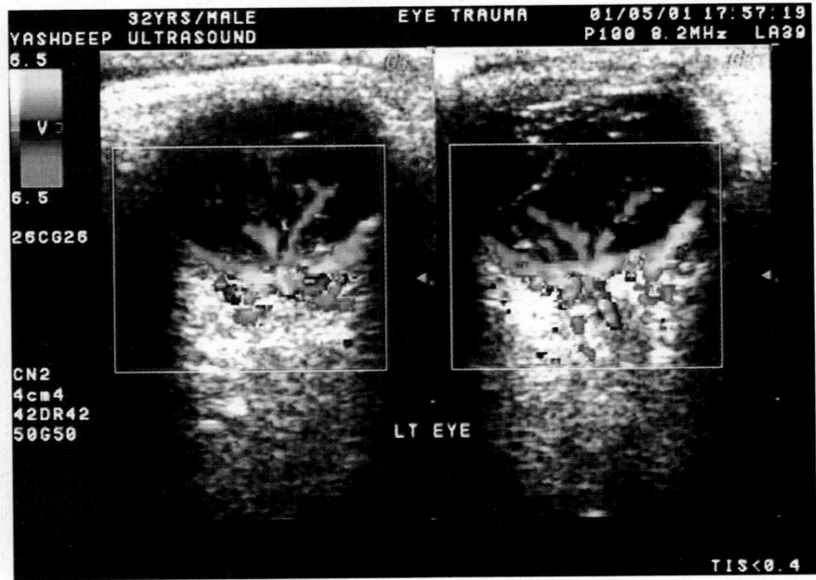

FIGURE 1.170: Color Doppler imaging of the same patient shows multiple blood vessels dragged by the membranous band in the vitreous cavity.

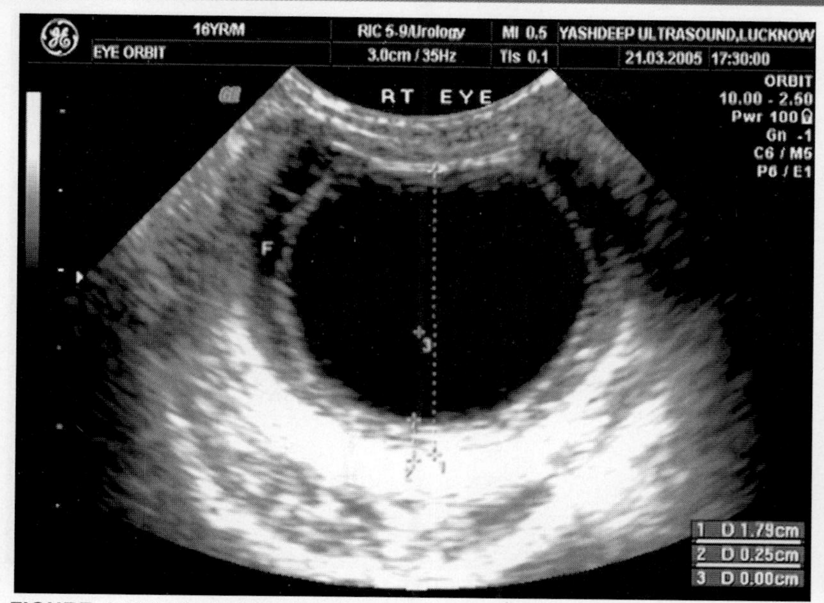

FIGURE 1.171: Complete detachment of the globe. HRSG shows complete detachment of the globe from the orbital wall in a patient with blunt trauma to the eye.

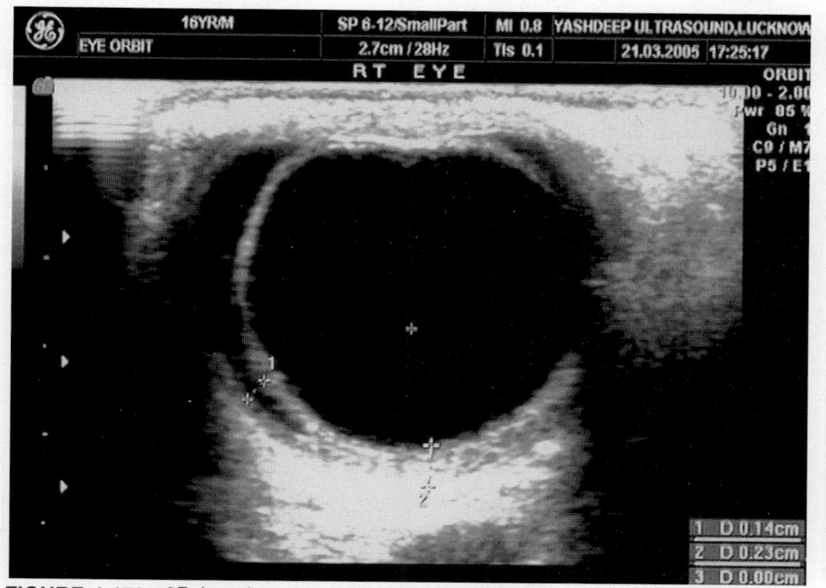

FIGURE 1.172: 3D imaging shows details of the separation of the globe with wall edema.

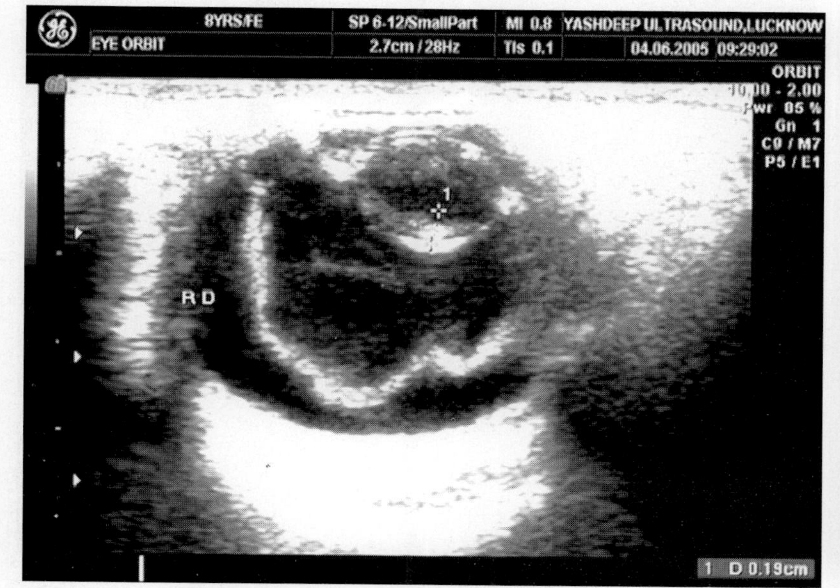

FIGURE 1.173: Blunt trauma of the eye shows traumatic cataract. Thickened calcified nucleus is seen in the lens. Posterior segment of the eye shows complete retinal detachment.

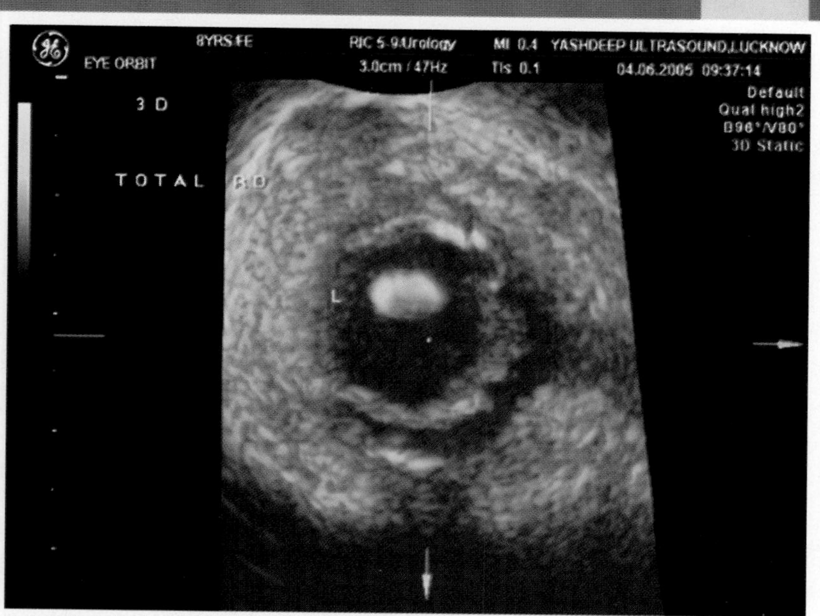

FIGURE 1.174: 3D imaging of the eye shows opaque lens with calcified nucleus and the retina is seen detached from it surface from all side.

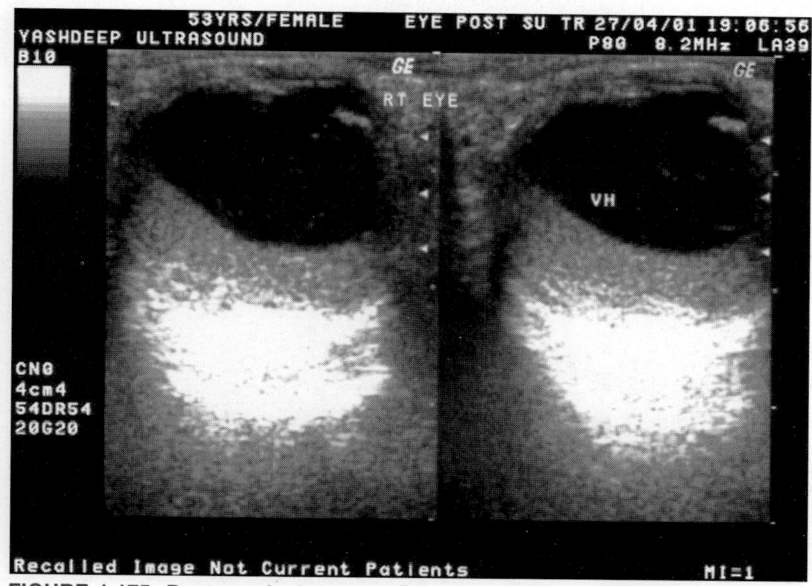

FIGURE 1.175: Post-surgical trauma. Dense intragel hemorrhage is seen in sub vitreal space with layering of the blood.

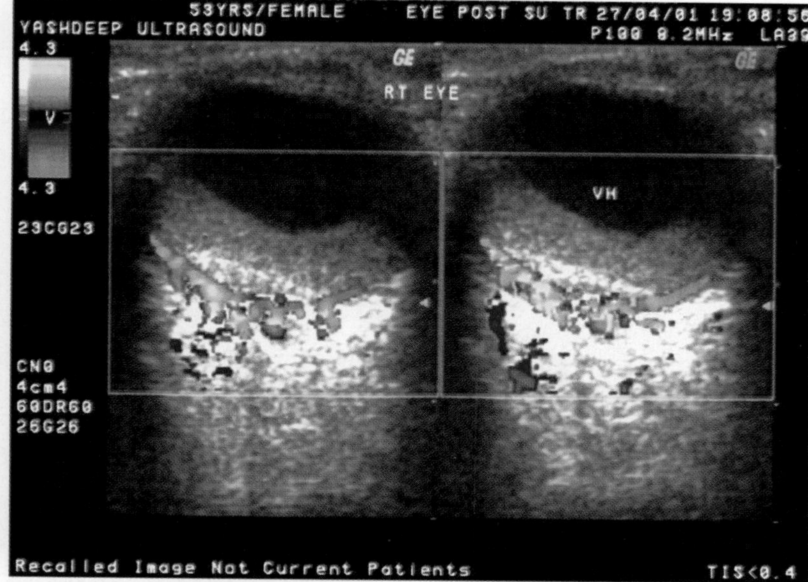

FIGURE 1.176: Post-surgical trauma. Color Doppler imaging of the same patient shows multiple vessels supplying the choroid suggestive of increased hyperemia.

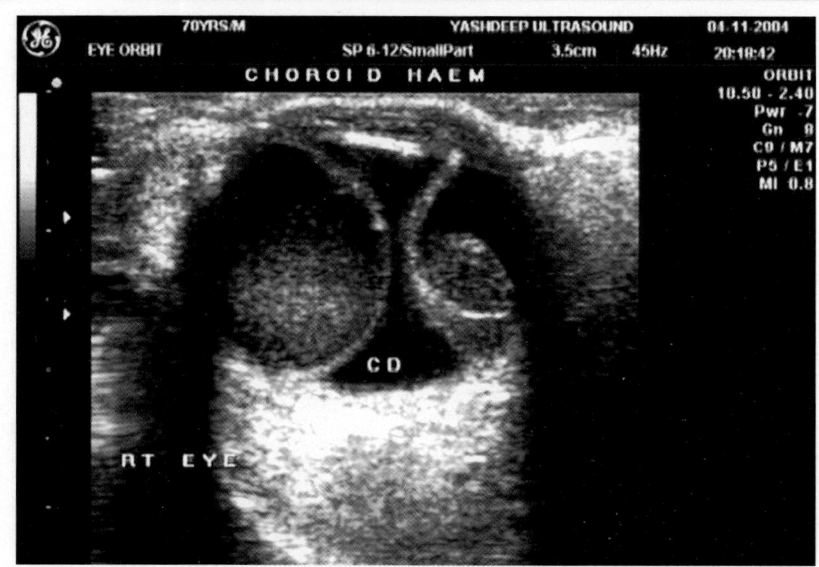

FIGURE 1.177: Post-surgical traumatic expulsive hemorrhage. HRSG shows dense subchoroidal hemorrhage with associated choroidal detachment after cataract surgery typical dome shape kissing bullae sign is positive with dense intragel collection typical of subchoroidal expulsive hemorrhage.

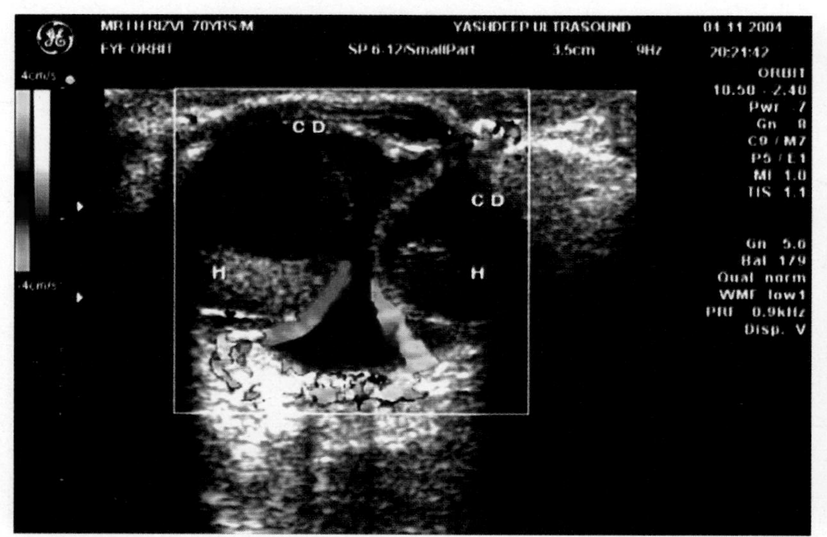

FIGURE 1.178: Color Doppler of the same patient shows dragging of the vessels with choroidal membrane.

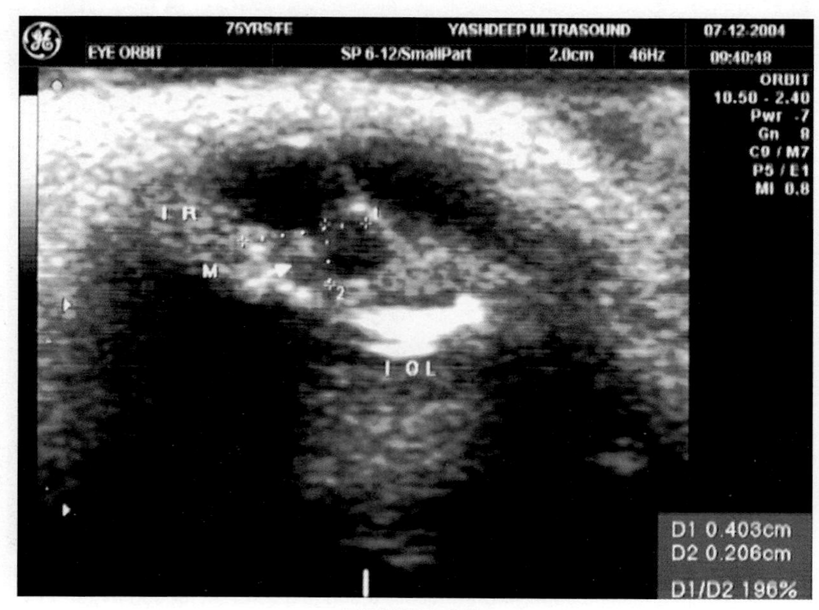

FIGURE 1.179

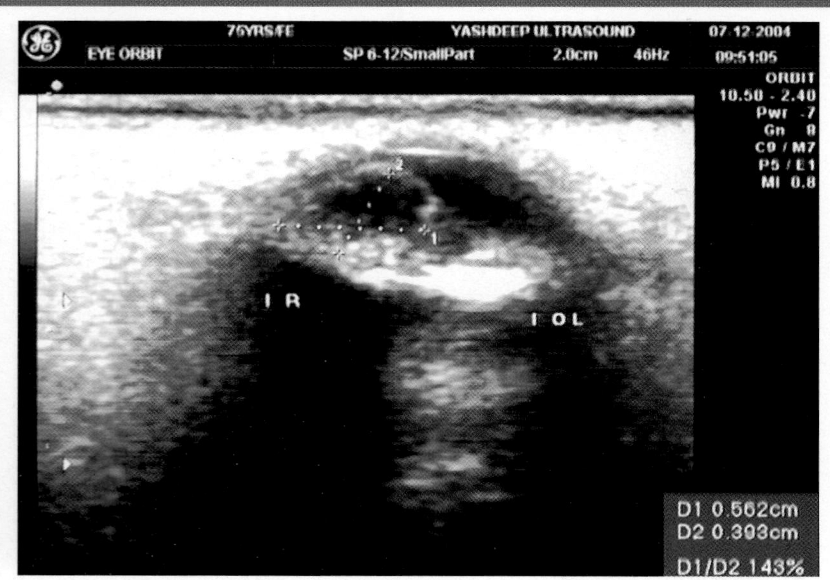

FIGURE 1.180

FIGURES 1.179 and 1.180: Post-surgical traumatic implantation cyst. HRSG shows a cystic mass coming out from the iris after cataract surgery. The iris is thickened and the cyst is seen in close relation with iris. IOL is also seen as a bright echogenic semilunar shadow.

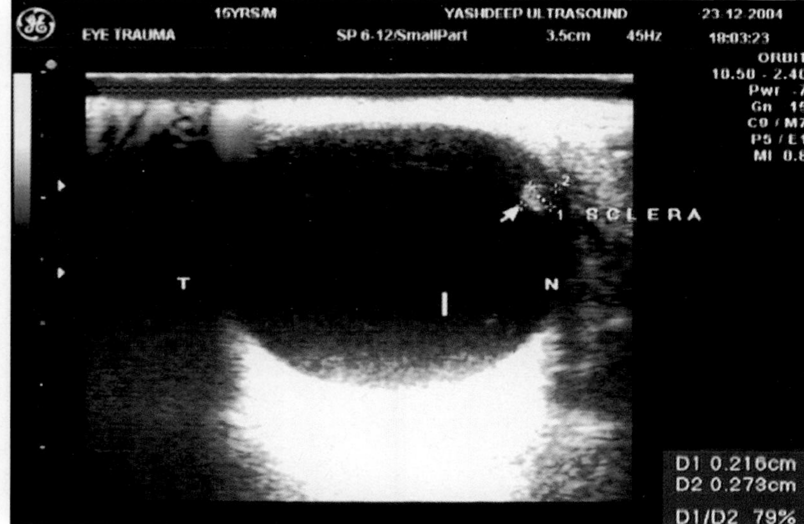

FIGURE 1.181

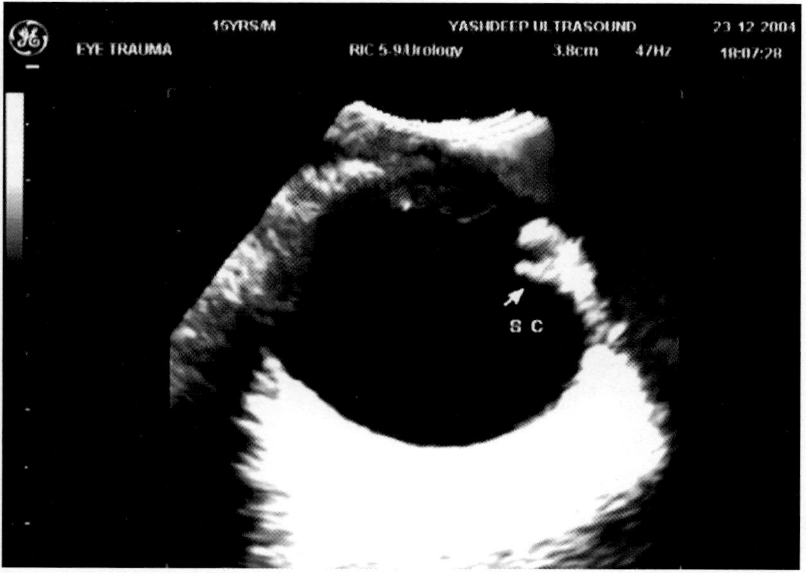

FIGURE 1.182

FIGURES 1.181 and 1.182: Post-surgical scleral fold. HRSG shows an echogenic membranous shadow lifted up from the sclera after the surgery. The 3D imaging clearly depicts the details of the scleral fold.

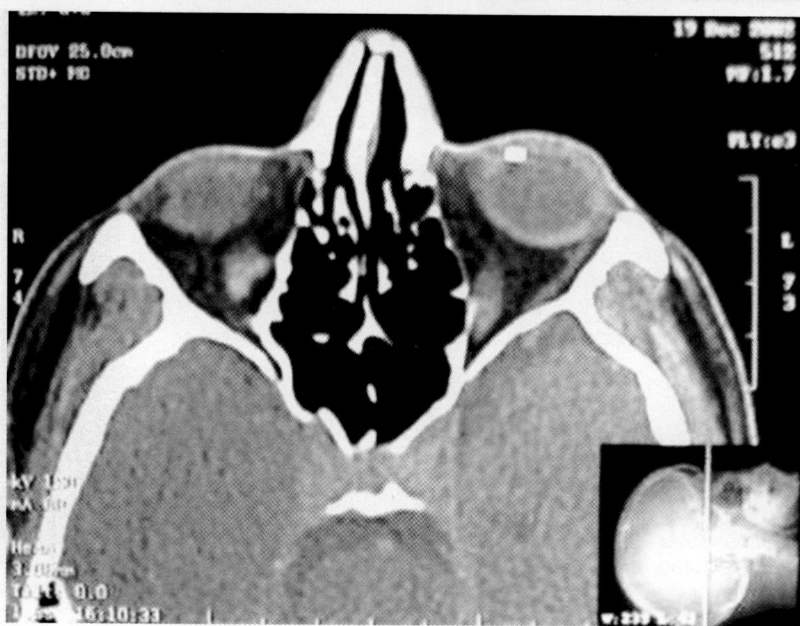

FIGURE 1.183A

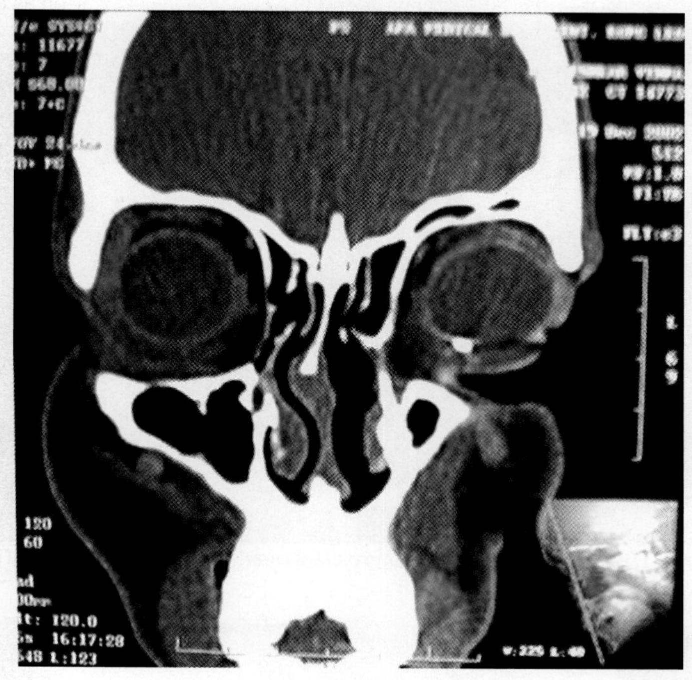

FIGURE 1.183B

FIGURES 1.183A and B: CT scan axial and coronal cut shows a hyperdense foreign body in the orbital socket. However, it is not confirmed whether it is lying in the eye or outside the eye.

intralenticular constituents and also accurately measures the thickness of cataract lens. Lens material may be lost in vitreous cavity during surgery. Dislocated lens may adhere to retina and does not move on dynamic scanning unlike a dislocated intact lens.

Intraocular Lens

Intraocular lens creates strong artifacts, which make evaluation of posterior segment difficult. The smooth very highly reflective surface of lens produces numerous signals producing strong artifacts. Therefore, for proper evaluation the probe should be kept peripheral to limbus.

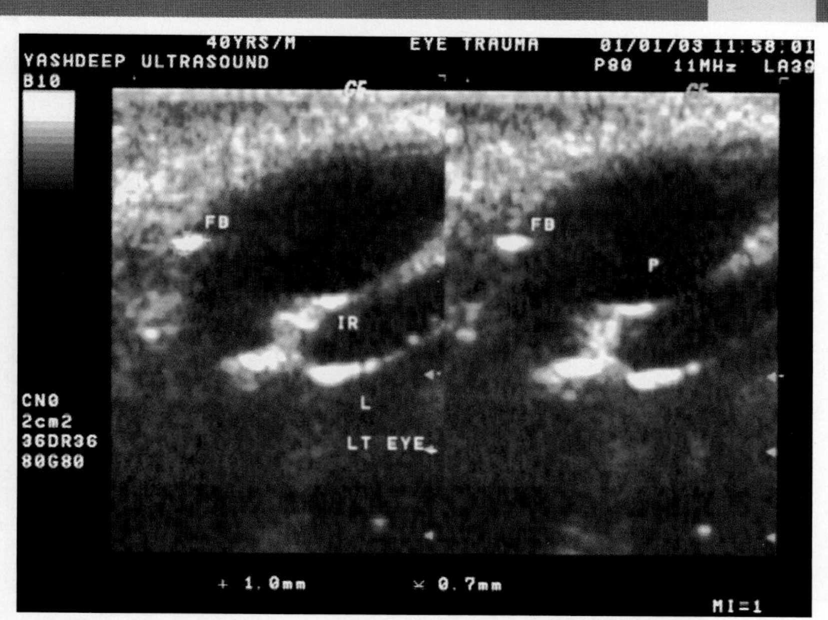

FIGURE 1.184: HRSG clearly shows that the foreign body is lying outside the eyeball. It is confined to the just in the conjunctiva.

Eye Trauma Foreign Bodies

Intraocular foreign bodies in most of the cases can be easily located on HRSG. Even when the foreign body (FB) is previously localized by X-ray or computed tomography (CT), Ultrasonographic examination should be performed for more precise location of FB and to determine the extent of intraocular damage. Further if FB is lying near to sclera, CT does not identify if it lies just within or just outside the globe. Most FB are detectable on HRSG. Small FB as metal, glass fragments, etc. traveling at high-speed cause penetrating injury to eye with metals, glass and stones are highly reflective masses and can picked up easily. They may cast acoustic shadowing. HRSG is highly sensitive even in picking up the track of traveling of FB outlining trajectory of the object. The entry of exist sites of FB may cause vitreous incarceration. Air bubbles are also seen in globe associated with FB. They stand out as highly reflective spots, which show rapid movements. Associated vitreous hemorrhage, retinal tears and scleral rupture can be well assessed on HRSG.

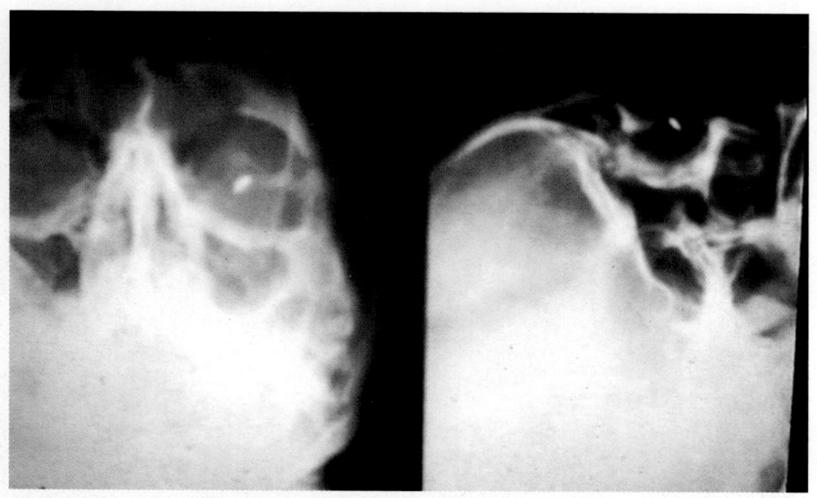

FIGURE 1.185: Foreign body in anterior chamber. X-ray of the patient shows a radiodense shadow in the orbit. However, exact location of the foreign body could not be determined.

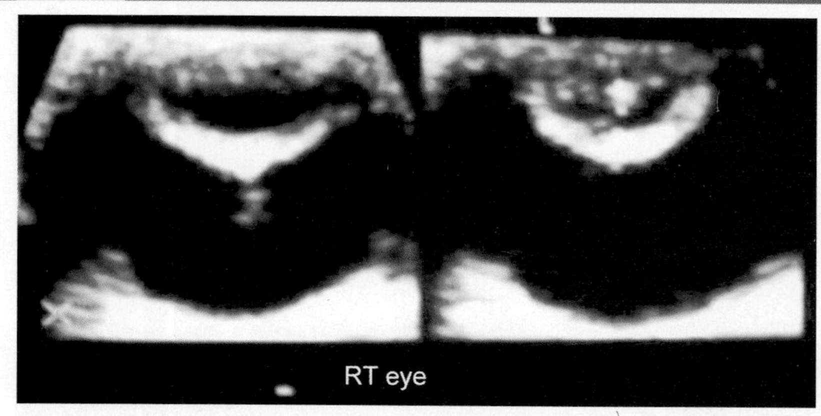

FIGURE 1.186: HRSG of the same patient shows an echogenic hyperreflective shadow in the anterior chamber. It is accompanied with acoustic shadowing.

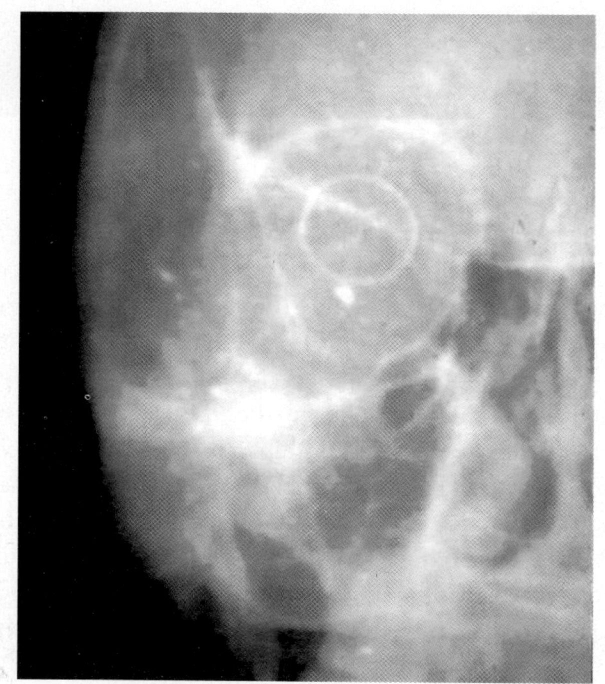

FIGURE 1.187: Foreign body in AC. X-ray of the patient shows radiodense opacity of metallic density in orbit.

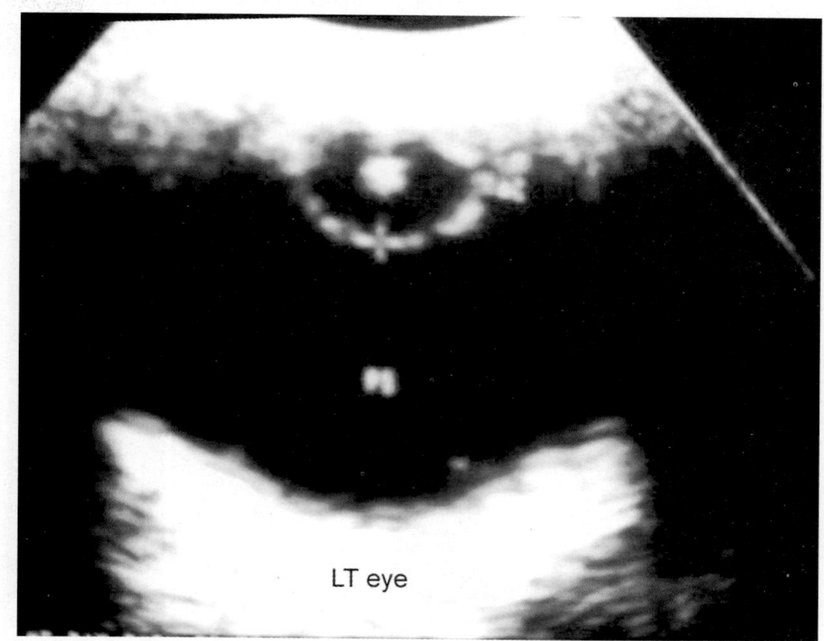

FIGURE 1.188: HRSG shows a hyperdense echogenic foreign body confined in the anterior chamber. Posterior segment of the eye is free.

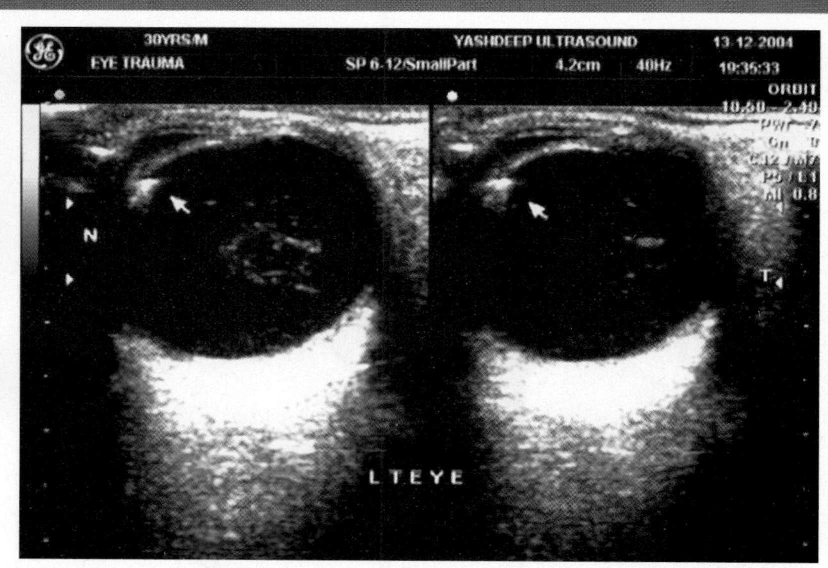

FIGURE 1.189: Foreign body trauma. HRSG shows a small echogenic shadow in the anterior chamber in just near to the lens. It is difficult to differentiate from the lens on 2D imaging.

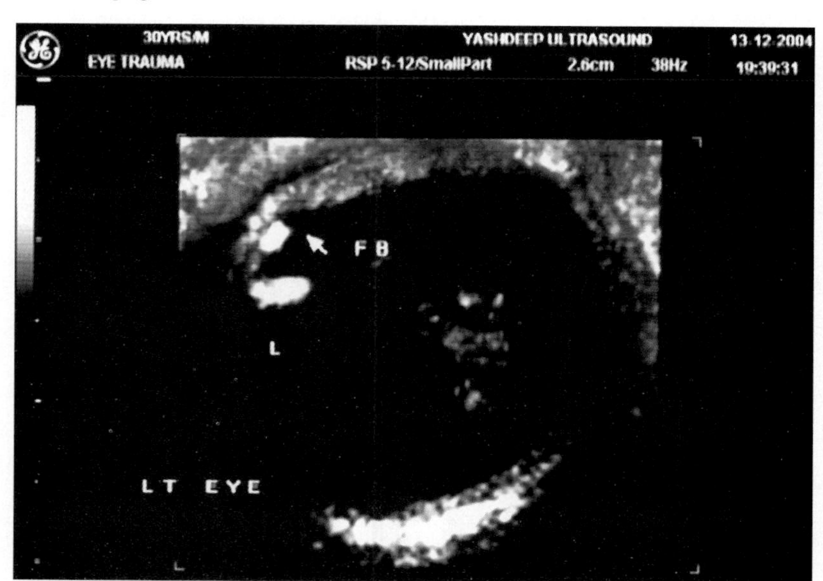

FIGURE 1.190: 3D imaging of the same patient clearly shows the foreign body is lying away from the lens.

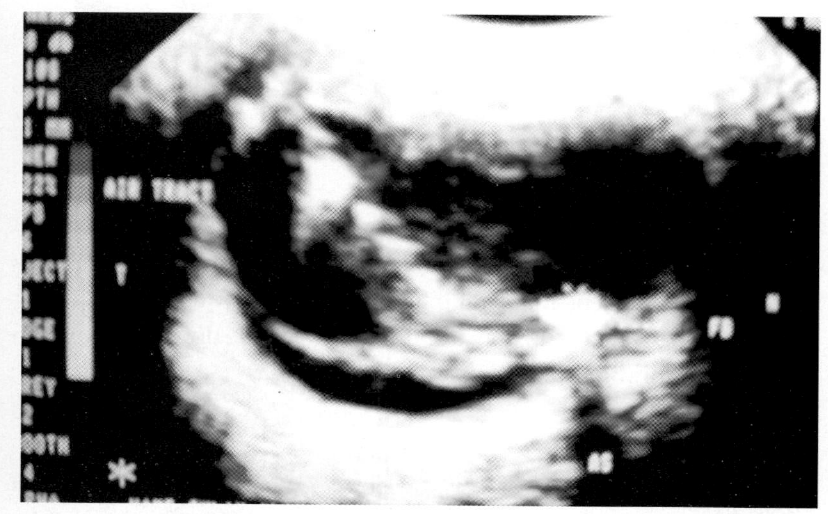

FIGURE 1.191: Penetrating injury to the eye. A dense echogenic shadow is seen lodged in the posterior segment just above the sclera. It is accompanied with acoustic shadowing. Multiple hyperdense floating echogenic shadows are also seen in the posterior segment suggestive of air bubble along the line of the trajectory of the foreign body. Associated intragel hemorrhage is also present.

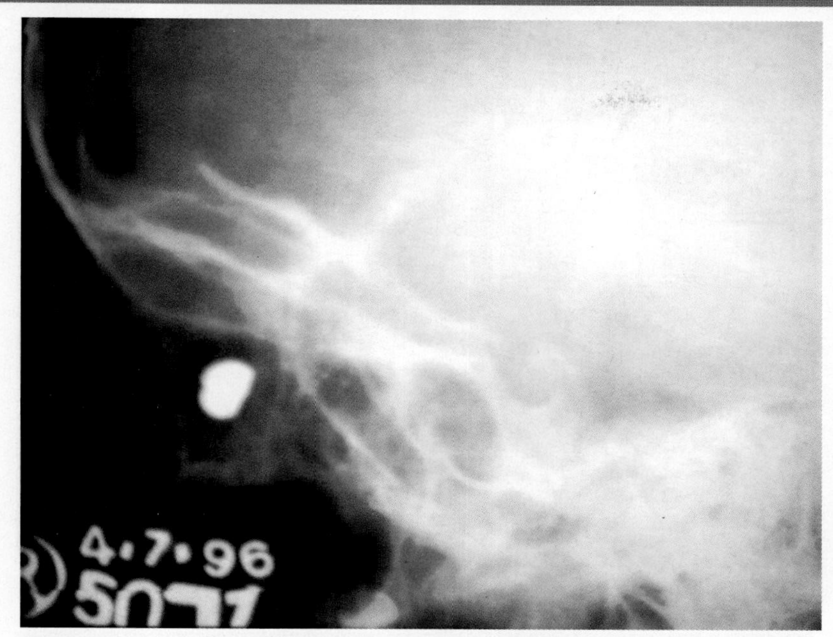

FIGURE 1.192: Penetrating injury to the eye. X-ray of the patient shows a hyperdense metallic shadow lodged in the orbital socket. The shape of the shadow is suggestive of a bullet.

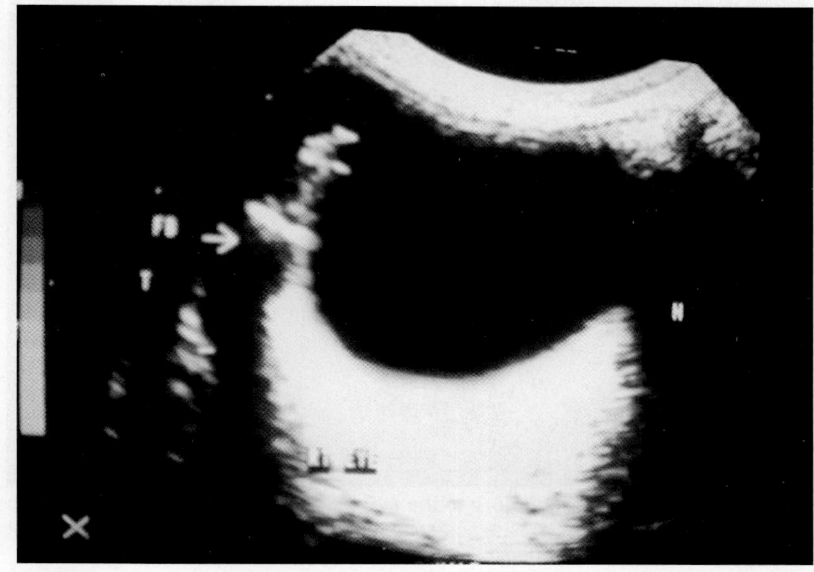

FIGURE 1.193: HRSG shows a hyperdense echogenic body lodged in the episcleral space. However, no intragel hemorrhage is seen. No retinal detachment is seen.

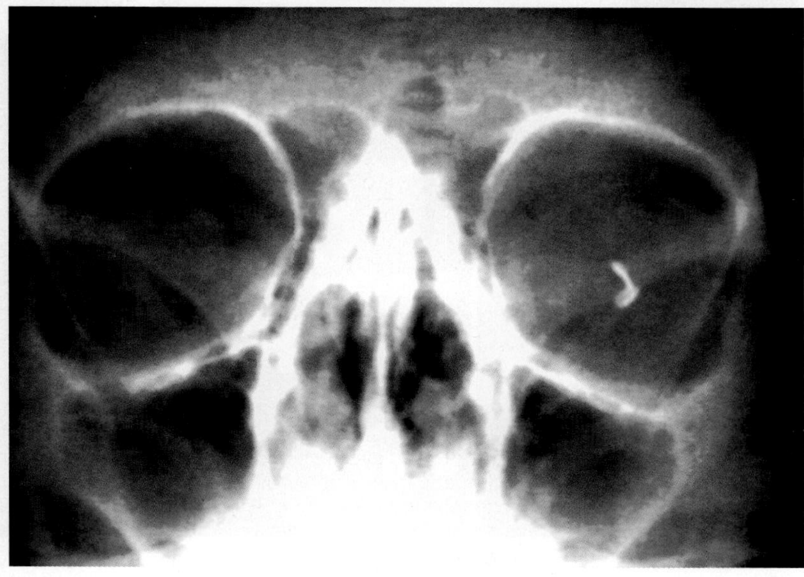

FIGURE 1.194

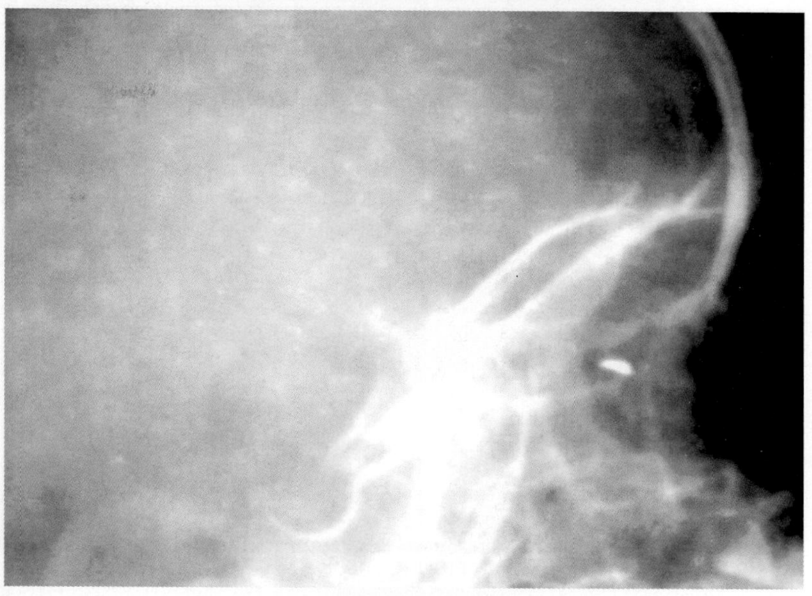

FIGURE 1.195

FIGURES 1.194 and 1.195: Penetrating injury to the eye. X-ray of the patient shows radiopaque foreign body of metallic density lodged in the posterior segment.:

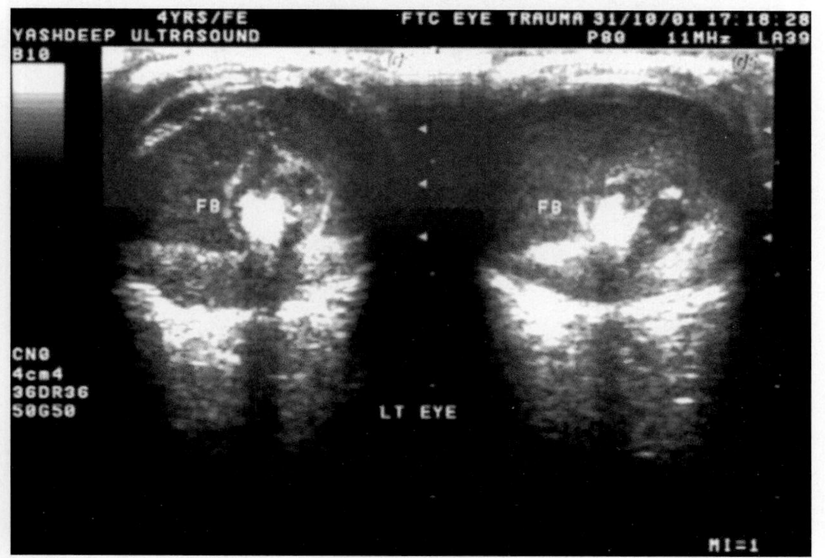

FIGURE 1.196

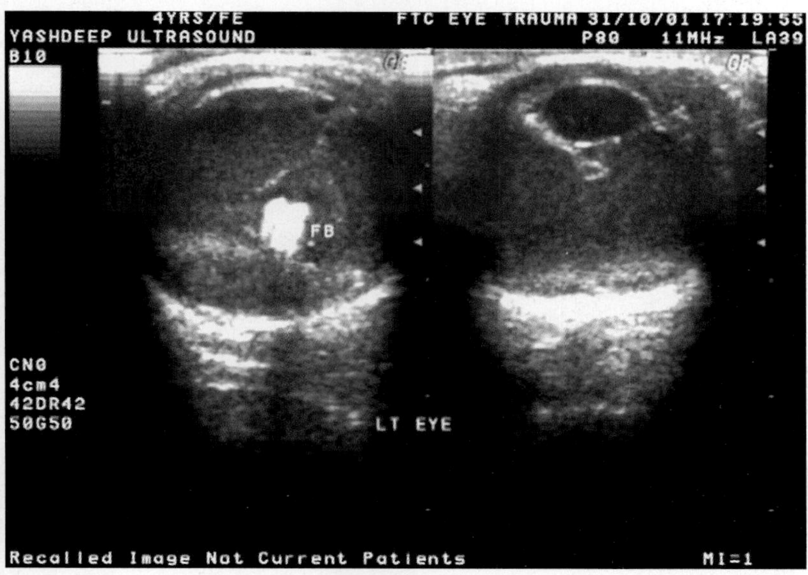

FIGURE 1.197

FIGURES 1.196 and 1.197: HRSG of the same patient shows a hyperdense echogenic foreign body lodged in the posterior segment with irregular shape. It is accompanied with acoustic shadowing. Associated dense intragel hemorrhage is also seen filling the posterior segment.

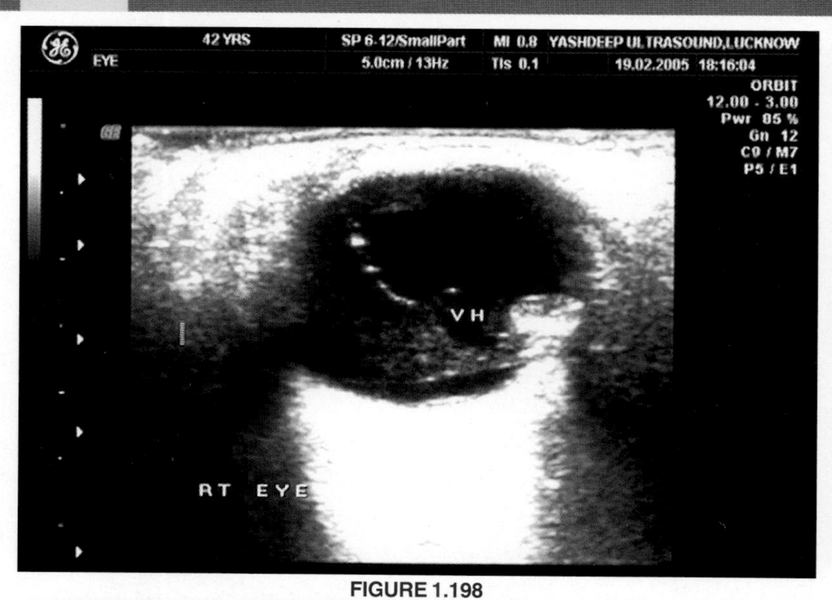

FIGURE 1.198

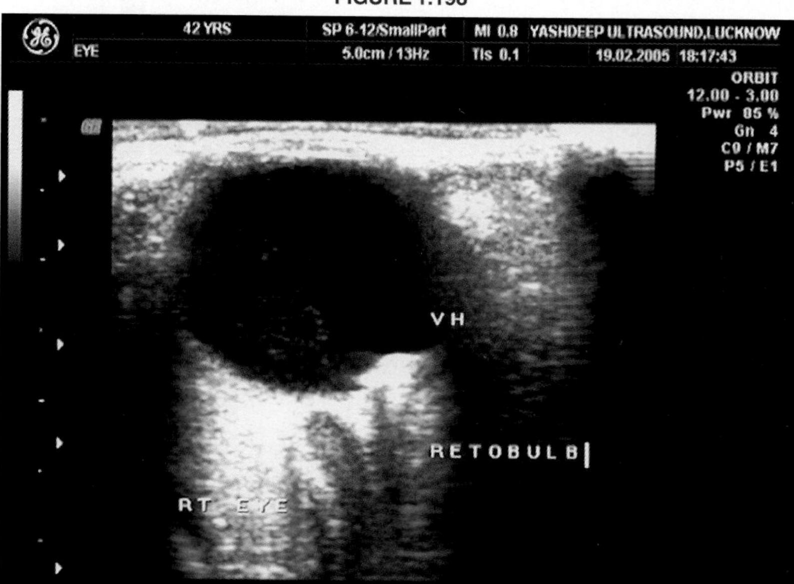

FIGURE 1.199

FIGURES 1.198 and 1.199: Subretinal foreign body. HRSG shows a big hyperechoic foreign body embedded in the subretinal space. Associated subretinal hemorrhage is also seen. Trajectory of foreign body is also seen with intragel hemorrhage.

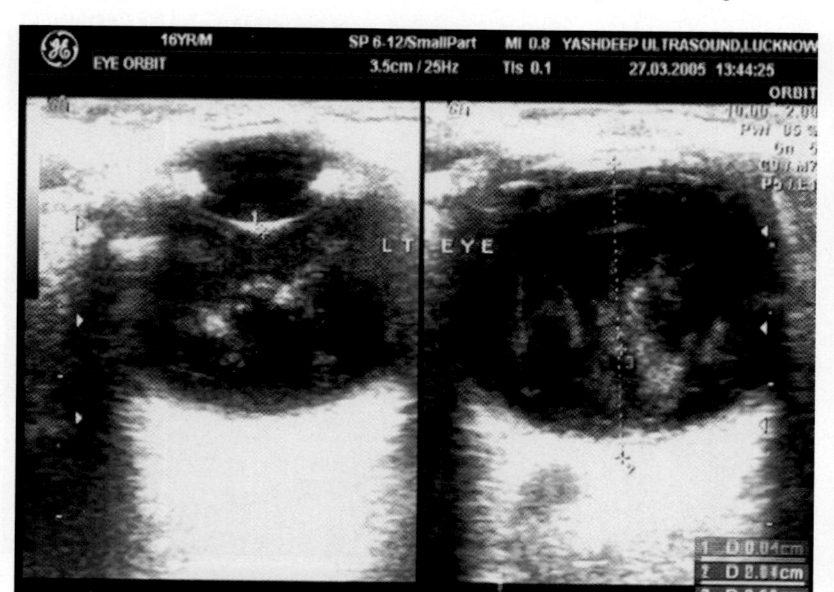

FIGURE 1.200: HRSG shows dense intragel hemorrhage in penetrating trauma. However, no foreign body is seen on 2D imaging.

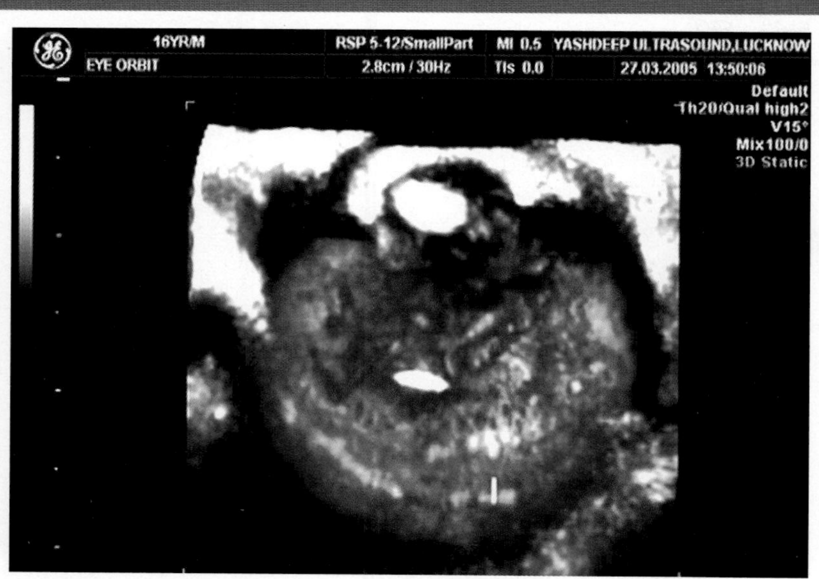

FIGURE 1.201: 3D imaging shows hyperreflective foreign body embedded in the hemorrhage, which was masked by the hemorrhage on 2D imaging.

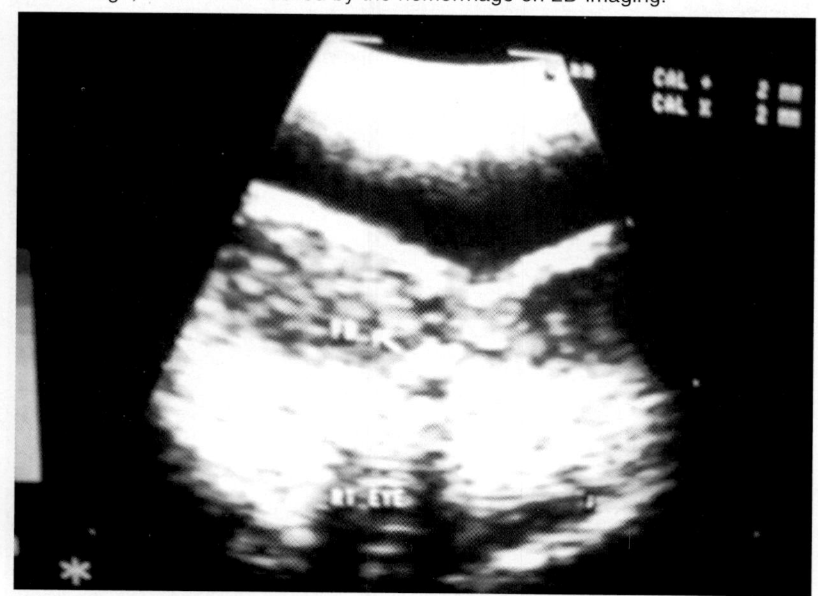

FIGURE 1.202: Penetrating injury to the eye. HRSG shows a hyperdense metallic density foreign body lodged in the posterior surface just above the sclera. Subvitreal hemorrhage with thick echogenic scleral fold buckling is also seen.

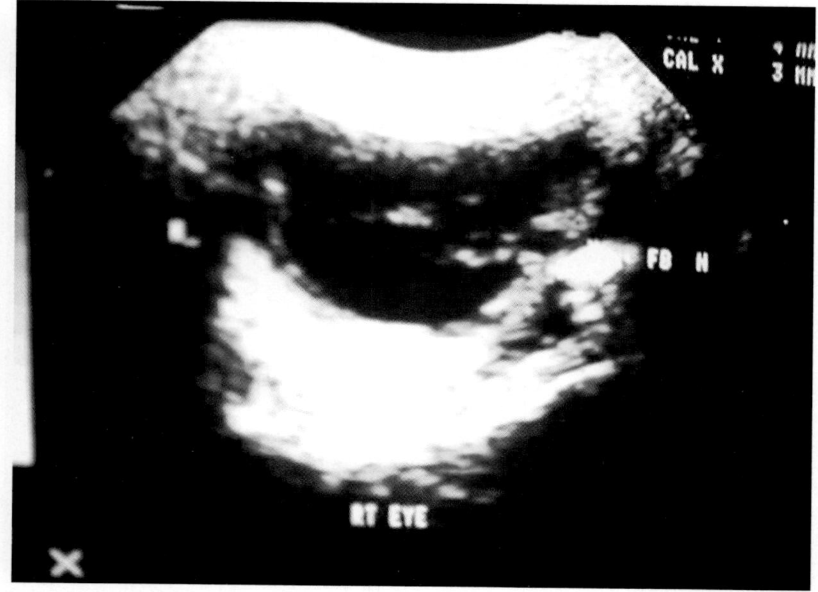

FIGURE 1.203: Scleral rupture. HRSG shows a dense echogenic foreign body lodged in the episcleral space after pearcing the sclera resulting into the rupture of the sclera. Collapsed of the eyeball is seen with intragel hemorrhage.

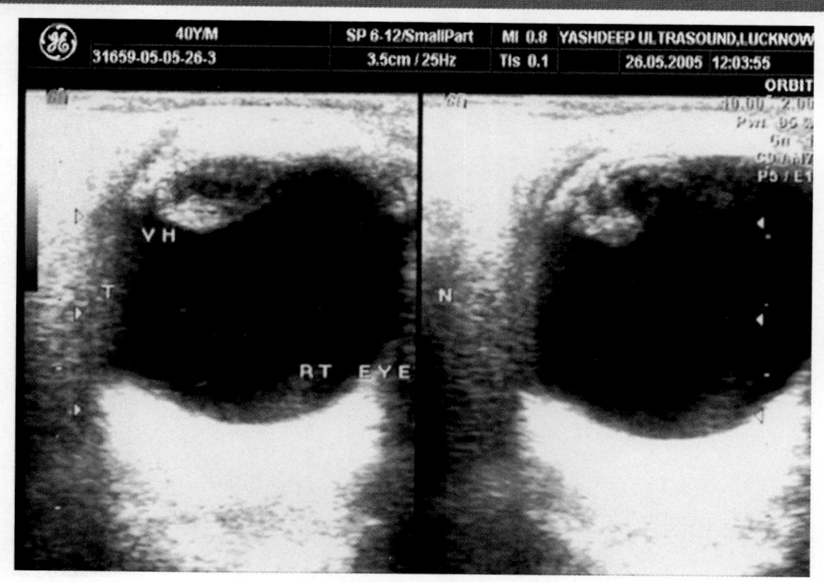

FIGURE 1.204

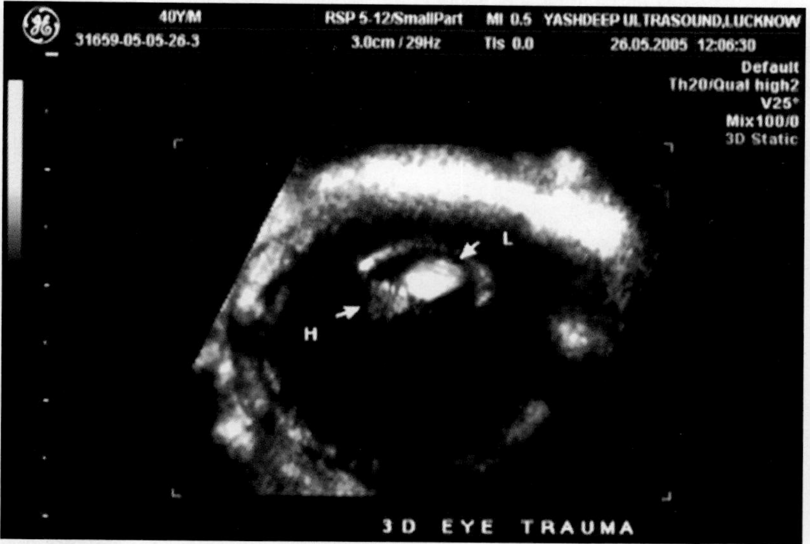

FIGURE 1.205

FIGURES 1.204 and 1.205: Organized intragel hemorrhage mimicking foreign body. HRSG shows organized intragel hemorrhage in the anterior chamber close to the lens mimicking as foreign body. 3D imaging clearly shows the organized hemorrhage instead of the foreign body.

Intraocular Tumors

High-resolution sonography (HRSG) has become one of the foremost investigation in evaluation of ocular tumors. It has also become the most important, noninvasive method to evaluate ocular tumors in the presence of clear ocular media. HRSG provides valuable assessment of size, shape, growth and regression of ocular tumor.

Sonographic Criteria

An intraocular tumor must have a minimum thickness or elevation before it can be picked up by HRSG. A lesion must have an elevation of minimum 8 mm before it can be picked up on HRSG. Choroidal tumors are more elevated than surrounding retinochoroid layer in order to be detected by HRSG. Ciliary body lesion need to be more elevated than choroidal lesion due to irregular surface of the ciliary body. A lesion as small as 3 mm can be easily picked up on HRSG. Other criteria includes reflectivity of lesions. Hypoechoic lesions are easily picked up than echogenic lesions. HRSG can tell about base, diameter and extraocular extension of the growth, solid and cystic nature of the mass.

Tumors of Anterior Chamber

HRSG is important investigation for evaluation of tumors of the anterior chamber. Like iris tumors, anterior ciliary body tumors, vascular masses of sclera.

Medulloepithelioma (Diktyoma)

Medulloepithelioma or diktyoma is a rare tumor of ciliary body. It is seen in children. The tumor presents as a whitish mass within the ciliary body. These tumors may be associated with iris cysts or free floating cysts within the anterior chamber or vitreous cavity. The tumors may be echogenic or shows moderate echogenicity. On color flow imaging moderate flow is seen in them. The teratoid variant of the tumor shows heterogeneous texture due to cartilage.

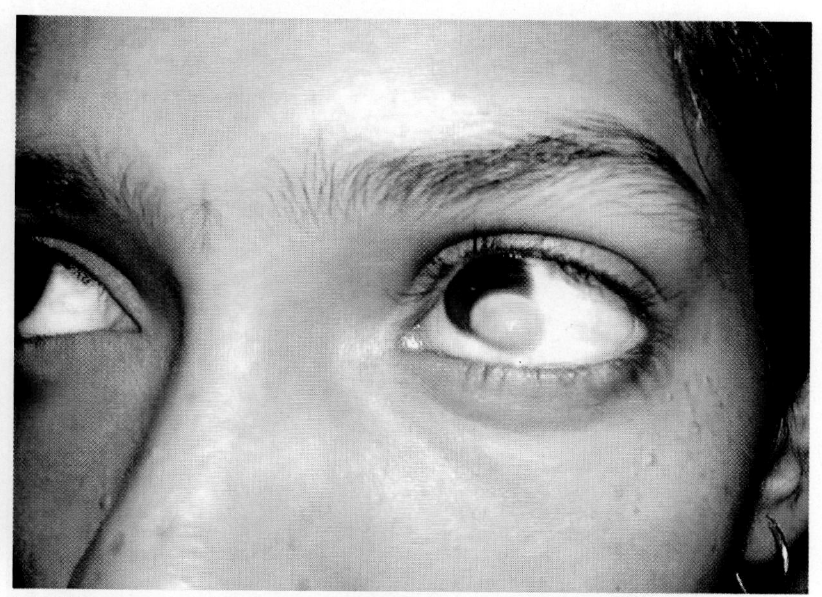

FIGURE 1.206: A well-defined mass with sharp seen over the left eye encircling the pupil and hampering vision at cornea.

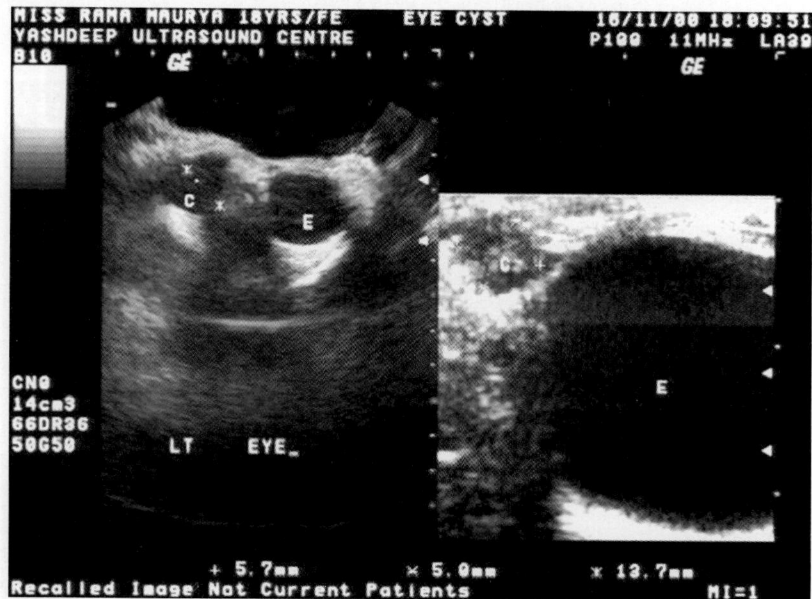

FIGURE 1.207: HRSG shows a well-defined cyst coming out from corneoscleral junction and also involving the iris. Multiple thick echoes are seen in it with thick wall due to keratinization.

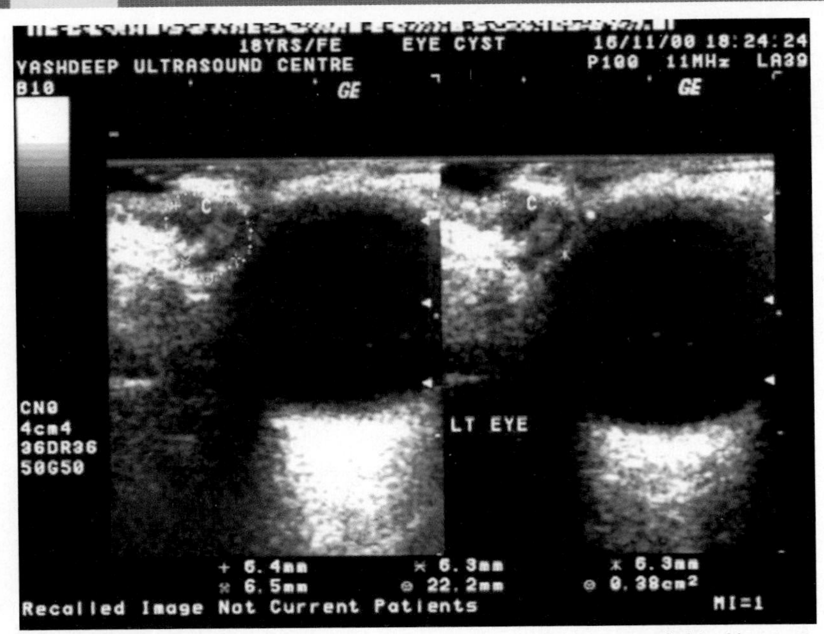

FIGURE 1.208: Part of the inner cyst is seen involving the iris on nasal side. However, the eyeball shows normal posterior segment.

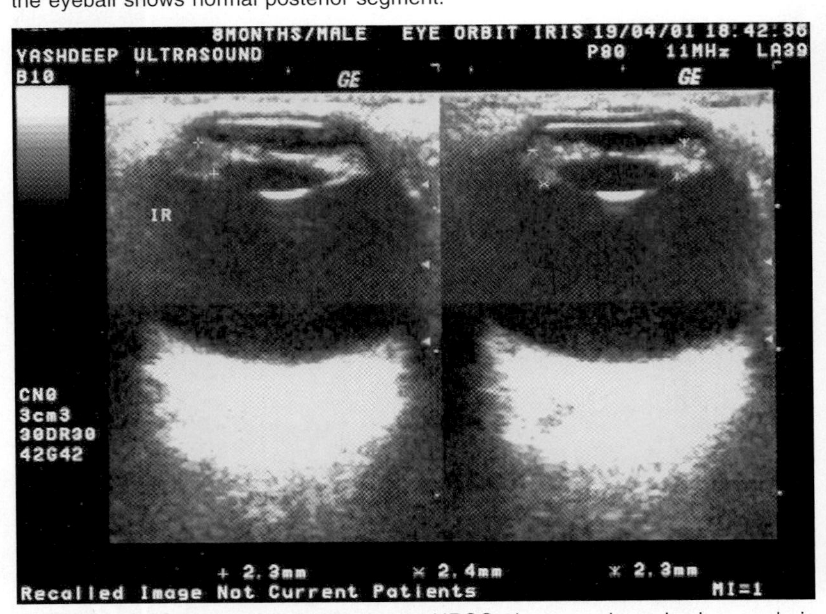

FIGURE 1.209: Iris mass: Epithelioma. HRSG shows an irregular hyperechoic mass in the iris in a four month old child.

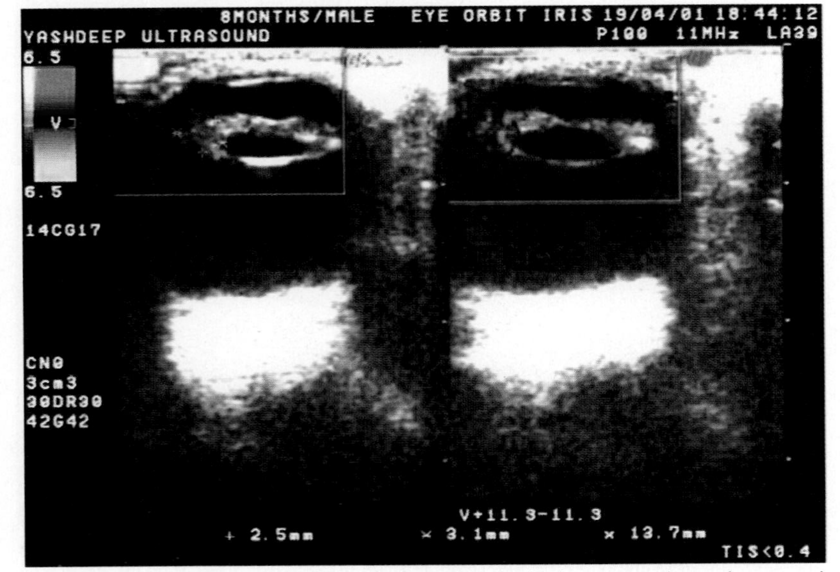

FIGURE 1.210: Iris mass. On color Doppler study the mass shows increased vascularity. Biopsy showed epithelioma.

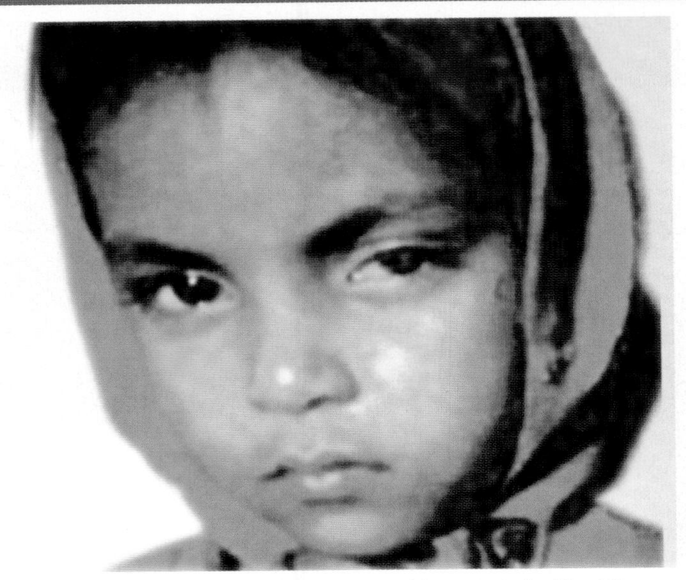

FIGURE 1.211: Diktyoma. A young girl presented with white eye reflex and a mass in anterior chamber on clinical examination.

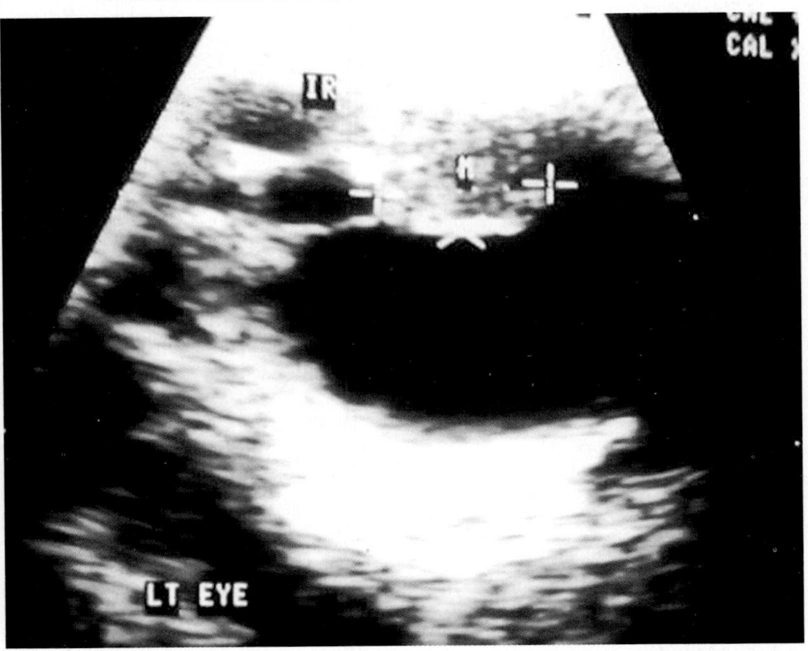

FIGURE 1.212: HRSG shows a well-defined homogeneous mass coming out from the ciliary body and confined to the anterior chamber. The sonographic findings are suggestive of diktyoma or medulloepithelioma.

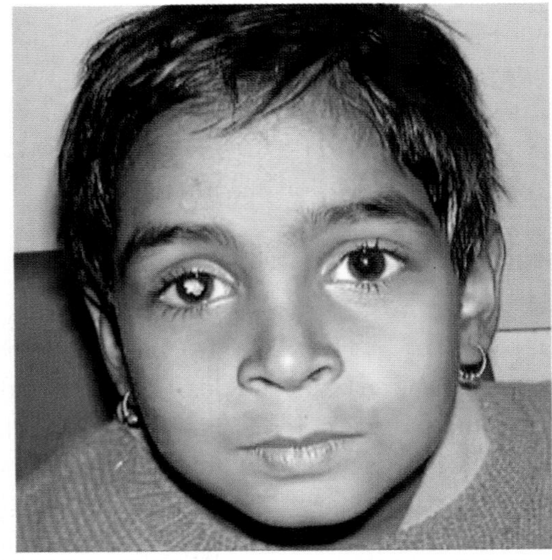

FIGURE 1.213: Diktyoma. A young girl presented with white eye reflex and diminished vision on clinical examination.

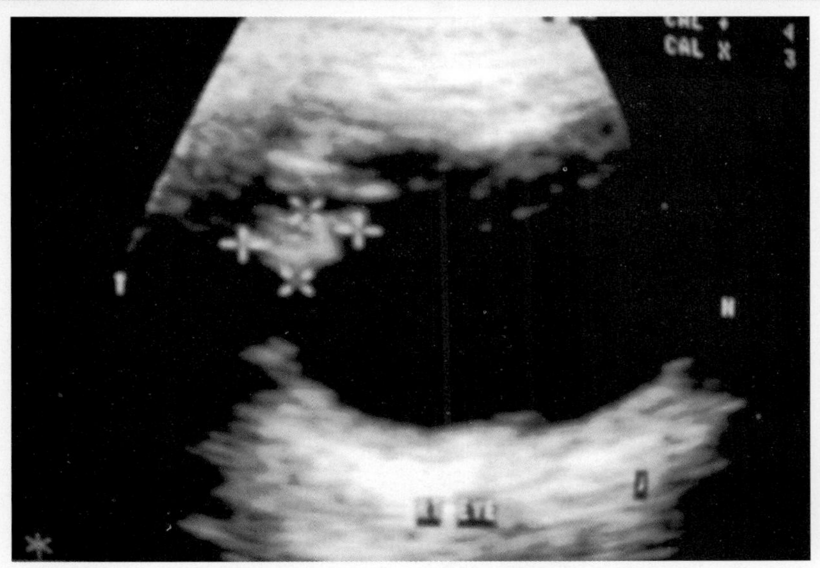

FIGURE 1.214: HRSG shows a well-defined echogenic mass coming out from the ciliary body and extending into the vitreous cavity. No cystic degeneration is seen in the mass. It came out to be diktyoma on Biopsy.

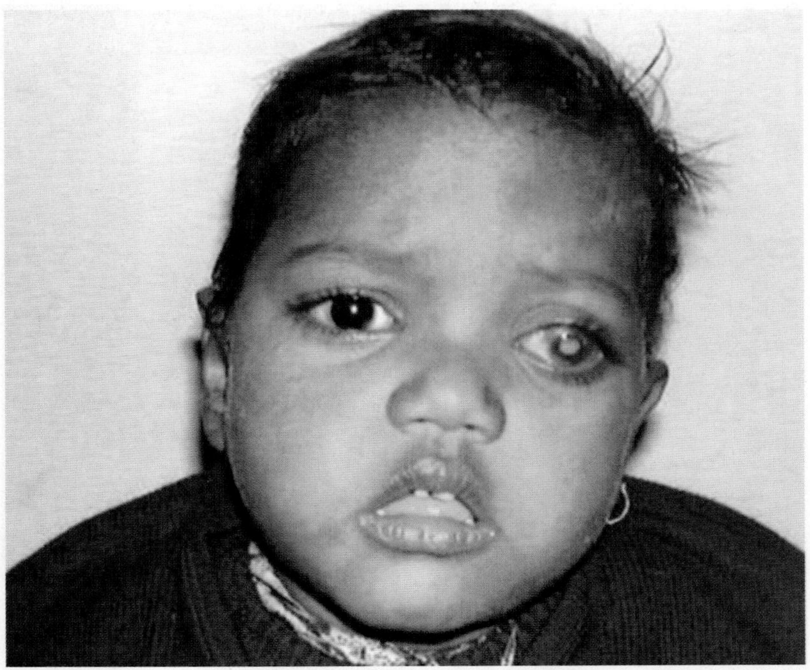

FIGURE 1.215: A young boy presented with Leukoria and loss of vision in Lt eye.

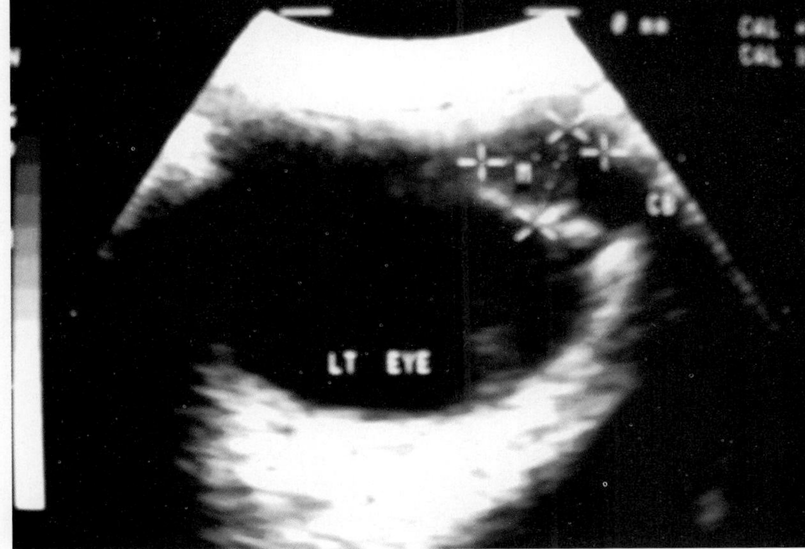

FIGURE 1.216

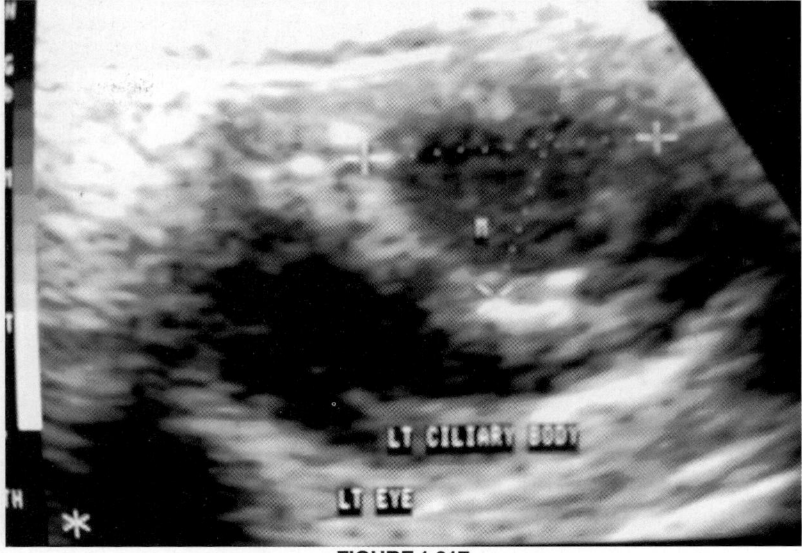

FIGURE 1.217

FIGURES 1.216 and 1.217: Diktyoma. HRSG shows a hypodense well-defined mass coming out from Lt ciliary body. A linear calcification is seen in the mass. The ciliary body has been destroyed by the mass.

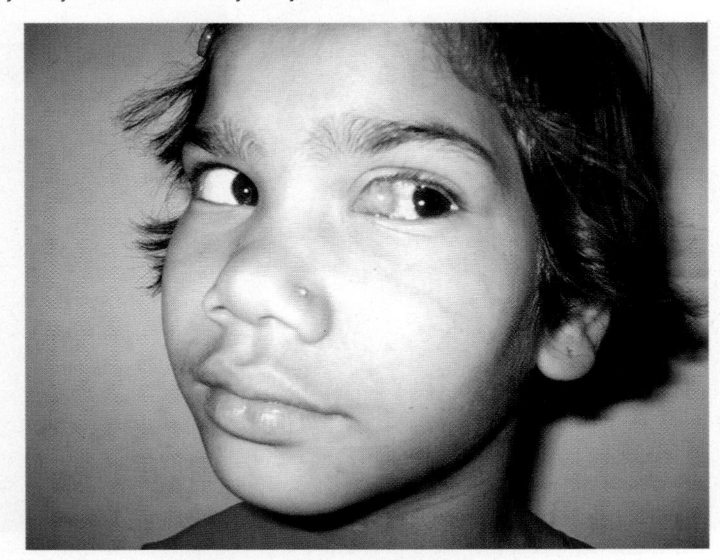

FIGURE 1.218: Medial canthus cysticercosis. A young girl presented with cystic mass over the medial canthus.

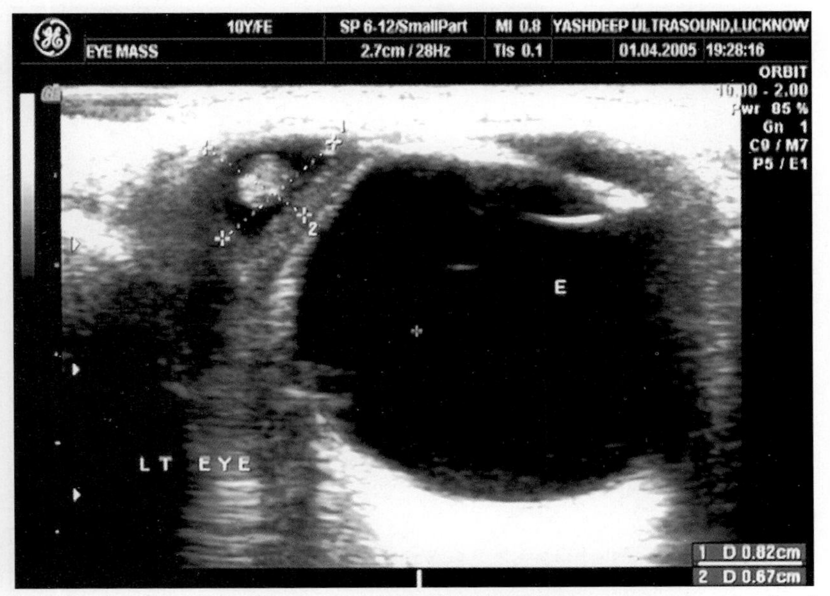

FIGURE 1.219: HRSG shows a well-defined cyst having an echogenic nidus fixed with inner wall of the cyst typical of cysticercus cyst.

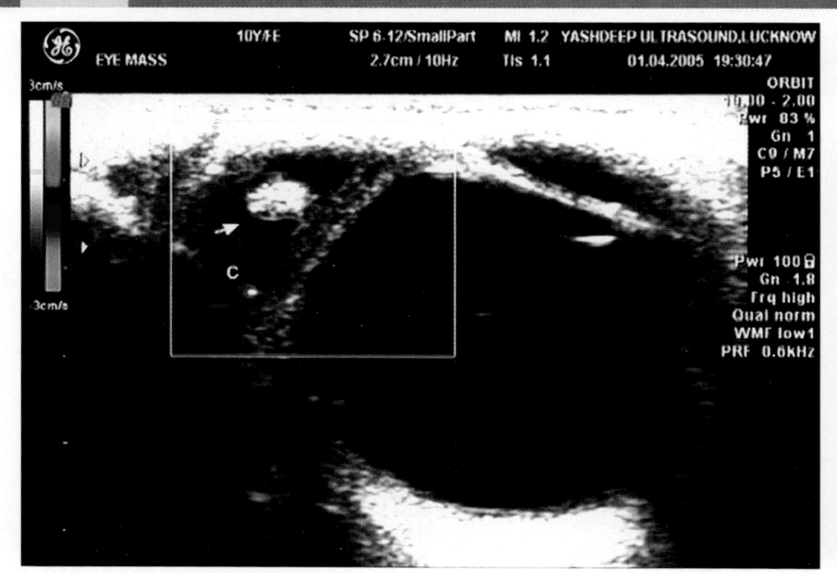

FIGURE 1.220: Color Doppler imaging shows no flow in the cyst.

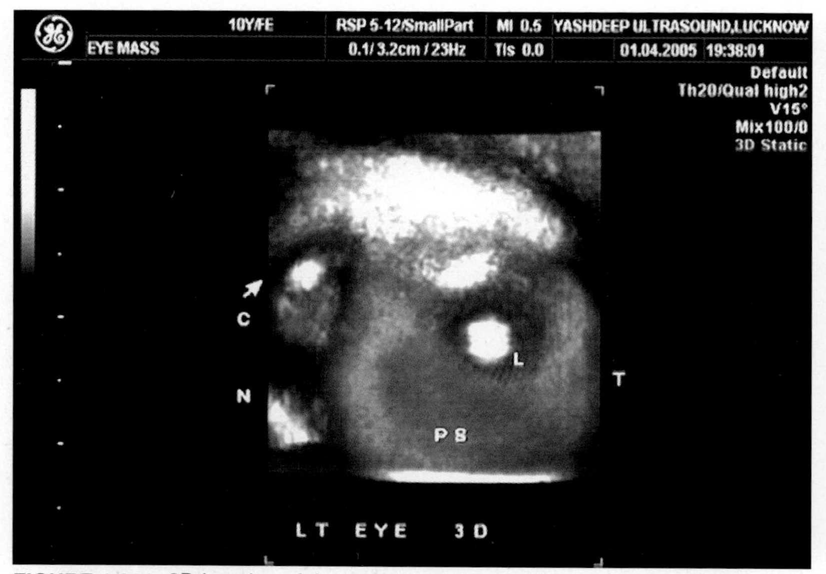

FIGURE 1.221: 3D imaging of the cyst shows the finer details of the cyst. It is seen confined to the sclera.

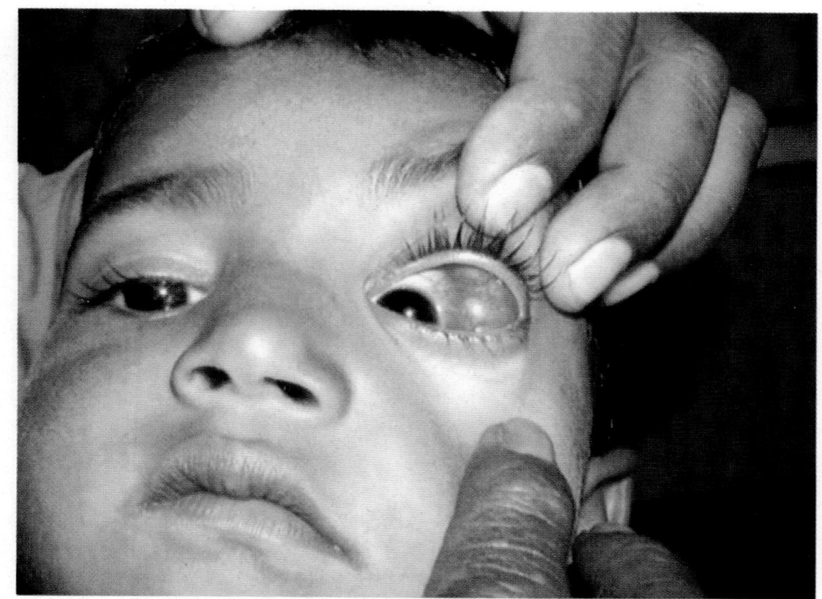

FIGURE 1.222: Lateral canthus cysticercus. A young child presented with left lateral canthus swelling.

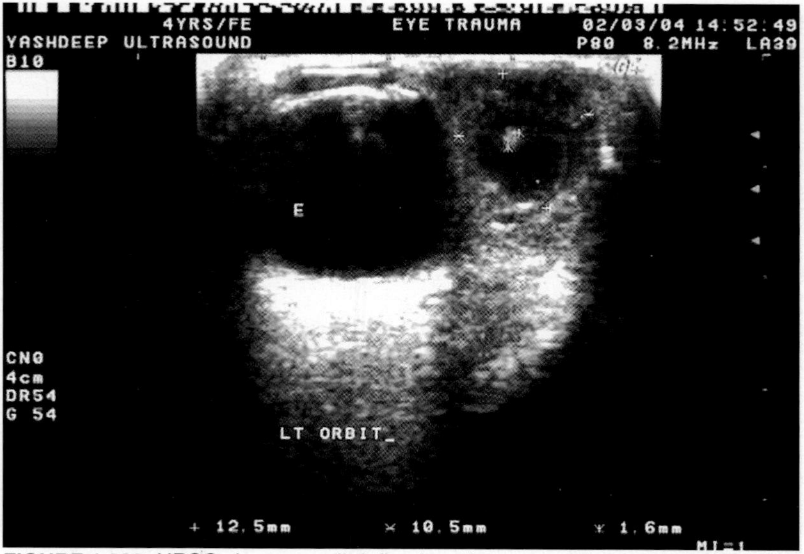

FIGURE 1.223: HRSG shows a well-defined cystic mass coming out from the lateral canthus. An echogenic nidus is seen fixed with the inner wall of the cyst typical of cysticercus cyst,

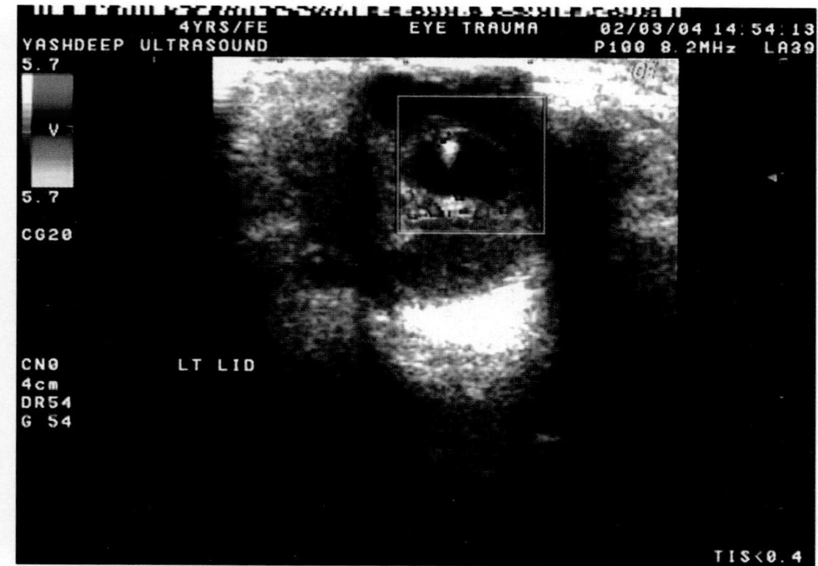

FIGURE 1.224: Color Doppler imaging of the cyst shows cyst wall hyperemia.

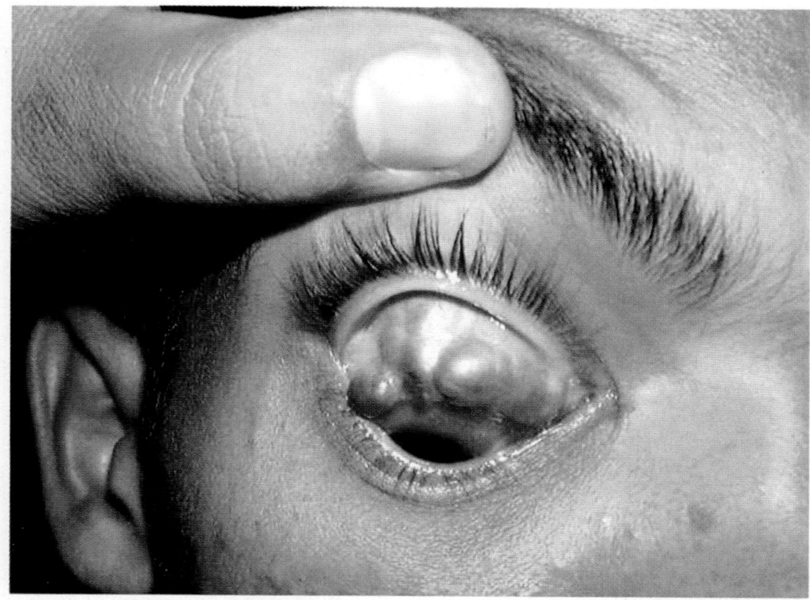

FIGURE 1.225: Cavernous hemangioma. A young man presented with mass having multiple dilated vessels in upper part of Rt eye.

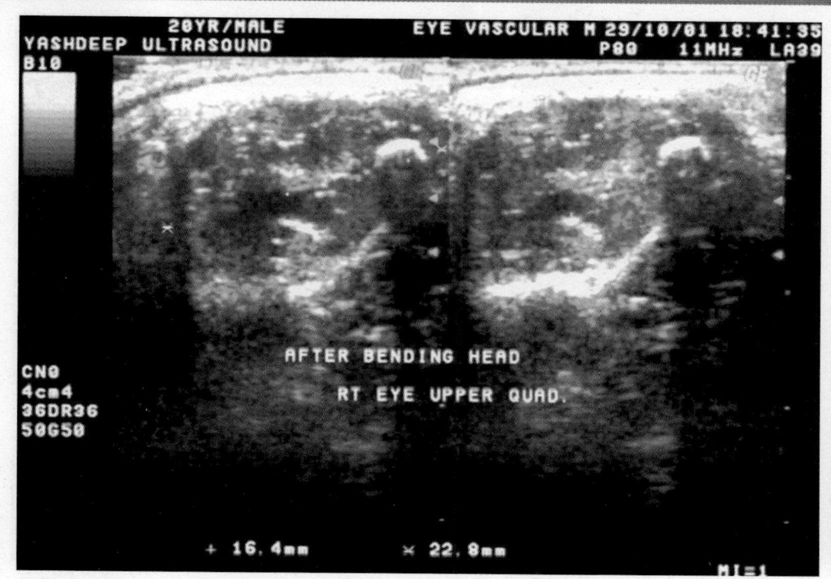

FIGURE 1.226: HRSG shows a soft tissue mass with multiple dilated vessels. Echogenic calcified shadows are also seen in the mass. They are associated with acoustic shadowing suggestive of phlebolith.

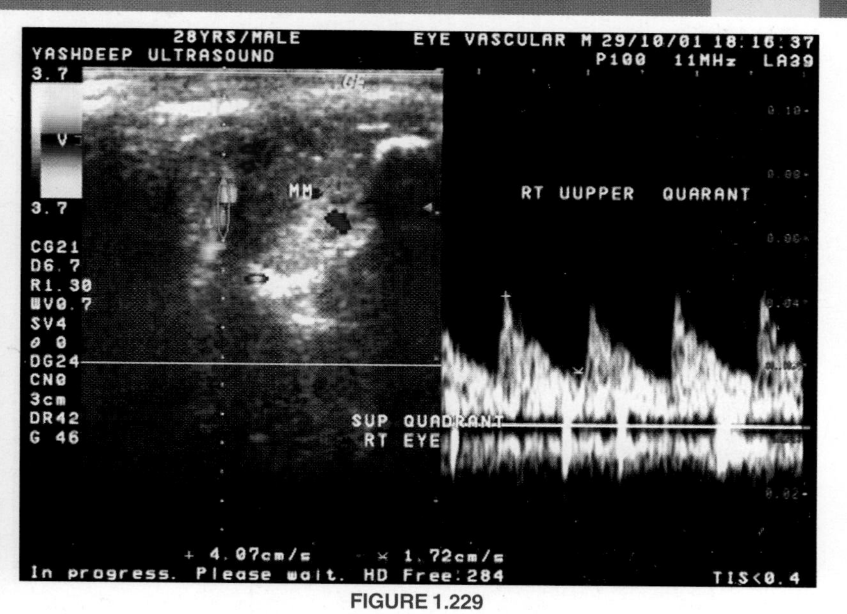

FIGURE 1.229

FIGURES 1.228 and 1.229: Color Doppler tracing shows venous and arterial flow in the mass.

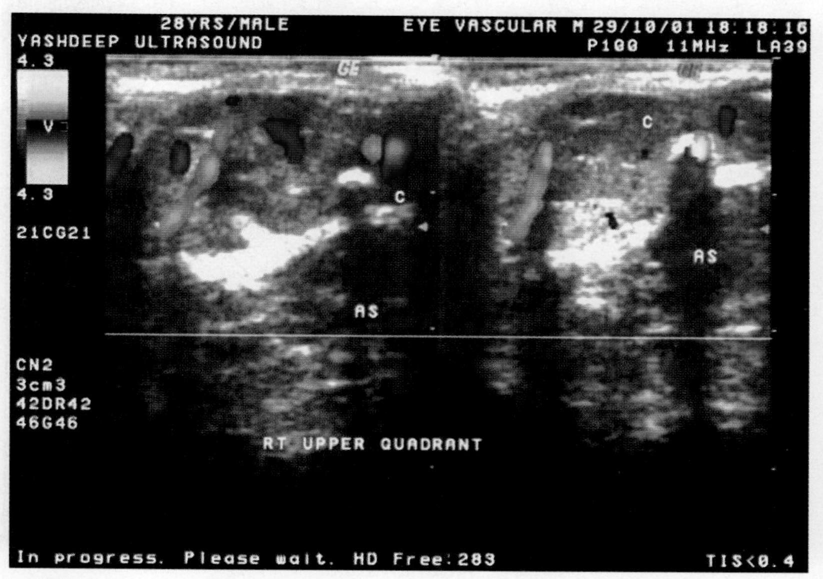

FIGURE 1.227: On color Doppler flow study, low flow is seen in the blood vessels with venous flow pattern.

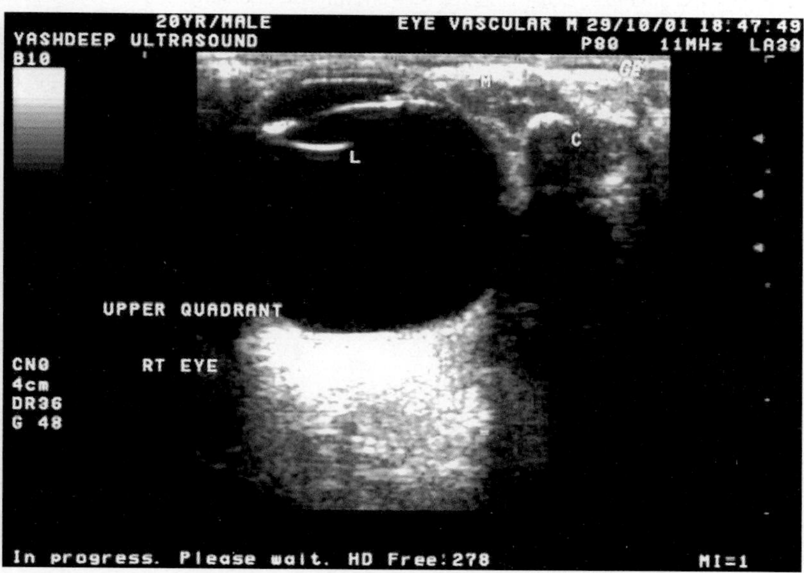

FIGURE 1.230: The mass is seen confined to the sclera anteriorly. It is not invading the eyeball posterior segment.

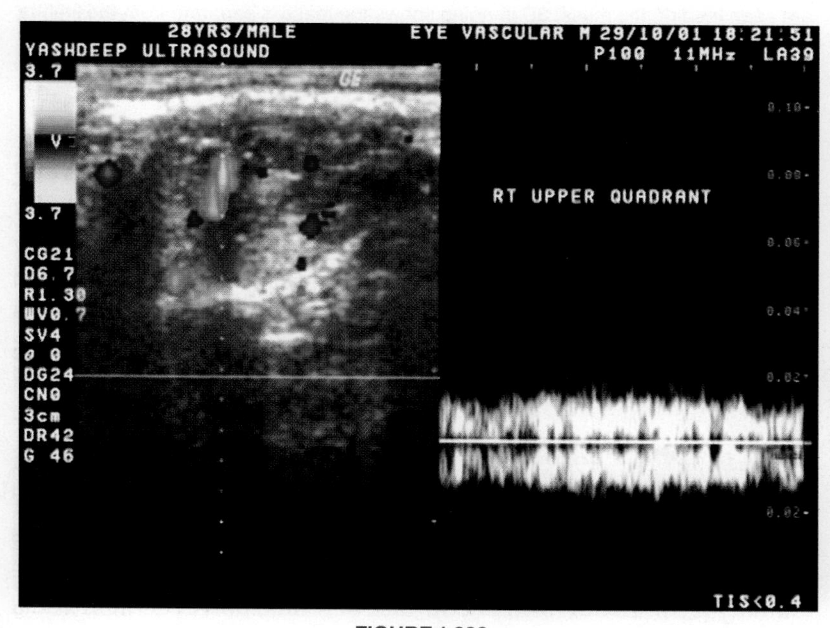

FIGURE 1.228

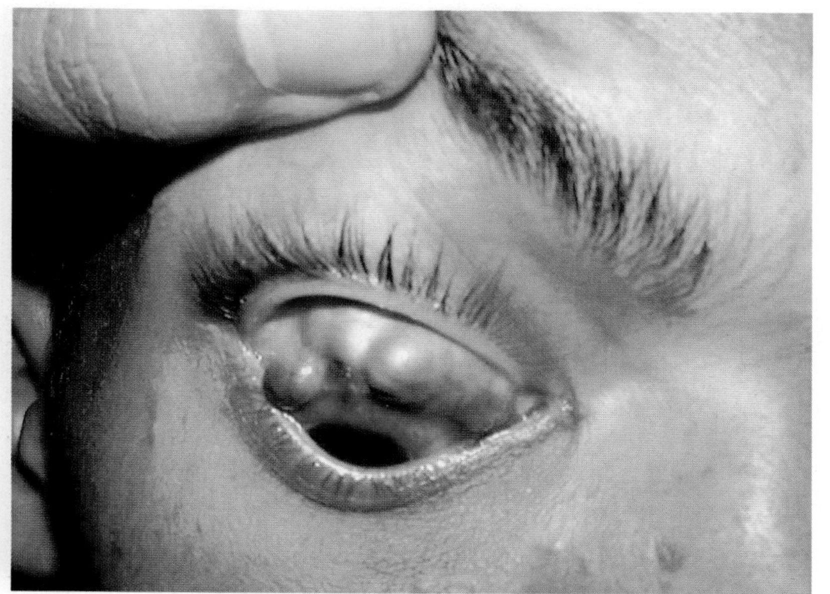

FIGURE 1.231A

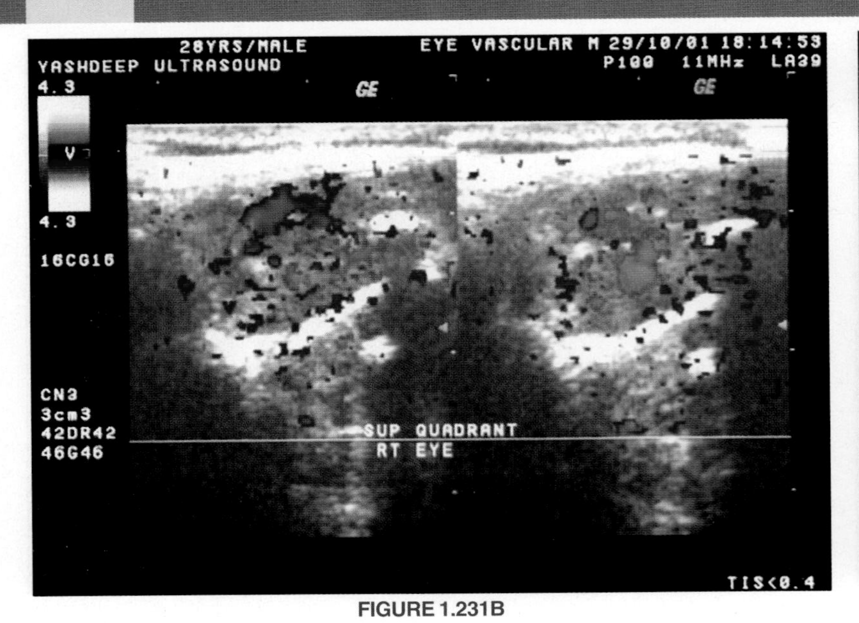

FIGURE 1.231B

FIGURES 1.231A and B: On persistent bending of head vessels show engorgement and increased in size. Color Doppler flow shows vessel enlargement.

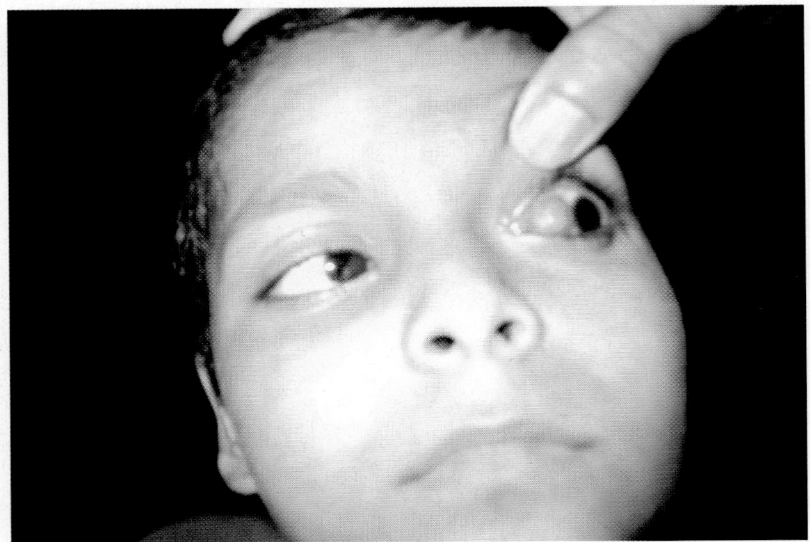

FIGURE 1.232: Soft tissue mass is coming out from medial side of sclera. A young child presented with a soft tissue mass coming out from the medial side of sclera reaching up to the cornea.

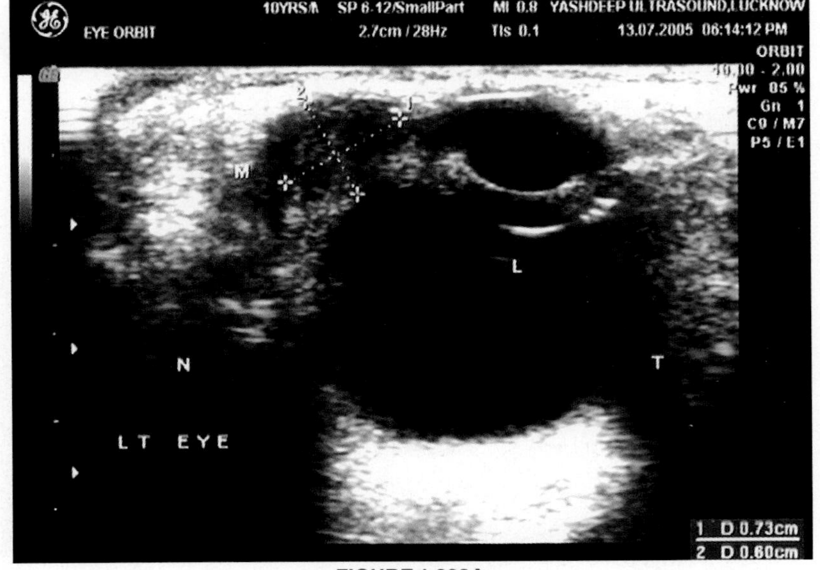

FIGURE 1.232A

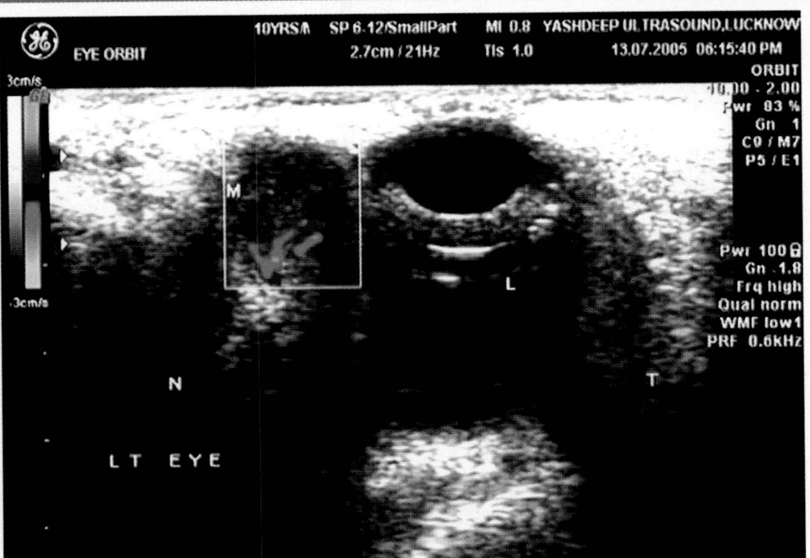

FIGURE 1.232B

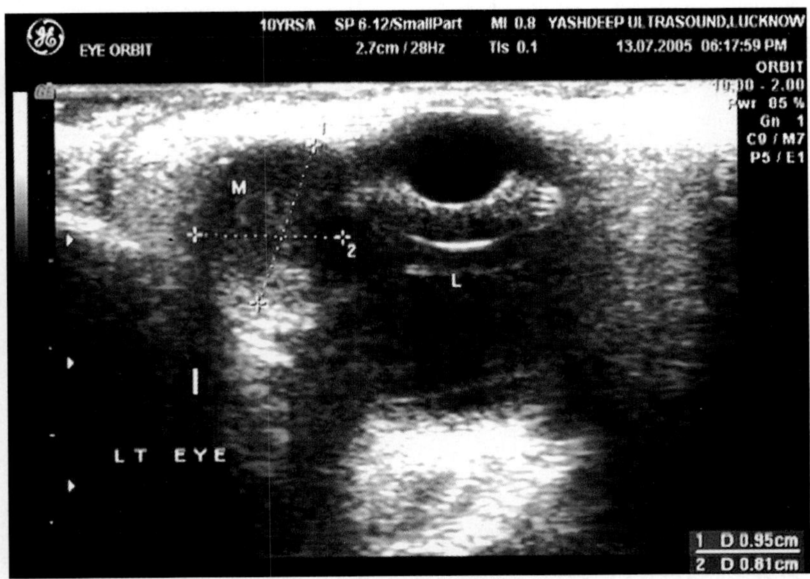

FIGURE 1.232C

FIGURES 1.232A to C: HRSG shows a medium level soft tissue mass coming out from the sclera. It is homogeneous in texture. On color flow imaging multiple vessels are seen feeding the mass. 3D imaging shows the mass is separate from the limbus. Biopsy shows a vascular mass.

Retinoblastoma

Retinoblastoma is the most common intraocular tumor in infant and children up to the age of 6 year. It comprises about 30% of all ocular tumors. It may be unilateral, bilateral, focal or multifocal and frequently presents as white reflex or cat eye reflex in the children. The tumor can grow either anteriorly from inner surface of the retina towards the vitreous (endophytic), or posteriorly form the posterior surface of retina towards the choroid (exophytic).

Sonographic Features

Sonographically the tumor can have smooth dome-shaped appearance when it is small in size. However, it is frequently having heterogeneous and irregular texture. Calcium deposit is the hallmark of retinoblastoma. Usually it comes out from one surface of retina, however, in unusual

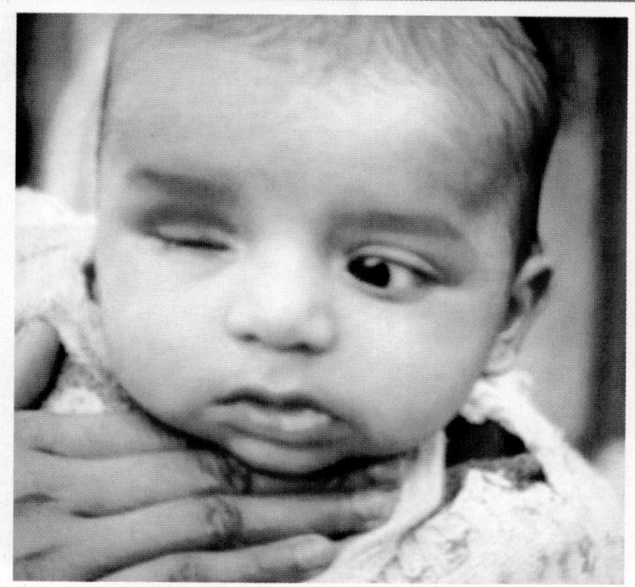

FIGURE 1.233: An infant presented with congenital retinoblastoma in Rt eye and it was operated. His father also had retinoblastoma in one eye. Another eye also showed white reflex after 3 month of his birth.

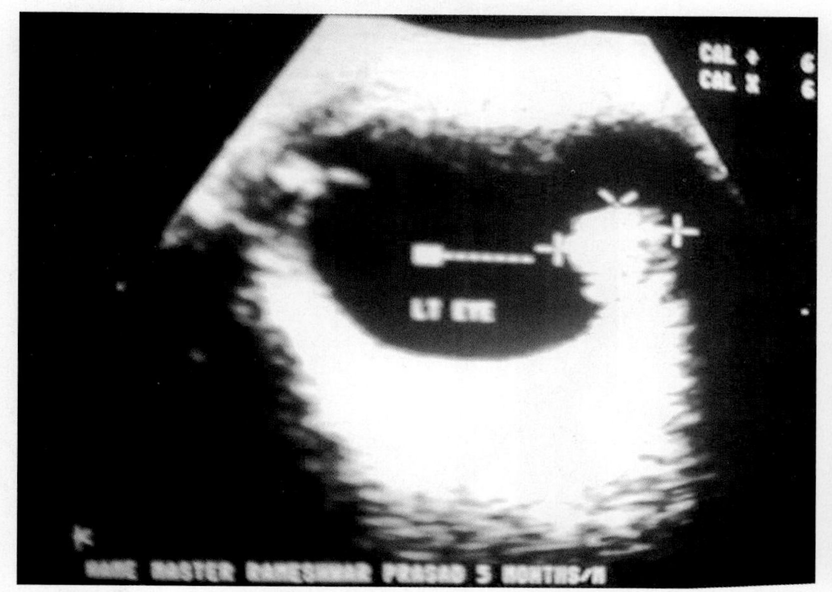

FIGURE 1.234: HRSG shows an echogenic mass on temporal side with wide base typical of retinoblastoma.

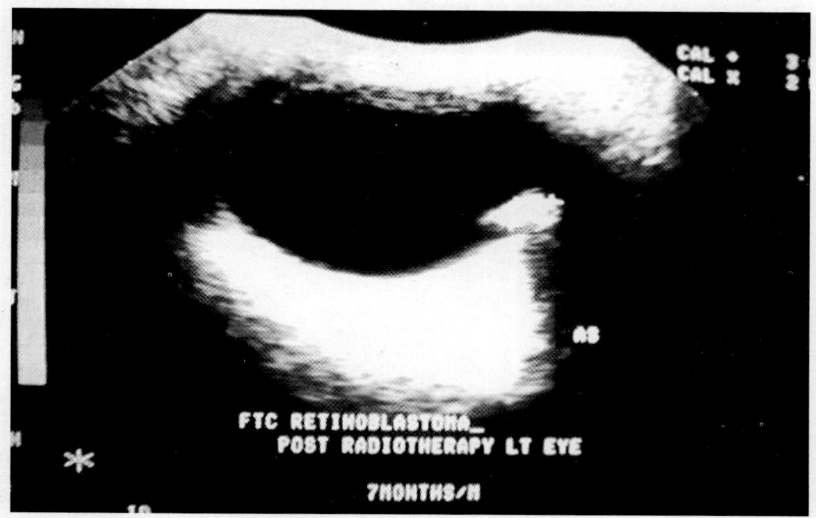

FIGURE 1.235: **Post tele radiotherapy.** The mass has regressed in size. Echogenic calcification is seen at the base of the mass. HRSG is an excellent modality to evaluate treatment response. It is non-invasive, cost effective and does not require any sedation.

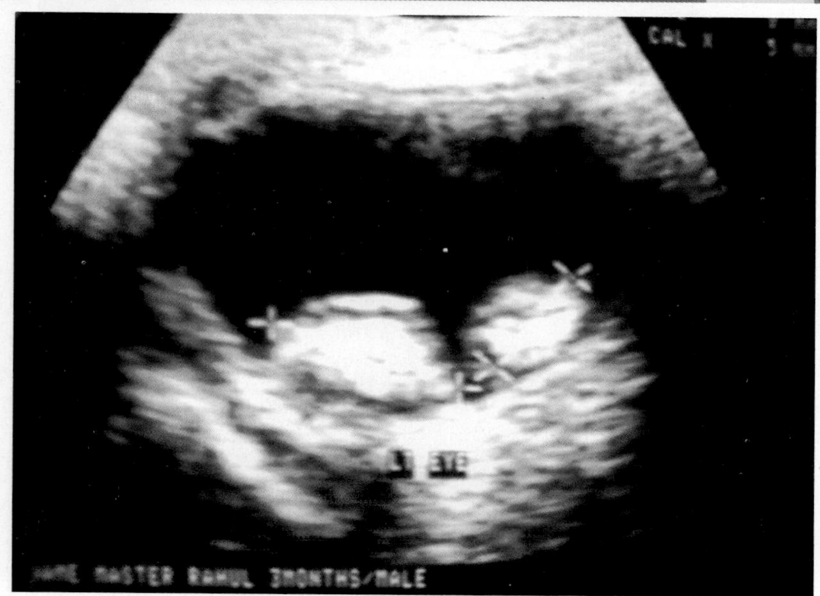

FIGURE 1.236: **Retinoblastoma.** 3 month old child presented with white eye reflex. HRSG shows a multifocal mass coming out from the retina on the posterior pole. Dense echogenic calcification is seen in the mass. No evidence of any extraocular extension is seen. No intragel hemorrhage is seen on HRSG.

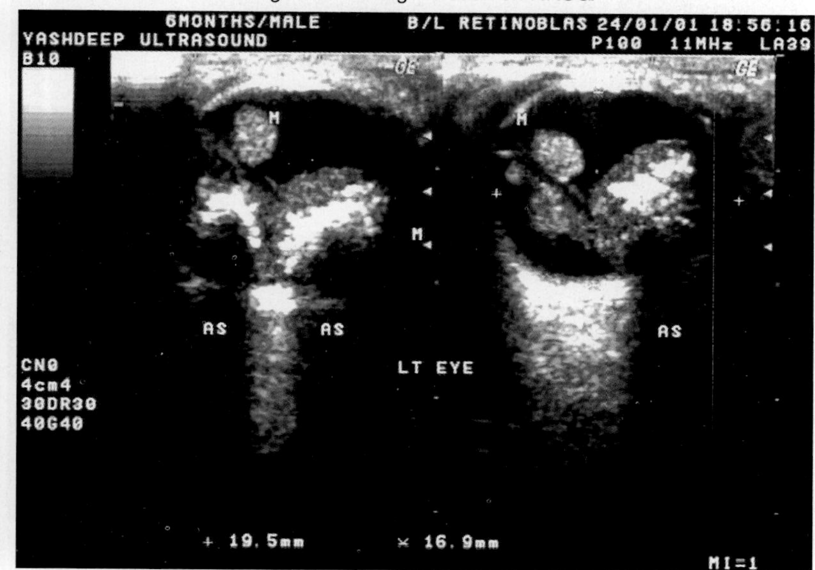

FIGURE 1.237: **Multifocal retroblastoma.** HRSG shows multifocal retinoblastoma invading whole of eyeball. Dense echogenic calcification is seen in the tumors.

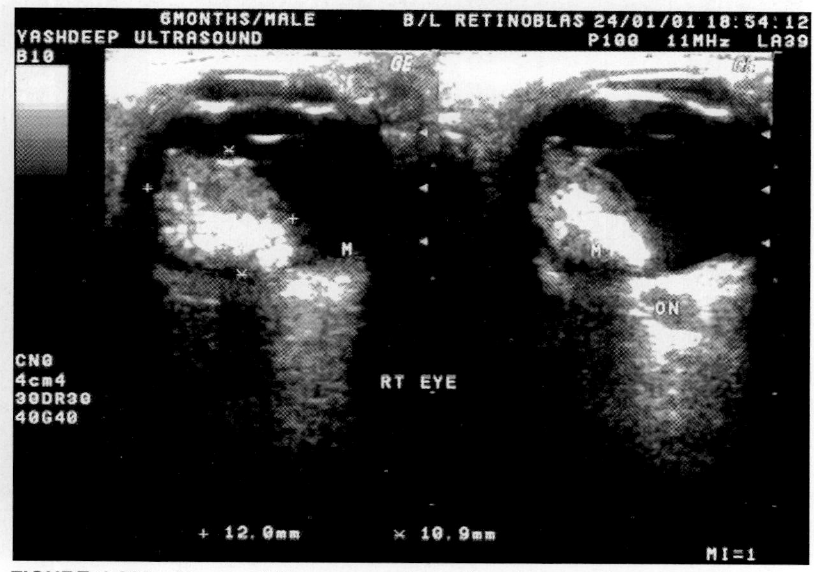

FIGURE 1.238: Rt eye of the same patient also shows a big tumor with dense calcification.

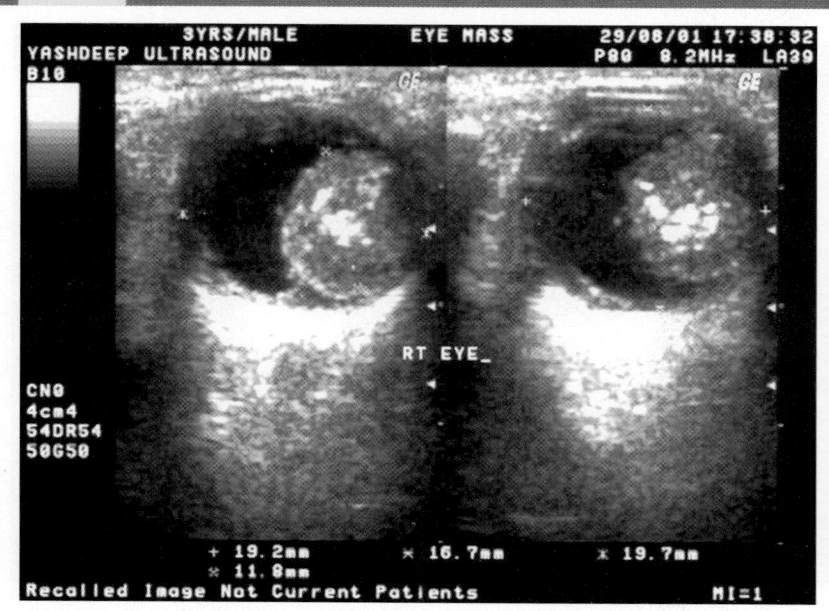

FIGURE 1.239: Lobulated retinoblastoma. A young child presented with Leukoria in both eyes with loss of vision in Rt eye.

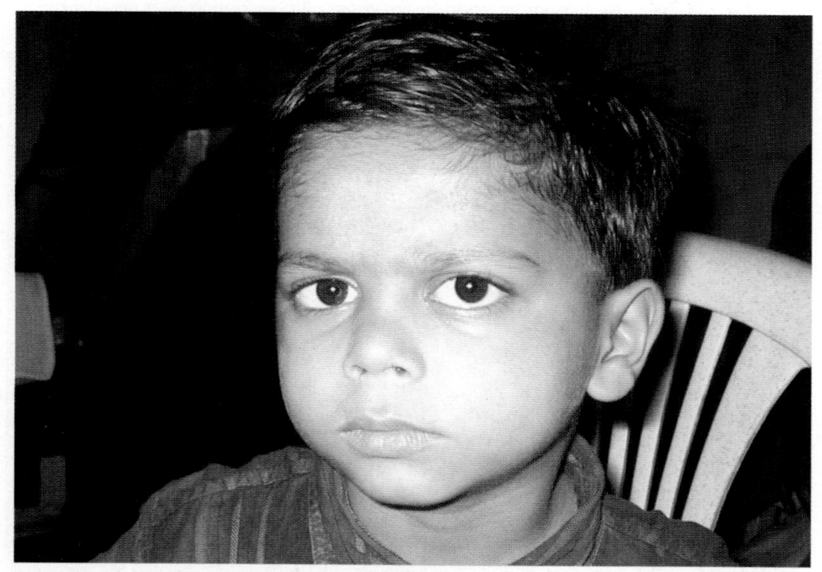

FIGURE 1.240: Bilateral retinoblastoma. A young child presented with bilateral cat eye reflex. It was more marked on Lt side.

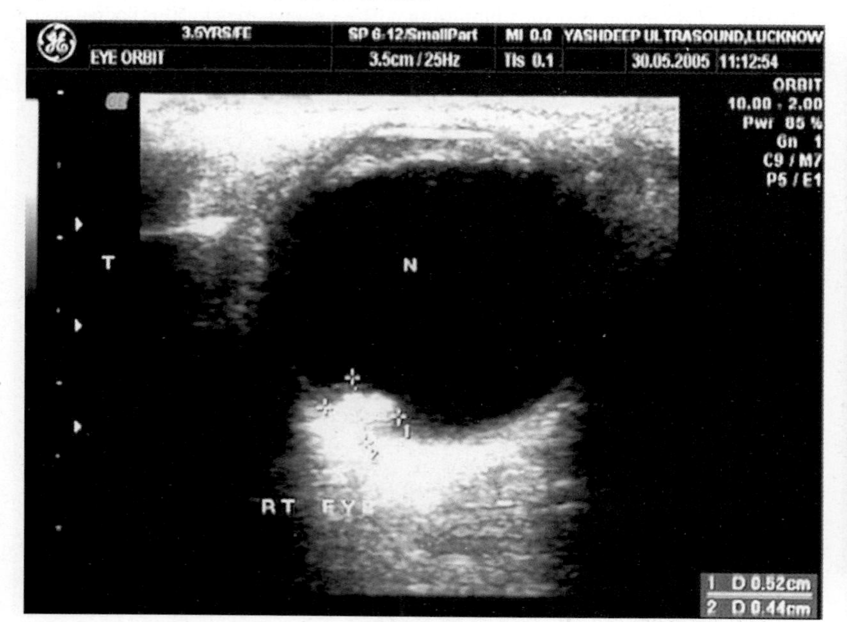

FIGURE 1.241: HRSG shows a small echogenic shadow fixed with the posterior pole of the eye with echogenic calcification suggestive of retinoblastoma.

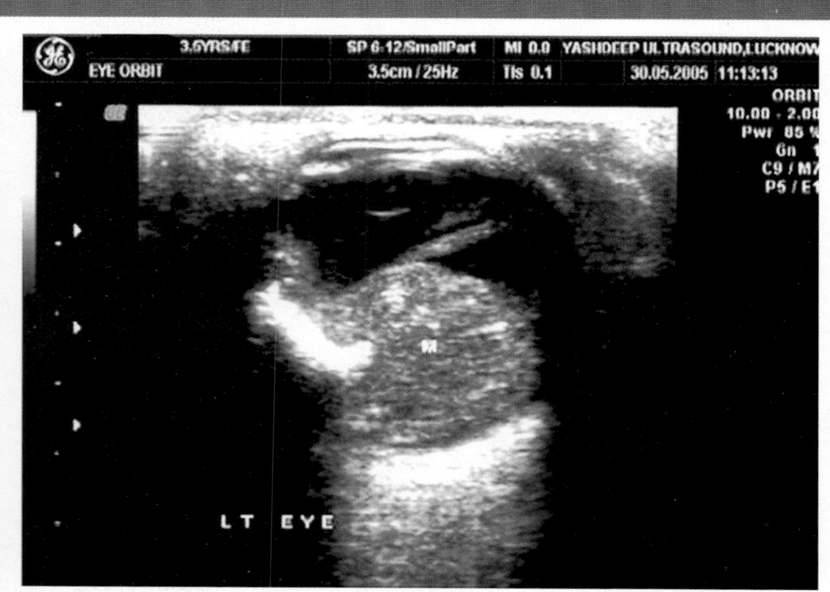

FIGURE 1.242: HRSG shows big tumor mass in Lt eye. It is irregular in outline. Dense calcification seen in the mass. However, it is confined to the retina. No choroidal excavation is seen.

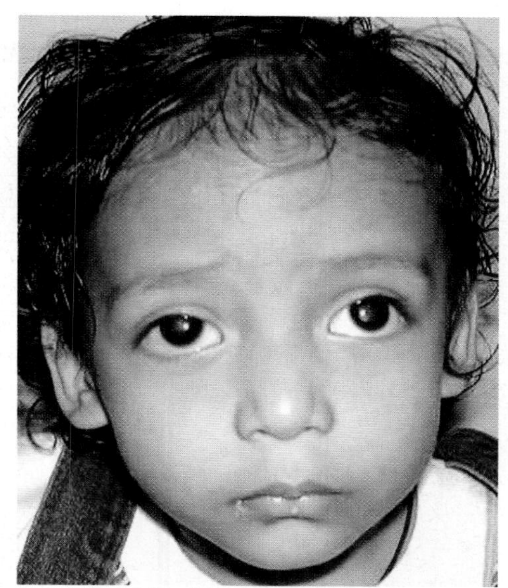

FIGURE 1.243: Bilateral retinoblastoma. A young child presented with white-eye reflex on either side.

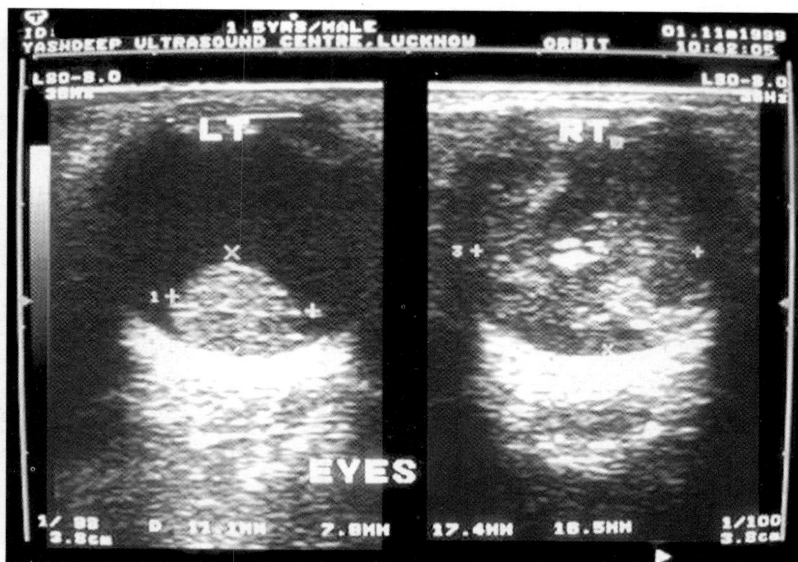

FIGURE 1.244: HRSG shows a tumor mass in the Rt eye with specks of calcification. Lt eye shows a big heterogeneous mass filling the posterior segment of the eye with dense calcification.

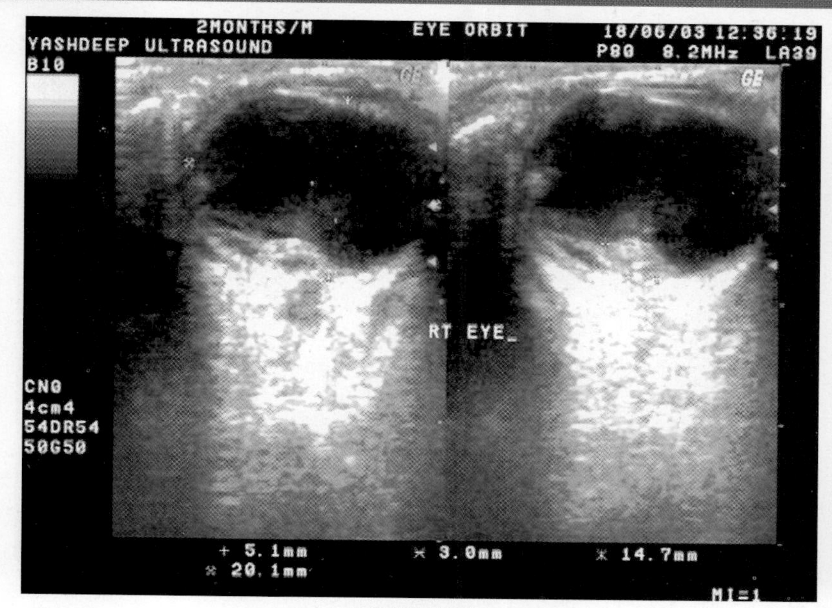

FIGURE 1.245

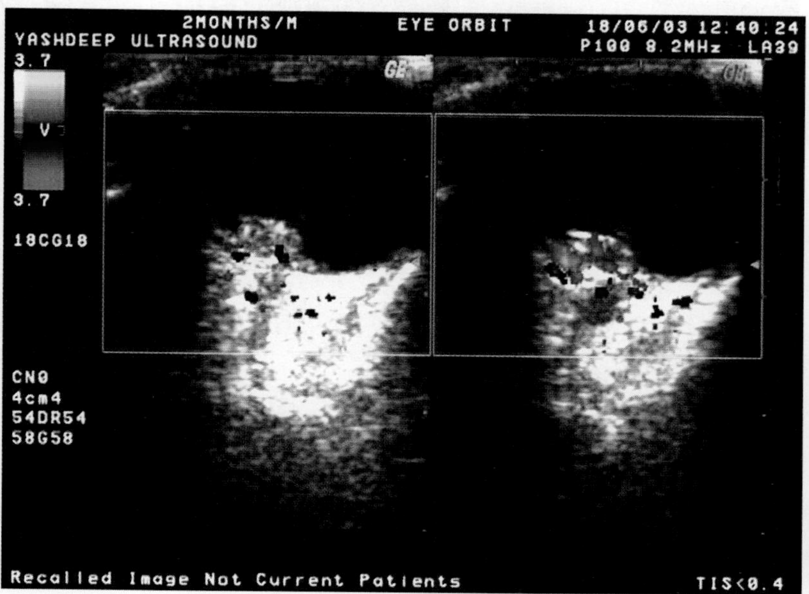

FIGURE 1.246

FIGURES 1.245 and 1.246: Small retinoblastoma. An infant was born with white-eye reflex in Rt eye. HRSG shows a small tumor fixed with the retina with fine calcification. **On color flow imaging** low flow is seen in the wall of the tumor.

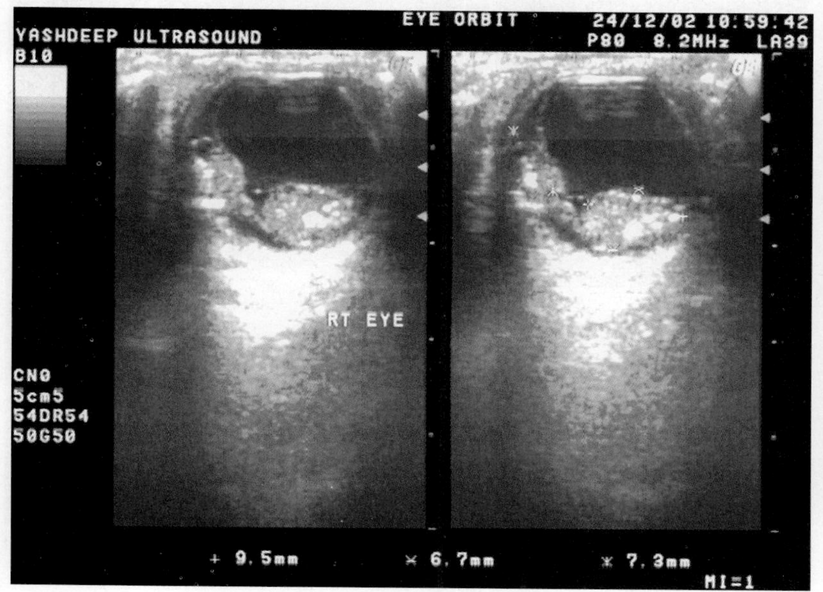

FIGURE 1.247

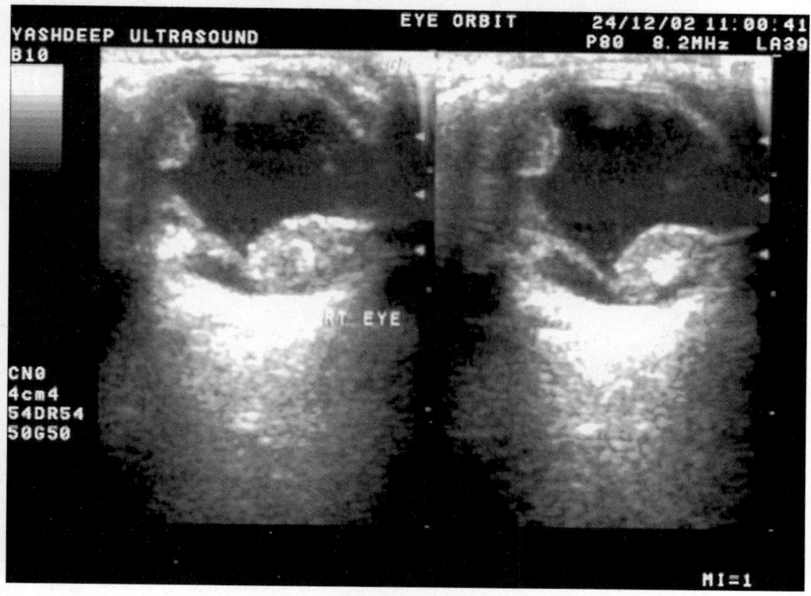

FIGURE 1.248

FIGURES 1.247 and 1.248: Multifocal retinoblastoma. HRSG shows multiple tumor masses coming out from the retina in the posterior segment of eye. The fine calcified specks are seen in the masses. Associated subretinal hemorrhage is also seen.

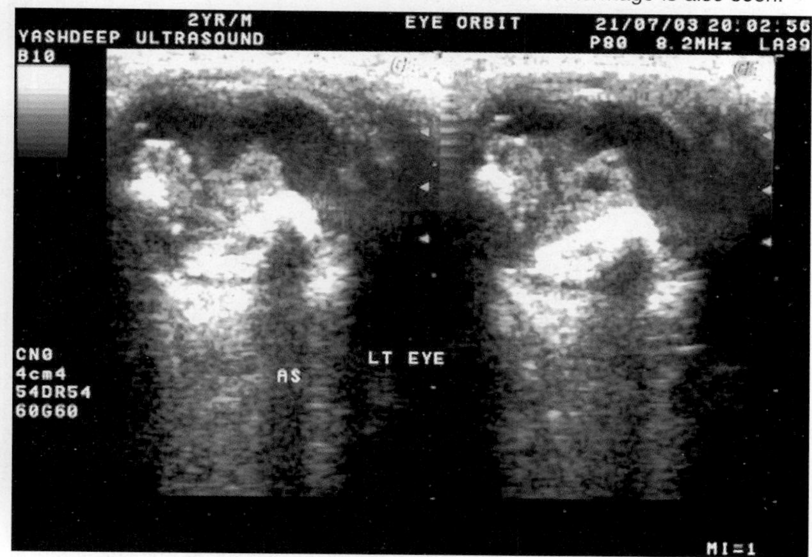

FIGURE 1.249A

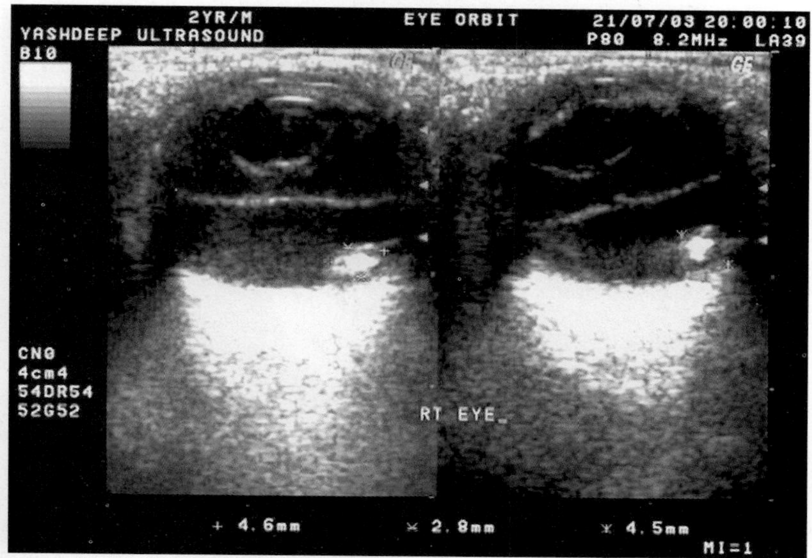

FIGURE 1.249B

FIGURES 1.249A and B: Bilateral retinoblastoma. HRSG shows a big heterogeneous tumor mass in the posterior segment of Lt eye. Dense echogenic calcification is seen in it. Rt eye also shows a small tumor confined to the eye coat.

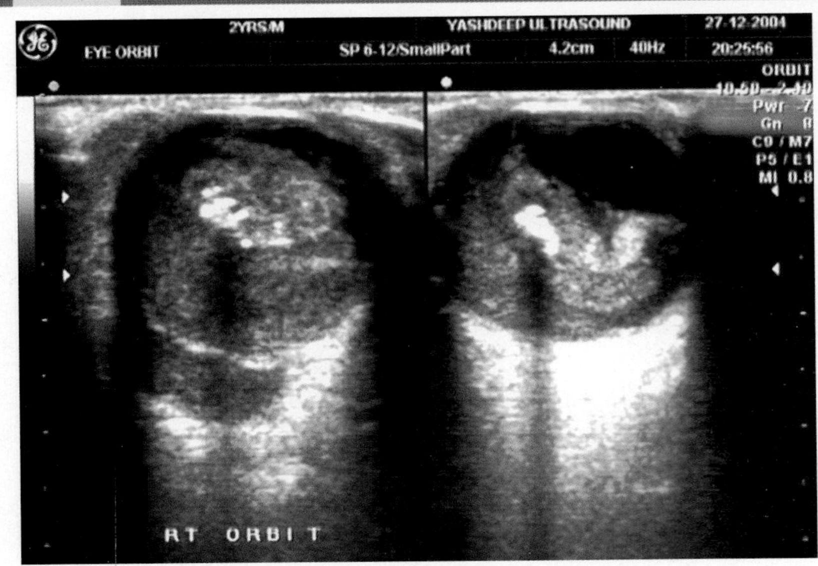

FIGURE 1.250

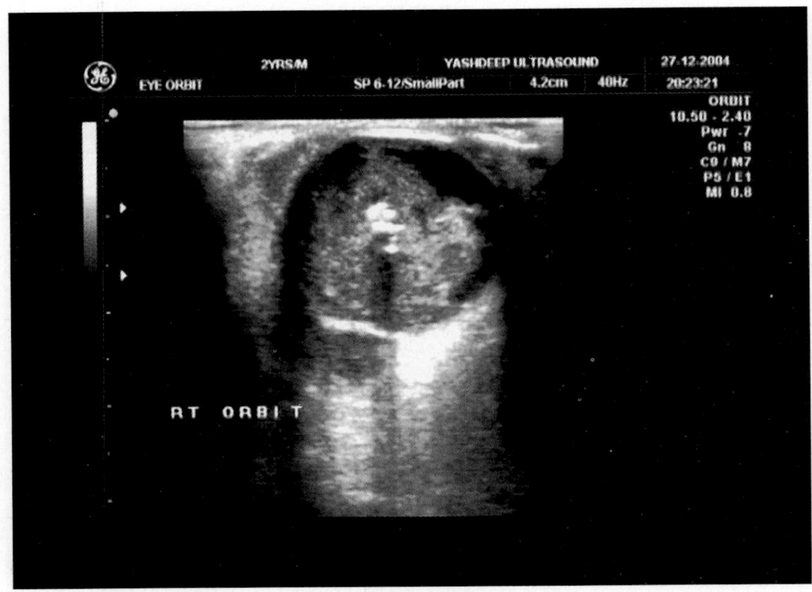

FIGURE 1.251

FIGURES 1.250 and 1.251: Retrobulbar extension of retinoblastoma. Big retinoblastoma is seen invading the sclera and also extending in retrobulbar space. The tumor has caused pressure over optic nerve.

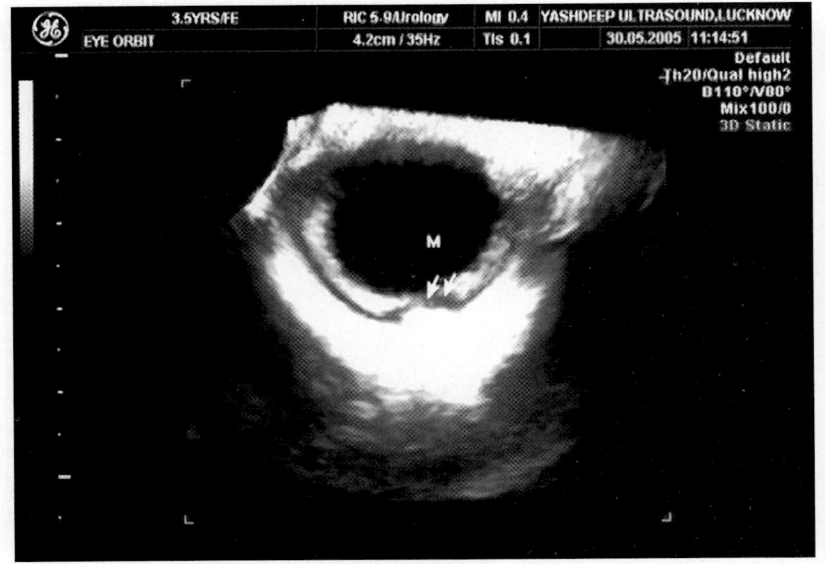

FIGURE 1.252: 3D of retinoblastoma. 3D imaging shows a small tumor mass seen confined in the subretinal space. It is seen lifting the retina at the time of study. However, no retinal breach is seen at the time of study.

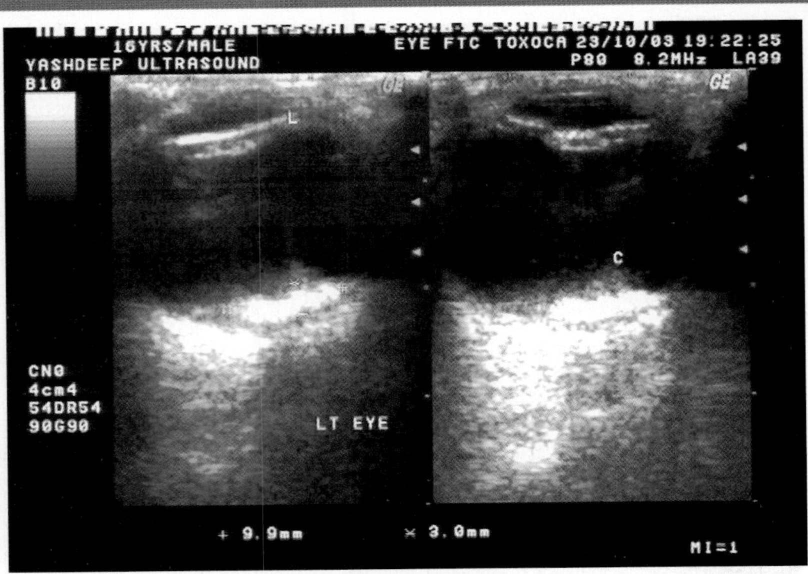

FIGURE 1.253

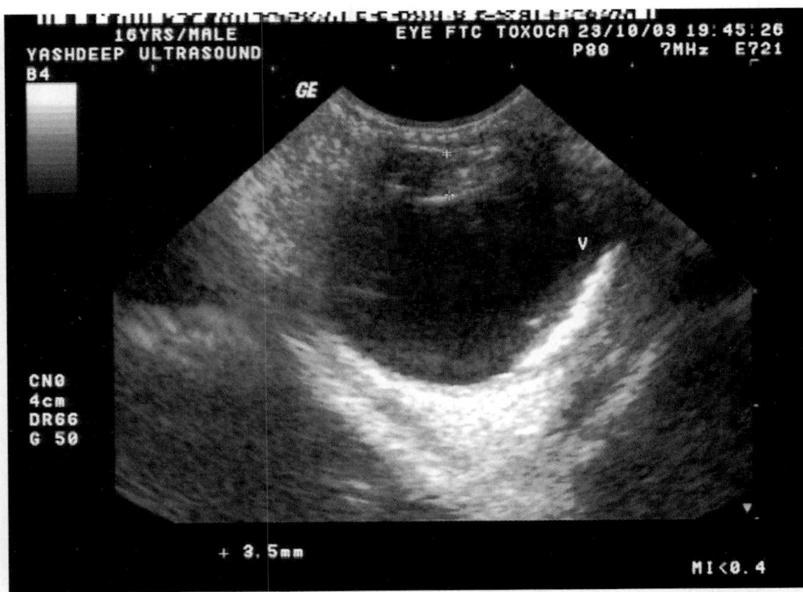

FIGURE 1.254

FIGURES 1.253 and 1.254: Differential diagnosis of retinoblastoma–Toxocara. A young child presented with white-eye reflex and painful red eye. There is evidence of linear plaque like calcification in the retina, which is different from clump like calcification seen in retinoblastoma. No tumor elevation is seen.

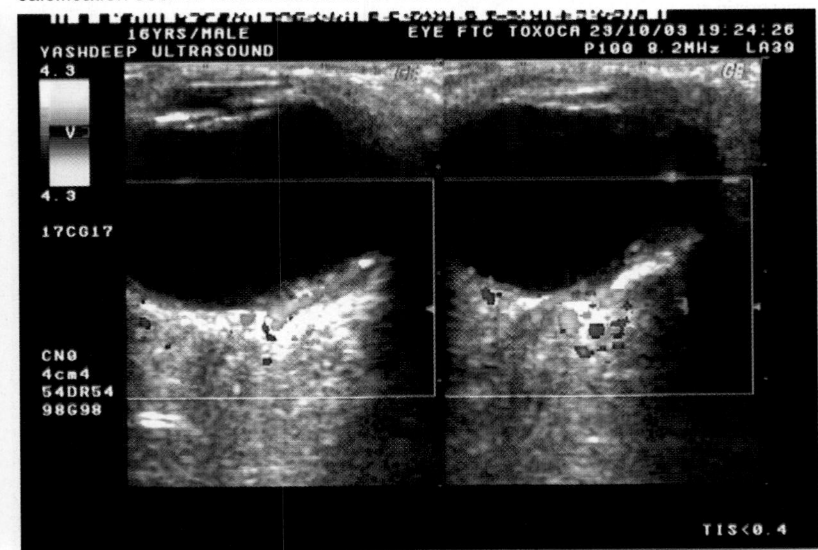

FIGURE 1.255: On color flow imaging increased flow is seen in the choroid and retina.

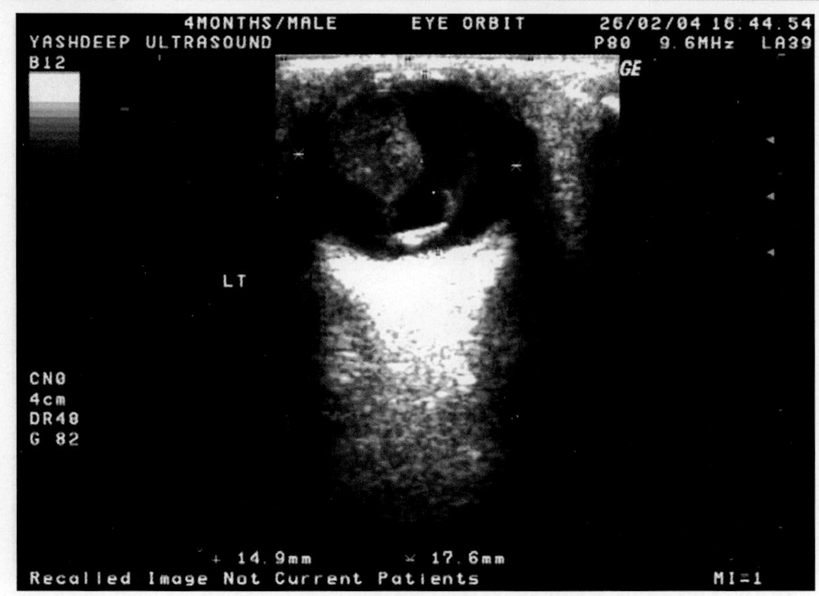

FIGURE 1.256

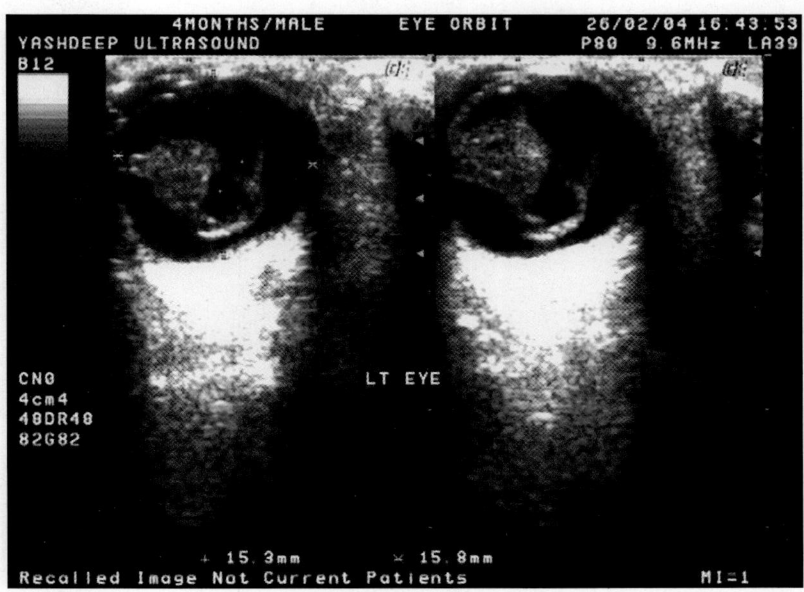

FIGURE 1.257

FIGURES 1.256 and 1.257: Differential diagnosis of retinoblastoma.Retinopathy of prematurity (ROP). An infant was born with white-eye reflex. HRSG shows echogenic membranous retina filling the posterior segment of the eye. It was shining through to the pupil giving the white-eye reflex suggestive of retinopathy of prematurity. No calcification is seen in the membrane.

case it involves retina from its entire surface known as retinoblastoma circumference. The tumor usually grows anteriorly and fills the vitreous cavity. However, it can invade the choroid and grows posteriorly in the retrobulbar space. HRSG can clearly assess the retrobulbar extension of tumor growth and its excavation in the choroid. HRSG is also very useful and noninvasive tool to monitor the treatment response of the tumor or to assess regression of tumor growth. The tumor is radiosensitive, and rapid regression takes place after radiotherapy and chemotherapy. The differential diagnosis of tumor are Coats disease or exudative retinal detachment, persistent hyperplastic primary vitreous (PHPV), retrolental fibroplasia and toxocara, a worm infestation caused by ingestion of intestinal worm. In ocular toxocara. The most common ocular lesions are ocular endophthalmitis or a posterior pole granuloma.

Ocular Melanoma

Ocular melanomas are the most common tumors in adults. It comprises about 40% of all ocular tumors. Eighty five percent of the melanomas arise from choroid and remaining 15% arises from ciliary body. Melanomas most commonly occur in fifth and sixth decades of life and rarely occur before third decade and after eight decade of life. The tumors are most common in white skin people and rare in blacks. HRSG is highly sensitive in evaluating the size, shape, growth and extent of the tumor, which are important prerequisites prior to surgery. Typical sonographic features are seen on HRSG in ocular melanomas.

Sonographic features: Sonographically, melanomas are typically dome-shaped, collar button appearance with smooth surface. They have low reflectivity with medium echoes. They are homogeneous.

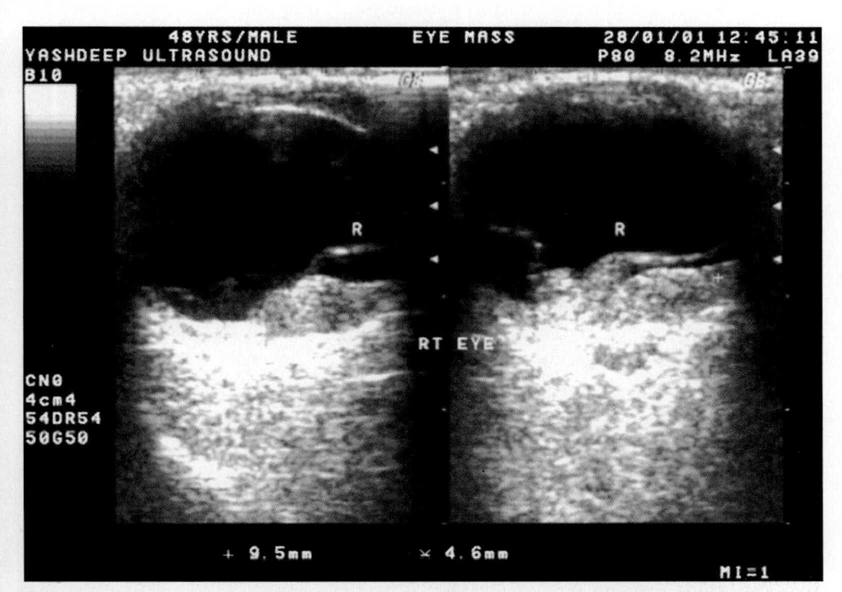

FIGURE 1.258: Disciform melanoma. HRSG shows well defined discoid mass over the posterior pole. It is homogeneous in texture. Associated retinal detachment is also seen.

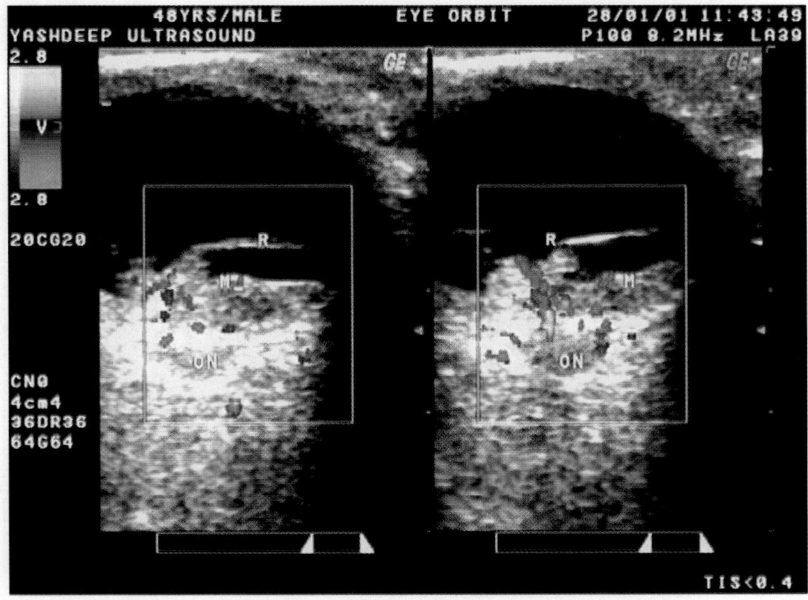

FIGURE 1.259: Color Doppler study shows moderate flow in the tumor.

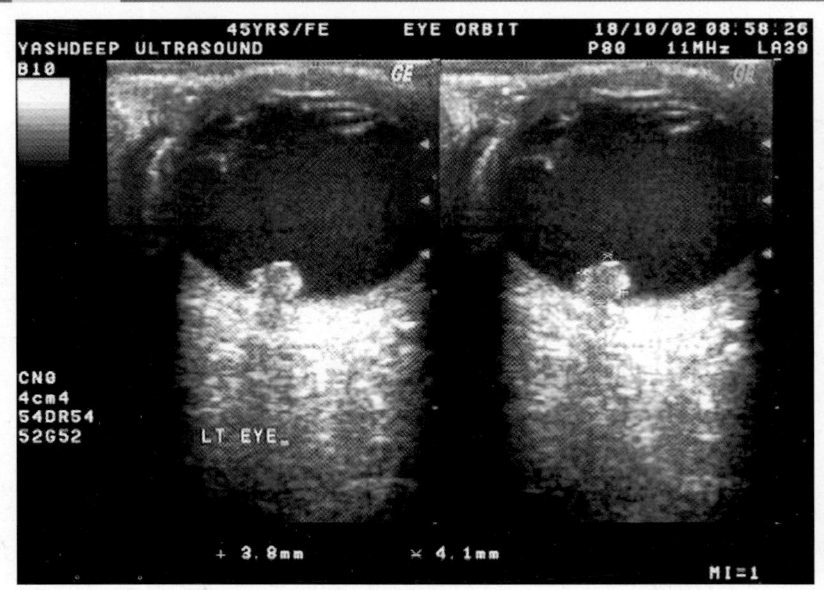

FIGURE 1.260: Small collar button melanoma. A well defined homogeneous soft tissue mass seen coming out from choroid.

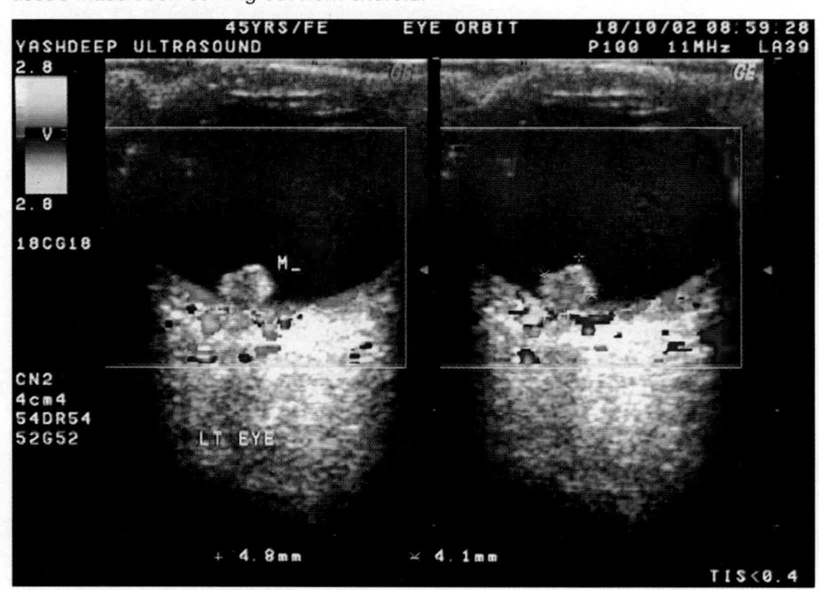

FIGURE 1.261: On color flow imaging a big vessel is seen feeding the tumor.

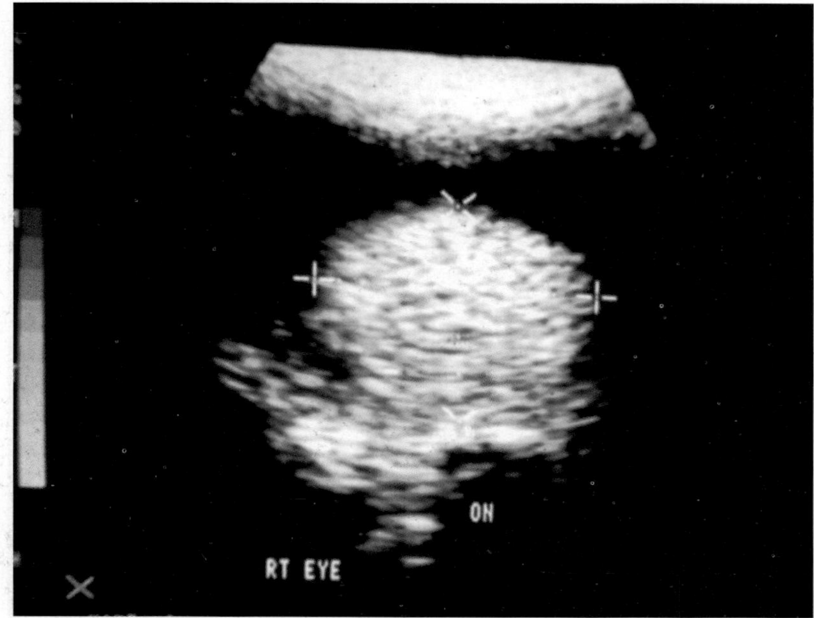

FIGURE 1.262: Collar button melanoma. HRSG shows well defined collar button shaped tumor with homogeneous texture. Fine interfaces are seen in the tumor mass suggestive of tightly packed tumor cells.

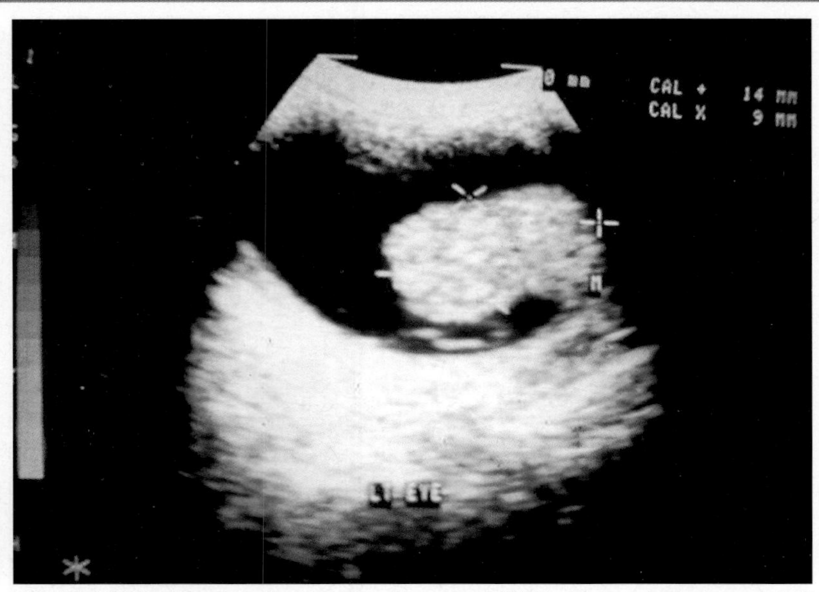

FIGURE 1.263: Dumb bell shaped melanoma. HRSG shows a typical dumb bell shaped cilio choroidal melanoma with a well-defined tumor neck. It is projecting in the posterior segment and shows typical homogeneous texture of tumor.

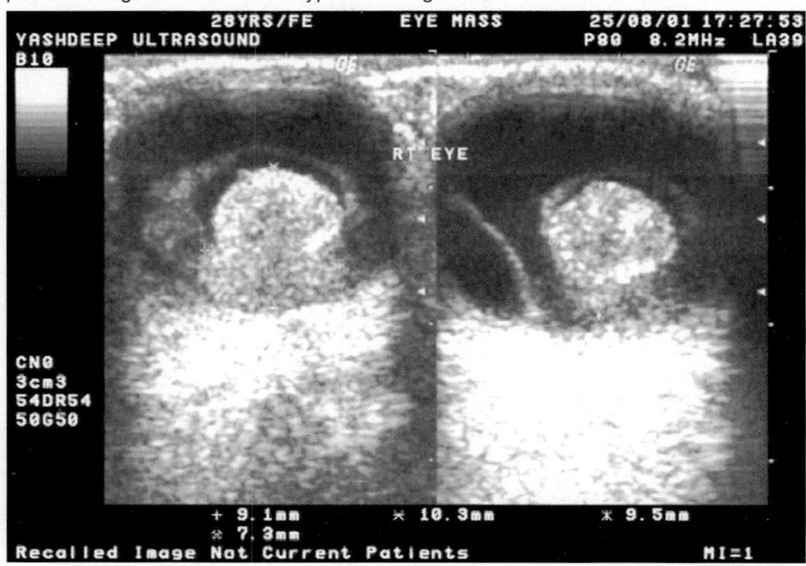

FIGURE 1.264: Collar button melanoma. Another case shows well defined homogeneous melanoma with wide base. The tumor is having smooth outline and confined to the choroid layer. No extra choroidal extension is seen.

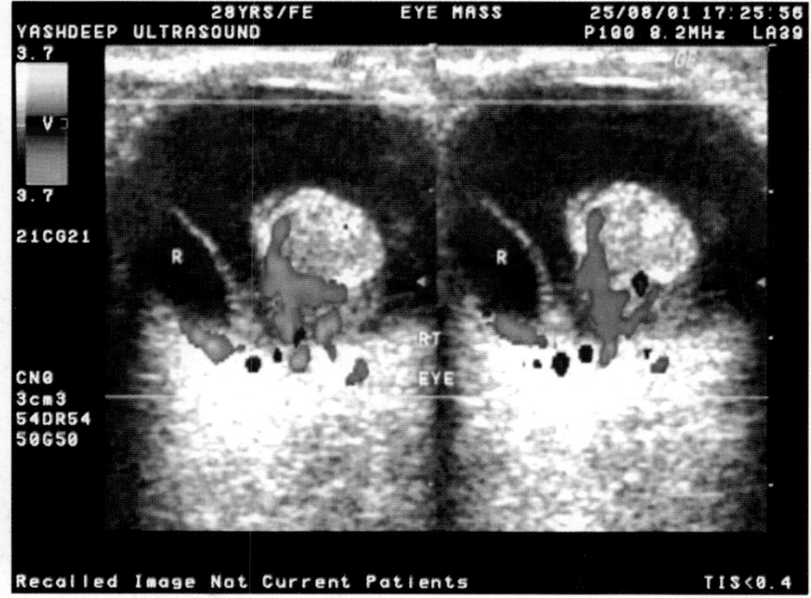

FIGURE 1.265: Color Doppler study shows high vascularity of the tumor with multiple feeding vessels. Associated retinal detachment is also seen.

FIGURE 1.266: A young child presented with blunt injury of eye while playing cricket and sustained loss of vision.

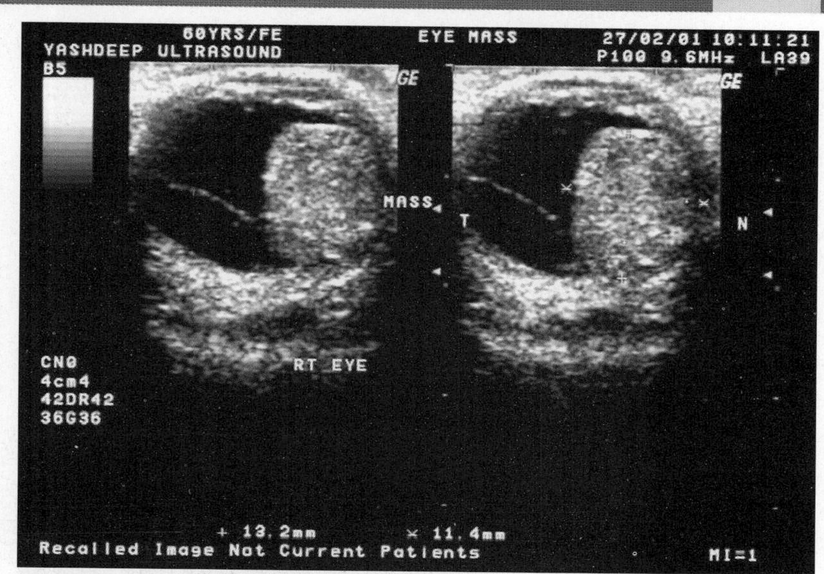

FIGURE 1.269: **Ciliary melanoma.** A big well defined lobulated mass coming out from the ciliary body. The tumor is by in large homogeneous in texture.

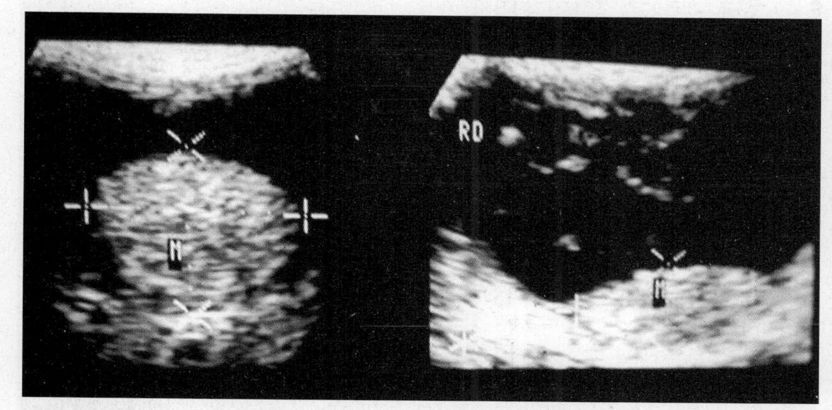

FIGURE 1.267: HRSG shows a big well-defined homogeneous tumor in the posterior segment. It shows typical collar button shape suggestive of melanoma. It is seen masked with the intragel hemorrhage.

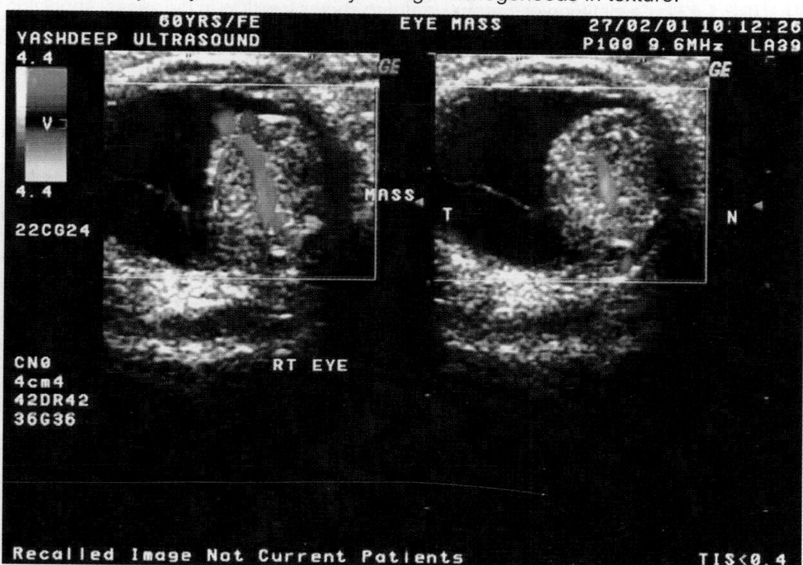

FIGURE 1.270: **Color Doppler study.** Color Doppler study shows a big vessel feeding the tumor.

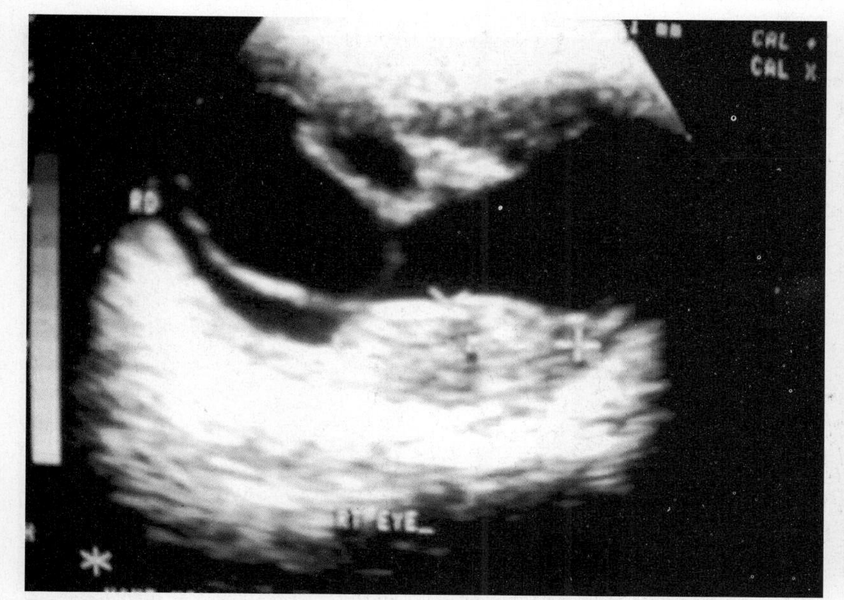

FIGURE 1.268: HRSG shows well-defined homogeneous soft tissue mass coming out from the choroid. It is also associated with intragel hemorrhage and retinal detachment.

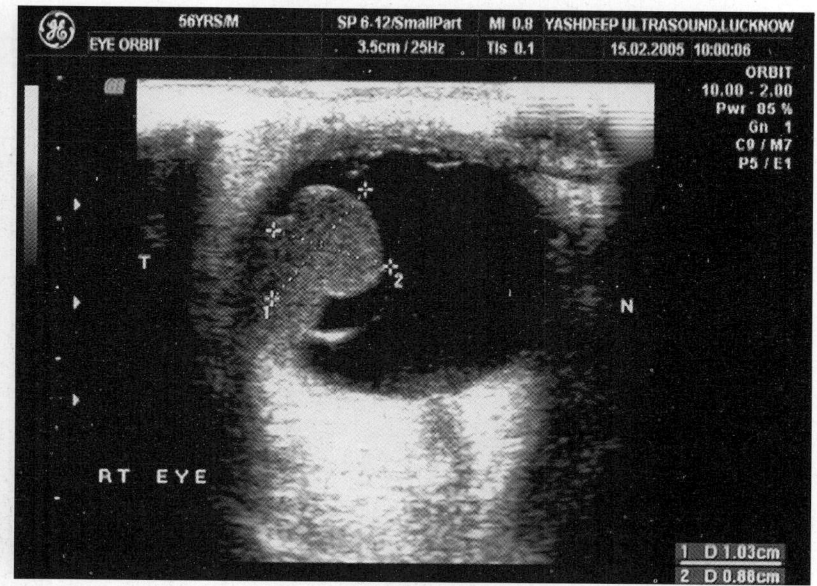

FIGURE 1.271: **Ciliary melanoma.** A big mass is seen coming out of the Rt ciliary body. It is club shaped and projecting in the posterior segment.

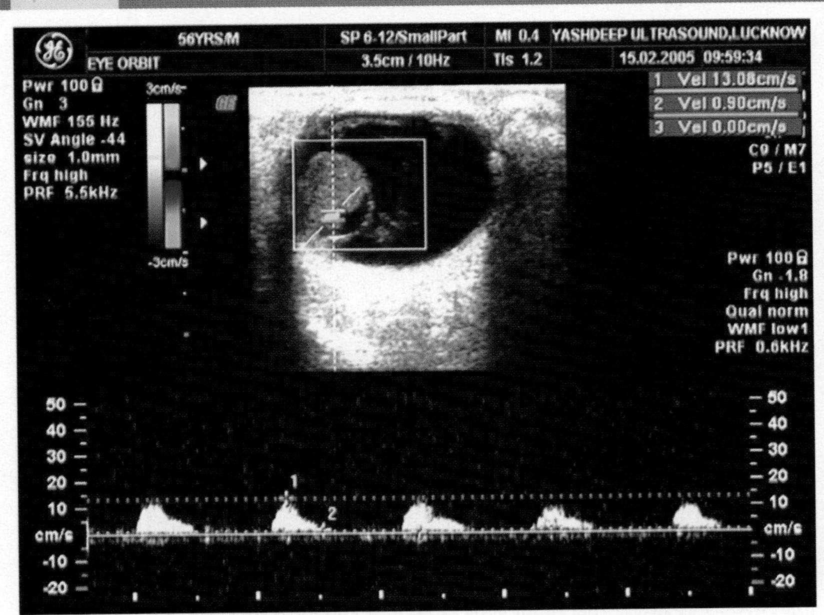

FIGURE 1.272: Color Doppler spectral imaging shows low flow in the vessel supplying the tumor mass.

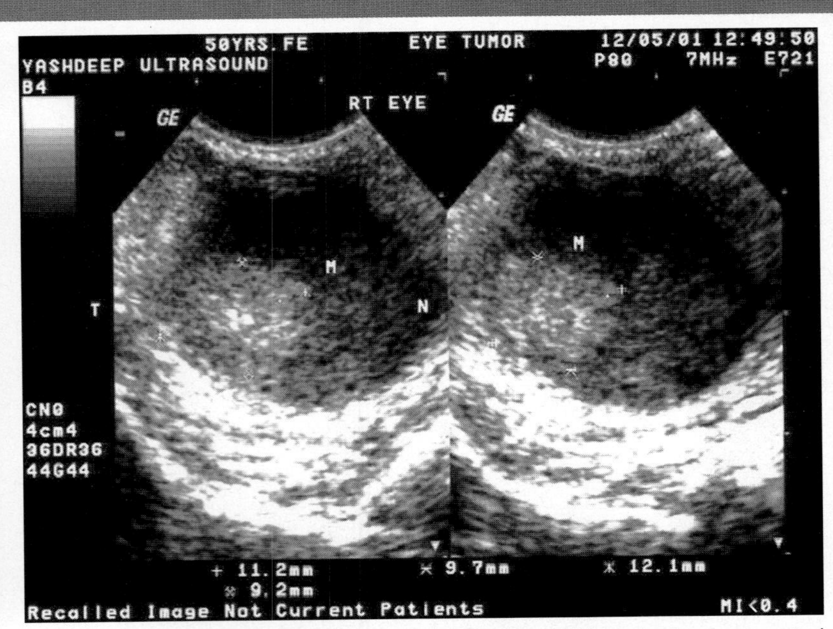

FIGURE 1.275: **Melanoma masked with hemorrhage.** This patient was presented with loss of vision direct ophthalmoscopy was not possible. HRSG shows a well-defined tumor fixed with temporal wall and shows smooth outline.

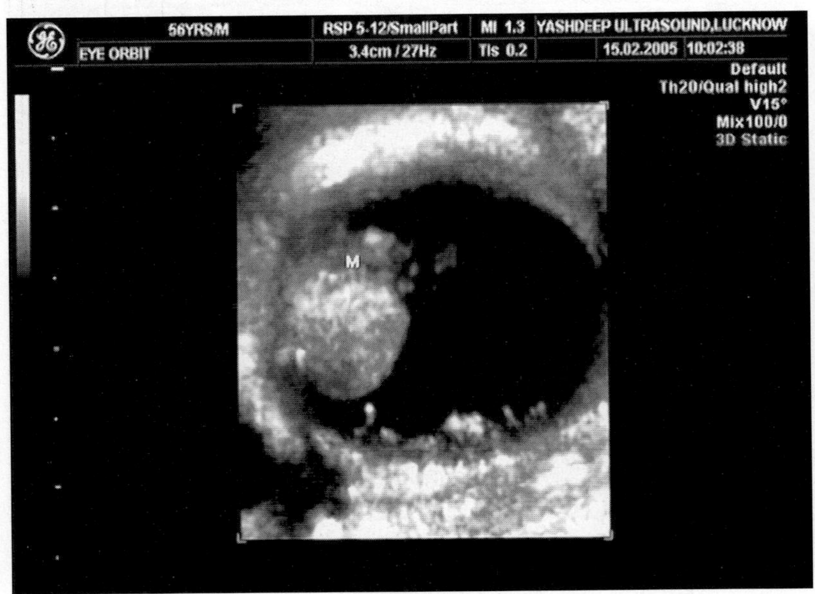

FIGURE 1.273

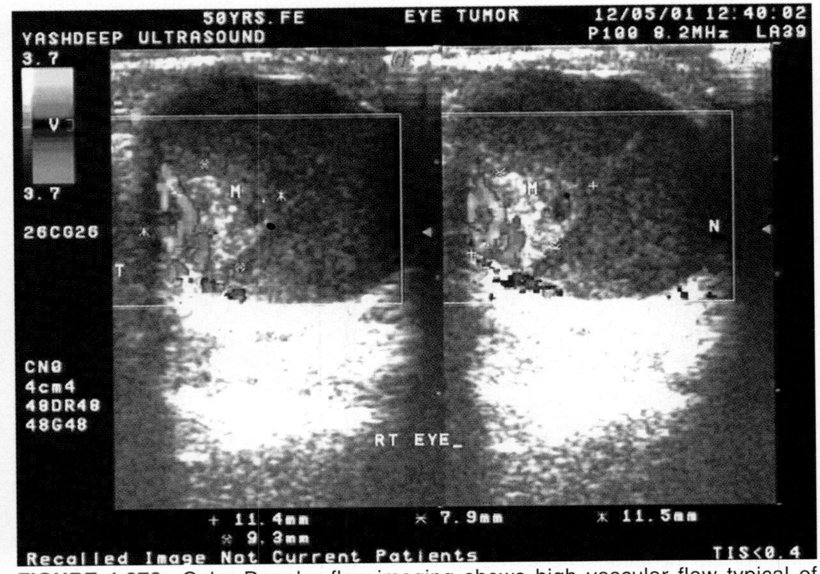

FIGURE 1.276: Color Doppler flow-imaging shows high vascular flow typical of melanoma.

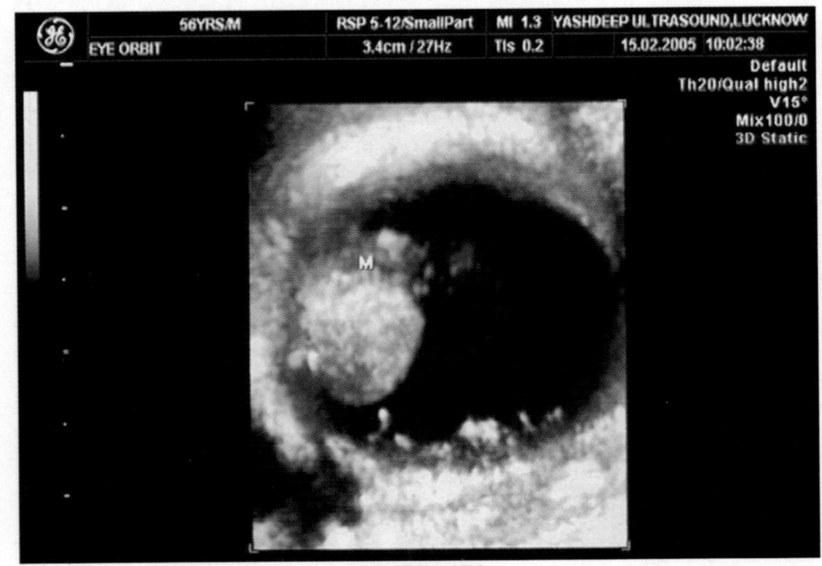

FIGURE 1.274

FIGURES 1.273 and 1.274: 3D imaging of the eye in the same patient shows the details of the tumor mass. Associated intragel hemorrhage is also seen.

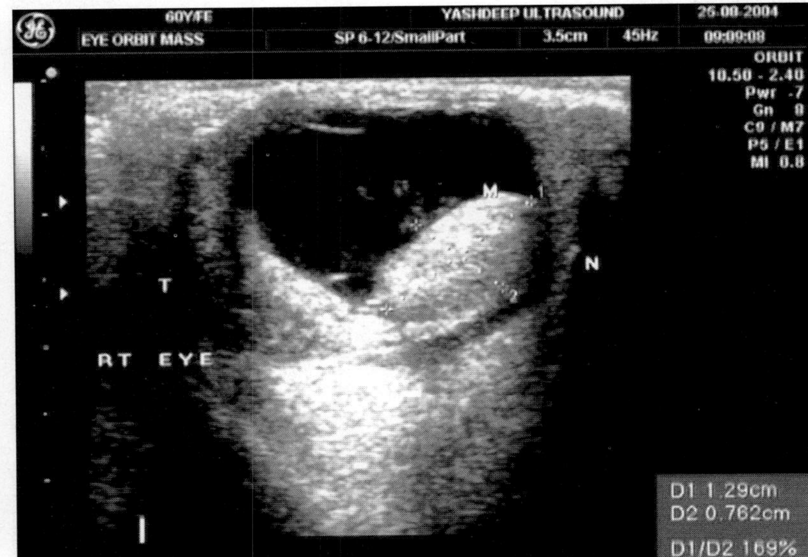

FIGURE 1.277: **Discoid melanoma.** Well-defined disc shaped homogenous mass is seen coming out from the choroid and bulging in the posterior segment. It shows wide base with smooth elevation.

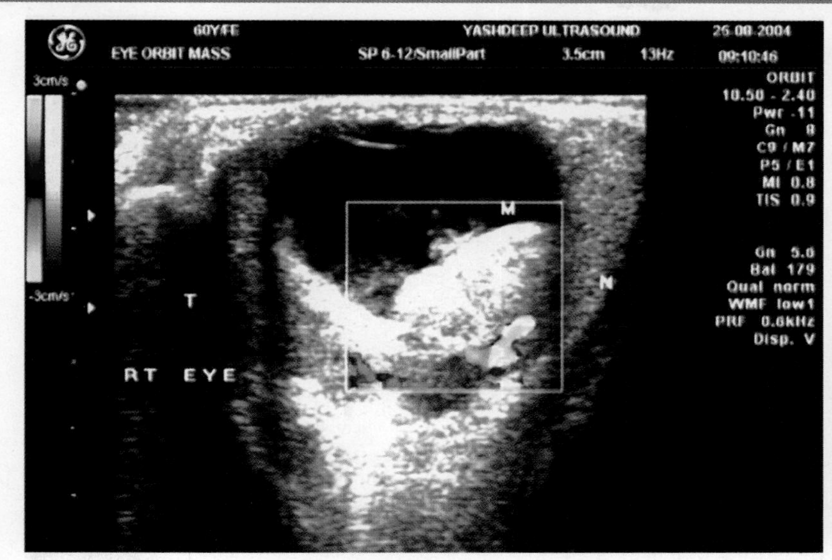

FIGURE 1.278

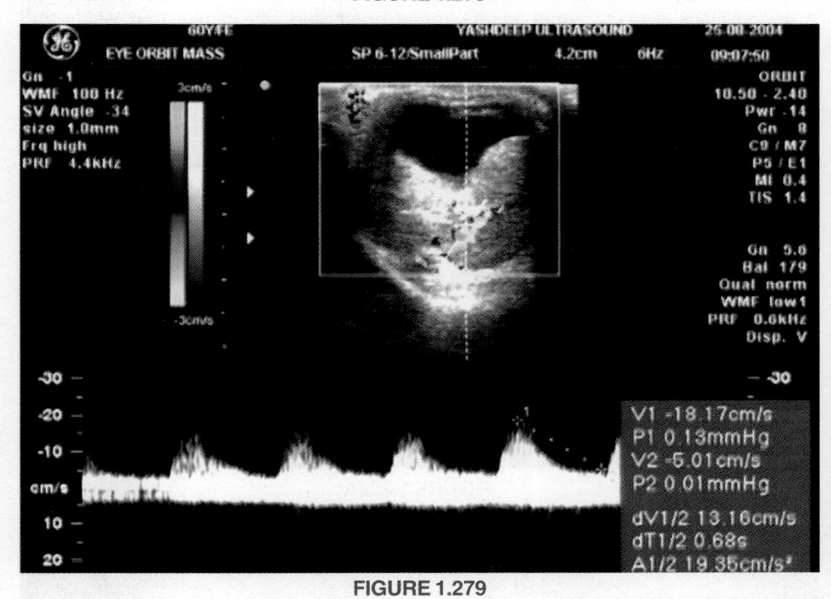

FIGURE 1.279

FIGURES 1.278 and 1.279: Color Doppler imaging shows the feeding vessels in the tumor and spectral Doppler tracing shows low flow in the vessels.

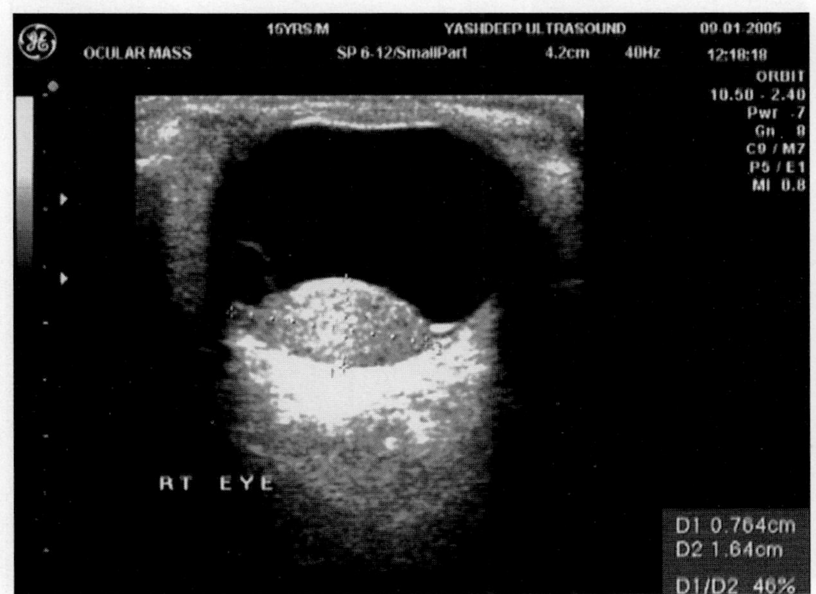

FIGURE 1.280

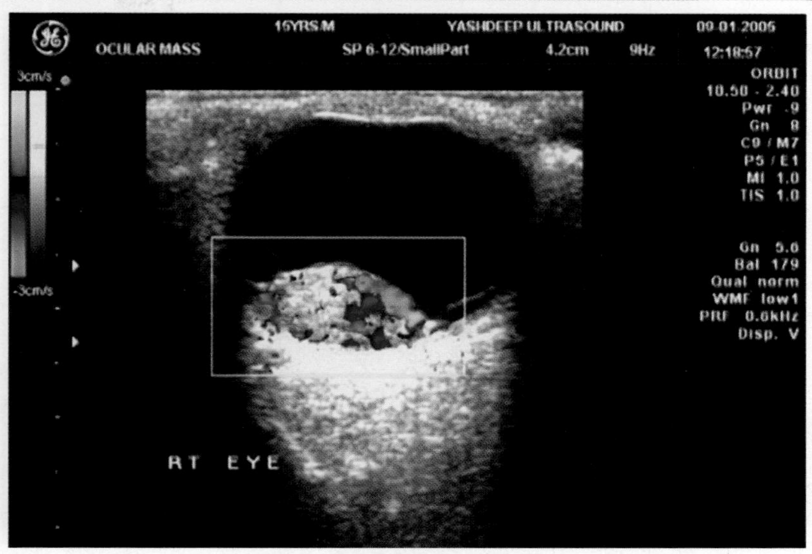

FIGURE 1.281

FIGURES 1.280 and 1.281: Discoid melanoma . HRSG shows a choroidal melanoma with smooth elevation and wide base. No evidence of any necrosis is seen in it. Multiple vessels are seen feeding the tumor.

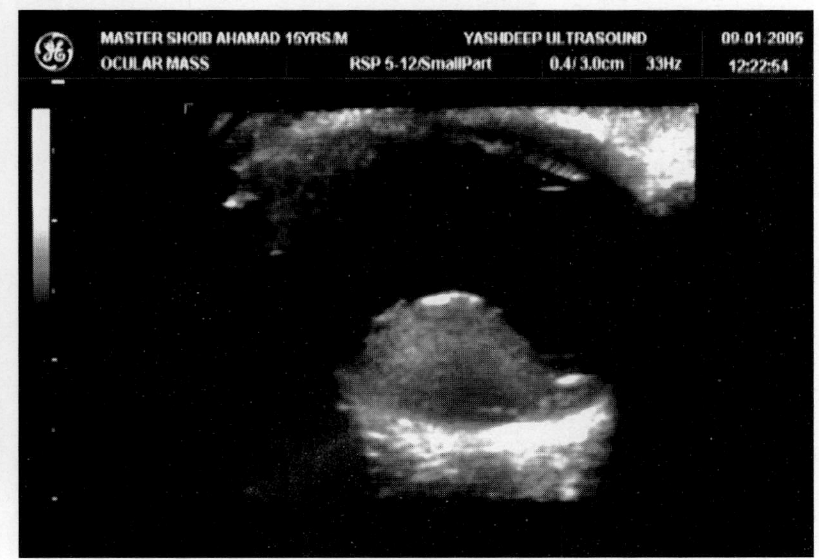

FIGURE 1.282

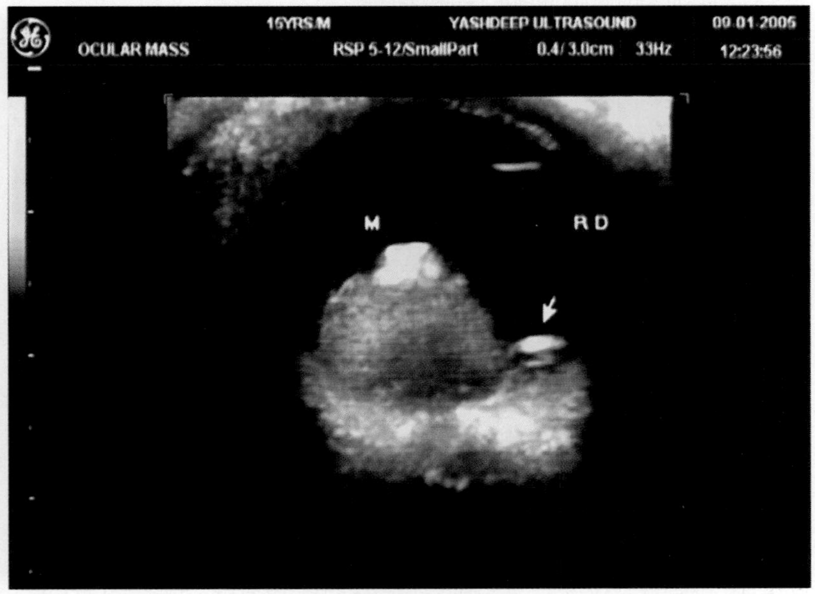

FIGURE 1.283

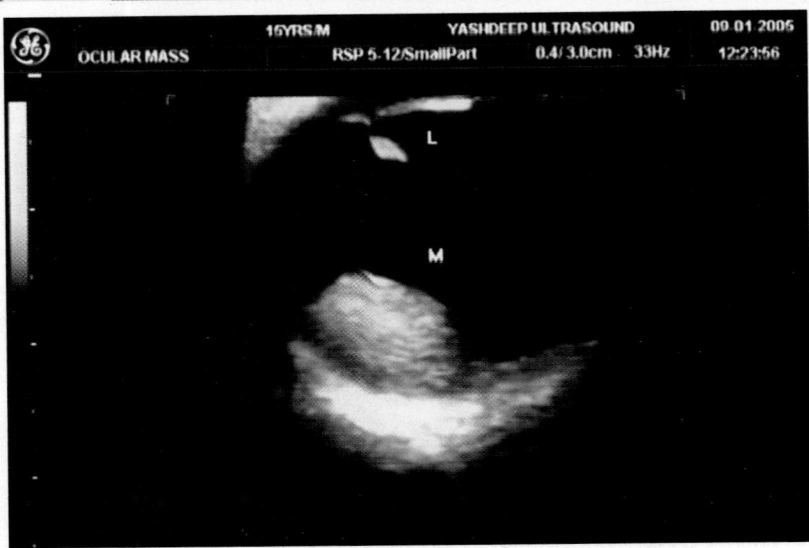

FIGURE 1.284

FIGURES 1.282 to 1.284: 3D imaging of the same tumor mass shows better details of the tumor. Associated intragel hemorrhage with focal retinal detachment is also seen.

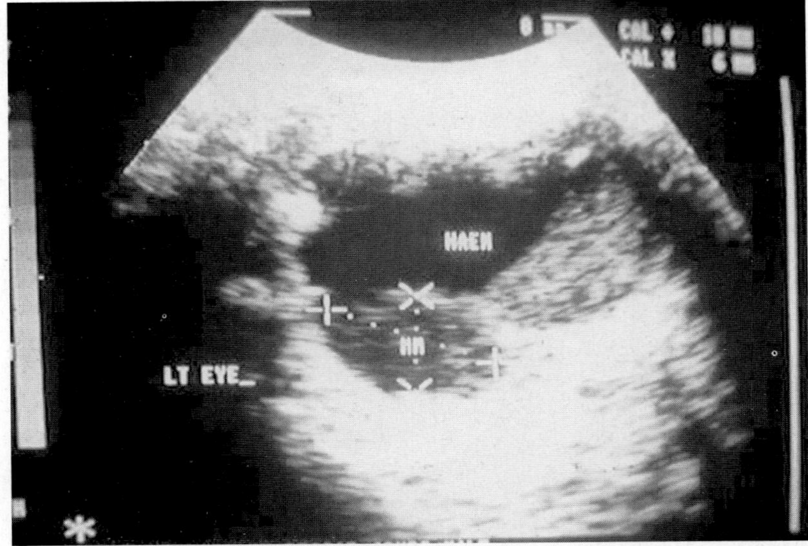

FIGURE 1.285: Melanoma with choroidal excavation. HRSG shows a choroidal melanoma invading the choroid and excavating it. Retrochoroidal extension is present. Associated vitreous hemorrhage is also present.

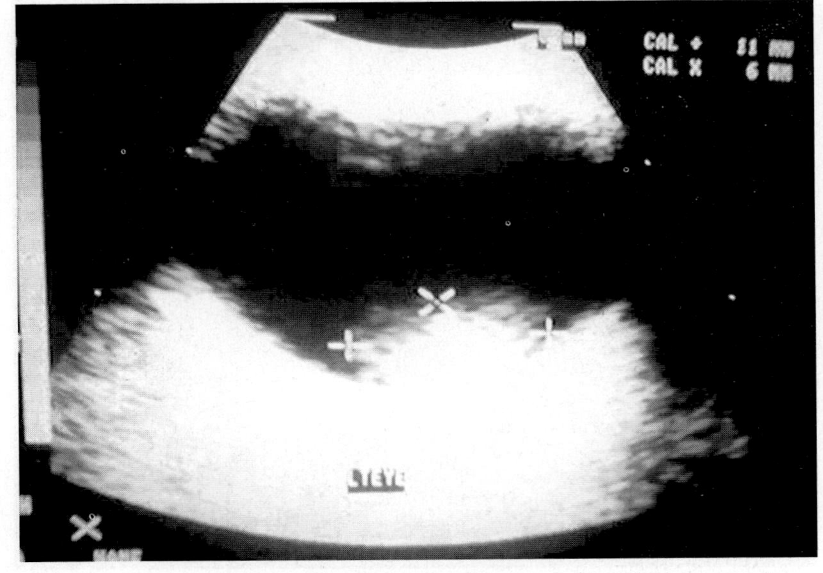

FIGURE 1.286: Differential diagnosis of melanoma Choroidal hemangioma. 3D Well defined homogeneous highly echogenic disc shaped mass seen in posterior segment suggestive of hemangioma. Hemangiomas are highly echogenic than melanomas, which show medium level echoes.

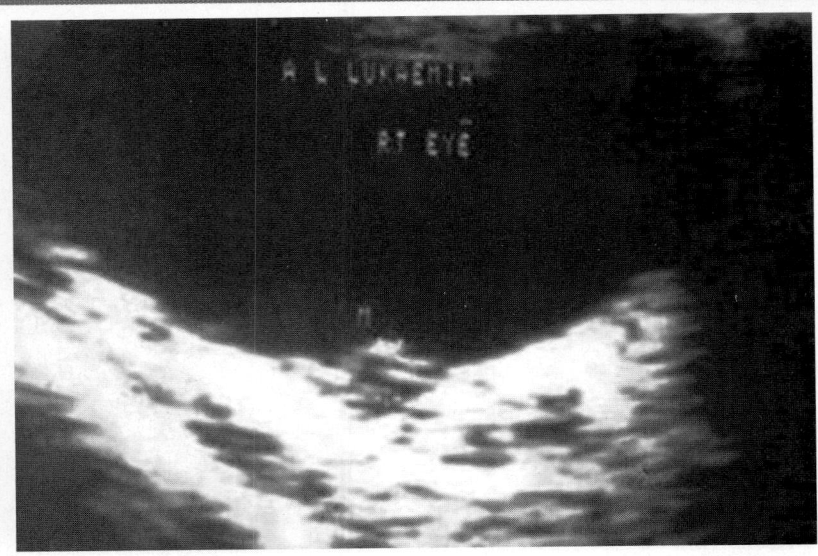

FIGURES 1.287: Differential diagnosis of melanoma—choroidal metastasis. A young child of acute lymphatic leukemia presented with diminished vision and painful eye. HRSG shows metastatic deposit in choroid. Associated retrobulbar deposit also seen as a hypodense irregular focus in retrobulbar fat.

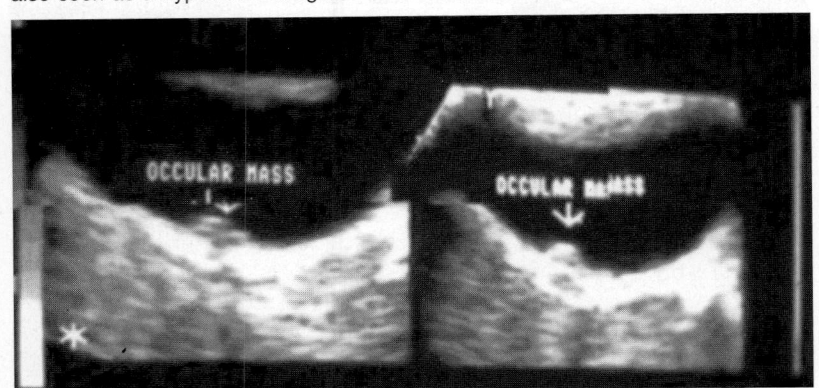

FIGURE 1.288A

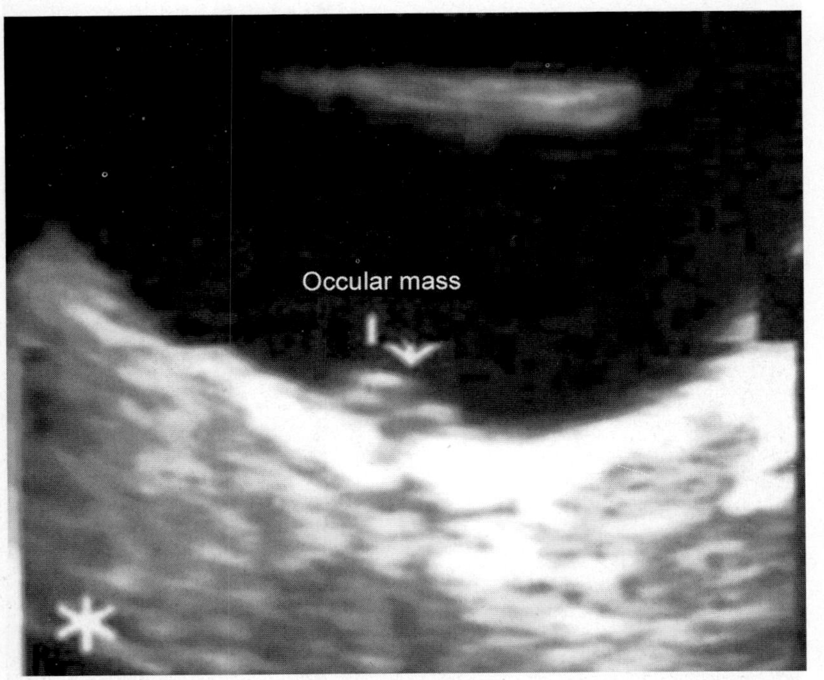

FIGURE 1.288B

FIGURES 1.288A and B: Differential diagnosis of melanoma—metastatic deposit choroid. A young child of known case of lymphoma presented with diminished vision and painful eye. HRSG shows metastatic deposit. The metastasis is seen as focal nodular shadow elevated from the sclera.

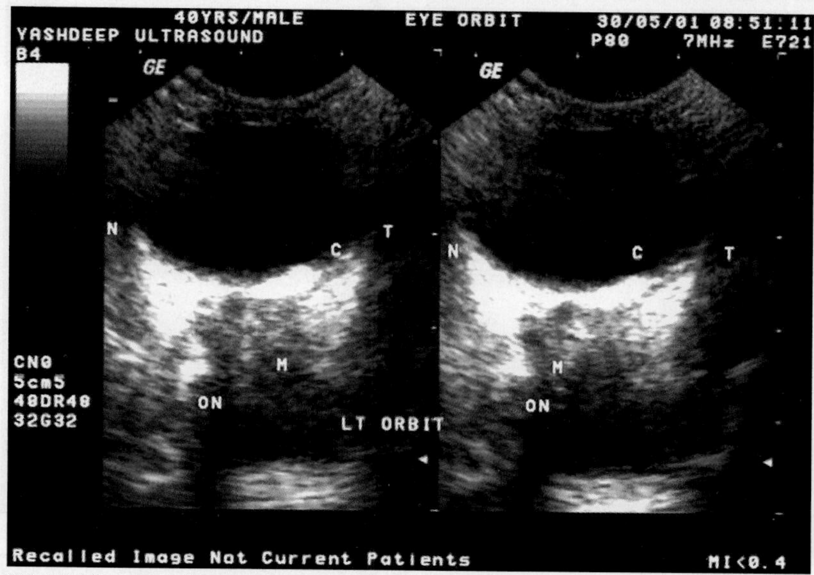

FIGURES 1.289: Differential diagnosis of melanoma—Choroidal osteoma. HRSG shows an elevated thickened mass with dense echogenic calcification. It is accompanied with acoustic shadowing. Typical of choroidal osteoma.

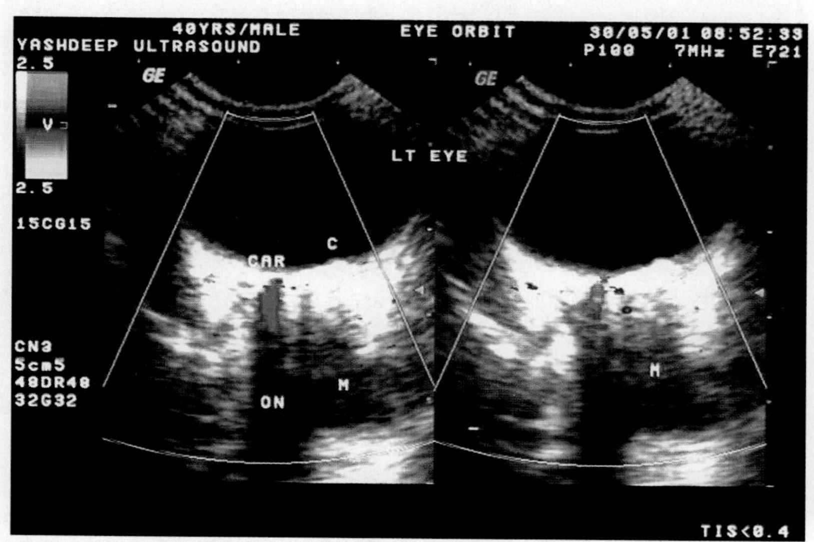

FIGURES 1.290: Color Doppler flow study shows poor flow in the choroidal osteoma.

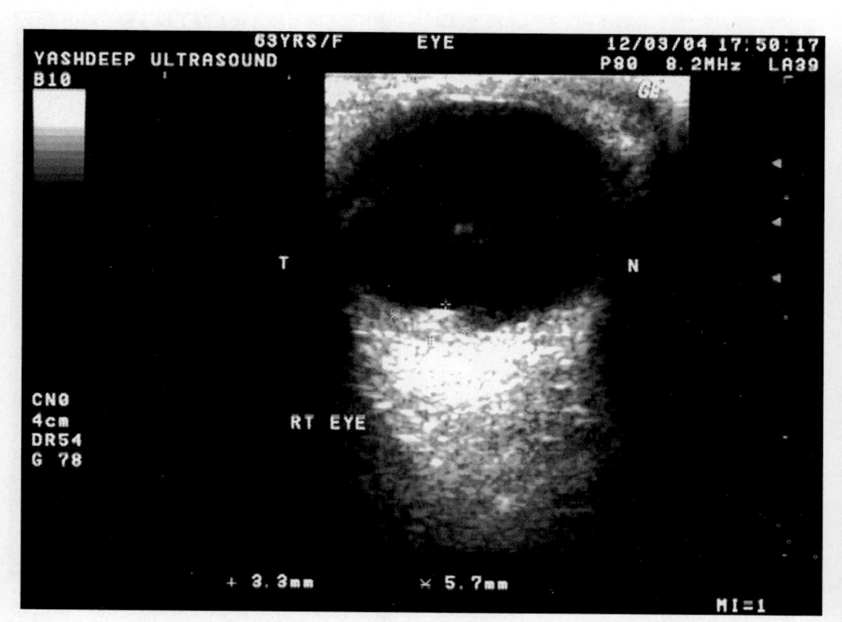

FIGURES 1.291: Differential diagnosis of melanoma—Choroidal angioma. A well defined homogeneous highly echogenic mass seen fixed with the choroidal layer.

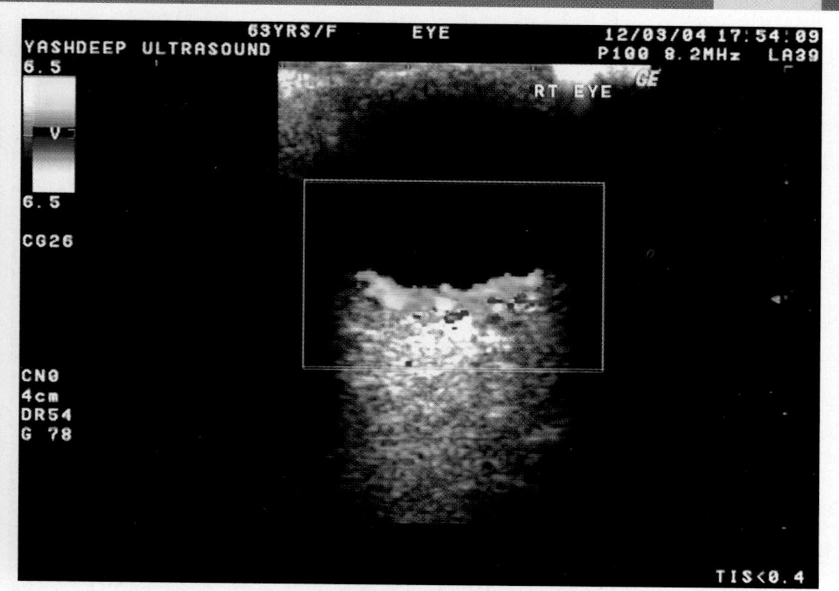

FIGURE 1.292

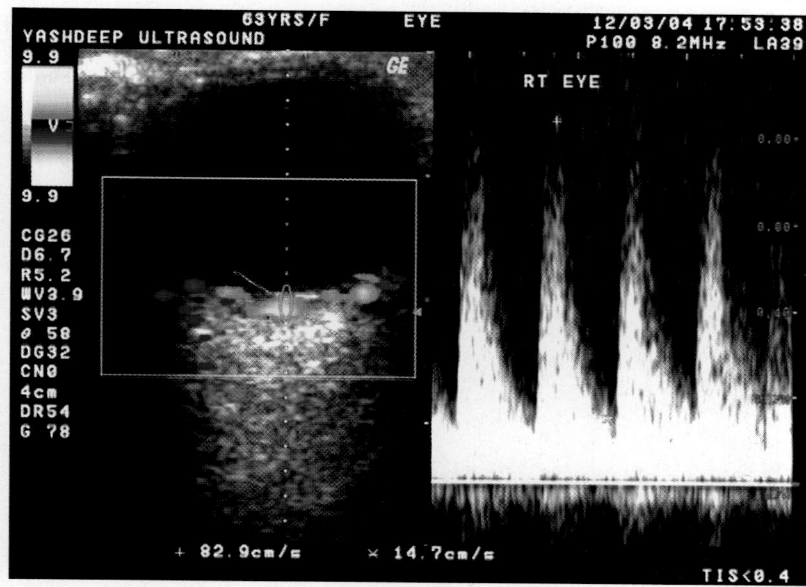

FIGURE 1.293

FIGURES 1.292 and 1.293: On color Doppler imaging it shows rich vascularity. On spectral Doppler tracing it shows high flow suggestive of high vascular tumor typical of choroidal angioma.

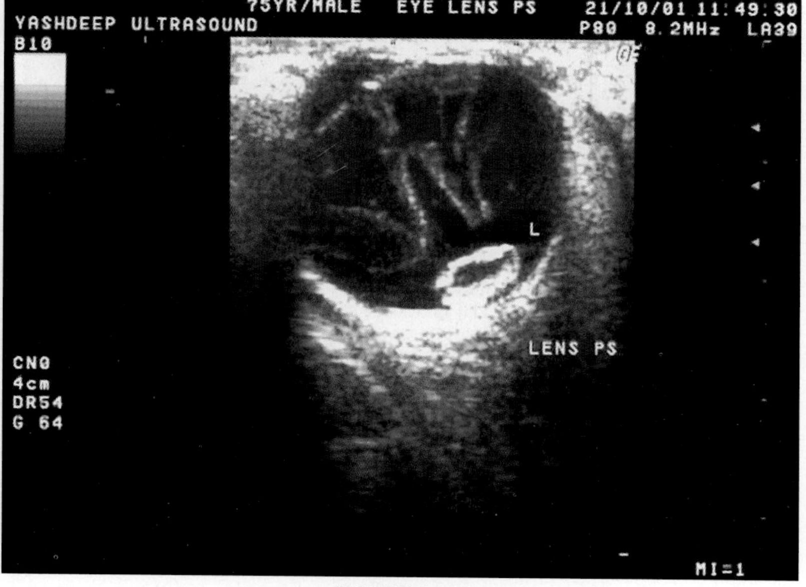

FIGURE 1.294

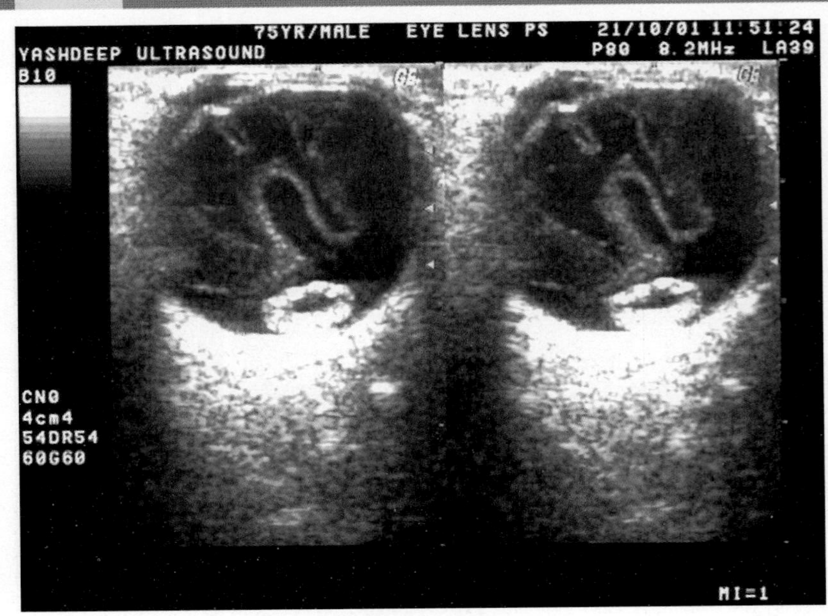

FIGURE 1.295

FIGURES 1.294 and 1.295: Differential diagnosis of melanoma. Dislocated lens in posterior segment. HRSG shows the thickened lens with associated intragel hemorrhage. A dislocated lens in posterior segment can be mistaken as melanoma.

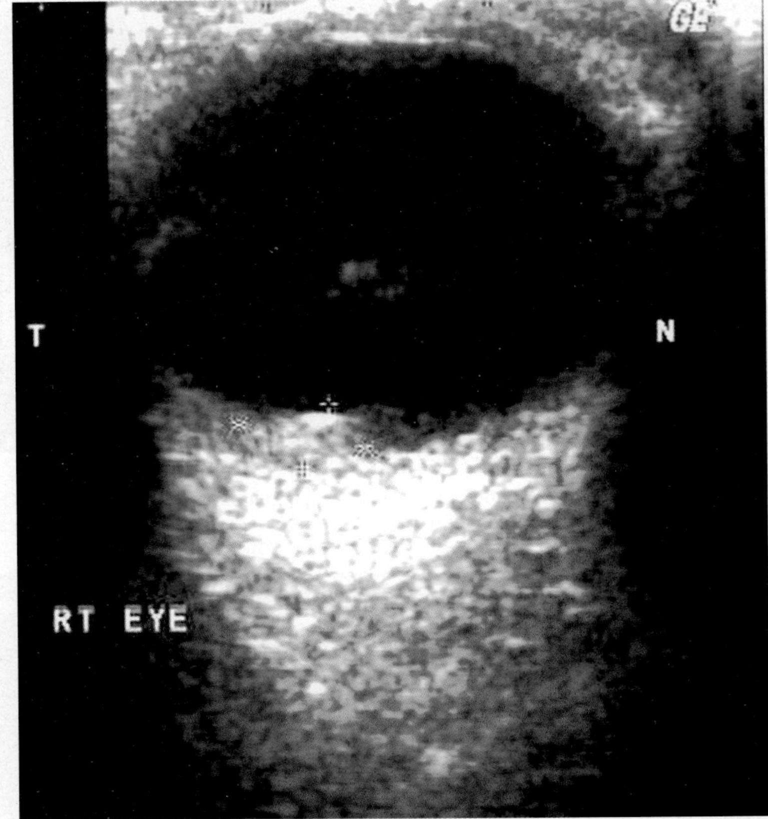

FIGURE 1.296: Differential diagnosis of melanoma—Choroidal metastasis in CA breast. A case of Carcinoma breast presented acute pain with loss of vision in the eye. HRSG shows a well defined mass lesion fixed with the choroid suggestive of metastatic deposits. Associated subchoroidal hemorrhage is also seen.

Posterior Segment Mass—Cyst

Other posterior segment masses can mimic as tumors of the choroid like cysticercosis, choroidal hemangioma or dislocated lens in the posterior segment. HRSG is an excellent modality to differentiate between benign pathological and malignant pathological conditions.

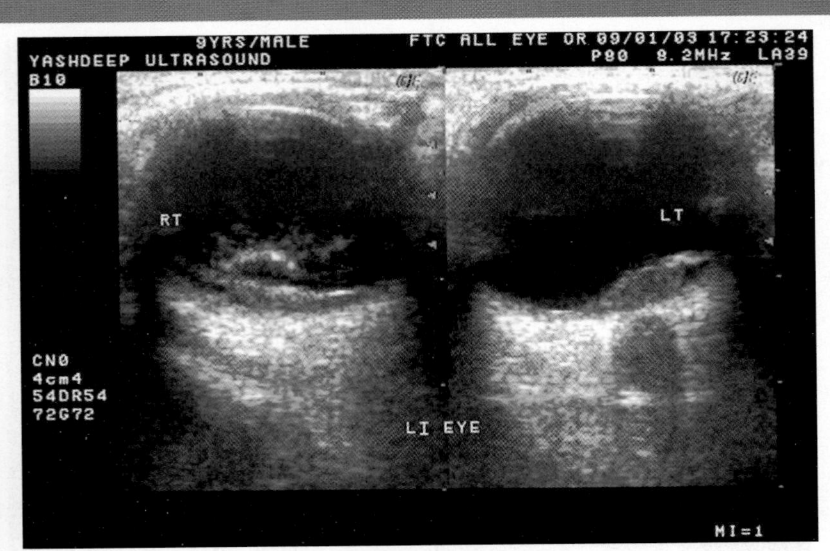

FIGURE 1.297

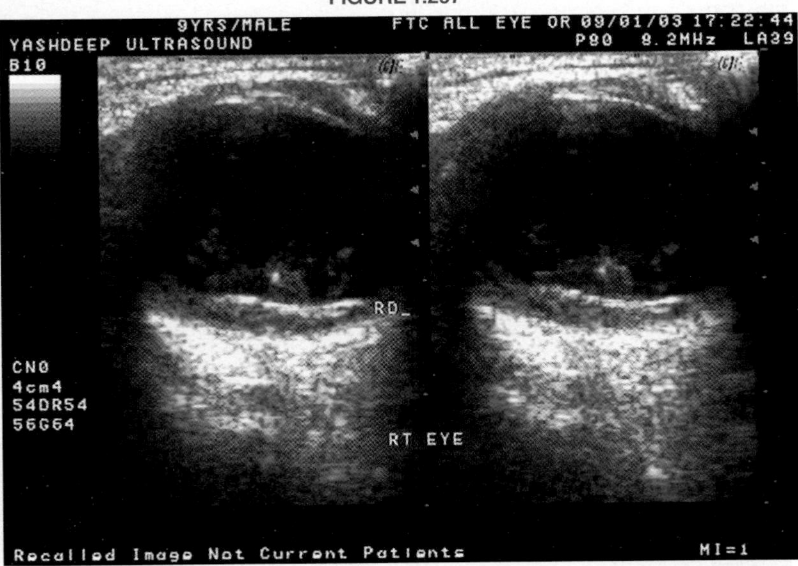

FIGURE 1.298

FIGURES 1.297 and 1.298: Differential diagnosis of melanoma—Choroidal metastasis in acute lymphoblastic leukemia. A young child known case of acute Lymphoblastic leukemia presented with acute pain in the eye with sudden loss of vision. HRSG shows subchoroidal hemorrhage with associated mass lesion suggestive of metastatic deposits. Associated retinal detachment is also seen.

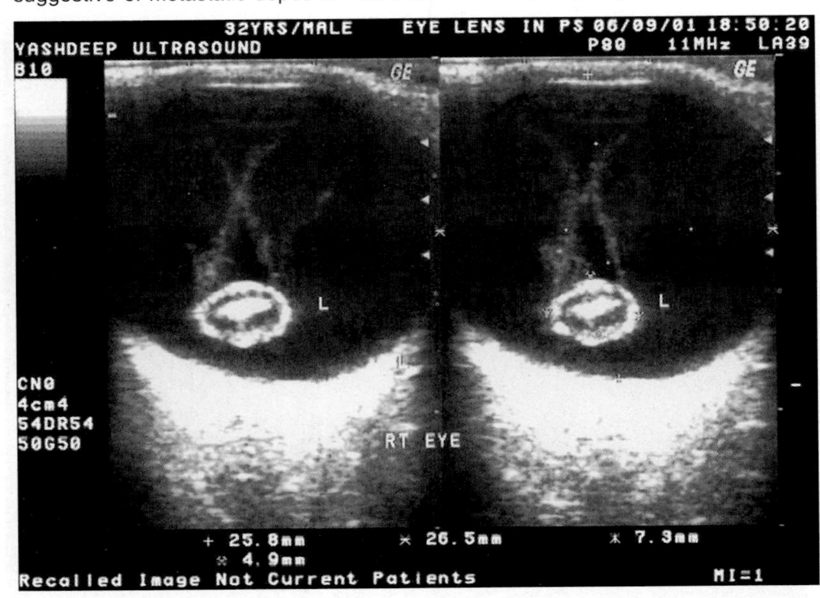

FIGURE 1.299: Lens in the posterior segment. HRSG shows thickened opaque lens seen in the posterior segment. Nucleus of the lens is calcified and the lens is seen suspended with the ligament.

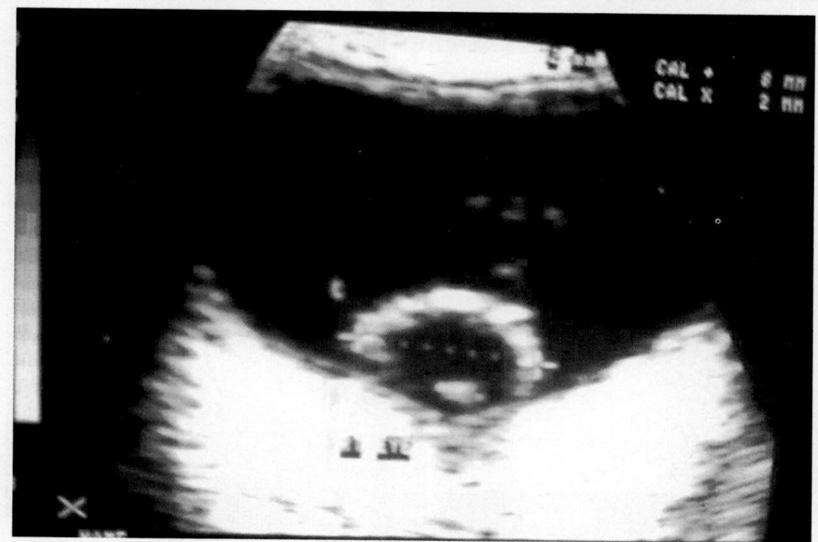

FIGURE 1.300: Cysticercosis cyst in posterior segment. HRSG shows well-defined thick walled cyst in the posterior segment just above the optic nerve head. An echogenic nidus is seen fixed with the inner wall of the cyst. Typical USG features of the cysticercus cyst.

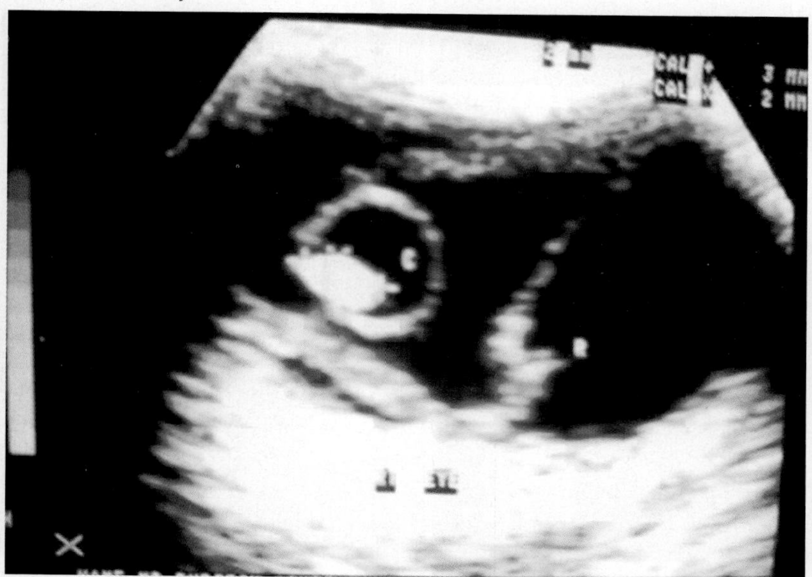

FIGURE 1.301: Ocular cysticercosis with retinal detachment. HRSG shows a well-defined cyst with echogenic nidus fixed with the inner wall of the cyst in Rt eye on temporal side. Thick echogenic membrane is also seen detached from its surface suggestive of total retinal detachment. The patient complained of sudden loss of vision.

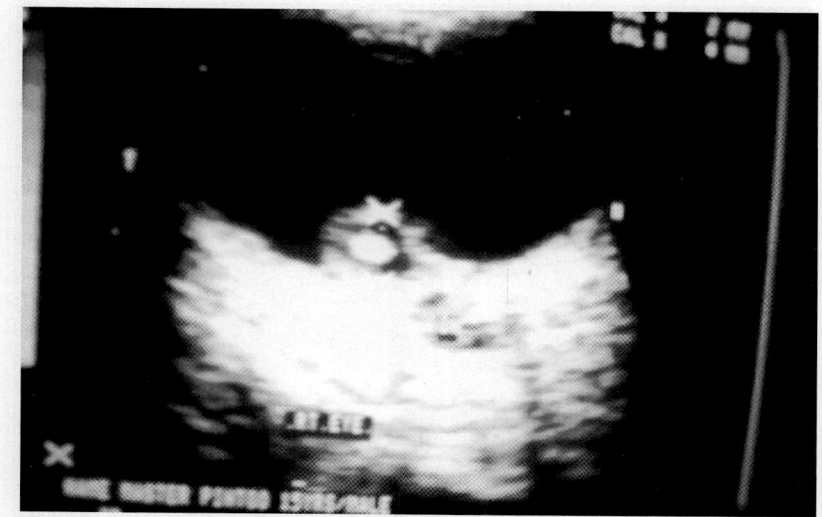

FIGURE 1.302: Intraocular cysticercosis. The patient came with the history of painful eye with blurred vision and dementia. HRSG shows a well-defined cyst fixed above the optic nerve head with an echogenic nidus fixed with the inner wall of the cyst, typical sonographic feature of cysticercus cyst.

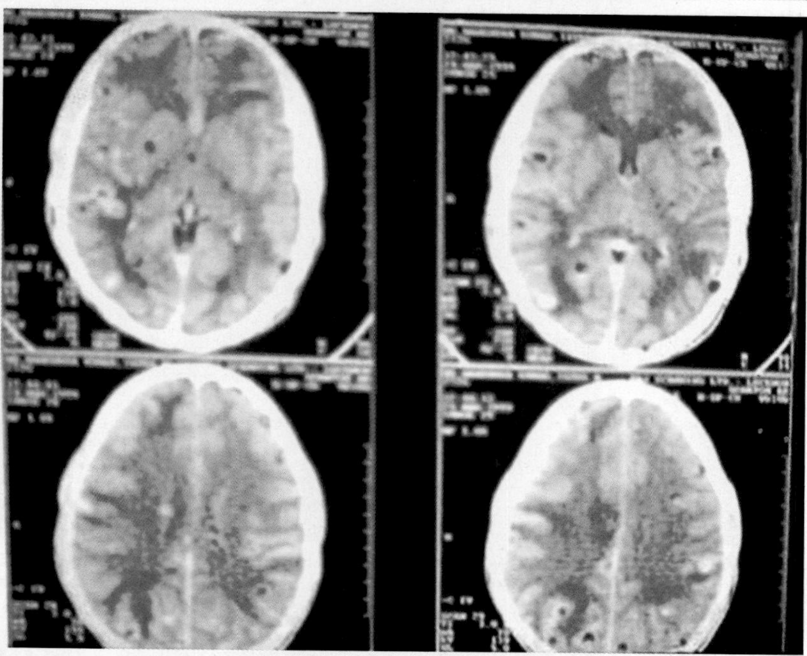

FIGURE 1.303: CT scan head of the same patient shows multiple hypodense dot like shadows in the brain showing diffuse intracranial cysticercus cysts and it was the cause of dementia.

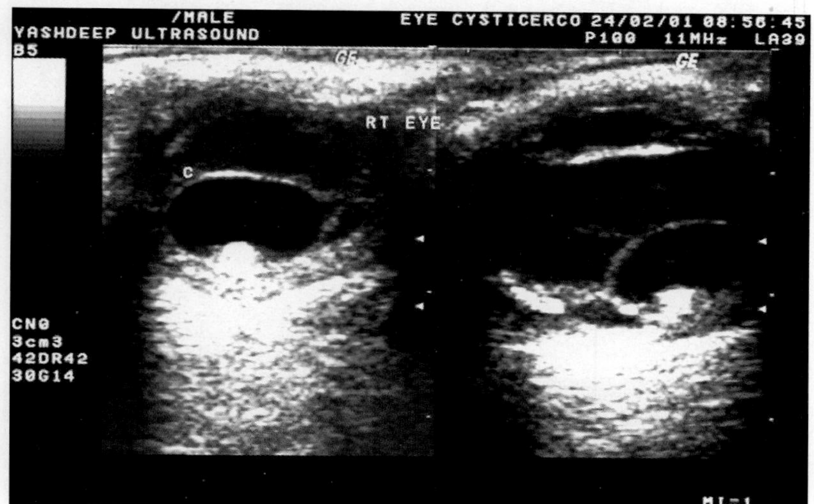

FIGURE 1.304: Intraocular cysticercosis with vitreous hemorrhage and RD. A patient came with sudden loss of vision with painful eye. HRSG shows a well-defined cyst in the vitreous cavity with an echogenic nidus suggestive of scolex with the inner wall of the cyst. Associated subvitreal and intragel hemorrhage is also seen in the eye. Partial lifting of the retina is also seen suggestive of retinal detachment.

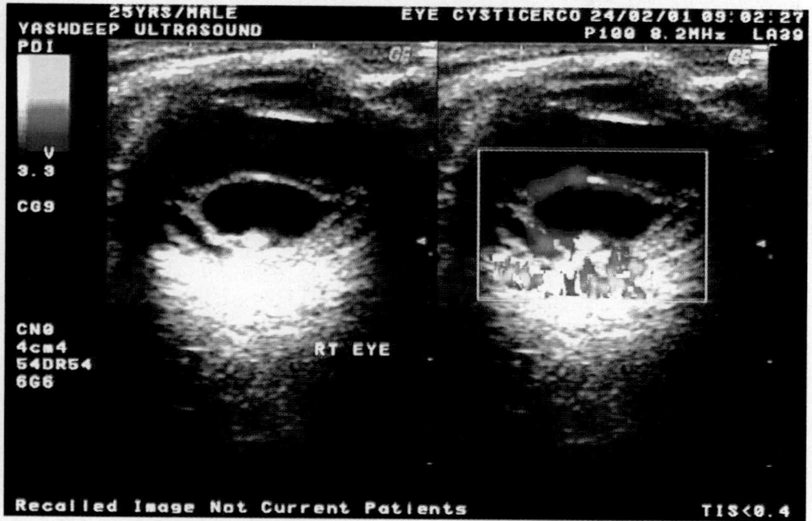

FIGURE 1.305: Color Doppler imaging of the patient shows increased hyperemia in the choroidal bed with cyst wall circulation.

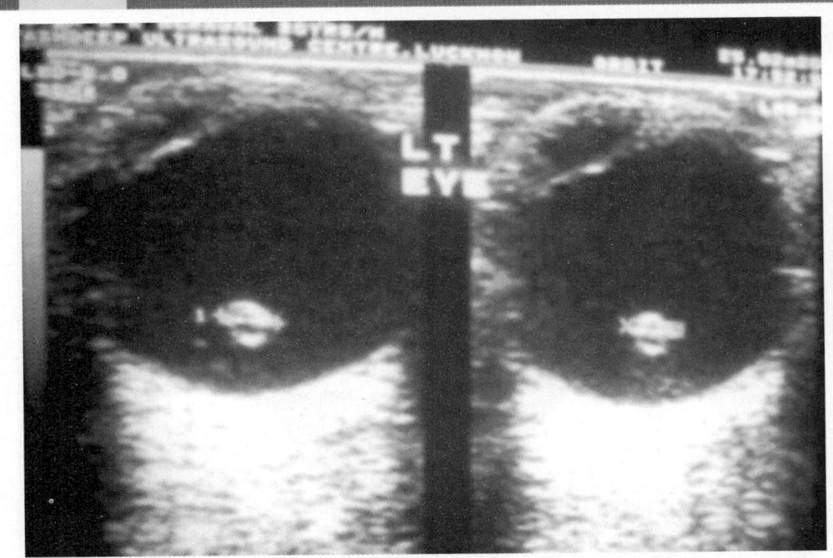

FIGURE 1.306: Rupture of the cysticercus cyst. Patient complains of sudden loss of vision with pain in the eye. HRSG shows rupture of the cysticercus cyst with free floating scolex in the posterior segment with intense vitreal reaction seen on HRSG.

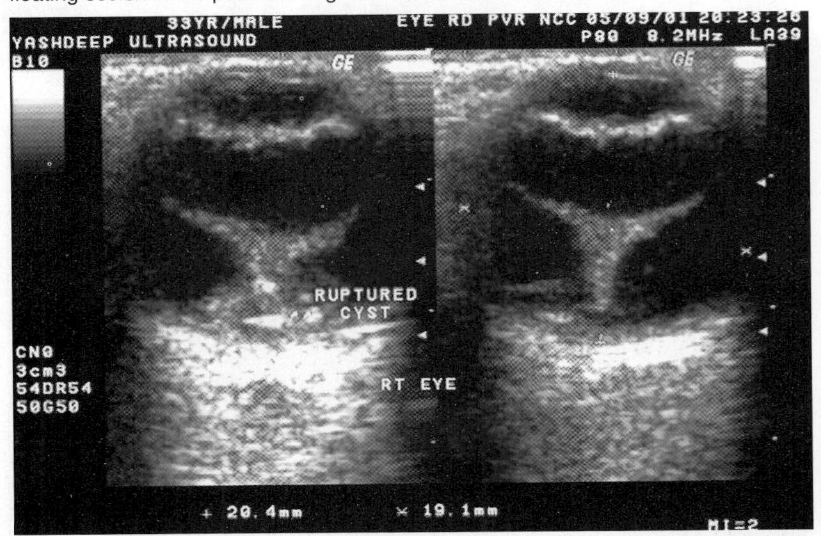

FIGURE 1.307: Rupture of the cysticercus cyst with PVR. HRSG shows rupture of the intraocular cysticercus cyst. Collapsed cyst is seen at the posterior pole with calcified echogenic nidus seen in the cyst. Associated proliferative vitreoretinopathy is also seen in the posterior segment. The patient was also having intracranial neuro-cysticercosis and was harboring the cyst in the eye. The vision was lost after the rupture and postrupture complication resulting into proliferative vitreoretinopathy.

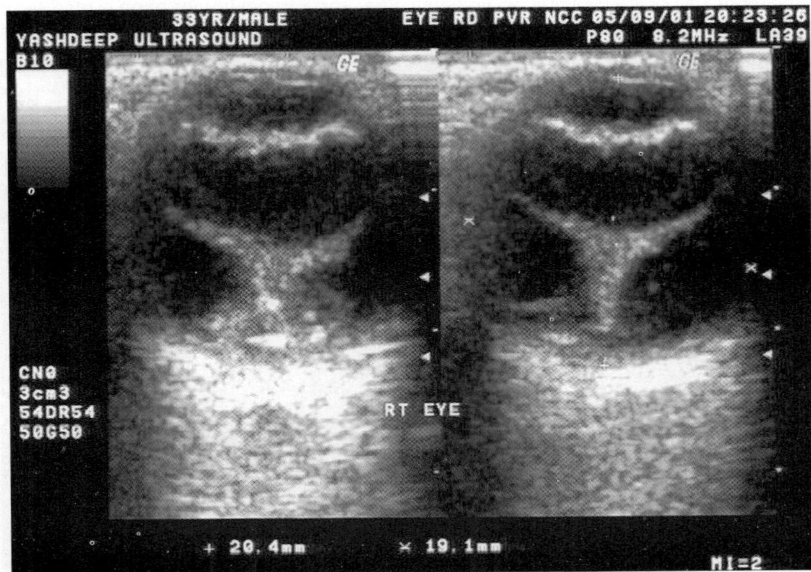

FIGURE 1.308: HRSG shows collapsed flattened cyst with vitreoretinal bands in the same patient.

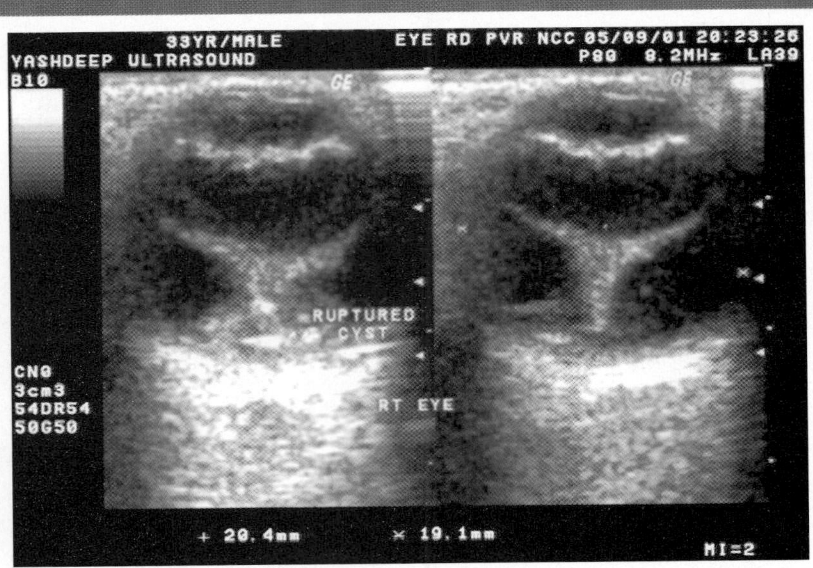

FIGURE 1.309: Color flow imaging shows dragging of the vessels along the retinal membranes.

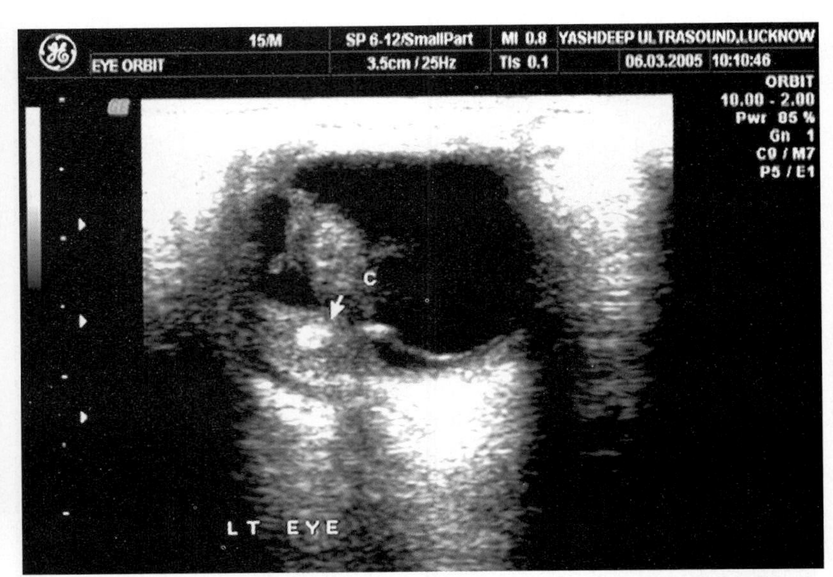

FIGURE 1.310: Ruptured cysticercus cyst and subvitreal space. A patient presented with sudden loss of vision with painful eye. HRSG shows a ruptured cysticercus cyst in subvitreal space with associated intragel hemorrhage.

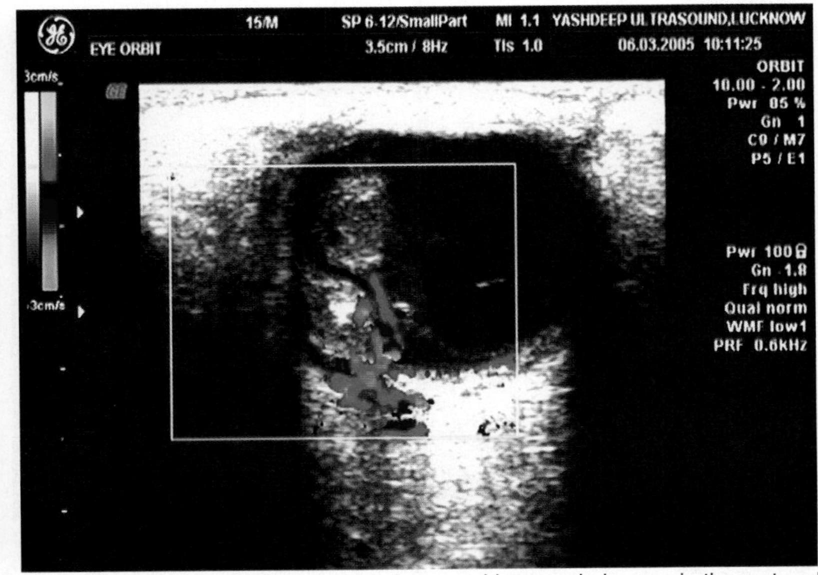

FIGURE 1.311: On color flow imaging increased hyperemia is seen in the ruptured cyst suggestive of recent rupture of the cyst.

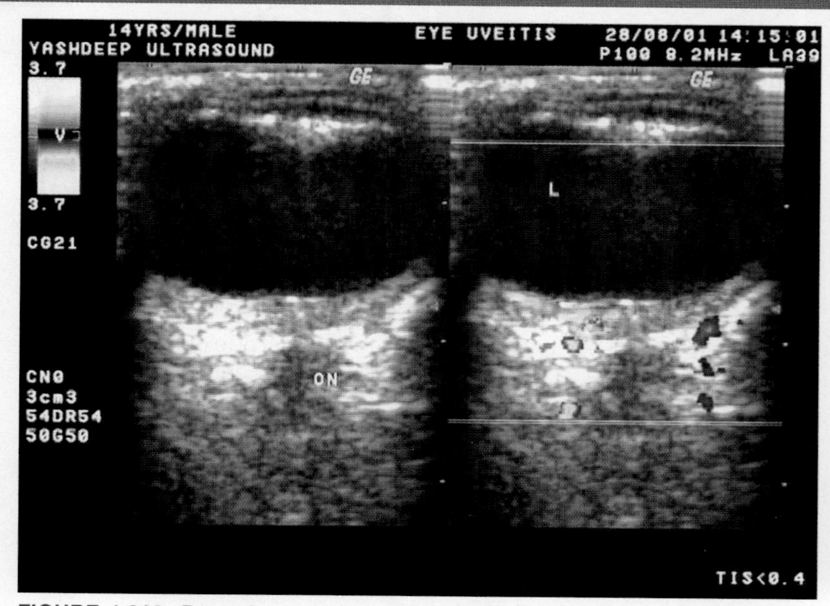

FIGURE 1.312: Posterior scleritis. HRSG shows thickening of the sclera with evidence of edematous hypoechoic sclera. The inflammatory fluid is seen seeping along the tenons space and also in the optic nerve sheath. Typical positive 'T' sign in posterior scleritis.

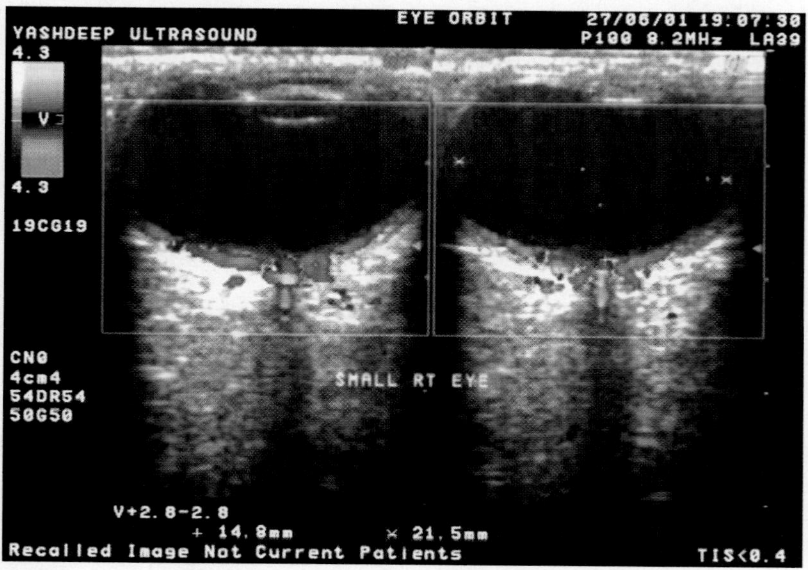

FIGURE 1.313: Color Doppler imaging of the same patient shows increased blood flow in the sclera due to hyperemia.

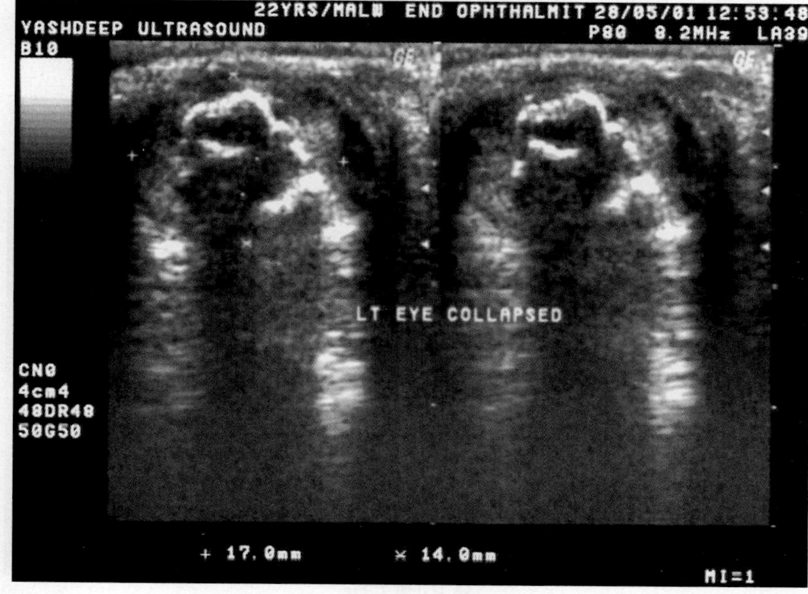

FIGURE 1.314

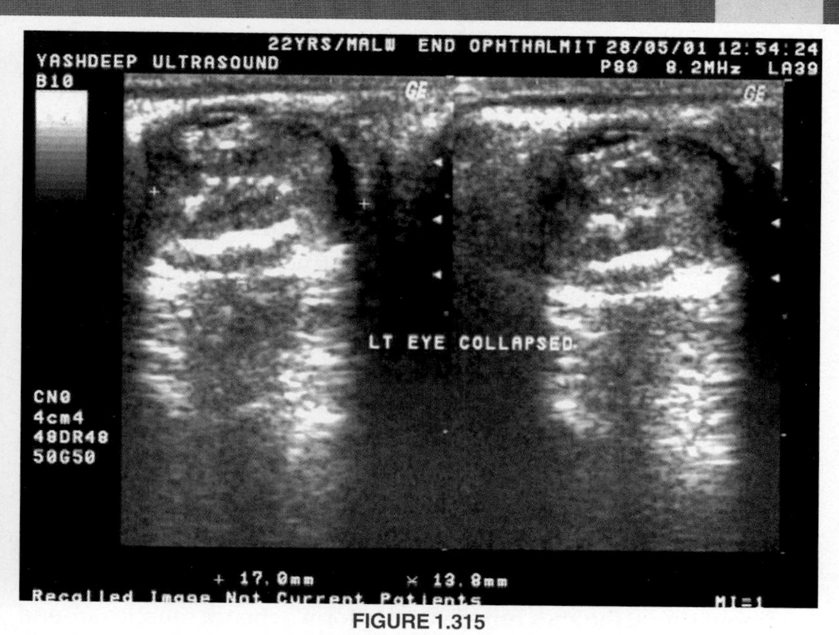

FIGURE 1.315

FIGURES 1.314 and 1.315: Endophthalmitis. HRSG shows small collapsed eye. Normal anatomy is destroyed. Dense echogenic collection is seen in the posterior segment. Multiple calcified specks are also seen. They are accompanied with acoustic shadowing. Antero posterior axis is small.

Examination of Orbit

Introduction

Standardized high-resolution sonography of orbit is highly sophisticated and accurate method for detecting and differentiating orbital lesions. HRSG is rapid noninvasive screening method to examine orbital lesions. Dynamic examination of lesion by HRSG is the added advantage in other imaging modalities. However, computed tomography (CT) is the investigation of choice for orbital lesions, as it produces excellent tissue details of surrounding structures with clear delineation of anatomy. But, the major disadvantage of CT is time-consuming; delivery of radiation dose to eye and costly investigation. Ultrasonography has advantage of rapidity and easy accessibility. With its ability to identify the orbital walls, retrobulbar fat, ocular muscles, optic nerve and orbital mass, it is a useful first line of investigation in orbital diseases and proptosis.

The aim of orbital imaging is to demonstrate an orbital lesion, determine its position and extent within the orbit. At times it becomes difficult to predict the pathological nature of lesion, as most of ultrasonographic features are nonspecific, therefore, clinical history is important.

Examination Technique

Short focus high-frequency transducers with frequency ranging from 5 mhz to 10 mhz are the ideal probes to examine the orbit. Direct contact method is the ideal technique. The examination is carried out in both transocular and paraocular approaches. The orbit is examined in transverse, axial and longitudinal views. Transocular method clearly delineates mid and posterior orbital masses. The paraocular approaches clearly demonstrate anterior lesions and relationship to the globe and orbital wall. The paraocular approach is used in both transverse and longitudinal directions. The orbit is examined in all four directions: superior, inferior, medial (nasal) and lateral (temporal).

Orbital Anatomy

The orbit is a bony socket, which contains eyeball, extrinsic muscles, and optic nerve embedded in fat, vessels and nerves innervating the eyeball. The two eyeballs are the dominant structures in the orbit, which are not lying along the axis of orbit. They face forward and lie parallel to each other. The optic nerve is echo poor in texture, which lies freely in the retrobulbar fat, which is highly echogenic. Intraorbital part of optic nerve is around 25 mm, whereas orbital length is around 18 to 20 mm. Optic nerve is moved with the movement of eye. Therefore, increased length of optic nerve does not make it stretched on eye movement. HRSG can show the whole length of optic nerve in the orbit with careful examination. It can be seen longitudinally and transversely as oval hypoechoic shadow. The four recti muscles arise from a tendinous ring at the apex broaden out to form a cone of muscles around the eyeball.

The orbital muscles are seen as thin, echo poor straps. These tendons are narrowed anteriorly and muscle belly is more fusiform in shape. The medial and lateral recti are seen best in horizontal planes and superior and inferior recti are seen in vertical planes. The inferior oblique muscle is seen behind the globe just below the macula. The superior oblique is seen in superomedial part of the orbit. Its sonoappearance is similar to the recti muscle appearance.

Proptosis

Anterior displacement of the eyeball due to retrobulbar pathology can be easily evaluated on HRSG. The retrobulbar masses, which can be detected well on HRSG are grouped as under.

i. Muscle hypertrophy in thyroid disease.
ii. Pseudo tumors—inflammatory orbital disease.
iii. Vascular tumors.
   1. Hemangioma
   2. A.V. malformations
   3. Orbital varix
   4. Lymphangioma
   5. Dilated superior ophthalmic vein.
iv. Parasitic infestations.
   1. Cysticercosis
   2. Hydatid cysts.
v. Optic nerve tumors—glioma
   1. Meningioma
   2. Optic nerve cyst
   3. Optic nerve glioma
vi. Rhabdomyo sarcoma
vii. Lymphoproliferative masses
viii. Lymphoma
ix. Metastatic deposits in the orbit
x. Orbital trauma.
   1. Orbital hematoma

Pseudoproptosis

Increased axial length of the eyeball congenital and acquired.

Vascular Tumors of Orbit

Many vascular tumors are found in the orbit. It is one of the most common causes of proptosis. HRSG is excellent modality to diagnose vascular tumors. Color doppler flow study can easily differentiate between venous or arterial nature of the tumor mass.

Hemangiomas

Orbital hemangioma (cavernous type) are common in adult. They are benign in nature. They occur in second to fifth decade of life. Clinically, they are characterized with slowly progressive swelling of eyeball with unilateral proptosis. HRSG shows a highly reflective echogenic mass in retrobulbar space, which is seen located in the muscle cone. The multiple interfaces of dilated capillary walls produce an echogenic mass. Doppler examination gives a good account of blood flow in the mass.

Arteriovenous Fistulae

These may develop after trauma or spontaneously. They are often missed clinically. Therefore, HRSG is useful investigation to diagnose them. Doppler examination is helpful in diagnosing small fistulae.

Orbital Varices

These are usually diagnosed clinically. They often increase in size with bending of head or Valsalva maneuver. These patients ultimately land up in endophthalmitis most likely due to fat necrosis.

Lymphangioma

These are mostly detected in children or young adults. These lesions grow slowly and causing proptosis. They may present as acute onset due to spontaneous secondary hemorrhage. These tumors can be small or large enough to fill the orbital space. On HRSG, multiple dilated septate lymph-filled spaces are seen in retrobulbar areas. A multiloculated multiseptate mass is seen with echogenic septa. If

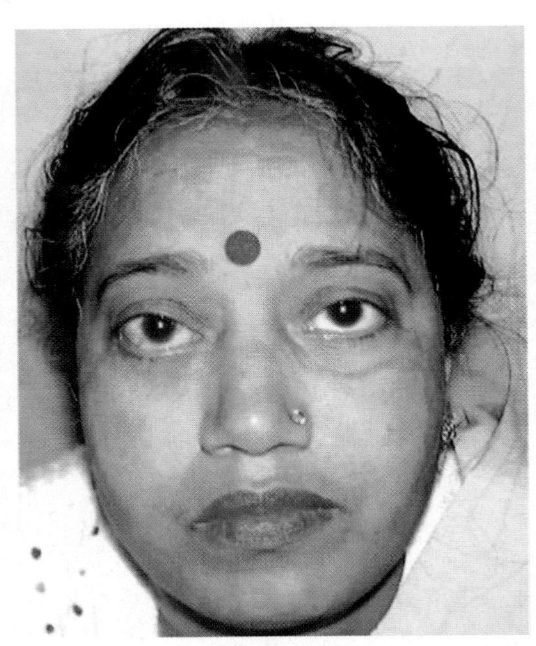

FIGURE 1.316

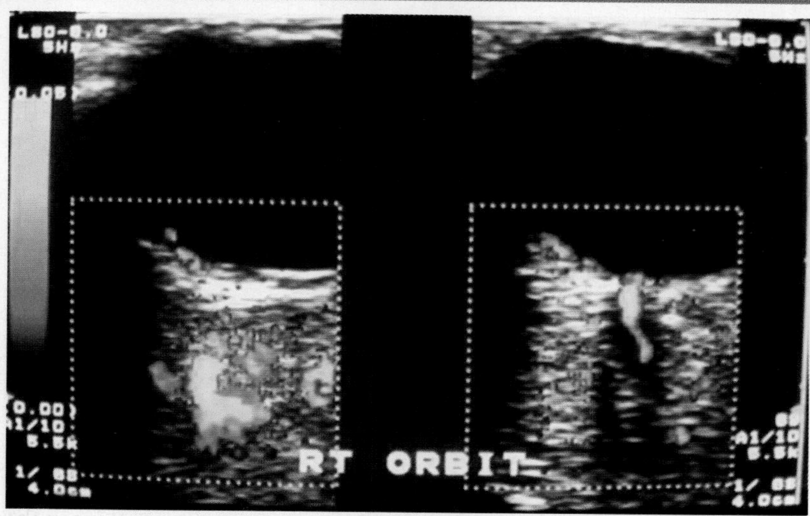

FIGURE 1.317

FIGURES 1.316 and 1.317: Orbital varix. Multiple detailed vessels are seen in retrobulbar intraconal area suggestive of orbital varix. On color Doppler flow, low venous flow is seen in them.

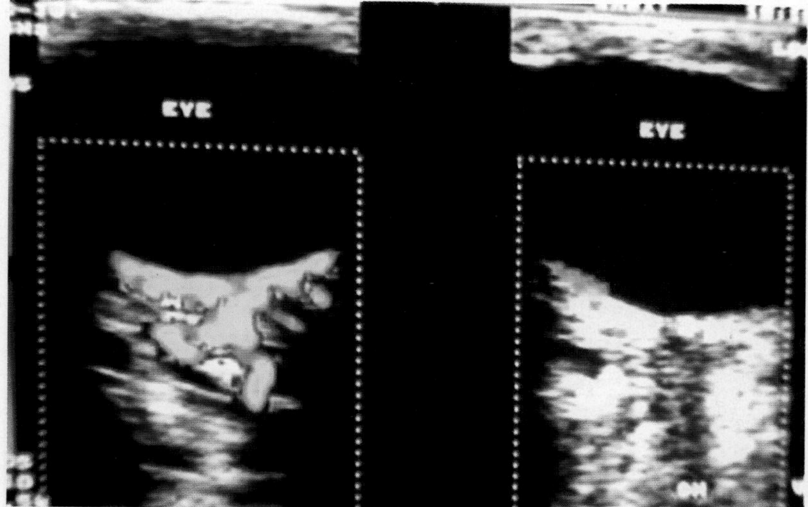

FIGURE 1.318

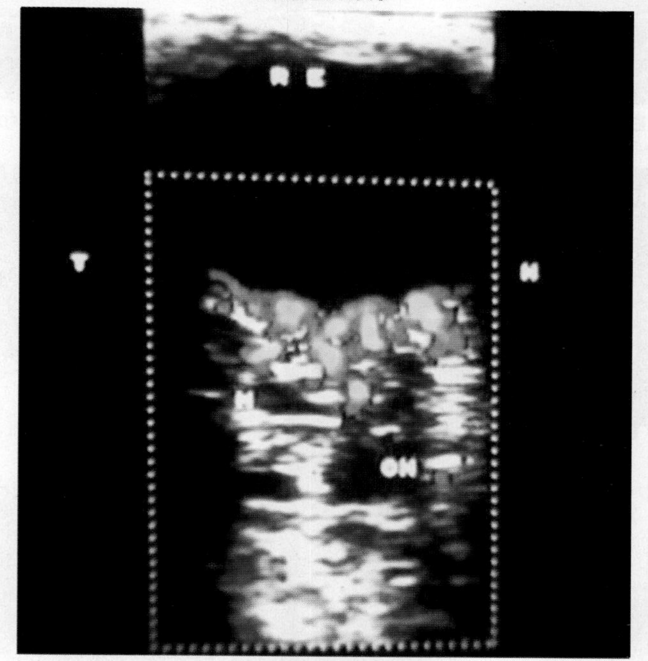

FIGURE 1.319

FIGURES 1.318 and 1.319: Orbital varices. Multiple dilated vessels are seen in Rt retrobulbar space in a patient with the history of proptosis on persistent bending of the head. Color Doppler shows multiple dilated vessels distended on drooping of the head. Confirms orbital varices.

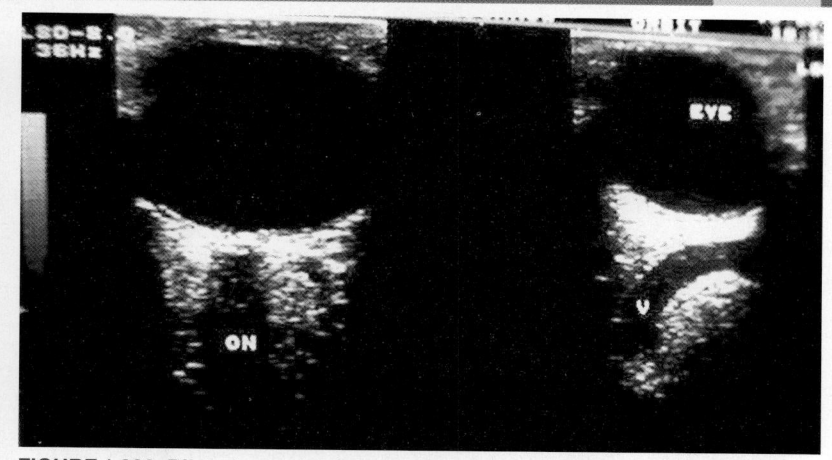

FIGURE 1.320: Dilated superior Ophthalmic vein. A patient presented with proptosis and optic disc edema. HRSG shows dilated superior ophthalmic vein congested in the intraconal part.

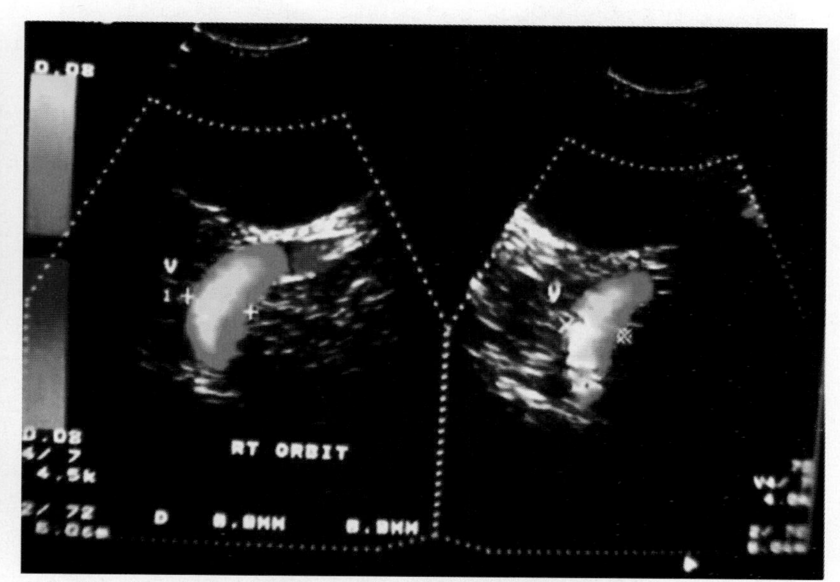

FIGURE 1.321: Color Doppler flow imaging shows monophasic venous flow in the vein.

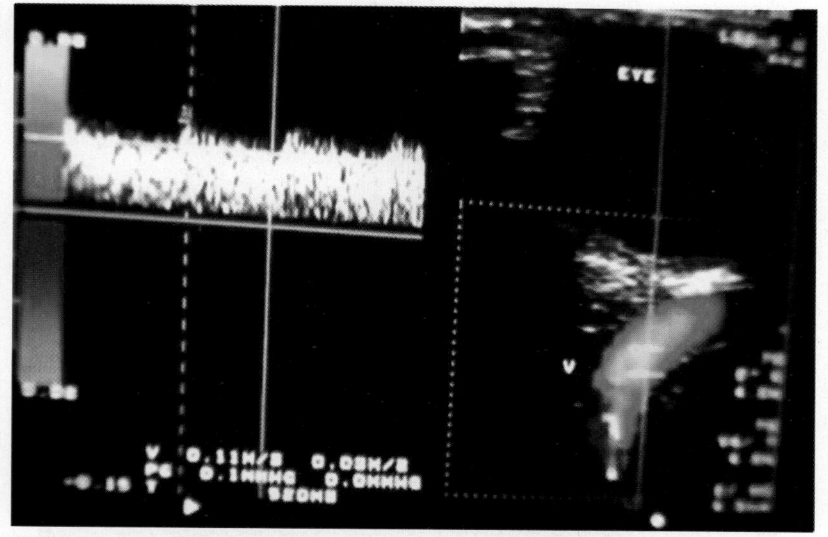

FIGURE 1.322: Spectral Doppler flow tracing confirms the venous flow pattern of dilated superior ophthalmic vein.

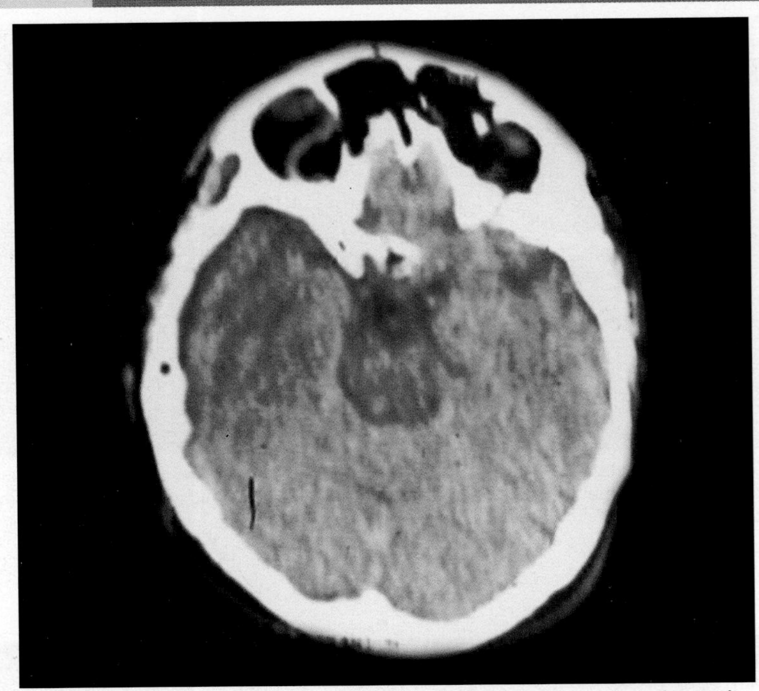

FIGURE 1.323: CT scan of the same patient shows tortuous dilated superior ophthalmic vein.

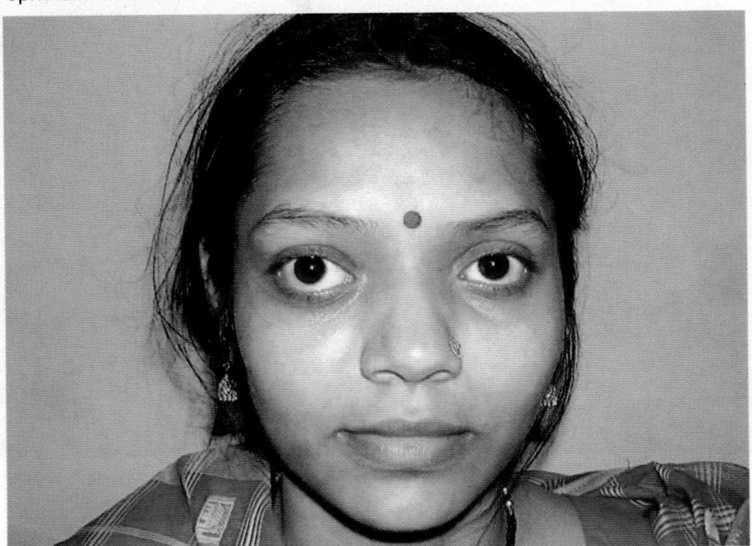

FIGURE 1.324: Cavernous hemangioma. A young lady came with the history of proptosis in Rt eye with blurring of vision.

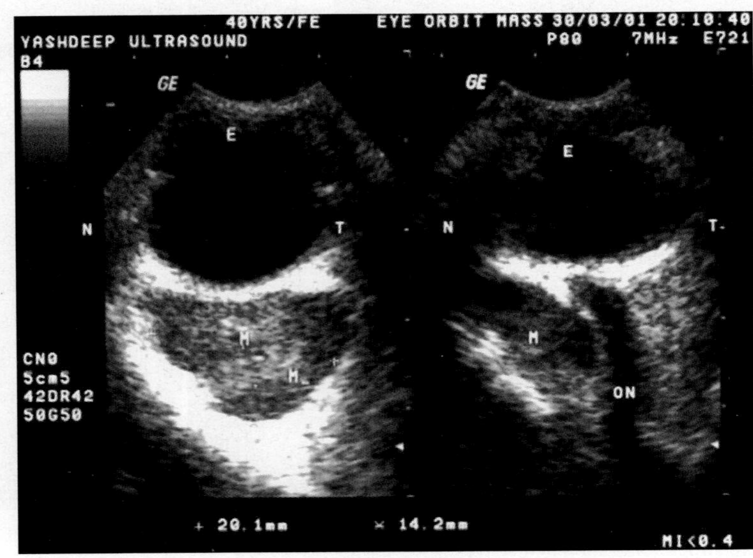

FIGURE 1.325

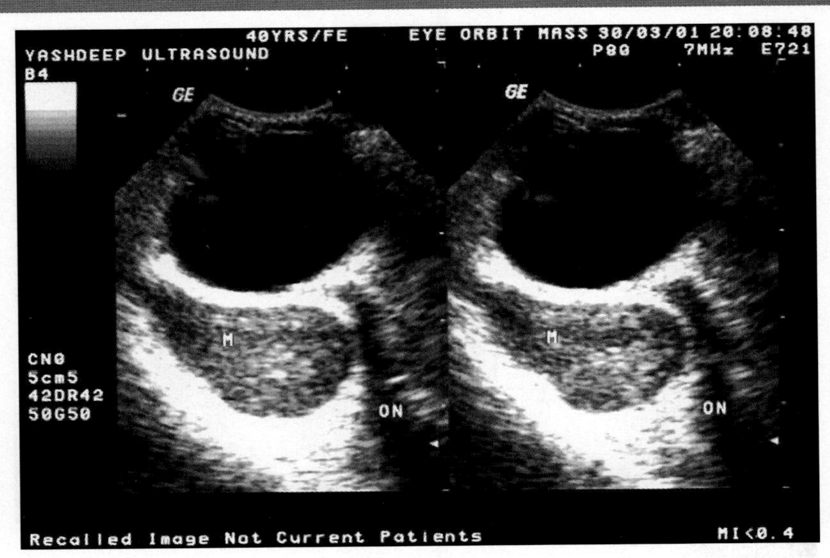

FIGURE 1.326

FIGURES 1.325 and 1.326: HRSG shows a low-level echo complex mass in intraconal part of Rt orbit. Multiple fine thin inter faces are seen in the mass. The mass is seen pressing over the optic nerve and displacing it to the opposite side. A smooth indentation is seen over the nerve. No evidence of any calcification is seen in the mass.

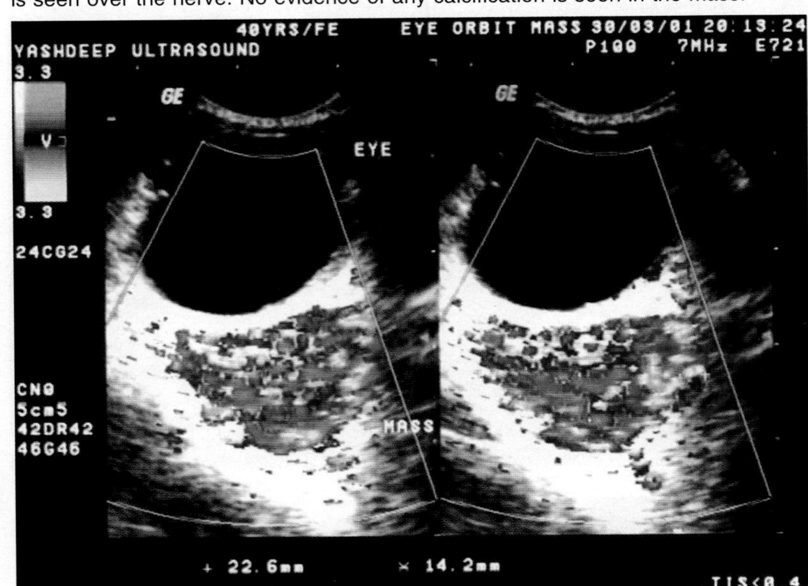

FIGURE 1.327: Color Doppler flow imaging show multiple dilated vessels seen in the mass with high flow pattern.

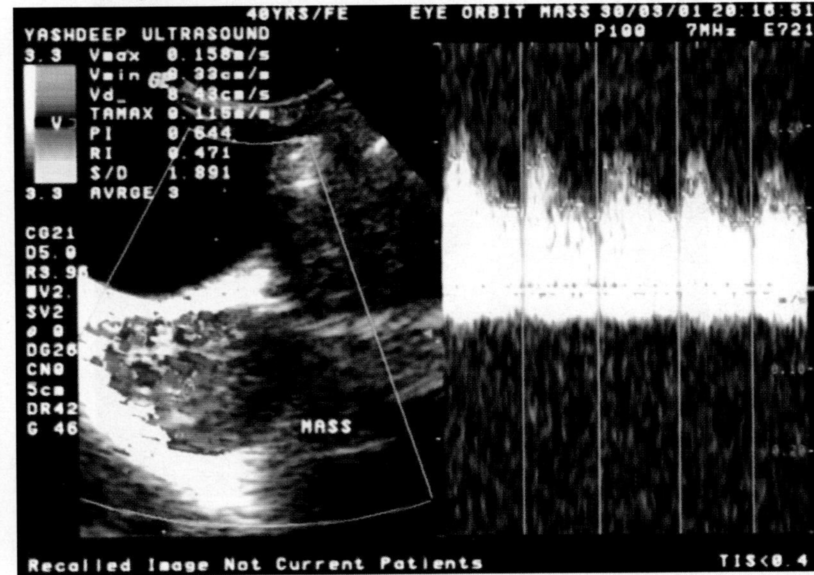

FIGURE 1.328: Spectral Doppler flow shows high flow with arterial pulsation in the mass suggestive of cavernous hemangioma.

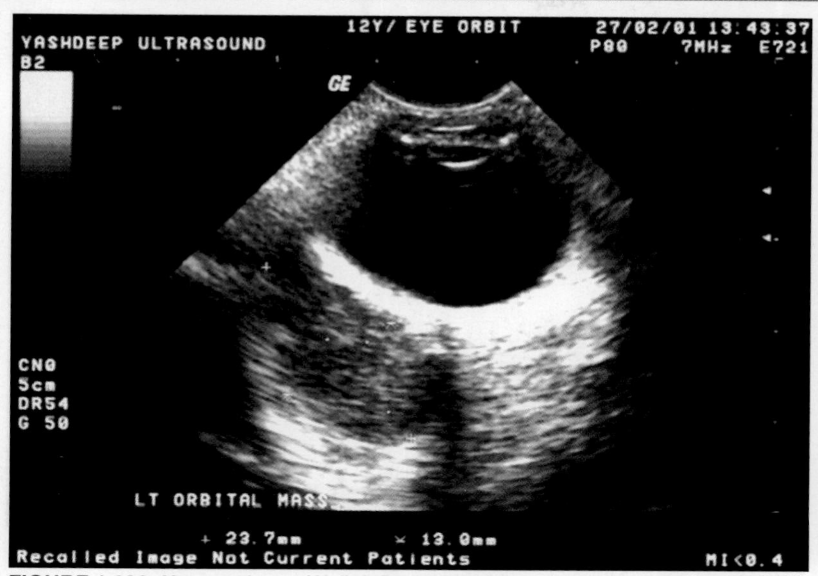

FIGURE 1.329: Hemangioma. Well defined homogeneous mass seen in retrobulbar area pressing over the optic nerve and anterior displacement of eyeball.

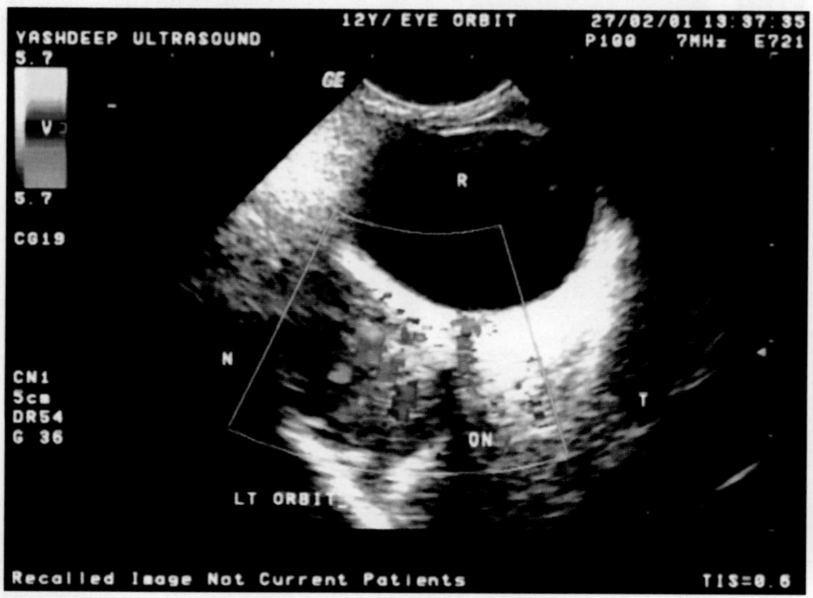

FIGURE 1.330: Color Doppler study shows moderately enhancing mass suggestive of hemangioma.

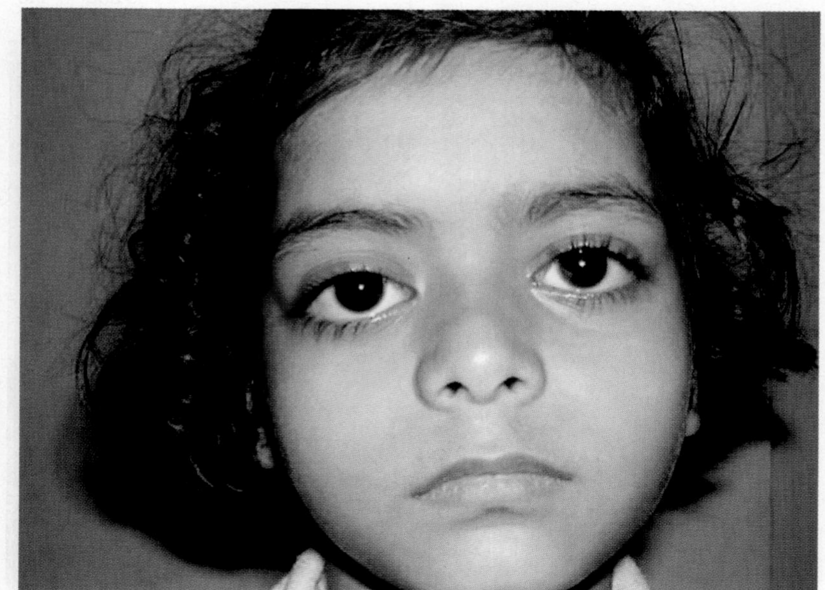

FIGURE 1.331: Young girl complains of proptosis Lt eye with no pain but diplopia.

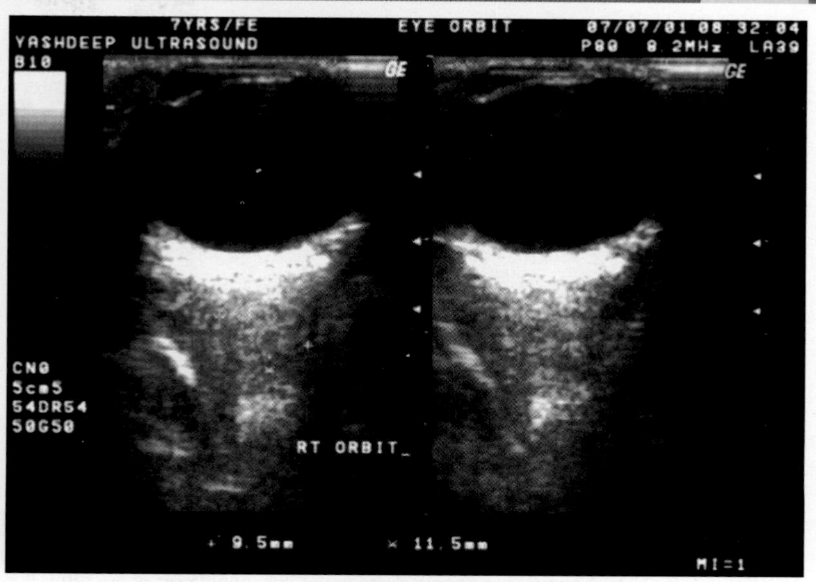

FIGURE 1.332

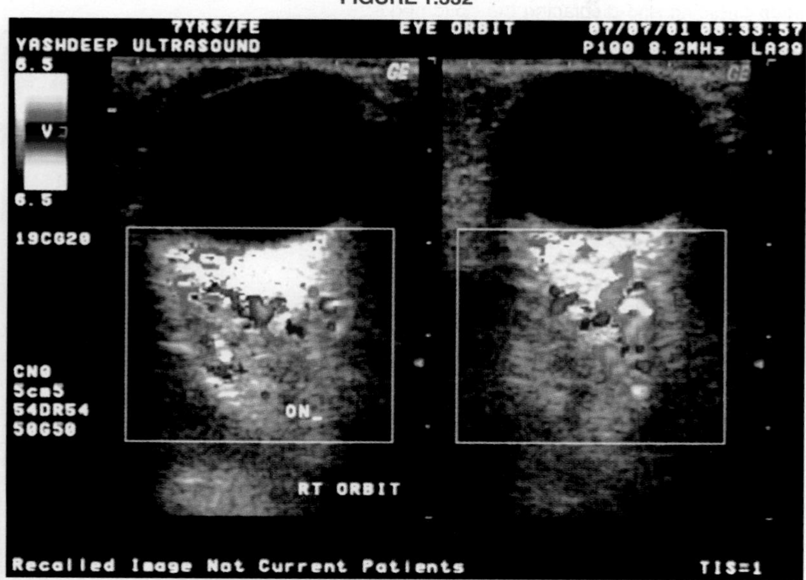

FIGURE 1.333

FIGURES 1.332 and 1.333: HRSG shows well defined intraconal mass with homogeneous texture. On color flow imaging moderate flow is seen in the mass suggestive of hemangioma

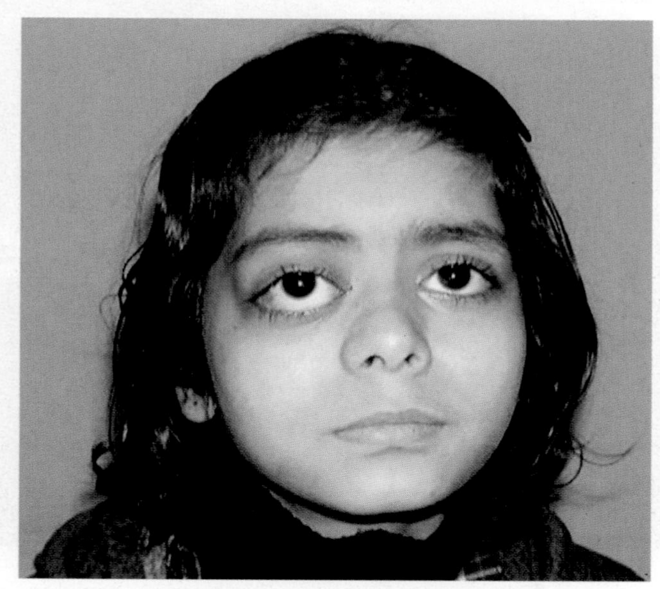

FIGURE 1.334: The same girl after 3 years presented with persistent proptosis with diplopia.

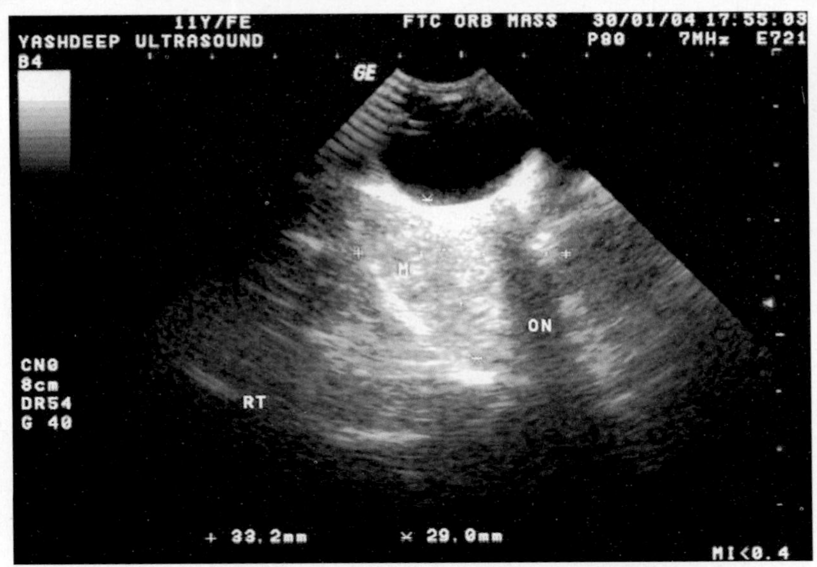

FIGURE 1.335: HRSG shows echogenic mass in the intraconal part of the orbit. It is seen pressing and displacing the optic nerve.

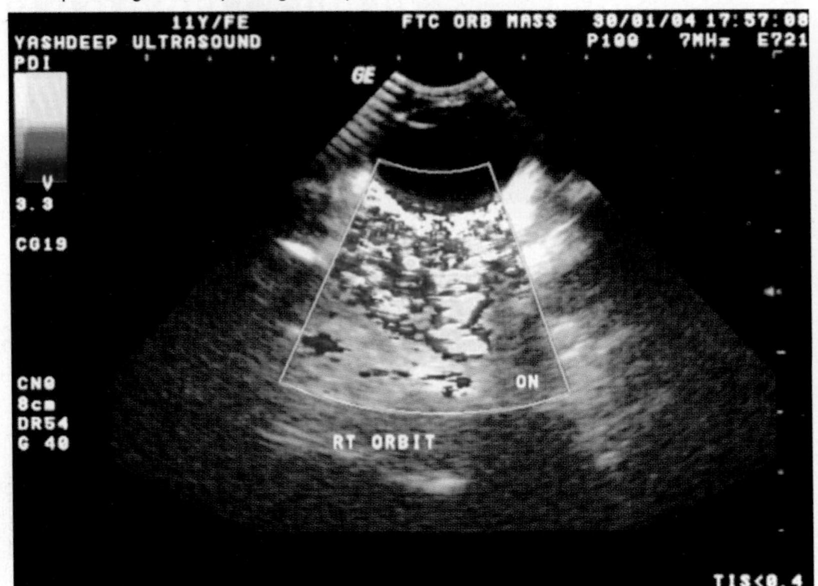

FIGURE 1.336

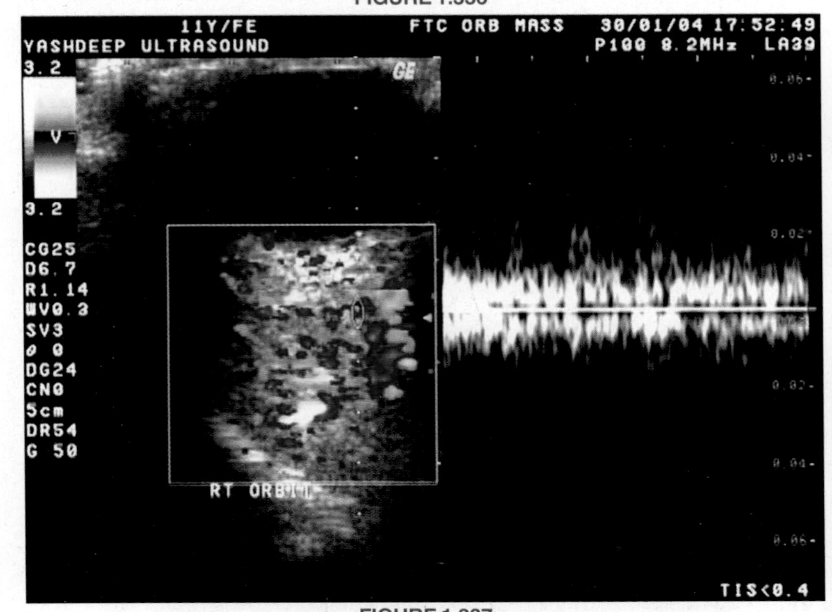

FIGURE 1.337

FIGURES 1.336 and 1.337: On color Doppler imaging multiple small vessels are seen in the mass suggestive of highly vascular mass. On spectral Doppler tracing low flow is seen in the mass.

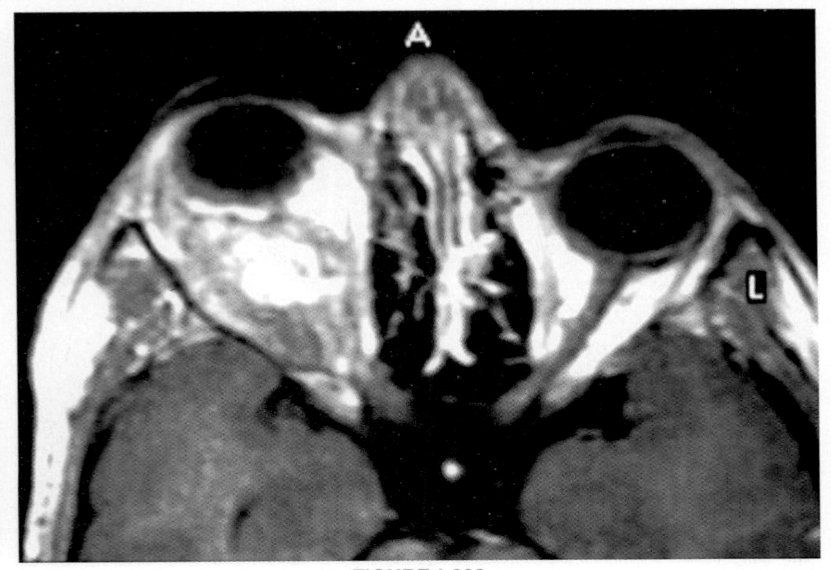

FIGURE 1.338

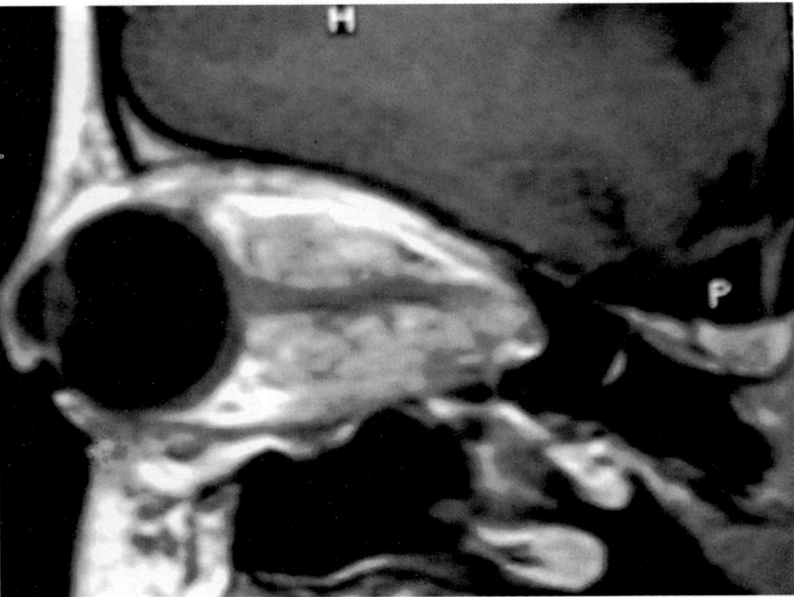

FIGURE 1.339

FIGURES 1.338 and 1.339: Contrast enhanced CT scan of the same patient shows intraconal mass. It is seen engulfing the optic nerve. It shows intense enhancement on contrast examination suggestive of highly vascular mass.

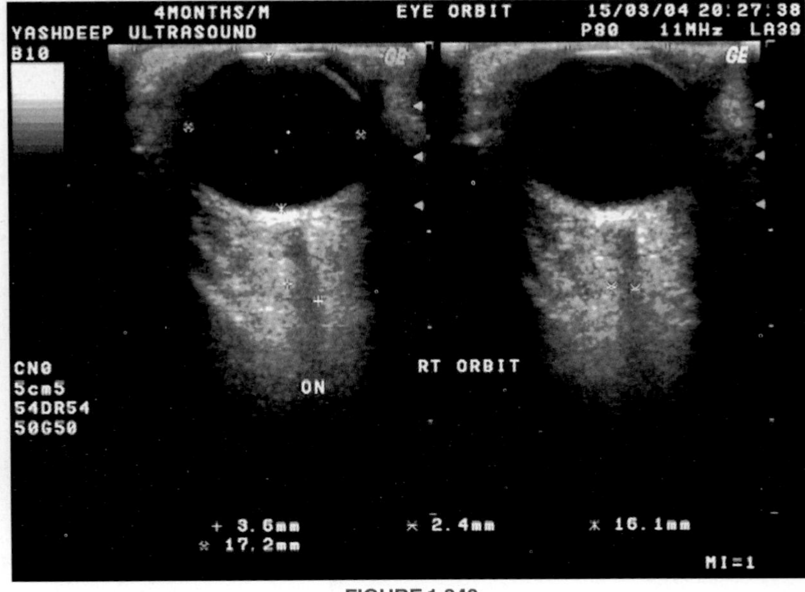

FIGURE 1.340

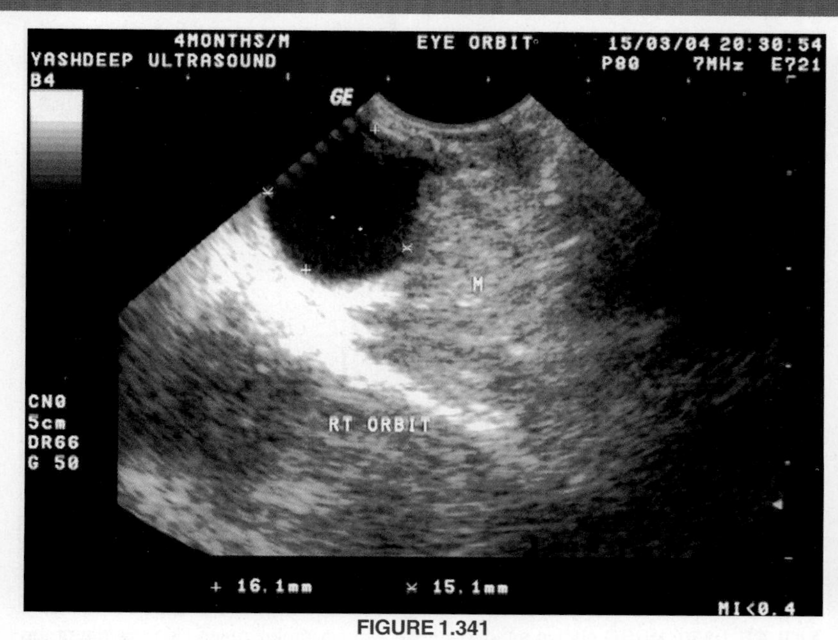

FIGURE 1.341

FIGURES 1.340 and 1.341: Vascular retrobulbar mass. An infant presented with proptosis with progressive bulging eye. The HRSG shows an echogenic mass filling the intraconal part of the orbit. The mass is seen covering the optic nerve and displacing it anteriorly.

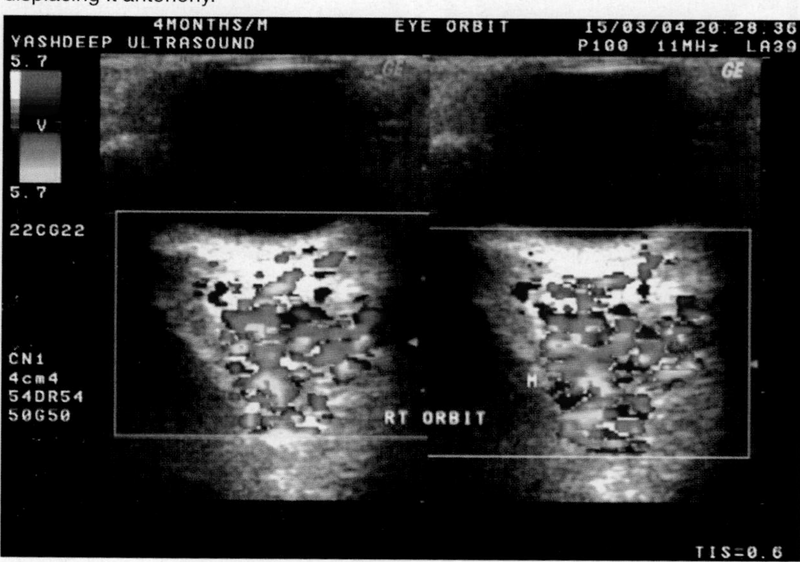

FIGURE 1.342

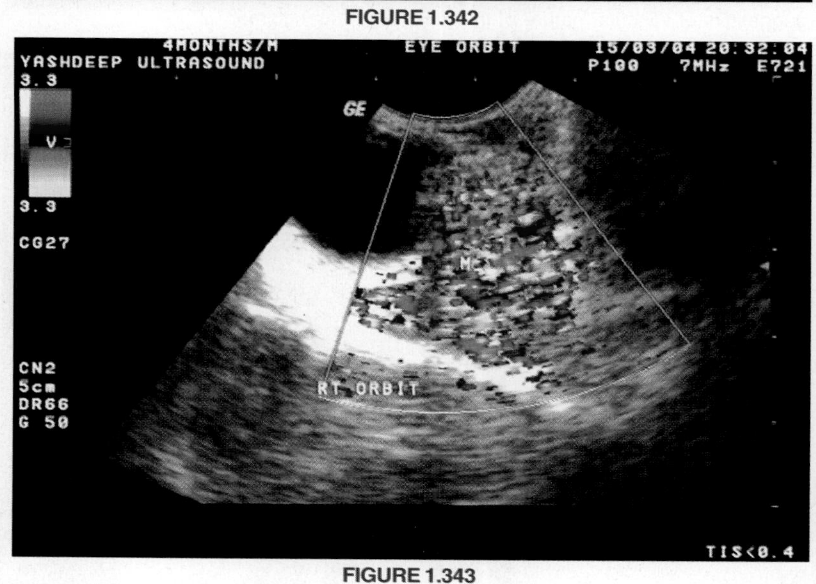

FIGURE 1.343

FIGURES 1.342 and 1.343: On color flow imaging multiple vessels are seen in the mass suggestive of highly vascular nature of the mass. On biopsy the mass turned out to be congenital hemangioma.

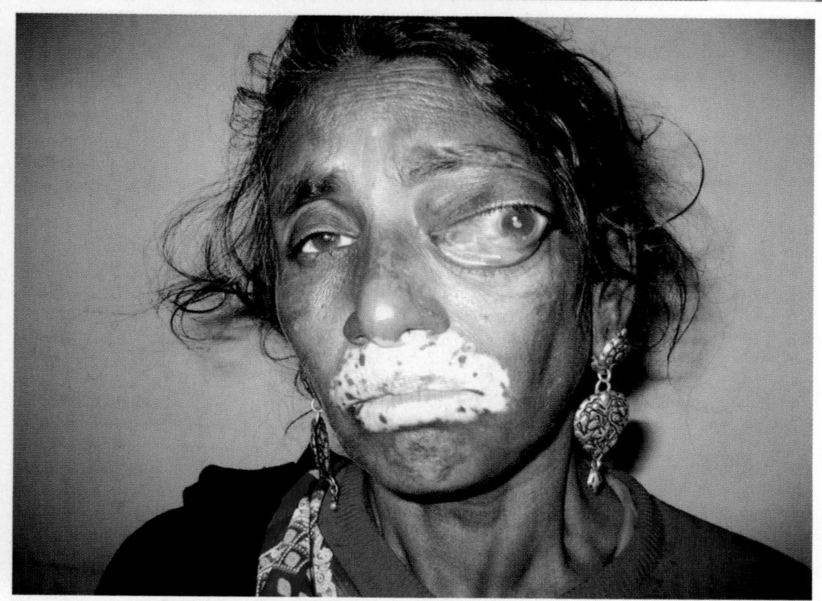

FIGURE 1.344: Big vascular mass. A middle aged lady presented with marked proptosis with lateral displacement of Lt eye.

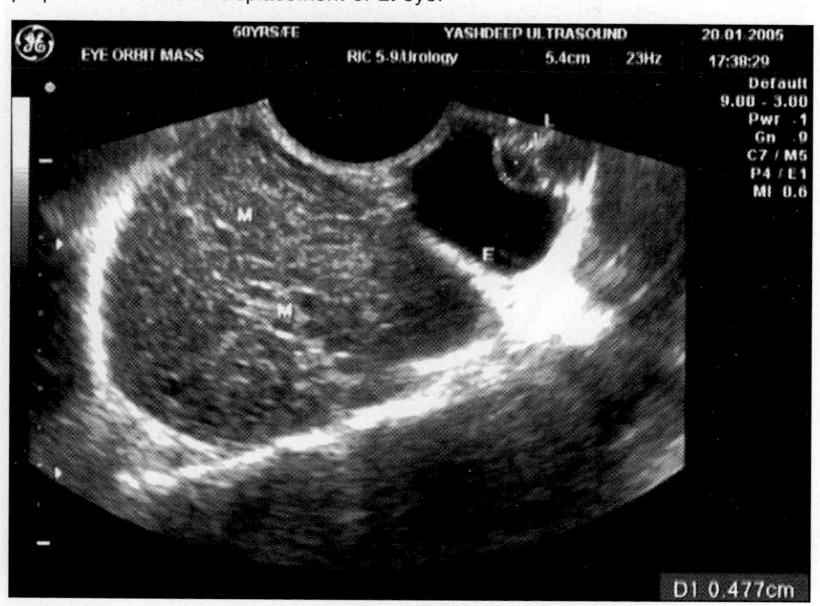

FIGURES 1.345: HRSG shows a big mass with medium level echoes in the retrobulbar space of the Lt eye. The mass is seen pressing and displacing of the eyeball. It is seen also extending into extra conal space. Multiple small cystic areas are seen in the mass.

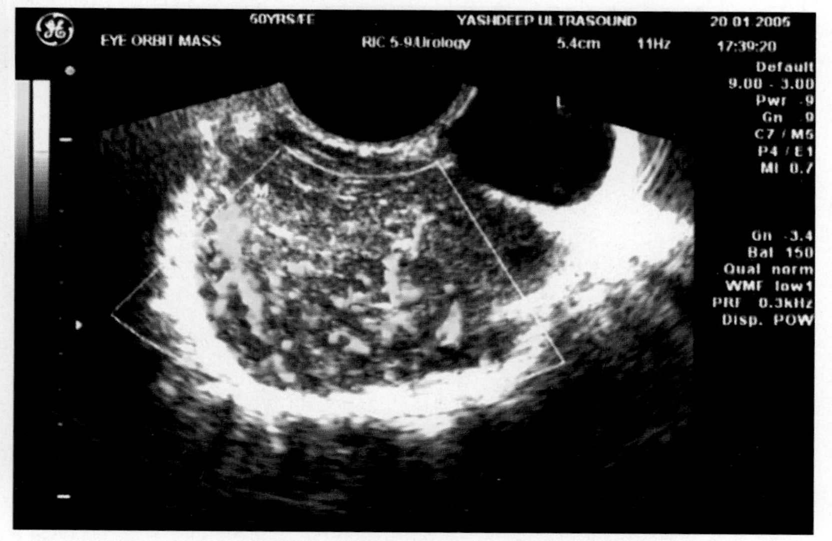

FIGURES 1.346: On color flow imaging multiple vessels are seen feeding the mass they are running to the mass suggestive of vascular mass.

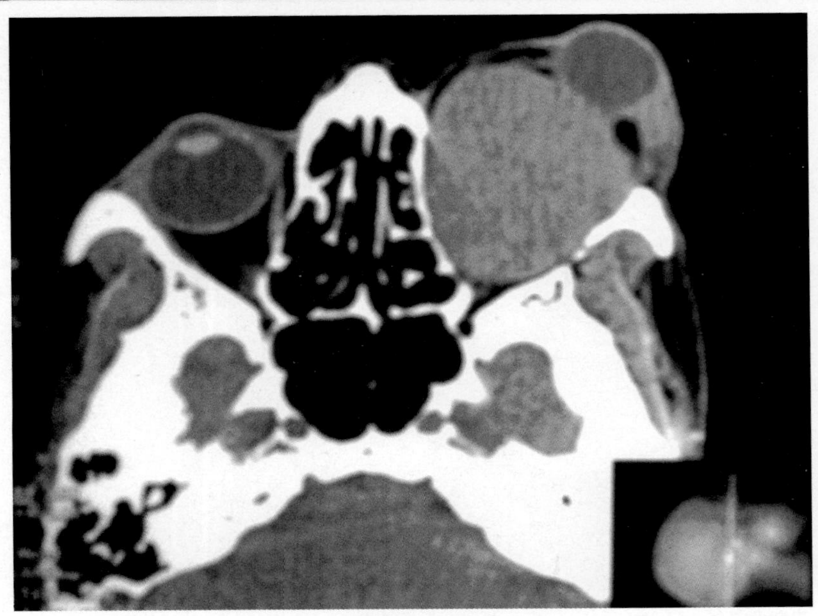

FIGURE 1.347

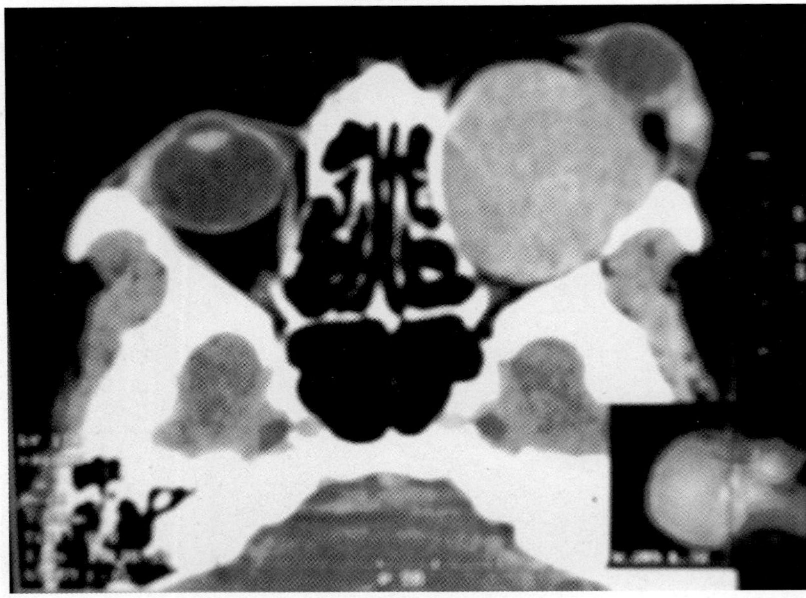

FIGURE 1.348

FIGURES 1.347 and 1.348: CT scan of the same patient shows a big homogeneous hypodense mass in the Lt retrobulbar space. The mass shows intense enhancement on contrast examination confirming the vascular nature of the mass.

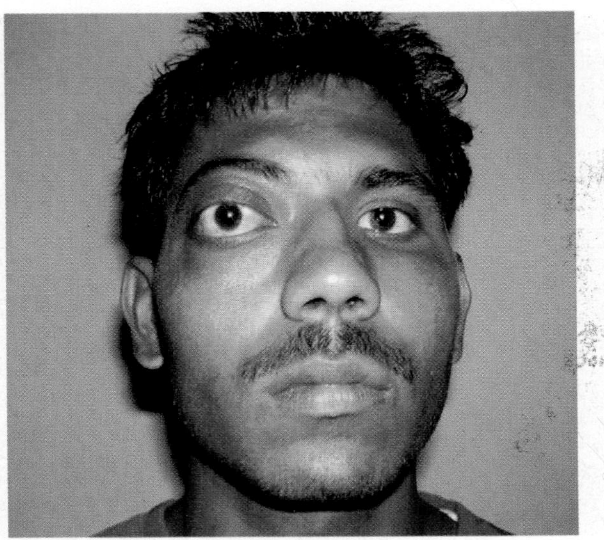

FIGURE 1.349: A young man presented with Rt sided proptosis with diplopia.

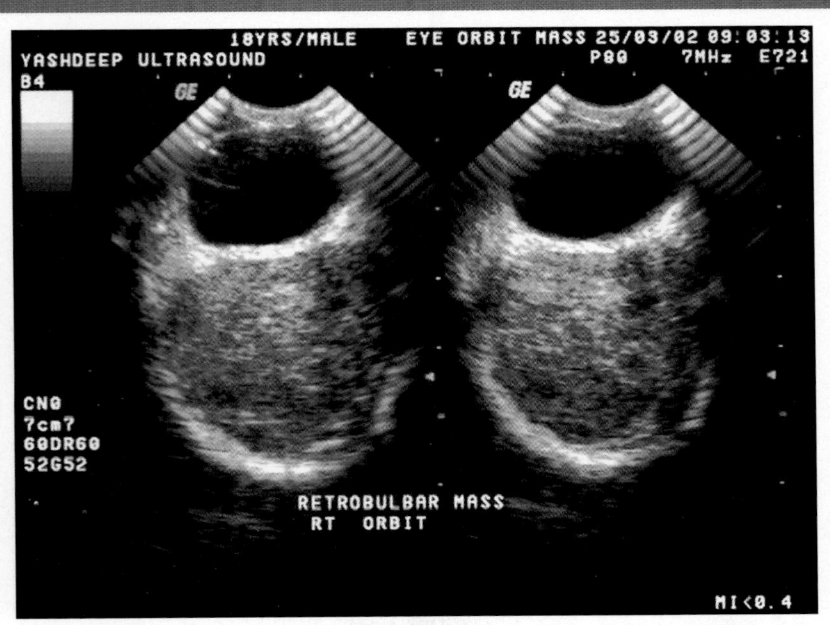

FIGURE 1.350: HRSG shows a big low-level echo complex mass filling whole of the retrobulbar space.

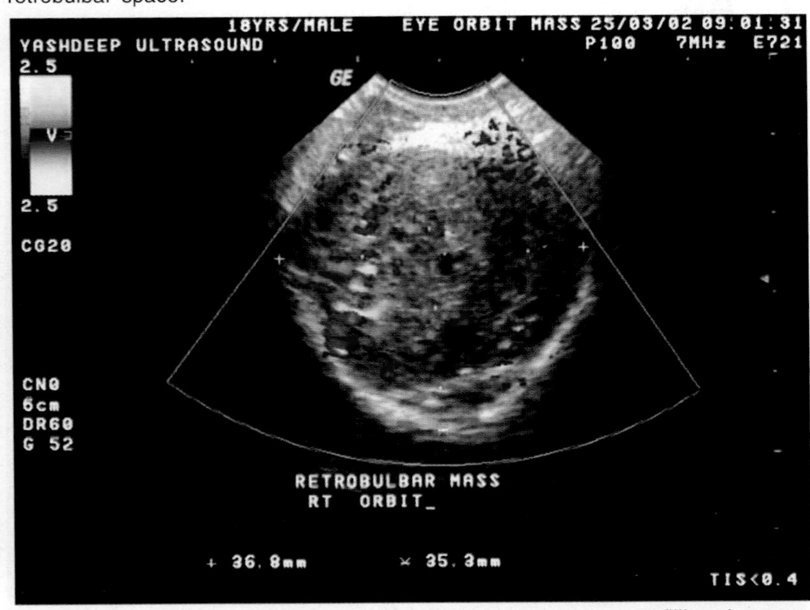

FIGURE 1.351: On color flow imaging multiple vessels are seen filling the mass suggestive of vascular nature of the mass big hemangioma.

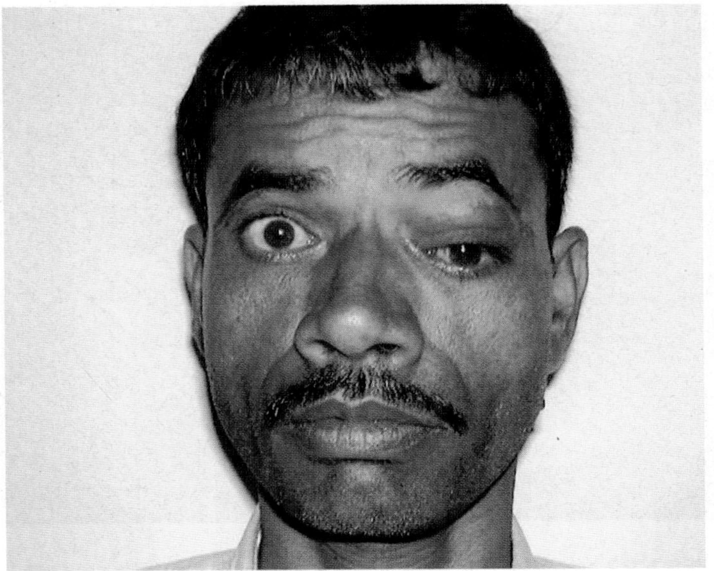

FIGURE 1.352: A man presented with soft tissue mass coming out from the superior part of the Lt orbit with downward displacement of the Lt eye.

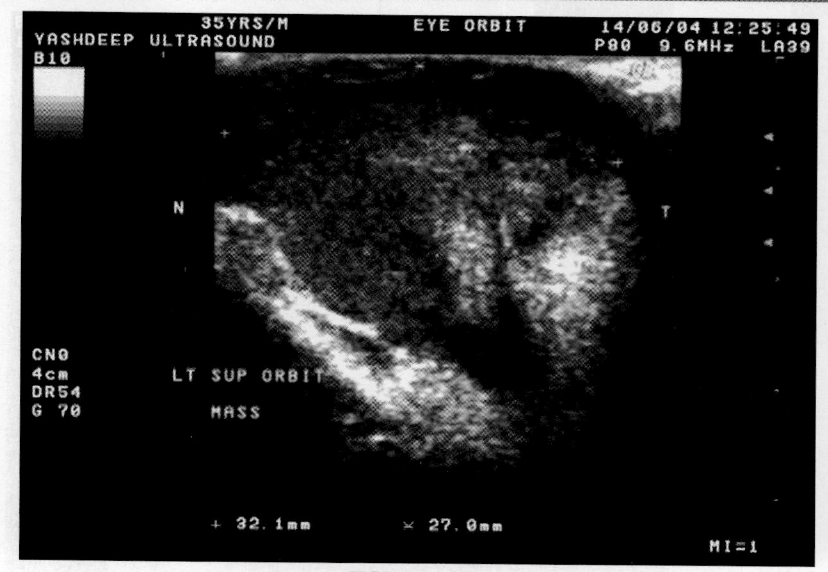

FIGURE 1.353

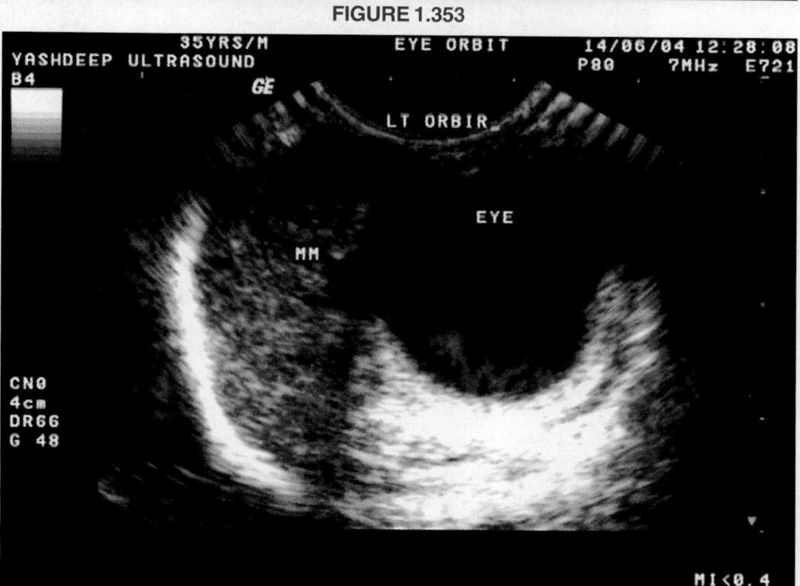

FIGURE 1.354

FIGURES 1.353 and 1.354: HRSG shows a low-level echo complex mass coming out from the superior quadrant of the Lt orbit. It shows homogeneous echoes and small tiny cysts.

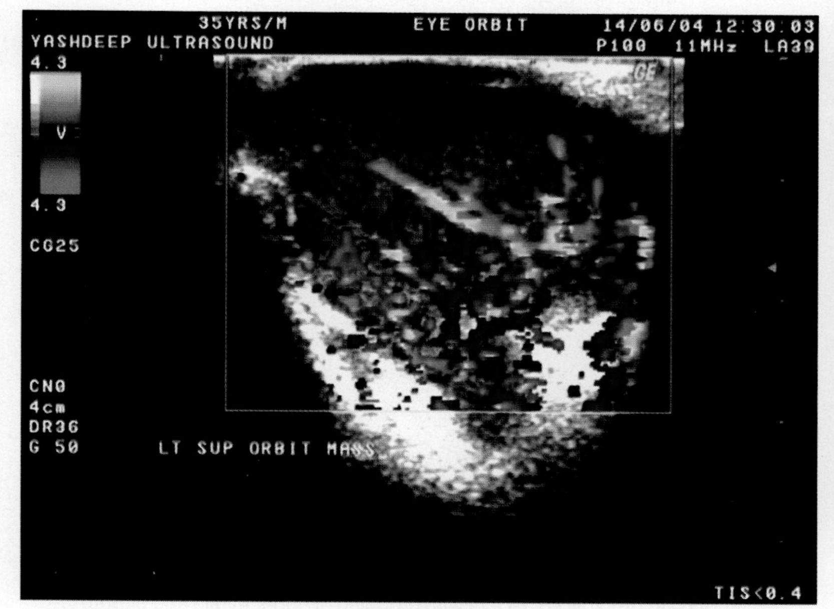

FIGURE 1.355

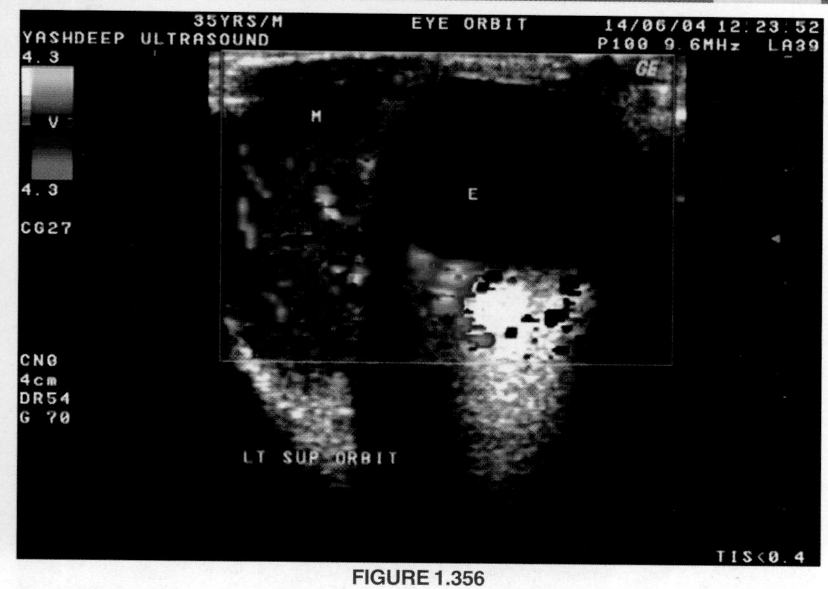

FIGURE 1.356

FIGURES 1.355 and 1.356: On color flow imaging, multiple vessels are seen feeding the mass suggestive of vascular nature of the mass.

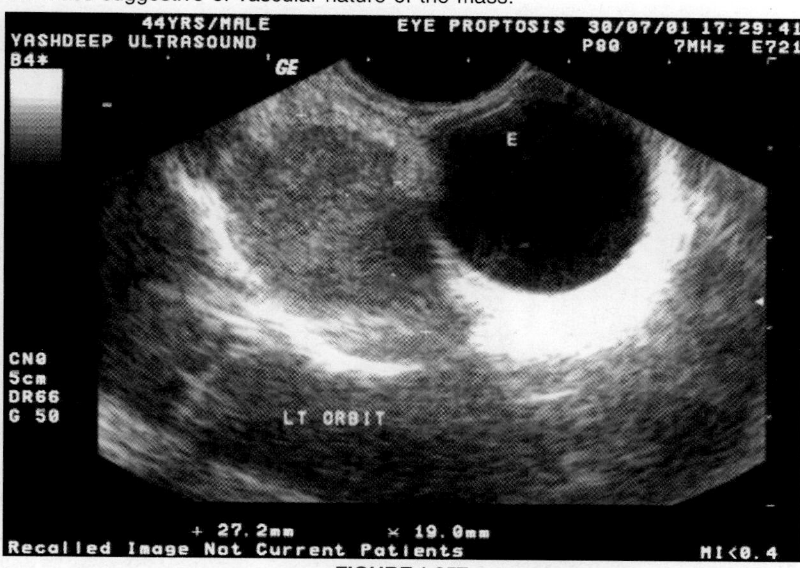

FIGURE 1.357

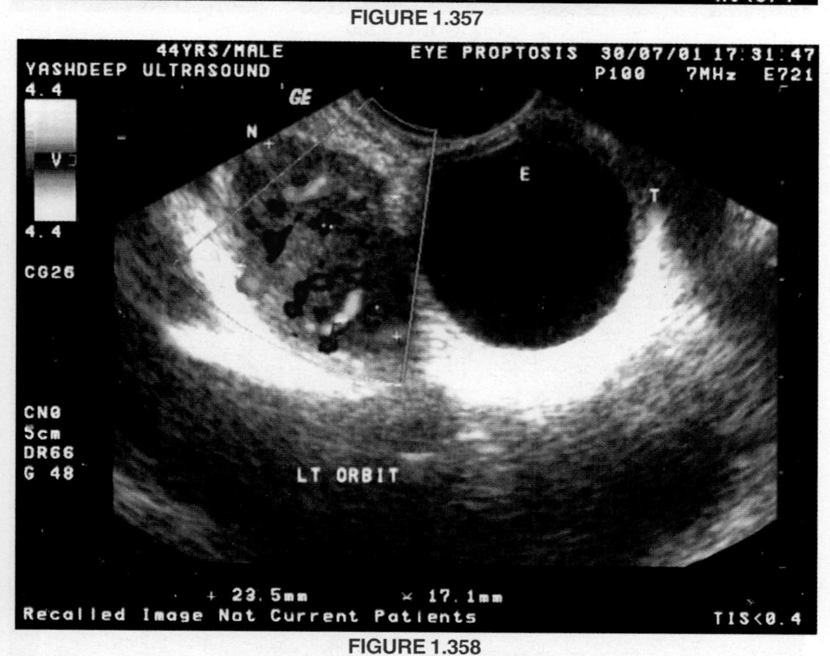

FIGURE 1.358

FIGURES 1.357 and 1.358: Retrobulbar vascular mass. HRSG shows well defined lobulated mass seen coming out from superior quadrant of Lt orbit with downward displacement of eyeball. On color flow imaging moderate flow is seen in the mass.

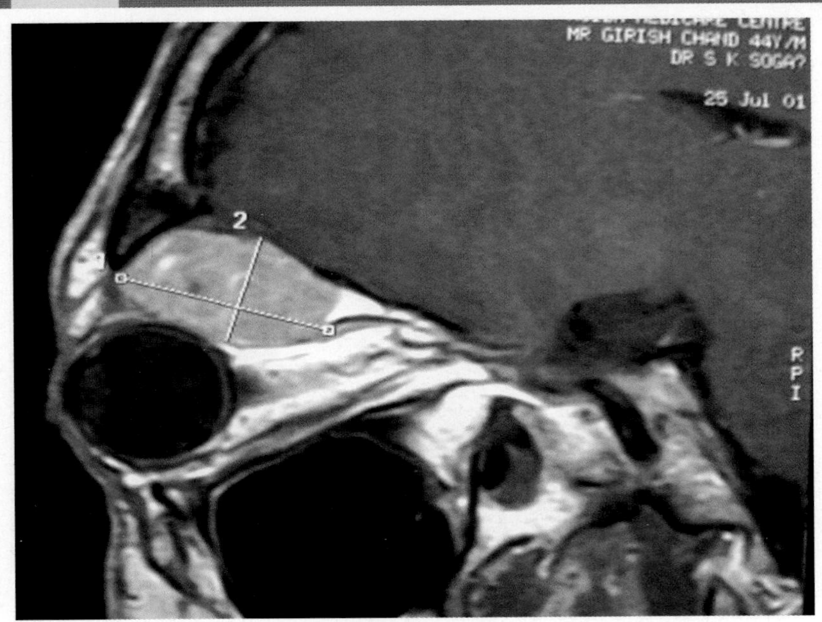

FIGURE 1.359

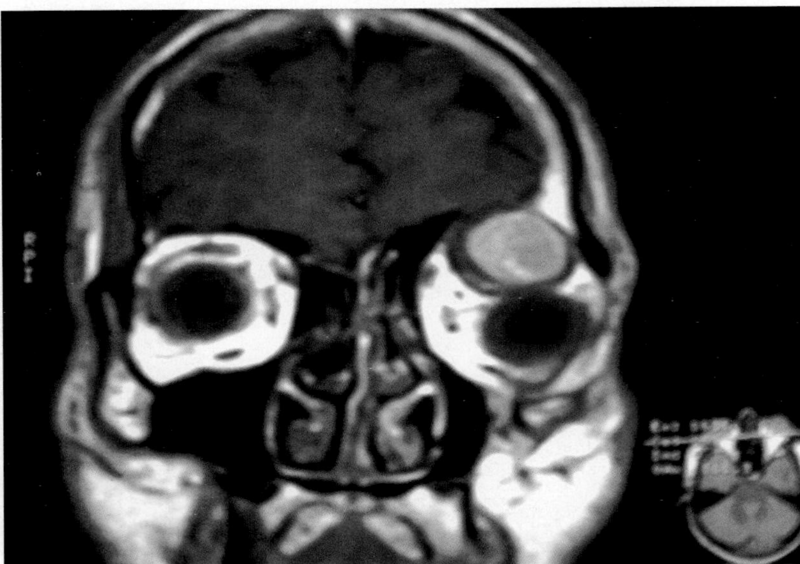

FIGURE 1.360

FIGURES 1.359 and 1.360: MRI of the same patient shows homogeneous mass in superior quadrant.

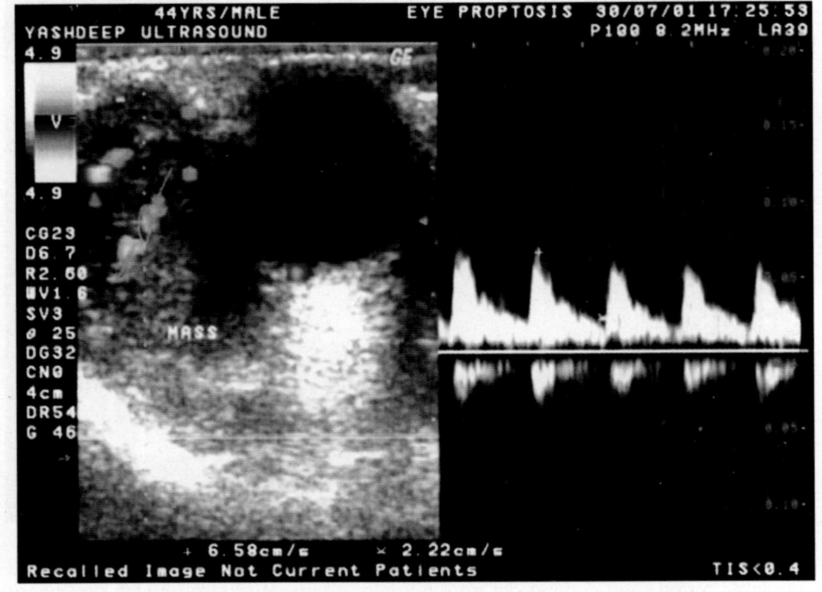

FIGURE 1.361: Spectral Doppler tracing shows arterial flow in the mass.

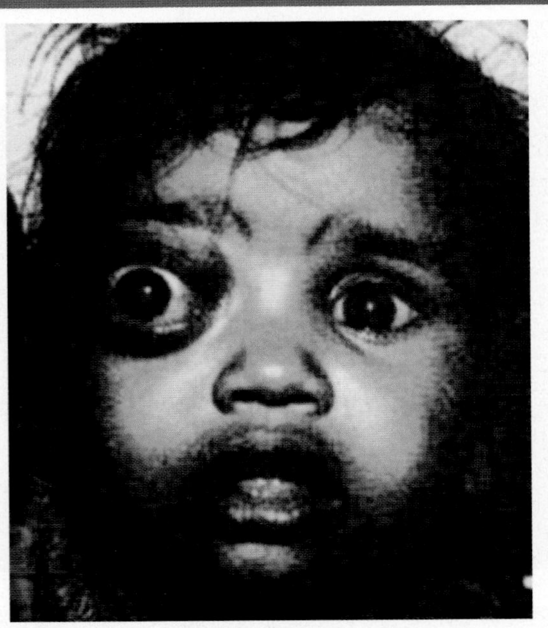

FIGURE 1.362: Lymphangioma. A young girl presented with Rt sided proptosis and diplopia.

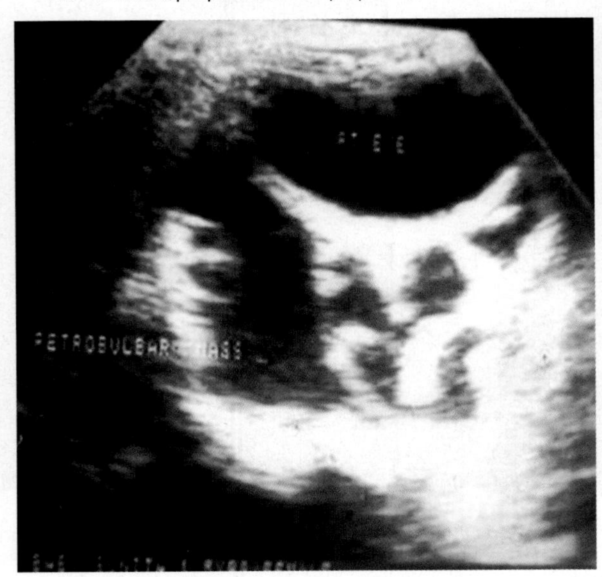

FIGURE 1.363: HRSG shows a big multiloculated cystic mass in retrobulbar area. Thick echogenic septa are seen in the cyst with low level echoes suggestive of lymphangioma. Biopsy confirms the sonographic findings.

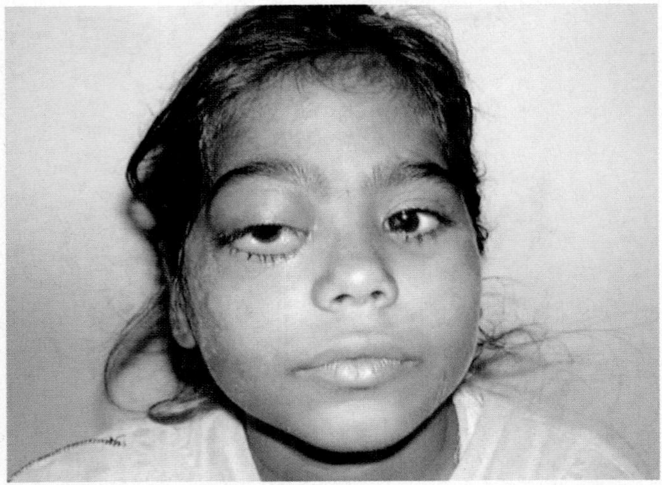

FIGURE 1.364: Lymphangioma. A young girl presented with slowly increasing proptosis of Rt eye with drooping of upper eyelid and blurred vision.

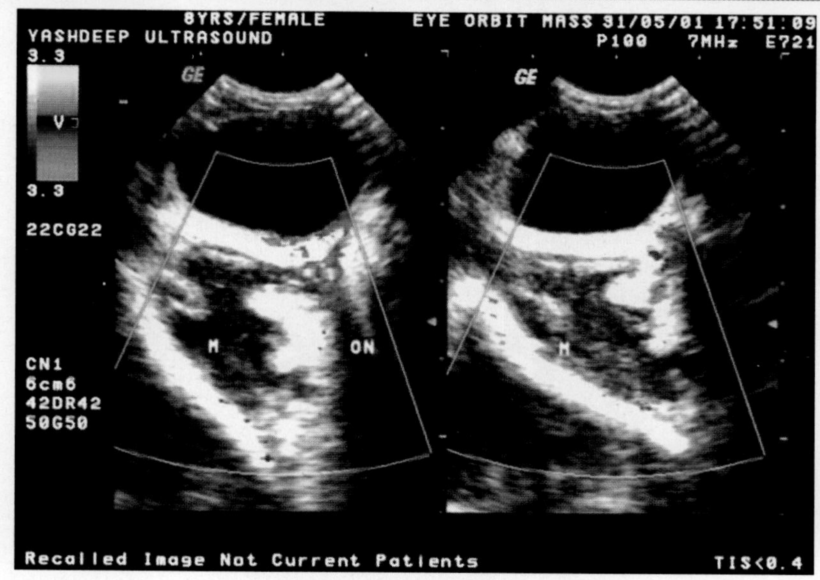

FIGURE 1.365

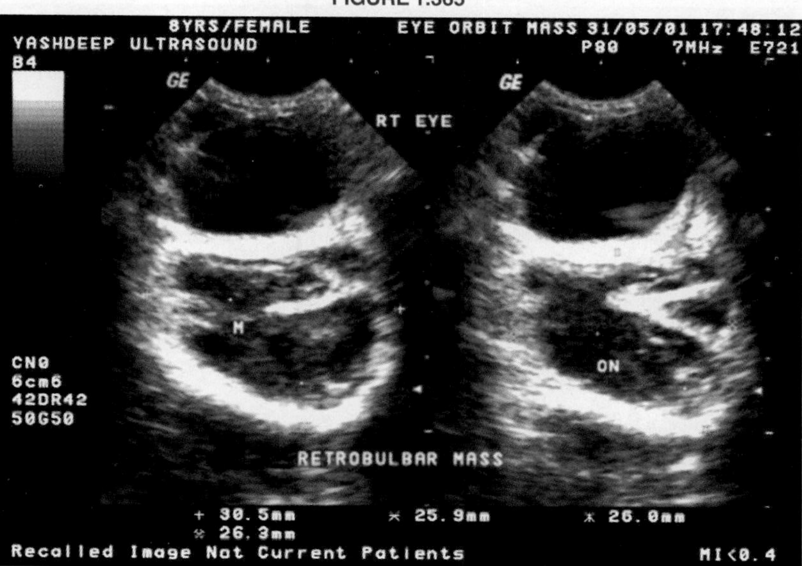

FIGURE 1.366

FIGURES 1.365 and 1.366: Lymphangioma. Cystic mass with thick echogenic septa is seen filling the retrobulbar space. Poor flow is seen in the mass on color flow. Biopsy proved lymphangioma.

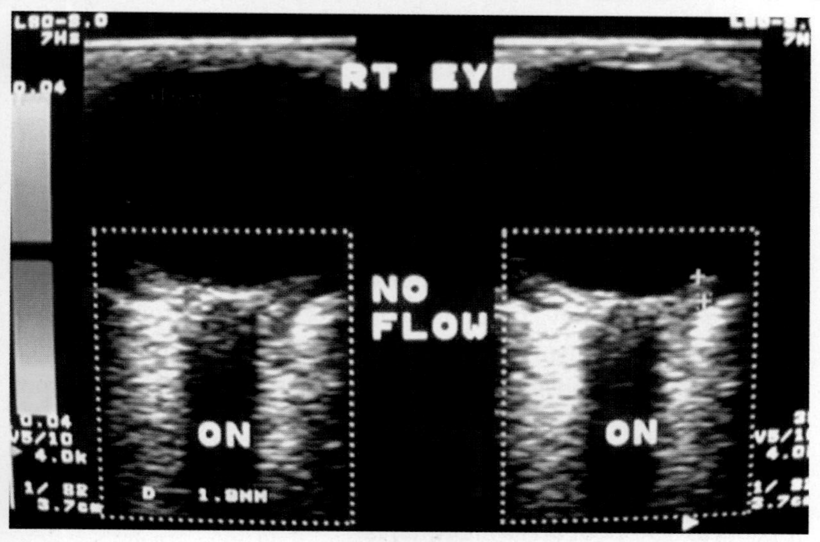

FIGURE 1.368

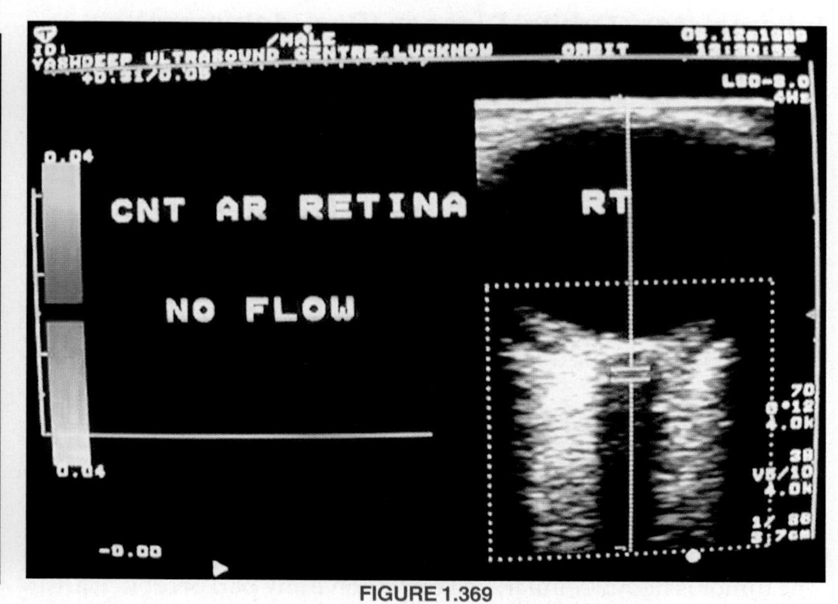

FIGURE 1.369

FIGURES 1.368 and 1.369: HRSG and color flow imaging show hypoechoic edematous optic nerve with no flow seen in central artery of retina suggestive of central artery of retina occlusion. It is confirmed on Duplex Doppler study.

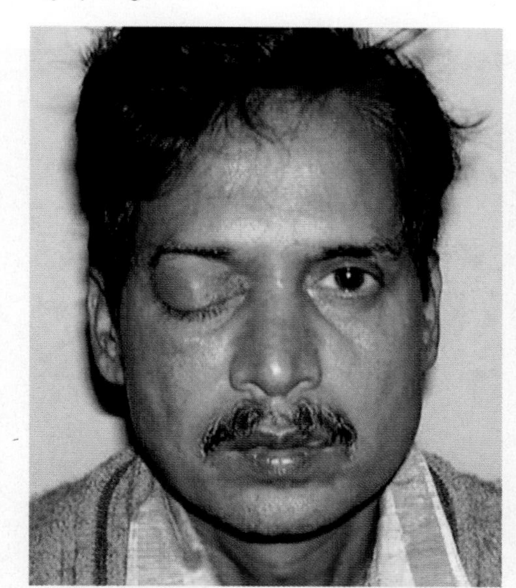

FIGURE 1.367: Central artery of retina thrombosis. A middle-aged man presented with sudden lose of vision and painful swollen Rt eye.

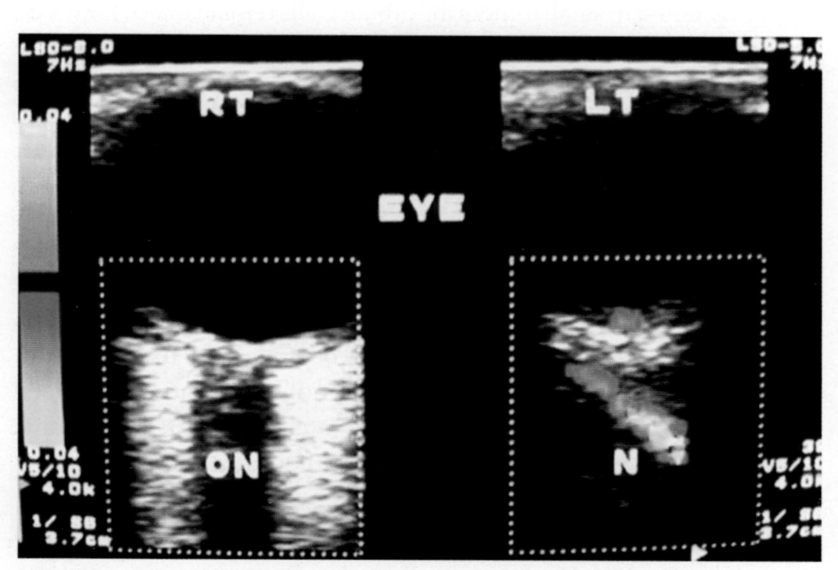

FIGURE 1.370: Color Doppler study shows no flow in Lt orbit, no flow in Rt Central artery of retina. However, normal flow is seen in Lt Central artery retina.

hemorrhage present in the tumor, low to medium level echoes are seen in the dilated spaces. Decompression of lymphangioma at times becomes necessary and "B" scan imaging is used to guide the aspiration needle.

Muscle Hypertrophy

In thyrotoxicosis (Graves' disease), ocular muscle involvement is common. However in 3 to 5% of the cases hypertrophied ocular muscles compress the optic nerve resulting into severe threat to the vision. HRSG is good noninvasive method to evaluate hypertrophy of muscles. Medial rectus muscle is taken as a standard and thickness of more than 4 mm is taken as hypertrophy of muscle suggestive of Graves' disease. Typically the enlargement takes place in muscle belly. Other orbital features include increased orbital fat and orbital edema. The edema appears as echopoor areas in orbital fat.

Inflammatory Orbital Disease (Pseudotumors)

Inflammatory orbital disease (pseudotumor) is the name given to a group of nonmalignant orbital tumors which involve orbital muscles. It includes muositis, periscleritis, perineuritis and pseudotumors. It results in proptosis, diplopia and at times painful eye. On HRSG thickening of the muscle belly is seen. It is echopoor in appearance with nodular appearance. The condition is unilateral and idiopathic in nature. However, it is difficult to differentiate from a malignant mass. But pseudotumors respond well to steroid therapy. In children, about one-third of the pseudotumors are bilateral.

Primary Tumor of Orbit

Rhabdomyosarcoma

Rhabdomyosarcoma is the most common primary tumor of orbit in childhood. It frequently presents as rapidly increasing exophthalmos. The tumor is highly cellular, and can involve any part of orbit. It arises from extraocular muscle. It is most commonly seen in the superonasal quadrant of orbit. Echographically, it is well circumscribed and medium to low echo complex mass. Connective tissue septa can be seen in tumors at times. In small tumors the orbital walls remain intact, however, in big tumors they may be eroded. Pseudotumors and lymphomas are the differential diagnosis of rhabdomyosarcomas.

Lymphoproliferative Masses

Orbital lymphoma is one of the three major causes of proptosis. It is usually of nonhodgkin type. HRSG shows a mixed echo complex mass in the retrobulbar space. It may elongated or oval. It shows good acoustic transmission determining the nature of the mass, whether it is benign or malignant. Usually it is bilateral, may be unilateral, focal or multi-focal in position. However, HRSG cannot differentiate between inflammatory or malignant lymphomas. Therefore, systemic examination is must in orbital lymphoproloferative diseases.

Orbital Metastasis

Orbital metastasis are common in 40% cases of neuroblastoma, they are also seen in cases of osteocarcinoma, Ewing's tumors and in rare cases of adenocarcinoma. They are hypoechoic nodular mass with heterogeneous texture. Infiltrating metastasis can destroy the bony walls or orbit.

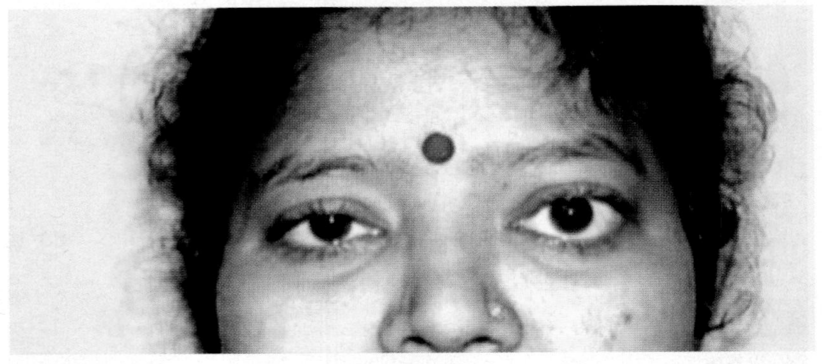

FIGURE 1.371

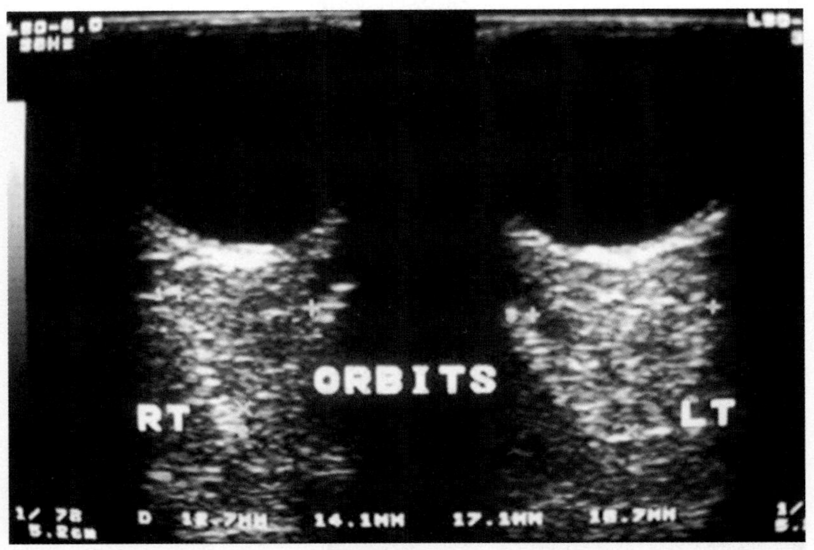

FIGURE 1.372

FIGURES 1.371 and 1.372: Thyrotoxicosis. Patient presented early proptosis Lt eye. HRSG shows increased amount of retrobulbar fat in Lt orbit coming early proptosis.

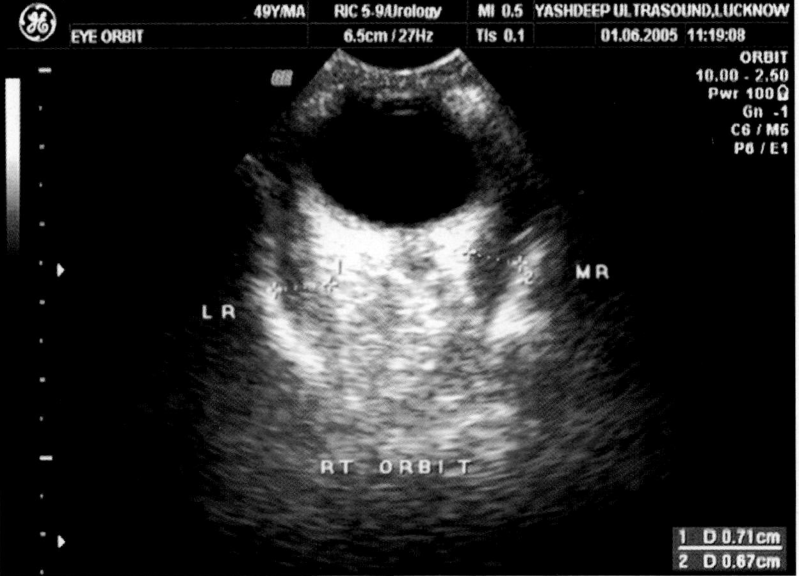

FIGURE 1.373

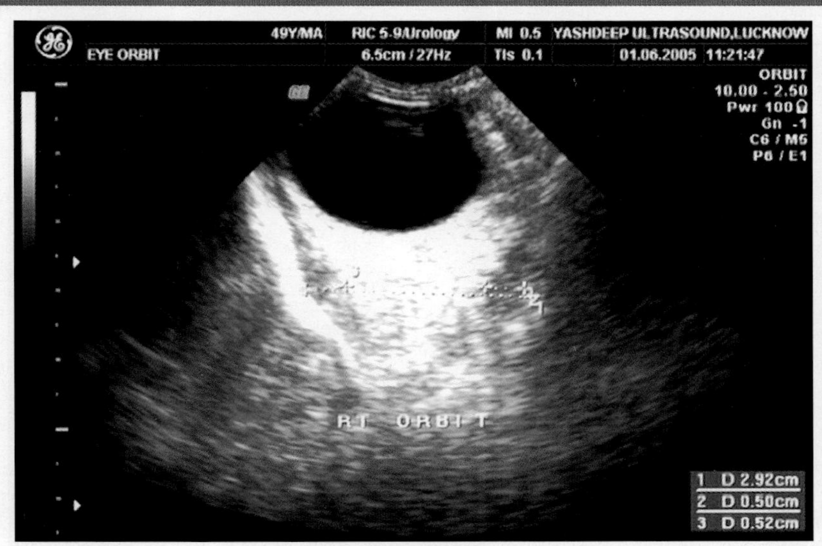

FIGURE 1.374

FIGURES 1.373 and 1.374: Muscle hypertrophy in thyrotoxicosis. A patient presented with bilateral orbital proptosis. HRSG shows marked thickening of the recti muscles in the orbit. The lateral and medial recti muscles are 7 and 6 mm in thickness (normal<3 mm). Increased amount of intraconal fat is also seen typical feature of thyrotoxicosis.

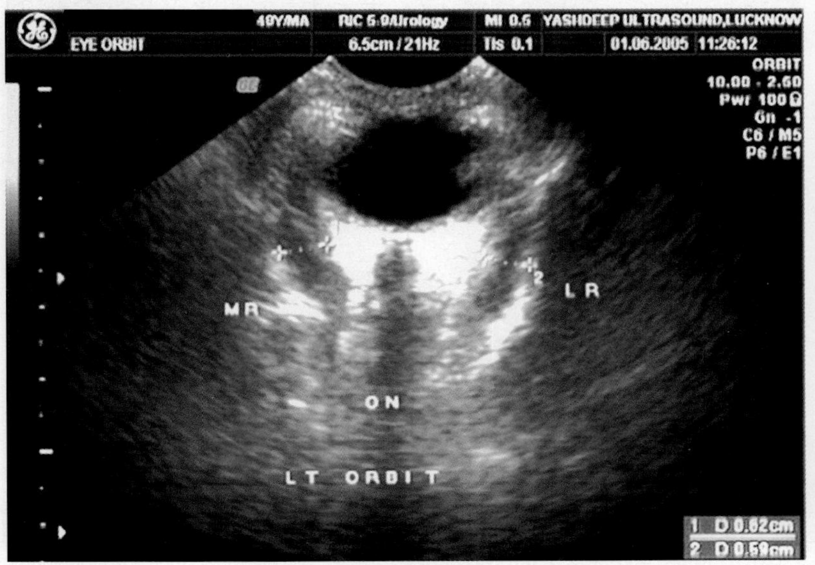

FIGURE 1.375

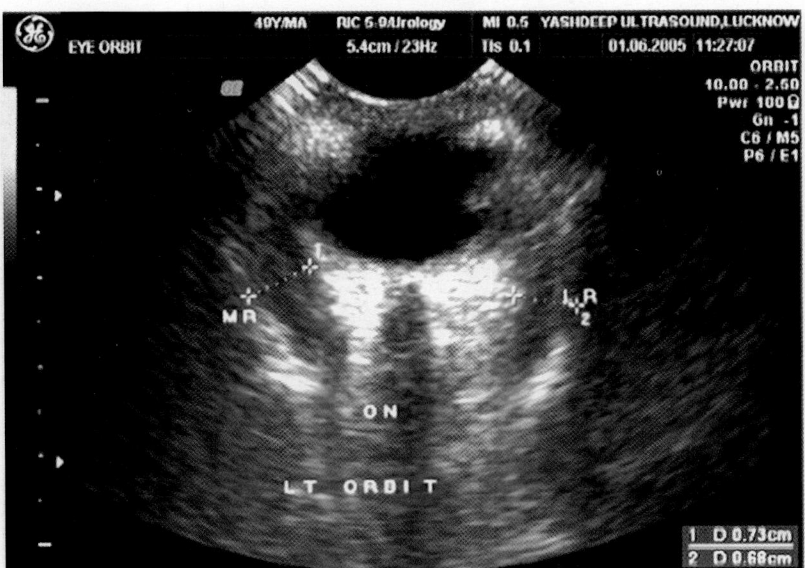

FIGURE 1.376

FIGURES 1.375 and 1.376: Lt eye also shows thickened recti muscles. They are 7 and 6 mm in thickness (normal <3 mm).

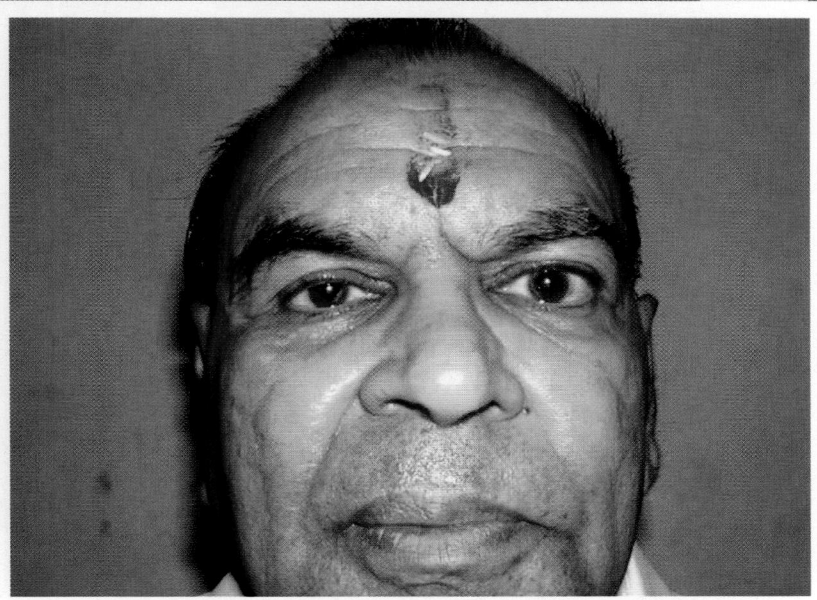

FIGURE 1.377: Myositis superior rectus muscle An old man presented with proptosis of the Lt eye.

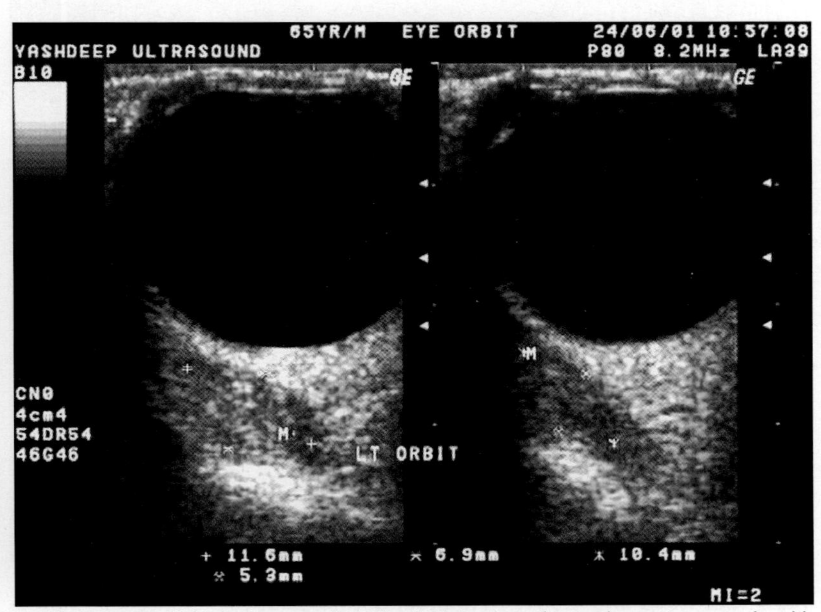

FIGURE 1.378: HRSG shows well-defined thickening of superior rectus muscle with homogeneous texture.

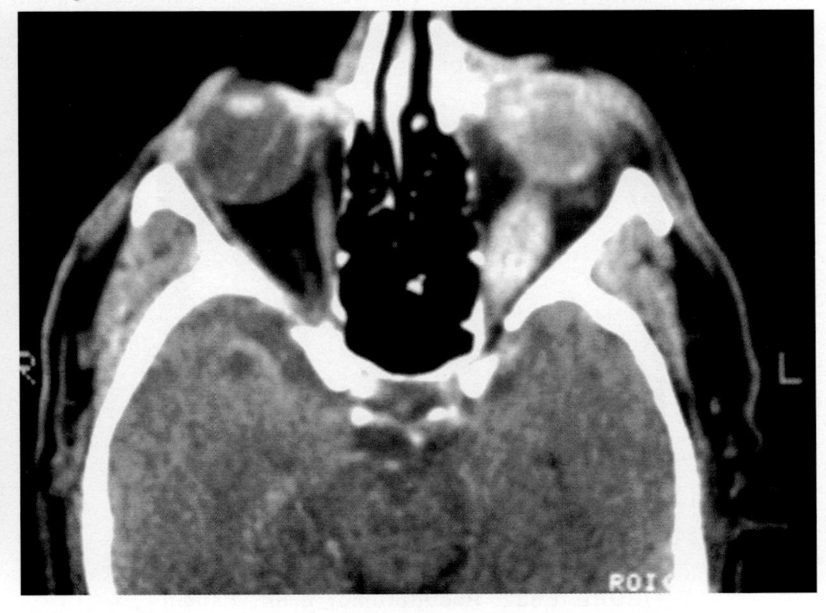

FIGURE 1.379: CT of the same patient also shows the same findings.

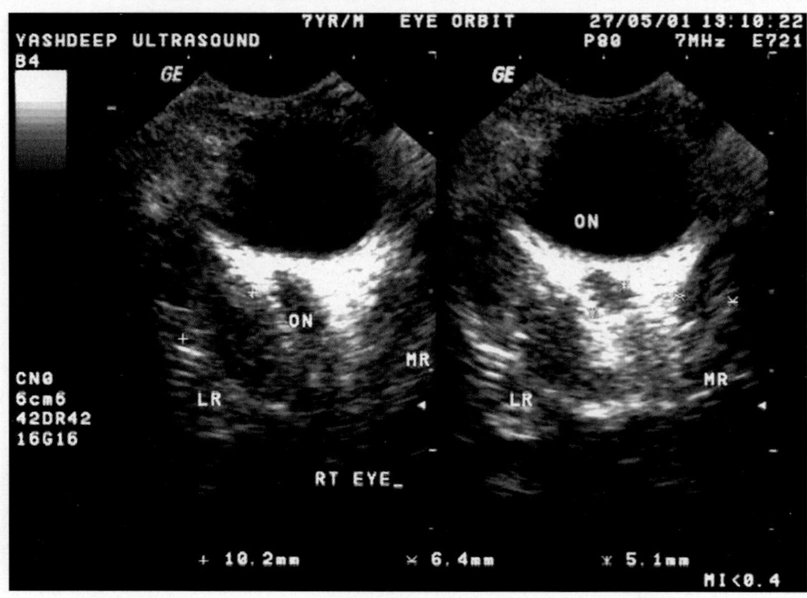

FIGURE 1.380

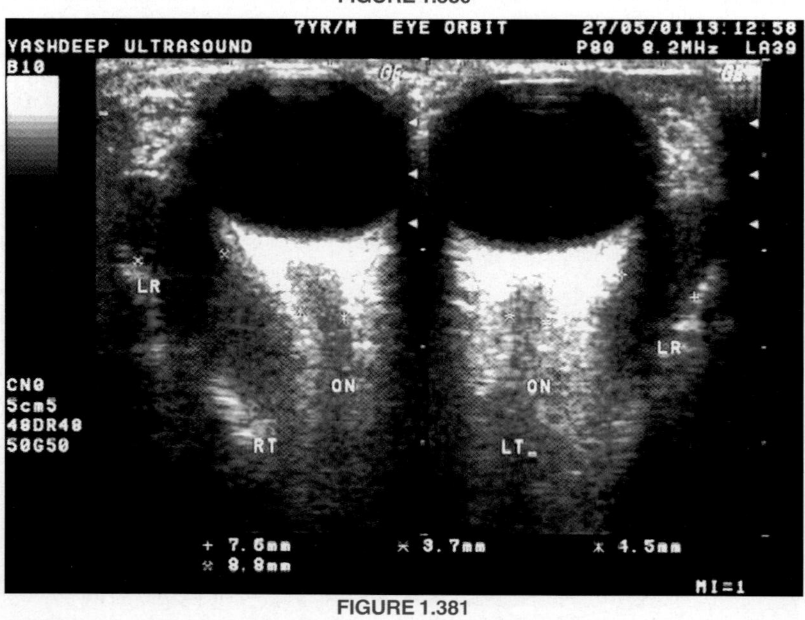

FIGURE 1.381

FIGURES 1.380 and 1.381: Reticuloendotheliosis. Bilateral rectus muscle enlargement in a case of reticuloendotheliosis. Smooth enlargement of medial and lateral rectus muscles is seen. Lateral rectus 10.2 mm and Medial rectus 6.4 mm (Normal <3 mm).

FIGURE 1.382: Pseudotumor orbit. Patient presented with unilateral proptosis with diplopia.

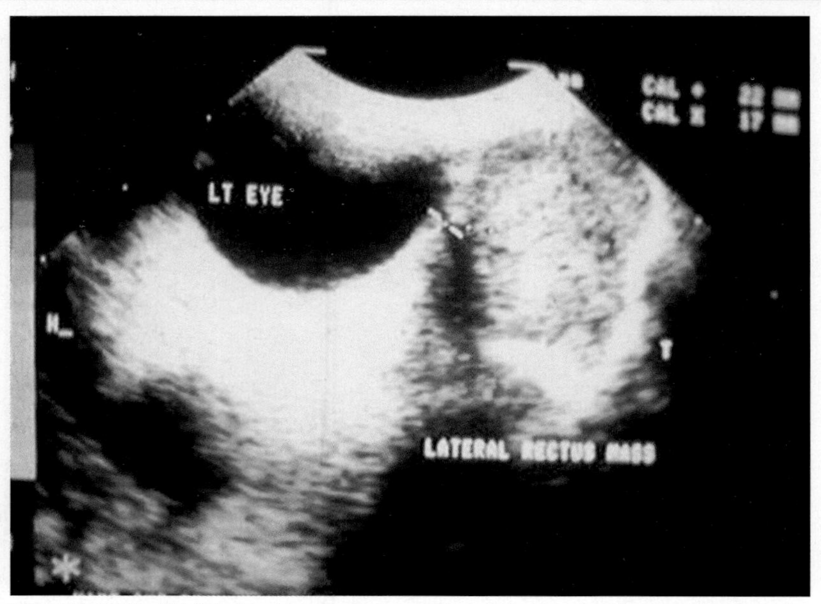

FIGURE 1.383: Well defined homogeneous lobulated mass in lateral rectus muscle. The mass regressed in size on corticosteroid therapy.

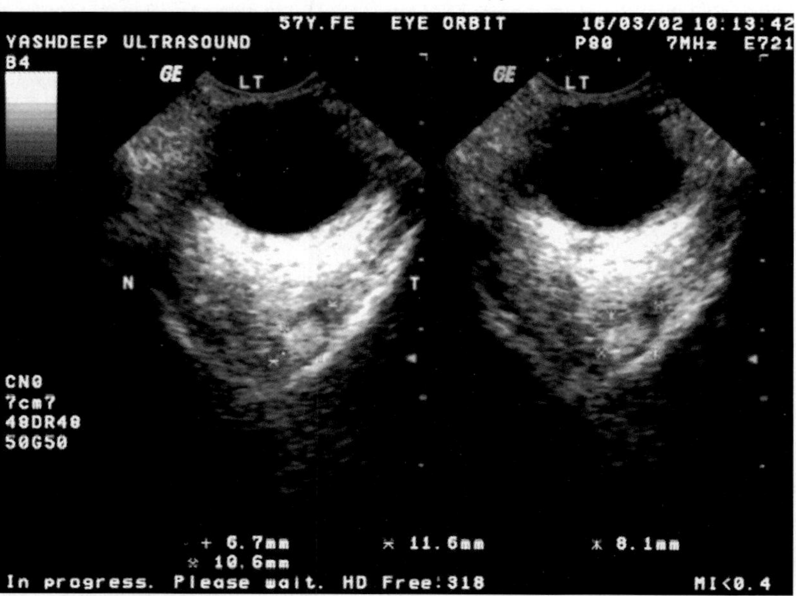

FIGURE 1.384: Myositis of lateral rectus muscle. HRSG shows localized thickening of lateral rectus muscle in a patient presented with proptosis.

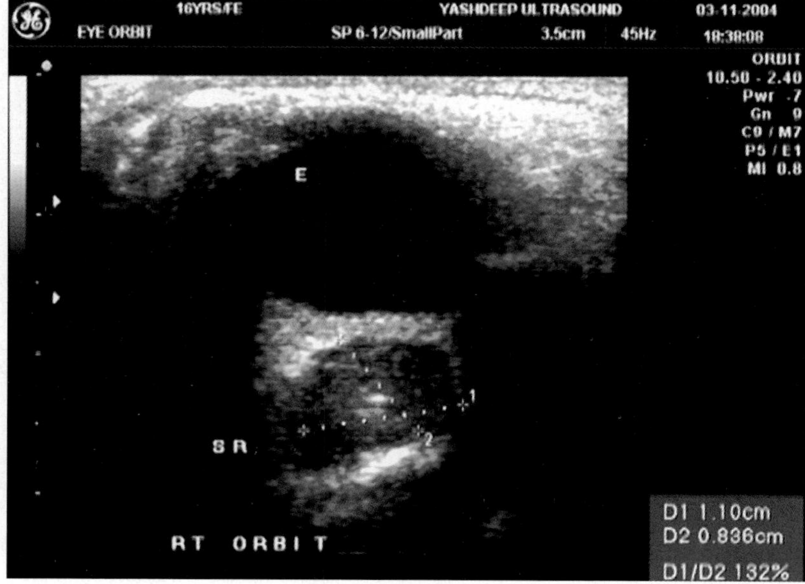

FIGURE 1.385

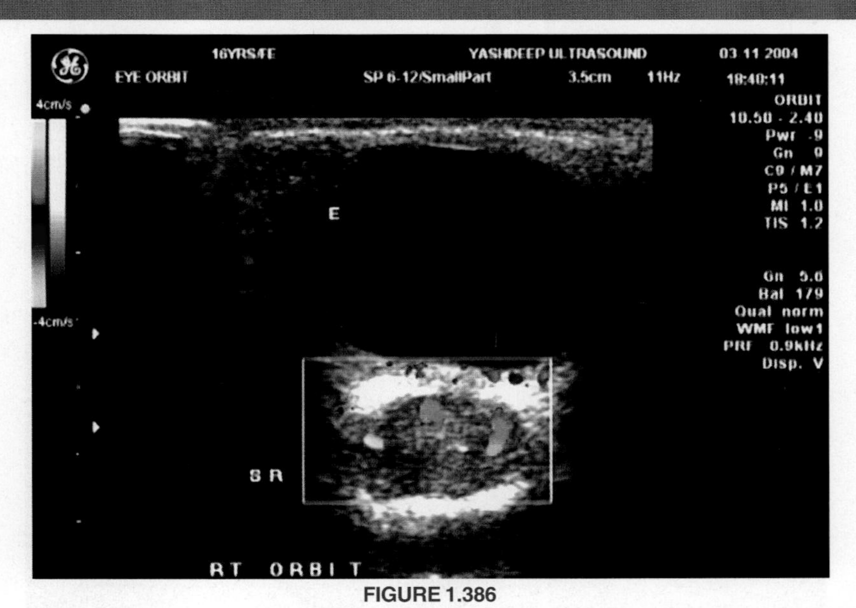

FIGURE 1.386

FIGURES 1.385 and 1.386: Myositis of superior rectus muscle. HRSG shows marked thickening of superior rectus muscle in a patient presented with proptosis. Color Doppler imaging shows increased flow in thickened muscle.

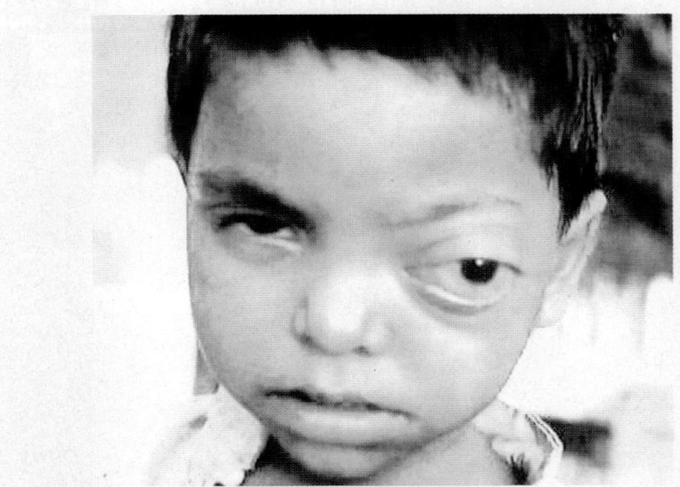

FIGURE 1.387

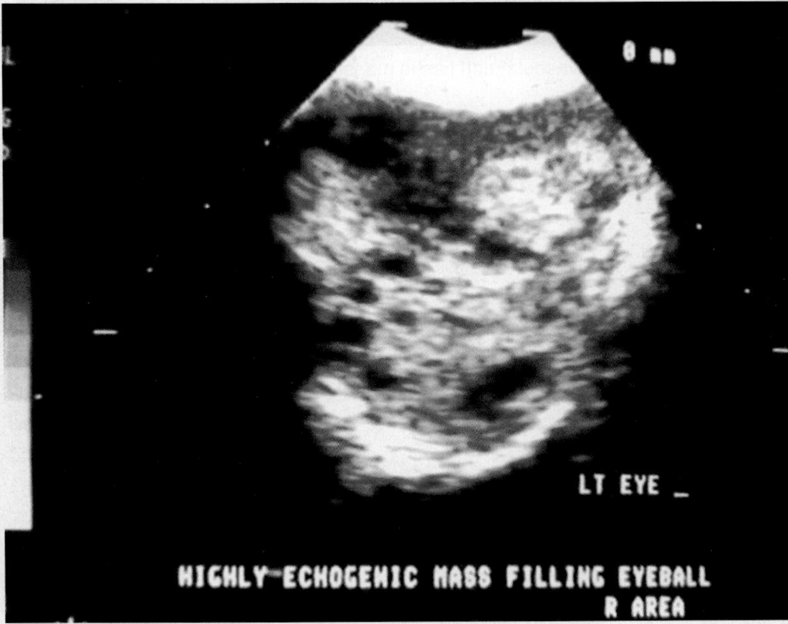

FIGURE 1.388

FIGURES 1.387 and 1.388: Leukemic deposits. Known case of acute lymphatic leukemia presented with proptosis and loss of vision. HRSG show highly heterogeneous mass filling the whole of retrobulbar area. It has destroyed normal anatomy and also invaded optic nerve areas of necrosis is also seen in the mass.

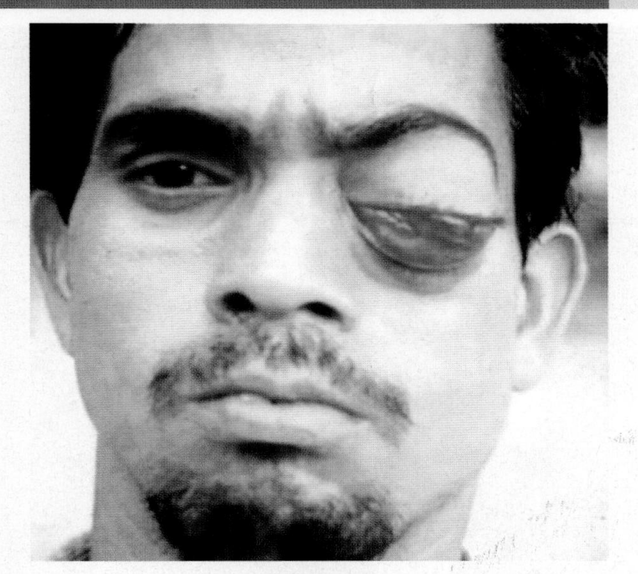

FIGURE 1.389

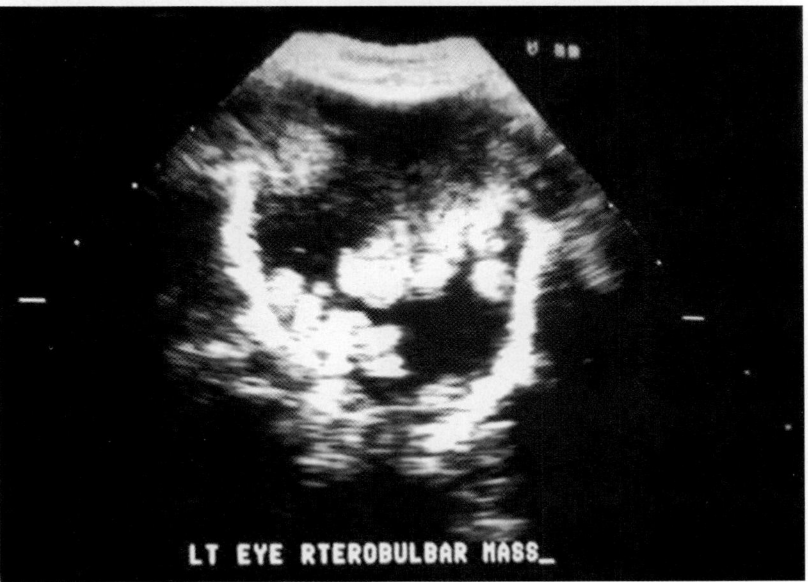

FIGURE 1.390

FIGURES 1.389 and 1.390: FTC operated lymphoma. A patient operated for lymphomatous deposits presented with recurrence and soft tissue mass filling the orbit suggestive of lymphoid tumor filling the orbit. Area of necrosis also seen in the mass.

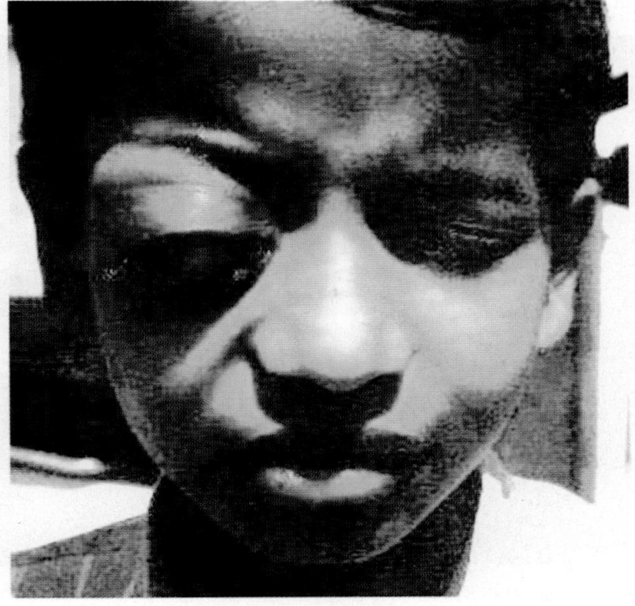

FIGURE 1.391

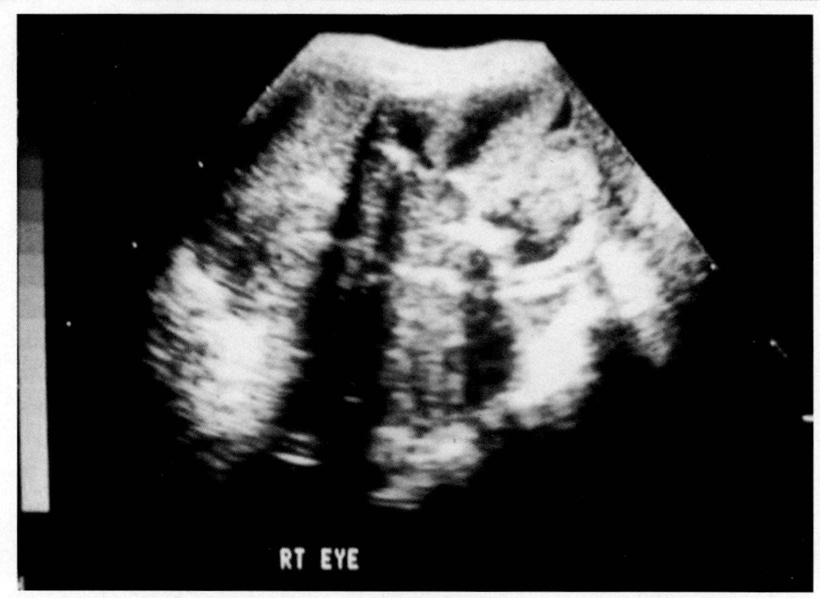

FIGURE 1.392

FIGURES 1.391 and 1.392: Orbital lymphoma. Heterogeneous mass is seen filling the orbit and also invaded the eyeball. The eye coat is broken by the mass and tumor is seen in the eyeball.

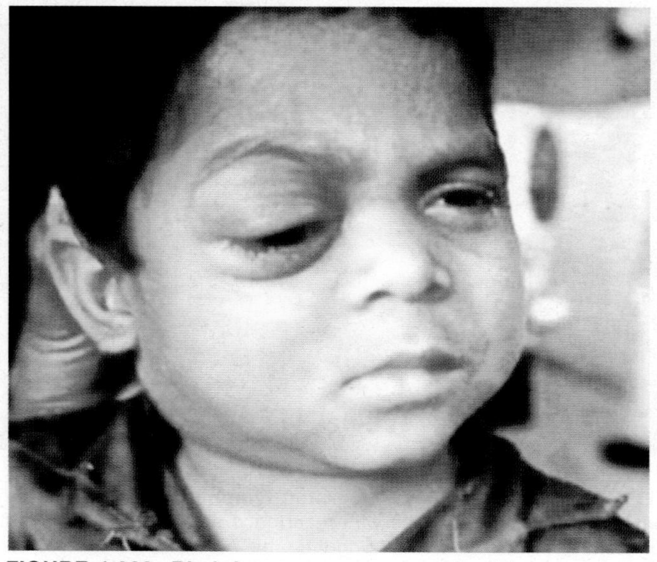

FIGURE 1.393: Rhabdomyosarcoma muscle. A boy presented with proptosis and loss of vision.

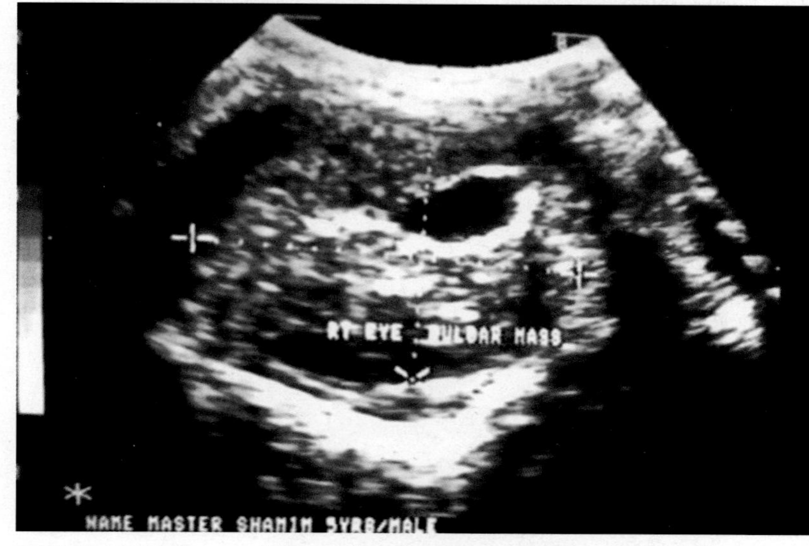

FIGURE 1.394: Big tumor is seen filling the orbit engulfing the eyeball and invaded it. Biopsy rhabdomyosarcoma.

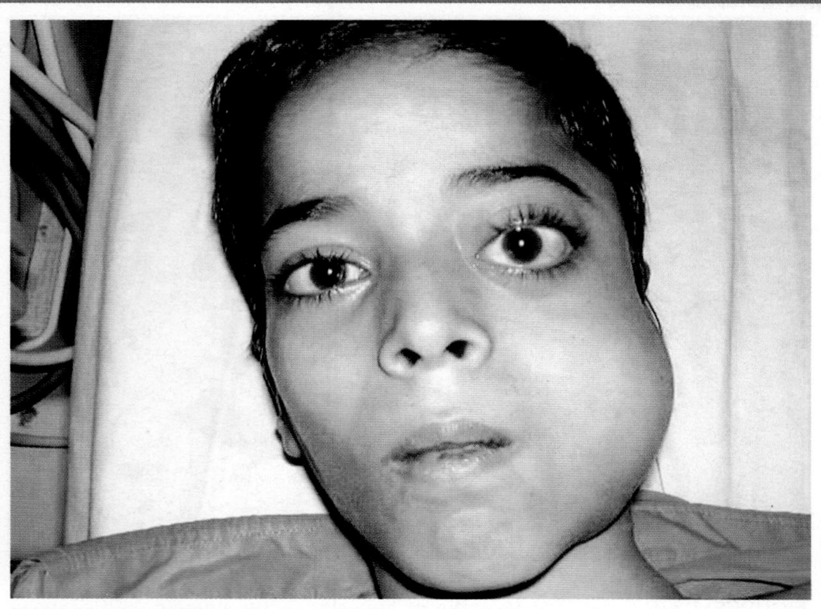

FIGURE 1.395: Metastatic deposits from neuroblastoma. A young boy presented Lt proptosis with swelling of the Lt jaw.

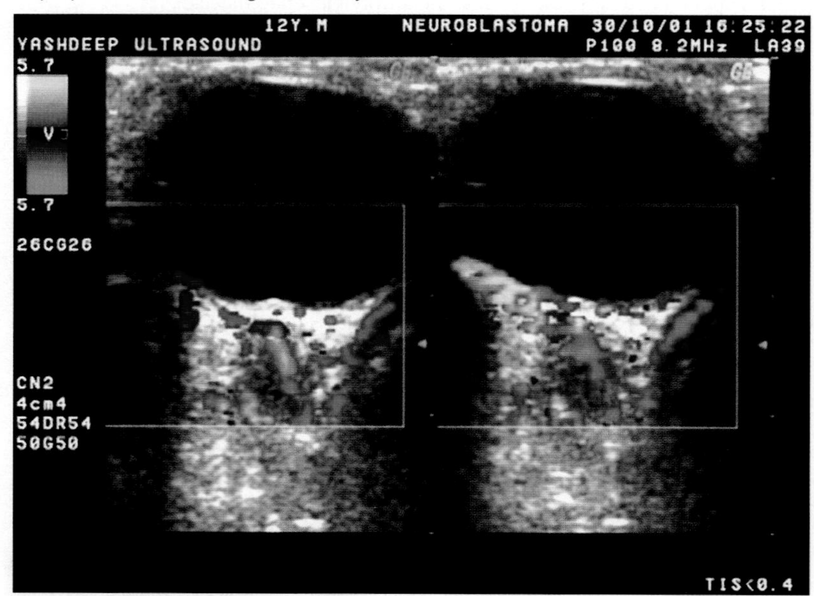

FIGURE 1.396: Highly vascular soft tissue mass in Lt orbit metastatic deposit from the tumor.

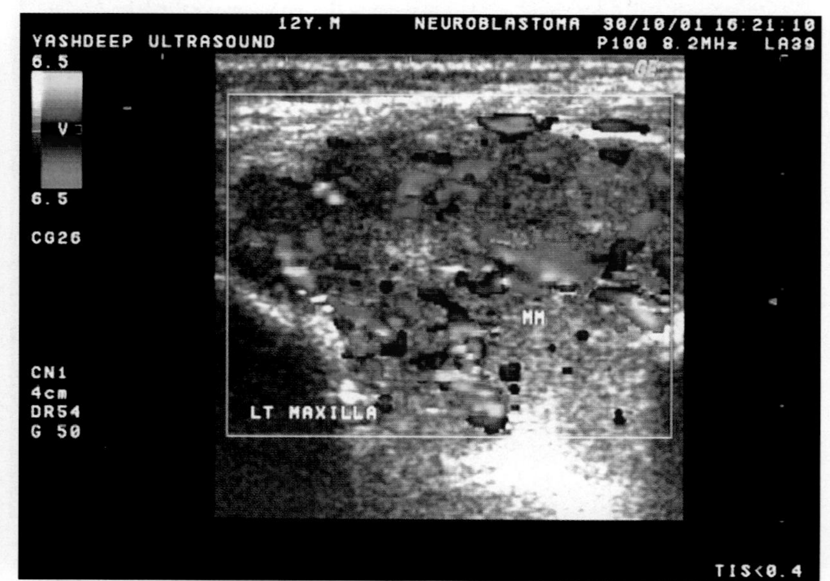

FIGURE 1.397: Highly vascular mass invading Lt maxilla and eaten up the bone metastatic deposit of Figure 1.395.

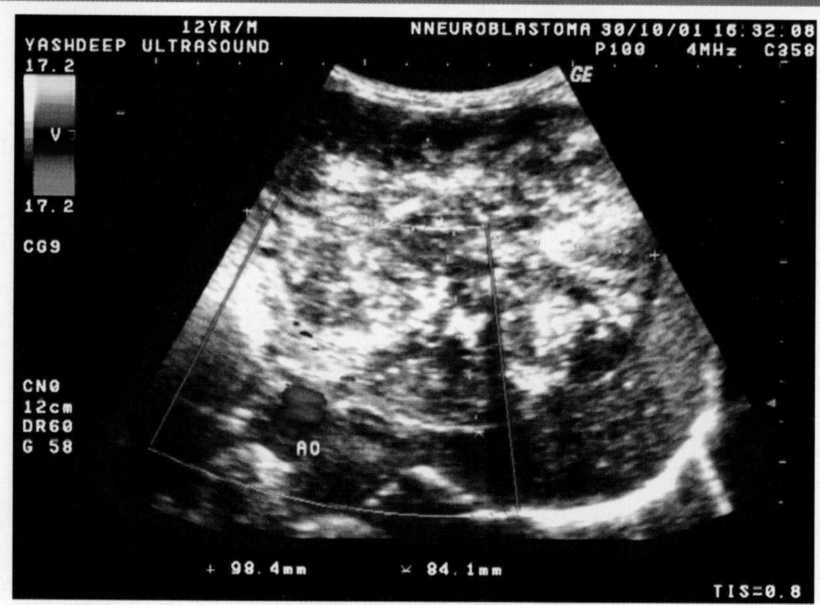

FIGURE 1.398: The primary tumor neuroblastoma in abdomen. A big heterogeneous mass is seen in Lt suprarenal area crossing the midline. HRSG shows the mass—Biopsy proved neuroblastoma.

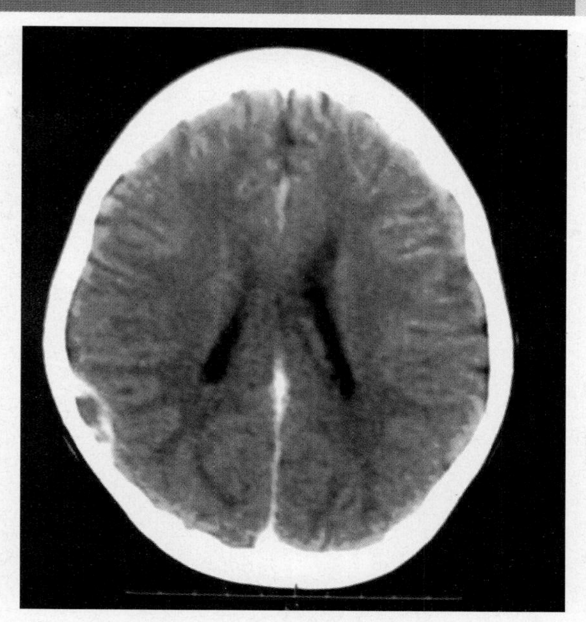

FIGURE 1.401: CT of the same patient shows bony metastasis of the parietal bone.

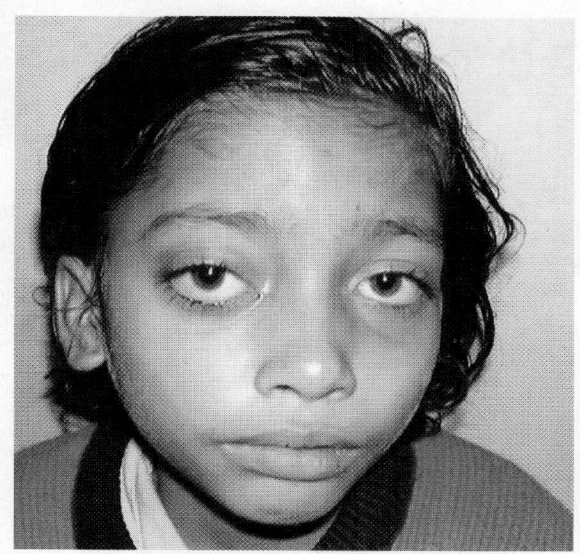

FIGURE 1.399: Primary neuroblastoma metastasis in the orbit. A young girl presented with proptosis Rt eye with painful eye.

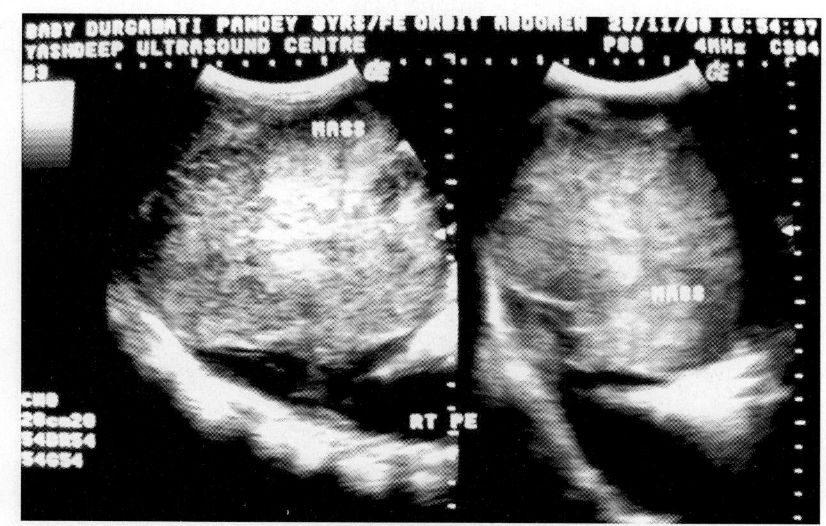

FIGURE 1.402

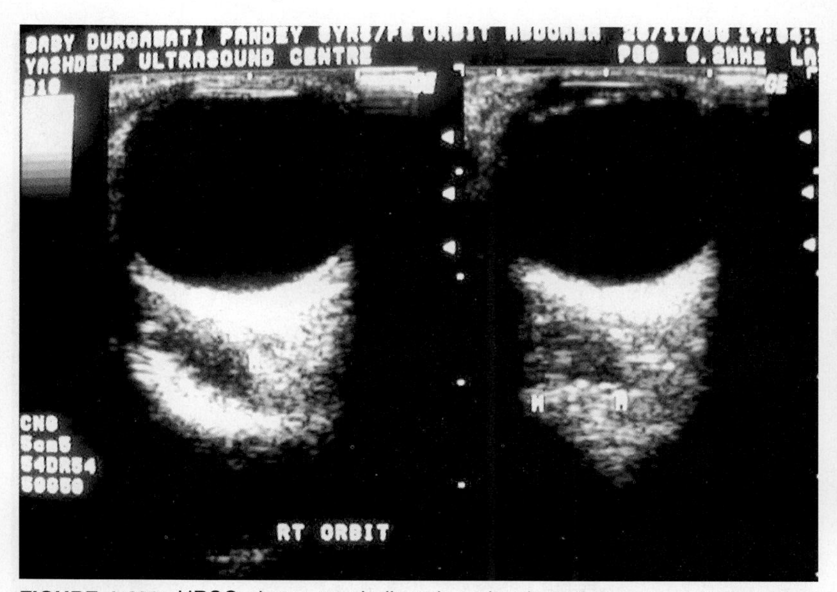

FIGURE 1.400: HRSG shows retrobulbar deposit a hypodense mass in intraconal region in case of Figure 1.399.

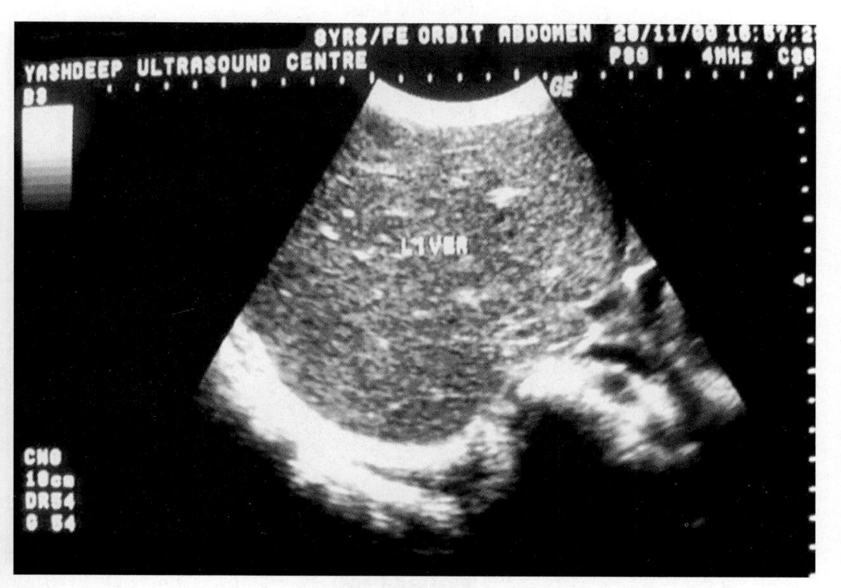

FIGURE 1.403

FIGURES 1.402 and 1.403: Abdominal USG shows heterogeneous mass in Rt suprarenal area with liver deposits and Rt sided pleural effusion.

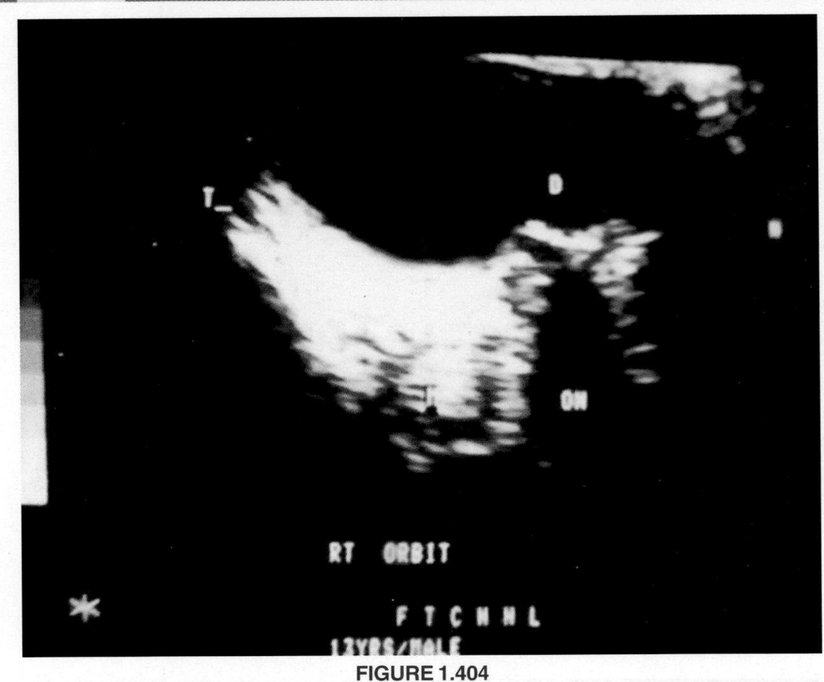

FIGURE 1.404

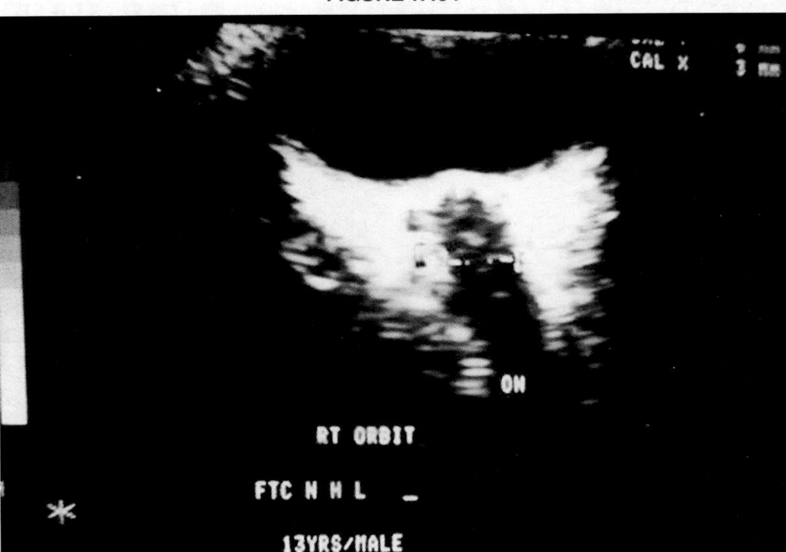

FIGURE 1.405

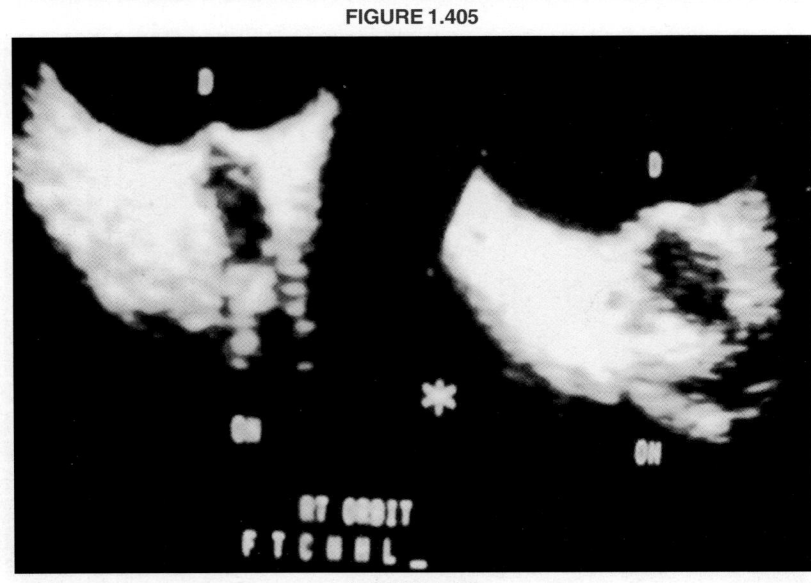

FIGURE 1.406

FIGURES 1.404 to 1.406: Metastasis of non-Hodgkin lymphoma. A known case of NHL presented loss of vision and pain in Rt eye. HRSG shows irregular mass invading optic nerve, optic nerve head. Calcification is also seen at optic nerve head accompanied with shadowing.

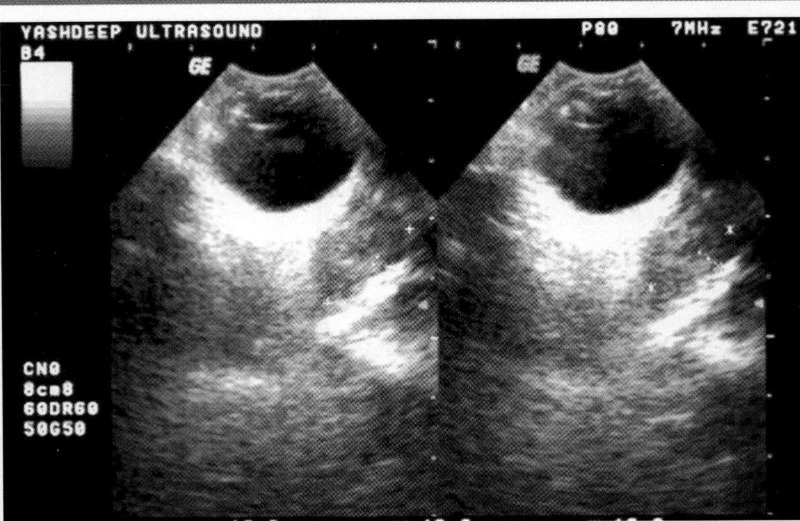

FIGURE 1.407: Metastatic deposit—Adenocarcinoma. Patient presented with swelling over lateral side of Lt orbit. HRSG showed irregular nodular mass in the lateral quadrant of orbit displacing the eyeball—open Biopsy—Metastatic deposit from adenocarcinoma.

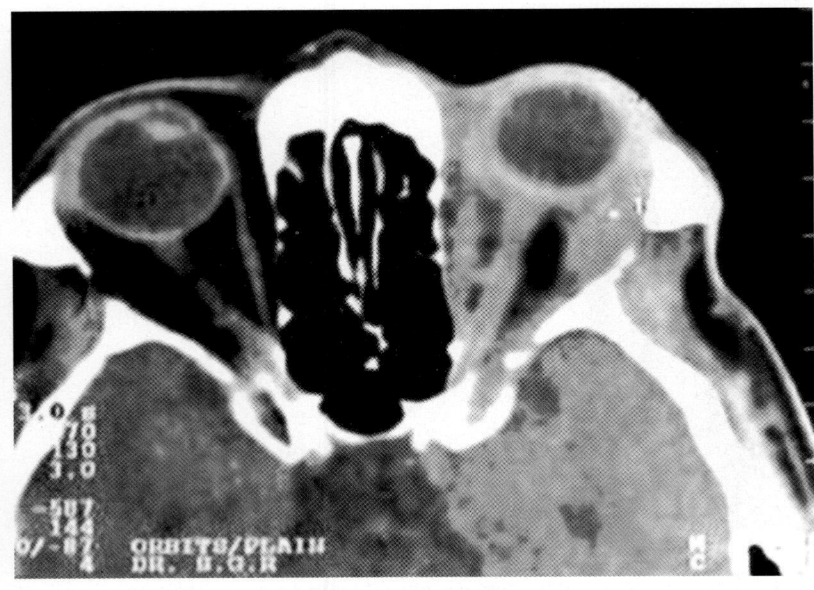

FIGURE 1.408: CT of the same patient shows the deposit mass in outer upper quadrant of Lt orbit.

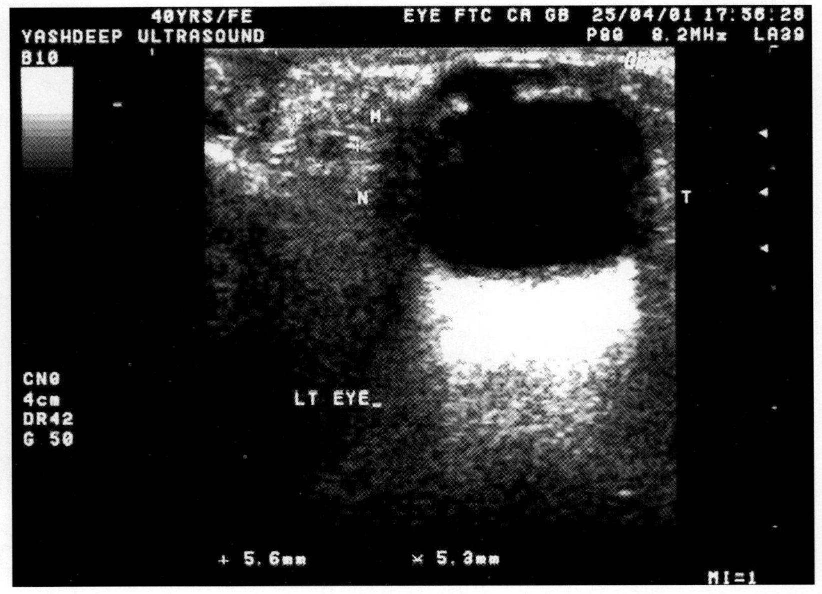

FIGURE 1.409

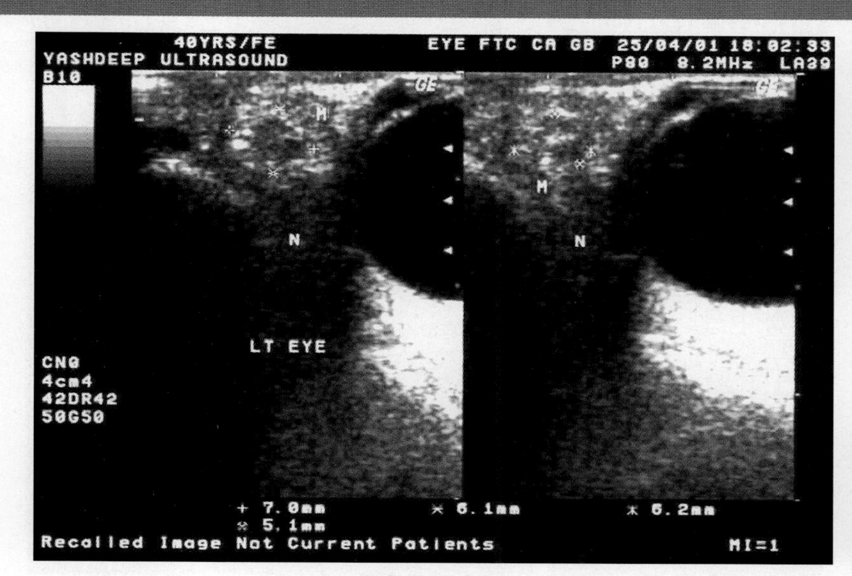

FIGURE 1.410

FIGURES 1.409 and 1.410: Metastasis from Carcinoma GB. A known case of carcinoma GB presented with mass in upper medial quadrant of Lt eye. HRSG shows heterogeneous nodular mass in upper medial part of Lt orbit. Biopsy proved metastasis Ca GB.

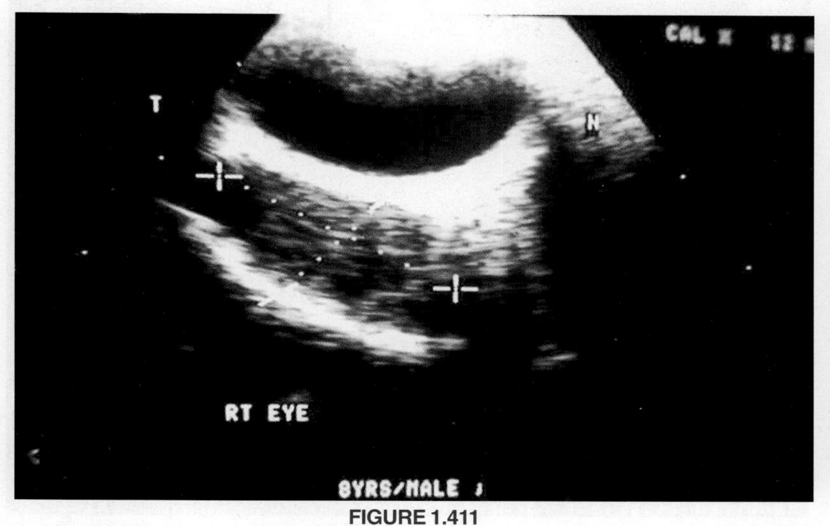

FIGURE 1.411

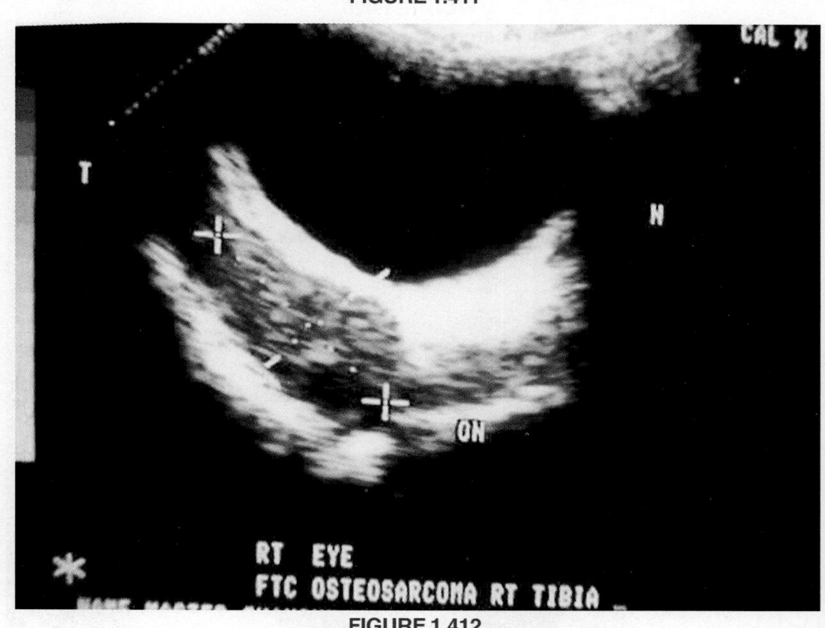

FIGURE 1.412

FIGURES 1.411 and 1.412: Metastatic deposits osteosarcoma tibia. A known case of osteosarcoma of tibia presented with Rt eye proptosis and painful eye. HRSG shows hypoechoic metastatic deposit in retrobulbar area. Osteosarcoma is not known to metastasize to orbit. This was an unusual finding.

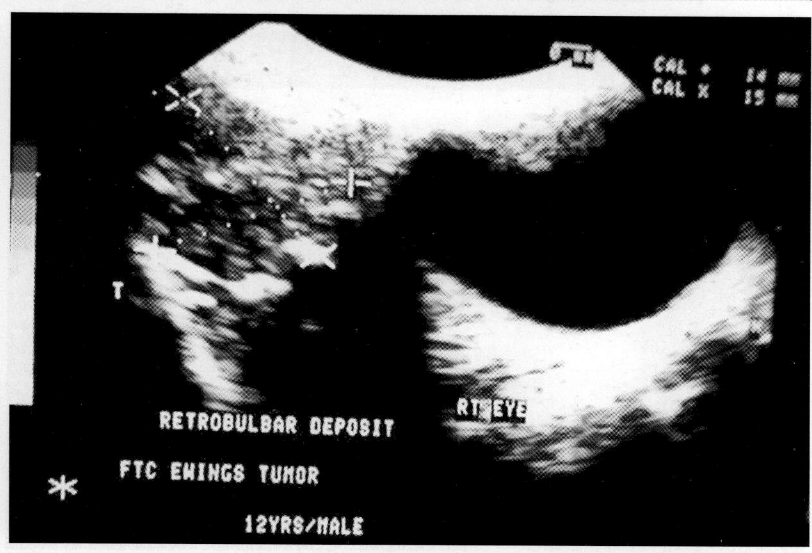

FIGURE 1.413: Metastatic deposit Ewing sarcoma. A young boy of 12 yrs of age known case of Ewing's tumor complained of diminished vision and Rt eye proptosis. HRSG shows a retrobulbar deposit in Rt upper outer quadrant. A heterogeneous mass was seen in upper outer quadrant.

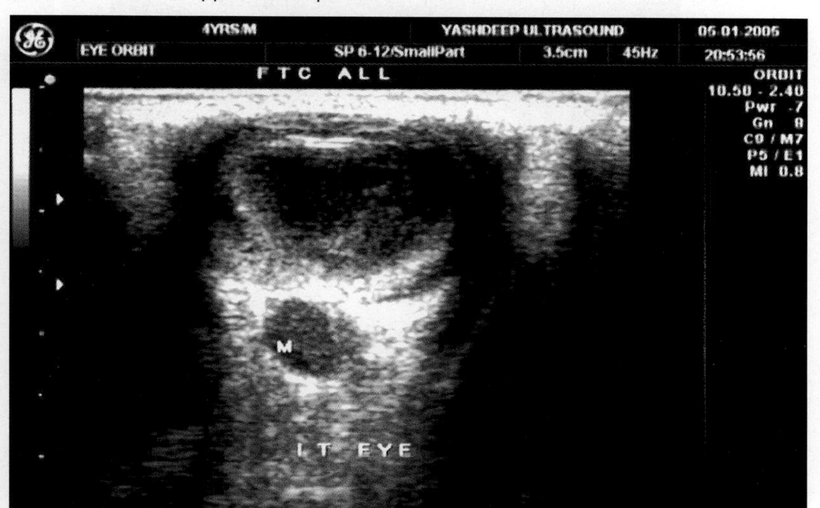

FIGURE 1.414: Metastasis deposit of acute lymphatic leukemia—(ALL). A young child known case of ALL presented with sudden loss of vision. HRSG shows a hypoechoic deposit in the retrobulbar space. Eye also shows associated subchoroidal hemorrhage with collapsed eye.

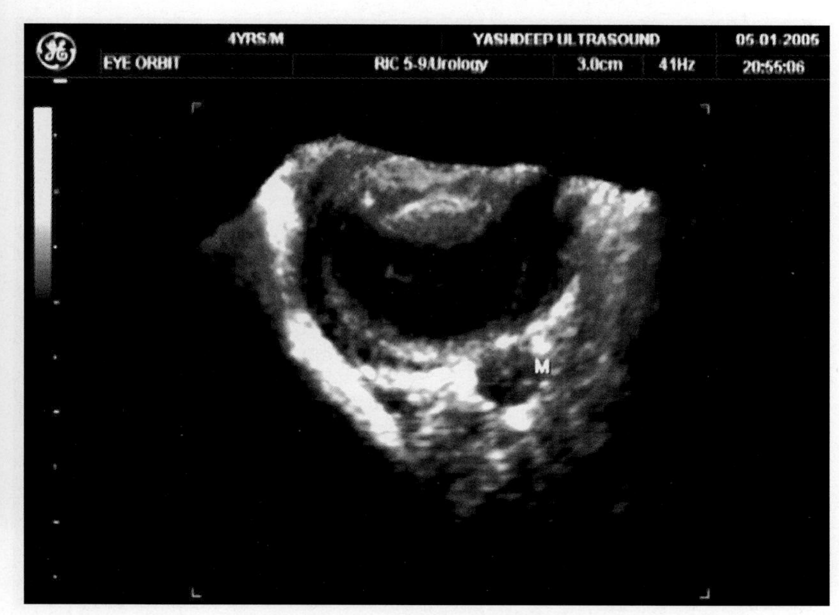

FIGURE 1.415: 3D imaging of the eye shows details of the deposit and subchoroidal hemorrhage.

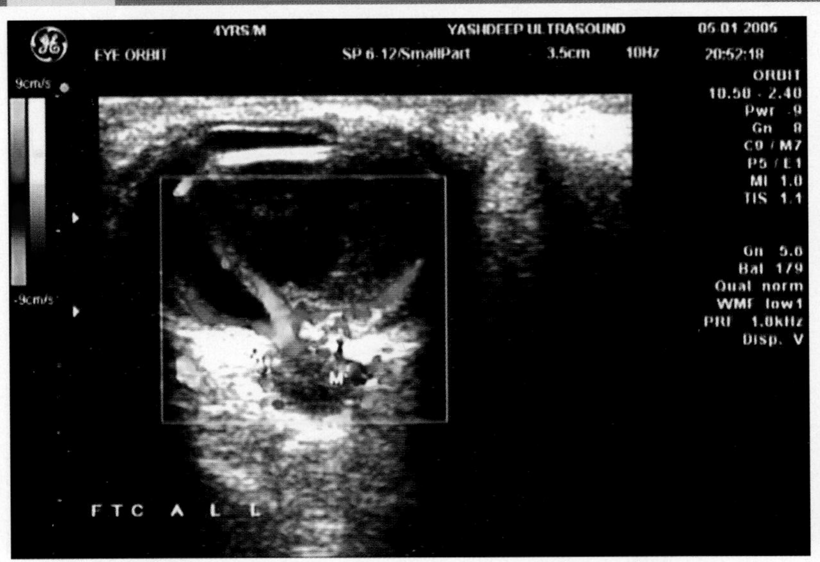

FIGURE 1.416: Color Doppler imaging shows increased flow in the metastasis and also in the membranes.

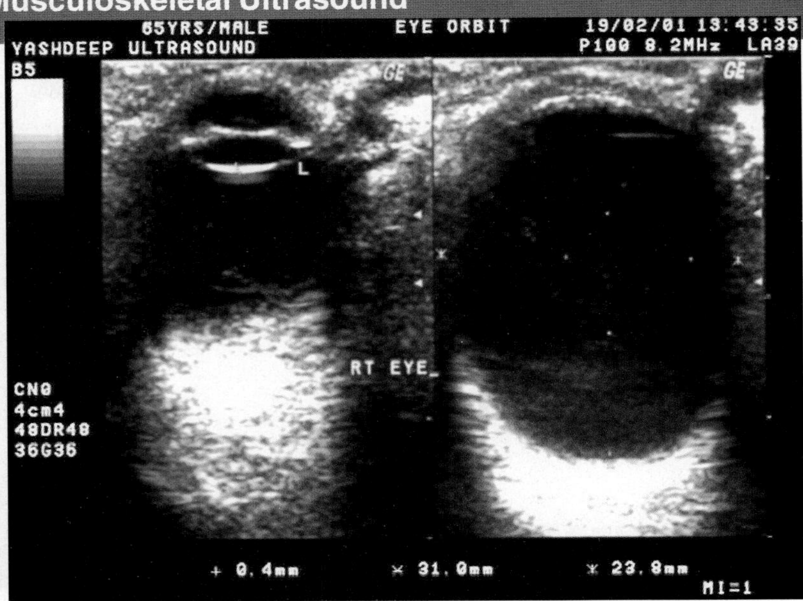

FIGURE 1.419

FIGURE 1.417: Pseudo Proptosis. A young boy presented with proptosis of Lt eye.

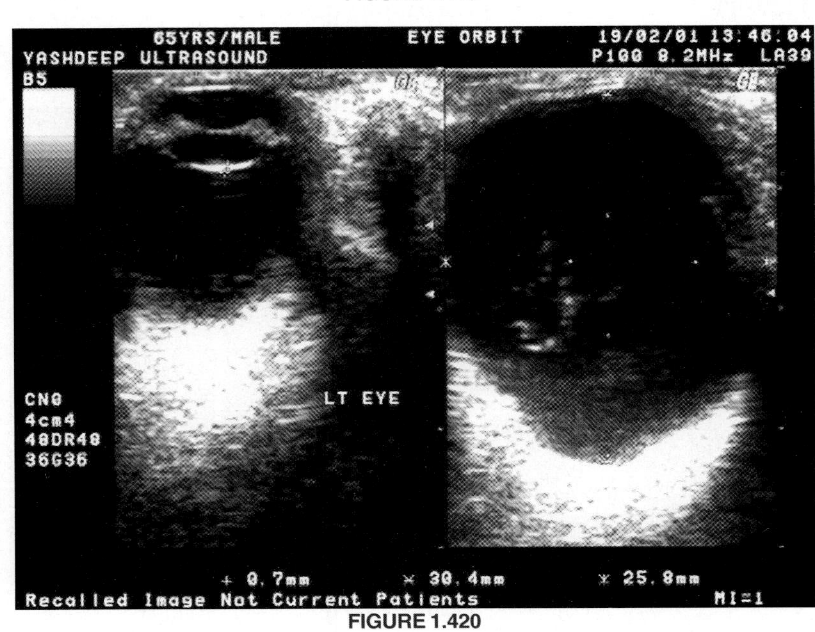

FIGURE 1.420

FIGURES 1.419 and 1.420: Posterior staphyloma. Both eyeballs are oblong in shape with increased axial length. Rt eye is 30 mm and Lt eye 30.4 mm. Presented as bilateral proptosis.

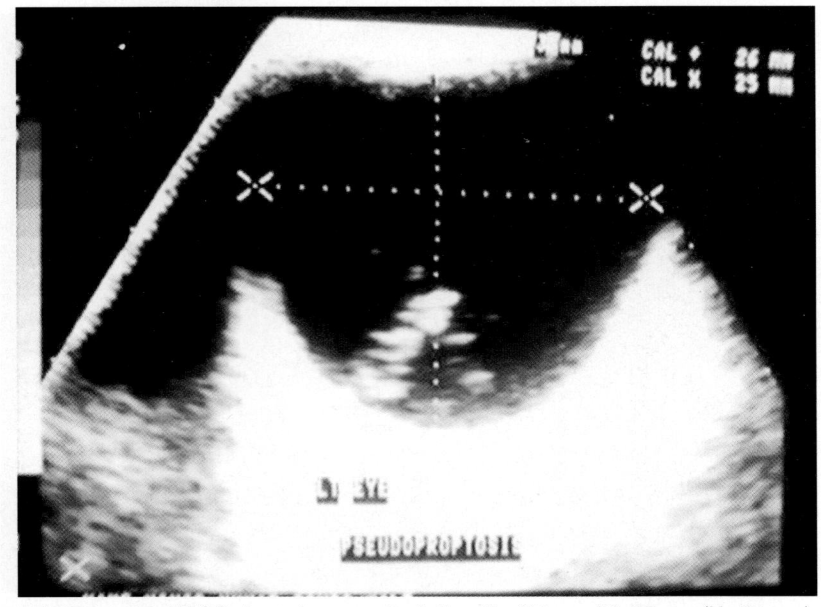

FIGURE 1.418: HRSG shows increased axial length of Lt eye. It is 26 mm (N <24 mm). No retrobulbar mass is seen.

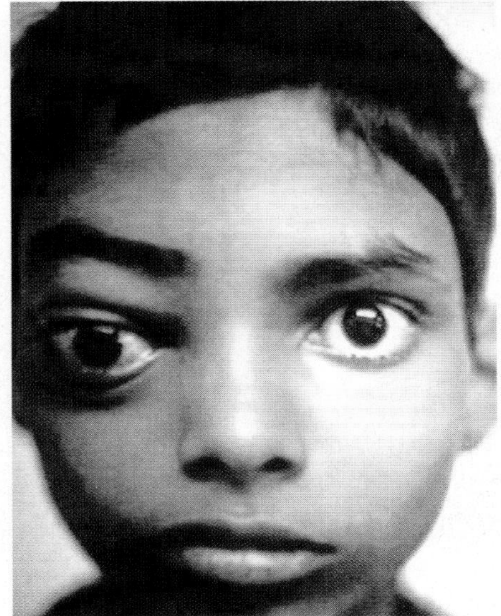

FIGURE 1.421: Retrobulbar hematoma. A young boy sustained blunt injury to eye and presented with proptosis with painful eye.

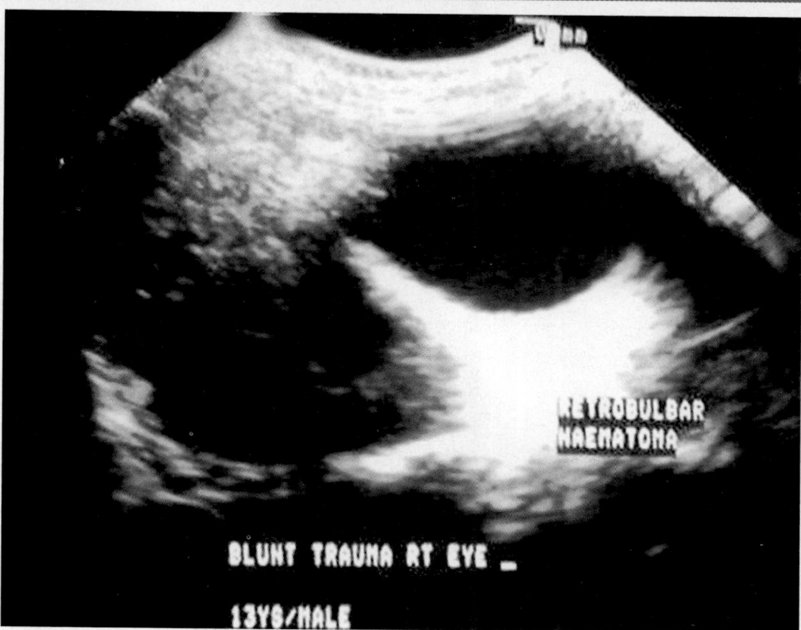

FIGURE 1.422: HRSG shows low-level echo complex mass in retrobulbar space with anterior displacement of eyeball. Few internal echoes are seen in it suggestive of retrobulbar hematoma. Eye was normal. Patient responded to compression bandage.

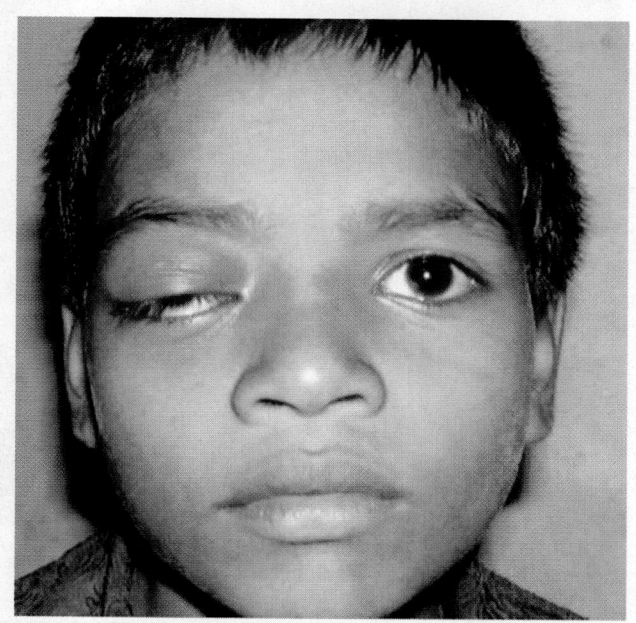

FIGURE 1.423: Retrobulbar abscess. A young boy presented with proptosis with painful eye and chymosis.

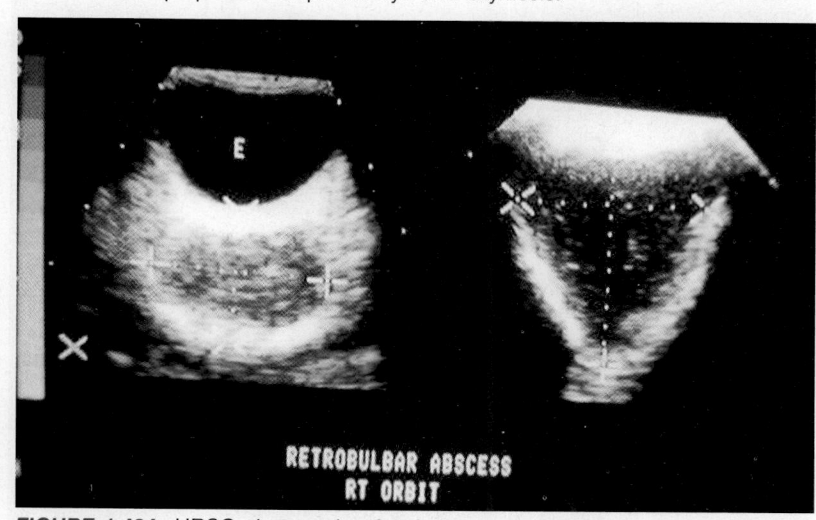

FIGURE 1.424: HRSG shows a low level echo complex mass in retrobulbar area layering of fluid is seen indicating a retrobulbar abscess.

Optic Nerve Tumors

Meningioma

Meningiomas are optic nerve sheath tumors. They present as slowly developing proptosis and unilateral in origin. Vision impairment is associated with the tumor when it enlarges and ruptures through the dura mater. It occupies the retrobulbar space. They are highly reflective masses, calcification may present in optic nerve sheath or in tumor.

Optic Nerve Glioma

Optic nerve glioma are smooth, fusiform or ovoid mass which replace the normal optic nerve. The lesion is poorly reflective and shows poor acoustic transmission. The thirty-degree test is negative in optic nerve glioma. CT is a better investigation for the diagnosis of optic nerve glioma.

Neurilemmoma

Neurilemmomas are rare tumors. They arise from proliferation of Schwann cells. They are hypoechoic nodular masses and seen in the superior orbit. They may be present in muscle cone in extraconal compartment.

Optic Neuritis

In optic neuritis, the optic nerve is enlarged due to fluid collection in perineural sheath. Therefore, the optic nerve gets thickened. The 30 degree test is the valuable test to differentiate between optic nerve tumor or inflammation. The test is carried out in fixed primary gaze position, and the patient is asked to rotate his eye to 30 degree towards to probe. In case of fluid collection, the fluid redistributes itself along the optic nerve sheath as the nerve stretches and the swelling subsides. However, the swelling persists on rotation of eye to 30 degree in case of tumor of optic nerve.

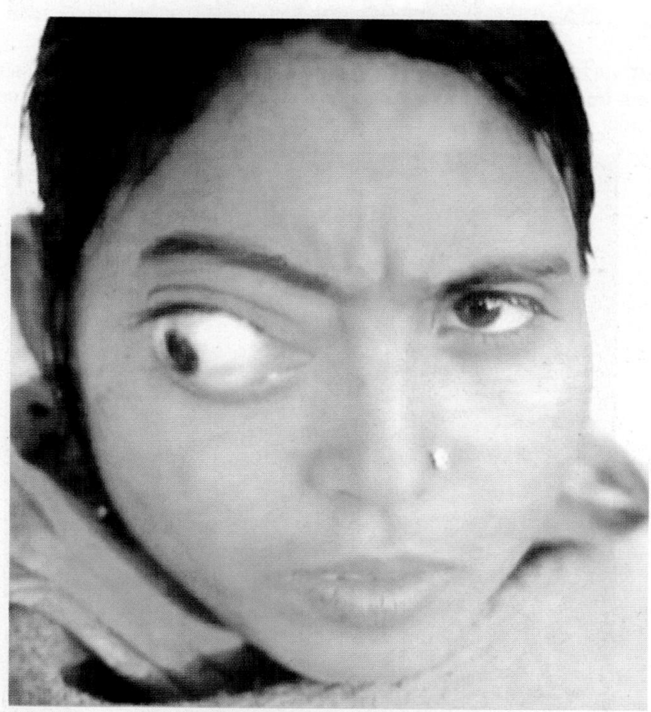

FIGURE 1.425: Meningioma optic nerve. A woman presented with proptosis and, diminished vision.

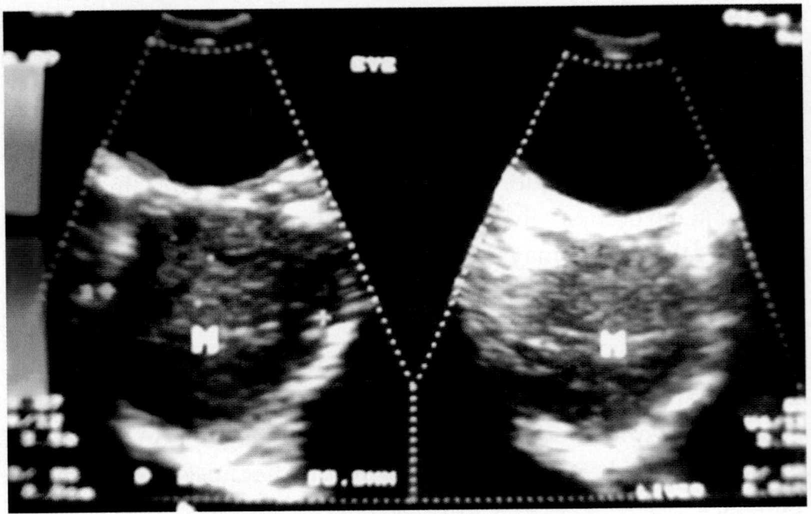

FIGURE 1.426

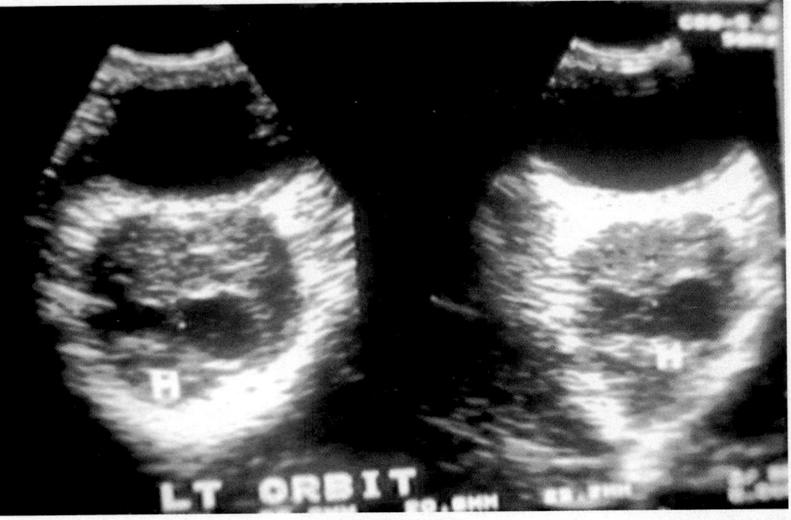

FIGURE 1.429

FIGURES 1.428 and 1.429: CT shows a heterogeneous intraconal mass with area of necrosis. HRSG shows a highly heterogeneous mass occupying whole intraconal space. No nerve is seen. Biopsy showed optic nerve glioma.

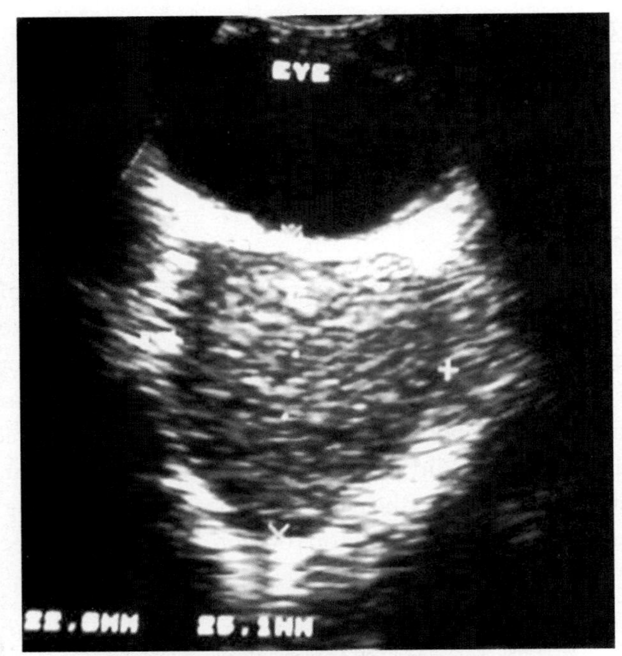

FIGURE 1.427

FIGURES 1.426 and 1.427: HRSG shows well-defined fusiform soft tissue mass coming out from the optic nerve. Echogenic calcification is seen in it. On color flow Doppler poor flow is seen in it. Biopsy shows optic nerve meningioma.

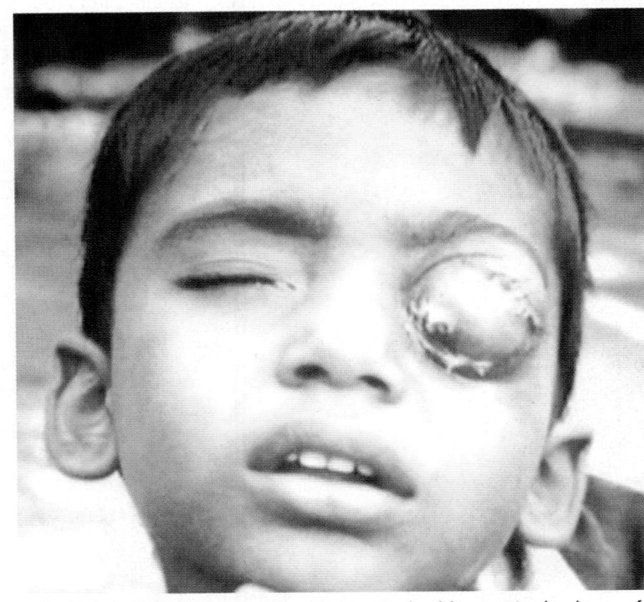

FIGURE 1.430: A young boy presented with proptosis, loss of vision with exposure keratitis.

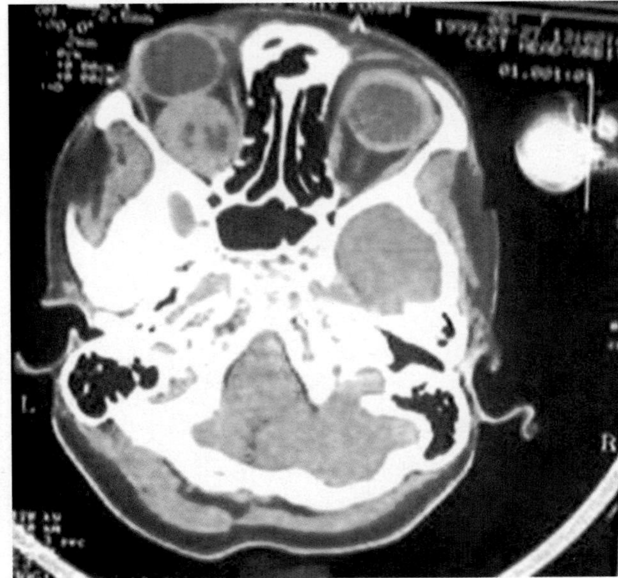

FIGURE 1.428

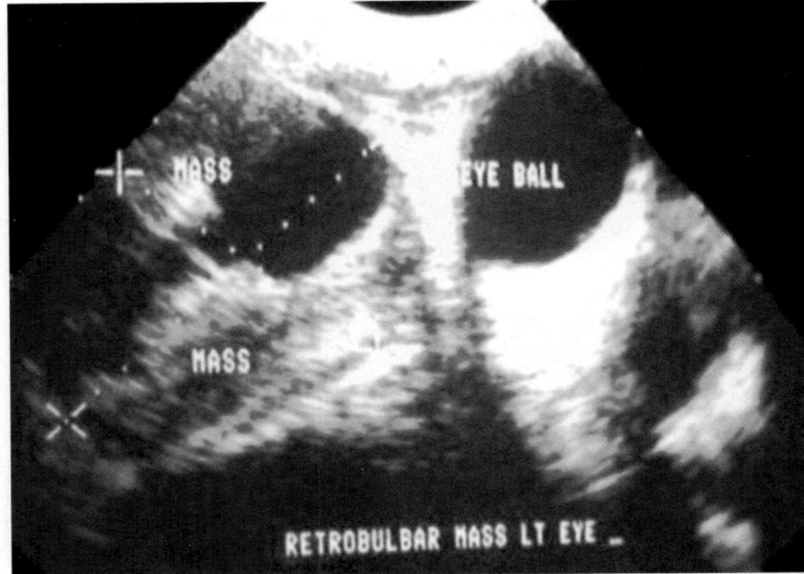

FIGURE 1.431: HRSG showed a heterogeneous cystic mass in retrobulbar space. Biopsy showed cystic glioma of optic nerve.

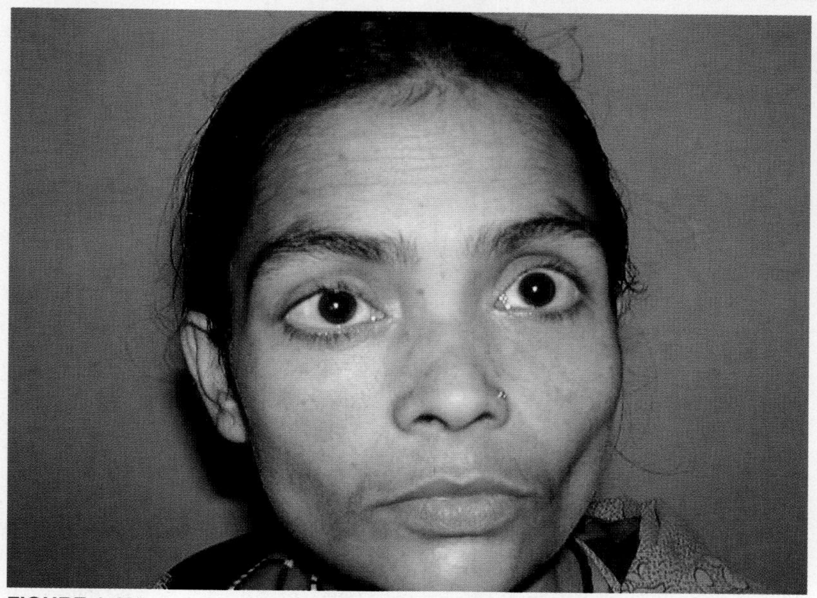

FIGURE 1.432: A young lady presented with loss of vision with proptosis Rt eye.

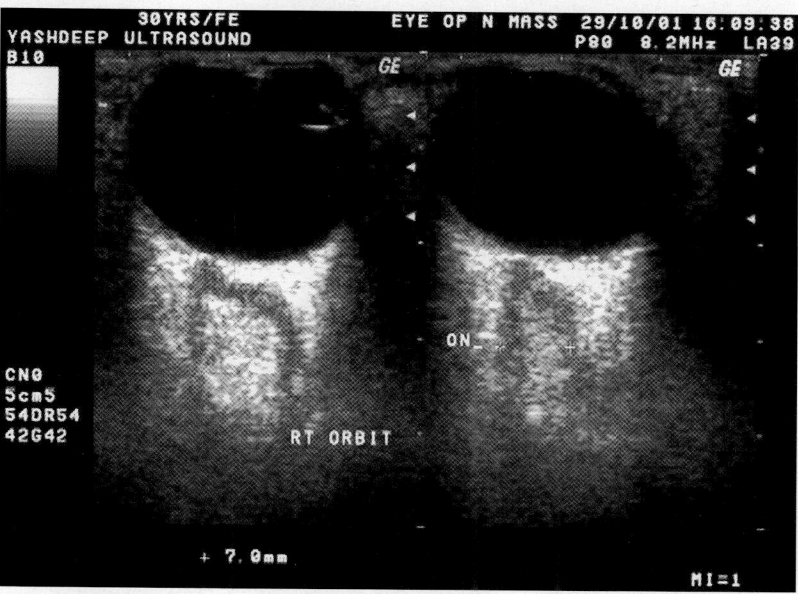

FIGURE 1.433

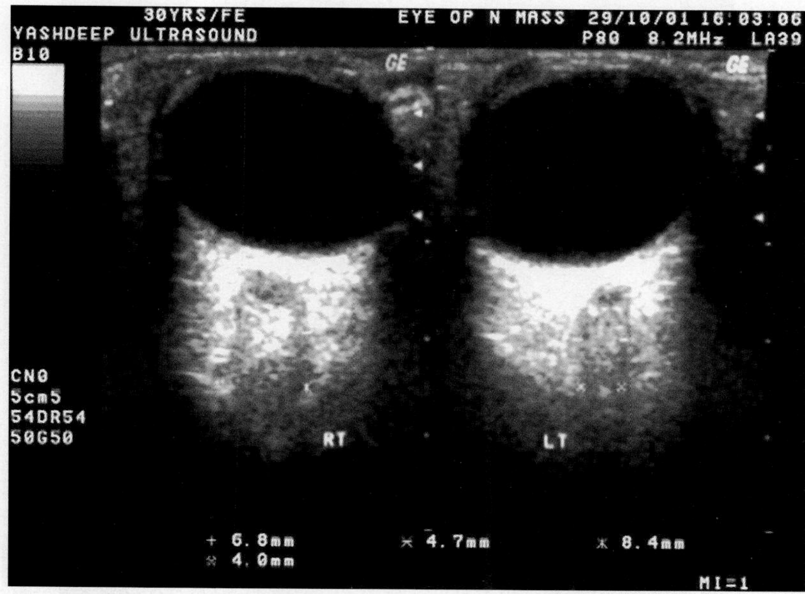

FIGURE 1.434

FIGURES 1.433 and 1.434: HRSG shows fusiform widening of optic nerve with echogenic mass in the nerve suggestive of optic nerve tumor.

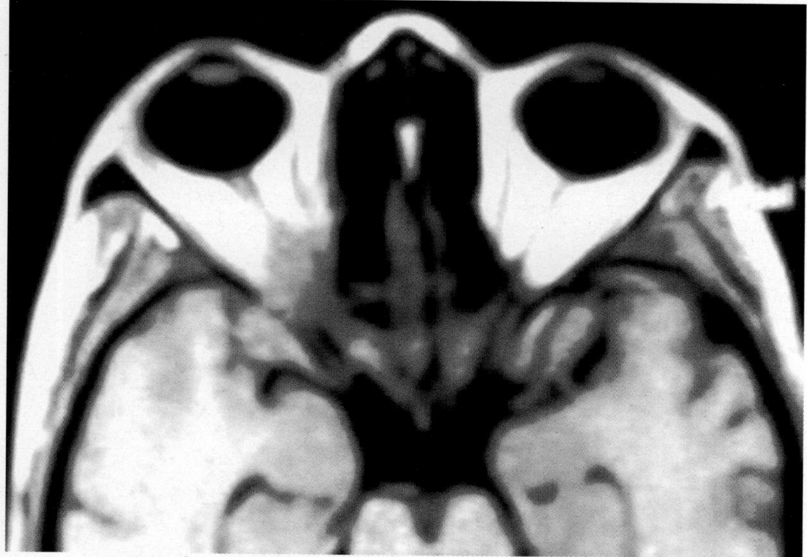

FIGURE 1.435: MRI of the same patient shows fusiform mass coming out from the optic nerve with expansion of apex of the cone. Biopsy showed optic nerve glioma.

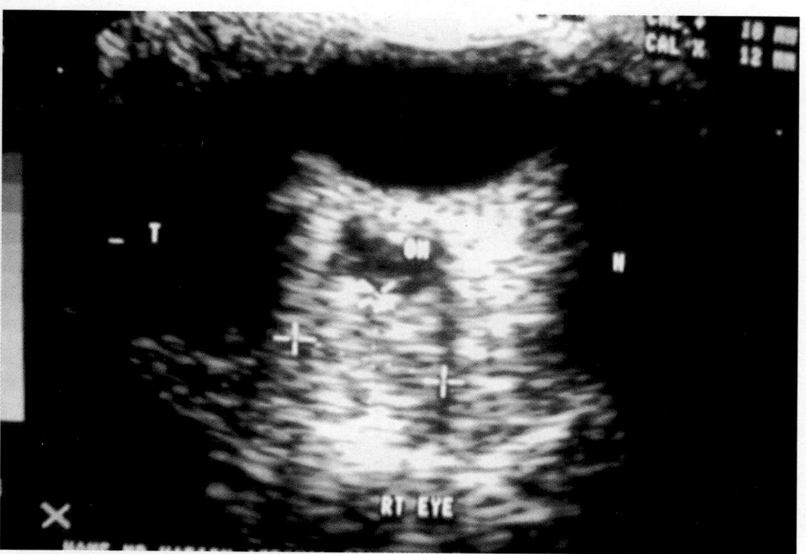

FIGURE 1.436: HRSG shows smooth widening of optic nerve. Expansion of the sheath is seen with a fusiform mass seen in the upper part of optic nerve. In a patient came with history of loss of vision. Sonographic features are suggestive of optic nerve tumor. Biopsy of the tumor show optic nerve glioma.

Parasitic Infestations

Orbital Cysticercosis

Orbital cysticercosis is one of the most common extracranial manifestation of cysticercus infestation. The disease mainly affects the extraocular muscles and mainly the recti. It presents as painful slowly progressing proptosis. The HRSG findings are classical. A well-defined cystic mass is seen in the belly of the muscle. A highly echogenic nidus is seen within the cyst. It is the scolex of the parasite. This scolex is echodense and present near the inner wall of the cyst. Echography is the choice of investigation in orbital or ocular cysticercosis. It is better than CT or MRI. It is also a very good tool to evaluate the post-therapy response of the disease after medical treatment, the parasite dies and cyst regresses in size and vanishes. The muscle belly shows localized thickening, which can be well appreciated on HRSG.

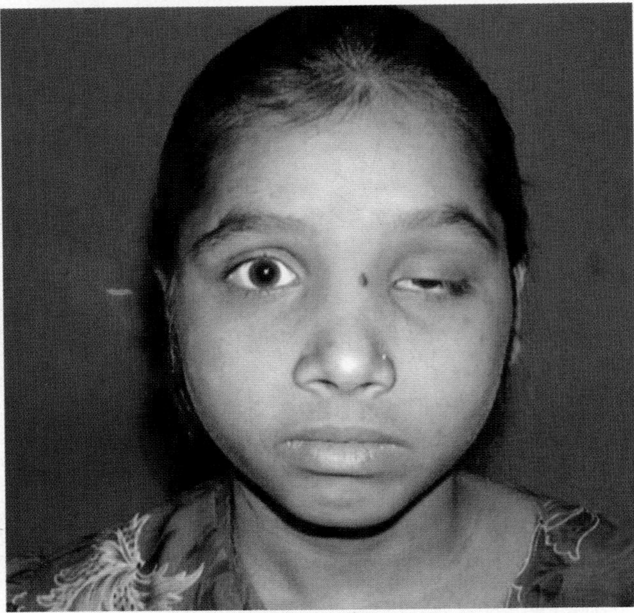

FIGURE 1.437: Cysticercosis superior rectus muscle. A young girl presented with proptosis and drooping of Lt eyelid and painful Lt eye.

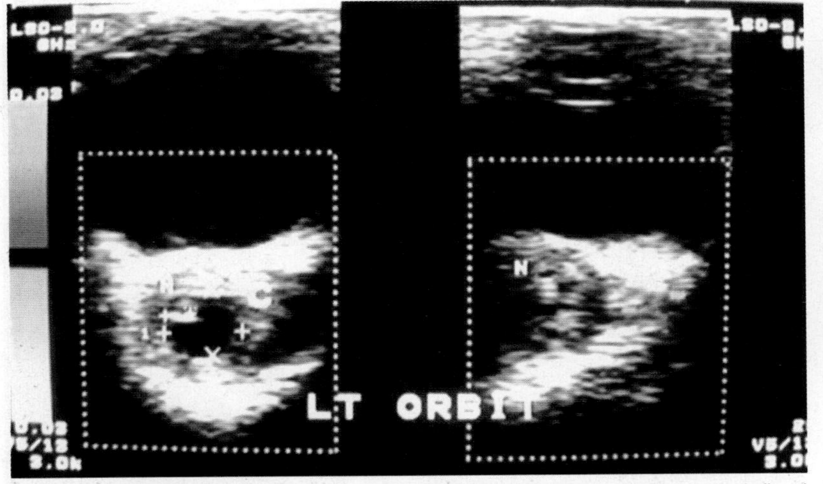

FIGURE 1.438: HRSG shows inflamed superior rectus muscle with a well-defined cyst. An echogenic nidus is seen fixed with inner wall of the cyst is scolex.

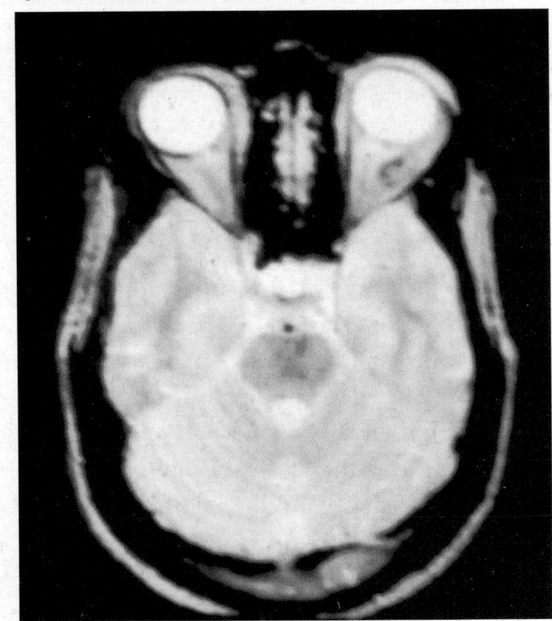

FIGURE 1.439. Lateral rectus cysticercosis: MRI shows an ill-defined hypodense mass in belly of lateral rectus muscle.

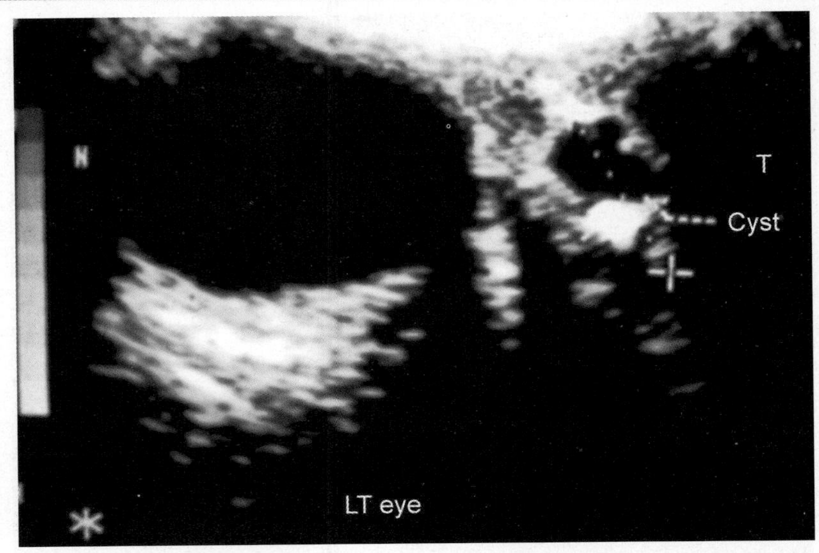

FIGURE 1.440: HRSG shows a well-defined cyst with an echogenic nidus fixed with inner wall of the cyst typical of cysticercus cyst.

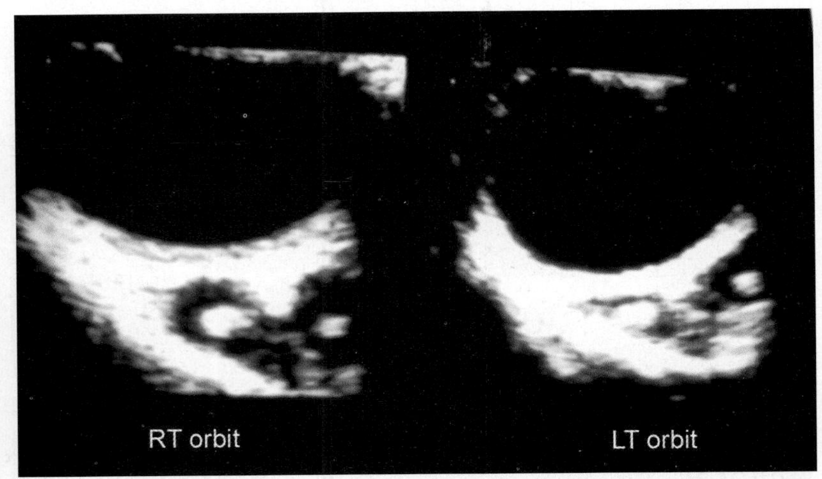

FIGURE 1.441. Bilateral cysticercosis. HRSG shows thick walled cysts in both side lateral recti with scolex fixed with inner wall of cyst as echogenic nidus.

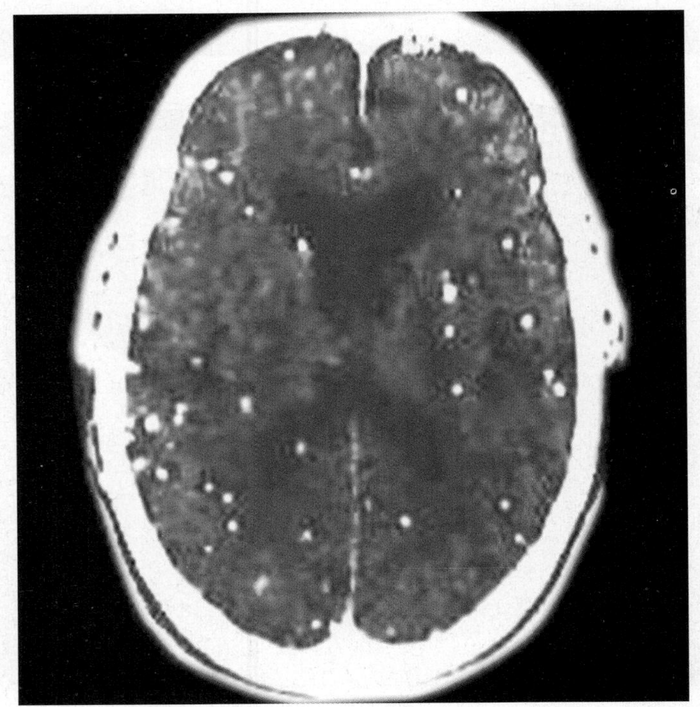

FIGURE 1.442: CT head of the same patient shows multiple hyperdense dot shadows diffuse in distribution in wide spread neurocysticercosis.

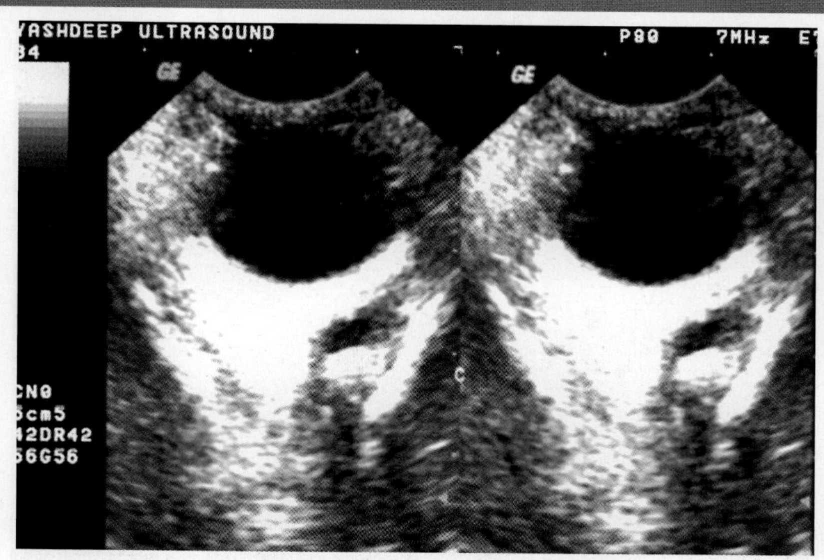

FIGURE 1.443: Lateral rectus cysticercosis. Well-defined cyst in lateral rectus with a hyperdense echogenic nidus fixed with inner wall of the cyst.

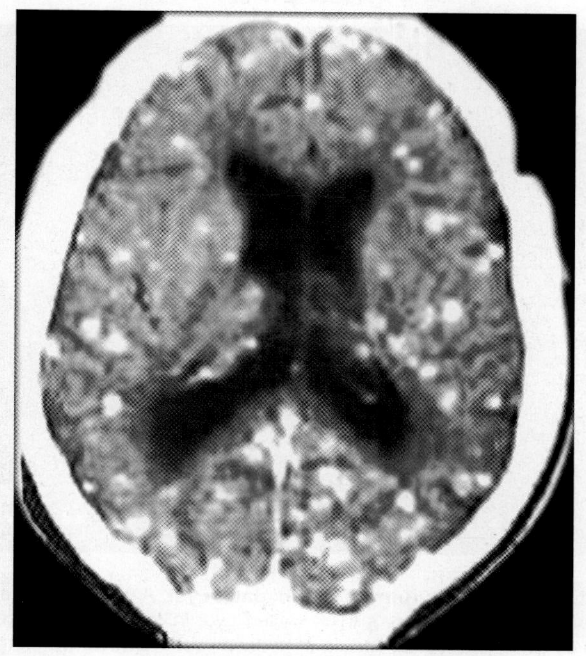

FIGURE 1.444: CT head of the same patient shows gross impregnation of brain parenchyma with neurocysticercosis.

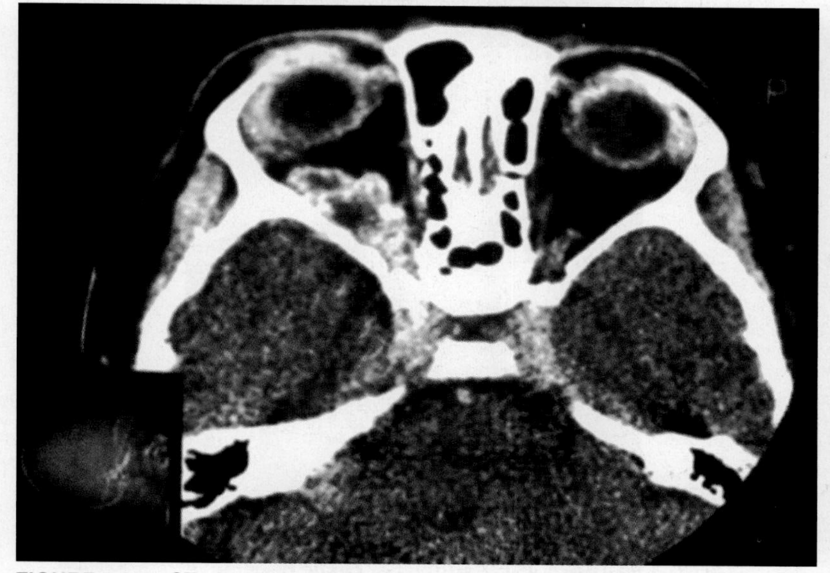

FIGURE 1.445: CT of a young girl shows irregular thickening of lateral rectus muscle with a hypodense area.

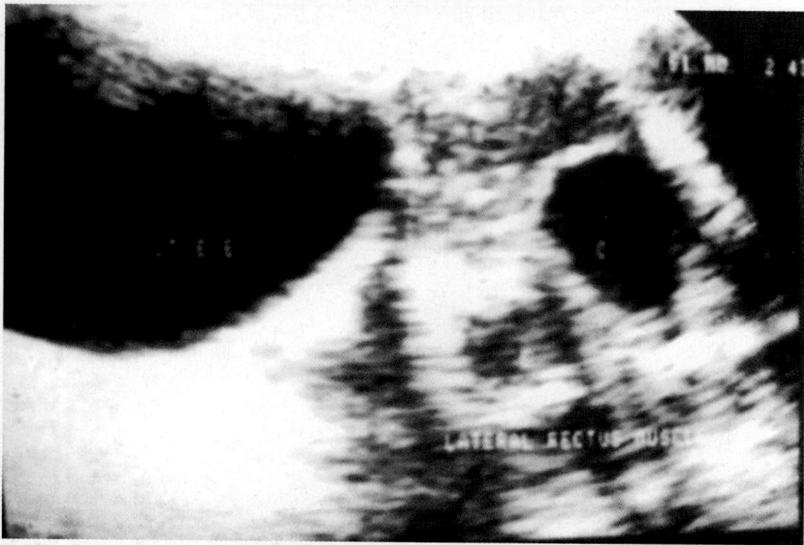

FIGURE 1.446: HRSG of the same patient shows well-defined cyst in lateral rectus with an echogenic nidus suggestive of cysticercus cyst.

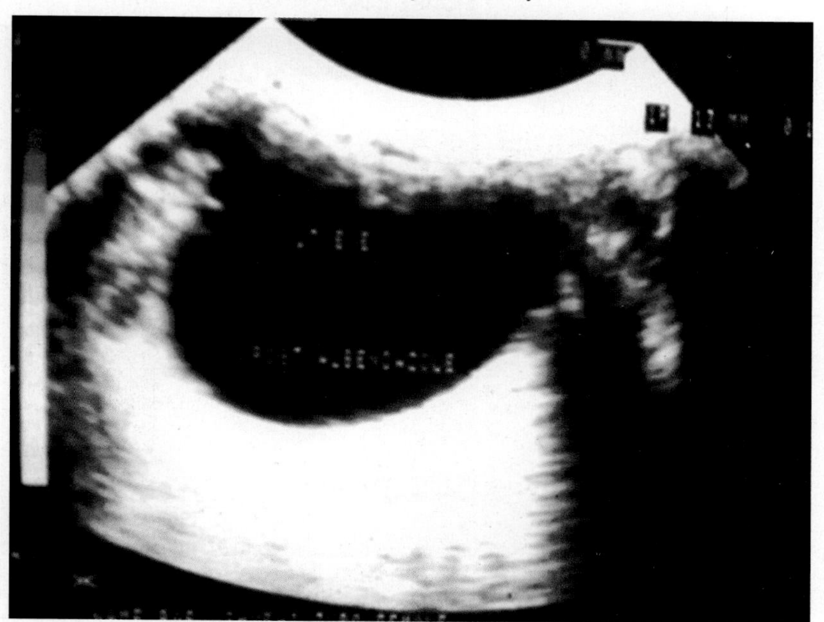

FIGURE 1.447: Post treatment. HRSG shows localized thickening of lateral rectus muscle with collapse of the cyst after medical treatment in the same patient Figure 1.446. HRSG is very good to assess the treatment response of the therapy.

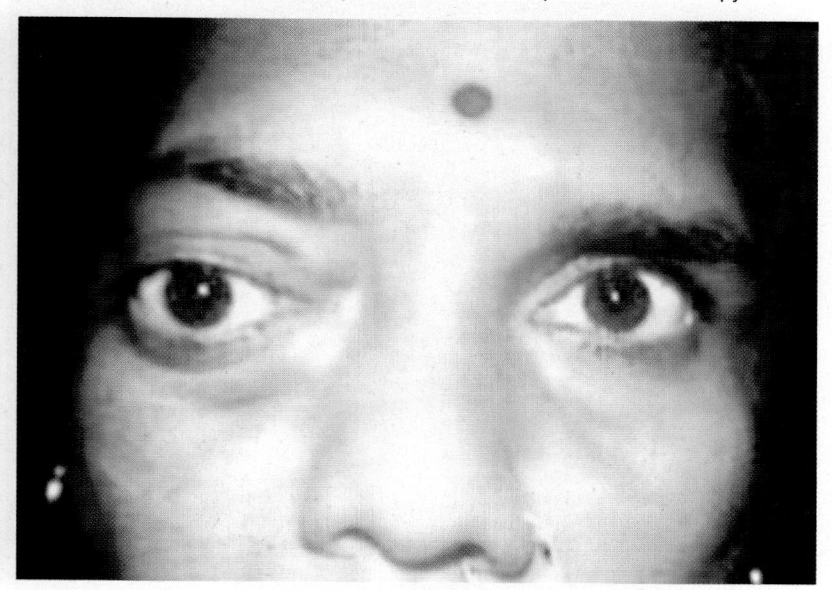

FIGURE 1.448: Hydatid cyst orbit. A middle-aged woman presented with Rt eye proptosis with diplopia.

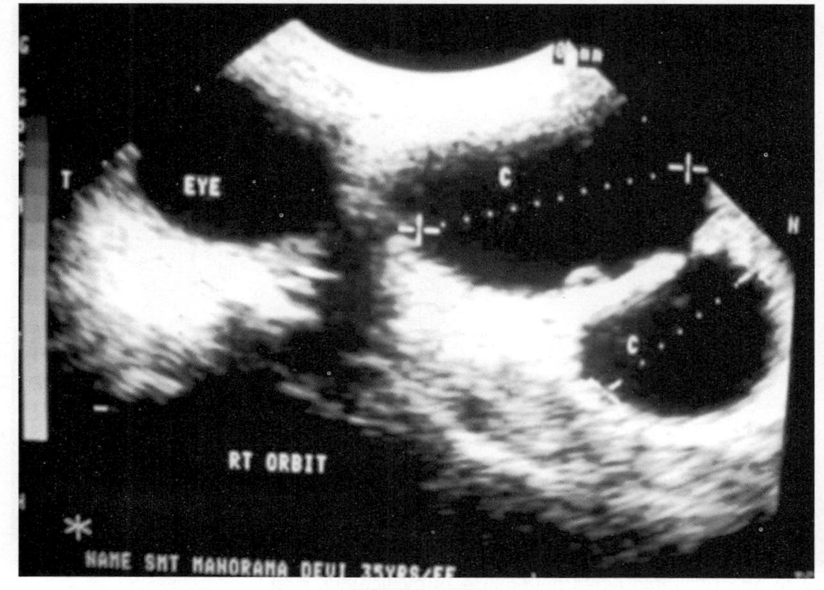

FIGURE 1.449

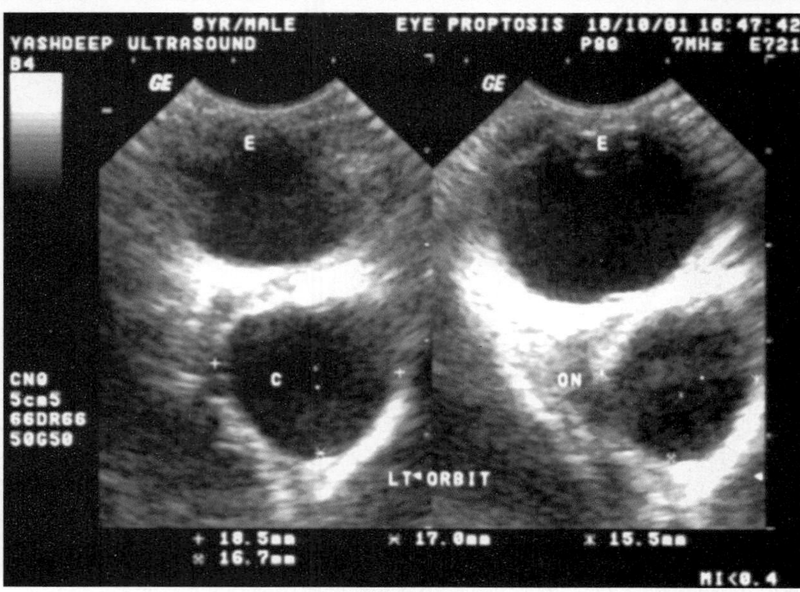

FIGURE 1.452

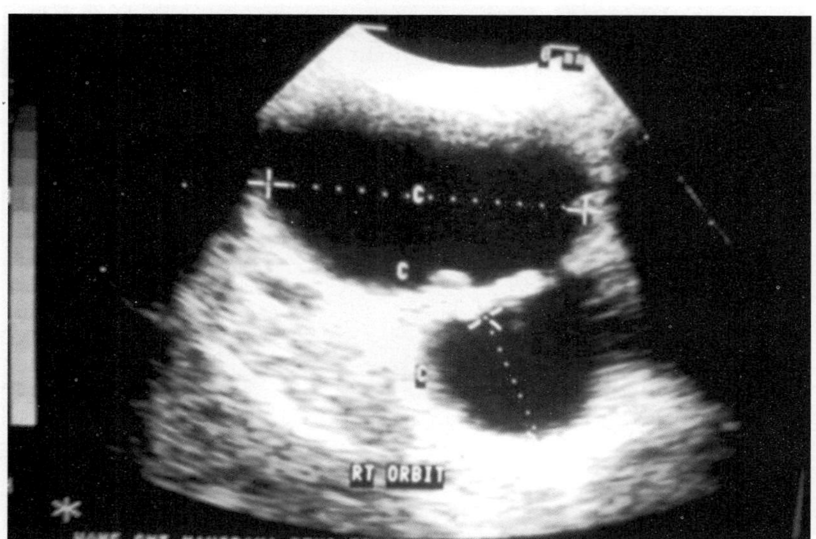

FIGURE 1.450

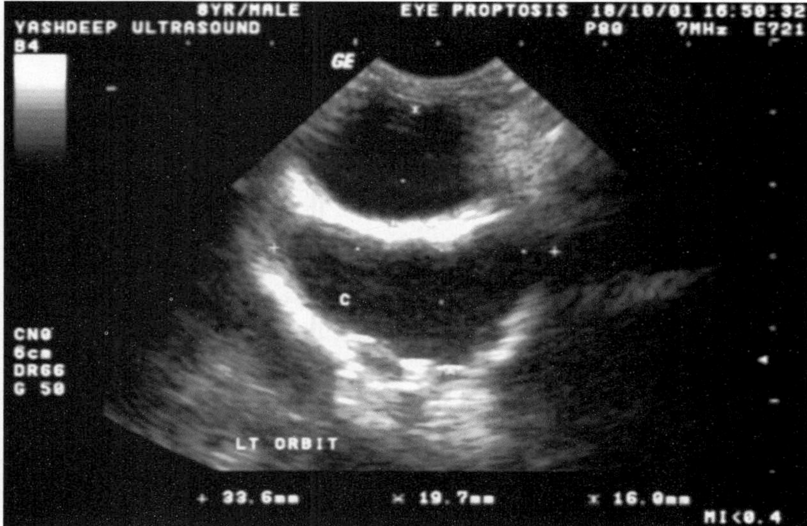

FIGURE 1.453

FIGURES 1.449 and 1.450: HRSG shows a bilobed cystic mass in retrobulbar space of Rt orbit. The big cyst shows another small daughter cyst fixed with its wall and anterior displacement of eyeball. Typical findings of the Hydatid cyst.

FIGURES 1.452 and 1.453: Retrobulbar hydatid cyst. A young boy presented with Lt eye proptosis. HRSG shows a thick walled cyst in retrobulbar space. Cyst wall also shows calcification, typical finding of hydatid cyst. However no daughter cyst is seen.

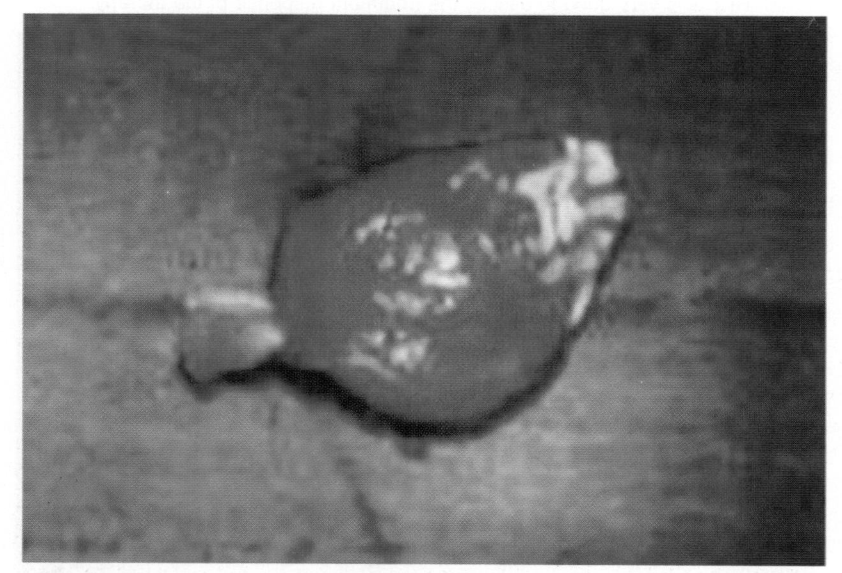

FIGURE 1.451: The gross postoperated specimen of the same patient shows the cyst in gross.

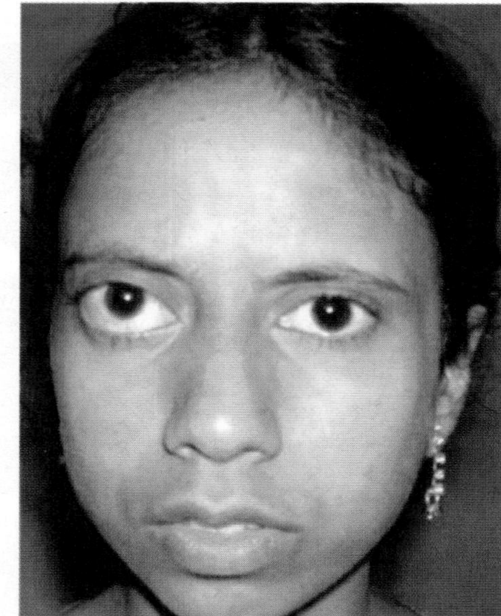

FIGURE 1.454: Retrobulbar cystic dermoid. A young girl presented with Rt sided proptosis.

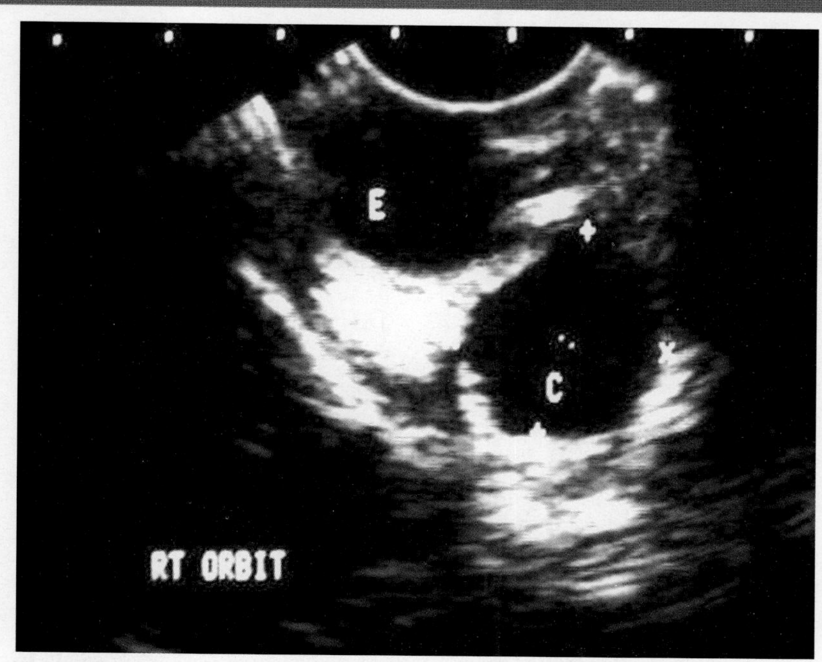

FIGURE 1.455: HRSG shows a thick walled cystic mass in the retrobulbar space. Echogenic calcification is seen in it.

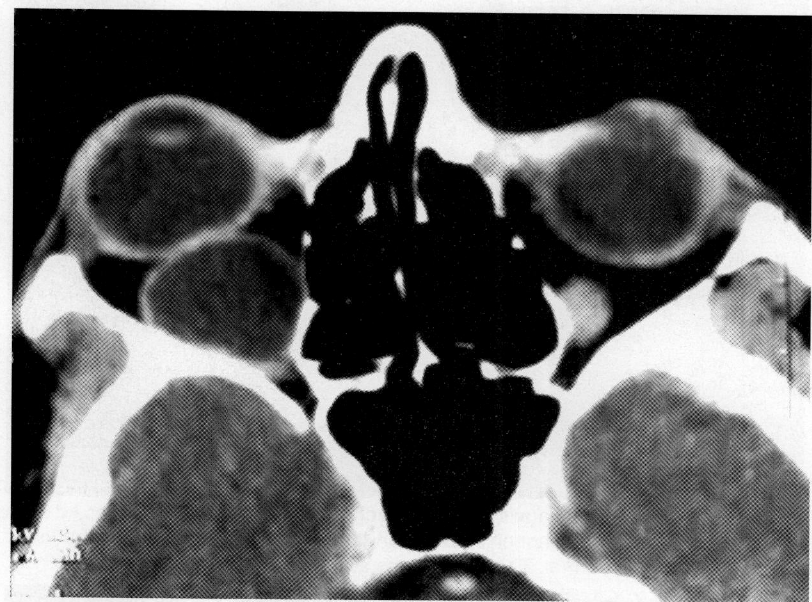

FIGURE 1.456: CT scan of the same patient shows well defined cyst in the retrobulbar intraconal space. Biopsy of the cyst was dermoid cyst.

Hydatid Cyst

Orbital hydatid cysts are not very common. They usually present in the retrobulbar area as a well-defined thin walled cystic mass. It stands out very clearly on HRSG. At times thin septa are seen in the cyst making the loculations and septations. The eye is displaced to opposite side by the pressure effect of the cyst, and associated proptosis is also present.

Periorbital Masses

HRSG is highly sensitive to evaluate periorbital masses like abscess, dermoid cyst, mucoceles, lid tumors and other periorbital pathologies.

Lacrimal Gland Tumors

HRSG is a good tool to evaluate tumors present in lacrimal fossa since large varieties of tumors can occur in lacrimal fossa, the differentiation is difficult. HRSG can very well identify the nature of the mass. Whether it is solid, cystic or mixed echo-complex in nature. The lesions most commonly found in this area includes inflammatory or lymphoid origin, primary lacrimal gland epithelial tumors and cyst. Normal lacrimal gland and surrounding soft tissues are highly reflective. Therefore, the gland cannot be separately delineated on HRSG. The most common lacrimal gland tumors are adenoid cystic carcinoma, mucoepidermoid carcinoma, pseudotumors and lymphoma.

Periorbital Abscess

A periorbital abscess is a painful swelling in the periorbital space which may push the eyeball to other side. HRSG shows a thick walled low-level collection with week sound transmission. It is also helpful in the guided needle aspiration of the abscess. In the thick abscess, areas of necrosis are also noted in the abscess cavity.

Dermoid Cyst

The most common orbital and periorbital cysts are dermoid cysts. They usually present superotemporally or superonasally. They may be present with the lids, at the orbital rim, near outer canthus, on the conjunctiva or within the orbit. They most commonly occur in infants and children. They are usually heterogeneous in texture, and different sound characteristics are seen in dermoids. Some attenuates the sound beam and other thoroughly transmits the sound. Homogeneous low-level echoes are seen in HRSG. They may contain keratins, sebaceous materials, hair follicles or inflammatory cells. When they are present deep in the orbit, it becomes difficult to differentiate them from the benign mixed tumors.

Mucoceles

Mucoceles are formed due to blockage in the sinus drainage openings. This leads to enlargement. This may bulge into the orbit. On HRSG, a mucocele looks like a low reflective, homogeneous nonvascular cystic mass, which is most commonly present in the supranasal quadrant (frontoethmoidal). This may cause outward and downward displacement of the globe.

Lid Tumors

Lid tumors arising from mebobain cells or tarsal glands are rare tumors. They are seen well-encapsulated homogeneous low-level echo complex mass on HRSG. The eyeball is pushed down and covered by the tumor mass. HRSG can clearly depict the involvement of eyeball or invasion of it by the tumor infiltration. If the tumor is well encapsulated and not involving the eyeball, it can be very well excised.

Lacrimal Ductal Cysts (Dacryops)

HRSG shows multiple small cysts in the upper lid involving palpebral part of lacrimal duct. Tiny echogenic calcified specks are also seen in them suggestive of ductal ectasia of lacrimal duct also known as dacryops. It is a rare condition of lacrimal duct caused due to obstruction and ectasia of lacrimal gland duct.

Carcinoma of Lid

Carcinoma of lid is the malignant tumor. It is highly heterogeneous in texture and rapidly growing tumor. The tumor can invade the eyeball and infiltrate the sclera which can be very well appreciated on HRSG.

The tumor is seen invading the conjunctiva, sclera and muscle coat of the eyeball, and the lid is firmly attached and it is not retractable. Therefore, direct examination of eye is not possible in adherent tumor of the lid.

Orbital Biometry

Ocular measurement can be accurately carried out by using standardized "A" scanning and 2D imaging. The most frequently measurements are required prior to cataract surgery for intraocular lens implantation. The measurements required are axial length of the eyeball and corneal curvature. The other measurements, which can be taken on HRSG, are thickness of cornea, retina, choroid, sclera or size, shape, elevation and base diameters of intraocular tumors. For anterior chamber depth measurement, immersion technique is used for proper evaluation.

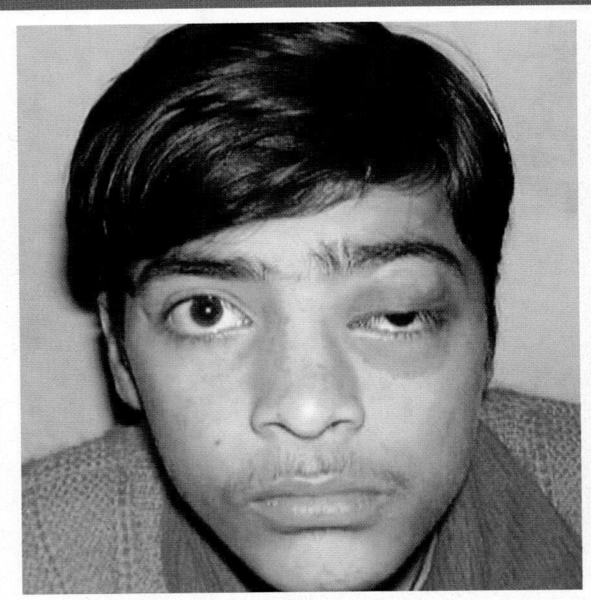

FIGURE 1.459: Upper lid abscess. A young boy presented with painful lid swelling with chymosis.

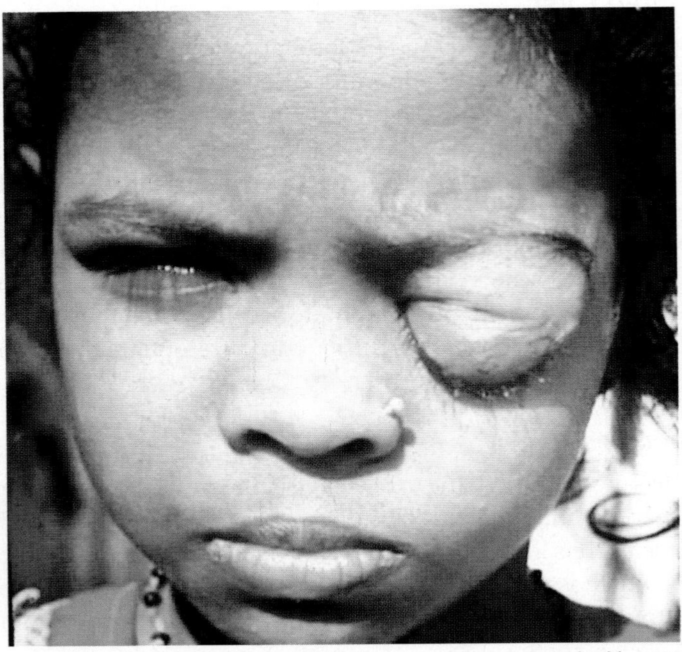

FIGURE 1.457: Neurofibroma lid. A young girl presented with very soft tissue mass over the Lt upper eyelid covering the eyeball.

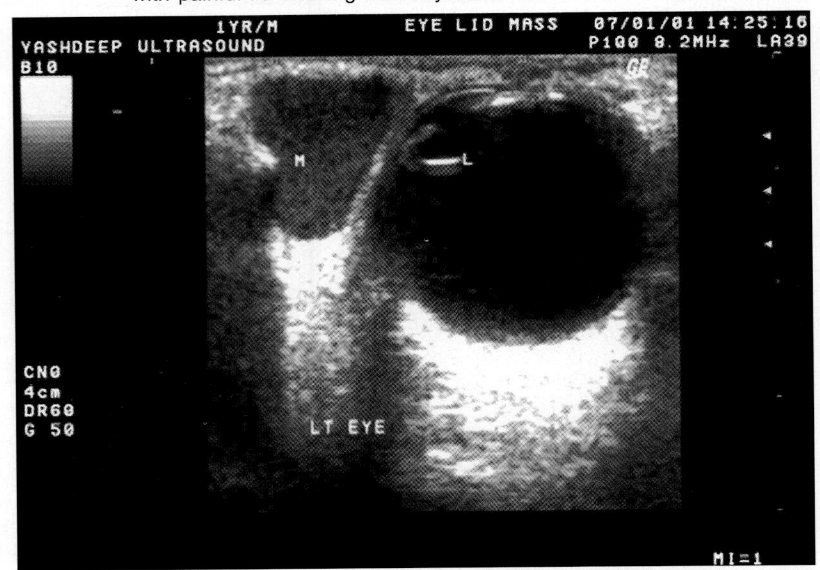

FIGURE 1.460: HRSG shows low-level echo complex mass in upper lid with homogeneous echoes. Echogenic collection is also seen sedimented at the lower side. The eyeball is normal. The findings are suggestive of lid abscess.

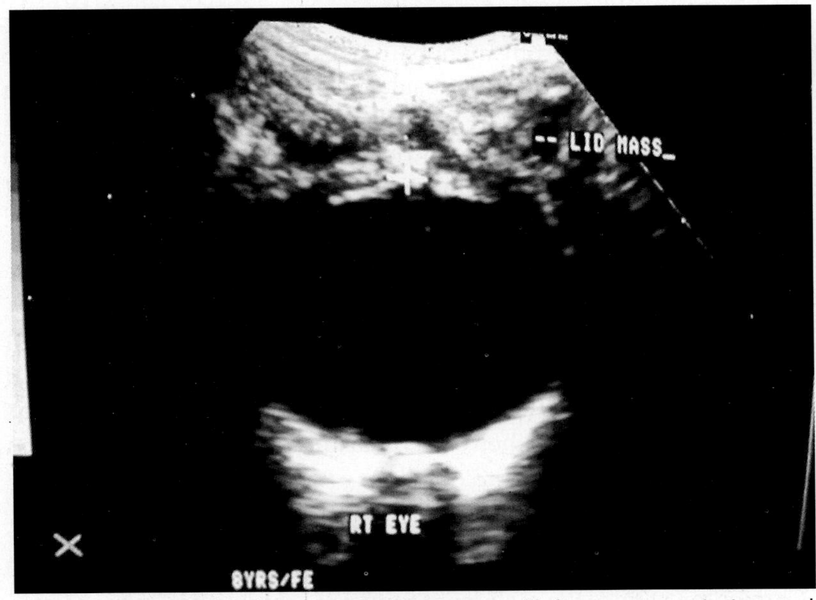

FIGURE 1.458: HRSG shows a soft tissue mass with homogeneous texture and covering the eyelid. It is not seen invading the eyeball. Eye is spared. The biopsy proved neurofibroma.

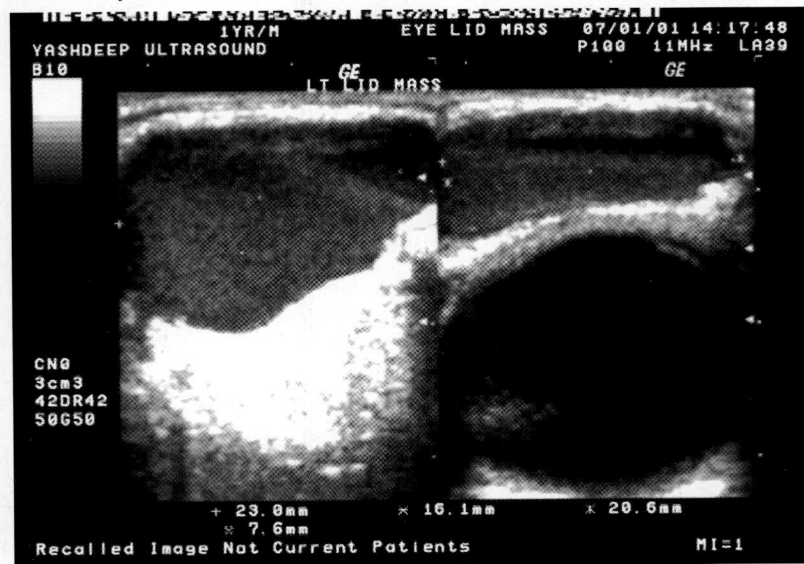

FIGURE 1.461: Same patient as in Figure 1.460 shows low-level echo collection in the lid with thick capsule suggestive of lid abscess.

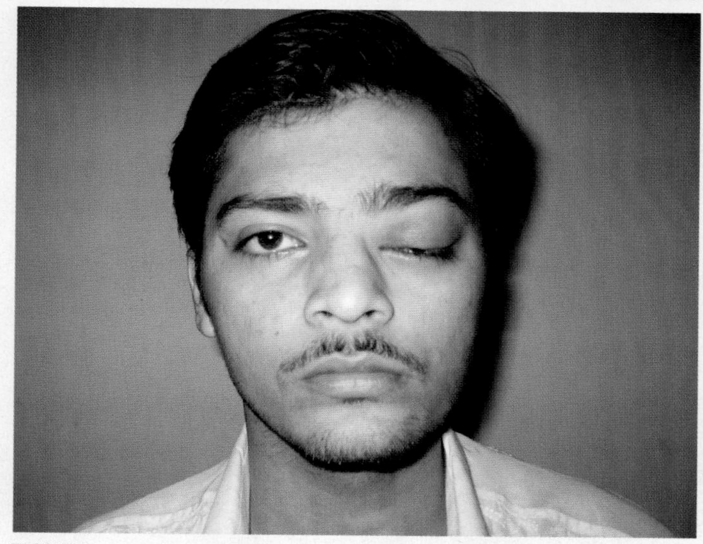

FIGURE 1.462: The same patient after 3 years, presented with drooping of the eyelid with lateral displacement of the eyeball.

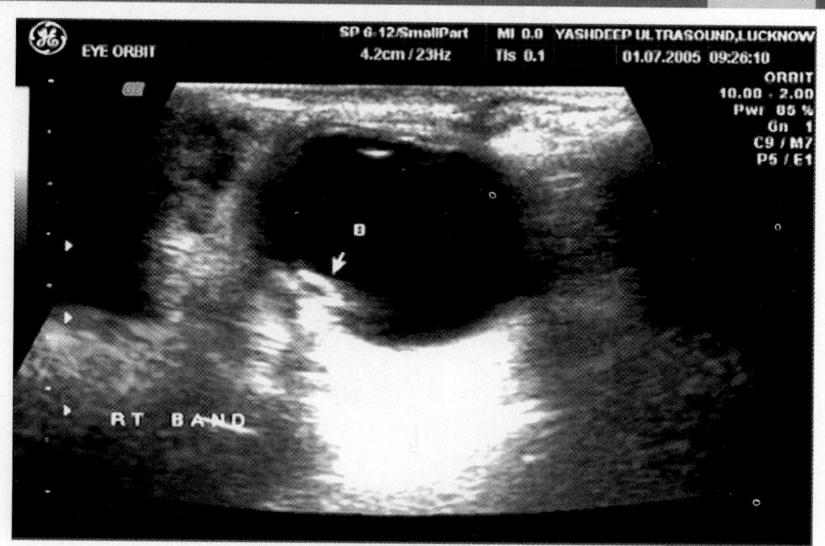

FIGURE 1.465

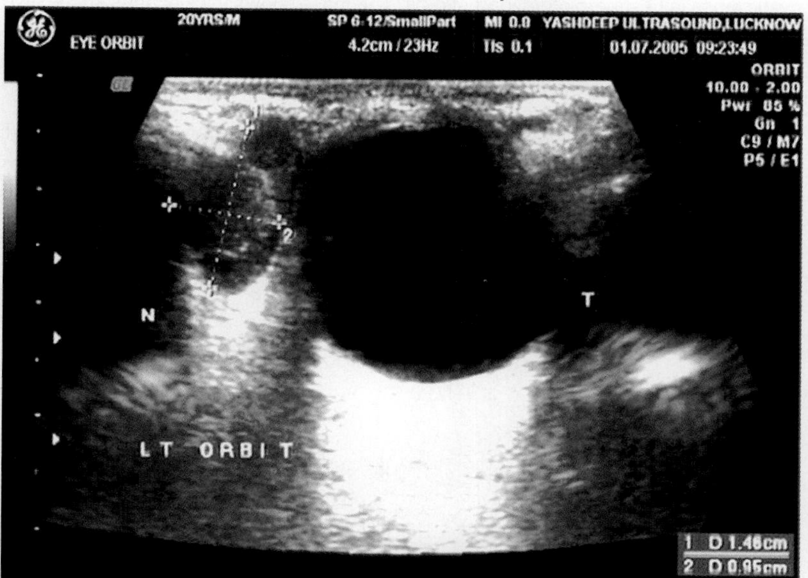

FIGURE 1.463

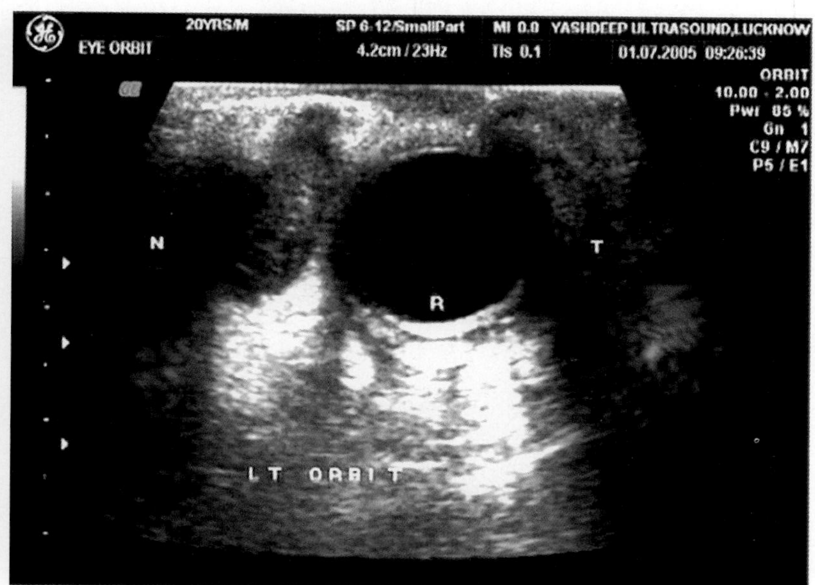

FIGURE 1.466

FIGURES 1.465 and 1.466: HRSG shows associated retinal buckle placed in position for associated retinal detachment. The retina is seen in place.

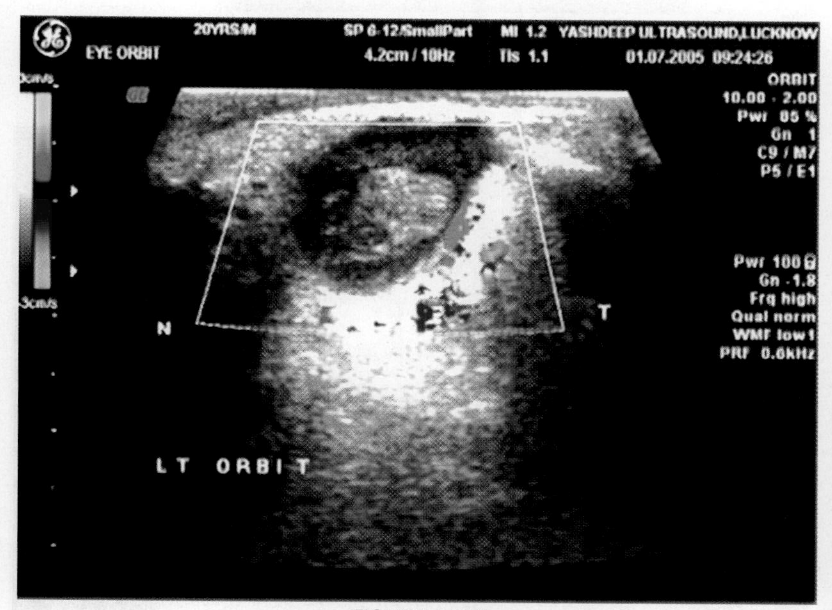

FIGURE 1.464

FIGURES 1.463 and 1.464: HRSG shows a thick walled low level echo complex mass coming out from medial side of the orbital wall. Thick echoes are seen in it suggestive of periorbital abscess.

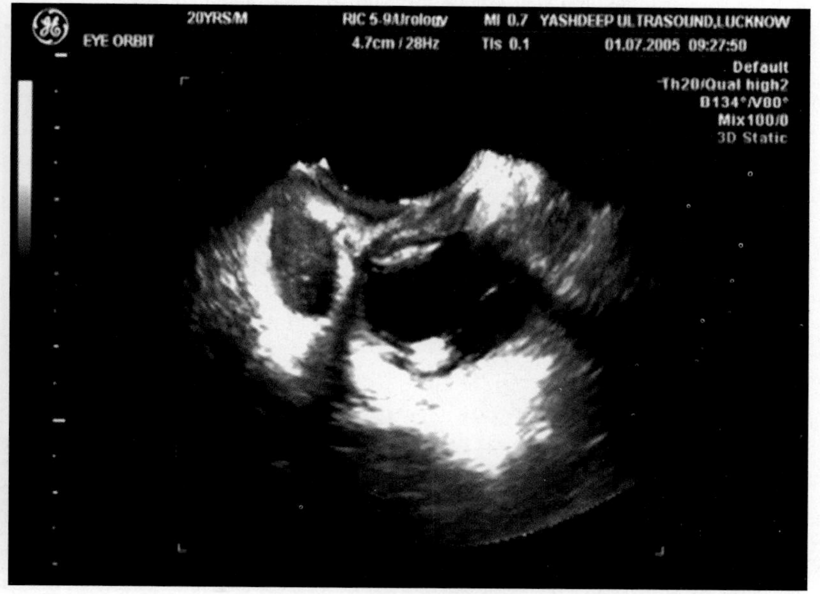

FIGURE 1.467

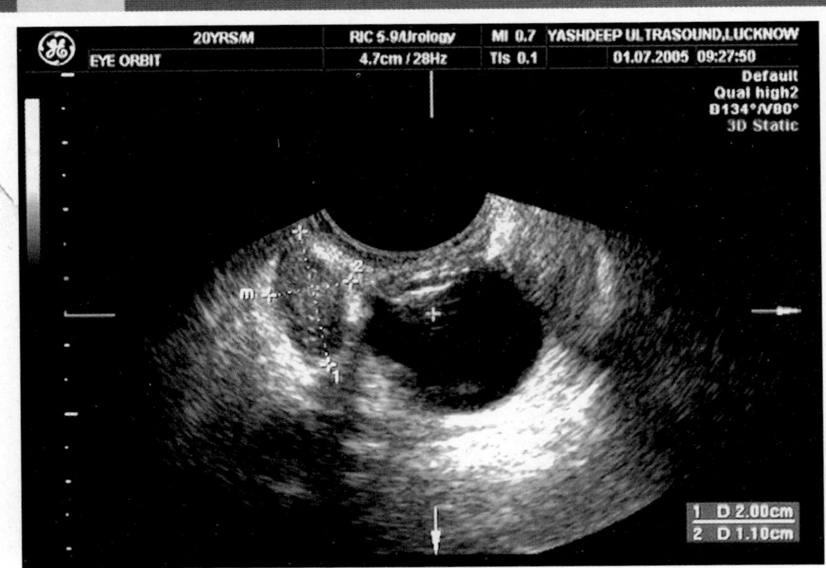

FIGURE 1.468

FIGURES 1.467 and 1.468: 3D of the same patient shows details of the periorbital abscess.

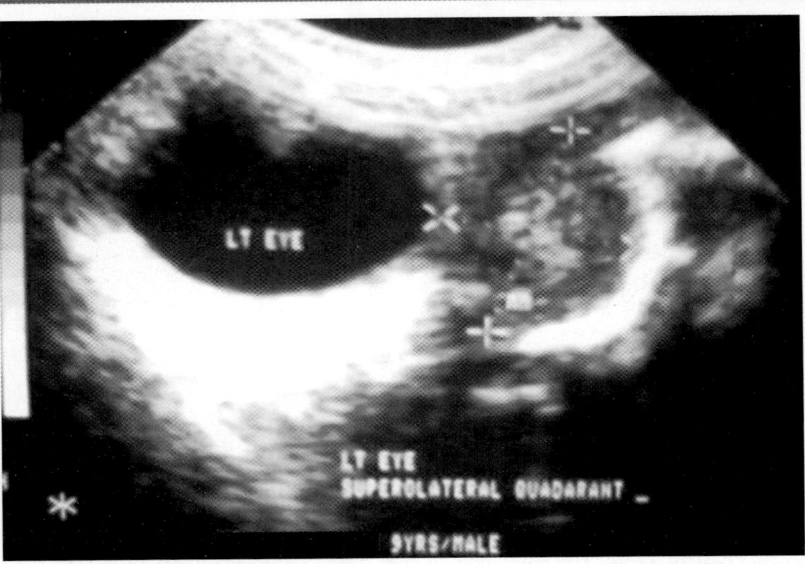

FIGURE 1.471: Periorbital abscess. HRSG shows well encapsulated thick walled cystic mass in Lt superolateral part of orbit. The mass shows thick echoes suggestive of pus. It was a lacrimal gland abscess. Eyeball is normal but pushed medially

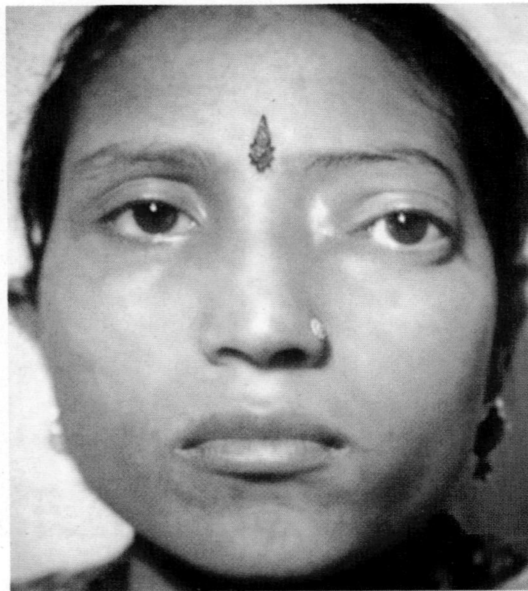

FIGURE 1.469: Lacrimal sac abscess. A patient presented with painful swelling over the upper medial quadrant of Lt eye. Purulent discharge through the nose was coming out.

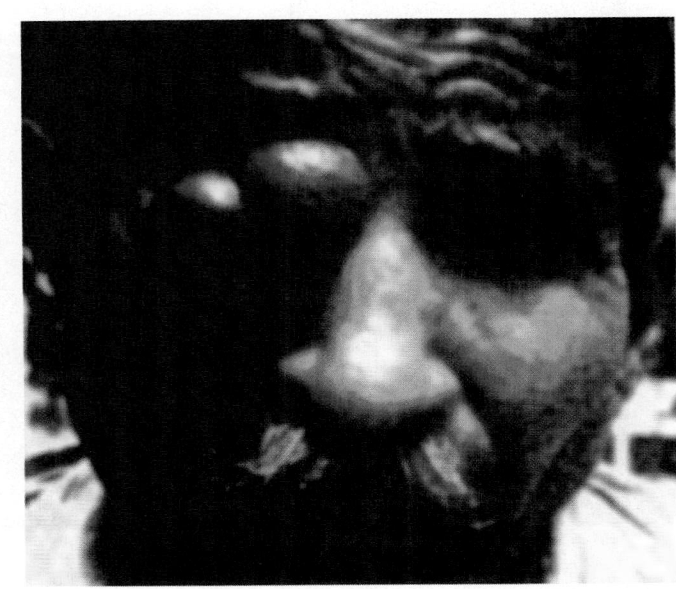

FIGURE 1.472: Lacrimal sac abscess. An old patient presented with semihard lobulated masses over the Rt eyeball. The lids could not be opened due to pressure by the masses and evaluation of eyeball was not possible.

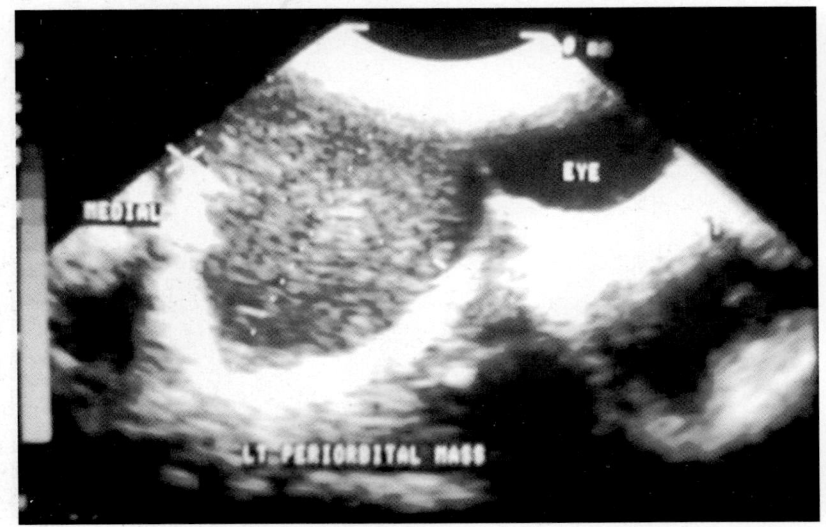

FIGURE 1.470: HRSG shows a thick walled low level echo complex mass seen in Lt upper medial quadrant of orbit. The mass is seen having homogenous collection pushing eyeball laterally. It is typical lacrimal sac abscess. The eyeball is normal.

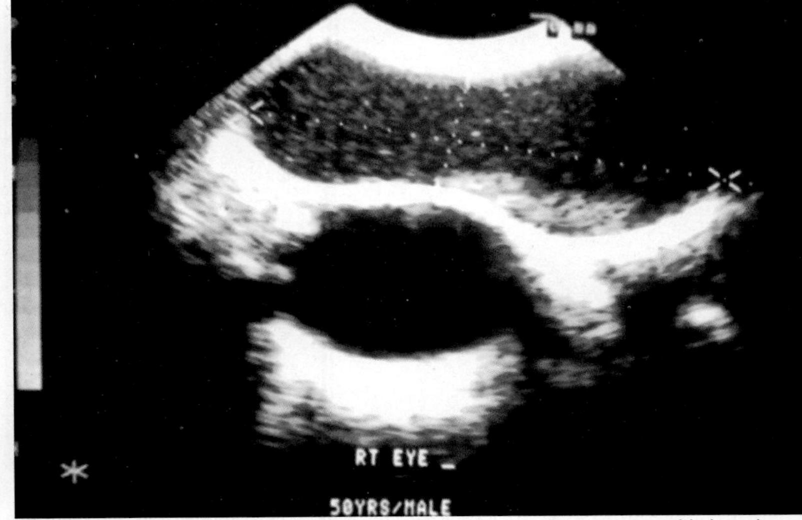

FIGURE 1.473: HRSG shows thick echo collection in the masses with layering of echogenic collection. It is showing multiple small echoes with thick wall. Eye is seen pushed down but appears to be normal.

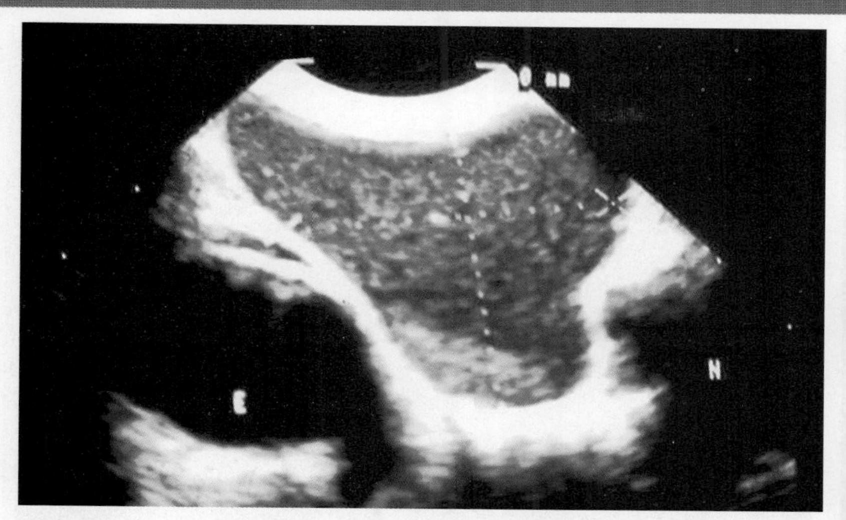

FIGURE 1.474: The collection is seen going down to the nasal side pushing the eyeball to lateral side. It came out to be a chronic abscess of lacrimal sac with thick walled. Eye was normal.

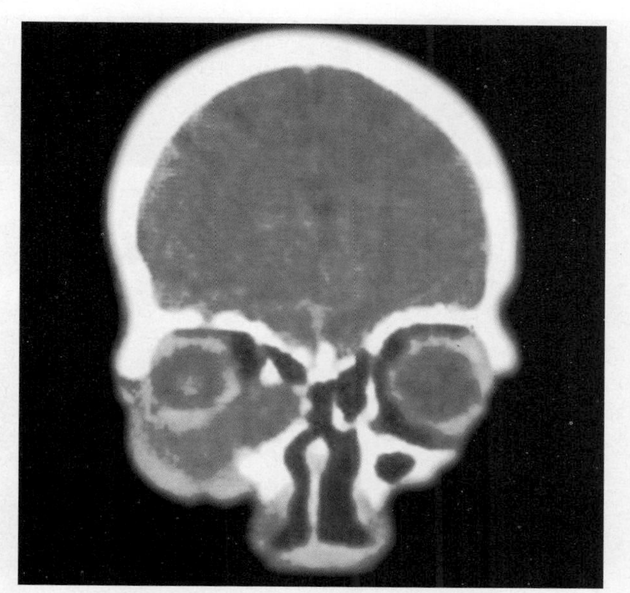

FIGURE 1.475: CT scan of same patient coronal views shows the abscess cavity coming out from the nasal side going down extending in the orbital space and engulfing the eyeball.

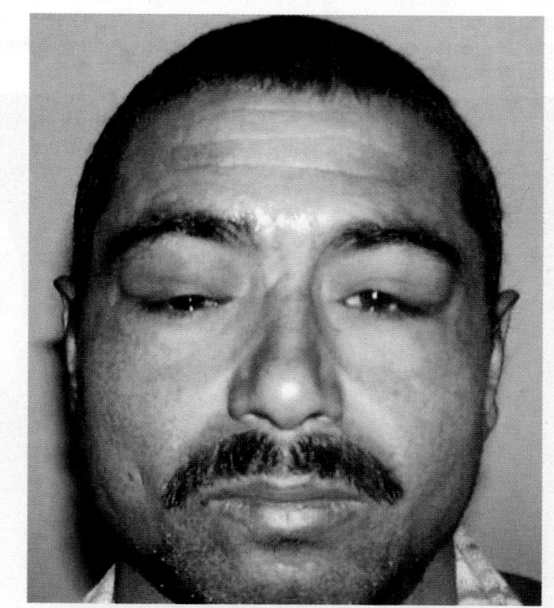

FIGURE 1.476: Neurofibroma of both lids. A middle aged man presented with bilateral swelling of lid with drooping eyelids. He finds difficulty in opening the lid.

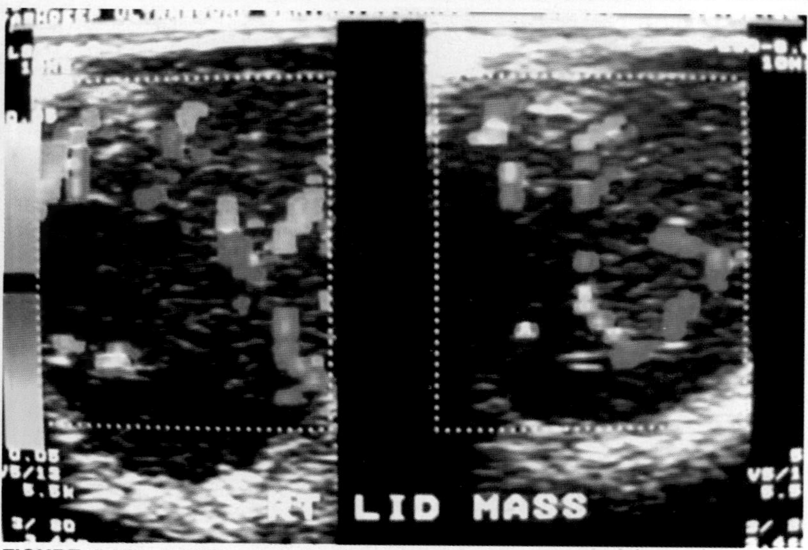

FIGURE 1.477: HRSG shows low-level echo masses in Rt lid. On color Doppler imaging multiple dilated vessels are seen in the mass.

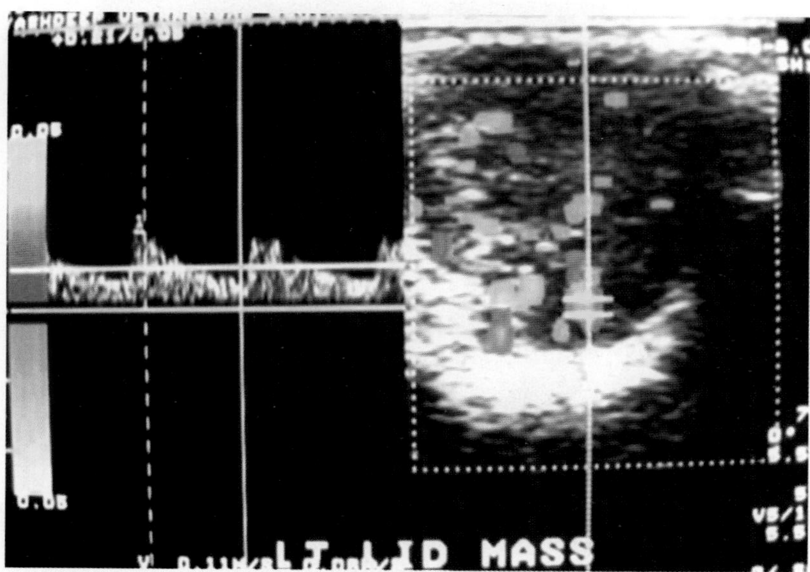

FIGURE 1.478: Spectral Doppler tracing of the mass shows low flow. The lid mass came out to be neovascularization in neurofibroma.

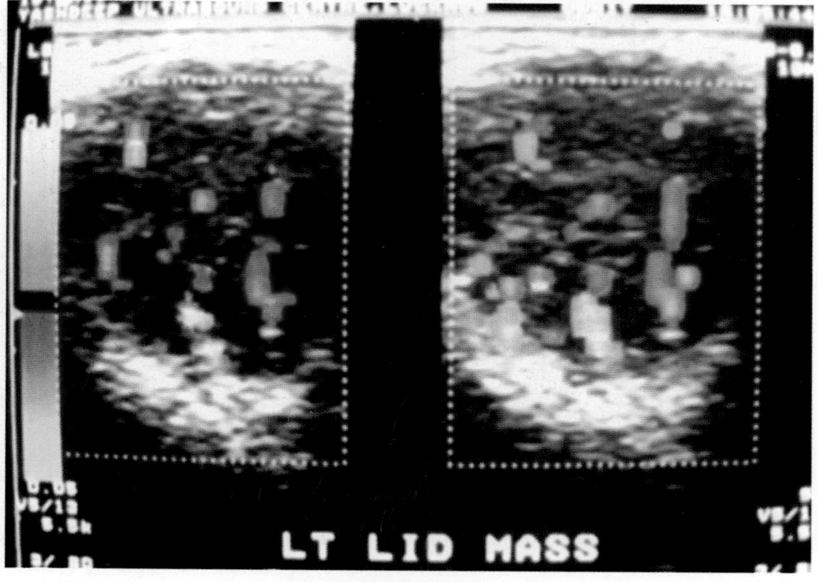

FIGURE 1.479

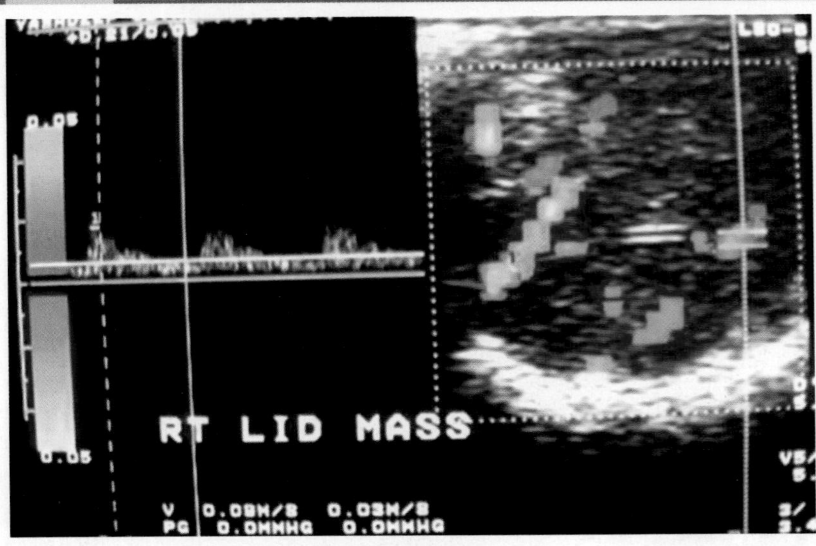

FIGURE 1.480

FIGURES 1.479 and 1.480: The same HRSG findings were noted in Lt lid mass with multiple dilated vessels. Neovascularization of neurofibroma in Lt lid.

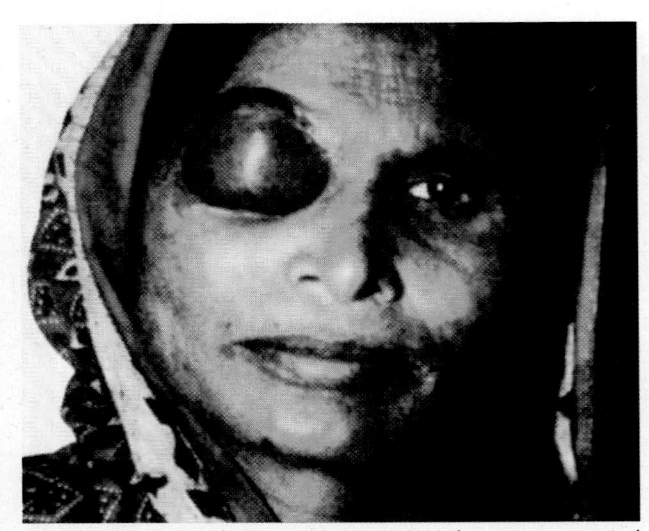

FIGURE 1.481: Mebobian cell tumor Rt lid. A woman presented with soft tissue mass over Rt eyelid. The mass was covering the eyeball and it was semihard in consistency.

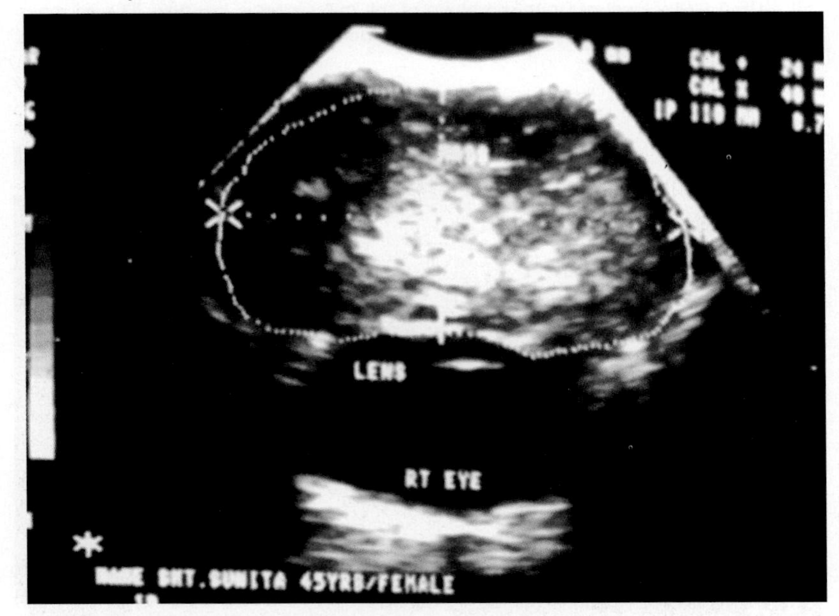

FIGURE 1.482: HRSG shows a well-defined echogenic mass over the eyelid covering the eyeball. Multiple bright echoes are seen in the mass. No cystic degeneration is seen. The mass is seen pushing the eyeball down. Lens is normal. Posterior segment of the eye is free. Biopsy of the man showed mebobian cell tumor of tarsal plate.

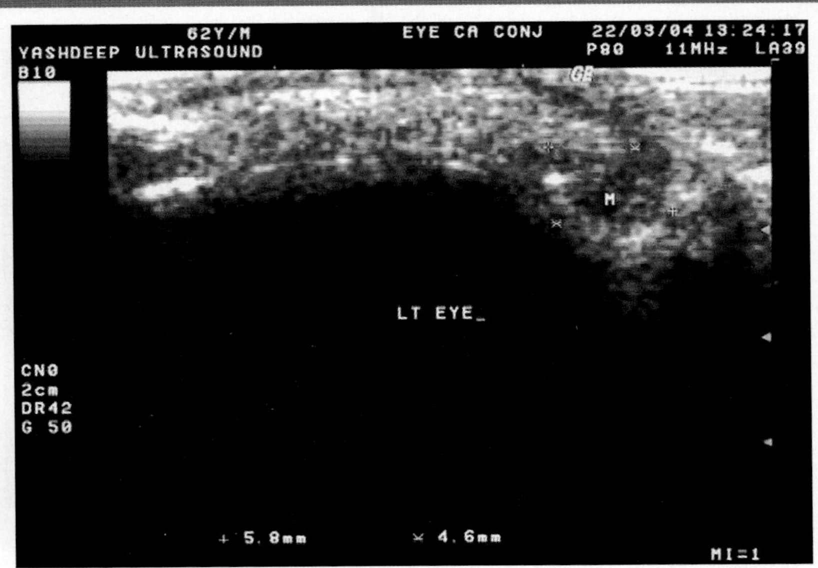

FIGURE 1.483

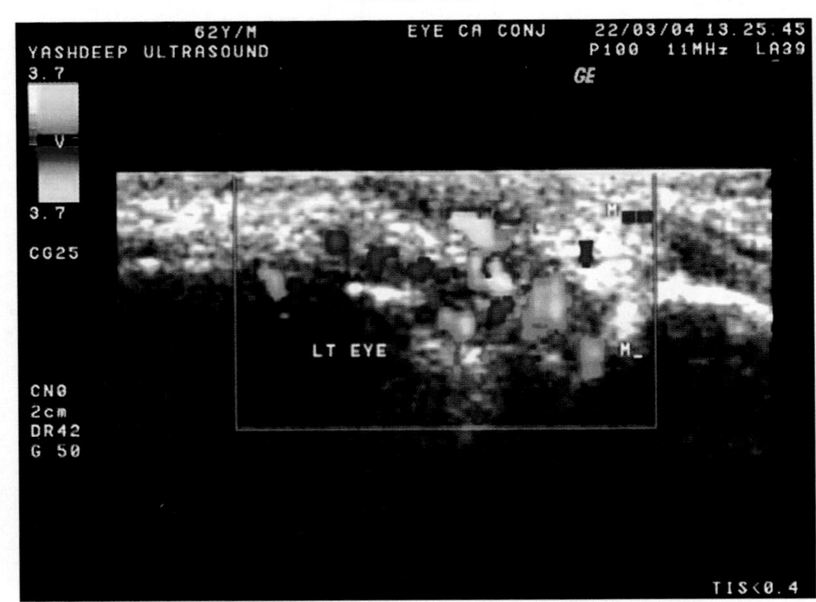

FIGURE 1.484

FIGURES 1.483 and 1.484: Carcinoma conjunctiva. A patient presented with a soft tissue mass over the lateral side of the eyeball coming out from conjunctiva. HRSG shows a heterogeneous irregular soft tissue mass on the lateral of the eye. It is seen invading the conjunctiva and also invading the sclera. On color flow imaging multiple vessels are seen invading the sclera.

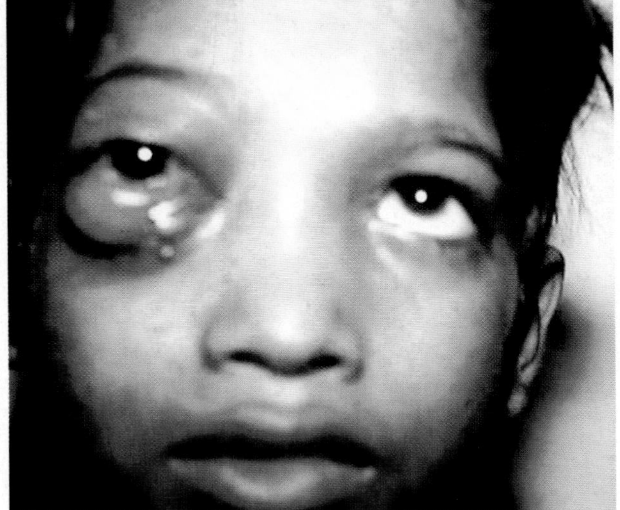

FIGURE 1.485: Carcinoma maxilla invading orbital floor. A young boy presented with soft tissue mass of lower lid with proptosed eyeball and diminished vision.

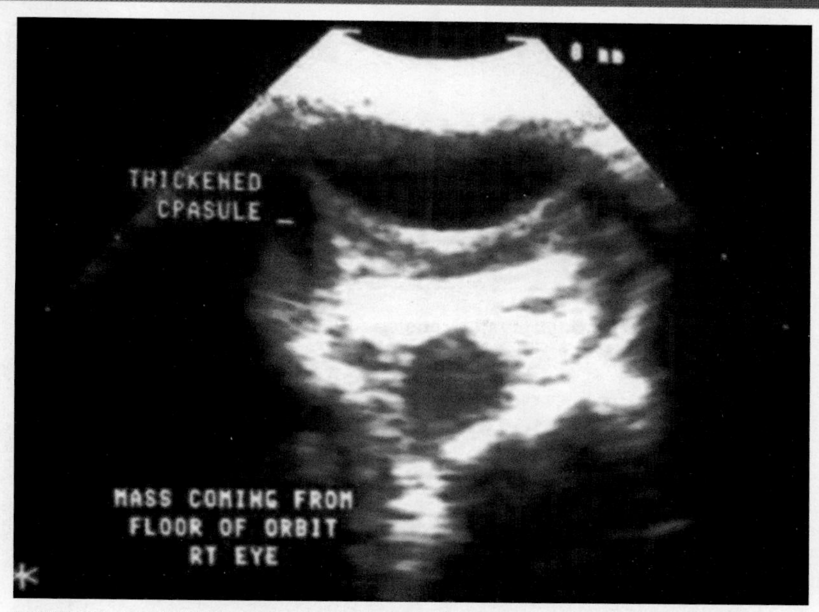

FIGURE 1.486: HRSG shows a soft tissue mass coming out from floor of orbit invading the sclera coat and pushing the eyeball anteriorly. Biopsy proved infiltrating carcinoma of Rt maxilla.

FIGURE 1.487: Carcinoma maxilla invading orbit. A young man presented with mild proptosis of Rt eye with diplopia. Fullness of Rt maxilla is also seen on clinical examination.

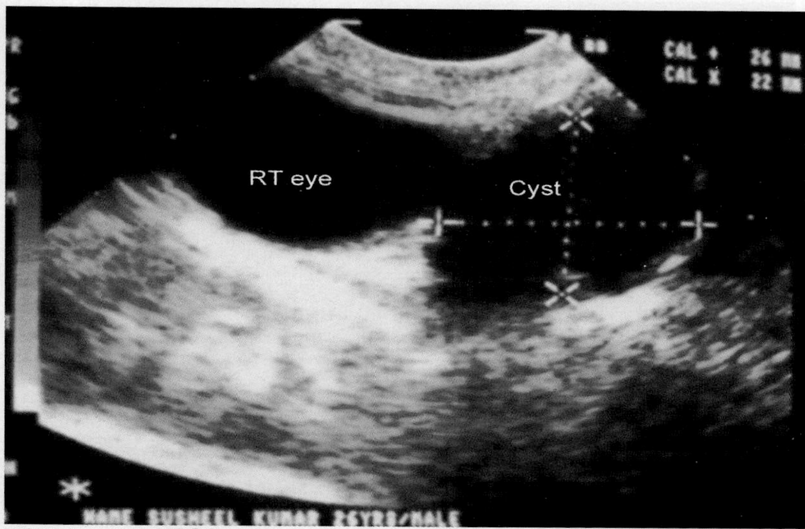

FIGURE 1.488: HRSG shows an irregular cystic mass in retrobulbar space coming out from the floor of orbit. The mass appears to be anechoic and pushing the eyeball anteriorly. However, eye appears to be spared. Biopsy proved maxillary carcinoma cystic variant.

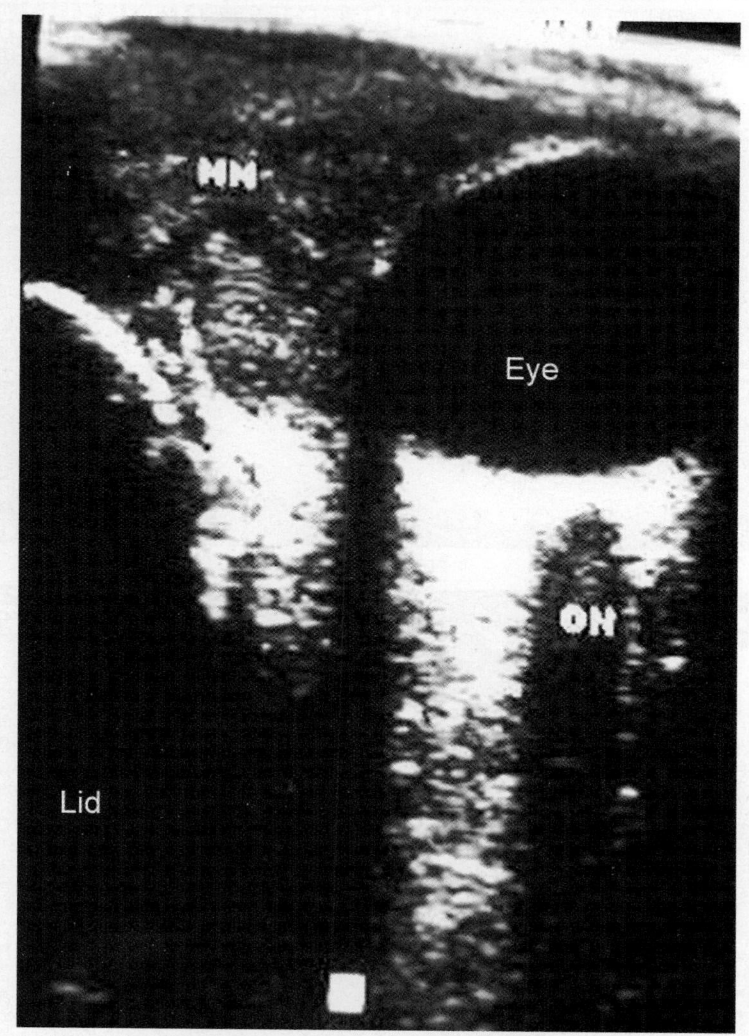

FIGURE 1.489: Squamous cell carcinoma of lid. HRSG shows an irregular mass over the Rt upper lid. The mass is seen invading the orbit and above the sclera into broken wall.

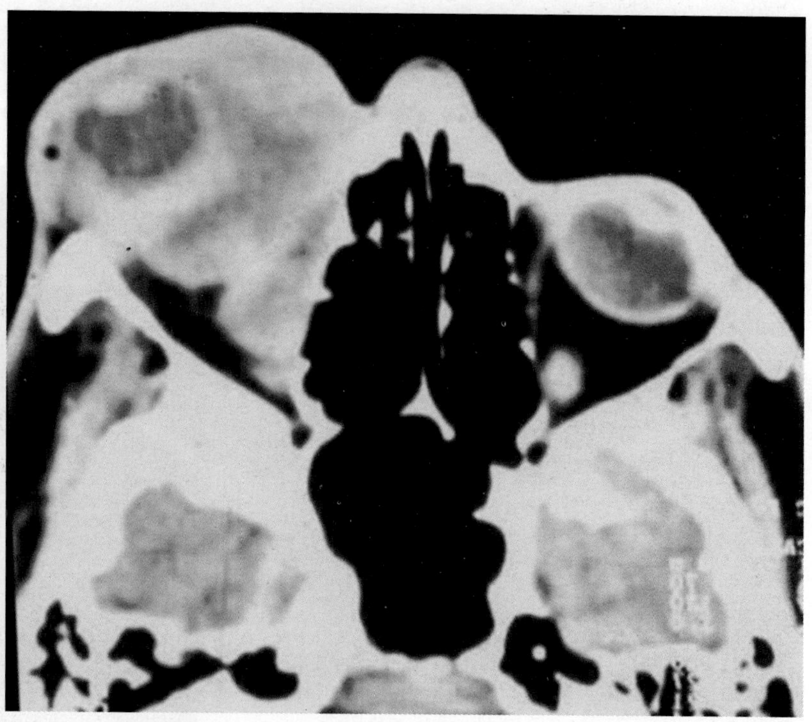

FIGURE 1.490: CT scan shows irregular heterogeneous mass over the Rt lid-invading orbit medially and also the eye coat and confirms the HRSG findings.

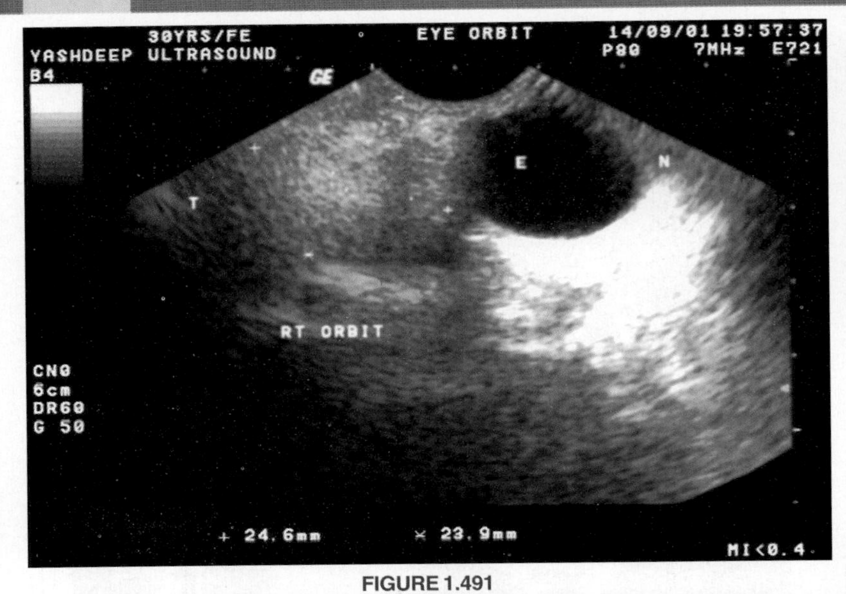

FIGURE 1.491

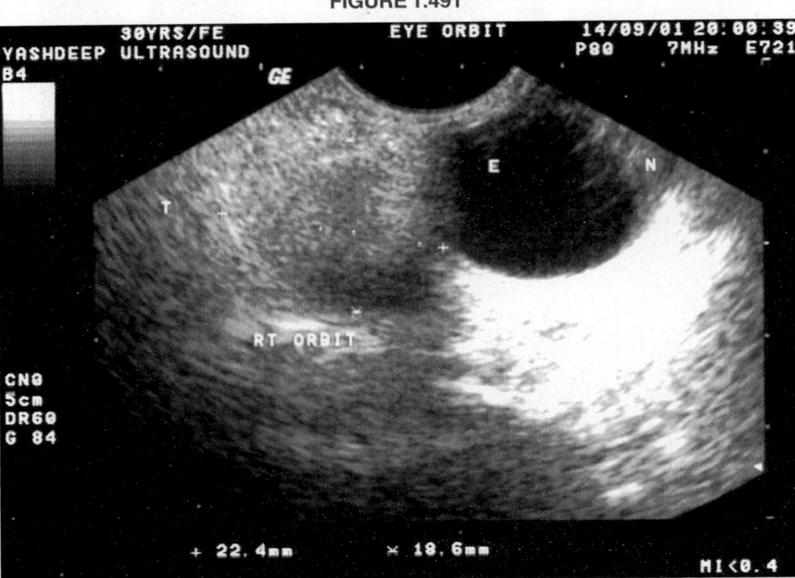

FIGURE 1.492

FIGURES 1.491 and 1.492: Dermoid cyst periorbital space. A patient presented with a soft tissue mass swelling over upper outer quadrant of orbit. HRSG shows a homogeneous well-defined soft tissue mass. Multiple fine homogeneous echoes are seen in the mass pushing the eye to medial side suggestive of a dermoid cyst.

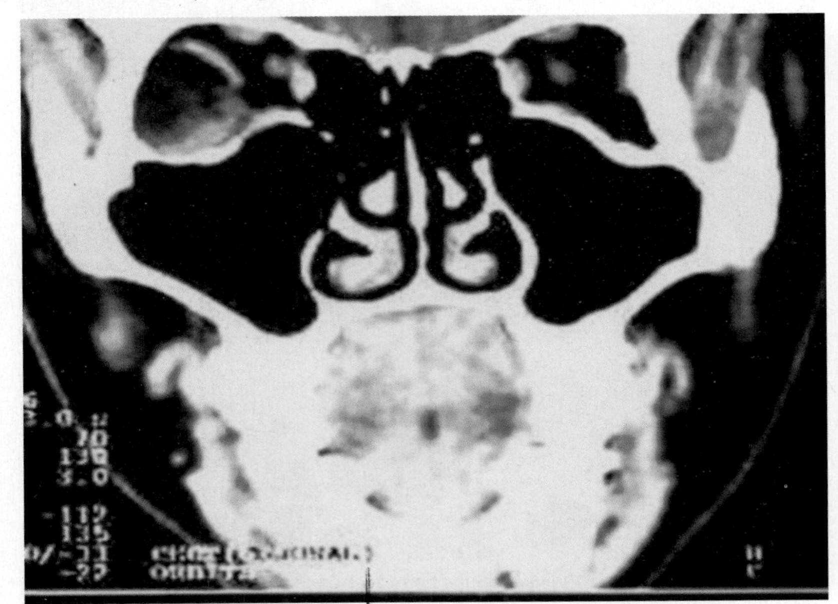

FIGURE 1.493: CT scan of the same patient shows a well defined soft tissue mass predominantly having fat contents confirms HRSG findings.

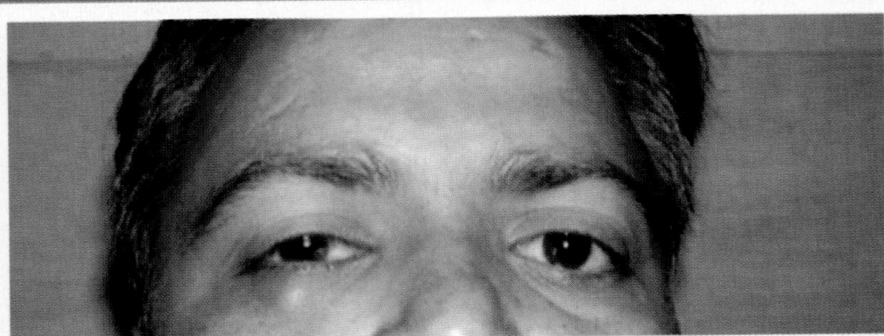

FIGURE 1.494: Infraorbital cysticercosis. A patient presented with soft tissue nodular shadow below the Rt lower orbital margin.

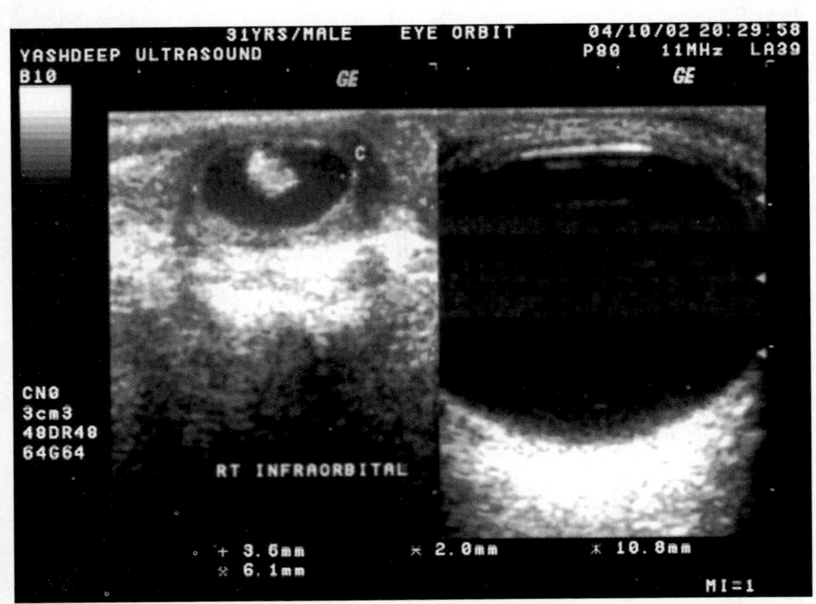

FIGURE 1.495: HRSG shows a well defined cystic mass. An echogenic nidus is seen fixed with inner wall of the cyst suggestive of cysticercosis.

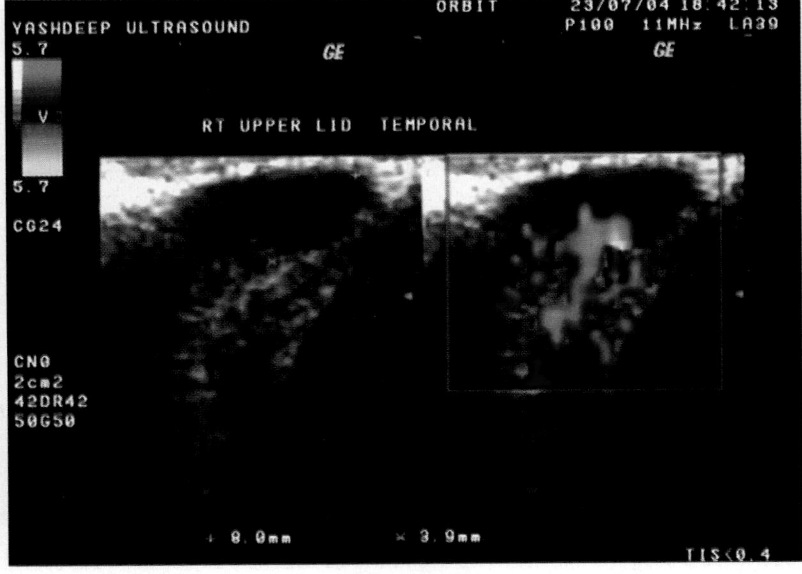

FIGURE 1.496

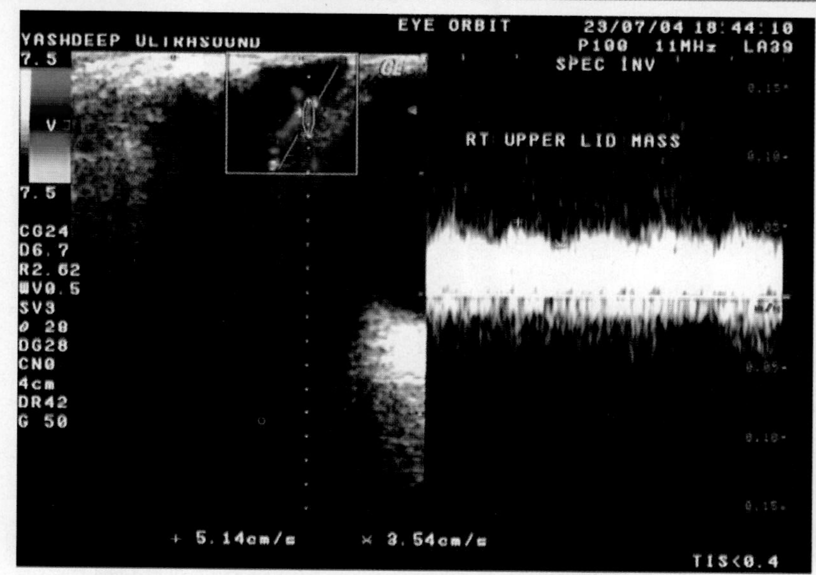

FIGURE 1.497

FIGURES 1.496 and 1.497: Vascular mass Rt upper lid temporal. A patient presented with soft tissue mass over the Rt upper lid on temporal side. HRSG shows a low level echo complex mass. No calcification is seen in it. On color flow imaging multiple vessels are seen in the mass. On spectral Doppler tracing low flow is seen in the mass.

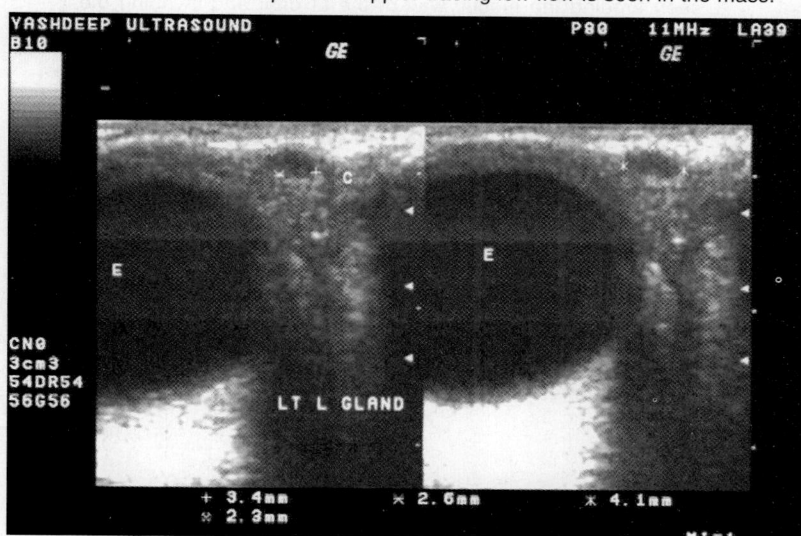

FIGURE 1.498

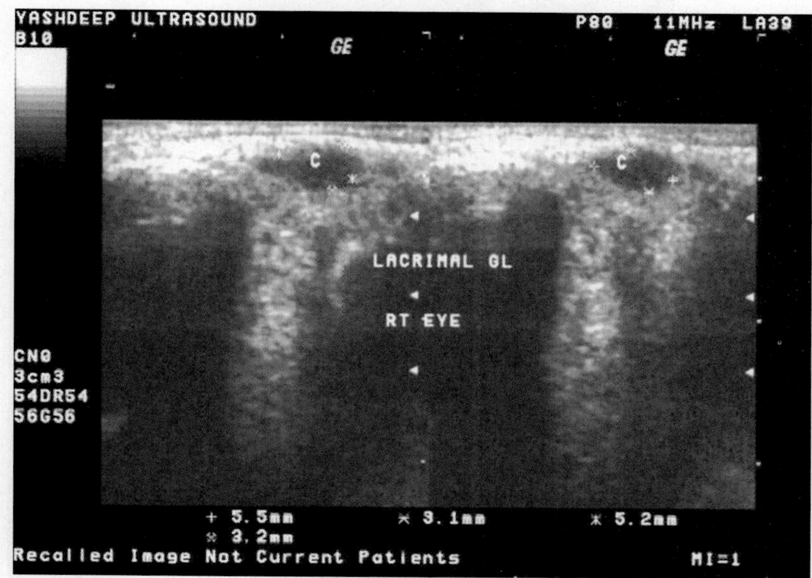

FIGURE 1.499

FIGURES 1.498 and 1.499: Lacrimal gland cysts. HRSG shows well-defined thin walled cysts in both lacrimal glands on either side. No internal debris is seen in the cyst. No septation is seen. Eyeballs are normal.

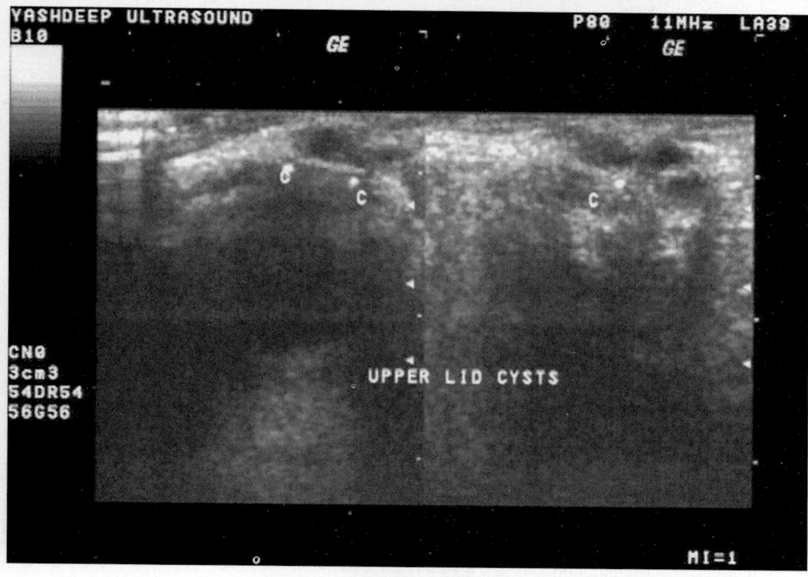

FIGURE 1.500

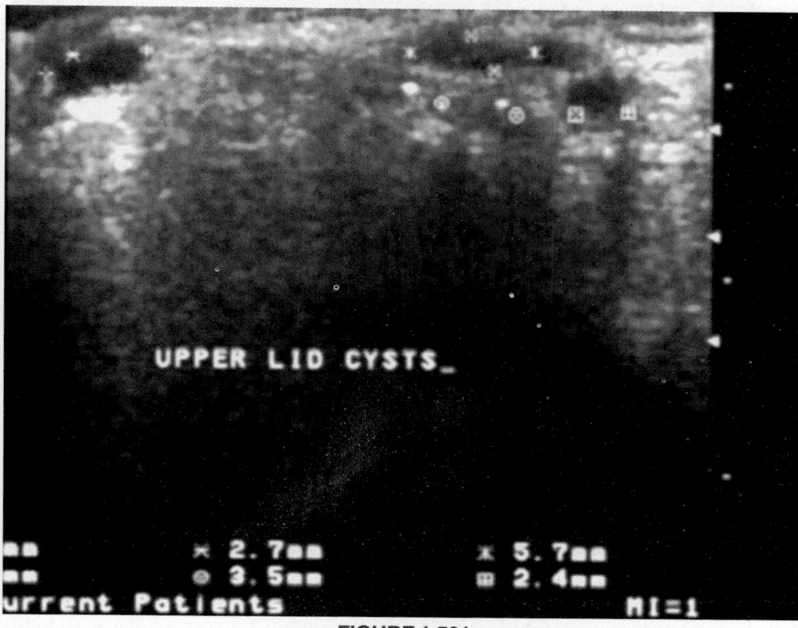

FIGURE 1.501

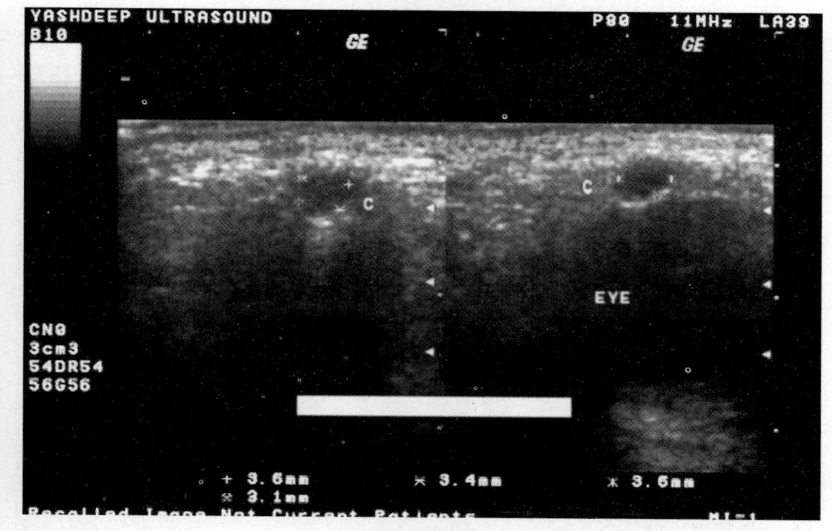

FIGURE 1.502

FIGURES 1.500 to 1.502: Lacrimal ductal cysts (Dacryops). HRSG shows multiple small cysts in the upper Lt lid involving palpebral part of lacrimal duct. Tiny echogenic calcified specks are also seen in them suggestive of ductal ectasia of lacrimal duct also known as dacryops. It is a rare condition of lacrimal duct caused due to obstruction and ectasia of lacrimal gland duct.

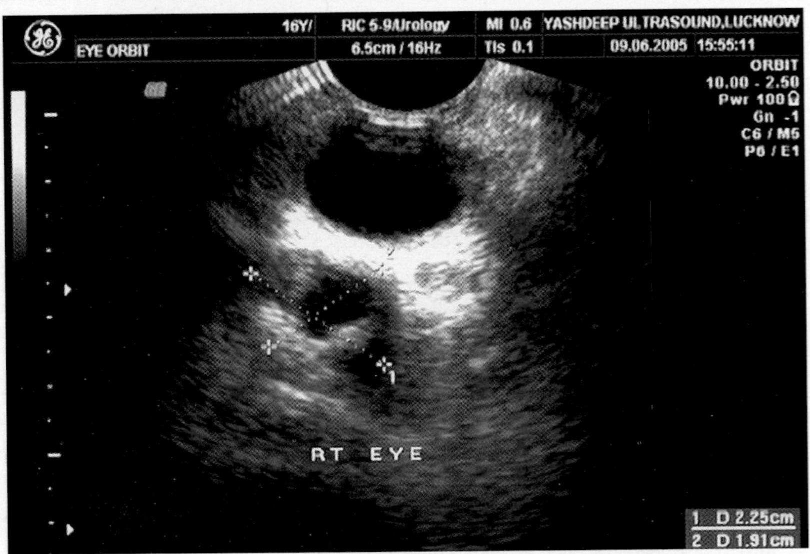

FIGURE 1.503

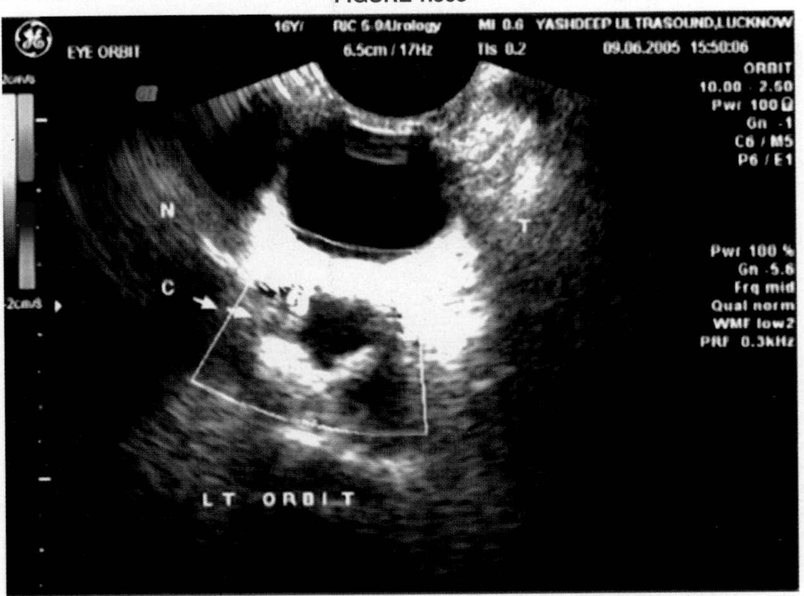

FIGURE 1.504

FIGURES 1.503 and 1.504: Nasal mass protruding in orbit. A young boy presented with lateral displacement of the eyeball with proptosis. HRSG shows a complex cystic mass coming out from the medial side of the orbit wall. Echogenic thick septa is seen in it. Multiple internal echoes are seen in it. The eye appears to be normal.

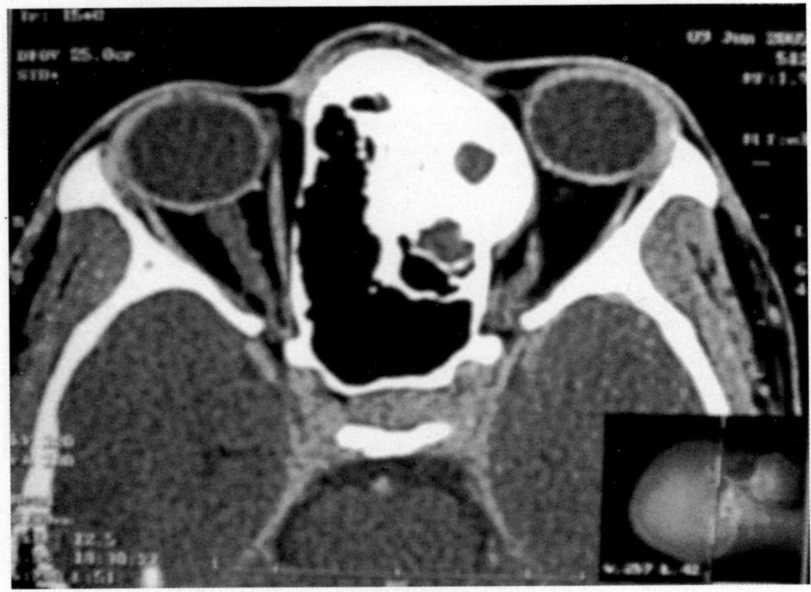

FIGURE 1.505

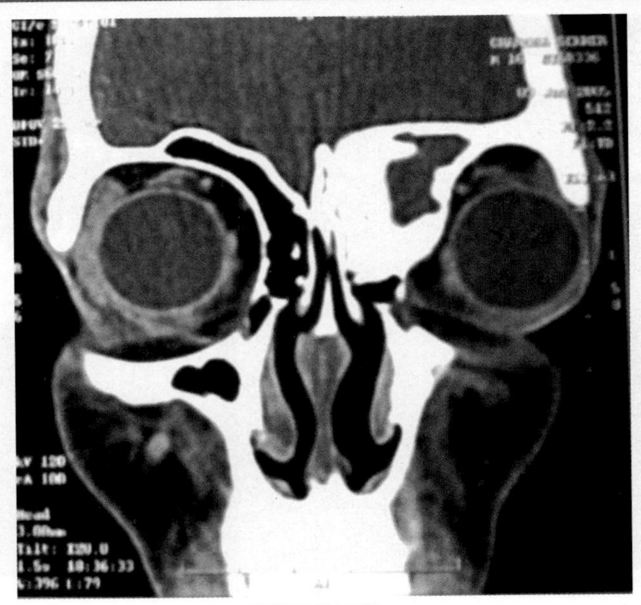

FIGURE 1.506

FIGURES 1.505 and 1.506: CT scan of the same patient shows highly enhancing mass seen coming out from the nasal cavity. It is seen eroding the medial wall of the Lt orbit and invading the retrobulbar space.

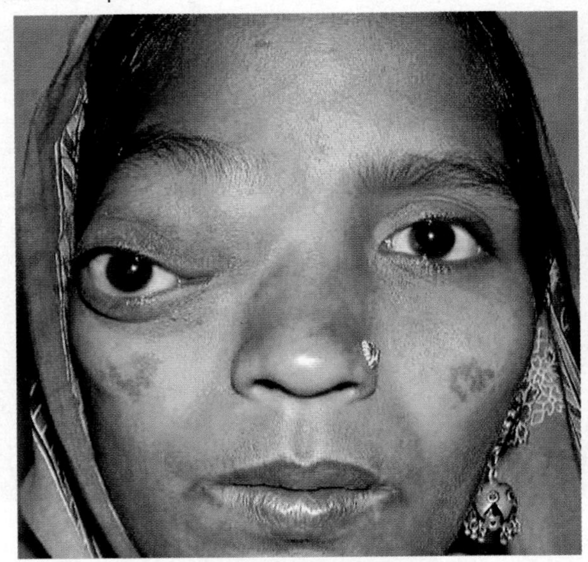

FIGURE 1.507: Mucocele invading the Rt orbit. A lady patient presented with swelling over the medial side of the eye with lateral displacement.

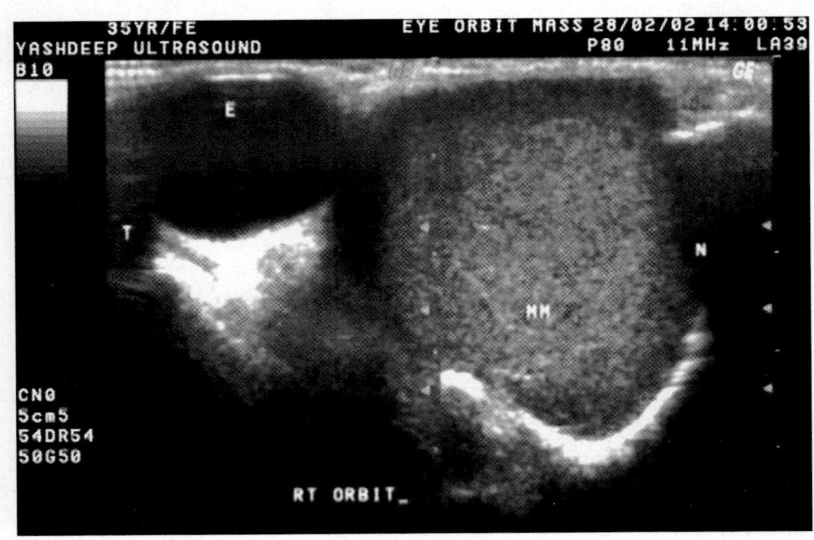

FIGURE 1.508

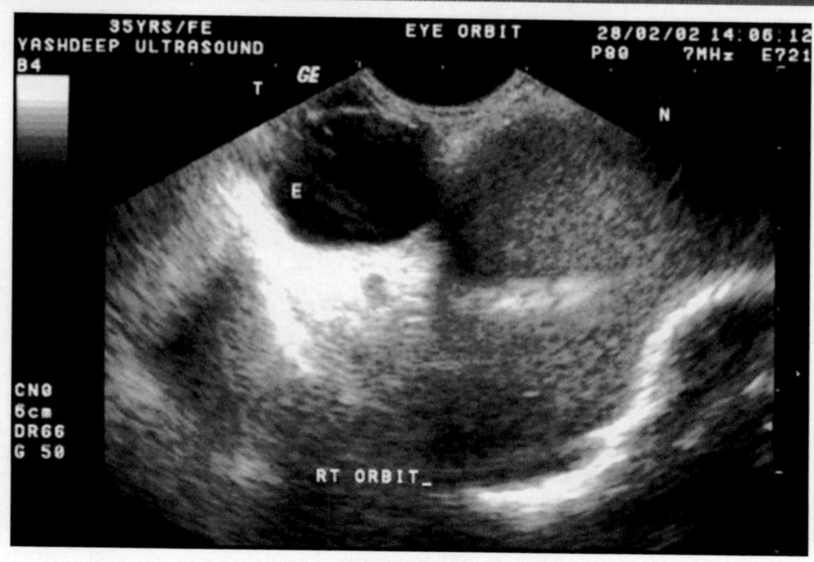

FIGURE 1.509

FIGURES 1.508 and 1.509: HRSG shows a thick echo complex mass coming out from the nasal cavity and pressing over the Rt orbit displacing the eyeball down. Thick echoes are seen in it suggestive of mucocele

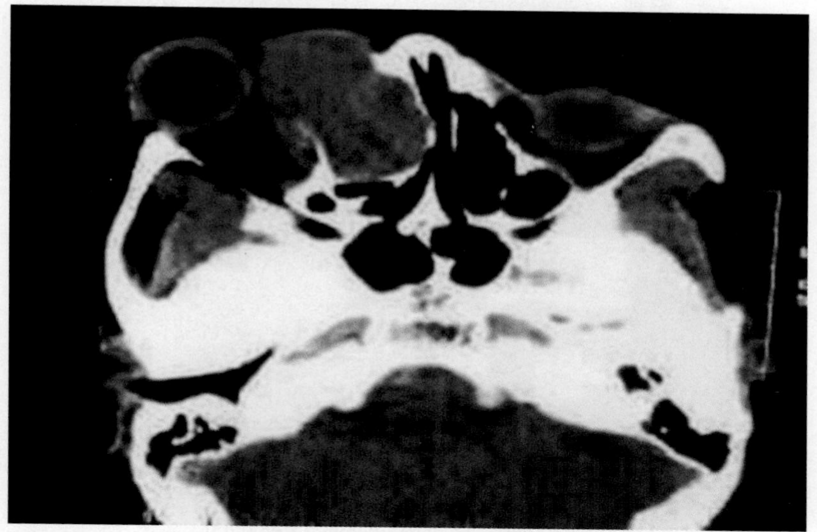

FIGURE 1.510: CT of the same patient shows a hypoechoic dense mass seen coming out from the Rt nasal cavity eroding the medial wall of the orbit and invading the retrobulbar space..

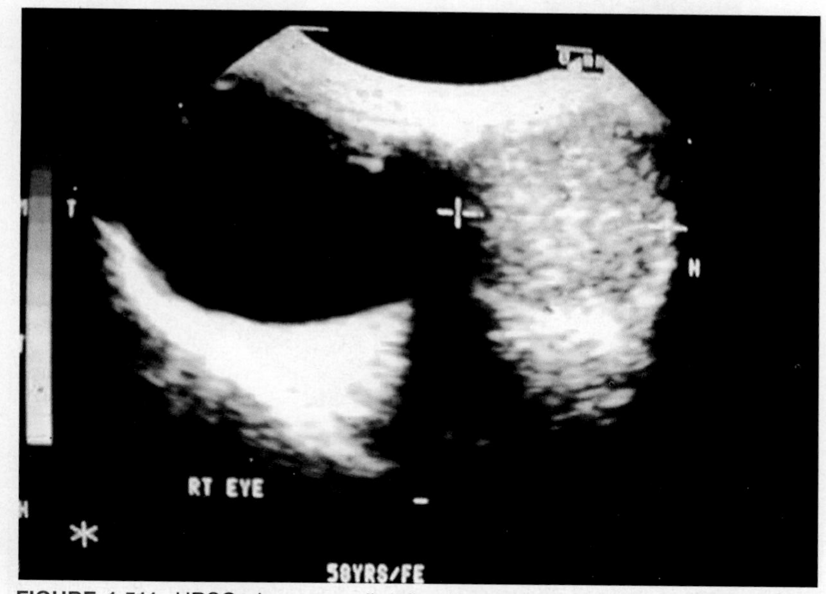

FIGURE 1.511: HRSG shows a well-defined echogenic mass in Rt upper medial quadrant. Fine echogenic interfaces are seen in the mass. Thick capsule is also seen. Biopsy shows dermoid cyst.

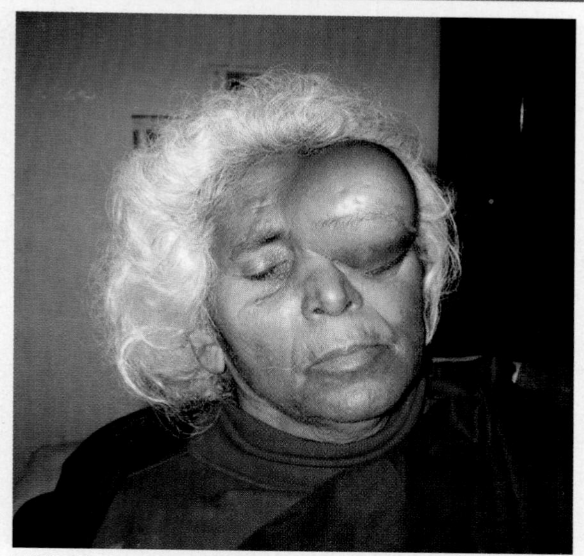

FIGURE 1.512: Metastasis from small cell carcinoma. A lady presented with a big soft tissue mass over the Lt forehead covering over the Lt orbit.

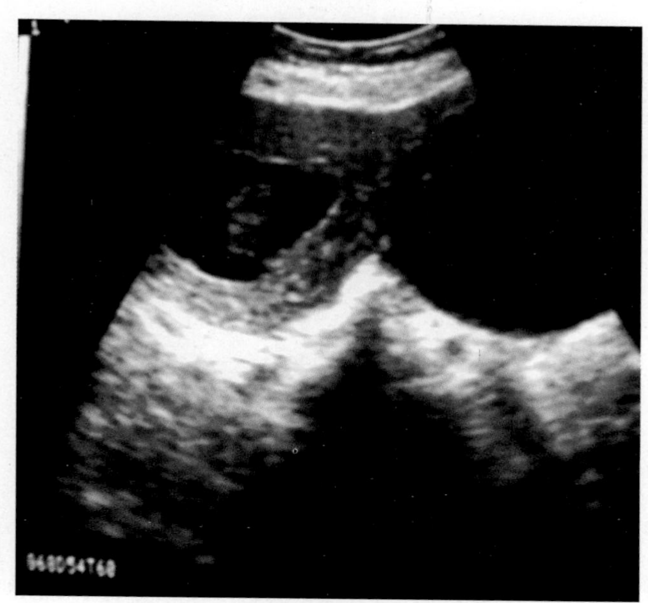

FIGURE 1.513

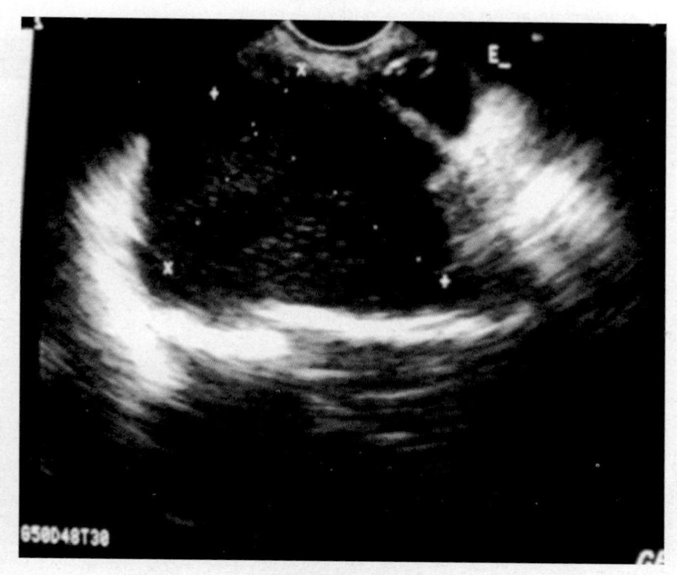

FIGURE 1.514

FIGURES 1.513 and 1.514: HRSG shows a big thick walled cystic mass over the Lt orbit. It is seen pressing and displacing the eyeball. However, the eyeball is separate from the mass and intact.

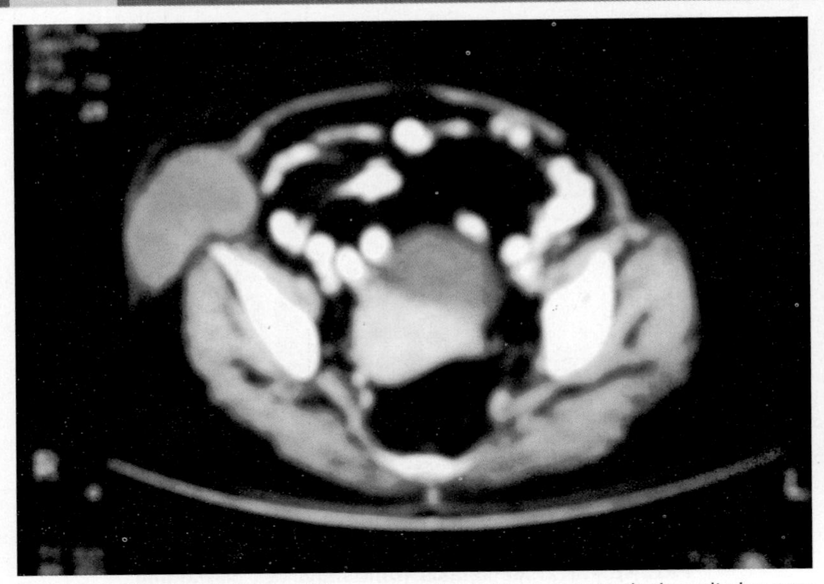

FIGURE 1.515: CT abdomen of the same patient shows metastatic deposit also over the Rt side of the anterior abdominal wall.

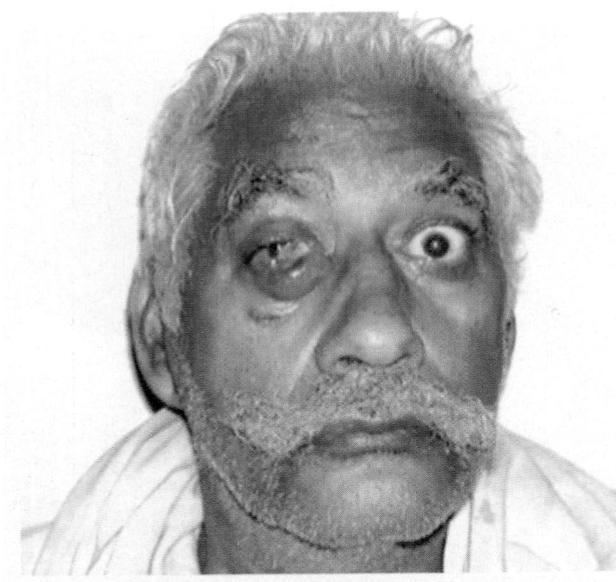

FIGURE 1.516: Lower lid carcinoma. An old man presented with soft tissue mass coming out from the Rt lower lid. The mass is seen invading the lower part of eyeball.

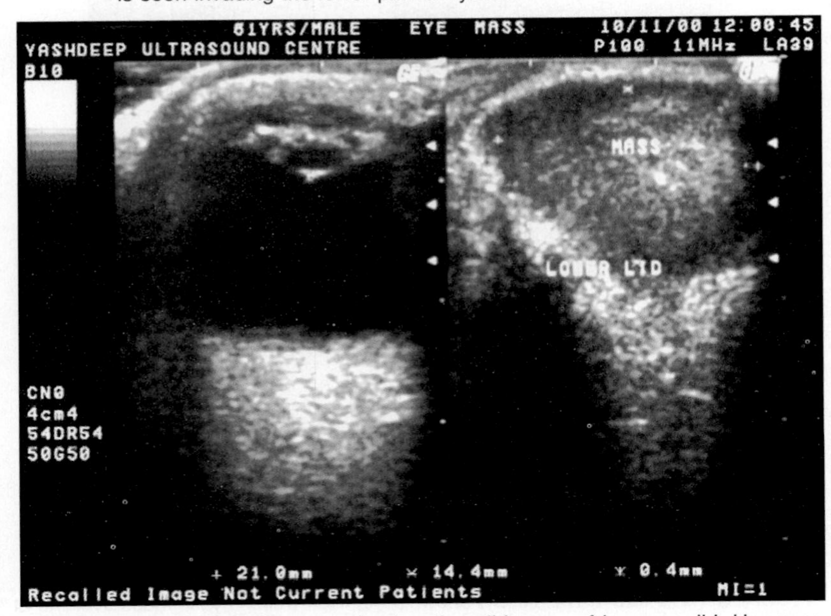

FIGURE 1.517: HRSG shows predominantly solid mass of lower eyelid. However, eyeball is seen spared at the time of study.

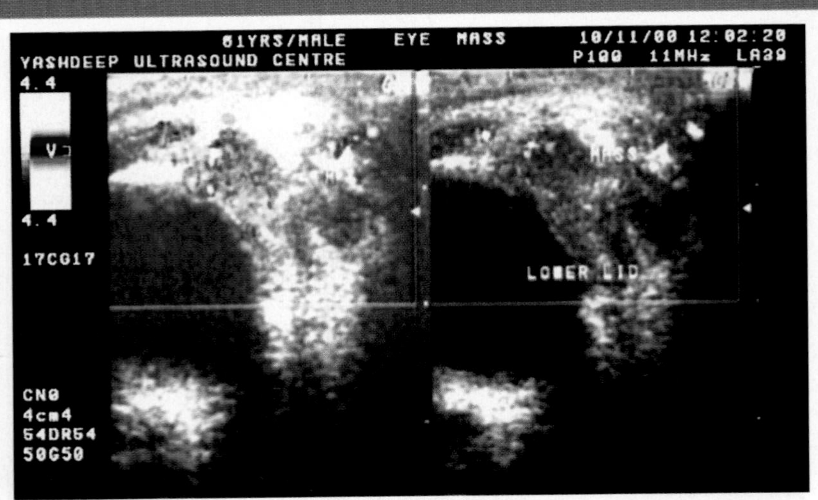

FIGURE 1.518: Color flow imaging of the mass shows poor flow. Biopsy of the mass came out to be carcinoma lid.

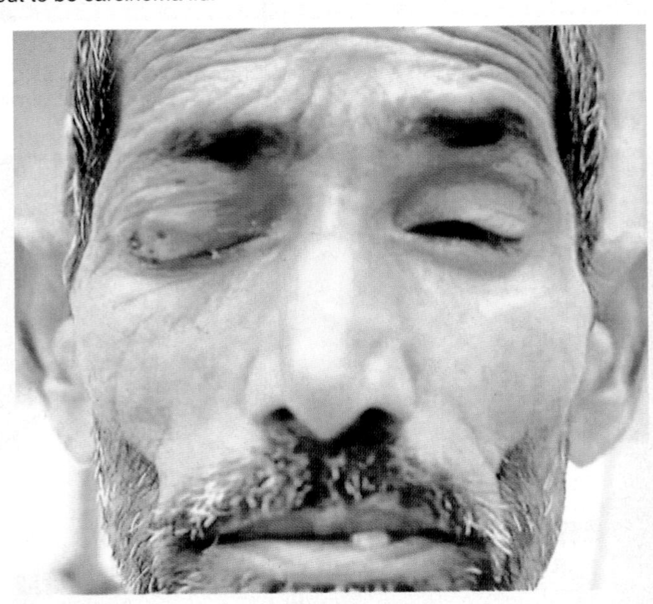

FIGURE 1.519: An old man presented with a stony hard soft tissue mass over Rt eyelid. Multiple nodules are also seen in the mass. Eyelid could not be opened and clinical examination was not possible.

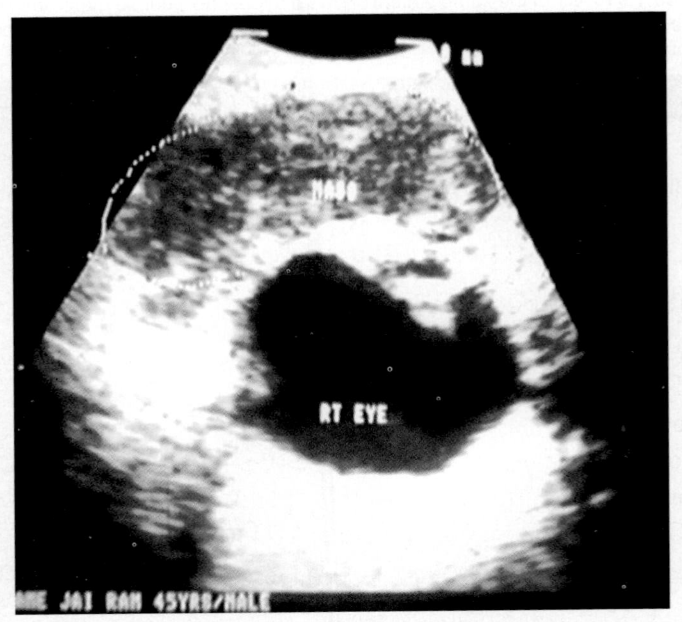

FIGURE 1.520: HRSG shows a heterogeneous irregular soft tissue mass of upper lid. The mass is seen invading the sclera and also eye coat. Lens is also thickened and opaque. Biopsy proved squamous cell carcinoma of upper lid with involvement of sclera.

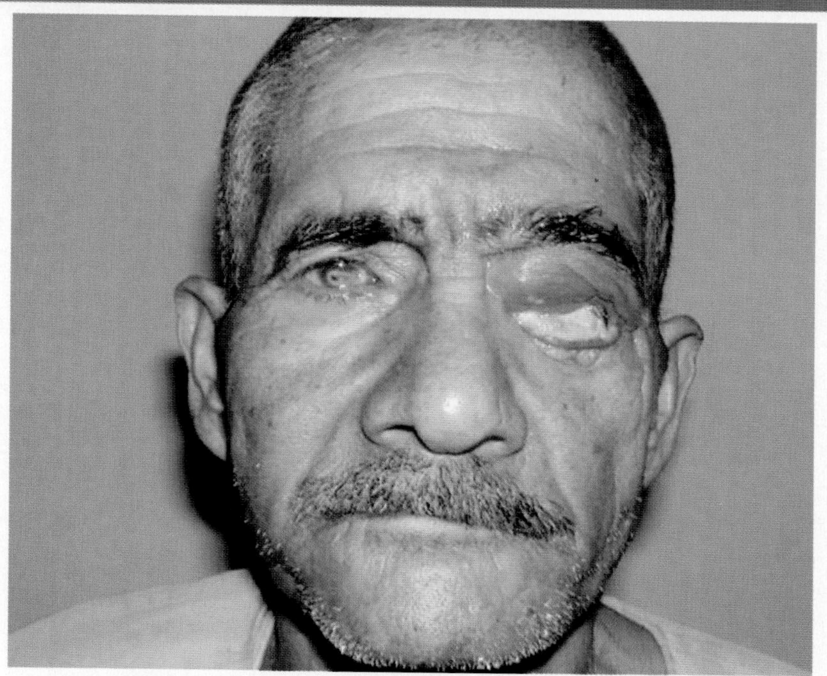

FIGURE 1.521: Carcinoma lower lid invading in the orbit. A patient presented with squamous cell carcinoma of lower lid of the Lt orbit. The lid was excised.

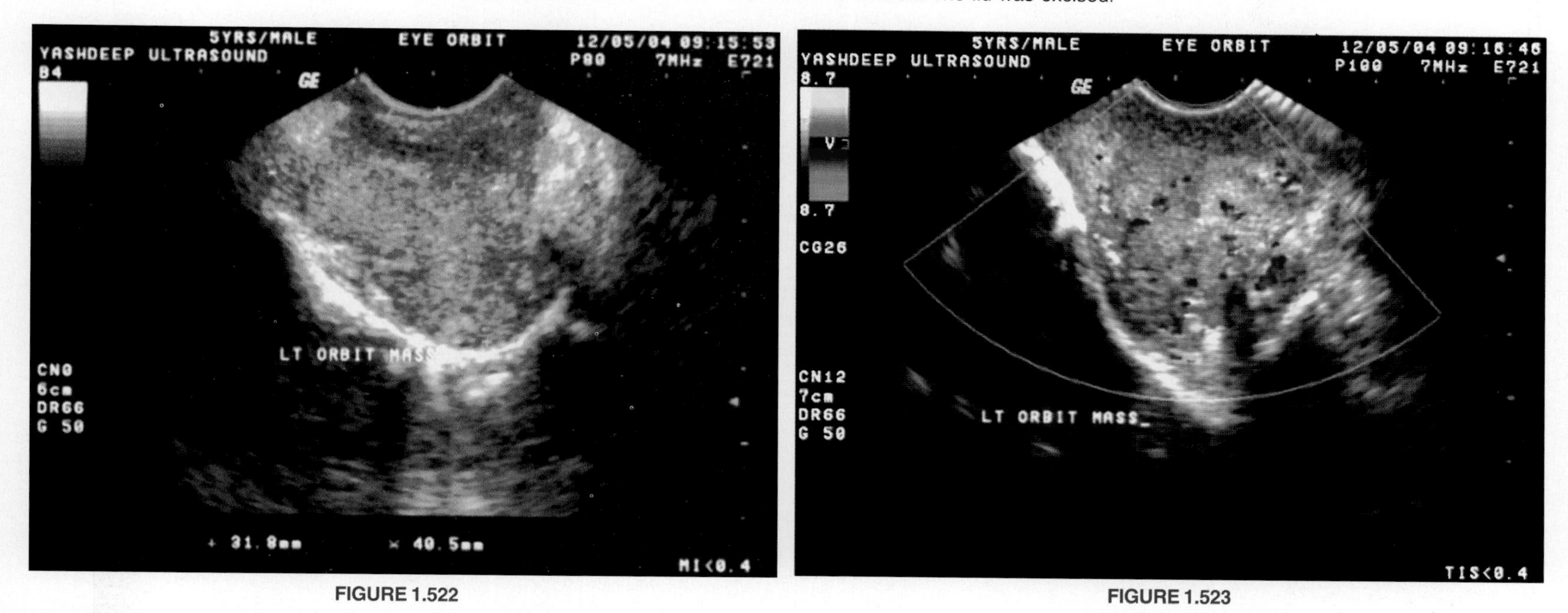

FIGURE 1.522

FIGURE 1.523

FIGURES 1.522 and 1.523: HRSG shows the mass invading the orbit. It is seen going posteriorly in the retrobulbar space and completely destroyed the normal orbital anatomy.

Conclusion

High-resolution sonography (HRSG) is noninvasive multi-planer, widely available imaging modality for evaluation of eye and orbit. The tissue characterization of mass is better than CT. It is the only practical method of examining the eye, when light conducting media are opaque. For orbit, CT is better technique as it provides the global view. But HRSG is better to differentiate between solid and cystic masses.

The limitation of HRSG is limited field of view and extra orbital extension of the disease in the cranium, where CT is definitely superior than USG. The biggest advantage of ultrasonography over CT is its wide availability, rapidity of investigation and cost-effectiveness. Therefore, it is first line of investigation in orbital proptosis.

2 Thyroid

Introduction

High-resolution sonography (HRSG) has become an established imaging modality to evaluate thyroid gland and its diseases. Since the gland is located superficially in anterior part of neck, it is an ideal organ to be imaged by HRSG. High-frequency transducers of 7.5 and 10 mhz are the ideal probes for thyroid imaging. However, in big thyroid mass, 5 mhz transducer is also required. Linear probes are preferred due to their wide contact surface.

Method

The patient is examined in supine position with extended neck with direct contact scanning technique. A thorough examination includes complete visualization of the thyroid gland, neck vessels, other adjoining structures and lymph node assessment, if they are enlarged. Subtle anatomic changes in thyroid gland and adjoining structures can be easily picked up on HRSG.

Normal Anatomy and Sono Appearance

The thyroid gland is situated in anterior part of the neck. The lateral lobes are situated on either side of trachea. They are connected with a band called isthmus. The lateral lobes are bounded laterally by great vessels, common carotid artery laterally and anteriorly. The recurrent laryngeal nerve lies in between carotid artery and jugular vein in a groove. Isthmus of gland lies anterior to trachea crossing it to both sides joining both lateral lobes. The normal gland measures more than 6 to 7 cm in length with thickness of 1 to 2 cm. The thickness of the gland is more reliable parameter than length and if it is more than 2 cm in thickness, it is suggestive of thyromegaly.

Sono Appearance

Normal ultrasonographic appearance of thyroid gland is characteristic. The gland shows homogeneous texture with medium level echoes. The echogenicity is more than the surrounding strap muscles. The gland is entrapped in thin highly reflective capsule. Vascular structures are seen piercing the glandular tissue mainly in upper parts. The strap muscles (sternohyoid and sternothyroid) are seen as sonolucent bands along the anterior surface of the thyroid gland. The air-filled in the midline gives a characteristic curvilinear reflecting with associated reverberation artifacts. At times echo free areas of 2 to 3 mm in size are seen in the gland substance suggestive of colloid collection. In elderly people, nodular calcifications or linear bands of fibrous tissue may be seen.

Indications for Thyroid Imaging

HRSG can reveal even minor changes in the thyroid gland anatomy due to pathological process. A routine examination of gland includes complete evaluation of thyroid gland, surrounding vasculature and lymph nodes evaluation if they are enlarged. When thyroid is partly present in retrosternal area, additional scans should be taken from suprasternal notch to see the retrosternal extension of the gland. Any ultrasonographic assessment of thyroid pathology should first determine the nature of glandular involvement, whether it is diffuse parenchymal disease, multinodular disease or focal or solitary thyroid nodule. However, any combination of these appearances can occur. The thyroid nodule should be further evaluated on the basis of its reflectivity, presence of cyst, calcification, degeneration, margins and presence of echo poor halo.

Diffuse Thyroid Disease

Diffuse thyroid disease includes thyroid hyperplasia which is most common pathology of thyroid gland. It may results due to iodine deficiency, may be familial or due to unilateral lobectomy. It may involve one of both lobes. When most of the gland is enlarged, it results into goiter. Diffuse hyperplasia may result in enlargement of one or both lobes. The gland size enlarges. The great vessels deviate laterally due to enlargement of gland. Pressure will be noted on trachea. However, tracheal walls are never infiltrated. The gland echo texture becomes inhomogeneous, but reflectivity remains the same. Nodular hyperplasia of thyroid gland shows multiple nodules in the gland. They vary in their size and numbers. They are seen studded in the normal parenchyma of the gland. HRSG can very well identify the multinodular status of the gland, which cannot be assessed clinically. HRSG can pick up nodule of even as small as 5 mm. A typical nodule has got the same reflectivity as normal gland. The nodules are isoechogenic. Twenty five percent of these nodules show colloid or hemorrhagic degeneration. Calcification may also present as bright as echogenic foci in the nodules. They are accompanied with acoustic shadowing. Benign calcification is coarse and is seen scattered throughout the parenchyma. However, all visible nodules must be evaluated for possible malignancy, as 10 to 30% of the nodules may be malignant.

Basedow's Disease

Diffuse hyperplasia of thyroid with thyrotoxicosis is characteristic of Basedow's disease. The gland shows generalized enlargement with low reflectivity. However, nodules are absent and veins are enlarged. The vessels status can be confirmed on color flow imaging.

Thyroiditis

Thyroiditis may be acute, sub-acute or chronic. In early thyroiditis, the gland gets enlarged. The enlargement may be partial or diffuse. Gland becomes hypoechoic and coarse. Parenchyma shows medium level echoes. Acute inflammation may lead to abscess formation. Chronic thyroiditis in hypertrophic stage shows glandular enlargement with lobulated irregular margins. The echogenicity is reduced. Typical lobulated pattern is seen which is due to fibrotic bands. Atrophic thyroiditis shows small shrunken gland. It becomes difficult to identify the gland, as its reflectivity merges with the surrounding tissue.

Hashimoto's Thyroiditis

Hashimoto's thyroiditis or chronic lymphocytic thyroiditis is an autoimmune disorder. There is painless, diffuse enlargement of the thyroid gland. It may involve one lobe or both lobes. The gland shows typical sonographic features. It becomes enlarged. Echo texture becomes hypoechoic and coarse. However, it is homogeneous in texture. No

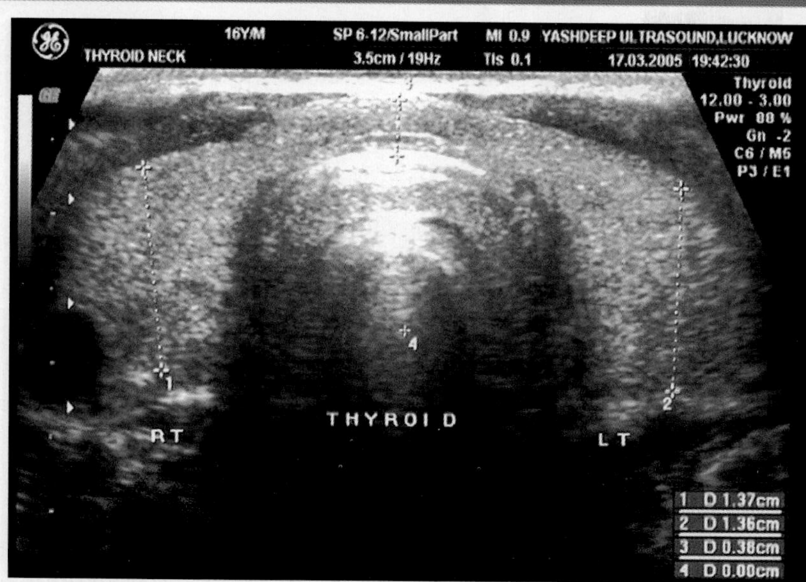

FIGURE 2.1: Normal thyroid. HRSG shows normal thyroid gland showing both the lobes, isthmus and the tracheal rings. The gland texture is homogeneous and shows medium level echoes. Highly reflective tracheal rings are seen as cresentric rings. Isthmus is seen a strap or band joint the lobes.

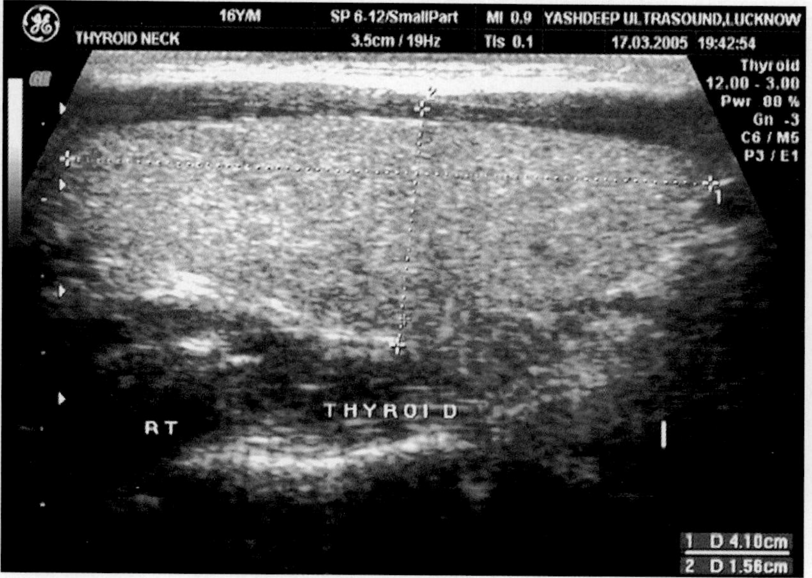

FIGURE 2.2A

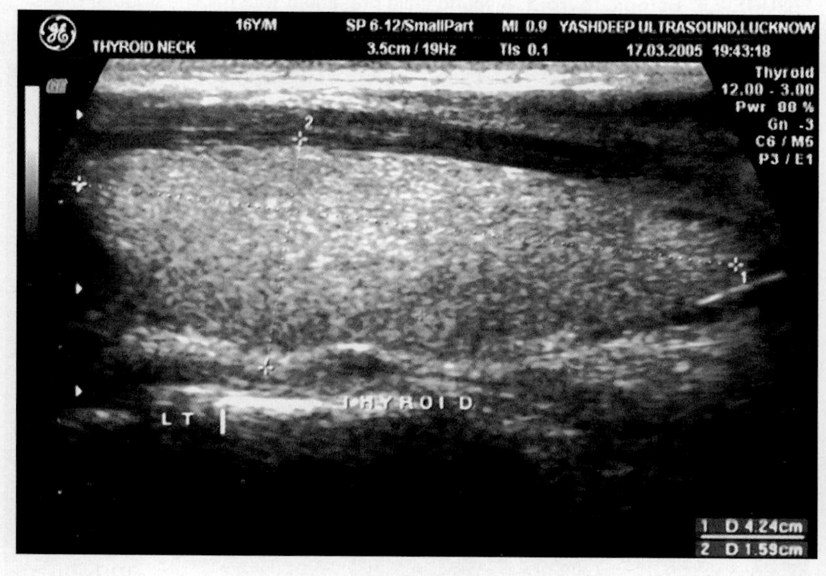

FIGURE 2.2B

FIGURES 2.2A and B: Normal thyroid gland. The longitudinal images of the thyroid lobes show normal homogeneous gland in long axis.

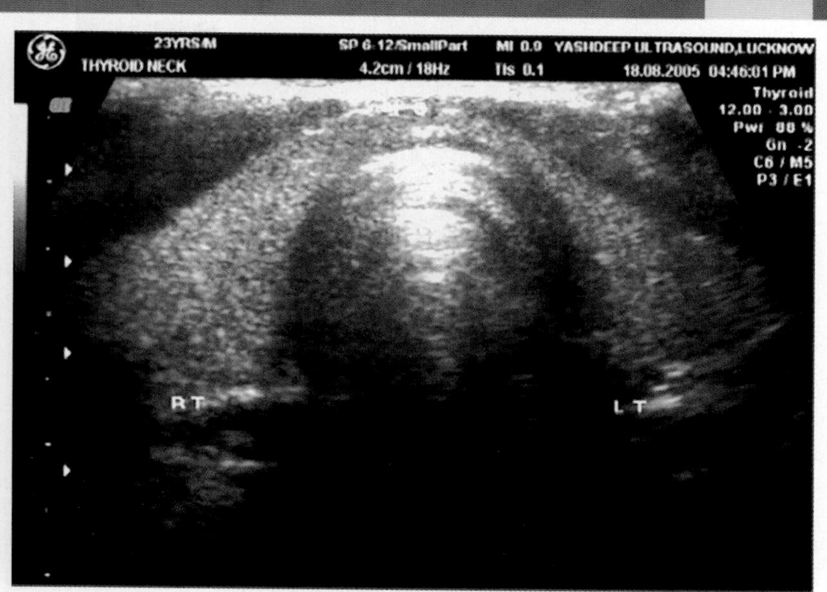

FIGURE 2.3A

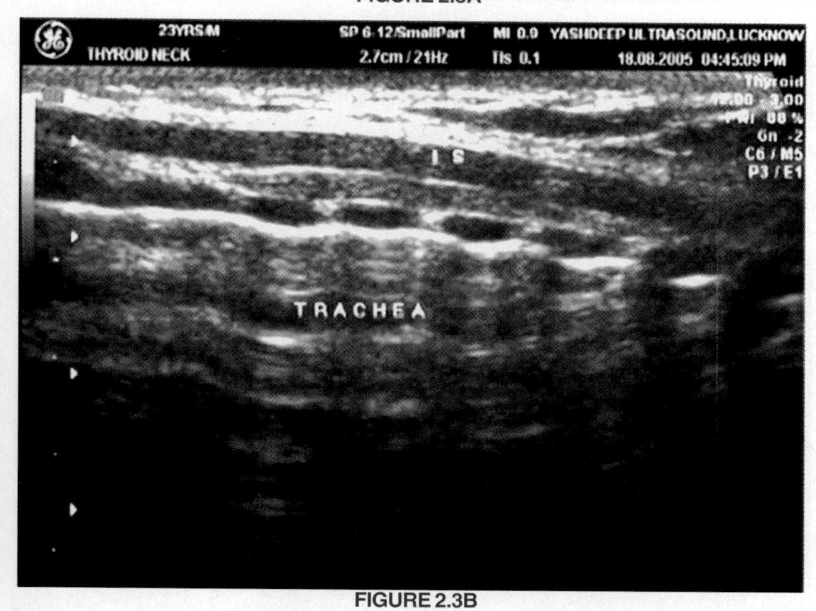

FIGURE 2.3B

FIGURES 2.3A and B: Normal isthmus. Isthmus is seen as a homogeneous band or strap joining the two lobes of thyroid. It shows normal echo texture similar to thyroid gland and crossing the trachea anteriorly. Normal thickness of the isthmus is 5 mm (upper limit).

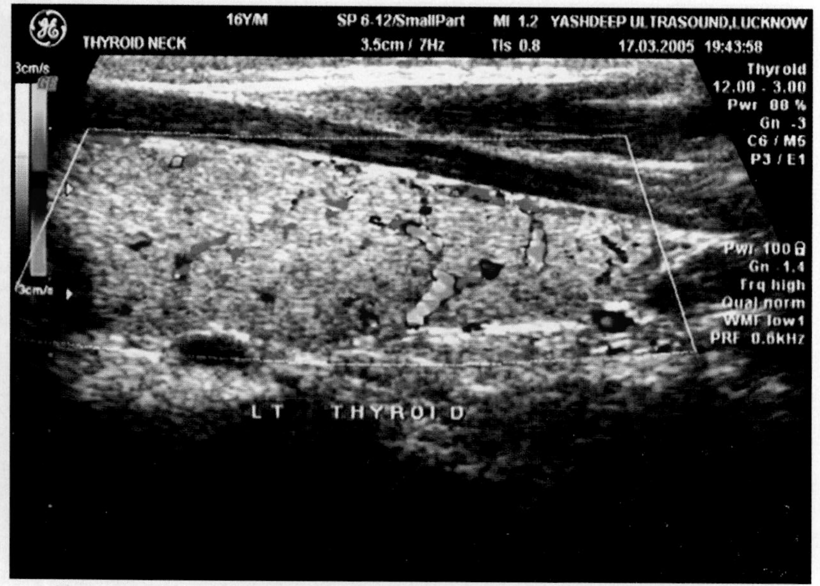

FIGURE 2.4: Color Doppler imaging shows high vascular nature of the gland. Multiple vessels are seen supplying the gland.

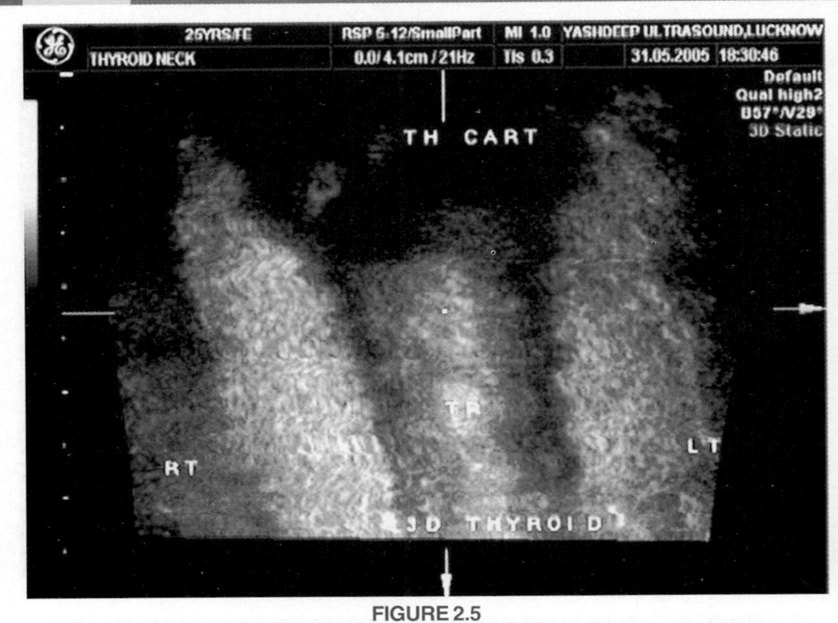

FIGURE 2.5

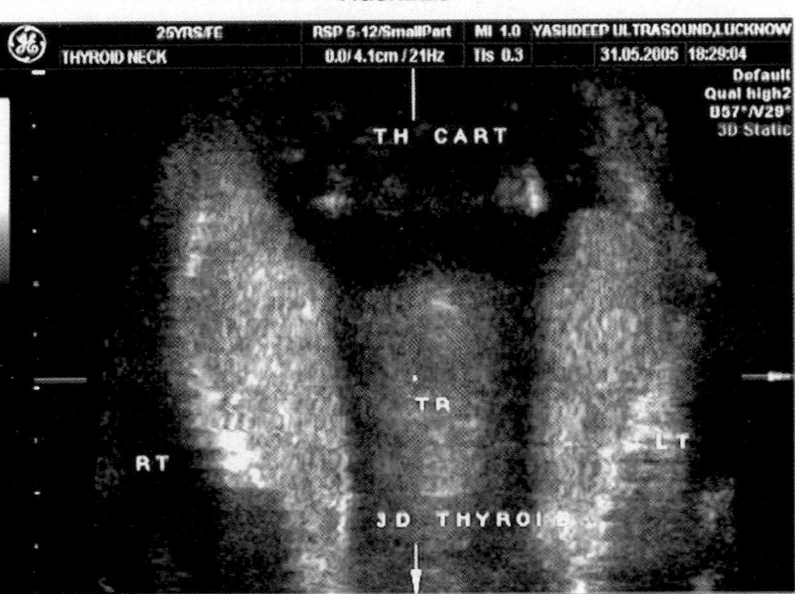

FIGURE 2.6

FIGURES 2.5 and 2.6: 3D imaging of the normal thyroid - 3D imaging of the thyroid gland in coronal plane shows the gland in relation to the trachea.

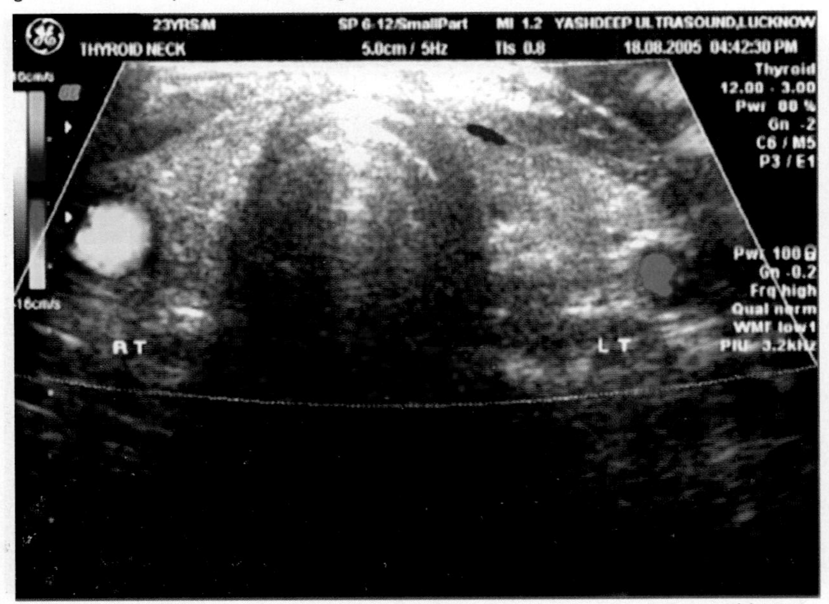

FIGURE 2.7: HRSG shows normal thyroid gland bounded by common carotid artery on either side. Homogeneous echoes are seen in the gland.

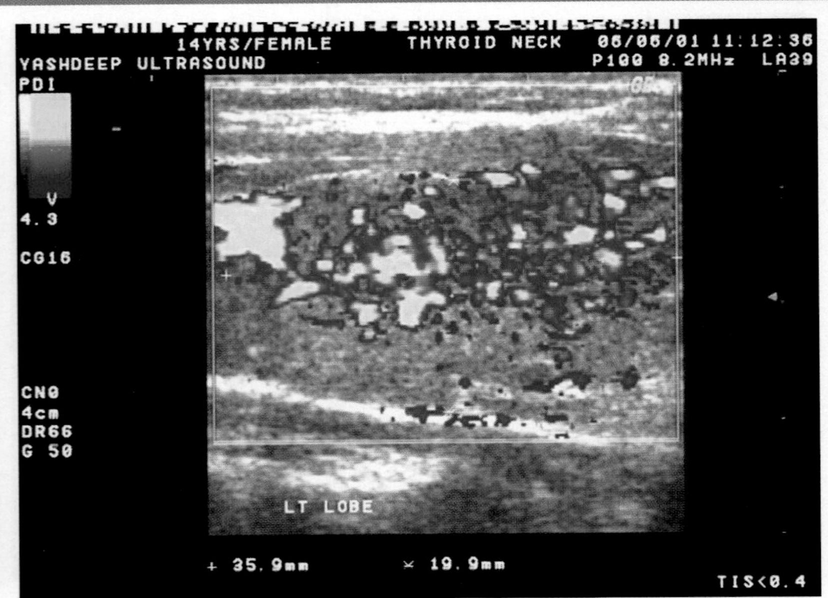

FIGURE 2.8: Power Doppler color flow imaging of normal gland. Power Doppler color flow imaging of thyroid gland shows high vascular nature of the gland. Multiple small vessels are seen traversing through the parenchyma.

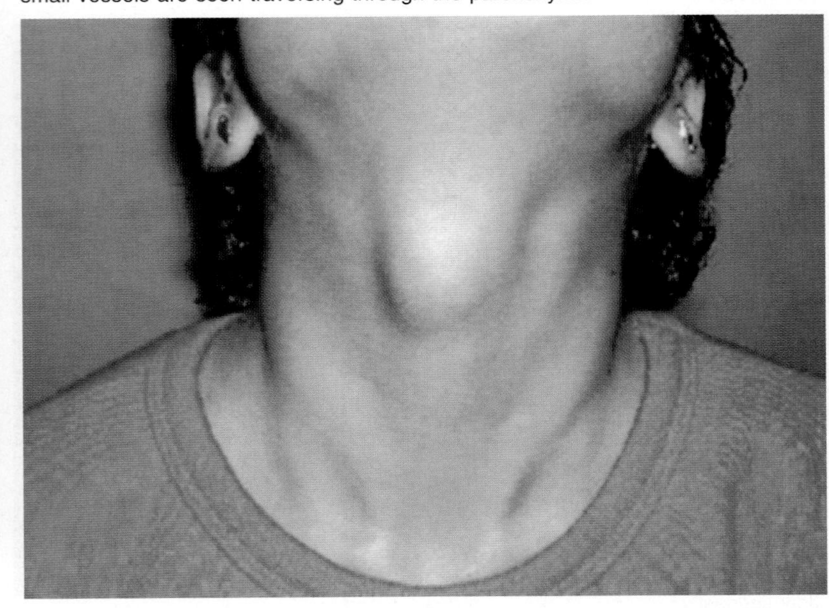

FIGURE 2.9: Ectopic thyroid. A young child presented with midline nodular mass.

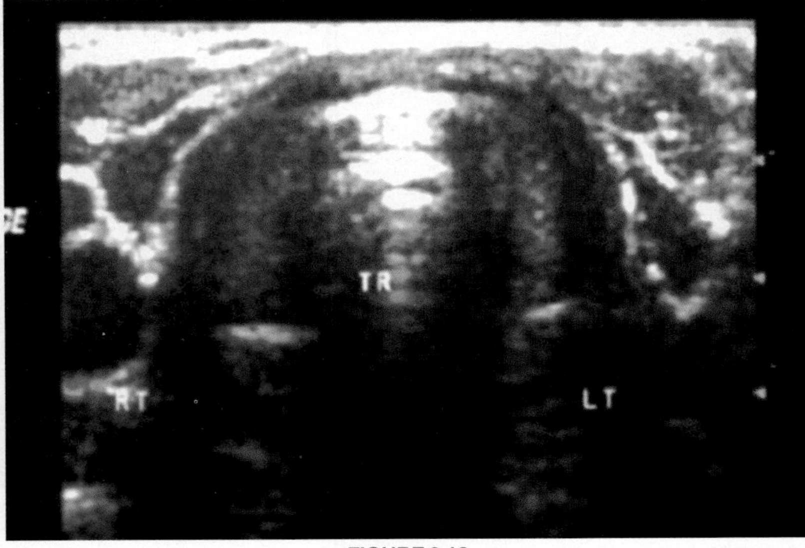

FIGURE 2.10

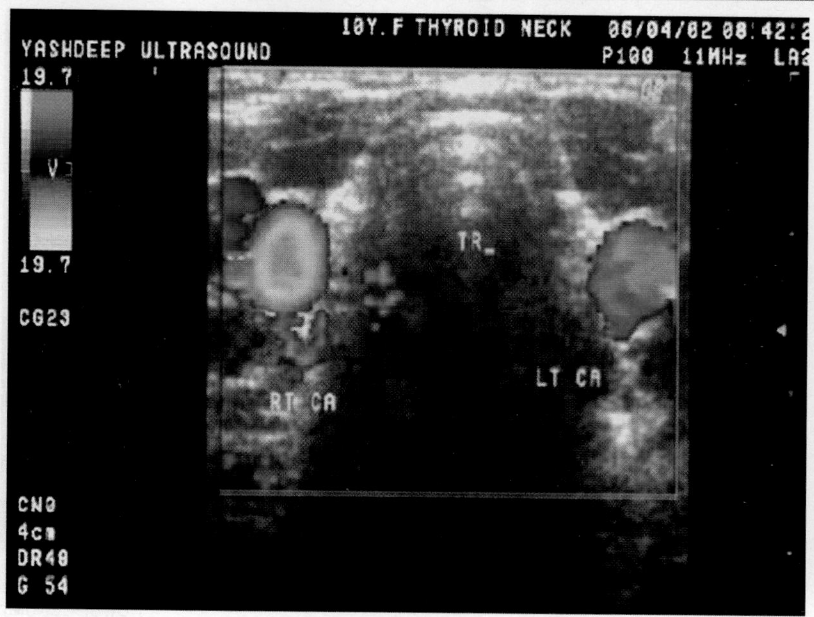

FIGURE 2.11

FIGURES 2.10 and 2.11: HRSG shows absence of thyroid gland on either side of the trachea in the thyroid fossa.

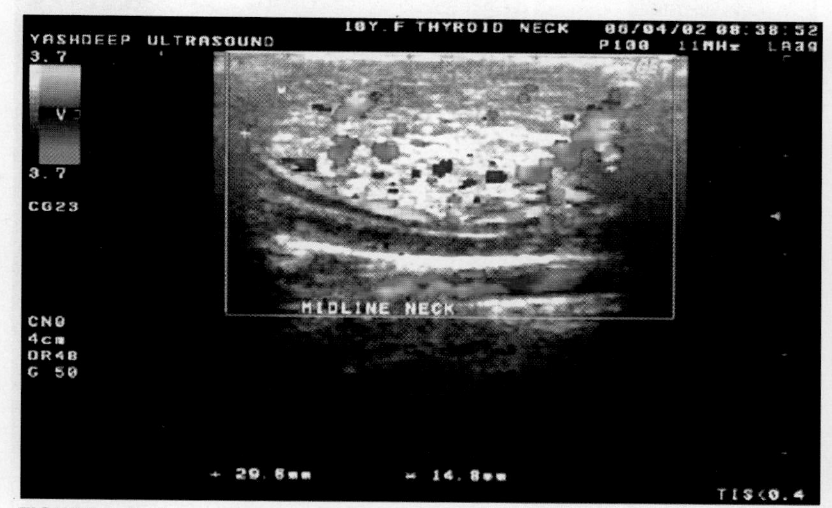

FIGURE 2.12: A homogeneous nodular shadow similar to the thyroid gland is seen in the midline suggestive of ectopic thyroid. On color flow imaging it shows increased flow suggestive of rich vascular supply of the ectopic thyroid.

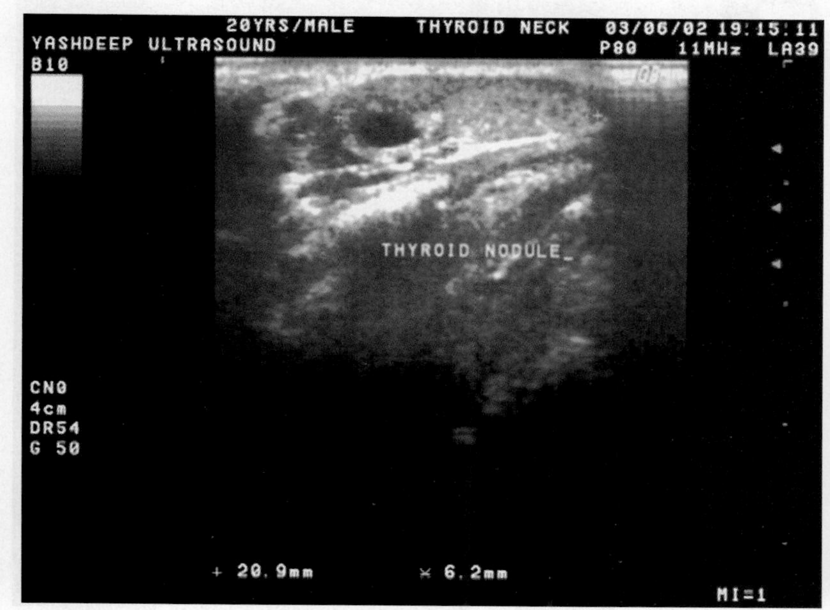

FIGURE 2.13

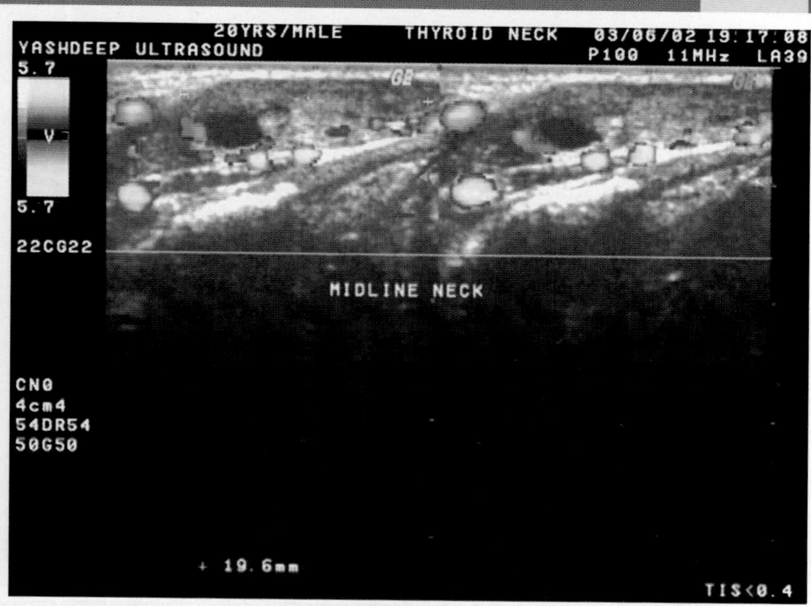

FIGURE 2.14

FIGURES 2.13 and 2.14: Ectopic thyroid. HRSG shows small normal thyroid glandular tissue in the Rt upper cervical region. On color flow imaging, it also shows rich flow in the gland.

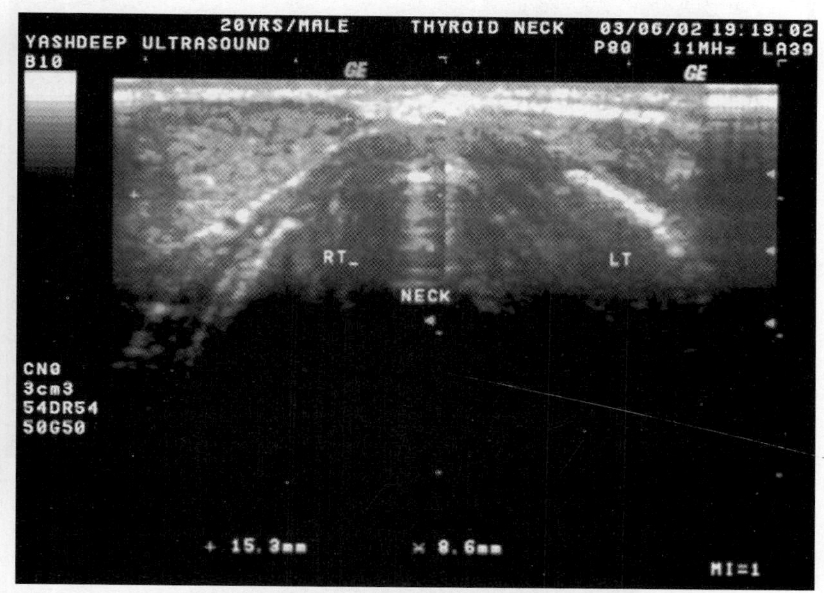

FIGURE 2.15: Normal thyroid gland is not seen in the thyroid fossa on either side.

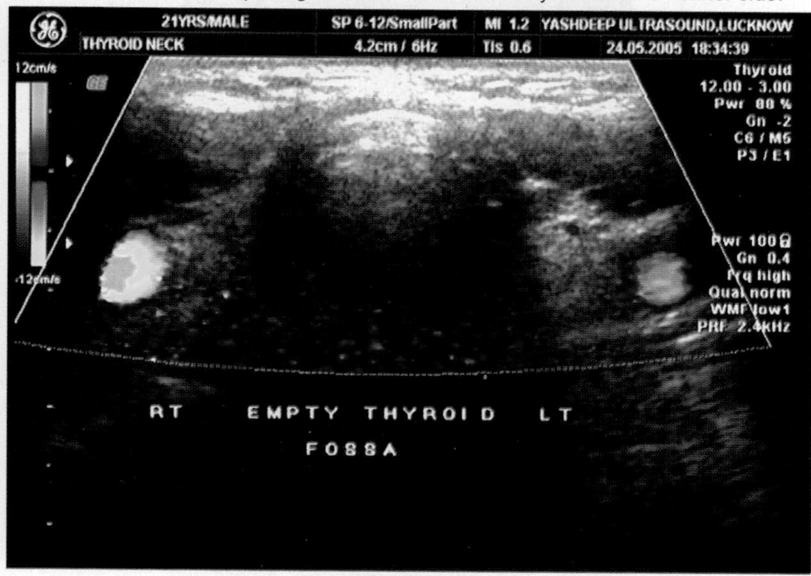

FIGURE 2.16

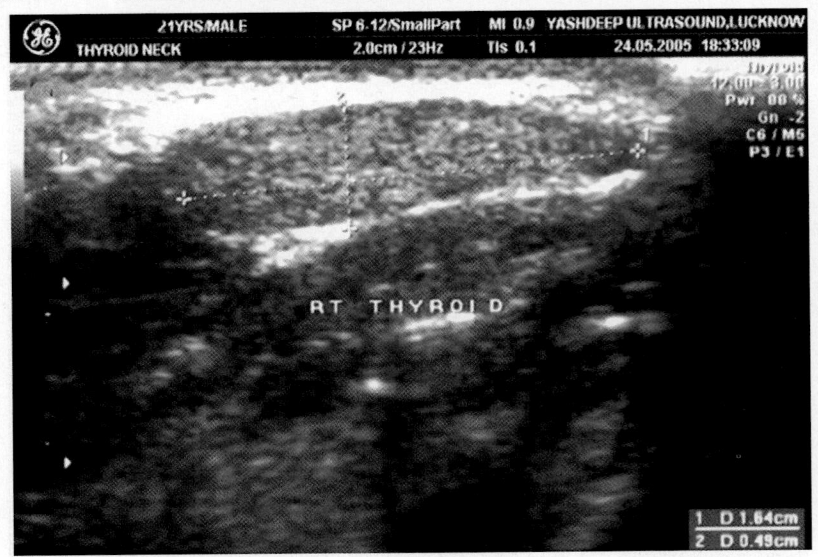

FIGURE 2.17

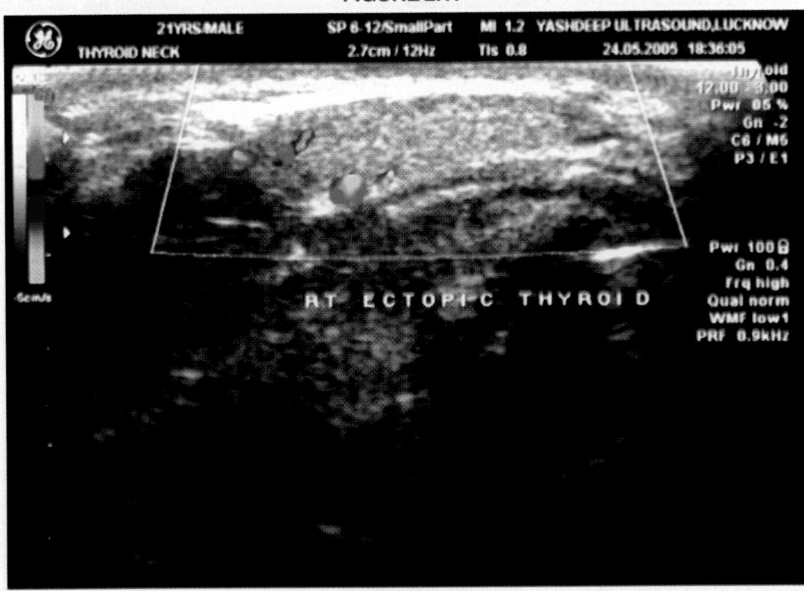

FIGURE 2.18

FIGURES 2.16 to 2.18: Same patient came after 2 years for repeat evaluation of the ectopic thyroid gland. HRSG shows the ectopic thyroid in the Rt upper cervical region as seen in the previous study. Normal thyroid gland was not seen in thyroid fossa.

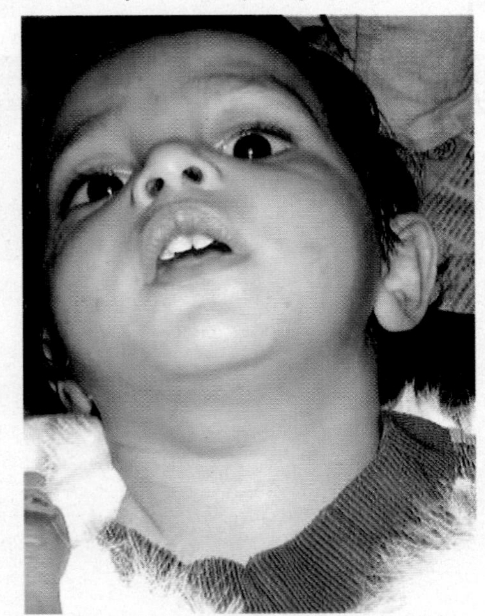

FIGURE 2.19: Thyroglossal cyst. A child was presented with midline cystic mass.

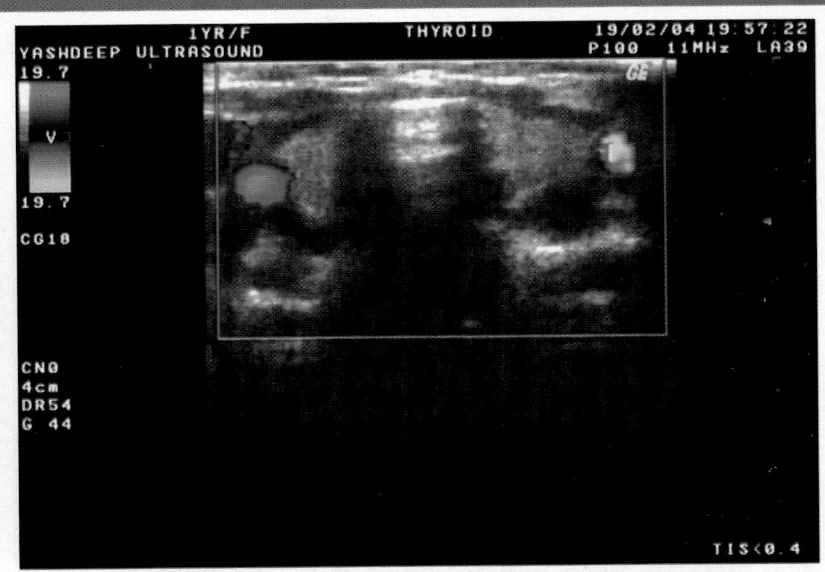

FIGURE 2.20: HRSG shows normal thyroid gland in the thyroid fossa on either side of the trachea. The cyst was not seen in relation to the thyroid.

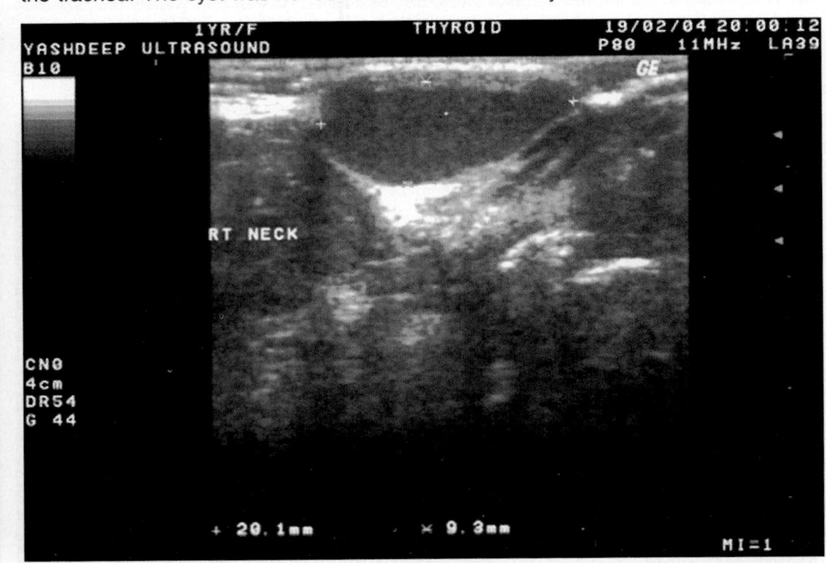

FIGURE 2.21

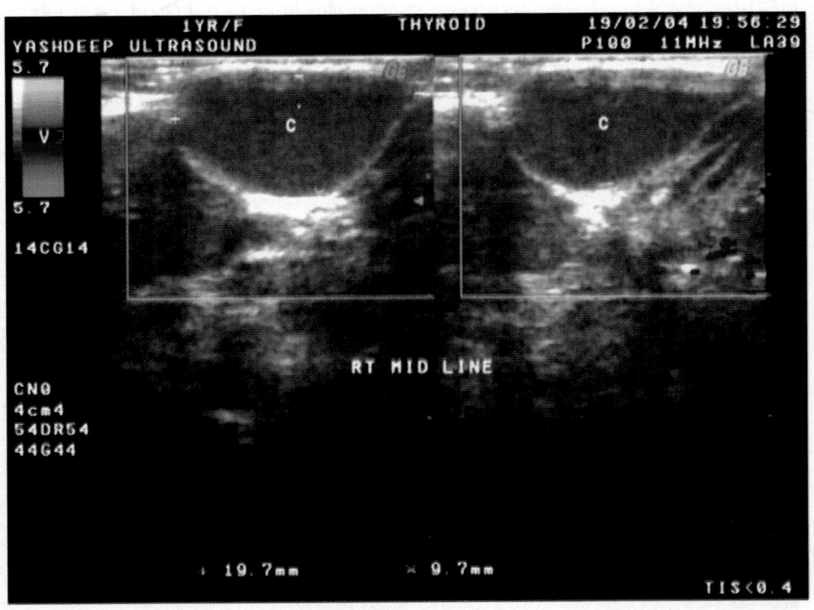

FIGURE 2.22

FIGURES 2.21 and 2.22: A well defined cystic mass is seen just above the thyroid gland in the midline. The cyst shows homogeneous anechoic texture. No evidence of any internal echo is seen in it suggestive of Thyroglossal cyst.

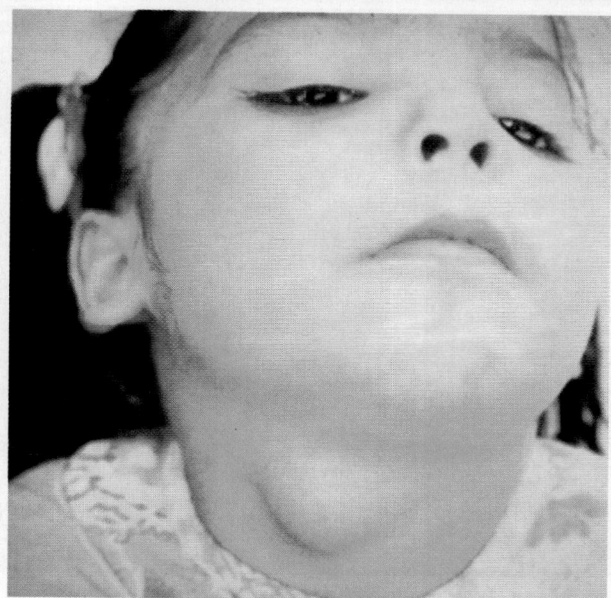

FIGURE 2.23: Thyroglossal cyst. A young girl was presented with the cystic swelling in the midline.

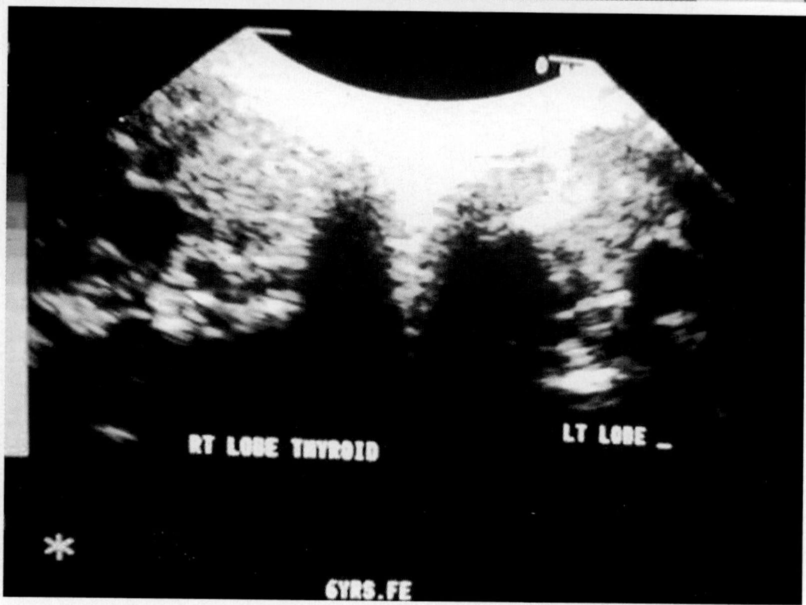

FIGURE 2.26: HRSG of the same patient shows normal thyroid confirming the Thyroglossal cyst.

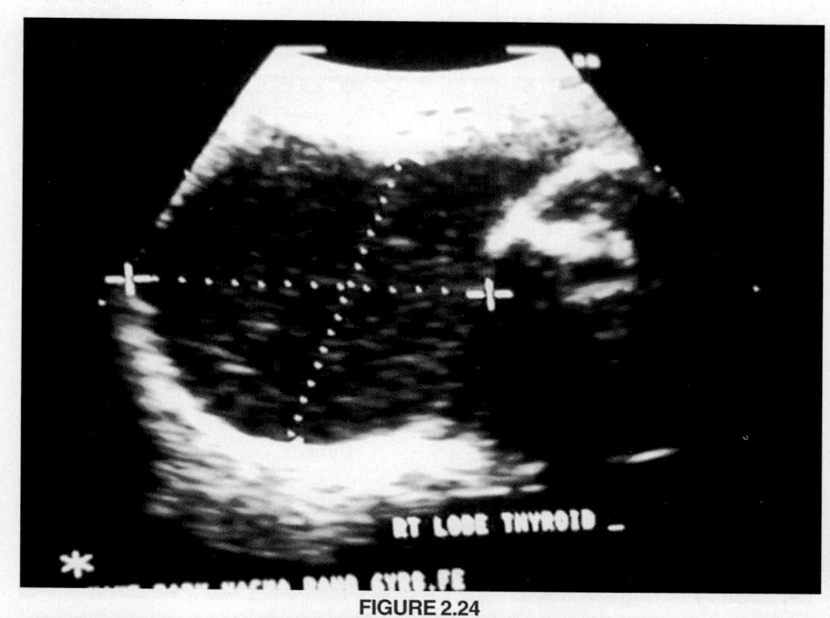

FIGURE 2.24

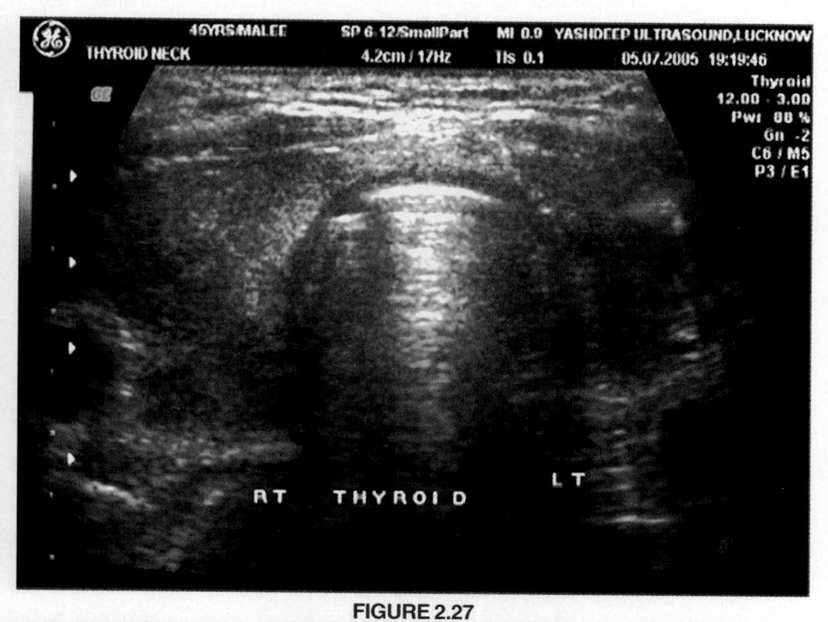

FIGURE 2.27

FIGURE 2.25

FIGURE 2.28

FIGURES 2.24 and 2.25: HRSG shows a well-defined thin walled cystic mass in the midline. No internal echo is seen in the cyst. No loculation or septation is seen (Typical of a Thyroglossal cyst).

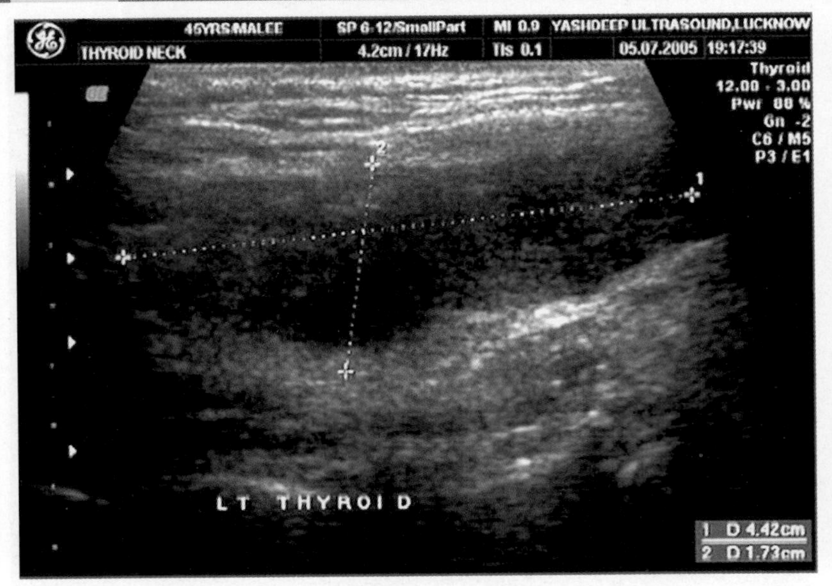

FIGURE 2.29

FIGURES 2.27 to 2.29: Acute thyroiditis. HRSG shows diffusely hypoechoic gland with evidence of multiple hypoechoic areas distributed in the parenchyma in acute thyroiditis. Subcapsular edema is also present in a patient presented with painful swelling of thyroid gland.

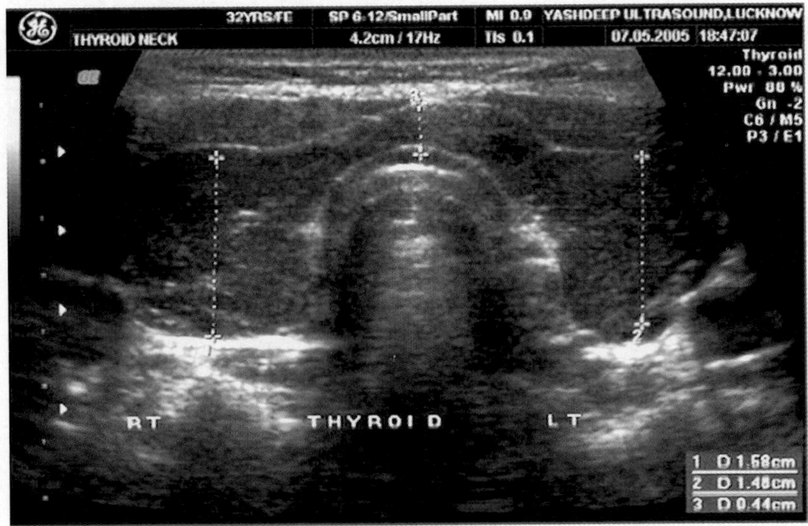

FIGURE 2.30

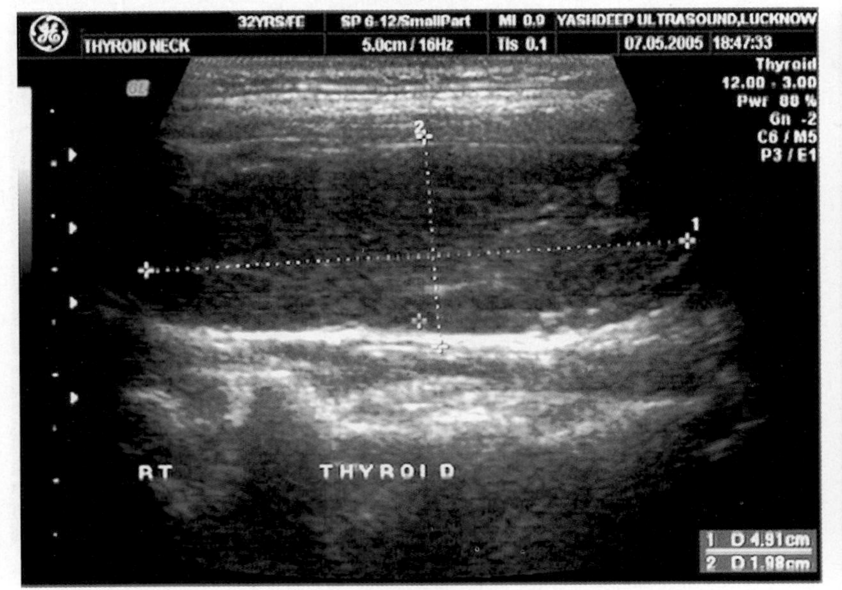

FIGURE 2.31

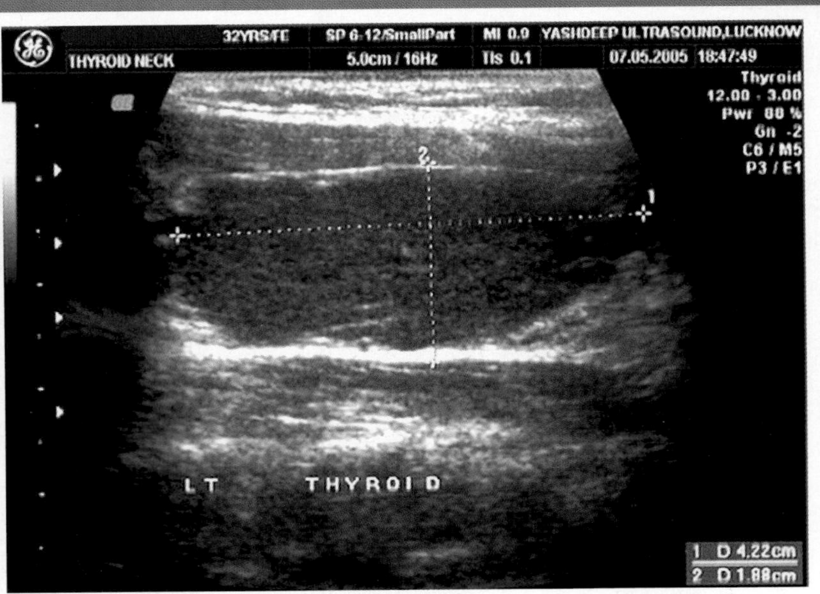

FIGURE 2.32

FIGURES 2.30 to 2.32: Acute thyroiditis. A patient presented with painful swelling in the neck. HRSG shows diffuse hypoechoic thyroid gland. Both lobes are hypoechoic. Normal glandular texture is lost. But no focal mass lesion is seen. Subcapsular edema is also seen.

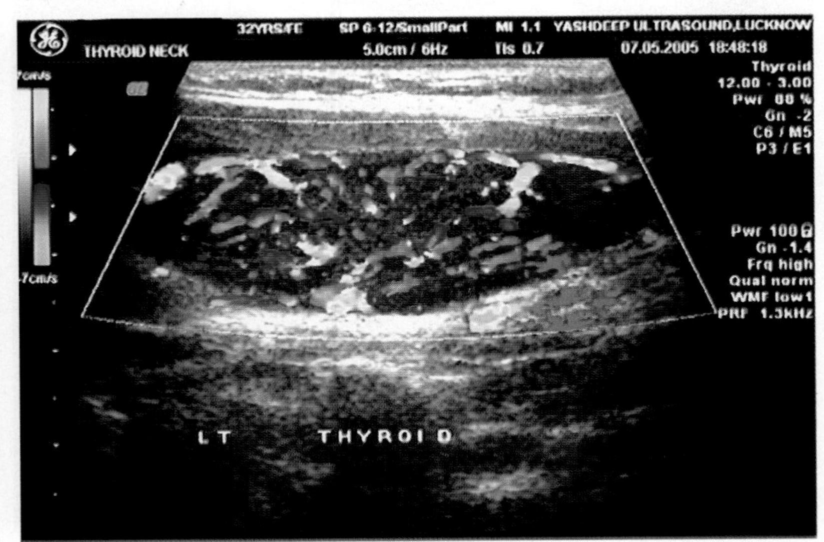

FIGURE 2.33

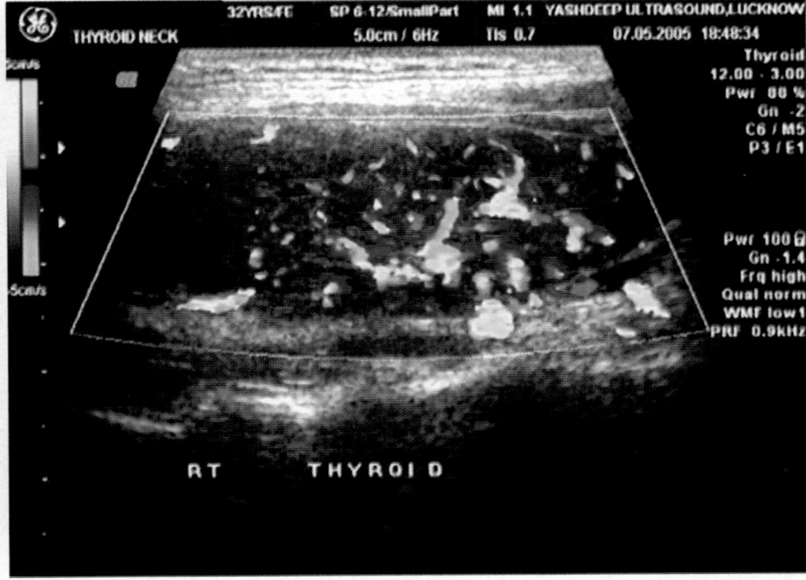

FIGURE 2.34

FIGURES 2.33 and 2.34: On color flow imaging very high flow is seen in the gland on both sides suggestive of acute hyperemia of the gland.

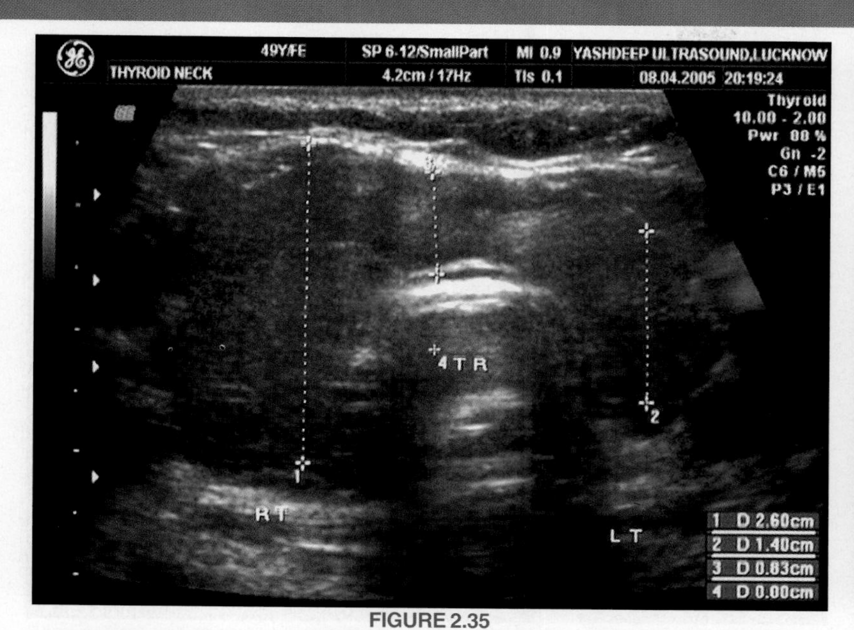

FIGURE 2.35

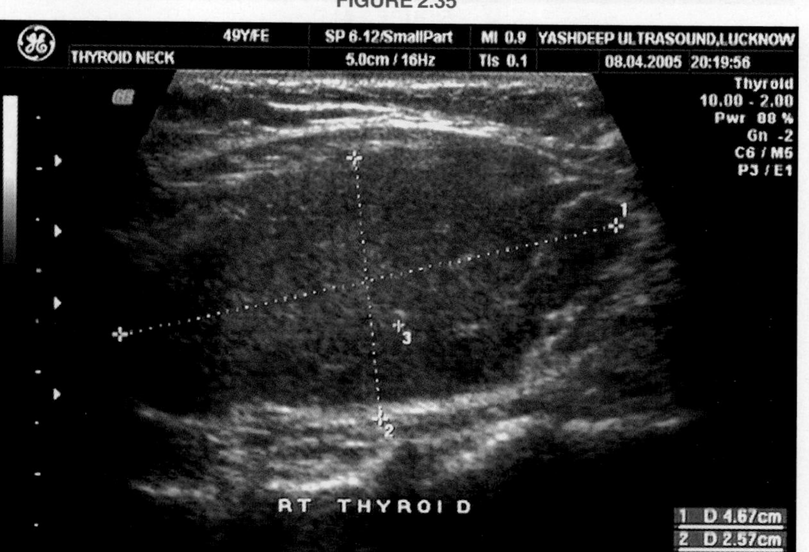

FIGURE 2.36

FIGURES 2.35 and 2.36: Acute thyroiditis with swollen edematous Rt lobe thyroid. A patient presented with acutely painful swelling in the Rt side of the neck. HRSG shows diffuse enlargement of Rt lobe of thyroid with distended capsule. Normal glandular texture is lost. Multiple hypoechoic foci are seen in it suggestive of microabscess.

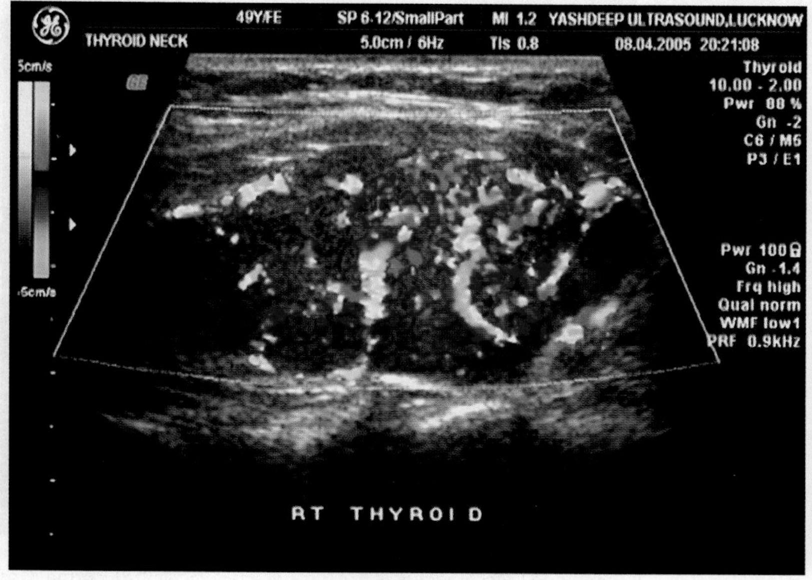

FIGURE 2.37

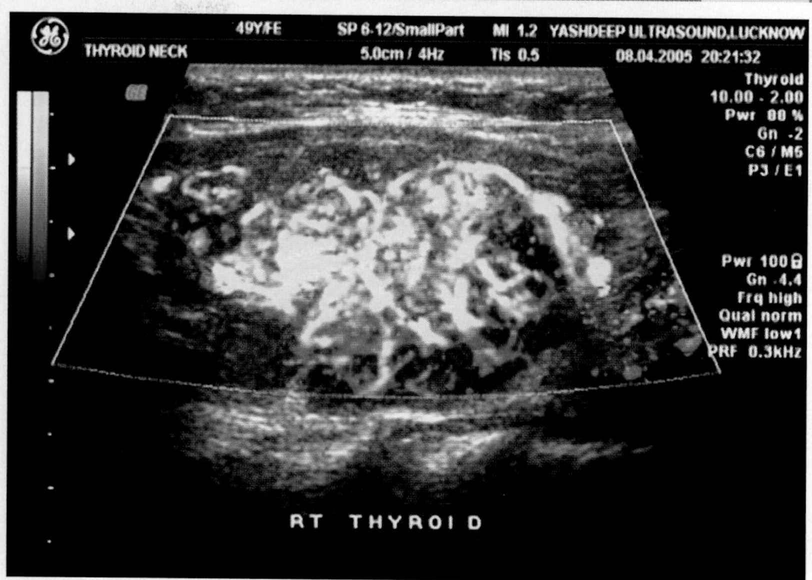

FIGURE 2.38

FIGURES 2.37 and 2.38: On color flow imaging very high flow is seen in the gland suggestive of acute hyperemia. **Thyroid inferno** sign is positive.

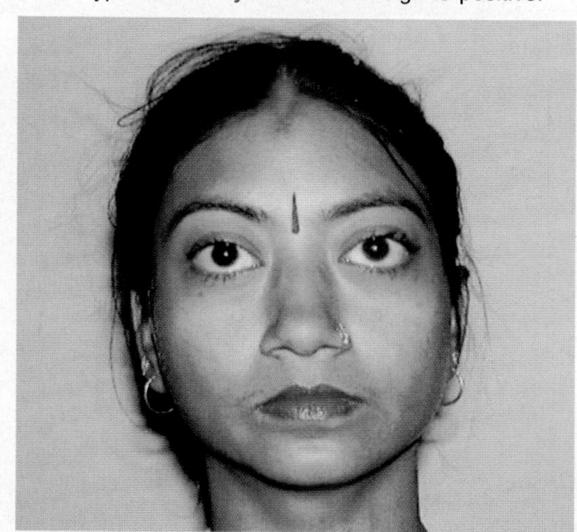

FIGURE 2.39: Acute thyrotoxicosis. A young lady presented with painful swelling in the neck with evidence bulging of eyeballs on both sides and pain in the eyes.

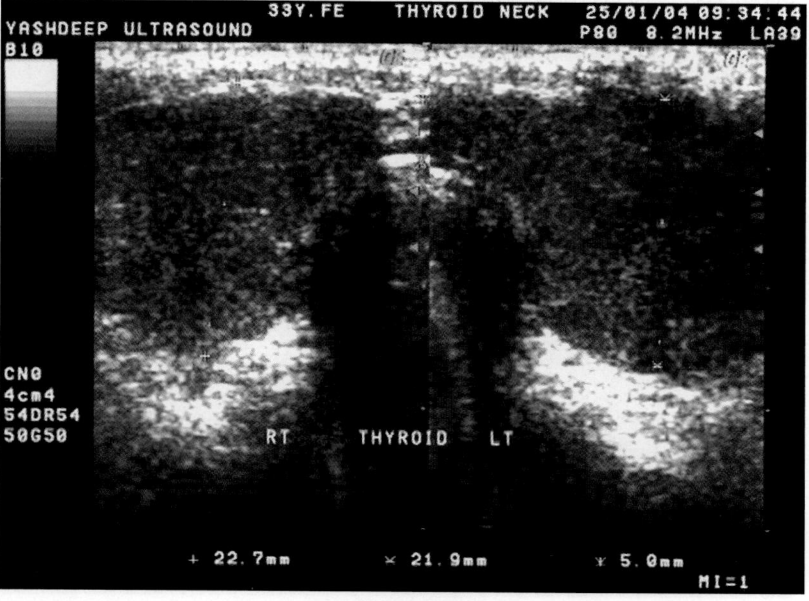

FIGURE 2.40

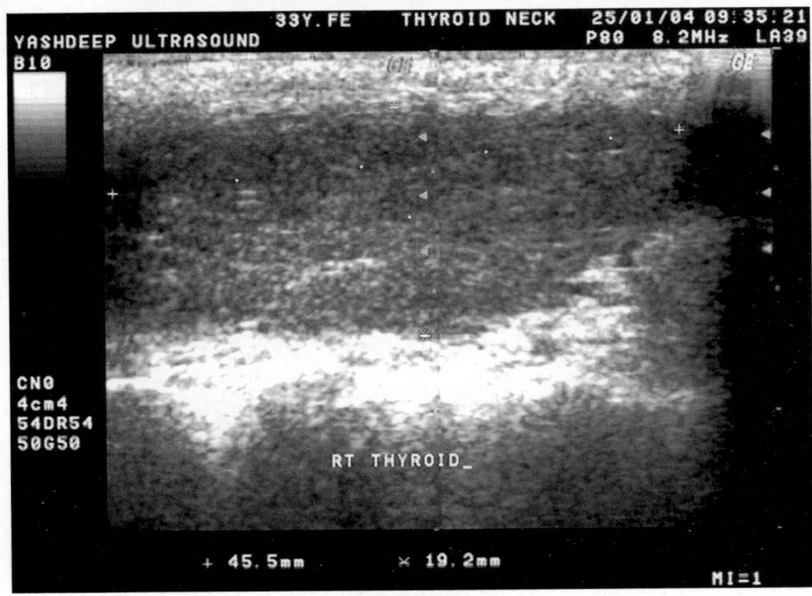

FIGURE 2.41

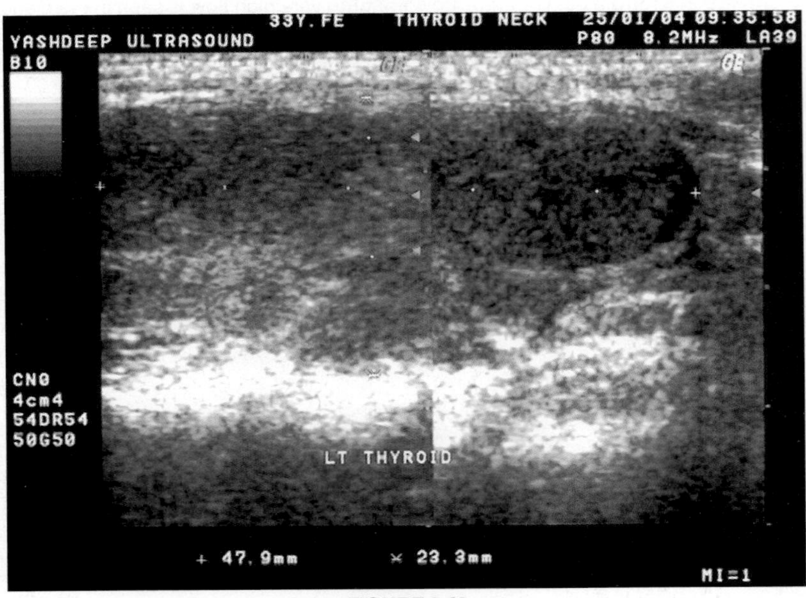

FIGURE 2.42

FIGURES 2.40 to 2.42: HRSG shows diffuse enlargement of both lobes of thyroid. The gland shows inhomogeneous texture with multiple hypoechoic areas. Subcapsular edema is also seen.

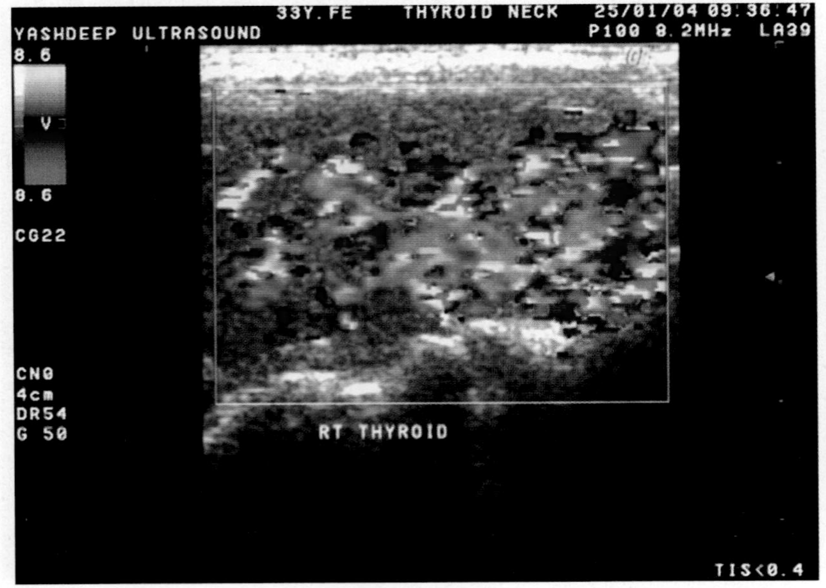

FIGURE 2.43

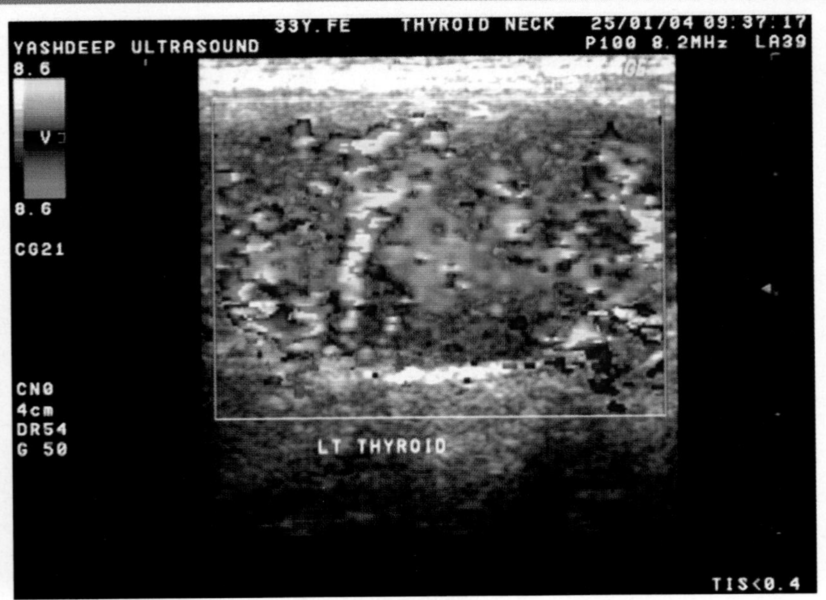

FIGURE 2.44

FIGURES 2.43 and 2.44: On color flow imaging high flow is seen in the gland suggestive of acutely inflamed gland.

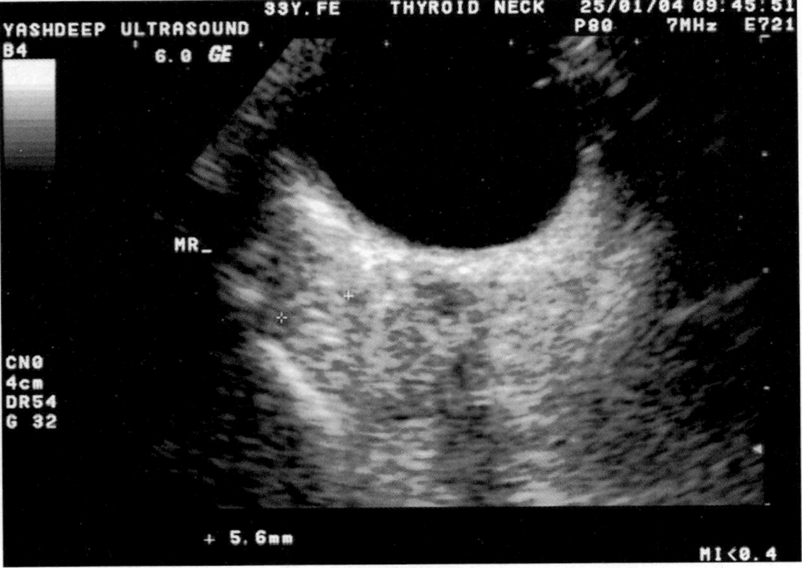

FIGURE 2.45

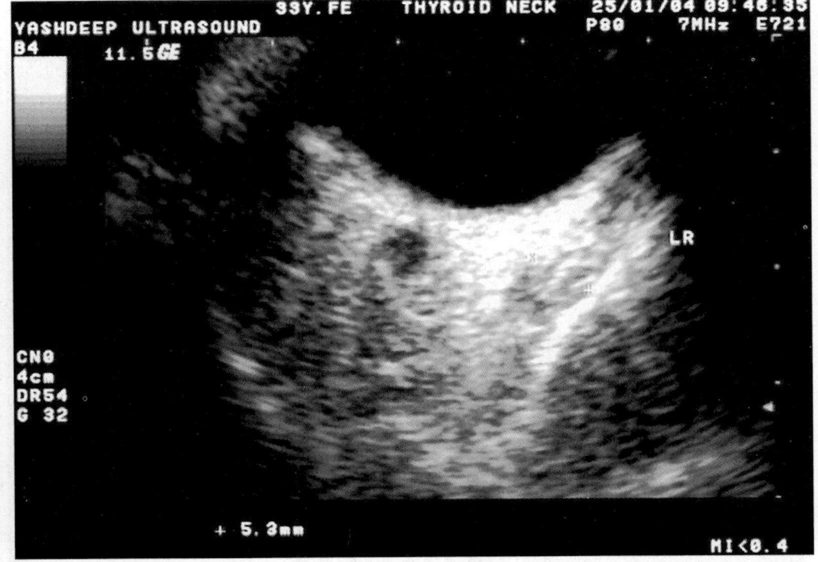

FIGURE 2.46

FIGURES 2.45 and 2.46: HRSG of the orbit shows thickened recti muscles. The medial rectus is 5.6 mm and Lateral rectus is 5.3 mm, which is more than normal limit (normal <3 mm). Retrobulbar fat is also increased.

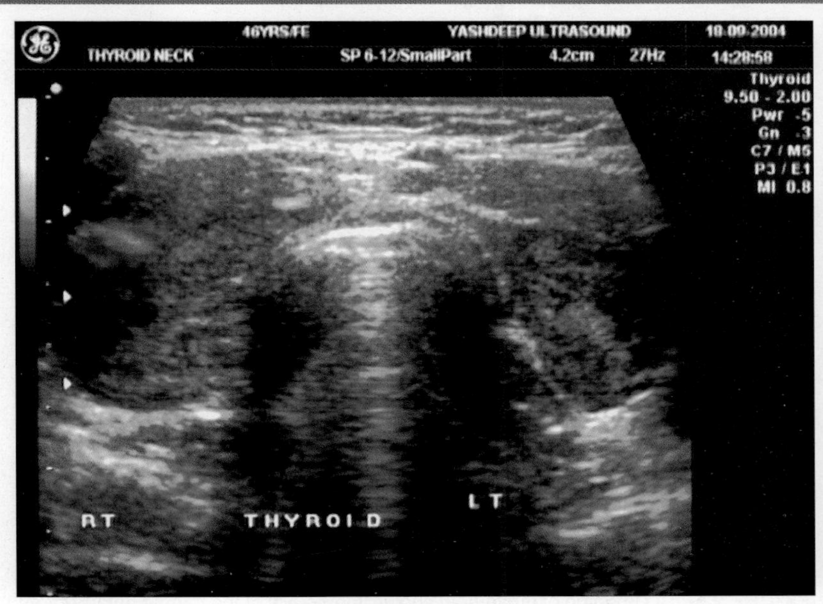

FIGURE 2.47

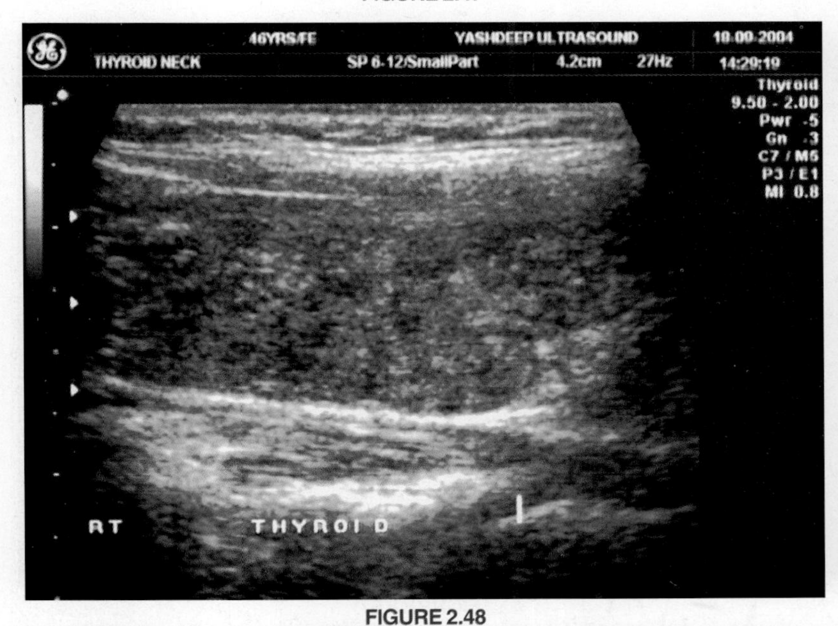

FIGURE 2.48

FIGURE 2.49

FIGURES 2.47 to 2.49: Basedow's disease. HRSG shows diffuse enlargement of the thyroid gland involving both the lobes. The gland shows overall diffuse low reflectivity. However, it shows homogeneous texture.

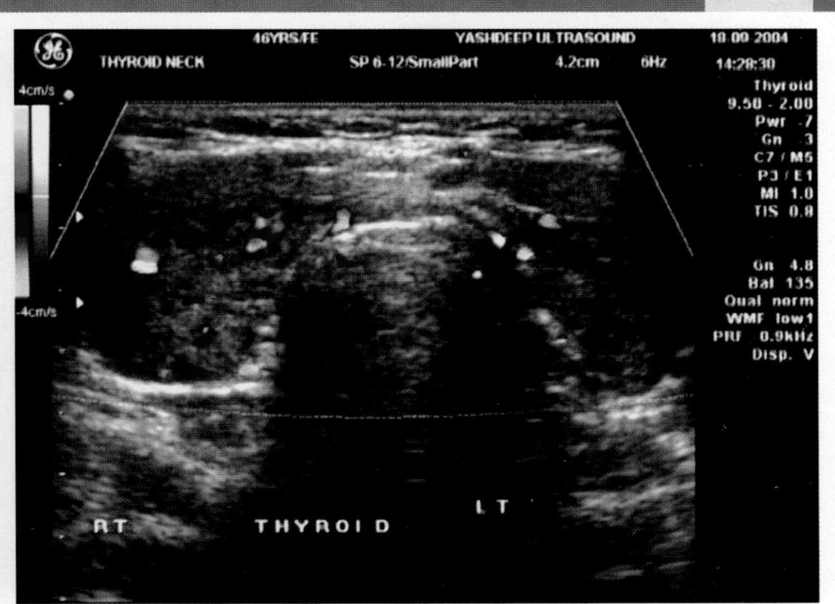

FIGURE 2.50: On color flow imaging increased flow is seen in the gland. The veins shows engorgement.

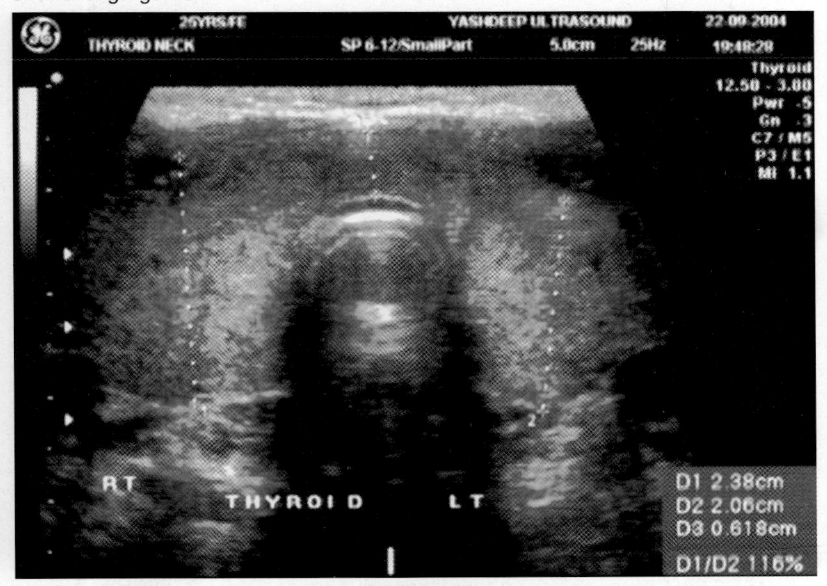

FIGURE 2.51

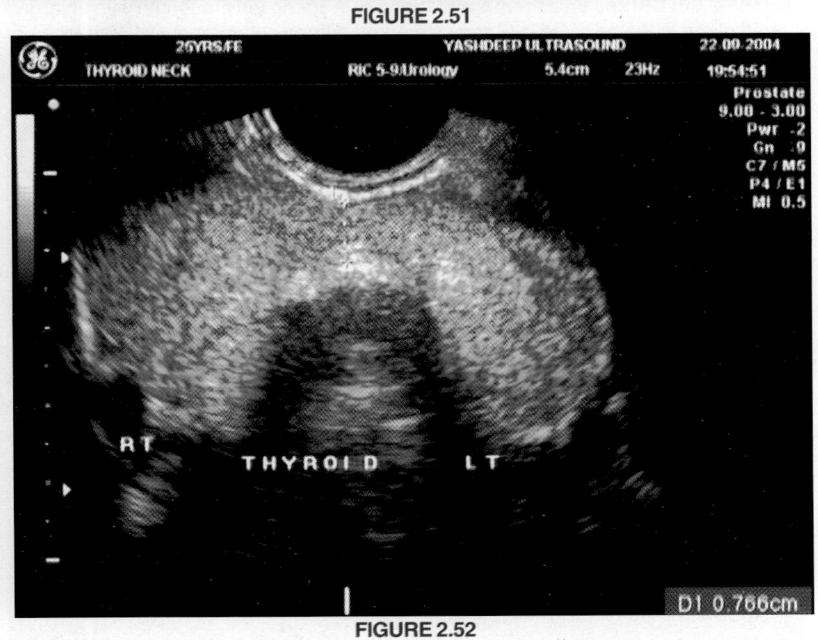

FIGURE 2.52

FIGURES 2.51 and 2.52: Diffuse thyroid hyperplasia. HRSG shows diffuse enlargement of thyroid gland. The gland shows homogeneous texture. No focal mass is seen in them.

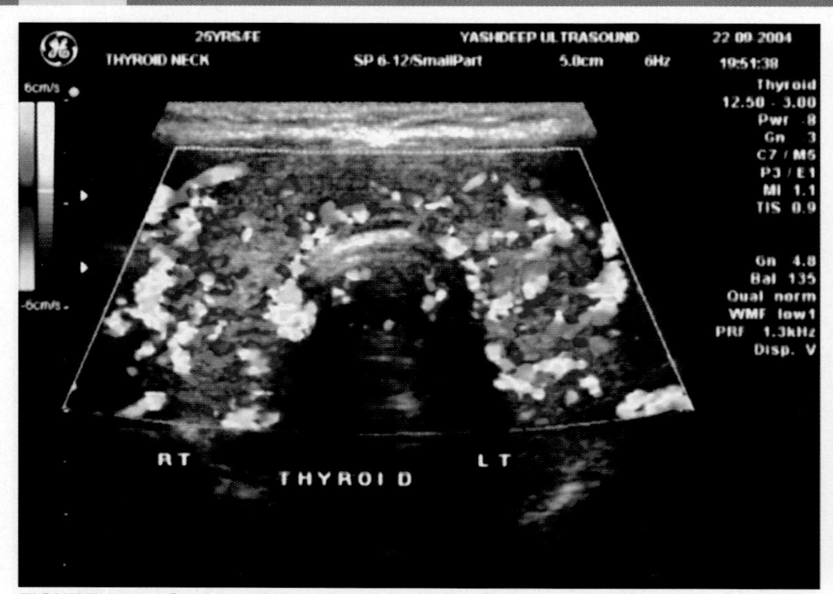

FIGURE 2.53: On color flow imaging high flow is seen in the gland suggestive of thyrotoxicosis.

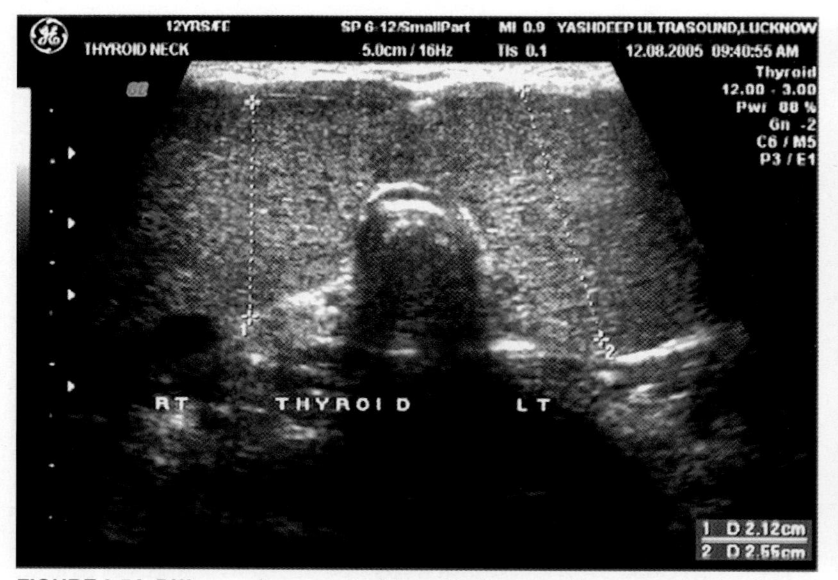

FIGURE 2.54: Diffuse enlargement of thyroid: A young girl presented with enlarged thyroid gland in the neck. HRSG shows marked diffuse enlargement of both lobes of thyroid. However, the gland texture is homogeneous. No calcification is seen in the gland. No necrosis is seen.

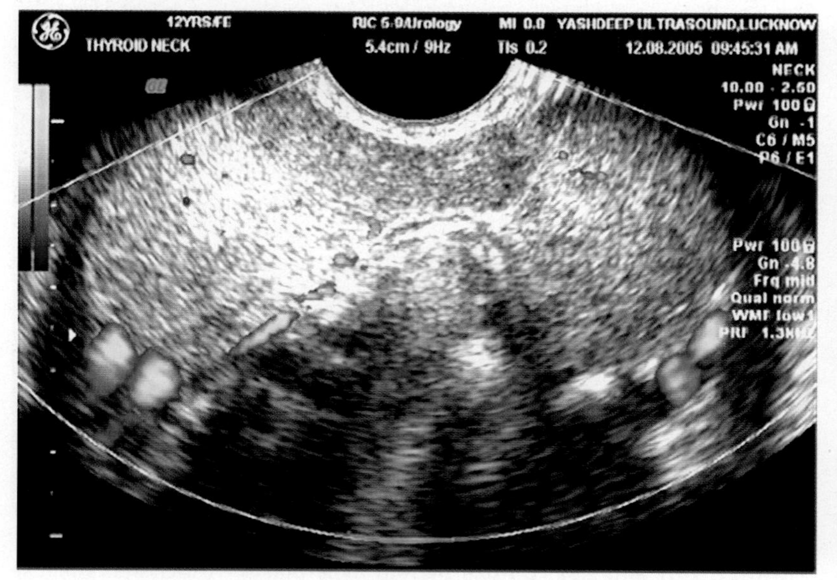

FIGURE 2.55: On color flow imaging the carotid artery is seen displaced laterly on either side by the enlarged thyroid gland.

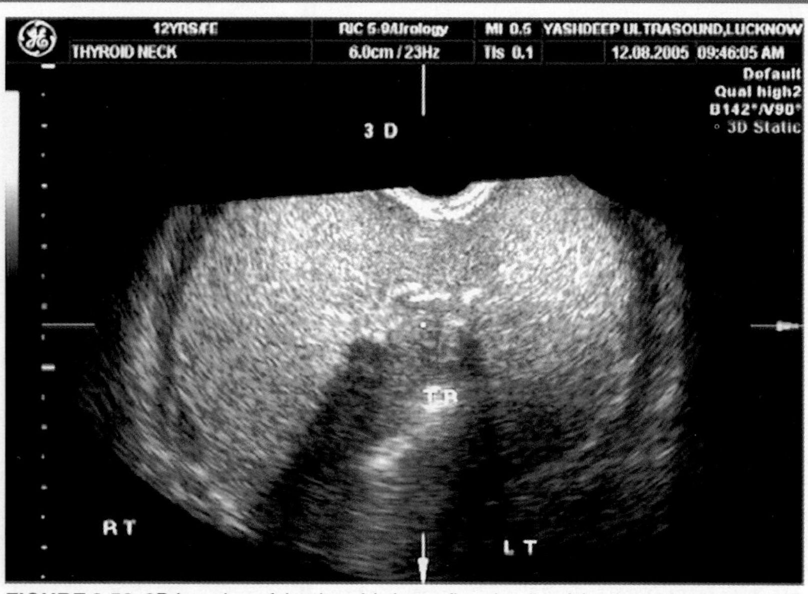

FIGURE 2.56: 3D Imaging of the thyroid shows fine details of the thyroid enlargement.

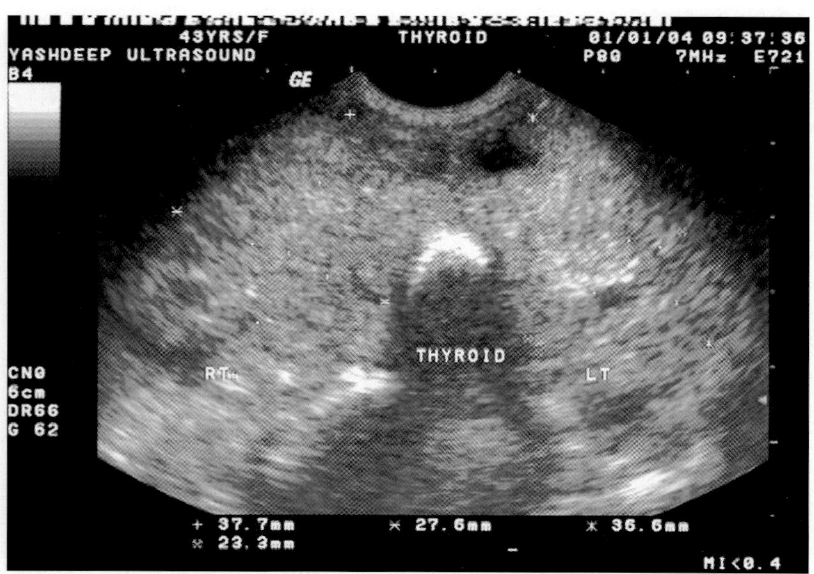

FIGURE 2.57

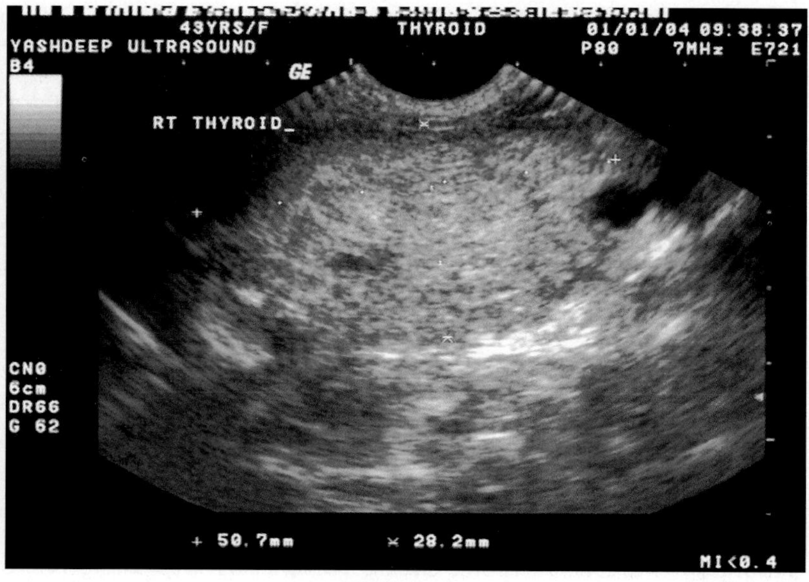

FIGURE 2.58

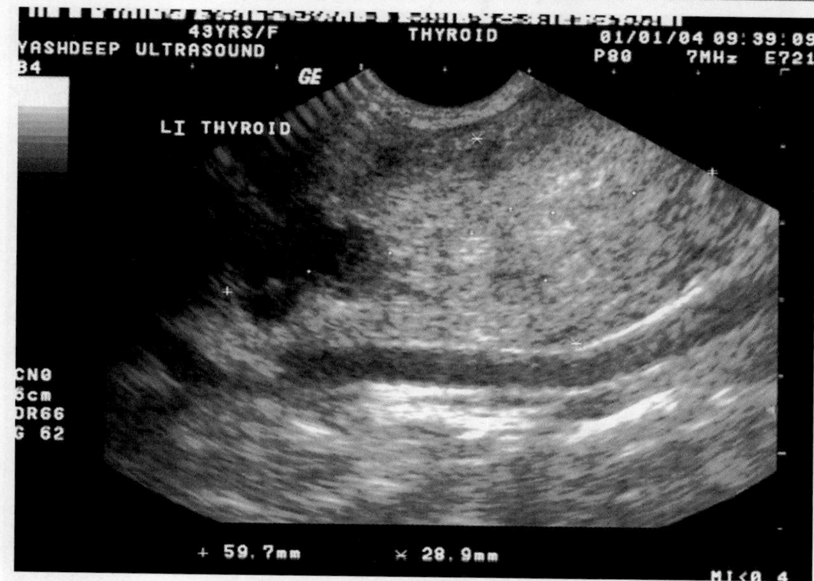

FIGURE 2.59

FIGURES 2.57 to 2.59: Nodular enlargement of thyroid Isoechogenic nodules are seen in the gland on both sides. They are seen distributed in the substance of the gland. Few nodules show cystic necrosis.

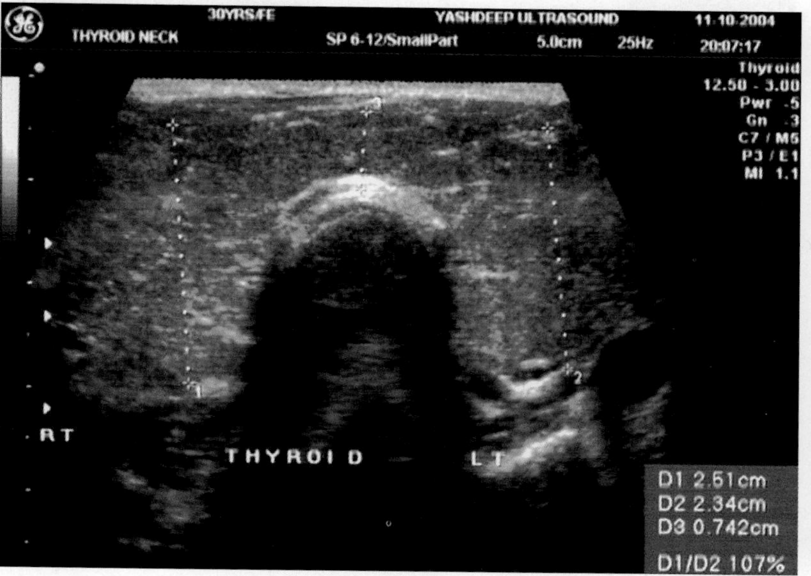

FIGURE 2.60

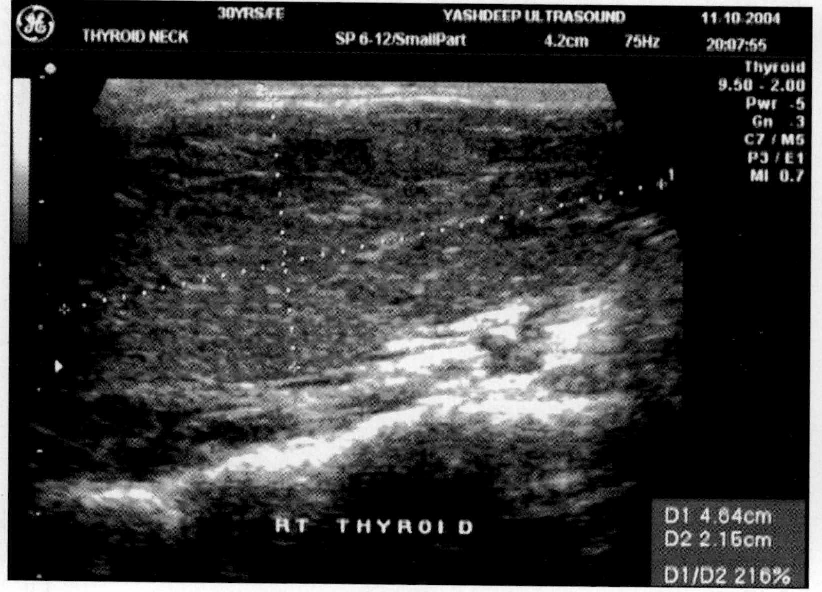

FIGURE 2.61

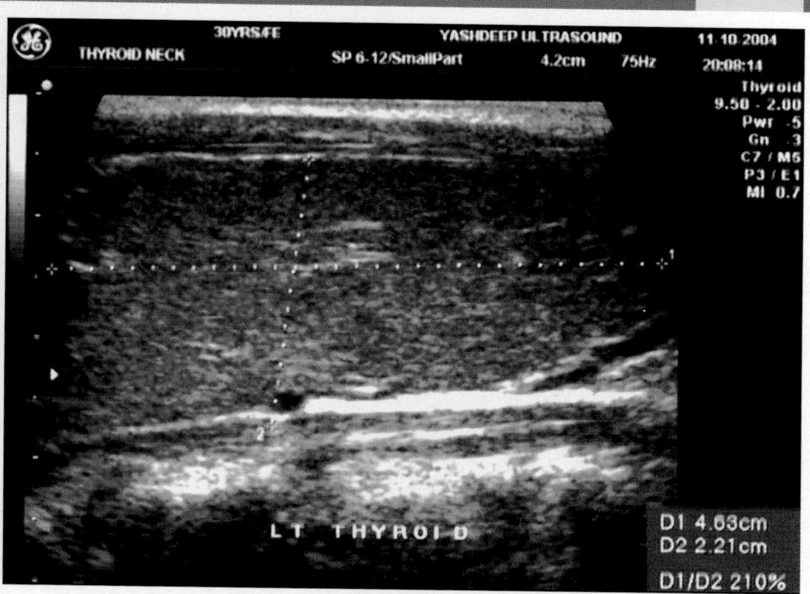

FIGURE 2.62

FIGURES 2.60 to 2.62: Diffuse enlargement of thyroid with hypoechoic texture. HRSG shows diffuse enlargement of the thyroid gland. The glandular texture is inhomogeneous and hypoechoic. Multiple low level areas are seen in the gland. Normal glandular texture is lost.

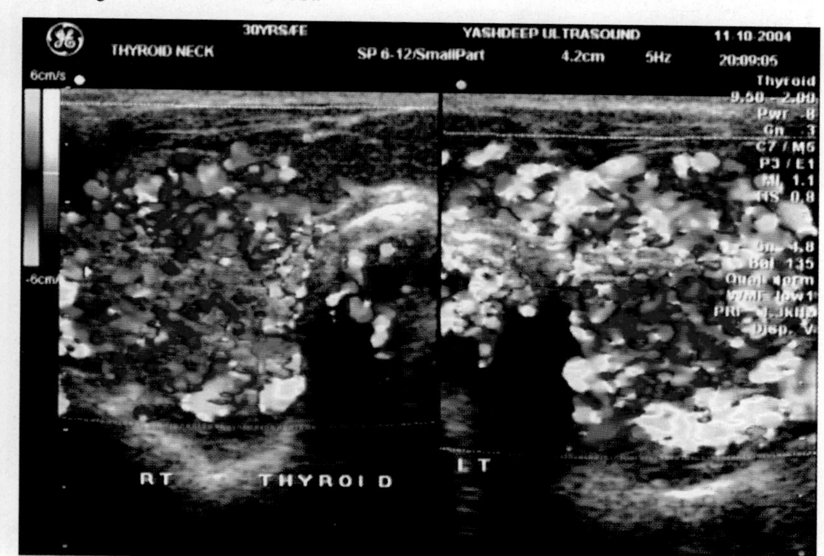

FIGURE 2.63

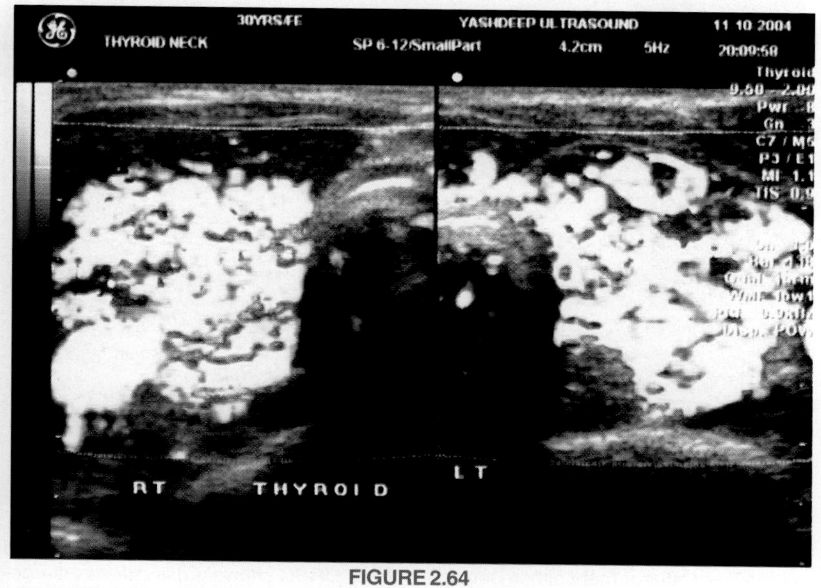

FIGURE 2.64

FIGURES 2.63 and 2.64: On color flow imaging very high flow is seen in the gland. **Thyroid Inferno** sign is positive suggestive of acute thyroiditis changes.

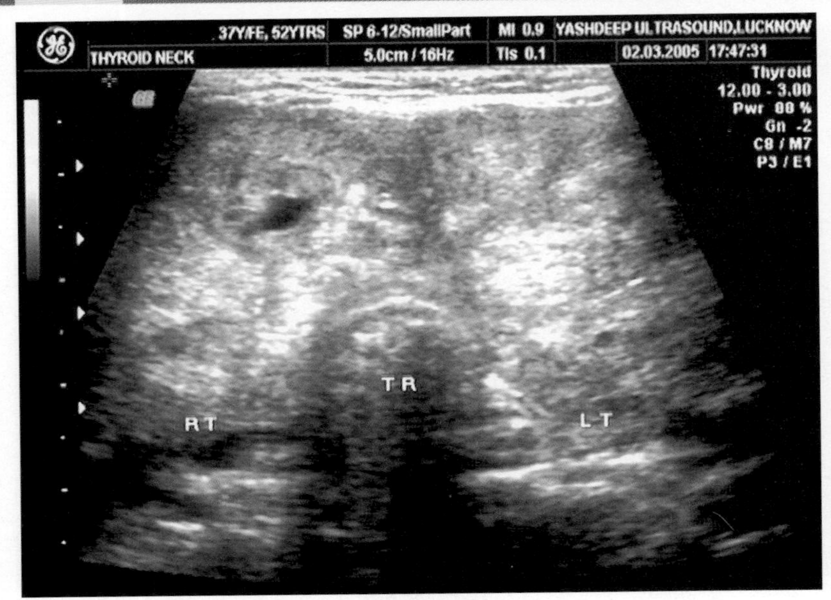

FIGURE 2.65

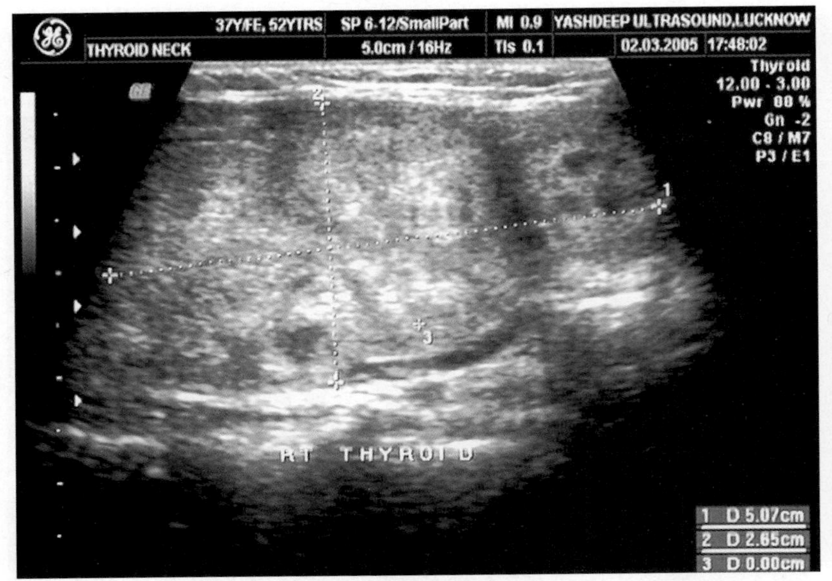

FIGURE 2.66

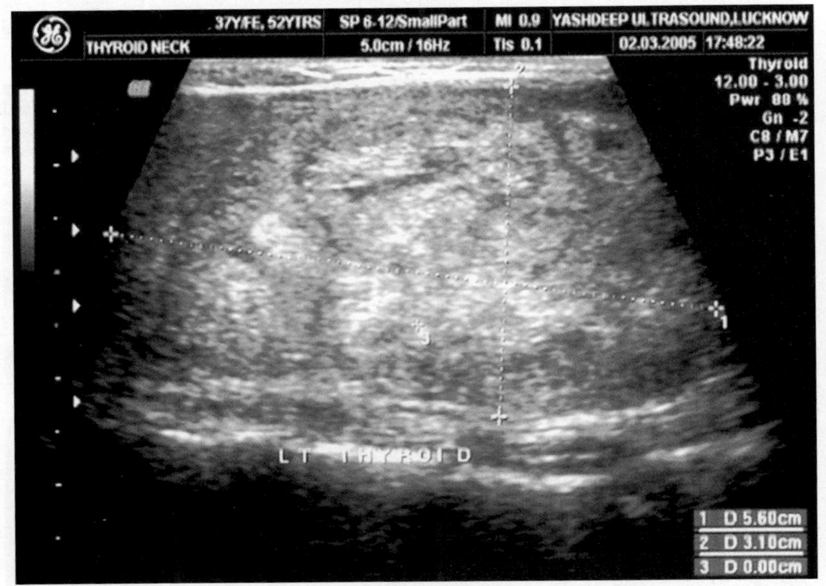

FIGURE 2.67

FIGURES 2.65 to 2.67: Multinodular hyperplasia of thyroid. A patient presented with diffuse enlargement of both lobes of thyroid. HRSG shows gross diffuse enlargement of the thyroid gland on both sides. Echogenic nodular shadows are seen in both the lobes. They are well defined. Few nodules show degenerative changes.

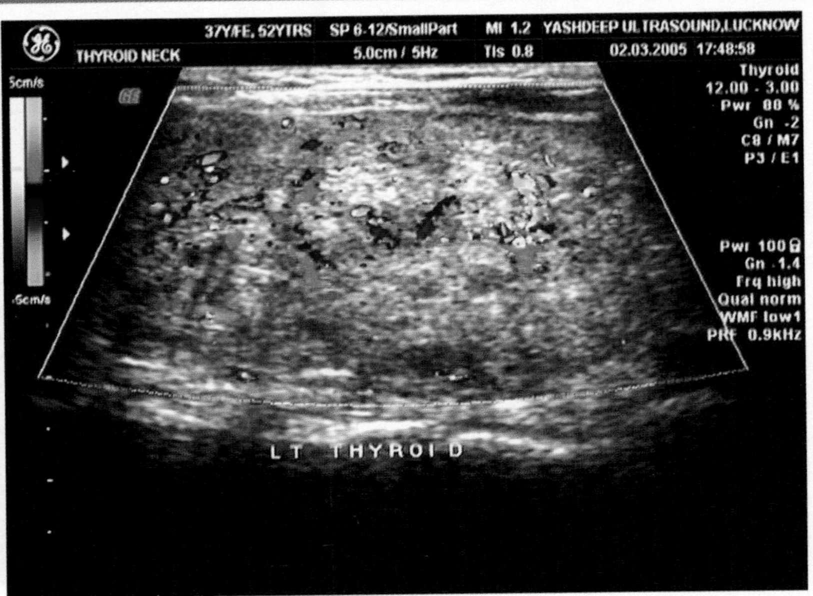

FIGURE 2.68

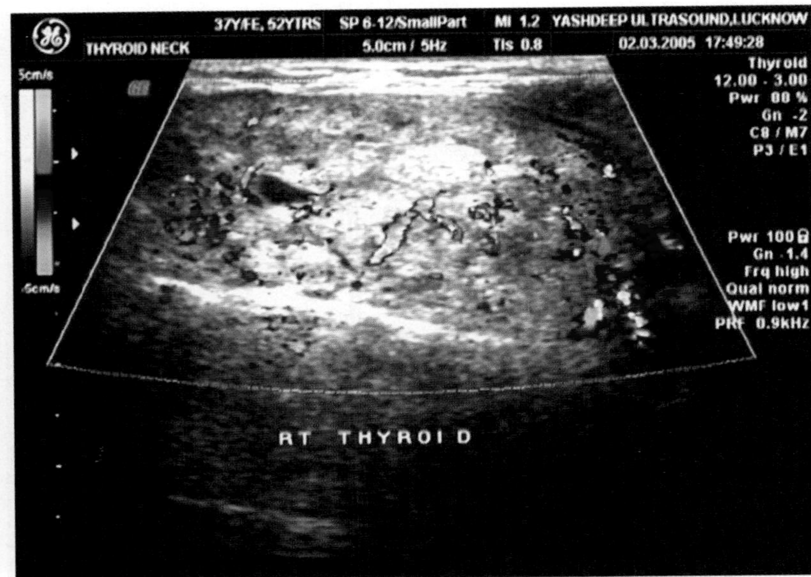

FIGURE 2.69

FIGURES 2.68 and 2.69: On color flow imaging moderate flow is seen around the nodules in both the lobes.

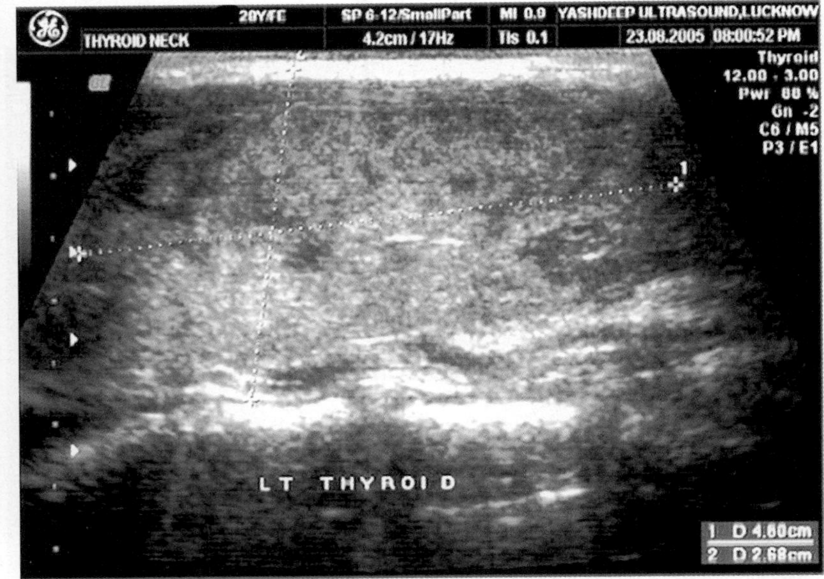

FIGURE 2.70: Multinodular hyperplasia. HRSG shows enlarged Lt lobe thyroid. Multiple well-defined nodular shadows are seen in both the lobes. They are homogeneous in texture.

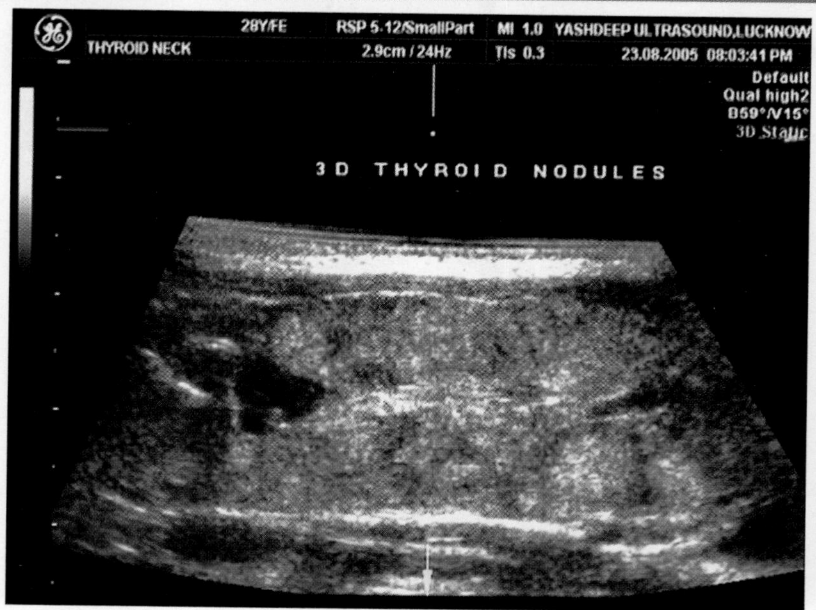

FIGURE 2.71A

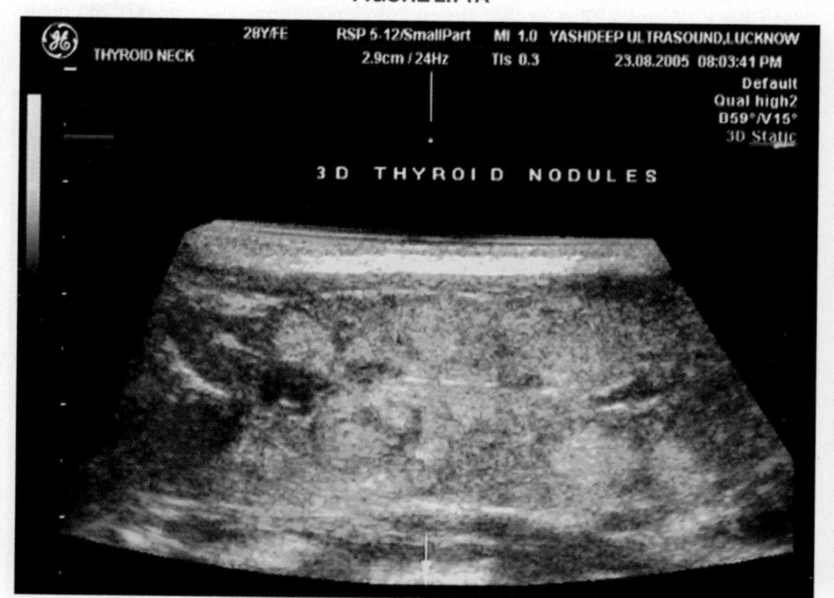

FIGURE 2.71B

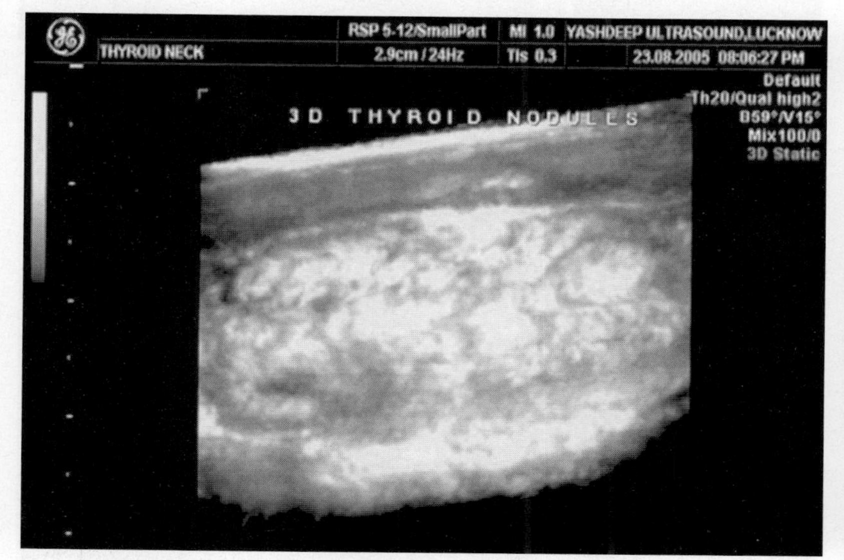

FIGURE 2.71C

FIGURES 2.71A to C: 3D images of the same patient shows better details of the nodules with sharp margins.

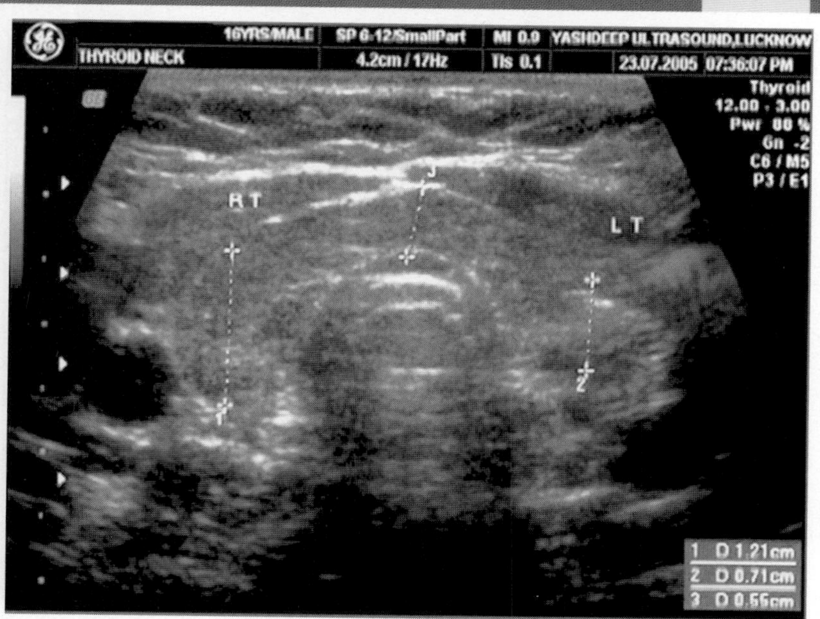

FIGURE 2.72

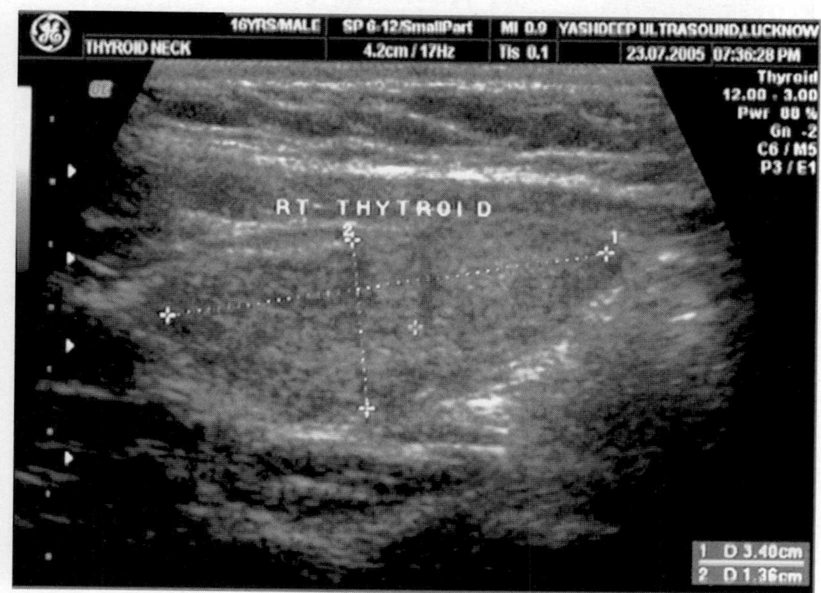

FIGURE 2.73

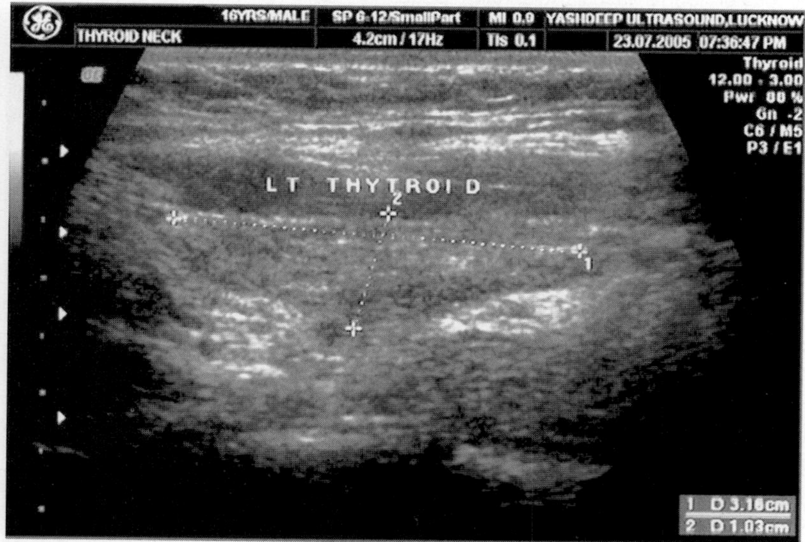

FIGURE 2.74

FIGURES 2.72 to 2.74: Chronic thyroiditis. HRSG shows a young child presented with chronic thyroid disease. HRSG shows small thyroid with inhomogeneous texture. Normal fine glandular texture is lost. HRSG shows diffuse parenchymal inhomogeneous texture with micronodularity. It shows typical “moth eaten” appearance.

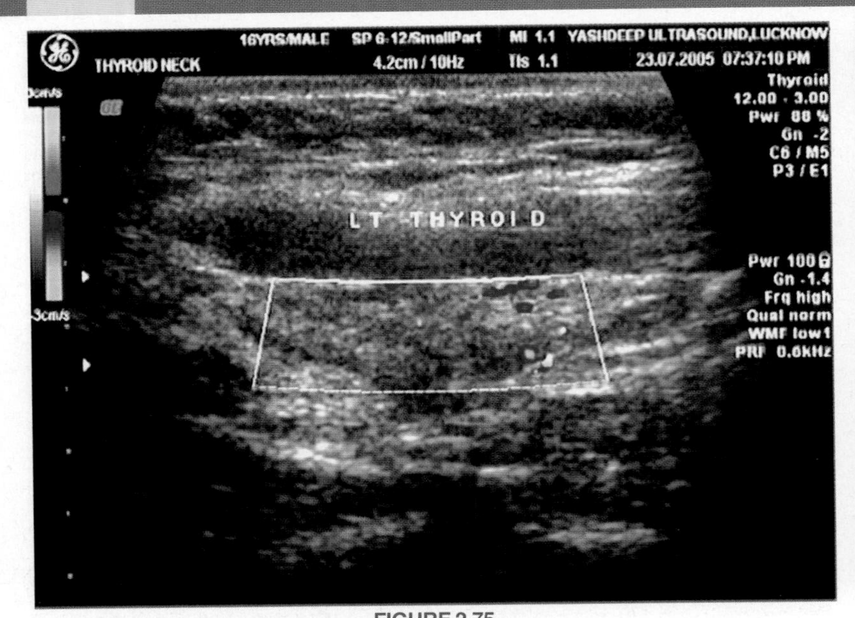

FIGURE 2.75

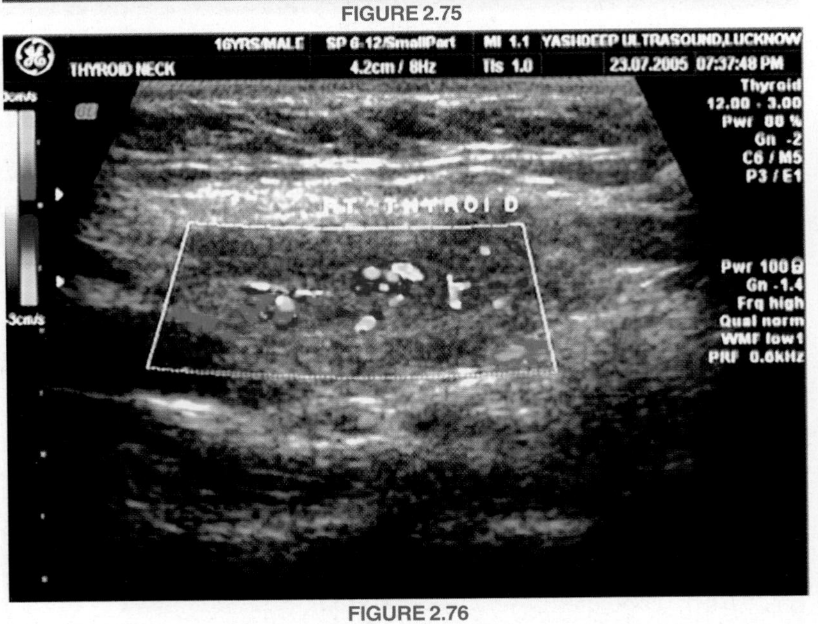

FIGURE 2.76

FIGURES 2.75 and 2.76: On color flow imaging the gland show low flow pattern due to loss of normal vascular glandular tissue.

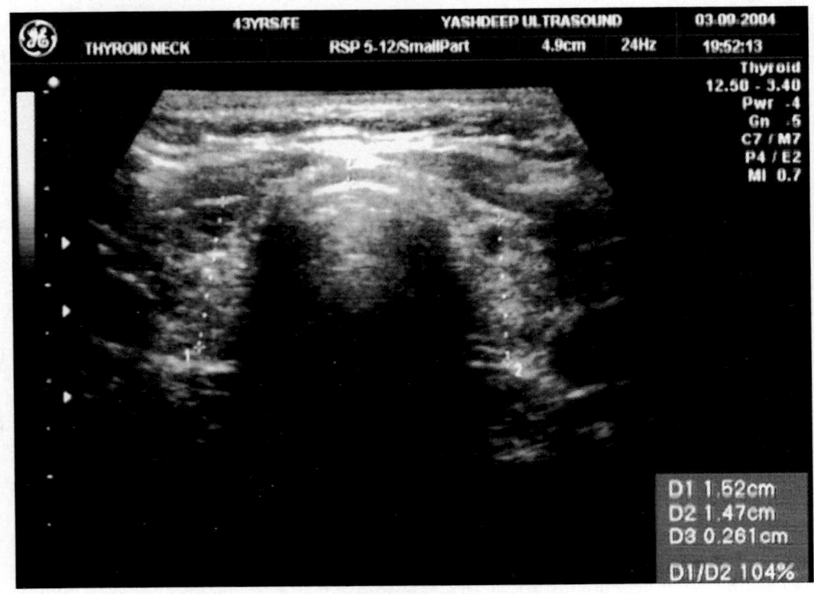

FIGURE 2.77

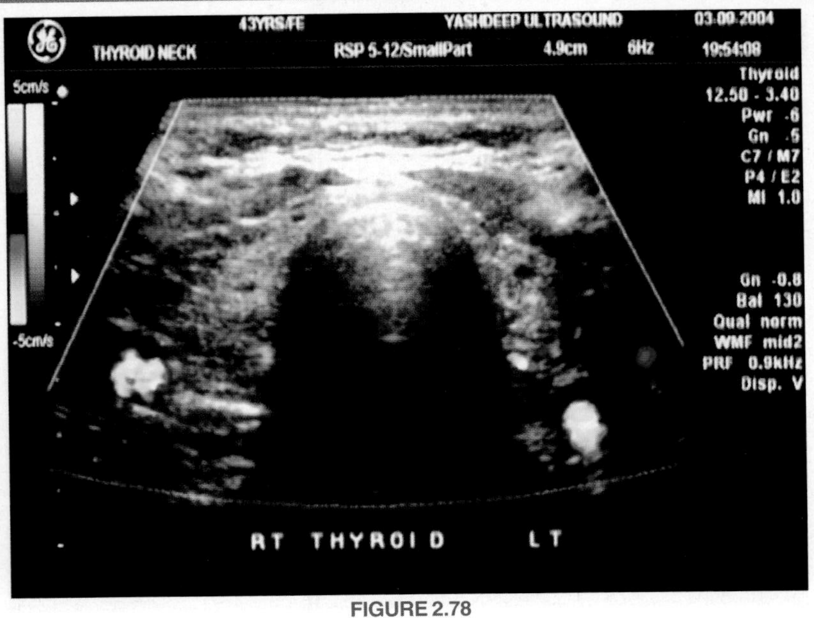

FIGURE 2.78

FIGURES 2.77 and 2.78: Chronic Hashimoto thyroiditis. The whole of the gland is small and grossly inhomogeneous in texture. Normal glandular texture is lost. The gland shows hyperplastic fibrous pattern. Color flow shows poor flow in the gland.

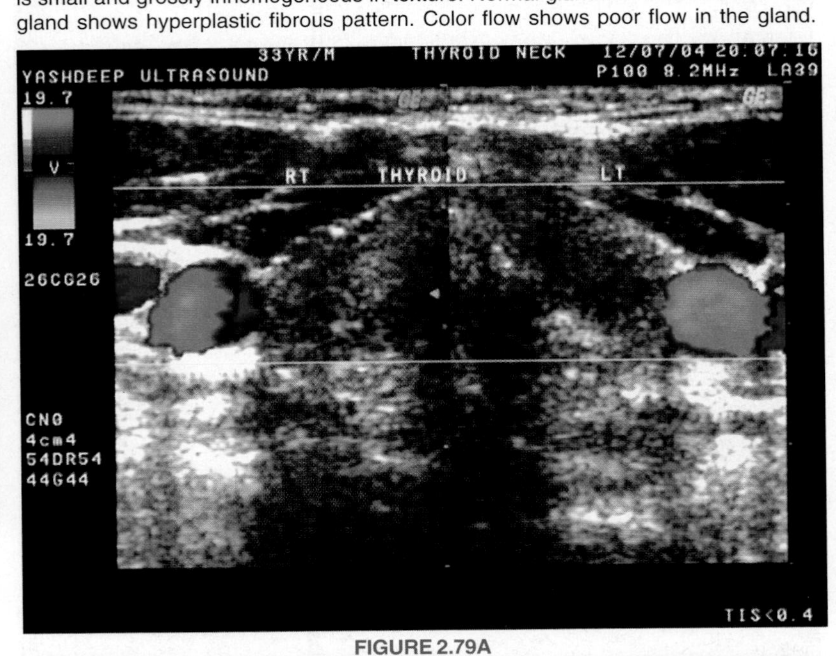

FIGURE 2.79A

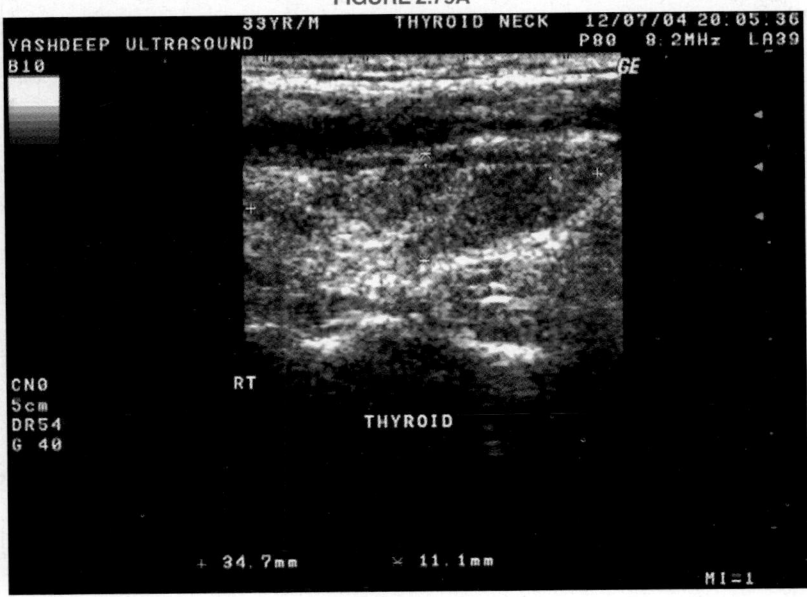

FIGURE 2.79B

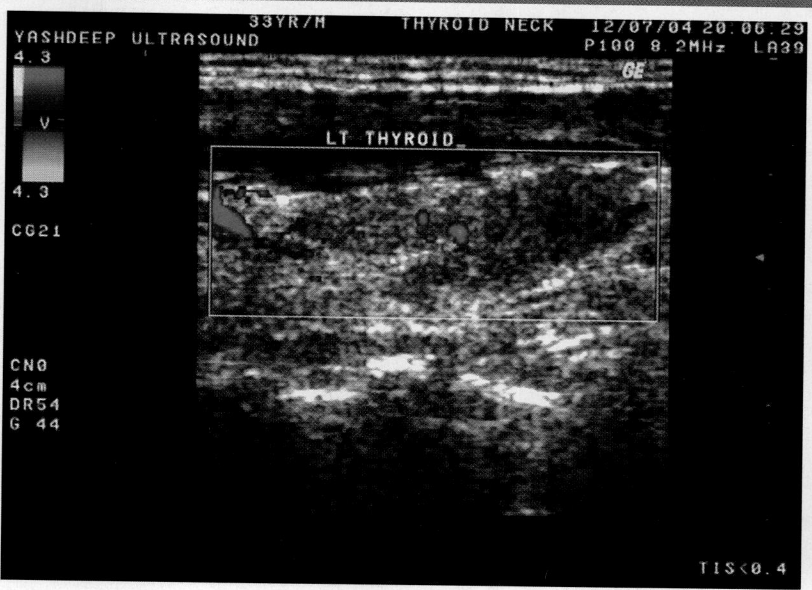

FIGURE 2.79C

FIGURES 2.79A to C: Chronic atrophic thyroiditis. HRSG grossly small shrunken gland on both sides. The echo texture is markedly inhomogeneous with multiple hypoechoic areas. On color flow imaging poor flow is seen

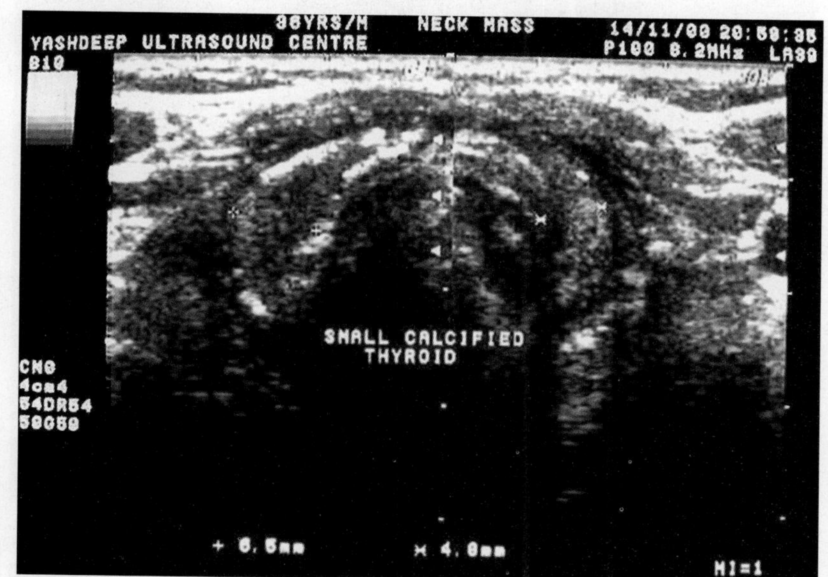

FIGURE 2.80: Chronic calcific thyroiditis. HRSG of the neck shows irregular small atrophic both lobes of the thyroid with multiple calcified specks.

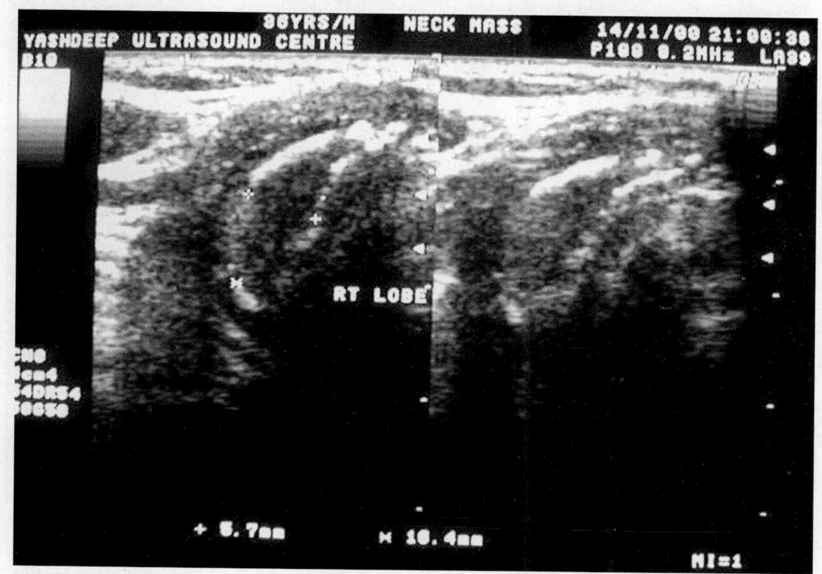

FIGURE 2.81A: The Rt lobe of the thyroid is small and measures 10 x 6 mm in size with multiple calcified shadows.

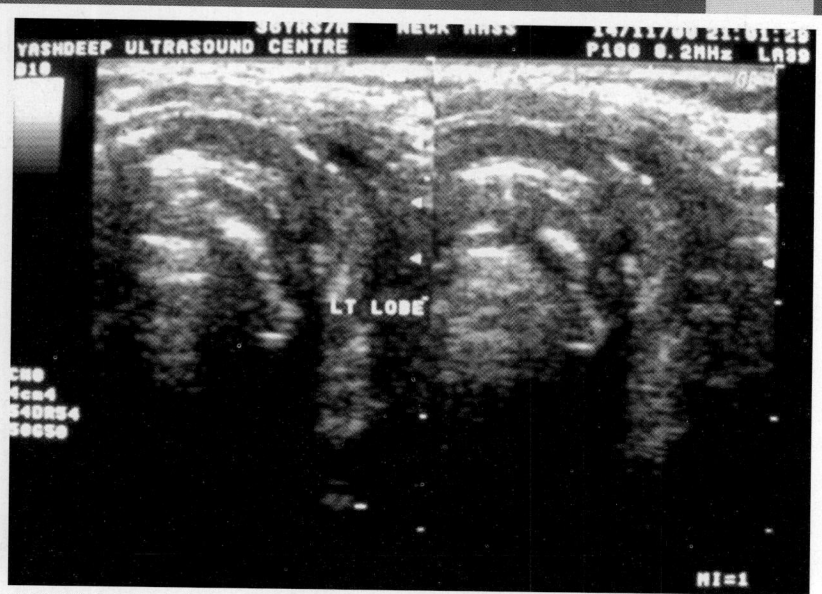

FIGURE 2.81B: Lt lobe of the thyroid also shows small lobe with hypoechoic texture. Multiple calcified shadows are seen. Biopsy of the gland shows atrophic calcific thyroiditis.

normal parenchymal pattern is identified. At times discrete nodules may be seen in Hashimoto's thyroiditis.

Solitary Thyroid Nodule

Solitary thyroid nodule is the most common manifestation of thyroid adenomas and carcinomas. Twenty five percent of the nodules show malignant changes. They should be evaluated for malignancy by illustrating other signs of malignancy on HRSG.

Benign Thyroid Nodules

Benign thyroid nodules include follicular and non-follicular adenomas.

Follicular Adenoma

Follicular adenomas are more common than non-follicular adenomas. Usually they are isoreflective. They present as homogeneous, well-defined isoechogenic nodule. They may undergo cystic degeneration or hemorrhage. The degenerative or hemorrhagic nodule shows low-level echoes on HRSG. The toxic adenomas usually are isoechogenic, and reflectivity of the nodule is same as of the gland. However, they are well demarcated by a peripheral "halo". These nodules usually undergo cystic degeneration on medical treatment. HRSG is good to evaluate the response of medical treatment. These nodules show hypervascularity on color flow imaging than the adjacent normal glandular tissue.

Thyroid Cyst

True cysts of thyroid are not common. However, they are encountered as well-marginated cystic swelling in anterior part of neck. HRSG shows classical sonoappearance of a cyst. They are well-defined with sharp margins, thin walled fluid-filled structures. No internal echo is seen in true cyst. No calcification is noted.

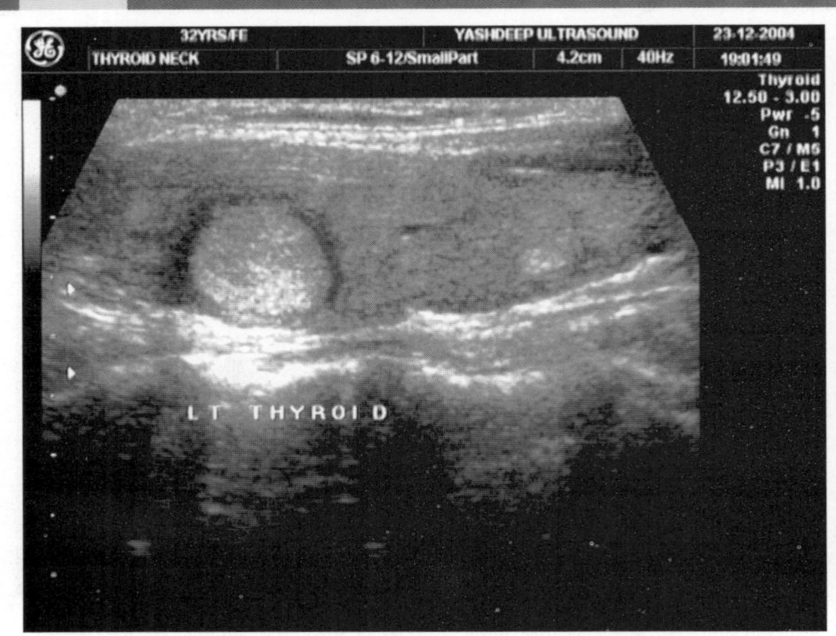

FIGURE 2.82: Solitary adenoma thyroid. HRSG shows well defined homogeneous echogenic nodular shadow in the Lt lobe of thyroid at the lower pole. Well-defined peripheral halo is also seen around the adenoma suggestive of benign nature of the adenoma.

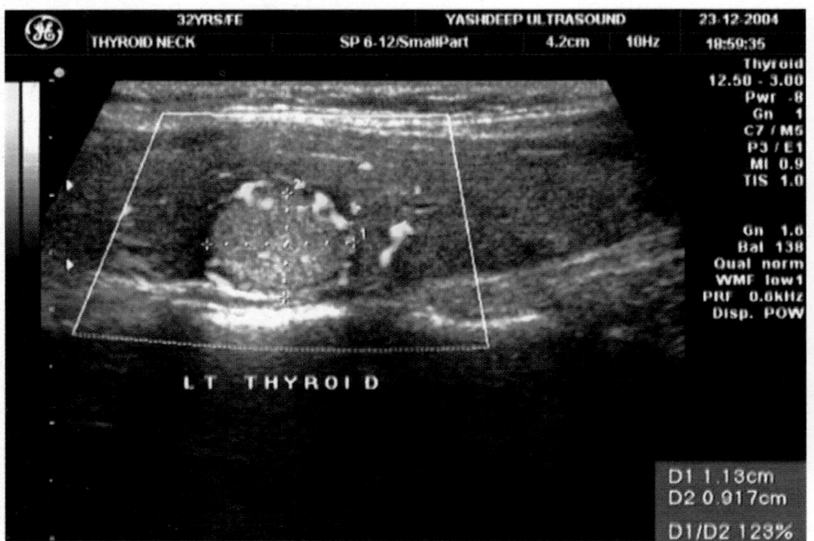

FIGURE 2.82A

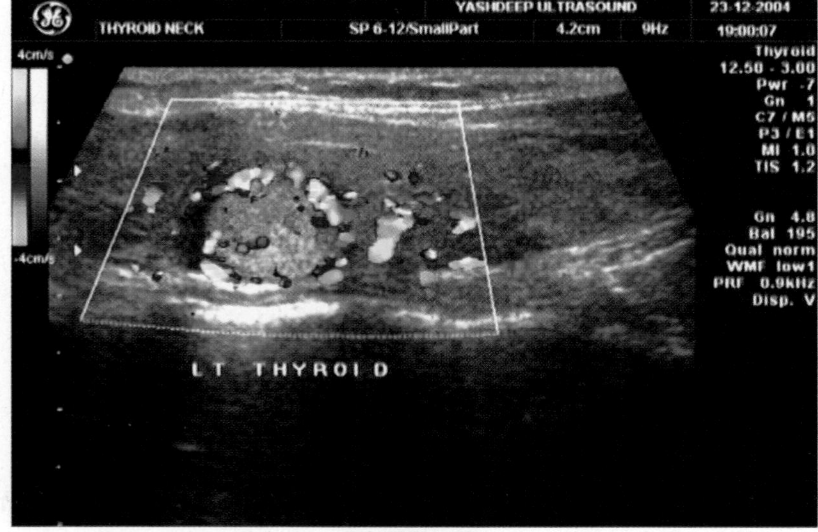

FIGURE 2.82B

FIGURES 2.82A and B: On color flow imaging smooth flow is seen around the adenoma. Few vessels are also seen feeding the adenoma suggestive of follicular nature of the adenoma.

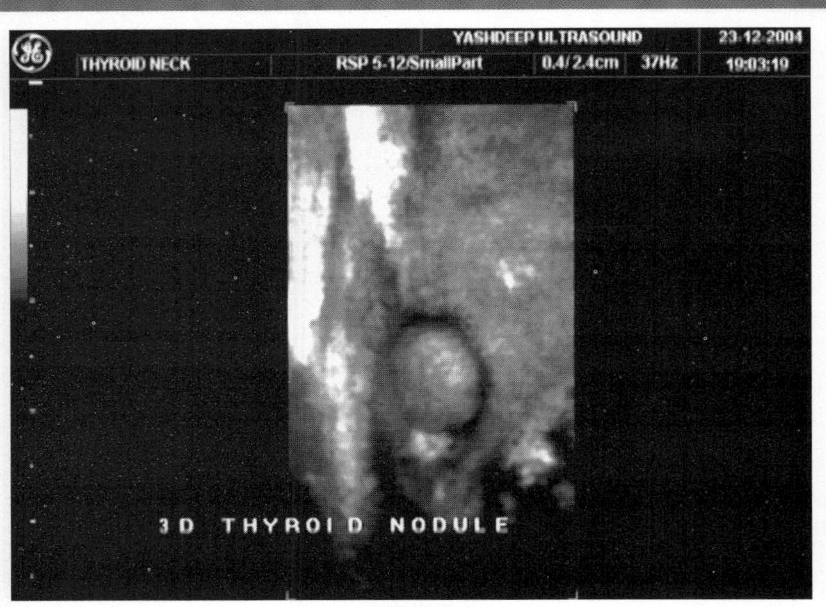

FIGURE 2.83

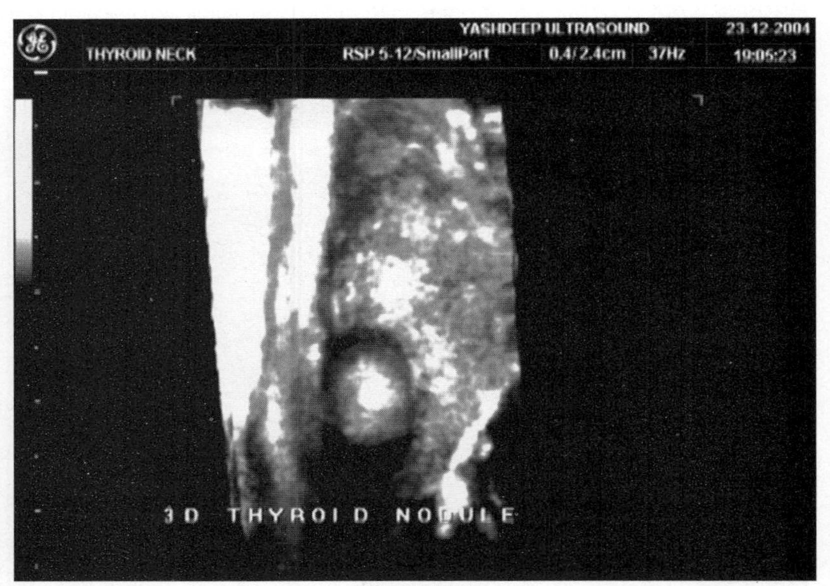

FIGURE 2.84

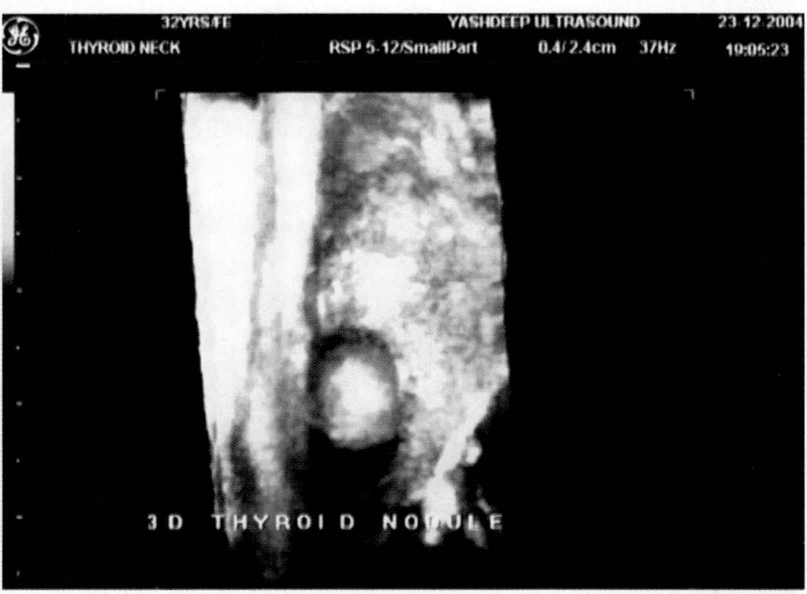

FIGURE 2.85

FIGURES 2.83 to 2.85: 3D imaging of thyroid adenoma of the same patient shows better sonographic details of the adenoma with finer details of anatomical relationship.

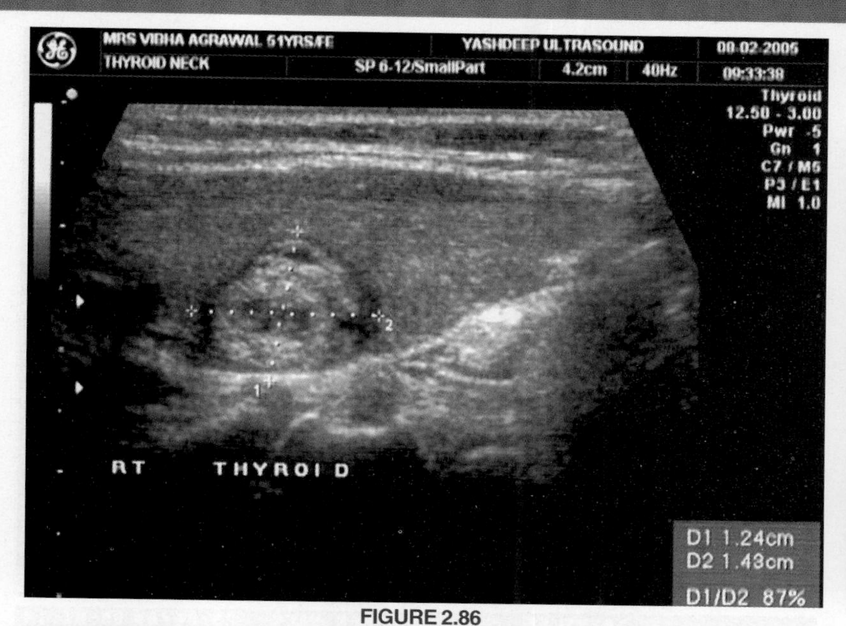

FIGURE 2.86

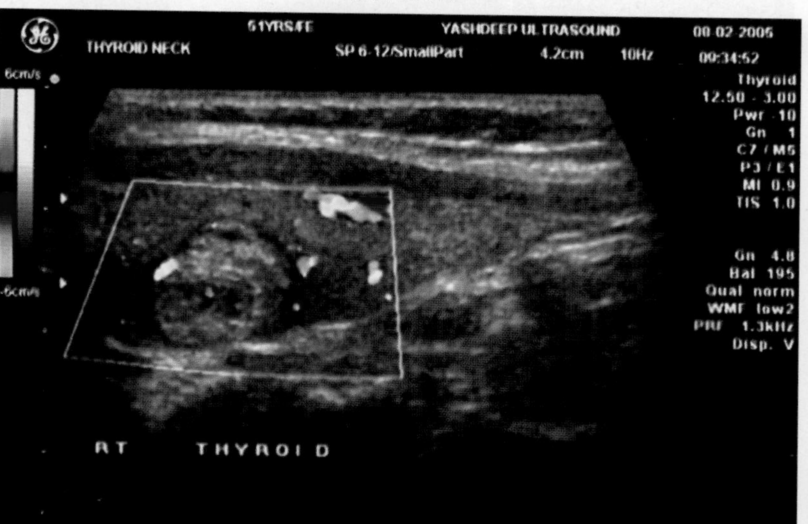

FIGURE 2.87

FIGURES 2.86 and 2.87: Solitary thyroid adenoma. HRSG shows well-defined homogeneous solitary thyroid adenoma in Rt lobe thyroid. It shows homogeneous peripheral halo around it. However, small area of necrosis is seen in the centre of adenoma. On color flow imaging it shows normal flow in the adenoma.

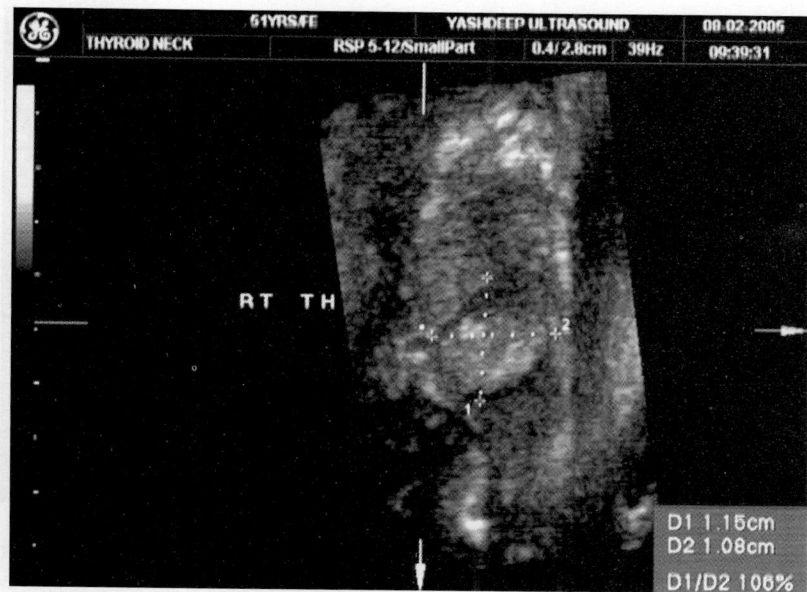

FIGURE 2.88: 3D imaging of the same patient shows better details of the adenoma.

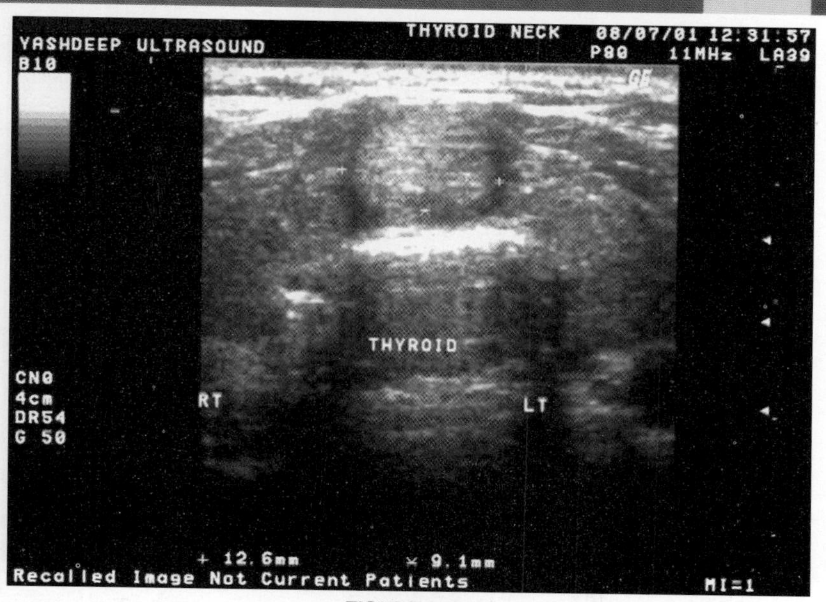

FIGURE 2.89

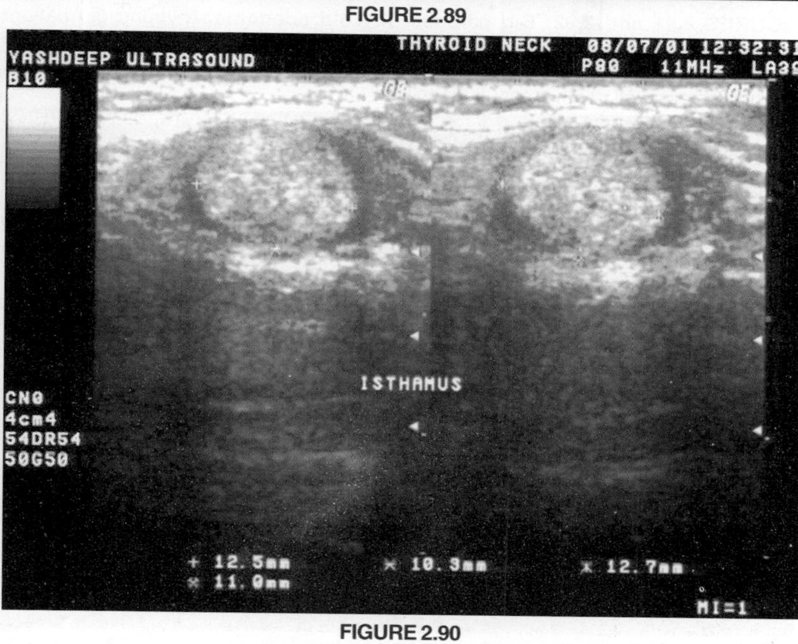

FIGURE 2.90

FIGURES 2.89 and 2.90: Solitary adenoma in the isthmus. HRSG shows well defined echogenic nodular shadows in the Isthmus of the gland. It is smooth in outline with homogeneous texture. Biopsy proved it a follicular adenoma.

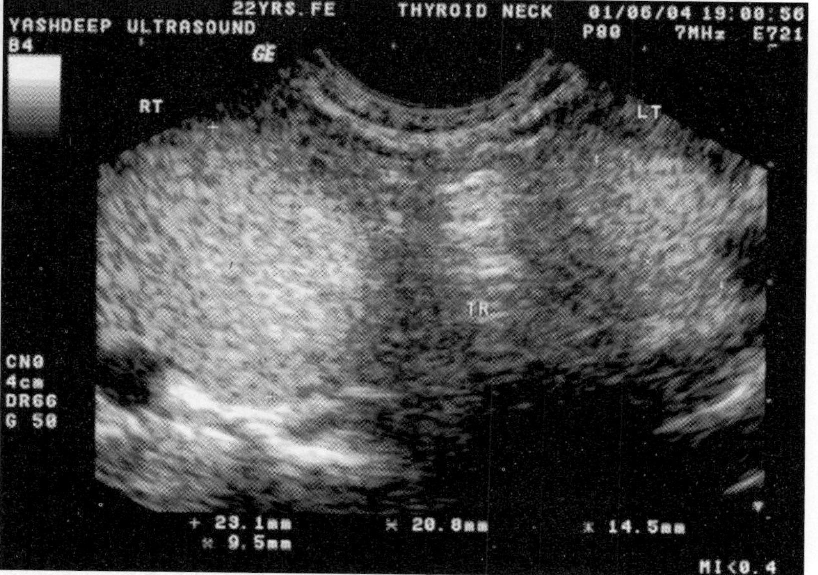

FIGURE 2.91

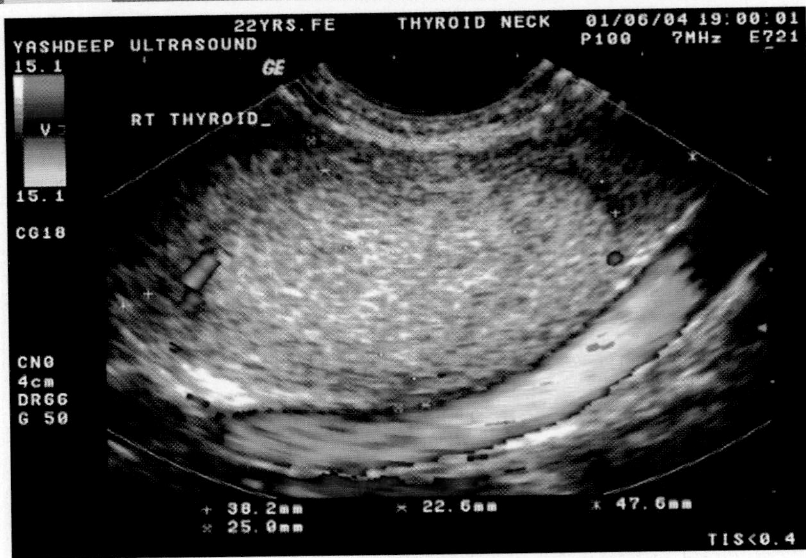

FIGURE 2.92

FIGURES 2.91 and 2.92: Big palpable thyroid adenoma. A young girl patient presented with a mass in the neck. HRSG shows well defined homogeneous echogenic mass in the Rt lobe of the thyroid. The mass is seen pressing over the neck vessels. But no vessels wall involvement is seen.

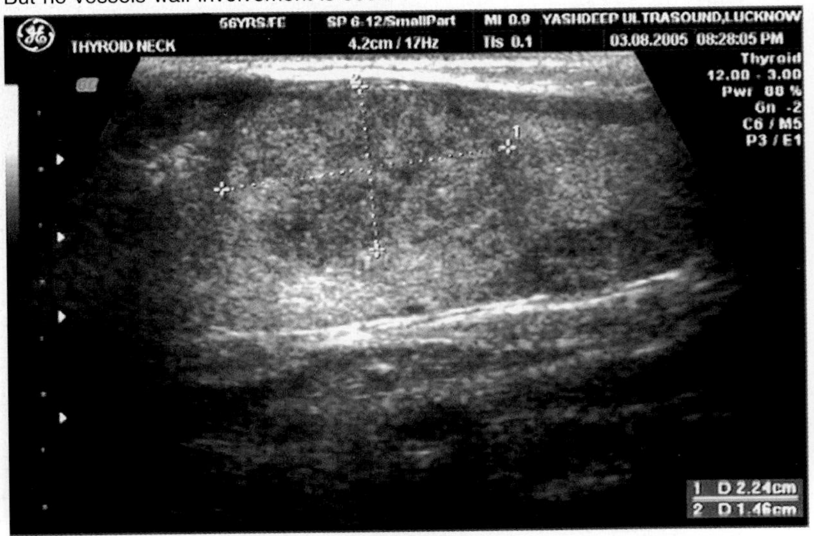

FIGURE 2.93

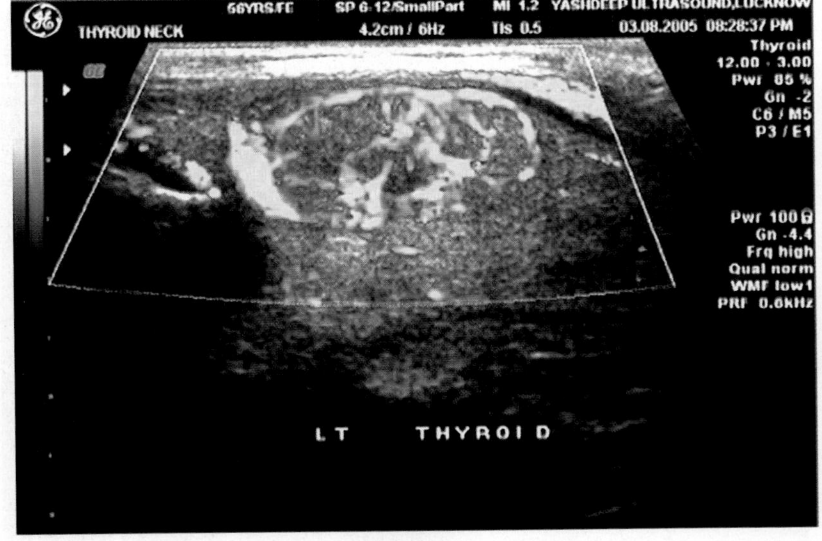

FIGURE 2.94

FIGURES 2.93 and 2.94: Follicular adenoma of thyroid. HRSG shows ill-defined homogeneous isoechogenic mass in the Lt lobe of thyroid. The mass is seen merging in the glandular tissue. On color flow imaging high flow is seen in the mass. Biopsy proved a follicular adenoma. This adenoma with very high vascularity has tendency to turn into malignant follicular adenomas.

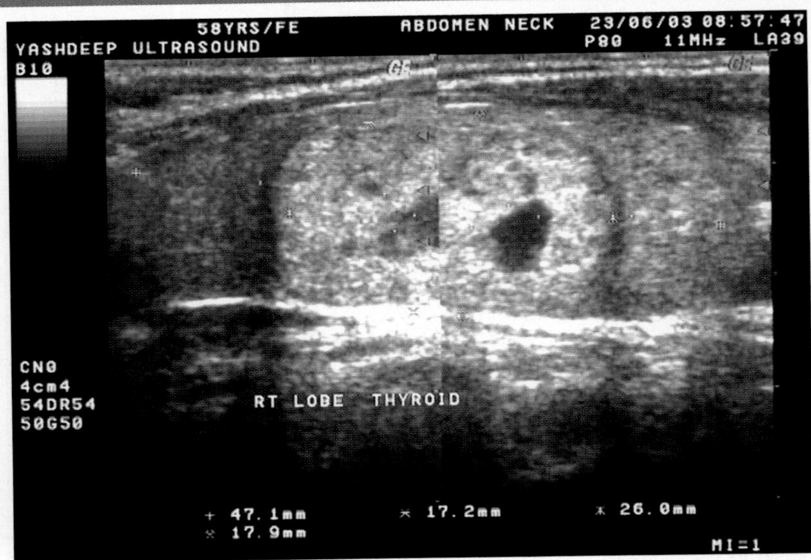

FIGURE 2.95

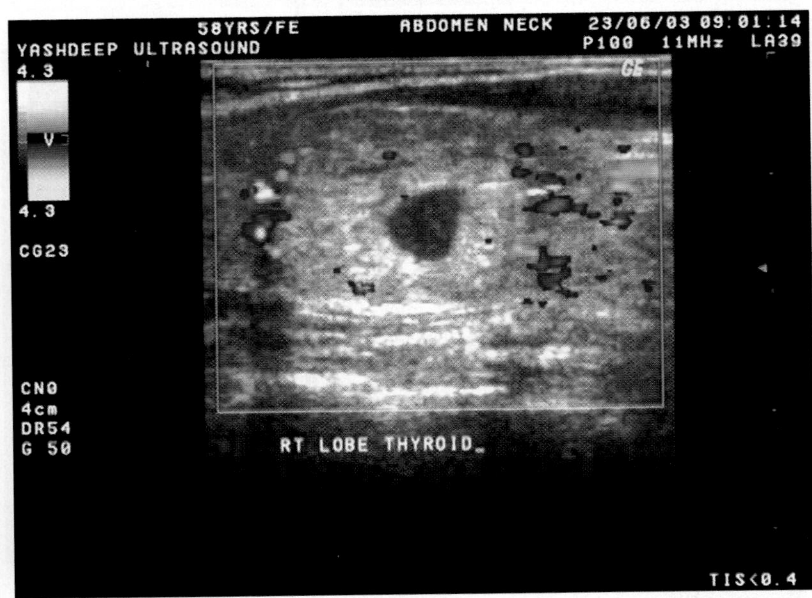

FIGURE 2.96

FIGURES 2.95 and 2.96: Follicular adenoma with cystic degeneration. HRSG shows well defined homogeneous echogenic mass in the gland. It shows central area of cystic degeneration. On color flow imaging high flow is seen in it. Ultrasound guided Biopsy proved follicular adenoma.

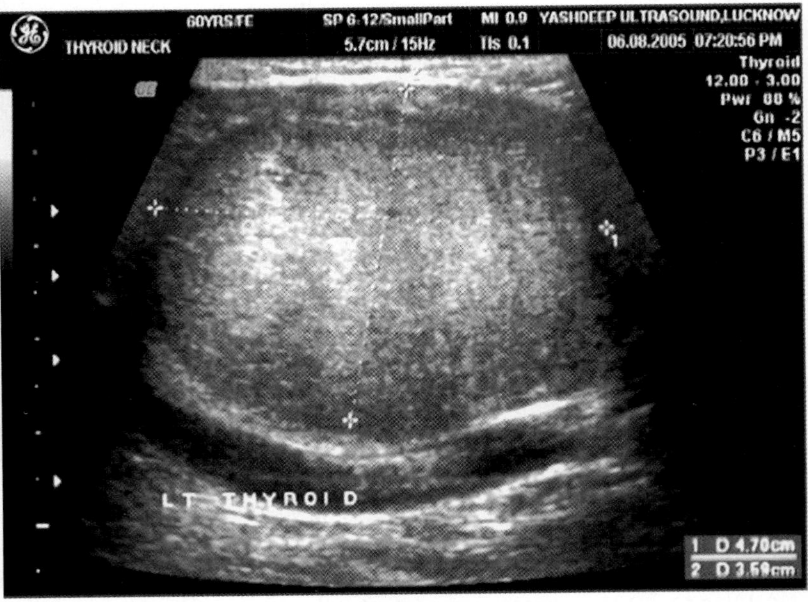

FIGURE 2.97

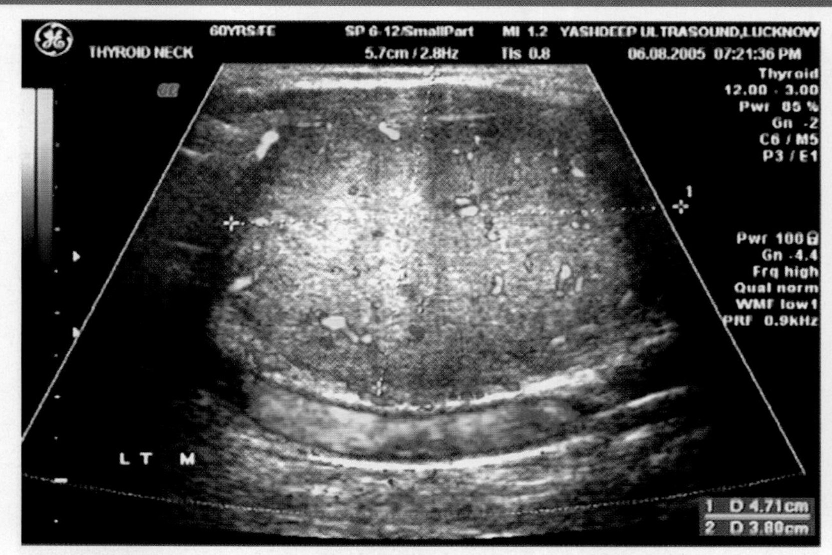

FIGURE 2.98

FIGURES 2.97 and 2.98: Big solid follicular adenoma. A patient presented with a big mass in the Lt lobe of thyroid. HRSG shows a big homogeneous solid mass in the Lt lobe of thyroid mid and lower part. The mass shows Isoechogenic texture similar to the thyroid. It is seen pressing over the neck vessels and displacing them.

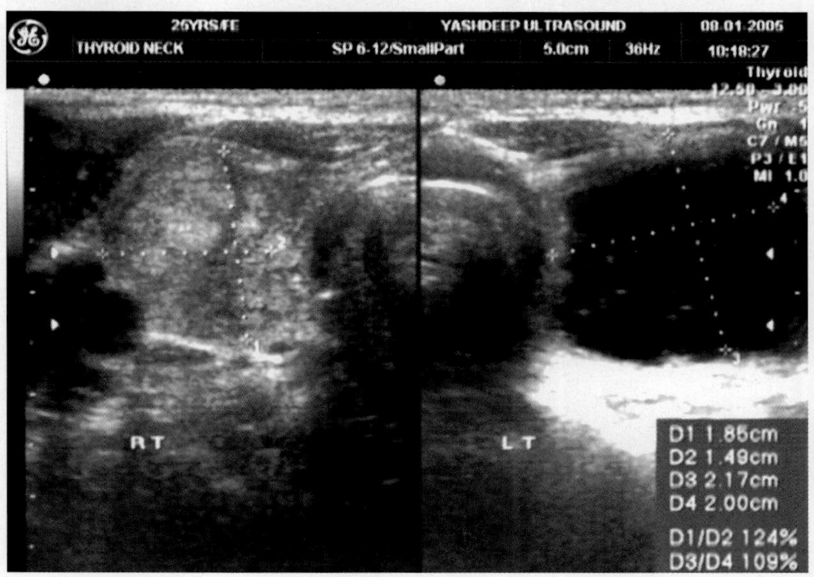

FIGURE 2.99: Cystic and solid adenomas thyroid. A patient presented with bilateral soft tissue mass in both lobes of thyroid. HRSG shows well defined homogeneous echogenic solid adenoma in the Rt lobe and a cystic adenoma in the Lt lobe. They show sharp margins suggestive of benign nature of the mass.

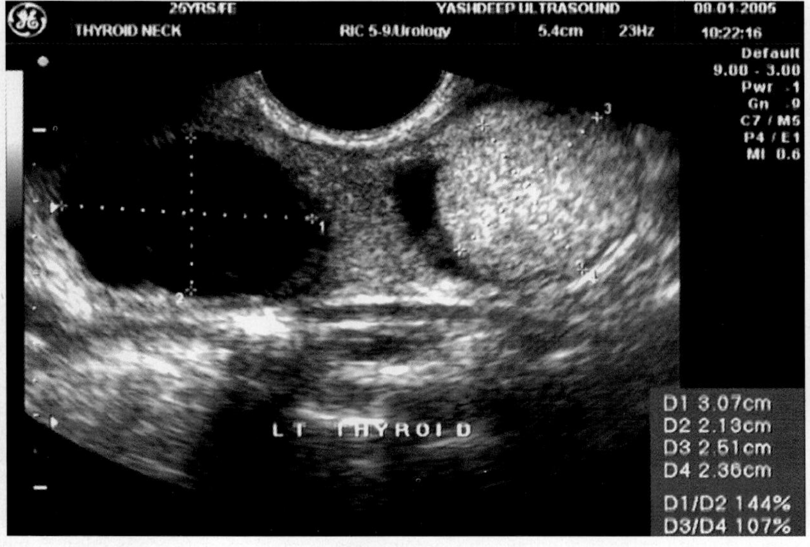

FIGURE 2.100

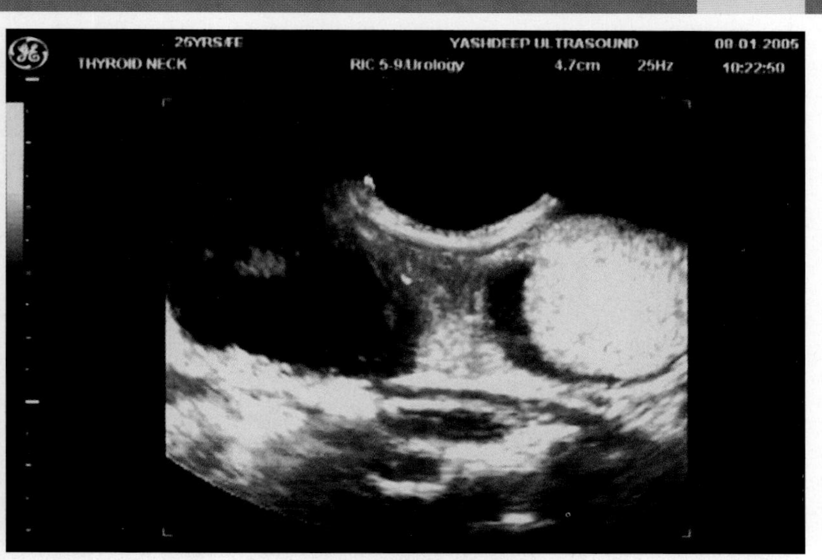

FIGURE 2.101

FIGURES 2.100 and 2.101: 3D imaging of the same patient shows the better and finer details of the adenomas.

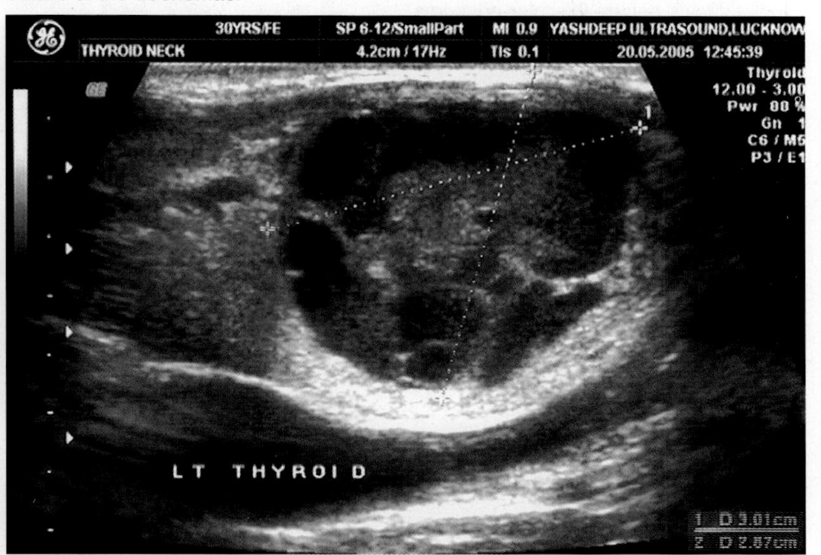

FIGURE 2.102

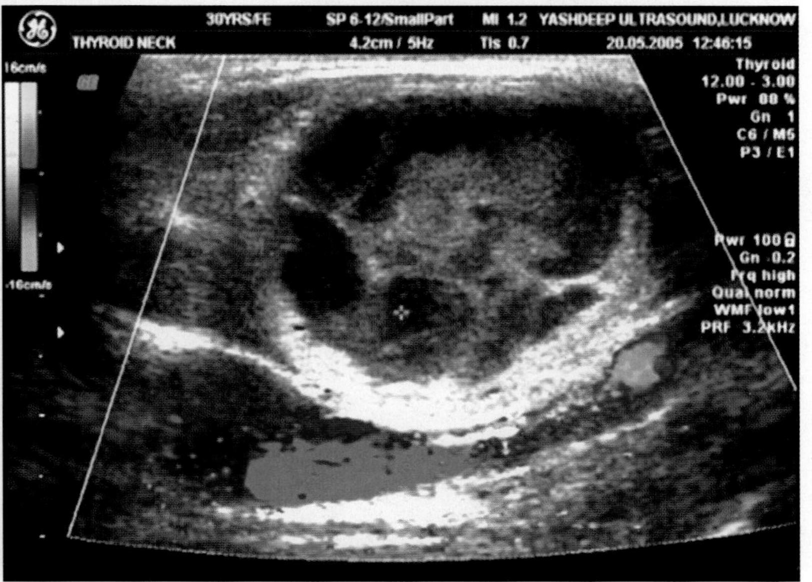

FIGURE 2.103

FIGURES 2.102 and 2.103: Complex adenoma thyroid. A patient presented with a soft tissue swelling in a thyroid. HRSG shows complex echo mass in the Lt lobe of thyroid. It shows solid cystic components. Margins are smooth in outline. No flow is seen on color flow imaging. No calcification is seen. Biopsy proved benign adenoma.

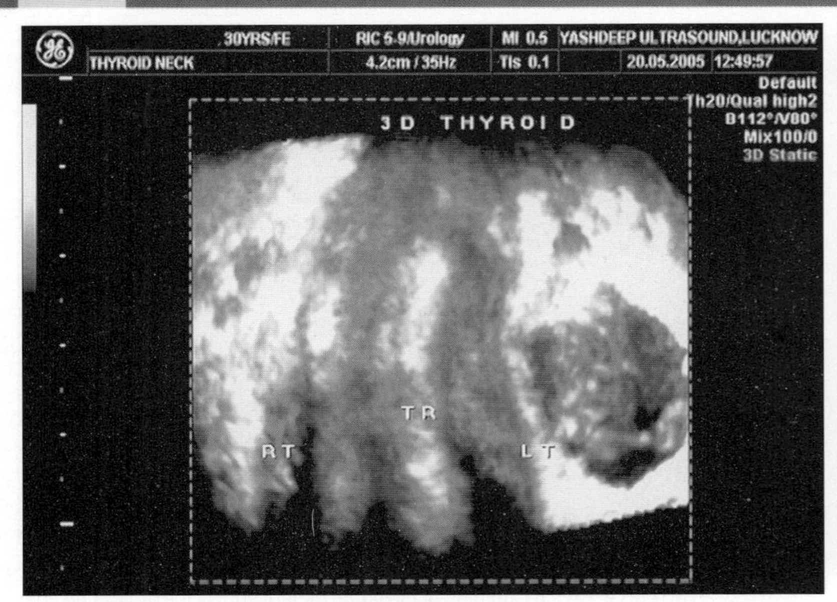

FIGURE 2.104: 3D Imaging shows details of the adenoma with multiple internal echoes. Biopsy shows benign follicular adenoma.

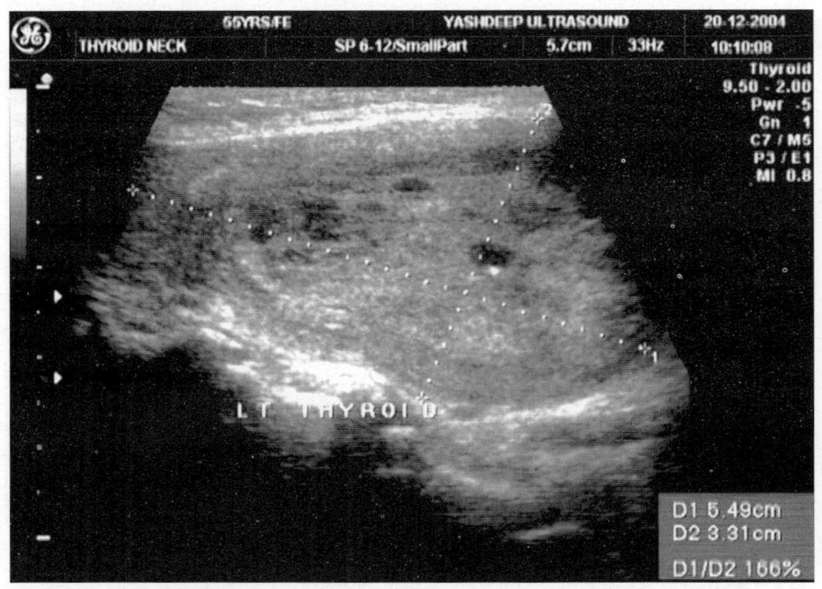

FIGURE 2.105: Benign adenoma. A patient presented with soft tissue mass in the Lt lobe of thyroid. The mass shows sharp margins. However, it shows heterogeneous texture.

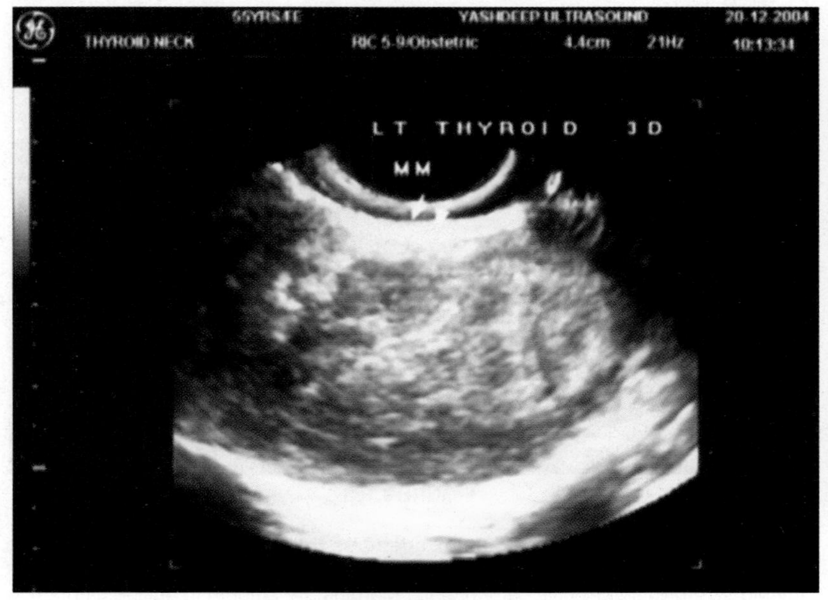

FIGURE 2.106

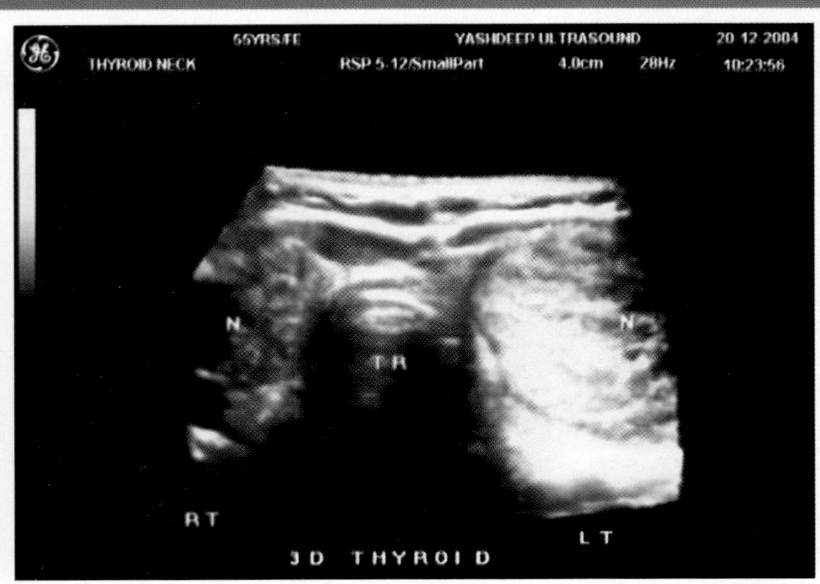

FIGURE 2.107

FIGURES 2.106 and 2.107: 3D of the mass show better anatomical details. The margins are sharp with smooth outline suggests benign nature of the mass.

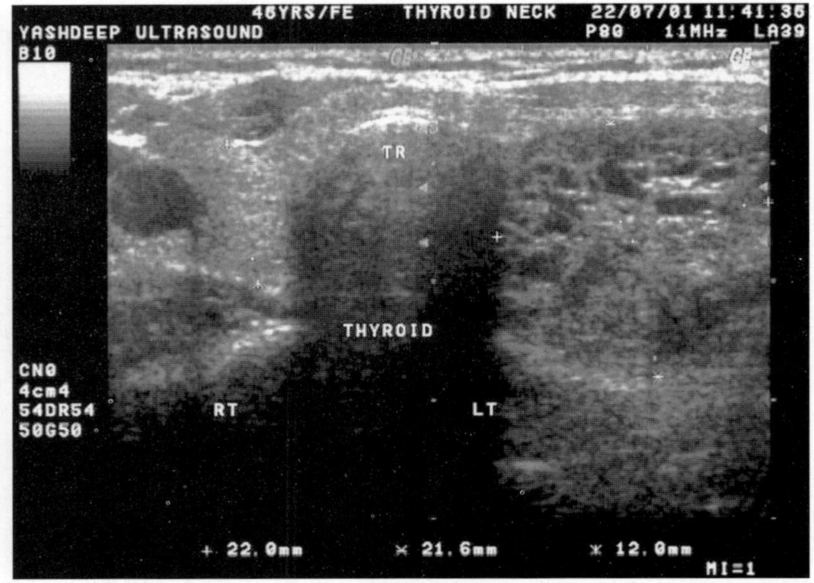

FIGURE 2.108

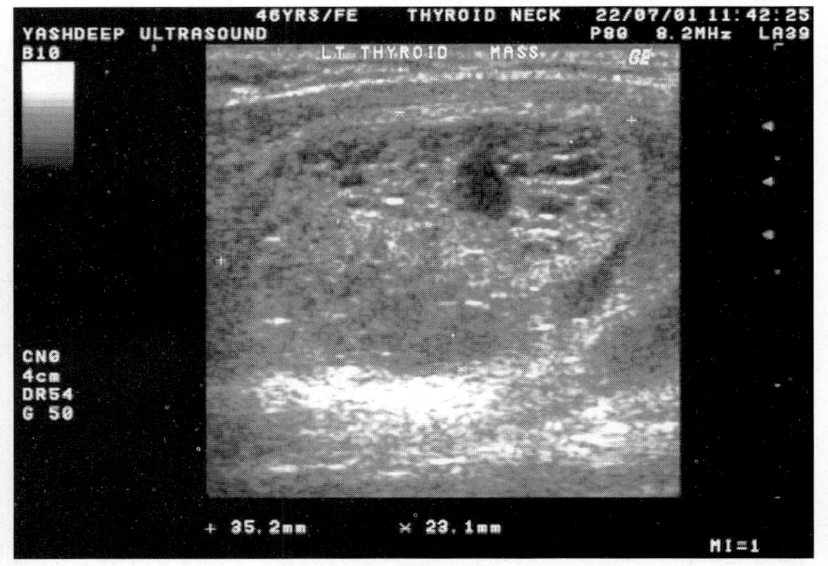

FIGURE 2.109

FIGURES 2.108 and 2.109: Complex adenoma with honeycomb pattern. HRSG shows benign mass in the thyroid. Complex honeycomb appearance is seen in the adenoma with cystic changes suggestive of benign nature of the mass.

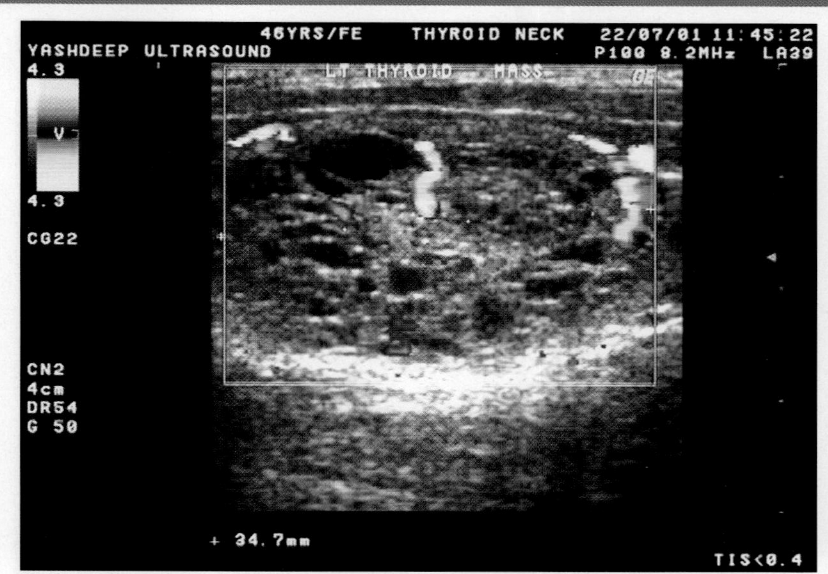

FIGURE 2.110: On color flow imaging moderate flow is seen in the adenoma.

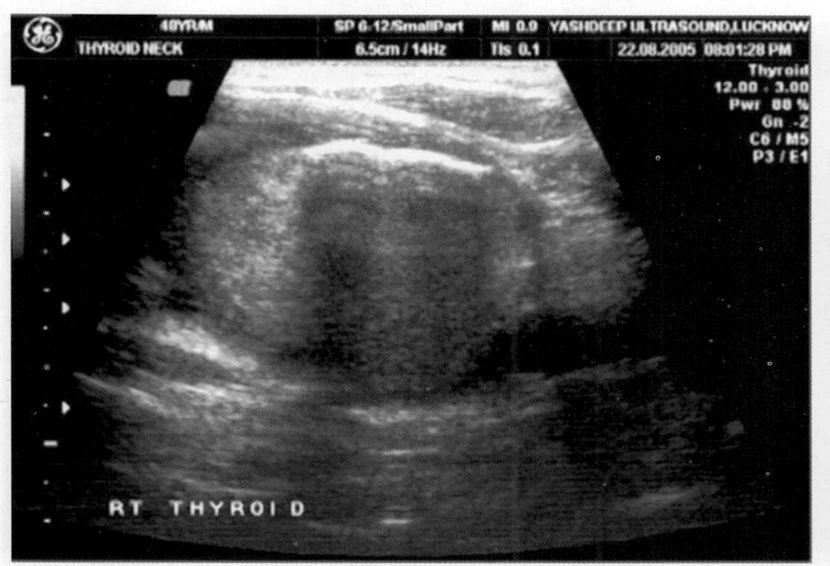

FIGURE 2.111

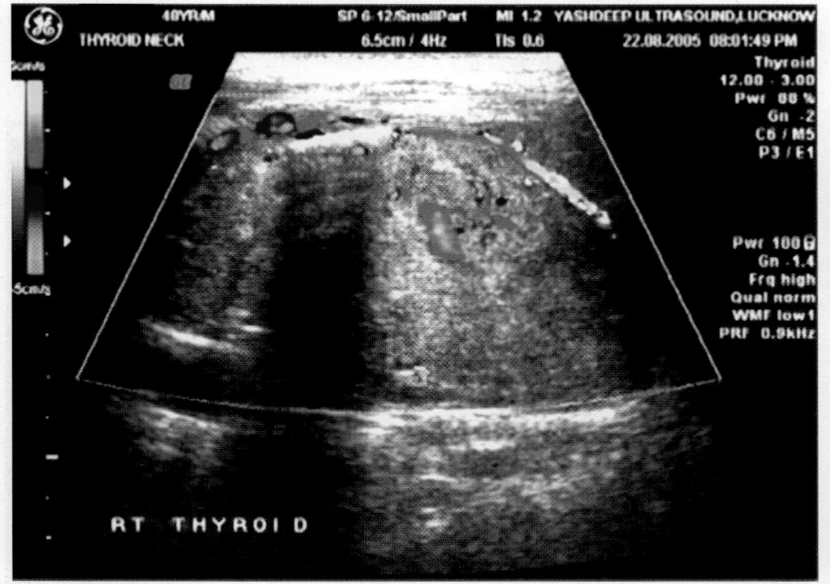

FIGURE 2.112

FIGURES 2.111 and 2.112: Benign adenoma with peripheral calcification. A patient presented with enlarged Rt lobe of thyroid. HRSG shows a big solid mass in the thyroid. Dense echogenic linear calcification is seen in it. It is accompanied with dense acoustic shadowing. The calcification looks like eggshell calcification suggestive of benign nature of the mass.

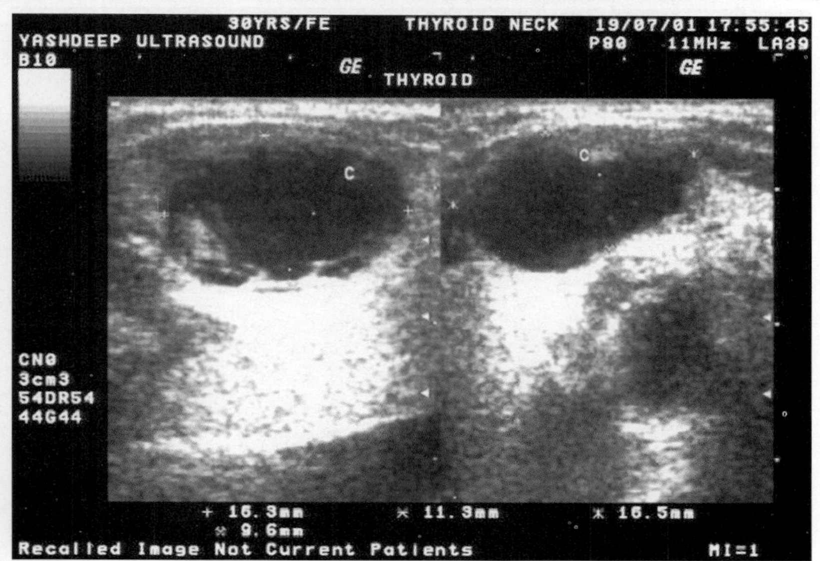

FIGURE 2.113

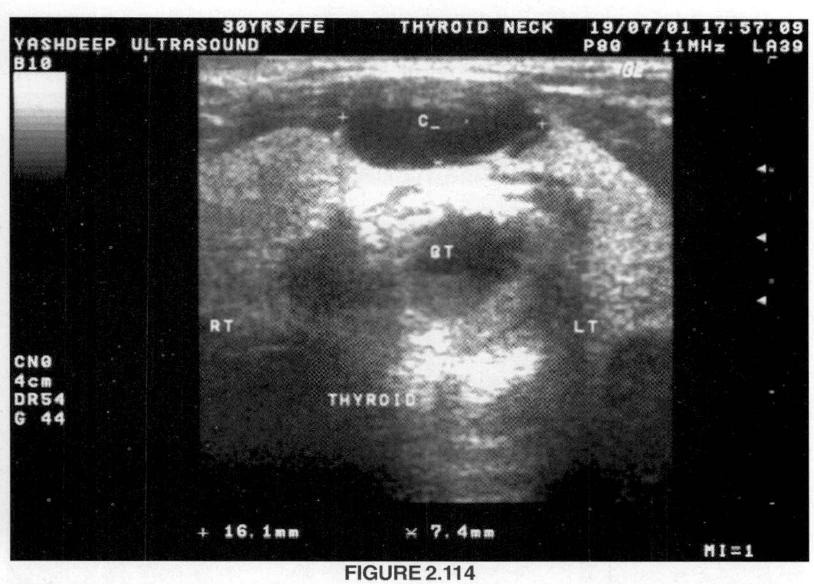

FIGURE 2.114

FIGURES 2.113 and 2.114: A patient presented with a cystic mass in the midline. HRSG shows a well-defined cyst in the Isthmus of the gland. No internal echo is seen in it.

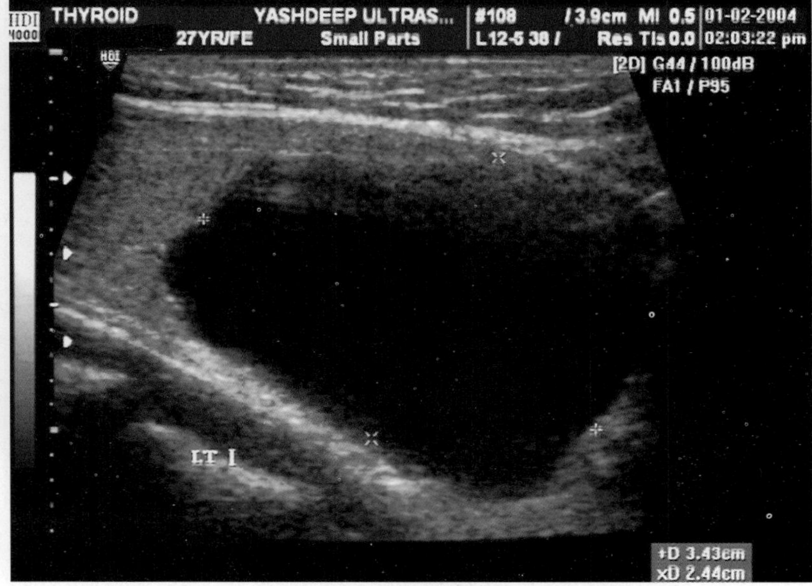

FIGURE 2.115

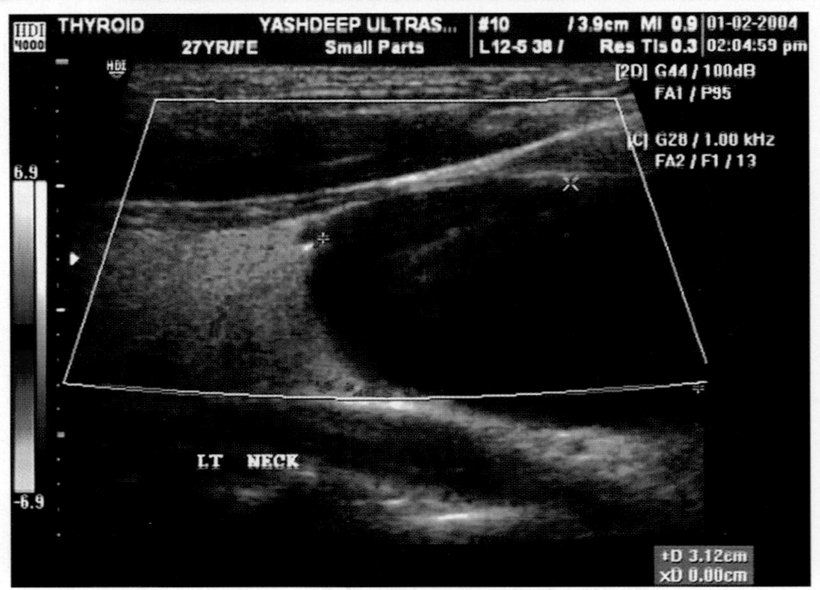

FIGURE 2.116

FIGURES 2.115 and 2.116: Big simple cyst in thyroid. HRSG shows a big cystic mass in the thyroid gland. The cyst is seen replacing whole of the glandular tissue. Aspiration of the cysts shows simple cyst. Simple cysts are uncommon in thyroid gland.

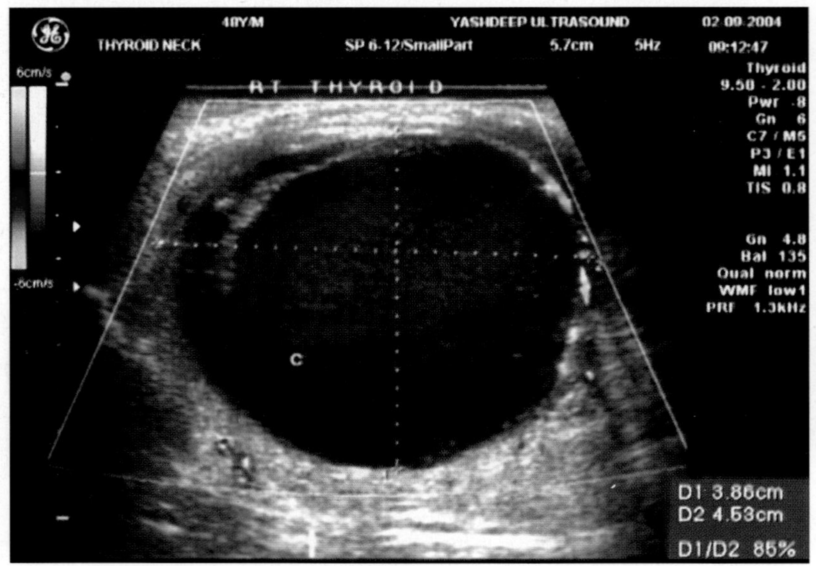

FIGURE 2.117

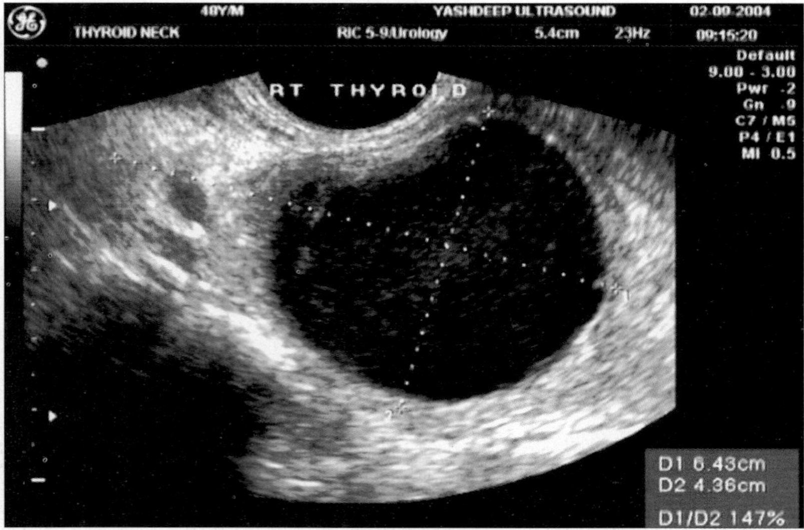

FIGURE 2.118

FIGURES 2.117 and 2.118: Simple cyst in the thyroid. Well defined cystic mass is seen in the Rt lobe of the thyroid. 3D shows no echoes in the cyst. Aspiration of the cyst suggests simple cyst.

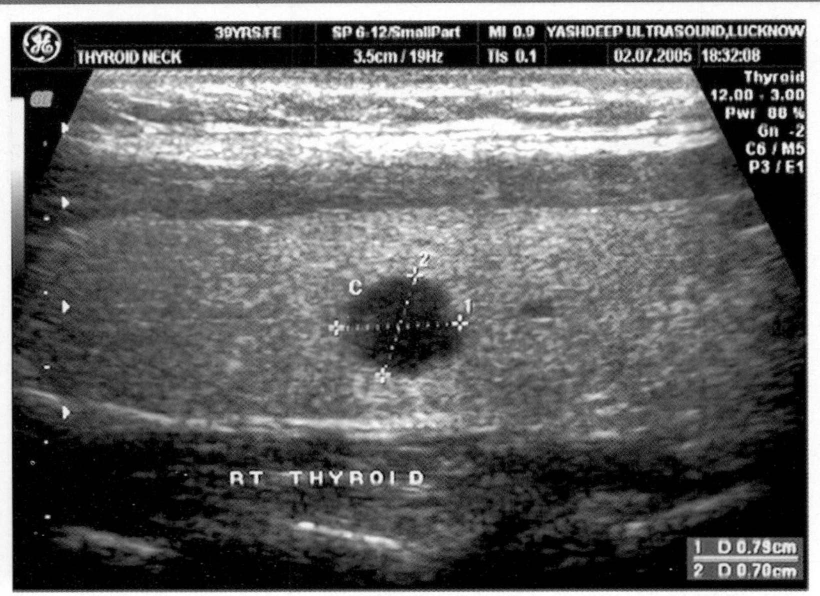

FIGURE 2.119

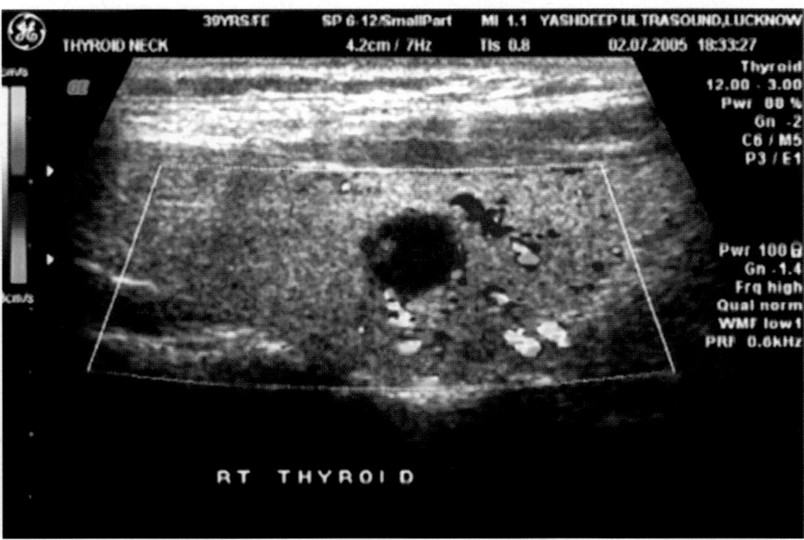

FIGURE 2.120

FIGURES 2.119 and 2.120: Non-palpable simple cyst. HRSG shows clinically non-palpable cyst in the Rt lobe of thyroid. No internal echoes seen in the cyst. Color flow imaging shows peripheral flow in the cyst.

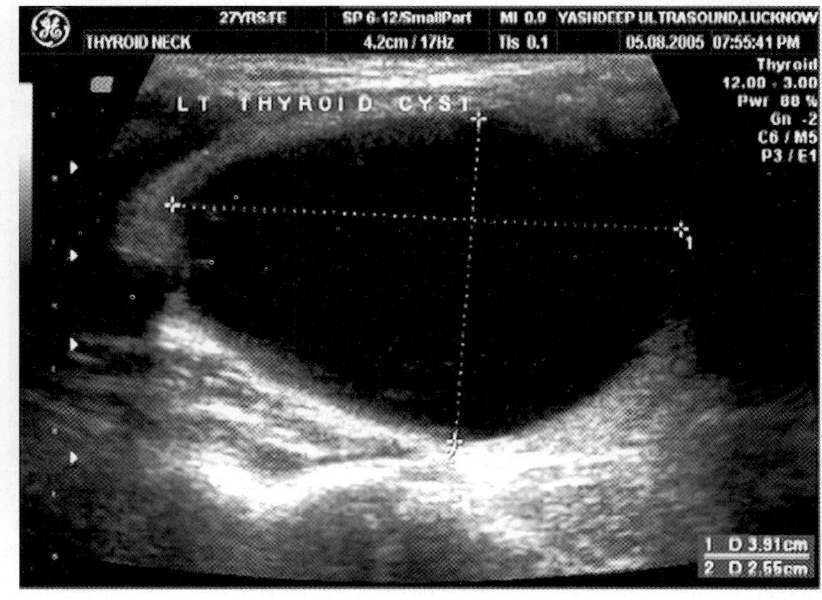

FIGURE 2.121

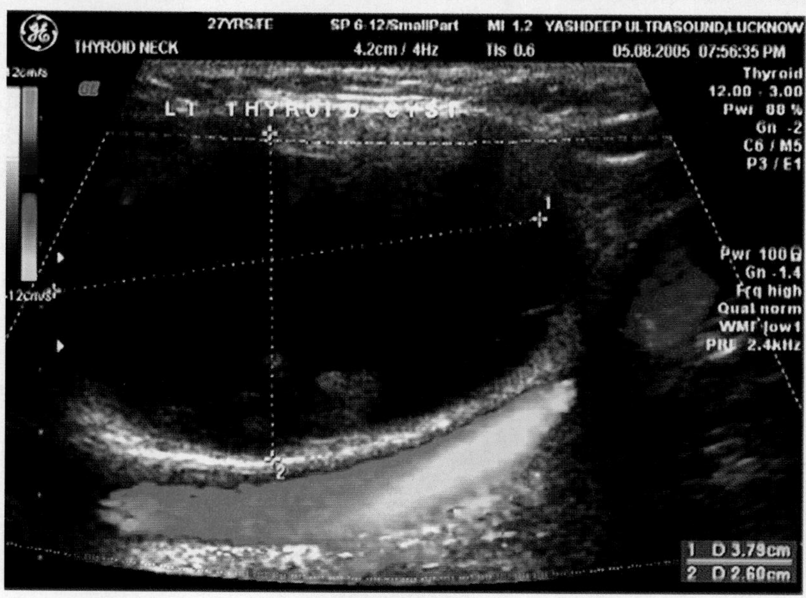

FIGURE 2.122

FIGURES 2.121 and 2.122: Simple cyst in thyroid with pressure displacement on the neck vessels. Big cystic mass is seen in the Lt lobe of thyroid. No internal echo is seen in the cyst. On color flow imaging it shows pressure over the CCA. Aliasing is seen due to pressure over the vessels.

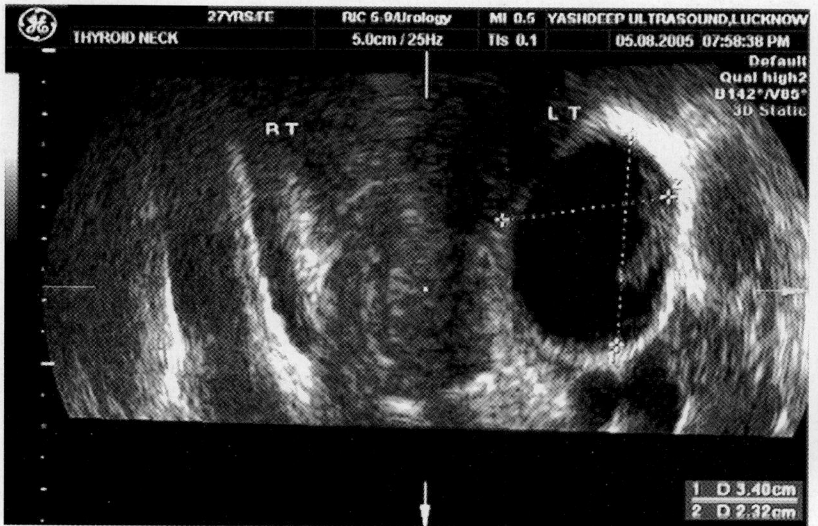

FIGURE 2.123

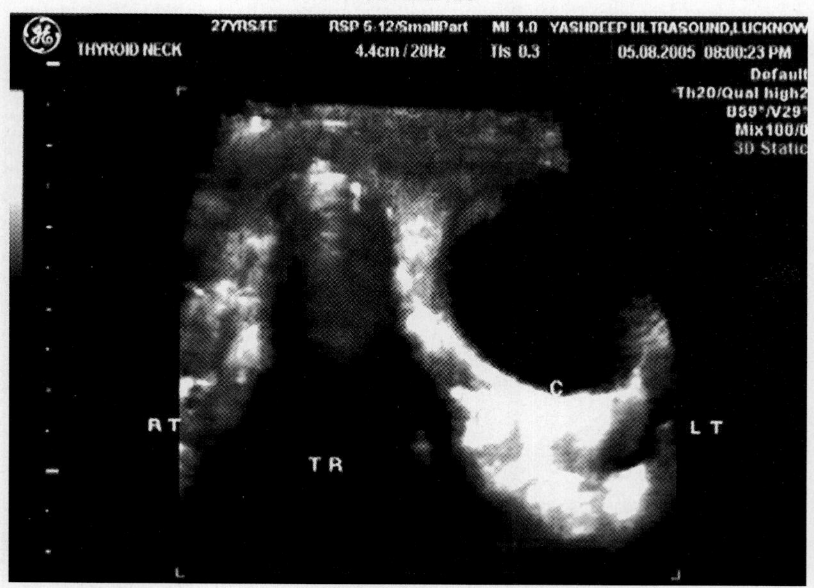

FIGURE 2.124

FIGURES 2.123 and 2.124: 3D imaging of the same patient shows better tissue details of the cyst.

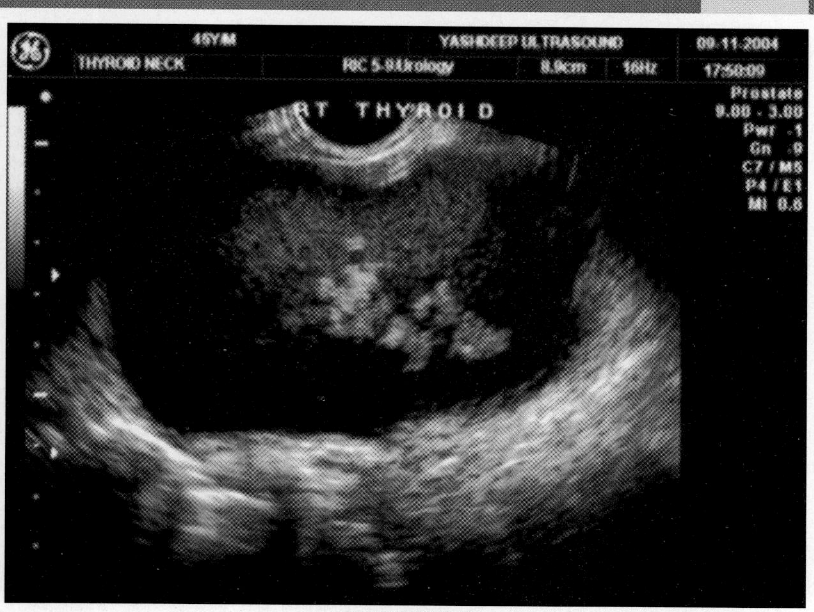

FIGURE 2.125

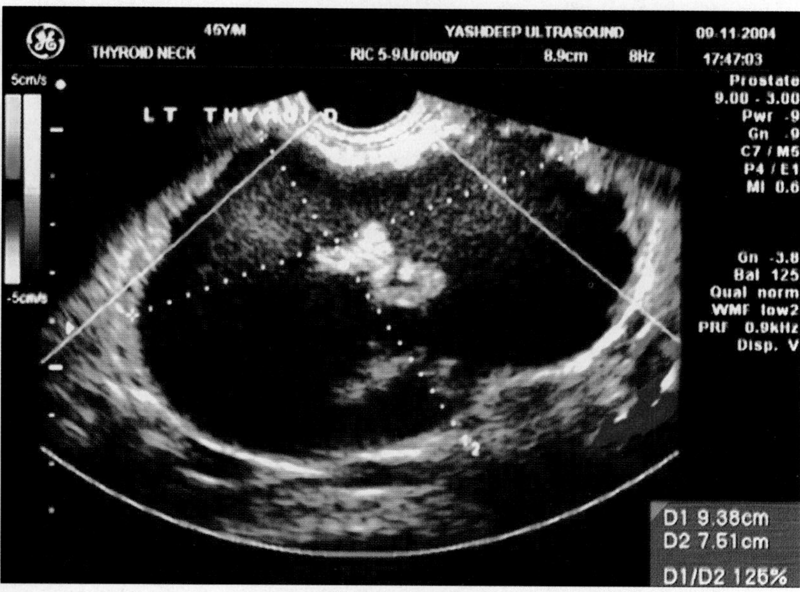

FIGURE 2.126

FIGURES 2.125 and 2.126: Cystic degeneration of the thyroid gland. HRSG shows a big cystic mass. It measures 93 x 75 mm in size. Small glandular tissue is also seen floating in the cyst. Biopsy shows cystic degeneration of the thyroid.

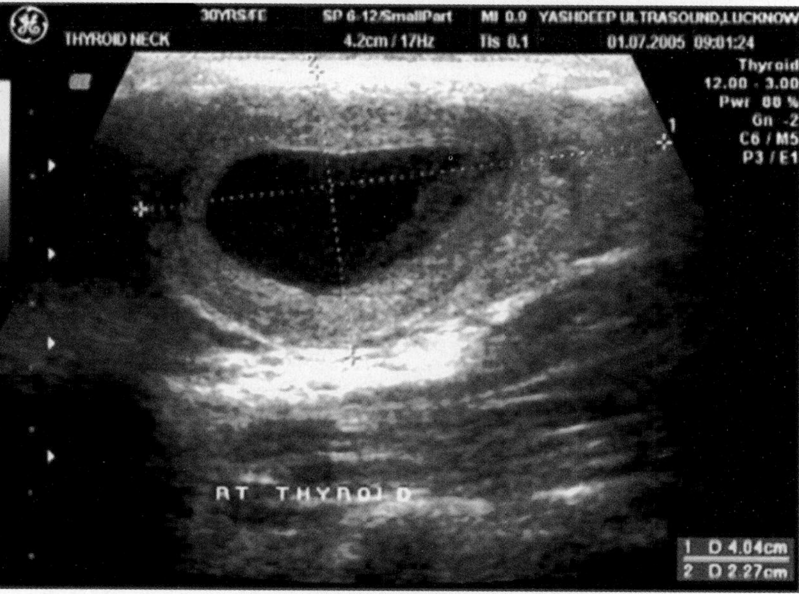

FIGURE 2.127

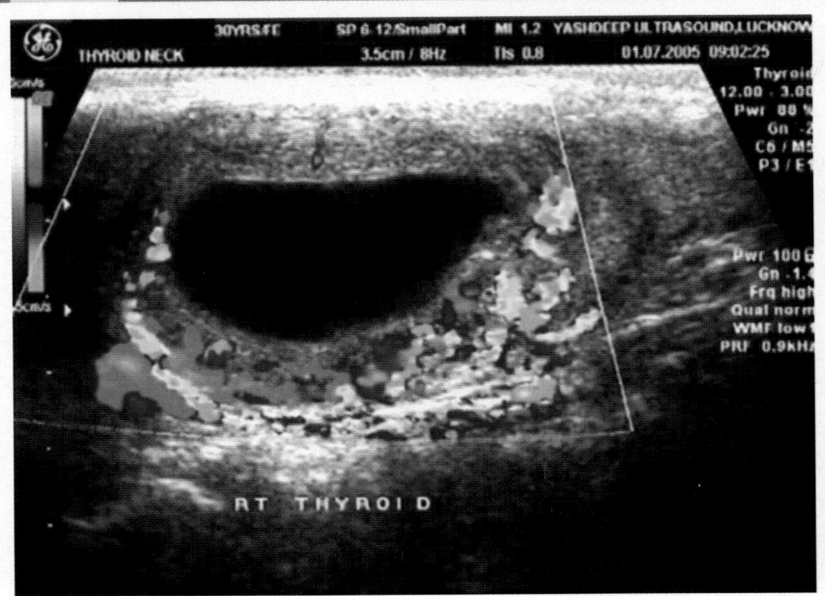

FIGURE 2.128

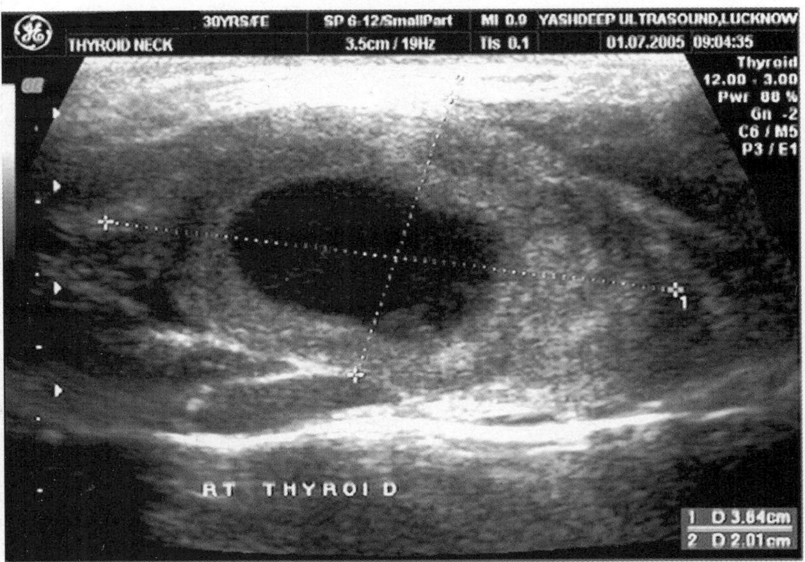

FIGURE 2.129

FIGURES 1.127 to 2.129: HRSG shows thick wall cystic mass in the Rt lobe of thyroid. No internal echo is seen in the cyst. Normal glandular tissue is seen around the cyst. 3D imaging of cyst also shows fine details of the cystic mass. On color flow imaging, high flow is seen in the normal glandular tissue.

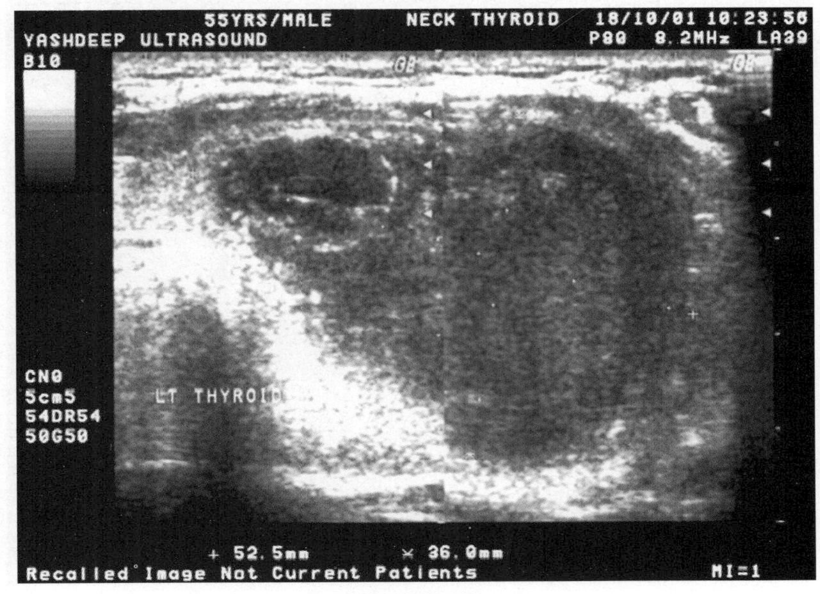

FIGURE 2.130

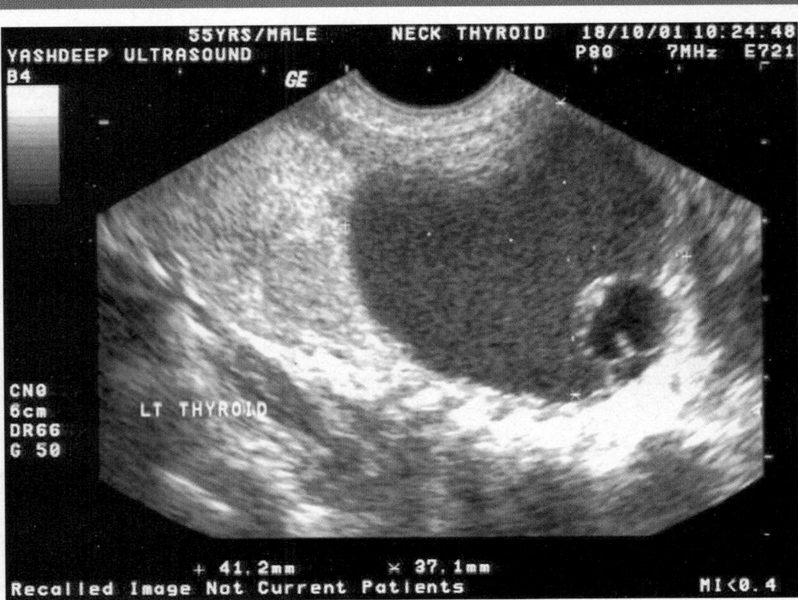

FIGURE 2.131

FIGURES 2.130 and 2.131: Big colloid cyst in the Lt lobe. A patient presented with a big mass in the Lt neck with dense cystic feeling on clinical examination. HRSG shows a big cystic mass—involving whole of the Lt lobe. It was seen extending in retrosternal area. Small cyst is also seen within the large cyst.

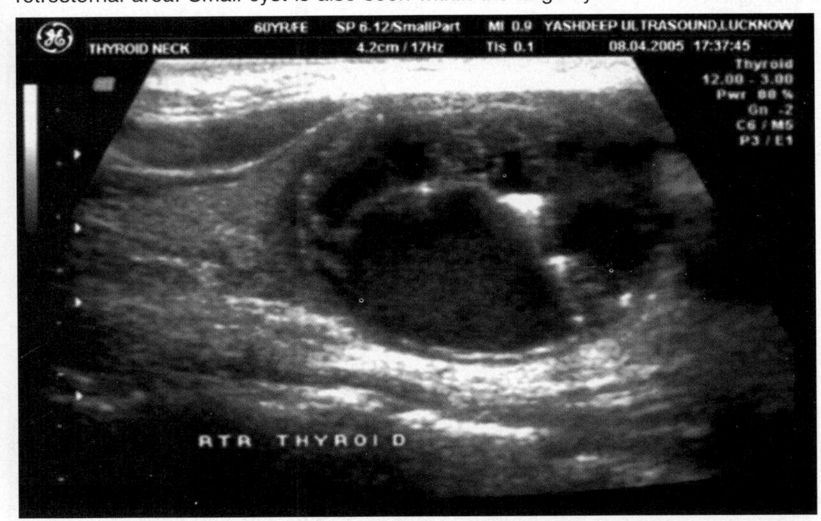

FIGURE 2.132: Colloid cyst with comet tail sign. HRSG shows a cystic mass in the thyroid. Multiple echogenic tiny foci are seen floating in the cyst suggestive of micro-colloid crystals. These floating shadows are seen associated with comet tail artifacts known as comet tail sign typical of colloid cyst.

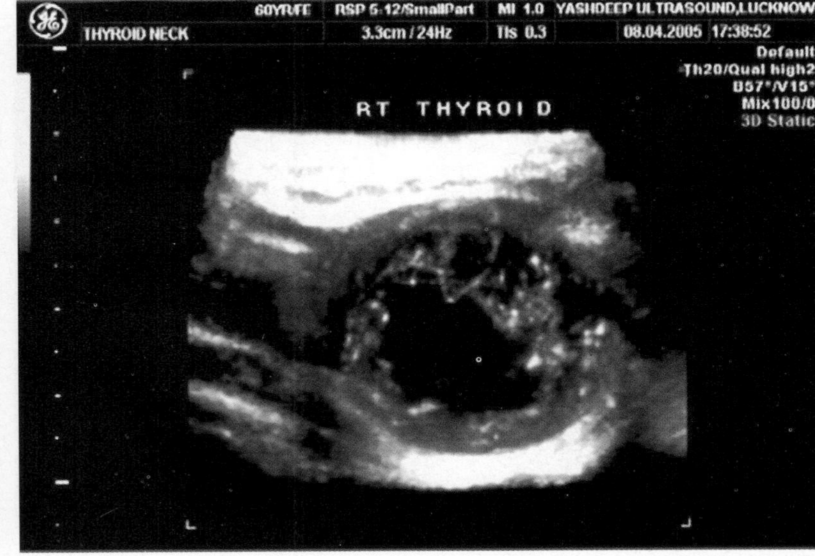

FIGURE 2.133: 3D imaging of the patient shows details of the floating microcrystals with evidence of thick jell like colloid in the cyst.

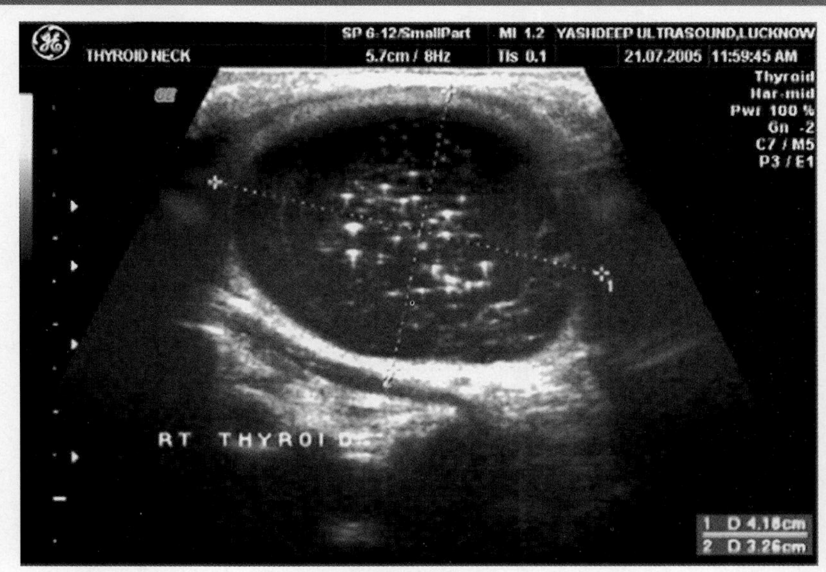

FIGURE 2.134

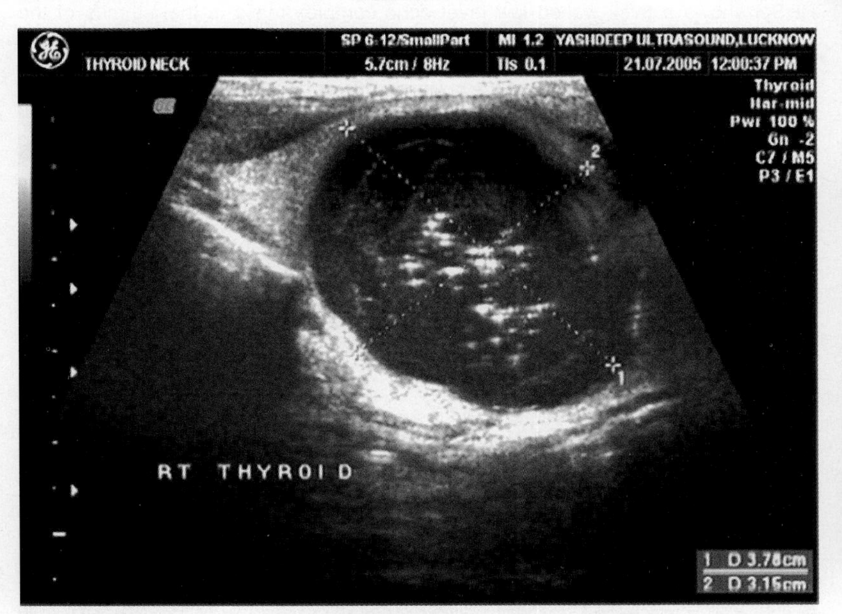

FIGURE 2.135

FIGURES 2.134 and 2.135: Colloid cyst with "Starry Sky" appearance. A thick wall cystic mass seen in the thyroid gland with multiple bright echogenic floating shadows in the cyst. They give **Starry Sky** appearance.

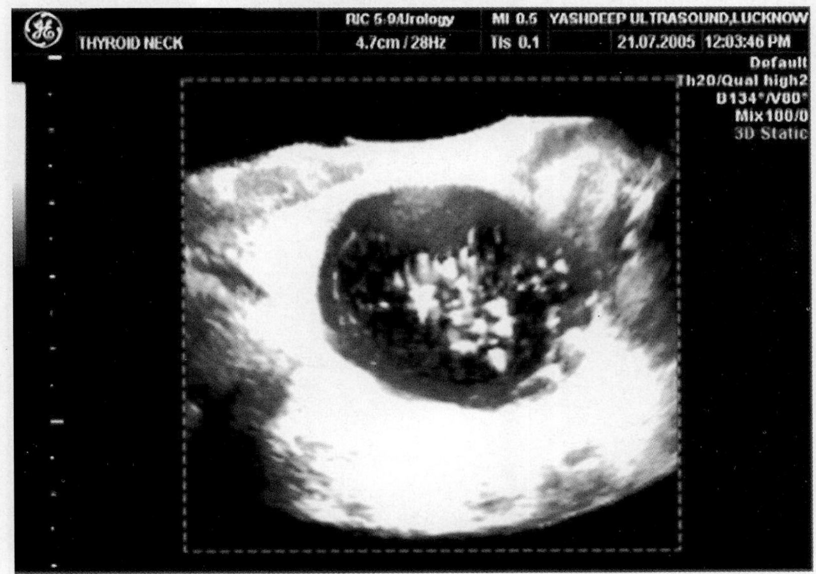

FIGURE 2.136: 3D of the same patient shows details of the floating crystals in the cyst.

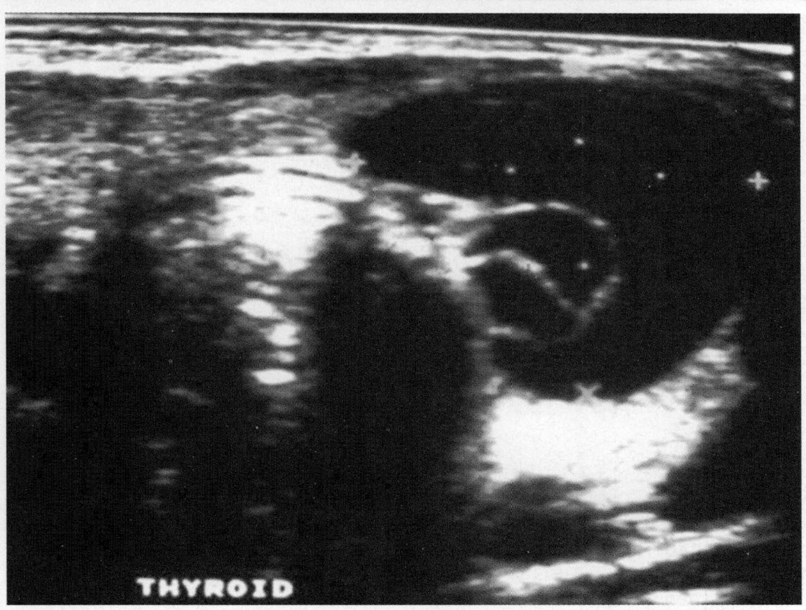

FIGURE 2.137: Hydatid cyst in the thyroid. HRSG shows a well defined cystic mass in the Lt lobe of the thyroid. There is also evidence of a multiloculated small cyst seen fixed with the inner wall of the cyst. Multiple small daughter cysts are seen in it typical feature of the hydatid cyst. Hydatid cysts are rare in thyroid.

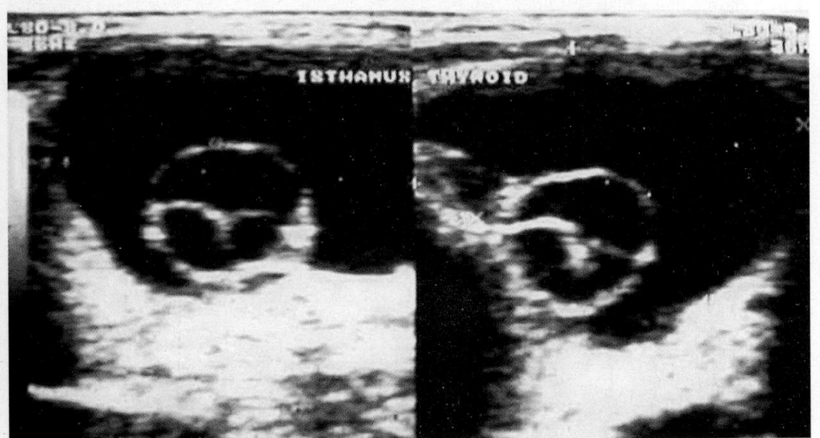

FIGURE 2.138

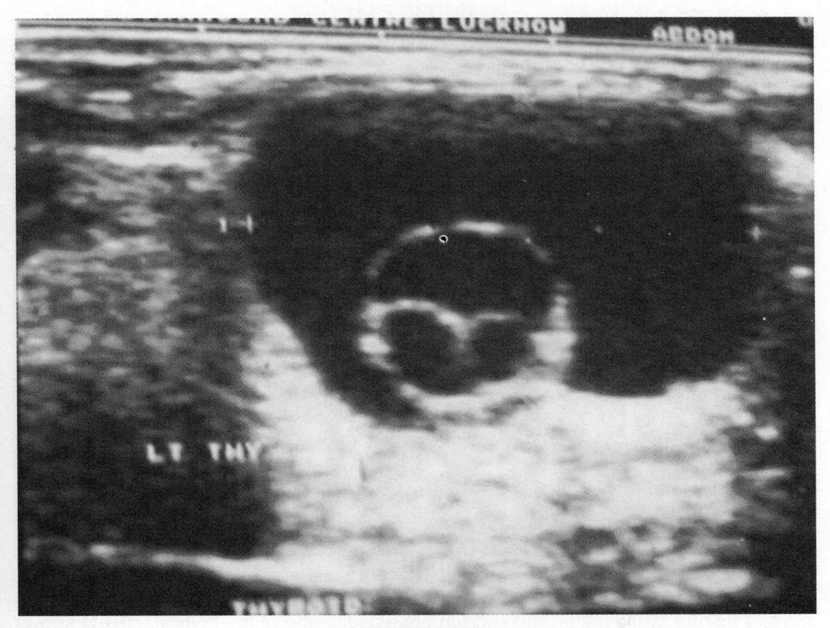

FIGURE 2.139

FIGURES 2.138 and 2.139: Hydatid cyst in thyroid. Well defined multiloculated cystic mass is seen in the Lt lobe of thyroid. Multiple small daughter cysts are seen in the big cyst typical of the hydatid cyst. Aspiration biopsy confirmed the hydatid cyst.

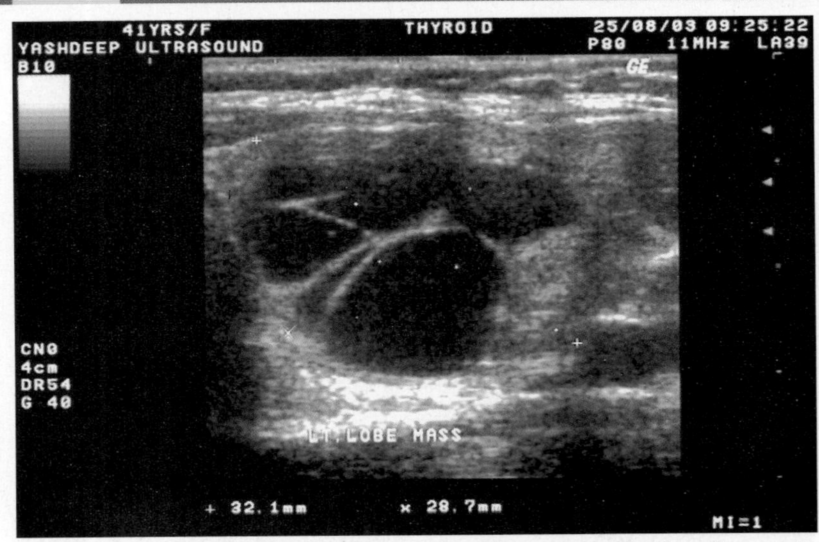

FIGURE 2.140: **Hydatid cyst in the thyroid.** A patient presented with cystic mass in the Lt lobe of thyroid. HRSG shows well-defined cyst in the thyroid. Fine echogenic septa are seen floating in the cyst suggestive of positive "Water Lily Sign". Aspiration confirmed the HRSG findings.

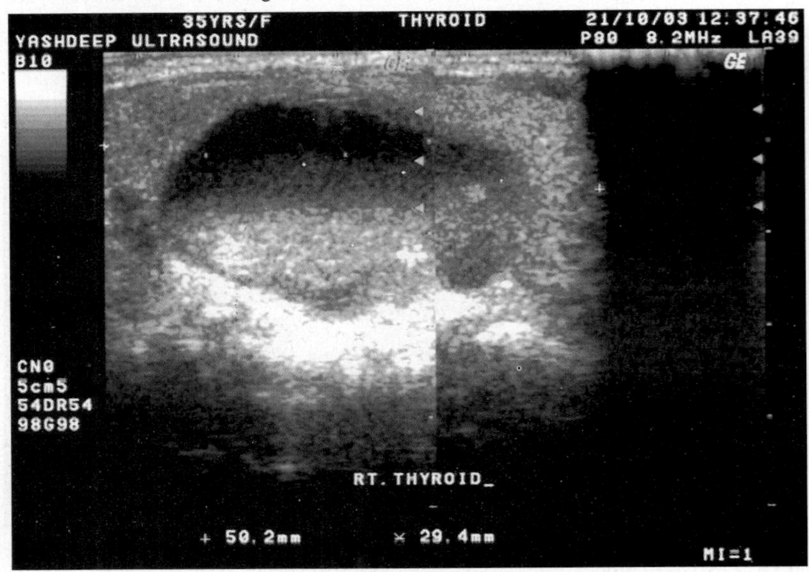

FIGURE 2.141

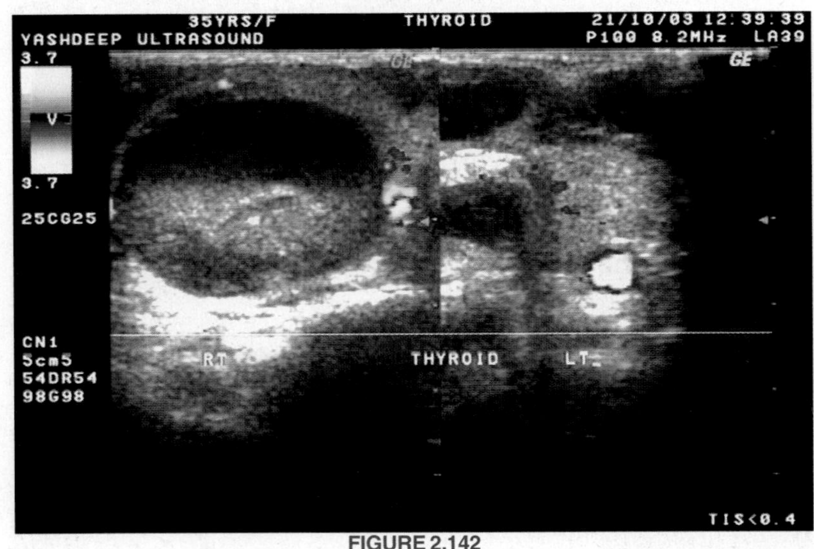

FIGURE 2.142

FIGURES 2.141 and 2.142: **Hemorrhagic cyst in thyroid.** A patient presented acutely painful cystic mass with sudden onset in Rt lobe of the thyroid. HRSG shows a cystic mass with echogenic collection in the cyst. The collection is seen settled at the base of the cyst suggestive of acute hemorrhage with fibrin deposits at the base. Fresh blood was aspirated from the cyst under USG guidance suggestive of acute hemorrhage in the cyst.

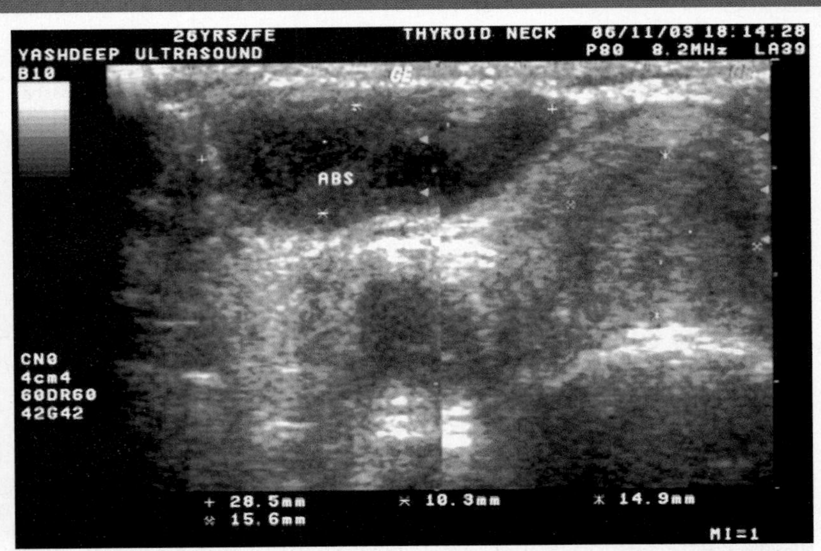

FIGURE 2.143: **Tubercular abscess in thyroid.** A patient presented with painless soft tissue mass in the midline and Rt side of the neck in the region of the thyroid. HRSG shows a thick walled low level echo complex mass in the isthmus of the thyroid. Aspiration of the fluid confirmed tubercular abscess in the thyroid. Multiple nodes are also seen enlarged in the cervical region on Rt side.

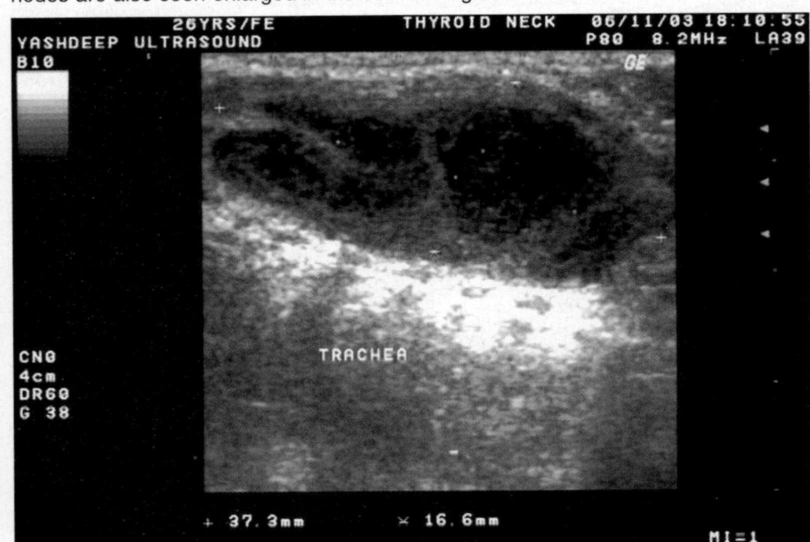

FIGURE 2.144: HRSG shows the abscess was limited only in the isthmus part of the gland. The tracheal cartilage was intact. They are seen as echogenic bright shadows.

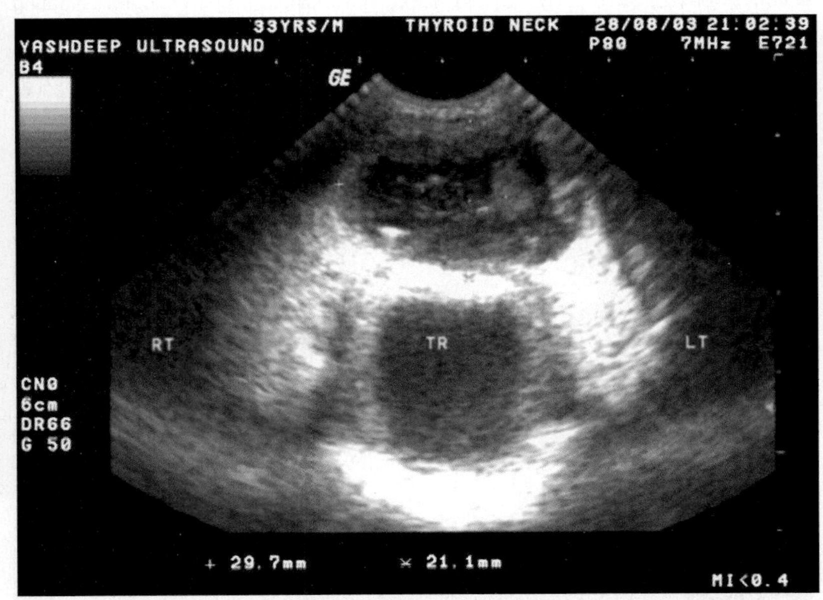

FIGURE 2.145: **Tubercular abscess in the thyroid.** Thick walled thick echo complex mass in the Isthmus of the gland. The mass shows multiple internal echoes. It was non-tender. Aspiration confirmed tubercular abscess.

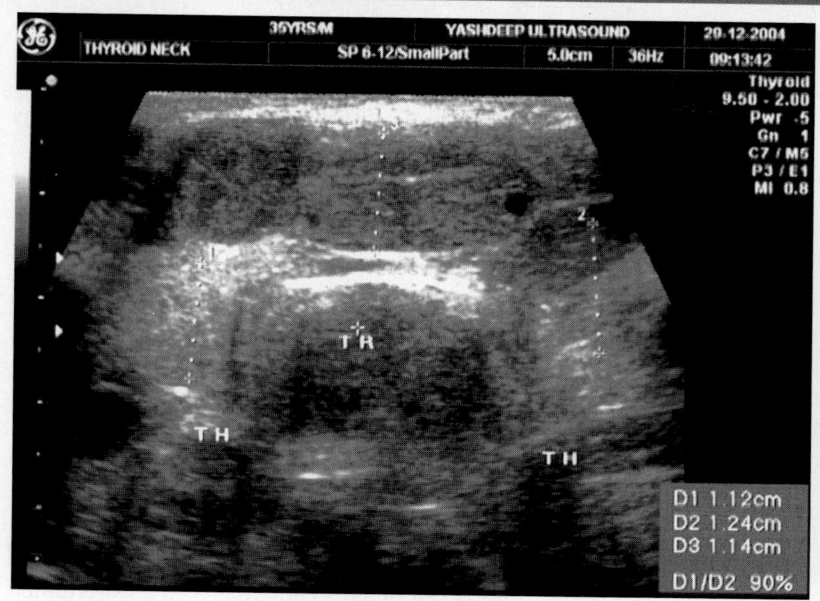

FIGURE 2.146

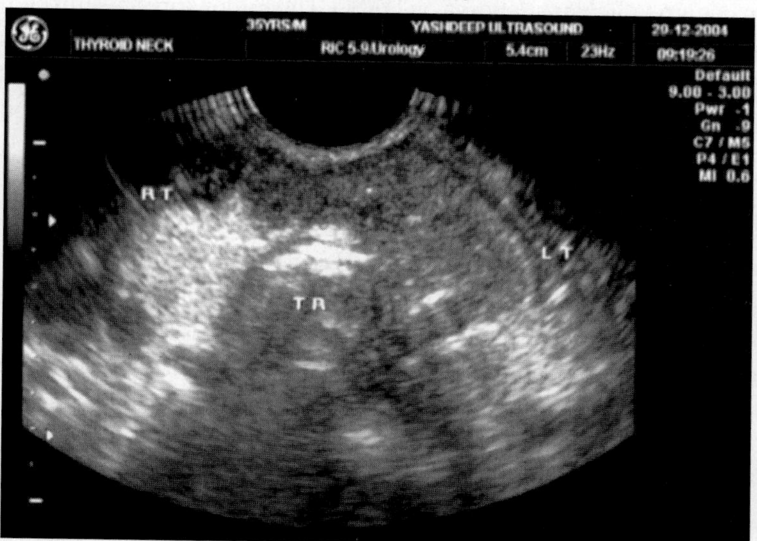

FIGURE 2.147

FIGURES 2.146 and 2.147: Acute pyogenic abscess in thyroid. A patient presented with acutely painful thyroid with high fever. HRSG shows distorted texture of the thyroid involving Isthmus and both upper lobes of the thyroid on either side. Diffuse necrosis of the gland is seen suggestive of acute pyogenic abscess.

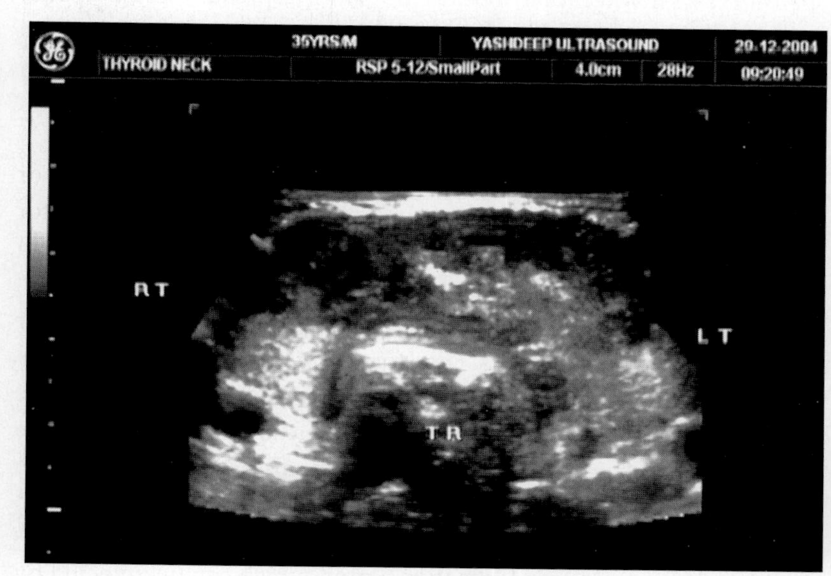

FIGURE 2.148

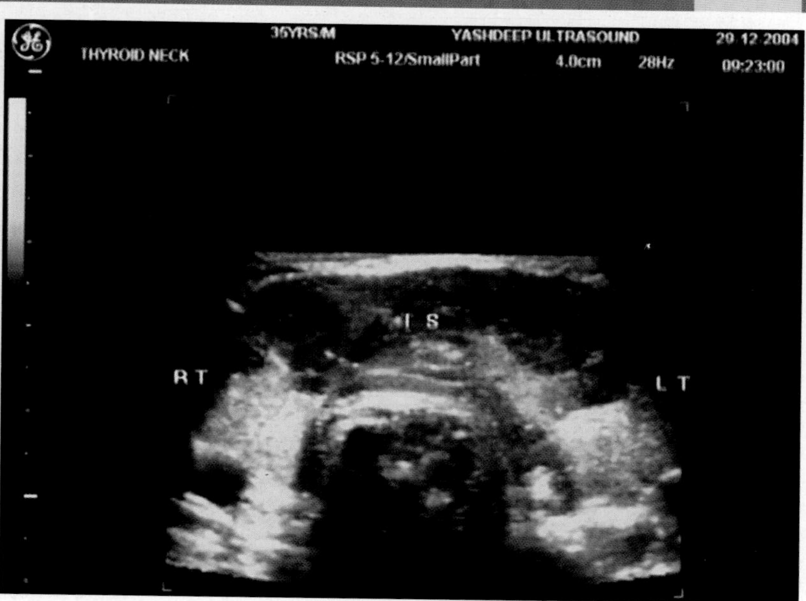

FIGURE 2.149

FIGURES 2.148 and 2.149: 3D Imaging of the same patient shows the finer details of the necrosis in the gland. The debris is also seen in the gland.

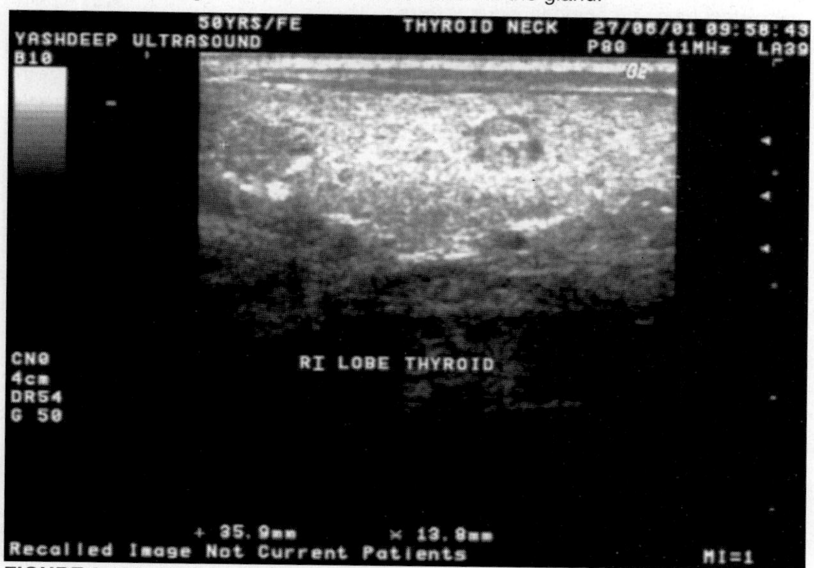

FIGURE 2.150: HRSG shows a small well-defined homogeneous solid nodule in the thyroid. It was clinically non-palpable. HRSG is highly sensitive in picking up sub-centimeter nodules in the thyroid.

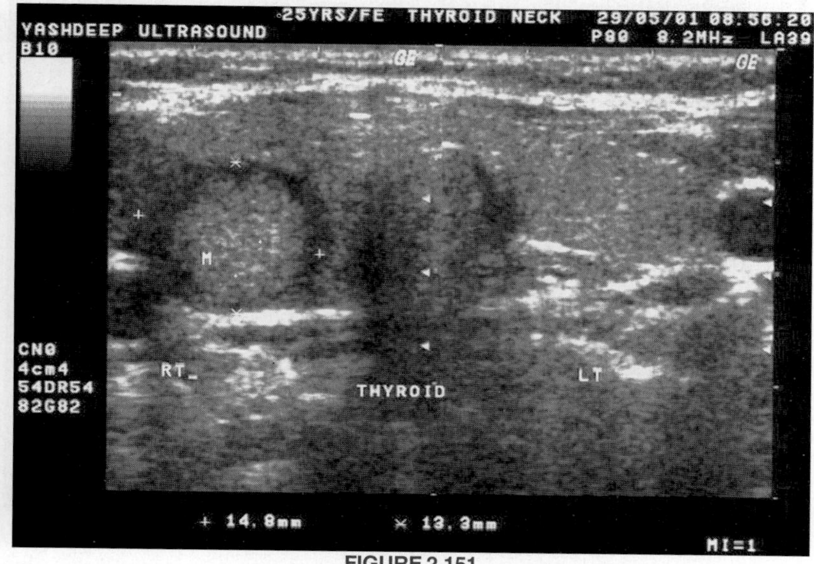

FIGURE 2.151

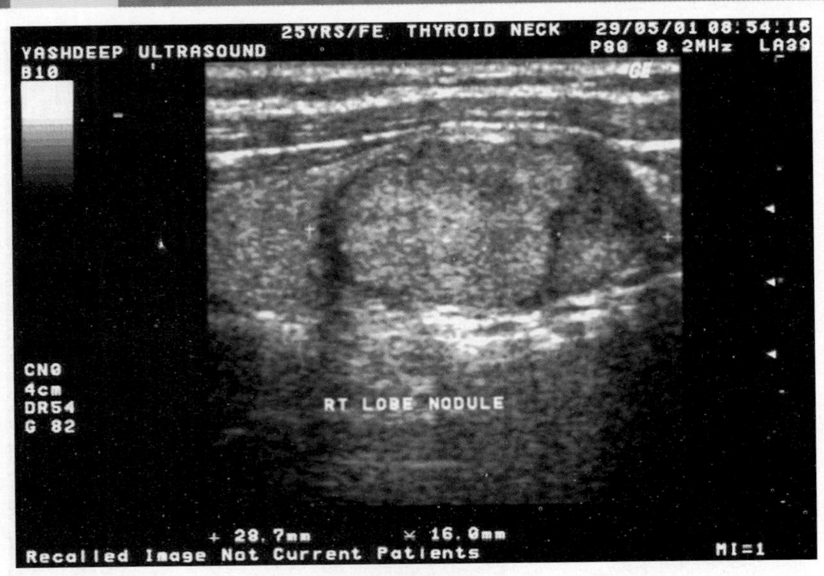

FIGURE 2.152

FIGURES 2.151 and 2.152: Follicular adenoma in thyroid. HRSG shows a well-defined echogenic nodular mass in the Rt lobe of the thyroid inferior pole with a typical peripheral halo. HRSG shows the details of the nodules in longitudinal plane with intact halo and homogeneous texture.

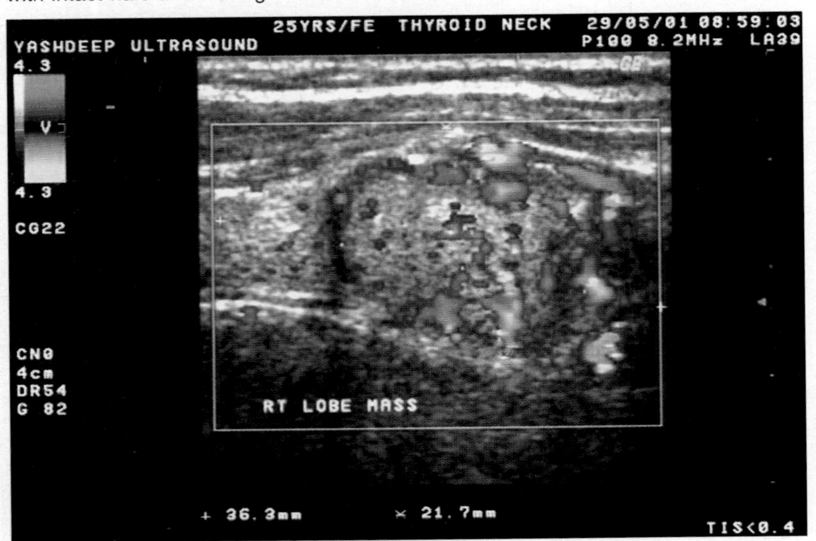

FIGURE 2.153

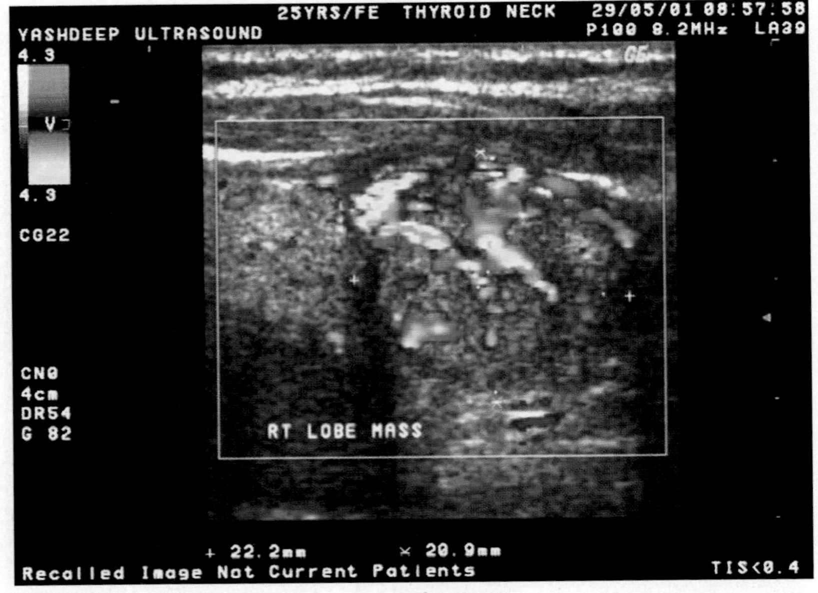

FIGURE 2.154

FIGURES 2.153 and 2.154: Color Doppler imaging shows increased flow in the nodule suggestive of increased hyperemia. The biopsy confirmed follicular adenoma.

Malignant Tumors of Thyroid

Malignant tumors of thyroid gland are not very common. They are present in 2 to 3 cases per 100,000. Thyroid neoplasms are grouped as follows.

Papillary Carcinoma

Papillary carcinoma is most common thyroid malignancy. It accounts for 60 to 65 percent of the total thyroid neoplasms. It is more common in females than males. The tumor incidence is high in young adults and elderly age group. This tumor spreads through regional lymphatics. However, lymph node involvement does not aggravate the prognosis. The prognosis is better in papillary neoplasms.

Follicular Carcinoma

Follicular carcinoma accounts for 15 to 20 percent of the thyroid neoplasms. It is more common in elder age group. It is well differentiated from benign follicular adenoma on HRSG. The tumor spread is through hematogenous route. Therefore, distant metastases to lung and bones are more common in follicular neoplasm. Typical cannon ball metastases are found in lungs. However at times, poorly differentiated follicular neoplasms are also encountered.

Anaplastic Carcinoma

Anaplastic carcinoma is the most aggressive variety of thyroid cancer. It is found in 10 to 15 percent of total thyroid malignancy. It has worse prognosis, and 5-year survival is found only in 3 to 4 percent of the cases. The tumor presents as rapidly growing mass with surrounding tissue invasion. Therefore, it often presents as an inoperable mass.

Medullary Carcinoma

Medullary carcinomas are present in 5 to 10% of the thyroid cancers. They are derived from parafollicular or C cells. They secrete calcitonin hormone. It is used as serum tumor marker. The tumor is commonly familial. It is also an essential part of multiple endocrine neoplasia (MEN) type II syndrome. The tumor is multicentric and in about 90 percent cases involves both lobes of the gland. It spreads through lymph node metastasis. Prognosis of the disease is considered worse than follicular neoplasm.

Sonographic Features

Most of the time, thyroid malignant tumor presents as an echo poor nodule. However, no specific ultrasonographic pattern is described for thyroid malignancy. The tumor may present as isoreflective nodule in 25% cases, and mixed echo complex mass in about 10% cases. However, echogenic pattern of tumor is very rare. Ninety percent of the echo poor nodules are malignant tumors. Therefore, all the thyroid nodules require fine-needle aspiration cytology (FNAC) for confirmation. It can be done under ultrasonographic guidance. Cystic degeneration is also encountered both in malignant and benign nodule. Coarse calcification is a feature of benign mass, but it can be seen in malignant tumor. Microcalcification is the typical feature of malignant neoplasm. It is highly echogenic focus of less than 2 mm, which can be detected by 10 mhz probe. Cystic degeneration is also encountered. The margins are irregular and ill-defined. A peripheral halo is seen. However, it is also seen in benign mass. Therefore, it is not a reliable sign, but in malignant tumor, it is irregular and incomplete. The most reliable indications of thyroid malignancy are surrounding tissue invasion and lymph node involvement. HRSG is very sensitive tool to detect them when they are non-palpable. Big tumors can destroy the normal glandular pattern and surrounding tissue relations are frequently encountered in anaplastic carcinoma. Dynamic sonography of the mass during swallowing can show the fixity of mass to strap muscles anteriorly or longus colli muscle posteriorly. Vascular invasion of the tumor mass can be easily identified on HRSG. The most common vessel involved is internal jugular vein. The vein can be either infiltrated or thrombosed due to metastatic compression. Recurrent laryngeal nerve

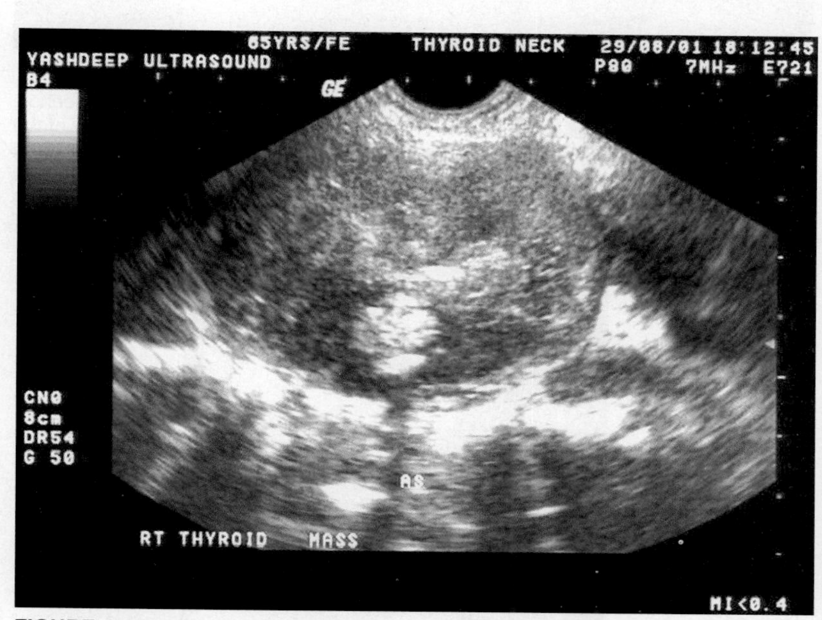

FIGURE 2.155: Heterogeneous mass in thyroid. A big irregular heterogeneous mass is seen involving Rt lobe of the thyroid. The mass has destroyed normal glandular texture. Echogenic calcification is also seen in the mass. It is accompanied with acoustic shadowing. The mass is seen involving and invading the tracheal ring. Biopsy confirmed a poorly differentiated carcinoma of thyroid.

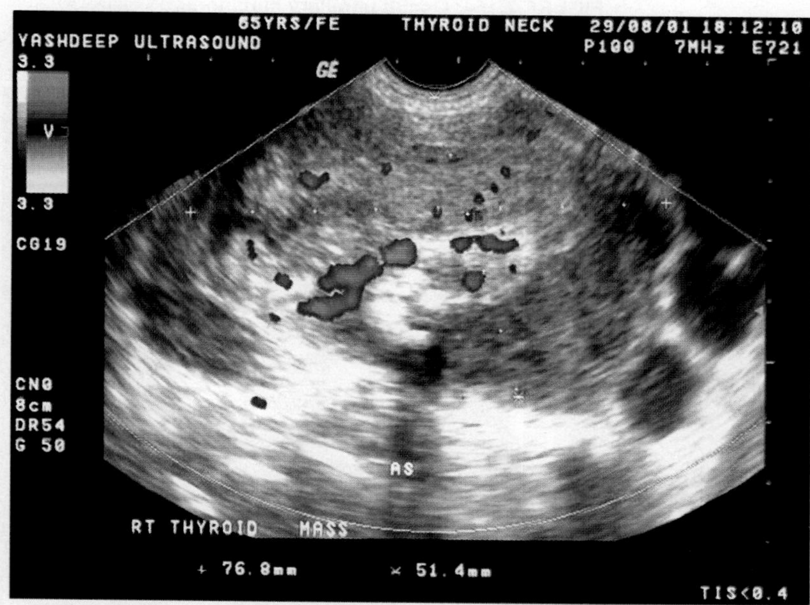

FIGURE 2.156: Color Doppler imaging shows low flow in the mass suggestive of hypovascular nature of the mass.

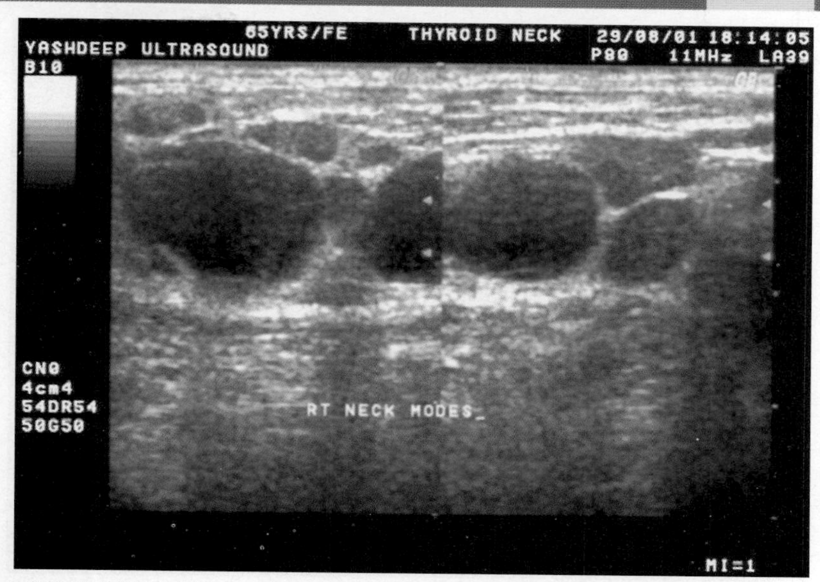

FIGURE 2.157: Multiple enlarged nodal shadows are seen in the neck suggestive of enlarged lymph nodes in the same patient. Normal nodal texture is lost with distorted hylum of the nodes. Biopsy confirmed metastatic nodes from adenocarcinoma.

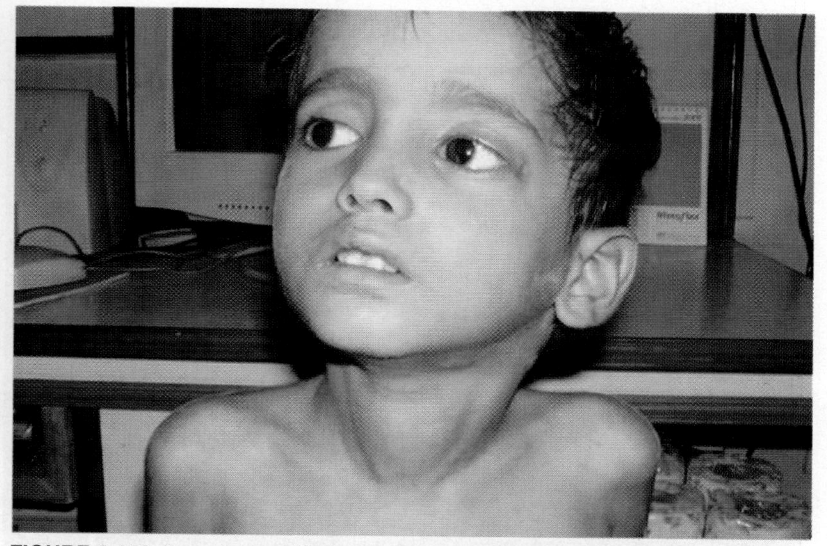

FIGURE 2.158: Invasive papillary carcinoma in a child. A young child presented with a soft mass swelling in the neck. The patient was presented with acute respiratory distress syndrome.

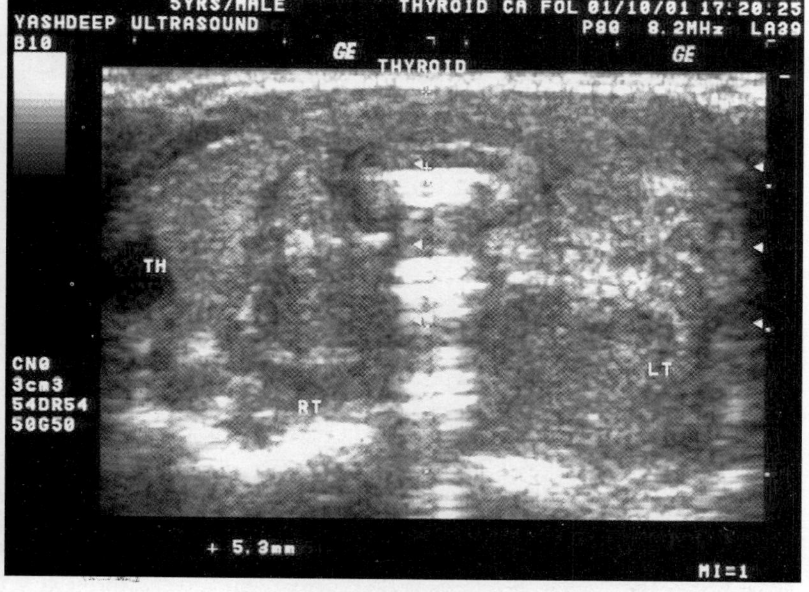

FIGURE 2.159

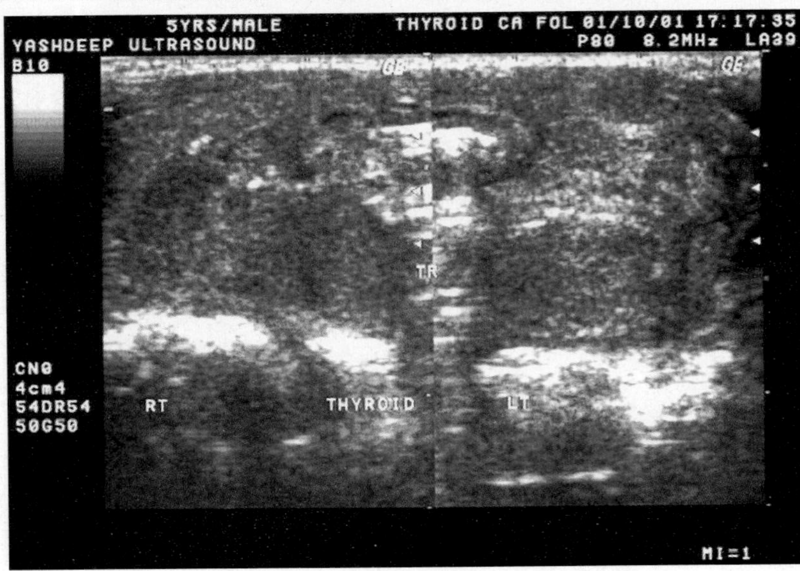

FIGURE 2.160

FIGURES 2.159 and 2.160: HRSG shows a big irregular heterogeneous mass in young child of 5 years in both lobes of the thyroid. The mass has destroyed normal thyroid anatomy. It is seen invading the muscle planes and also the tracheal rings and part of trachea. Echogenic calcification is also seen in the mass. But no cystic degeneration is seen. Normal anatomy was destroyed.

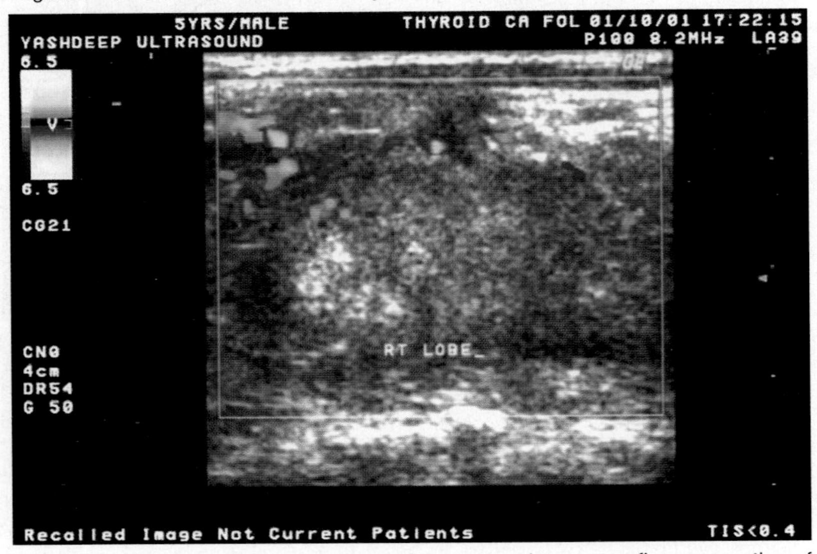

FIGURE 2.161: Color Doppler imaging of the mass shows poor flow suggestive of low vascularity, which was only confined to the periphery.

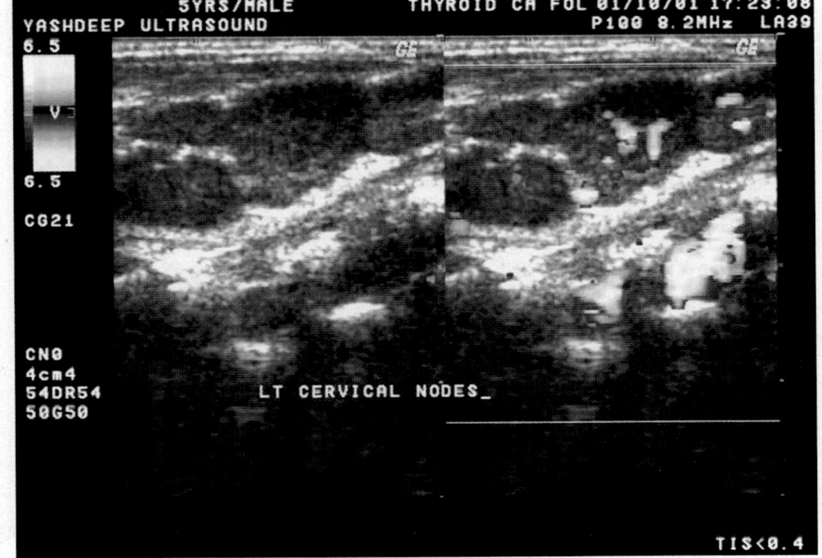

FIGURE 2.162

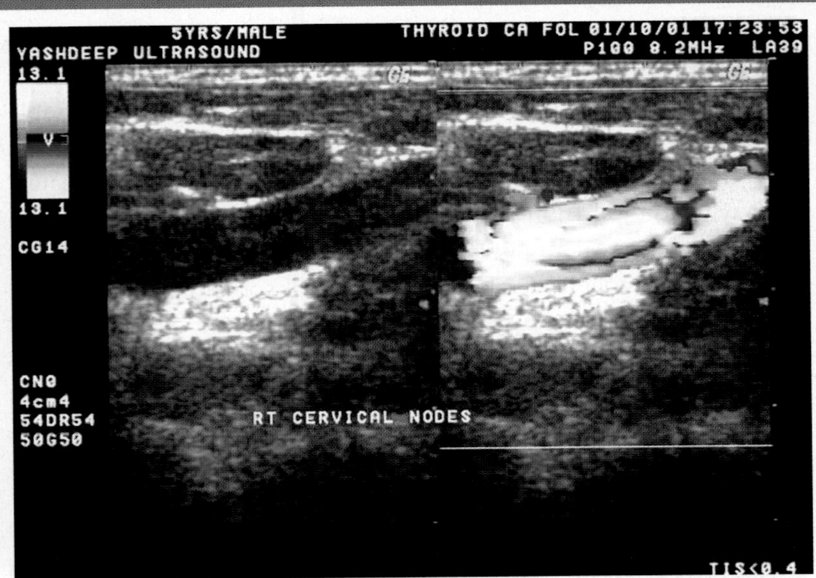

FIGURE 2.163

FIGURES 2.162 and 2.163: HRSG also shows multiple enlarged irregular hypoechoic lymph node masses on either side of the neck with lost normal texture. Biopsy of the mass shows invasive papillary carcinoma with follicular predominance. Carcinoma of thyroid in children is rare and not reported below the age of 7 years. However, this was a rare case when invasive papillary carcinoma with follicular predominance was found in a young child of 5 years of age with local spread.

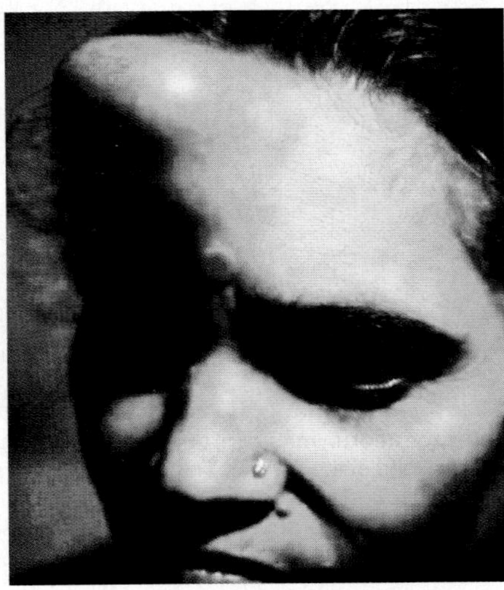

FIGURE 2.164: Occult papillary carcinoma in thyroid. A young lady was presented with a soft tissue mass over the frontal area.

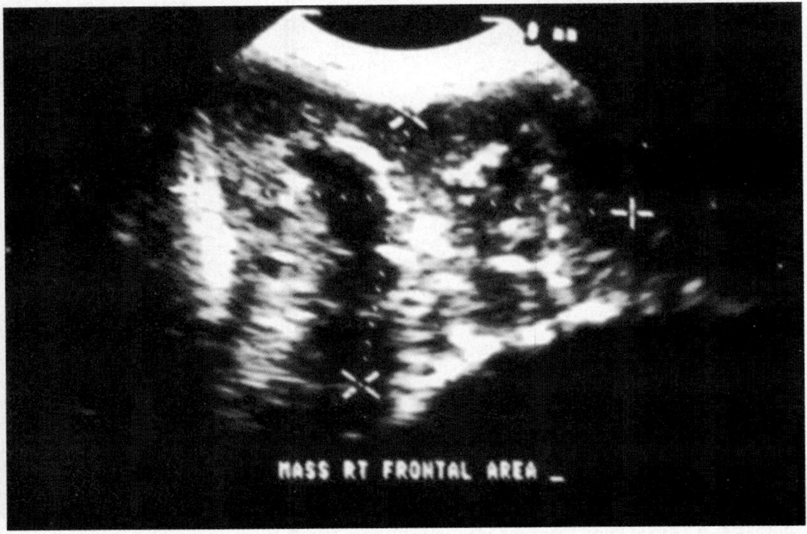

FIGURE 2.165: HRSG of the mass shows irregular heterogeneous mass with multiple hypoechoic areas. Biopsy of the mass shows metastatic deposits from adeno carcinoma.

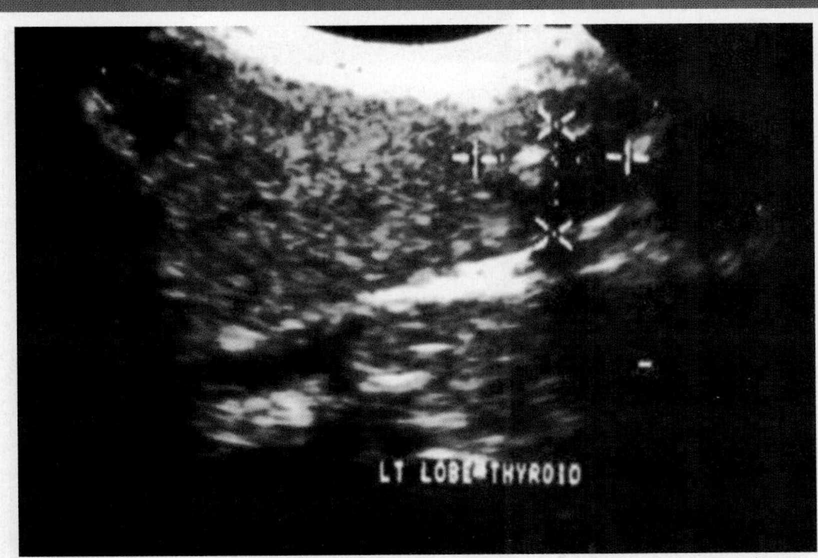

FIGURE 2.166: HRSG of the thyroid gland of the same patient shows small non-palpable irregular nodular mass in the Lt lobe of the thyroid measures 10 x 8 mm in size. Ultrasound guided FNACA of the mass shows papillary carcinoma.

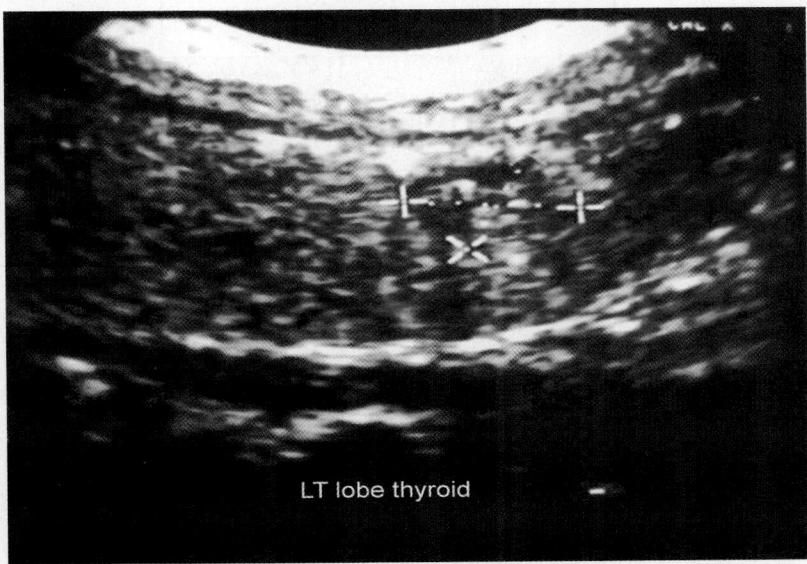

FIGURE 2.167

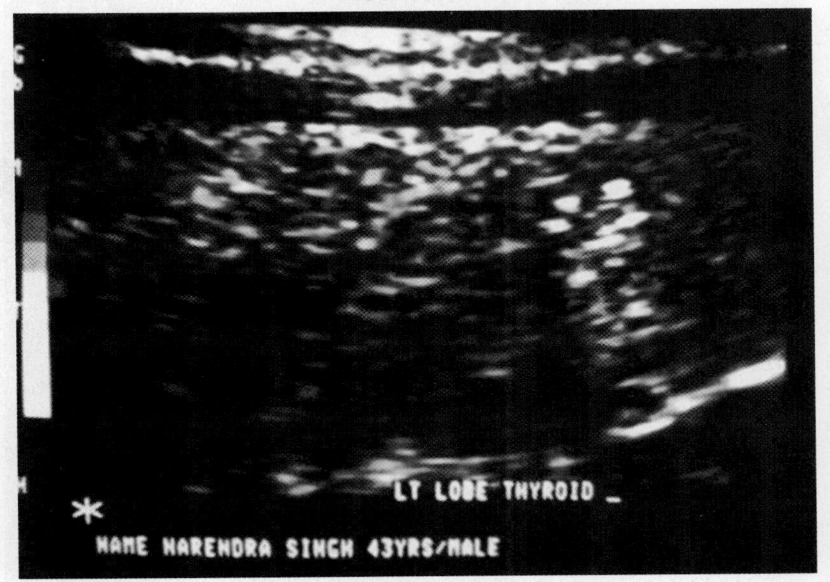

FIGURE 2.168

FIGURES 2.167 and 2.168: Occult primary carcinoma of thyroid. A middle-aged man was presented with multiple metastatic nodes in the Lt side of the neck. Biopsy through metastasis of the adenocarcinoma HRSG of thyroid shows irregular heterogeneous nodular mass in the Lt lobe. Which was non-palpable in clinical examination. Ultrasound guided biopsy confirmed occult primary in thyroid.

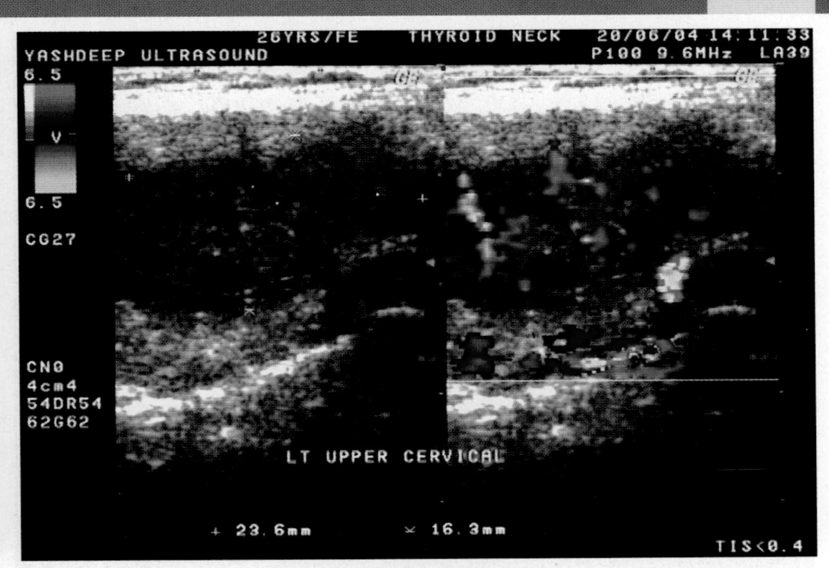

FIGURE 2.169

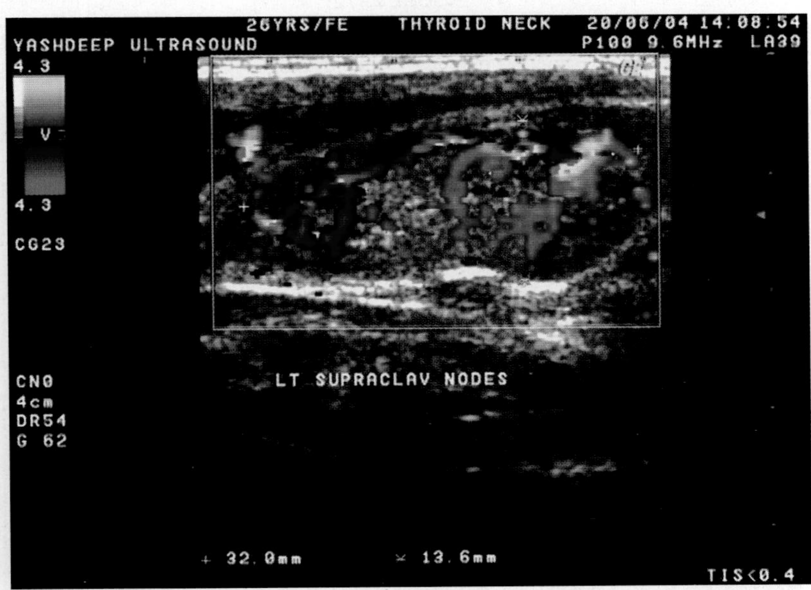

FIGURE 2.170

FIGURES 2.169 and 2.170: Invasive non-palpable papillary carcinoma thyroid. A young girl presented with multiple enlarged lymph nodes in the neck. HRSG shows irregular heterogeneous nodal masses in the neck. Biopsy of the nodes proved metastasis from adenocarcinoma. Color flow imaging shows high flow in the nodes.

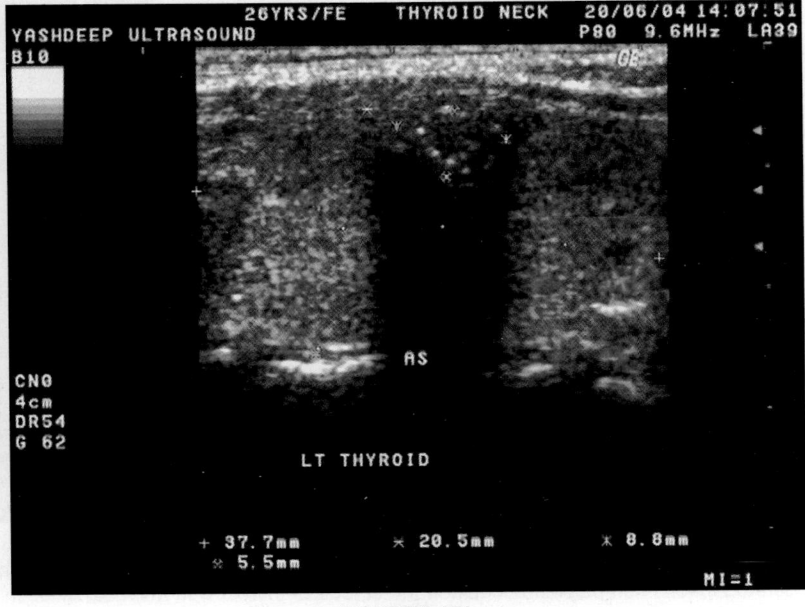

FIGURE 2.171

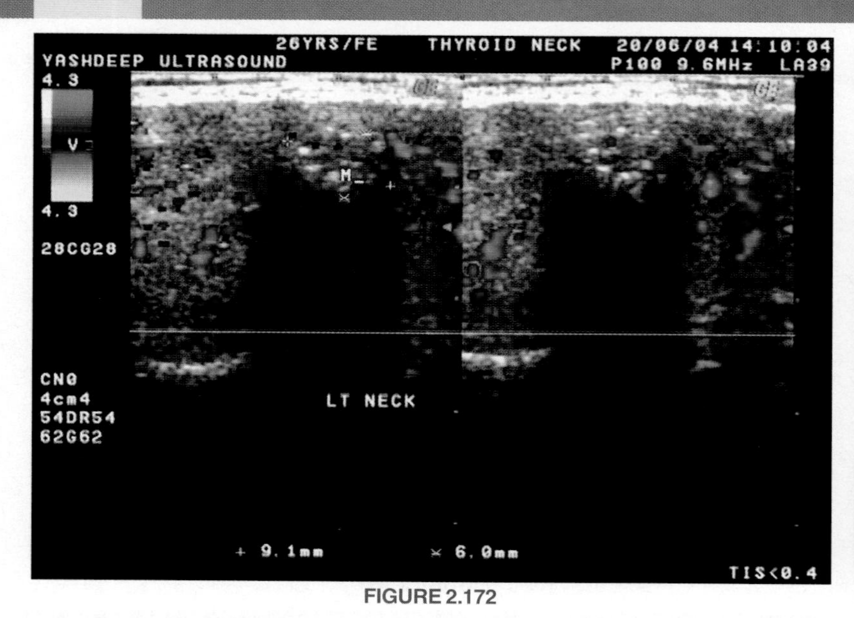

FIGURE 2.172

FIGURES 2.171 and 2.172: HRSG shows small heterogeneous nodular mass in the Lt lobe of thyroid gland. Echogenic punctate calcified specks are seen in it. Biopsy from the nodule confirmed invasive papillary carcinoma.

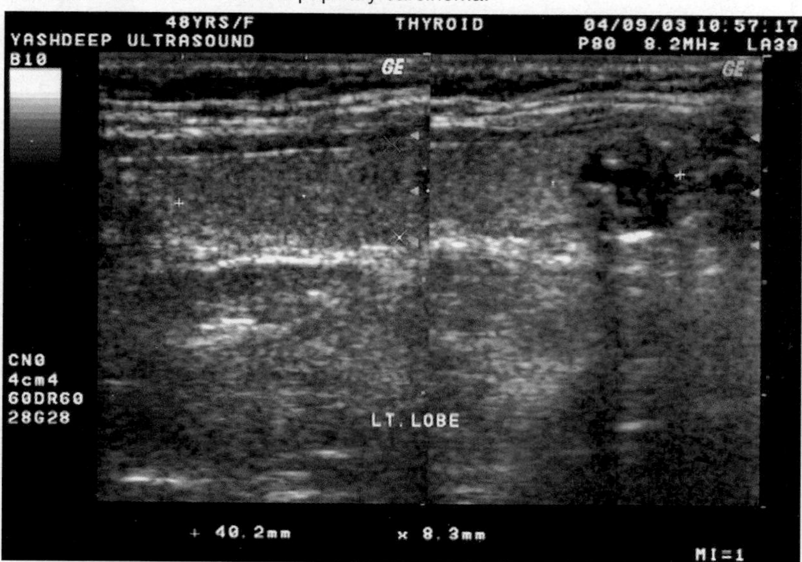

FIGURE 2.173

FIGURE 2.174

FIGURES 2.173 and 2.174: Papillary carcinoma thyroid. A patient presented with heterogeneous non-palpable nodule in the Lt lobe of thyroid with microcalcification. HRSG guided FNAC confirmed papillary adenocarcinoma of thyroid.

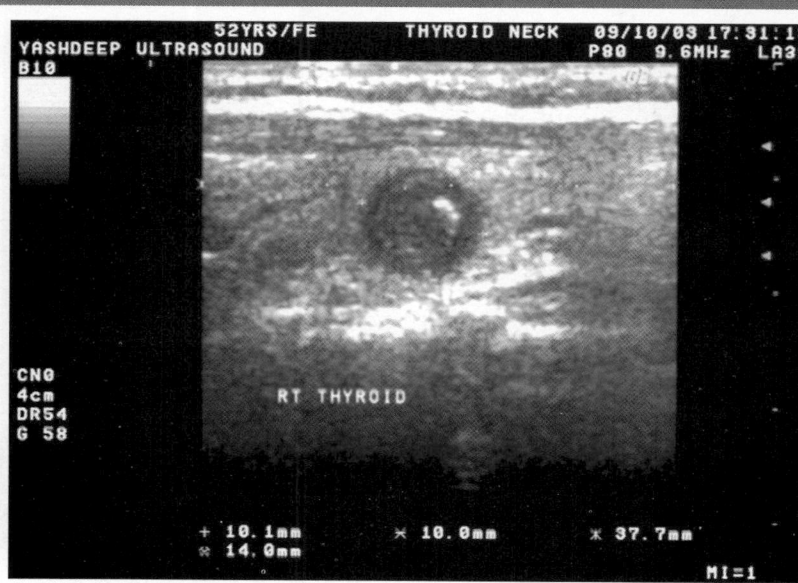

FIGURE 2.175: Papillary adenocarcinoma thyroid. Longitudinal image of thyroid shows a hypoechoic nodular mass in the Rt lobe of the gland. Echogenic punctate calcification is seen in it. Biopsy showed papillary adenocarcinoma.

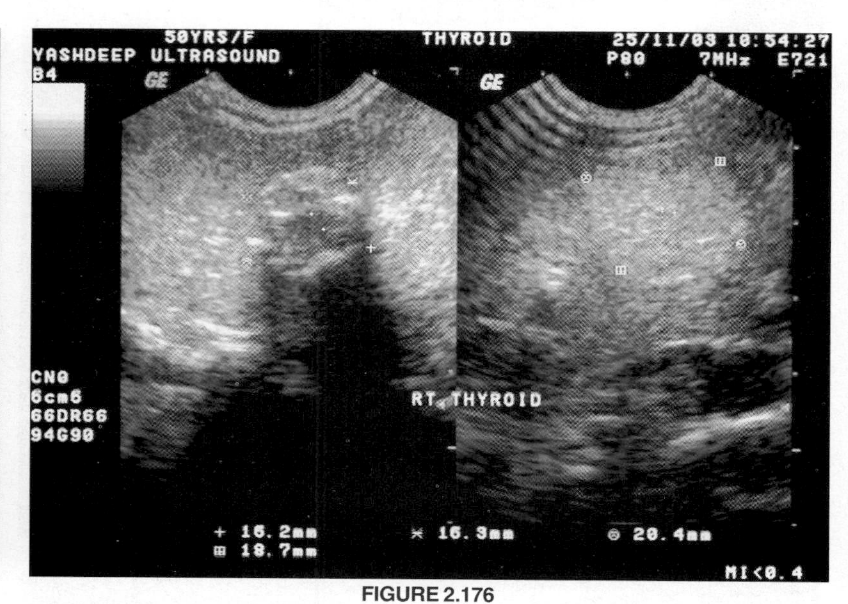

FIGURE 2.176

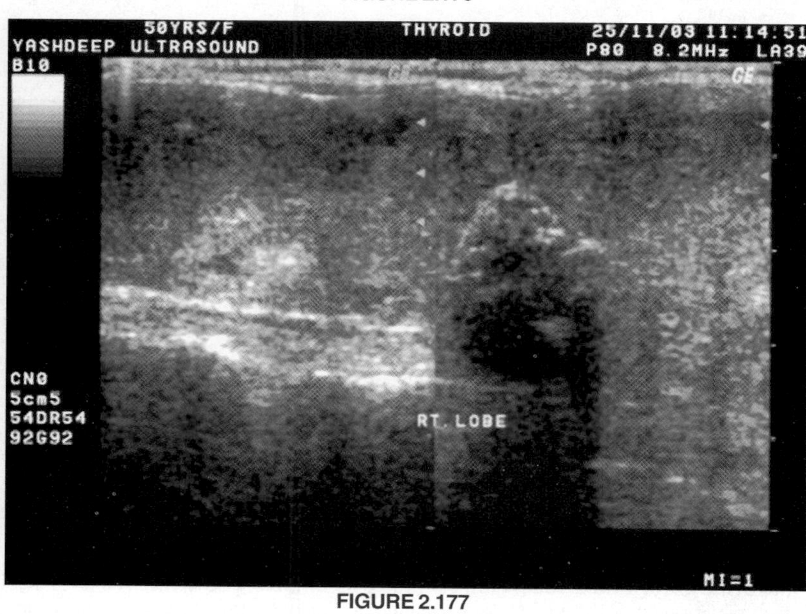

FIGURE 2.177

FIGURES 2.176 and 2.177: Papillary carcinoma. A patient presented with multiple nodes in the neck. HRSG shows an irregular mass in the Rt lobe of the thyroid. Calcified rim is seen around the mass. Punctate calcification is also seen in the mass. No halo is seen. Biopsy proved papillary adenocarcinoma.

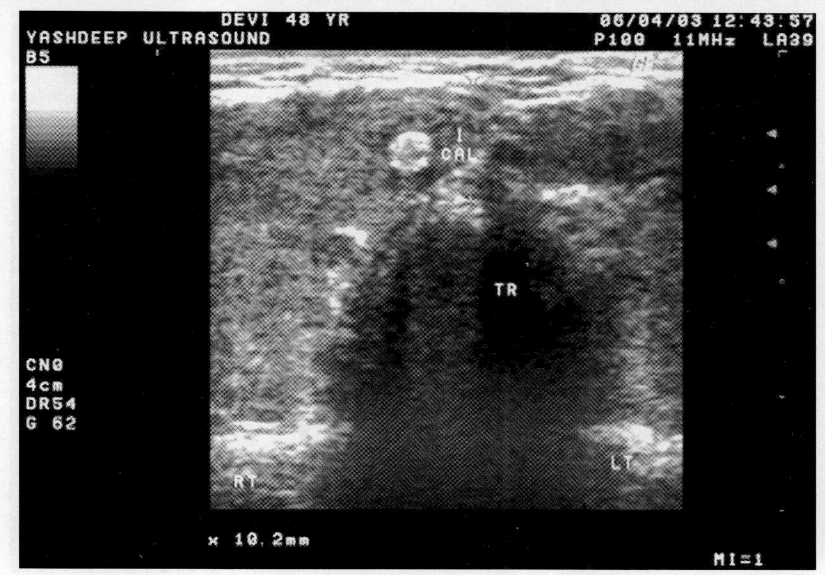

FIGURE 2.178

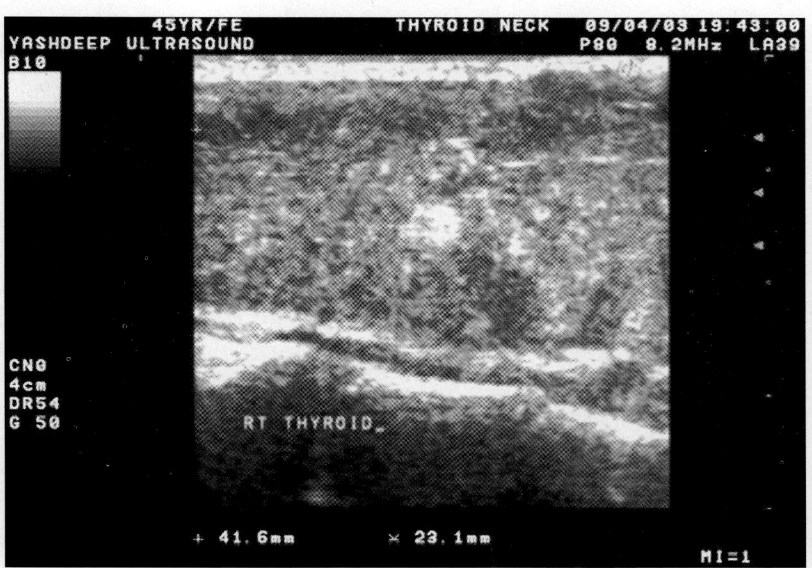

FIGURE 2.179

FIGURES 2.178 and 2.179: Benign adenoma with macrocalcification. A hyperechoic non-palpable nodular mass is seen in the Rt lobe of thyroid. Dense echogenic macrocalcification is seen in it. Biopsy shows benign adenoma. This macro calcification is different from the punctate microcalcification found in papillary adenocarcinoma.

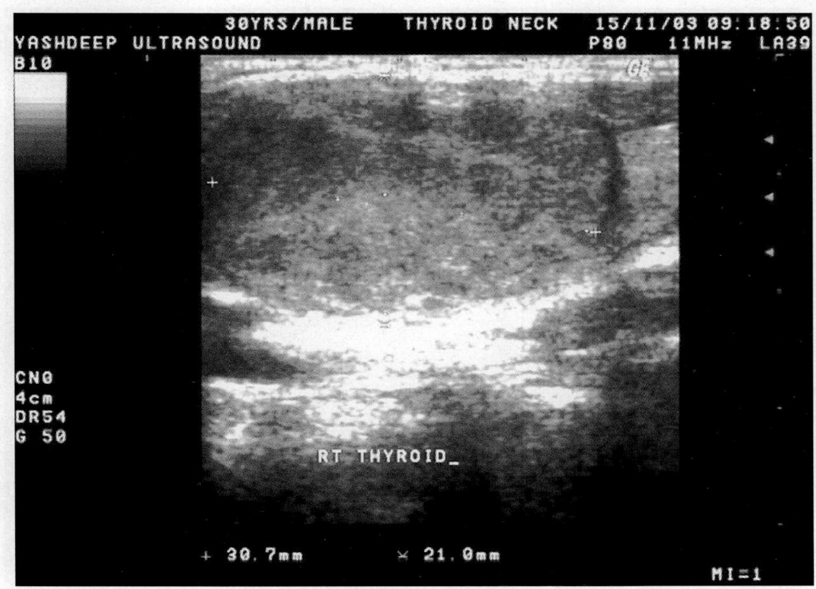

FIGURE 2.180

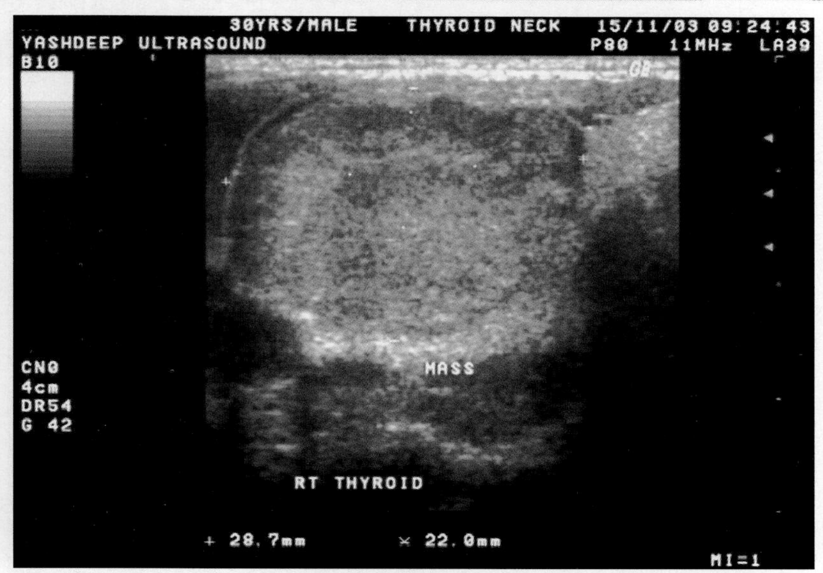

FIGURE 2.181

FIGURES 2.180 and 2.181: Follicular adenocarcinoma. HRSG shows a hypoechoic ill defined mass in the Rt lobe of thyroid. The mass shows bulging margins. No halo is seen around the mass. It is merging with the gland. However, it is homogeneous in texture. No calcification or necrosis is seen in the mass.

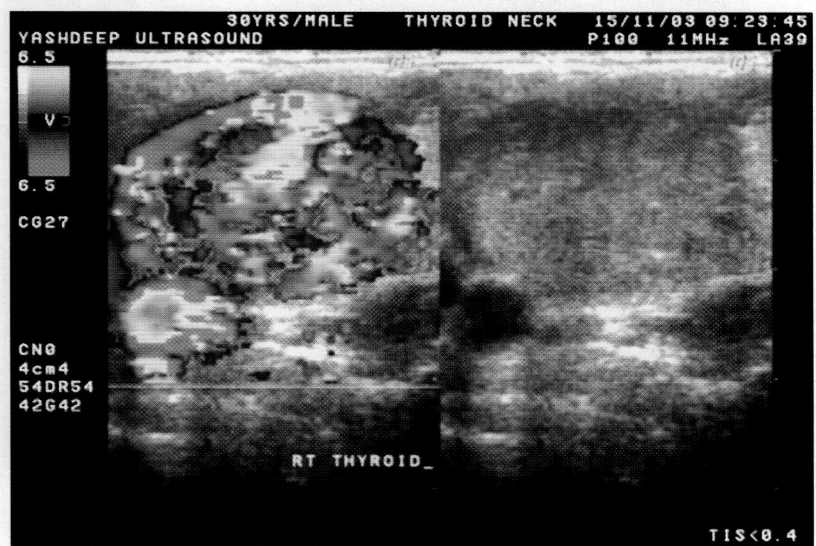

FIGURE 2.182

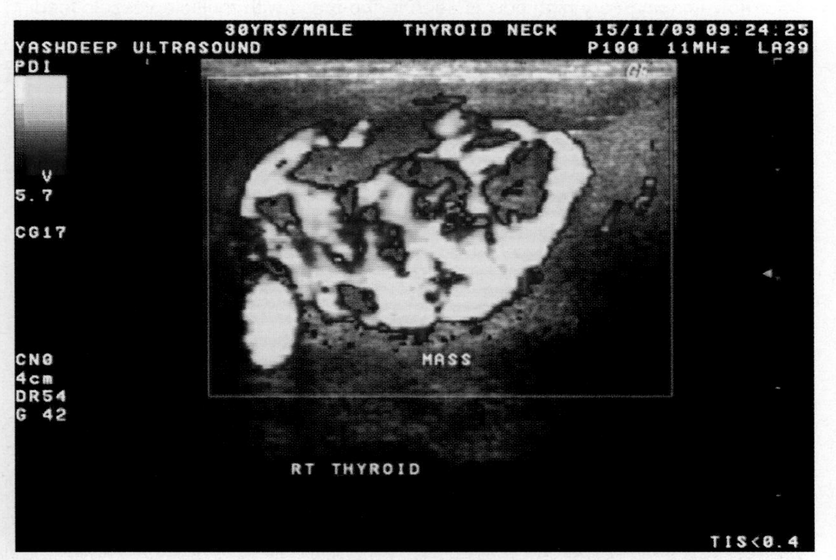

FIGURE 2.183

FIGURES 2.182 and 2.183: On color flow imaging very high flow is seen in the mass. Multiple vessels are seen feeding the mass in bizarre fashion. Biopsy proved follicular adenocarcinoma.

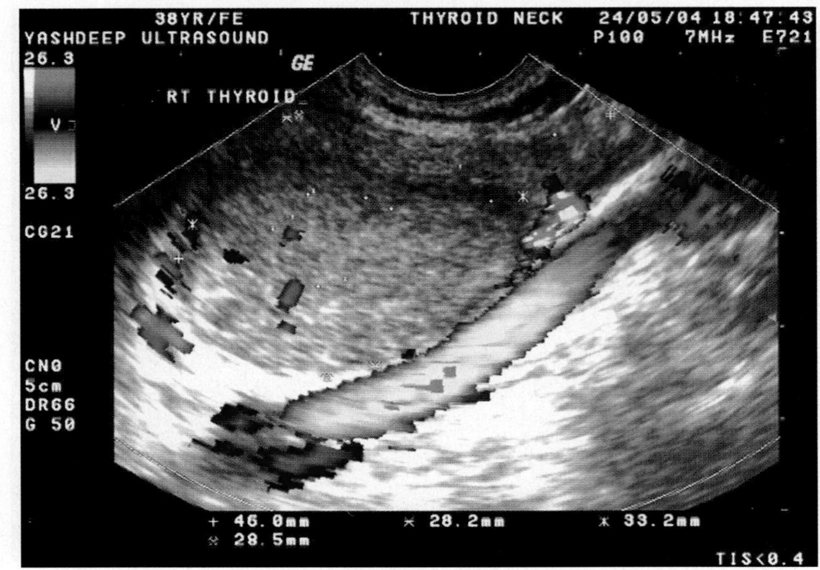

FIGURE 2.184

FIGURE 2.185

FIGURES 2.184 and 2.185: Follicular adenocarcinoma. An irregular hypoechoic mass is seen in the Rt lobe of the thyroid. The mass shows homogeneous texture. On color flow imaging very high flow is seen in the mass with multiple vessels feeding the mass. Biopsy shows follicular adenocarcinoma.

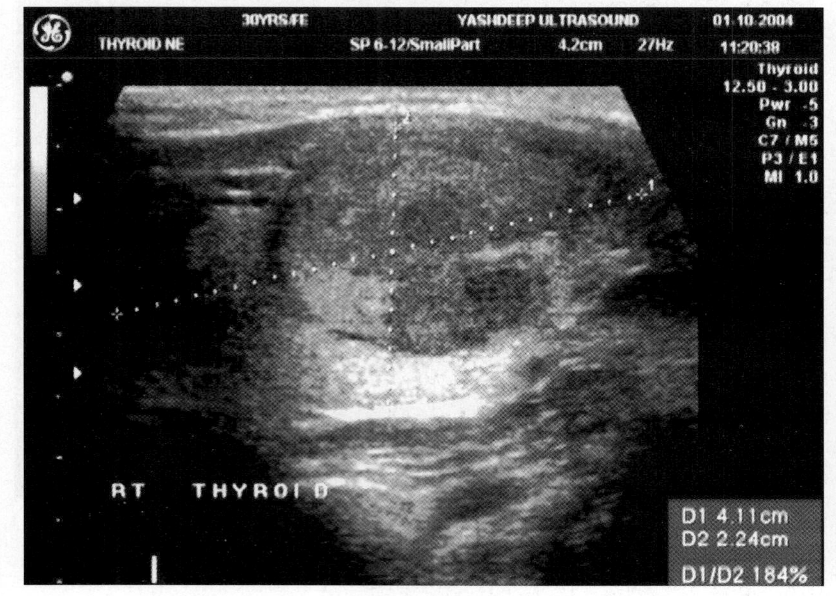

FIGURE 2.186

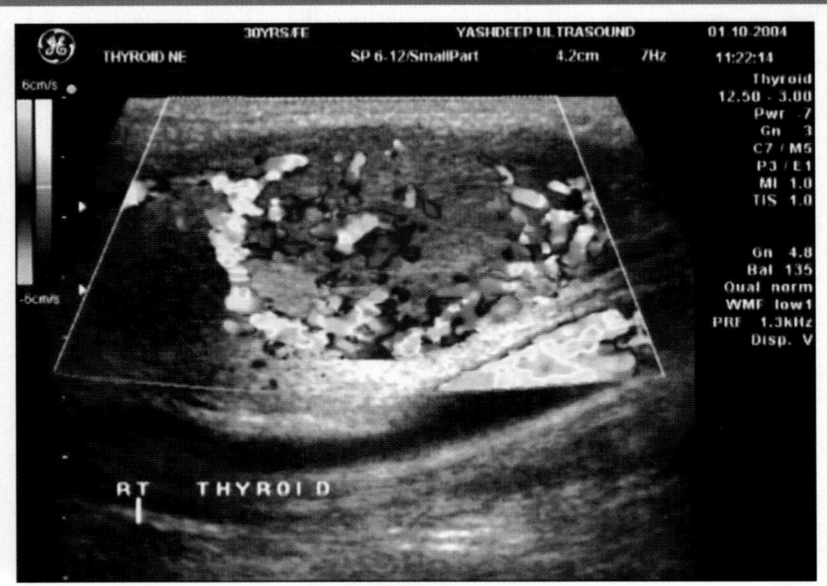

FIGURE 2.187

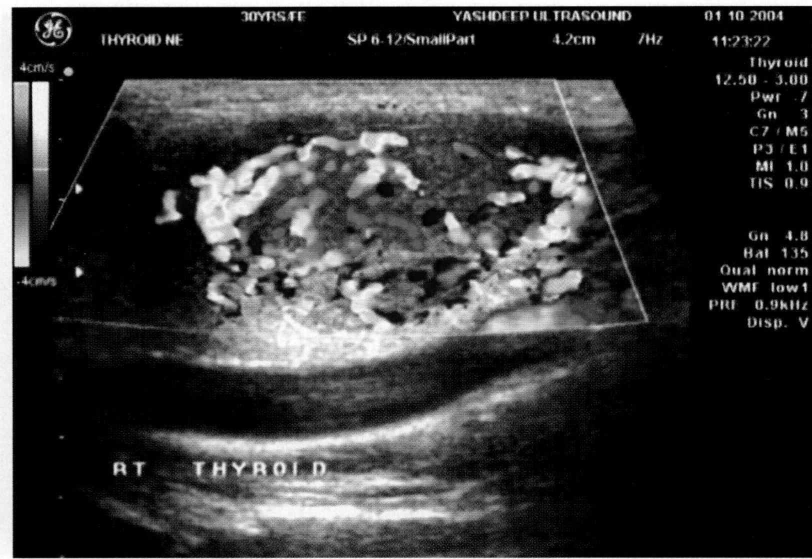

FIGURE 2.188

FIGURES 2.186 to 2.188: Heterogeneous follicular adenocarcinoma thyroid. A hypoechoic mass is seen in the Rt lobe of the thyroid. Margins are irregular and ill-defined. It shows heterogeneous texture. On color flow imaging very high flow is seen in the mass. Biopsy proved adenocarcinoma.

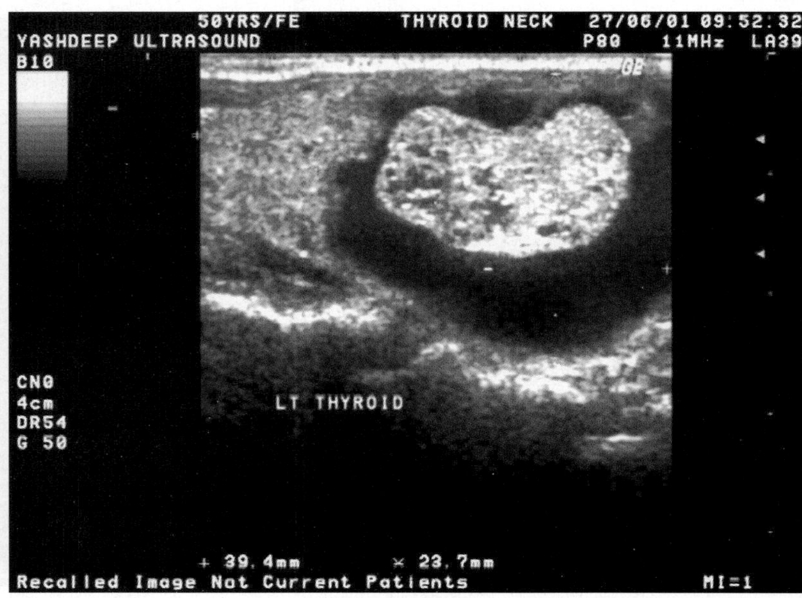

FIGURE 2.189

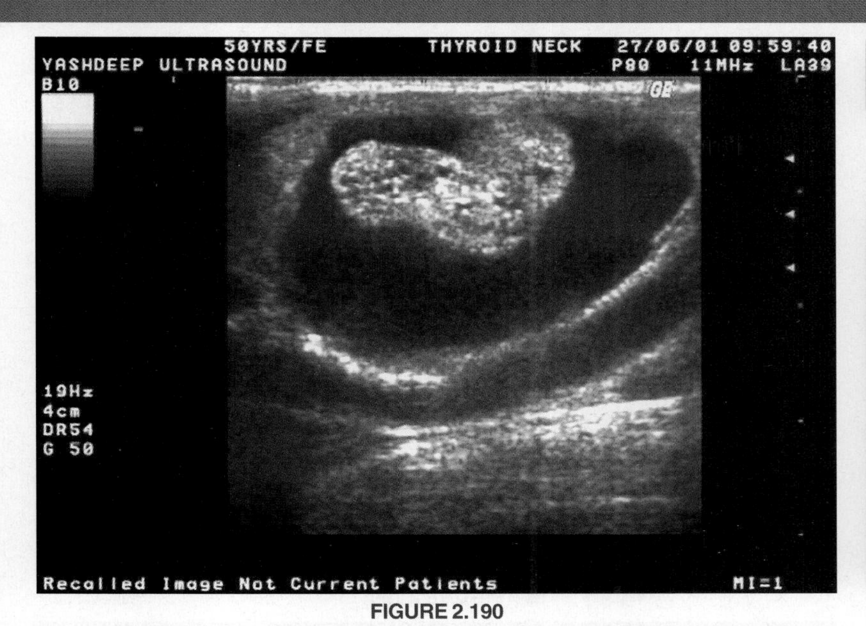

FIGURE 2.190

FIGURES 2.189 and 2.190: Malignant cystic mass in thyroid. HRSG shows a heterogeneous cystic mass in the Lt lobe of the thyroid. The cyst was having an irregular heterogeneous soft tissue mass, which was fixed with the upper wall of the cyst. Multiple hypoechoic areas are seen in the mass. Irregular calcification was also seen in the solid mass.

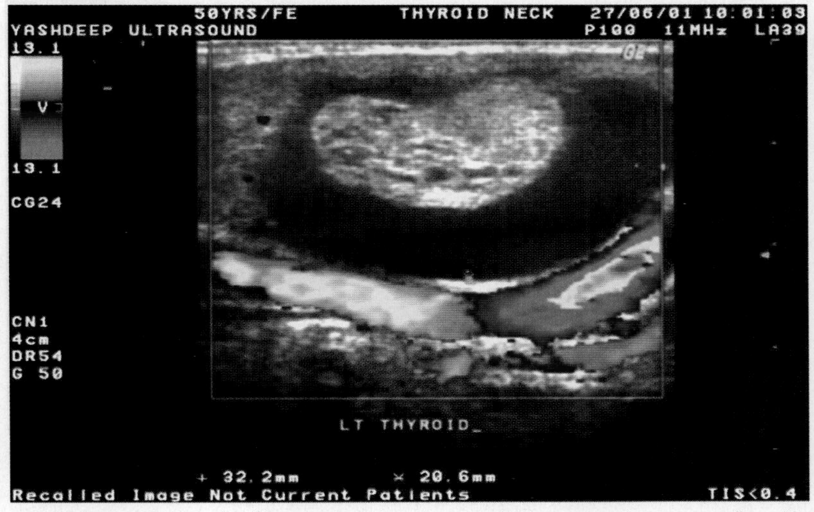

FIGURE 2.191: On color Doppler imaging the cyst in the mass shows poor flow. No evidence of any flow was seen in the soft tissue mass. Ultrasound guided FNAC and biopsy confirmed poorly differentiated adenocarcinoma.

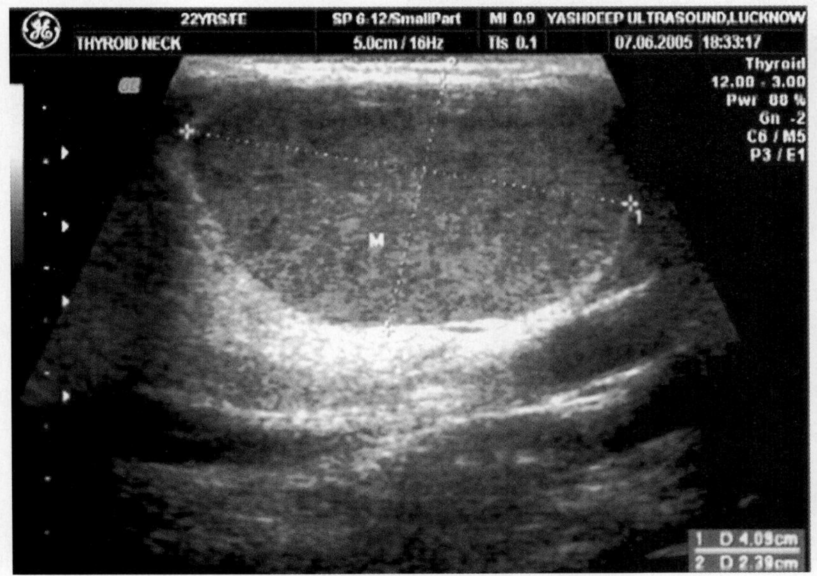

FIGURE 2.192

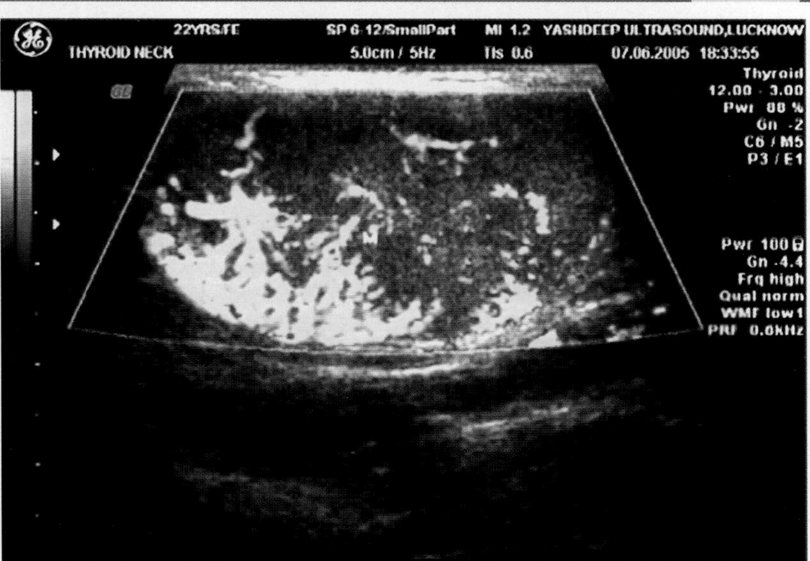

FIGURE 2.193

FIGURES 2.192 and 2.193: Follicular adenocarcinoma in the isthmus. A hypoechoic mass is seen in the isthmus of the gland. The mass shows homogeneous texture with scalloping of the margins. No peripheral halo is seen in the mass. On color flow imaging very high flow is seen in it with bizarre pattern. FNAC proved follicular adenocarcinoma.

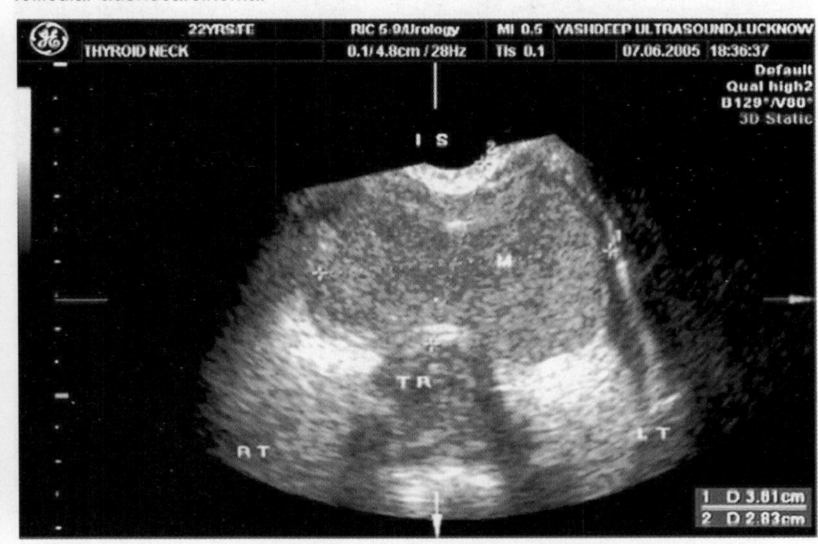

FIGURE 2.194: 3D of the same patient shows invasion of the tracheal ring by the mass.

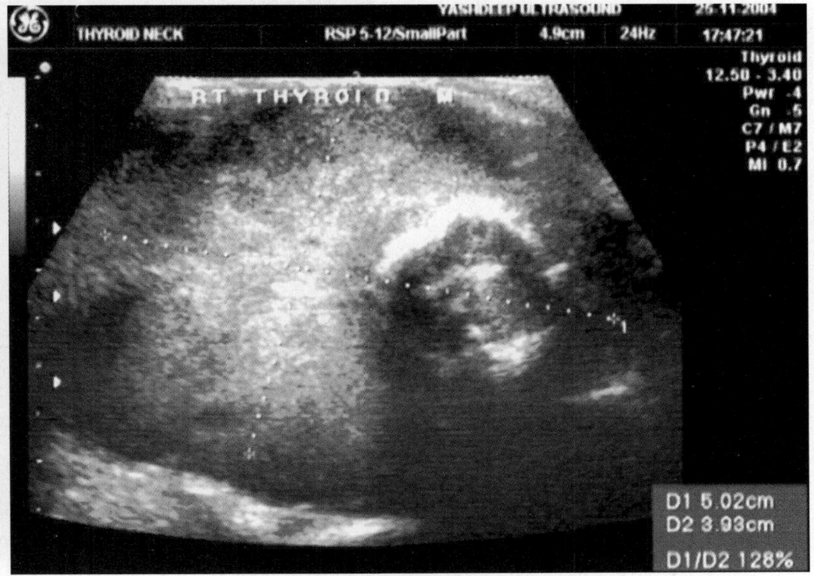

FIGURE 2.195

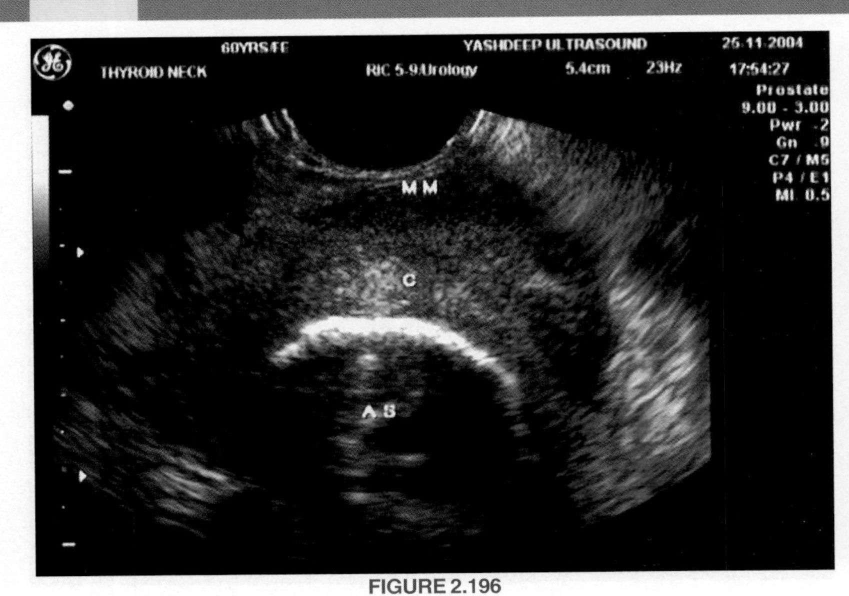

FIGURE 2.196

FIGURES 2.195 and 2.196: Invasive adenocarcinoma thyroid. A big heterogeneous mass is seen in the Rt lobe of thyroid. Normal glandular texture is lost. The mass shows dense calcification. It is seen invading the surrounding tissues. Normal anatomy is distorted.

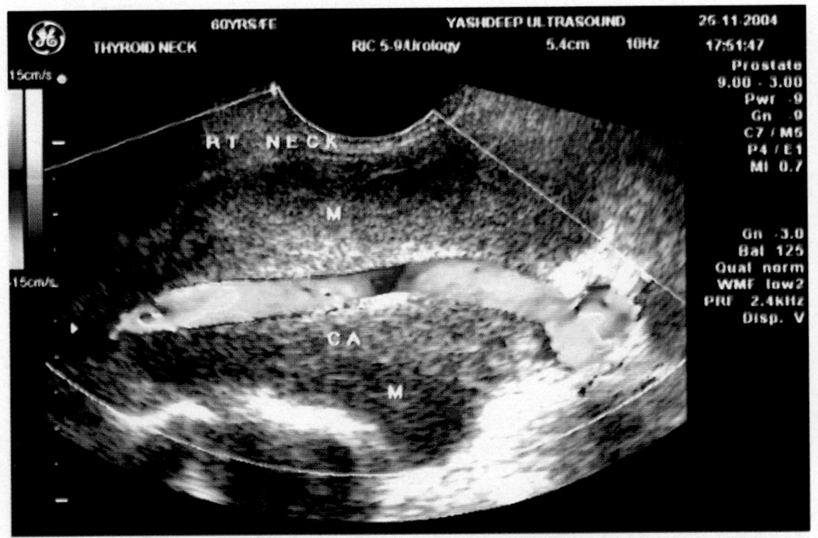

FIGURE 2.197: On color flow imaging the mass is seen invading the neck vessels and going around the Rt common carotid artery. It is seen encasing the artery.

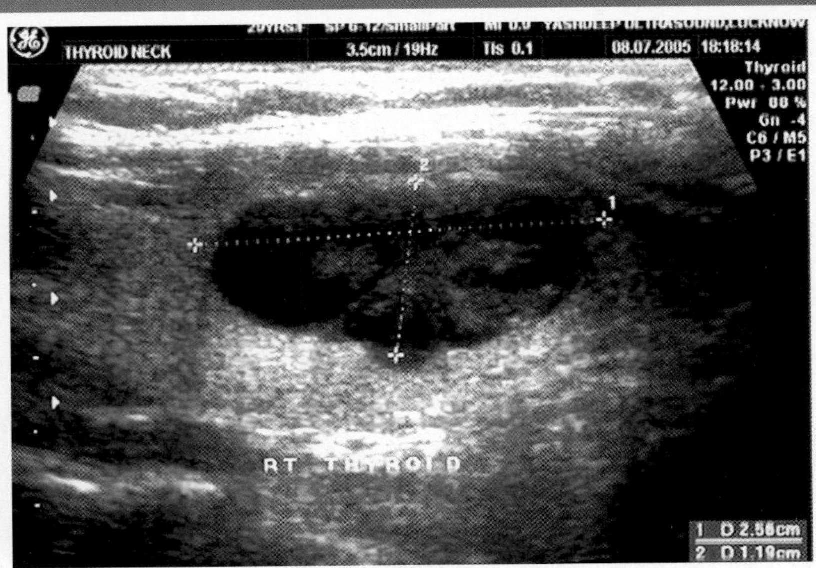

FIGURE 2.199

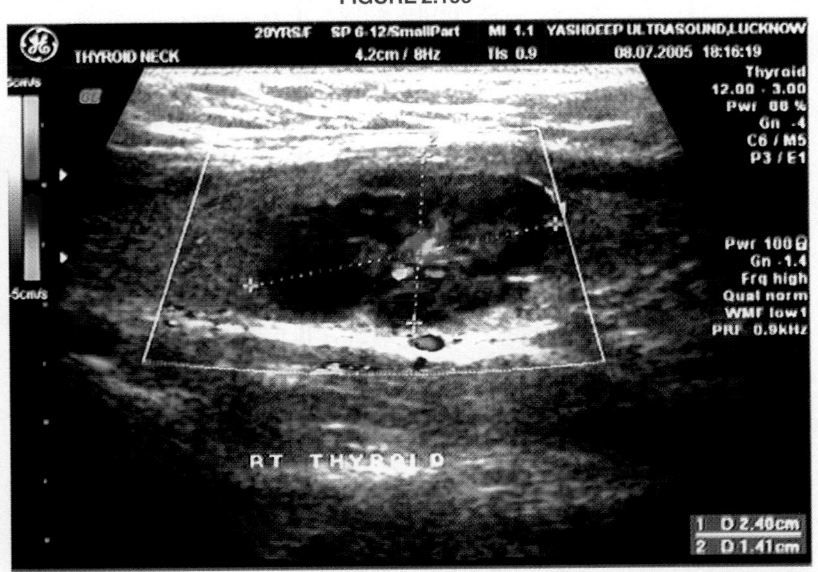

FIGURE 2.200

FIGURES 2.199 and 2.200: Metastatic mass in thyroid. An irregular hypoechoic mass is seen in the Rt lobe of thyroid. The mass does not show any halo. FNAC of the mass shows metastasis from carcinoma breast. The patient was known case of Carcinoma breast operated in the past.

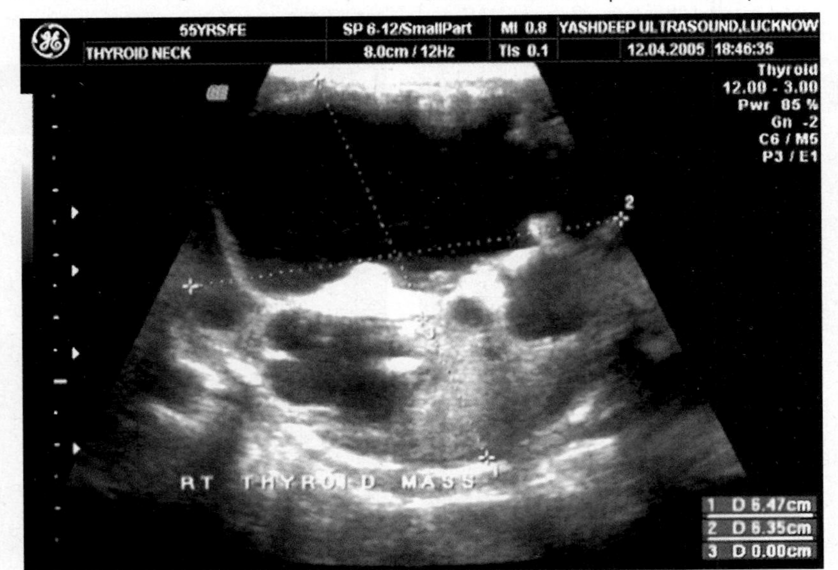

FIGURE 2.198: Invasive anaplastic adenocarcinoma. A big heterogeneous mass is seen in the Rt lobe of the thyroid. The capsule is broken. Mass shows cystic degeneration. Normal anatomy is distorted. Calcified specks are also seen in the mass. Biopsy proved anaplastic adenocarcinoma.

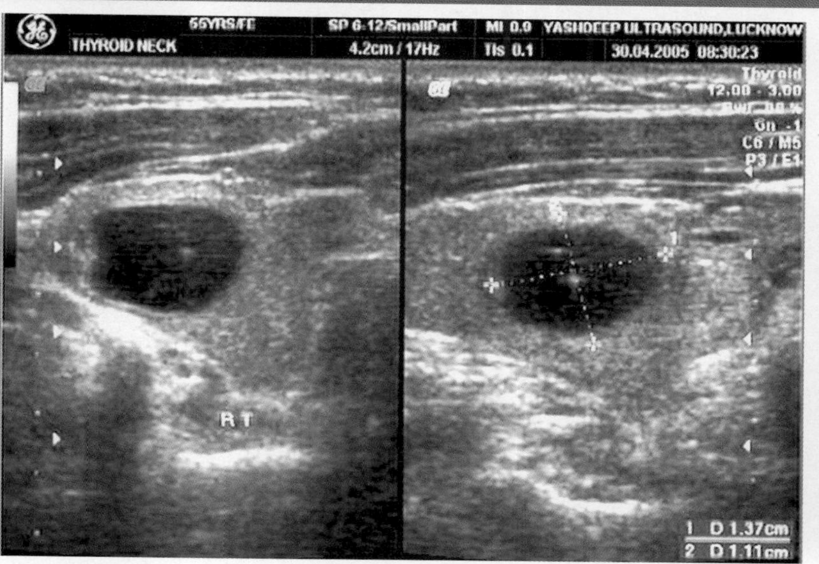

FIGURE 2.201

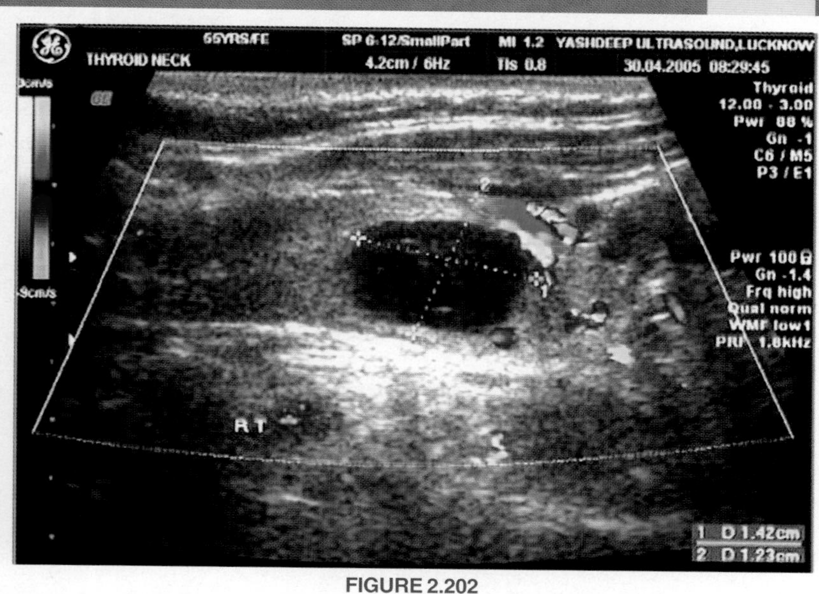

FIGURE 2.202

FIGURES 2.201 and 2.202: Metastatic deposit of carcinoma bronchus. A patient presented with nodular mass in the thyroid. A hypoechoic mass is seen in the Rt lobe. FNAC of the mass shows metastasis deposit from carcinoma bronchus.

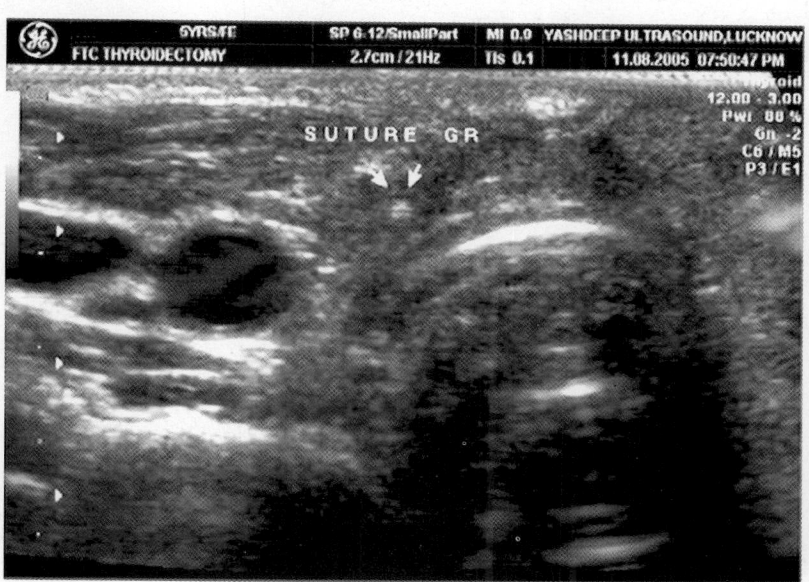

FIGURE 2.203

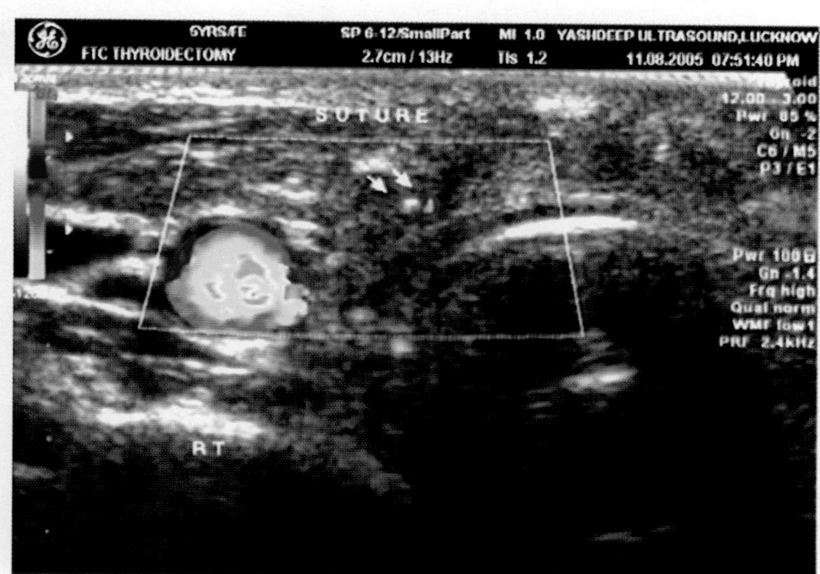

FIGURE 2.204

FIGURES 2.203 and 2.204: Suture granuloma. A patient presented with palpable nodule in the Rt side of the neck after total thyroidectomy. HRSG shows the suture knot, echogenic fibrous tissue is seen around it suggestive of suture granuloma.

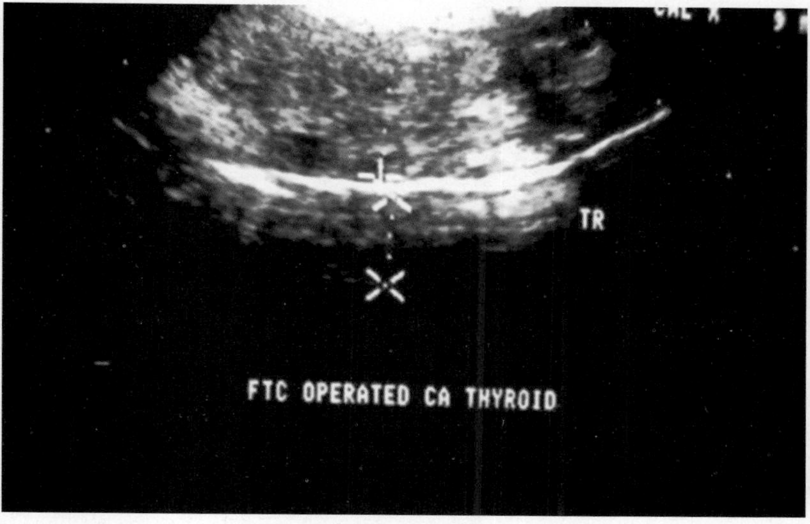

FIGURE 2.205

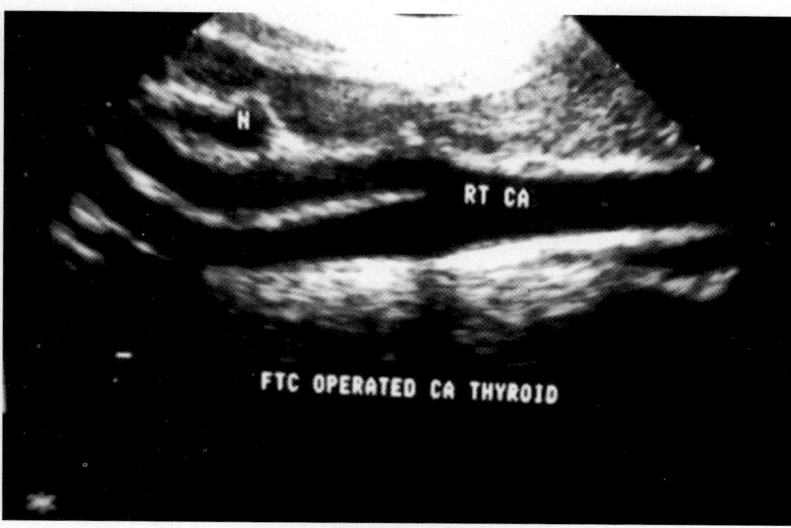

FIGURE 2.206

FIGURES 2.205 and 2.206: Recurrence of the mass in operated case of CA thyroid. HRSG shows irregular heterogeneous mass in the neck anterior to the neck vessels in postoperated CA thyroid. There is also suggestion of involvement of external carotid artery seen by the mass. HRSG is an excellent modality for evaluation of treatment response and recurrence the growth.

can be assessed on HRSG in 5 to 7% of the cases. Cervical lymph node involvement is found mainly in papillary or medullary carcinoma. HRSG is also valuable to assess post-surgery or post-irradiation status of thyroid malignancy. Recurrence of tumor or lymph node involvement is most easily picked up on HRSG, as surgical scarring and post-irradiation fibrosis make clinical examination difficult.

3

Neck

Introduction

High-resolution sonography (HRSG) is an excellent modality to examine the neck and neck masses. The excellent tissue details and anatomical landmarks in neck like thyroid cartilage; trachea, strap muscles and neck vessels have made assessment of neck masses a practical proposition. The neck masses have been divided into the following heads for convenience besides thyroid and parathyroid glands:

1. Cervical masses
2. Lymph node mass
3. Salivary glands.

Cervical Masses

Besides thyroid and parathyroid gland pathologies, other cervical masses are mainly present below hyoid bone in midcervical or lower cervical regions. They are mainly confined to anterior triangle of the neck; however, few masses are also seen in the posterior triangle of the neck like nerve tumors and abscess. The cervical masses are grouped as follows:

1. Congenital masses
2. Inflammatory masses
3. Vascular masses
4. Benign tumors
5. Malignant tumors.

Congenital Masses

Congenital masses are found in children usually as developmental anomalies. They are grouped as: (i) thyroglossal duct cysts, (ii) branchial cysts, and (iii) cystic hygromas.

Thyroglossal Duct Cysts

These are formed due to remnant of the duct, which is filled due to secretion of epithelial, cells rests. In normal embryological development, the thyroid gland descends down along thyroglossal duct to its normal position. The duct undergoes progressive atrophy. However, when it remains patent is filled with clear fluid filled cystic mass in the midline anterior to the thyroid gland. Small cyst is difficult to differentiate from the reactive lymph node enlargement, which undergoes cystic necrosis. However, high frequency transducers of 13 to 15 mhz can differentiate them. Hemorrhage is common in the cyst. It can be seen as highly reflective mobile shadow in the dependent part of the cyst.

Branchial Cysts

These cysts arise from the branchial cleft usually from the second cleft. It is lateral in position. It is situated in the anterior triangle of neck along the anterior margins of sternocleidomastoid muscle. It is situated between suprahyoid and carotid spaces. These cysts are present in the childhood. The cyst is seen as a well-defined thin walled echo free mass on HRSG. It is round with sharp clear margins. It is clearly separated and demarcated with sternocleidomastoid muscle border. Secondary infection may take place in the cyst and low level echoes are noted on HRSG.

Cystic Hygromas

These are serous benign cystic masses arising from the lymphoid tissue. They are non-capsulated and usually confined to posterior triangle of the neck. When they are big they are seen reaching up to anterior part of neck or upper part of the superior mediastinum. They may also be seen reaching up to axilla on HRSG; it appears to be multilocular cystic masses with echogenic fibrous septa with thin regular, smooth margins. The internal structures are usually free and anechoic. However, flow amplitude echoes may be seen in the loculae representing fine jelly like consistency of cystic hygroma. Secondary hemorrhage is also encountered in cystic hygroma.

Sublingual Dermoid

It is a benign congenital condition. It is seen a medium to hard consistency mass with sharp margins on clinical examination. It presses over the floor of the mouth, and pressure is seen on the tongue.

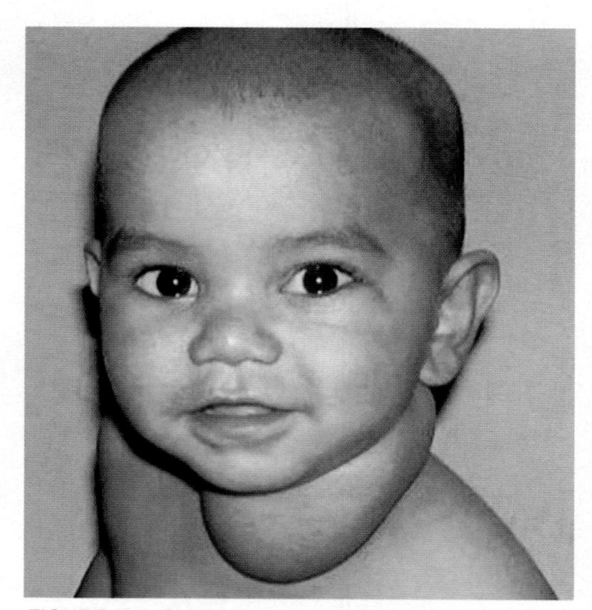

FIGURE 3.1: Cystic hygroma. A young child presented with a soft tissue swelling in the neck.

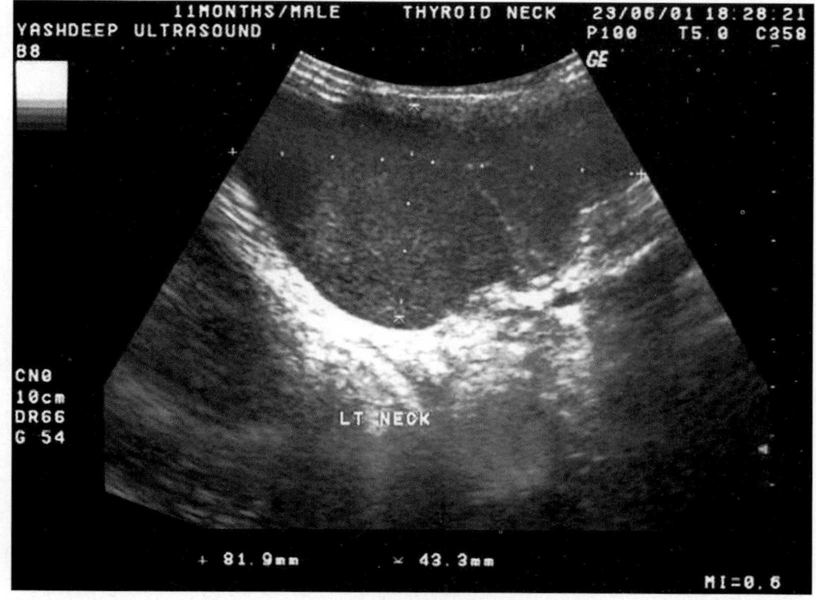

FIGURE 3.2

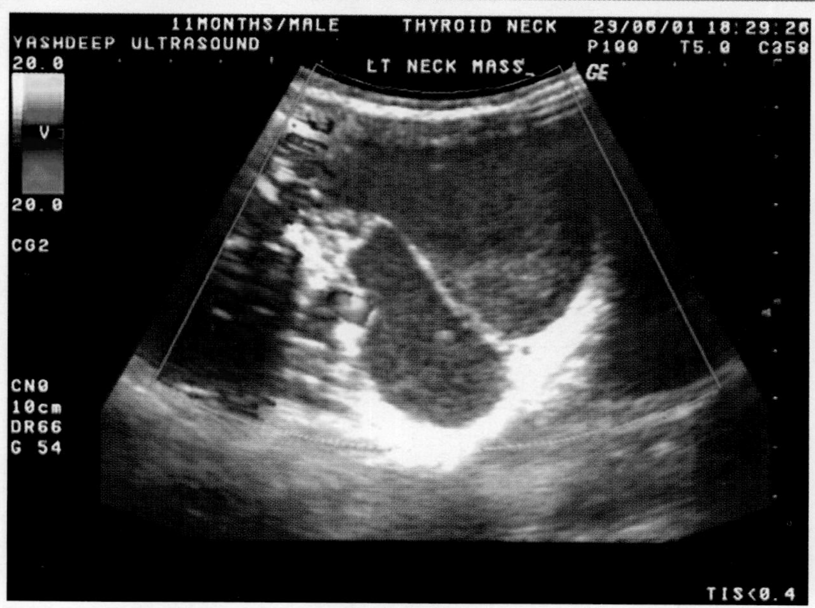

FIGURE 3.3

FIGURES 3.2 and 3.3: HRSG shows a multiloculated well defined cystic mass seen in the Lt side of the neck. Echogenic septa are also seen in the mass. Low-level echo collection is seen in the cyst suggestive of serous collection. The margins are smooth, typical findings of a cystic hygroma.

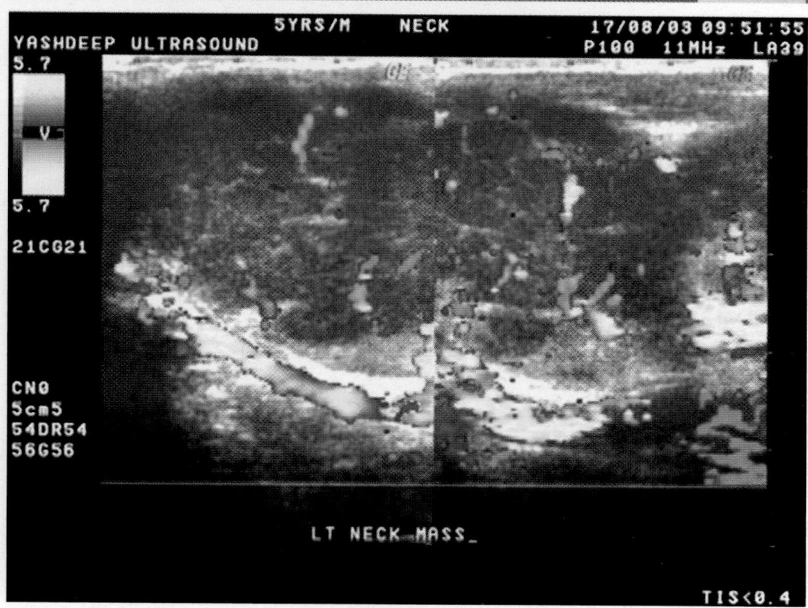

FIGURE 3.6

FIGURES 3.5 and 3.6: HRSG shows a well-defined low-level echo complex mass in the Lt neck. The mass shows low-level thick echoes. Thin echogenic septa are also seen in the mass. On color flow imaging multiple vessels are seen feeding the mass. FNAC confirmed lymphangioma.

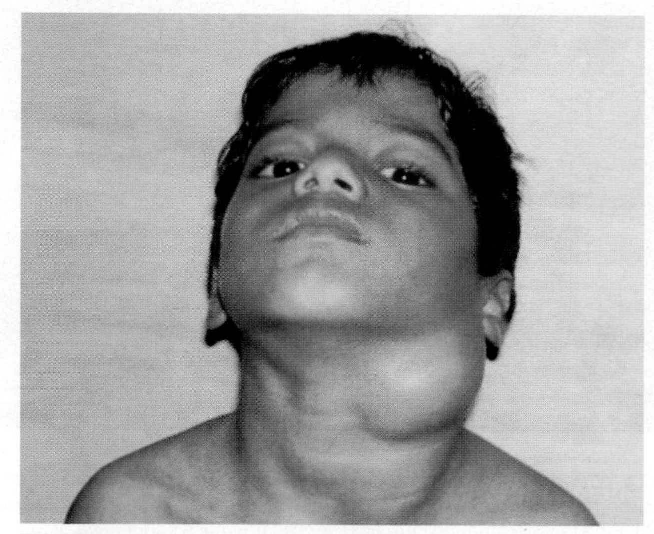

FIGURE 3.4: Lymph angioma neck. A young child presented with cystic mass in the Lt upper cervical region.

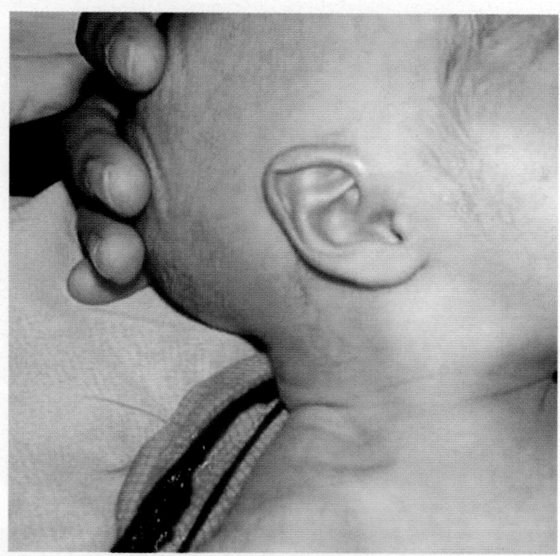

FIGURE 3.7: Sternocleidomastoid tumor (SCM Tumor). A young child was born with soft tissue swelling in the Rt side of the neck.

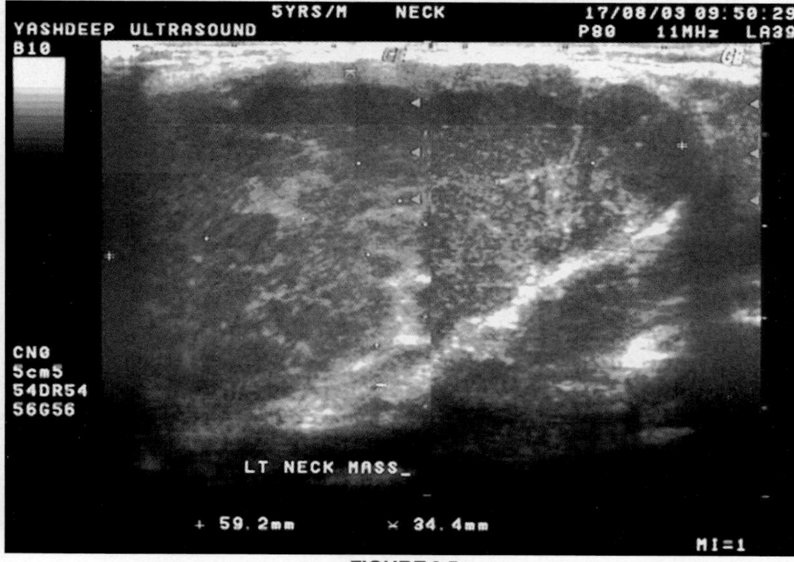

FIGURE 3.5

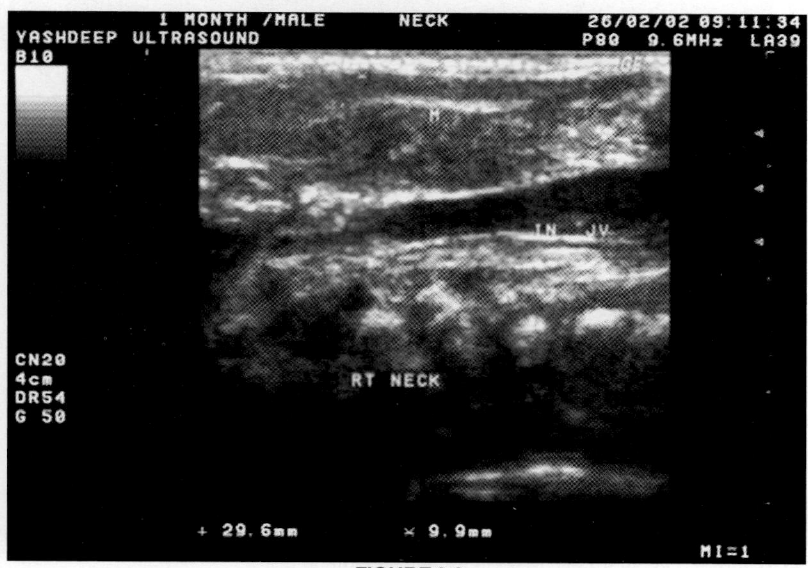

FIGURE 3.8

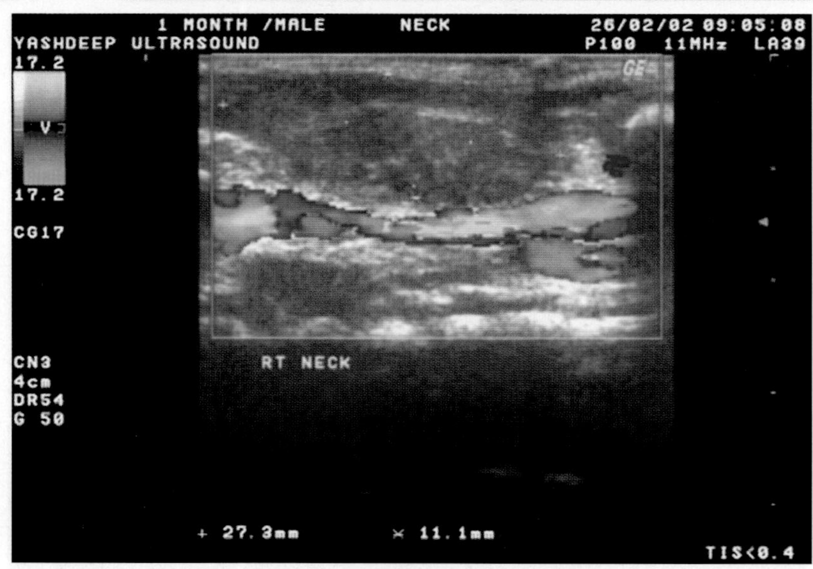

FIGURE 3.9

FIGURES 3.8 and 3.9: HRSG shows a well-defined fusiform mass in the sterno-cleidomastoid muscle. It is homogeneous in texture. No calcification or necrosis is seen in it. On color flow imaging no flow is seen in it.

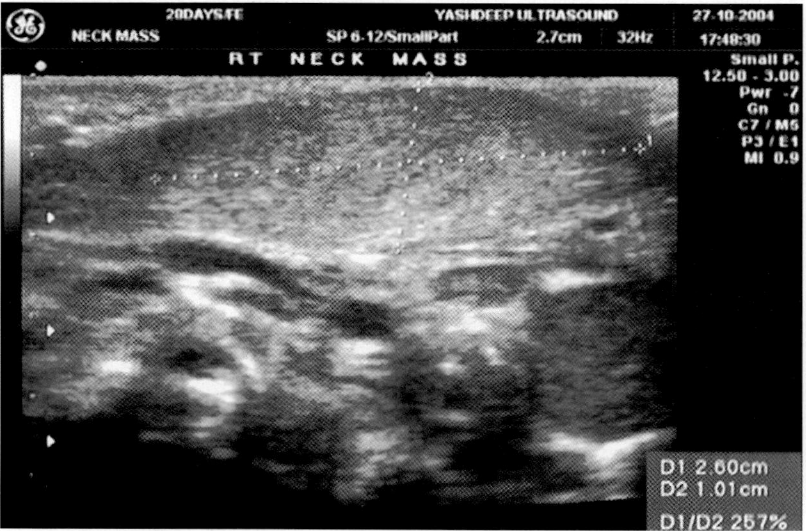

FIGURE 3.10

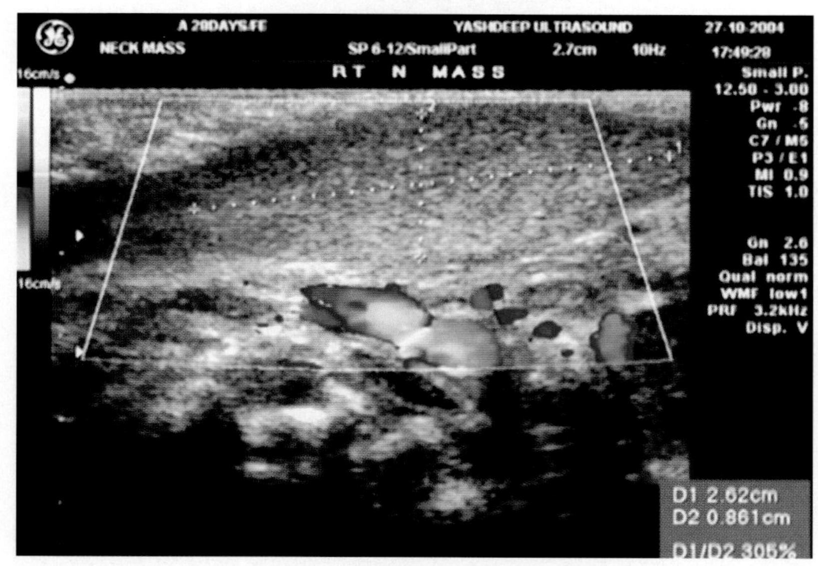

FIGURE 3.11

FIGURES 3.10 and 3.11: Sternocleidomastoid tumor. An Infant girl was born with soft tissue mass in the neck. HRSG shows a well-defined mass in the sterno-cleidomastoid muscle. Fine muscle striations are also seen in the mass suggestive of SCM tumor. On color flow imaging no flow is seen in it.

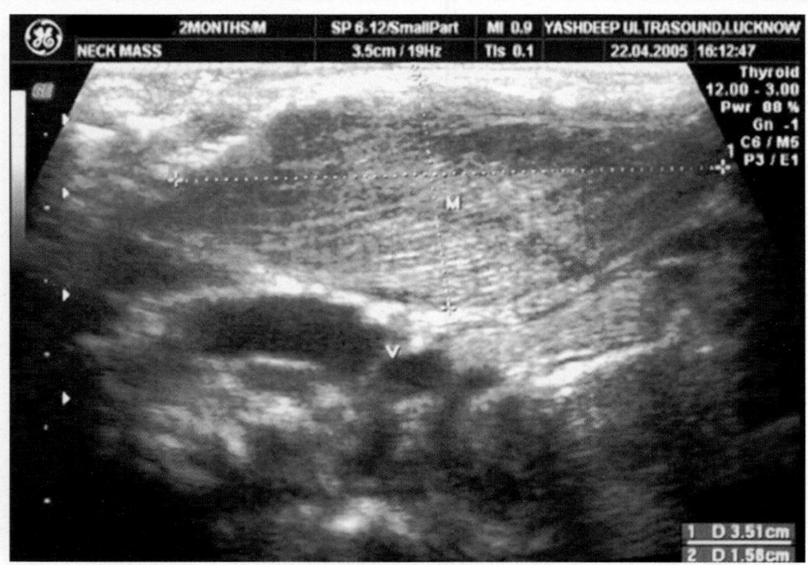

FIGURE 3.12

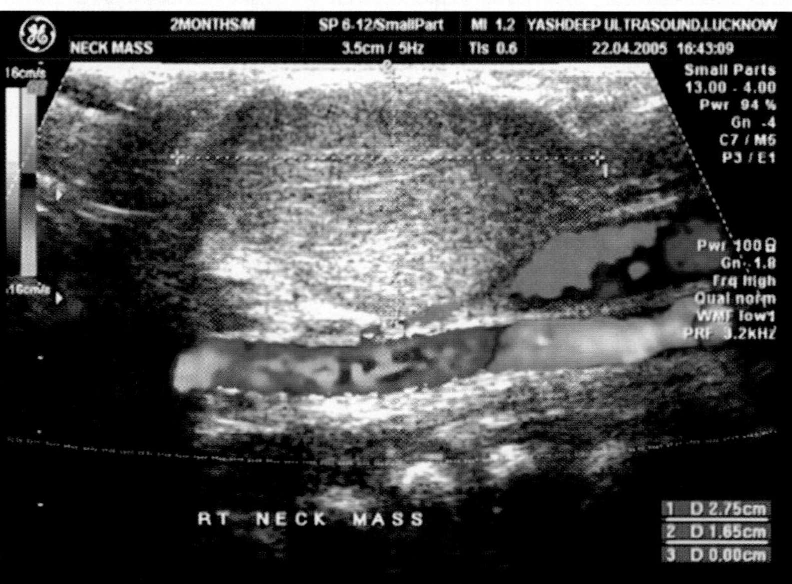

FIGURE 3.13

FIGURES 3.12 and 3.13: SCM tumor. A well defined soft tissue mass is seen in the Rt side of the neck of a newborn baby. HRSG shows well-defined soft tissue mass in the SCM. Fine muscle striations are also seen in the mass suggestive of SCM tumor.

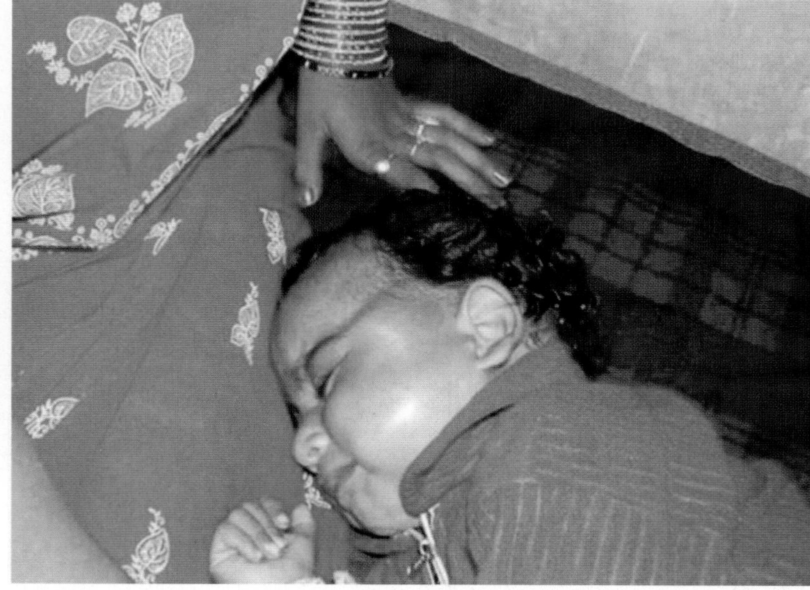

FIGURE 3.14

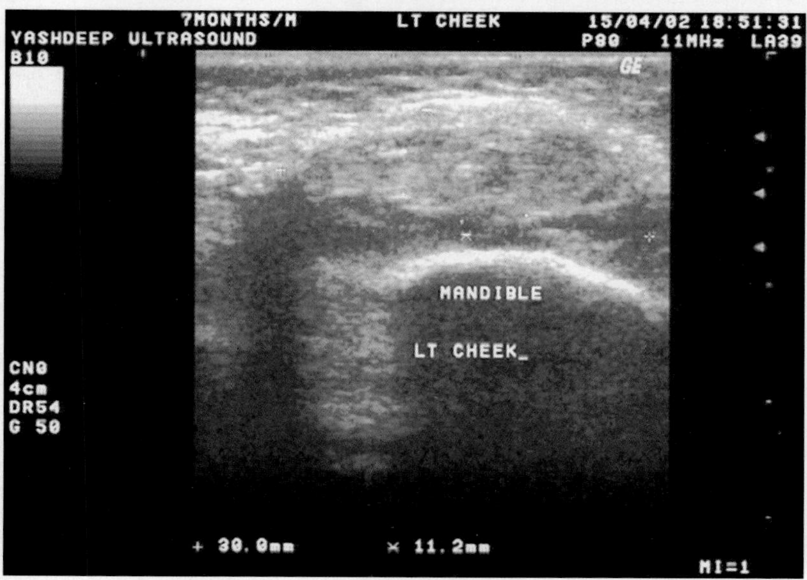

FIGURE 3.15

FIGURES 3.14 and 3.15: Tumor of the masseter muscle. A child presented with soft tissue mass over the Lt masseter muscle. It is irregular in outline. No calcification is seen in it. No cystic degeneration is seen. FNAC confirmed masseter muscle mass. The mass is not seen invading the bone. FNAC confirmed a myosarcoma.

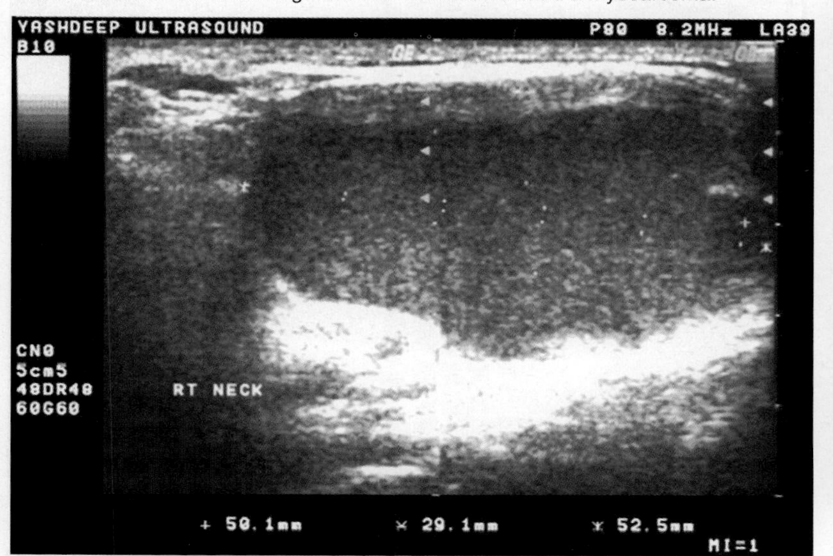

FIGURE 3.16: Branchial cyst. A young child presented with a well-defined cystic mass in the Rt supraclavicular area. Margins are smooth. No septation is seen. No loculation is seen.

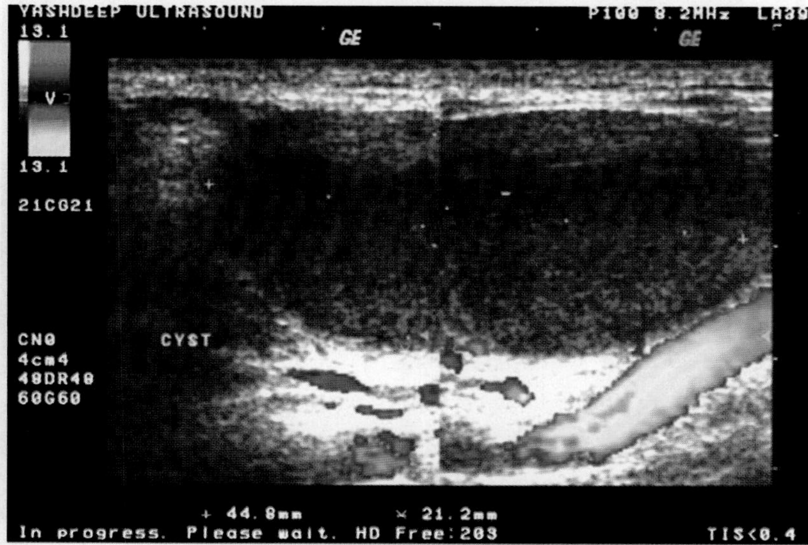

FIGURE 3.17: Color Doppler imaging shows no flow in the cyst. Neck vessels are seen displaced down with the cyst. Biopsy shows branchial cyst.

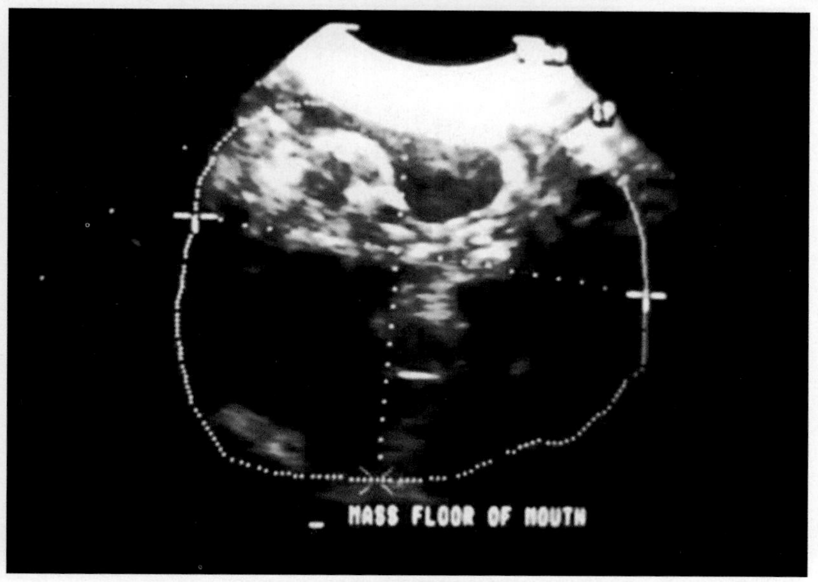

FIGURE 3.18

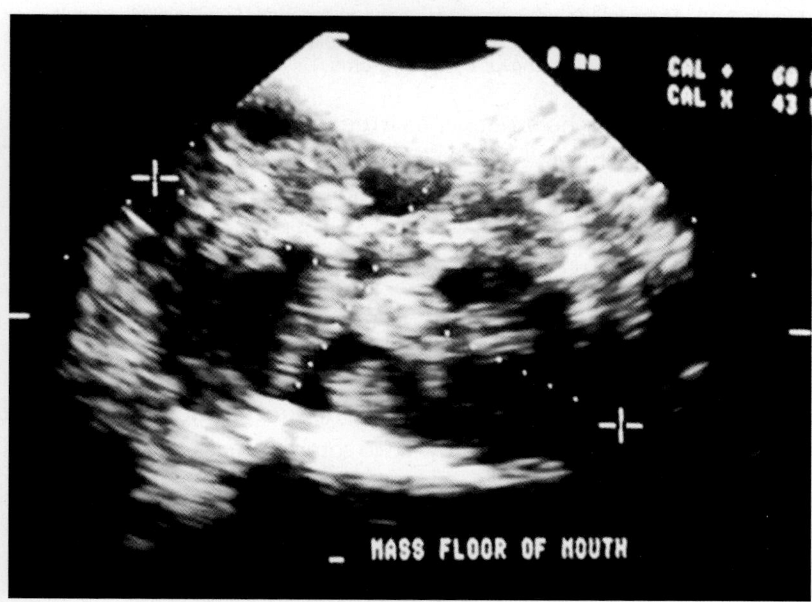

FIGURE 3.19

FIGURES 3.18 and 3.19: Sublingual dermoid. An infant was presented with a soft tissue mass in the submental area. HRSG shows a heterogeneous mass in the sublingual area with irregular margins. Multiple low level echoes are seen in it. Few calcified specks are also seen. Biopsy of the mass showed sublingual dermoid.

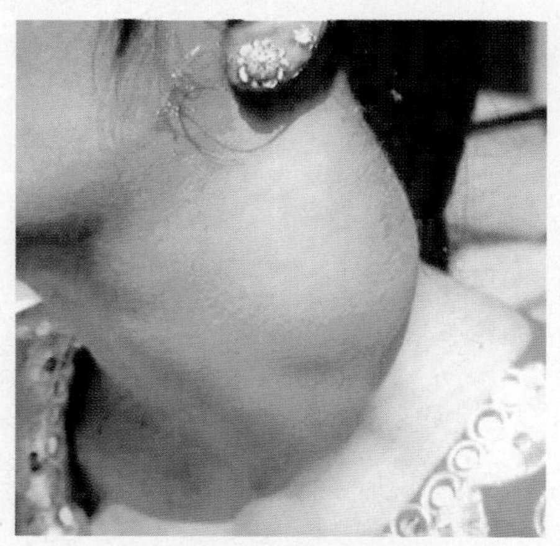

FIGURE 3.20: Infected cystic hygroma. A young lady presented with a soft tissue mass swelling in the Lt side of the neck present since the birth. There was the history of intervention done in the cystic mass and serous fluid was aspirated.

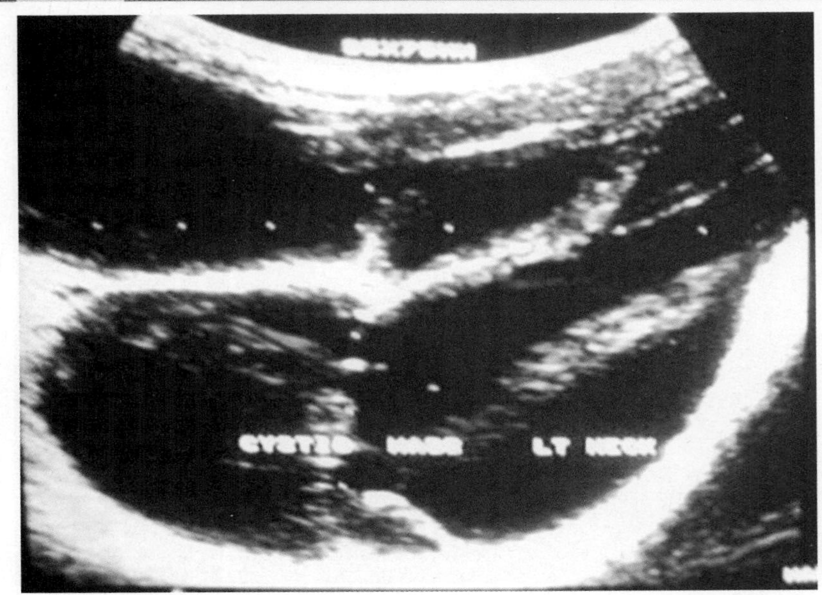

FIGURE 3.21: HRSG shows a big multiloculated cystic mass in the same patient of Figure 3.12. Multiple internal echoes are seen in the cyst. Few calcified specks are also seen. The cyst shows low level echogenic collection at the base suggestive of debris. Biopsy of the cyst shows cystic hygroma.

On HRSG, a highly heterogeneous irregular mass with poor margins with variable texture is seen. Areas of low-level echoes are seen in it, suggestive of cystic degeneration. Echogenic shadows are also seen which represent fibrous element of the cyst. At times, echogenic foci are also seen suggestive of calcification. They are accompanied with acoustic shadowing.

Inflammatory Masses

Inflammatory or cervical abscesses are not uncommon pathology in neck. They show thick walled low echo complex mass on HRSG. Walls are irregular. Central area of necrosis is seen in the abscess. Echogenic debris may be seen in the abscess on the dependent part.

Cold Abscess

Cold or tubercular abscess are also seen in the neck mainly in the posterior triangle of the neck. These are usually present in painful

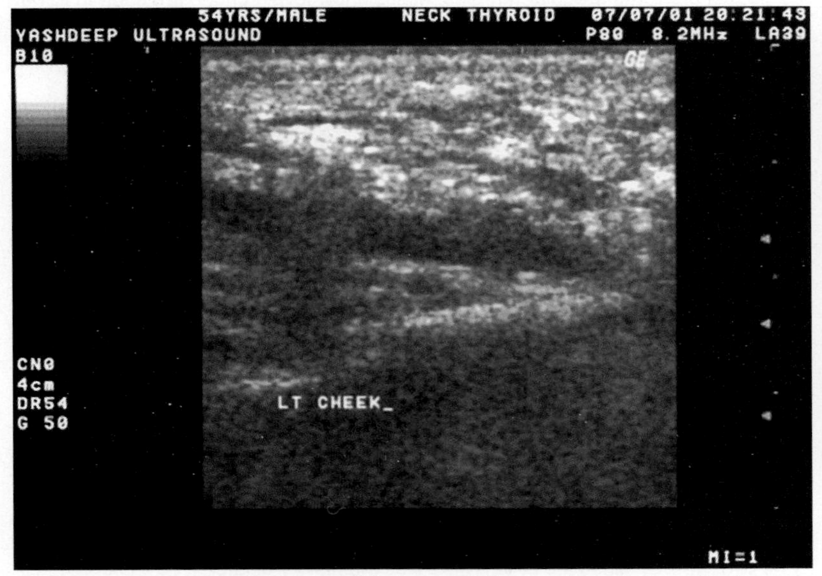

FIGURE 3.22: Acute cellulitis. HRSG show soft tissue edema in the subcutaneous planes. Low-level collection is seen in it. Pressure is seen on the neck vessels.

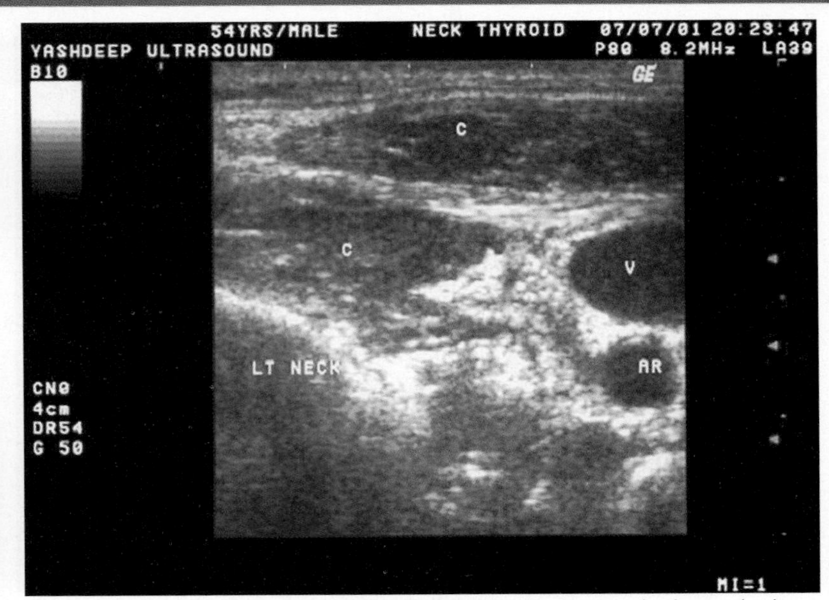

FIGURE 3.23: Acute cellulitis. Hypoechoic collection is seen in the neck planes anterior to the carotid artery. Soft tissue edema is also present in the muscles.

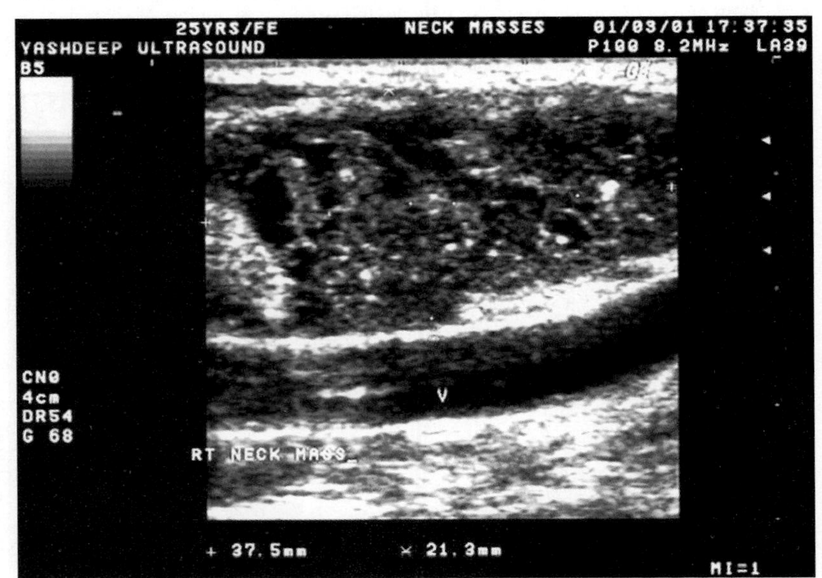

FIGURE 3.24

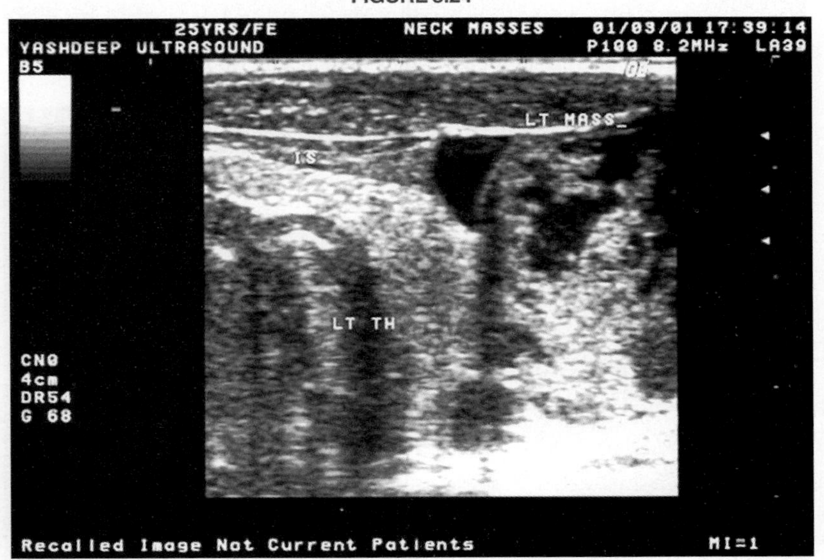

FIGURE 3.25

FIGURES 3.24 and 3.25: Pyogenic neck abscess: A young lady presented with Lt sided neck swelling. HRSG shows thick low level echo collection mass in the anterior part of the neck. Multiple internal echoes are seen in it, suggestive of debris. The abscess is seen separate from the thyroid gland.

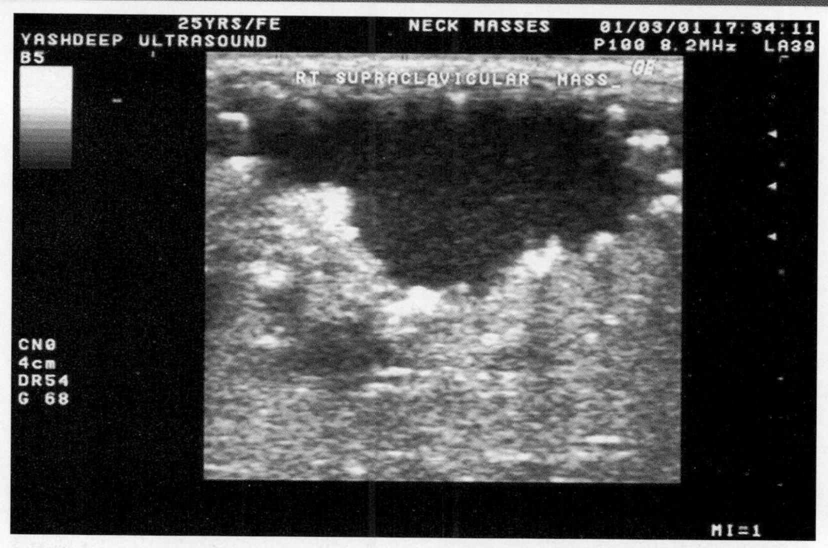

FIGURE 3.26: A big area of necrosis is also seen in the abscess suggestive of cystic degeneration.

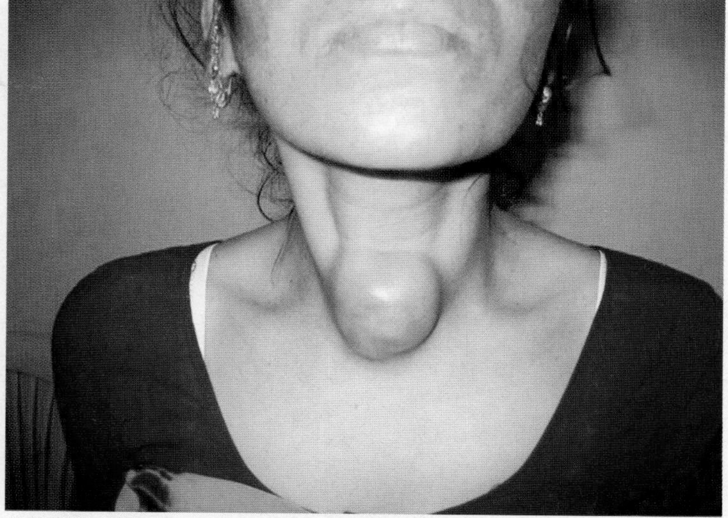

FIGURE 3.29: **Pyogenic suprasternal abscess.** A lady presented with well defined red and hot mass in the suprasternal notch.

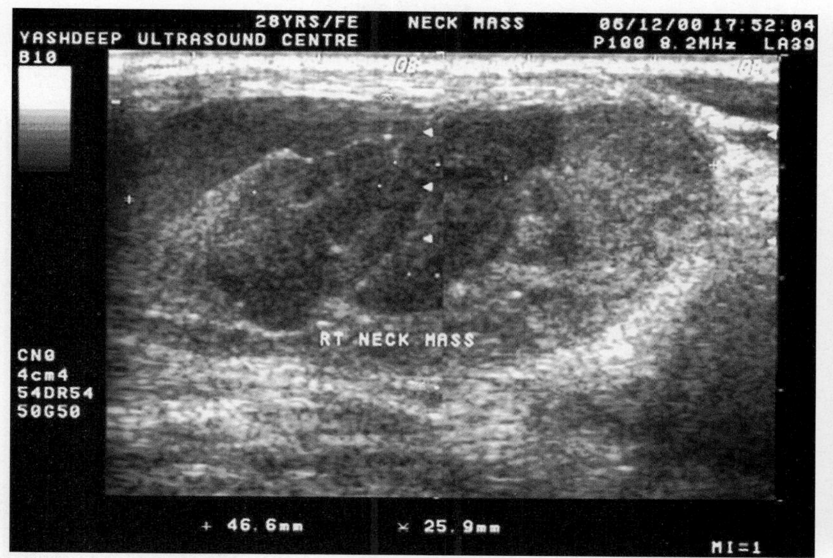

FIGURE 3.27

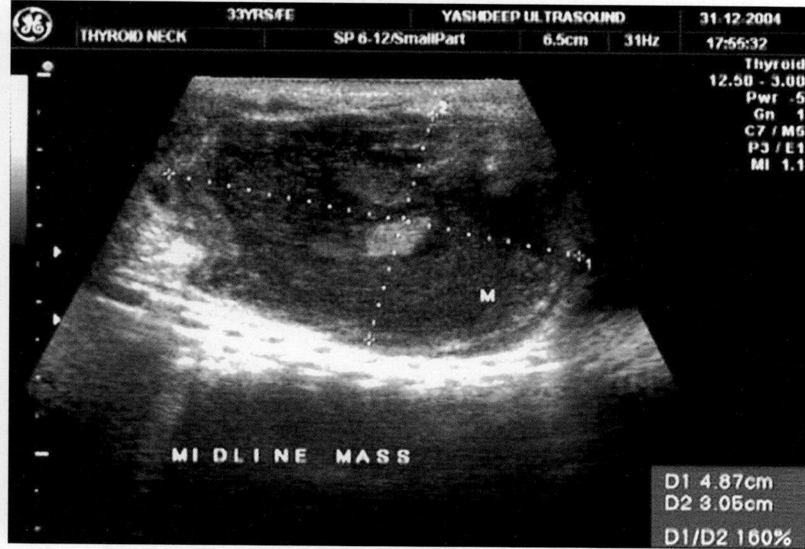

FIGURE 3.30

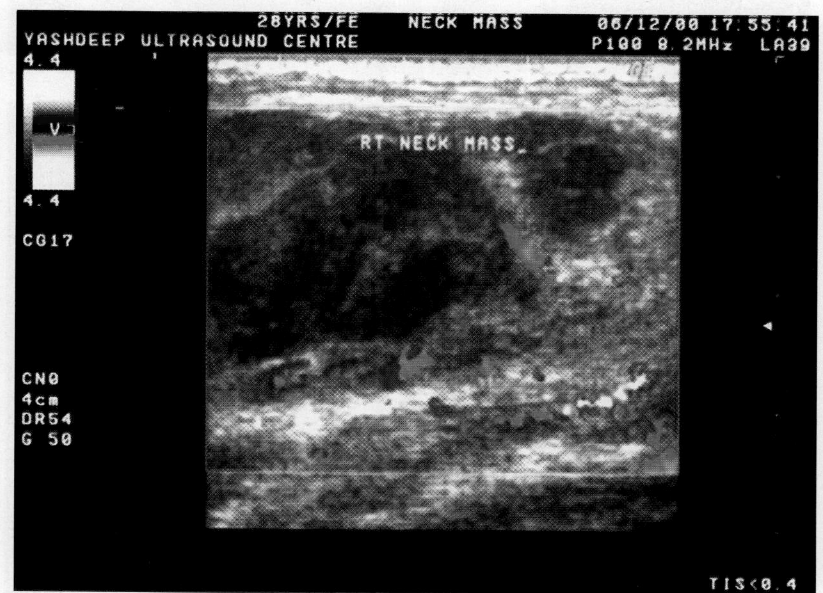

FIGURE 3.28

FIGURES 3.27 and 3.28: **Pyogenic neck abscess.** A young lady was presented with a big Rt sided neck mass. It was tender and hot on clinical examination. HRSG shows a big thick wall low level echo complex mass central area of necrosis is seen in it on color flow imaging, moderate flow is seen in the septa, suggestive of a big pyogenic abscess.

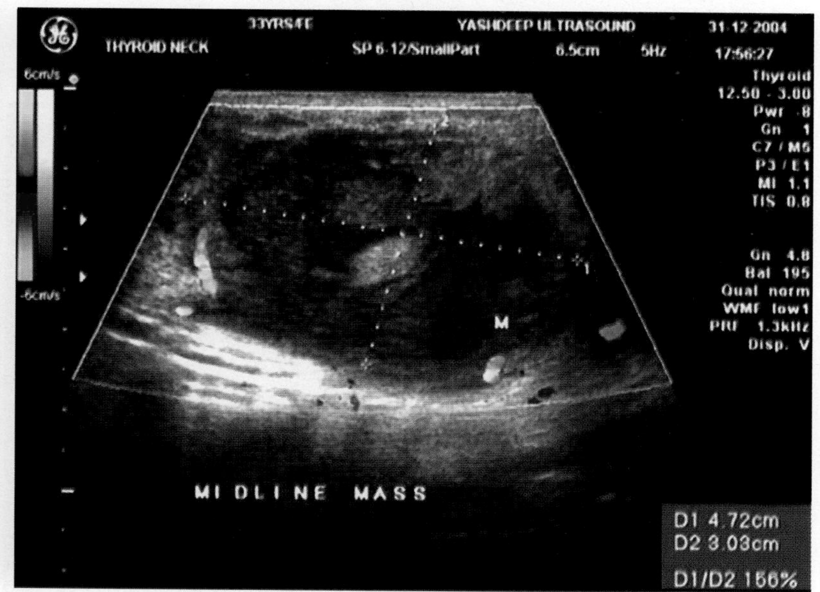

FIGURE 3.31

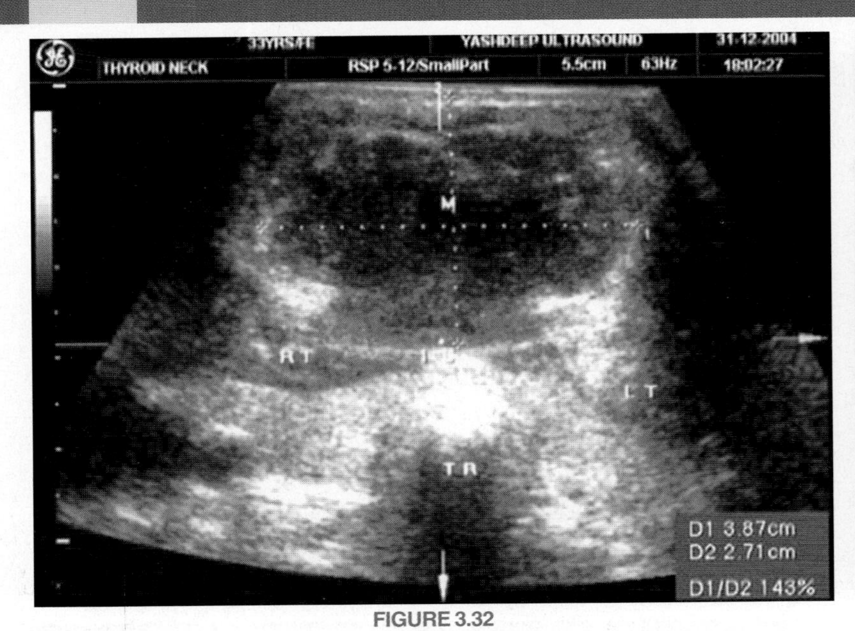

FIGURE 3.32

FIGURES 3.30 to 3.32: HRSG shows a thick walled, thick echo mass in the suprasternal area. Echogenic debris is also seen in it. On color flow imaging low flow is seen. Aspiration showed a thick pyogenic abscess.

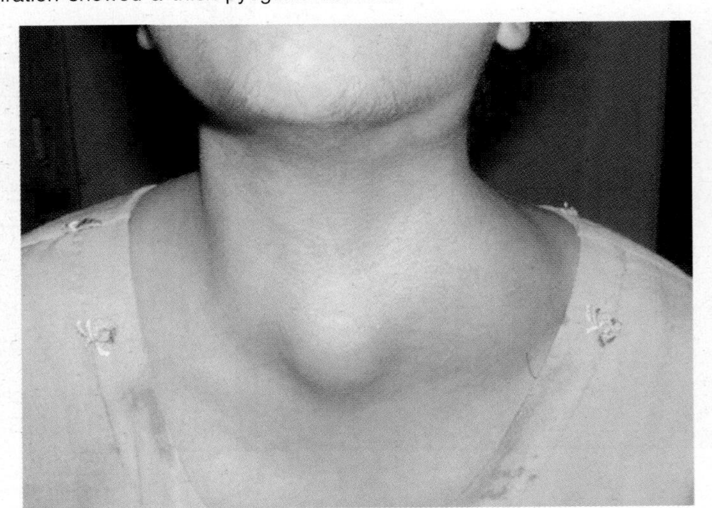

FIGURE 3.33: Localized pyogenic abscess (AntiBioma). A young girl presented with cystic mass in the midline of the neck in suprasternal area.

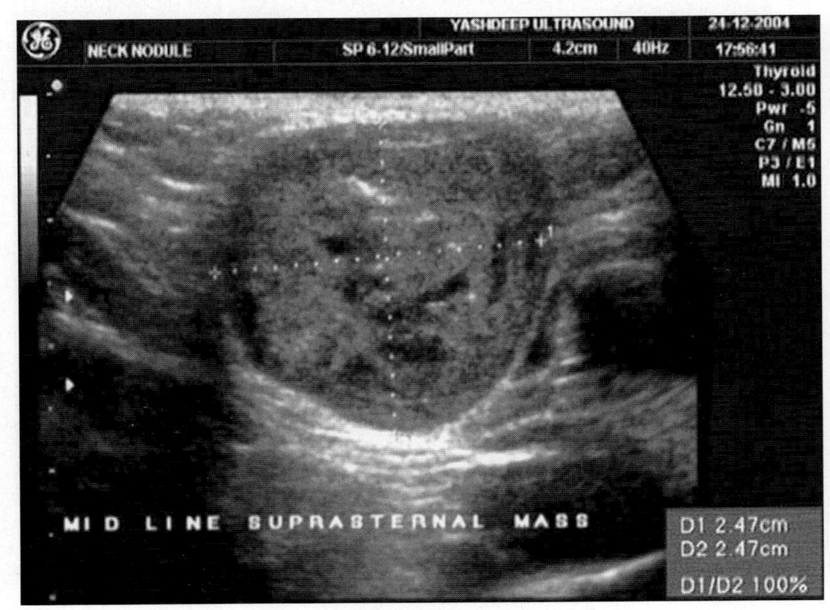

FIGURE 3.34

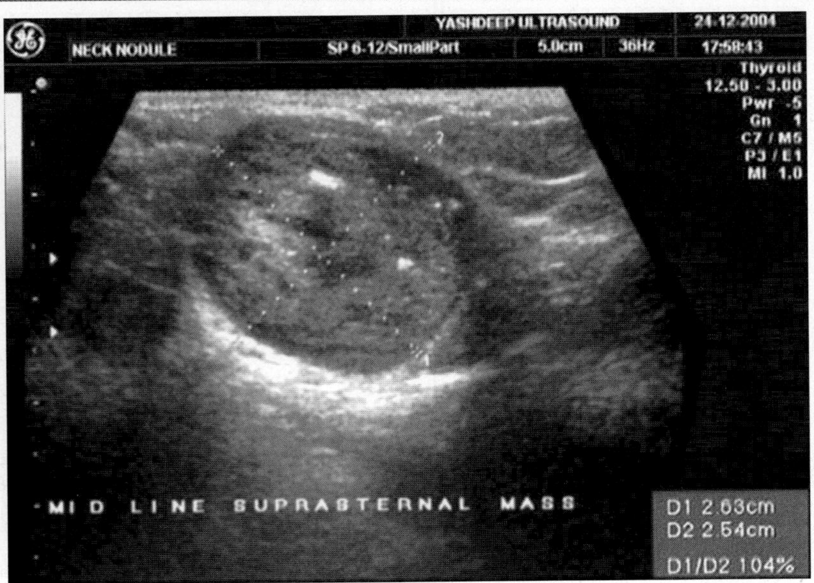

FIGURE 3.35

FIGURES 3.34 and 3.35: HRSG shows a thick jelly like mass in the suprasternal area. Few cystic areas are seen in it suggestive of degeneration.

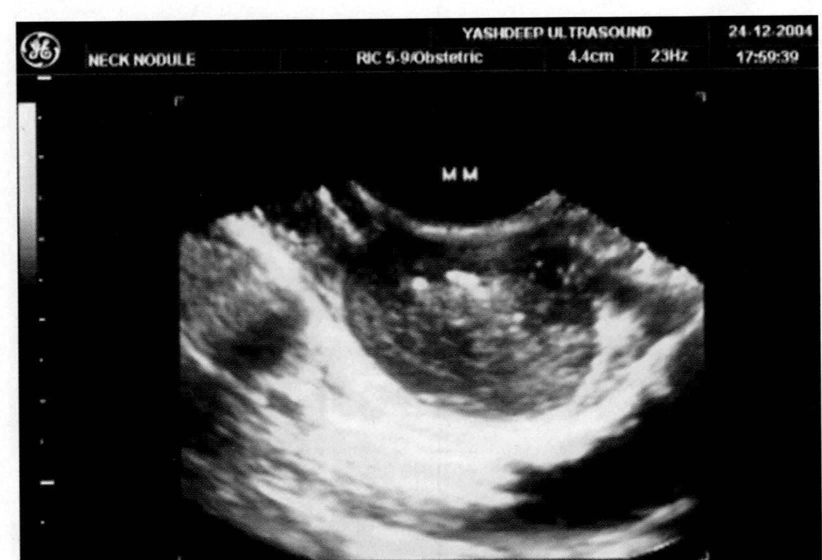

FIGURE 3.36

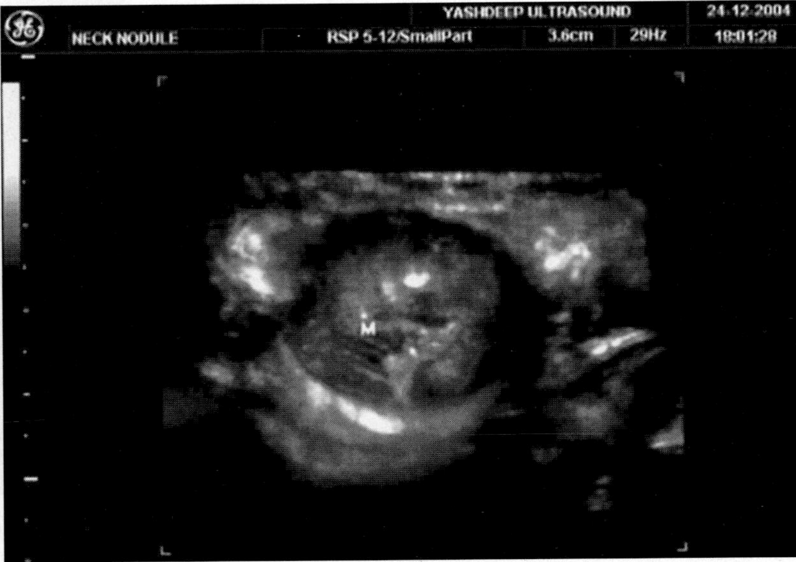

FIGURE 3.37

FIGURES 3.36 and 3.37: 3D imaging of the same patient shows the details of the mass. Biopsy confirmed an AntiBioma.

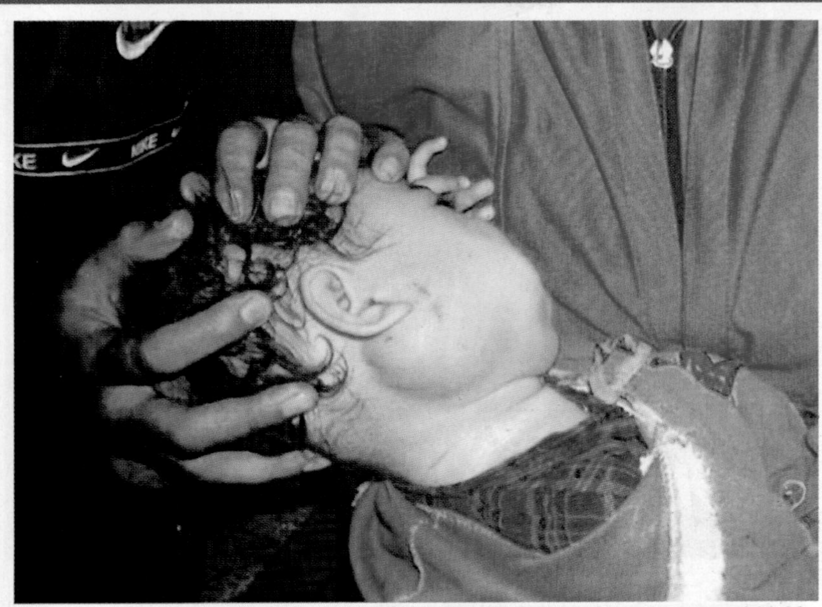

FIGURE 3.38: **Acute abscess in the Rt neck.** A newborn baby presented highly tender mass in Rt upper cervical and postauricular area.

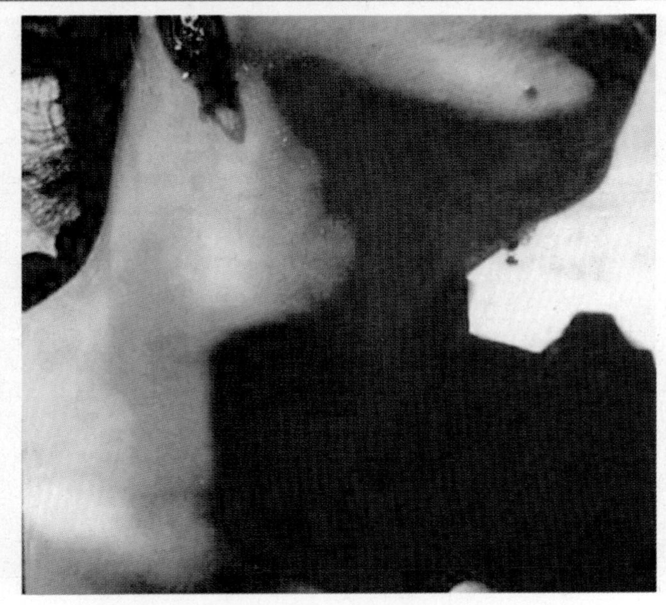

FIGURE 3.41: **Tubercular paravertebral abscess extending in the neck.** A young lady presented with a nontender swelling in the Rt side of neck.

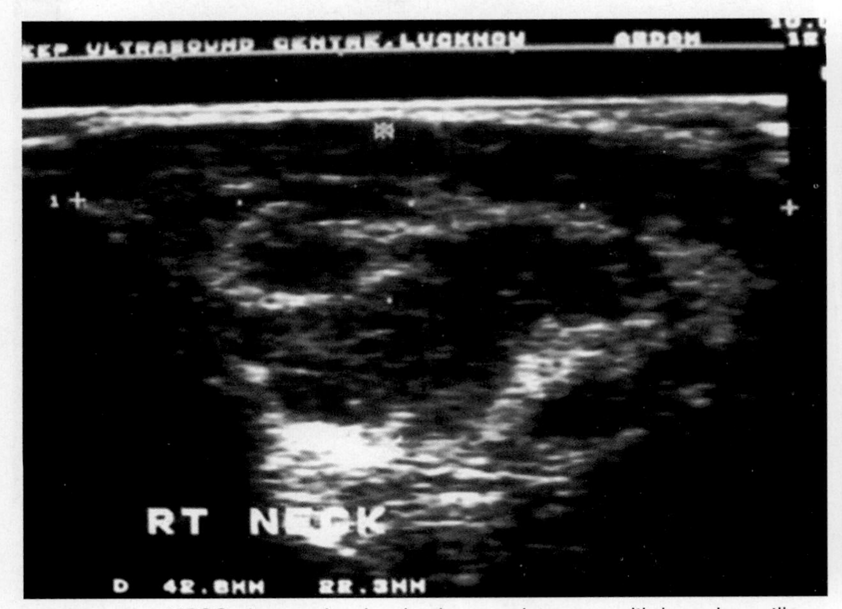

FIGURE 3.39: HRSG shows a low-level echo complex mass with irregular outlines and necrosis.

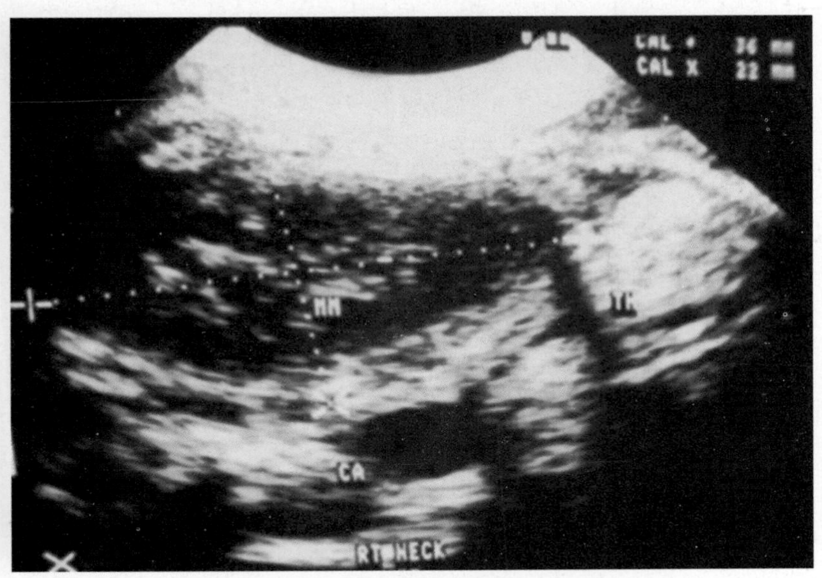

FIGURE 3.42: HRSG shows a big thick walled low-level echo complex mass in the midcervical area. Multiple internal echoes are seen in it suggestive of an abscess.

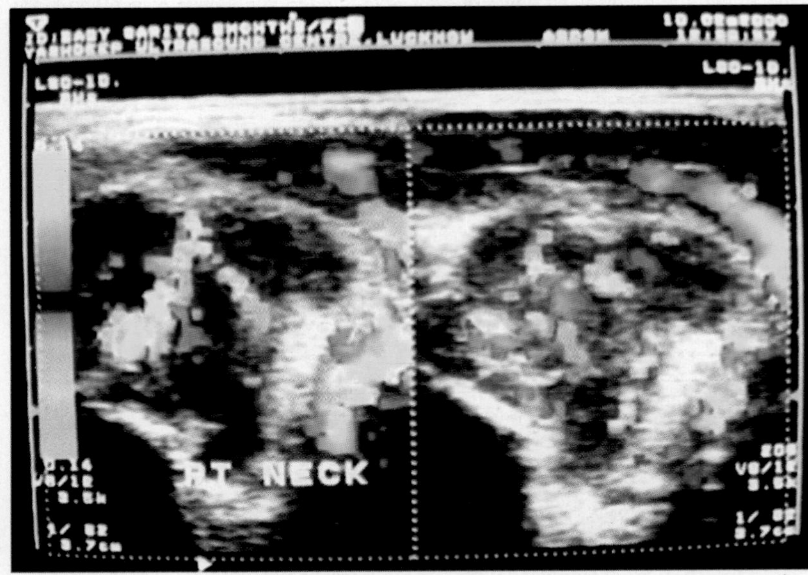

FIGURE 3.40: Color Doppler imaging shows increased flow in the mass, suggestive of hyperemia. Findings are suggestive of an abscess.

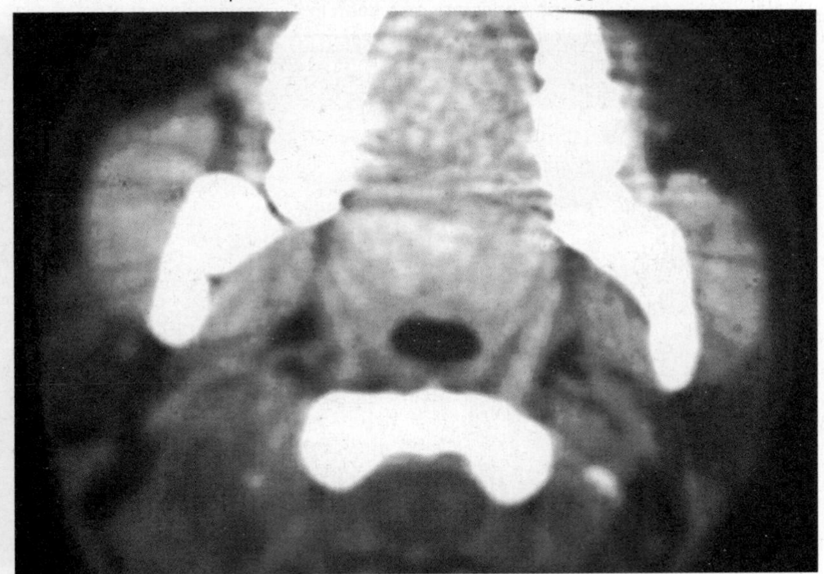

FIGURE 3.43: CT scan of the same patient shows a hypodense mass in the paravertebral and parapharyngeal area, known case of tubercular abscess.

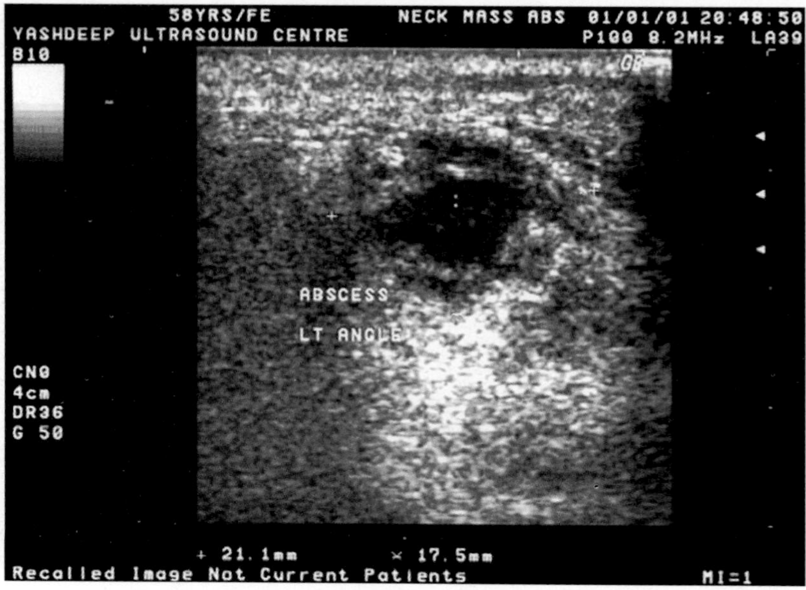

FIGURE 3.44

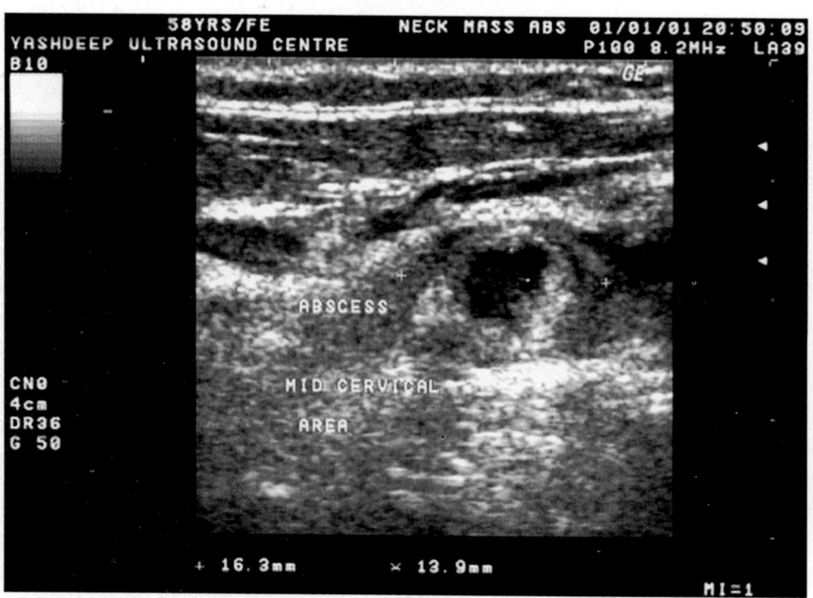

FIGURE 3.45

FIGURES 3.44 and 3.45: Pyogenic abscess in the Lt side of the neck. HRSG shows a thick walled low-level echo complex mass in central area of necrosis in Lt upper cervical region. Few enlarged lymph nodes are also seen in the upper cervical area.

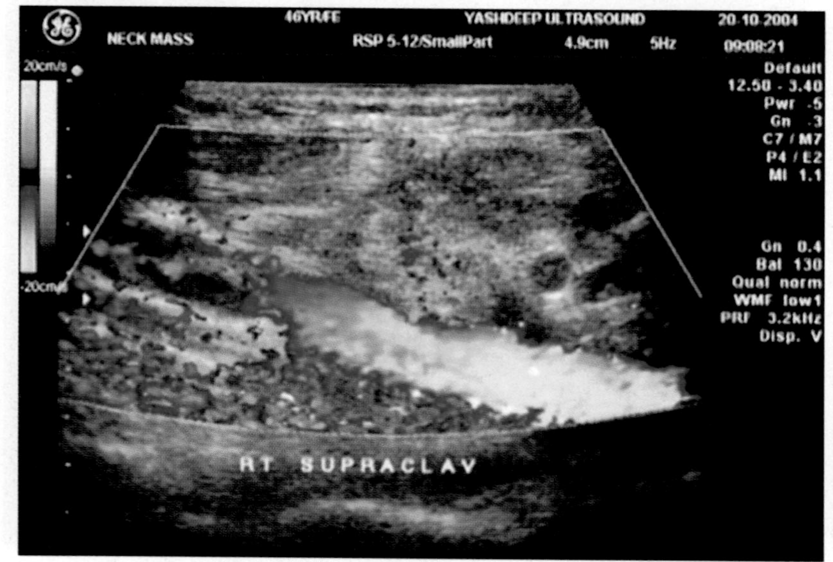

FIGURE 3.46

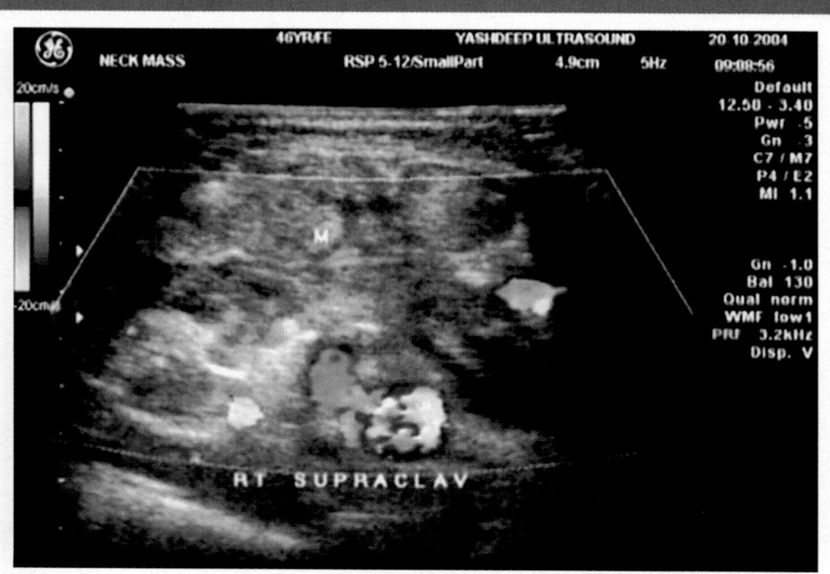

FIGURE 3.47

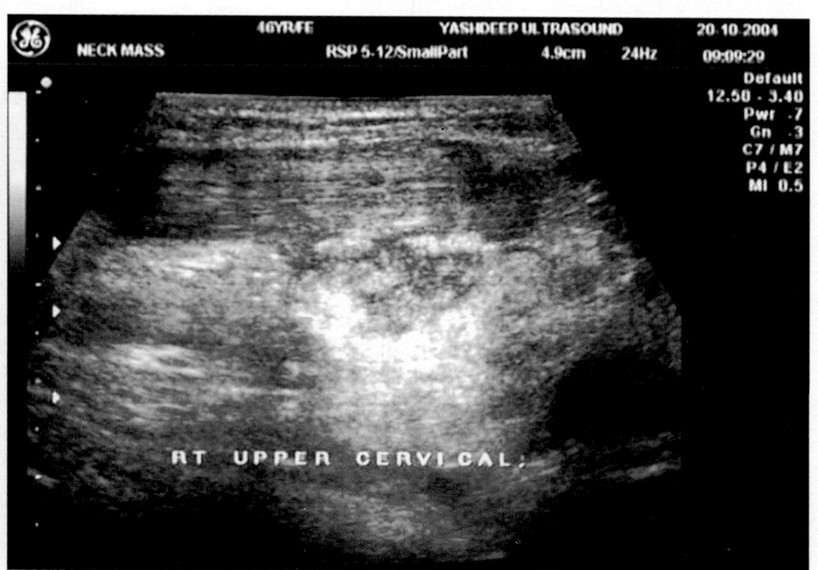

FIGURE 3.48

FIGURES 3.46 to 3.48: Acute cellulitis. A patient presented with acute swelling in the Rt supraclavicular and midcervical region. HRSG shows dermis and hypodermis edema. Muscle edema is also present with distorted tissue planes. Low-level echoes are seen in the muscles with microabscess formations. On color flow imaging low flow is seen in it. Breaking down of the tissue is seen suggestive of acute cellulitis.

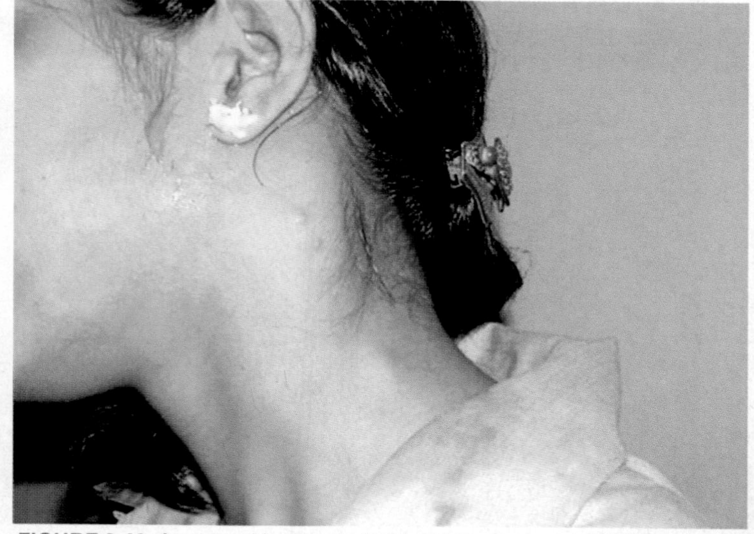

FIGURE 3.49: A young girl presented with soft tissue mass in Lt cervical area. No tenderness was present in the mass.

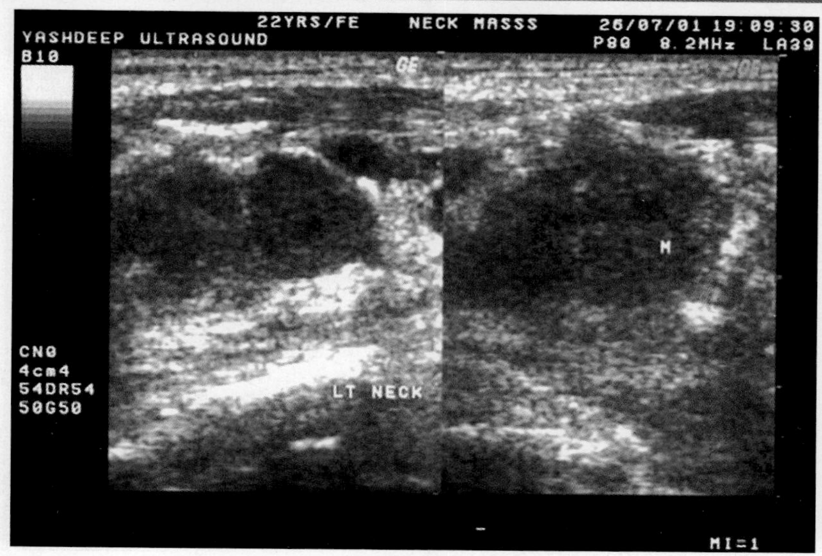

FIGURE 3.50

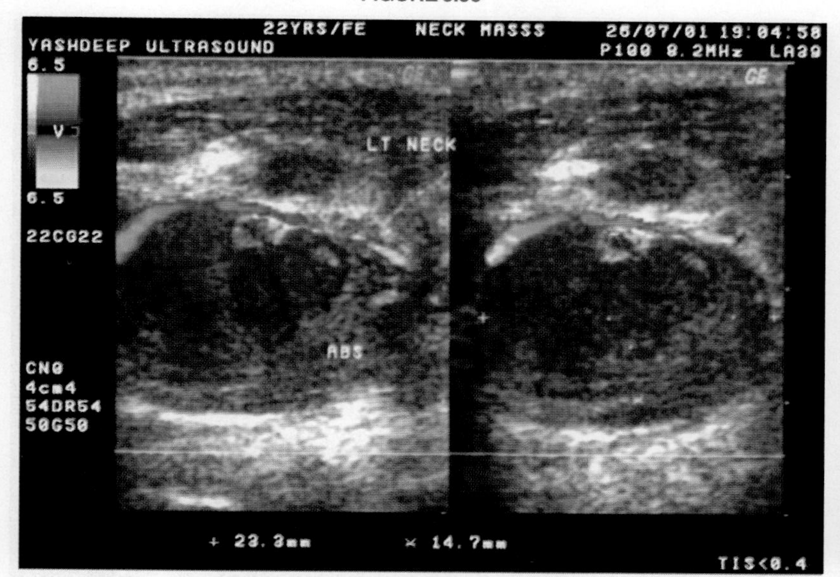

FIGURE 3.51

FIGURES 3.50 and 3.51: Tubercular abscess. HRSG shows well-defined thick wall low-level echo complex masses in Rt cervical area. Central area of necrosis is seen in them on color Doppler imaging poor flow is in the masses. However capsular flow is present.

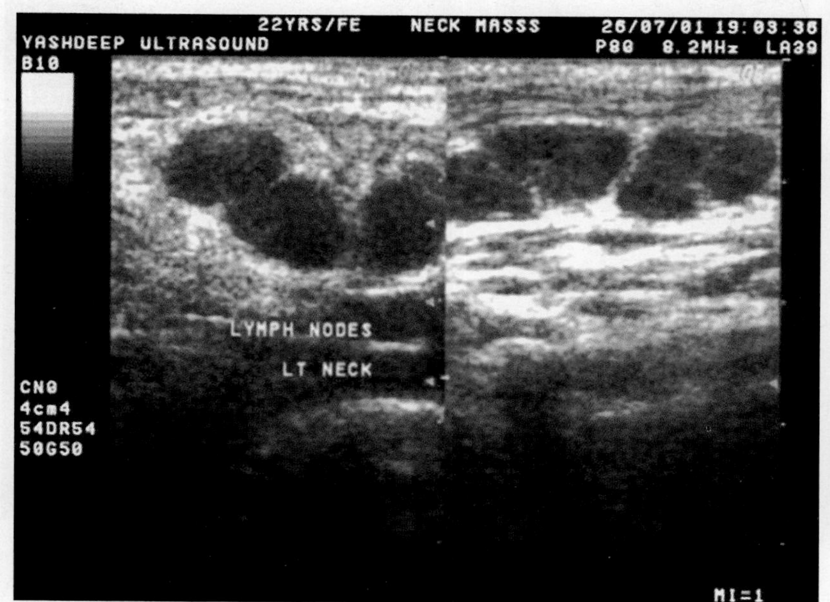

FIGURE 3.52: Multiple enlarged lymph nodes are seen in upper cervical area in the same case of the Figure 3.4, suggestive of tubercular lymphadenitis.

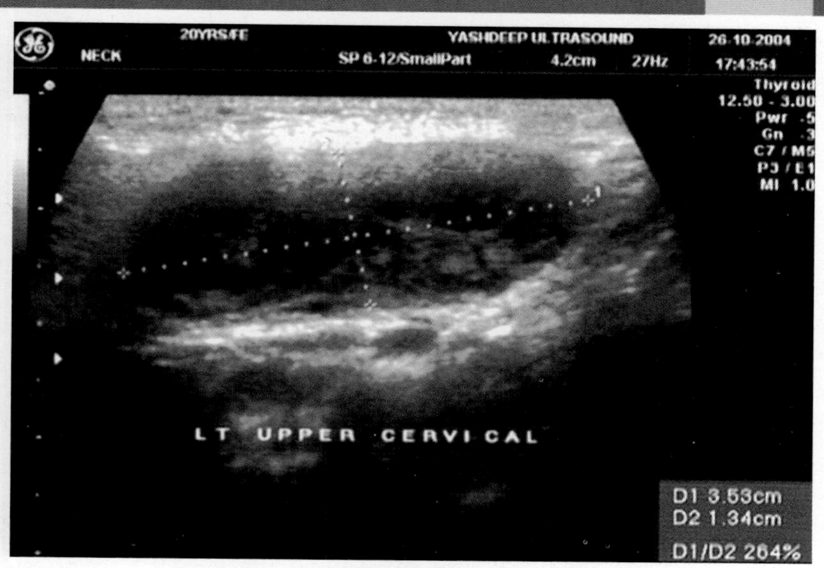

FIGURE 3.53A

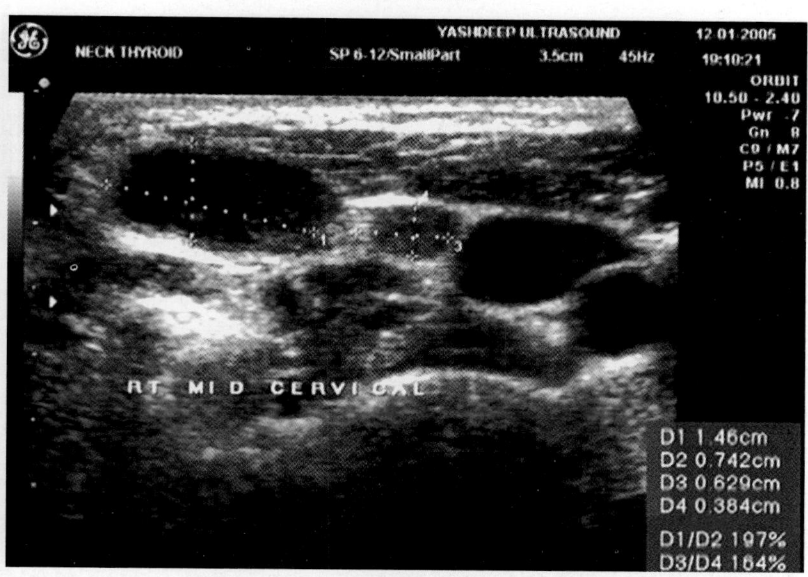

FIGURE 3.53B

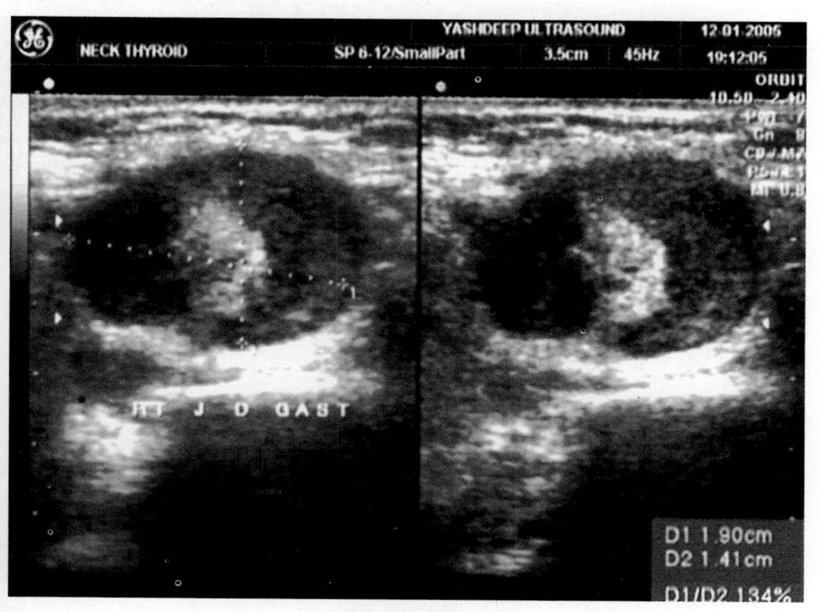

FIGURE 3.54

FIGURES 3.53A and B and 3.54: Tubercular abscess. A young girl presented with multiple nodes on either side of the neck. Big abscess is also seen in the Lt upper cervical region. FNAC confirmed tubercular abscess.

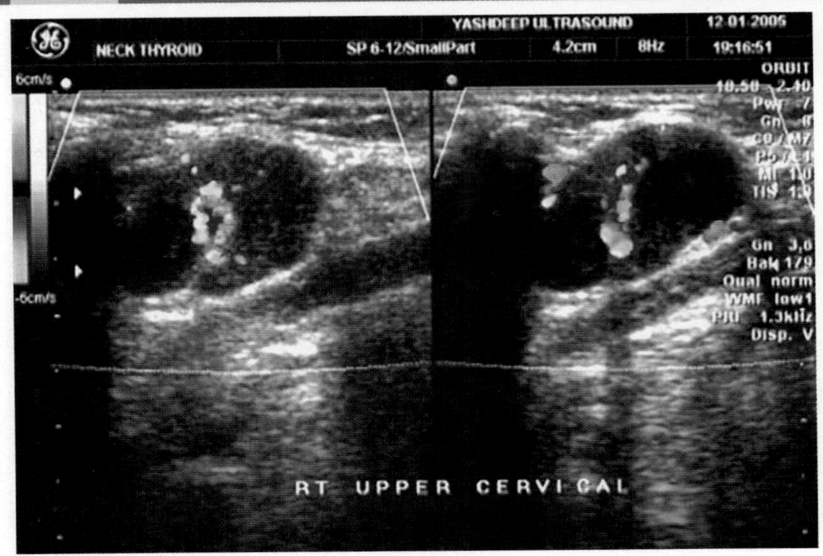

FIGURE 3.55

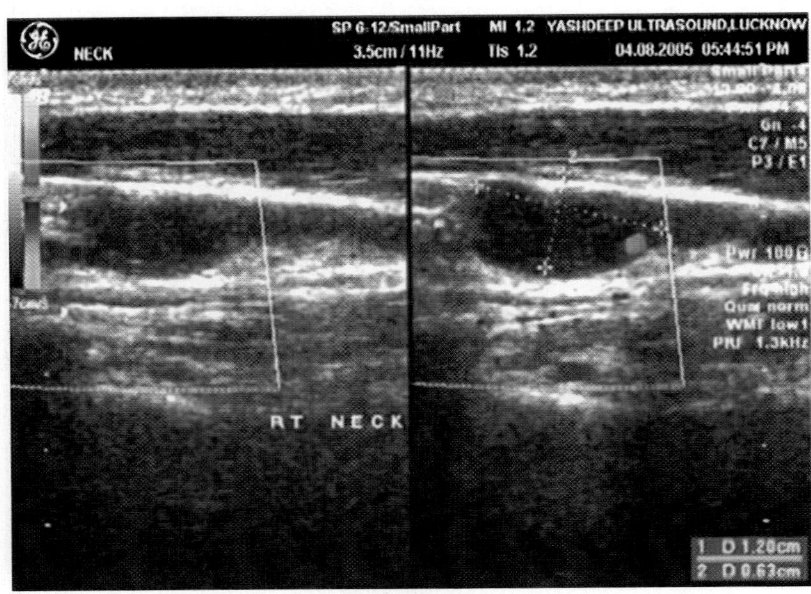

FIGURE 3.56

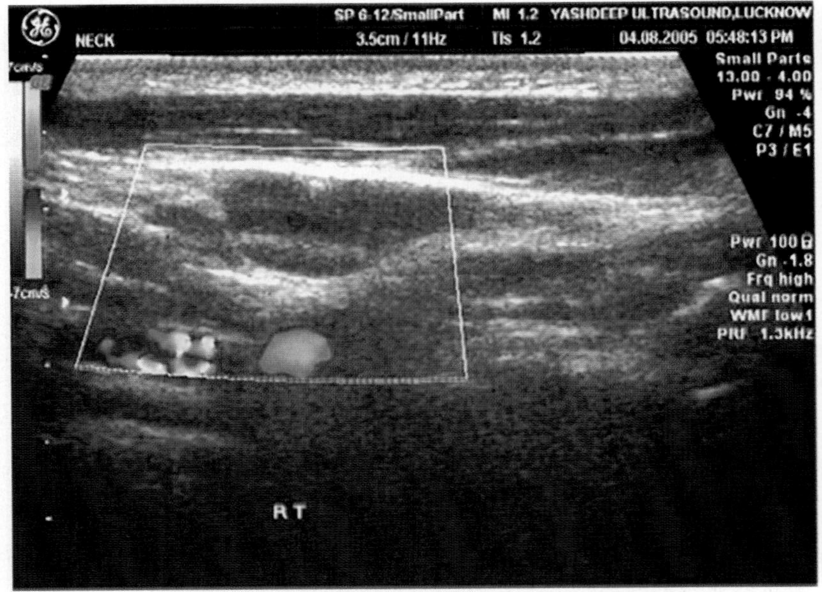

FIGURE 3.57

FIGURES 3.55 to 3.57: The same patient seen after one year of the treatment. HRSG shows disappearance of the tubercular abscess. The nodes are also seen markedly regressed in size.

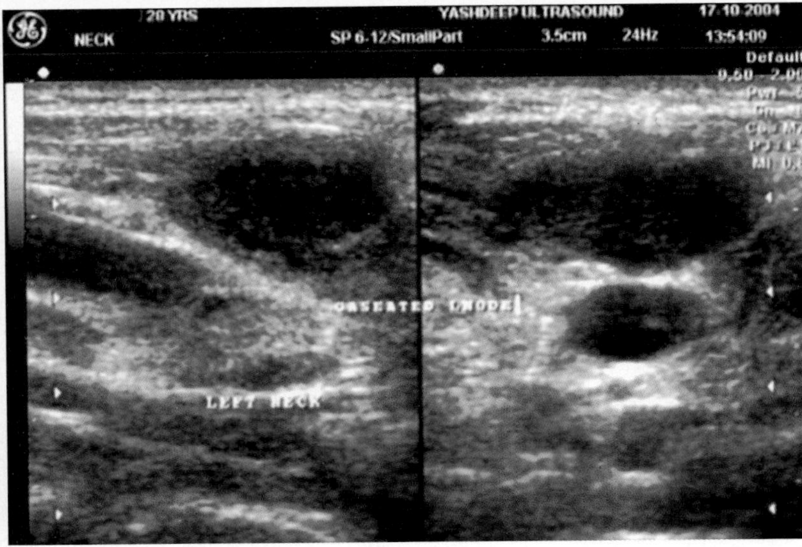

FIGURE 3.58

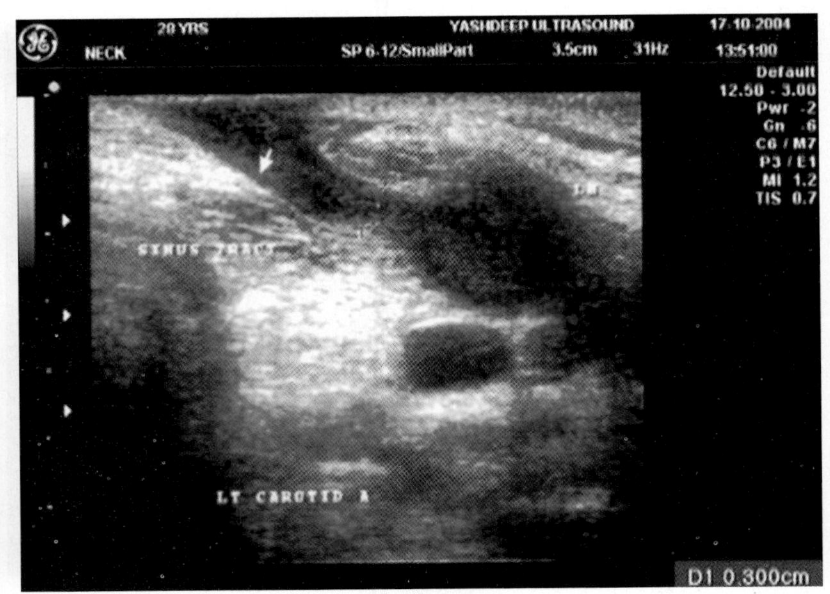

FIGURE 3.59

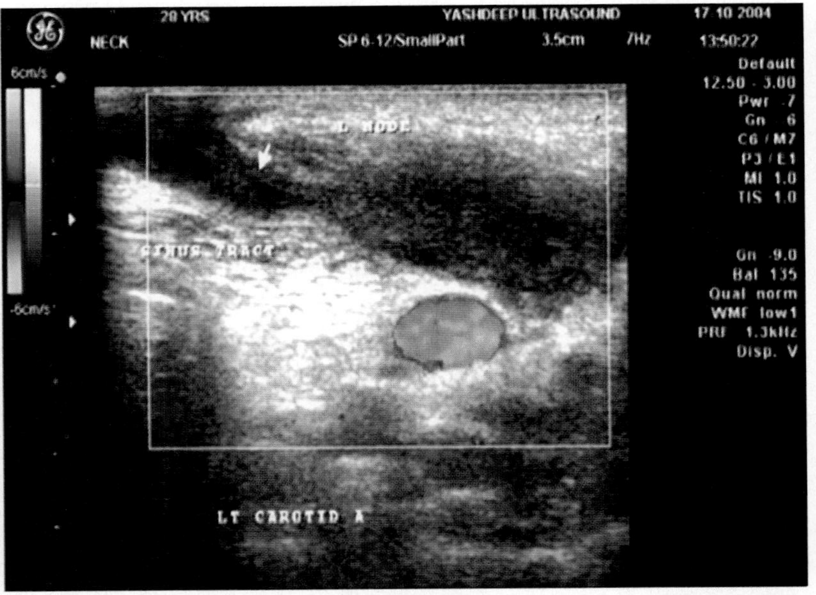

FIGURE 3.60

FIGURES 3.58 to 3.60: Tubercular sinus with caseating nodes in the neck. A young lady presented with discharging sinus in the neck. HRSG shows the sinus tract. Multiple enlarged nodes are also seen in the upper cervical region. Central area of necrosis is seen in them. The sinus tract is seen communicating with the caseating nodes.

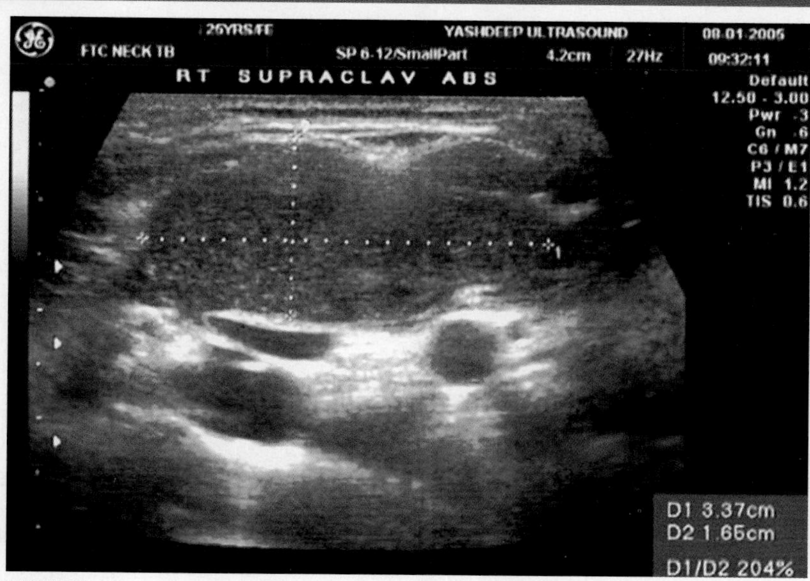

FIGURE 3.61: Big tubercular abscess in the supraclavicular region. A young lady presented with soft tissue non-tender mass in the Rt supraclavicular region. HRSG shows big low-level echo complex mass with homogeneous echo pattern suggestive of tubercular abscess.

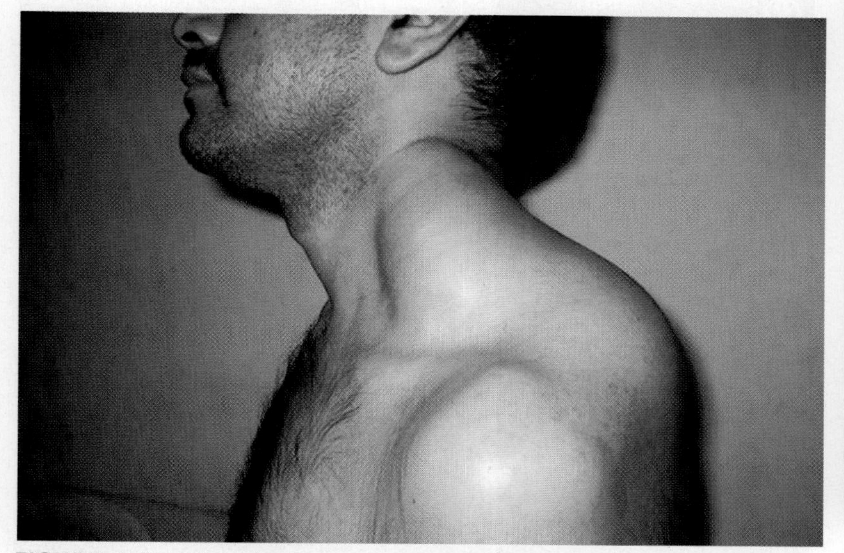

FIGURE 3.62: Big tubercular abscess in the shoulder muscles and suprascapular region. A young man presented with soft tissue swelling over the Lt shoulder and suprascapular region. Bulging nontender mass is also seen in the posterior triangle of the neck.

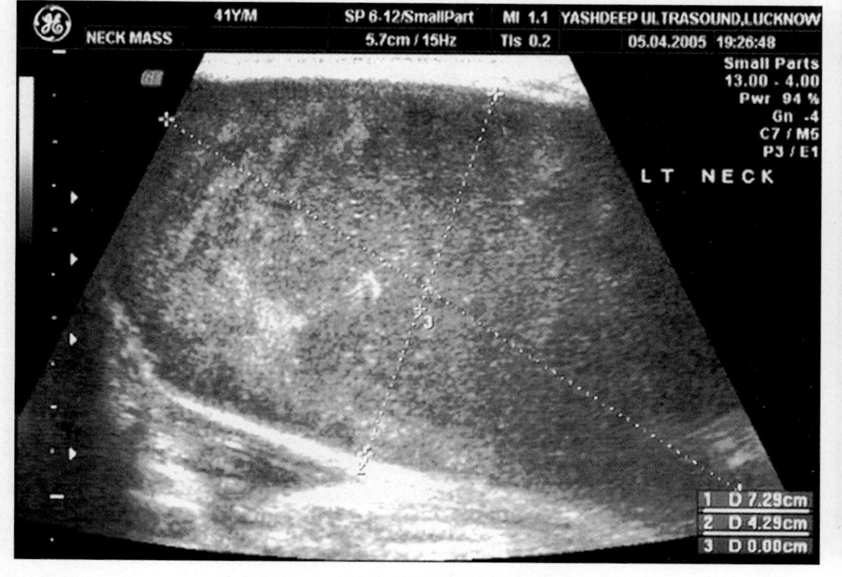

FIGURE 3.63

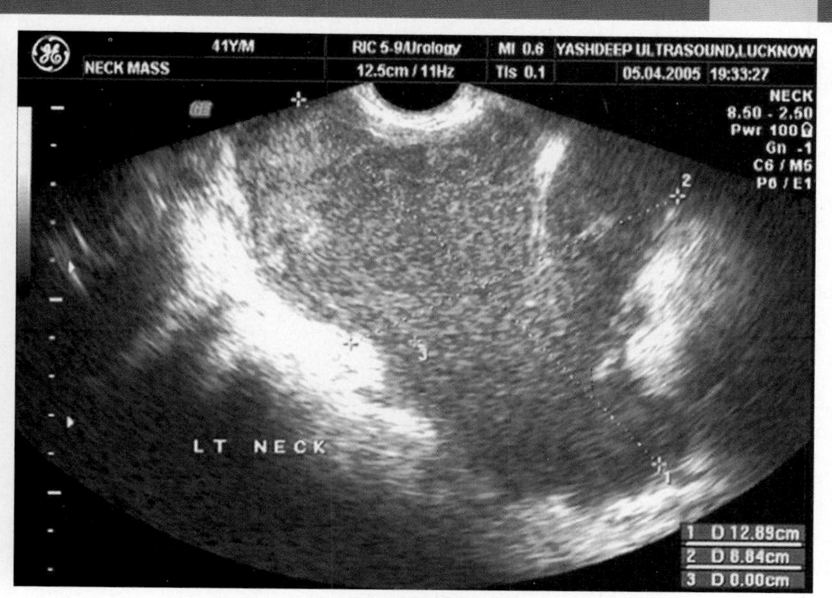

FIGURE 3.64

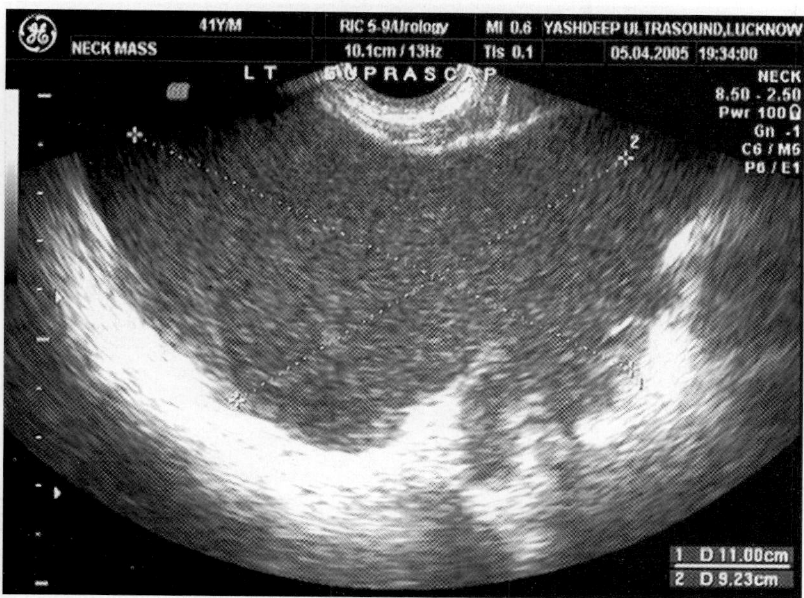

FIGURE 3.65

FIGURES 3.63 to 3.65: HRSG shows big low-level echo collection in the shoulder muscles posterior triangle of the neck. It is also seen going down in the Lt subscapular, suprascapular and in the region of axilla. USG guided aspiration confirmed tubercular collection.

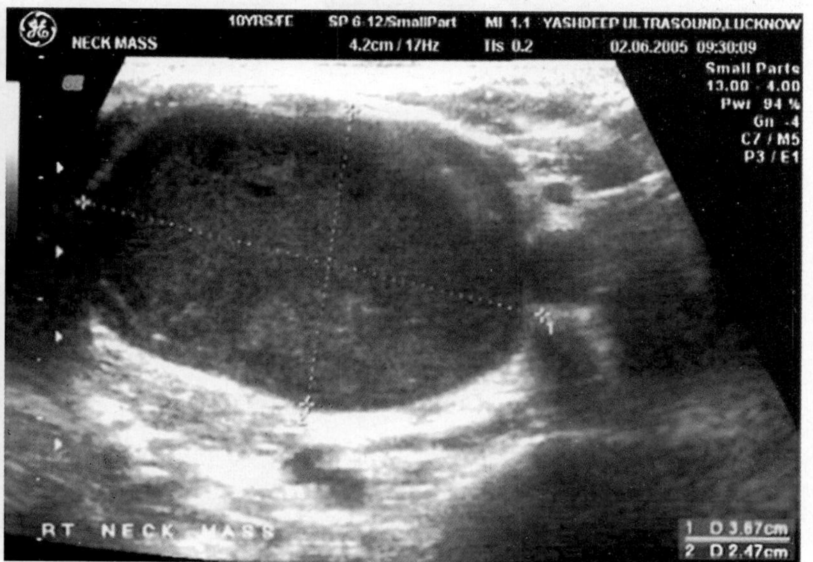

FIGURE 3.66

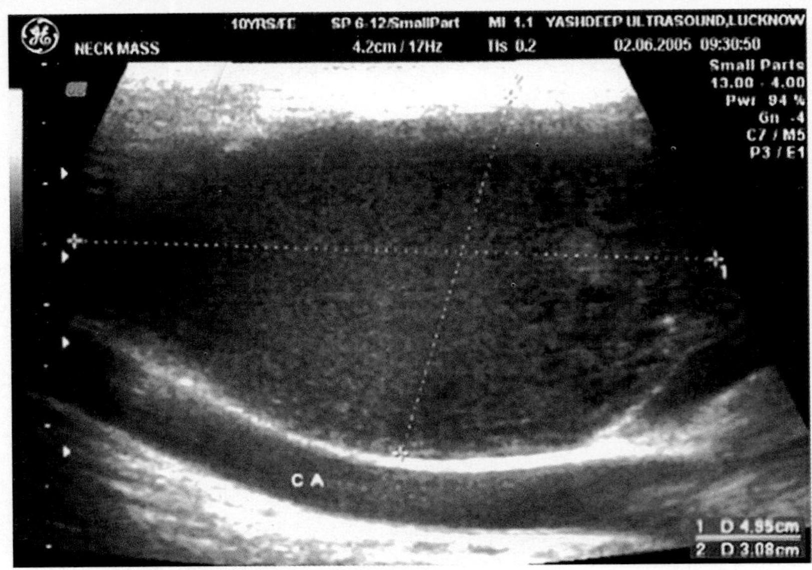

FIGURE 3.67

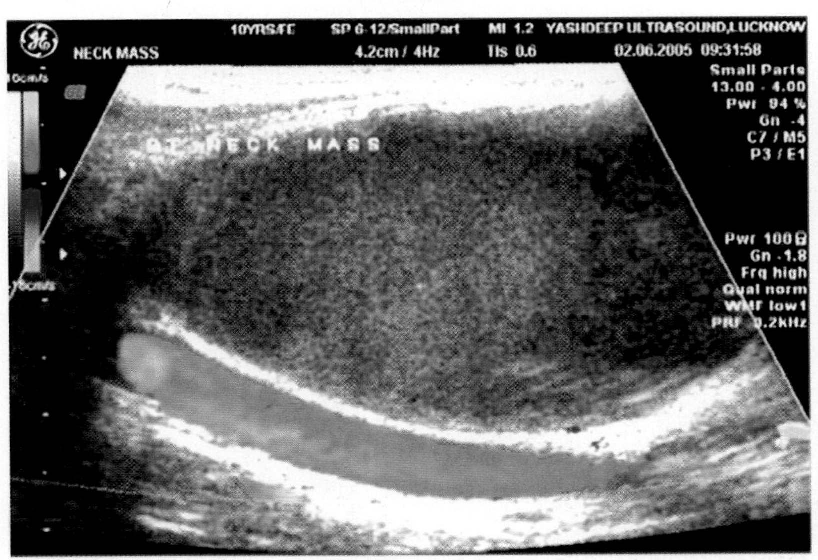

FIGURE 3.68

FIGURES 3.66 to 3.68: Supraclavicular tubercular abscess. A young girl presented with a big thick echo complex mass in the Rt supraclavicular region. Homogeneous echoes are seen in it.

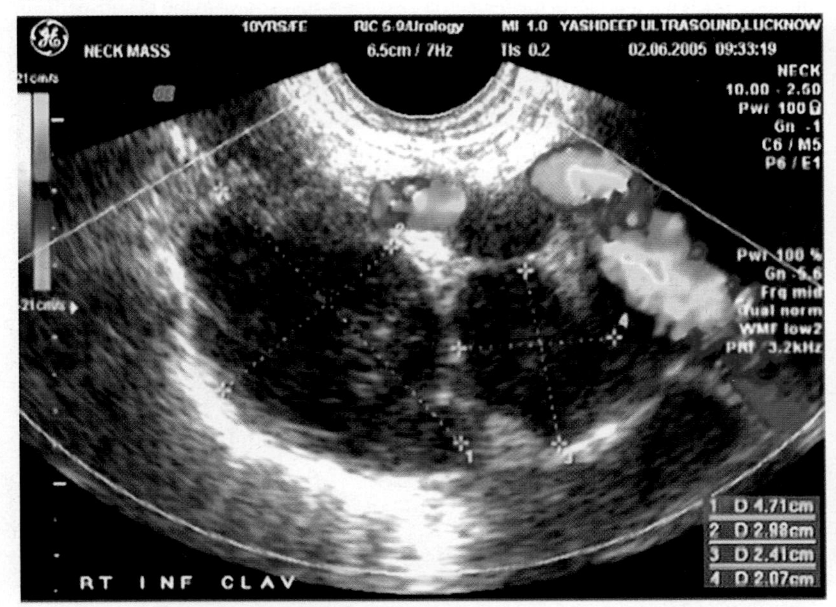

FIGURE 3.69: Multiple enlarged nodes are seen in the neck matting of the nodes is also seen suggestive of tubercular adenitis.

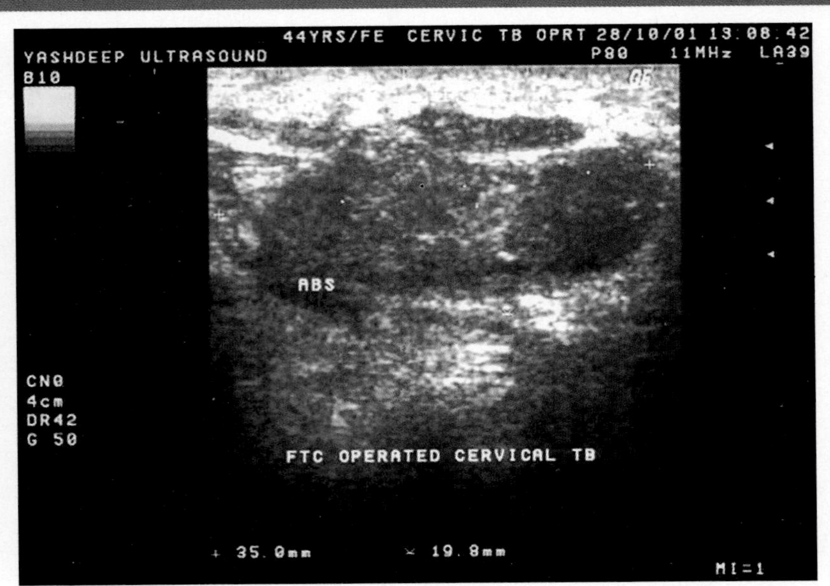

FIGURE 3.70: FTC tubercular abscess in the neck. An operated patient of cervical paravertebral tubercular abscess presented with a mass lesion in Rt upper cervical and mid cervical area. HRSG shows low-level echo complex collection, suggestive of tubercular abscess in the upper and mid cervical area.

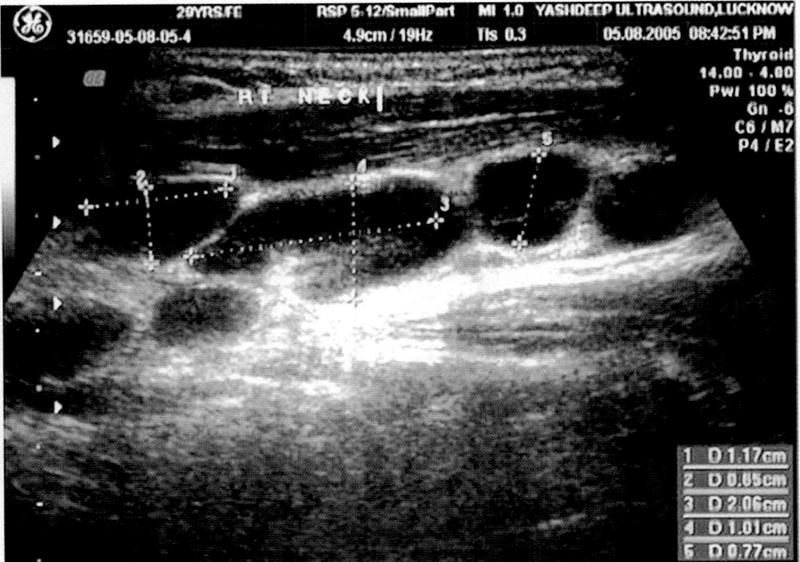

FIGURE 3.71

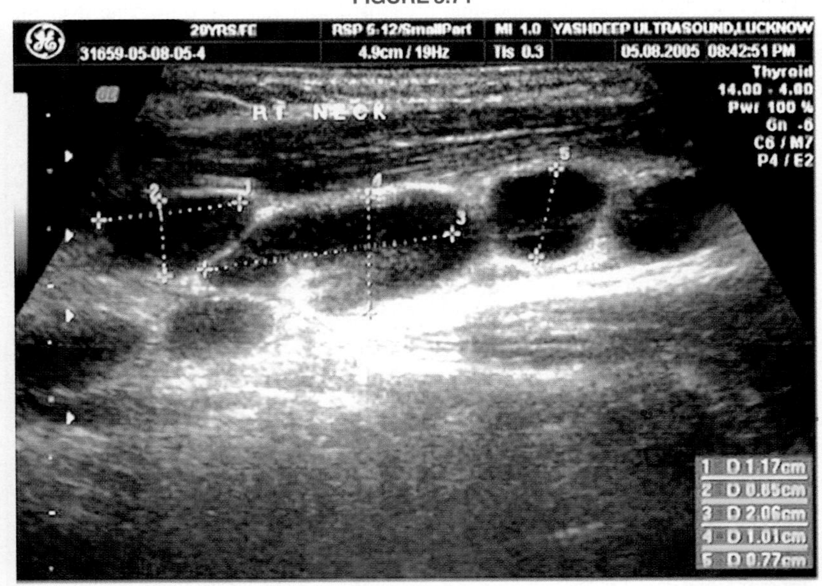

FIGURE 3.72

FIGURES 3.71 and 3.72: Tubercular lymphadenitis. A patient presented with multiple enlarged nodes in the neck. Nodes are seen in the chain. FNAC confirmed tubercular lymphadenitis.

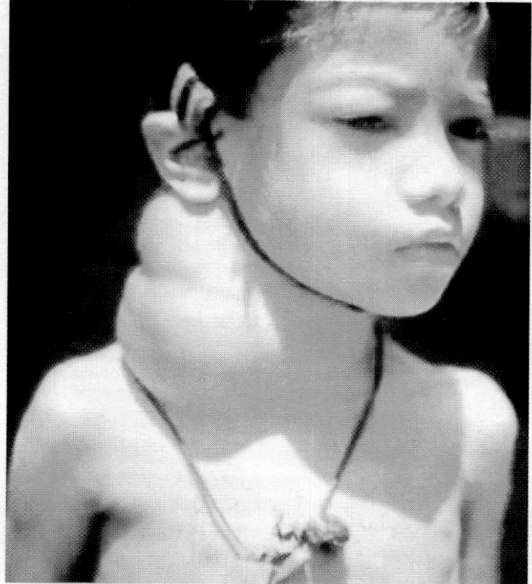

FIGURE 3.73: A young boy presented with multinodular soft tissue mass involving whole Rt side of the neck with wavy outline.

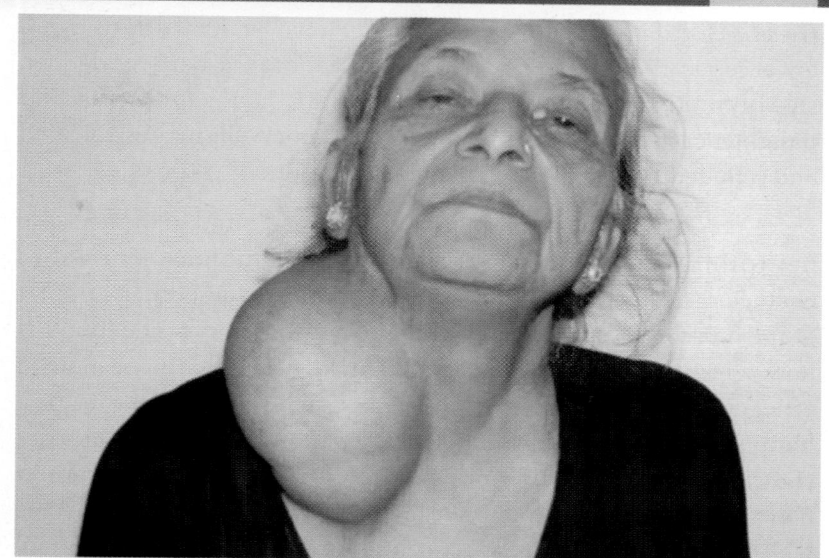

FIGURE 3.75: An old lady presented with big soft tissue mass in the Rt side of the neck. The mass is coming out from anterior triangle of the neck.

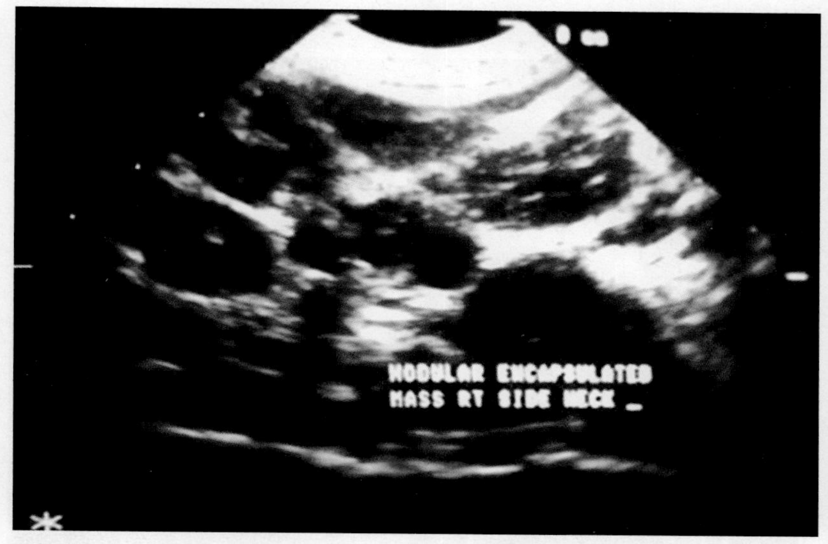

FIGURE 3.74: HRSG of the same patient shows multiple hypoechoic nodular masses involving Rt side of the neck. They are smooth in outline. No necrosis is seen. Biopsy shows non-specific lymphadenitis.

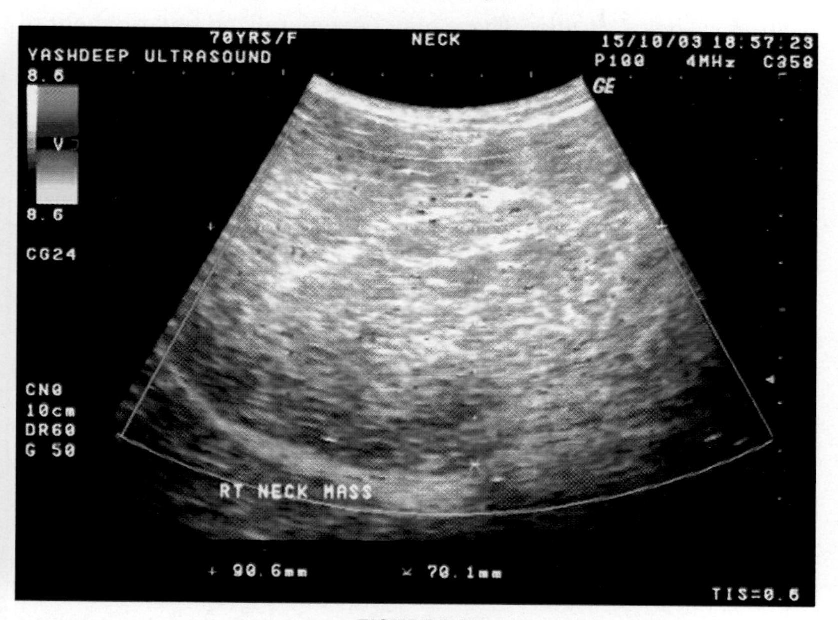

FIGURE 3.76

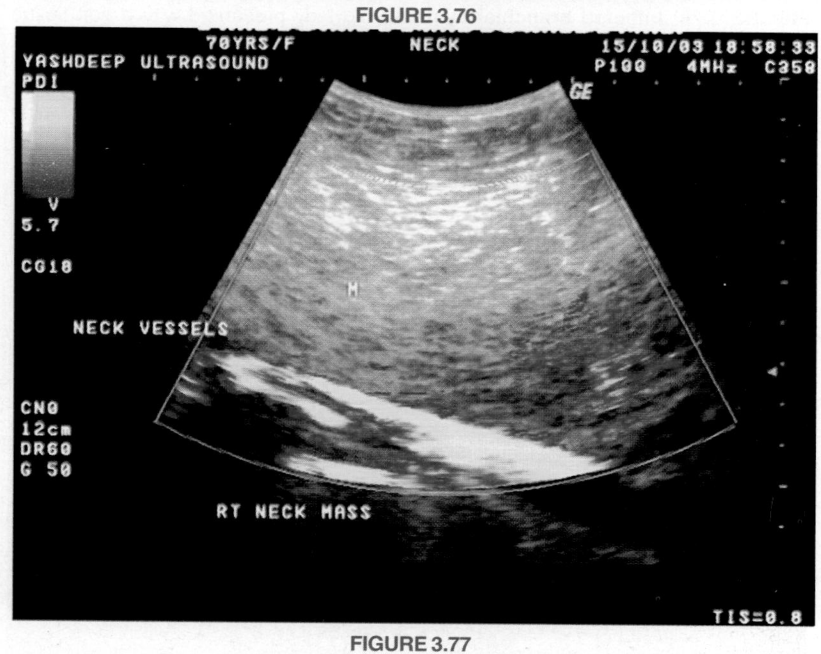

FIGURE 3.77

FIGURES 3.76 and 3.77: HRSG shows a big homogeneous soft tissue mass in the neck. The mass shows homogeneous echoes. No calcification is seen in the mass. No cystic degeneration is seen. The mass is not seen pressing over the neck vessels.

swelling with medium to hard consistency. They are usually extension of paravertebral abscess in the neck and bulge in the posterior triangle or paravertebral space. Heterogeneous low amplitude echo complex mass is seen on HRSG with thick walls. Echogenic calcified foci may be seen in the abscess. However, no perifocal edema is present. Extension of abscess to paravertebral space is well evaluated on CT examination, and exact size of the abscess cavity can be assessed. HRSG is a good tool to evaluate and assess the treatment response of abscess. Associated reactive lymph node enlargement can also be seen on HRSG.

Neck Masses—Benign

Benign Tumors

The most common benign tumors found in neck are hemangiomas, fibromas, lipomas and tumors of nerves as neurofibromas. Sublingual dermoids are also seen as a congenital benign mass. The lymph node enlargement and lymph node masses are commonly found in the neck as primary lymph node disease or metastatic lymph node disease.

Hydatid cyst: It is uncommon in the neck. However, it may be found in lower cervical and superior mediastinum. It appears to be a well defined sharply marginated cystic mass on HRSG. Echogenic septa and daughter cysts are also seen on ultra sonogram. Walls may be calcified and echogenic. On real time imaging, water lily sign is positive.

Nerve tumors: The most common nerve tumors found in the neck are neurofibroma, ganglioneuroma and neurinomas. These are characteristically found along the nerve route, and associated atrophy of the muscle is found supplied by the involved nerve. The tumor is homogeneous in texture and well defined. They are hypoechoic in echogenicity. The margins are echogenic. Usually hyperplastic reactive lymph nodes are seen associated with the tumor. The tumor does not show any flow or little flow on color flow imaging. However, carotid body tumors chemodectomas are highly vascular. They are echogenic well defined tumors, which are seen at the carotid artery bifurcation. The tumor may be seen adherent with the carotid sheath.

Nonspecific lymphadenitis: There may be generalized enlargement of lymph nodes presenting as a mass in the neck. Ultrasonogram shows

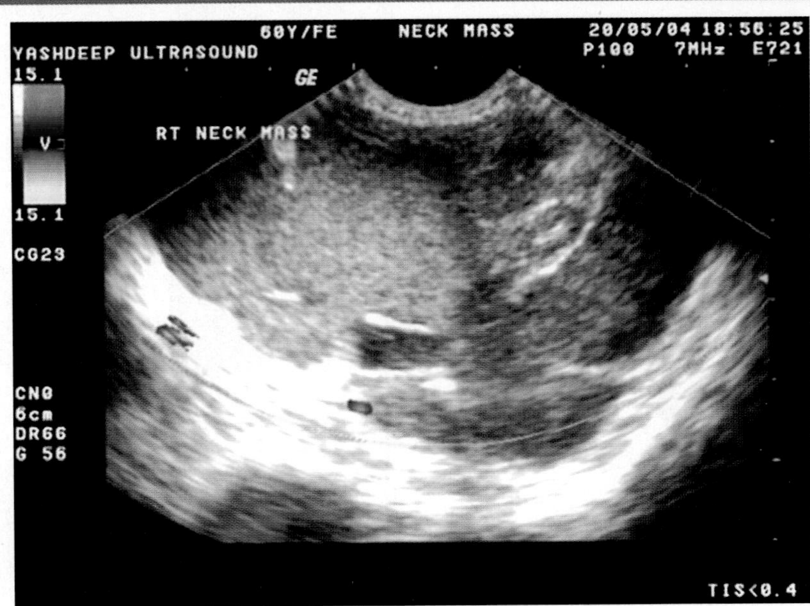

FIGURE 3.80

FIGURES 3.79 and 3.80: HRSG shows a thick echo mass. No calcification is seen in it. No cystic degeneration is seen. Few echogenic septa are seen in the mass.

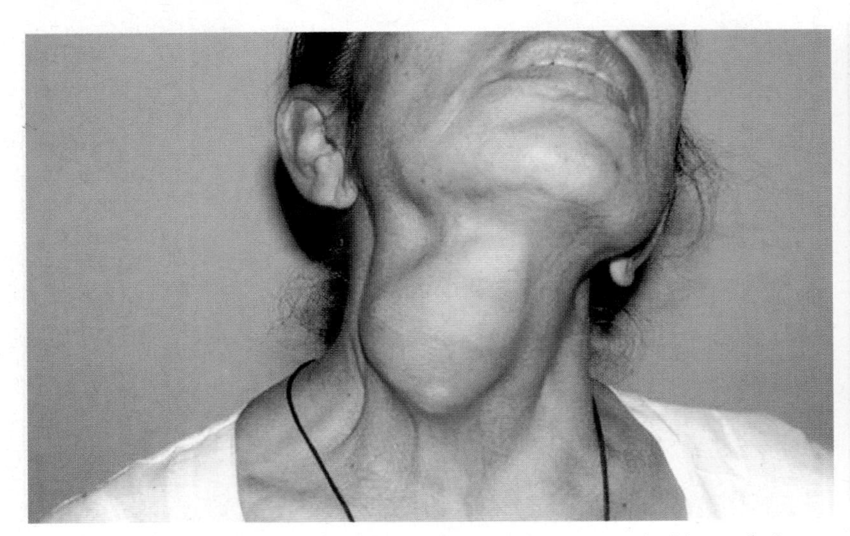

FIGURE 3.78: Infected branchial cyst. An old lady presented with a soft tissue mass in the neck. The mass is seen in the anterior triangle.

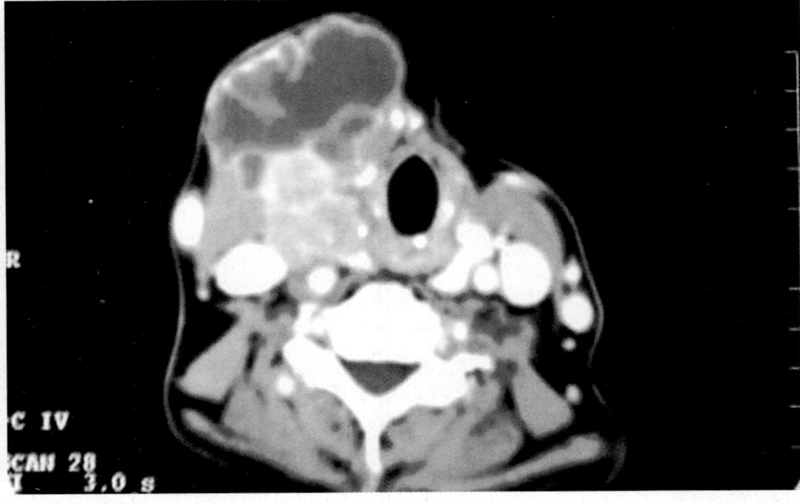

FIGURE 3.81: CT of the same patient shows detail of the multiloculated cystic mass. FNAC confirmed branchial cyst.

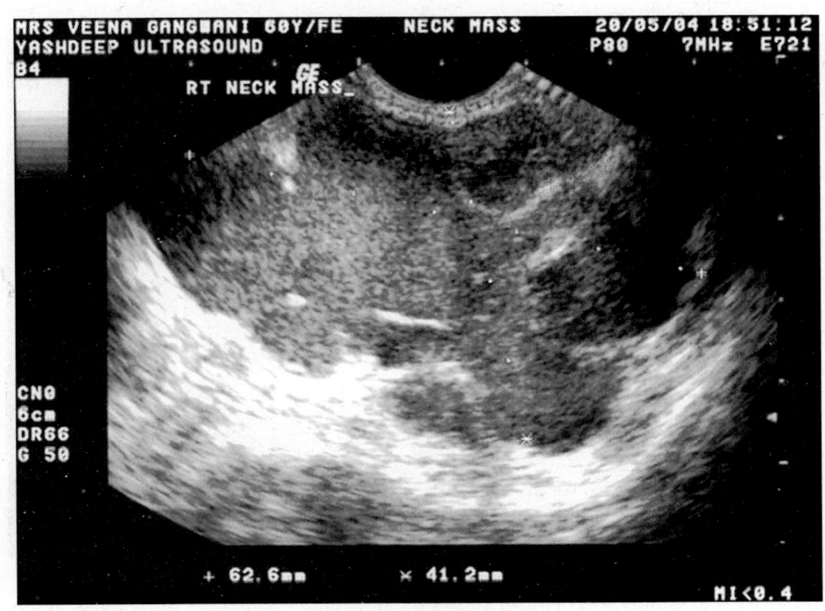

FIGURE 3.79

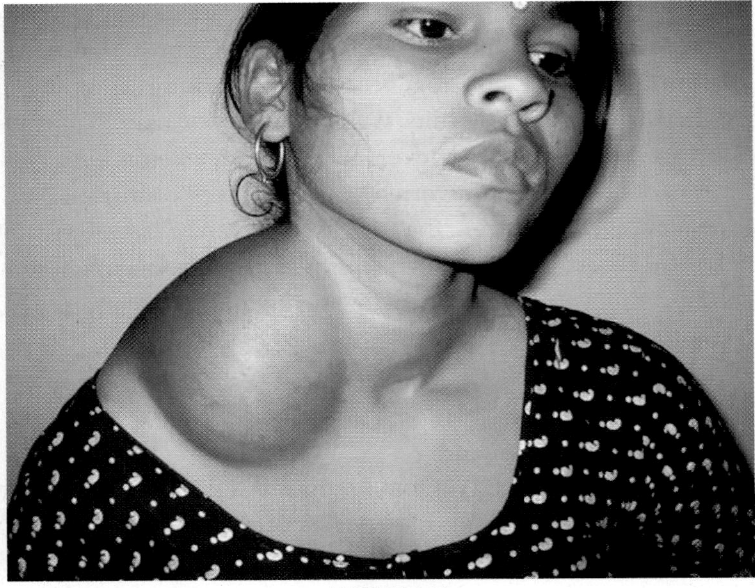

FIGURE 3.82

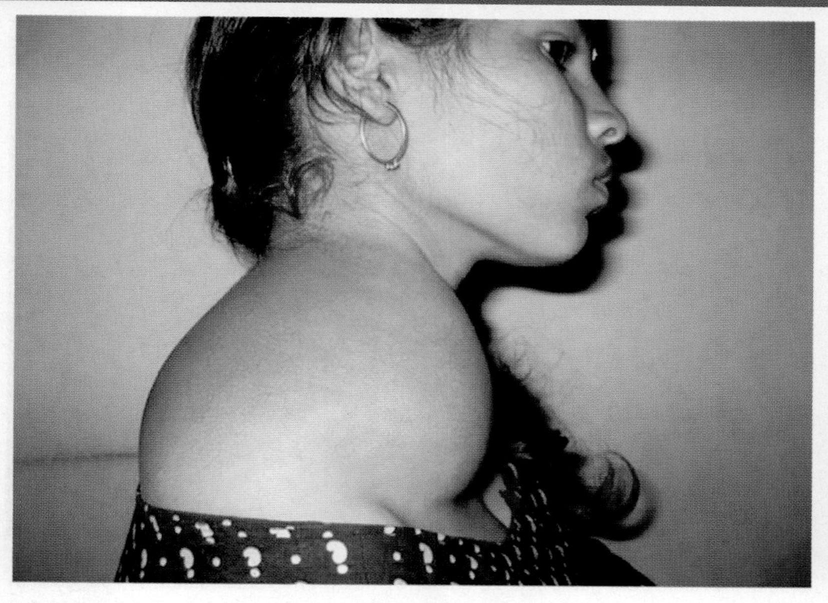

FIGURE 3.83

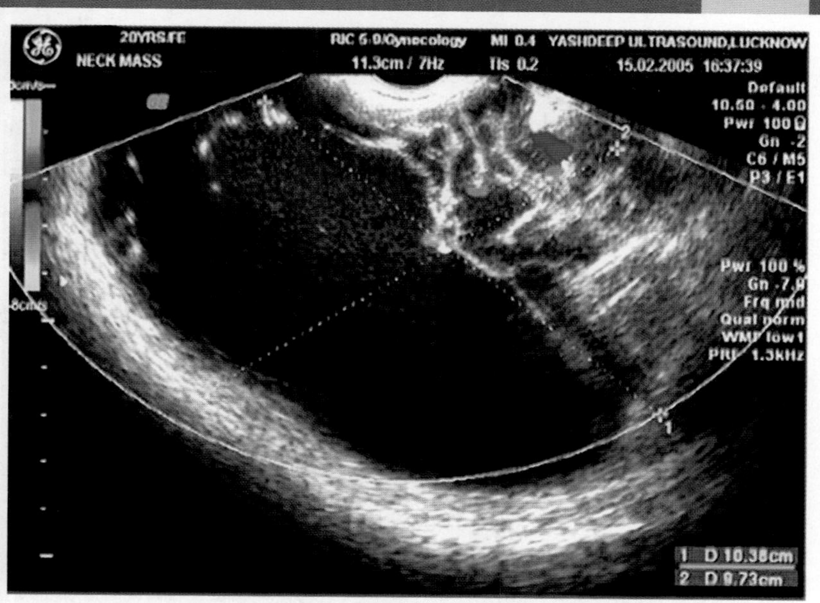

FIGURE 3.86

FIGURES 3.82 and 3.83: Cystic hygroma. A young lady presented with soft tissue cystic mass in the side of the neck. The mass is seen coming out anteriorly.

FIGURES 3.84 to 3.86: HRSG shows a big multiloculated cystic mass in the neck. Thick echogenic septa are seen in it. On color flow imaging few vessels are seen in the septa.

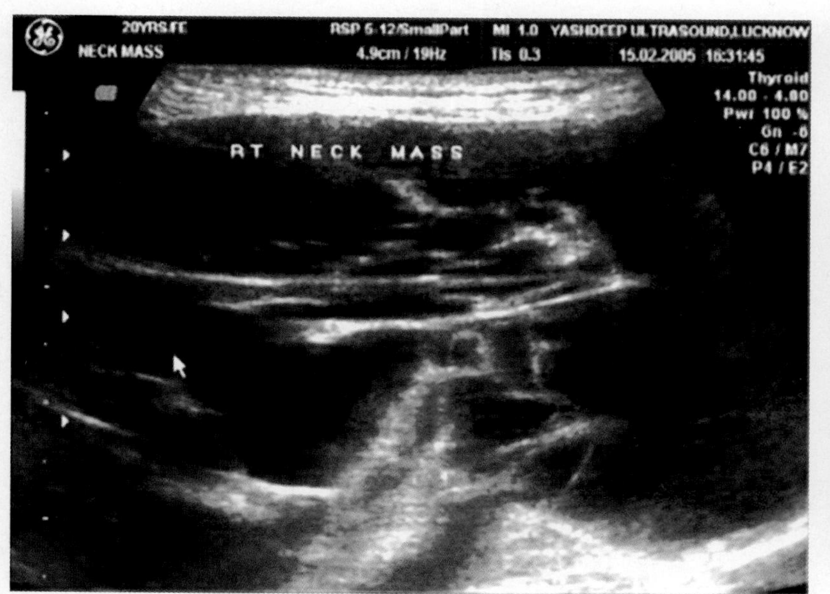

FIGURE 3.84

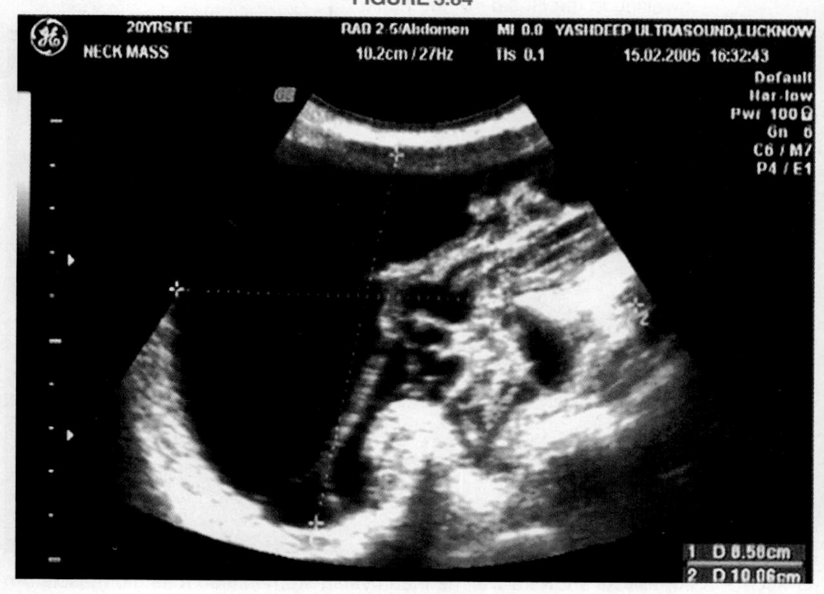

FIGURE 3.85

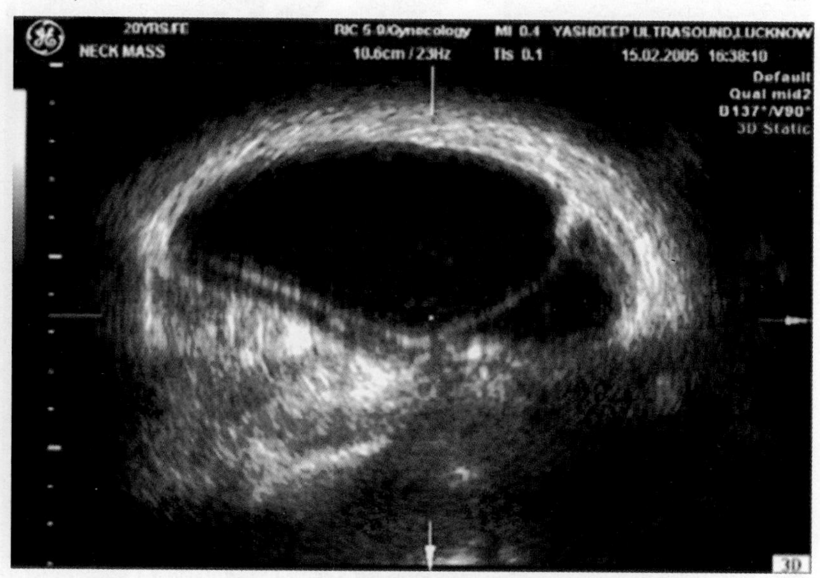

FIGURE 3.87

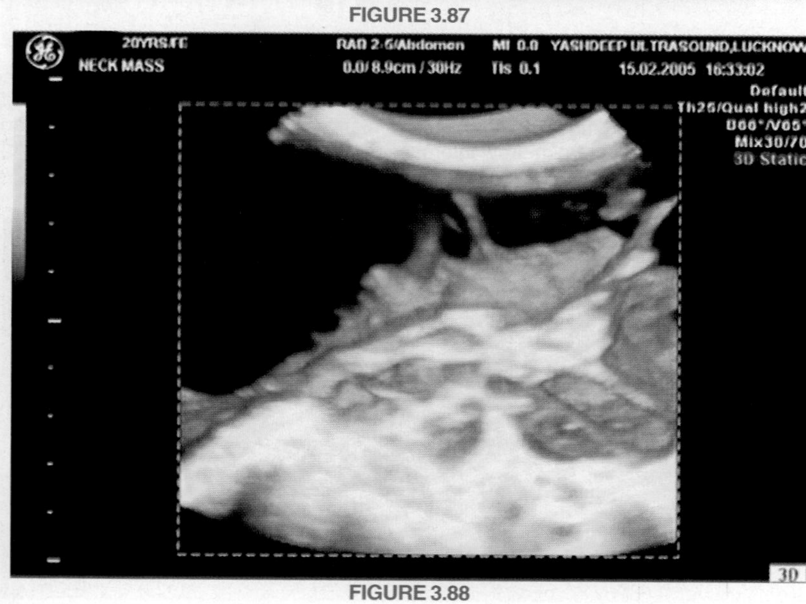

FIGURE 3.88

FIGURES 3.87 and 3.88: 3D of the same patient mass shows better details of the cyst. The septa stand out clearly. Biopsy confirmed cystic hygroma.

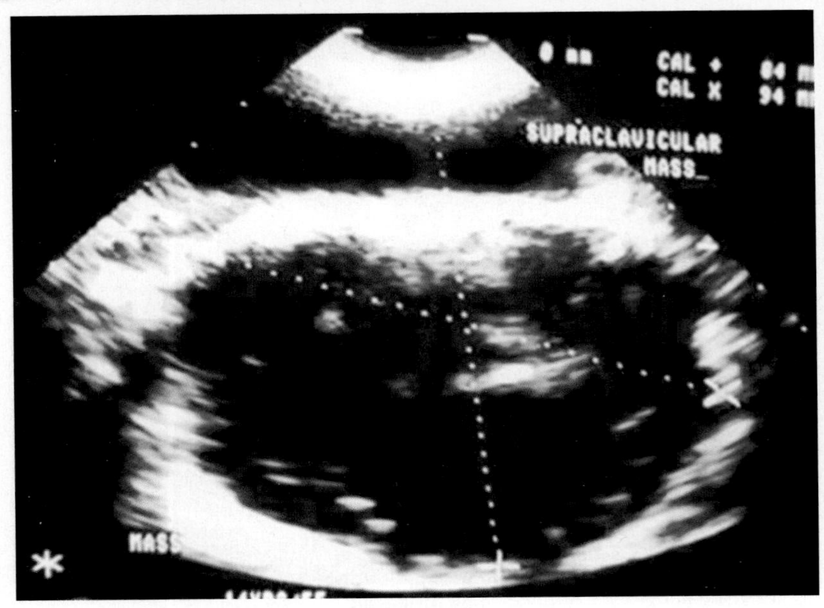

FIGURE 3.89

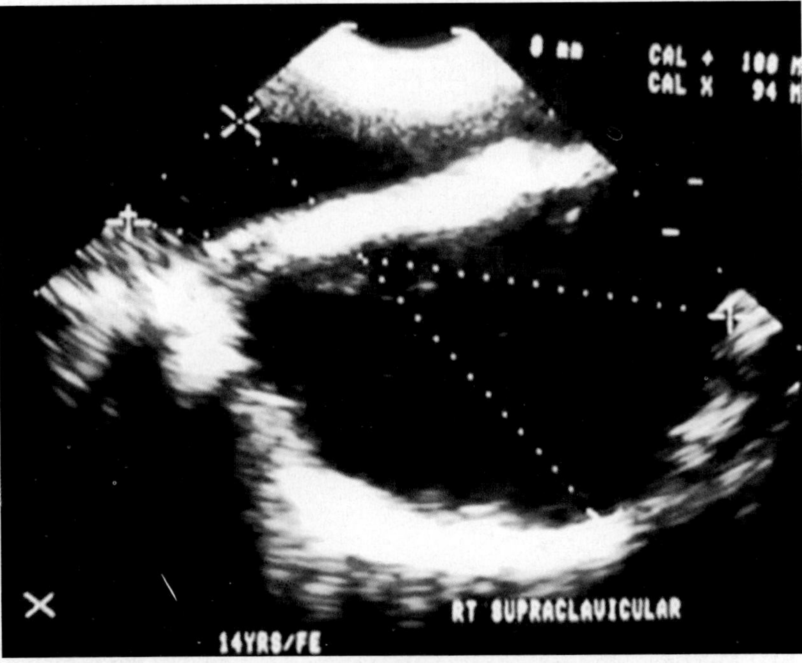

FIGURE 3.90

FIGURES 3.89 and 3.90: Hydatid cyst in the neck. HRSG shows a thick walled multiloculated cystic mass in Rt supraclavicular area. The septa are thickened and calcification is seen in them. Few internal echoes are seen in the cyst suggestive of hydatid cyst.

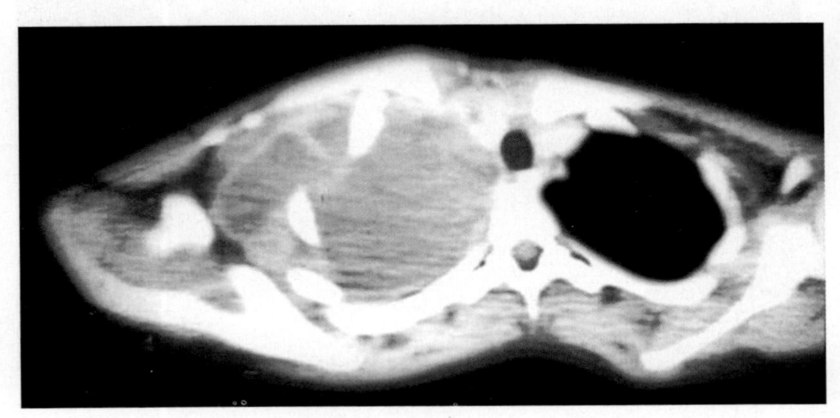

FIGURE 3.91: CT scan of the same patient shows hypodense mass with homogeneous echoes. Biopsy confirmed hydatid cyst.

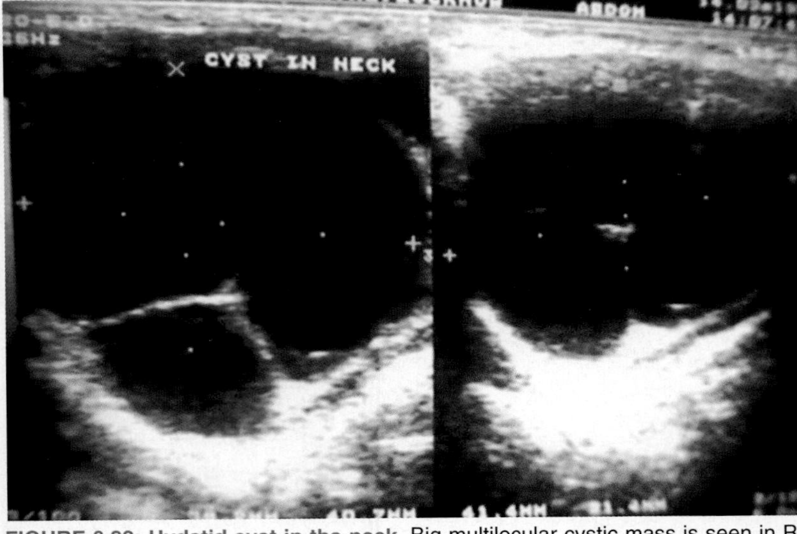

FIGURE 3.92: Hydatid cyst in the neck. Big multilocular cystic mass is seen in Rt side of the neck. Echogenic septa are seen in the cyst with calcification. No suggestion of any debris is seen in the cyst. The septa shows movements. Sonographic features are suggestive of hydatid cyst.

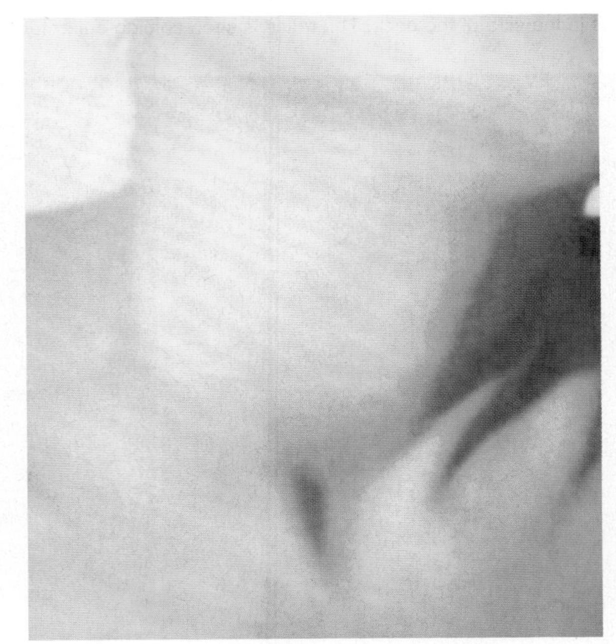

FIGURE 3.93: A young lady presented with a soft tissue mass swelling in the Rt side of the neck.

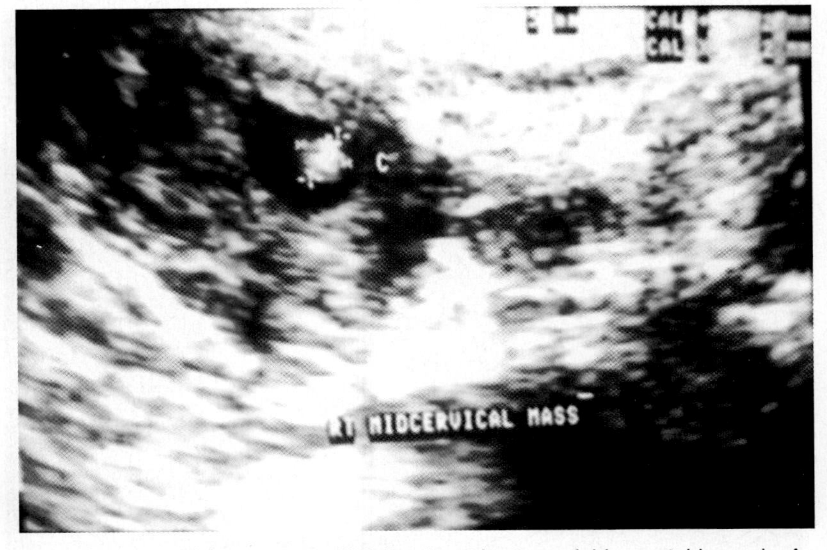

FIGURE 3.94: HRSG shows a well-defined cyst in sternocleido mastoid muscle. An echogenic nidus is seen fixed with the inner wall of the cyst suggestive of cysticercus cyst. Typical USG features on HRSG. The echogenic nidus is the scolex of the cyst.

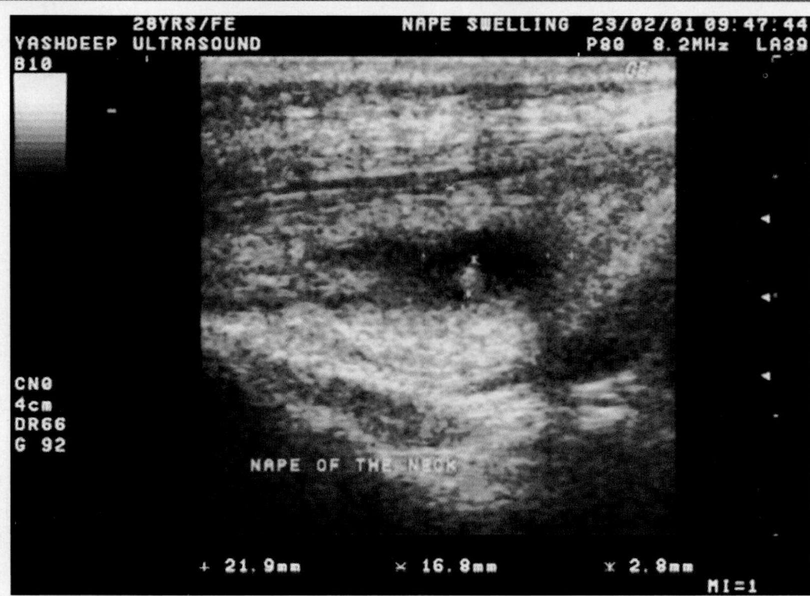

FIGURE 3.95: **Cysticercus cyst in the occiput.** HRSG shows a well-defined cyst in the muscle of the occiput. An echogenic nidus is seen fixed with the inner wall of the cyst, typical feature of the cysticercus cyst.

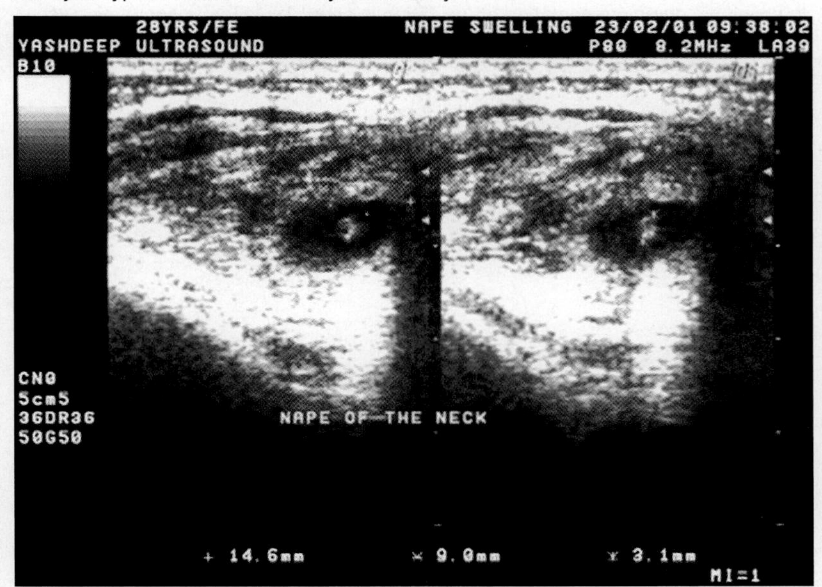

FIGURE 3.96: **Cysticercus cyst.** HRSG shows a well-defined cyst at the nape of the neck. An echogenic nidus is seen in the cyst.

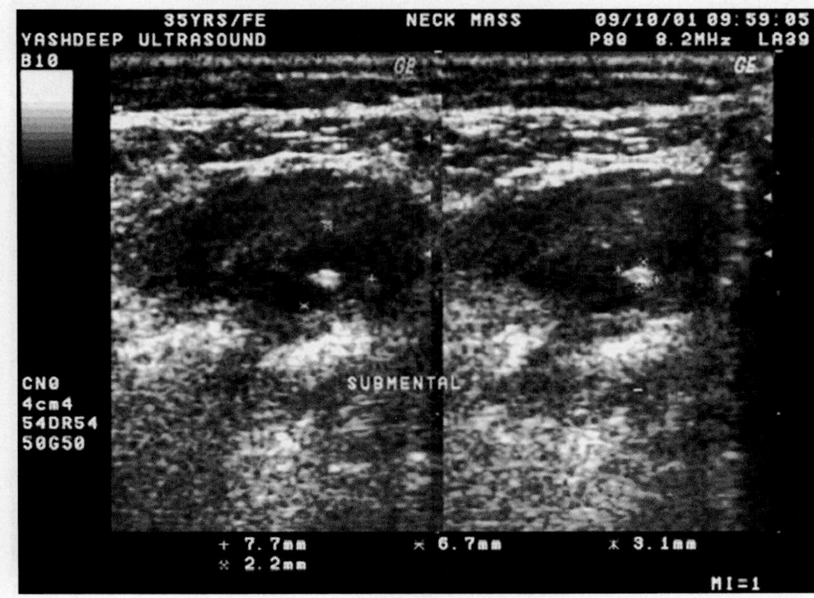

FIGURE 3.97

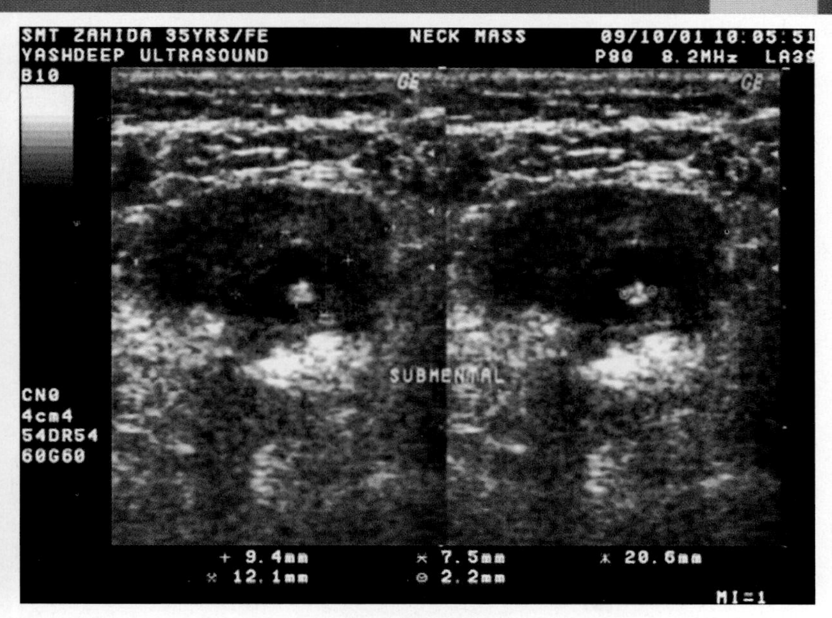

FIGURE 3.98

FIGURES 3.97 and 3.98: **Submental cysticercosis.** HRSG shows a well-defined thick wall cyst in the submental area. An echogenic nidus is seen fixed with the inner wall of the cyst, suggestive of scolex typical USG features of cysticercosis.

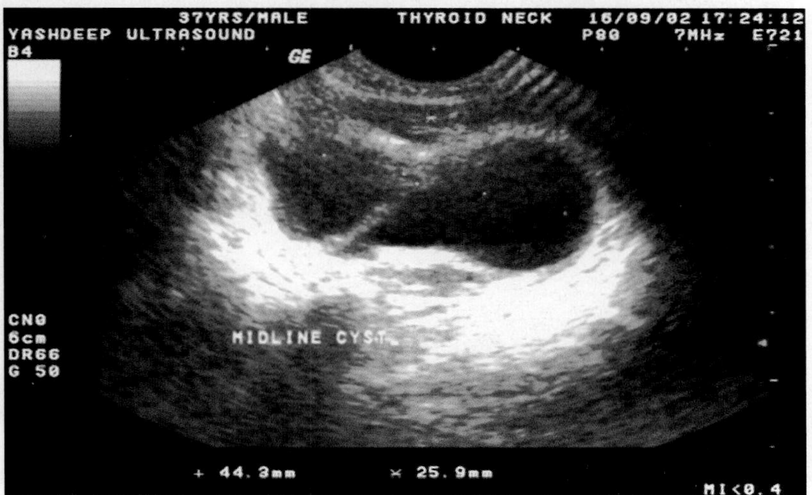

FIGURE 3.99

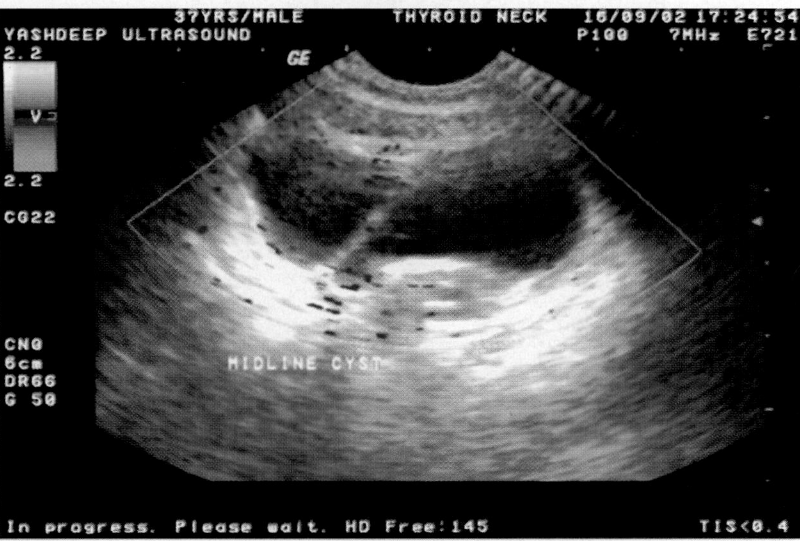

FIGURE 3.100

FIGURES 3.99 and 3.100: **Ranula of sublingual gland.** A patient presented with cystic swelling of the floor of the mouth. HRSG shows well-defined cystic mass in the floor of the mouth, which shows thick echogenic septum. Echogenic debris is also seen in it. But no soft tissue mass is seen. On color flow imaging low flow seen in the wall of the cyst. Biopsy of the cyst confirmed sublingual gland cyst suggestive of ranula of submandibular gland.

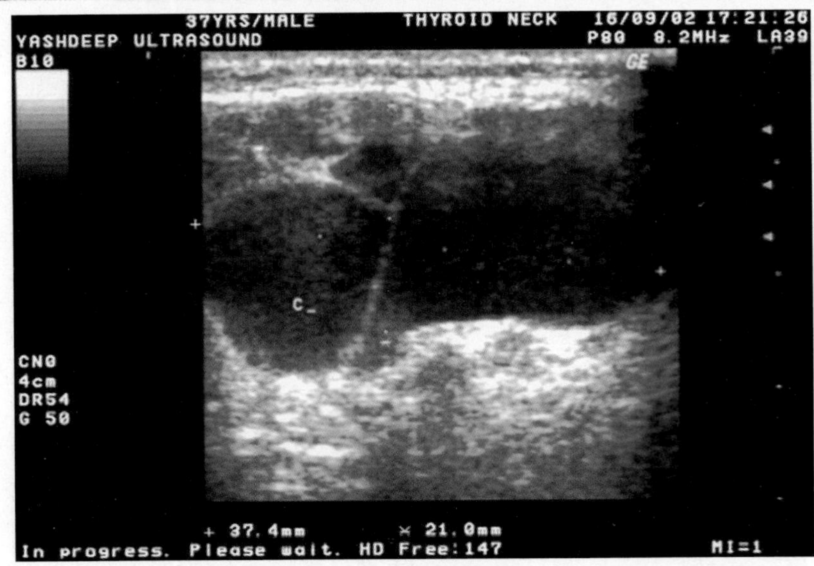

FIGURE 3.102

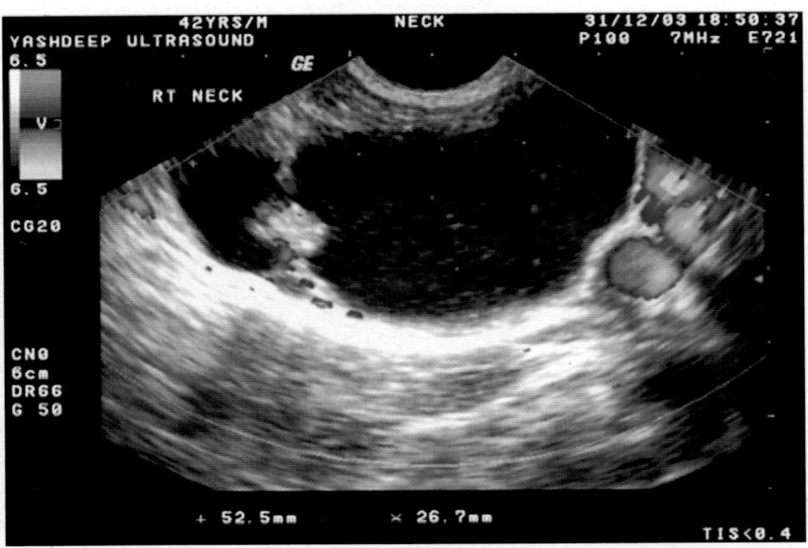

FIGURE 3.103

FIGURES 3.102 and 3.103: Lymphatic cyst in the neck. A patient presented with a big cystic mass in the neck. HRSG shows big multiloculated cystic mass. Thick echogenic septum is seen in it. FNAC confirmed lymphatic cyst.

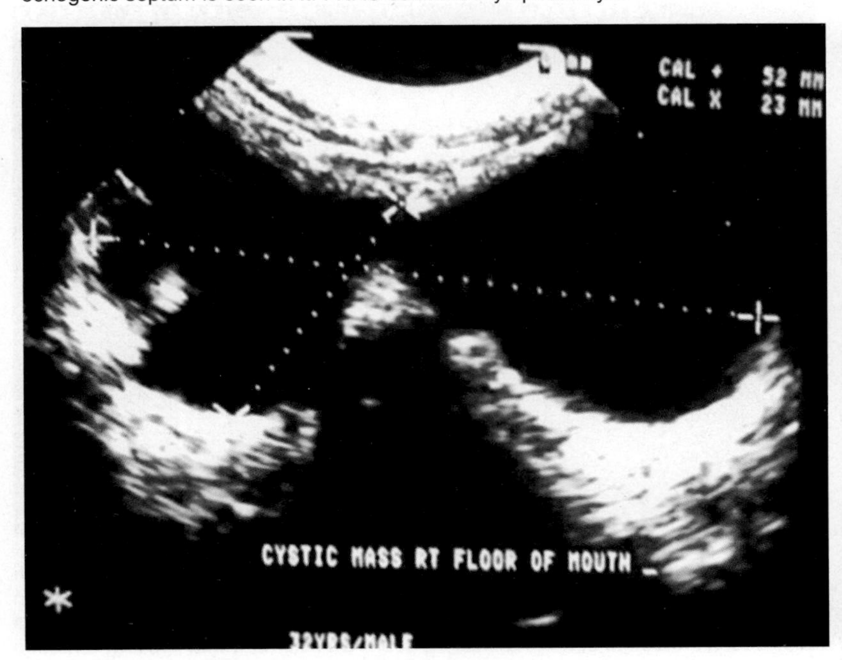

FIGURE 3.104

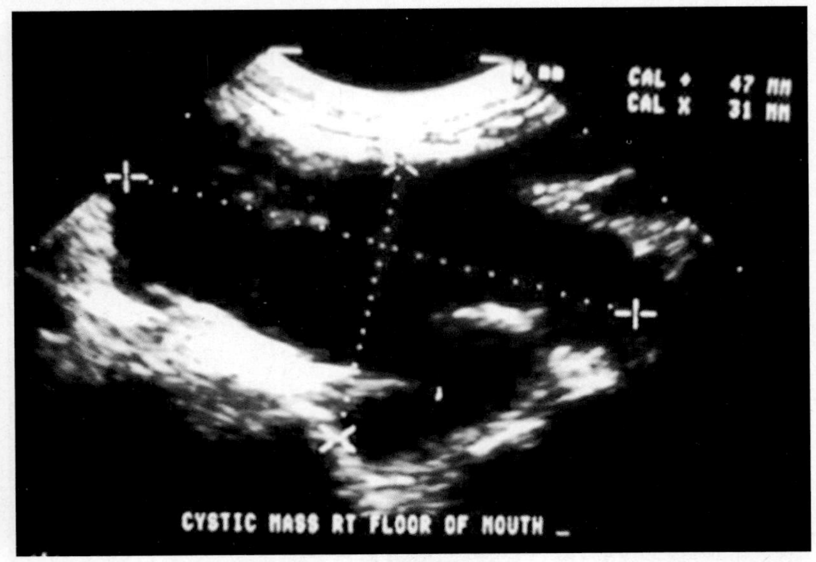

FIGURE 3.105

FIGURES 3.104 and 3.105: Lymphatic cyst. A young man presented with a cystic mass in Rt side of the neck and in the floor of the mouth. HRSG shows a big unilocular cyst with partial thickened echogenic septa. Few calcified specks are also seen in the cyst. No complete septation or loculation is seen, Typical of a lymphatic cyst.

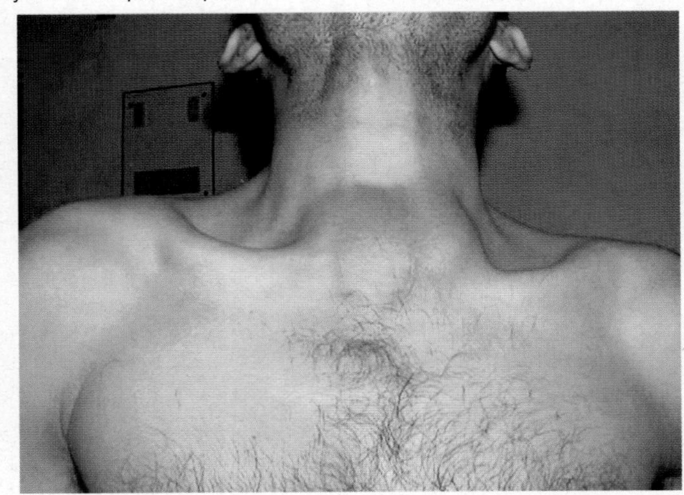

FIGURE 3.106: Lymphangioma. A patient presented with a soft tissue mass in the suprasternal area.

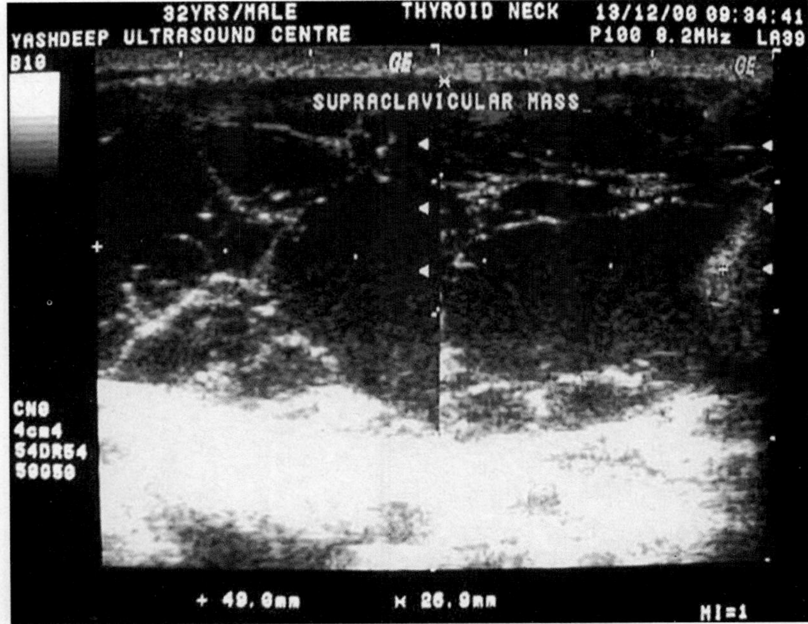

FIGURE 3.107

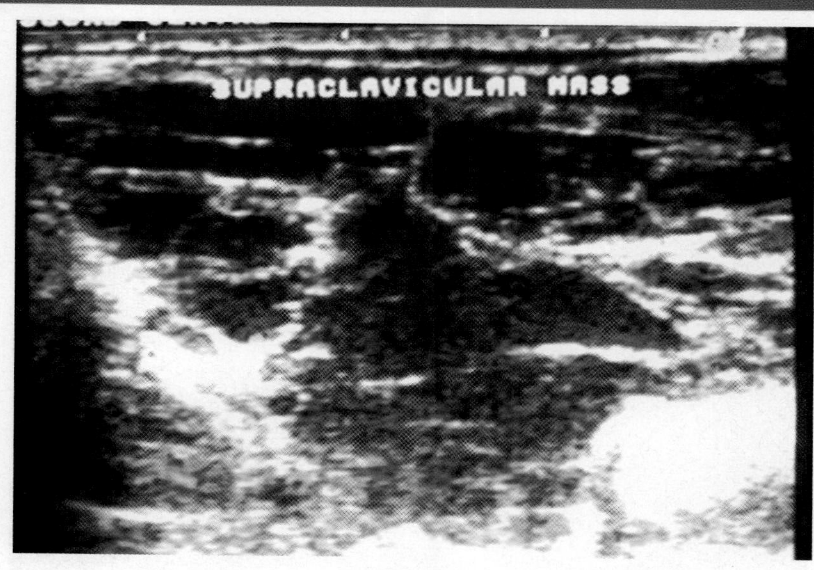

FIGURE 3.108

FIGURES 3.107 and 3.108: HRSG shows a multiloculated cystic mass. Multiple thinned echogenic septa are seen in the mass. Few internal echoes are seen in the cyst, suggestive of debris. Color Doppler imaging shows poor flow in the mass. Biopsy of the mass was lymph angioma.

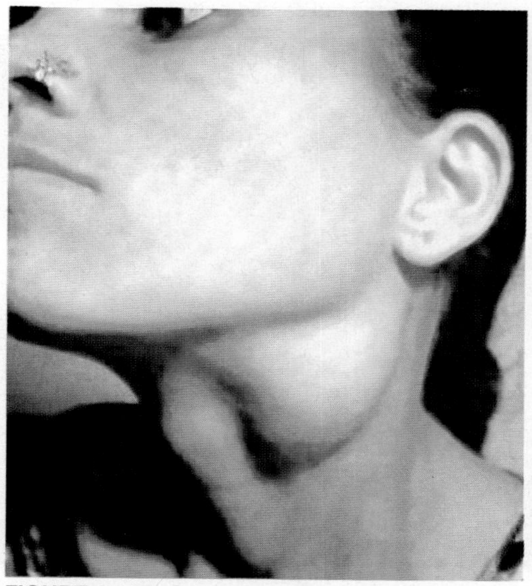

FIGURE 3.109: Carotid body tumor. A young lady presented with a soft tissue mass in the Lt side of the neck in its upper part. It was non-tender.

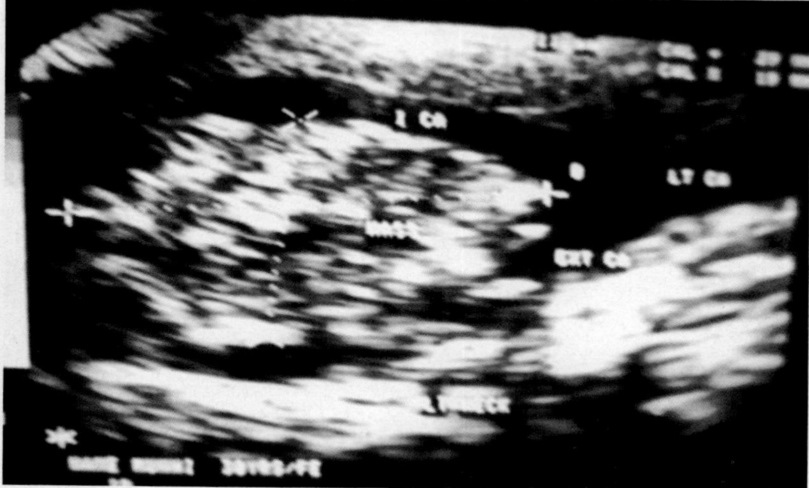

FIGURE 3.110: HRSG shows a well-defined echogenic tumor mass seen in between the carotid bifurcation. The tumor is seen splaying in the internal and external carotid artery and was adherent with carotid artery sheath. Biopsy shows carotid body tumor or chemodectomas.

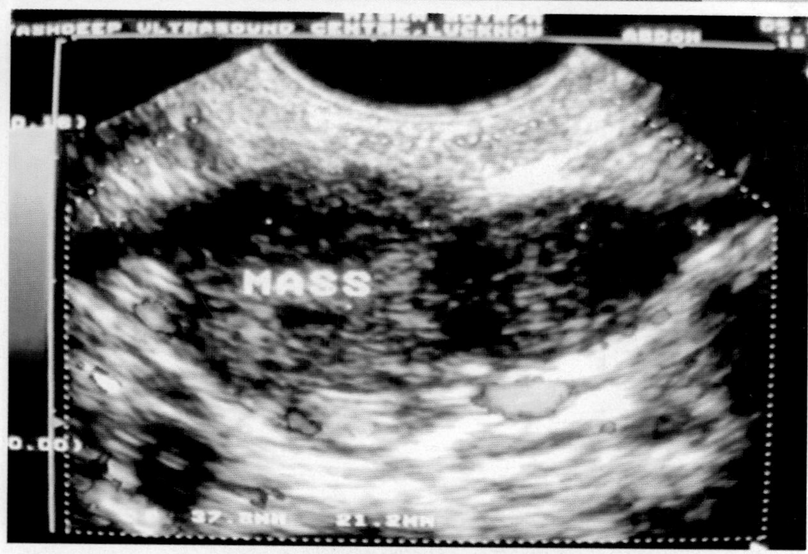

FIGURE 3.111: Carotid body tumor. An old man presented with a soft tissue mass in the Lt upper cervical area. HRSG shows a well-defined lobulated potato shaped tumor mass. It is homogeneous in texture and seen in between the internal and external carotid arteries and compressing them.

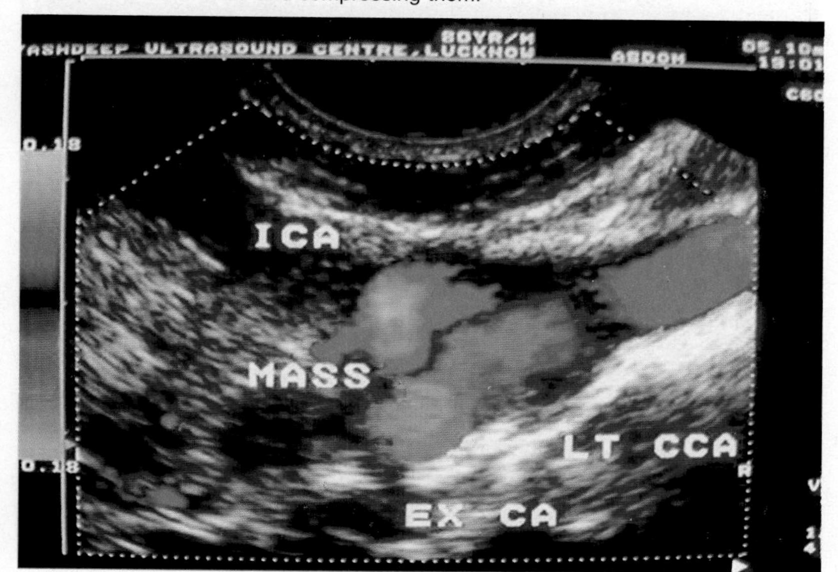

FIGURE 3.112

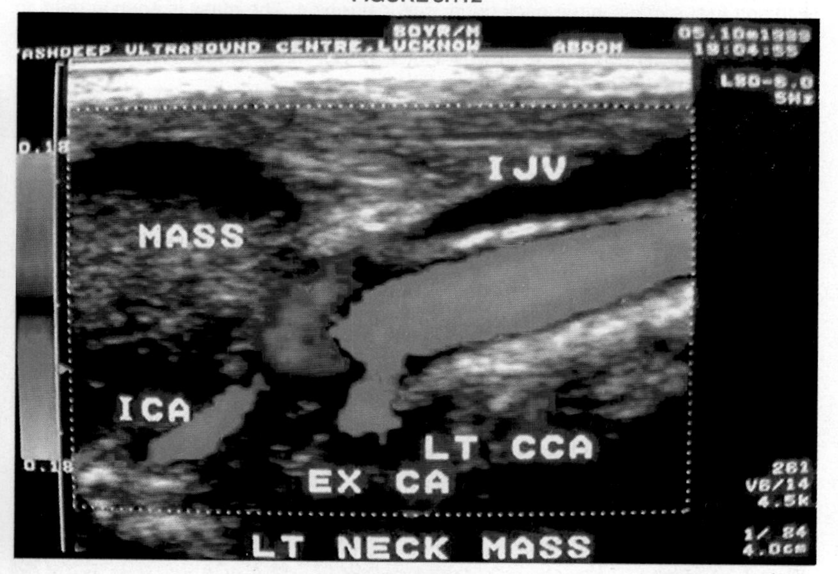

FIGURE 3.113

FIGURE 3.112 and 3.113: Color Doppler flow study shows poor blood flow in the tumor and the tumor was seen pressing over external and internal carotid arteries. The tumor was separate from carotid artery and its branches.

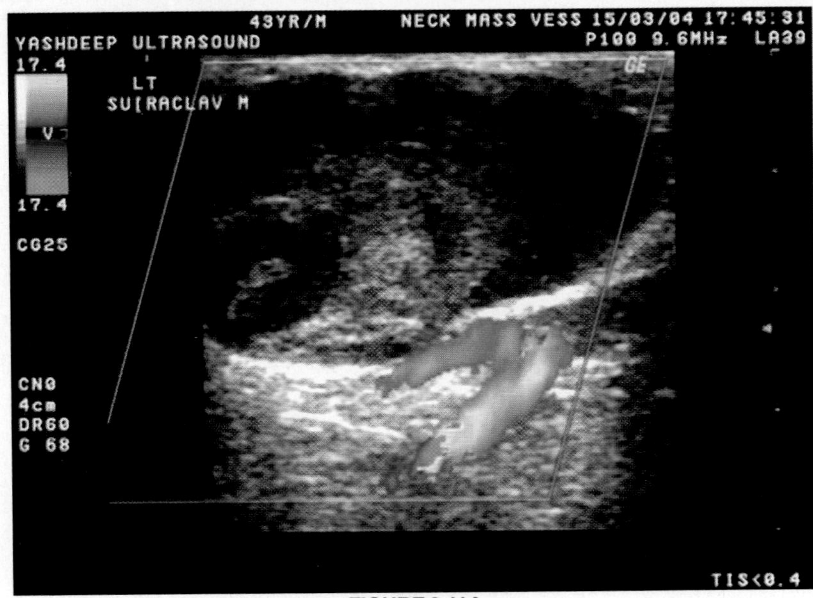

FIGURE 3.114

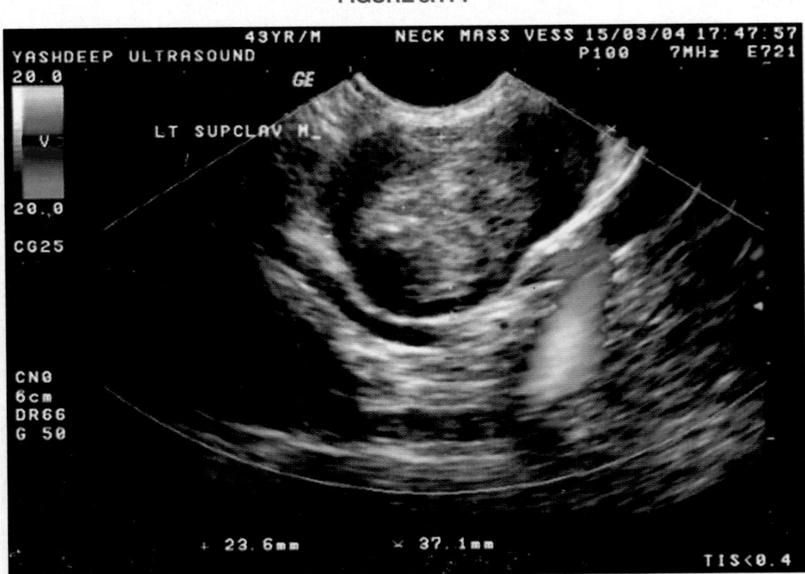

FIGURE 3.115

FIGURES 3.114 and 3.115: Carotid body tumor (potato tumor): A patient presented with soft tissue mass in the neck upper cervical region. The mass is seen in relation to the carotid vessels. It is homogeneous in texture. No necrosis is seen in it. No cystic degeneration is seen. However, it was not fixed with the carotid sheath. Biopsy confirmed carotid body tumor.

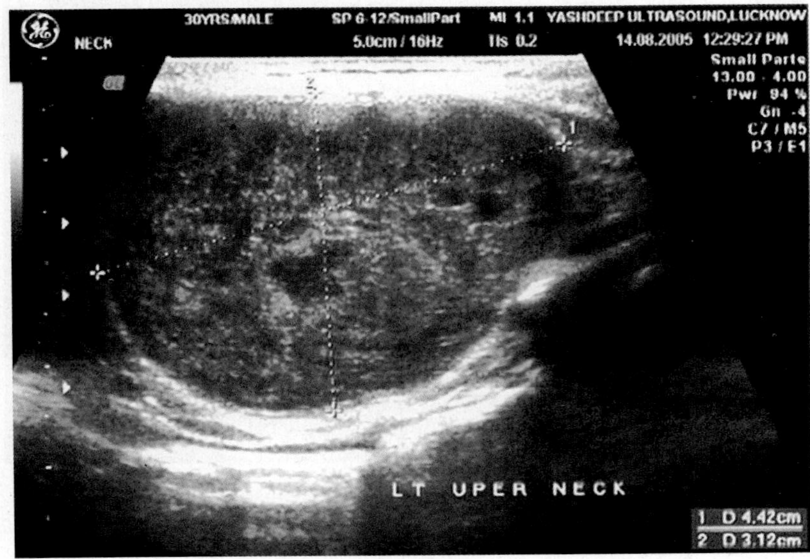

FIGURE 3.116

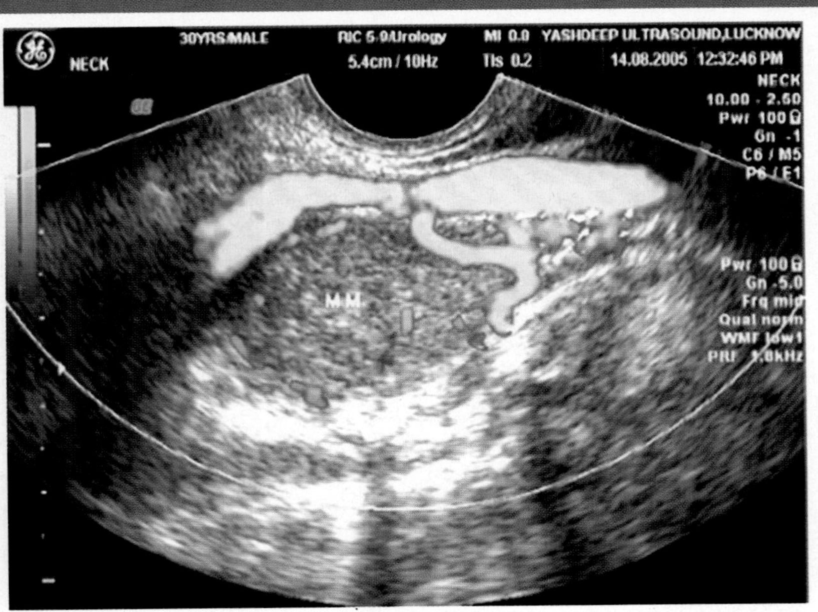

FIGURE 3.117

FIGURES 3.116 and 3.117: Carotid body tumor. A patient presented with well defined soft tissue mass in the neck. The mass is seen in relation to the carotid vessels. It is seen fixed with the carotid sheath. Color Doppler shows a feeding vessel in the mass. Biopsy confirmed carotid body tumor.

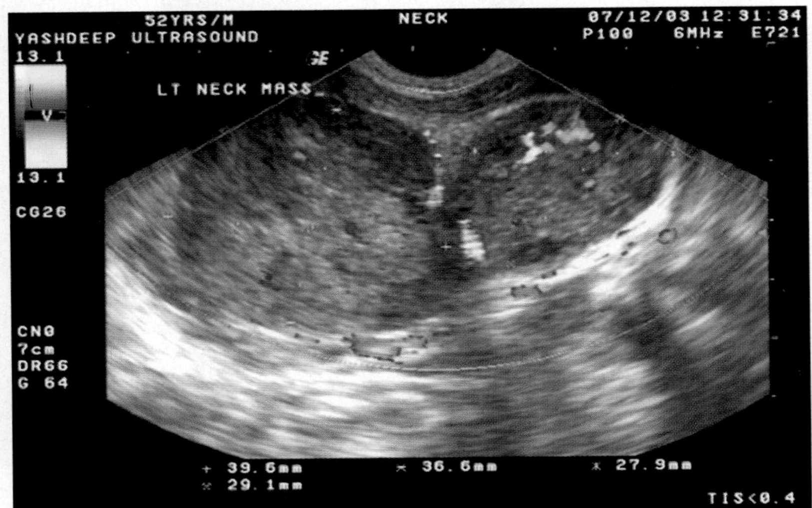

FIGURE 3.118

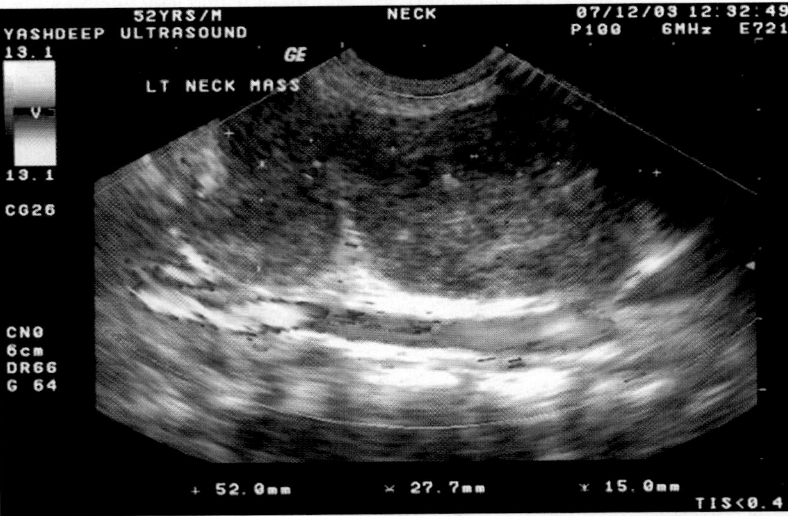

FIGURE 3.119

FIGURES 3.118 and 3.119: Nerve tumors (Schwann cell neuromas). Well defined homogeneous lobulated mass seen in the Lt side of the neck. It shows sharp margins with homogeneous texture. The mass is separate from the carotid vessels and not pressing in them. On color flow imaging low flow is seen in the mass. Biopsy showed nerve cell tumor.

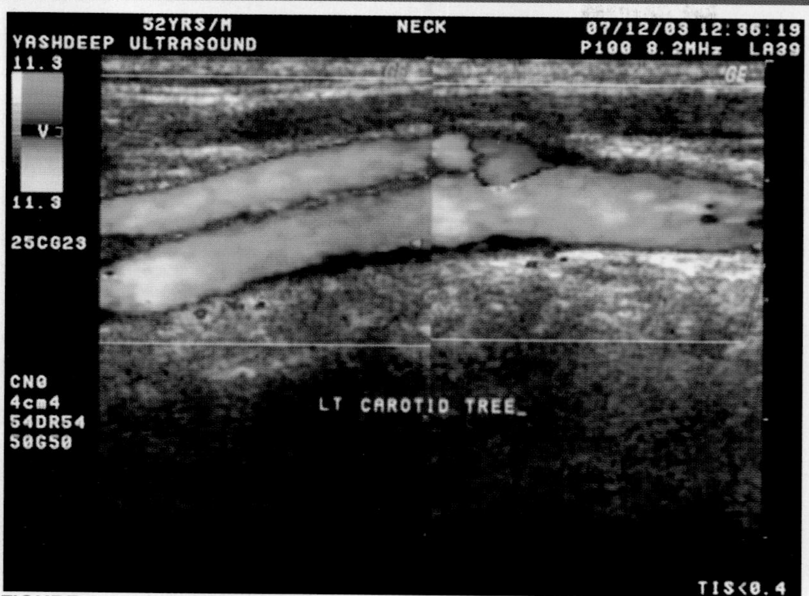

FIGURE 3.120: The tumor is not seen pressing over the carotid tree. The carotid vessels are seen normal in sonoappearance.

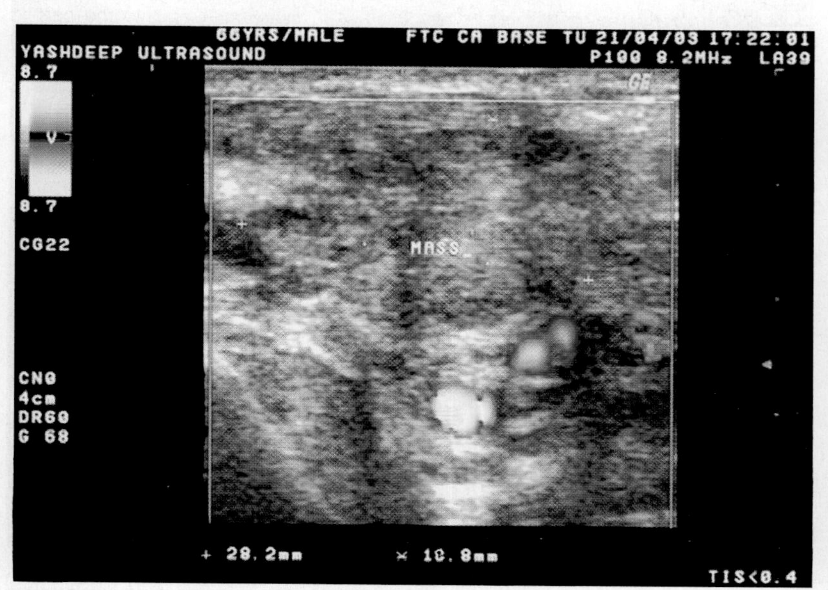

FIGURE 3.121

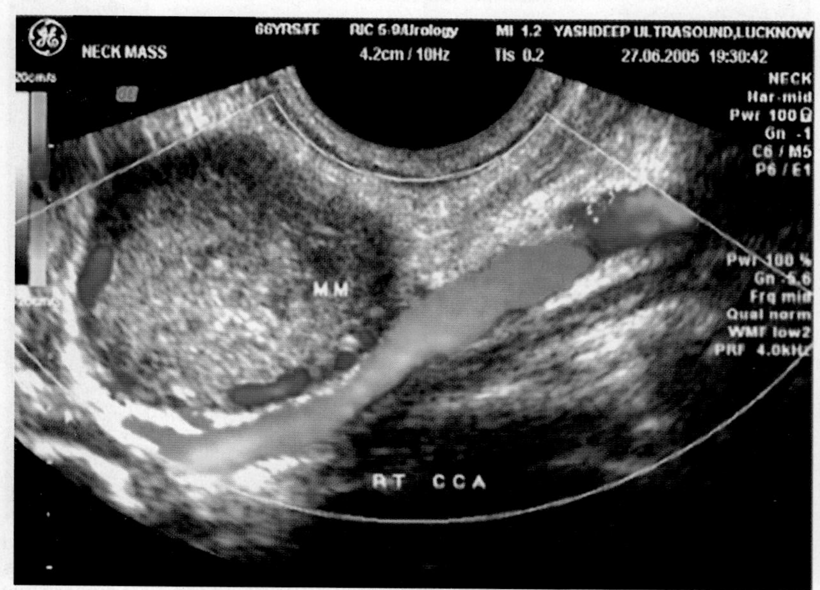

FIGURE 3.122

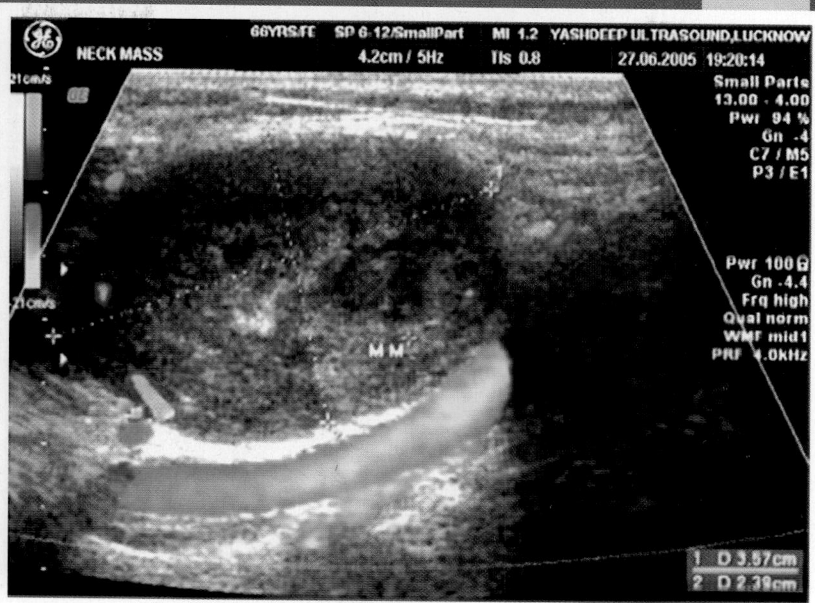

FIGURE 3.123

FIGURES 3.121 to 3.123: Shwan cell neuromas. An old lady presented with soft tissue mass in Rt side of the neck. The mass is homogeneous in texture with well-defined margins. No necrosis is seen in it. On color flow imaging low flow is seen in the mass. It is separate from the neck vessels.

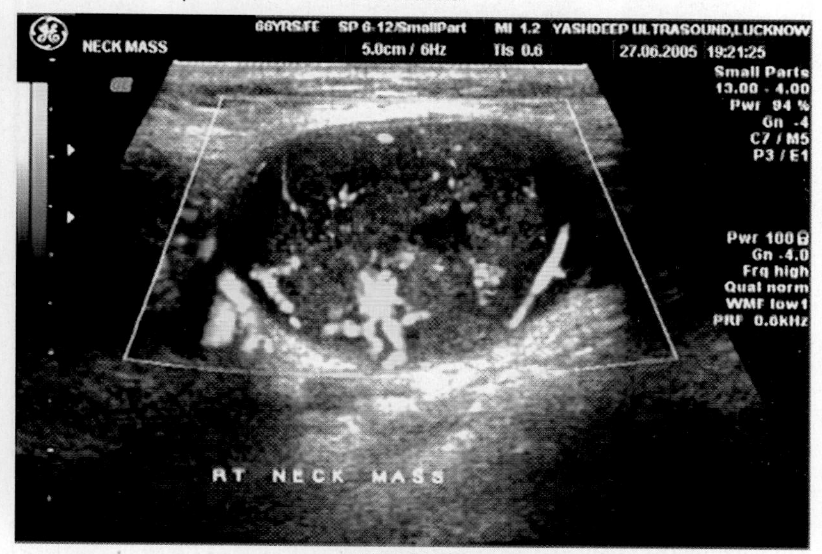

FIGURE 3.124

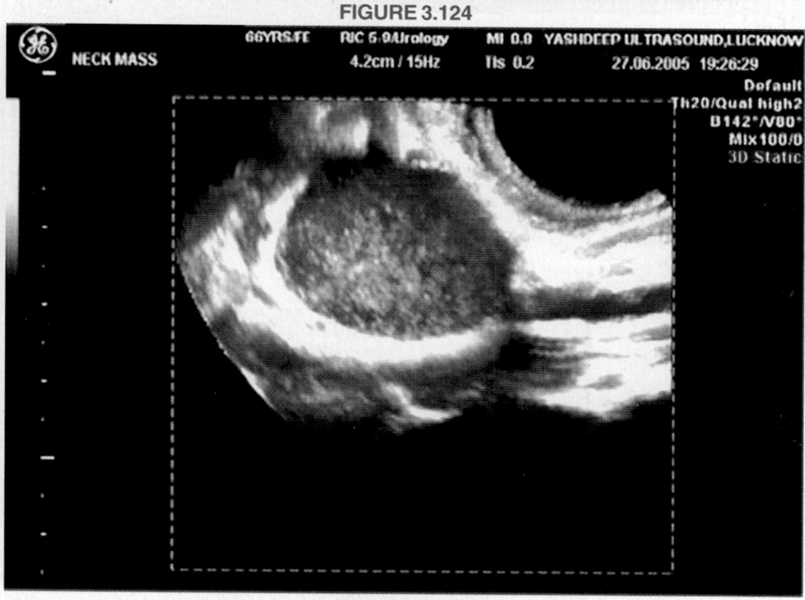

FIGURE 3.125

FIGURES 3.124 and 3.125: 3D imaging of the mass shows finer details with relationship with the neck vessels. Biopsy shows Schwann cell neuromas.

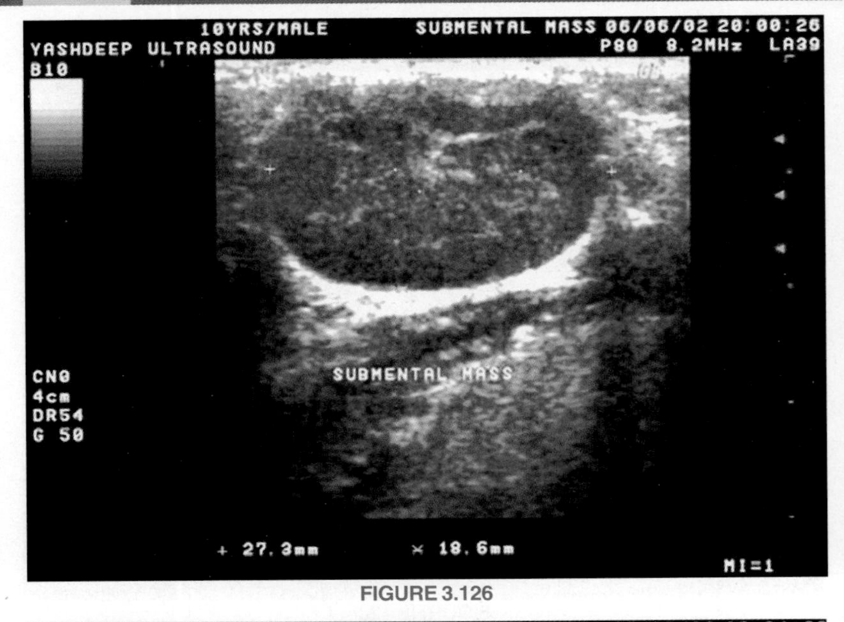

FIGURE 3.126

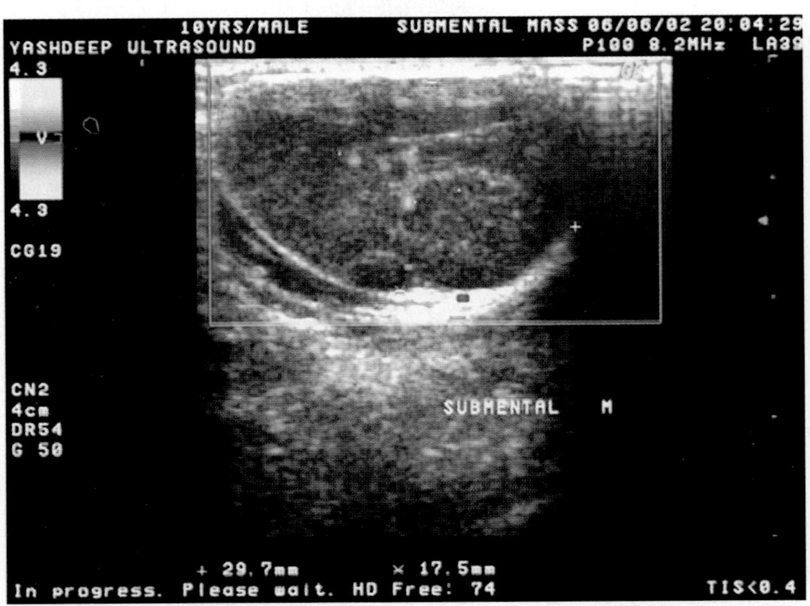

FIGURE 3.127

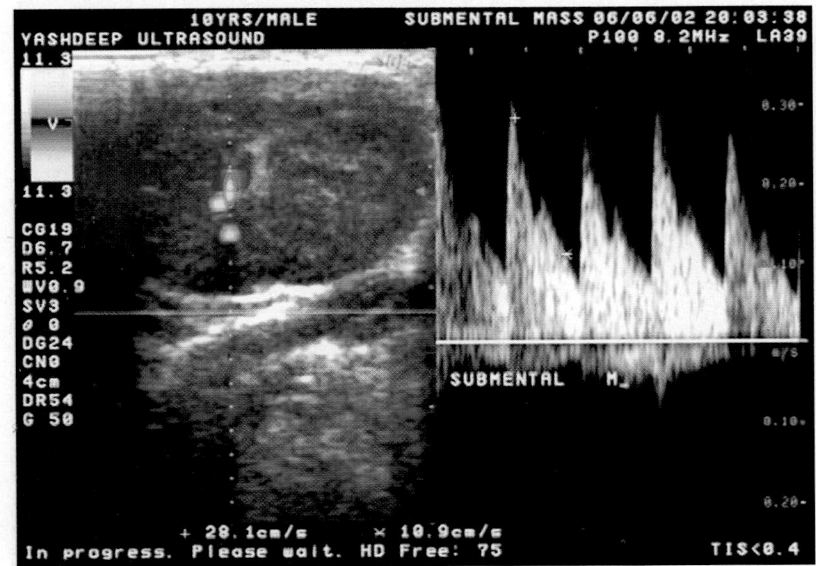

FIGURE 3.128

FIGURES 3.126 to 3.128: Submental dermoid. A young boy presented with soft tissue mass in the submental region. HRSG shows well defined thick echo mass in the submental region. No calcification is seen in it. No cystic degeneration is seen. On color flow imaging moderate flow is seen in the cyst. Biopsy shows a submental dermoid.

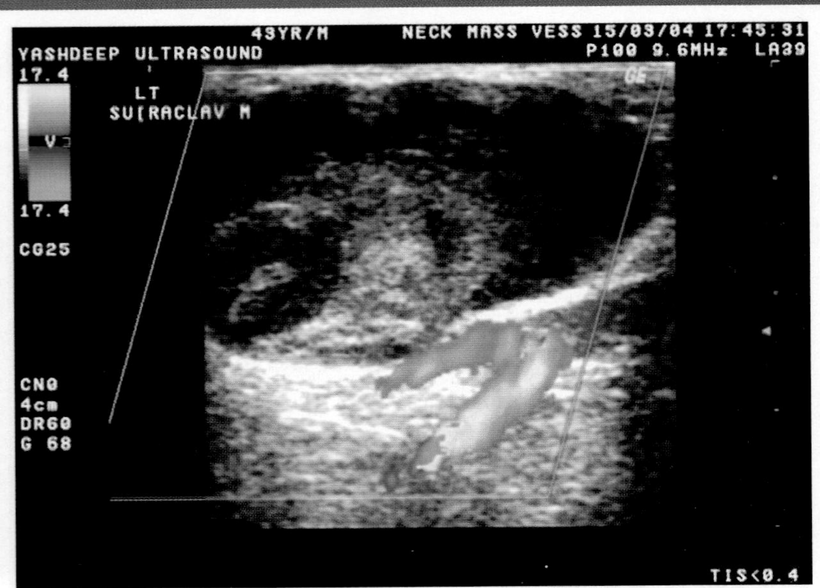

FIGURE 3.129

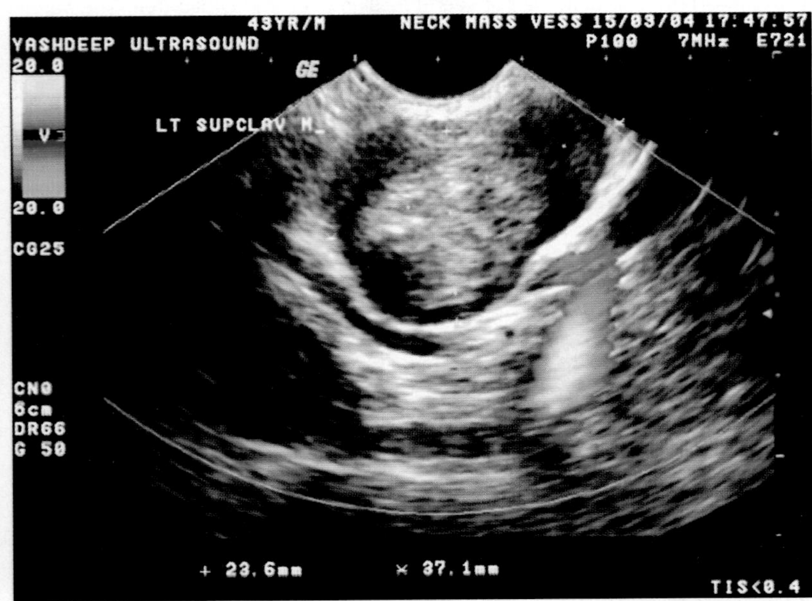

FIGURE 3.130

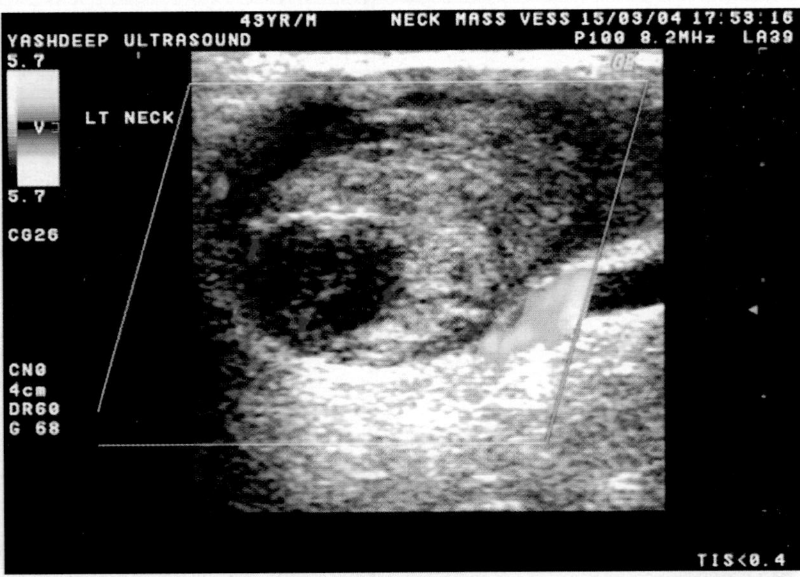

FIGURE 3.131

FIGURES 3.129 to 3.131: Hematoma neck. A patient sustained blunt injury in the neck. He presented with soft tissue mass in the neck. HRSG shows a big mass in the neck. The mass shows heterogeneous texture with central area of necrosis. No evidence of any bleeding vessel is seen in it. However, It was seen pressing over the neck vessels. Biopsy shows organised hematoma.

well-defined lobulated pattern of the mass. It shows homogeneous texture. No calcification or necrosis is seen in the mass. Pressure may be seen on the neck vessels due to large size of the mass. FNAC (fine-needle aspiration cytology) is the final investigation to confirm the diagnosis in these cases.

Vascular Masses

HRSG is very sensitive and accurate in detecting vascular masses in the neck. Hemangiomas, cavernous venous malformations and arteriovenous malformations can be easily differentiated on HRSG. Color flow imaging is mandatory in deciding the nature of the mass.

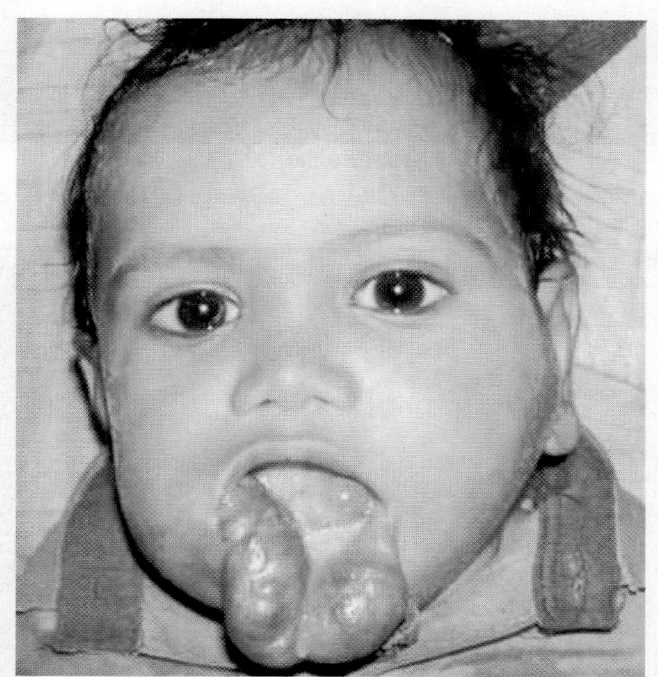

FIGURE 3.132: Arteriovenous malformation. A young child presented with marked lip hypertrophy with multiple bluish masses seen over the tongue and fullness of both cheek and the floor of the mouth.

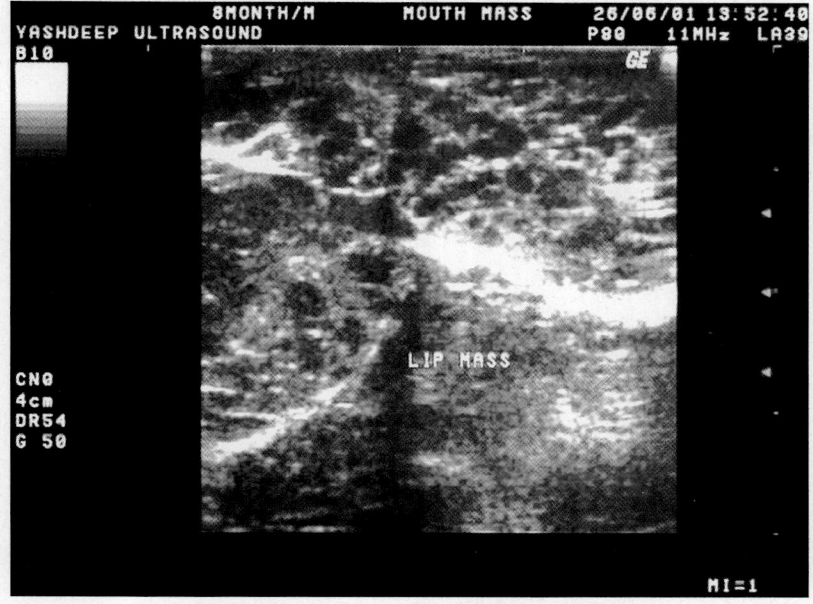

FIGURE 3.133: HRSG shows markedly hypertrophied lips with multiple dilated vessels are seen in the lower lip. Few calcified specks are also seen in it.

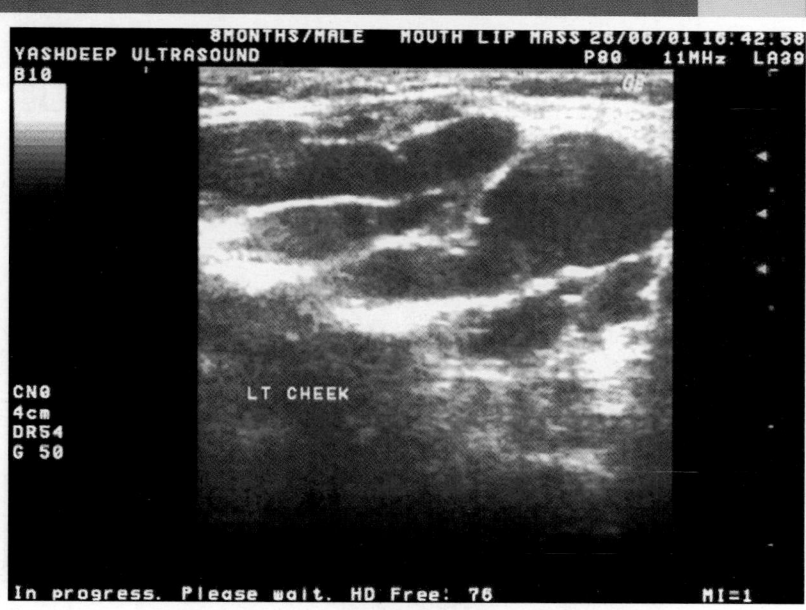

FIGURE 3.134

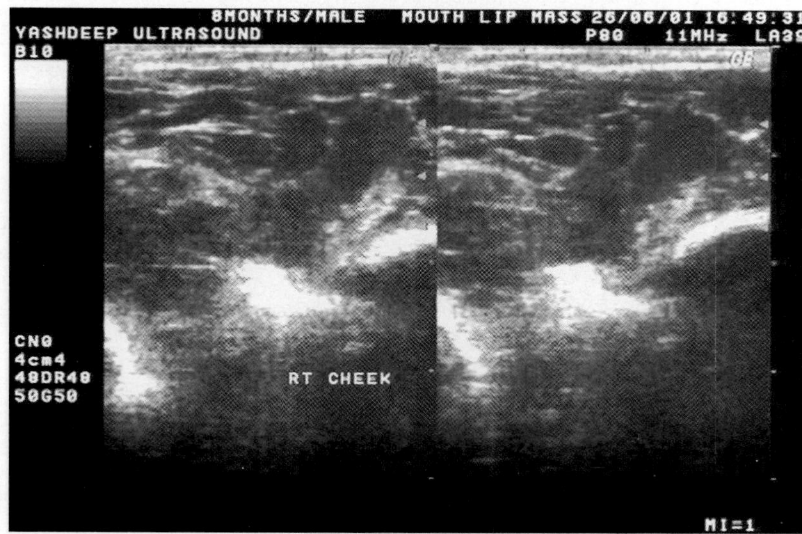

FIGURE 3.135

FIGURES 3.134 and 3.135: HRSG shows grossly dilated vessels in Rt and Lt cheeks.

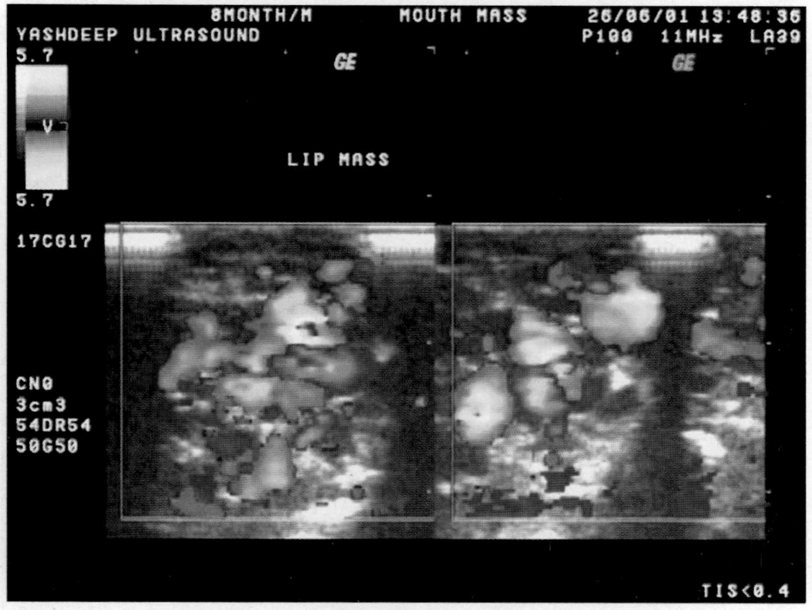

FIGURE 3.136

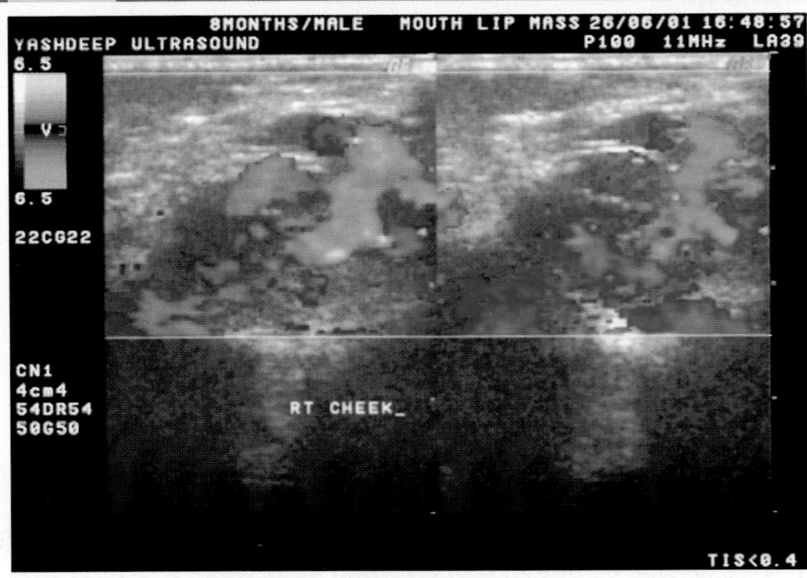

FIGURE 3.137

FIGURES 3.136 and 3.137: Color Doppler imaging shows high flow in the lip vessels with irregular pattern.

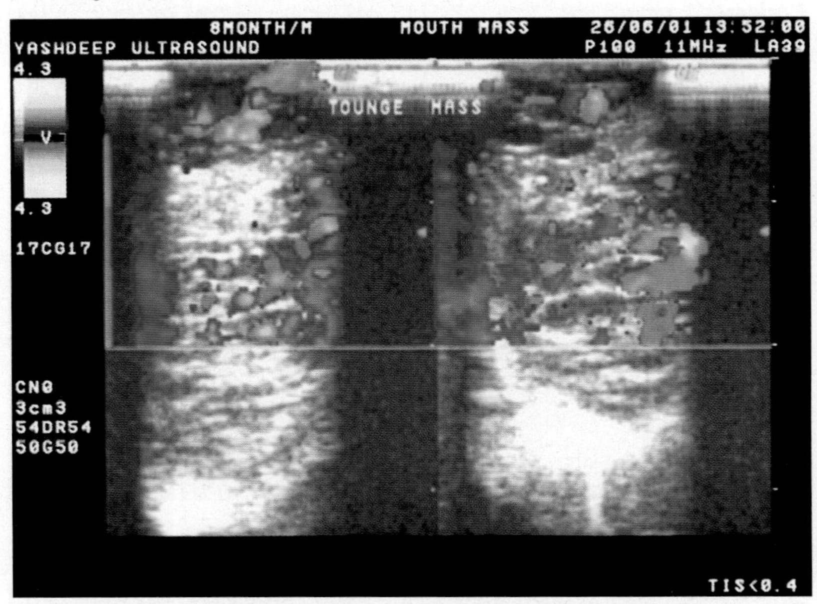

FIGURE 3.138. HRSG shows multiple dilated vessels on the under surface of the tongue. Color Doppler imaging shows high flow in these vessels.

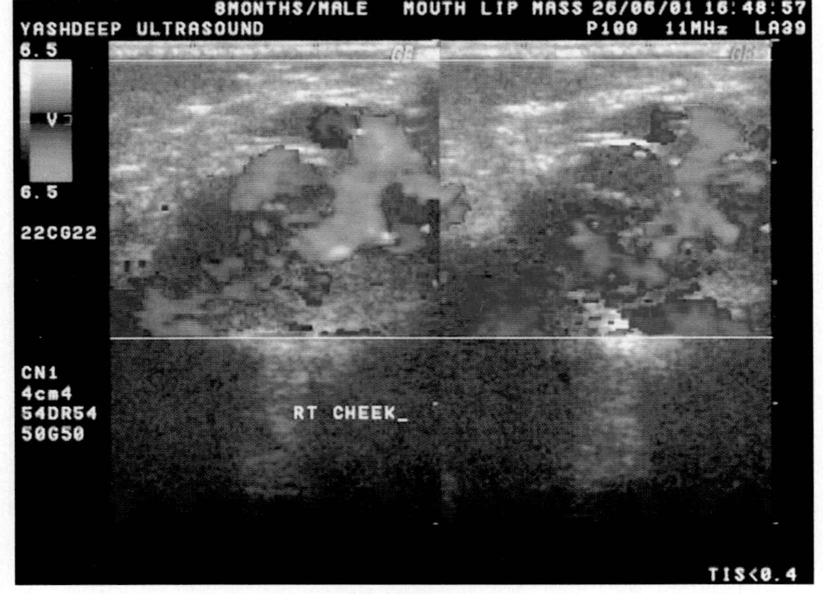

FIGURE 3.139

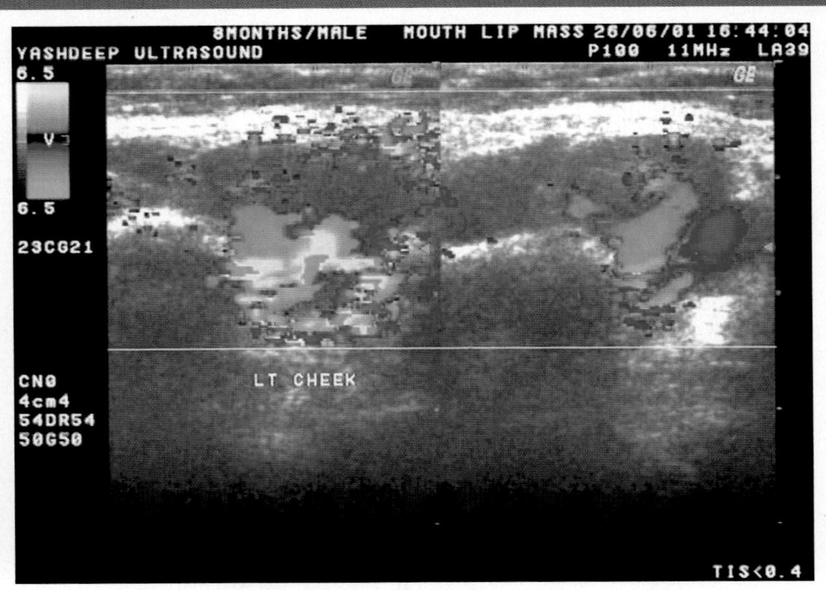

FIGURE 3.140

FIGURES 3.139 and 3.140: Color flow imaging shows increased flow in the cheek vessels on either side.

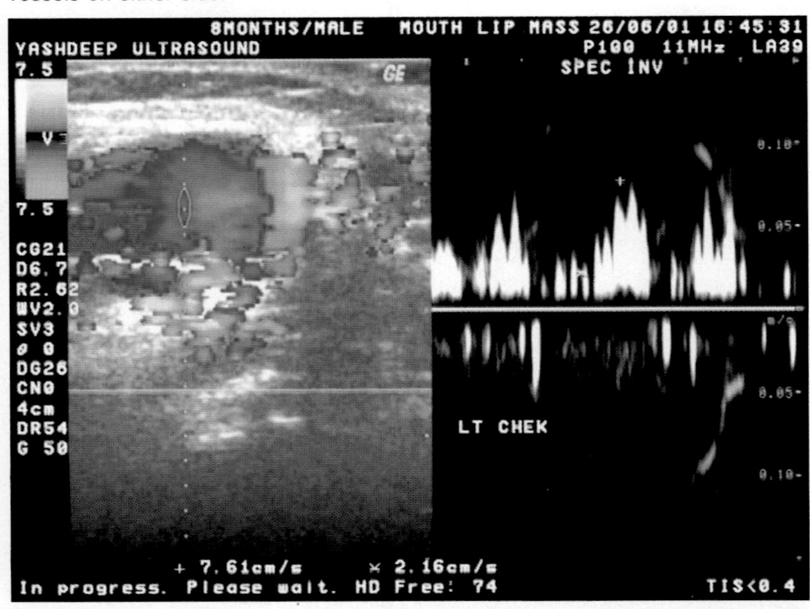

FIGURE 3.141: Spectral Doppler flow-tracing shows high flow in the vessels with arterial and venous flow pattern suggestive of arteriovenous malformation involving tongue, lower lip, both cheek and floor of the mouth.

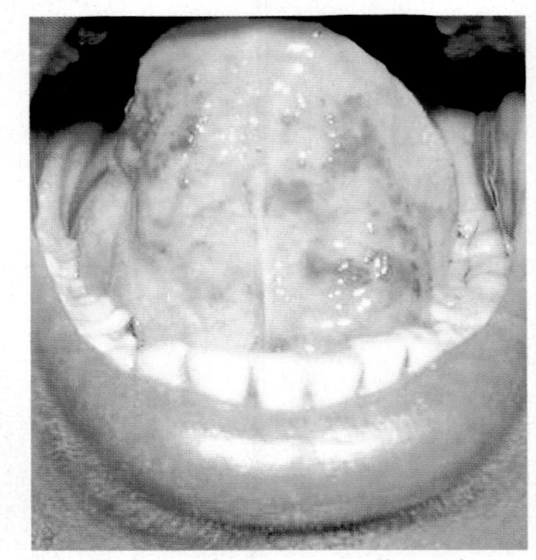

FIGURE 3.142: Cavernous malformation on the under surface of fongue. A young man presented with multiple dilated vessels over under surface of tongue more marked on Lt side.

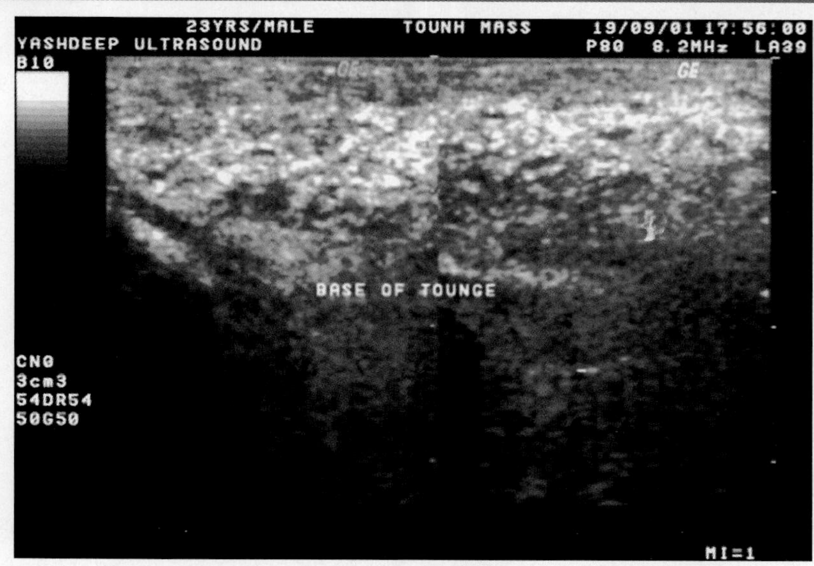

FIGURE 3.143: HRSG shows multiple dilated vessels on the tongue surface. They are seen running in zigzag session.

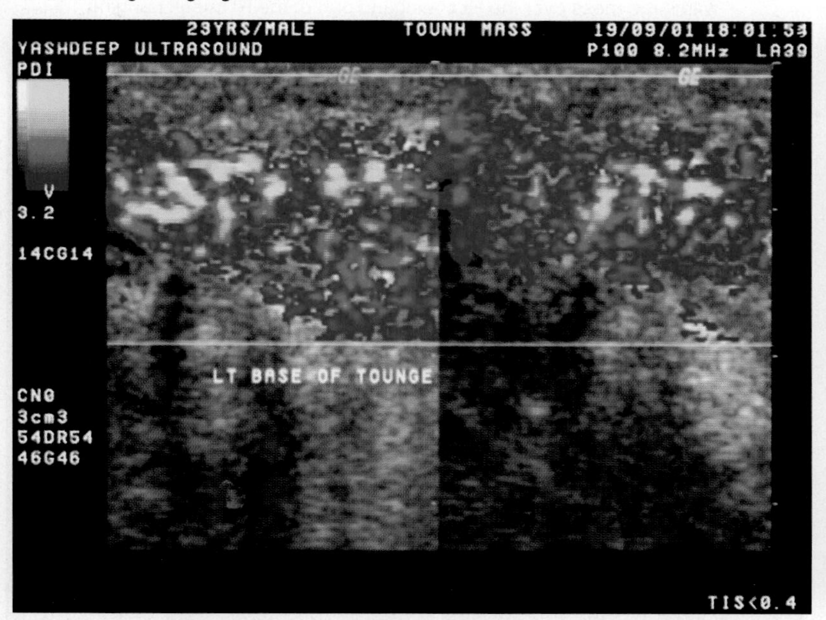

FIGURE 3.144: Color flow imaging shows blood flow in the vessels.

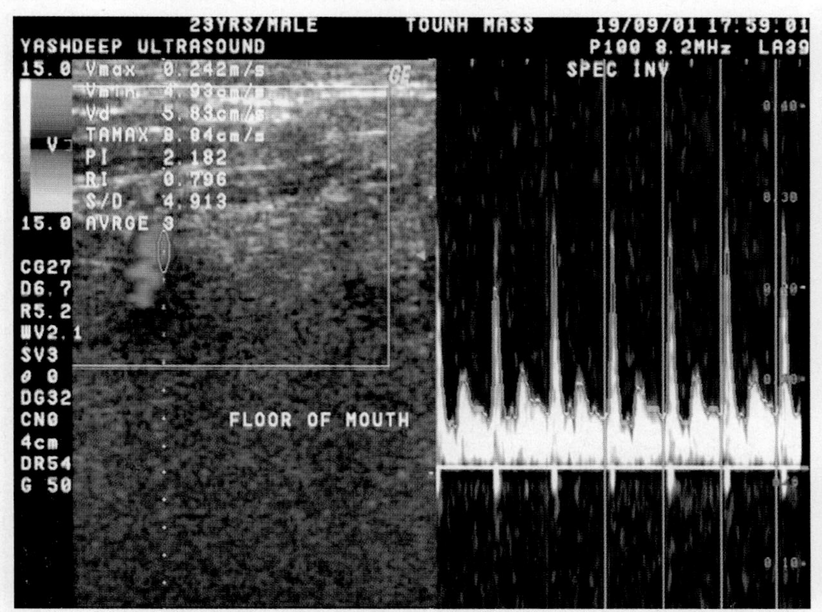

FIGURE 3.145

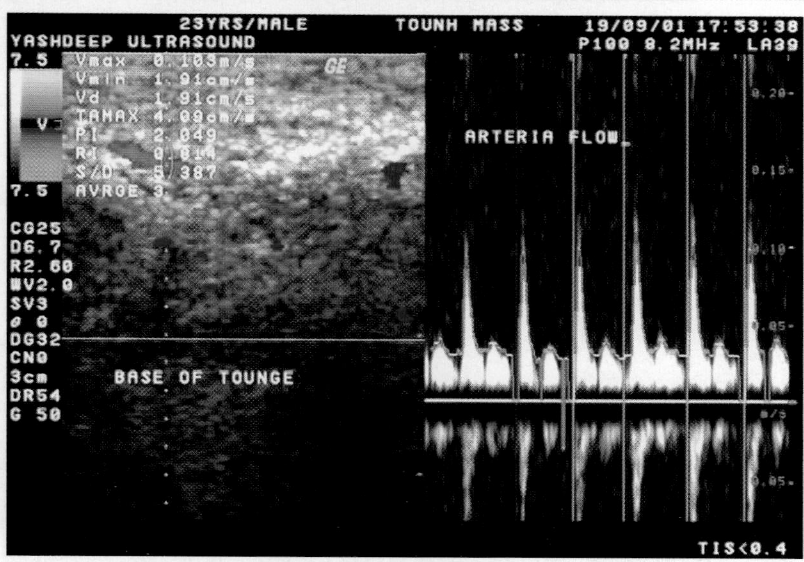

FIGURE 3.146

FIGURES 3.145 and 3.146: Spectral Doppler tracing shows arterial waveform in the vessels with venous flow pattern suggestive of cavernous malformation.

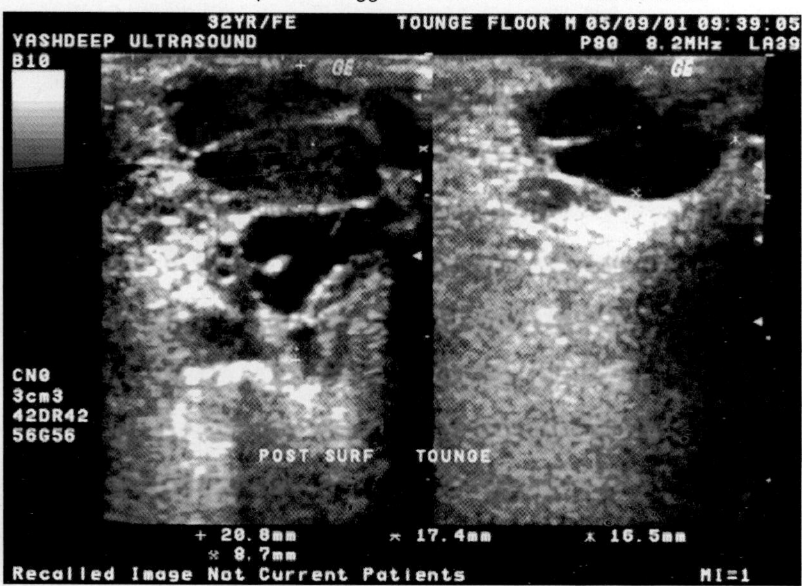

FIGURE 3.147

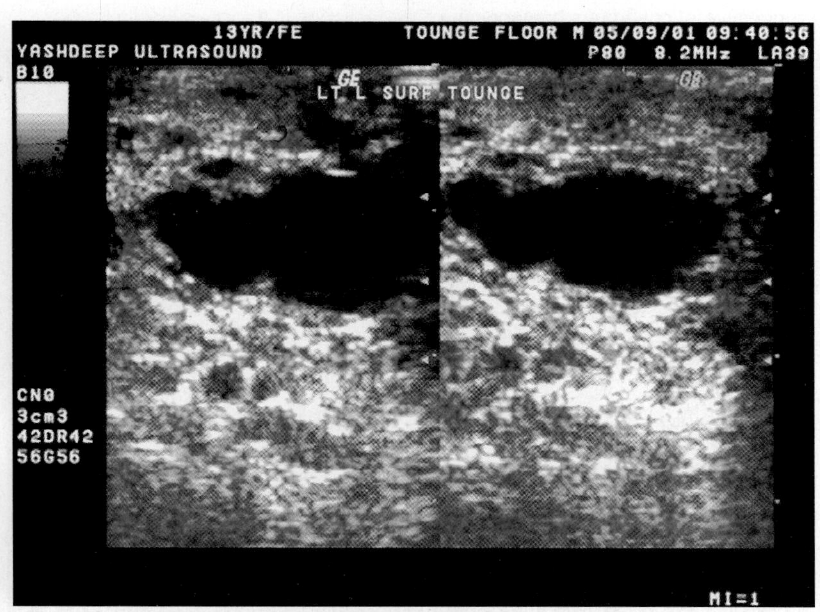

FIGURE 3.148

FIGURES 3.147 and 3.148: A young lady came with a soft tissue mass over the under surface of the tongue. HRSG shows multiple dilated vessels in the mass.

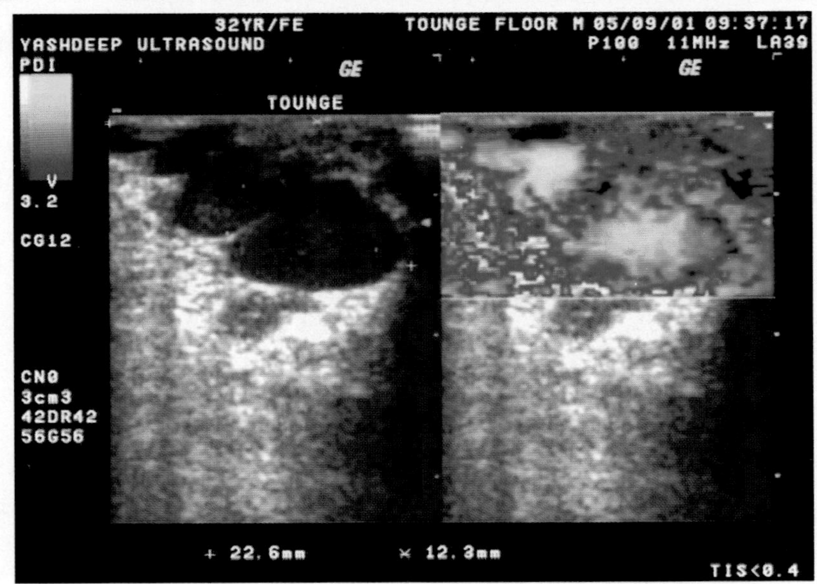

FIGURE 3.149: Color Doppler imaging shows large vessels running in the mass and show slow blood flow.

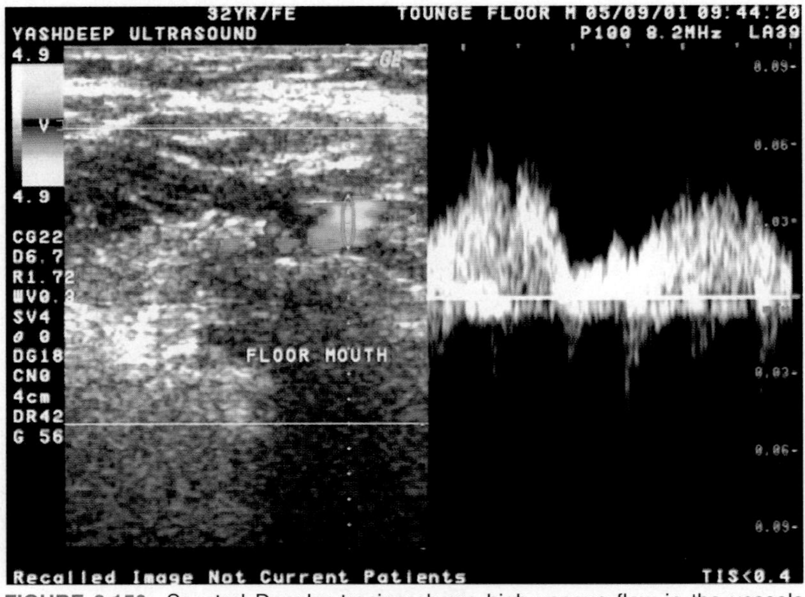

FIGURE 3.150: Spectral Doppler tracing shows high venous flow in the vessels suggestive of cavernous hemangioma.

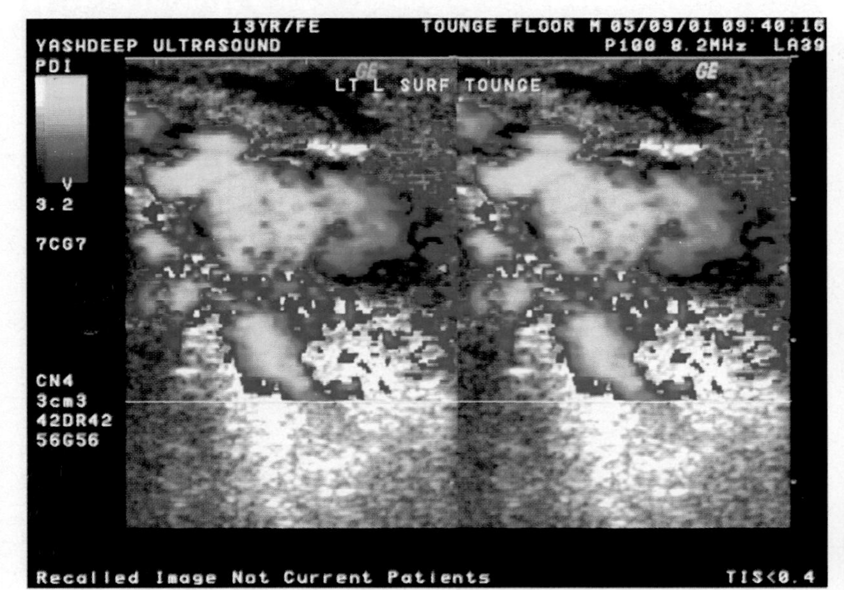

FIGURE 3.151: Power Doppler color flow imaging shows dilated vessels filled with blood. The vessels stand out very clearly on power Doppler imaging.

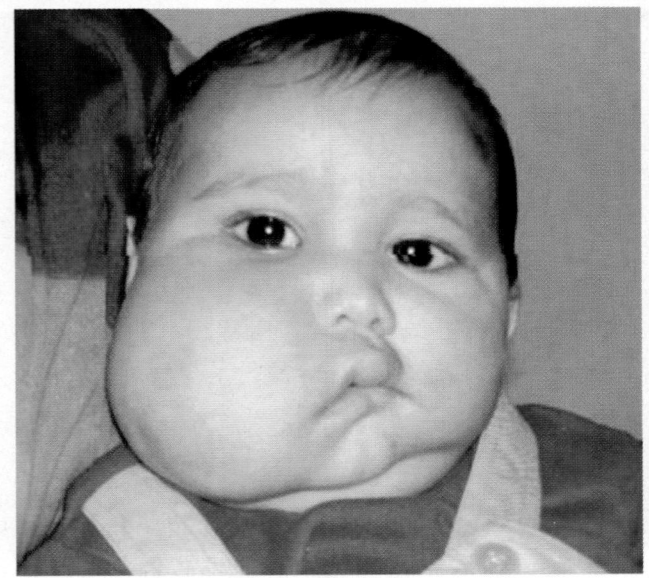

FIGURE 3.152: **Lymphangioma.** A four months old boy presented with large mass over the Rt cheek and part of the Rt cheek hanging down involving part of the chin.

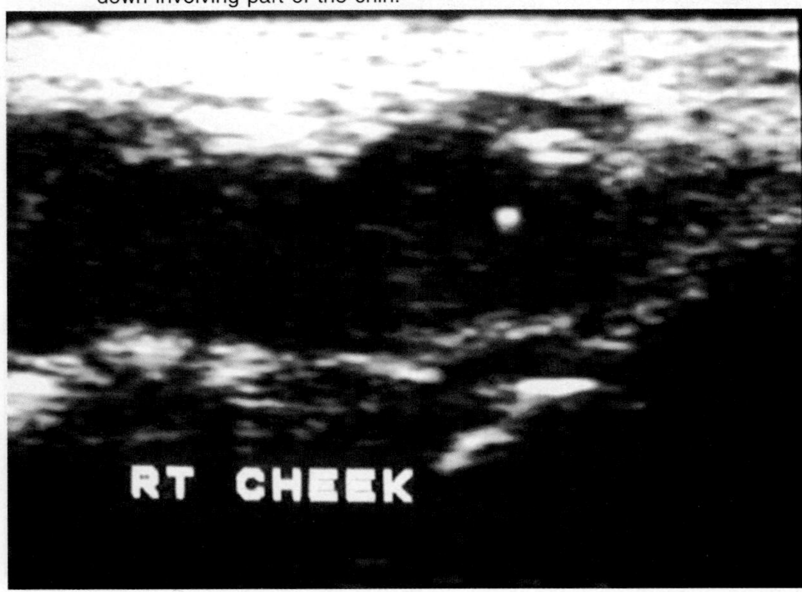

FIGURE 3.153: HRSG shows a multiloculated cystic mass. Dilated vessels are also seen in it.

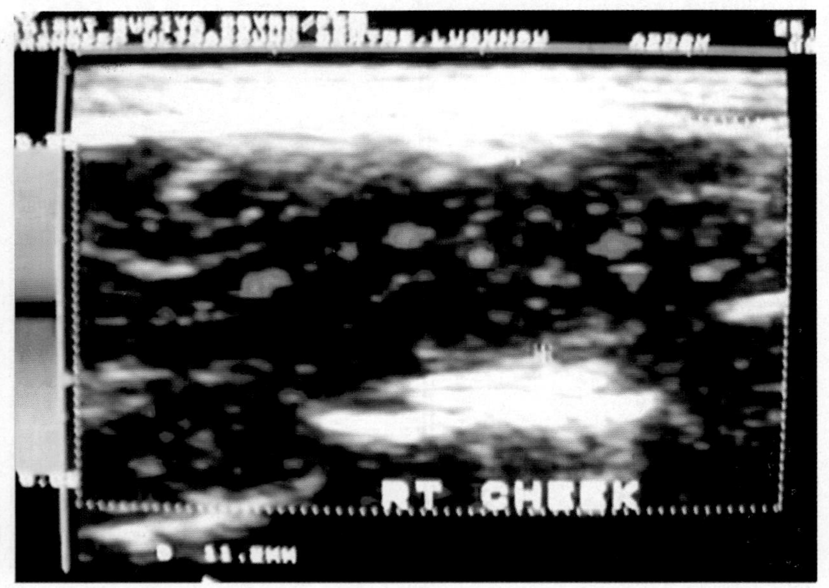

FIGURE 3.154: Color Doppler flow study shows little flow in the mass. However, few blood vessels are seen feeding the mass.

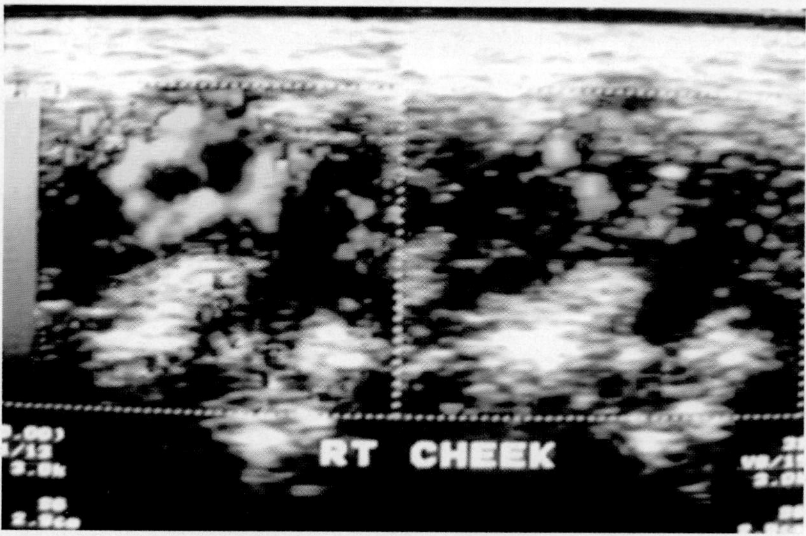

FIGURE 3.155: Power Doppler imaging shows better visualization of the vessels showing blood flow in the vessels. Postoperative findings confirmed lymphangioma.

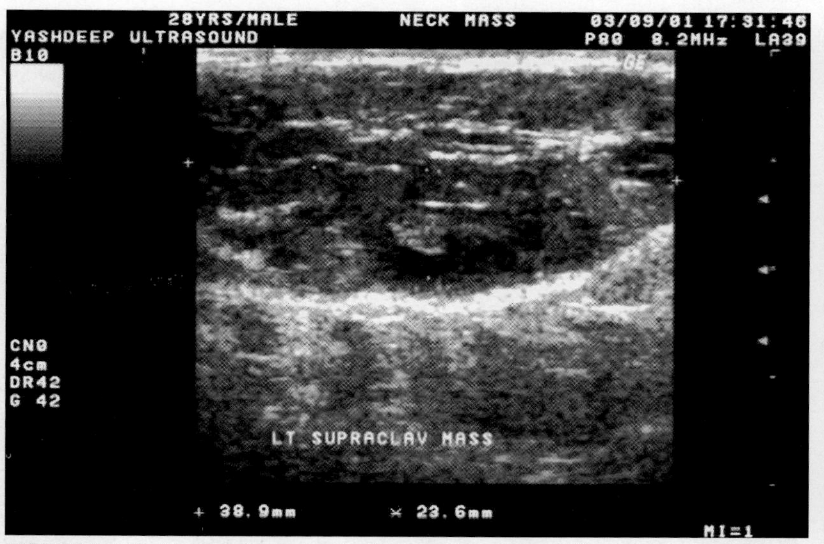

FIGURE 3.156: Hemangioma neck. A young boy came with a soft tissue mass in the Lt supraclavicular area. HRSG shows a low-level echo complex mass over the Lt supraclavicular area. The mass shows multiple vessels.

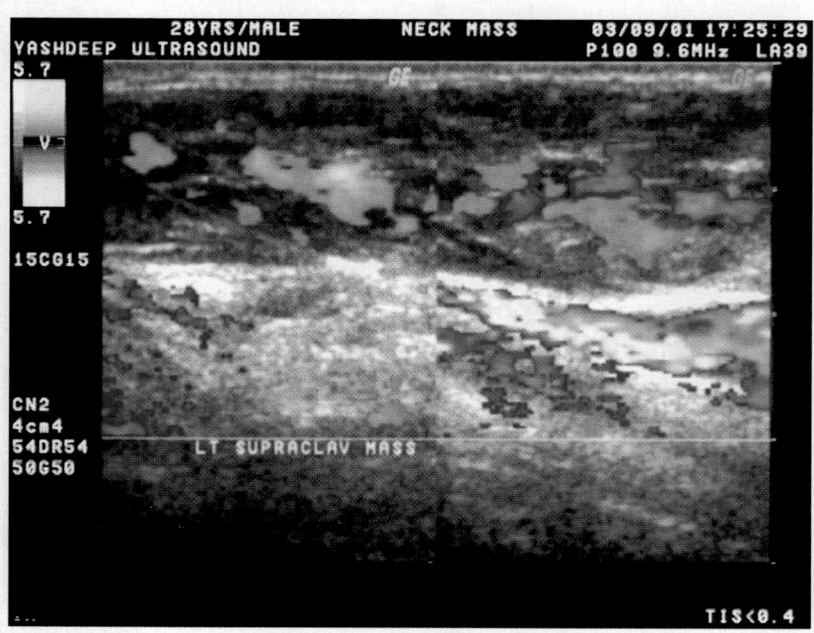

FIGURE 3.157

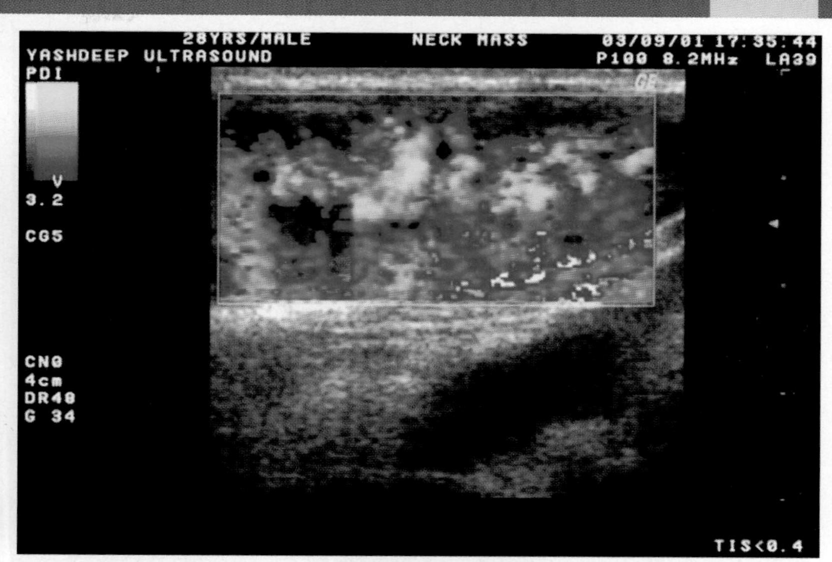

FIGURE 3.158

FIGURES 3.157 and 3.158: Color Doppler imaging shows multiple vessels feeding the mass with low flow pattern suggestive of hemangioma.

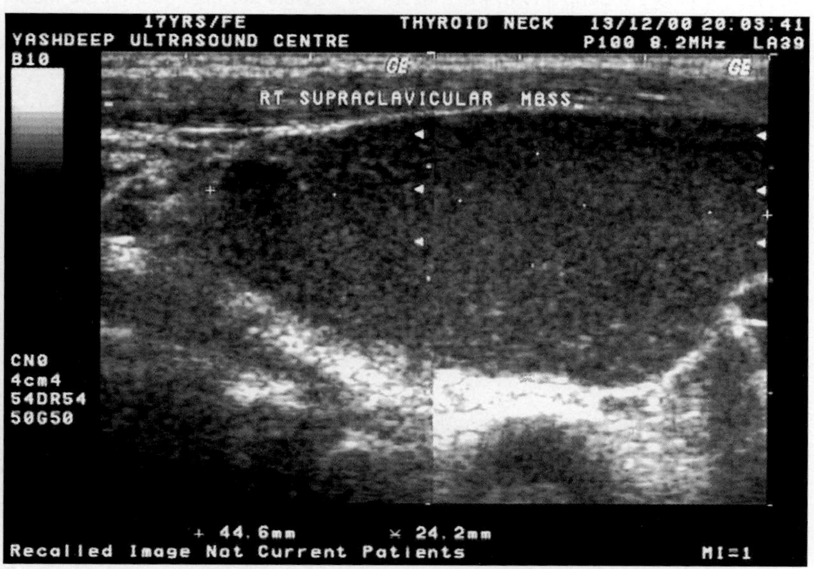

FIGURE 3.159: A young girl presented with a well-defined low echo complex mass in Rt supraclavicular area. HRSG shows a hypoechoic low-level echo complex mass in the Rt supraclavicular area.

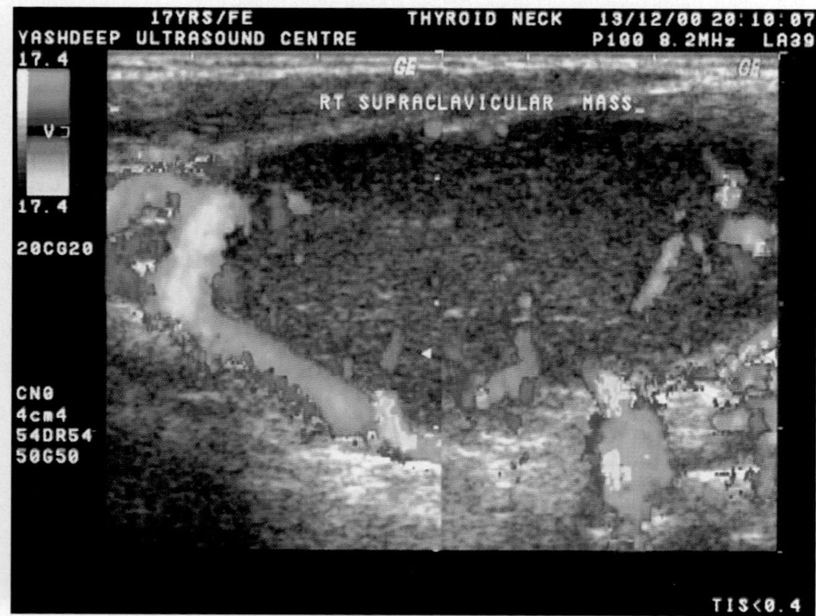

FIGURE 3.160

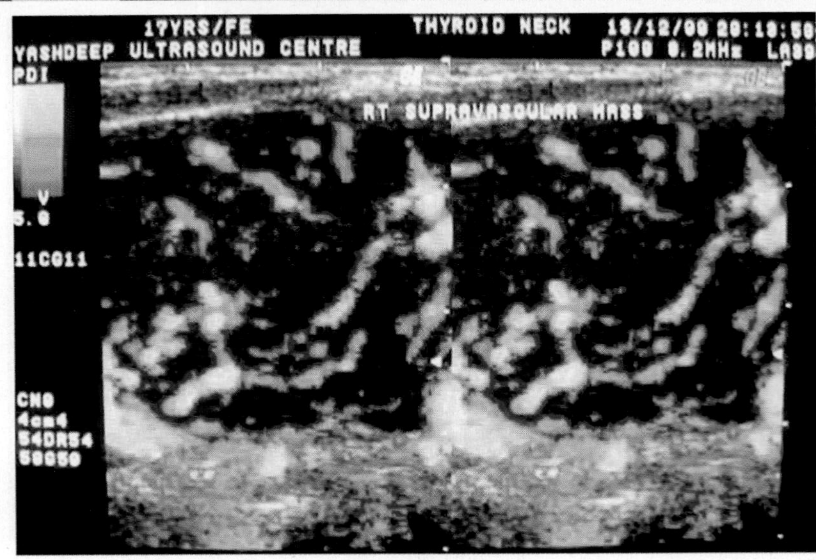

FIGURE 3.161

FIGURE 3.160 and 3.161: Lymphangioma. Color Doppler imaging shows well defined vessels around the periphery of the mass. Few vessels are also seen in the mass. Power Doppler imaging shows multiple feeding vessels in the mass cris-crossing the mass. Biopsy of the mass shows lymphangioma.

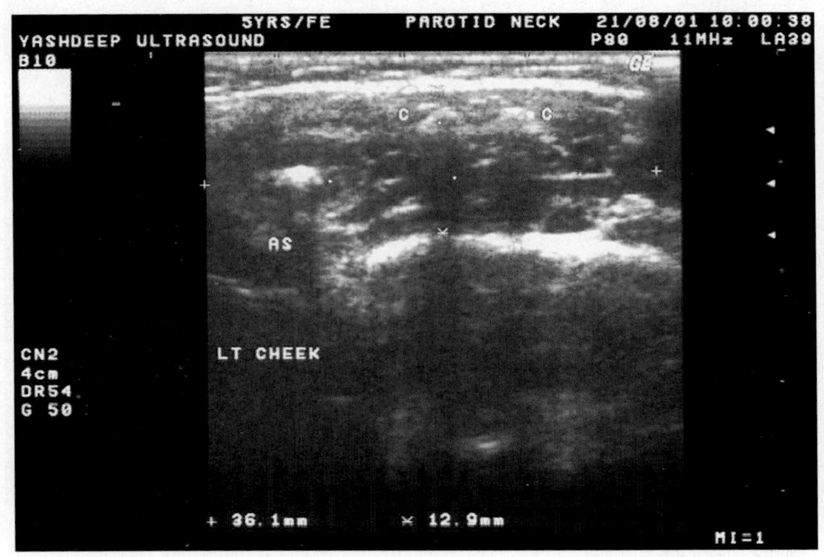

FIGURE 3.162: Cavernous venous malformation. A young girl presented with a soft tissue mass over the Lt cheek. HRSG shows hypoechoic low level echo complex mass. Multiple dilated vessels are seen running in the mass.

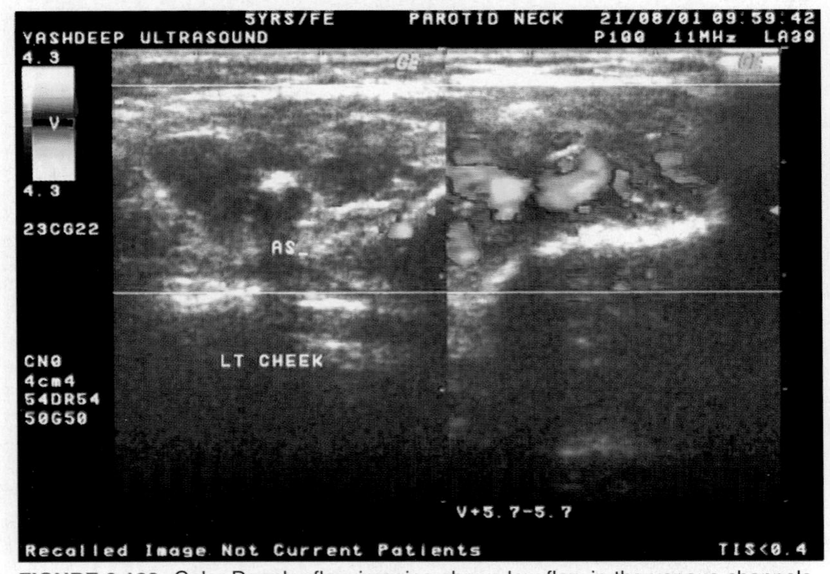

FIGURE 3.163: Color Doppler flow-imaging shows low flow in the venous channels. Echogenic calcified shadow is phlebolith.

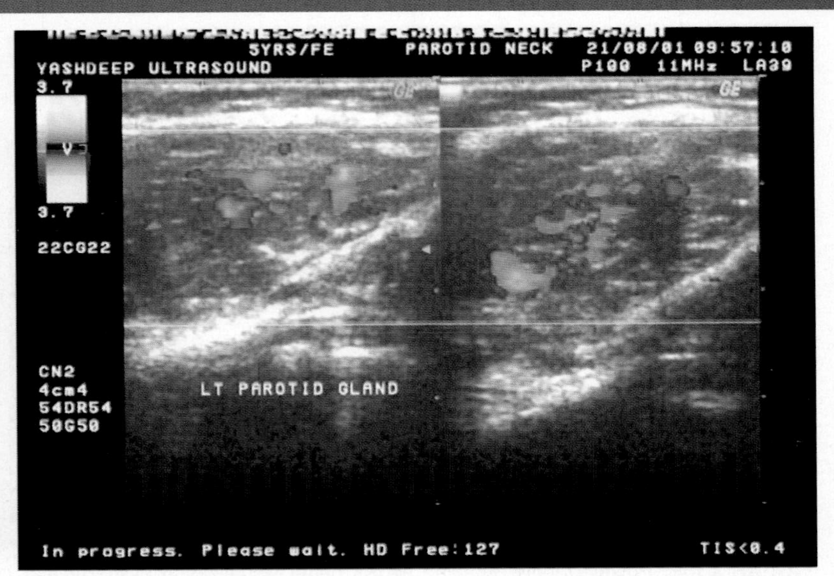

FIGURE 3.164: Dilated vessels are also seen in the Lt parotid gland in the same patient.

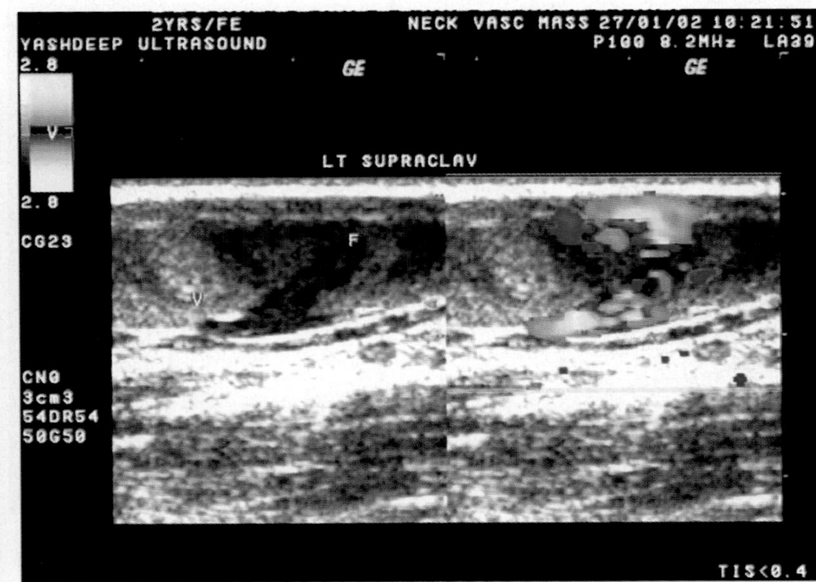

FIGURE 3.165

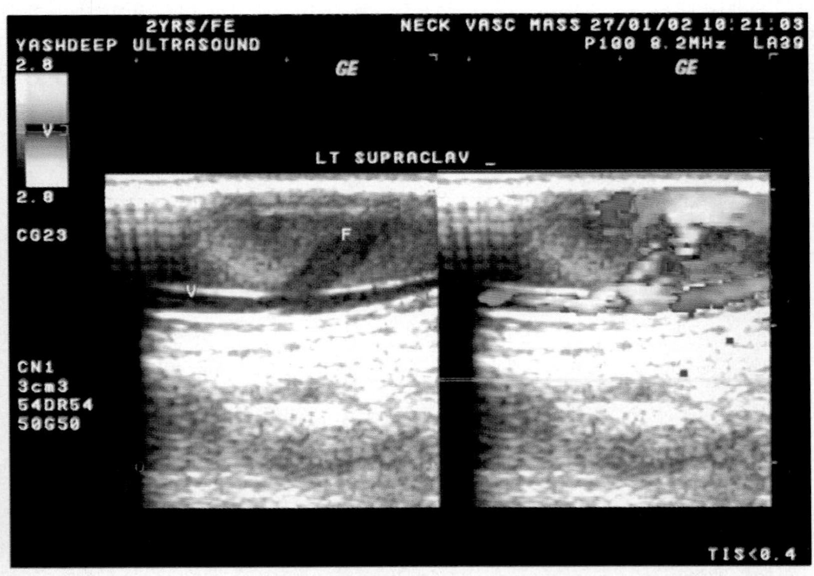

FIGURE 3.166

FIGURES 3.165 and 3.166: Bleeding vessel hematoma. A patient presented with increasing swelling in the Lt supraclavicular region. HRSG shows a bleeding vessel in the neck. Color flow shows blood coming out from the bleeding vessel.

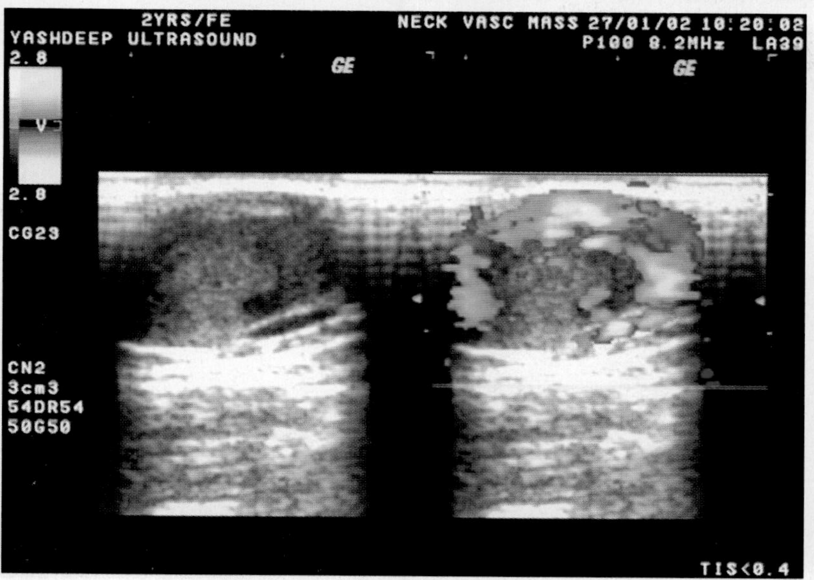

FIGURE 3.167

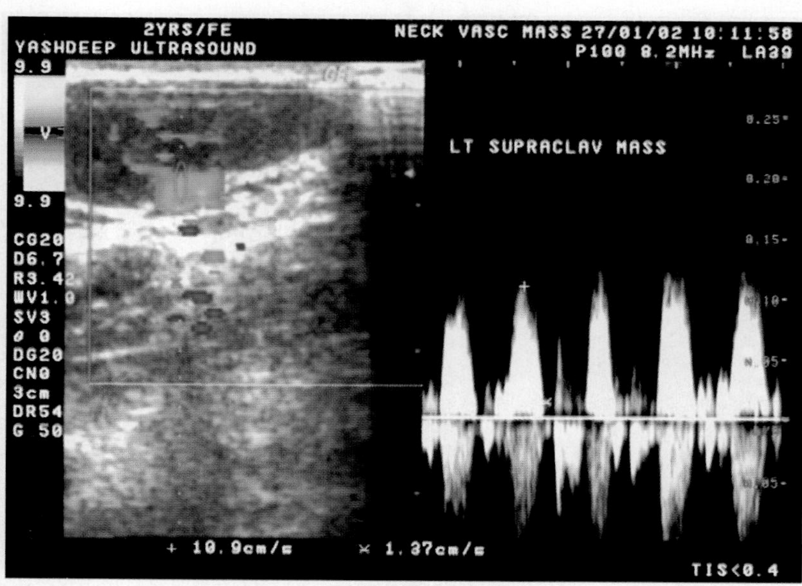

FIGURE 3.168

FIGURES 3.167 and 3.168: Color flow imaging and spectral Doppler imaging shows blood flow in the hematoma with evidence of high flow noted on spectral Doppler tracing.

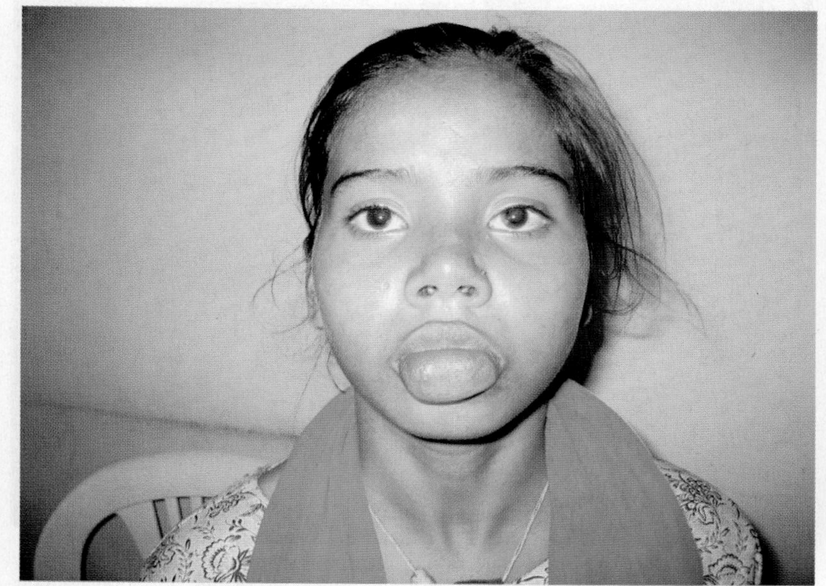

FIGURE 3.169: Traumatic AVM lower lip. A young girl presented with slowly increasing lower lip mass after sustaining trauma to the lip after a fall.

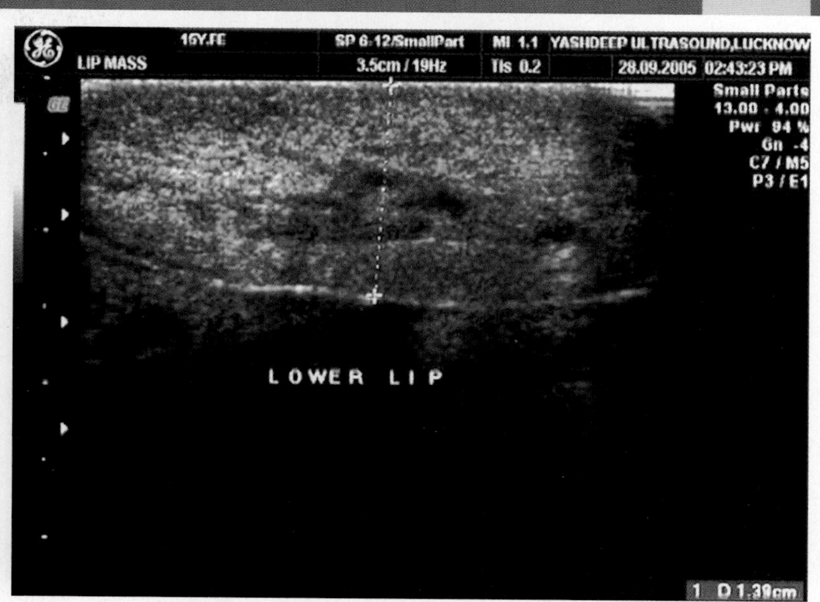

FIGURE 3.170: HRSG shows soft tissue mass over the lower lip multiple dilated vessels are seen in the mass.

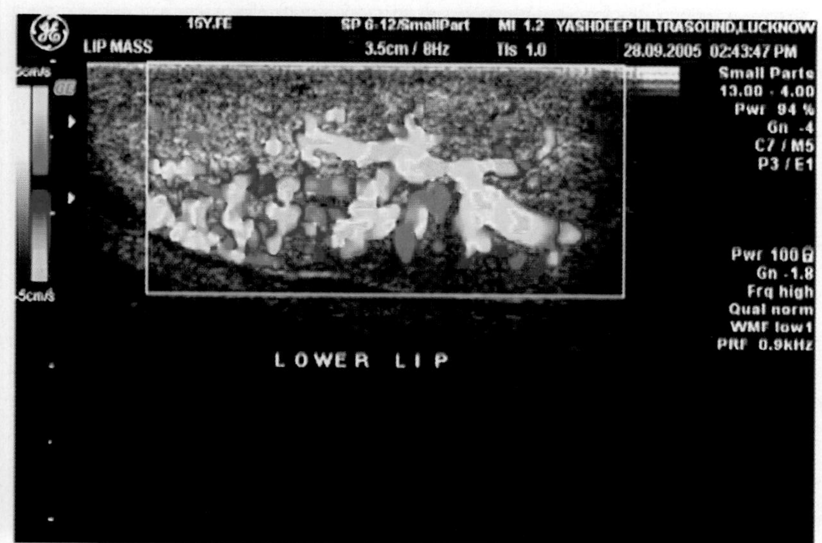

FIGURE 3.171

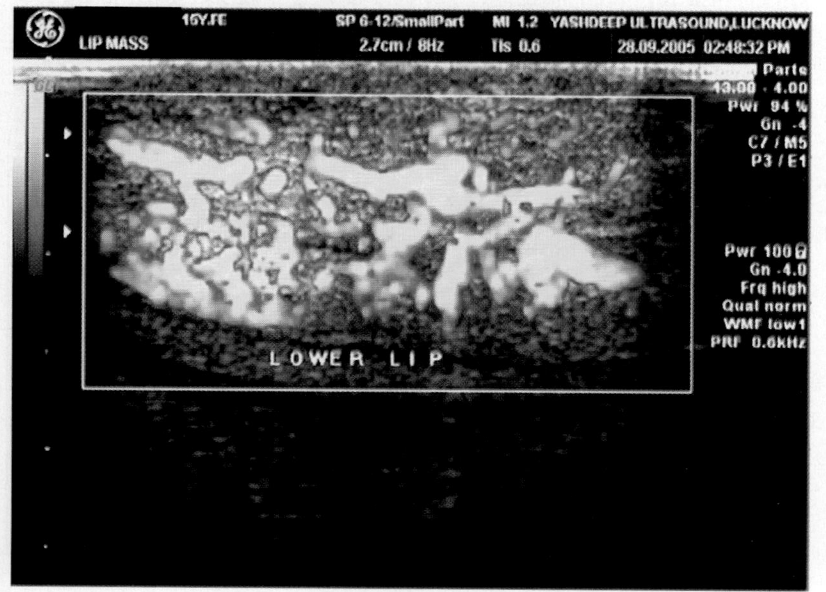

FIGURE 3.172

FIGURES 3.171 and 3.172: On color flow imaging multiple vessels are seen in the mass. Power Doppler also shows multiple vessels feeding the mass.

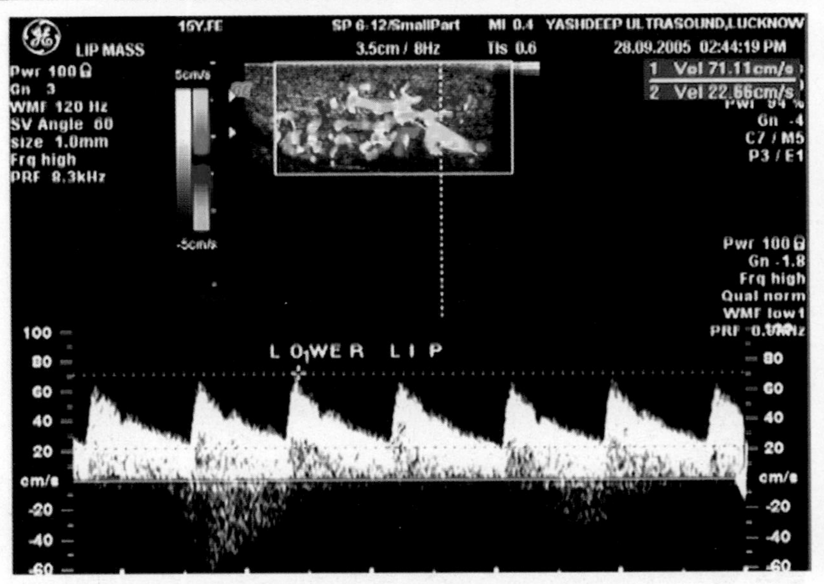

FIGURE 3.173: Spectral Doppler tracing shows predominantly arterial flow in the vessels. Sonographic findings were consistent with AVM.

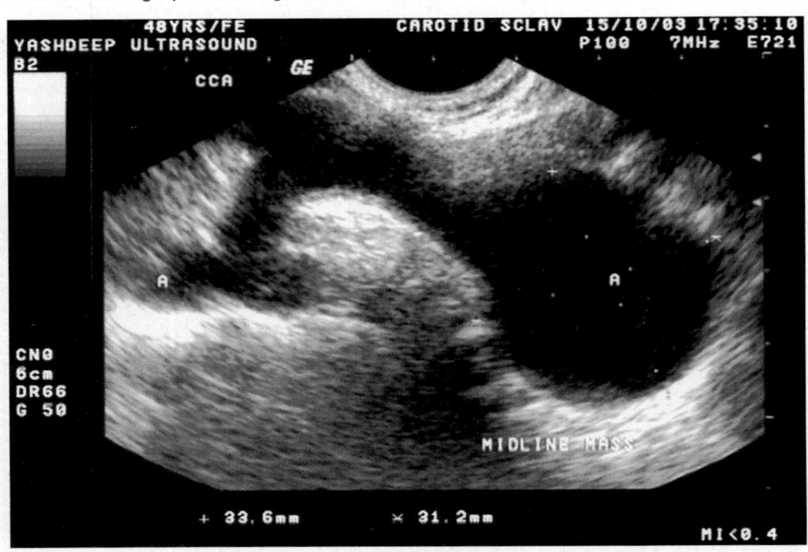

FIGURE 3.174

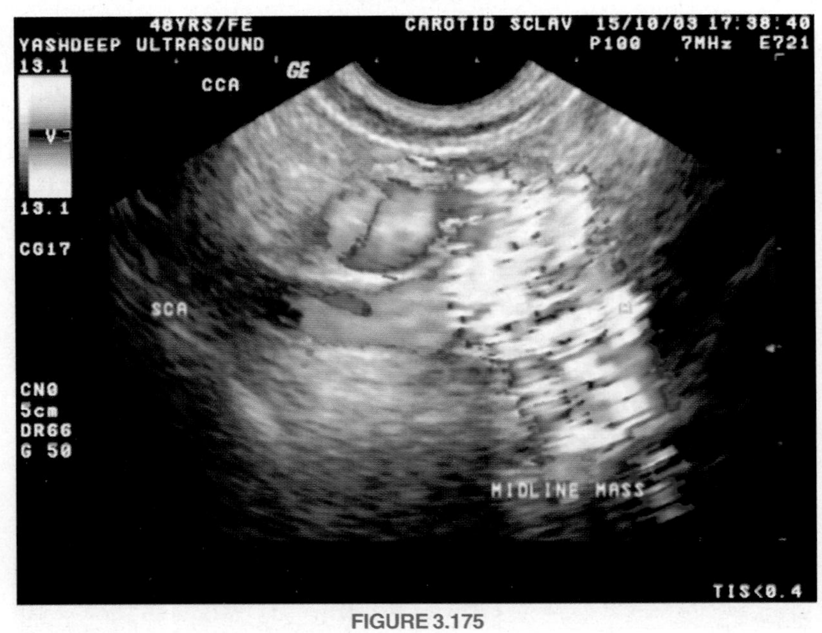

FIGURE 3.175

FIGURES 3.174 and 3.175: Aneurysm of Rt carotid artery. A patient presented with soft pulsatile swelling in the Rt supraclavicular area. Color Doppler imaging shows aneurysmal dilatation of proximal common carotid artery. The blood is seen flowing in the aneurysm in whirlpool fashion. No calcification is seen in the aneurysm.

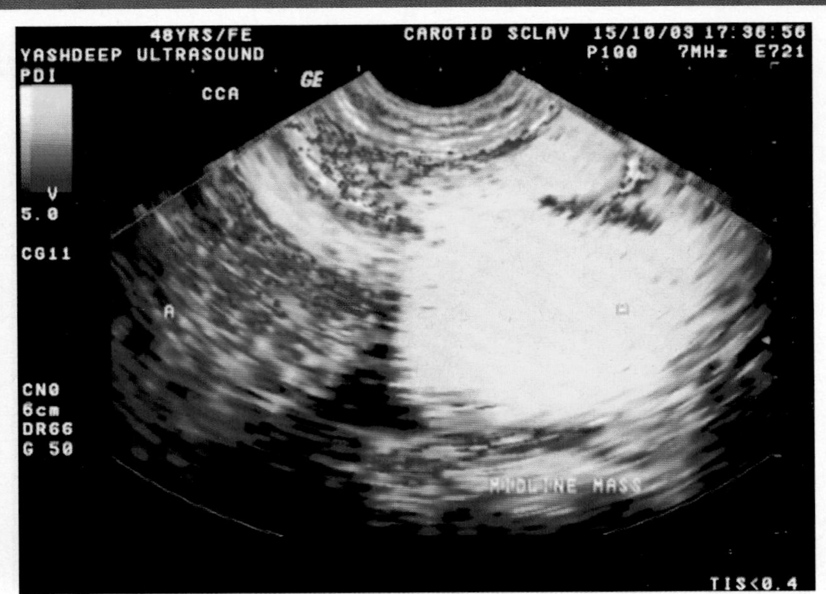

FIGURE 3.176

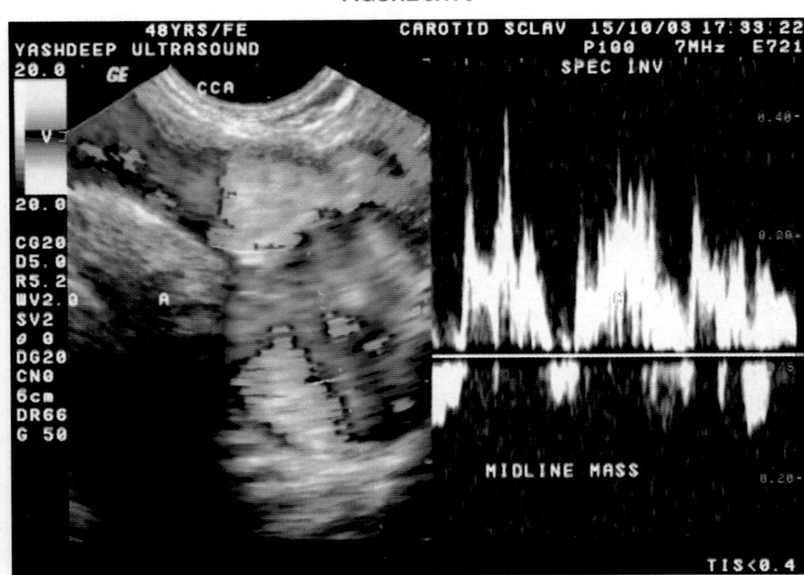

FIGURE 3.177

FIGURES 3.176 and 3.177: Power Doppler imaging shows pooling of blood in the aneurysm. Spectral Doppler shows bizarre flow pattern in the aneurysm.

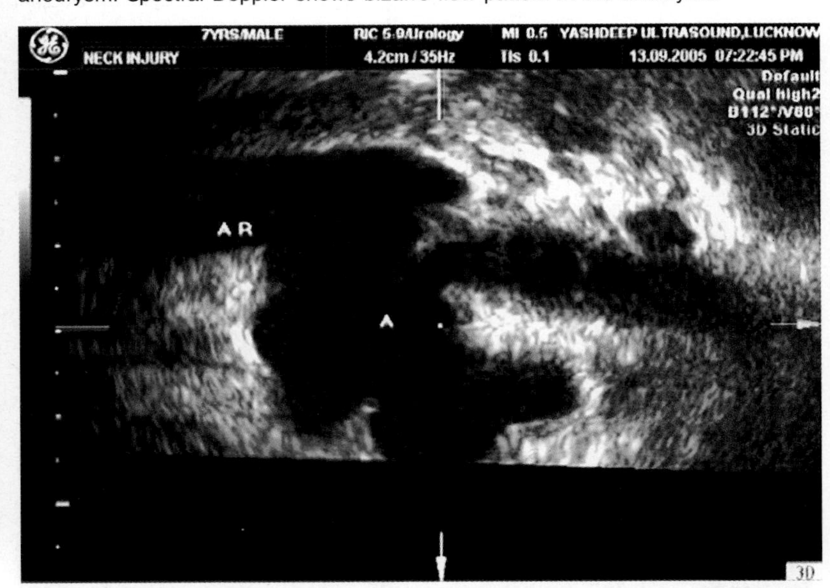

FIGURES 3.178: 3D traumatic aneurysm Lt brachiocephalic trunk. A young boy sustained sharp knife injury in the Lt side of the neck. No carotid artery pulsation is seen on the Lt side. Soft tissue mass was seen in the neck. HRSG shows traumatic aneurysm of Lt brachiocephalic trunk.3D imaging shows details of the aneurysm.

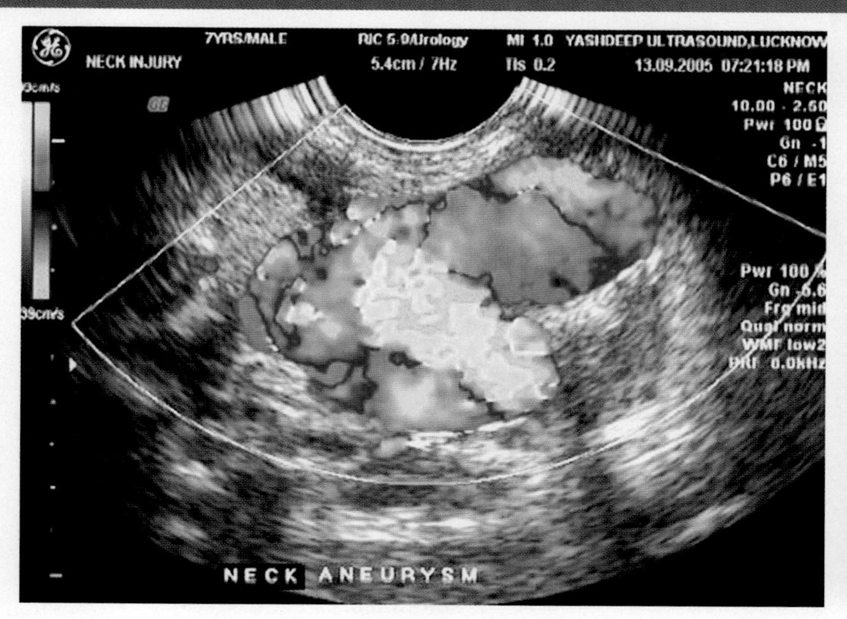

FIGURE 3.179

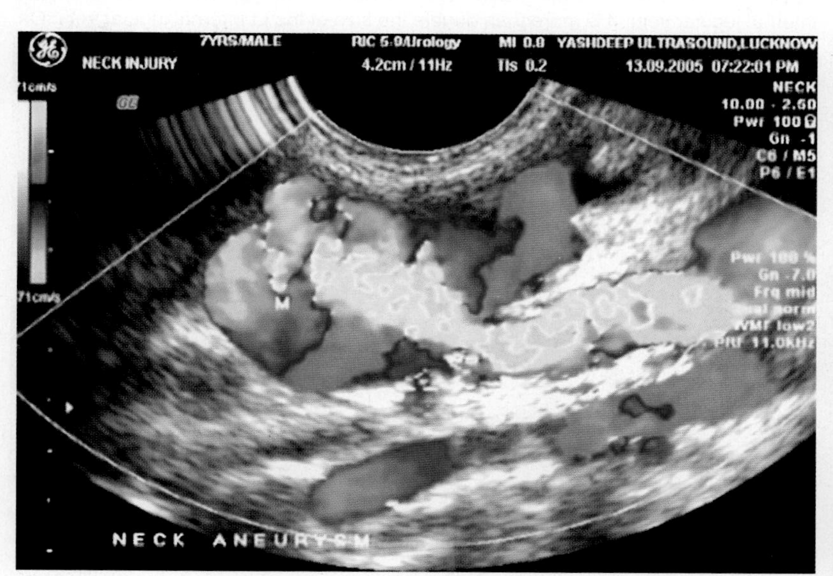

FIGURE 3.180

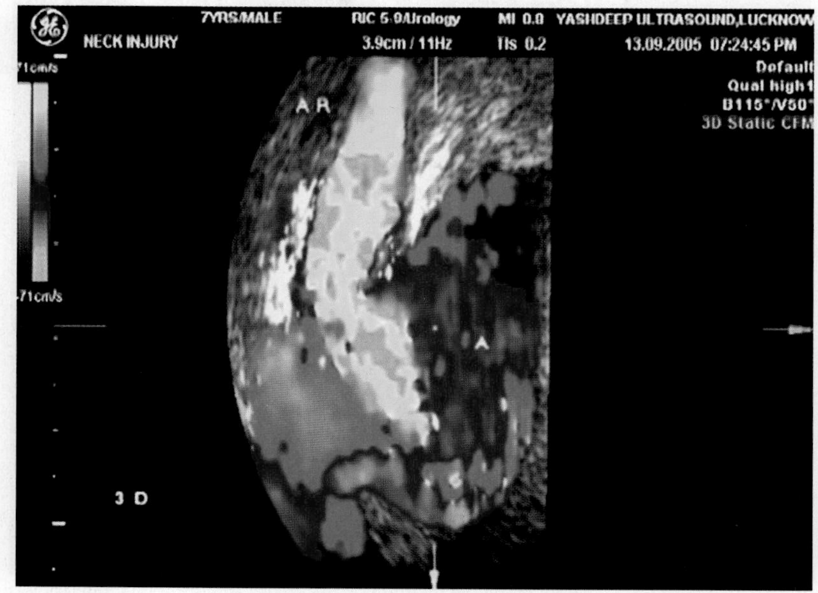

FIGURE 3.181

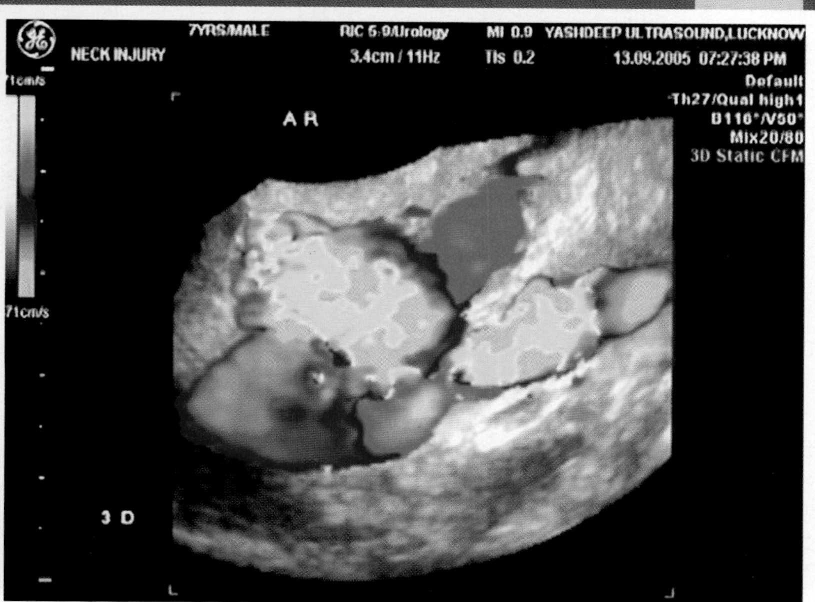

FIGURE 3.182

FIGURES 3.179 to 3.182: 3D imaging of the aneurysm with color flow imaging shows filling of the aneurysm with the blood with bizarre fashion. The proximal brachiocephalic trunk is seen feeding the aneurysm.

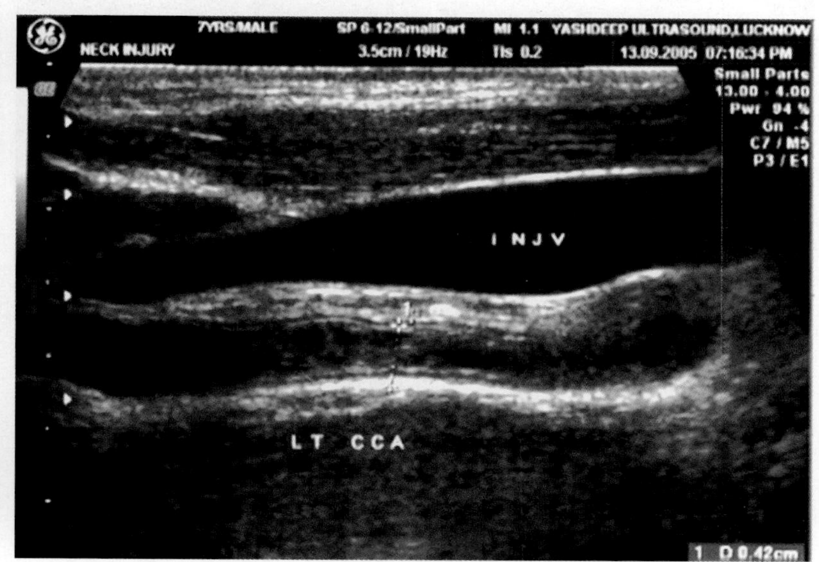

FIGURE 3.183

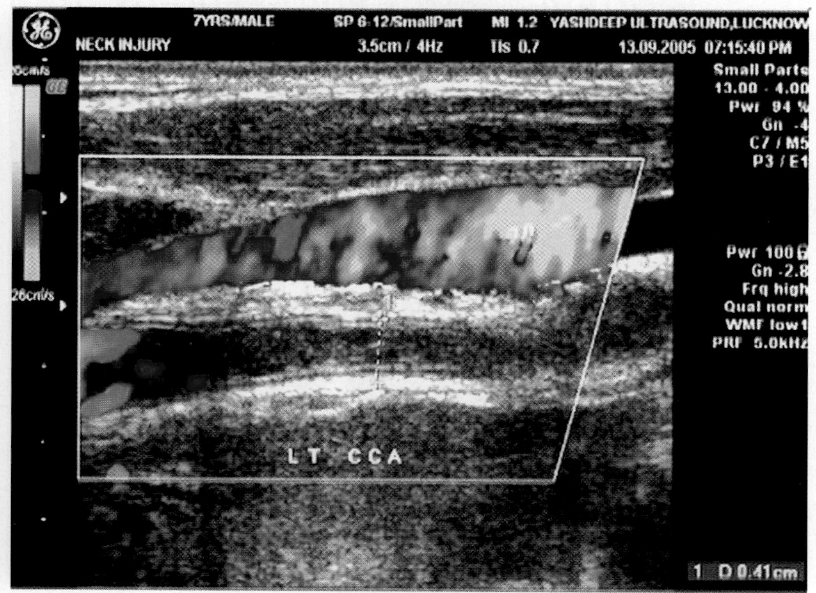

FIGURE 3.184

FIGURES 3.183 and 3.184: The Lt carotid artery is thickened. No flow is seen in the Lt carotid artery. However, the flow is seen in internal jugular vein.

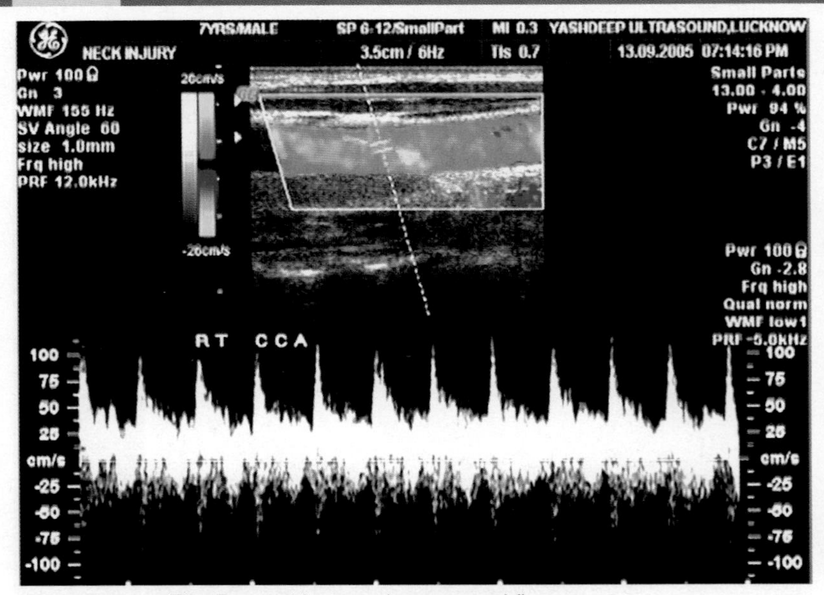

FIGURE 3.185: The Rt carotid artery shows normal flow.

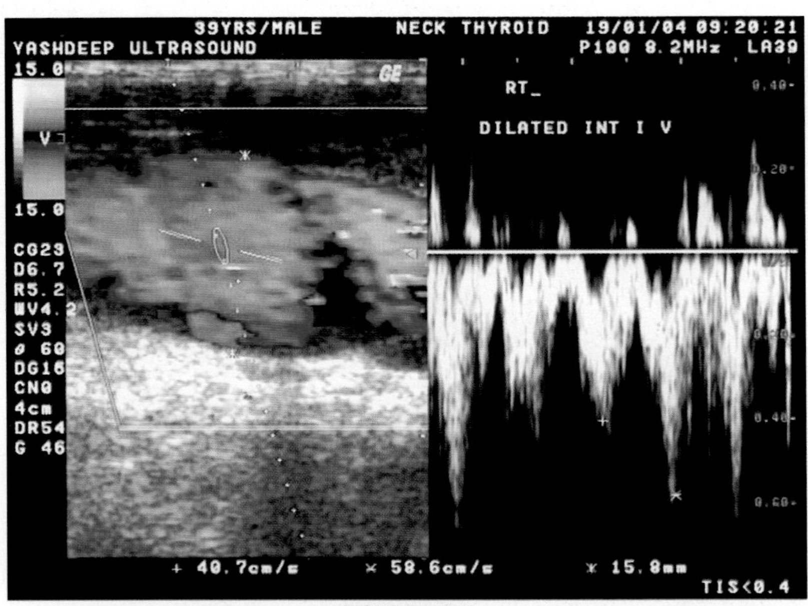

FIGURE 3.186

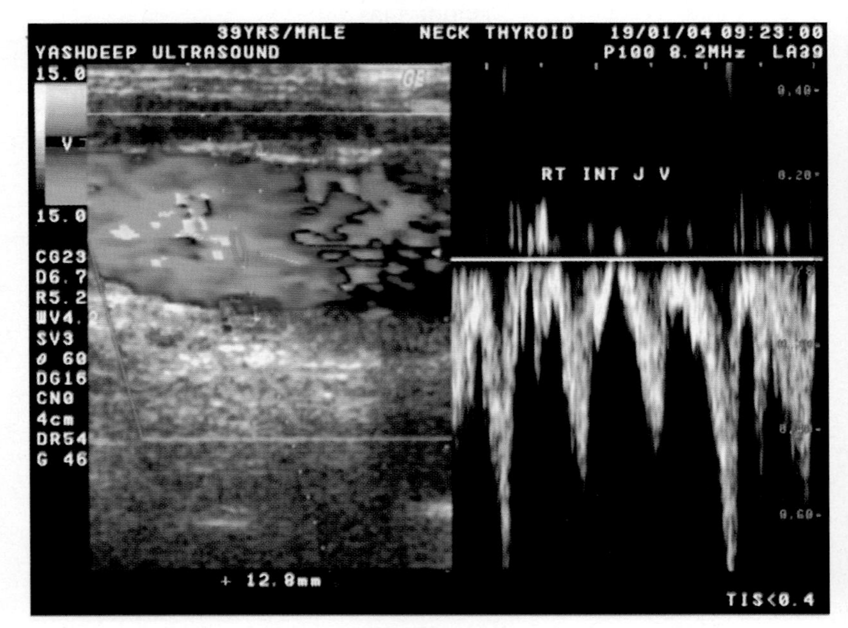

FIGURE 3.187

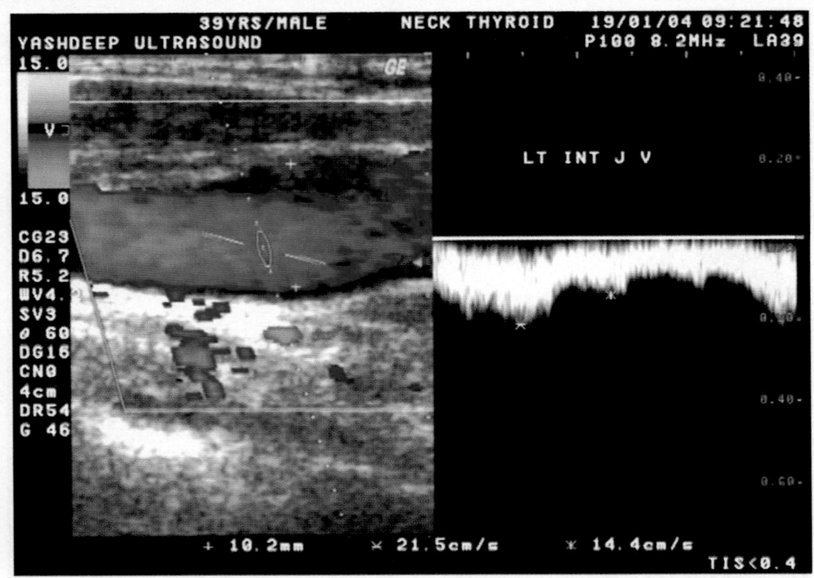

FIGURE 3.188

FIGURES 3.186 to 3.188: Dilated Rt internal jugular vein with pulsatile arterial flow. A patient presented with pulsation in the neck on Rt side. HRSG shows dilated internal jugular vein. It is more than double the size of the Lt internal jugular vein. On spectral Doppler tracing high pulsatile flow is seen in the Rt IJV. The Lt IJV vein shows normal monophasic flow pattern.

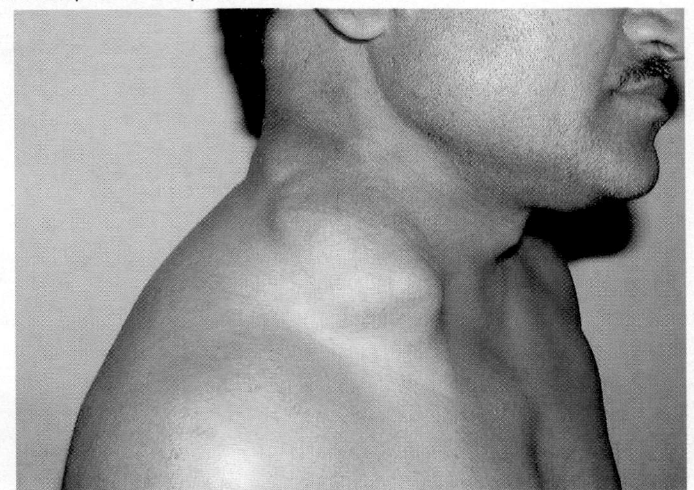

FIGURE 3.189: Localized AVM Rt supraclavicular region. A patient presented with cystic mass in the Rt supraclavicular region.

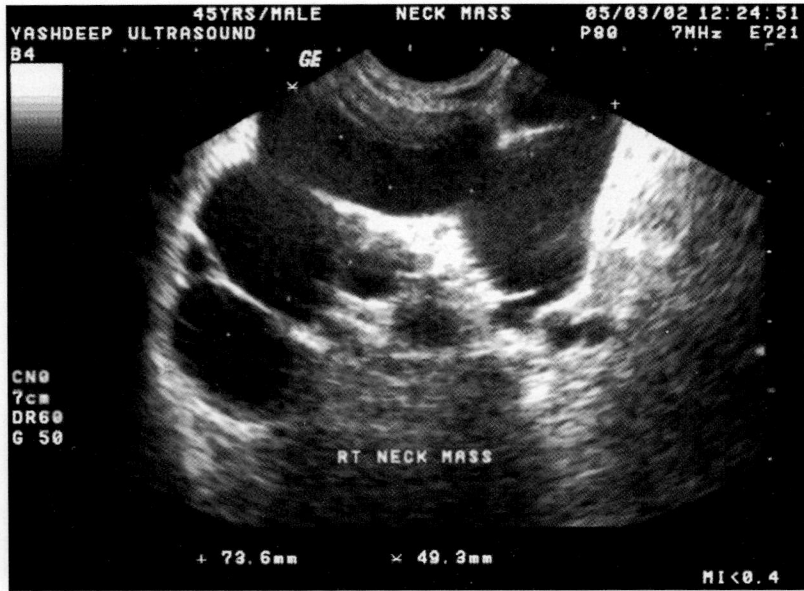

FIGURE 3.190: HRSG shows multiple dilated vessels in the Rt supraclavicular region and upper cervical region.

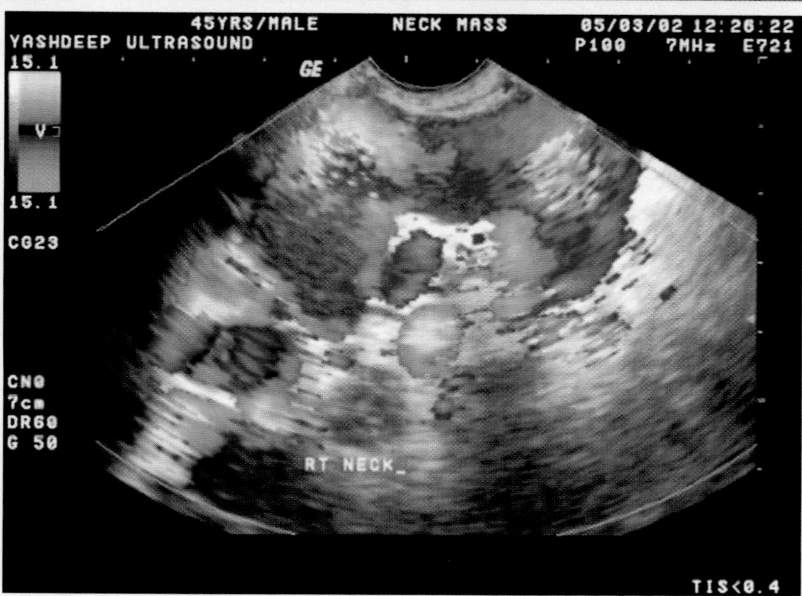

FIGURE 3.191

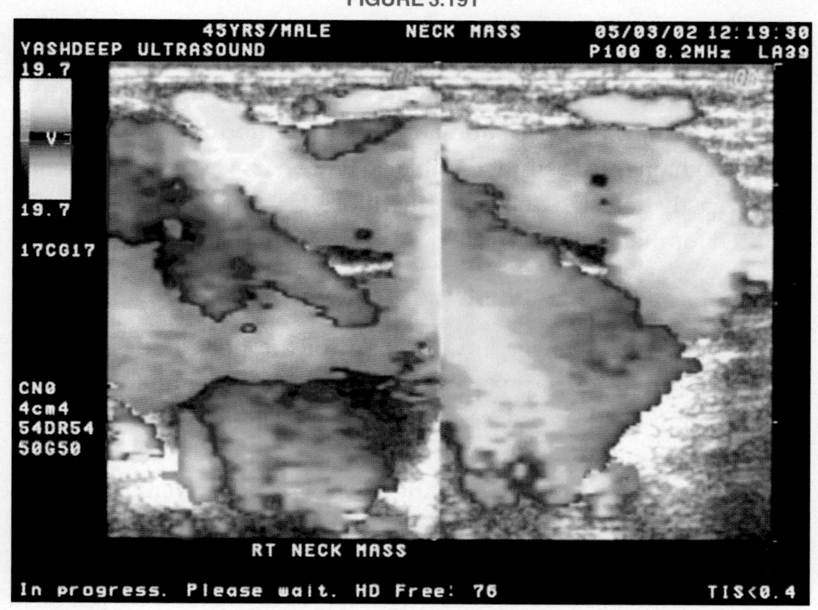

FIGURE 3.192

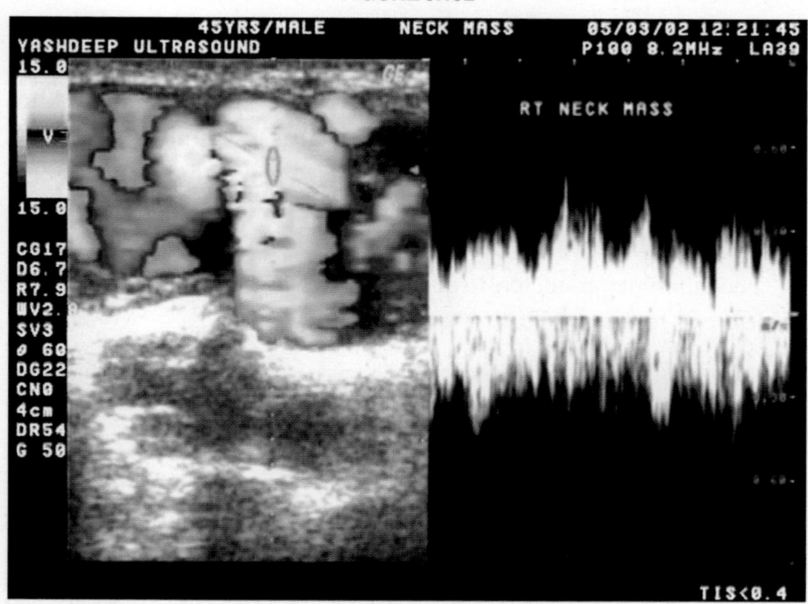

FIGURE 3.193

FIGURES 3.191 to 3.193: On color flow imaging high flow is seen in the vessels. On spectral Doppler tracing bizarre flow pattern seen in it suggestive of AVM.

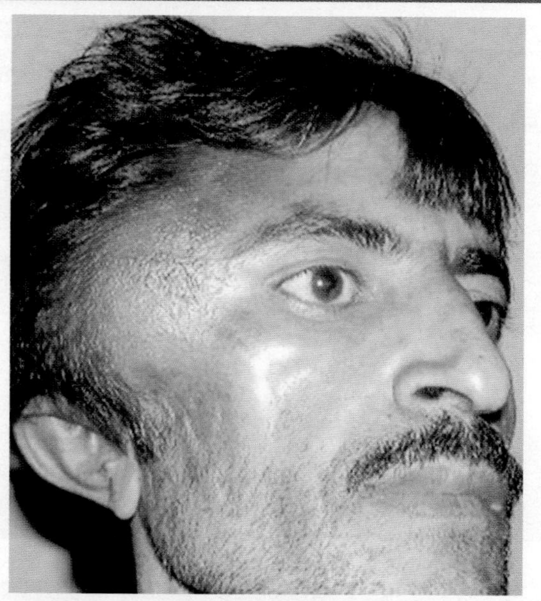

FIGURE 3.194: AVM Rt temporal region. A patient presented with increasing swelling over the Rt temporal region.

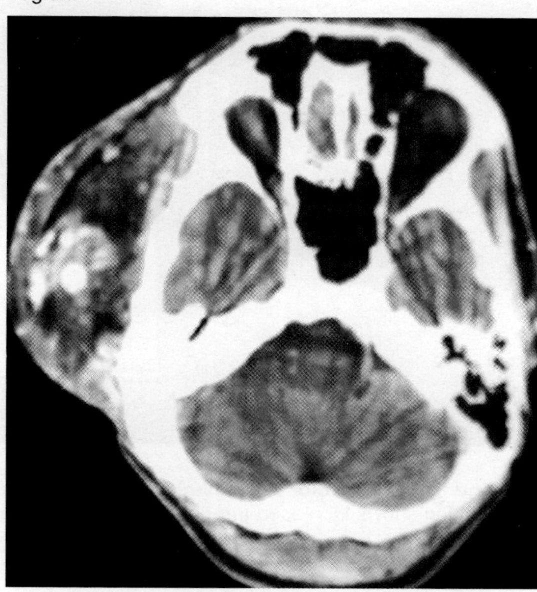

FIGURE 3.195: CT scan of the same patient shows enhancing soft tissue mass in the Rt temporal region.

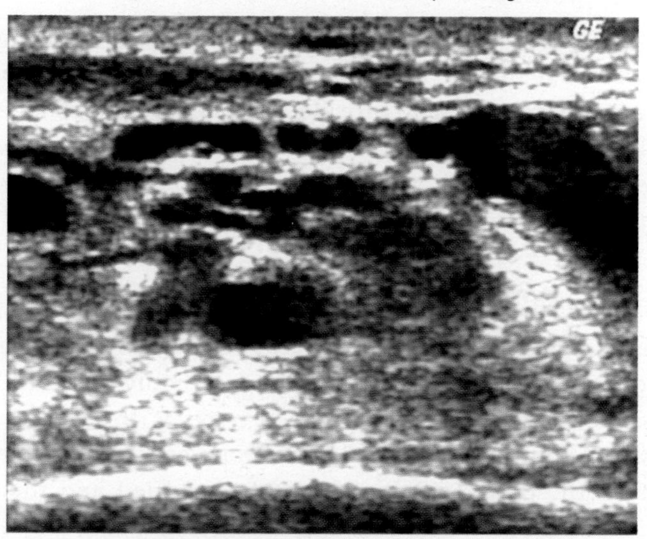

FIGURE 3.196: HRSG shows multiple dilated vessels in the Rt temporal region.

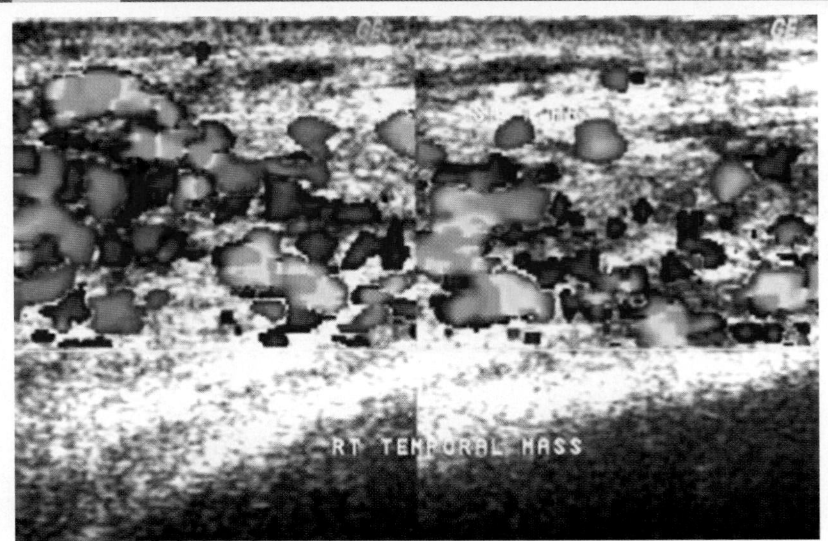

FIGURE 3.197

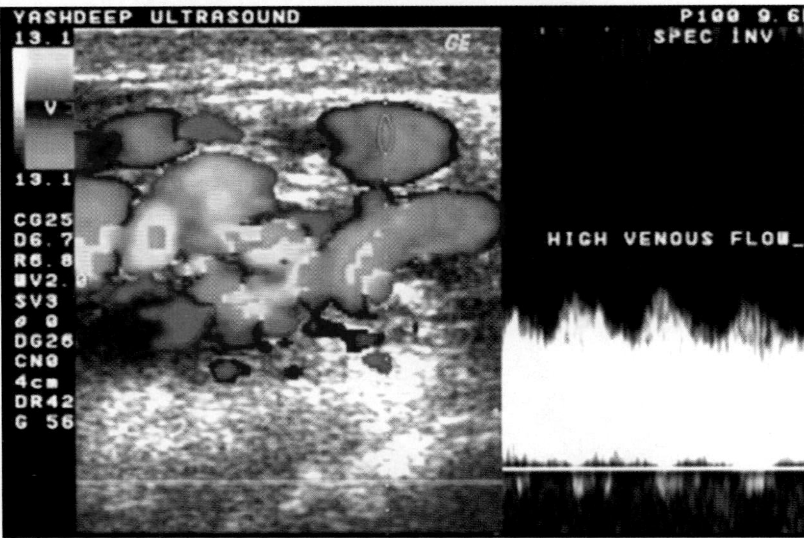

FIGURE 3.198

FIGURES 3.197 and 3.198: On color flow imaging high flow seen in the vessels. Spectral Doppler tracing shows increased arterial flow pattern suggestive of AVM of superficial temporal artery.

Hemangiomas

These are soft, cystic and mobile mass present superficially or deep in the soft tissue planes. They may be pulsatile. Ultrasonogram shows low amplitude mass with jelly like consistency. Fine echogenic septa are seen in the mass. Echogenic hemorrhagic collection is seen when there is secondary Hemorrhage in the hemangioma. Color flow imaging is an added advantage in hemangioma.

Aneurysmal Dilatation

Aneurysmal dilatation of great vessels may bulge in the lower cervical region on either side or in the midline. It is highly pulsatile throbbing mass. It is seen as a well-defined mass with anechoic lumen. Ultrasonogram shows blood flow on Doppler color flow imaging and confirm the diagnosis. Walls may be calcified and thickened in long standing aneurysm. Echogenic thrombus may be seen in the aneurysm.

Neck Malignant Masses and Lymph Nodes

Lymph Node Mass

Normally lymph nodes of the neck are not resolved on ultrasonographic examination because the reflectivity of lymph nodes is similar to the subcutaneous fat. But when they are enlarged or involved in a disease process, they become hypoechoic and enlarged, therefore they can be very well picked up on HRSG. Seven main chains of lymph nodes are present in the neck. They are as follows:

1. Internal jugular
2. Anterior jugular
3. Parotid chain
4. Submandibular chain
5. Anterior juxtavisceral
6. Retropharyngeal
7. Spinal accessory chain.

From clinical point of view, internal jugular, anterior jugular, parotid and submandibular chains are important. They are further grouped as superficial and deep chains. For convenience, they are further grouped as low cervical, mid cervical and high cervical groups. Internal jugular, anterior jugular, parotid, submandibular, jugulodigastric and post-auricular chains are easily accessible on HRSG. However, other chains are not easily detected on HRSG, therefore, CT of MRI examination is required for their evaluation.

Clinical Assessment

In head and neck cancer, involvement of the neck nodes decides the staging of the disease, which is very important for further management of the disease. Clinical assessment of lymph node metastasis becomes difficult when the lymph nodes are small in size and non-palpable. When they are less than 1 cm in size or when they are deeply present as posterior to sternocleidomastoid muscles they are Nonpalpable. Clinical assessment is also difficult in post-surgical or post-irradiated neck due to extensive fibrosis or in a large head and neck cancer mass. Low accuracy of clinical assessment of disease in head and neck cancer is estimated up to 30%, which is very high in the group. Staging, management and treatment of disease become controversial. Therefore, different imaging modalities were tried to improve the results of clinical assessment like CT, MRI and HRSG. Among all of them, HRSG has been found to be highest sensitive. It can pick up nodes as small as 2 mm in size in superficial group.

The different parameters to distinguish reactive node from metastatic node are as follows:

Lymph node size: It is seen that normal lymph node has an axial diameter between 2 to 5 mm. In reactive enlargement of lymph node as in tonsil infections, the jugulodigastric nodes become enlarged. The size increases more than 2 cm in length and 1 cm in thickness. However, in metastatic lymph node enlargement, the size does not increase more than 2 cm and it remains under 1 cm in length and in thickness. But lymph node size and axial diameter has got limited value in the differential diagnosis of metastatic lymph node enlargement.

Lymph node shape: It has been observed that metastatic nodes have a tendency to be of round shape. Reactive lymph node enlargement appears to be more enlarged. Sakai et al have defined the numerical index for lymph node shape. It is known as rounded index (RI). The longitudinal diameter is divided by axial diameter. It is seen that metastatic nodes have a RI of less than 1.5 cm, whereas reactive lymph nodes have RI of more than 2 cm.

Echo pattern: Normally the lymph nodes are hypoechoic than the adjacent thyroid parenchyma. However, no specific echo pattern is reported for lymph node pathology. The metastatic nodes are less hypoechoic. When they are small, they have sharp margins. However, when they are large, they are heterogeneous in texture and show tumor necrosis. Micro-calcification may be found in the metastatic nodes. Cystic changes are also found in them mainly in papillary carcinoma of thyroid.

Reactive nodes are homogeneous in texture. They have sharp margins and regular smooth shape. Macrocalcification is seen in them mainly on the granulomatous disease.

Lymph node hilum: In reactive lymph node enlargement, there is a highly reflective band seen in the central part of node or in lymph node sinus. It is seen running through the substance of node. It is considered to be a sign of non-malignancy. Presence of lymph node hilum combined with RI increases the significance of this sign for the detection of metastatic lymph node disease. It is seen that presence of lymph node hilum with RI of more than 2, the chances of reactive lymph node enlargement are high.

Extra Nodal Tumor Extension

Extra nodal tumor extension is a poor prognostic sign for patient. HRSG is very sensitive tool to assess extra nodal tumor spread to muscles, jugular vein or carotid artery.

Post-radiotherapy Evaluation

Clinical assessment is difficult in post-irradiated neck due to extensive fibrosis. HRSG is very sensitive to detect presence of recurrence of disease in post-irradiated neck. Nodes become well-defined, hypoechoic and homogeneous in texture after irradiation. Nodal necrosis is also encountered. Recurrence of primary growth or disease can be very well assessed by HRSG.

Malignant Tumors

Malignant cervical tumors are very rare. They include lip sarcoma, malignant chemodectomas, malignant epithelioma and rhabdomyosarcoma. However, tumors of larynx and laryngopharynx when enlarge, they bulge as a mass in the neck. The extension of the growth can be assessed by HRSG. In carcinoma of larynx and laryngopharynx, associated lymph node enlargement and assessment of non-palpable lymph node can be done on HRSG. Cervical lymphomas are also very common and big lobulated masses are seen in the neck with heterogeneous texture. Retrosternal extension of the mass can be assessed from the suprasternal notch. Echo texture, lymph node enlargement and vessel involvement can be assessed on HRSG.

HRSG is also a good tool to assess the treatment response and recurrence of growth in carcinoma of larynx or laryngopharynx. However, no specific sonographic feature is found in cervical malignancy. But the signs are suggestive of malignancy found elsewhere in body like irregular margins, heterogeneous texture, lobulated pattern and surrounding tissue invasion and infiltration and vessel involvements are the features that suggest the malignant nature of the mass. Biggest advantage of HRSG is to do the guided fine-needle aspiration biopsy of the tumor for histopathological examination to confirm the diagnosis.

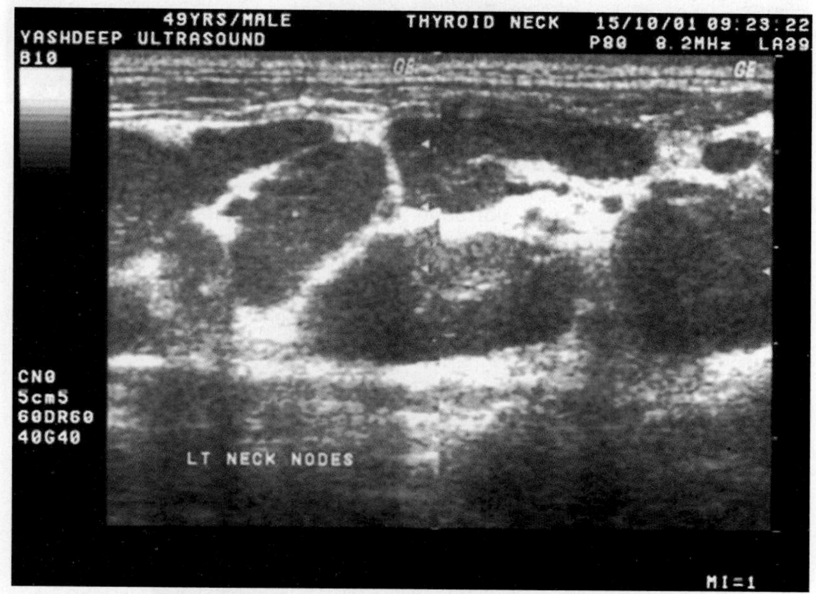

FIGURE 3.199: Malignant lymph node enlargement. A middle aged man presented with a lymph node nodular mass in the Lt side of the neck. HRSG shows multiple nodular masses. They are irregular in outline. Hilum is destroyed. No necrosis is seen in them. Biopsy shows lymphoma.

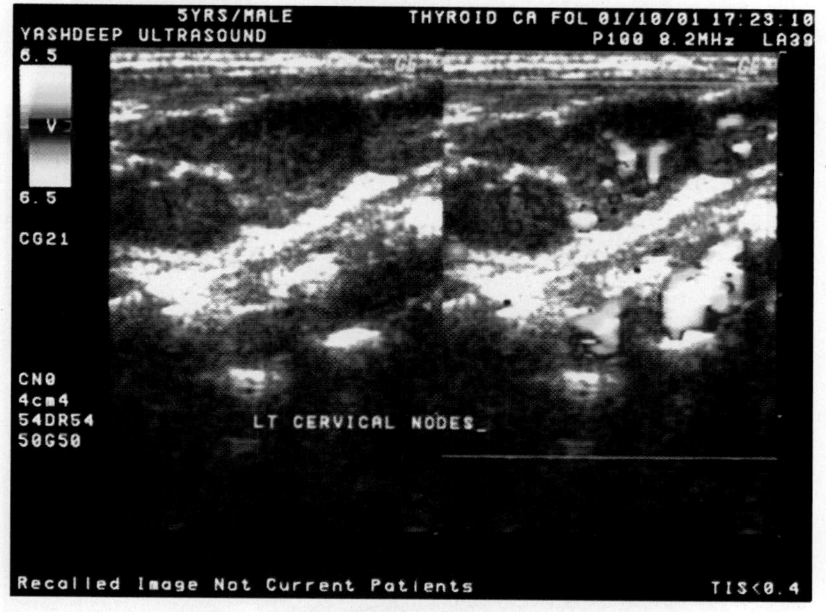

FIGURE 3.200

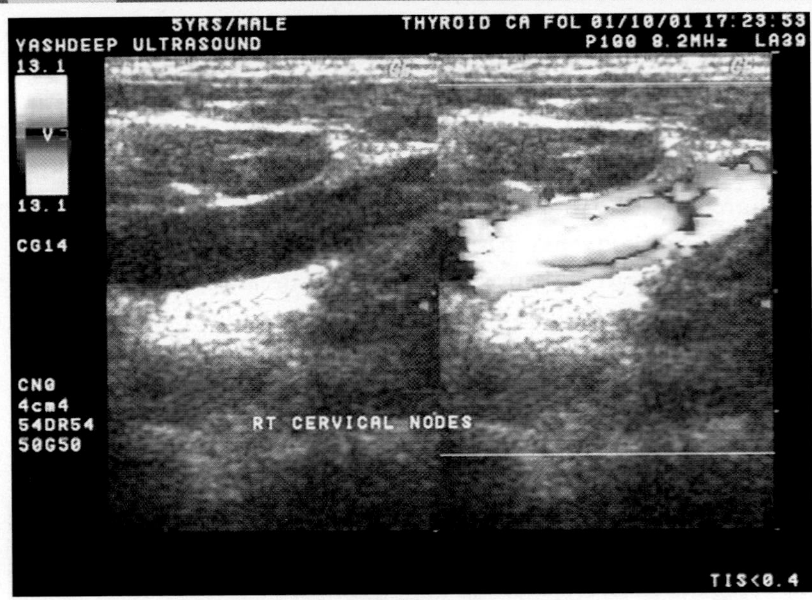

FIGURE 3.201

FIGURES 3.200 and 3.201: A young child presented with multiple hypoechoic enlarged nodular lymph node masses in both cervical areas. Normal nodal texture is lost. Hilum is destroyed. Necrosis was not seen. Biopsy shows metastatic lymph node enlargement from adenocarcinoma. The patient was having papillary adenocarcinoma of thyroid. Color Doppler imaging of the nodal mass shows moderate flow.

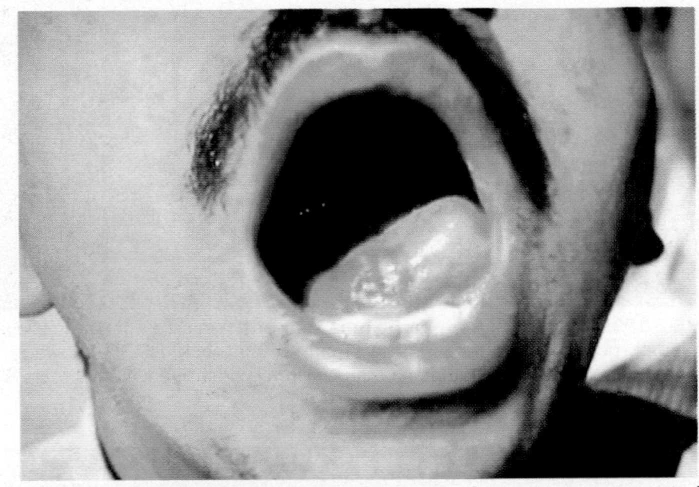

FIGURE 3.202: Carcinoma base of tongue. A young man presented with an ulcerative growth over the Rt side of the tongue.

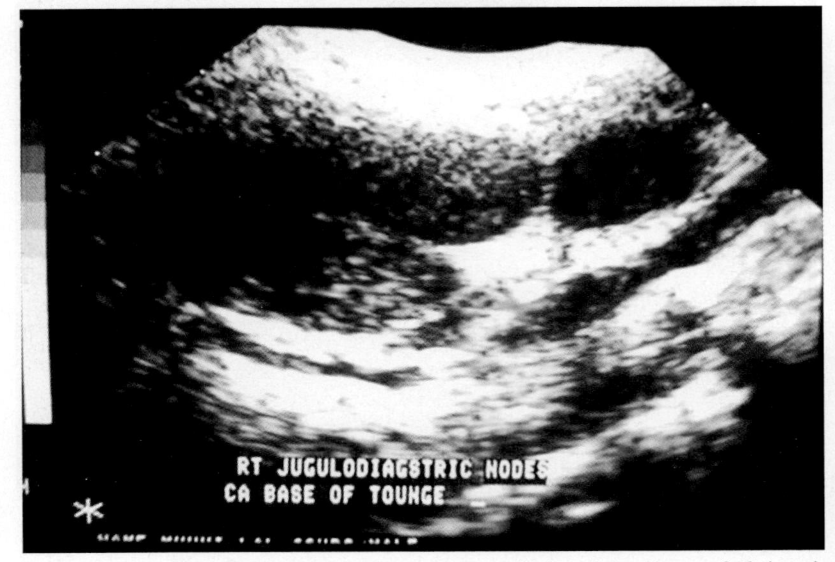

FIGURE 3.203: HRSG of the same patient shows multiple enlarged hypoechoic lymph node masses in jugulodiagastrics. Normal nodal borders are lost and the node forms a mass suggestive of metastatic deposits from Carcinoma base of tongue.

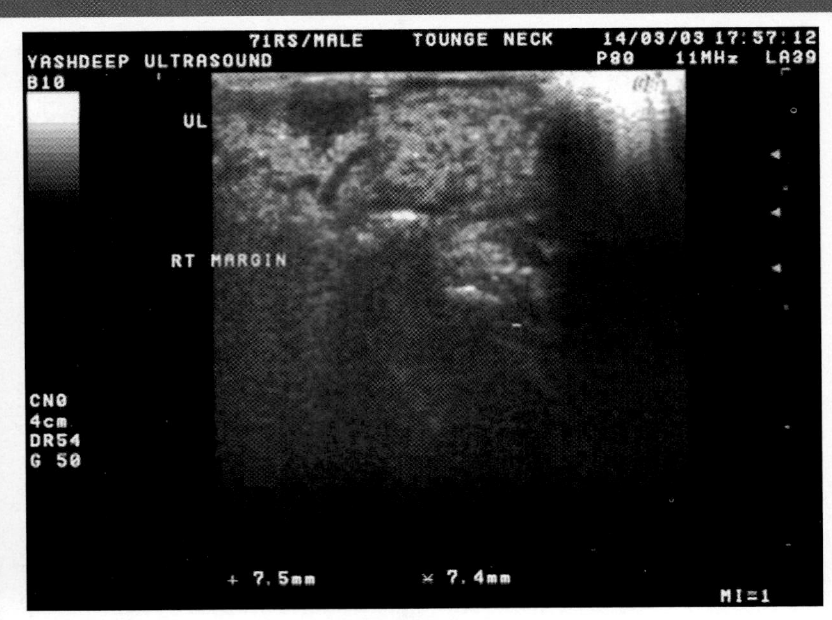

FIGURE 3.204

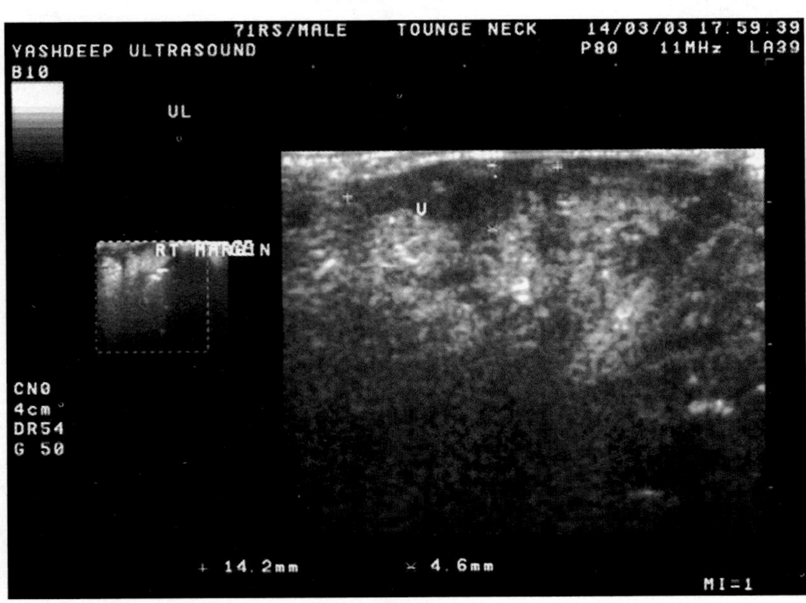

FIGURE 3.205

FIGURES 3.204 and 3.205: CA lateral border of tongue. An old patient presented with ulcerative growth over the lateral border of the tongue. HRSG shows soft tissue heterogeneous mass involving the Rt lateral border of the tongue. It is seen invading the deep layer of the tongue.

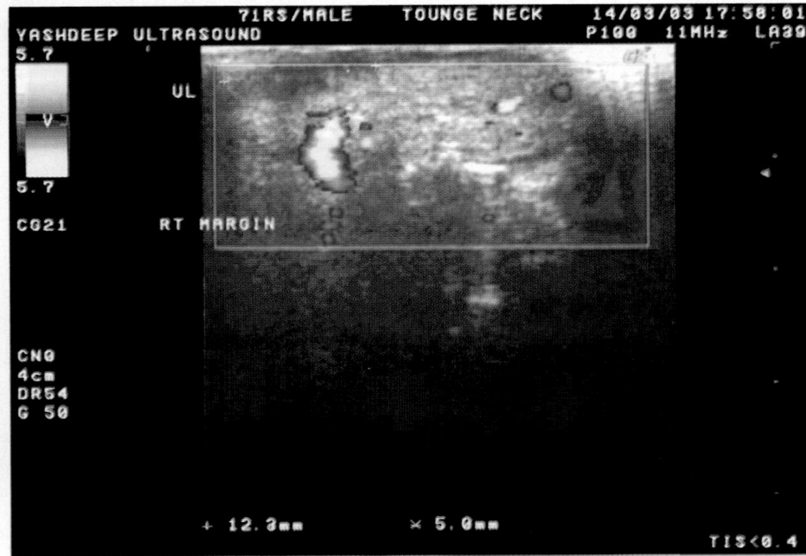

FIGURE 3.206: Color flow imaging shows multiple vessels feeding the mass.

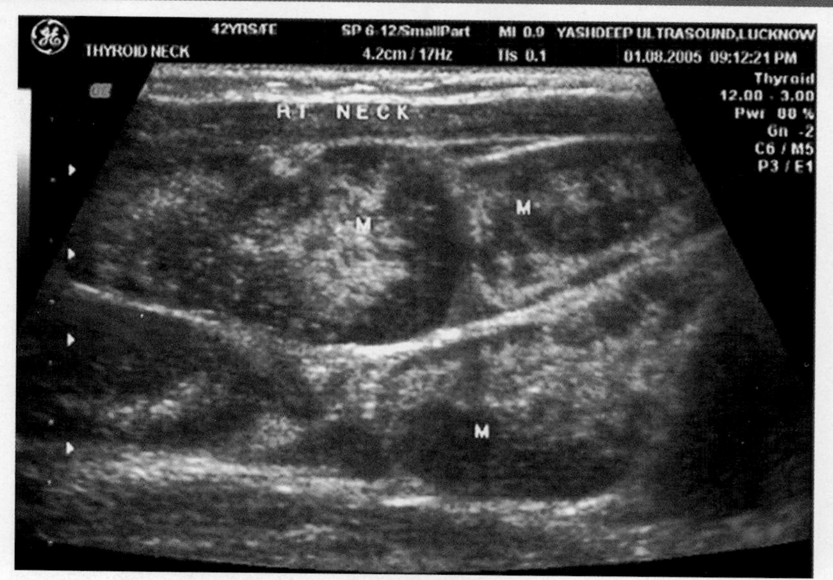

FIGURE 3.207

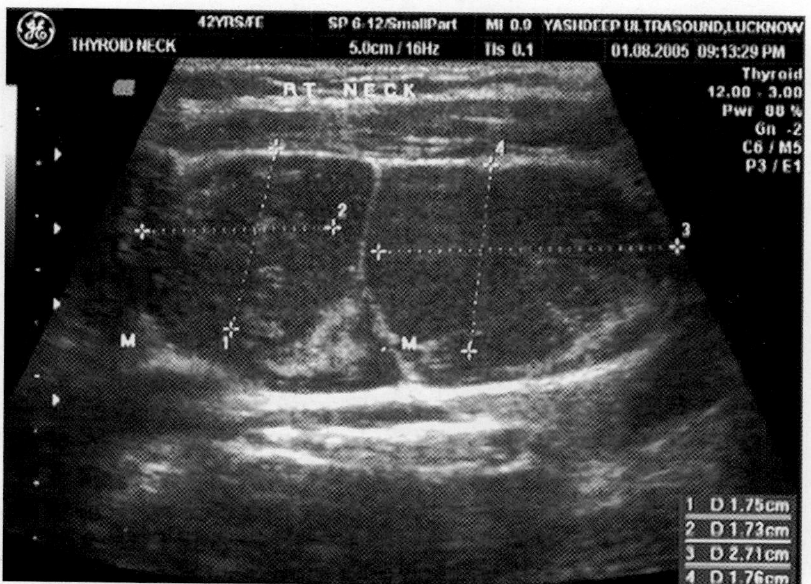

FIGURE 3.208

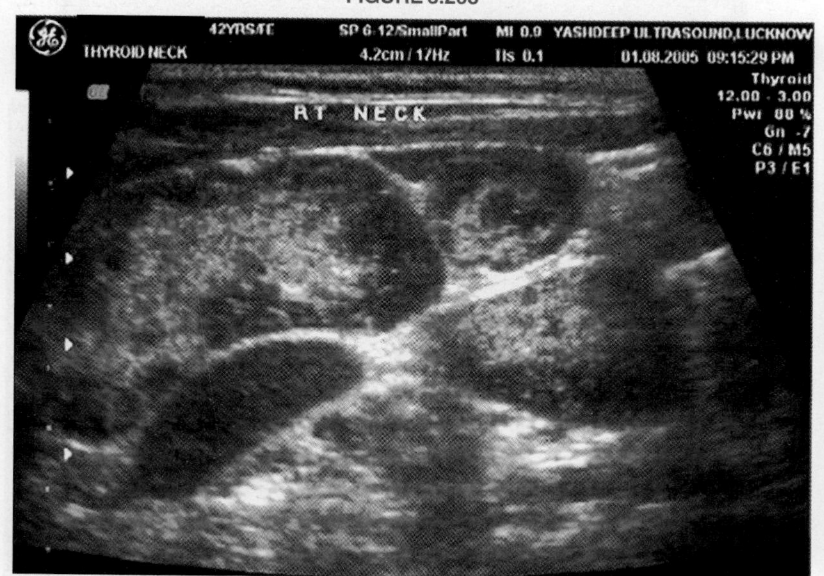

FIGURE 3.209

FIGURES 3.207 to 3.209: Malignant enlarged lymph nodes in the neck. An old lady presented with multiple enlarged nodes in the Rt side of the neck. HRSG shows heterogeneous nodal enlargements. The nodes have lost the normal texture. The lymph node hilum is distorted. The ovoid shape of the node is lost. But no necrosis is seen in them.

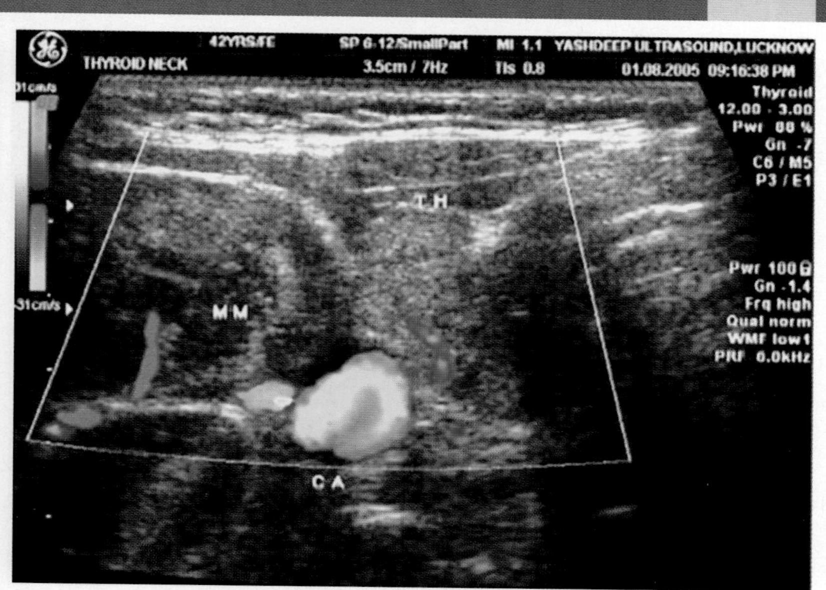

FIGURE 3.210: The color flow imaging shows the nodes are not invading the neck vessels. They are separate from the thyroid gland. Biopsy of the node shows metastatic nodal mass.

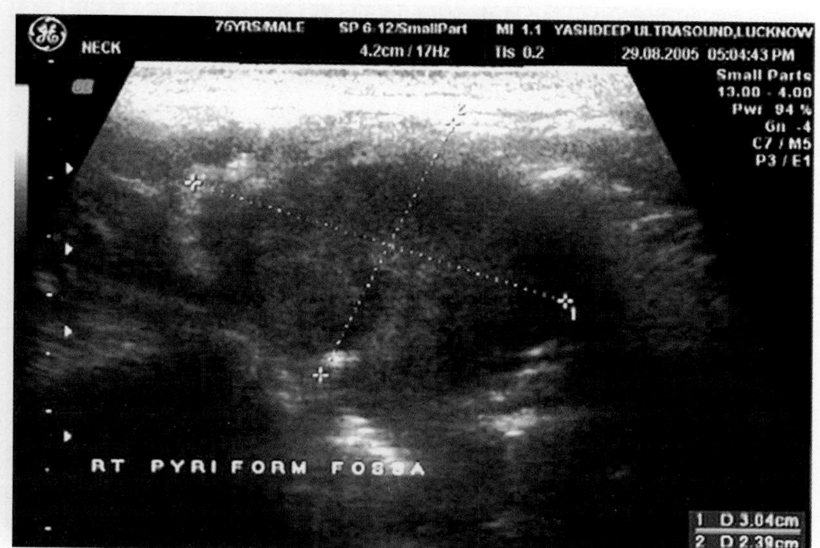

FIGURE 3.211

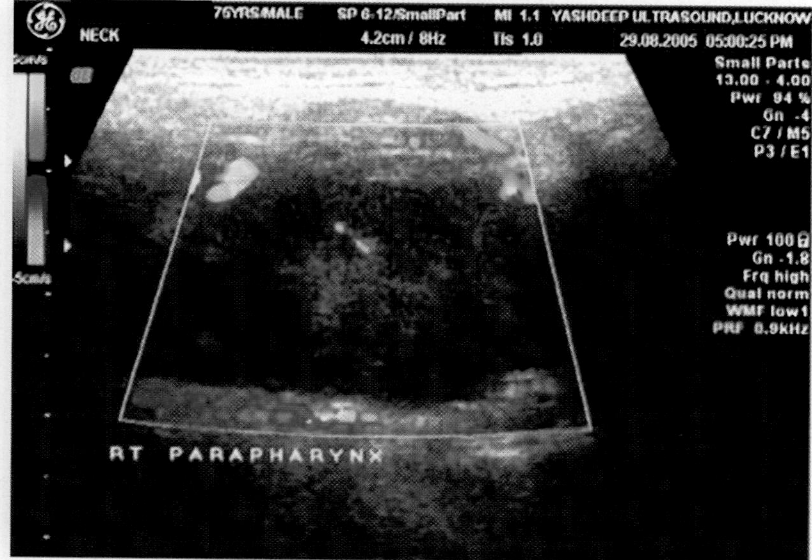

FIGURE 3.212

FIGURES 3.211 and 3.212: Parapharyngeal malignant mass. An old man presented with mass and fullness in the midline of the neck. HRSG shows a big heterogeneous hypoechoic mass in the parapharyngeal space. Biopsy of the mass showed malignant squamous cell carcinoma.

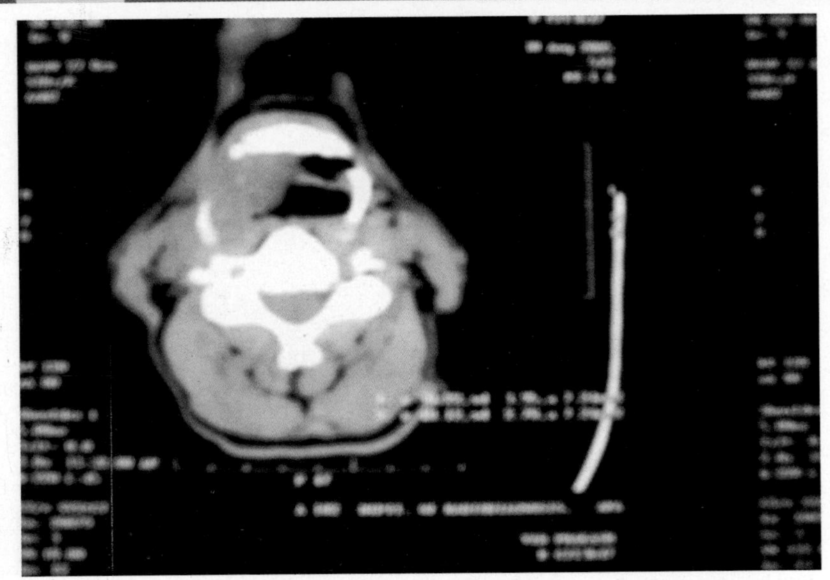

FIGURE 3.213: CT scan of the same patient confirmed the HRSG findings. The mass is seen invading Rt parapharyngeal space and also the pyriform fossa.

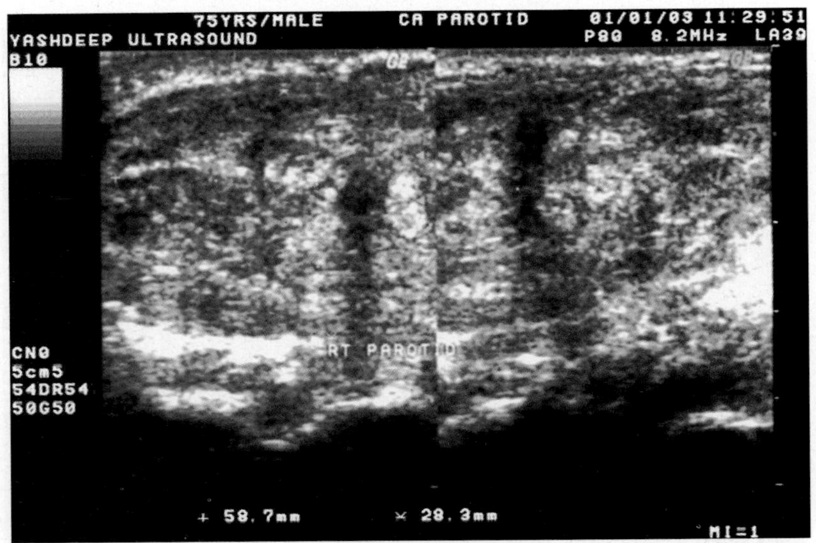

FIGURE 3.214

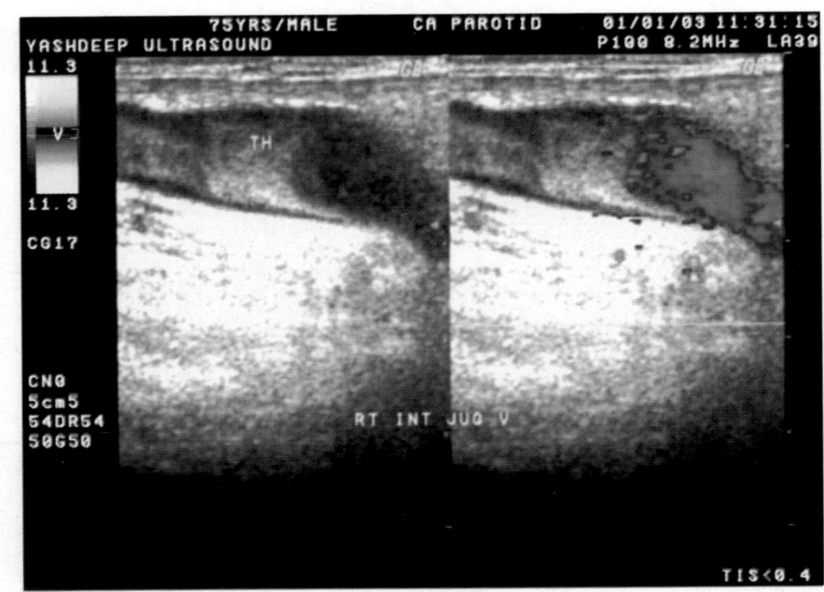

FIGURE 3.215

FIGURES 3.214 and 3.215: Malignant parotid mass. A patient presented with a big soft tissue mass in the upper cervical region. HRSG shows heterogeneous mass in the parotid region. Biopsy confirmed malignant mass. The mass was seen pressing over the internal jugular vein. An echogenic thrombus is seen filling the vein and obstructing the flow.

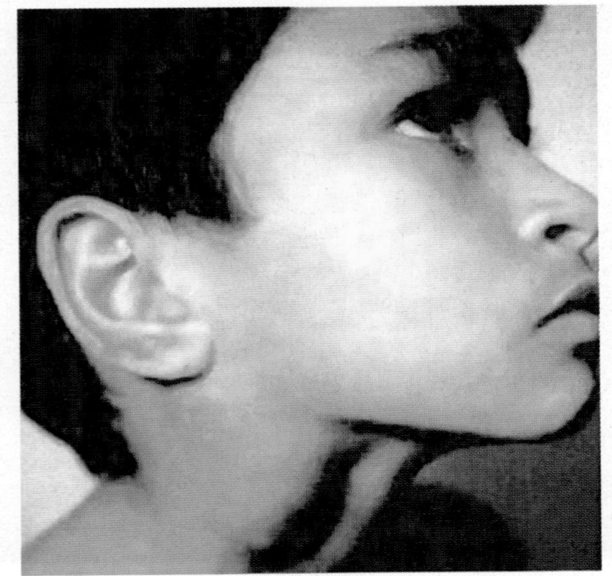

FIGURE 3.216: A young boy presented with a mass shadow in the Rt side of the neck in posterior triangle. The mass is seen pressing over the carotid artery. However, no flow obstruction is seen. It is predominantly homogeneous. Few calcified specks are seen in the mass. But no cystic necrosis is seen.

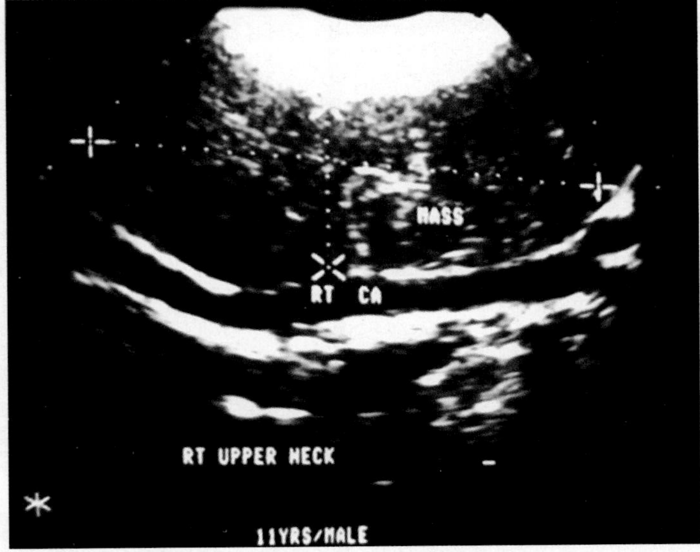

FIGURE 3.217: Biopsy shows primary muscle tumor.

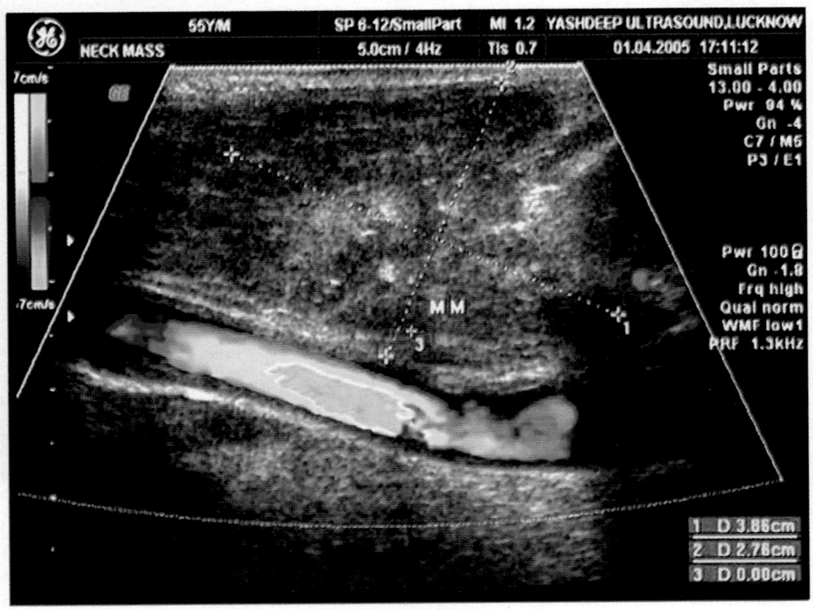

FIGURE 3.218

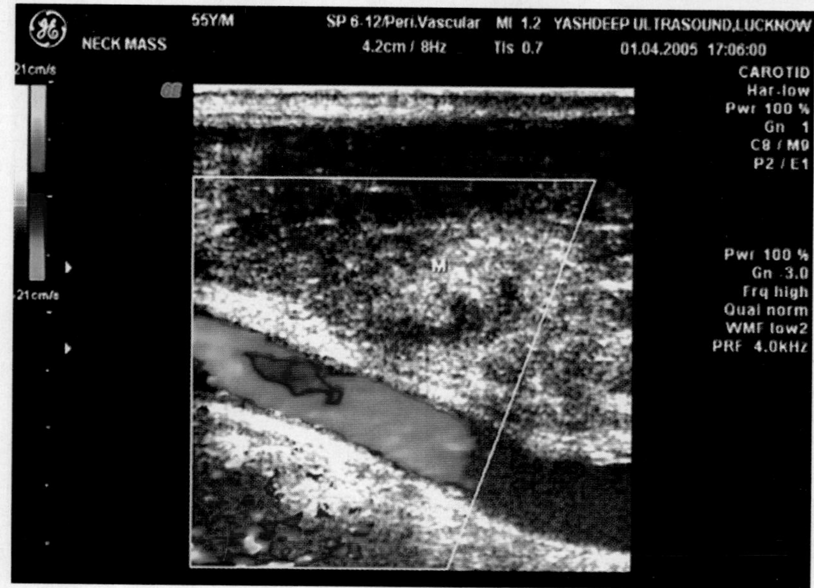

FIGURE 3.219

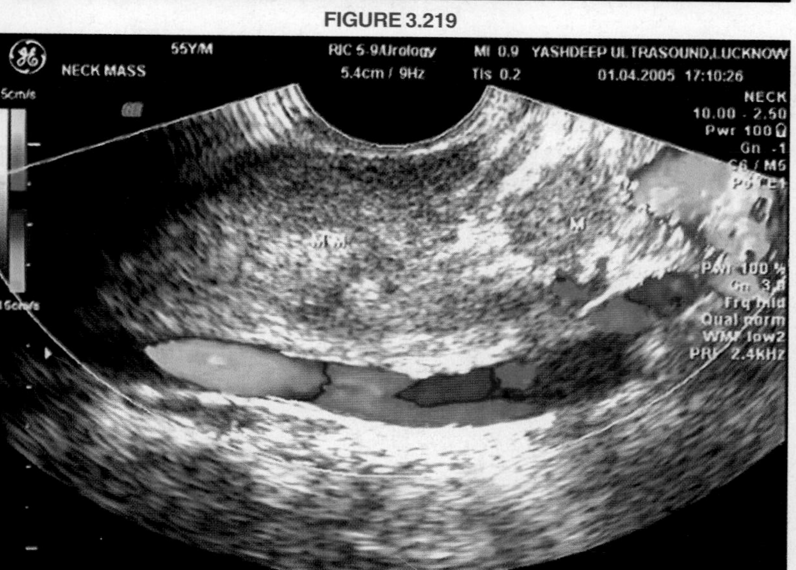

FIGURE 3.220

FIGURES 3.218 to 3.220: Recurrence malignant growth invading the carotid sheath. A known patient of CA tonsil presented with recurrence in the neck. HRSG shows a big irregular heterogeneous mass in the upper cervical region. The mass is seen fixed with the carotid sheath of Rt carotid artery. However, it was not seen obstructing the flow of the vessel. Biopsy shows metastatic nodal mass adherent with the carotid sheath.

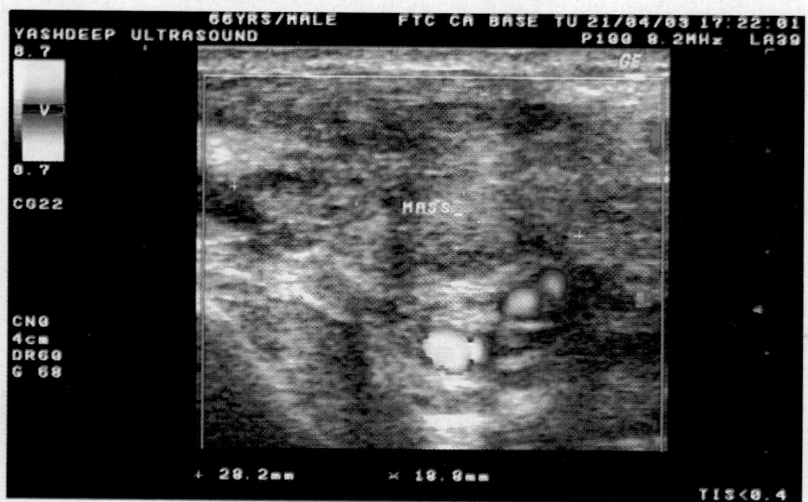

FIGURE 3.221: Recurrence of growth in CA base of tongue. HRSG shows a big irregular heterogeneous mass in the Rt upper cervical region in a known case of CA base of tongue. Biopsy confirmed the recurrence of the growth.

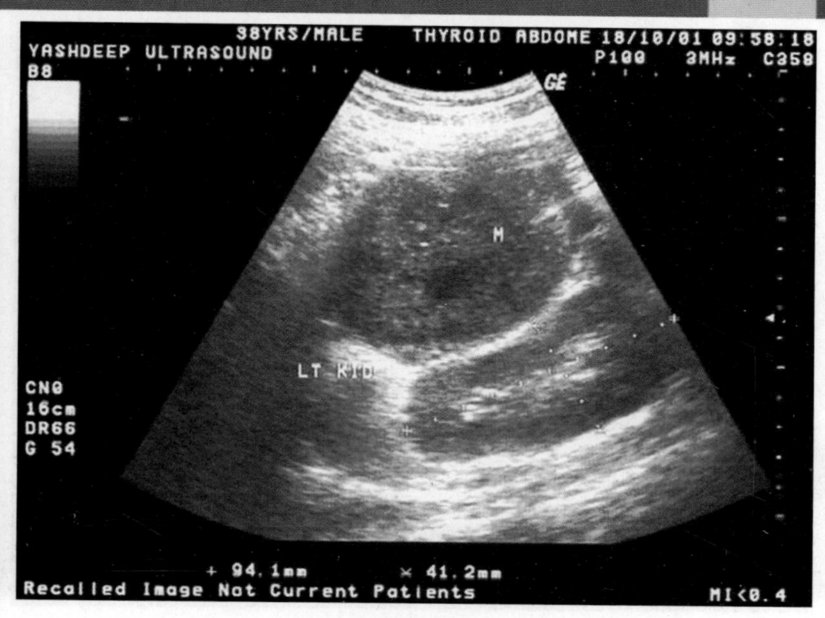

FIGURE 3.222

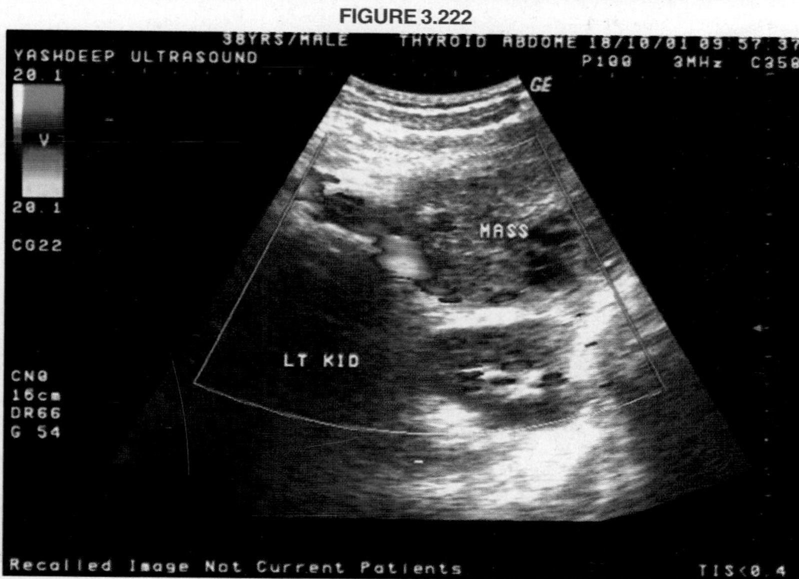

FIGURE 3.223

FIGURES 3.222 and 3.223: Adrenal tumor metastasizing to scalp. HRSG shows a big irregular hypoechoic mass lesion in the Lt adrenal area. A well-defined hypoechoic mass is seen anterior to the Lt kidney. It has pushed the kidney down. However, renal outline is maintained. Biopsy of the mass shows adrenal tumor.

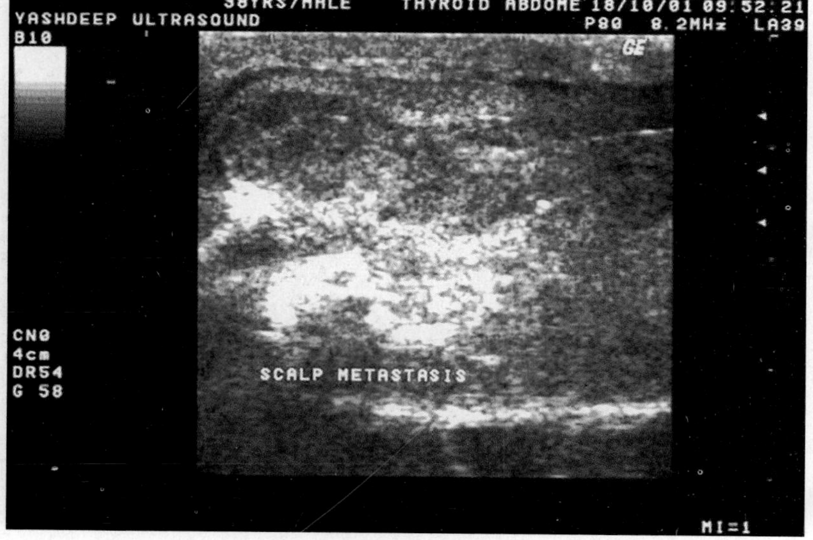

FIGURE 3.224

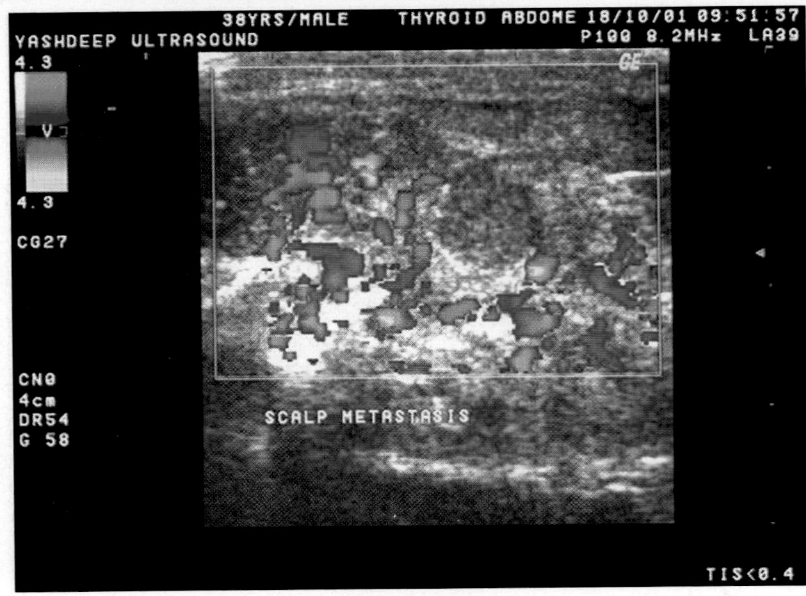

FIGURE 3.225

FIGURES 3.224 and 3.225: Scalp metastasis. HRSG of scalp of the same patient shows an irregular hypodense mass over the Lt scalp. It is irregular in outline with multiple hypodense areas. On color flow imaging increased flow is seen in the mass. Biopsy of the mass shows metastatic deposits from the adrenal tumor.

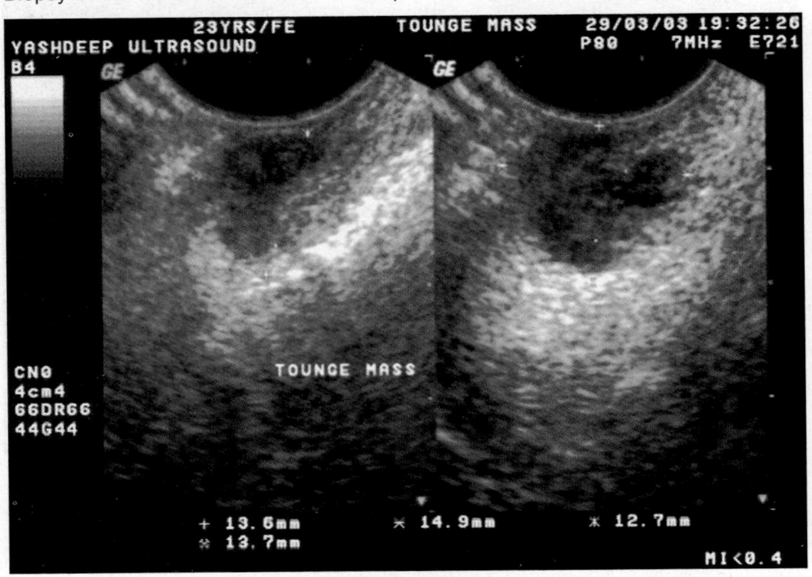

FIGURE 3.226

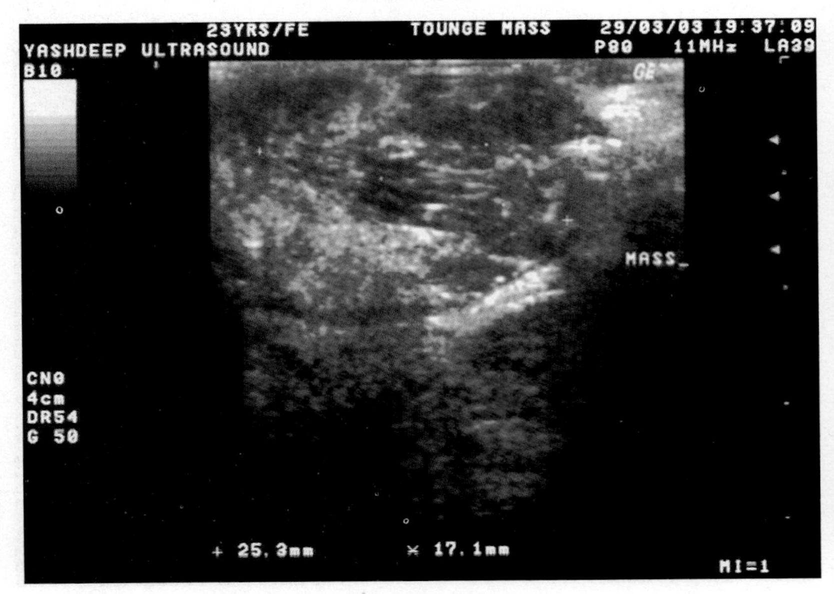

FIGURE 3.227A

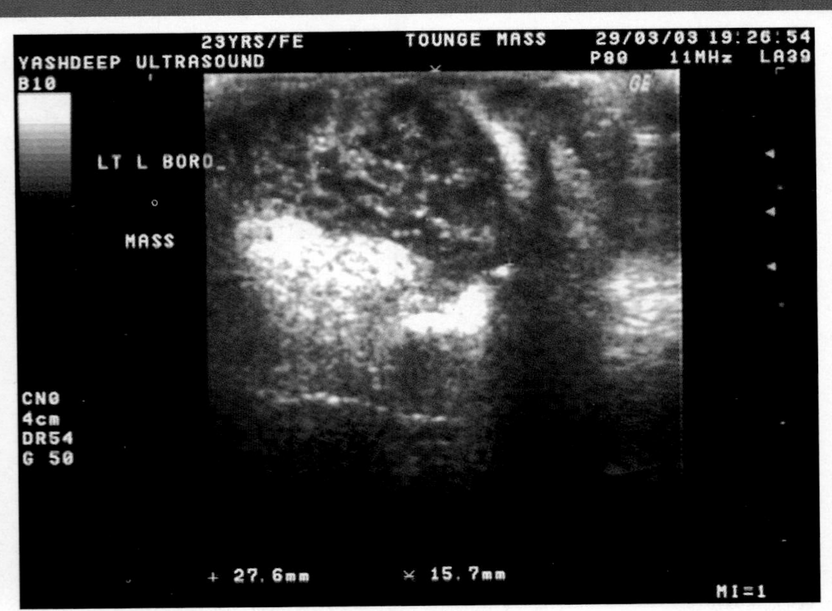

FIGURE 3.227B

FIGURES 3.226, 227A and B: CA tongue. A patient presented with growth over the lateral border of the tongue. HRSG shows a big irregular heterogeneous mass invading the tongue. The mass is seen invading the deeper layers of the tongue. It is distorting the normal texture of the tongue.

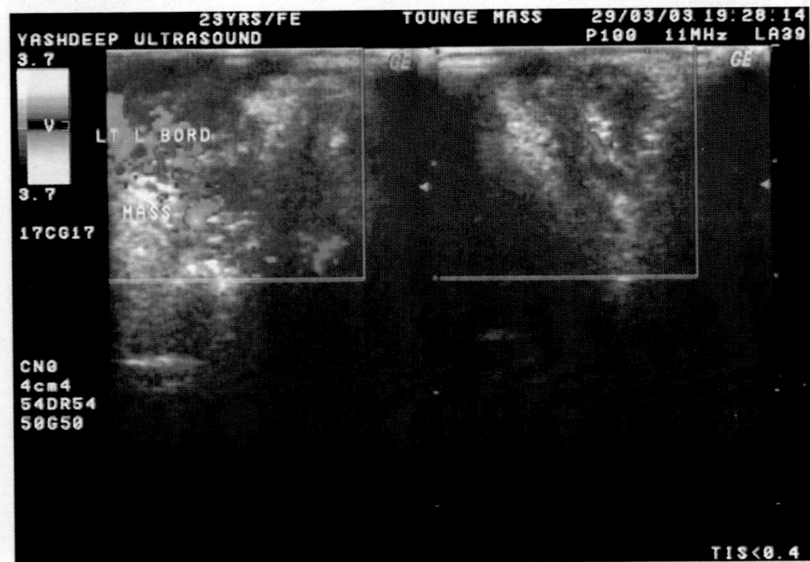

FIGURE 3.228A

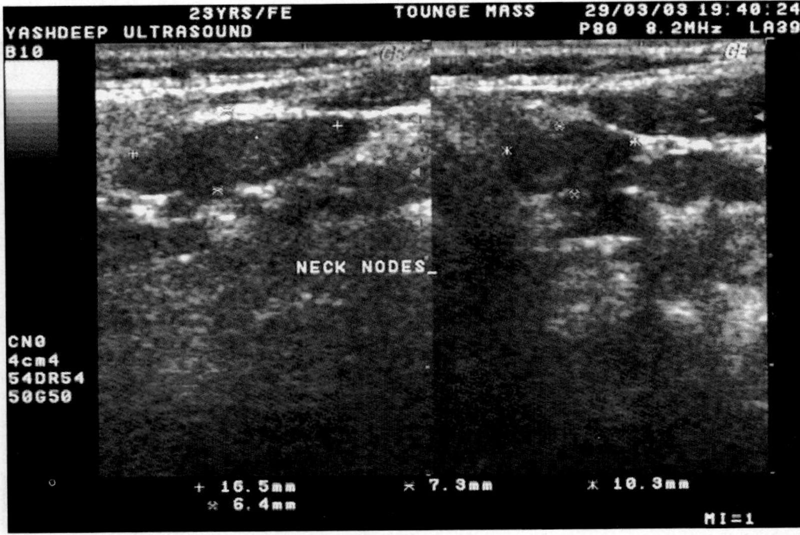

FIGURE 3.228B

FIGURES 3.228A and B: On color flow imaging increased flow in the mass. Multiple enlarged nodes are also seen in the floor of the mouth with malignant features.

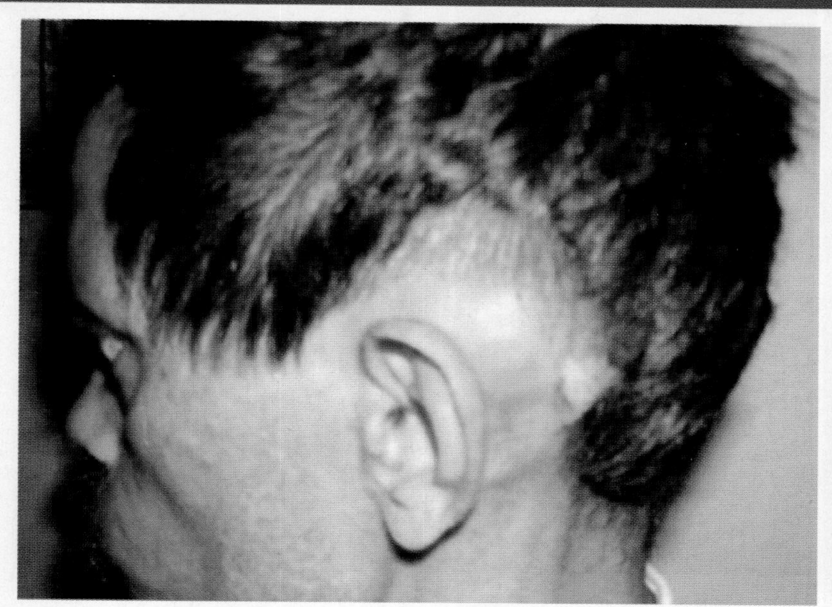

FIGURE 3.229: A young man came with pulsatile mass in the mastoid region.

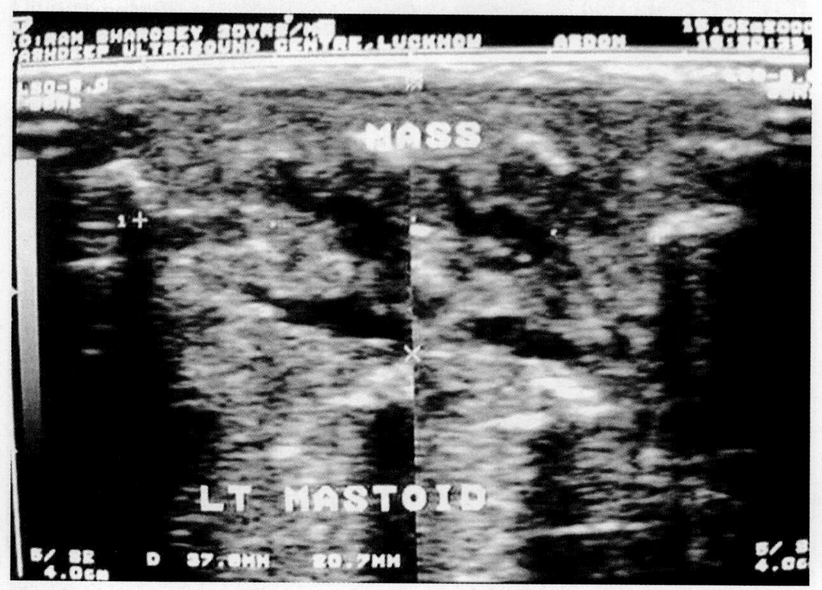

FIGURE 3.230: HRSG shows an irregular hypodense mass in the mastoid region with multiple dilated vessels.

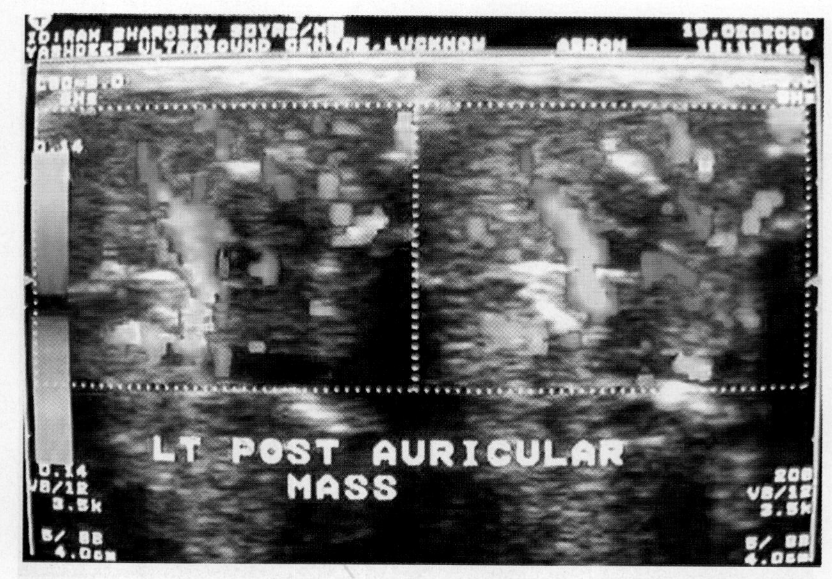

FIGURE 3.231: Metastatic deposit. Color flow imaging of the mass shows multiple dilated vessels in the mass suggestive of high vascular nature of the mass. Biopsy of the mass shows metastatic deposits.

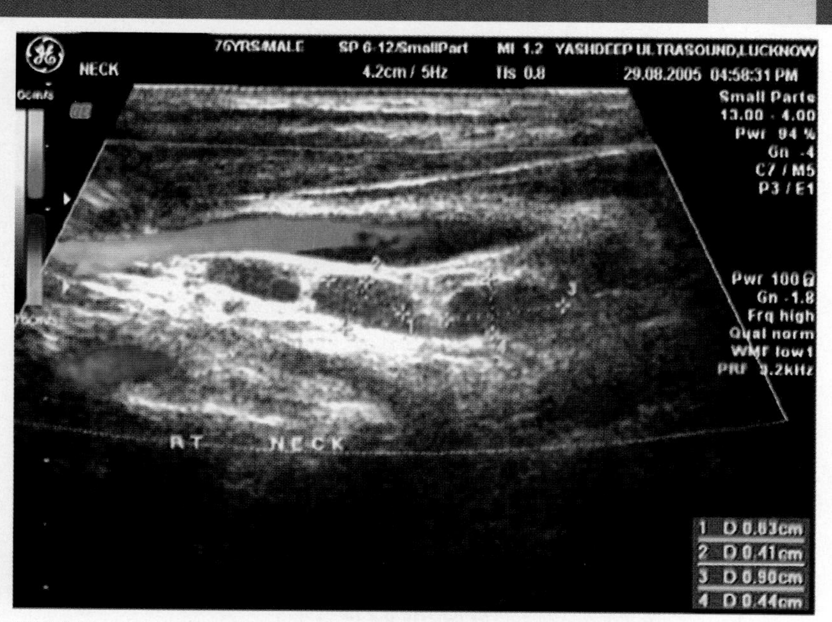

FIGURE 3.232: Post radiotherapy lymph node. Small irregular flattened hypoechoic lymph node seen in a case of carcinoma cheek after radiotherapy. Normal texture is lost. HRSG is very sensitive in detecting the residual lymph node disease in post-irradiated neck, because clinical examination of neck was not possible due to fibrosis.

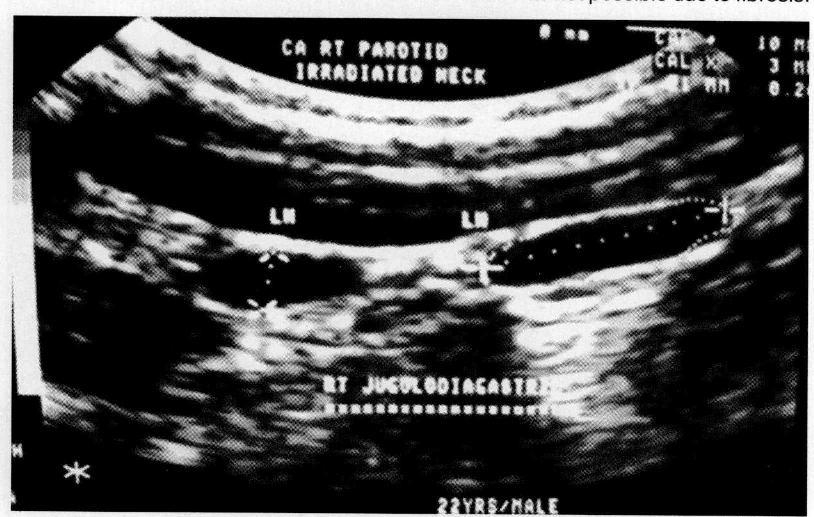

FIGURE 3.233: Postirradiated neck. HRSG shows multiple small hypoechoic nodes in a case of carcinoma, parotid gland after radiotherapy. Normal nodal texture is lost. However, nodes are small in size and flattened.

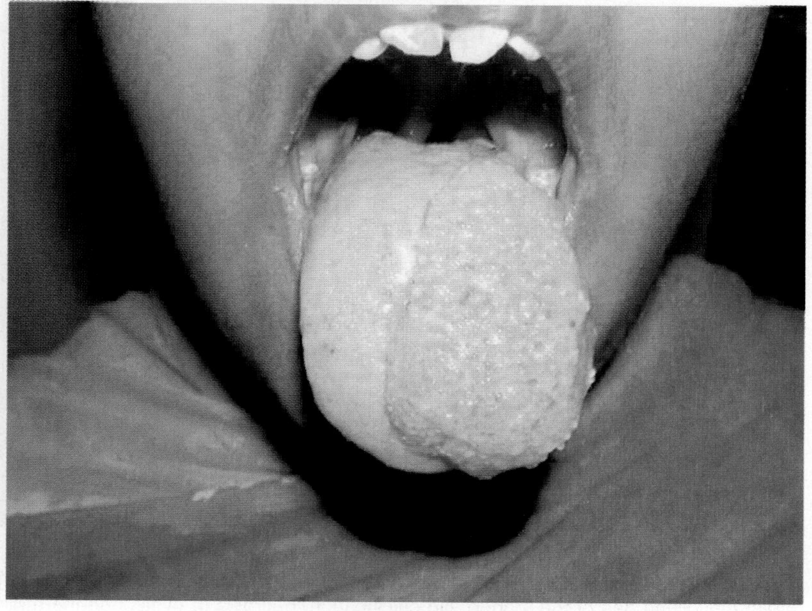

FIGURE 3.234

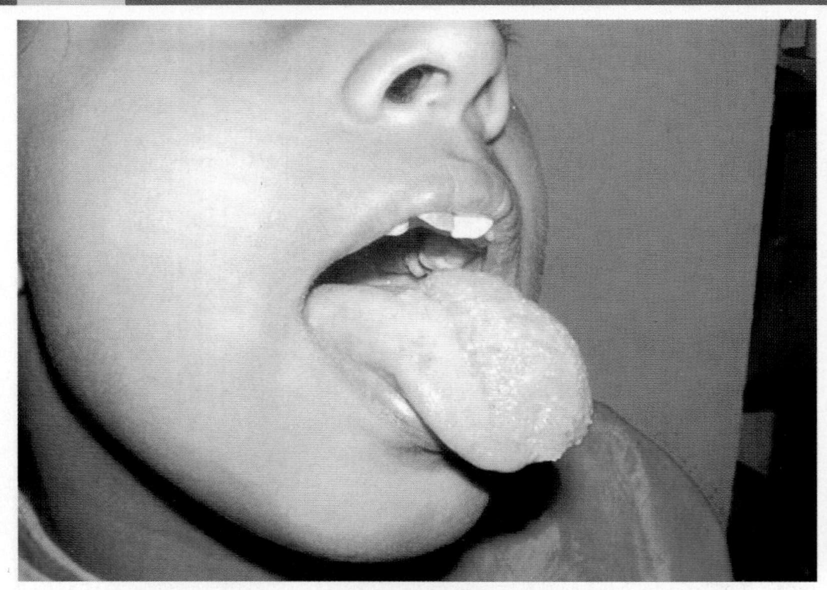

FIGURE 3.235

FIGURES 3.234 and 3.235: Papilloma tongue. A young girl presented with a well defined homogeneous soft tissue mass over one half of the tongue.

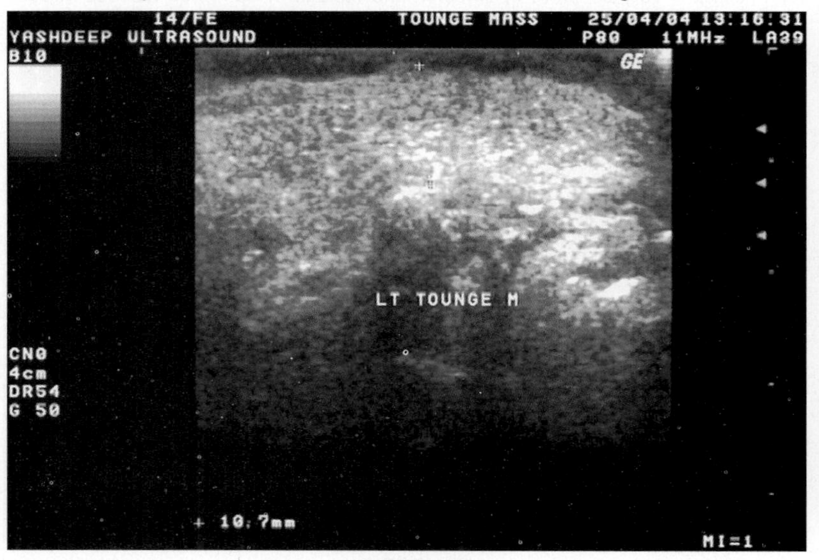

FIGURE 3.236

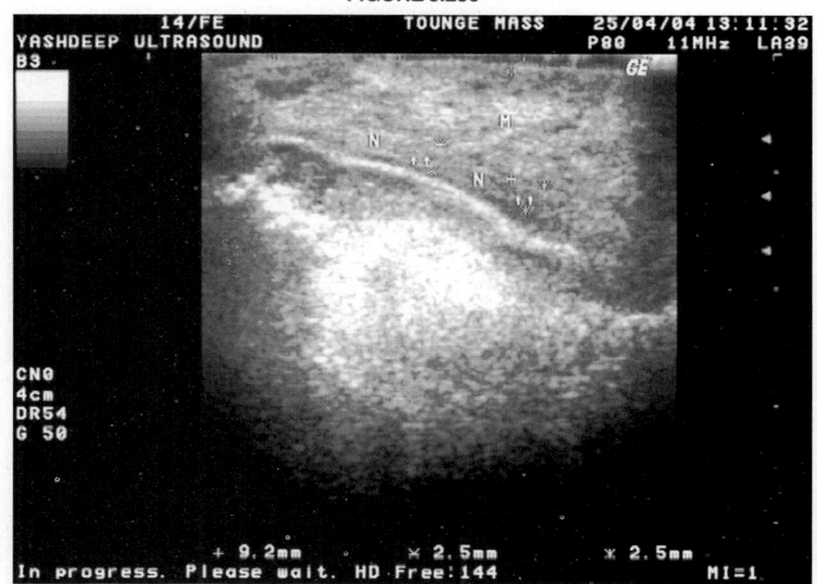

FIGURE 3.237

FIGURES 3.236 and 3.237: HRSG shows homogeneous soft tissue mass with smooth outline involving half of the tongue. No calcification, necrosis or cystic degeneration is seen in the mass.

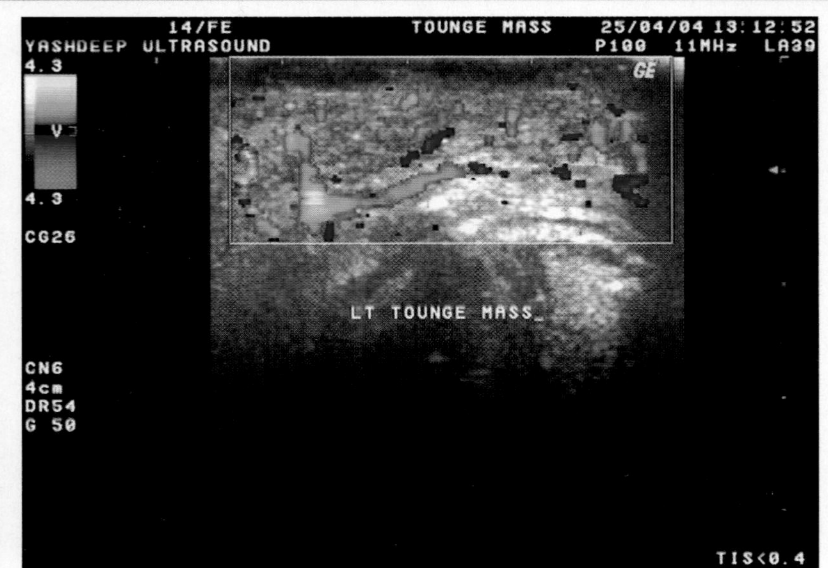

FIGURE 3.238

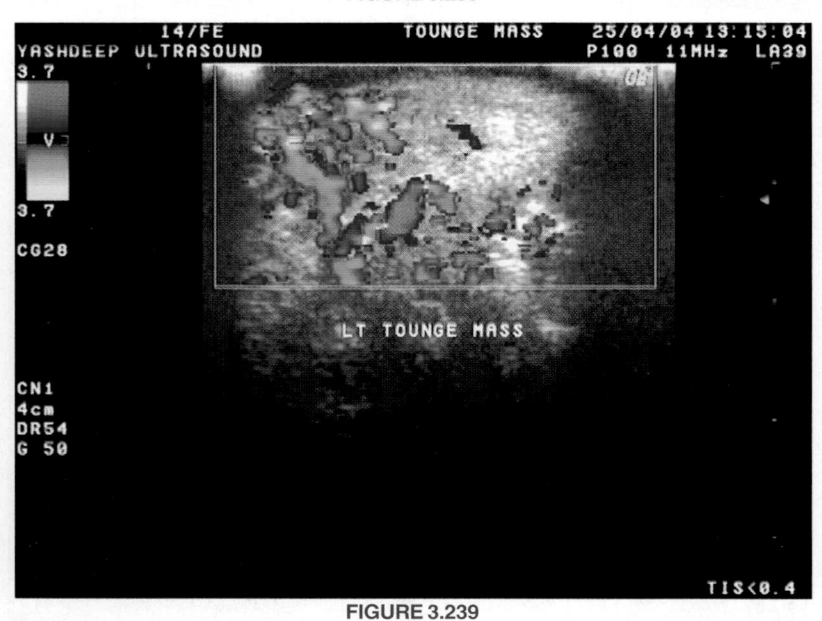

FIGURE 3.239

FIGURES 3.238 and 3.239: On color flow imaging multiple vessels are seen feeding the mass. Biopsy confirmed papilloma of the tongue.

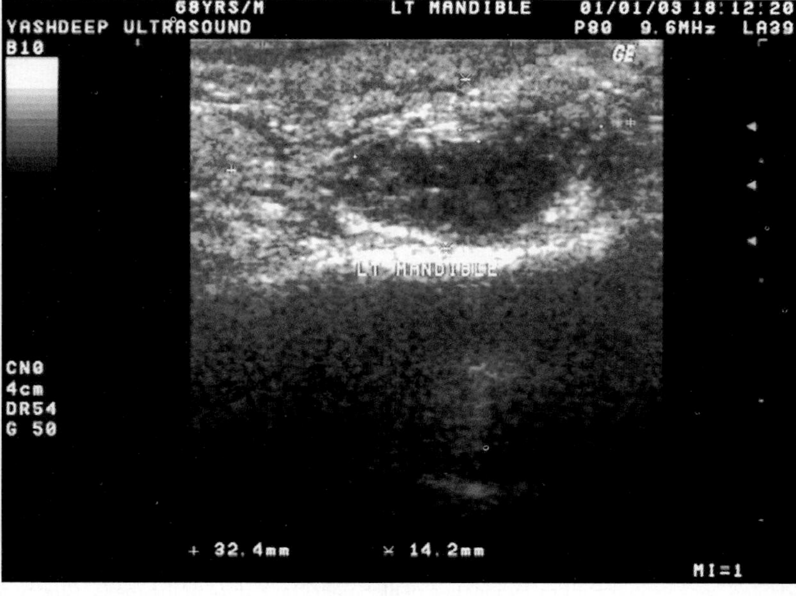

FIGURE 3.240

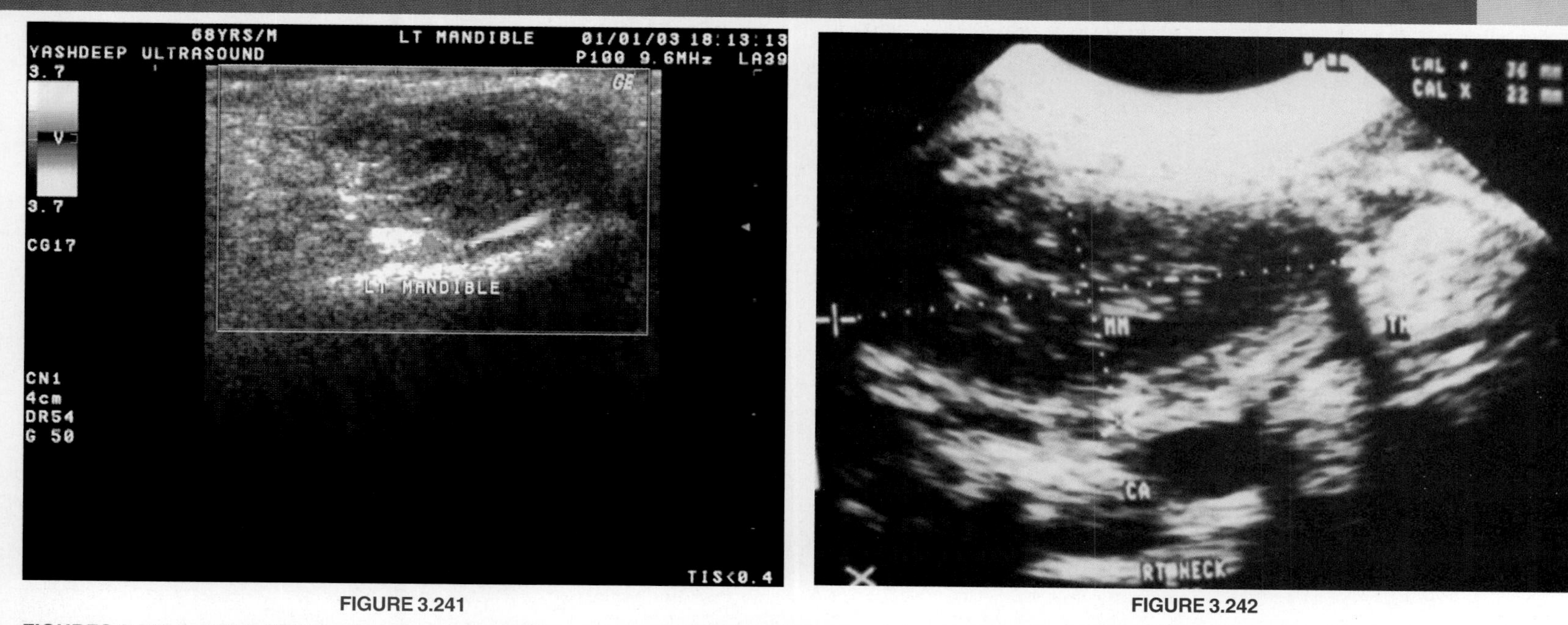

FIGURE 3.241

FIGURE 3.242

FIGURES 3.240 to 3.242: Malignant mass coming out from the ramus of the mandible. An old patient presented soft tissue swelling over the ramus of the Lt mandible. HRSG shows an irregular hypoechoic soft tissue mass coming out from the ramus of the mandible. On color flow imaging low flow is seen in the mass. Biopsy showed malignant mass.

4 Salivary Glands

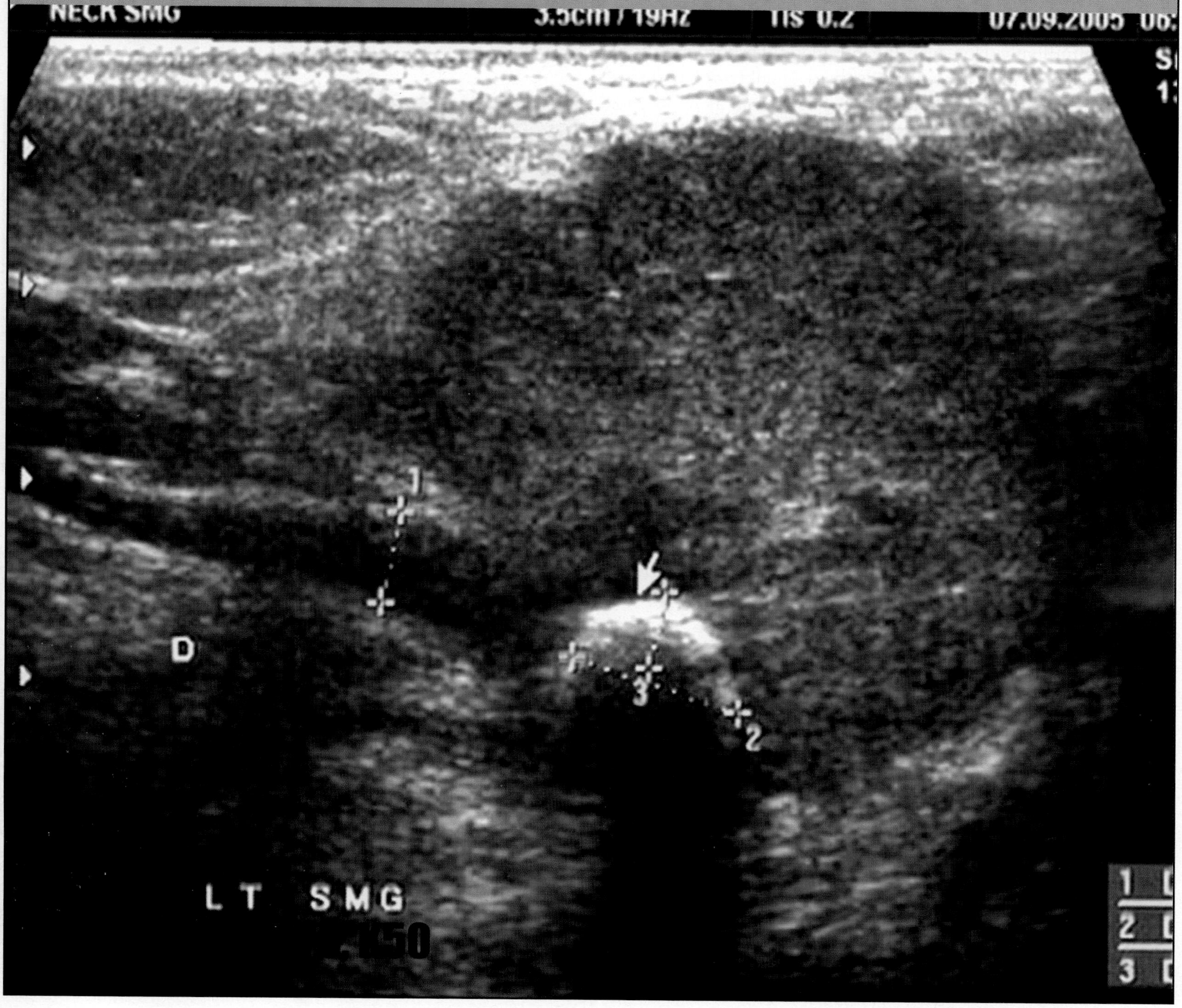

Introduction

High-resolution ultrasonography is capable to image all the three major salivary glands, i.e. parotid, submandibular and sublingual glands. It can detect most of the lesions affecting them. A 7.5 or 10 mhz transducer is the ideal transducer to detect salivary glands. The gland can be imaged in multiple planes like axial, longitudinal or cross-section or oblique planes.

The parotid glands are the largest salivary glands. They lie over the ramus of mandible on each side. Each gland has got two lobes. One superficial lobe which rests over ramus of mandible and another deep lobe which is seen behind the ramus in between the mastoid and stern mastoid muscle and in front of the styloid process. The gland shows a homogeneous texture. It is more echogenic than the surrounding muscles. The margins of the gland are not well defined, and the retro pharyngeal portions of the gland cannot be evaluated on HRSG. The parotid duct can be resolved on HRSG running in the substance of the gland longitudinally. However, the intraglandular portion of facial nerve is seen only in 30 percent of the cases. Small lymph nodes are commonly present in the gland, which can be seen as hypoechoic nodular shadow on HRSG. The submandibular gland is seen below the jaw and sublingual glands are seen in the floor of the mouth. They have got the similar echotexture to that of parotid gland. Sublingual glands are difficult to be resolved on HRSG. They can be seen as small triangular solid echogenic nodules in the floor of the mouth in axial scanning.

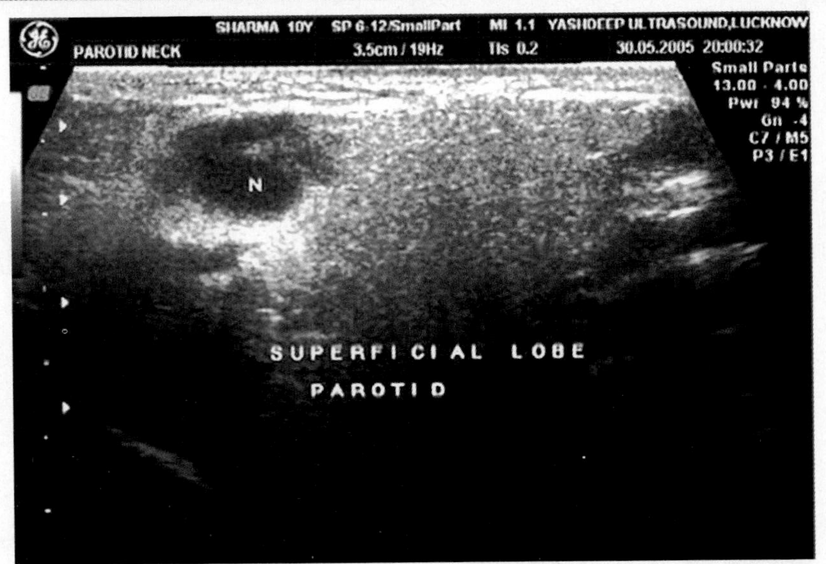

FIGURE 4.1: HRSG shows normal parotid gland. Fine glandular texture of the gland is seen. Hypoechoic nodular shadow is also seen in the gland suggestive of sentinel node.

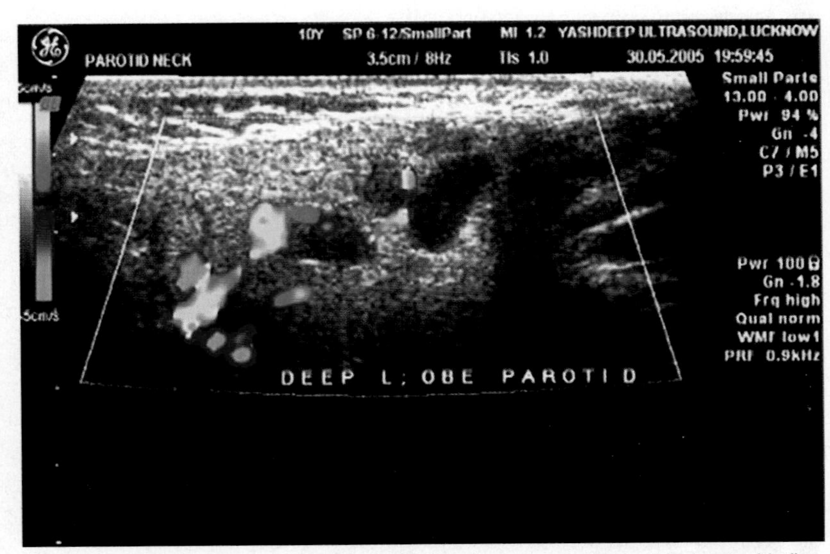

FIGURE 4.2: Normal parotid gland: HRSG shows normal parotid gland with fine well-defined globules. Few vessels are also seen feeding the gland.

Pathology

Acute Inflammation

In acute inflammation of parotid gland, it becomes enlarged and hypoechoic in texture and shows inhomogeneous pattern. Multiple hypoechoic pockets are seen in the gland suggestive of micro abscess. Breaking down of glandular tissue may lead to large abscess formation.

It shows thick wall with low-level echo complex necrotic area. Ultrasonographic-guided aspiration of the pus can be done. Stone in the parotid duct can be easily picked up on HRSG. It is better than sialography and noninvasive in nature. The calculi are seen as a bright echogenic foci with posterior acoustic shadowing.

Tumors

Tumors of salivary gland can be easily picked up on HRSG. The aims of ultrasonographic imaging in a palpable mass in salivary gland is to confirm the presence of the mass, to differentiate between intra and extra glandular mass, and the nature of the mass whether it is solid or cystic. Detection of parotid mass is very accurate and its sensitivity is 100 percent as it can detect subclinical nodules, although identification of nature of mass is more difficult and FNAC can confirm the diagnosis. Pleomorphic adenomas, mixed parotid gland tumor are more common and they remain silent for many years. By and large, all the benign tumors of parotid gland are echo poor and homogeneous in texture. They are sharply demarcated. Malignant tumors are grossly heterogeneous and irregular in outline. Irregular pockets of collection

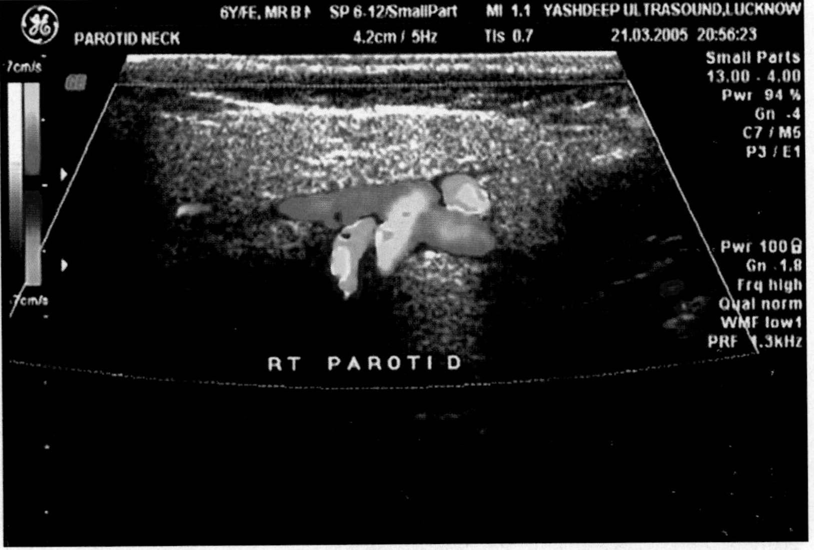

FIGURE 4.3

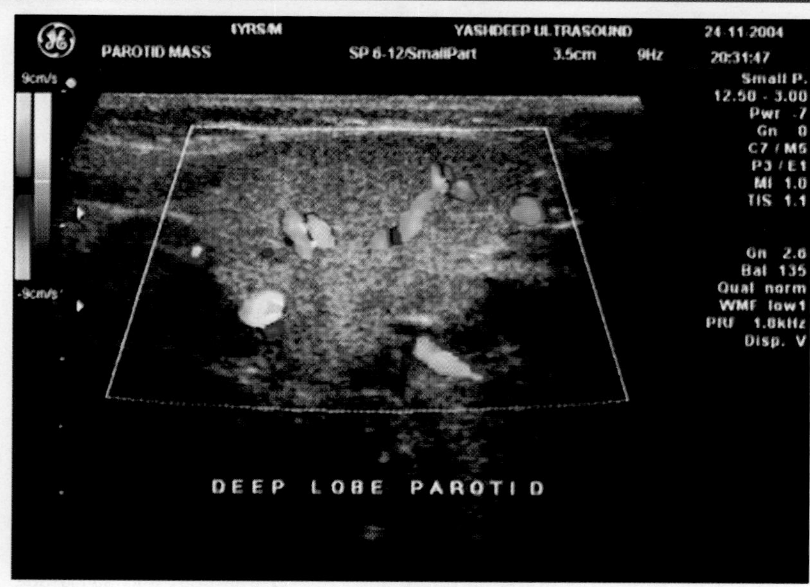

FIGURE 4.4

FIGURES 4.3 and 4.4: Normal parotid gland: Color Doppler imaging shows normal gland. A vessel is seen traversing through the gland.

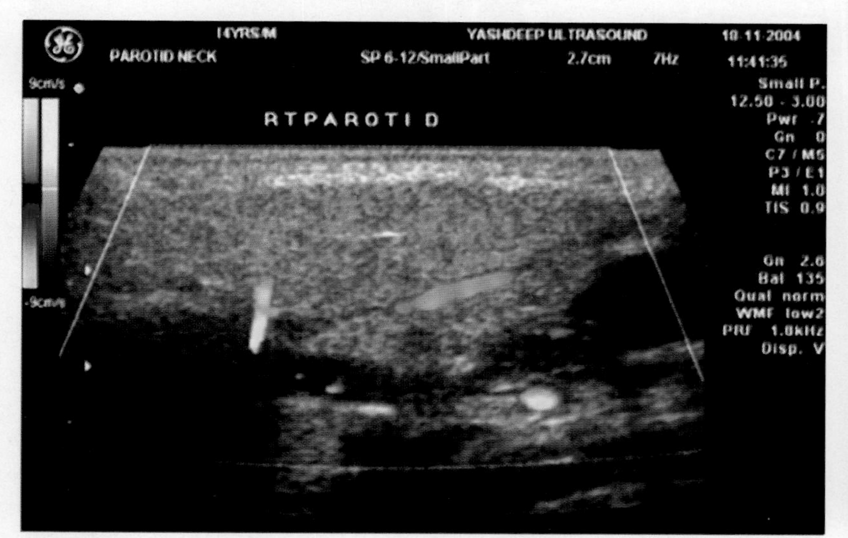

FIGURE 4.5: Normal parotid gland: HRSG shows normal gland. A fine echogenic line is seen suggestive of normal duct.

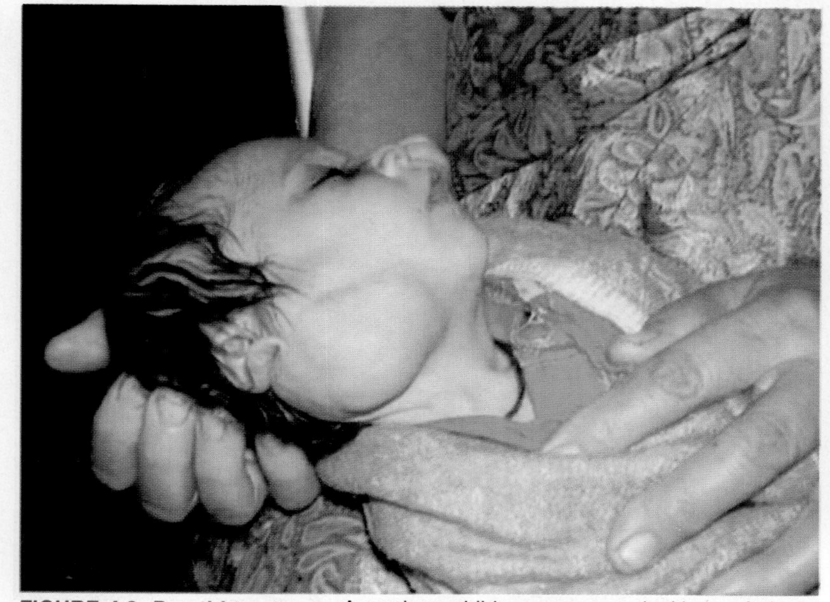

FIGURE 4.6: Parotid teratoma: A newborn-child was presented with a soft tissue mass over the Rt parotid region and lower neck. The mass is lobulated and reaching the upper cervical region going posteriorly.

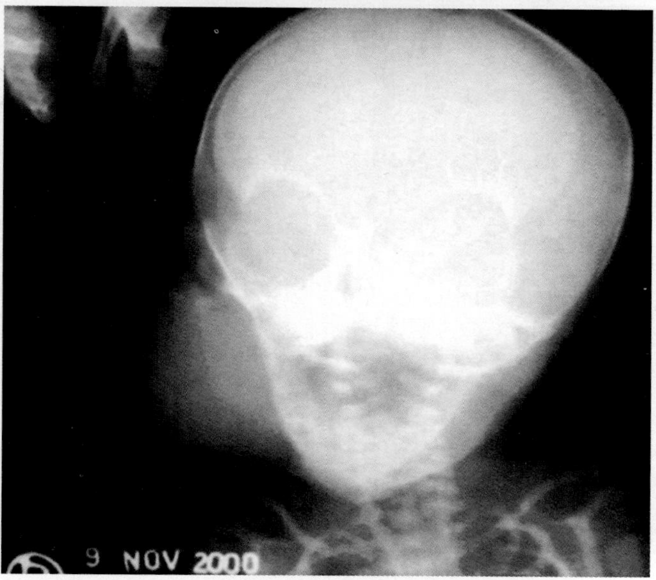

FIGURE 4.7: Plain X-ray of the head shows soft tissue mass over the Rt side of the neck. Skull bones are normal.

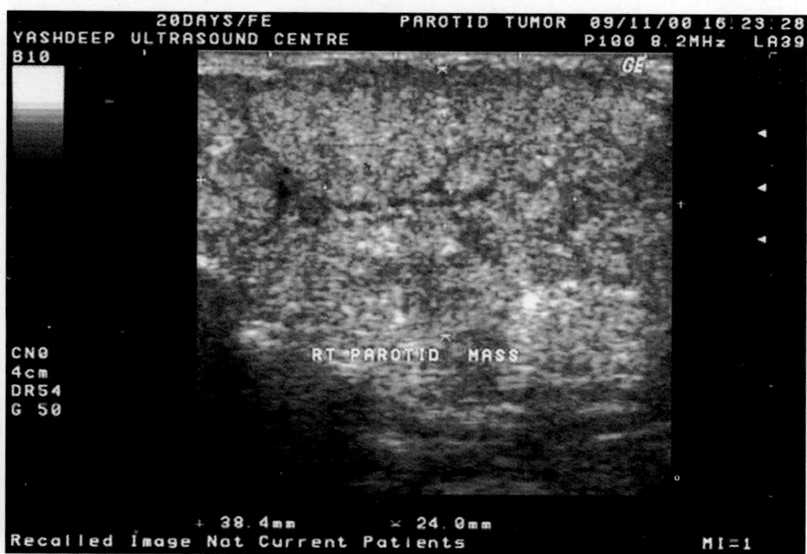

FIGURE 4.8

FIGURE 4.9

FIGURES 4.8 and 4.9: HRSG shows a predominantly multi-lobulated solid mass in the Rt parotid region and upper cervical region. Few well-defined solid lobules are seen in the mass. No cystic degeneration is seen.

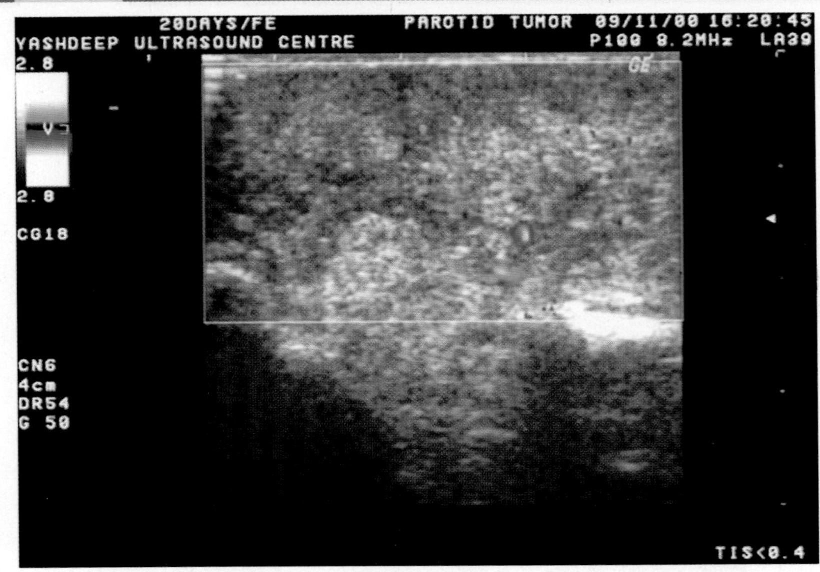

FIGURE 4.10

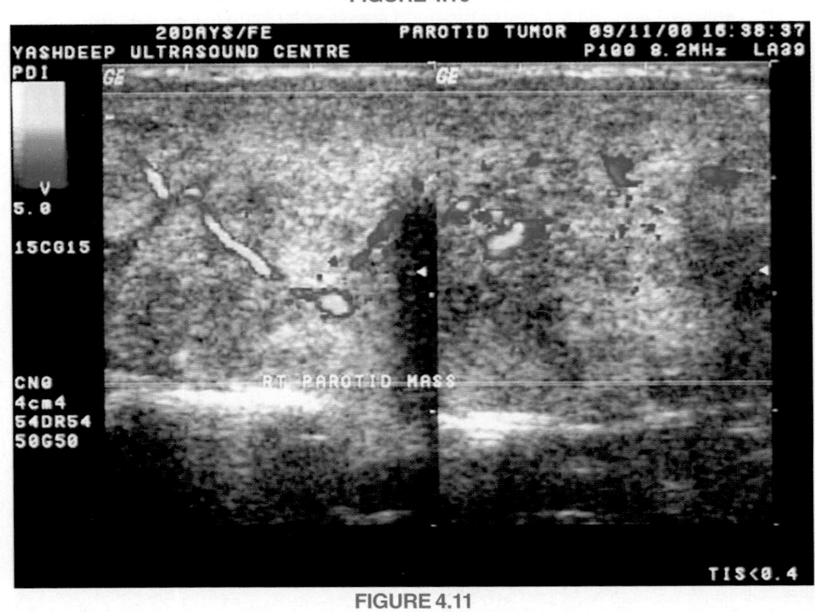

FIGURE 4.11

FIGURES 4.10 and 4.11: HRSG shows the mass is going through the upper cervical area with multiple hypoechoic septa running in the mass. Color Doppler imaging shows little flow in the mass.

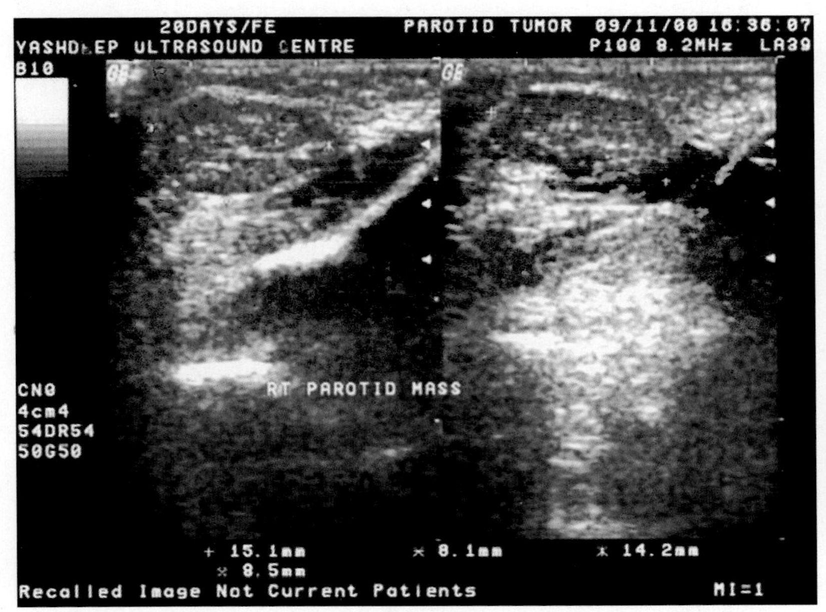

FIGURE 4.12: Part of the normal glandular tissue is seen in the parotid mass.

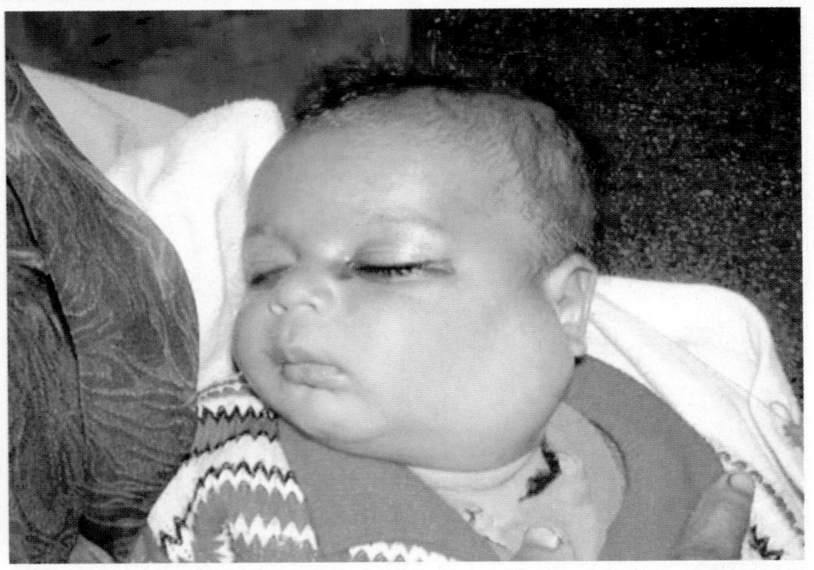

FIGURE 4.13: **Angiolymphoma parotid gland.** A child was born with soft tissue mass over the parotid region.

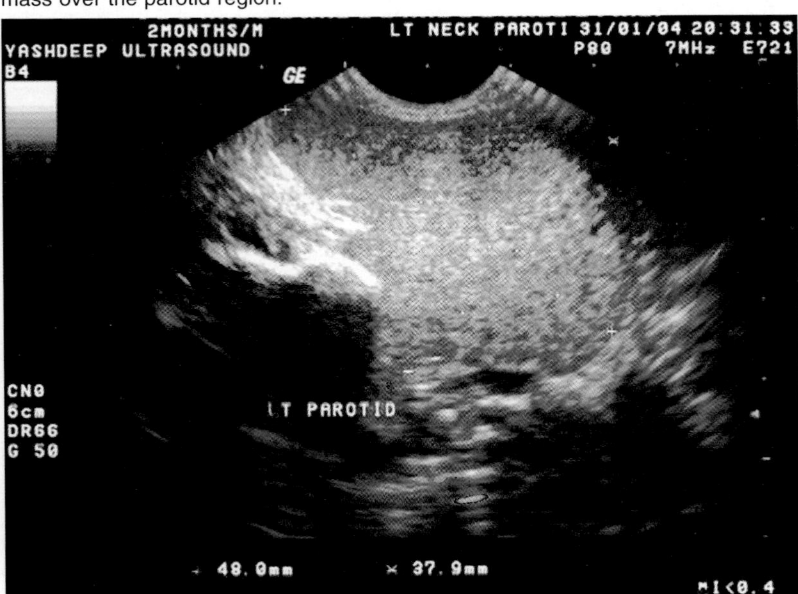

FIGURE 4.14

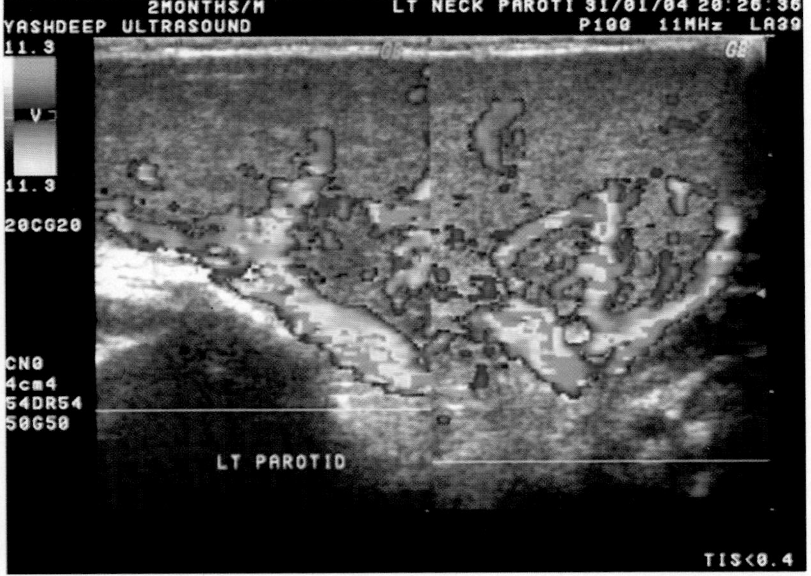

FIGURE 4.15

FIGURES 4.14 and 4.15: HRSG shows homogeneous soft tissue mass in the parotid region. Normal gland is not seen. On color flow imaging multiple vessels are seen feeding the mass. Biopsy confirmed angiolymphoma.

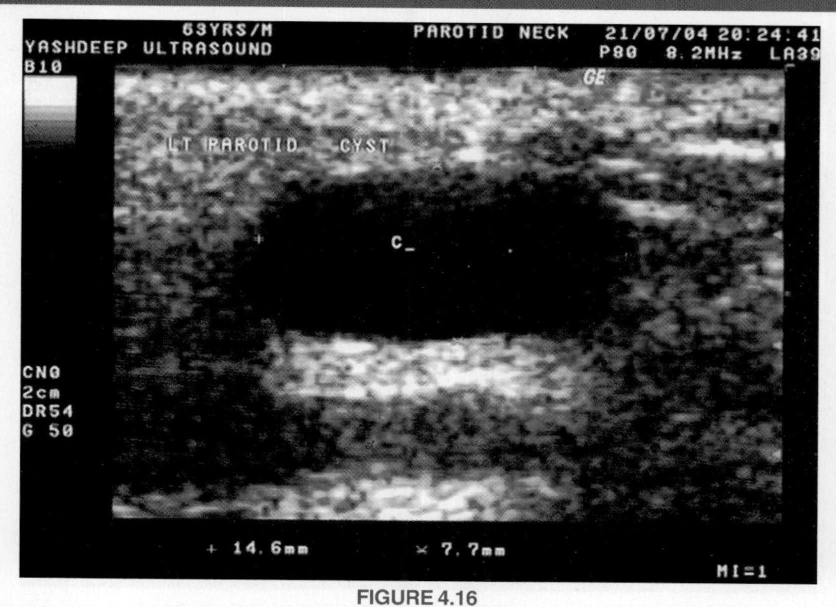

FIGURE 4.16

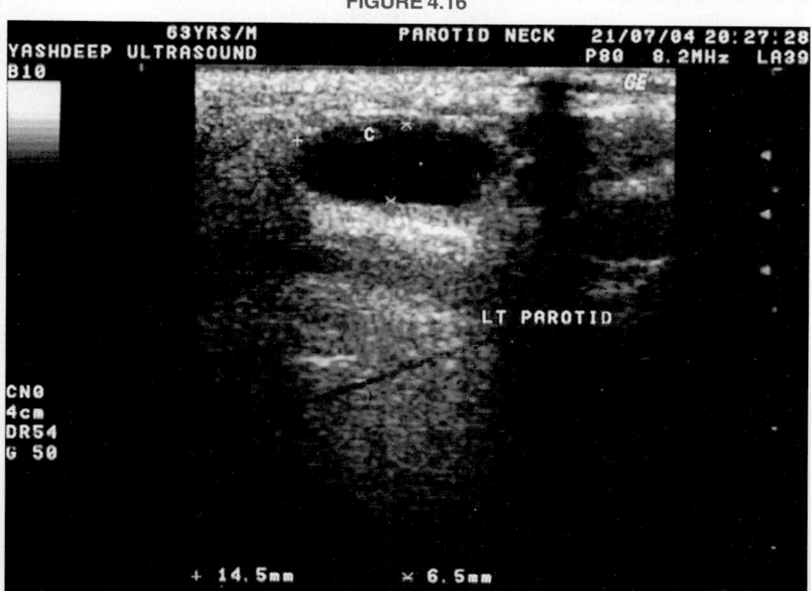

FIGURE 4.17

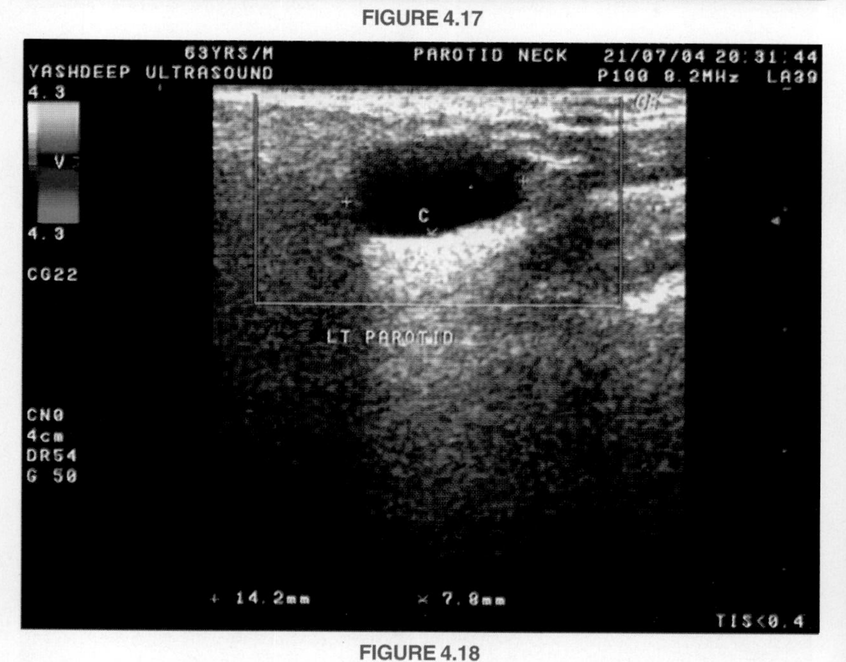

FIGURE 4.18

FIGURES 4.16 to 4.18: Simple parotid cyst. A patient presented with a cystic mass in the Lt parotid gland. HRSG shows a well-defined thin walled cyst in the superficial lobe of the gland. No internal echo is seen in the cyst. Post cystic enhancement is also seen.

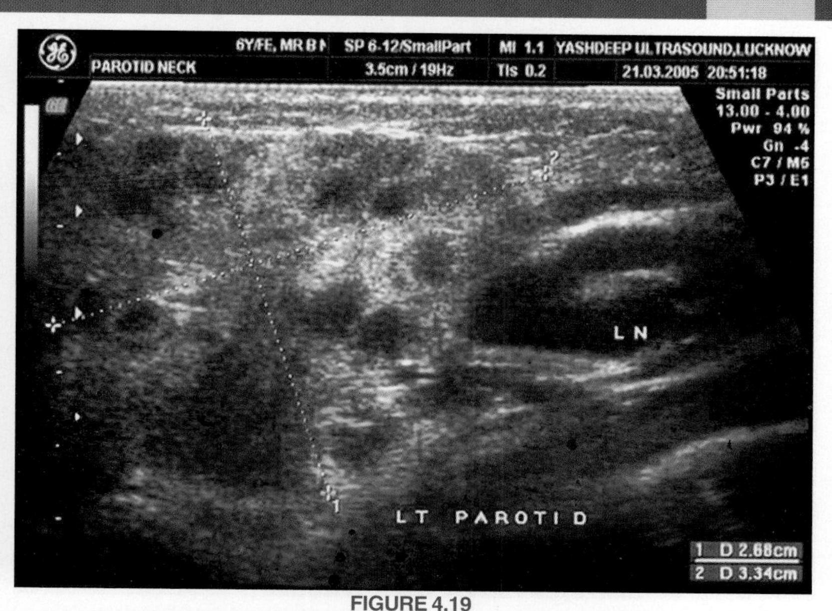

FIGURE 4.19

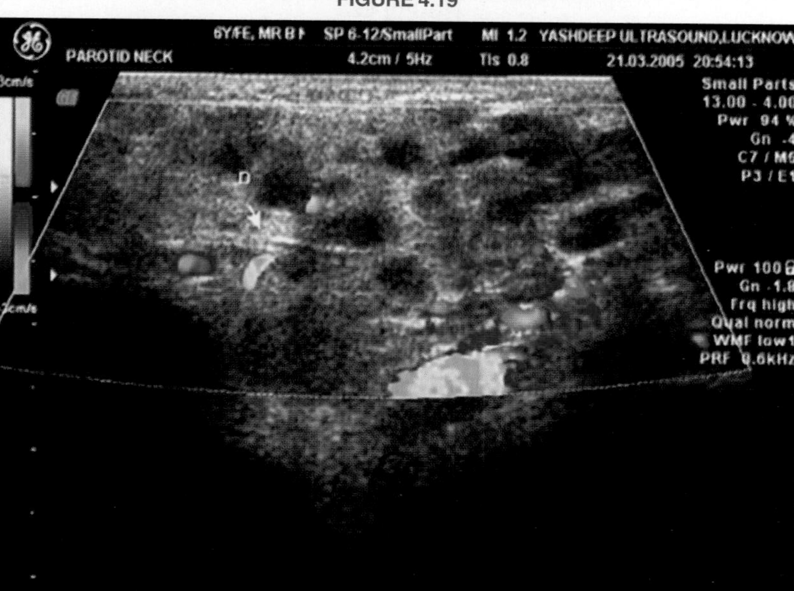

FIGURE 4.20

FIGURES 4.19 and 4.20: Acute inflammation in parotid gland. HRSG shows edematous hypoechoic enlargement of the Lt parotid gland. Margins are blurred in acute inflammation of the gland. Multiple hypoechoic foci are seen within the gland suggestive of small abscesses.

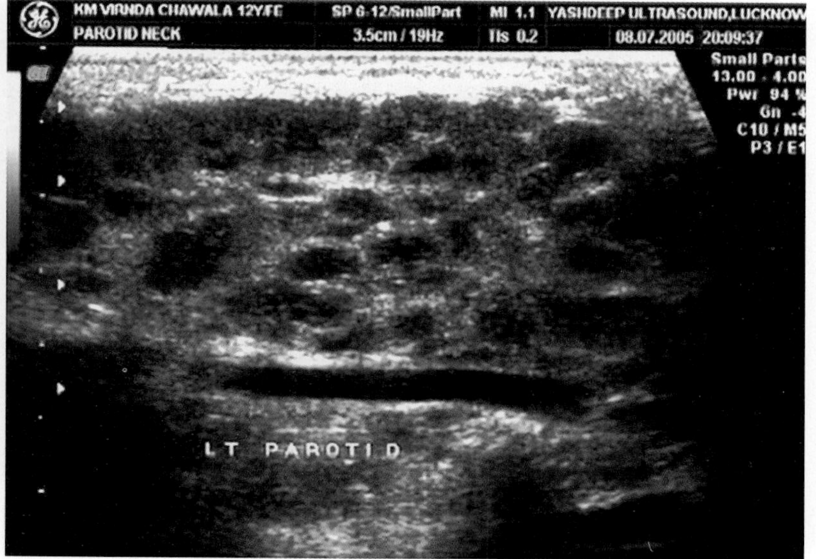

FIGURE 4.21

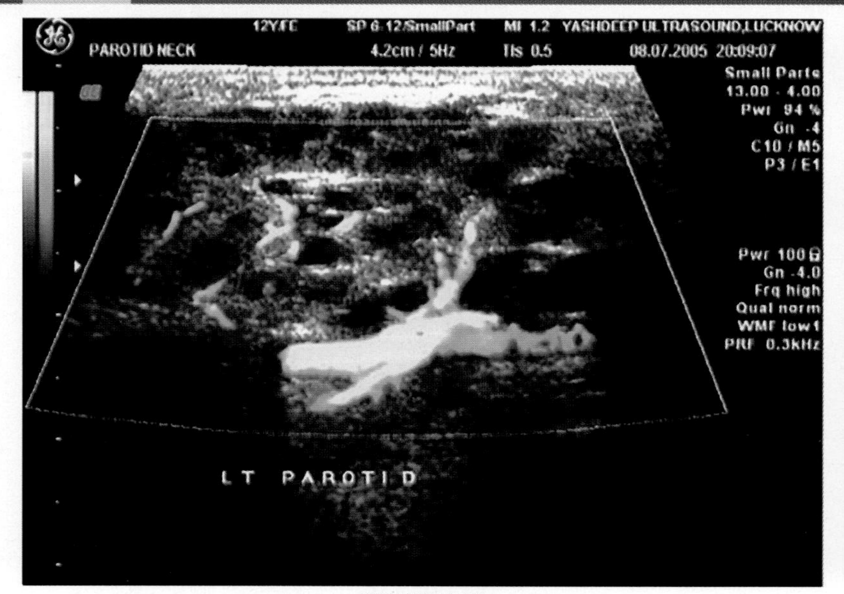

FIGURE 4.22

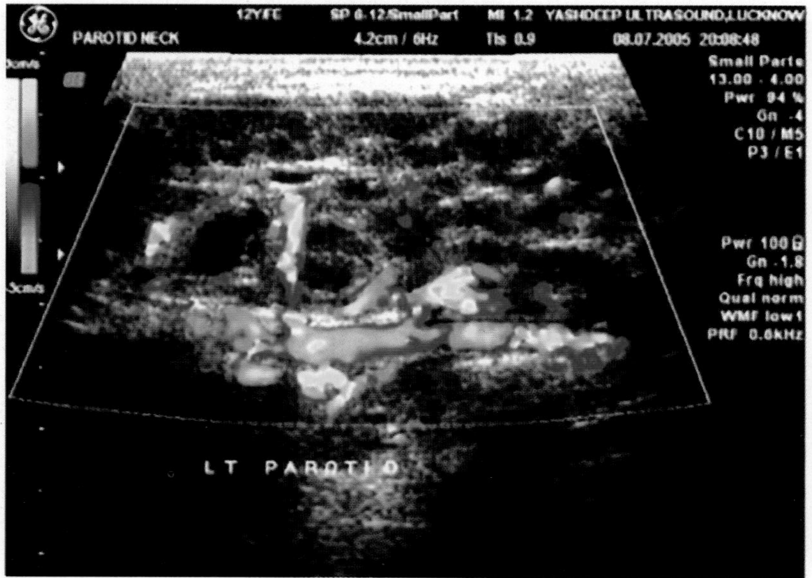

FIGURE 4.23

FIGURES 4.21 to 4.23: Acute Inflammation. HRSG shows heterogeneous enlargement of the Lt parotid gland with blurred outlines. Multiple small micro abscesses are seen in the gland. Color Doppler imaging shows increased flow of the gland suggestive of hyperemia.

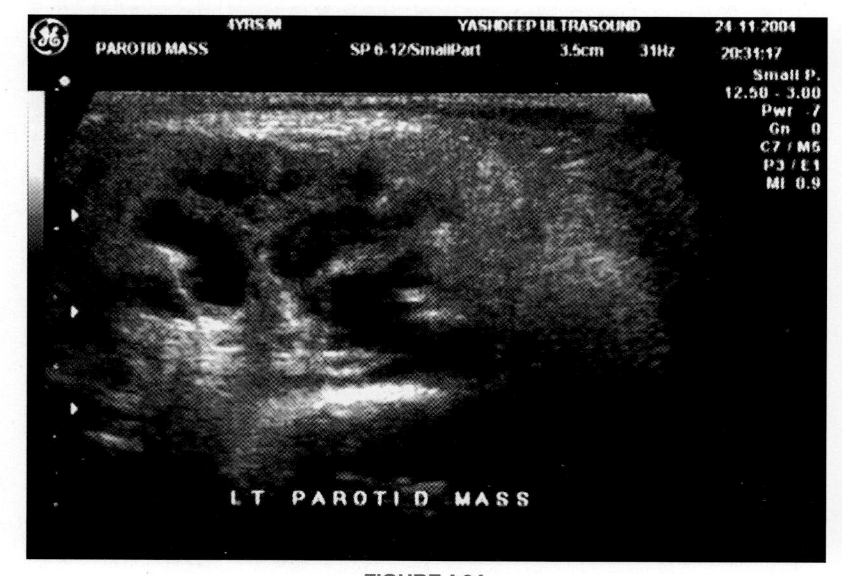

FIGURE 4.24

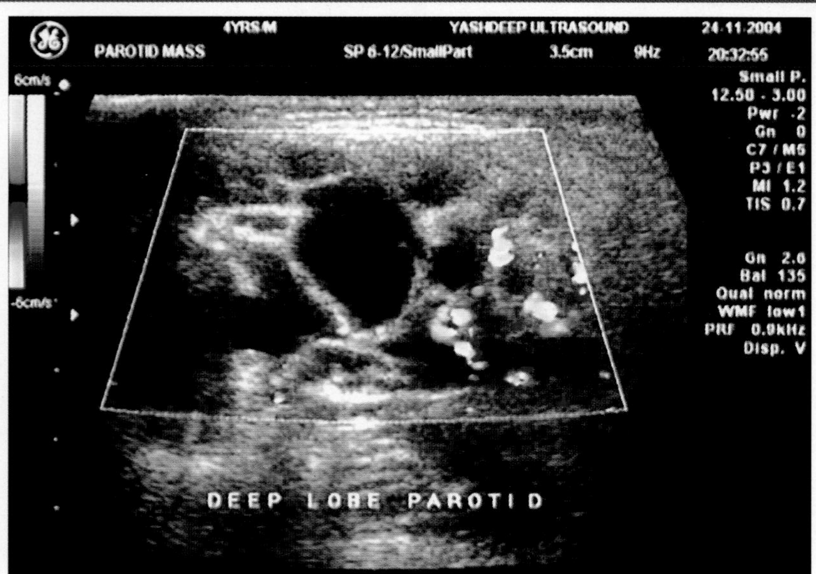

FIGURE 4.25

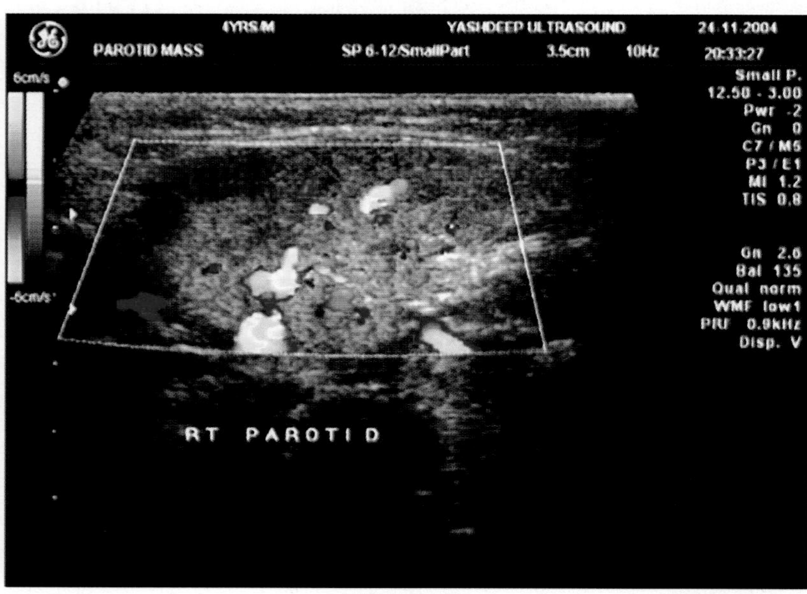

FIGURE 4.26

FIGURES 4.24 to 4.26: Acute parotitis. A young child presented with acute swelling of the Lt parotid gland. HRSG shows edematous enlarged parotid gland. Big abscess is seen in the gland. On color flow imaging increased flow is seen in the gland. Lt parotid is normal in appearance.

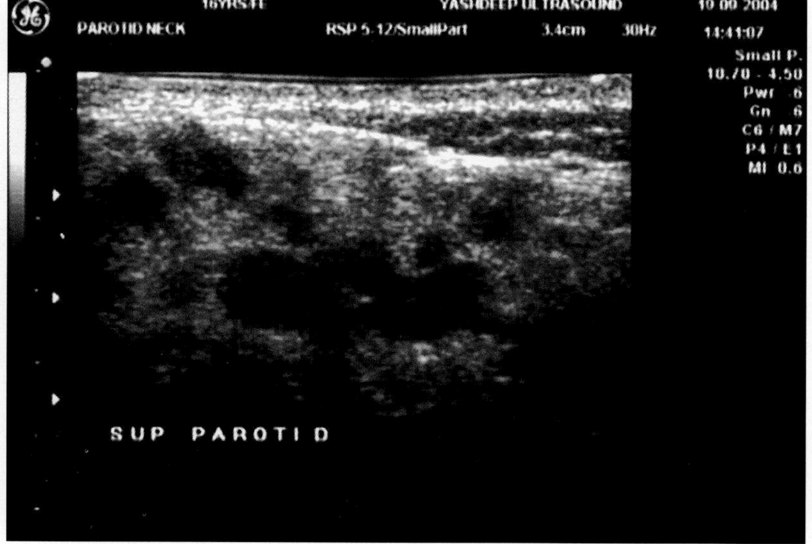

FIGURE 4.27

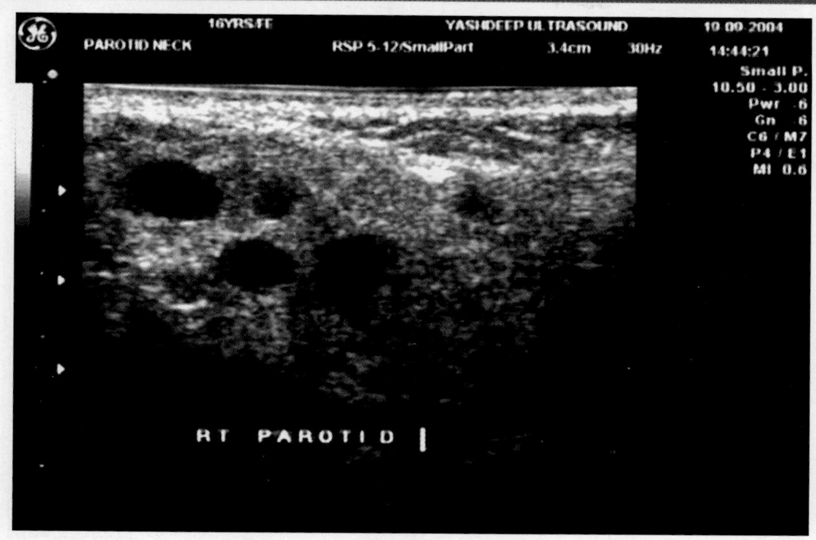

FIGURE 4.28

FIGURES 4.27 and 4.28: Small abscess in the superficial lobe of parotid. A young child presented with fever and painful swelling in the Rt parotid region. HRSG shows enlarged inflamed parotid gland. Multiple small abscesses are also seen in the superficial lobe of the gland.

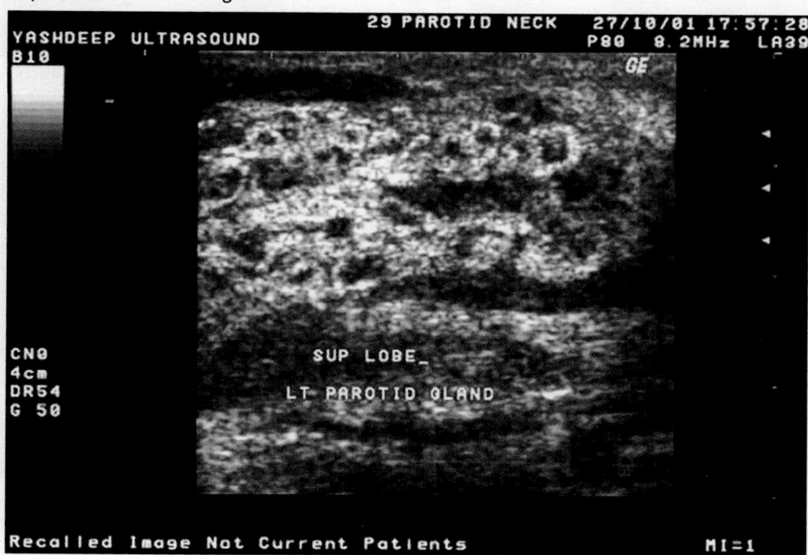

FIGURE 4.29

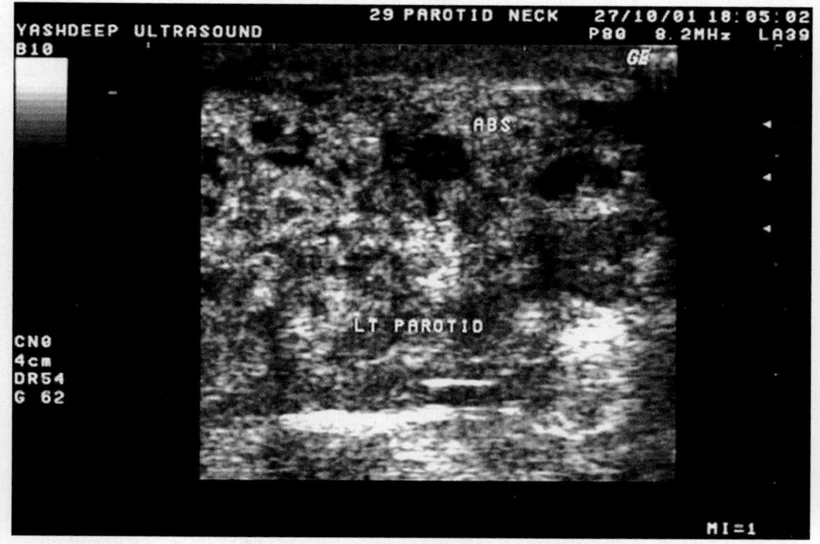

FIGURE 4.30

FIGURES 4.29 and 4.30: Acute fulminating parotitis. A young man presented with markedly tender swelling over the Lt parotid gland. HRSG shows grossly enlarged irregular parotid gland. The margins are blurred and irregular in outline. Inter glandular fluid collection is seen suggestive of edema.

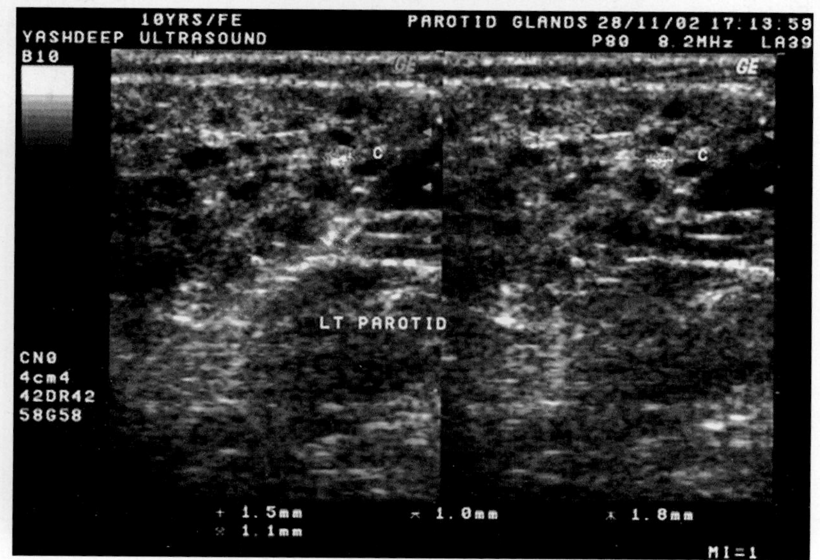

FIGURE 4.31

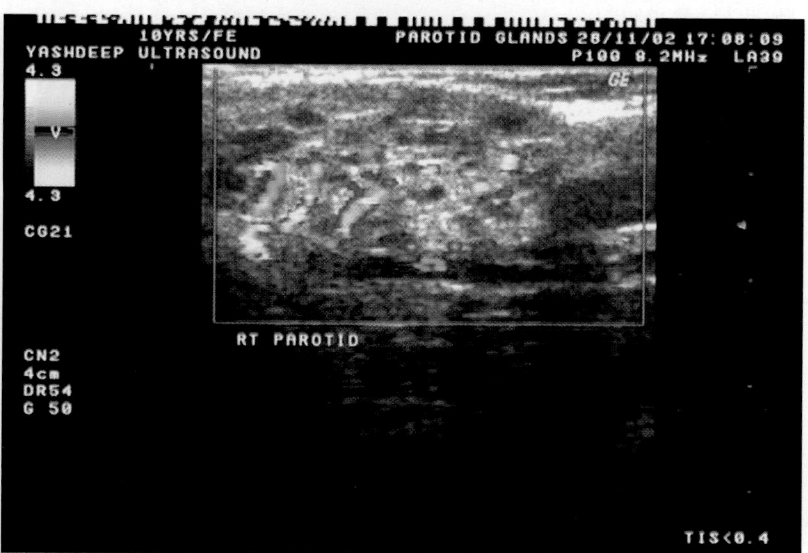

FIGURE 4.32

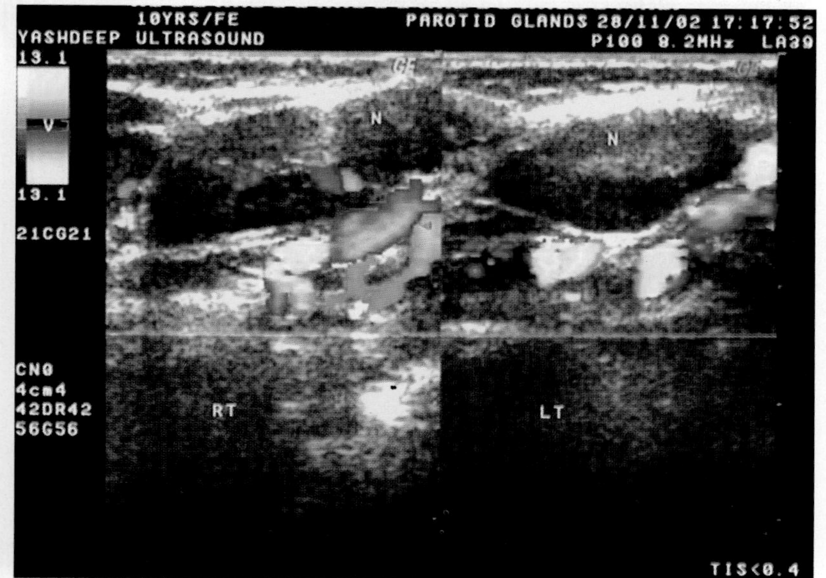

FIGURE 4.33

FIGURES 4.31 to 4.33: Acute recurrent sialectatic parotitis. A young girl presented with recurrent infection of the Lt parotid gland with swelling. HRSG shows multiple irregular hypodense areas in the gland. The gland is more heterogeneous. Small echogenic punctate calcified specks are seen in the gland. Enlarged intraglandular lymph nodes are also seen. On color flow imaging poor flow is seen in the gland on power Doppler imaging. No intraductal dilatation is seen.

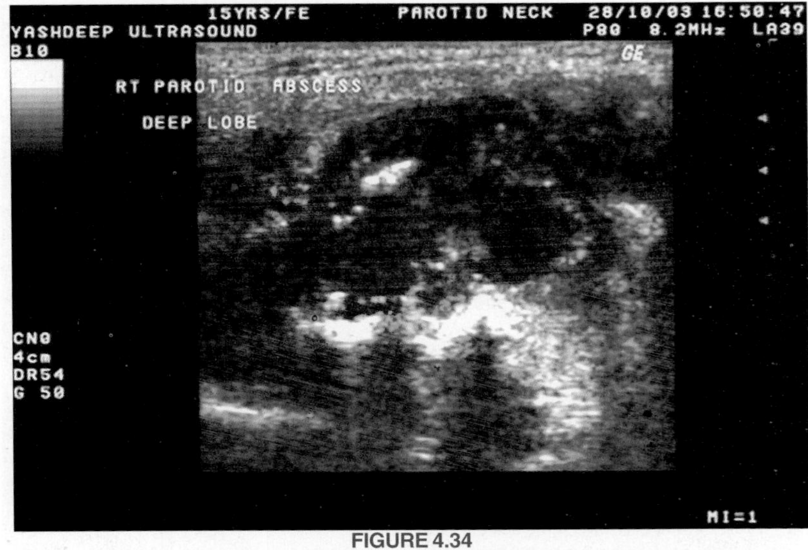

FIGURE 4.34

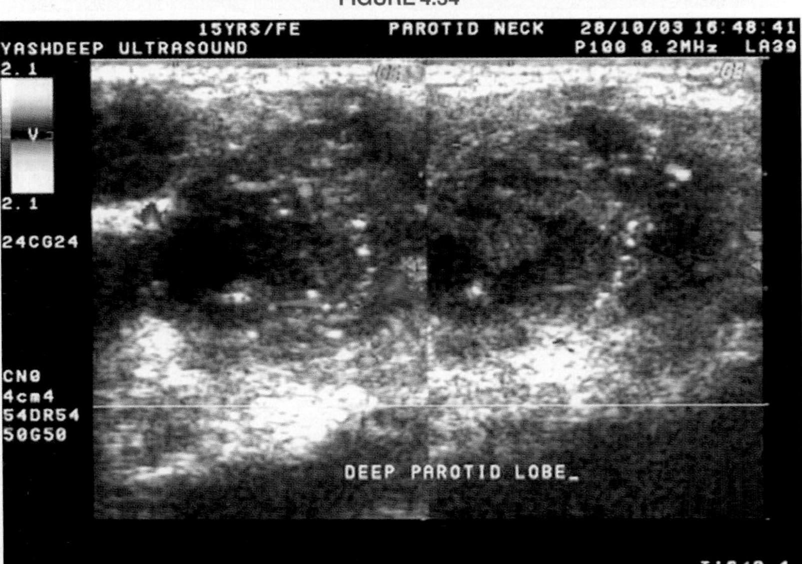

FIGURE 4.35

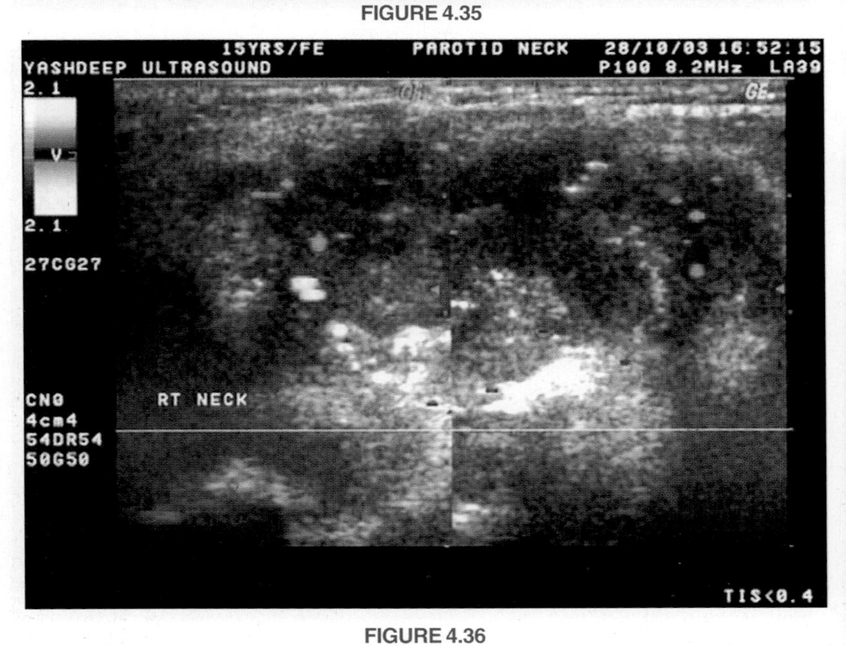

FIGURE 4.36

FIGURES 4.34 to 4.36: Acute necrotizing parotitis. A young girl presented with acutely painful swelling in Rt side of the neck with high fever. HRSG shows irregular hypoechoic enlarged deep lobe of parotid. Normal glandular texture is lost. It is grossly inhomogeneous. Multiple abscess are seen in the gland with necrosis suggestive of acute necrotizing parotitis. On color flow imaging low flow is seen in the gland.

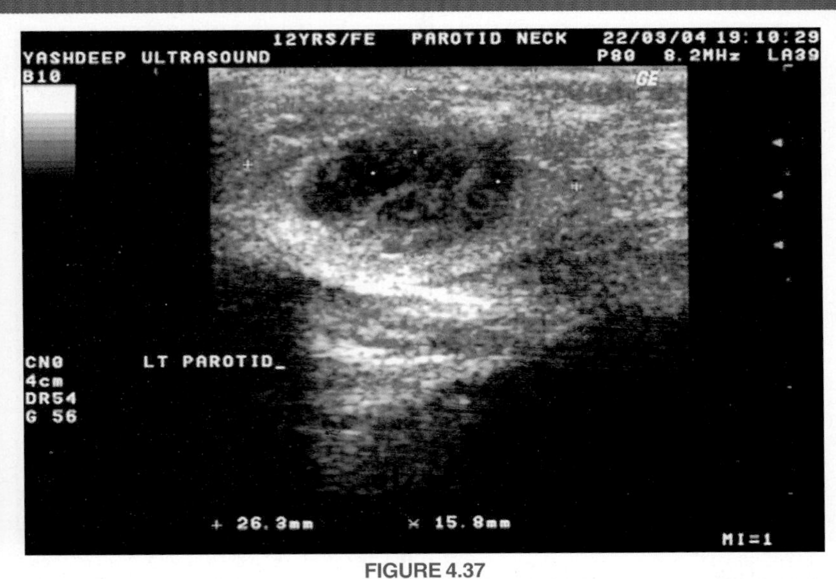

FIGURE 4.37

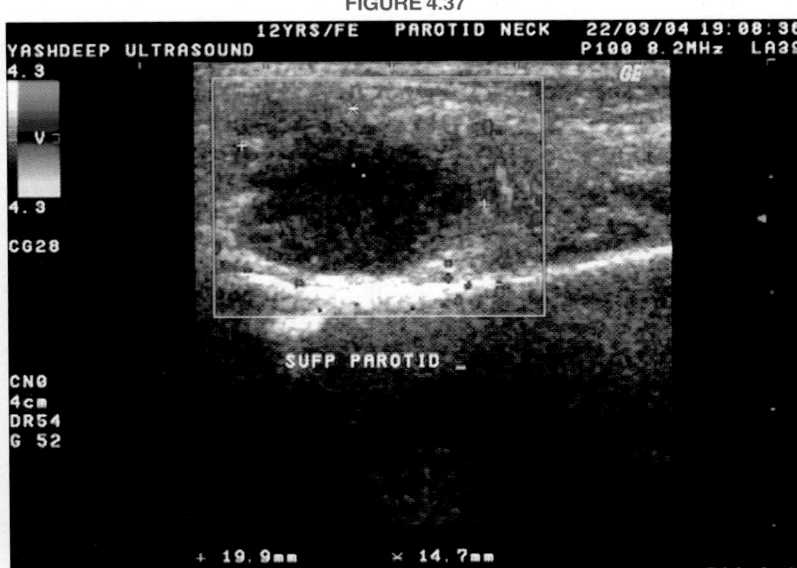

FIGURE 4.38

FIGURES 4.37 to 4.38 Subacute parotid abscess. A young girl presented with soft tissue swelling in the superficial lobe of Lt parotid gland. HRSG shows a thick walled low level echocomplex collection in the superficial lobe suggestive of chronic abscess. On color flow imaging low flow is seen in it.

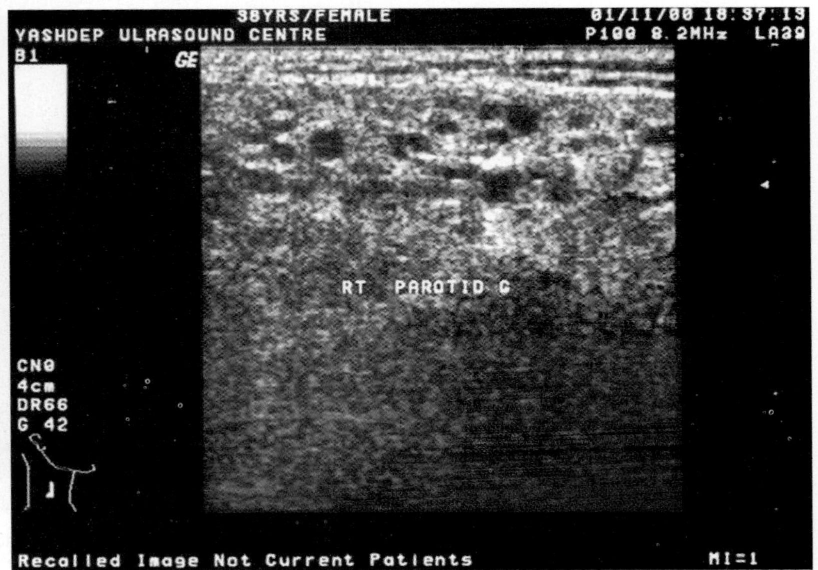

FIGURE 4.39: Chronic recurrent sialectasia with lethiasis. A middle aged women presented with chronic swelling of the Rt parotid gland. HRSG shows enlargement of the gland with multiple hypoechoic areas. Few fine calcified specks are seen in the substance of the gland suggestive of micro lethiasis. But no abscess formation is seen.

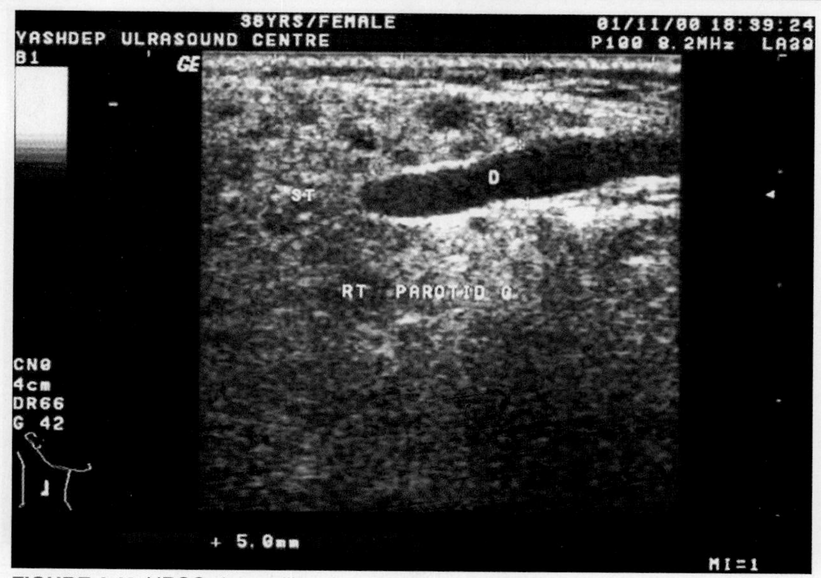

FIGURE 4.40: HRSG shows dilated parotid duct with blocking its mouth by an echogenic calculus.

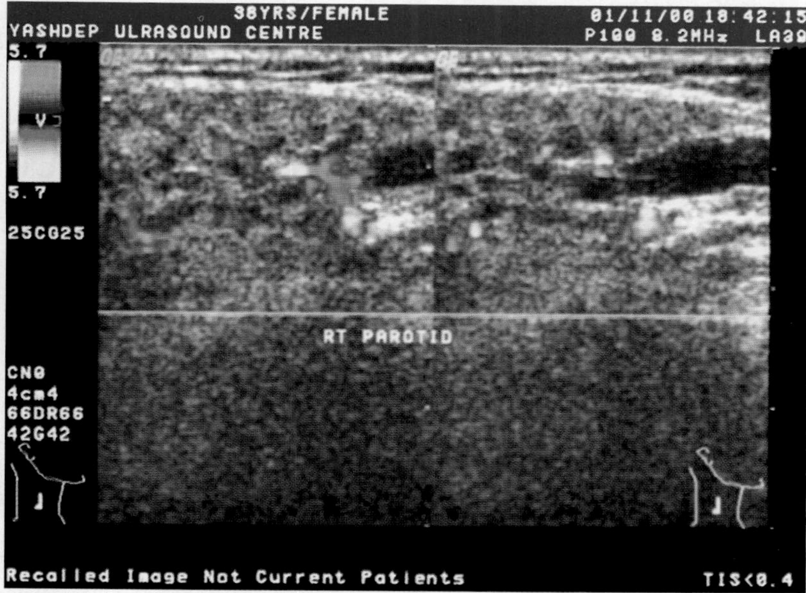

FIGURE 4.41: Color Doppler imaging shows little or poor flow in the gland suggestive of chronic inflammation.

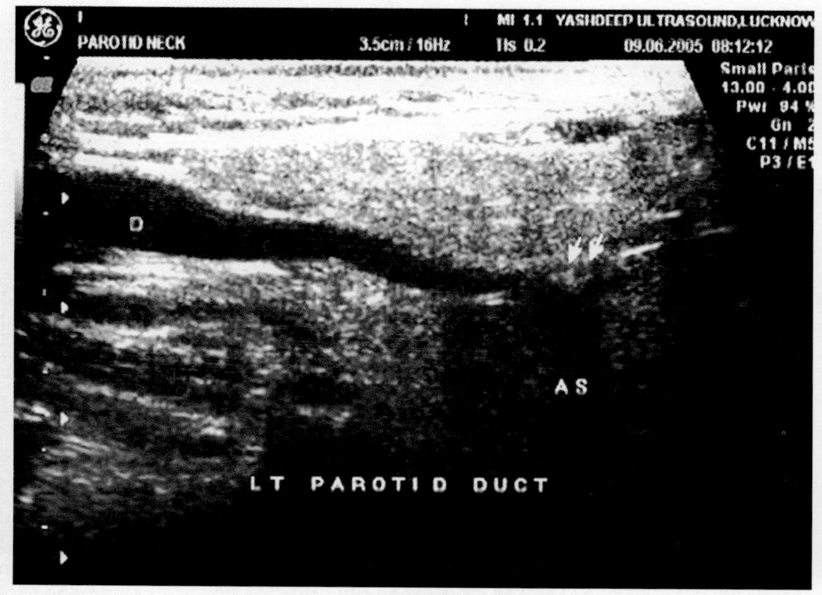

FIGURE 4.42

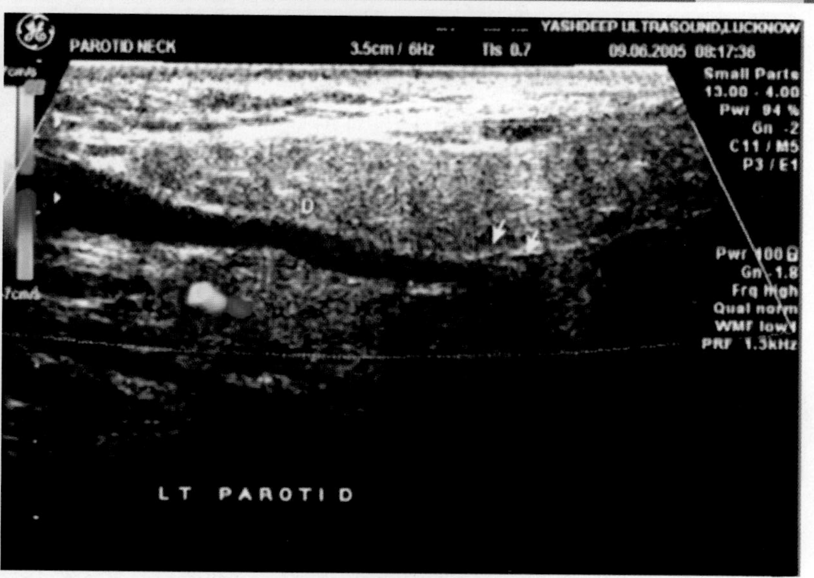

FIGURE 4.43

FIGURES 4.42 and 4.43: Dilated parotid duct with impacted stone at the mouth of the duct. A patient presented with recurrent painful swelling in the Lt parotid region. HRSG shows markedly dilated parotid duct. An echogenic shadow is seen impacted at the mouth of the duct. It is accompanied with acoustic shadowing suggestive of calculus.

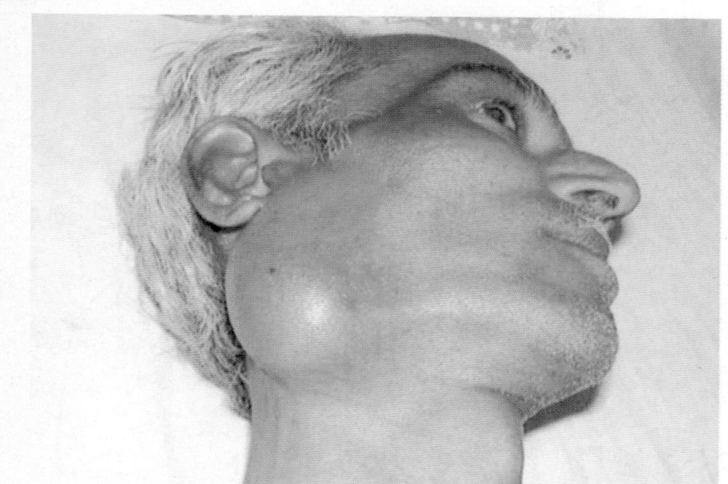

FIGURE 4.44: Parotid abscess. A man presented with soft tissue tender mass over the Rt parotid area.

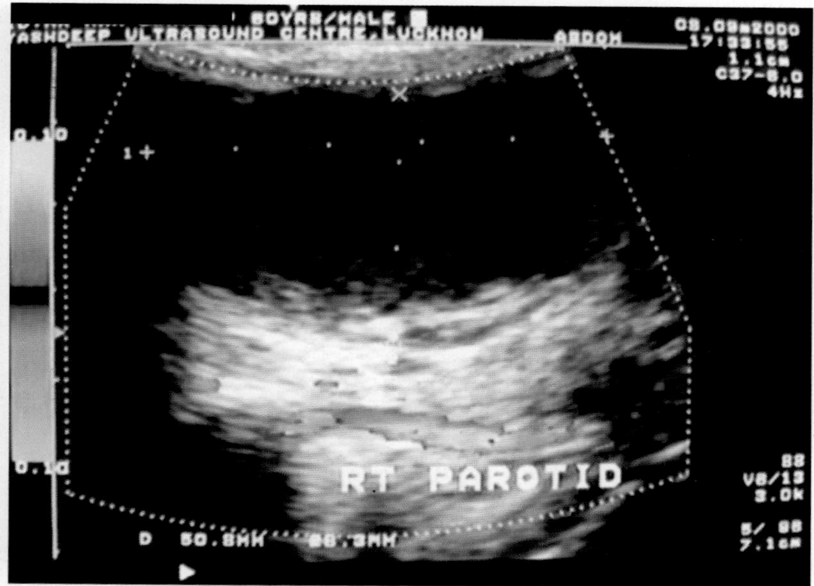

FIGURE 4.45: HRSG shows a low-level echo complex mass in the parotid gland. Central area of necrosis is seen in it. Color flow-imaging shows poor flow in the mass suggestive of parotid abscess.

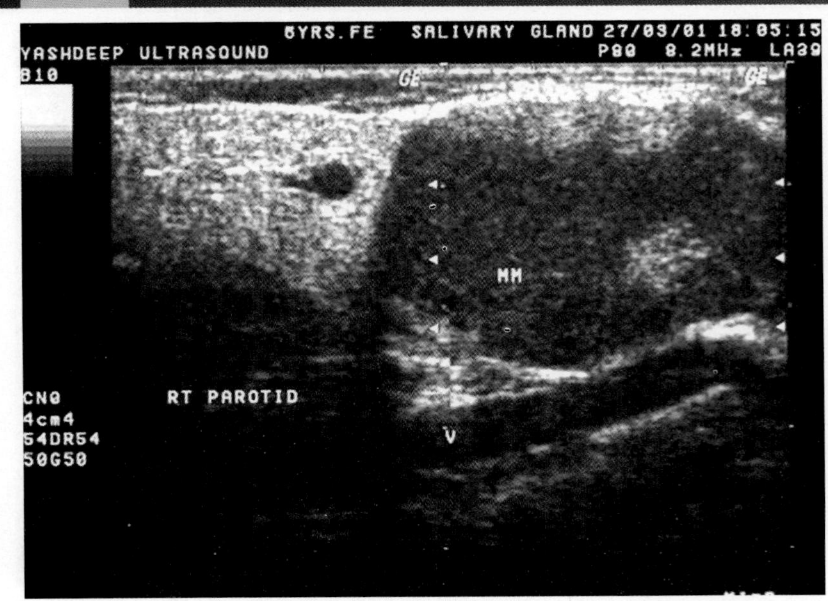

FIGURE 4.46 Parotid abscess. A young girl presented with painful cystic mass. HRSG shows a well-defined echo complex mass in the parotid gland suggestive of an abscess. The Lt parotid gland is normal.

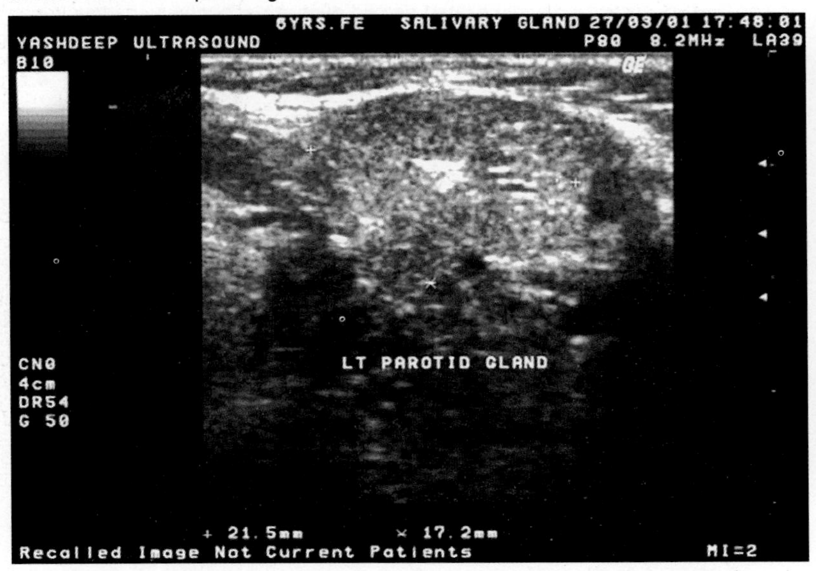

FIGURE 4.47: HRSG of the same patient shows that the abscess is not pressing over the parotid duct. The duct is seen normally. No evidence of any calculus is seen.

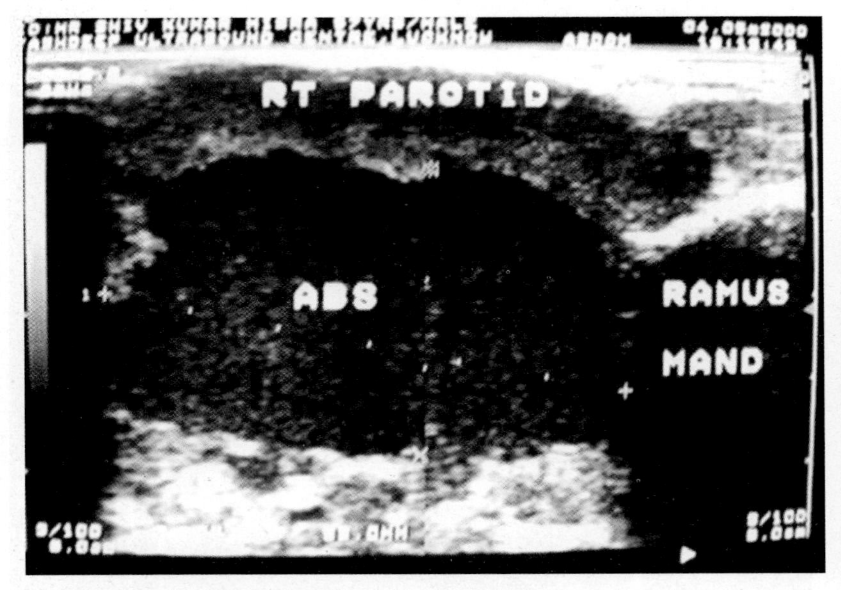

FIGURE 4.48: Parotid abscess. A man presented with painful cystic mass in the parotid region on Rt side. HRSG shows low-level echo complex mass with central area of necrosis in the parotid gland suggestive of parotid abscess.

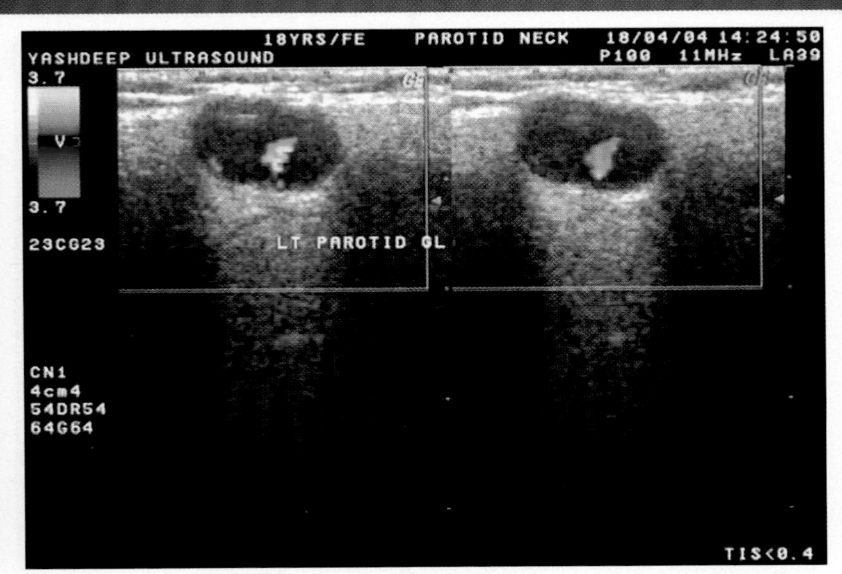

FIGURE 4.49

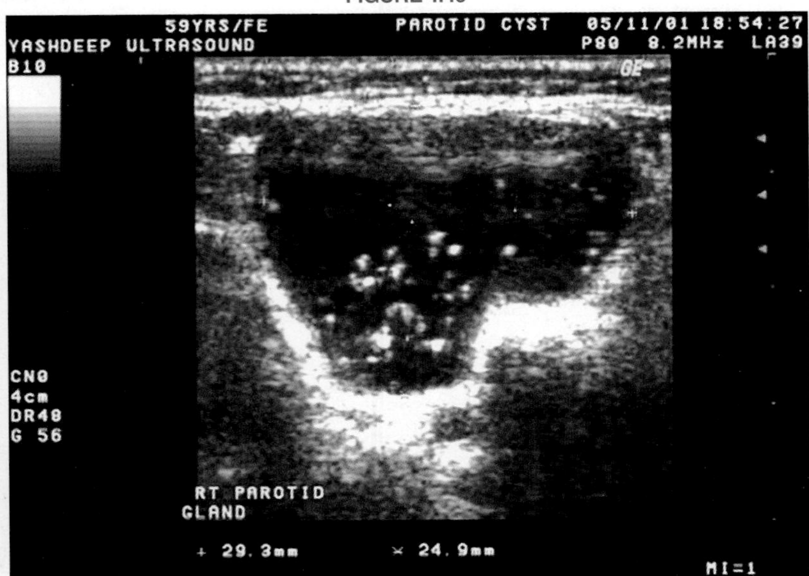

FIGURE 4.50

FIGURES 4.49 and 4.50: Mucoid degeneration in the parotid gland. A old lady presented with a cystic swelling of the Rt parotid gland. HRSG shows a well-defined cystic mass in the parotid gland. The cyst has completely destroyed the normal glandular texture. However, it is having sharp margins. No septation or loculation is seen in the cyst. Multiple bright echogenic specks are seen floating in the cyst. Biopsy of the cyst shows mucoid degeneration.

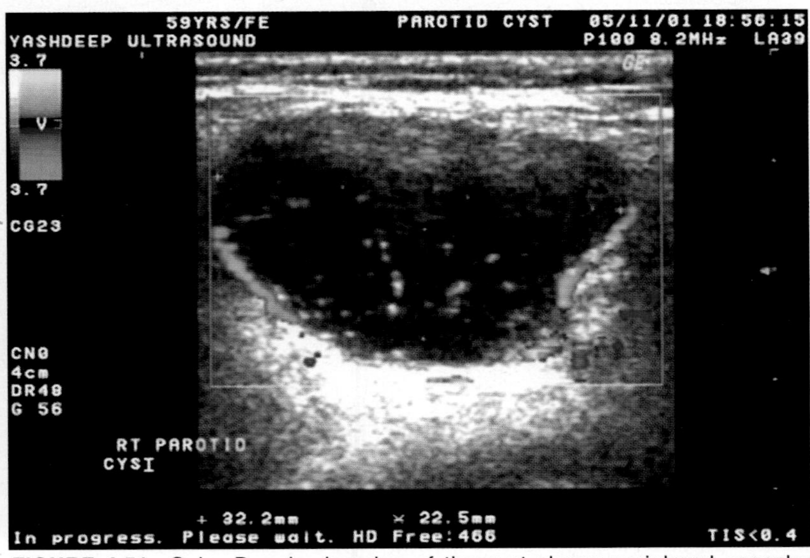

FIGURE 4.51: Color Doppler imaging of the cyst shows peripheral capsular enhancement of the cyst.

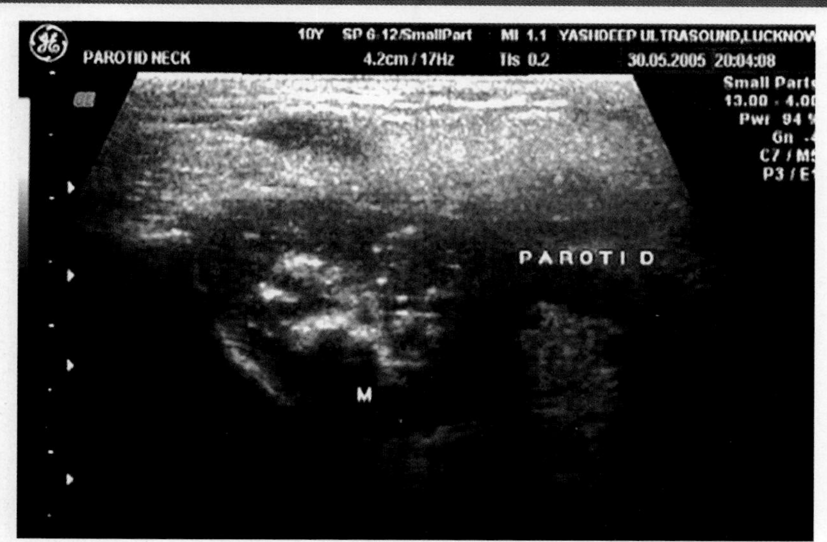

FIGURE 4.52

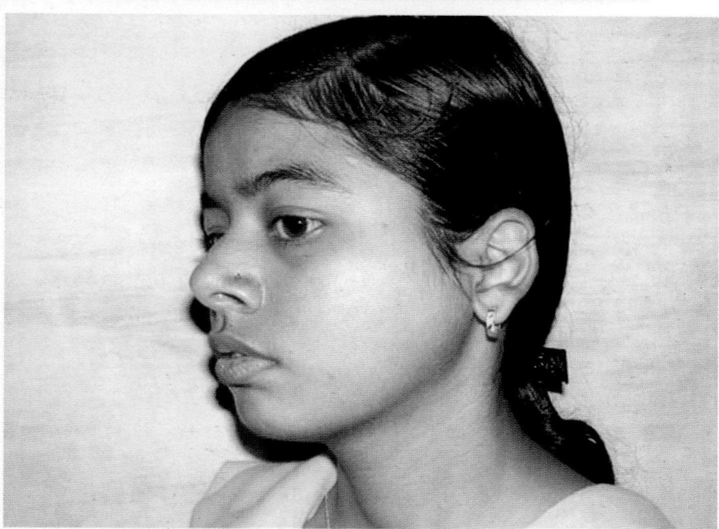

FIGURE 4.55: **Cysticercosis in the cheek muscle.** A young girl presented soft tissue swelling over the Lt cheek muscle.

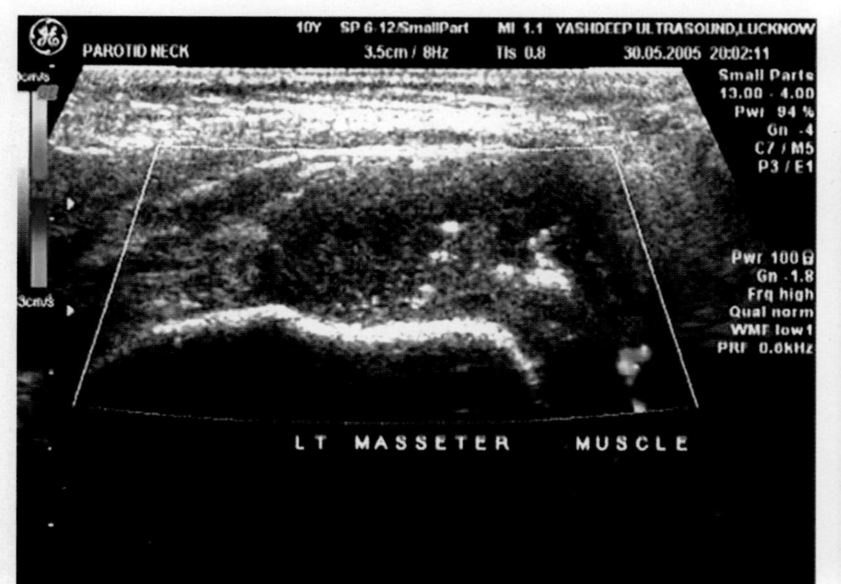

FIGURE 4.53

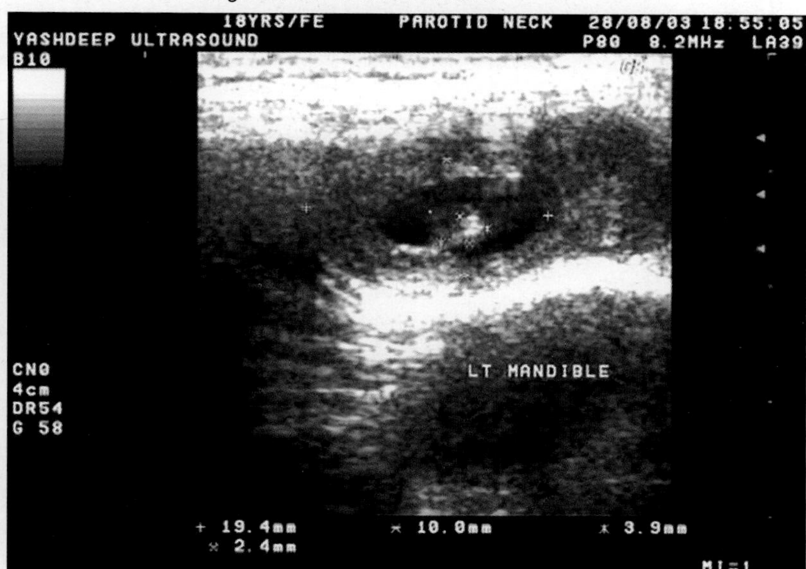

FIGURE 4.56

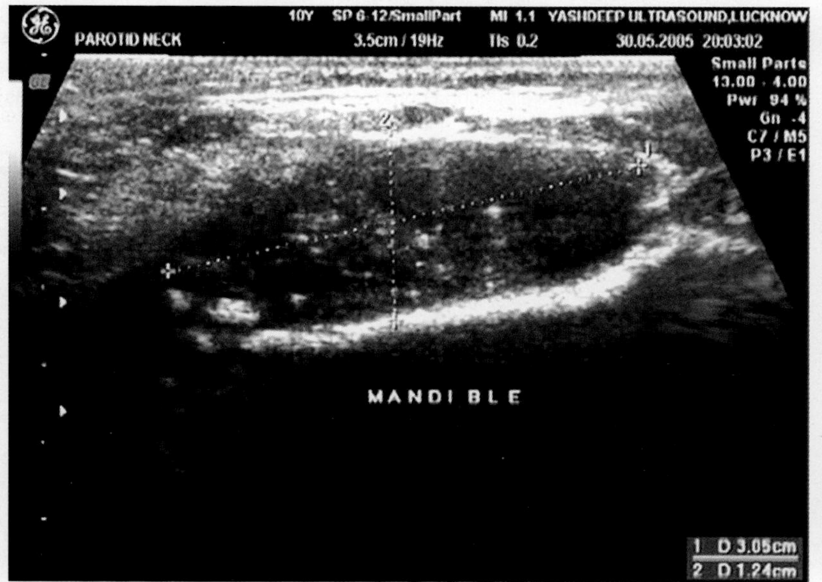

FIGURE 4.54

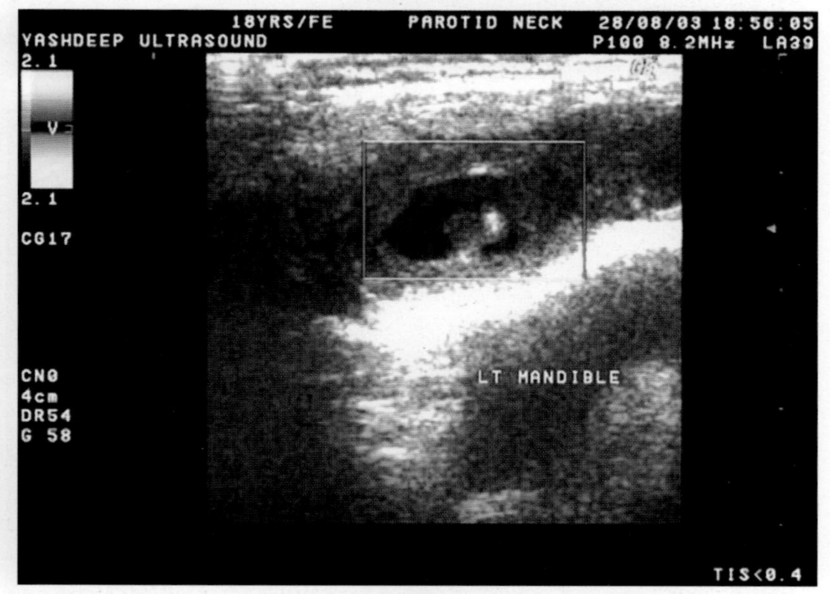

FIGURE 4.57

FIGURES 4.52 to 4.54: Tubercular abscess in the parotid. A young girl presented with chronic non-tender swelling involving the Lt parotid gland. HRSG shows mixed echocomplex mass in the parotid echogenic specks are also seen in the mass. It is seen extending in the Lt masseter muscle. On color flow imaging poor flow is seen in the abscess. Biopsy of the mass confirmed chronic tubercular abscess.

FIGURES 4.56 and 4.57: HRSG shows a well-defined thick wall cystic mass in the masseter muscle. Echogenic nidus is seen in the cyst fixed with the inner wall suggestive of cysticercus cyst.

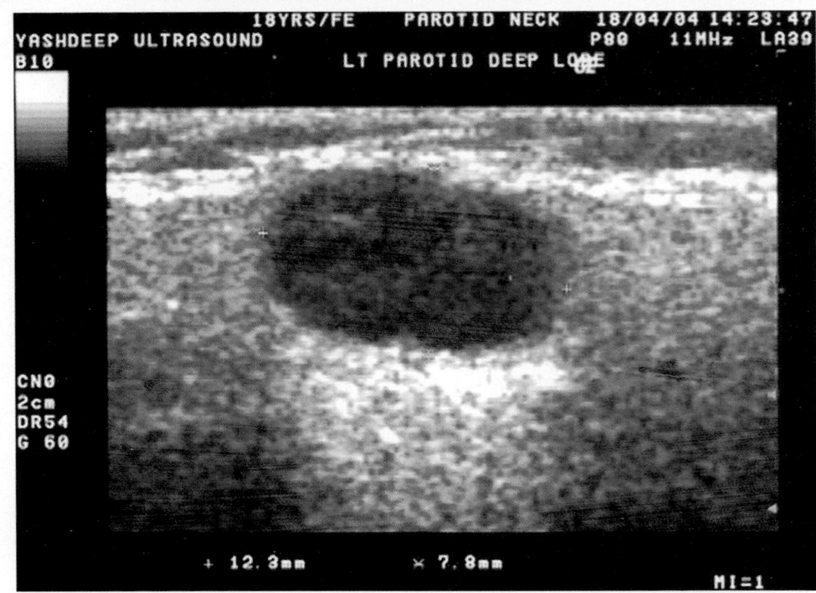

FIGURE 4.58: Benign adenoma parotid. A patient presented with palpable mass in the deep lobe of Lt parotid gland. HRSG shows well-defined homogeneous hypoechoic mass in the gland. Biopsy of the mass showed benign adenoma.

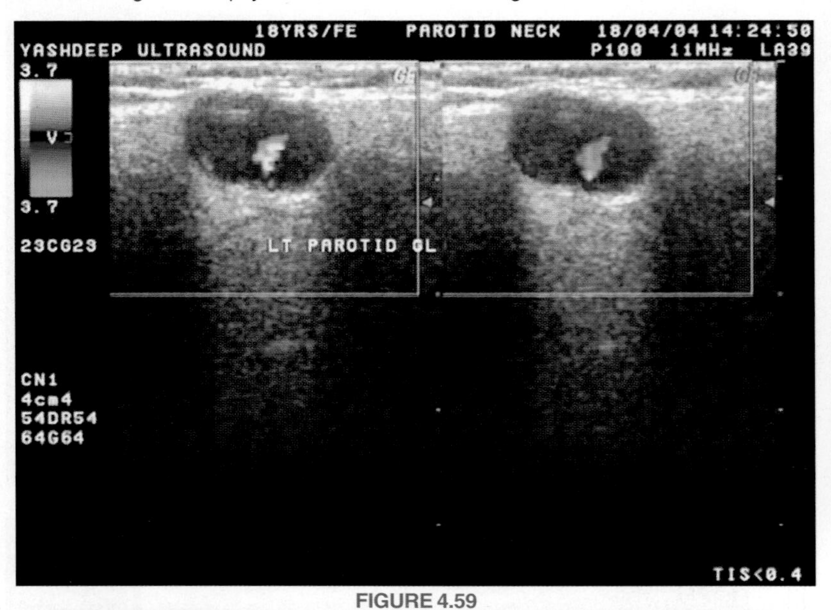

FIGURE 4.59

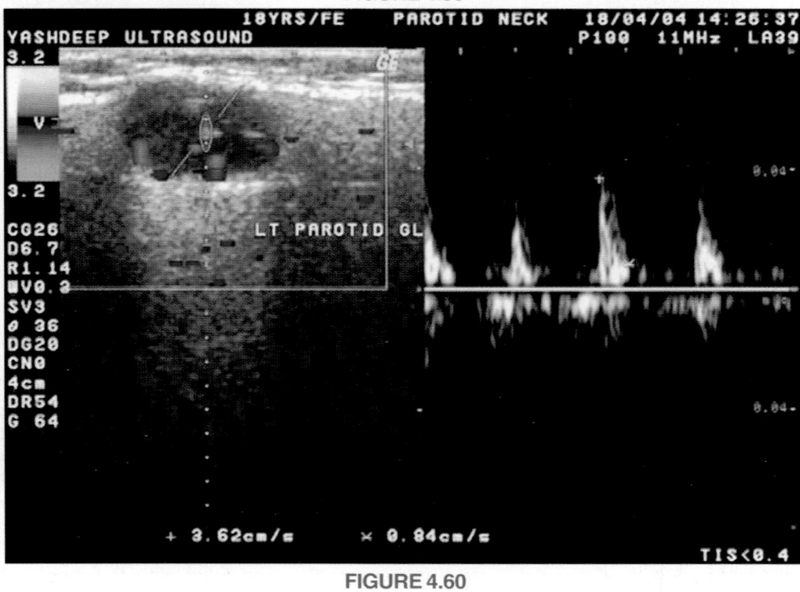

FIGURE 4.60

FIGURES 4.59 and 4.60: There is also evidence of few enlarged nodes seen in the upper cervical region. On color flow imaging normal flow is seen in them.

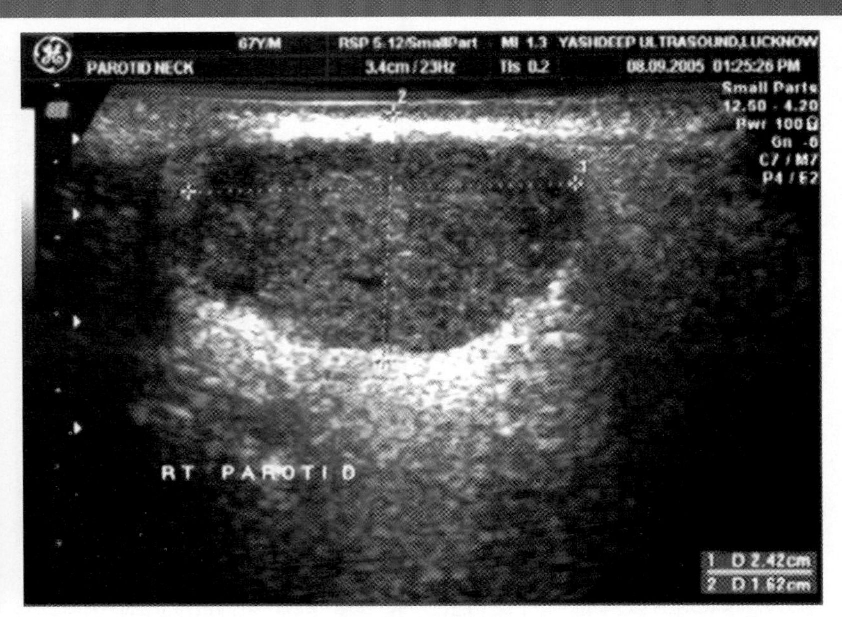

FIGURE 4.61

FIGURE 4.62

FIGURES 4.61 and 4.62: A lady presented with soft mass in the Rt parotid gland. HRSG of the gland shows well-defined homogeneous mass. It shows sharp margins. No necrosis or calcification is seen in it.

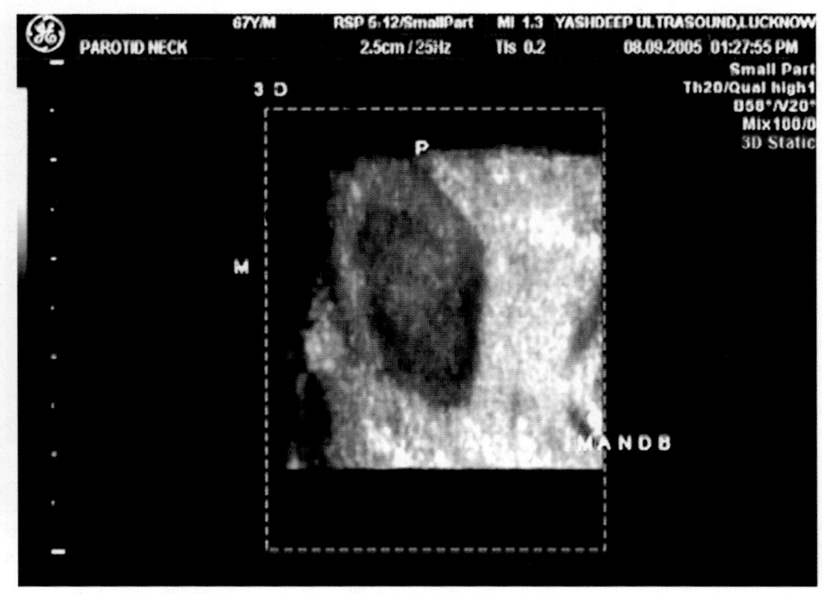

FIGURE 4.63: 3D of the same mass shows better details of the mass.

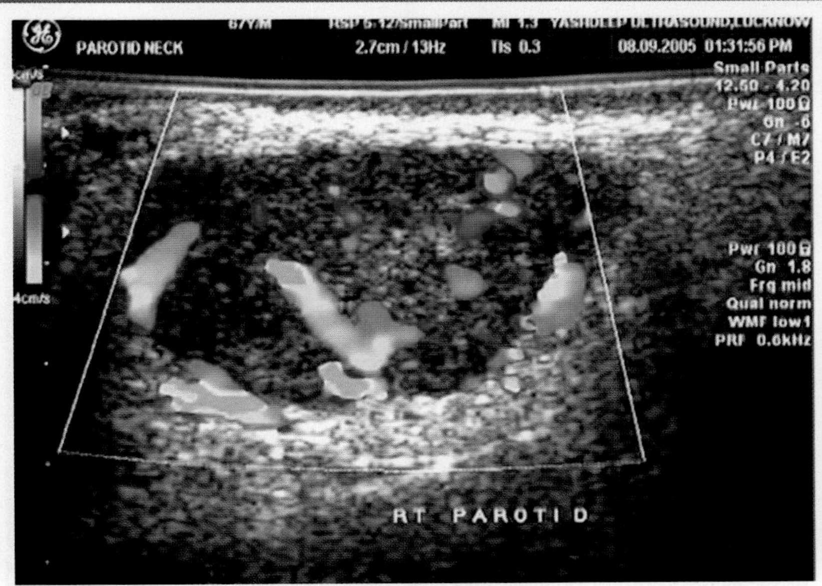

FIGURE 4.64

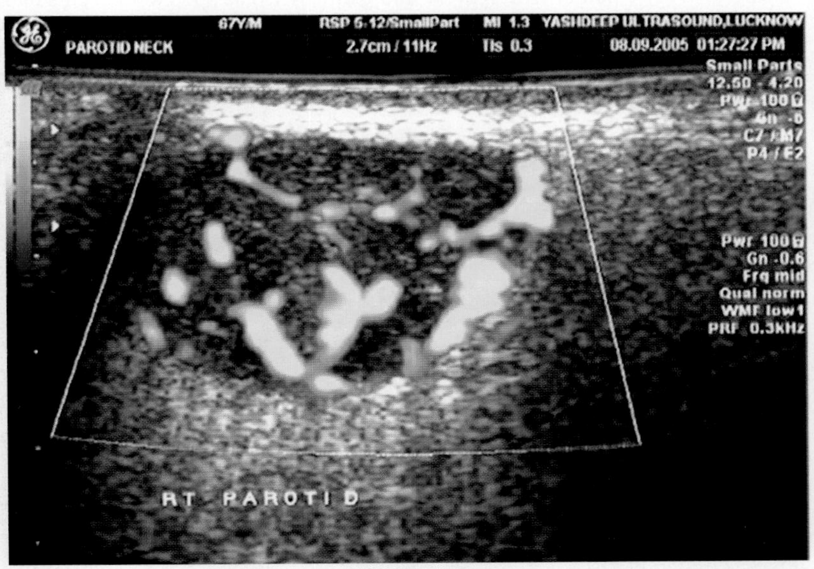

FIGURE 4.65

FIGURES 4.64 and 4.65: Color flow imaging of the mass shows increased flow in the mass. Biopsy of the mass proved benign pleomorphic adenoma.

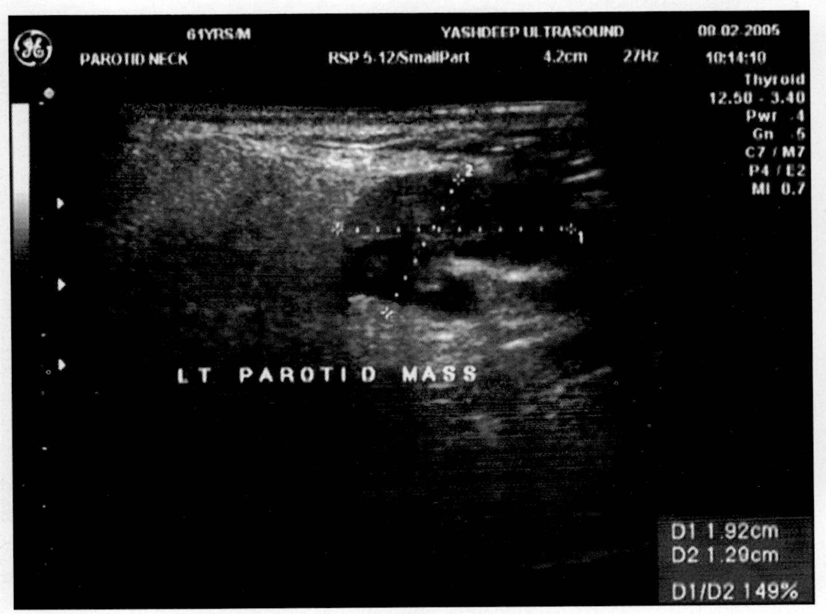

FIGURE 4.66

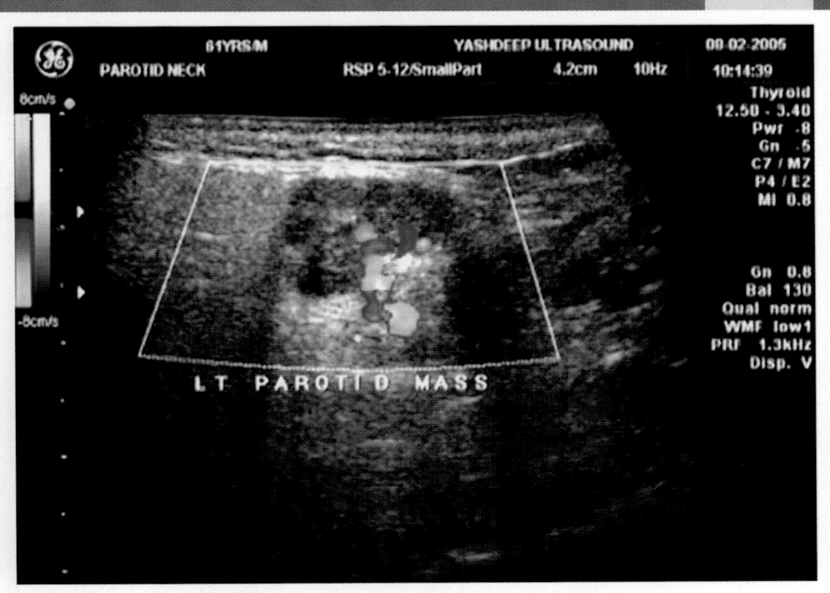

FIGURE 4.67

FIGURES 4.66 and 4.67: Benign adenoma. Homogeneous soft tissue mass seen in the Lt parotid gland with sharp margins. Color flow imaging shows moderate flow in the mass. Biopsy shows mixed tumor of the parotid gland.

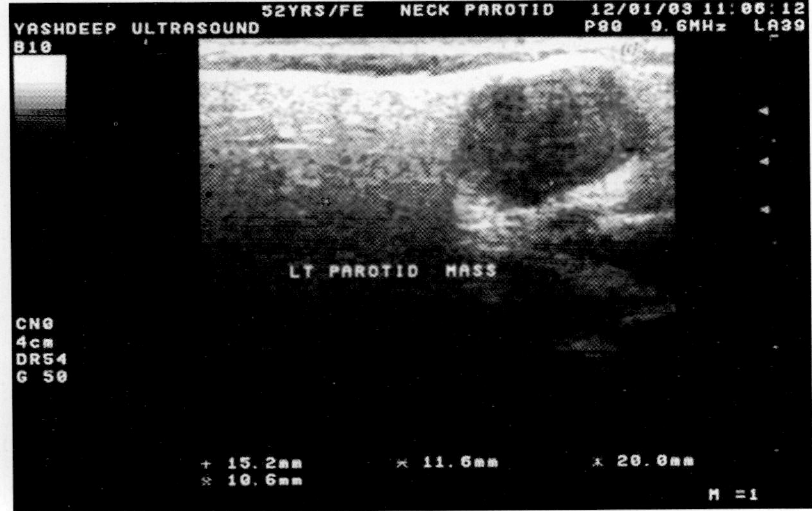

FIGURE 4.68

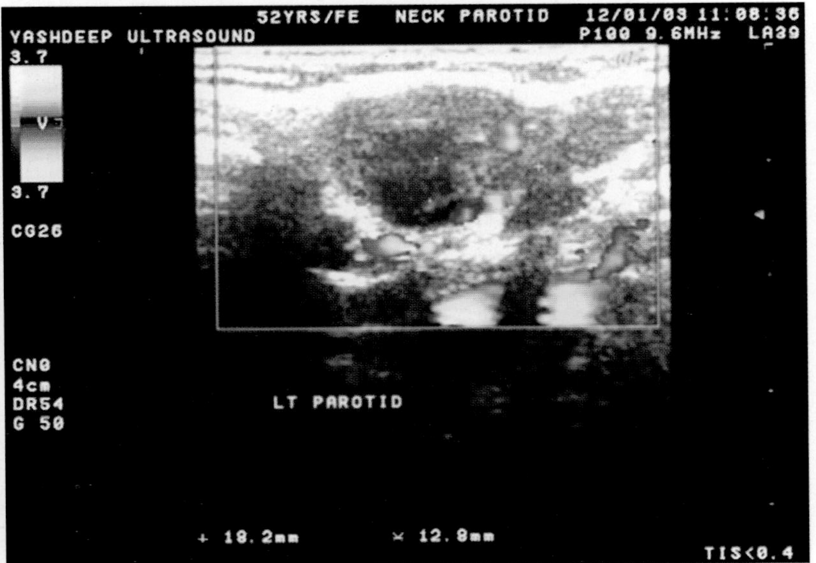

FIGURE 4.69

FIGURES 4.68 and 4.69: A middle aged lady presented with slowly growing mass in the Lt parotid gland. HRSG shows homogeneous soft tissue mass involving the parotid gland. No necrosis is seen in the mass. On color flow imaging moderate flow is seen in the mass. Biopsy shows mixed tumor of the parotid gland.

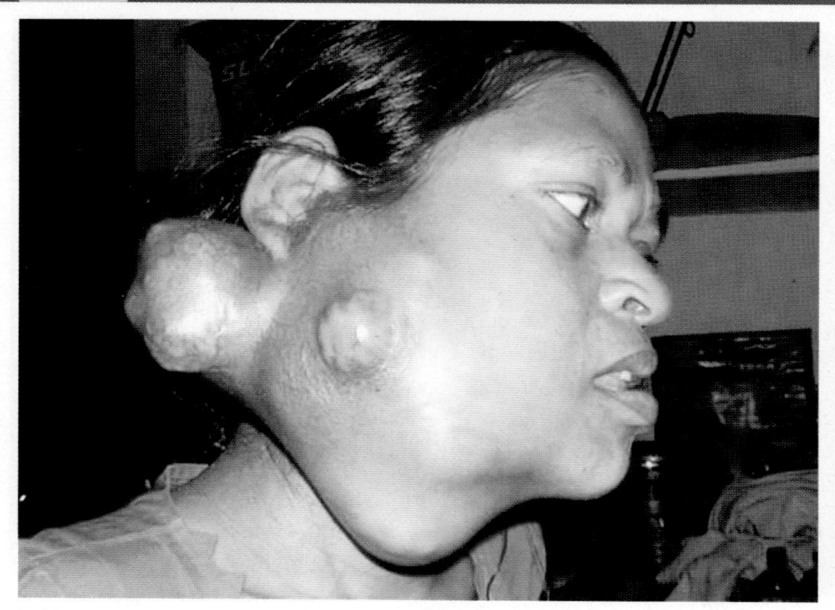

FIGURE 4.70: A lady presented with big lobulated nodular masses involving the Rt parotid gland.

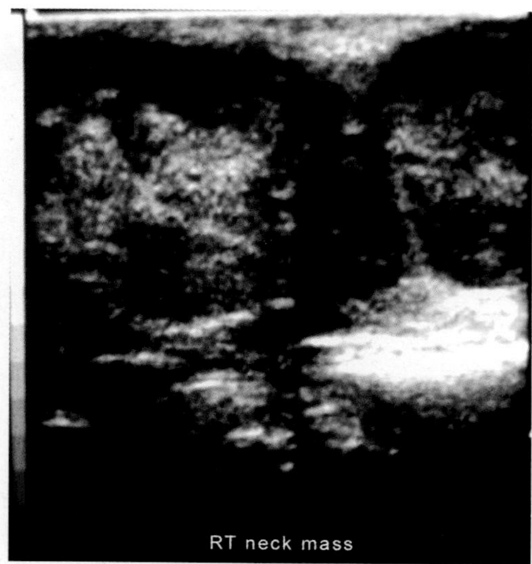

FIGURE 4.71

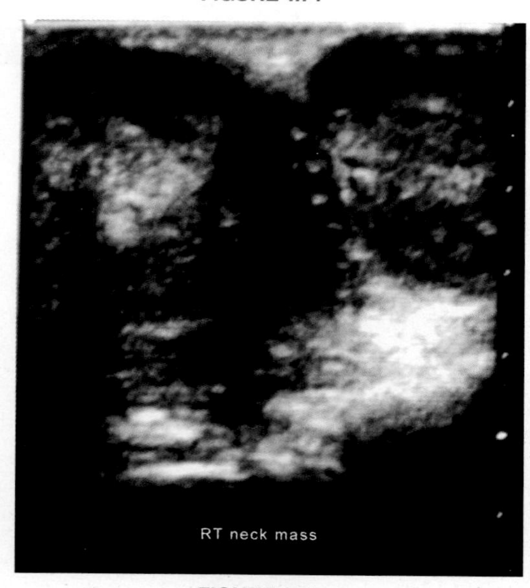

FIGURE 4.72

FIGURES 4.71 and 4.72: HRSG shows multilobulated soft tissue mass involving the whole of parotid gland. No necrosis is seen in them. No cystic degeneration is seen. Biopsy showed pleomorphic adenoma.

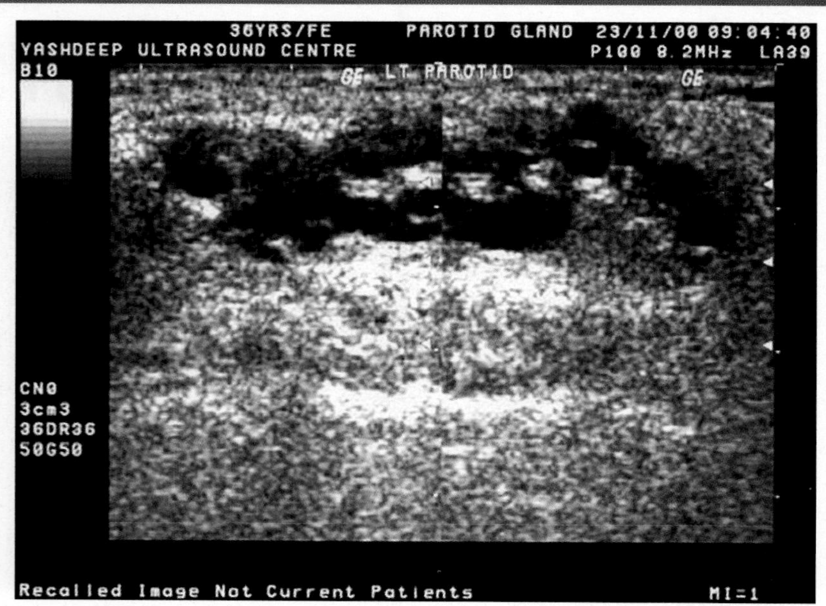

FIGURE 4.73: **AV malformation of parotid gland.** A middle aged women presented with a pulsatile mass swelling in the Lt parotid gland. HRSG shows multiple dilated vessels in the gland.

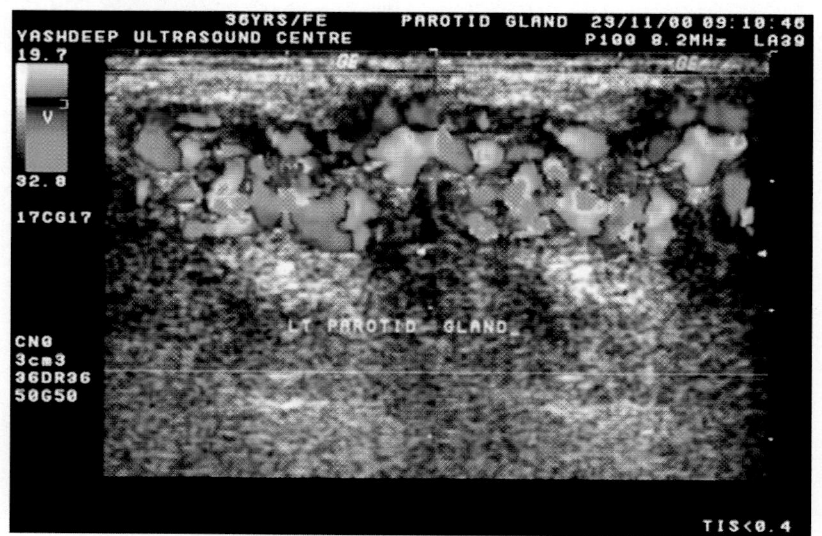

FIGURE 4.74

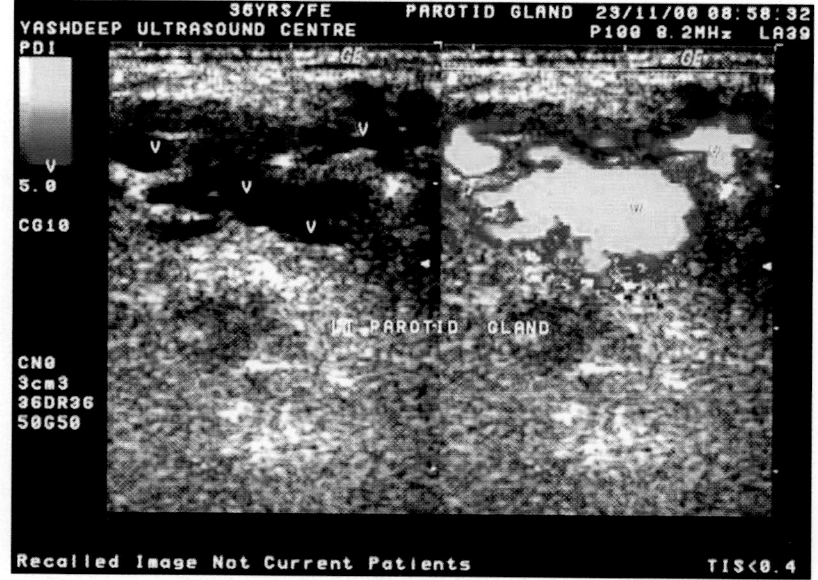

FIGURE 4.75

FIGURES 4.74 and 4.75: Color Doppler imaging of the same patient shows high blood flow in the vessels filling whole of the gland. Power Doppler imaging shows irregular filling of the sinusoides.

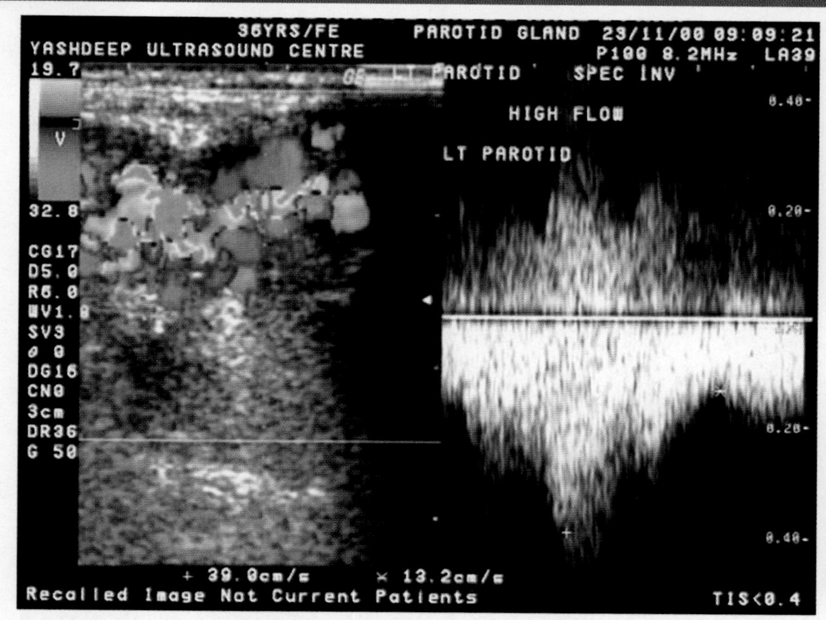

FIGURE 4.76: Spectral Doppler imaging shows high flow in the vessels running through the gland with arterial and venous flow pattern. Doppler findings are highly suggestive of AV malformation in the parotid gland.

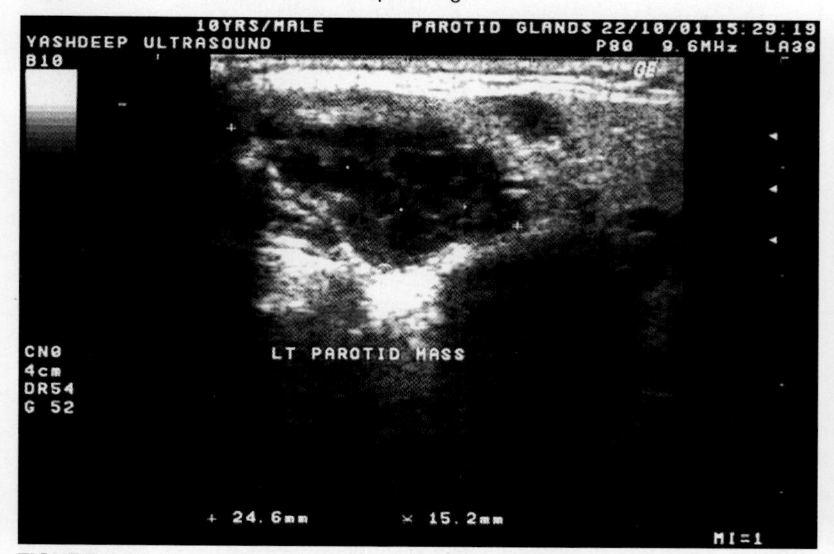

FIGURE 4.77: Cavernous venous malformation in the parotid gland. A young boy presented with a cystic mass in the Lt parotid gland. HRSG shows multiple dilated vessels in the gland. Normal glandular texture is lost.

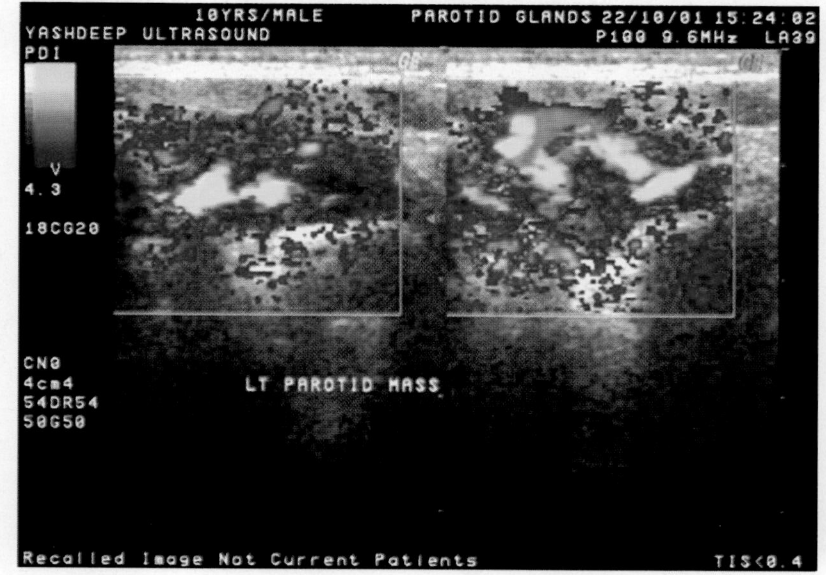

FIGURE 4.78: Power Doppler imaging shows low flow in the blood vessels suggestive of venous flow.

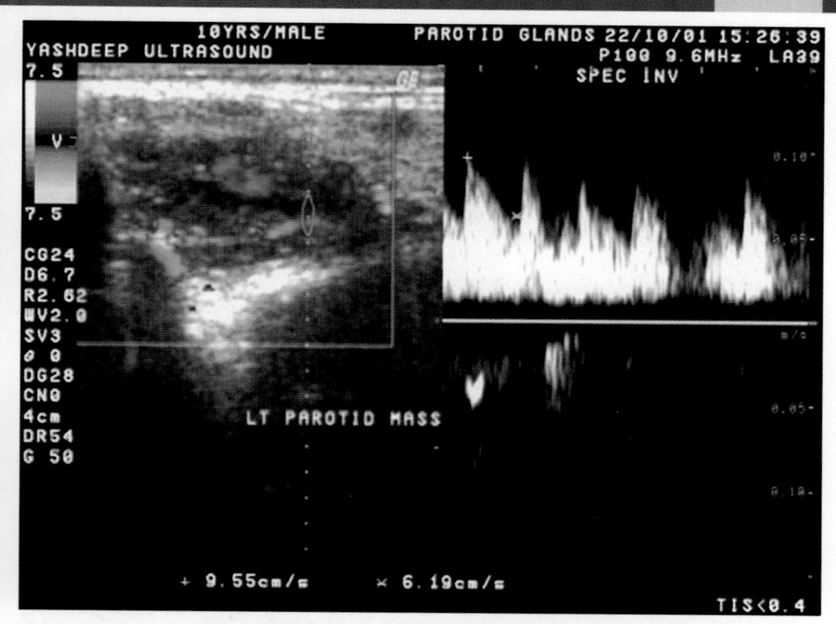

FIGURE 4.79: Spectral Doppler flow-imaging shows pulsatile venous flow in the mass. Biopsy shows cavernous venous malformation.

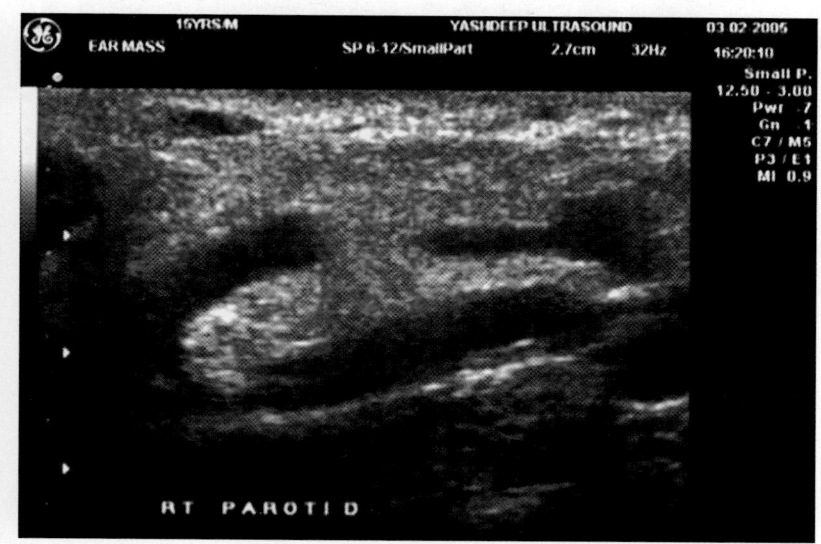

FIGURE 4.80: Parotid AVM. A patient presented with swelling over the Rt parotid region and also thickened pinna of the Rt ear. HRSG shows smooth enlargement of Rt parotid gland. Multiple dilated vessels are seen in the gland.

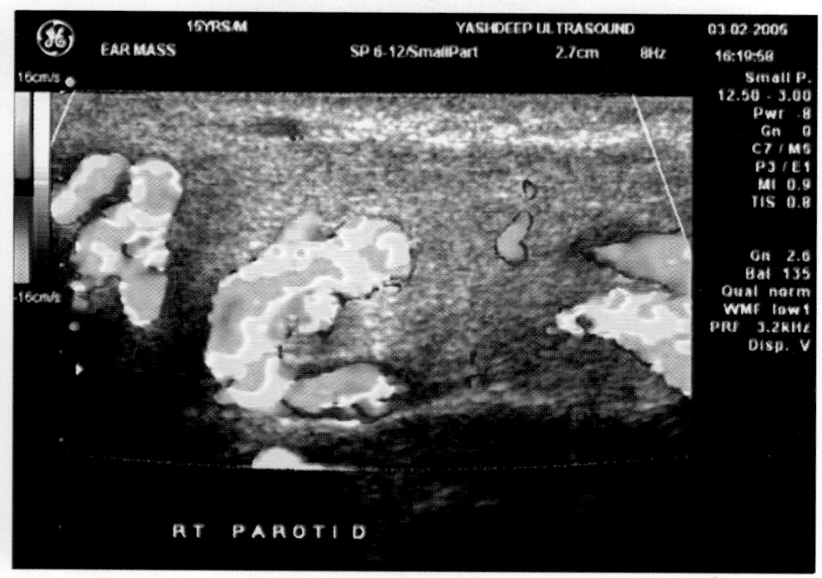

FIGURE 4.81

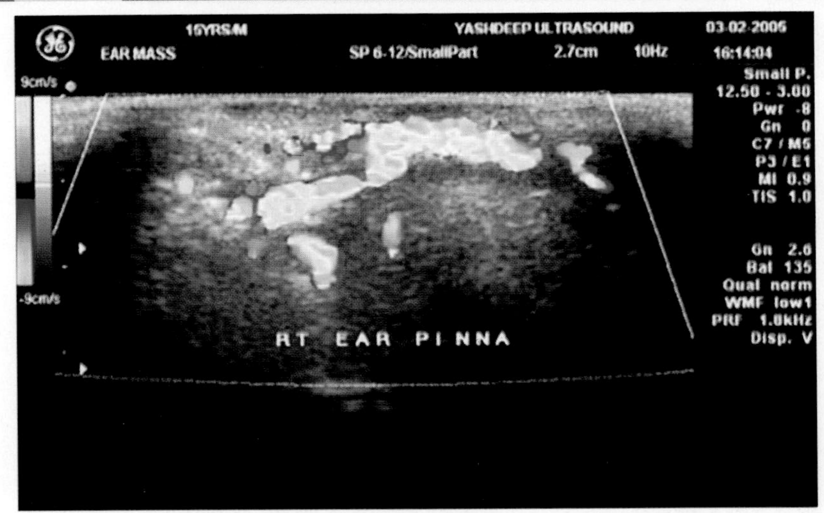

FIGURE 4.82

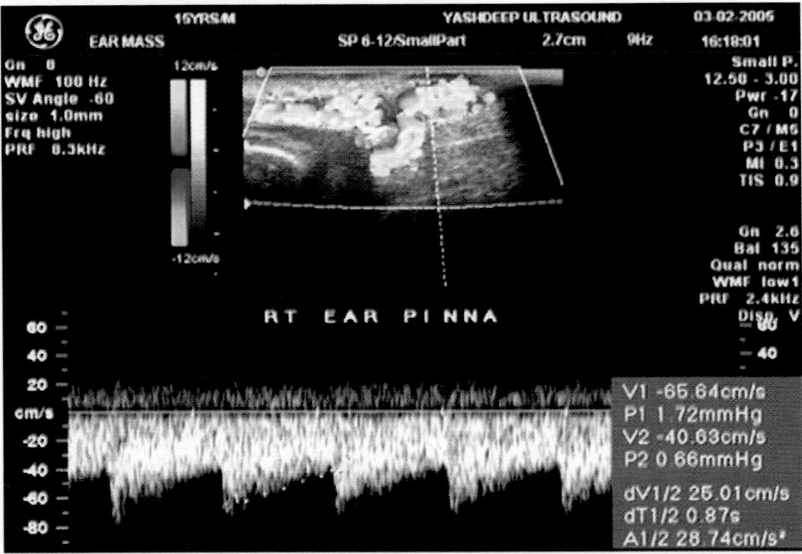

FIGURE 4.83

FIGURES 4.81 to 4.83: Color Doppler imaging shows dilated vessels in the parotid gland. The vessels are also seen feeding pinna of the ear. On spectral Doppler tracing increased flow seen in the vessels confirmed the finding of AVM.

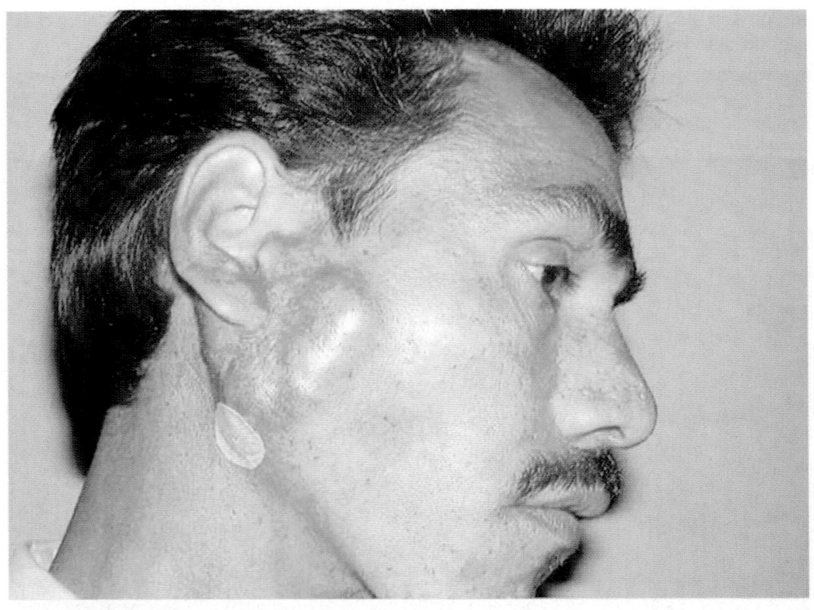

FIGURE 4.84: Adenocarcinoma of Parotid. A patient presented with irregular soft tissue mass over the Rt parotid area.

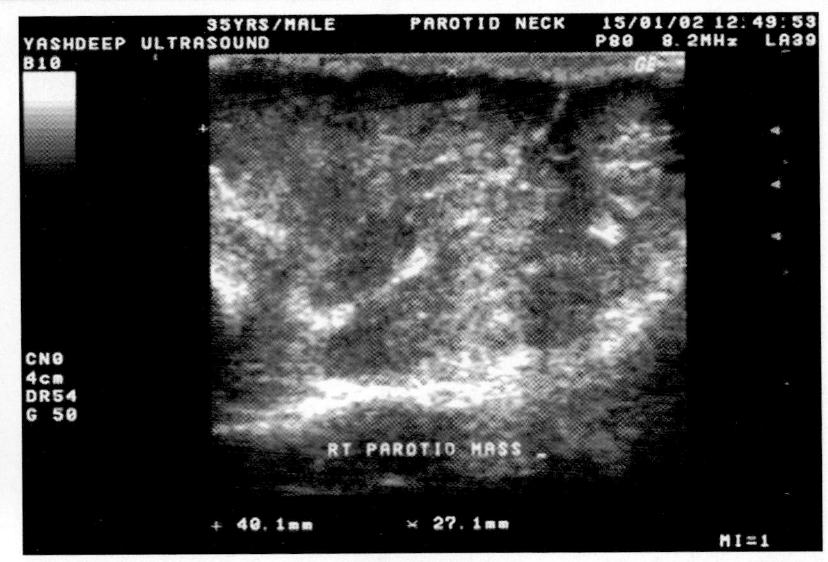

FIGURE 4.85A

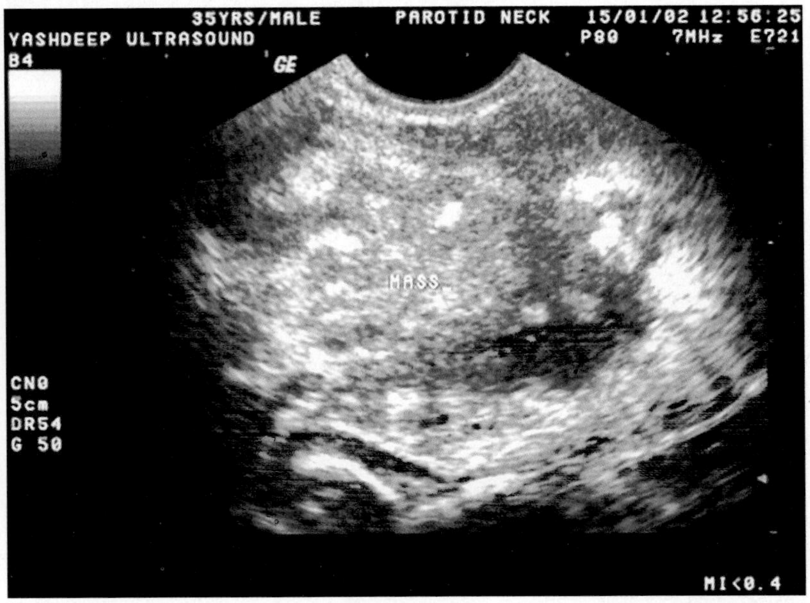

FIGURE 4.85B

FIGURES 4.85A and B: HRSG shows a big irregular heterogeneous mass invading the parotid gland. The mass is seen invading the deep lobe of parotid. The capsule of the gland was also broken. Biopsy confirmed invasive adenocarcinoma.

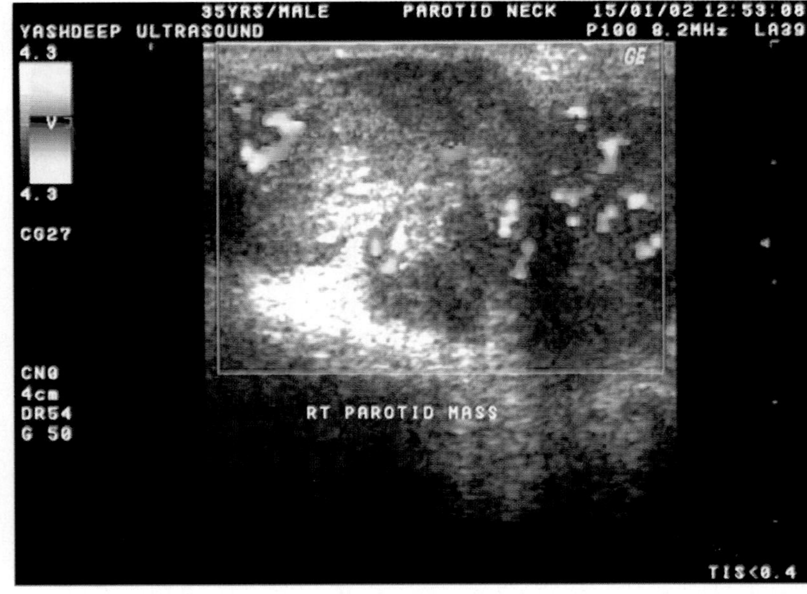

FIGURE 4.86

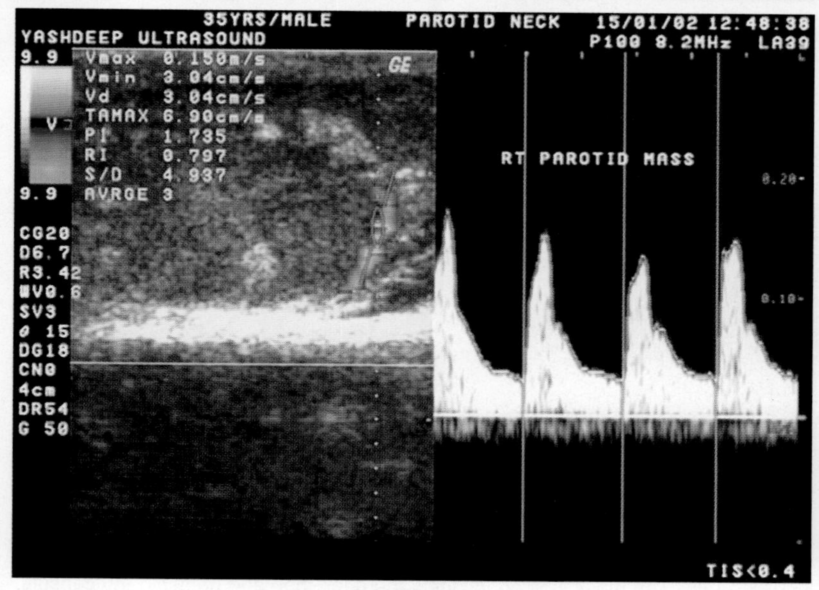

FIGURE 4.87

FIGURES 4.86 and 4.87: On color flow imaging increased vascularity of the mass is seen. On spectral Doppler tracing high flow is seen in the mass.

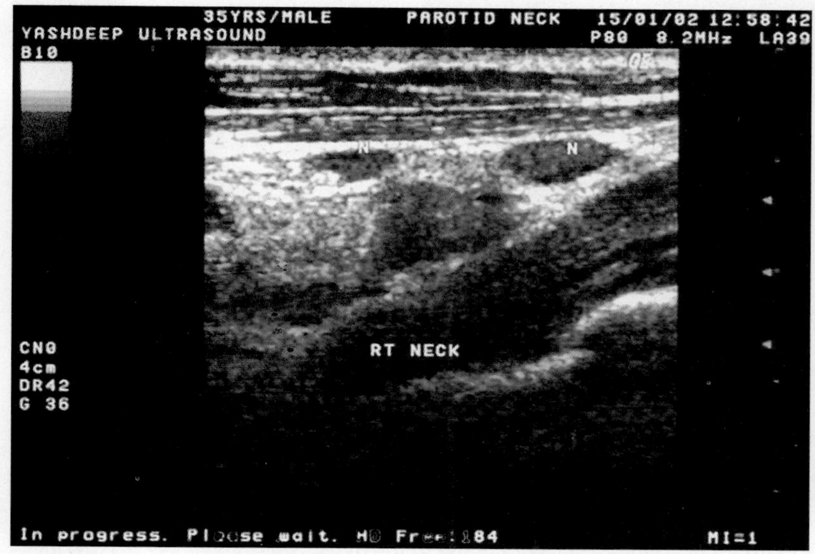

FIGURE 4.88: Multiple enlarged nodes are also seen in the mid and upper cervical region. The lymph nodes hylum is lost. Normal lymph nodes texture is also distorted.

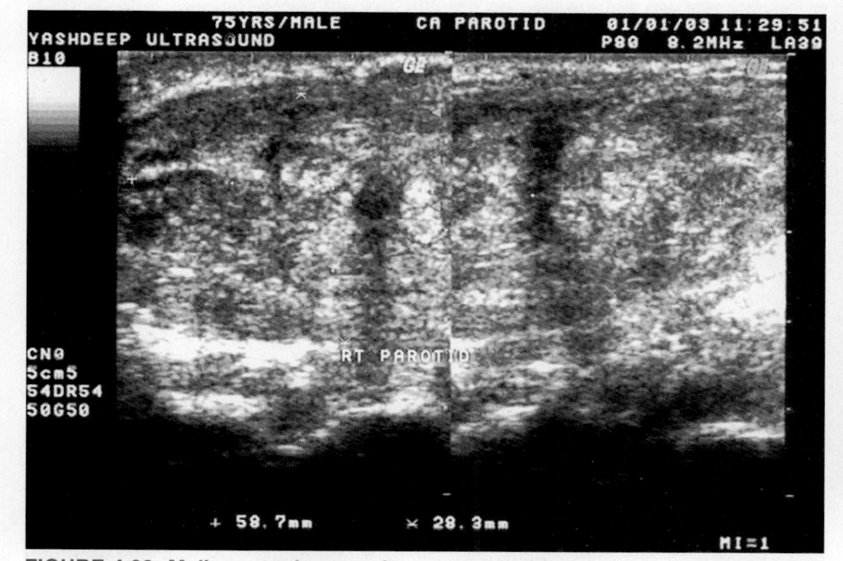

FIGURE 4.89: Malignant adenocarcinoma parotid. A patient presented with a big mass in the Rt parotid gland. The mass has totally destroyed the gland. The capsule is broken and it is seen invading the adjacent tissue planes. Biopsy confirmed invasive adenocarcinoma.

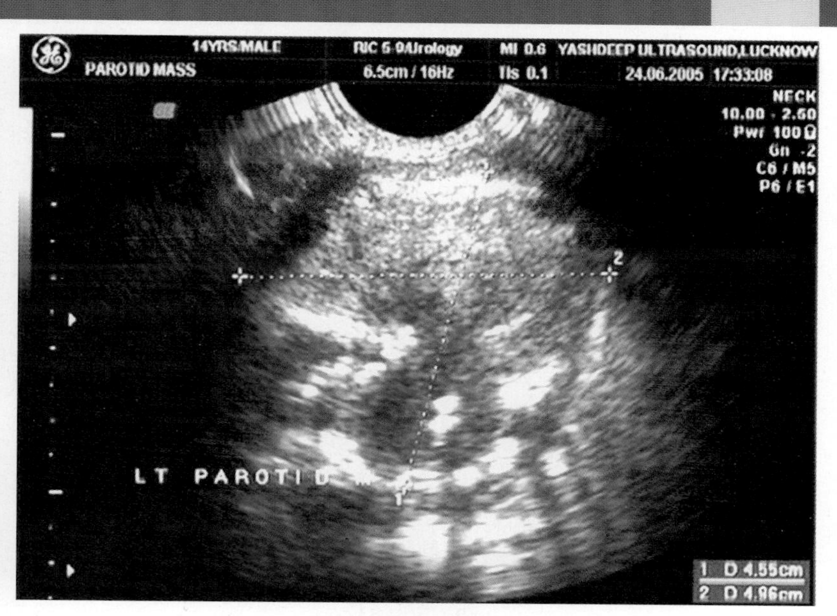

FIGURE 4.90

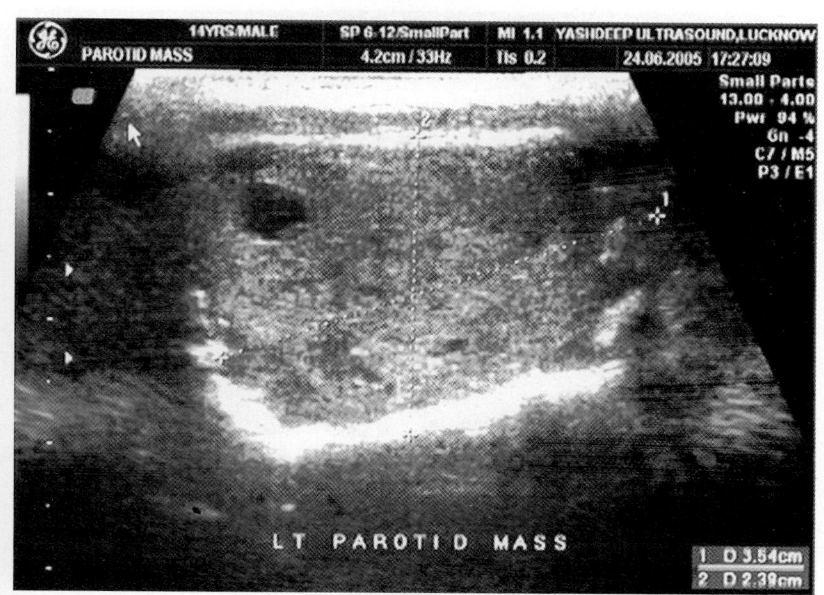

FIGURE 4.91

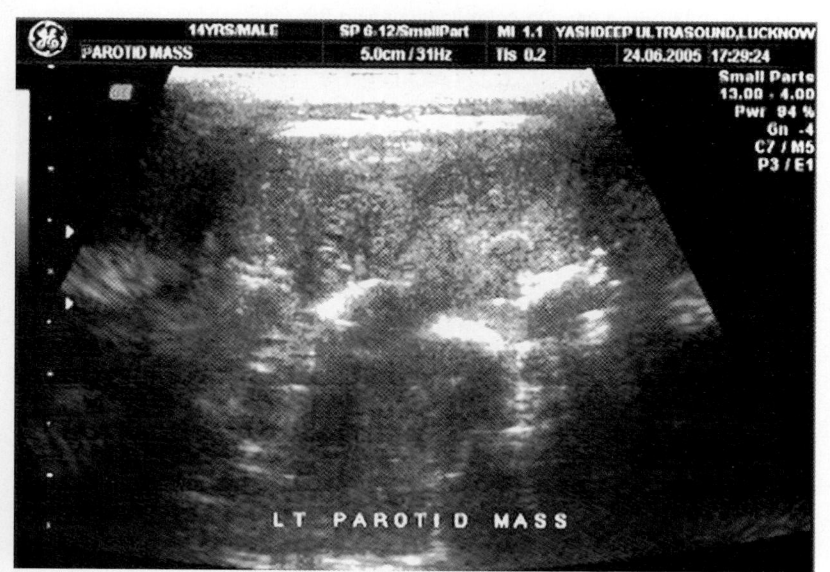

FIGURE 4.92

FIGURES 4.90 to 4.92: Invasive adenocarcinoma. A young child presented with a big soft tissue mass in the Lt parotid region. The mass is coming out from the parotid gland. It is seen broken the capsule of the gland and invading the adjacent structure. Normal glandular texture is lost. Echogenic calcification is also seen in the mass. Biopsy showed invasive adenocarcinoma.

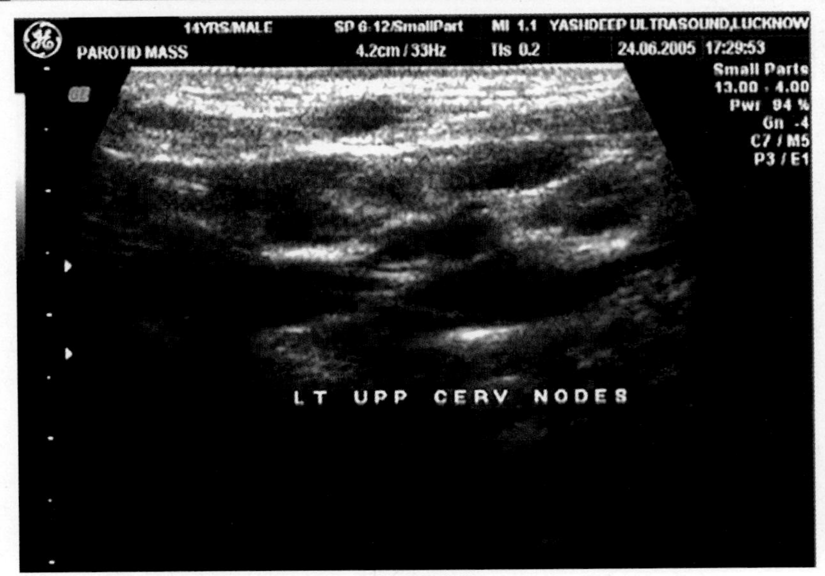

FIGURE 4.93

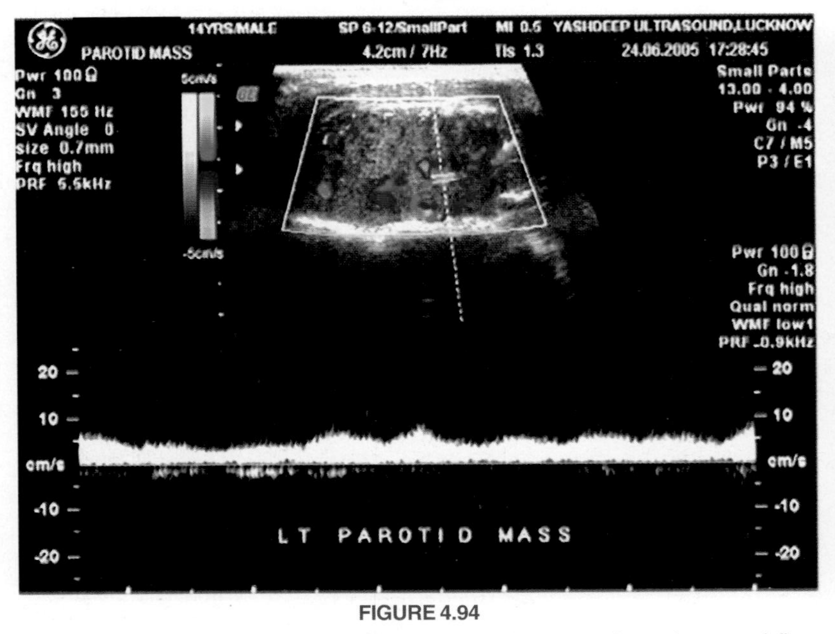

FIGURE 4.94

FIGURES 4.93 and 4.94: On color flow imaging the mass shows increased flow. Multiple enlarged nodes are also seen in the upper cervical region.

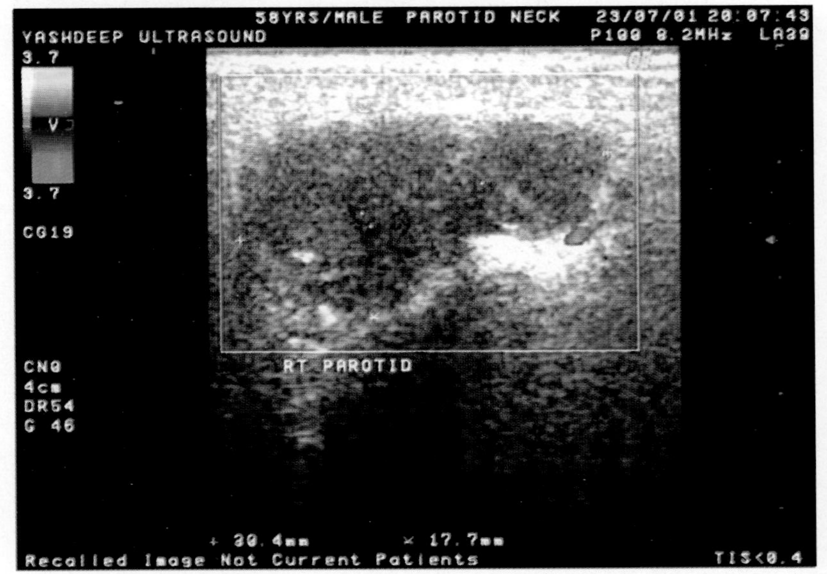

FIGURE 4.95: Adenocarcinoma of parotid gland. An old man presented an irregular soft tissue mass in the Rt parotid gland. Normal glandular texture is lost. Capsule is broken. It is seen invading the deep lobe of the gland. Color Doppler imaging of the mass shows little or no flow in the mass.

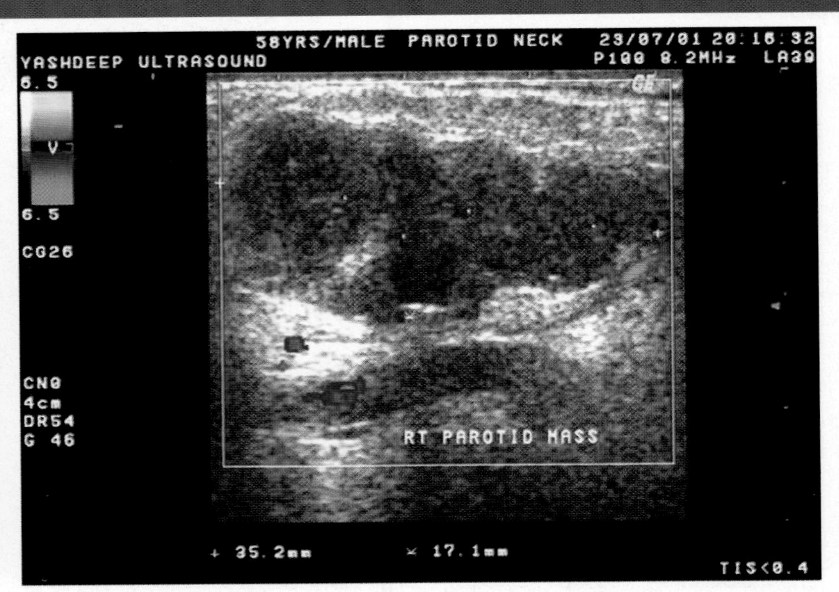

FIGURE 4.96: The mass is seen invading the deep lobe and breaking the capsule. Biopsy of the mass shows adenocarcinoma.

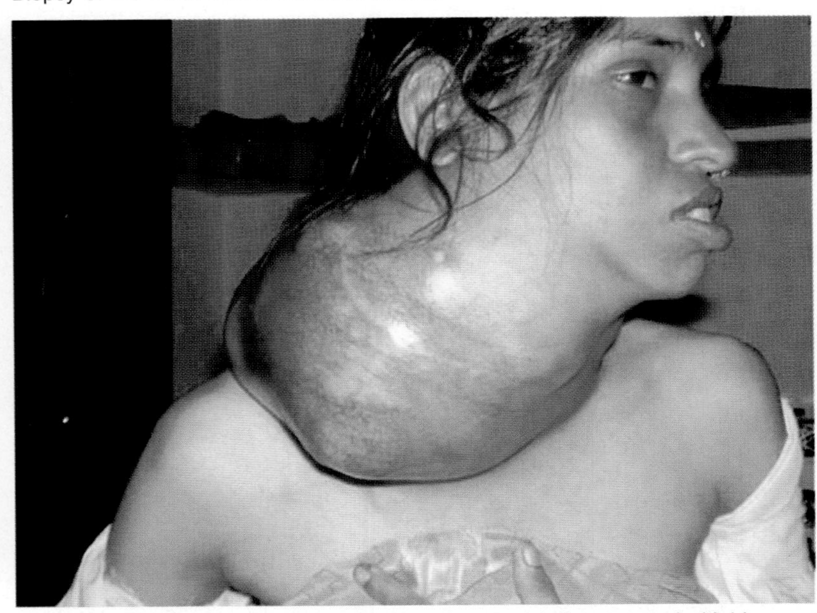

FIGURE 4.97: Spindle cell carcinoma of the neck. A lady presented with big mass in the neck region.

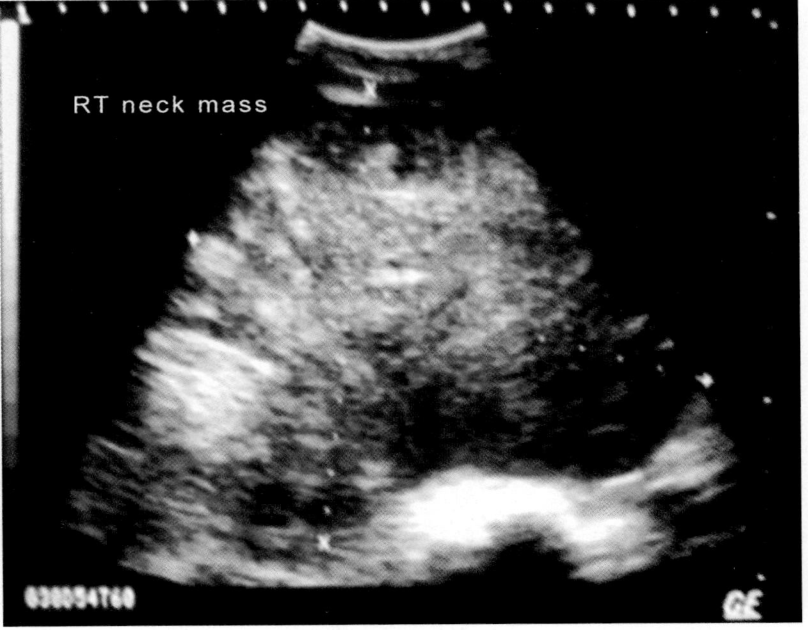

FIGURE 4.98

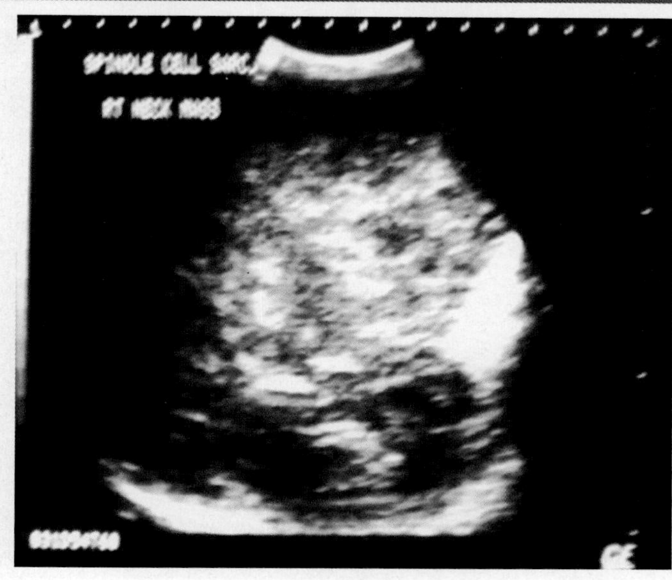

FIGURE 4.99

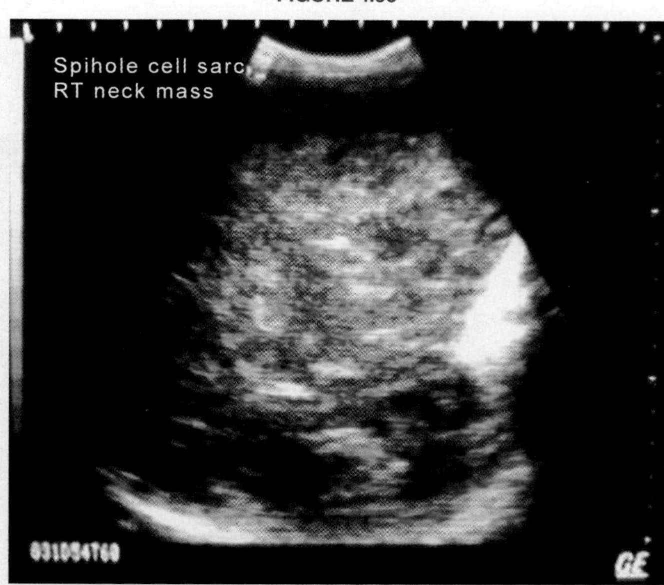

FIGURE 4.100

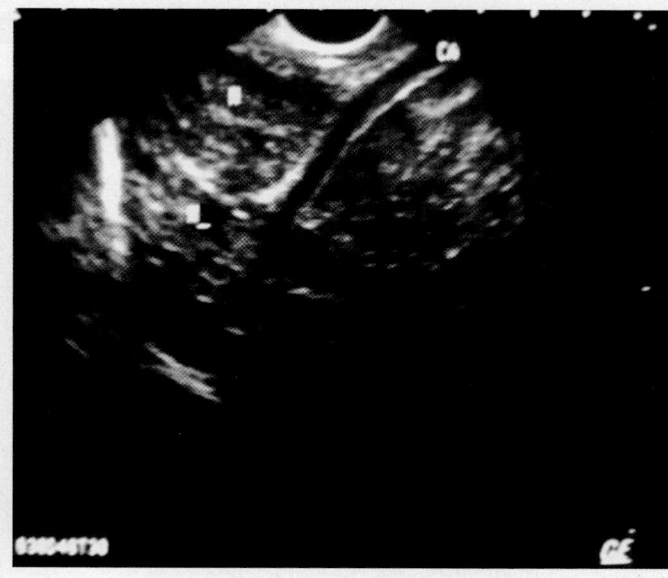

FIGURE 4.101

FIGURES 4.98 to 4.101: HRSG shows a big heterogeneous mass invading the neck. Normal anatomy is destroyed. Biopsy of the mass shows spindle cell carcinoma. However, the mass is not seen invading the neck vessels. Though it is seen pressing over the neck vessels. However, they are not invaded by the mass.

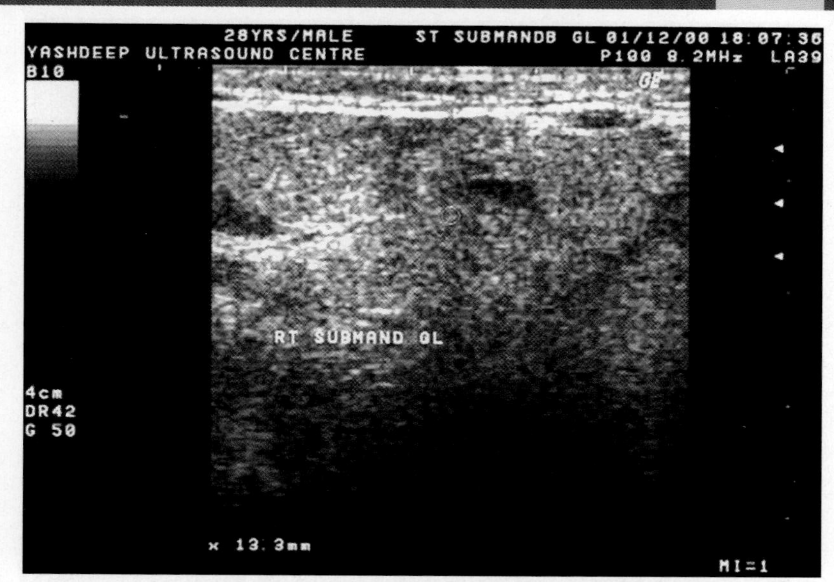

FIGURE 4.102: Normal submandibular gland. HRSG shows well-defined homogeneous granular texture of the submandibular gland. Few hypodense areas are seen in the substance of the gland. No evidence of any calcification is seen.

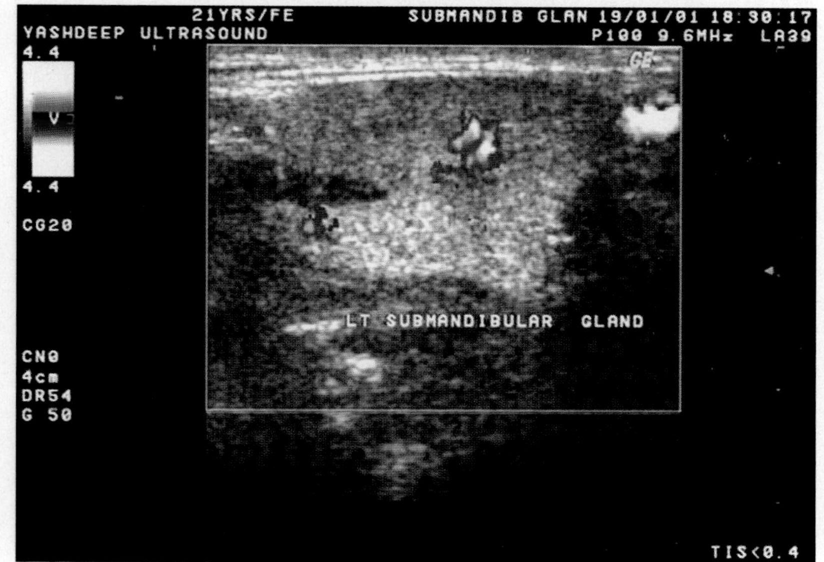

FIGURE 4.103

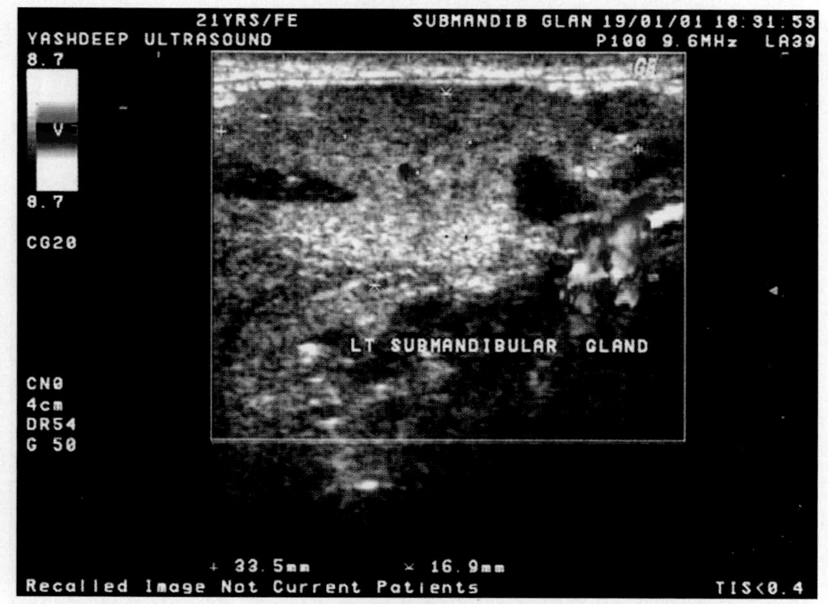

FIGURE 4.104

FIGURE 4.103 and 4.104: Normal submandibular gland. Color Doppler imaging shows low flow in the gland.

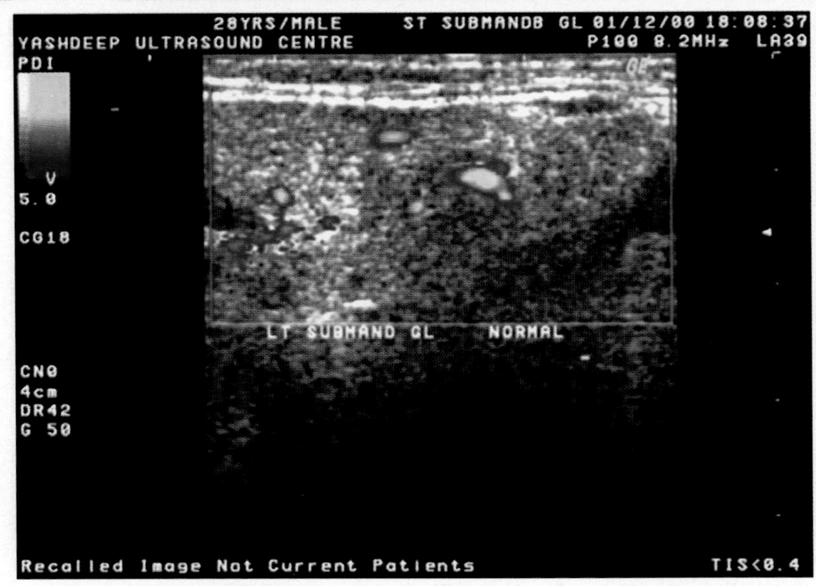

FIGURE 4.105: Power Doppler imaging of the patient shows better depiction of the flow in the substance of the gland.

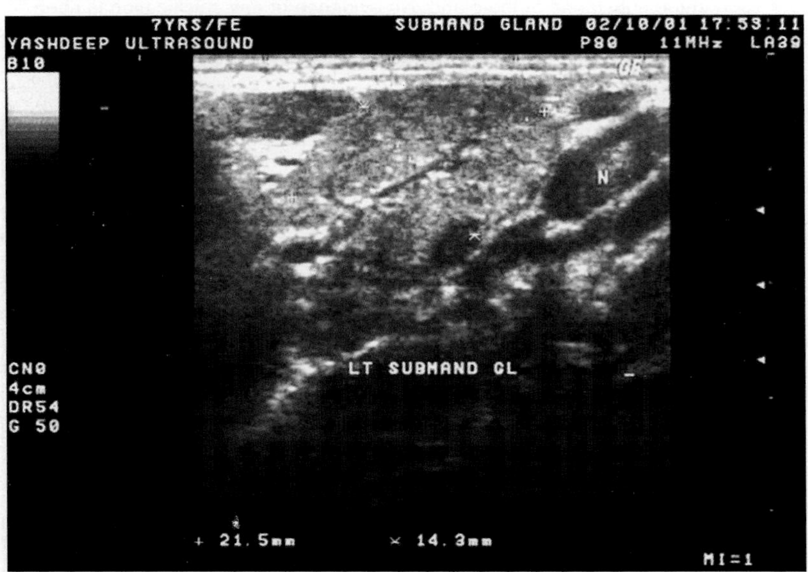

FIGURE 4.106: HRSG shows edematous inflamed submandibular gland. The duct is dilated. Small-calcified calculi are seen in the duct suggestive of sialectasis. Enlarged lymph nodes are also seen in the submental area.

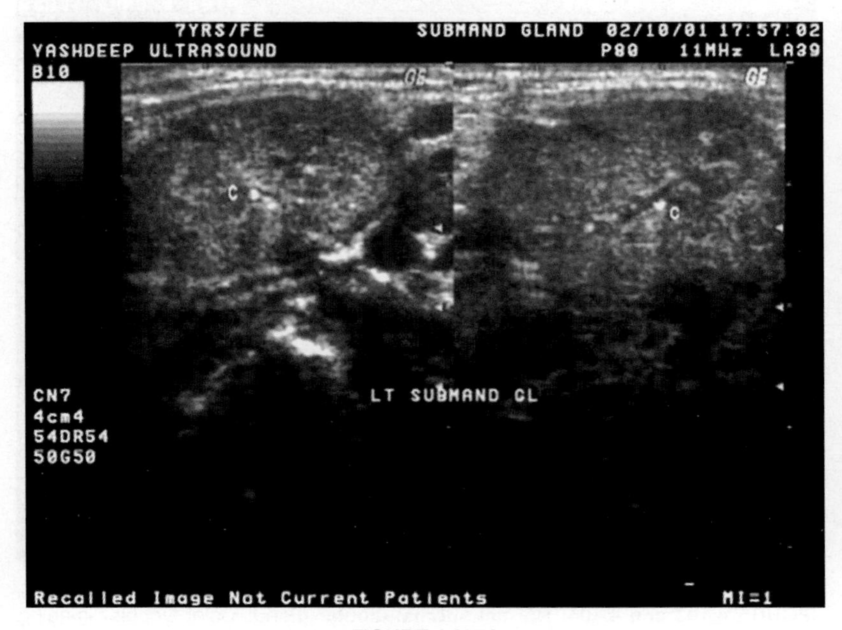

FIGURE 4.107A

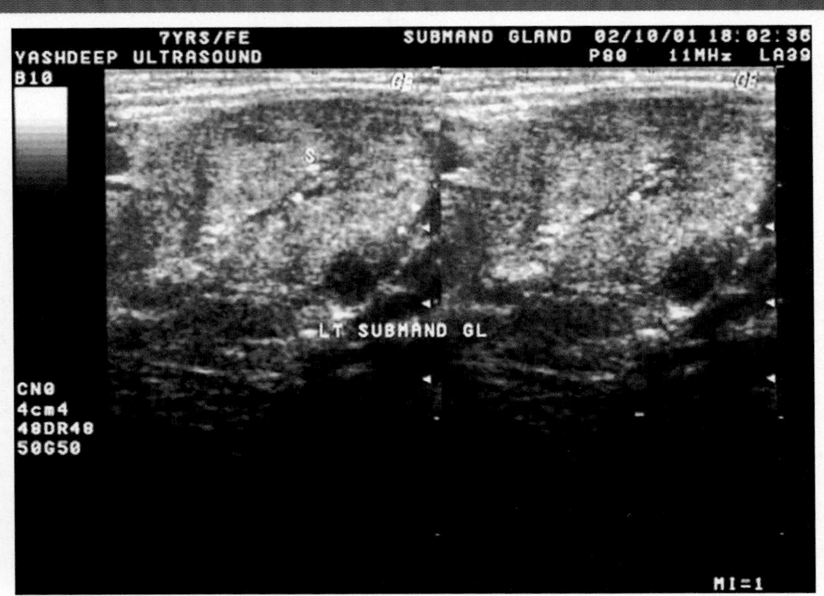

FIGURE 4.107B

FIGURES 4.107A and 4.107B: Acute inflammation of the gland. A young girl presented with swelling in the Lt submandibular area. The gland is enlarged and increased in size. It is lobulated. However, the margins are maintained. The duct is dilated. Few calcified small calculi are also seen in the duct of the gland.

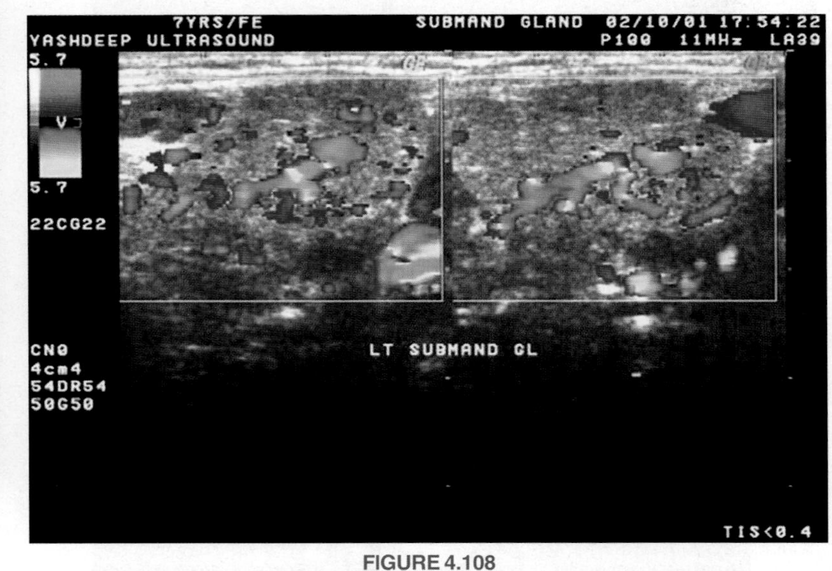

FIGURE 4.108

FIGURE 4.109

FIGURES 4.108 and 4.109: Color Doppler imaging shows increased flow in the gland with increased vascularity in the intra glandular tissue. There is also evidence of hypoechoic lymph node enlargement seen in the upper cervical area. No evidence of any necrosis is seen in the lymph node.

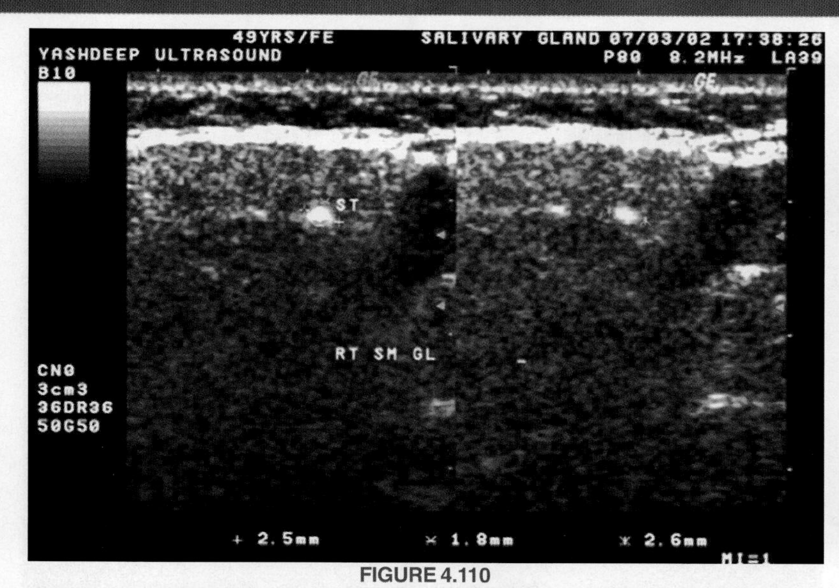

FIGURE 4.110

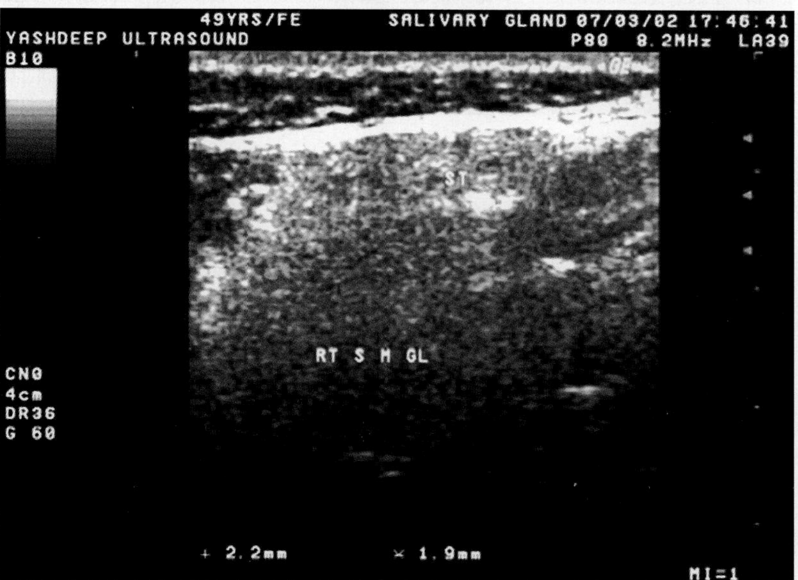

FIGURE 4.111

FIGURES 4.110 and 4.111: Small calculi in SMG: A patient present recurrent swelling of the submandibular gland. HRSG shows coarse glandular texture. Small echogenic calculi are seen in the gland suggestive of sialectasia.

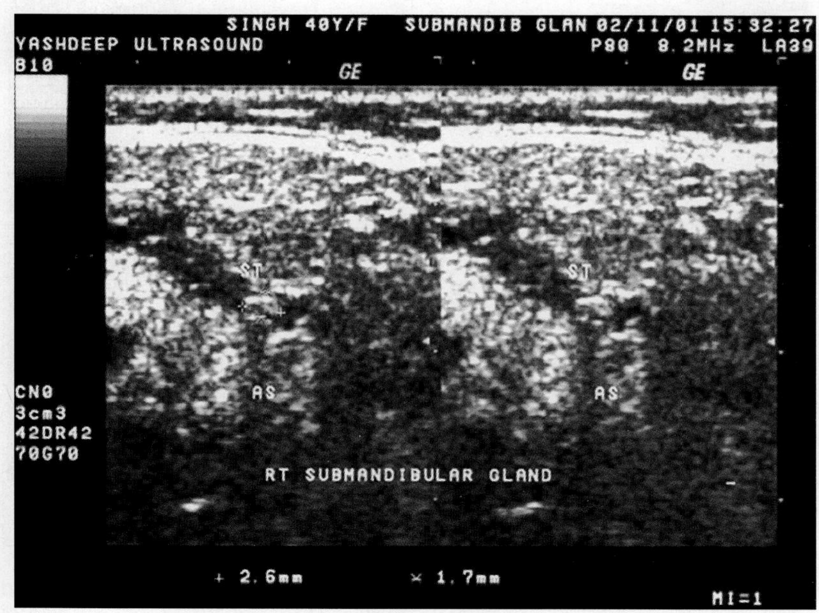

FIGURE 4.112

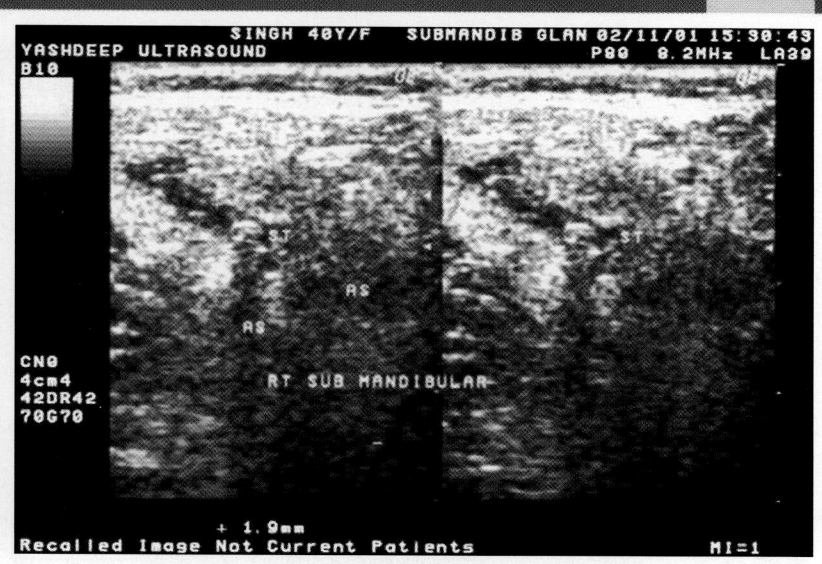

FIGURE 4.113

FIGURES 4.112 and 4.113: Chronic sialectasis with lethiasis. HRSG shows enlarged Rt submandibular gland. The lobules are increased in size. The main duct is dilated. Echogenic calcified foci are seen in the main duct suggestive of calculi. They are accompanied with acoustic shadowing. But no evidence of any cystic degeneration is seen.

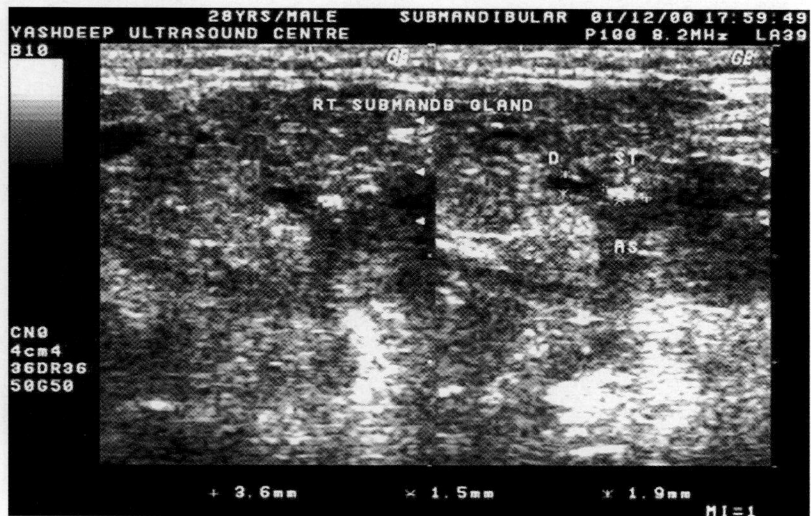

FIGURE 4.114

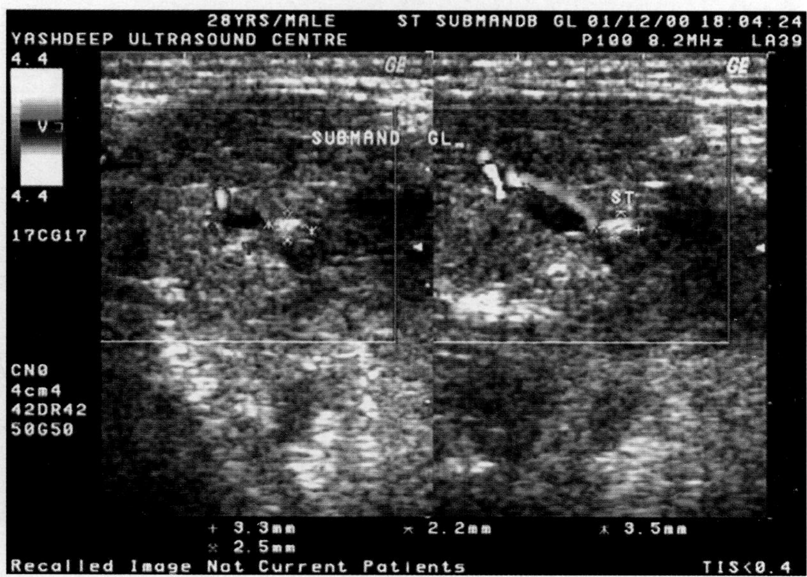

FIGURE 4.115

FIGURES 4.114 and 4.115: Chronic sialectasia with lethiasis. HRSG shows small abscess formation in the case of chronic recurrent sialectasis of the submandibular gland. On color Doppler imaging little flow is seen in the substance of the gland.

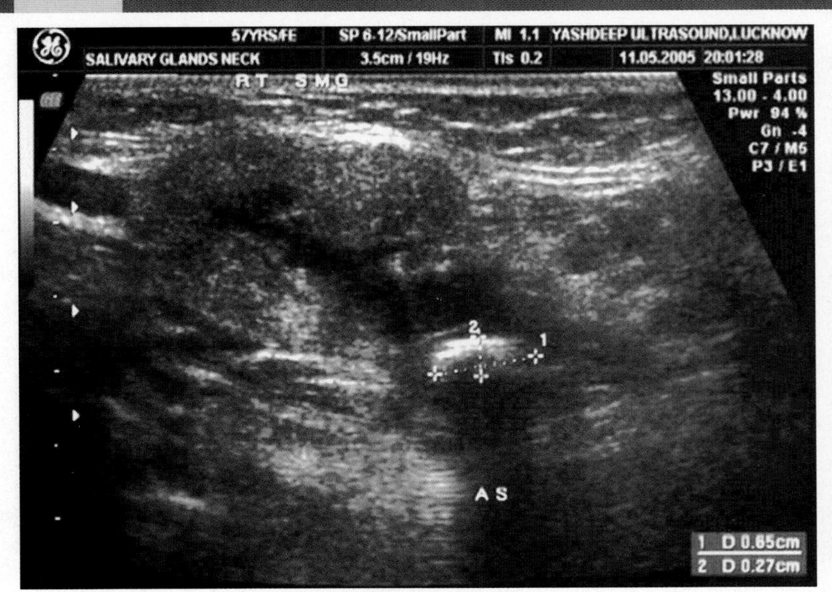

FIGURE 4.116

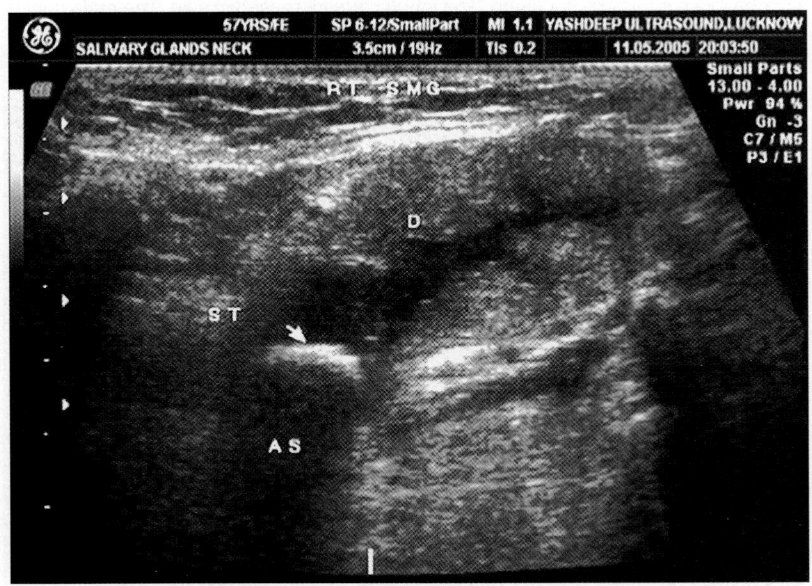

FIGURE 4.117

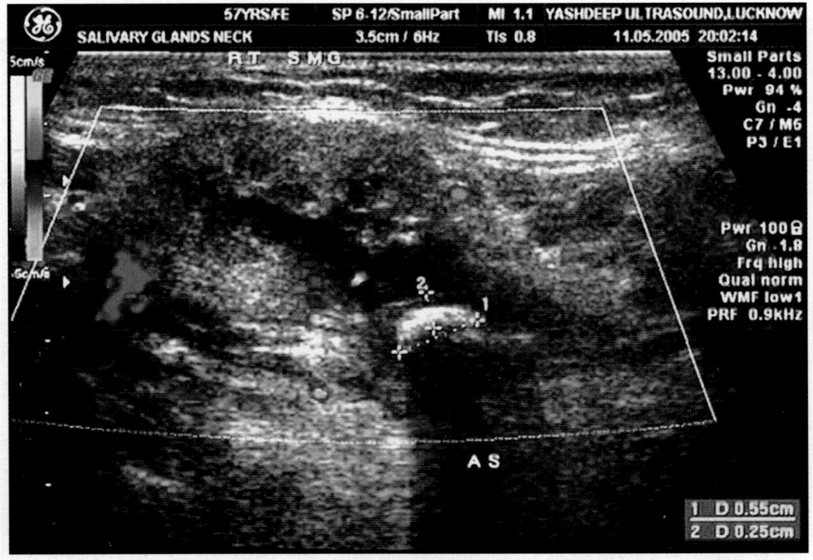

FIGURE 4.118

FIGURES 4.116 to 4.118: Big submandibular calculus. A patient presented with a palpable nodular mass in Rt submandibular gland. HRSG shows a big echogenic shadow seen in the duct of the gland. It is accompanied with acoustic shadowing suggestive of a big calculus. The gland is also enlarged and hypoechoic in texture. The ductal branches are also dilated.

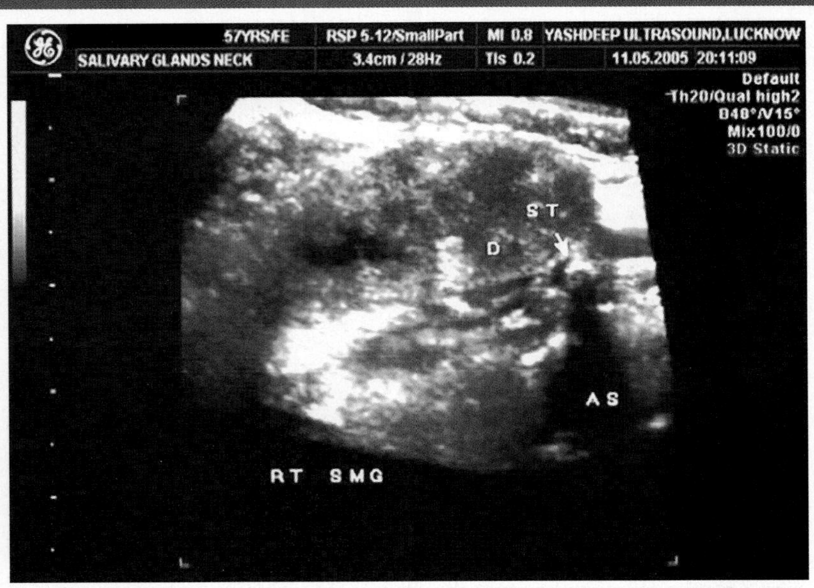

FIGURE 4.119: 3D imaging of the gland shows details of the ductal dilatation with impacted stone at the mouth of the duct.

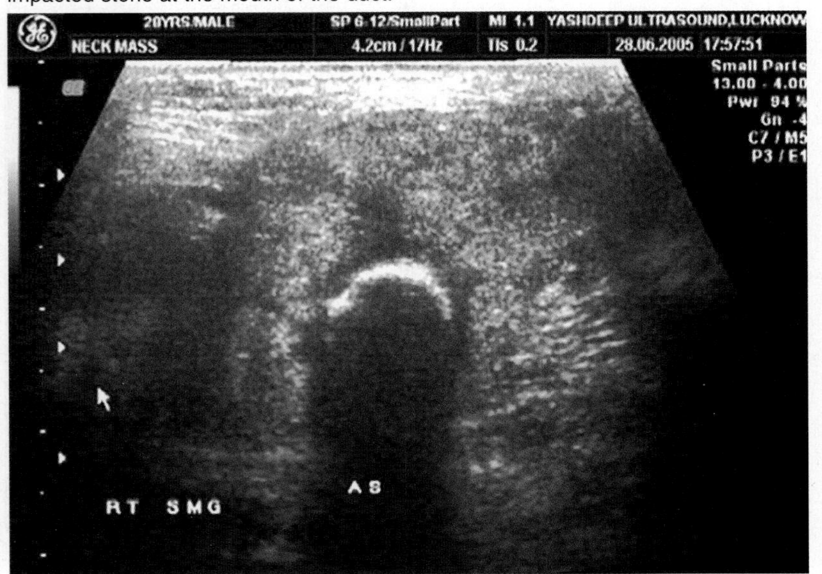

FIGURE 4.120

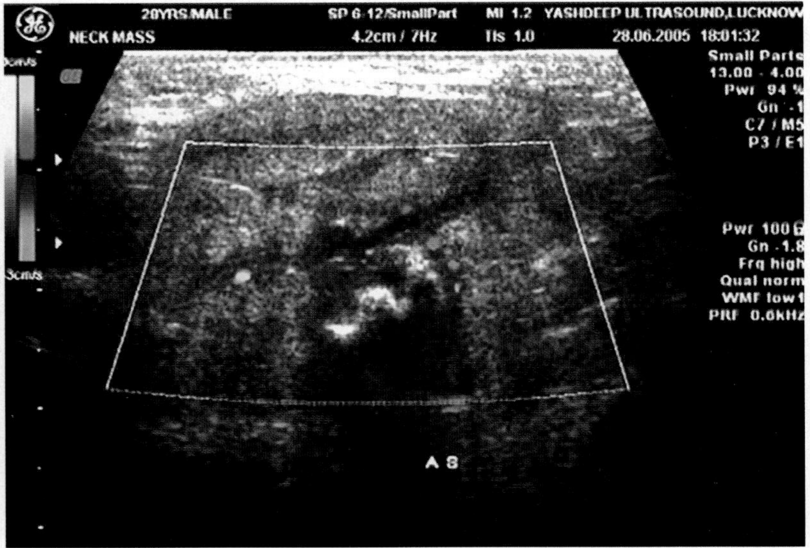

FIGURE 4.121

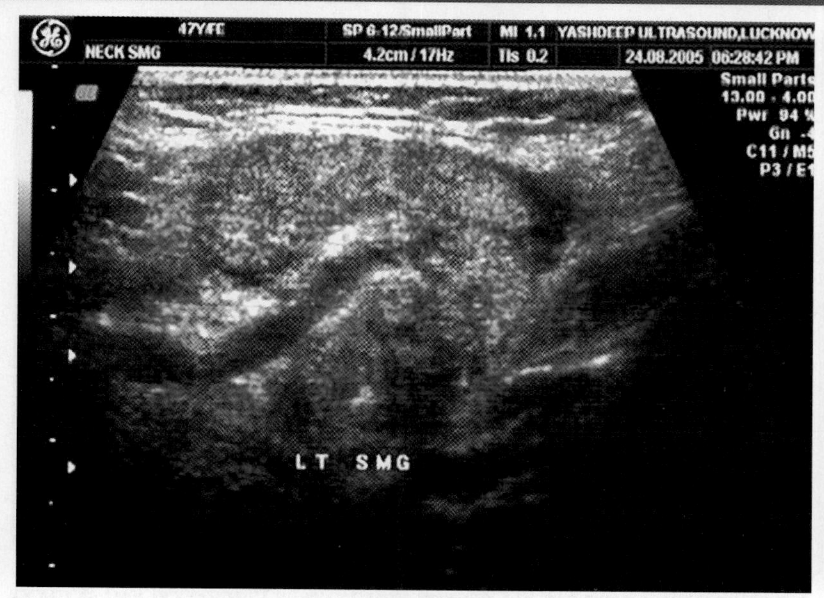

FIGURE 4.122

FIGURES 4.120 to 4.122: Big submandibular calculus. A patient presented with recurrent pain and swelling in the Rt SMG. HRSG shows a big calculus impacted in the duct of the gland. Small calculi are also seen in the ductal branches. The big calculus shows acoustic shadowing.

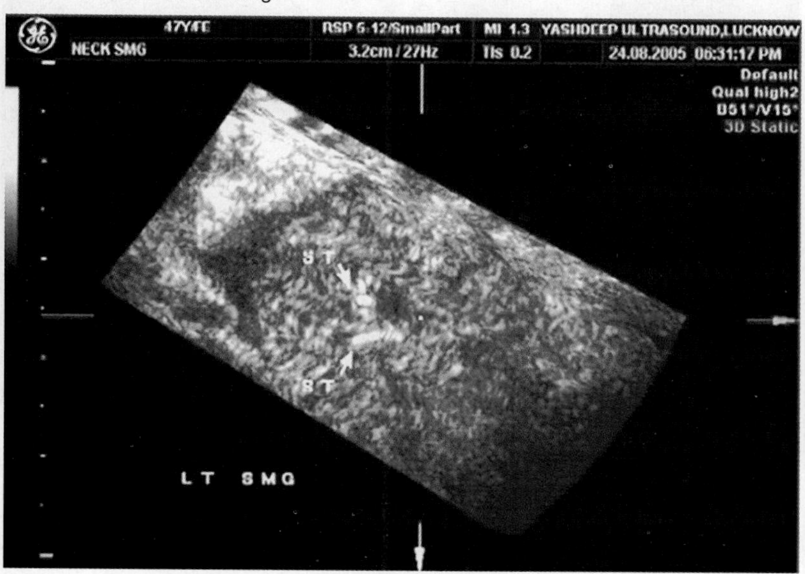

FIGURE 4.123

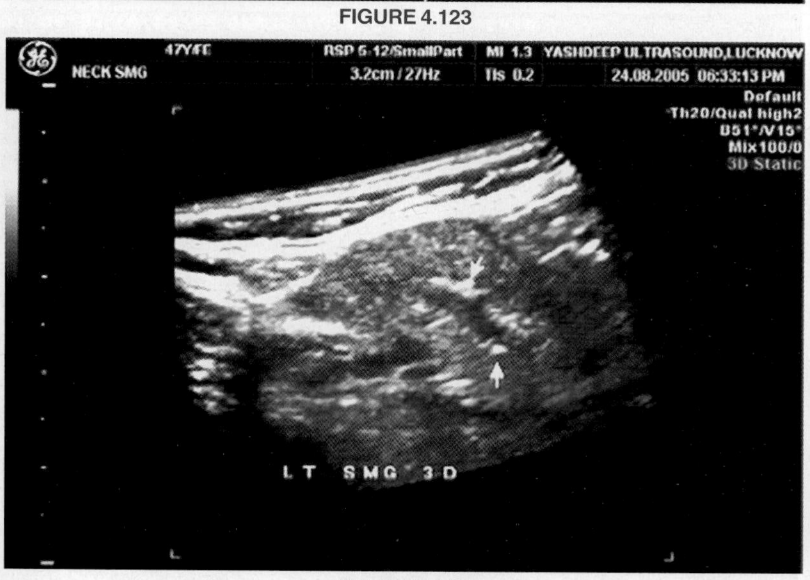

FIGURE 4.124

FIGURES 4.123 and 4.124: 3D imaging of the gland shows small calculi impacted in the ductal branches.

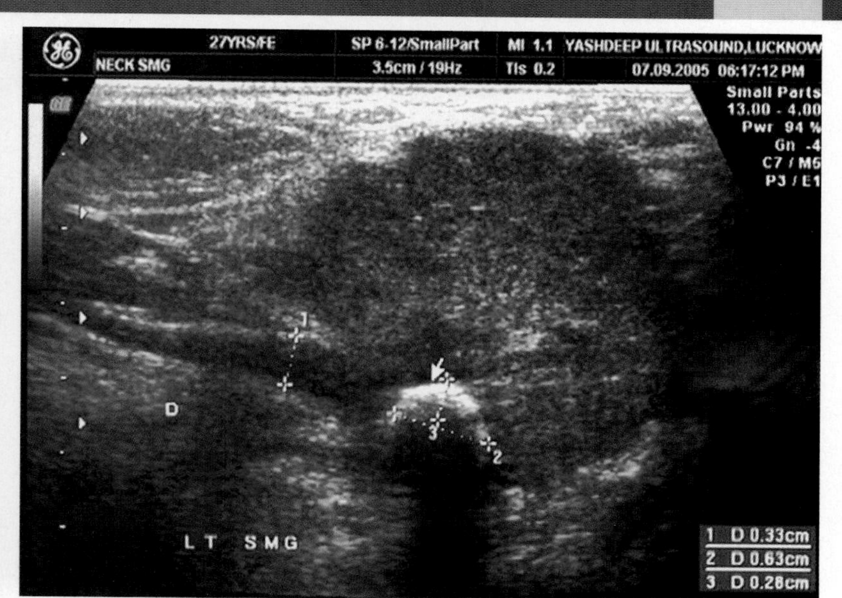

FIGURE 4.125

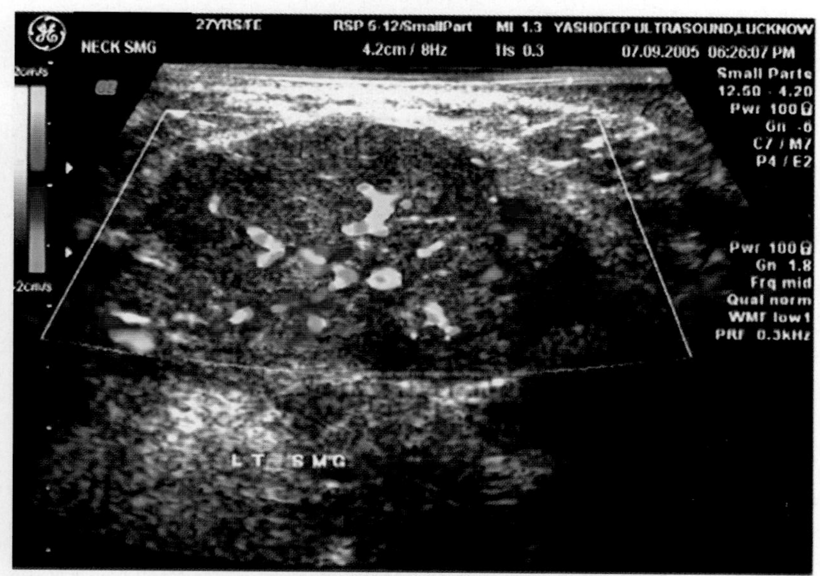

FIGURE 4.126

FIGURES 4.125 and 4.126: Impacted submandibular gland calculus. A patient presented with painful swelling of Rt submandibular gland with fever. HRSG shows enlarged swollen hypoechoic gland. On color flow imaging increased flow is seen in the gland. The main duct of the gland is dilated. A big echogenic calculus is seen impacted at the mouth of the duct. It is accompanied with acoustic shadowing.

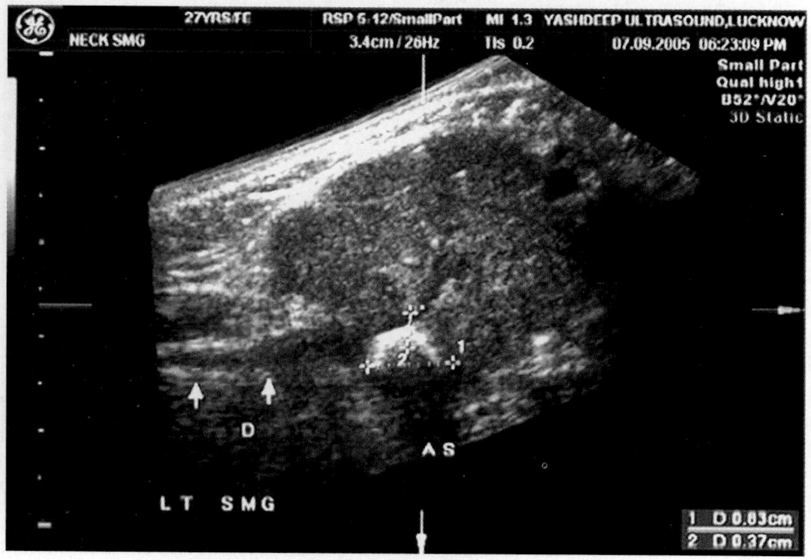

FIGURE 4.127

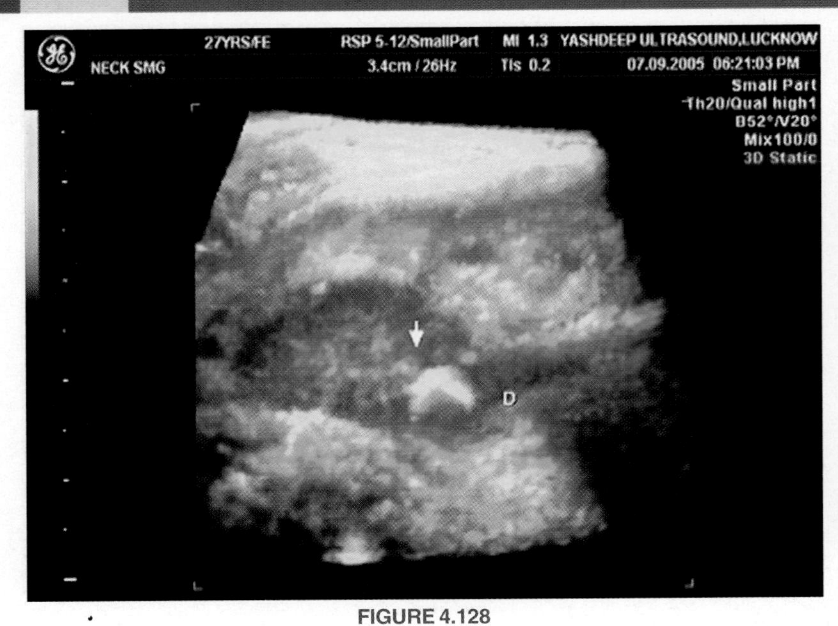

FIGURE 4.128

FIGURES 4.127 and 4.128: 3D imaging of the gland shows details of the glandular tissue with impacted stone in the mouth of the duct.

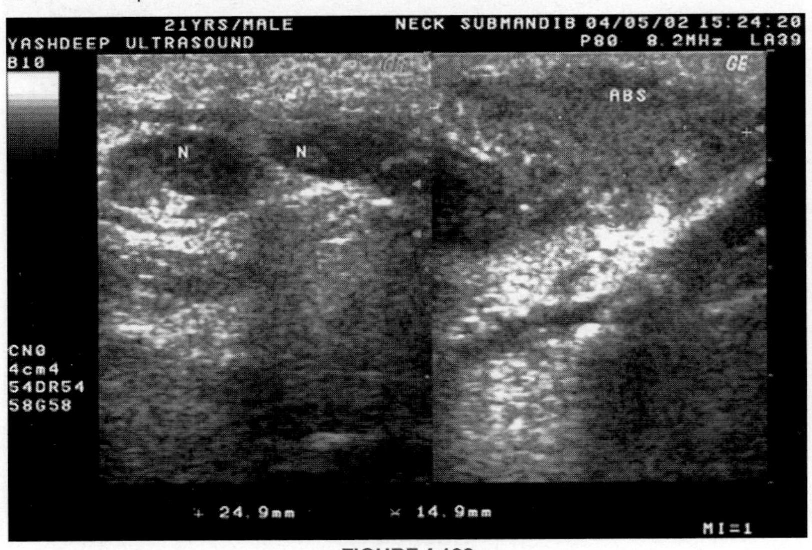

FIGURE 4.129

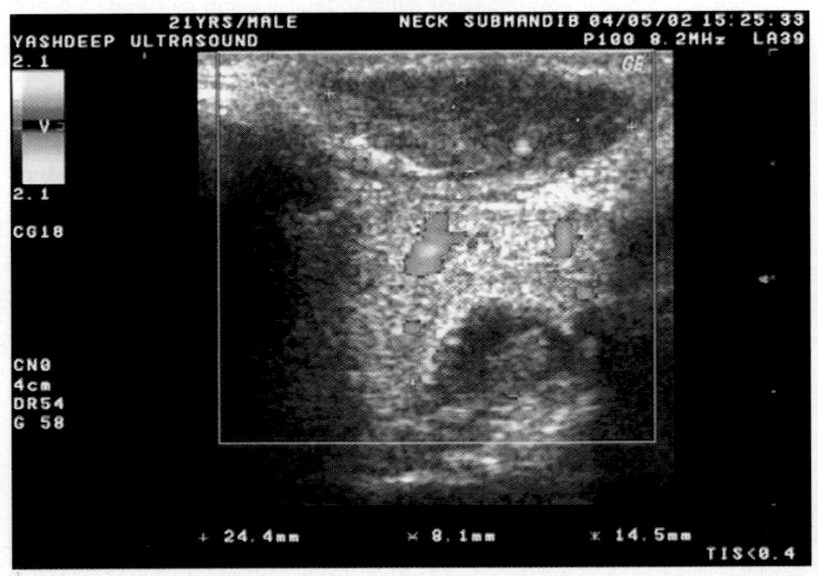

FIGURE 4.130

FIGURES 4.129 and 4.130: Pyogenic submandibular gland abscess. A patient presented with painful swelling in the Rt submandibular gland. HRSG shows thick low level echo complex mass in the gland. Multiple enlarged nodes are also seen in the upper cervical region. Biopsy confirmed an abscess.

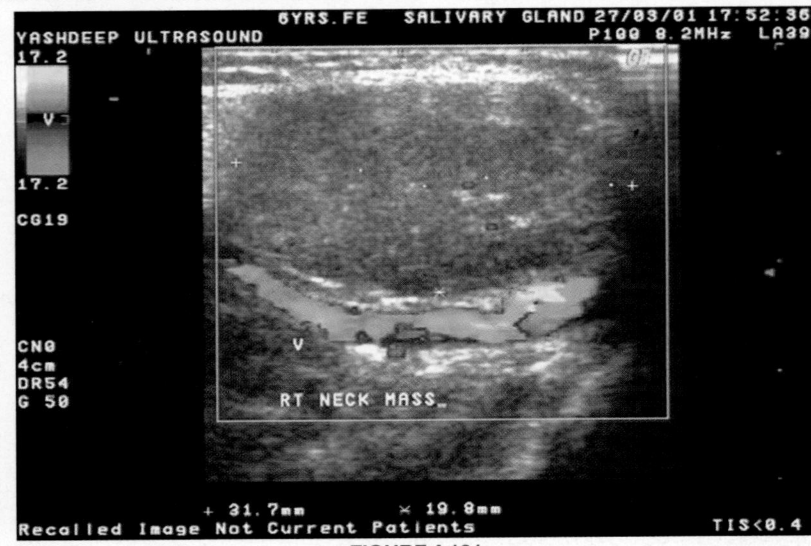

FIGURE 4.131

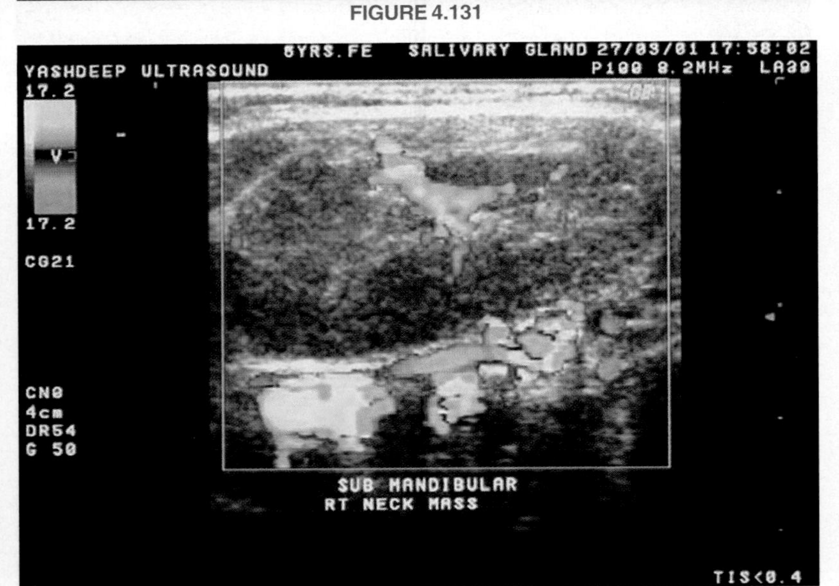

FIGURE 4.132

FIGURES 4.131 and 4.132: Submandibular gland vascular mass. A young girl presented with a painful swelling over the Rt submandibular area. HRSG shows a well-defined homogeneous low level echo complex mass in the Rt submandibular gland. On color Doppler imaging moderate flow is seen in the mass.

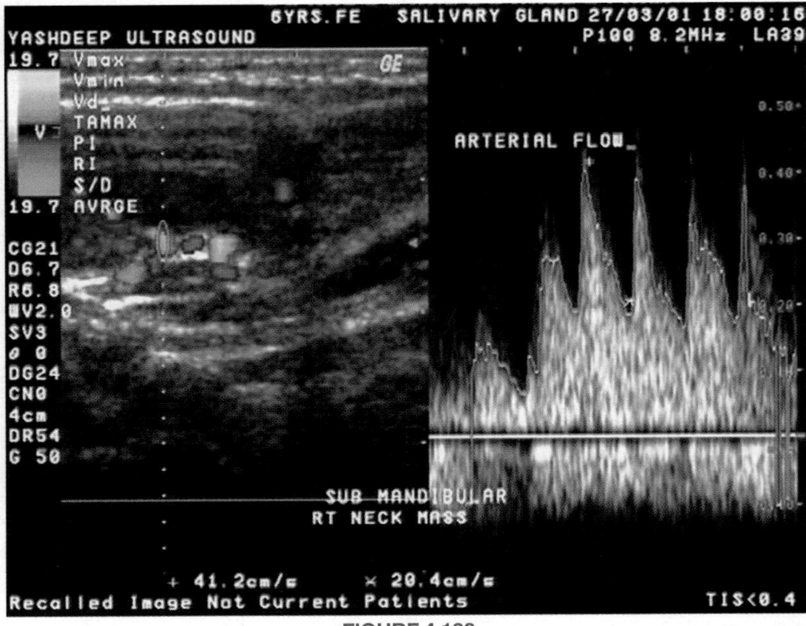

FIGURE 4.133

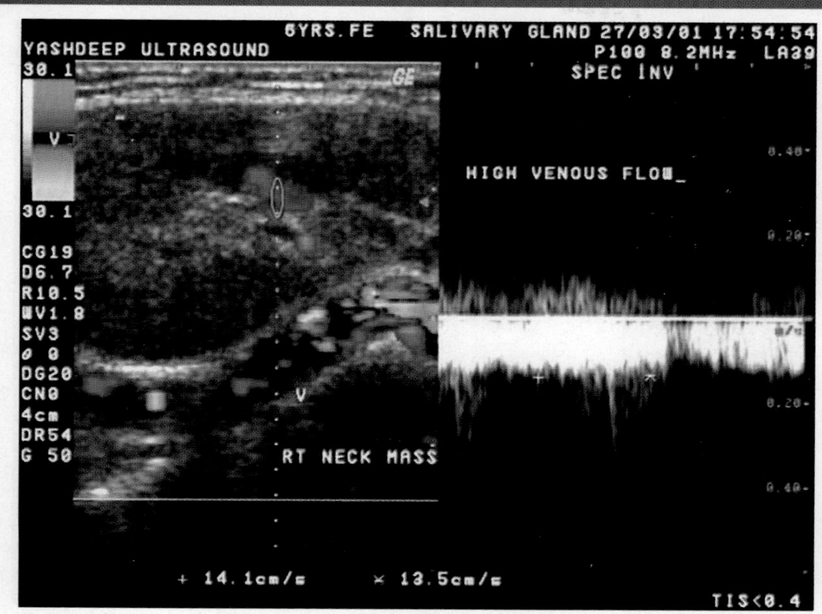

FIGURE 4.134

FIGURES 4.133 and 4.134: Vascular mass. Spectral Doppler flow imaging shows arterial and venous flow in the mass in the same patient suggestive of vascular malformation.

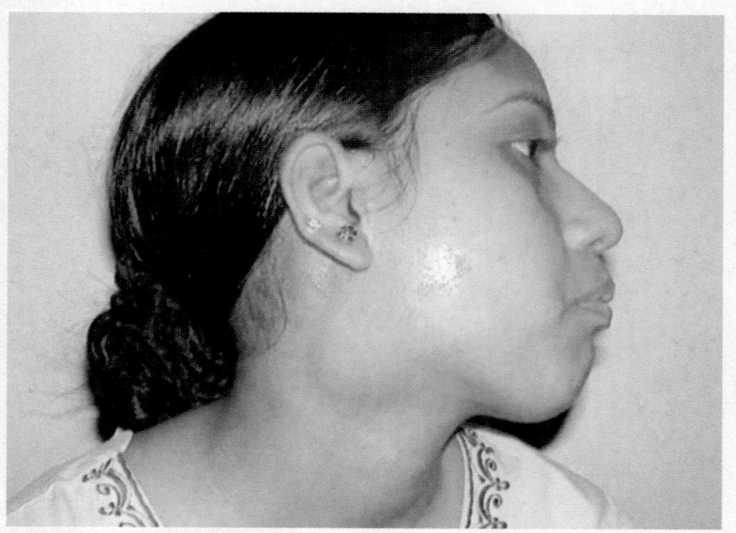

FIGURE 4.135: Pyogenic abscess. A young lady presented with a soft tissue mass in Rt submandibular area.

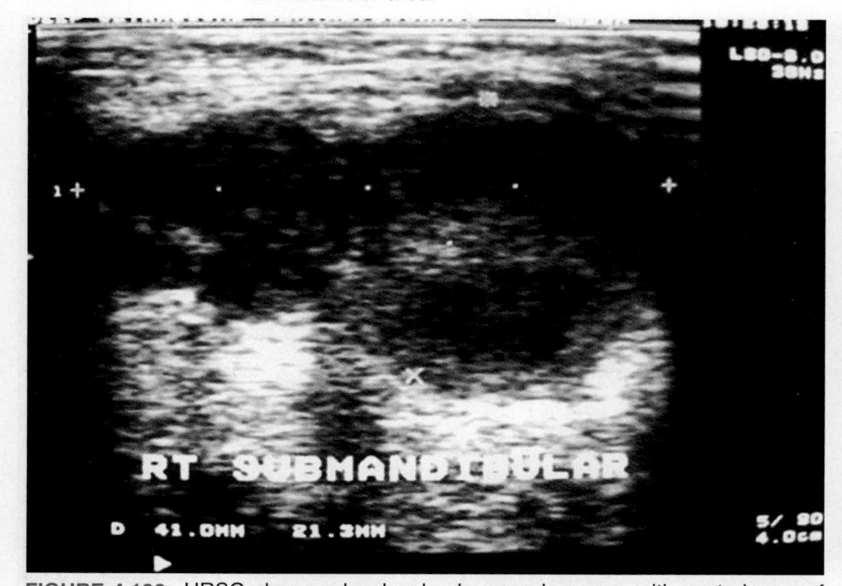

FIGURE 4.136: HRSG shows a low-level echo complex mass with central area of necrosis. Echogenic debris is also seen in the mass. The mass was bilobed. Biopsy shows pyogenic abscess.

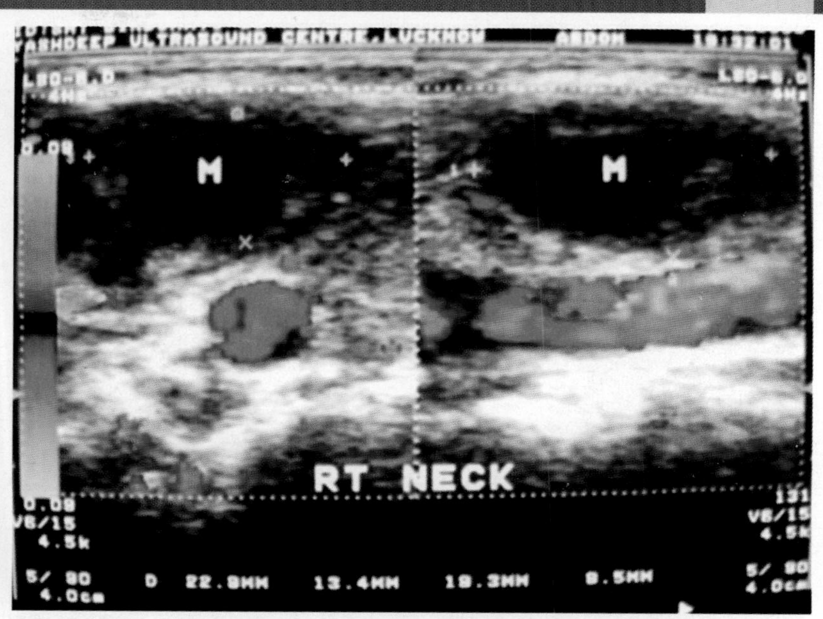

FIGURE 4.137: Multiple enlarged lymph nodes are also seen in the upper cervical area suggestive of reactive lymph node enlargement.

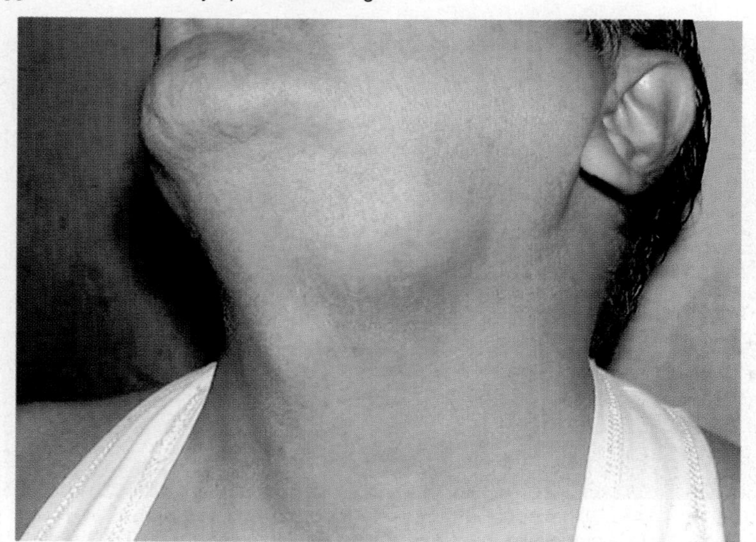

FIGURE 4.138: Submandibular gland pleomorphic adenoma. A patient presented with a soft tissue mass swelling over the Lt submandibular area.

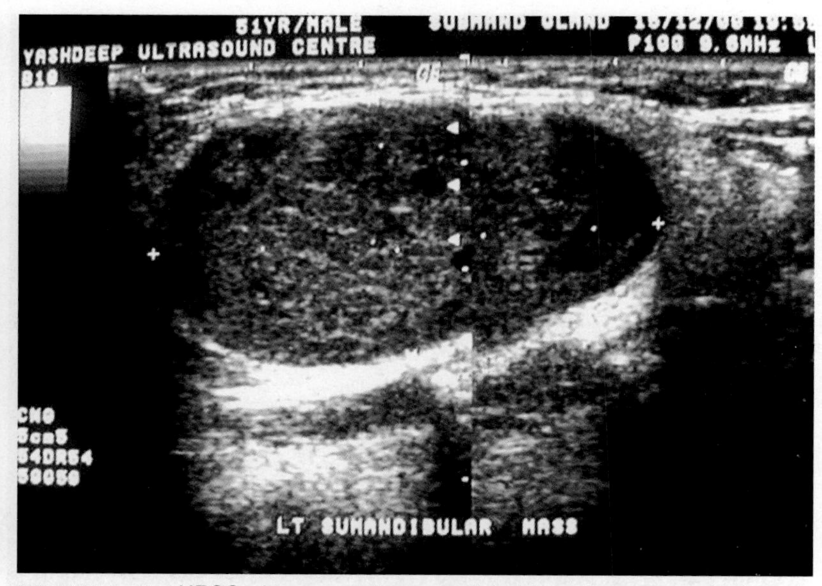

FIGURE 4.139: HRSG shows a well-encapsulated hypodense mass in the sub mandibular area. The mass shows medium level echoes. No cystic degeneration is seen in the mass. No perifocal edema is present.

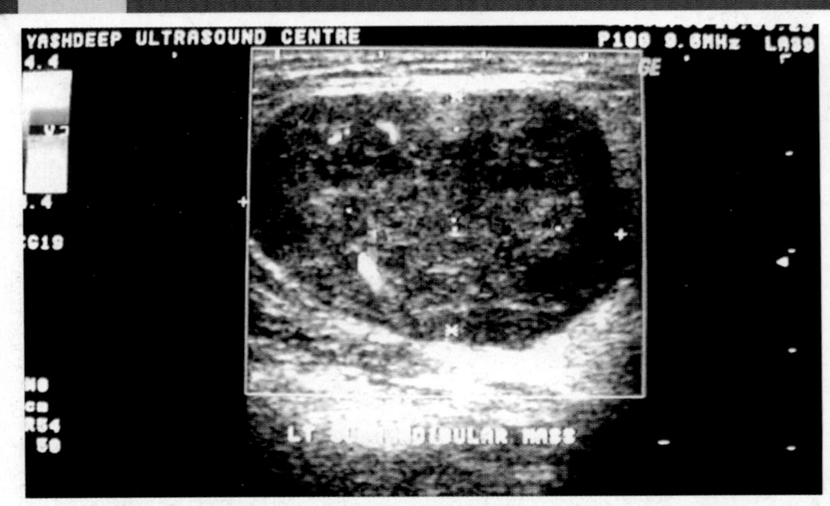

FIGURE 4.140: Color Doppler imaging of the mass shows poor flow. Biopsy of the mass shows pleomorphic adenoma.

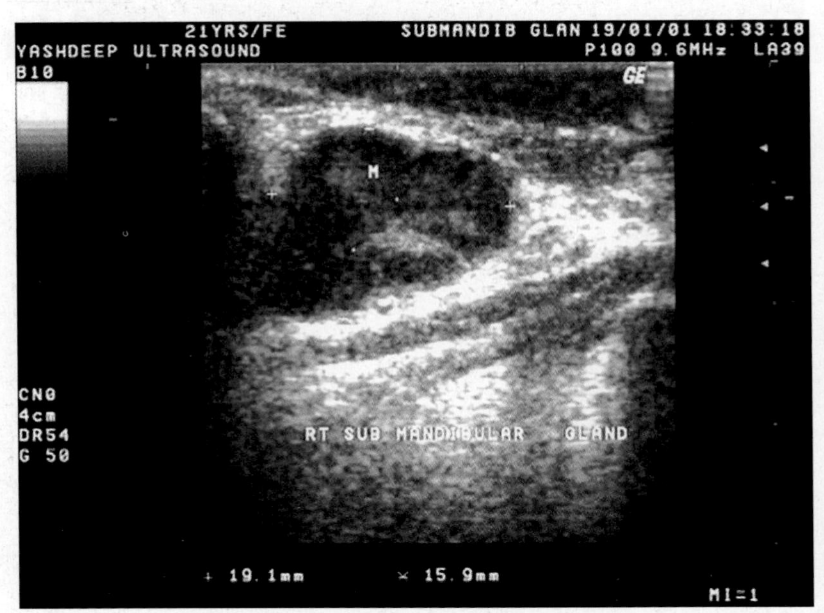

FIGURE 4.141: Tubercular abscess in submandibular gland. A young girl presented with a soft cystic mass in the Rt submandibular area. HRSG shows low-level echo complex mass in the sub mandibular gland. Normal glandular texture is lost. No perifocal edema is present. Biopsy shows tubercular abscess. Patient was also having pulmonary tuberculosis.

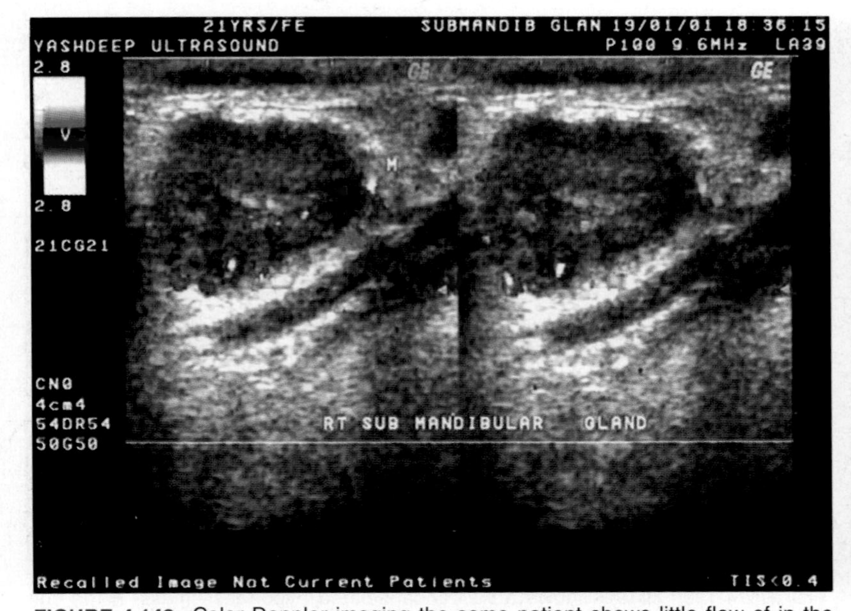

FIGURE 4.142: Color Doppler imaging the same patient shows little flow of in the abscess cavity. However, capsular vascularity is seen.

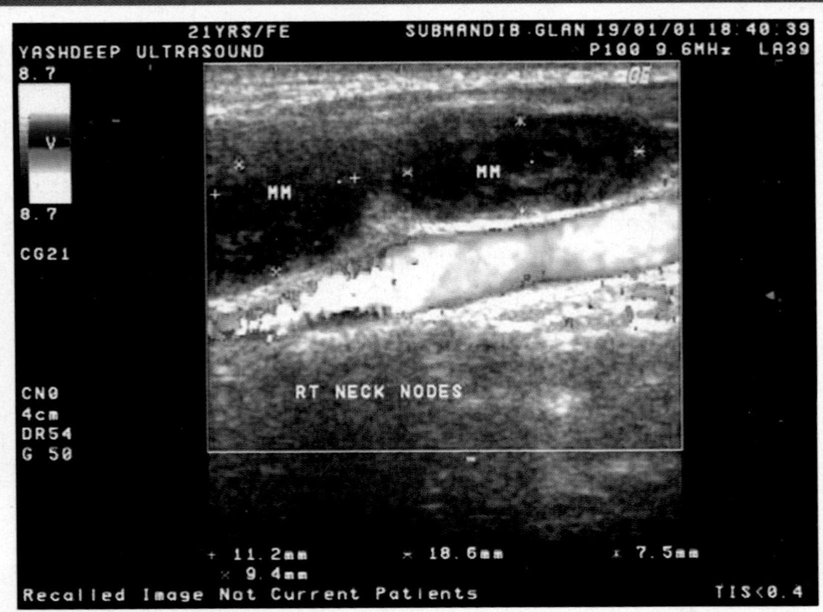

FIGURE 4.143: Multiple enlarged lymph nodes are seen in the upper cervical area in the same patient suggestive of tubercular lymphadenitis.

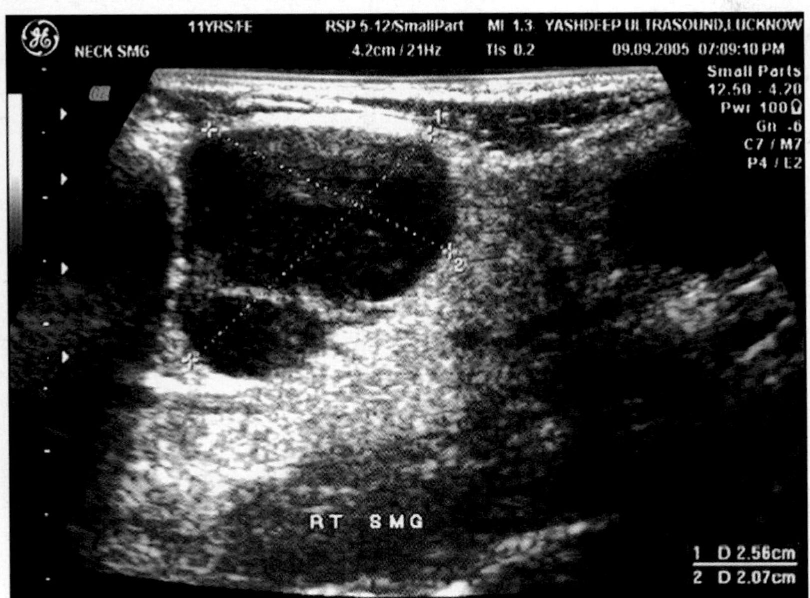

FIGURE 4.144

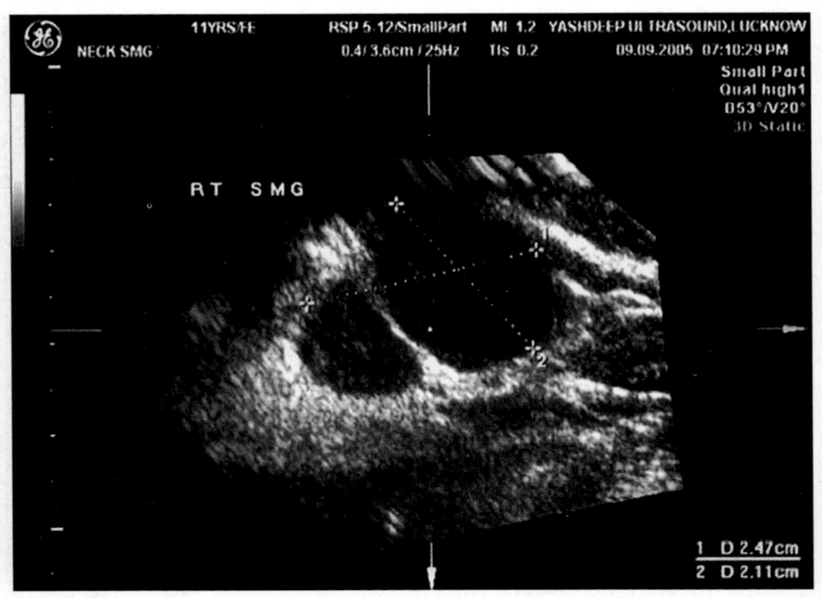

FIGURE 4.145

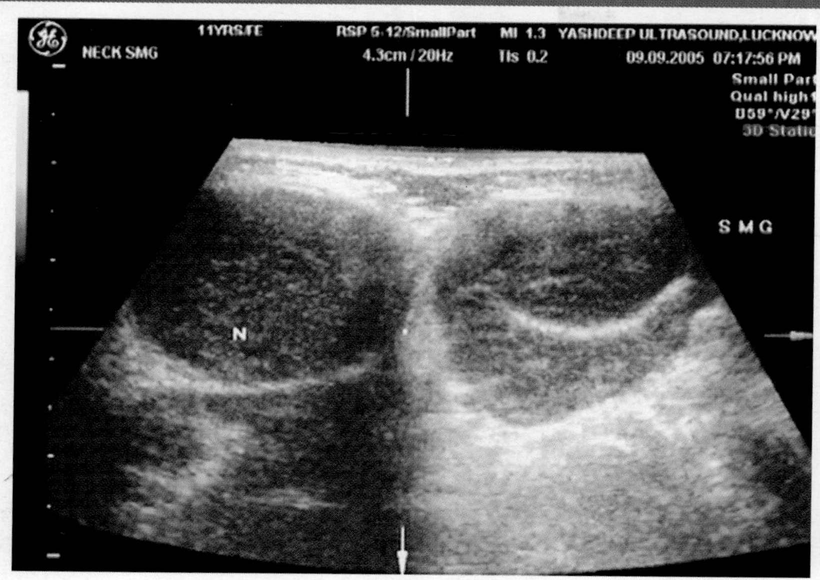

FIGURE 4.146

FIGURES 4.144 to 4.146: SMG tuberculosis. A young girl presented with swelling in the Rt submandibular gland. HRSG shows thick walled loculated cystic mass in the SMG. Echogenic septa are also seen in it. Thick collection is seen in the gland. 3D imaging of the gland shows details of the cystic cavity. Aspirate of the cyst came out to be thick pus suggestive of chronic infection.

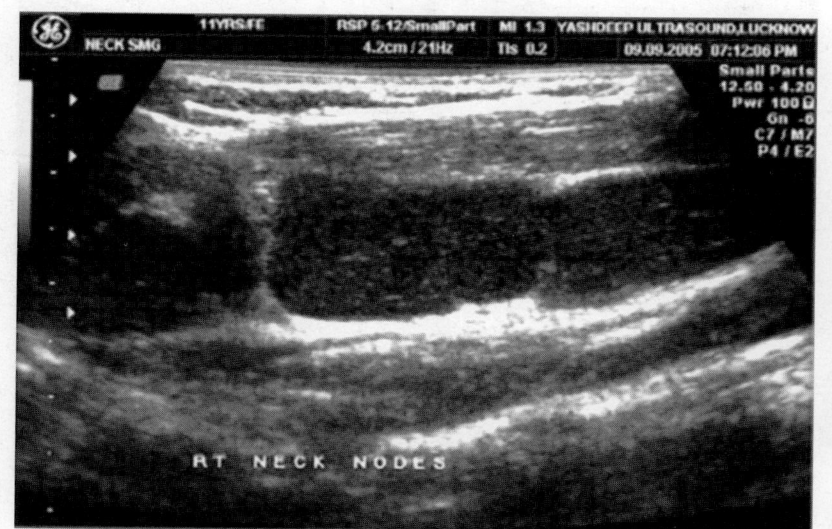

FIGURE 4.147

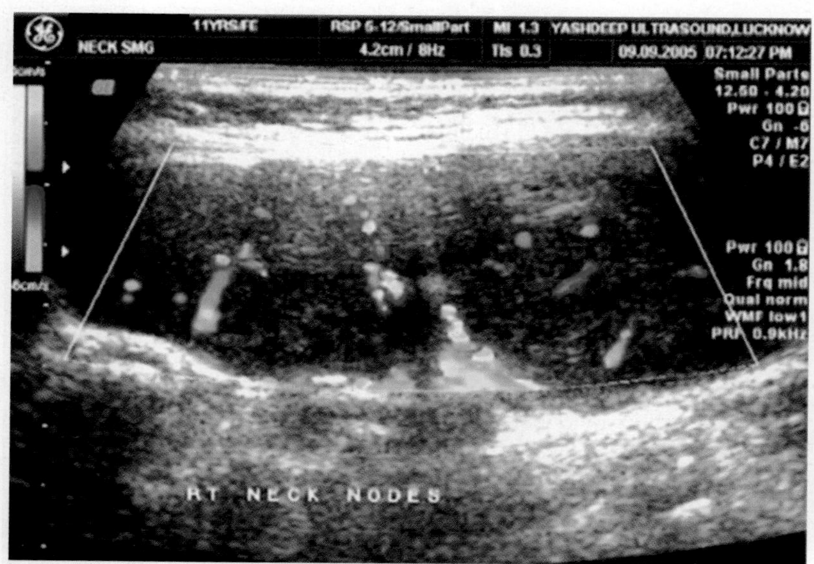

FIGURE 4.148

FIGURES 4.147 and 4.148: Multiple enlarged nodes are also seen in the Rt upper cervical and mid cervical region. Increased flow is seen in them on color flow imaging. Biopsy confirmed tubercular abscess.

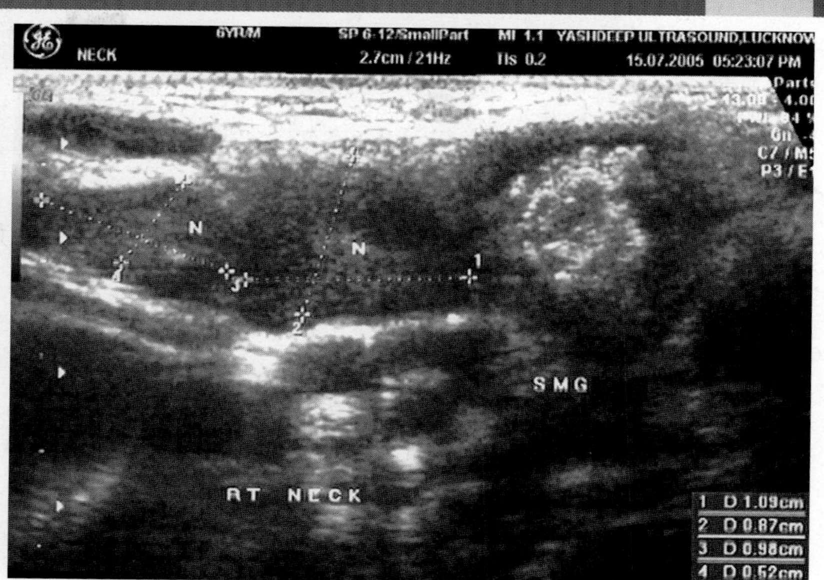

FIGURE 4.149

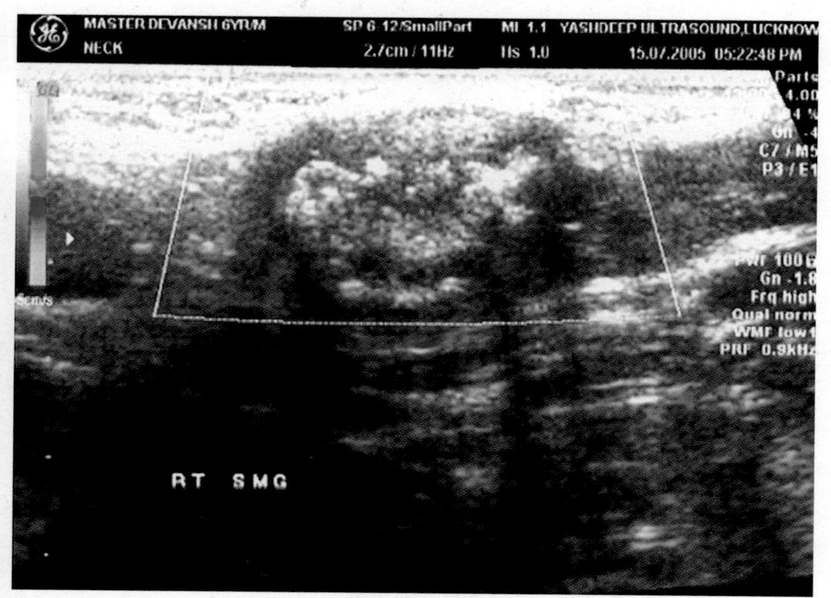

FIGURE 4.150

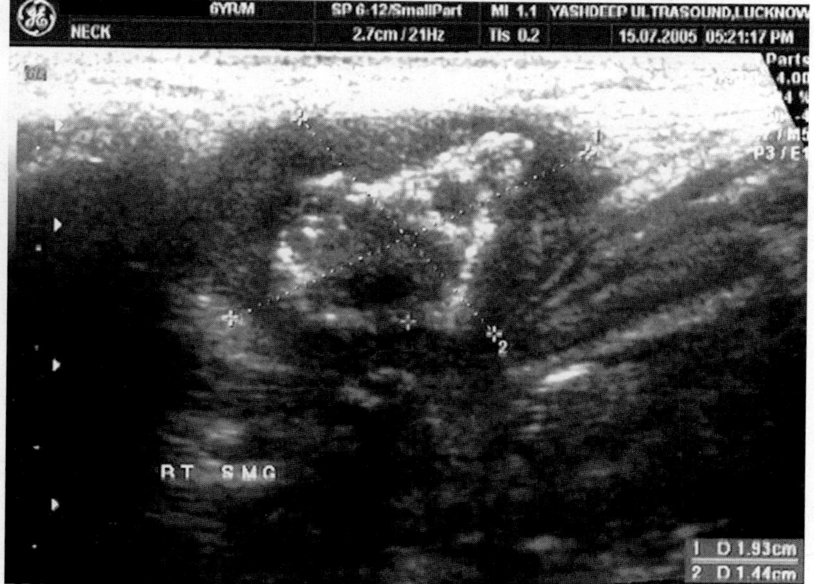

FIGURE 4.151

FIGURES 4.149 to 4.151: Chronic tubercular abscess. A young child presented with mass with non tender mass in the Rt submandibular gland. HRSG shows thick echogenic collection in the submandibular gland. Echogenic calcified specks are also seen in it. Multiple enlarged nodes are also seen in the neck. Biopsy confirmed chronic tubercular abscess.

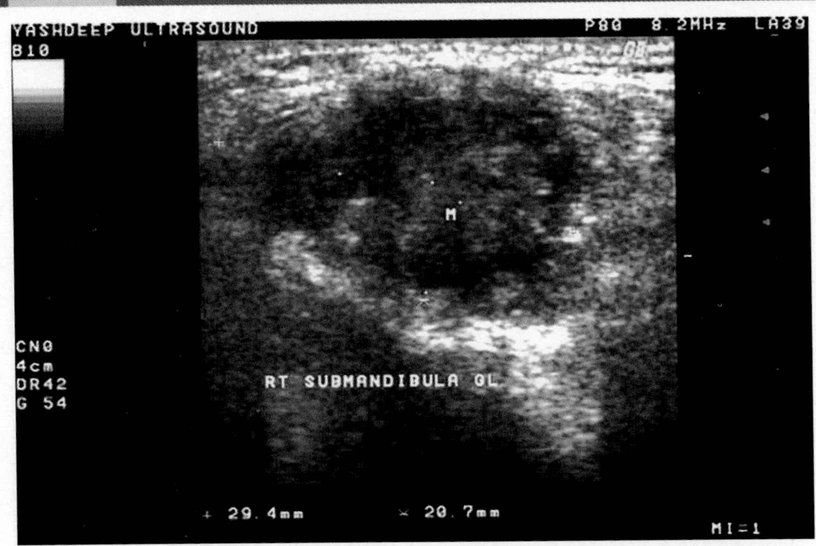

FIGURE 4.152: Malignant mass at the submandibular gland. HRSG shows an irregular heterogeneous mass in the Rt submandibular area. Margins are irregular. Capsule is broken.

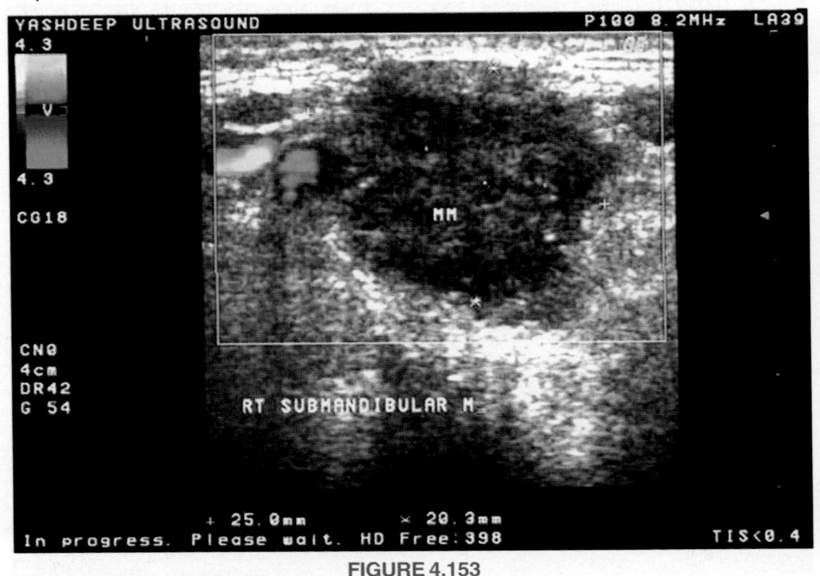

FIGURE 4.153

62YRS/MALE SALIVARY GLAND 09/11/01 17:40:04
YASHDEEP ULTRASOUND P80 8.2MHz LA39
B10
MM
CN0
4cm
DR42
G 58
RT SUBMANDIBULAR M
MI=1

FIGURE 4.154

FIGURES 4.153 and 4.154: Color Doppler imaging shows poor flow in the mass. There is also suggestion of finger like projections seen coming out from the mass invading the fascial planes. No cystic degeneration is seen in the mass. Biopsy of the mass shows malignant epidermoid carcinoma.

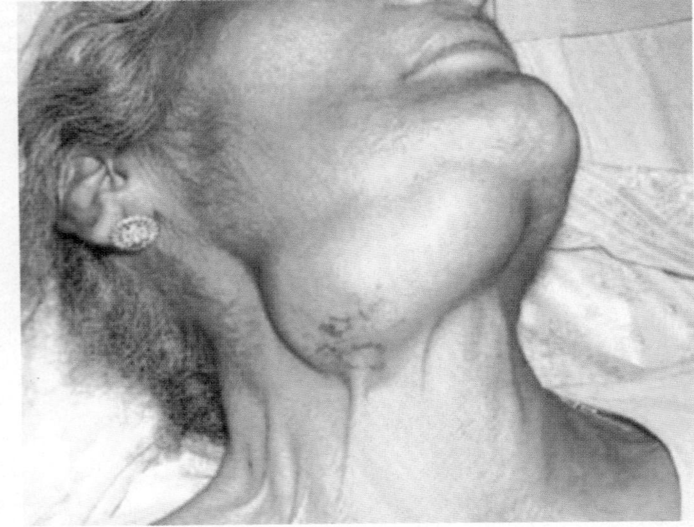

FIGURE 4.155: Malignant mass in the neck. Old lady presented with a solid mass in the Rt submandibular region.

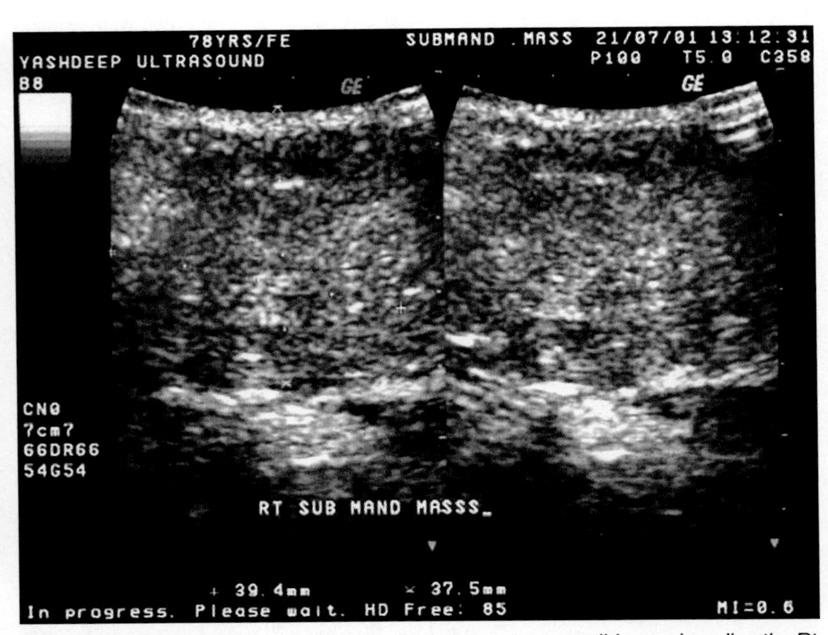

FIGURE 4.156: HRSG shows a big irregular heterogeneous solid mass invading the Rt parotid gland. No necrosis is seen in it. No cystic degeneration is seen.

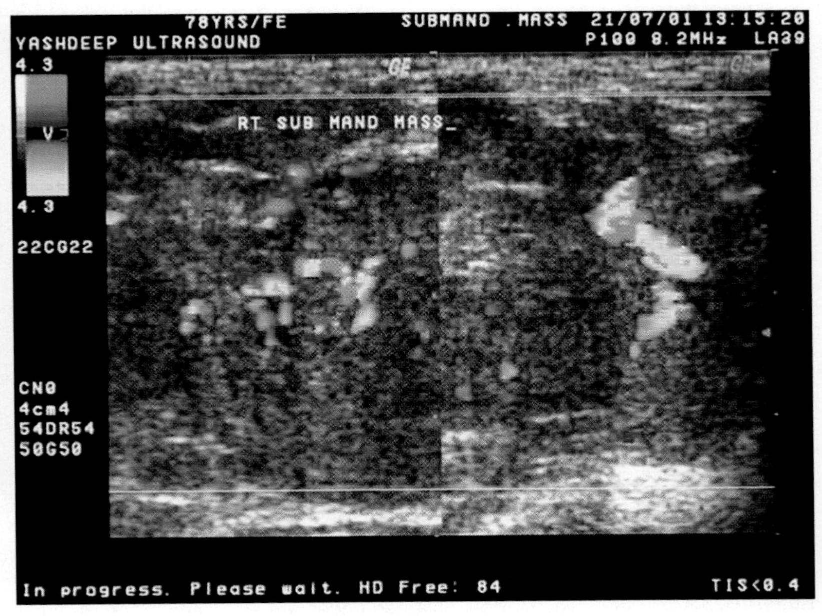

FIGURE 4.157

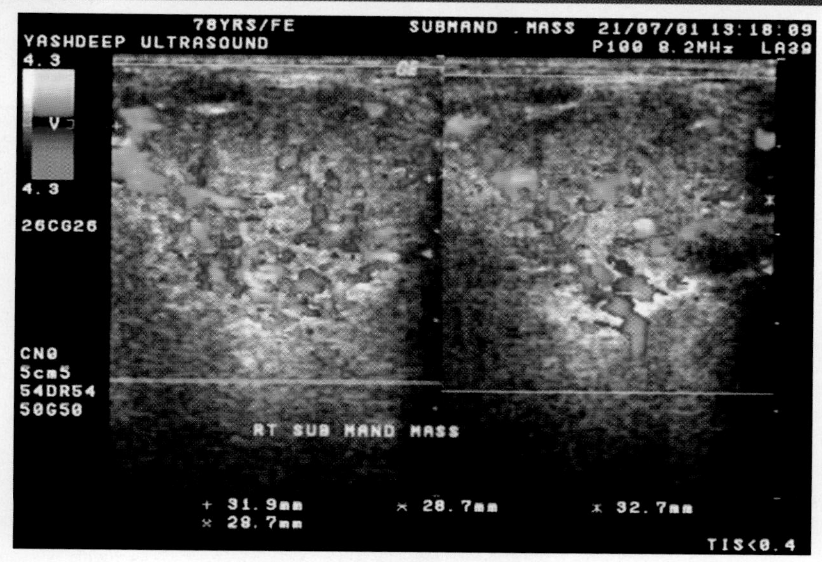

FIGURE 4.158

FIGURES 4.157 and 4.158: Color Doppler of the mass shows increased vascularity in the mass. However, the flow pattern is bizarre. Biopsy of the mass shows carcinoma.

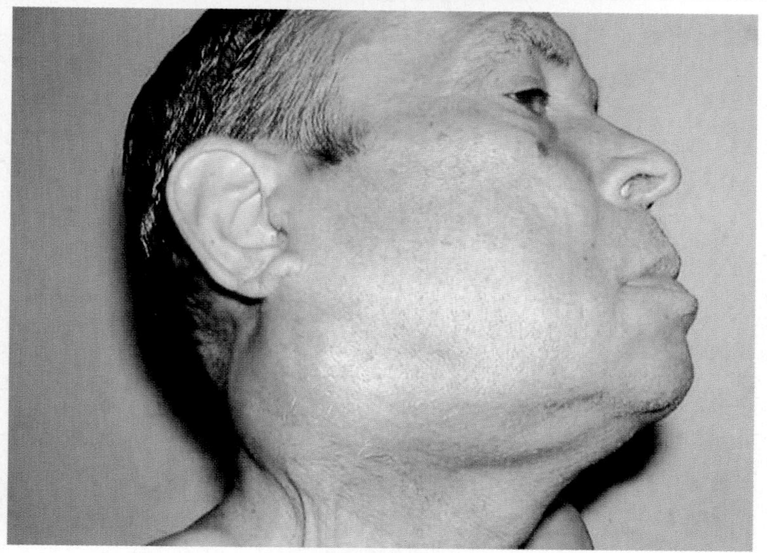

FIGURE 4.161

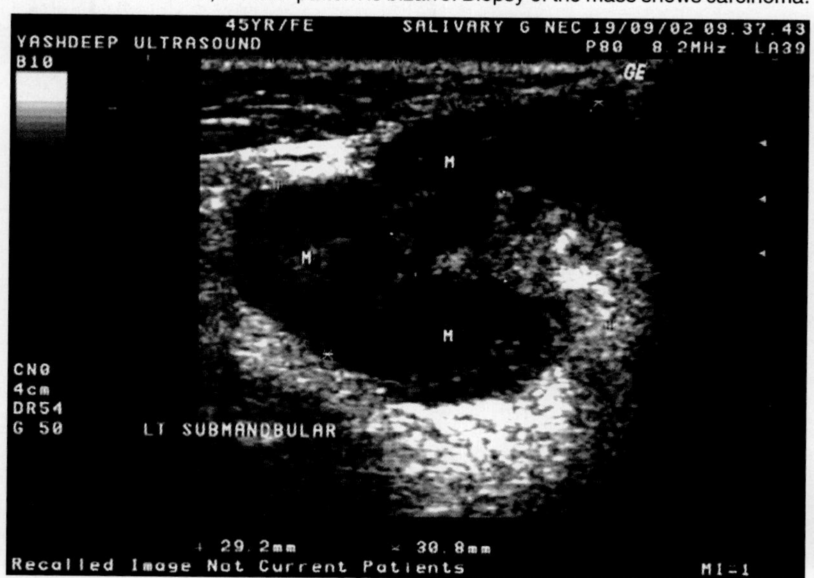

FIGURE 4.159

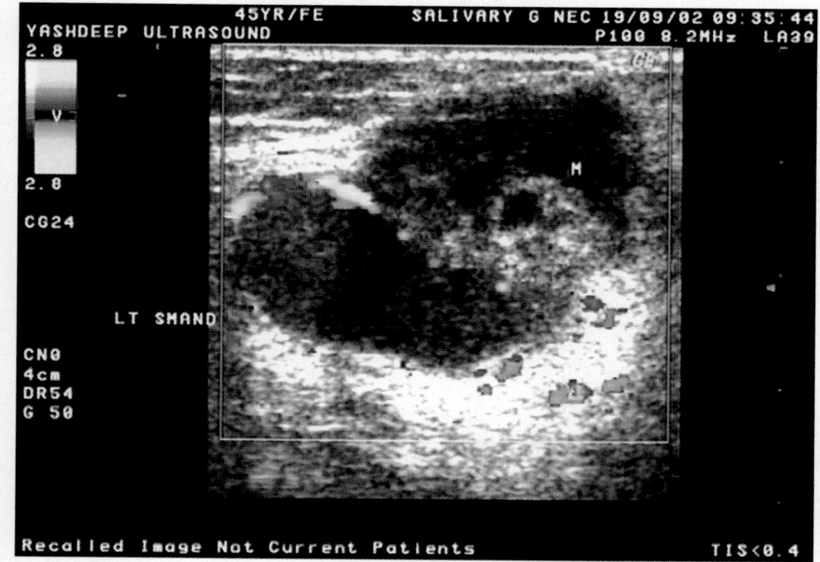

FIGURE 4.160

FIGURES 4.159 and 4.160: Malignant mass in the submandibuar gland. A middle aged man presented with heterogeneous mass in the Lt submandibular gland. Normal glandular texture is lost. The capsule is broken. The mass is seen invading the soft tissue planes. Biopsy confirmed invasive adenocarcinoma.

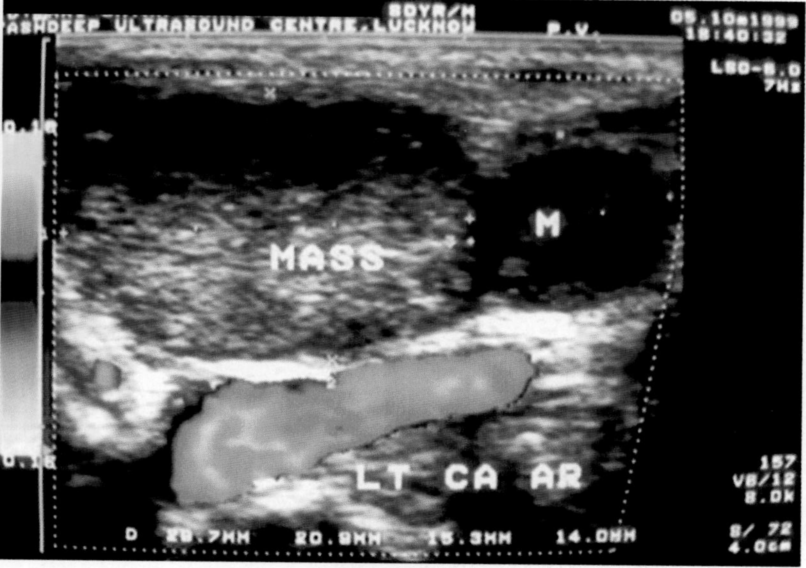

FIGURE 4.162

FIGURES 4.161 and 4.162: Malignant mass. An old man presented with a solid cystic mass over the Rt submandibular area. HRSG shows heterogenous mass in SMG. Biopsy of mass confirmed adenocarcinoma.

are seen in the tumor suggestive of hemorrhage. Associated lymph node enlargement around the lesion is suggestive of malignancy. However, HRSG guided biopsy is ultimate to confirm the diagnosis. HRSG is also helpful in staging of the disease for proper treatment and management.

CONCLUSION

High resolution sonography is an excellent non-invasive multi planter imaging modality for evaluation of salivary glands. It does not require any sedation or specific preparation for evaluation of salivary glands. Small ductal stones can be easily picked up on HRSG. The benign masses are easily evaluated on HRSG and it is gold standard in differentiating between solid and cystic masses. Vascular masses are also easily picked up on HRSG and color Doppler imaging improves the diagnostic accuracy. In indeterminate lesion ultrasound guided FNAC is of immense value in arriving to the correct diagnosis.

[illegible] An old man presented with a solid [illegible] HRSG shows heterogenous mass in [illegible] carcinoma

[illegible] the tumor suggestive of hemorrhage. Associated lymph-[illegible] the lesion is suggestive of malignancy. [illegible] biopsy is ultimate to confirm the diagnosis. [illegible] also helpful in staging of the disease for proper treatment and management.

High [illegible] sonography is an excellent non-invasive [illegible] imaging [illegible] evaluation of salivary glands. It does not require any [illegible] preparation for evaluation of salivary glands. Small [illegible] easily picked up on HRSG. The benign masses [illegible] evaluated on HRSG and it is gold standard in differentiating [illegible] solid and cystic masses. Vascular masses are also easily [illegible] HRSG and color Doppler imaging [illegible] the [illegible] lesion ultrasound-guided FNAC [illegible] the correct diagnosis.

5 Neurosonography of Neonatal Brain

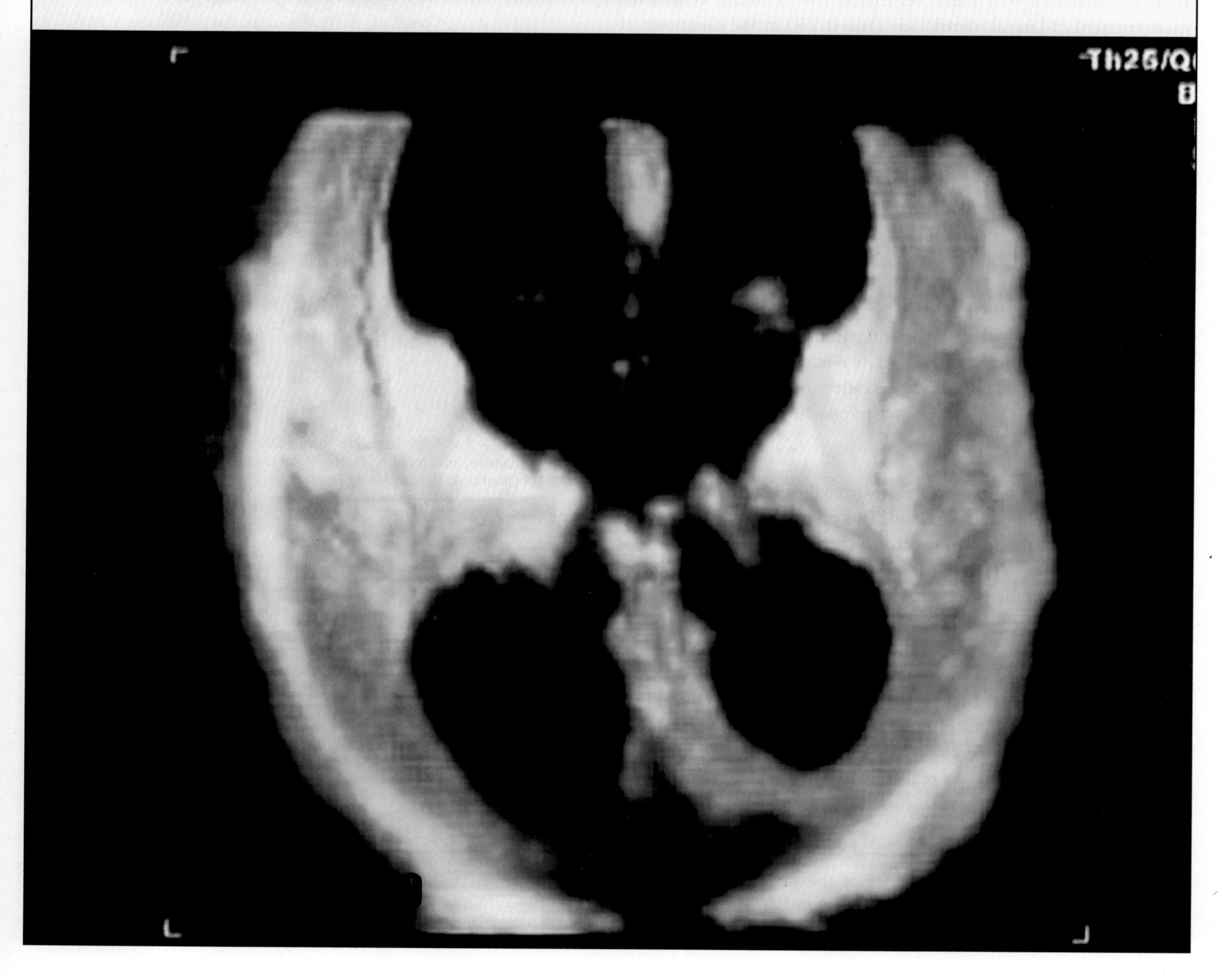

Introduction

With the advent of high-resolution sonography and lately addition of color Doppler have paved the way for detailed evaluation of neonatal brain. The fact that it can be taken to the bedside makes it an ideal imaging tool in the evaluation of critically ill premature newborn babies.

Technique and Instrumentation

To obtain undisturbed images, mild sedation is helpful. The newborn suspected to have suffered anoxia/birth injury is seen in the main department. The pre-term baby is examined at bedside with portable machine having high resolution probe without disturbing the life-support. The anterior fontanelle is used as acoustic window and multiple coronals, sagittal and parasagittal scans are taken using 5.0/7.5 Mhz sector transducers. Positioning the transducer just above the external auditory meatus does axial scanning. By this technique, we get excellent view of lateral ventricles. The technique is used for measurement of lateral ventricular ratio (LVR) in the diagnosis of hydrocephalus.

Normal Anatomy

Coronal Scans

Placing the probe transversely in anterior fontanelle and sweeping the probe from front to back to obtain the image in coronal plane. It gives excellent details of ventricular system and adjoining brain parenchyma as described by Shuman et al.

Plane Through Frontal Horns

The frontal horns are seen as crescentic fluid-filled spaces separated by cavum septum pellucidum. Caudate nucleus occupies the lateral concavity of the frontal horns. Corpus callosum, which forms the roof of lateral ventricles is seen as concave hypoechoic band.

Plane Through Thalamus

With the slight shift of probe posteriorly from the plane of frontal horns, section through thalami is obtained. This plane, the lateral ventricles are seen as arcuate fluid-filled structures, thalami appear as hypoechoic structures separated by slit-like third ventricle. The echogenic choroid plexus is seen prominently occupying the body of lateral ventricles.

Plane Through Quadrigeminal Cistern

The cerebellum and quadrigeminal cistern appear as an echogenic tree. Body of lateral ventricle is seen superiorly, and temporal horn is seen inferiorly. The tentorium cerebelli appear as thick echogenic band on either side of cerebellum. The middle cerebral artery can be recognized pulsating on real-time in echogenic Sylvain fissure laterally.

Plane Through Trigone of Lateral Ventricle

The choroid plexus occupies most of the lateral ventricle. The parietal lobe is seen superiorly. The tentorium forms the inverted V posteriorly.

Sagittal Scans

The sagittal scans are obtained by placing the probe parallel to the anterior fontanelle. The first plane obtained is in the midline. The cavum septum pellucidum is seen as comma-shaped structure just above the third ventricle, corpus callosum is seen as hypoechoic band running just above, whereas fourth ventricle and echogenic cerebellar vermis are appreciated posterior to cavum septum pellucidum. By slight angulations laterally, one visualizes caudate nucleus, shallow lateral ventricle and cerebellar hemispheres posteriorly. By further angulations laterally, the body of lateral ventricle with choroid plexus resting on its floor and occipital horns can be appreciated.

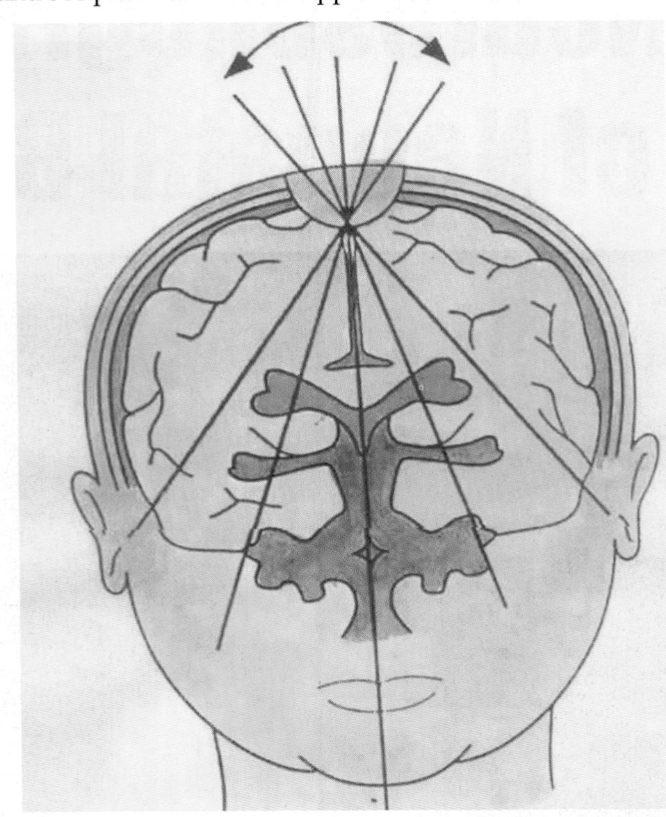

FIGURE 5.1: Line diagram shows scanning plane through anterior fontanelle, in coronal and transorbital planes (*Courtesy* — Textbook of Diagnostic Bltrasound by Rumac).

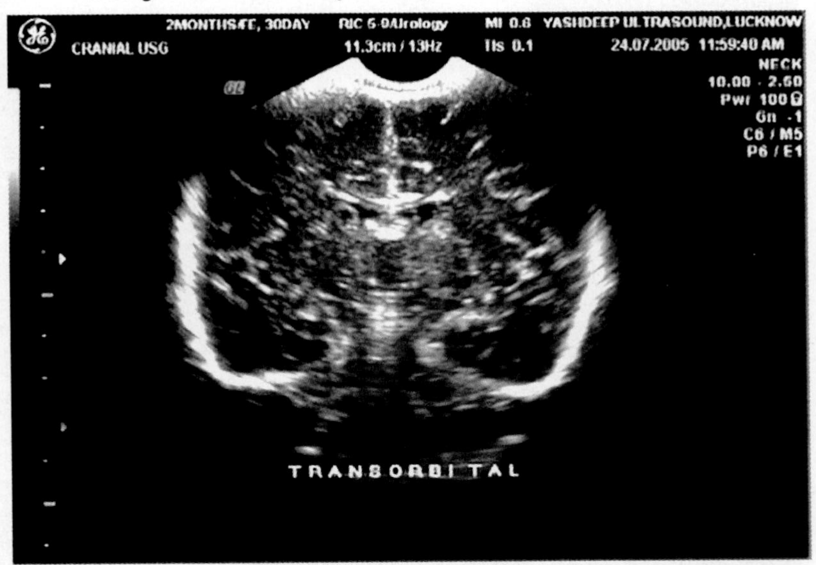

FIGURE 5.2: Normal brain anatomy: Transorbital view. The coronal plane shows normal brain anatomy. Anterior horn, third ventricle and both thalami are seen in the transorbital views.

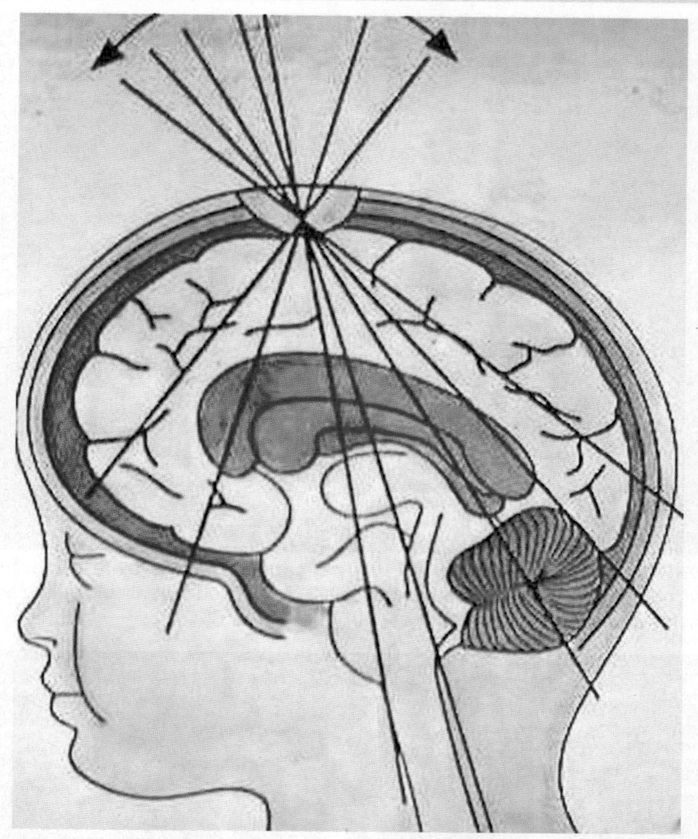

FIGURE 5.3: **Line diagram-parasagittal view.** The line diagram shows scanning plaines in parasagittal axis. Rocking movement of transducer can scan whole of the brain in parasagittal view. (*Courtesy*—Textbook of Diagnostic Bltrasound by Rumac).

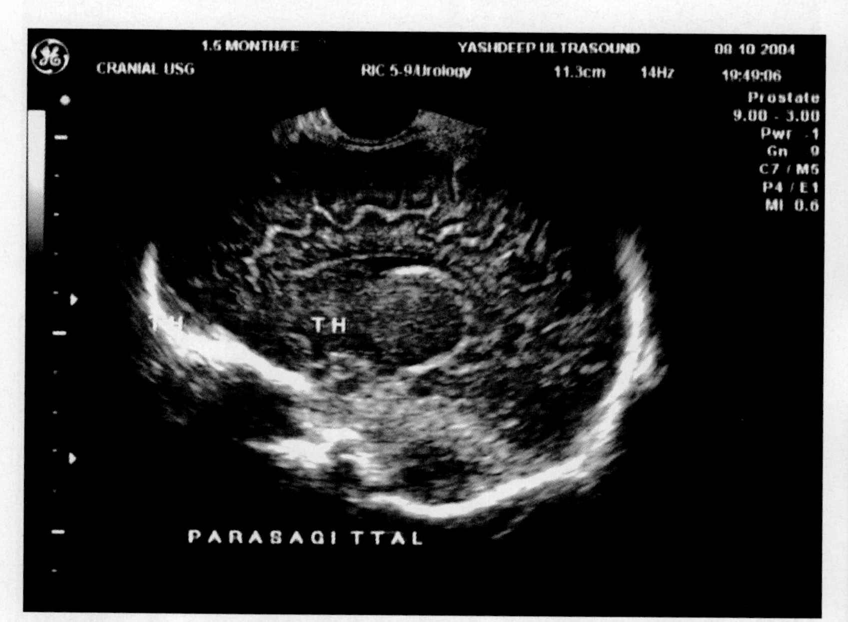

FIGURE 5.4: Normal parasagittal plane. HRSG shows normal infant brain anatomy in parasagittal planes. Cavum septum pellucidum is seen a small collapsed fluid filled shadow. The thalamus is seen well-defined rounded body. Collapsed fourth ventricle is seen as a streak. Normal salci and gyri are seen.

Congenital Anomalies

Congenital malformations are quite common and with the availability of high-resolution sonography most of the disorder can be diagnosed pre-natally. Brain malformations are classified based on the various stages of development of the brain. De Myer had suggested the classification in early seventies, which had been modified by Harwood-

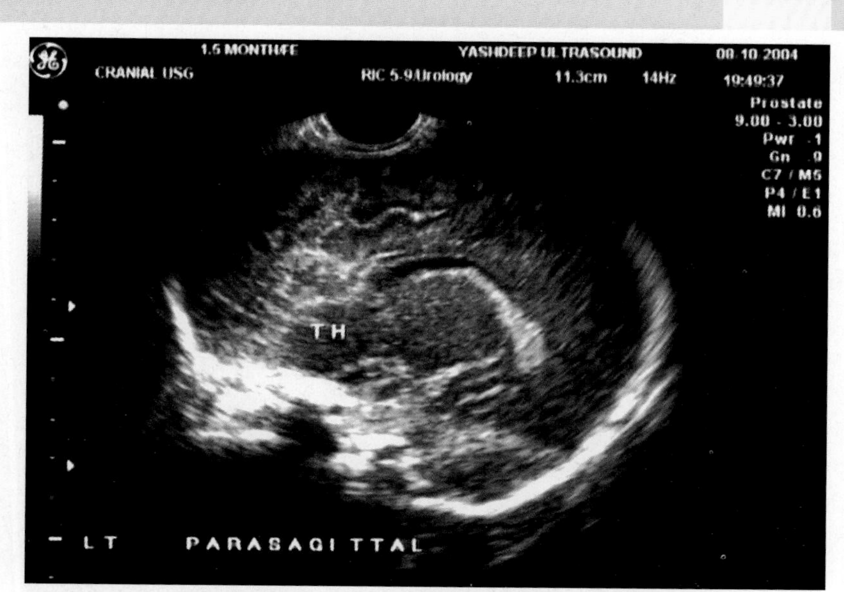

FIGURE 5.5: Normal parasagittal view. HRSG shows normal infant brain in para sagittal planes. Body of the third ventricle is seen and echogenic choroid plexus is seen filling the body at the atrium. Head of the caudate nucleus is also seen adjacent to the thalamus.

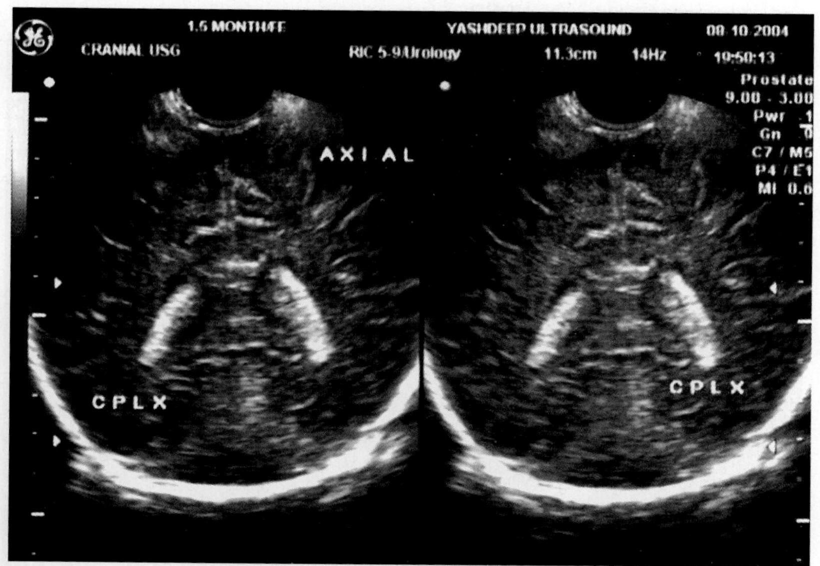

FIGURE 5.6: Normal axial view. HRSG shows normal axial view of the brain parenchyma. Body of the lateral ventricles is seen. Echogenic choroid plexus fill the lateral ventricles. Parietal lobes show normal sulci and gyri.

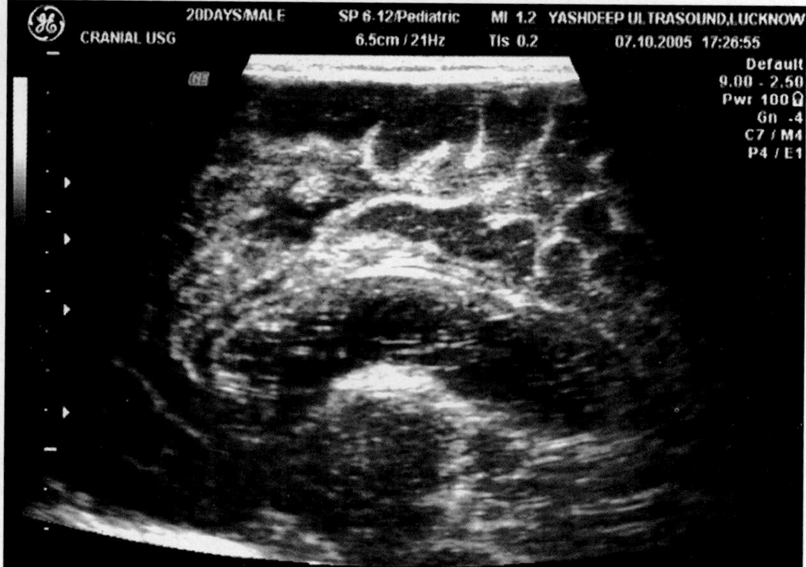

FIGURE 5.6A

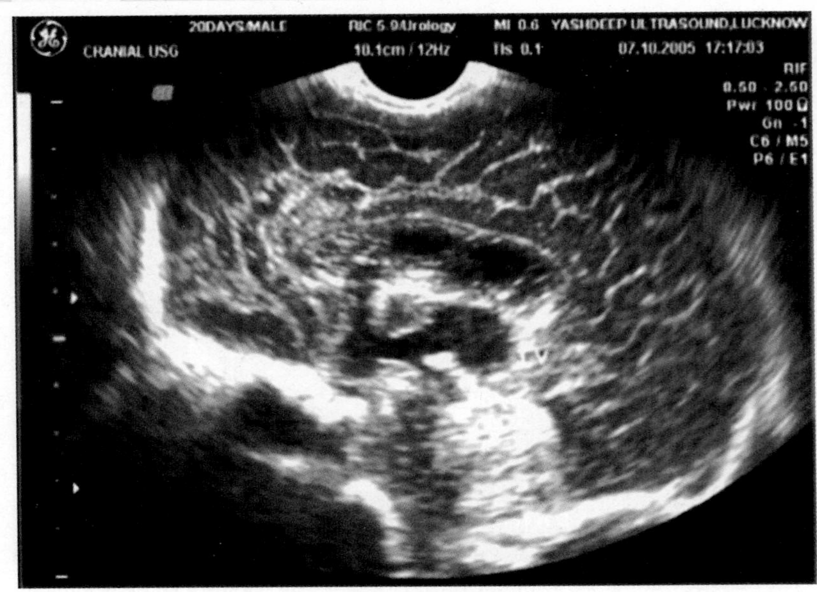

FIGURE 5.6B

FIGURES 5.6A and B: Normal sulci and gyri on high resolution. High resolution ultrasound shows the details of the sulci and gyri of the brain parenchyma. The sulci and gyri stand out clearly on HRSG.

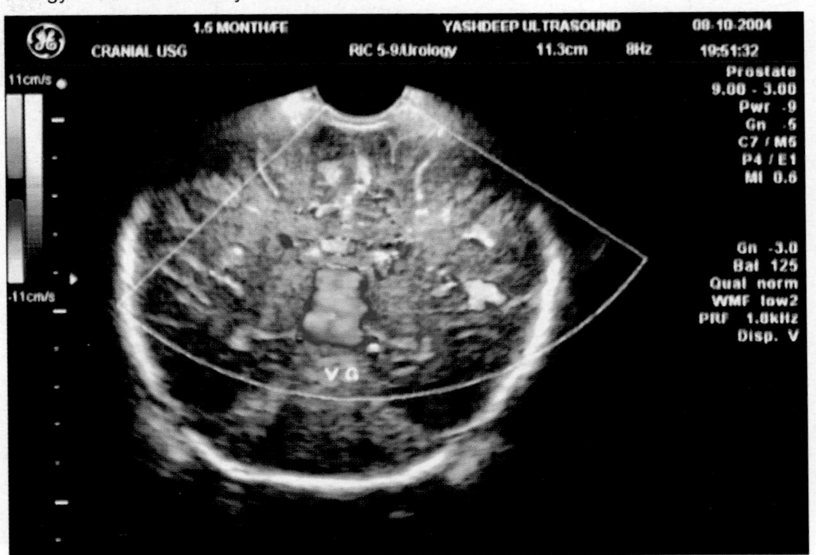

FIGURE 5.7

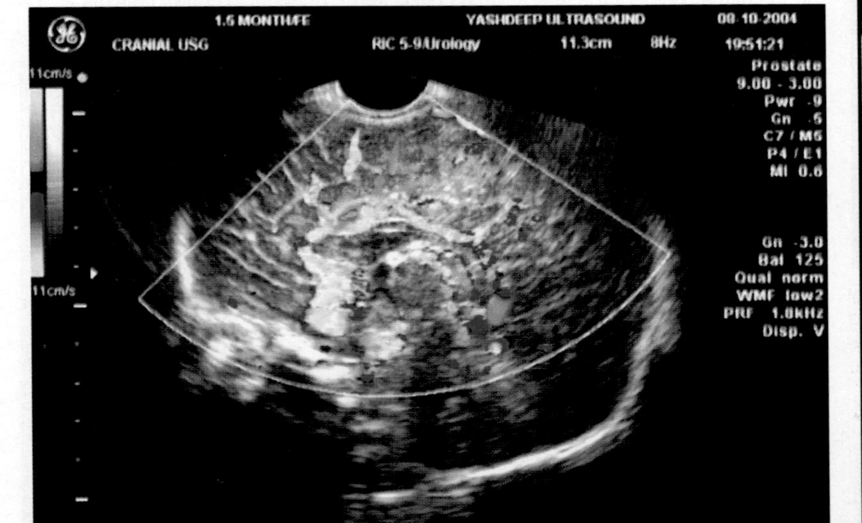

FIGURE 5.8

FIGURES 5.7 and 5.8: Normal brain parenchymal flow. Color flow imaging shows normal parenchymal flow in the brain parenchyma. The vein of Galen is seen as the dominant vessel.

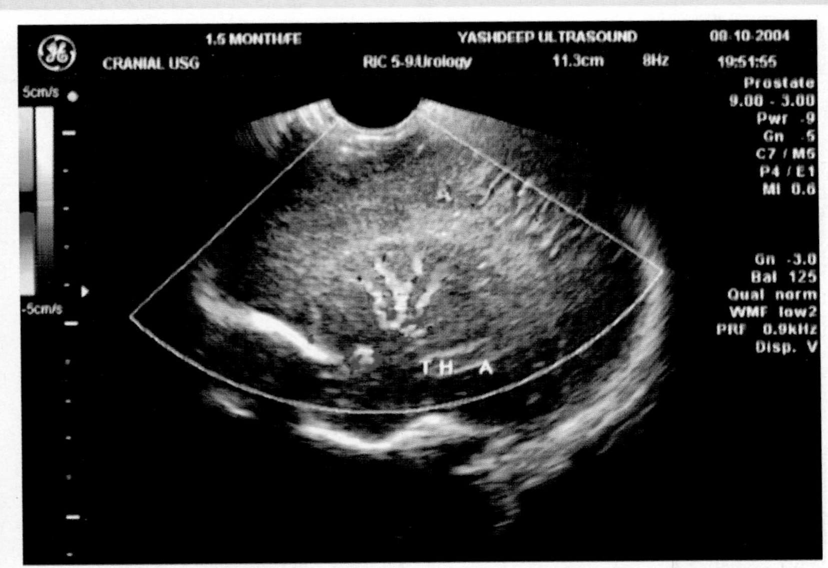

FIGURE 5.9: Normal thalamic and internal capsular flow. Color Doppler imaging shows normal internal capsular and thalamic flow supplied by lenticulo striate arteries, branches of middle cerebral artery.

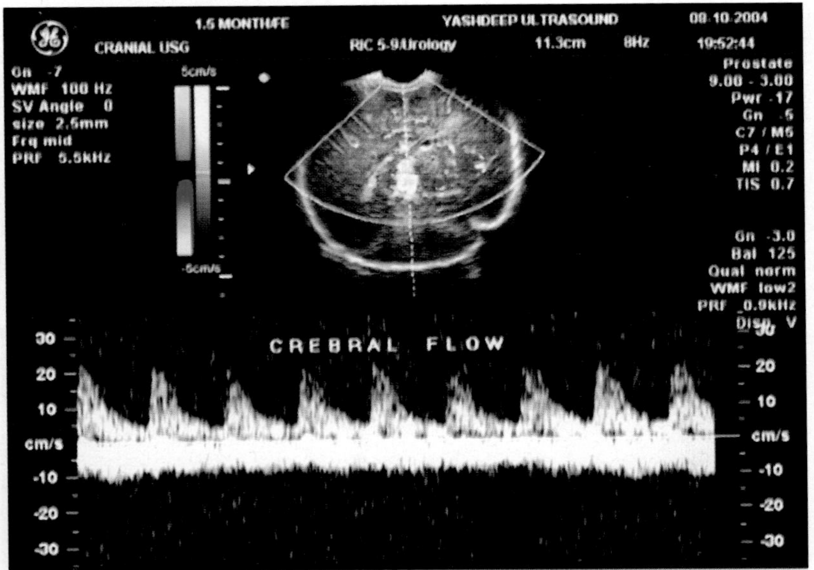

FIGURE 5.10

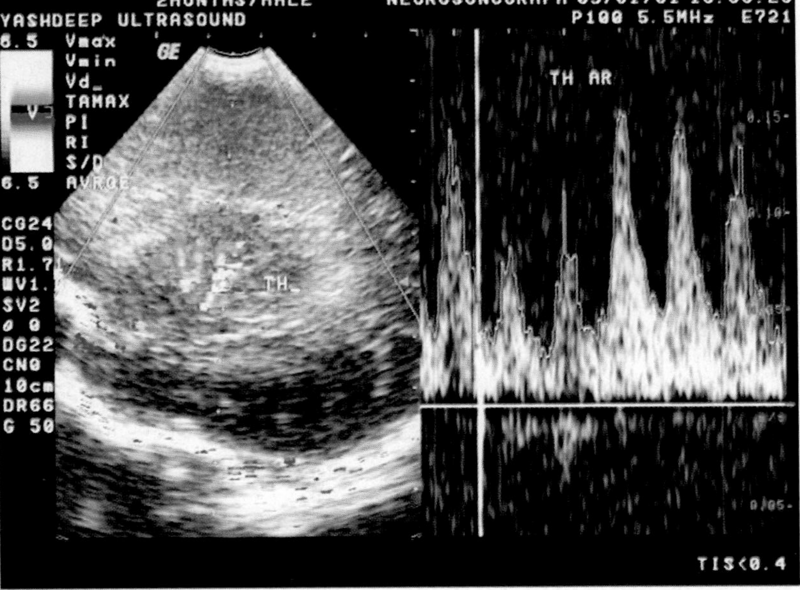

FIGURE 5.11

FIGURES 5.10 and 5.11: Spectral Doppler flow in the cerebral arteries. Spectral Doppler tracing flow shows arterial flow pattern in the intracerebral arteries.

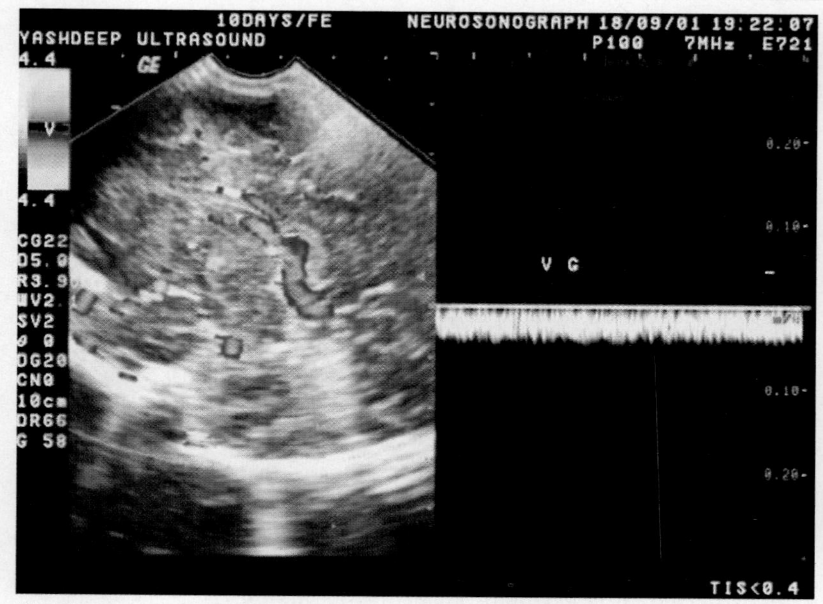

FIGURE 5.12: **Spectral Doppler flow in vein of galen.** Spectral Doppler tracing flow shows monophasic flow pattern of vein of Galen.

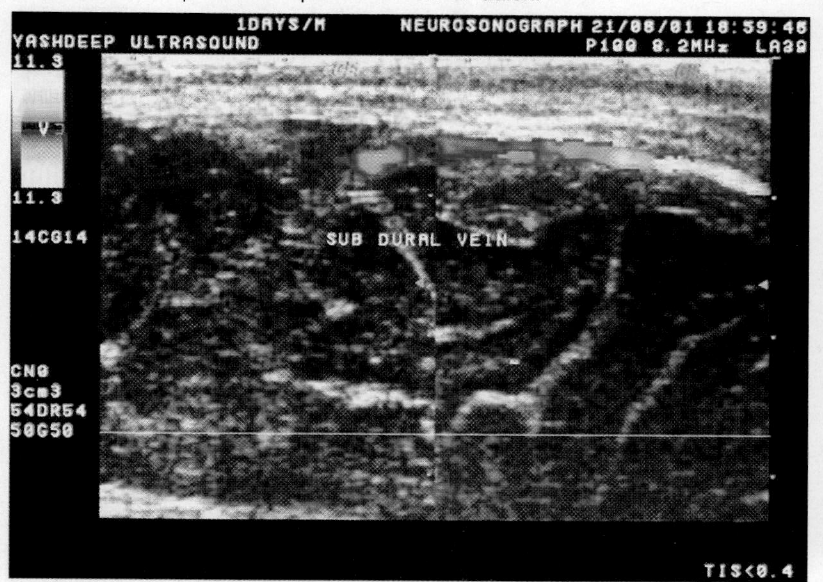

FIGURE 5.13

FIGURE 5.14

FIGURES 5.13 and 5.14: Dural sinus. HRSG shows normal dural sinus running in the superior surface of the vein. Color Doppler imaging shows normal venous flow in the dural sinus.

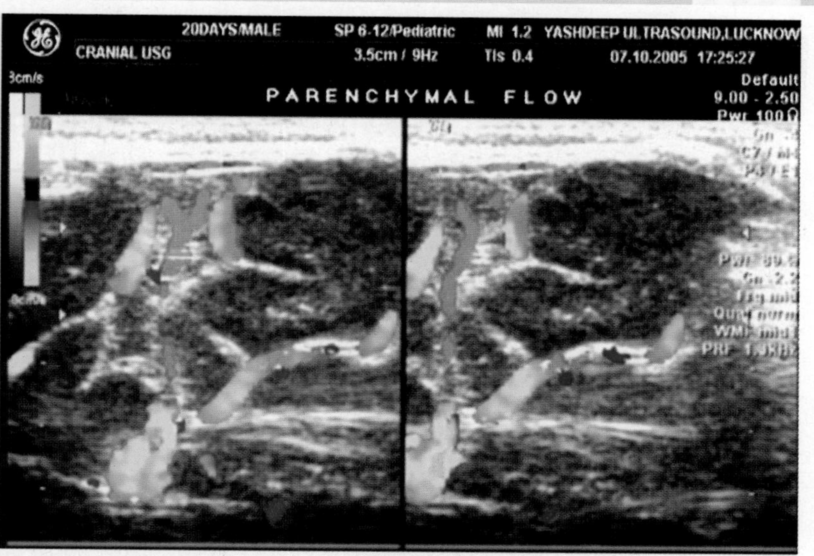

FIGURE 5.15

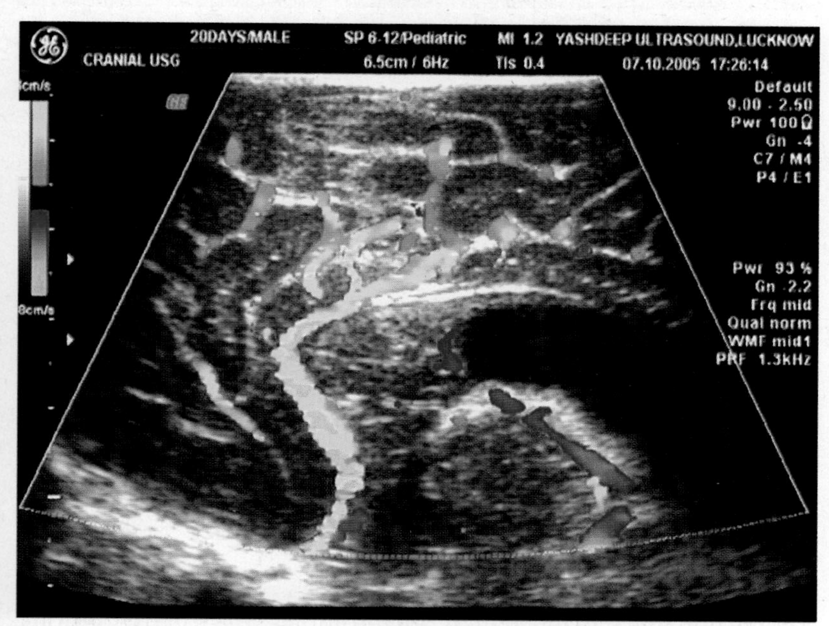

FIGURE 5.16

FIGURES 5.15 and 5.16: Normal brain parenchymal flow. High resolution color flow imaging shows normal parenchymal flow in neonatal brain.

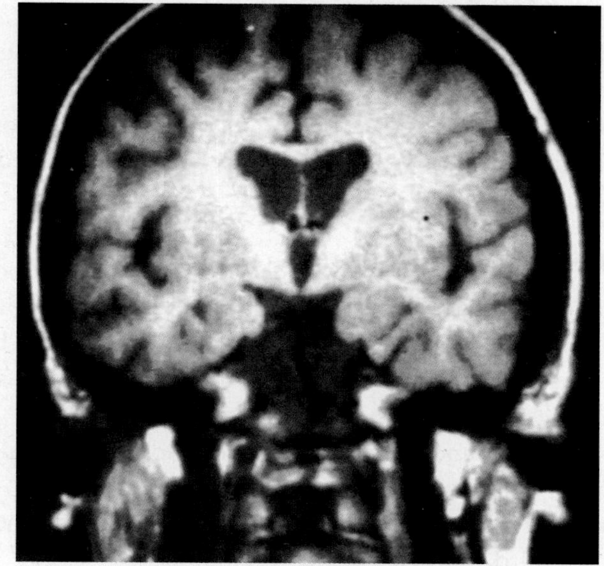

FIGURE 5.17: **MRI image of the normal brain.** MRI image of the normal brain shows normal brain anatomy. Which is comparable to the normal sonographic anatomy of the brain.

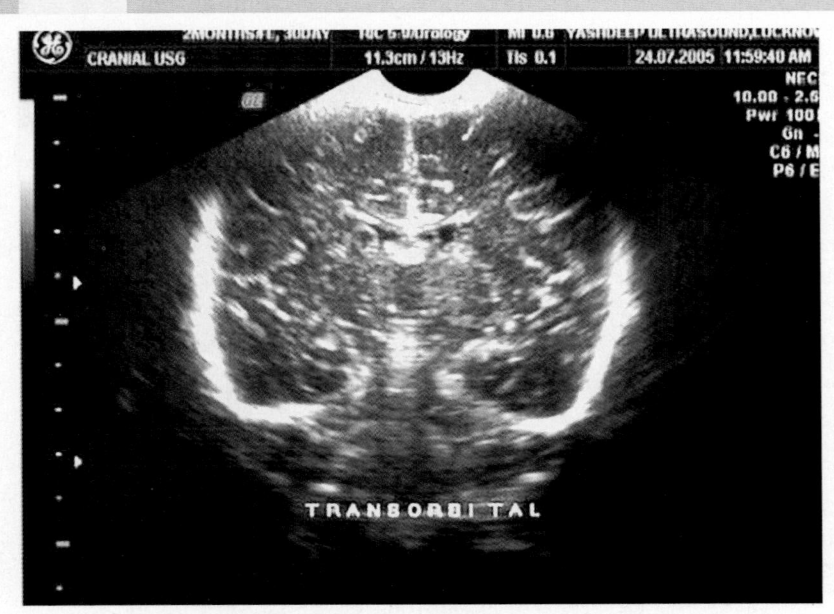

FIGURE 5.18: Normal sonographic image of infant brain. HRSG shows normal image of brain parenchyma in coronal plane. The image is comparable to the MRI image of the same patient.

Nash and Rumack and Johnson, accordingly the brain malformations can be classified as follows: -

Disorders of Organogenesis

Disorders of Neural Tube Closure

Chairi malformation, Dandy-Walker malformation, agenesis of the corpus callosum, lipoma of the corpus callosum, teratoma disorders of diverticulation Septo-optic dysplaisa, holoprosencephaly, aventricular cerebrum.

Disorders of Proliferation— Microcephaly

Disorders of Sulcation and Gyration

Lissencephaly, schizencephaly, heterotopias, destructive lesions, disorders of organization, disorders of myelination.

Disorders of Histogenesis

Phakomatosis (tuberous sclerosis, neurofibromatosis), neoplasia, vascular lesions.

Among the malformations, disorders of neural tube closure are frequently encountered in our day-to-day practice and need further attention.

Chiari Malformation

Chairi malformations have been classified into Chiari I to IV, however, Chiari II is most frequently seen, and its early recognition becomes important, as invariably it is associated with meningomyelocele.

The salient sonographic features are as follows: -

- Batwing appearance of frontal horns due to anterior and inferior pointing.
- Cavum septum pellucidum completely or partially absent.
- Enlarged third ventricle
- Enlargement of mass intermedia.
- Non-visualization of fourth ventricle.

The above changes can be explained on account of downward herniation of cerebellar tonsils into upper spinal canal.

Dandy-Walker Malformation

The salient sonographic features are as follows: -

- Large posterior fossa cyst community with fourth ventricle.
- Hypoplastic cerebellar hemispheres.
- Associated hydrocephalus.

Associated with other anomalies in 68% cases, i.e. agenesis of corpus callosum, encephalocele, aqueductal stenosis.

Dandy-Walker variant

- Small posterior fossa cyst.
- Mild hydrocephalus

Differential diagnosis of Dandy-Walker malformation

- Arachnoid cyst—no communication with fourth ventricle.
- Large cisterna magna – no hydrocephalus.

Agenesis of Corpus Callosum

Agenesis of corpus callosum may present as complete agenesis, partial agenesis or hypoplasia, 80% of the cases are associated with other anomalies such as Dandy-Walker, Chiari malformation and midline cysts.

The salient sonographic features are as follows: -

- Distorted sulci/gyri anatomy.
- Superior extension of third ventricle.
- Parallel lateral ventricles
- Absent cavum septum pellucidum.
- Absent cingulate gyrus.

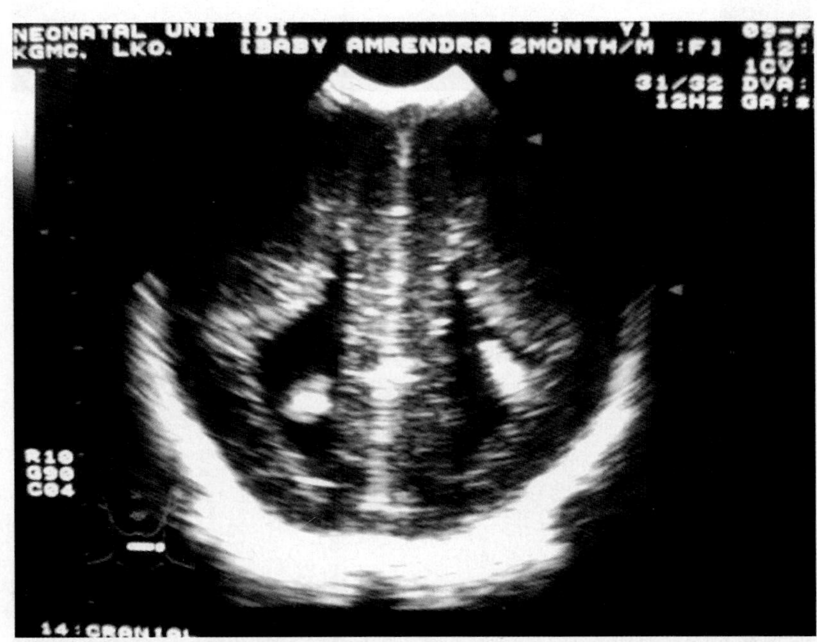

FIGURE 5.19: Agenesis of corpus callosum. HRSG shows parallel placed lateral ventricles suggestive of agenesis of corpus callosum. The ventricles are seen widely separated. The sulci and the gyri are seen running in radial fashion. Typical sonographic findings are agenesis of corpus callosum.

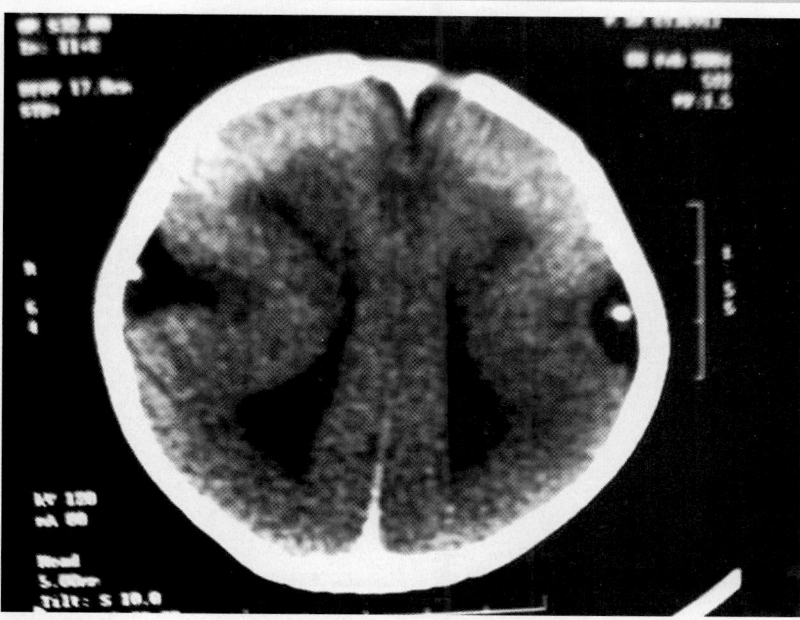

FIGURE 5.20: CT scan of the same patient also shows parallel lateral ventricles. Thus confirms the sonographic findings.

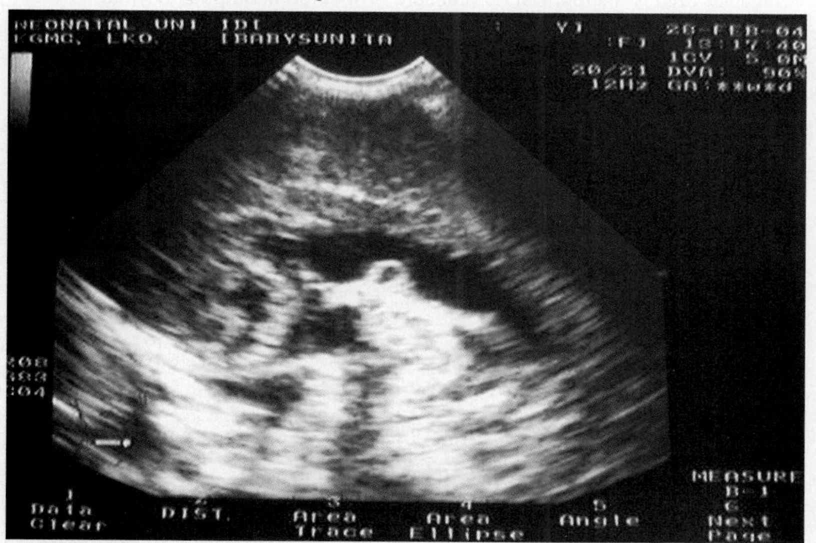

FIGURE 5.21

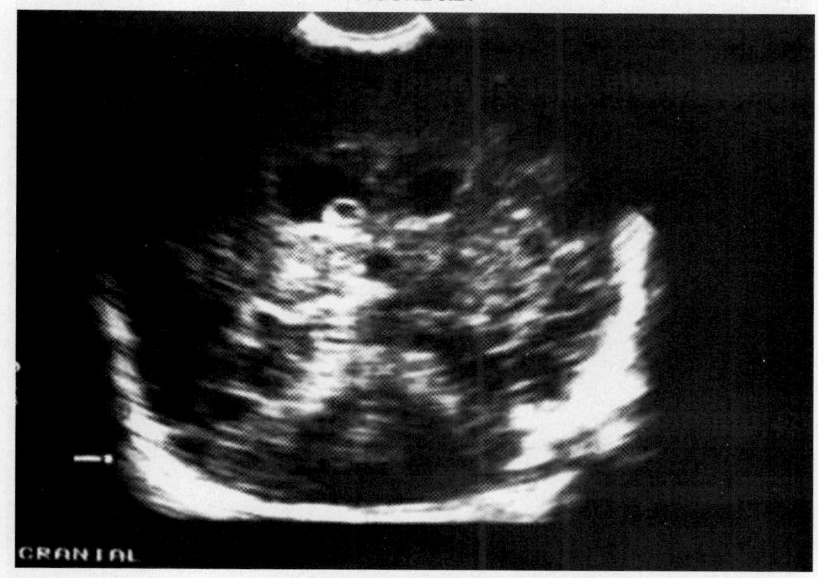

FIGURE 5.22

FIGURES 5.21 and 5.22: Choroid plexus cyst. An infant child presented with fullness of the fontanelle. HRSG shows mild dilatation of the ventricles. A well defined thin walled cyst is seen attached with the choroid plexus suggestive of choroid plexus cyst.

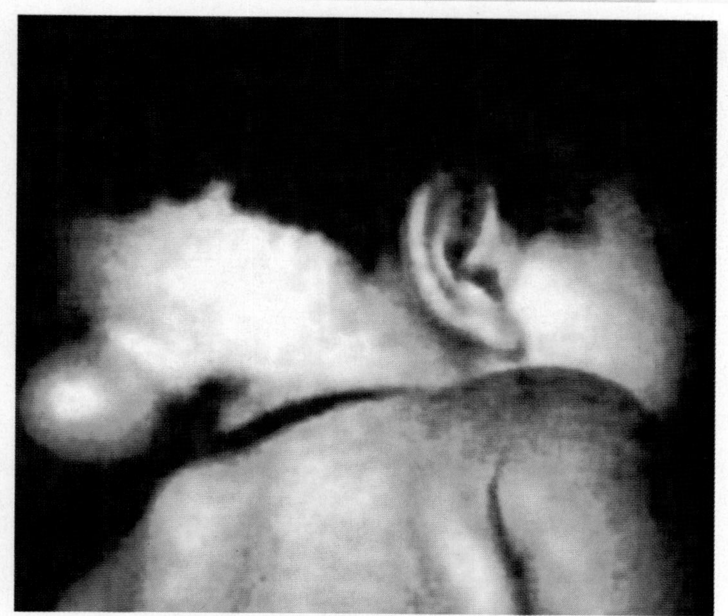

FIGURE 5.23: Occipital meningocele. A young boy presented with a cystic mass in the occiput.

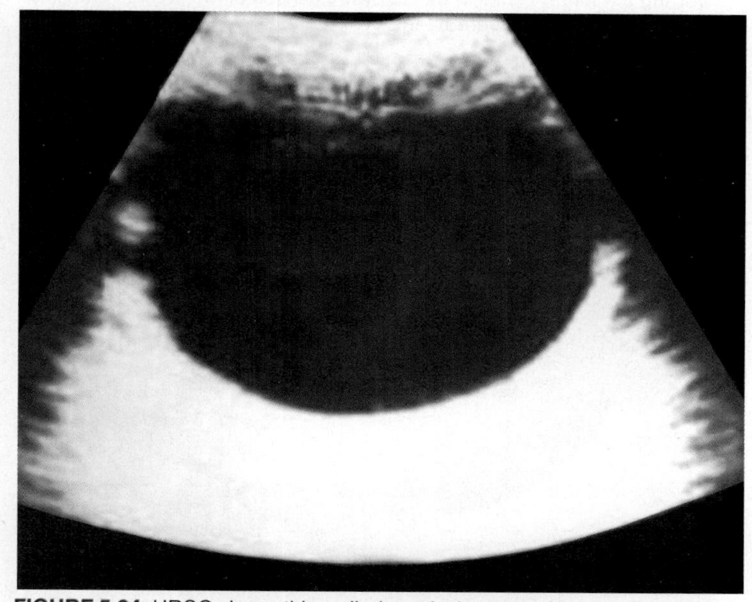

FIGURE 5.24: HRSG shows thin walled cyst in the mass. No brain herniation is seen in the cyst. No septation, calcification or any membranous shadow is seen in the cyst.

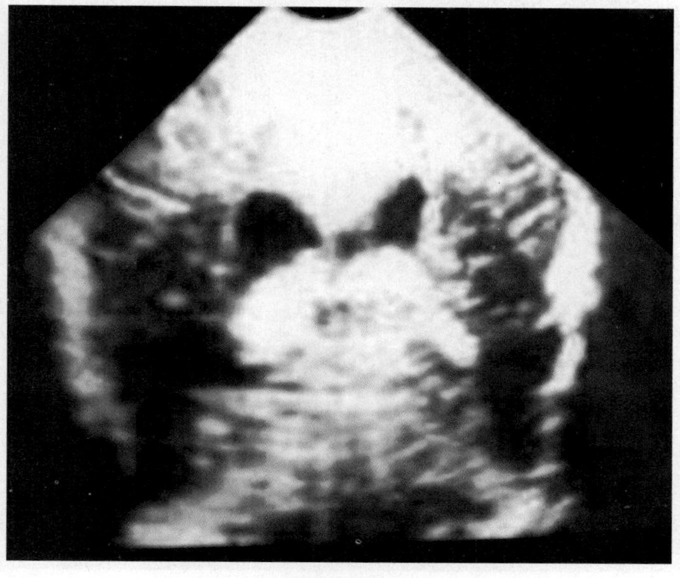

FIGURE 5.25

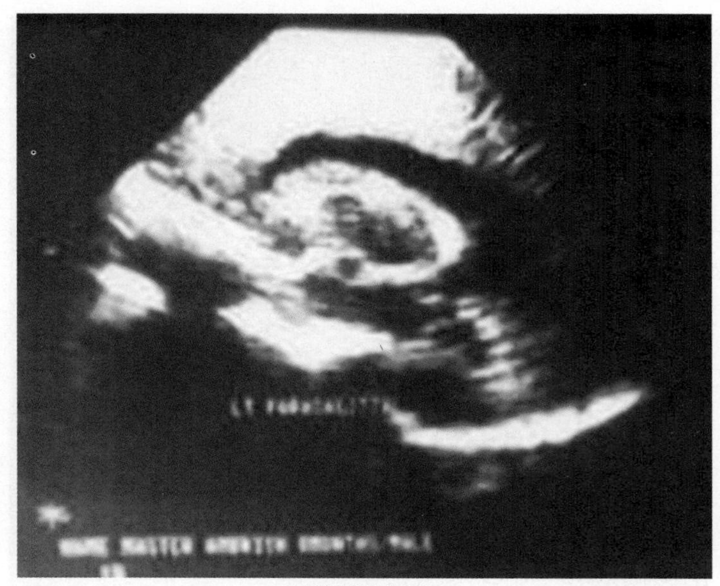

FIGURE 5.26

FIGURES 5.25 and 5.26: HRSG shows mild pressure on the outflow tract involving the fourth ventricle by the cyst. Fullness of the lateral ventricles is seen due to the pressure by the cyst.

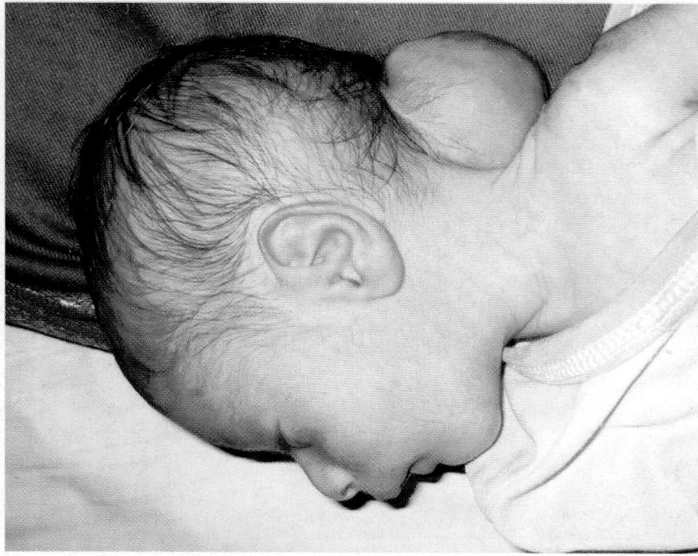

FIGURE 5.27: Occipital encephalocele. An infant is born with soft tissue swelling over the occiput.

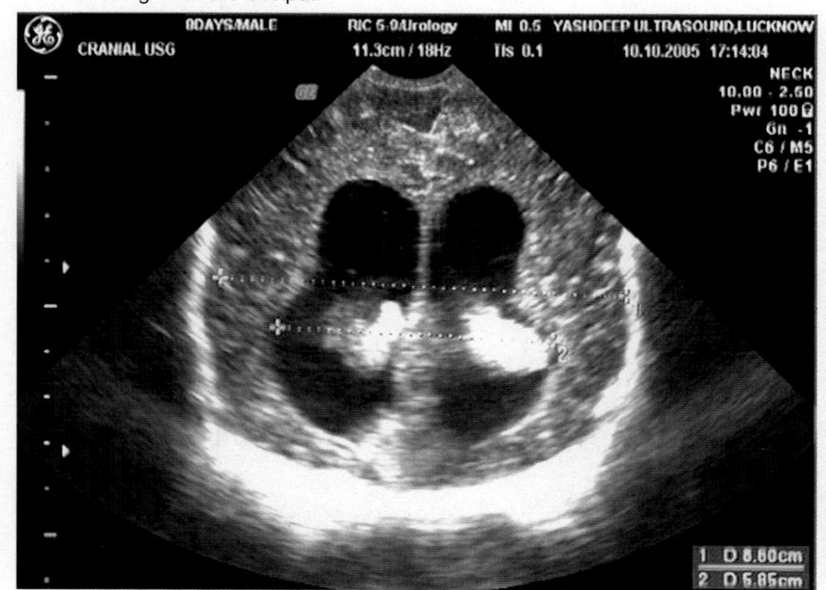

FIGURE 5.28

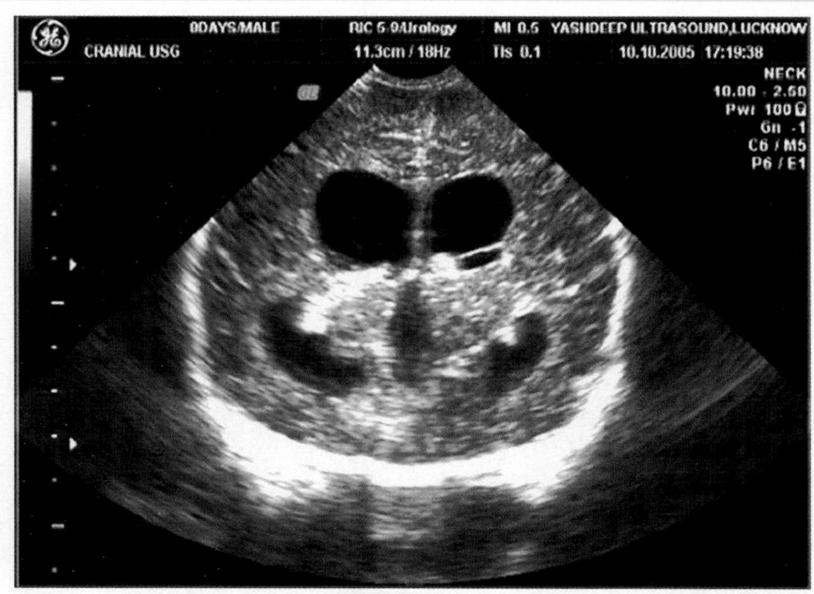

FIGURE 5.29

FIGURES 5.28 and 5.29: Cranial ultrasound shows dilated ventricles. Cephalo ventricular ratio is altered. But no pressure is seen on brain parenchyma.

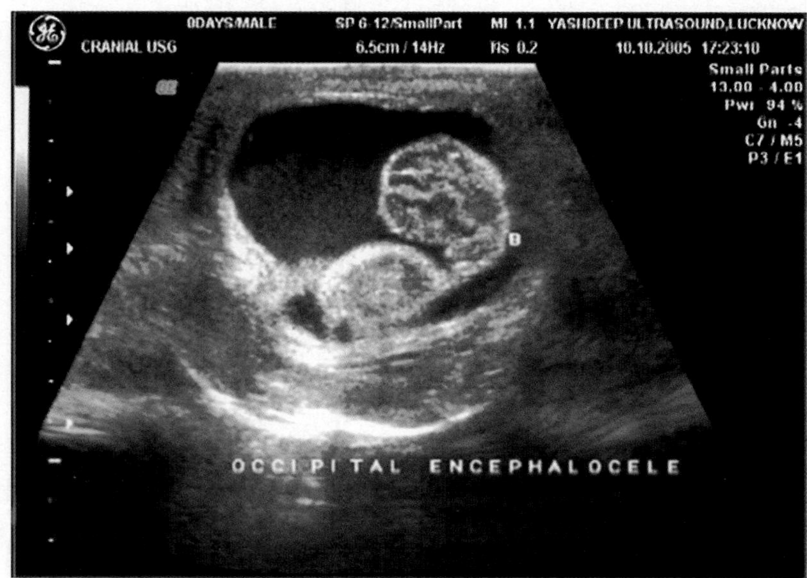

FIGURE 5.30

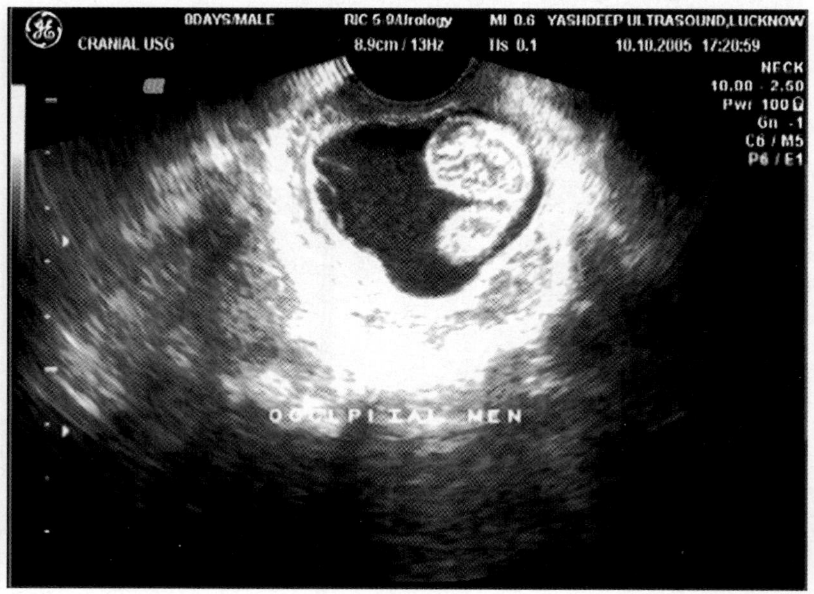

FIGURE 5.31

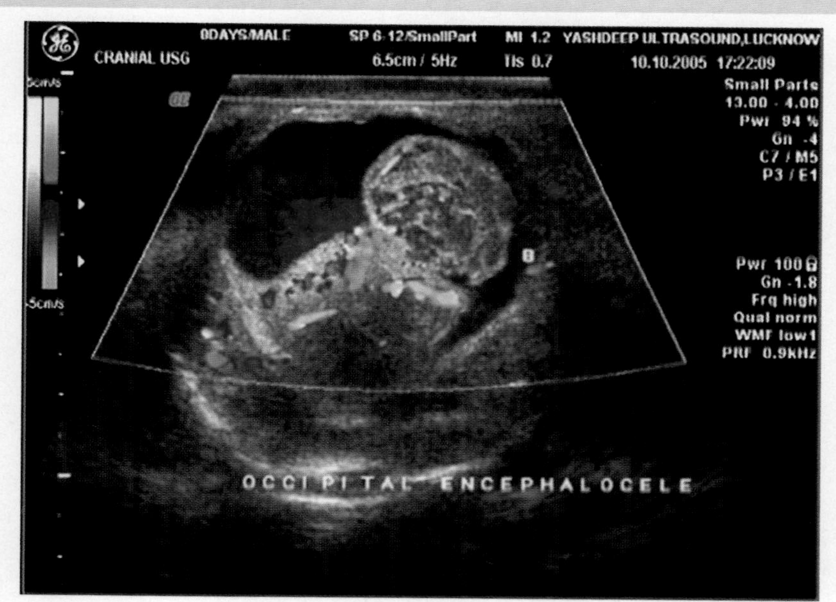

FIGURE 5.32

FIGURES 5.30 to 5.32: High-resolution ultrasound of the occipital cystic mass shows herniation of the brain tissue in the cyst suggestive of encephaloceles part of the cerebellum is seen in the cyst. On color flow imaging vessels are seen in the brain tissue.

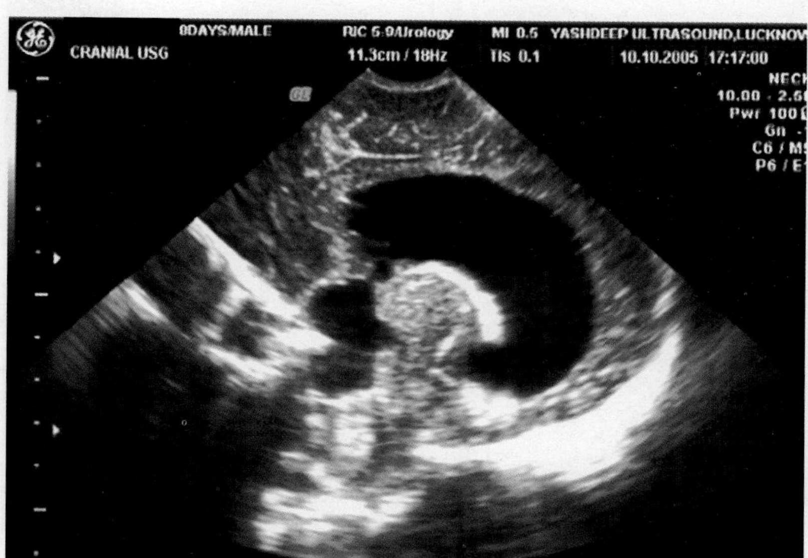

FIGURE 5.33: 3D imaging of the same patient shows defect in the occiput. The brain is seen herniating through the defect in the cyst.

FIGURE 5.34: **Anterior meningocele.** A young boy presented with a cystic mass over the nasion.

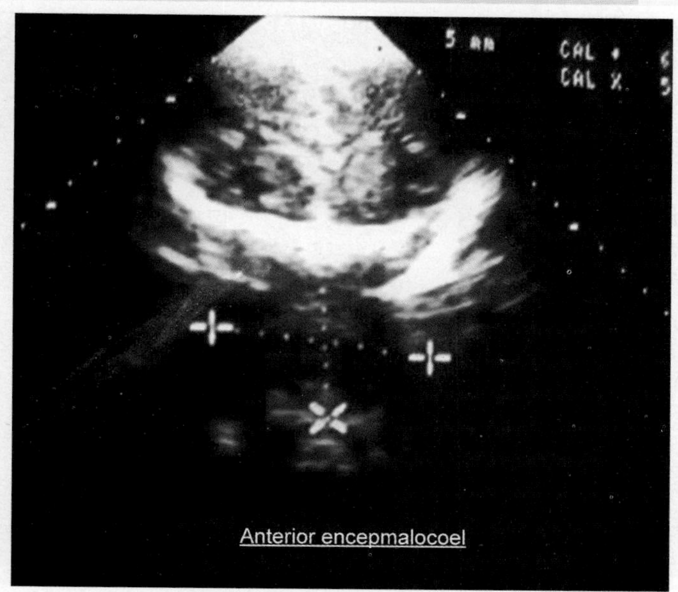

FIGURE 5.35: A well-defined thick walled cystic mass is seen in the mass. Homogeneous low-level collection is seen in it. No septation or loculation is seen suggestive of anterior meningocele.

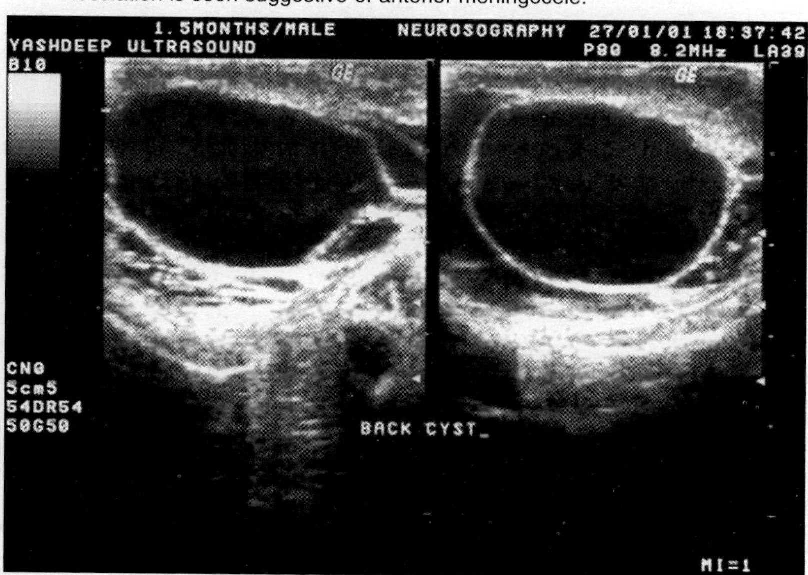

FIGURE 5.36

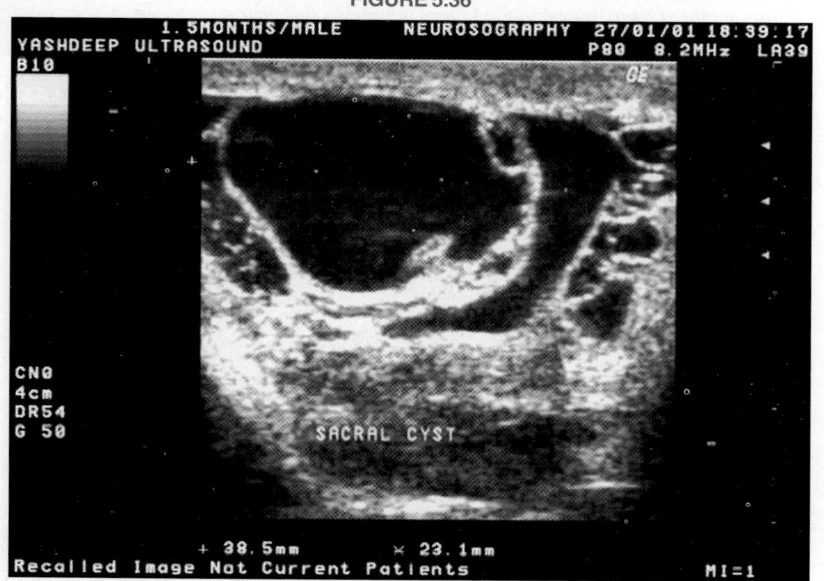

FIGURE 5.37

FIGURES 5.36 and 5.37: **Sacral meningocele.** HRSG shows a cystic mass in the sacral region in an infant. Echogenic septa are seen in the cyst. But no soft tissue shadow is seen suggestive of a meningocele.

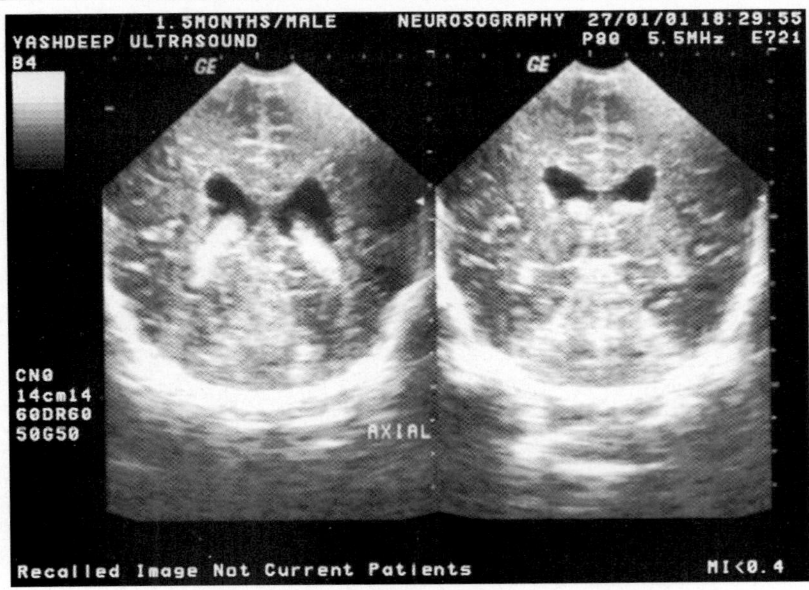

FIGURE 5.38

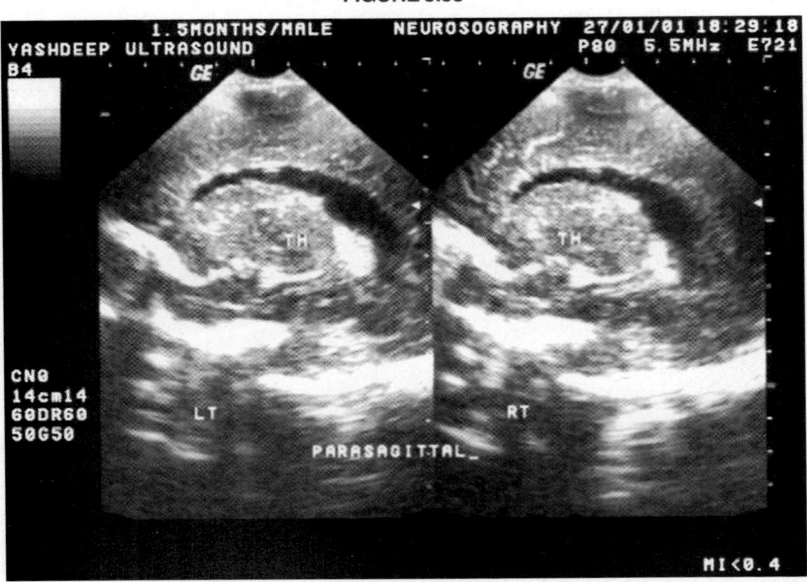

FIGURE 5.39

FIGURES 5.38 and 5.39: Cranial ultrasound of the same patient shows mild fullness of ventricular system involving the lateral ventricles due to the pressure by the sacral cyst. But no periventricular leak is seen.

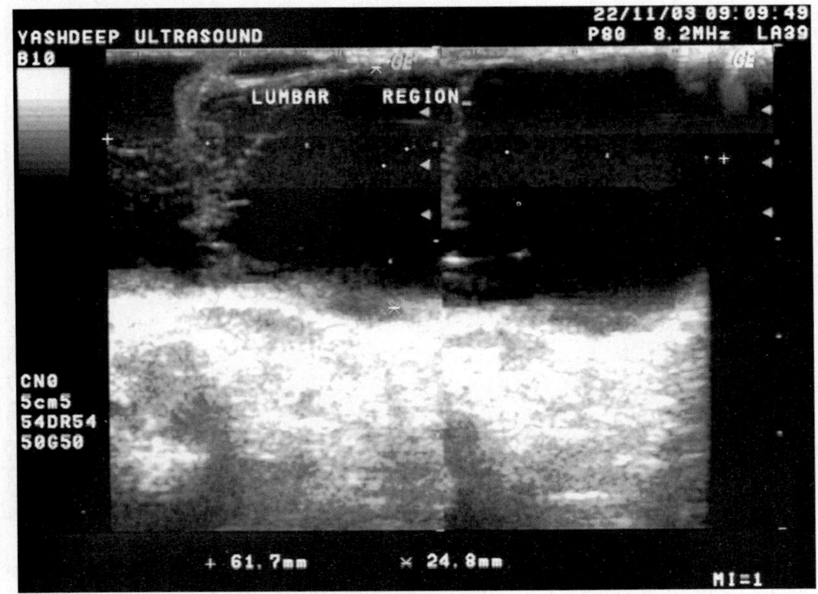

FIGURE 5.40

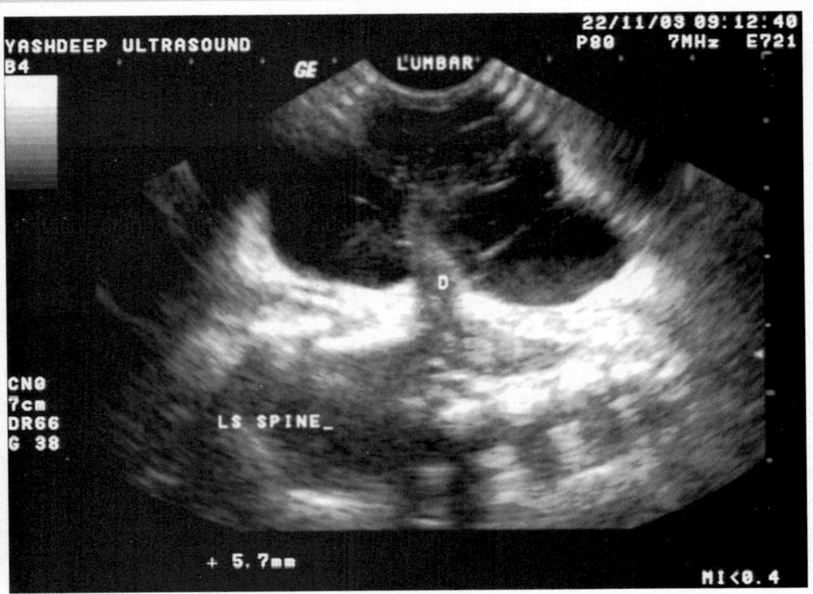

FIGURE 5.41

FIGURES 5.40 and 5.41: Sacral meningocele. A child is born with a big cystic mass in the sacral area. HRSG shows well-defined cyst. No evidence of any septation or loculation is seen in the cyst suggestive of pure meningocele.

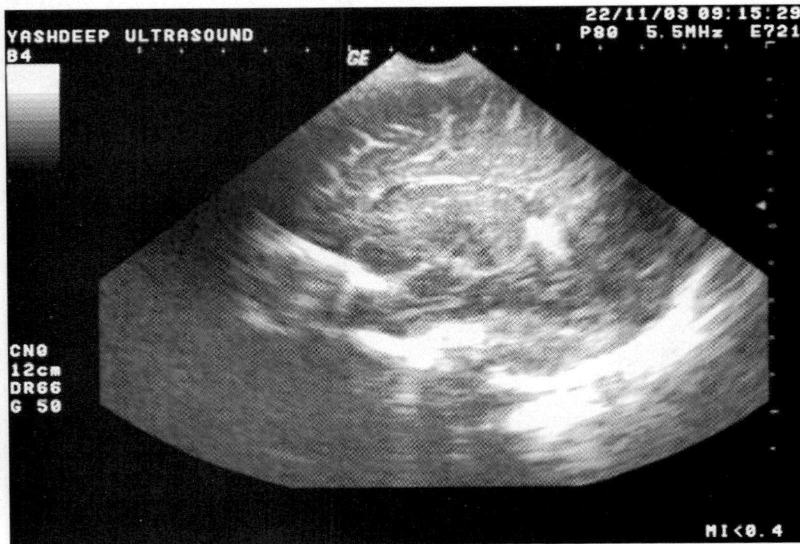

FIGURE 5.42

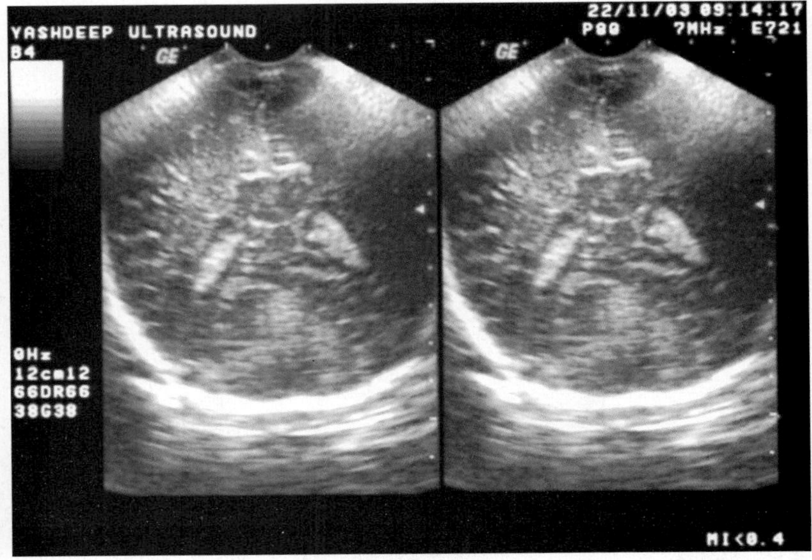

FIGURE 5.43

FIGURES 5.42 and 5.43: Cranial ultrasound of the same patient does not show any dilatation of the ventricles. No evidence of any brain parenchyma pathology is seen.

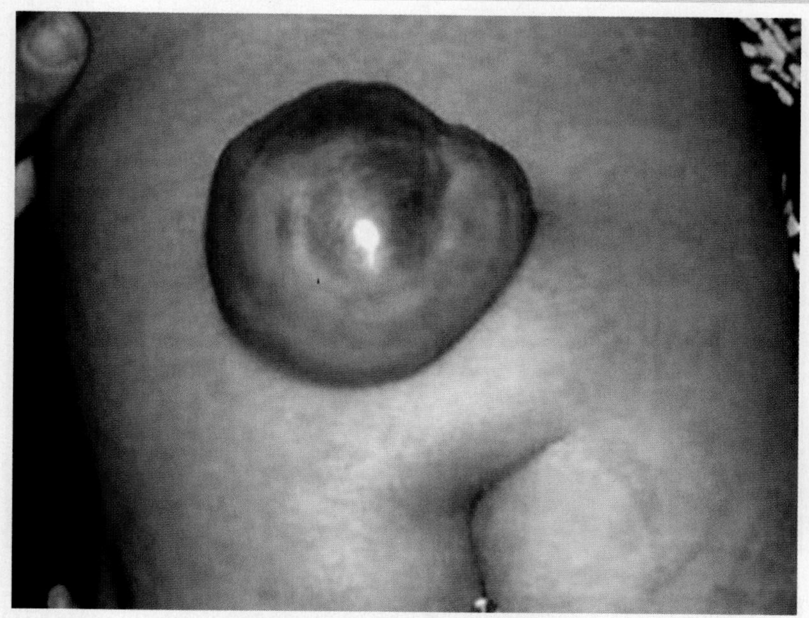

FIGURE 5.44: Sacral meningocele. An infant was born with a soft tissue mass over the sacral area.

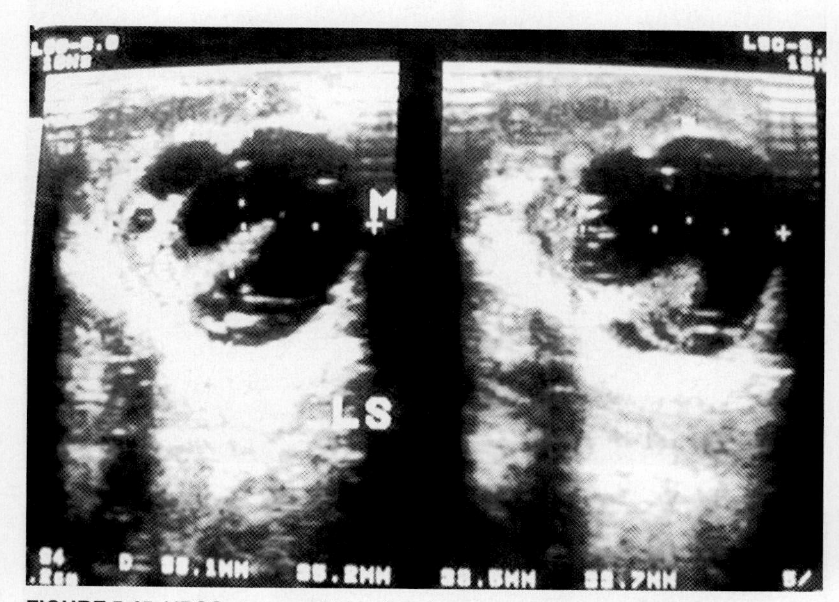

FIGURE 5.45: HRSG shows multi-loculated cystic mass with multiple echogenic septa suggestive of sacral meningocele. No membranous herniation is seen in it.

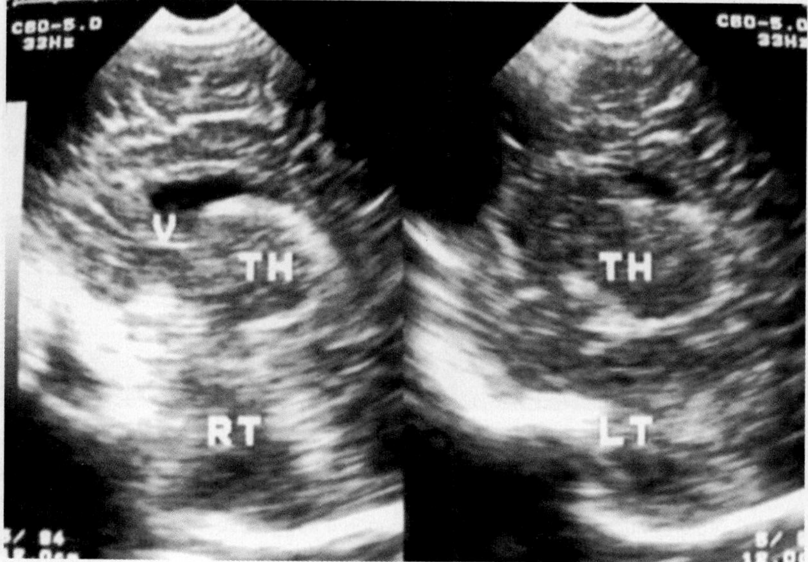

FIGURE 5.46

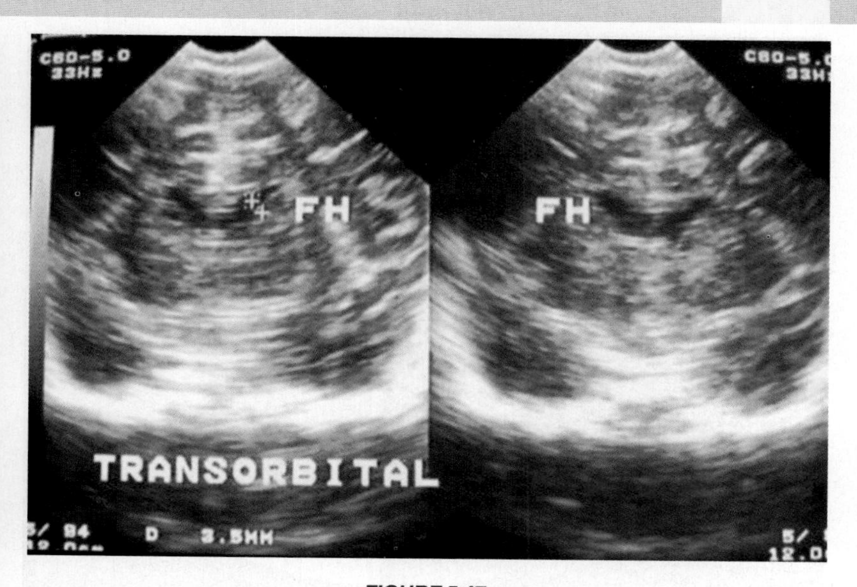

FIGURE 5.47

FIGURES 5.46 and 5.47: Cranial ultrasound of the same patient shows normal ventricular system. No backpressure is seen on the ventricles by the meningocele.

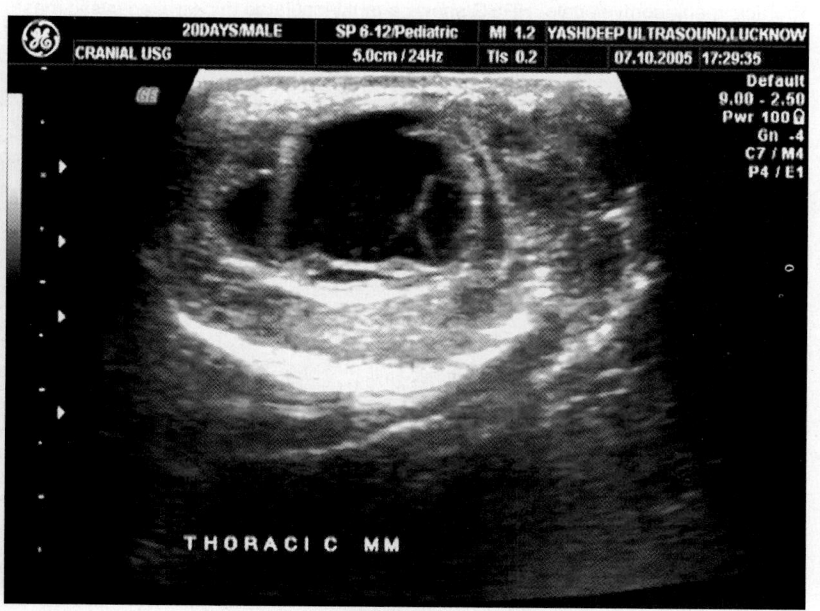

FIGURE 5.48

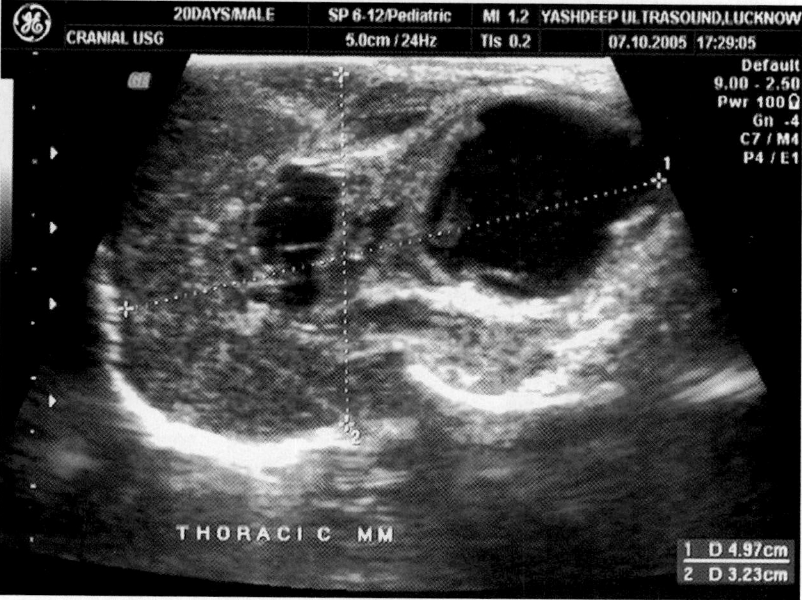

FIGURE 5.49

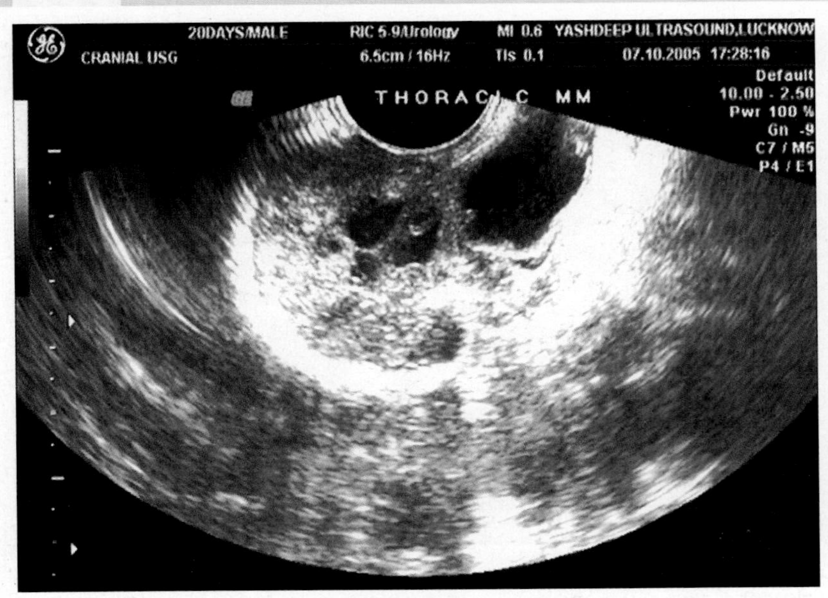

FIGURE 5.50

FIGURES 5.48 to 5.50: Thoracic meningomyelocele. A child was born with big thoracic meningomyelocele. HRSG shows a multi-loculated thick walled cystic mass. The mass shows herniation of the membranes.

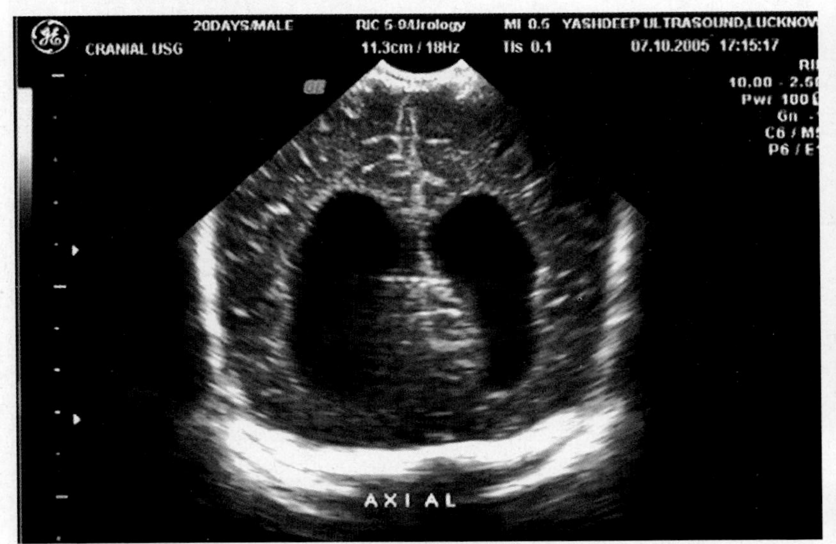

FIGURE 5.51

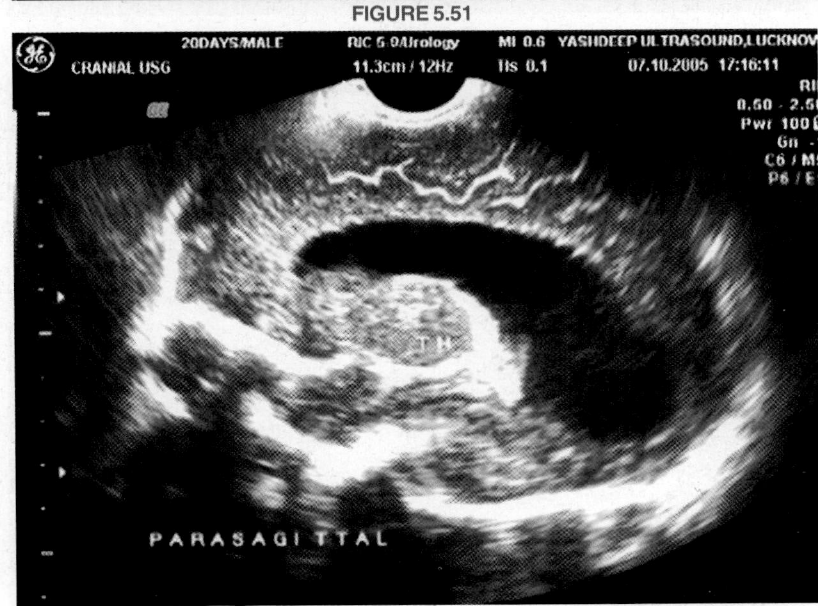

FIGURE 5.52

FIGURES 5.51 and 5.52: The cranial ultrasound of the same patient shows dilated ventricles due to obstructive pathology.

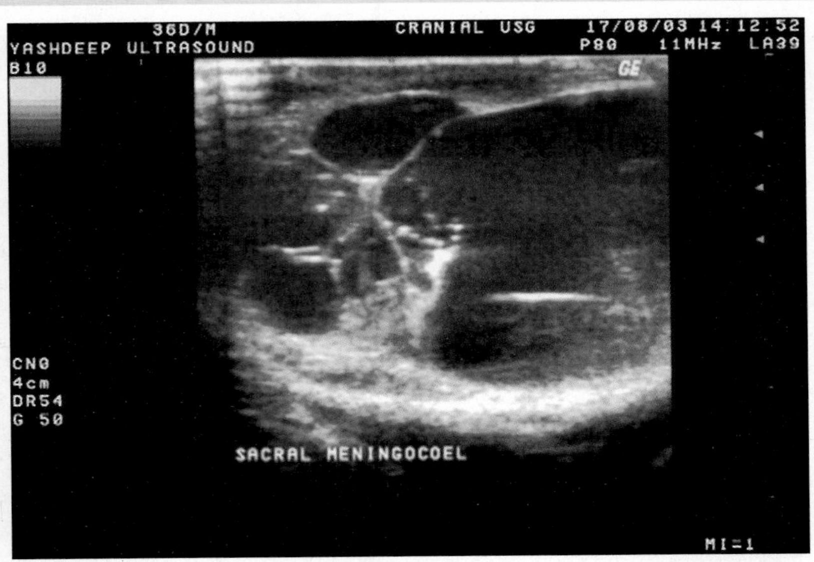

FIGURE 5.53

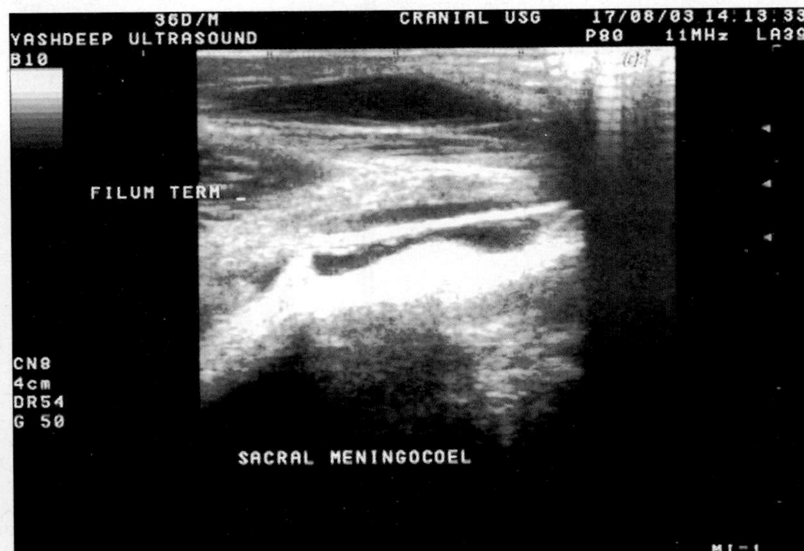

FIGURE 5.54

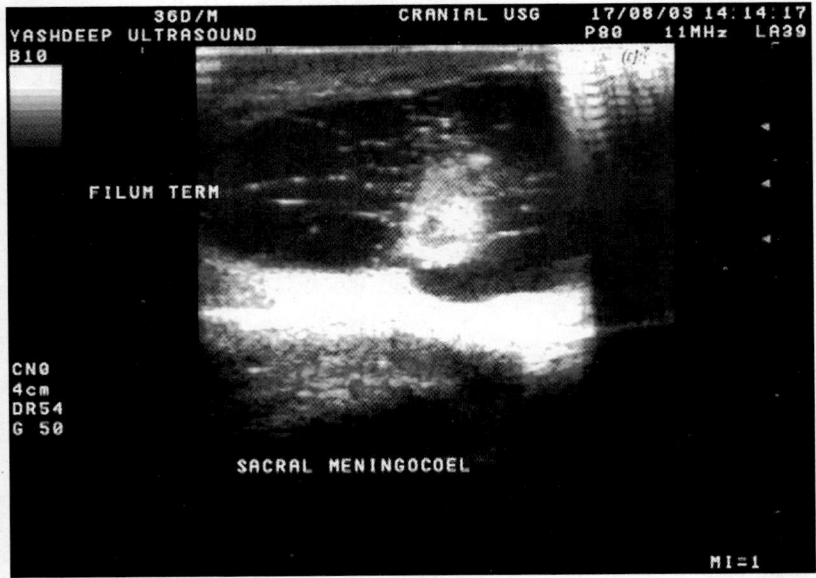

FIGURE 5.55

FIGURES 5.53 to 5.55: Lumbosacral meningomyelocele. An infant is born with a big cystic mass in the lumbosacral region. HRSG shows a complex cystic mass in the LS region. Multiple membranes are seen in it suggestive of meningomyelocele. The Filum Terminal is seen floating in the cyst. The fibril pattern of Filum Terminal is well appreciated on HRSG.

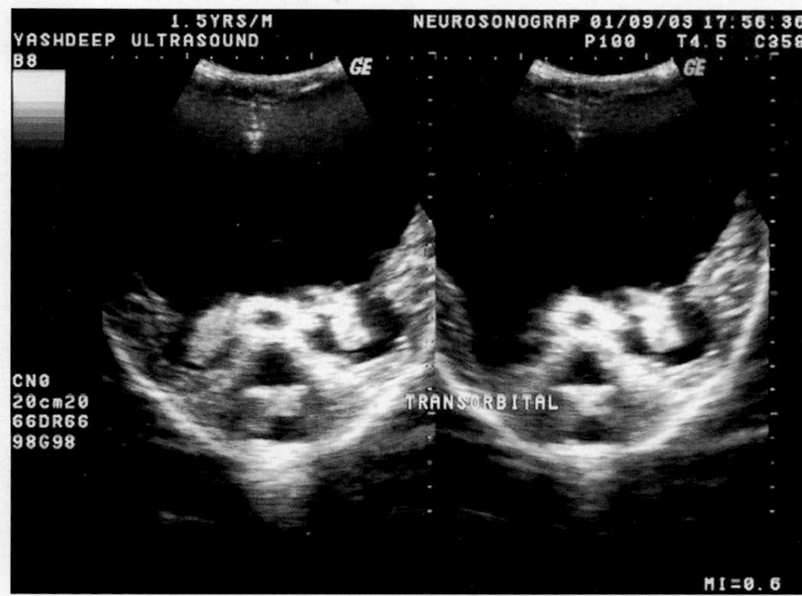

FIGURE 5.56

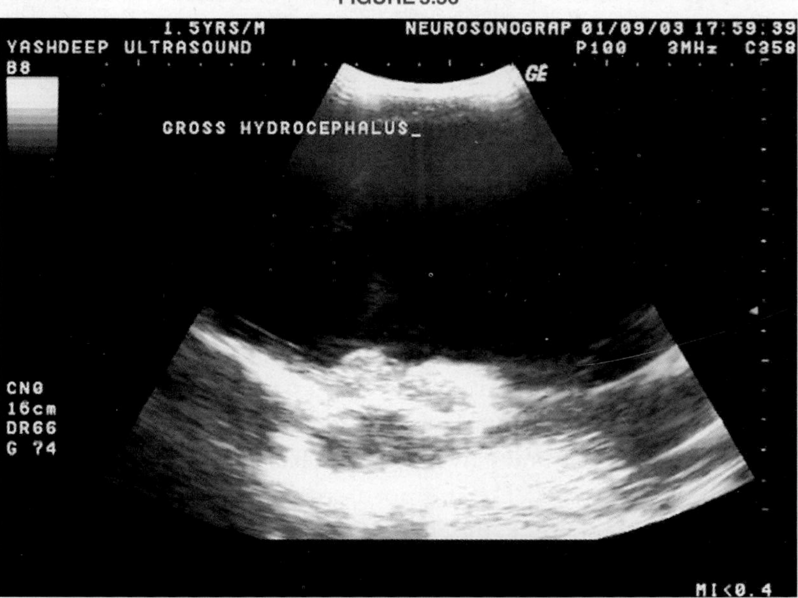

FIGURE 5.57

FIGURES 5.56 and 5.57: Cranial ultrasound of the same patient shows gross hydrocephalous. The dilated ventricles thin the brain parenchyma.

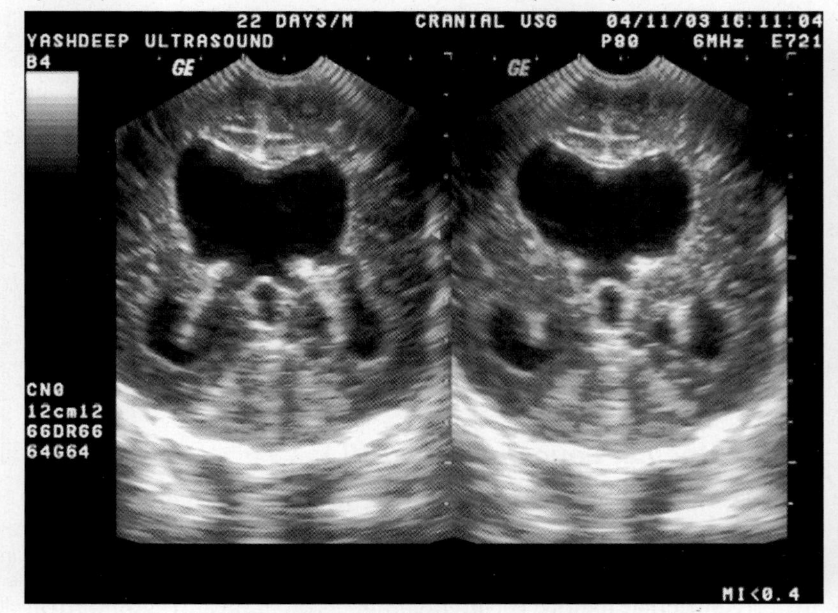

FIGURE 5.58

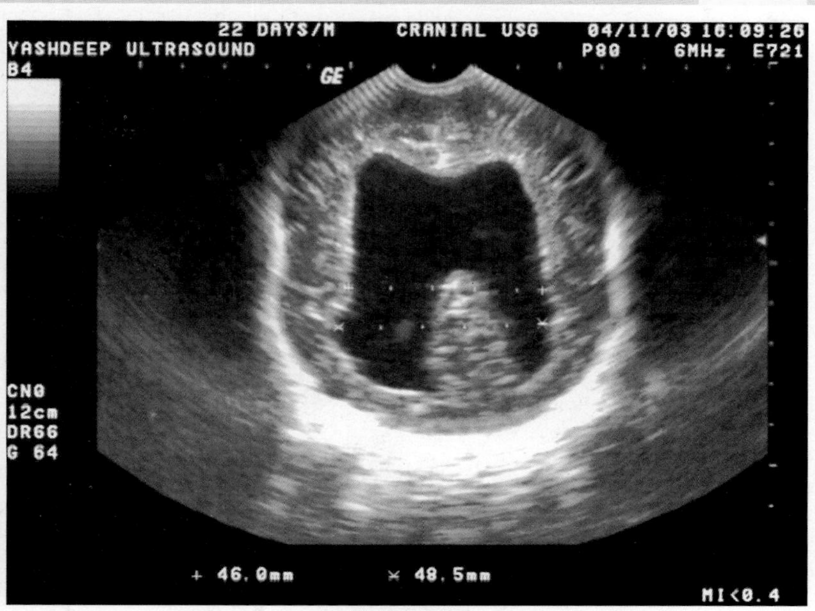

FIGURE 5.59

FIGURES 5.58 and 5.59: Septo optic dysplasia with lobar holoprosencephaly. HRSG shows flattening of the frontal horns with evidence of absence of septum pellucidum. The anterior horns are fused and squared shaped. However, the occipital horn is separate.

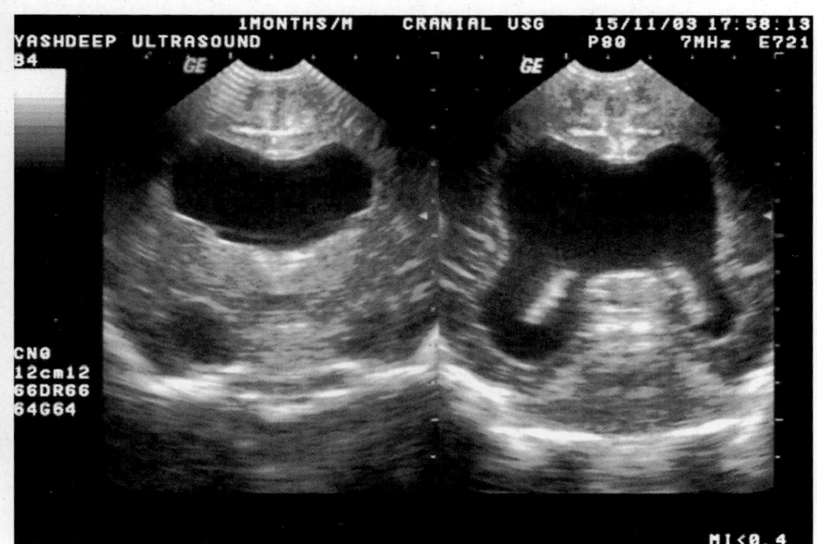

FIGURE 5.60

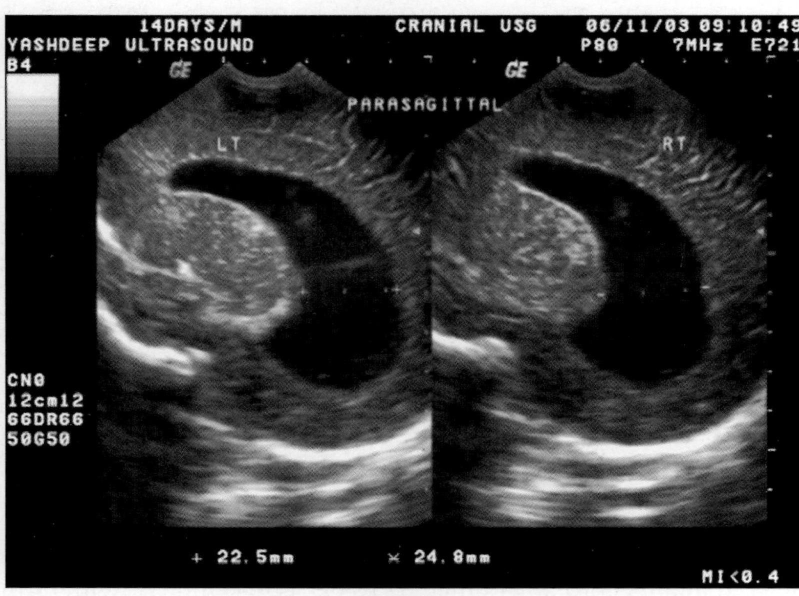

FIGURE 5.61

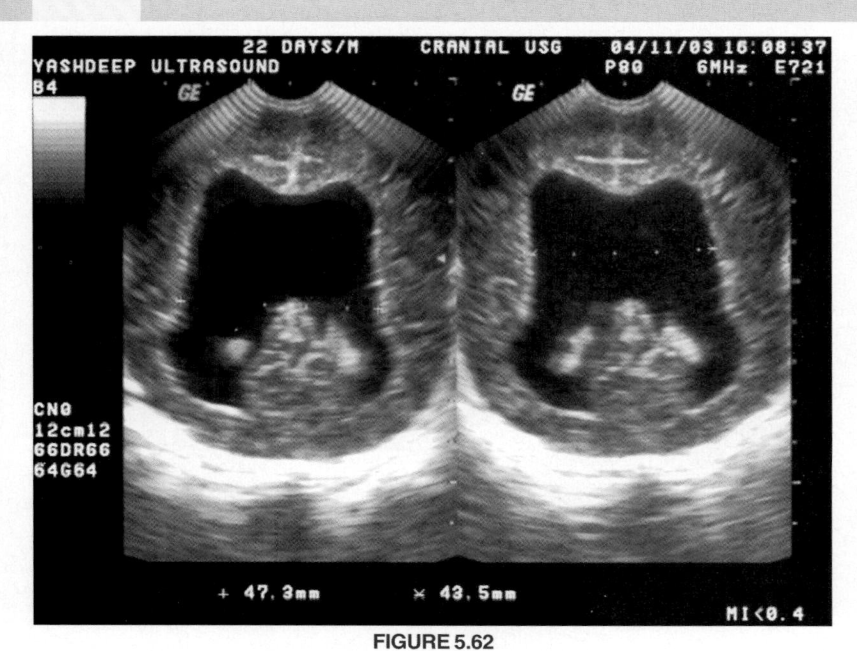

FIGURE 5.62

FIGURES 5.60 to 5.62: Lobar holoprosencephaly. HRSG shows details of the Holoprosencephaly. The frontal horns are fused. However, the occipital horns are separated. The ventricles are dilated.

Hydrocephalus

Hydrocephalus is the most common sequelae of hemorrhage in preterm infants leading to obstruction at the level of third of fourth ventricle. The other possible cause may be interference in absorption of CSF due to inflammatory ependymitis or basilar arachnoiditis due to hemorrhage/infection. Rarely, over production of CSF from choroid plexus papilloma may be responsible for production of hydrocephalus (non-obstructive).

Ultrasonographic Diagnosis of Hydrocephalus

Various qualitative and quantitative methods have been described for diagnosis of hydrocephalus. The lateral ventricular ratio (LVR) method describer by Rumack and Jhonson has been commonly used. The measurements are made from midline echo to the inner surface of lateral ventricle and from midline to inner table of skull on axial ultrasonography. The normal LVR is up to 35 percent, LVR of 35 to 40% is considered mild hydrocephalus, LVR of 41 to 50% is considered moderate, and LVR of more than 50% is severe hydrocephalus. In fetus, the ventricular measurement at the level of atrium is found to be more accurate, wherein less than 10 mm is considered to be normal. However, with sonographer gaining experience, the ventricular dilatation can be appreciated on routine coronal and sagittal scans without any critical measurements.

Ventricular dilatation can be documented weeks before clinical signs of hydrocephalus are noticed. Serial sonograms on weekly intervals are recommended to look for progress of hydrocephalus.

Intracranial Hemorrhage (ICH)

Intracranial hemorrhage and periventricular leucomalacia (PVL) are the most common CNS problems in premature infants. Prevalence of ICH in infants born at less than 32 weeks is gestation or with birth weight less than 1500 gm is 30 to 55%. All such infants should be screened routinely on days 4 to 7 after birth. Infants with ICH should be followed weekly to detect early hydrocephalus.

Pathophysiology

The germinal matrix, which contains rich vascular stroma is a fragile structure present in the region of caudothalamic groove in the inferolateral wall of lateral ventricle, beyond 32 weeks germinal matrix involutes. However, before 32 weeks it is highly vascular and fragile in the event of increased cerebral blood flow and raised blood pressure.

Papile et al have nicely graded the extent of hemorrhage as follows:

Grade I Subependymal hemorrhage related to germinal matrix.
Grade II Intraventricular extension without hydrocephalus.
Grade III IVH + Hydrocephalus.
Grade IV IVH + Parenchymal extension

Ultrasonography is the most effective method for detection and follow-up of various grades of hemorrhage, mostly newborns are affected within 7 days, hence, first sonographic evaluation is recommended after 7 days of birth.

Subependymal Hemorrhage (SEH)

SEH is seen as an area of increased echogenecity in the inferolateral wall of lateral ventricle, acute Hemorrhage is hyper echogenic and is similar in echo genecity to choroid plexus. In few cases, the SEH liquefy centrally resulting in cyst formation, which may persist for months.

Intraventricular Hemorrhage (IVH)

The hemorrhage may rupture through ependymal lining into the ventricles. Sonography reveals an echogenic clot filling and expending the pathways of CSF and may reach up to basal cisterns. At times it is difficult to diagnosis IVH if it is not associated with hydrocephalus, clot commonly adheres to the choroid plexus and is indistinguishable form it. However, with due course of time, the clot becomes hypoechoic in the center due to liquefaction, the clot contracts and gradually decreases in size.

Intraparenchymal Hemorrhage (IPH)

The SEH may extend into surrounding parenchyma. The intraparenchymal hemorrhage is commonly seen (approximately in 8%) in the fetus born with birth-weight of less than 1500 gms. Though it is commonly believed that intraparenchymal hemorrhage is extension of SEH, however, few studies have revealed that in neonates, the intraparenchymal hemorrhage may be manifestation of secondary hemorrhage within an area of periventricular infarction. It shows all the features of hemorrhage seen elsewhere in the body being echogenic in the beginning and leading to formation of porencephalic cyst as the clot resorbs.

Periventricular Leucomalacia (PVL)

PVL develops as a consequence of ischemia in the periventricle white matter leading to gliosis or cystic changes depending upon the severity. Sonographically, PVL appears as focal areas of increased echogenicity in periventricular region. The involvement is bilateral and symmetrical.

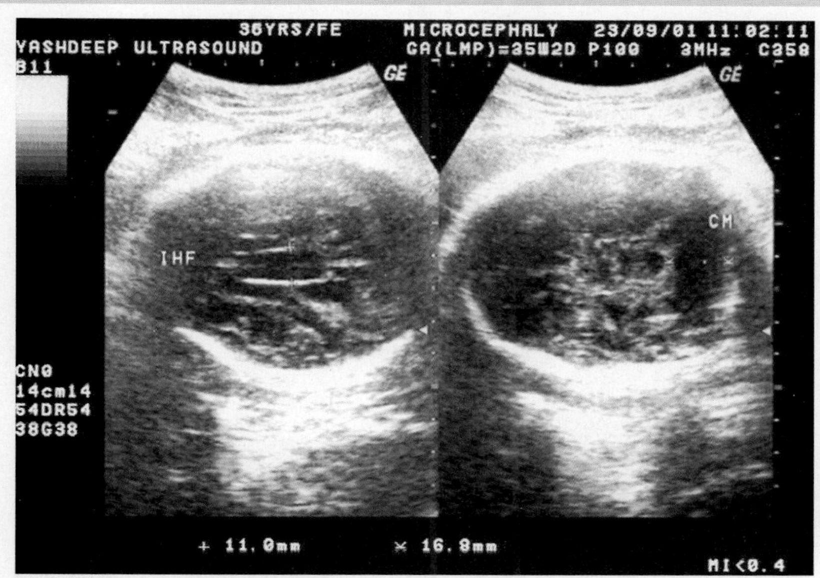

FIGURES 5.63: Early hydrocephalus. Fetal head shows dilated cisterna magna. It is 16 mm in caliber (normal upper limit 10 mm). The lateral ventricles are also dilated and it is 11 mm at the atrium (normal upper limit 8 mm) suggestive of early hydrocephalus.

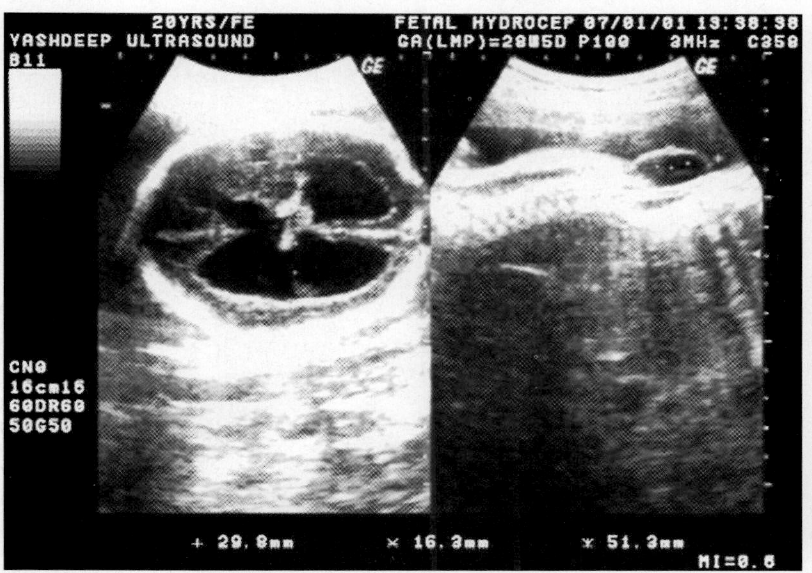

FIGURE 5.64: Fetal hydrocephalus with sacral meningocele. HRSG shows grossly dilated ventricles in the fetal head. Pressure is seen on the brain parenchyma. The fetal spine also shows curvature defect with a cystic mass coming out from the sacrum suggestive of sacral meningocele.

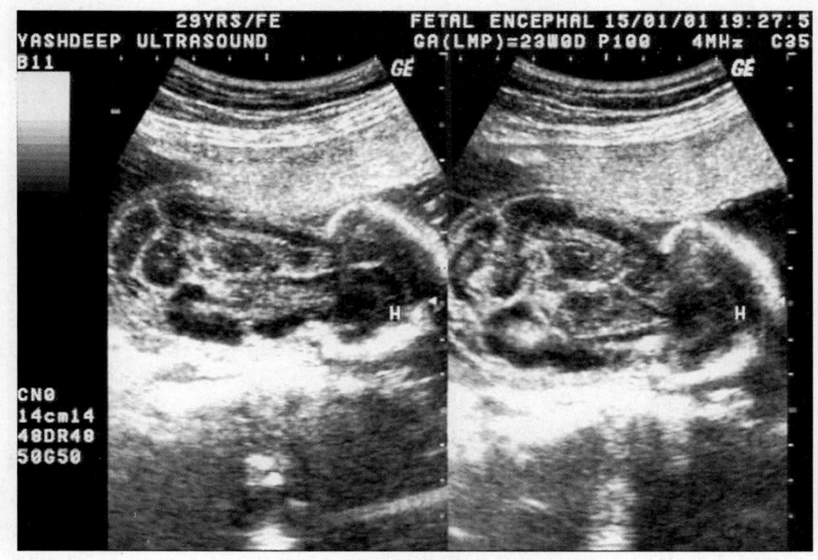

FIGURE 5.65

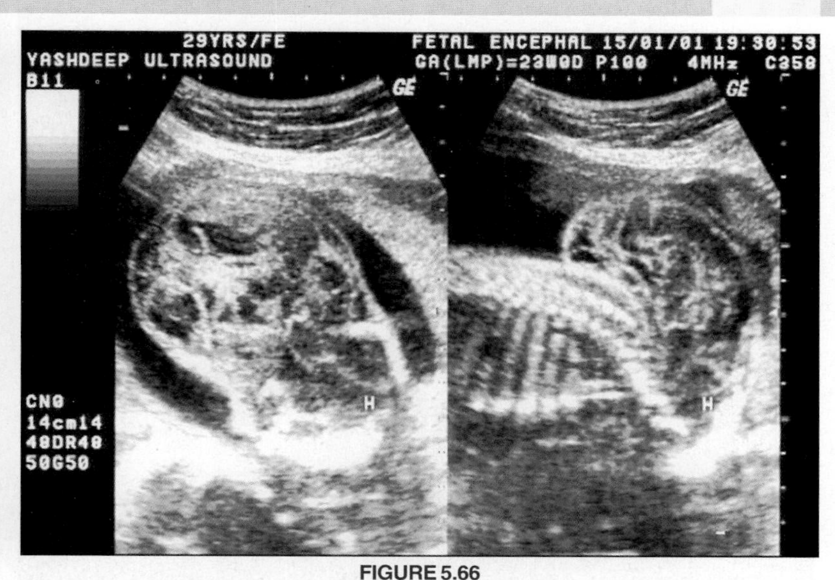

FIGURE 5.66

FIGURES 5.65 and 5.66: Fetal encephalocele. The fetal brain is seen herniated from the skull in a cystic mass. A big defect is seen in the occiput resulting into herniation in the fetal brain detected on HRSG.

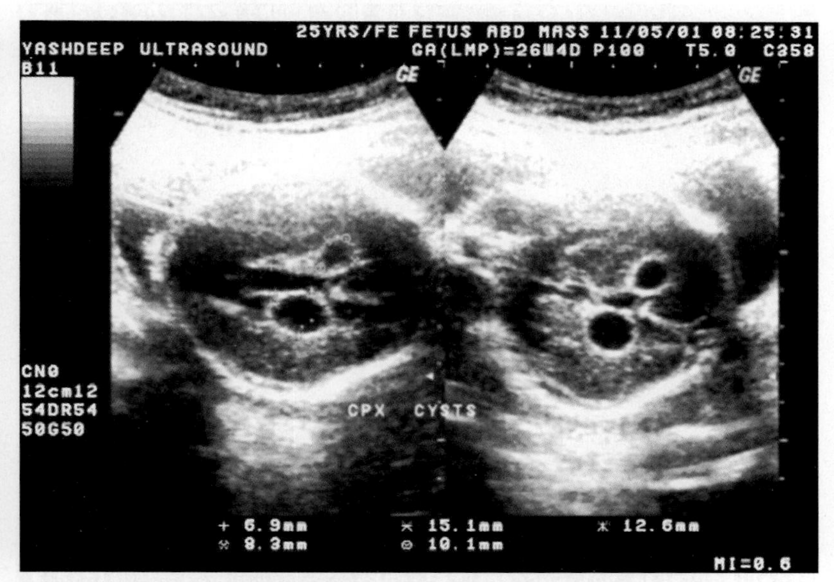

FIGURE 5.67: Choroid plexus cyst. Fetal brain shows well-defined thin walled cyst in both choroid plexus suggestive of choroid plexus cyst. The baby was also having other congenital anomalies and was trisomy 21.

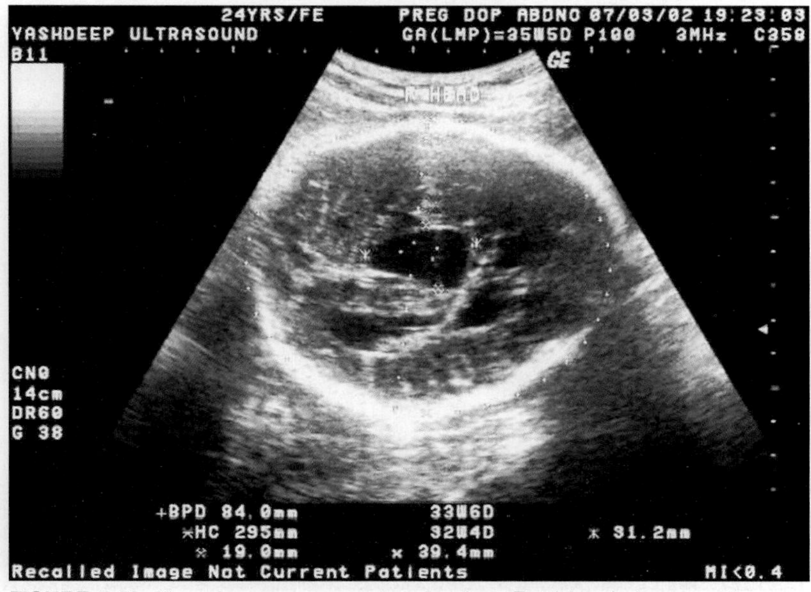

FIGURE 5.68: Vascular-aneurysm vein of galen: Fetal head shows cystic mass in the paraventricular region mistaken as dilated ventricle.

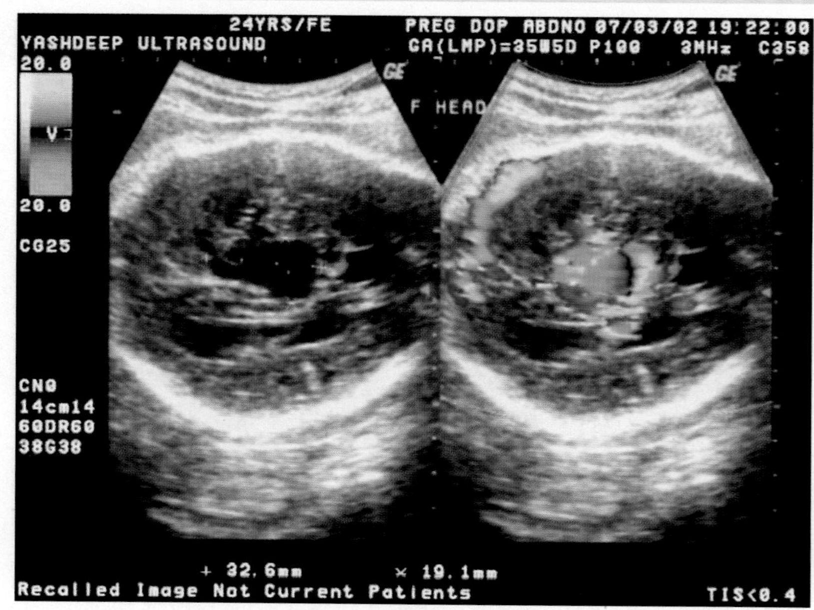

FIGURE 5.69

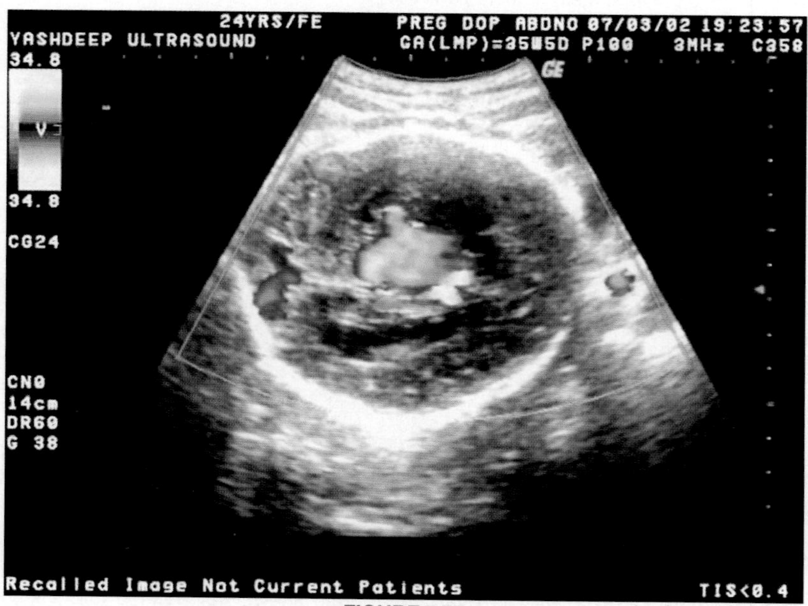

FIGURE 5.70

FIGURES 5.69 and 5.70: On color Doppler imaging it turns out to be a dilated vessel not the ventricle suggestive of aneurysm of vein of Galen.

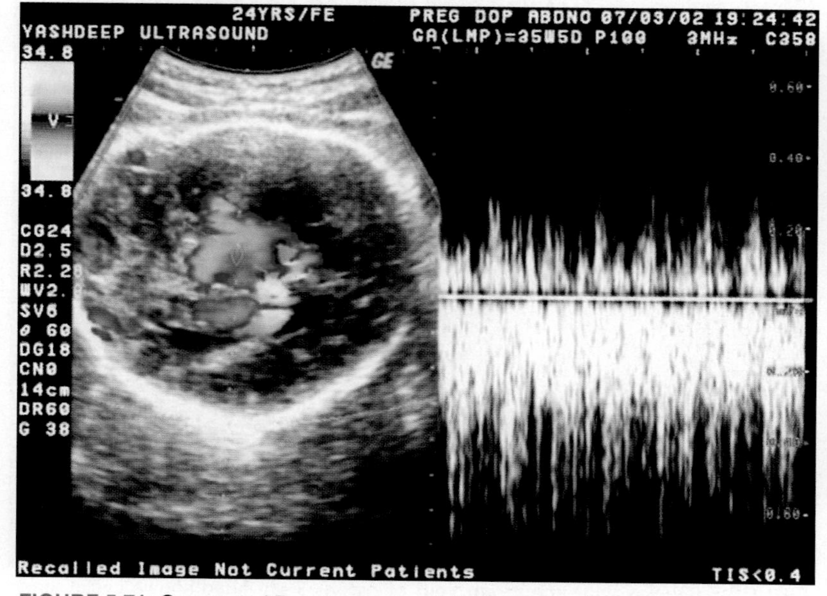

FIGURE 5.71: On spectral Doppler tracing high bizarre flow is seen in the aneurysm.

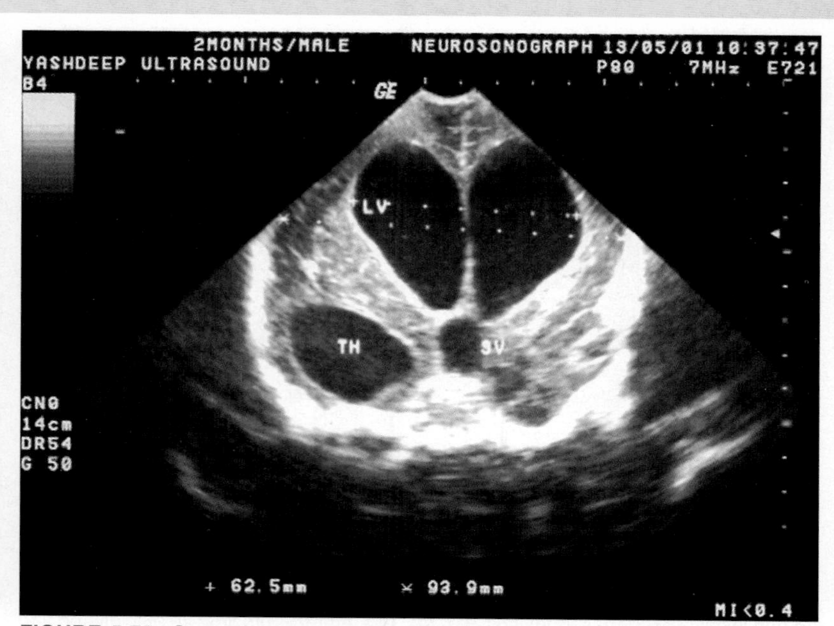

FIGURE 5.72: **Gross hydrocephalus.** Neurosonography shows grossly dilated ventricles involving all the four ventricles with marked pressure on the brain parenchyma.

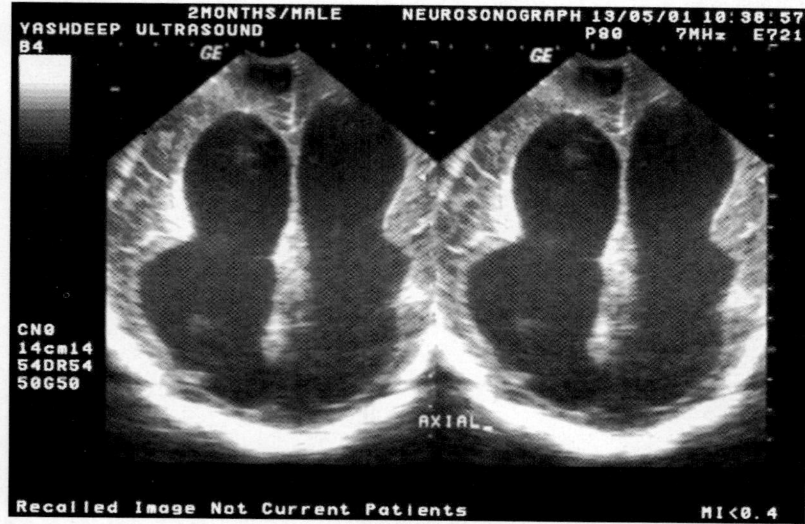

FIGURE 5.73

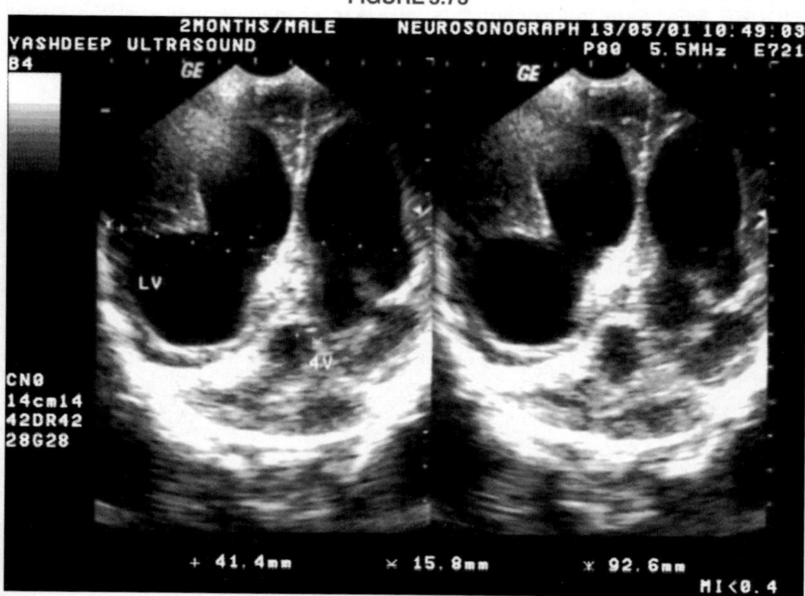

FIGURE 5.74

FIGURES 5.73 and 5.74: **Gross hydrocephalus.** The lateral ventricles are markedly dilated. Fourth ventricle is also dilated in brain parenchyma suggestive of gross hydrocephalus.

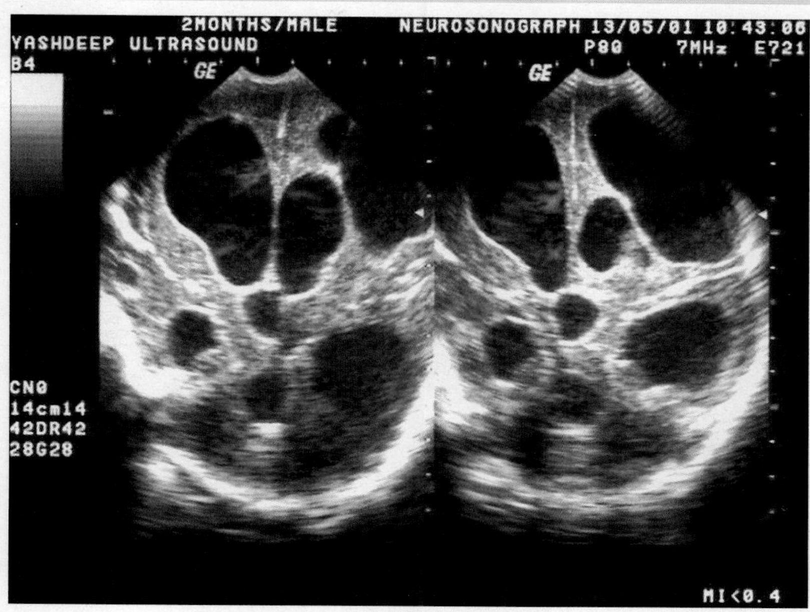

FIGURE 5.75

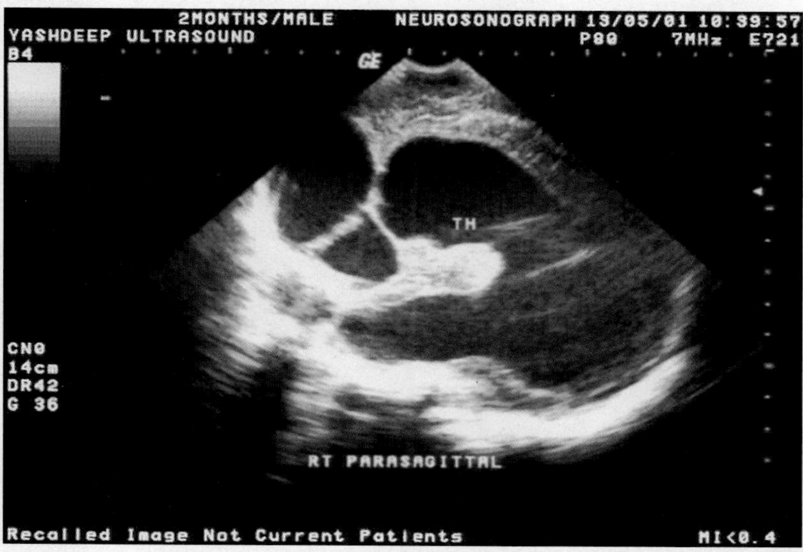

FIGURE 5.76

FIGURES 5.75 and 5.76: Axial and the parasagittal planes also show gross dilatation of all the four ventricles. The occipital horn, temporal horn and fourth ventricles are grossly dilated. Marked pressure is seen in brain parenchyma.

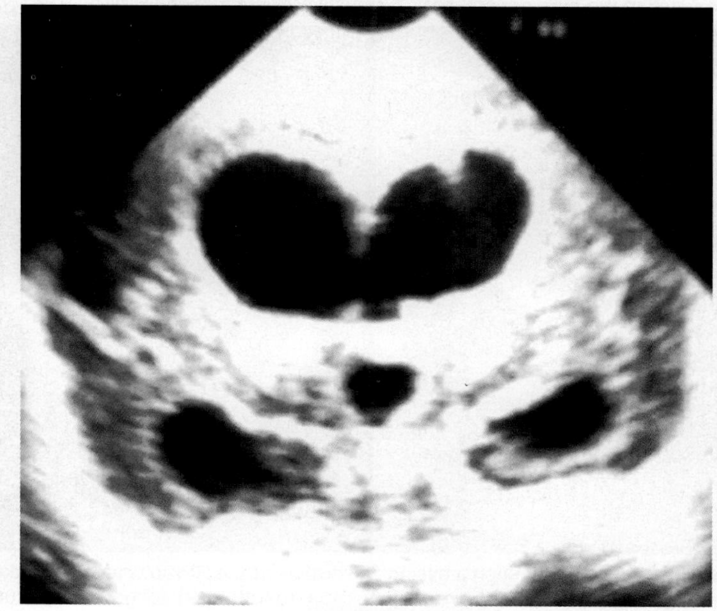

FIGURE 5.77

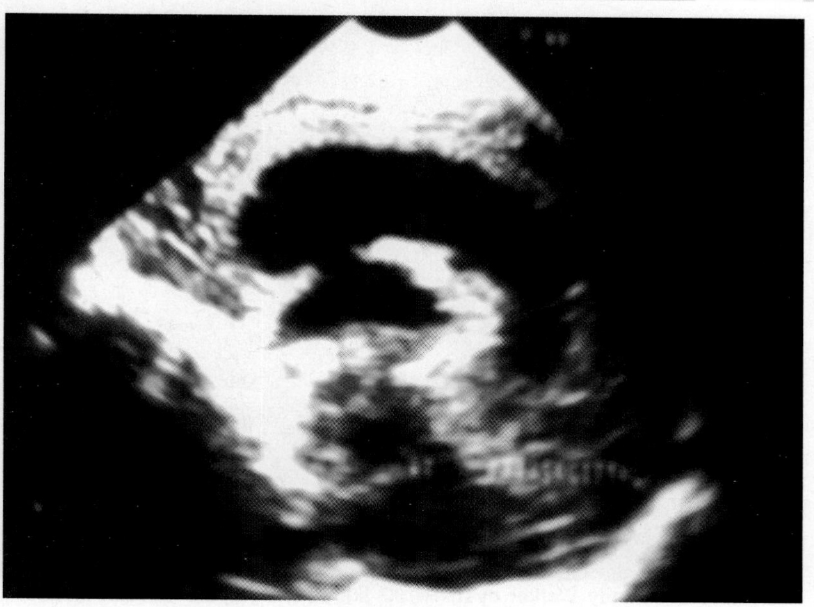

FIGURE 5.78

FIGURES 5.77 and 5.78: Obstructed hydrocephalus. HRSG shows dilated ventricles involving all the four ventricles in a case of obstructed hydrocephalus due to basal arachanoditis due to brain infection.

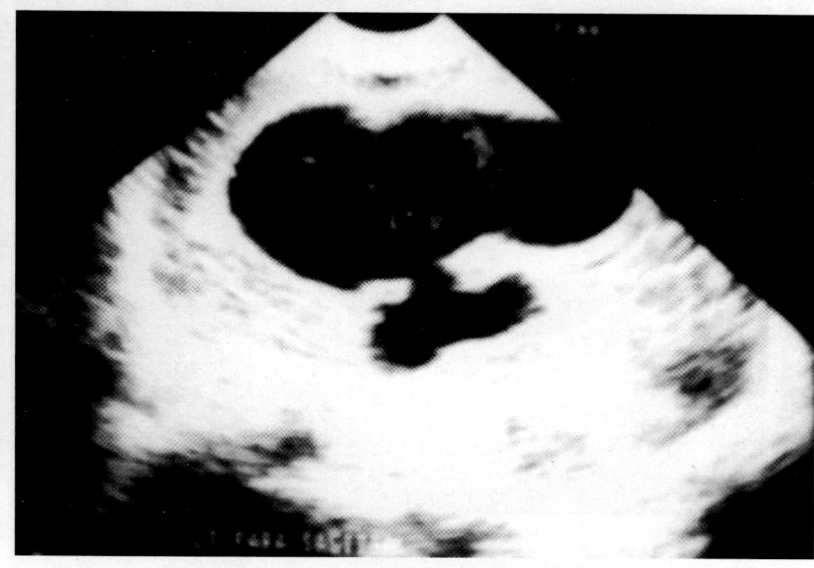

FIGURE 5.79: Aqueduct stenosis. HRSG shows obstructed hydrocephalus due to aqueduct stenosis. The third ventricle and lateral ventricles are grossly dilated. Fourth ventricle is not dilated.

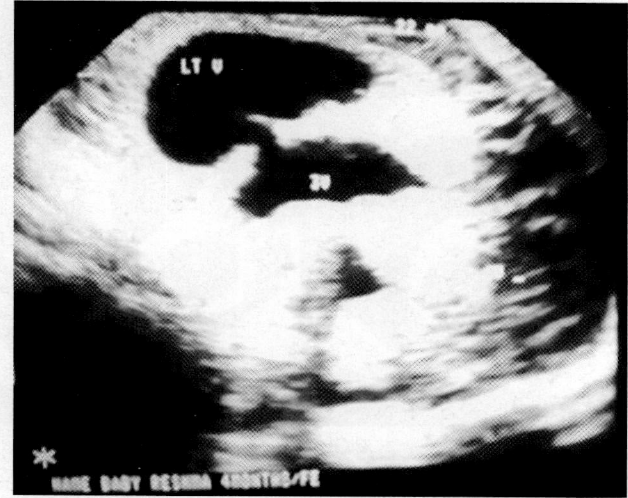

FIGURE 5.80: Aqueduct stenosis. The aqueduct of sylvius is seen obstructed resulting into dilated third ventricle, lateral ventricle. The foramen of monro is seen distended.

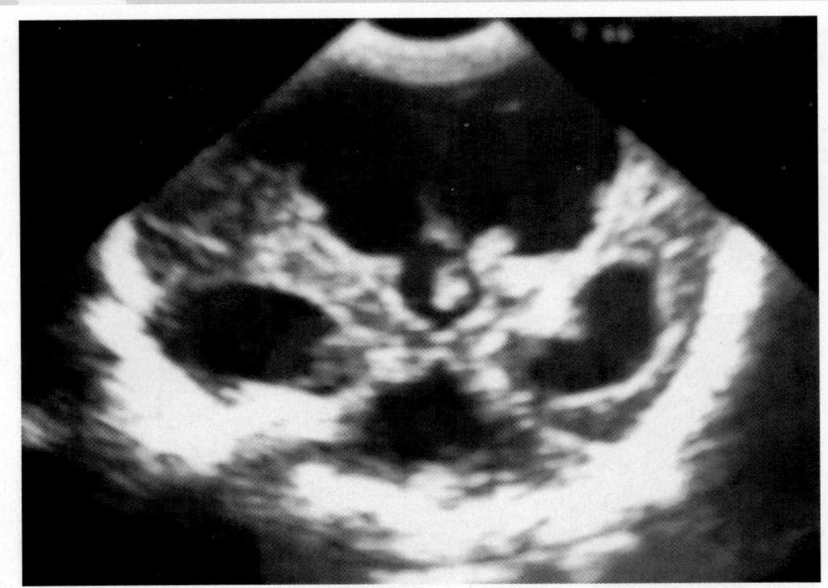

FIGURE 5.81: Dandy Walker cyst. HRSG shows a big cystic mass in the fourth ventricle with a Dandy Walker cyst. It has obstructed the outflow tract resulting into obstructed hydrocephalus.

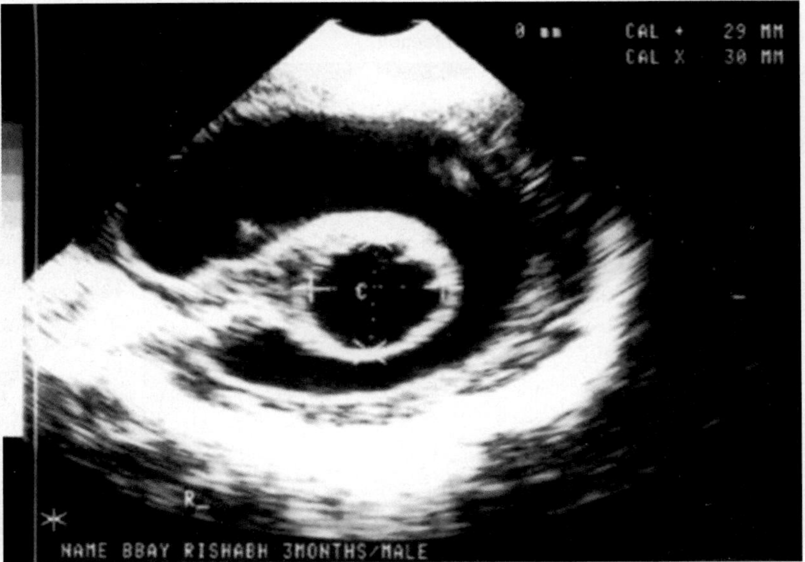

FIGURE 5.82

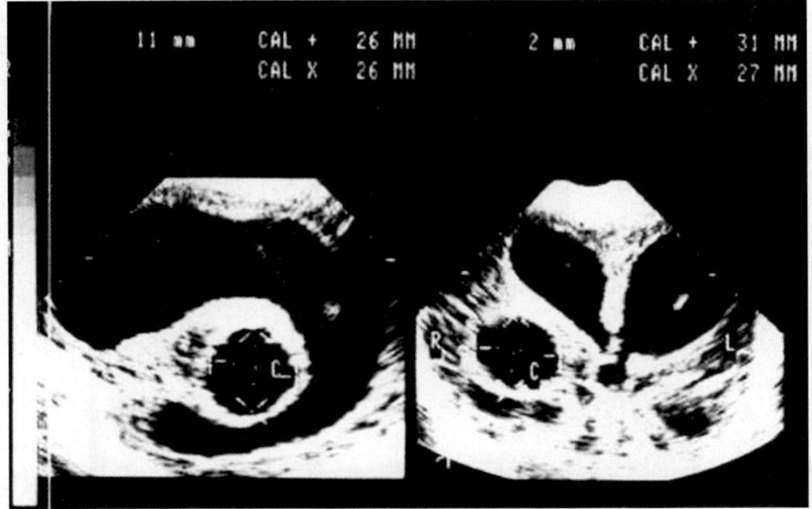

FIGURE 5.83

FIGURES 5.82 and 5.83: Gross hydrocephalus with thalamic cyst. HRSG shows gross hydrocephalus involving all the ventricles. Pressure thinning of the brain parenchyma is also seen. There is also evidence of a well-defined thin walled cyst seen in the thalamus. No internal echo is seen in the cyst.

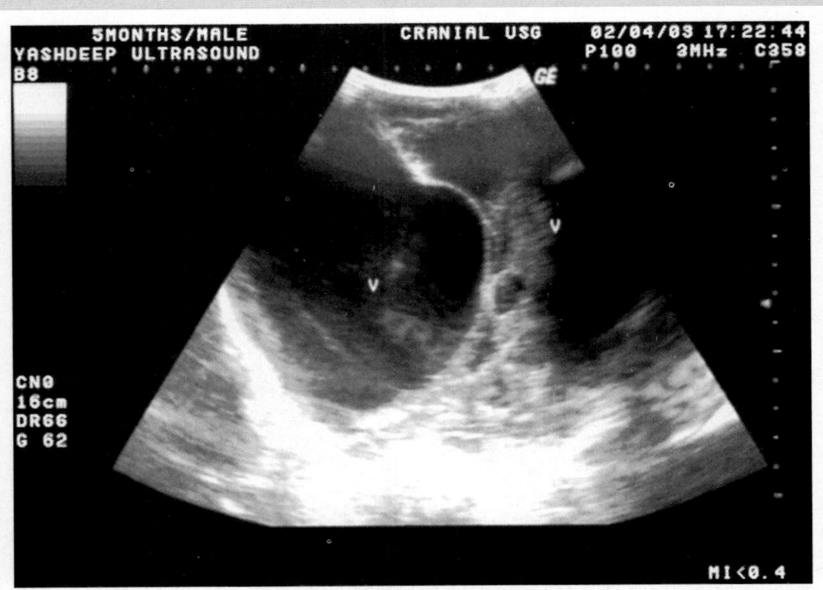

FIGURE 5.84

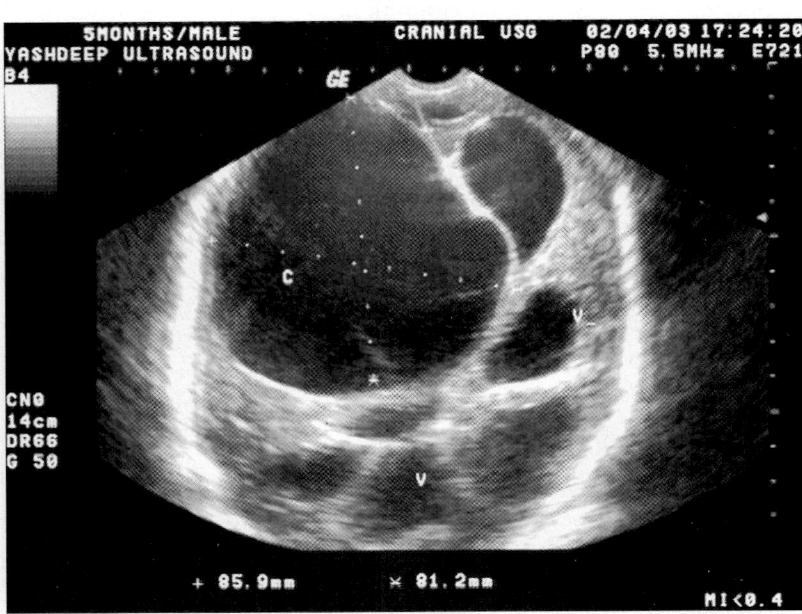

FIGURE 5.85

FIGURES 5.84 and 5.85: Porencephalic cyst. HRSG shows grossly dilated lateral ventricle in parasagittal plane. The dilated ventricle is seen communicating with a cystic mass in the high parietal area suggestive of porencephalic cyst.

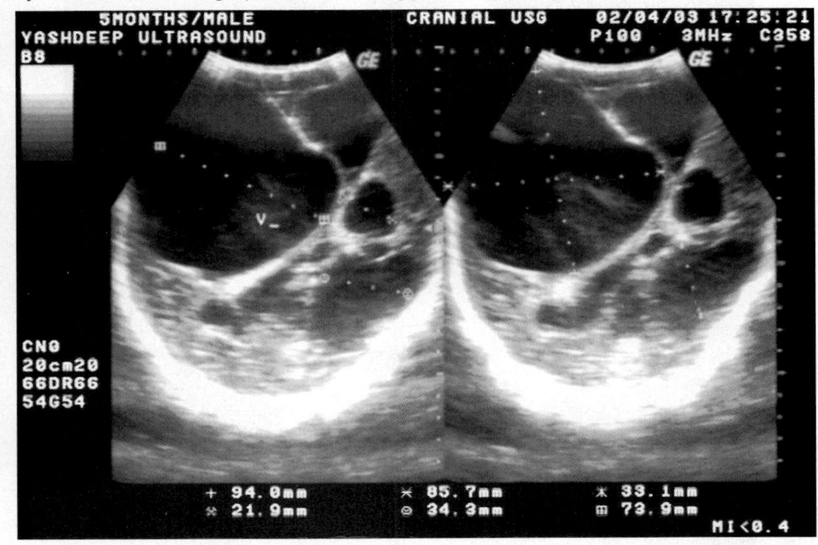

FIGURE 5.86: The cyst is seen pressing and displacing the Rt ventricle to the opposite. It is seen pressing over the out flow tract resulting in to gross obstructed hydrocephalus.

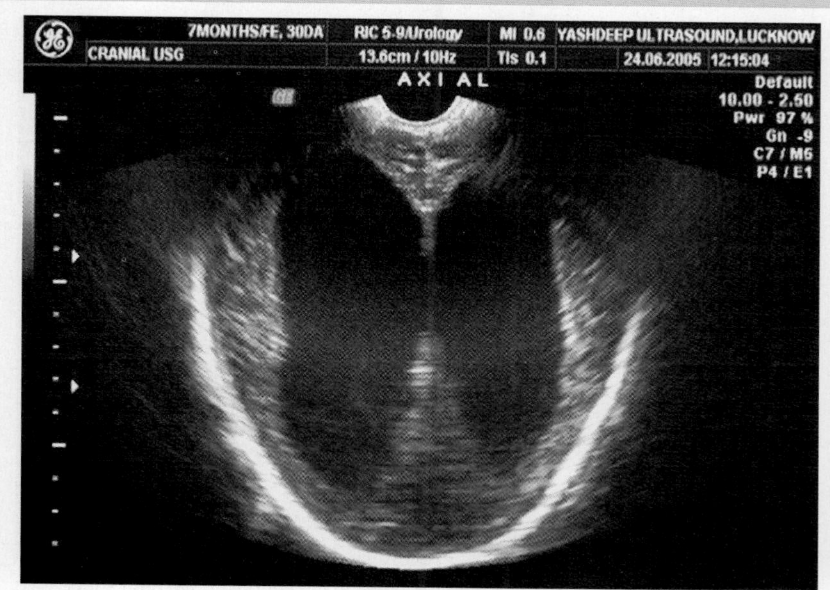

FIGURE 5.87

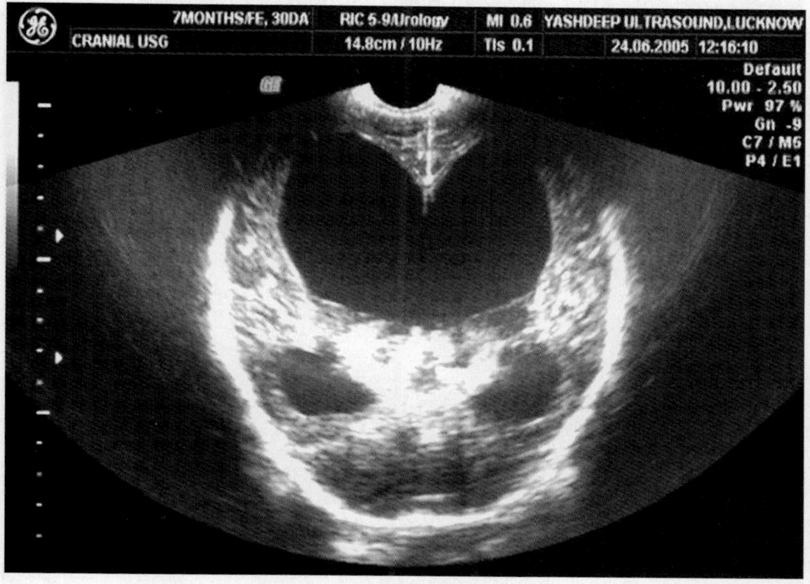

FIGURE 5.88

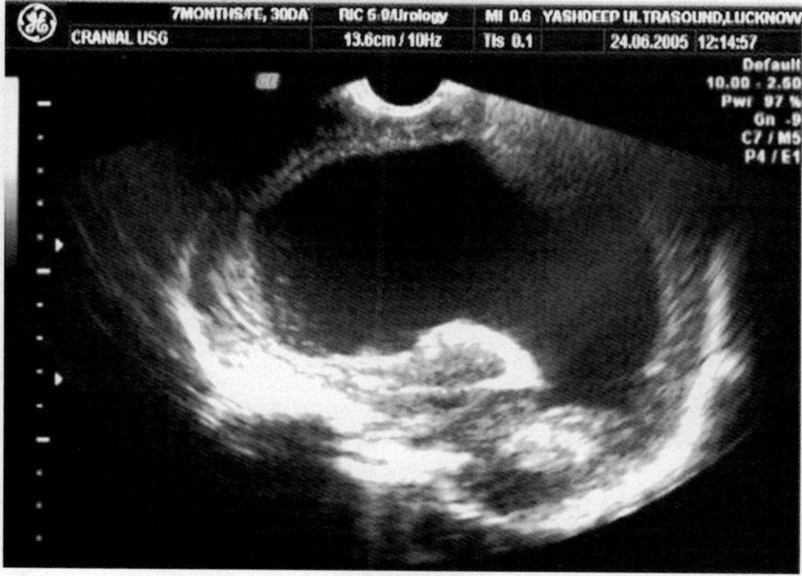

FIGURE 5.89

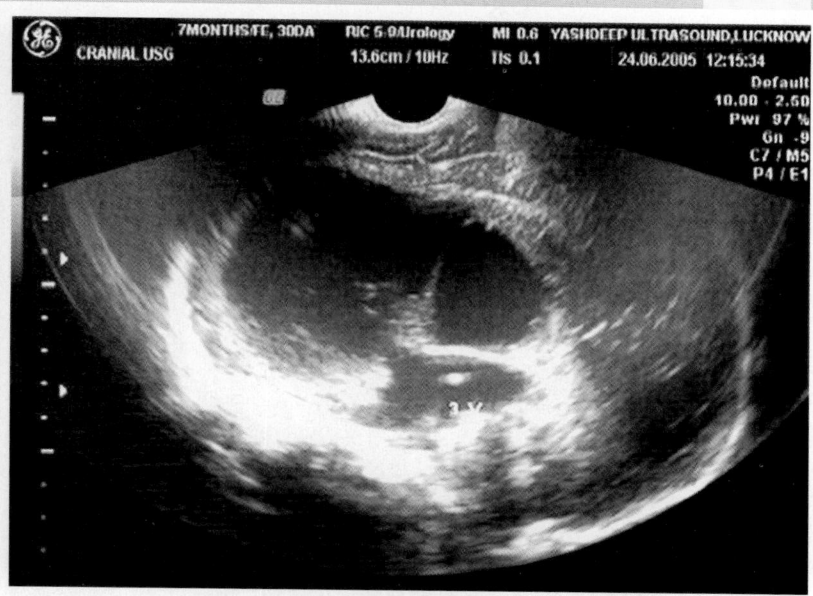

FIGURE 5.90

FIGURES 5.87 to 5.90: Gross hydrocephalus. HRSG shows gross dilatation of the ventricles. All the ventricles are dilated. Brain parenchyma is seen compressed by the fluid pressure. Third ventricle is also well seen on HRSG.

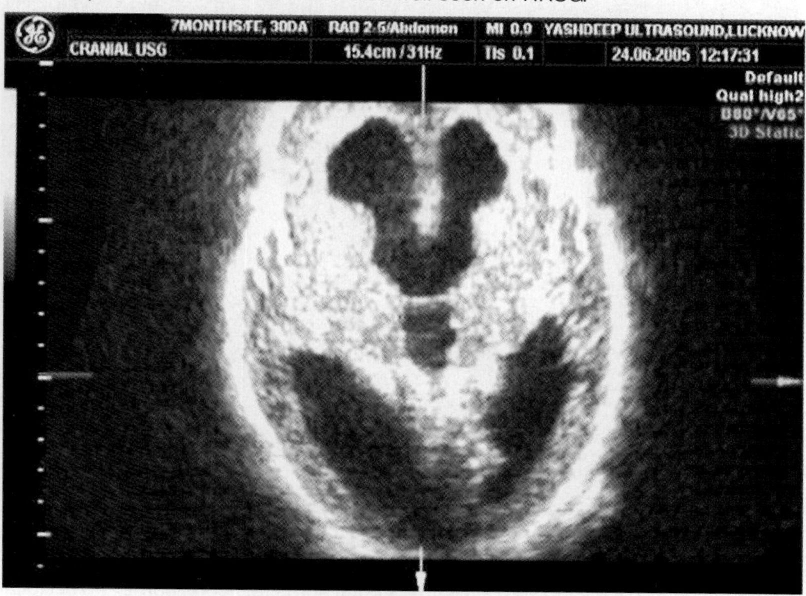

FIGURE 5.91

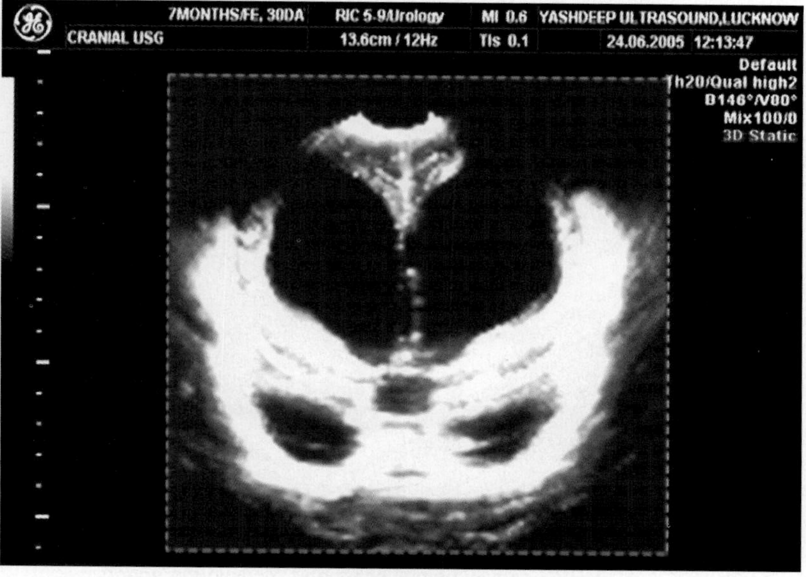

FIGURE 5.92

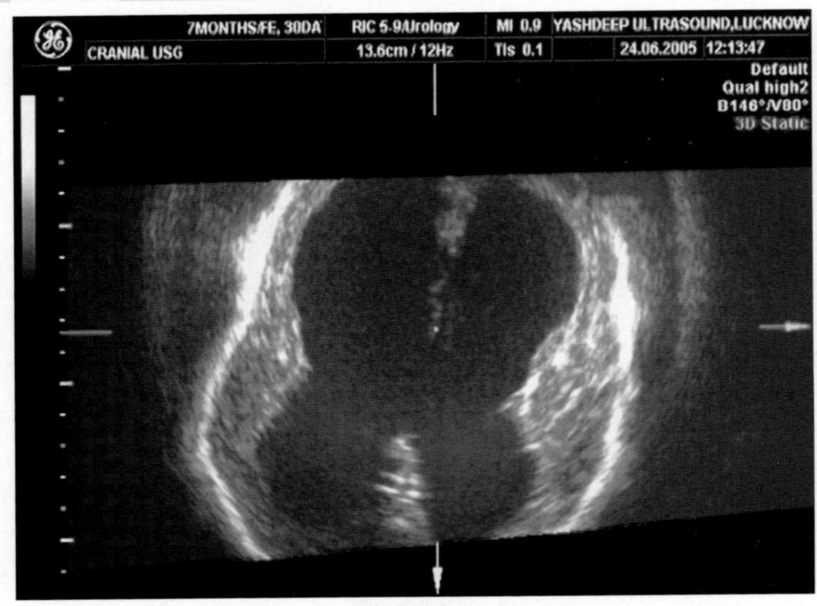

FIGURE 5.93

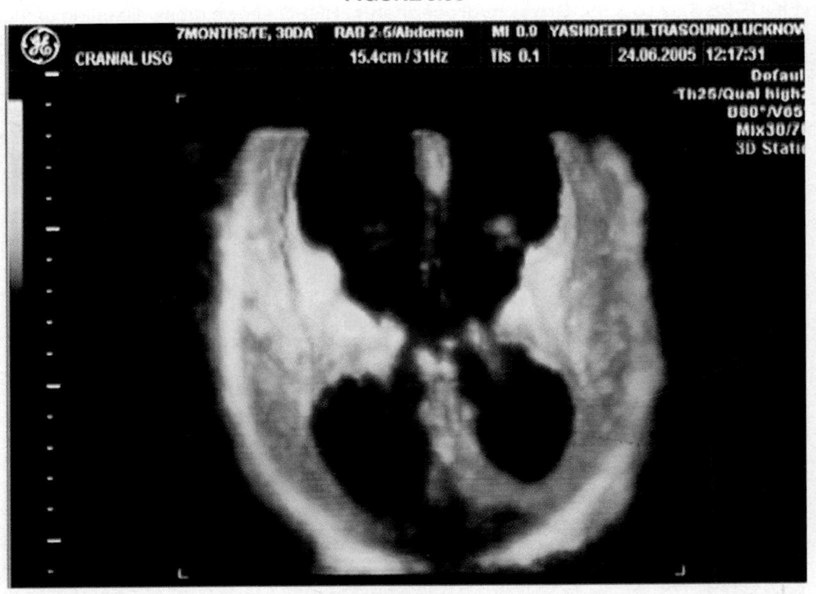

FIGURE 5.94

FIGURES 5.91 to 5.94: 3D of hydrocephalus. 3D ultrasound of the same patient shows the details of dilated ventricles. The frontal horns, temporal horns and occipital horns are well visualized. 3D images are well comparable to MRI images.

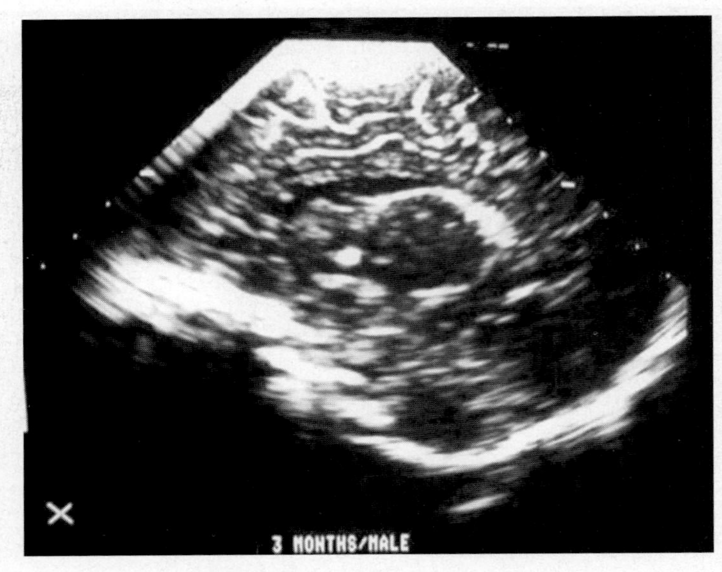

FIGURE 5.95: HRSG shows a small echogenic focus in the ependymal region in para sagittal axis.

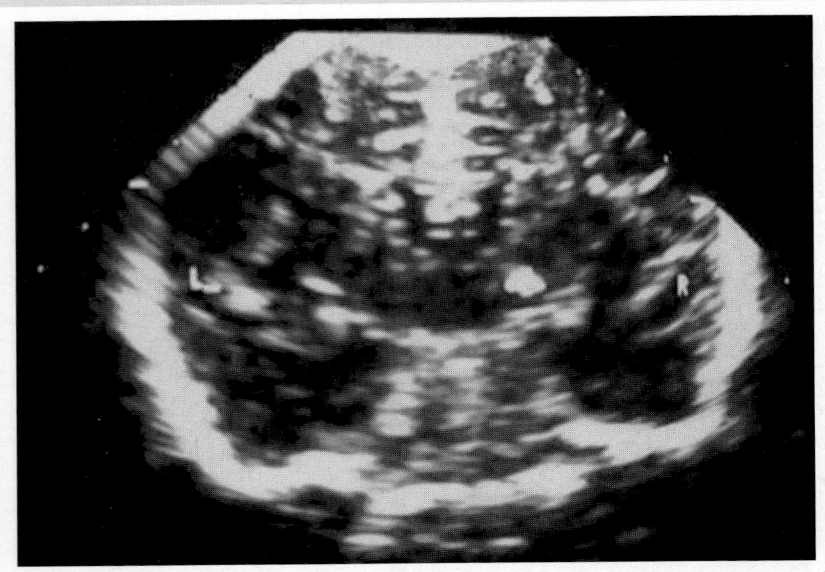

FIGURE 5.96: The coronal plane shows the bleed in the floor of frontal horn suggestive of sub ependymal bleed.

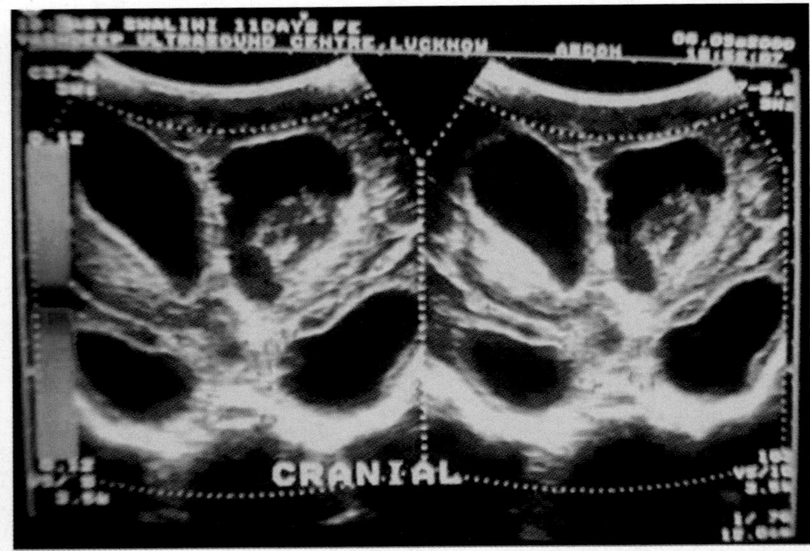

FIGURE 5.97

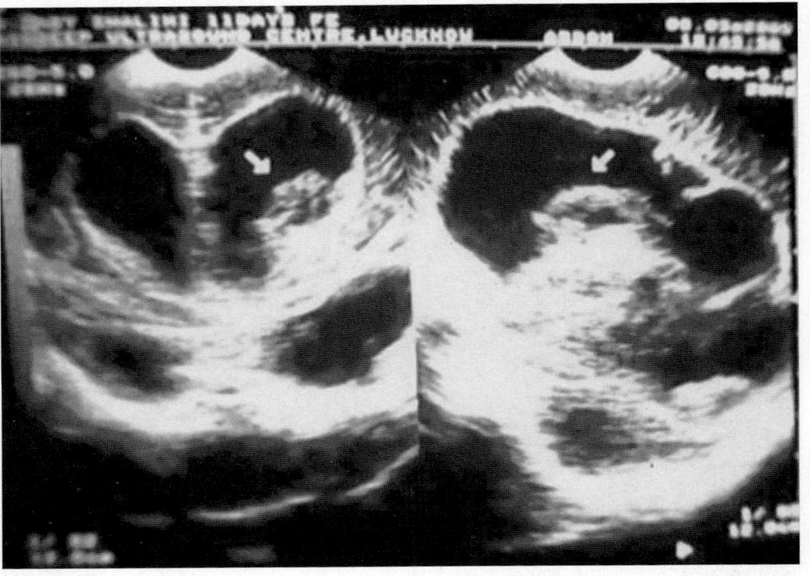

FIGURE 5.98

FIGURES 5.97 and 5.98: Intraventricular hemorrhage. HRSG shows an echogenic clot in the frontal horn of the ventricle. The ventricle wall is irregular. Echogenic thin strands are seen in the ventricles. On color flow imaging increased flow is seen in the ventricular wall suggestive of ventriculitis.

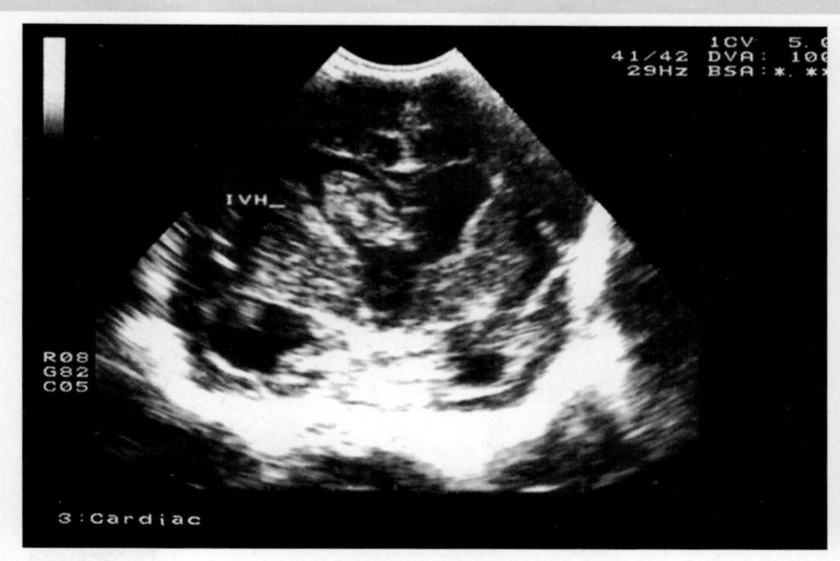

FIGURE 5.99

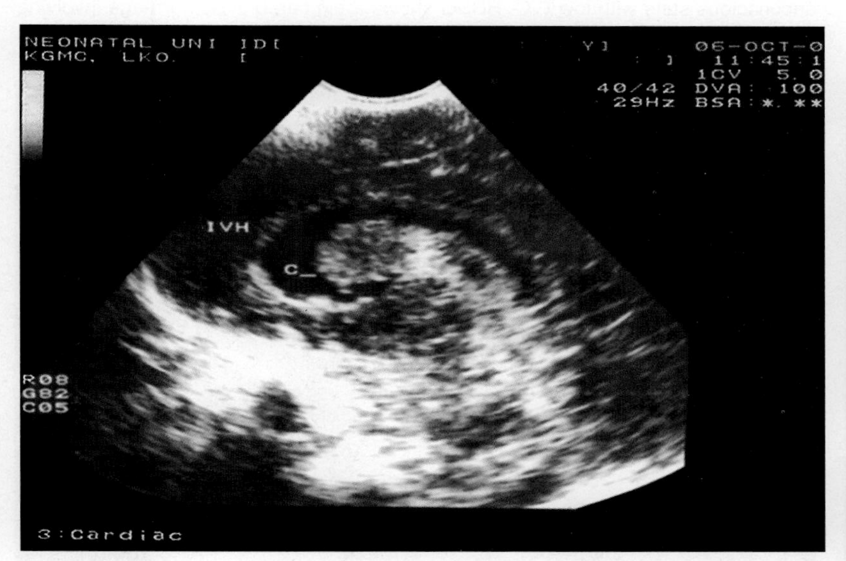

FIGURE 5.100

FIGURES 5.99 and 5.100: Intraventricular hemorrhage. HRSG shows a big echogenic hemorrhagic clot in the lateral ventricle. The clot is seen obstructing the foramen resulting into obstructed hydrocephalus.

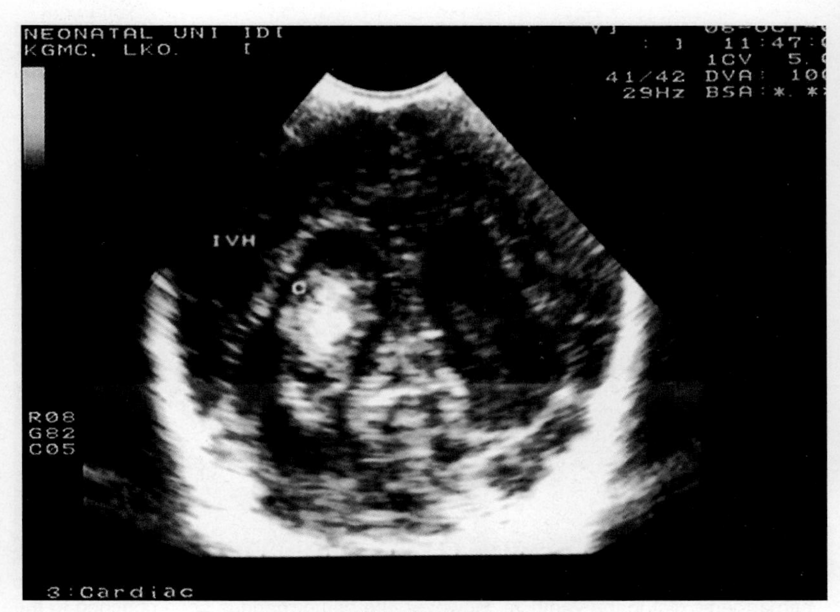

FIGURE 5.101

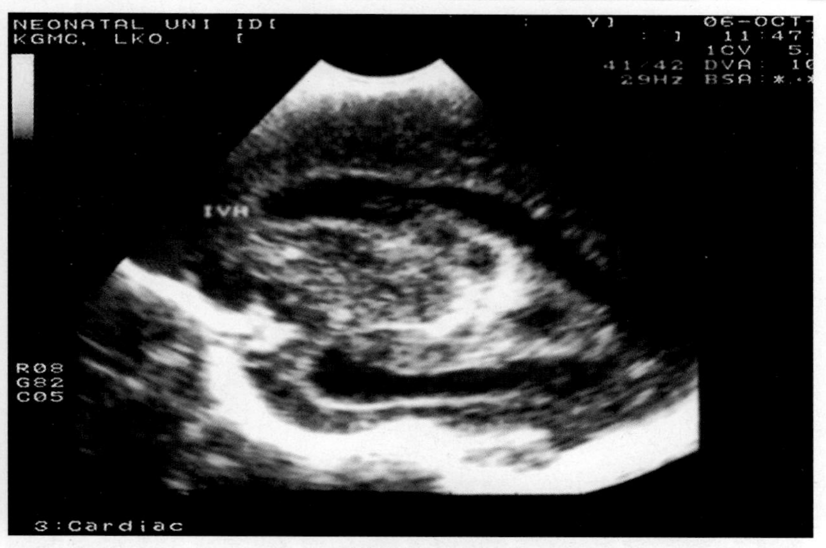

FIGURE 5.102

FIGURES 5.101 and 5.102 Intraventricular hemorrhage. Big echogenic hemorrhage is seen filling the ventricular cavity. It is also pressing over the foramen causing obstructed hydrocephalus.

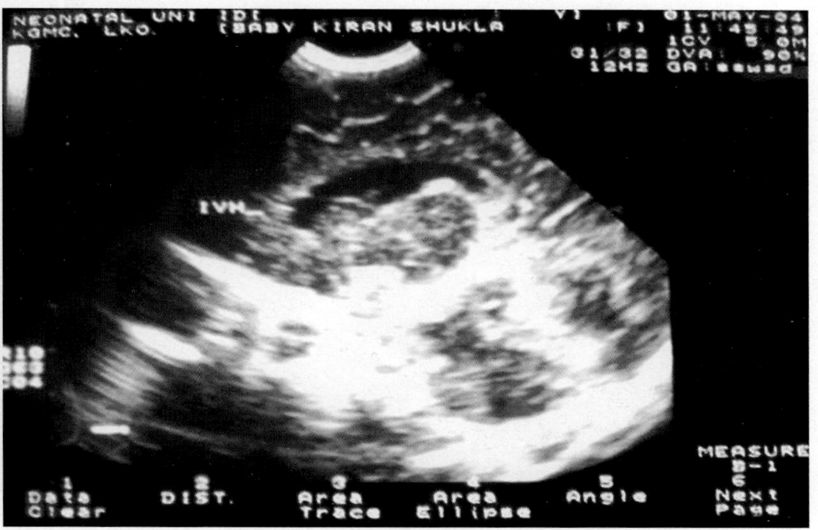

FIGURE 5.103

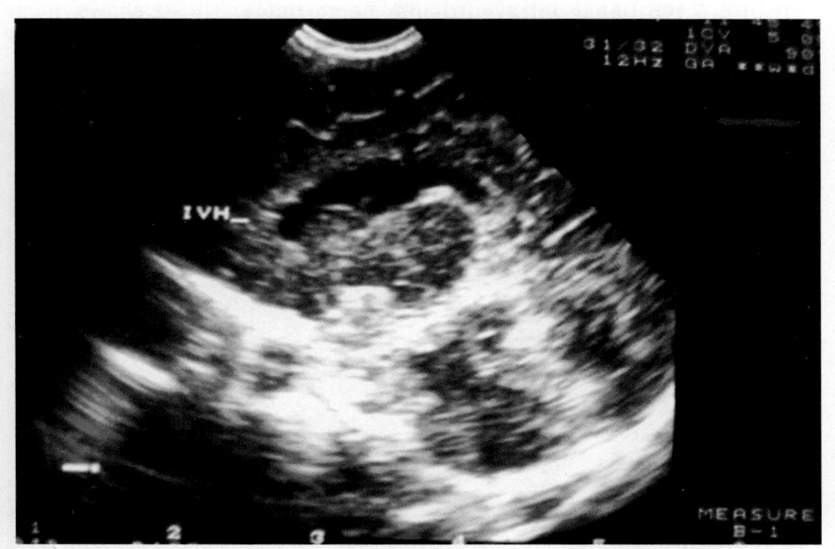

FIGURE 5.104

FIGURES 5.103 and 5.104: Intraventricular hemorrhage. Small echogenic hemorrhage is seen in the ventricle. However, no evidence of any calcification is seen in it.

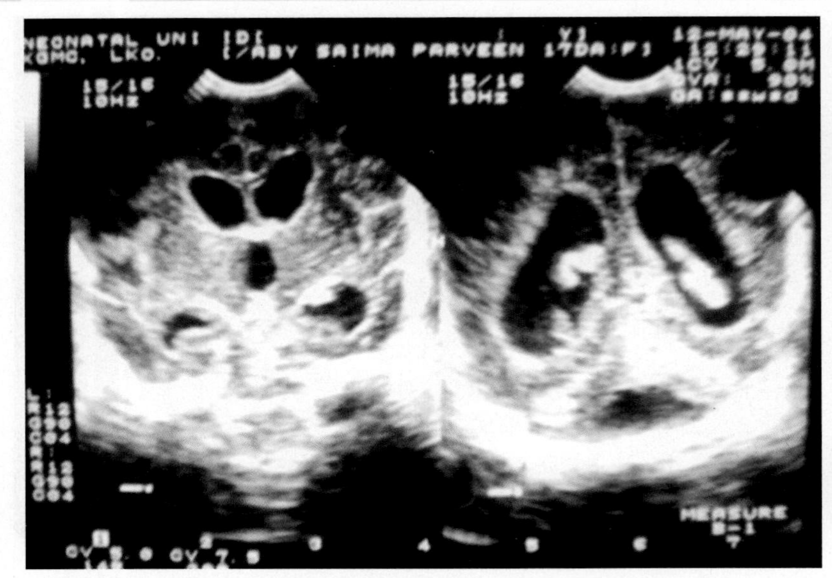

FIGURE 5.105: Intraventricular hemorrhage. HRSG shows echogenic-floating clot in the lateral ventricle associated hydrocephalus is also seen.

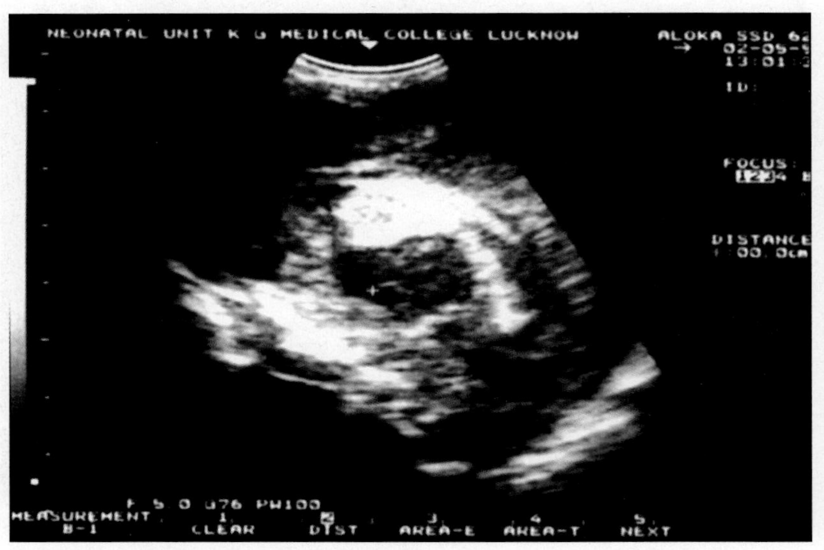

FIGURE 5.106 Dense intraventricular hemorrhage. HRSG shows dense echogenic hemorrhagic collection filling whole of the lateral ventricle. Chalky white collection is seen in the hemorrhage.

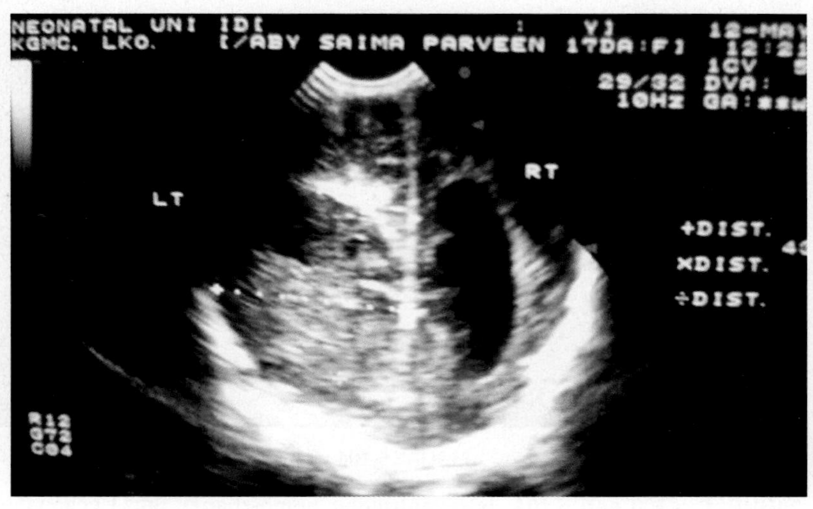

FIGURE 5.107

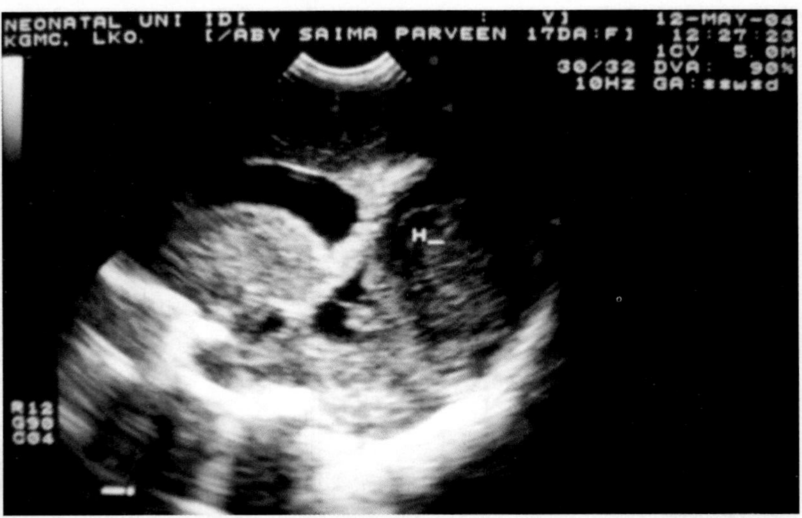

FIGURE 5.108

FIGURES 5.107 and 5.108: Cerebral hemorrhage. A patient presented with semi unconscious state with low G.C. HRSG shows a big cerebral hemorrhage involving the Lt cerebral cortex. The hemorrhage is seen causing mass effect and mid line shift. Obstructed hydrocephalus is also seen.

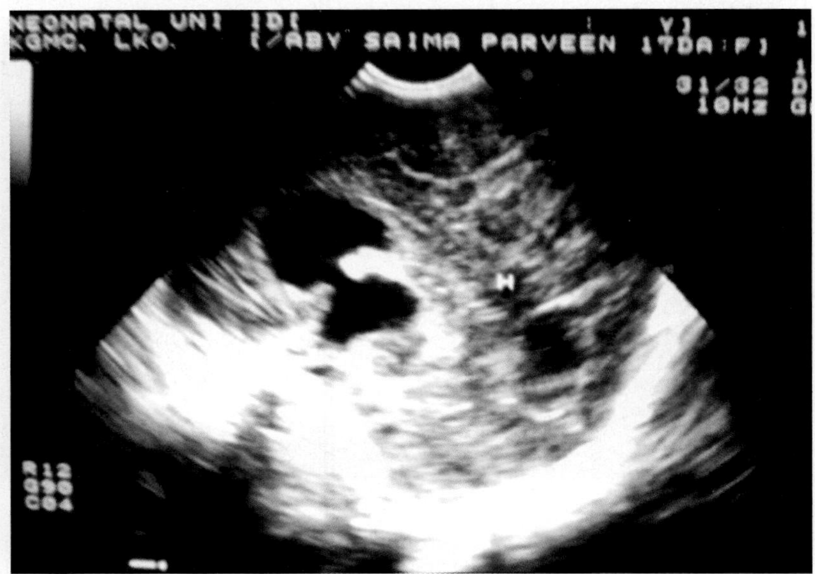

FIGURE 5.109: The parasagittal view shows intracerebral hemorrhage pressing and displacing the ventricle to the opposite side.

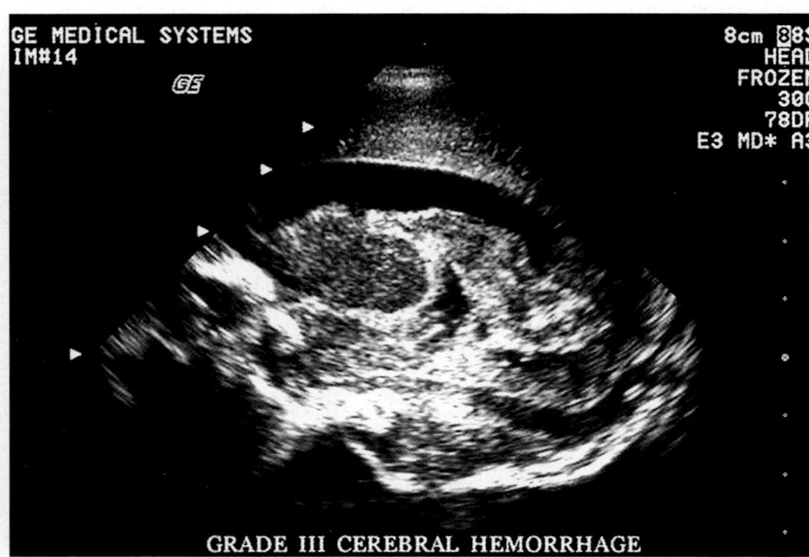

FIGURE 5.110: Intraventricular hemorrhage. HRSG shows a big echogenic hemorrhagic clot in the ventricle. However, the hemorrhage is seen confined to the ventricular cavity suggestive of grade III hemorrhage.

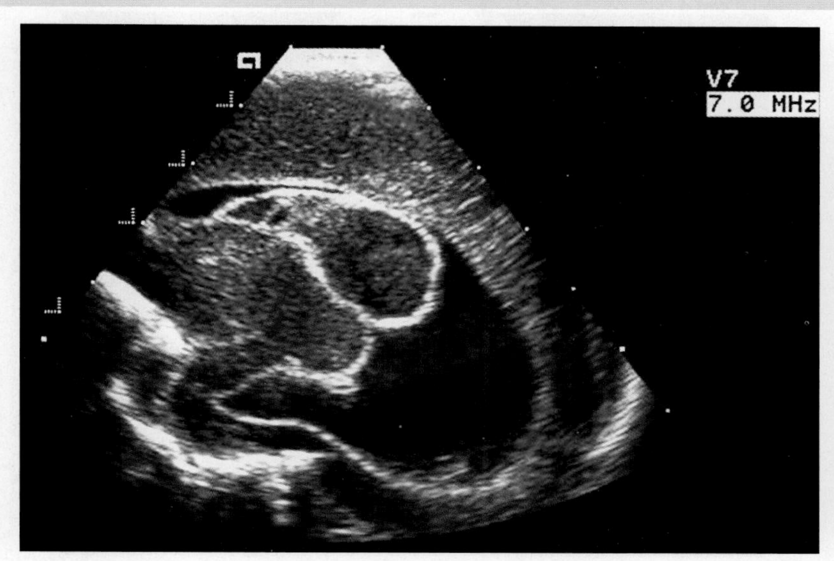

FIGURE 5.111: **Retracting hemorrhagic clot.** HRSG shows retracting clot filling the ventricular cavity.

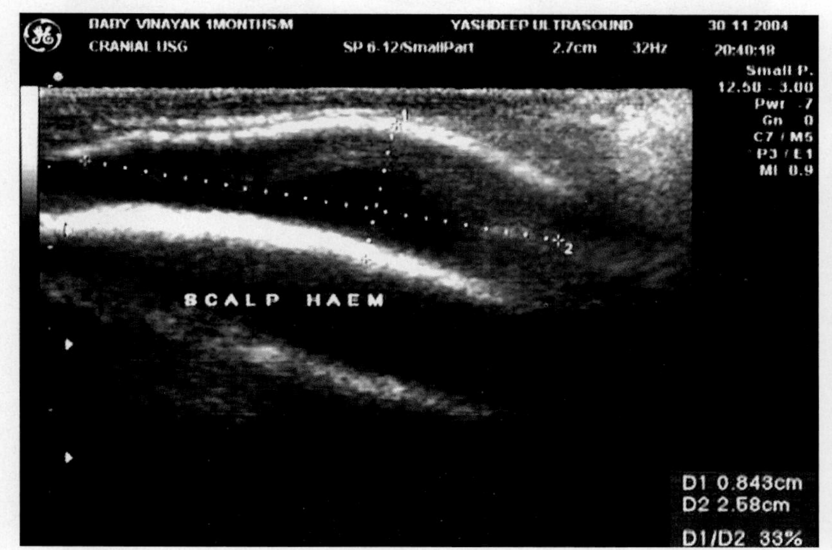

FIGURE 5.112

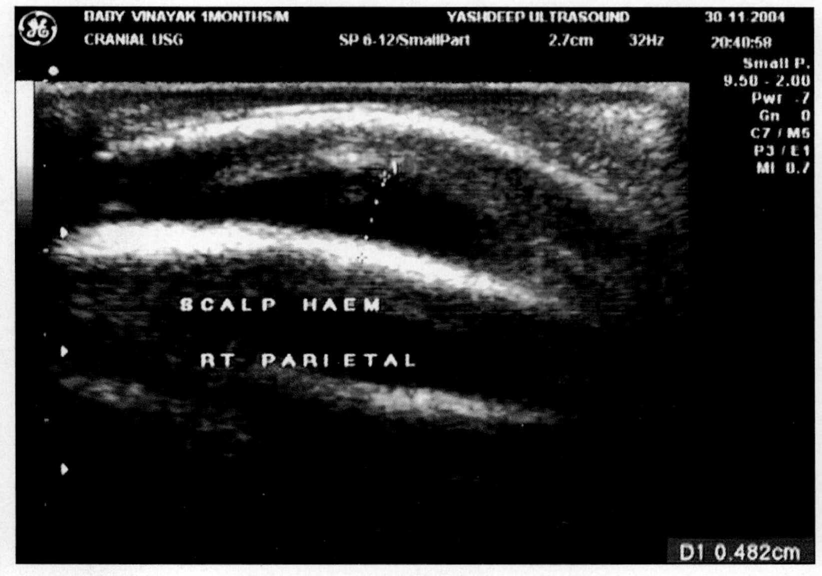

FIGURE 5.113

FIGURES 5.112 and 5.113: **Scalp hematoma.** A patient presented a soft tissue swelling over the parietal scalp after a fall injury. HRSG shows big low-level collection in the scalp layer suggestive of scalp hematoma. However, the bones appear to be intact.

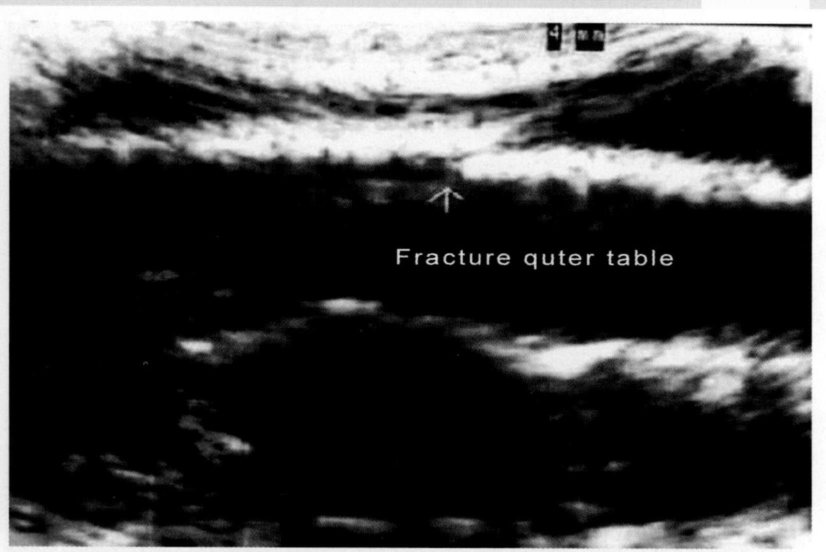

FIGURE 5.114: **Parietal bone fracture.** A baby sustained injury to the head after had a fall and presented with soft swelling over the scalp. HRSG shows fracture of the outer table of the parietal bone. The bony fragments are seen well in apposition on HRSG.

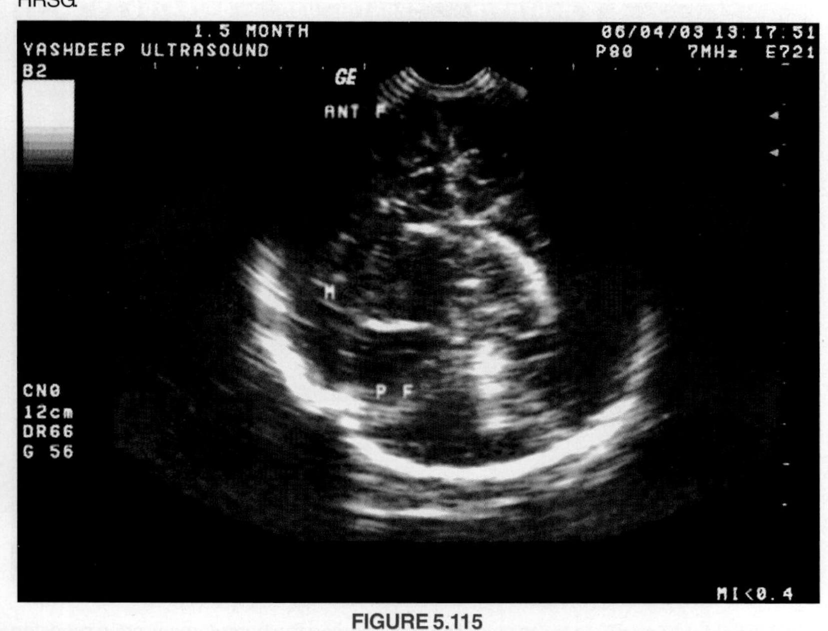

FIGURE 5.115

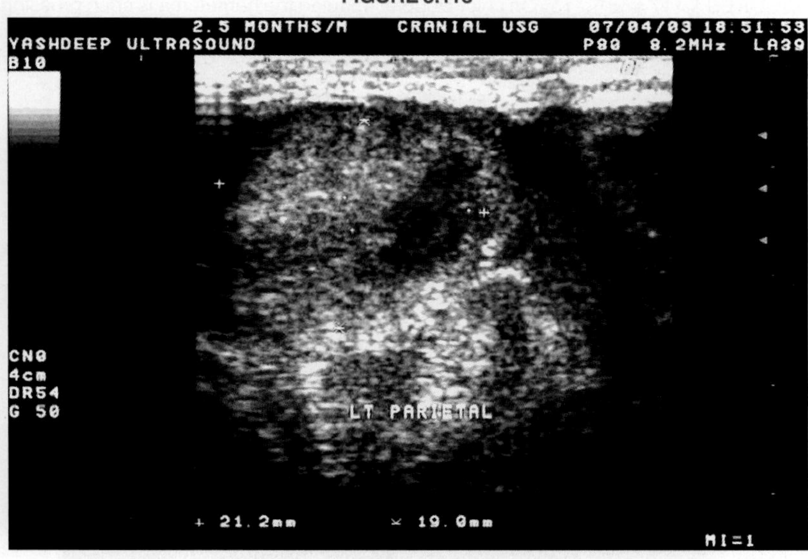

FIGURE 5.116

FIGURES 5.115 and 5.116: **Big parietal hematoma.** A patient presented in unconscious state after had a fall. HRSG shows big parietal hematoma with associated sub dural hematoma. It is seen pressing and displacing the contralateral brain.

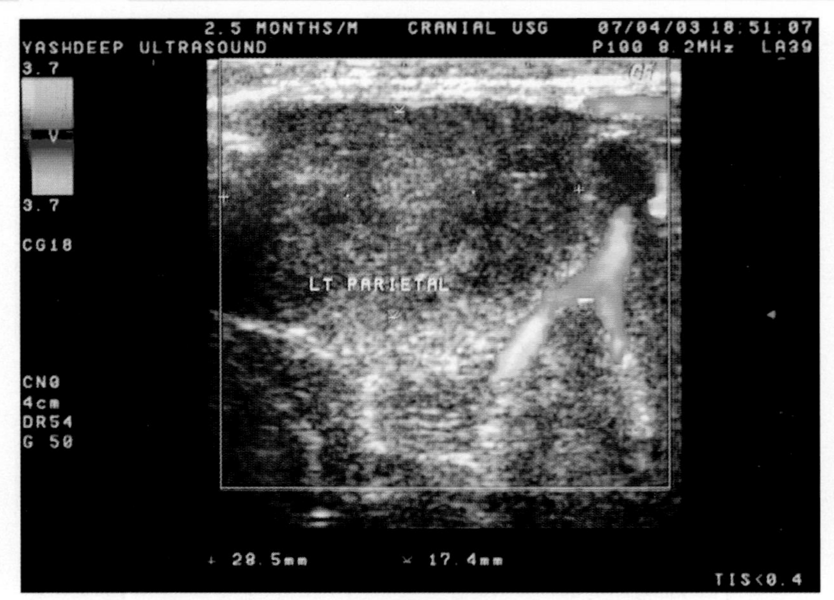

FIGURE 5.117

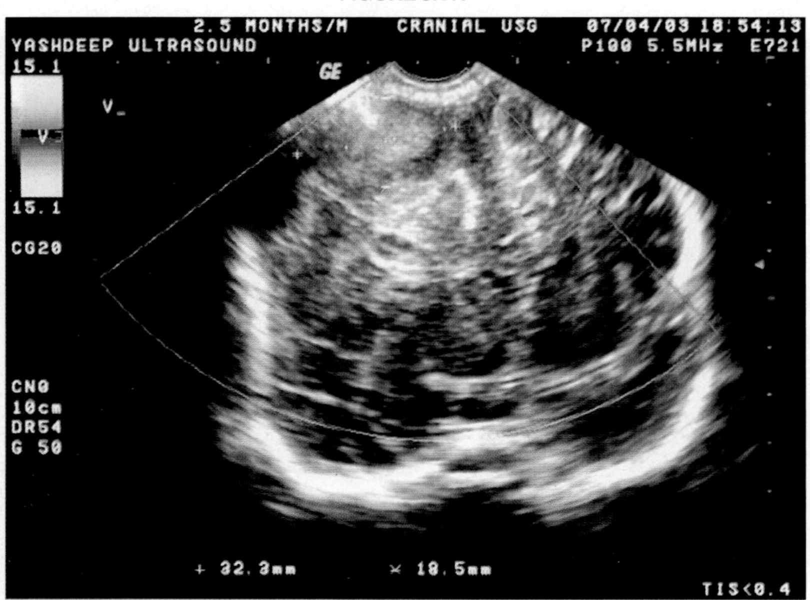

FIGURE 5.118

FIGURES 5.117 and 5.118: On color Doppler imaging the hematoma is seen pressing and displacing the vessels.

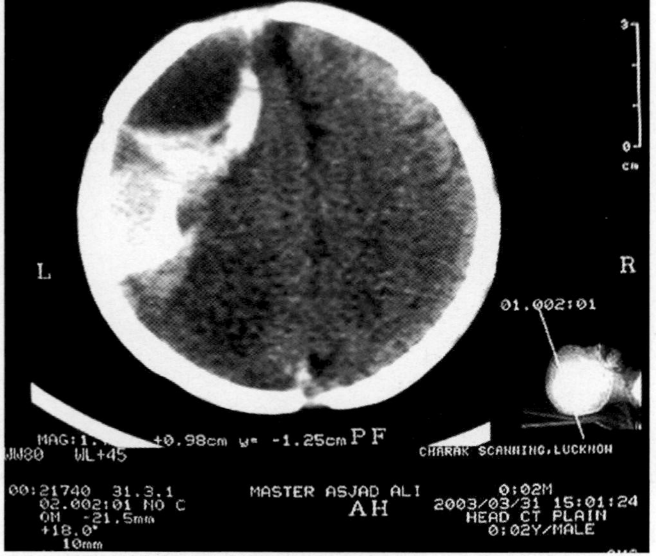

FIGURE 5.119: CT scan of the same patient shows the hematoma with mass effects and confirms the sonographic findings.

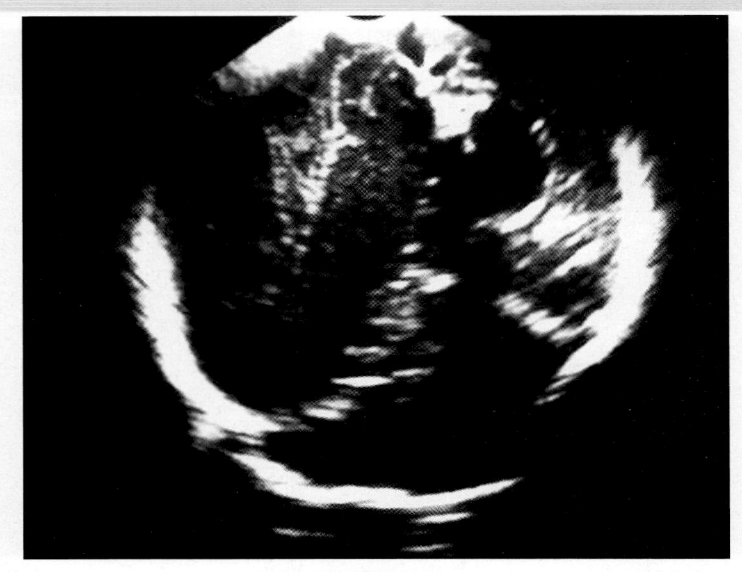

FIGURE 5.120

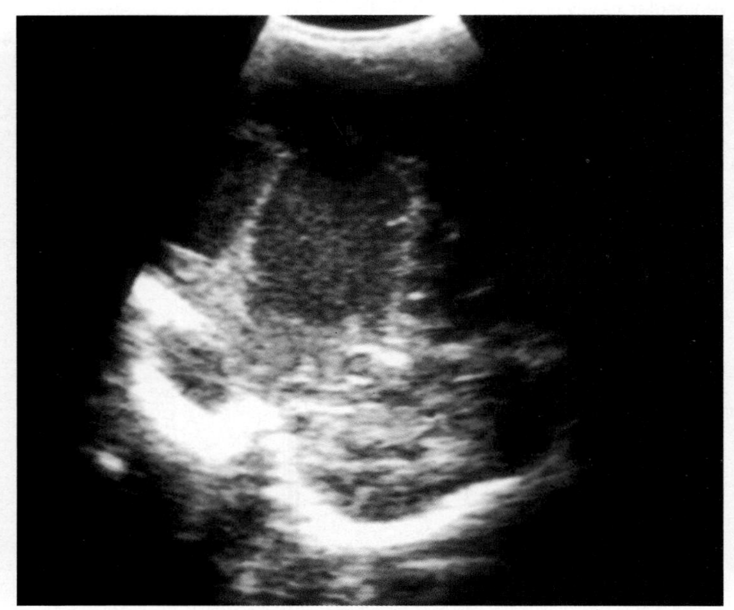

FIGURE 5.121

FIGURES 5.120 and 5.121: Chronic subdural hematoma. A patient presented in unconscious state. HRSG shows a big low level echo collection in the subdural space. It is seen pressing and displacing the brain parenchyma. Marked thinning of the brain parenchyma is seen. Aspiration of the fluid confirmed chronic subdural hematoma.

In several cases, it appears as periventricular cystic formation, however, since sensitivity of ultrasonography is 30% only. In cases of strong clinical suspicion with negative sonography, MRI is indicated.

Intracranial Infections

The intracranial infections can be divided into: -

i. Antenatal acquired.
ii. Postnatal acquired.

Antenatal Infections

Most common infections acquired in utero are due to cytomegalovirus, herpes simplex, rubella or toxoplasma. Cytomegalovirus typically presents with periventricular calcification. The typical features are present at or immediately after birth. Though initially they may be

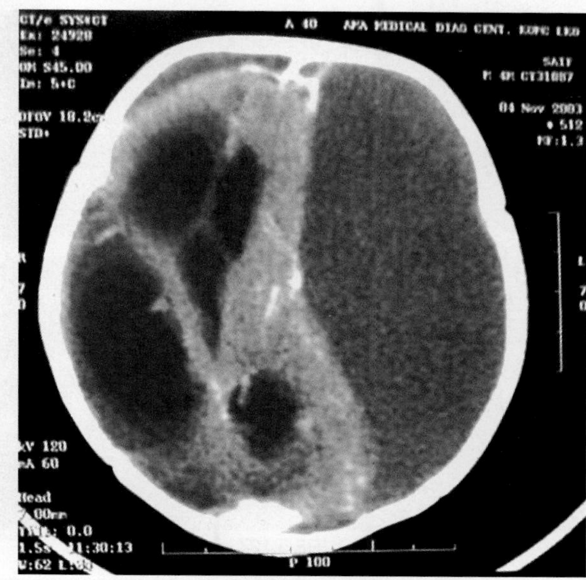

FIGURE 5.122

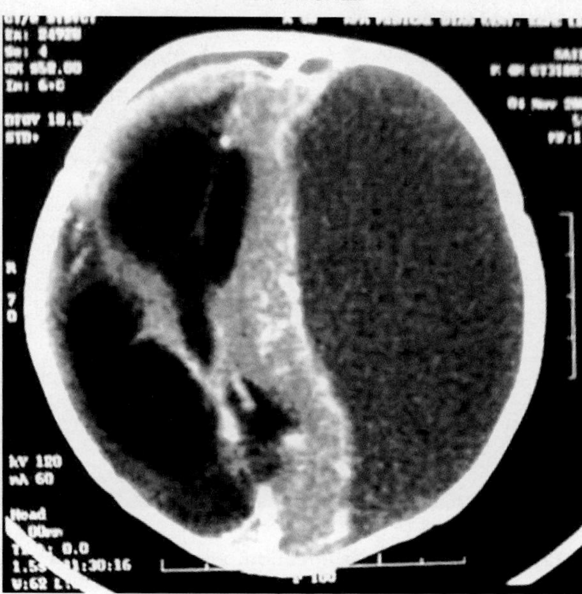

FIGURE 5.123

FIGURES 5.122 and 5.123: Chronic subdural hematoma. CT scan of the same patient shows details of the subdural hematoma. The hematoma is seen completely displacing the brain parenchyma with marked thinning.

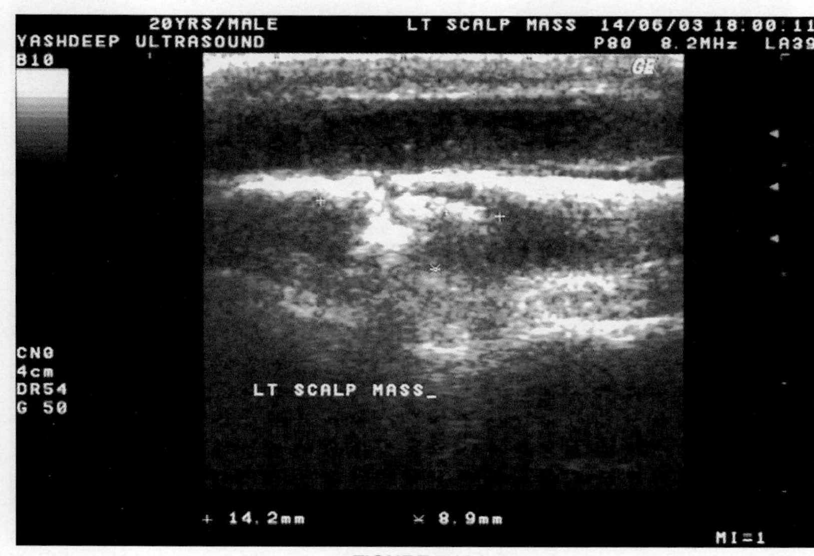

FIGURE 5.124

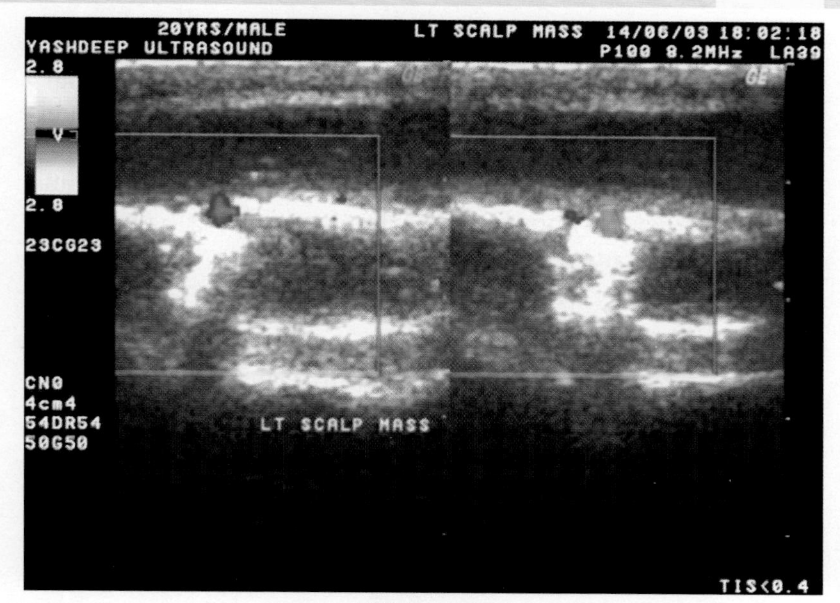

FIGURE 5.125

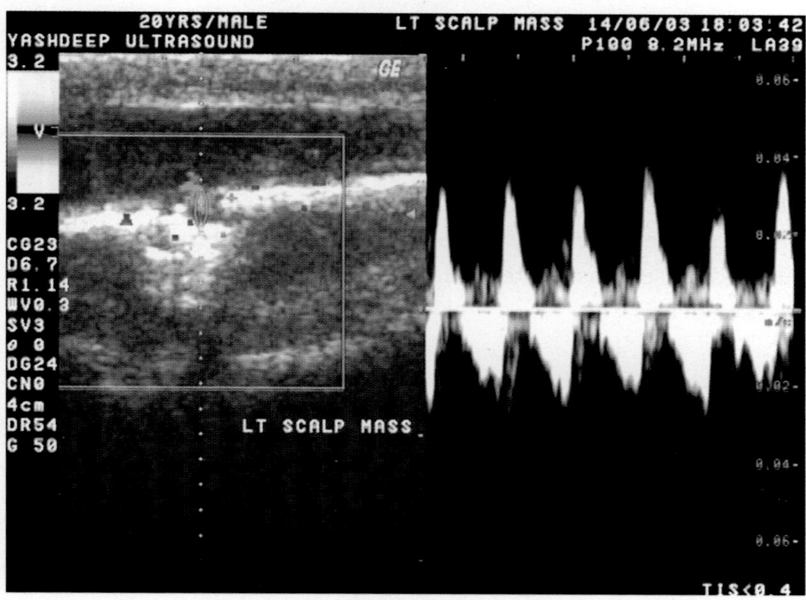

FIGURE 5.126

FIGURES 5.124 to 5.126: Scalp mass. A patient presented with soft tissue mass over the scalp. HRSG shows soft tissue mass over the scalp. Bony erosion is also seen on color flow imaging increased flow is seen in it. Biopsy showed metastatic deposit.

small enough not to cause shadowing their persistence after 1 to 2 weeks of birth differentiates them from subependymal hemorrhage. Toxoplasmosis is the second most common infection. Unlike periventricular distribution is seen in CMV, toxoplasmosis presents as scattered calcification specially involving the basal ganglia. Severe cases may present as destructive lesions presenting as multiple cystic encephalomalacia.

Postnatal Infections

Bilateral meningitis, ventriculitis and encephalitis are common brain infection encountered frequently in neonates. Thirty percent of the survivors develop hydrocephalus. Neonates are usually affected by β-hemolytic streptococci or *Escherichia coli*, whereas *Haemophilus influenzae* generally affects the infants.

Meningitis

It has nonspecific sonographic appearances. It may present as an area of diffuse increased or decreased echogenicity of brain parenchyma representing edema, cerebritis, or evolving infarction. Inflammatory exudates may present as increased echogenicity and widening of sulci and fissures. Few may present with extra/axial fluid collection, which resolves without any sequelae.

Ventriculitis

It is a serious complication and presents with characteristic sonographic findings. The findings include floating debris in ventricles with septation and ventricular dilatation. The development of intraventricular septa poses a great problem in effective shunting.

Multicystic Encephalomalacia (MCE)

It is a serious complication of bacterial meningitis with poor outcome. Ultrasonography shows multiple bilateral cystic lesions involving the brain parenchyma with or without ventriculomegaly. Early detection helps in malignant and family counseling.

Abscess and Infarctions

These are commonly seen in neonates with poor resistance. Focal areas of increased echogenicity represent vasculitis, cerebritis or infarction. Abscess presents as an area of cavitations with thickened wall with or without debris.

Intracranial Masses

The brain tumors in the neonatal age group are uncommon and mostly are congenital. Most tumors are present as hydrocephalus, which can easily be picked up by ultrasonography. Ultrasonography is found to be excellent screening modality, however, for detailed delineation contrast. CT of MRI is preferred method of imaging. In many cases, these tumors may present as hemorrhage, and one should be aware about their presentations. The commonly occurring tumors in this age group are astrocytomas choroid plexus papilloma and ependymoma.

Cystic Tumors

Cystic tumors are easily identified on sonography. Arachnoid cysts are frequently encountered in neonates. The other common cystic lesions are porencephalic cyst, which are seen invariable in continuation with ventricular system. Choroid plexus cyst presents as cystic mass within choroid plexus. Subependymal cysts are formed as a consequence of SEH and are found in the region of caudothalamic groove involving the lining of ventricles.

Galenic venous malformation commonly describes as vein of Galen aneurysm seen as midline cystic mass above the third ventricle. Doppler ultrasonography helps in confirming the diagnosis.

Color Doppler USG

Though duplex Doppler and color Doppler are being extensively used in various clinical conditions, however, in cranial ultrasonography, it

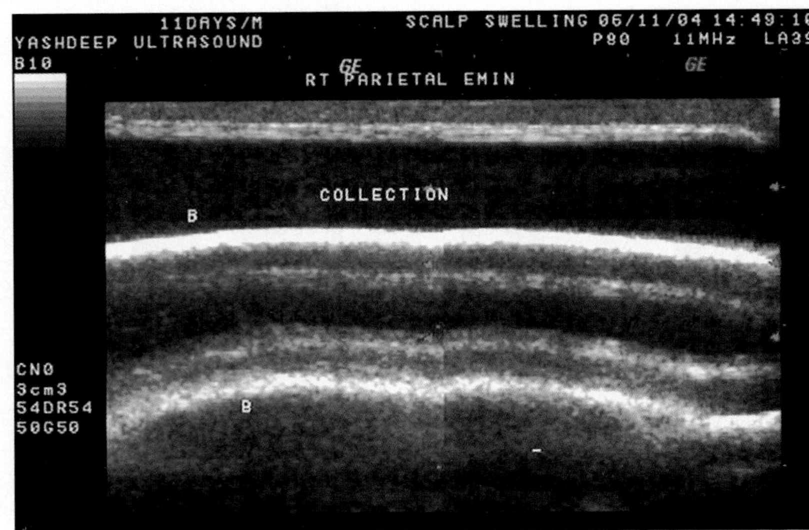

FIGURE 5.127

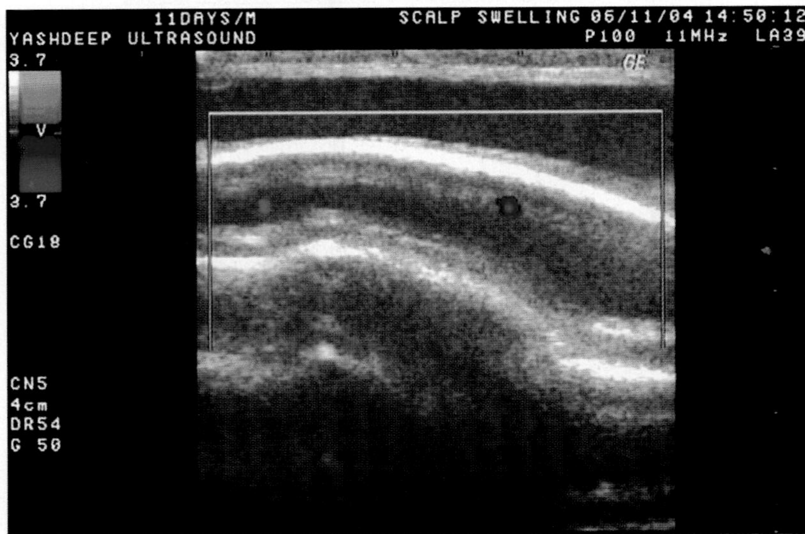

FIGURE 5.128

FIGURES 5.127 and 5.128: Scalp infection. A child presented with tender hot swelling over the parietal region. HRSG shows low-level echo collection in the scalp. However, the collection was limited only to the scalp layer. No subdural collection was seen. Aspiration confirmed pyogenic collection.

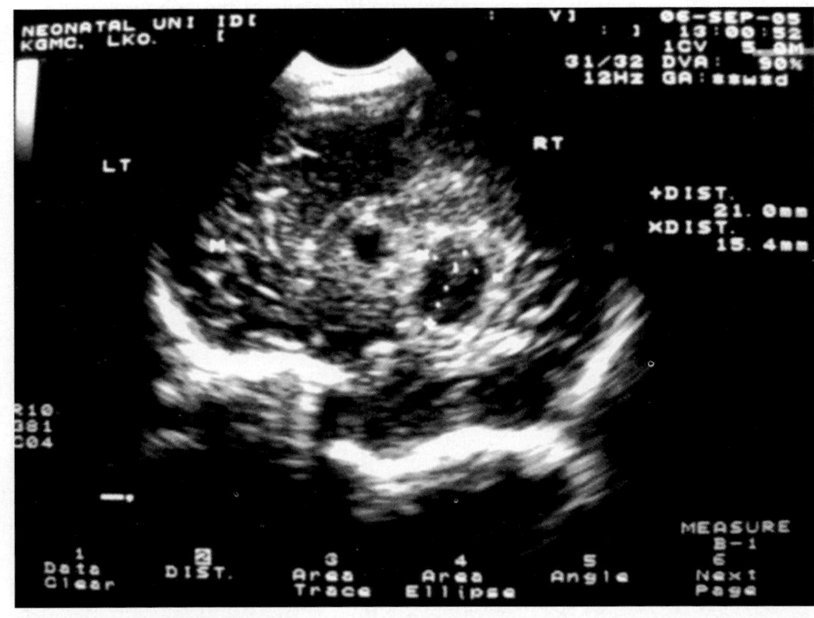

FIGURE 5.129: Brain abscess. A patient presented with high fever. HRSG shows irregular low-level echo complex mass in the Rt parietal lobe suggestive of brain abscess.

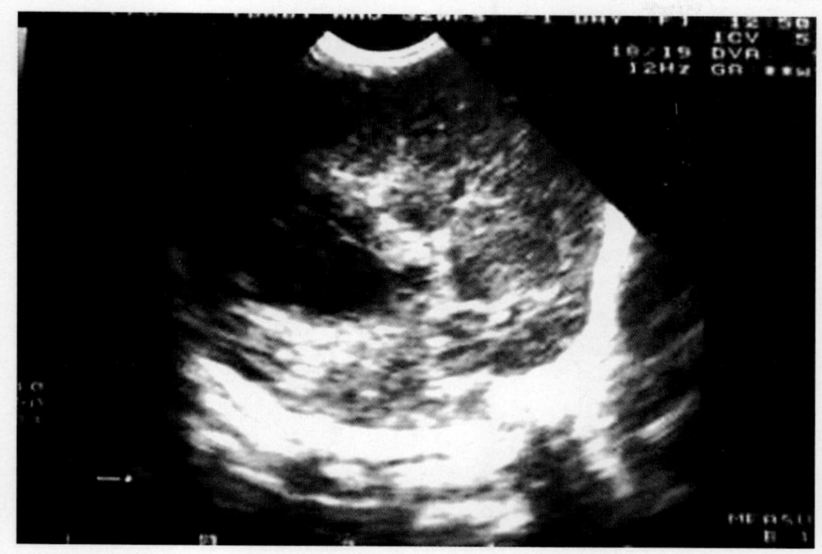

FIGURE 5.130: **Brain abscess.** HRSG shows irregular hypoechoic low-level echo complex mass in the Lt parietal lobe. The mass is seen involving the thalamic region.

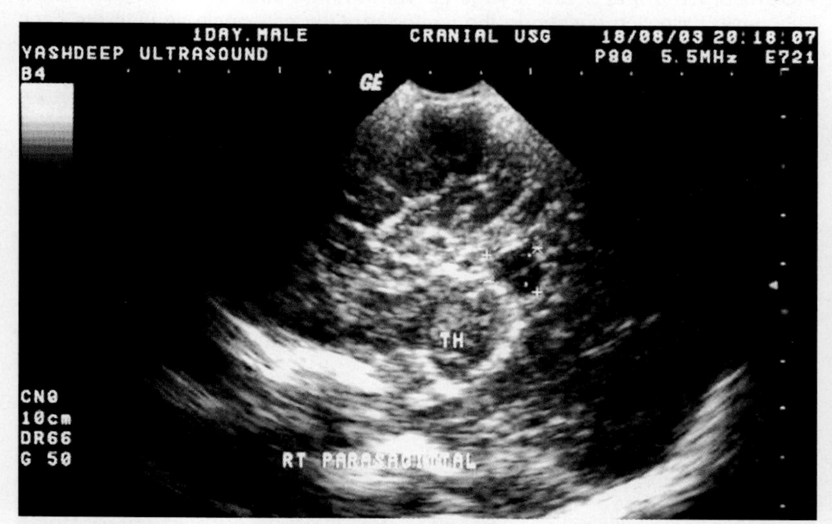

FIGURE 5.131: **Brain abscess with parietal cyst.** A patient presented with high fever with fullness of the ventricles. HRSG shows low-level echo complex mass in the parathalamic region suggestive of an abscess.

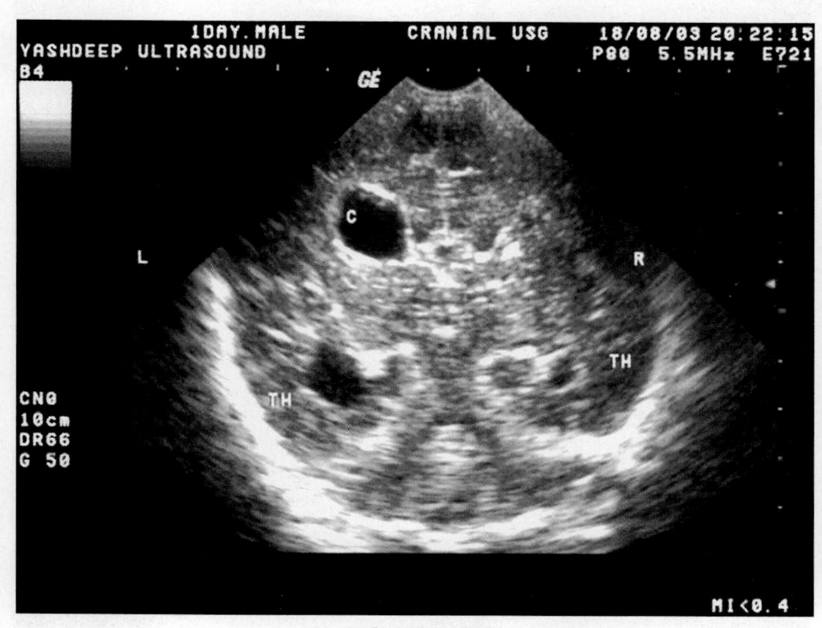

FIGURE 5.132

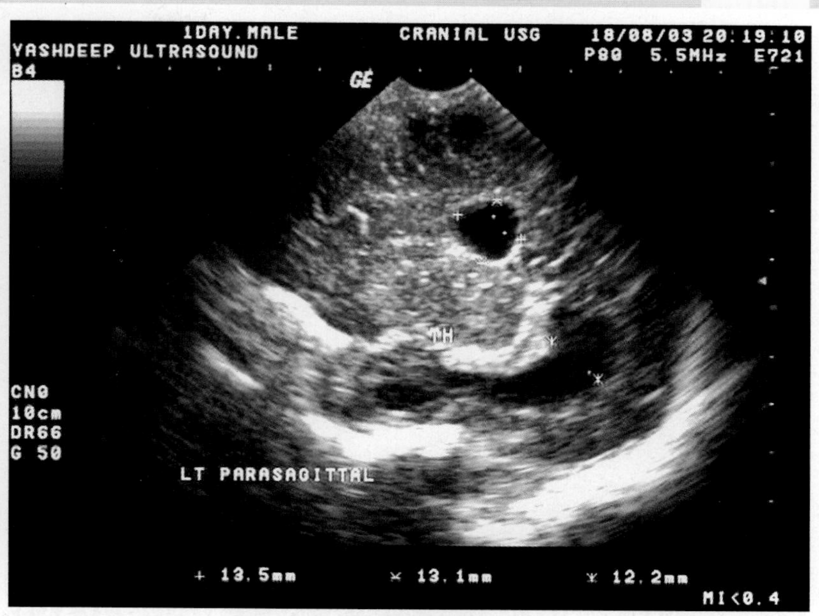

FIGURE 5.133

FIGURES 5.132 and 5.133: There is also evidence of well-defined thin walled cyst seen in the Lt parietal lobe.

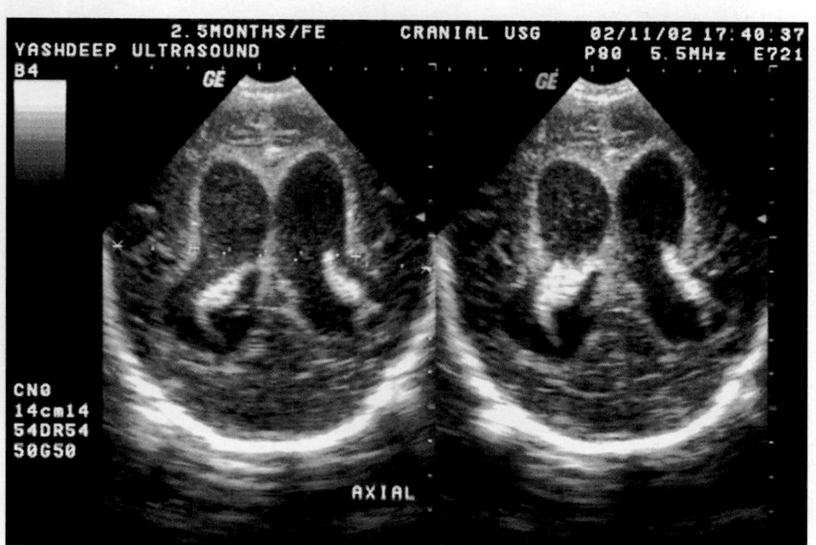

FIGURE 5.134

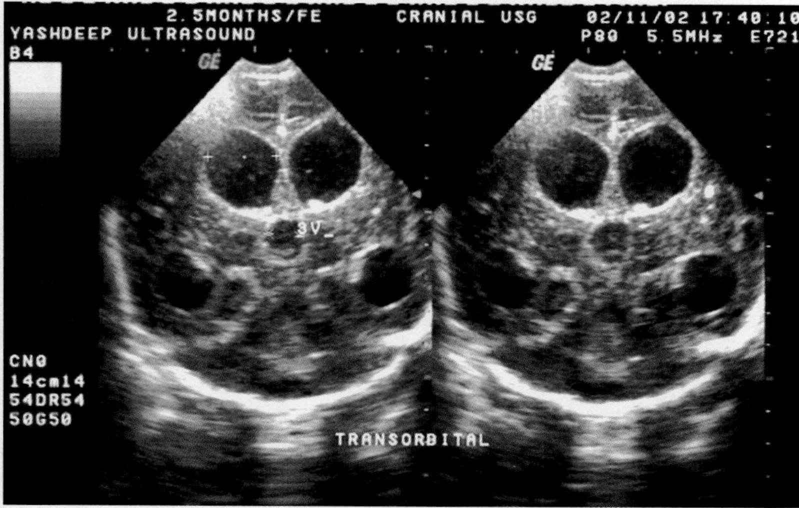

FIGURE 5.135

FIGURES 5.134 and 5.135. Ventriculitis with hydrocephalus. A child presented with high fever and fullness of the fontanelle. HRSG shows dilated ventricles. Echogenic collection is seen in the ventricles. The walls of the ventricles are echogenic. Sonographic findings are suggestive of ventriculitis. Spinal fluid examination confirmed the pyogenic meningitis.

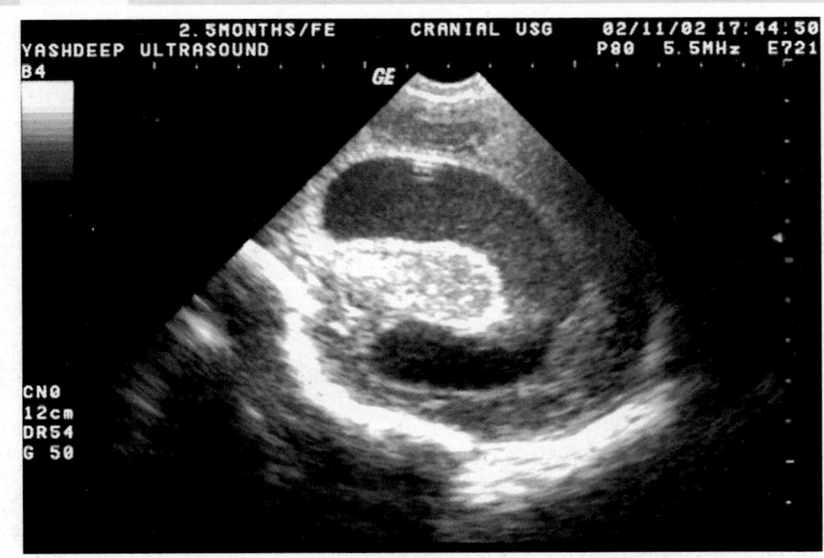

FIGURE 5.136

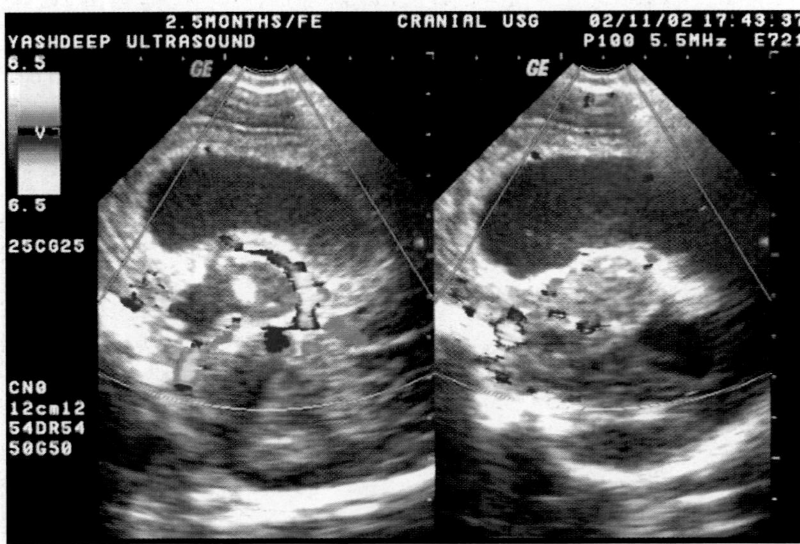

FIGURE 5.137

FIGURES 5.136 and 5.137: Ventriculitis with hydrocephalus. Echogenic collection is seen filling the lateral ventricles. On color flow imaging high flow is seen in periventricular vessels.

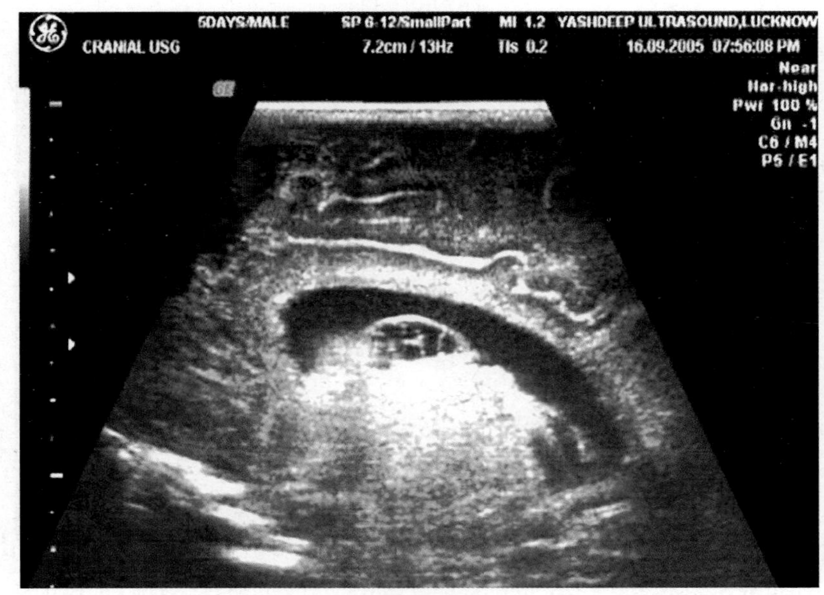

FIGURE 5.138

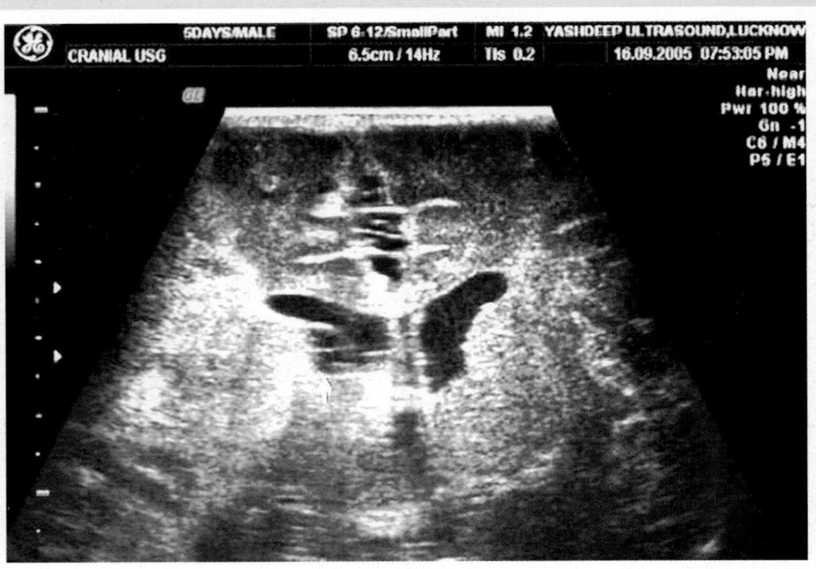

FIGURE 5.139

FIGURES 5.138 and 5.139: Pyogenic ventriculitis-shows dilated ventricles. Echogenic collection is seen in the ventricles. Fine echogenic strands are seen running in the ventricles suggestive of protein bands in a case of pyogenic meningitis.

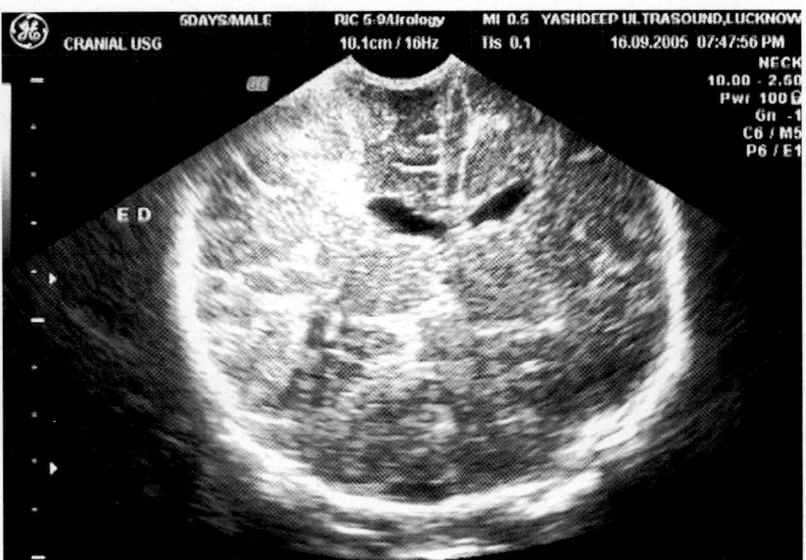

FIGURE 5.140

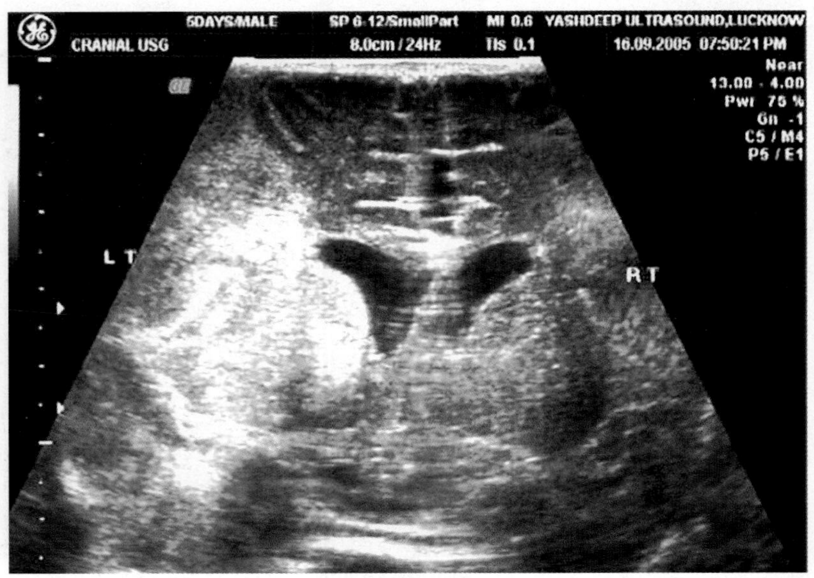

FIGURE 5.141

FIGURES 5.140 and 5.141: Pyogenic ventriculitis with brain edema. There is also evidence of echogenic brain parenchyma seen in the same patient suggestive of associated brain edema. The sulci and gyri are effaced by the brain edema.

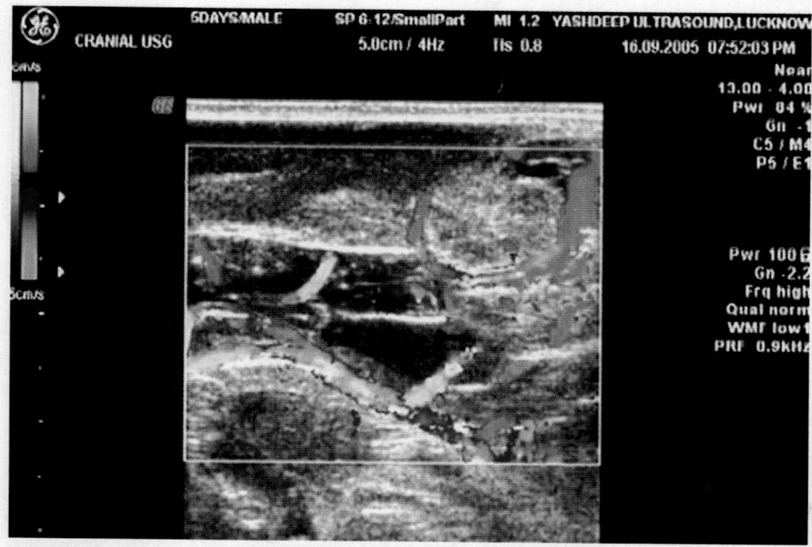

FIGURE 5.142: On color flow imaging increased flow is seen in the brain parenchyma due to hyperemia.

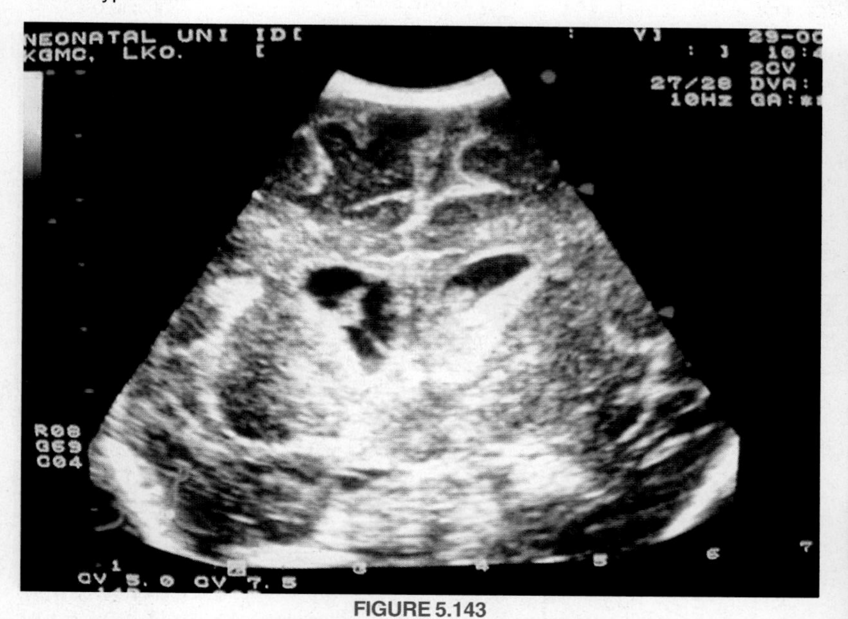

FIGURE 5.143

FIGURE 5.144

FIGURES 5.143 and 5.144: Meningitis with ventriculitis. A patient presented with high fever with fullness of the ventricles. HRSG shows mild dilatation of the ventricles. Ventricular walls are echogenic. Fine echogenic strands are seen floating in the ventricles suggestive of ventriculitis with meningitis.

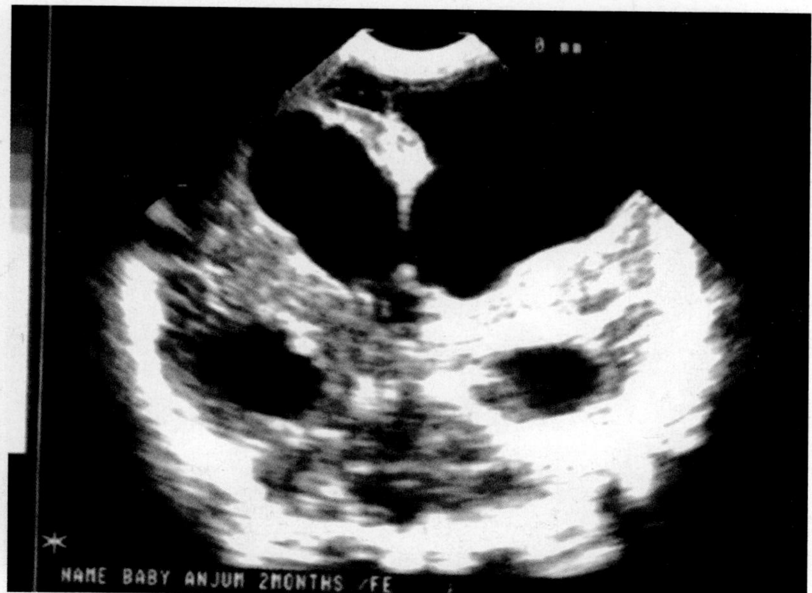

FIGURE 5.145

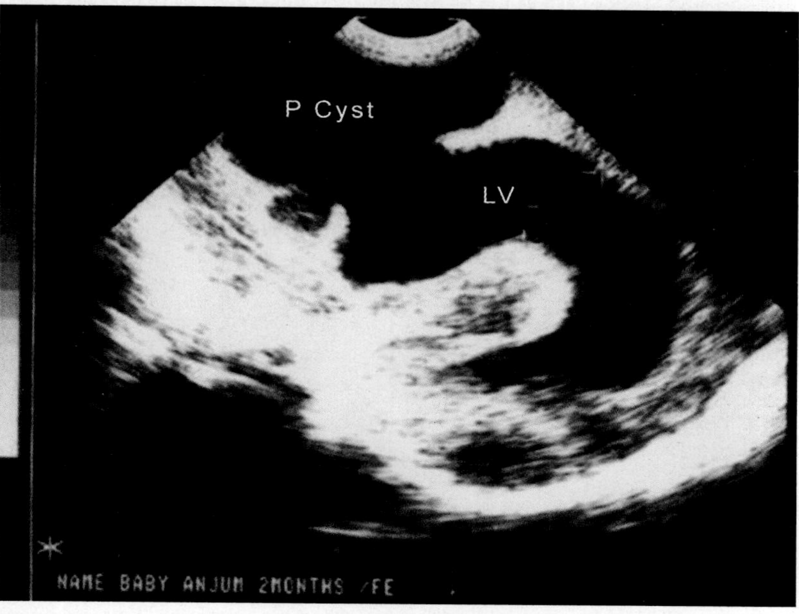

FIGURE 5.146

FIGURES 5.145 and 5.146: Porencephalic cyst. HRSG shows well-defined cystic mass communicating with lateral ventricle on the Lt side suggestive of a big porencephalic cyst. The contralateral frontal horn and temporal horns are also dilated.:

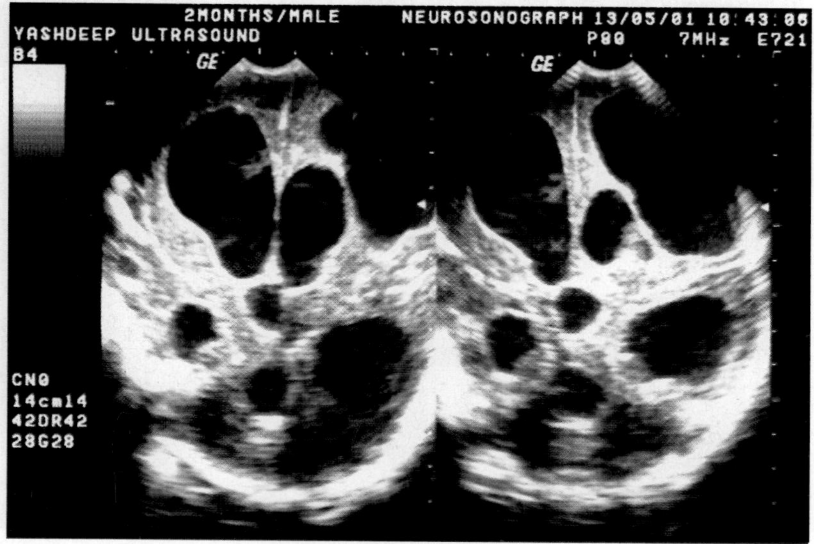

FIGURE 5.147

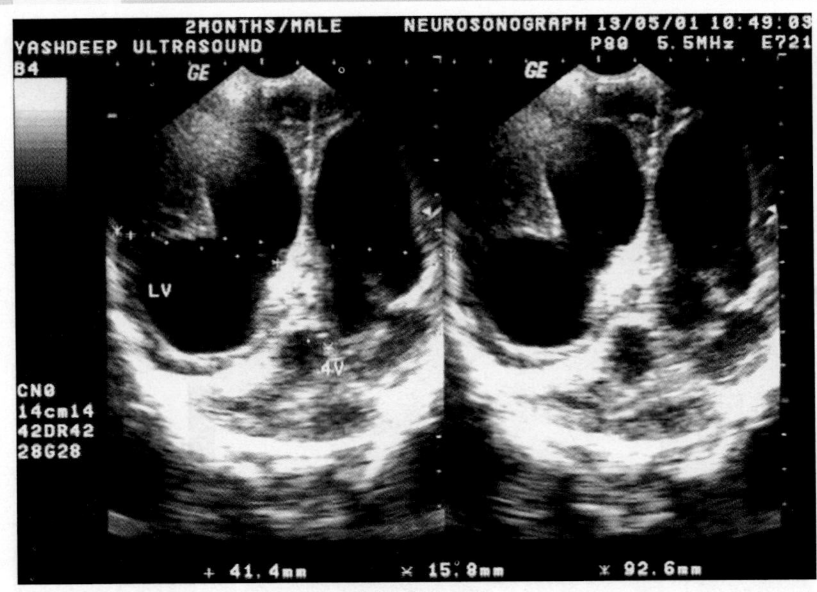

FIGURE 5.148

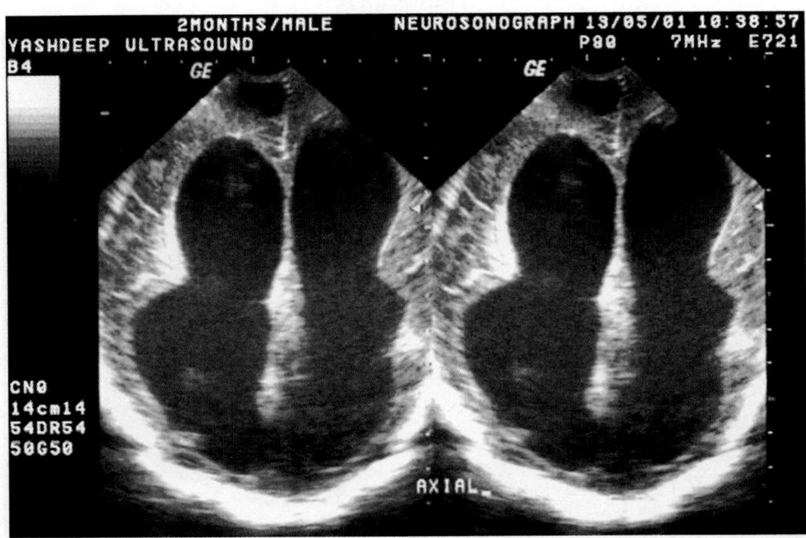

FIGURE 5.149

FIGURES 5.147 to 5.149: Ventriculitis with hydrocephalus. A patient of pyogenic meningitis presented with high fever with fullness of fontanelle. HRSG shows dilated ventricles. Echogenic cerebral fluid is seen filling the ventricles. Fine echogenic strands are seen in the ventricles. The ventricular walls are also irregular in outline. Obstructed hydrocephalus is seen due to basal arachnoiditis.

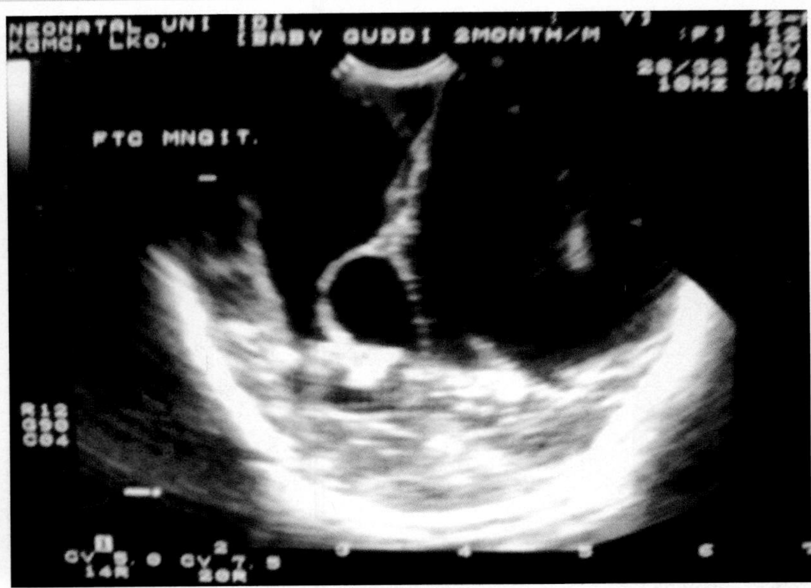

FIGURE 5.151

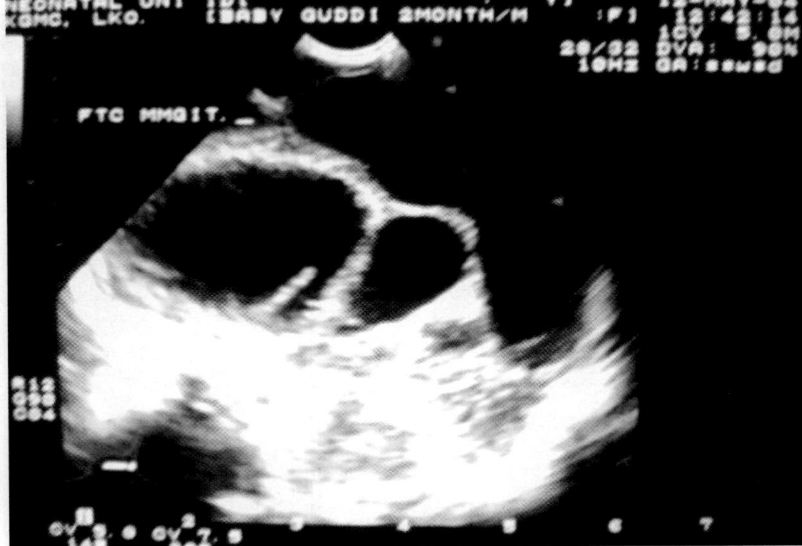

FIGURE 5.152

FIGURES 5.150 to 5.152: Postmeningitis hydrocephalus with gliosis. A patient presented with gross hydrocephalus after meningitis. HRSG shows distorted brain anatomy. Marked fluid collection is seen in the ventricles. Brain parenchyma is thinned out. Echogenic strands are also seen in the ventricles. Ultrasound is very good for evaluation of sequelae of cranial infections.

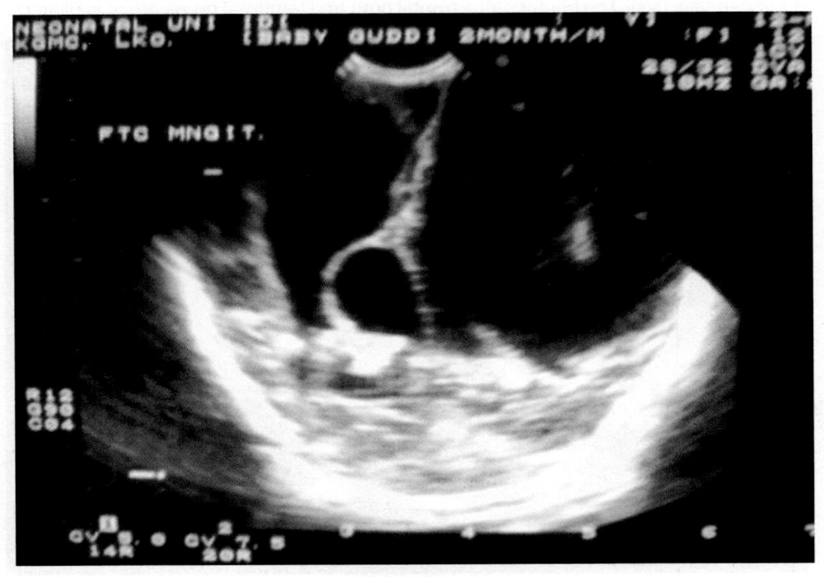

FIGURE 5.150

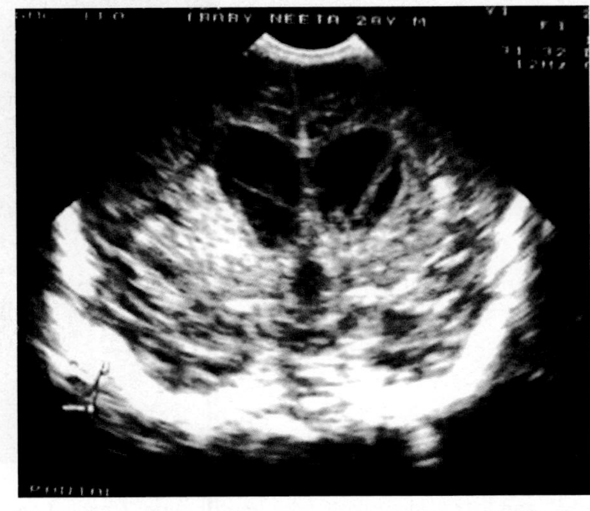

FIGURE 5.153: Ventriculitis with hydrocephalus. HRSG shows dilated frontal horns. Echogenic collection is seen in the ventricles. Fine echogenic strands are seen running in the ventricles.

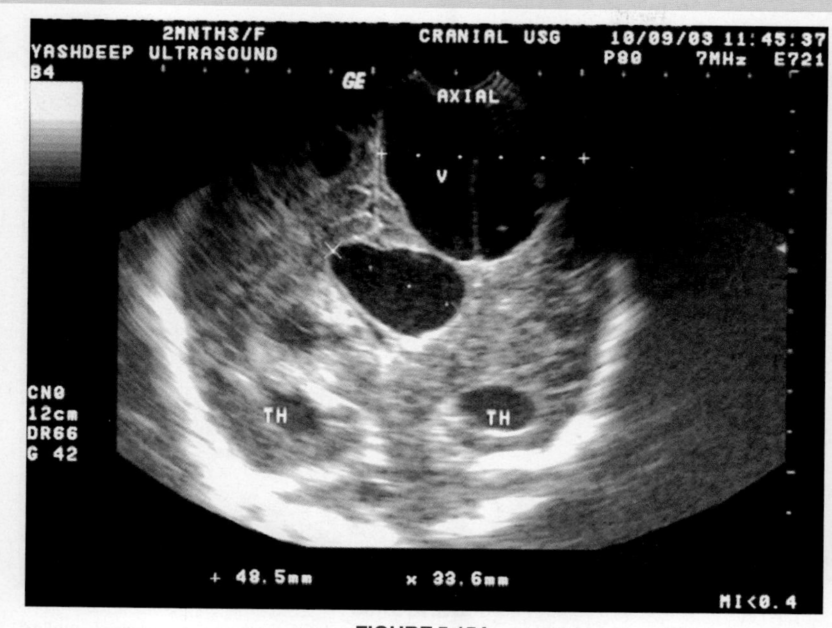

FIGURE 5.154

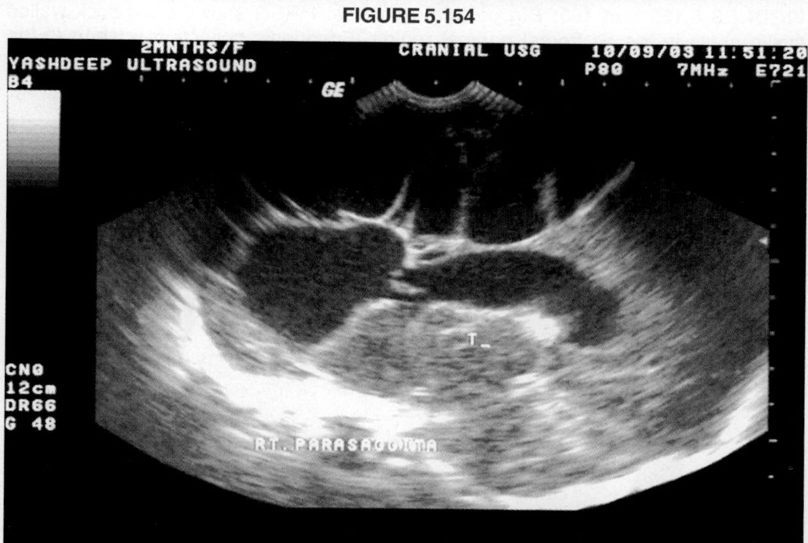

FIGURE 5.155

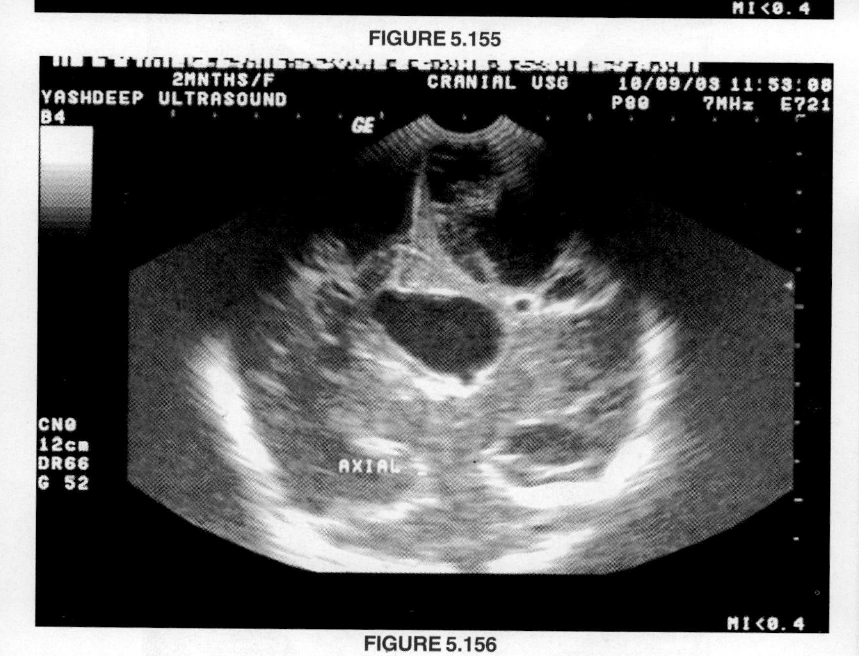

FIGURE 5.156

FIGURES 5.154 to 5.156: Postmeningitis porencephalic cyst with hydrocephalus. A patient presented with progressively increasing hydrocephalus after meningitis. HRSG shows marked dilatation of the ventricles. A big cyst is seen communicating with the Rt frontal horn suggestive of porencephalic cyst. Echogenic strands are seen floating in the cyst suggestive of pyogenic exudative collection.

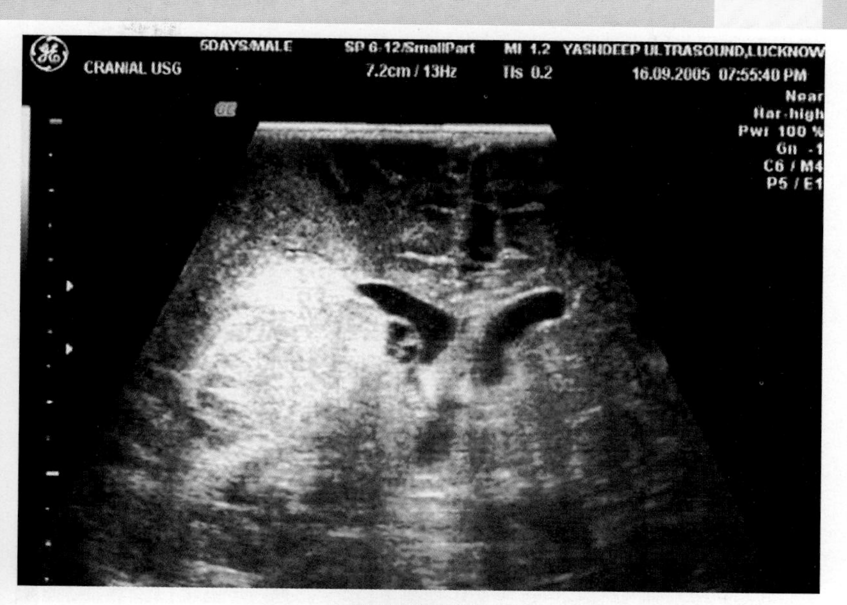

FIGURE 5.157

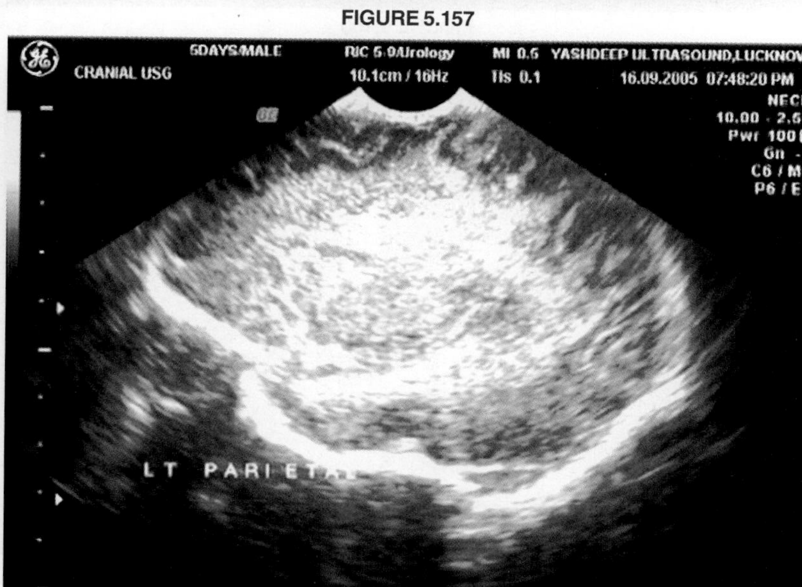

FIGURE 5.158

FIGURES 5.157 and 5.158: Gross brain edema with ventriculitis. HRSG shows gross brain edema. The edematous parenchyma is wide. Sulci and gyri are effaced. Ventricles are dilated with echogenic collection. Fibrous bands are also seen in the ventricles.

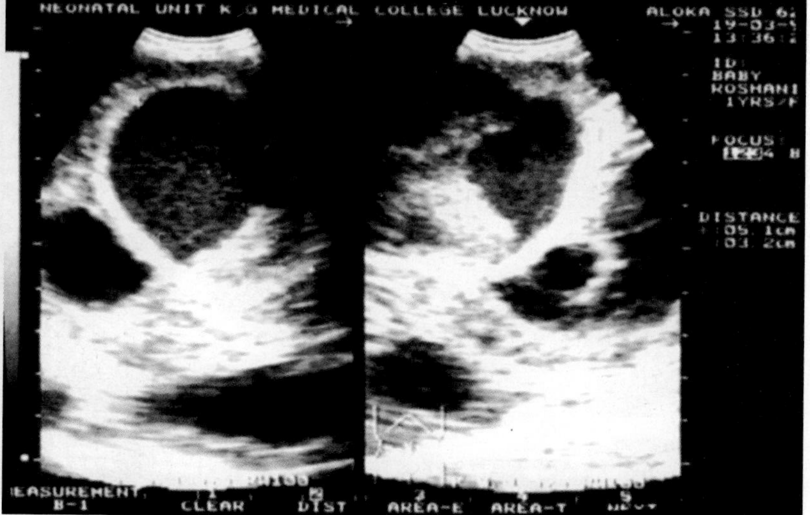

FIGURE 5.159

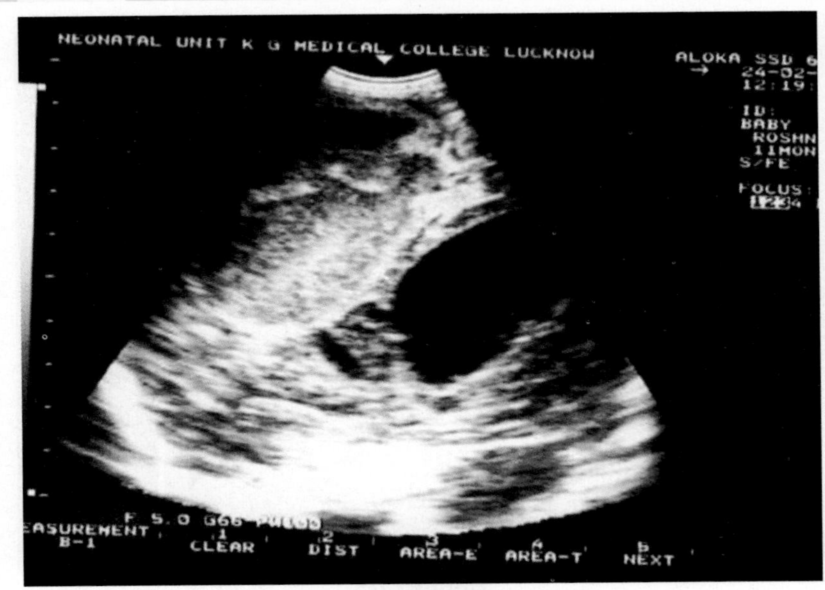

FIGURE 5.160

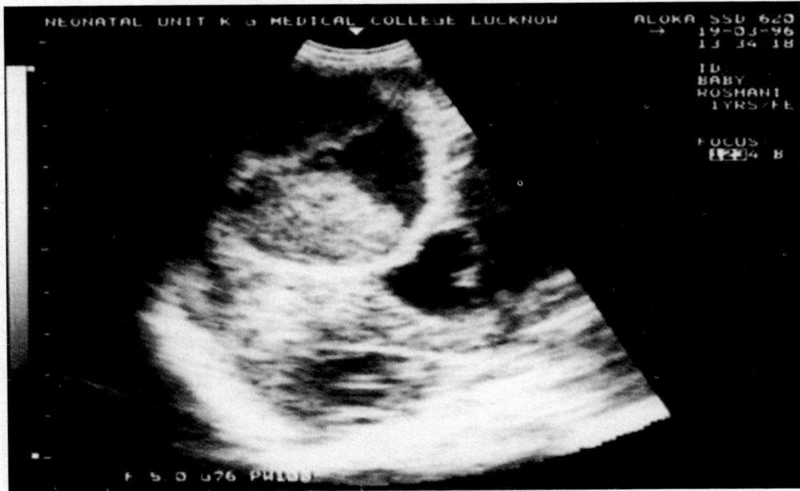

FIGURE 5.161

FIGURES 5.159 to 5.161: Big parietal lobe abscess. Neurosonography shows multiple thick walled low level echo complex masses in the frontoparietal area in a child presented with septicemia. Echogenic debris is also seen in the abscess. Low-level collection is seen filling the abscess cavity. It is seen pressing over the ventricles resulting into obstructed hydrocephalus associated mass effect is also seen.

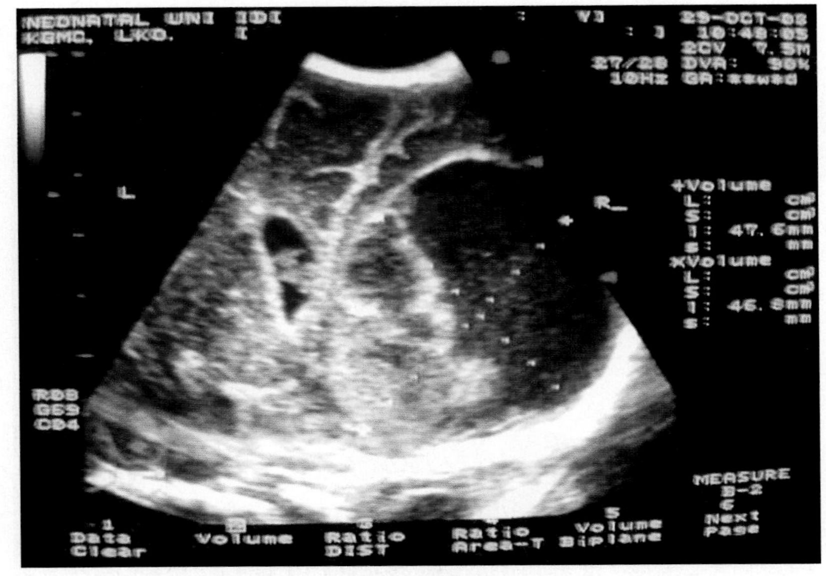

FIGURE 5.162

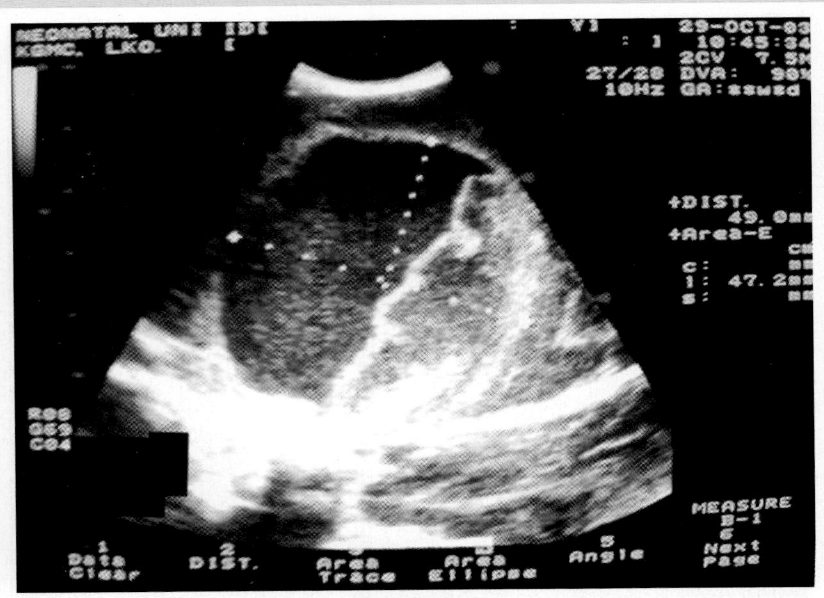

FIGURE 5.163

FIGURES 5.162 and 163: Big parietal abscess. HRSG shows a big thick walled cystic mass in the Rt frontal parietal region. Echogenic collection is seen in the abscess. Layering sign is positive. Marked perifocal edema is present with mass effect suggestive of big brain abscess.

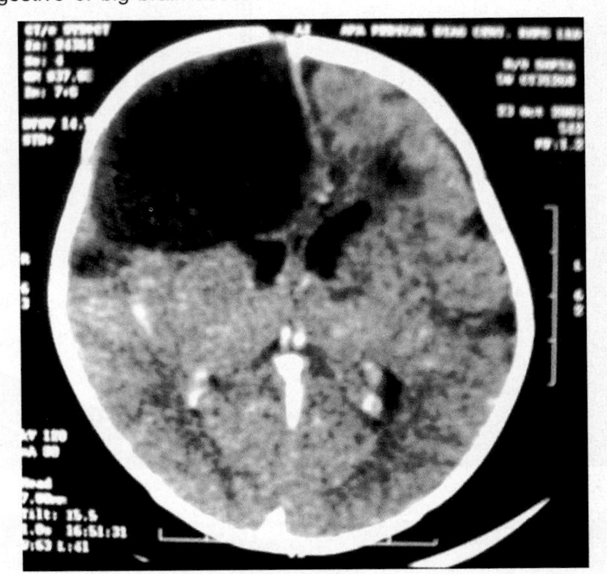

FIGURE 5.164

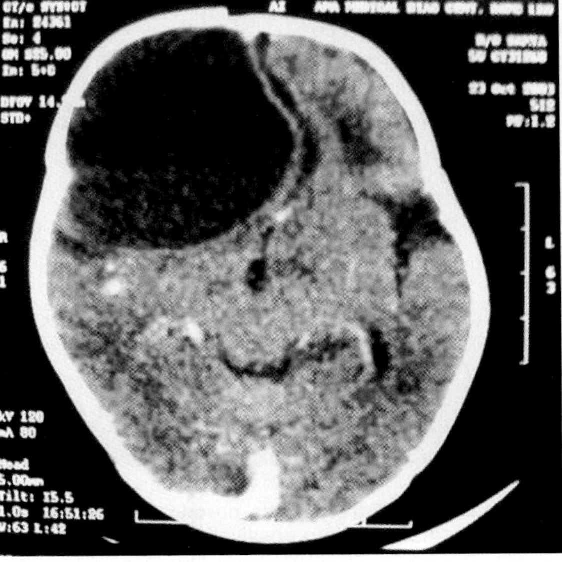

FIGURE 5.165

FIGURES 5.164 and 5.165: CT scan of the same patient shows big abscess in the Rt frontal parietal region with mass effect.

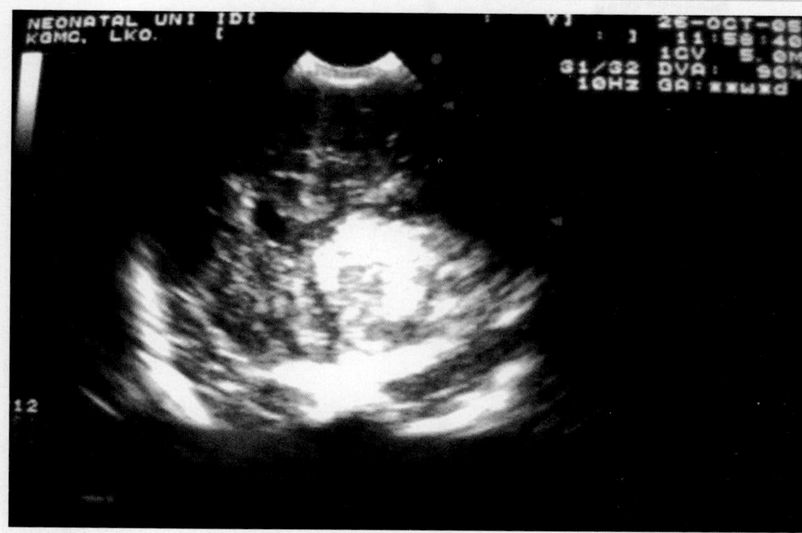

FIGURE 5.166

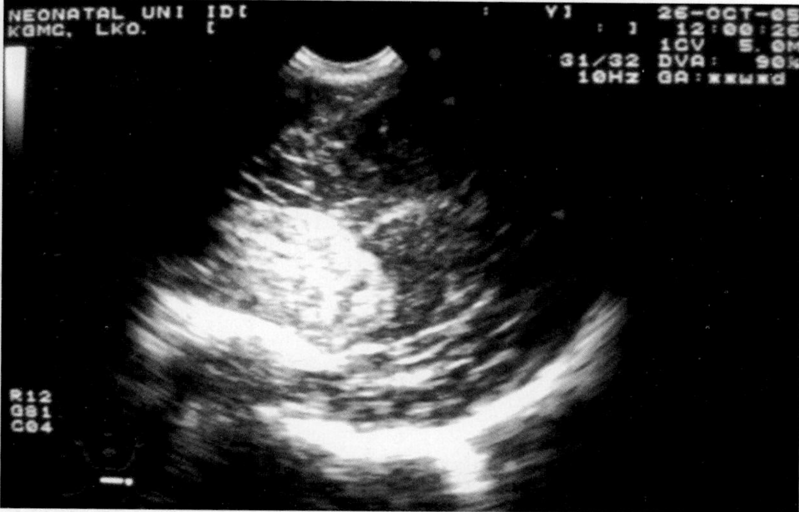

FIGURE 5.167

FIGURES 5.166 and 5.167: Intracranial tumor-glioma. A two months old child presented with fall and subsequent comma. HRSG shows a big irregular highly echogenic mass in the Rt parietal lobe. The mass is seen pressing over the Rt lateral ventricle with mass effect.

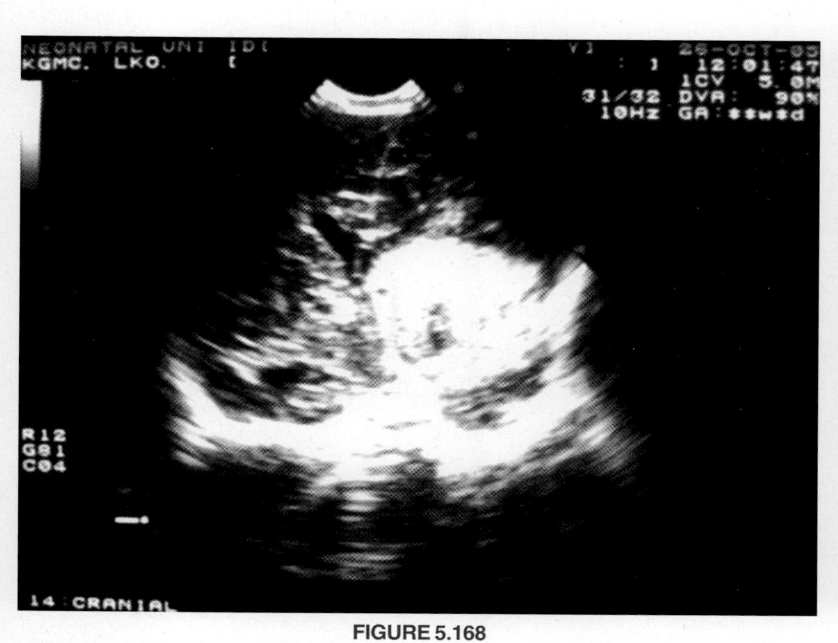

FIGURE 5.168

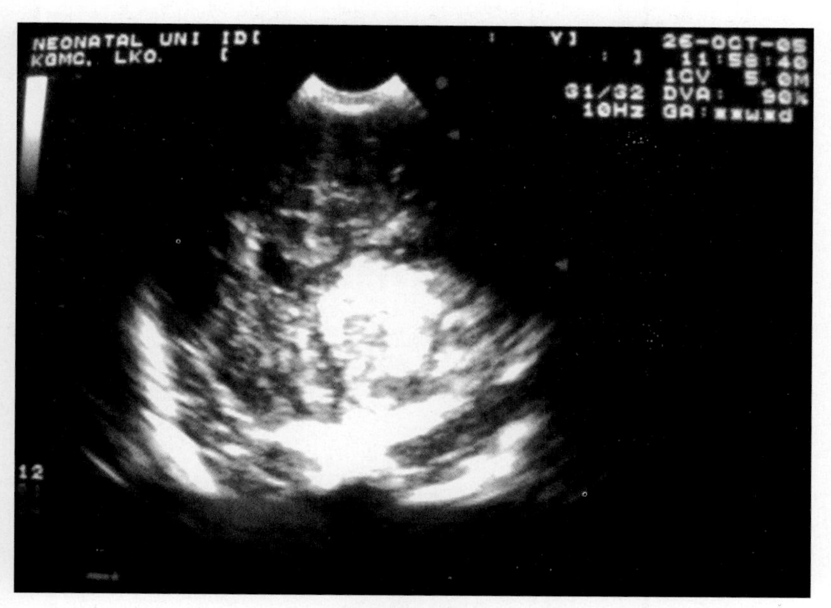

FIGURE 5.169

FIGURES 5.168 and 5.169: The mass is seen pressing over the ventricles. Obliteration of the ventricle is seen. Contralateral shift is also seen. Biopsy of the mass confirmed glioma.

is still struggling to find its place. It has been found to be extremely useful to differentiate between vein of Galen malformation and arachnoid cyst. It has also been use to detect patency of intracerebral vessels. Sagittal sinus thrombosis has been accurately diagnosed using color Doppler technique.

In severe respiratory disease where extra corporeal membrane oxygenation is increasingly being used, color Doppler has been found to be useful to study the collaterals vessel pathways that maintain blood flow during and after extra corporeal membrane oxygenation. Lately transcranial Doppler (TCD) has been used in diagnosis of increased intracranial pressure leading to vascular spasm. TCD utilizes high frequency (2.25 Mhs) probe. In near future with development of technology, it may prove to be an effective tool for evaluation of intracerebral circulation.

6

Peripheral Chest

Introduction

Ultrasonography has emerged as a useful tool in evaluating a wide range of perplexing clinical problems of the chest. The presence of fluid in the pleural space, tumor, consolidation or atelectasis in the lung parenchyma provides a window for the ultrasound to penetrate. The image, so derived can well diagnose or at least limit the differential diagnosis of the condition under consideration. Since the instrumentation is portable, it allows for a bedside examination of the critically ill patient. The procedure not only offers direct and constant visualization of pleural, parenchyma or mediastinal abnormalities, but can also facilitate and guide the safe and accurate placement of needles for diagnostic or therapeutic aspiration of pleural, parenchyma and mediastinal abnormalities, and the role and methodology of ultrasonographic-aided invasive interventions in the diagnosis and treatment.

Peripheral Chest

HRSG is excellent modality for evaluation of peripheral chest, masses like abscess formation, Hematoma, Parasitic cyst in the chest wall, vascular masses like hemangioma, cavernous venous malformation, Arteriovenous malformation, metastatic tumors or deposits and early rib infections like osteomylitis and bony tumors of the rib. HRSG can tell local involvement of the chest wall or distant involvement of other organ in the same sitting. Congenital anomalies like cystic hygromas can be very well evaluated on HRSG.

Pleural Space

The pleural space is best visualized by either direct (thoracic) approach using high frequency (5 MHz) linear transducer or indirect (abdominal) approach using lower frequency (3.5 MHz) sector transducer. In the normal lung an examination through thoracic approach while holding transducer perpendicular to the intercostals spaces, the ribs show as rounded echogenecities with prominent acoustic shadowing, whereas parietal and visceral pleura are seen as thin bright echogenic lines separated by a dark line of normal pleural fluid. The air filled lung causes reflection of the sound beam producing a pattern of bright echoes which diminish with increasing distance from a transducer – an indicator of the air filled lung. In the abdominal (indirect) approach the normal air filled lung above the bright echogenic diaphragm is an indicator of absence of pleural fluid. A static scan encompassing large region of interest involving lower thorax and upper abdomen showing normal appearances.

A pleural effusion on ultrasonography is recognized as an area of echolucency separating the parietal and visceral pleura. With the abdominal approach, the presence of hypoechoic fluid above the diaphragm, the absence of the mirror image reflection of the liver or spleen above the diaphragm or visualizations of lung through the window created by fluid collection, are the findings that indicate the presence of a pleural effusion. As opposed to the criteria for fluid masses in the abdomen, posterior echo accentuation does not serve as an indicator of fluid in the chest because of air around the lesion. Lymphoma and leukemic involvement may show as anechoic lesion like fluid, hence warranting an optical gray scale setting using varying frequency transducers. On the contrary the complex pleural fluid shows up as echogenic, making it difficult to differentiate from solid tissue. However, the ultrasonographic examination is often able to detect prominent septation of pleural spaces, which is usually missed on chest radiography or CT examination. Discrimination of pleural fluid to be transudates or exudates is usually not possible on ultrasonographic appearance alone and requires a diagnostic thoracocentesis. However, the ultrasonography can determine the appropriate site and approach for aspiration of fluid from the pleural space.

This modality also offers the means to characterize the focal pleural lesions, which may be a loculated effusion, thickened fibrous pleura of a solid pleural tumor. Presence of fluid in a lesion is suggested by moving septation or floating echogenic shadows in the lesions and change in shape with respiration. Linear plaque like calcification of pleural thickening of punctate internal calcification of the tumor further

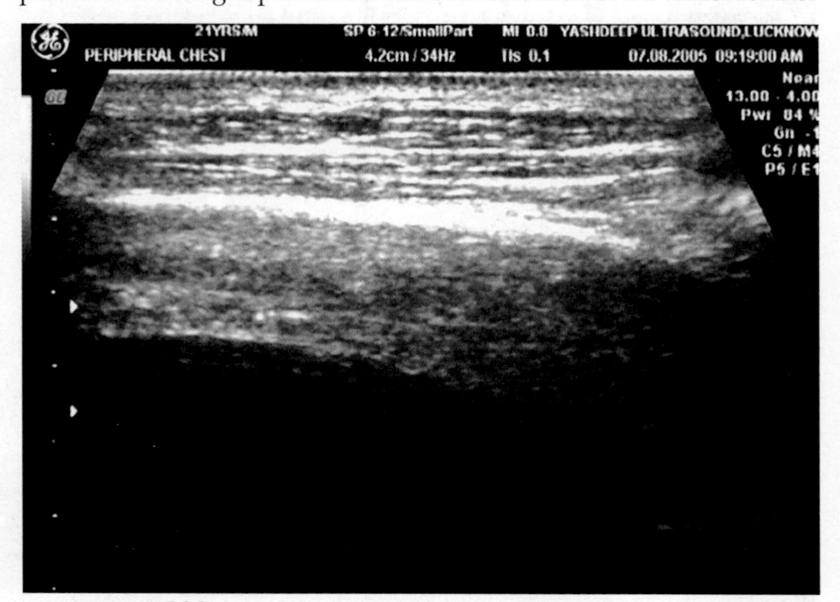

FIGURE 6.1: HRSG shows bright echogenic line representing the pleura. Parietal and visceral layers are not separated in normal conditions. The peripheral lung is seen bright soft tissue interface on high-resolution ultrasound.

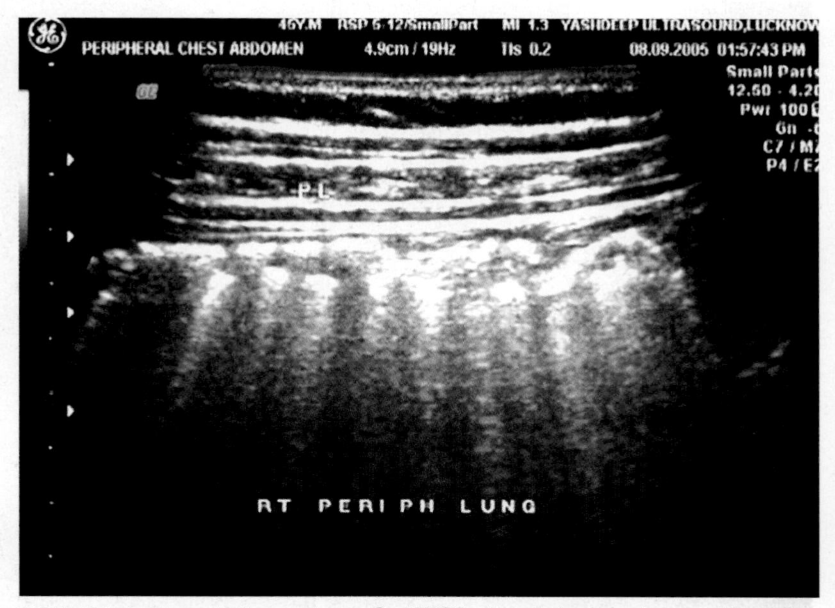

FIGURE 6.2: Normal pleura chest. On HRSG pleural shadow shows artifacts due to interposing ribs. Rib shadow casting the acoustic shadowing.

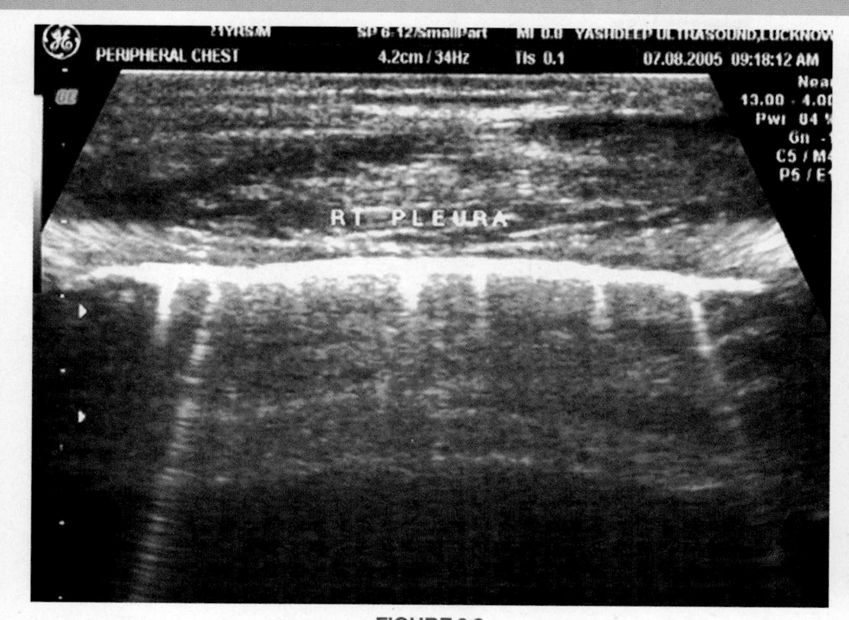

FIGURE 6.3

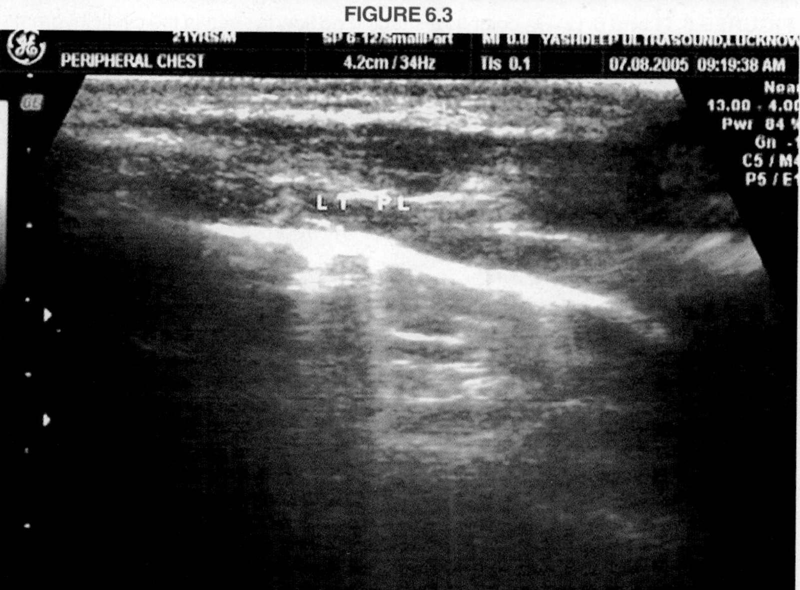

FIGURE 6.4

FIGURES 6.3 and 6.4: Normal peripheral lung. HRSG shows bright interface of the peripheral lung. Bright echogenic shadows are seen in the peripheral lung with comet tail artifacts due to air filled lung.

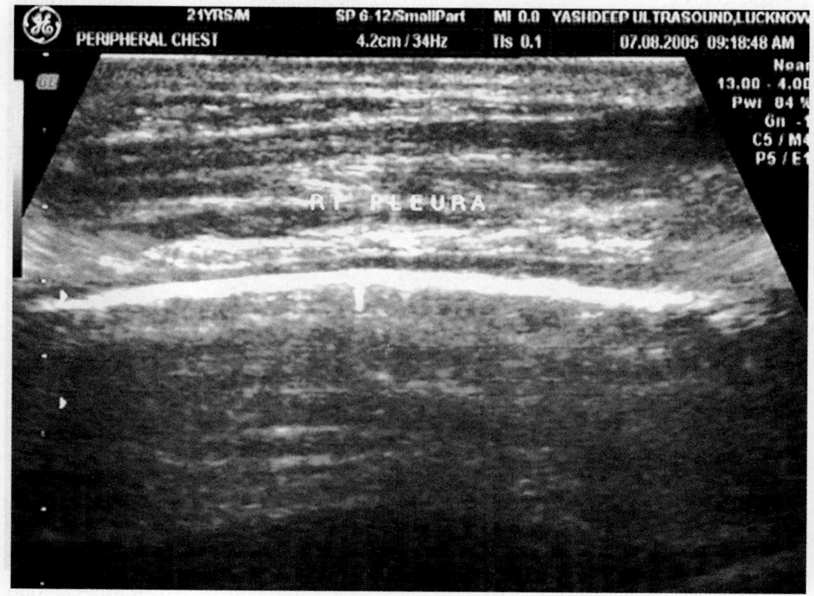

FIGURE 6.5

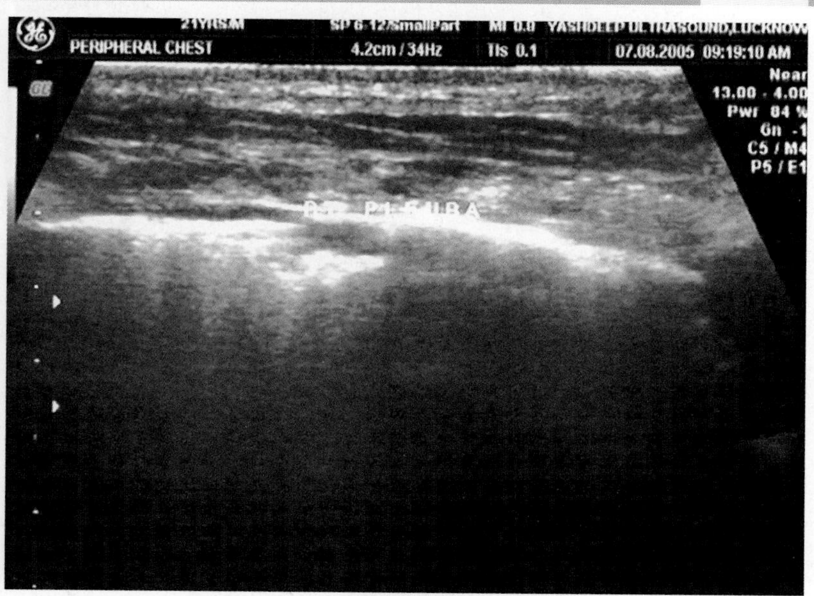

FIGURE 6.6

FIGURES 6.5 and 6.6: Normal chest. Subcutaneous tissue is well visualized. The echogenic strip of bright echoes shows the pleural layer. Reverberation artifacts are seen posterior to the pleural shadow.

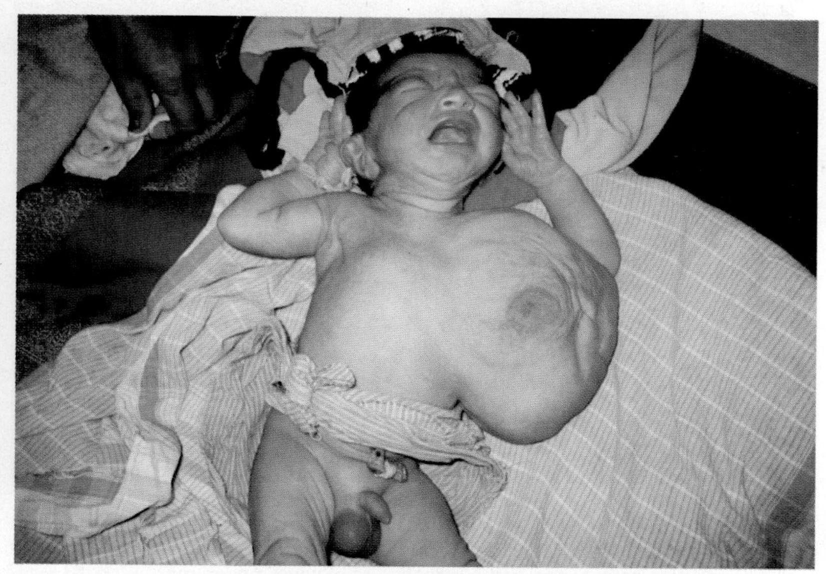

FIGURE 6.7: Cystic hygroma fixed with the chest wall. A child is born with big soft tissue mass hanging from the Lt anterior chest wall.

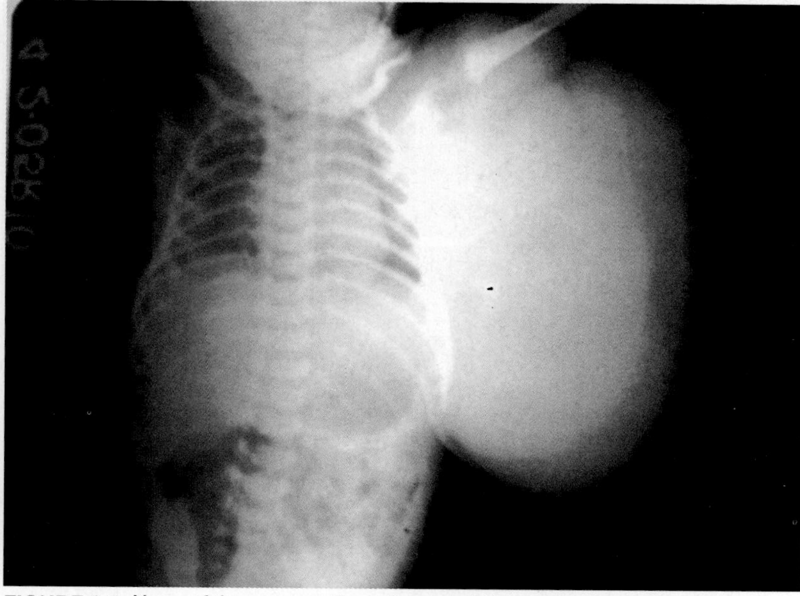

FIGURE 6.8: X-ray of the same patient shows soft tissue shadow over the Lt thorax.

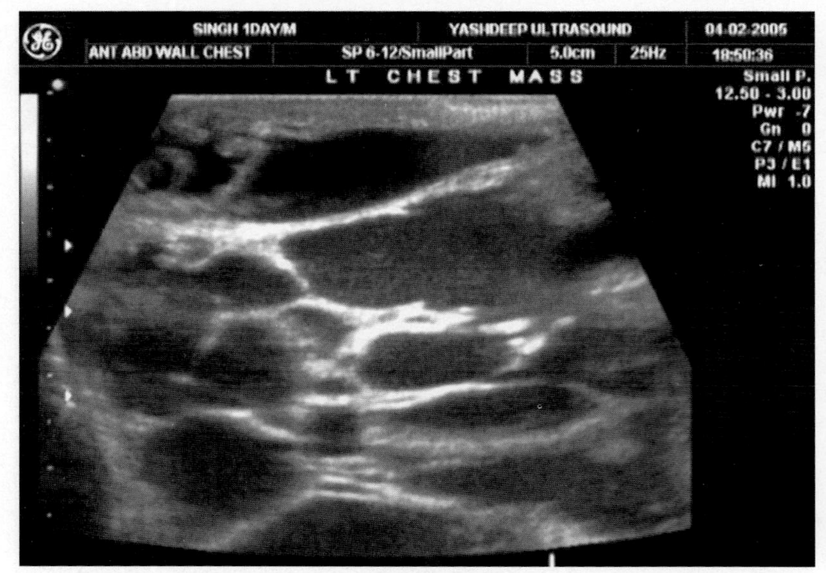

FIGURE 6.9

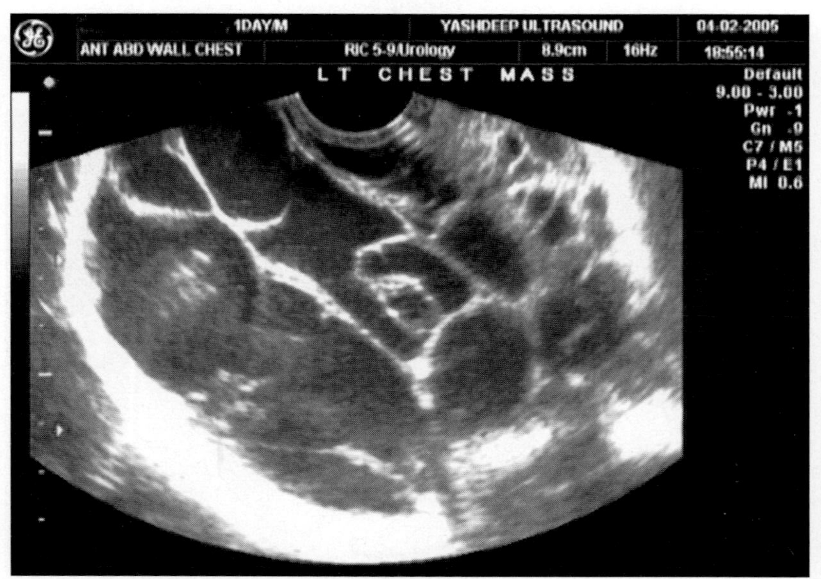

FIGURE 6.10

FIGURES 6.9 and 6.10: HRSG of the same patient multi septate cystic mass fixed with the Lt anterior chest wall. Bright echogenic septa are seen in the cyst. Thick low-level collection is also seen in the cyst suggestive of big cystic hygroma.

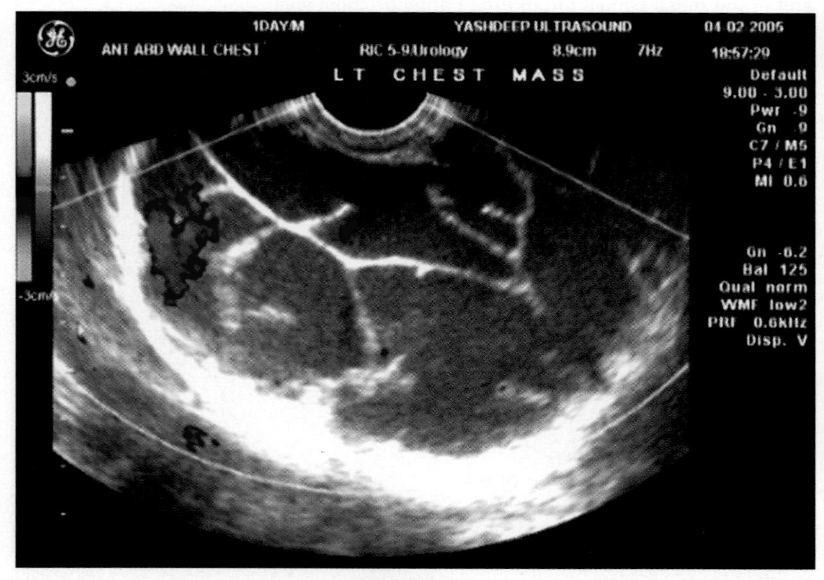

FIGURE 6.11

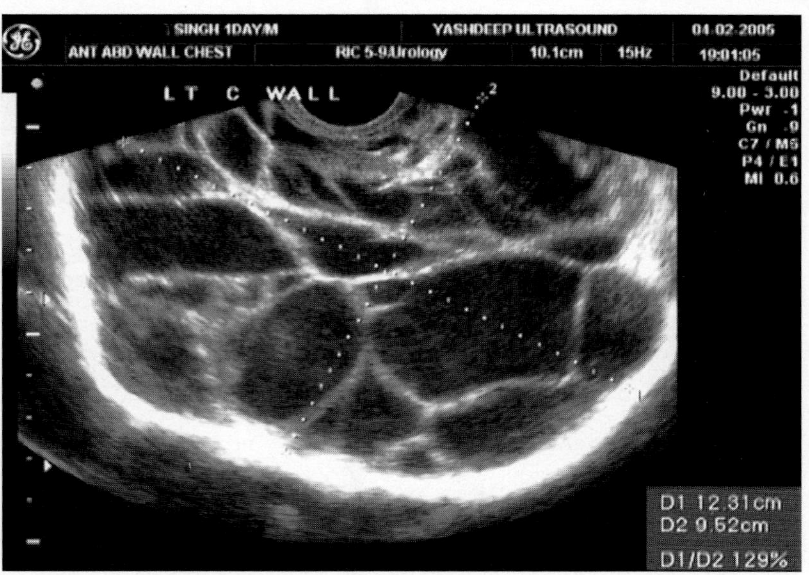

FIGURE 6.12

FIGURES 6.11 and 6.12: On color flow imaging no flow is seen in the cystic hygroma.

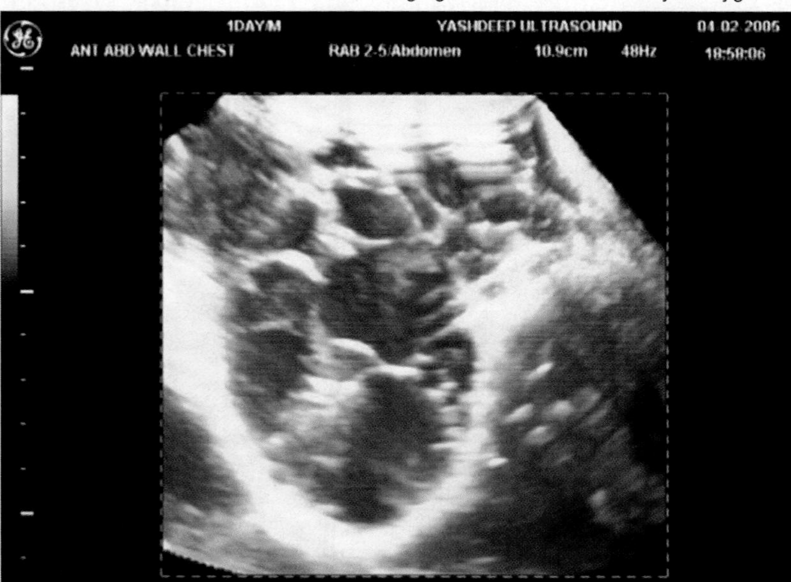

FIGURE 6.13

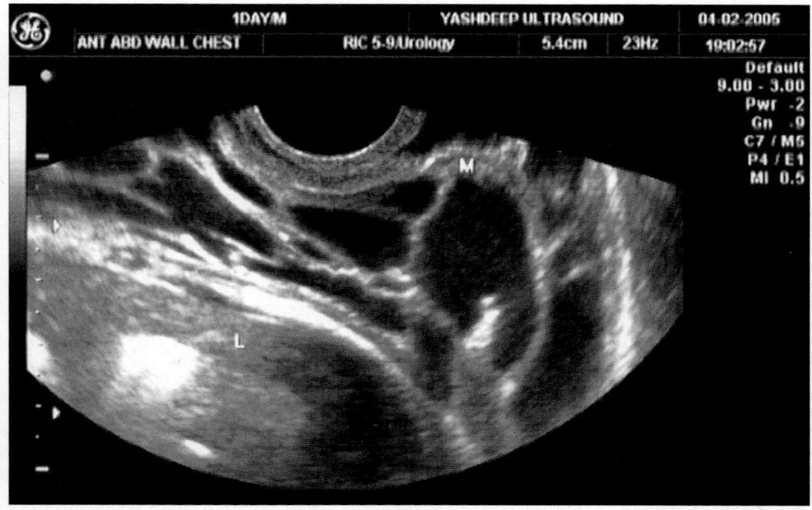

FIGURE 6.14

FIGURES 6.13 and 6.14: 3D imaging of the same patient shows details of the cystic hygroma. It is not seen fixed with the chest wall muscles. The peripheral lung appears to be normal.

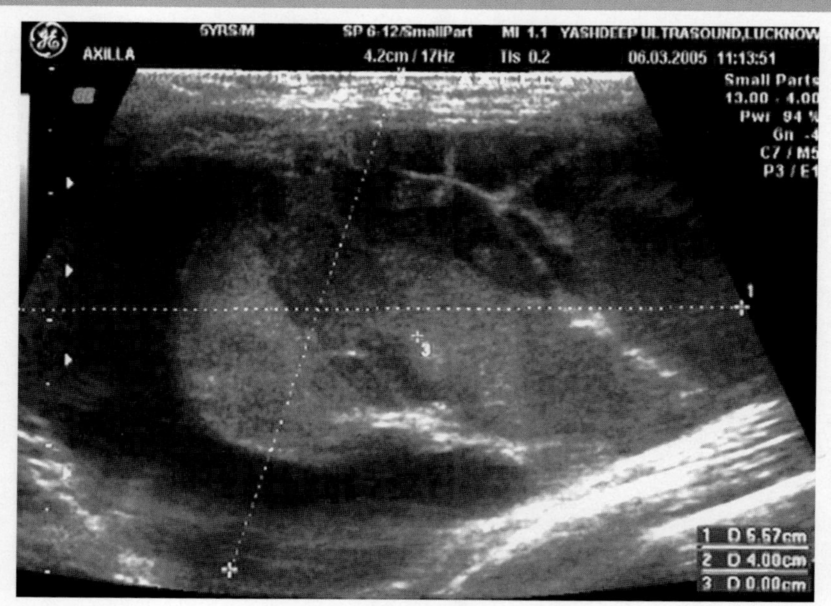

FIGURE 6.15

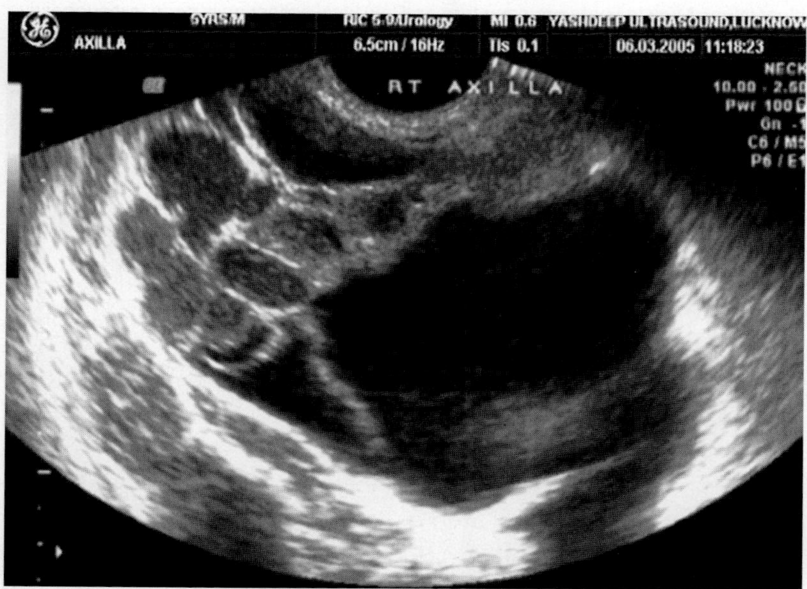

FIGURE 6.16

FIGURES 6.15 and 6.16: Cystic hygroma Rt chest wall and axilla. A child presented with soft tissue mass over the Rt axilla and chest wall. HRSG shows a big multi septate cystic mass fixed with anterior chest wall. Thick jelly like fluid is seen filling the cyst suggestive of a big cystic hygroma.

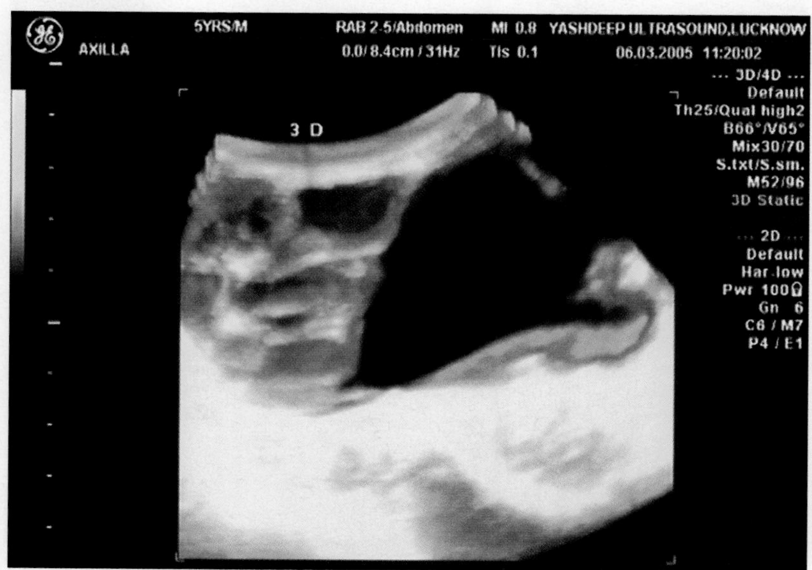

FIGURE 6.17: 3D imaging of the cyst mass shows details of the cystic hygroma.

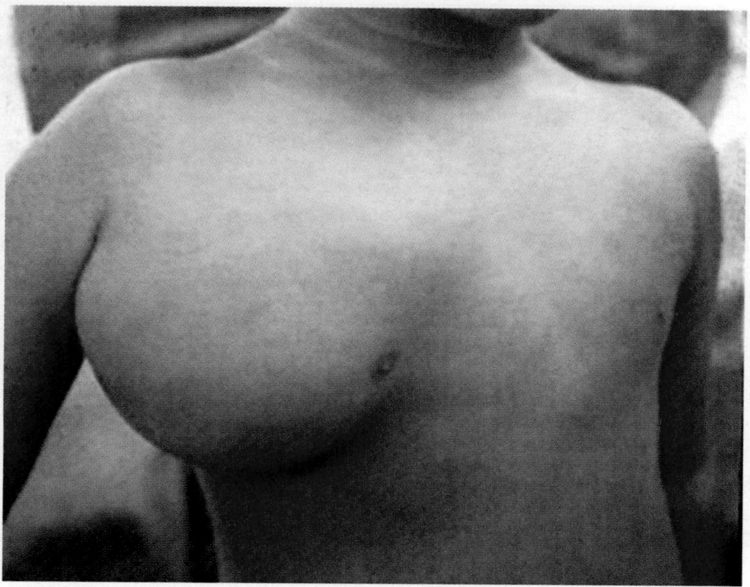

FIGURE 6.18: Cystic hygroma anterior chest wall. A young boy presented with a soft tissue cystic mass in the Rt axilla and anterior chest wall.

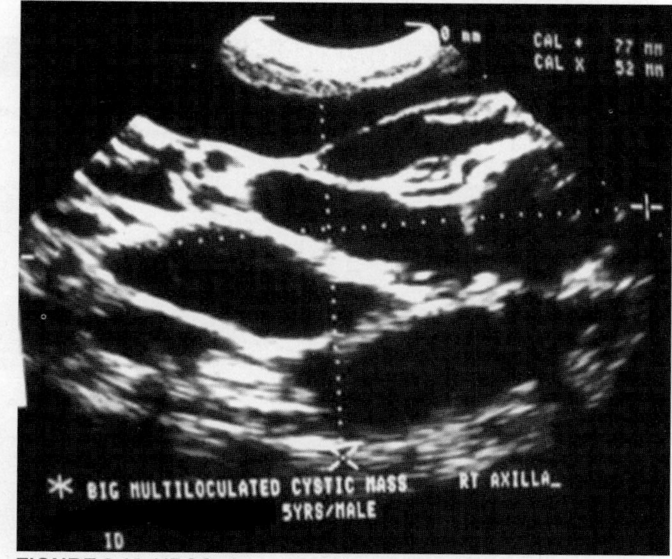

FIGURE 6.19: HRSG shows a multi septate multi loculated soft tissue mass. The septa are well defined. Locullae are anechoic. However, few locullae show internal echoes, typical finding of cystic hygroma.

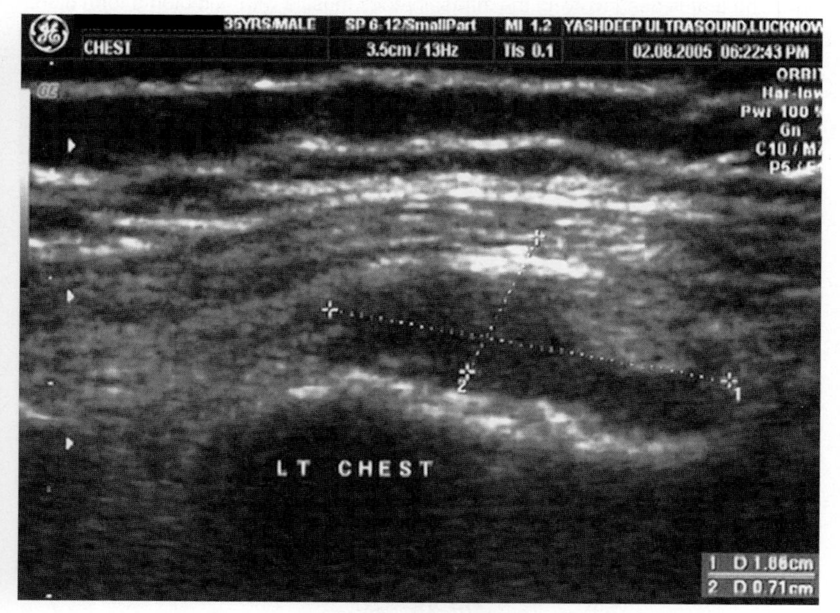

FIGURE 6.20A

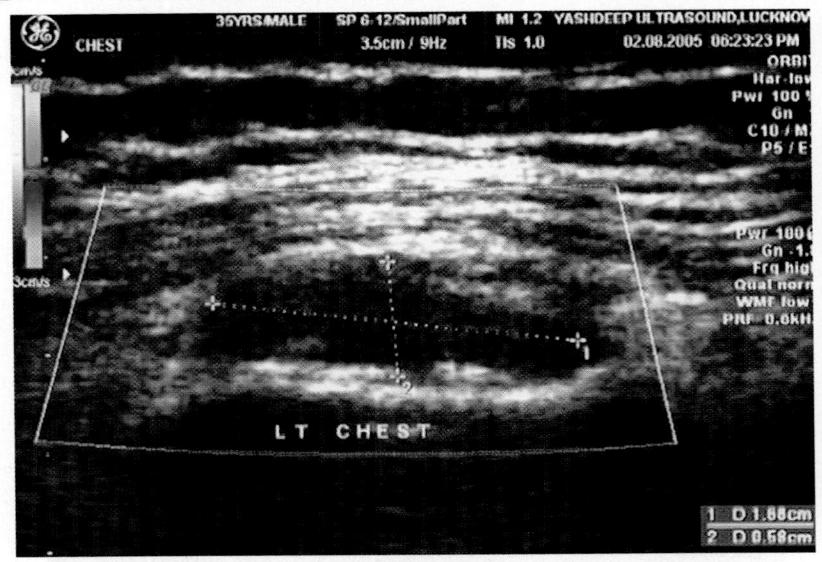

FIGURE 6.20B

FIGURES 6.20A and B: Anterior chest wall hematoma with tear. HRSG shows a hypoechoic tear with low level collection in a patient sustained blunt injury to the anterior chest wall. A big hypoechoic cleft seen in the chest wall muscle. On color flow imaging no flow seen in it.

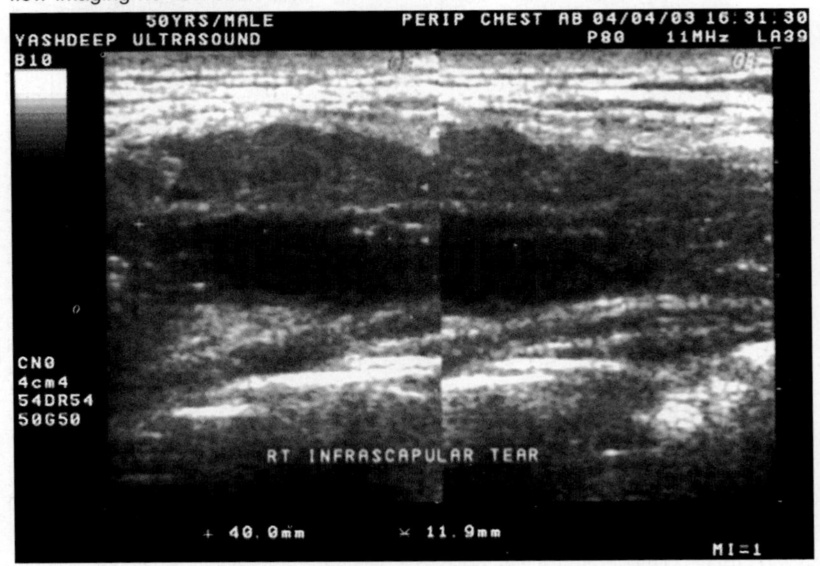

FIGURE 6.21: Big tear in the infrascapular region. A big hypoechoic cleft is seen in the Infrascapular region. Muscle fibers disruption with thick collection is seen in the tear. The tear is see involving the infraspinatous muscle with hematoma formation.

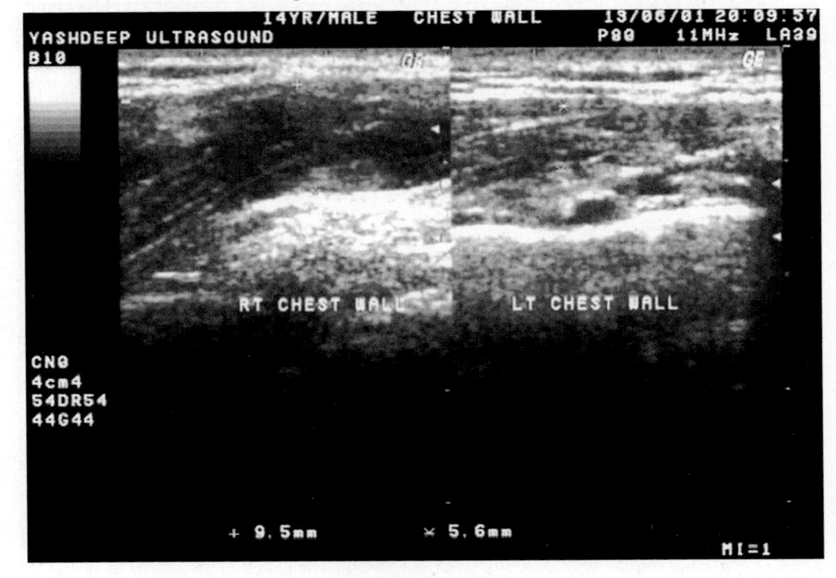

FIGURE 6.22A

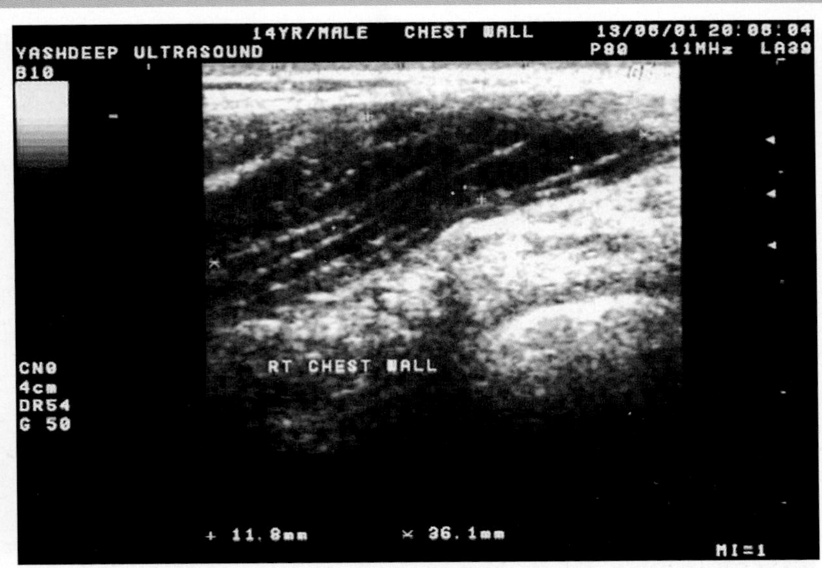

FIGURE 6.22B

FIGURES 6.22A and B: Chest wall tear. HRSG shows chest wall muscle tear in a blunt injury. Muscle fibers are seen torn and low level collection is seen in it suggestive of hematoma formation. Lt chest wall muscle shows normal texture.

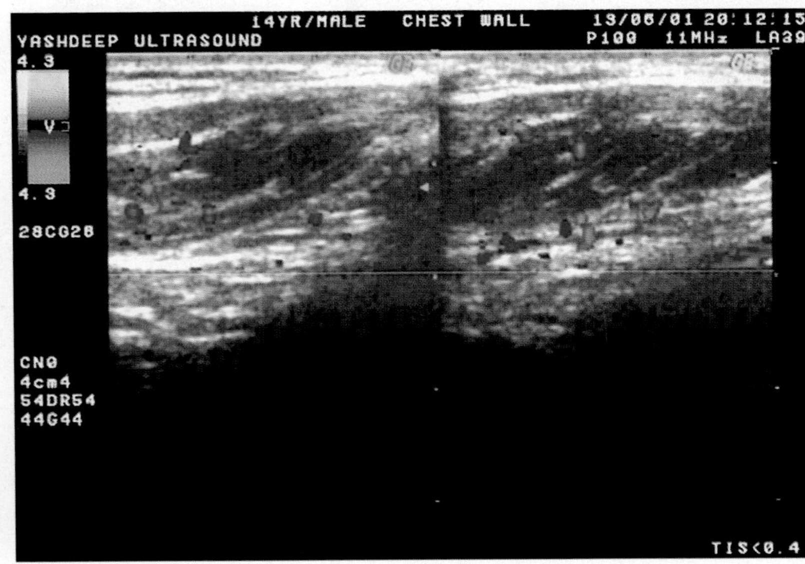

FIGURE 6.23: Color flow imaging does not show any flow in the hematoma.

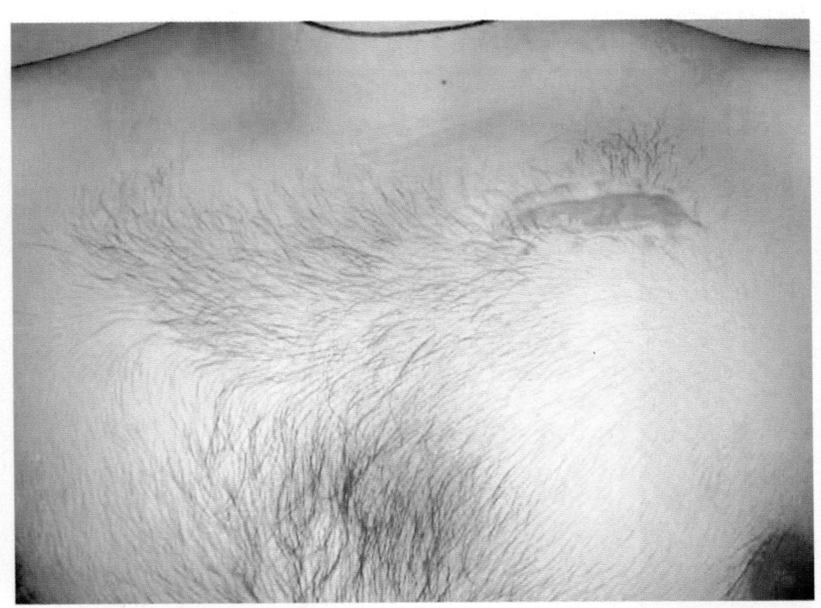

FIGURE 6.24: Tubercular abscess anterior chest wall muscle. A young boy presented with soft tissue swelling over the Lt anterior chest wall.

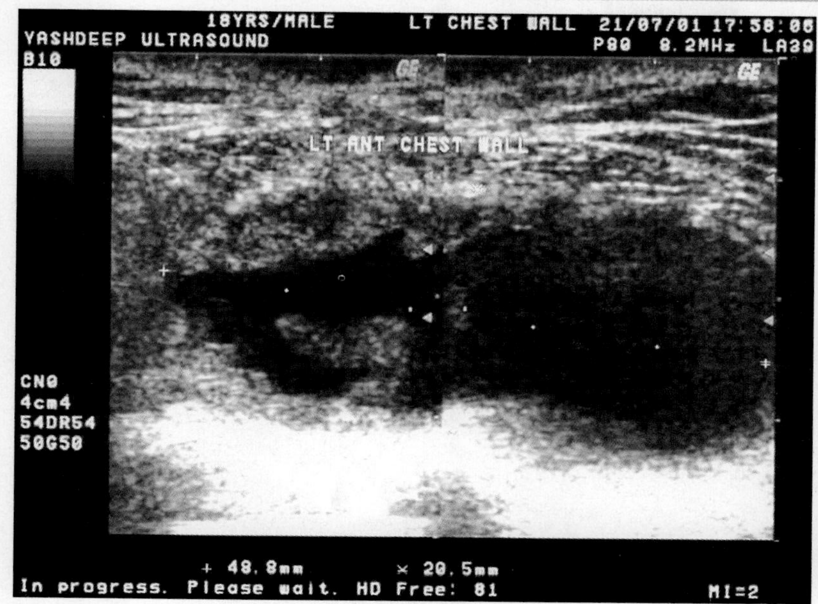

FIGURE 6.25A

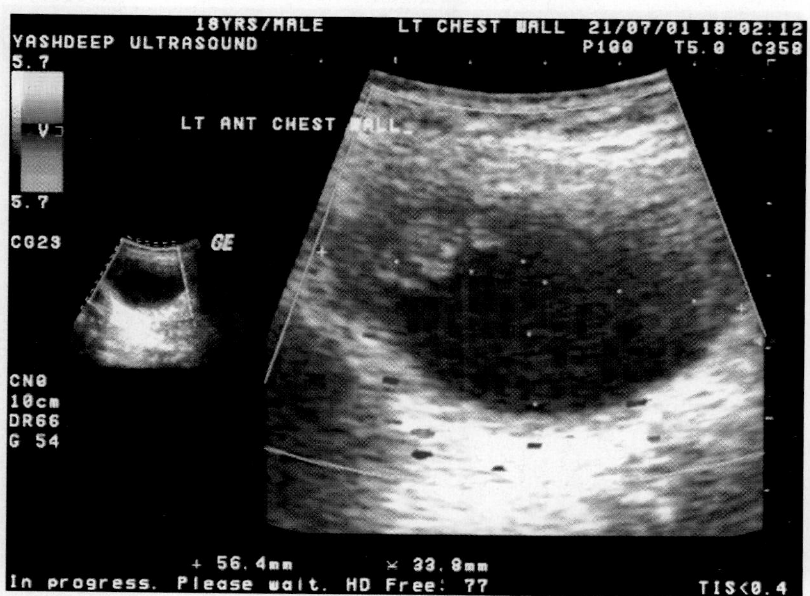

FIGURE 6.25B

FIGURES 6.25A and B: HRSG shows low-level echo collection in the muscle. Thick echogenic wall is seen around it. Debris is also seen. No perifocal edema is present indicating a cold abscess.

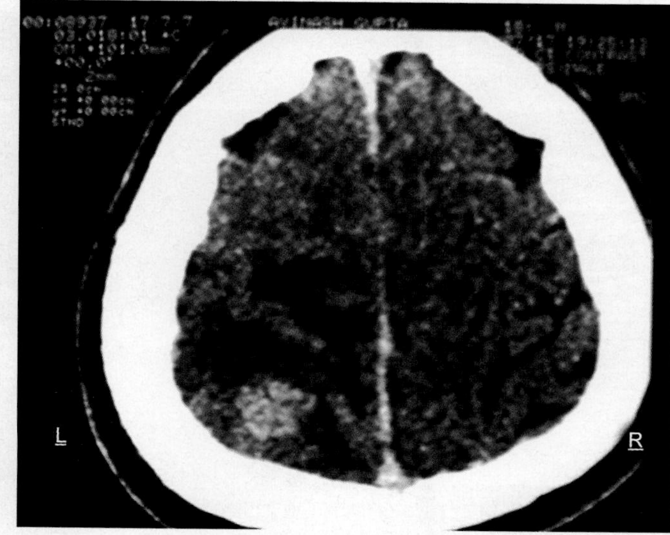

FIGURE 6.26: Cranial CT of the same patient shows a hypodense focus in the Lt parietal lobe with perifocal edema suggestive of tuberculoma.

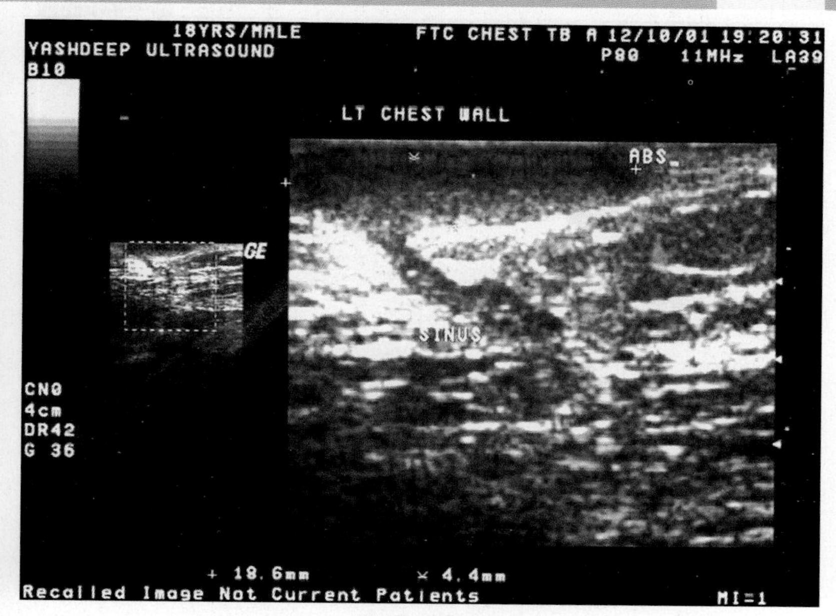

FIGURE 6.27A

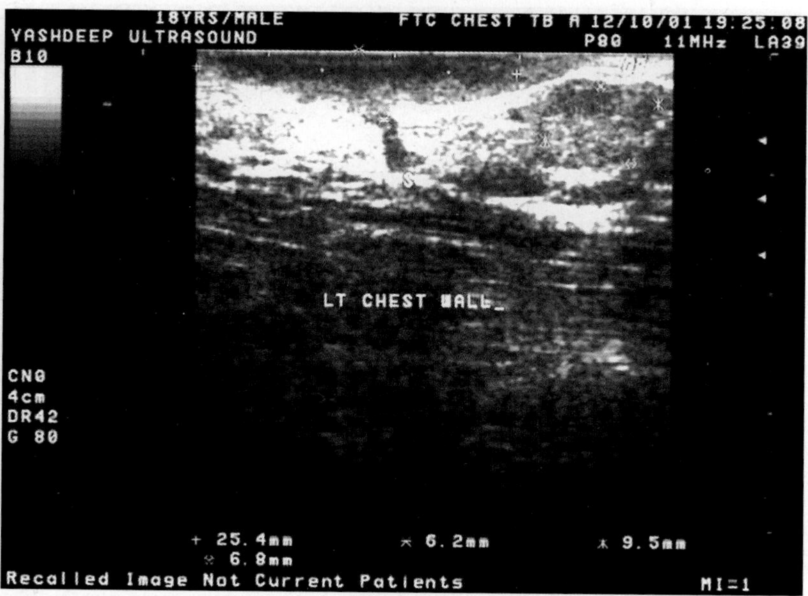

FIGURE 6.27B

FIGURES 6.27A and B: FTC of tubercular abscess anterior chest wall muscle. The abscess cavity in the same patient has reduced in size after three months of antitubercular treatment. HRSG shows thick walled cavity reduced in size with echogenic debris in it suggestive of organization of the abscess. The abscess cavity is showing a sinus tract going to the chest wall from the superficial abscess.

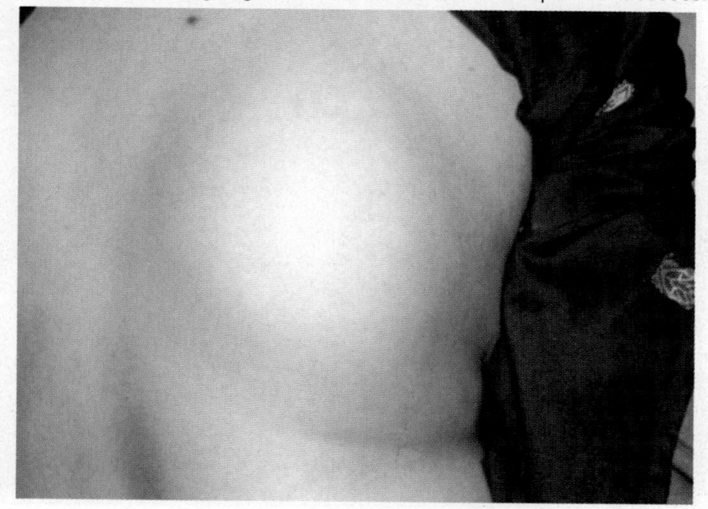
FIGURE 6.28: Tubercular abscess over the Rt scapular region. A young girl presented with soft tissue non-tender mass lesion over the Rt Infrascapular region over the posterolateral chest wall.

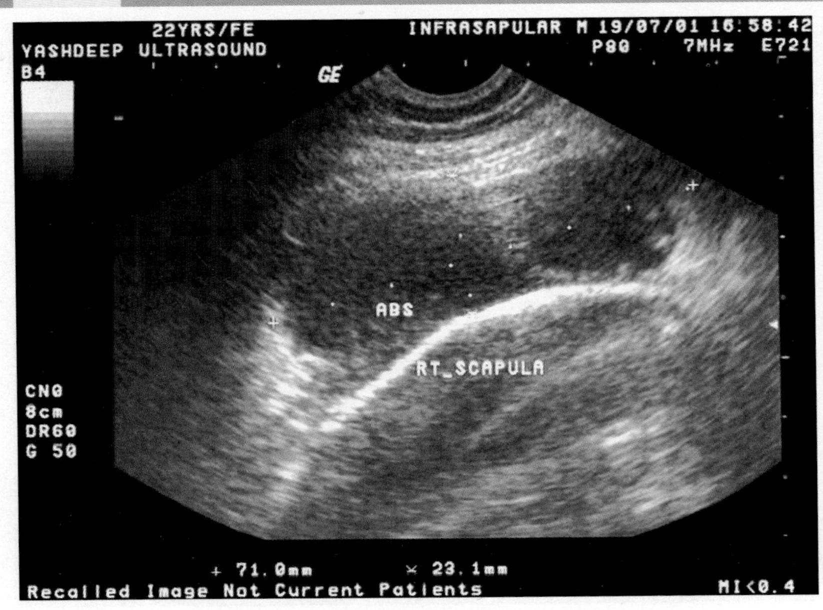

FIGURE 6.29

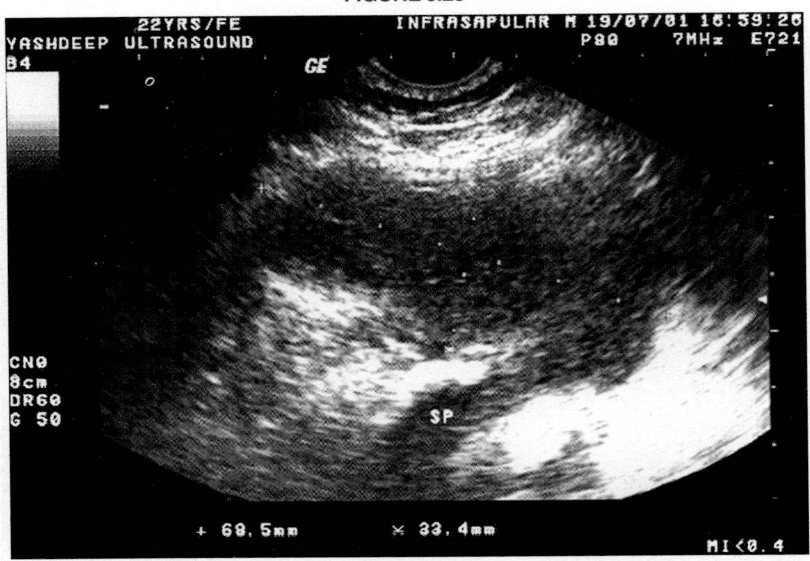

FIGURE 6.30

FIGURES 6.29 and 6.30: HRSG shows a low level echo complex mass over the infrascapular region. No perifocal edema is present. The inferior border of the scapula shows bony destruction. The collection turns out to be chronic inflammatory abscess suggestive of tubercular abscess.

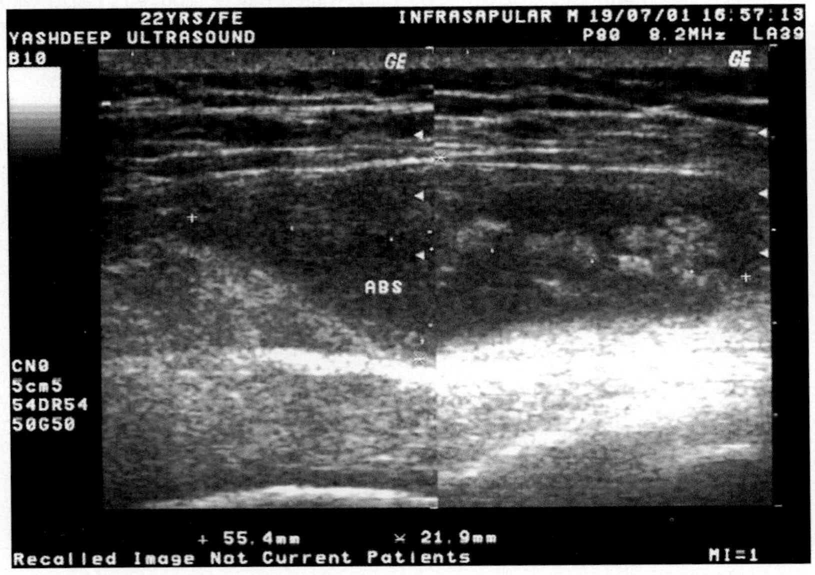

FIGURE 6.31: The abscess is also seen extending in to the Rt paraspinal muscles. Central area of necrosis is seen in it. But no perifocal edema is present.

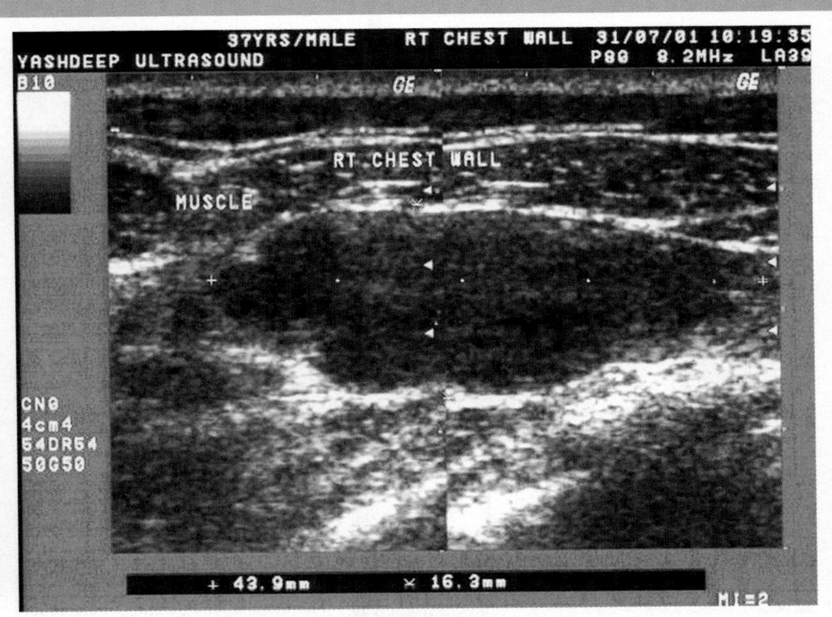

FIGURE 6.32

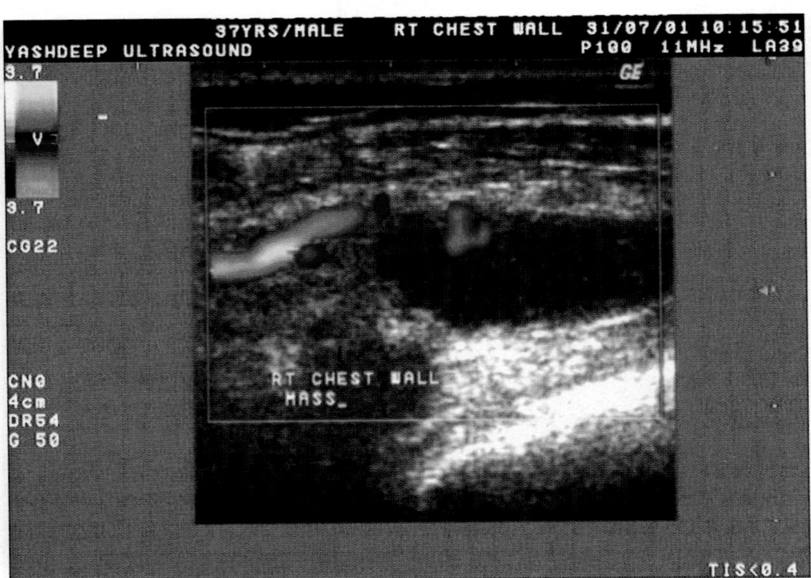

FIGURE 6.33

FIGURES 6.32 and 6.33: Tubercular abscess Rt anterior chest wall. HRSG shows a well-defined low-level echo complex mass confined to the Rt anterior chest wall in infraclavicular region. Central area of necrosis is also seen in it. On color Doppler imaging poor flow is seen in the abscess. A peripheral feeding vessel is seen in it. Biopsy of the mass shows tubercular abscess.

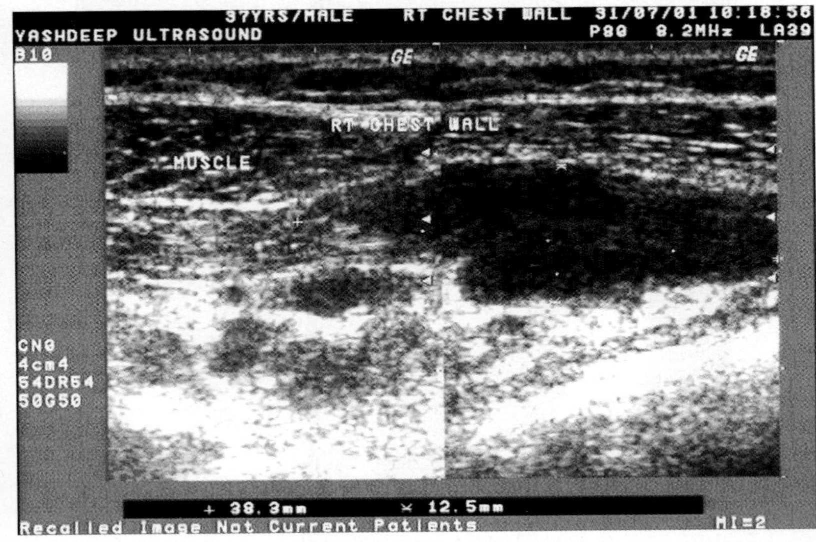

FIGURE 6.34: HRSG shows the abscess is confined to the muscle belly. Adjacent normal muscle is also seen having normal bipinnate pattern.

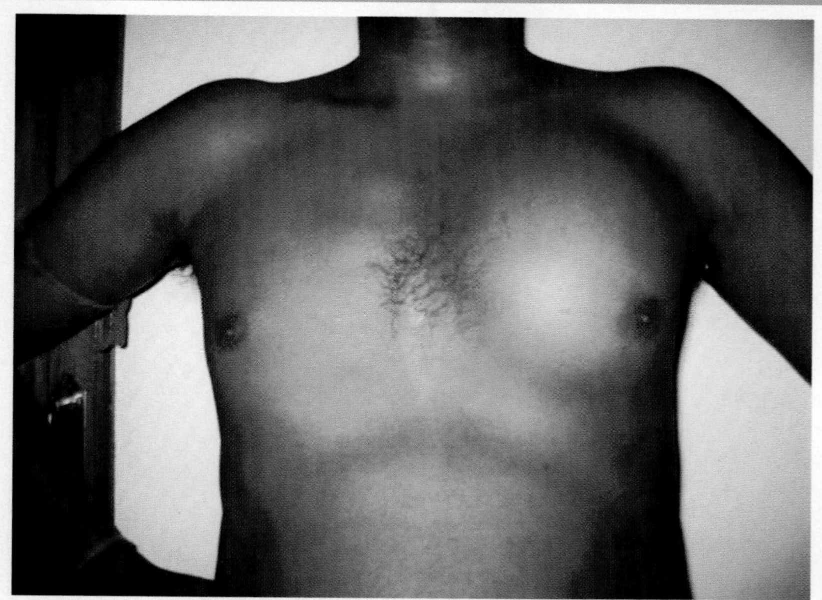
FIGURE 6.35: **Tubercular abscess anterior chest wall.** A young man presented with painless non-tender soft tissue mass over the Lt anterior chest wall.

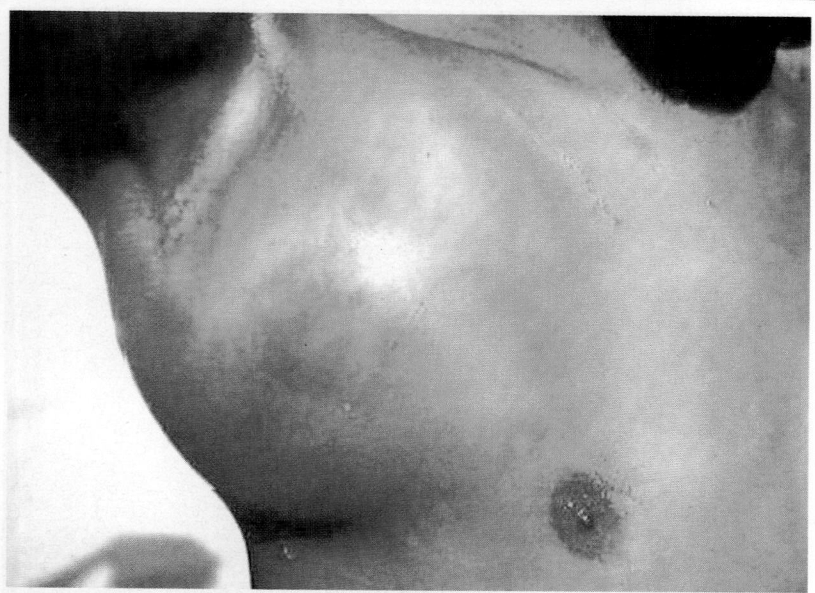
FIGURE 6.37: **Pyogenic abscess.** A young man presented with a big tender over the Rt anterior chest wall and extending to the axilla.

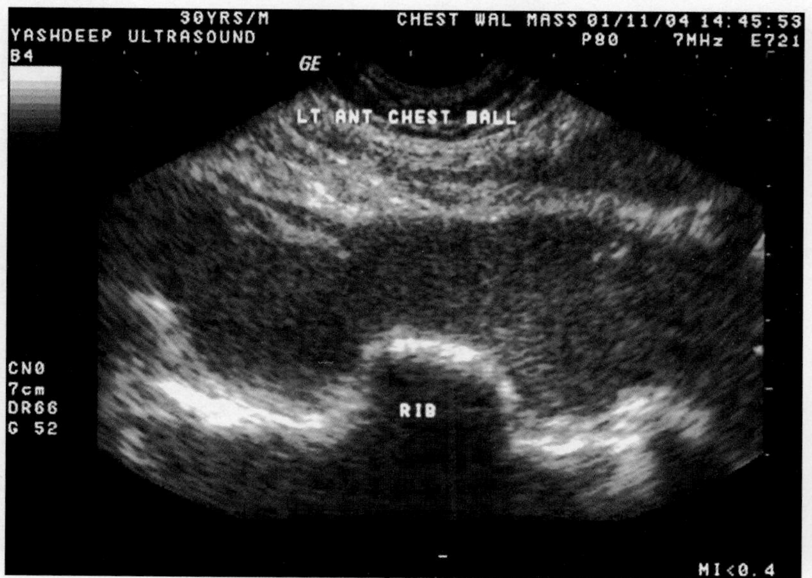

FIGURE 6.36A

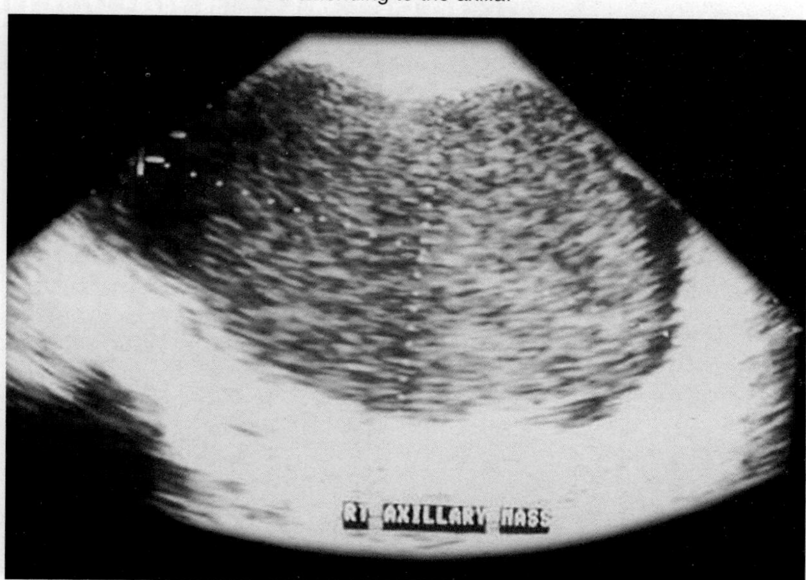

FIGURE 6.38: HRSG shows a low level echo complex mass. Homogeneous collection is seen in it. Layering sign is positive. Perifocal edema is present. No septation or loculation is seen in the mass suggestive of a big pyogenic abscess.

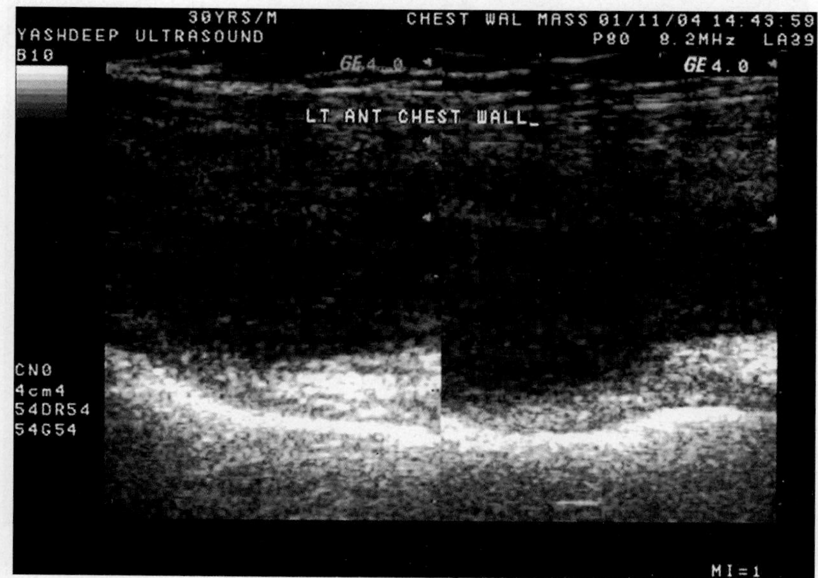

FIGURE 6.36B

FIGURES 6.36A and B: HRSG shows thick low-level echo cystic mass over the anterior chest wall. Thick collection is seen in it. However, the collection is not seen eroding the rib or the bony cage. Aspiration confirmed tubercular abscess.

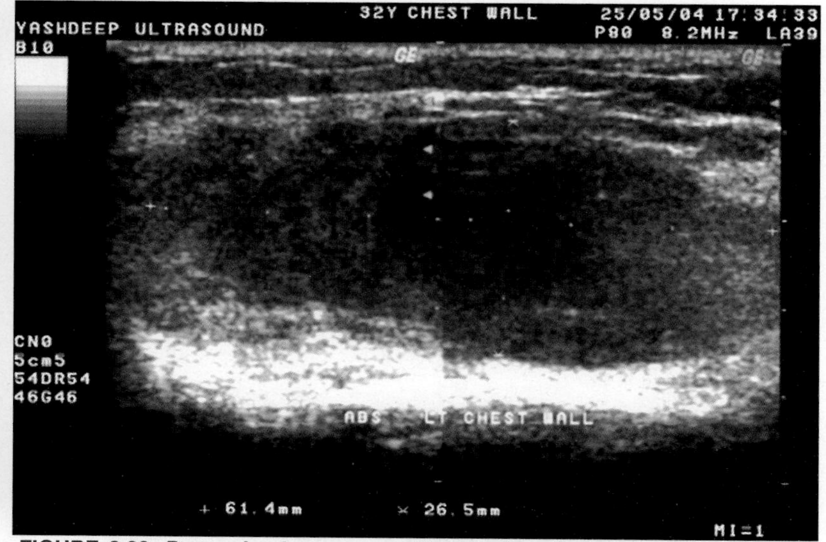

FIGURE 6.39: **Pyogenic abscess over the chest wall.** A patient presented with red hot swelling over the chest wall. HRSG shows low-level echo collection in the muscles. Perifocal edema is also present. Internal debris is seen in the collection suggestive of pyogenic collection.

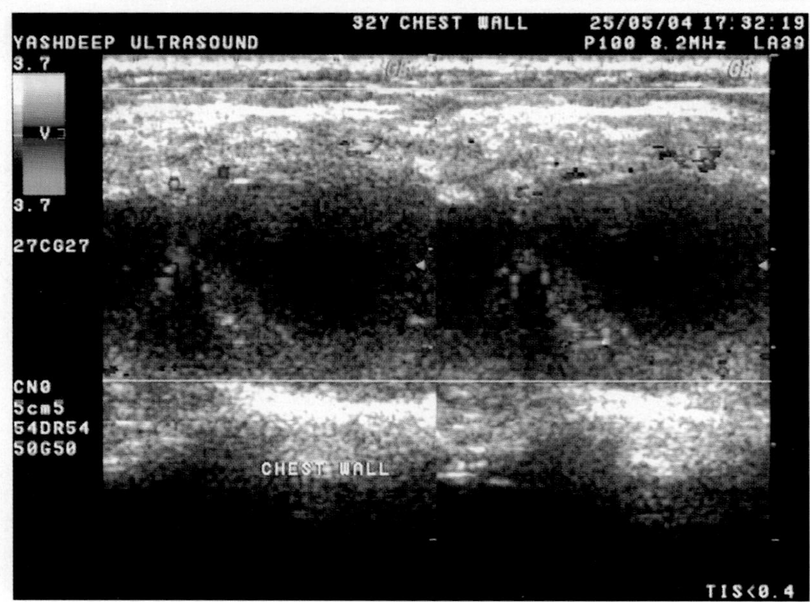

FIGURE 6.40A

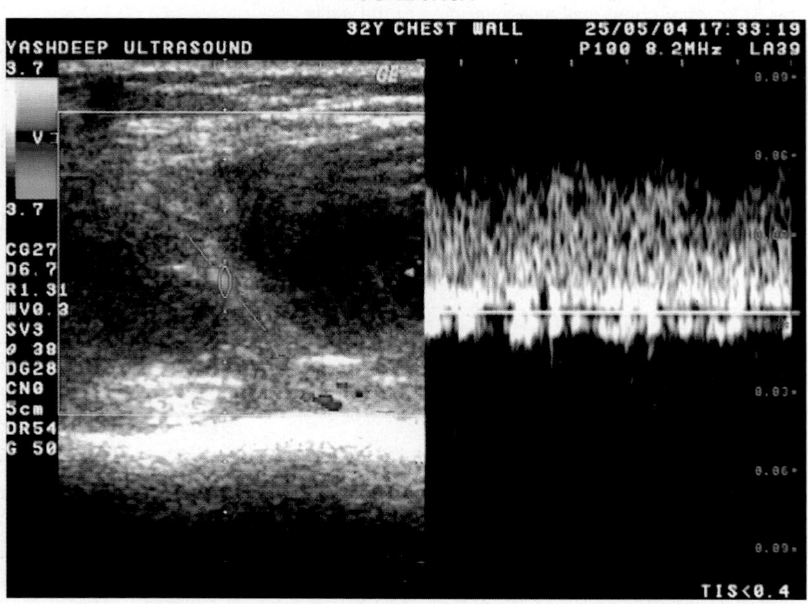

FIGURE 6.40B

FIGURES 6.40A and B. On color flow imaging capsular flow is seen in the collection. On spectral Doppler tracing high flow is seen in the vessels suggestive of hyperemia.

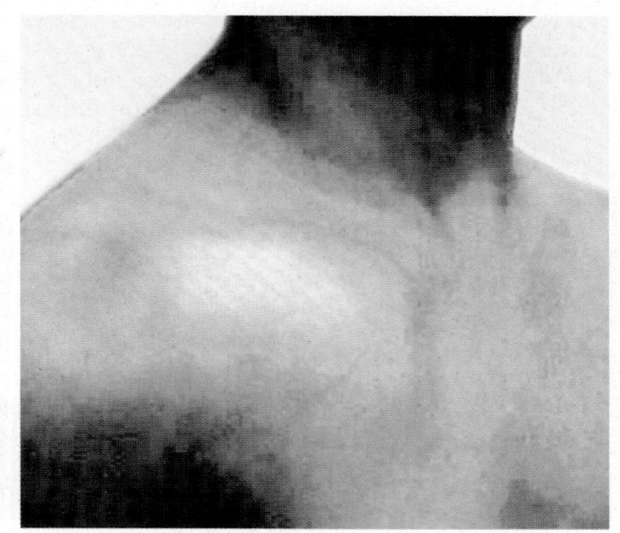

FIGURE 6.41: Hydatid cyst in the infraclavicular region. A young man presented with a soft tissue mass over the Rt infraclavicular region.

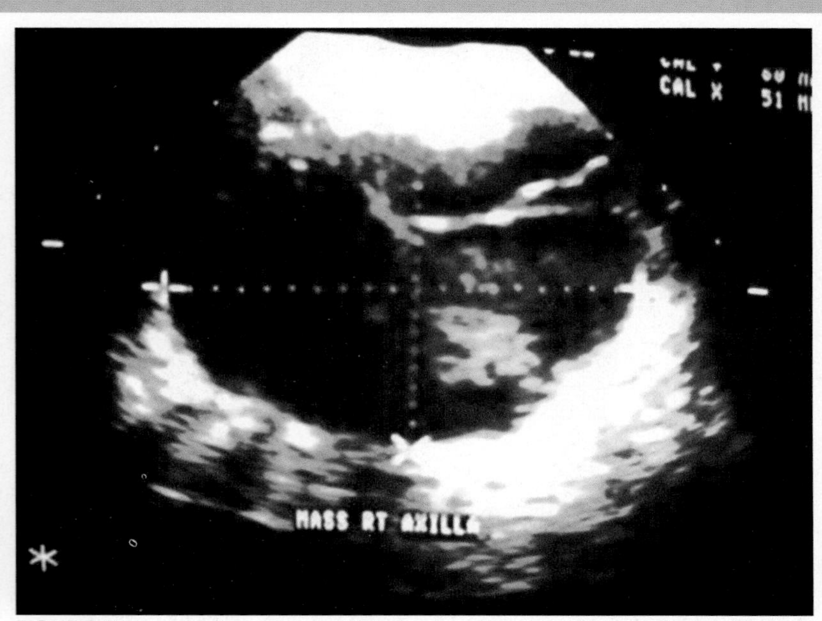

FIGURE 6.42: HRSG shows a well-defined cystic mass. Echogenic membrane is seen floating in the cyst suggestive of hydatid cyst. Low-level echocollection is also seen in it.

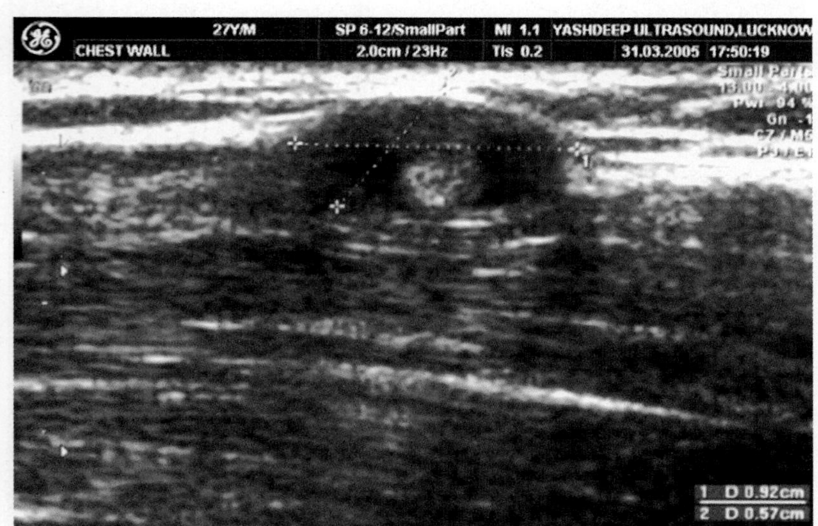

FIGURE 6.43

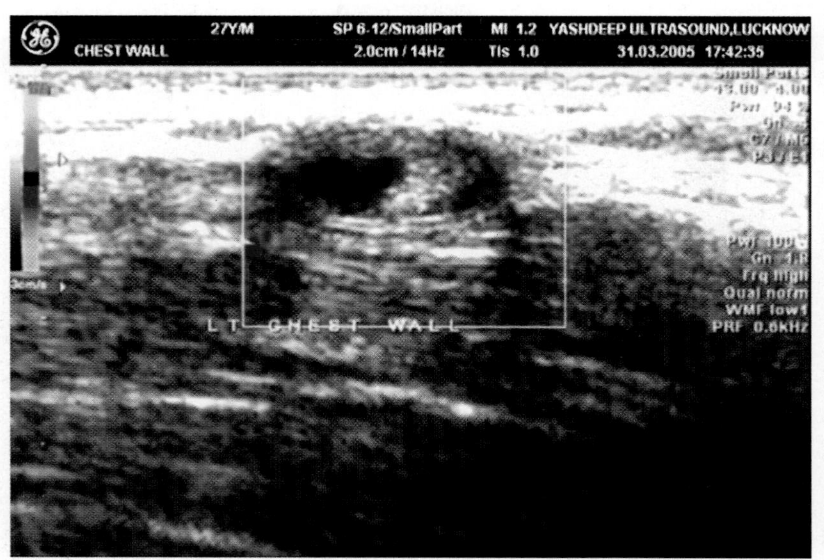

FIGURE 6.44

FIGURES 6.43 and 6.44: Cysticercus cyst in Lt anterior chest wall. A patient presented with nodular mass over the Lt anterior chest wall. HRSG shows a well-defined thick wall cystic mass in the muscle. Echogenic nodule is also seen in it fixed with the inner wall of the cyst suggestive of cysticercus cyst. On color flow imaging no flow is seen in it.

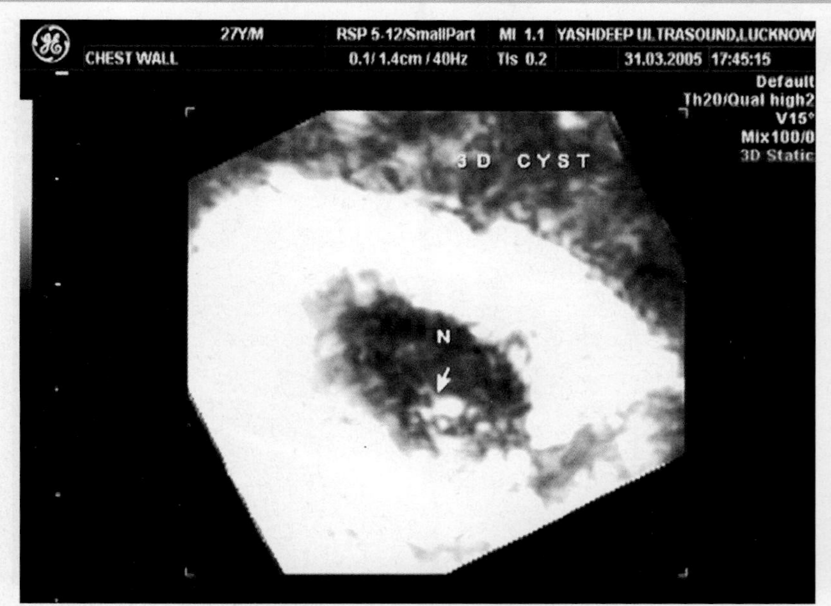

FIGURE 6.45: 3D imaging of the same patient shows better details of the cysticercus cyst with echogenic nidus fixed with the inner wall of the cyst.

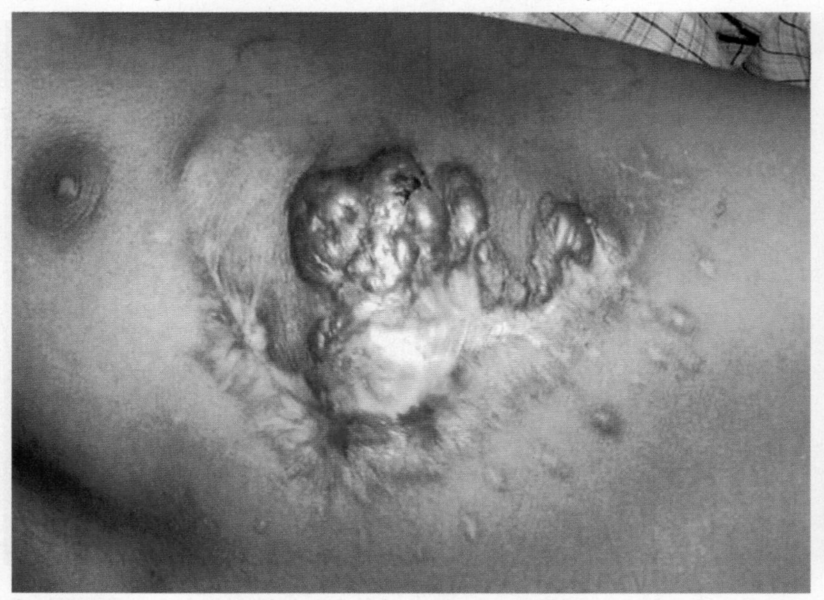

FIGURE 6.46: **Vascular mass over the anterior chest wall.** A young man presented with a vascular mass over the Lt anterior chest wall. Multiple dilated vessels are seen in the mass.

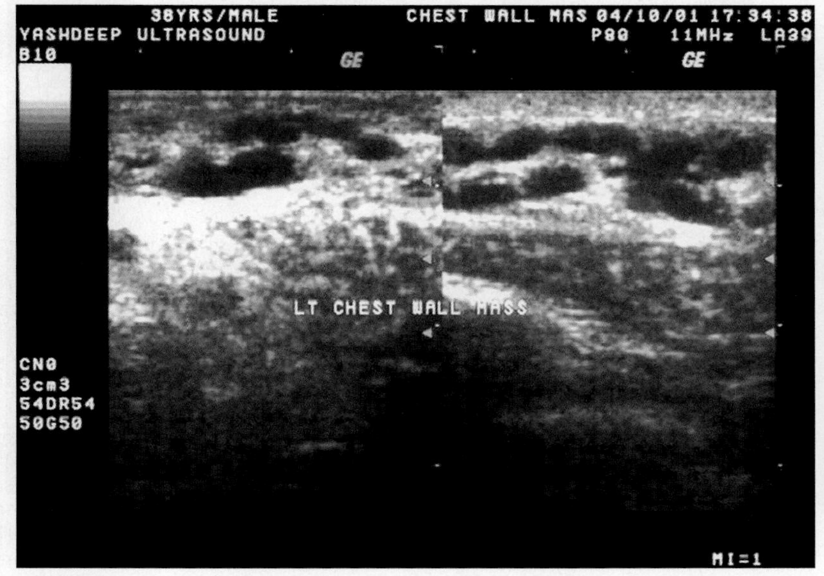

FIGURE 6.47: HRSG shows multiple dilated channels in the mass suggestive of vessels.

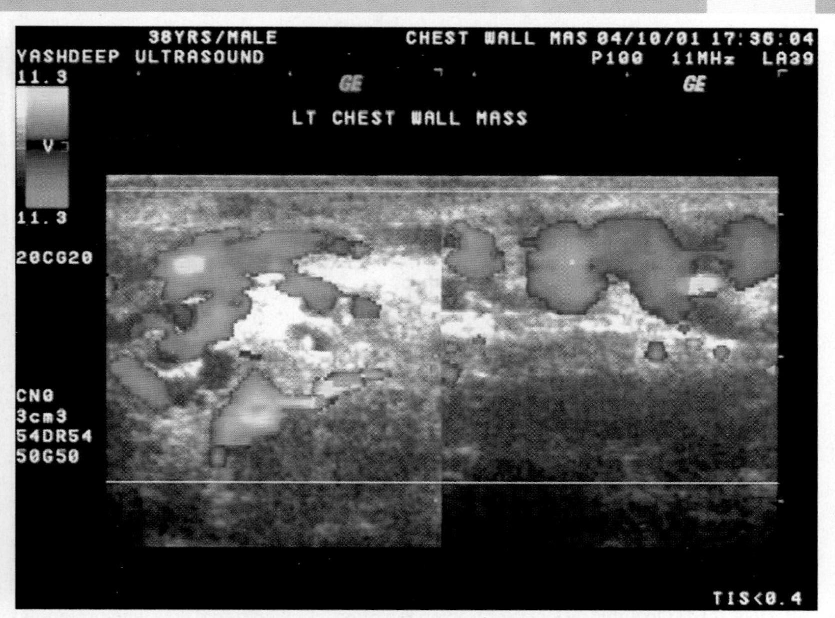

FIGURE 6.48: On color flow imaging low blood flow is seen in the vessels.

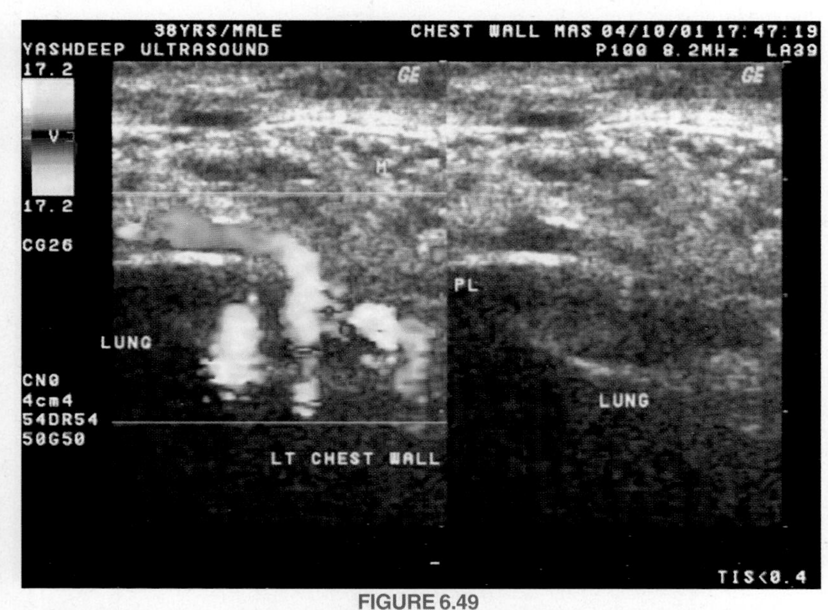

FIGURE 6.49

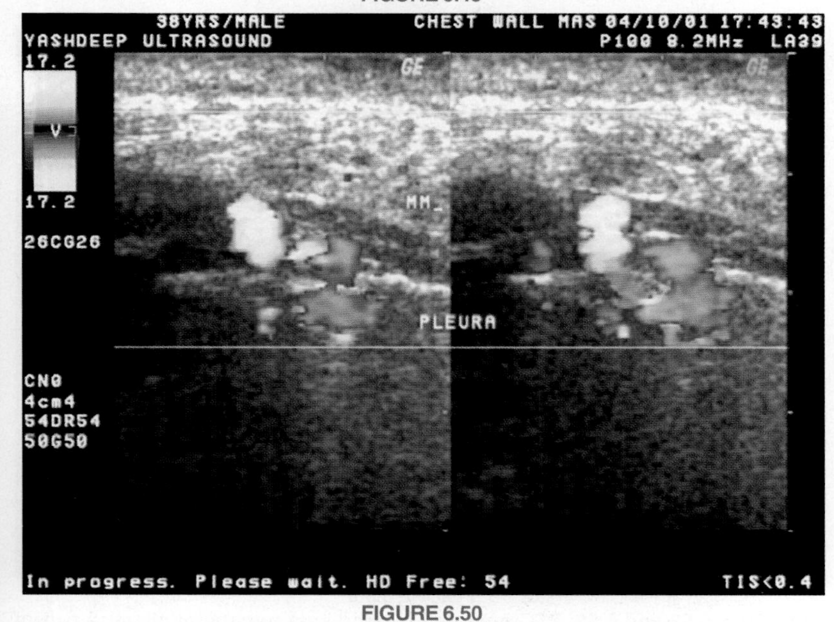

FIGURE 6.50

FIGURES 6.49 and 6.50: HRSG shows the vessels are seen invading the chest wall and also perforated the pleura. They are seen going to the peripheral lung. Therefore, it is suggestive of deeper invasion of the vascular mass.

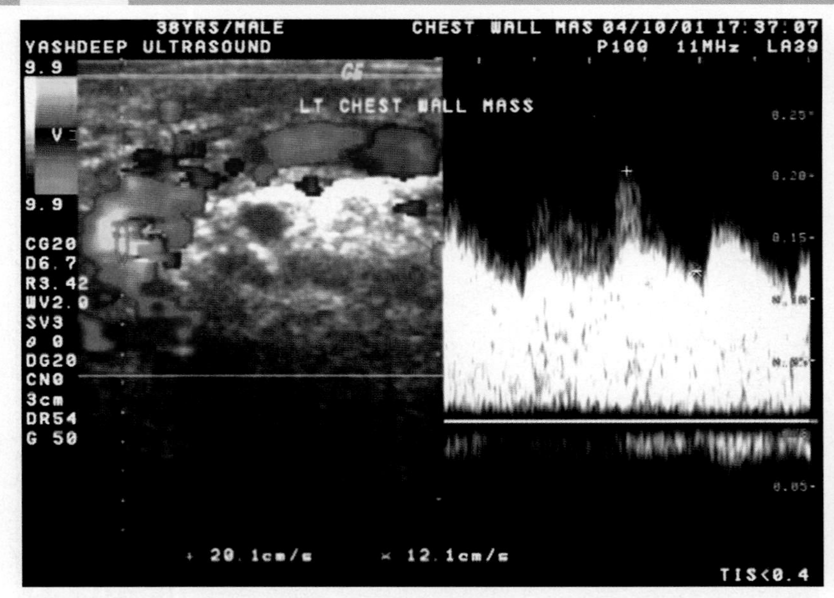

FIGURE 6.51: On spectral Doppler tracing the mass shows arterial flow with prolonged diastolic flow. Venous flow was also seen in the vessels suggestive of AV fistula in the mass.

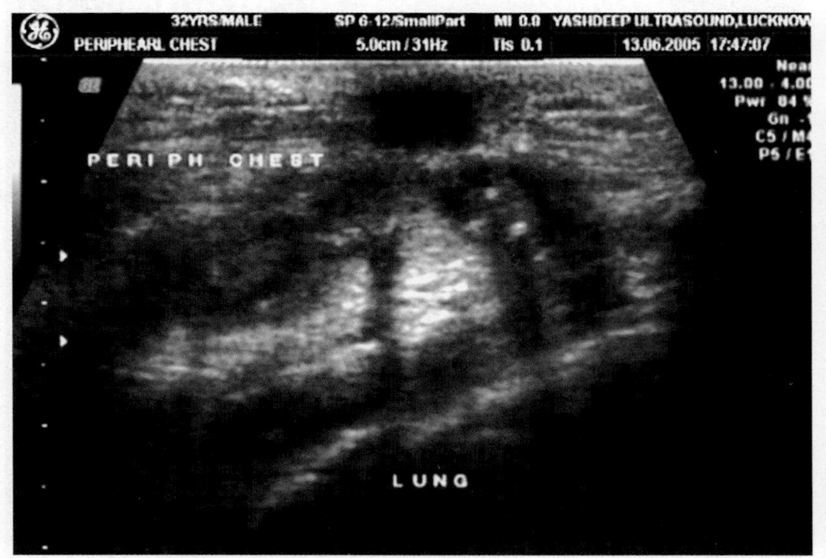

FIGURE 6.52

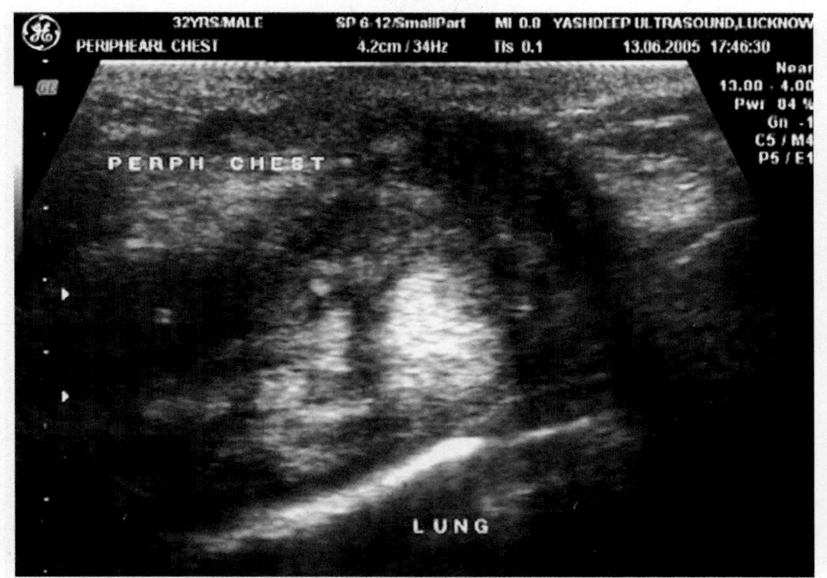

FIGURE 6.53

FIGURES 6.52 and 6.53: FTC operated hydatid cyst peripheral chest. A patient operated for hydatid cyst peripheral chest presented with pain and swelling over the scar line. HRSG shows irregular hypoechoic areas over the scar tissue suggestive of granulomatous tissue with secondary infection.

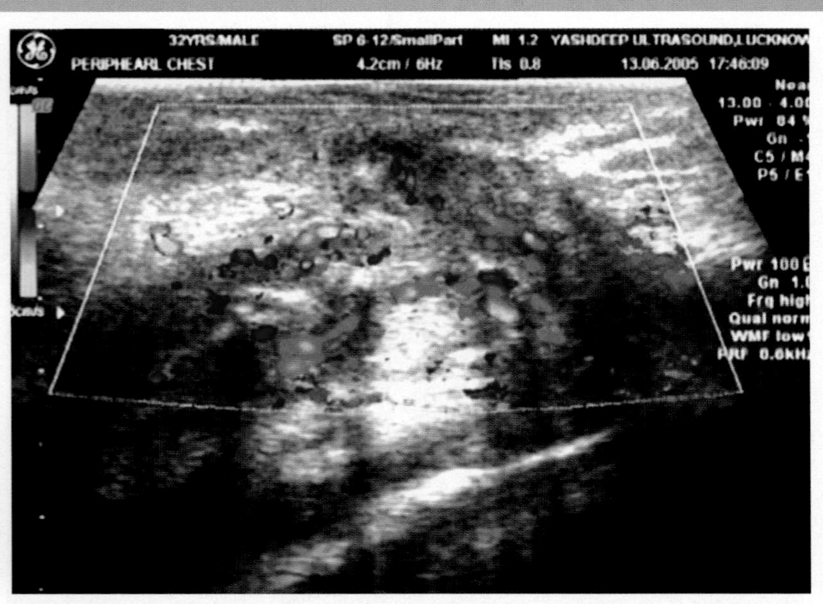

FIGURE 6.54: On color flow imaging increased flow is seen around the scar suggestive of secondary infection.

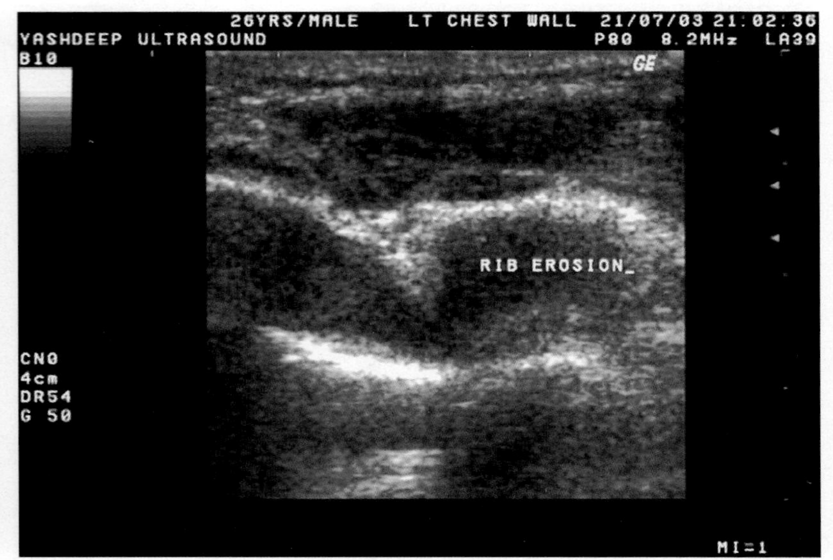

FIGURE 6.55

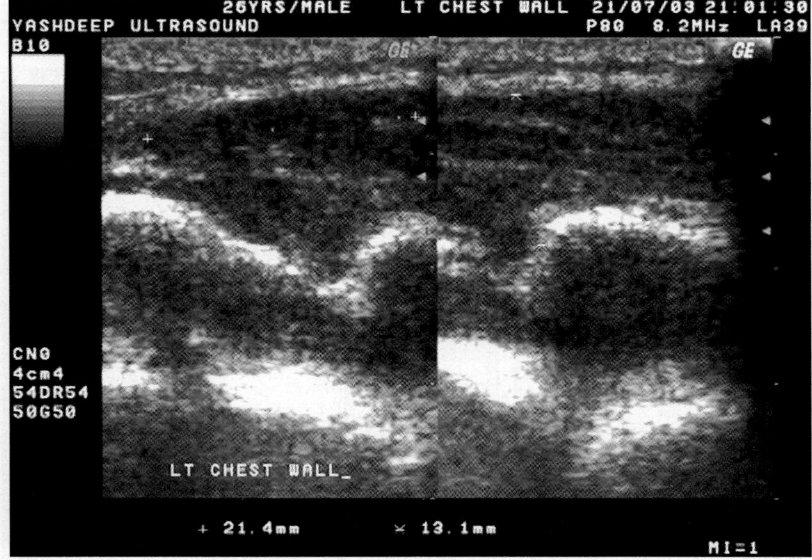

FIGURE 6.56

FIGURES 6.55 and 6.56: Rib tumor invading the chest wall. A patient presented with persistent soft tissue mass over the chest wall, which was hard. On clinical examination HRSG shows a soft tissue mass coming out from the rib. It is seen invading the chest wall muscle. Rib erosion is also seen suggestive of malignant tumor.

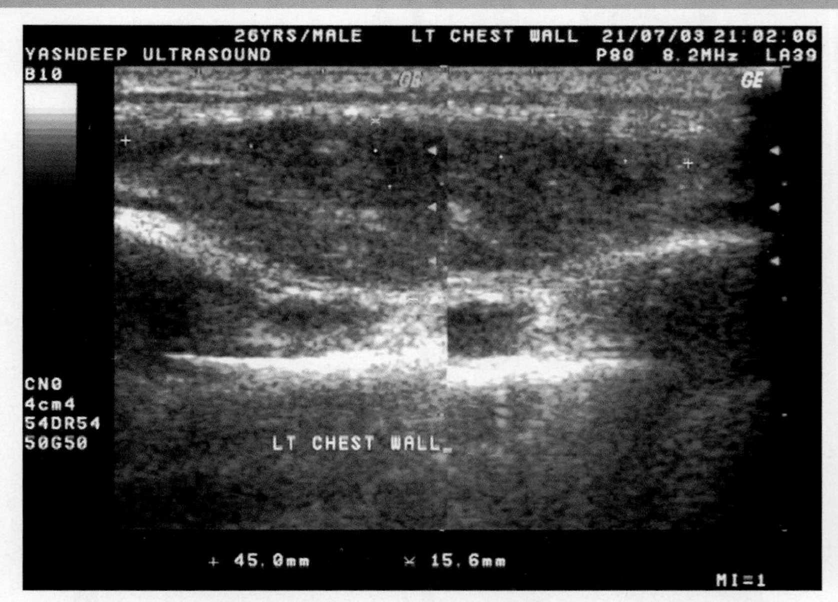

FIGURE 6.57: High-resolution ultrasound shows soft tissue component of the tumor. Biopsy of the tumor confirmed chondrosarcoma.

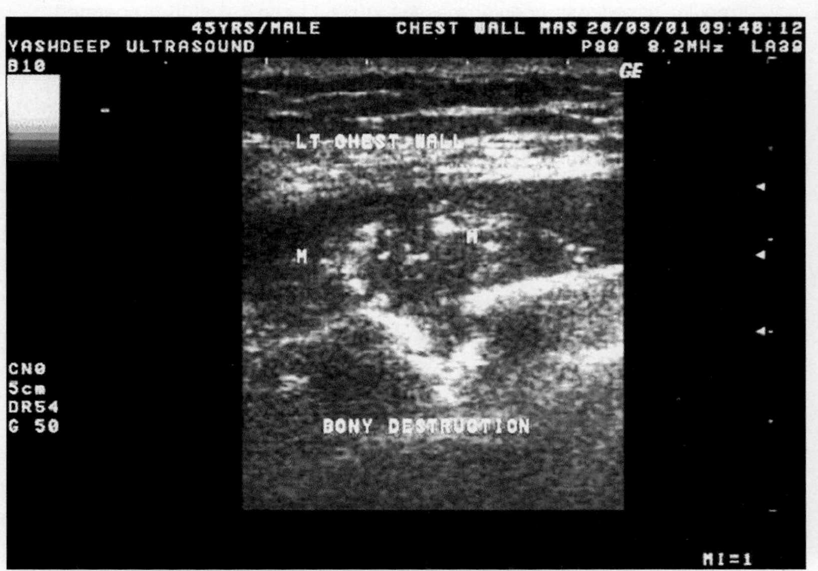

FIGURE 6.58

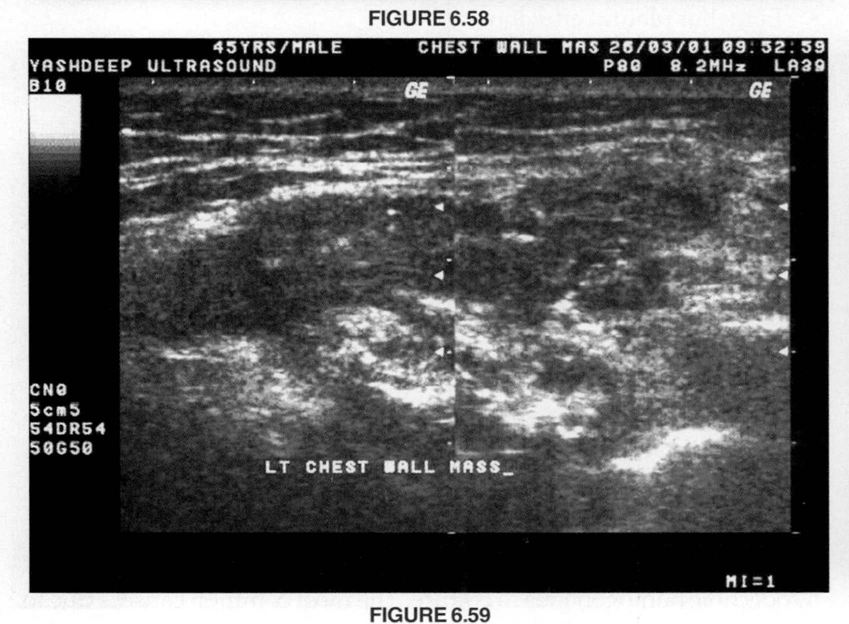

FIGURE 6.59

FIGURES 6.58 and 6.59: Tumor mass invading chest wall and the rib. HRSG shows an irregular heterogeneous mass invading the rib of the Lt chest wall and the mass has destroyed the rib. It is also seen invading the muscles of the chest wall. Biopsy confirmed invasive chondrosarcoma.

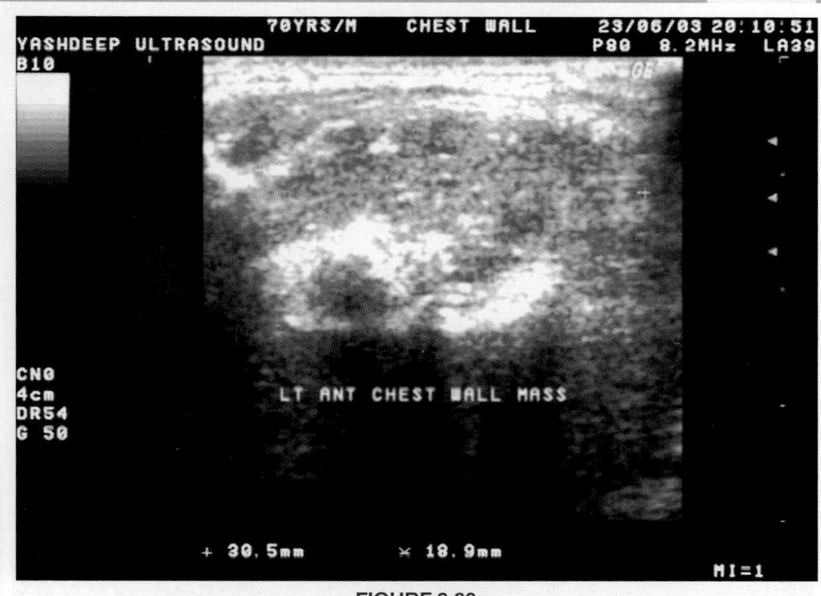

FIGURE 6.60

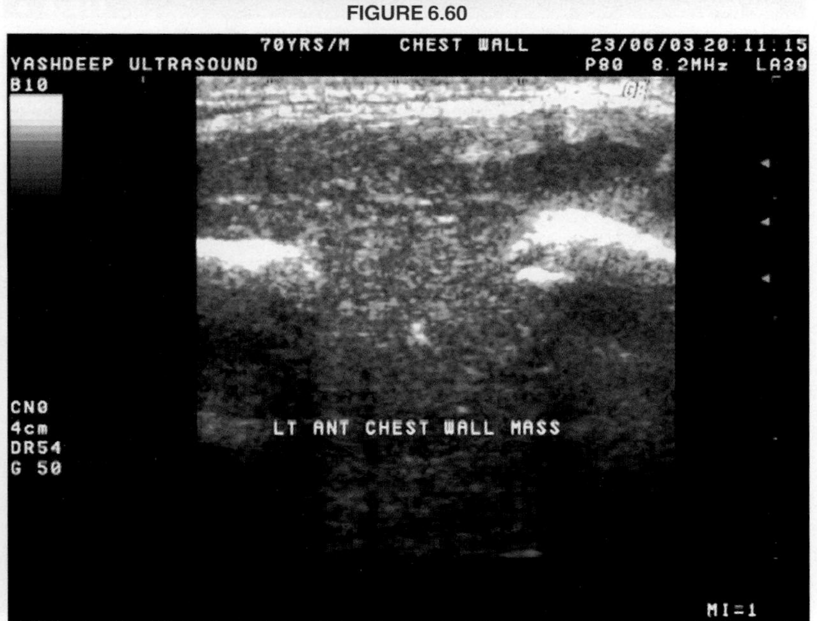

FIGURE 6.61

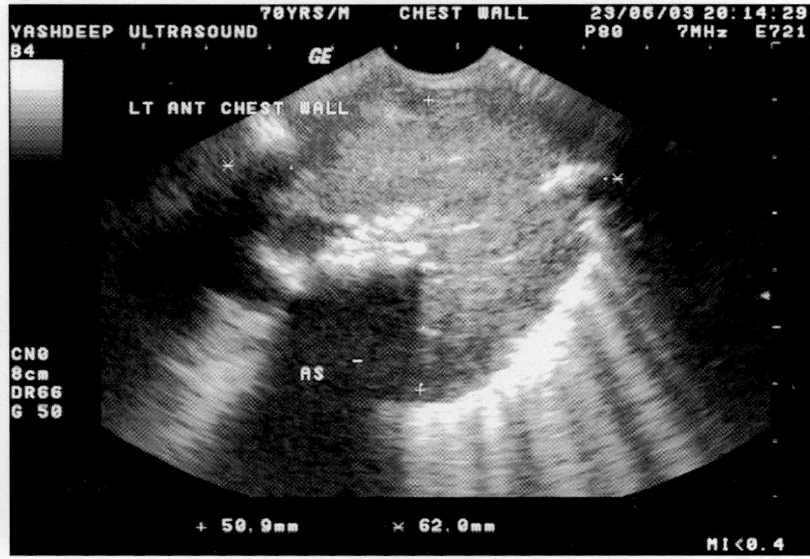

FIGURE 6.62

FIGURES 6.60 to 6.62: Big heterogeneous malignant mass invading the chest wall and pleura and peripheral lung. A patient presented with big heterogeneous soft tissue mass fixed with the chest wall. The mass is seen invading the peripheral chest. It is also seen going down deep and invading the rib cage with the bony destruction. Pleural invasion with invasion of peripheral lung is also seen.

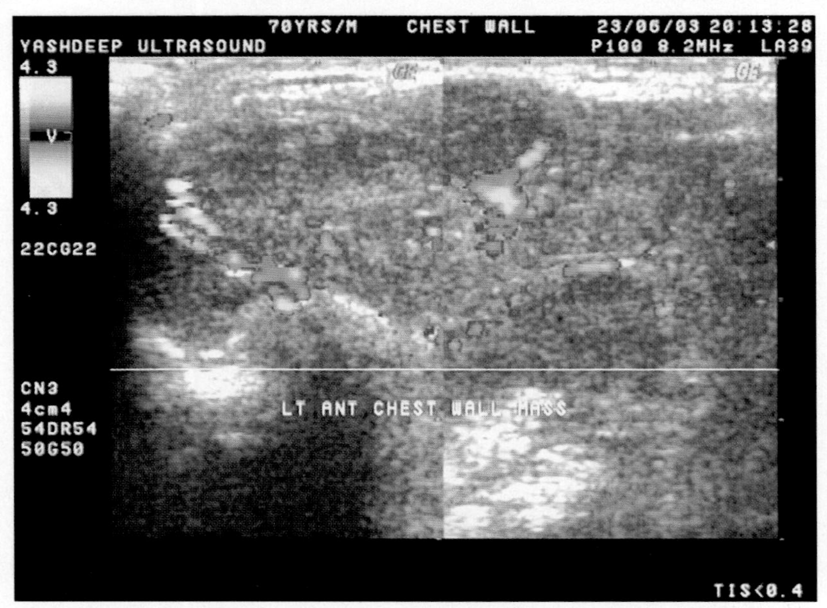

FIGURE 6.63: On color flow imaging low flow seen in the mass. Biopsy of the mass shows malignant rhabdomyosarcoma.

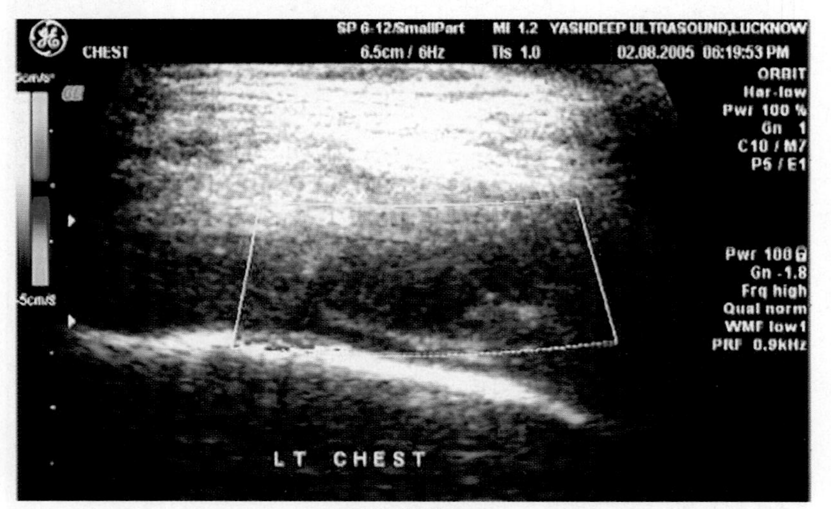

FIGURE 6.64

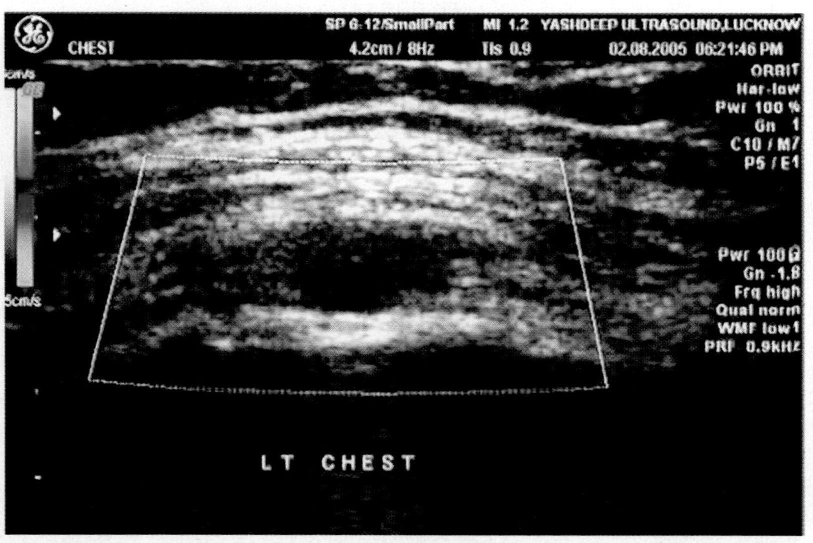

FIGURE 6.65

FIGURES 6.64 and 6.65: Metastatic deposit. A patient of Ca breast presented with soft tissue mass over the chest wall. HRSG shows a hypoechoic irregular mass over the chest wall suggestive of metastatic deposit.

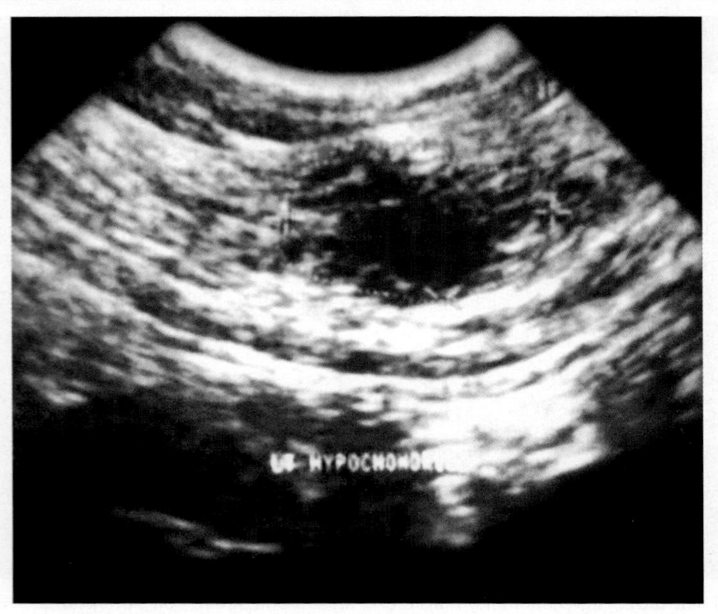

FIGURE 6.66: Metastatic deposits from CA breast. HRSG shows irregular hypodense nodular masses in the Lt chest wall in a operated case of Ca breast. They are well defined smooth in outline suggestive of enlarged lymph nodes. Irregular hypoechoic mass is also seen in the chest wall. Biopsy of the mass shows metastatic deposits.

helps in the characterization of lesion. Ultrasonography can also identify area of solid tumor within the pleural effusion or a focal area of pleural thickening suitable for guided needle biopsy. Presence of air in pleural space is appreciated as bright echo with prominent reverberation artifact caused by near complete reflection of the ultrasound beam at the interface between air and tissue. The juxta diaphragmatic lesions like amoebic hepatic abscess with thoracic complications are well seen through the liver.

Indications for High Resolution Sonography

- Simple pleural effusion
- Encysted pleural effusion
- Lamellar pleural effusion
- Pleural calcification
- Empyema
- Hemothorax
- Pleural thickening
- Pleural tumors
  - Primary tumors are rare
  - Benign tumors
  - Mesothelioma, fibroma
  - Lymphoma
  - Metastatic tumors—Most common cause is carcinoma bronchus.

Large collection of the fluid in the pleural space may cause hemothorax resulting into collapse of the lung toward hylum. It results in to inversion of diaphragm. It can be best evaluated on real time. HRSG shows paradoxical movements on real time imaging.

Mesothelioma is benign tumors. They are poorly echogenic or hypoechoic homogeneous in texture. The most common cause is due to prolonged exposure asbestos. Homogeneous nodular pleural thickening is seen on HRSG.

Lung Parenchyma

Ultrasonographic examination of the lung parenchyma to determine the nature of lesion is possible only when the barrier of air in lung periphery is broken by collapse of lung or filling in by fluid and inflammatory cells (in consolidation). Consolidated lung is hypoechoic as compared to aerated lung, liver or spleen. Air within bronchi surrounded by consolidated lung produces highly reflective linear branching echoes that can be recognized as sonographic air bronchogram. Aerated alveoli surrounded by consolidated lung produce highly reflective globular echoes that can be recognized as sonographic air alveolograms. In the consolidated lung, fluid filled bronchi or pulmonary vessels can be seen as anechoic tubular branching structures. Differentiation between the two can be recognized as observing pulsation in vessels or using Doppler. The identification of sonographic air bronchogram, air alveologram, fluid bronchogram and/or pulmonarary vasculature helps to differentiate consolidated lung from parenchyma masses and pleural lesions. Besides ultrasonographic examination is useful to differentiate pneumonia alone from pneumonia with pleural effusion of empyema. Atelectatic lung appears as wedge shaped density moving synchronously with respiration through the pleural fluid and have crowded fluid or air filled bronchi within the collapsed portion of the lung.

Lung abscesses have thick irregular walls with echogenic debris and air within the internal fluid and area are often associated with surrounded area of consolidations. An empyema is confined within the pleural space and tends to have smooth walls of uniform thickness. It also compresses and displaces the surrounding lung parenchyma. The radiographic dilemma of differentiating lung abscess from empyema is resolved to large extent by ultrasonography. With the real time sonography, a lung abscess during inspiration, demonstrates expansion of its entire circumference, whereas in case of empyema only the visceral pleural surface show motion. Hydatid cyst has characteristics appearance of daughter cysts, lifting up of membrane or sonographic water lily sign. They may have associated with hepatic involvement or some other pathology.

Lung tumors, both peripheral and central, depending on size and relation to chest wall can usually be visualized by ultrasonography. They usually have well-defined outlines and stand out in presence of surrounding air filled lung, consolidation or collapse. Although most of the lesions that are smaller than 5 cm are hypoechoic compared to aerated lung. The increased echogenecity of larger lesions may be caused by internal echolucency with hyperechoic walls while calcification is seen as echogenic speck with or without posterior shadow. For centrally located lung tumors associated consolidation or collapse is required as a window to permit visualization. Tumors that are surrounded by consolidation lung appear as hyperechoic masses within the hypoechoic fluid filled lung. Ultrasonography can help in diagnosing invasion of the chest wall by using lung tumor. Besides this modality permit us to identify various pathologies present in case of opaque hemithorax.

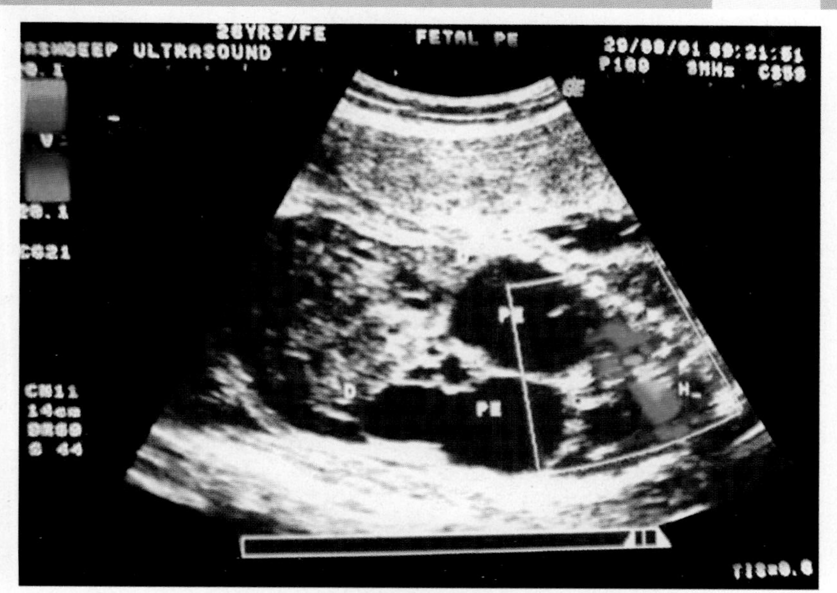

FIGURE 6.67

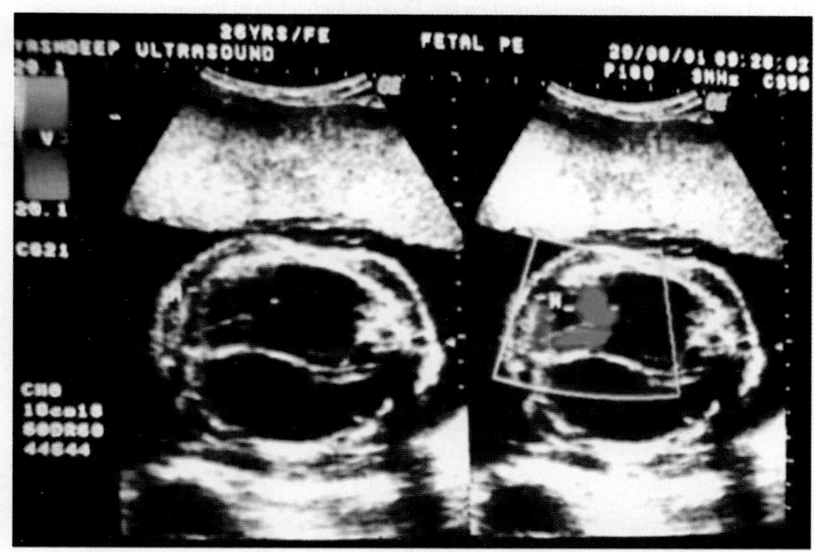

FIGURE 6.68

FIGURES 6.67 and 6.68: Bilateral fetal pleural effusion. HRSG shows bilateral pleural effusion in both fetal thorax. The lungs are seen collapsed on either side. Heart is also seen pushed and compressed by the pleural effusion.

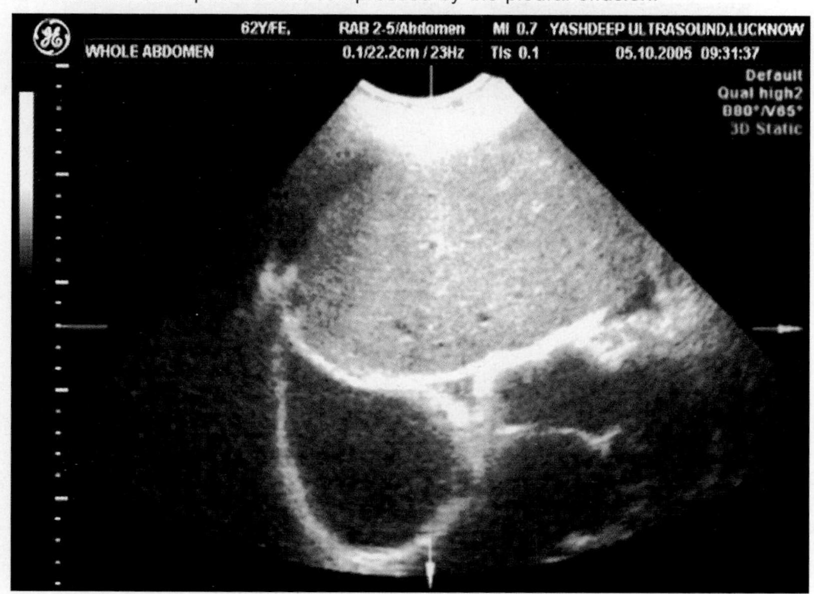

FIGURE 6.69: Rt sided pleural effusion. HRSG shows moderate amount of fluid collection in the Rt pleural cavity. No septation is seen in the collection.

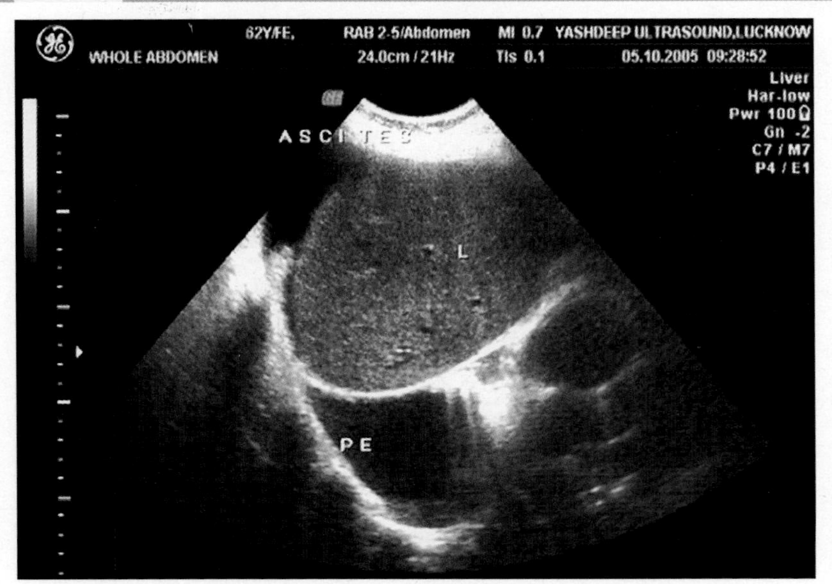

FIGURE 6.70: Simple pleural effusion with ascites. HRSG shows small amount of fluid collection in the Rt pleural cavity. Fluid is also seen in the peritoneal cavity.

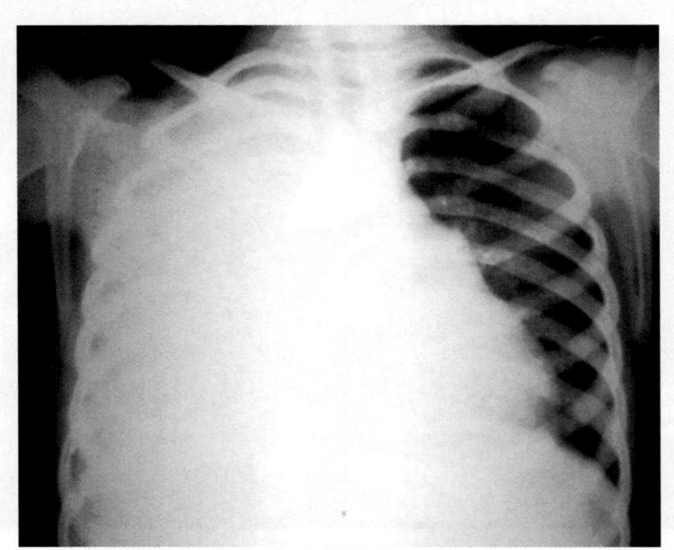

FIGURE 6.71: Massive pleural effusion. Plane X-ray chest of the patient shows Rt sided hemithorax with radio dense opacities filling whole of the Rt thorax of a young child.

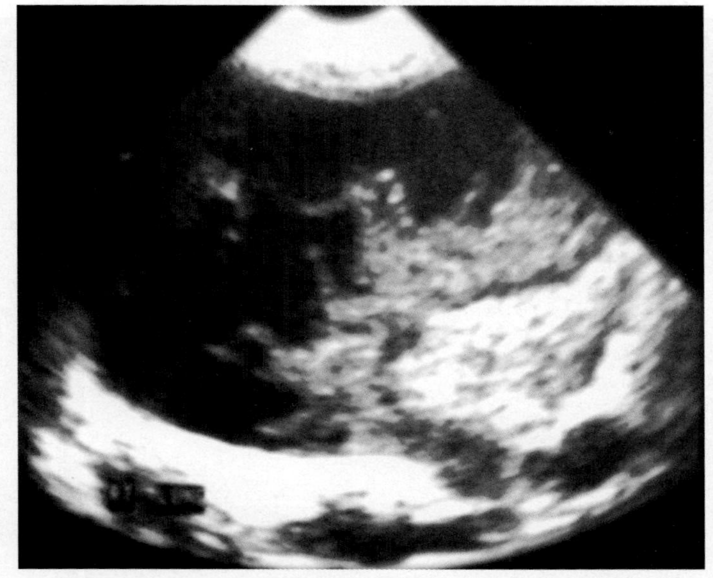

FIGURE 6.72: HRSG shows massive fluid collection in the Rt pleural cavity. The lung is seen collapsed and floating in the fluid collection. Lung is seen collapsed by the collection. Multiple fibrous bands are seen in the collection. Echogenic debris is also seen in the collection.

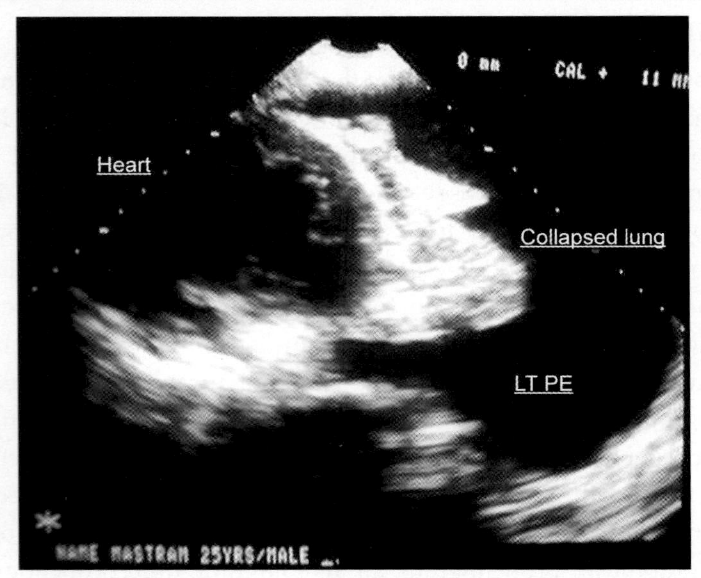

FIGURE 6.73: Massive pleural effusion. HRSG shows gross amount of fluid collection in the Lt pleural cavity. The lung is collapsed and compressed by the pressure of the fluid. Heart is also seen shifted to the Rt side.

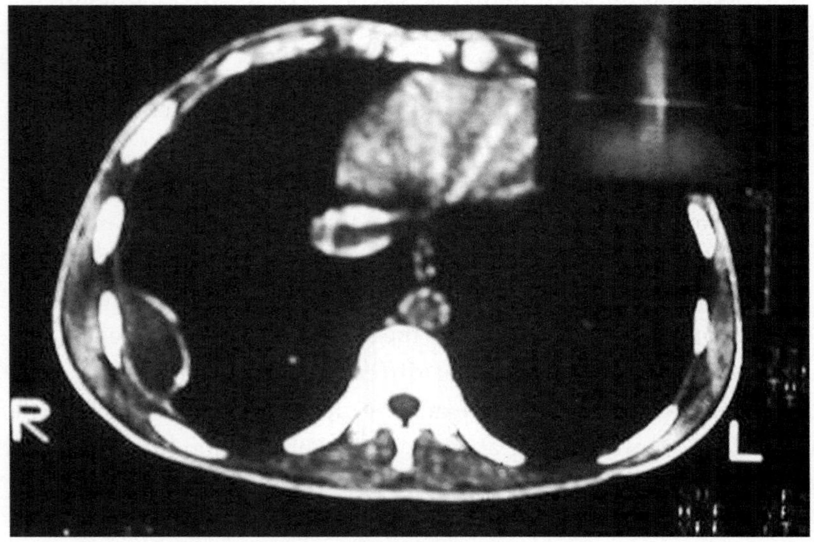

FIGURE 6.74: Loculated anterior pleural effusion. CT scan of the patient shows a hypodense shadow in the anterior part of the Rt lung.

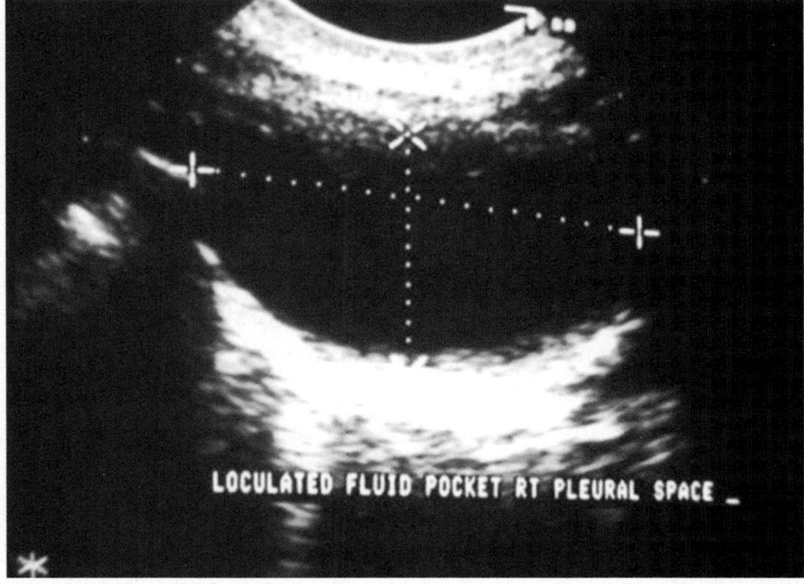

FIGURE 6.75: HRSG of the same patient shows encysted fluid collection in the anterior part of the Rt lung.

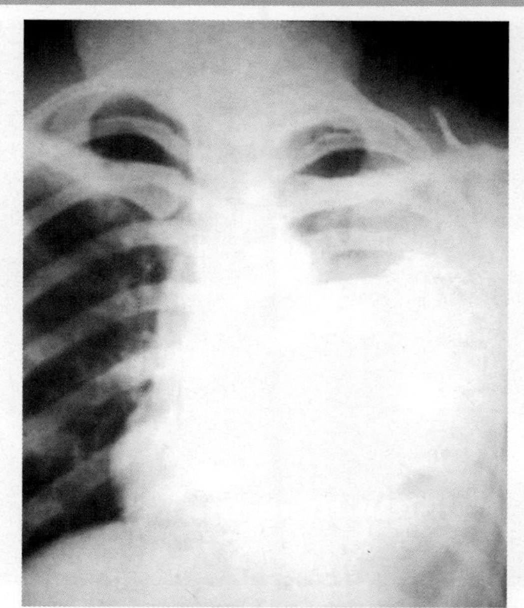

FIGURE 6.76: Plain X-ray of chest of the patient shows Lt sided radio opacity with obliterated CP angle.

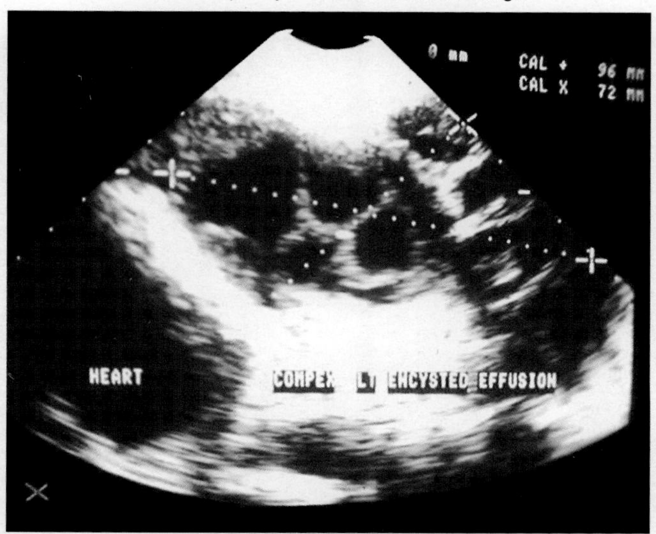

FIGURE 6.77: HRSG of the same patient shows multiloculated Lt sided pleural effusion. Multiple echogenic septa are seen in the collection.

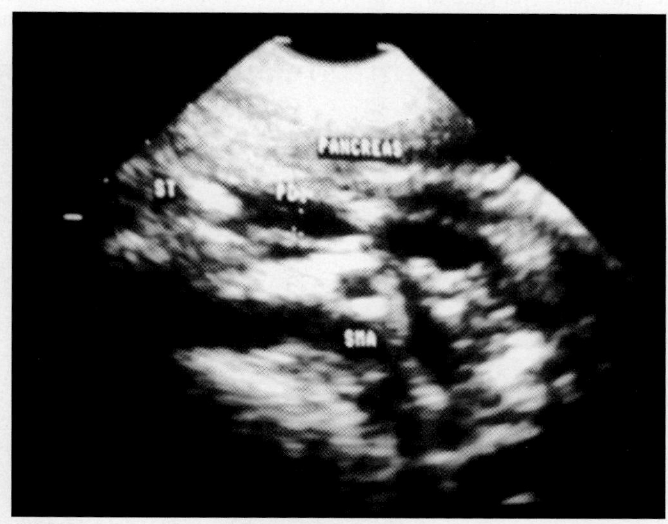

FIGURE 6.78: Abdominal sonography of the same patient shows chronic calcific pancreatitis with dilated pancreatic duct. Multiple calculi are seen in the pancreatic duct. Tail of the pancreas is enlarged hypoechoic irregular in outline. The pancreatitis is the cause of recurrent Lt sided pleural effusion in the same patient. Pancreatitis involving the tail of the pancreas is known to cause reactive pleural effusion on the Lt side.

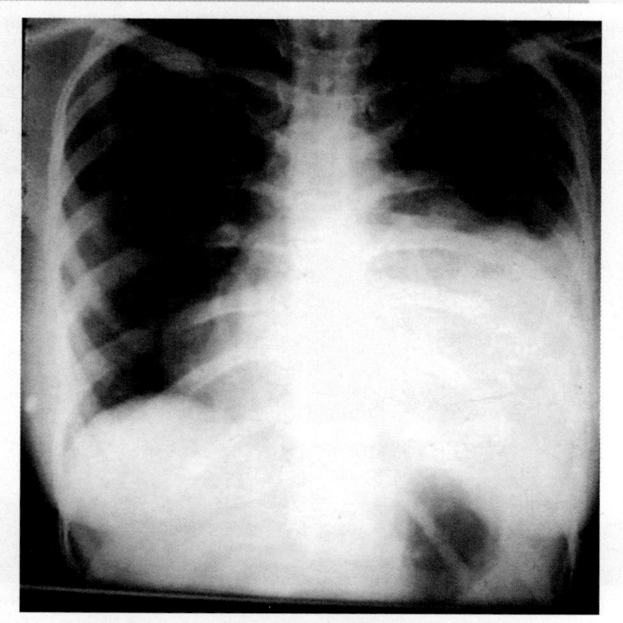

FIGURE 6.79: Lt sided pleural effusion. PlainX-ray of the patient shows obliterated Lt costophrenic angle with opacity in the lower part of the Lt chest.

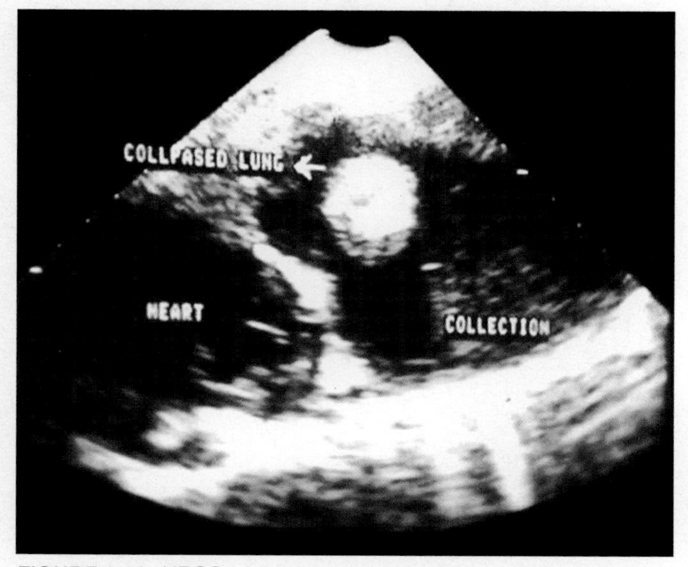

FIGURE 6.80: HRSG of the same patient shows collection in the Lt pleural cavity. Associated collapsed of the lung is also seen in the chest. Which is seen as an echogenic shadow.

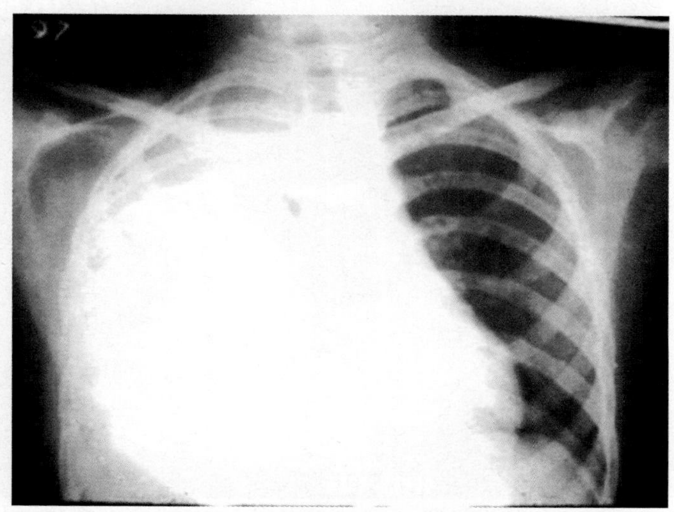

FIGURE 6.81: Empyema thorax. Plain X-ray of the patient shows Rt sided hemithorax radiodense opaque shadow is seen filling the Rt pleural cavity and opicify the thorax.

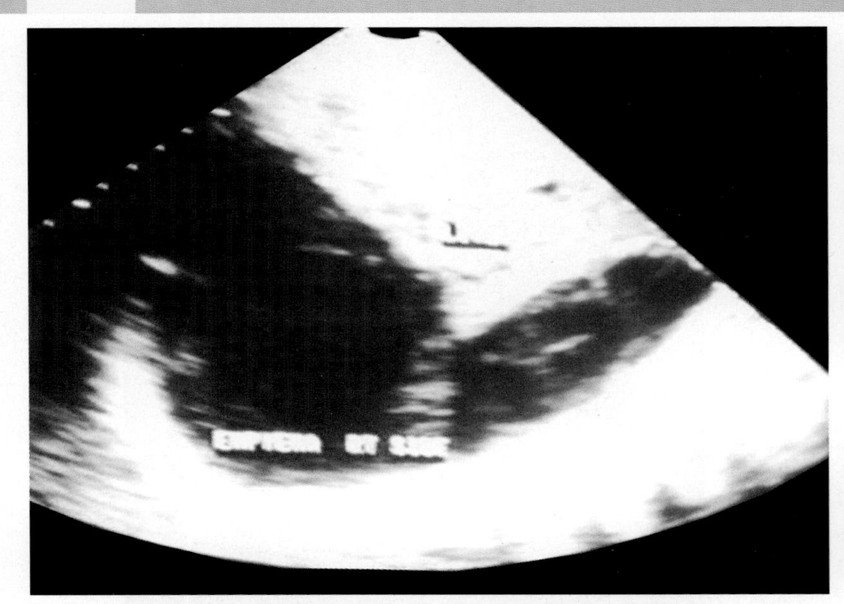

FIGURE 6.82: HRSG of the same patient shows big thick low level echo collection filling the pleural cavity. The lung is seen collapsed and sharp borders of the collapsed lung are seen. On aspiration of the fluid pyogenic collection was seen coming out from the fluid.

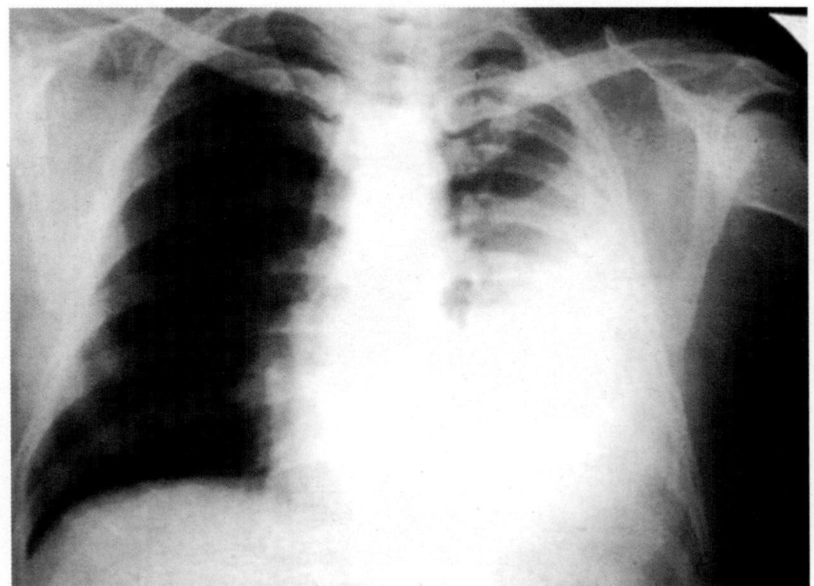

FIGURE 6.83: X-ray chest shows Lt thoracic opacity.

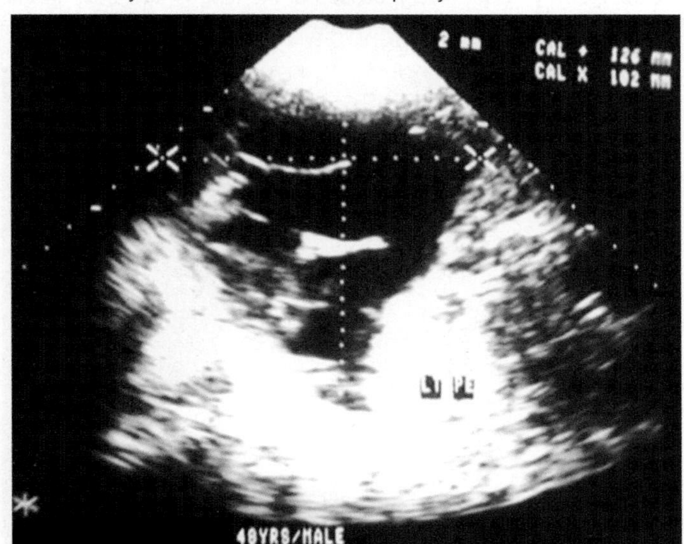

FIGURE 6.84: HRSG of the same patient shows Lt sided pleural effusion. Multiple locullae are seen formed due to the fibrotic bands in the collection. Multiple thin septa are seen in the collection.

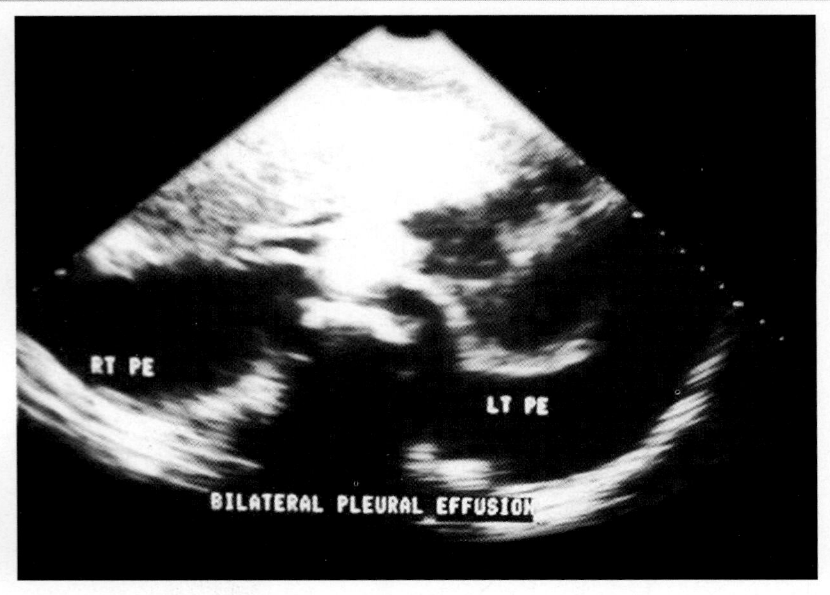

FIGURE 6.85: **Bilateral pleural effusion.** HRSG of the patient shows bilateral simple pleural effusion. Collection is seen in both pleural cavities on either side.

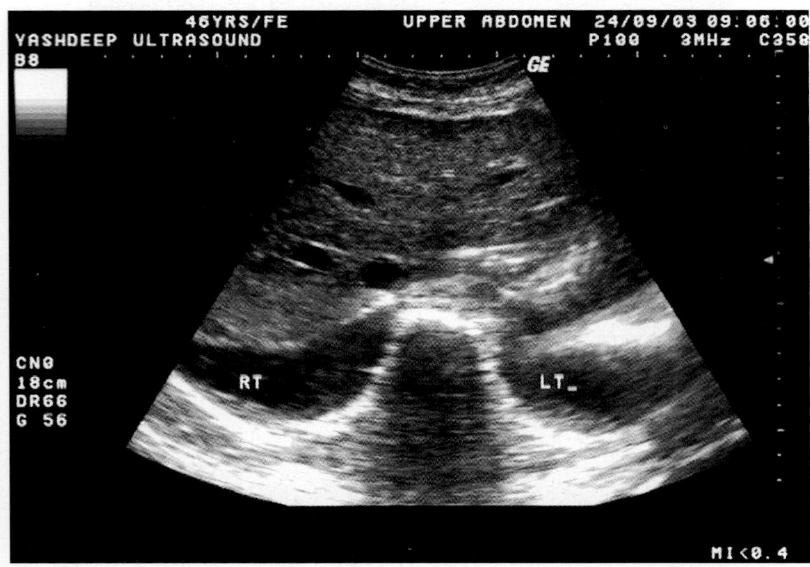

FIGURE 6.86: **Bilateral pleural effusion.** A patient presented with breathlessness. HRSG shows bilateral pleural effusion. Lt lung is collapsed and it is seen floating in the Lt pleural effusion as an echogenic shadow.

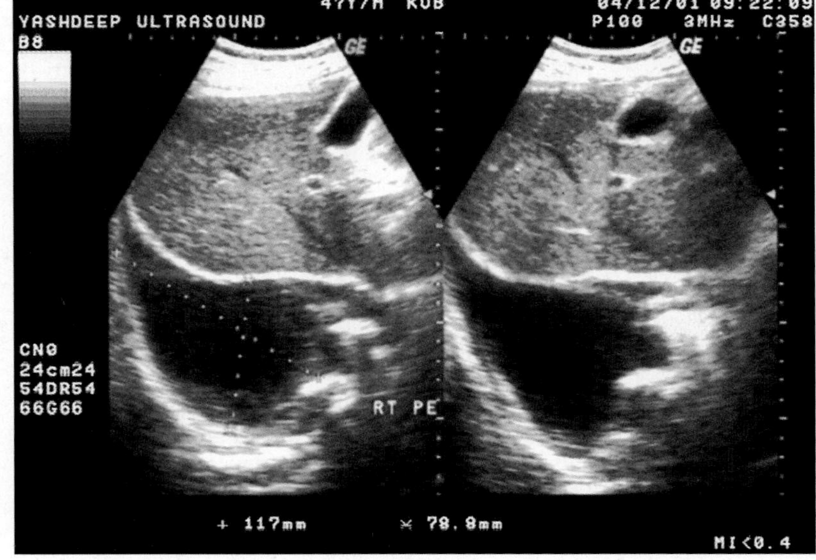

FIGURE 6.87

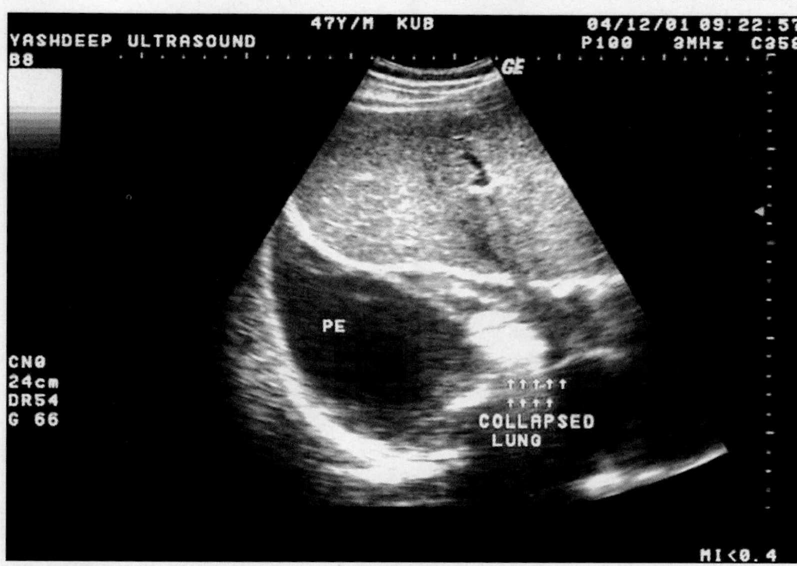

FIGURE 6.88

FIGURES 6.87 and 6.88: Simple pleural effusion with partial collapse of the lung. HRSG of the patient shows simple fluid collection in the Rt pleural cavity. Collapsed part of the lung is seen as an echogenic shadow.

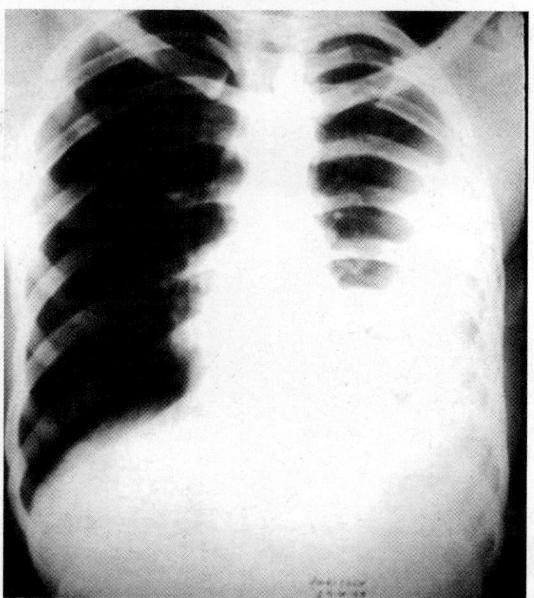

FIGURE 6.91: Peripheral lung mass. Plain X-ray of the cyst shows a radiodense opacity in the Lt thorax with obliterated costophrenic angle.

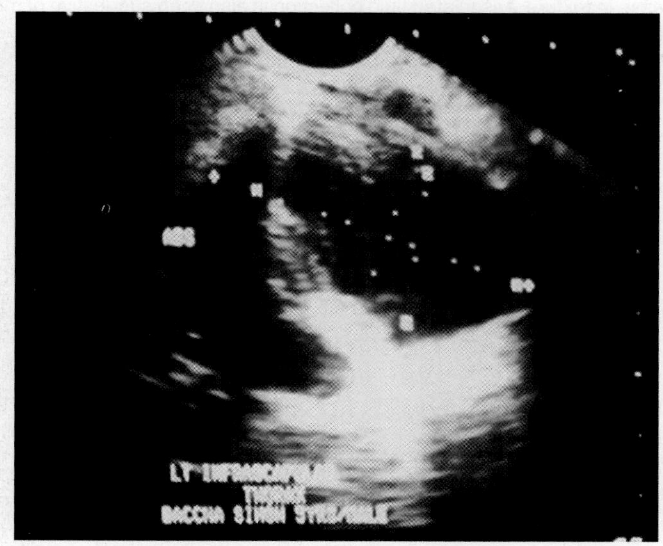

FIGURE 6.89: Peripheral lung abscess. HRSG of the patient shows a thick low-level echo complex mass in the posterior upper part of the Lt lung. Multiple internal echoes are seen in it suggestive of peripheral lung abscess.

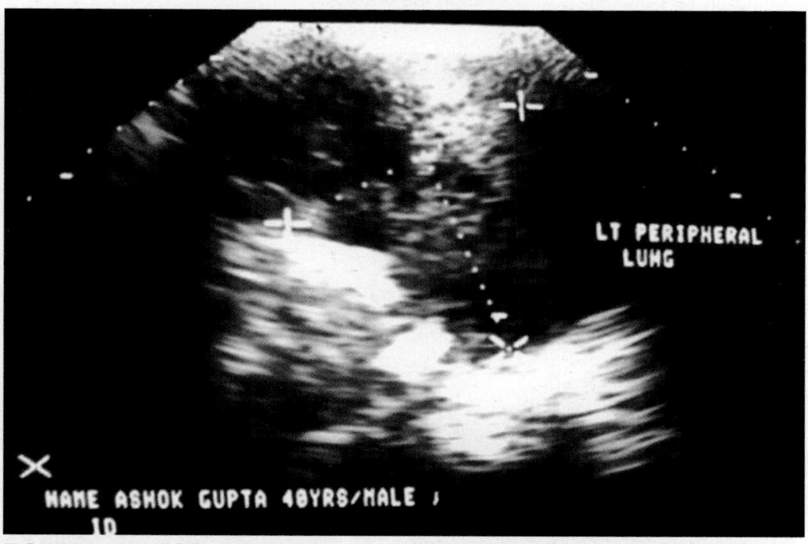

FIGURE 6.92: HRSG of the same patient shows a soft tissue mass in the Lt mid zone and lower zone. The mass shows irregular margin. Irregular hypoechoic areas are also seen in it suggestive of an abscess. USG guided biopsy of the mass shows bronchogenic carcinoma.

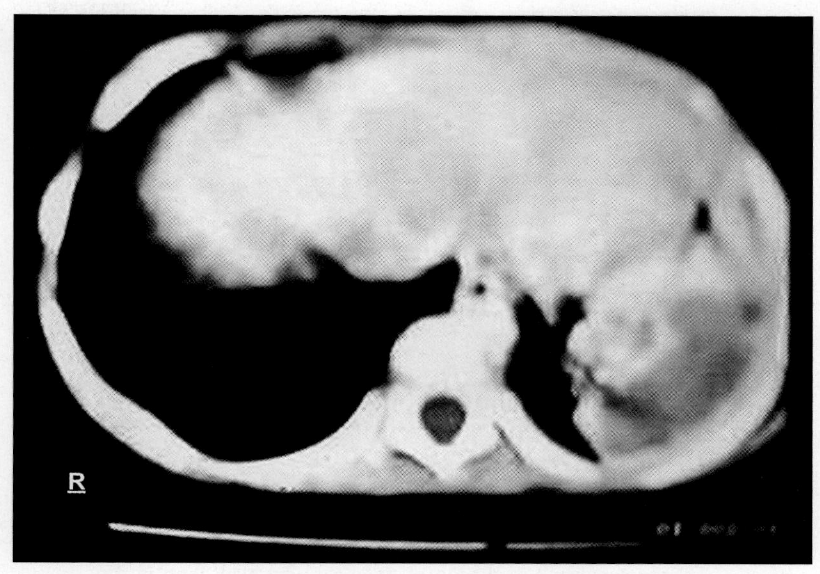

FIGURE 6.90: CT scan of the same patient shows a thick abscess cavity in the upper posterior part of the Lt lung.

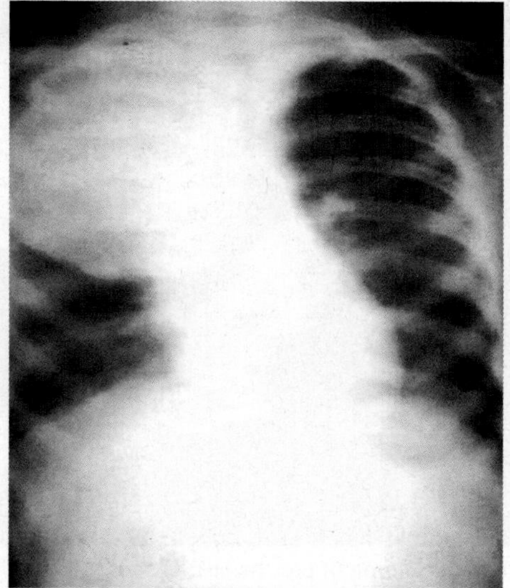

FIGURE 6.93: Hydatid cyst of the lung. X-ray of the young girl shows a well-defined homogeneous mass in the Rt upper zone. The mass shows sharp margins.

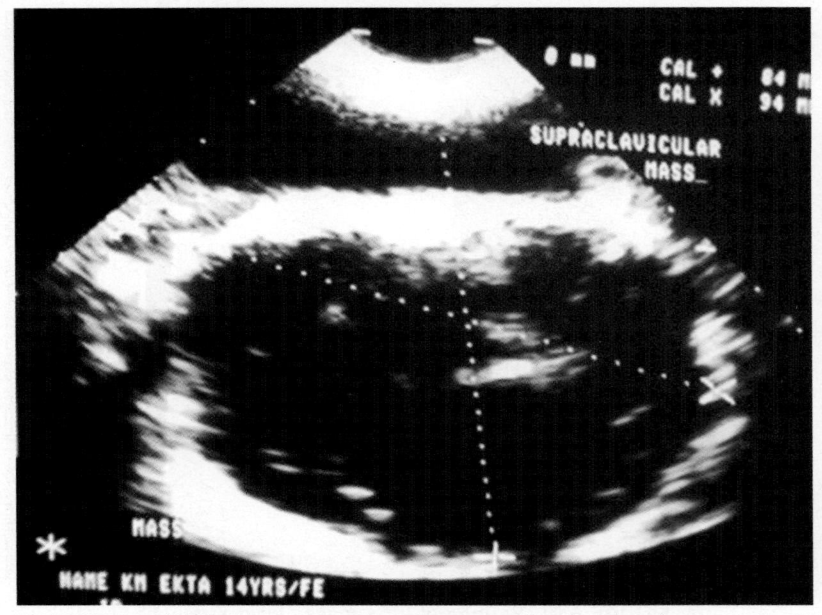

FIGURE 6.94: HRSG of the same patient shows a multiloculated cystic mass in the Rt supraclavicular and infraclavicular area going to the upper part of the Rt thorax. Echogenic calcification is also seen in the mass.

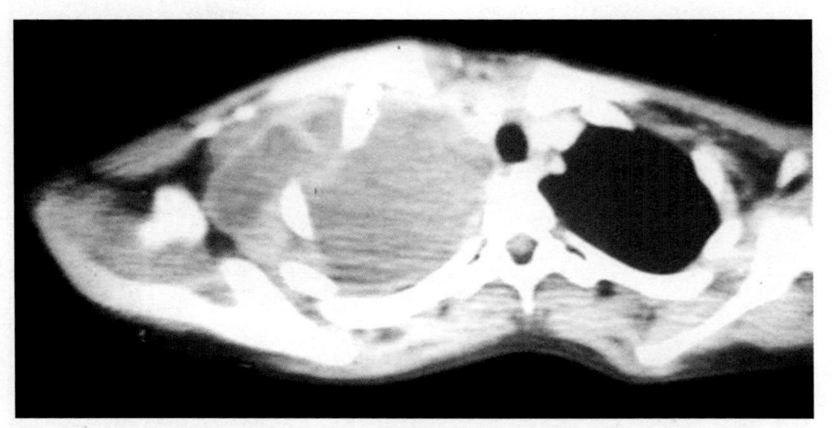

FIGURE 6.95: CT scan of the same patient shows homogeneous soft tissue mass with thick septa. Biopsy of the mass shows hydatid cyst.

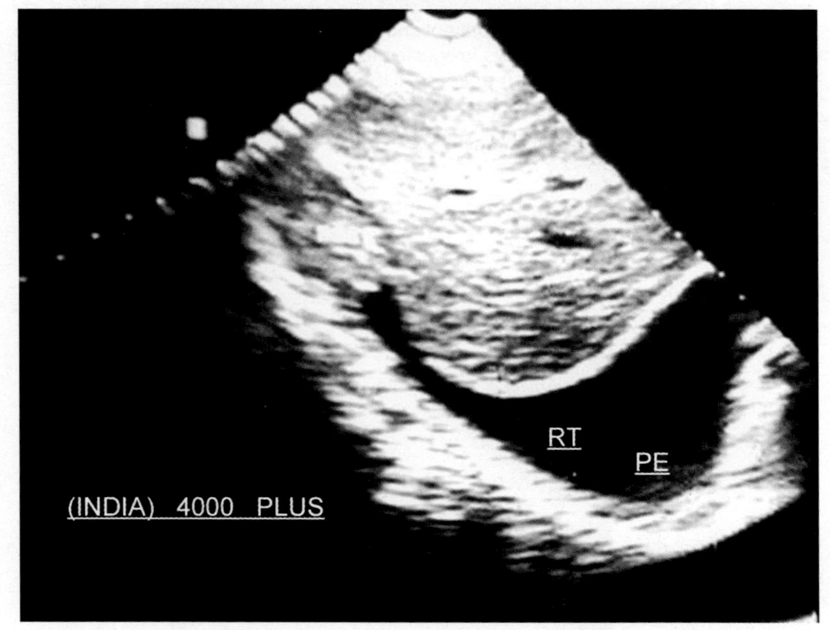

FIGURE 6.96: **Mesothelioma of pleura.** HRSG shows a well-defined homogeneous hyperechoic nodular mass coming out from the pleura on the Rt side on the anterior recess. No calcification is seen in the mass. No cystic degeneration is seen. Associated pleural effusion is also seen. Biopsy of the mass shows mesothelioma.

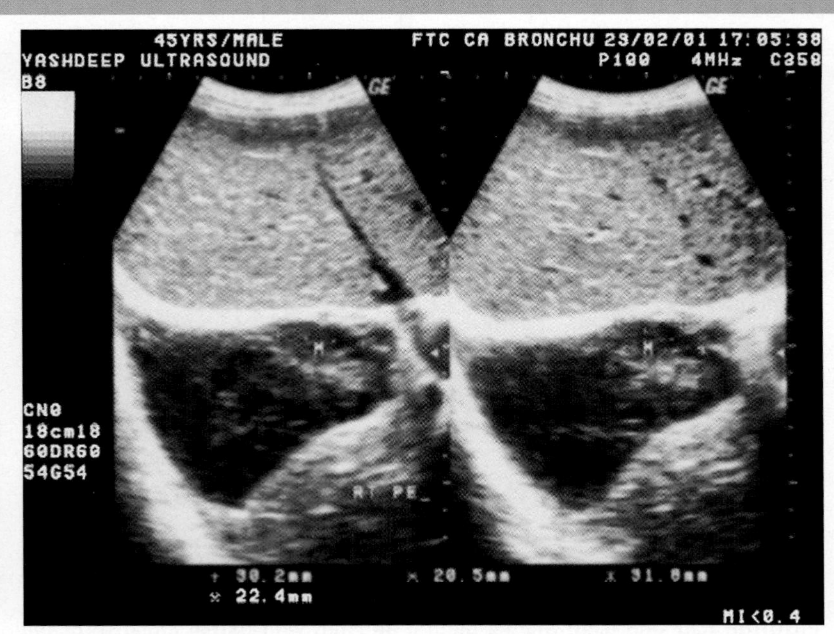

FIGURE 6.97: **Metastatic deposits from Ca bronchus.** A soft tissue mass shadow is seen fixed with the Rt pleura in a case of Ca bronchus. The mass is well-defined, hypodense in texture. Associated fluid collection is also seen suggestive of metastatic deposits from Ca bronchus.

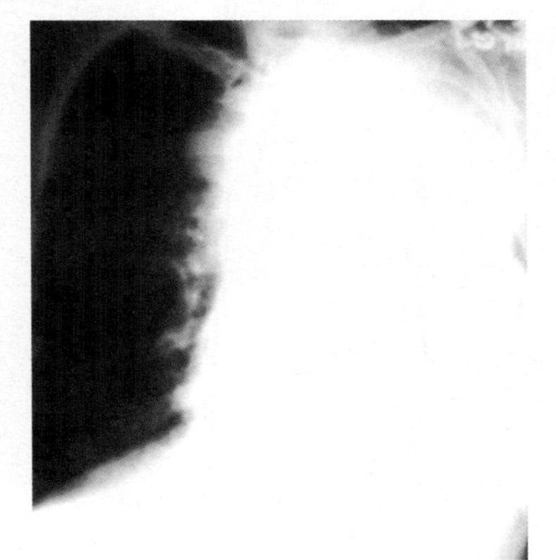

FIGURE 6.98: X-ray of the patient shows complete Lt sided hemithorax with radiodense opacity filling whole of the Lt thorax.

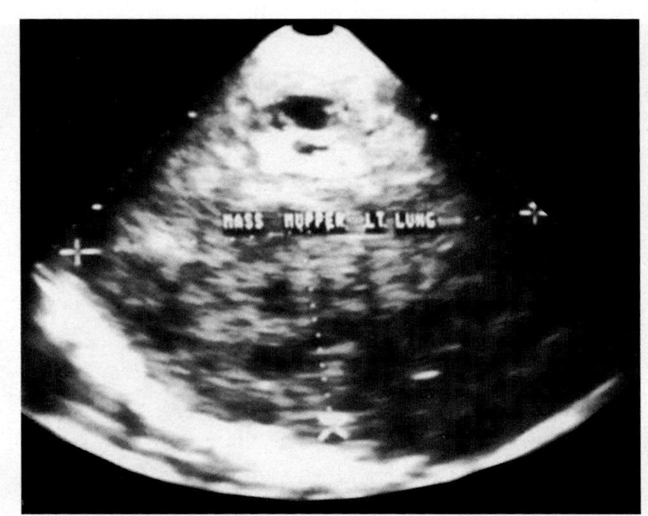

FIGURE 6.99: HRSG shows a big soft tissue mass involving the whole of the lung and filling the Lt lung. Biopsy of the mass shows malignant mass. Small area of necrosis was also seen in the mass.

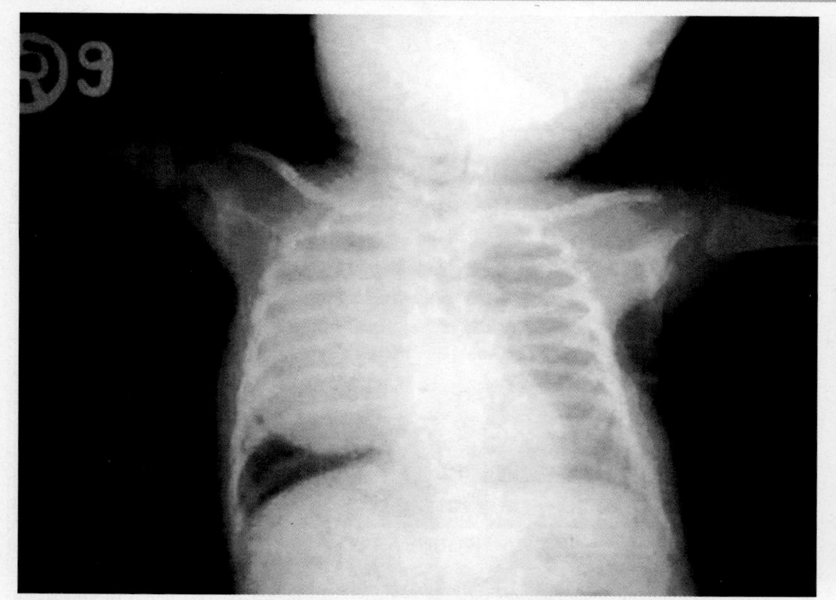

FIGURE 6.100: **Cystic adenomatoid malformation of lung.** A newborn child presented with dyspnea. X-ray chest shows well define homogeneous opacity in the Rt lung.

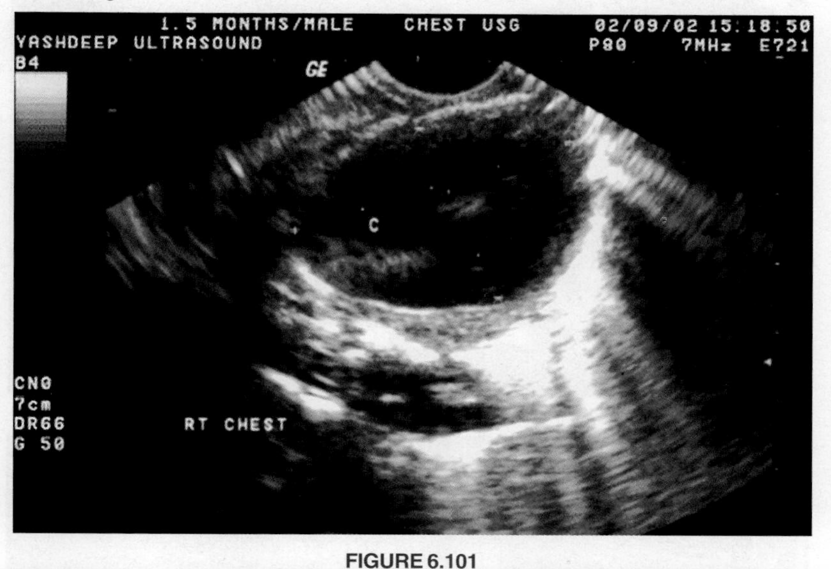

FIGURE 6.101

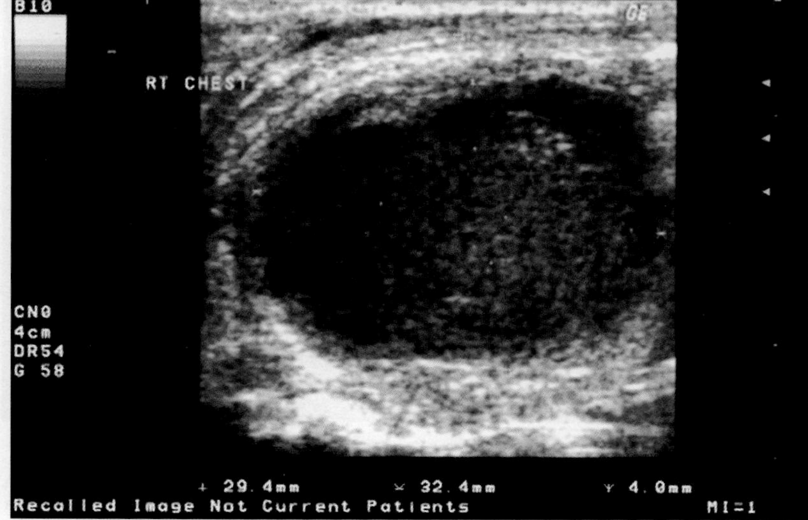

FIGURE 6.102

FIGURES 6.101 and 102: HRSG of the same patient shows well-defined cyst in the Rt lung. No septation is seen in it.

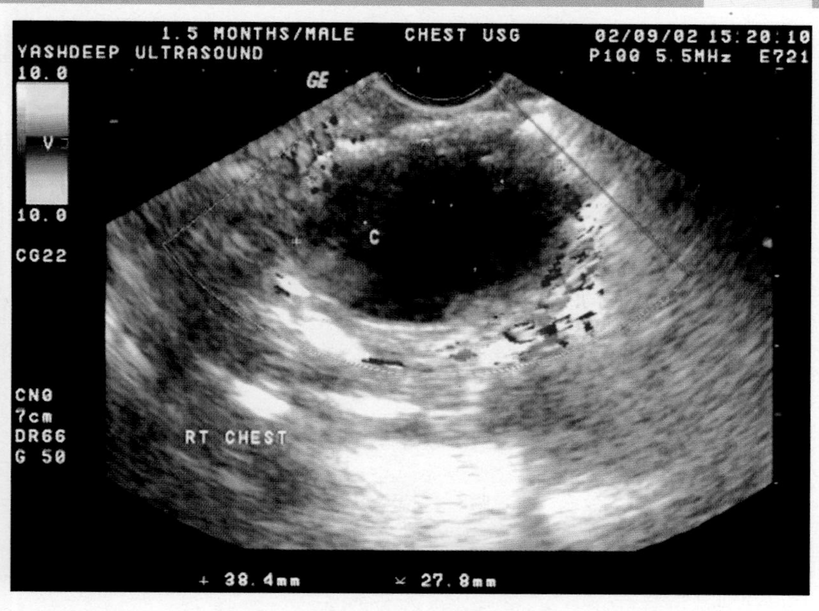

FIGURE 6.103: On color flow imaging peripheral flow seen in the cyst. Biopsy of the cyst confirmed cystic adenomatoid malformation.

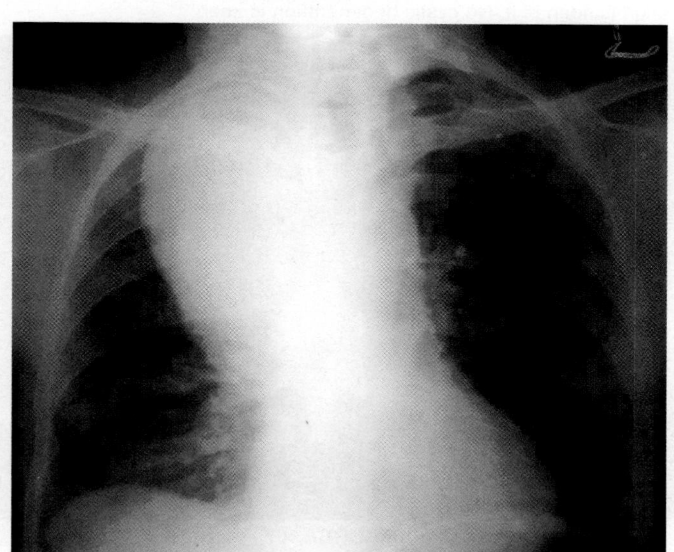

FIGURE 6.104: A patient presented with short breadth. X-ray chest shows big opacity in the Rt thorax.

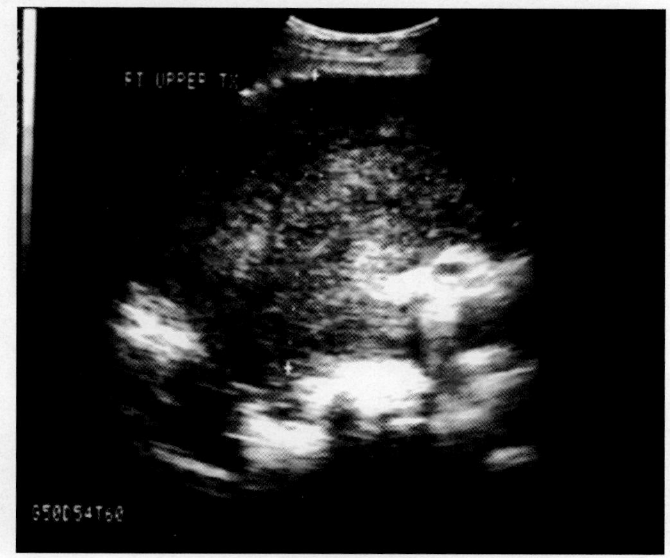

FIGURE 6.105

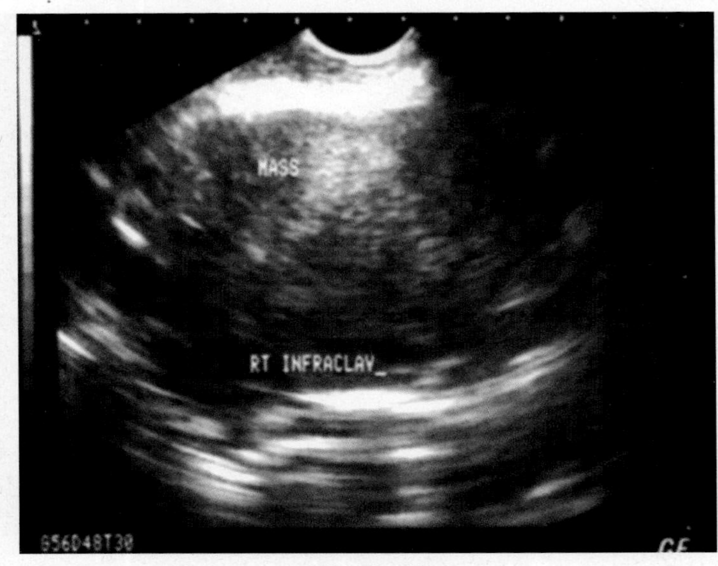

FIGURE 6.106

FIGURES 6.105 and 6.106: HRSG shows big soft tissue mass in the Rt peripheral lung and also seen in upper and mid lobe region. It is homogeneous in texture. No calcification is seen in it. No cystic degeneration is seen.

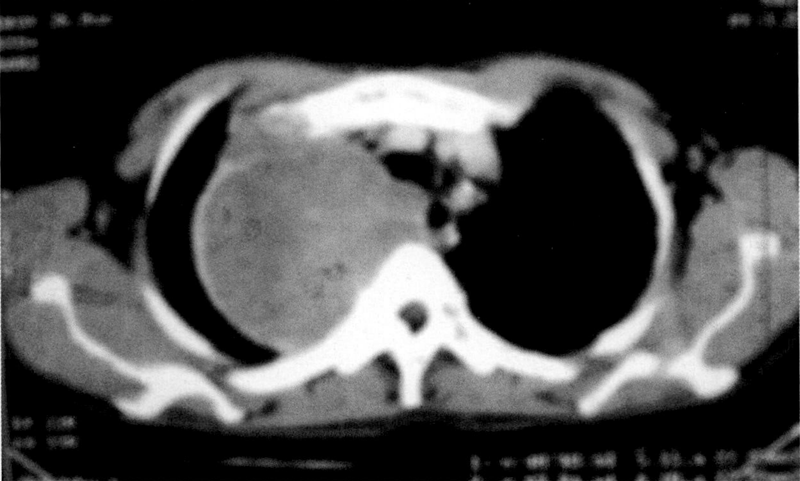

FIGURE 6.107

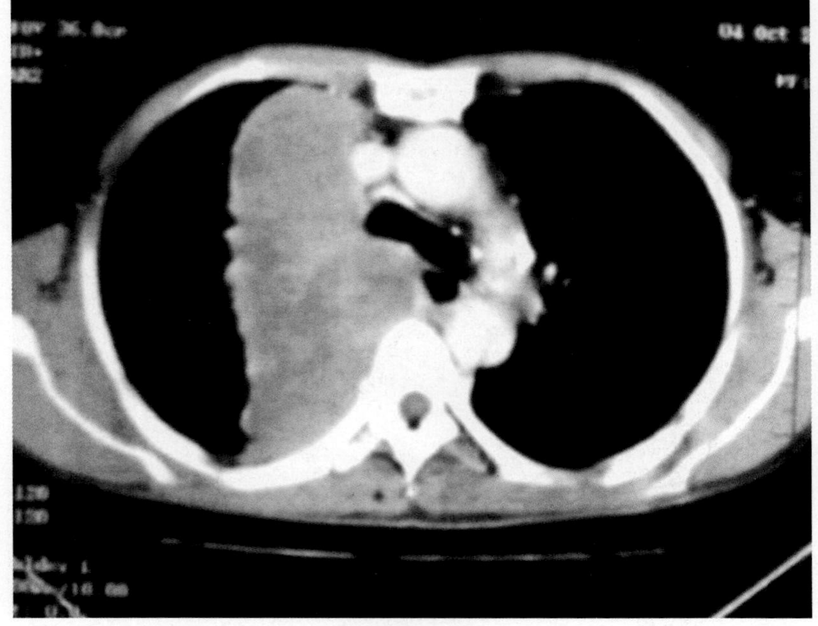

FIGURE 6.108

FIGURES 6.107 and 6.108: CT scan of the same patient conformed the biopsy of the mass showed bronchogenic carcinoma.

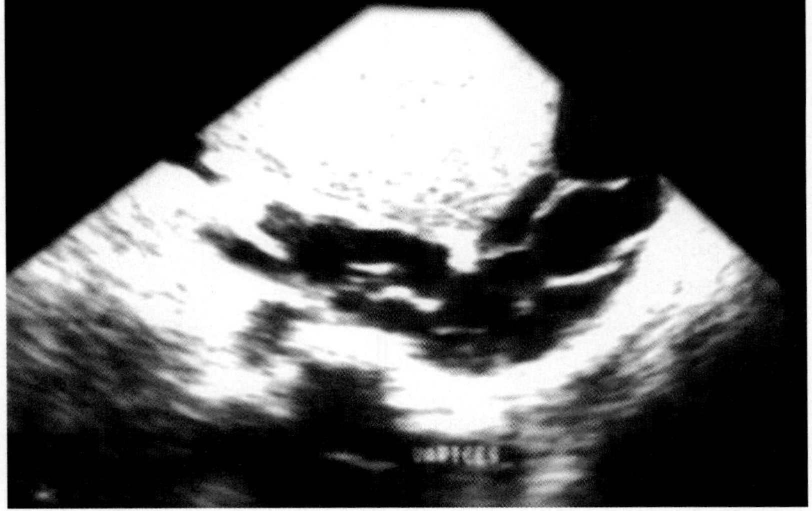

FIGURE 6.109

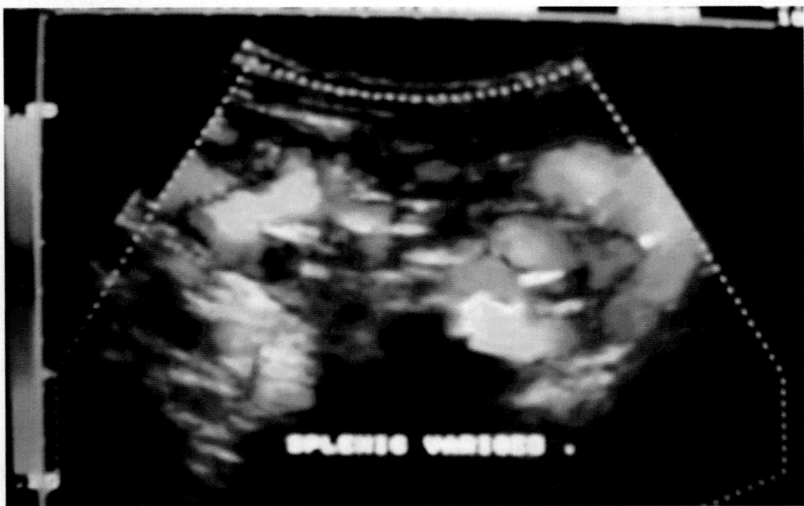

FIGURE 6.110

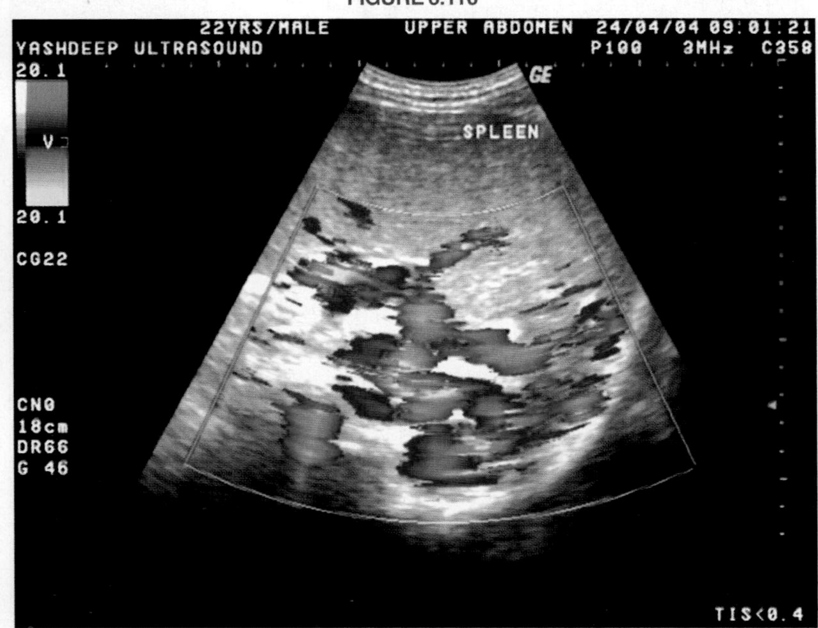

FIGURE 6.111

FIGURES 6.109 to 6.111: Splenic varices presented as pleural effusion. A patient presented with pain on the Lt side of the chest. Plan X-ray of the patient shows obliterated Lt costophrenic angle with soft tissue shadows and suspected Lt sided pleural effusion. Aspiration of the collection was done and frank blood was seen coming out HRSG was done to find out the nature of the collection and multiple dilated vessels were seen in the splenic bed. Color Doppler imaging of the same patient shows multiple splenic varices presented as Lt sided pleural effusion.

Mediastinum

Although shadowing bone and the reflective lung surrounds mediastinum, most of its area can still be effectively examined with careful attention to technique and patient positioning. The upper mediastinum is accessible to sonographic examination by the use of a suprasternal approach. Patients are examined in a supine position with a pillow placed beneath the shoulders and the neck extended. The transducer is placed at the base of the neck and angled caudally behind the manubrium. Visualization is done in oblique sagittal and coronal planes. Parasternal scanning of the mediastinum is aided by placing the patient in the appropriate lateral decubitus position. The ascending aorta, anterior mediastinum and subcarinal region is best imaged from a right parasternal approach with patient in a left lateral position. The pulmonary trunk and left side of the anterior mediastinum are best imaged with a left parasternal approach with patient in right lateral position. Posterior masses are imaged from a posterior paravertebral approach. Further for large masses, a direct look through intercostal space can be undertaken. Mediastinal masses can be precisely characteristic cystic, vascular or calcified. By sonographic Doppler imaging precise localization of solid tissue in tumor achieved.

Diaphragm

Diaphragm is a fibromuscular band, which limits the thorax from abdomen. It is highly echogenic curvilinear band on USG. It shows normal movements. Ultrasound allows multiple scanning planes to visualize the diaphragm. Rt dome of the diaphragm is higher than the Lt dome of diaphragm due to heart depresses the Lt dome and not the liver the Rt dome pushes up. Rt dome of the diaphragm can be imaged to liver window and Lt through spleen. The biggest advantage of ultrasound is real time imaging.

Indications for Evaluation of Diaphragm

1. Sub-diaphragmatic fluid collection of abscess.
2. Rupture liver abscess in the sub-diaphragmatic space.
3. Traumatic rupture of the diaphragm.
4. Malignant metastatic deposits and invasion: Tumors of the diaphragm are rare. Benign tumors are lipoma, Neurofibroma, fibroma and the cyst. Pleural effusion is associated with sarcomas can present as tumor.
5. Diaphragmatic hernia.
   i. *Bochdelek hernia:* This hernia arises through patent pleuro-peritoneal canal. It is usually congenital and present at the time of birth. 90% hernias are Lt sided. The hernia may have bowel loop fat, omentum, spleen or kidneys. It results ipsilateral hypoplasia of lung with mediastinal shift to the Rt side.
6. *Hernia of Morgagni:* It is seen on the Rt side. It is rare due to maldevelopment of septum transversum. It is usually associated with pericardial defect. It appears as triangular mass in Rt cardiophrenic angle, anteriorly and medially.
7. *Eventration of diaphragm:* The Lt diaphragm is elevated with marked mediastinal displacement to the Rt side. It is due to inherent muscle weakness, which is thin and reduced in thickness.

Ultrasonography in Invasive Procedures of Thorax

Ultrasonographic delineation of structures helps in invasive procedures for diagnosis or treatment of condition related to pleura, parenchyma or mediastinum. It adds accuracy and safety to the procedures without increasing cost significantly or radiation burden. It is particularly useful when it is difficult to move the patient or decide if pleural effusion is subpulmonic or subphrenic. Vital structures like heart and aorta can be identified and avoided. A safe site for thoracocentesis is chosen, based upon careful diagnostic ultrasound examination that uses the direct approach. Although the optimal position for diagnostic thoracocentesis is the erect sitting position with the patient's arms resting comfortably on a bedside table but if required lateral decubitus or supine position can be used if patient is unable to sit.

The puncture site in intercostal space be so chosen that the needle crosses the top of the rib and avoids the neurovascular bundle that courses along the undersurface of the rib. After local asepsis and anesthesia the needle puncture should be perpendicular to the chest wall while entering pleural space. A characteristic 'pop' can usually be felt as the needle pierces the parietal pleura. While a 22-gauze needle is usually recommended in literature, our own experience suggests that an 18 or a 20 needle is more appropriate. Aspirate may then be subjected to a physical, biochemical, and microbiological and/or pathological assessment as the case may be. In therapeutic drainage of effusion, it must be remembered that not more than one liter fluid be removed in a drainage of empyema, catheters can be removed when less than 10 ml of fluid drains from pleural spaces in 24 hours. A bronchopleural fistula should be suspected in cases that show presence of both air and fluid in the pleural space, or are slow to resolve even after catheter placement.

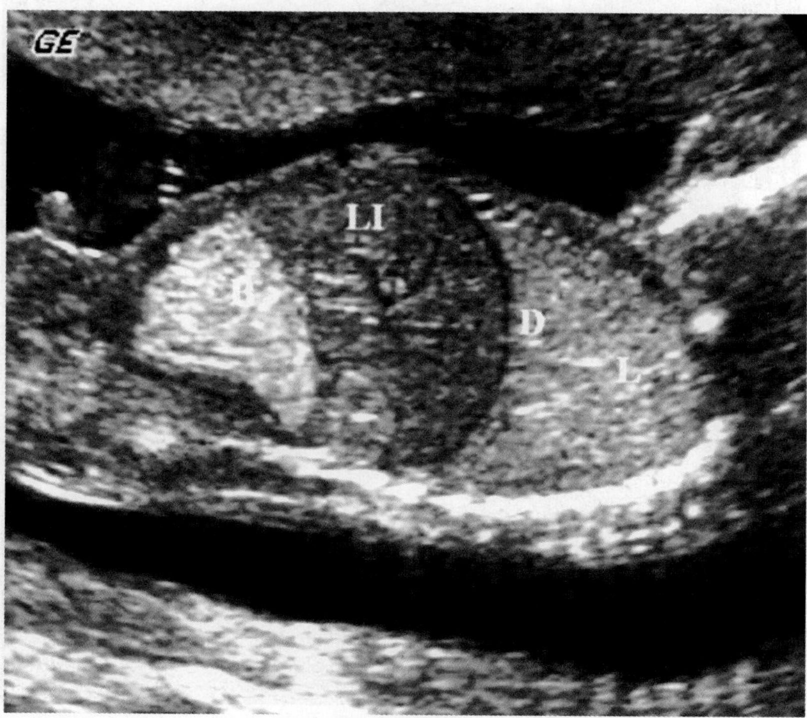

FIGURE 6.112: Fetal thorax. HRSG shows fetal thorax, abdomen and the diaphragm. The diaphragm is seen as a hypoechoic curvilinear band separating the abdominal cavity and the thorax.

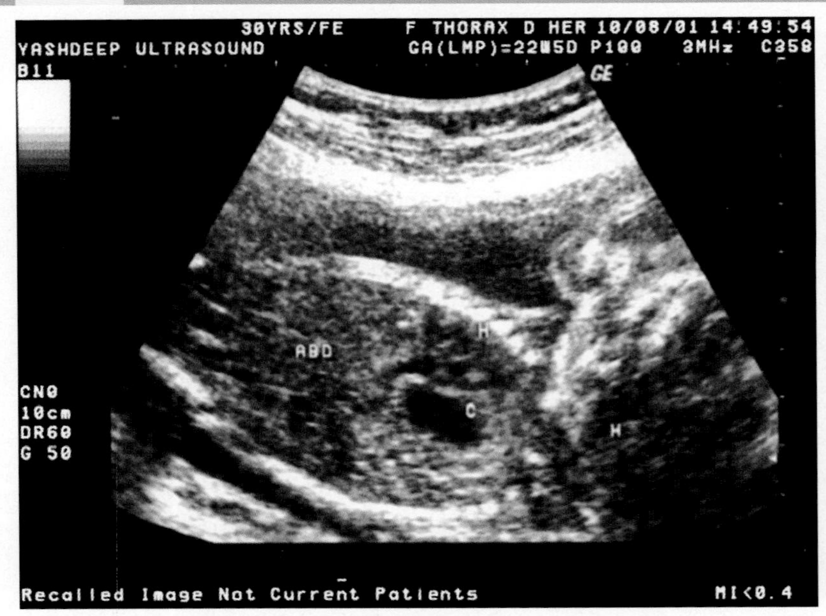

FIGURE 6.113: Diaphragmatic hernia. HRSG shows a fetal thorax having a cystic mass in the thorax suggestive of herniated stomach in the fetal thorax.

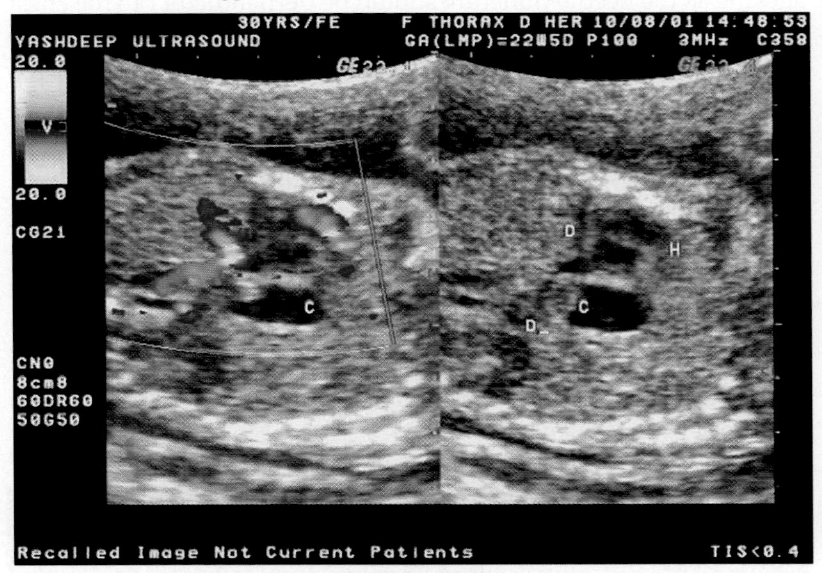

FIGURE 6.114

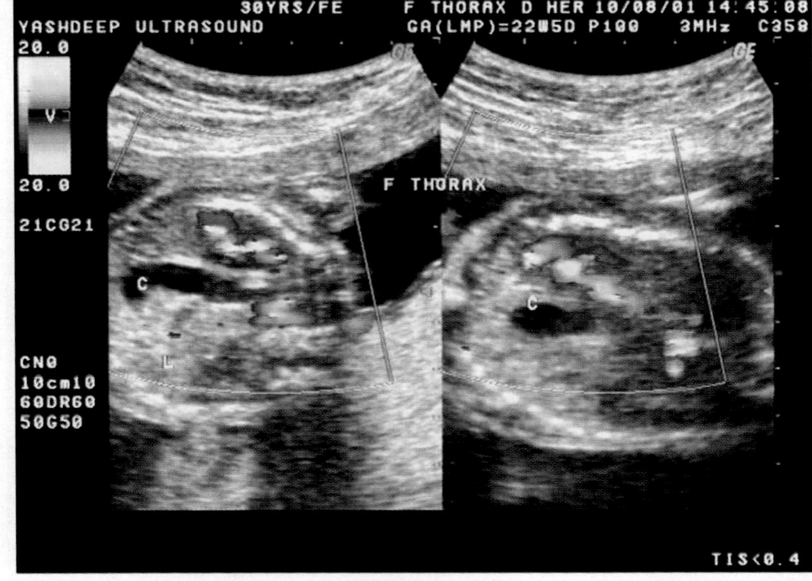

FIGURE 6.115

FIGURES 6.114 and 6.115: Color Doppler imaging of the patient shows shifting of the heart to the Rt side. Herniated stomach is seen as a non-vascular cystic structure.

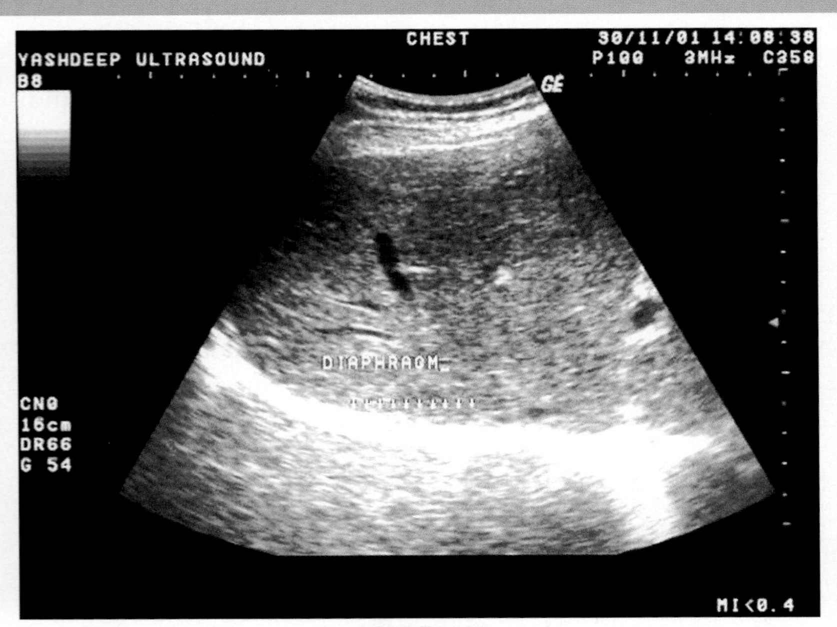

FIGURE 6.116

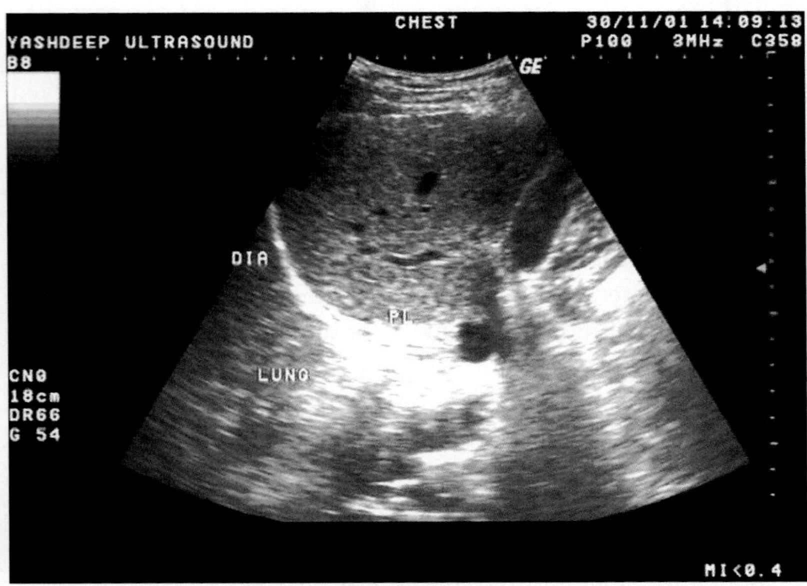

FIGURE 6.117

FIGURES 6.116 and 6.117: Normal diaphragm. Normal diaphragm is seen as a hyper-reflective echogenic curvilinear band limiting the inferior surface of the liver. The pleural shadow is seen posterior to the diaphragm casting strong reverberation artifacts.

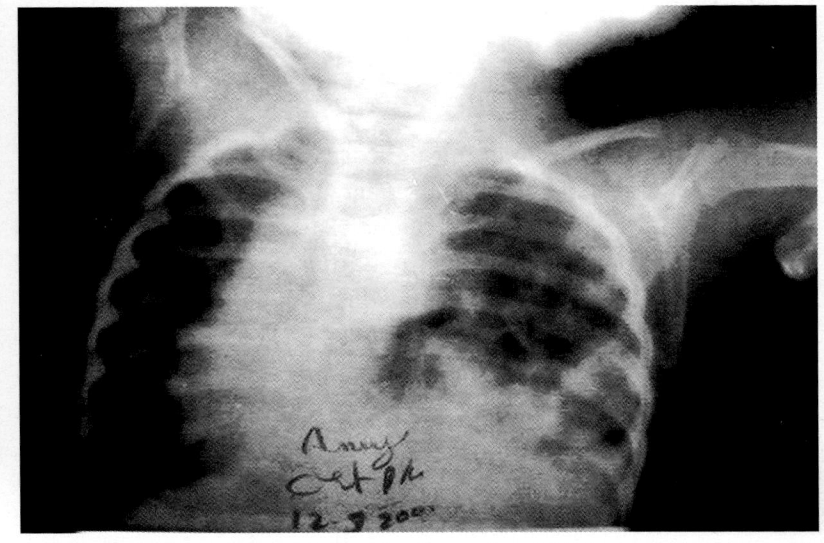

FIGURE 6.118: Diaphragmatic hernia. X-ray of a child shows a loculated cystic mass in the Lt thorax. The heart is seen shifted in the Rt thorax.

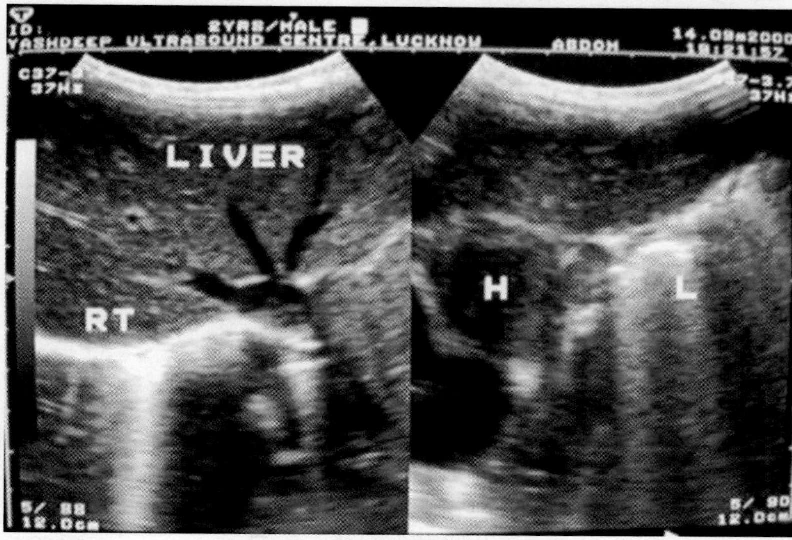

FIGURE 6.119

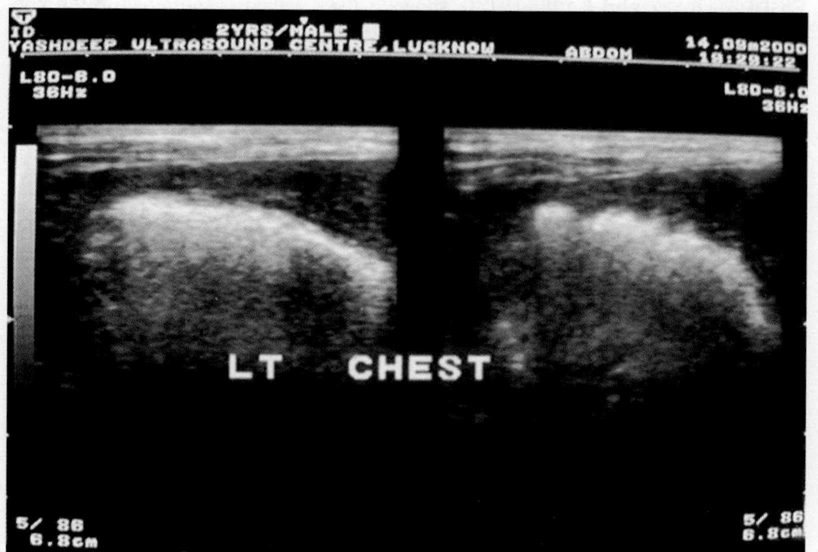

FIGURE 6.120

FIGURES 6.119 and 6.120. HRSG shows a herniated bowel loop in the thorax. The bowel loop is filled with air and seen as an echogenic shadow in the thorax. Heart is also seen pushed to the Rt side.

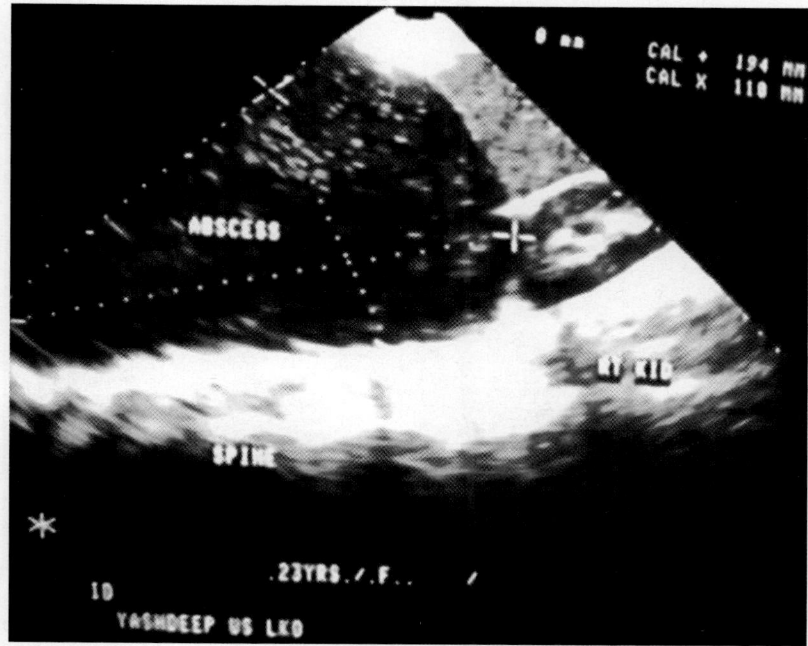

FIGURE 6.121

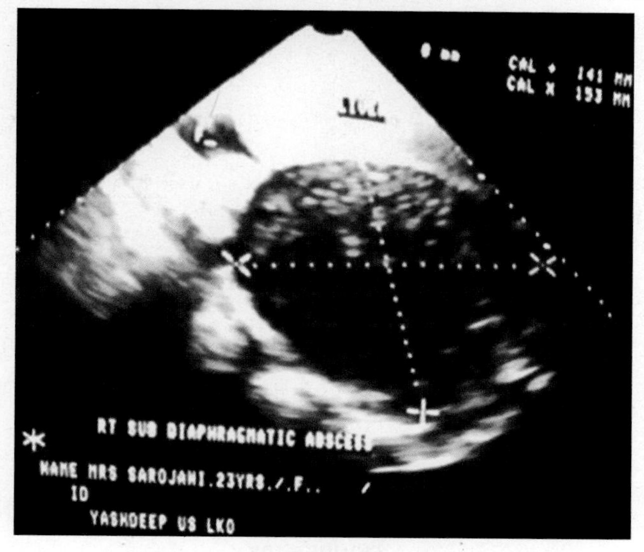

FIGURE 6.122

FIGURES 6.121 and 6.122: Sub-diaphragmatic abscess. HRSG shows thick low level echo collection in the Rt sub-diaphragmatic space. The collection is seen well defined and seen in the posterior recess suggestive of a big sub diaphragmatic abscess.

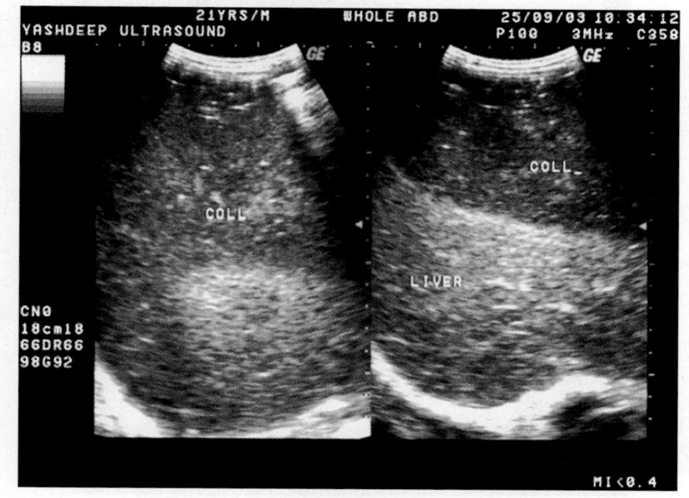

FIGURE 6.123

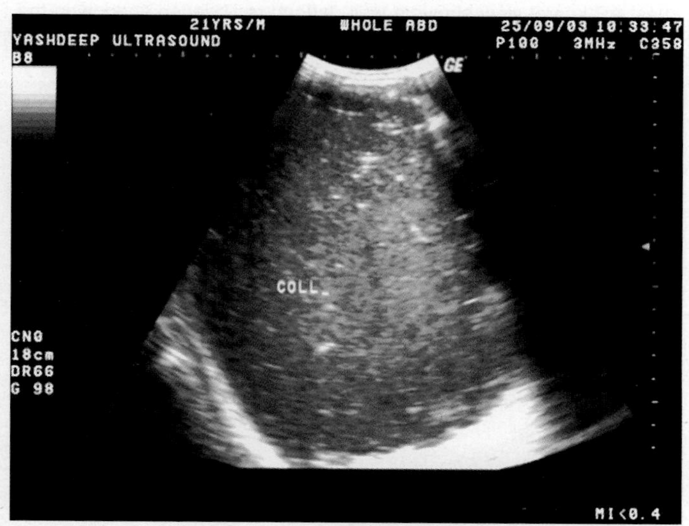

FIGURE 6.124

FIGURES 6.123 and 6.124: Big sub-diaphragmatic abscess. A patient presented with high fever with dyspnea. HRSG shows big collection in the Rt sub-diaphragmatic space. Thick echoes are seen in it. Aspiration of the collection came out to be pyogenic collection.

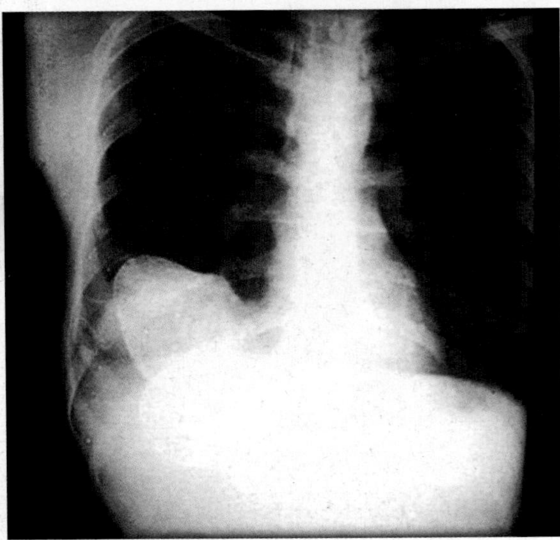

FIGURE 6.125

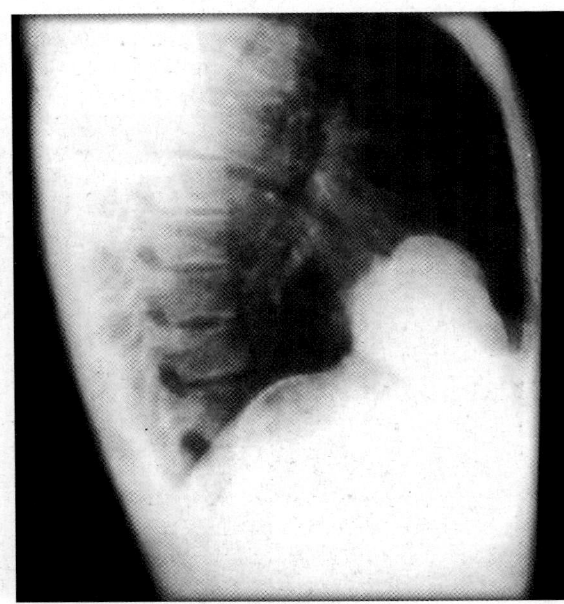

FIGURE 6.126

FIGURES 6.125 and 6.126: Hydatid cyst in the sub-diaphragmatic area. Plain X-ray chest AP lateral view shows a well-defined radiodense shadow in the Rt sub-diaphragmatic space. It is having sharp margins with wavy outline. Diaphragm lifting is also seen.

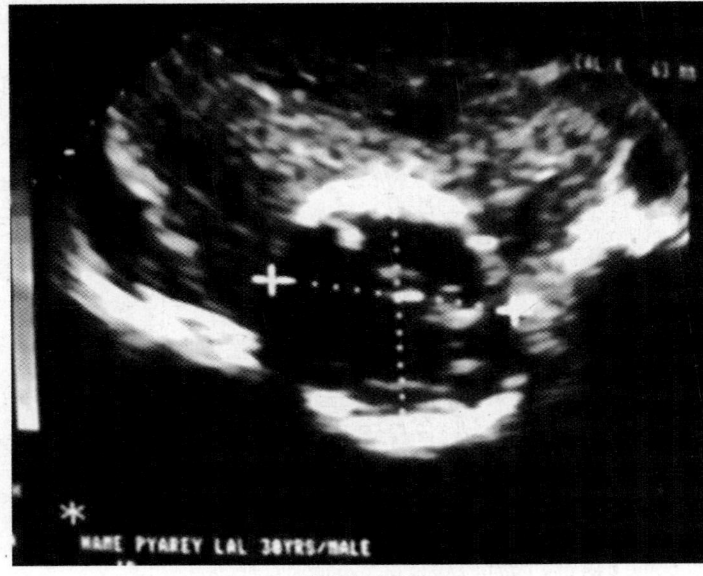

FIGURE 6.127

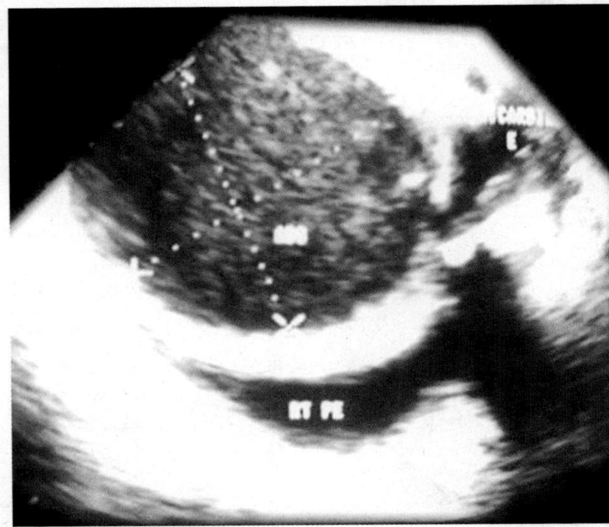

FIGURE 6.128

FIGURES 6.127 and 6.128: HRSG shows a thick walled cystic mass in the sub-diaphragmatic space. Echogenic calcification is seen in the mass. Calcified specks are also seen in it suggestive of hydatid cyst. Small associated pleural fluid is also seen in the Rt pleural cavity.

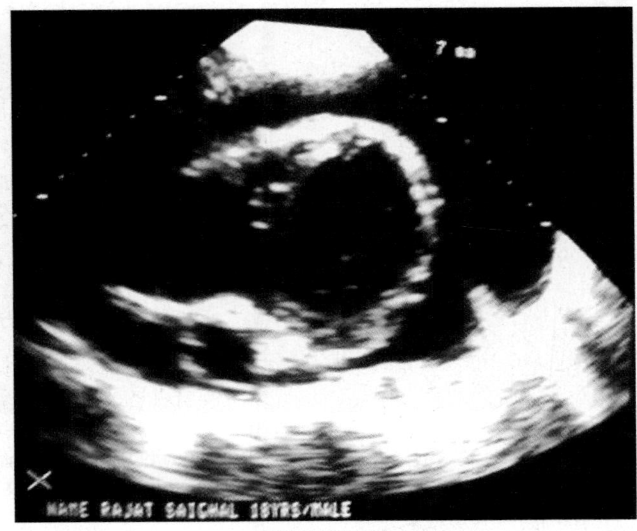

FIGURE 6.129

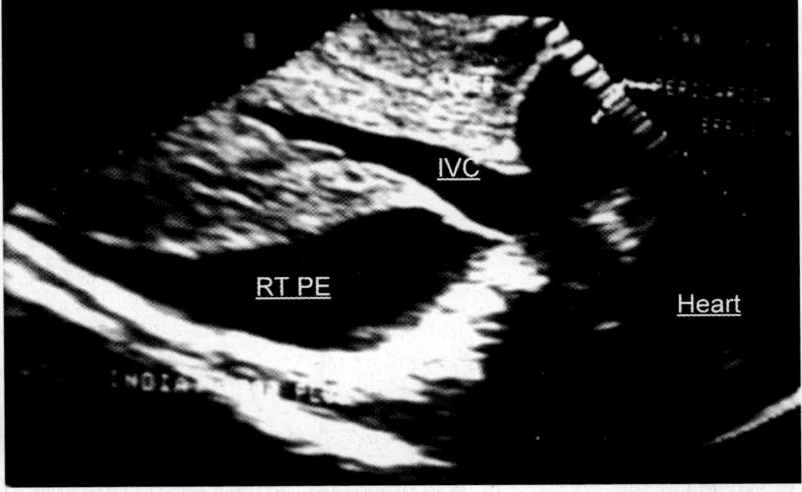

FIGURE 6.130

FIGURES 6.129 and 6.130: Massive pericardial effusion. HRSG shows a large pericardial effusion. The heart is seen compressed. The liver shows congested hepatic veins and inferior vena cava in a patient of cardiac temponad.

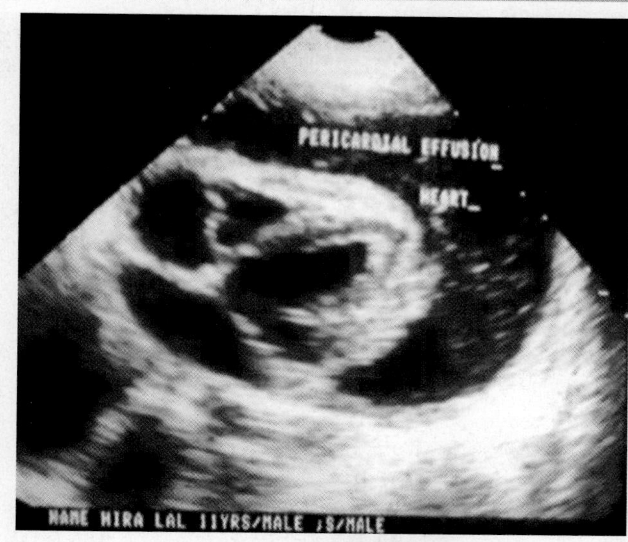

FIGURE 6.131: **Pericardial effusion.** HRSG shows thick cheesy collection in the pericardium. The heart is compressed due to the collection. Multiple internal echoes are seen in the collection in a case of tubercular pericarditis.

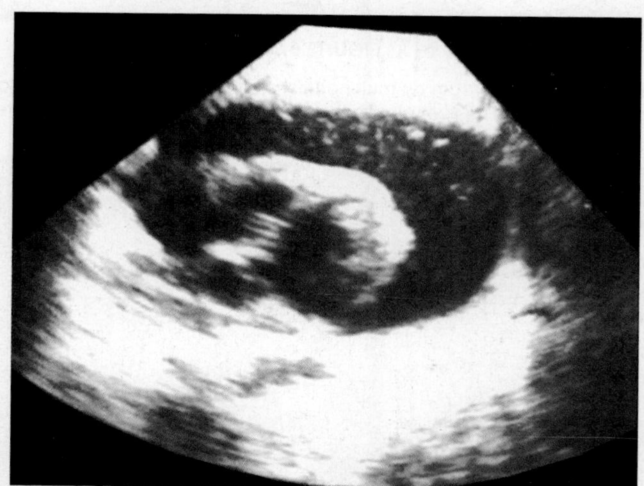

FIGURE 6.132: **Tubercular pericarditis.** HRSG shows thick echogenic collection in the pericardium in a case of tubercular pericarditis. The heart is shrunken and compressed in a case of tubercular constrictive pericarditis.

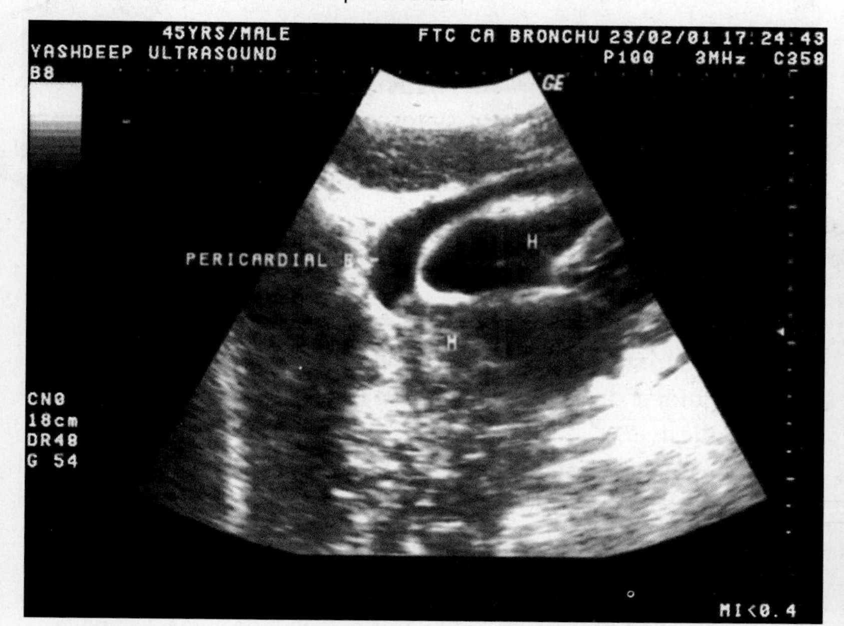

FIGURE 6.133: **Metastatic deposit in the pericardium.** HRSG shows a soft tissue mass shadow invading the pericardium in a case of Ca bronchus. Small amount of fluid collection is also seen in the pericardium.

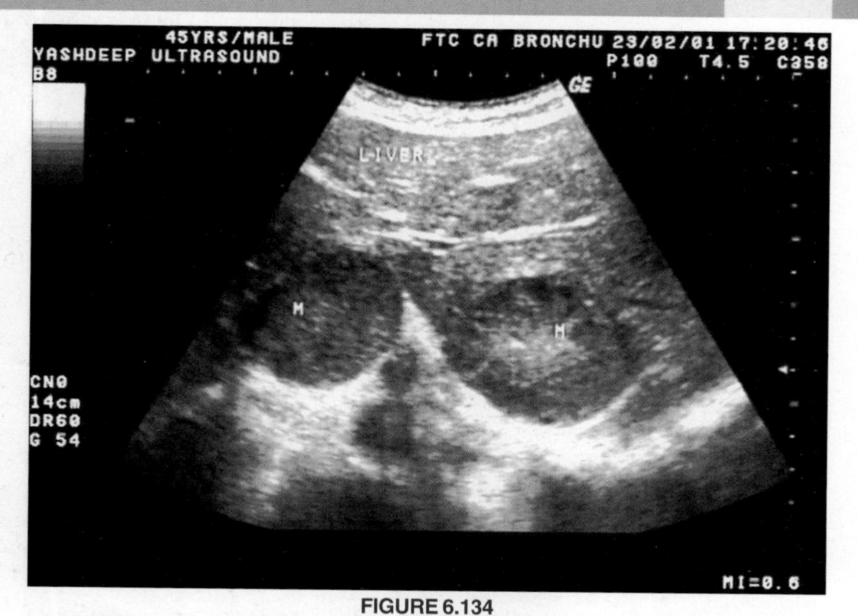

FIGURE 6.134

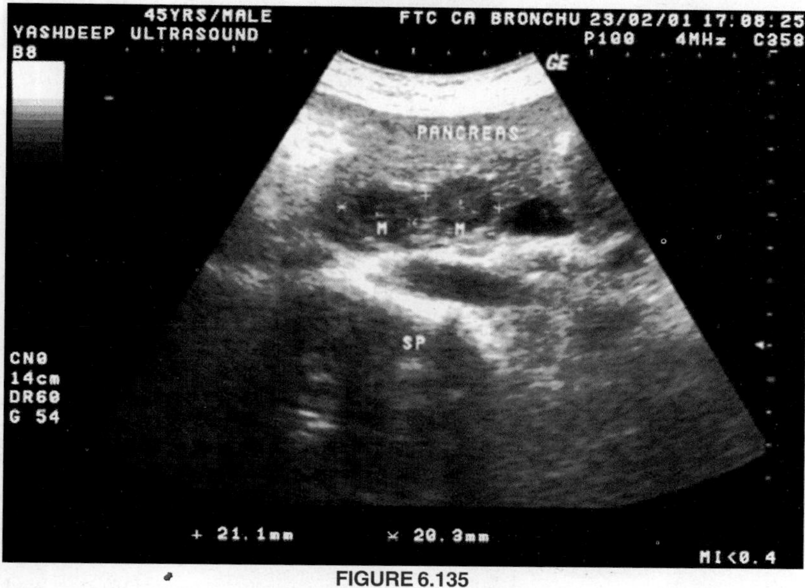

FIGURE 6.135

FIGURES 6.134 and 6.135: Metastatic deposits are also seen in both lobes of the liver in the same patient. Hypoechoic nodular masses are seen in the pancreas in the same patient of Ca bronchus. Biopsy of the masses shows metastasis of adeno-carcinoma.

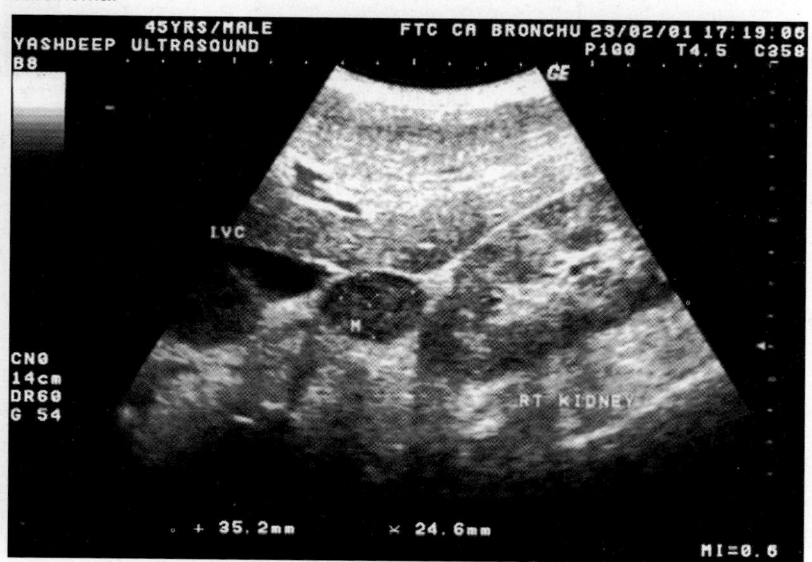

FIGURE 6.136: HRSG also shows Rt adrenal metastasis in the same patient of Ca bronchus.

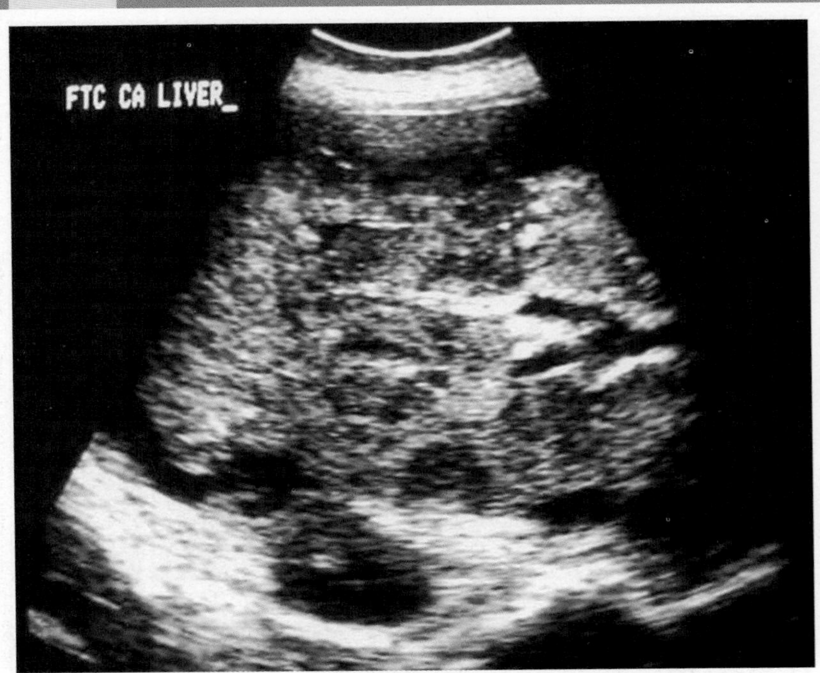

FIGURE 6.137

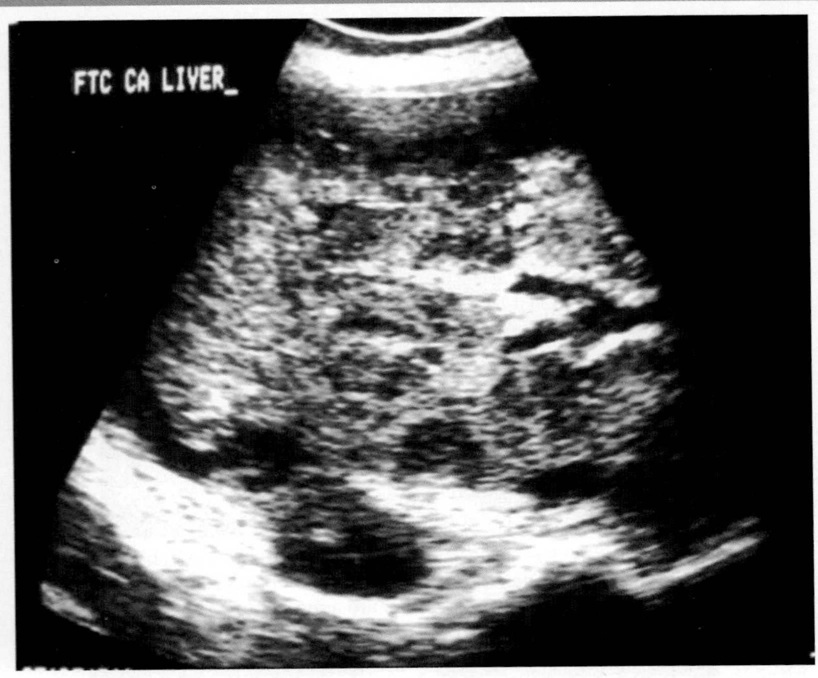

FIGURE 6.138

FIGURES 6.137 and 6.138: Metastatic invasion of diaphragm. HRSG shows destruction of Rt dome of diaphragm by multinodular heterogeneous mass in Rt lobe of the liver. The diaphragmatic sleeve is destroyed and eaten up by the mass in a case of acute lymphatic leukemia. The liver shows the leukemic infiltrate in the liver parenchyma.

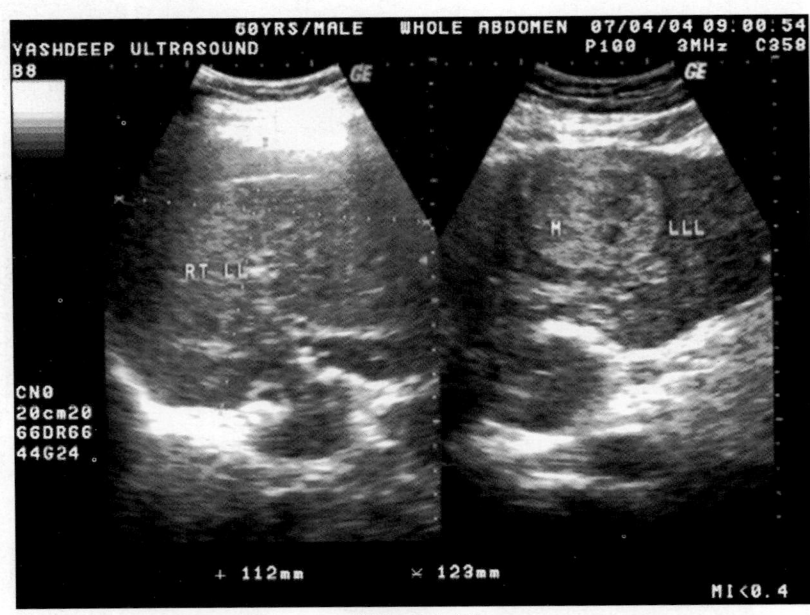

FIGURE 6.139

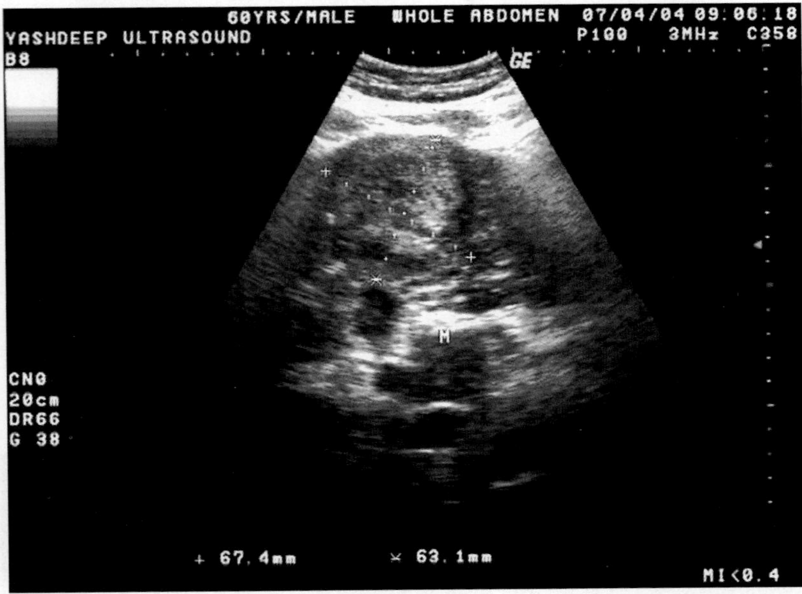

FIGURE 6.140

FIGURES 6.139 and 6.140: Metastatic mass in the Rt dome of the diaphragm. HRSG shows a hypoechoic mass invading the Rt lobe of the liver posterior segment in a case of carcinoma bronchus breast. The mass is also seen invading the Rt dome of the diaphragm.

Ultrasonographic guidance for biopsy of pleural, parenchymal or mediastinal masses not only ascertain proper placement of needle tip but also helps to avoid puncturing of aerated lung or vasculature. Ultrasonography is particularly useful in guiding biopsy of peripheral lung tumors obscured by pleural effusion. For this procedure the patient is asked to suspend respiration while the needle is being advanced in to the lesion. The patient may then resume shallow respiration while the needle is allowed to swing freely. The patient is again asked to stop breathing while the biopsy is taken. The needle is usually guided directly into abnormal area or can also be passed through the liver.

Of the mediastinal lesions, large lesions that about the chest wall and those in the anterior mediastinum are the most amenable to ultrasound-directed biopsy.

Breast Ultrasound

7

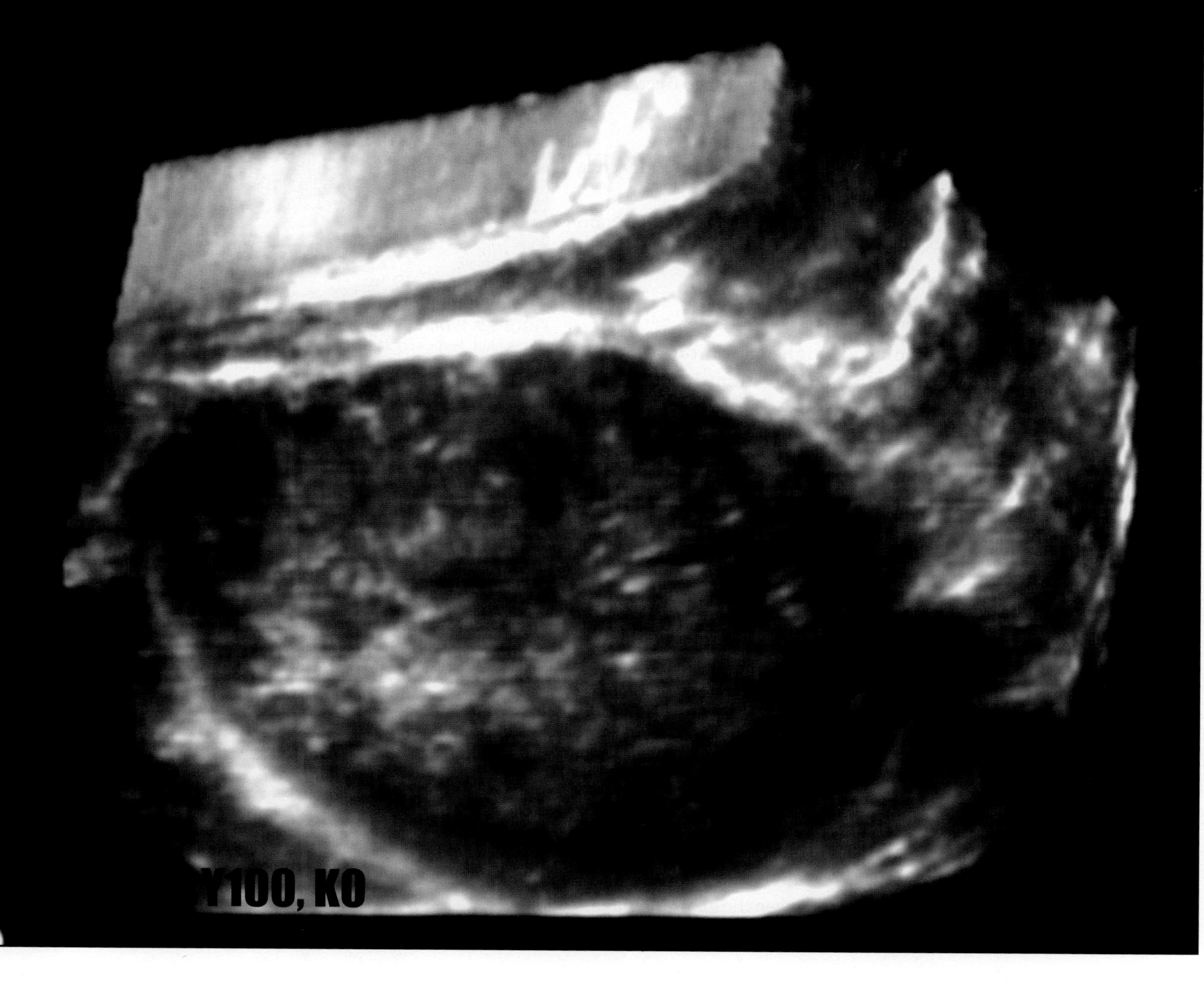

Y100, KO

Introduction

The breast or mammary glands are modified sweat glands. They are exocrine organs. Their main function is the secretion of milk during pregnancy (lactation). Breast being superficial organs is ideal for HRSG evaluation. The other techniques like CT scanning, MRI or mammography has got limitations. Sonomammography due to its versatility has become a choice of investigation for breast imaging. Though X-ray mammography is still preferred for mass screening program for detection of breast cancer, but with use of sonomammography the nature of lesion, its surrounding tissue invasion and lymph node enlargement can be better evaluated.

Normal Anatomy

Breast is superficially placed anterior to pectoralis major, serratus anterior and external oblique muscles and the sixth rib. It is bounded medially by sternum and laterally by the margins of axilla. Second or third ribs and inferior border form the superior border by the seventh costal cartilage.

The size of normal breast varies depending on the age, functional state of the breast and amount of fibro adipose tissue in the breast. There is increase in the size in puberty due to stimulation by estrogen hormone and decrease in the size after menopause due to decrease in hormonal stimulation. The breast has got three layers.

1. *Subcutaneous layer*—It contains skin and subcutaneous fat.
2. *Mammary layer*—It contains glandular tissues, ducts and connective tissue.
3. *Retromammary layer*—It contains retromammary fat, muscle and deep connective tissue.

Scanning Method

Since breast is superficially placed, high-resolution real-time transducers of 7.5 and 10 MHz are the requisite transducers. Linear transducer is preferred over sector and convex probes. The patient is examined in sitting position. It allows a full thickness study of breast tissue.

The biggest advantage of breast ultrasonographic imaging is to assess the mobility of the breast tissue and any other lesion in the breast. Use of high frequency transducers as 10 MHz can give additional information about cancer nidus or micro calcification and reactive halo around the nidus. Doppler ultrasonography can give better information about benign or malignant mass in the breast.

Normal Sonoappearance of Breast

Normal ultrasonographic appearance of breast depends on many factors like age of patient, size of breast and functional state of the breast. Breast tissue is seen in three distinct layers. Most anterior is skin and subcutaneous fat. It appears to be a highly reflective layer on HRSG. The middle layer contains the glandular tissue, which is seen sandwiched between subcutaneous layer and retro mammary layer. Rounded fat lobules are seen distributed between the glandular layers. They are uniformly reflective in young breast. The ducts are seen as tubular structures running towards the nipple or areola. The maximum normal caliber of ducts is 3 mm. In postmenopausal woman, glandular tissue decreases in amount and fatty tissue predominates and it becomes less echogenic. The ducts are also not seen in this breast. The young breasts are denser due to high percentage of glandular tissue. Dense breast is difficult to be seen on X-ray mammography. Therefore, sonomammography is better investigation to evaluate young breast. Nipple and areola attenuate the sound beam. This casts the shadowing, which is seen as a hypoechoic band posterior to nipple and areola. Glandular tissue predominates during pregnancy and ducts are dilated and radiating towards nipple from all directions. They can be seen well on HRSG. The axilla contains fibro fatty layer. It is having mixed echo texture. Normal lymph nodes are not identified on HRSG.

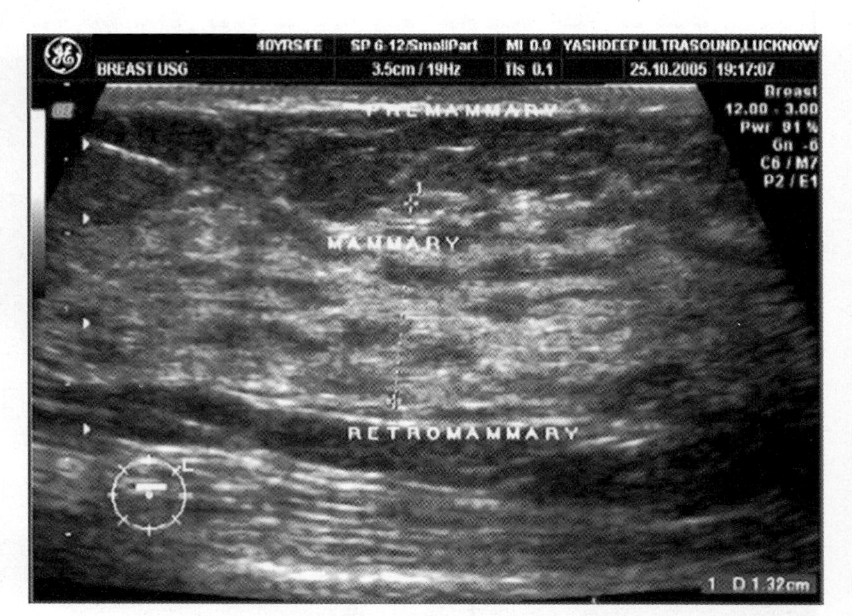

FIGURE 7.1: Normal breast. HRSG shows the normal breast appearance. Echogenic line is skin. Hypoechoic layer posterior to the skin is subcutaneous fat. A hypoechoic layer after the subcutaneous fat is premammary layer. The mammary layer shows main glandular components of the breast. Multiple fibroglandular lobules are seen dispersed in the glandular tissue. The retroperitoneum layer is seen posterior to the mammary chest wall and a hyper-reflective layer is seen after the chest wall is the pleura.

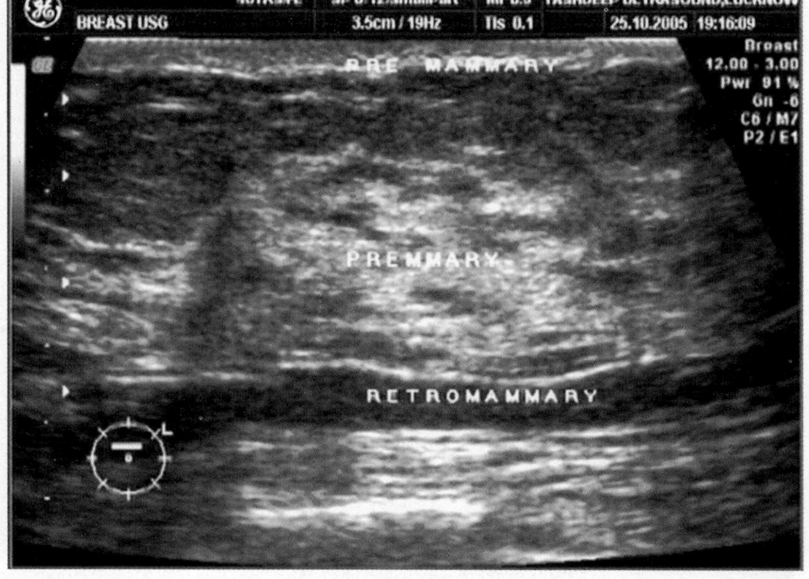

FIGURE 7.2

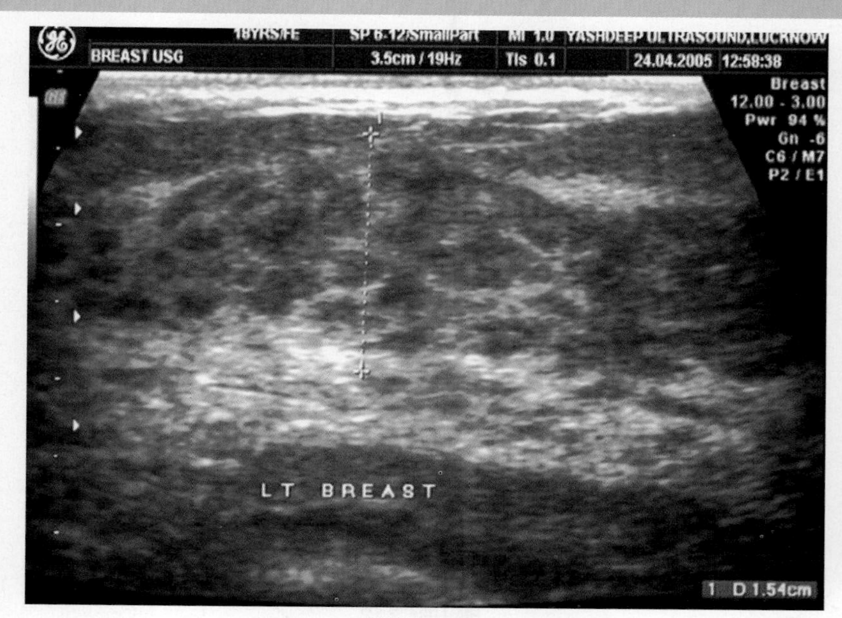

FIGURE 7.3

FIGURES 7.2 and 7.3: Normal breast. HRSG shows normal breast. The skin, subcutaneous fat, mammary layer and retromammary layers are well visualized.

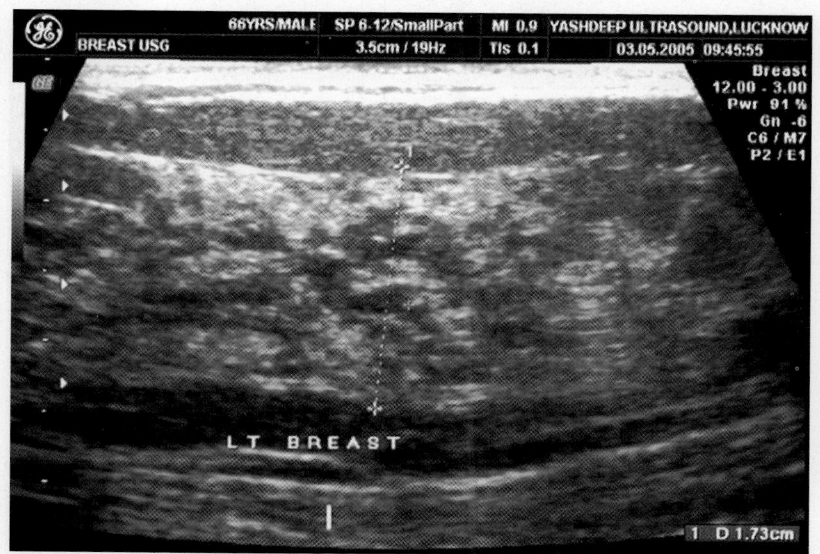

FIGURE 7.4: Periareolar ducts. HRSG shows few ducts seen running to the areola. Normal duct diameter should not exceed 3 mm.

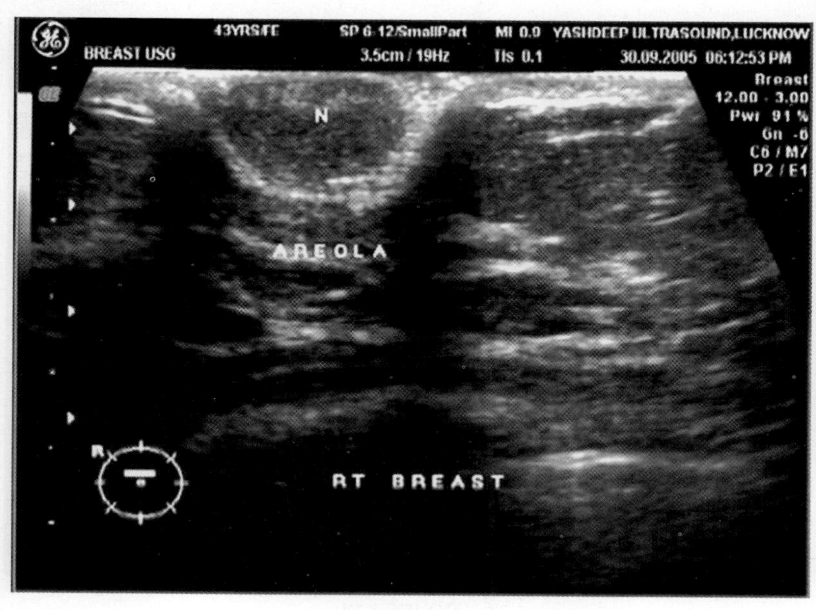

FIGURE 7.5

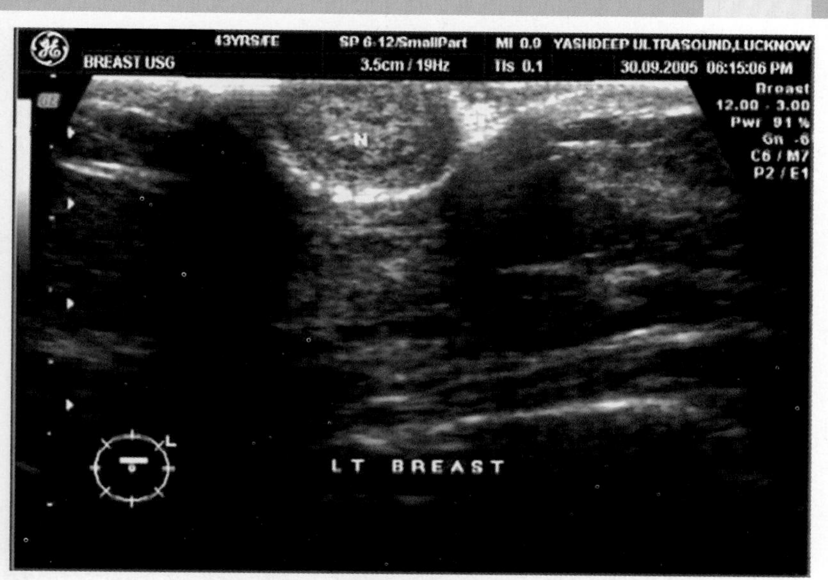

FIGURE 7.6

FIGURES 7.5 and 7.6: Nipple areola complex. HRSG shows nipple areola complex. Nipple is seen a hypoechoic well-defined oval shadow. Echogenic border of the nipple is seen. Normal nipple cast shadowing.

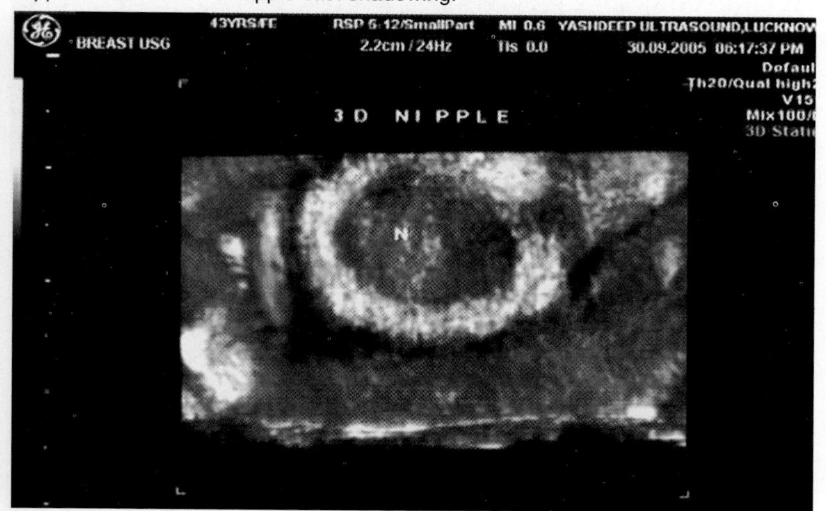

FIGURE 7.7

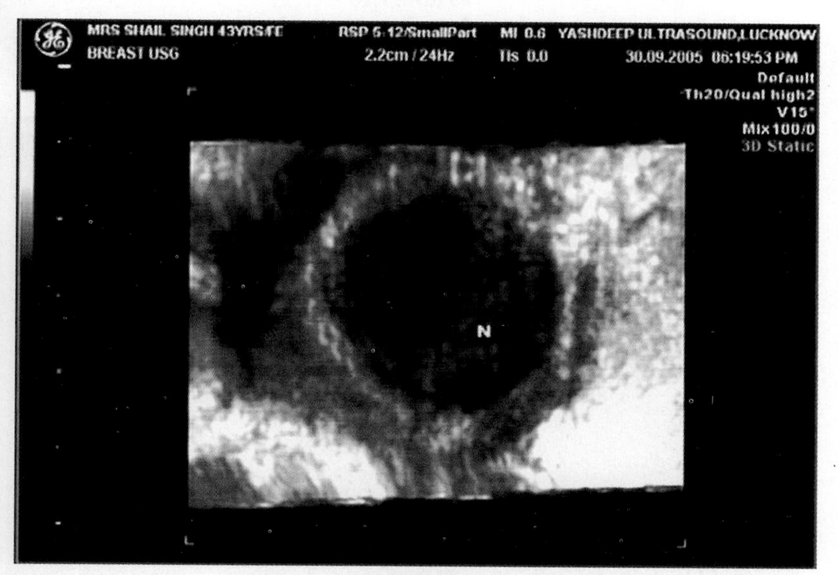

FIGURE 7.8

FIGURES 7.7 and 7.8: 3D imaging of the nipple areola complex. 3D imaging of the nipple areola: Complex shows details of the nipple. The retroareolar shadowing is removed from the scanning plane. Therefore any lesion in the retroperiareolar region, which is usually masked by the shadowing, can be easily evaluated on 3D imaging.

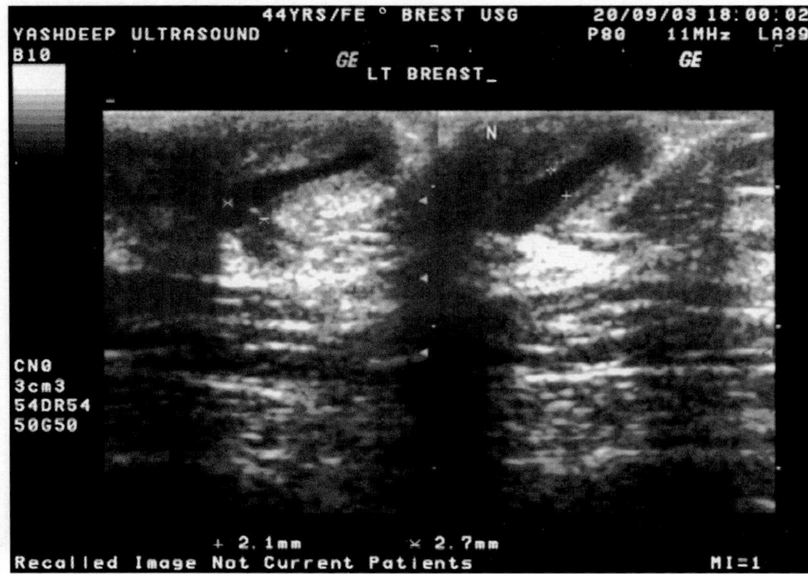

FIGURE 7.9

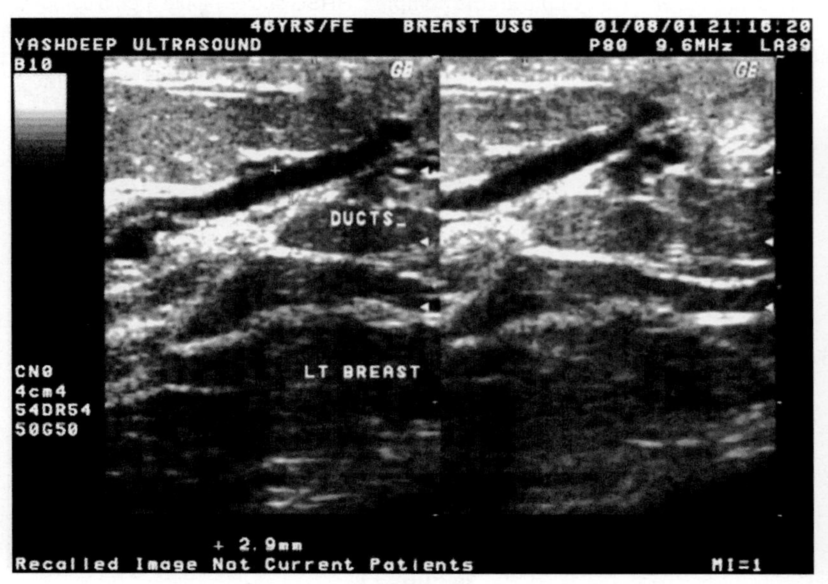

FIGURE 7.10

FIGURES 7.9 and 7.10: Normal ducts. HRSG shows normal lactiferous ducts running to areola. A single duct is seen running through the nipple. Normal duct diameter should not exceed more than 3 mm in non-lactating breast.

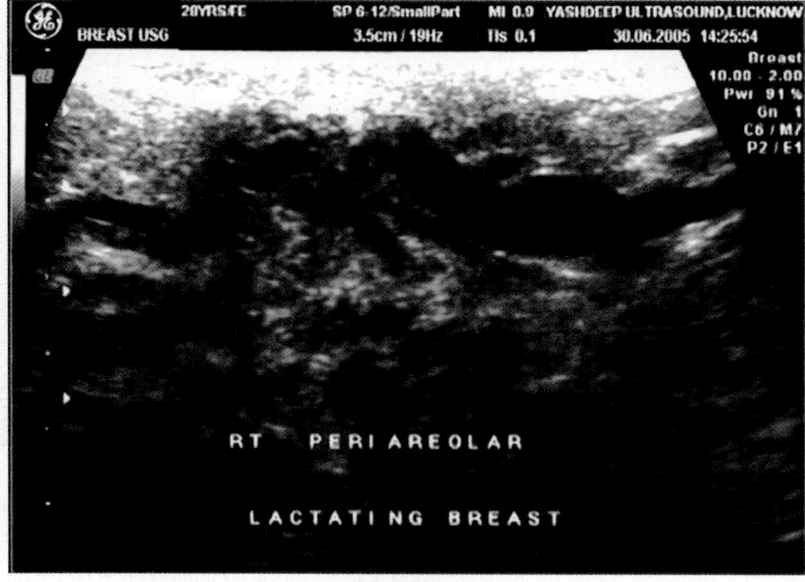

FIGURE 7.11

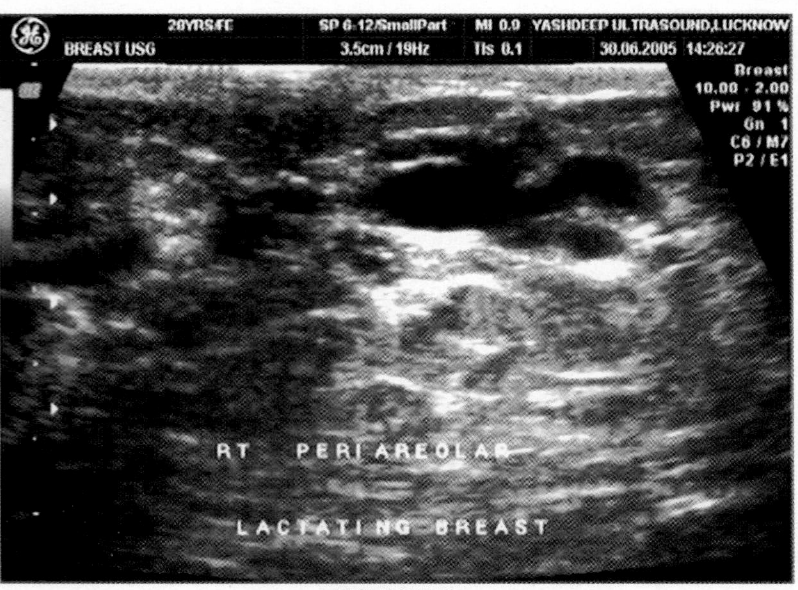

FIGURE 7.12

FIGURES 7.11 and 7.12: Dilating ducts in a lactating mother. Lactating breast shows multiple dilated ducts. They are running to the areola from all the directions in the spider leg fashion. The ducts are engorged and should not be mistaken as ductal ectasia.

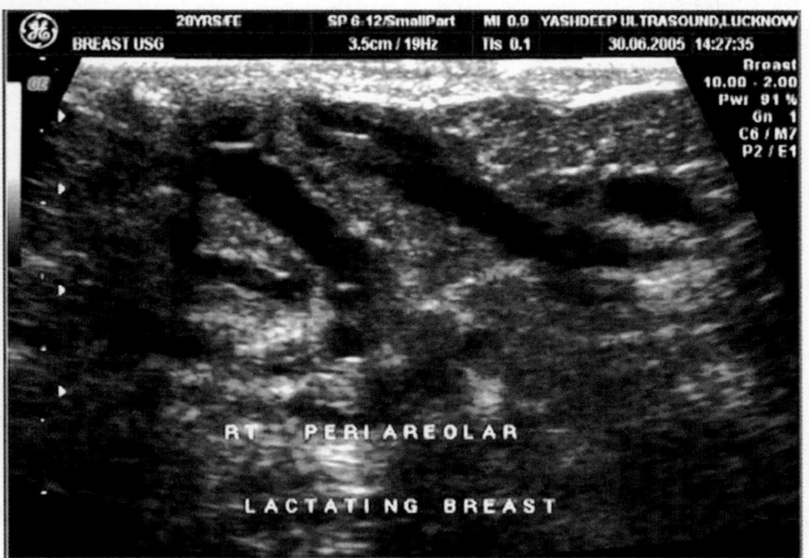

FIGURE 7.13

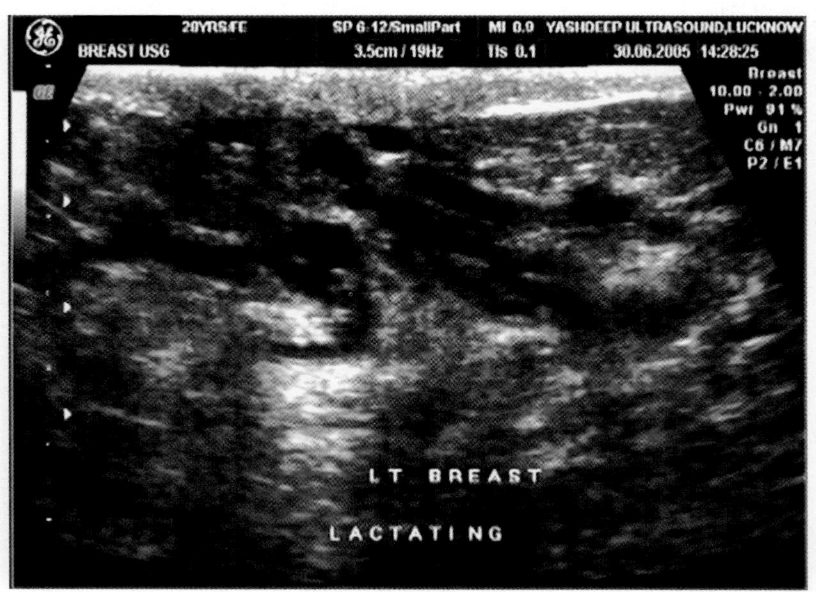

FIGURE 7.14

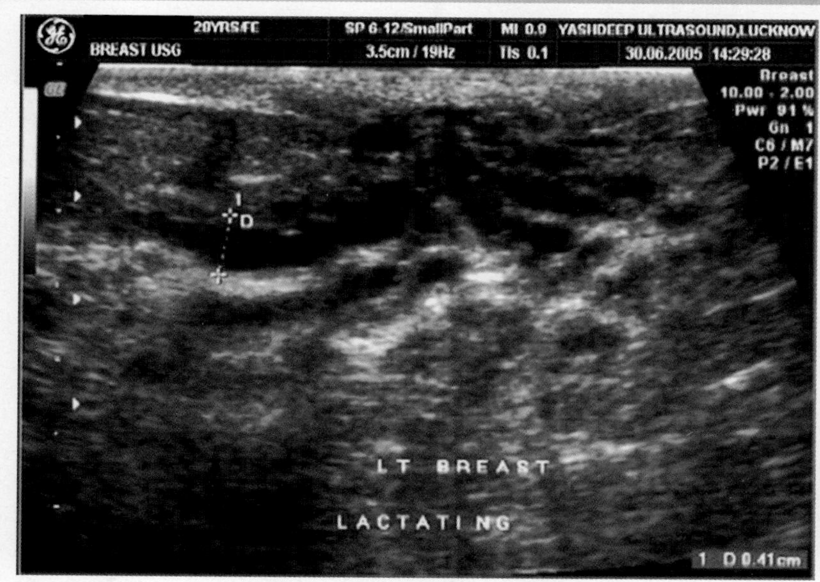

FIGURE 7.15

FIGURES 7.13 to 7.15: Dilated ducts in lactating breast. Multiple dilated ducts are seen running from all the directions to areola in lactating breast in the same patient.

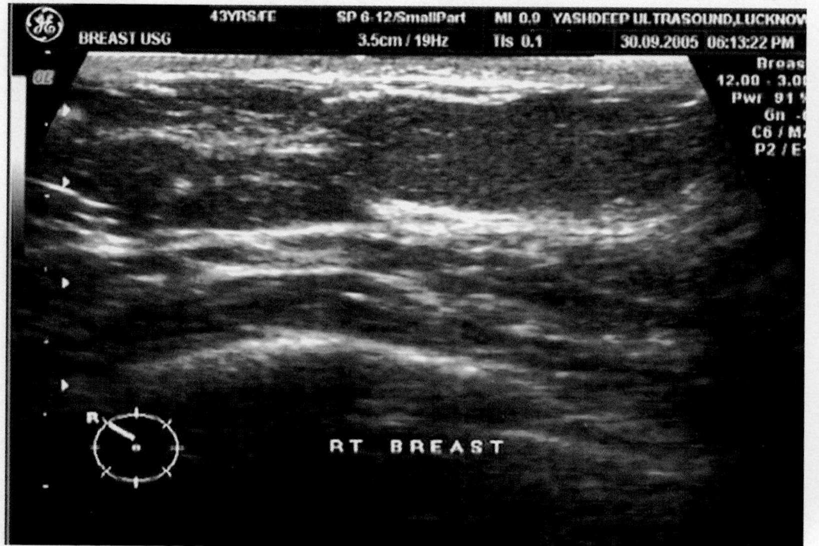

FIGURE 7.16

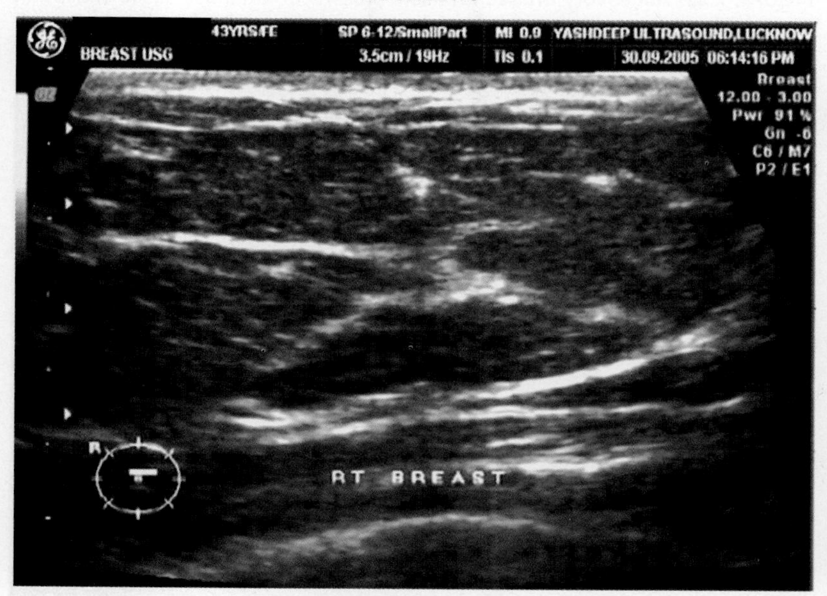

FIGURE 7.17

FIGURES 7.16 and 7.17: Normal breast in middle aged woman. HRSG shows normal breast appearance in middle-aged woman. The amount of glandular tissue is reduced in this age group. The fatty tissue starts getting dominant in the breast.

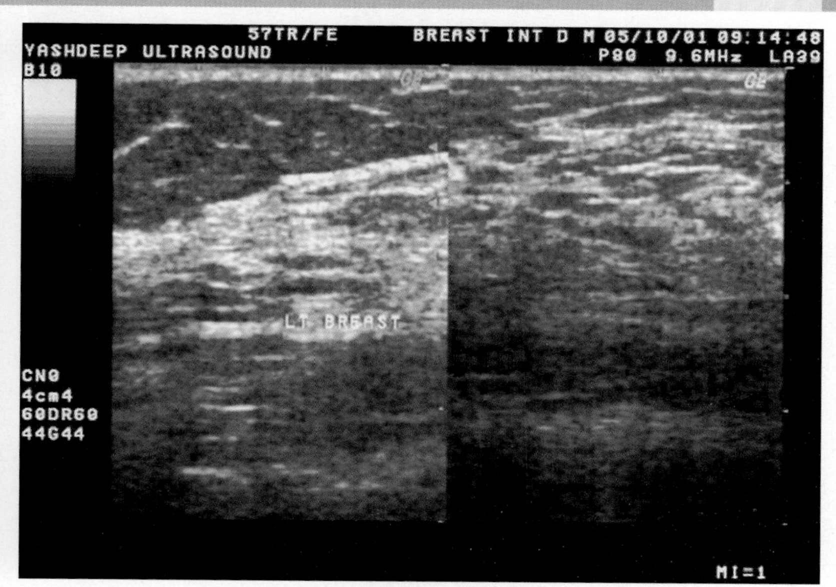

FIGURE 7.18

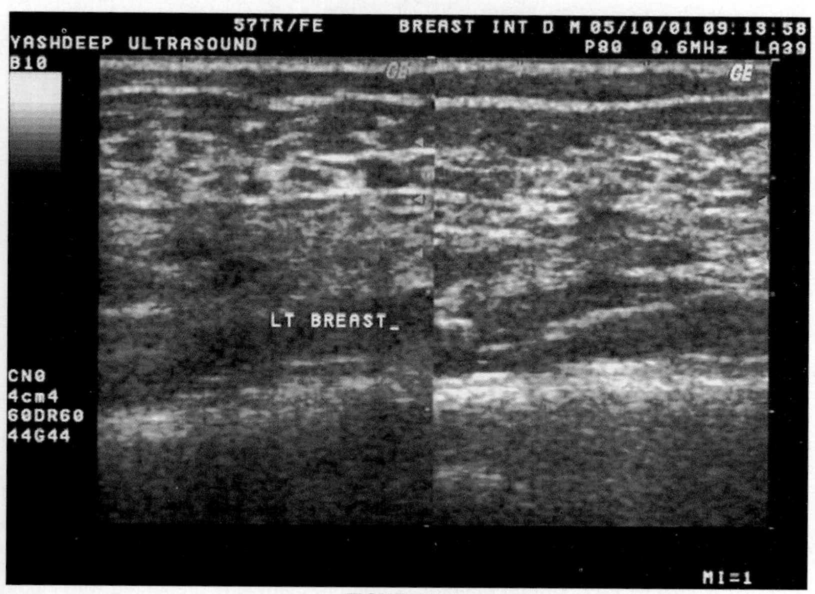

FIGURE 7.19

FIGURES 7.18 and 7.19: Postmenopausal breast. HRSG shows postmenopausal breast. The glandular tissue is decreased n amount and fatty tissue predominates in postmenopausal breast.

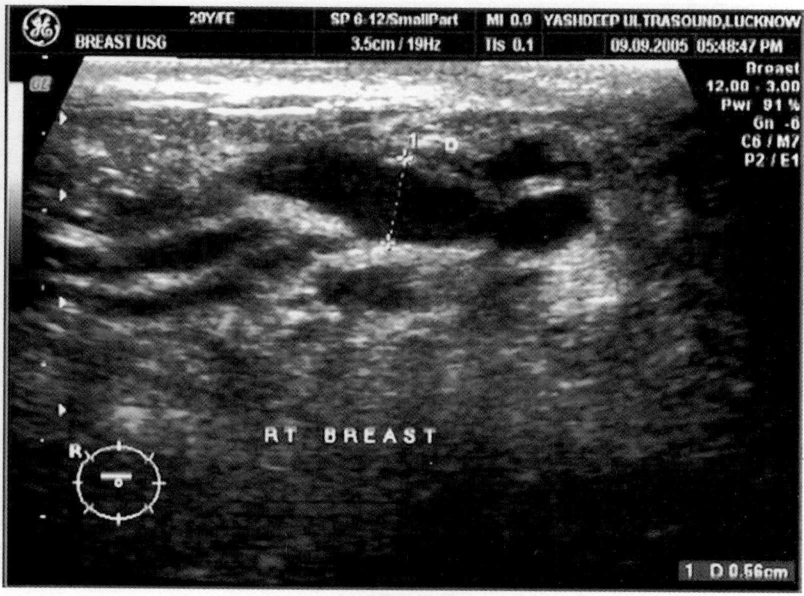

FIGURE 7.20

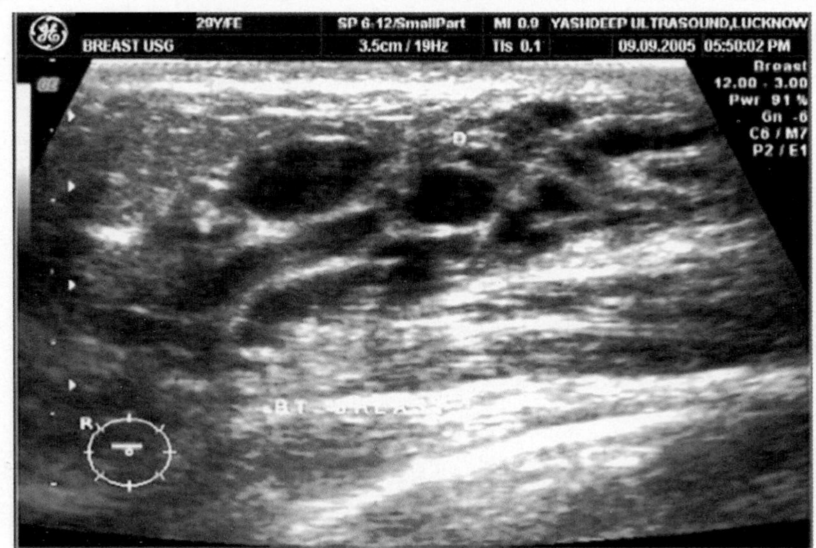

FIGURE 7.21

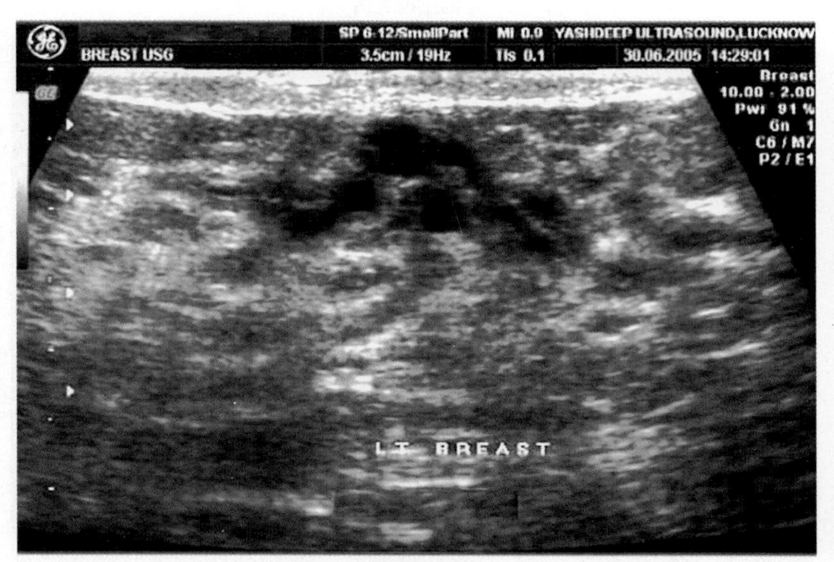

FIGURE 7.22

FIGURES 7.20 to 7.22: Ductal ectasia. A non-pregnant lady presented with nipple discharge from both the breasts. HRSG shows dilated ducts in both the breasts. No intraductal pathology is seen. The duct size was more than 3 mm sonographic findings are ductal ectasia.

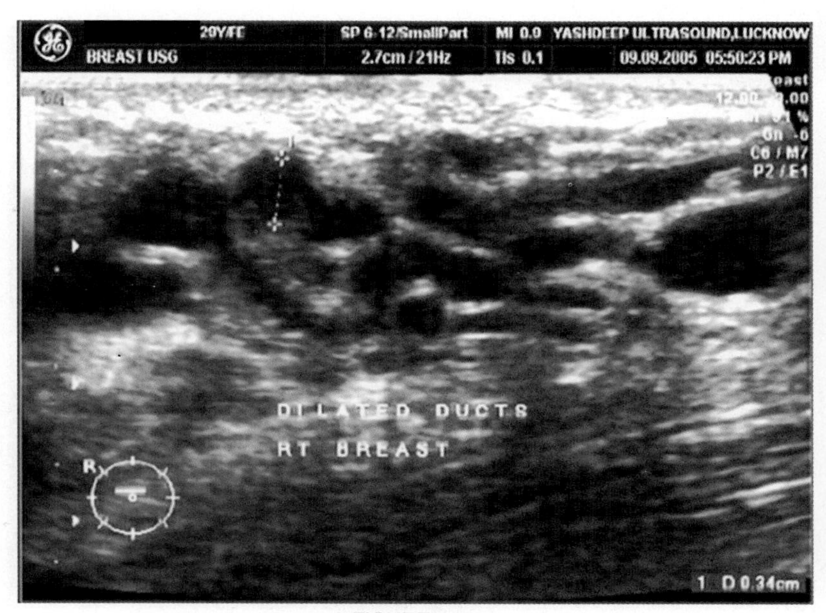

FIGURE 7.23

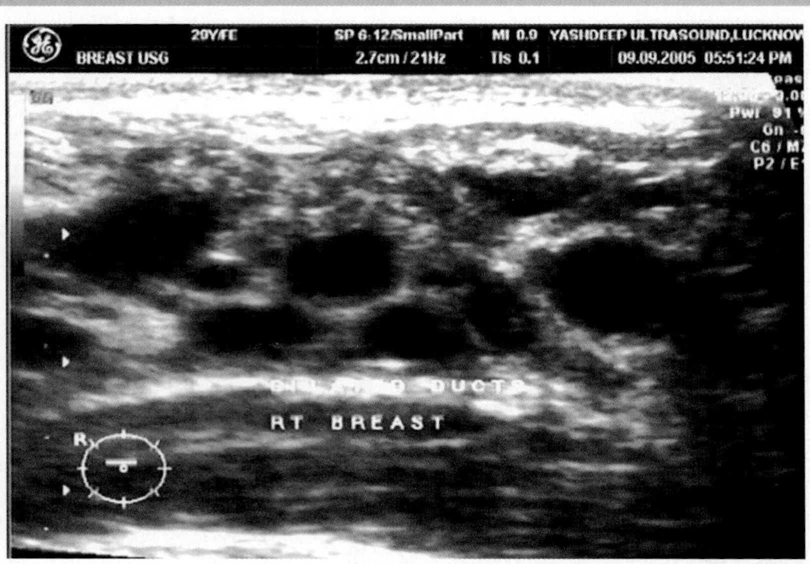

FIGURE 7.24

FIGURES 7.23 and 7.24: The magnified and cross sectional view of the breast show the details of the dilated ducts.

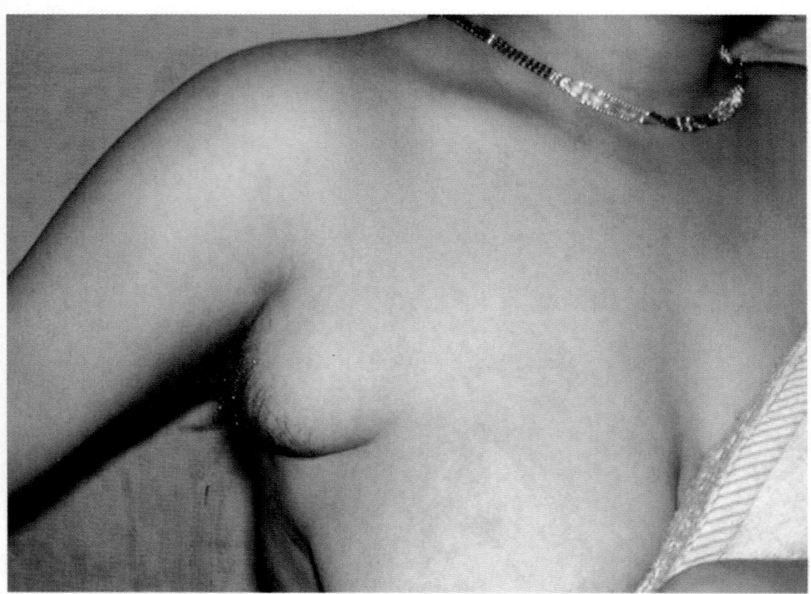

FIGURE 7.25: Accessory breast. A patient presented with a soft tissue shadow in the Rt axilla.

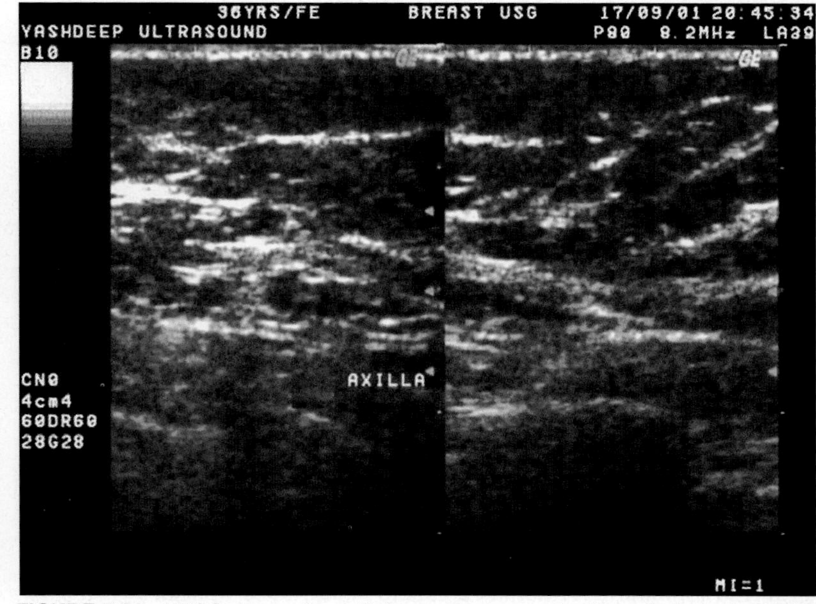

FIGURE 7.26: HRSG shows the soft tissue shadow in the Rt axilla is having echo texture similar to the breast suggestive of accessory breast.

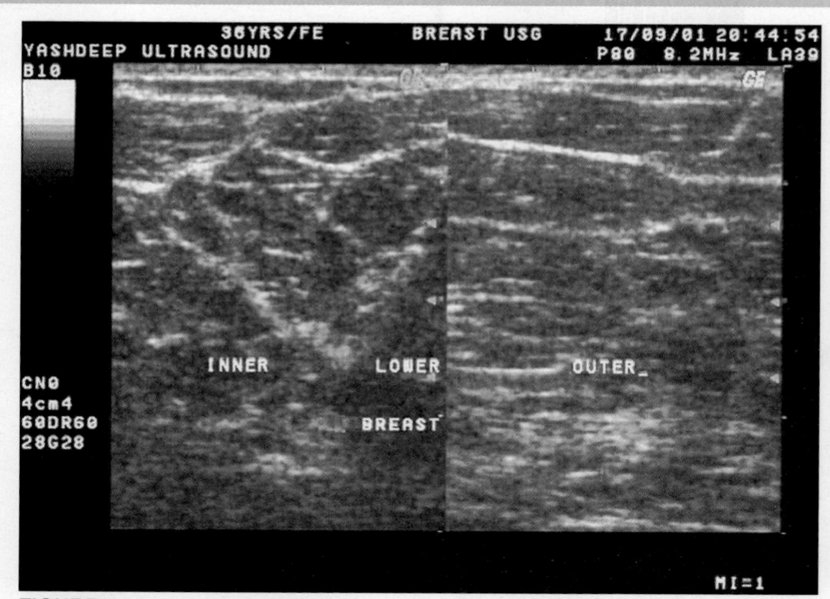

FIGURE 7.27: The normal breast tissue is seen in the upper inner and outer quadrant of the breast, which is similar to the soft tissue texture seen in Rt axilla.

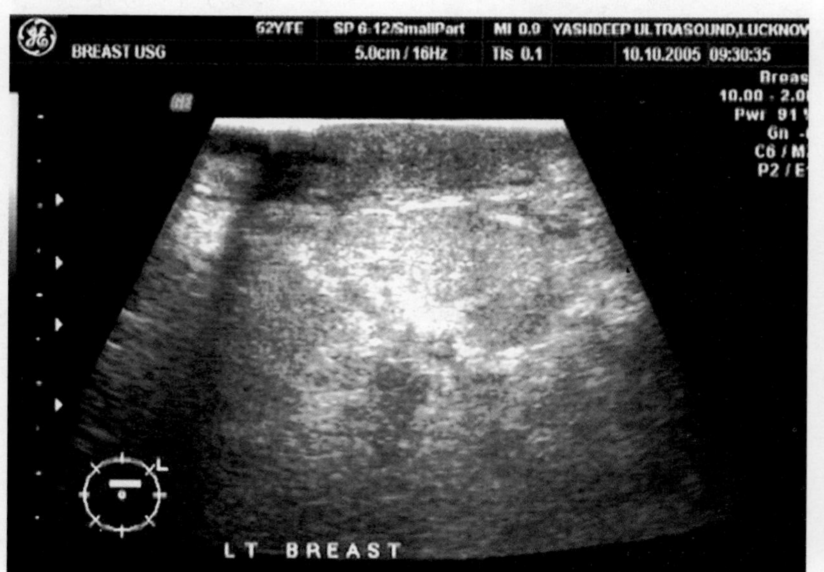

FIGURE 7.28

FIGURE 7.29

FIGURES 7.28 and 7.29: Diffuse breast edema. A patient presented with markedly swollen Lt breast. HRSG shows diffuse breast edema. No evidence of any focal mass is seen in the breast tissue.

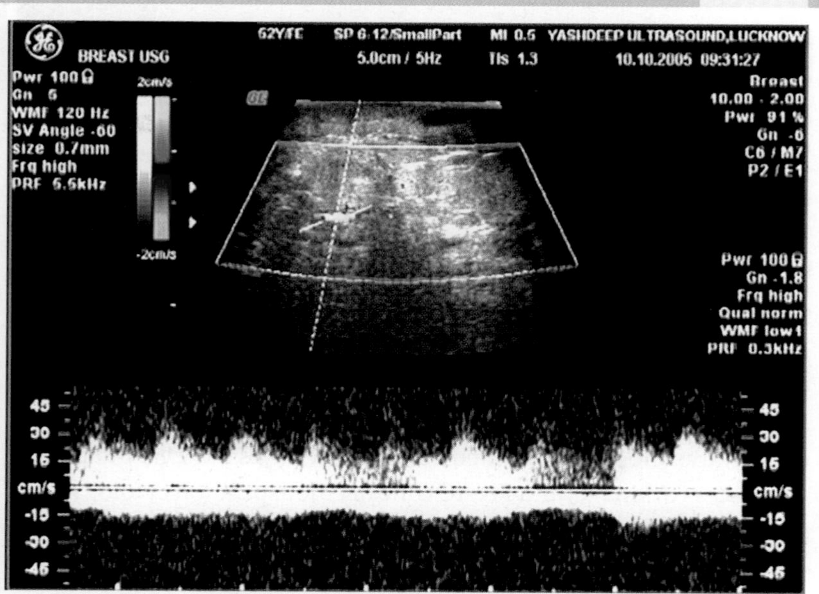

FIGURE 7.30: On color flow imaging shows dilated vessels around the breast. Spectral Doppler shows increased flow in the dilated vessels.

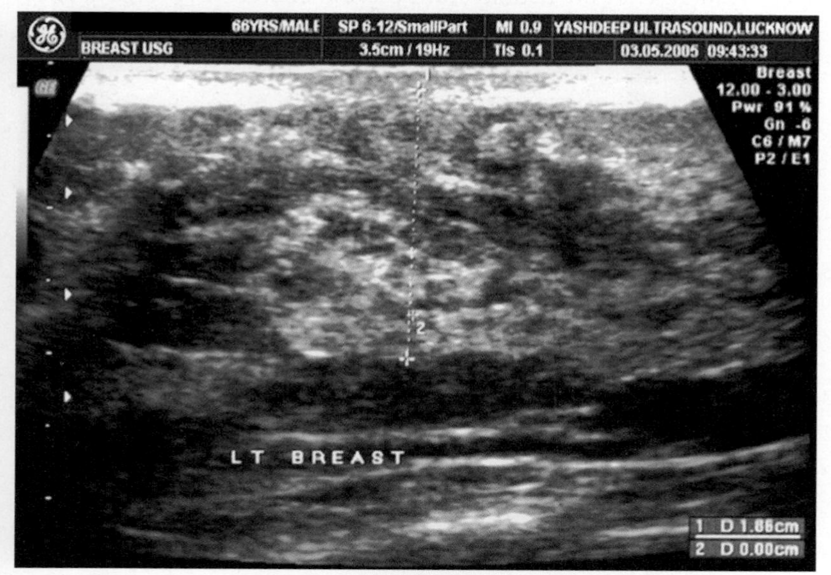

FIGURE 7.31

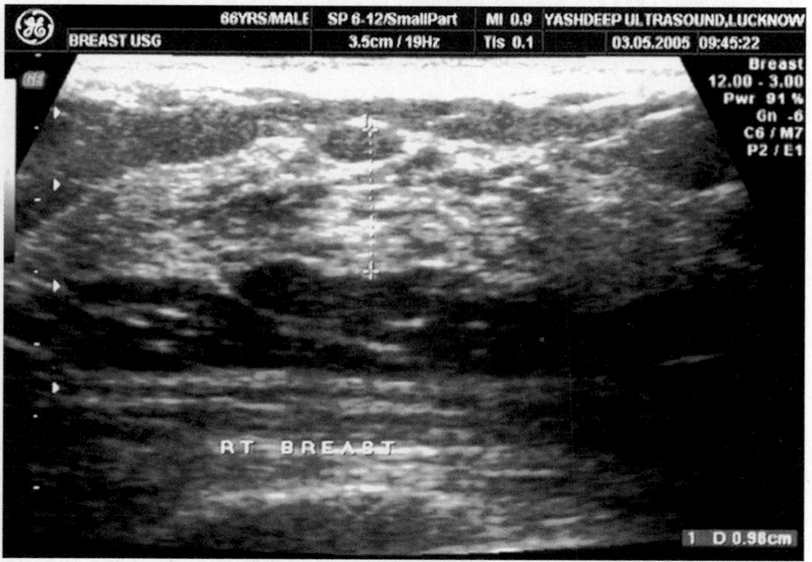

FIGURE 7.32

FIGURES 7.31 and 7.32: Gynecomastia in Lt male breast. A patient presented with enlargement of the Lt breast. HRSG shows increased amount of normal glandular tissue in the Lt breast suggestive of gynecomastia. No evidence of any focal mass is seen in the breast tissue. The Lt breast tissue is thickened in comparison to the Rt breast tissue.

However, when they are enlarged they can be easily picked up on HRSG.

Indications of Sonomammography

The main indications for breast sonography are as follows:

1. To identify the nature of palpable mass in the breast.
2. To evaluate a palpable mass in young females.
3. To evaluate a non-palpable mass in breast where X-ray mammography findings are uncertain.
4. Simultaneous comparison of both breasts.
5. In full thickness breast tissue study.
6. In ultrasonographic guided breast procedures like cyst punctures, abscess drainage, and FNAC (fine needle aspiration cytology), etc.

Breast Cyst

Breast cysts are very common and frequently encountered in middle-aged women. They are usually multiple and bilateral in position. They are more commonly seen in the women who are on hormonal replacement therapy. Ultrasonographic features include well-defined sharply marginated, thin walled anechoic fluid filled masses with post cystic enhancement of echoes. They are usually round but may be flattened due to surrounding tissue pressure. Ultrasonographic sensitivity is 100% in cyst diagnosis. However, when a cyst contains

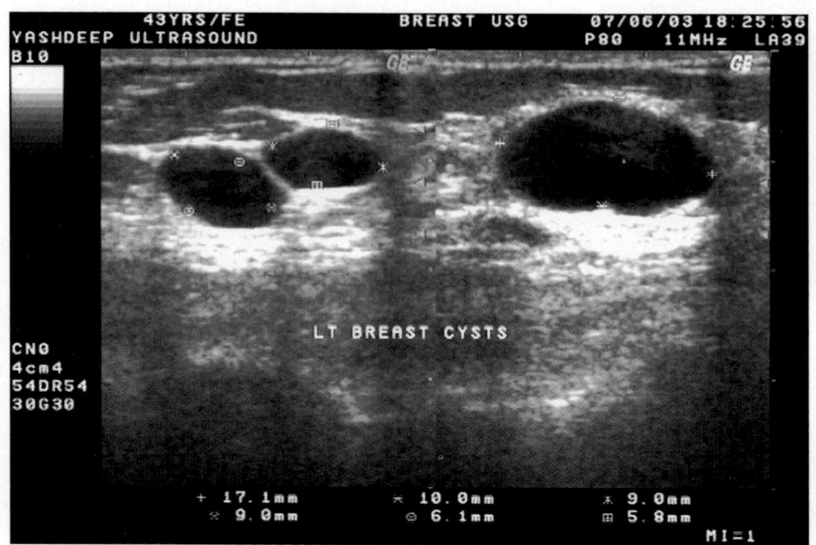

FIGURE 7.33A

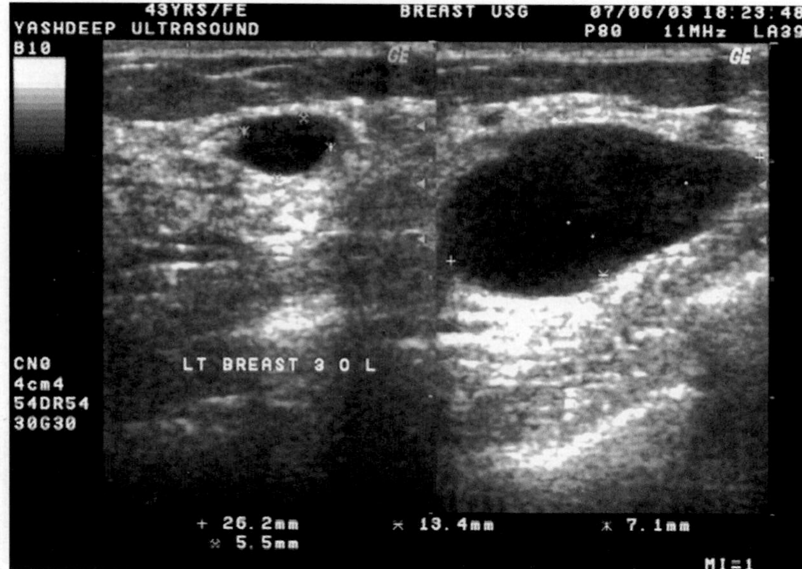

FIGURE 7.33B

FIGURES 7.33A and B: HRSG shows multiple well-defined thin walled cysts in the Lt breast. The cysts show thin echogenic sharp margins. They are anechoic. Postcystic enhancement is seen posterior to the cysts suggestive of clear fluid in the cysts. HRSG diagnosis is 100% in cystic lesion.

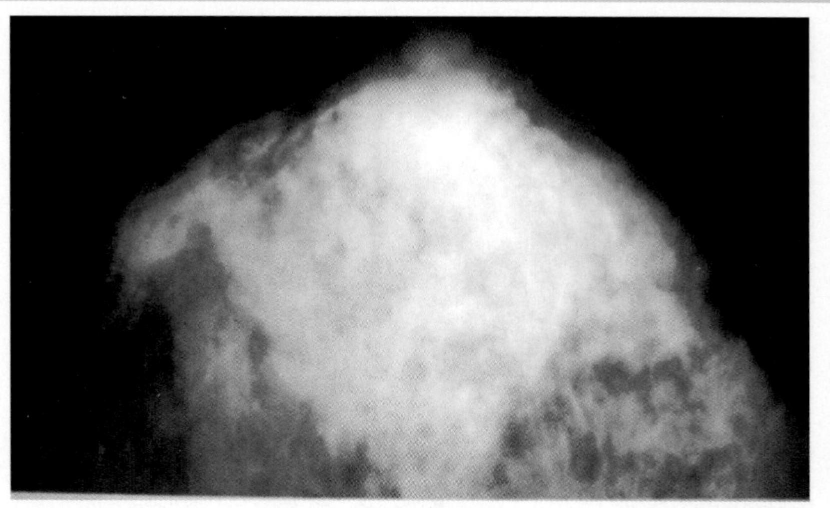

FIGURE 7.34

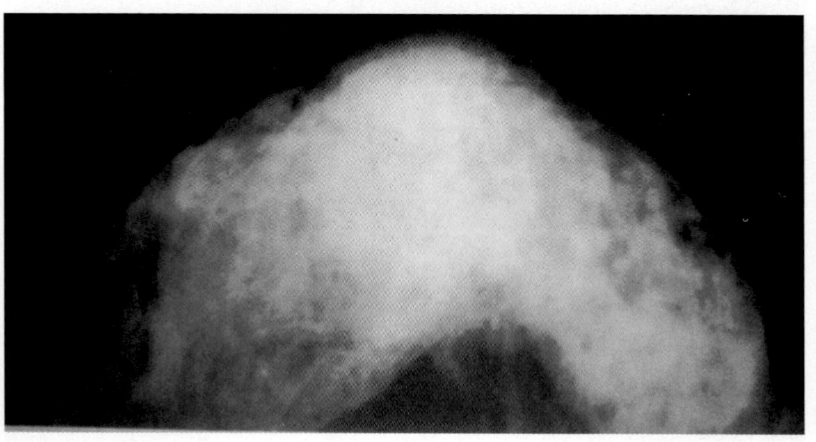

FIGURE 7.35

FIGURES 7.34 and 7.35: A patient presented with lump in Rt breast. X-ray mammogram shows ill-defined opacities in the breast and it is inconclusive.

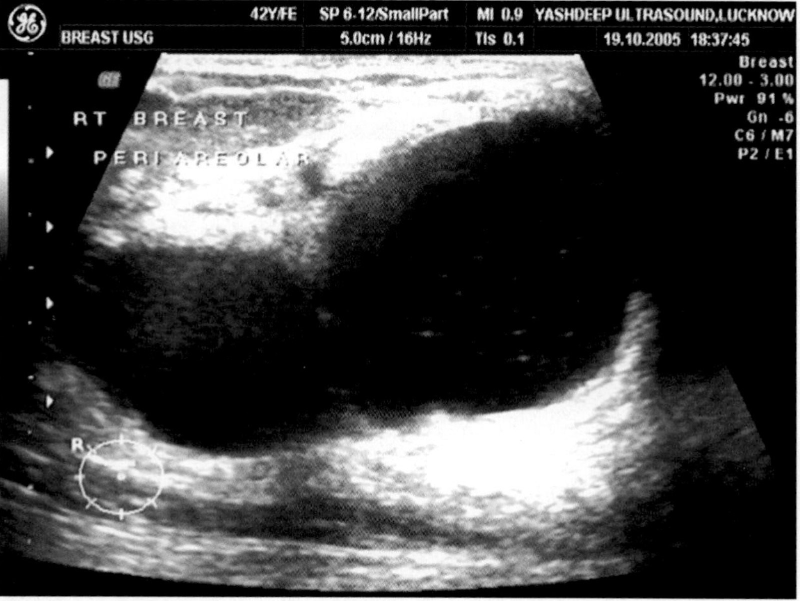

FIGURE 7.36

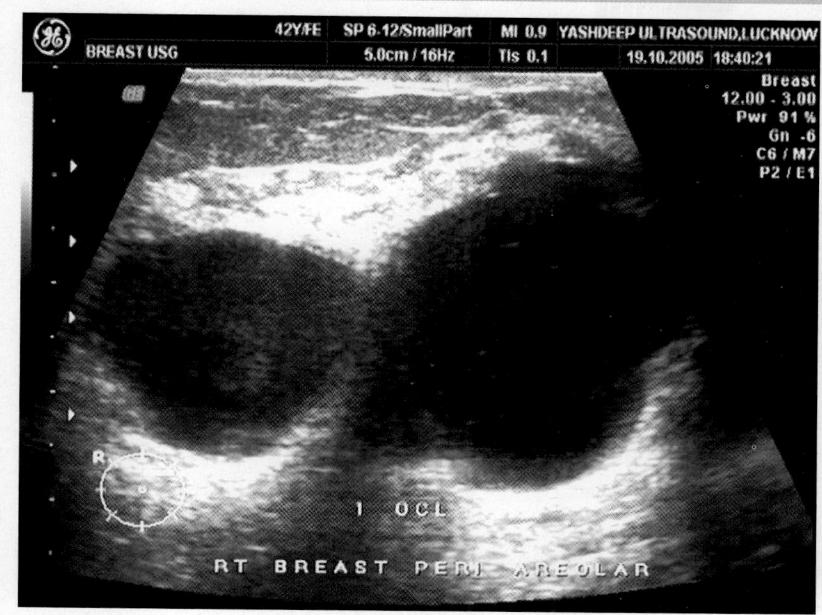

FIGURE 7.37

FIGURES 7.36 and 7.37: HRSG shows well-defined cystic masses in the breasts. They are anechoic. Post cystic enhancement is seen.

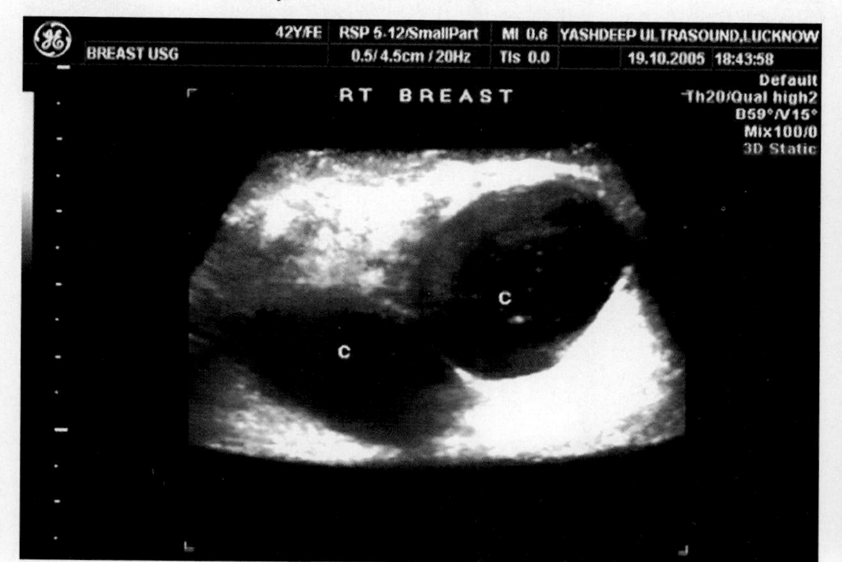

FIGURE 7.38

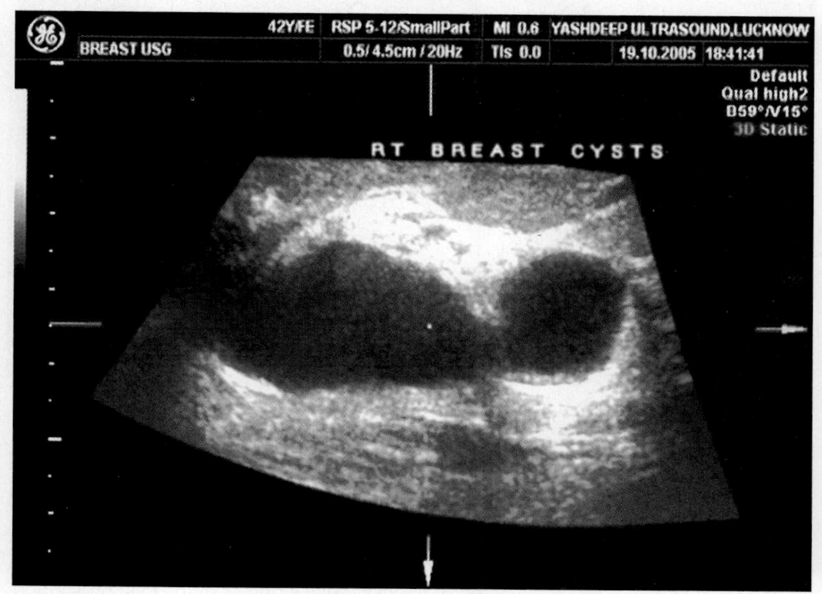

FIGURE 7.39

FIGURES 7.38 and 7.39: 3D imaging of the breast shows details of the cyst.

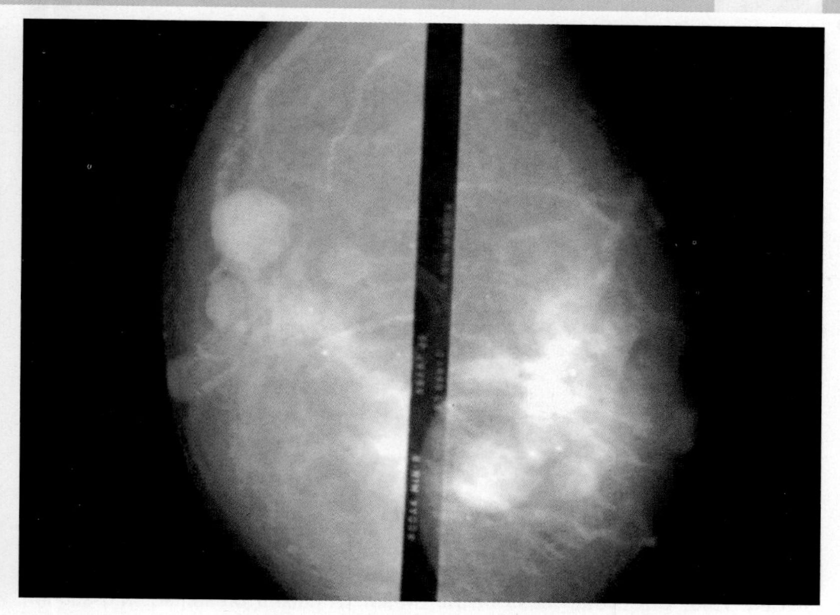

FIGURE 7.40: X-ray mammogram shows well-defined radio dense opacity in the breast.

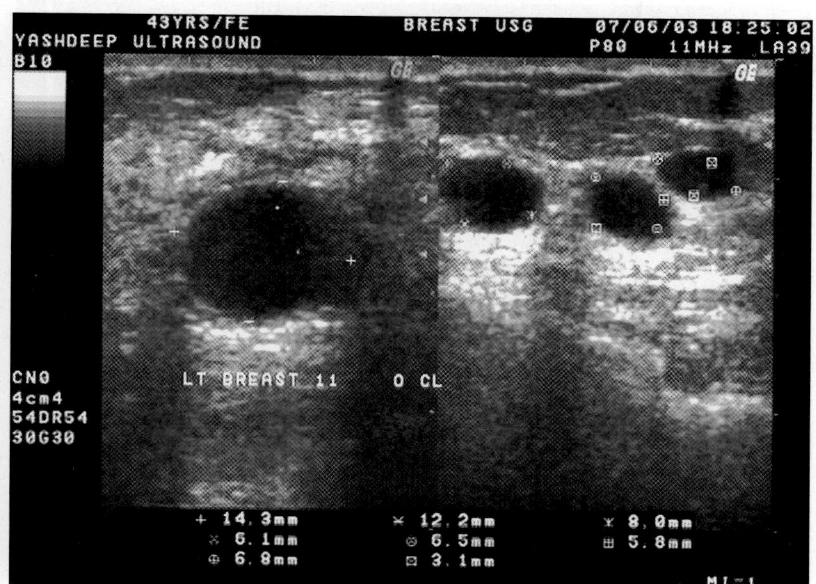

FIGURE 7.41

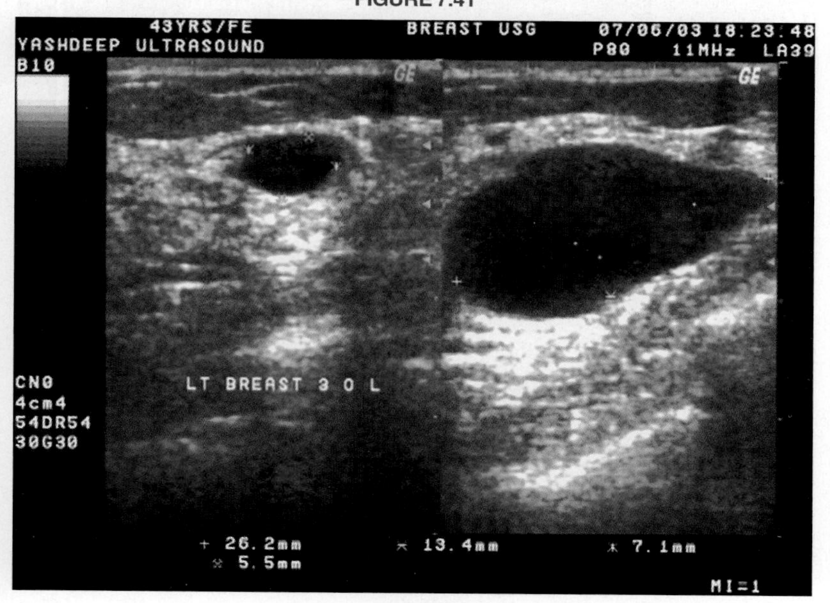

FIGURE 7.42

FIGURES 7.41 and 7.42: Ultrasound of the same patient shows well defined cysts in the breast. They are anechoic with no internal echoes.

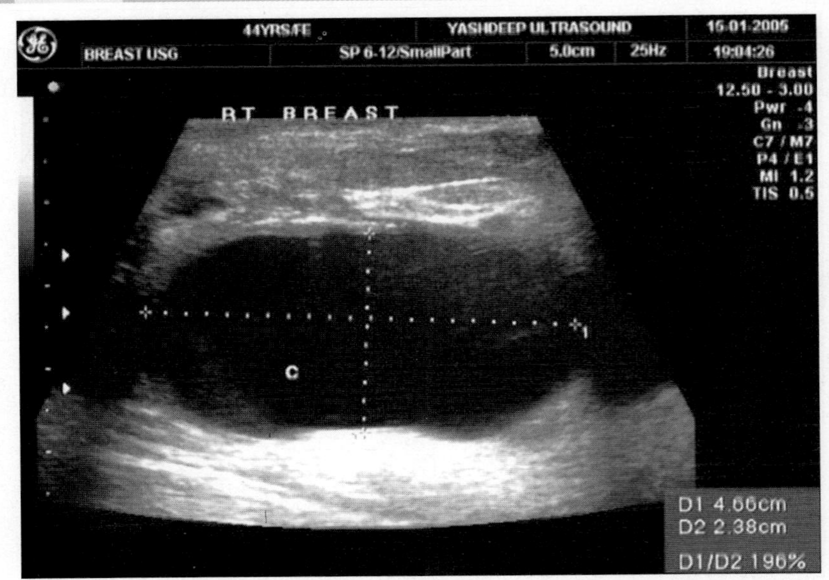

FIGURE 7.43: **Big cystic mass in the breast.** A patient presented with sudden appearance of the lump in the breast. HRSG shows well define cystic mass in the breast.

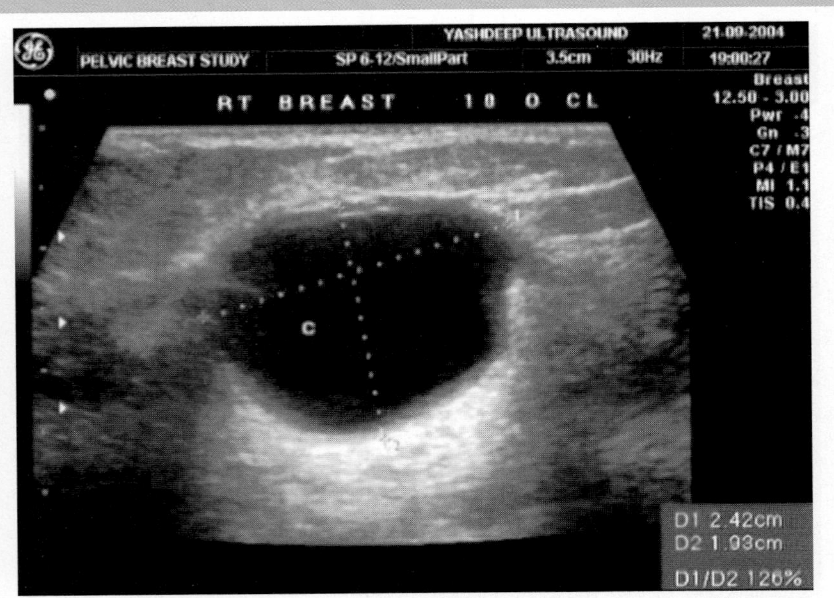

FIGURE 7.46: **Big cyst in the breast.** HRSG shows a big cystic mass in the breast. Walls are sharp. No internal echo is seen.

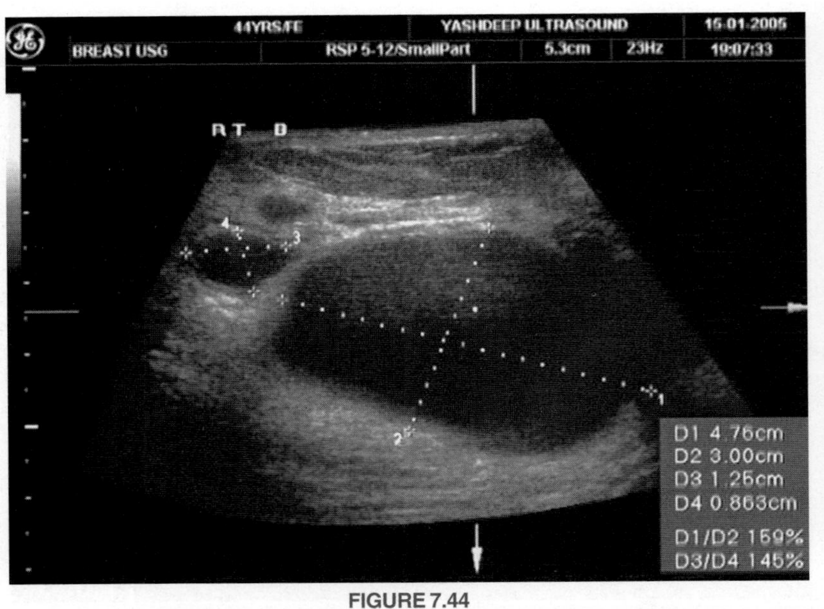

FIGURE 7.44

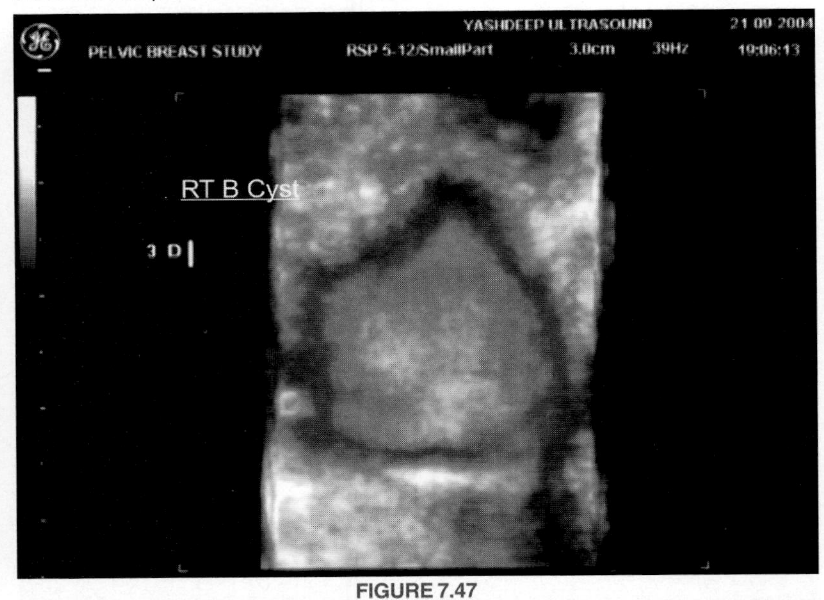

FIGURE 7.47

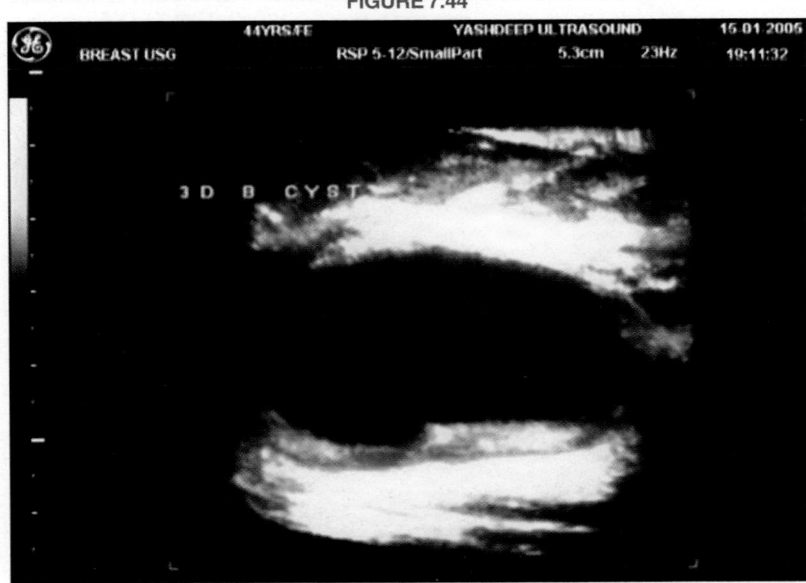

FIGURE 7.45

FIGURES 7.44 and 7.45: 3D imaging of the same breast shows details of the cyst in better view. Small satellite cysts are also seen around the big cyst in 3D imaging.

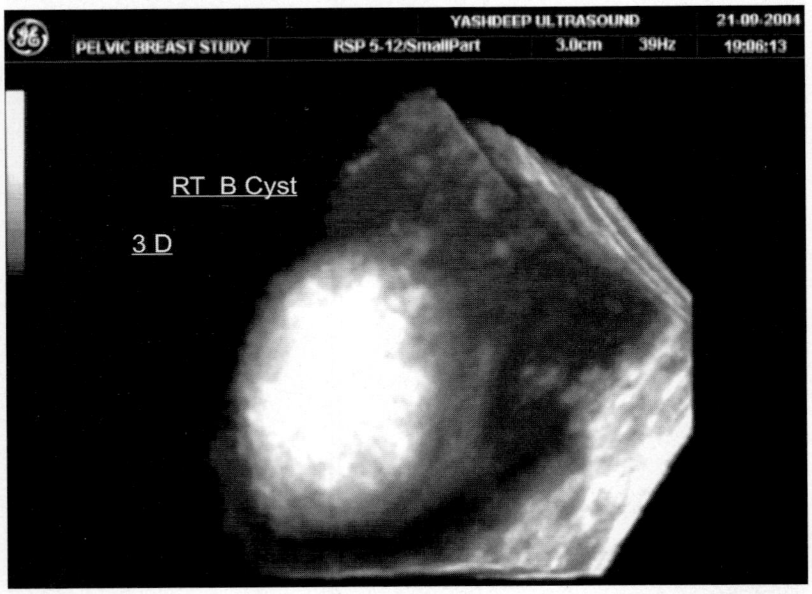

FIGURE 7.48

FIGURES 7.47 and 7.48: 3D imaging of the cyst shows details of the cyst on surface rendering. Reverse 3D imaging shows the cyst is lying in the superficial plane of the breast tissue.

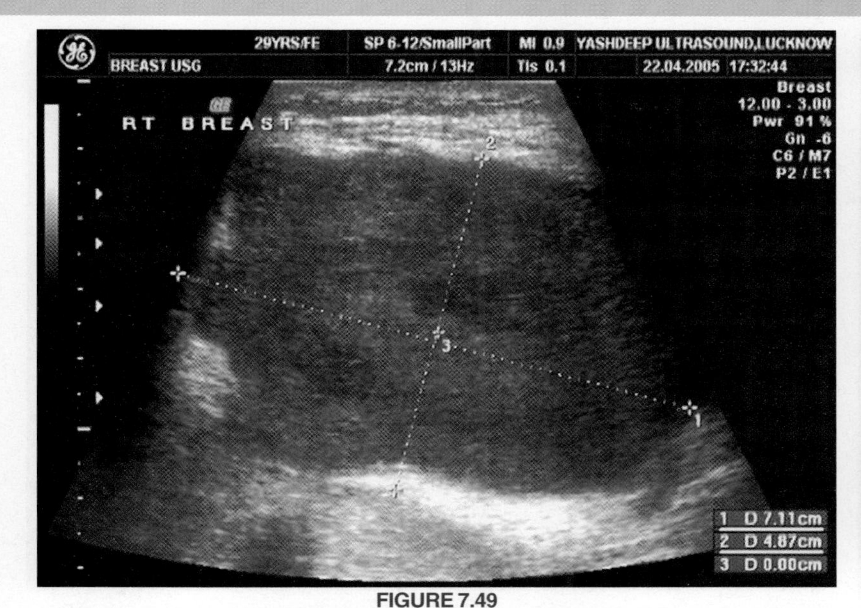

FIGURE 7.49

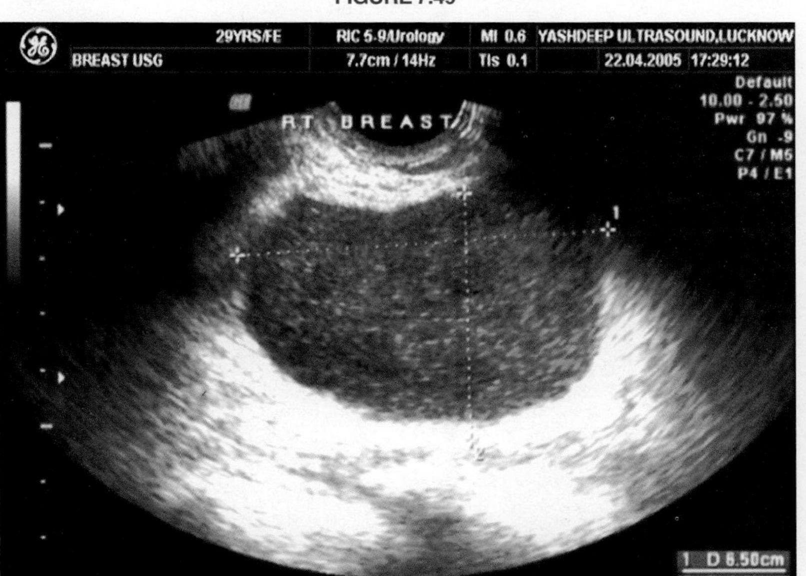

FIGURE 7.50

FIGURES 7.49 and 7.50: Galactocele in the breast. A lactating mother presented with big mass in the breast upper outer quadrant. HRSG shows a big low-level echo complex mass in the breast. Thick echoes are seen in it. But no septation is seen. No mass lesion is seen. Aspiration of the cyst shows milk suggestive of bib galactocele.

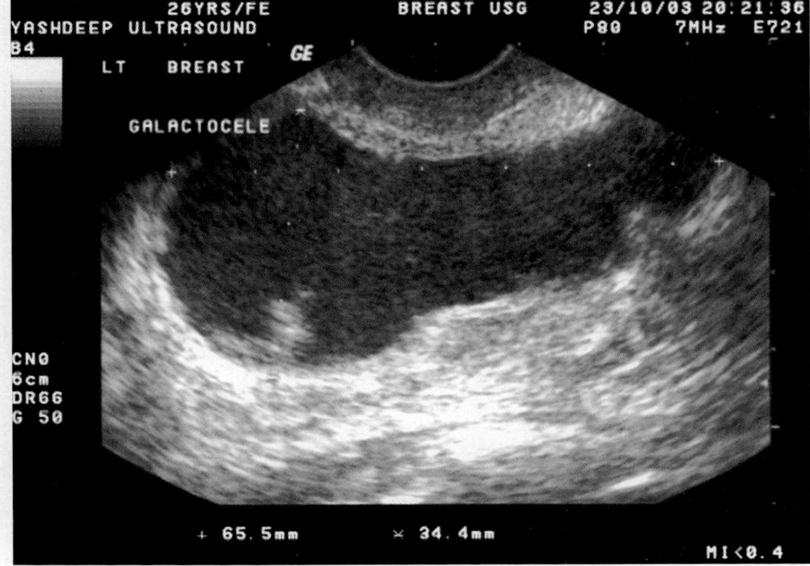

FIGURE 7.51

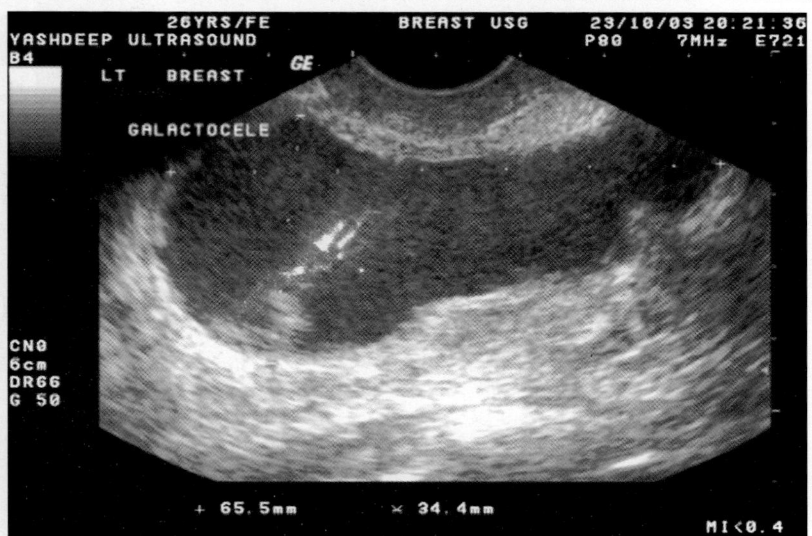

FIGURE 7.52

FIGURES 7.51 and 7.52: Galactocele in the breast. HRSG shows a big low-level echo complex mass in the Lt breast. The mass shows multiple echoes with homogenous echo texture. A tounge of the mass is seen going towards the axilla. Aspiration confirmed a galactocele.

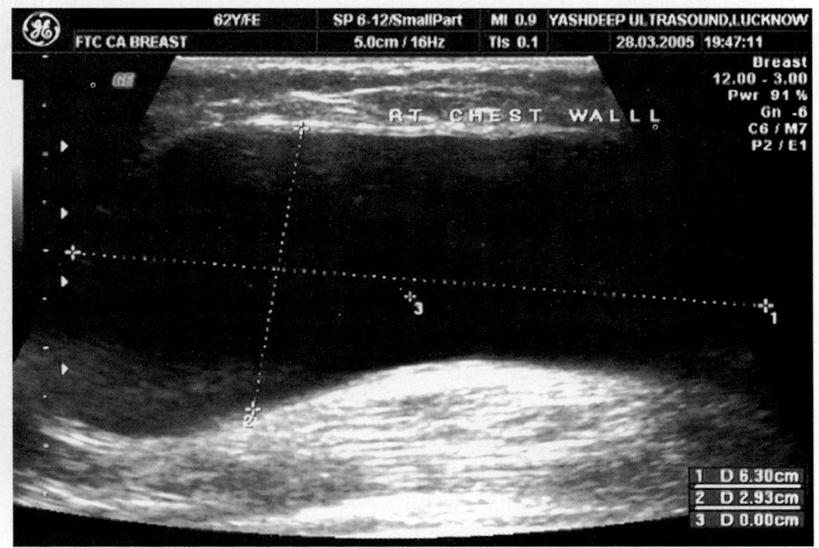

FIGURE 7.53

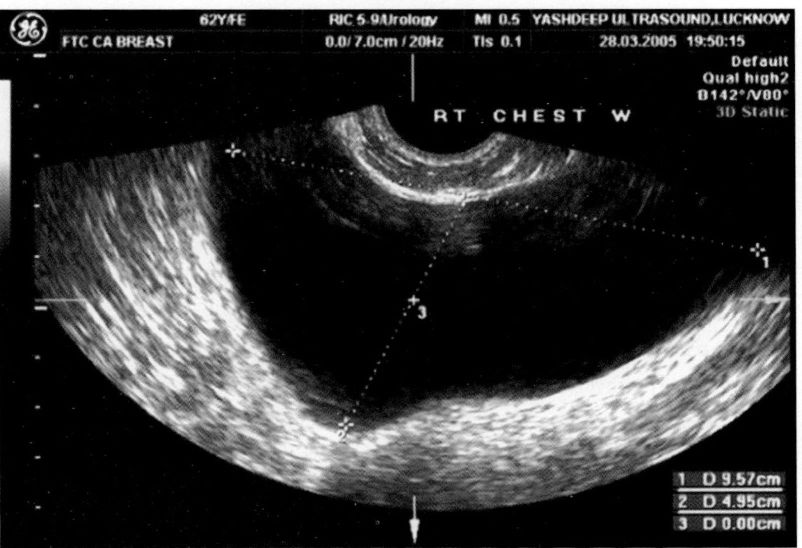

FIGURE 7.54

FIGURES 7.53 and 7.54: Postoperative seroma. a patient presented with big lump in the chest wall after surgery for CA breast. HRSG shows big anechoic collection in the chest wall. The collection is seen going to the axilla. It is also seen confined to anterior chest wall.

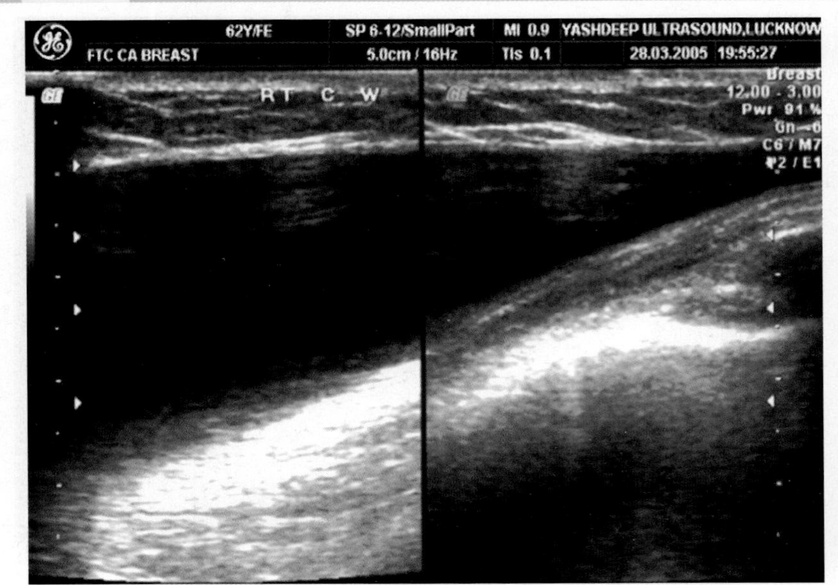

FIGURE 7.55

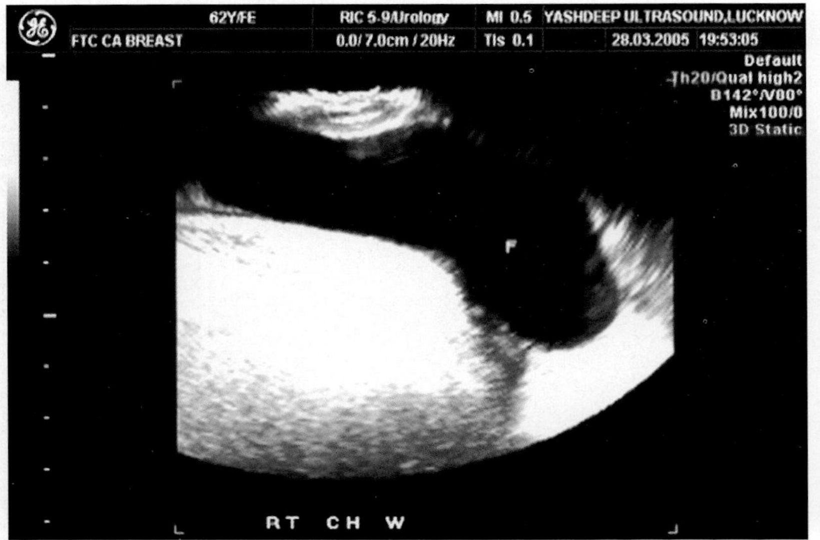

FIGURE 7.56

FIGURES 7.55 and 7.56: 3D imaging the same patient shows details of the fluid collection going into the axillary tail.

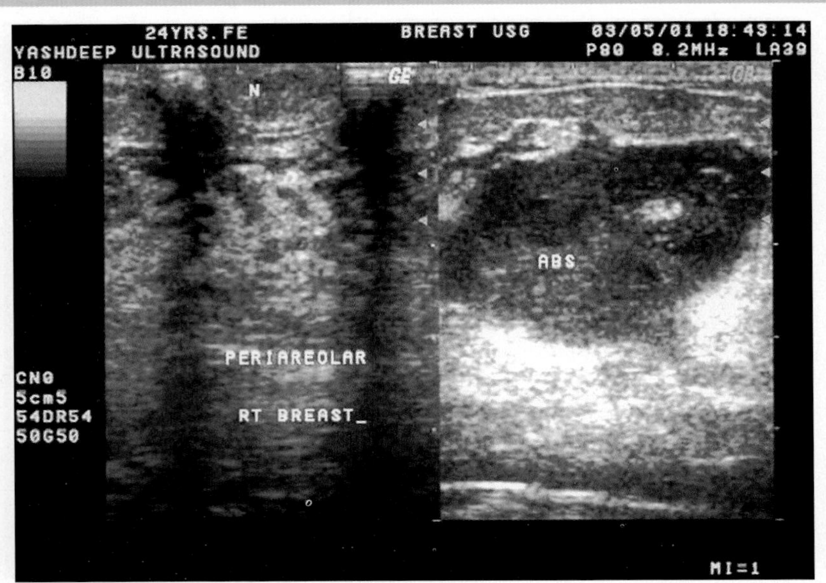

FIGURE 7.57

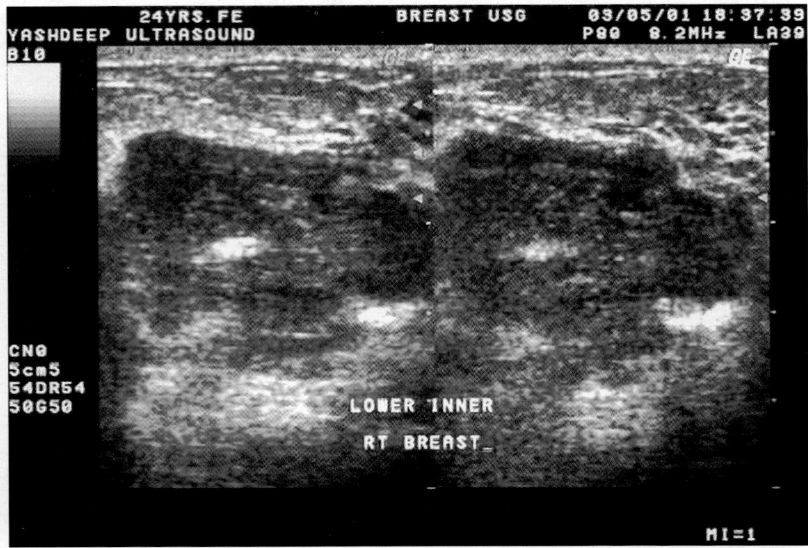

FIGURE 7.58

FIGURES 7. 57 and 7.58: HRSG shows a big low-level echo complex mass in the Rt breast in the lower inner quadrant in a 5-month-old pregnant lady. Central area of necrosis is seen in it. Perifocal edema is present. It is seen pressing over the ducts resulting into dilatation of periareolar ducts.

turbid fluid needs further evaluation. In such setting, USG guided aspiration is done. Intracystic carcinomas are rare. However, malignant changes in cyst are also encountered, but they are rare.

Breast Abscess

A breast abscess may result due to infected galactocele or due to secondary infection in the cyst. It appears to be a thick walled, echo poor low level echo complex mass. It shows heterogeneous texture with areas of attenuation and debris. The diagnosis of breast abscess can be confirmed by aspiration of the pus from it under USG guidance. However, a breast abscess can mimic a narcotizing tumor or fat necrosis.

Fat Necrosis

It is difficult condition to be diagnosed on USG. As it causes sound beam attenuation and due to fibrosis, aspiration biopsy is also not very fruitful. Therefore, at times open biopsy is required. It resembles as carcinoma and it become difficult to exclude breast carcinoma.

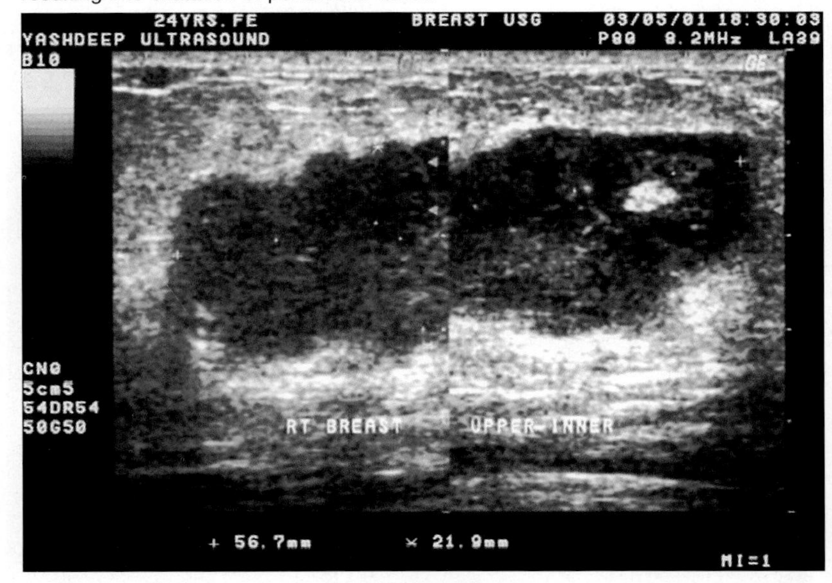

FIGURE 7.59A

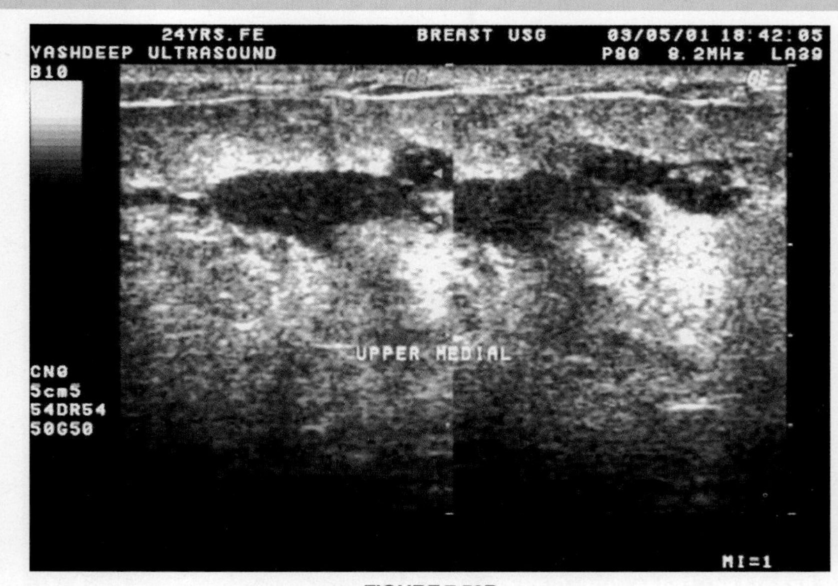

FIGURE 7.59B

FIGURES 7.59A and B: The abscess is also seen occupying the upper medial quadrant of the breast and extending to the outer side. The periareolar ducts are dilated. On aspiration it turns out to be a big pyogenic abscess.

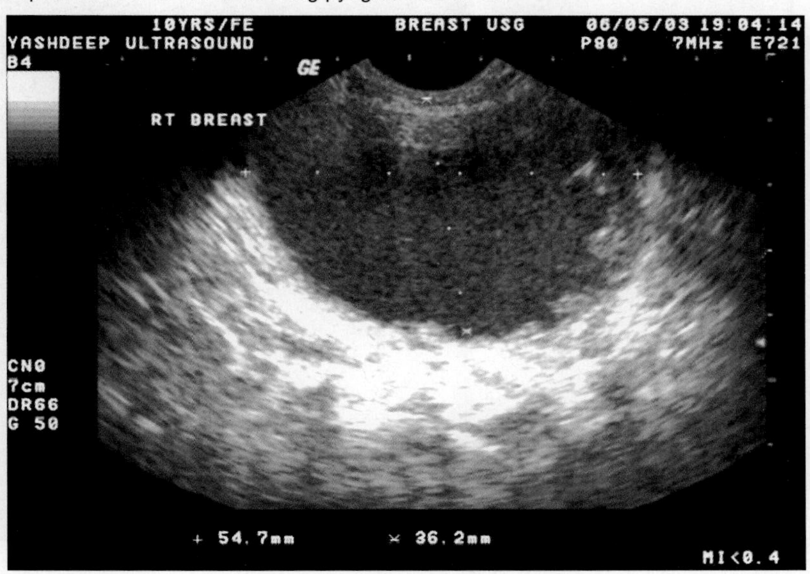

FIGURE 7.60

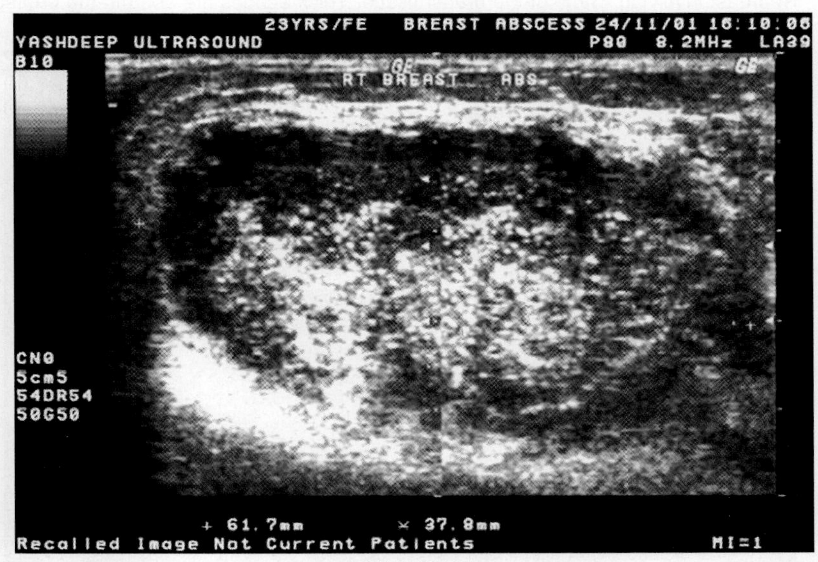

FIGURE 7.61

FIGURES 7.60 and 7.61: HRSG shows a huge pyogenic abscess in the breast. Central area of necrosis is seen in it. Layering sign is positive. Echogenic debris is seen in the abscess on the dependent part.

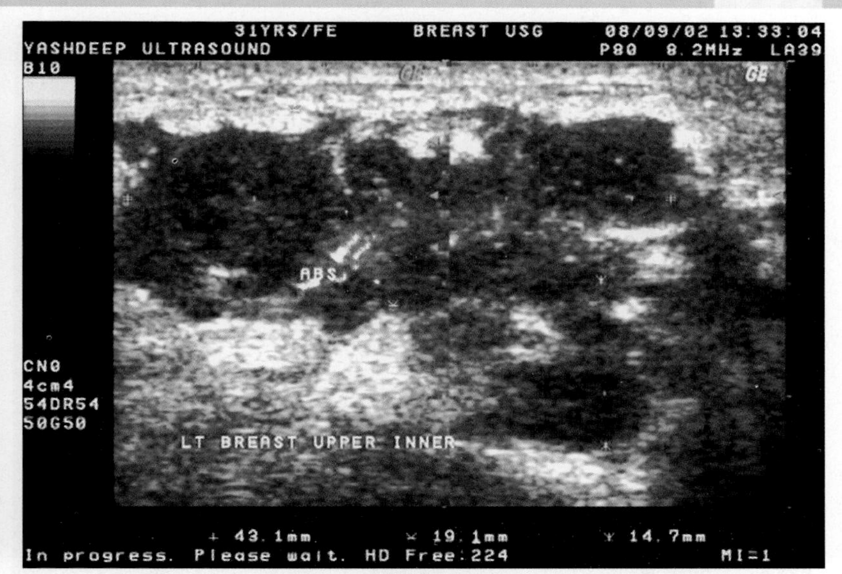

FIGURE 7.62

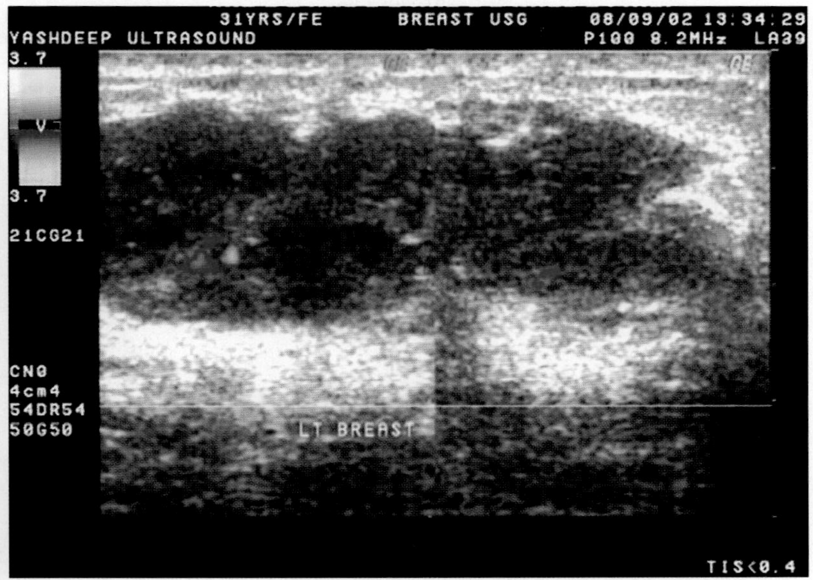

FIGURE 7.63

FIGURES 7.62 and 7.63: Big breast abscess. A patient presented with acutely inflamed Rt breast with highly tender mass in the breast. HRSG shows big low attenuating complex mass involving whole of the upper quadrant of the breast. Thick echoes are seen in it. Perifocal edema is also present. On color flow imaging no flow is seen in it suggestive of big necrotizing abscess in the breast.

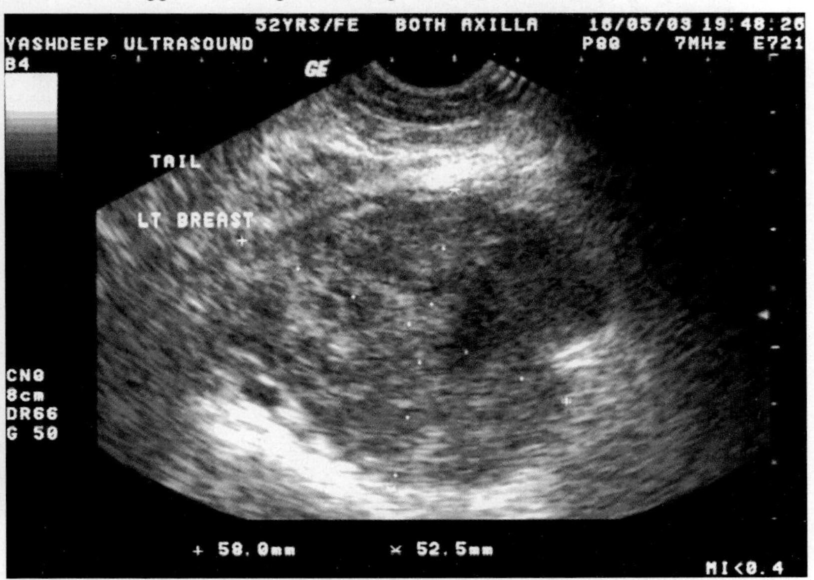

FIGURE 7.64

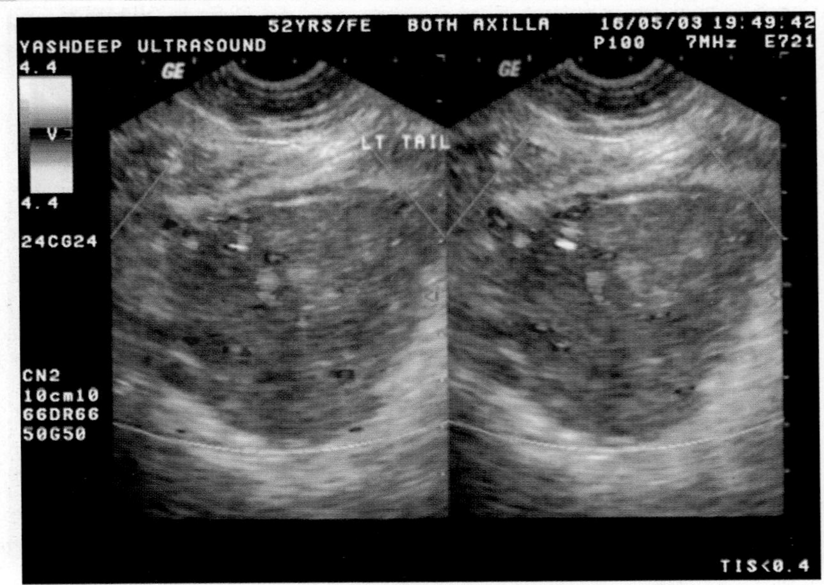

FIGURE 7.65

FIGURES 7.64 and 7.65: Big breast abscess. A lady presented with pyogenic abscess in the breast. Thick low levels echo complex mass is seen involving whole of the breast with central area of necrosis. On color flow imaging low flow is seen in the abscess.

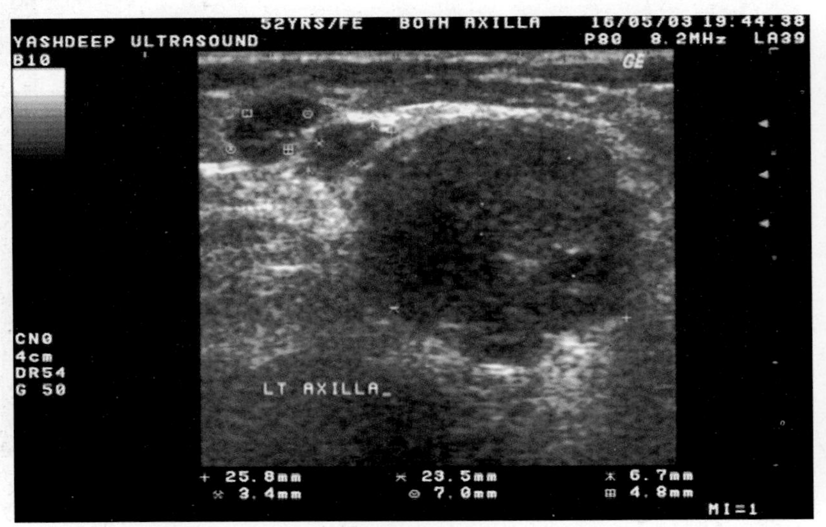

FIGURE 7.66

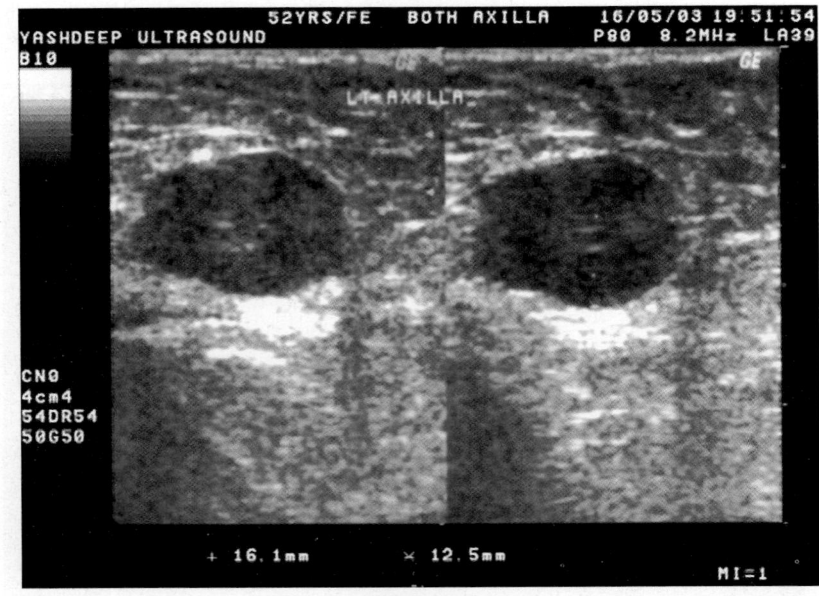

FIGURE 7.67

FIGURES 7.66 and 7.67: The abscess is seen extending to the Lt. axilla. Enlarged nodes are also seen in the axilla. Biopsy confirmed pyogenic abscess.

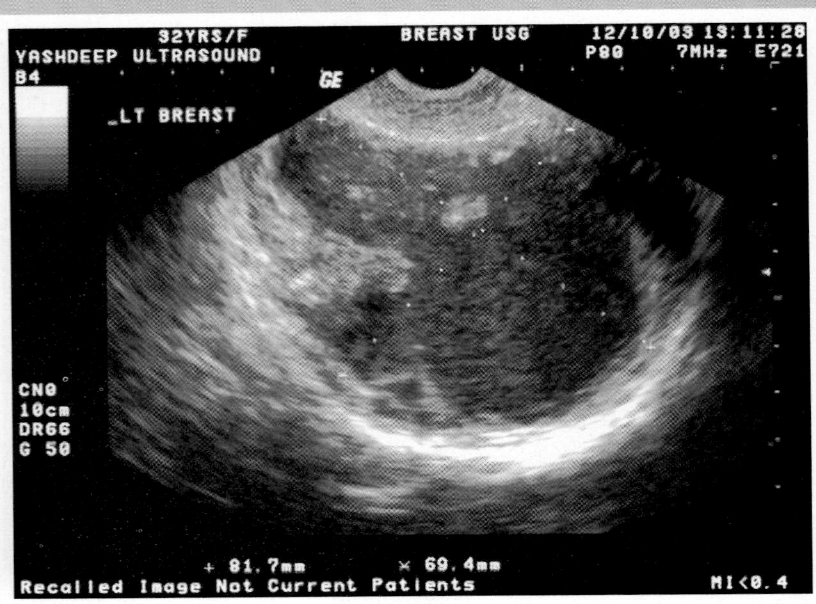

FIGURE 7.68

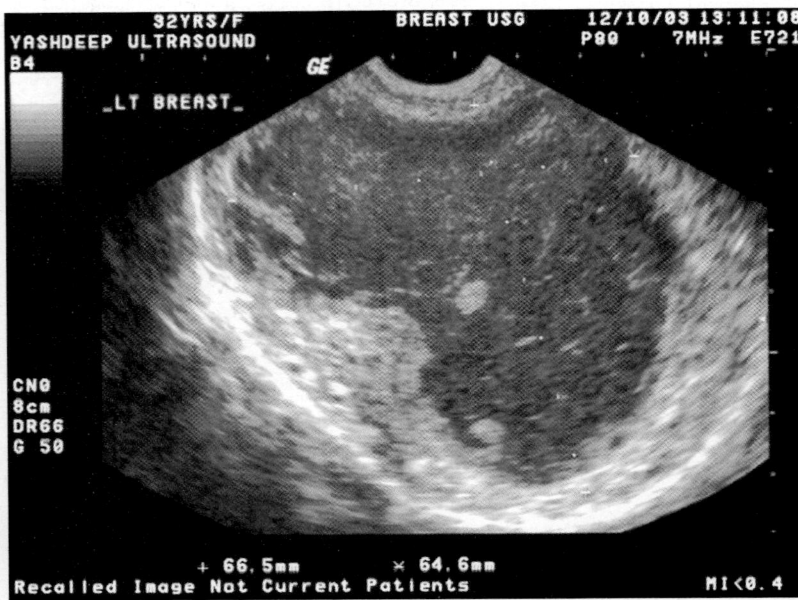

FIGURE 7.69

FIGURES 7.68 and 7.69: Big abscess in the breast. A big abscess is seen in the Rt breast involving whole of the breast tissue. Thick echoes are seen in the breast cavity. Central area of necrosis is also seen in it.

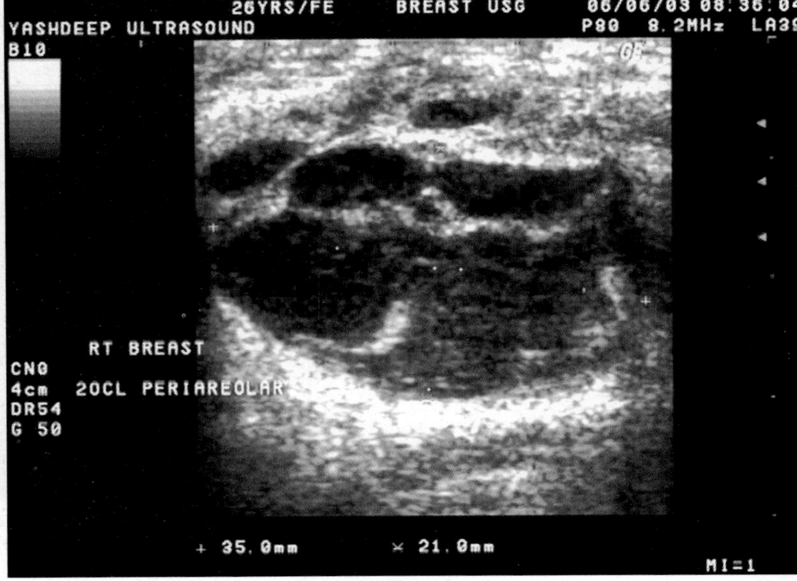

FIGURE 7.70

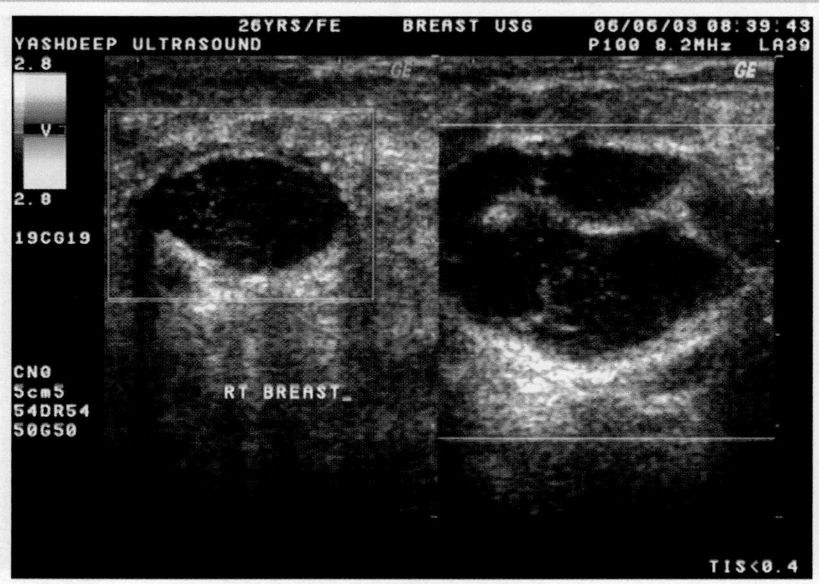

FIGURE 7.71

FIGURES 7.70 and 7.71: Loculated breast abscess. A patient presented with multi loculated thick echo complex mass in the breast. Thick echogenic septa are also seen in it. On color flow imaging low flow is seen in it.

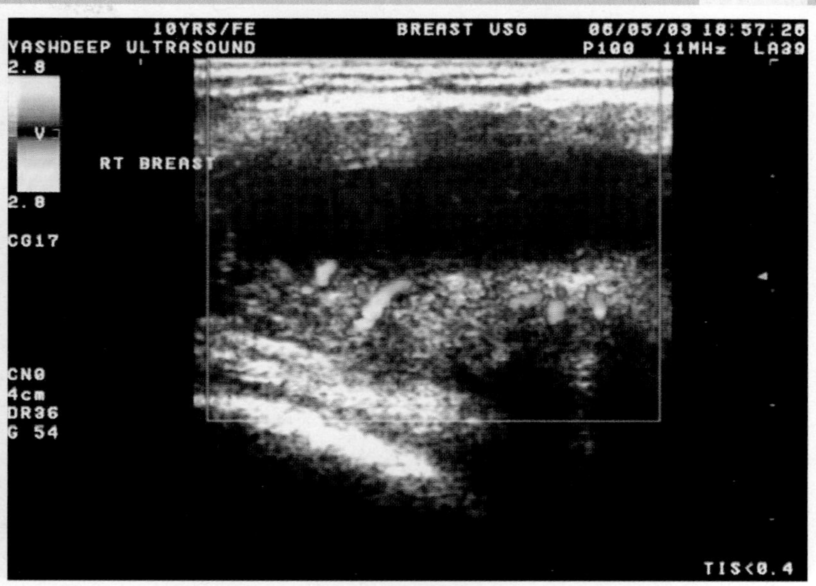

FIGURE 7.74

FIGURES 7.73 and 7.74: HRSG shows a thick low-level echo complex mass in the breast. Walls are thickened. On color flow imaging peripheral wall flow is seen in it. Aspiration of the abscess confirmed tubercular abscess.

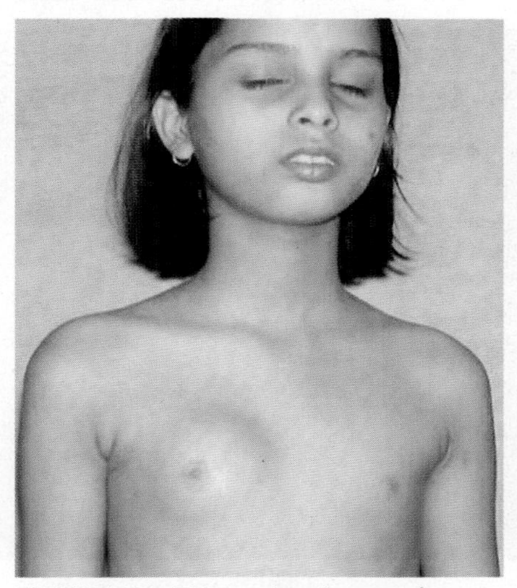

FIGURE 7.72: Tubercular abscess. A young girl presented with slowly growing mass in the Rt breast. It was non-tender. Patient presented with low-grade fever.

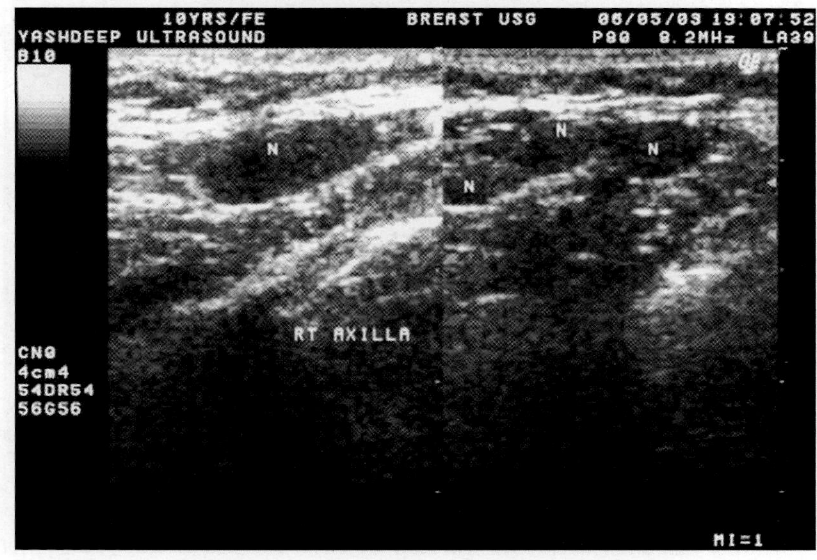

FIGURE 7.75: Multiple enlarged nodes are also seen in the Rt axilla. Nodal biopsy also confirmed tubercular abscess.

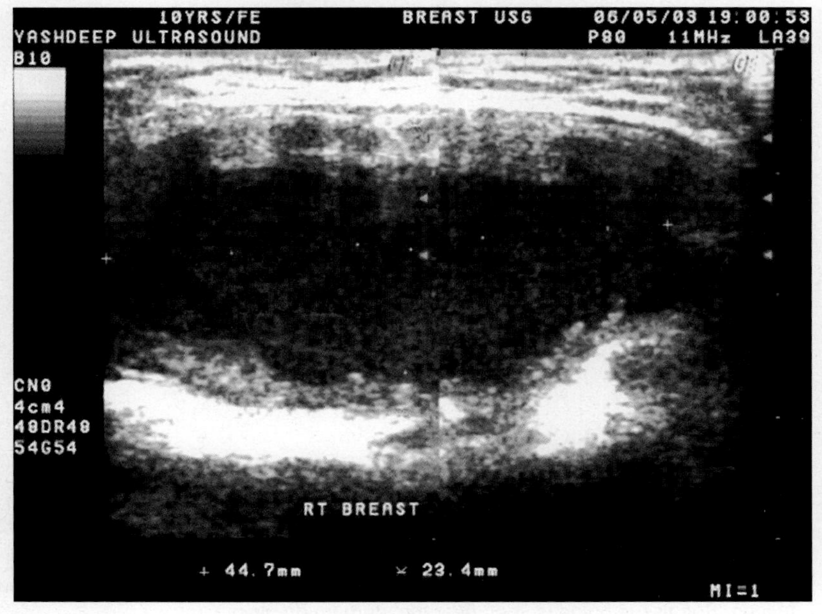

FIGURE 7.73

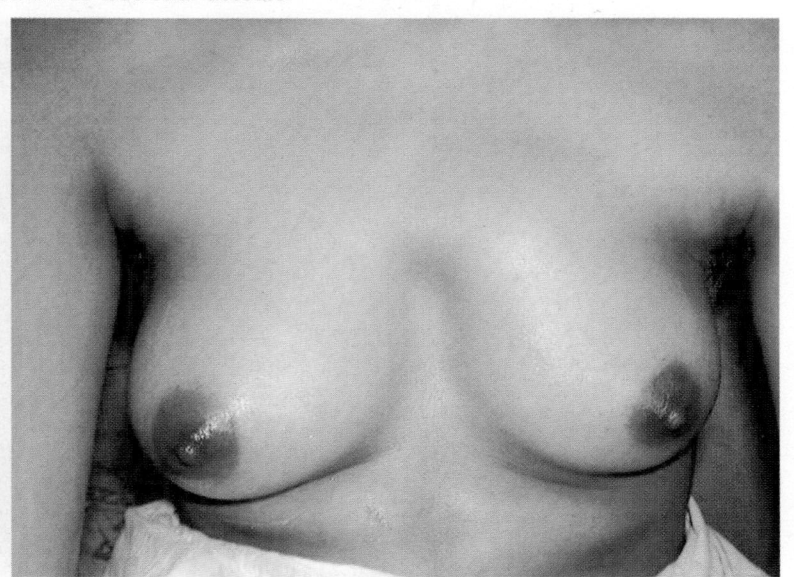

FIGURE 7.76: Tubercular abscess Lt breast. A young girl presented with soft tissue swelling over the upper medial quadrant of Lt breast. It was not very tender on clinical examination and semihard in consistency.

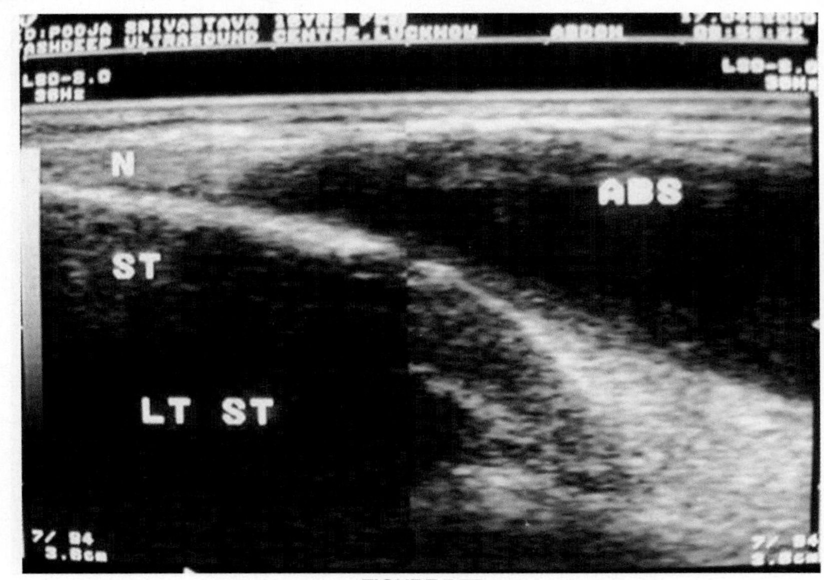

FIGURE 7.77

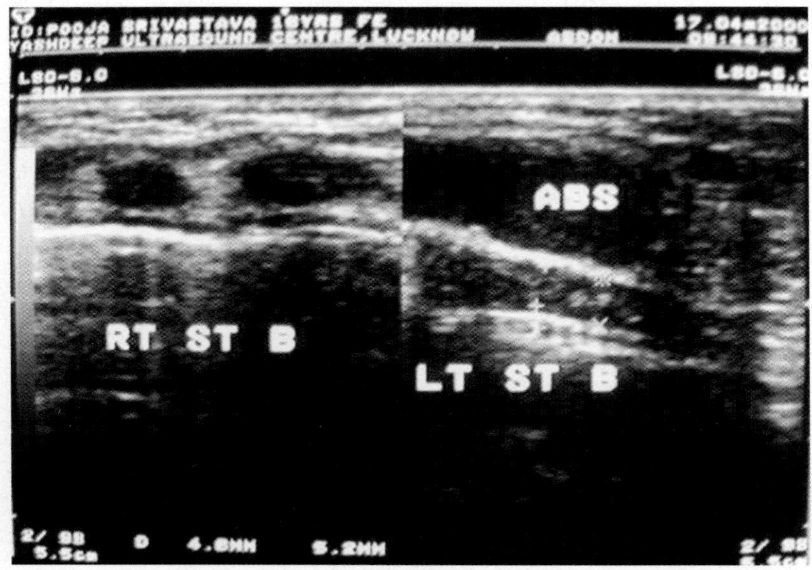

FIGURE 7.78

FIGURES 7.77 and 7.78: HRSG shows low-level echo complex collection coming out from the Lt border of sternum and going to the upper medial quadrant to the Lt breast.

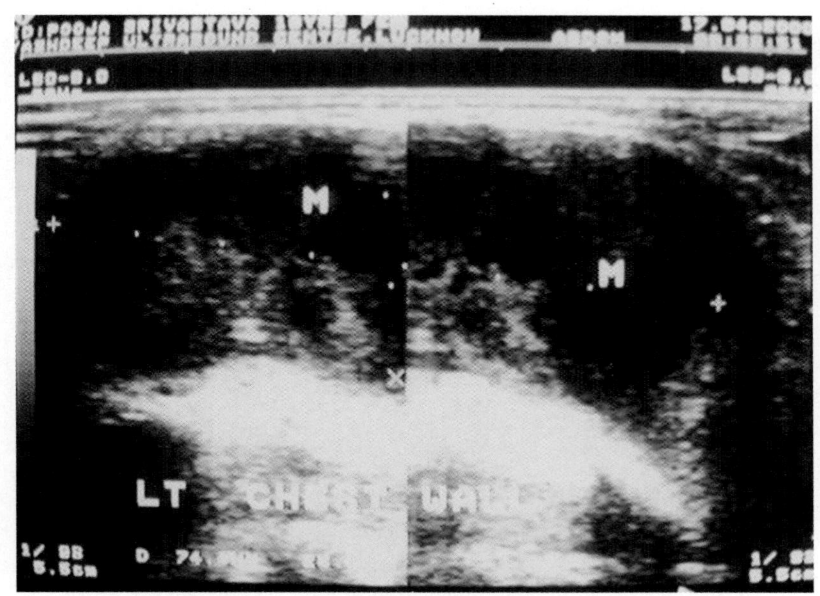

FIGURE 7.79

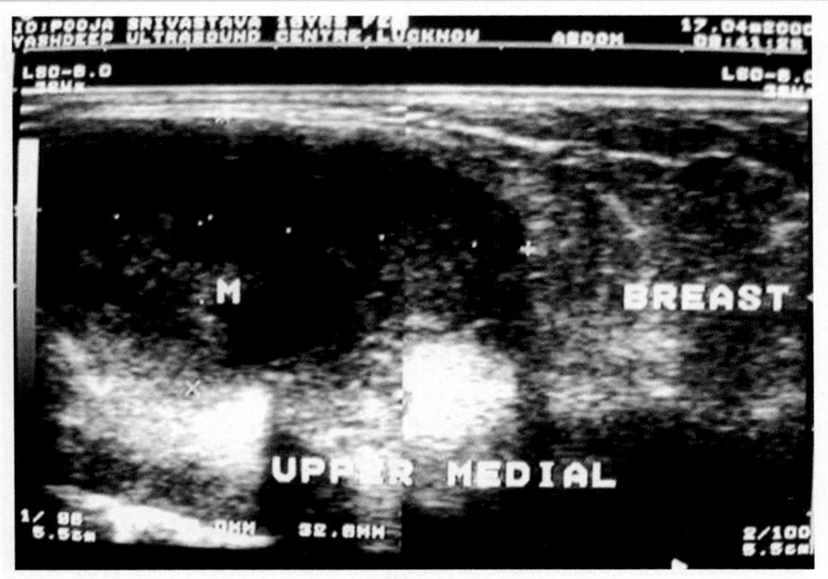

FIGURE 7.80

FIGURES 7.79 and 7.80: HRSG shows thick low-level echo collection involving the upper medial quadrant of the Lt breast suggestive of an abscess. On aspiration thick cheesy material was taken out suggestive of tubercular abscess.

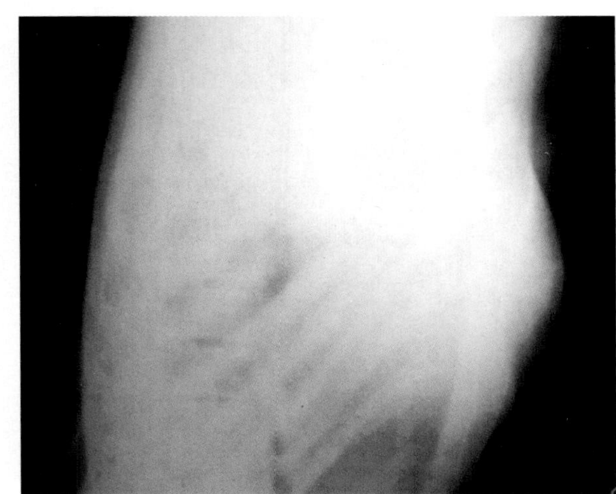

FIGURE 7.81

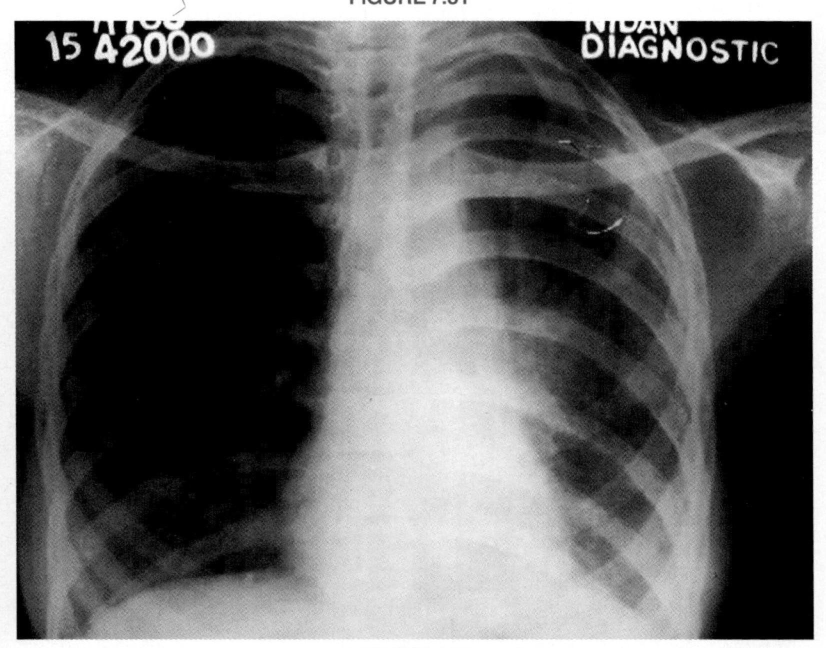

FIGURE 7.82

FIGURES 7.81 and 7.82: P/A, lateral view shows tubercular infiltration involving the Lt lung. Soft tissue bulge is seen over the sternum on plain X-ray.

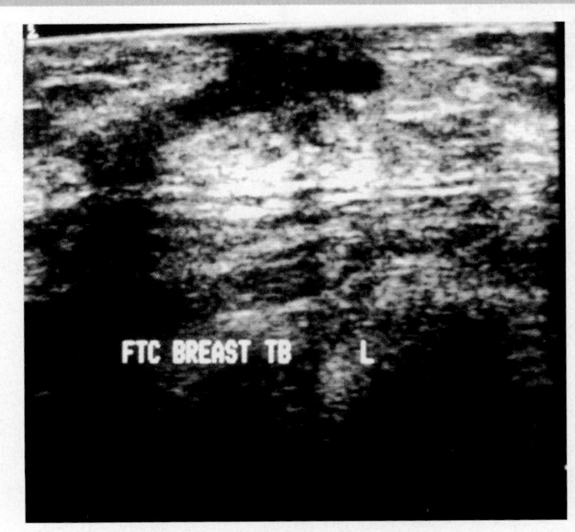

FIGURE 7.83

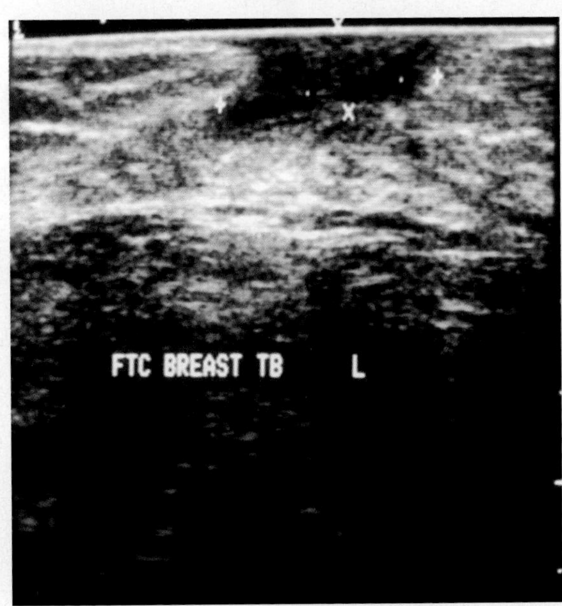

FIGURE 7.84

FIGURES 7.83 and 7.84: FTC of tubercular abscess breast. HRSG shows residual breast abscess in a case of tubercular abscess breast. After 6 months of antitubercular treatment, small abscess cavity is seen in the superficial part of the breast.

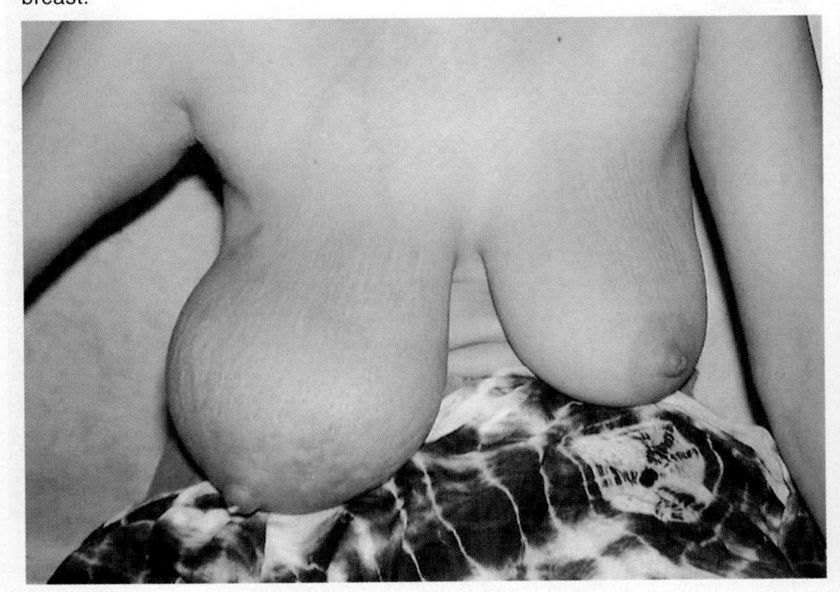

FIGURE 7.85: Filarial breast. A patient presented with markedly enlarged tender, hot. Inflamed Rt breast with acute mastalgia. The breast is almost doubles the size of the Lt breast.

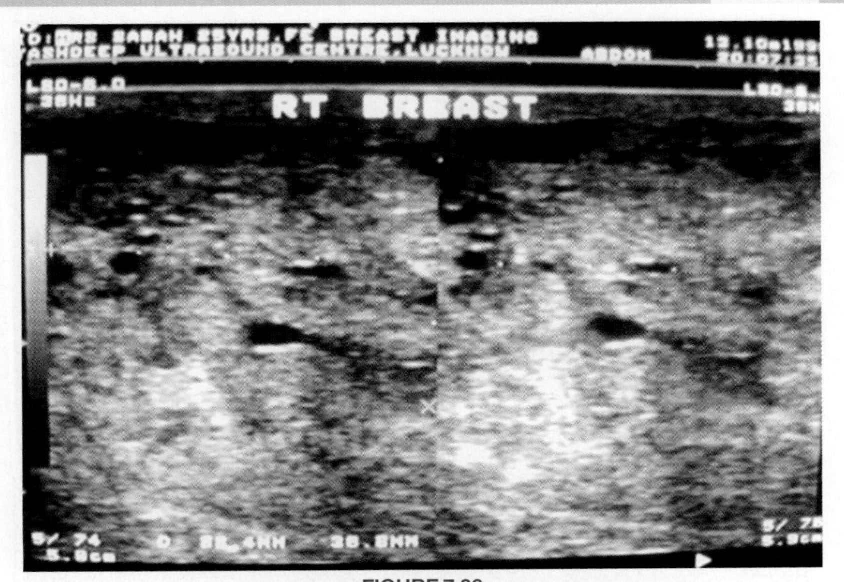

FIGURE 7.86

FIGURE 7.87

FIGURES 7.86 and 7.87: HRSG shows diffuse enlargement of the breast, involving whole of the breast. The ducts were dilated and engorged. Marked soft tissue edema is present. But no focal mass lesion is seen. The patient was having positive history of filarial and it was acute filarial involvement of Rt breast.

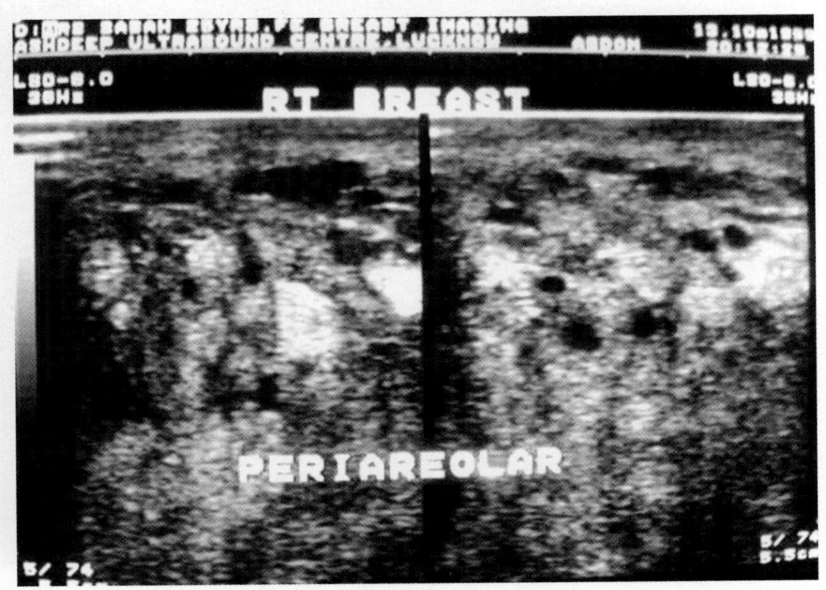

FIGURE 7.88: HRSG shows grossly edematous Rt breast with multiple dilated tubules in the same patient.

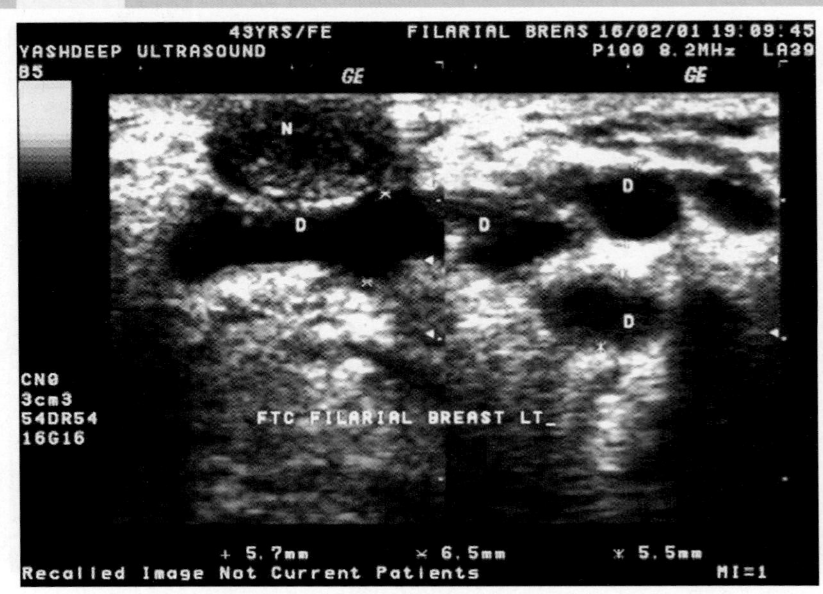

FIGURE 7.89: Bilateral filarial breast. A known case of bilateral filarial breast infestation presented with enlarged engorged breast with pain. HRSG shows markedly dilated tubules. They are running to the areola. The ducts are engorged.

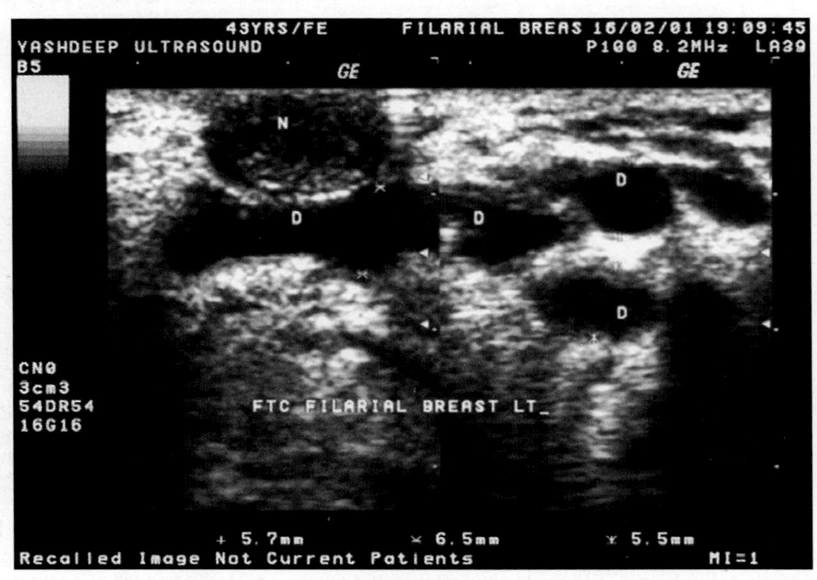

FIGURE 7.90

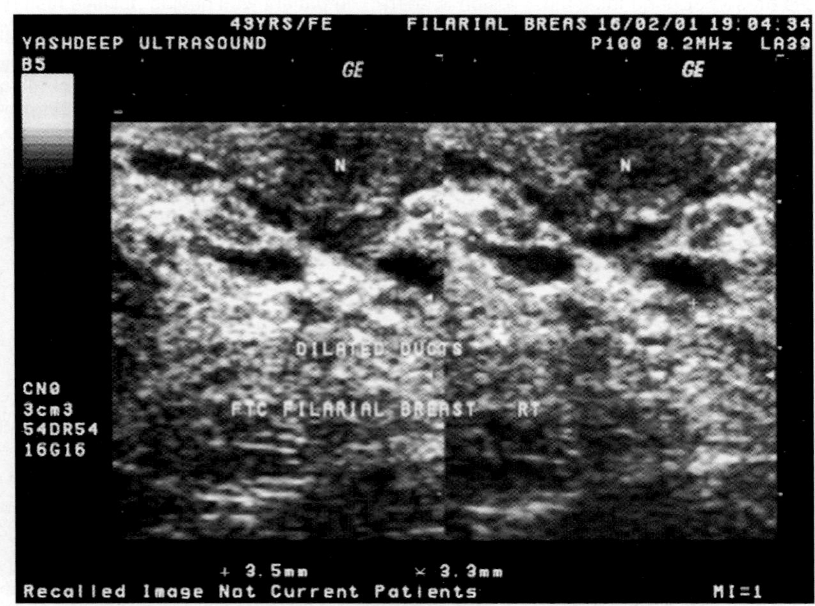

FIGURE 7.91

FIGURES 7.90 and 7.91: Rt breast also shows dilated ducts running to the tubules.

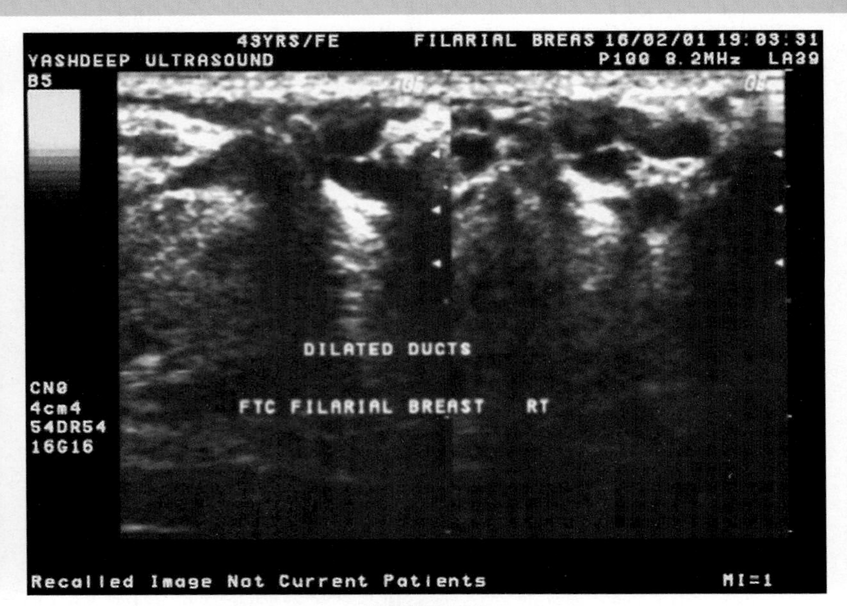

FIGURE 7.92

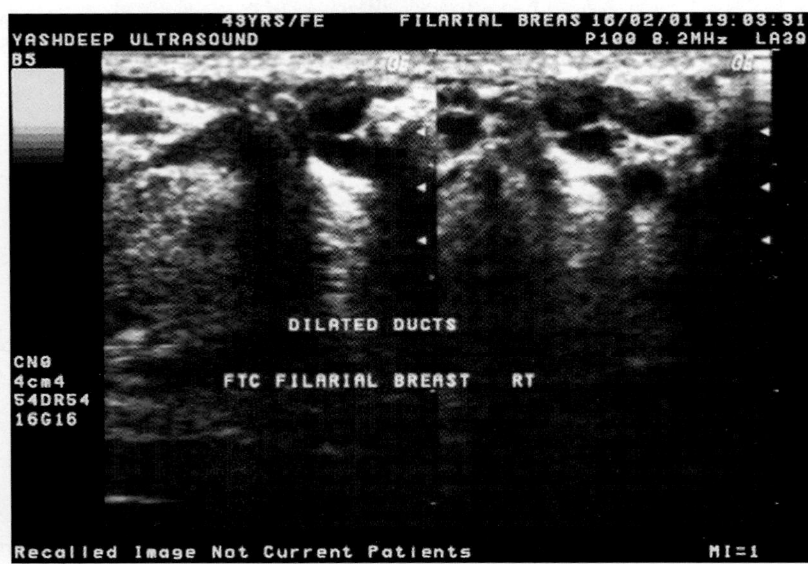

FIGURE 7.93A

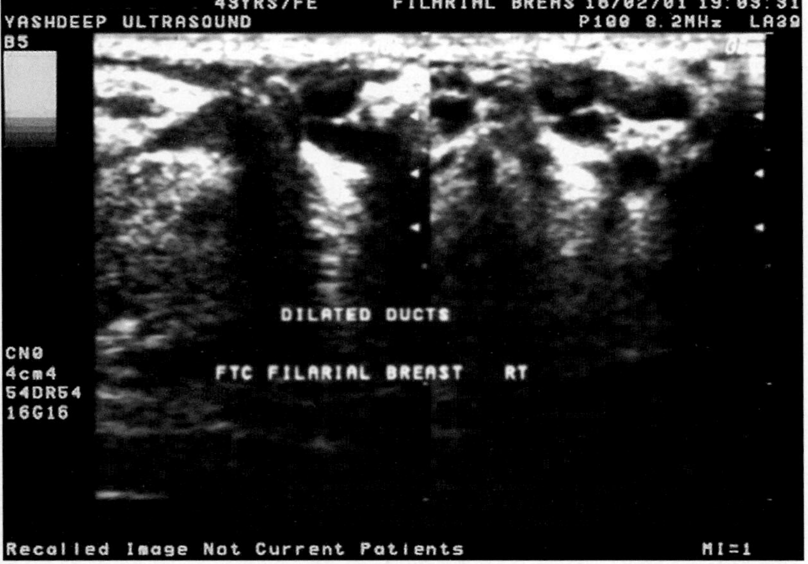

FIGURE 7.93B

FIGURES 7.92 and 7.93: Gross dilatation of the ducts is seen involving both the breast suggestive of filarial ductal dilatation with engorgement. Patient was giving history of while milk like fluid coming out from the both breast in large quantity suggestive of chyle.

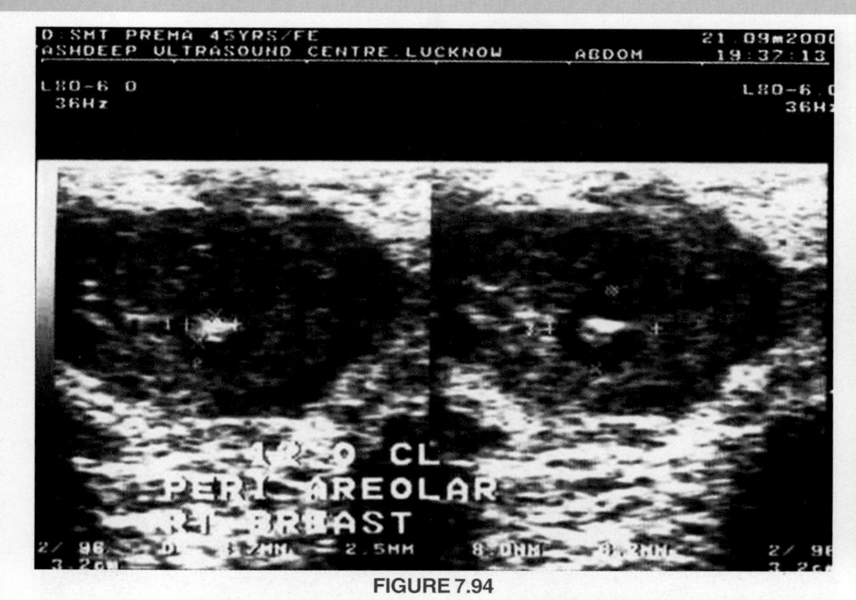

FIGURE 7.94

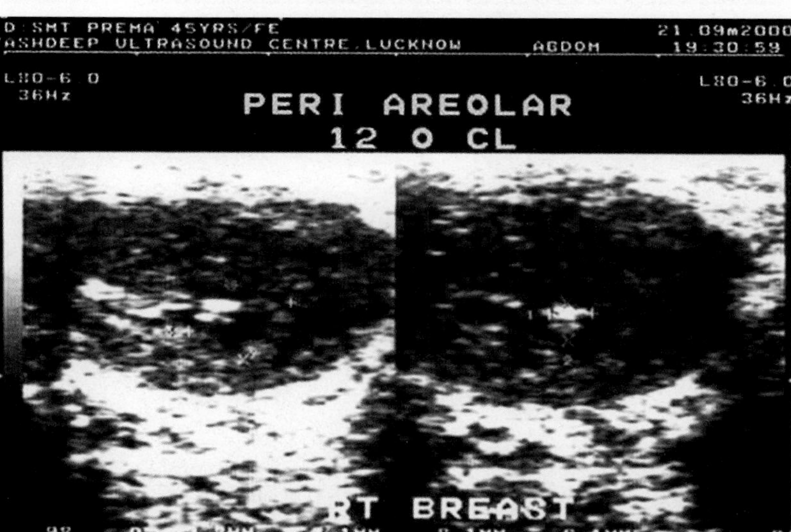

FIGURE 7.95

FIGURES 7.94 and 7.95: Cysticercus cyst in the breast. HRSG shows a homogeneous mass shadow in the breast in a patient presented with a lump. A well-defined cyst is seen in the mass. An echogenic nidus is seen fixed with the inner wall of the cyst suggestive of scolex. This is a typical sonographic appearance of cysticercus cyst. It is an unusual finding and not commonly seen in breast.

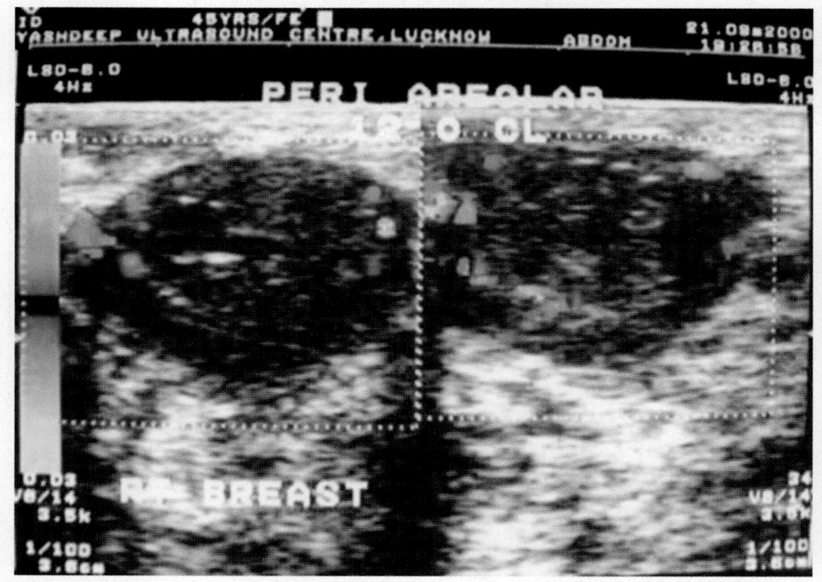

FIGURE 7.96: On color Doppler imaging low flow is seen in the cyst suggestive of hyperemia.

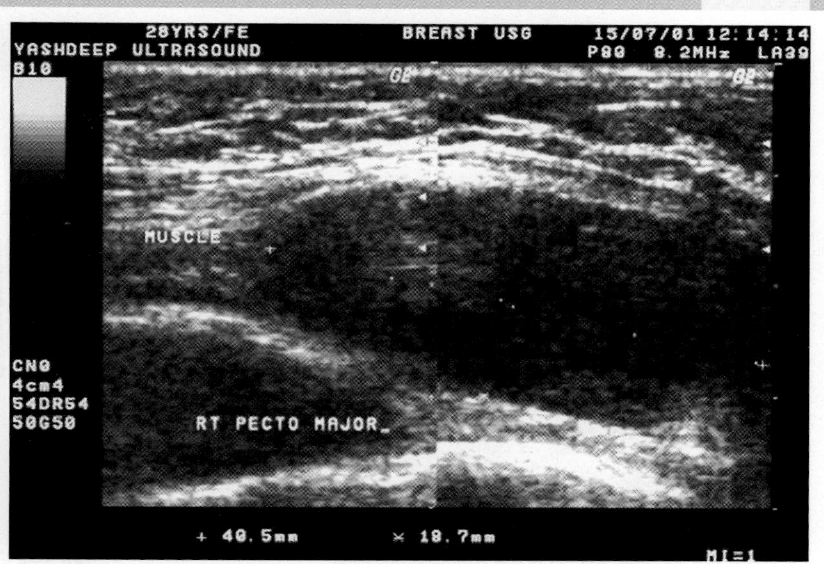

FIGURE 7.97

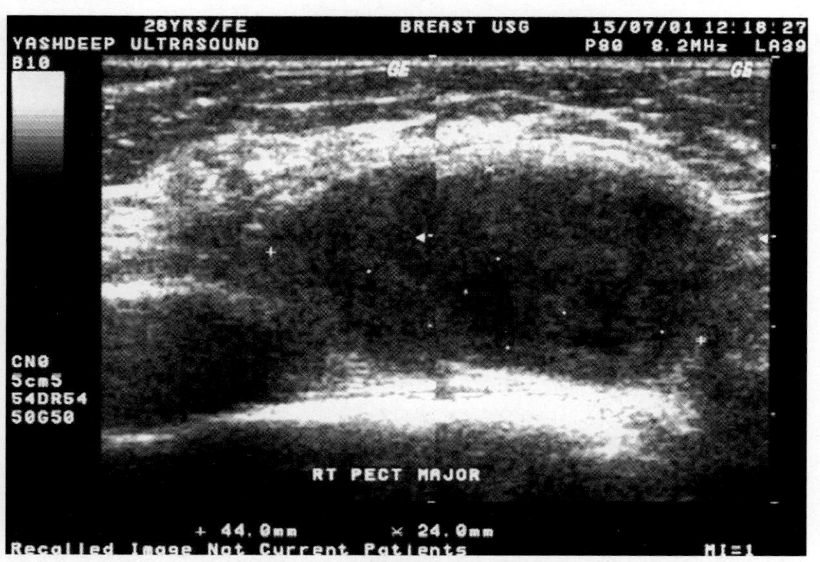

FIGURE 7.98

FIGURES 7.97 and 7.98: Pectoralis major muscle abscess. HRSG shows a big low-level echo complex mass in the Rt pectoralis major muscle suggestive of an abscess. The patient presented as a breast lump. However, the breast was absolutely normal and abscess was seen in the muscle.

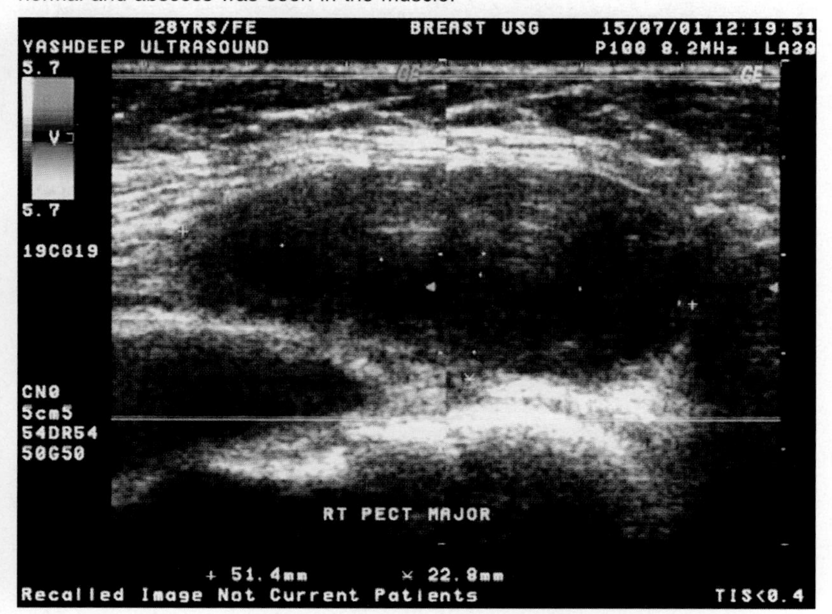

FIGURE 7.99: On color Doppler imaging no flow is seen in the abscess cavity.

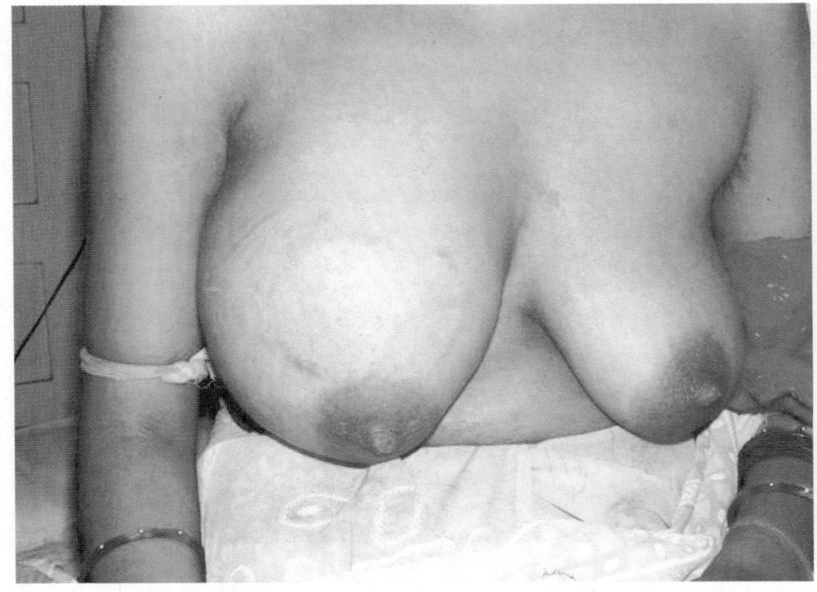

FIGURE 7.100: Ruptured breast. A patient presented with acute swelling of the breast after sustaining blunt trauma to the breast.

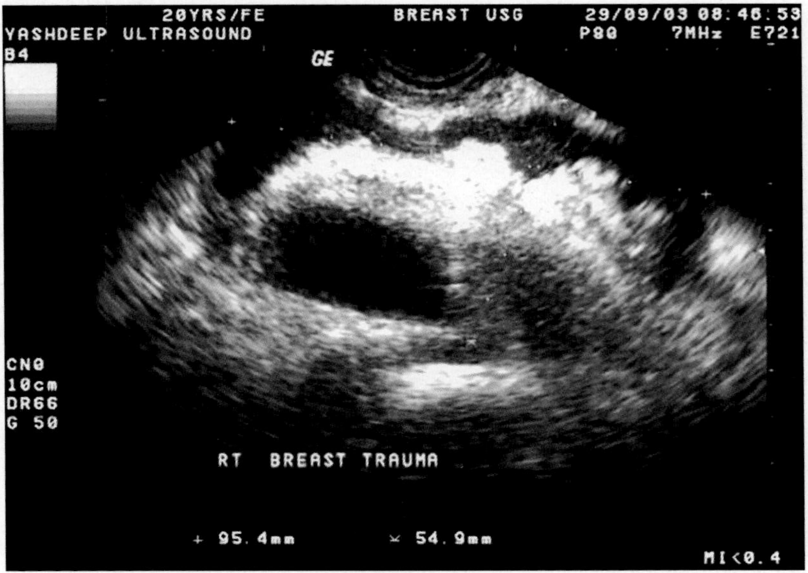

FIGURE 7.101

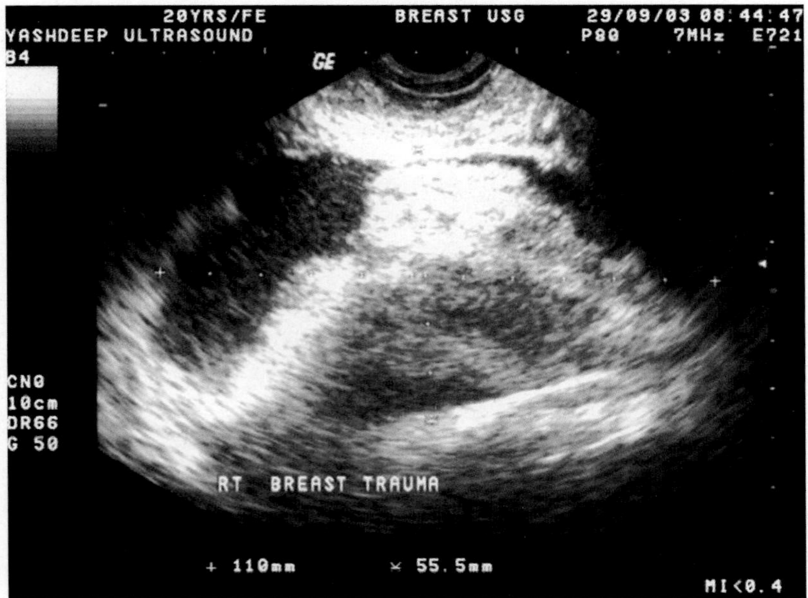

FIGURE 7.102

FIGURES 7.101 and 7.102: HRSG shows rupture of the breast capsule with hematoma formation. The floating echogenic shadows are suggestive of fresh blood.

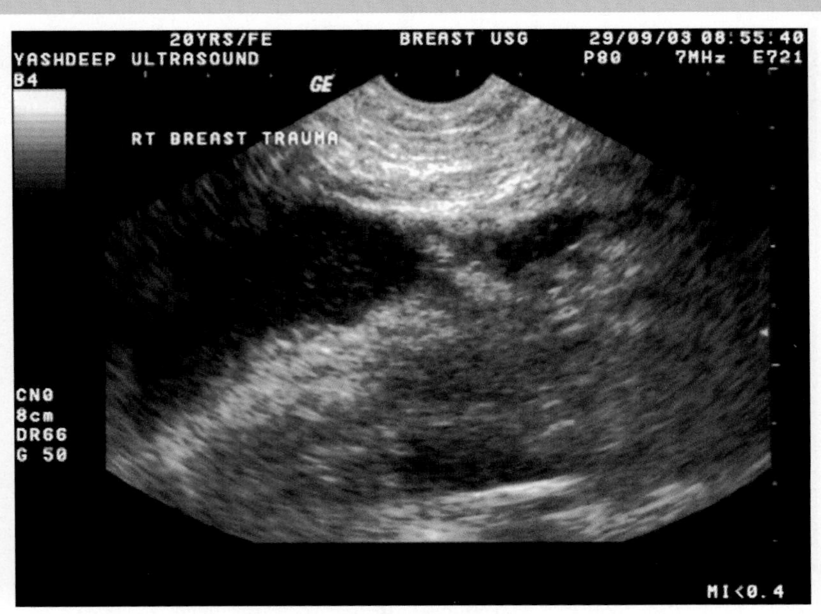

FIGURE 7.103

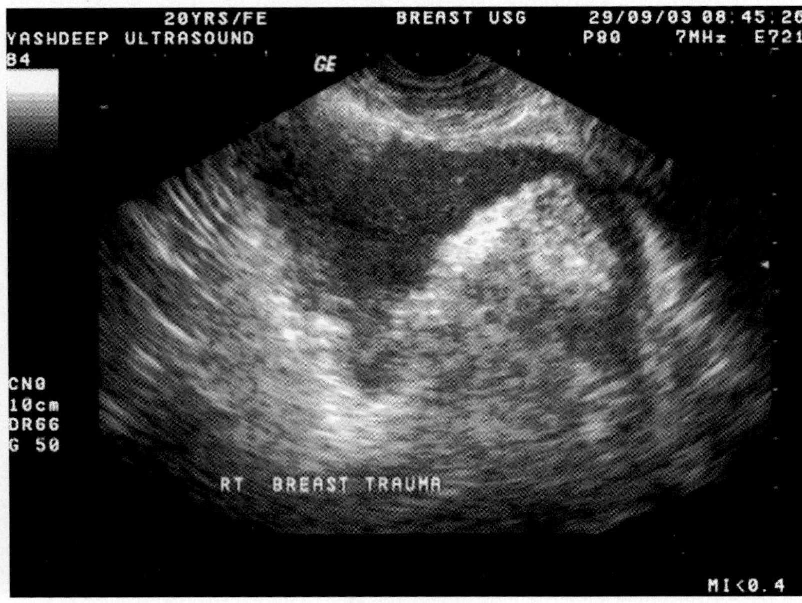

FIGURE 7.104

FIGURES 7.103 and 7.104: The collected blood is seen gravitating on the dependent part. The ruptured breast tissue is seen floating in the hemorrhagic collection.

Tuberculosis of the Breast

Tuberculosis of the breast is uncommon condition. However, the disease can attack the breast from surrounding tubercular focus as in sternum, rib, chest wall or systemic focus elsewhere in body. A patient presents with soft tissue swelling in the breast, which is most of the time, is not painful. HRSG shows low-level echo collection with multiple internal echoes suggestive of debris. Liquefaction is rare. Inflammatory nodes are seen in axilla.

Benign Breast Masses

The benign breast masses are grouped as follows:

1. Fibroadenoma
2. Adenomata and papillomata
3. Breast cyst
4. Hematoma

Fibroadenoma

Fibroadenomas are well-defined, lobulated solid, encapsulated masses present anywhere in the breast. They are homogeneous in texture. They are highly mobile and clinically termed as "breast mouse" due to their frequent changing position. When they are present near to skin, they are sharply marginated on HRSG. The capsule is echogenic. The mass is incompressible. No retrotumoral effect or shadowing is seen in fibroadenoma. At times there may be many fibroadenoma in one breast and they are more common in young breast. Presence of fibroadenoma in both breasts is also common. Macro calcification is seen in few fibroadenoma. It may cast acoustic shadowing. At times areas of cystic necrosis are also seen in the tumors. Usually the fibroadenoma is small to medium size. But giant fibroadenoma is also reported. They are irregular in outline and heterogeneous in texture. Margins are ill-defined. It becomes difficult to differentiate it from a malignant breast lump. They overlap in ultrasound image of fibroadenoma and malignant breast lump is well established. Therefore, other features of fibroadenoma like mobility of mass, its compressibility, homogeneous texture and its tendency to lie along the tissue planes. The long axis of adenoma is more than the transverse axis and known as LAP (Longitudinal/Anteroposterior) ratio. It should be more than 1.4 in fibroadenoma. However, wherever there is any doubt about the diagnosis, a biopsy is must to establish the diagnosis.

Adenomata and Papillomata

Adenomata and papillomata are the other benign masses. They are difficult to be differentiated from fibroadenoma on ultrasonographic features alone. However, they tend to appear near periareolar area and usually surrounded by a rim of fluid. Hamartomas are usually sharply marginated and well demarcated. An internal structure is heterogeneous with calcification. Lipoma is well-defined echo poor mass with homogeneous texture.

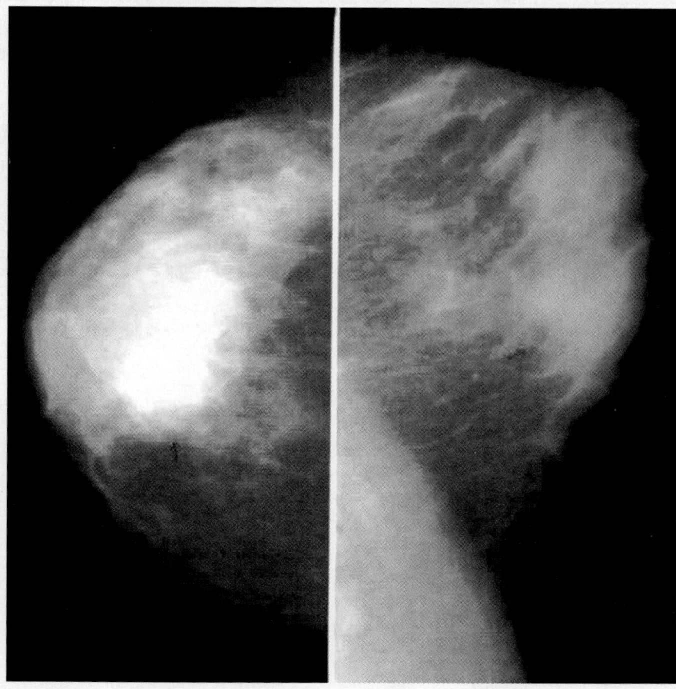

FIGURE 7.105: **Fibroadenoma in the breast.** A patient presented with palpable mass in the Rt breast. X-ray mammogram shows a radiodense shadow in the breast.

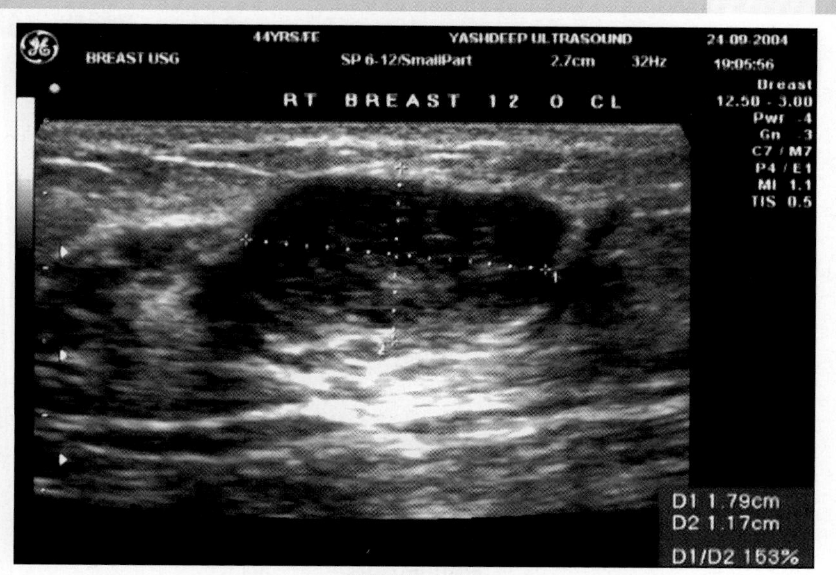

FIGURE 7.106

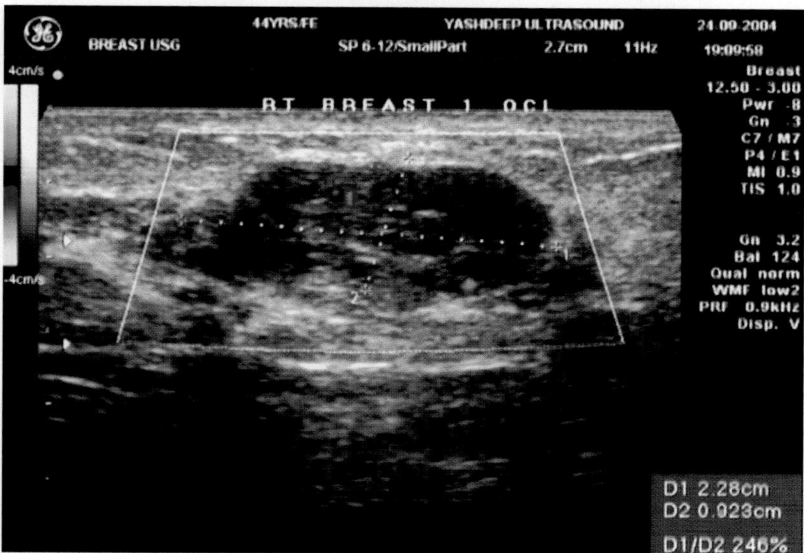

FIGURE 7.107

FIGURES 7.106 and 7.107: HRSG shows well-defined homogeneous mass at 12 O' clock position. Margins are smooth. It is homogeneous in texture. No calcification is seen in it. Findings are suggestive of fibroadenoma breast.

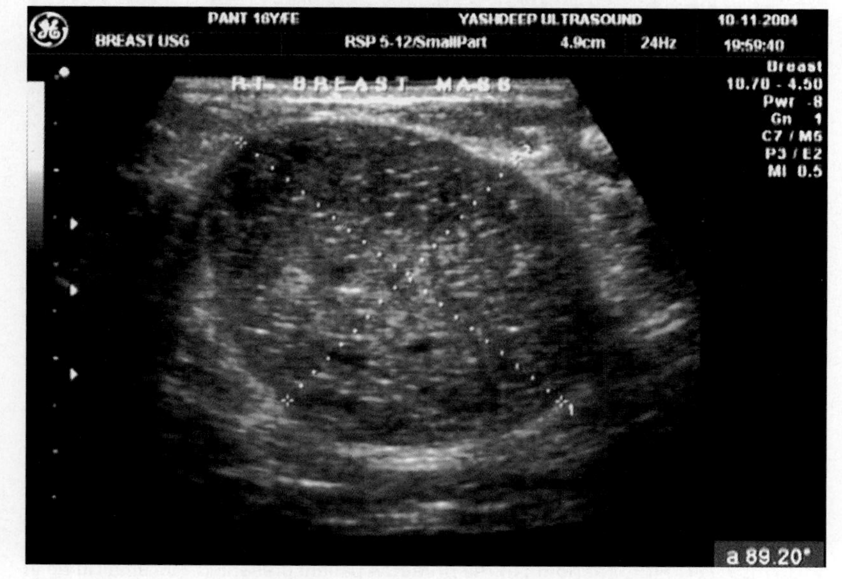

FIGURE 7.108: **Fibroadenoma in the breast.** Well-defined homogeneous mass seen in the breast. The mass shows sharp margins. Homogeneous echoes are seen in it. No calcification is seen. No cystic degeneration is seen suggestive of fibroadenoma.

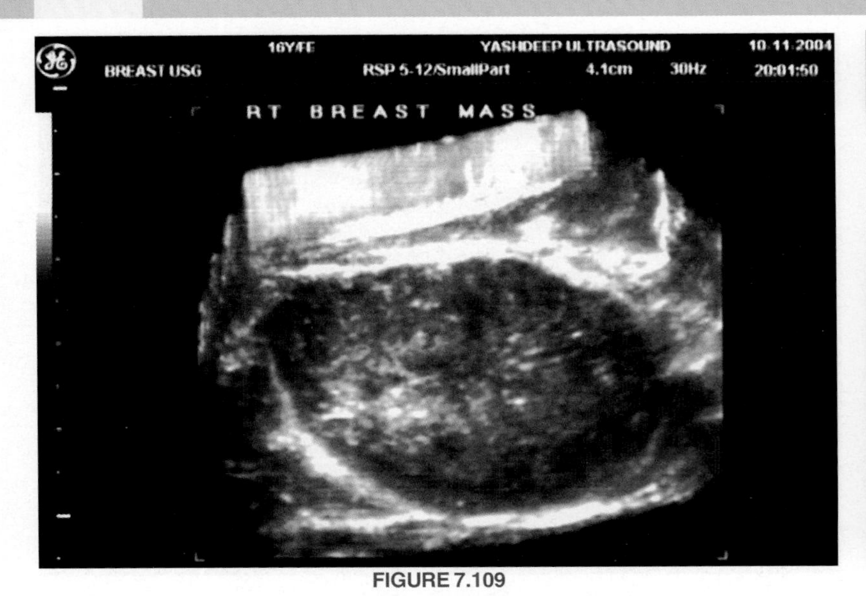

FIGURE 7.109

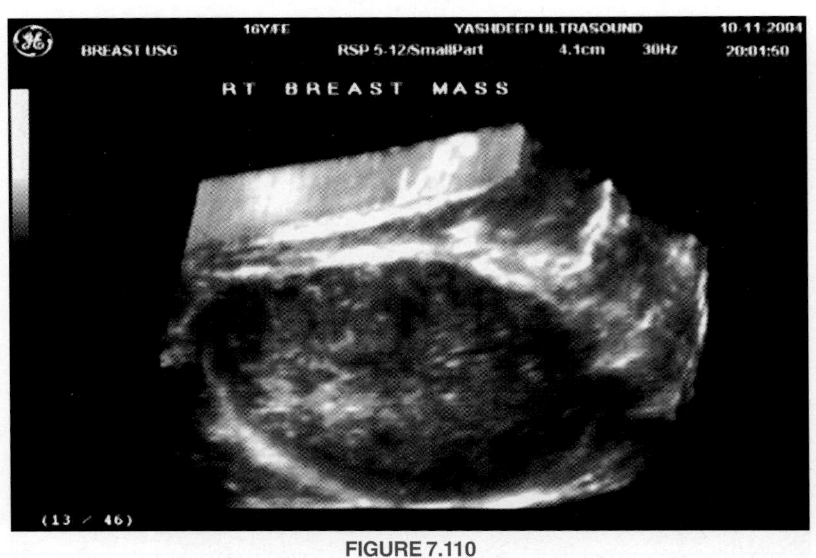

FIGURE 7.110

FIGURES 7.109 and 7.110: 3D imaging of the same patient shows details of the fibroadenoma.

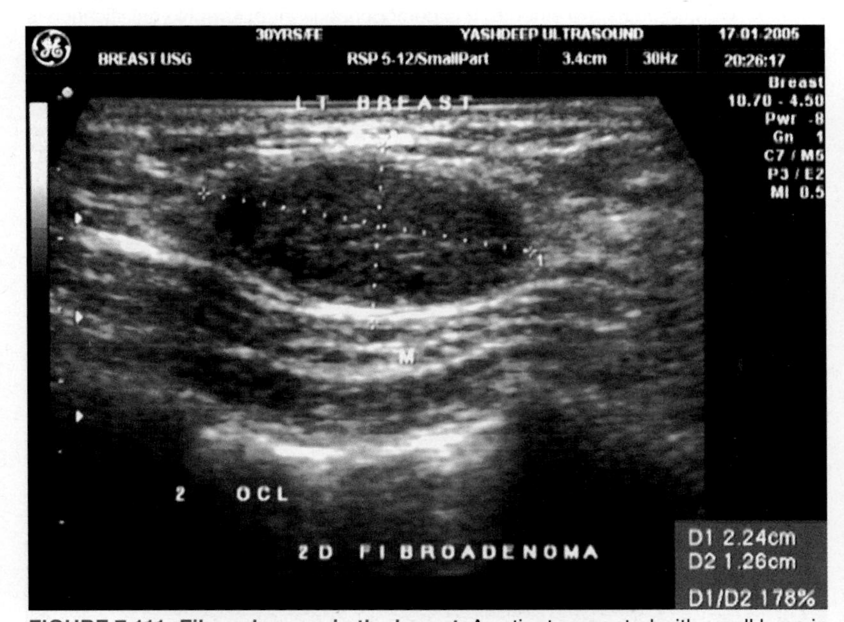

FIGURE 7.111: Fibroadenoma in the breast. A patient presented with small lump in the breast. HRSG shows well-defined homogeneous soft tissue mass in the Rt breast at 2 O'-clock position. The anteroposterior diameter is less than longitudinal diameter. LAP sign is positive; it is 2.2 (L/AP diameter). Positive LAP sign suggest benign nature of the mass.

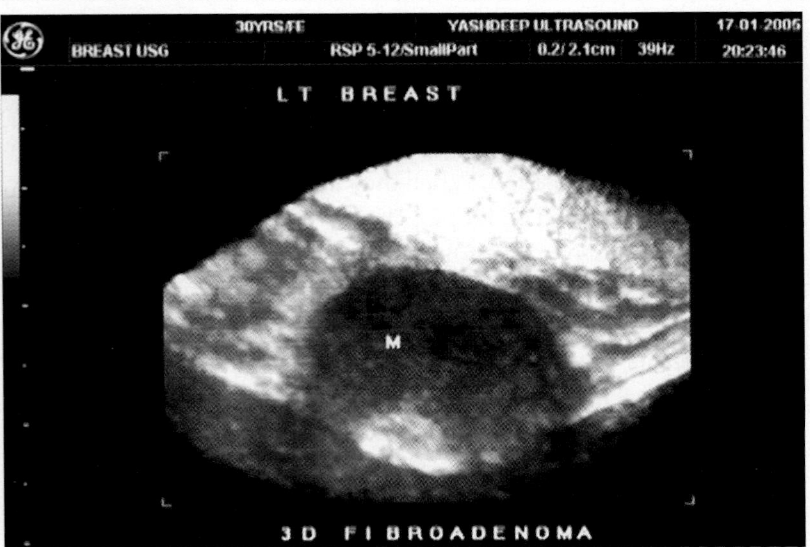

FIGURE 7.112

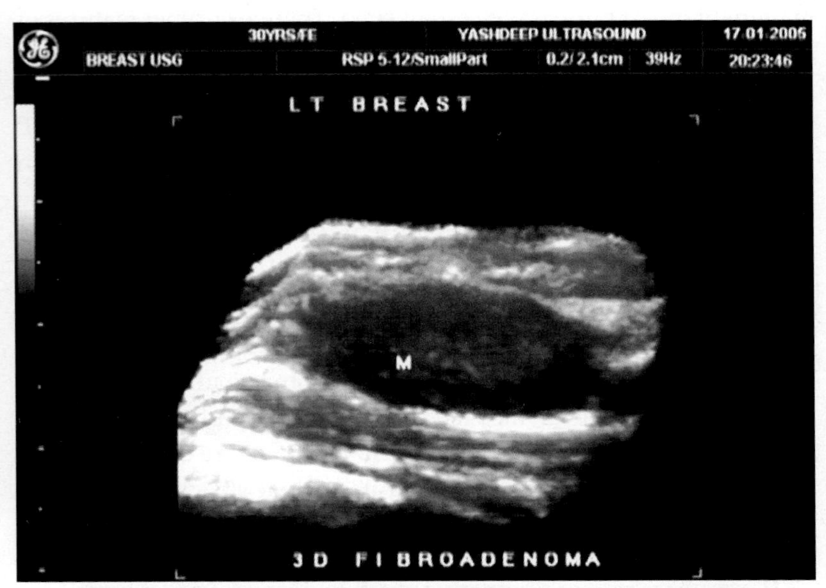

FIGURE 7.113

FIGURES 7.112 and 7.113: 3D imaging of the mass shows details of the fibroadenoma with homogeneous texture.

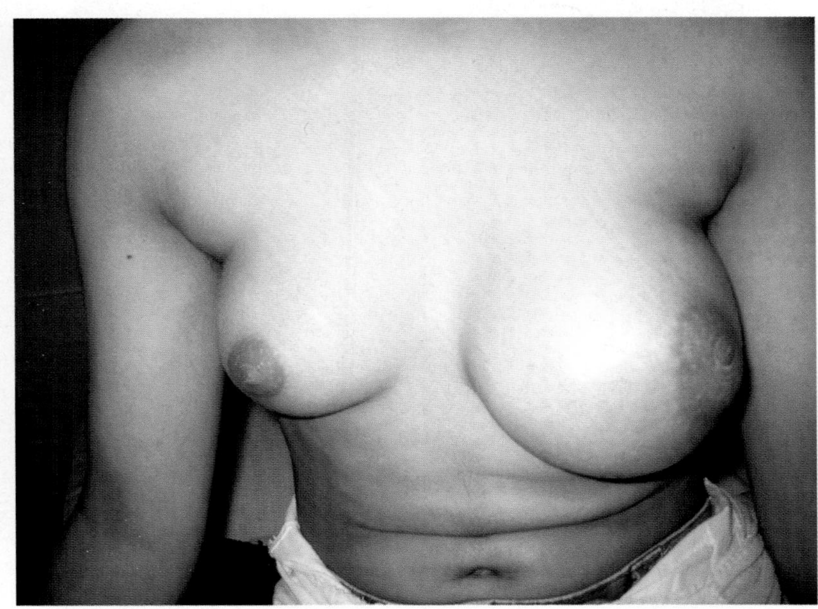

FIGURE 7.114: Big fibroadenoma breast. A young girl presented with big soft tissue mass in the Lt breast. Diffuse enlargement of the breast is seen.

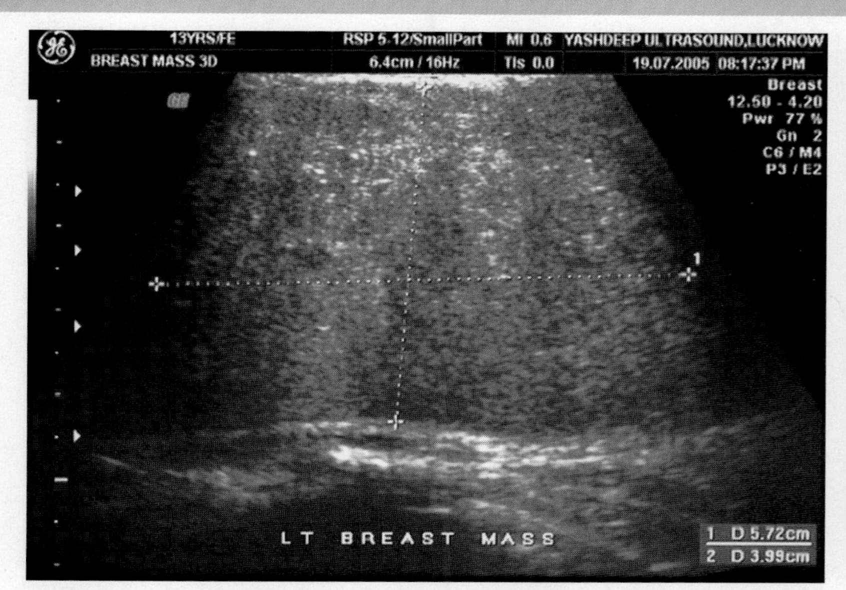

FIGURE 7.115: HRSG shows a well-defined encapsulated soft tissue mass in the Lt breast. The mass shows sharp margins with homogeneous texture. No calcification or cystic necrosis is seen in the mass suggestive of huge fibroadenoma.

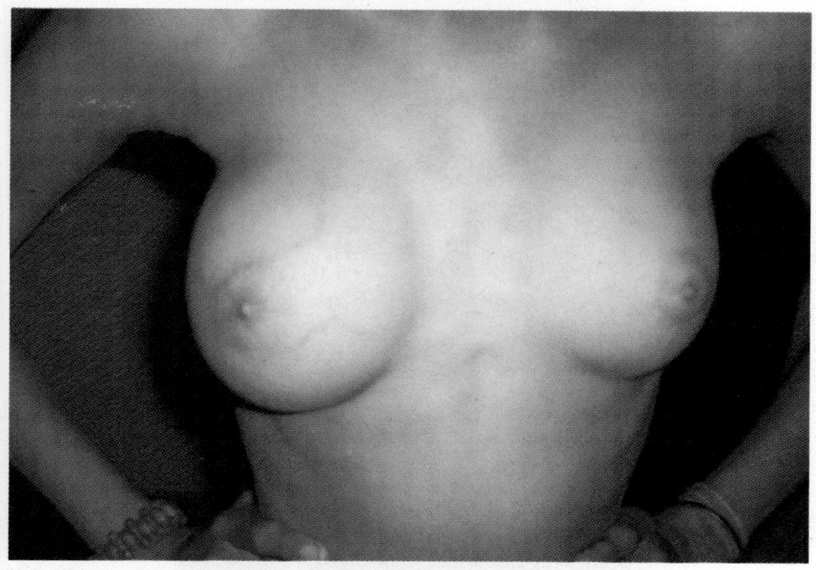

FIGURE 7.118 Big fibroadenoma breast. A young girl presented with big mass in the Rt breast. The mass was non-tender.

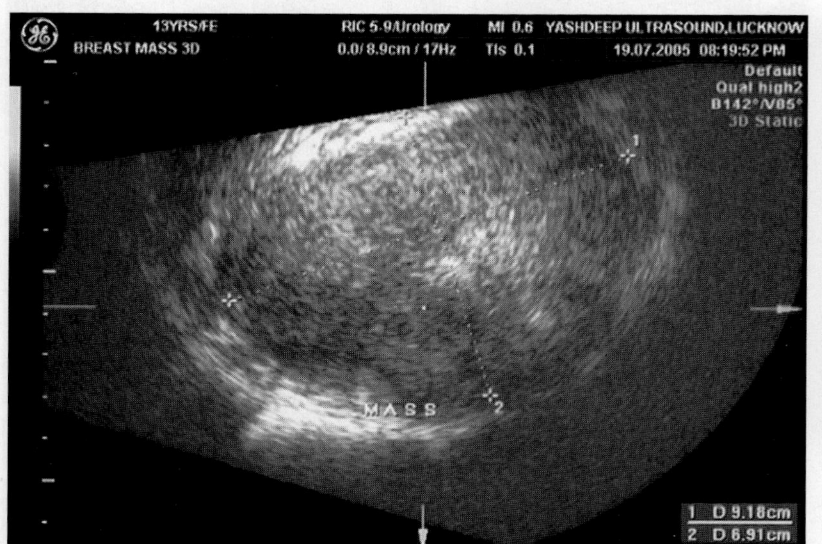

FIGURE 7.116

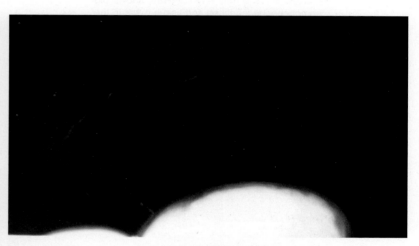

FIGURE 7.119: X-ray mammogram shows radiodense opacity in the Rt breast. No mass details are seen on X-ray mammogram.

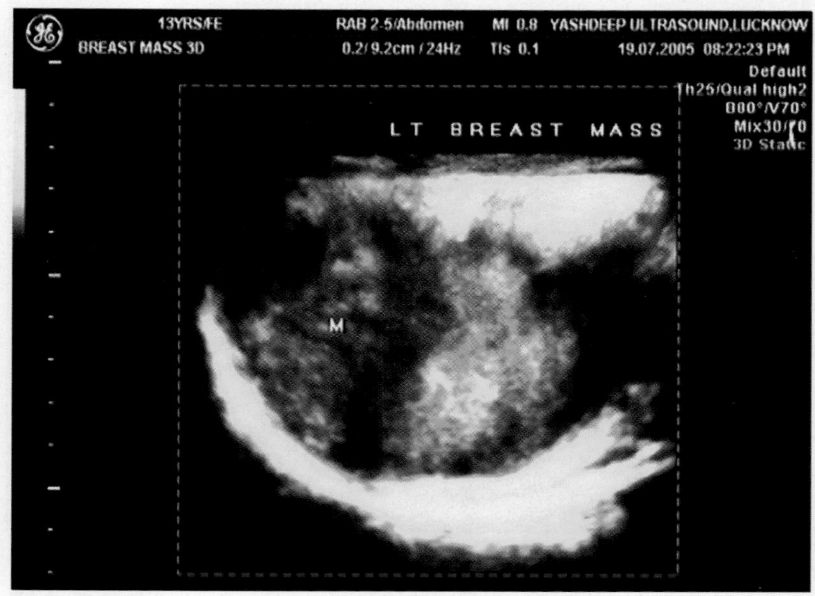

FIGURE 7.117

FIGURES 7.116 and 7.117: 3D imaging of the mass shows the details of the mass. Depth of the mass is well appreciated on 3D imaging.

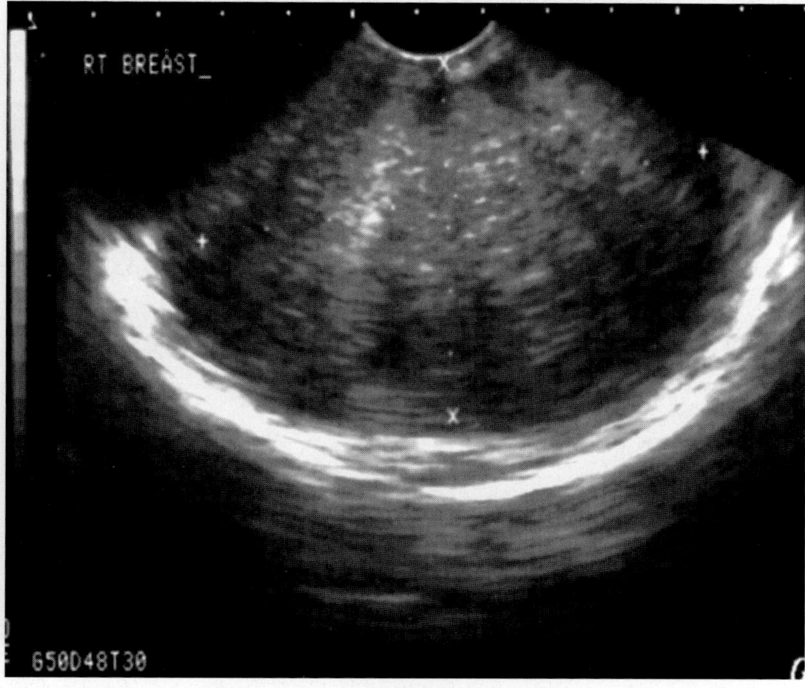

FIGURE 7.120

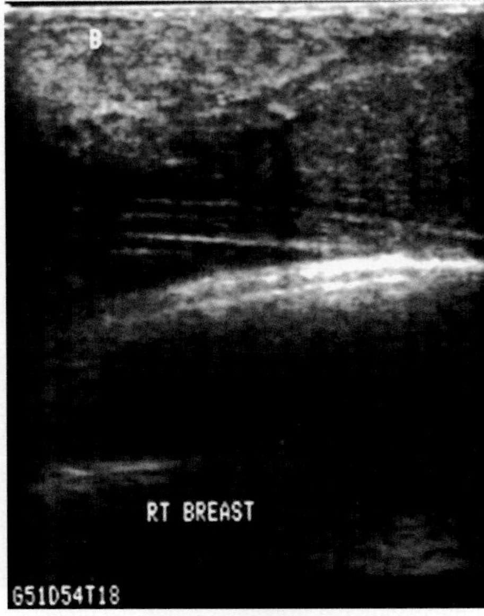

FIGURE 7.121

FIGURES 7.120 and 7.121: HRSG shows big homogeneous lobulated soft tissue mass in the Rt breast. The mass is well encapsulated. Normal breast tissue is seen out side of the mass in the Rt upper outer quadrant. Biopsy confirmed a big fibroadenoma.

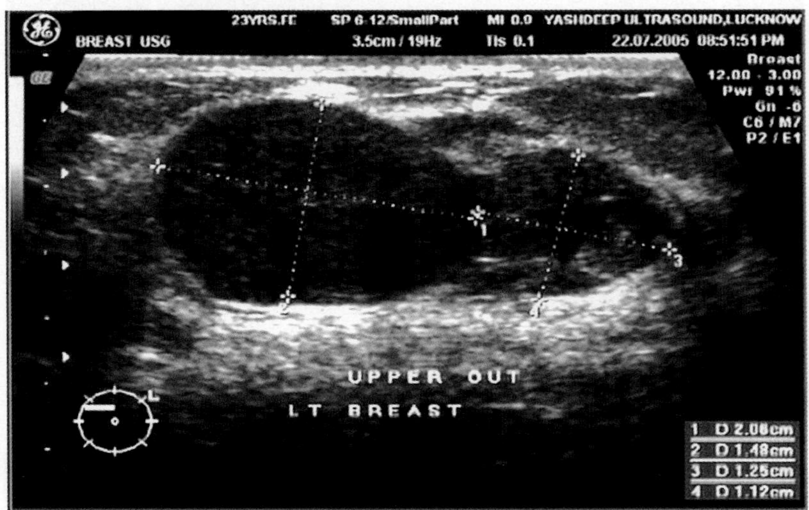

FIGURE 7.122

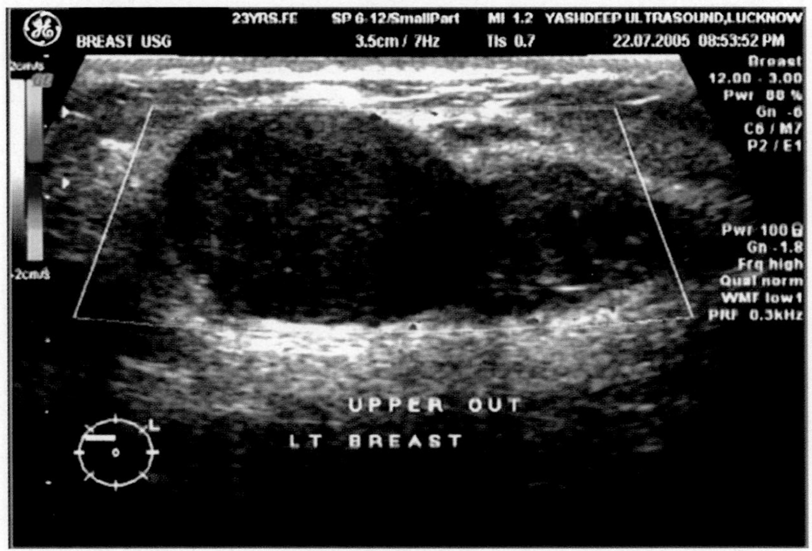

FIGURE 7.123

FIGURES 7.122 and 7.123: Lobulated fibroadenoma. HRSG shows big lobulated fibroadenoma in the Lt breast. Well-defined homogeneous lobules are seen in the mass with sharp margins.

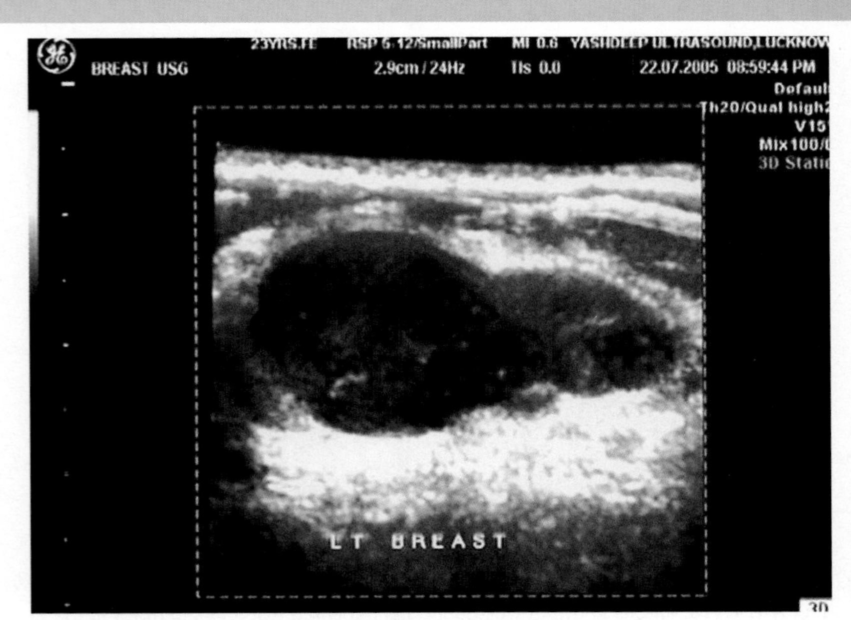

FIGURE 7.124: 3D imaging of the mass shows details of the lobules.

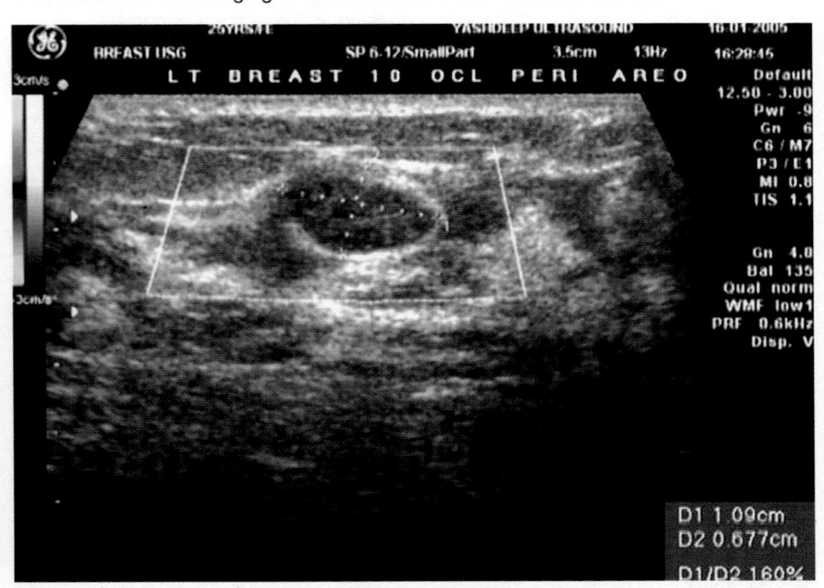

FIGURE 7.125A

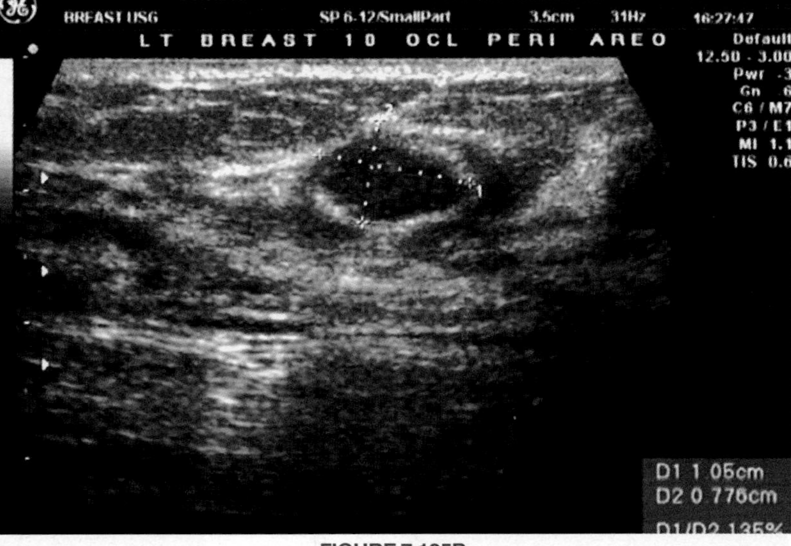

FIGURE 7.125B

FIGURES 7.125A and B: Small nonpalpable fibroadenoma. HRSG shows well-defined homogeneous shadows in the Lt breast 10 O'-clock position Periareolar area. The shadow shows sharp margins. It is homogeneous in texture. No calcification is seen in them suggestive of small fibroadenoma. Clinically small fibroadenomas are Nonpalpable and they are incidental findings on HRSG.

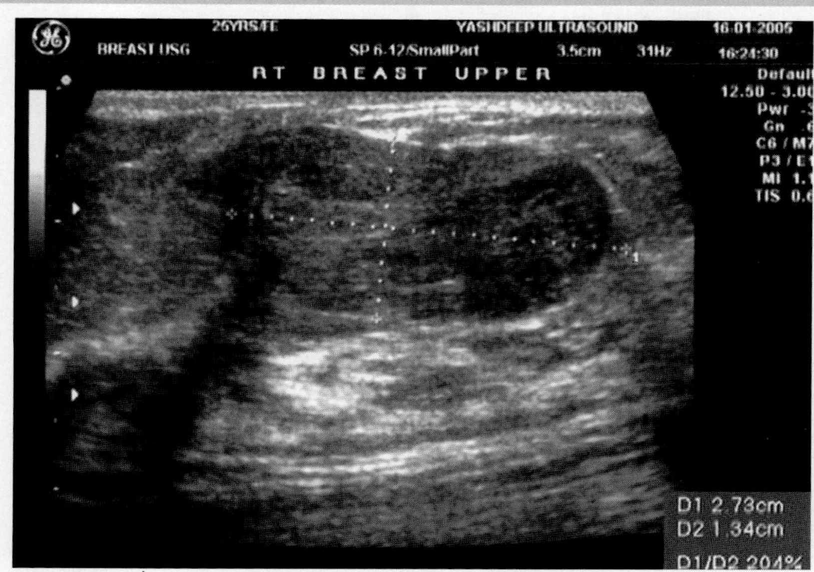

FIGURE 7.126: **Solitary fibroadenoma.** Well-defined homogeneous soft tissue mass is seen in the Rt breast suggestive of fibroadenoma. The mass shows sharp margins.

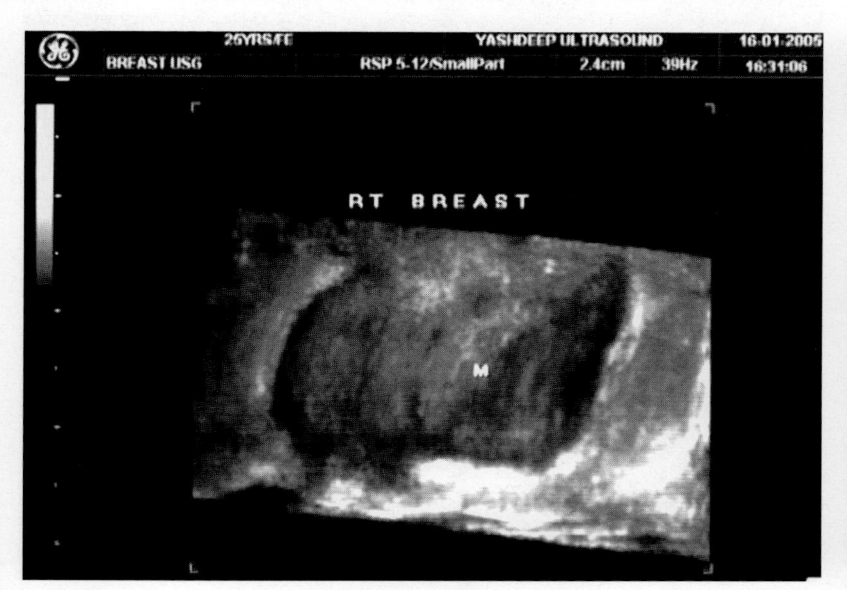

FIGURE 7.127A

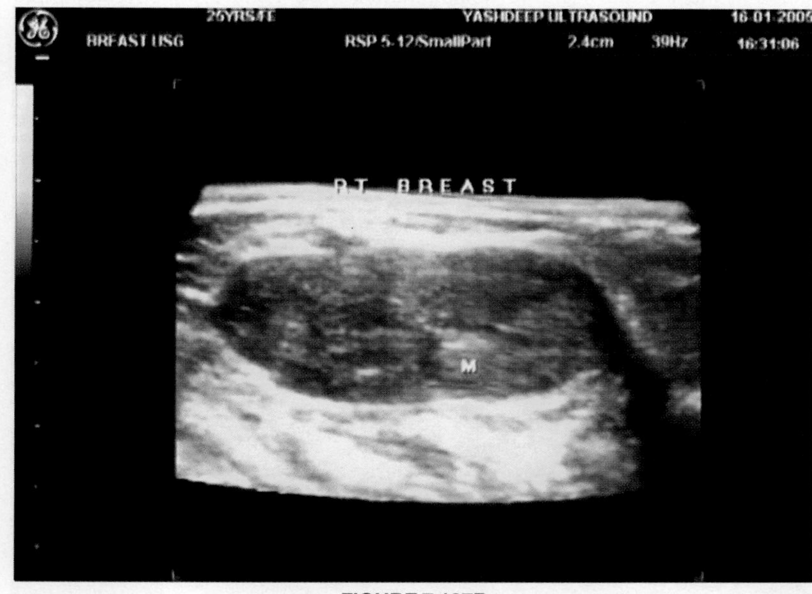

FIGURE 7.127B

FIGURES 7.127A and B: 3D imaging of the mass shows details of the mass. Depth of the mass is well appreciated on 3D imaging.

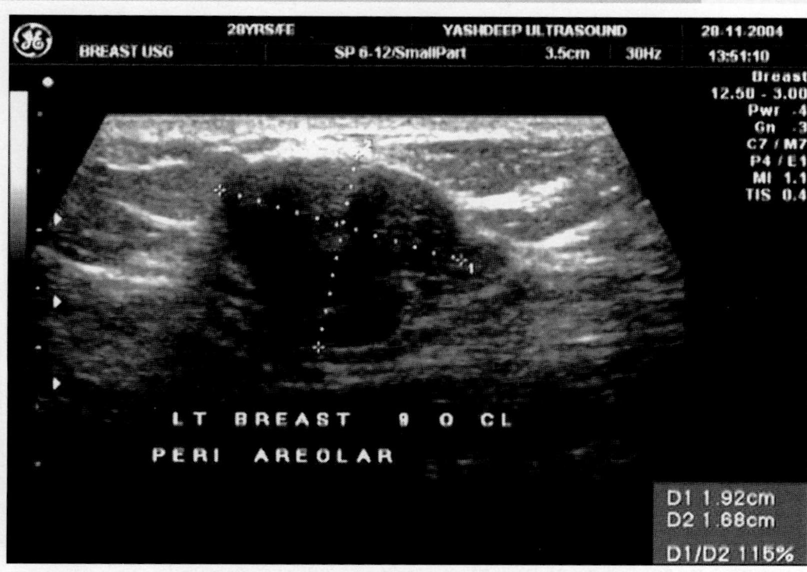

FIGURE 7.128: **Hypoechoic fibroadenoma.** HRSG shows a well-defined homogeneous mass in the breast. The mass shows hypoechoic texture. But no evidence of any cystic degeneration is seen in the mass. No calcification is seen in it.

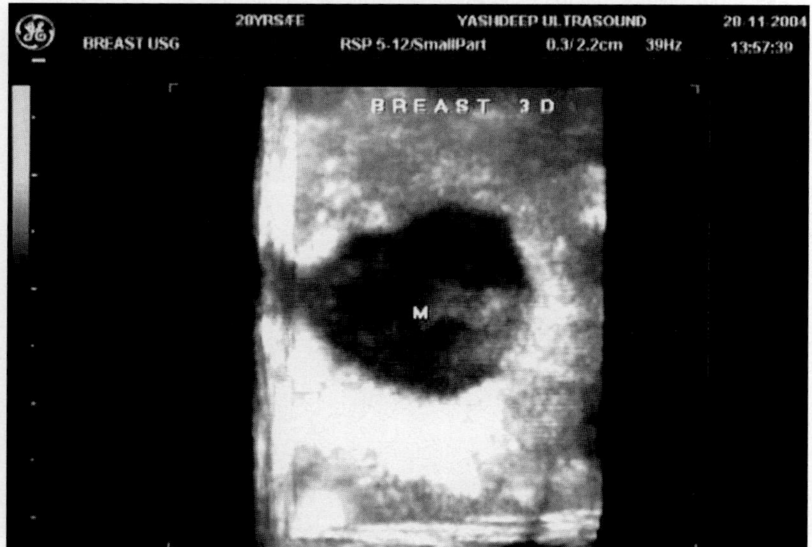

FIGURE 7.129: 3D imaging of the mass shows details of the mass. No evidence of any calcification is seen in it.

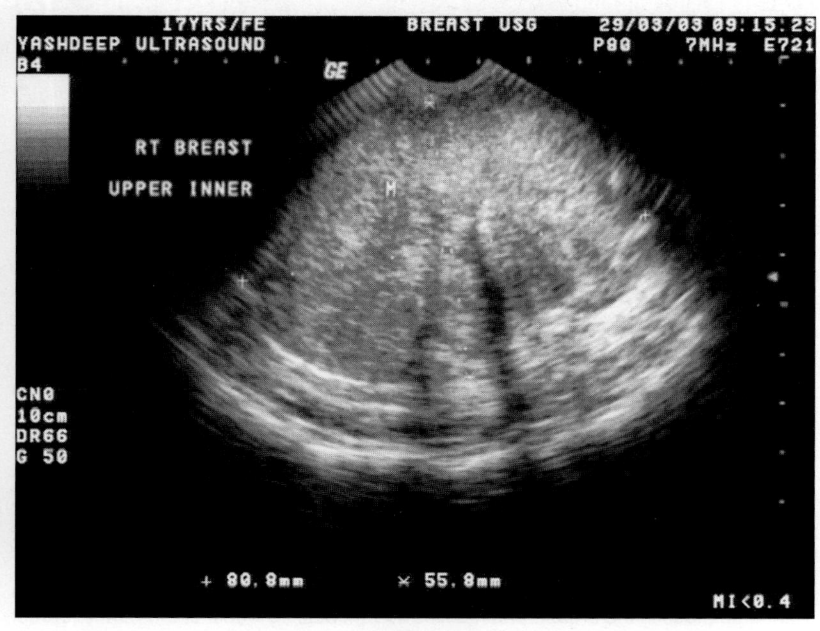

FIGURE 7.130

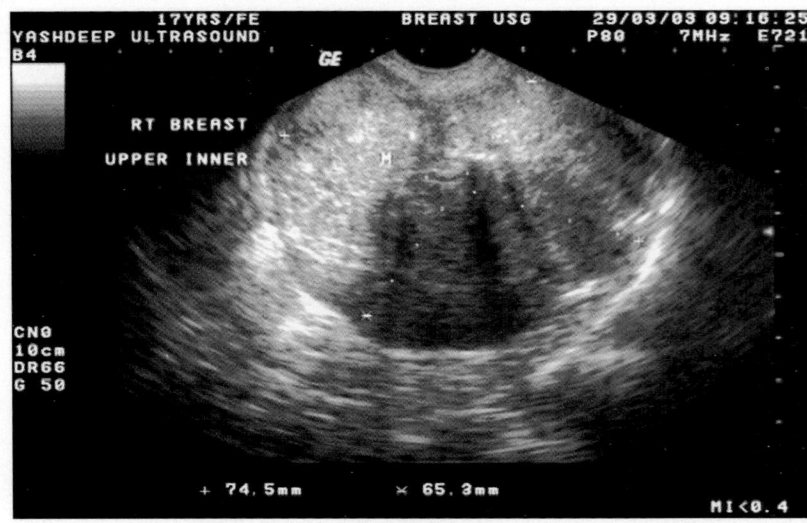

FIGURE 7.131

FIGURES 7.130 and 7.131: Big fibroadenoma with macro calcification. A young girl presented with soft tissue mass in the breast. HRSG shows a big mass in the breast in upper inner quadrant. Echogenic calcified shadows are also seen in the mass suggestive of macro calcification. They are accompanied with dense acoustic shadowing.

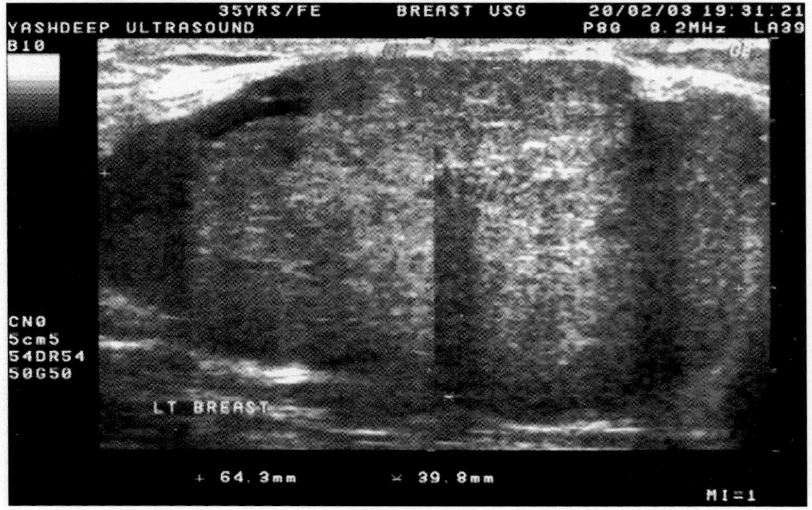

FIGURE 7.132

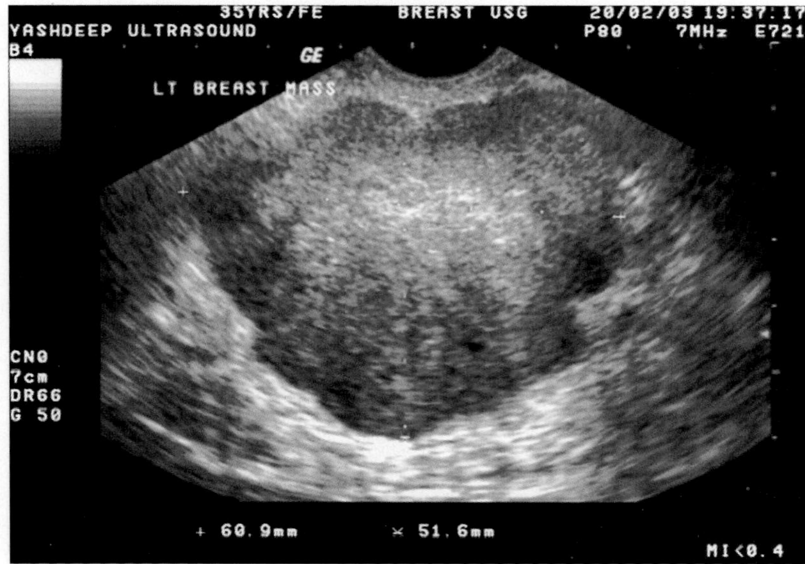

FIGURE 7.133

FIGURES 7.132 and 7.133: Big fibroadenomata in the breast. A lady presented with huge mass in the breast. The mass is seen replacing whole of the normal breast tissue. It is occupying whole of the breast. It is homogeneous in texture. Calcified specks are seen in the mass, which are accompanied with acoustic shadowing.

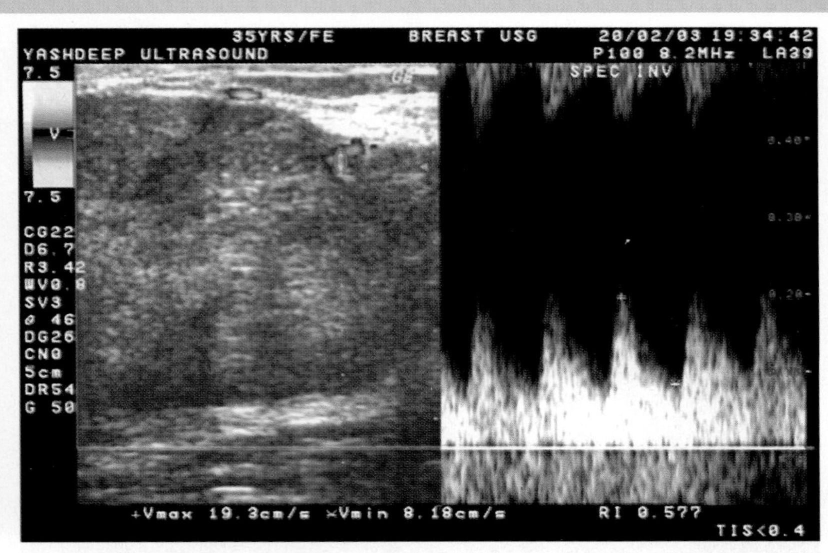

FIGURE 7.134: On color imaging peripheral vessels are seen in the capsule of the mass. On spectral Doppler tracing it shows low arterial flow.

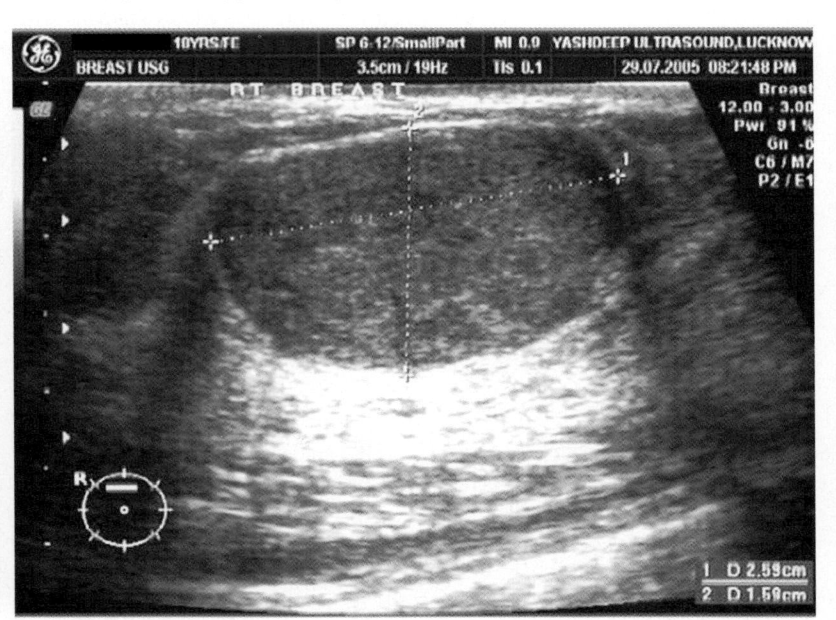

FIGURE 7.135: Lobulated mass in the breast. HRSG shows well-defined lobulated mass in the breast. It shows homogeneous echo texture. Biopsy confirmed fibroadenoma.

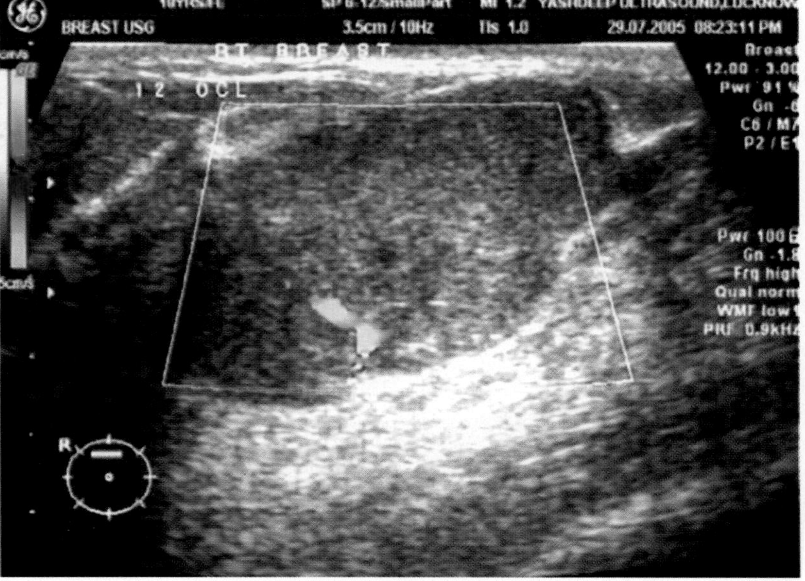

FIGURE 7.136

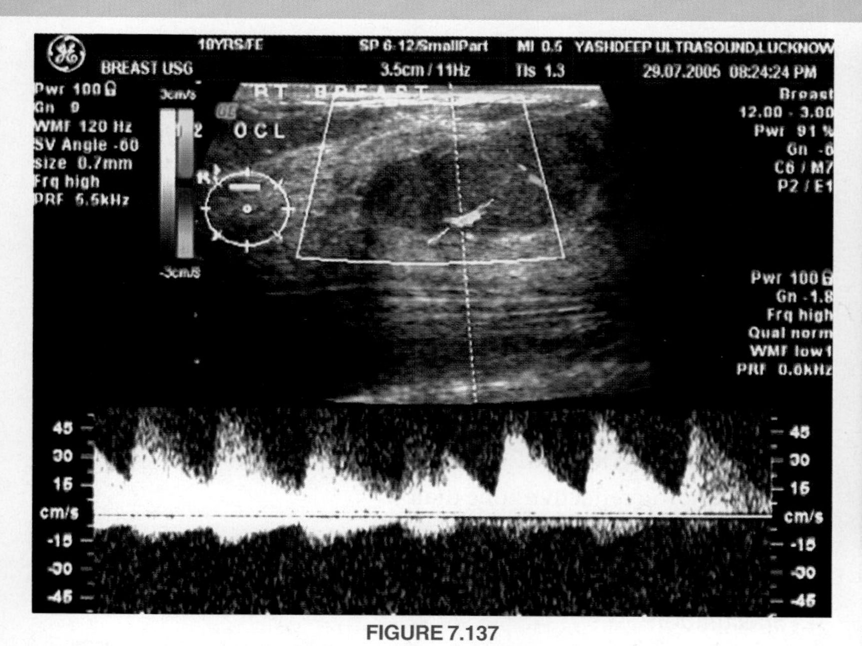

FIGURE 7.137

FIGURES 7.136 and 7.137: On color flow imaging low flow is seen in the mass.

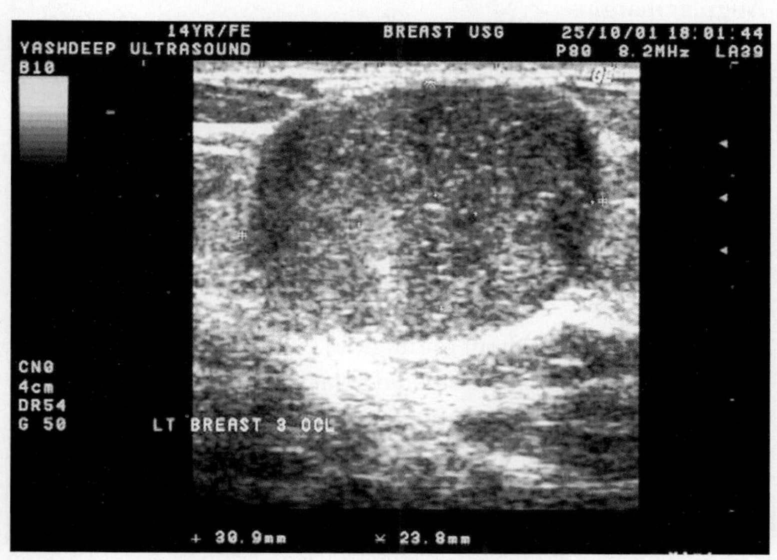

FIGURE 7.138: Big fibroadenoma in the breast. HRSG shows a big hypoechoic ovoid mass in the Lt breast at 13 O'-clock position. It is homogeneous in texture and smooth in outline. No necrosis or calcification is seen in it.

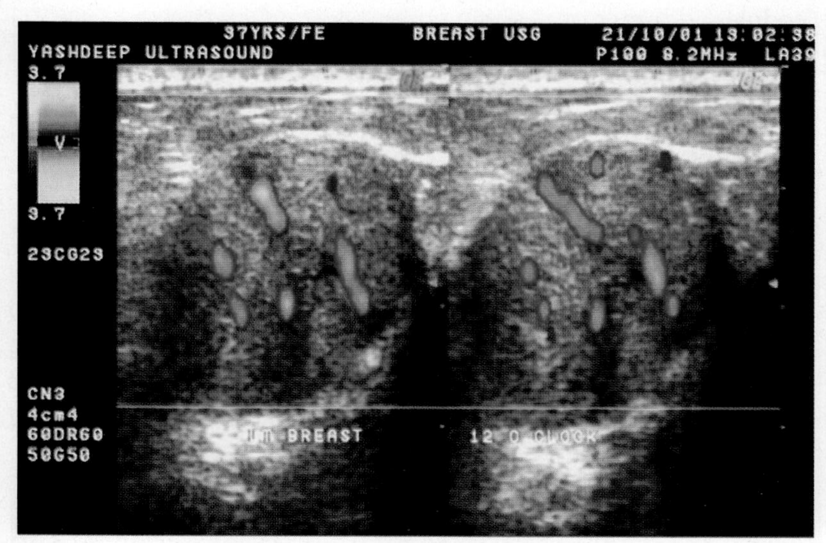

FIGURE 7.139

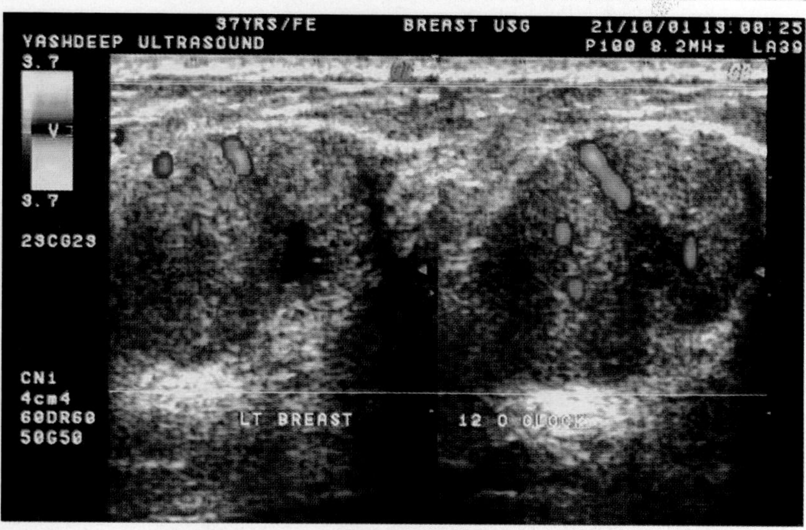

FIGURE 7.140

FIGURES 7.139 and 7.140: On color Doppler imaging moderate flow is seen in the fibroadenoma.

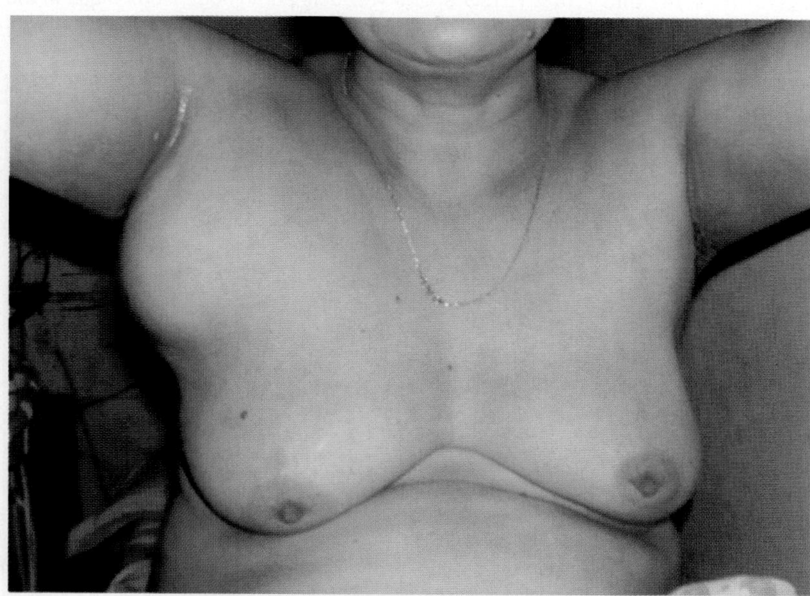

FIGURE 7.141: Lipoma breast. A patient presented with soft tissue mass in the Rt axilla and also reaching up to Rt upper outer quadrant.

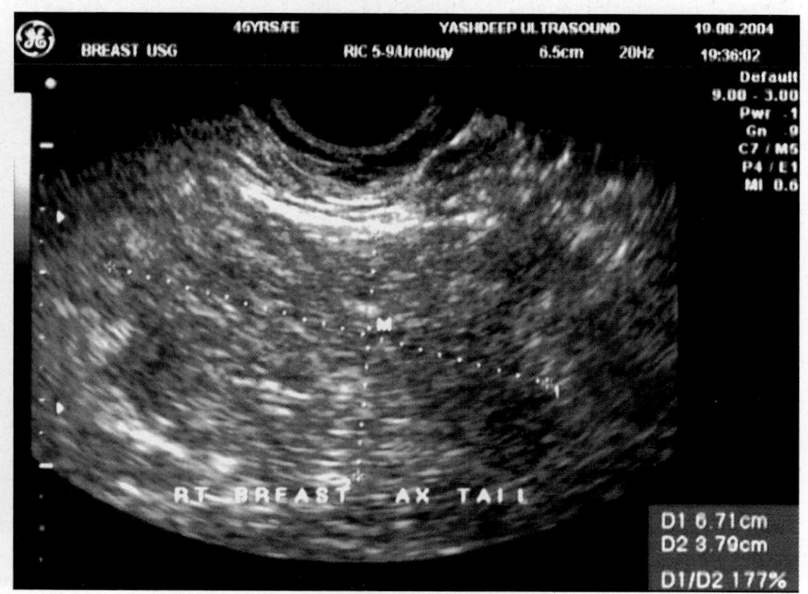

FIGURE 7.142

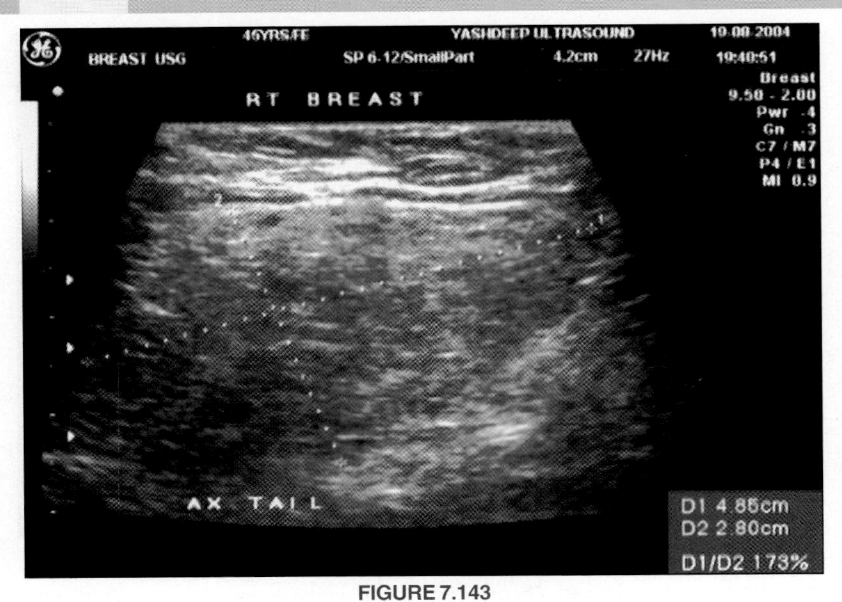

FIGURE 7.143

FIGURES 7.142 and 7.143: HRSG shows well-defined homogeneous mass in the breast. The mass shows medium level echoes. No calcification is seen in it. Biopsy of the mass shows lipoma of the breast.

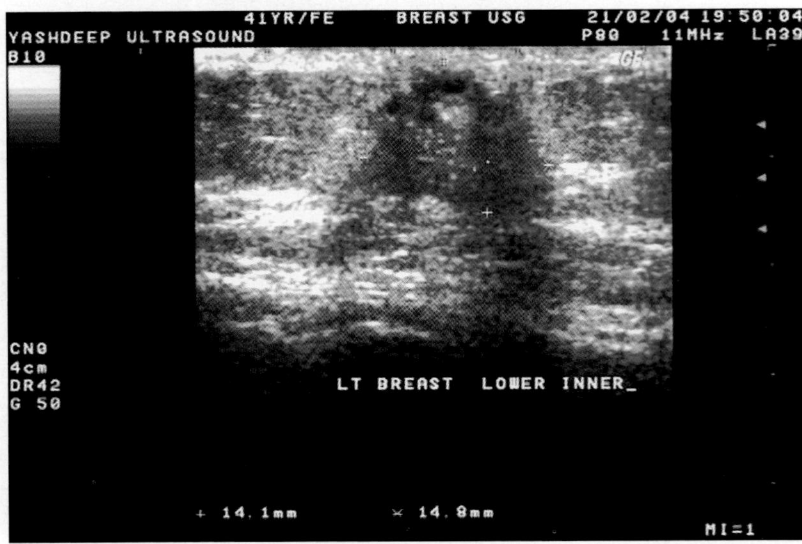

FIGURE 7.144

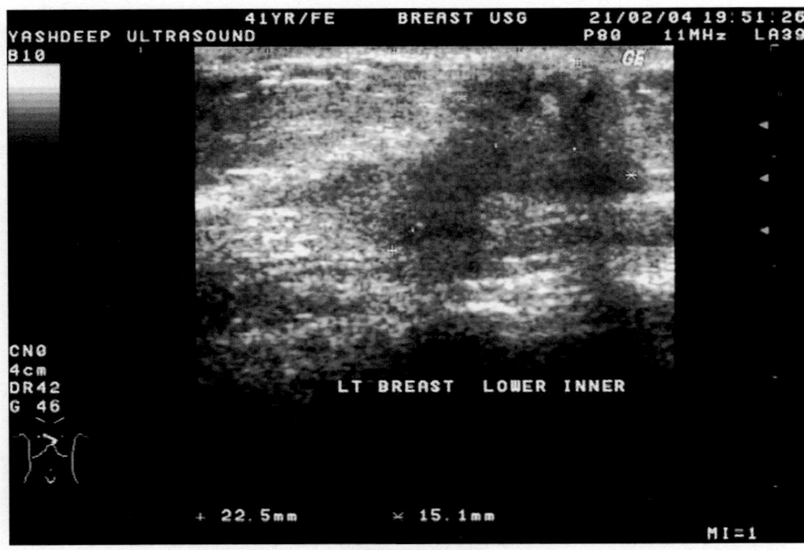

FIGURE 7.145

FIGURES 7.144 and 7.145: Fat necrosis. A patient presented with painful lump in the breast lower inner quadrant. HRSG shows diffuse irregular hypoechoic mass in the Lt inner quadrant. Margins are ill-defined. Perifocal edema is present. But no calcification is seen in it. Biopsy from the lesion confirmed fat necrosis.

Malignant Breast Lesions

Breast carcinoma is one of the most common malignancies in women. The incidence of carcinoma of breast is on increase, and early detection offers better treatment and survival rate. Although X-ray mammography is still choice of investigation for mass screening program for detection of carcinoma breast, yet ultrasonography contributes significantly in making the diagnosis and disease infiltration in the local tissue. Ultrasonographic feature of typical breast carcinomas are as follows:

1. Echo poor or hypoechoic texture of the mass.
2. Irregular border and outlines.
3. Posterior acoustic shadowing.
4. Heterogeneous texture.
5. Surrounding reflective zone (halo).

The sonographic criteria's for breast malignancy described by **Stavros** and his colleagues are as follows:

A. Speculation
B. Taller than wide (larger AP than transverse dimensions)
C. Angular margins
D. Markedly hypoechoic (composed to fat)
E. Shadowing
F. Calcifications
G. Duct extension
H. Branch pattern
I. Microlobulation

Individual criteria for benign lesions (smooth and well-circumscribed) are:

A. Absence of any malignant findings
B. Markedly hyperechoic (compared to fat)
C. Three or more gentle lobulations
D. Thin, echogenic capsule.

Malignant Findings

Speculation

This consists of alternating hypoechoic and relatively hyperechoic straight lines radiating out perpendicularly from the surface of the nodule.

Taller than Wide

It indicates that a part or the entire solid nodule is greater in its AP dimension than it is in either the craniocaudal or transverse dimensions. This suggests growth across the tissue planes.

Angular Margins

Angular margins refer to the contour of the junction between the hypoechoic or isoechoic solid nodule and the surrounding tissues. These may be acute, obtuse or nearly 90°. This finding has the highest sensitivity of any of the individual findings approaching 75%.

Markedly Hypoechoic

The central part of the majority of nodules is very hypoechoic. This suggests malignant change, but the converse is not true.

Shadowing

Some breast malignancies attenuate the sound beam and cause acoustic shadowing.

Calcification

Sonography is not as sensitive as mammography to detect punctate calcification. However, when detected are highly suggestive of malignancy.

Branch Pattern

A malignant breast nodule may have projections from the surface of the nodule extending in radial fashion within a duct toward the nipple and / or within duct away from the nipple. These findings suggest the extension of the nodule into or along the ductal system.

Microlobulation

Presence of small lobulations on the surface of a solid breast nodule is worrisome for malignancy. Microlobulations are often associated with angular margins.

Other additional features like skin thickening, disturbed trabeculae pattern, distorted tissue planes and microcalcification are also encountered on HRSG. But typical combination of **nidus, halo and posterior shadowing** is found in 40 to 50% of the breast cancers. The tumors nidus is echo poor and ill-defined in outlines. It is heterogeneous in texture. It may be uniform when it is small or under 1 cm in size. However, in large mass, it becomes grossly heterogeneous with distorted and disrupted normal echo pattern.

Posterior acoustic shadowing: It is caused due to the fibrosis caused by the tumor. It is seen on 40 to 60% of cases. However, it is also seen in fat necrosis, post-surgical cases and nonspecific findings.

Irregular borders: These are also one of the features of carcinoma of breast. Thus it is also an unreliable sign and may be found in many benign conditions such as fibroadenoma, abscess, hematomas and fat necrosis.

Skin thickening and edema: Localized thickening of skin and skin edema are the features of inflammation in breast cancer. It also occurs in radiation therapy. HRSG can reveal the thickening of skin. The thickened dermis becomes more echogenic than the normal pattern due to tumor infiltration, fibrosis or edema. Engorged lymphatics and interstitial fluid collection can be noticed on HRSG. Asymmetrical ductal dilatation is the feature of intraductal carcinoma. However, it may be seen in few benign conditions as intraductal papilloma.

Microcalcifications: These are the most important X-ray mammographic markers for non-palpable breast cancers, especially intraductal carcinomas. Occasionally with HRSG, microcalcification can be picked up specially when it is present within the mass. They are seen as tiny echogenic specks in the breast parenchyma and can be seen with high-gain setting. The size of microcalcification detected on X-ray mammogram and soft tissue abnormalities picked up on HRSG are the good indicators for breast carcinoma. However, detection of microcalcification on currently available ultrasonographic scanners is an unreliable parameter for confirmation of breast cancer. Therefore, other associated features should be looked for the final diagnosis.

Size of tumor: When carcinoma is diffuse it disrupts the normal breast tissue pattern. It is replaced by a poorly reflective mass, which does not cause shadowing. It becomes difficult to distinguish it from the benign mass, and FNAC is required to establish the diagnosis.

HRSG can detect non-palpable nodes in axilla and also metastatic deposits in the liver in the same sitting, thus, helps in the staging of disease.

Ultrasonographic Guided Biopsy

Localization of breast lump for biopsy can be easily done with ultrasonographic imaging. Free-hand technique is the best to do USG guided FNAC. The lump is identified on USG and making a mark on the skin does localization. The depth of lesion can be found out on HRSG. Needle tip is guided into substance of the mass under USG and guided biopsy is done. It is simple, rapid and non-invasive technique for doing biopsy of the breast mass

BIRADS Nomenclature and Categories

American College of Radiology (ACR) is trying to develop standard reporting patterns for mammographic imaging of breast lesions. It is known as Breast, Imaging, Reporting and Data System (BIRADS). Though it is not yet complete yet it is useful to classify the breast lesion under BIRADS risk categories for every mammogram. Though BIRADS is developed for mammogram, but same principles can be applied for Sono mammograms. Since most sonomammograms are targeted after clinical or mammographic suspicion of breast lesion, therefore, BIRADS is very useful in sonomammograms. As per the ACR the different categories of BIRADS are as follows:

BIRADS category	*Description*	*Risk of malignancy (%)*	*Management*
1	Suspicious clinical presentation with doubtful X-ray mammogram Normal sono mammogram	0%	Clinical follow-up
2	Benign breast lesions on ultrasound like cysts, Ductal Ectasia, Reactive lymph node enlargement, Benign solid breast masses—lipomas, fibroadenomas	0%	Clinical follow-up Periodic screening
3	Probably benign lesions—Chances of malignancy <2% Complex cysts, intraductal papilloma, subset of fibro adenomas	≤ 2%	Patient choice for follow-up and biopsy
4	Large group of breast lesions with suspicion of malignancy. This group is devided into two sub categories:	3-49%	Biopsy
4a	Mildly suspicious of breast malignancy	2-<50%	Additional imaging - MRI
4b	Moderately suspicious	50-89%	Biopsy
5	Malignant masses	90 or >90%	Biopsy

Advantage of Sonomammography over X-ray Mammography

Sonomammography has got distinctive advantage over X-ray mammography. They are as follows:

1. Sensitivity of sonomammography in carcinoma breast is high up to 95%.

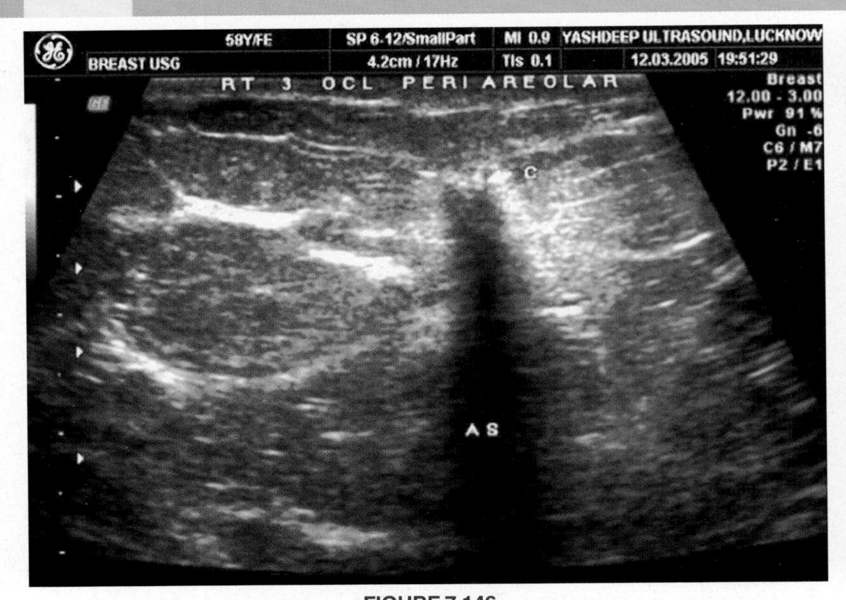

FIGURE 7.146

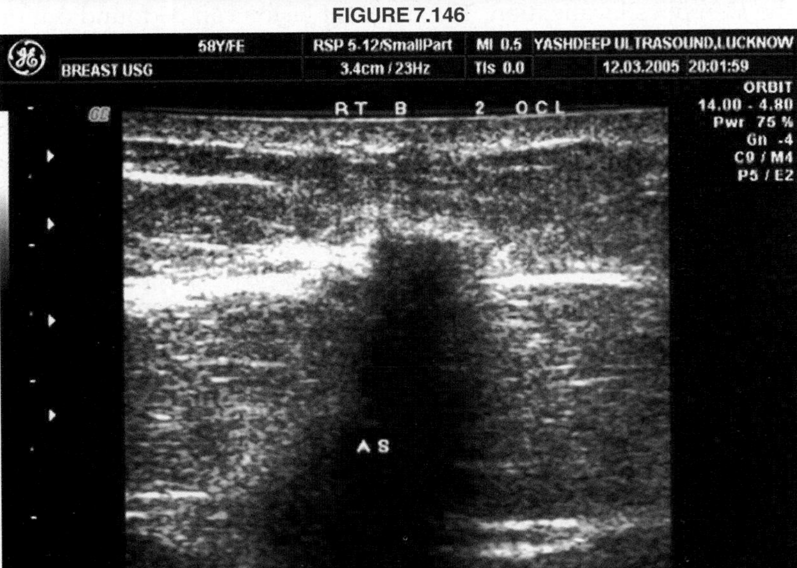

FIGURE 7.147

FIGURES 7.146 and 7.147: Nonpalpable malignant tumor breast. A patient presented with irregular nodularity in the breast. HRSG shows a small irregular heterogeneous focus at 3 O'-clock position in the breast. Fine calcification is seen in the mass. Dense acoustic shadowing is seen posterior to the tumor nidus. Malignant tumors of the breast are known to have characteristic retrotumoral acoustic shadowing in 70% of cases.

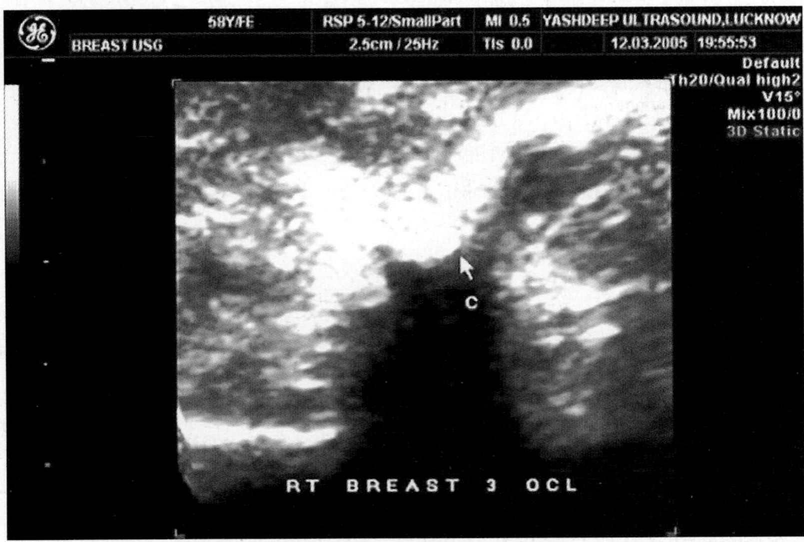

FIGURE 7.148: 3D imaging of the same patient shows calcified focus in the tumor, which was responsible for dense acoustic shadowing.

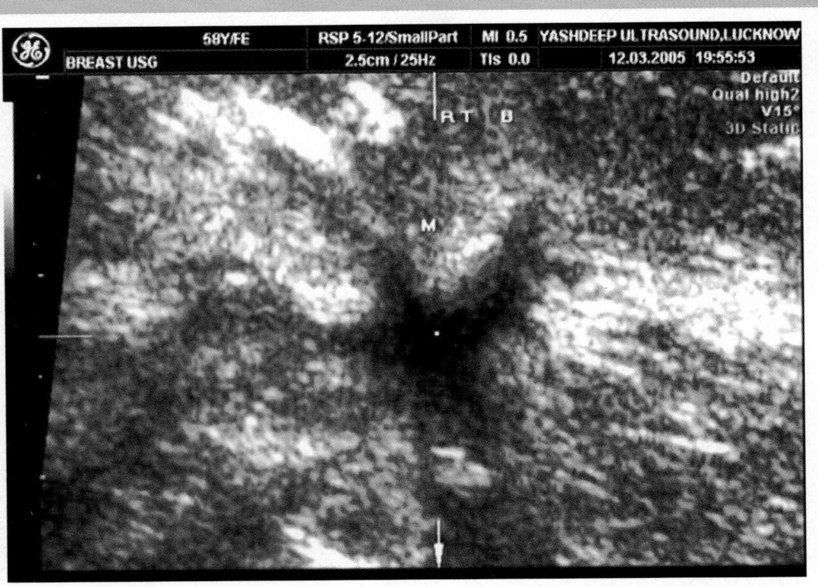

FIGURE 7.149: 3D imaging of the tumor shows tumor nidus. Branching pattern of the mass is well appreciated on 3D imaging. The calcified shadow is also seen clearly in the tumor.

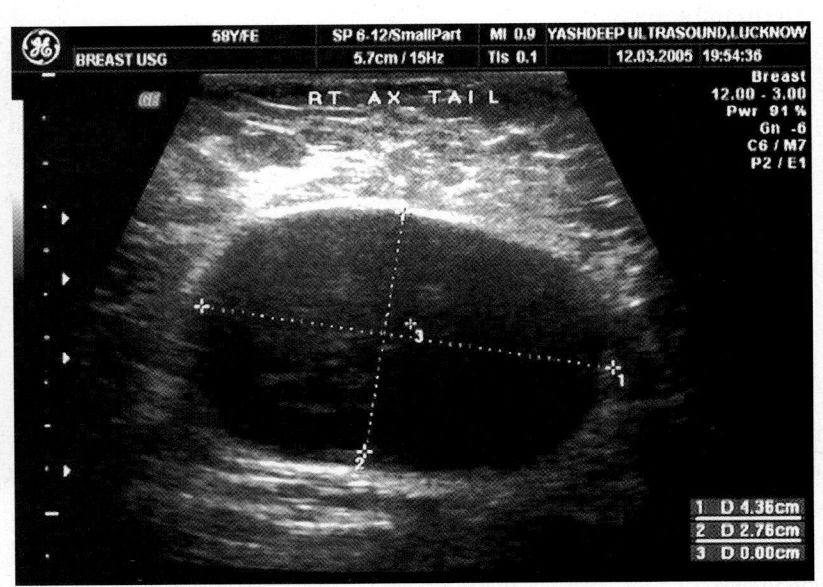

FIGURE 7.150: Enlarged node is seen in the axilla. The node is round in shape with hylum distorted suggestive of malignant node in the same patient.

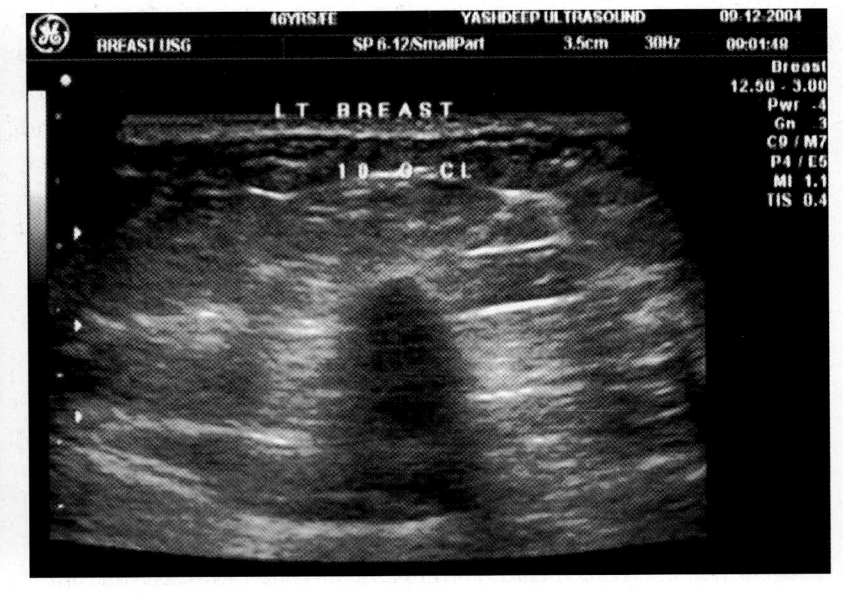

FIGURE 7.151

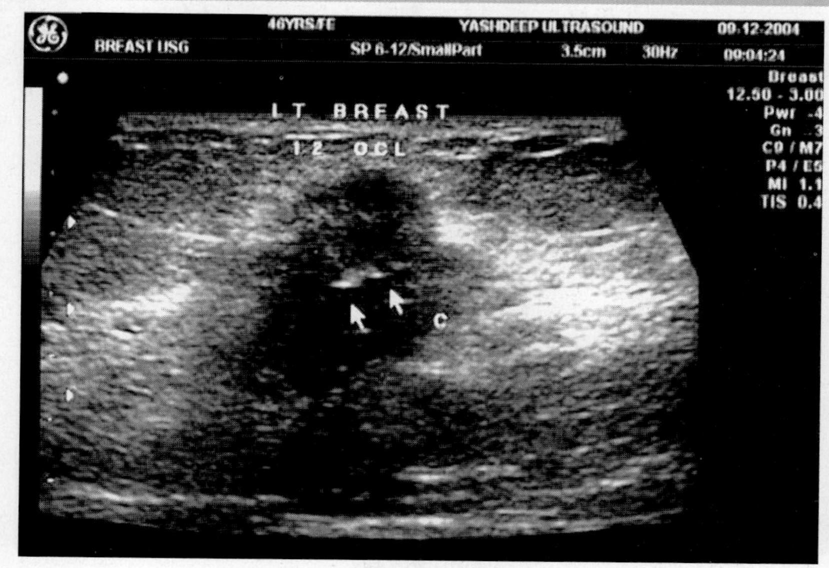

FIGURE 7.152

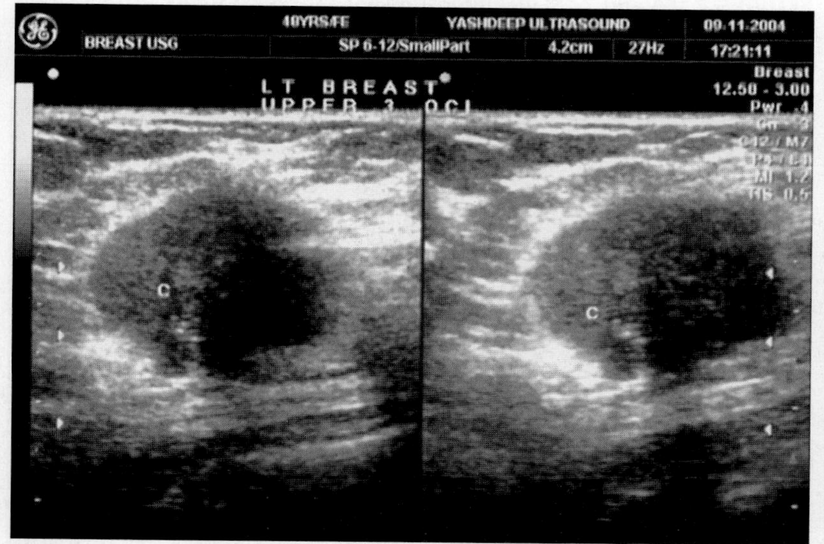

FIGURE 7.153

FIGURES 7.151 to 7.153: Heterogeneous mass with microcalcification in the breast. HRSG shows an irregular heterogeneous mass in the breast. The mass shows dense acoustic shadowing. Microcalcification is seen in the mass. The calcified specks are highlighted with image compression technique. They are seen bright echogenic specks in the mass.

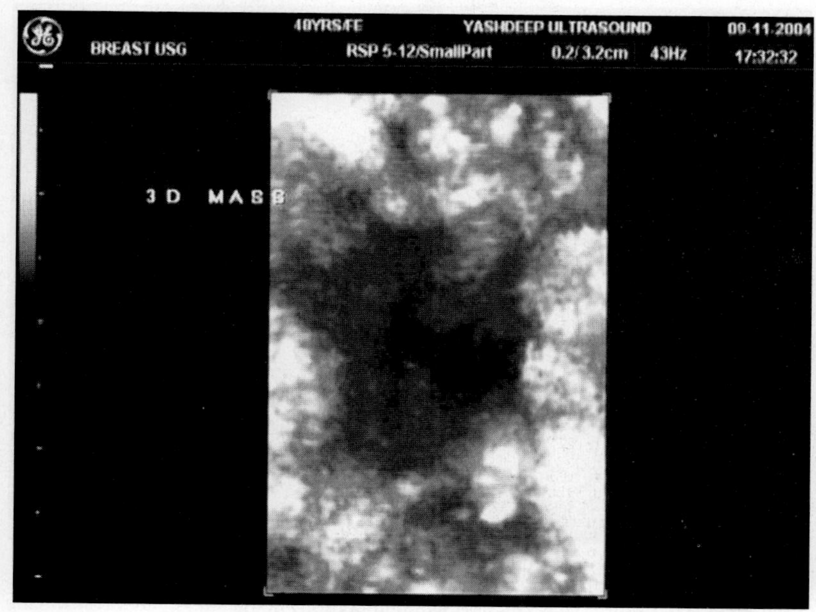

FIGURE 7.154

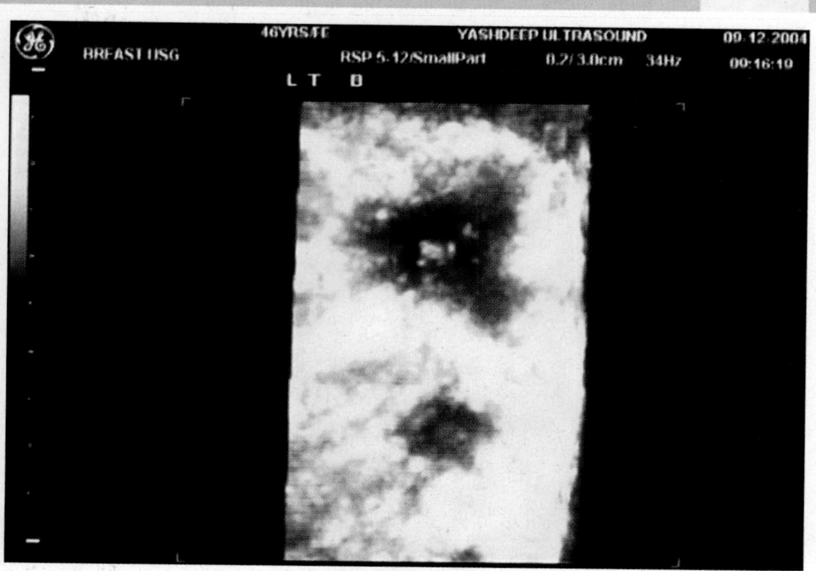

FIGURE 7.155

FIGURES 7.154 and 7.155: 3D imaging of the mass shows star appearance with branching pattern typical of malignant nature of the mass. Microcalcification is seen embedded in the tumor bed.

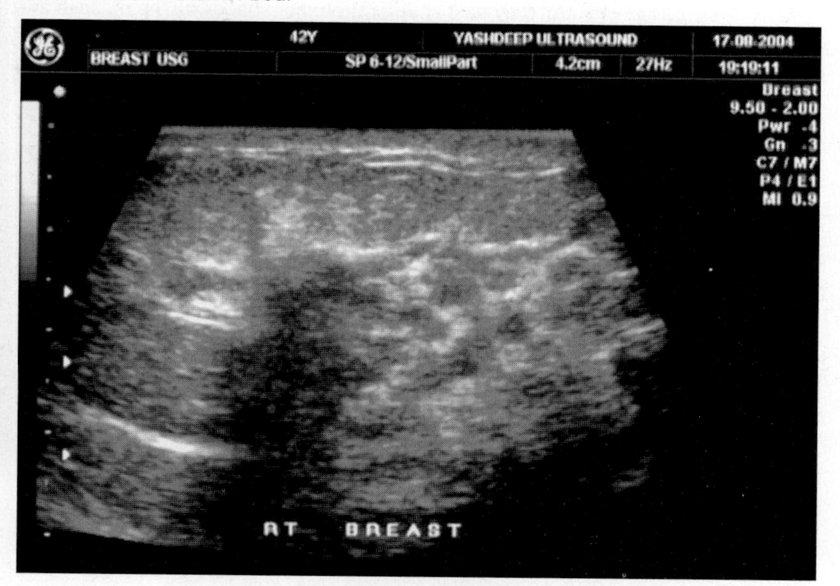

FIGURE 7.156

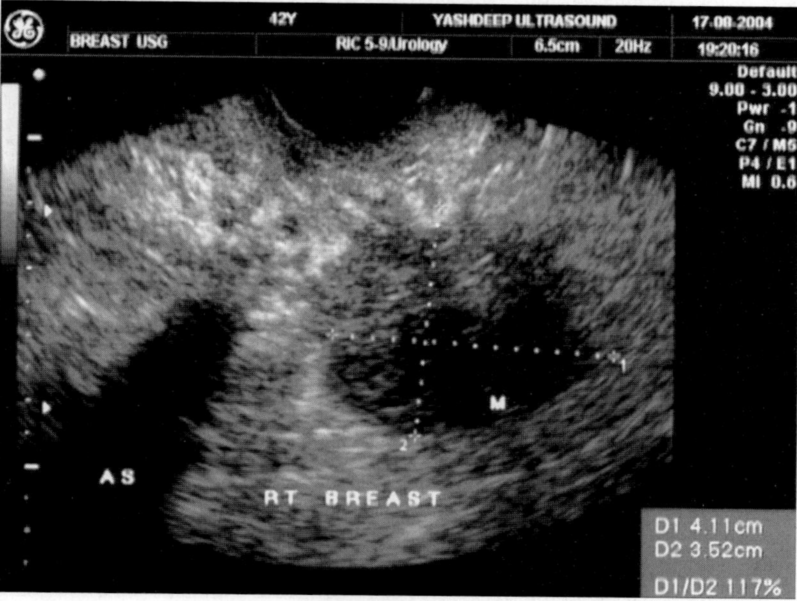

FIGURE 7.157

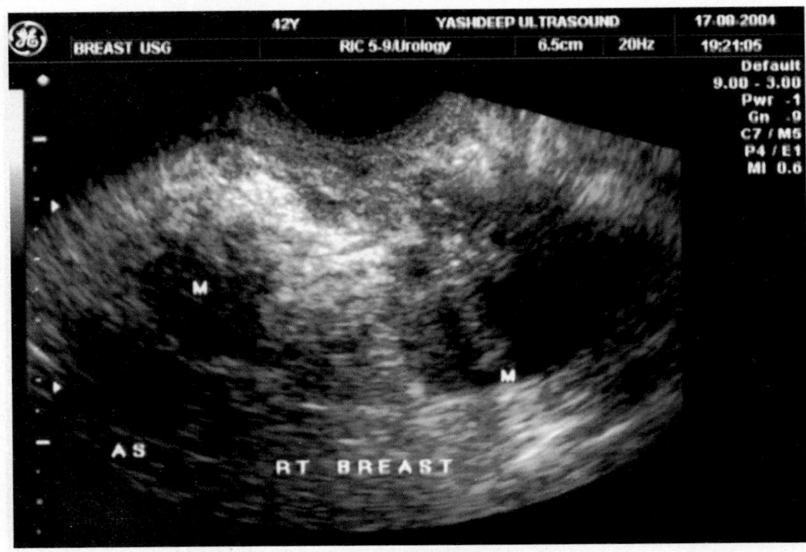

FIGURE 7.158

FIGURES 7.156 to 7.158: Multifocal malignant masses. A patient presented with palpable lump in the Rt breast. HRSG shows multiple irregular heterogeneous masses in the breast. They are accompanied with dense acoustic shadowing. The tumor niduses are small in size. But they cast strong retrotumor shadows. The masses are seen confined in the mammary zone of the breast.

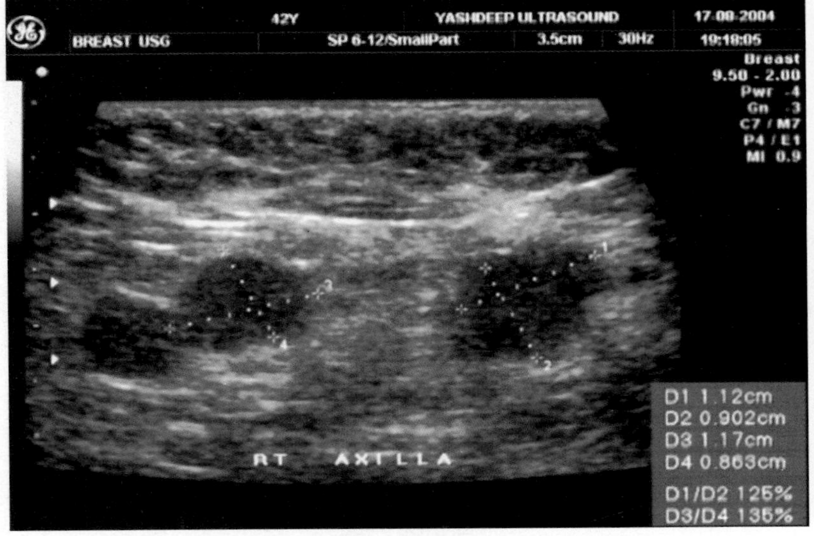

FIGURE 7.159: The Rt axilla also shows multiple nodes. They are irregular in outline with distorted normal texture of the lymph nodes suggestive of malignant nature.

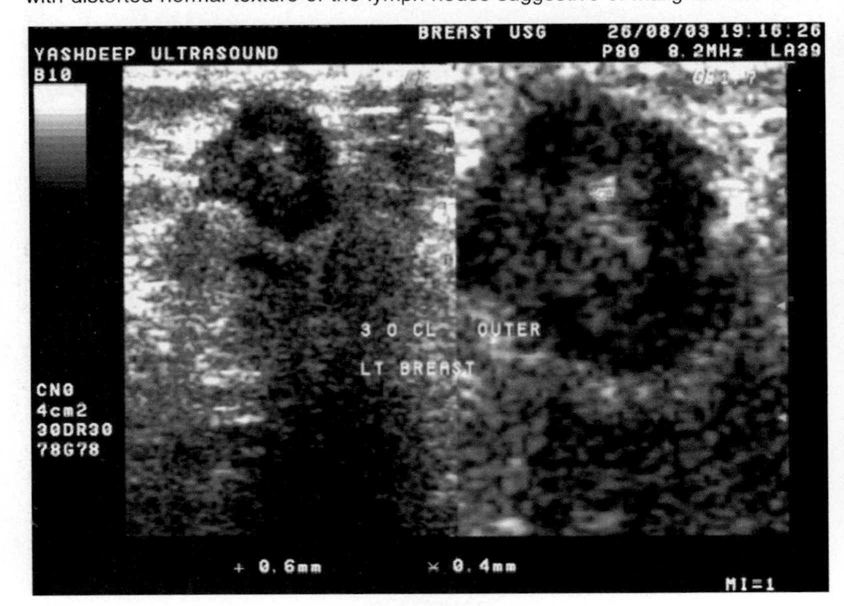

FIGURE 7.160

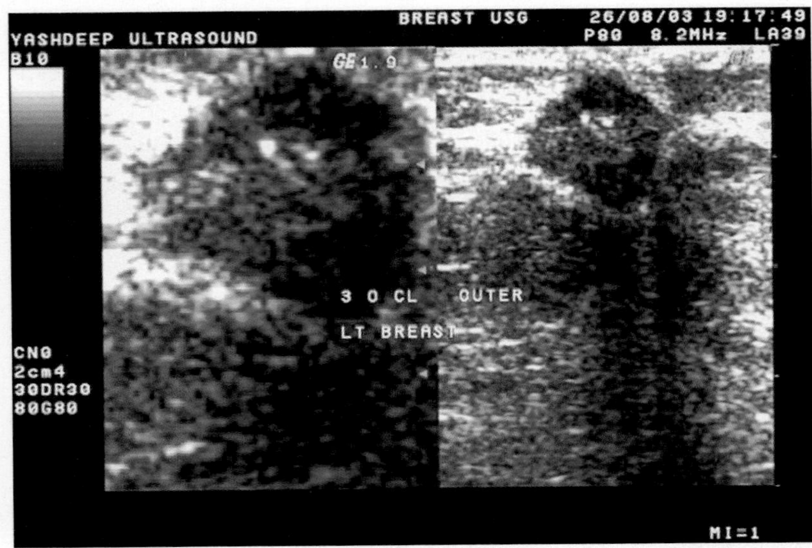

FIGURE 7.161

FIGURES 7.160 and 7.161: Multifocal malignant breast tumor. HRSG shows irregular heterogeneous masses in the Lt breast. The masses show microlobulations coming out from the periphery. Small echogenic calcified specks are also seen in the masses suggestive of microcalcification.

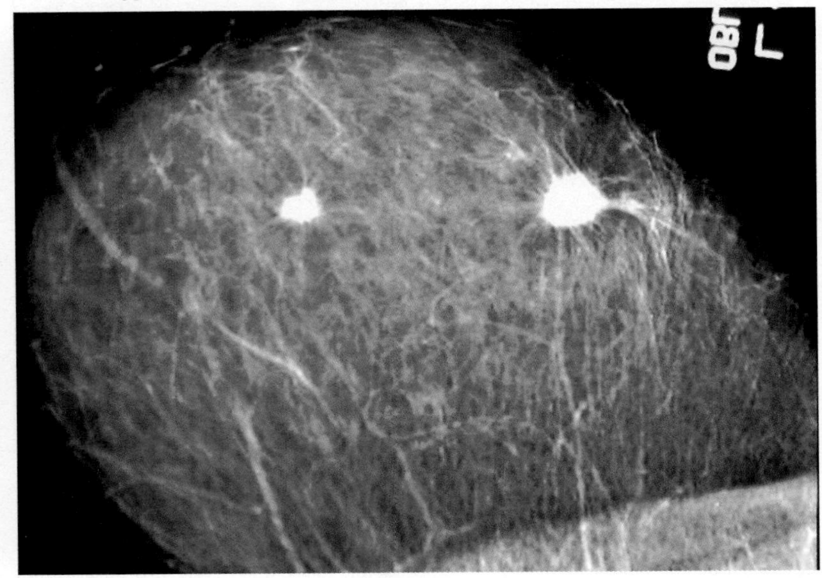

FIGURE 7.162

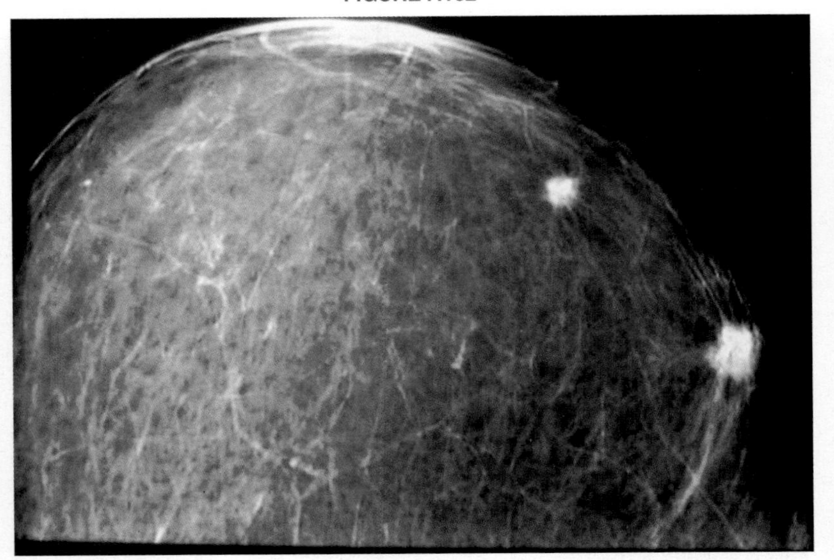

FIGURE 7.163

FIGURES 7.162 and 7.163: The digital mammography of the same patient shows the masses as described on USG and confirmed the USG findings.

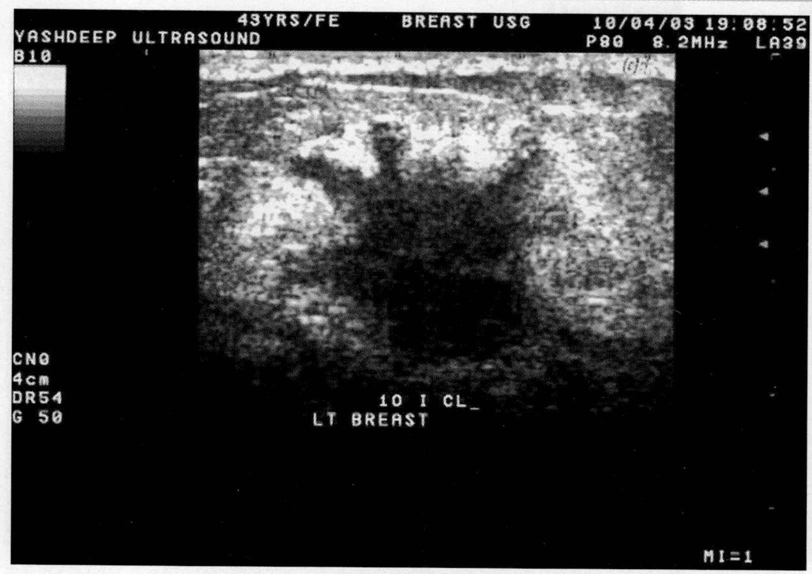

FIGURE 7.164

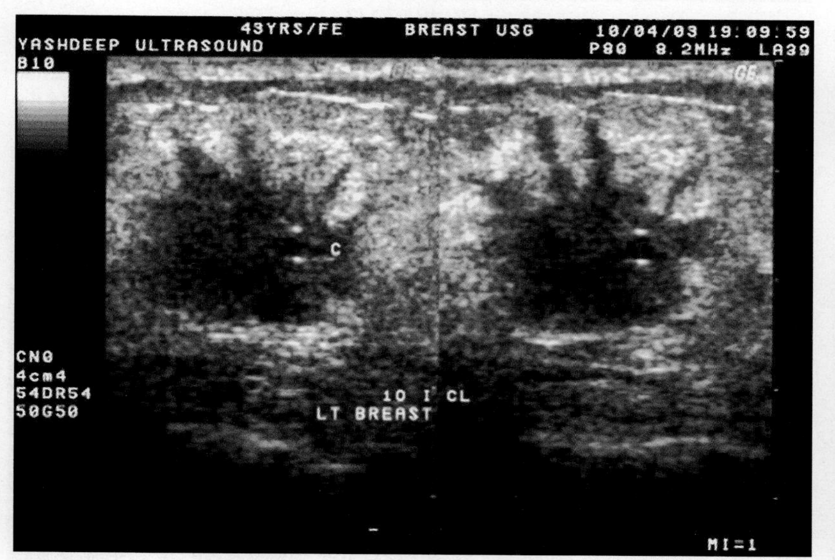

FIGURE 7.165

FIGURES 7.164 and 7.165: Heterogeneous mass with spiculated pattern. HRSG shows irregular heterogeneous mass in the breast. Fine spicules are seen coming out from the mass with "Sunray appearance", typical of the malignant nature of the mass. Echogenic microcalcification is also seen in the mass.

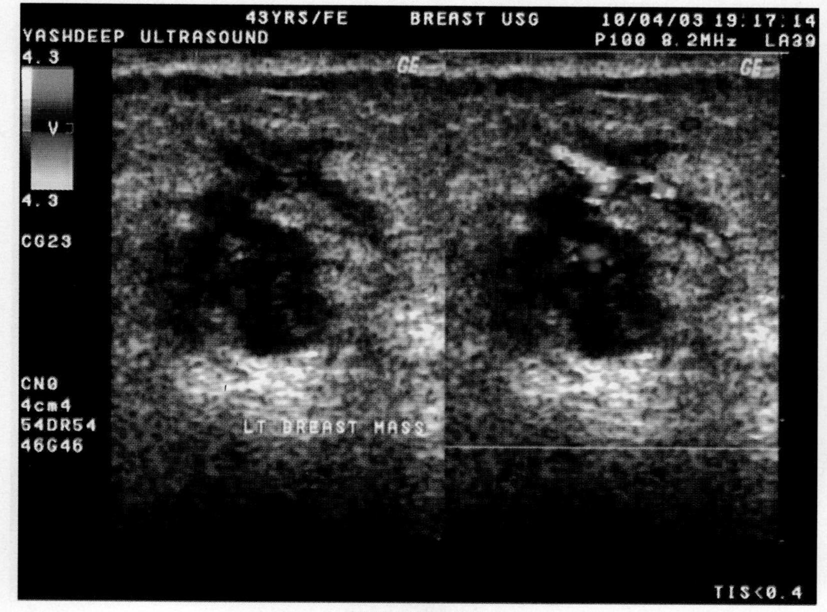

FIGURE 7.166

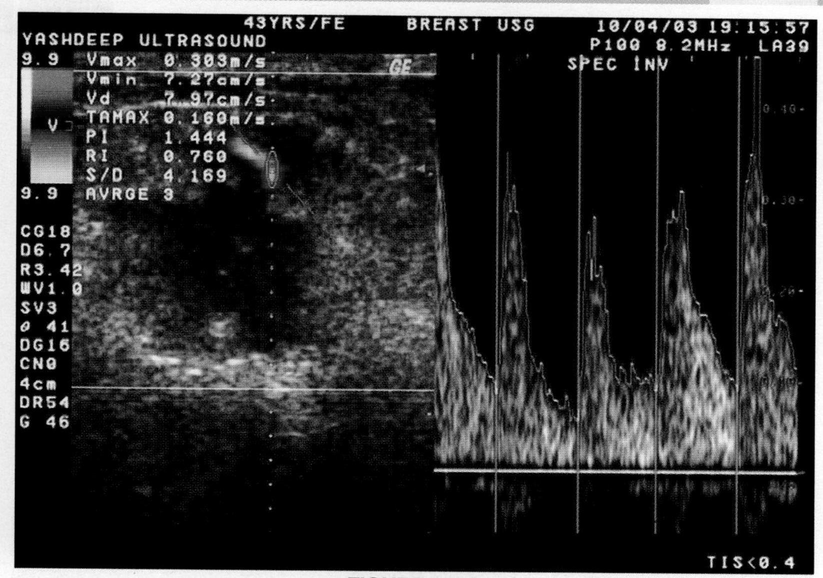

FIGURE 7.167

FIGURES 7.166 and 7.167: On color flow imaging a big vessel is seen feeding the tumor. High flow is seen in the vessels on spectral Doppler tracing.

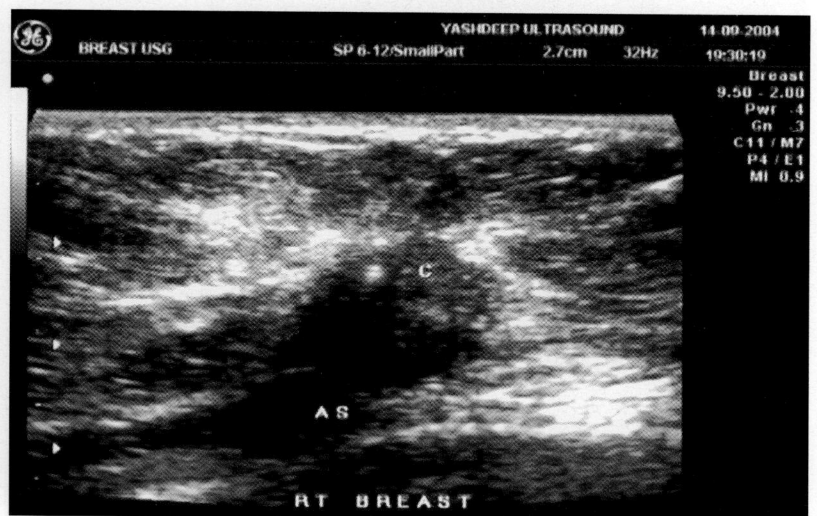

FIGURE 7.168

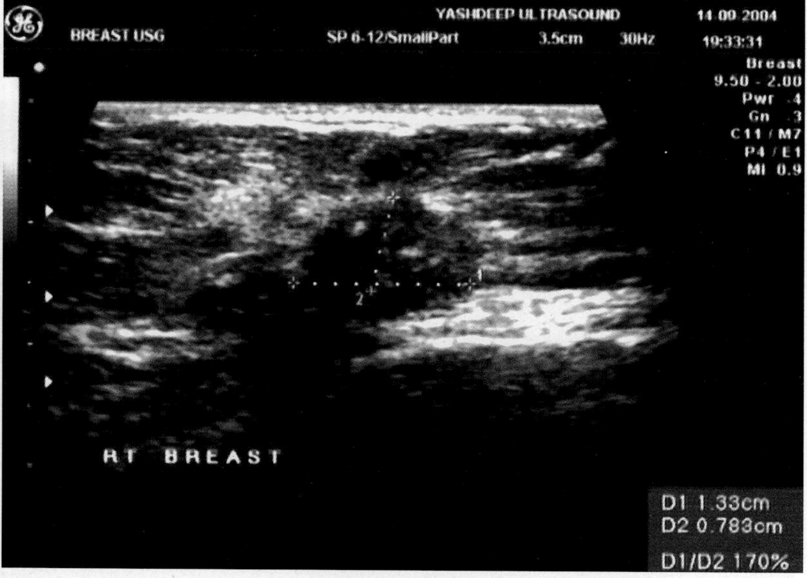

FIGURE 7.169

FIGURES 7.168 and 7.169: Nonpalpable carcinoma breast. A patient presented with slight skin puckering in the outer quadrant of the breast. HRSG shows irregular heterogeneous mass. The mass was clinically not palpable. It shows adherence with skin. Echogenic calcification was also seen in the mass with acoustic shadowing.

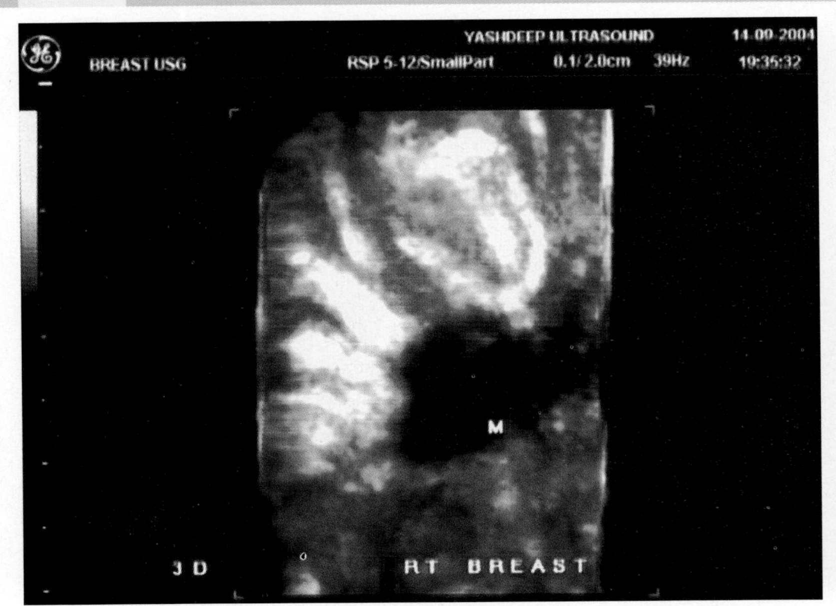

FIGURE 7.170

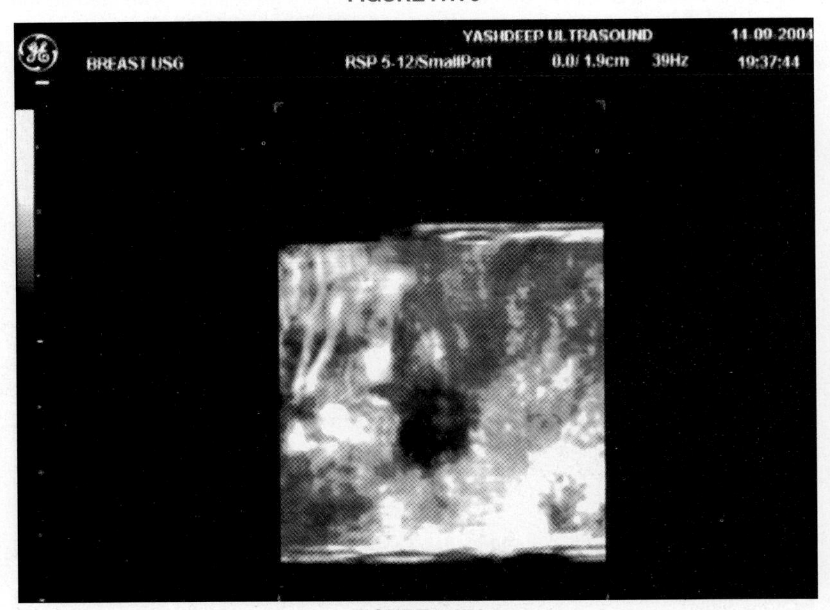

FIGURE 7.171

FIGURES 7.170 and 7.171: 3D imaging of the mass shows the details of the tumor typical branching pattern of the malignant tumor is well appreciated on 3D imaging.

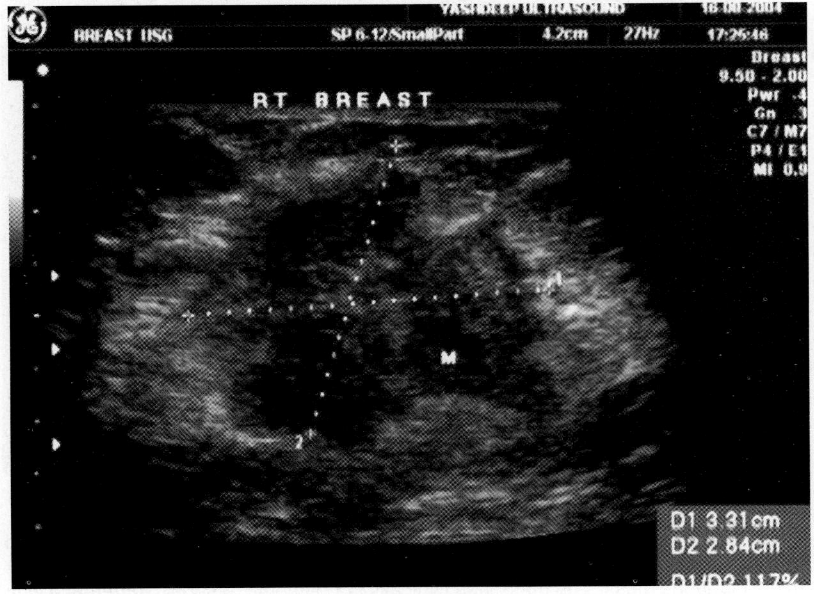

FIGURE 7.172

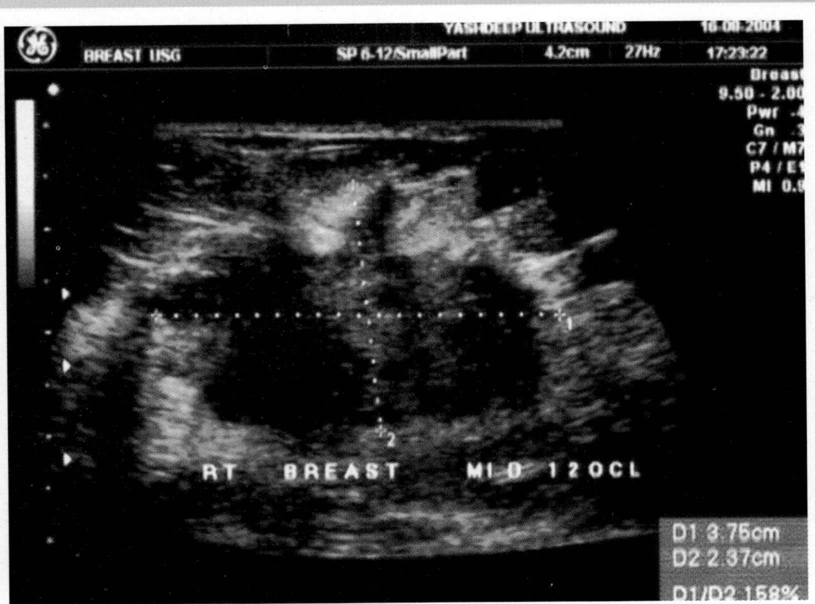

FIGURE 7.173

FIGURES 7.172 and 7.173: Big heterogeneous mass with retromammary invasion. HRSG shows in the breast. The mass shows multiple lobules. Associated ray pattern is also seen in the mass. The mass is seen in the mammary layer of the breast. However, it is seen invading the retromammary layer of the breast.

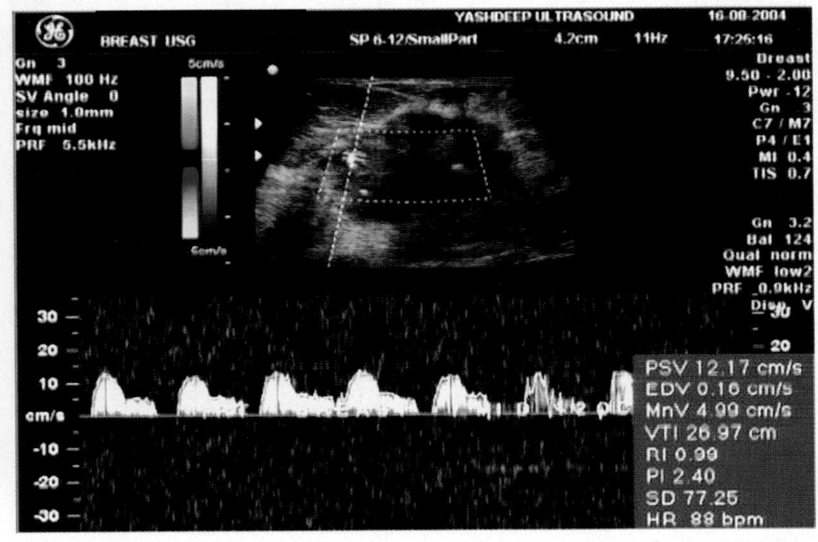

FIGURE 7.174: On color flow imaging multiple vessels are seen going into the mass with low flow pattern.

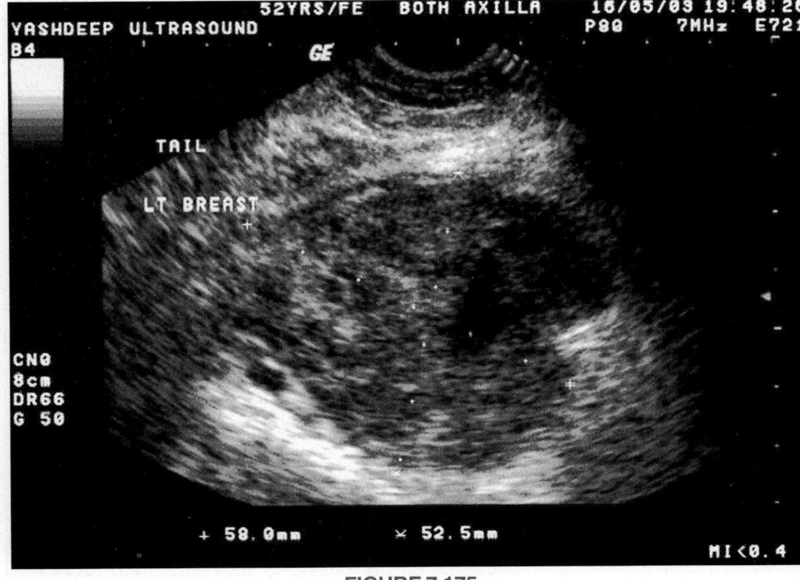

FIGURE 7.175

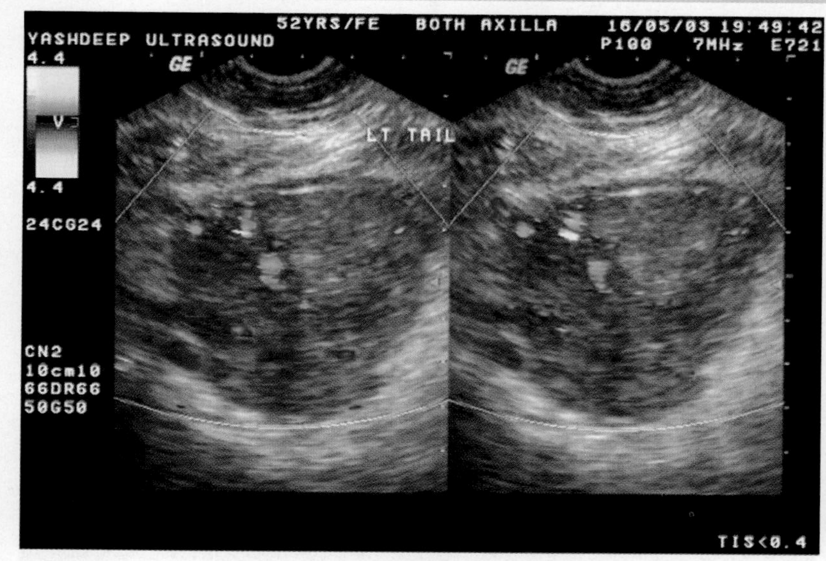

FIGURE 7.176

FIGURES 7.175 and 7.176: Big heterogeneous mass with retromammary invasion. A patient presented with big mass in the breast. HRSG shows big heterogeneous mass in the breast. The mass is seen invading the retromammary layer. It is also seen invading the axillary tail. On color flow imaging multiple vessels are seen feeding the mass.

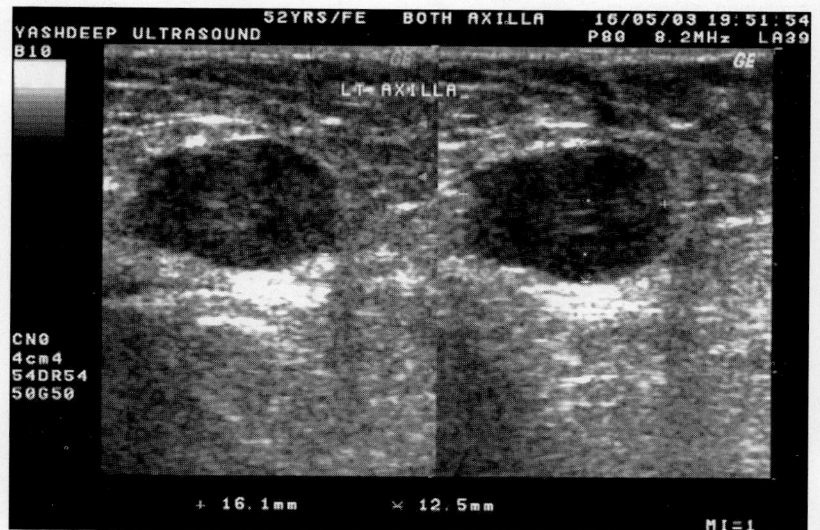

FIGURE 7.177

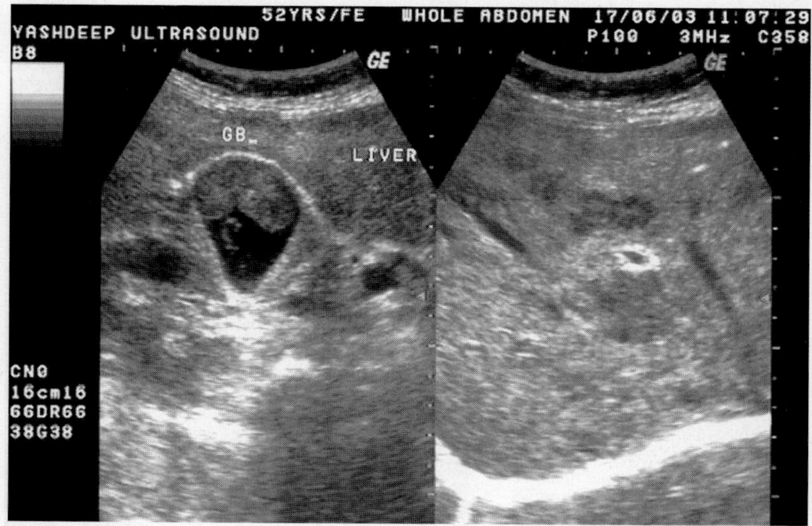

FIGURE 7.178

FIGURES 7.177 and 7.178: HRSG shows enlarged nodes in the axilla with malignant changes in the nodes. They are round in shape. Lymph node hylum is destroyed. The liver also shows multiple irregular hypoechoic masses suggestive of metastatic deposits.

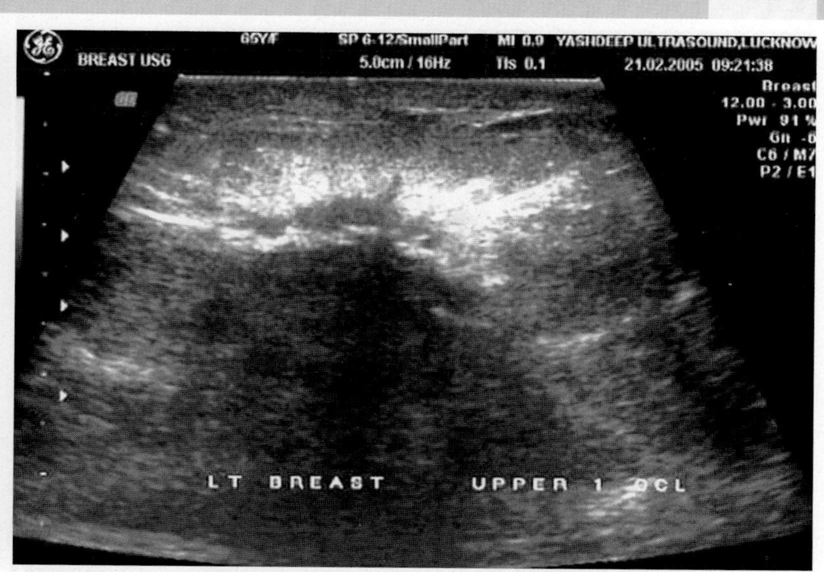

FIGURE 7.179: Malignant mass in the breast. HRSG shows irregular heterogeneous mass in the breast. The mass is not clearly defined as it is masked with acoustic shadowing.

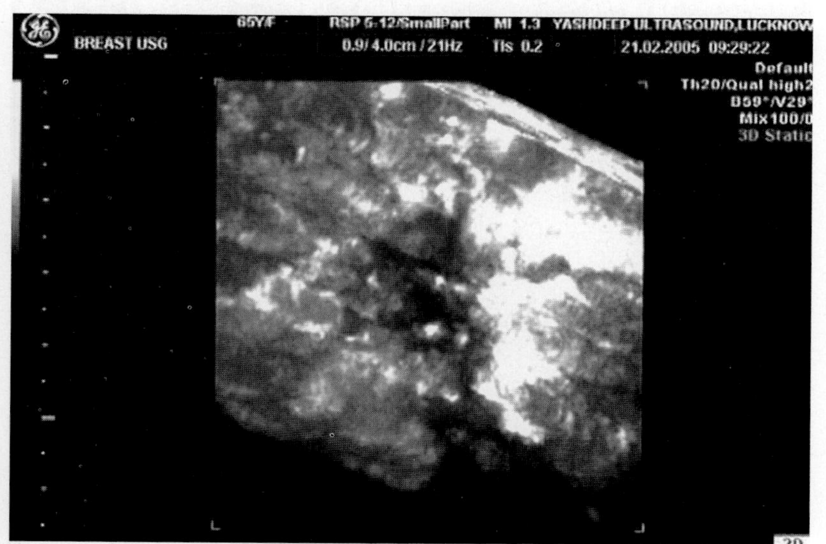

FIGURE 7.180

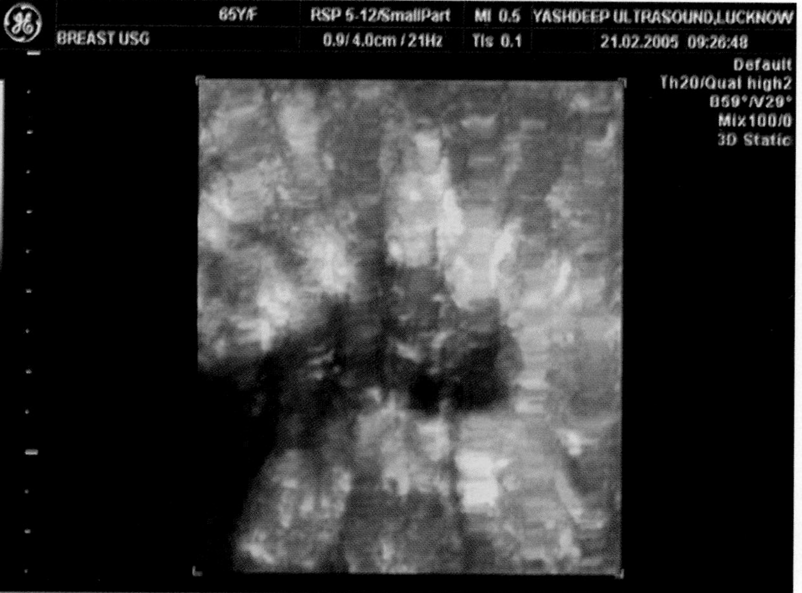

FIGURE 7.181

FIGURES 7.180 and 7.181: 3D imaging of the mass shows typical branching pattern of malignant mass. Multiple branches are seen coming out from the mass and going in all the directions with "Star Like Appearance" echogenic calcification is also seen in the mass suggestive of microcalcification.

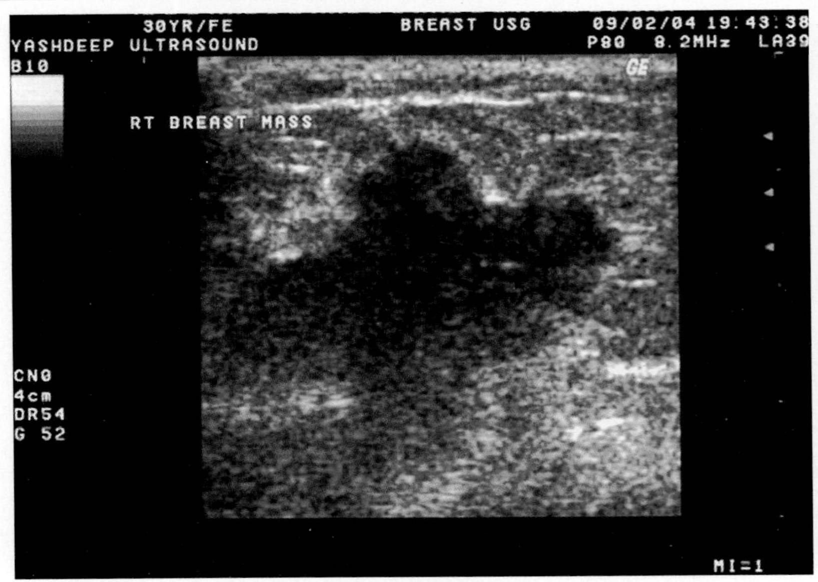

FIGURE 7.182

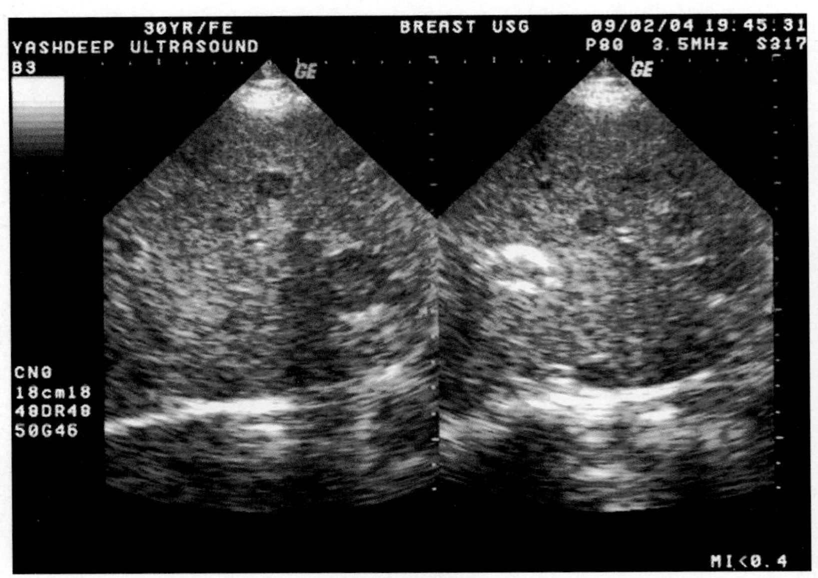

FIGURE 7.183

FIGURES 7.182 and 7.183: Malignant breast mass with liver metastasis. HRSG shows big heterogeneous soft tissue mass in the breast. The mass shows multiple lobulations. It is seen confined to the mammary layer. Liver shows multiple hypodense irregular deposits suggestive of metastasis. HRSG is very good for staging of the disease.

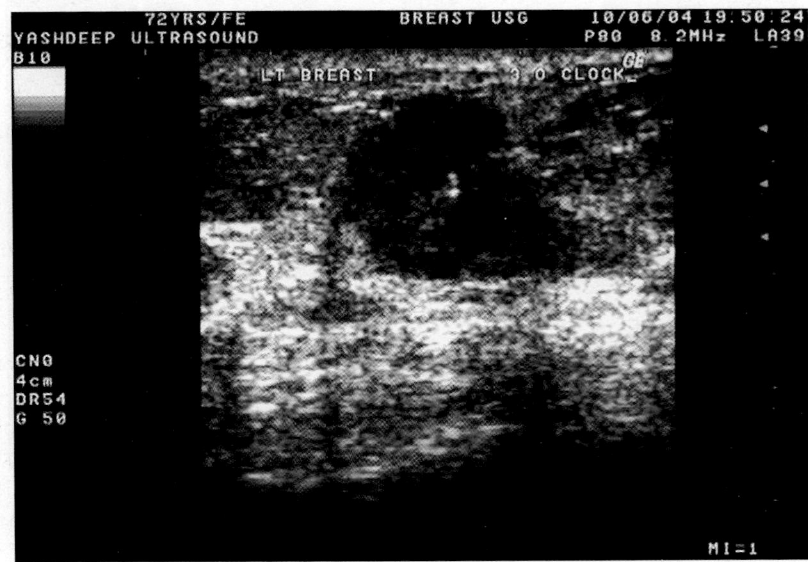

FIGURE 7.184

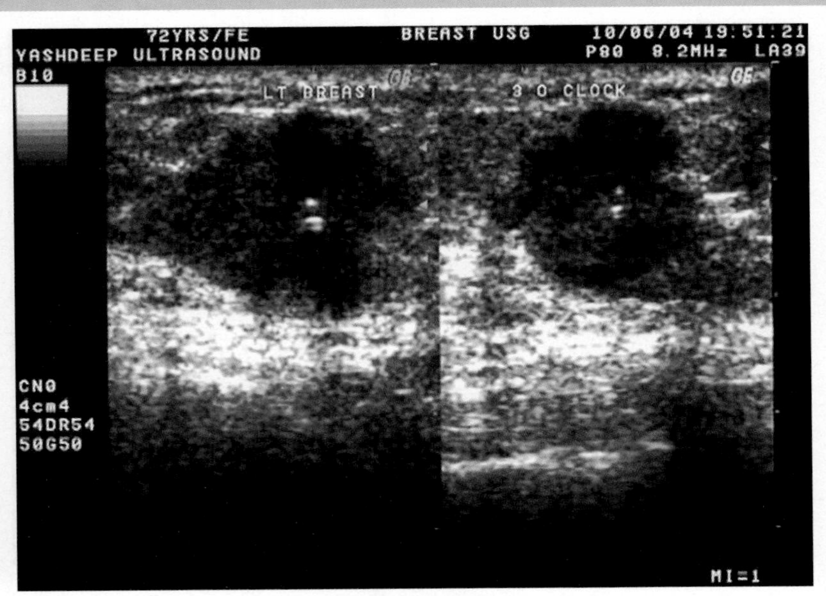

FIGURE 7.185

FIGURES 7.184 and 7.185: Malignant mass with microcalcification. HRSG shows big irregular heterogeneous mass in the mammary zone. Echogenic calcification is seen in the mass indicating microcalcification. However, the mass shows smooth margins. Though marginal microlobulation are seen in the mass. Biopsy of the mass confirmed malignancy.

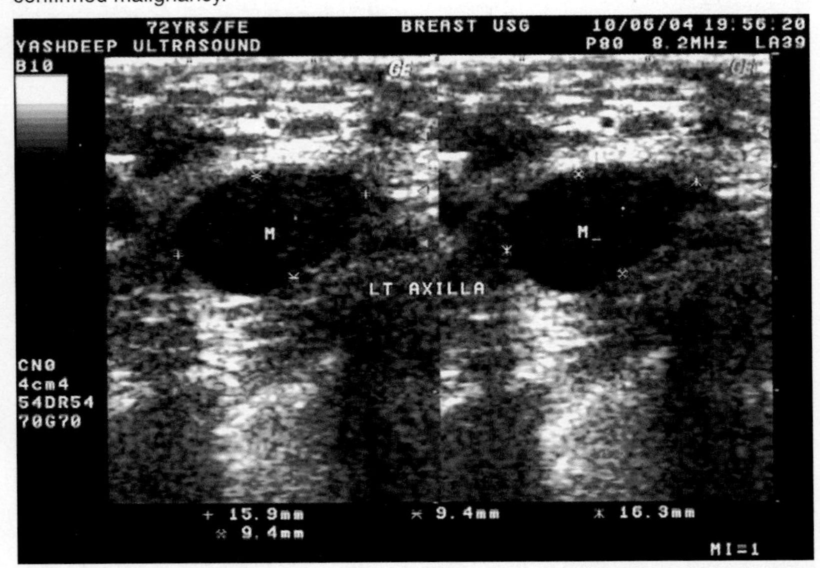

FIGURE 7.186: Malignant nodes are also seen in the axilla. They show distorted texture.

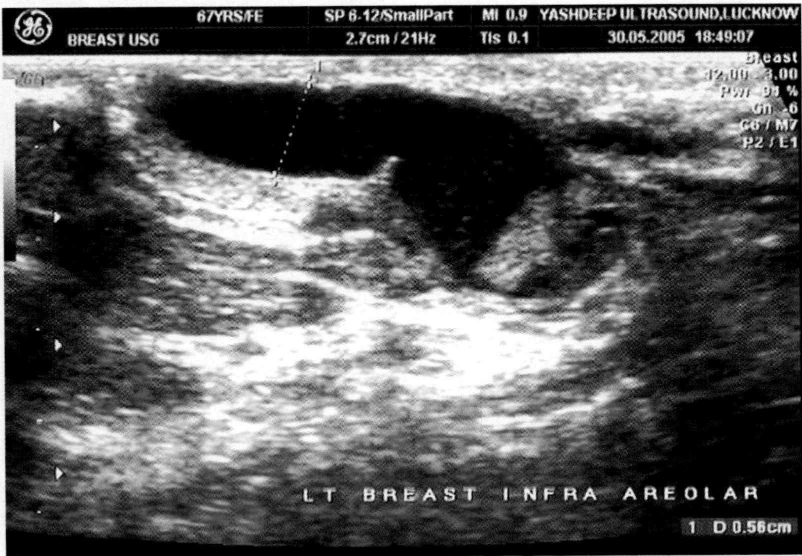

FIGURE 7.187

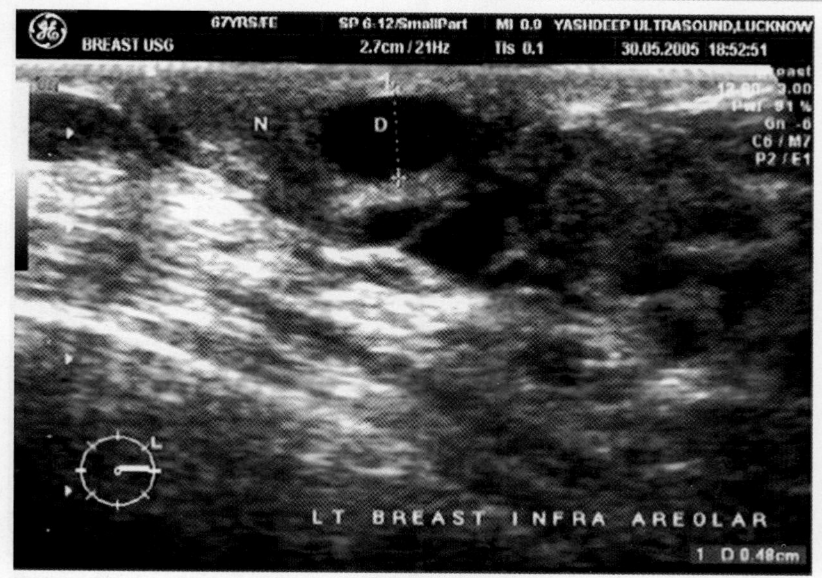

FIGURE 7.188

FIGURES 7.187 and 7.188: Intraductal Ca. A patient presented with blood discharge from the nipple. HRSG shows grossly dilated ducts in the breast. The soft tissue mass is seen in the duct. The mass is seen invading the duct and blocking the mouth of the duct.

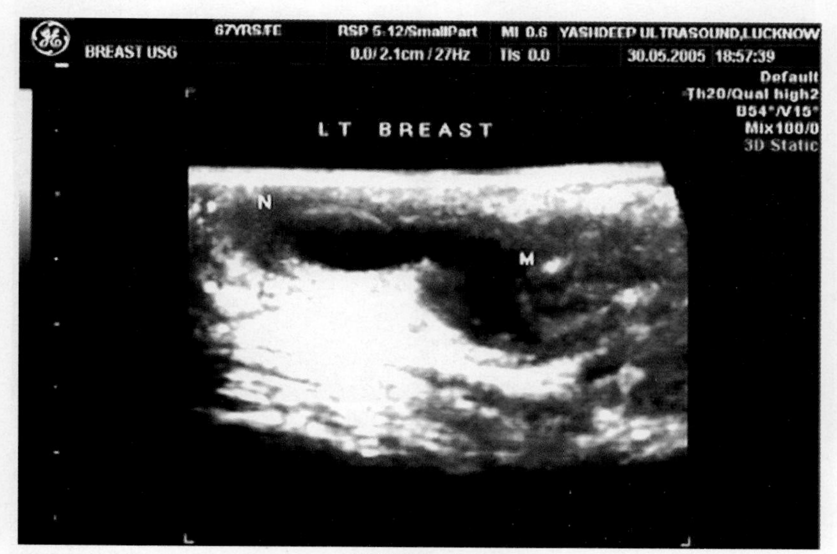

FIGURE 7.189

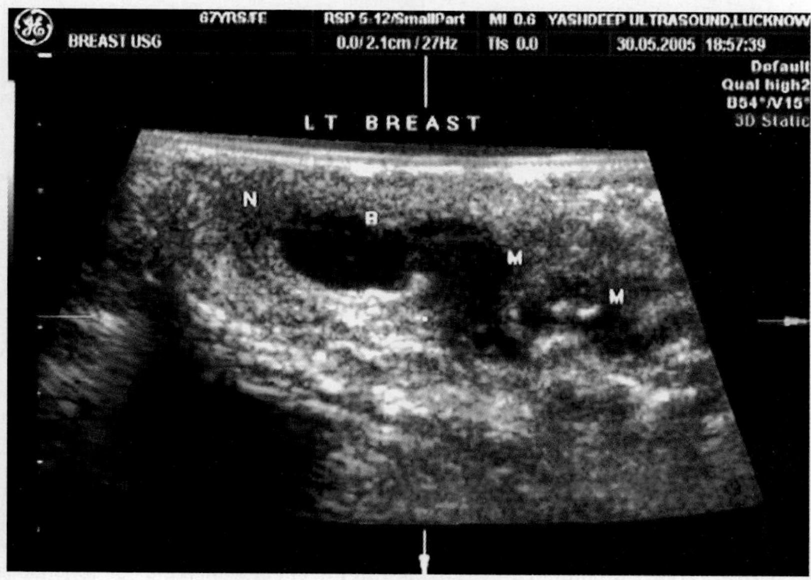

FIGURE 7.190

FIGURES 7.189 and 7.190: 3D imaging of the mass shows details of the mass invading the duct.

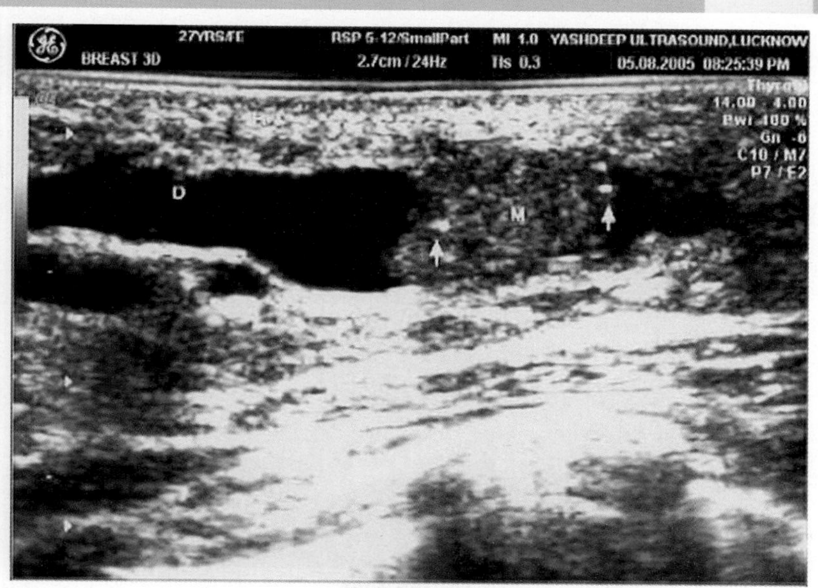

FIGURE 7.191

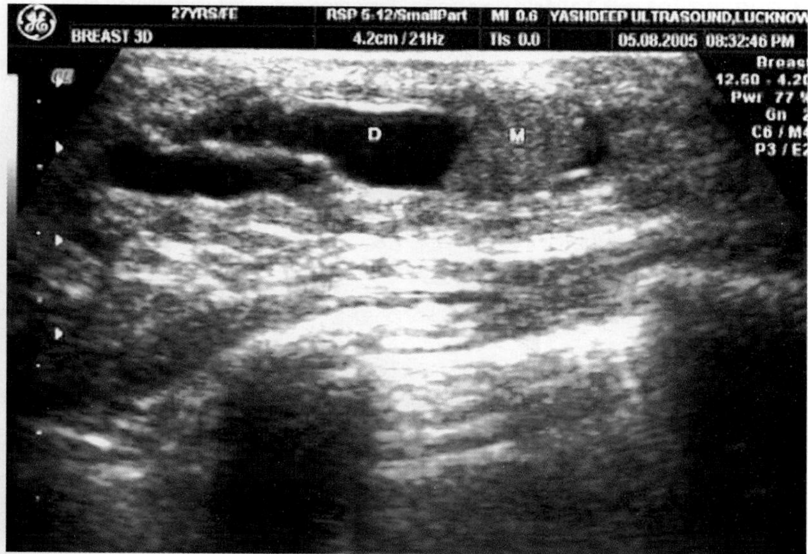

FIGURE 7.192

FIGURES 7.191 and 192: Intraductal Ca. A patient presented with blood from the nipple with a palpable lump in periareolar area. HRSG shows dilated ducts in the periareolar area. A big soft issue mass is seen in one of the ducts. The duct is markedly dilated. Microcalcification is also seen in the mass. Biopsy confirmed Intraductal carcinoma.

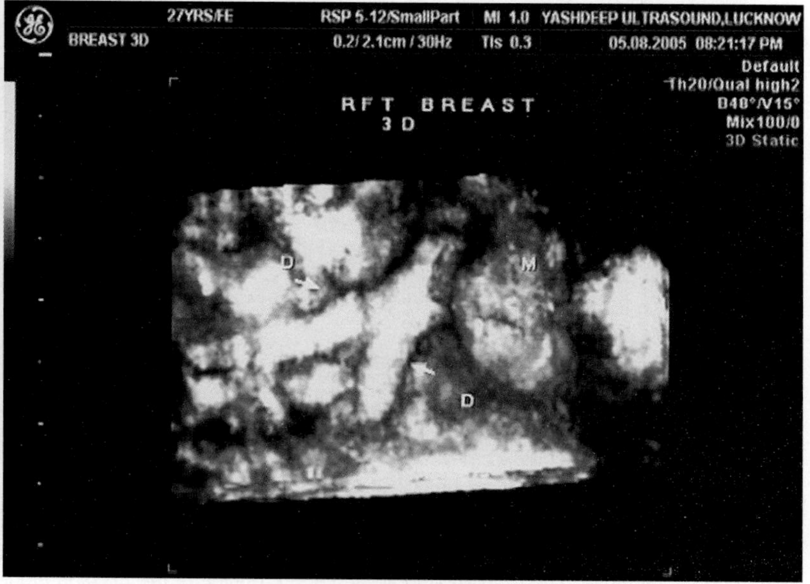

FIGURE 7.193

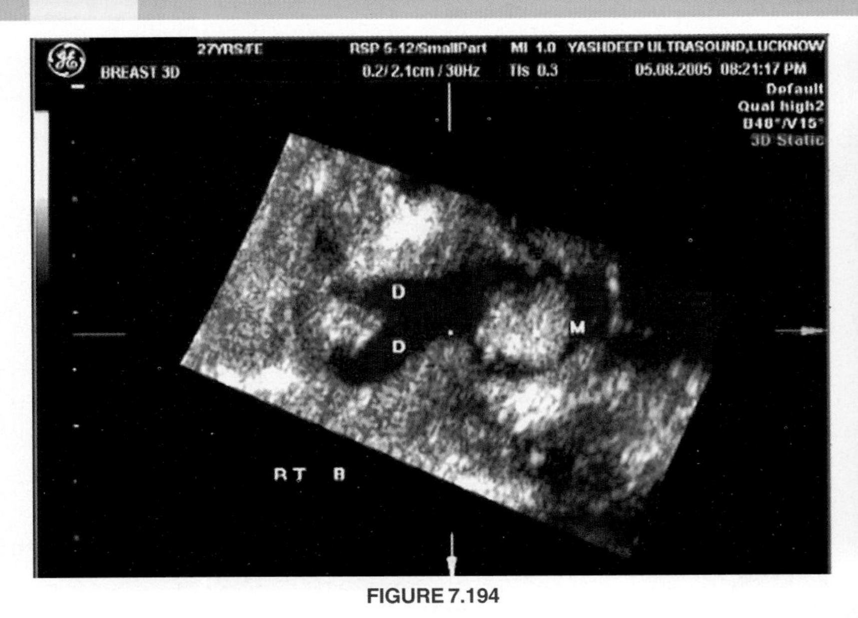

FIGURE 7.194

FIGURES 7.193 and 7.194: 3D imaging of the same patient shows dilated ducts. The mass is seen blocking mouth of two ducts. It is clearly delineated on 3D imaging.

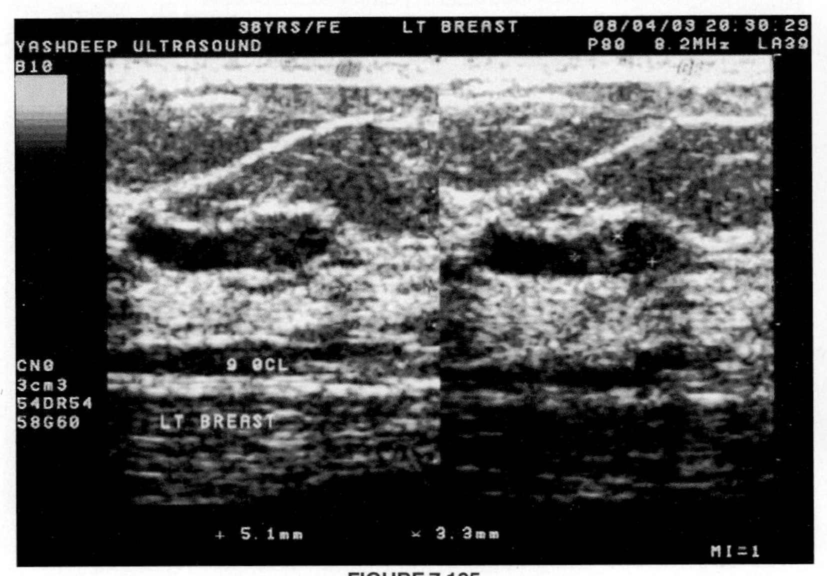

FIGURE 7.195

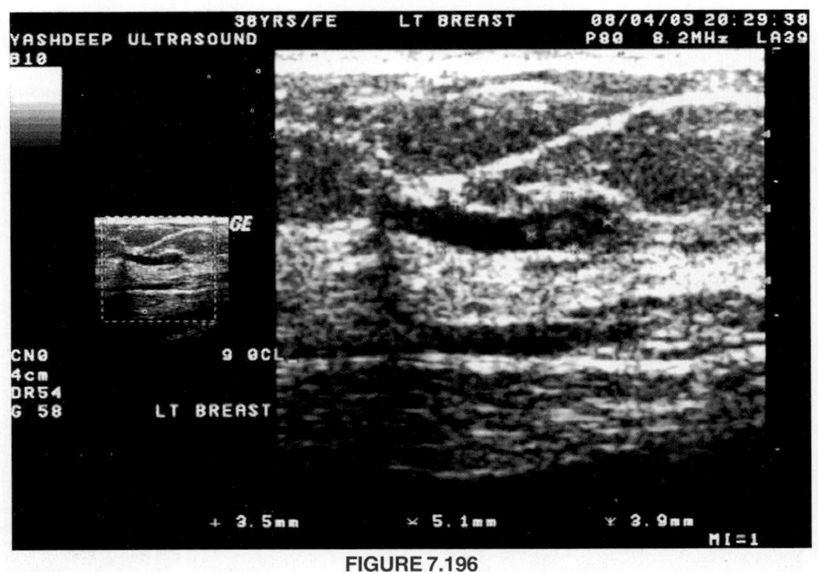

FIGURE 7.196

FIGURES 7.195 and 7.196: Intraductal papilloma. Well-defined homogeneous soft tissue mass is seen in dilated duct. The mass is smooth in outline. No evidence of any calcification is seen in it. Biopsy of the mass confirmed intraductal papilloma. Papilloma is benign ductal tumor. The regular, smooth and sharp outline of the mass is highly suggestive of benign nature of the mass.

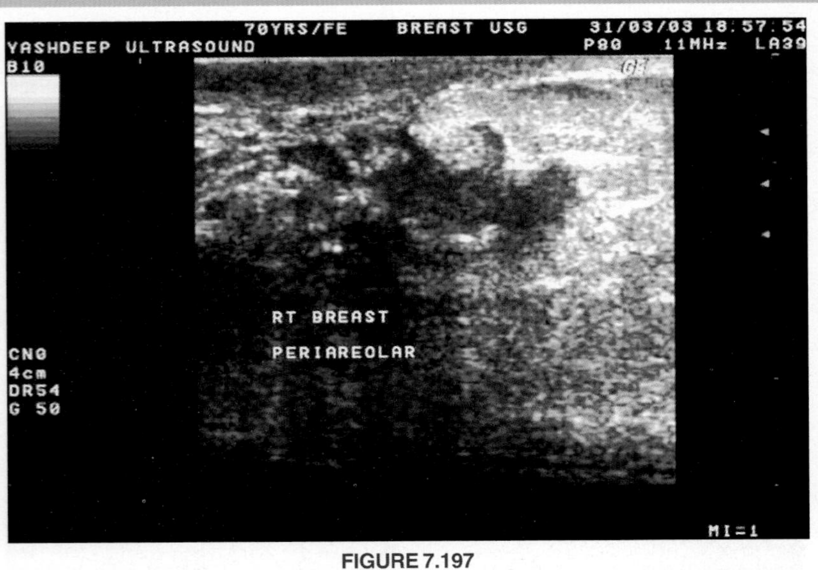

FIGURE 7.197

FIGURE 7.198

FIGURES 7.197 and 7.198: Infiltrating ductal carcinoma. A patient presented with palpable lump in the breast. HRSG shows big irregular heterogeneous mass in the periareolar area. The mass is seen obstructing the ducts. The nipple is seen retracted. Biopsy confirmed infiltrating ductal carcinoma.

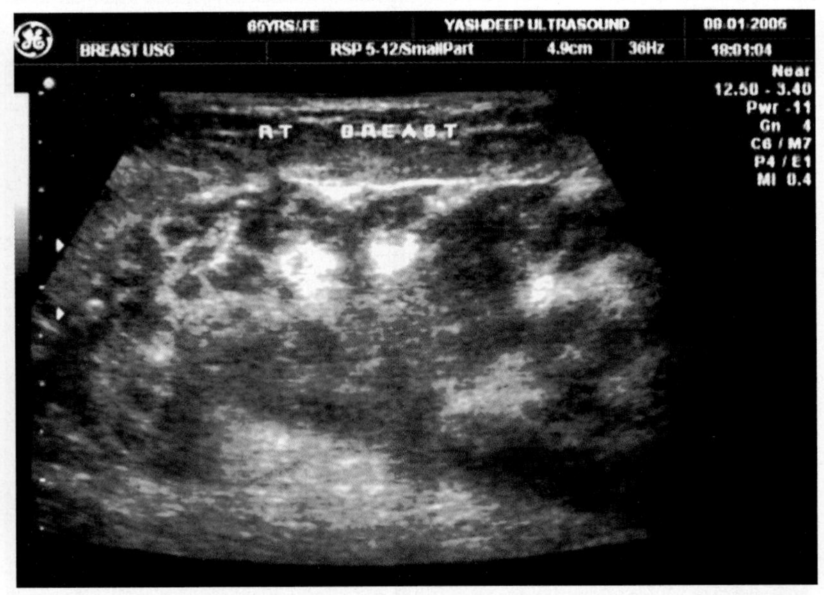

FIGURE 7.199: Carcinoma breast fixed with the chest wall. A big heterogeneous soft tissue mass is seen in the Rt breast. The mass has completely destroyed the breast tissue. It is seen invading the mammary and retromammary layers. It is also seen invading muscles of the chest wall and fixed with the chest wall.

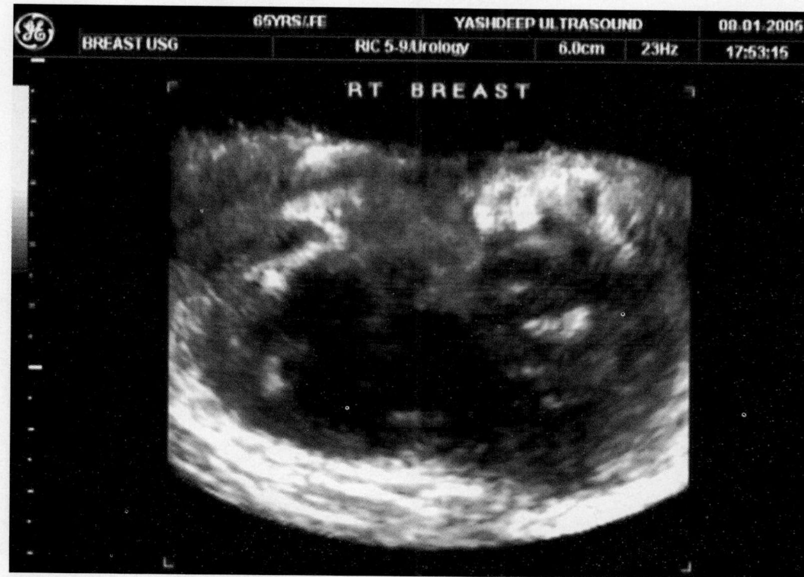

FIGURE 7.200

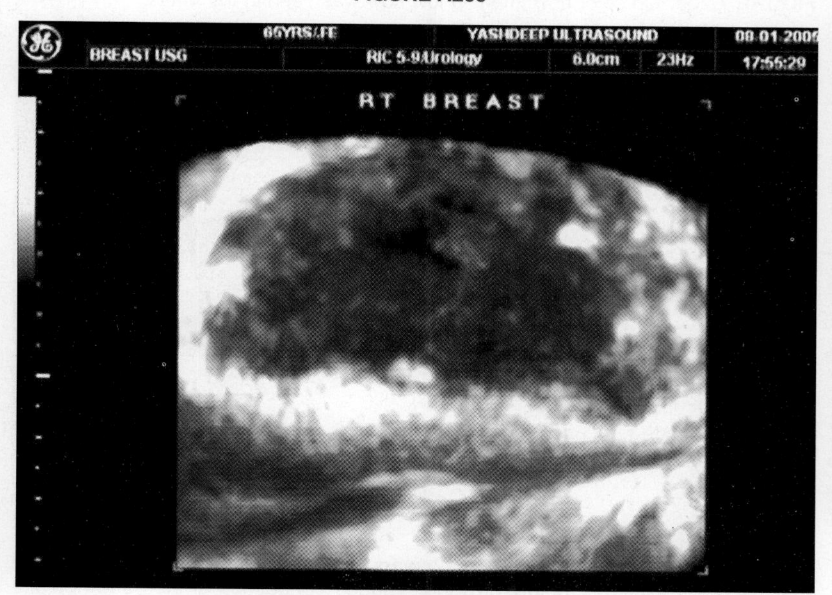

FIGURE 7.201

FIGURES 7.200 and 7.201: 3D imaging of the same patient shows the tumor invading the deeper layers of the breast and it is also seen invading the chest wall.

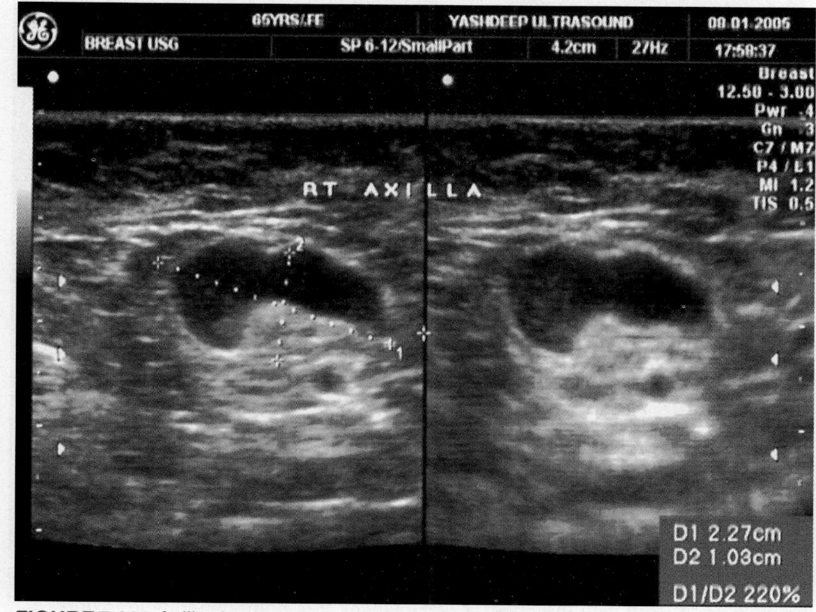

FIGURE 7.202: Axilla shows multiple enlarged nodes. The lymph nodes hylum is distorted.

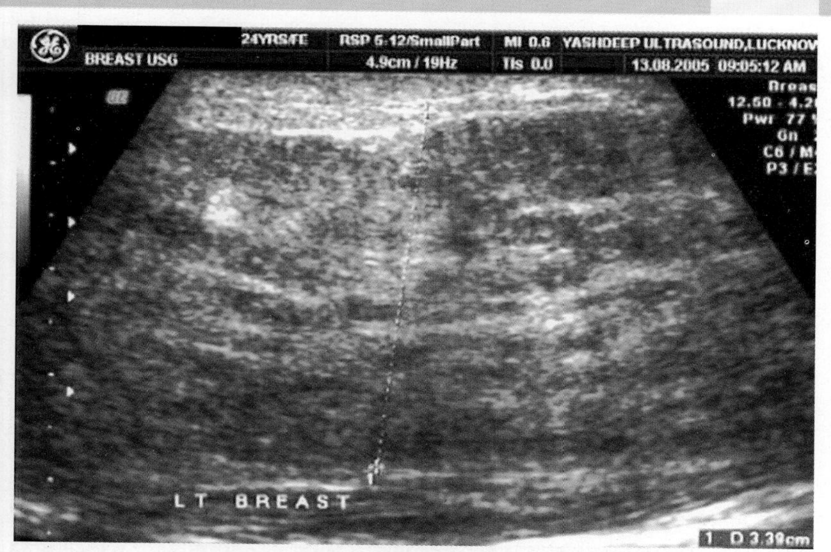

FIGURE 7.203

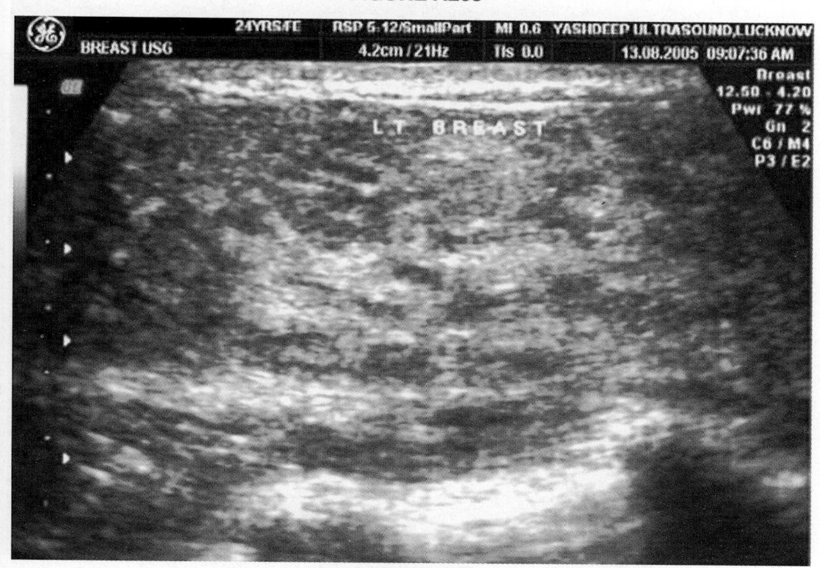

FIGURE 7.204

FIGURES 7.203 and 7.204: Inflammatory carcinoma of the breast. A lady presented with diffuse painful swelling of the breast. HRSG shows markedly thickened skin with enlargement of the breast. The breast tissue is edematous. Normal distinction between the fibrous and fatty tissue is lost. The whole of the breast is inhomogeneous with gross breast edema with exudative effusion in the breast tissue. Biopsy confirmed inflammatory carcinoma breast.

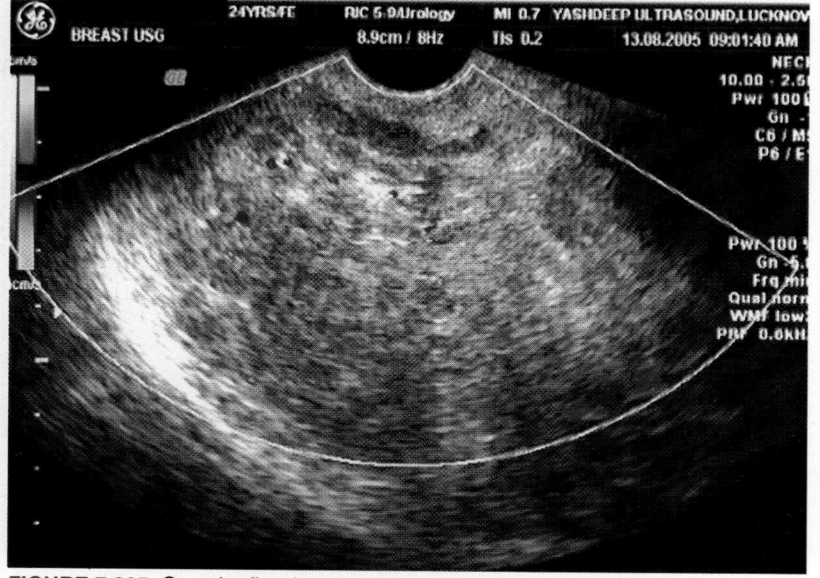

FIGURE 7.205: On color flow imaging little or poor flow is seen in the mass.

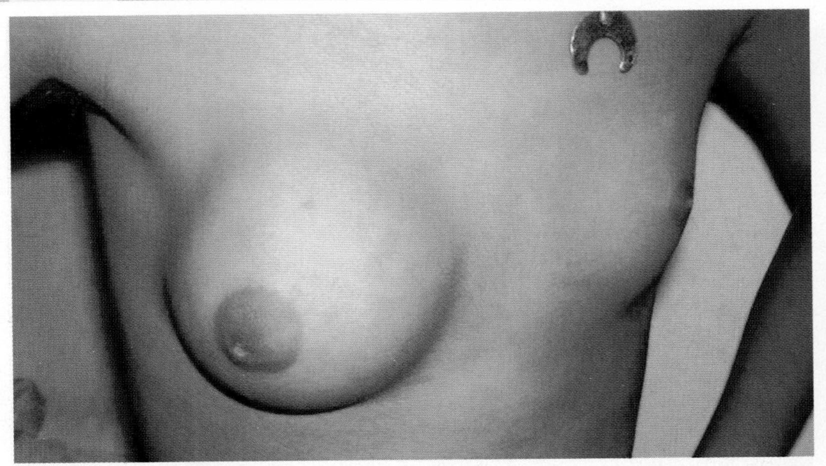

FIGURE 7.206: Breast sarcoma. A 13 years old young girl presented with soft tissue mass in the Rt breast. The mass was slowly growing and it was not painful.

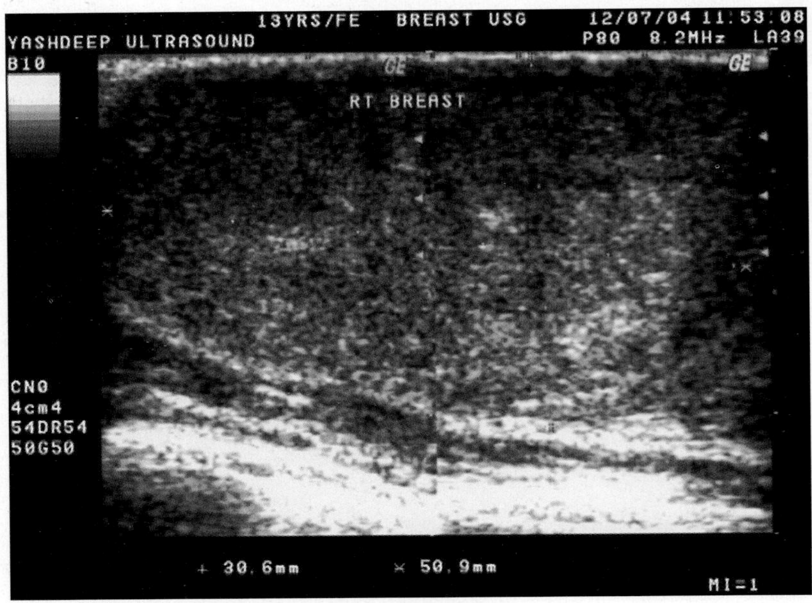

FIGURE 7.207

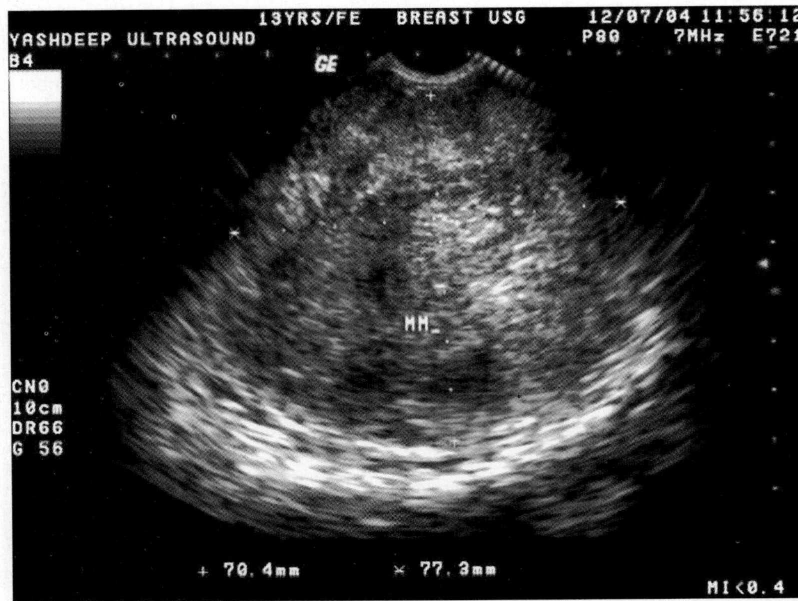

FIGURE 7.208

FIGURES 7.207 and 7.208: HRSG shows a big soft tissue mass in the breast. Normal texture of the breast is lost. However, It is homogeneous in texture. No calcification is seen in it. No cystic degeneration is seen. Biopsy of the mass confirmed breast sarcoma.

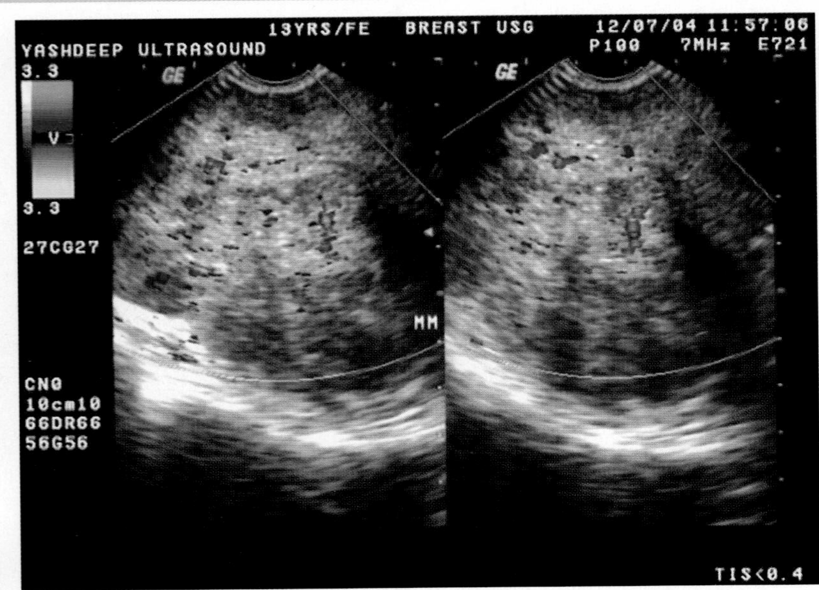

FIGURE 7.209: On color flow imaging low flow was seen in the mass.

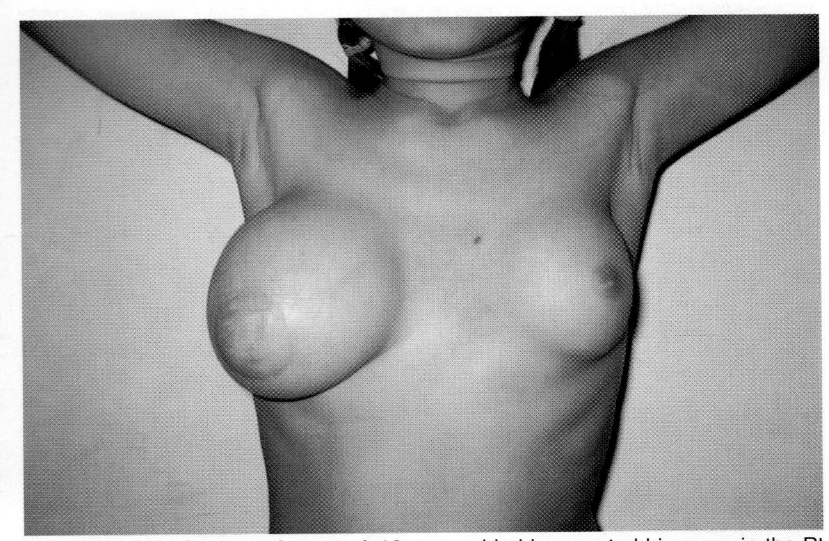

FIGURE 7.210: Sarcoma breast. A 12 years old girl presented big mass in the Rt breast. The mass was rapidly growing. Venous congestion is also seen on the surface of the breast.

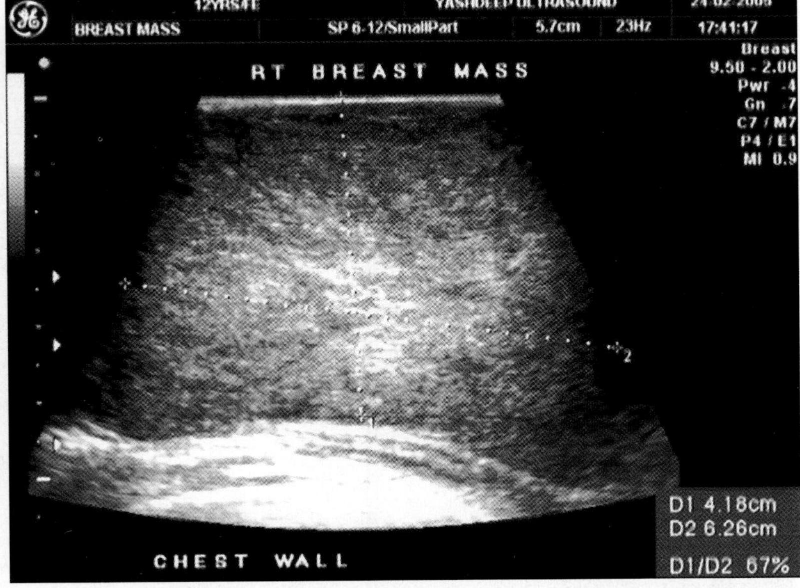

FIGURE 7.211: HRSG shows big mass in the breast. Normal breast texture is lost. The mass is seen fixed with the chest wall. However, it is homogeneous in texture.

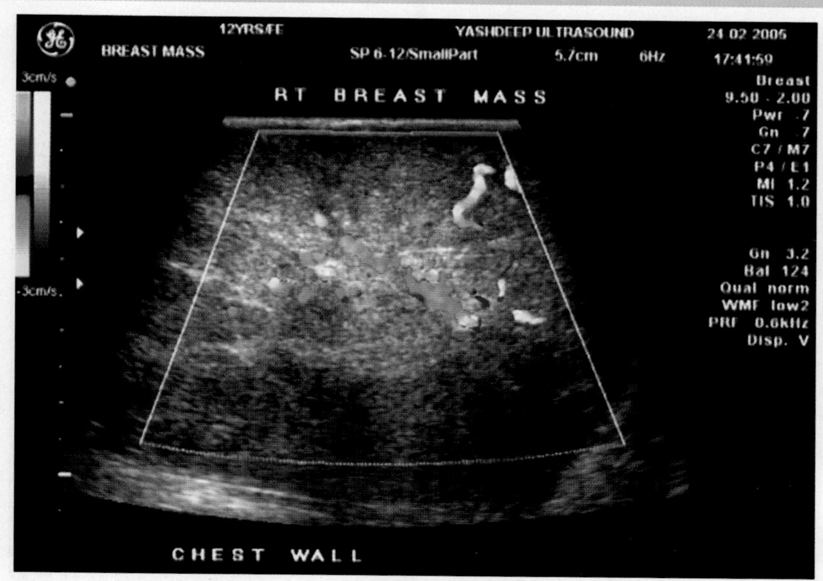

FIGURE 7.212

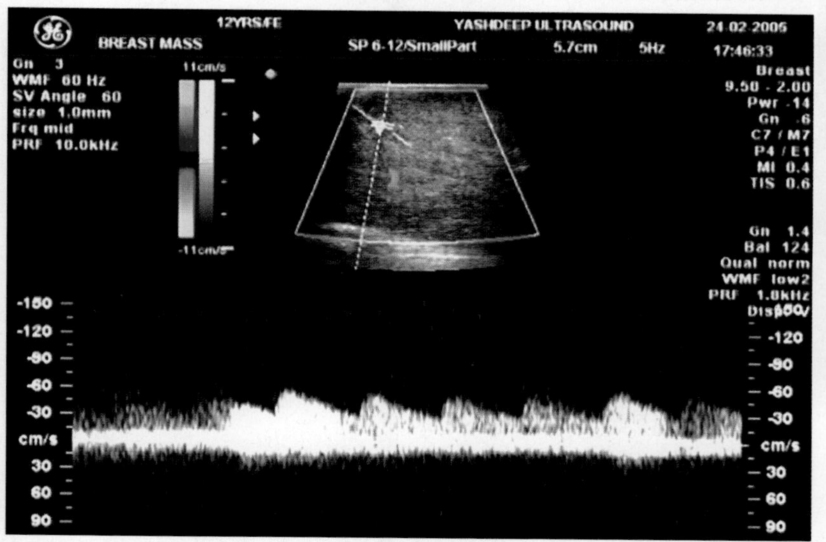

FIGURE 7.213

FIGURES 7.212 and 7.213: On color flow imaging multiple vessels are seen feeding the mass. Multiple vessels are seen in the mass. On spectral Doppler tracing moderate flow is seen in the mass.

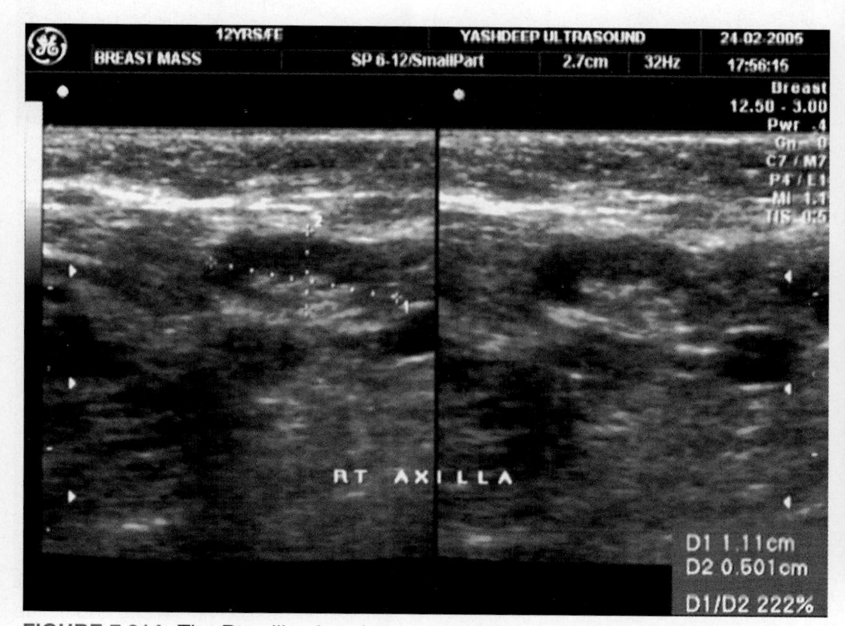

FIGURE 7.214: The Rt axilla also shows enlarged nodes with malignant changes in the nodes.

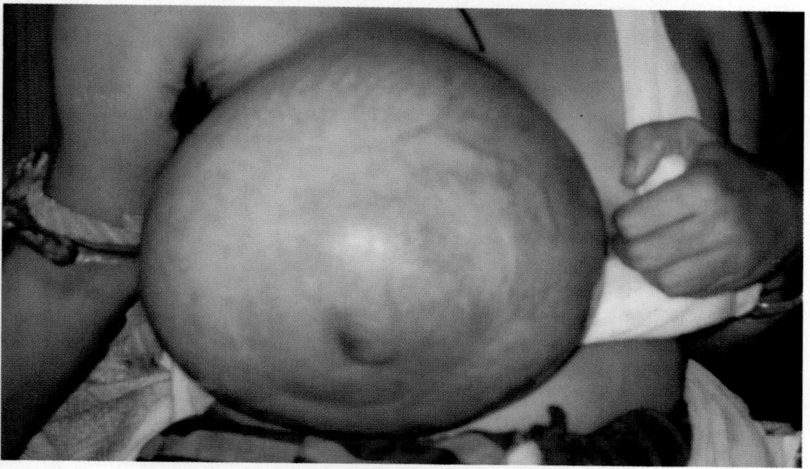

FIGURE 7.215

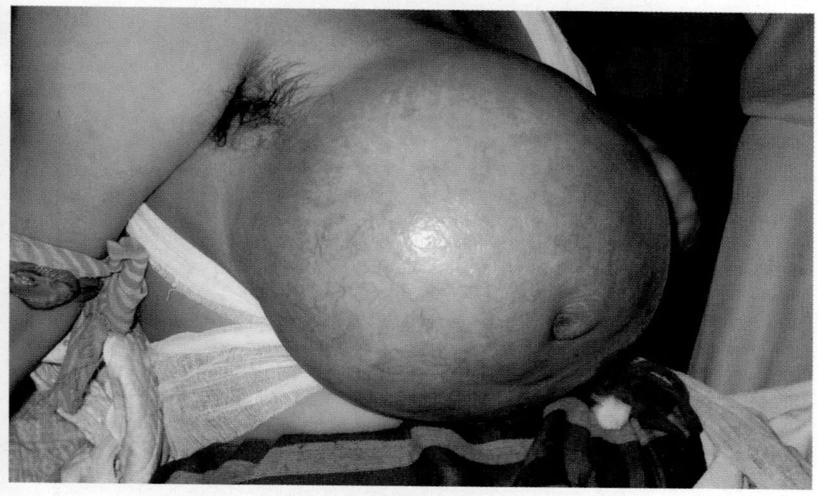

FIGURE 7.216

FIGURE 7.215 and 7.216: Huge sarcoma breast. A 22 years old girl presented with rapidly increasing mass in the breast. The mass is seen involving whole of the anterior chest wall. It is seen grossly enlarged. The surface of the mass shows multiple dilated veins.

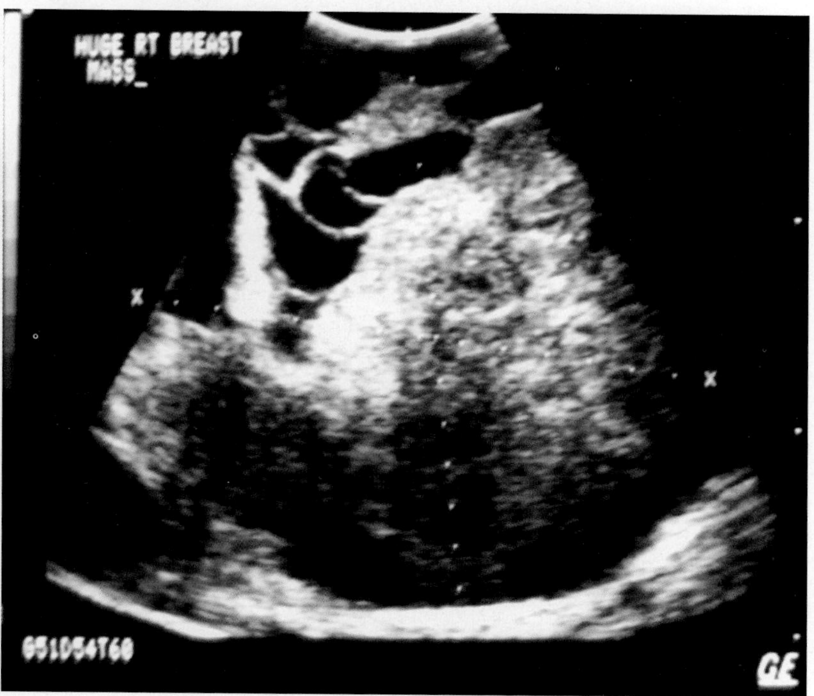

FIGURE 7.217: HRSG shows a big mass in the breast with predominantly solid texture. Few cystic areas are also seen in the mass with echogenic septa due to big size of the mass, lower frequency transducer was used for evaluation of tumor mass.

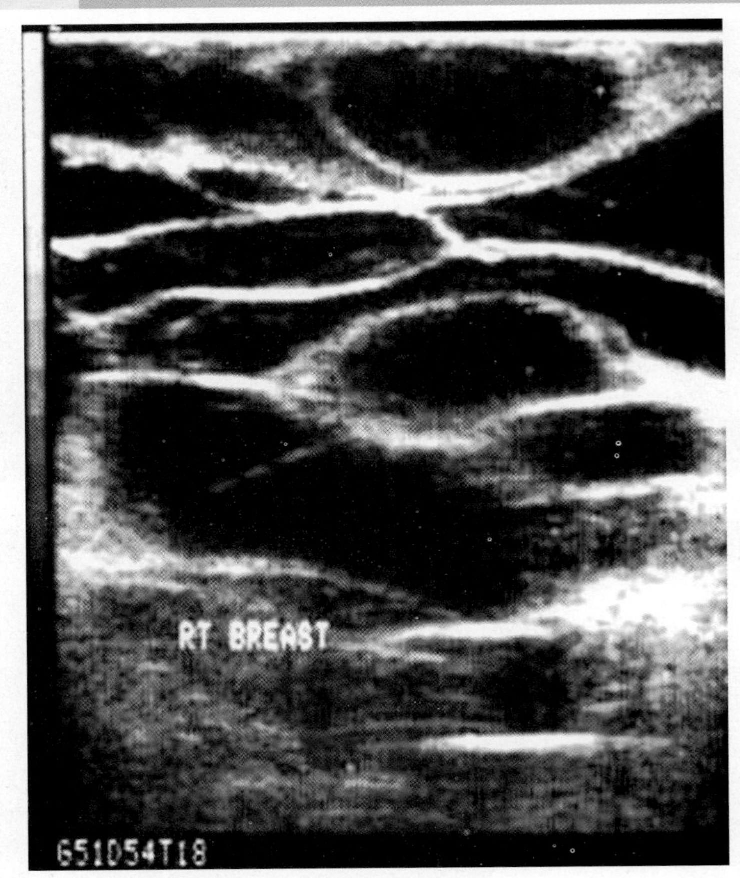

FIGURE 7.218

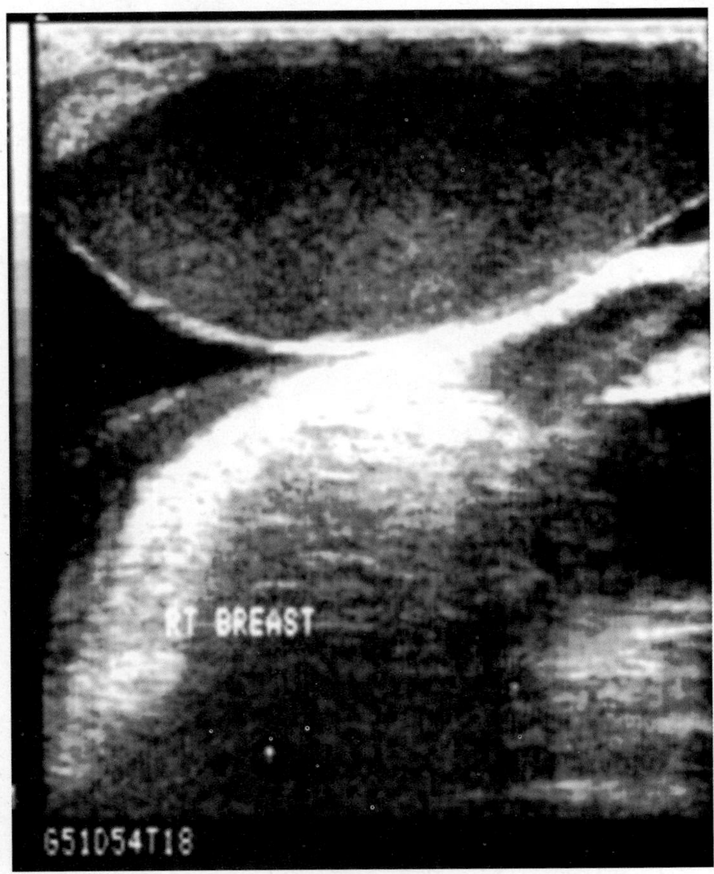

FIGURE 7.219

FIGURES 7.218 and 7.219: High-resolution ultrasound shows the cystic areas in much clear detail. Multiple cysts are seen in the solid tumor with fine echogenic septa. Thick collection was seen in the cysts. Biopsy of the tumor confirmed malignant sarcoma.

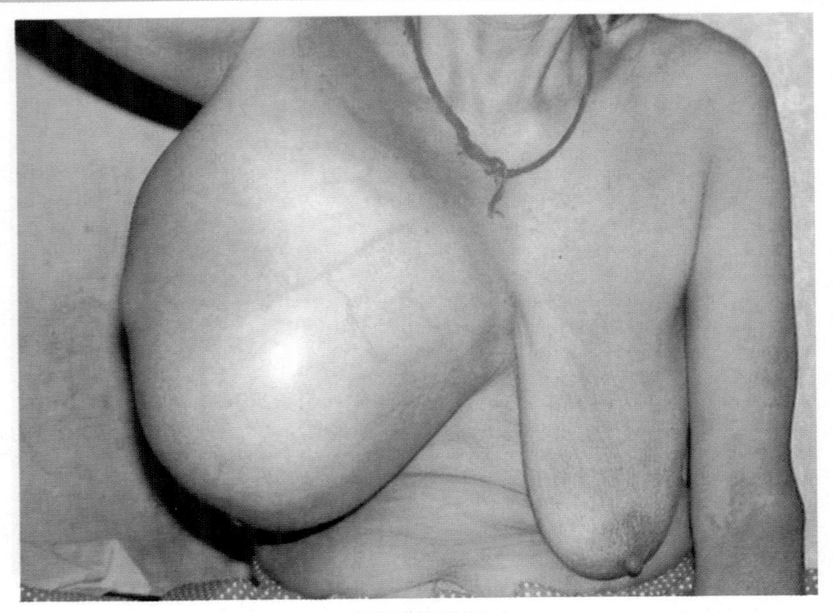

FIGURE 7.220

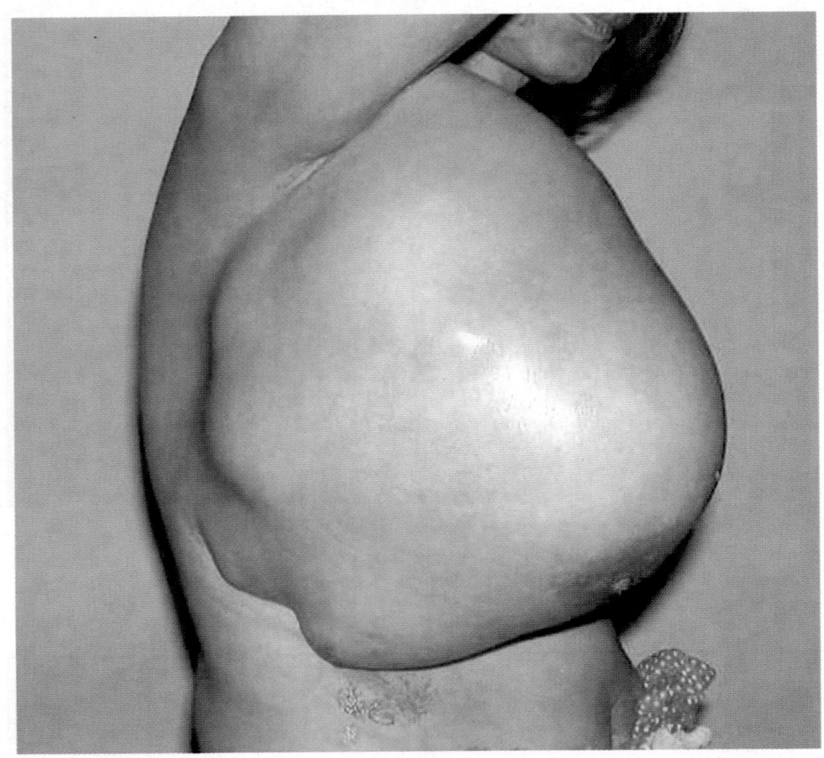

FIGURE 7.221

FIGURES 7.220 and 7.221: Phylloides tumor. A lady presented with huge mass coming out from the Rt breast. The mass shows lobulated surface. It is seen extending anteroposteriorly and also going on the backside.

2. Better differentiation between solid and cystic masses. It is superior to X-ray in radiologically dense breast as in young females, and breast density below 35years of age is usually a limitation in X-ray mammography.
3. Ultrasonographic-guided tumor biopsy can be done from the tumor substance.
4. In mastalgia X-ray mammography is contraindicated.
5. In pregnancy sonomammography is preferred.
6. In massive breast lesions.
7. In severe physical disability.

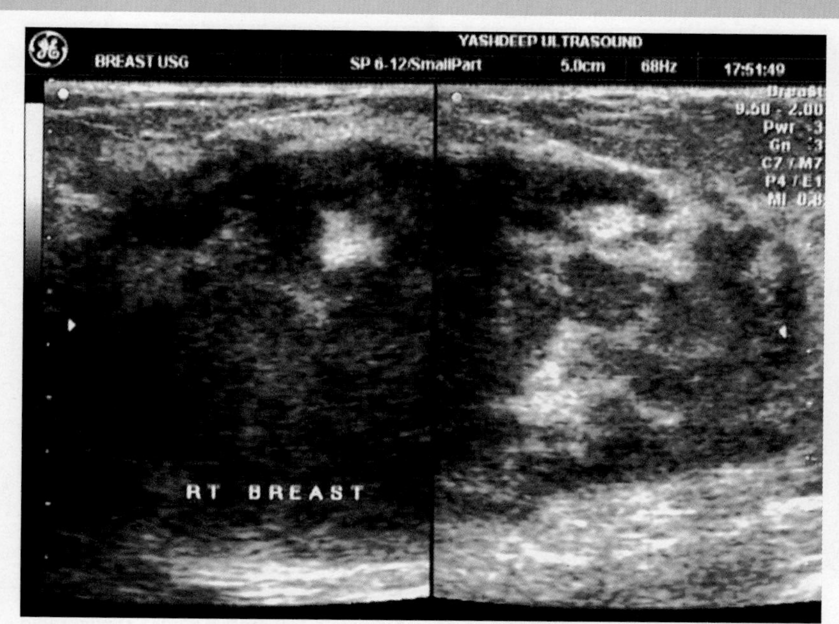

FIGURE 7.222: HRSG shows big mass in the breast. The mass could not be wholly assed on ultrasound, as it was not properly covered even on 3 MHz transducer. The details of the mass were well appreciated on HRSG. It was predominantly solid with variegated texture. Few areas show degeneration. Biopsy of the mass confirmed malignant Phylloides tumor.

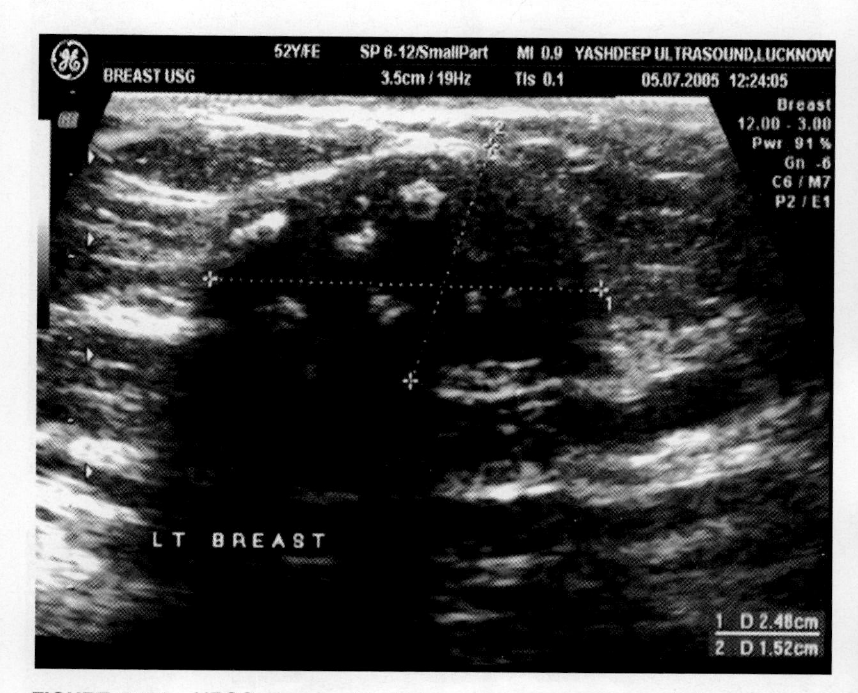

FIGURE 7.223: HRSG shows a big heterogeneous mass in the breast. It shows lobulations. Echogenic calcified shadows are also seen in the mass. They are accompanied with acoustic shadowing.

Limitations

The insensitivity of currently available ultrasonography equipment to detect microcalcification does not allow HRSG to be used in mass screening program for cancer detection. The modality is highly operator dependent and requires operator skill. Therefore, the best result can be achieved when it is used in conjunction with the X-ray and the combined sensitivity of sono and X-ray mammography can go up to 97% in detection of the breast cancer in preoperative patients.

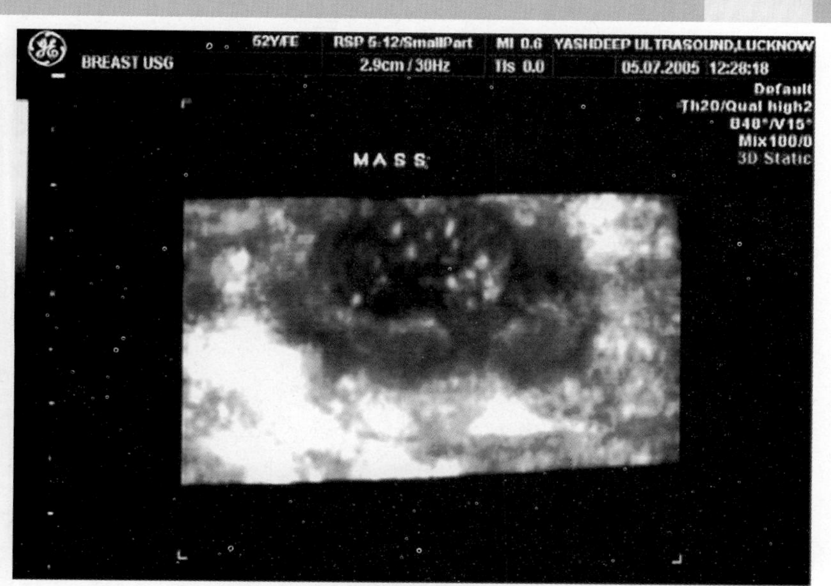

FIGURE 7.224

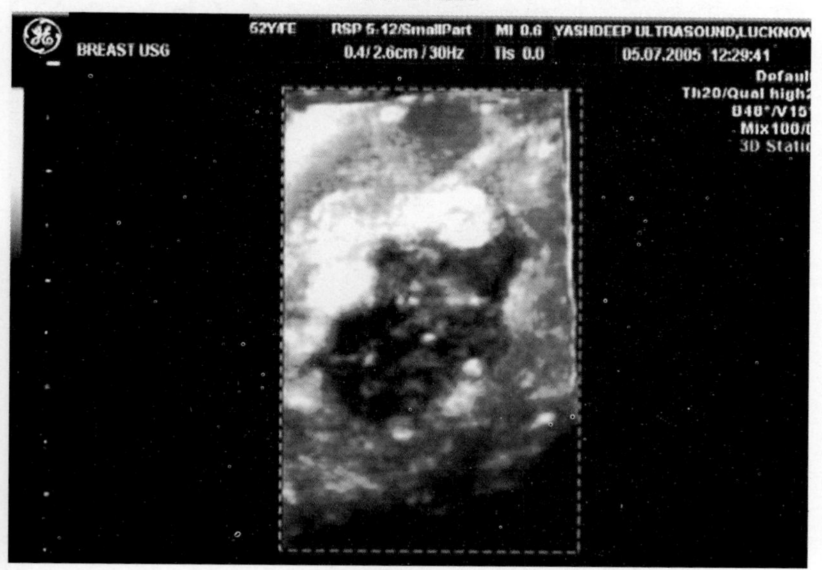

FIGURE 7.225

FIGURES 7.224 and 7.225: 3D imaging of the same mass shows details of the mass with multiple calcified shadows embedded in the mass. The mass shows crenated margins and it is confined to the mammary zone.

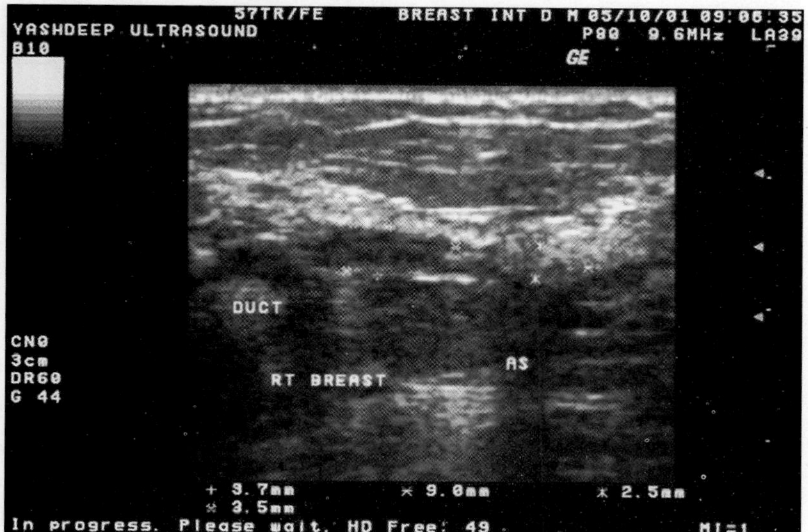

FIGURE 7.226: **Intraductal carcinoma.** An old patient presented with browning discharge from the nipple. HRSG shows dilated periareolar ducts. An echogenic mass is seen blocking the mouth of the duct. Biopsy of the mass showed intraductal carcinoma.

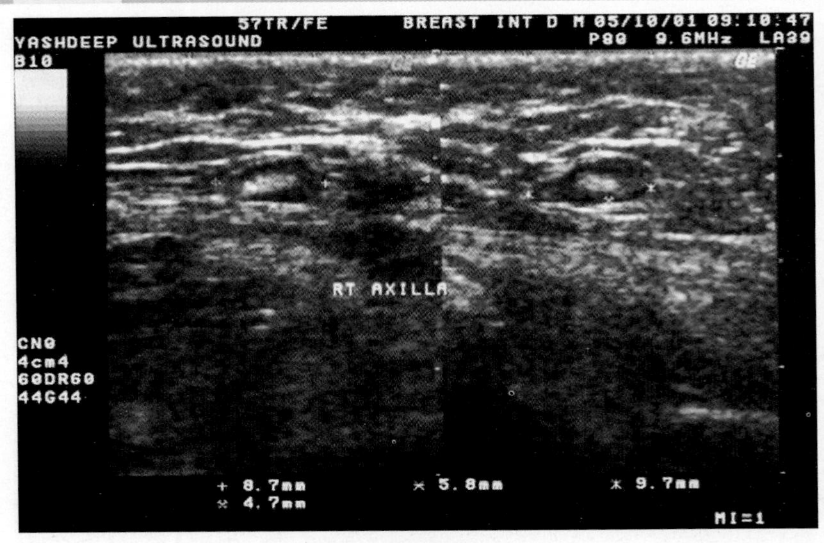

FIGURE 7.227

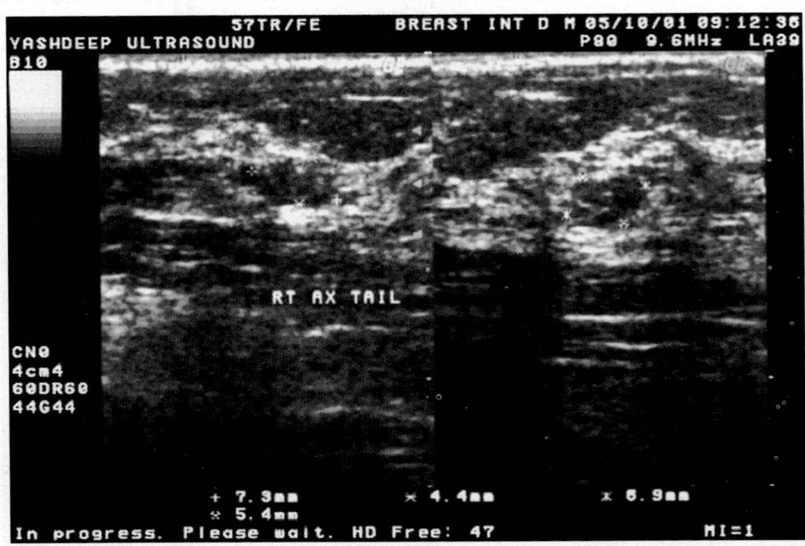

FIGURE 7.228

FIGURES 7.227 and 7.228: HRSG also shows hypoechoic-enlarged lymph nodes in axilla in same patient. Node biopsy was positive for malignancy.

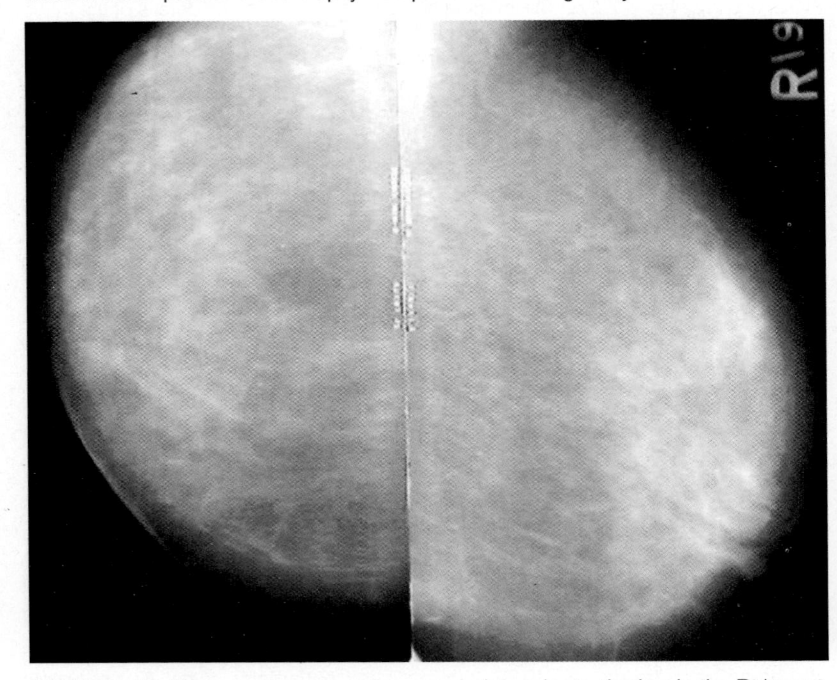

FIGURE 7.29: X-ray mammogram shows an indeterminate shadow in the Rt breast, which was not conclusive.

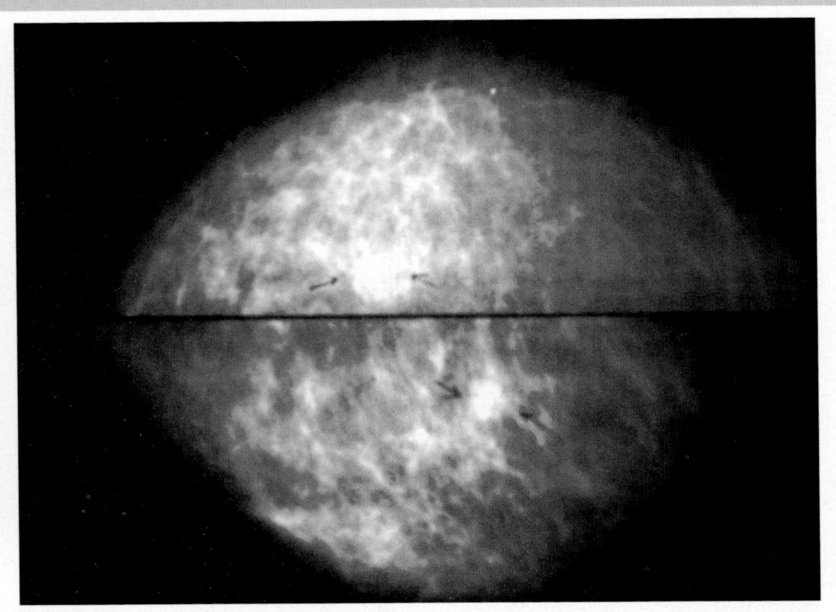

FIGURE 7.230: Macrocalcification in the breast. X-ray mammogram shows radio dense calcification in the breast. But no spiculation is seen.

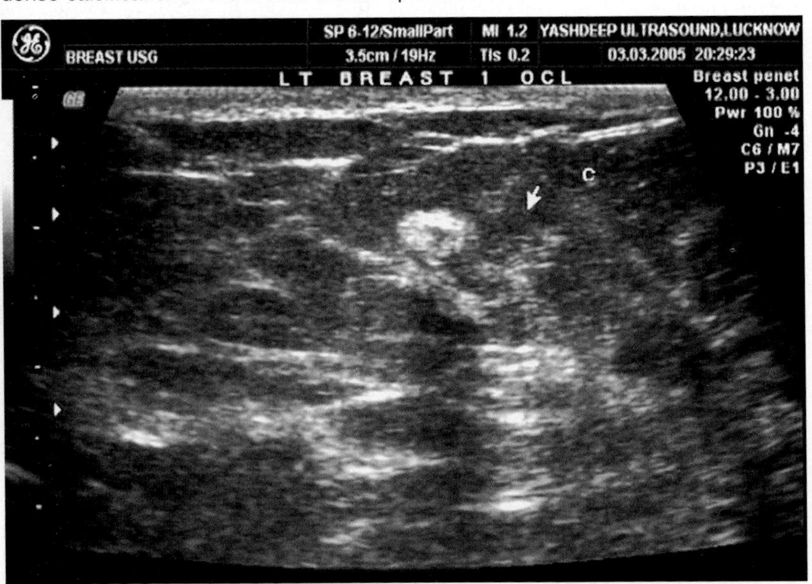

FIGURE 7.231: HRSG shows dense echogenic calcified focus in the breast. It is accompanied with acoustic shadowing. No evidence of any spiculation or satellite lesion is seen suggestive of benign calcification.

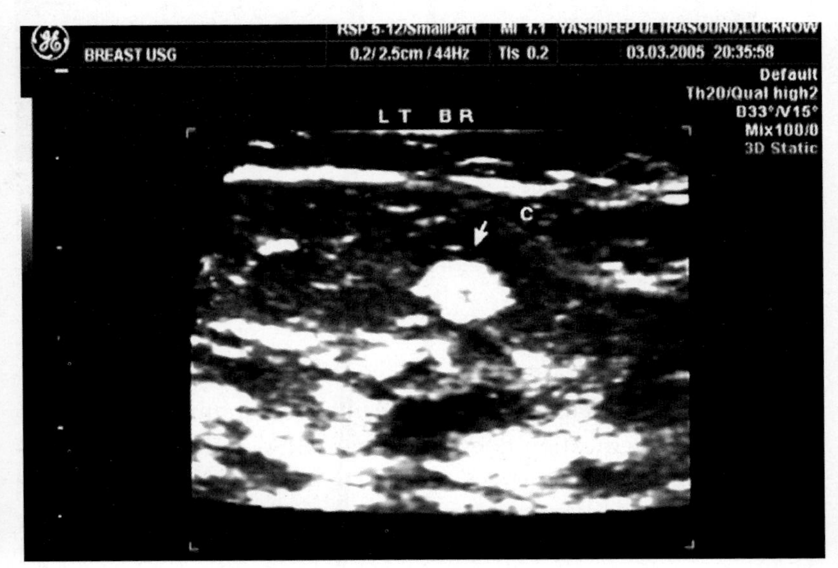

FIGURE 7.232

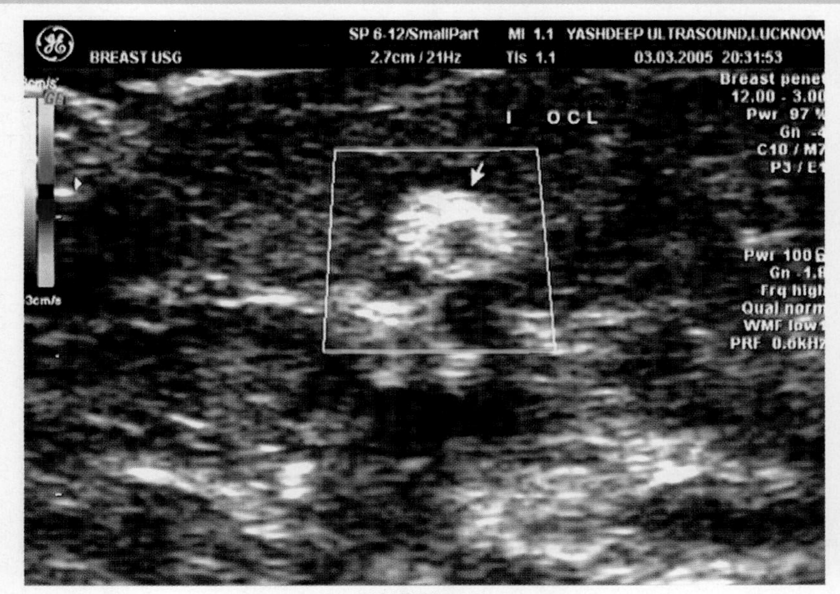

FIGURE 7.233

FIGURES 7.232 and 7.233: 3D imaging shows calcified focus clearly. On color flow imaging no flow seen in it. Biopsy confirmed benign lesion.

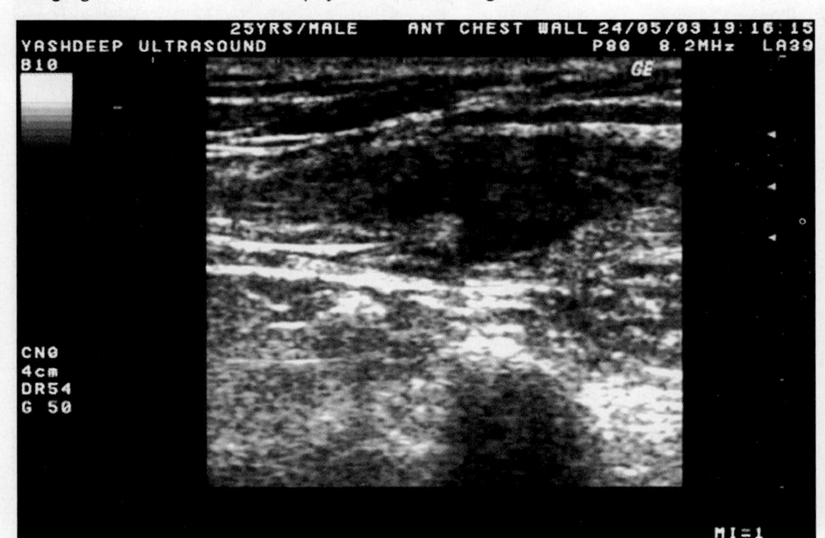

FIGURE 7.234

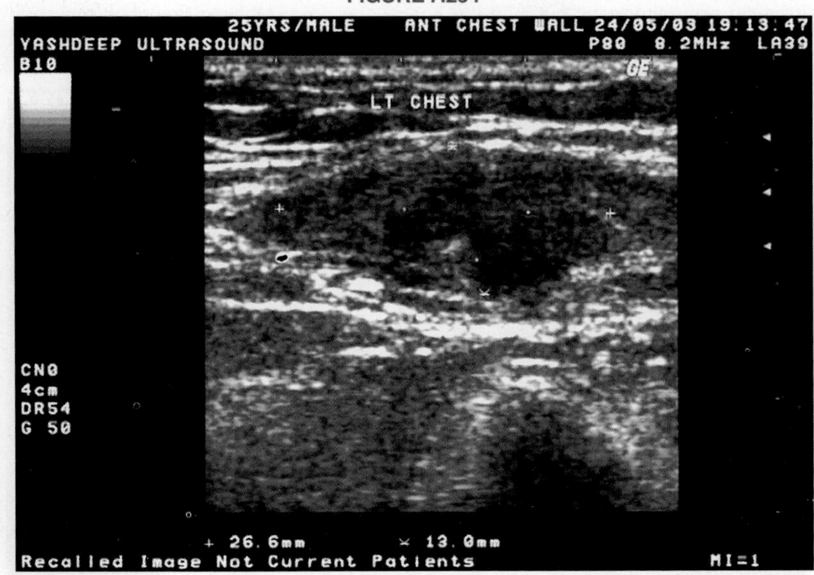

FIGURE 7.235

FIGURES 7.234 and 7.235: Male breast carcinoma. A patient presented with irregular lump in the periareolar area. HRSG shows heterogeneous irregular mass in the periareolar area.

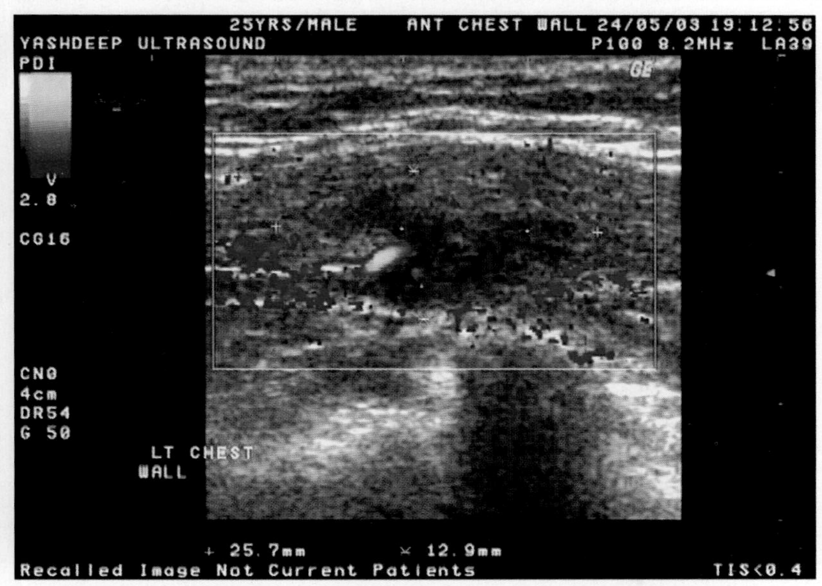

FIGURE 7.236: On color flow imaging vessel is seen feeding the mass. Biopsy of the mass confirmed carcinoma breast.

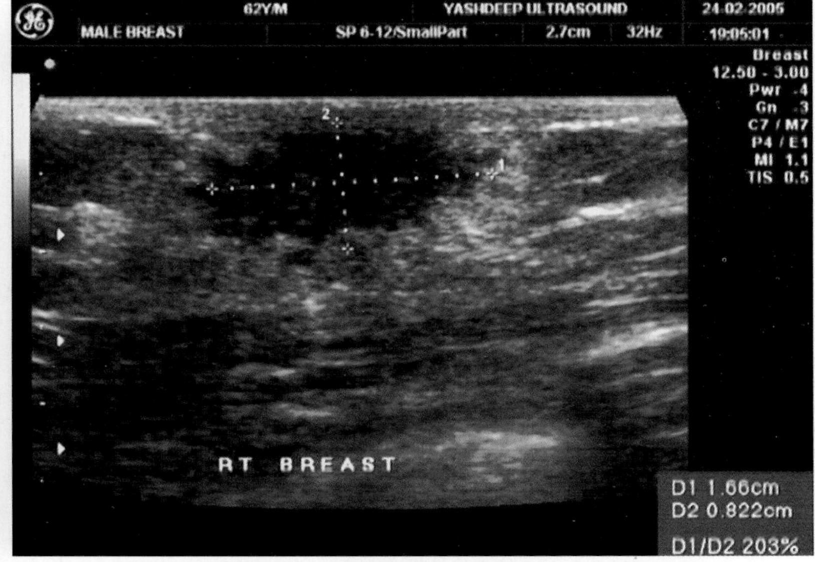

FIGURE 7.237

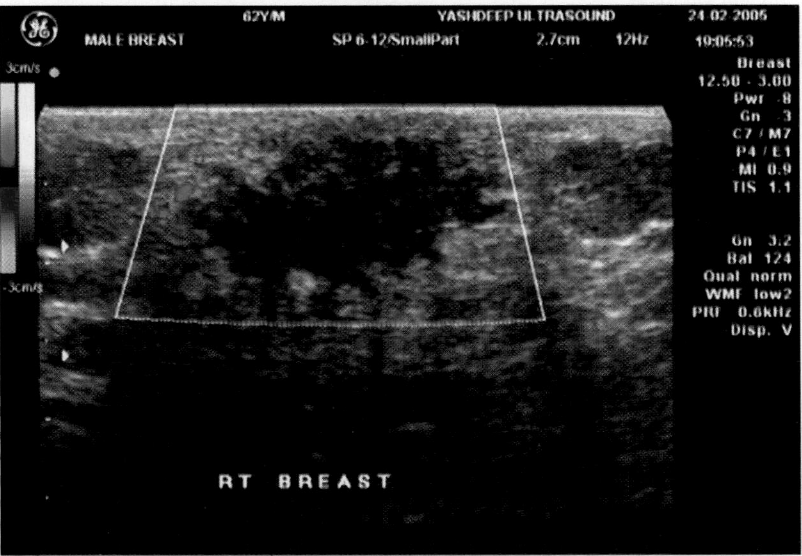

FIGURE 7.238

FIGURES 7.237 and 7.238: Carcinoma male breast. An old man presented with heterogeneous mass in the Rt breast Periareolar area. HRSG shows irregular heterogeneous mass in the sub-areolar region. The mass shows branching pattern running in all the directions. Biopsy confirmed carcinoma.

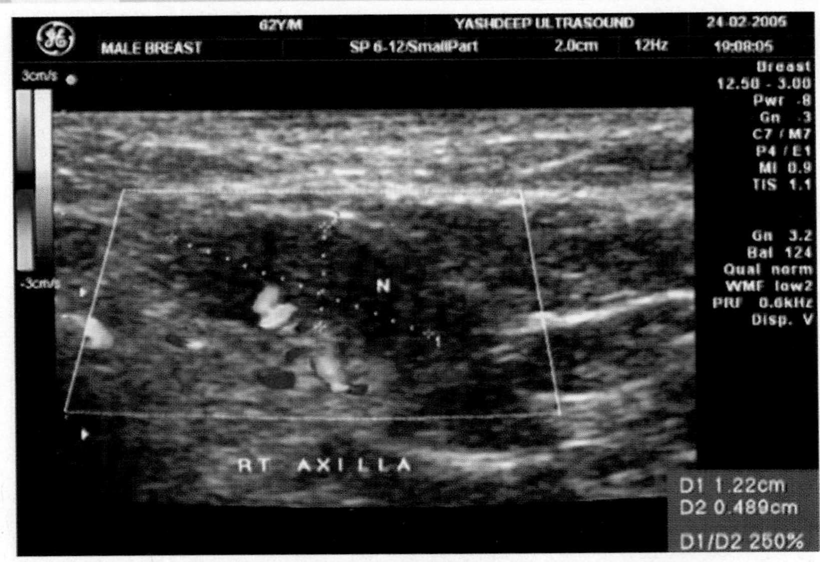

FIGURE 7.239

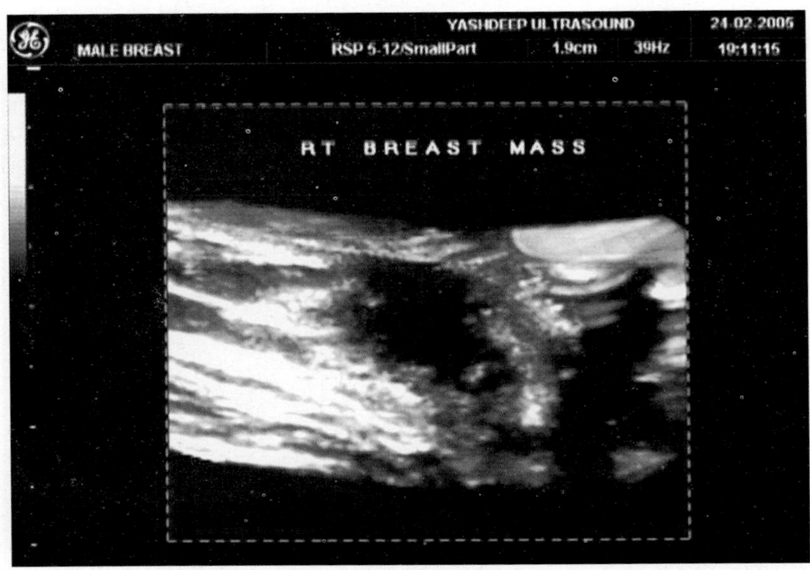

FIGURE 7.240

FIGURES 7.239 and 7.240: 3D imaging of the mass shows details of the invasion of the mass in the chest muscle. On color flow imaging multiple vessels are seen feeding the mass.

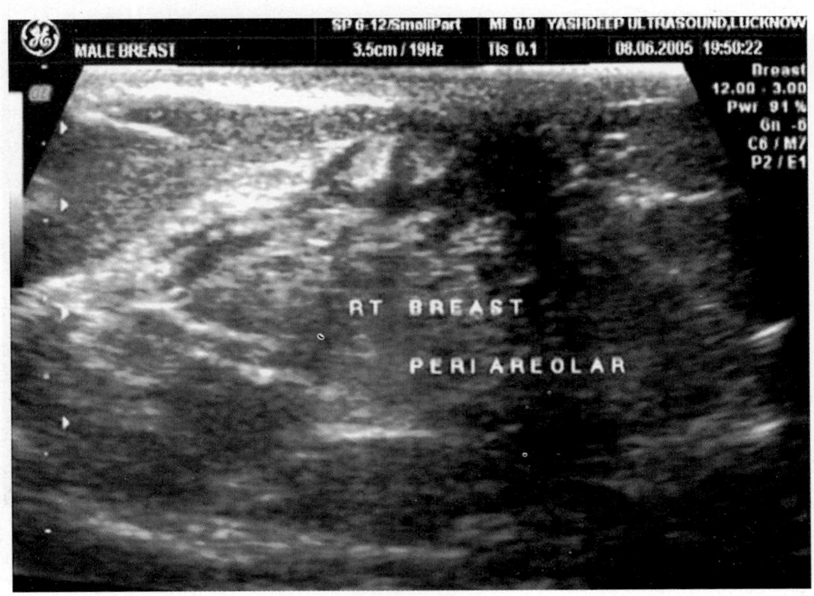

FIGURE 7.241

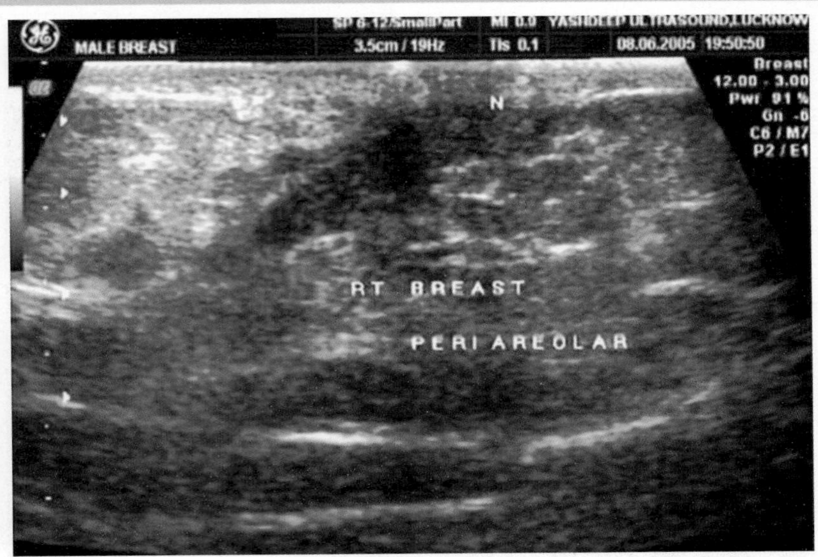

FIGURE 7.242

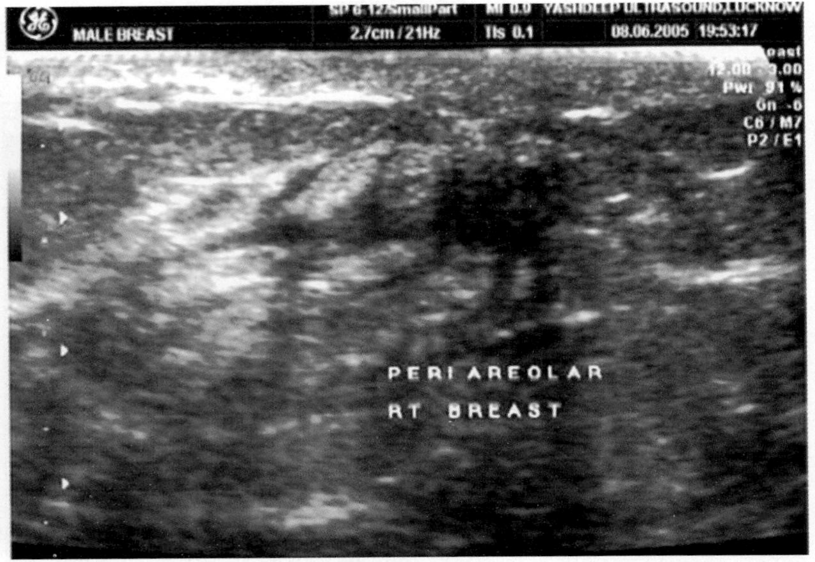

FIGURE 7.243

FIGURES 7.241 to 7.243: Carcinoma male breast. An old patient presented with irregular heterogeneous mass in the Rt breast in periareolar area. HRSG shows irregular heterogeneous mass in the periareolar area. Multiple branches are seen coming out from the mass. The mass is seen invading areolar tissue and the chest wall muscle.

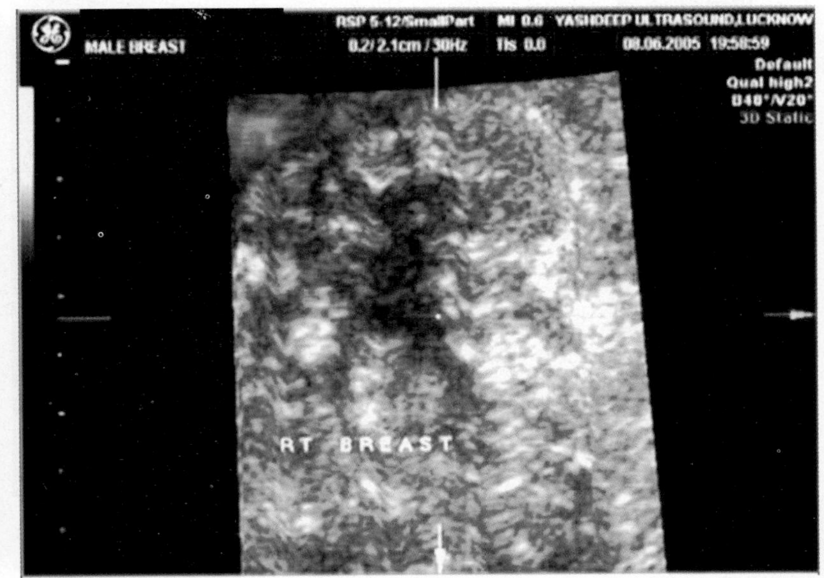

FIGURE 7.244: 3D imaging of the mass shows typical branching pattern of malignant mass.

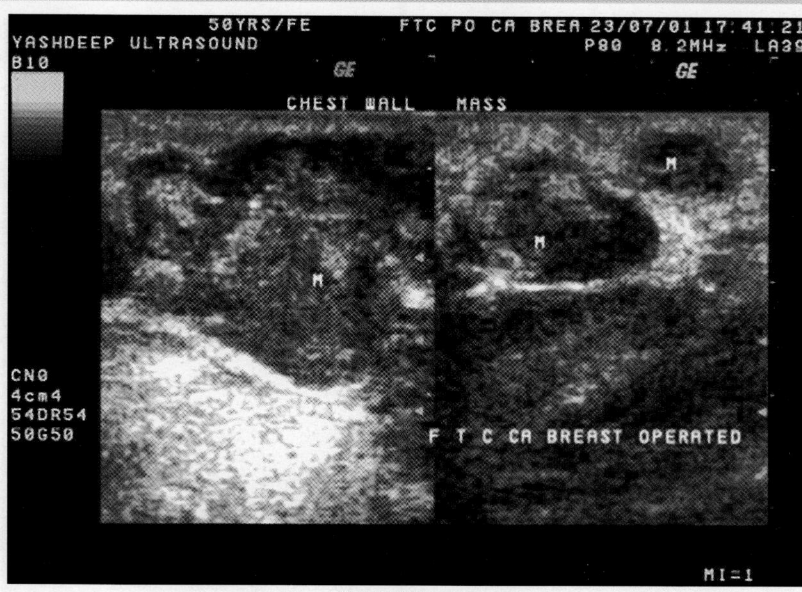

FIGURE 7.245

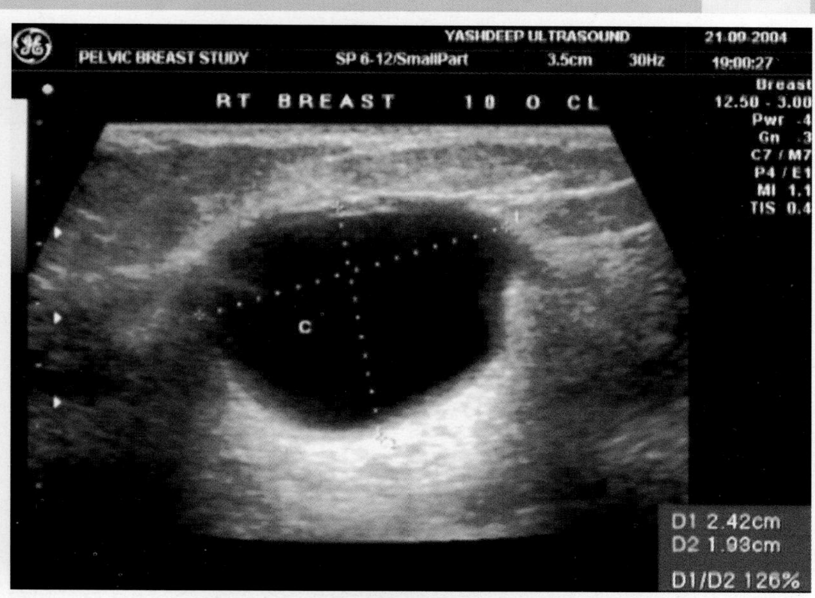

FIGURE 7.246

FIGURES 7.245 and 7.246: Recurrence of operated carcinoma breast. HRSG shows multiple irregular heterogeneous mass over the Rt chest wall in an operated case of carcinoma breast. Chest wall muscle invasion is also seen.

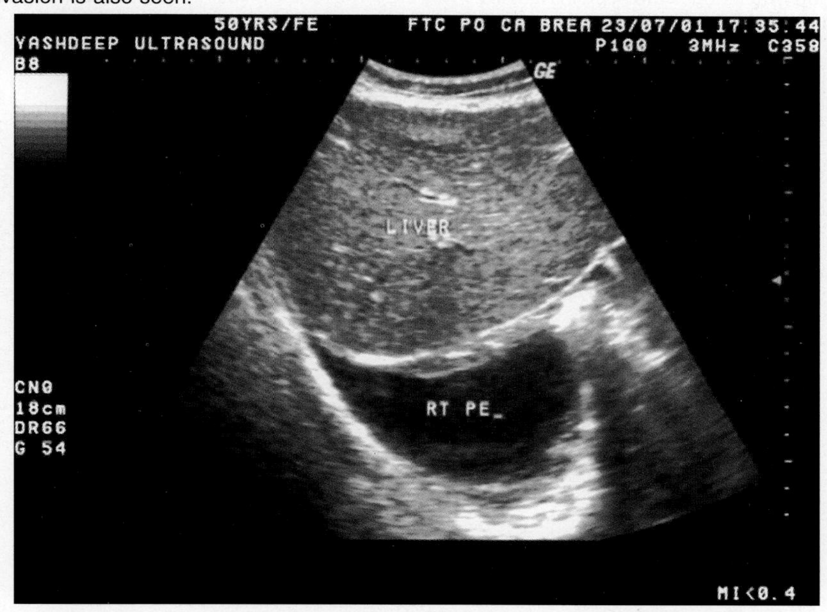

FIGURE 7.247: Abdominal ultrasound shows associated Rt sided pleural effusion.

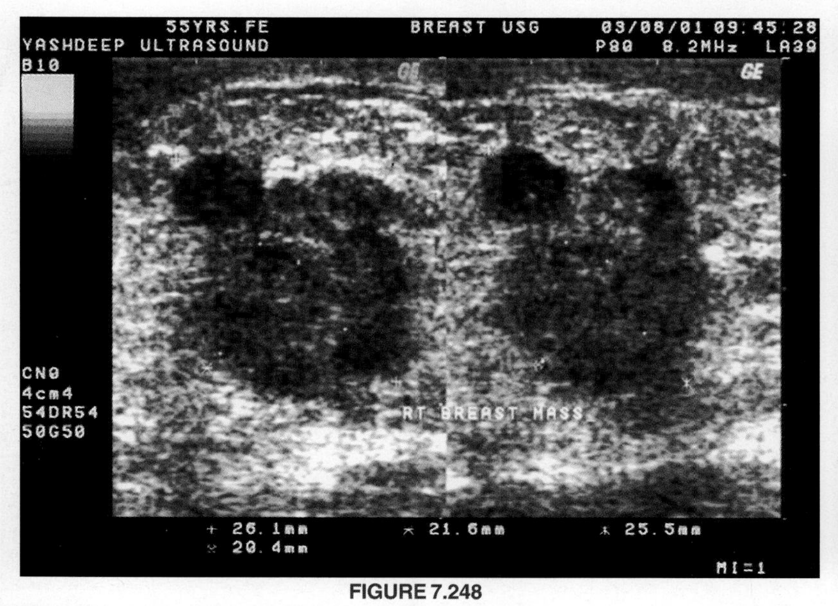

FIGURE 7.248

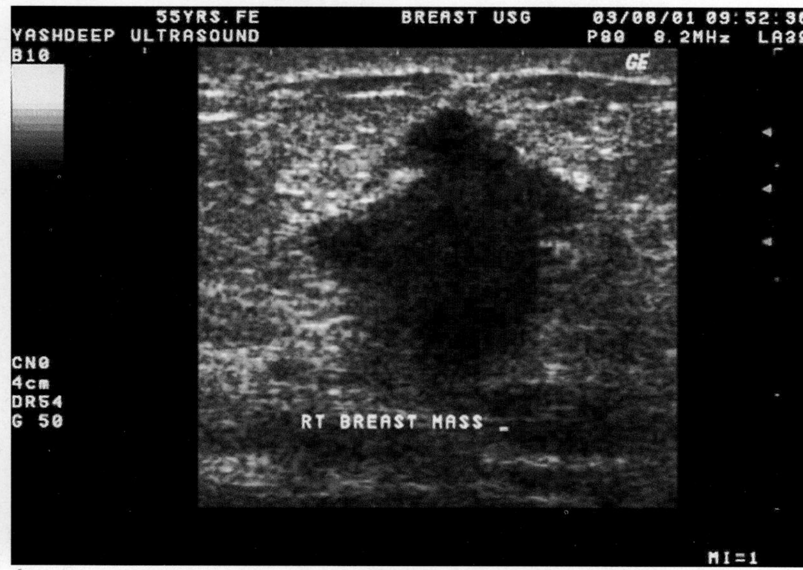

FIGURE 7.249

FIGURES 7.248 and 7.249: HRSG shows a big heterogeneous mass in the mammary zone. The mass is seen growing vertically. It is taller than wide. It shows multiple lobulations and confined to the mammary zone.

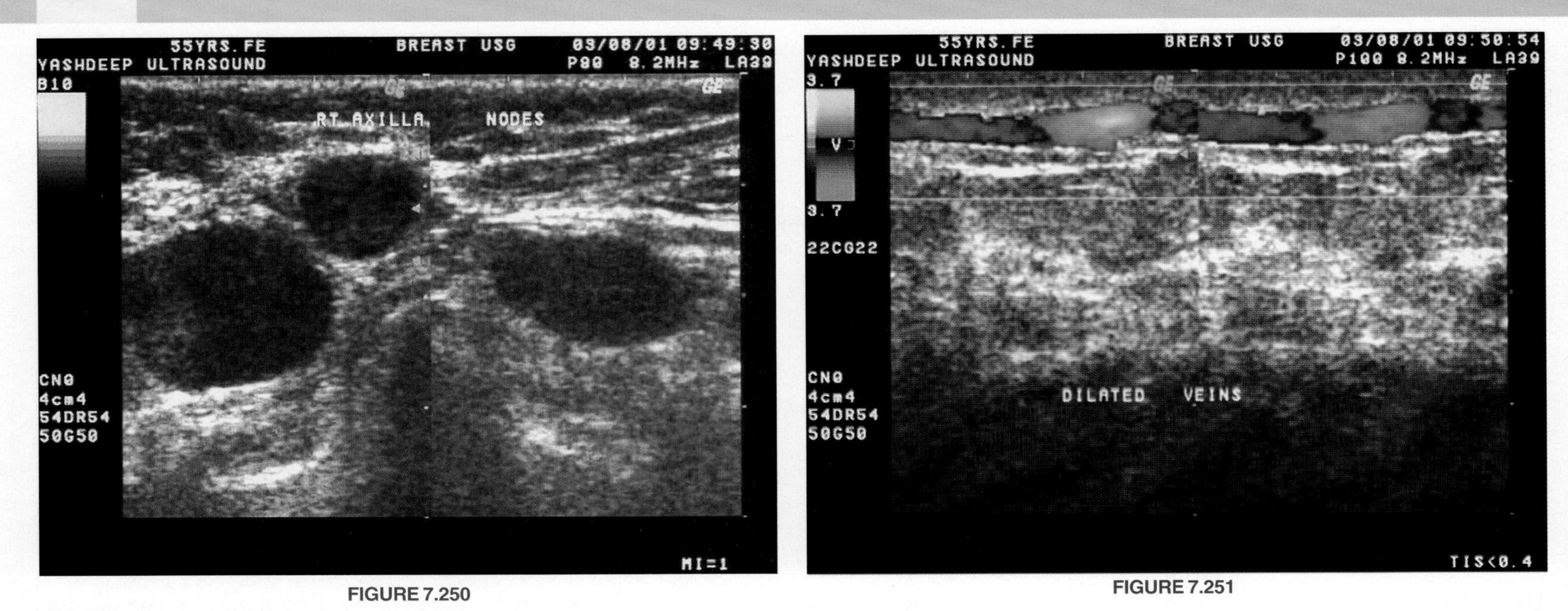

FIGURE 7.250

FIGURE 7.251

FIGURES 7.250 and 7.251: Multiple enlarged lymph nodes are seen in Rt axilla. They are hypoechoic in texture and round in shape. Lymph node hylum is destroyed. On color flow imaging vessels are seen congested.

Gastrointestinal System

8

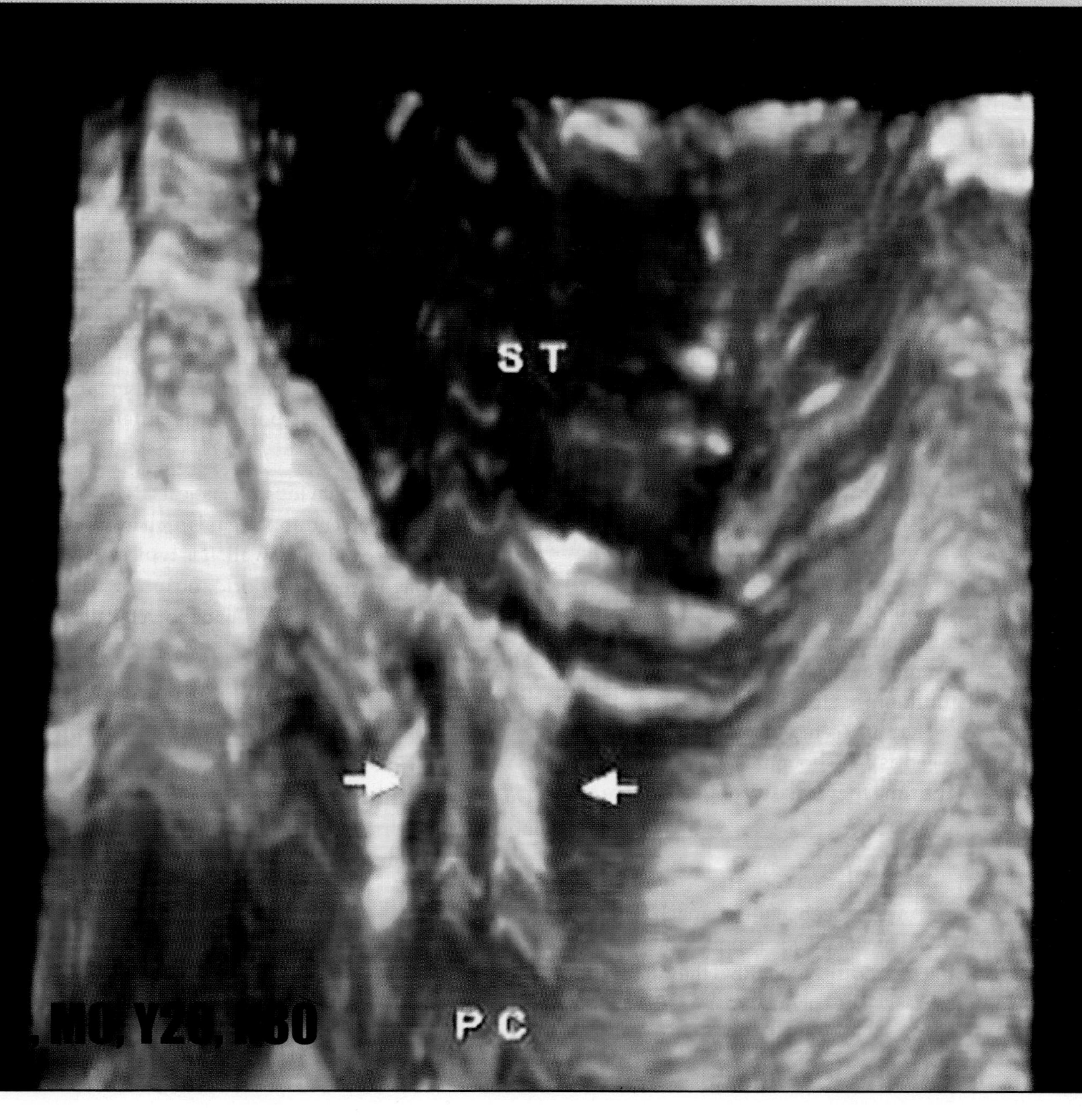

Introduction

High-resolution sonography plays an important role in evaluating and diagnosing many pathological conditions of gastrointestinal tract.

Stomach can be very well visualized by transabdominal approach by filling water in to it. By turning the patient to the Rt oblique position and displacing the gas from one part to other, stomach wall and mass adhered to it can be easily evaluated. Duodenum usually is difficult to see, as it is gas filled. But giving a water load and turning the patient to Rt lateral position, one can easily see first and second part of duodenum. Duodenal cap is seen lying posterior to gall bladder when scanning done through liver and seen contracting and relaxing.

Sonographic Features of Gut Wall

The gut wall is a multi-layered structure. It is seen as many layers of echogenic and hypoechoic layers parallel to mucosal and serosal surfaces. The five-layer structure is seen in good scanning technique. However, most of the time only three layered gut wall structure is seen, 1. Mucosal surface, 2. Middle echo poor muscularis mucosa, 3. Outer echogenic serosal surface.

Pathological Conditions

Stomach

Normal stomach wall is 3 to 5 mm in thickness when distended. If it is more than 10 mm in thickness, malignancy should be suspected. However, inflammatory pathology can also lead to thickening. Majority of carcinoma stomach can be detected on HRSG. The malignancy can be picked up as a localized thickening of stomach wall with echo poor texture. When the tumor spreads there is loss of normal layered pattern of the stomach wall. Lymph node enlargement can be picked easily on HRSG. About 80% of carcinoma stomach are presented with lymph node enlargement, when clinically detected. Endoscopic ultrasound is very useful in picking up early gastric lymphoma when barium meal and endoscopic findings are in conclusive.

Benign gastric tumors are much less common than the malignant tumors. Adenomatous polyp is thought to be a premalignant condition. It arises from mucosal or sub mucosal layer. Hyperplastic polyps are not true polyp but are more common than adenomatous polyp. They appear as localized thickening and usually multiple. Carcinoid tumors, hamartoma are other tumors of stomach. The most common benign tumor is leiomyomas. They are from muscle layer and seen as polypoidal growth. They are seen as echo poor mass coming out from muscularis propria. Most of the time they have ulcerated surface and seen on HRSG.

Gastric Bezoars

Bezoars are the foreign material masses seen in the stomach. They are trichobezoars (hair) phyto bezoars (vegetables) or concretions like cow milk in infants and antacid tablets in adult. On HRSG they are seen as highly reflective masses in stomach. The trichobezoar is seen as solid echogenic mass with total acoustic shadowing suggestive of calcified surface.

Hypertrophic Pyloric Stenosis

Hypertrophic pyloric stenosis is usually seen in male infants at birth or up to 6 month of age. HRSG is choice of investigation for diagnosing it. It shows the pyloric muscle thickening. If it is more than 4 mm in thickness, it is abnormal. The total thickness of pyloric muscle from one surface to other surface should be more than 15 mm and length of pyloric canal is less than 17 mm in pyloric stenosis.

Gastric Diaphragm

The gastric diaphragm is the congenital membrane. It is also known as web. It is seen extending into the gastric antrum and partially obstructing it. Most of the time it is 2 cm before the pylorus. When it is complete it is thought to be one of the types of gastric atresia. However, most of the diaphragms are incomplete and results into various form of obstruction.

Duodenal Atresia

Typical double bubble sign is seen in duodenal atresia on HRSG. The common causes of duodenal obstruction are bands, atresia and stenosis. The other less common causes are annular pancreas, preduodenal and portal vein. The distal small bowel obstruction or volvulus can also result in duodenal obstruction.

The duodenal hematoma can occur due to blunt trauma to abdomen. It is seen a hypoechoic or echo free mass indenting the wall of duodenum. It is to be differentiated from pseudopancreatic cyst.

Bowel

Bowel Pattern

Normal bowel shows patterns on HRSG. It depends upon the part of bowel seen on HRSG, amount of fluid in bowel, gas in the bowel, and peristalsis present in bowel. The different bowel patterns are as follows:

Mucous pattern: It is having the classical target pattern. Highly reflective core of mucous with entrapped gas is seen in the center. It is surrounded by echo poor halo of bowel wall. It is typical of a sphincter region like gastroesophageal region, pyloric antrum or in duodenum. HRSG can show the multi-layered structure of bowel wall.

Fluid pattern: The bowel is seen as tubular organ on long axis when filled with fluid. Different portion of bowel can be seen on HRSG. The jejunum with valvulae conniventes usually give an appearance of a ladder.

Gas pattern: The typical gas pattern of bowel and its distribution in the bowel may help in tracking down the disease. The typical ring down artifact often tapers and described as "comet tail pattern".

Colon pseudotumor pattern: At time colonic contents trapped in gas can give rise pseudotumor pattern. It may be caused due to fecalith.

Bowel Pathology

Variation for normal bowel pattern is known as bowel pathology. Therefore, it is mandatory to assess the normal bowel pattern, distribution of gas, amount of fluid in bowel and evaluation of bowel wall thickness and study of peristalsis and man movement.

Pseudo Kidney Sign

Normal bowel wall thickness is around 5 mm in non-distended state and 3 mm in distended state. If it is greater than the normal limits it should be treated as abnormal. When bowel wall is thickened the mucous pattern changes to thickened irregular halo of bowel wall, central echogenic complex thinned and compressed and become irregular, pathological changes should be suspected and known as pseudokidney sign. The various conditions which may show pseudo kidney signs are as listed below:

Inflammatory disease

1. Tuberculosis of bowel
2. Ruptured appendix with appendiceal mass
3. Chronic granulomatous disease
4. Crohn's disease
5. AntiBioma
6. Diverticulosis

Tumors

1. Adenocarcinoma
2. Lymphoma
3. Leiomyosarcoma
4. Carcinoid tumor
5. Metastatic deposits

Intussusception: HRSG shows typical crescentric triple layer appearance of intussusceptum in transverse axis. The longitudinal scan shows trident sign representing intussusceptum and intussuscipiens.

Mucoviscidosis: Mucoviscidosis or cystic fibrosis is uncommon condition involving small bowel loops, pancreas, kidney and the lungs. The bowel loops are seen dilated. Thick low-level collection is seen filling the bowel loop. It is thick jelly like in consistency. The thick echogenic bands are seen crisscrossing the bowel loops and cause the intestinal obstruction. Specks of calcification are also seen in kidneys.

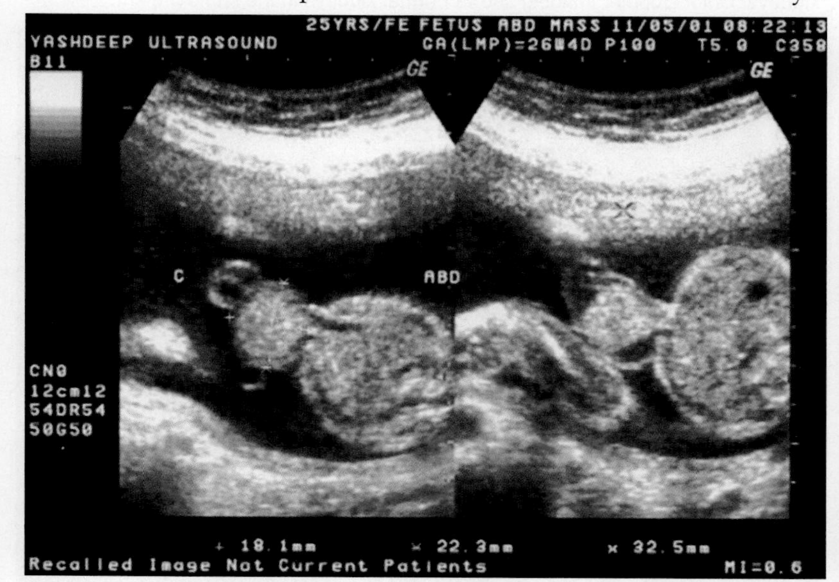

FIGURE 8.1: Congenital omphalocele. HRSG shows herniation of bowel loops from the defect of anterior abdominal wall in a fetal abdomen. The bowel loops are seen floating in the amniotic cavity.

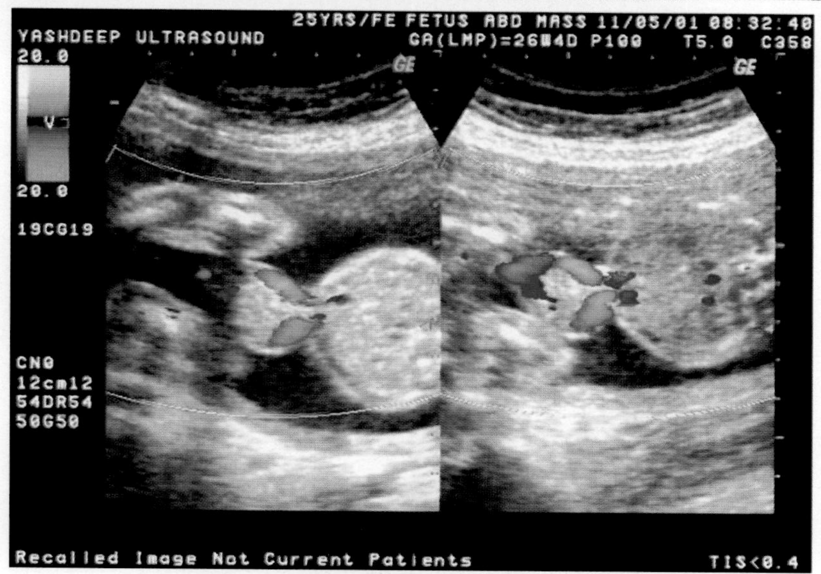

FIGURE 8.2

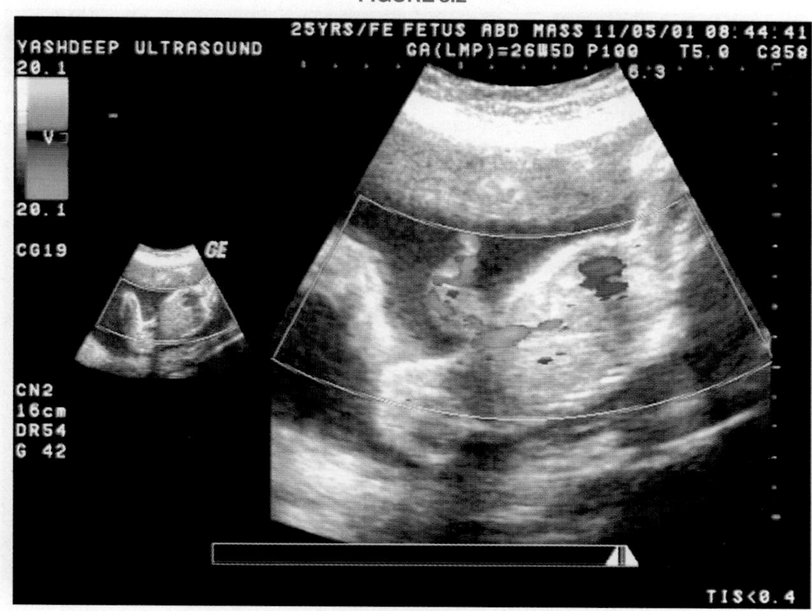

FIGURE 8.3

FIGURES 8.2 and 8.3: Omphalocele. Color Doppler imaging shows herniated bowel loop floating in the amniotic cavity. A single umbilical artery is also seen entering into the fetal abdomen at the umbilicus.

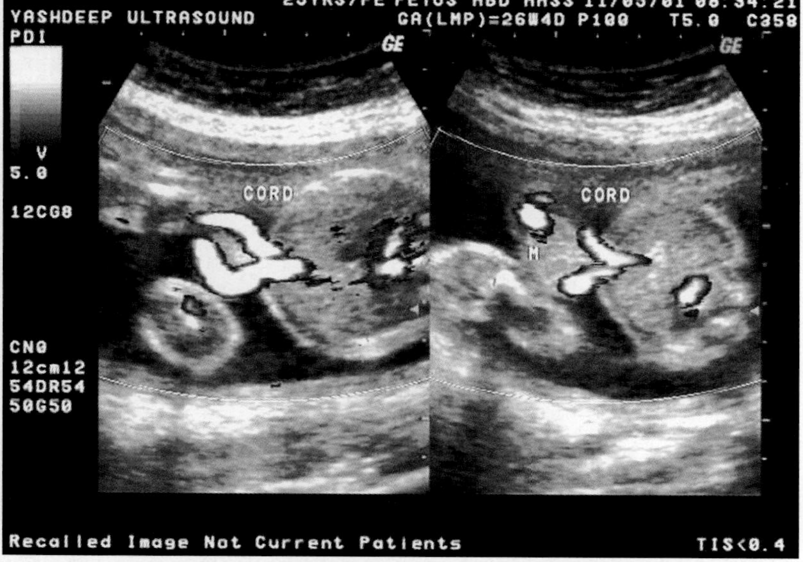

FIGURE 8.4: Power Doppler imaging shows the umbilical artery encircling the bowel loop.

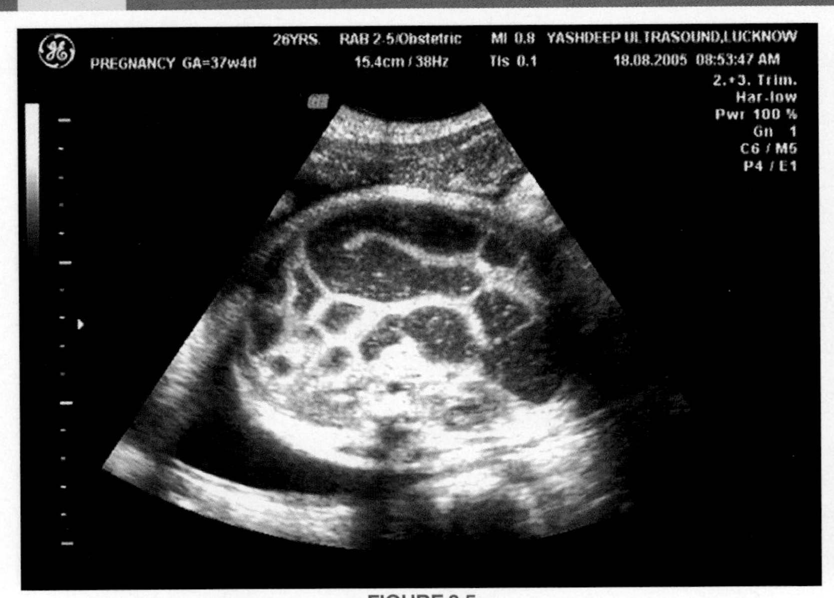

FIGURE 8.5

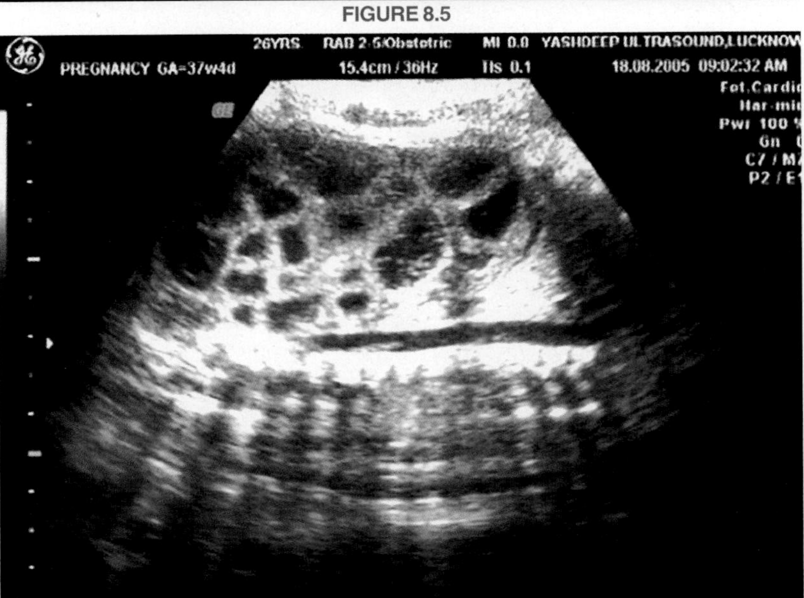

FIGURE 8.6

FIGURES 8.5 and 8.6: Small bowel obstruction. HRSG shows multiple dilated bowel loops in the fetal abdomen. Coiling of the loops is seen. Echogenic collection is seen in the loops suggesting stagnant meconium.

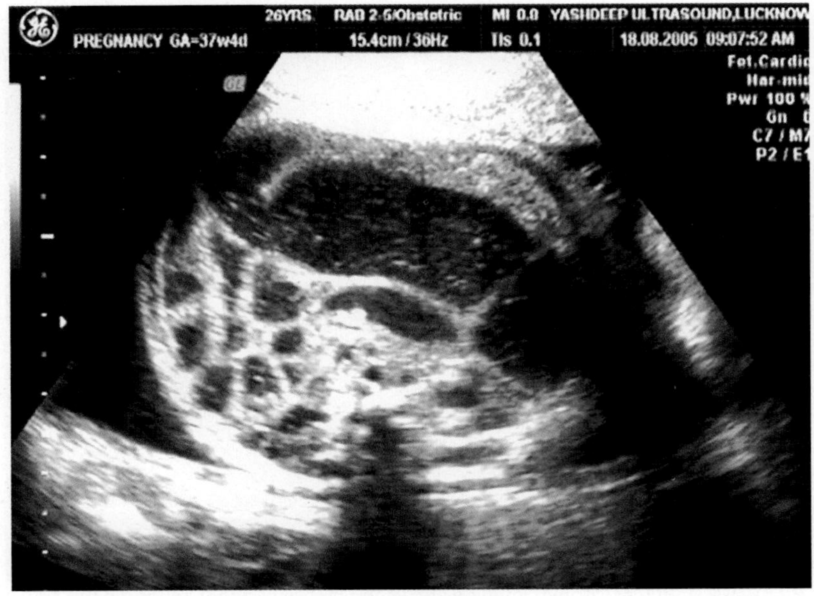

FIGURE 8.7

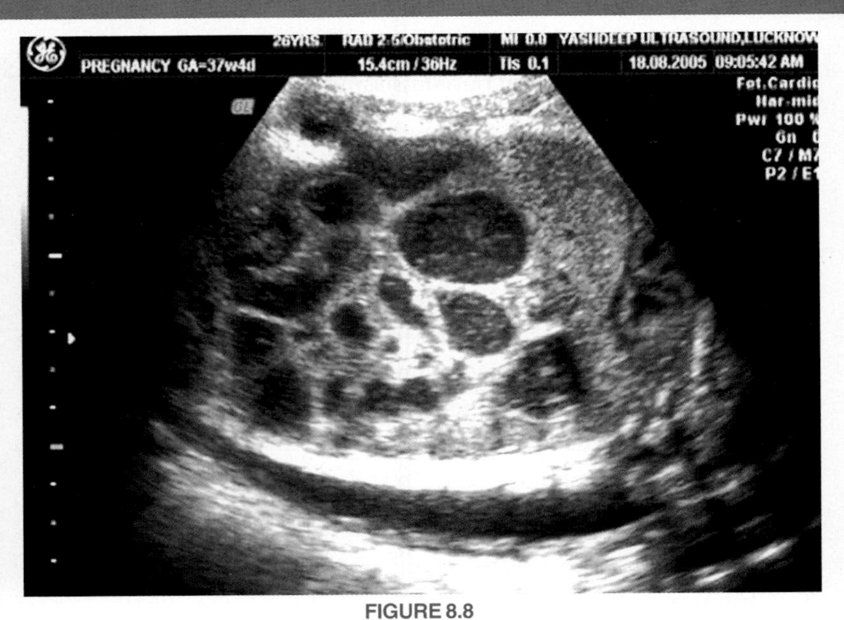

FIGURE 8.8

FIGURES 8.7 and 8.8: Distended stomach is also seen in the fetal abdomen with fluid stagnation. The distended bowel loops are well appreciated with fine echogenic bands in the loops.

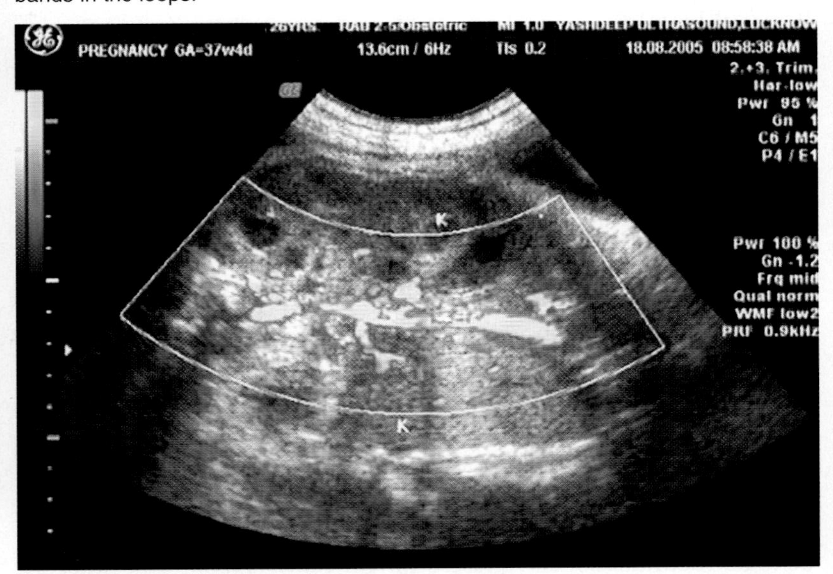

FIGURE 8.9: The dilated bowel loops at times are confused with hydronephrotic kidneys. Color Doppler imaging shows normal kidneys in the same patient.

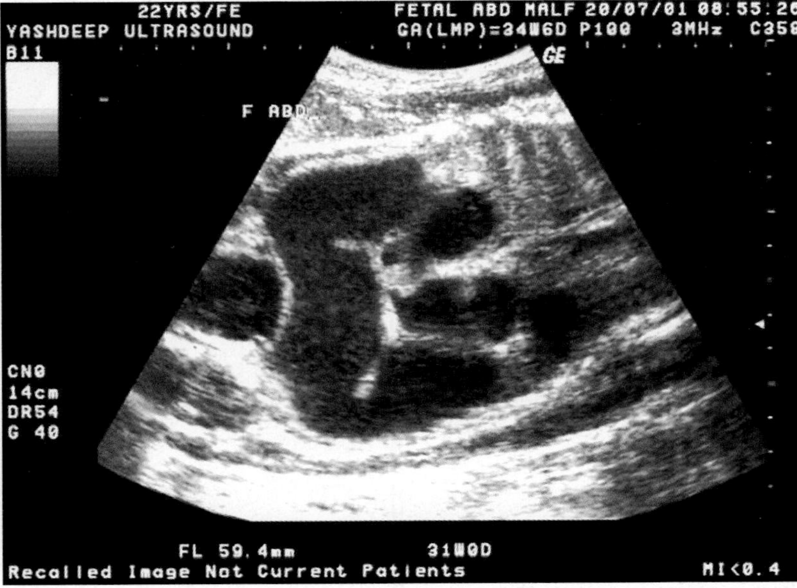

FIGURE 8.10

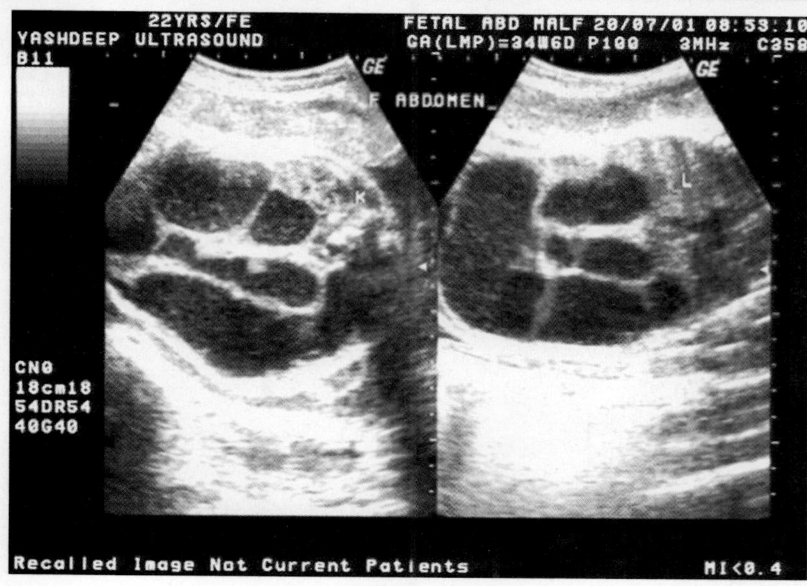

FIGURE 8.11

FIGURES 8.10 and 8.11: Anorectal atresia. HRSG shows grossly dilated large bowel loop in fetal abdomen. The distended bowel loops are seen occupying the 2/3rd part of fetal abdomen in a case of anorectal atresia.

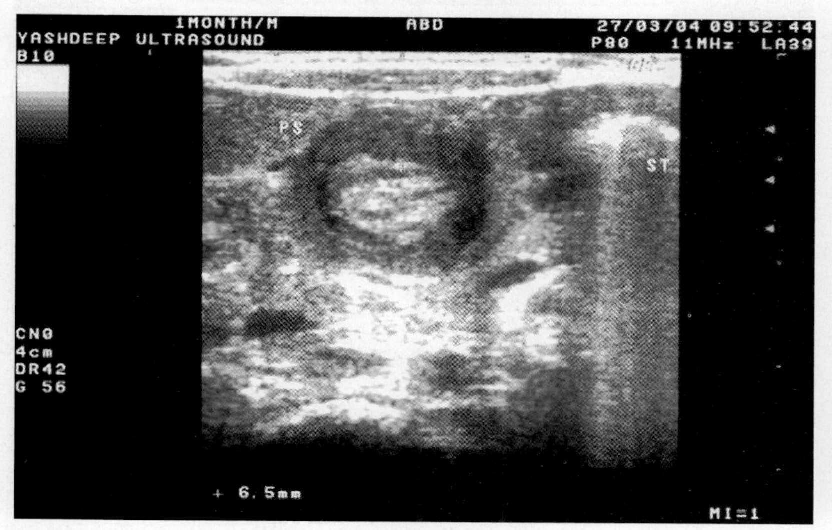

FIGURE 8.12

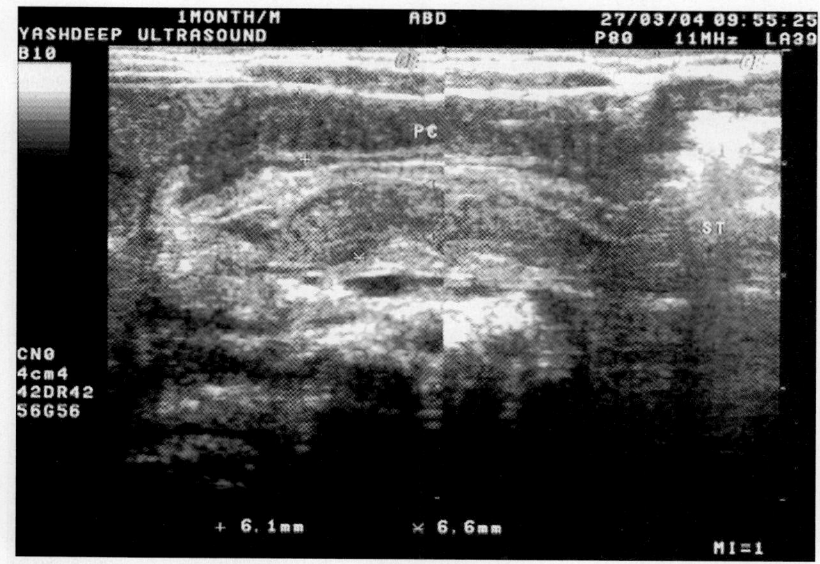

FIGURE 8.13

FIGURES 8.12 and 8.13: Hypertrophic pyloric stenosis. HRSG shows markedly thickened pyloric muscle in a 1-month-old infant. The pyloric muscle is 6 mm in thickness. Pyloric canal is collapsed and a thick mucous line is seen in the pyloric canal. In the transverse axis, typical target sign is seen on HRSG.

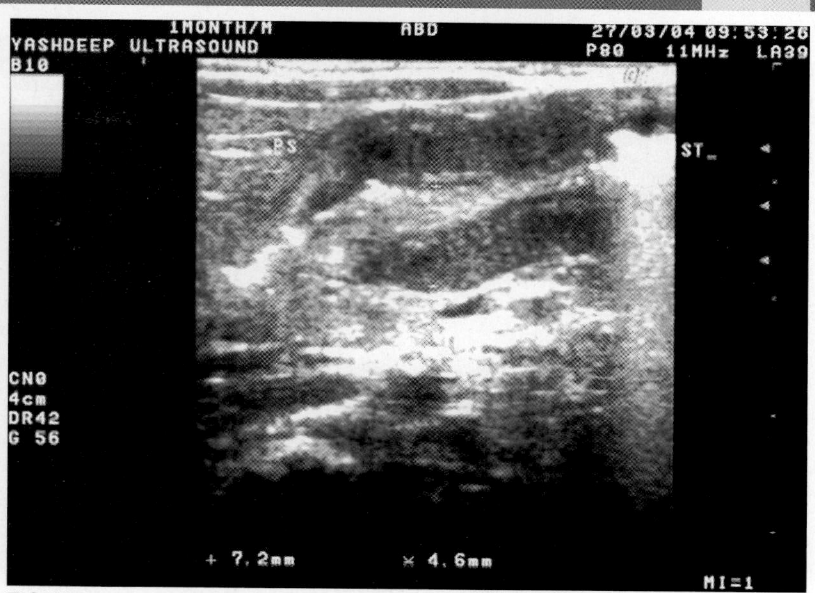

FIGURE 8.14: Longitudinal scan shows the collapsed pyloric canal. A column of air is seen entrapped in the obliterated canal. The pyloric muscle is markedly thickened.

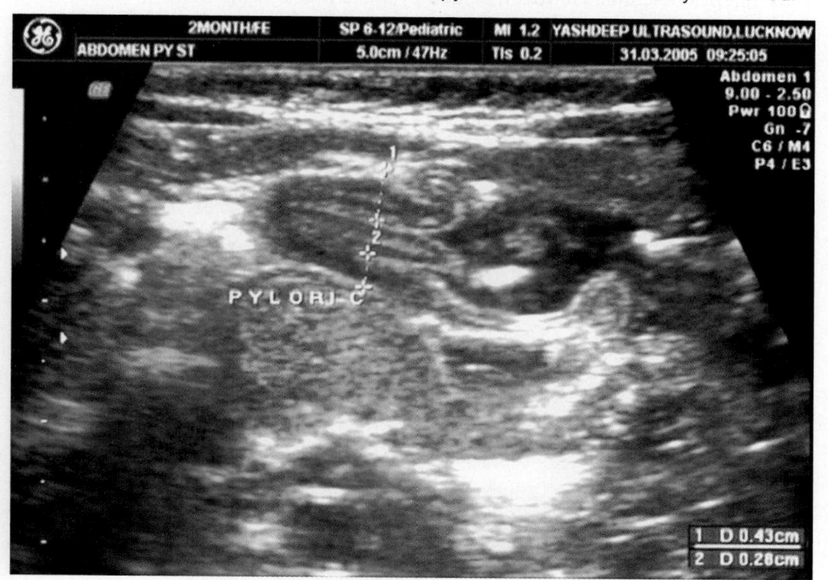

FIGURE 8.15

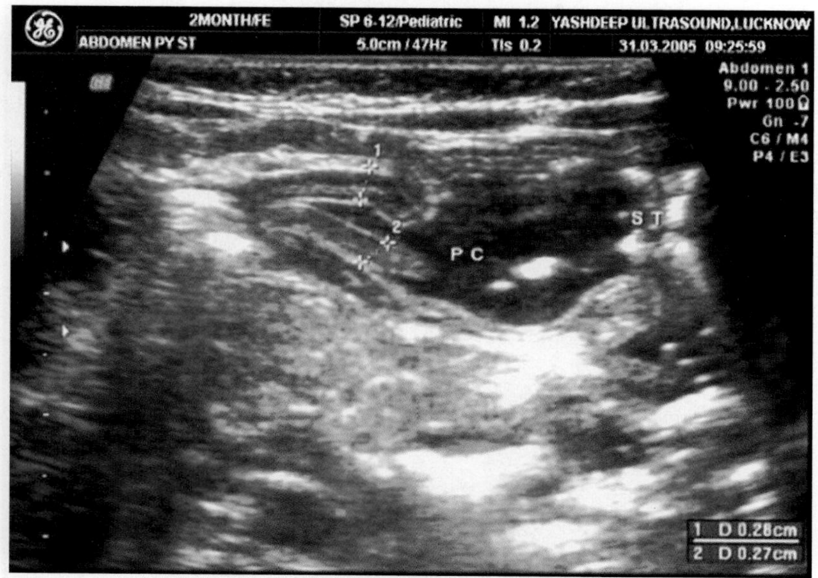

FIGURE 8.16

FIGURES 8.15 and 8.16: Moderate muscular hypertrophy with pylorospasm. HRSG shows moderate pyloric muscle hypertrophy with persistent pyloric spasm partial opening of the pyloric canal is seen. Small fluid is seen in the canal.

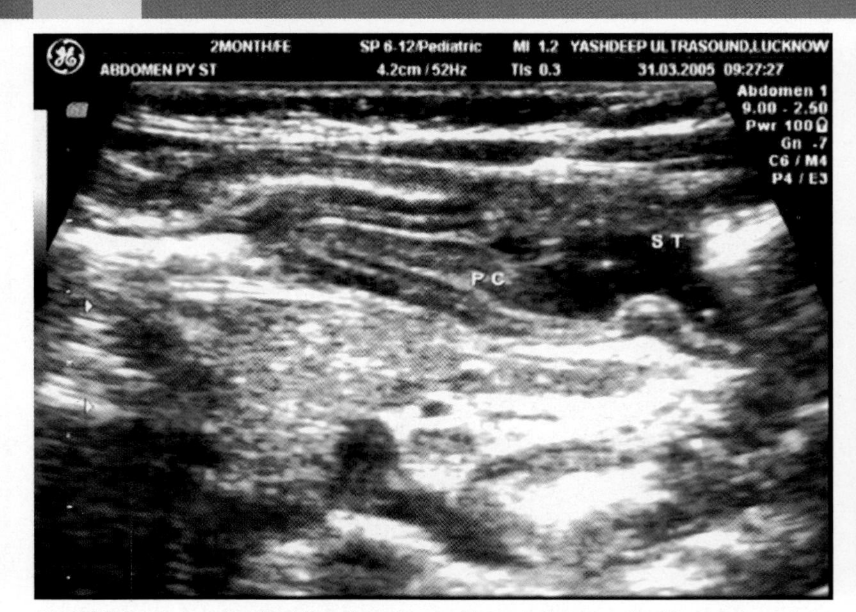

FIGURE 8.17: Longitudinal imaging shows distended stomach with opening of the proximal pyloric canal. However, distant pyloric canal is obliterated.

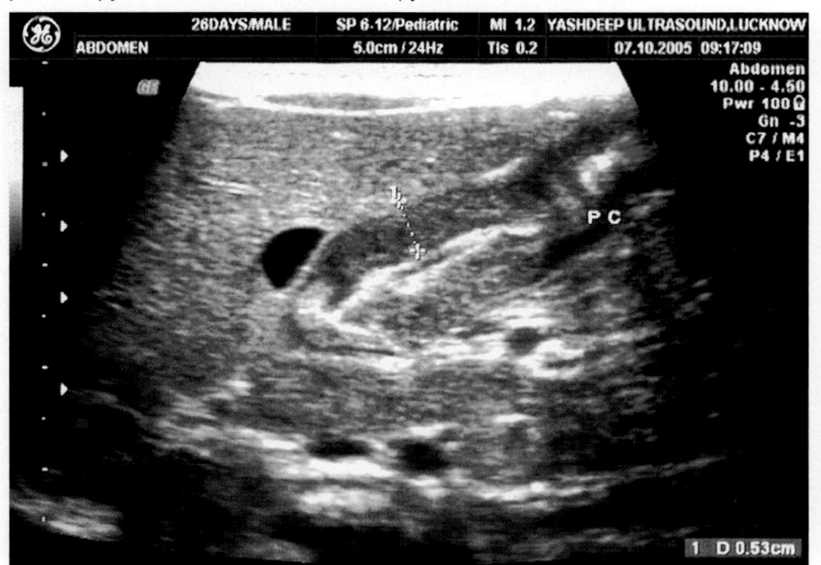

FIGURE 8.18

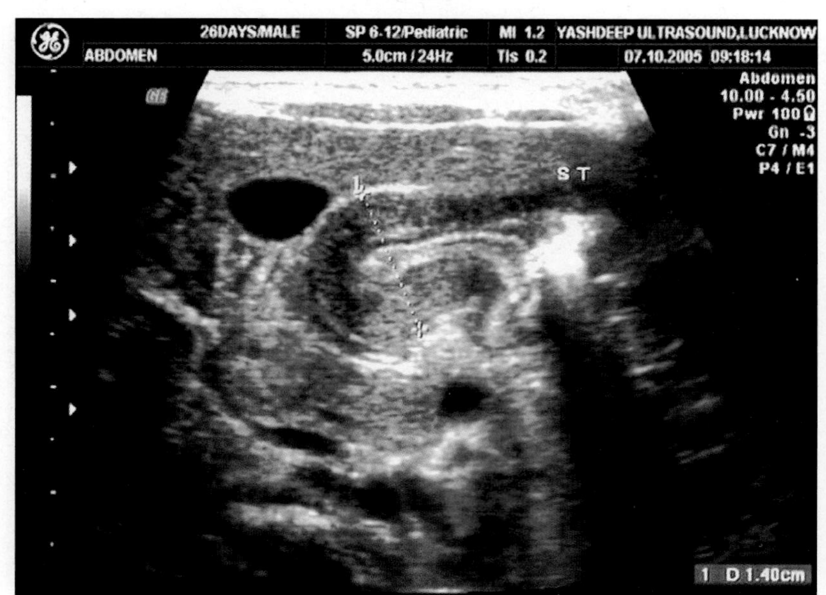

FIGURE 8.19

FIGURES 8.18 and 8.19: Hypertrophic pyloric stenosis. Longitudinal and transverse scan show mark thickening of the pyloric muscle. Transverse image shows typical doughnut sign.

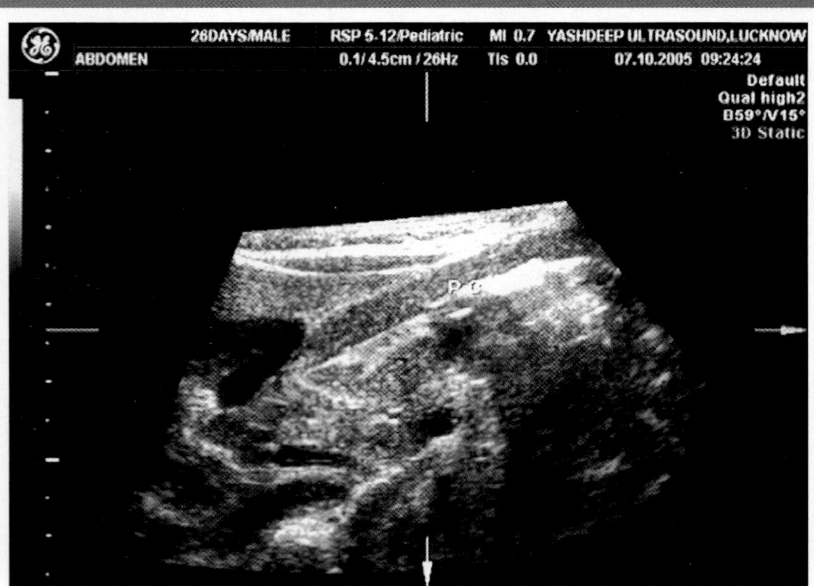

FIGURE 8.20: 3D imaging of the pyloric canal shows details of the hypertrophy of the pyloric muscles.

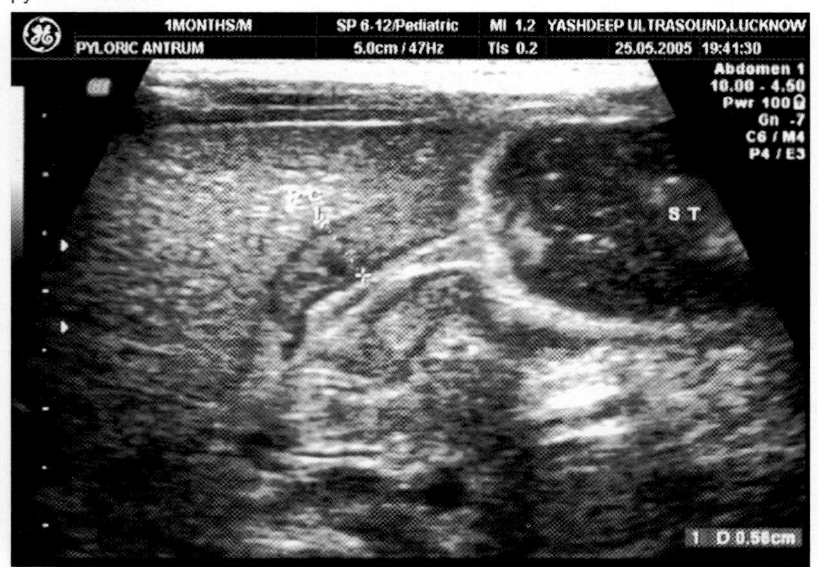

FIGURE 8.21: Gross pyloric muscle hypertrophy. HRSG shows gross hypertrophy of the muscle. The pyloric canal is seen obliterated in it proximal part. The stomach is seen distended with sudden obliteration of the proximal part of pyloric canal with marked beaking.

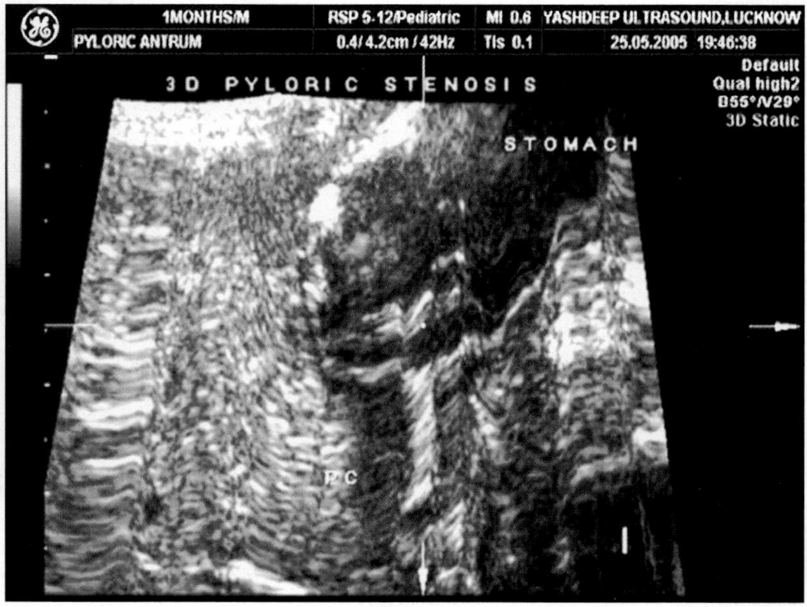

FIGURE 8.22

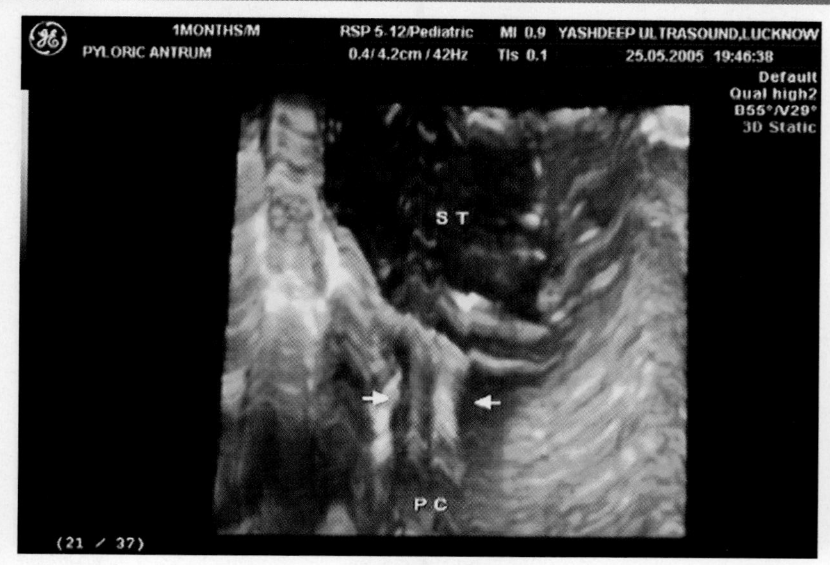

FIGURE 8.23

FIGURES 8.22 and 8.23: 3D imaging through the pyloric canal and stomach shows the details of the hypertrophy of the pyloric muscle and stomach. The distended stomach show thickened mucosal folds. The pyloric muscle hypertrophy is well appreciated on 3D. The obliterated canal is also seen beautifully in the cut section on 3D imaging.

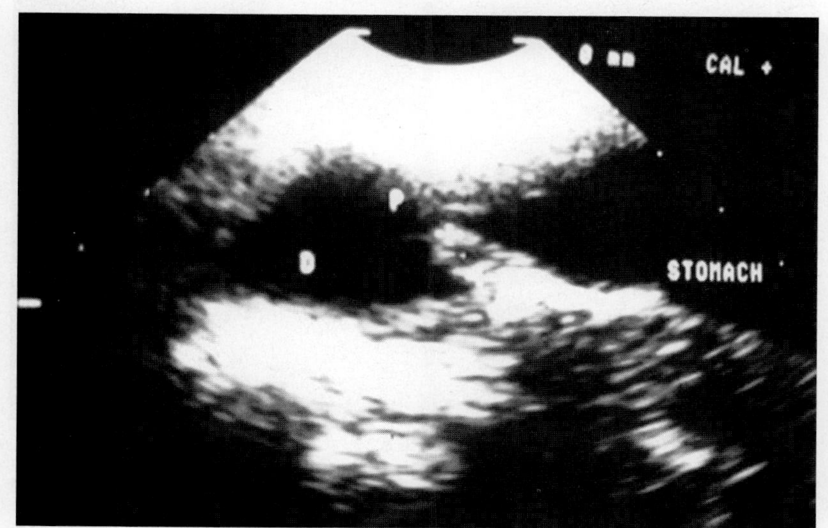

FIGURE 8.24: Gastric diaphragm. Echogenic bulging membrane is seen in the antrum before the pylorus suggestive of a gastric diaphragm. The fold is seen partially obliterating the pyloric canal. First part of duodenum is seen distended.

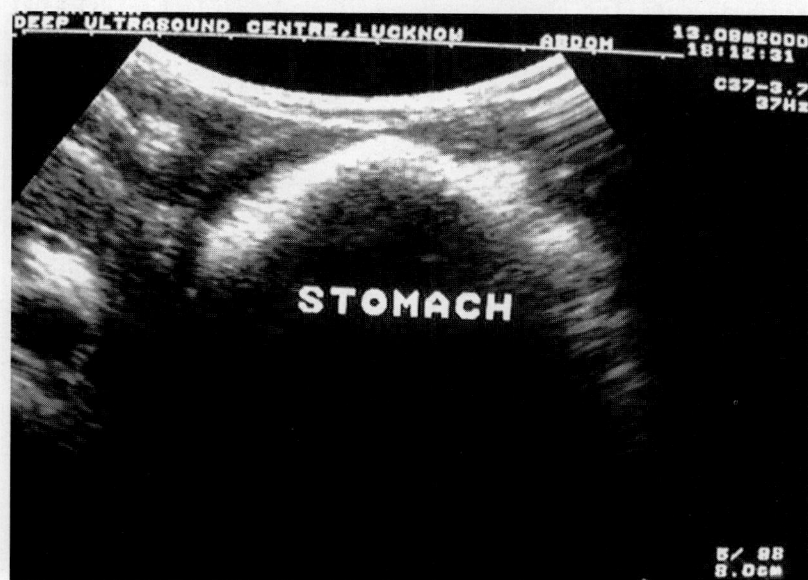

FIGURE 8.25

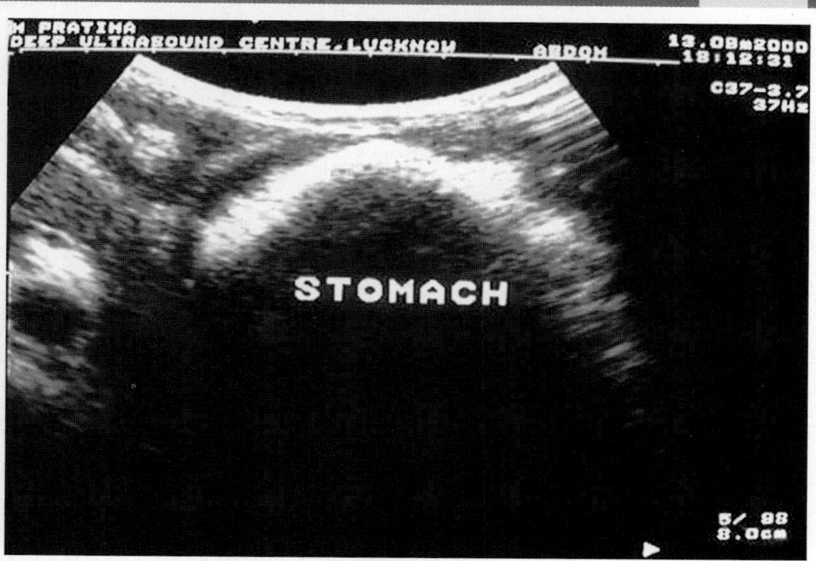

FIGURE 8.26

FIGURE 8.25 and 8.26: Trichobezoar. A young girl presented with history of loss of appetite and vomiting. HRSG shows a dense echogenic mass filling the stomach cavity. Strong acoustic shadowing is seen posterior to the mass and masking the posterior wall of the stomach suggestive of calcification over the mass. The mass was a Trichobezoar on endoscopic evacuation.

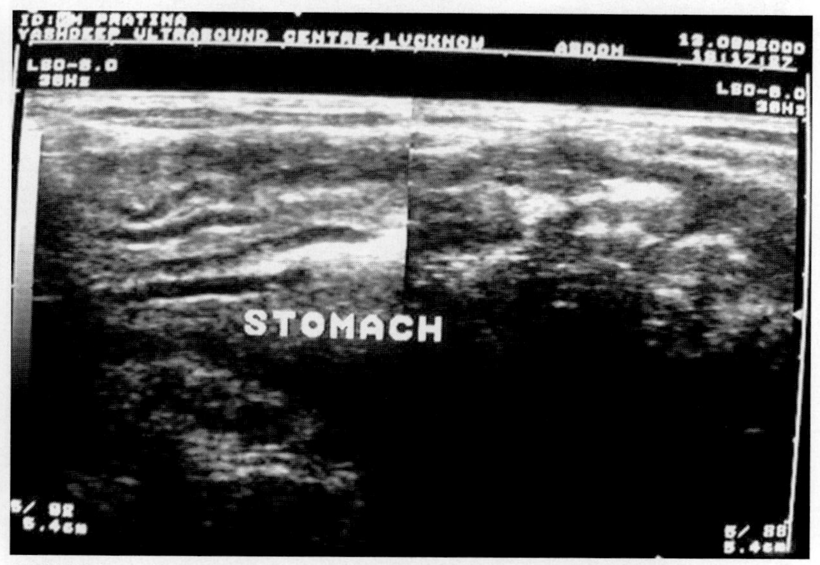

FIGURE 8.27: Trichobezoar. There is evidence of displacement of gas seen in the fundal part of the stomach. Irregular thickening of the muscle wall is also seen due to the chronic inflammation, in the same case.

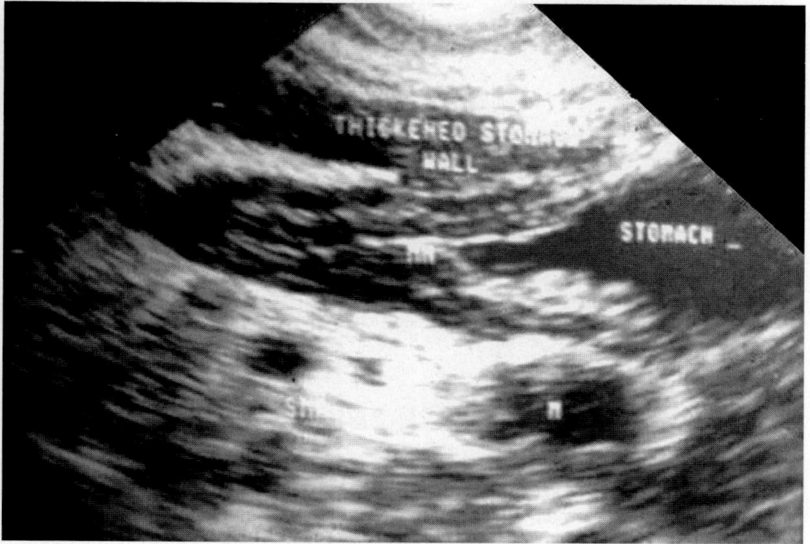

FIGURE 8.28

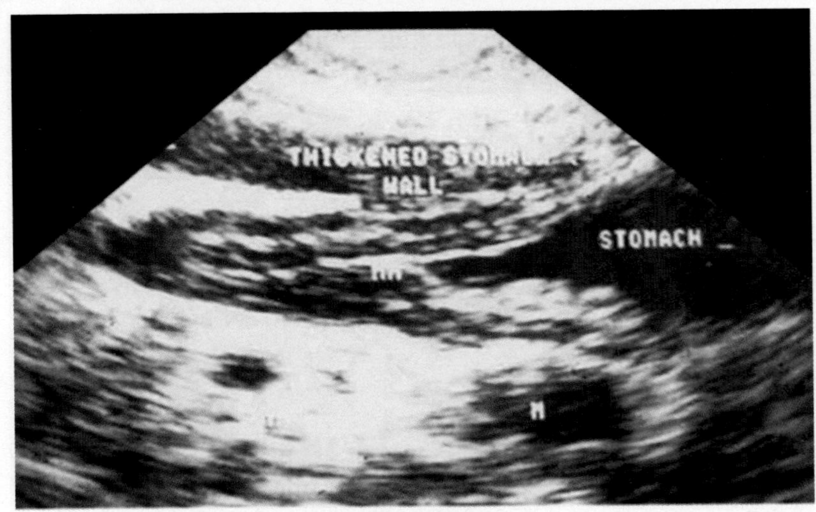

FIGURE 8.29

FIGURES 8.28 and 8.29: Carcinoma stomach. HRSG shows marked irregular thickening of the stomach wall seen involving the lower part of the stomach. The stomach cavity is reduced suggestive of malignancy. A hypoechoic nodular shadow is also seen in the stomach bed suggestive of enlarged lymph nodes.

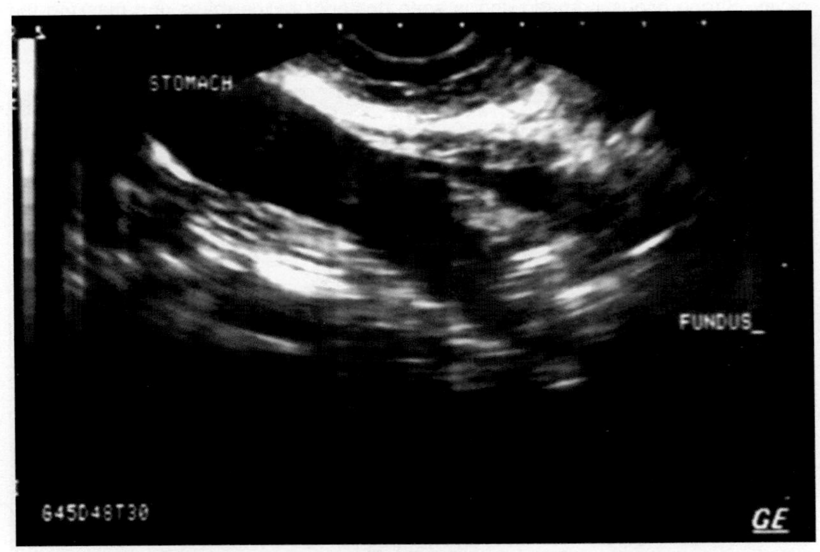

FIGURE 8.30

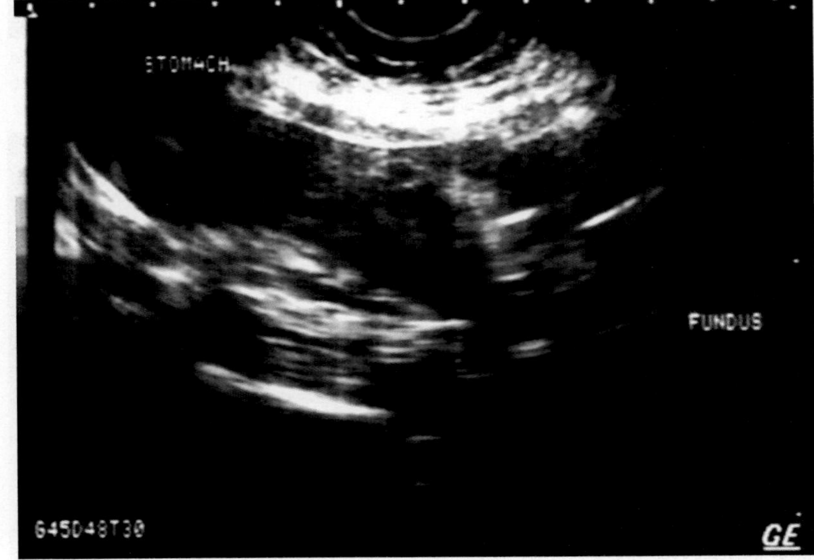

FIGURE 8.31

FIGURES 8.30 and 8.31: Carcinoma stomach linitis plastica. HRSG shows markedly thickened stomach wall. The cavity is obliterated. Irregular thickening of the stomach wall is seen with grossly hypoechoic texture. The patient complained of marked reduction in the appetite. Biopsy confirmed linitis plastica.

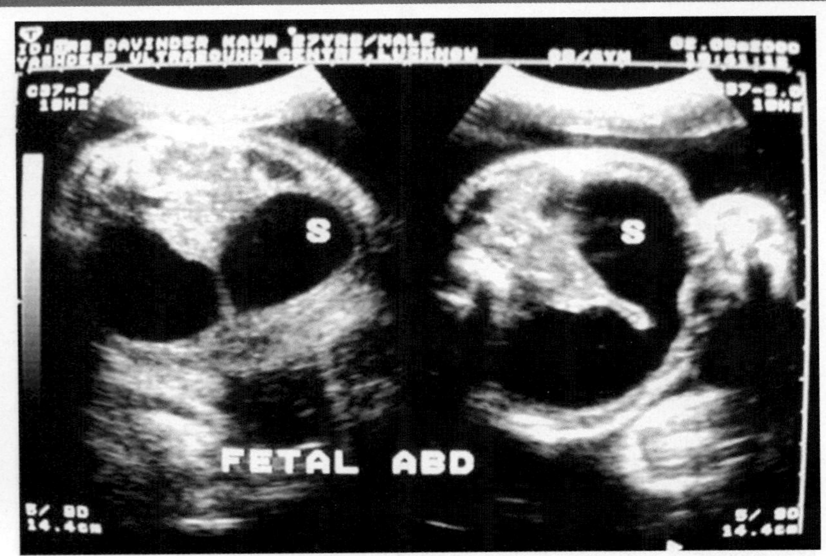

FIGURE 8.32

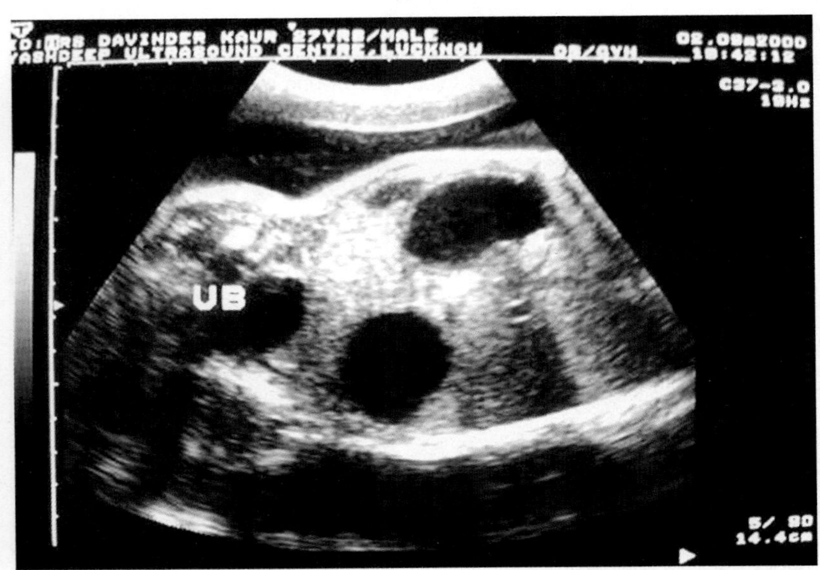

FIGURE 8.33

FIGURES 8.32 and 8.33: Duodenal atresia. HRSG shows double bubble sign in fetal abdomen suggestive of duodenal atresia. The stomach and first part of the duodenum are seen distended with fluid filled structure, typical features of duodenal atresia.

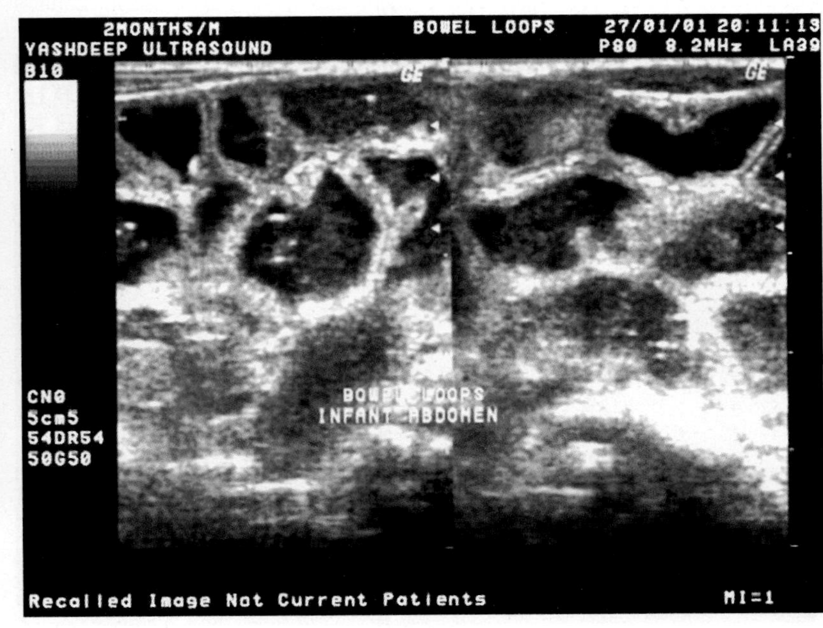

FIGURE 8.34

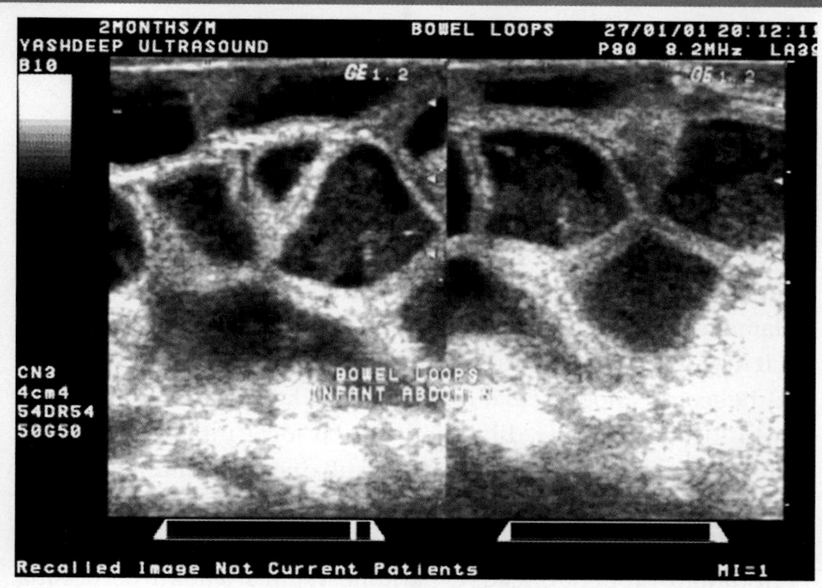

FIGURE 8.35

FIGURES 8.34 and 8.35: Mucoviscidosis. A three months old child presented with persistent vomiting and failure to thrive. HRSG shows dilated bowel loops with multiple echogenic bands crisscrossing the loops giving fisherman net appearance. Thick echogenic collection is also seen in bowel loops suggestive of cystic fibrosis (Mucoviscidosis) of the bowel. The echogenic bands are fibrosed septa. The sweat test of the same patient was positive.

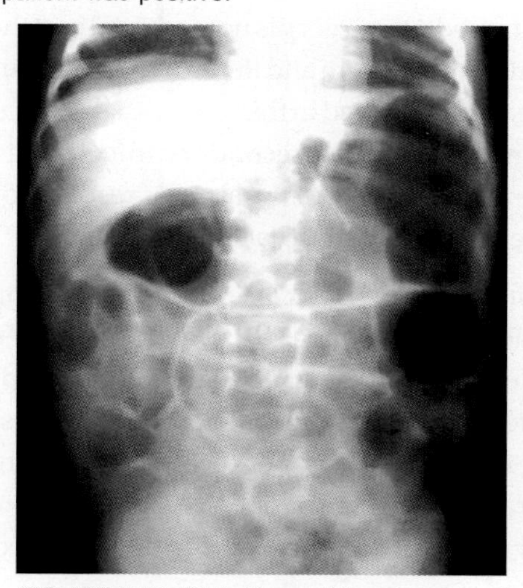

FIGURE 8.36: Plain X-ray of the abdomen of the same patient in erect position shows multiple radio opaque bands in the abdominal cavity. No fluid level is seen in the bowel loops.

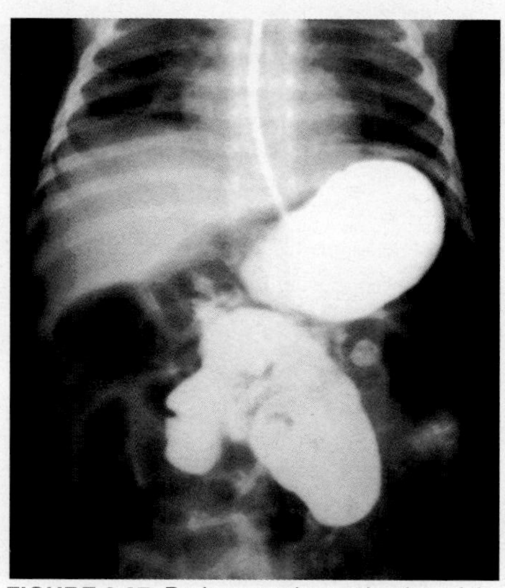

FIGURE 8.37: Barium meal examination of the patient distended stomach and duodenum.

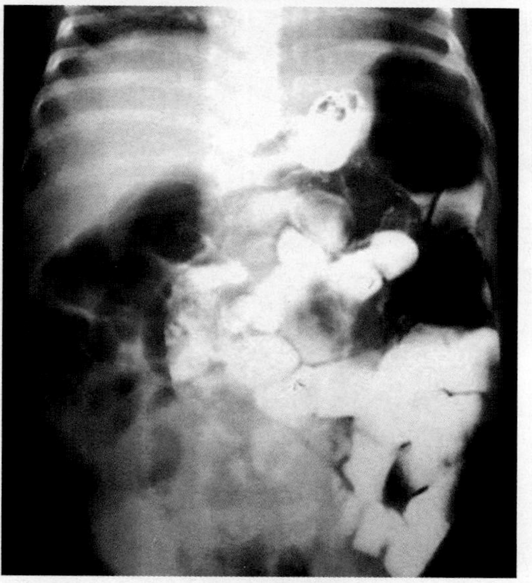

FIGURE 8.38

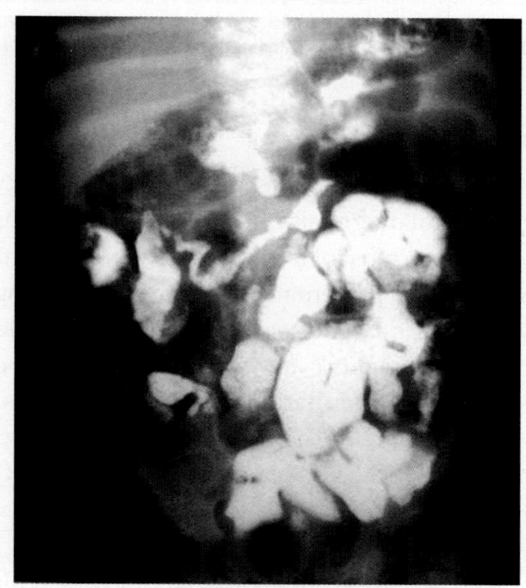

FIGURE 8.39

FIGURES 8.38 and 8.39: Barium meal follow through examination shows multiple bands in the bowel loops causing the indentation and constriction of bowel loops.

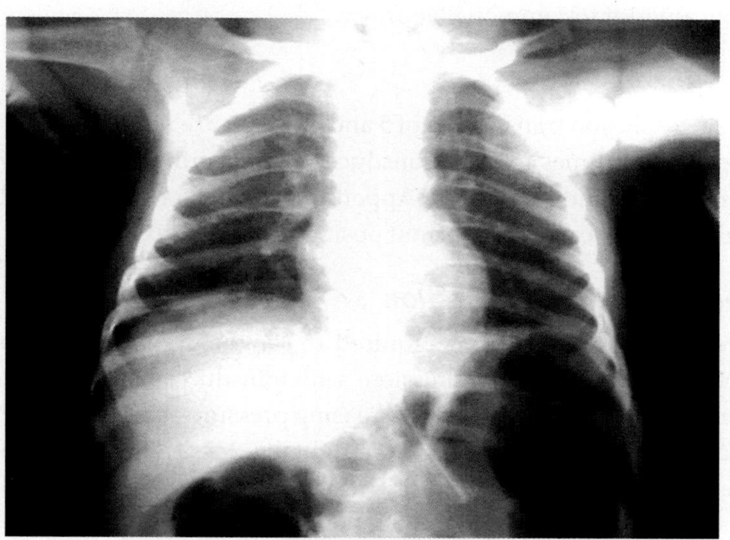

FIGURE 8.40: Plain X-ray chest of the patient does not show any sign of cystic fibrosis in the lung.

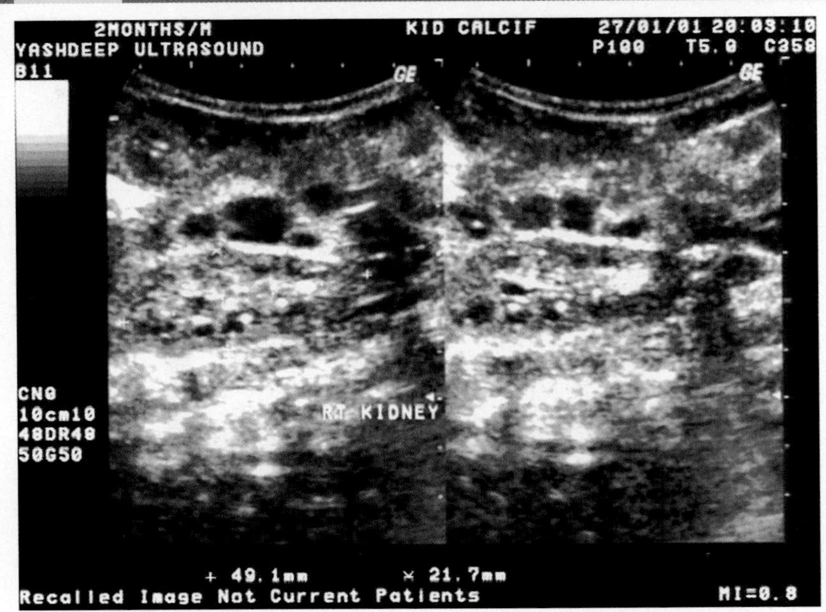

FIGURE 8.41: Abdominal ultrasound of the kidney shows multiple calcified specks in the medulla suggestive of medullary calcification in the same patient.

Appendix

Acute appendicitis is one of the most common causes of acute abdomen. Although the clinical diagnosis of classical acute appendicitis can be made easily, however, most of the time the patient presents with vague symptoms and atypical presentation. In such settings, high-resolution sonography (HRSG) is a good tool to study appendicular pathology. The hallmark of the modality is to visualize inflamed appendix directly and also to detect any other pathology, which mimicks appendicitis clinically. Appendix can be visualized in almost every position except when it is present retrocaecal, it cannot be seen on HRSG.

Normal Anatomy

Appendix is seen as a blind ended aperistaltic tube with typical "gut signature" in appendicular fossa. The normal appendix is usually not visualized. However, in few patients, it can be seen as a thin compressible blind-ended tubular structure. Its wall obliterate with graded compression, as it is easily compressible, mobile tube. It is seen lying anterior to common iliac artery.

Equipment

High-resolution transducers of 5 and 7.5 MHz are required to see the appendix. At times 10 MHz transducer is also required to see it in very thin patients and in children. Appendix can be seen in every position except in retrocaecal or aberrant position of appendix.

Technique and Preparation

No specific preparation is required to see the appendix. A special technique of "graded compression" with transducer head is applied to see the appendix with slowly increasing pressure applied on anterior abdominal wall in appendicular or right iliac fossa. The fat and bowel are displaced and compressed. This reduces the artifacts from bowel gas and its contents. It also brings transducer near to the appendix. Full urinary bladder displaces bowel loops in appendicular fossa; therefore, urinary bladder should be empty prior to examination. Gentle graded compression should be applied to avoid pain.

Acute Appendicitis

The biggest advantage of HRSG is the direct visualization of inflamed appendix, which appears to be a thick-walled incompressible, sausage-like tube with concentric layers on HRSG. It is an elongated tube with one end blind. When wall thickness exceeds 6 mm, it is suggestive of inflammation. Many a times, fecalith is seen in the lumen of appendix.

It may the cause of obstruction. A small amount of fluid collection may be seen in appendicular fossa due to inflammation. The adjacent bowel loop may show dilatation due to local inflammation and irritation. Local adynamic ileus may result due to inflammation.

Appendiceal Mass

Perforation of appendix leads to peritoneal spill. It leads to peritonitis. The spill of pus may be limited due to sealing of perforation by omentum, mesentery or bowel loops. They all form a mass due to adhesion. HRSG shows a non-compressible mass. It is predominantly echo poor with heterogeneous texture. However, secondary infection may lead to an abscess formation with fluid collection and debris. When appendiceal mass is formed, immediate surgery is not required and these patients are treated conservatively. Caecum and ileum also show muscle thickening and they can be easily identified in thin patients. An abscess can develop in appendiceal mass following secondary infection. HRSG can very well identify the abscess. Most of the abscess resolves on conservative treatment, or they may drain in adjacent bowel loops.

Differential diagnosis: Many clinical conditions may present as acute appendicitis. They may include right ureteric colic, perforated peptic ulcer, right sided twisted ovarian cyst, right sided ectopic tubal pregnancy, acute cholecystitis or right iliac fossa abscess. Therefore, in 50% cases presented as acute appendicitis, the reason is found to be something else rather than acute appendicitis. Therefore, HRSG is also helpful in diagnosing the above conditions by excluding acute appendicitis.

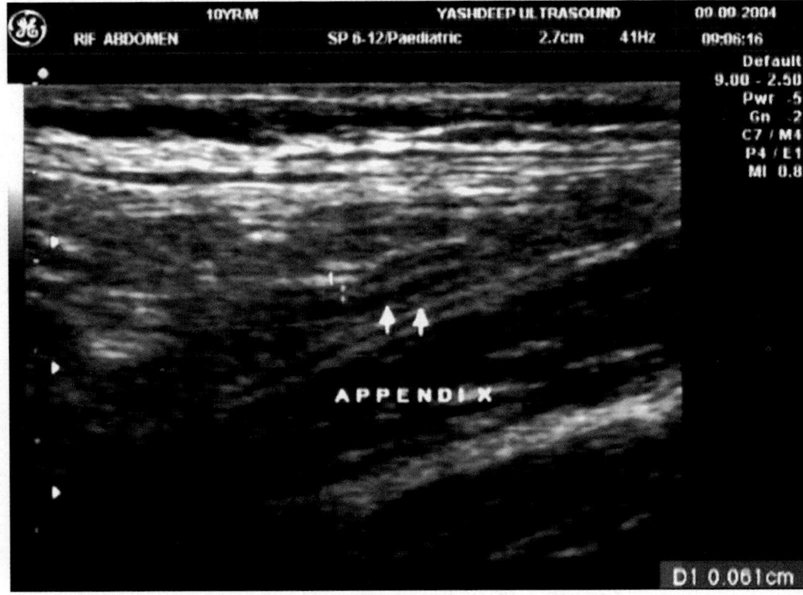

FIGURE 8.42

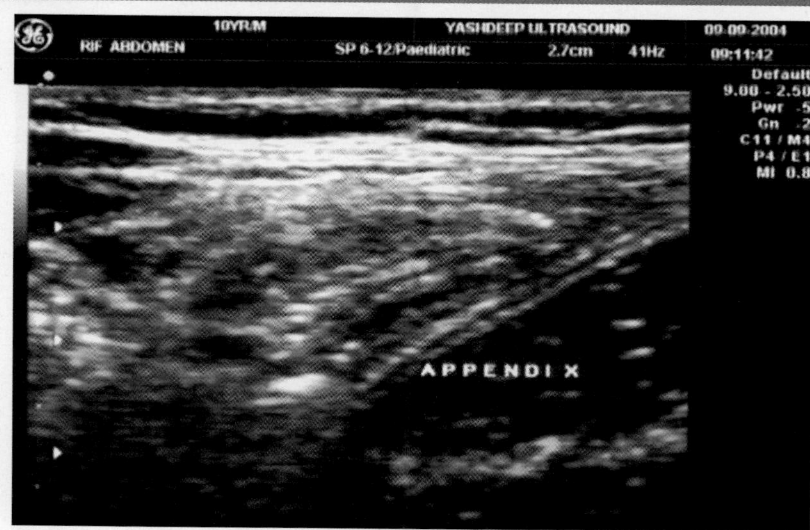

FIGURE 8.43

FIGURE 8.42 and 8.43: Normal appendix. HRSG shows a normal appendix in appendicular fossa. Appendix is seen as an echogenic tubular shadow having blind one end. The lumen is anechoic. Normal wall thickness is 3 mm.

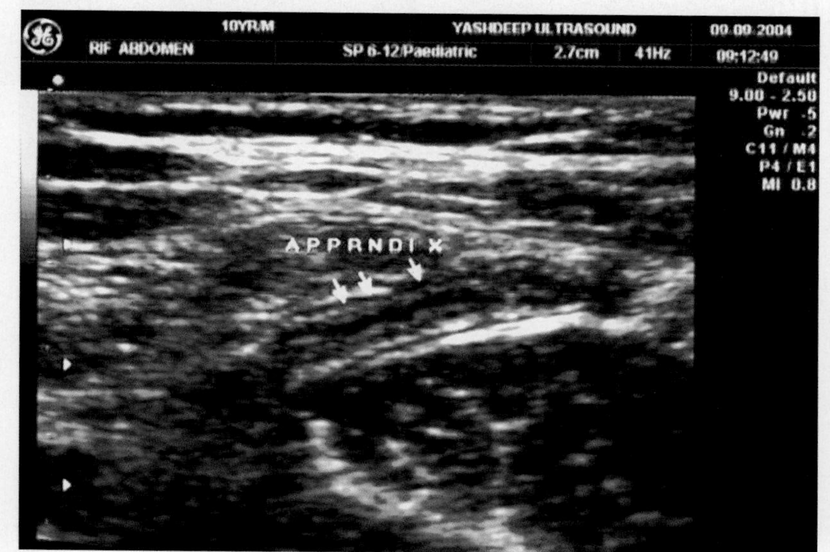

FIGURE 8.44: Normal appendix. HRSG shows normal appendix in the Rt iliac fossa. It is seen anterior to the iliac vessels. Walls are normal. Lumen is free. No internal echo is seen in the lumen.

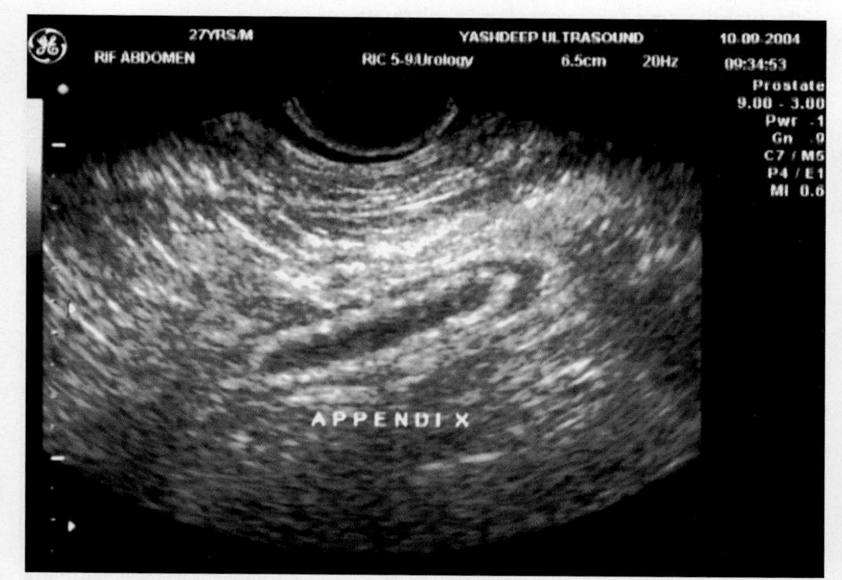

FIGURE 8.45: Acute appendicitis. HRSG shows inflamed thickened edematous appendix anterior to the iliac vessels. The appendix is seen as thickened echogenic tubular shadow. Small amount of fluid is also seen in appendicular fossa suggestive of reactionary fluid collection.

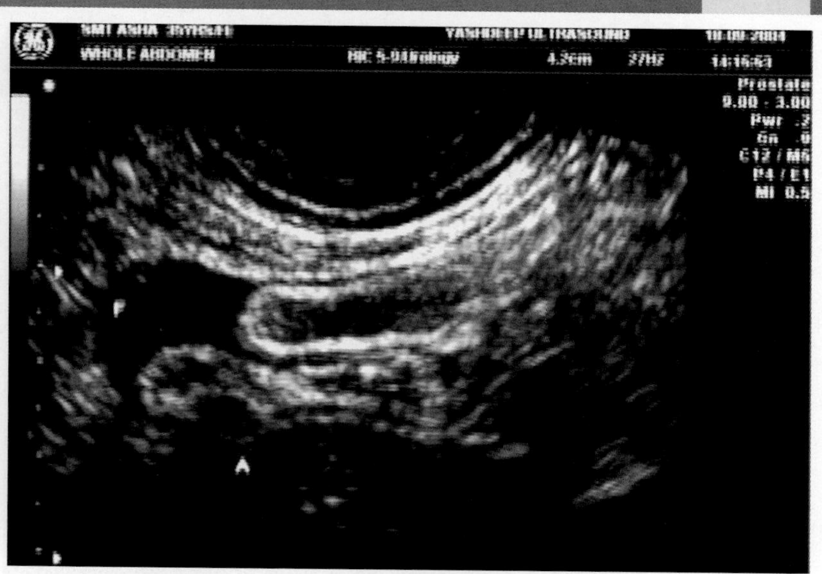

FIGURE 8.46

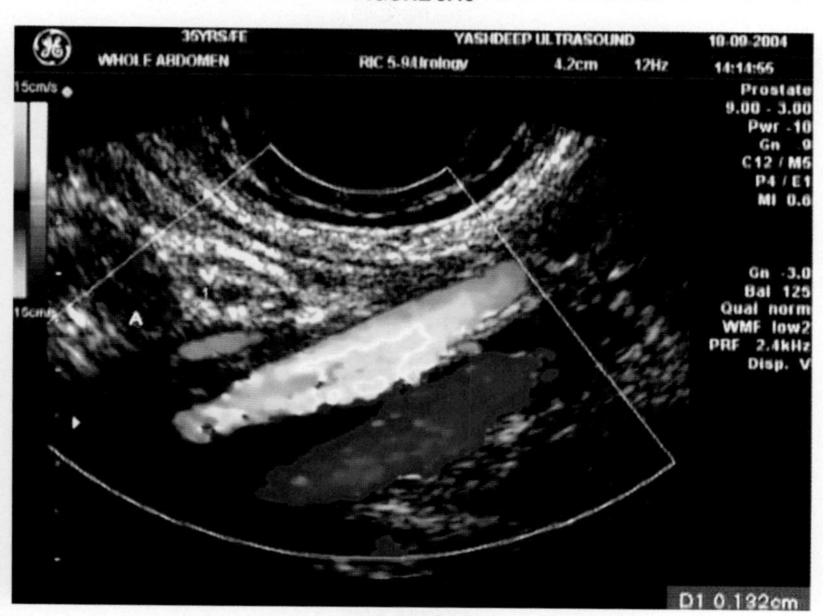

FIGURE 8.47

FIGURES 8.46 and 8.47: Acute appendicitis. The appendix is seen as an inflamed edematous tubular shadow in appendicular fossa in a young girl. Wall edema is also present. Perifocal edema is present. Low-level echo collection is seen in the lumen of the appendix. Small amount of fluid collection is seen in the appendicular fossa. Appendix is seen to the iliac vessels.

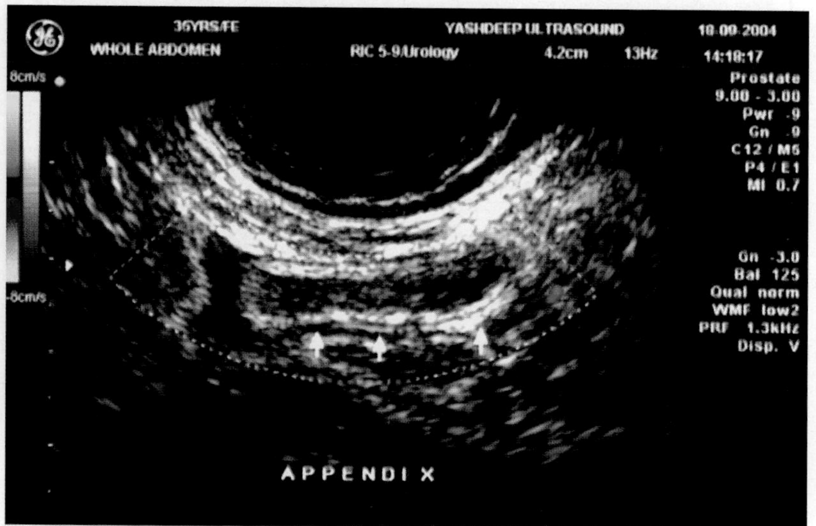

FIGURE 8.48: HRSG also shows subserosal edema in the wall of appendix.

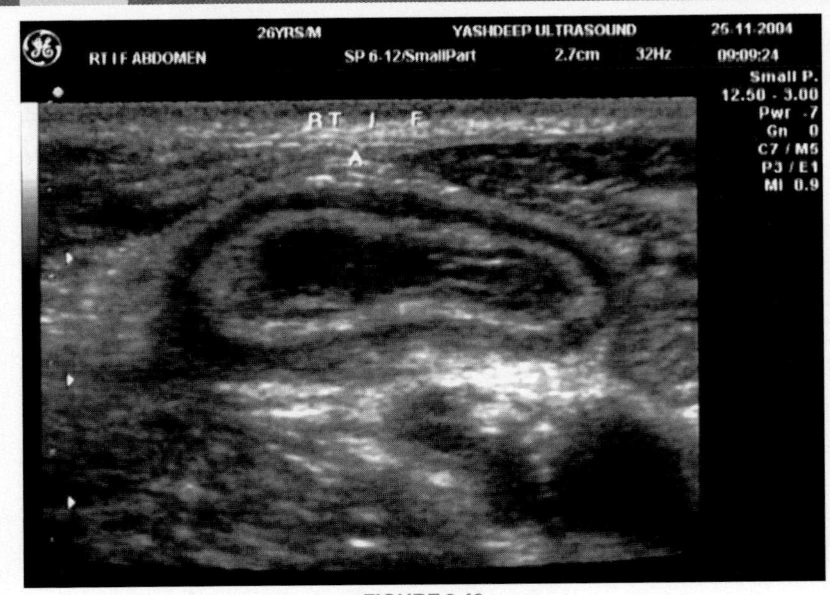

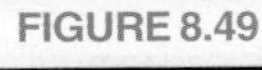
FIGURE 8.49

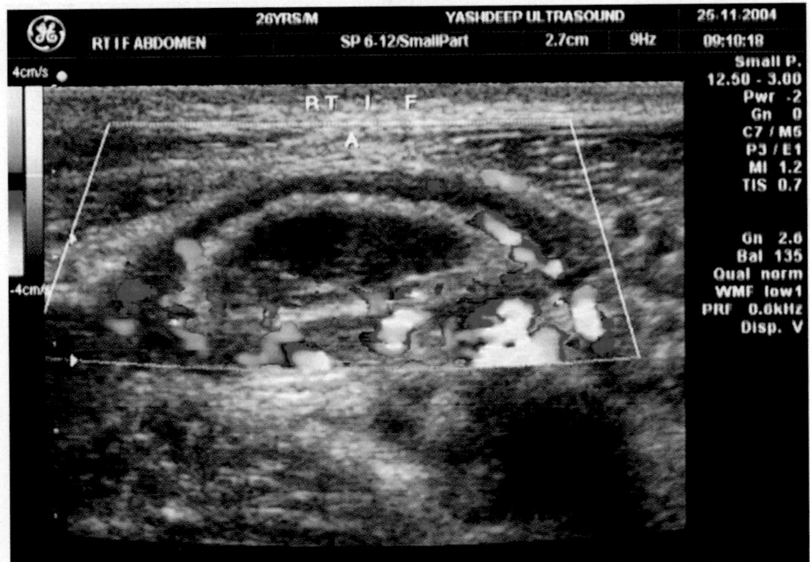

FIGURE 8.50

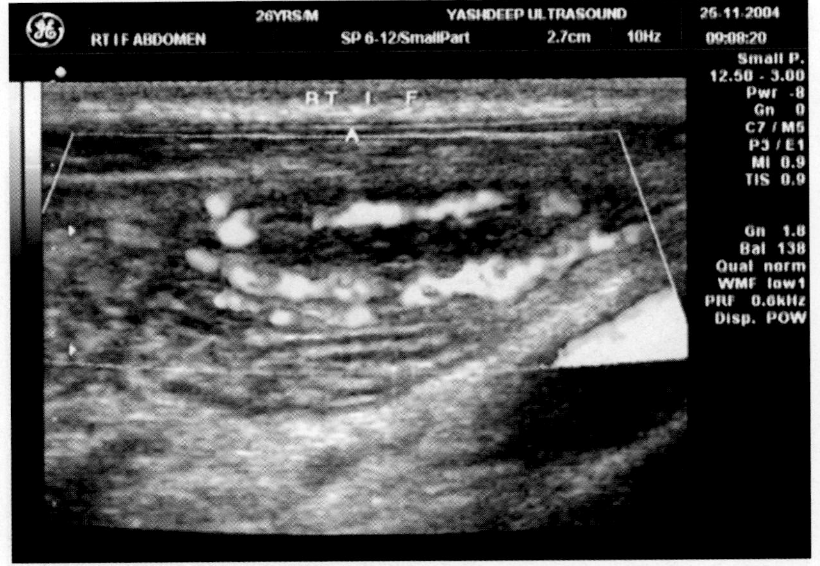

FIGURE 8.51

FIGURES 8.49 to 8.51: Acute appendicitis. HRSG shows acutely inflamed appendix. Appendicular walls are edematous. Small amount of collection is seen in appendicular fossa. On color flow imaging increased flow is seen in appendicular wall suggestive of hyperemia. Power Doppler imaging shows appendicular wall flow. It is seen as ring of fire suggestive of acute inflammation.

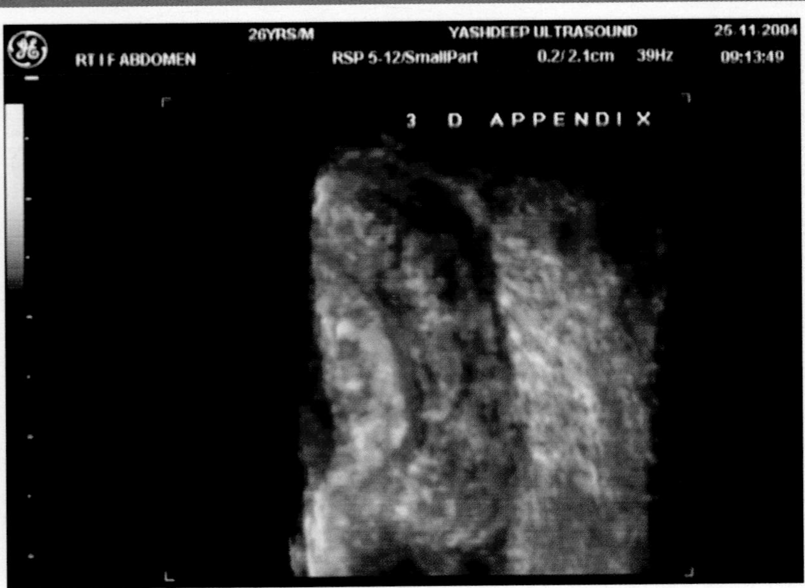

FIGURE 8.52

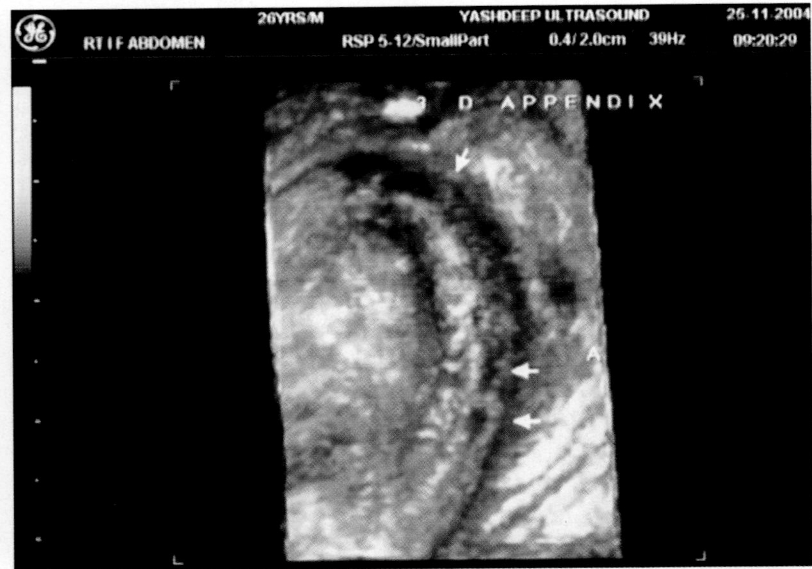

FIGURE 8.53

FIGURES 8.52 and 8.53: 3D surface rendering imaging of the appendix shows grossly edematous appendicular walls. Multiple hypoechoic areas are seen on the surface of the appendix suggestive of micro abscesses.

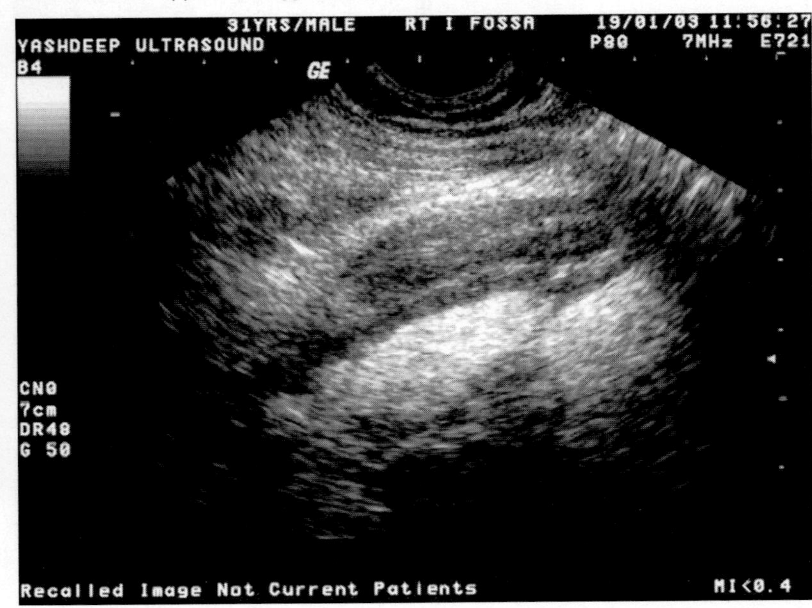

FIGURE 8.54: Acute appendicitis. HRSG shows thickened edematous tubular shadow in the appendicular fossa.

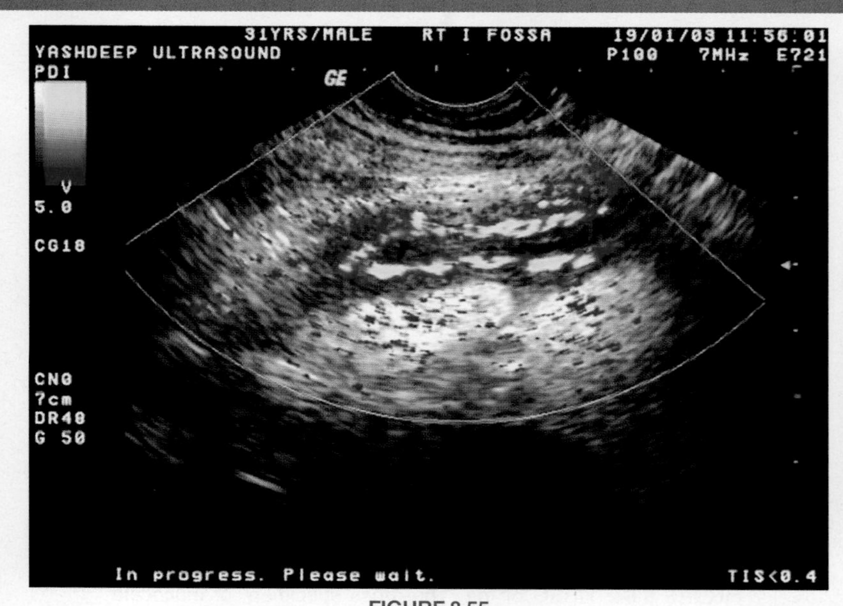

FIGURE 8.55

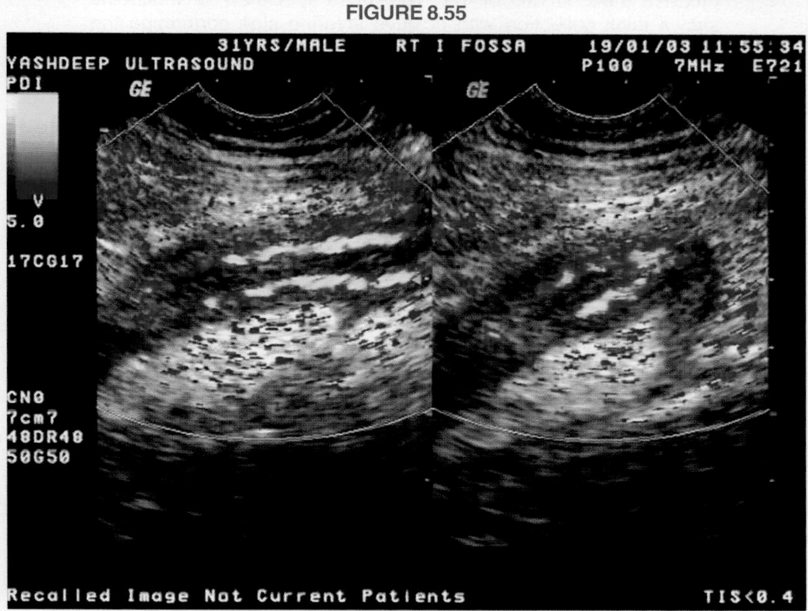

FIGURE 8.56

FIGURES 8.55 and 8.56: On color flow imaging wall hyperemia is seen. Power Doppler imaging shows positive ring of fire sign.

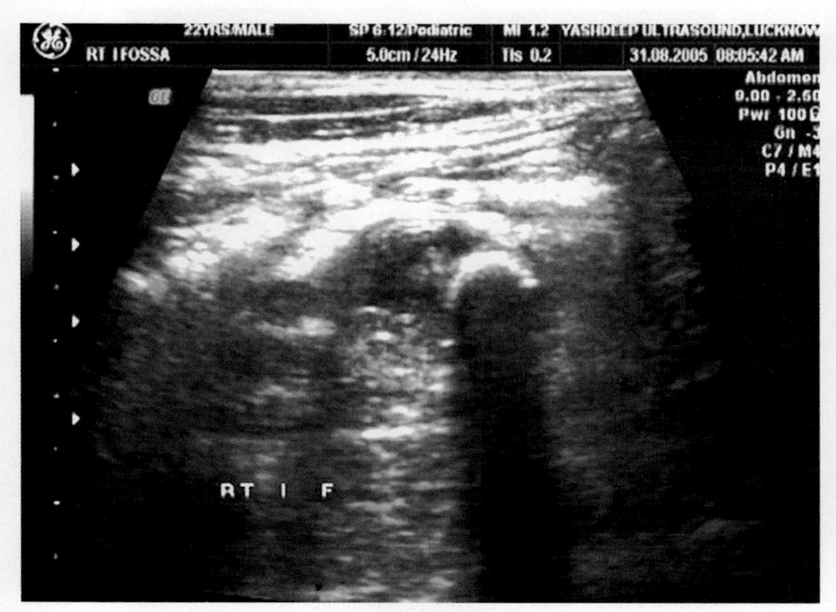

FIGURE 8.57

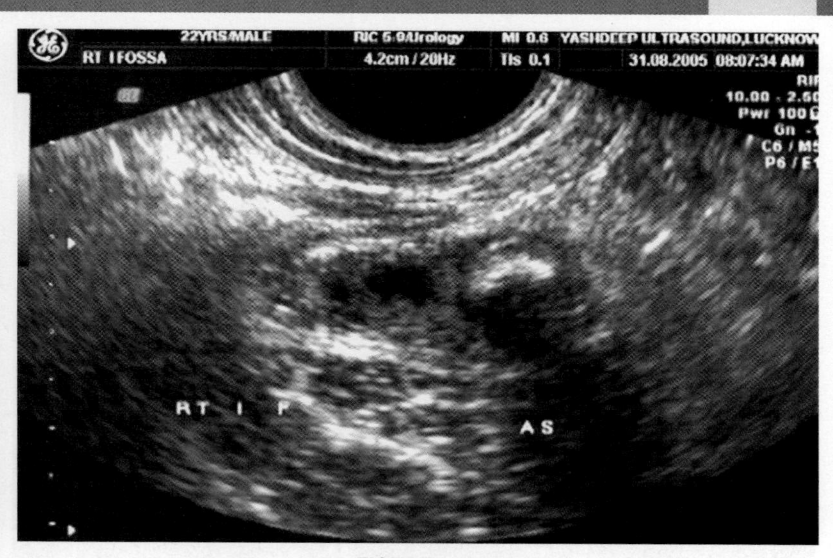

FIGURE 8.58

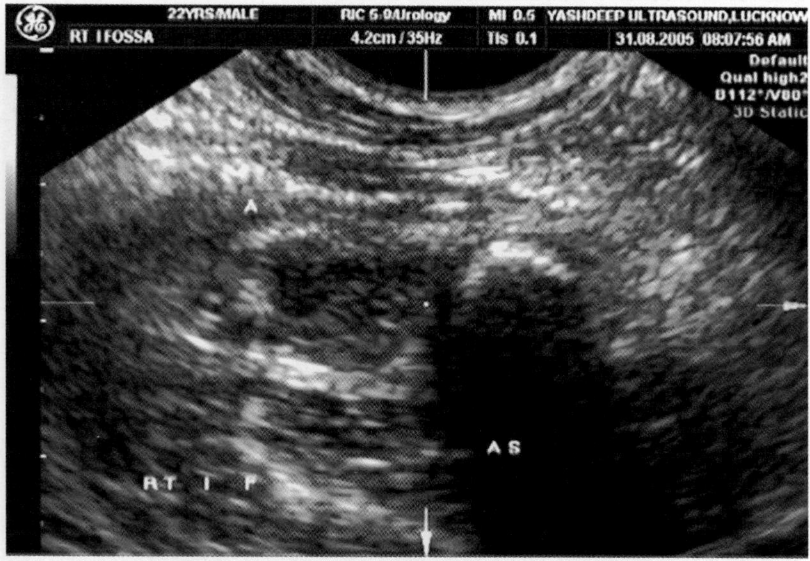

FIGURE 8.59

FIGURES 8.57 to 8.59: Obstructed appendix. HRSG shows edematous thickened appendix. The lumen is distended. Low-level echoes are seen in the lumen of appendix. Big echogenic shadow is seen blocking of the mouth of the appendix. It is accompanied with acoustic shadowing suggestive of appendicolith. Perifocal edema is also seen. Walls of appendix also show edema.

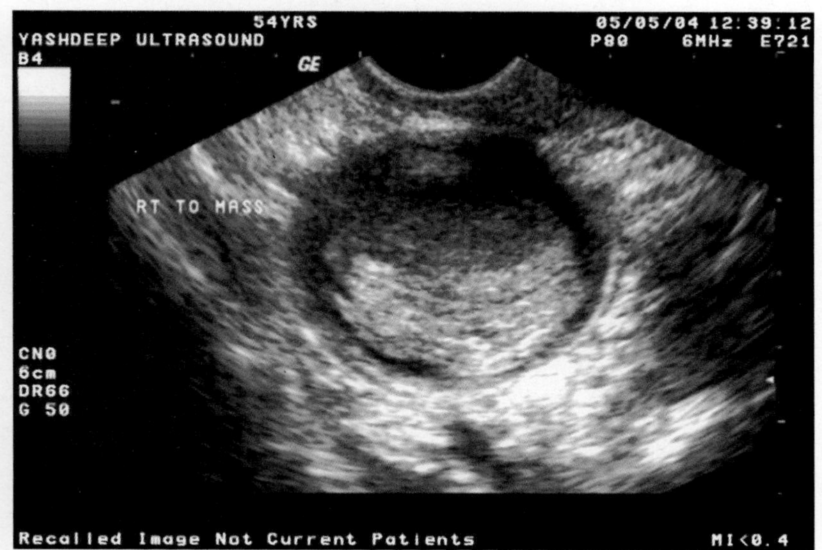

FIGURE 8.60: Mucocele of appendix. A patient presented with palpable lump in the Rt iliac fossa. It was mistaken as ovarian cyst. HRSG shows a thick walled cystic mass in cross section. Multiple internal echoes are seen in it with positive layering sign.

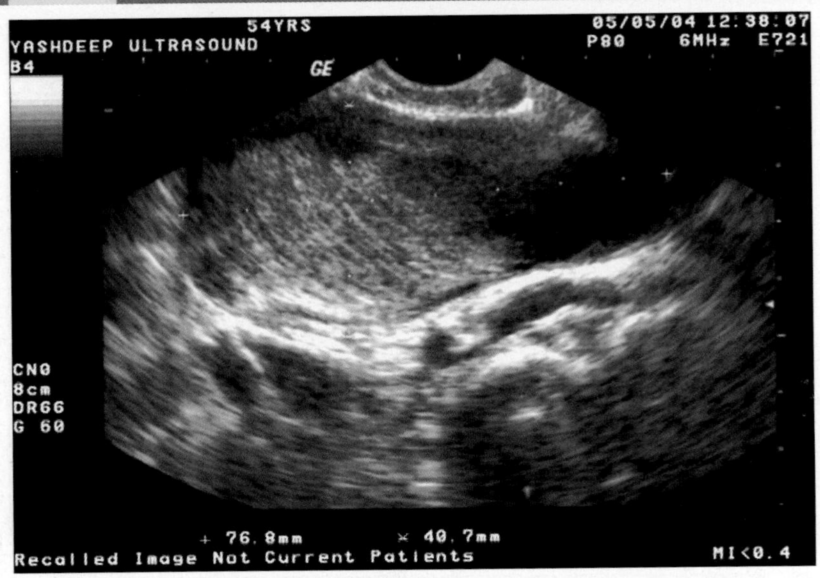

FIGURE 8.61

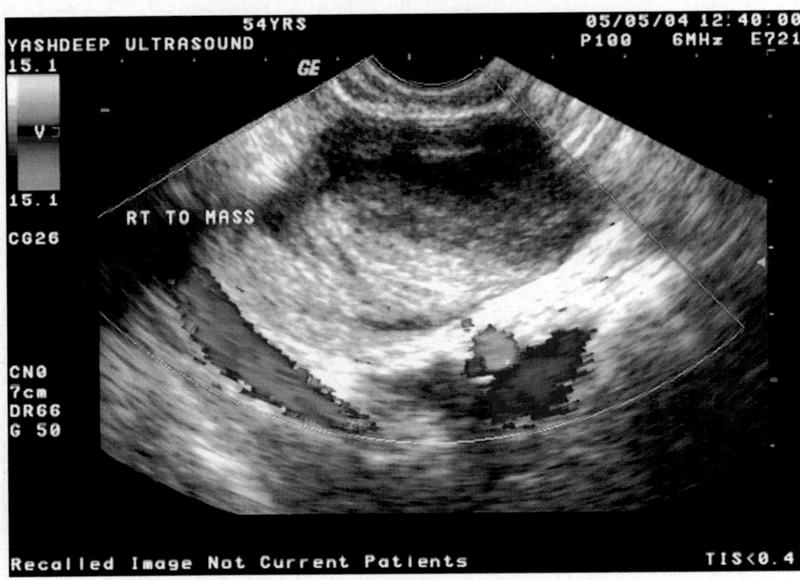

FIGURE 8.62

FIGURES 8.61 and 8.62: The longitudinal imaging shows a tubular fusiform shadow in the Rt iliac fossa. Thick debris is seen filling the lumen of it suggestive of mucocele of appendix. On color flow imaging poor flow is seen in the wall of appendix.

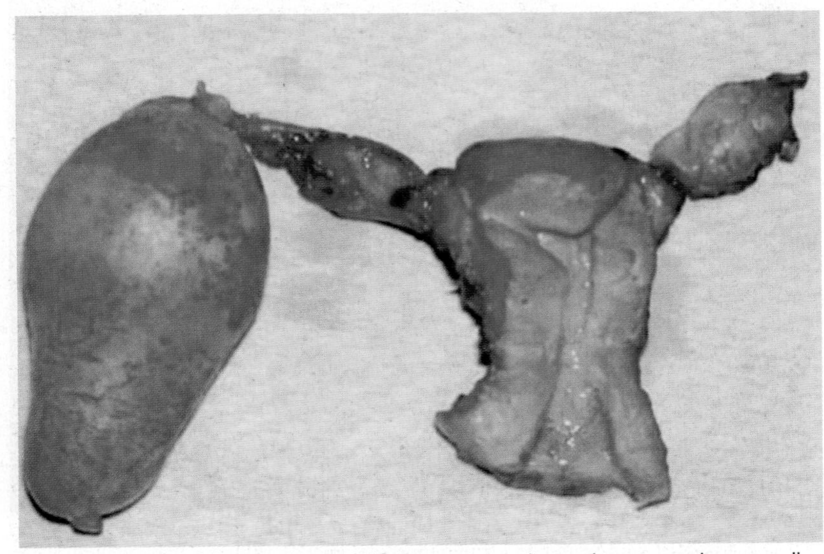

FIGURE 8.63: "The gross specimen" after surgery shows the mucocele appendix. The Rt ovary is seen normal. Uterus and Lt ovary are also seen with the specimen.

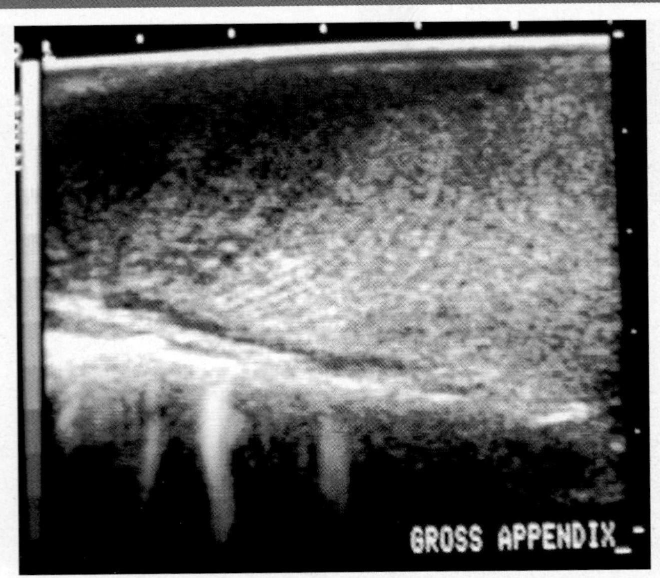

FIGURE 8.64: HRSG of the "The gross specimen" of mucocele shows thick collection with positive layering sign confirming the findings of HRSG of the patient.

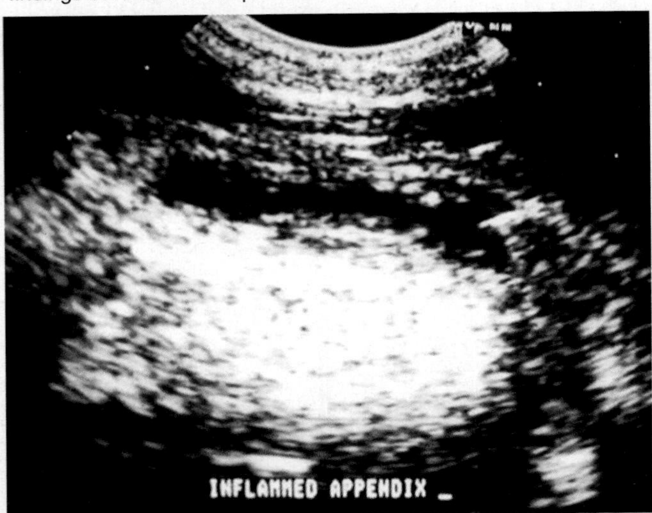

FIGURE 8.65: Obstructed appendix. HRSG shows an acutely inflamed appendix with thickened echogenic wall. An echogenic-calcified shadow is seen blocking the mouth of appendix suggestive of appendicolith. It is accompanied with acoustic shadowing. Perifocal edema is also present.

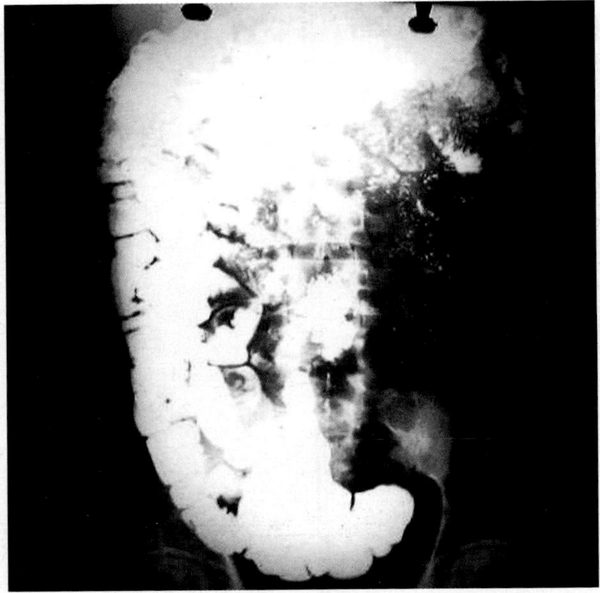

FIGURE 8.66: Barium meal follow through examination of the same patient does not show the appendix.

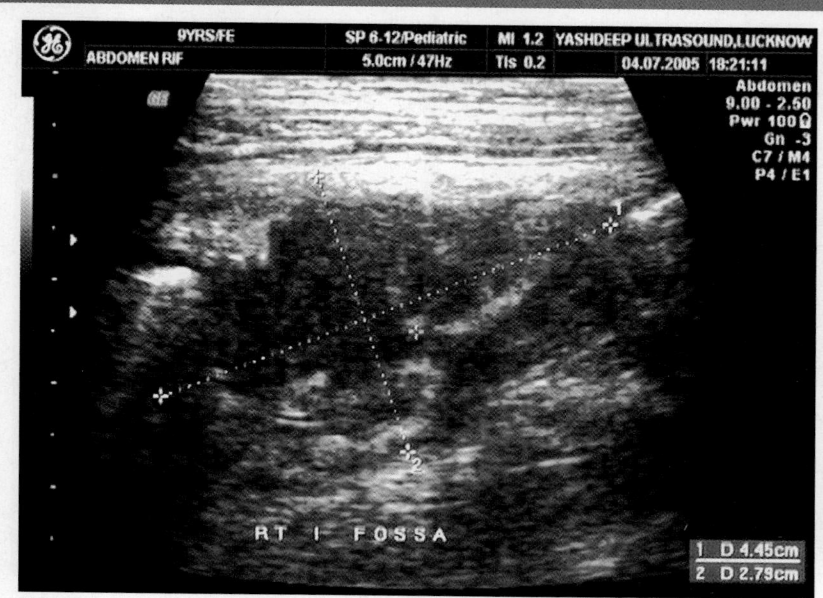

FIGURE 8.67: Abscess formation in appendicular fossa. A patient operated for acute appendicitis presented with a lump in Rt iliac fossa. HRSG shows an irregular heterogeneous low-level echo complex mass in the Rt iliac fossa suggestive of an abscess formation. Adjacent bowel loops are also seen adherent with the mass.

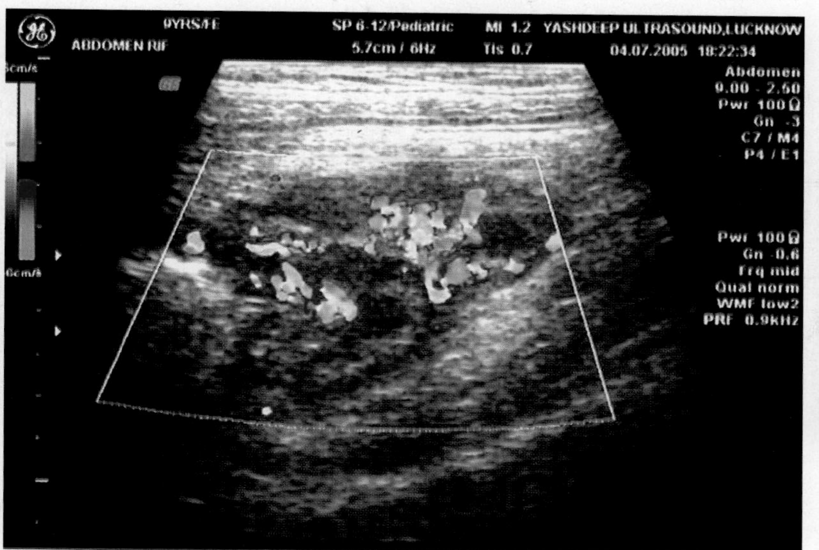

FIGURE 8.68

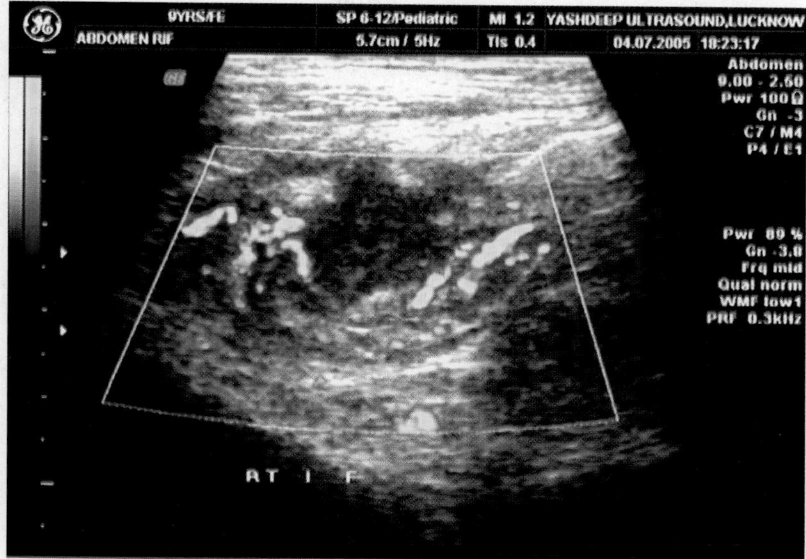

FIGURE 8.69

FIGURES 8.68 and 8.69: On color flow imaging, the abscess wall shows increased flow around it suggestive of acute hyperemia. Internal debris is also seen in the abscess.

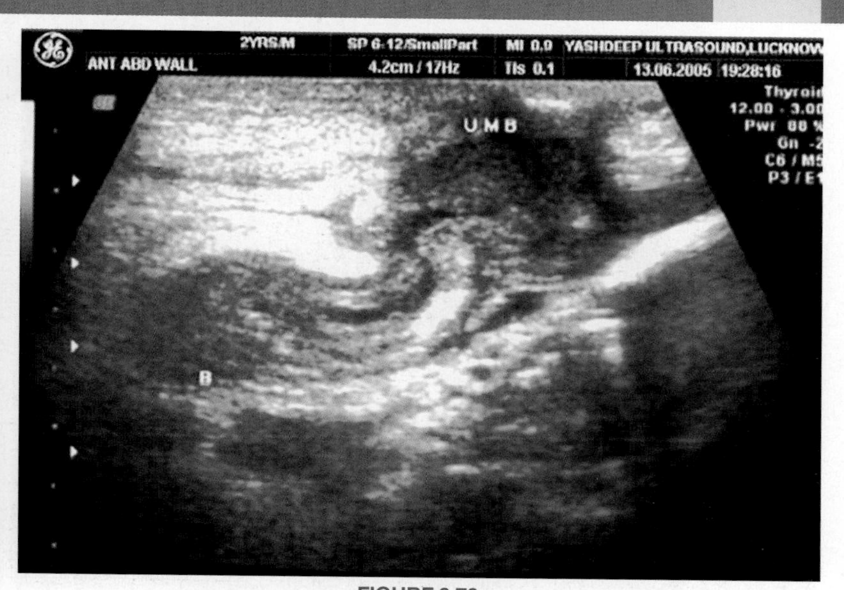

FIGURE 8.70

FIGURE 8.71

FIGURES 8.70 and 8.71: Obstructed Meckel's diverticulum. A young child presented with pain in the abdomen. HRSG shows an elongated edematous tubular shadow running in the paraumbilical area and going towards the umbilicus. The neck of the diverticulum is seen torsed. Marked perifocal edema is present. Fluid is also seen in the peridiverticulum area suggestive of obstruction of the diverticulum.

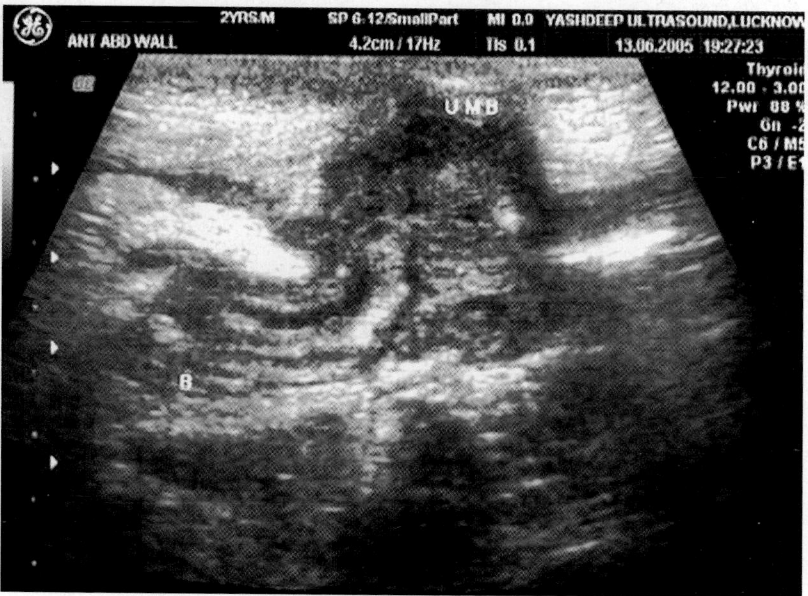

FIGURE 8.72

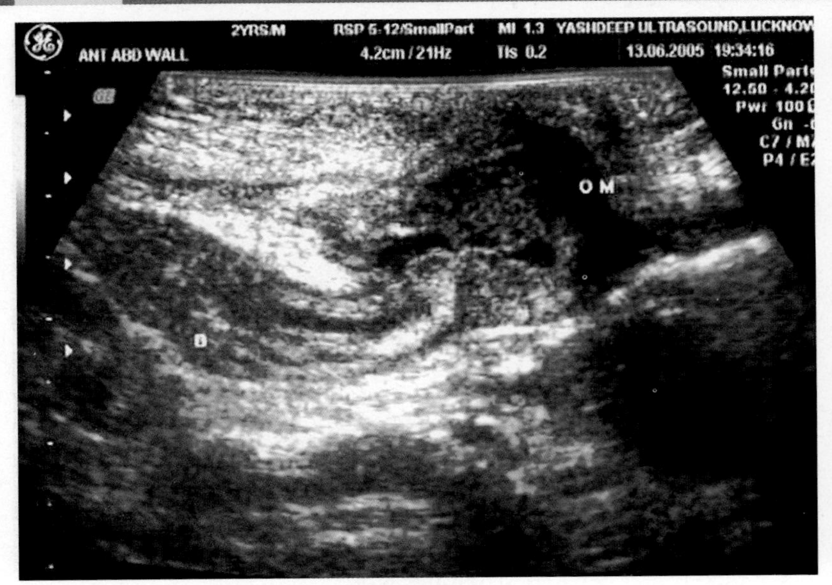

FIGURE 8.73

FIGURES 8.72 and 8.73: HRSG shows clearly the edematous mucosal folds of the diverticulum. Small abscess is also seen in the floor of the diverticulum. Partial herniation is seen in the umbilical region. Postoperative findings confirmed obstructed torsed Meckel's diverticulum.

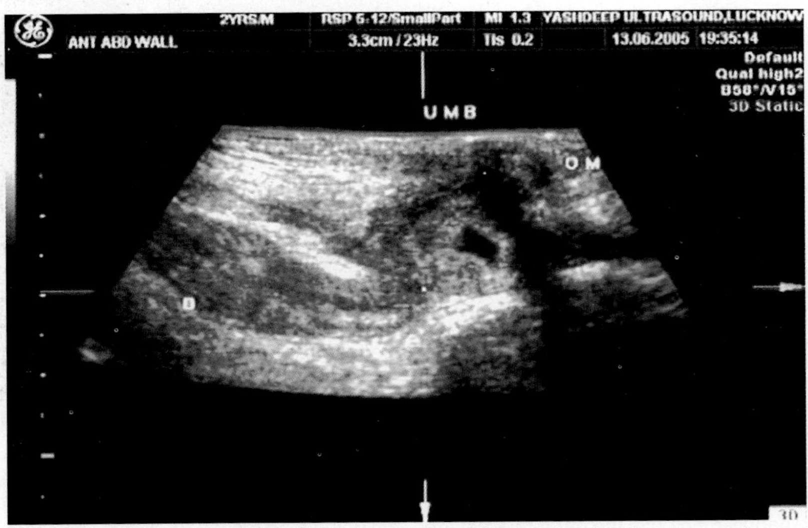

FIGURE 8.74: 3D imaging of the same patient gives the details of the inflamed diverticulum.

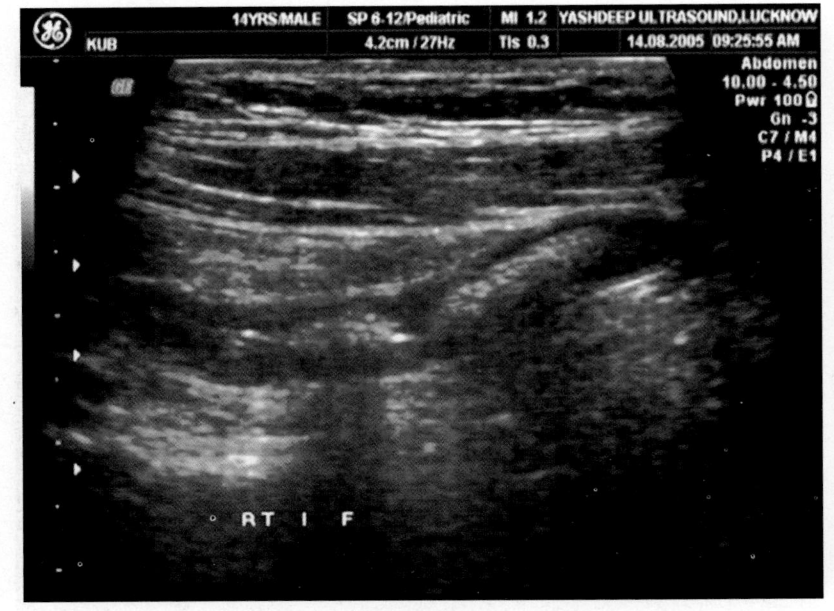

FIGURE 8.75

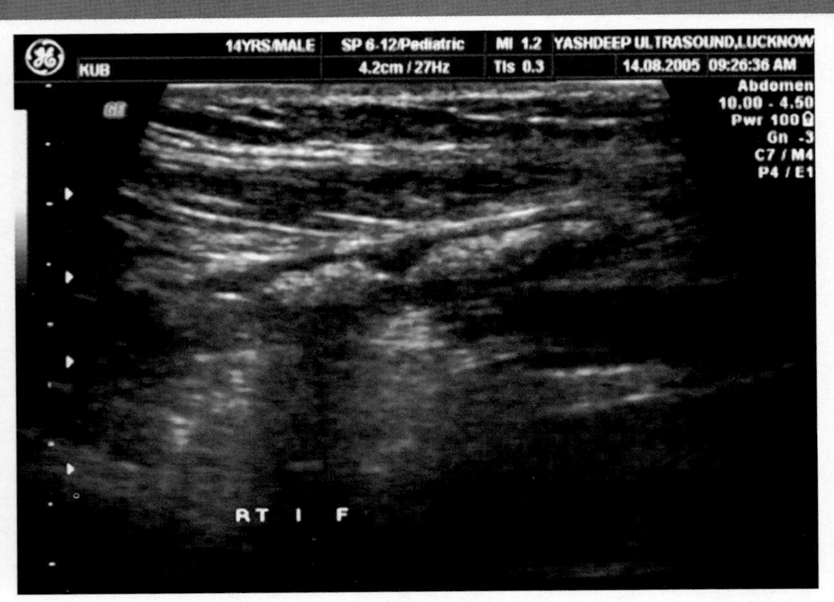

FIGURE 8.76

FIGURES 8.75 and 8.76: Inflamed Meckel's diverticulum. A young patient presented with pain in the Rt iliac fossa. HRSG shows thickened edematous tube. It was rigid and non compressible on transducer pressure. Echogenic shadows are seen filling the lumen with gas entrapment. It was mistaken as inflamed appendix. However, postoperative findings confirmed inflamed Meckel's diverticulum.

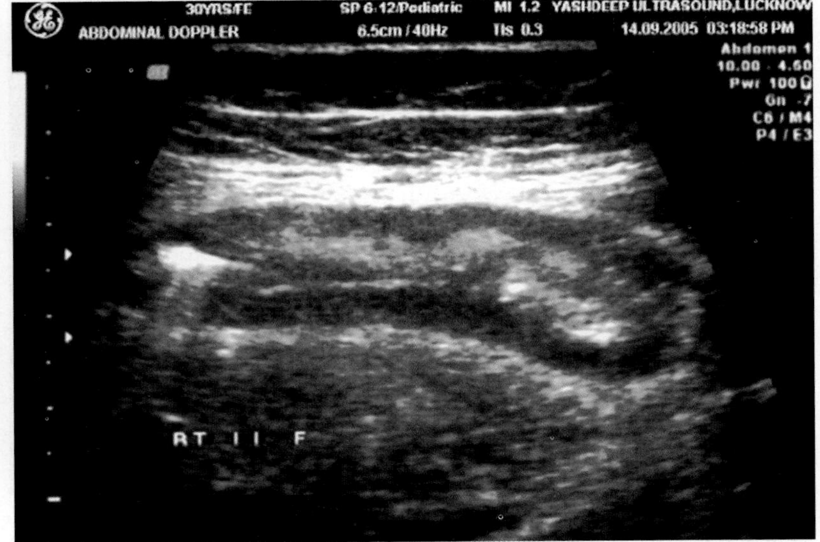

FIGURE 8.77

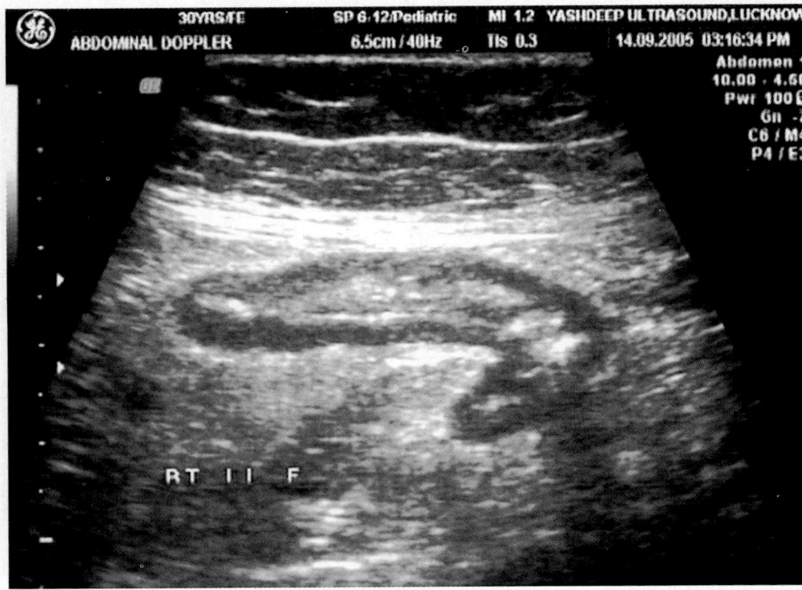

FIGURE 8.78

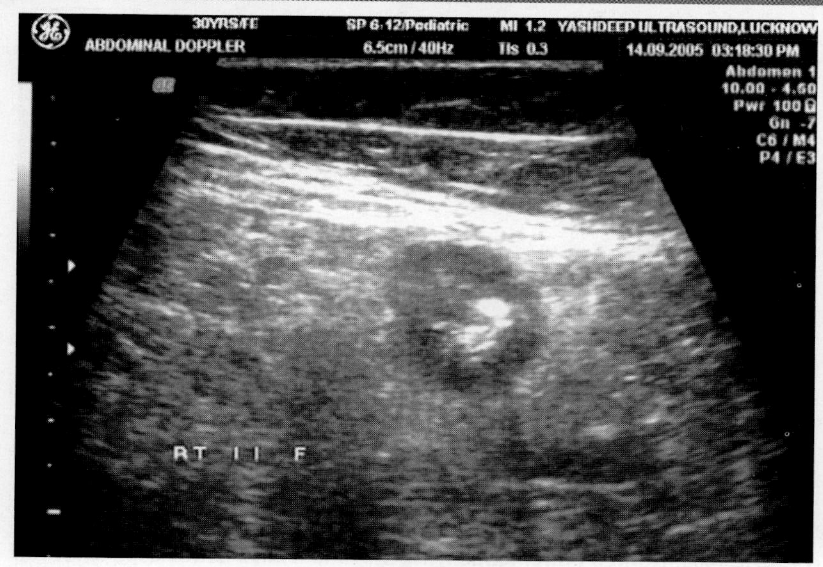

FIGURE 8.79

FIGURES 8.77 to 8.79: Grossly thickened ascending colon. A patient presented with recurrent pain in Rt iliac fossa. HRSG shows markedly thickened bowel loop in the Rt iliac fossa. Walls are edematous. Lumen is obliterated. Gas entrapment is seen.

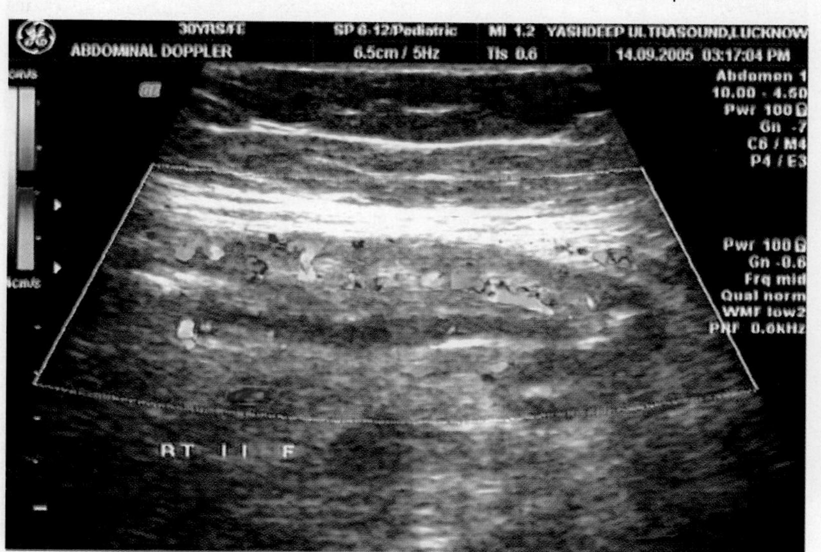

FIGURE 8.80

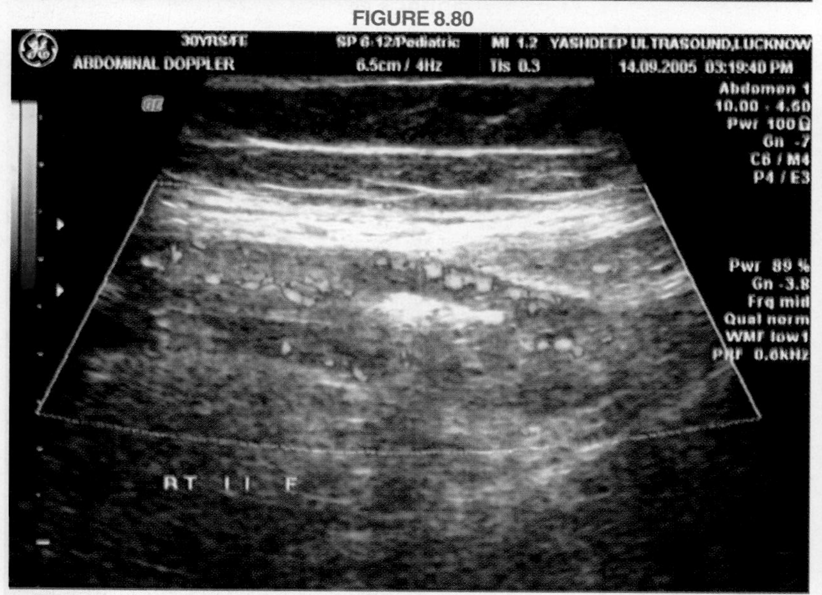

FIGURE 8.81

FIGURES 8.80 and 8.81: On color flow imaging moderate flow is seen in the wall of the thickened bowel loop. Power Doppler imaging shows blood flow in much better way suggestive of chronic inflammation. No evidence of any fluid collection seen in the lumen of the loop.

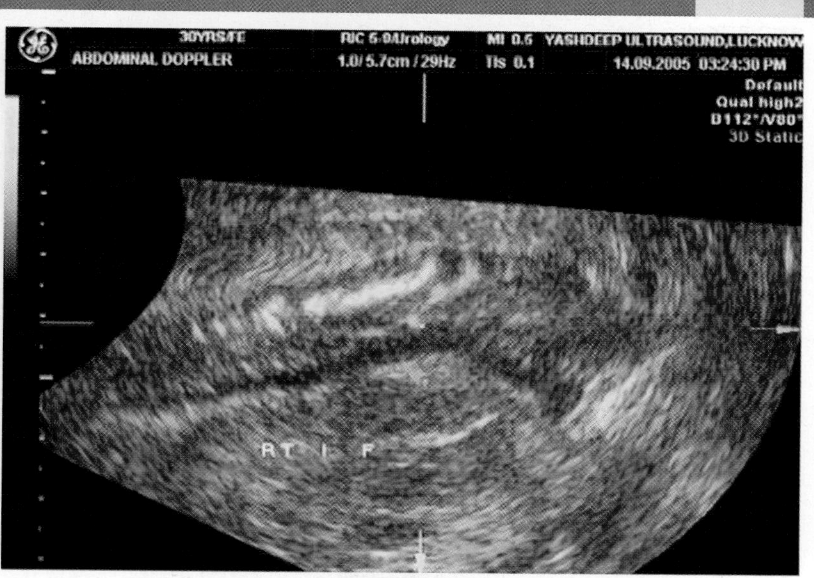

FIGURE 8.82

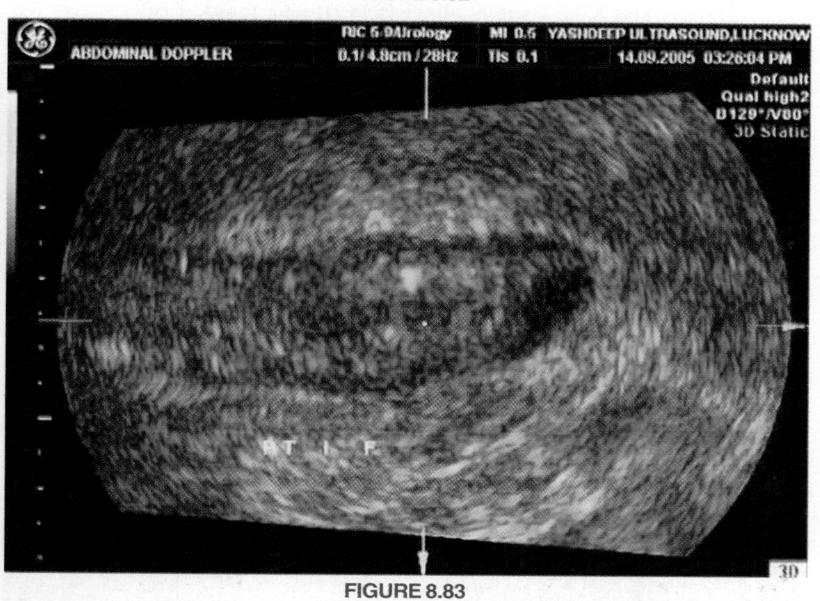

FIGURE 8.83

FIGURES 8.82 and 8.83: 3D surface rendering imaging of the bowel shows marked thickening of the wall of the bowel. Postoperative findings confirmed chronic inflammatory disease.

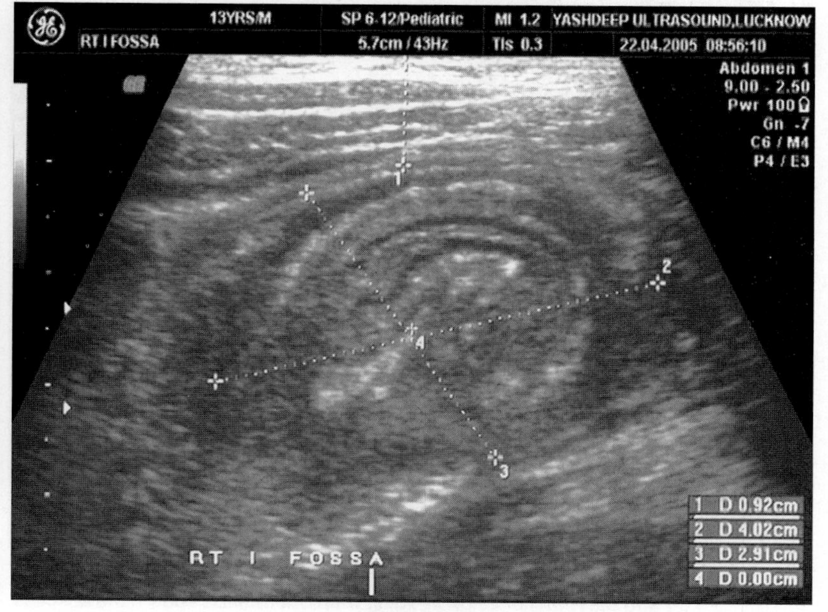

FIGURE 8.84

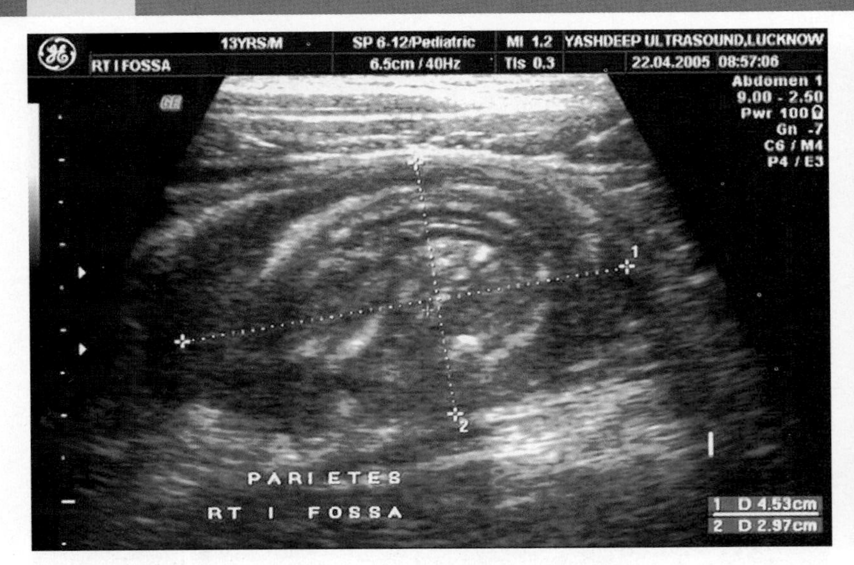

FIGURE 8.85

FIGURES 8.84 and 8.85: Parietal abscess in Rt iliac fossa. HRSG shows well defined parietal abscess in the Rt iliac fossa. The abscess is seen confined to the parietes. Layering sign is positive. Perifocal edema is present. It was mistaken as appendicitis on clinical examination.

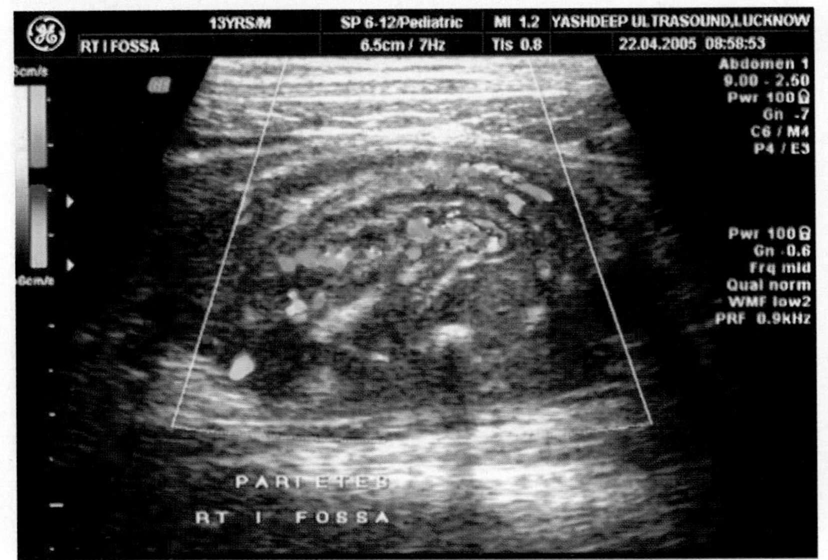

FIGURE 8.86

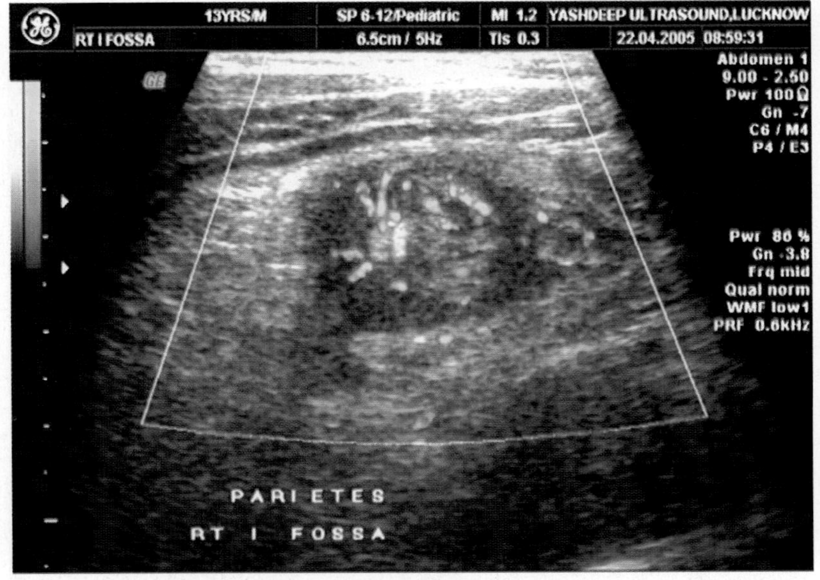

FIGURE 8.87

FIGURES 8.86 and 8.87: On color flow imaging wall hyperemia is seen in the abscess. Multiple vessels are seen feeding the abscess.

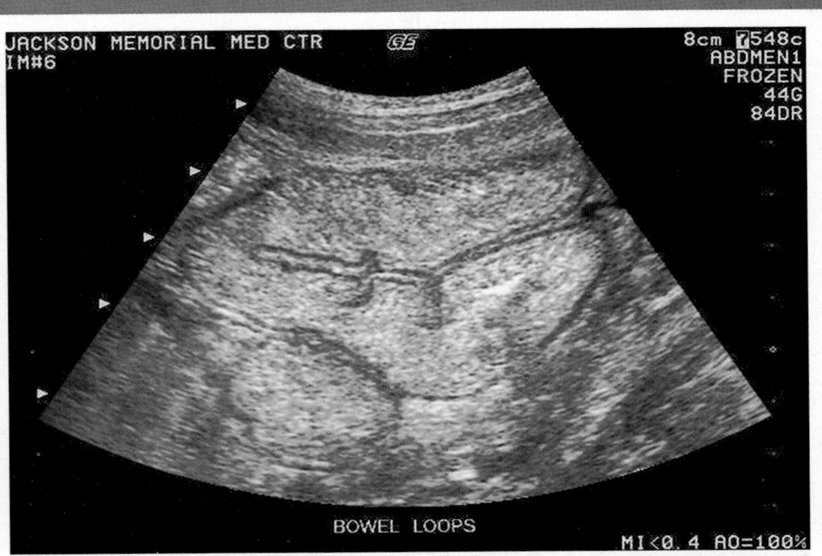

FIGURE 8.88: Normal bowel loops. HRSG shows normal bowel loops. They are seen as smooth well-defined coiled tubular shadows filled with fluid. Intervening bands are seen in it.

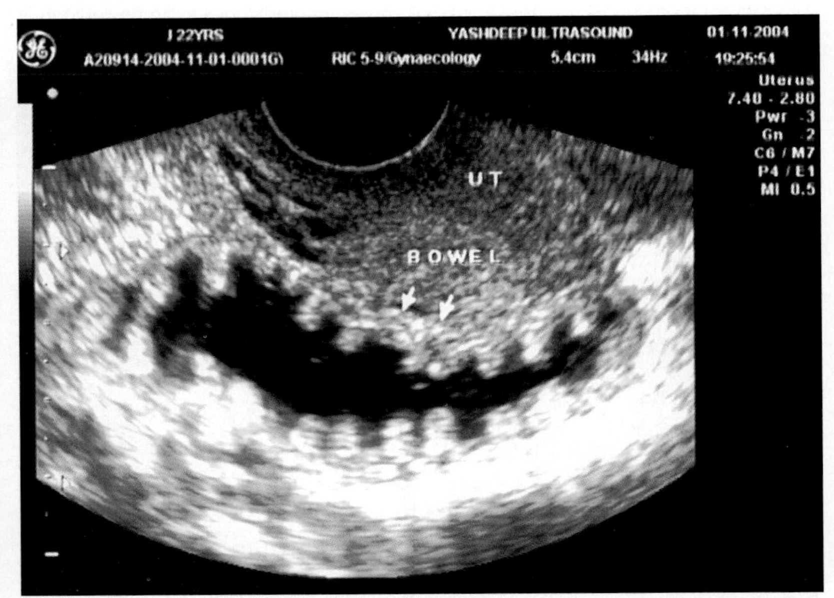

FIGURE 8.89

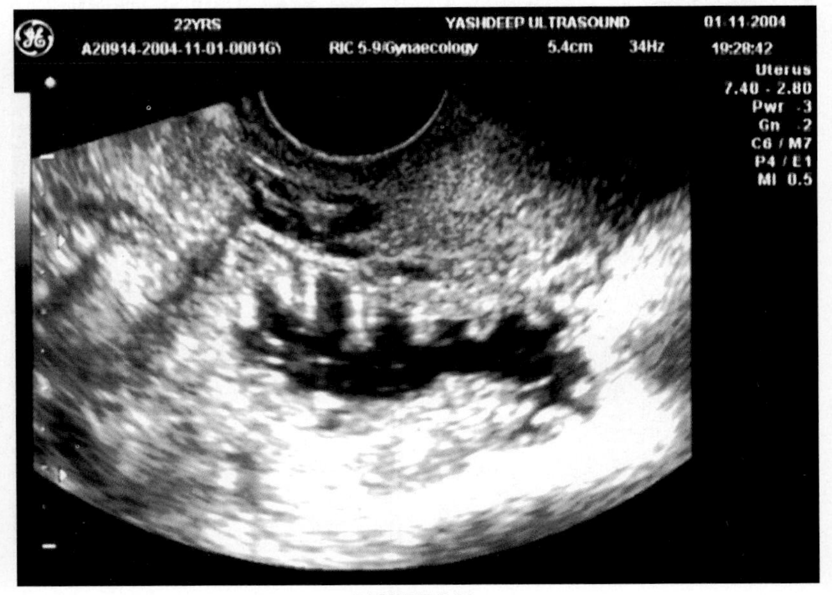

FIGURE 8.90

FIGURES 8.89 and 8.90: Dilated jejunal loops. HRSG shows normal dilated fluid filled jejunal loops. The valculae conniventes are seen clearly in the loops.

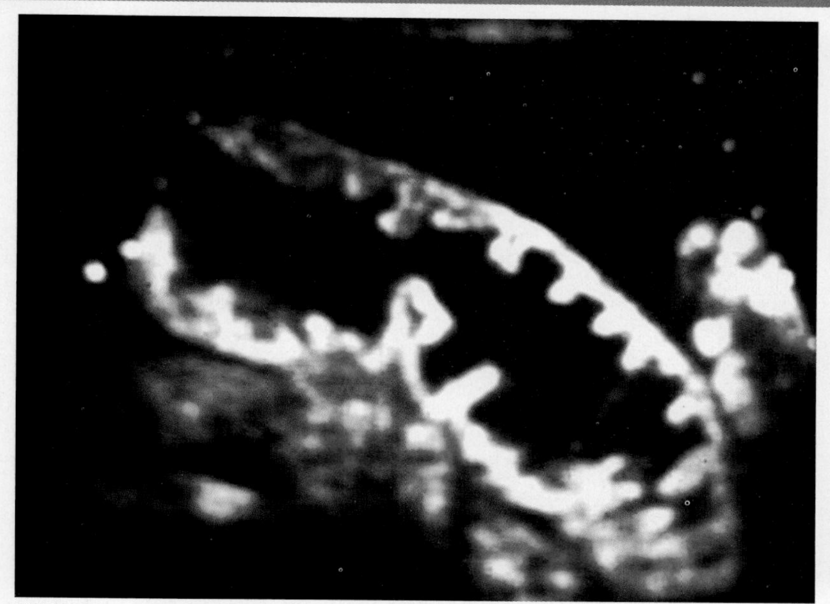

FIGURE 8.91: Dilated jejunal loops. HRSG shows dilated jejunal loops with prominent valculae conniventes.

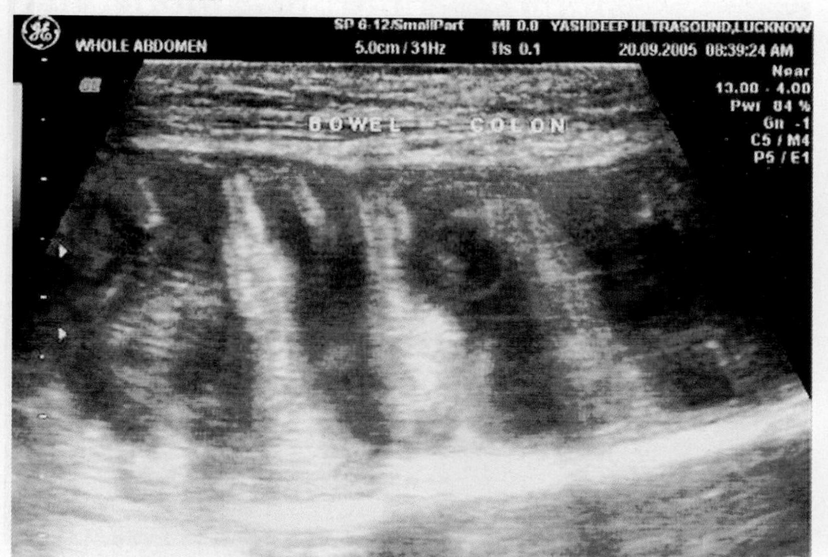

FIGURE 8.92

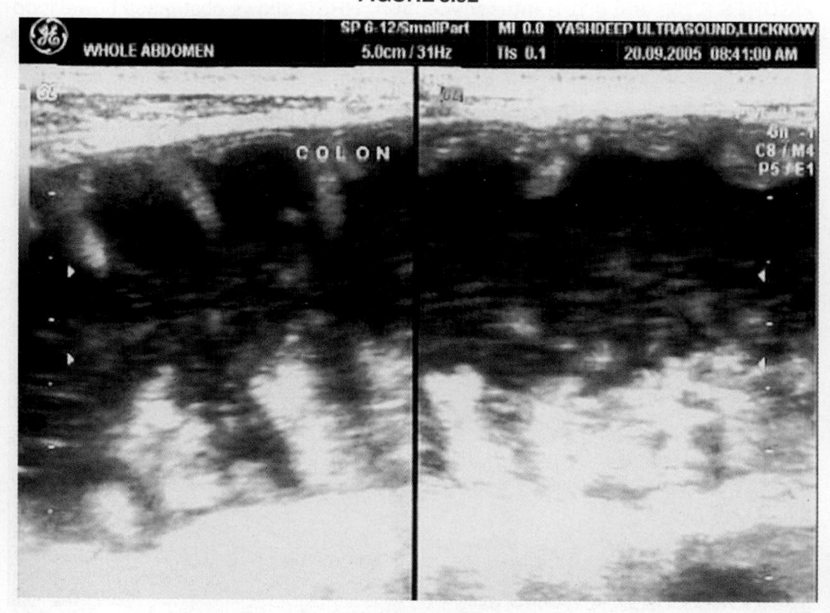

FIGURE 8.93

FIGURES 8.92 and 8.93: Dilated large bowel. HRSG shows dilated large bowel loops prominent haustrations are seen running to the lumen of the loops.

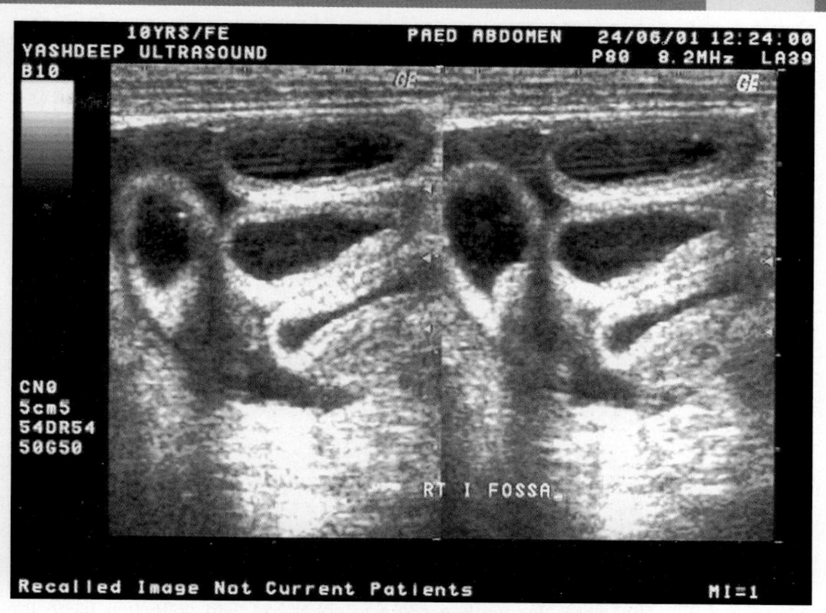

FIGURE 8.94

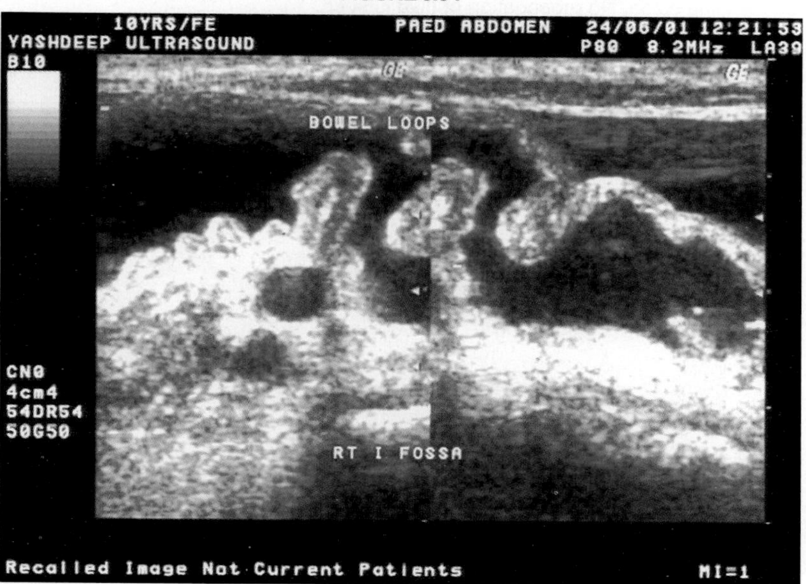

FIGURE 8.95

FIGURES 8.94 and 8.95: Subacute intestinal obstruction. A young girl presented with pain in abdomen and vomiting. HRSG shows multiple dilated bowel loops in the abdomen cavity. Coiling of the bowel loops is seen.

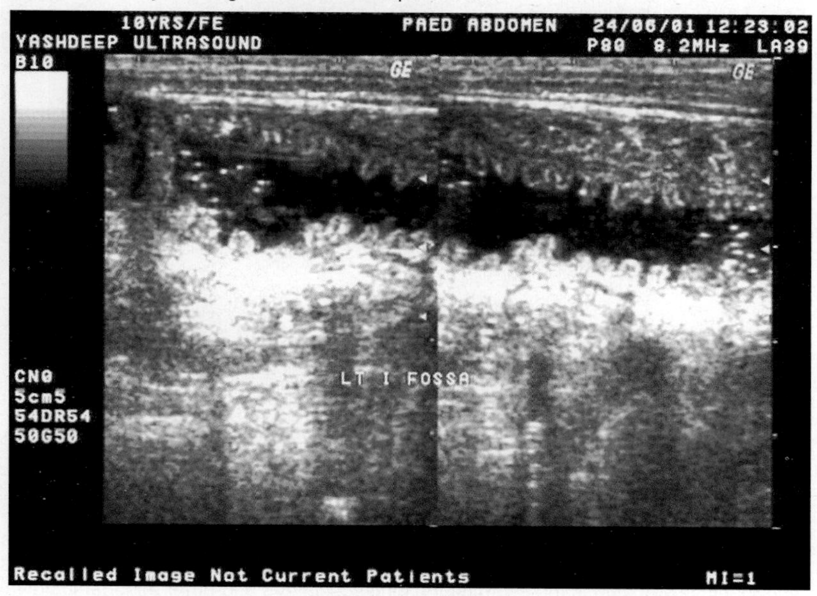

FIGURE 8.96: Proximal jejunal loops are also dilated.

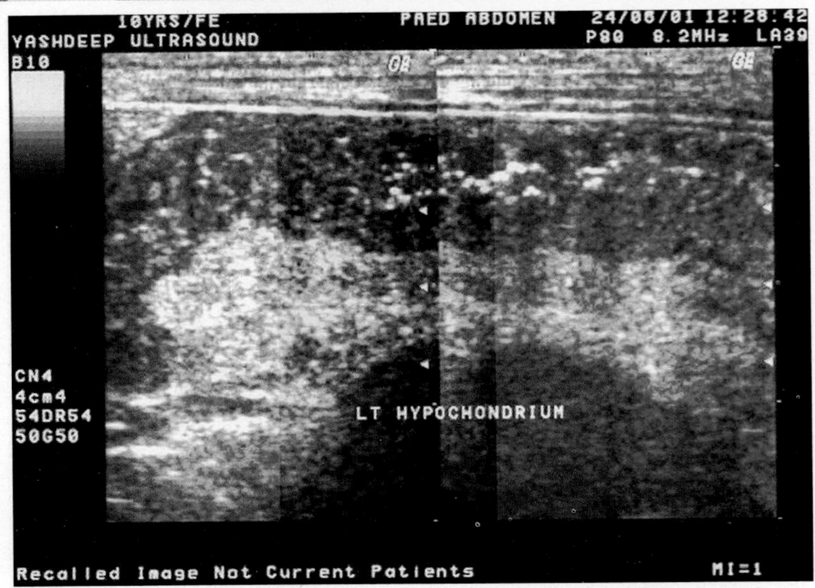

FIGURE 8.97: The distal descending colon is collapsed with collection filling the descending colon. No propagation of fluid is seen.

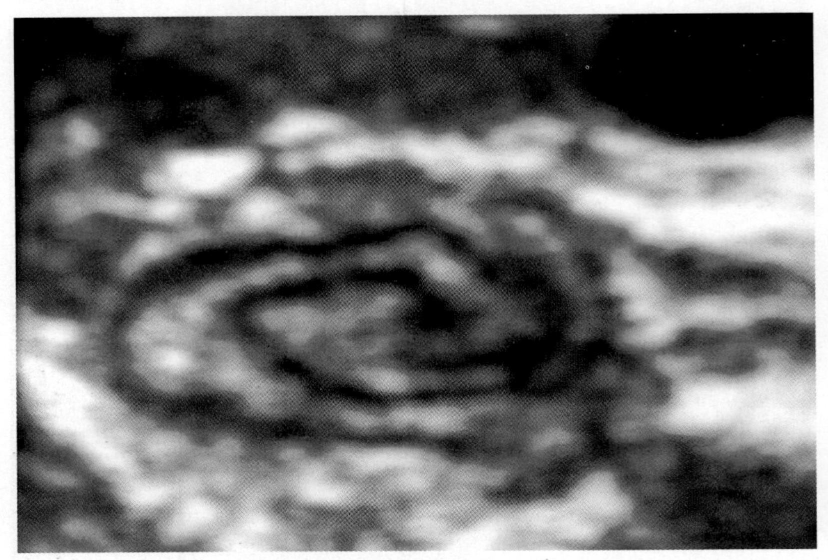

FIGURE 8.98

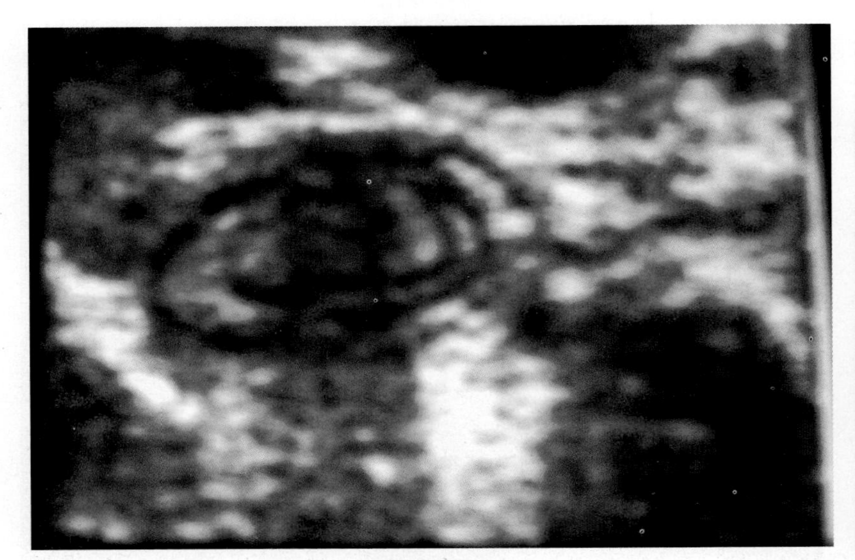

FIGURE 8.99

FIGURES 8.98 and 8.99: Intussusception. HRSG shows an intussusception in the abdomen; typical crescentric triple layer bowel appearance is seen. The inner collapsed echogenic bowel loop is the invaginating intussuscipiens and outer echogenic bowel loop is the intussusceptum.

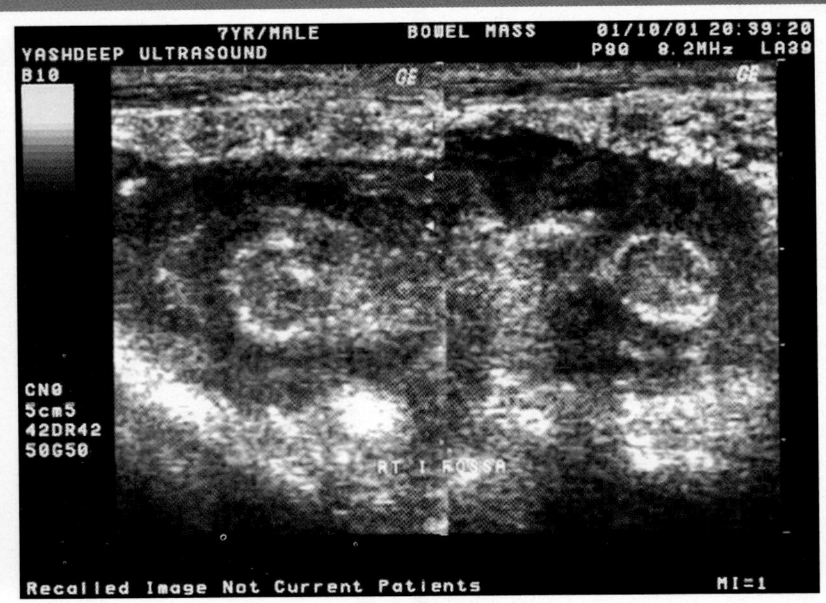

FIGURE 8.100

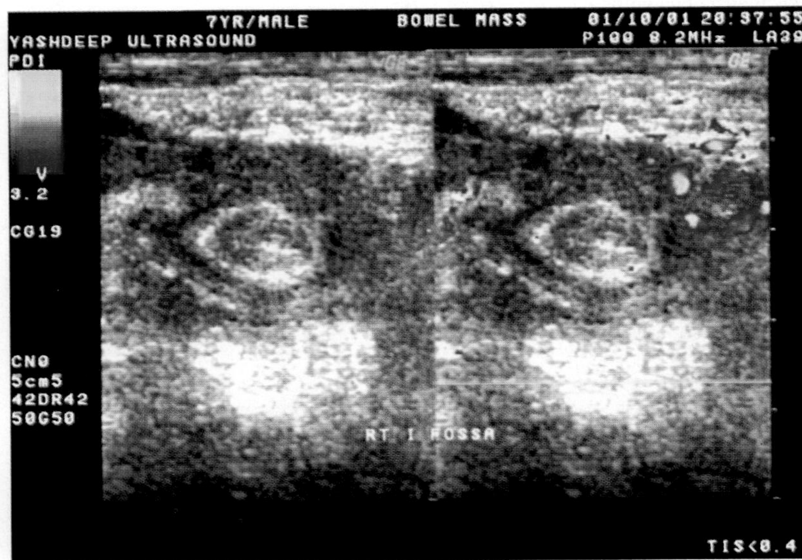

FIGURE 8.101

FIGURES 8.100 and 8.101: Intussusception. HRSG shows obstructed bowel loop in the Rt iliac fossa. Echogenic crescentric loops are seen with hypoechoic rim suggestive of intussusception. The inner collapsed echogenic loop is the intussuscipiens and outer loop is the intussusceptum, typical doughnut sign is seen.

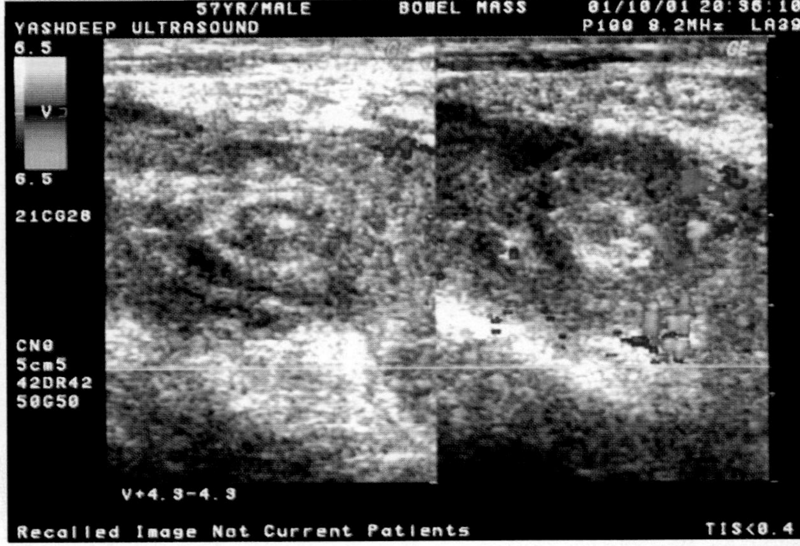

FIGURE 8.102: On color Doppler imaging increased flow is seen in the bowel wall due to inflammation.

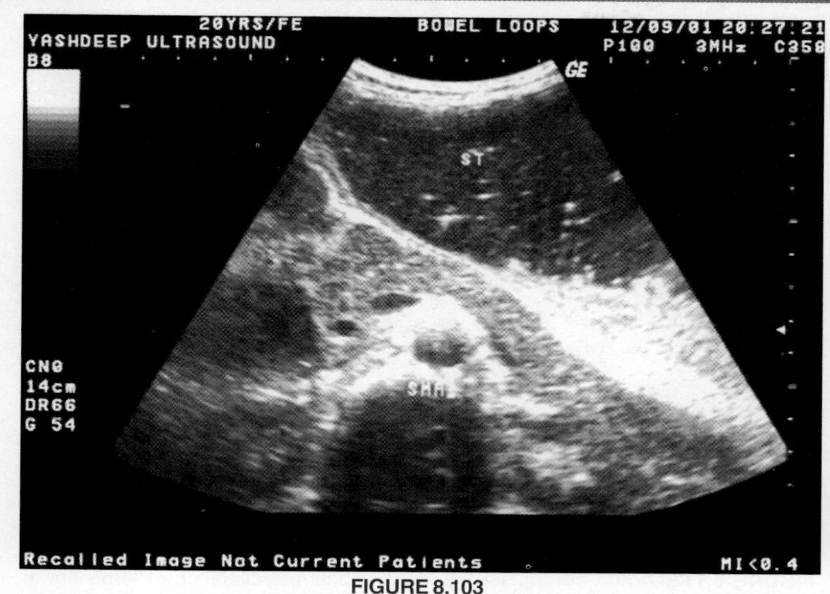

FIGURE 8.103

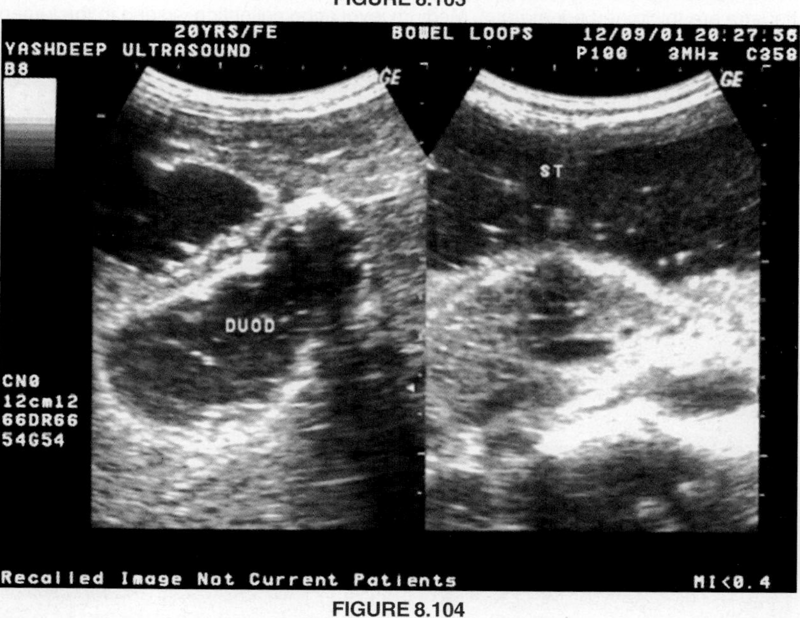

FIGURE 8.104

FIGURES 8.103 and 8.104: Intestinal obstruction volvulus. A young patient presented with acute pain in the abdomen with persistent vomiting. The stomach is markedly distended. Fluid collection is seen in the stomach. The duodenum and proximal bowel loops are also dilated.

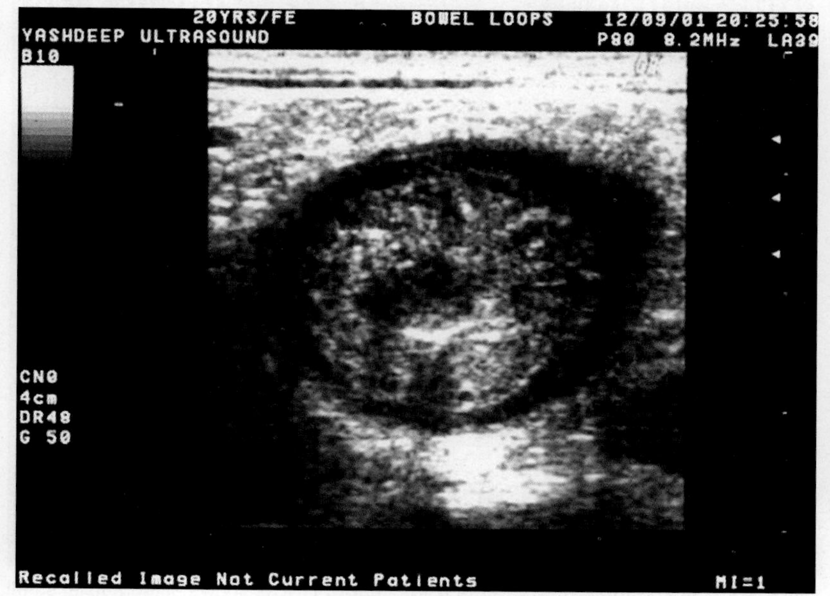

FIGURE 8.105

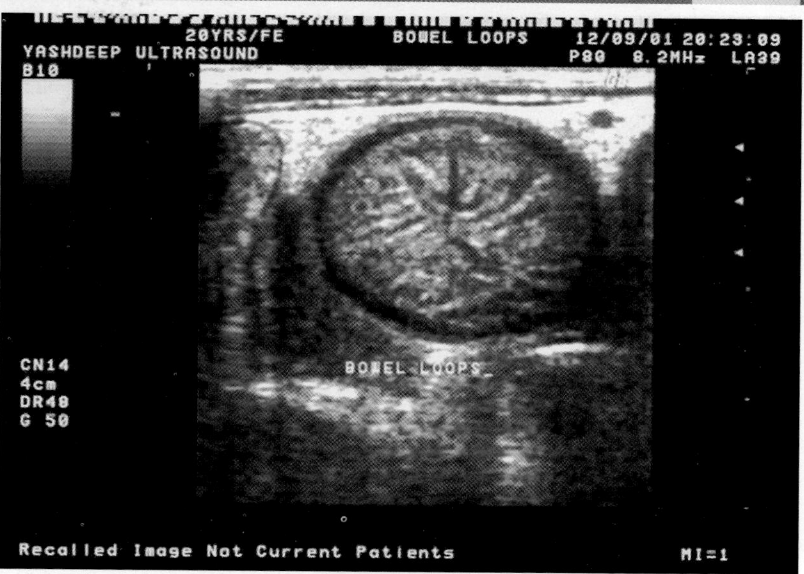

FIGURE 8.106

FIGURES 8.105 and 8.106: HRSG shows the volvulus of small intestine. Hypoechoic perifocal edema is seen around the obstructed bowel loops. Hypoechoic central area is also seen.

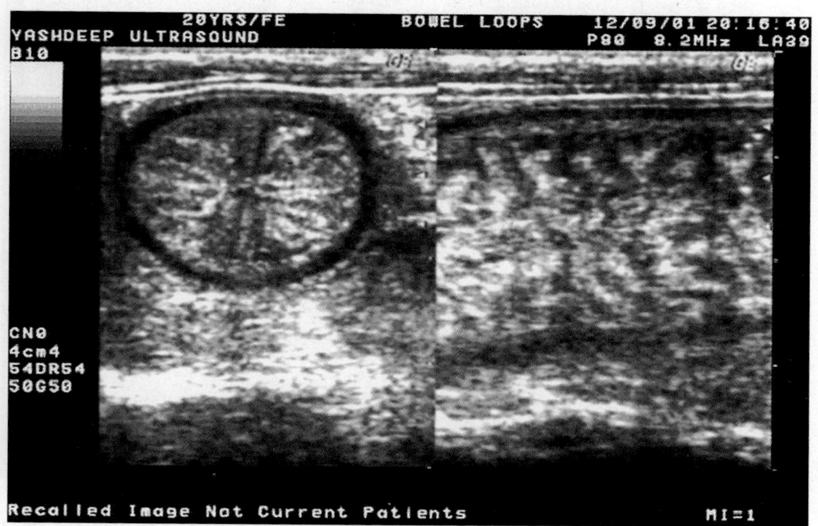

FIGURE 8.107

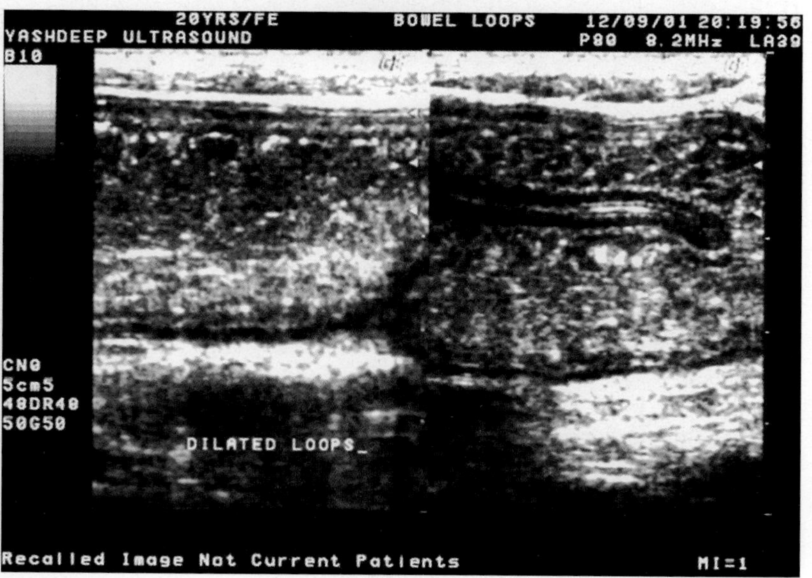

FIGURE 8.108

FIGURES 8.107 and 8.108: HRSG shows edematous obstructed bowel loops in the longitudinal axis. The bowel wall is edematous and hypoechoic. Echogenic collection is seen in the lumen. No peristalsis is seen.

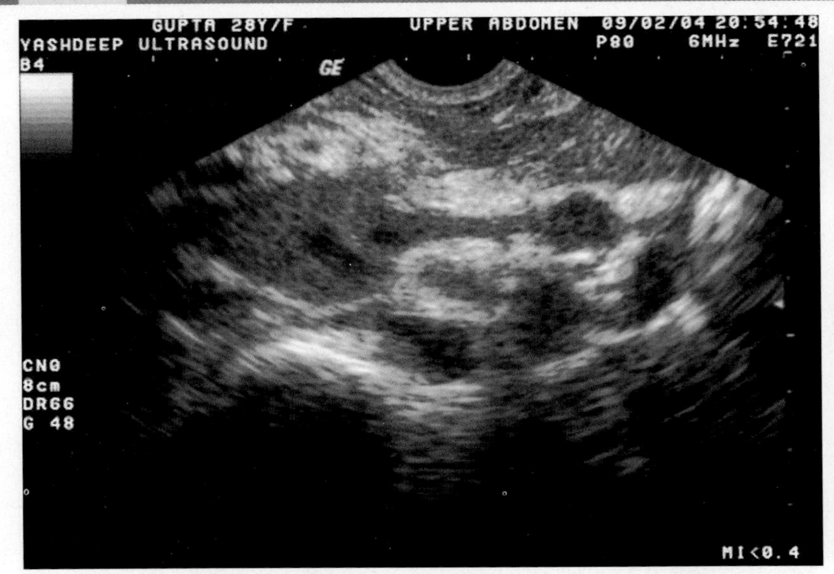

FIGURE 8.109: **Small intestine volvulus.** HRSG shows volvulus of small intestine. The bowel loop is edematous and hypoechoic. Perifocal edema is present.

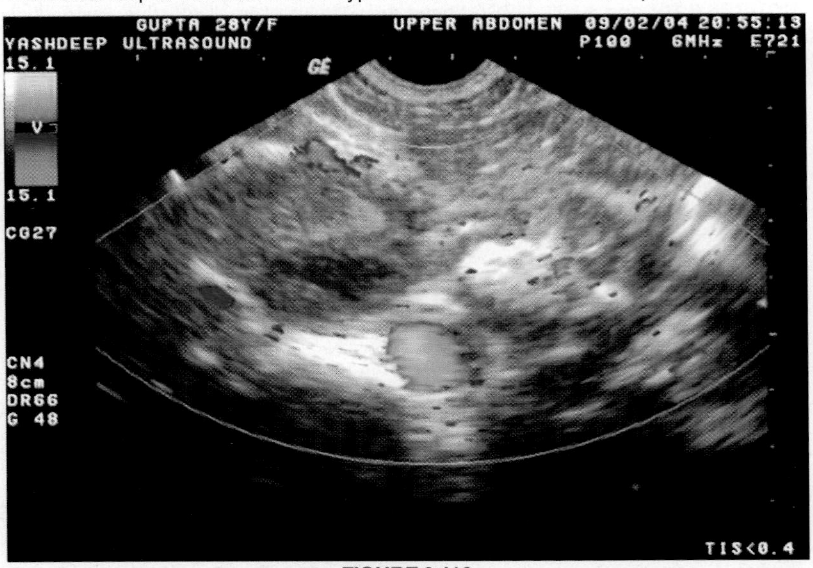

FIGURE 8.110

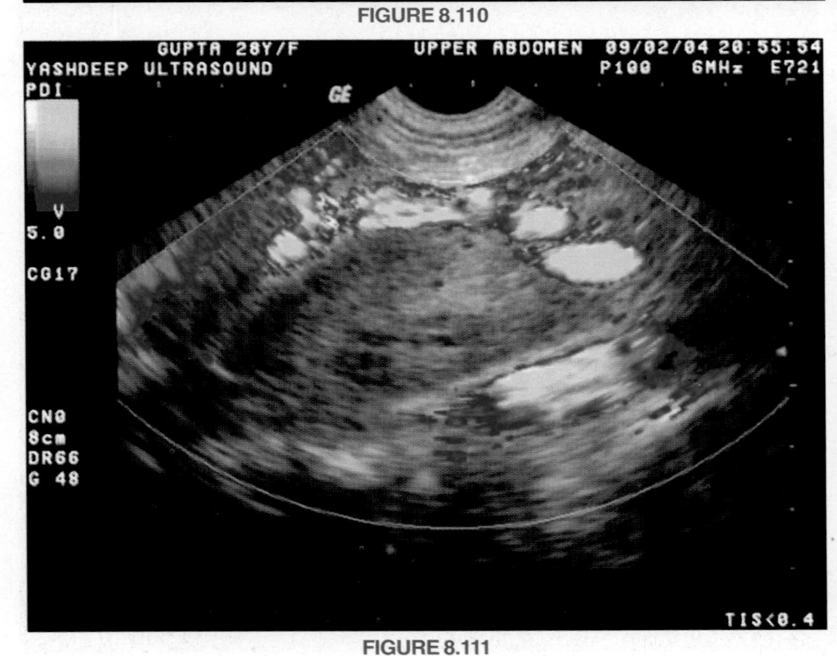

FIGURE 8.111

FIGURES 8.110 and 8.111: On color flow imaging peripheral flow is seen around the bowel loop. A vascular band is running around the volvulus. It is seen as a ring around the constricting bowel loop. Power Doppler imaging shows the vascular band beautifully.

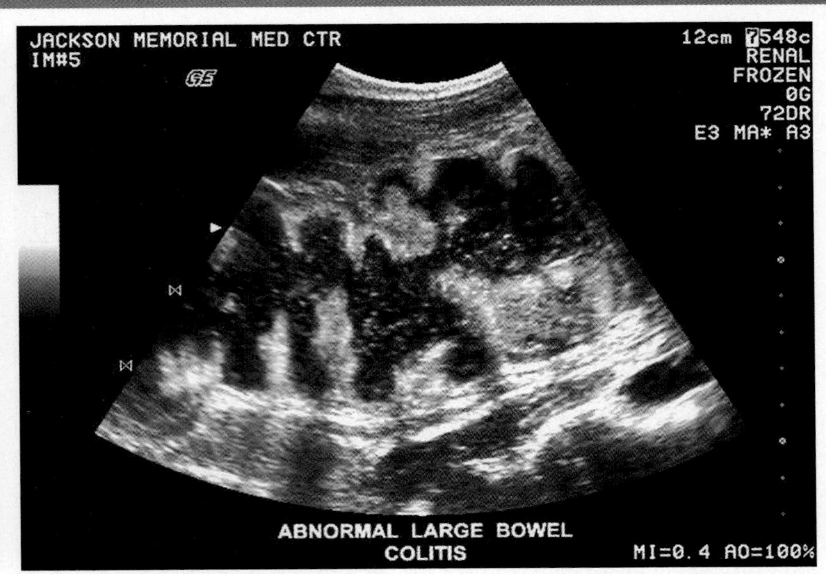

FIGURE 8.112: **Acute colitis.** HRSG shows abnormal dilatation of large bowel, haustra are thickened edematous. Medium level echo collection is seen in the lumen suggestive of stagnation in a case of acute colitis.

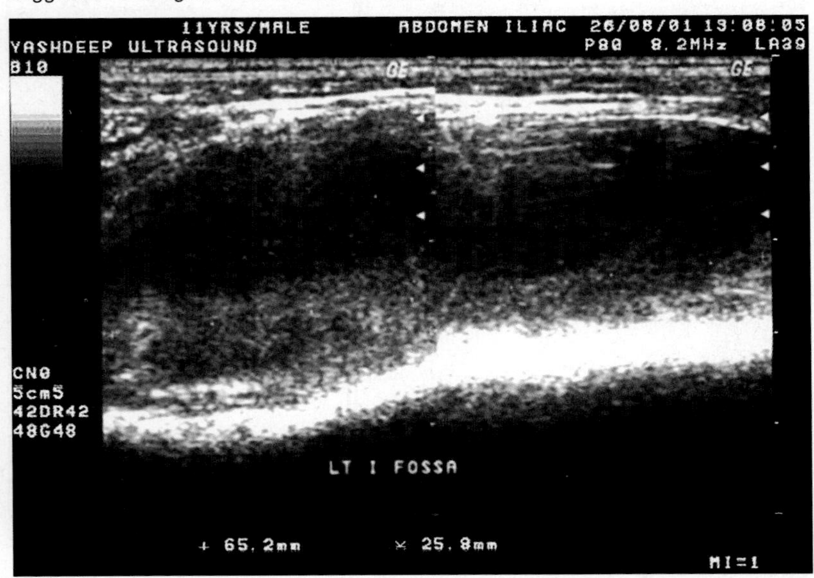

FIGURE 8.113: **Large bowel dilatation.** Large bowel dilatation is seen in case of acute inflammation. Collection is seen on the dependant part of bowel loop.

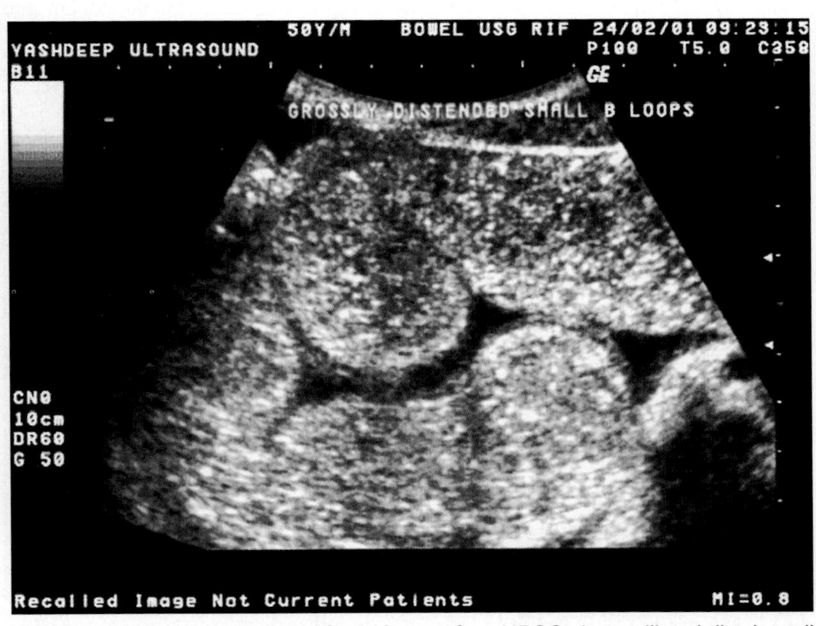

FIGURE 8.114: **Subacute intestinal obstruction.** HRSG shows dilated distal small bowel loops with echogenic collection.

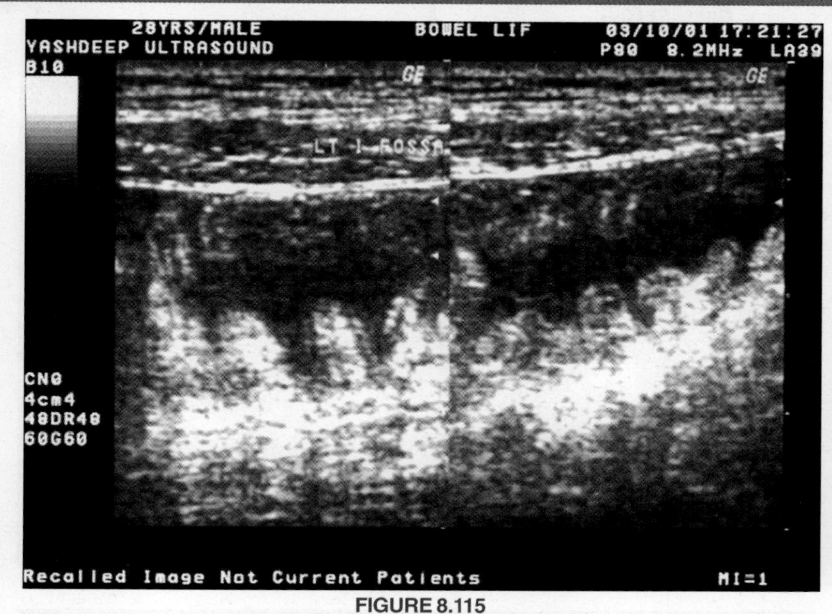

FIGURE 8.115

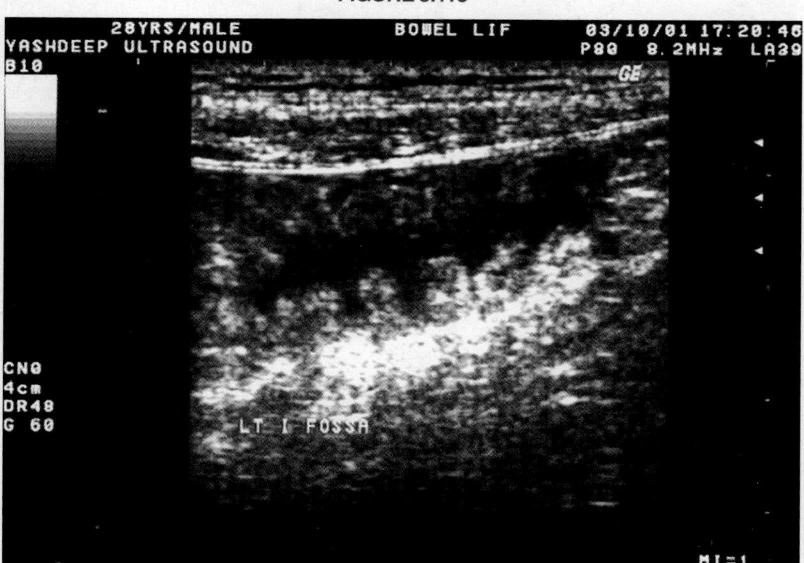

FIGURE 8.116

FIGURES 8.115 and 8.116. The proximal jejunal loops are dilated in a case of sub acute intestinal obstruction.

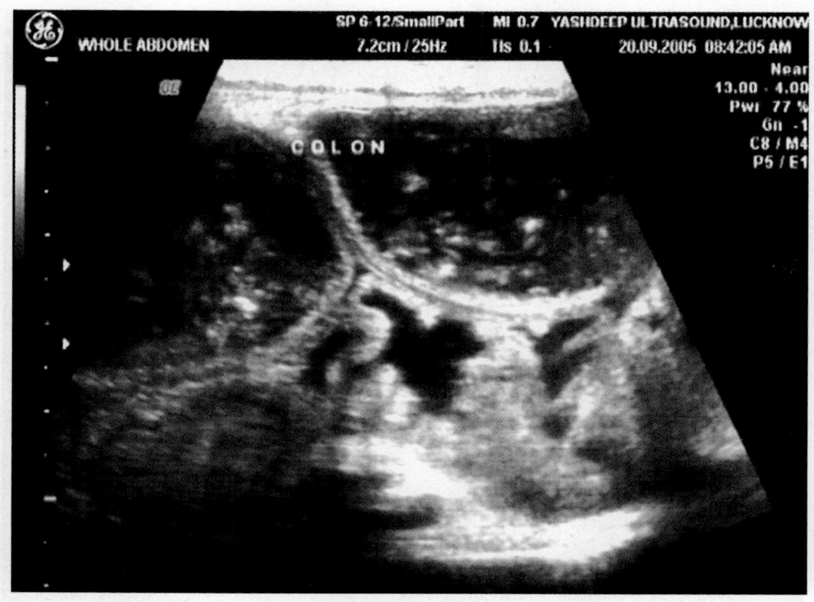

FIGURE 8.117

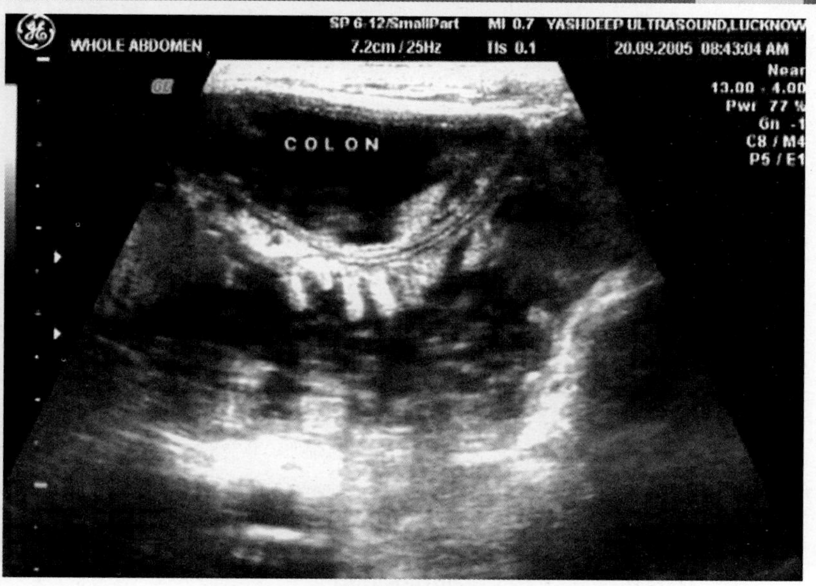

FIGURE 8.118

FIGURES 8.117 and 8.118: Gross dilatation ascending and transverse colon. HRSG shows gross dilatation of ascending and transverse colon in the case of intestinal obstruction. The haustrations are seen standing out clearly in the fluid background. No peristalsis is seen.

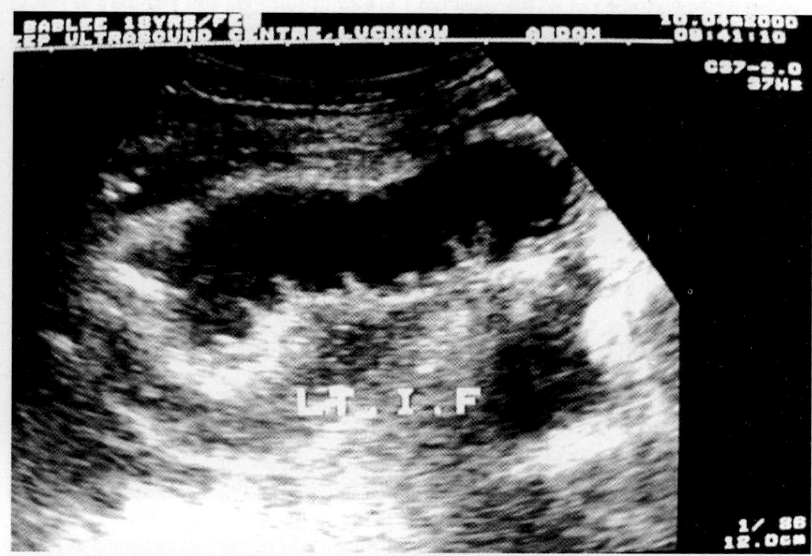

FIGURE 8.119: Dilated bowel loops in the Lt iliac fossa. HRSG shows dilated large bowel loop in the Lt iliac fossa in the case of intestinal obstruction.

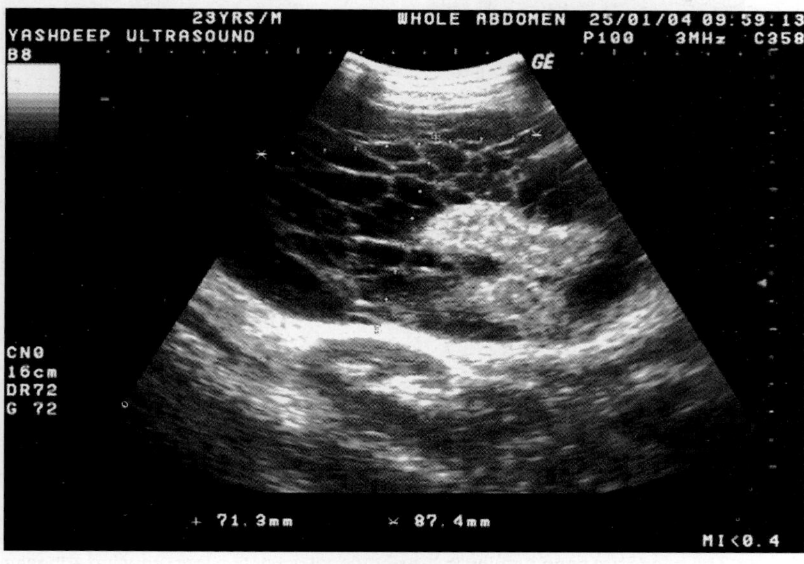

FIGURE 8.120

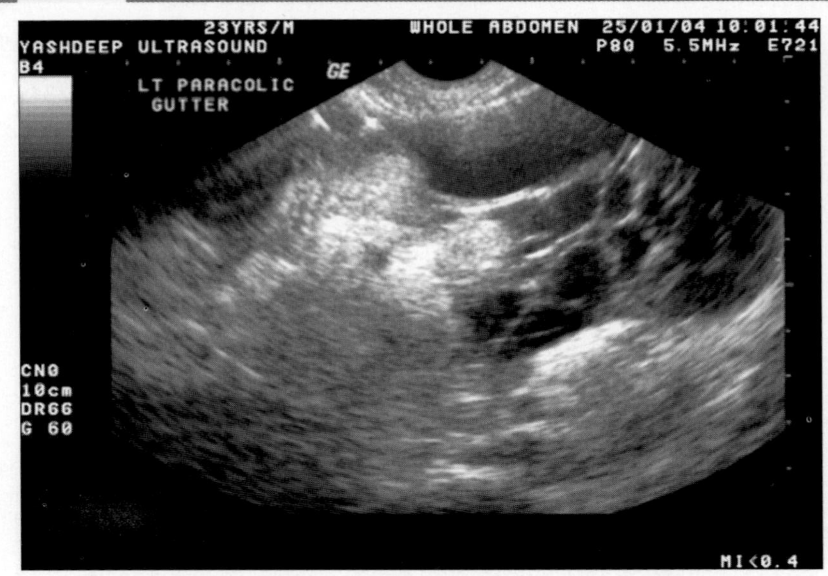

FIGURE 8.121

FIGURES 8.120 and 8.121: Tubercular bowel with peritonitis. HRSG shows thick collection in the peritoneal cavity. Multiple echogenic bands are seen in the collection. Bowel loops are seen adherent due adhesive bands forming a jumbled mass.

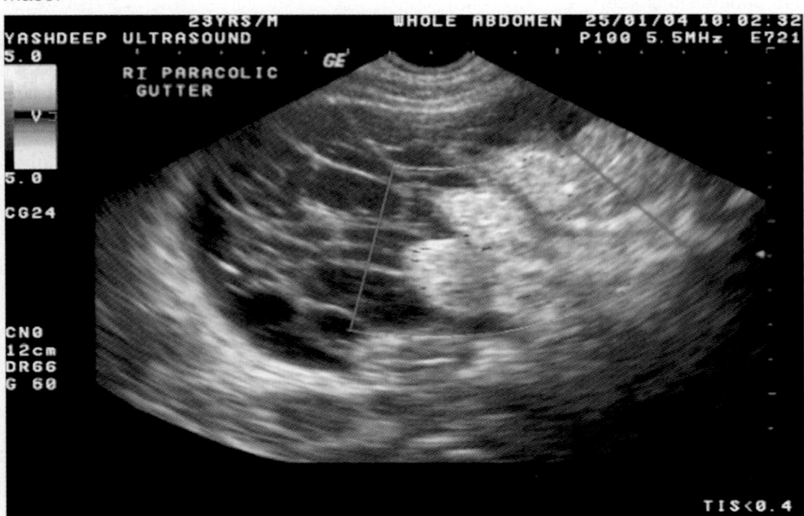

FIGURE 8.122

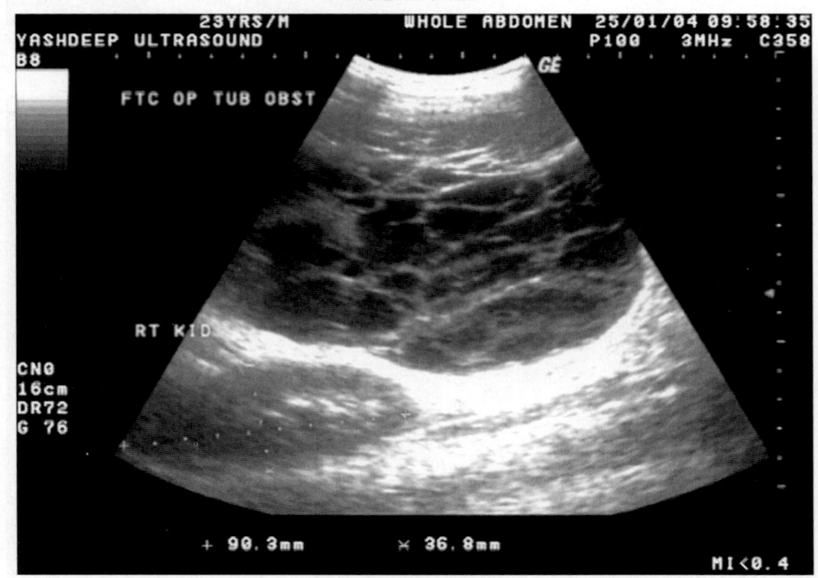

FIGURE 8.123

FIGURES 8.122 and 8.123: The bowel loops are seen held together with multiple bands. The bands are seen running in the peritoneal collection form causing adhesive plastic peritonitis. On color flow imaging no flow seen in the loops.

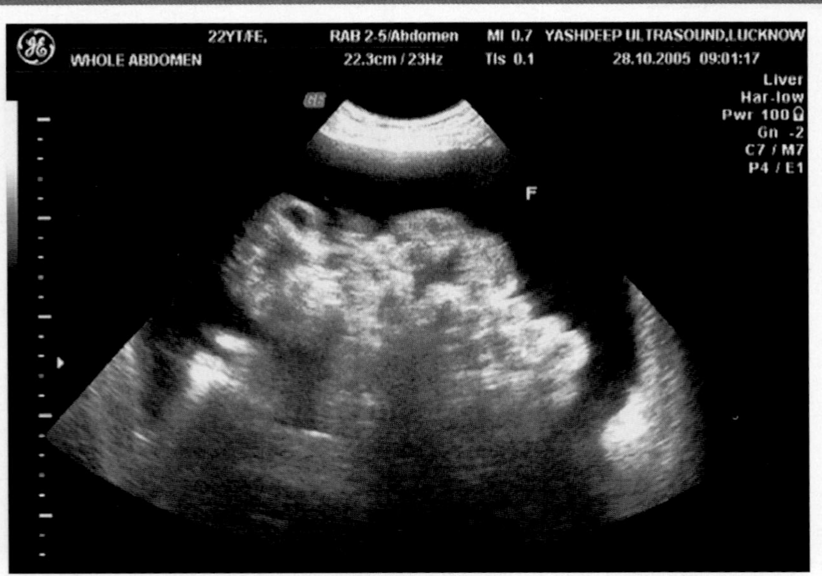

FIGURE 8.124

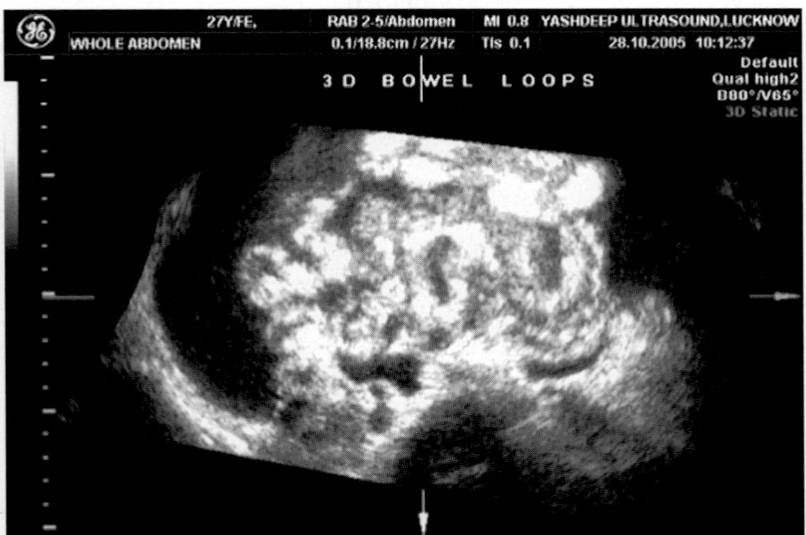

FIGURE 8.125

FIGURES 8.124 and 8.125: Tubercular bowel (Cocoon ball). HRSG shows tubercular bowel loops clumped together forming a cocoon due the adhesive tubercular capsulitis. This cocoon ball keeps on moving in the abdomen in the peritoneal fluid. Therefore the patient complains of some ball moving in the abdomen. 3D imaging shows clumped bowel loops in a much greater detail.

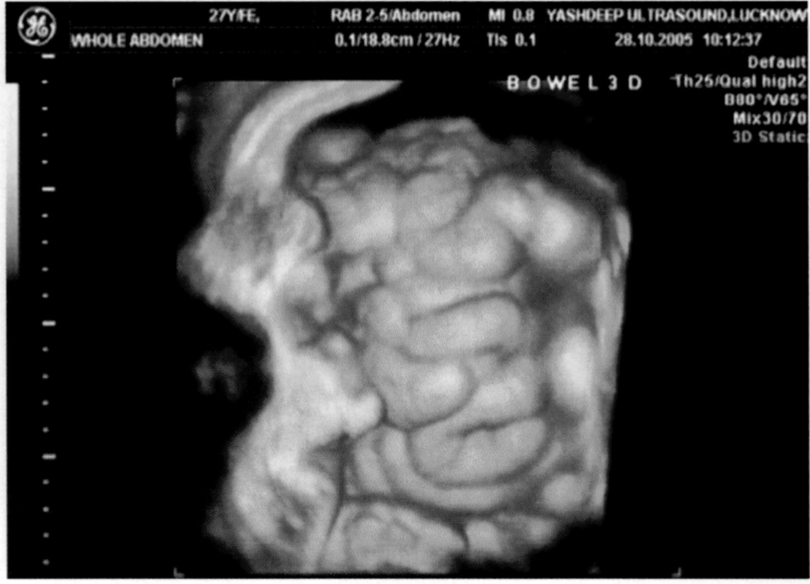

FIGURE 8.126

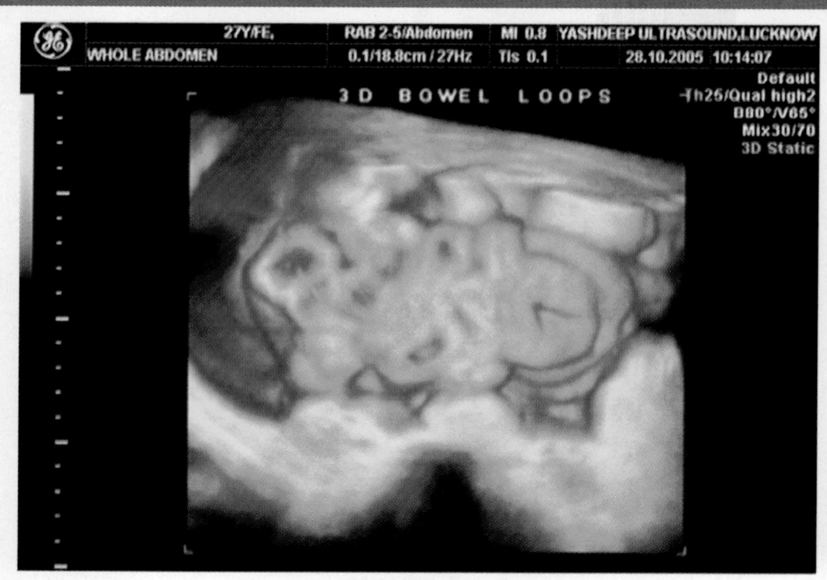

FIGURE 8.127

FIGURES 8.126 and 8.127: 3D surface rendering of clump bowel loops show the well form cocoon due to tubercular adhesive capsulitis around the bowel loops.

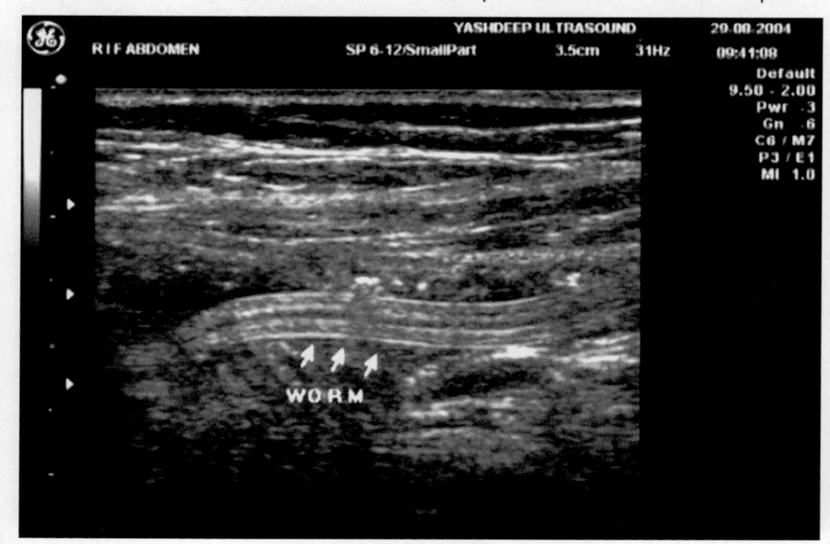

FIGURE 8.128

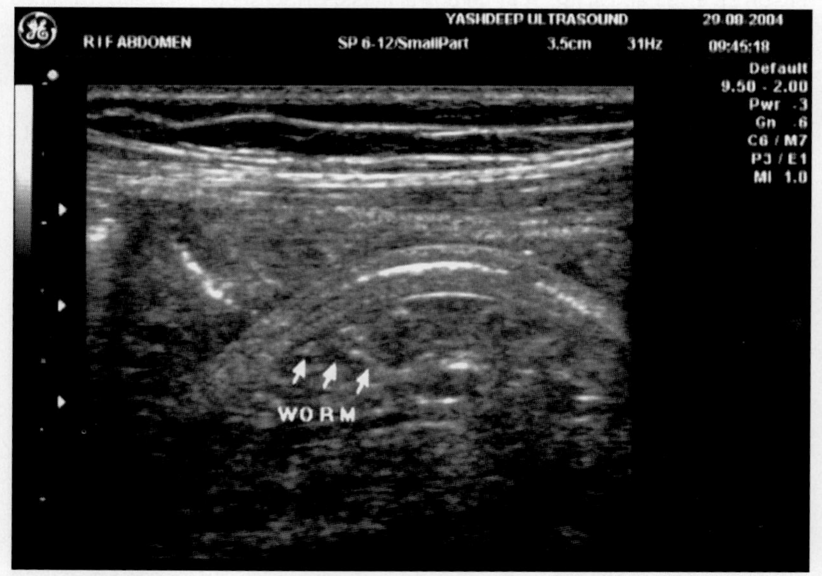

FIGURE 8.129

FIGURES 8.128 and 8.129: Ascaris in the bowel loops. A young child presented with recurrent pain in the abdomen. HRSG show moving tapeworm (Ascaris). The details of the worm are beautifully resolved on HRSG.

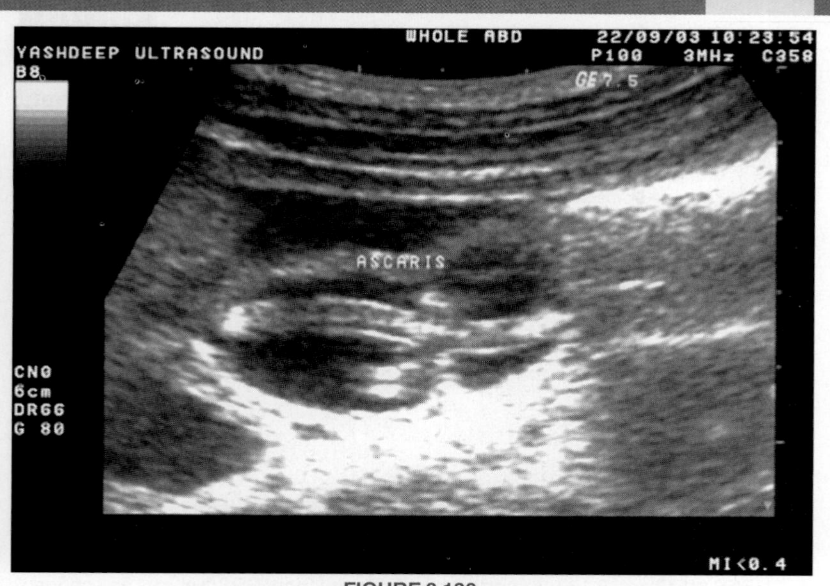

FIGURE 8.130

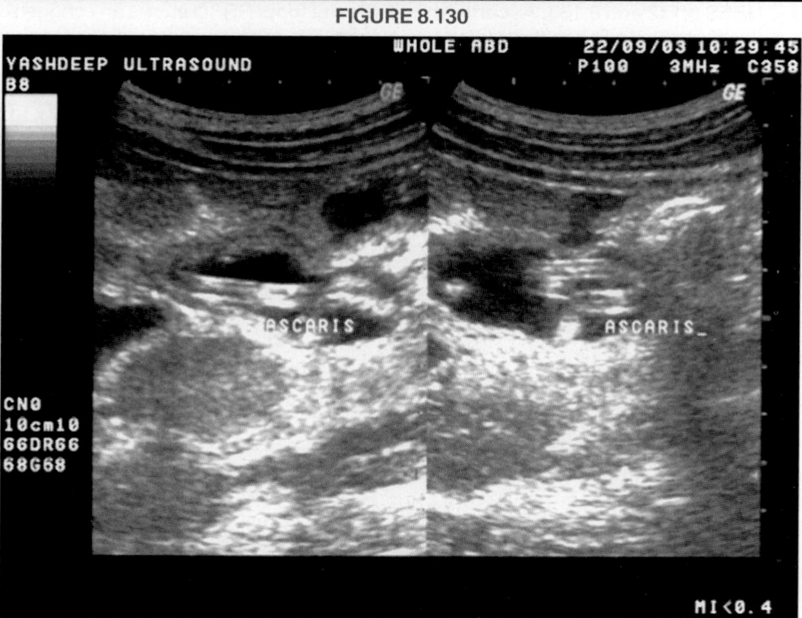

FIGURE 8.131

FIGURES 8.130 and 8.131: Ascaris in the dilated loops. HRSG shows floating Ascaris in the dilated ascending colon in a patient presented with pain in abdomen. Fluid is also seen.

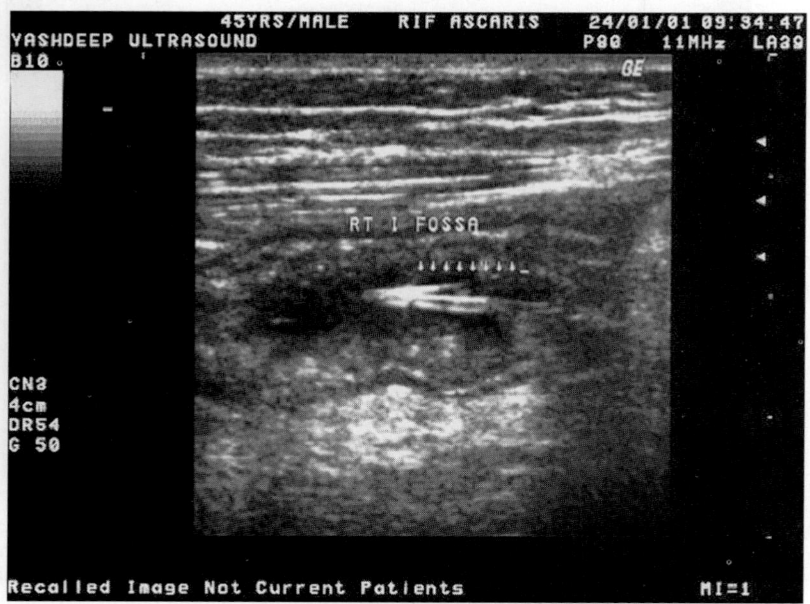

FIGURE 8.132

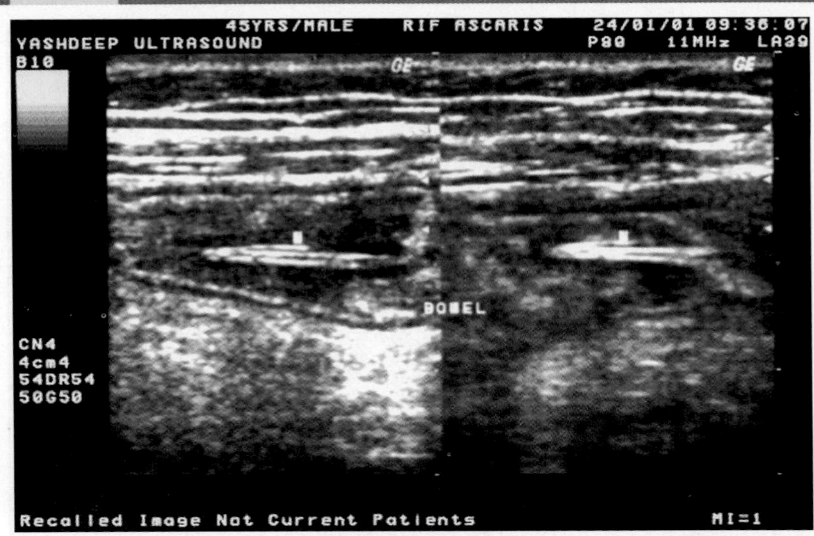

FIGURE 8.133

FIGURES 8.132 and 8.133: Ascaris worm in bowel loop. HRSG shows an echogenic tubular shadow moving in the dilated bowel loop. It is having typical triple layer sign showing the wall and the lumen of worm suggestive of ascaris in the large bowel.

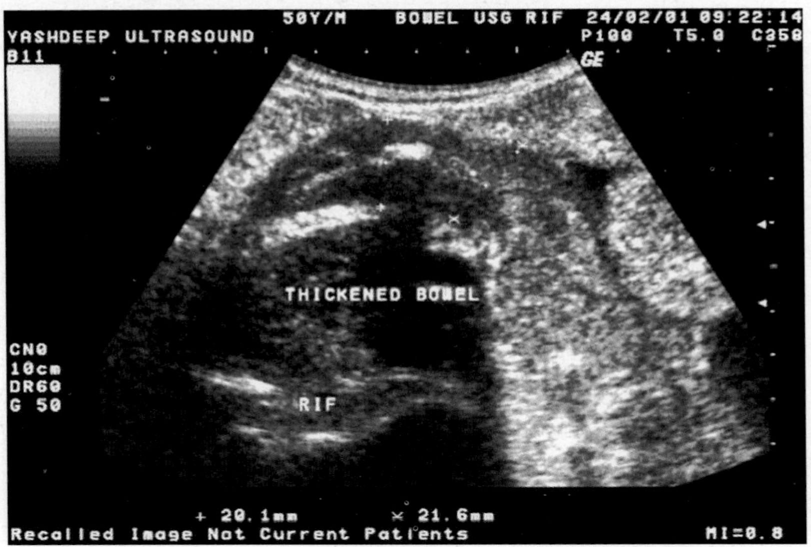

FIGURE 8.134

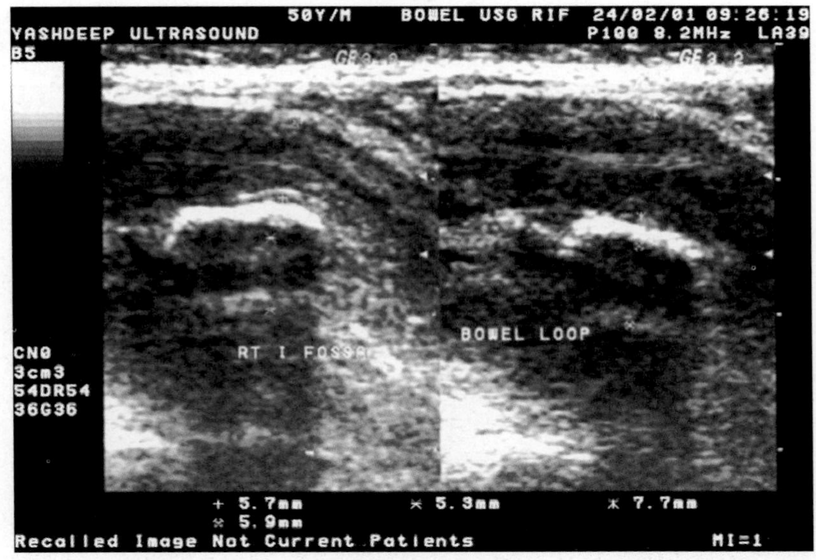

FIGURE 8.135

FIGURES 8.134 and 8.135: Large bowel mass. HRSG shows markedly thickened bowel wall involving the ascending colon. The echogenic gas is seen entrapped in the center giving positive pseudo kidney sign appearance. The proximal small bowel loops are dilated. The lumen of the large bowel loop is obliterated suggestive of a bowel malignancy. The biopsy shows carcinoma colon.

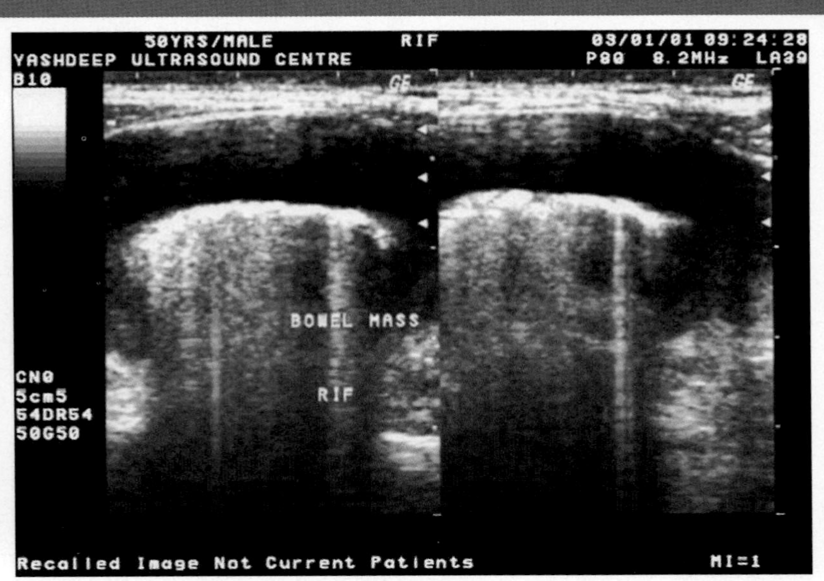

FIGURE 8.136

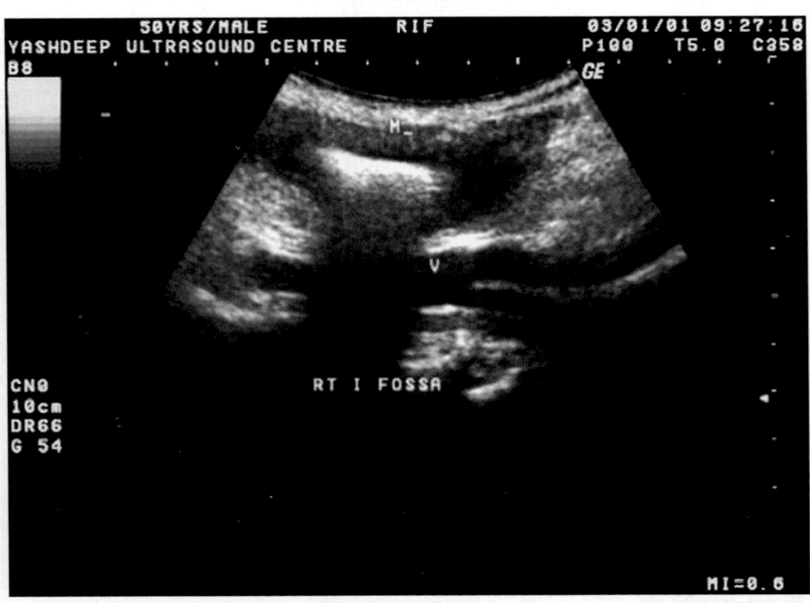

FIGURE 8.137

FIGURES 8.136 and 8.137: A large bowel mass. HRSG shows markedly thickened ascending colon wall with obliterated lumen. Air in the lumen is compressed and it is causing strong reverberation artifacts masking the posterior wall of the bowel. Biopsy of the mass shows adenocarcinoma.

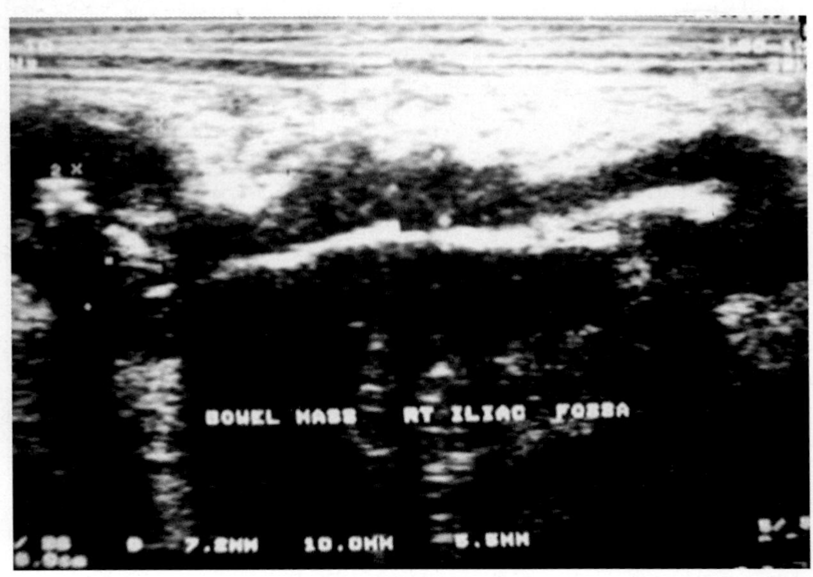

FIGURE 8.138

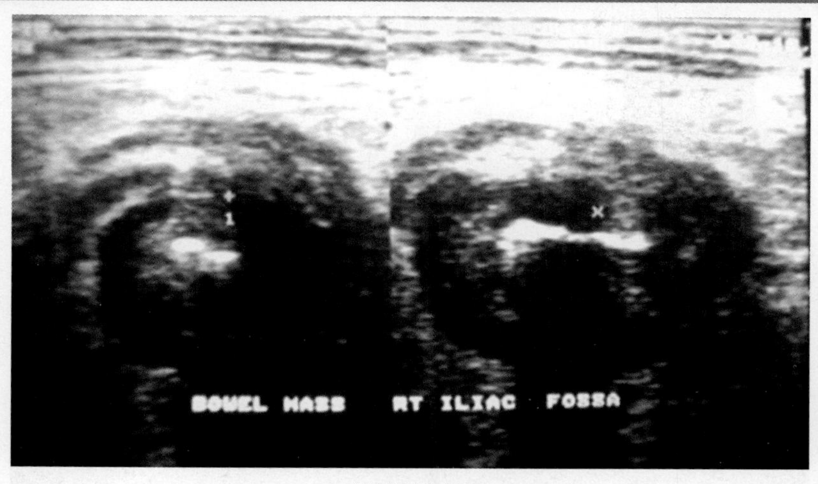

FIGURE 8.139

FIGURES 8.138 and 8.139: Lymphomatous infiltration of large bowel. HRSG shows markedly thickened large bowel loops. It is hypoechoic in texture. Normal muscular wall pattern is lost. Pseudo kidney sign is positive. On cross section typical target sign is seen. Biopsy of the mass shows lymphomatous infiltration of the bowel.

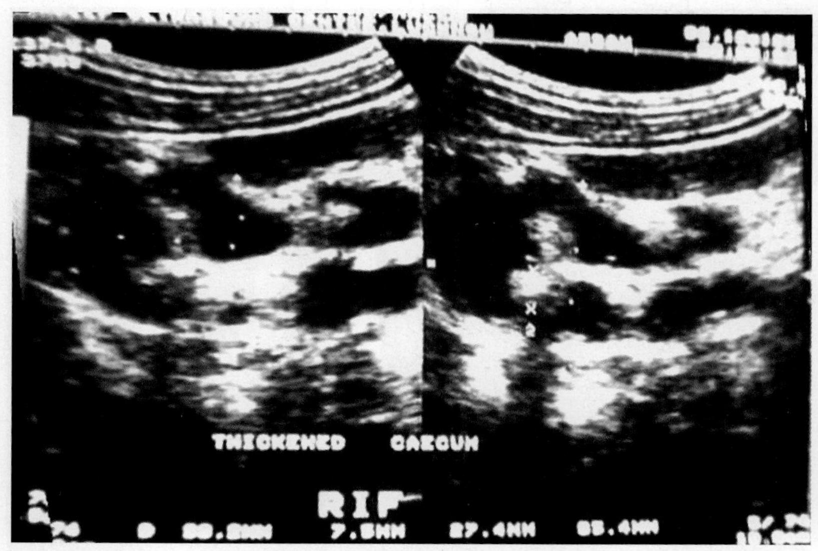

FIGURE 8.140

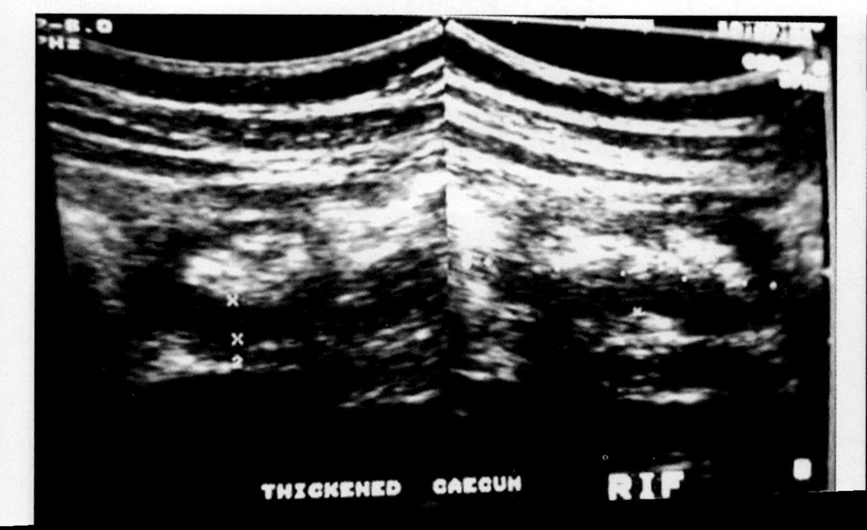

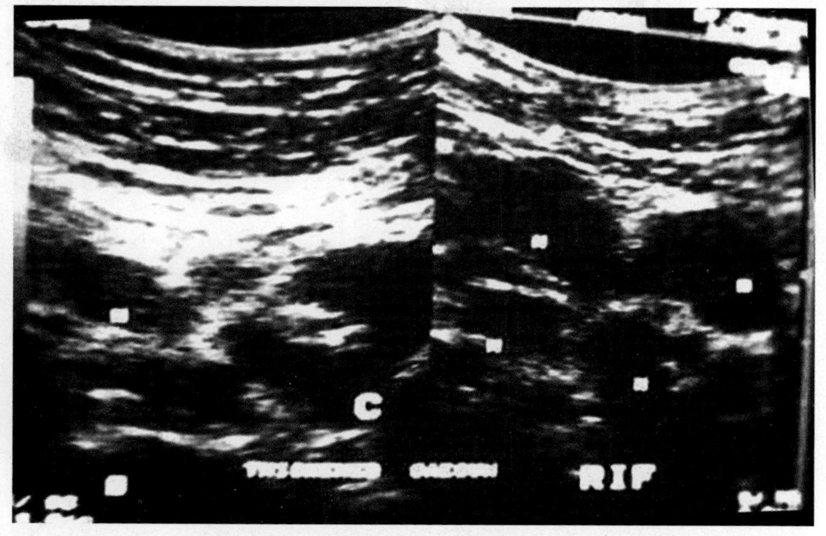

FIGURE 8.142: Multiple enlarged lymph nodes are also seen in the iliac fossa. Biopsy of the mass shows ileocaecal tuberculosis.

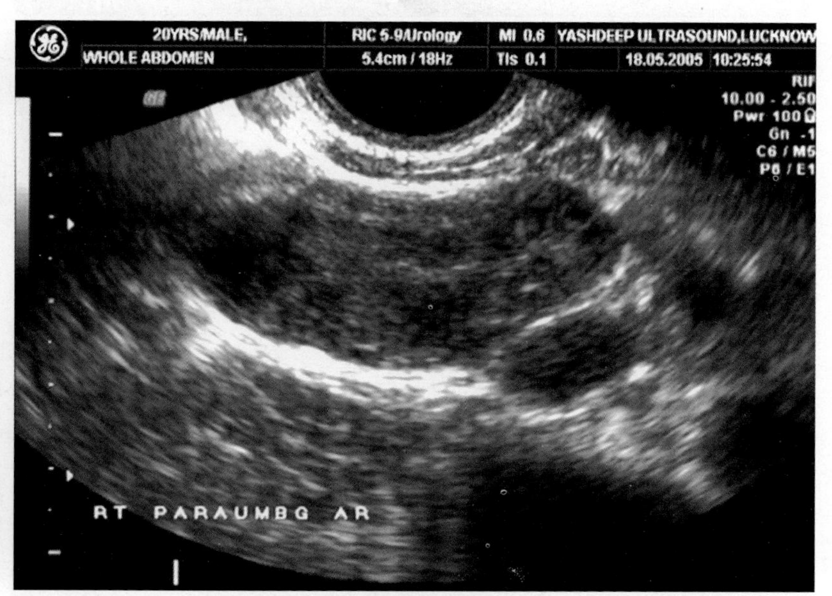

FIGURE 8.143

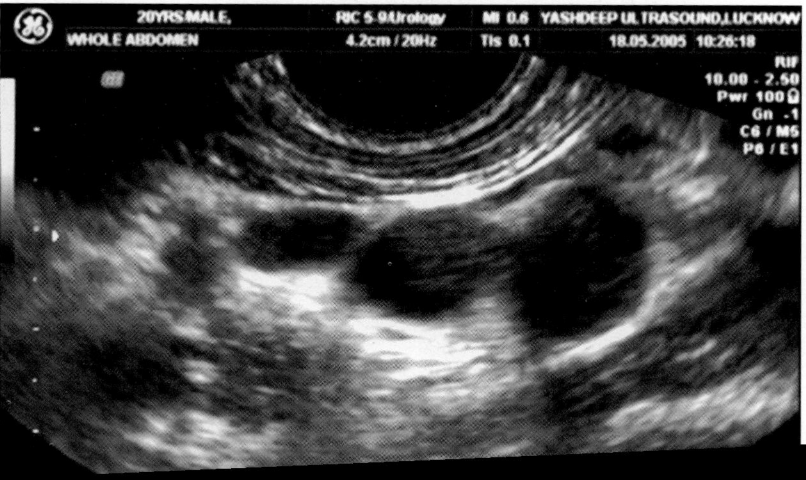

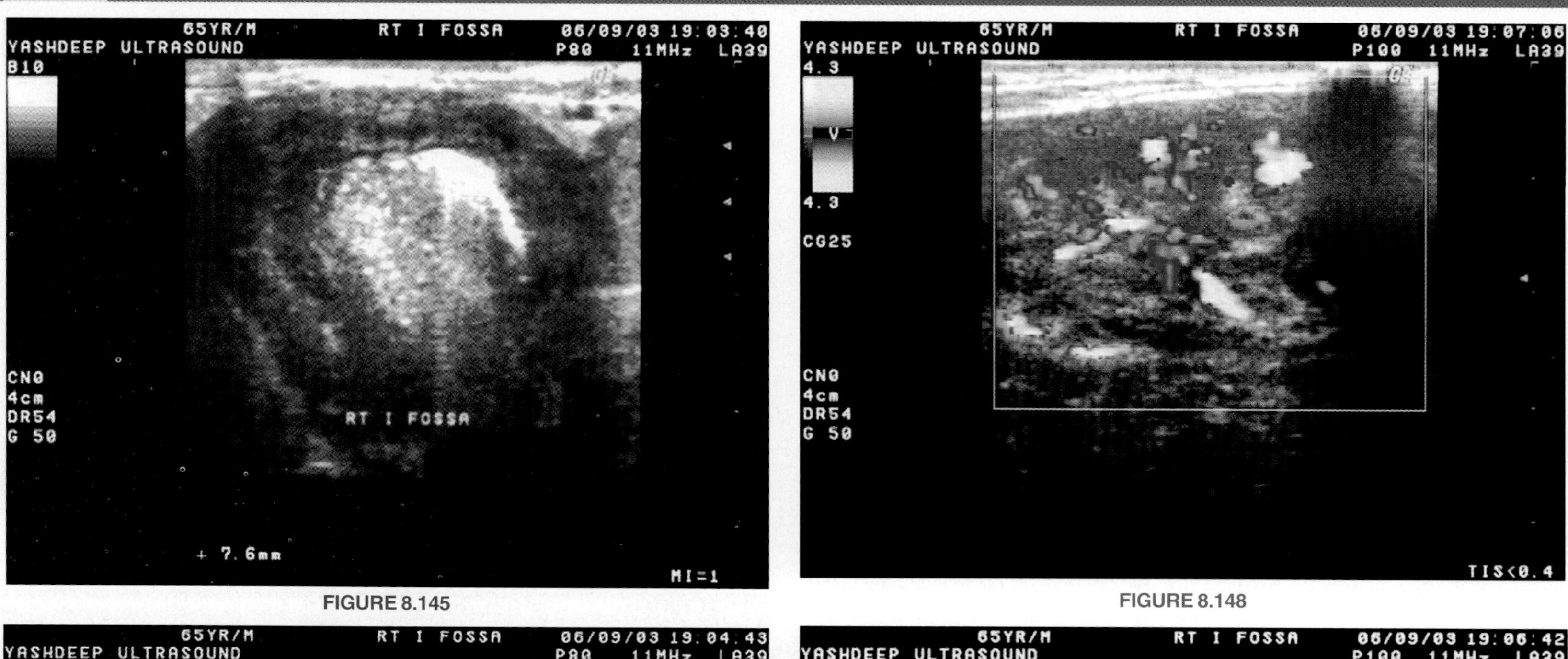

FIGURE 8.145

FIGURE 8.148

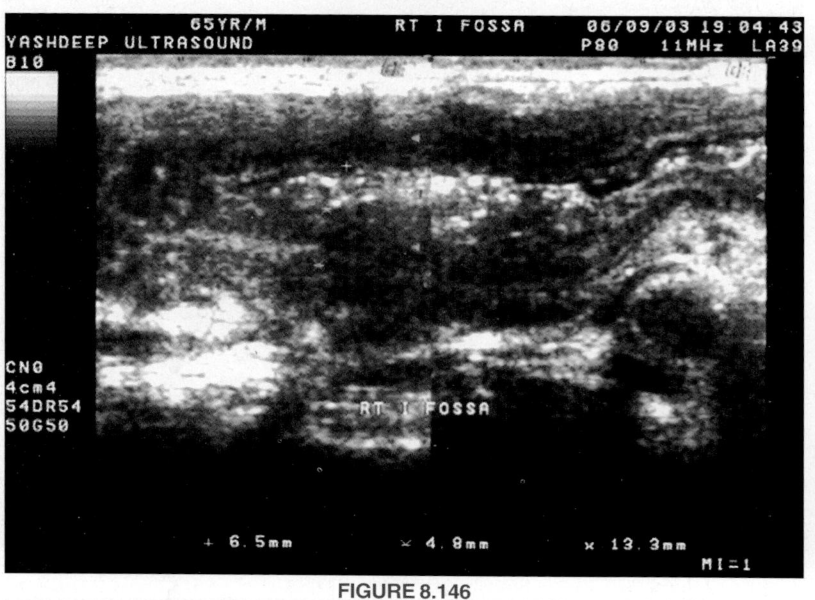

FIGURE 8.146

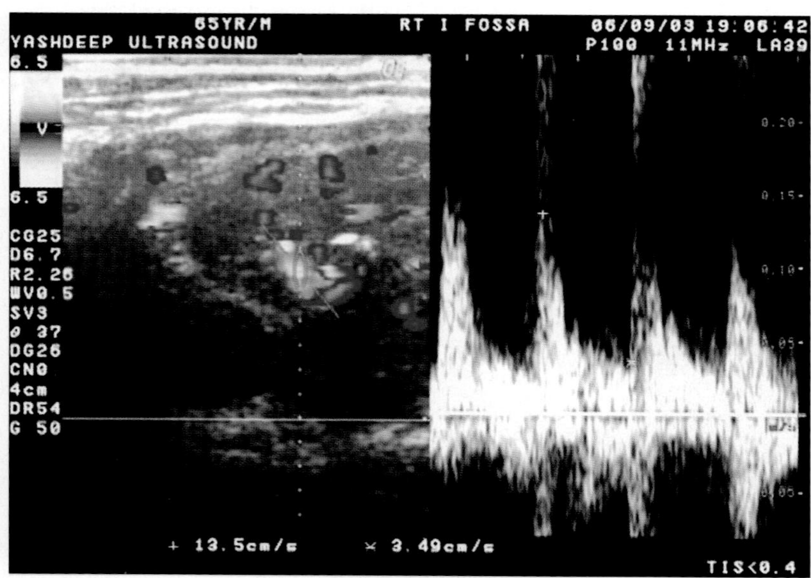

FIGURE 8.149

FIGURES 8.148 and 8.149: On color flow imaging few vessels are seen feeding the mass. On spectral Doppler tracing high flow seen in the mass. Biopsy of the mass confirmed lymphoma of the bowel.

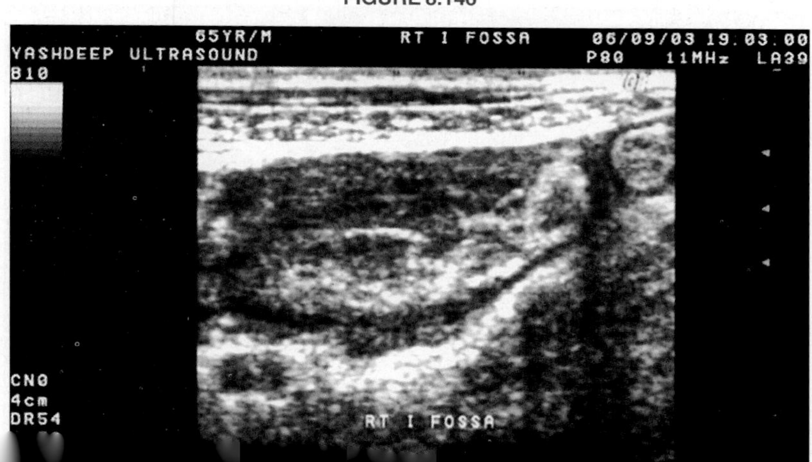

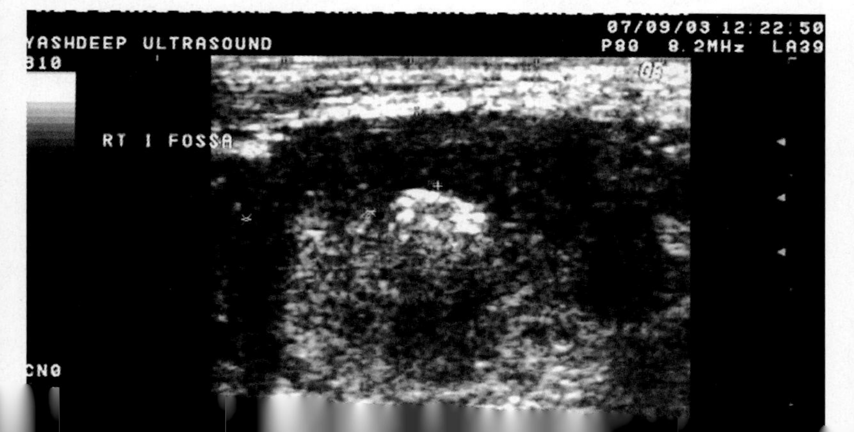

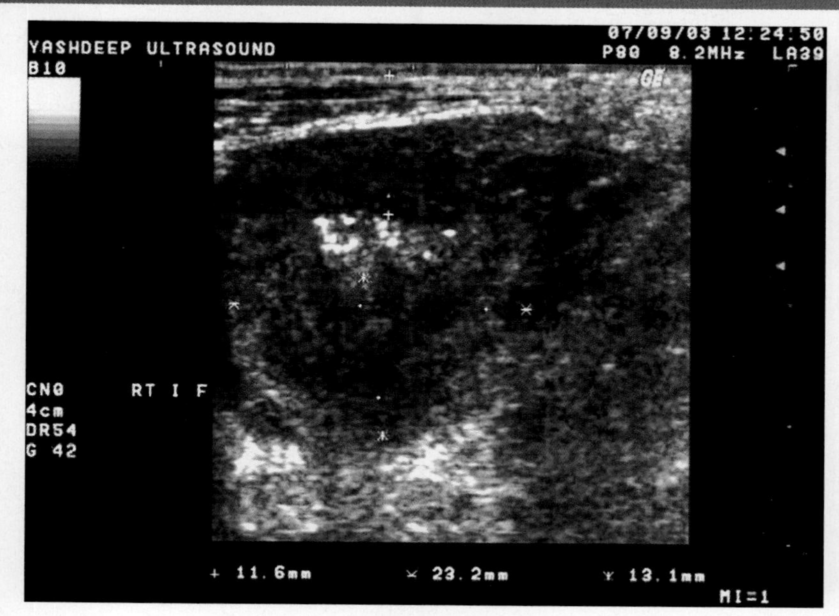

FIGURE 8.151

FIGURES 8.150 and 8.151: HRSG shows grossly thickened bowel loops. Lumen is obliterated. Thin column of air is seen in the lumen. Target sign is positive. Biopsy of the mass confirmed lymphoma of bowel.

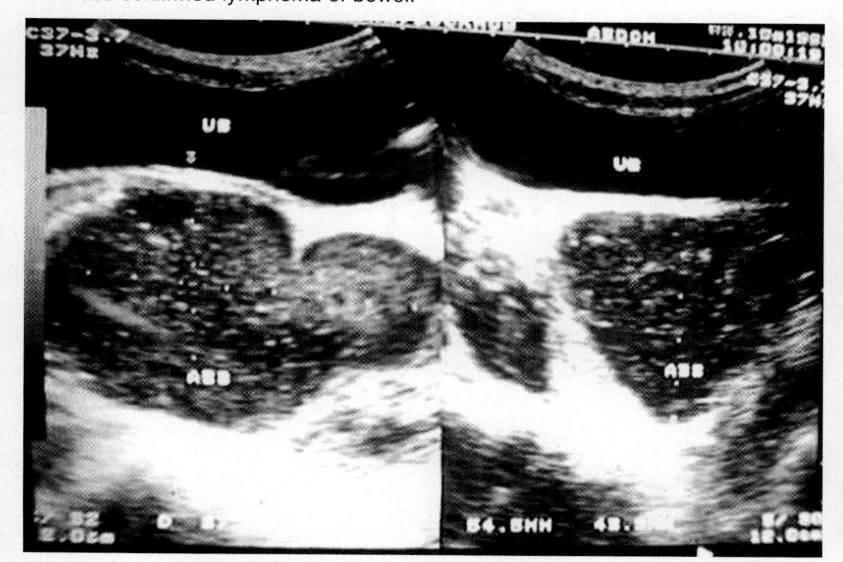

FIGURE 8.152: Rectal abscess. HRSG shows a big low-level echo complex mass in the retrovesicle pouch suggestive of an abscess. Homogeneous low-level echoes are seen in it. Layering is also seen, typical feature of an abscess.

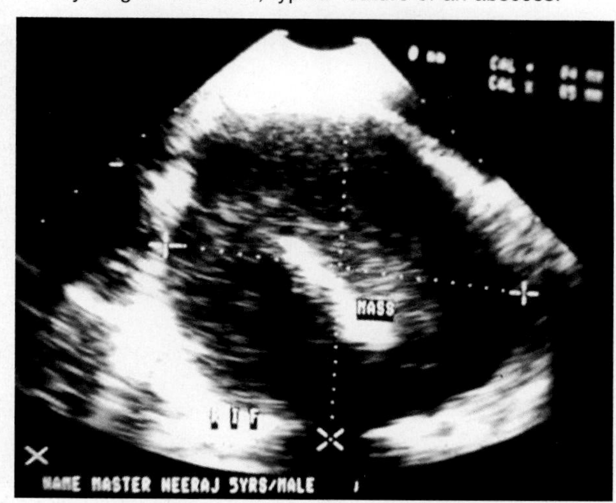

FIGURE 8.153: Rectal abscess. A big low level echo complex mass is seen in the rectum. The mass shows homogeneous echoes. The gas entrapment is seen in it in the collapsed rectal lumen.

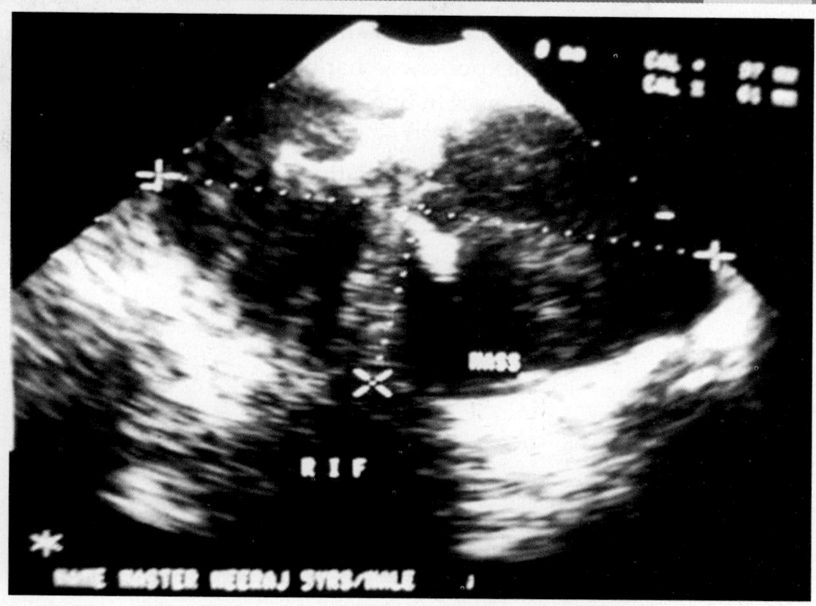

FIGURE 8.154: The abscess is seen extending in the ischiorectal fossa on the Rt side. Irregular margins are seen. Ultrasound guided aspiration of the abscess was done.

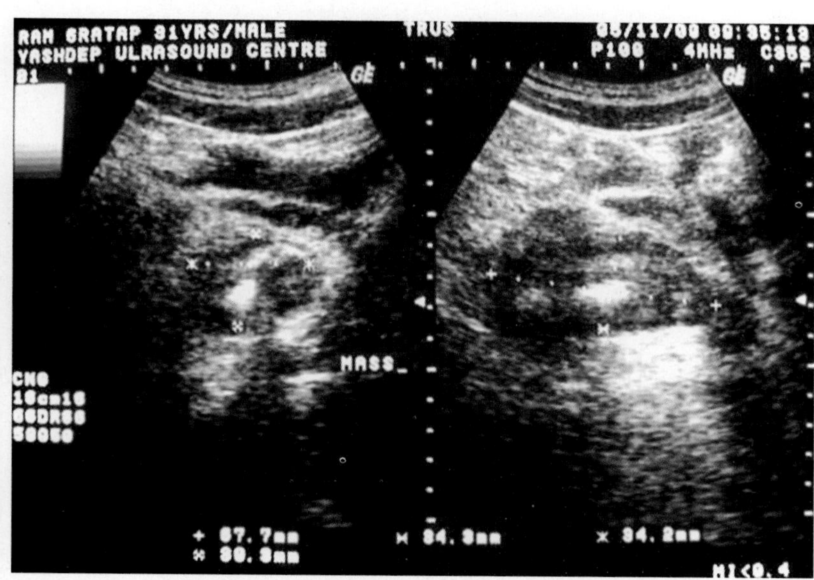

FIGURE 8.155: Rectal mass. A patient presented with soft tissue mass in the rectum. The mass was suspected as enlarged prostate. However, HRSG shows the mass was separate from the prostate and it is seen involving the rectal wall. Transrectal ultrasound shows muscle infiltration of the rectum by the mass.

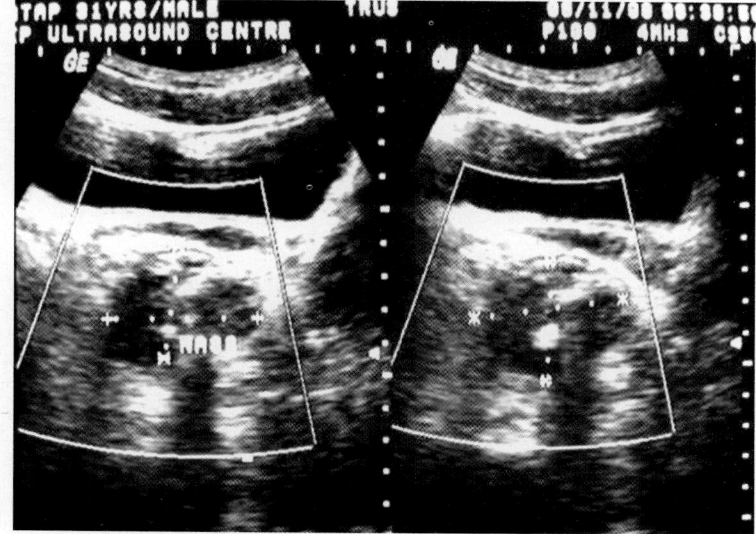

FIGURE 8.156: Color Doppler imaging of the mass shows little flow.

Limitations: Failure to visualize inflamed appendix may lead to erroneous diagnosis in inexperienced hands. An inflamed appendix can be seen in 90% of the cases in non-perforated cases, 80 to 85% in appendiceal mass and 50 to 55% in free perforation of appendix. However, in over 80% of the cases, a positive diagnosis of appendicitis can be made on HRSG. But non-visualization of appendix does not rule out appendicitis particularly when it is aberrantly placed or retrocaecal in position. In these situations, indirect signs of local ileus, small fluid in appendicular fossa or an abscess formation may lead to diagnosis.

Mukund Joshi
PK Srivastava

Prostate and Seminal Vesicles

9

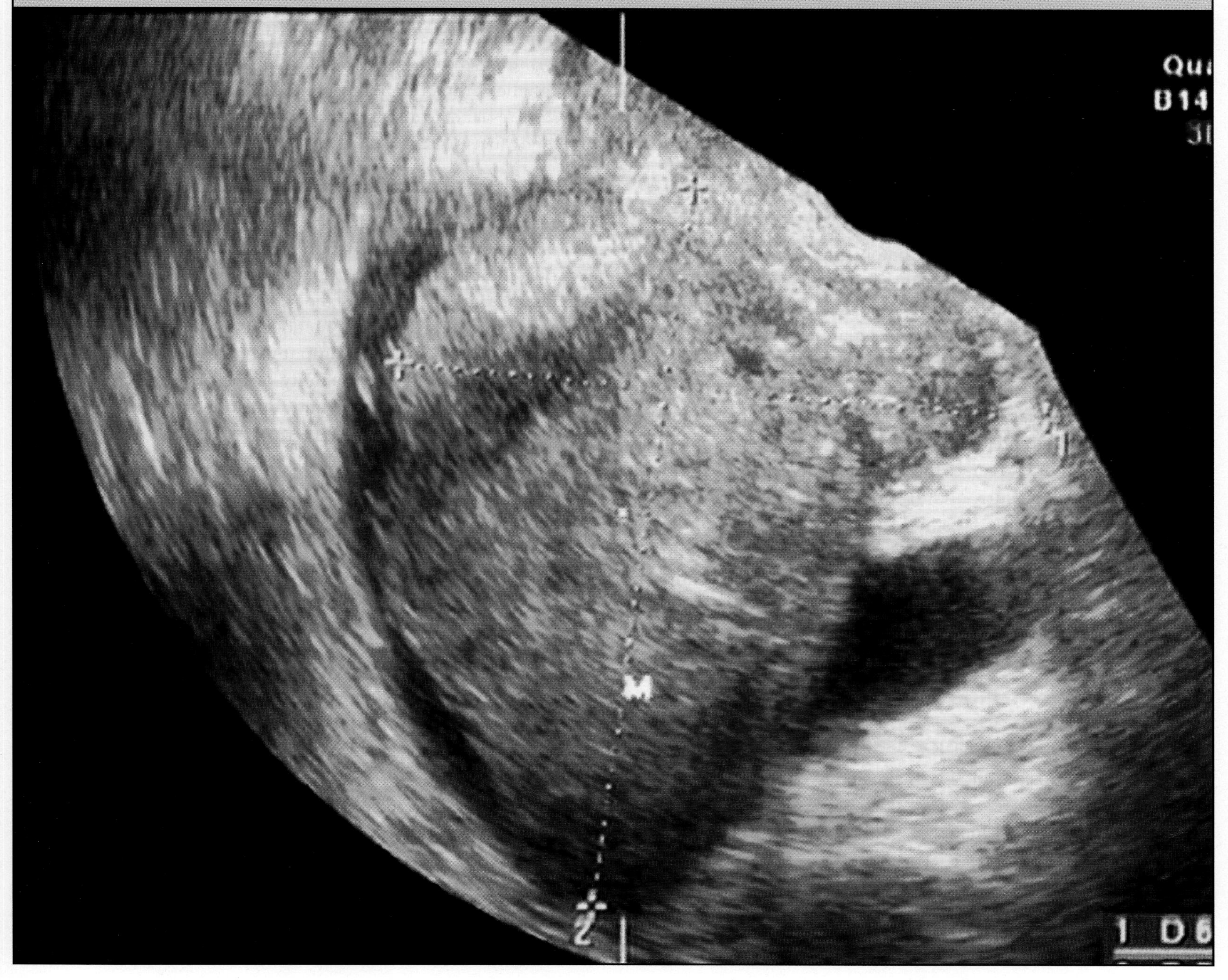

Introduction

The approach to the diagnosis of prostatic disease has changed over the last two decades. Apart from digital rectal examination (DRE), transabdominal ultrasound and the development of serum assays, the availability of transrectal ultrasound with biopsy has revolutionized the thought process in prostatic pathology. Transrectal US (TRUS) of the prostate is a dynamic study and needs to be carried out systematically. In order to understand the pathologies, it is quite essential to know the internal US anatomy of the prostate in great details. The availability of excellent quality color Doppler facilities has added a new dimension to the diagnosis of prostatic cancer.

Gross Anatomy

The normal prostate in a young adult has a triangular or conical shape and a weight of about 12-20 gms. The gland progressively enlarges with age. The average measurements of a normal prostate are 4.0-4.5 cm (transverse) × 2.5-3.0 cm (AP) × 3.0-4.0 (cephalo caudad). The prostate is surrounded by a thin "pseudo-capsule" which consists of dense fibrous tissue and smooth muscle, which is connected with the muscular layers of the prostatic urethra. This pseudocapsule has a few areas of weakness.

- At the insertion of the neuro vascular bundle
- At the insertion of the ejaculatory duct
- At the insertion of the internal sphincter
- At the apex and external sphincter

The prostate is situated immediately superior to the urogenital diaphragm. Its anterior relation is the symphysis pubis and the pubis bone. Laterally, the obturator internus and levator ani border the gland. Posteriorly, the 'Denonvilliers' fascia separates the prostate and its superior margin are in close contact with the urinary bladder.

The seminal vesicles are paired structures situated postero-superior to the prostate. The vas deferens is also paired structures originating from the epididymis and joining the seminal vesicles anteromedially. The prostatic urethra traverses the prostate. There are two important groups of vessels that surround the prostate. The Santorini's plexus is situated anterior to the prostate. The neuro-vascular bundles are situated postero-laterally in the prostate and course from the apex of the gland towards the seminal vesicles. The importance of neurovascular bundles is that they can be involved by tumor extension. They are also important during surgical interference for prostatic cancer. The branches from the neuro-vascular bundles perforate the capsule and enter the prostate.

Lobar Anatomy (Lowsley's Classification)

This divides the prostate into five major and two minor lobes:

- The anterior lobe
- The median or middle lobe
- The posterior lobe
- Two lateral lobes
- Two minor accessory lobes viz., the sub-cervical and sub trigonal lobes.

However, the newer classification depending upon the histopathology of various prostatic disorders is now the accepted standard and not lobar anatomy.

Zonal Anatomy of Prostate

McNeal in 1968, first introduced the concept of zonal prostatic anatomy. He defined three glandular zones, which are covered anteriorly by a fibromuscular stroma. The urethra and the ejaculatory ducts form the network for these glandular zones. The three zones described are:

- The central zone
- The transition zone
- The peripheral zone

Each zone has significant implications related to the origin of the various disease processes.

Central Zone (CZ)

The central zone is pyramidal in shape and contains approximately 25% of glandular tissue at the base of the prostate. Its apex is at the verumontenum. Ducts from the CZ radiate form the base of the gland and terminate into the proximal urethra. The ejaculatory ducts pass through the CZ and terminate into the urethra at the verumontenum. The CZ is resistant to disease process. It is the site of origin of 10% of all prostatic cancers.

Peripheral Zone (PZ)

The peripheral zone is the largest of the glandular zones and contains 70% of prostatic glandular tissue and surrounds the distal urethra. It occupies the posterolateral and apical regions of the prostate and extends anteriorly. The ducts of the PZ enter the distal urethra, distal of the verumontenum. The PZ is homogeneous in echo texture. About 70% of prostatic cancers originate in the PZ. The majority of these are located in close proximity of the prostatic capsule. They are separated from the central and transition zones by the surgical capsule.

Transition Zone (TZ)

The transition zone in the earlier years of the life contains approximately 5% of prostatic glandular tissue. In the ageing prostate, the TZ shows marked hyperplasia and constitutes the majority of overall prostatic glandular elements. It is located on both sides of the urethra. Prostatic calcification within the periurethral glands in the proximal urethra and verumontenum produces typical "Eiffel tower" effect. About 20% of prostatic cancers arise within the transition zone. These are embedded in the longitudinal smooth muscle of the proximal urethra, which is also known as the internal prostatic sphincter.

Anterior Fibromuscular Stroma (AFMS)

The AFMS is non-glandular tissue, which forms the anterior surface of the prostate. It is situated anterior surface of the prostate. It is situated anterior to the urethra and is composed of smooth muscle that is continuous with the detrusor muscle fibers. It is thickest just distal to the verumontanum where it mainly consists of fibrous tissue. It becomes thinner as it reaches the apex of the prostate.

Prostatic Calcifications

Calcifications within the gland may form denovo or represent sequelae secondary to inflammatory conditions. Pathologically, they usually represent calcifications of the corpora amylacea, a proteinaceous material present within the gland. Dahnert et al showed in a sonographic-xeroradiographic-histopathologic correlation study that calcifications within the gland and especially within the corpora amylacea cause bright reflections with or without acoustic shadowing. Calcification in this study was not associated with cancer of the prostate. The distribution of calcifications was either in the central portion of the gland immediately adjacent to the urethra or at the margins of the internal and external gland in an arc/wing-shaped configuration.

A more recent study, however, reported that occasionally tumors may contain echogenic foci, which at a pathological level correspond either to high-grade tumors with extensive central comedonecrosis and calcifications, or unusual deposits of small intraluminal crystalloid within tumor glands. These however, were never as bright or large as calcified corpora amylacea and usually were central in location within the tumor.

Clinical Indications for TRUS

The following are the indications for the use of transrectal sonography:

1. *Patients with abnormal DRE:* Sonography should be performed when the DRE is abnormal. Sonography can differentiate a benign process (e.g. a cyst or calculus) from a neoplastic process. A neoplastic lesion would require further evaluation by biopsy. However, not every palpable lesion (seen on DRE) can be identified using sonography.
2. *Biopsy guidance:* TRUS is an accurate, safe approach for obtaining tissue from a focal abnormality in the gland especially from nonpalpable lesions.
3. *Abnormal laboratory studies:* Men who have normal DRE may still harbor unsuspected prostatic cancer. Abnormally elevated levels of the prostate specific antigen (PSA) and value, the more likely it is that malignancy exists. TRUS can evaluate such occult neoplasms and a transrectal biopsy can be easily performed.
4. *Prostatic inflammation:* US of the prostate is generally not required in acute prostatitis. However, in patients who do not respond to antibiotics, US may delineate a fluid collection/abscess.
5. *Infertility:* Occasionally, male infertility may be due to agenesis and atresia of the seminal vesicles or ejaculatory duct obstruction. TRUS can be used to identify these anatomic abnormalities.
6. *Staging of prostate cancer:* TRUS can only detect local extension of the disease. Once cancer is diagnosed, MRI performs staging.
7. *Deflation of a Foley's catheter*
8. Evaluation of carcinoma of cervix
9. Brachytherapy
10. Screening.

Cancer of prostate is one of the most prevalent malignancies in the world and is the second most common cause of related morbidity in the United States. Should screening be used to detect prostate cancer, earlier than is now possible by DRE, and if so, should sonography be the technique used? The statistics of prostatic cancer suggest that early diagnosis is beneficial. It has been shown that the cancer becomes more aggressive with associated increased metastatic potential when the tumor screening tests with TRUS have shown a small but significant yield of carcinoma in patients without clinical suspicion. US is complementary to DRE, but cannot stand alone as a screening tool. When combined with DRE, and routing PSA/PAP, TRUS helps in detecting more cancers at an earlier stage in high-risk age groups.

Procedure (Technique of TRUS)

When Watanabe and co-workers first introduced TRUS in Japan in 1967, the patients were examined in the lithotomy position following a routine urologic examination. In the early 1970s however, a sitting position with a chair mounted 3.5 transrectal probe was used for more efficient examination. Most investigators today prefer equipment using hand-held transducers with frequencies ranging from 4.5 to 7.5 MHz. The patient is usually examined in a left side down decubitus position. In this fashion, the operator has more control over the endorectal probe position. An enema prior to the examination ensures adequate evacuation of the rectum. A rectal examination done just before the study gives the sonologist a "feel" of the prostatic nodularity, its firmness and the size of the prostate.

With the advancement of technology multiplanar imaging is now possible with accurate localization of abnormalities and extent of the disease. Three major planes are imaged-axial (transverse), sagittal and oblique coronal. The transaxial plane encompasses views made transversely across the gland. Moving the probe cephalocaudally and angling the transducer posteriorly allows all of the prostate and seminal vesicles to be visualized in multiple obtained by moving the probe form side to side within the rectum. Transverse images yield more information concerning the lateral margins and symmetry of the prostate, while longitudinal images show the base and apex of the gland more clearly. The transverse images are easier to understand than the longitudinal images. The sonographer must adopts a standardized examination technique in order to evaluate the prostate and the periprostatic issue completely. By moving the transrectal probe in a cephalocaudally direction, the entire prostate can be visualized. The apex, the base and the intervening can be well delineated, but the bladder is not well evaluated in this plane. The seminal vesicles appear as symmetrically paired structures (bow-tie appearance) at the base of the prostate. On the longitudinal section, the bladder neck and the urethra are well identified. The ejaculatory duct is then seen to traverse through the prostate to the verumontanum.

Sonographic Appearances of Normal and Abnormal Prostate

Normal Prostate

On transrectal sonography, the normal prostate appears as a symmetrical, triangular, ellipsoid structure that is covered by a continuous prostatic capsule. The internal texture of the prostate is composed of multiple, fine, diffuse, homogeneous echoes. The true

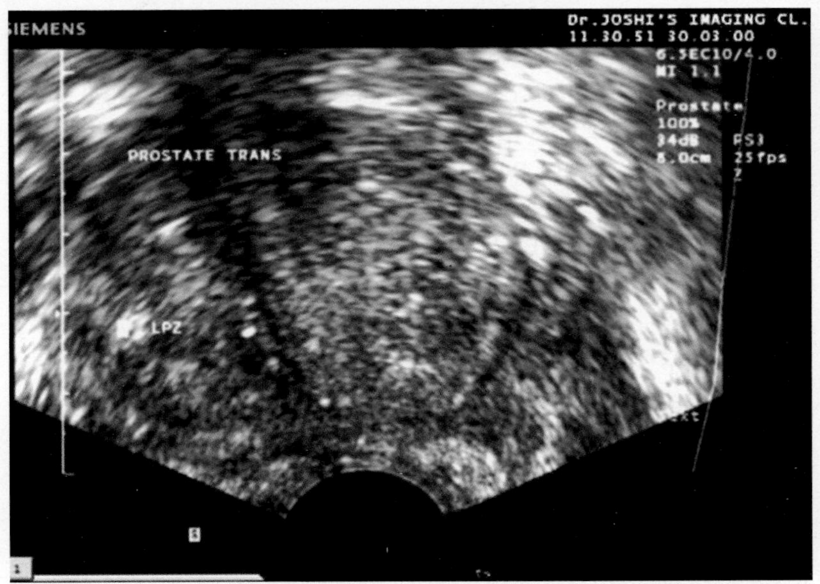

FIGURE 9.1: Normal prostate. TRUS shows normal prostate gland. The lateral lobes are seen as a medium level echoes. The Lt and Rt peripheral zones are seen.

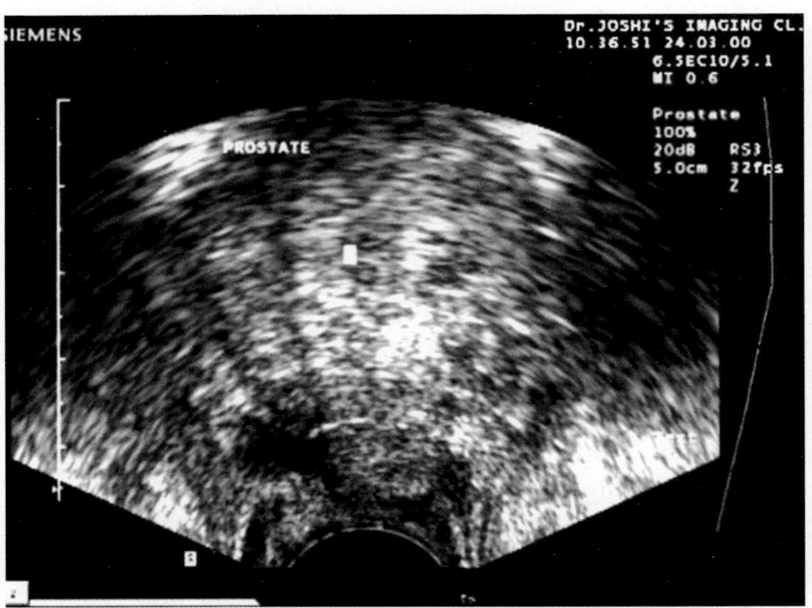

FIGURE 9.2: Normal prostate. TRUS shows normal prostate gland. The central zone and peripheral zones are seen very clearly on HRSG.

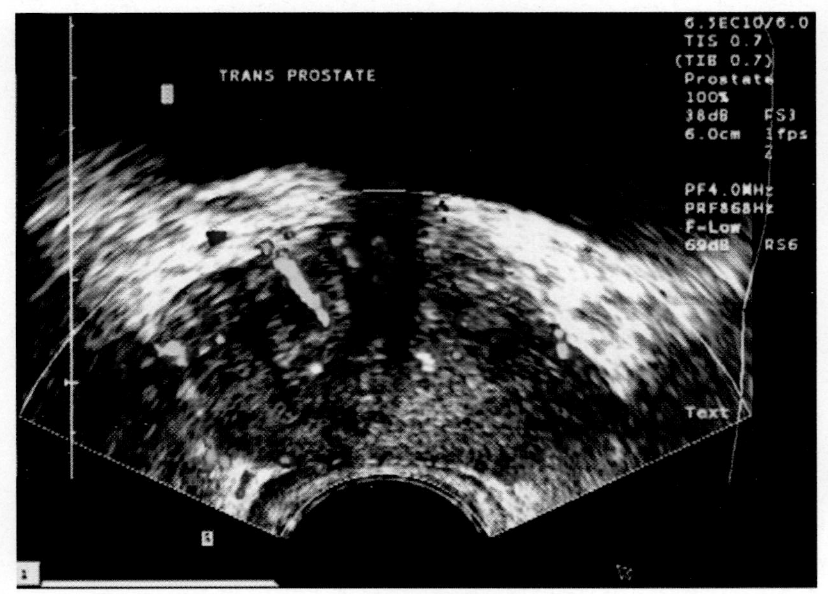

FIGURE 9.3

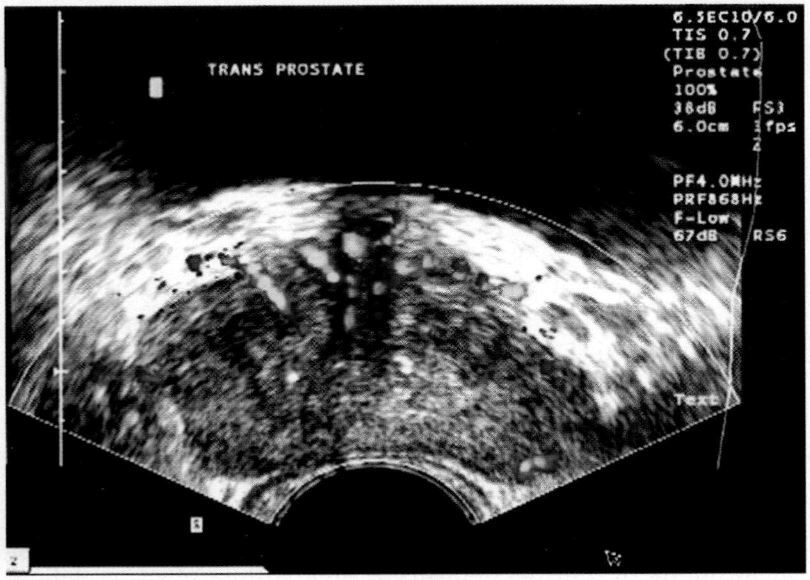

FIGURE 9.4

FIGURES 9.3 and 9.4: Normal flow of the gland. HRSG shows normal flow of the prostate gland. The vessels show low flow.

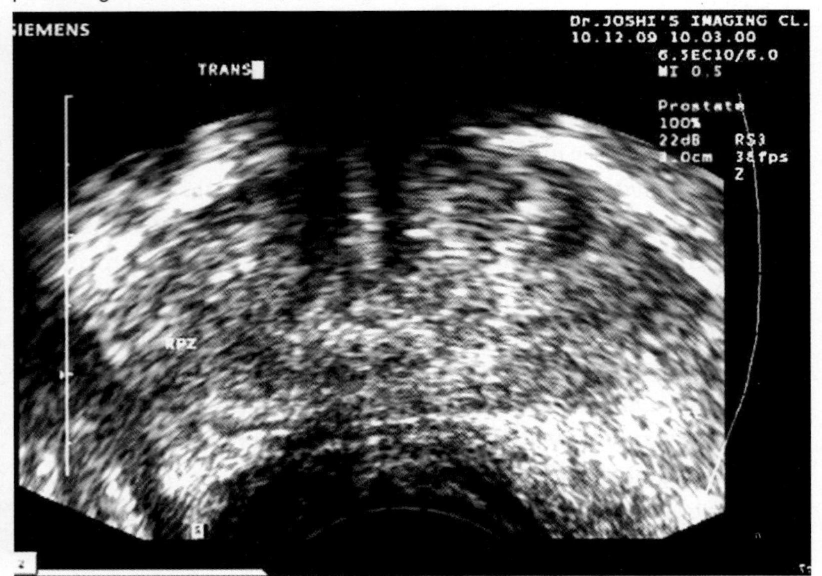

FIGURE 9.5

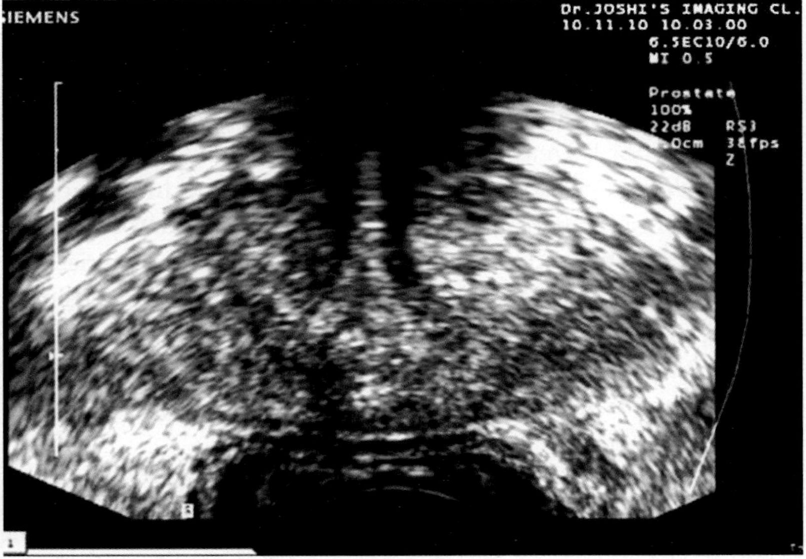

FIGURE 9.6

FIGURES 9.5 and 9.6: Normal Eiffel tower. HRSG shows transition zone in the gland. Typical Eiffel tower appearance of gland is seen in transition zone due to urethra and periurethral glands.

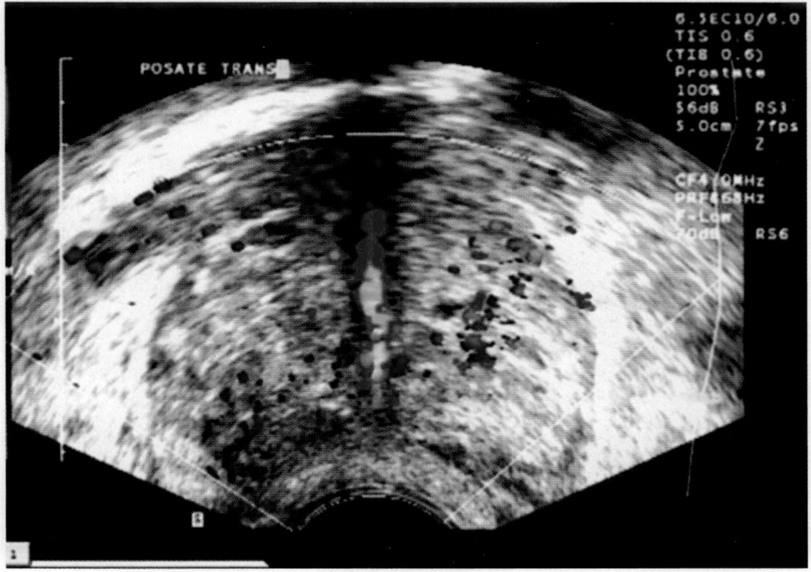

FIGURE 9.7

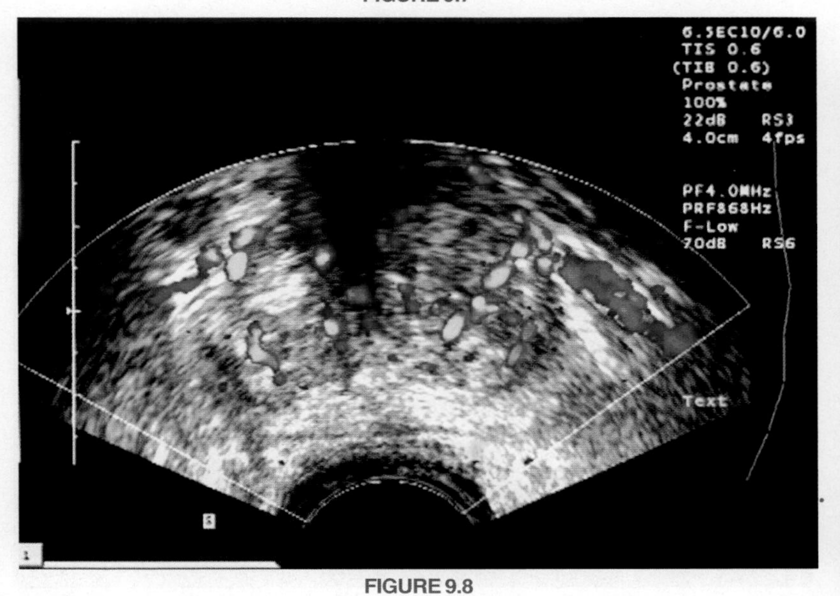

FIGURE 9.8

FIGURES 9.7 and 9.8: HRSG and color flow imaging show normal flow in the peripheral zone on both sides. Low flow is also seen in the central zone.

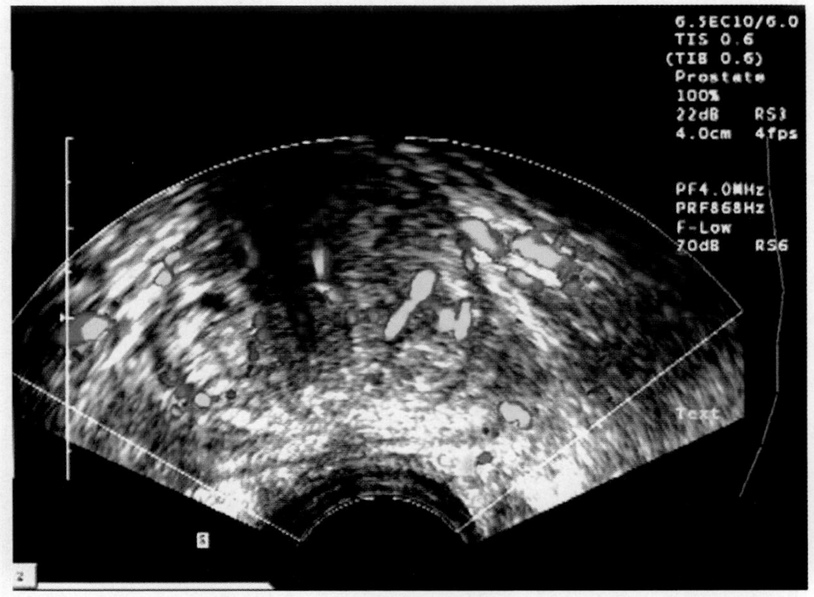

FIGURE 9.9

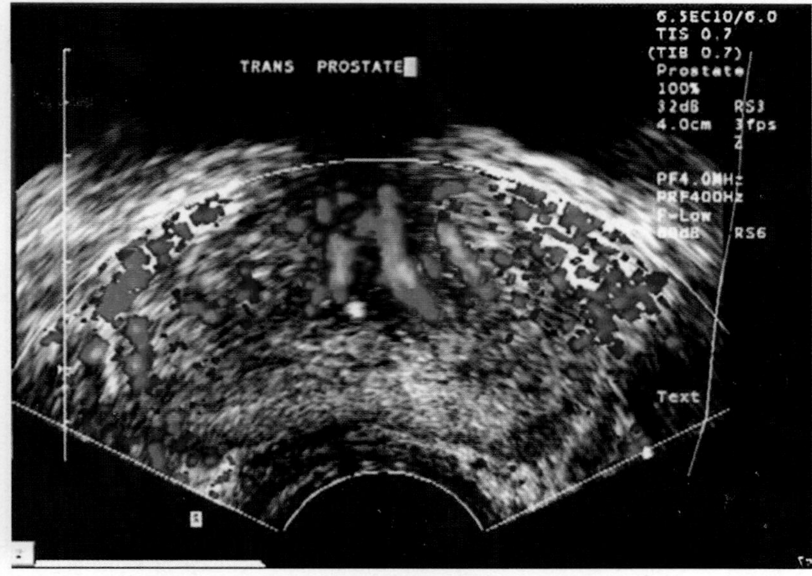

FIGURE 9.10

FIGURES 9.9 and 9.10: Normal flow. Normal flow is seen in the gland in the peripheral zone. Power Doppler imaging shows rich flow.

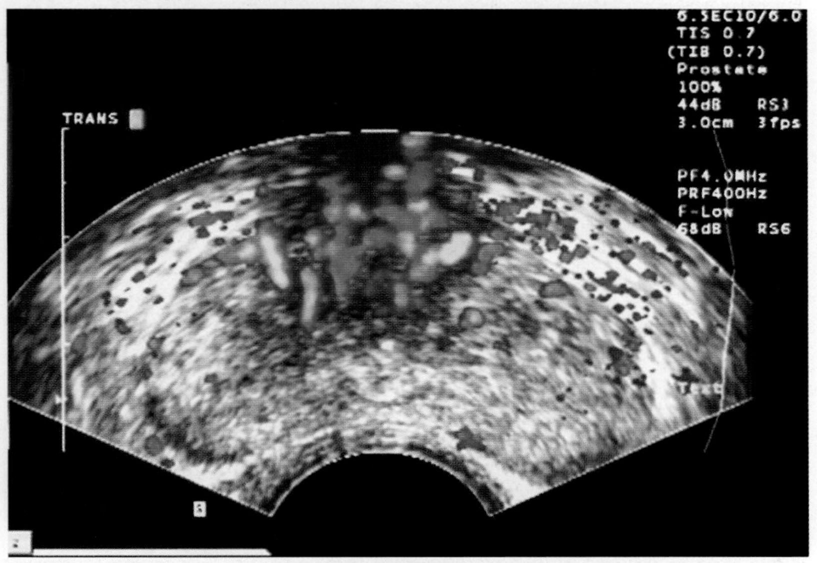

FIGURE 9.11

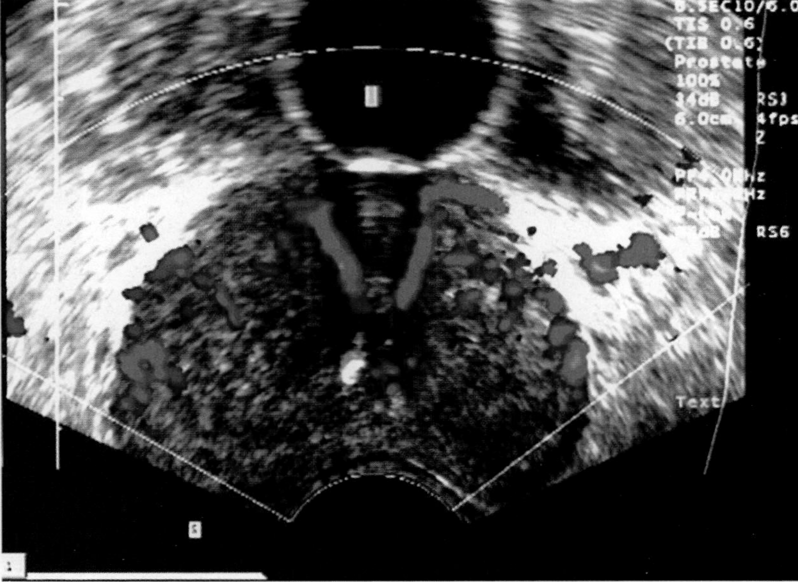

FIGURE 9.12

FIGURES 9.11 and 9.12: Normal flow. Power Doppler imaging shows multiple vessels feeding the central zone. The pericapsular vessels are also seen.

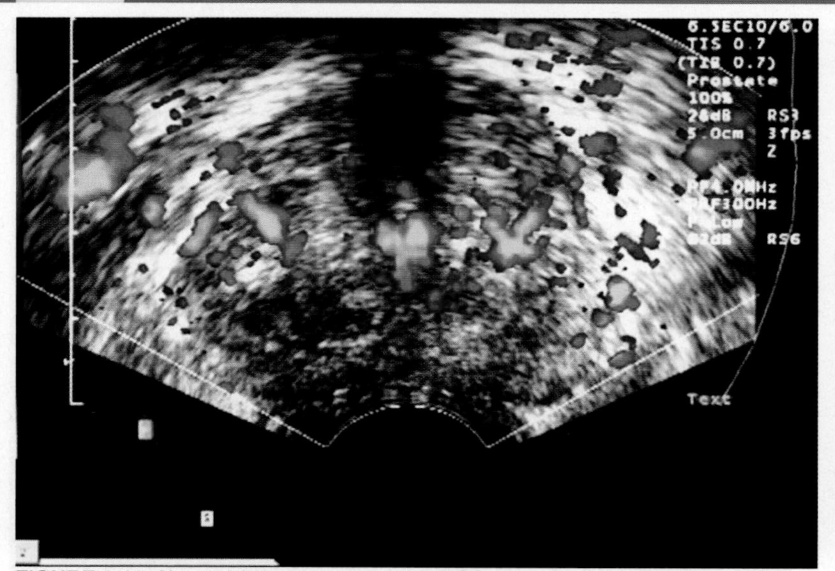

FIGURE 9.13: **Normal flow in the gland.** HRSG shows normal flow in the gland in central and peripheral zones. Multiple periprostatic vessels are also seen.

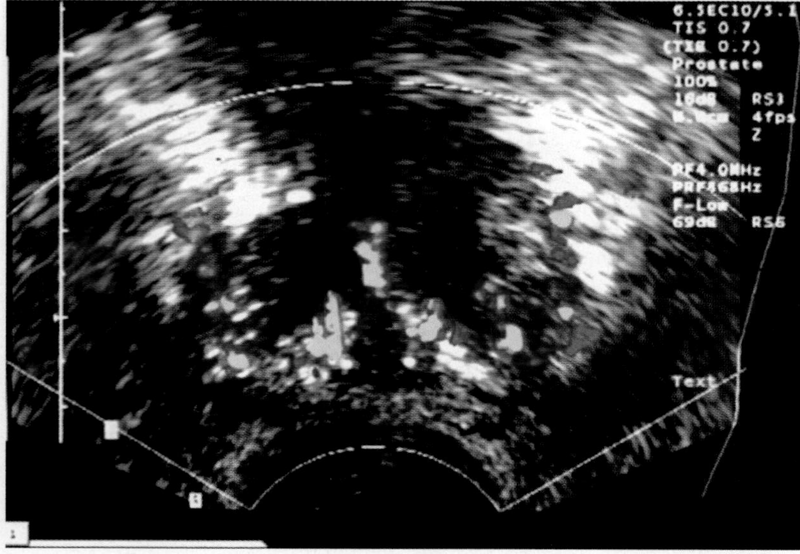

FIGURE 9.16

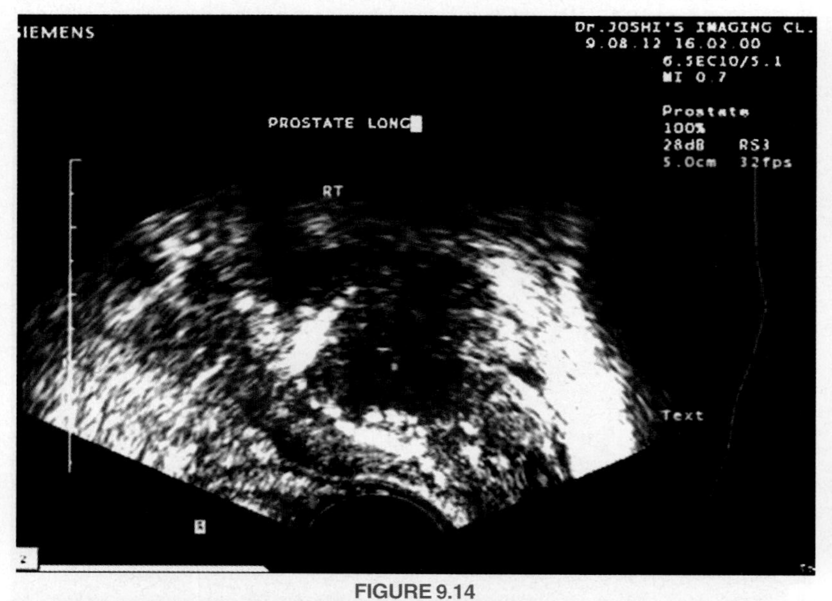

FIGURE 9.14

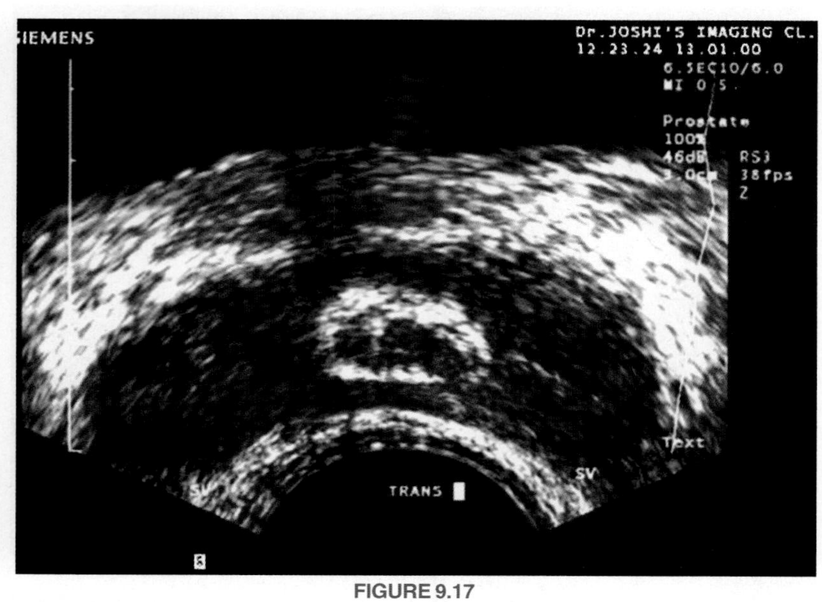

FIGURE 9.17

FIGURES 9.16 and 9.17: **Corpora amylacea.** HRSG shows corpora amylacea calcification. Specks of calcification with calcified rim are seen. Color flow imaging shows normal flow.

FIGURE 9.15

FIGURES 9.14 and 9.15: **Corpora amylacea.** HRSG shows echogenic-calcified specks between the peripheral and central zones of the gland. It is curvilinear calcification between peripheral and central zone and presents corpora amylacea calcification.

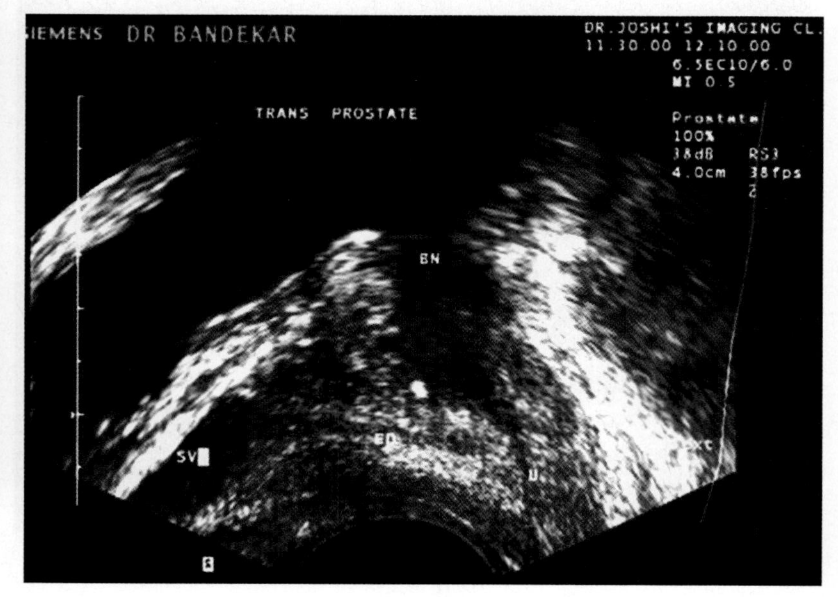

FIGURE 9.18

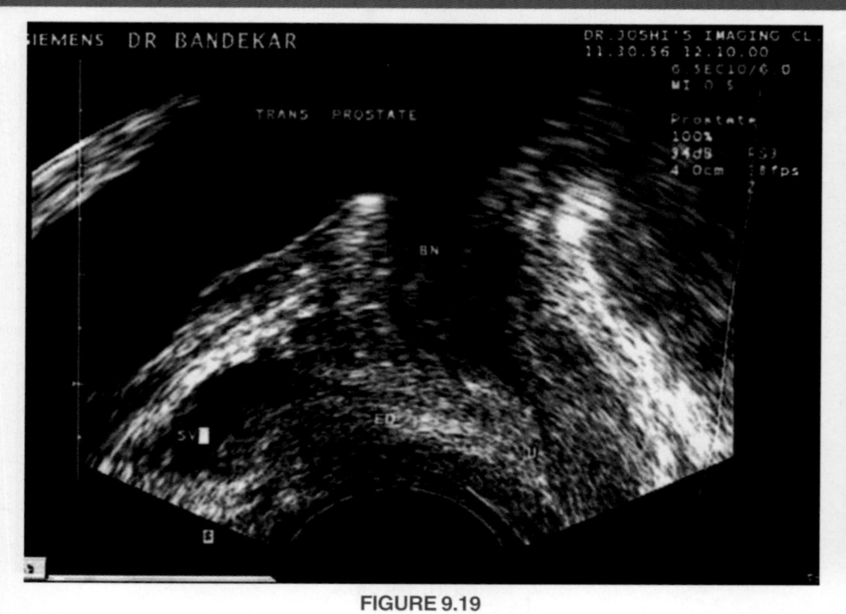

FIGURE 9.19

FIGURES 9.18 and 9.19: HRSG shows normal gland. The seminal vesicles are seen partially on the Lt side. The ejaculatory duct and urethra are also seen.

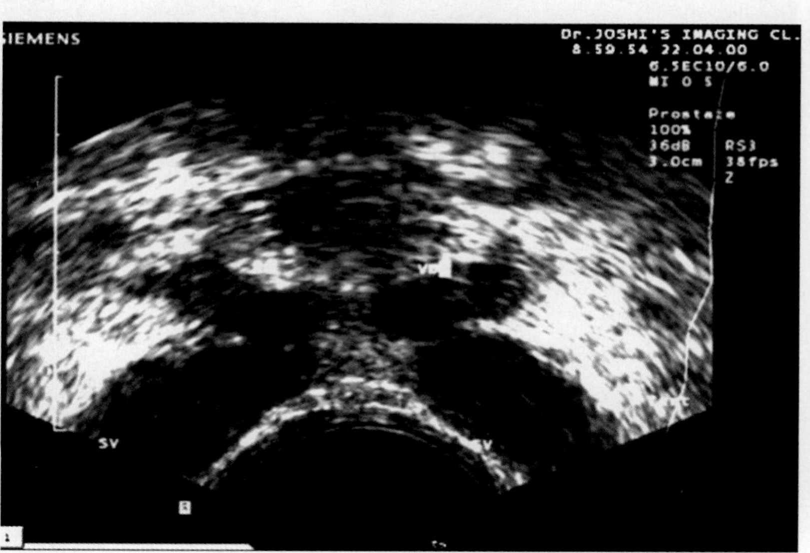

FIGURE 9.20

FIGURE 9.21

FIGURES 9.20 and 9.21: Normal seminal vesicle and vas deferentia. HRSG shows normal seminal vesicle. The vas deferentia is seen anterior to the seminal vesicle on either side. They have similar appearance of seminal vesicle. The ejaculatory duct is also seen running in central zone and opening the verumontenum.

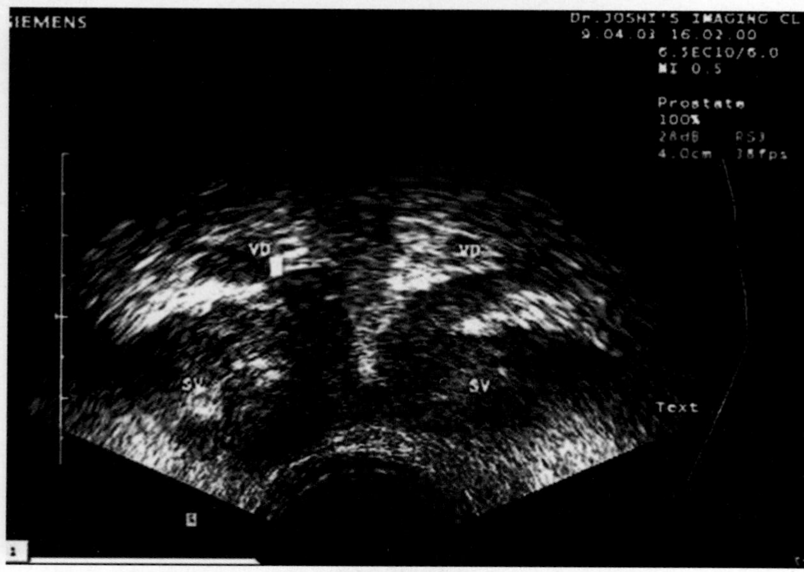

FIGURE 9.22

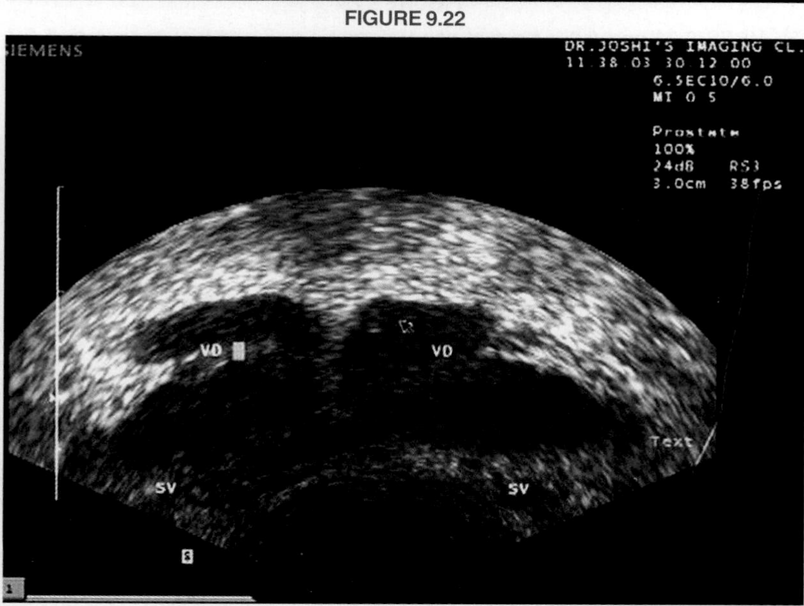

FIGURE 9.23

FIGURES 9.22 and 9.23: Normal seminal vesicle and vas deferentia in transverse plane. HRSG shows normal vas deferentia and seminal vesicle in transverse plane. Ejaculatory duct is also seen.

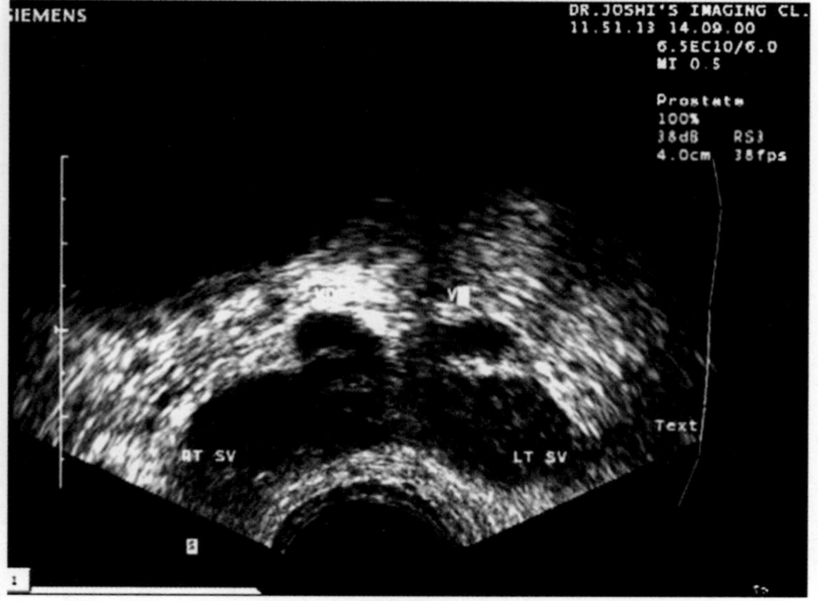

FIGURE 9.24

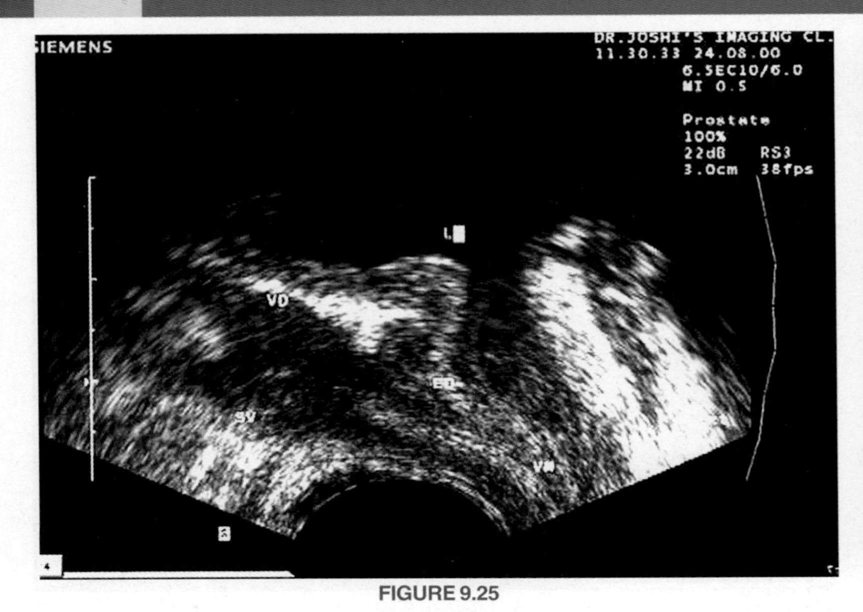

FIGURE 9.25

FIGURES 9.24 and 9.25: Normal vas deferentia and seminal vesicle and ED. HRSG shows normal seminal vesicles on either side. The vas deferentia are seen anterior to the seminal vesicles. Ejaculatory duct is also seen in the central zone and terminating at verumontenum.

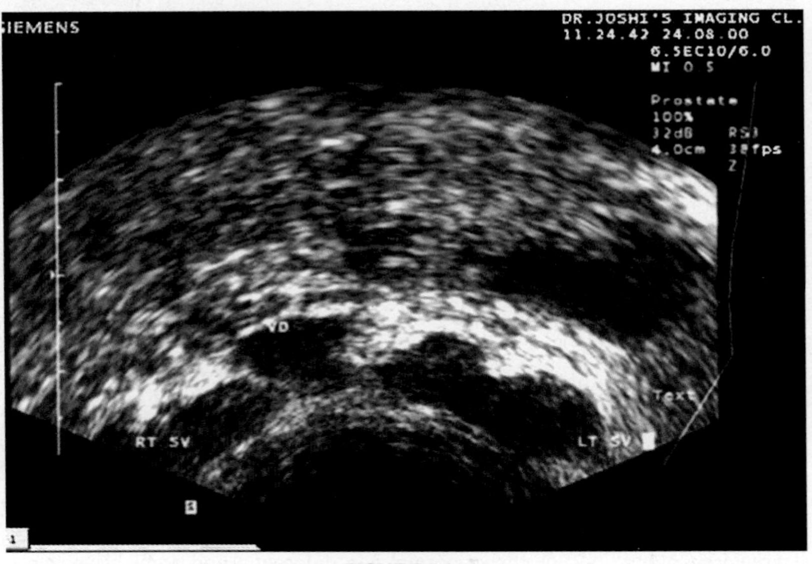

FIGURE 9.26

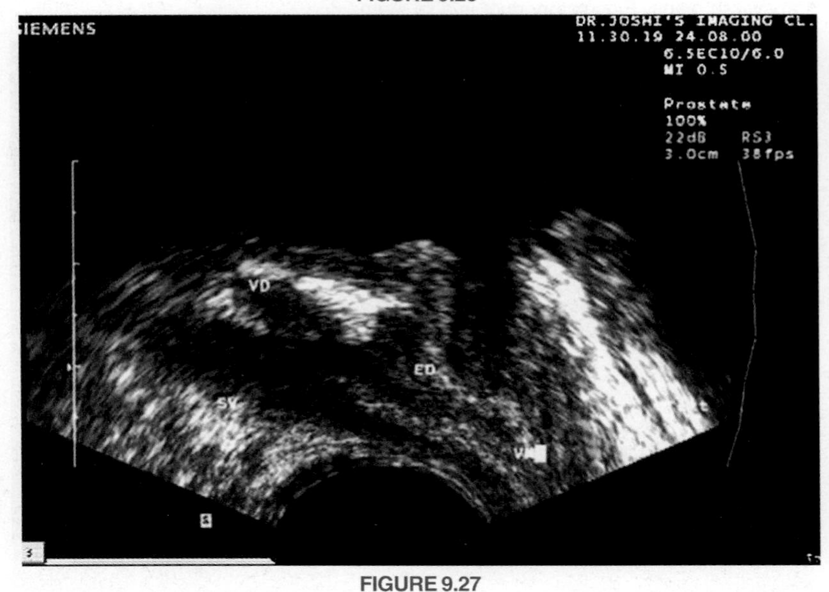

FIGURE 9.27

FIGURES 9.26 and 9.27: HRSG shows normal seminal vesicle, vas deferens and ejaculatory duct.

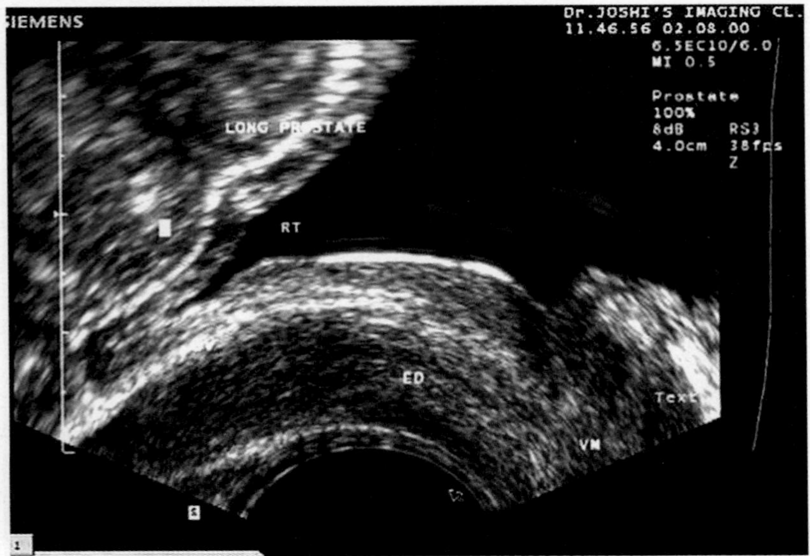

FIGURE 9.28

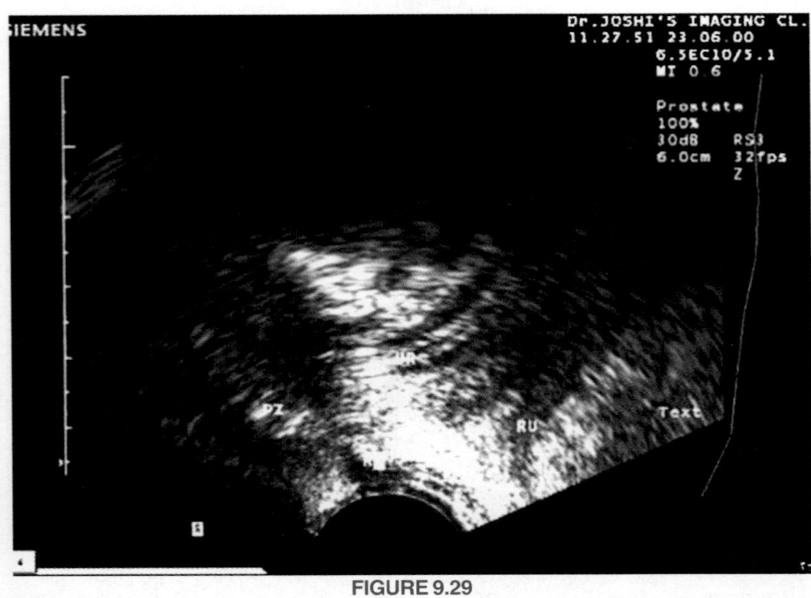

FIGURE 9.29

FIGURES 9.28 and 9.29: HRSG shows normal ejaculatory duct running in central zone and opening in the urethra at verumontenum. Normal upper limit of ejaculatory duct is 2 mm.

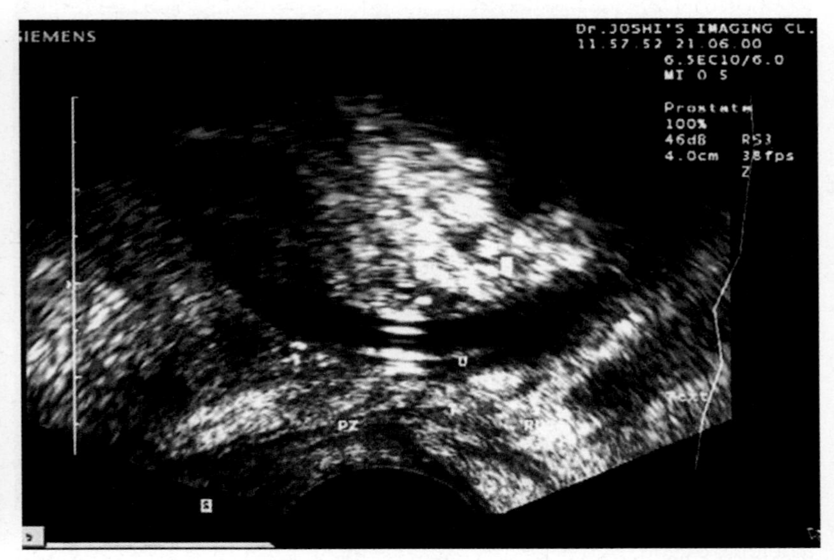

FIGURE 9.30: Normal trapezoid area. HRSG shows normal trapezoid area. The catheters tube is seen in position passing through prostate.

capsule of the prostate is not identified on US. However, the demarcations between the edge of the prostate and the surrounding periprostatic tissue are usually well visualized on TRUS. The surgical capsule of the prostate is comprised of the peripheral and central zones, which becomes compressed into a thin layer by benign hyperplasia of the transition zones. The surgical capsule represents the demarcation between the transition zone and the two outer zones viz. the central and peripheral zones.

Benign Prostatic Hypertrophy (BPH)

Most men with BPH see an urologist because of obstructive and irritative symptoms. These manifestations have been grouped together as 'prostatism'. DRE demonstrates a smooth firm and soft enlargement of the prostate. However, depending upon the anatomical site of the adenoma, the gland may feel normal. Abnormal uroflowmetery suggests decrease in stream and force of the urine stream.

TRUS is used very commonly for the evaluation of the prostate. Periurethral enlargement is noticed as intravesical extension of the gland. The AP and the lateral dimensions of the gland appear normal, but there is cephalad extension of the prostate. Approximately, 95% of cases of BPH arise from the transition zone and 5% from periurethral glandular tissue. Distinct nodules or diffuse enlargement can be present. The US appearance of BPH varies depending upon histopathologic changes. Typically, the central gland enlarges of BPH. Often with BPH the inner gland remains hypoechoic. On TRUS, it is frequently seen as a hypoechoic or heterogeneous, well-delineated mass involving the periurethral regions. Occasionally, BPH may compress the central zone as well as the peripheral zone to only a few millimeters in thickness. The surgical capsule is the line of demarcation between the central zone and the peripheral zone. Often, a halo and /or calcifications may be seen. In normal young men, there is usually no clear differentiation between the central and peripheral glands. However, when BPH develops, those two areas can usually be identified separately. Focal BPH nodules can bulge the capsule of the gland or compress the lateral margins, but they should not disrupt the capsule or periprostatic fat. If diffuse involvement with BPH is present, the zonal anatomy of the gland can no longer be discerned.

Prostatitis—Prostatic Abscess

Prostatitis is an acute inflammatory condition that can involve either the peripheral zones or the central, periurethral areas. It can be seen in men of all ages. Prostatitis, when not due to surgical manipulation, develops nearly exclusively in the acinar prostate. Therefore, the distribution of hematogeneous spread of infection is predominantly in the peripheral zone resembling prostate cancer. When prostatitis is secondary to post-surgical instrumentation or manipulation, the central gland can be infected. Unless abscess formation occurs, the US features of prostatitis are rather non-specific. Focal or diffuse hypo-or hyperechoic areas may be seen, often mimicking cancers of the gland. In chronic prostatitis, the TRUS findings are variable, usually showing a heterogeneous echotexture with an irregular appearance resulting in difficulties with the diagnosis. Dystrophic calcifications can develop in the peripheral gland due to chronic inflammation (chronic prostatitis). Additional US features than can be seen in chronic prostatitis include focal masses of different echogenicities that, if hypoechoic can mimic cancer, ejaculatory duct calcifications, capsular thickening or periurethral glandular irregularities. Occasionally, enlarged blood vessels creating a translucent zone around the prostate have been described in patients with chronic prostatitis. Color Doppler US may help in the detection of increased vascularity in the inflammatory phase of prostatitis. Complications of prostatitis, i.e. abscess formation may be seen as hypoechoic or fluid filled areas within the central or peripheral gland. US guided aspiration and instillation of antibiotic into the abscess can be performed via a transperineal or transrectal route.

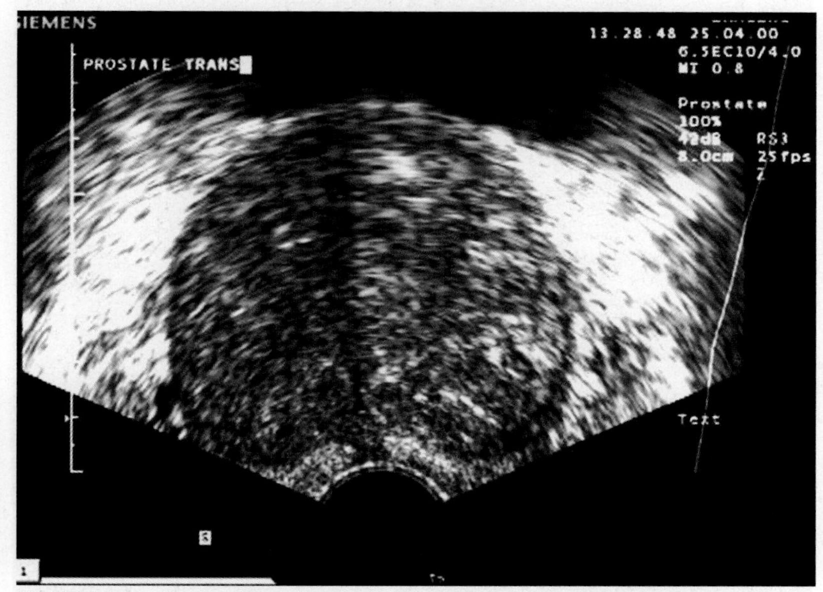

FIGURE 9.31: Benign hypertrophy. HRSG shows diffuse enlargement of prostate gland. However, the glandular texture is homogeneous. No calcification is seen.

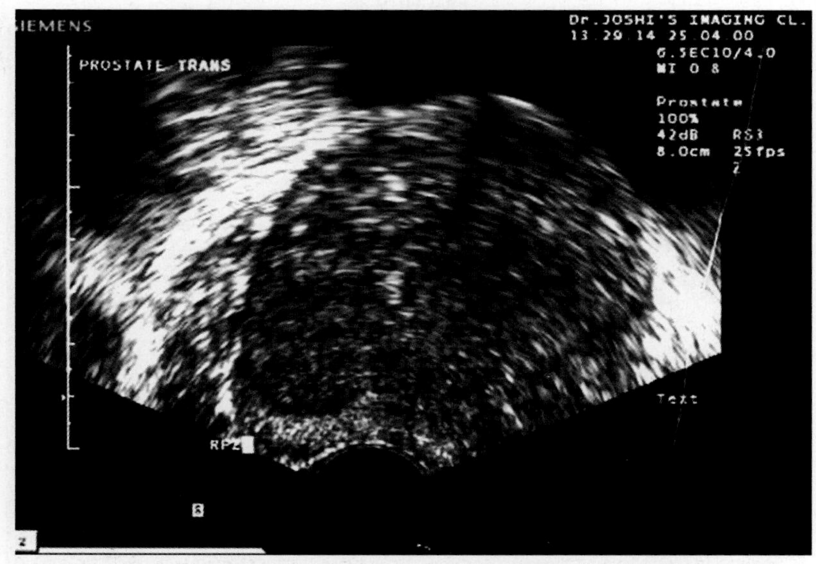

FIGURE 9.32: Benign prostatic enlargement in the transverse plane. HRSG shows diffuse enlargement of the prostate gland. Homogeneous enlargement of the peripheral zone is seen.

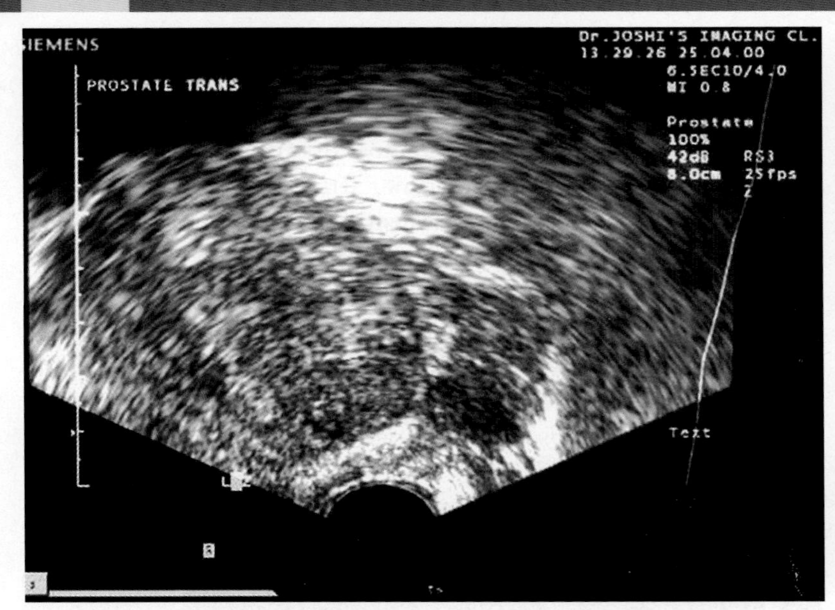

FIGURE 9.33: Benign prostate enlargement in transverse plane. HRSG shows diffuse enlargement of Lt lateral lobe of the prostate. Homogeneous echoes are seen in the lobe.

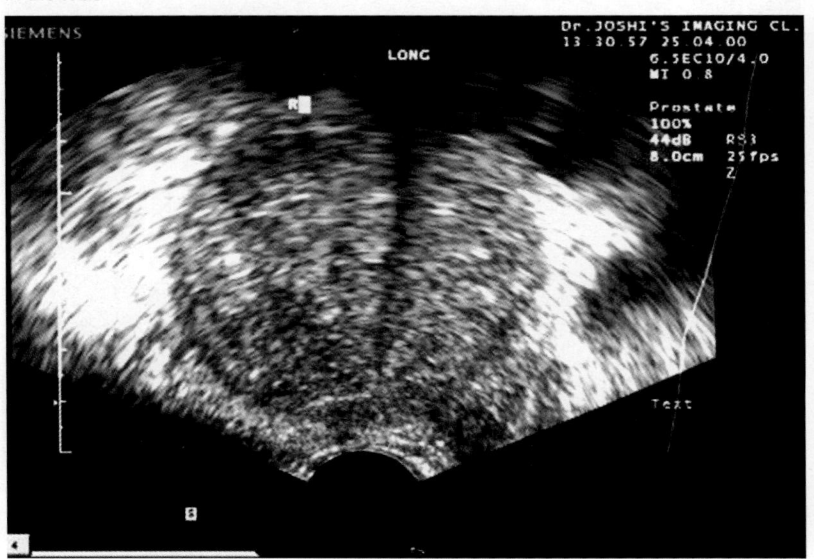

FIGURE 9.34

FIGURE 9.35

FIGURES 9.34 and 9.35: Homogeneous enlargement of the gland. HRSG shows homogeneous glandular enlargement involving both lobes of the prostate. No necrosis or calcification is seen in it.

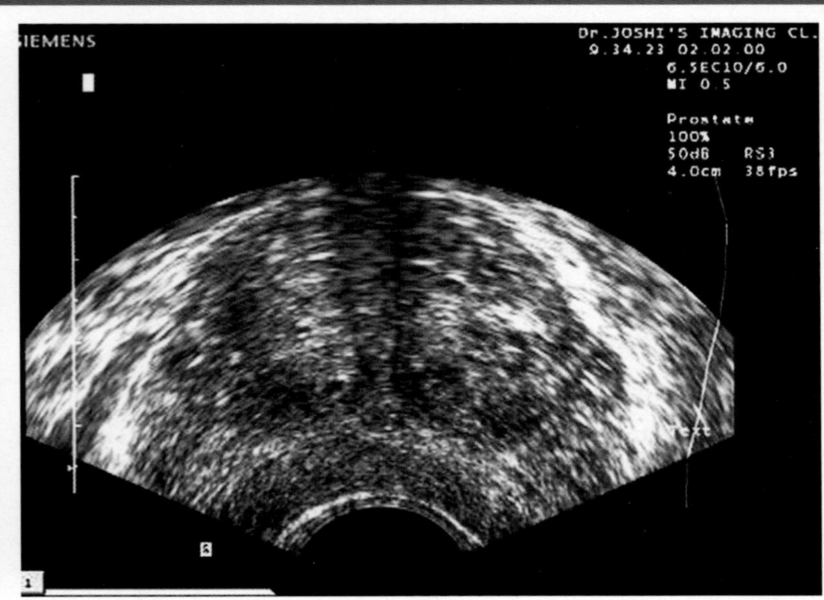

FIGURE 9.36

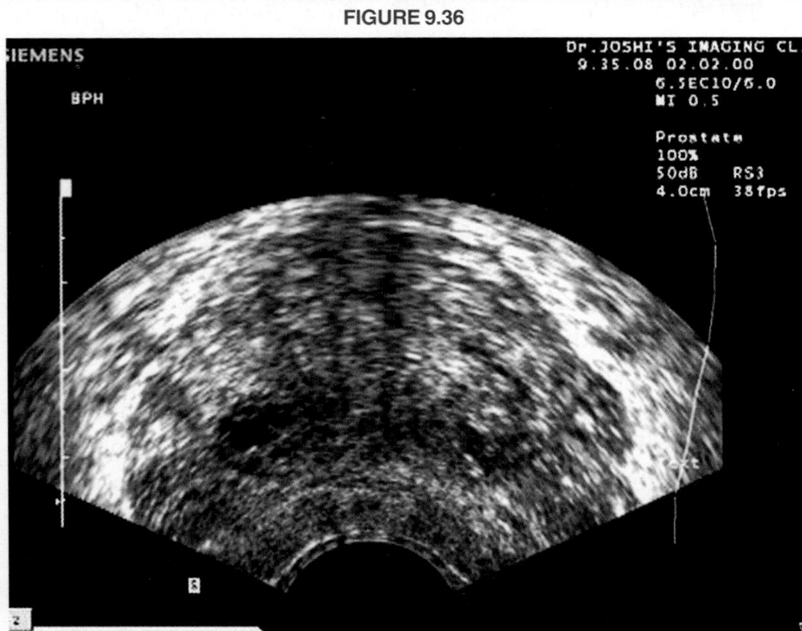

FIGURE 9.37

FIGURES 9.36 and 9.37: Benign prostatic enlargement. HRSG shows benign prostatic enlargement involving both lateral lobes. Few hypoechoic areas are also seen in the gland.

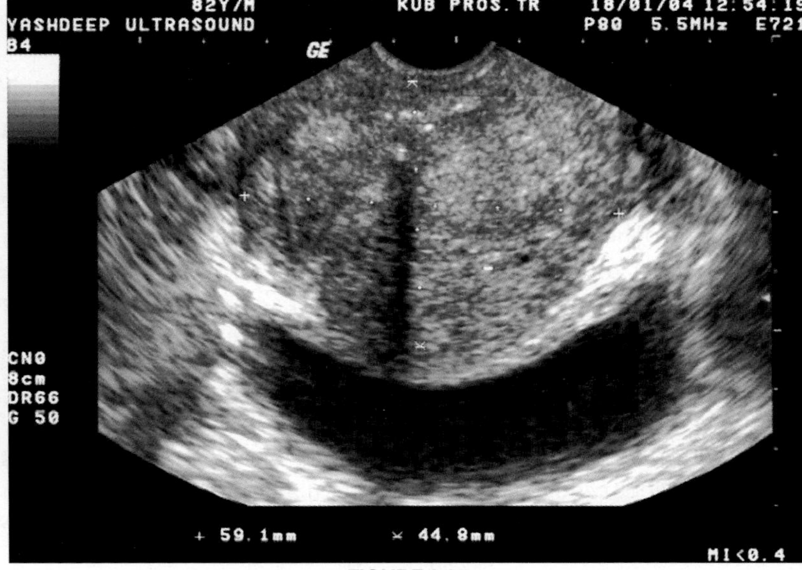

FIGURE 9.38

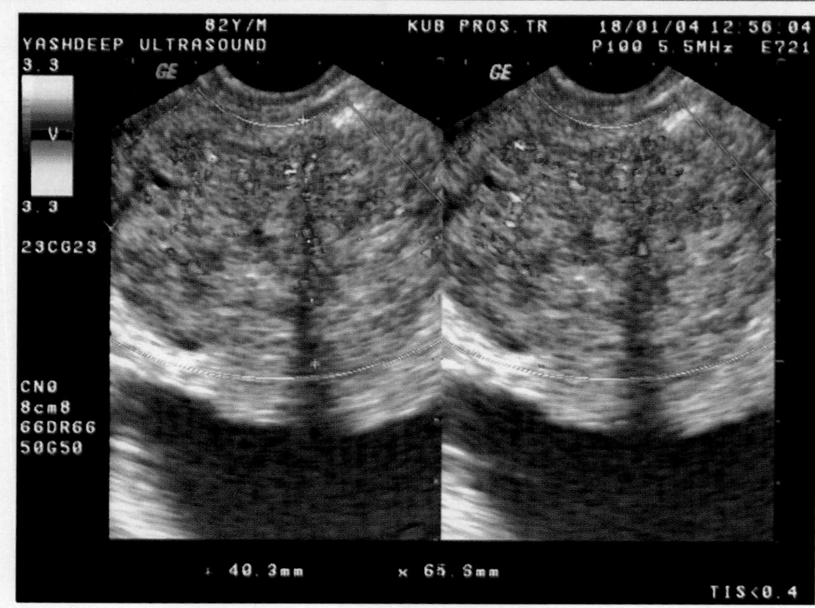

FIGURE 9.39

FIGURES 9.38 and 9.39: Diffuse enlargement of prostate. HRSG shows diffuse prostatic enlargement involving all the lobes. However, the Rt lateral lobe more enlarged than the Lt lateral lobe. No evidence of any calcification or necrosis is seen in it. On color flow imaging normal flow seen in the gland.

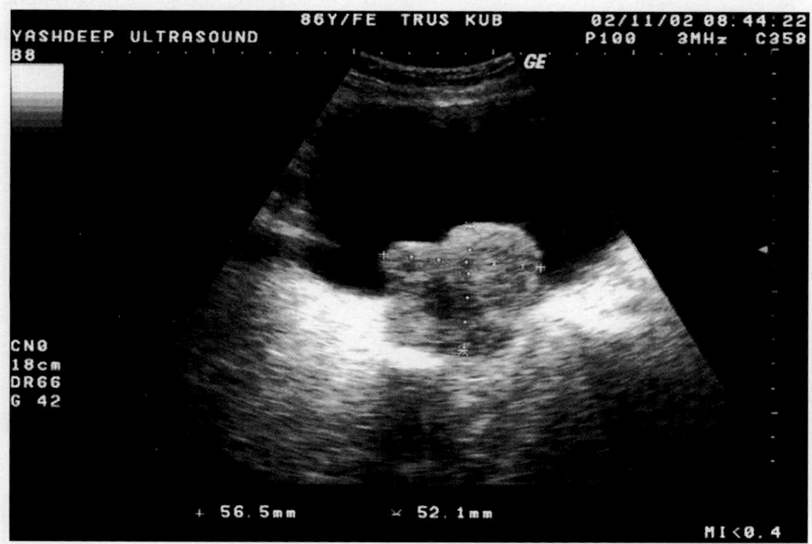

FIGURE 9.40

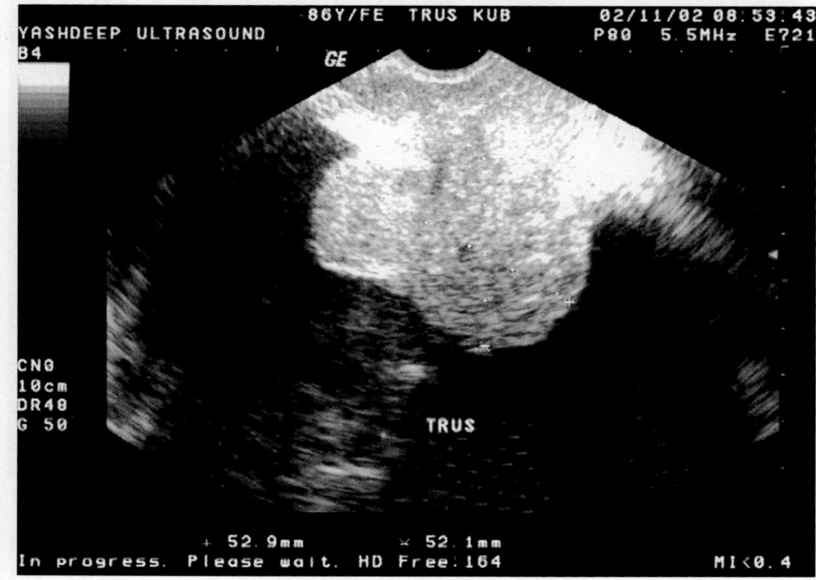

FIGURE 9.41

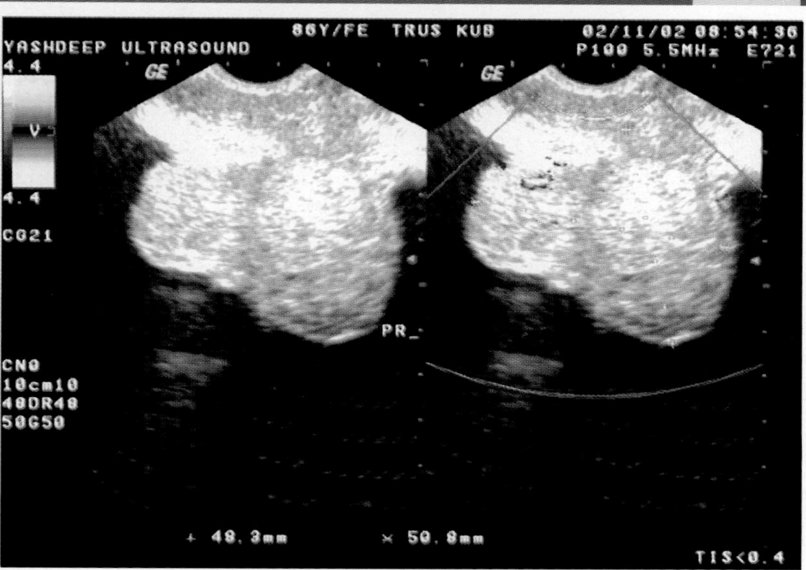

FIGURE 9.42

FIGURES 9.40 to 9.42: Marked median lobe enlargement. HRSG shows marked enlargement of the median lobe of the gland. It is seen bulging in the posterior wall of urinary bladder. However, the lobe shows smooth bulging. TRUS shows well-defined lobulated structure of the enlarged lobe with smooth outline. The lateral lobes are not significantly enlarged.

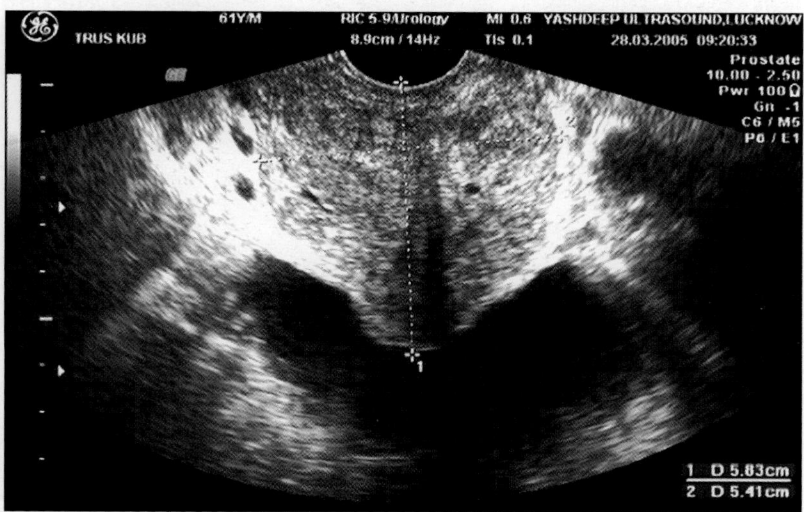

FIGURE 9.43: Diffuse enlargement of prostate with prominent median lobe. HRSG shows diffuse enlargement of the prostate gland involving all the lobes. However, the median lobe is more marked than the lateral lobes.

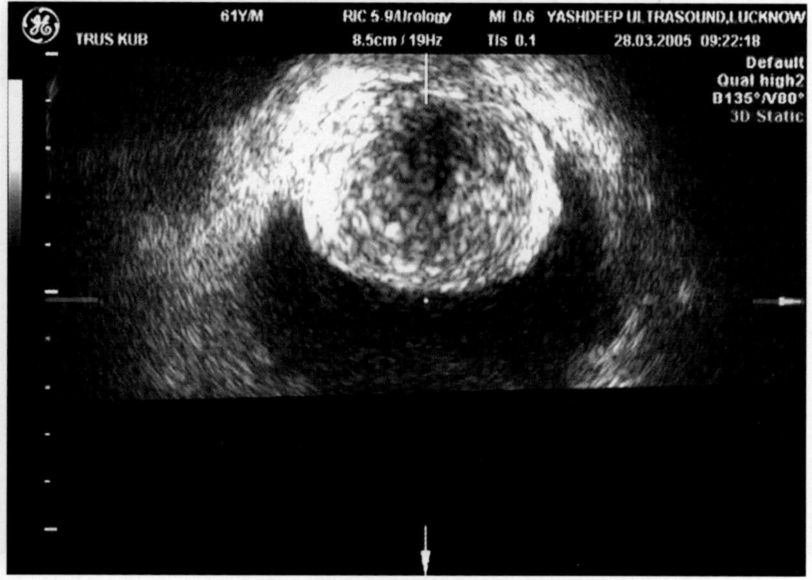

FIGURE 9.44

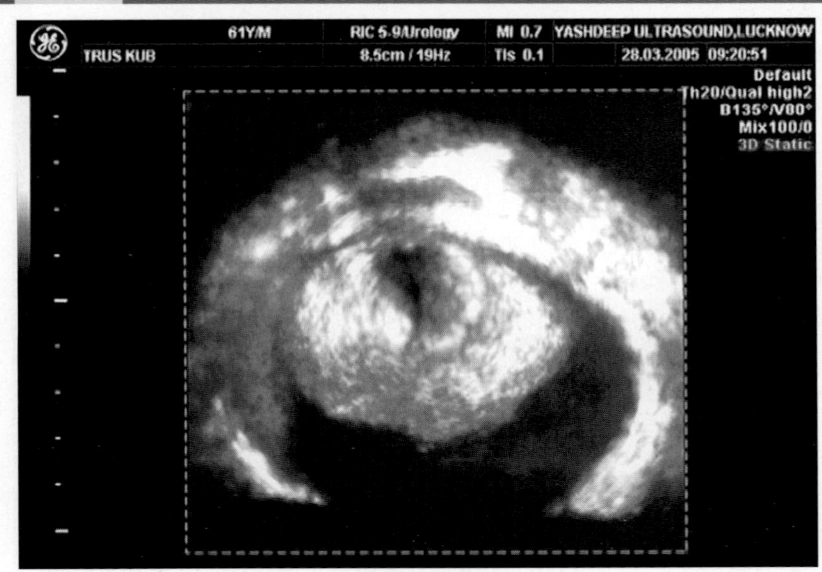

FIGURE 9.45

FIGURES 9.44 and 9.45: 3D imaging of the enlarged median lobe shows better anatomic details of the median lobe. No evidence of any mass lesion is seen in the lobe. It is seen projecting in the posterior wall of UB.

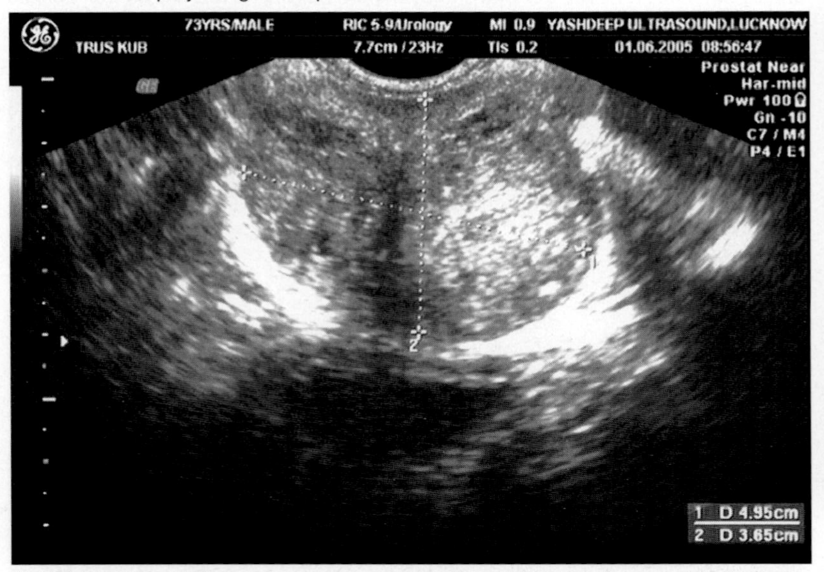

FIGURE 9.46: Diffuse enlargement of the lateral lobes. HRSG shows homogeneous smooth enlargement of both lateral lobes of the gland. The median lobe is not enlarged. No evidence of any necrosis is seen in the lobes.

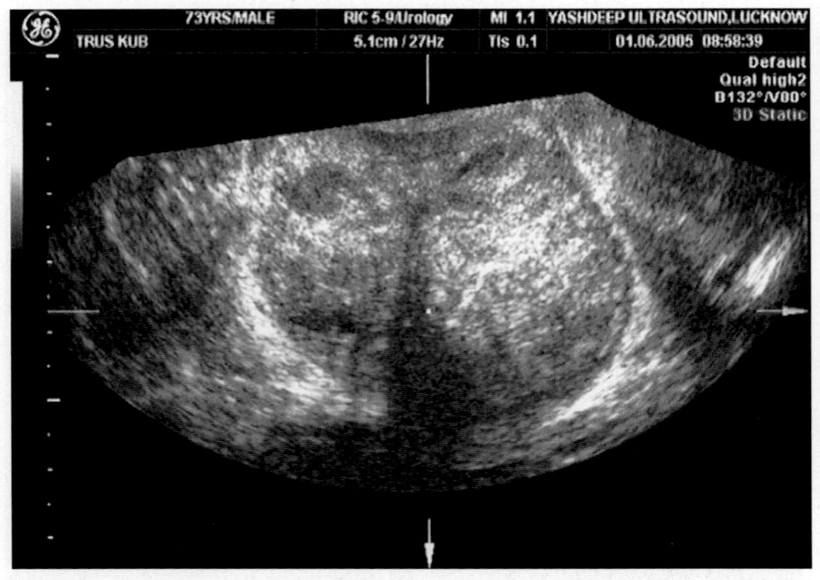

FIGURE 9.47: 3D imaging of the gland shows better anatomic details of the enlarged lobes the Lt lobe is more enlarged than the Rt lobe.

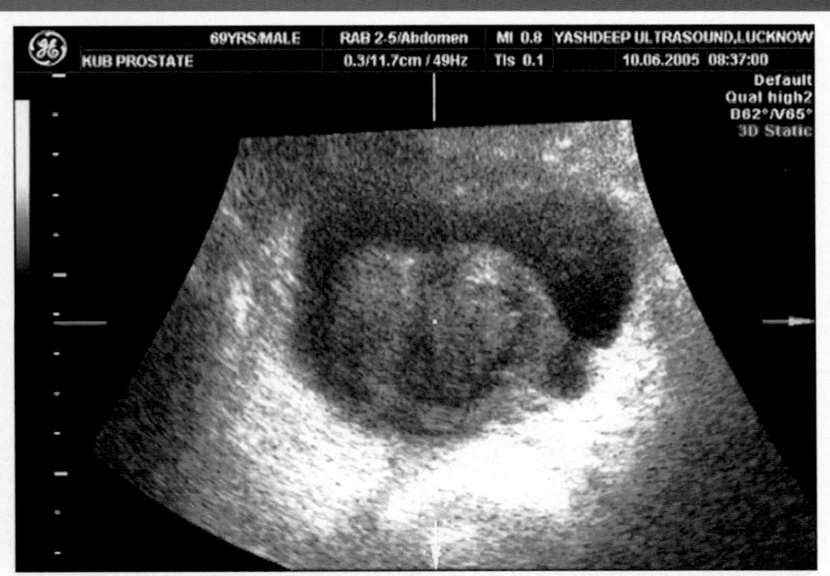

FIGURE 9.48: Diffuse enlargement of the gland. 3D imaging of the prostate shows diffuse lobulated enlargement of the gland with homogeneous in texture.

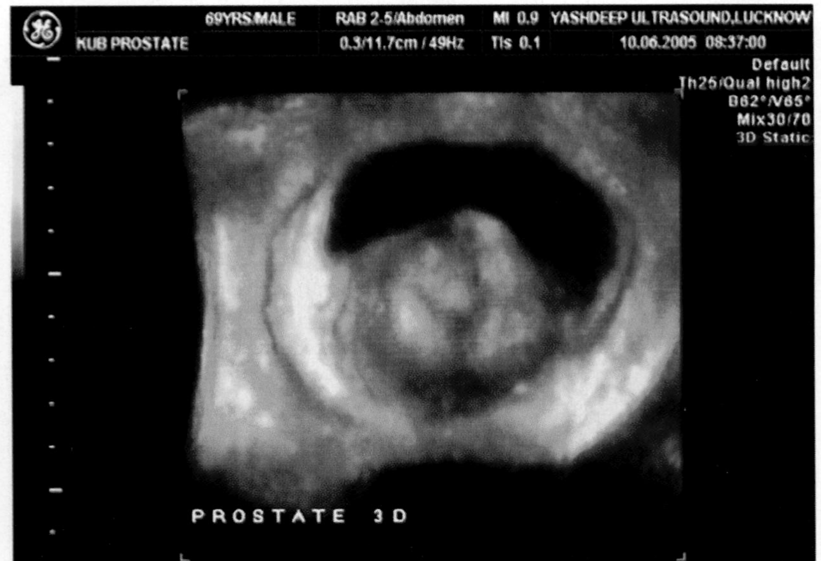

FIGURE 9.49

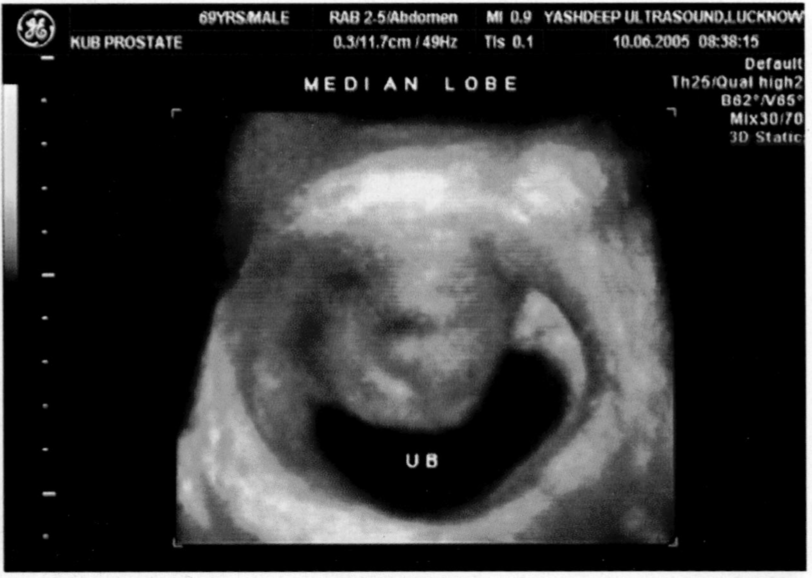

FIGURE 9.50

FIGURES 9.49 and 9.50: 3D surface rendering of the gland shows well-defined lobulated appearance of the gland. No evidence of any mass lesion is seen in the gland.

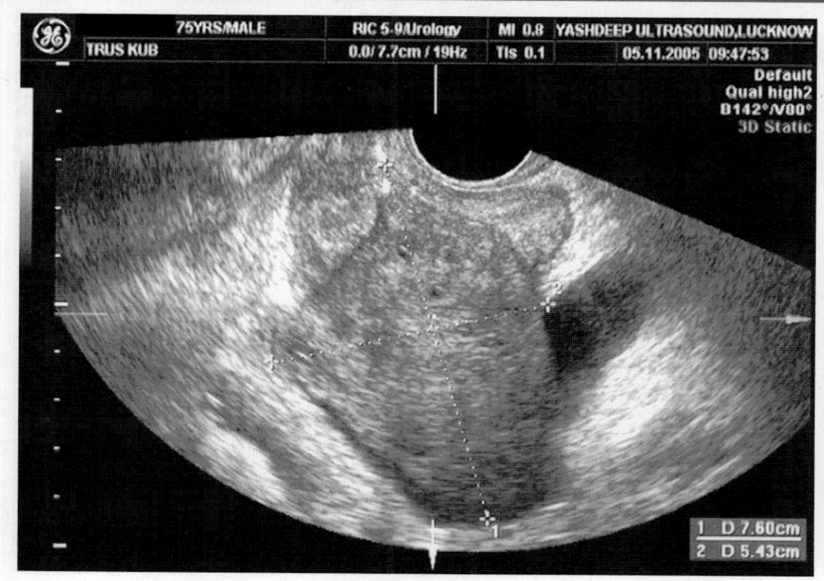

FIGURE 9.51

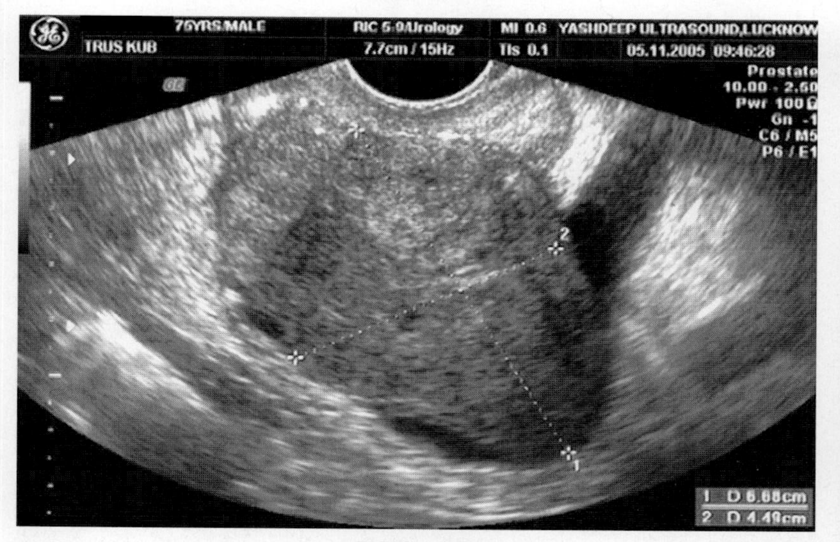

FIGURE 9.52

FIGURES 9.51 and 9.52: Marked enlargement of median lobe. HRSG shows marked conical enlargement of median lobe of the gland. The median lobe is seen coming out from the gland displacing the lateral lobe. However, the facial planes are seen intact. It is seen projecting in the UB wall. The lobe shows homogeneous texture. However, few small cysts are seen in the lobe.

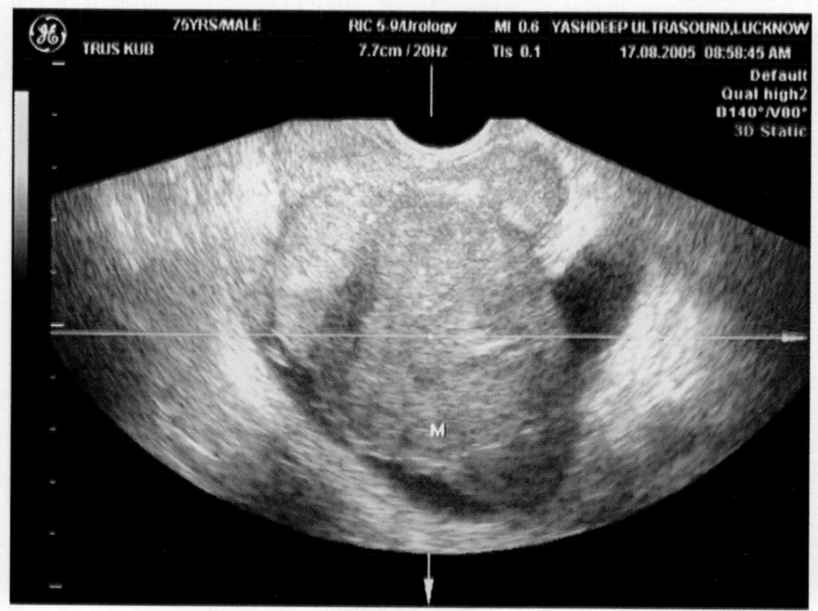

FIGURE 9.53

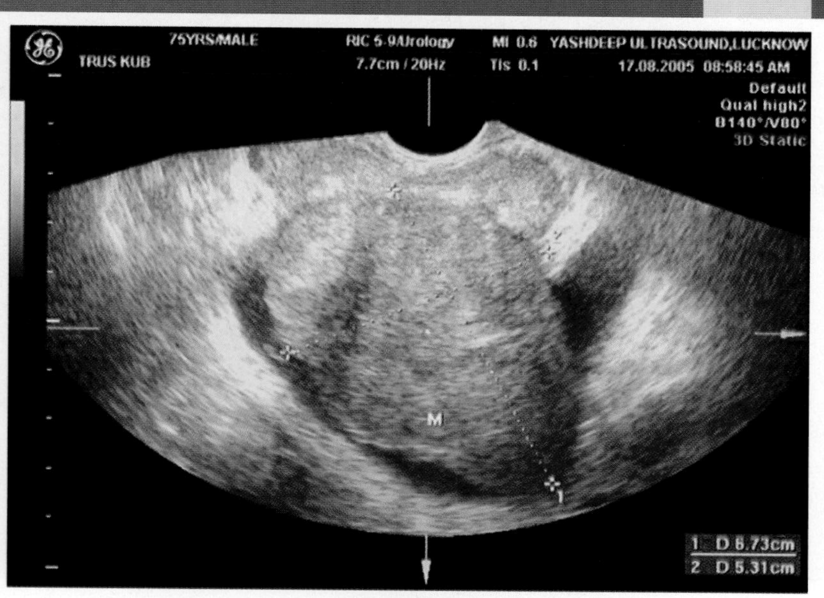

FIGURE 9.54

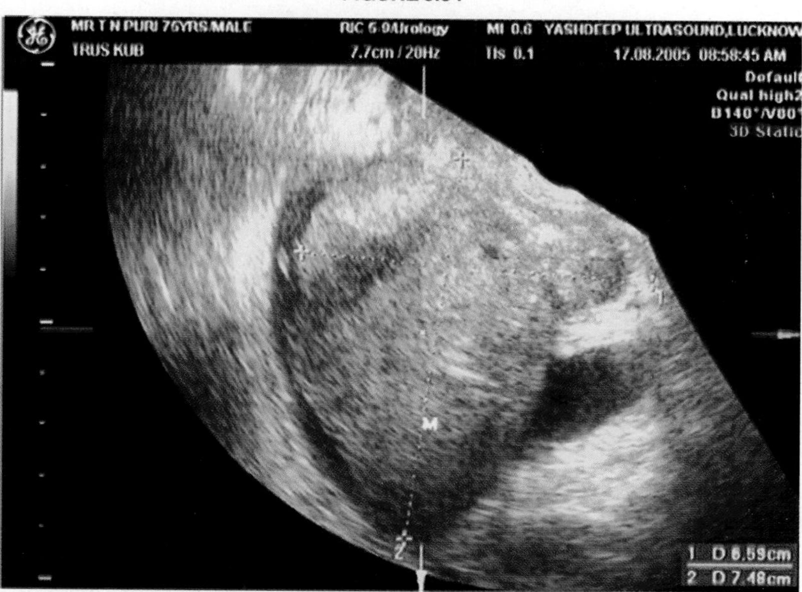

FIGURE 9.55

FIGURES 9.53 to 9.55: 3D surface rendering of the lobes show better clinical details of the lobe. The margins are well defined with smooth outline. However, macro lobulation are seen at the margins suggestive of macro nodular texture of the lobe.

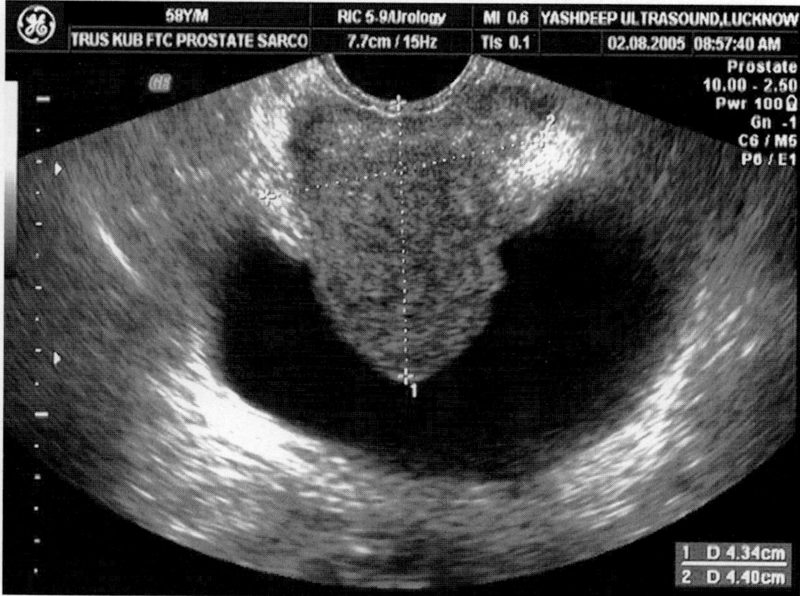

FIGURE 9.56

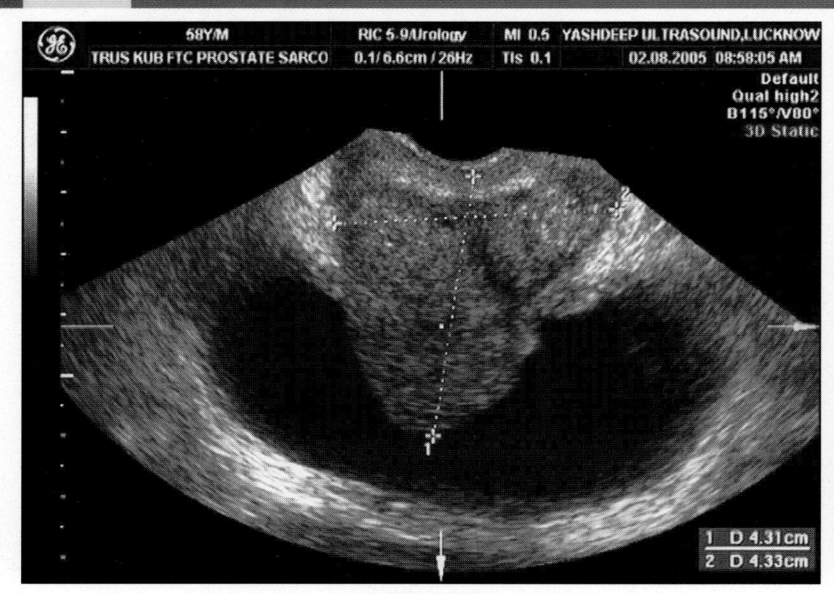

FIGURE 9.57

FIGURES 9.56 and 9.57: Prostate sarcoidosis. A known patient of sarcoidosis presented with obstructed uropathy. HRSG shows marked enlargement of the gland. The lobe is hypoechoic in texture with smooth outline. The median lobe is seen as cone. No necrosis is seen in it. The lateral lobes are also enlarged.

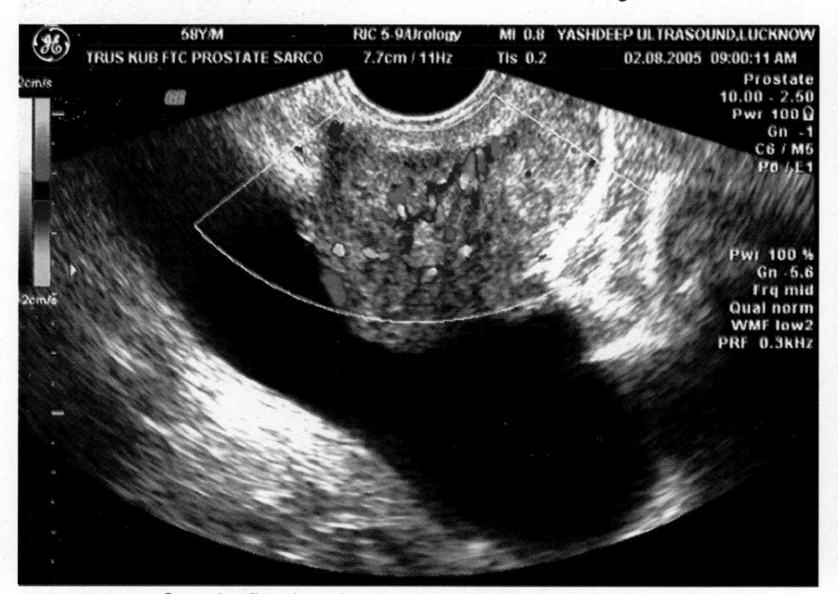

FIGURE 9.58: On color flow imaging increased flow is seen in the gland.

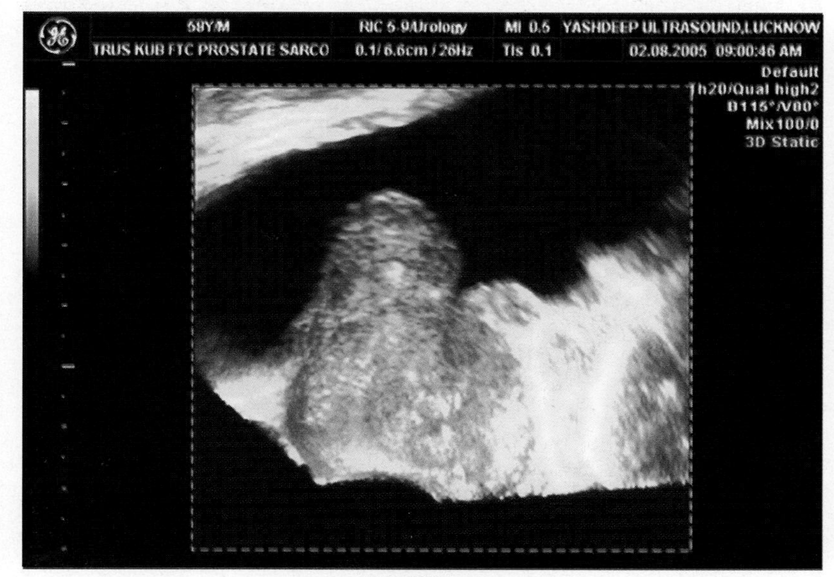

FIGURE 9.59

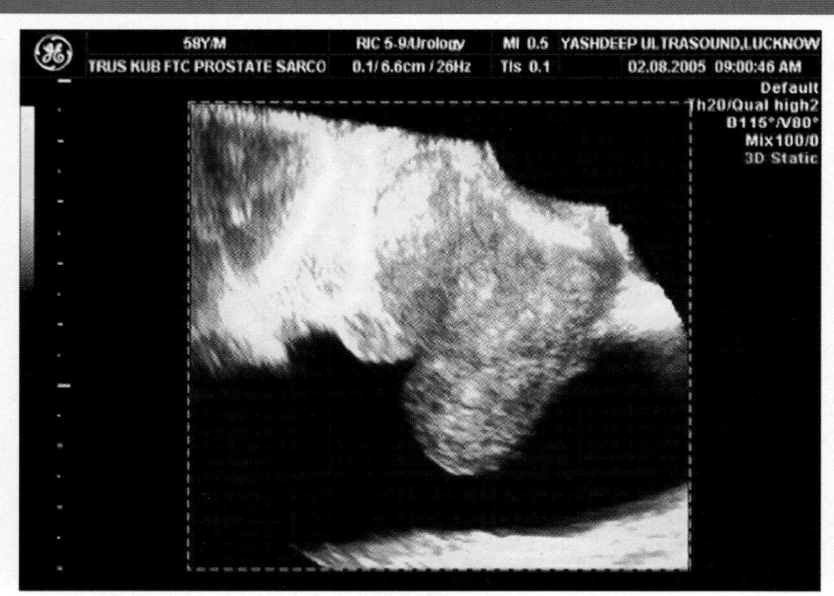

FIGURE 9.60

FIGURES 9.59 and 9.60: 3D surface rendering of the enlarged lobes shows better clinical details of the gland. The cone shape enlarged median lobe is well appreciated on 3D imaging. Biopsy of the enlarged lobe confirmed sarcoidosis.

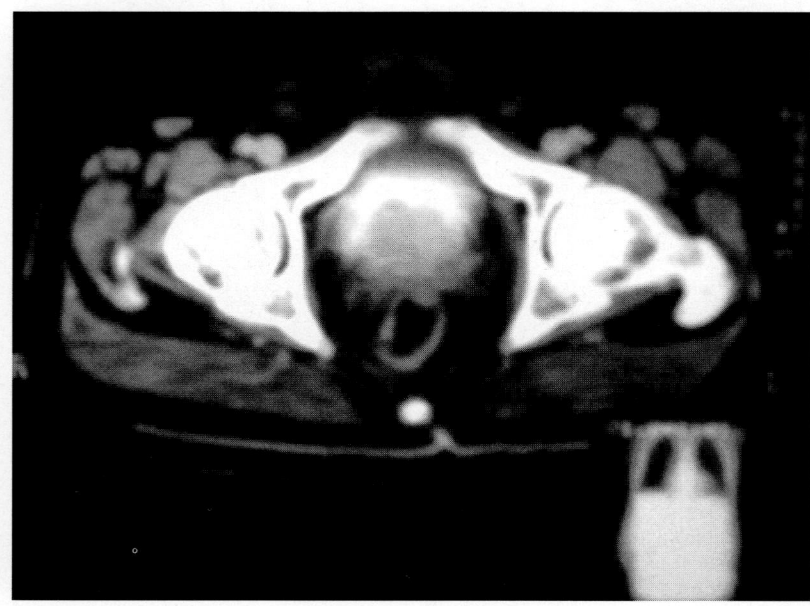

FIGURE 9.61: CT of the pelvis of the same patient shows enlarged median lobe indenting over UB wall.

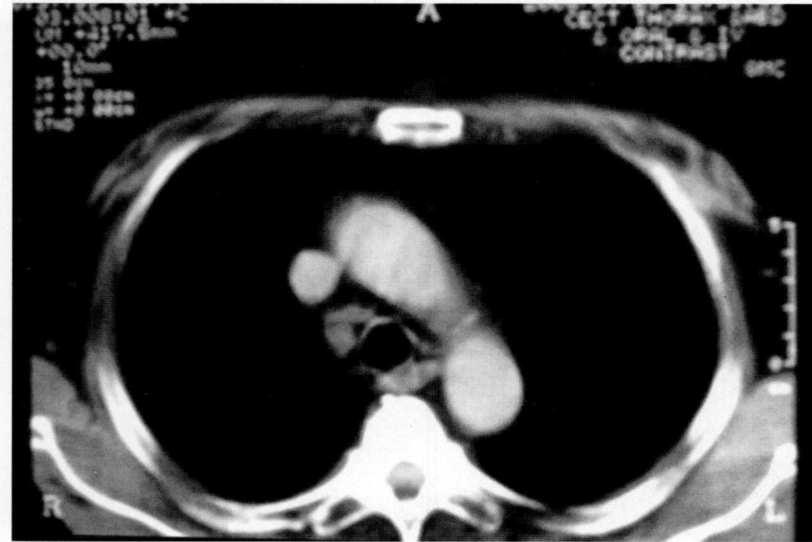

FIGURE 9.62

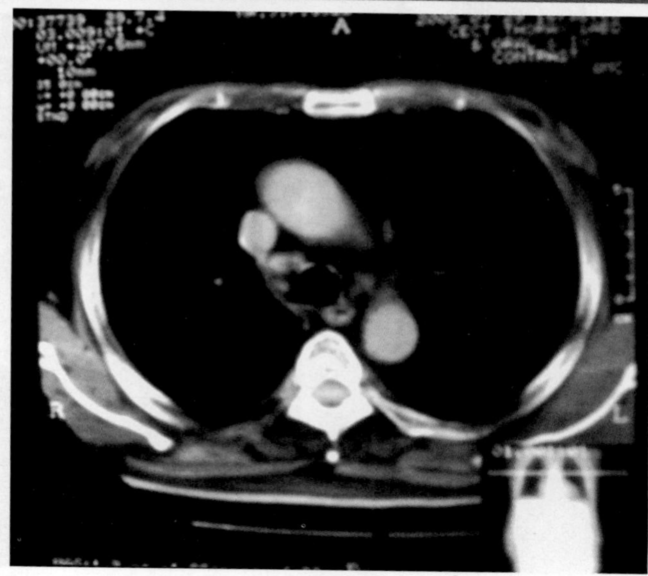

FIGURE 9.63

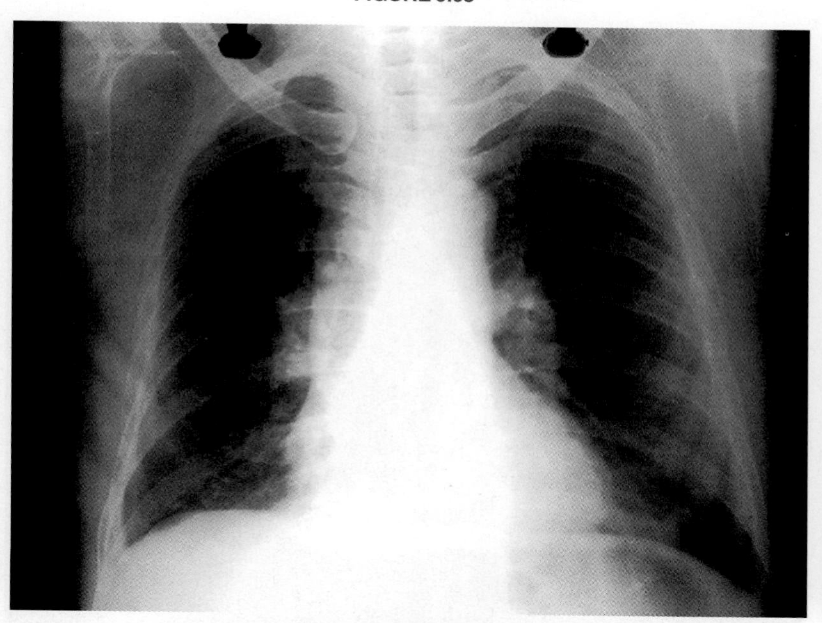

FIGURE 9.64

FIGURES 9.62 to 9.64: X-ray and CT of the same patient shows enlarged hilar shadows suggestive of hilar lymphadenopathy.

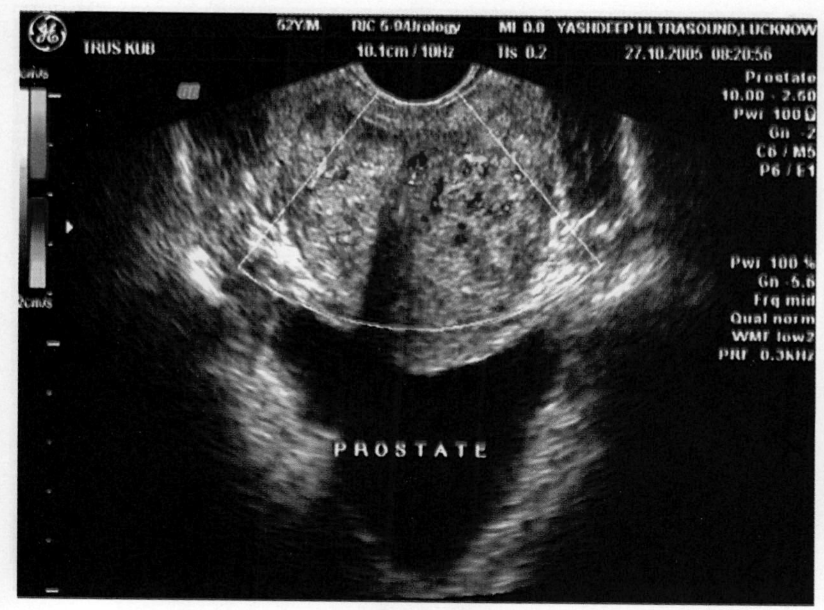

FIGURE 9.65

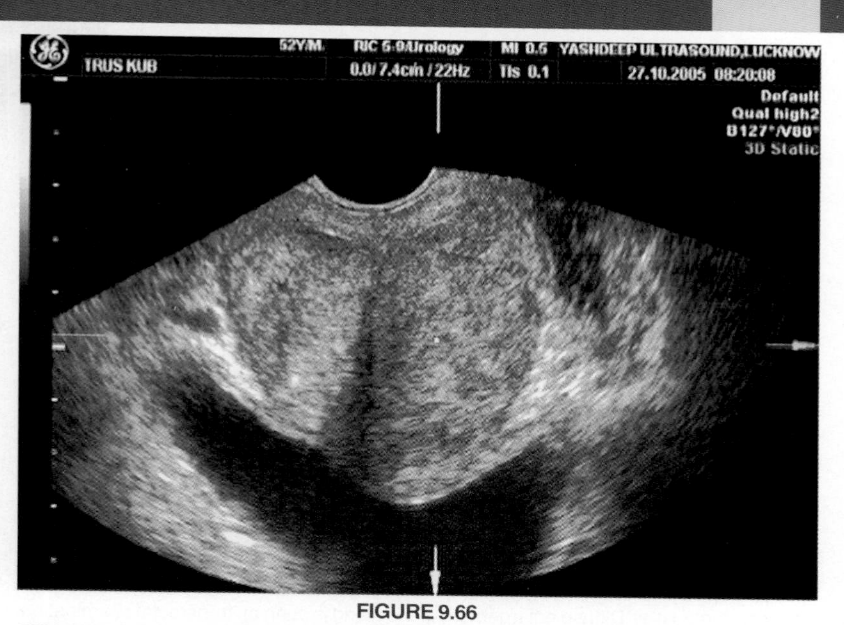

FIGURE 9.66

FIGURES 9.65 and 9.66: Lateral lobe enlargement. HRSG shows both lateral lobes enlargement of the gland. The Lt lobe is more marked than the Rt lobe. 3D imaging of the same patient shows better details of the lobe enlargement.

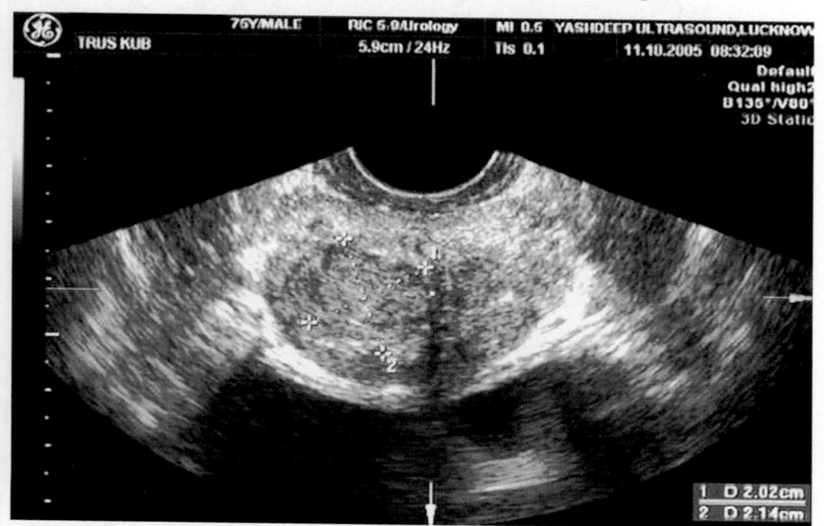

FIGURE 9.67: Unilateral enlargement of the lobe. HRSG shows unilateral enlargement of Rt lobe of prostate. The lobe is smooth in outline. No calcification or necrosis is seen in it. The Lt lobe is normal.

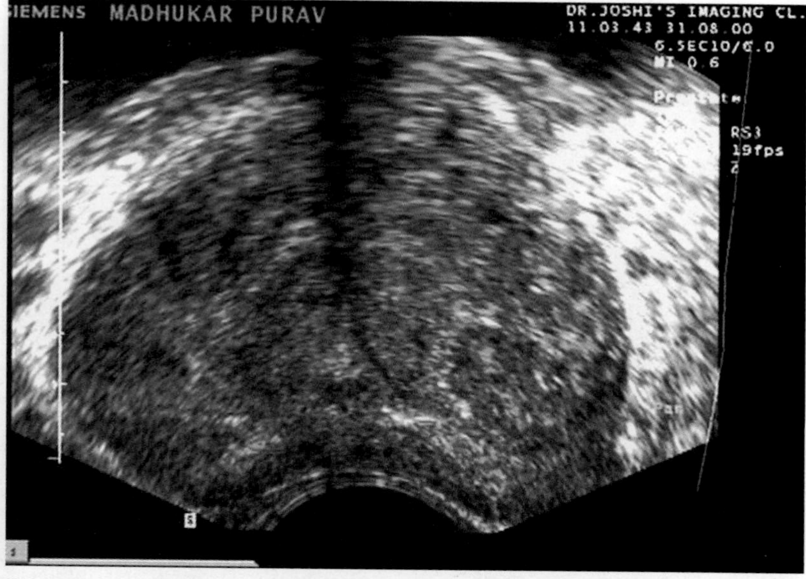

FIGURE 9.68: Marked enlargement of lobes. HRSG shows marked enlargement of the gland involving both lateral lobes. Compression of the central zone is seen. However, the glandular texture is homogeneous. No focal mass is seen in it.

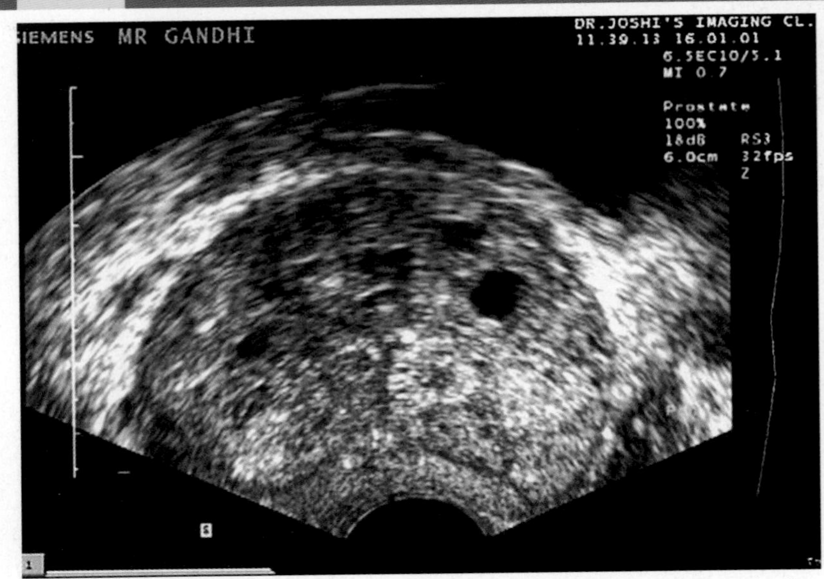

FIGURE 9.69: BPH. Diffuse enlargement of the gland is seen on transrectal examination. A small cyst is seen on the Rt side in the transition zone.

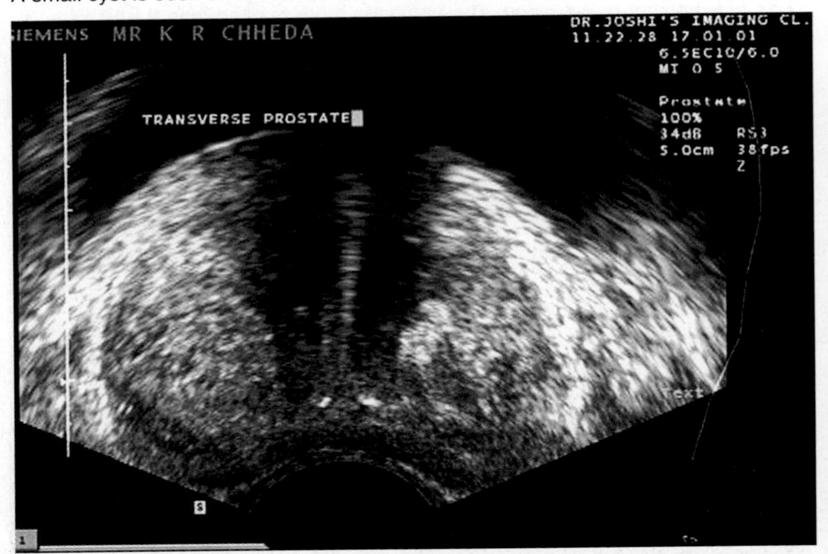

FIGURE 9.70: BPH with normal Eiffel tower. HRSG shows diffuse enlargement of prostate gland involving the transition zone. The “Eiffel Tower” is seen normally in the center of gland.

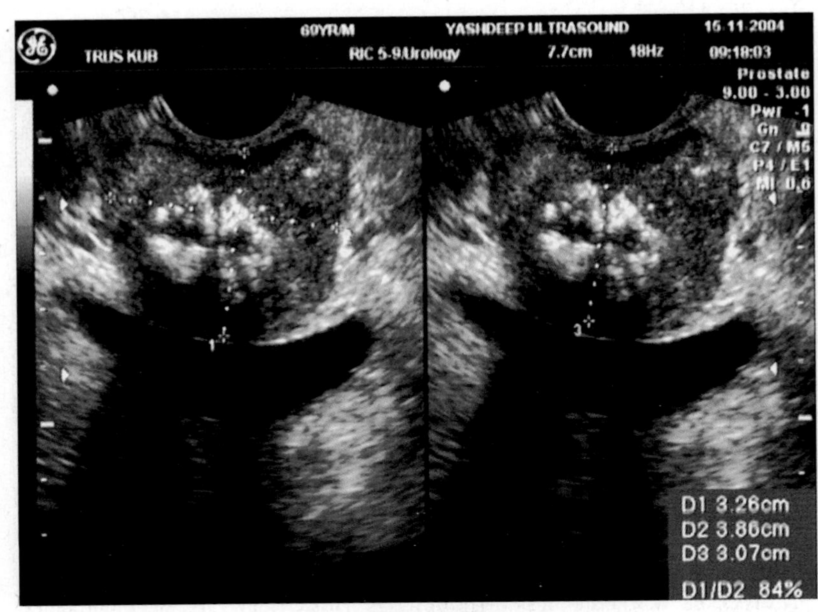

FIGURE 9.71

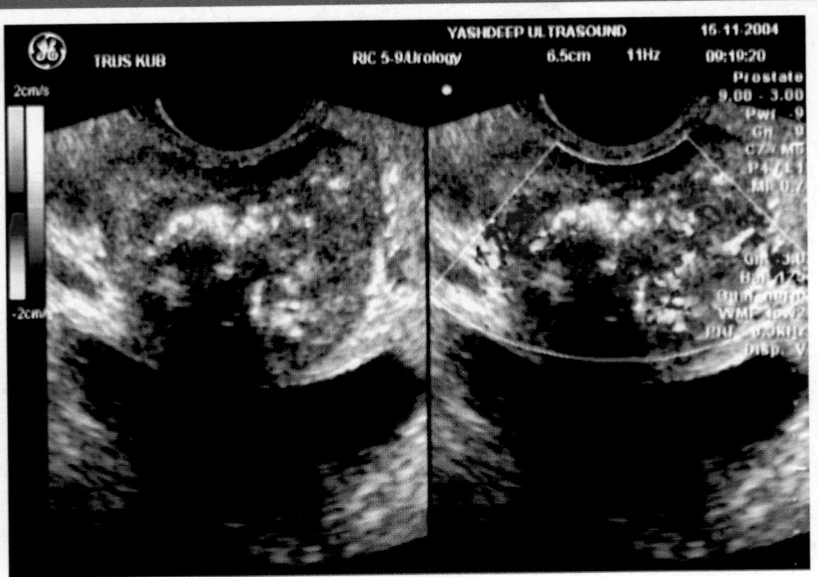

FIGURE 9.72

FIGURES 9.71 and 9.72: BPH with amylacea. HRSG shows diffuse enlargement of the prostate gland. Echogenic calcification is seen between peripheral and central zone suggestive of corpus amylacea calcification.

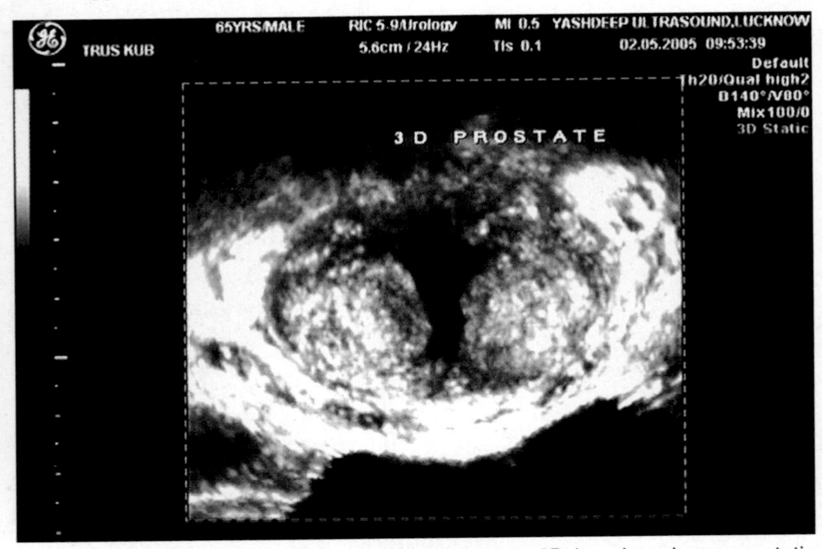

FIGURE 9.73: Patent prostatic tunnel after TUR. 3D imaging shows prostatic tunnel made after TUR in a case of BPH. Enlarged lateral lobes are also seen on either side of the tunnel.

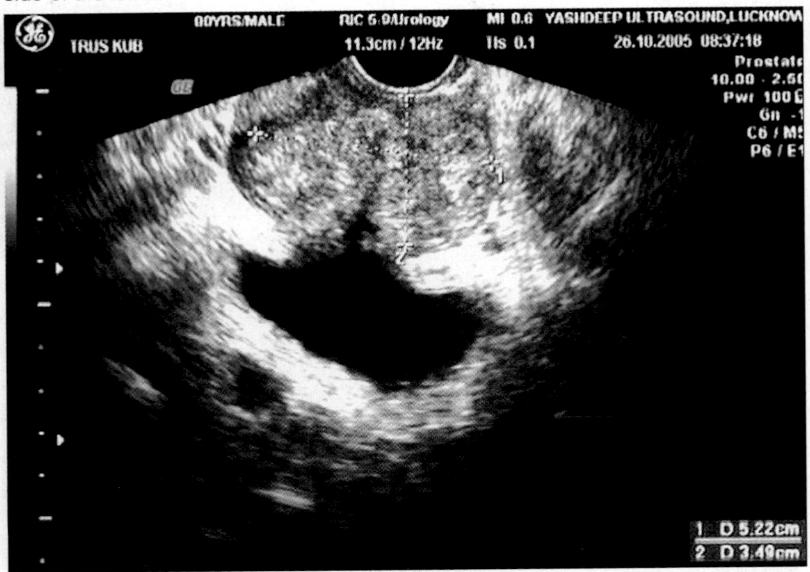

FIGURE 9.74

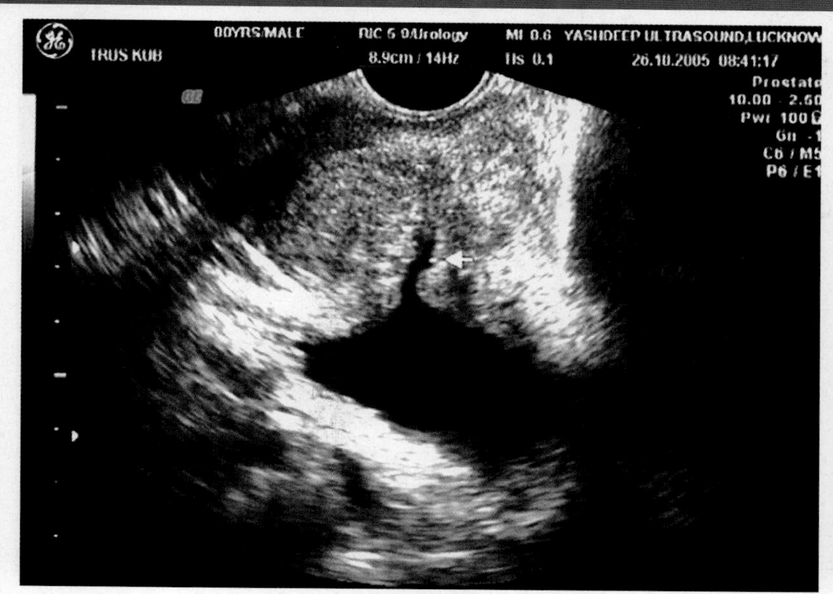

FIGURE 9.75

FIGURES 9.74 and 9.75: Obliterated prostatic tunnel. 3D HRSG shows obliterated prostatic tunnel in a patient came with retention of urine after having TUR operation for BPH. The both lobes are seen pressing over the tunnel and obliterating it.

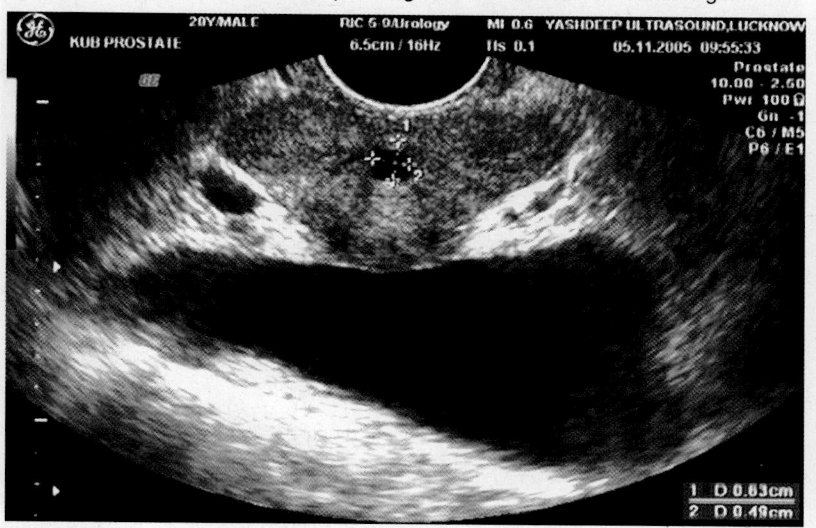

FIGURE 9.76

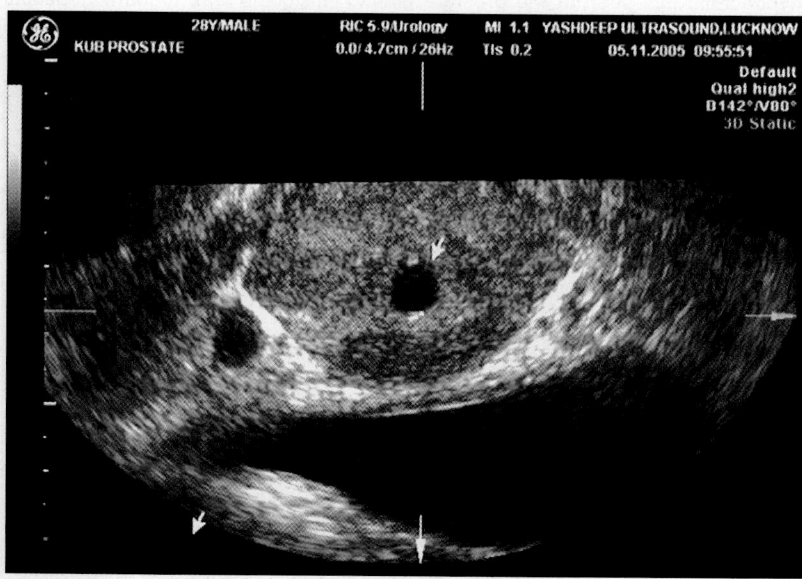

FIGURE 9.77

FIGURES 9.76 and 9.77: Small prostatic cyst. HRSG shows well-defined thin walled cyst in the prostate. 3D imaging shows better details of cyst. Prostatic cysts are utricle cyst, which are benign in nature.

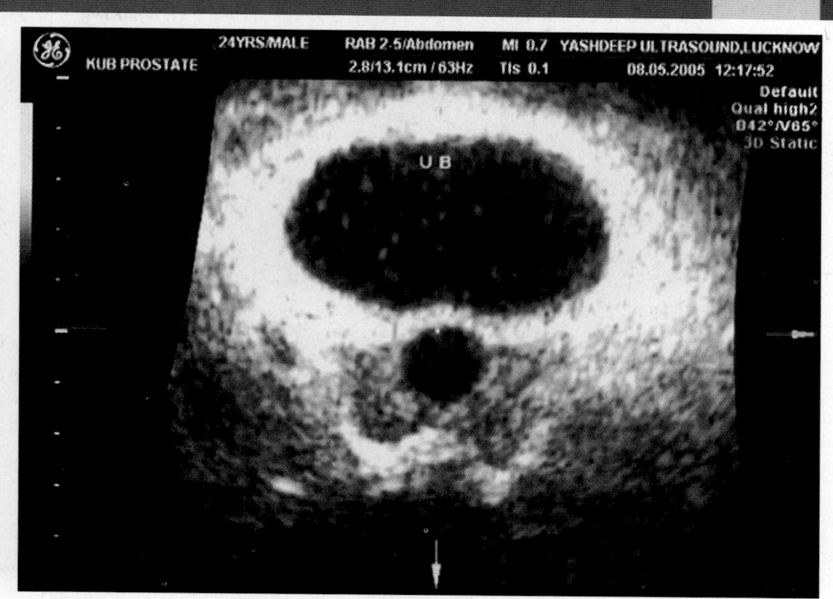

FIGURE 9.78: 3D imaging of utricle cyst. HRSG shows well define thin walled cyst in the prostate. The cyst wall is sharp. No evidence of any internal echo seen in the cyst.

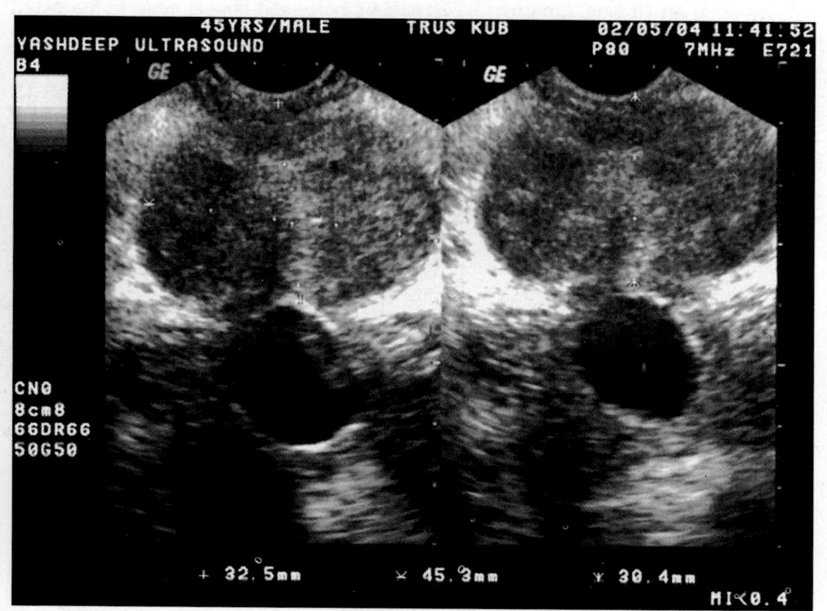

FIGURE 9.79

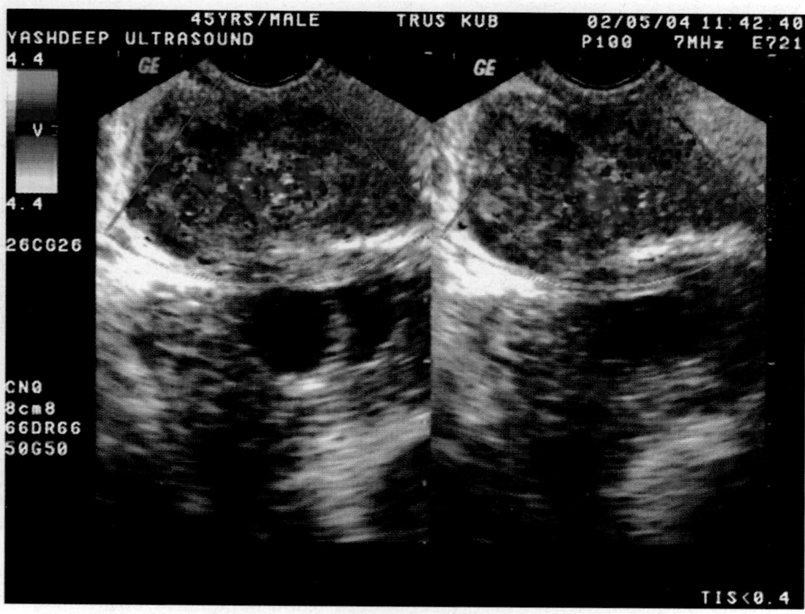

FIGURE 9.80

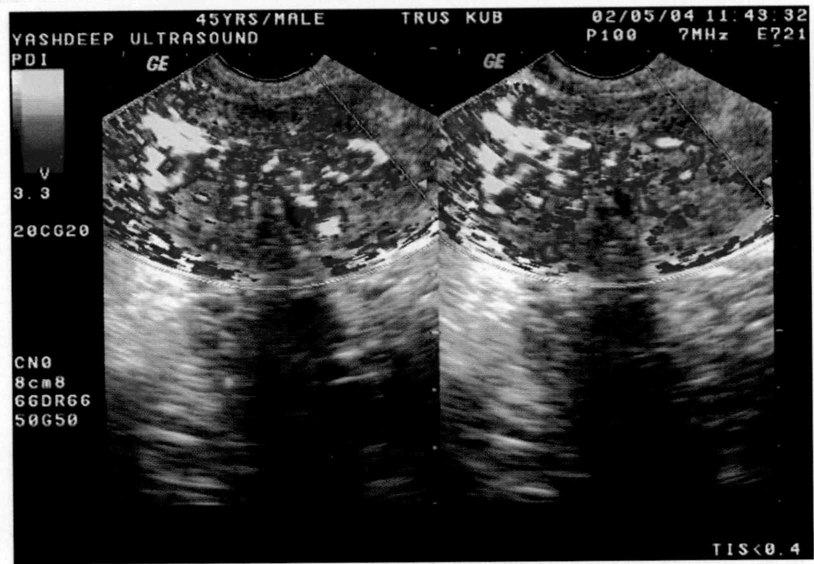

FIGURE 9.81

FIGURES 9.79 to 9.81: BPH with prostatitis. HRSG shows benign prostatic enlargement. However, multiple hypoechoic areas are seen in the gland. Small abscess is also seen on Rt side. On color flow imaging increased flow is seen in the gland suggestive of hyperemia indicating prostatitis. On power Doppler imaging marked increased flow is seen in the inflamed prostate.

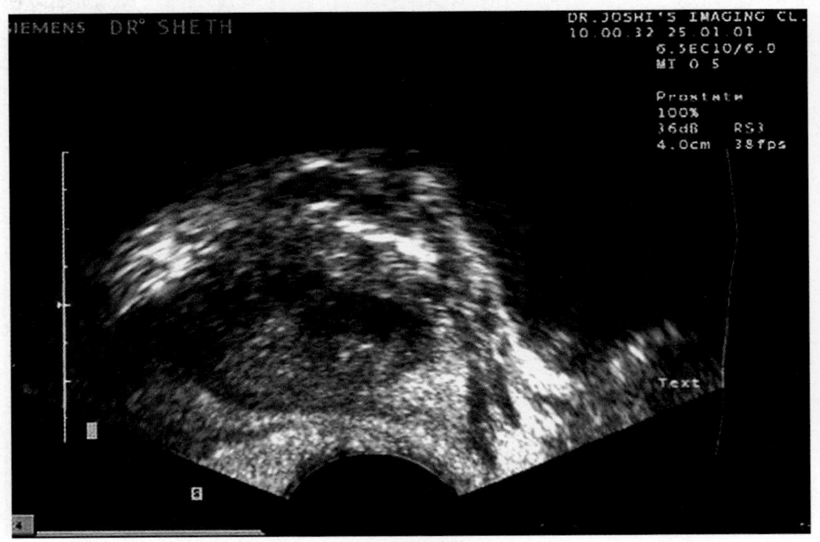

FIGURE 9.82

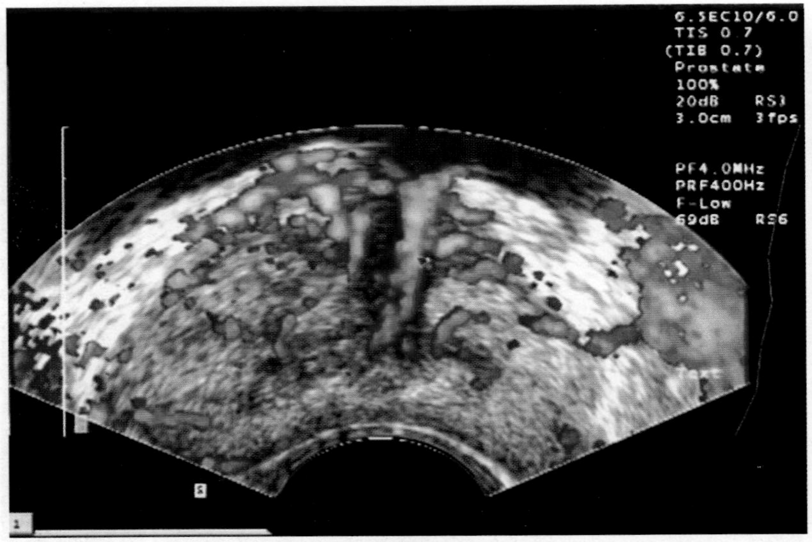

FIGURE 9.83

FIGURES 9.82 and 9.83: Acute prostatitis. HRSG shows irregular hypoechoic low-level echo complex mass in the prostate gland suggestive of an abscess formation. On color flow imaging high flow is seen in the gland suggestive of hyperemia.

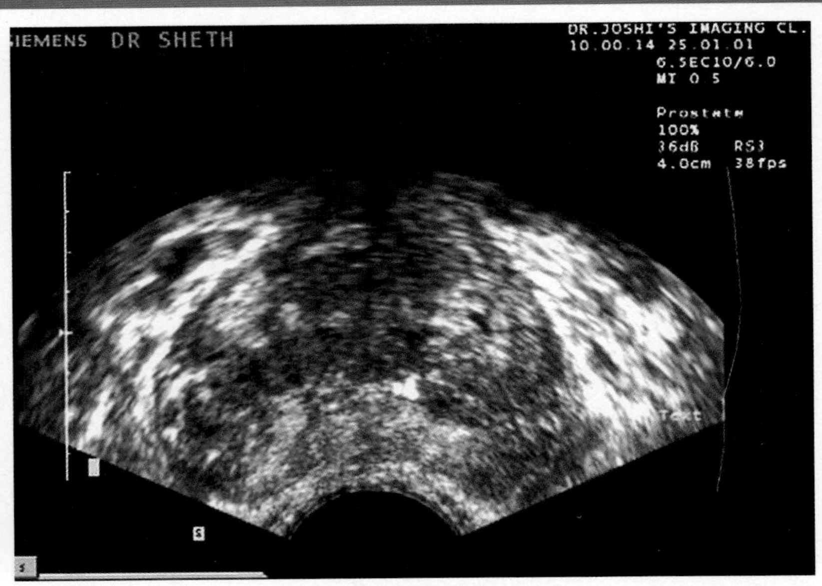

FIGURE 9.84

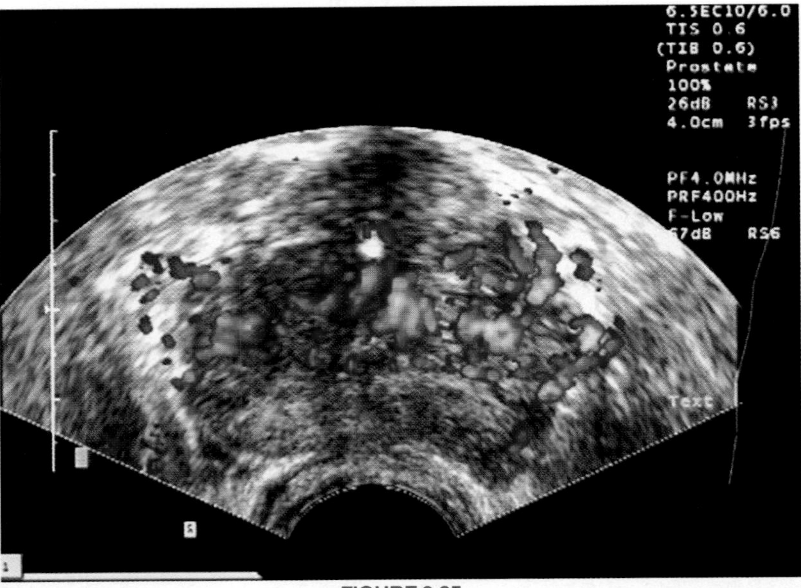

FIGURE 9.85

FIGURES 9.84 and 9.85: Acute prostatitis. HRSG shows multiple irregular hypoechoic foci seen in the prostate gland. The gland was tender on clinical examination suggestive of prostatitis. On color flow imaging increased flow is seen in the gland.

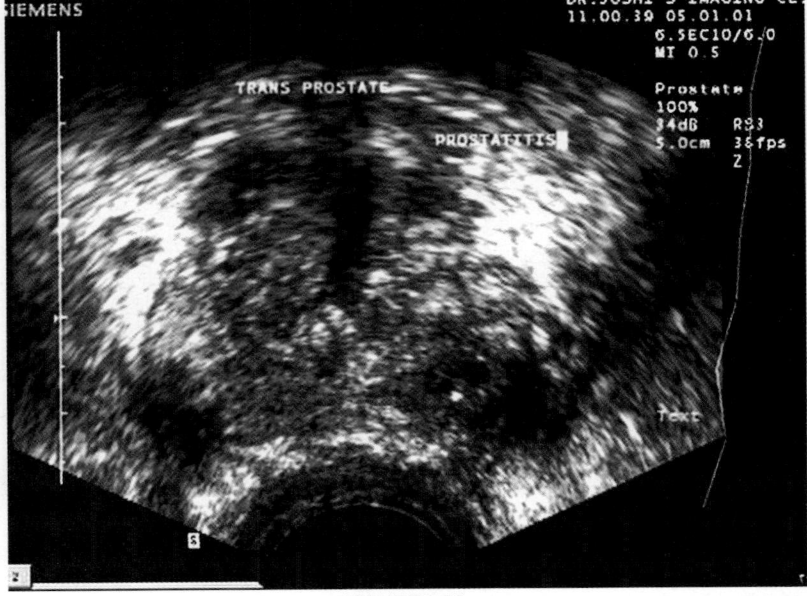

FIGURE 9.86

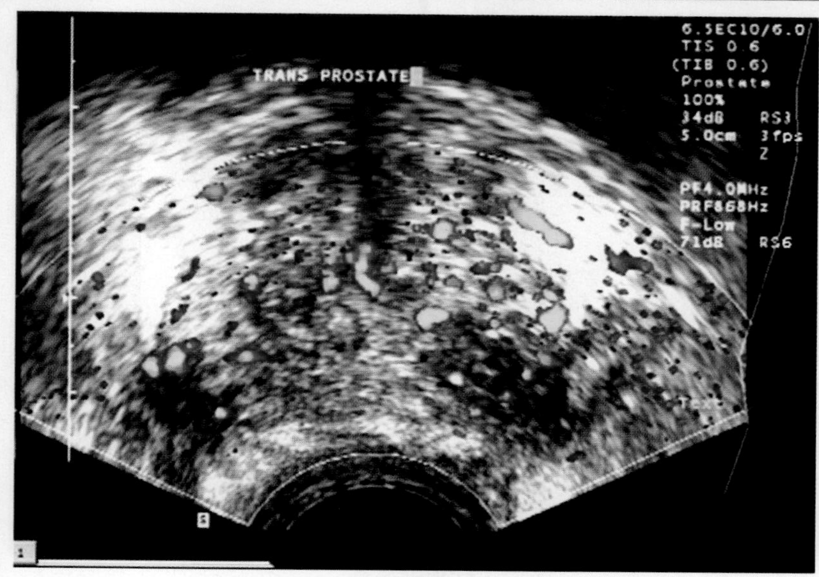

FIGURE 9.87

FIGURES 9.86 and 9.87: Chronic prostatitis. HRSG shows multiple irregular hypoechoic foci in the prostate gland involving both the lobes. They are discrete in distribution. However, no abscess formation is seen in them. On color Doppler imaging moderate flow is seen in the gland.

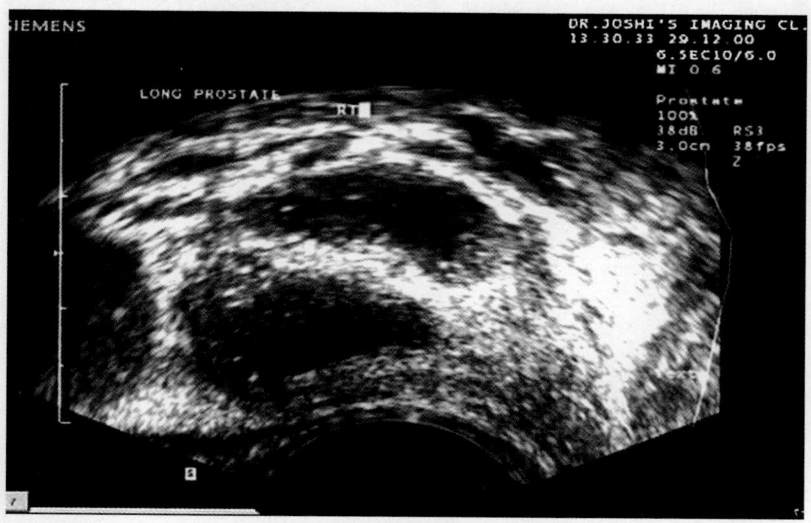

FIGURE 9.88

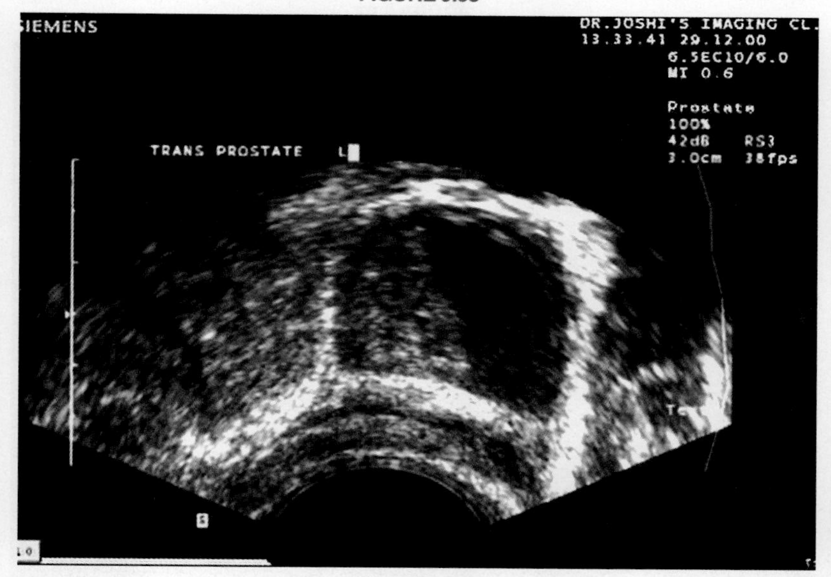

FIGURE 9.89

FIGURES 9.88 and 9.89: Abscess formation in the prostate gland. HRSG shows a thick walled low-level echo complex mass on the Lt side of the gland. Central area of necrosis is seen in the gland. Echogenic debris is also seen suggestive of an abscess formation.

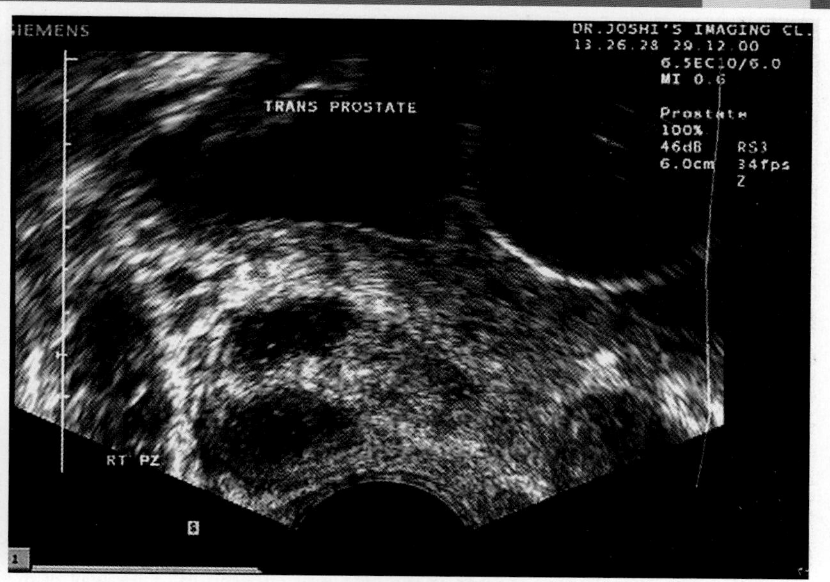

FIGURE 9.90

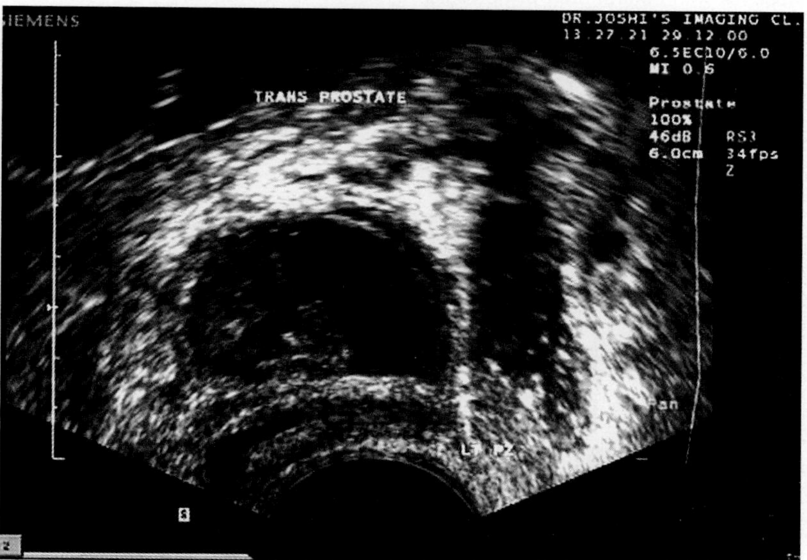

FIGURE 9.91

FIGURES 9.90 and 9.91: Abscess formation in the prostate. HRSG shows a big abscess in the prostate gland on Lt side. Few small abscesses are also seen in the peripheral zone.

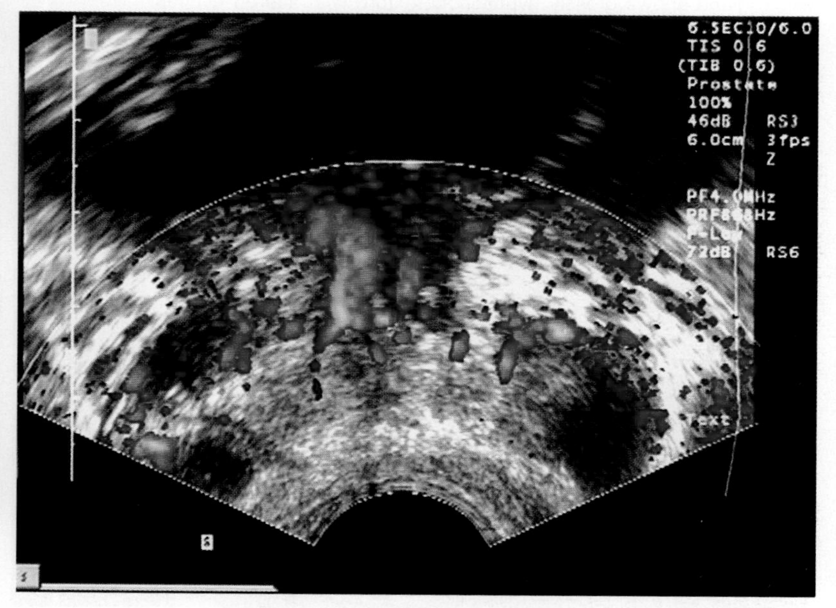

FIGURE 9.92: On color Doppler imaging increased flow is seen around the wall of abscess.

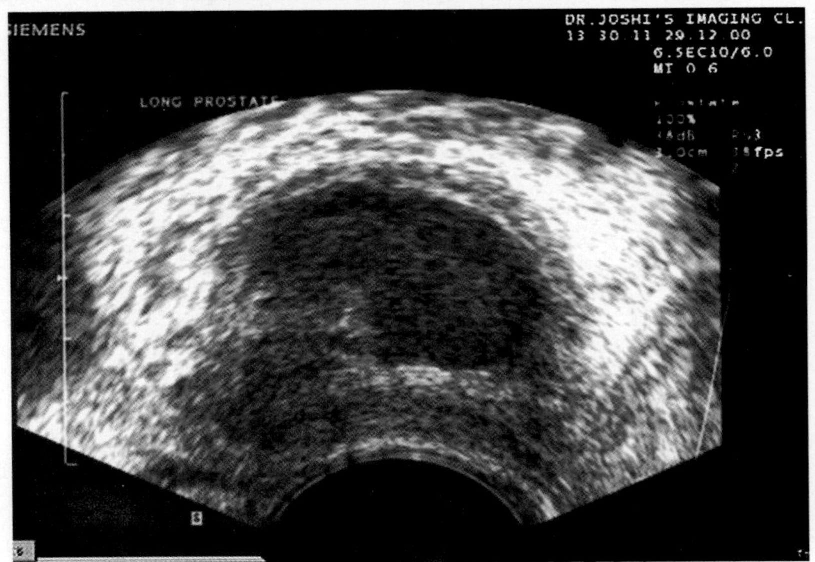

FIGURE 9.93: The abscess is seen in the long axis showing the thick wall of the abscess with central area of necrosis.

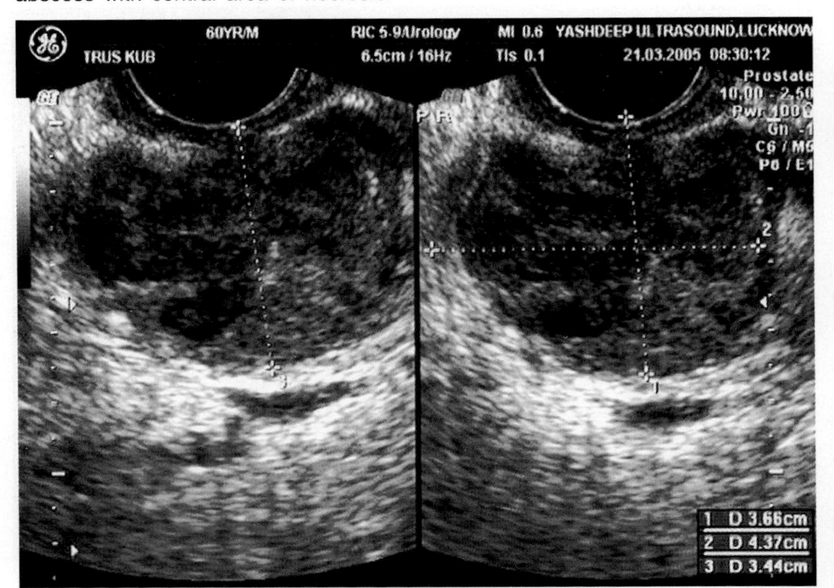

FIGURE 9.94: Multiple abscesses in the prostate. A patient presented with pyouria and acute pain in the perineum with high fever. HRSG shows diffuse enlargement of the gland with multiple abscess in both the lobes in the gland. The whole gland shows grossly hypoechoic texture with altered echoes.

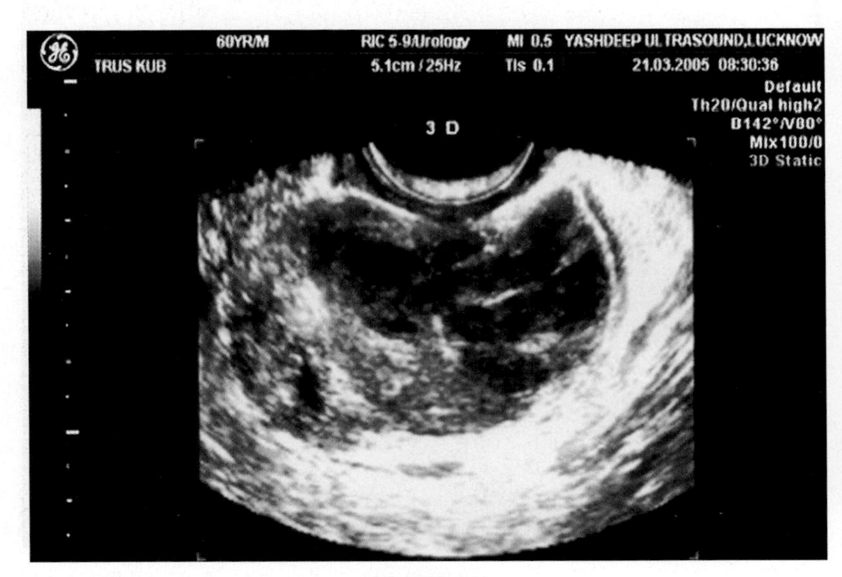

FIGURE 9.95

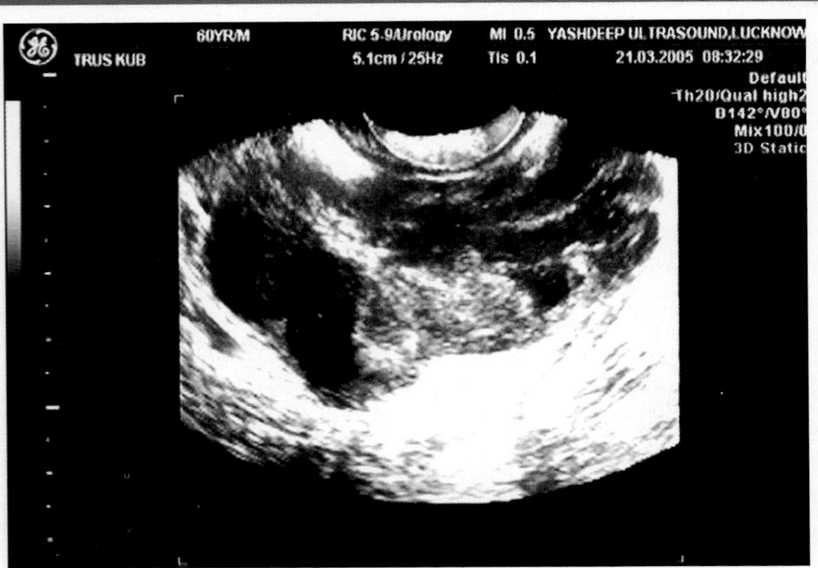

FIGURE 9.96

FIGURES 9.95 and 9.96: 3D imaging of the same patient shows the abscess in much clinical details. Thin echogenic septa are seen running in the abscess cavity. Internal debris is also seen in the abscess.

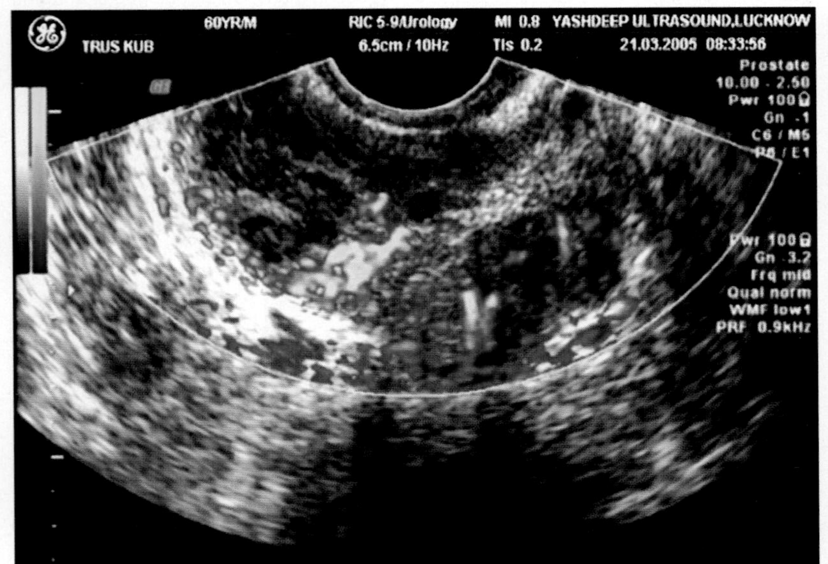

FIGURE 9.97

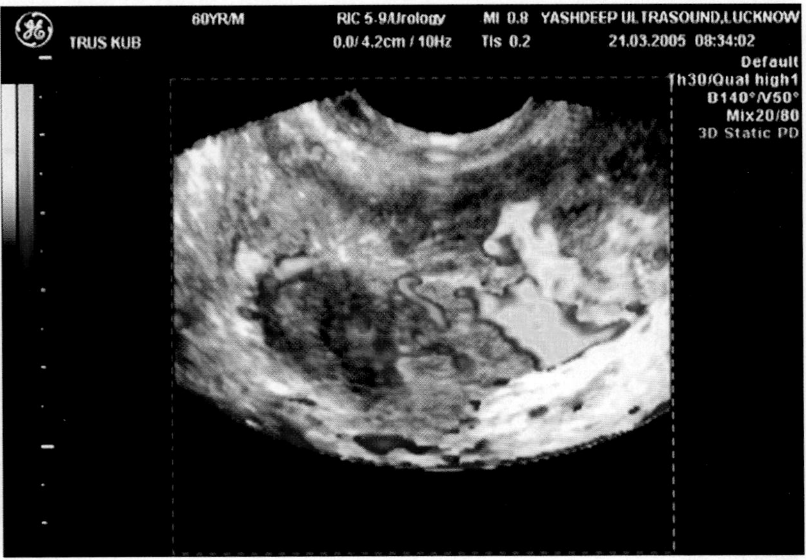

FIGURE 9.98

FIGURES 9.97 and 9.98: On color flow imaging high flow seen in the capsule of the abscess. Multiple vessels are seen feeding the abscess.

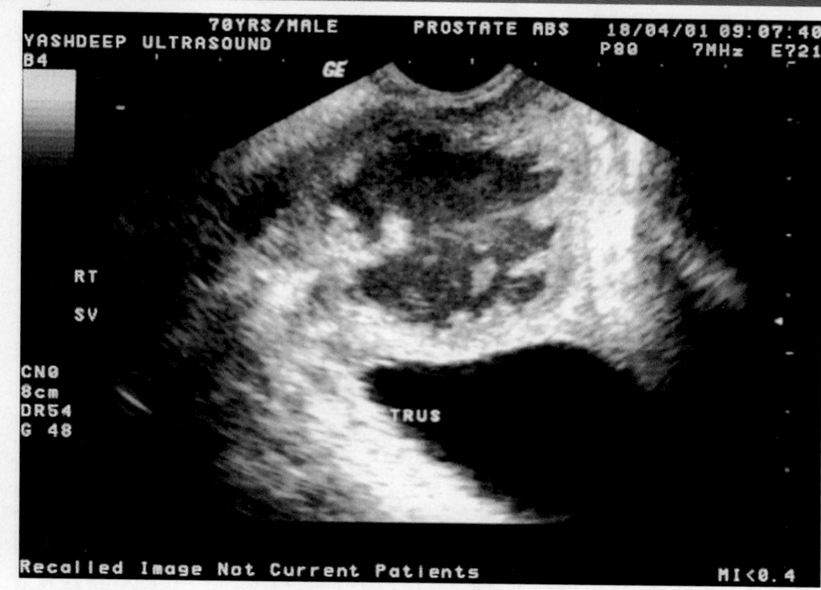

FIGURE 9.99

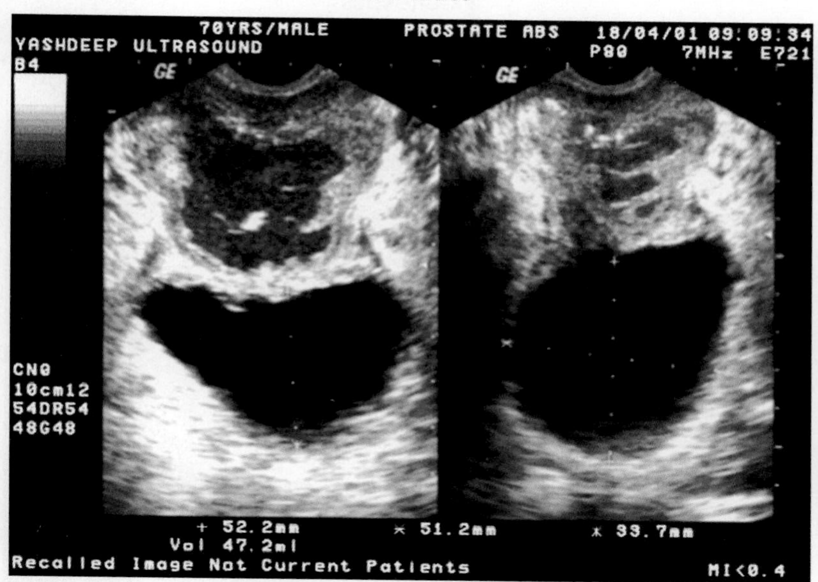

FIGURE 9.100

FIGURES 9.99 and 9.100: Pyogenic abscess in the prostate. HRSG shows an irregular heterogeneous low-level echo complex mass in the substance of gland. Central area of necrosis is also seen. Normal prostatic gland texture is lost suggestive of abscess. Patient was having severe pyouria.

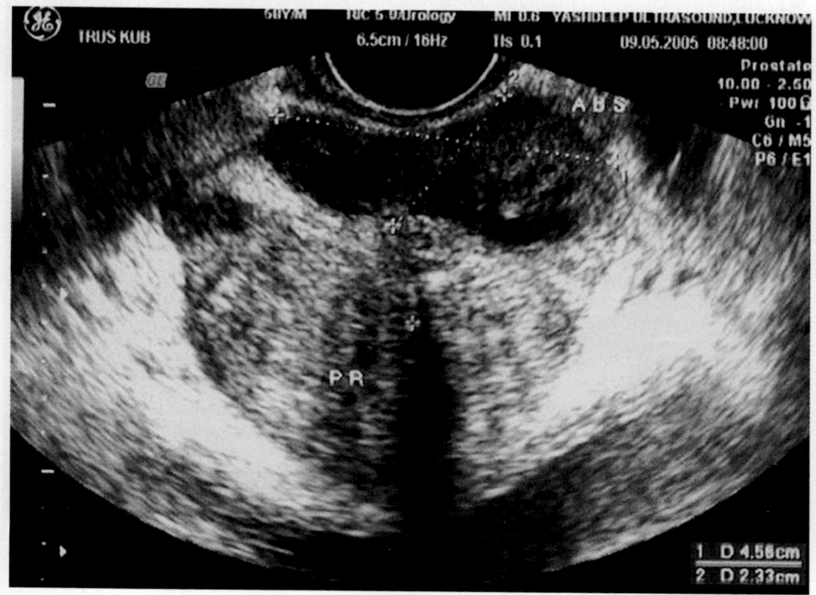

FIGURE 9.101

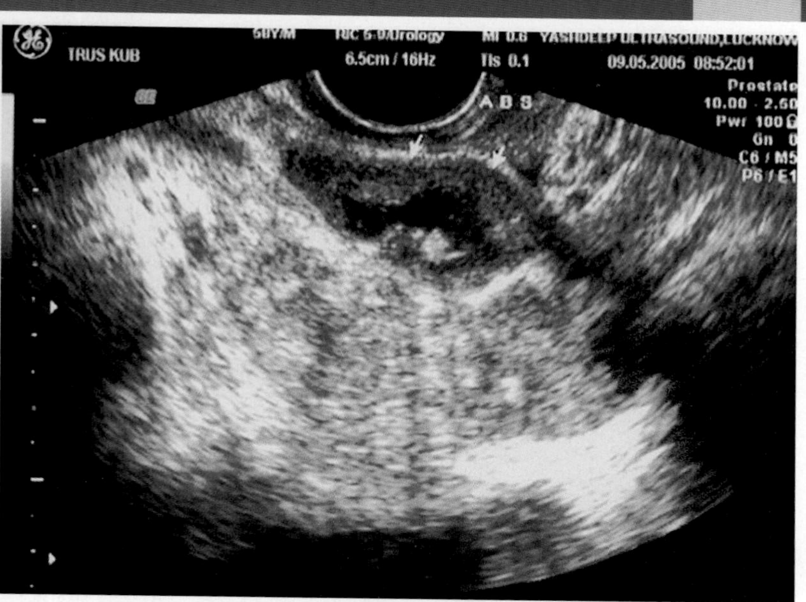

FIGURE 9.102

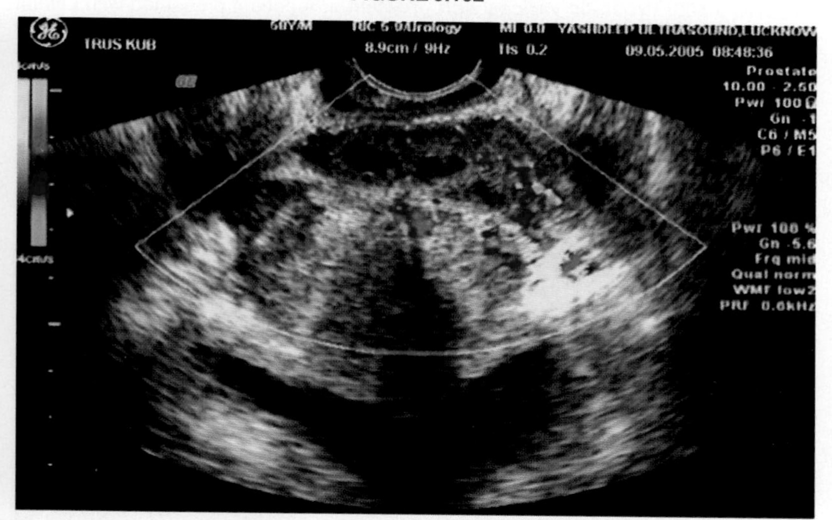

FIGURE 9.103

FIGURES 9.101 to 9.103: Abscess in the prostatic bed. Patient presented with high fever with chills and riggers and tender perineum. HRSG shows low-level echo complex mass in the prostatic bed. Thick echoes are seen in it. The abscess is seen pressing over the prostate. However, the prostatic capsule is intact. No evidence of any breaking down of the capsule is seen. On color flow imaging multiple vessels are seen around the capsule of the abscess.

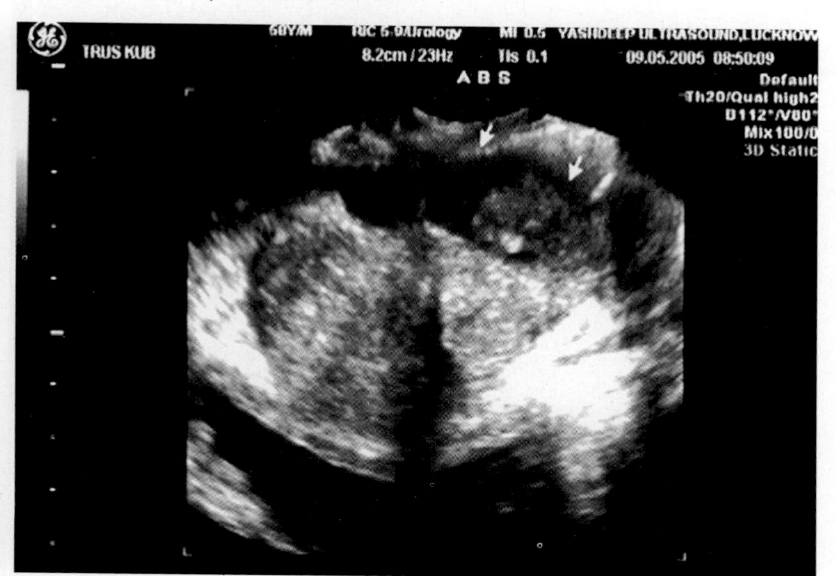

FIGURE 9.104: 3D imaging of the abscess clearly demarcated the abscess separate from the prostate. Echogenic debris is also seen.

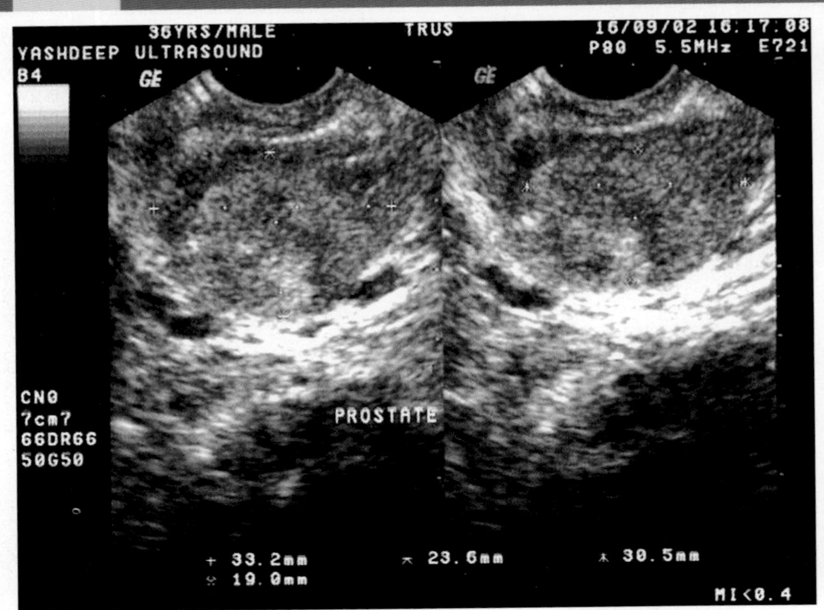

FIGURE 9.105: Prostatic varices with prostatitis. A patient presented with painful urination with infertility. HRSG shows in homogeneous prostate with hypoechoic texture. Multiple dilated vessels are seen around the prostate.

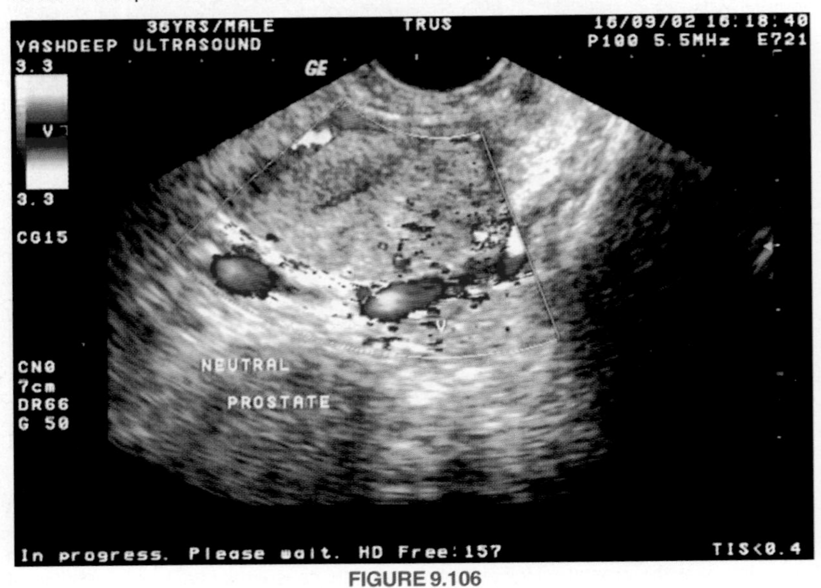

FIGURE 9.106

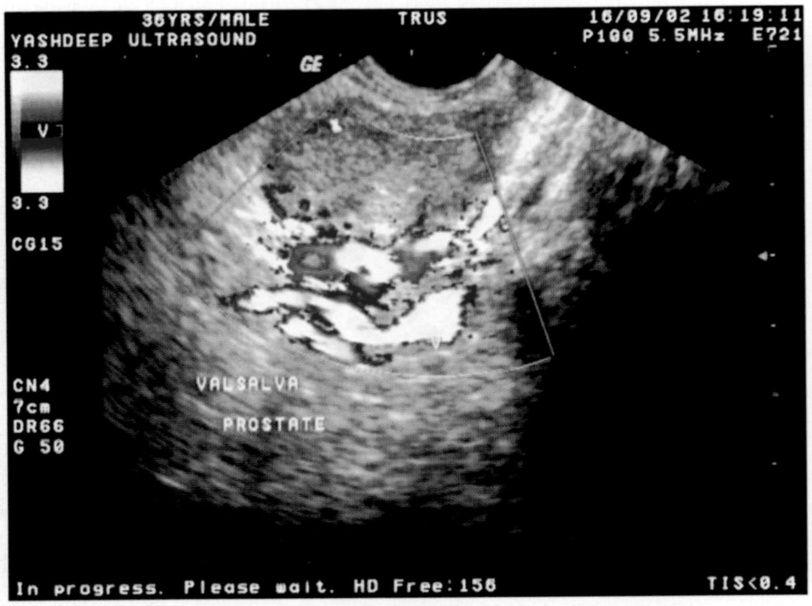

FIGURE 9.107

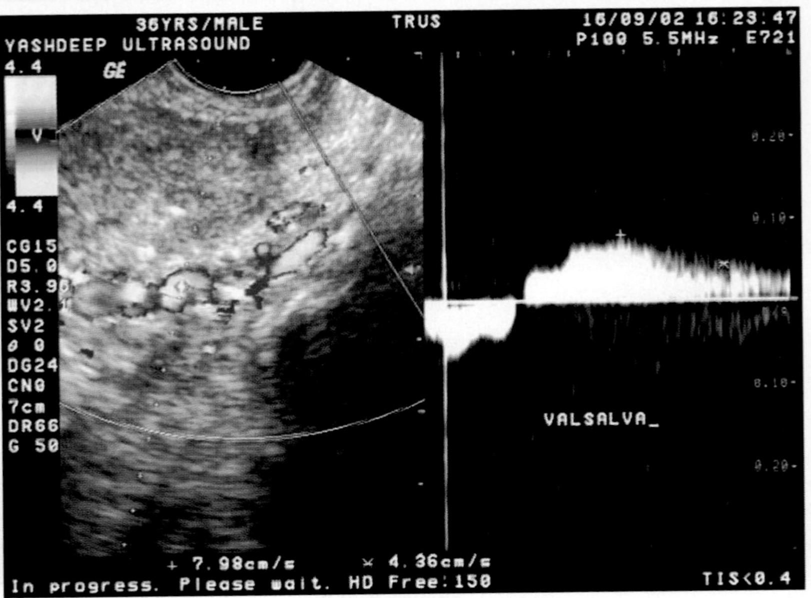

FIGURE 9.108

FIGURES 9.106 to 9.108: On color flow imaging dilated vessels are confirmed around the capsule. The vessels distended on Valsalva maneuver suggestive of varices. On spectral Doppler tracing high flow seen in the prostatic varices.

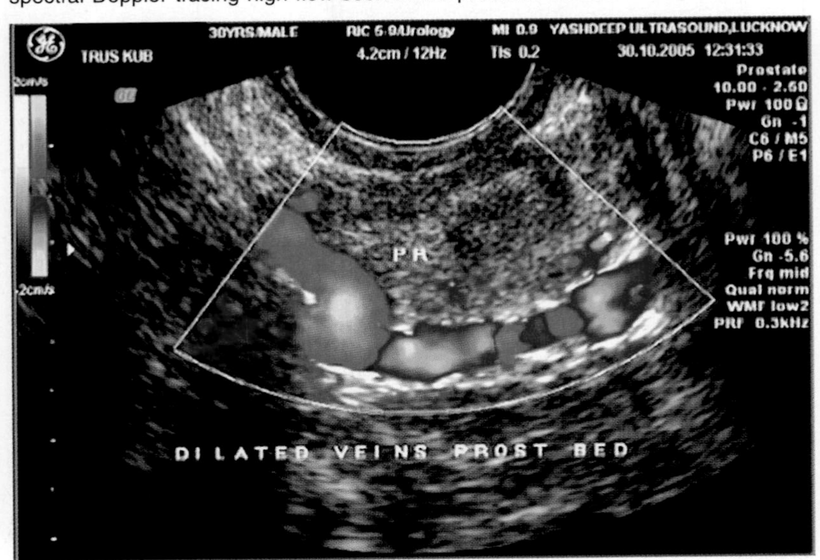

FIGURE 9.109

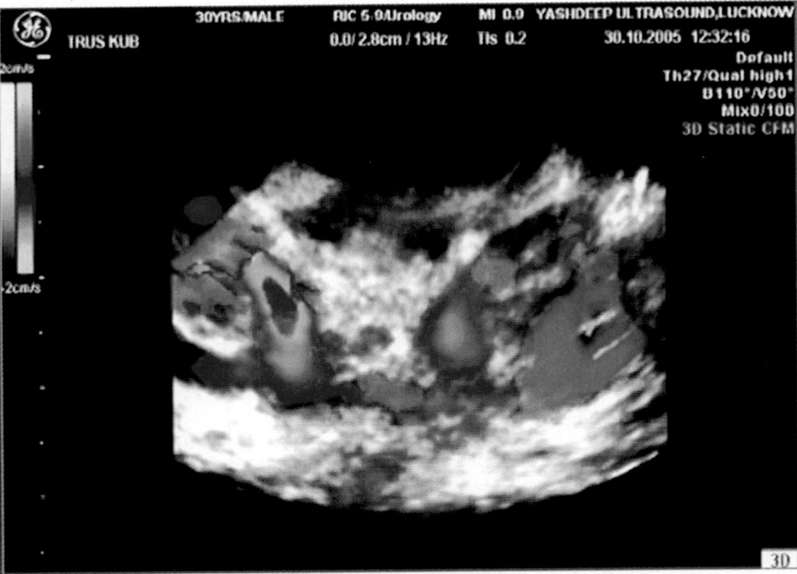

FIGURE 9.110

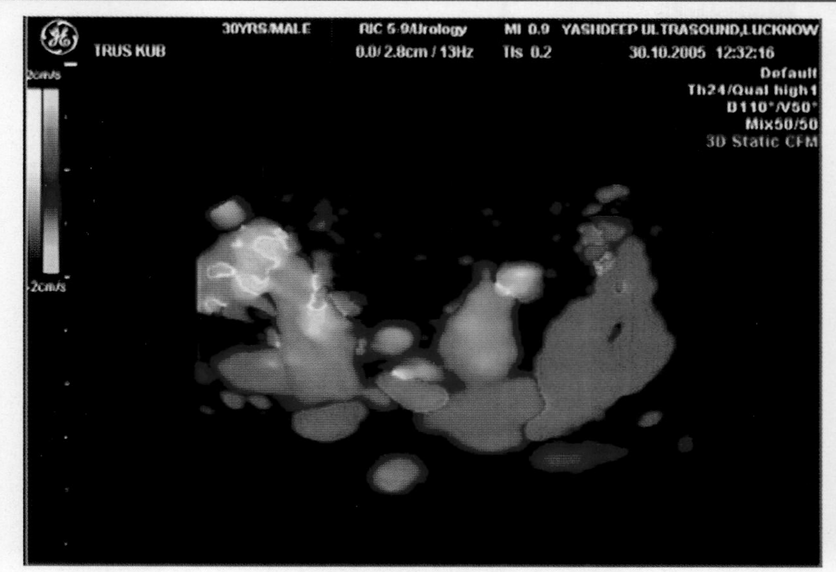

FIGURE 9.111

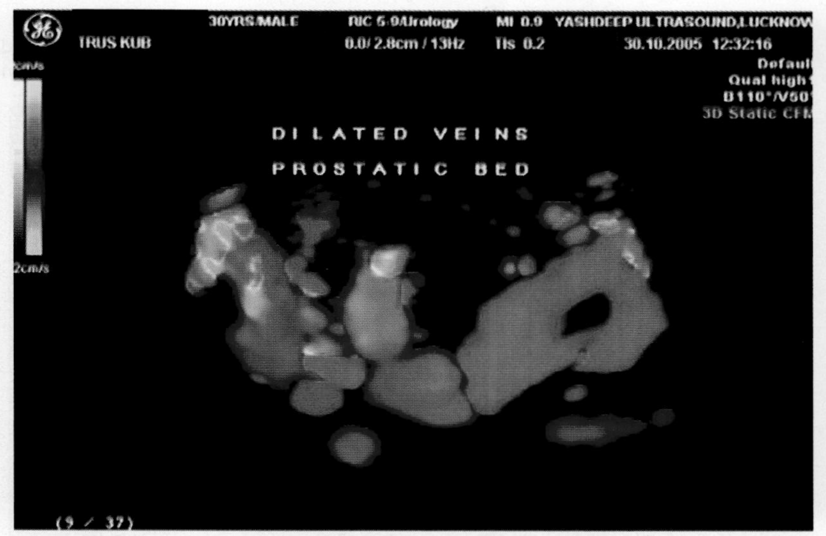

FIGURE 9.112

FIGURES 9.109: to 9.112: Prostatic varices 3D imaging. A patient presented with a infertility. HRSG shows multiple dilated veins in the prostatic bed. They are grossly dilated and surround the prostate from all side. 3D imaging with tissue subtraction technique highlight the veins and gives the better view the dilated vein.

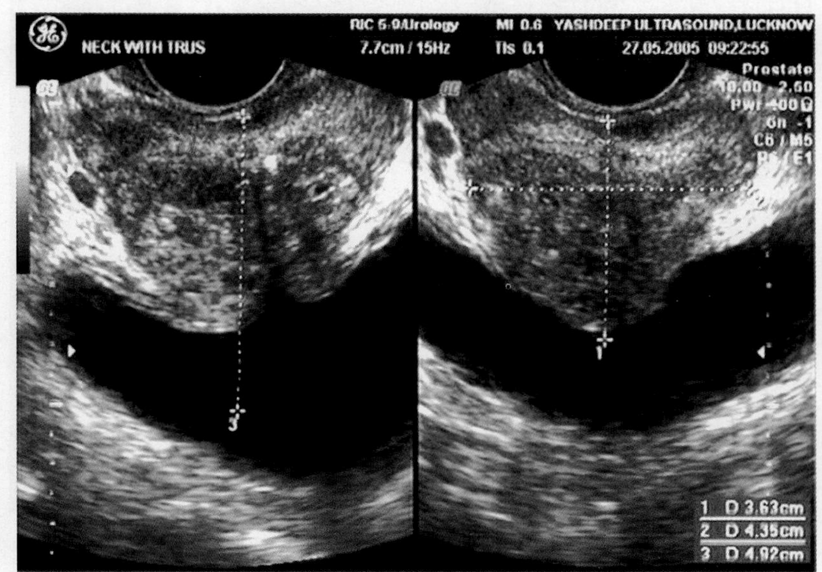

FIGURE 9.113: Tubercular prostatitis. An old man presented with gradual loss of weight and signs of prostatism. HRSG shows enlarged inhomogeneous gland. The glandular texture is hypoechoic small nodules are seen in the gland. They are discrete in position.

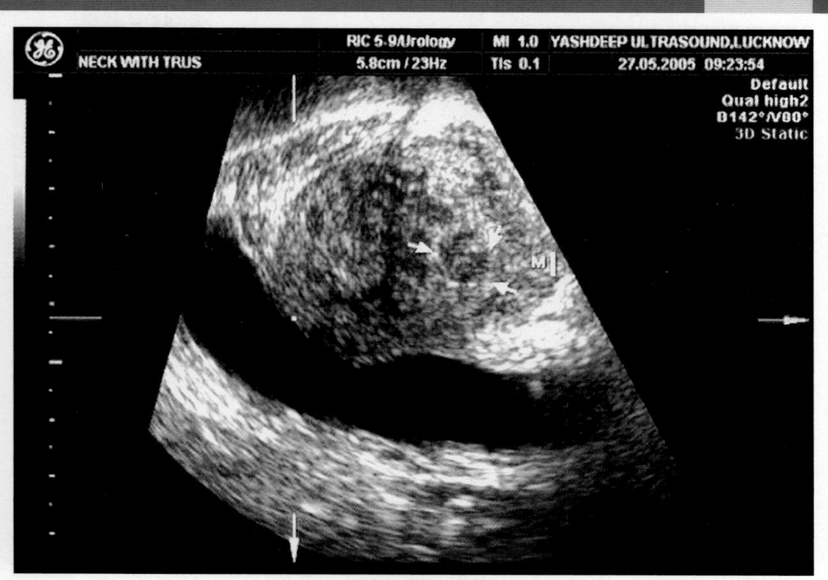

FIGURE 9.114

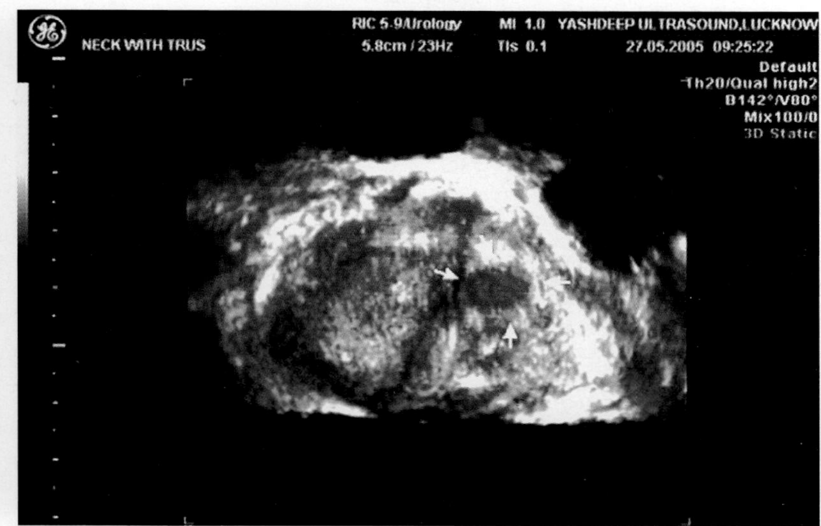

FIGURE 9.115

FIGURES 9.114 and 9.115: 3D imaging of the gland shows the nodules in the central zone of the gland. Biopsy of the nodules confirmed tuberculosis.

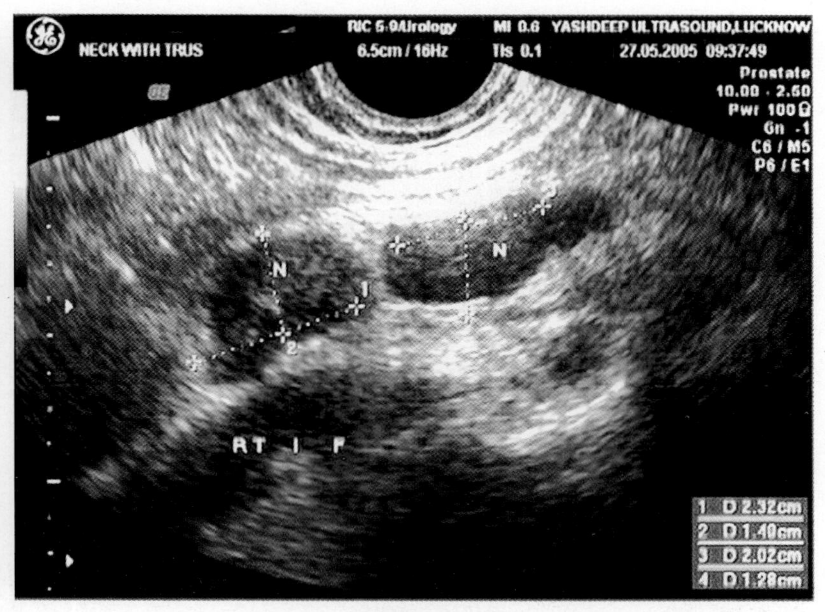

FIGURE 9.116

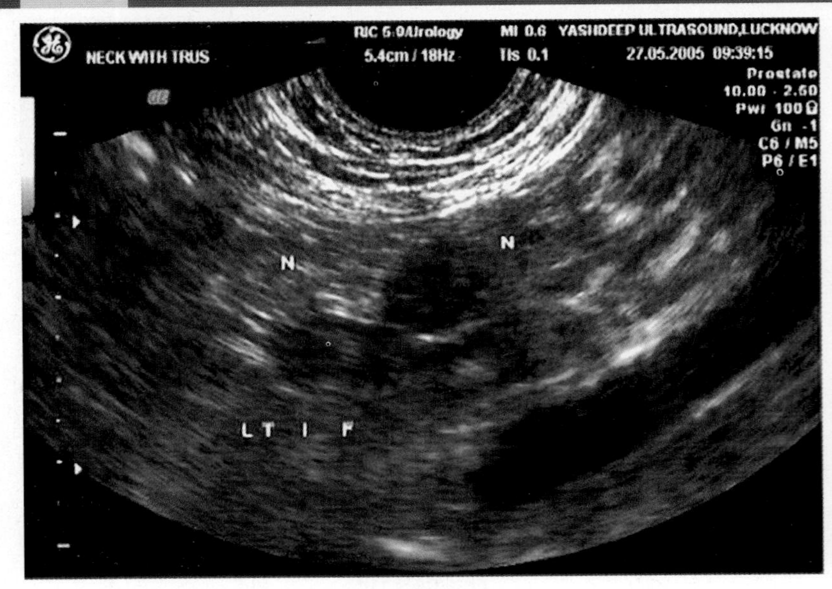

FIGURE 9.117

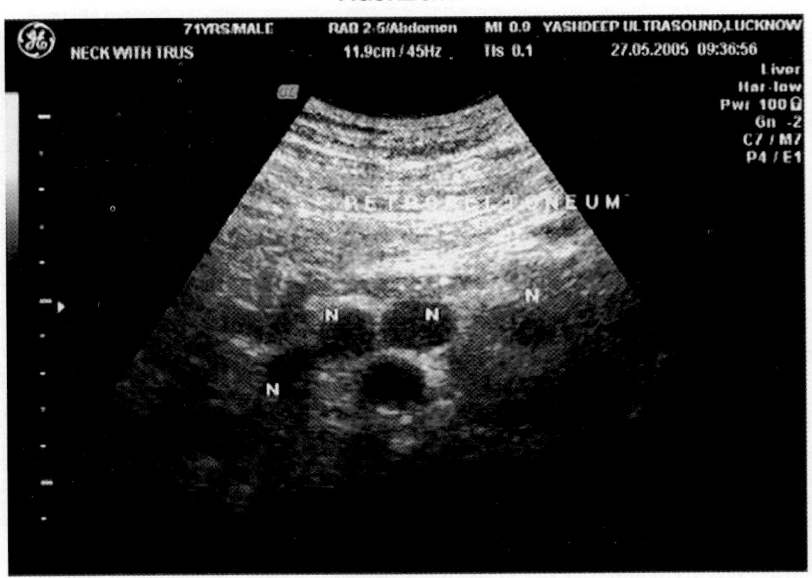

FIGURE 9.118

FIGURES 9.116 to 9.118: Abdominal imaging shows multiple enlarged nodes in both iliac fossa. Enlarged nodes are also seen in the retroperitoneum of the same patient.

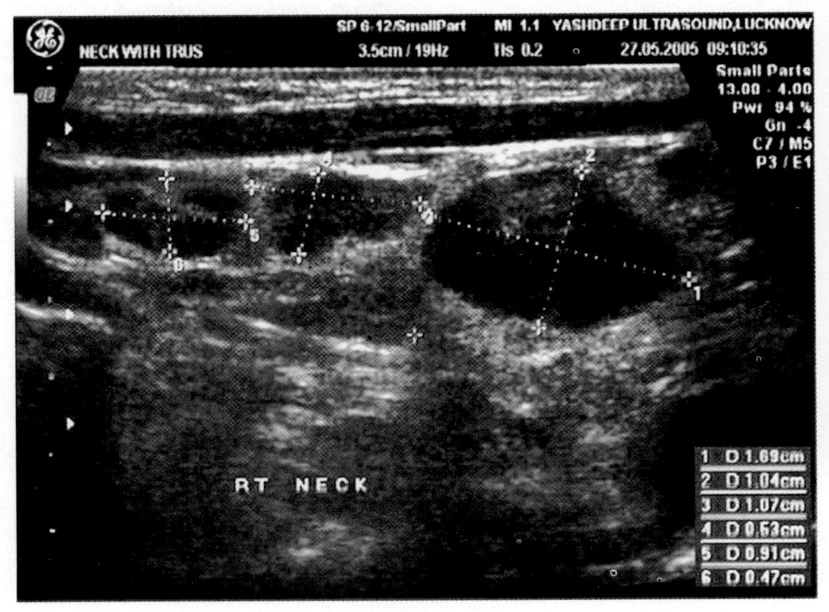

FIGURE 9.119

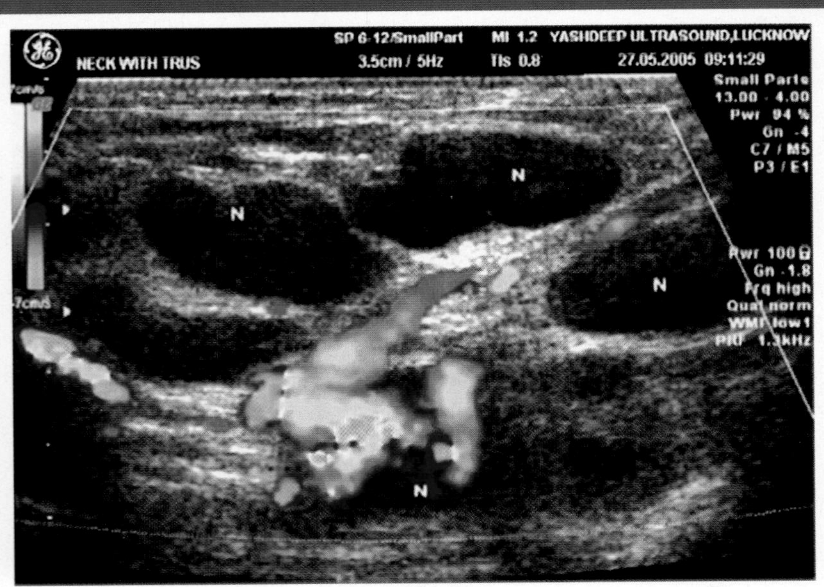

FIGURE 9.120

FIGURES 9.119 and 9.120: HRSG of the neck of the same patient shows multiple enlarged nodes in the supraclavicular regions. Biopsy of the nodes confirmed tuberculosis.

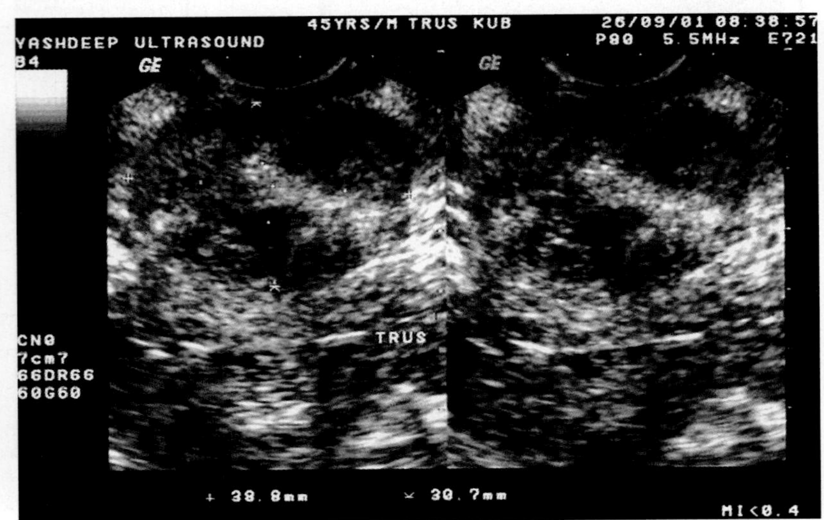

FIGURE 9.121

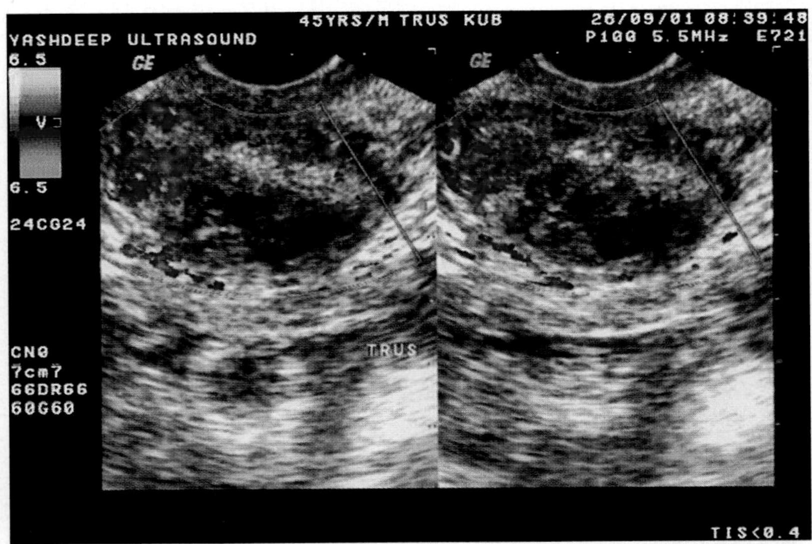

FIGURE 9.122

FIGURES 9.121 and 9.122: Abscess anterior to the prostate. A patient presented with marked tender perineum. Clinical examination showed a mass and suspected prostatic mass. Patient was having severe dysuria. TRUS showed a irregular hypoechoic low level echo complex mass anterior to the prostate. The prostate gland is seen pushed down and displaced.

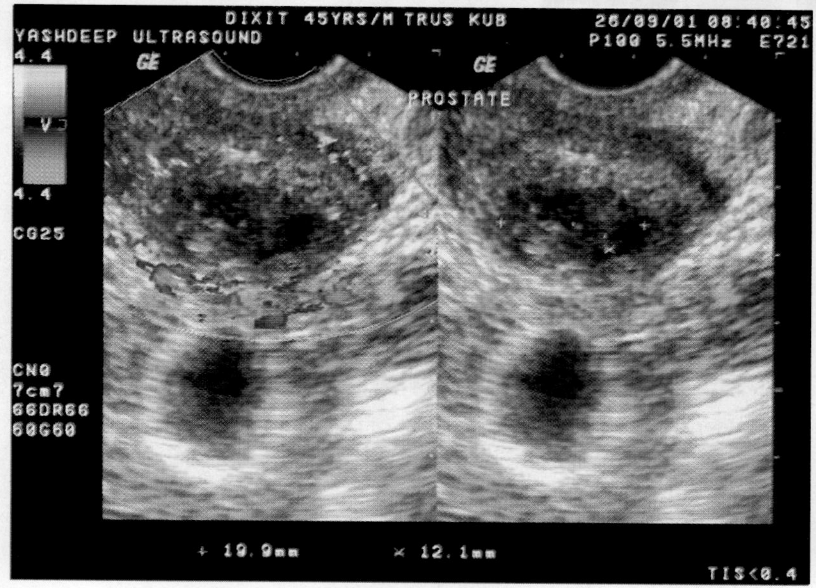

FIGURE 9.123

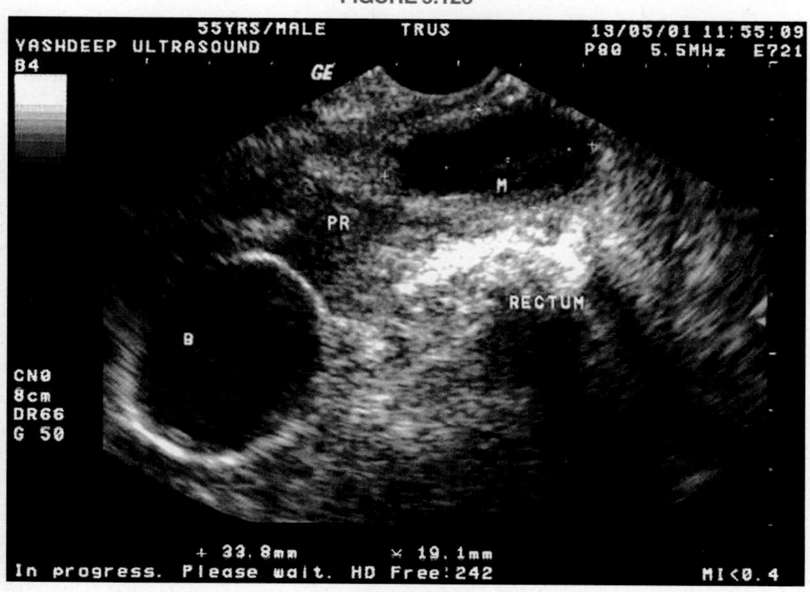

FIGURE 9.124

FIGURES 9.123 and 9.124: On color flow imaging moderate flow is seen in the mass. Wall circulation is also seen. Prostate is seen down and displaced.

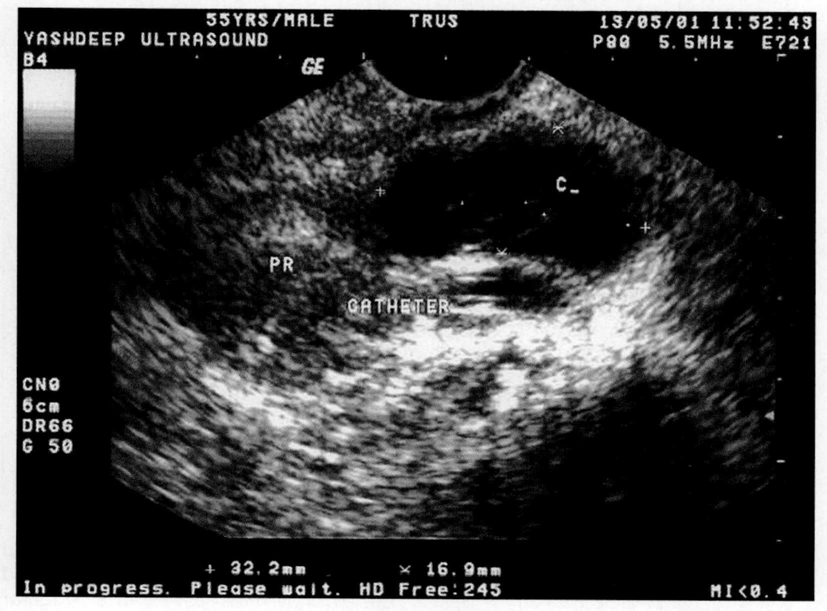

FIGURE 9.125

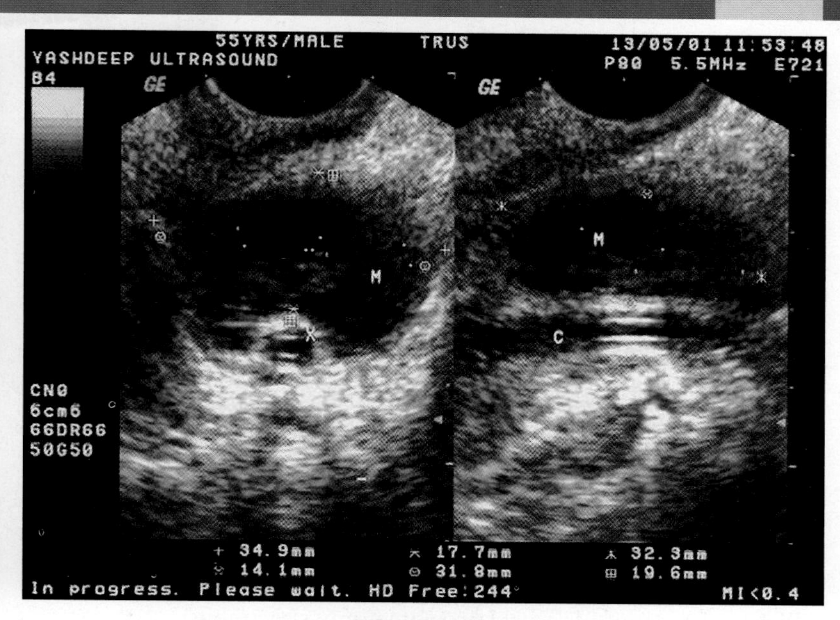

FIGURE 9.126

FIGURES 9.125 and 9.126: HRSG clearly shows the abscess anterior to the prostate gland. Prostate is normal in texture. The prostatic urethra shows catheter in position. Aspiration showed an abscess.

Carcinoma of the Prostate

Localized prostatic cancer classically presents as a hypoechoic lesion with ill-defined or well-defined borders in the peripheral zone of the prostate. Contrast between the echotexture of the tumor and that of the normal, evenly echogenic surrounding peripheral zone allows easy detectability. Sonographic-pathologic correlation studies of radical prostatectomy specimens have shown that approximately 70-75% of tumors have this classical US appearance, while; about 30% of cancers are isoechoic and flush with the surrounding tissue. These cannot be identified by TRUS unless secondary sonographic signs of cancer, such as a bulge of the capsule or distortion of the periprostatic fat are present. A smaller number of cancers have a heterogeneous echotexture with subtle areas of hyperechogenicity or stippled small echogenic foci scattered within the hypoechoic lesion.

The reasons for the variable sonographic appearance of prostate carcinoma are incompletely understood. Well-differentiated tumors tend to be hypoechoic. Higher-grade lesions are more heterogeneous perhaps due to a higher degree of stromal fibrosis and desmoplastic reaction or invasion of areas of prostate hyperplasia by larger cancers. Hyperechogenic foci within the tumor correspond pathologically to comedo calcification within highly undifferentiated anaplastic tumors. Echogenic areas located in the periphery of the tumor are usually calcified corpora amylacea is adjacent benign prostatic tissue.

Factors which Affect the Detectability

Two factors affect the detectability of a tumor on TRUS.

Tumor echogenicity: Hypoechoic tumors are readily visible, whereas isoechoic lesions blend with the surrounding tissues.

Tumor size: Small cancers (less than 5 mm) are often undetected. Very extensive cancers involving the entire peripheral zone are difficult to identify because comparison with normal tissue is impossible. Almost 50% of cancers are found in the anterior half of the prostate, and either arises from the central or transition zone, or from the anterolateral portion of the peripheral zone. These tumors are often located beyond the focal zone of the transducers, or blend in with areas of prostatic hyperplasia. Stage A cancers (cancers not palpable clinically), are specifically difficult to detect, even in retrospect unless they extend into the posterior portion of the peripheral zone, or appear as irregular hypoechoic extend into the posterior portion of the peripheral zone, or appear as irregular hypoechoic mass adjacent to the TURP defect.

Accuracy of Diagnosis

Estimation of tumor volume by TRUS: It is often inaccurate for two reasons: Tumor size is usually underestimated particularly in large cancers, which tend to infiltrate into and spread through the gland to invade areas of benign prostate hyperplasia. Prostatic cancer is often multifocal. Only approximately half of contra lateral tumor foci were prospectively identified by TRUS in one study comparing pre-operative TRUS and sections of radical prostatectomy specimens.

Accuracy of TRUS in the diagnosis of prostatic cancer: The US appearance of early prostate cancer is not specific. The positive predictive value (PPV) for a hypoechoic lesion to be cancer increases with the size of the lesion, presence of a palpable nodule, and elevated PSA. Overall, the use of TRUS can also determine glandular volume accurately. The most commonly used method to measure prostatic or tumor volume is the formula for a prostatic ellipse (length x width x height x 0.523). Volumes can then be easily converted to weight because of 1 mass lesion of prostatic tissue is equivalent to 1 g.

Prostate Specific Antigen (PSA)

Prostate specific antigen is a low molecular weight glycoprotein, which is produced exclusively by the prostate and to a smaller extent by the seminal vesicular tissue. It is normally present in small amounts in all post-pubertal males and is absent in women. The two techniques for measuring PSA are:

- Polyclonal assay (Yang method)
- Monoclonal assay (Hybritech method)

The monoclonal assay is the most commonly used method the world over.

PSA levels have been evaluated as normal borderline and abnormal values. The following are the accepted levels.

- < 4 ng/cc– normal
- 4.0-10.0ng/cc—borderline
- >10 ng/cc—abnormal

However, the diagnosis of prostatic malignancy is not specific or absolute with PSA studies. A normal PSA does not necessarily mean that the patient has no cancer. Similarly, abnormal values are not definitive for the presence of prostatic malignancy. There are several entities, which can increase the PSA values. These are:

- Prostate cancer
- Benign prostatic hypertrophy
- Prostatic inflammation
- Prostatic infarct
- Post digital rectal examination

In view of the limitations of the role of PSA in detecting prostatic cancer, newer parameters are being evolved. Some of these are as follows:

Free Prostate Specific Antigen

This is one of the parameters, which is now being advocated since the absolute PSA is not definitive of prostatic cancer. Evaluation of free versus bound PSA may provide additional information for the evaluation of prostatic cancer. However, further work is needed to confirm the role of free PSA in prostatic pathology.

Prostate Specific Antigen Density

It has been suggested that a measure of the volume of the prostate may correlate with PSA level to a specific disease. It has now been shown that men who have PSA levels >4 ng/cc with no palpable lesion on digital rectal examination, but have a PSAD>0.10 and 0.12/0.15 have cancer.

Prostate Specific Antigen Velocity

It had been suggested that the changing levels of PSA especially increasing values, even if within the normal range may be significant and require further evaluation. A PSA level that doubles or triples in the course of a single year should raise suspicion that the patient may be harboring prostatic cancer.

Age-related Prostate Specific Antigen

As men age, their prostate also enlarge due to BPH. It has been suggested that the normal PSA values may relate to the age of the individual and the range may vary with the age. The following parameters may be useful for "normal age-related PSA".

40-50 years	0.0-2.5ng/cc
50-60 years	0.0-3.5ng/cc
60-70 years	0.0-4.5ng/cc
>70 years	0.0-6.5ng/cc

Prostatic Biopsy

The following are the indications for undertaking transrectal prostatic biopsies.

- A suspicious lesion, i.e. a peripheral, ill-defined lesion in the PZ.
- Focal bulge in the prostate
- Capsular or pericapsular irregularity
- Rising PSA levels

Biopsies of the prostate gland can be performed either by the transperineal route or the transrectal approach. The transrectal biopsy is a relatively simple procedure carried out under antibiotic cover. A true-cut 18-gauge spring-loaded needle is used. The position of the transducer is adjusted until the puncture guideline, superimposed on the image, transects the area of biopsy. The tip of the needle is visualized as it penetrates the rectal wall and enters the prostate. Transrectal biopsies are performed without anesthesia and are well tolerated in most patients.

Hypoechoic lesions that are larger and exist with significantly elevated PSA levels have a higher risk of malignancy. Lesions that are smaller and are in patients with lower or normal PSA levels have a lower risk of being malignant. Lesions in the TZ and the CZ or other gland have much lower rates of malignancy than those with similar lesions in the PZ.

Color Doppler of the Prostate

In general, cancer anywhere in the body has the following characteristics on color Doppler:

- An increased number of visualized vessels
- An increase in the size of the vessels
- Increase in the irregularity and contours of the vessels
- Increased diastolic flow indicating a low resistance system.

The normal prostate has abundant surrounding vasculature. The two largest collections of vessels are the Santorini's plexus and the neurovascular bundles.

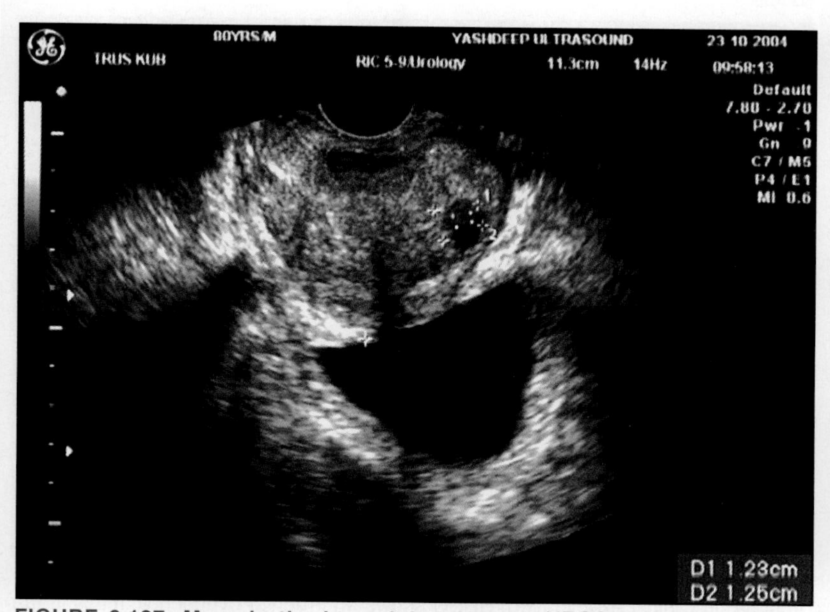

FIGURE 9.127: Mass in the Lt peripheral zone. HRSG shows an irregular hypoechoic mass in the Lt peripheral zone. The margins are irregular in outline. No necrosis is seen in the mass. Biopsy of the mass confirmed adenocarcinoma.

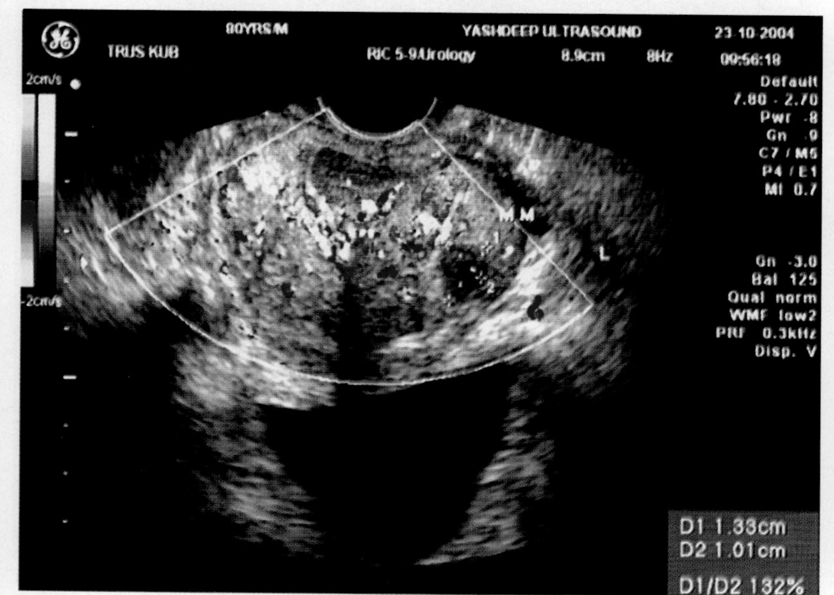

FIGURE 9.128: On color flow imaging there is evidence of increased flow is seen around the nodule.

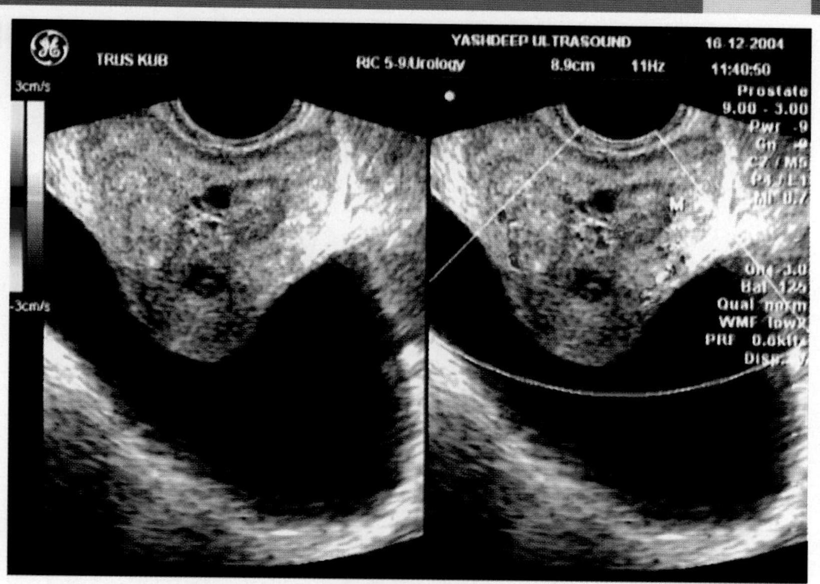

FIGURE 9.129

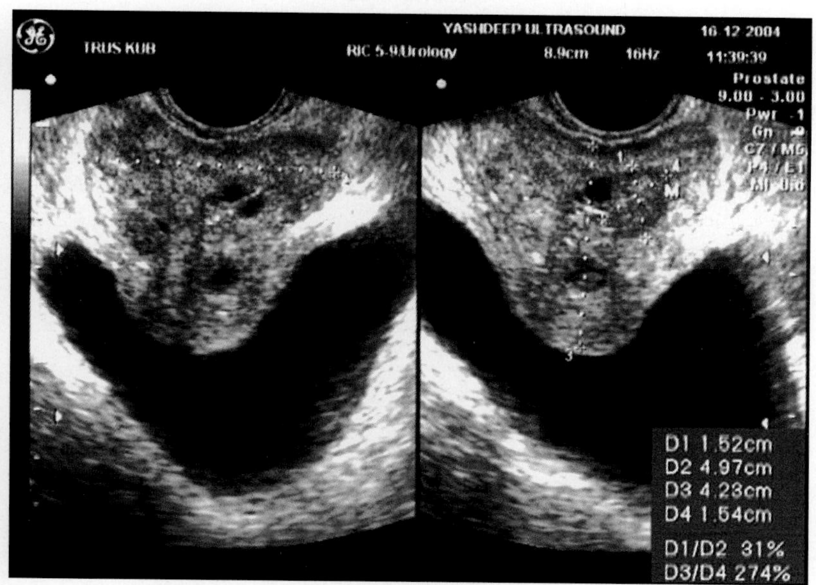

FIGURE 9.130

FIGURES 9.129 and 9.130: Irregular hypoechoic mass in the Lt peripheral zone. HRSG shows irregular hypoechoic nodular mass in the Lt peripheral zone. The mass is seen encroaching the central zone. Few cystic areas are seen in the mass. But no calcification is seen. Bulging of the lateral margins of the lobe is also seen.

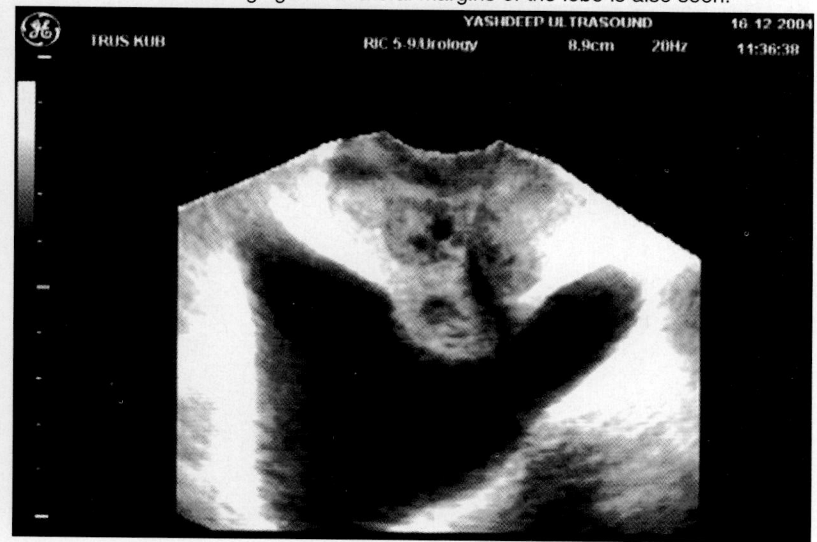

FIGURE 9.131: On 3D Imaging multiple nodules are seen in the gland. Another hypoechoic nodule is also seen in the median lobe. 3D imaging is good technique to look for hidden nodules in the prostate gland.

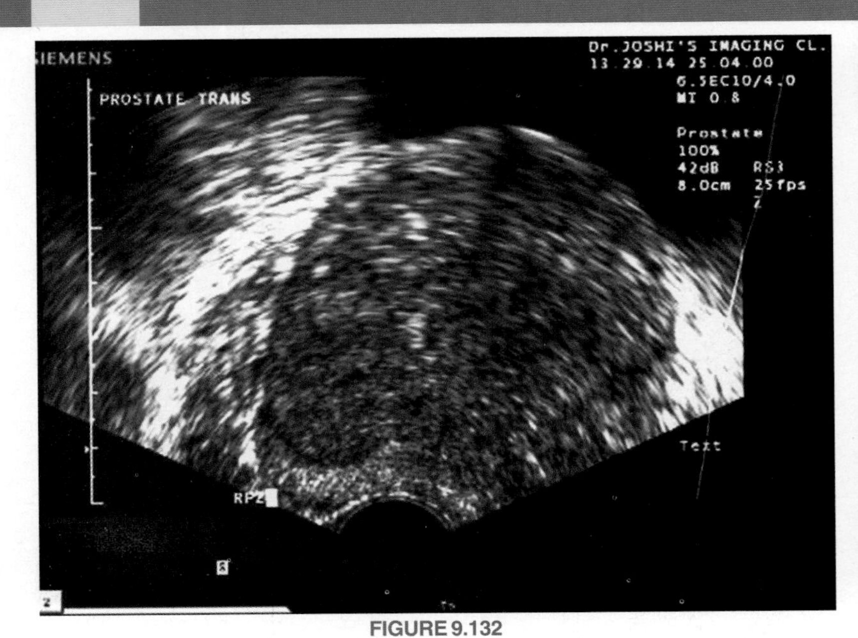

FIGURE 9.132

FIGURE 9.133

FIGURES 9.132 and 9.133: Hypoechoic nodular mass in the Rt peripheral zone. HRSG shows well-defined hypoechoic mass in the Rt peripheral zone. The mass shows variegated texture. No calcification is seen in it. No cystic necrosis is seen in it.

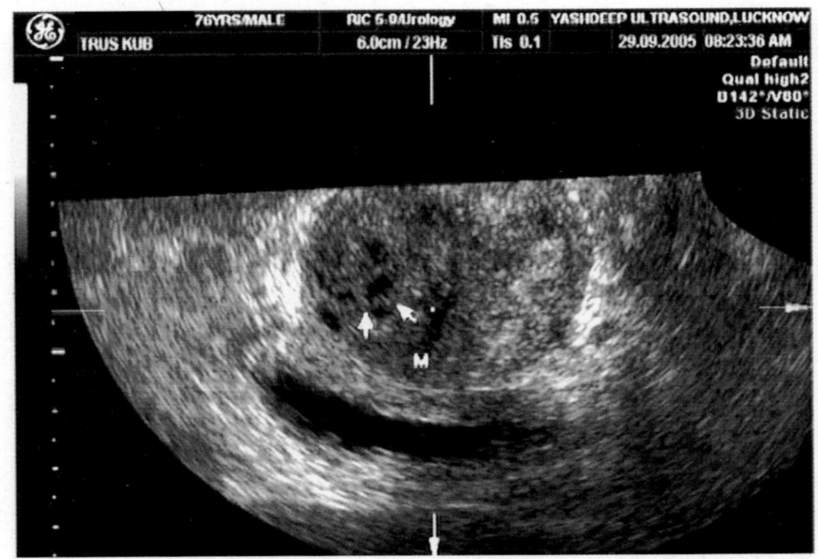

FIGURE 9.134: 3D imaging of the prostate shows irregular heterogeneous texture of the mass in much better view. Multiple hypoechoic areas are seen in the mass. The margins are ill-defined. Biopsy of the mass confirmed adenocarcinoma.

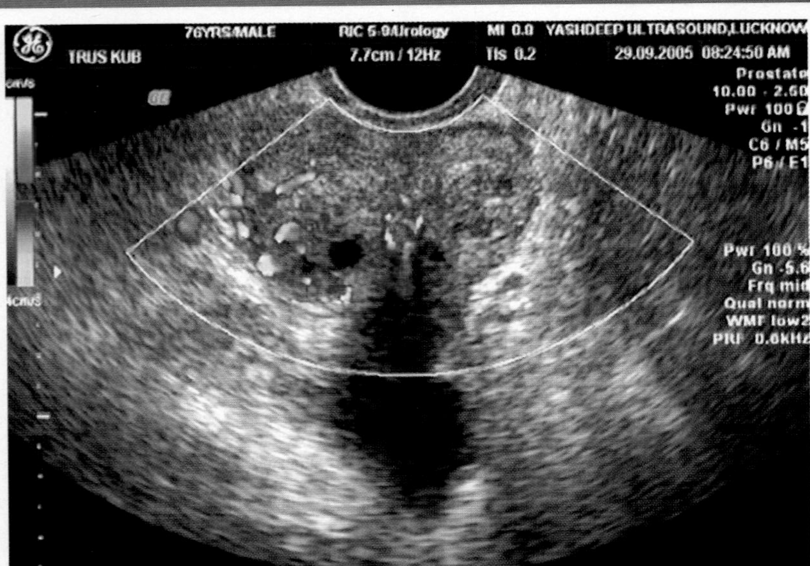

FIGURE 9.135: On color flow imaging of the mass shows increased flow around the nodule.

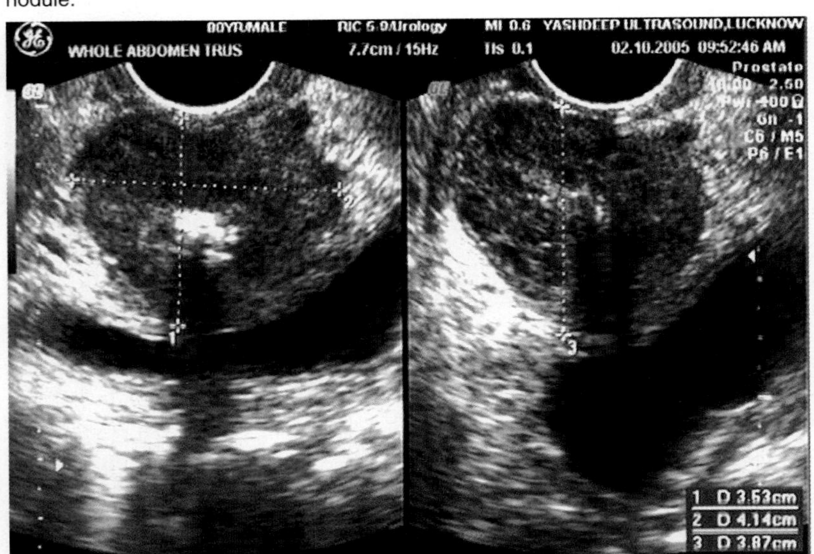

FIGURE 9.136: An old man presented with markedly increased PSA with nodular prostate on clinical examination. HRSG shows an irregular hypoechoic mass in the transition zone. The mass is seen bulging toward the Rt lateral lobe.

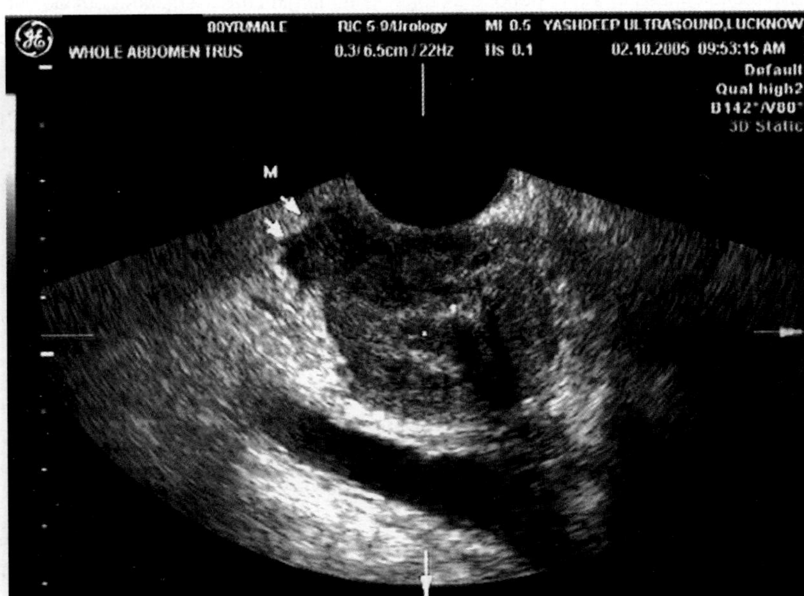

FIGURE 9.137: 3D imaging of the mass shows broken capsule of the gland in the Rt lateral lobe. The mass is seen growing beyond the capsule and invading the periphrastic area.

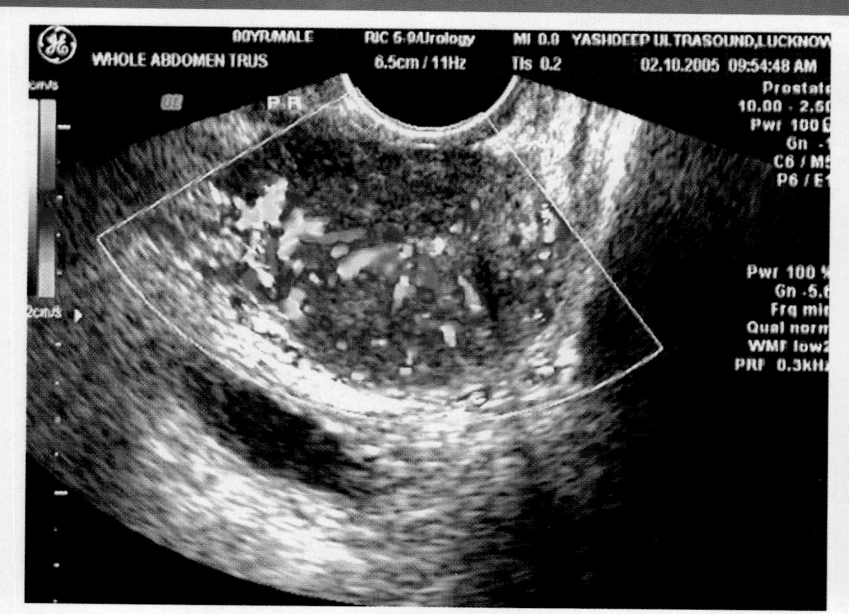

FIGURE 9.138

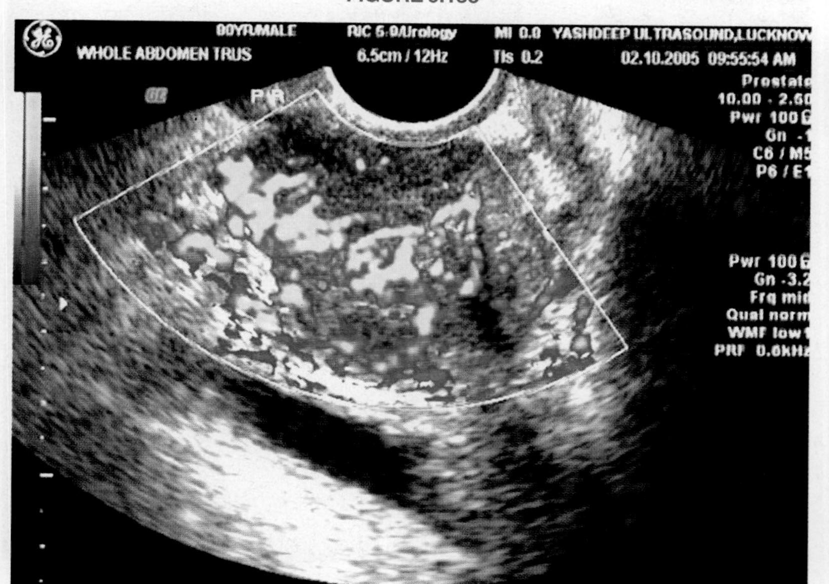

FIGURE 9.139

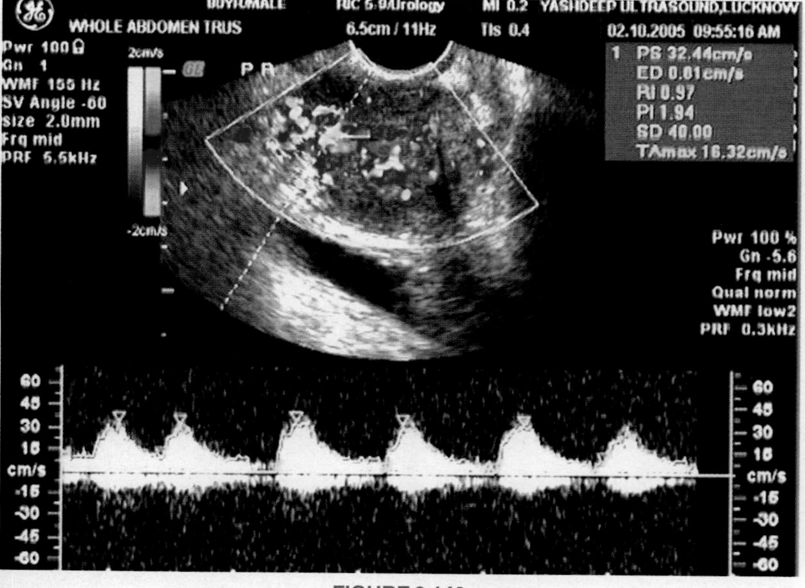

FIGURE 9.140

FIGURES 9.138 to 9.140: On color flow imaging high flow seen in the gland. Marked increased flow seen in the Lt peripheral zone on power Doppler imaging. On spectral Doppler tracing high flow is seen in the mass.

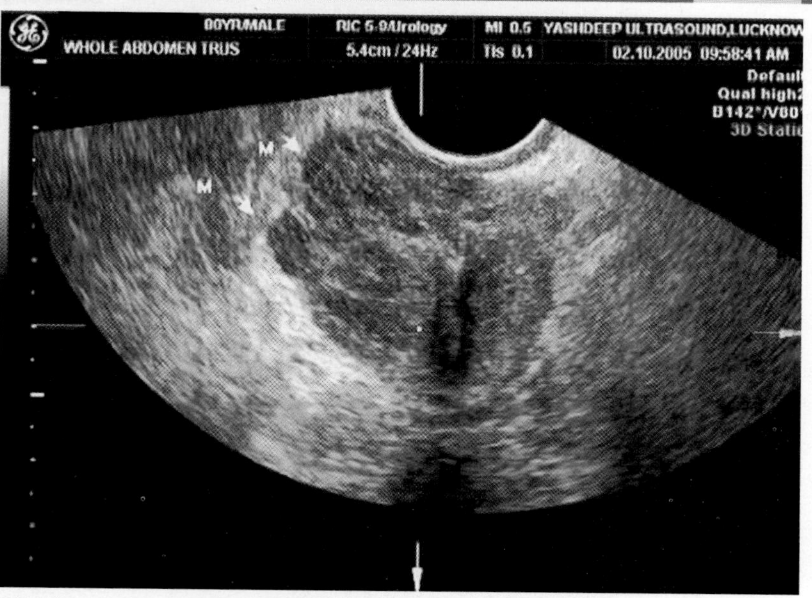

FIGURE 9.141: 3D imaging of the prostate shows multiple heterogeneous nodules in the Rt peripheral zone. They are seen bulging out.

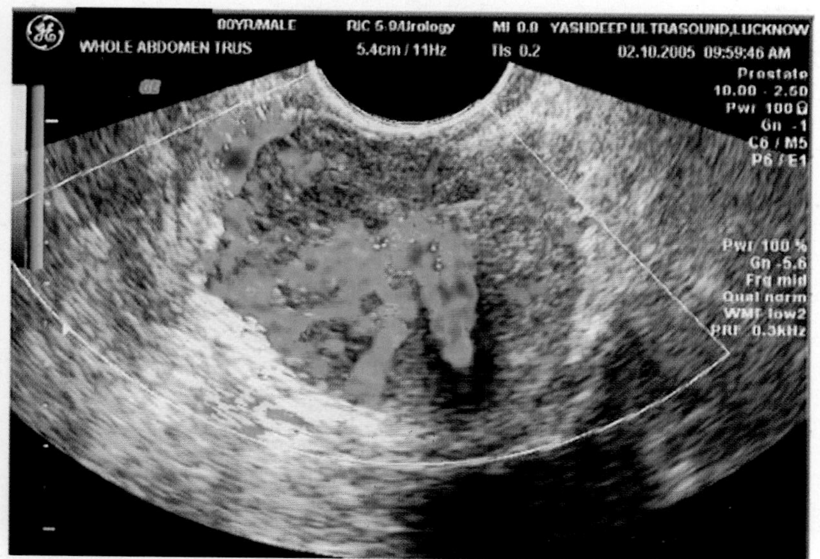

FIGURE 9.142

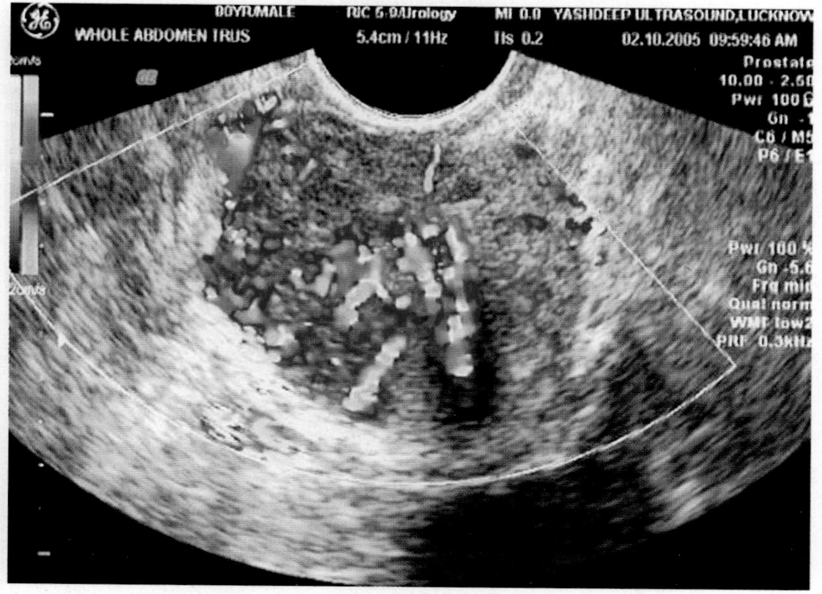

FIGURE 9.143

FIGURES 9.142 and 9.143: On directional power Doppler imaging high flow is seen around the mass. Biopsy of the mass confirmed adenocarcinoma prostate.

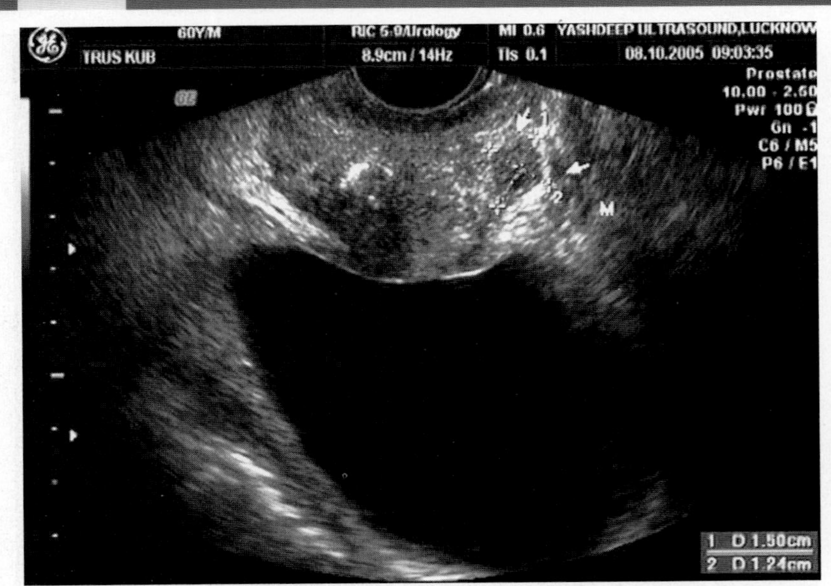

FIGURE 9.144: Nodular mass in the Lt peripheral zone. HRSG shows irregular hypoechoic mass in the Lt peripheral zone. The mass is seen pressing over the capsule. Capsular bulge is seen. But no evidence of any breaking of the capsule is seen.

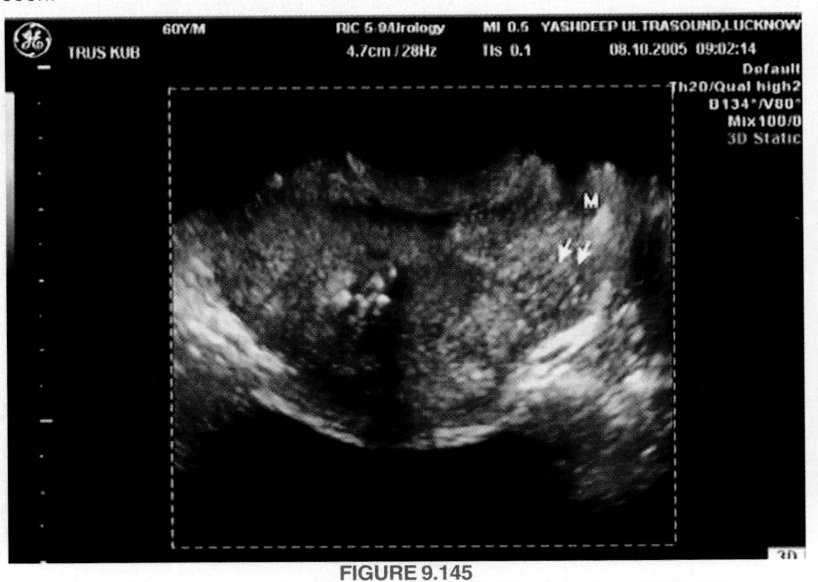

FIGURE 9.145

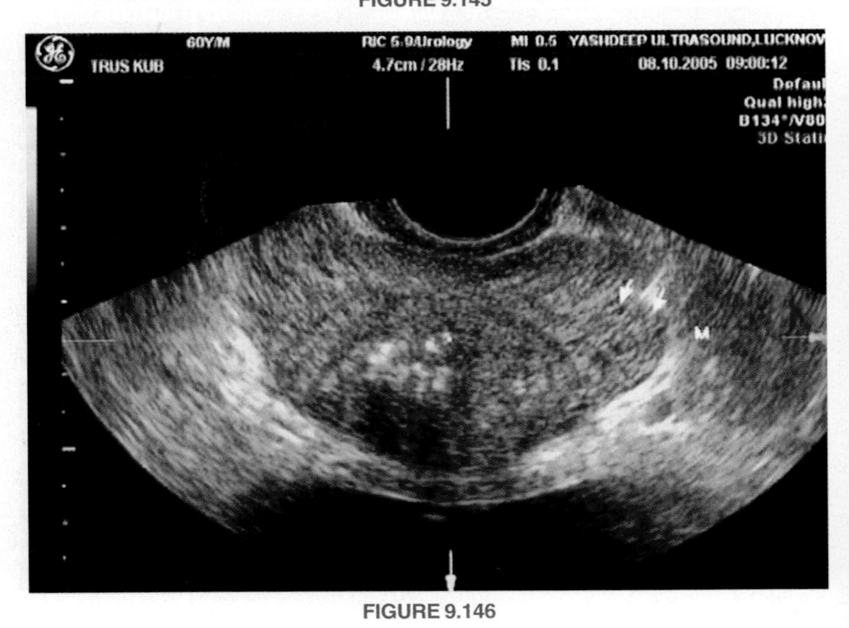

FIGURE 9.146

FIGURES 9.145 and 9.146: 3D surface rendering of the prostate shows the mass in much clear details with capsule bulge.

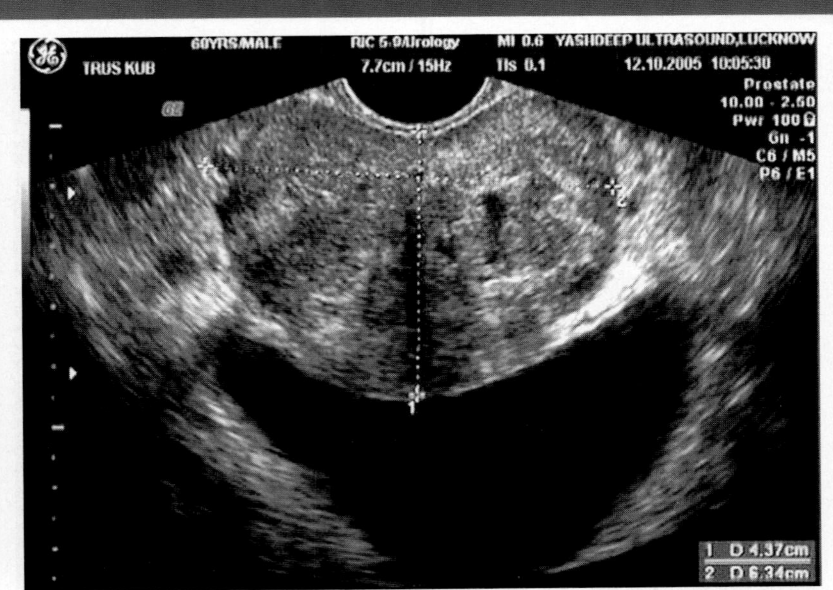

FIGURE 9.147

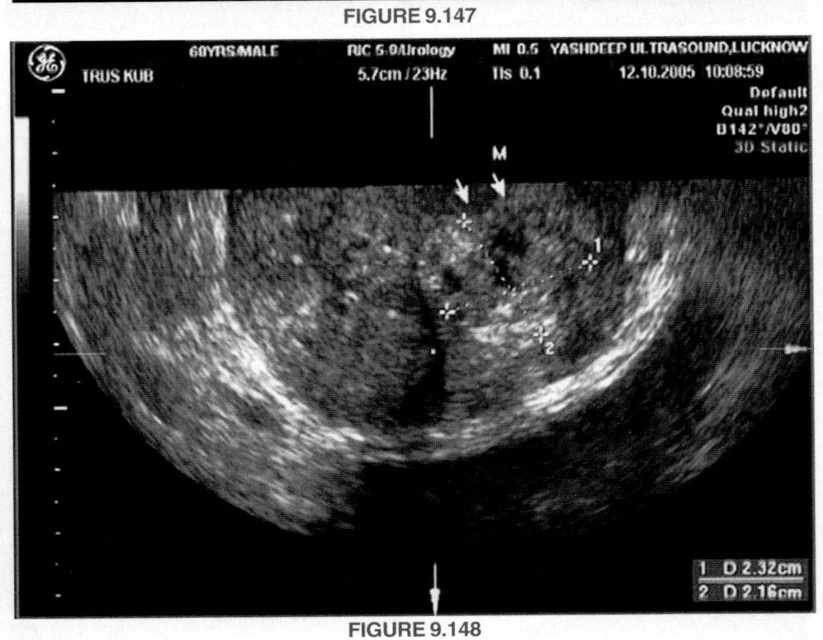

FIGURE 9.148

FIGURES 9.147 and 9.148: Heterogeneous mass in the Lt peripheral zone. Irregular heterogeneous mass is seen in the Lt peripheral zone. The mass is confined to the zone. No pressure is seen on capsule. 3D imaging of the prostate shows the mass in much better clearity.

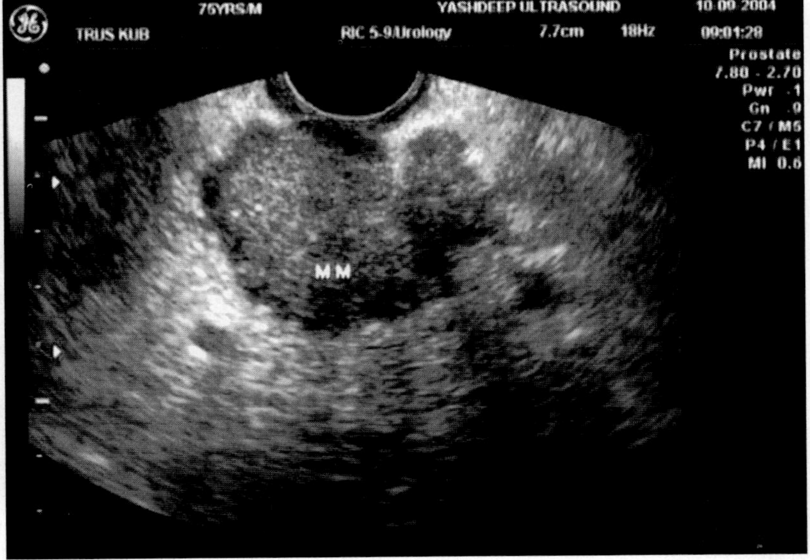

FIGURE 9.149

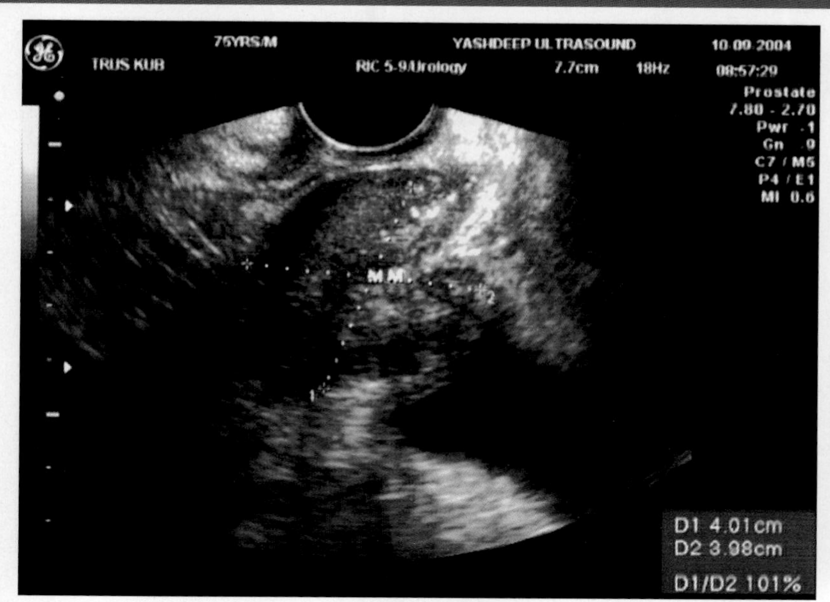

FIGURE 9.150

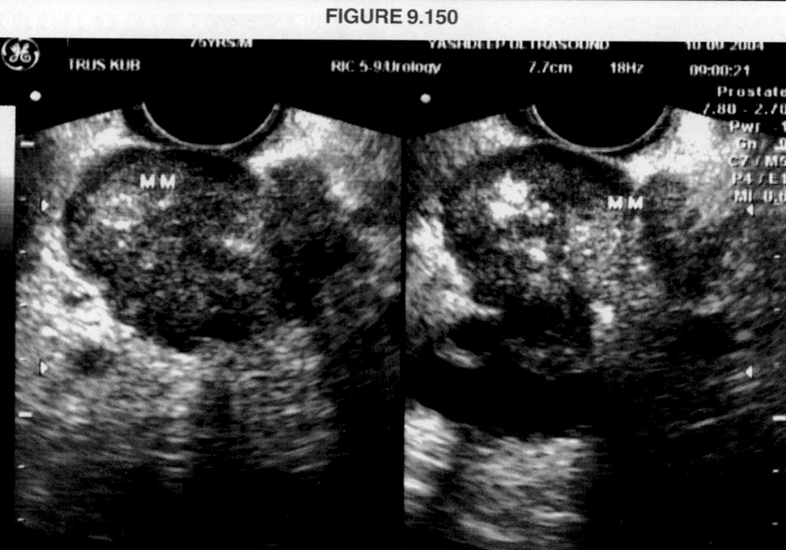

FIGURE 9.151

FIGURES 9.149 to 9.151: Heterogeneous mass in the prostate with broken capsule and local invasion. HRSG shows big heterogeneous mass in the gland involving the Lt peripheral zone. The mass is seen breaking the capsule of the gland and invading the surrounding tissue. The inner mass of the gland is also seen invaded by the mass.

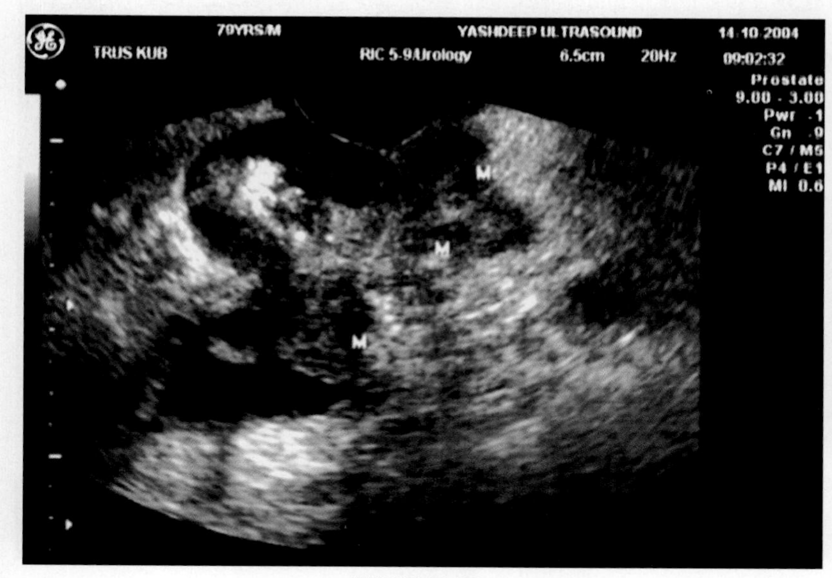

FIGURE 9.152

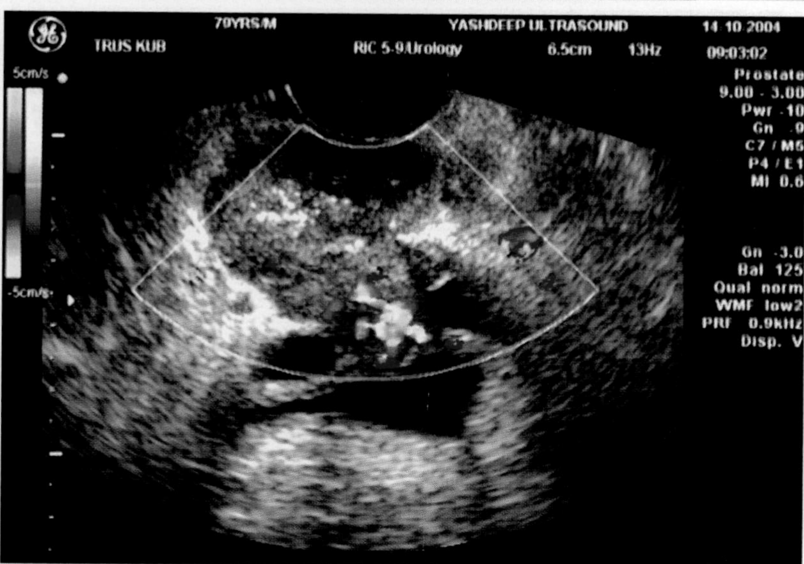

FIGURE 9.153

FIGURES 9.152 and 9.153: Heterogeneous malignant mass lateral and median lobe of the gland. HRSG shows big heterogeneous mass in the Lt peripheral zone and also the median lobe of the gland. Normal glandular texture is lost. Irregular hypoechoic nodular masses are seen in the gland. Capsule of the gland is broken. On color flow imaging high flow seen in the mass. Biopsy of the mass confirmed CA prostate.

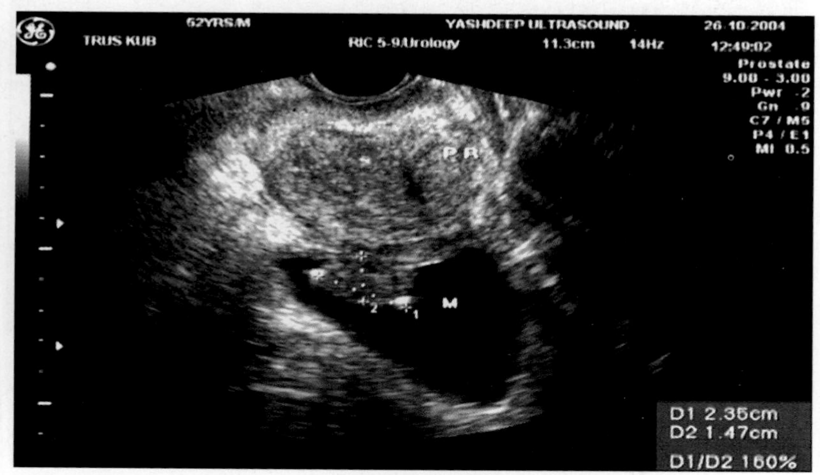

FIGURE 9.154

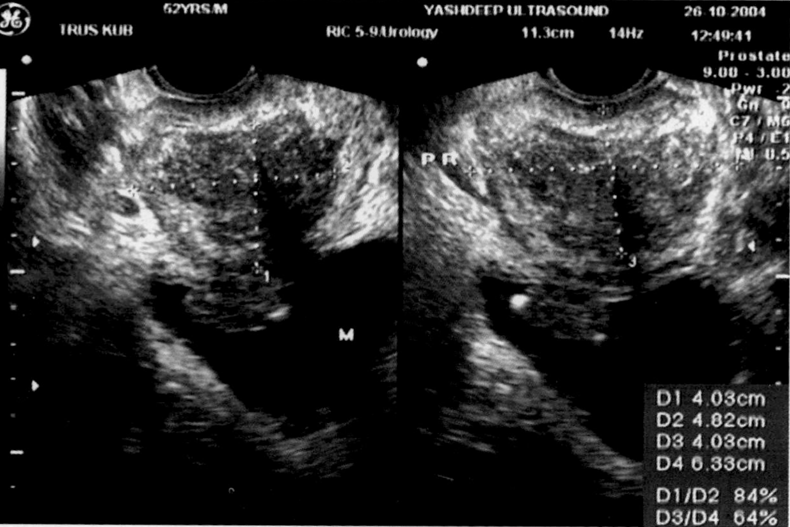

FIGURE 9.155

FIGURES 9.154 and 9.155: Malignant mass of the median lobe. HRSG shows irregular hypoechoic mass coming out from the median lobe of the gland. The mass shows heterogeneous texture. No calcification is seen in the mass. Another mass is also seen in the inner part of the gland and more confined to the Rt side. Biopsy of the mass confirmed malignancy.

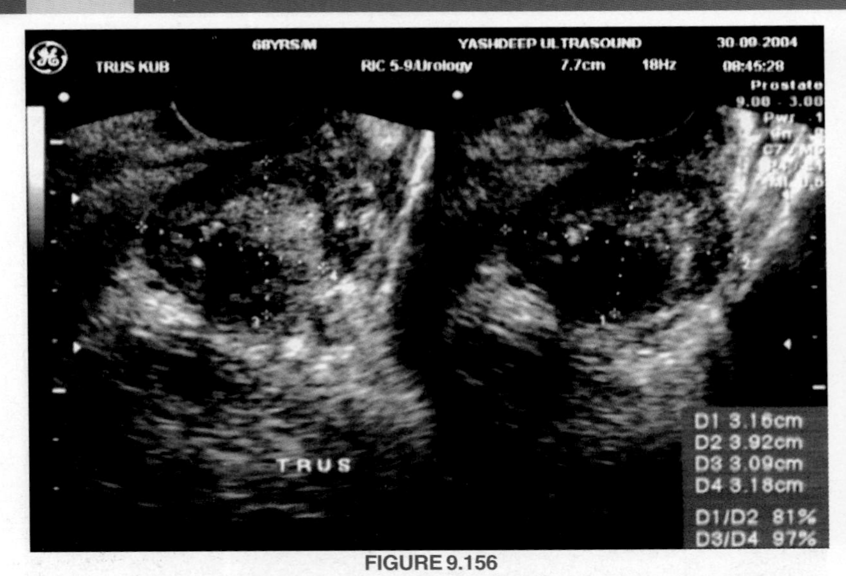

FIGURE 9.156

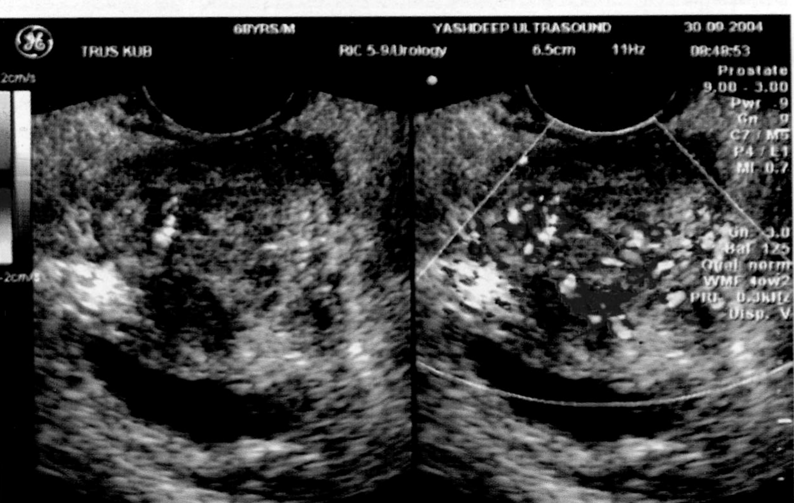

FIGURE 9.157

FIGURES 9.156 and 9.157: CA prostate Rt peripheral zone. HRSG shows grossly distorted prostate. Irregular heterogeneous mass is seen in the Rt peripheral zone. The mass is seen invading the inner gland mass. Normal texture is lost. On color flow imaging increased bizarre flow low pattern is seen in the mass. Biopsy of the mass turned out to be adenocarcinoma.

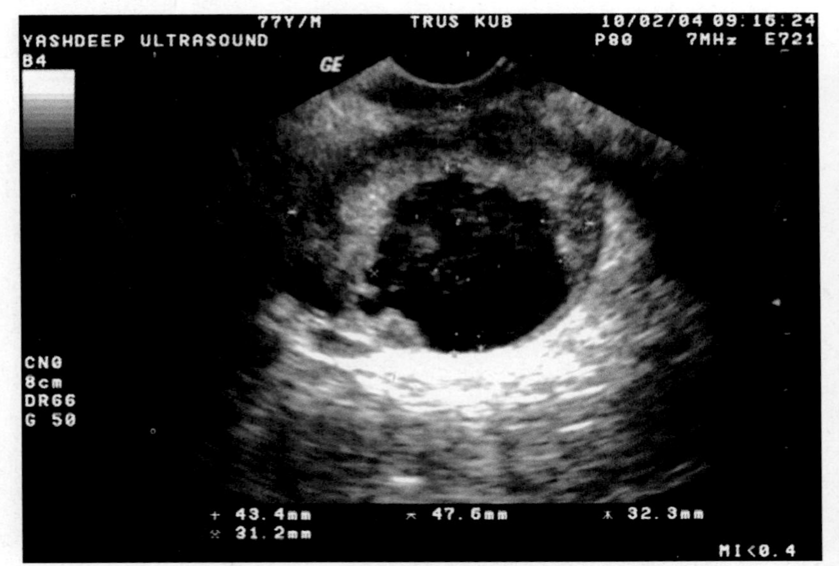

FIGURE 9.158: Adenocarcinoma prostate with cystic degeneration. An old patient presented with heterogeneous prostomegaly. The PSA titer was very high. HRSG shows grossly heterogeneous prostate gland. Irregular cystic degeneration was also seen in the gland. The big mass is seen in the inner gland. Central area of necrosis is seen in it.

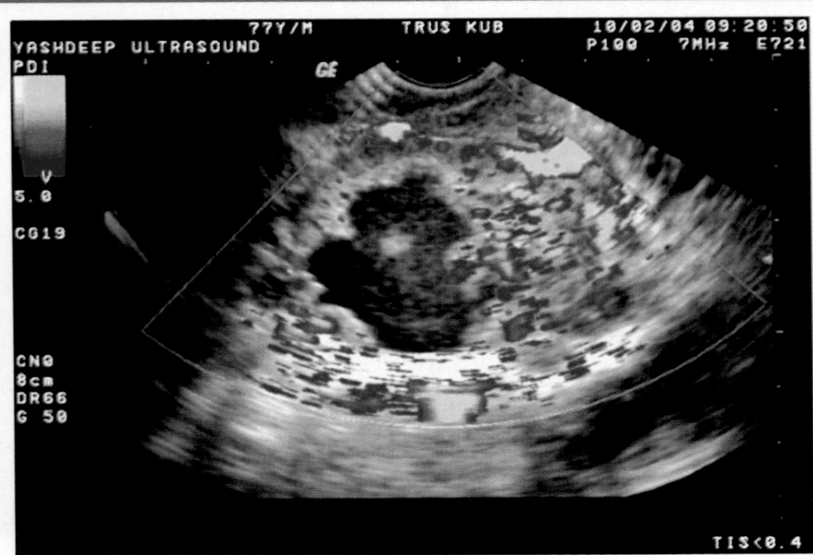

FIGURE 9.159

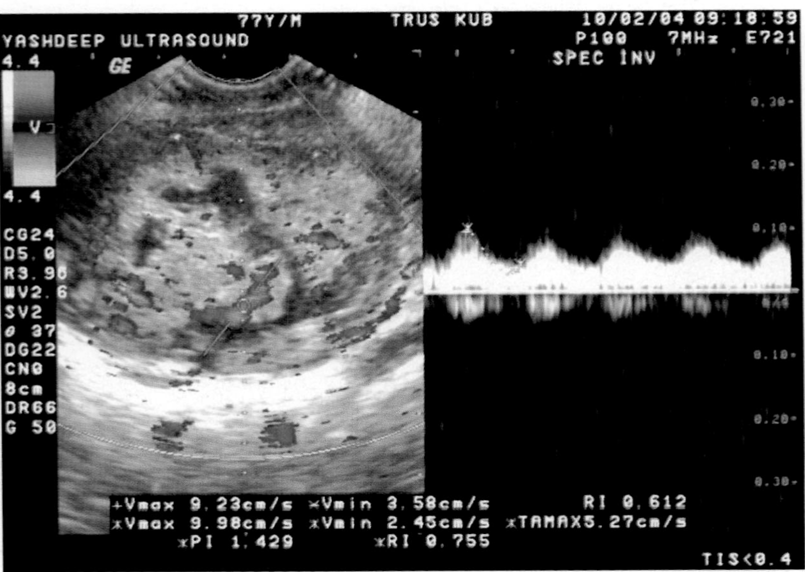

FIGURE 9.160

FIGURES 9.159 and 9.160: On color flow imaging high flow is seen in the gland. On spectral Doppler tracing increased arterial flow seen in the gland. Biopsy of the mass confirmed adenocarcinoma.

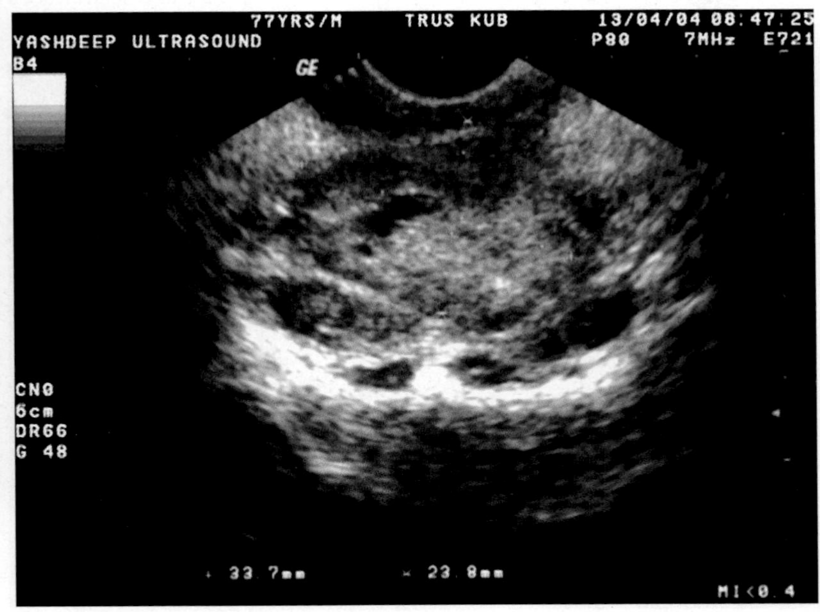

FIGURE 9.161

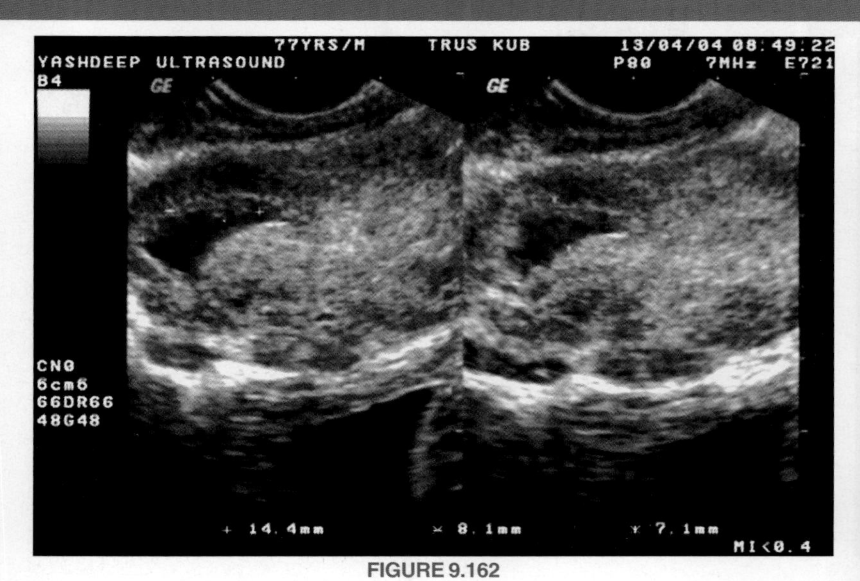

FIGURE 9.162

FIGURES 9.161 and 9.162: Adenocarcinoma of the gland after hormone therapy. The same patient was put after hormone therapy for the CA prostate. A repeat TRUS examination after 2 months shows reduction in the size of the gland. The cystic degeneration also reduced in size. The gland has become shrunken and homogeneous in texture. However, focal capsular nodularity was persistent with the gland. HRSG is very good for evaluating the post therapy response of the treatment.

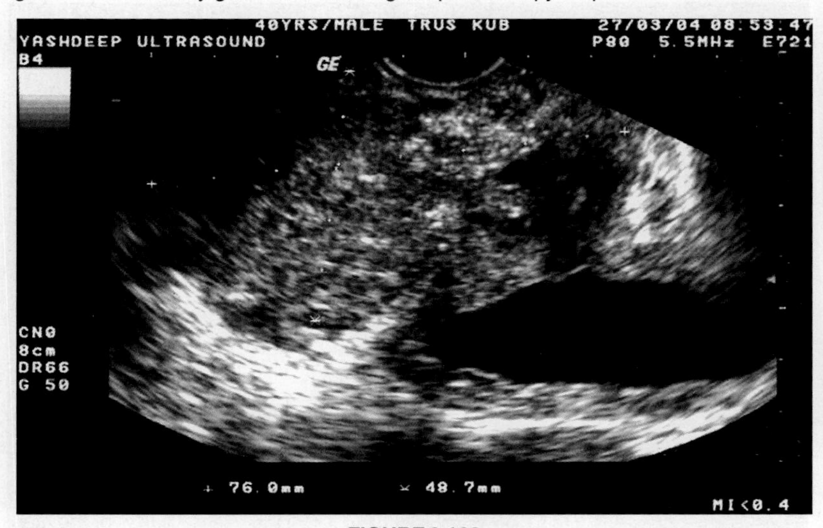

FIGURE 9.163

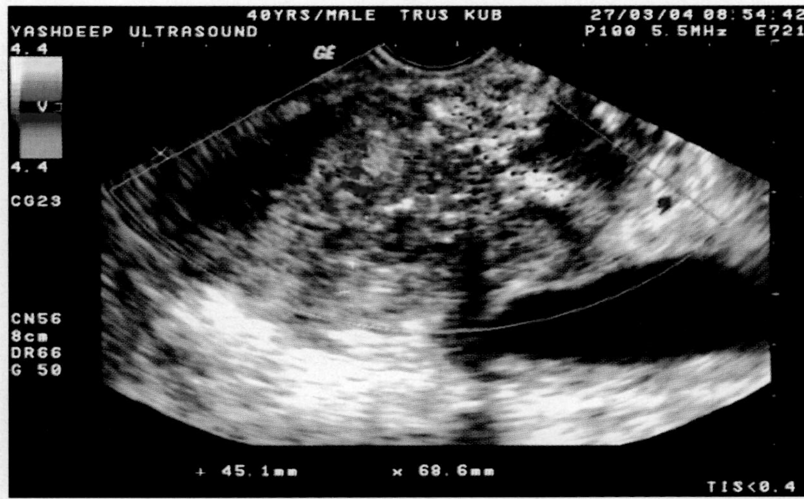

FIGURE 9.164

FIGURES 9.163 and 9.164: Adenocarcinoma of the gland with gross destruction. HRSG shows grossly enlarged prostate with heterogeneous texture in middle-aged man of 40 yrs. The patient presented with retention of urine. Normal prostate texture is lost. The capsule is broken. The mass is seen invading the surrounding tissue areas. However, the color flow imaging poor flow of the mass. Biopsy of the mass confirmed invasive carcinoma.

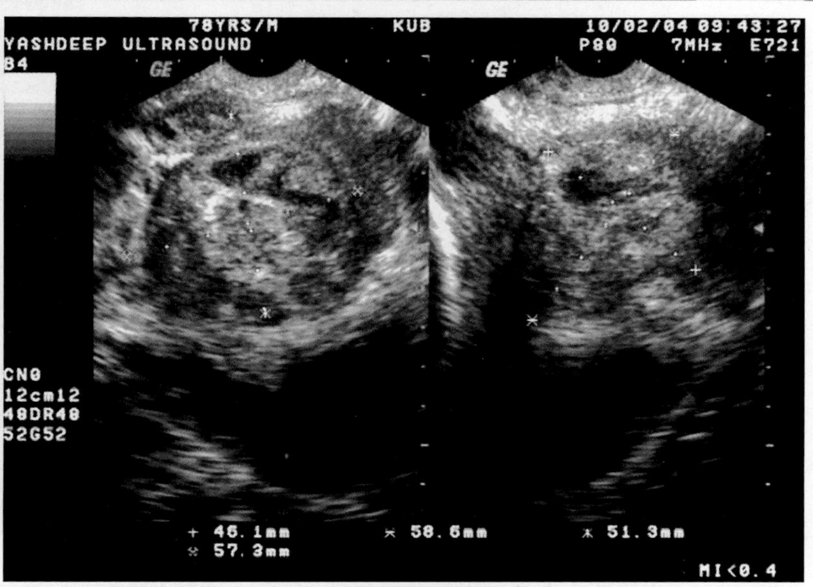

FIGURE 9.165

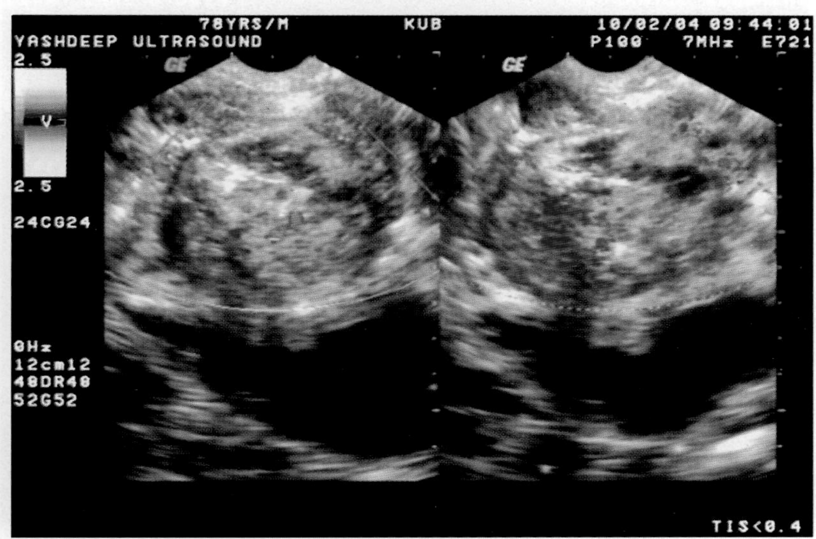

FIGURE 9.166

FIGURES 9.165 and 9.166: Invasive adenocarcinoma of the prostate. HRSG shows grossly heterogeneous gland. Multiple irregular hypodense areas are seen in the gland. Cystic degeneration is also seen in them. On color flow imaging moderate flow is seen in the prostate. Biopsy of the gland shows adenocarcinoma.

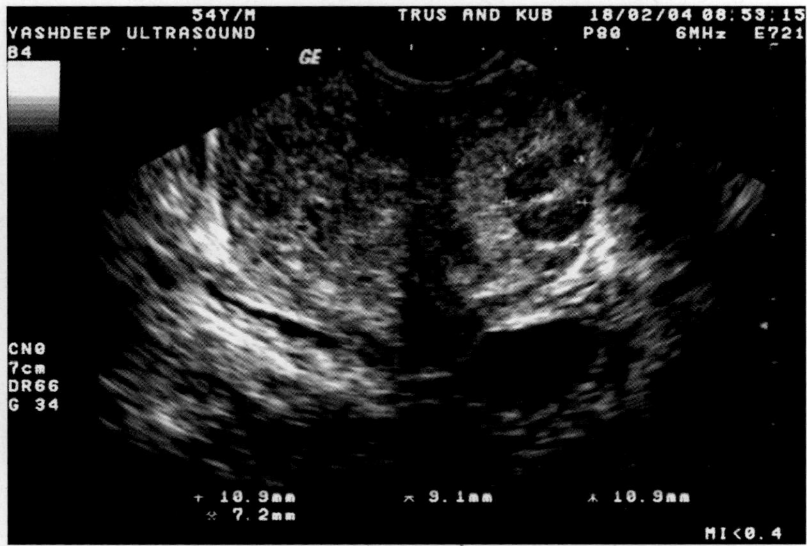

FIGURE 9.167: Malignant adenonodule in the prostate. HRSG shows irregular hypoechoic nodular mass in the Lt peripheral zone. The mass shows sharp margin. No necrosis seen in it. However, biopsy of the mass turned out to be adenocarcinoma.

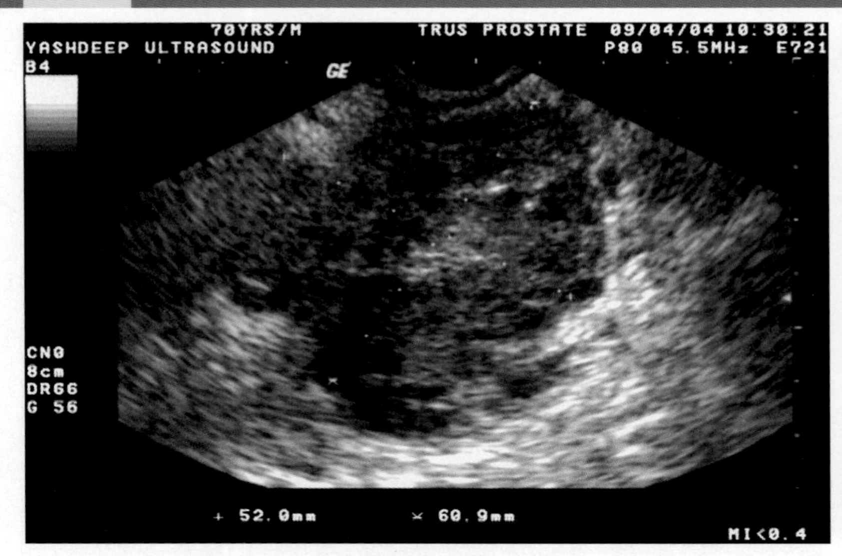

FIGURE 9.168

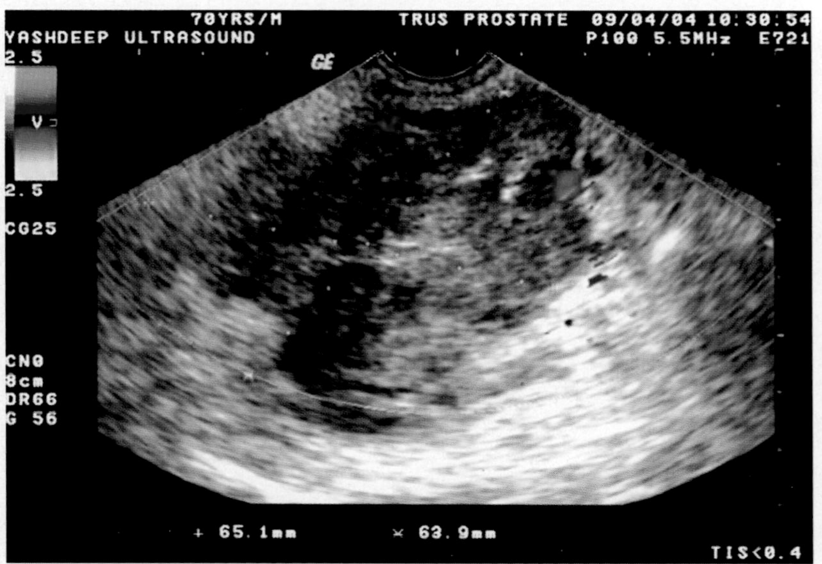

FIGURE 9.169

FIGURES 9.168 and 9.169: Invasive adenocarcinoma prostate. HRSG shows big gross prostomegaly. The mass is seen invading the floor of the prostate. Normal glandular texture is lost. Capsule is seen broken from all side. Biopsy of the mass confirmed invasive carcinoma.

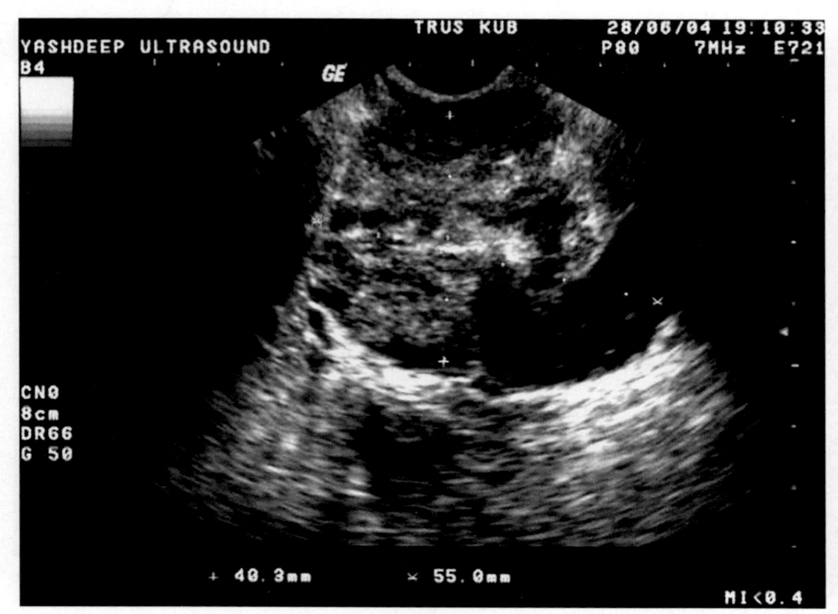

FIGURE 9.170

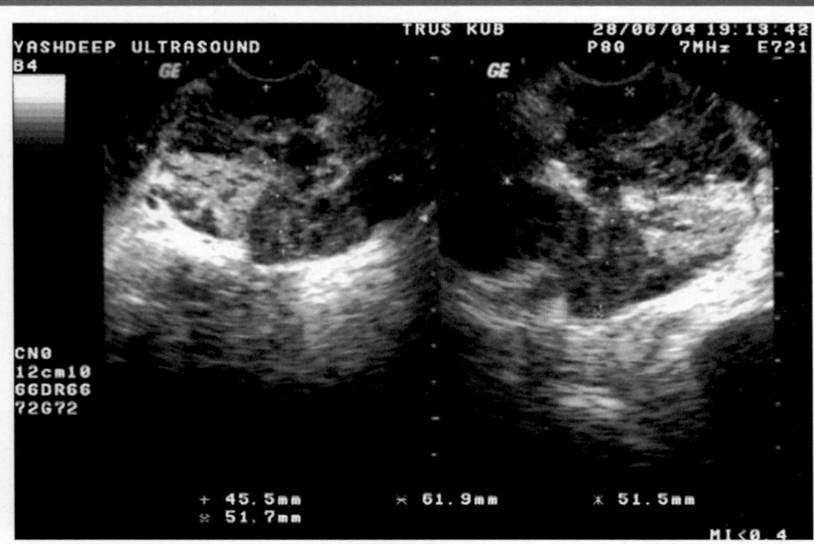

FIGURE 9.171

FIGURES 9.170 and 9.171: Invasive adenocarcinoma of prostate. A middle aged patient presented with markedly enlarged PSA with fever and discomfort in urination. HRSG shows enlarged prostate whole of the gland shows grossly distorted texture. Multiple irregular hypoechoic areas are seen in the gland. Cystic degeneration is seen in it. Capsule was seen broken. The gland was seen invading the surrounding tissues.

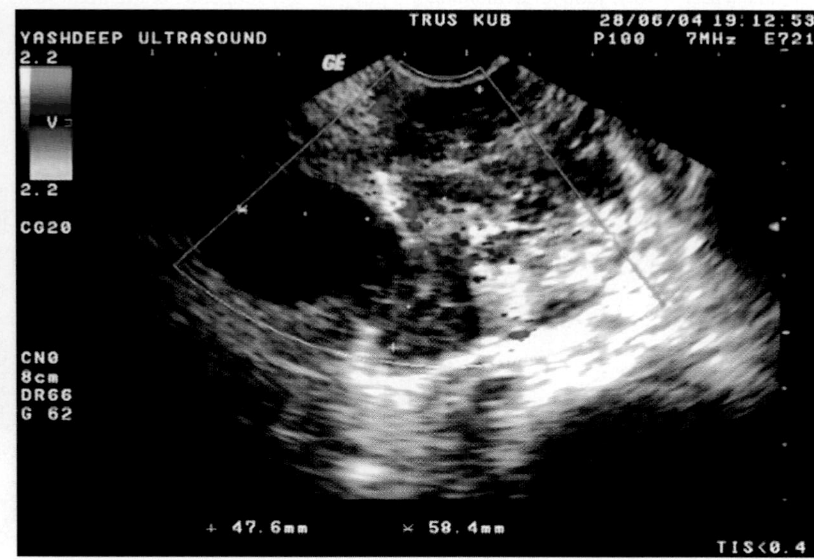

FIGURE 9.172: On color flow imaging poor flow was seen in the gland.

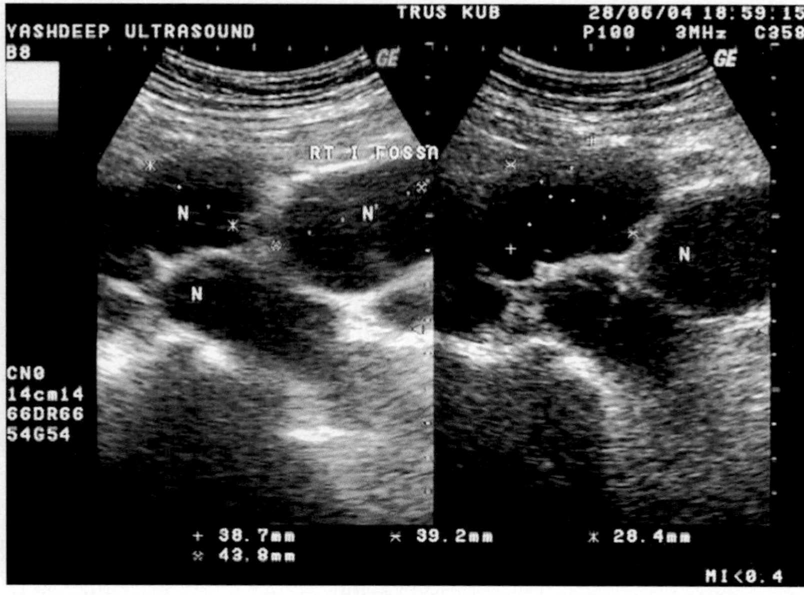

FIGURE 9.173

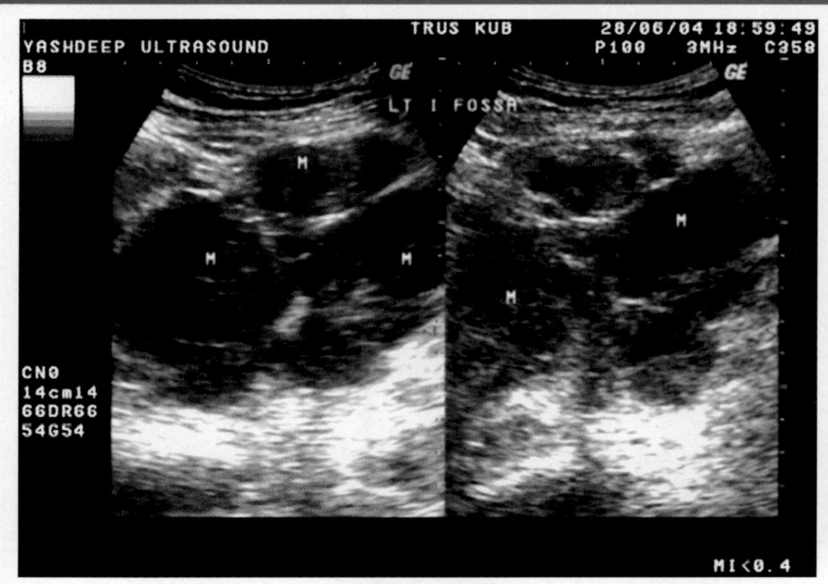

FIGURE 9.174

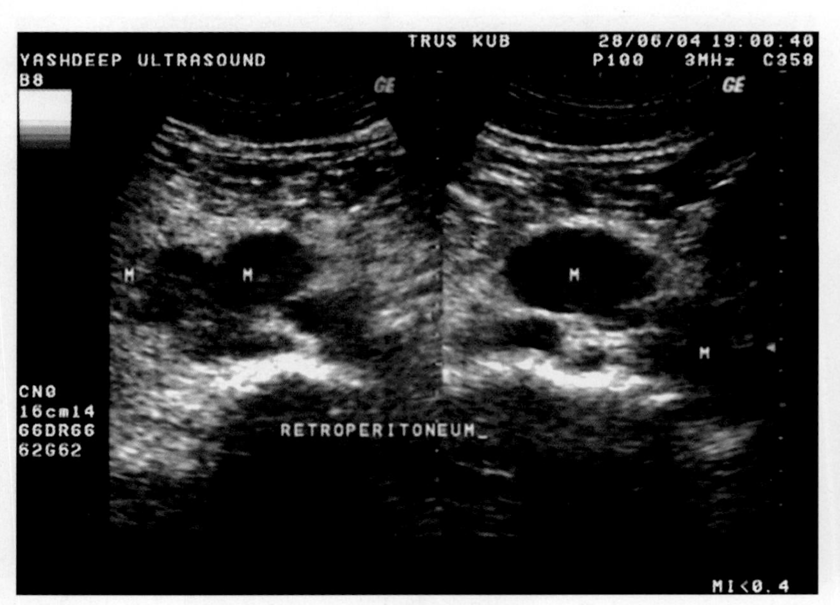

FIGURE 9.175

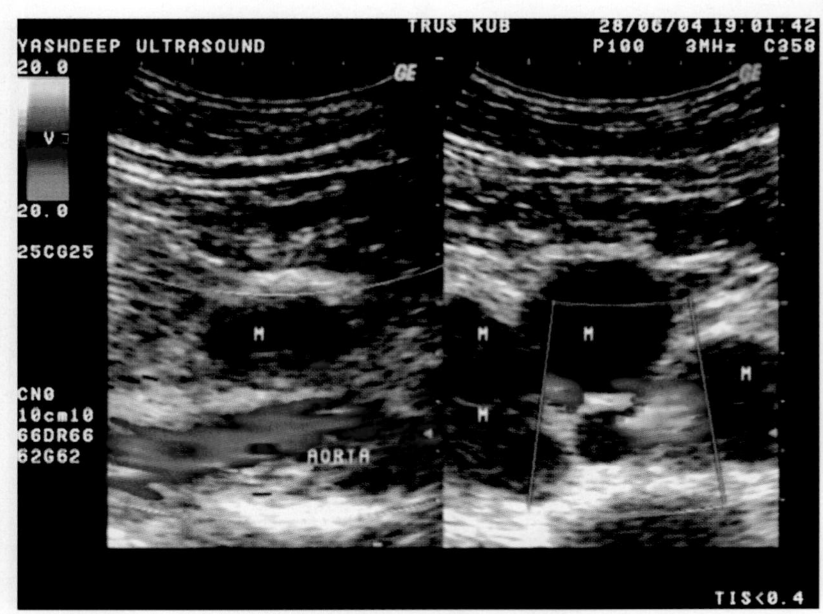

FIGURE 9.176

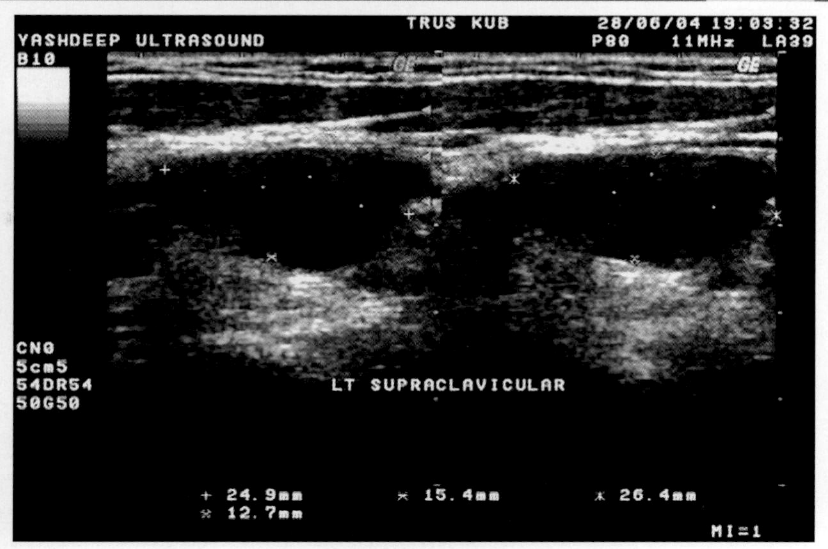

FIGURE 9.177

FIGURES 9.173 to 9.177: Abdominal and neck imaging of the same patient showed multiple enlarged lymph nodes in both iliac fossae and retroperitoneum. Lymph node hilum was seen destroyed. Normal texture was lost. Multiple enlarged nodes were also seen in the supraclavicular region. Biopsy of the nodes confirmed invasive adenocarcinoma.

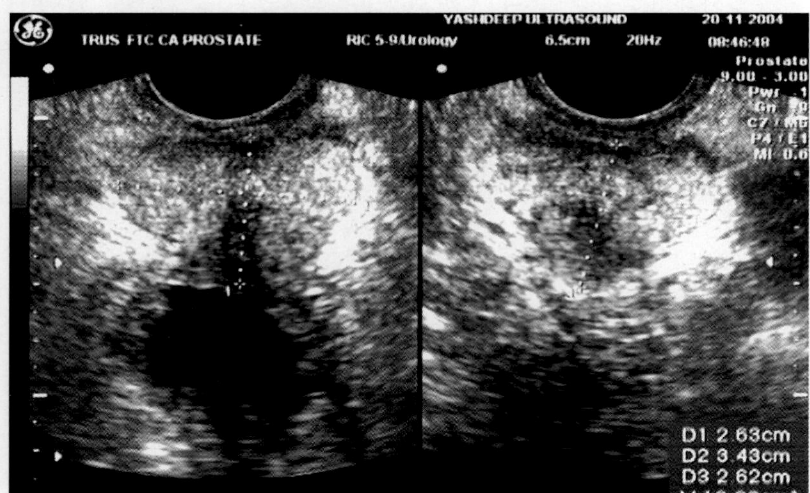

FIGURE 9.178

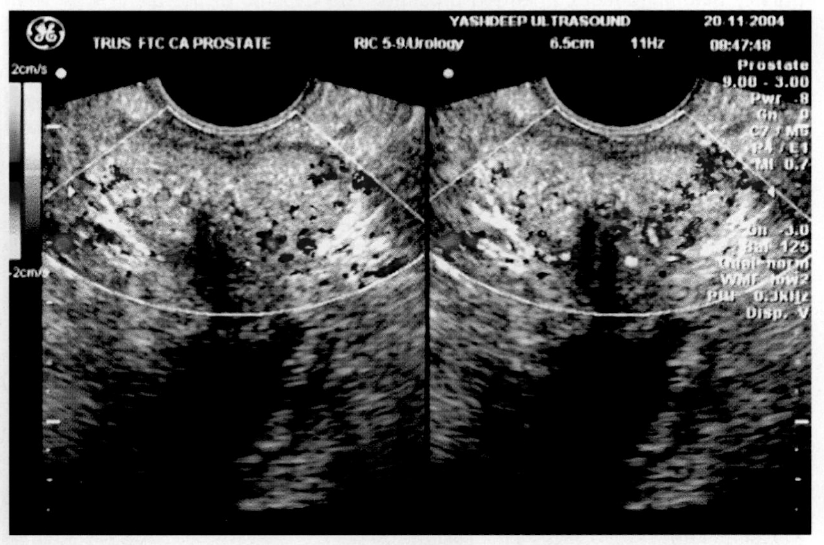

FIGURE 9.179

FIGURES 9.178 and 9.179: The same patient after 6 months of hormone therapy for CA prostate was examined. The prostate gland was seen reduced in size. Echo texture has become almost homogeneous. The gland got shrunk in size. The iliac, retroperitoneal and supraclavicular nodes also regressed in size. Patient responded well to the treatment.

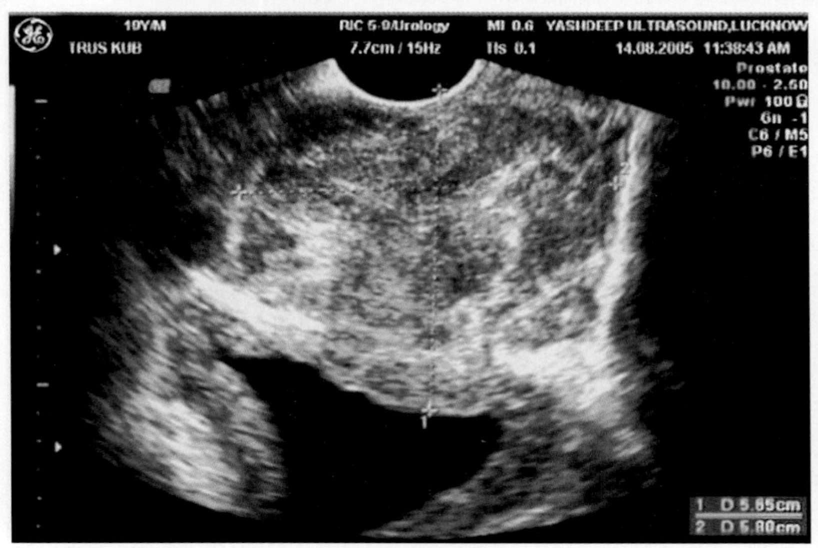

FIGURE 9.180

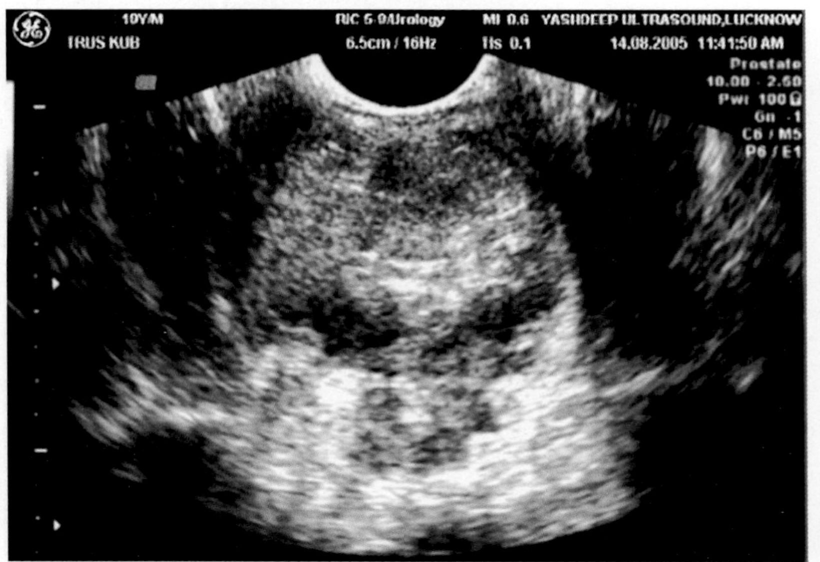

FIGURE 9.181

FIGURES 9.180 and 9.181: Rhabdomayosarcoma of prostate. A young boy of 19 years of age presented with retention of urine with pain in perineum. TRUS showed grossly inhomogeneous enlarged prostate. The capsule of the gland was seen broken. Multiple hypoechoic areas are seen in the gland. Biopsy of the mass confirmed rhabdomayosarcoma prostate.

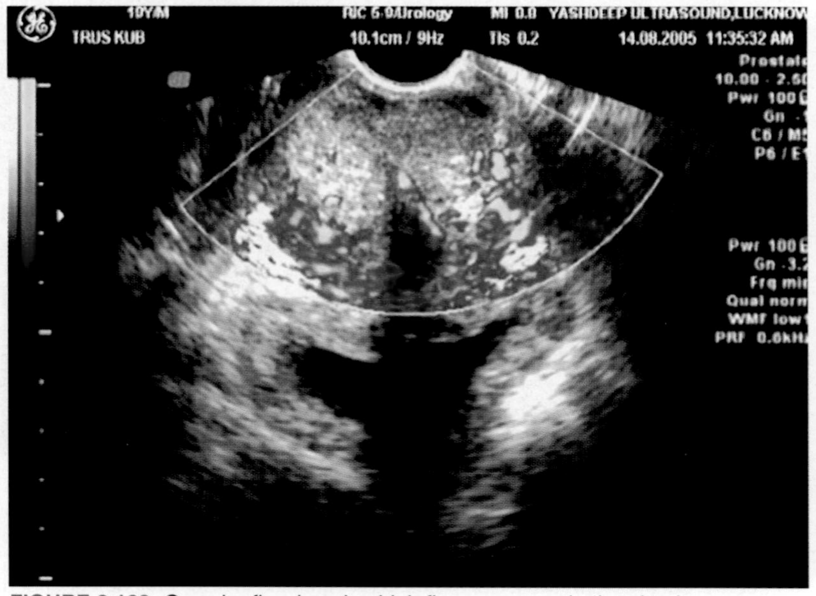

FIGURE 9.182: On color flow imaging high flow was seen in the gland.

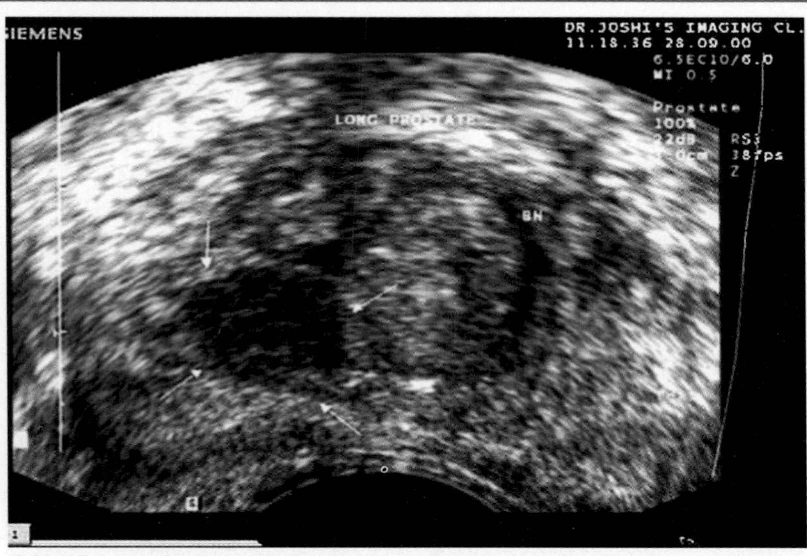

FIGURE 9.183: Mass in the Rt inner gland. A hypoechoic low level echo complex mass is seen in the Rt inner gland. No calcification is seen in the mass. Biopsy of the mass showed malignancy.

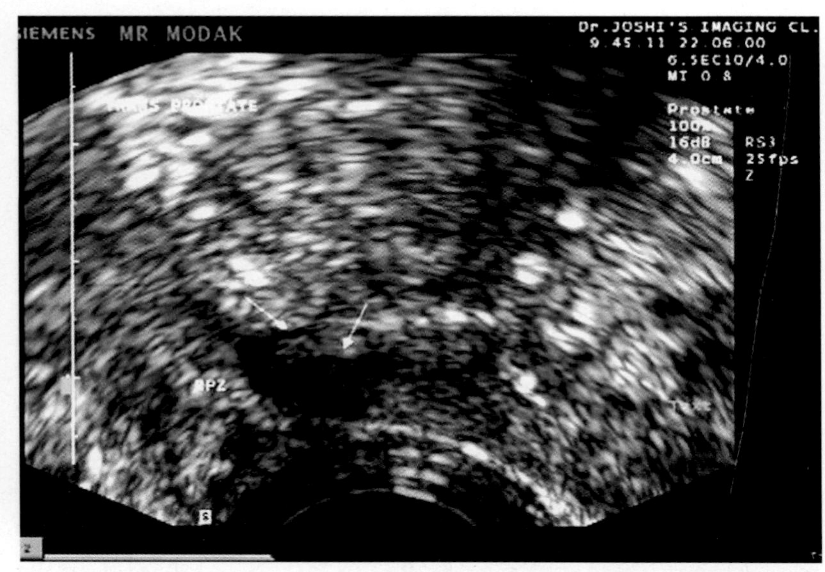

FIGURE 9.184: Rt peripheral zone mass. HRSG shows irregular heterogeneous soft tissue mass in the Rt peripheral zone. It is hypoechoic in texture. No necrosis is seen in the mass. Biopsy of the mass confirmed carcinoma prostate.

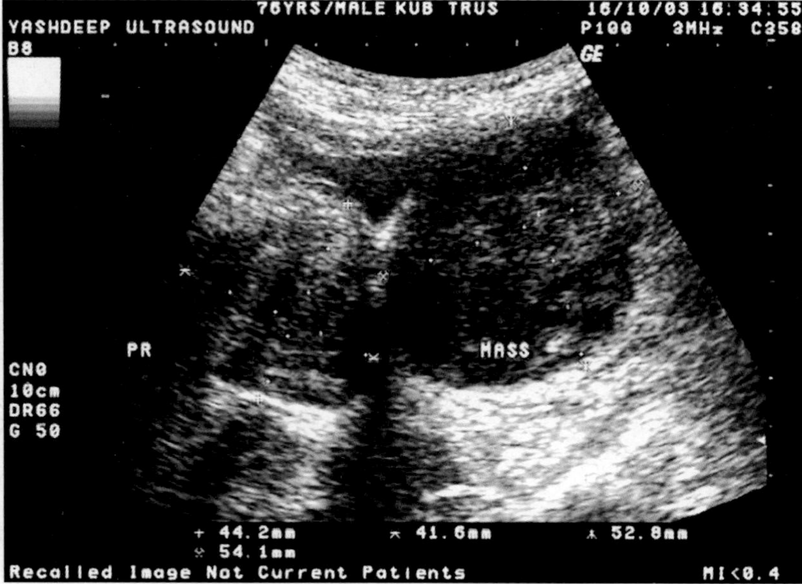

FIGURE 9.185

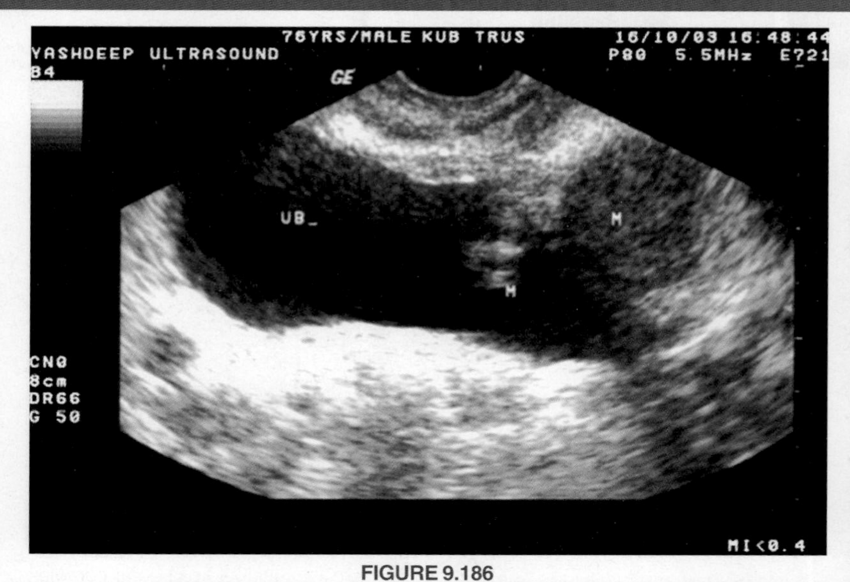

FIGURE 9.186

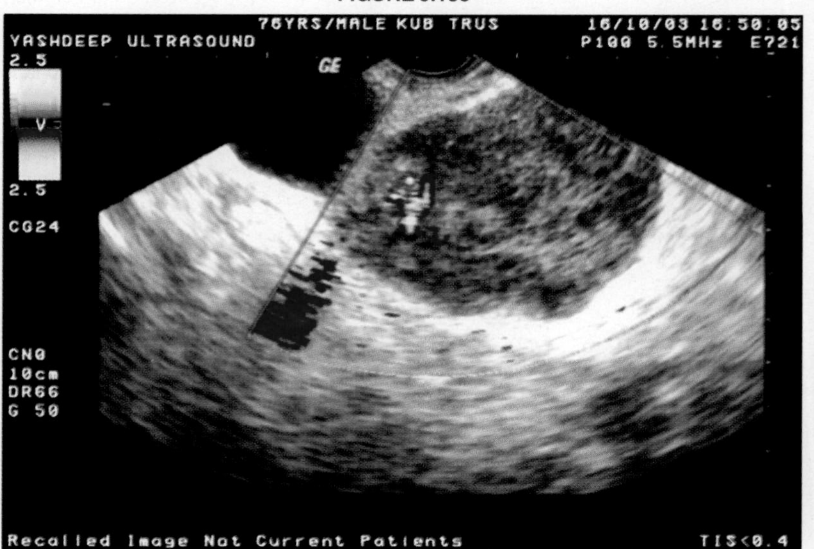

FIGURE 9.187

FIGURES 9.185 to 9.187: CA urinary bladder presented with retention of urine. A patient presented with retention of urine. HRSG shows enlarged prostate gland. However, the gland was homogeneous in texture. No evidence of any focal mass was seen in the gland. But there was a growth seen hanging from the apex of the urinary bladder. It was irregular in outline. The growth is seen invading the UB wall. A big hypoechoic mass was also seen in the extra vesicle space. It was irregular in outline. However, it was separate from prostate gland. The biopsy of the mass confirmed carcinoma urinary bladder with enlarged nodal spread to the pelvis.

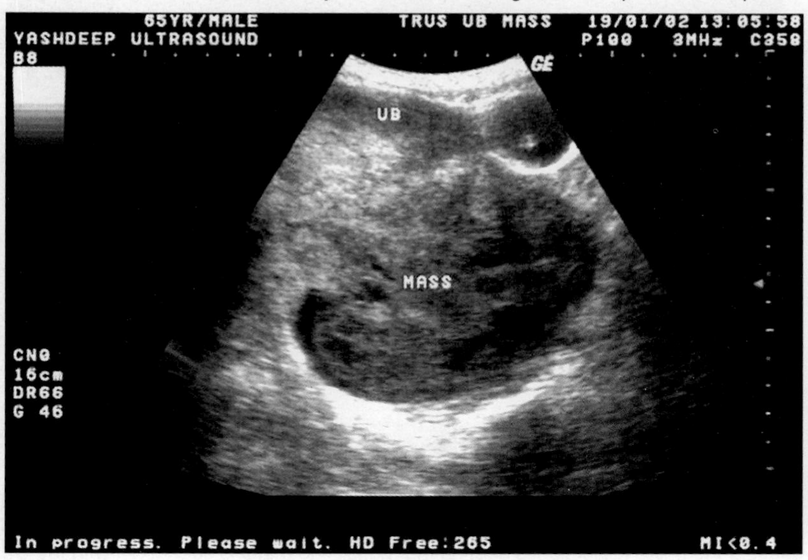

FIGURE 9.188

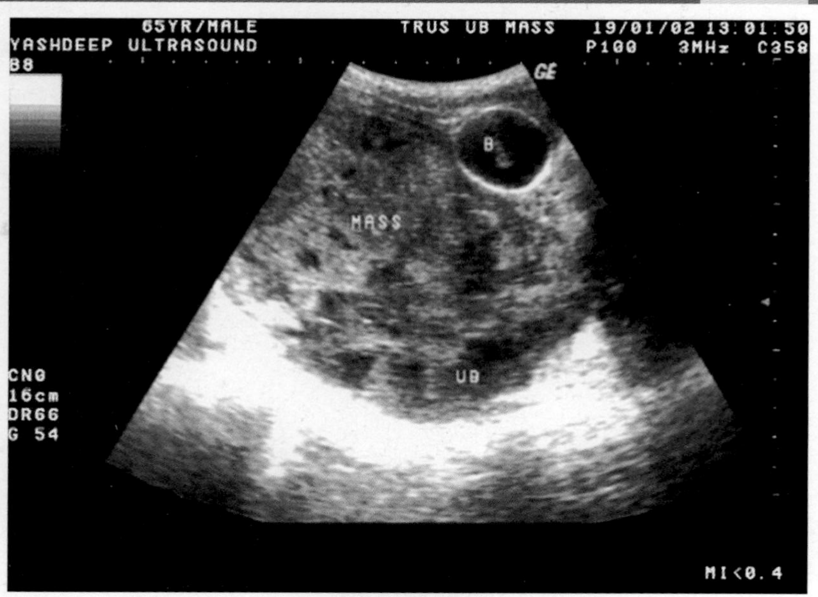

FIGURE 9.189

FIGURES 9.188 to 9.189: Huge mass in urinary bladder. HRSG shows a huge mass in the urinary bladder the mass is seen occupying whole of the urinary bladder and filling the cavity. Clinically it was thought to be prostate mass. However, HRSG settled the diagnosis to be CA urinary bladder.

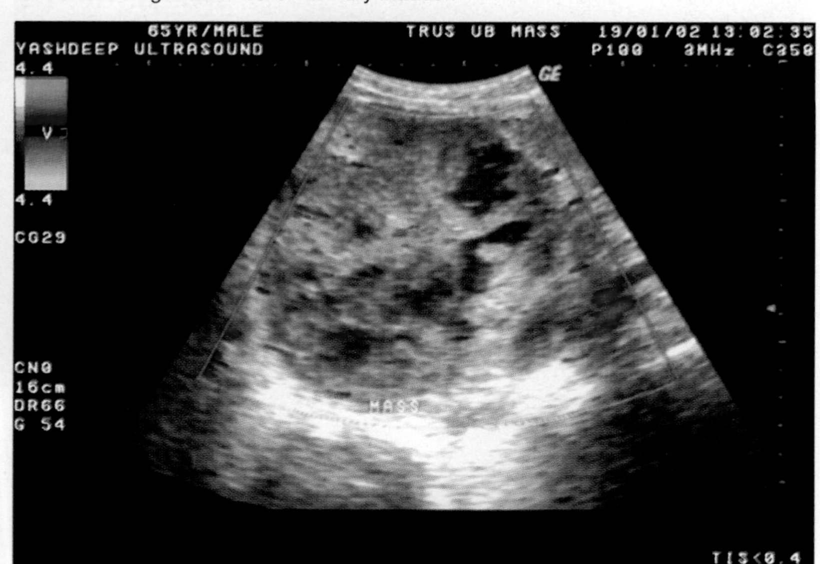

FIGURE 9.190: On color flow imaging poor flow was seen in the mass.

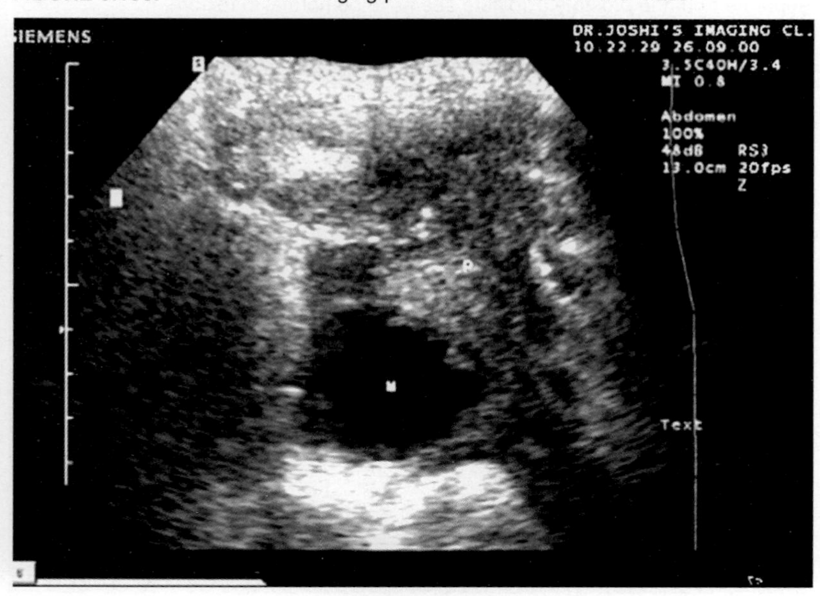

FIGURE 9.191

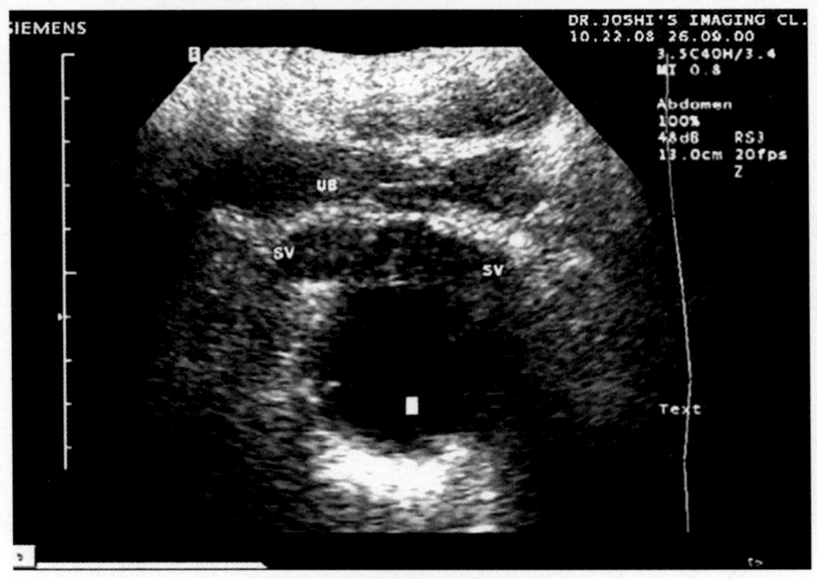

FIGURE 9.192

FIGURES 9.191 and 9.192: Periprostatic fluid collection. HRSG shows low-level echo collection in periprostatic area. No septation or loculation is seen in collection. It is seen pressing over the prostate.

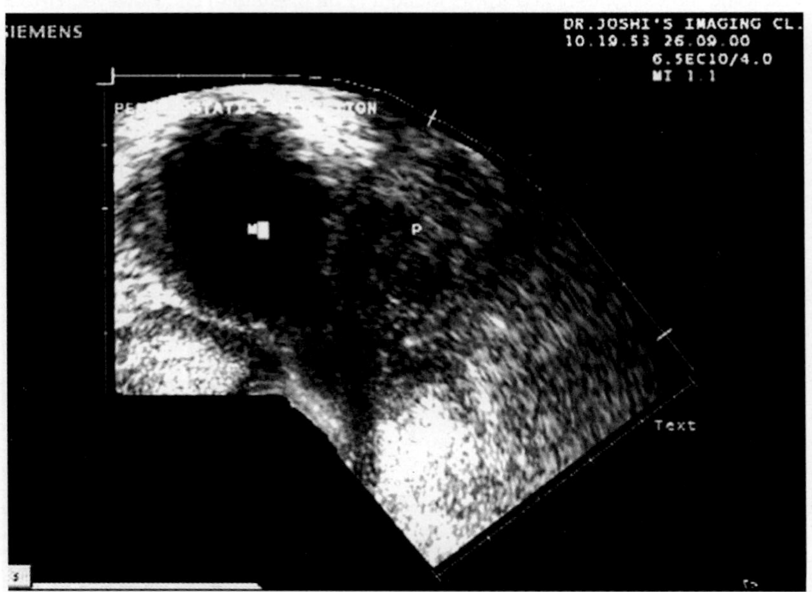

FIGURE 9.193: The prostate gland is well visualized. The fluid collection is seen separate from the prostate. Prostate seems to be normal in texture.

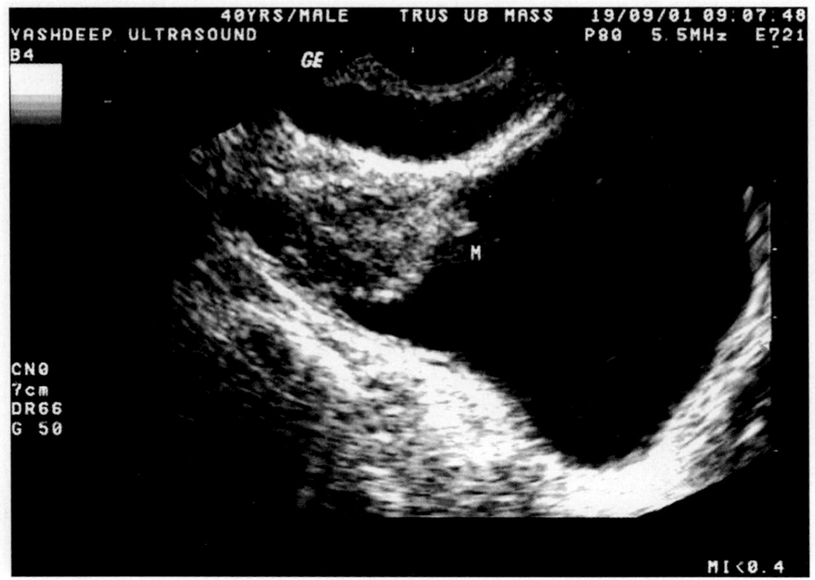

FIGURE 9.194

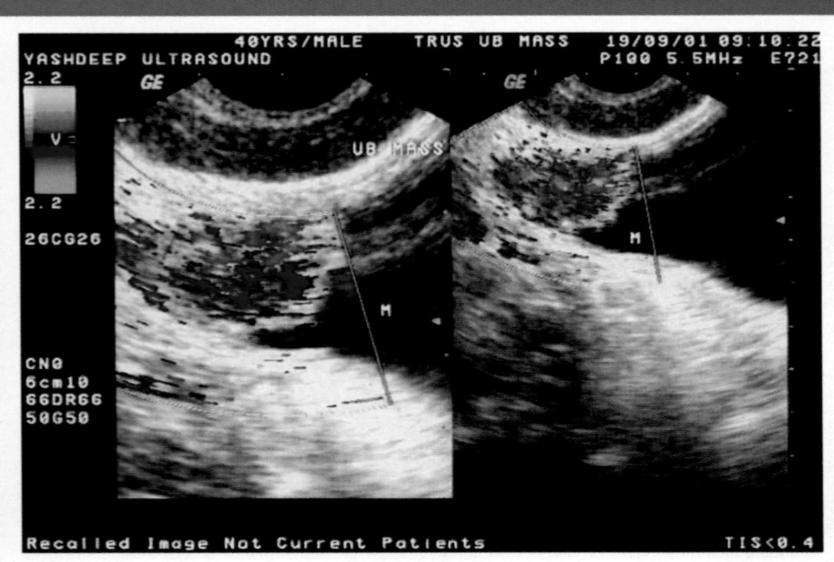

FIGURE 9.195

FIGURES 9.194 and 9.195: Urinary bladder mass. High resolution TRUS shows mass hanging from the superior wall of urinary bladder. It is homogeneous in texture and well define in shape. No extra vesicle spread is seen. On color flow imaging moderate flow is seen in the mass. Biopsy of the mass shows carcinoma urinary bladder.

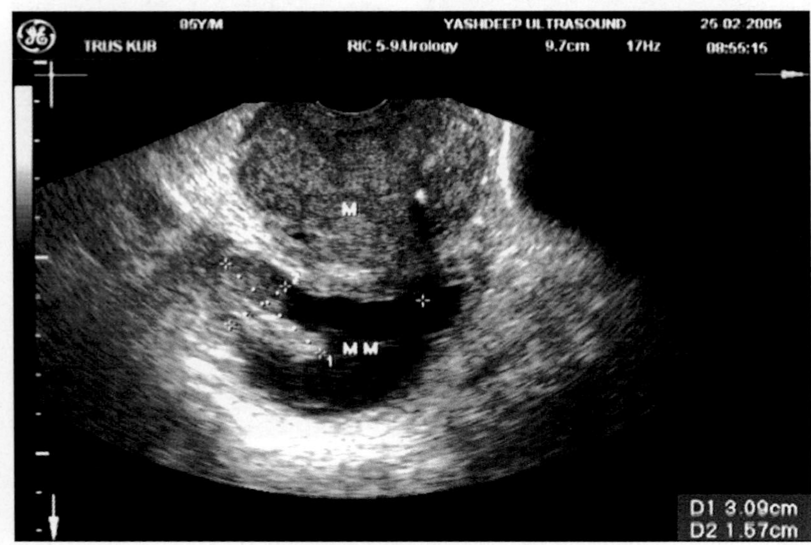

FIGURE 9.196

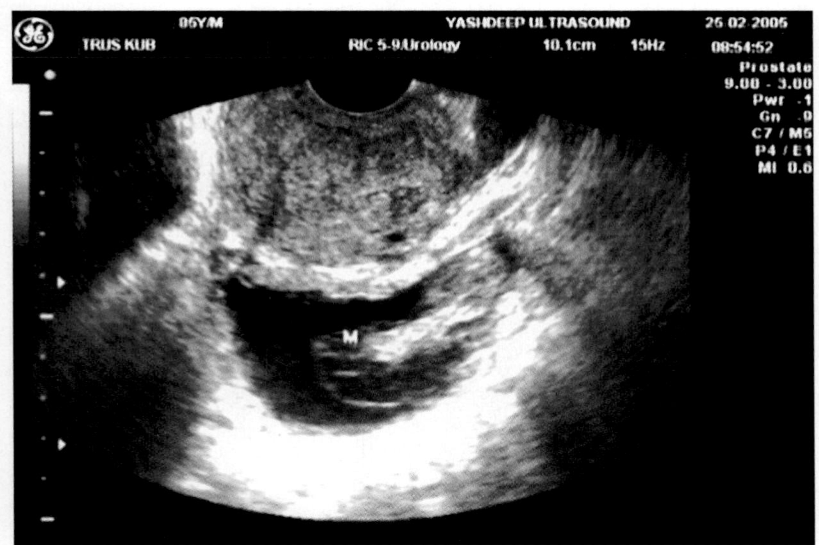

FIGURE 9.197

FIGURES 9.196 and 9.197: UB mass. Patient presented with hematuria and prostomegaly. HRSG shows an irregular heterogeneous mass fixed with the UB mass. Finger like projection is seen in the mass. The mass is seen fixed with the UB wall. No calcification necrosis is seen in the mass.

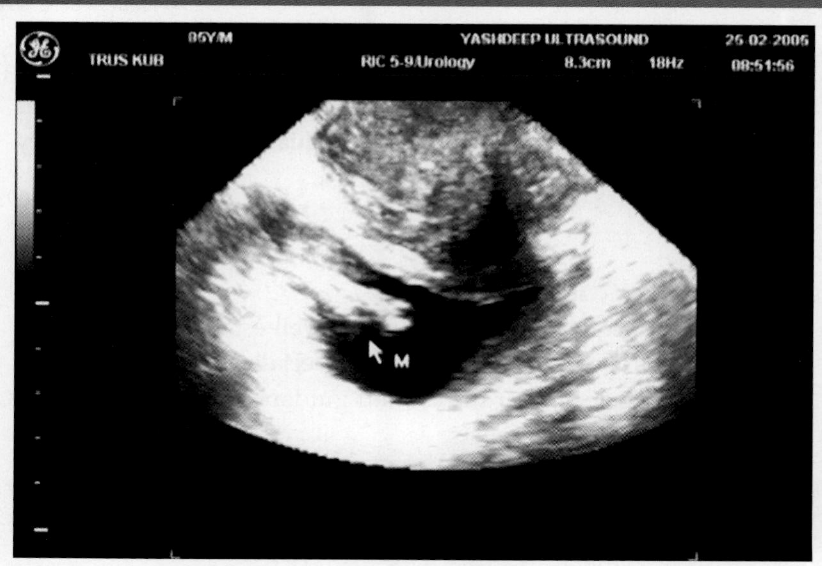

FIGURE 9.198: 3D imaging of the mass shows it separate from the prostate. Prostate shows homogeneous enlargement. No focal mass is seen in the prostate.

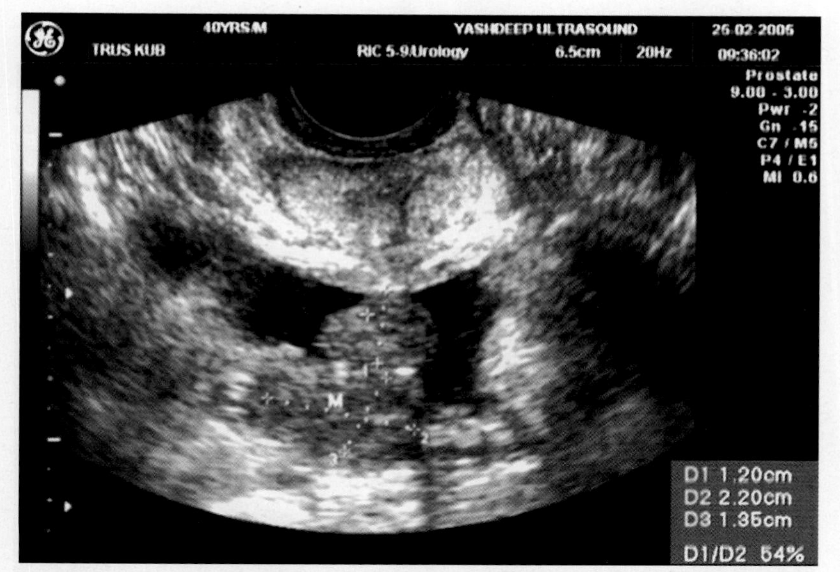

FIGURE 9.199: UB Mass with bladder wall invasion. HRSG shows a polypoidal growth hanging in the UB. The mass is seen invading the UB wall and shows extra vesicle spread. No calcification or necrosis is seen in the mass.

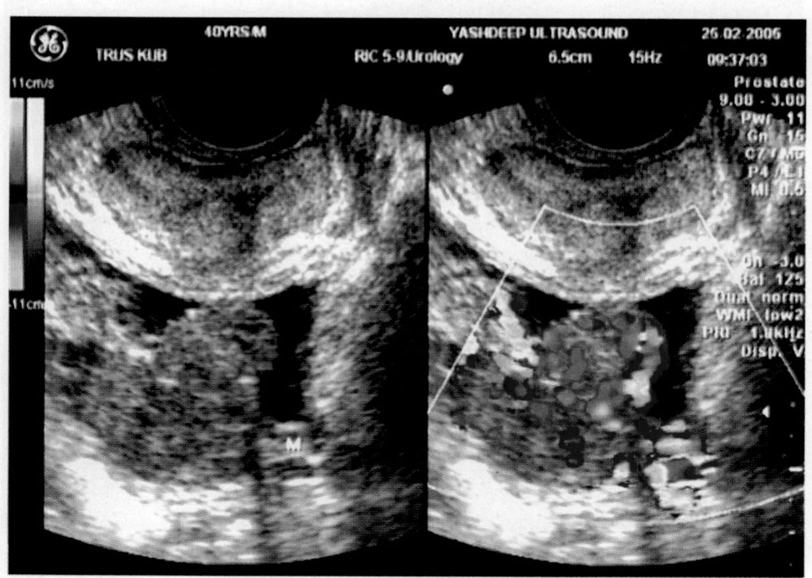

FIGURE 9.200

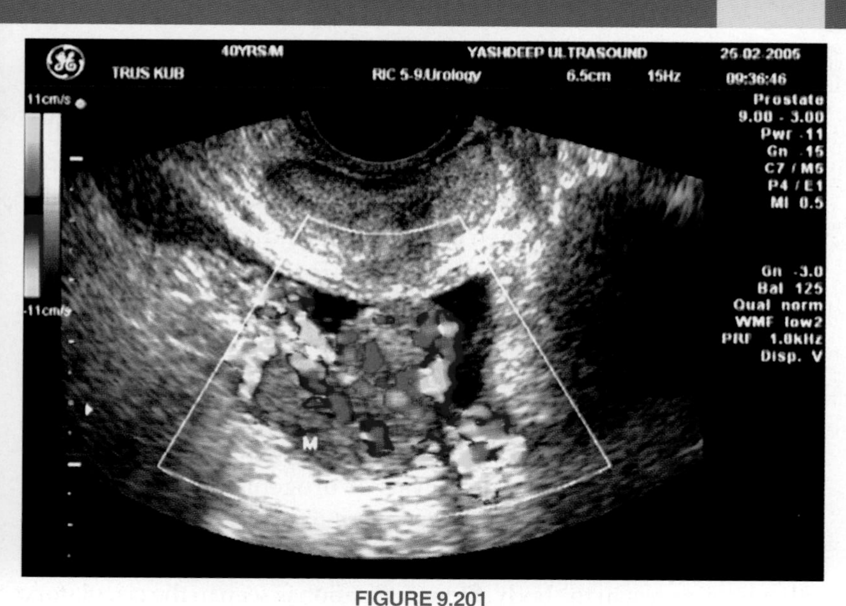

FIGURE 9.201

FIGURES 9.200 and 9.201: On color flow imaging high flow seen in the mass.

In BPH, there is increased vascularity within the inner gland and the surgical capsule. However, the flow may vary from patient to patient. In prostatitis, there is increased high velocity and low resistance flow. In diffuse prostatitis, there is diffuse increase in prostatic vascularity throughout the gland. Early studies in prostatic cancer have suggested increased blood flow in cancers greater than 5 mm in size. However, recent studies have demonstrated that only 50% of patients with cancer have increased flow. Three different patterns have been identified. All have low resistance patterns; diffuse flow, focal flow or surrounding flow. The most frequently encountered pattern is diffuse flow within the lesion. When there is focal flow it is not pathognomonic of cancer. In general, color Doppler has not been of great help in distinguishing between benign and malignant lesions.

Seminal Vesicles

The normal seminal vesicles are paired, well defined, saccular and elongated structures situated cephalad to the prostate and posterior to the urinary bladder. They present a classical "bow-tie" appearance in transverse imaging. The seminal vesicles (SV) have internal septations giving rise to the saccular convoluted structure. There are marked variations in the size, shape and degree of distension of the seminal vesicles. They are usually symmetric and measure 3.0-1.5 cms in size. When the seminal vesicles are not identified on more than one occasion with abstinence, the diagnosis of absent seminal vesicles is made. If the seminal vesicles are smaller than 30% of normal volume (13-17 ccs), the diagnosis of hypoplastic seminal vesicles is made.

The vas differentia including their ampula is normally well visualized on a transrectal study, cephalad to the seminal vesicles. They have appearances similar to those of the seminal vesicles and are usually bilaterally symmetrical. The confluence of the seminal vesicles and the ampullary portion of the vas deferens join together to form the ejaculatory duct, which crosses the prostate through the central zone and terminates at the level of the verumontanum. On an average the normal ejaculatory ducts have a lumen of about 2 mms. With good technique and meticulous examination, the ejaculatory ducts can be visualized in most of the patients.

US have become a major aid in the evaluation of male infertility. Seminal vesicles abnormalities are reported to occur in 90% of patients with agenesis of the vas deferens. Congenital unilateral absence of the vas deferens occurs in 1-7men per thousand individuals. Congenital bilateral absence of the vas deferens occurs in approximately 1% of patients. A good few of these patients also have associated renal anomalies such as renal agenesis and crossed renal ectopic pelvic kidney commonly seen on the left side.

Seminal vesicle cysts may be either congenital or acquired in association with obstructive lesions. The congenital variety is rare in midline cysts or Müllerian duct cysts or urogenital sinus cysts. The orifices of the ejaculatory ducts enter into the cysts. These cysts contain sperms. On US, they are easy to identify and have the characteristics of other benign cyst in the body. Stone formation is seen in the ejaculatory duct as a sequelae to inflammation and are well visualized on the trans rectal study.

In all congenital anomalies of the seminal vesicles, the vas deferens or the ejaculatory ducts it is mandatory to evaluate the rest of the urinary tract to rule out other congenital anomalies.

Summary

Transrectal ultrasound of prostate is now a well-established procedure in the diagnosis of prostatic pathologies, especially for prostatic cancer. Transrectal biopsies under ultrasound guidance have been of great value in establishing diagnosis.

Evaluation of the seminal vesicles by TR sonography has enabled to evaluate patients with infertility. However, it is vital to understand the limitations of this procedure especially for diagnosing prostatic lesions.

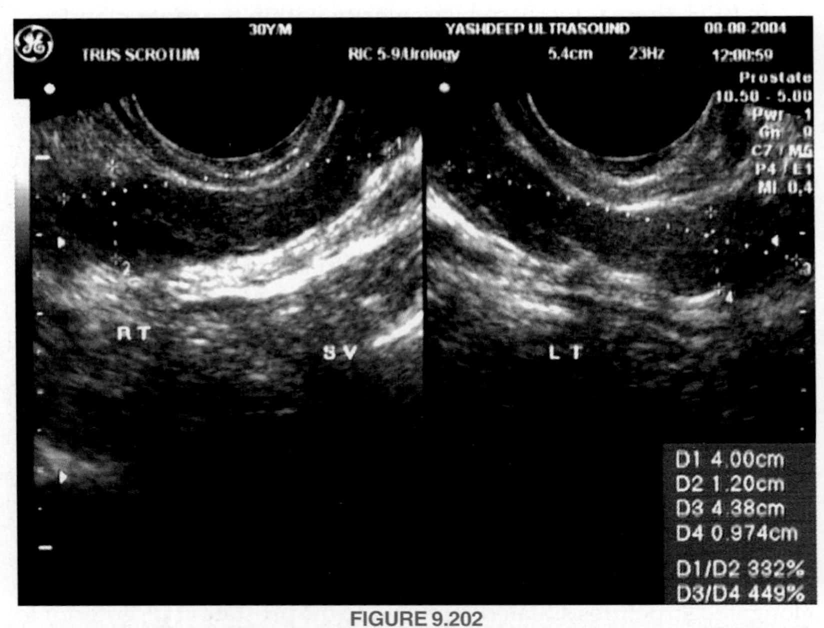

FIGURE 9.202

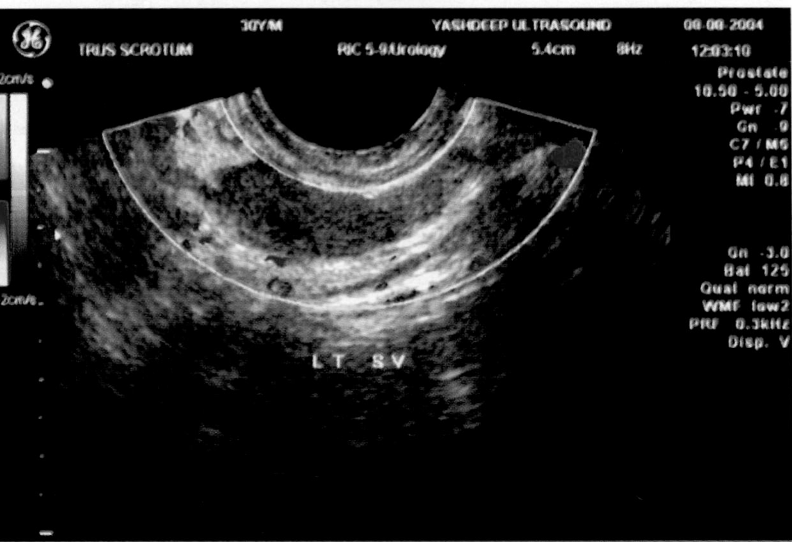

FIGURE 9.203

FIGURES 9.202 and 9.203: Normal. HRSG shows normal seminal vesicles on both sides. They are seen as hypoechoic convoluted saccular tubal shadows. They give typical "Bow Tie" appearance. On color flow imaging low or no flow is seen in them.

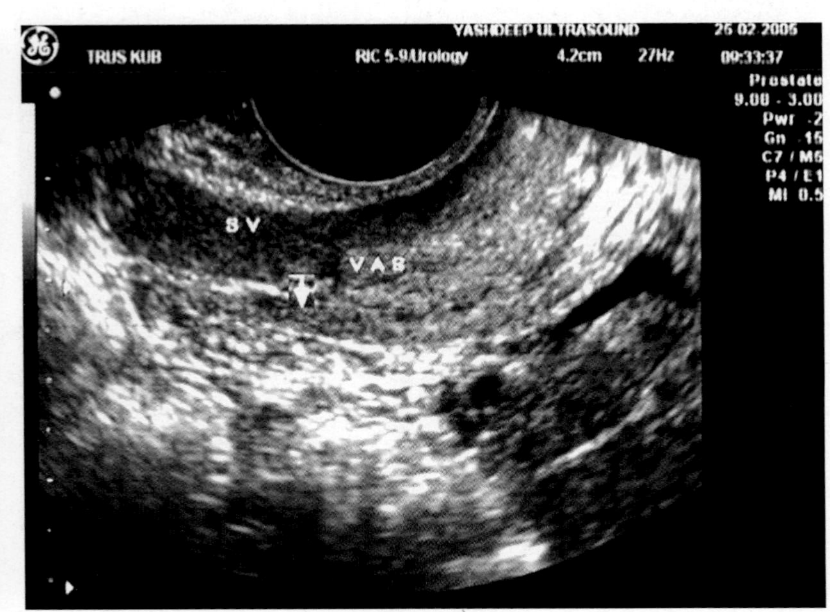

FIGURE 9.204

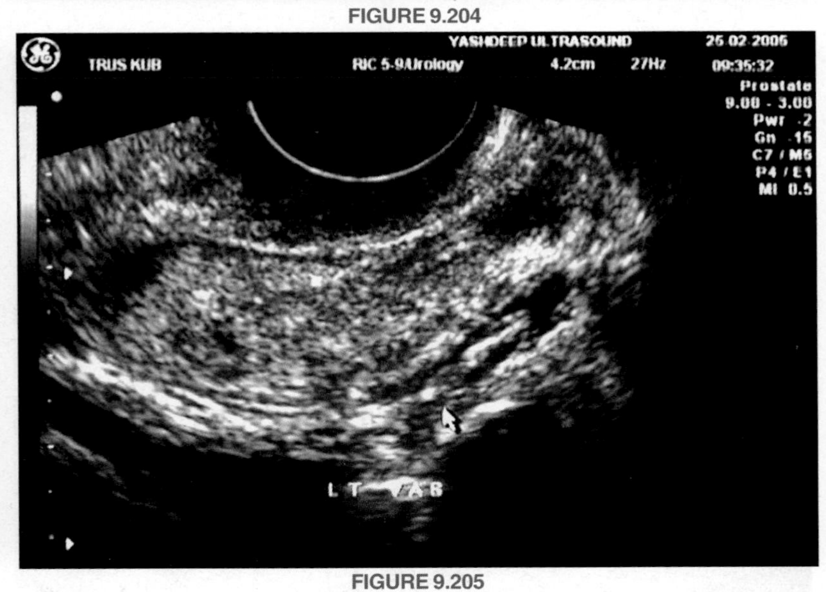

FIGURE 9.205

FIGURES 9.204 and 9.205: Normal vas deferens. Normal vas deferens are seen as hypoechoic tubular shadows running posterior to the seminal vesicle in parallel fashion.

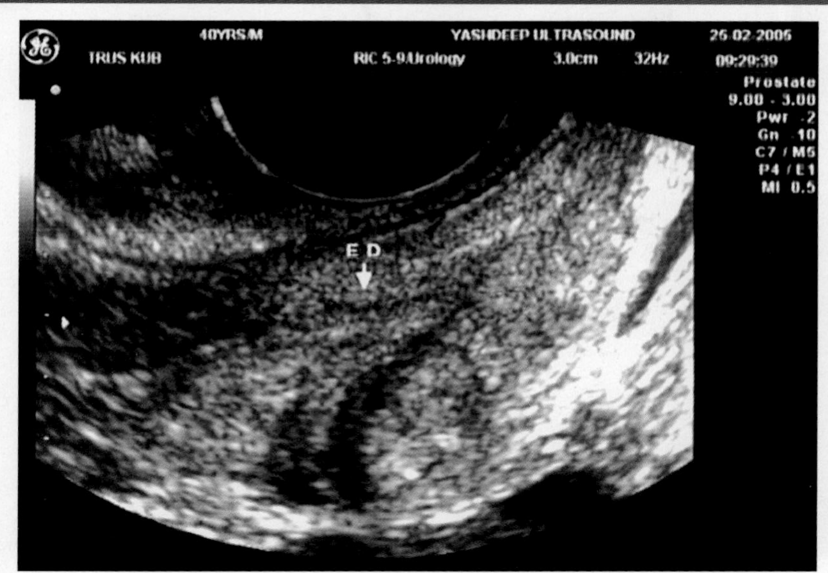

FIGURE 9.206

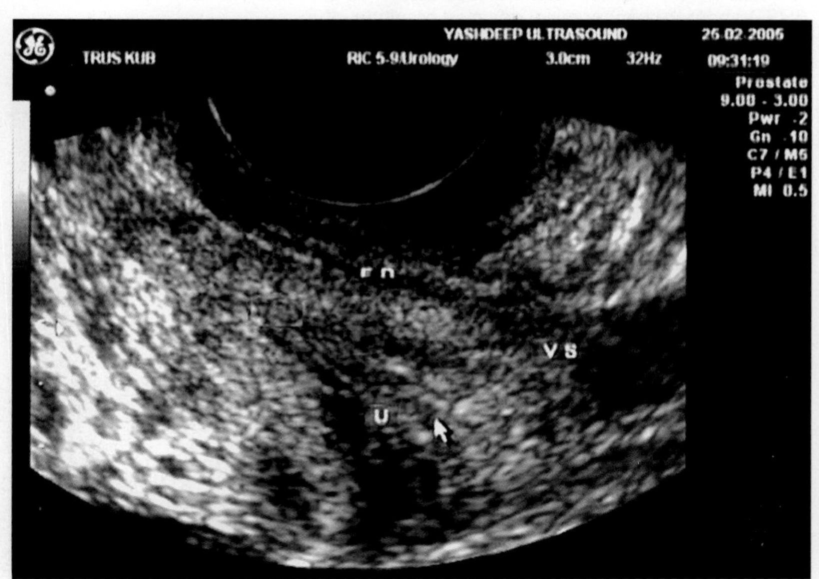

FIGURE 9.207

FIGURES 9.206 and 9.207: Normal ejaculatory ducts. HRSG shows normal ejaculatory ducts in longitudinal section of the prostate. It is seen as thin hypoechoic line in the running obliquely and opening in the prostatic part of urethra.

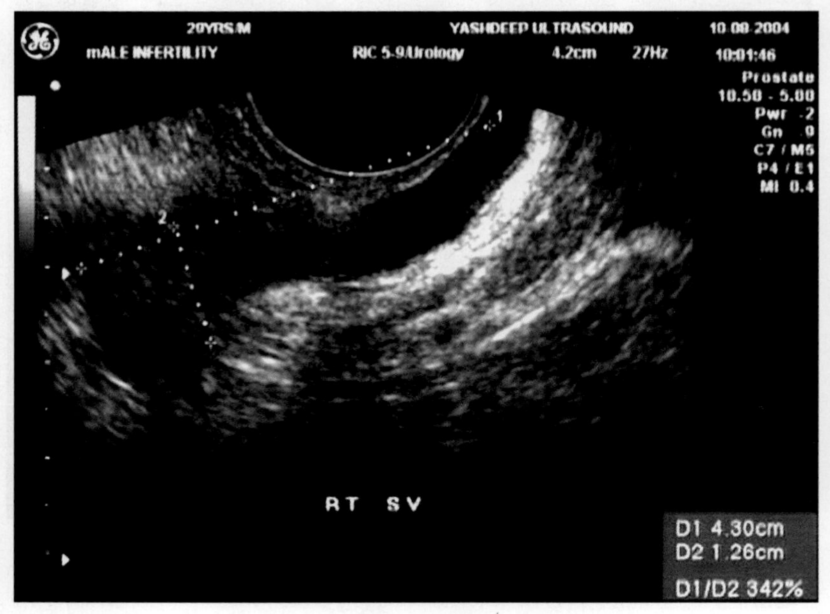

FIGURE 9.208

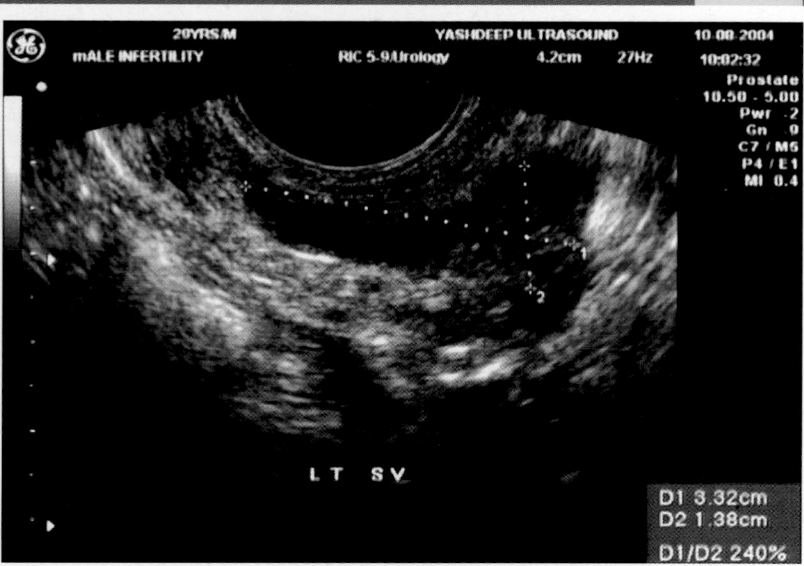

FIGURE 9.209

FIGURES 9.208 and 9.209: Normal ampula of seminal vesicle. The ampula of seminal vesicles are seen at distal end of the seminal vesicle. They are seen as flared reservoirs. Thick echo low-level collection is seen in them suggestive of seminal vesicle fluid.

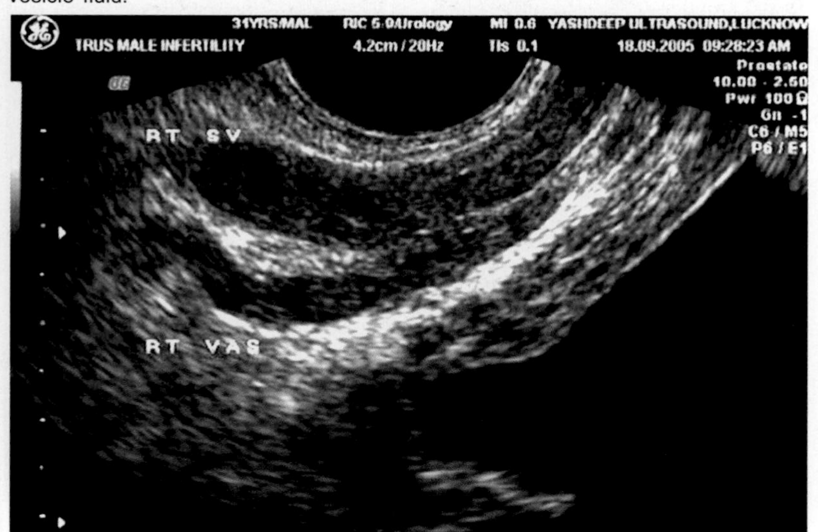

FIGURE 9.210

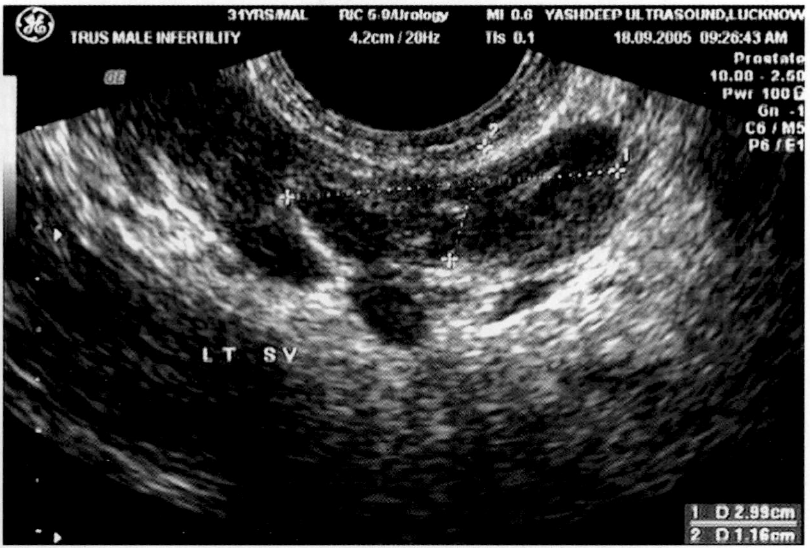

FIGURE 9.211

FIGURE 9.210 and 9.211: Aberrant origin of Lt vas deferens. A patient of male infertility presented with oligospermia. HRSG shows aberrant origin of Lt vas deferens. The Lt vas is seen coming out vertically from the Lt SV and runs down in an abnormal fashion. The Rt vas is seen normal in position.

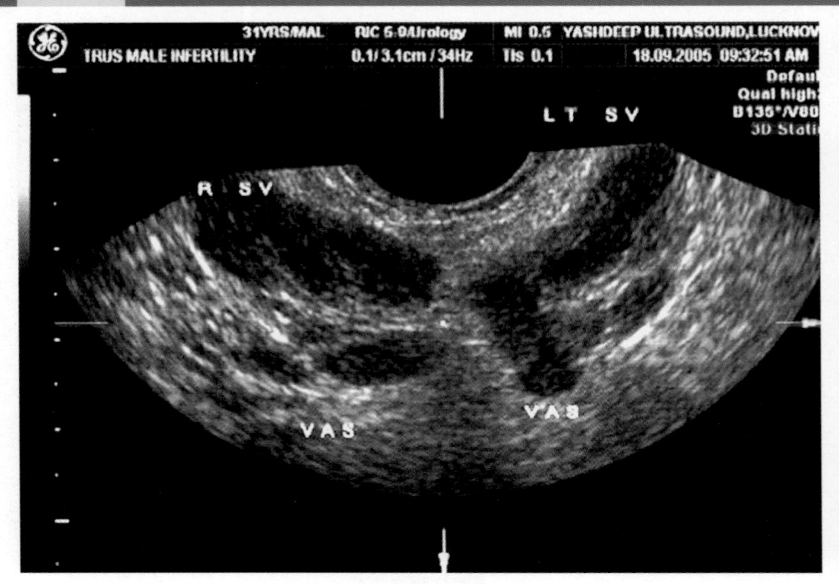

FIGURE 9.212

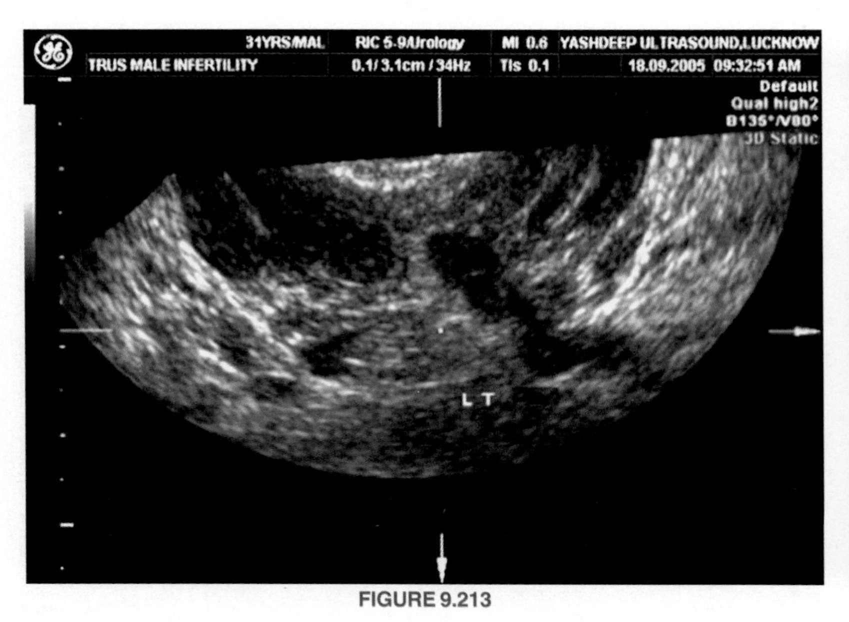

FIGURE 9.213

FIGURES 9.212 and 9.213: 3D imaging of the vas shows abnormal origin of Lt vas in better details.

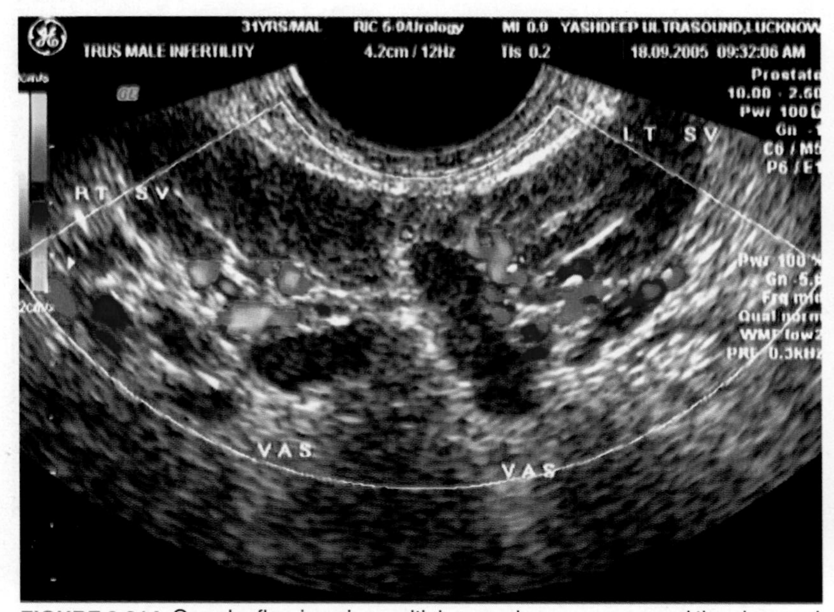

FIGURE 9.214: On color flow imaging multiple vessels are seen around the abnormal origin of the vas and periseminal vesicle area.

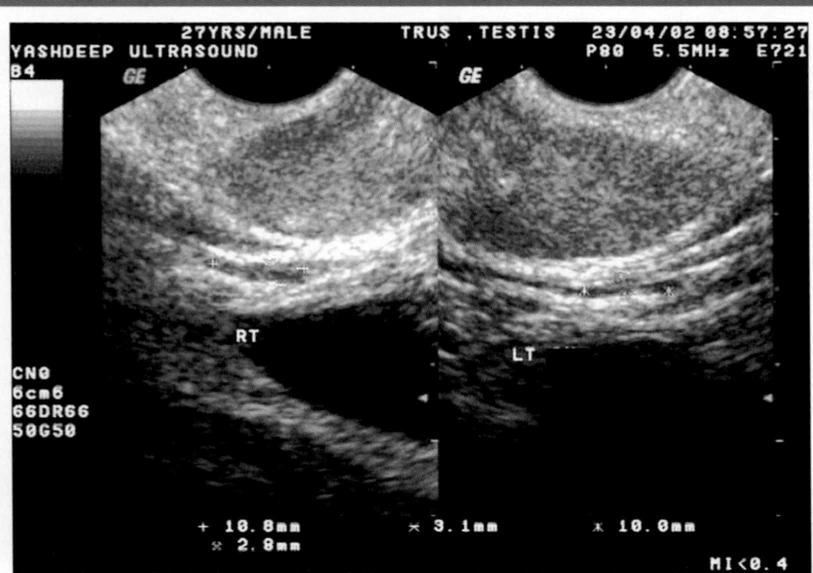

FIGURE 9.215: Absent seminal vesicles. A patient presented with azospermia. HRSG shows absent seminal vesicles on either side.

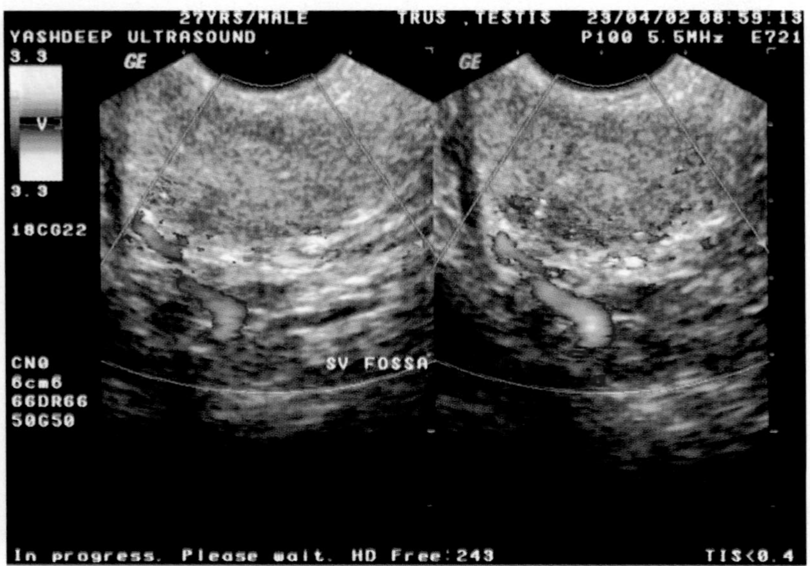

FIGURE 9.216

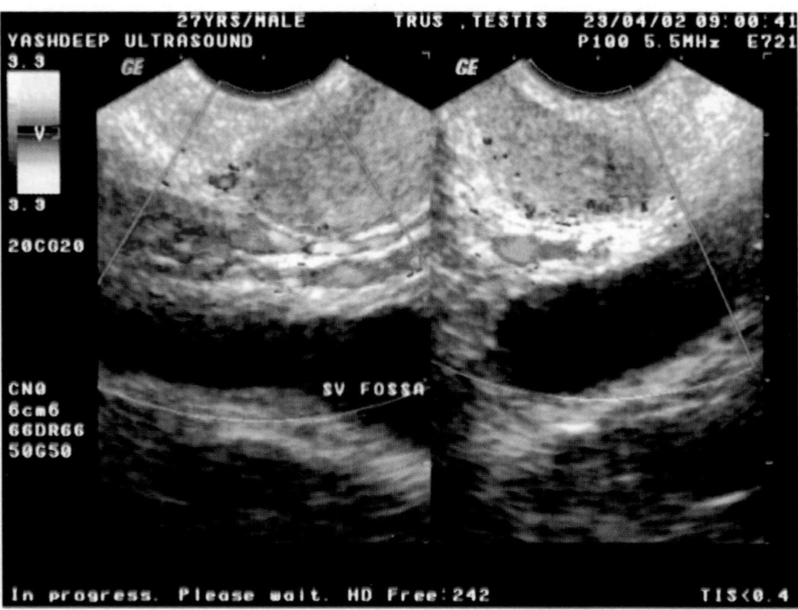

FIGURE 9.217

FIGURES 9.216 and 9.217: On color flow imaging dilated vessels are seen filling the seminal vesicle fossa. They are confirmed on color Doppler imaging.

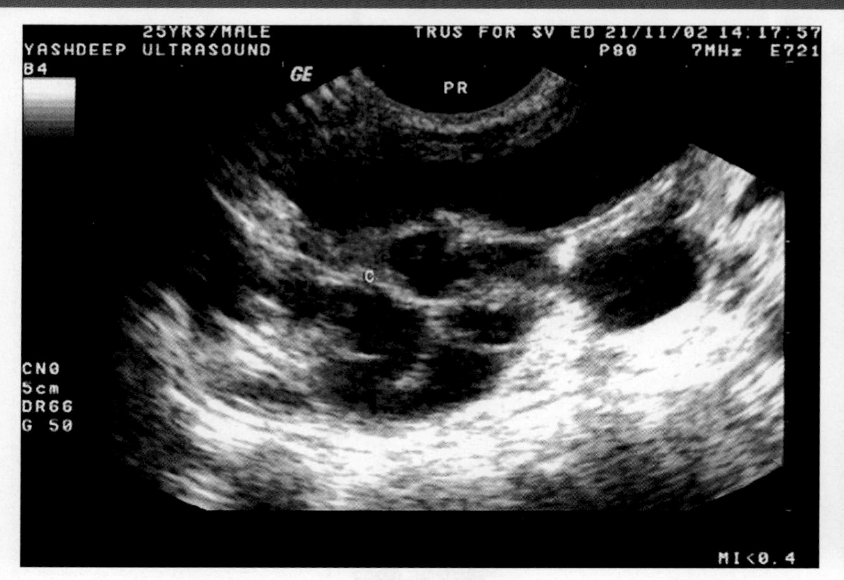

FIGURE 9.218

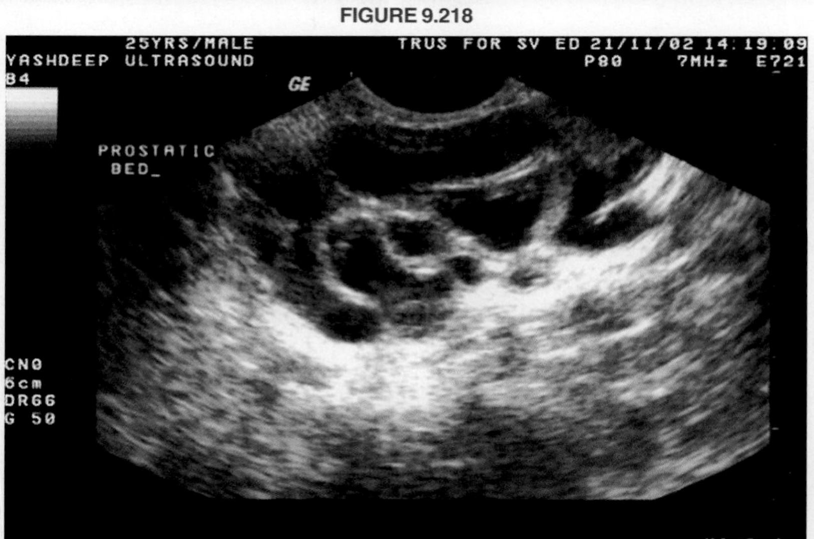

FIGURE 9.219

FIGURES 9.218 and 9.219: Abnormal seminal vesicle cysts. A patient of male infertility presented with azospermia. TRUS shows abnormal multiple cysts in both SV fossae. The cysts are non-communicating. Thick echogenic septa are seen in the cysts.

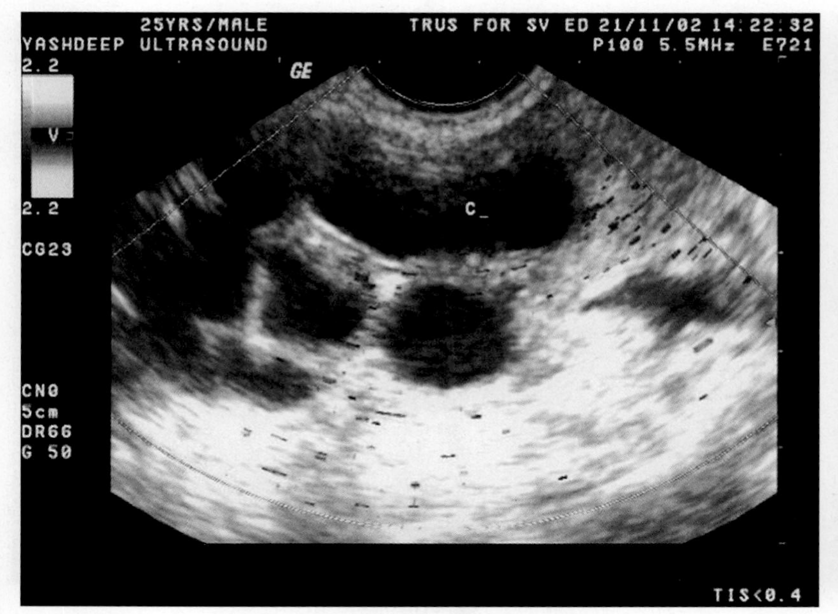

FIGURE 9.220

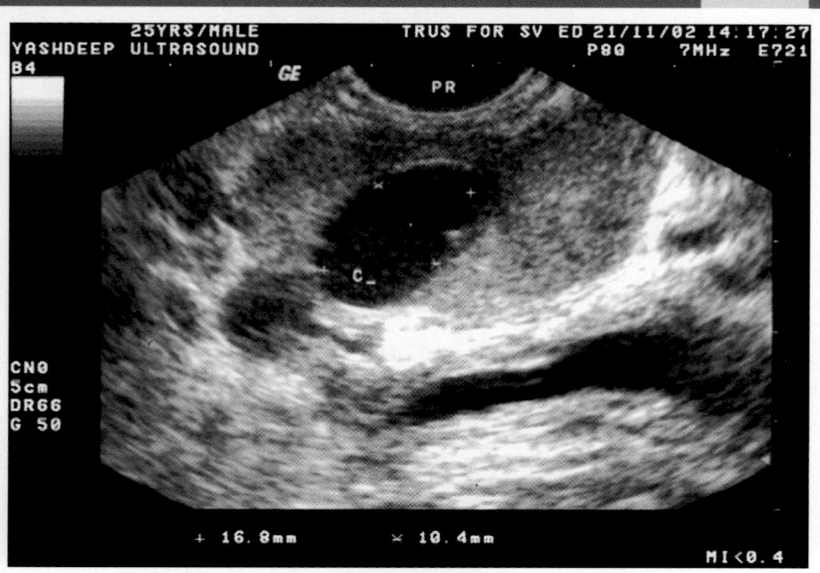

FIGURE 9.221

FIGURES 9.220 and 9.221: Multiple cysts are seen in the prostatic bed in the same patient. A big cyst is seen in the substance of prostate suggestive of a big utricular cyst. Biopsy confirmed congenital cyst.

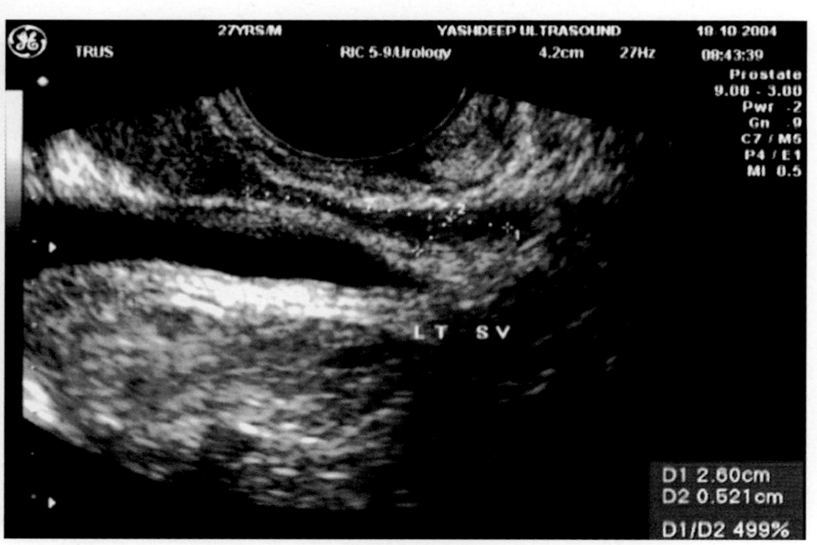

FIGURE 9.222

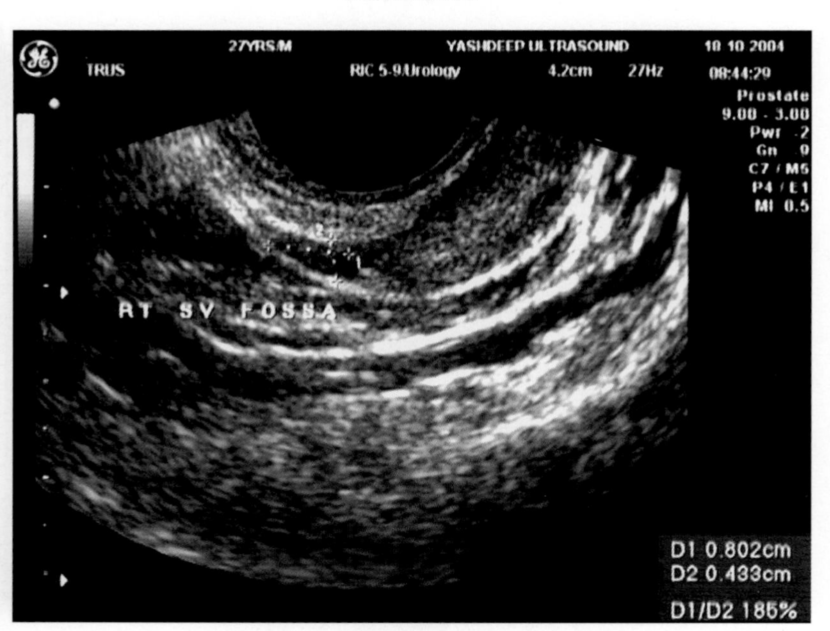

FIGURE 9.223

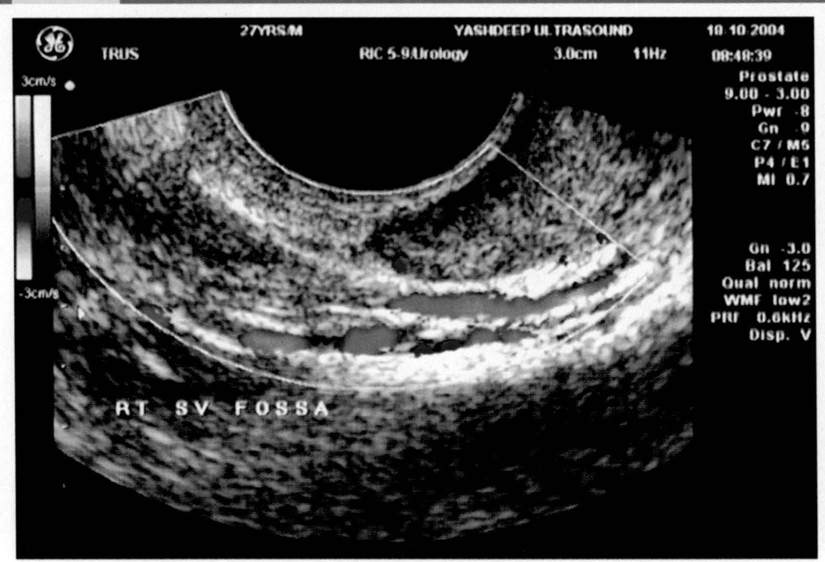

FIGURE 9.224

FIGURES 9.222 to 9.224: Absent Rt seminal vesicle and small Lt seminal vesicle. HRSG shows absent Rt seminal vesicle. However, Lt seminal vesicle is small in size. On color Doppler imaging vessels are seen in the Rt seminal vesicle fossa. The patient presented as azospermia.

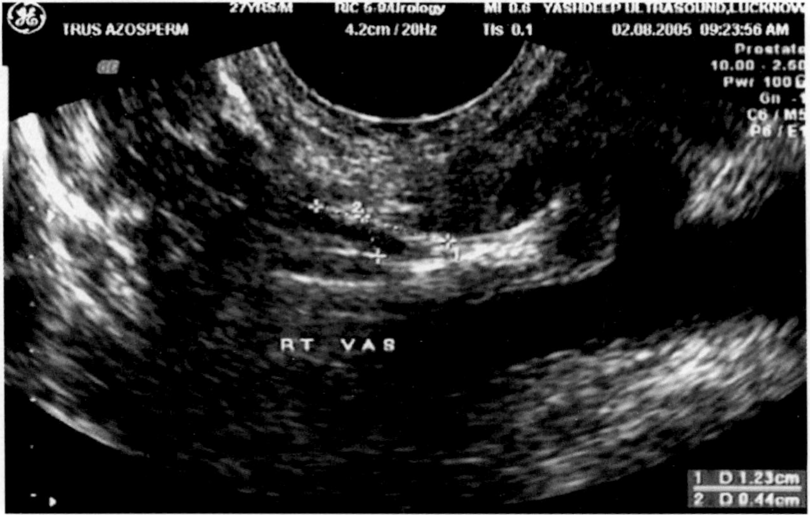

FIGURE 9.225

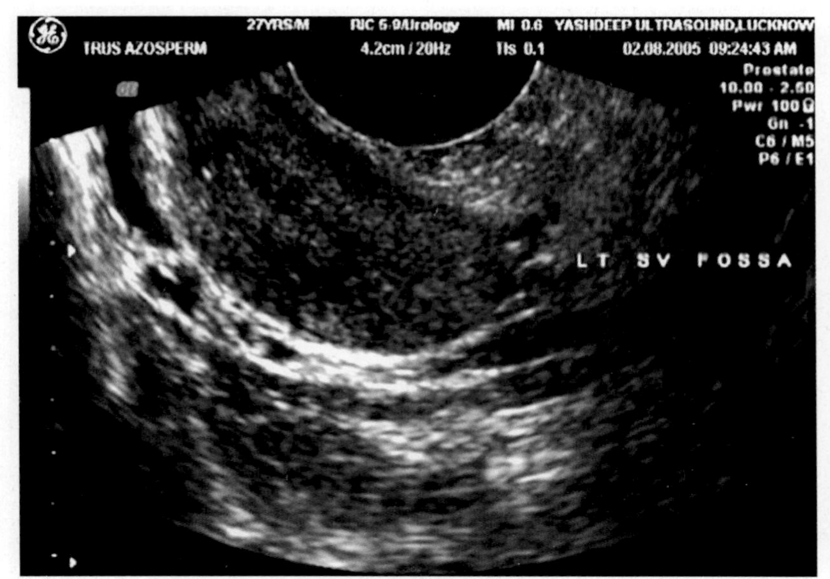

FIGURE 9.226

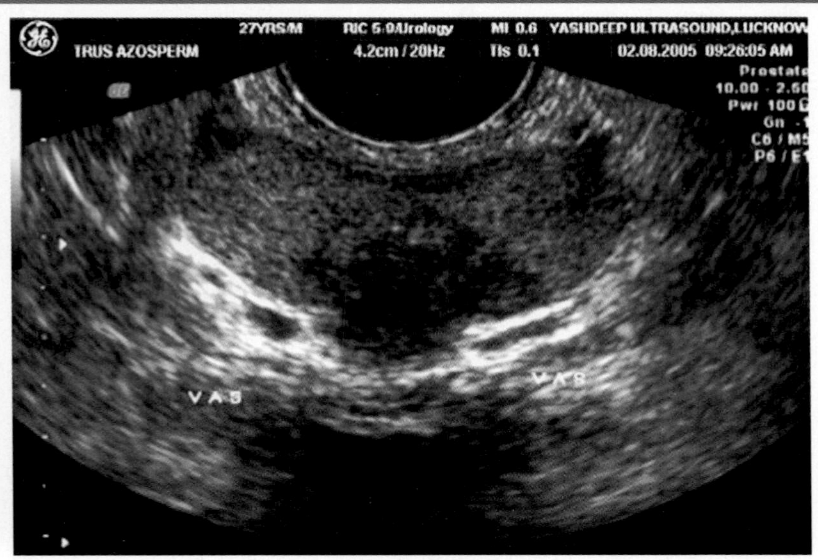

FIGURE 9.227

FIGURES 9.225 to 9.227: Bilateral absent seminal vesicles. HRSG shows absent seminal vesicle on either side in a patient presented with azospermia. The vas deferens is also seen small in size suggestive of hypoplasia.

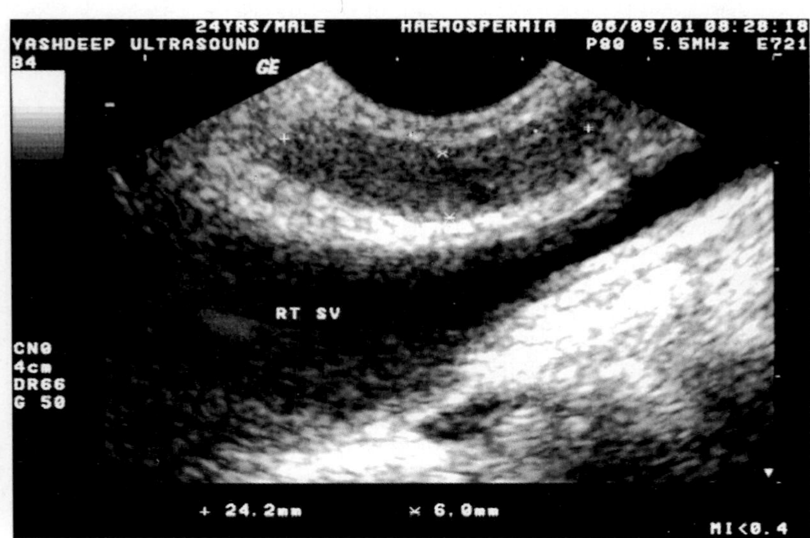

FIGURE 9.228

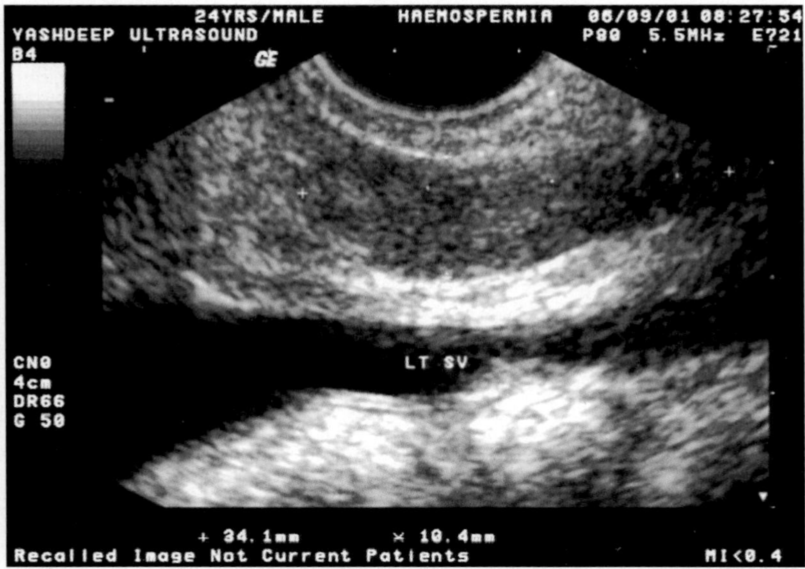

FIGURE 9.229

FIGURES 9.228 and 9.229: Bilateral seminal vesiculitis. HRSG shows low-level echo collection in a patient presented with pyospermia. The collection shows multiple homogeneous echoes suggestive of pyogenic collection. Peri focal edema is also seen.

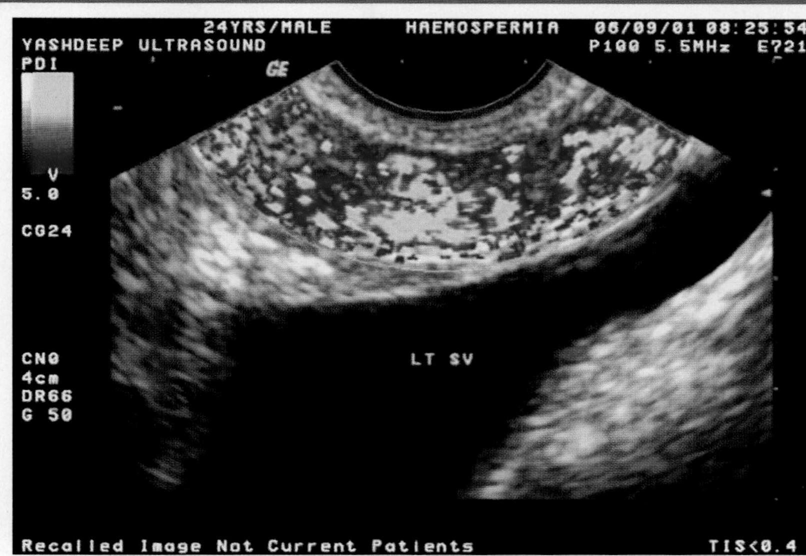

FIGURE 9.230

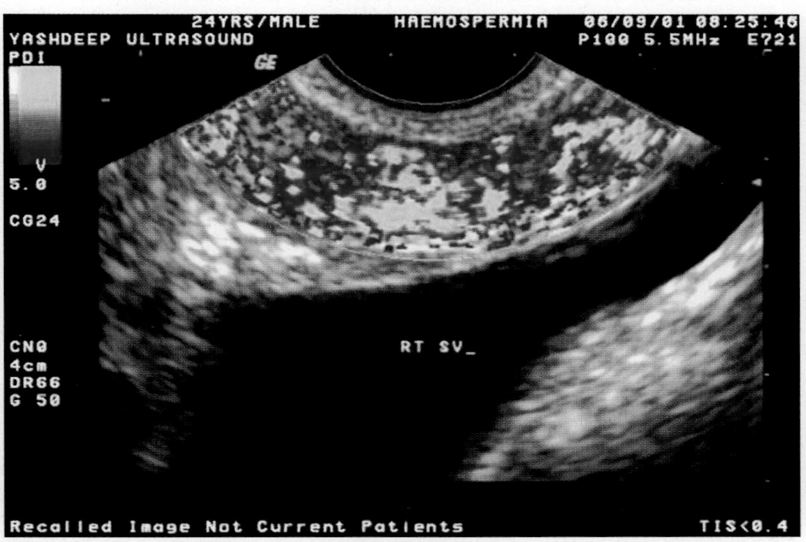

FIGURE 9.231

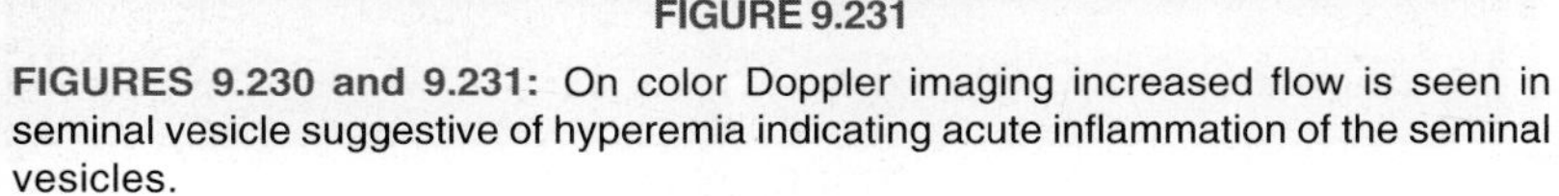

FIGURES 9.230 and 9.231: On color Doppler imaging increased flow is seen in seminal vesicle suggestive of hyperemia indicating acute inflammation of the seminal vesicles.

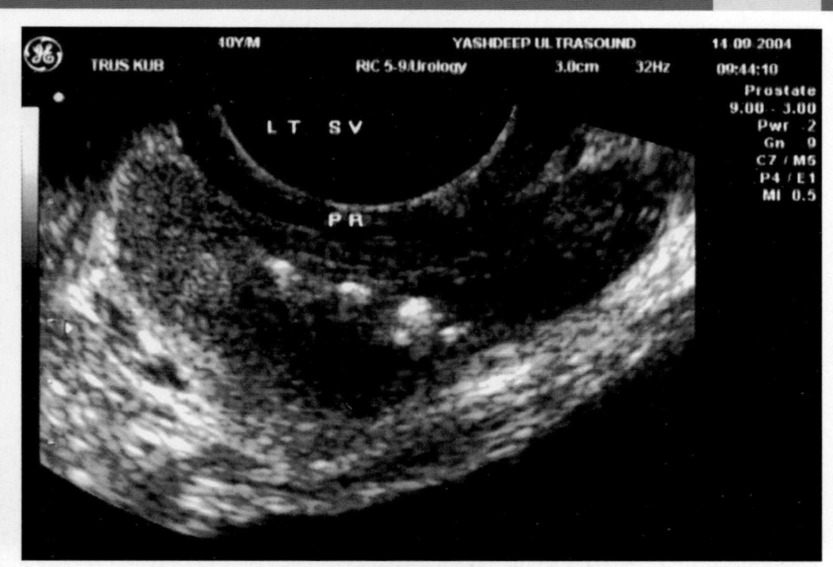

FIGURE 9.233

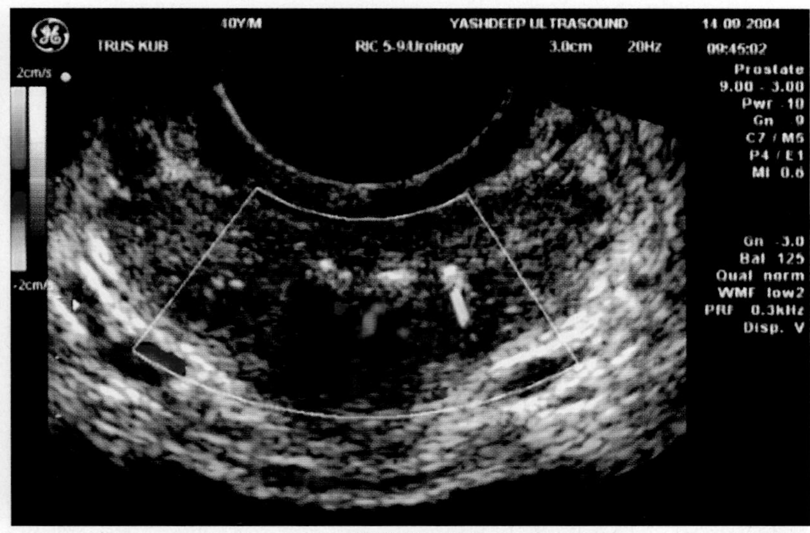

FIGURE 9.234

FIGURES 9.232 to 9.234: Chronic seminal vesiculitis. A patient presented with pyospermia. HRSG shows hypoechoic distended seminal vesicle. Thick collection is seen in the seminal vesicle. Echogenic calcified specks are also seen in the wall of the SV. Echogenic calcified shadows are also seen in the prostate.

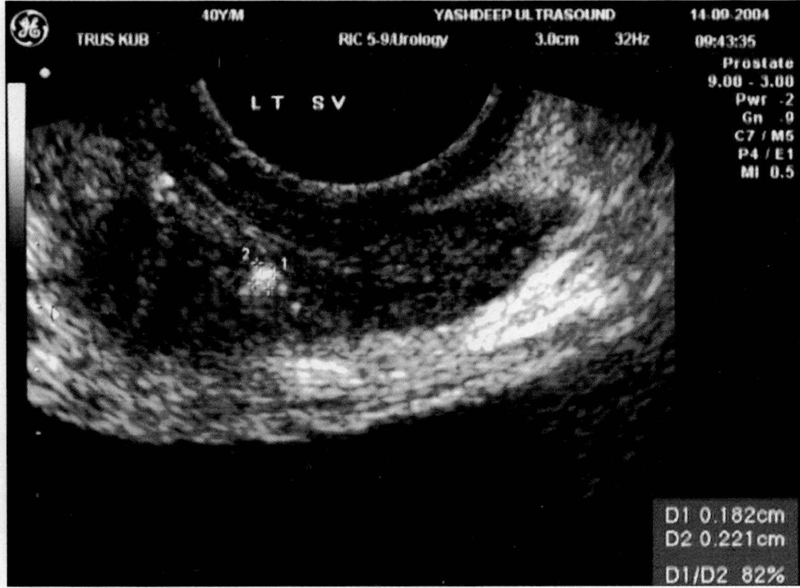

FIGURE 9.232

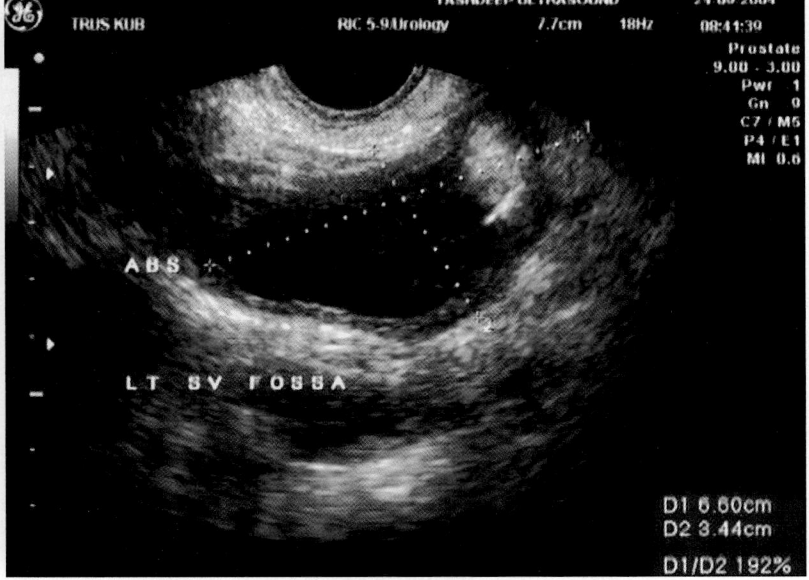

FIGURE 9.235

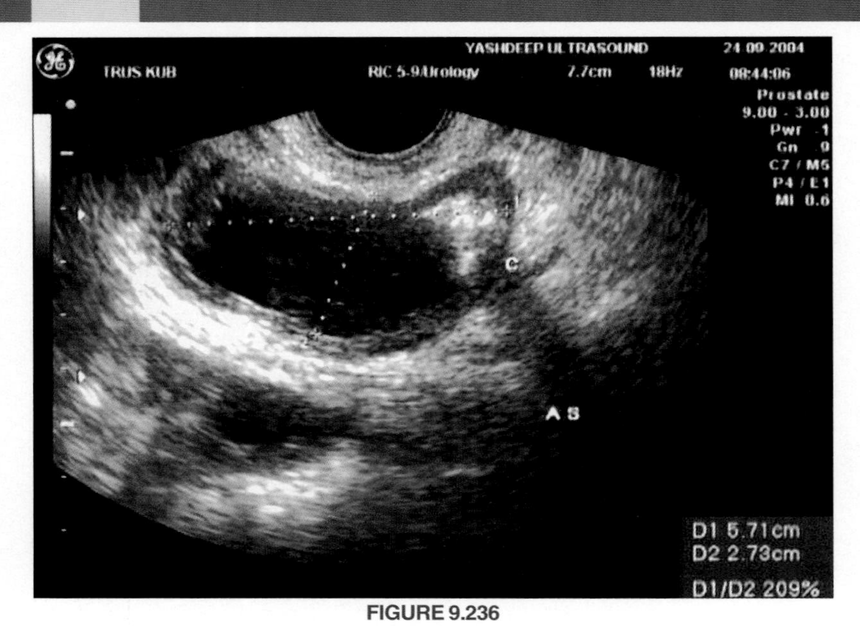

FIGURE 9.236

FIGURES 9.235 and 9.236: Seminal vesicles abscess. A patient presented with pyospermia with fever. HRSG shows thick low-level echo collection in the Lt SV. Echogenic floating shadows are seen in it suggestive of abscess. Echogenic debris is also seen in the seminal vesicle. Calcification is seen in it suggestive of chronic abscess.

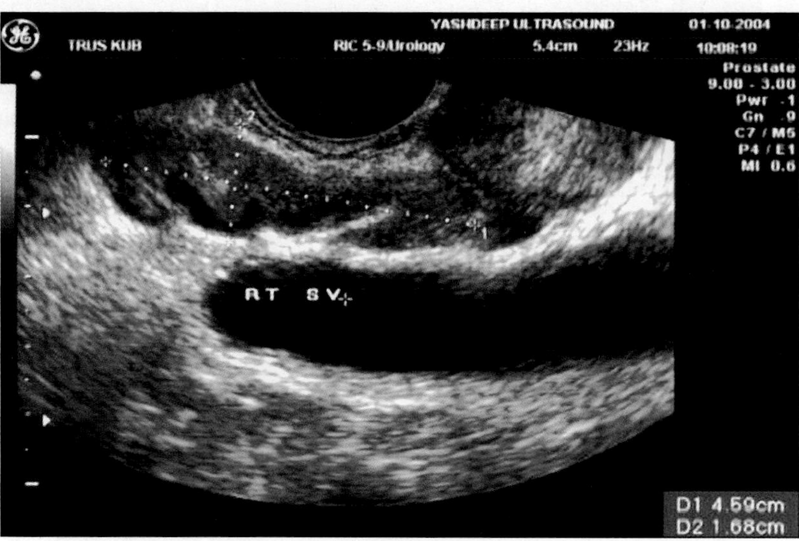

FIGURE 9.237

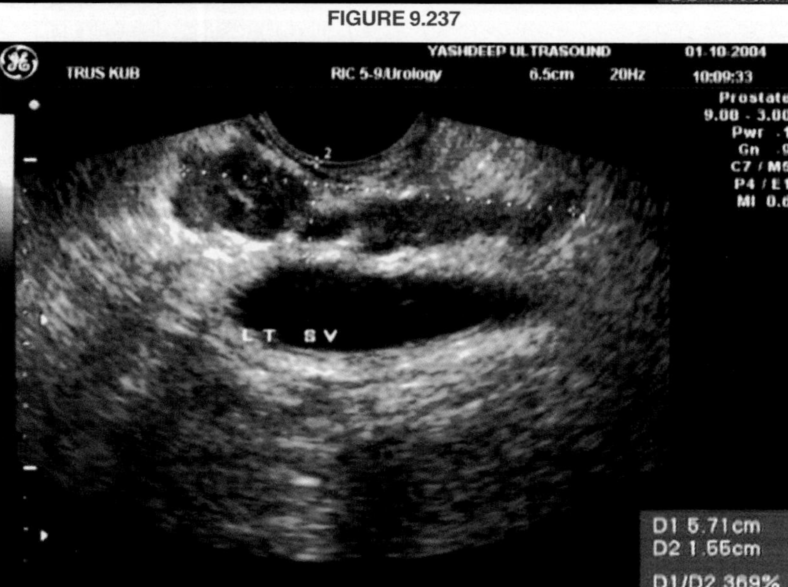

FIGURE 9.238

FIGURES 9.237 and 9.238: Rt SV also shows echogenic collection. The SV walls are irregular in outline.

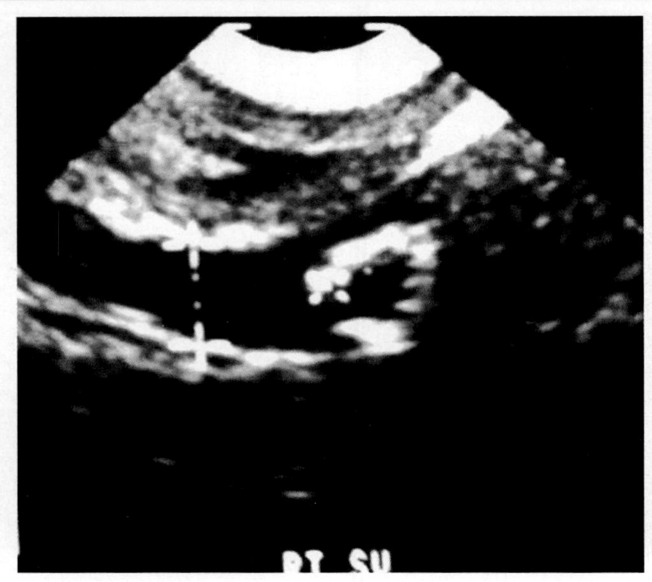

FIGURE 9.239: Rt seminal vesicle calculus. HRSG shows a well-defined echogenic calcified ring like shadow blocking the mouth of the Rt seminal vesicle. It is accompanied with acoustic shadowing. The seminal vesicle is dilated and engorged.

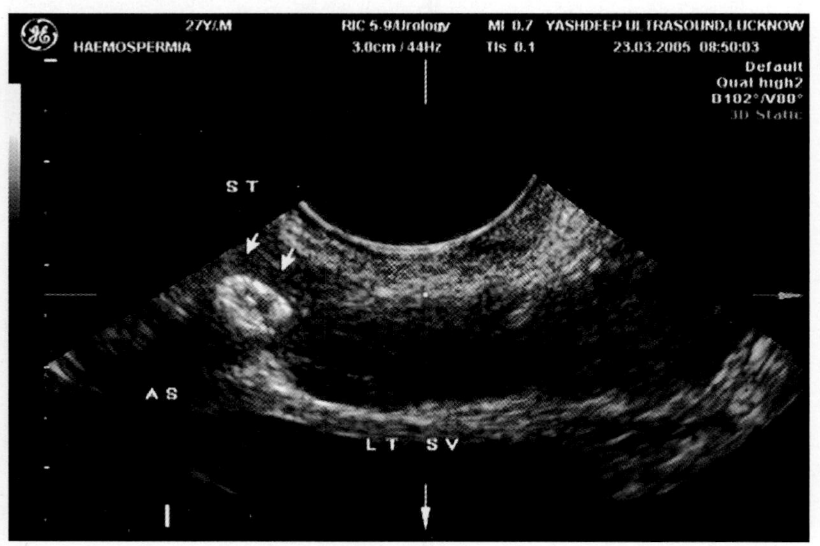

FIGURE 9.240

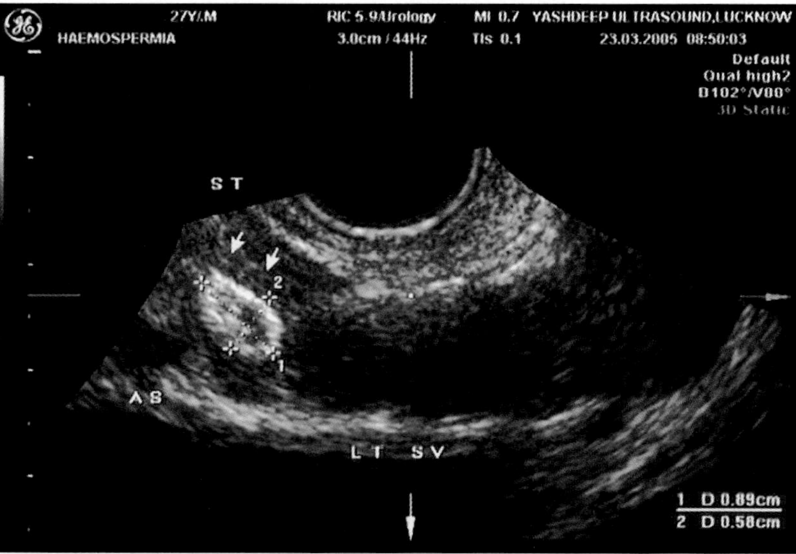

FIGURE 9.241

FIGURES 9.240 and 9.241: Lt seminal vesicle calculus. A patient presented with oligospermia. HRSG shows dilated Lt seminal vesicle. A big echogenic calculus shadow is seen impacted at the mouth of seminal vesicle. It is well-defined and accompanied with acoustic shadowing.

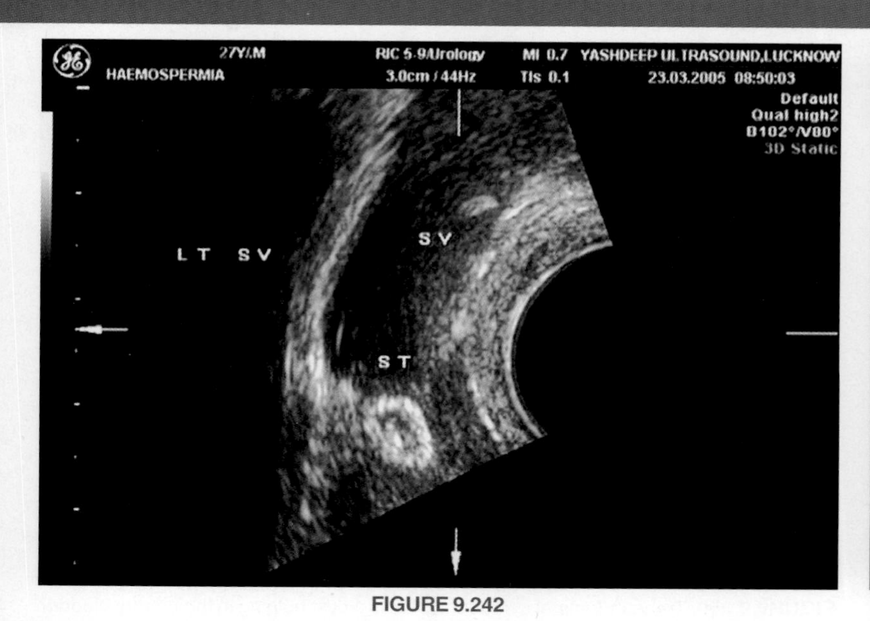

FIGURE 9.242

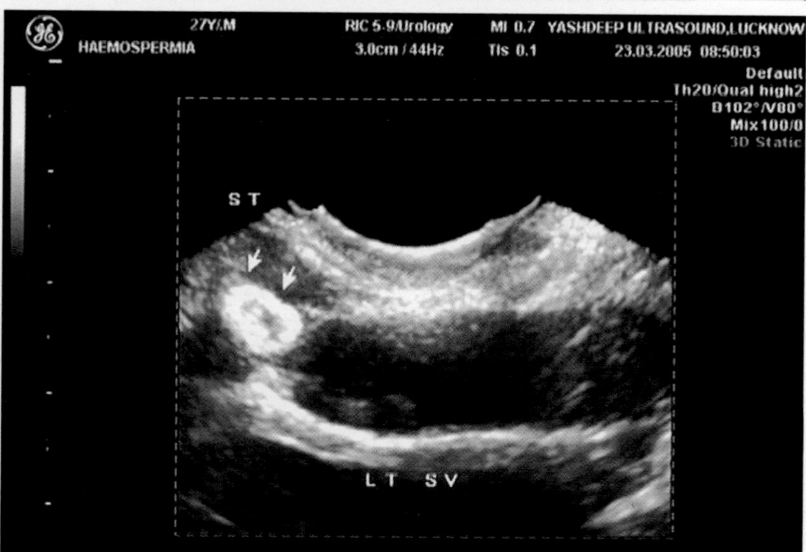

FIGURE 9.243

FIGURES 9.242 and 9.243: 3D surface rendering imaging shows the calculus in much better view. Echogenic floating collection is also seen in the seminal vesicle.

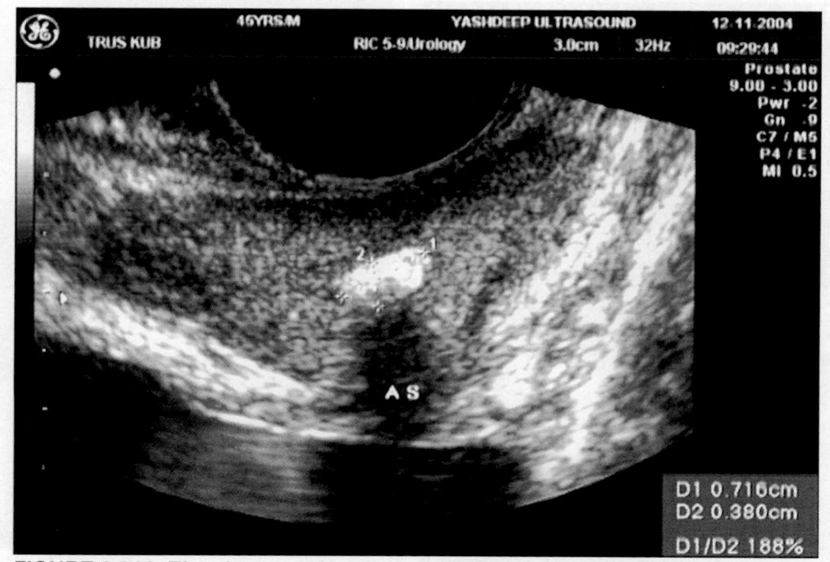

FIGURE 9.244: Ejaculatory duct calculus. HRSG shows a big echogenic calculus shadow at the confluence of ejaculatory and prostatic urethra in a patient of azospermia. It is accompanied with acoustic shadowing. The calculus is seen blocking the mouth of the duct.

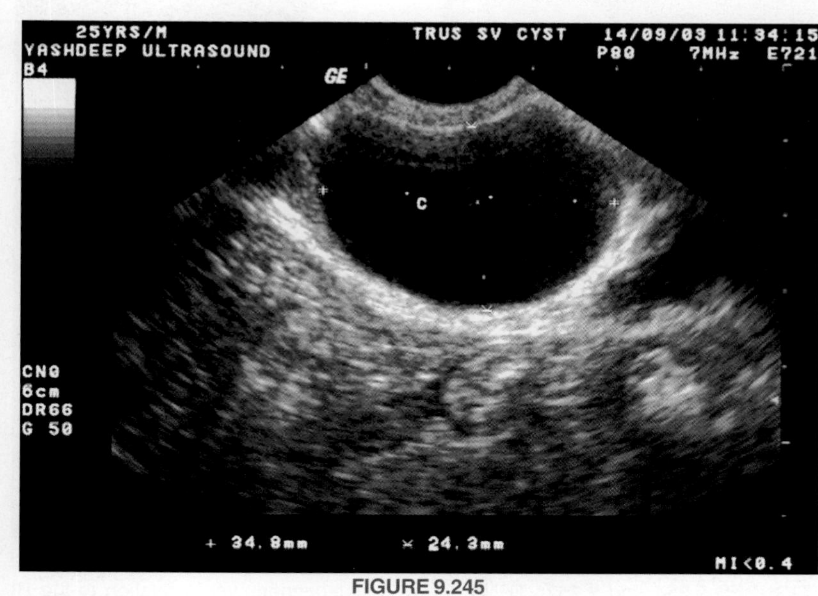

FIGURE 9.245

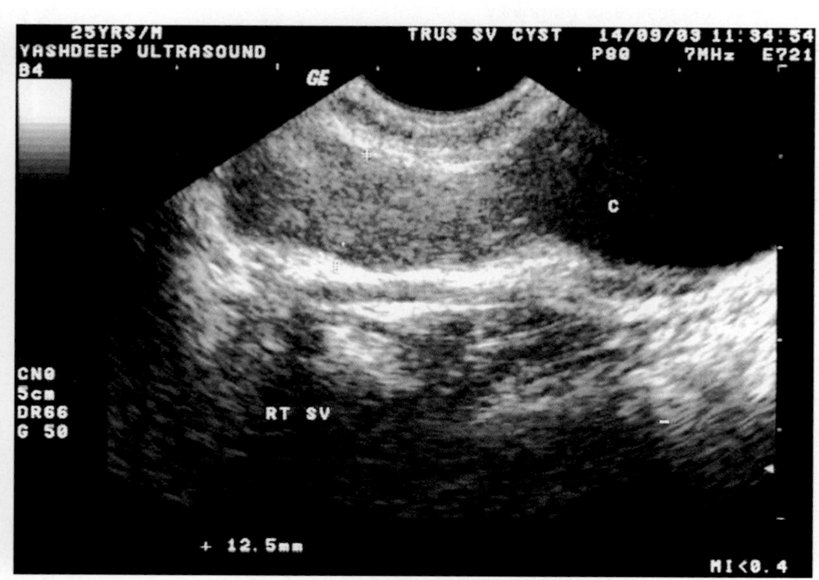

FIGURE 9.246

FIGURES 9.245 and 9.246: Rt seminal vesicle cyst. HRSG shows a big cystic mass in relation to the Rt seminal vesicles in a patient of oligospermia. The cyst is seen blocking the mouth of the seminal vesicle.

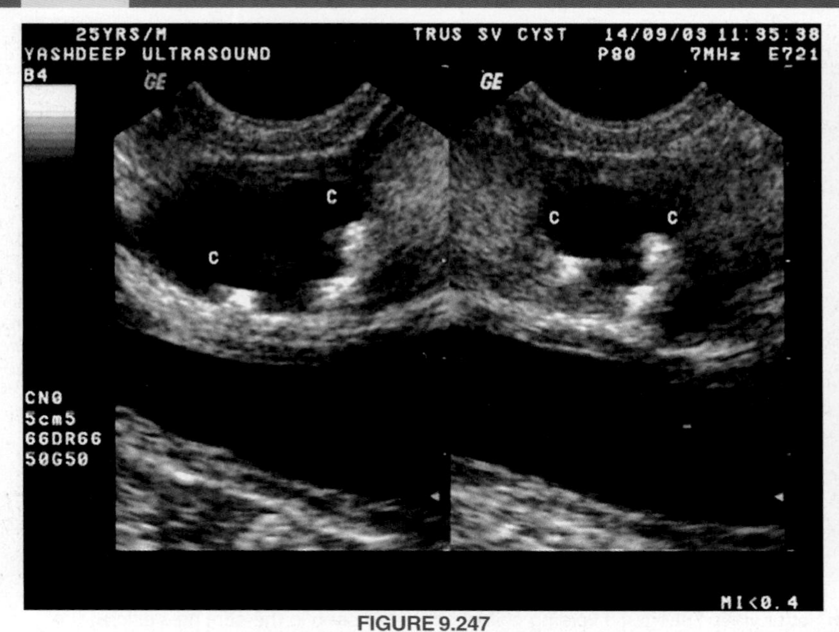

FIGURE 9.247

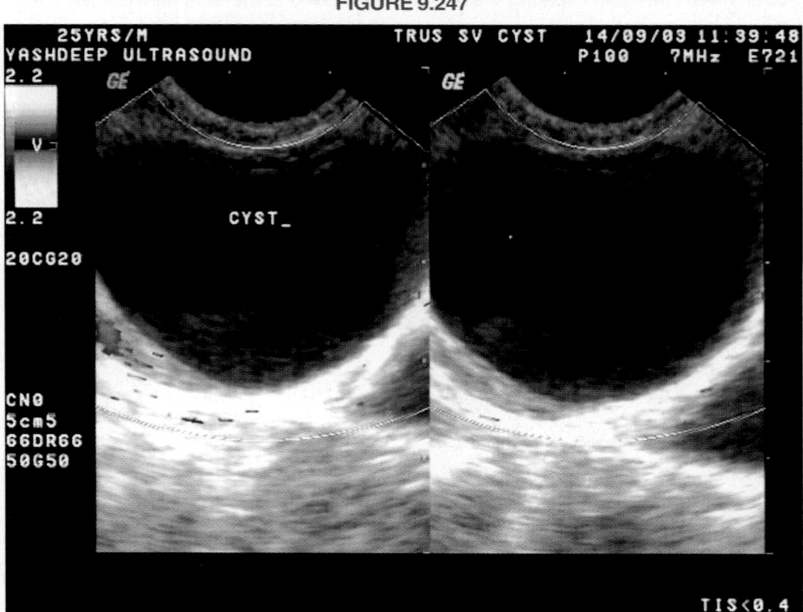

FIGURE 9.248

FIGURES 9.247 and 9.248: Echogenic calcified specks are also seen fixed with the cyst wall suggestive of chronicity.

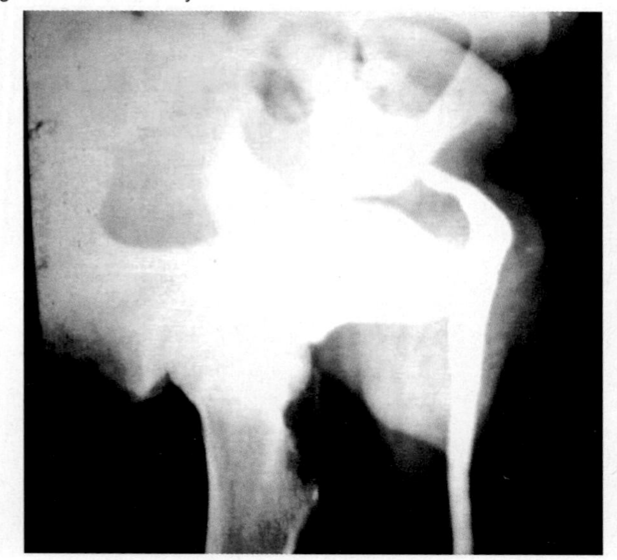

FIGURE 9.249: Seminal vesicle cyst. A patient presented with pyo-oligospermia. Retrograde urethrogram shows pressure over the prostatic part of the urethra with indentation.

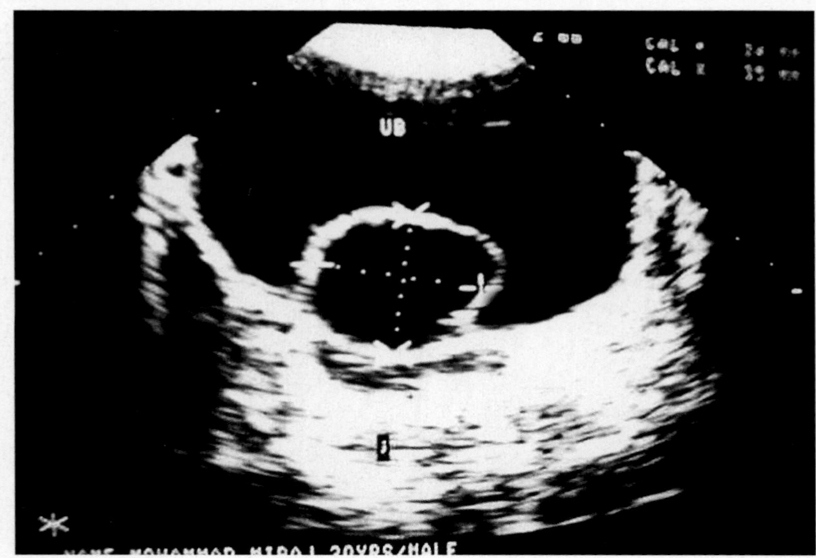

FIGURE 9.250: Transabdominal ultrasound shows a cyst bulging in the urinary bladder.

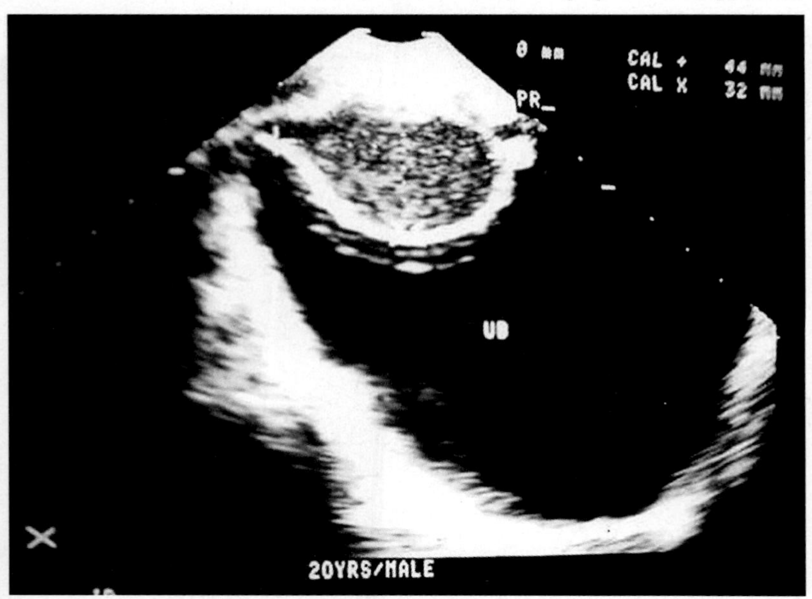

FIGURE 9.251

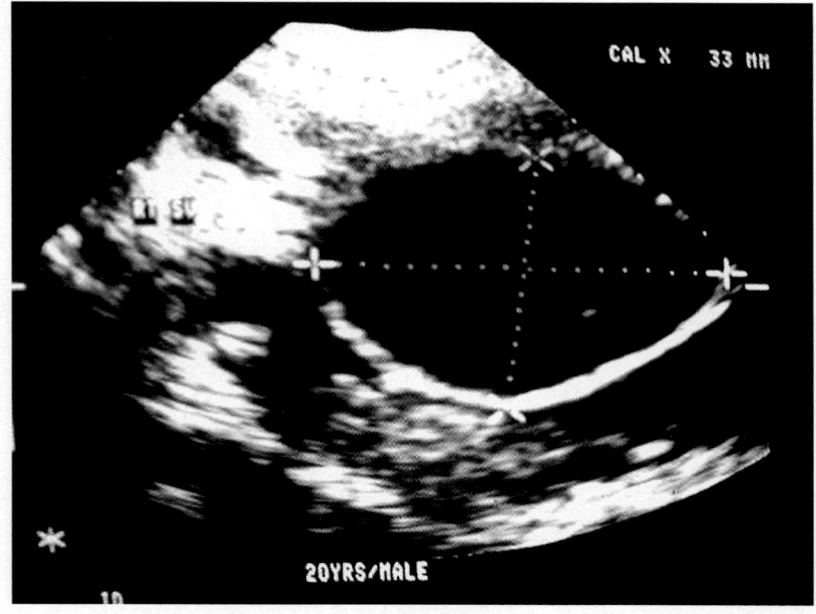

FIGURE 9.252

FIGURES 9.251 and 9.252: TRUS shows a well-defined cyst in relation to the Rt seminal vesicle. Multiple internal echoes are seen in it. USG guided aspiration showed an infected seminal vesicle cyst.

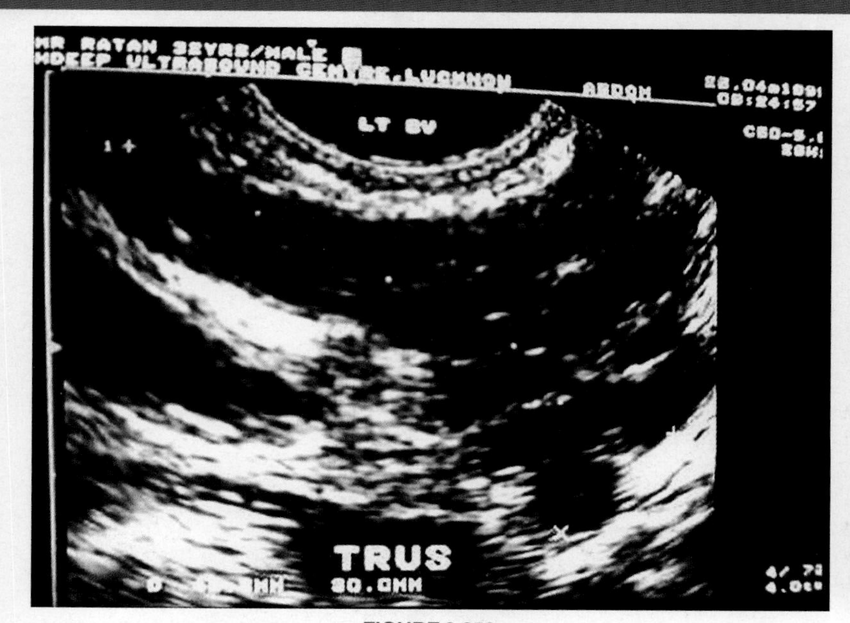

FIGURE 9.253

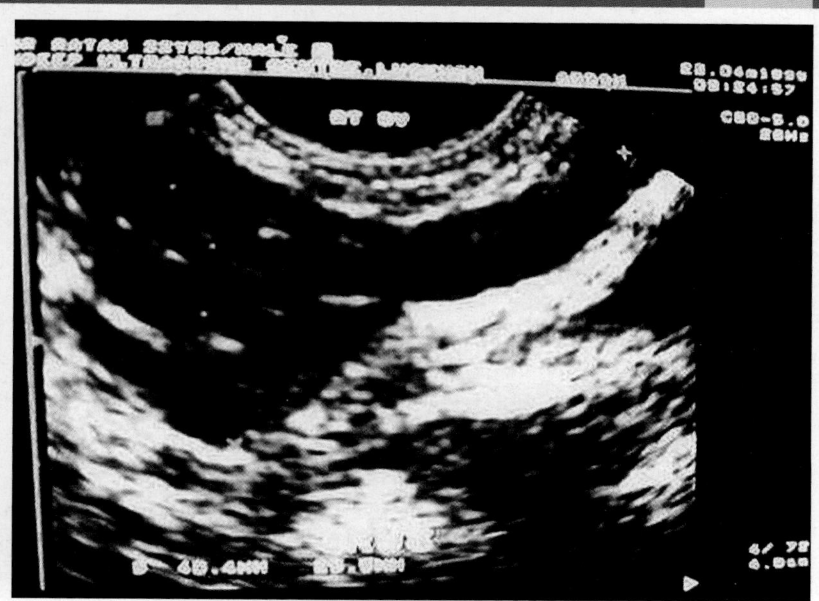

FIGURE 9.254

FIGURES 9.253 and 9.254: Tubercular seminal vesiculitis. HRSG shows bilateral dilated seminal vesicles. Multiple echogenic septa are seen in the seminal vesicles. Septation and honeycombing is seen on both sides. Low-level echo collection is seen in them. The patient presented with oligospermia. On FNAC chronic granulomatous infection is seen in both the SV suggestive of tubercular infection.

10 Scrotum and Testis

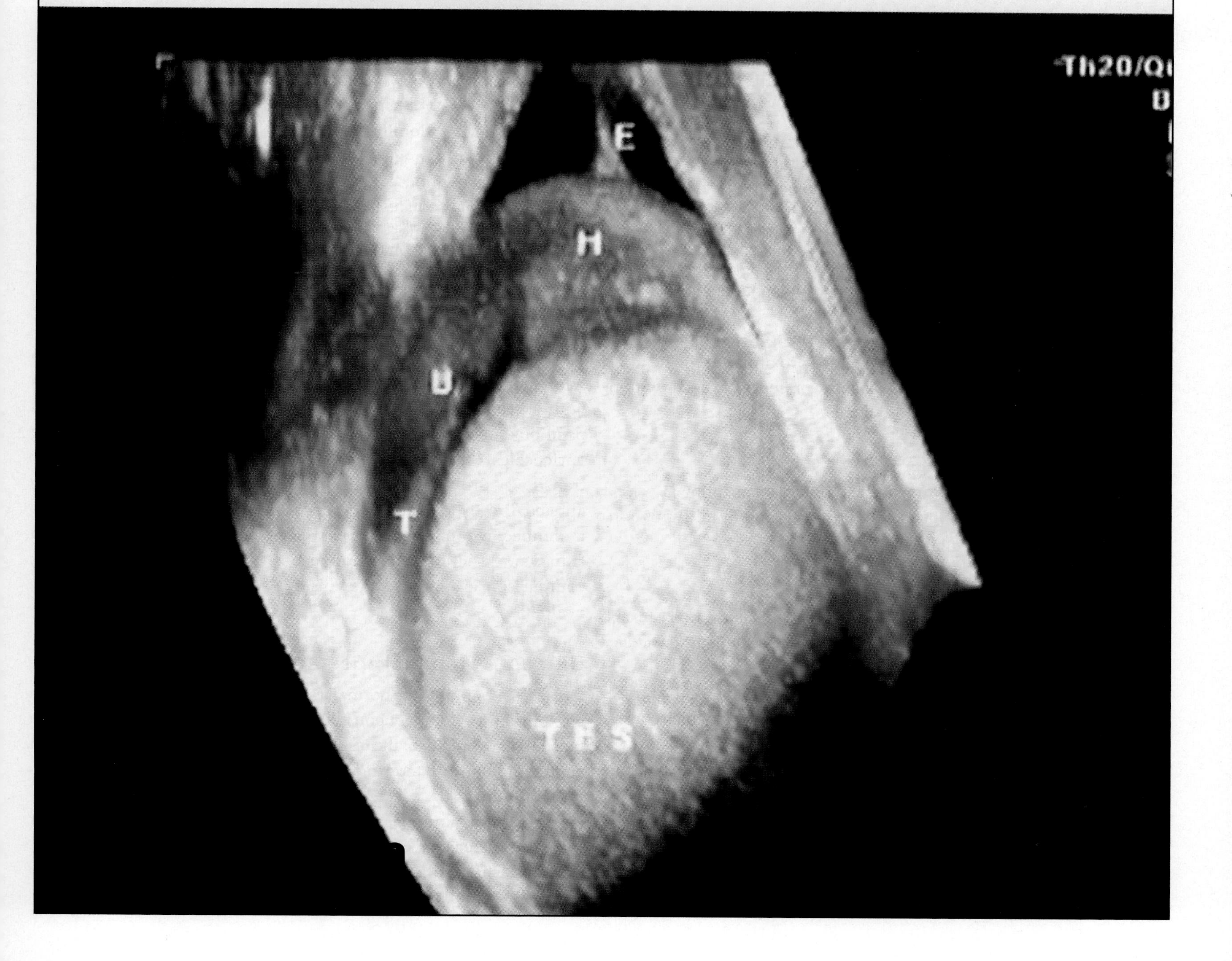

Introduction

At one time, diagnosis of scrotal pathologies was dependent not only on clinical manifestation but to a large extent on the palpatory examination of the scrotum. Today the advent of newer high-resolution ultrasonographic machines has revolutionized imaging of the scrotum. Further addition of color Doppler imaging has increased the efficiency of diagnosis of scrotal pathologies especially in acute scrotum. In fact probably the single most important advantage of ultrasonography is its ability of distinguish between intra and extratesticular mass lesions.

In general, the initial confusion of overlapping signs and symptoms in many conditions is now significantly made easy due to the availability of sonography. Ultrasonography allows the clinician to treat a patient early with no delays and confirms the clinical diagnosis with relative ease, and yields adequate information in most cases.

Normal Anatomy

The testis is symmetrical, ovoid structures measuring about 2 to 3 cm in diameter and 3 to 5 cm in length. Each testis is encased in a dense white membrane called tunica albuginea. The Epididymis is a tortuous tubular structure located superolaterally over the posterior aspect of the testis. It is round or triangular in shape with two expansions, the head (globus major) at the lower pole. The testis, Epididymis and proximal segment of spermatic cord are invested in the visceral layer of tunica vaginalis, which covers the entire testis except for a small area posteriorly where it folds on itself. This bare area, the mediastinum testis, is where the spermatic cord and its contents, i.e. the blood and lymphatic vessels, the nerves and epididymal ducts join the testicle. The parietal layer of tunica vaginalis lines the inner wall of the scrotum.

The spermatic cord is visualized as a heterogeneous structure (predominantly echogenic) arising in close proximity of the head of the Epididymis. Color Doppler studies show presence of vessels within it.

Blood Supply

The blood supply to the scrotum and its contents is as follows:

1. *Testicular artery:* It supplies mainly the testis and Epididymis and arises from the abdominal aorta.
2. *Differential artery:* It arises from the superior or inferior vesicle artery and supplies to the vas deferens.
3. *Cremasteric artery:* It arising from the inferior epigastric artery forms a network over the tunica vaginalis to anastomose with the testicular artery at the mediastinum.

The testicular artery divides into branches, which pierce the tunica albuginea. These branches form the centripetal branches, which course along the margins of the testis and converge on to the mediastinum. From the mediastinum, recurrent ramie are formed, which traverse the testis centrifugally and finally branch into arterioles and capillaries. This has been best studied by color Doppler sonography. Color Doppler sonography can easily identify the capsular as well as the intratesticular arteries. On spectral analysis, these vessels have a low resistance pattern with high diastolic flow.

Venous Drainage

The pampiniform plexus exits from the mediastinum and converges into the testicular vein, which drains into the renal vein on the left side and inferior vena cava on the right. The deferential vein drains into the pelvic veins, and cremasteric vein drains into tributaries of the inferior epigastric and deep pudendal veins.

Indications for Scrotal Sonography

1. Evaluation of acute scrotal disorders
   - Torsion
   - Inflammation
   - Trauma
2. Evaluation of scrotal fluid collections
   - Hydrocele
   - Pyocele/hematocele
3. Evaluation of a scrotal mass
   - Extratesticular
   - Intratesticular
4. Evaluation of metastatic disease
   - Retroperitoneal lymphadenopathy
   - Testicular involvement—lymphoma, leukemia
5. Unexplained abdominal mass
   - Malignant change in an undescended testis
6. Evaluation of undescended testis
7. Exclusion of occult testicular neoplasm
   - Leydig cell tumor
8. Evaluation of infertility
   - Varicocele.

Technique

The examination of the scrotum is normally done in supine position with the patient's scrotum supported by towels or other means with spread-out legs. The examination may also be done in erect position especially in the evaluation of male infertility for visualizing varicoceles. A real-time high-resolution (5-10 mhz) linear array transducer provides excellent images. The transducer is normally applied directly to the scrotal surface through a coupling gel. The testis is scanned in both transverse as well as sagittal planes. If necessary a transverse section to visualize both testis simultaneously for comparison is undertaken. Valsalva's maneuver is useful in the evaluation of varicocele.

Ultrasonographic Appearances

On ultrasonography, the normal testis is an ovoid structure with smooth contours and a homogeneous echo texture—medium level echoes. In fact the echogenicity of the testis simulates that of the thyroid. A bright echogenic band extending craniocaudally in the testis indicates mediastinum testis.

The tunica albuginea is not well identified normally while linear hypoechoic areas traversing through the testis indicate testicular septa. The rete testis, not very commonly seen, presents as a hypoechoic area in close proximity of the epididymal head.

The epididymis is an elongated structure measuring about 6 to 7 cm in length. It is situated posterolateral to the testis. The head of the Epididymis is the largest portion of the Epididymis and is situated in close proximity of the upper pole of the testis. The body of the Epididymis is thinner and lies along the lateral surface of the testis. The tail is situated in the region of the lower pole of the testis but is not usually visualized. The Appendicular testis is a remnant of the mullerian duct and is best visualized in the presence of hydrocele. On an average, the Epididymis is relatively hypoechoic compared to the testis.

With the use of color Doppler sonography, arterial flow is normally demonstrated in the testis and in the spermatic cord. However, the arterial flow in the Epididymis is seldom or never seen because of the small size of the vessels and very slow flow (any detectable flow in the Epididymis is usually considered abnormal) within the testis the capsular, centripetal and centrifugal arteries are well identified.

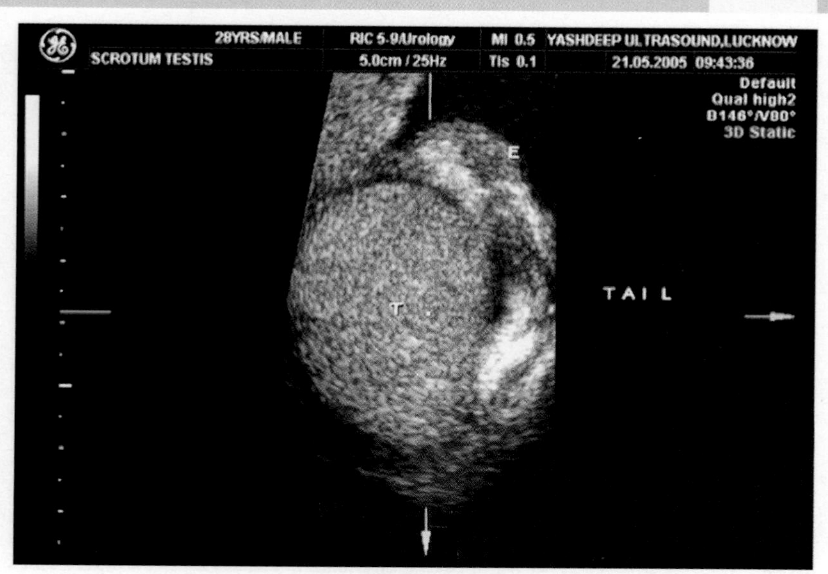

FIGURE 10.3A

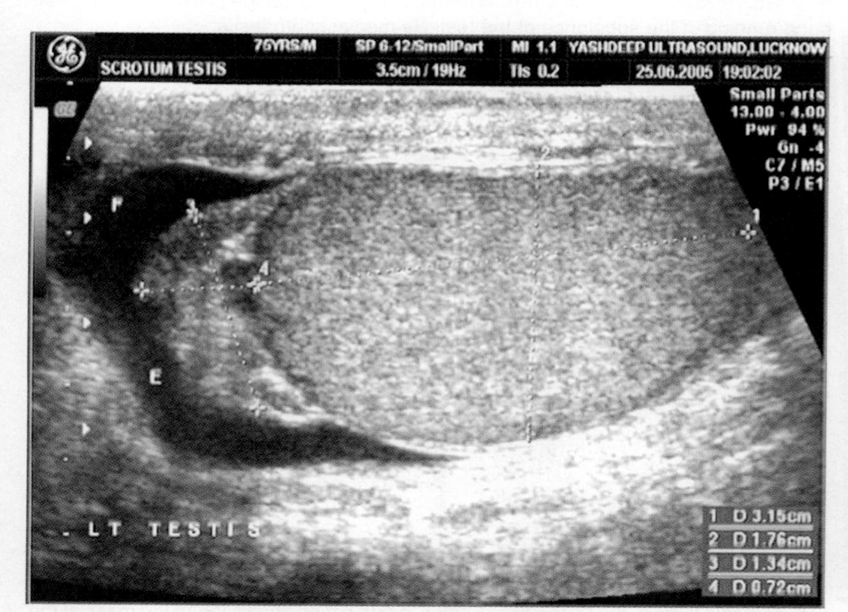

FIGURE 10.1: Normal testis—HRSG shows normal testis. The testis is ovoid in shape with fine granular echoes and homogeneous texture. Head of the epididymis is seen in proximity to the upper pole of the testis. It is less echogenic than the testis.

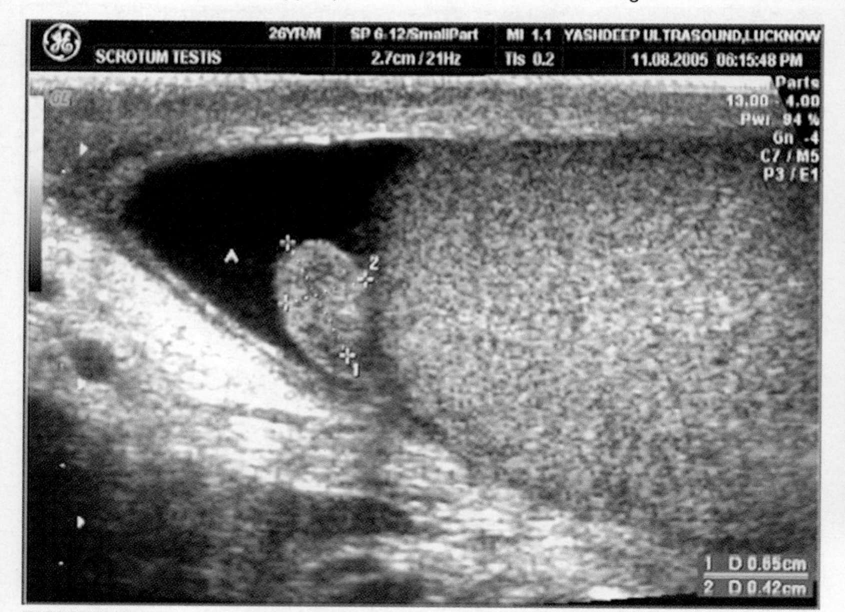

FIGURE 10.2: Appendix testis—HRSG shows a small well-defined Iso echogenic nodular shadow fixed with upper pole of the testis is appendix testis. Small amount of fluid is seen in the scrotal sac and the appendix is seen clearly in fluid background.

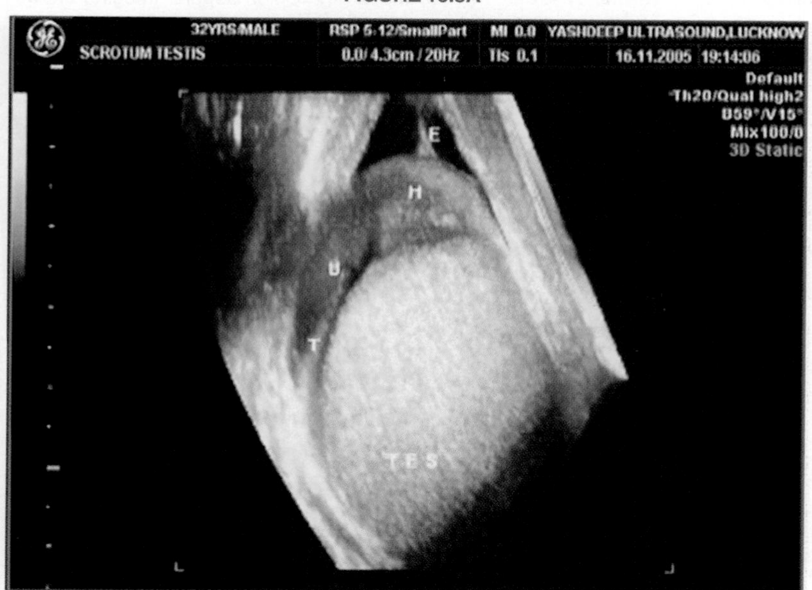

FIGURE 10.3B

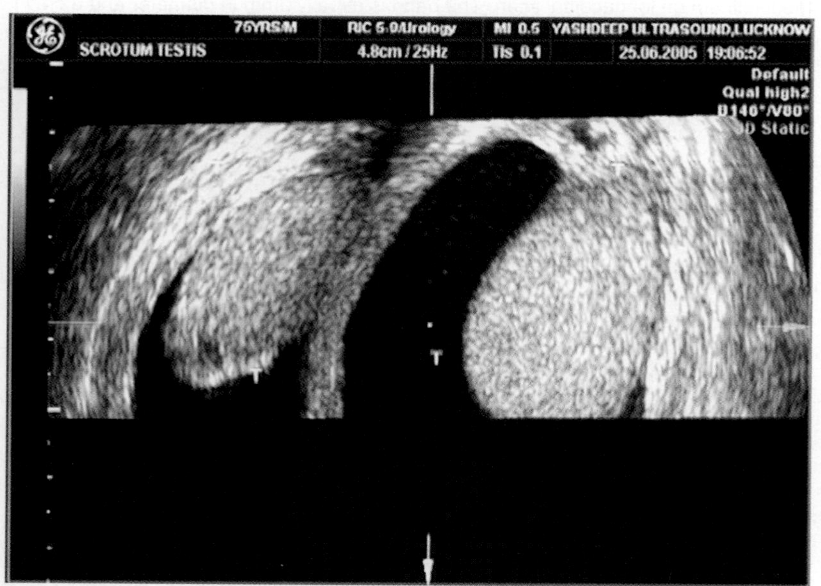

FIGURE 10.3C

FIGURES 10.3A to 3C: 3D imaging of the normal testis: 3D surface rendering of the testis clearly shows the testis. Head of epididymis body and tail of the epididymis are well appreciated on 3D imaging. 3D view of the scrotum shows both testis the raphe is seen dividing the scrotal cavity into two. Each testis is well visualized in each scrotal sac.

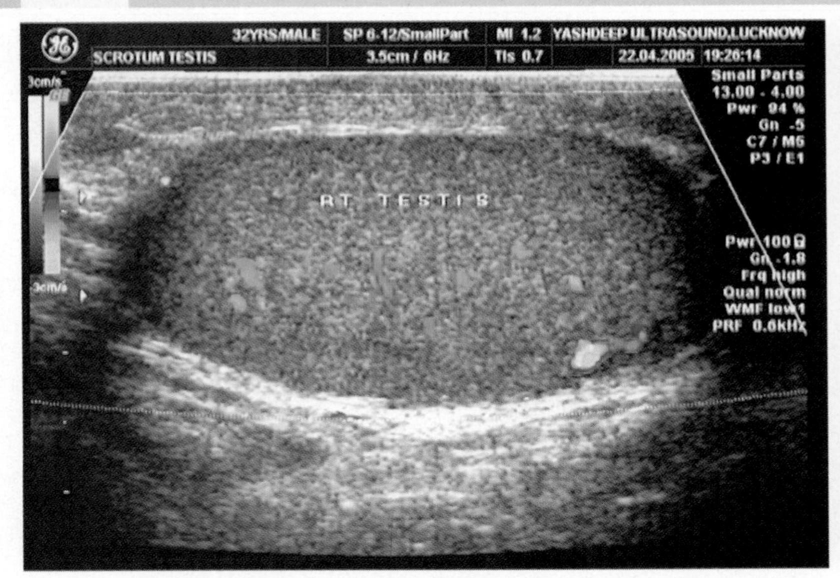

FIGURE 10.4: Normal vascular flow of the testis. Color Doppler imaging shows normal capsular testicular flow.

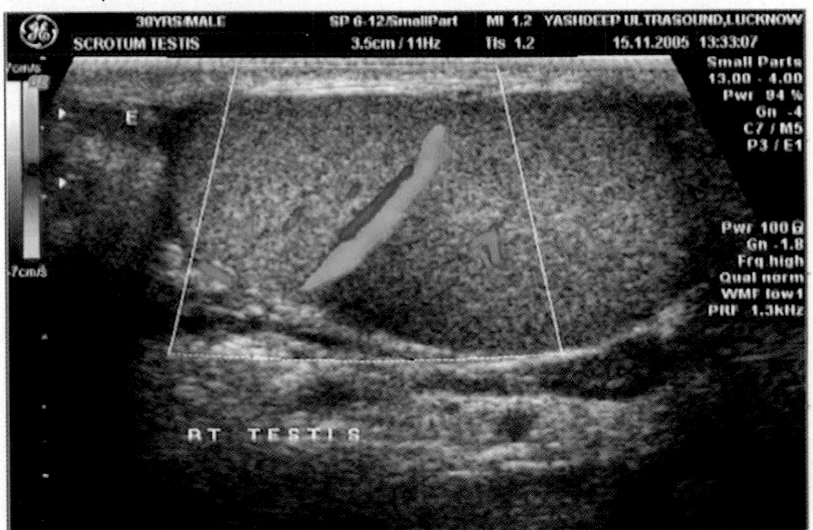

FIGURE 10.5: Vascular Band. HRSG color flow imaging shows well-defined intra testicular vascular band traversing through the testis. It is almost perpendicular to the mediastinum on longitudinal scan. The band is usually 3mm in diameter and 3 cm in length. Color Doppler imaging shows low arterial flow in the band.

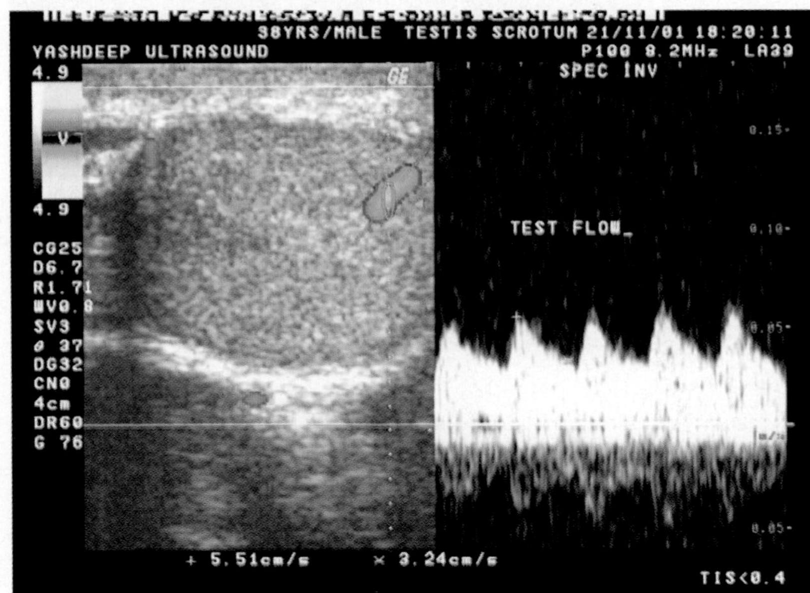

FIGURE 10.6: Normal testicular arterial flow. High Resolution color flow imaging shows low resistance flow in the testicular artery with prolonged diastolic phase characteristic of low flow waveform.

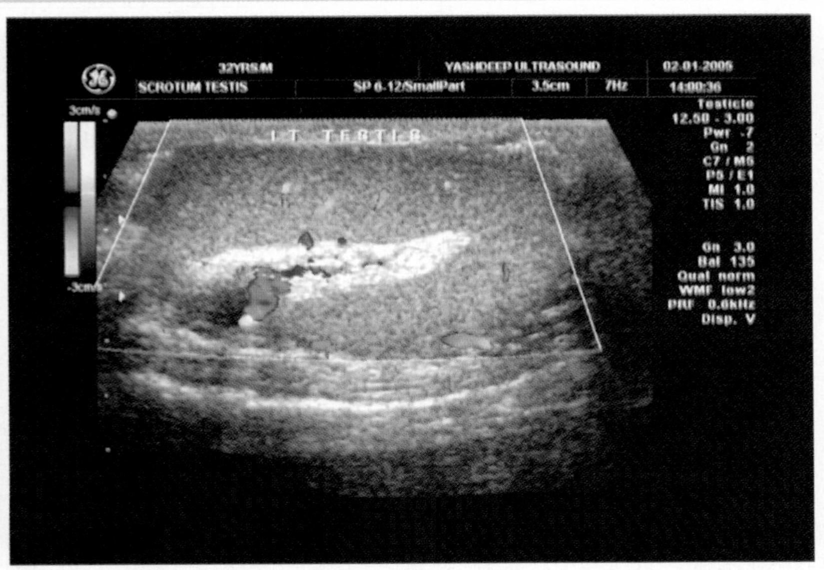

FIGURE 10.7: Normal mediastinum testis. Longitudinal image shows an echogenic band running in the substance of the testis is mediastinum testis.

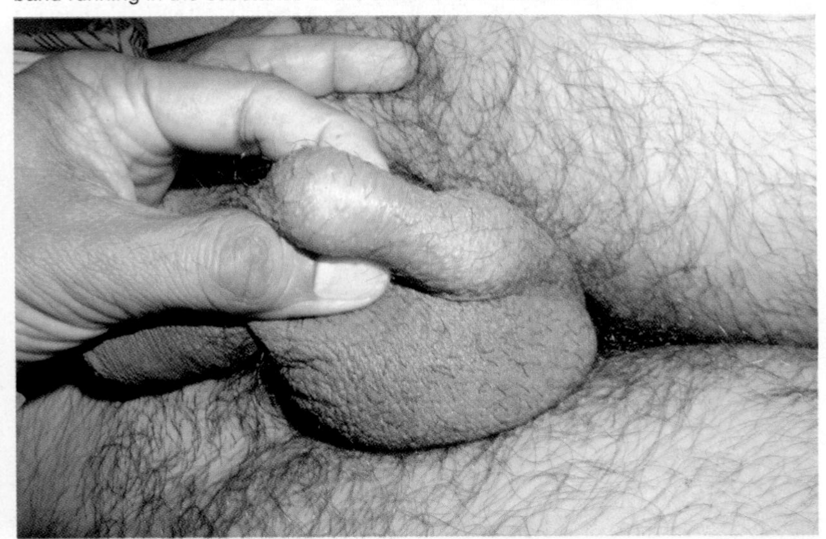

FIGURE 10.8: Triorchatism (Supernumerary Testis). A young patient presented with palpable lump in the Lt scrotum. Clinical examination shows lobulated mass at the upper pole of the Lt testis.

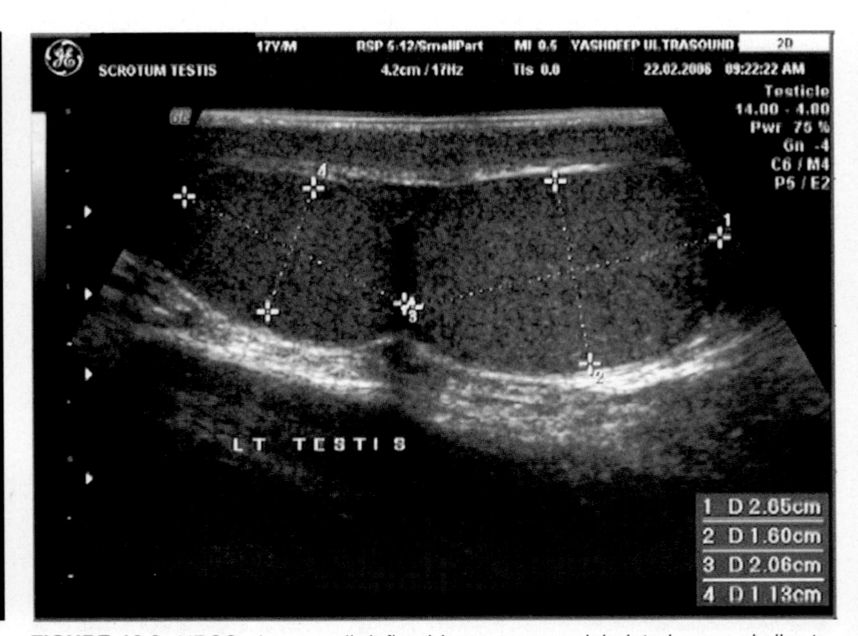

FIGURE 10.9: HRSG shows well-defined homogeneous lobulated mass similar to testicular texture in relation to the Lt testis. It appears to be another testis in the Lt scrotum suggestive of supernumerary testis.

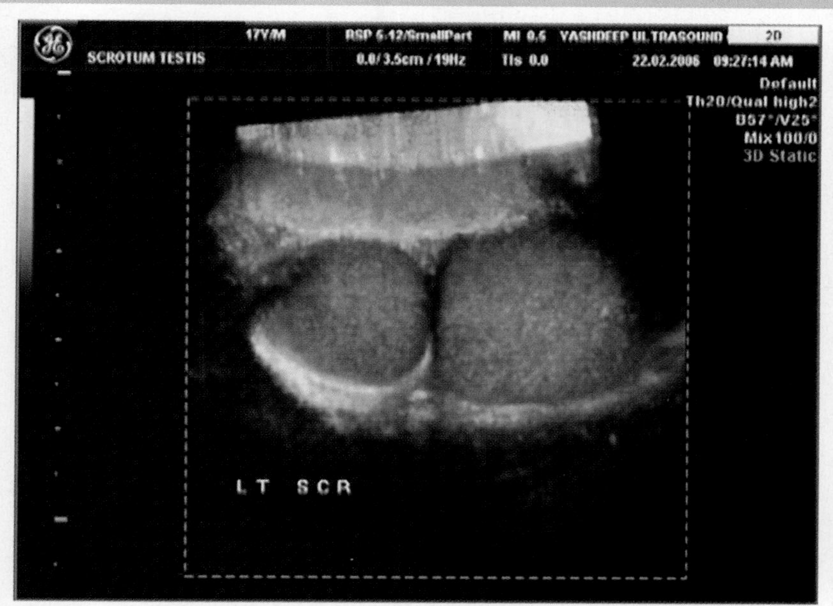

FIGURE 10.10: **3D surfaces rendering** imaging shows the supernumerary testis and the normal testis with better sonographic details. The testis shows the same texture. A single epididymis is seen in the Lt scrotum.

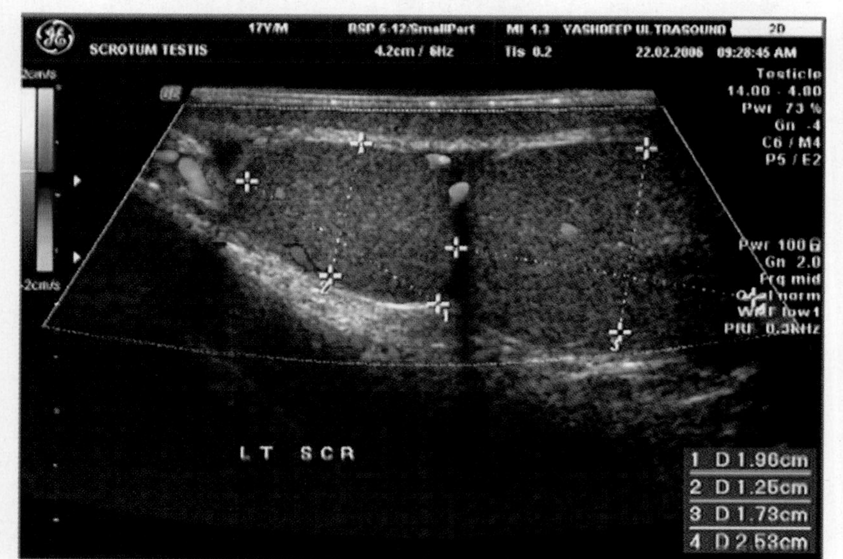

FIGURE 10.11: **On color doppler imaging** normal flow is seen in the normal testis and supernumerary testis. Normal intratesticular flow is seen.

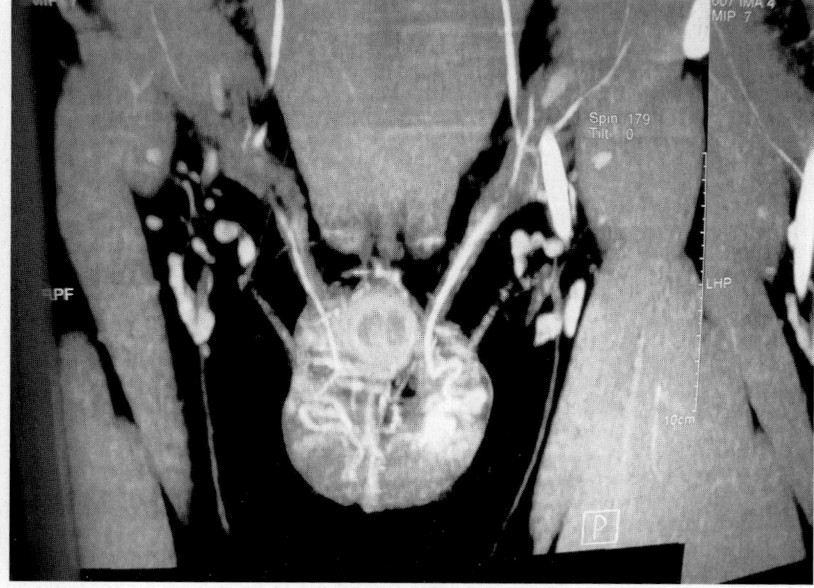

FIGURE 10.12

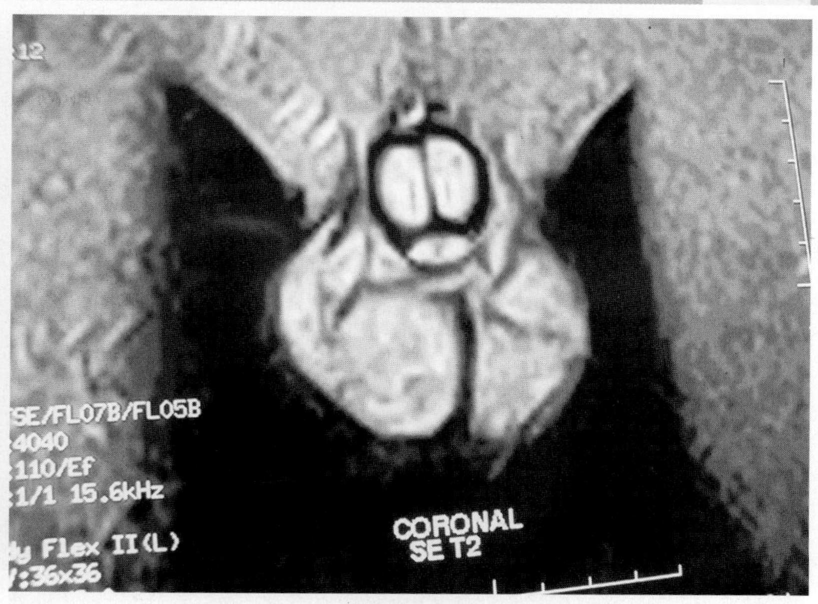

FIGURE 10.13

FIGURES 10.12 and 10.13: **CT Angio and MRI** of the same case shows single testicular artery supplying the both testis. MRI image shows normal signals in both the testis confirming the sonographic findings. Triorchatism is a rare phenomenon and only hundred cases are reported in world literature. Modern imaging techniques confirmed the Triorchatism without any interventional procedures.

Congenital Anomalies

HRSG is an excellent modality in evaluation of developmental anomalies of testis and scrotum. Many congenital conditions like Ambiguous genitalia, hermaphrodites and undescended testis can be easily evaluated on HRSG.

Undescended Testis

Undescended testis is a congenital anomaly as a result of abnormal descent during antenatal life. It is found in 4% of all full term infants. It has been noted that the incidence of undescended testis is increasing. The testis may be seen to be mobile entering the scrotum and retracting. They may be in the inguinal canal, superficial inguinal region or in the pelvis. Chances of infertility and malignancy increase significantly in these patients. Orchipexy in the early years reduces the risk of infertility; the risk of a malignant change is not eliminated.

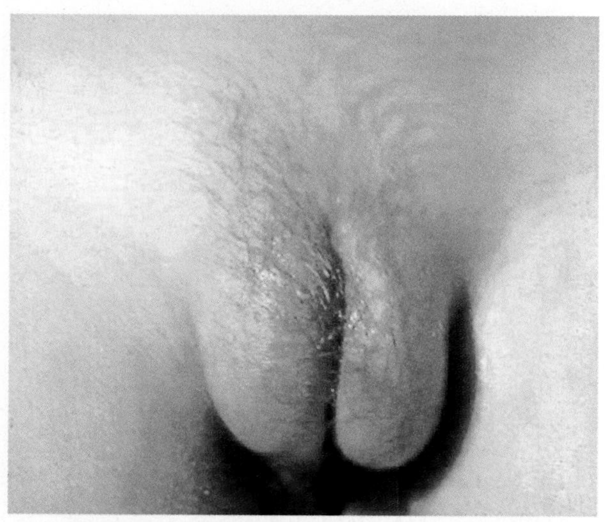

FIGURE 10.14: **Labial testis.** A young child was born with bilateral labial masses. No phallus was seen.

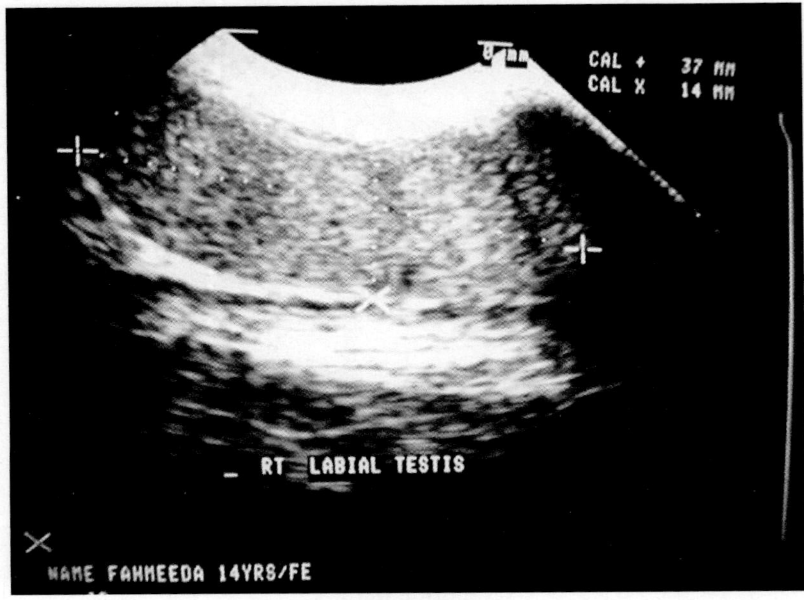

FIGURE 10.15

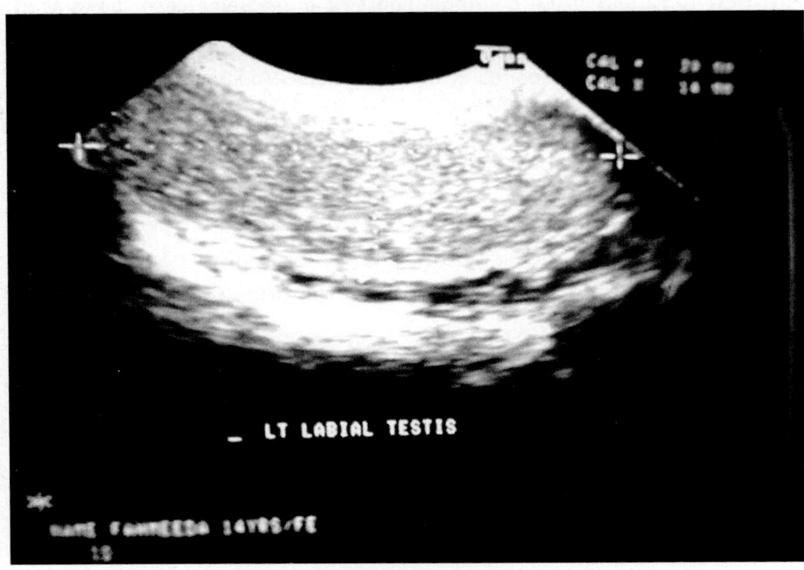

FIGURE 10.16

FIGURES 10.15 and 10.16: HRSG shows well-defined homogeneous ovoid shadows in Rt and Lt labia with fine granular texture similar to the testicular shadows suggestive of bilateral labial testis. However, the patient was not having any uterine or ovarian shadow.

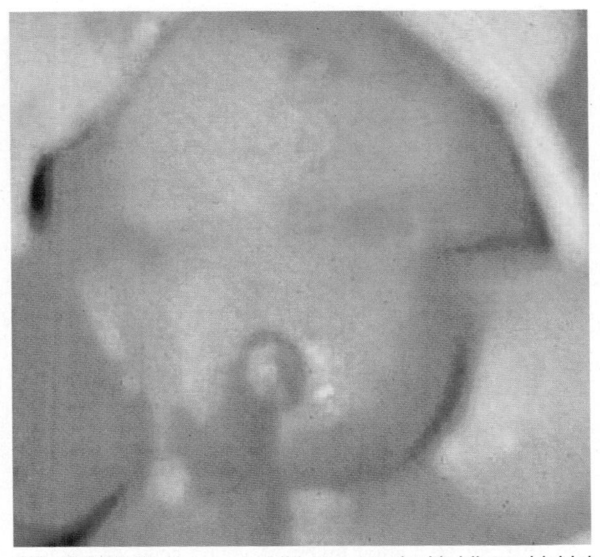

FIGURE 10.17: A young child presented with bilateral labial masses. Hypertrophied clitoris is also seen.

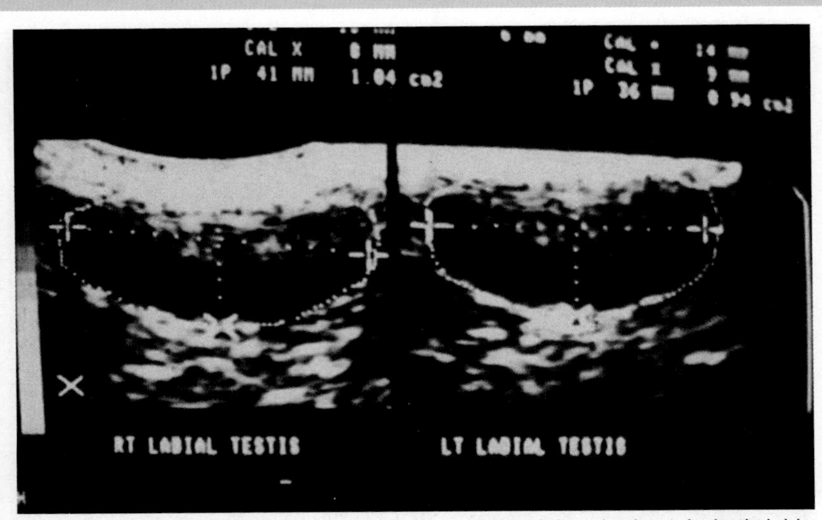

FIGURE 10.18: HRSG shows hypoechoic ovoid nodular shadows in both labia suggestive of labial testis. No uterine or ovarian shadow seen in the pelvis.

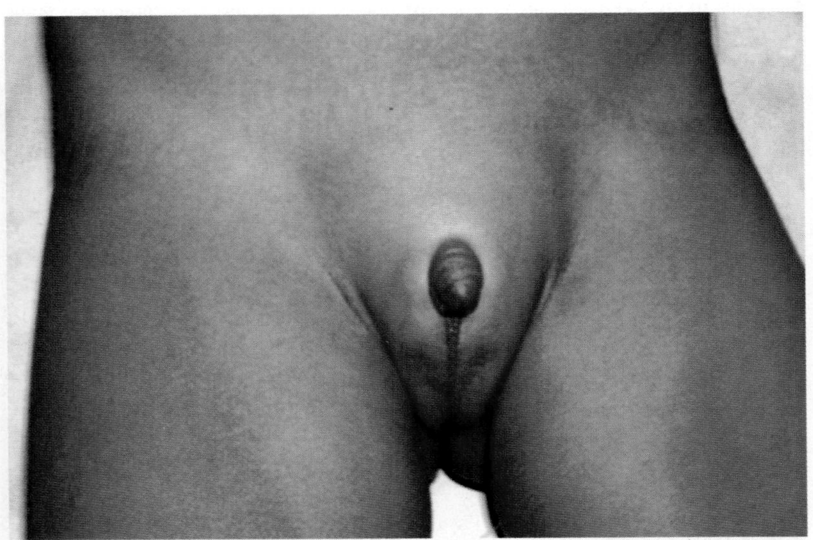

FIGURE 10.19: Hydrometro colpos. A young child presented with rudimentary penis.

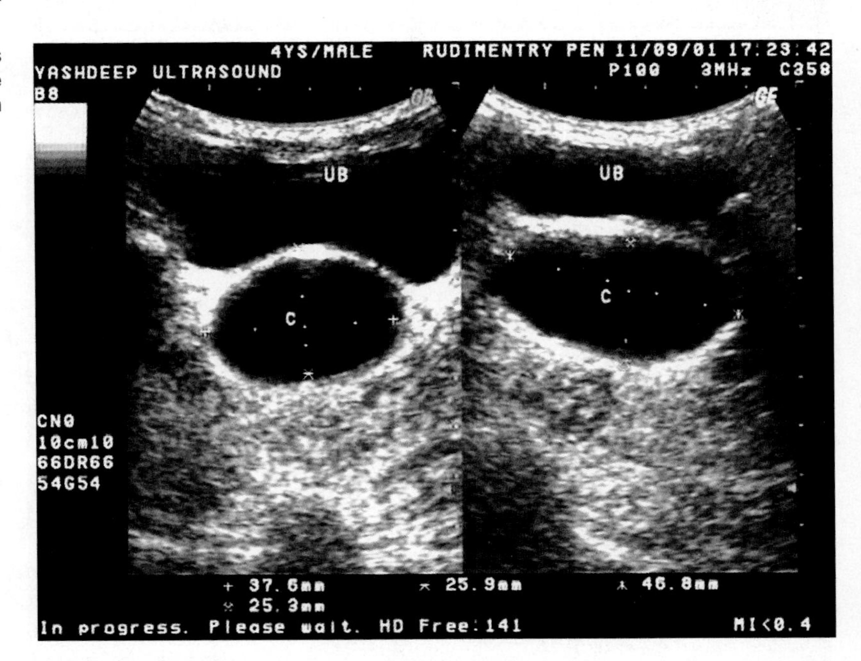

FIGURE 10.20: HRSG shows a cystic mass in the pelvis. However, ovarian shadows are not seen. The cystic mass turned out to be a hydrometrocolpos on laparoscopy.

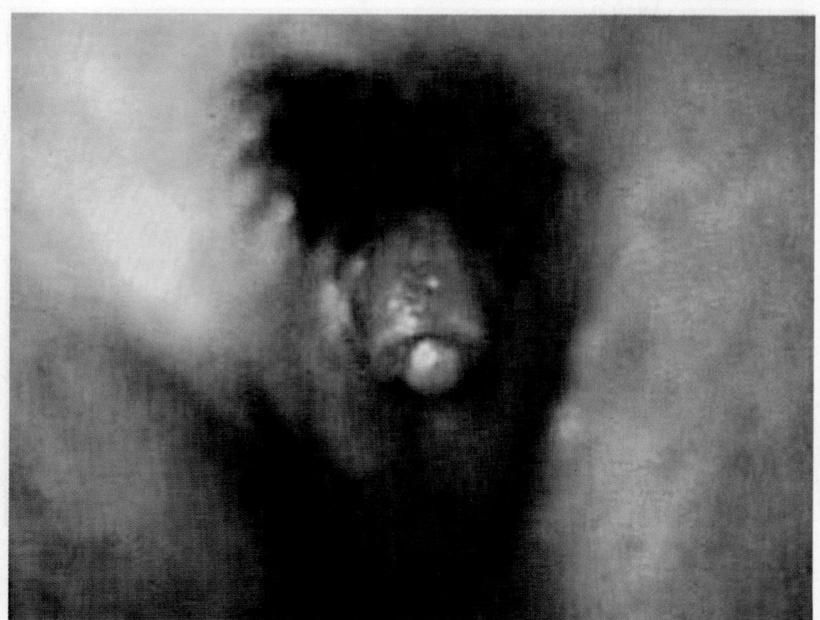

FIGURE 10.21: **Hermaphrodite.** A teen aged patient presented with scrotal shadows with small penile shadow. The scrotum was empty on either side. No testicular shadow was seen in the scrotum.

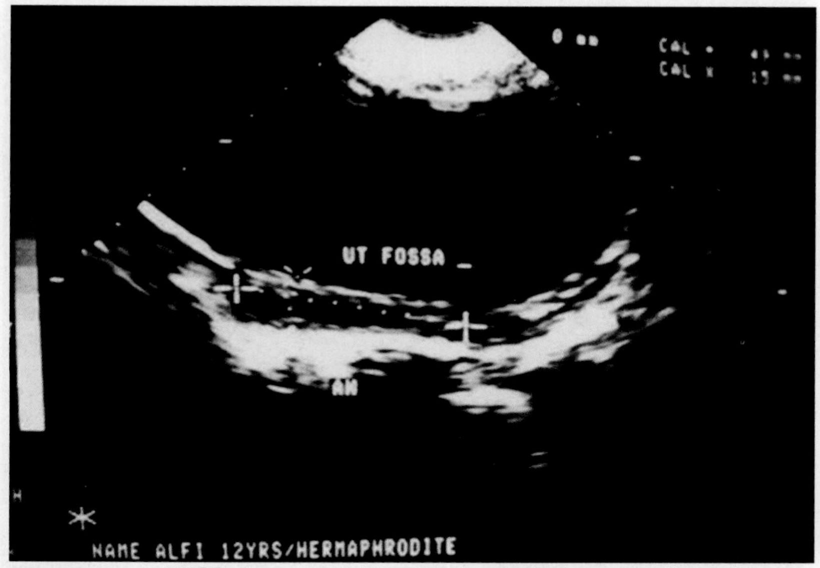

FIGURE 10.22: HRSG shows a small uterine shadow in the uterine fossa. Lt ovary was also seen anterior to the uterus. Rt ovary was not seen suggestive of hermaphrodite.

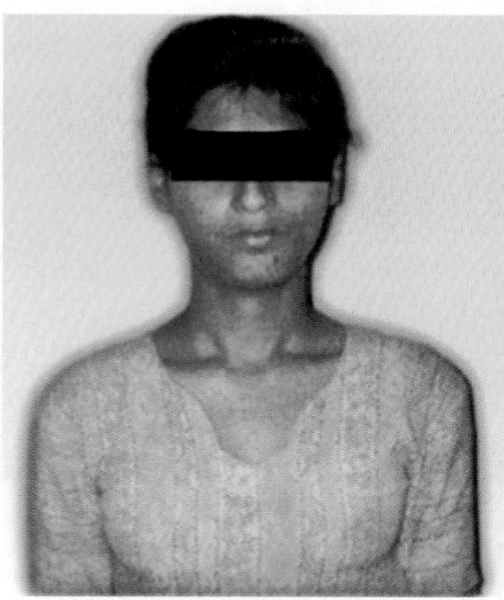

FIGURE 10.23

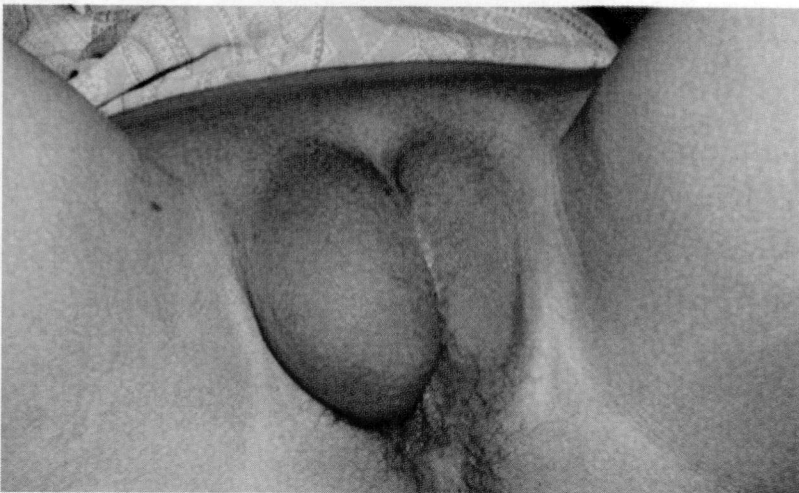

FIGURE 10.24

FIGURES 10.23 and 10.24: **True hermaphrodite.** A young girl presented with bilateral labial masses. The serum testosterone was very high. Then girl was having well-developed breast.

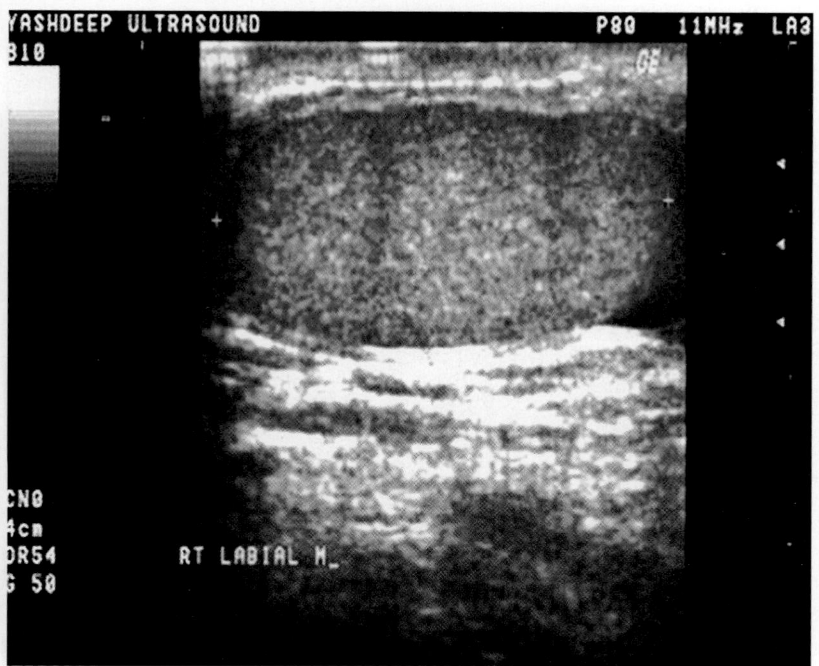

FIGURE 10.25

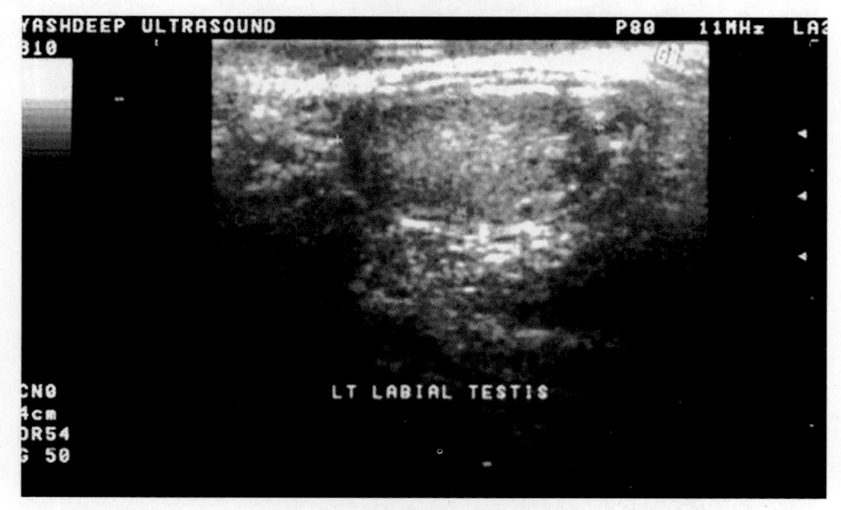

FIGURE 10.26

FIGURES 10.25 and 10.26: HRSG shows well-defined homogeneous ovoid masses in both labia. They show fine granular texture similar to the testis. Biopsy of the mass confirmed labial testis. After surgery of the testis the serum testosterone came to normal range.

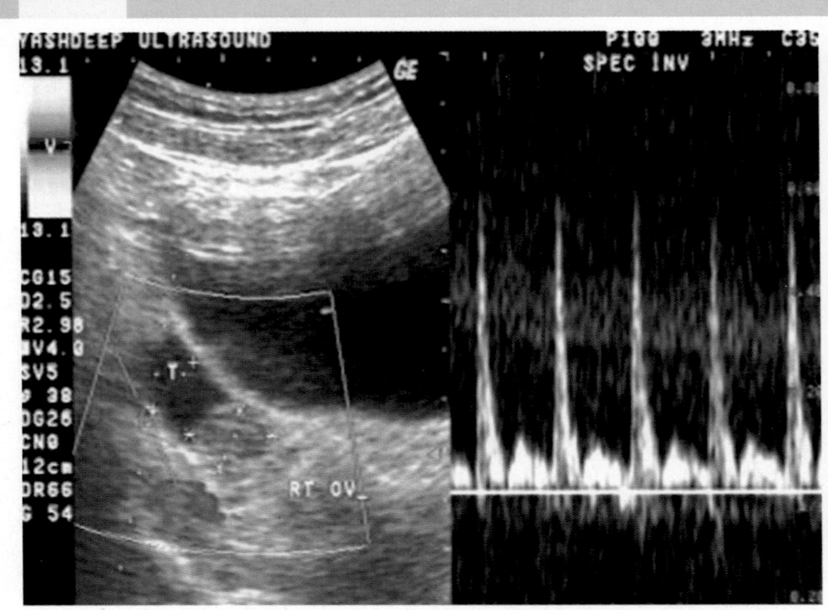

FIGURE 10.27: The patient was also having well-developed ovaries. Therefore it was true hermaphrodite.

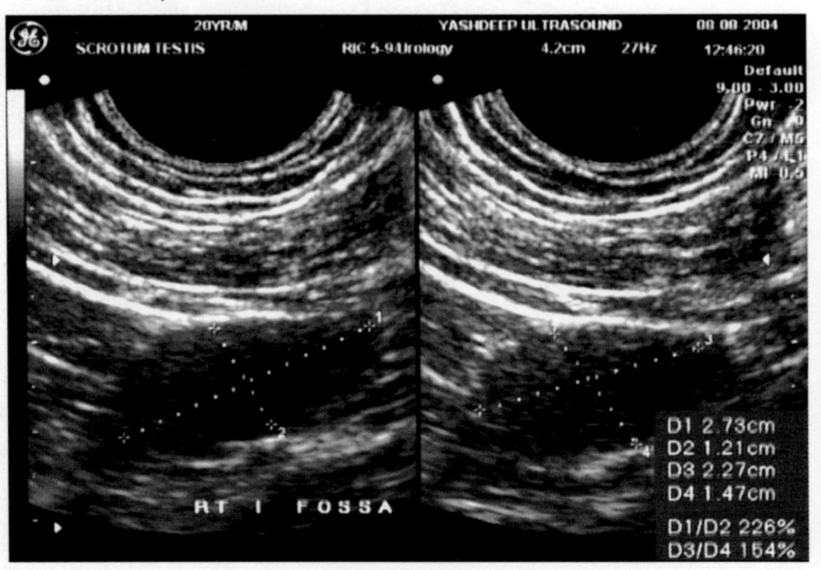

FIGURE 10.28

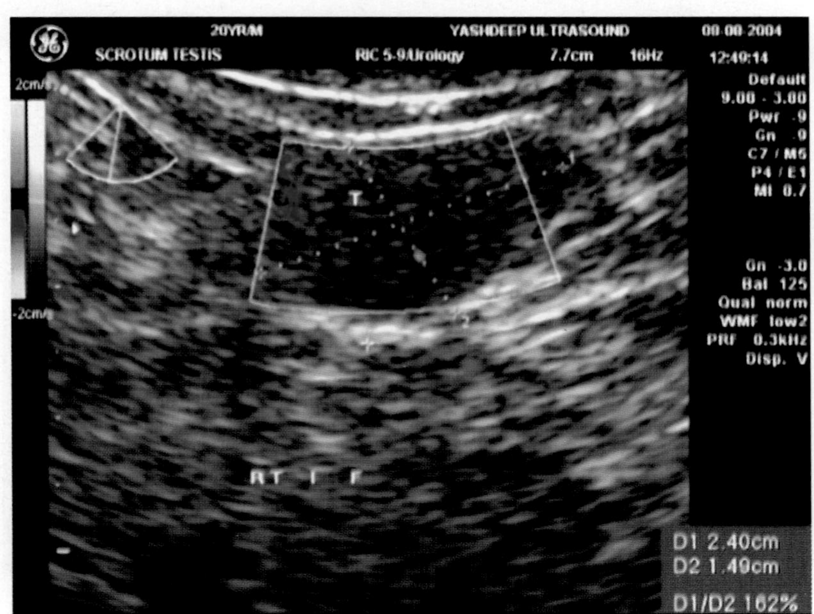

FIGURE 10.29

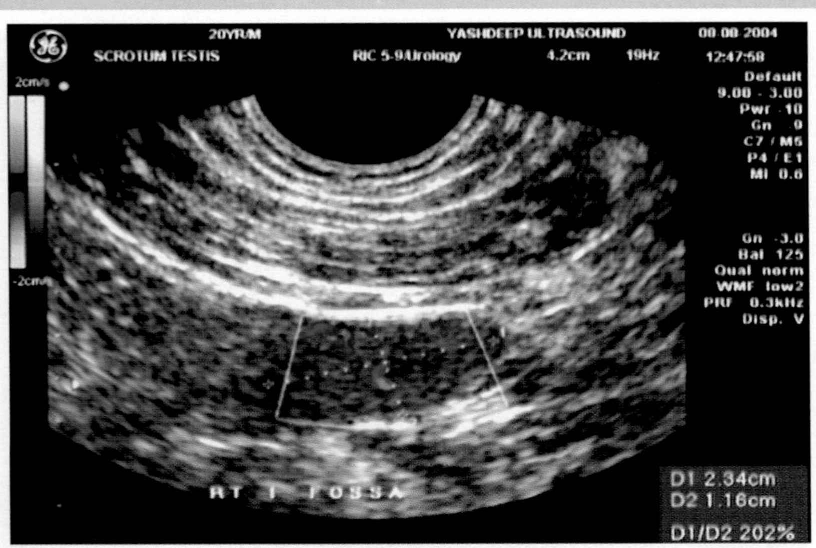

FIGURE 10.30

FIGURES 10.28 to 10.30: Ectopic testis in the Rt iliac fossa. A 20 years old patient presented Rt sided undescended testis. HRSG shows well-defined hypoechoic shadow in Rt iliac fossa. It is homogeneous in texture. On color flow imaging low flow is seen in the testis suggestive of arrest of the testis in the inguinal region.

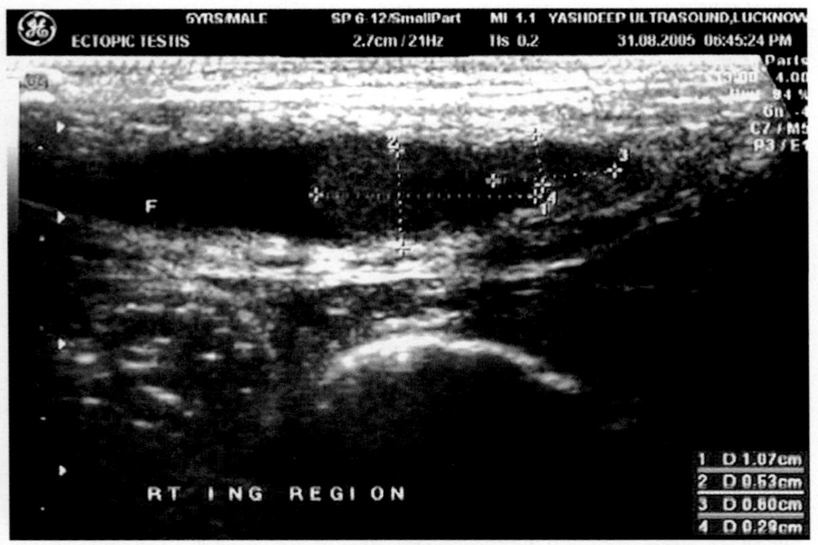

FIGURE 10.31

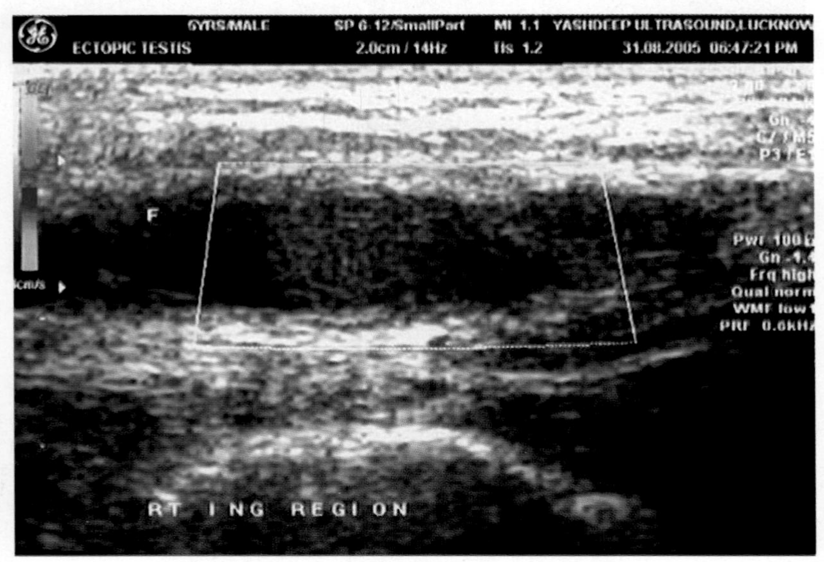

FIGURE 10.32

FIGURES 10.31 and 10.32: Ectopic inguinal testis. A young boy presented with Rt empty scrotum. HRSG shows well-defined hypoechoic ovoid shadow in Rt inguinal canal with fine granular echo pattern. Small amount of fluid pattern is also seen in the inguinal canal. On color flows imaging low flow is seen in the testis.

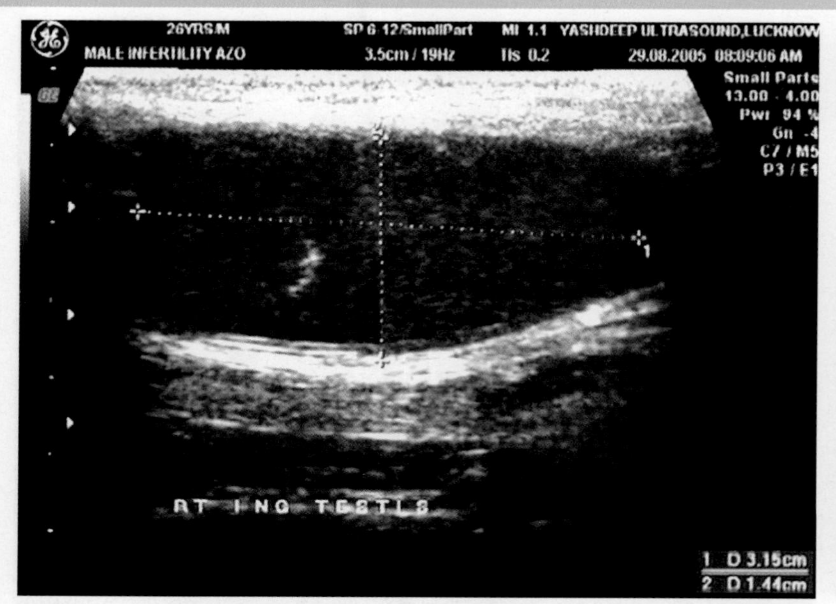

FIGURE 10.33: **Rt inguinal testis.** A 20 years old boy presented with absent Rt testis in the scrotum. HRSG shows well defined hypoechoic nodular shadow at the root of the scrotum suggestive of arrest of the testis in the inguinal canal.

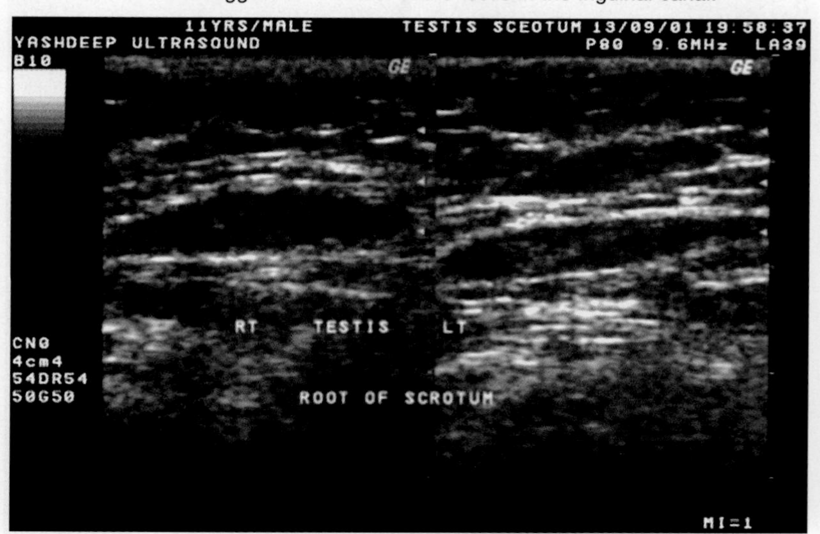

FIGURE 10.34A

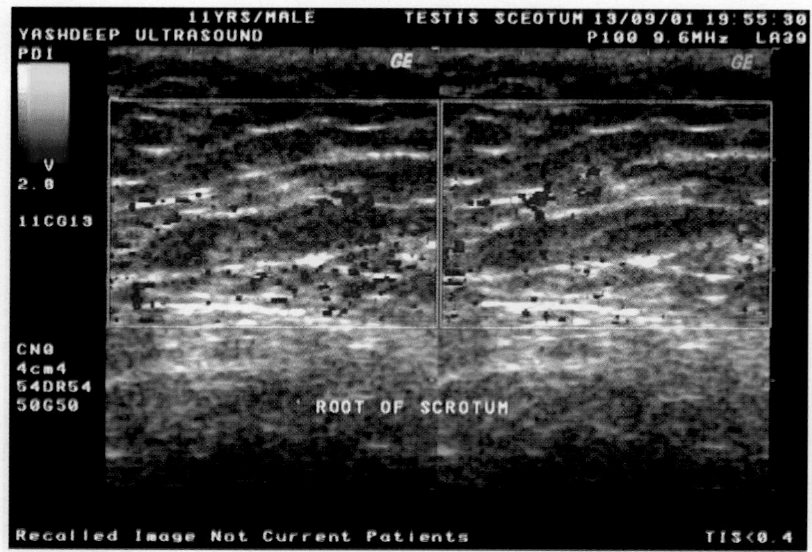

FIGURE 10.34B

FIGURES 10.34A and 10.34B: **Bilateral atrophic inguinal testis.** A 11 years old child presented with congenital absence of scrotal testis on either side. HRSG shows hypoplastic flat testis in both inguinal canal. The testicular echoes are not normal. The testicular size was very small. On color power Doppler imaging very little flow was seen in the testis. Biopsy shows bilateral hypoplastic testis in both inguinal canal.

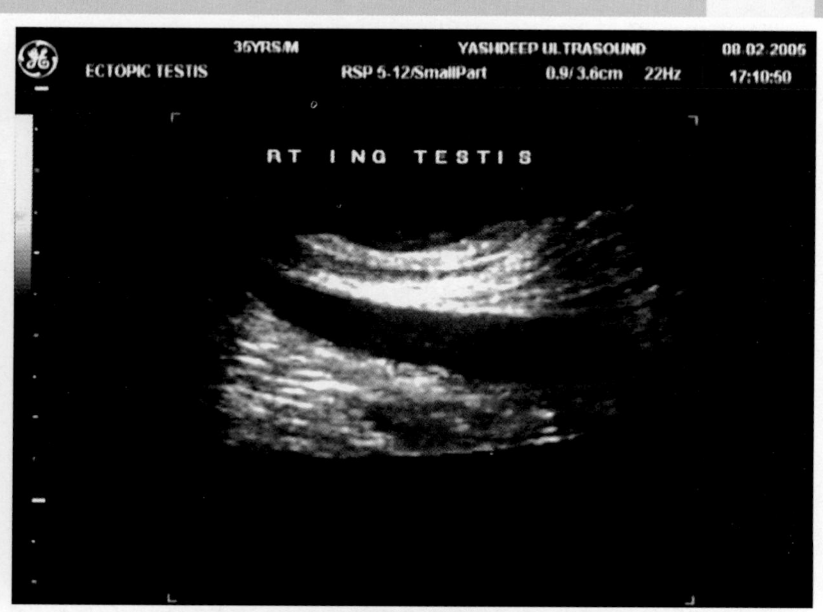

FIGURE 10.35

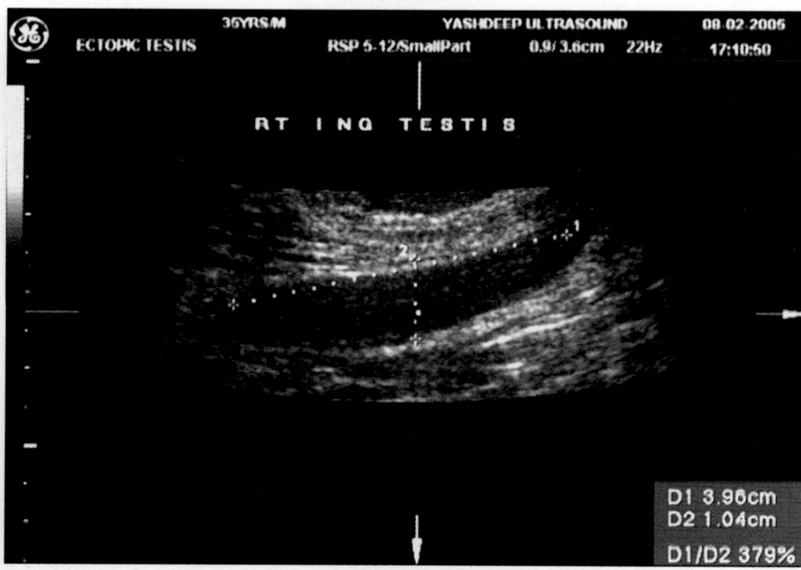

FIGURE 10.36

FIGURES 10.35 and 10.36: **Atrophic inguinal testis with bilateral scrotal testis implants.** A patient presented with bilateral small atrophic inguinal testis. The testis is seen as hypoechoic flat band like shadows in the inguinal canal.

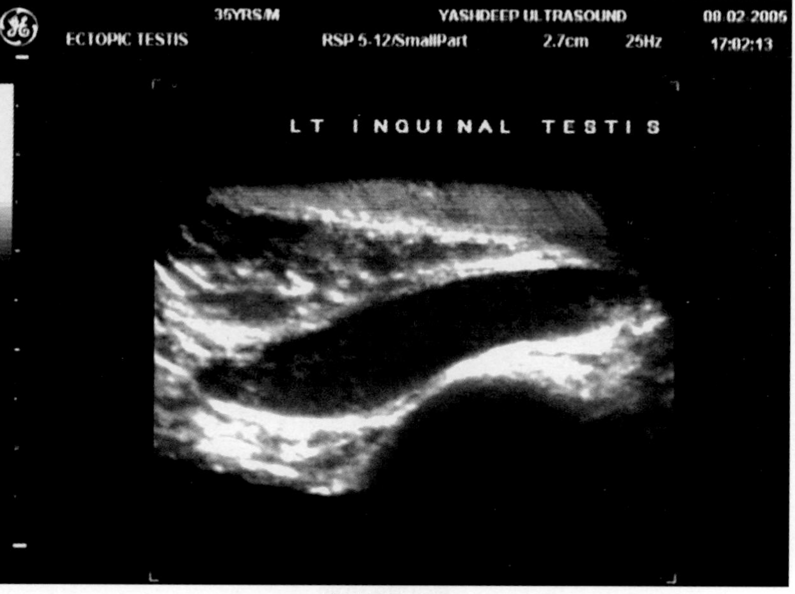

FIGURE 10.37

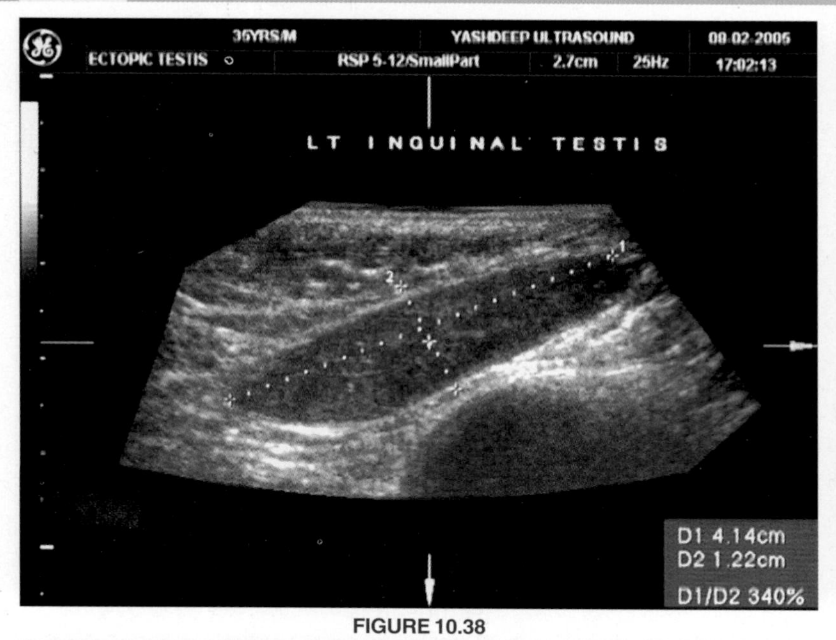

FIGURE 10.38

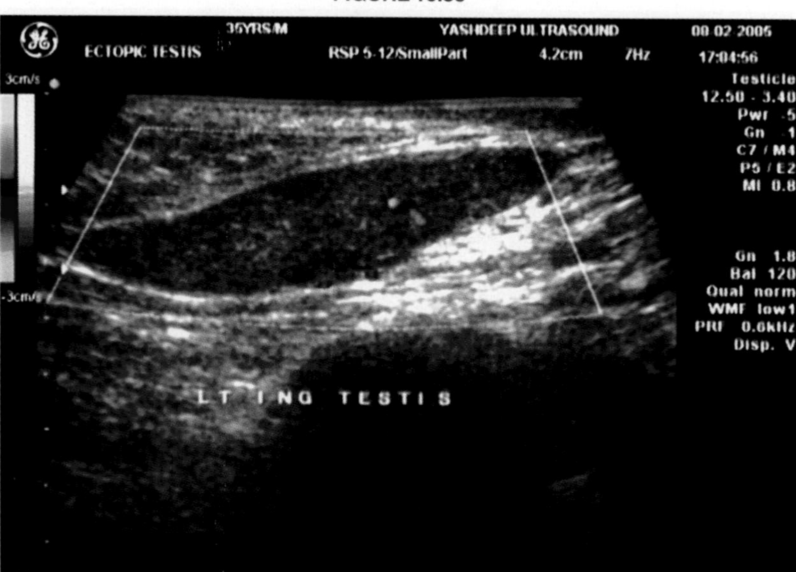

FIGURE 10.39

FIGURES 10.37 to 10.39: 3D imaging of the testis shows the testis in much clinical details. On color Doppler imaging low flow seen in the testis.

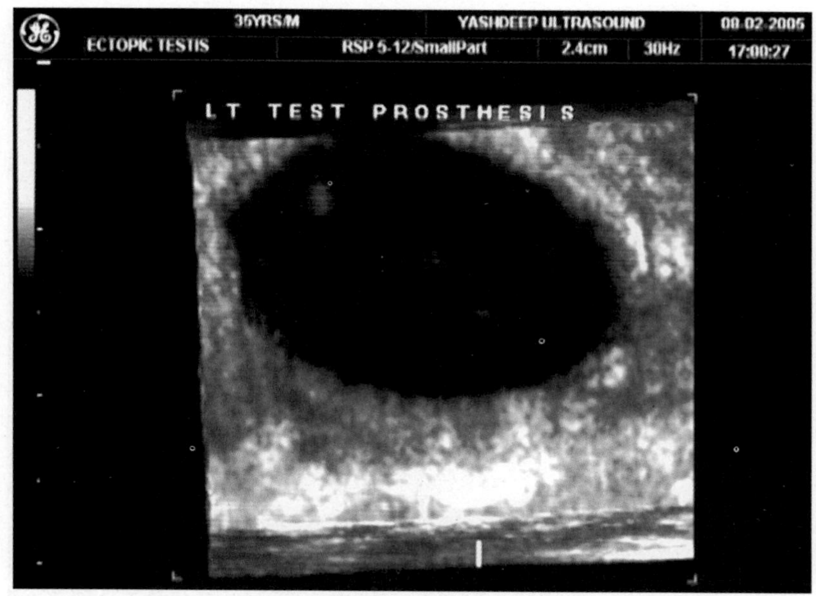

FIGURE 10.40

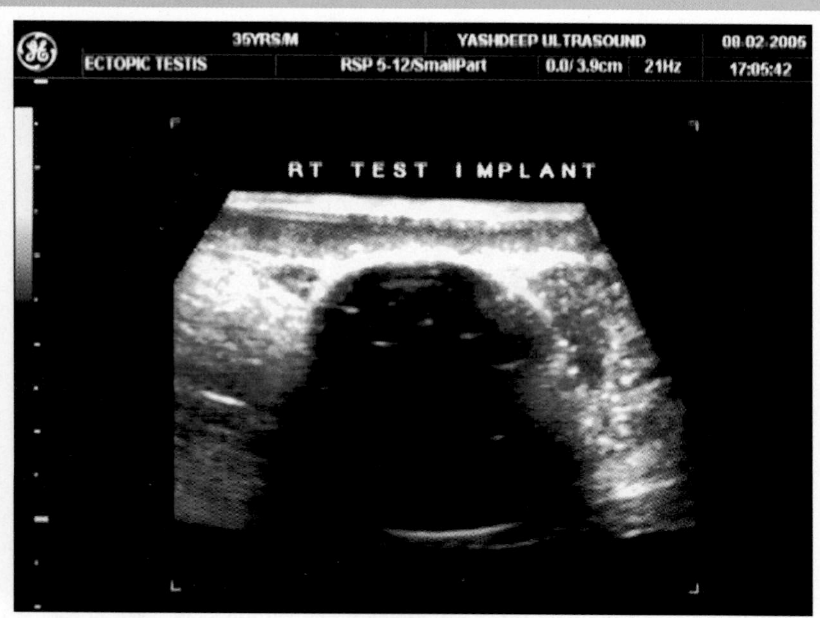

FIGURE 10.41

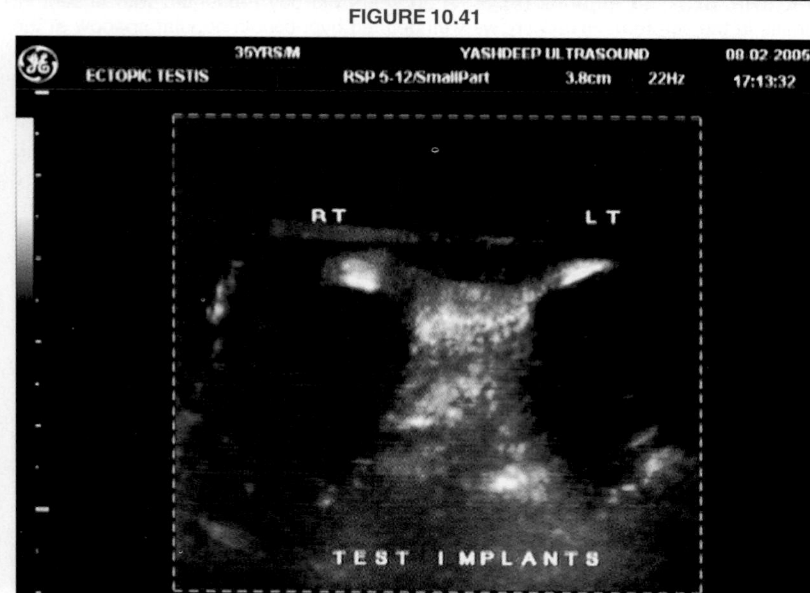

FIGURE 10.42

FIGURES 10.40 to 10.42: 3D imaging of the both scrotum show the testicular implant. They are seen as well defined homogeneous anechoic masses. The implants attenuate the sound beam making them anechoic on HRSG.

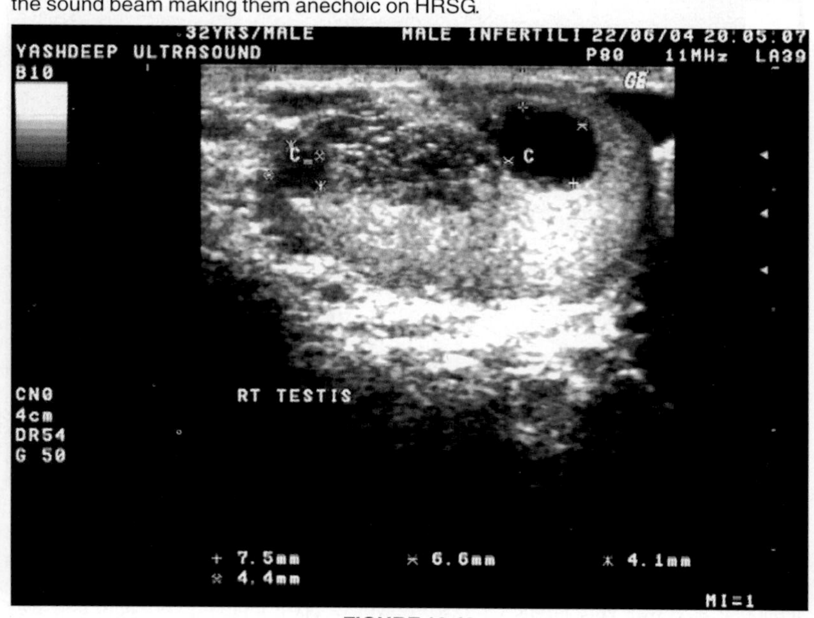

FIGURE 10.43

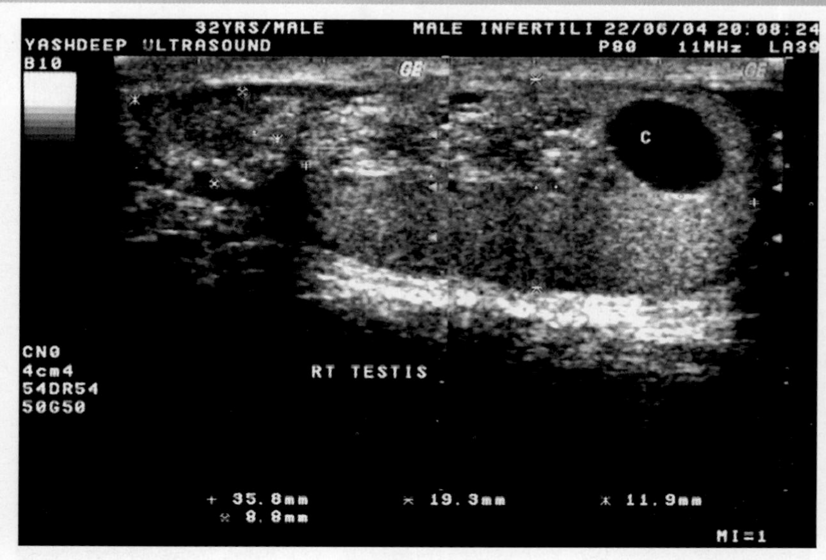

FIGURE 10.44

FIGURES 10.43 and 10.44: Cysts of rete testis. A patient presented with male infertility azospermia. HRSG of testis show multiple cysts running through the mediastinum testis in the substance of the testis. Few medium size cysts are also seen in the testis. They are disperse in the testicular substance suggestive of cysts of rete testis.

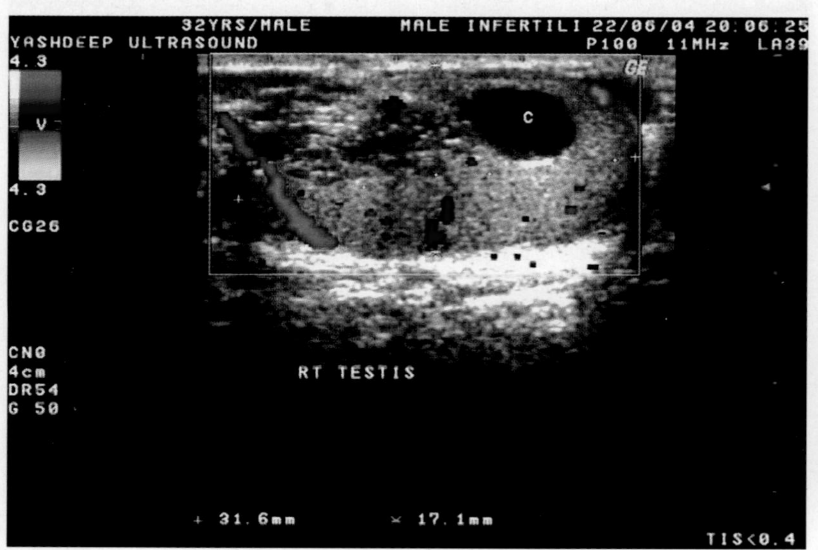

FIGURE 10.45

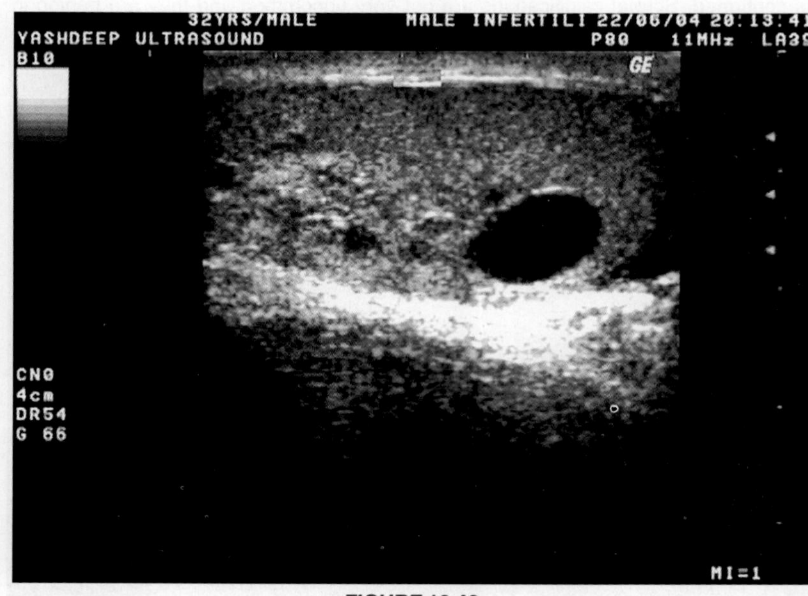

FIGURE 10.46

FIGURES 10.45 and 10.46: On color flow imaging no flow seen in the cysts.

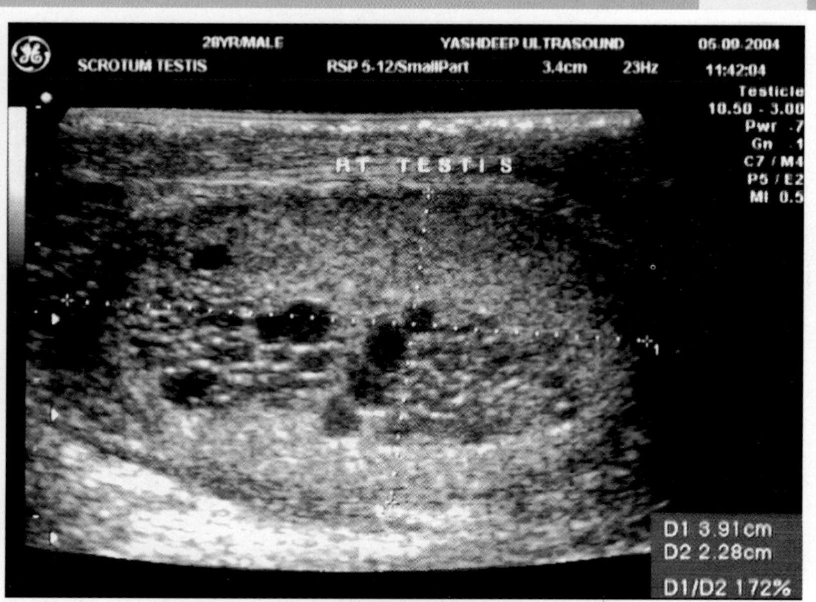

FIGURE 10.47

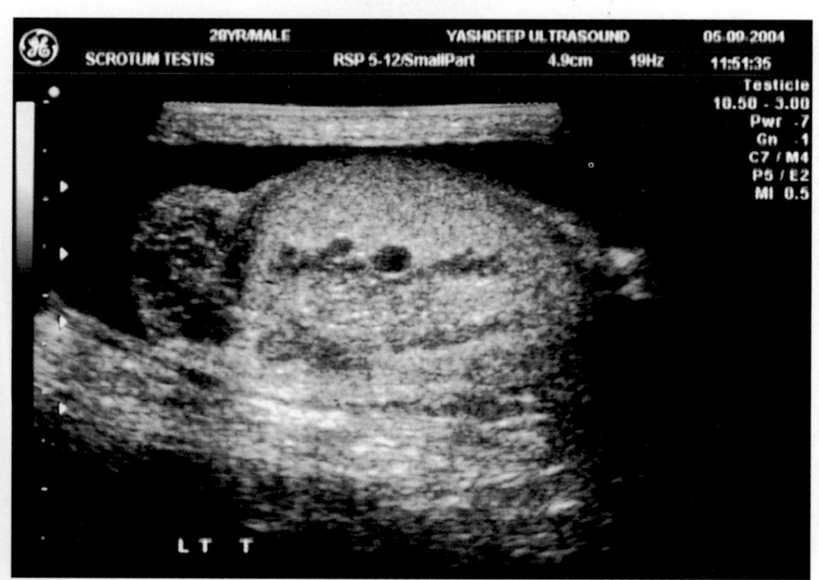

FIGURE 10.48

FIGURES 10.47 and 10.48: Bilateral rete testis cysts. A patient of male infertility shows bilateral cysts in the testis. The cysts are diffuse in distribution suggestive of rete testis cysts.

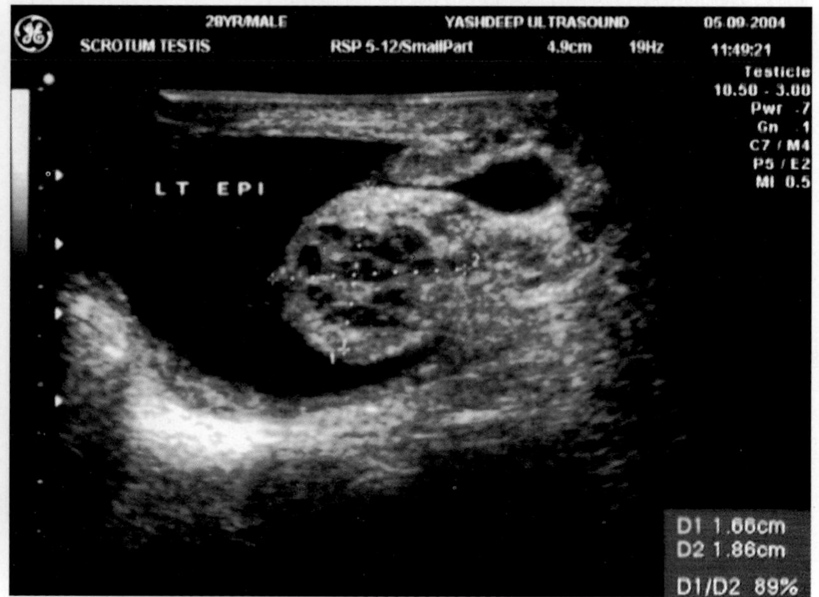

FIGURE 10.49

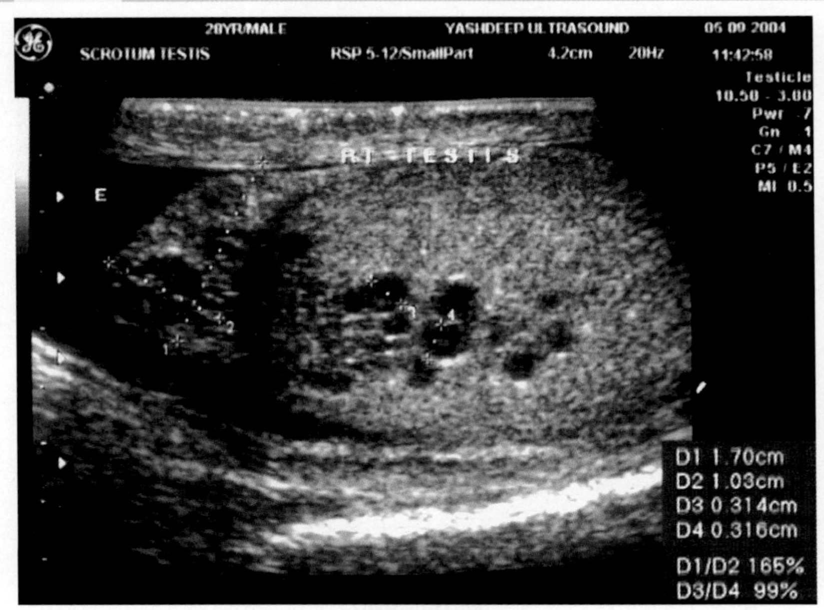

FIGURE 10.50

FIGURES 10.49 and 10.50: The cysts are also seen in the epididymis of both testis.

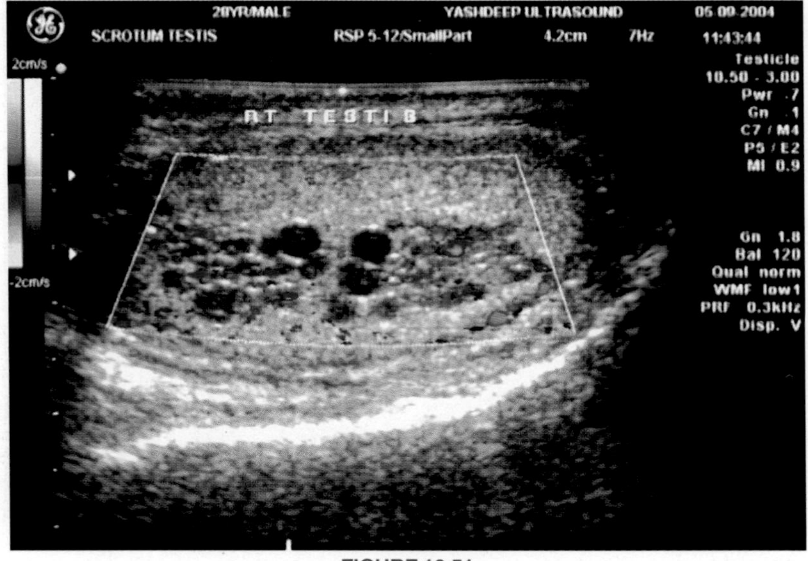

FIGURE 10.51

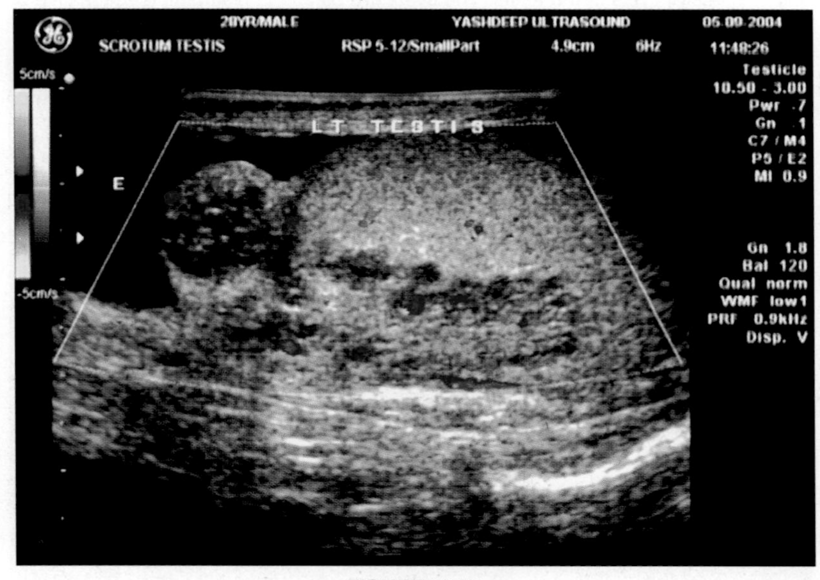

FIGURE 10.52

FIGURES 10.51 and 10.52: On color flow imaging no flow seen in the cyst.

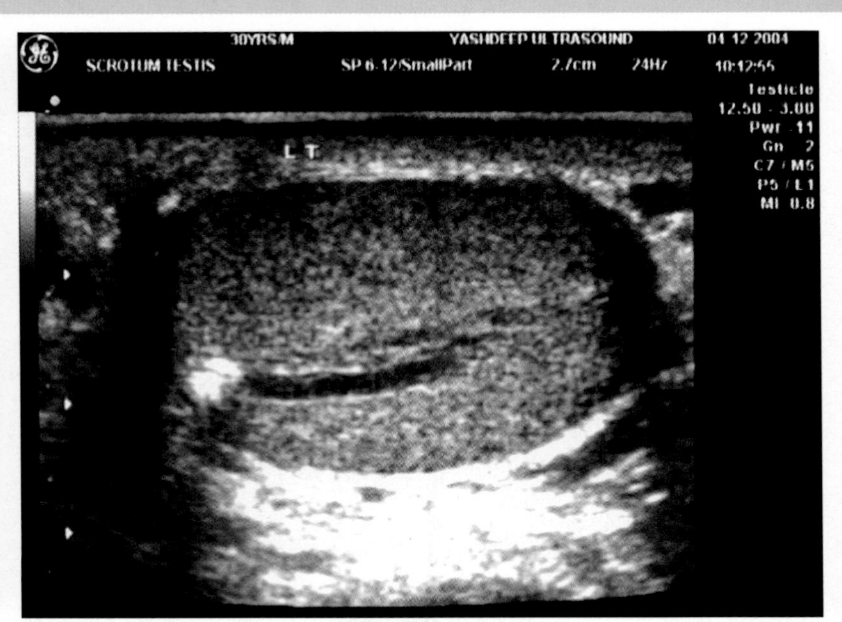

FIGURE 10.53

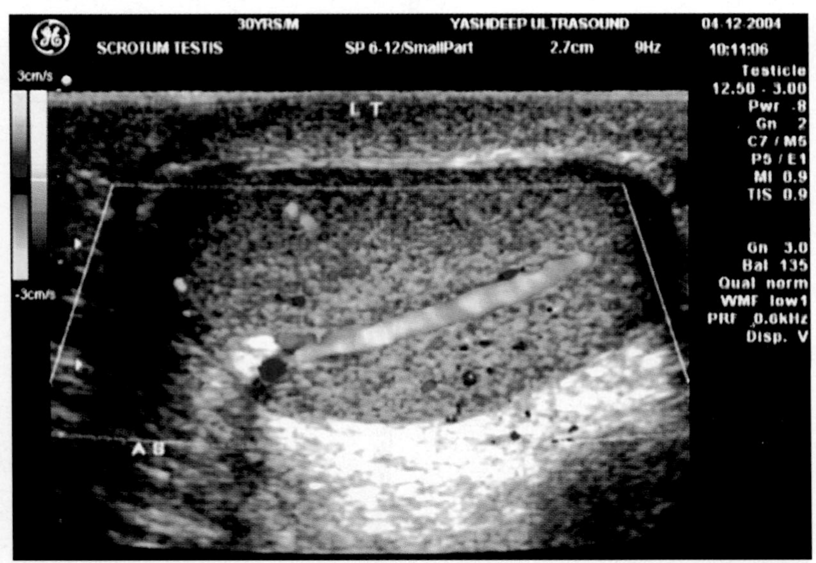

FIGURE 10.54

FIGURES 10.53 and 10.54: Scrotal calcification. HRSG shows well defined calcified shadow in the testis. It is seen along the vascular band.: On color flow imaging vessel is confirmed. Scrotal calcifications are not very uncommon and they are benign in nature.

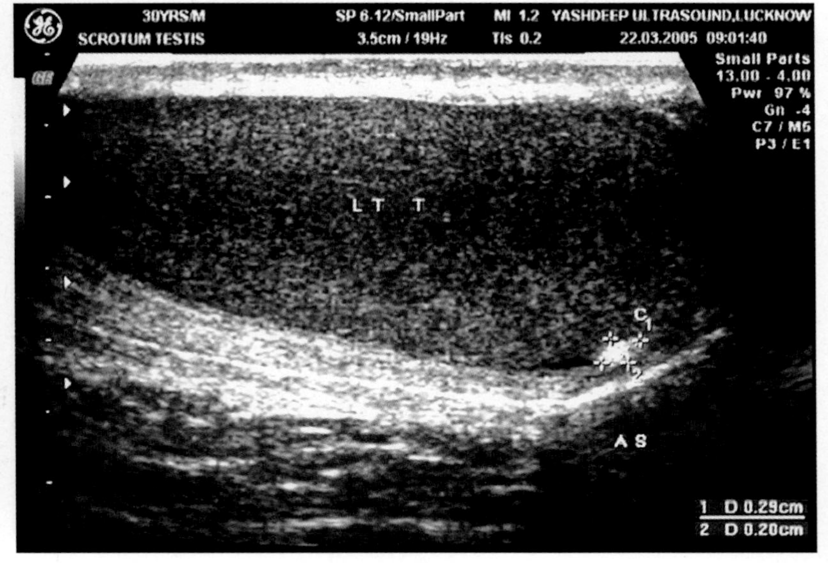

FIGURE 10.55

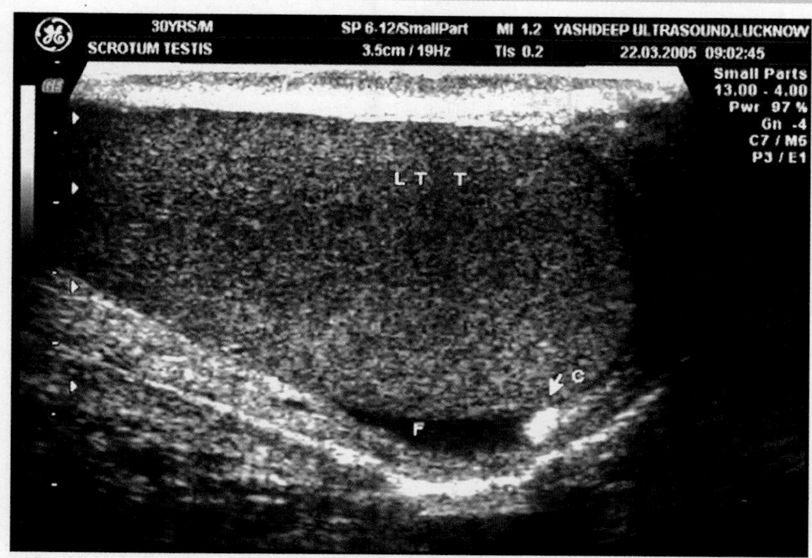

FIGURE 10.56

FIGURES 10.55 and 10.56: Scrotal mouse. Scrotal mouse are calcified nodules, which is seen in tunica albeugenia. They move on palpation that's why known as scrotal mouse. On HRSG they are seen echogenic calcified shadow between tunica and the testis.

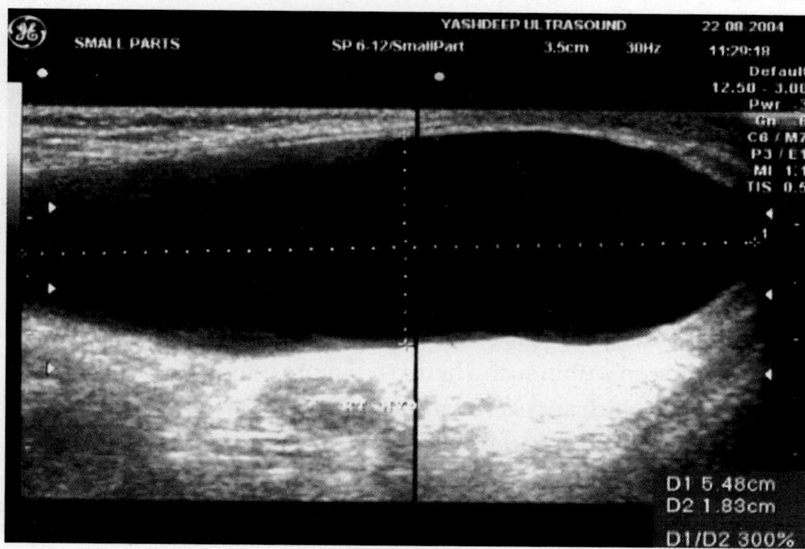

FIGURE 10.59

FIGURES 10.58 and 10.59: Inguinal hydrocele. A boy presented with swelling in the Rt scrotal inguinal region. HRSG shows well defined fluid collection in Rt inguinal canal. No septation is seen. However, the collection is seen only in the inguinal canal and not going down in the scrotum suggestive of inguinal hydrocele.

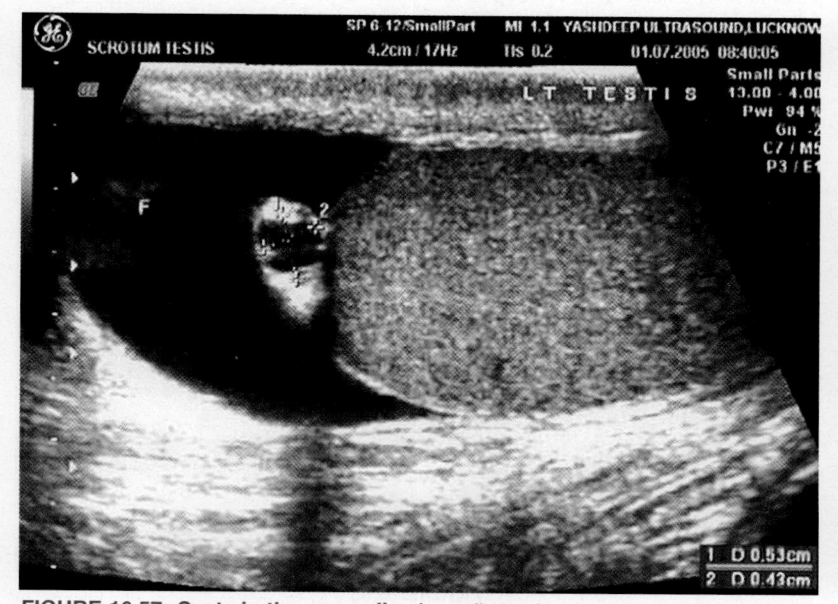

FIGURE 10.57: Cysts in the appendix. A small cyst in appendix testis. The cyst is well defined with sharp margins.

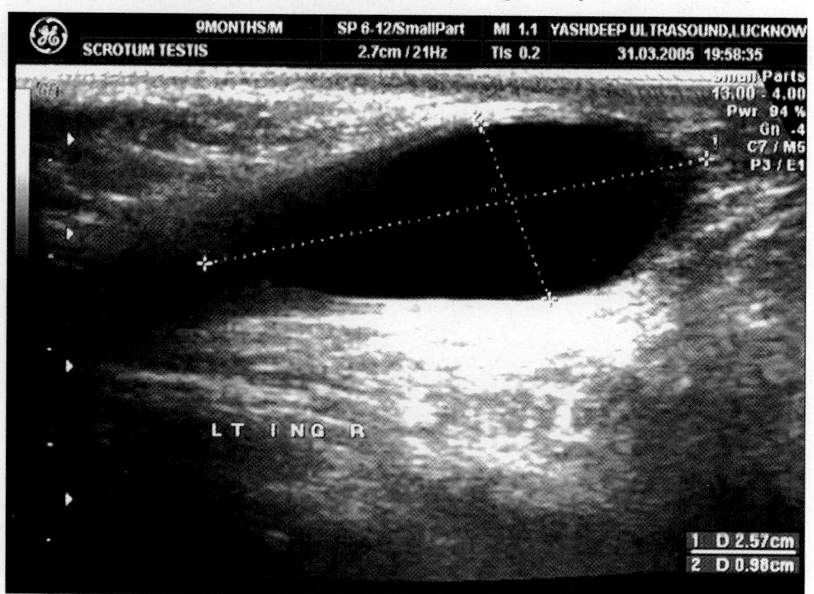

FIGURE 10.60

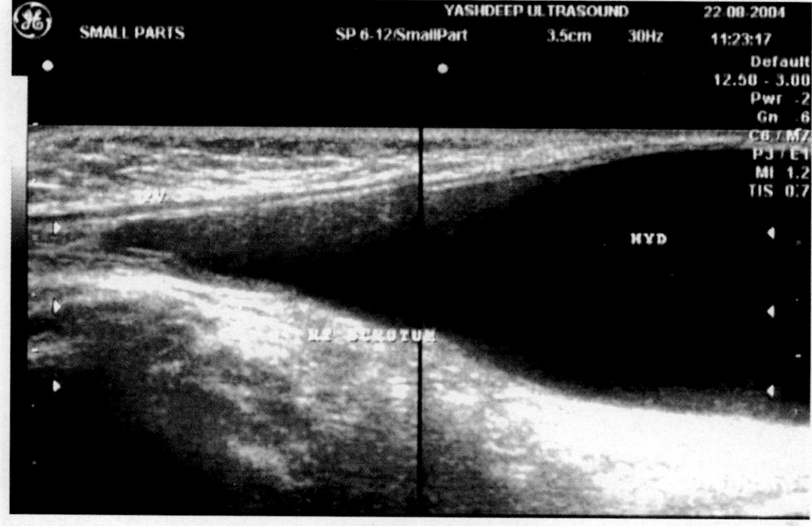

FIGURE 10.58

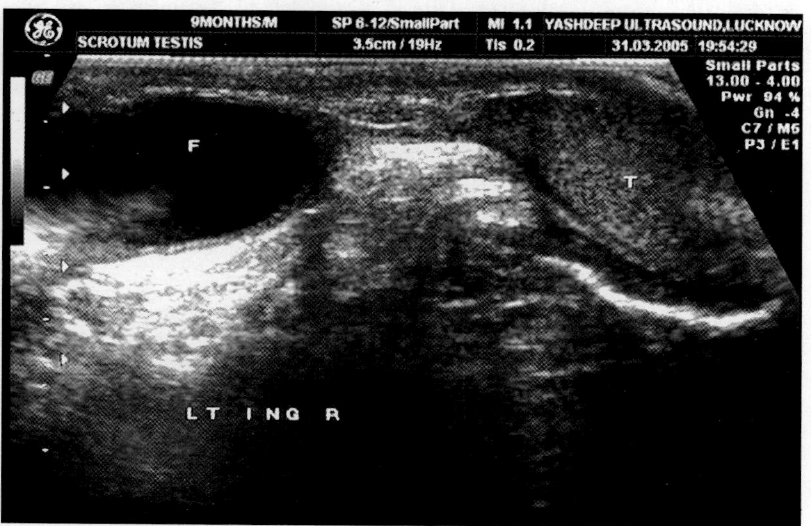

FIGURE 10.61

FIGURES 10.60 and 10.61: Inguinal hydrocele. HRSG shows well defined fluid collection in the inguinal canal. The collection is confined to the inguinal canal. No suggestion of any septation or loculation is seen in it. The Rt testis is well visualized and it is normal in appearance.

In sonography, the testis is homogeneous and less echogenic than the normal testis. The Epididymis is not well identified. Color flow imaging reveals poor vascularity. The testis is atrophied or often not identified in adults.

Scrotal Hernia

Majority of scrotal hernias are correctly diagnosed by clinical examination. However, if the hernia is incarcerated and indurated, the patient may be referred for an ultrasonographic study. Rarely bowel loop will be seen within such a hernia but more commonly only omental fat is present, which appears as an amorphous mass (not corresponding to any recognizable testicular structure) in a scrotal sac.

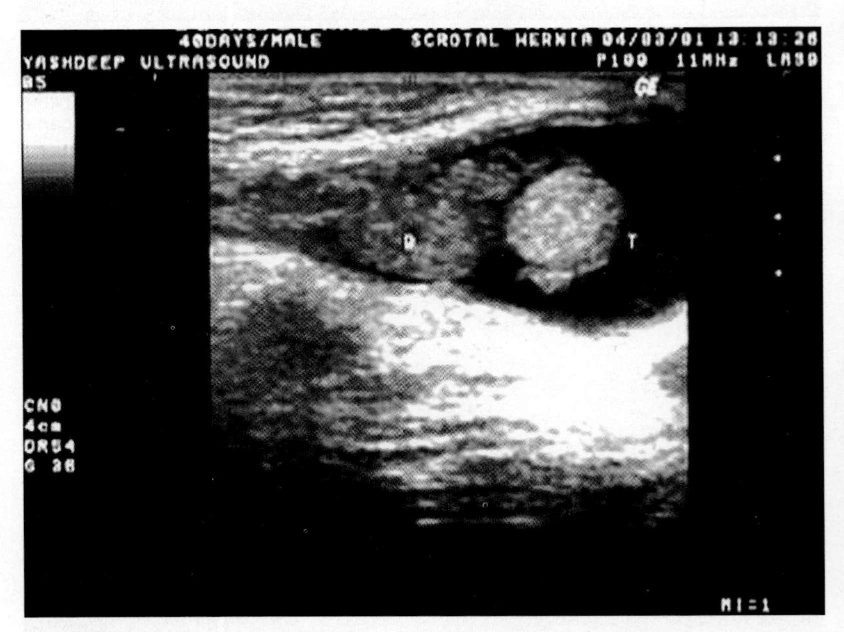

FIGURE 10.62: Scrotal hernia. A bowel loop is seen herniated in scrotum in an infant presented with a scrotal swelling. Testis appears to be normal. Associated hydrocele is also seen.

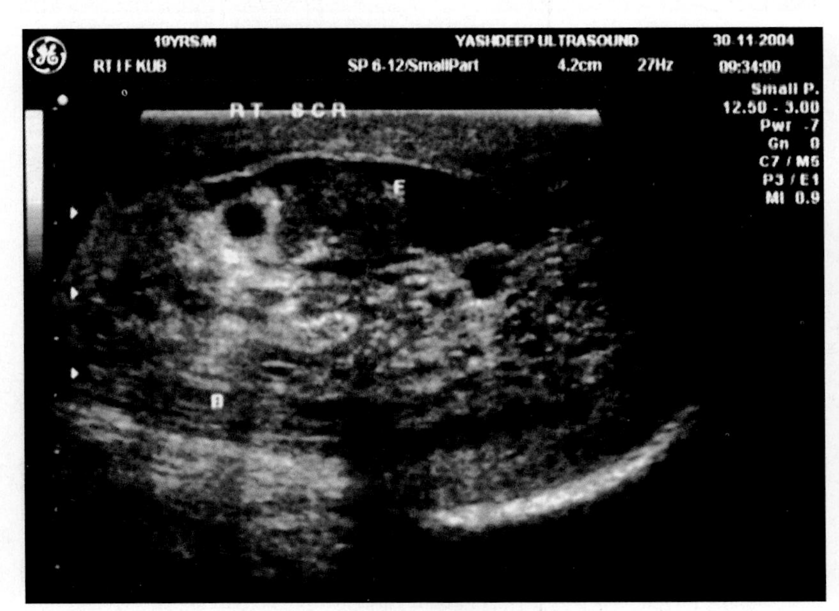

FIGURE 10.63

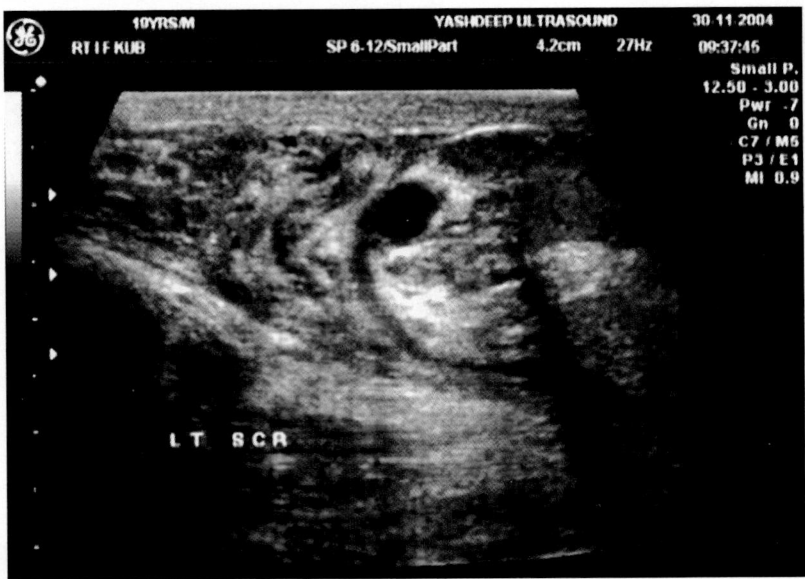

FIGURE 10.64

FIGURES 10.63 and 10.64: Scrotal hernia. A patient presented with Rt sided inguino-scrotal swelling. HRSG shows a bowel loop herniating in the scrotal sac along with omentum. The floating bowel loop is well appreciated in the scrotal sac against the fluid background.

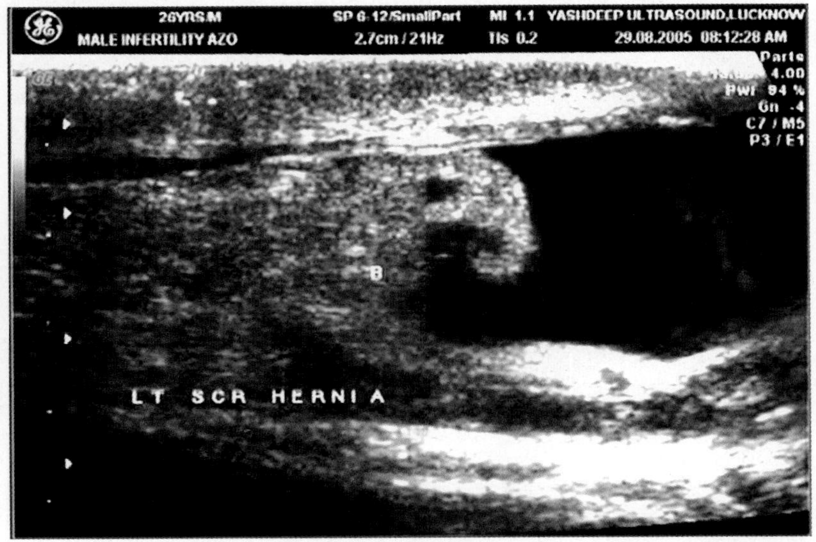

FIGURE 10.65

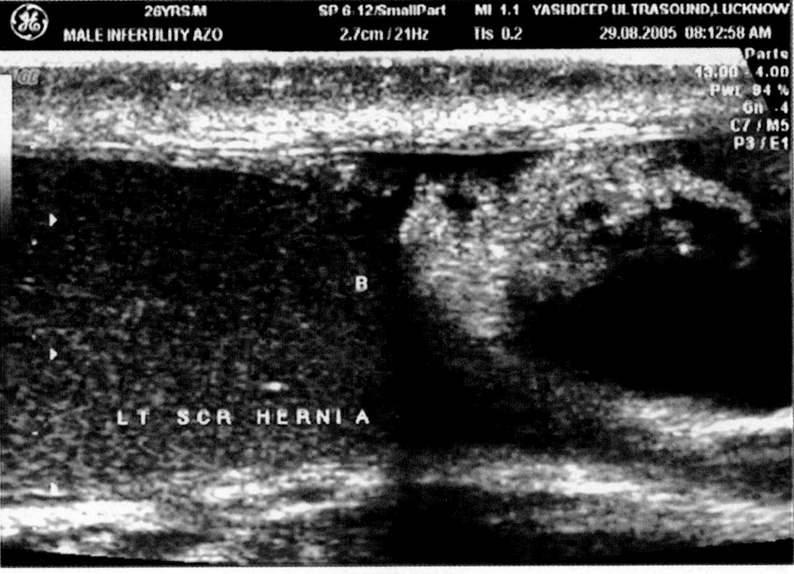

FIGURE 10.66

FIGURES 10.65 and 10.66: Lt scrotal hernia. HRSG shows herniating bowel loop in the Lt scrotal sac. The bowel loop is well appreciated along with normal testis.

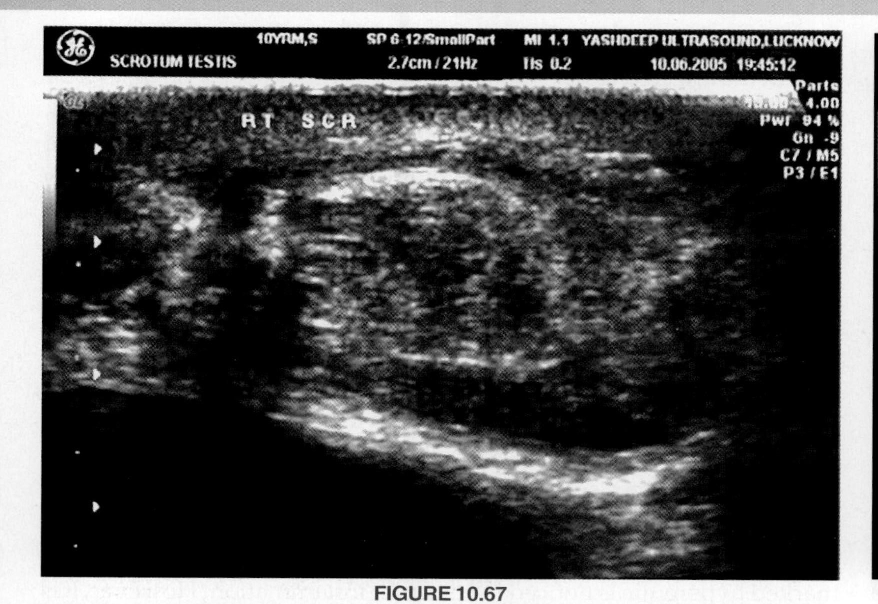

FIGURE 10.67

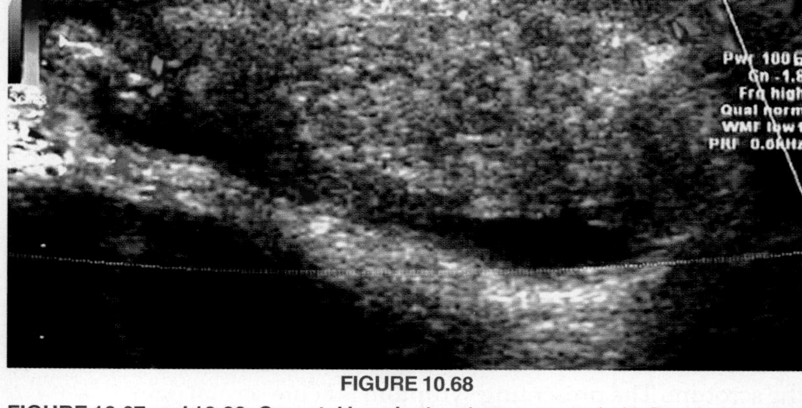

FIGURE 10.68

FIGURE 10.67 and 10.68: Omental hernia. A patient presented with Rt sided inguino-scrotal swelling. HRSG shows echogenic omentum herniating in the inguinal canal and also flowing down in the scrotum. No bowel loop is seen coming out along with the omentum. On color flow imaging peripheral flow is seen along the herniating omentum.

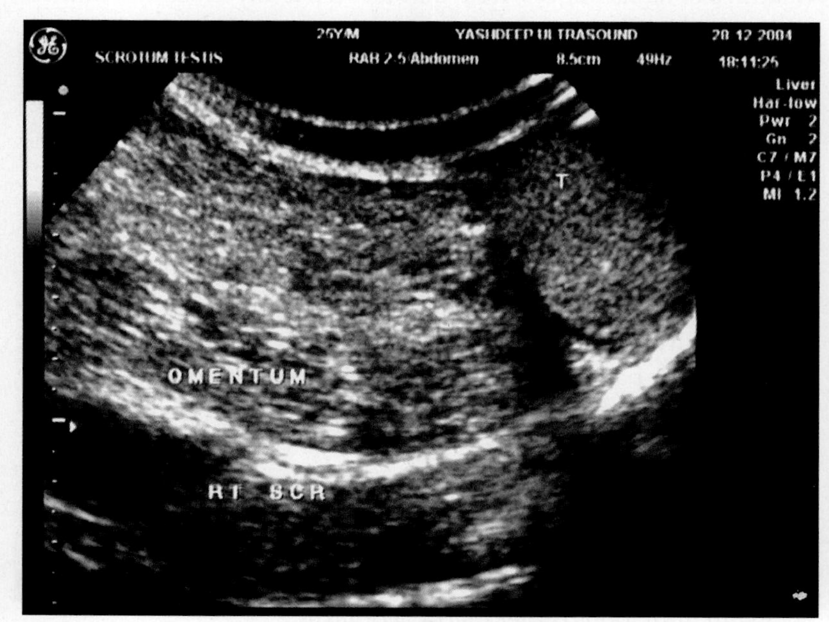

FIGURE 10.69

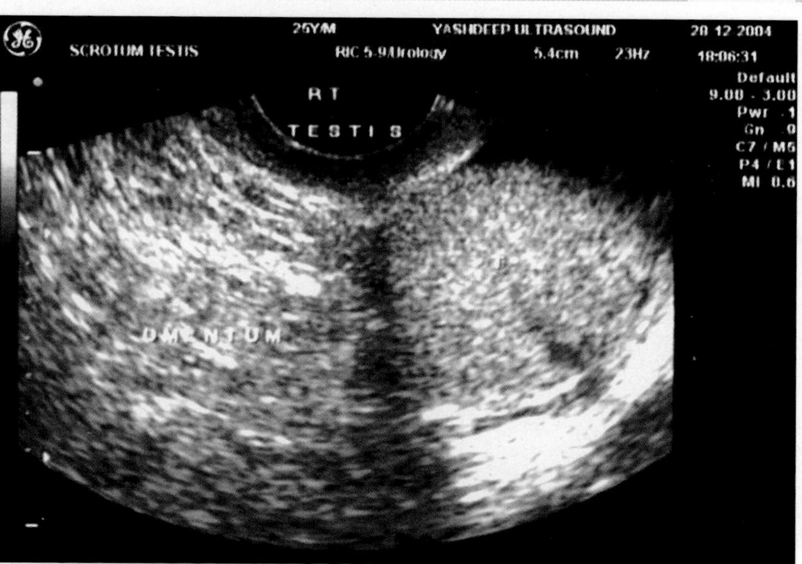

FIGURE 10.70

FIGURES 10.69 and 10.70: Herniating omentum in the scrotal sac. A patient presented with diffuse scrotal swelling on the Rt side. Echogenic omentum is seen herniating in the scrotal sac. The omentum is seen pushing the testis to one side due to fatty nature of omentum. It is difficult to differentiate the omentum from the scrotal contents. A proper gain setting is must for the omental visualization.

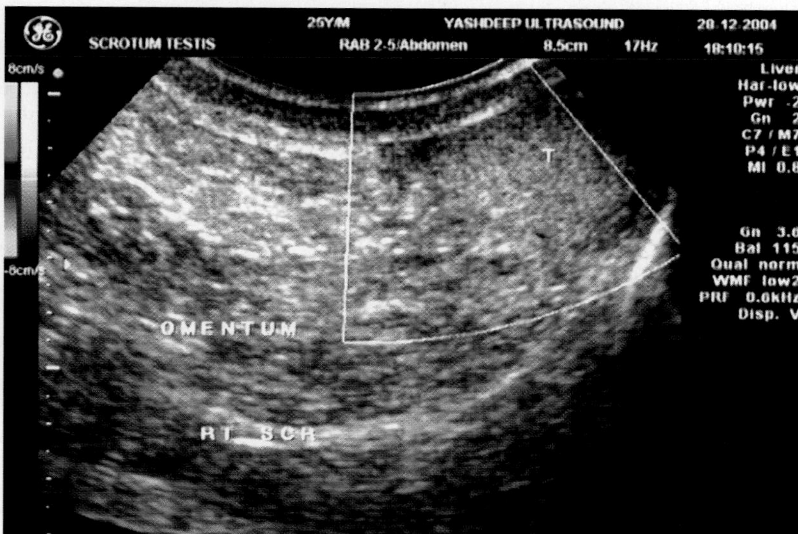

FIGURE 10.71: On color flow imaging normal flow is seen in the testis. However, no flow is seen in the omentum.

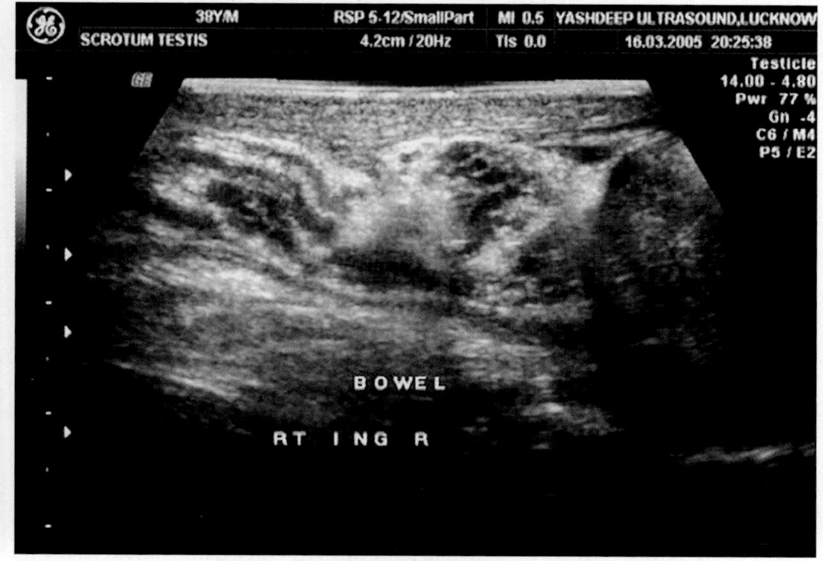

FIGURE 10.72

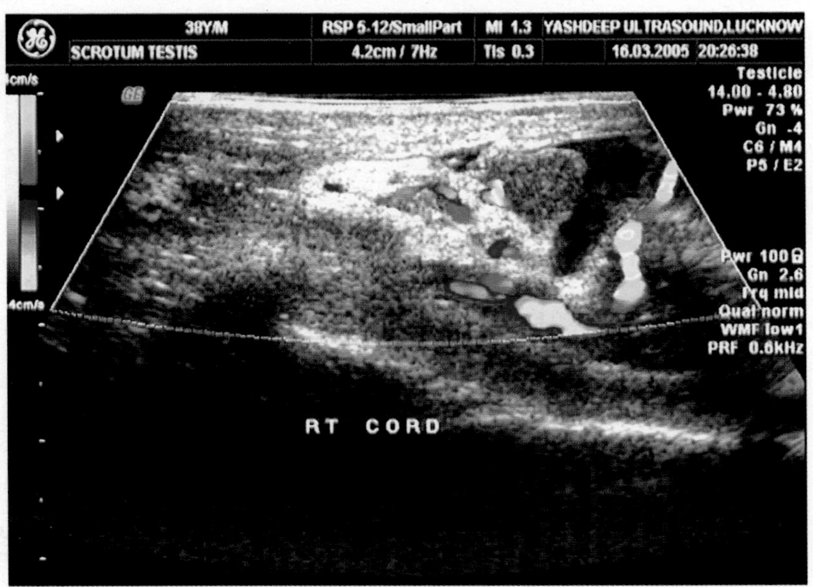

FIGURE 10.73

FIGURES 10.72 and 10.73: Bowel herniation. HRSG shows herniating bowel loop in the scrotal sac on the Rt side. Color Doppler shows the vessels along with the bowel. Color Doppler imaging is very informative and must to rule out strangulation of hernia. On Doppler imaging no flow seen in strangulated bowel loops.

Acute Scrotal Pathologies

Acute scrotum is a general term used to indicate various pathologies which present with a triad of symptoms and signs, viz. (i) acute pain, (ii) swelling and (iii) redness. Ultrasonographic studies help significantly in aiding the clinician's decision of "surgery or no surgery". Some of the major causes of acute scrotum are as follows:

1. Testicular torsion
2. Testicular inflammation (epididymitis, orchitis)
3. Testicular trauma
4. Torsion of the appendix, Epididymis and testis
5. Strangulated hernia
6. Testicular vasculitis and infarction

Testicular (Spermatic Cord) Torsion

Testicular torsion is defined as "rotation of the testis" on the longitudinal axis of the spermatic cord. The events that take place when the spermatic cord is twisted are as follows.

1. Blocked venous drain (edema and hemorrhage)
2. Impaired arterial flow (hemorrhage necrosis of testis)

Depending upon the site of twisting, torsion has been classified as extravaginal, intravaginal and mesorchial.

Intravaginal torsion is more common and occurs at 12 to 18 years of age. It occurs in patients with absent or short mesenteric attachment of the testis to the scrotal sac, allowing excess mobility of the testis and cord. This in turn results in twisting of the vascular pedicle, which results in complete, incomplete or recurrent obstruction to venous flow and eventually the arterial flow. Initially thin walled drainage veins are compressed by the twisted cord leading to congestion and edema, thereby, elevating pressure in the cord. Subsequently, it causes occlusion of the arteries leading to ischemia and infarction of the testis. In the early phase (with 4 hours), there is ischemia of the testis, a little reactive hydrocele and hyperemia of the scrotum. Sonography may be normal at this stage. However, color Doppler studies will show no detectable flow in the affected testis.

In the mid phase (5 to 24 hours), there is increasing congestion, edema and hemorrhage in the testis. Ultrasonography in this phase shows an abnormal testis that is enlarged and hypoechoic. As congestion, hemorrhage and ischemia progress, the testicle becomes heterogeneous with areas of decreased echogenicity. Color Doppler shows no flow in the affected testis, but flow is seen in the peritesticular area and in the scrotal wall due to reactive hyperemia.

In the late phase (more than 24 hours), infarction sets in. The testis is enlarged and shows heterogeneous texture. The epididymis is thickened, there is reactive hydrocele and thickened scrotal walls. If an abscess develops it is seen as a focal, complex, fluid-filled collection.

Color Doppler sonography has been of immense value in distinguishing between testicular torsion and inflammation. While there is no testicular flow in torsion with associated peritesticular flow, marked hyperemia is noticed in testicular inflammation. However, it is often difficult to differentiate between a spontaneously detorsed testis and testicular inflammation, when both pathologies show evidence of increased intratesticular vascularity.

Detorsion of the cord must be achieved as early as possible. If surgery is performed within six hours, the testicular viability is significantly high. If performed within 24 hours, the salvage rate is 60 to 70%. It fails to be less than 20% when surgery is undertaken after 24 hours. Scintigraphy is very useful in patients with torsion when reduced or absent perfusion is demonstrated (94 to 90%).

It may also be remembered that the sensitivity of the machine for detecting low flows is vital in visualizing intratesticular flows. With a situation like a detorsed testis or a poor sensitivity machine, the surgeon cannot depend only on color flow imaging to decide the line of therapy.

Testicular Inflammation

Epididymitis and orchitis are the most common infections involving the scrotum. The presenting symptom is acute scrotal pain.

There is considerable overlap of clinical findings between testicular torsion and acute inflammatory disorders. Color Doppler is of immense value in distinguishing one from the other especially since the treatment is completely different.

Scrotal infections are usually secondary to prostatitis or bladder infections. Orchitis can also be the result of viral infection sonographic features.

1. Enlargement of the testis.
2. Hypoechoic testis with involvement being diffuse or focal.
3. Abscess formation may show a totally heterogeneous epididymis.
4. Color Doppler studies reveal hypervascular pattern.

These vessels have a low vascular resistance and more venous flow indicating hyperemia. Normally the epididymis is avascular, but increased flow pattern in the epididymis is diagnostic of inflammatory changes. The common pitfalls in the diagnosis of acute epididymo-orchitis are as follows:

1. Diffuse increase in vascularity is often seen in infiltrative lesions like leukemia or lymphoma.
2. A distorted testis may also have increased vascular pattern. Commonly a hydrocele and thickening of the scrotal wall are noted in inflammations rather than a neoplasm.
3. Focal orchitis may be mistaken for an occult neoplasm.

Testicular Trauma

The common causes of testicular trauma are sport injuries and vehicular accidents.

Sonographic Features

Several sonographic features are seen. There is thickening of the scrotal skin as a result of edema. The thickness can also be due to hemorrhage and extravasations. It there is associated bladder rupture; a multilayered pattern of the scrotal wall is seen. Hemorrhage is generally echo poor, and low-level mobile echoes are seen within the fluid.

Hematocele may have a complex or echogenic pattern. In the presence of hematoceles, the testis must be thoroughly scanned for an intratesticular hematoma or a fracture.

Acute hematomas in any location are heterogeneous masses often with increased echogenicity. As hematoma resolves in 1 to 2 weeks, it becomes smaller and anechoic cystic mass with septations replacing the heterogeneous lesion.

Intratesticular hematomas have an appearance similar to extra testicular ones, but because of their position they may mimic a tumor. These cases require close observation as hematomas will become cystic and change their appearance rapidly over the course of several days, whereas a tumor will remain unchanged for a long time.

Testicular rupture leads to loss of normal smooth contours of testis with interruption of bright capsule of the tunica albuginea and protrusion of parenchyma. Echo poor and heterogeneous areas will be seen in the testicle as a result of hemorrhage and parenchymal disruption.

Testicular contusion or fracture without tunical rupture will have a smooth and remain surrounded by bright tunica albuginea. The fracture will be seen as a linear hypoechoic area within the testis. There are associated changes in the epididymis in the form of enlargement, reduced echogenicity or abnormal position in relation to the testis.

Fluid Collections

Hydrocele is the most common cause of an enlarged scrotum for a patient presenting to the surgeon. It is an abnormal fluid collection between the two layers of the tunic vaginalis.

Most commonly hydrocele are idiopathic. In the neonates, this can be the result of the processes vaginalis being open. It invariably closes by about 2 years. In the adults, the other causes of hydrocele are inflammation, torsion and trauma. Hydroceles are not commonly encountered with testicular tumors.

Sonographic Features

Hydroceles are present as hypoechoic/anechoic fluid collections surrounding the testis. The testis as often displaced posteriorly and medially.

Septations may be present and indicate old hemorrhage or infections. Thick septations are often seen in tuberculosis. Diagnosis of hydrocele is usually straightforward. There may be fine debris within the fluid in long-standing hydroceles; these indicate cholesterol crystals, some of these patients where the hydroceles are tense may even suggest solid lesion clinically. This may be associated with thickening of the scrotal wall in longstanding hydroceles. Parietal calcifications are also noted.

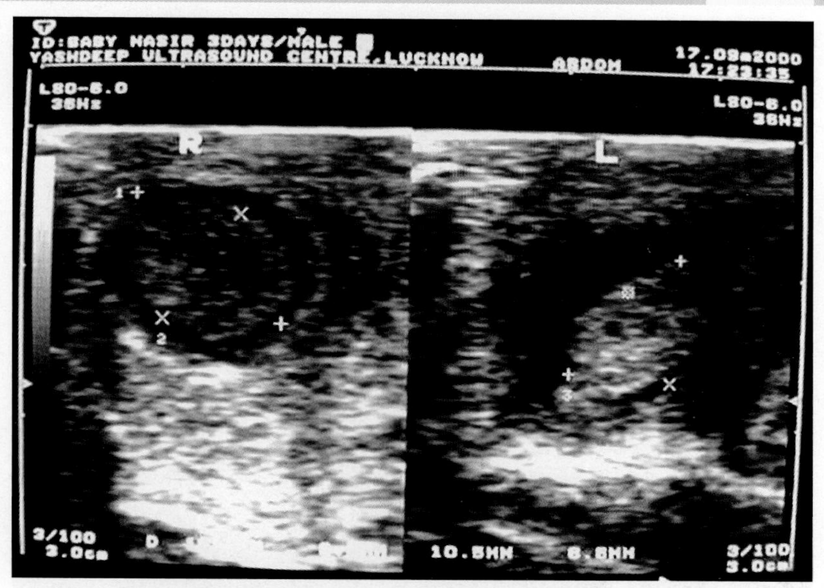

FIGURE 10.74: Neonatal torsion. The Rt testis is enlarged. It is irregular in outline and hypoechoic in texture. Perifocal edema is present. Fluid collection is also seen in the scrotal sac suggestive of torsion.

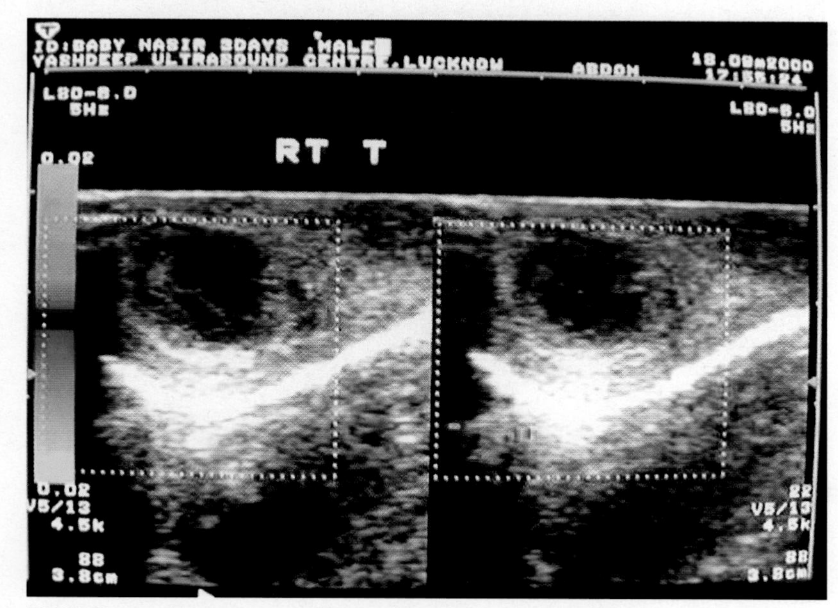

FIGURE 10.75: No flow is seen in the testis. The Lt testis shows normal flow in the testicular artery.

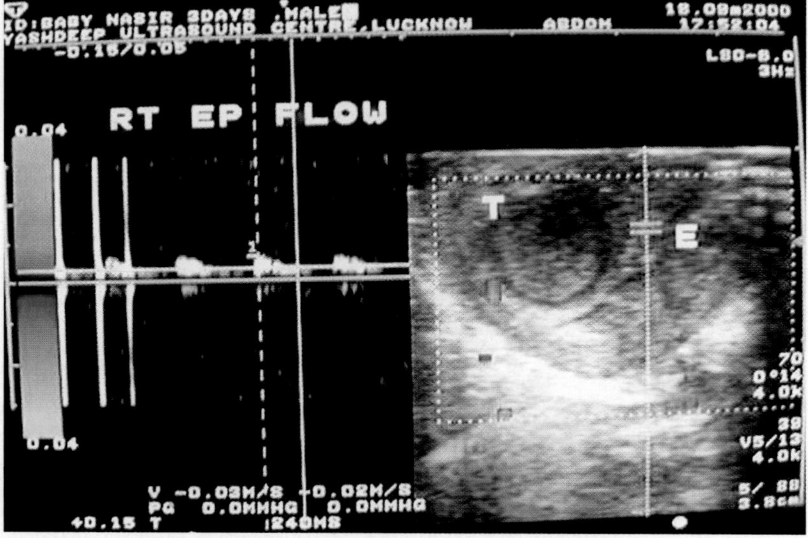

FIGURE 10.76: No testicular flow is seen on color flow imaging. However, flow is seen in the epididymis of the Rt testis.

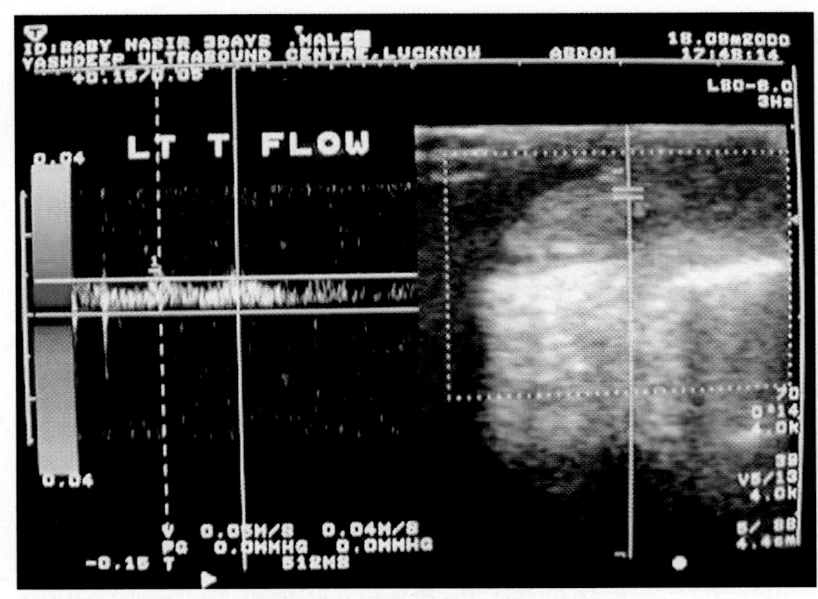

FIGURE 10.77: Lt testis of the same patient shows normal flow.

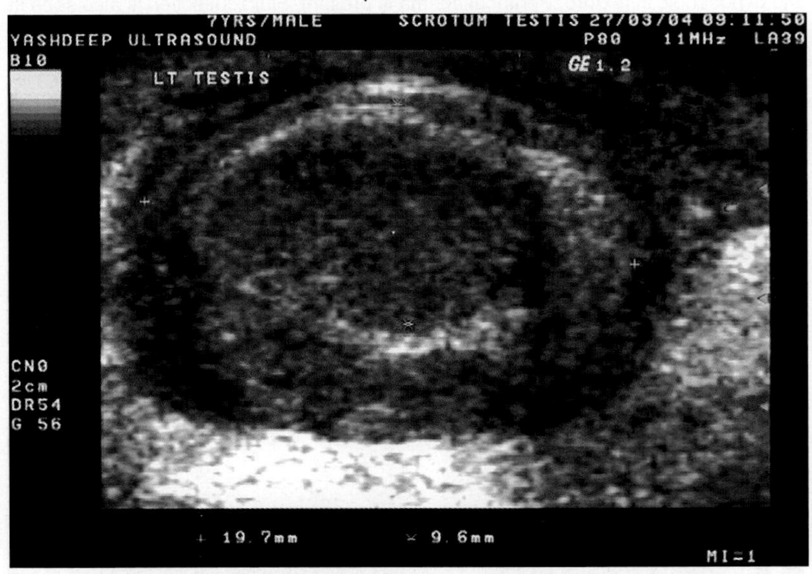

FIGURE 10.78: Acute torsion testis. A young boy presented with acutely painful Lt scrotum. HRSG shows enlarged edematous hypoechoic Lt testis. Marked testicular edema is present.

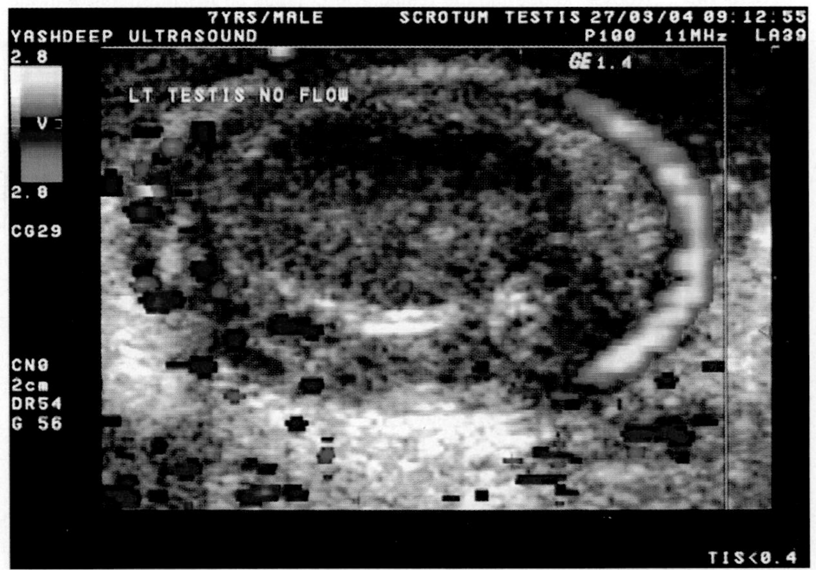

FIGURE 10.79: On color flow imaging no flow is seen in the testis. However, increased peripheral flow seen in the peritesticular region. But no suggestion of any flow is seen in the intratesticular arteries suggestive of acute torsion.

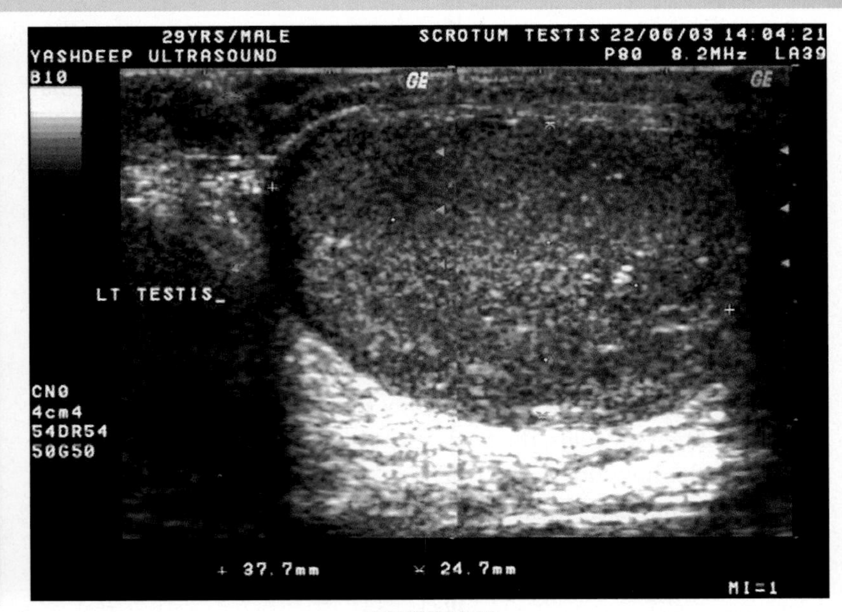

FIGURE 10.80

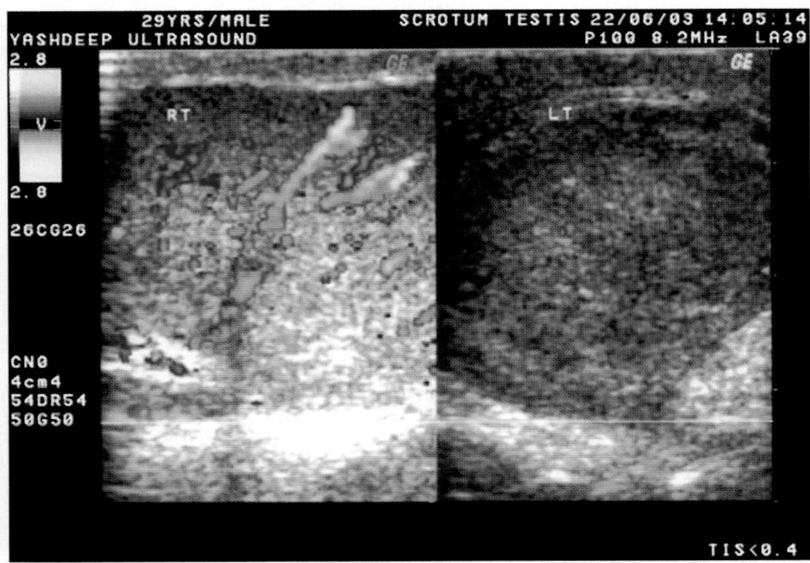

FIGURE 10.81

FIGURES 10.80 and 10.81: Testicular torsion. HRSG shows hypoechoic enlarged edematous swollen Lt testis. On color flow imaging no flow is seen in the testis. The Rt testis shows normal flow. The Lt testis shows grossly hypoechoic texture in comparison to the Rt testis.

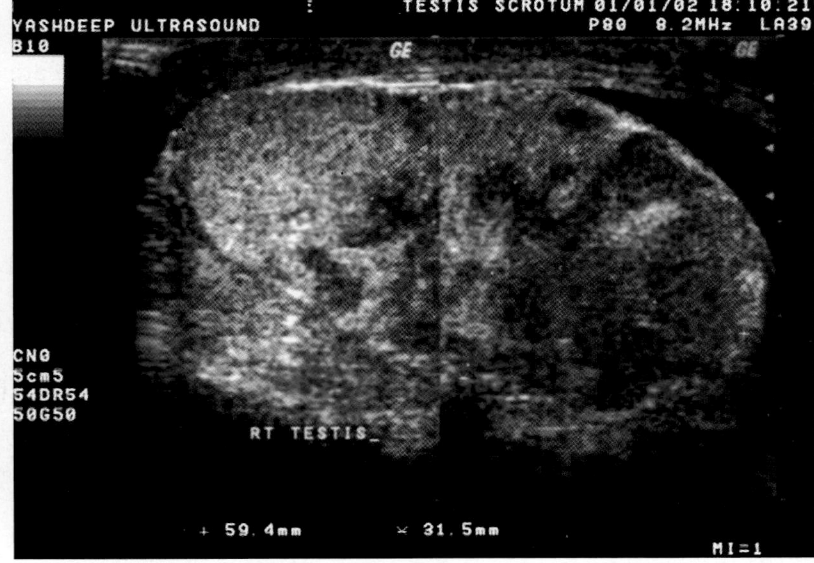

FIGURE 10.82

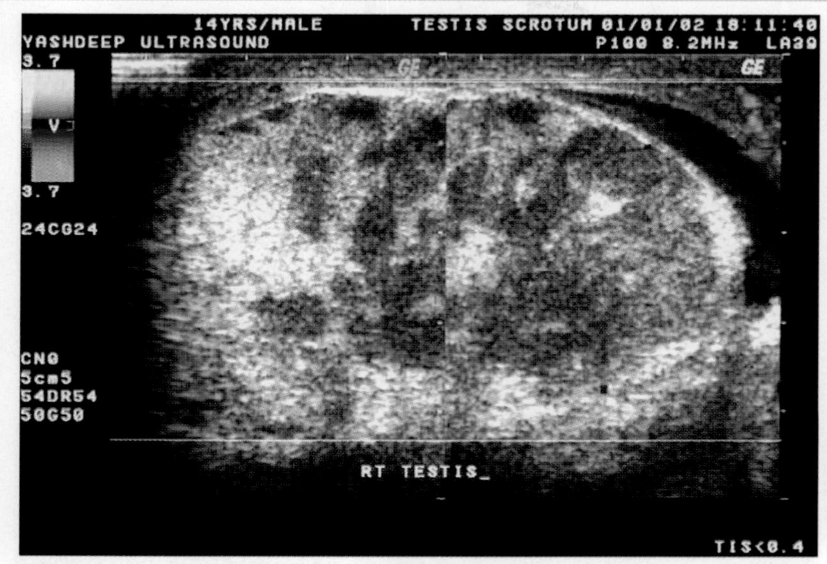

FIGURE 10.83

FIGURES 10.82 and 10.83: Torsion with necrosis. A young boy presented with painful swelling in the Rt scrotum. However, it was not properly attended and ultrasound was delayed. The delayed ultrasound examination shows enlarged heterogeneous swelled testis. Multiple hypoechoic areas are seen in the testis. They are irregular in outline suggestive of necrosis. On color flow imaging no flow seen in the testis.

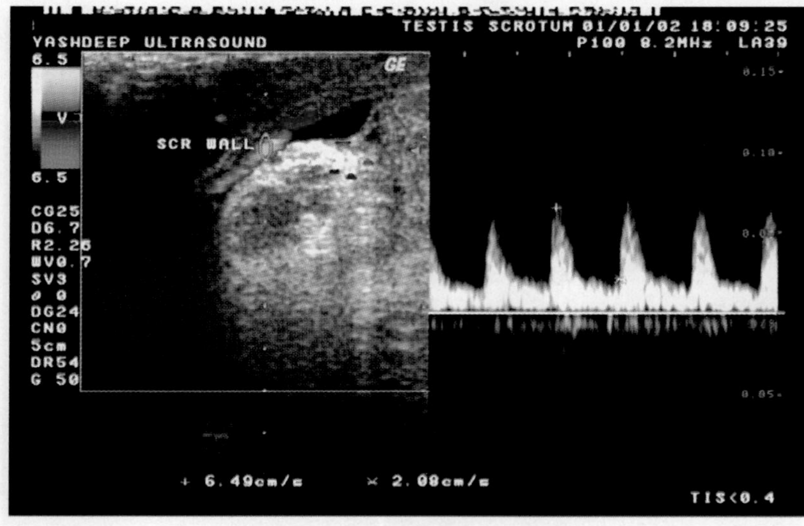

FIGURE 10.84

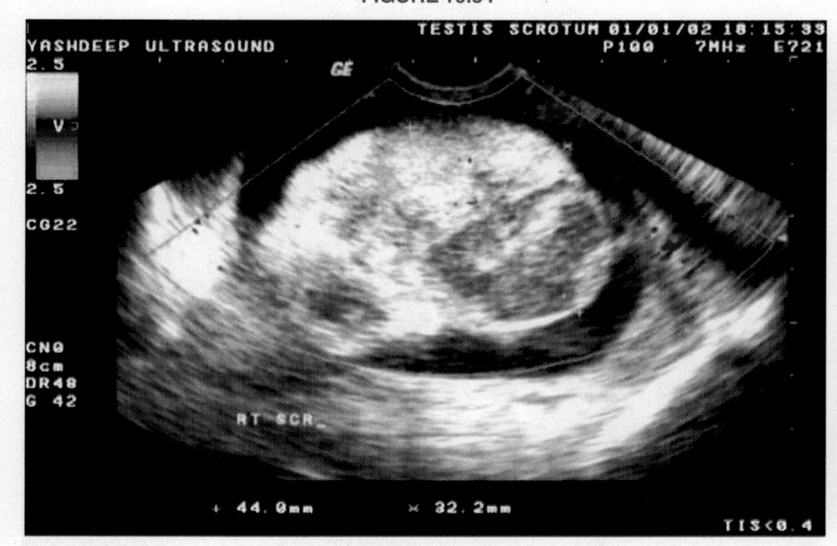

FIGURE 10.85

FIGURE 10.84 and 10.85: The epididymis is also swelled up and heterogeneous in texture. It is irregular in outline. Peripheral arterial flow is seen in the scrotal wall. The scrotal wall was edematous. There is also evidence of big echogenic floating hemorrhagic collection seen in the scrotal sac suggestive of hematocele.

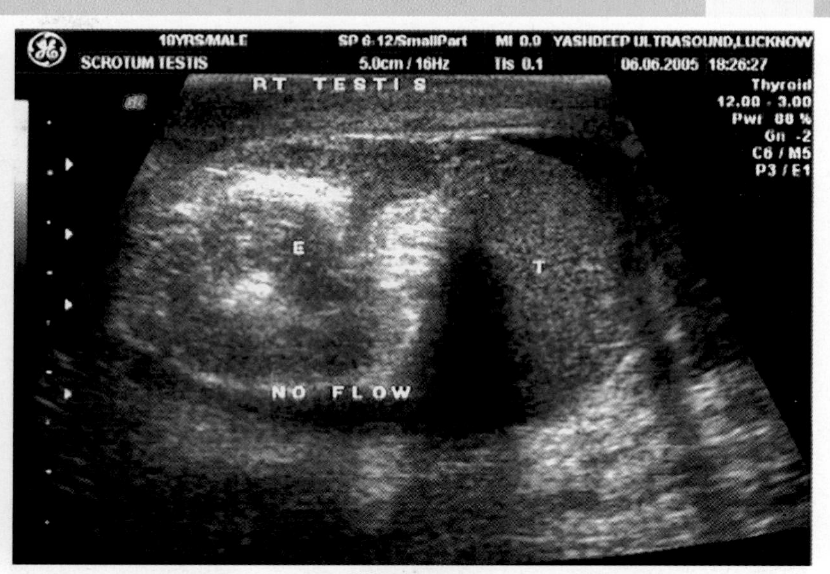

FIGURE 10.86

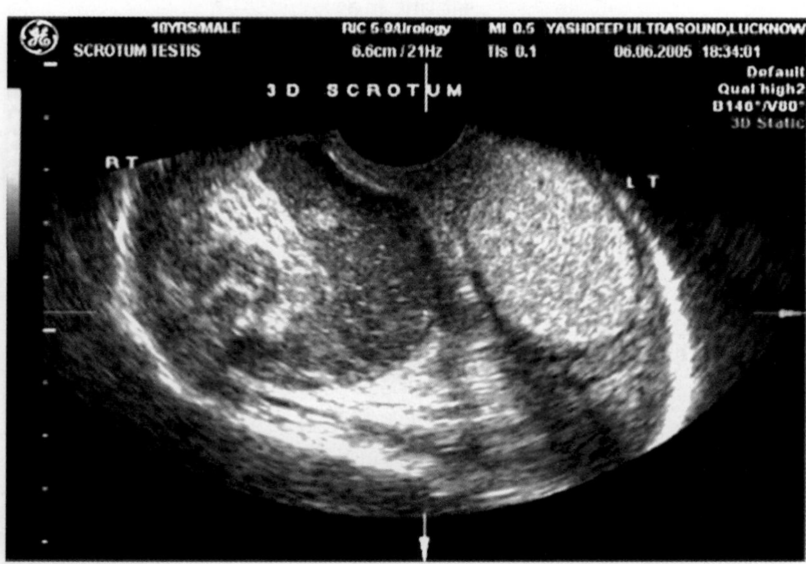

FIGURE 10.87

FIGURES 10.86 and 10.87: Acute torsion with hemorrhagic epididymis. A young boy presented with acutely painful swollen Rt scrotum. HRSG shows edematous hypoechoic Rt testis. The epididymis is also enlarged and swollen. Echogenic collection is seen in the scrotal sac suggestive of hematocele.

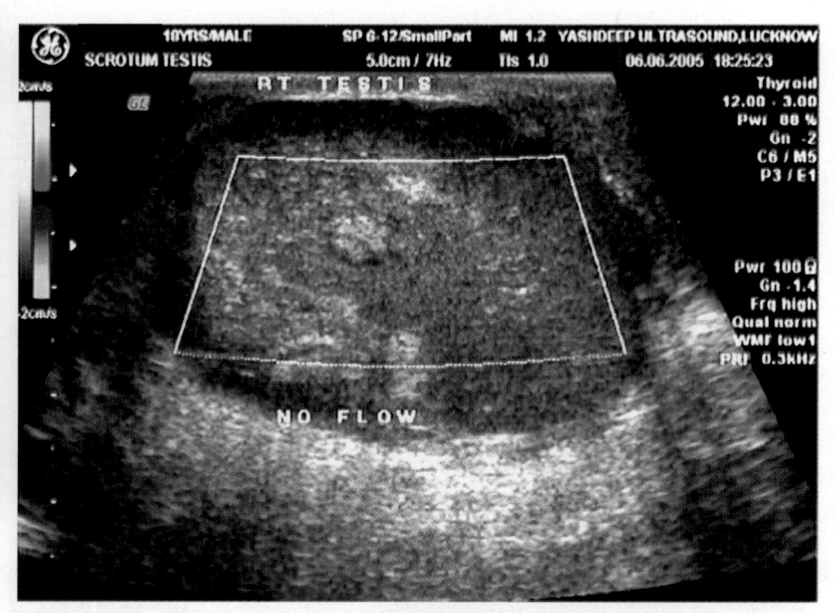

FIGURE 10.88: The testis is enlarged and swollen. It is grossly hypoechoic. On color flow imaging no flow is seen in the testis suggestive of acute torsion.

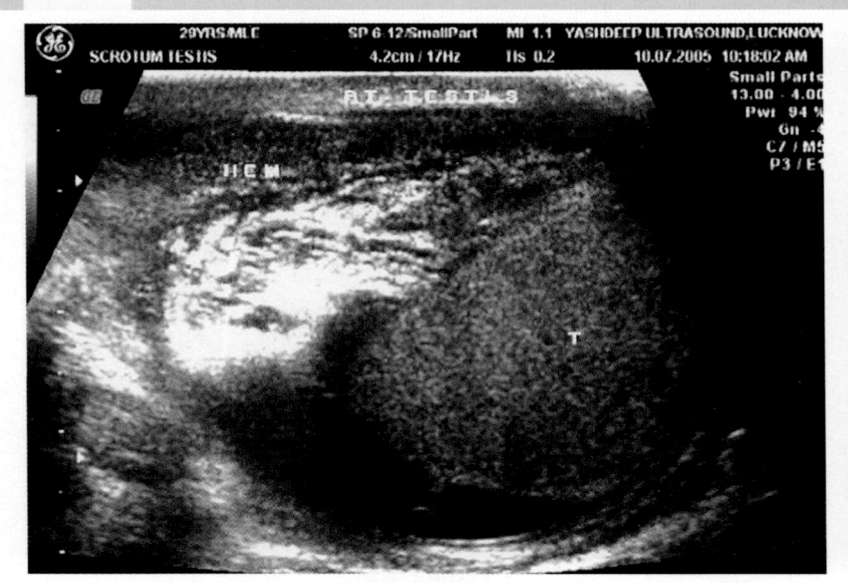

FIGURE 10.89

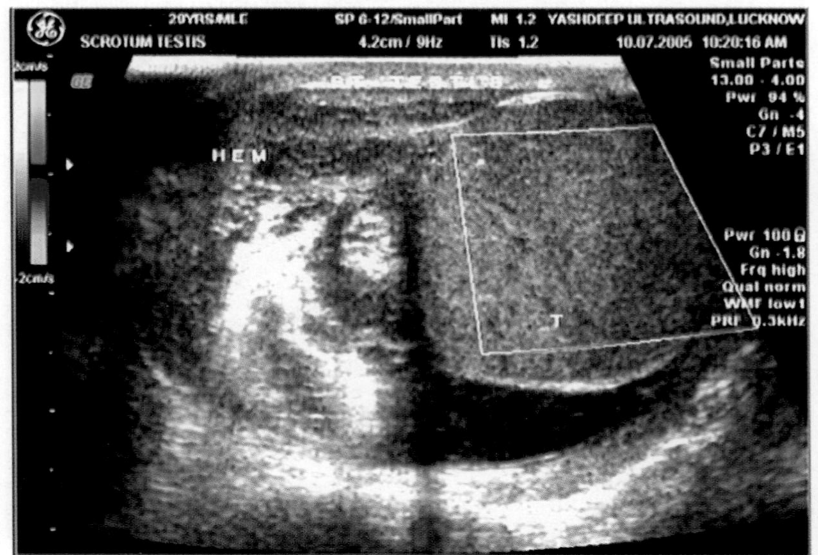

FIGURE 10.90

FIGURES 10.89 and 10.90: 3D imaging of the same patient shows the better sonographic details of the torsion. The hemorrhage is seen as echogenic shadows floating in the scrotum. The testis is grossly hypoechoic. But no flow is seen in the testis.

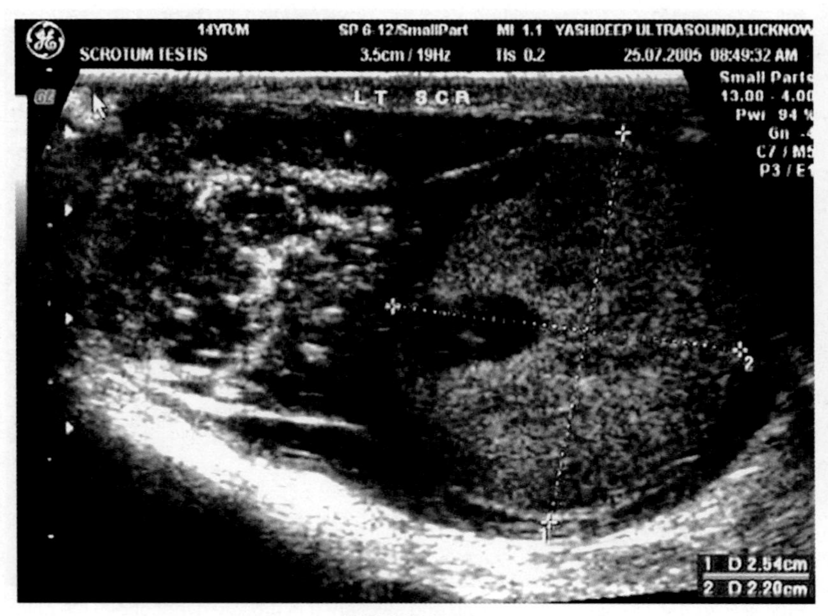

FIGURE 10.91

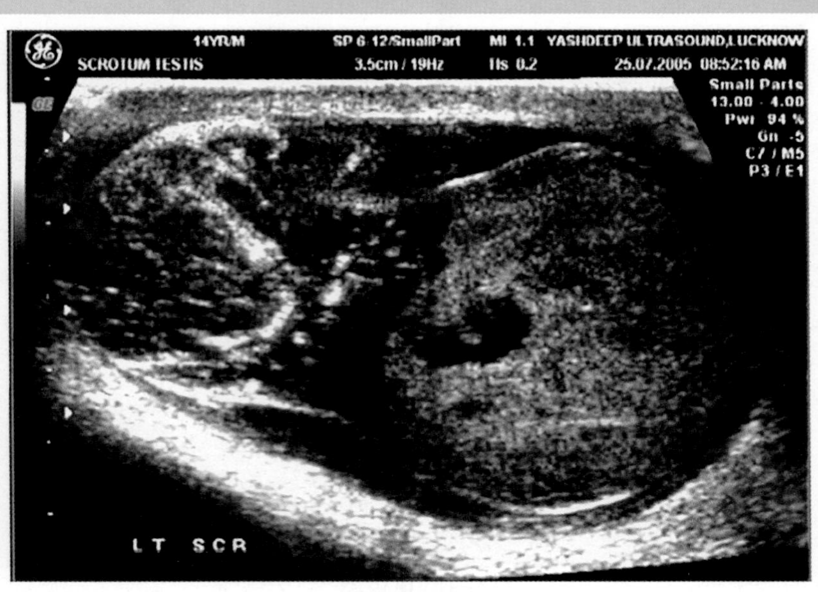

FIGURE 10.92

FIGURES 10.91 and 10.92: Torsion with testicular necrosis. A patient presented with acutely painful swelling in the Lt scrotum. HRSG shows enlarged swollen hypoechoic testis. Irregular hypoechoic necrotic area is also seen in the testis. The epididymis is also enlarged and irregular in outline. Multiple hypodense shadows are seen floating in the scrotum suggestive of hematocele.

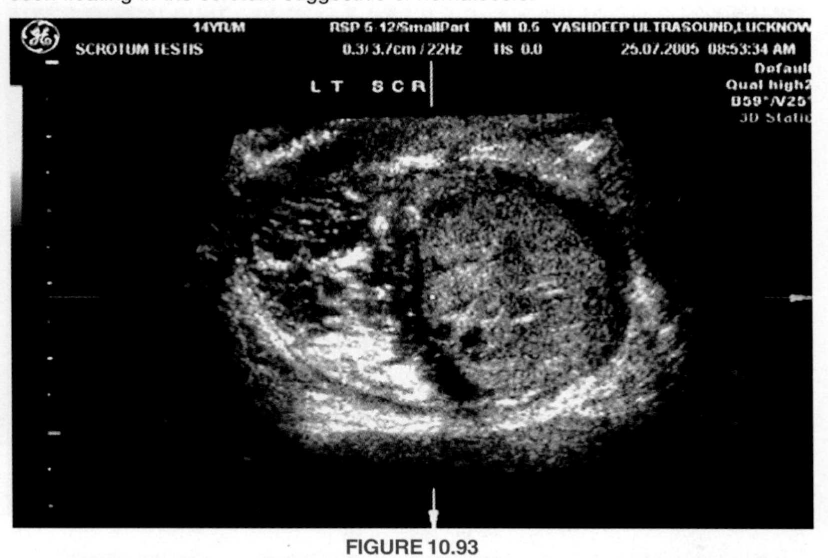

FIGURE 10.93

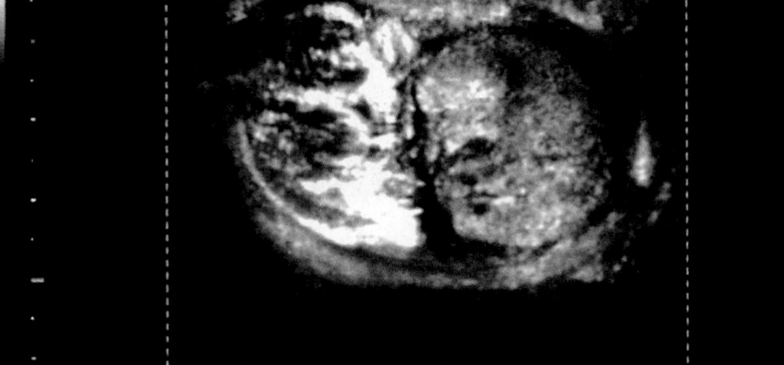

FIGURE 10.94

FIGURES 10.93 and 10.94: 3D surface rendering of the patient shows global view of the torsed testis. The whole of the testis is well visualized with multiple hypoechoic areas. A big hematocele is also seen in the scrotal sac.

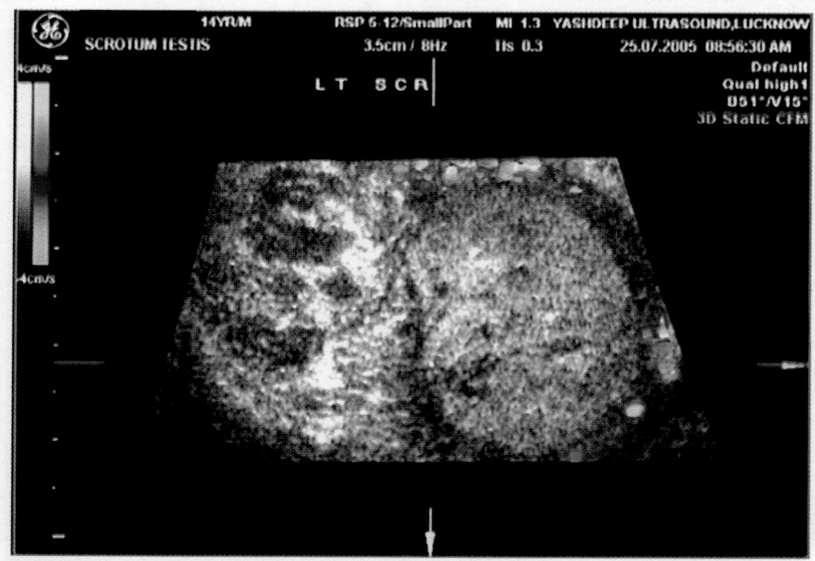

FIGURE 10.95: Marked scrotal edema is seen. On color flow imaging edematous scrotal wall shows hyperemic flow.

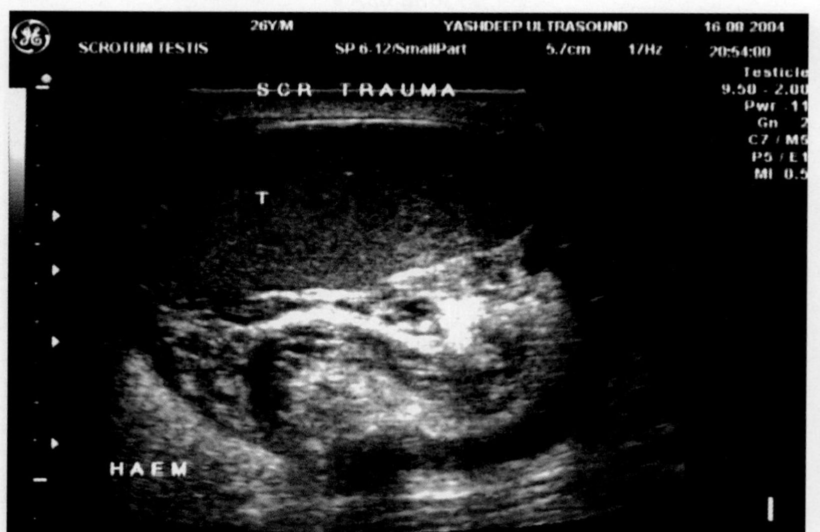

FIGURE 10.96

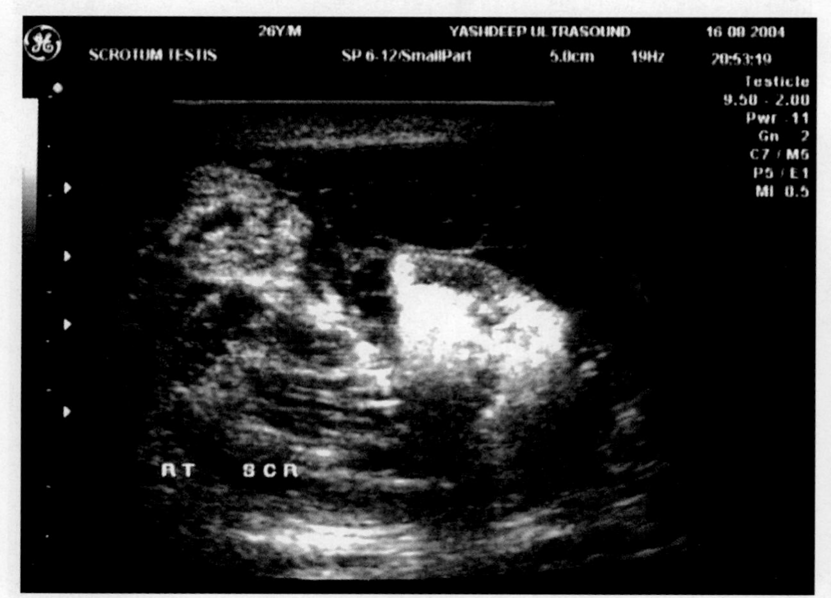

FIGURE 10.97

FIGURES 10.96 and 10.97: Fracture of testis. A patient presented with acutely painful swollen Rt testis after sustaining blunt trauma. HRSG shows the Rt testis enlarged swollen. The testis shows a sharp line running through the substance of the testis dividing the testis into two suggestive of fracture of the testis. Hemorrhagic collection is seen in the scrotum with big hematocele formation.

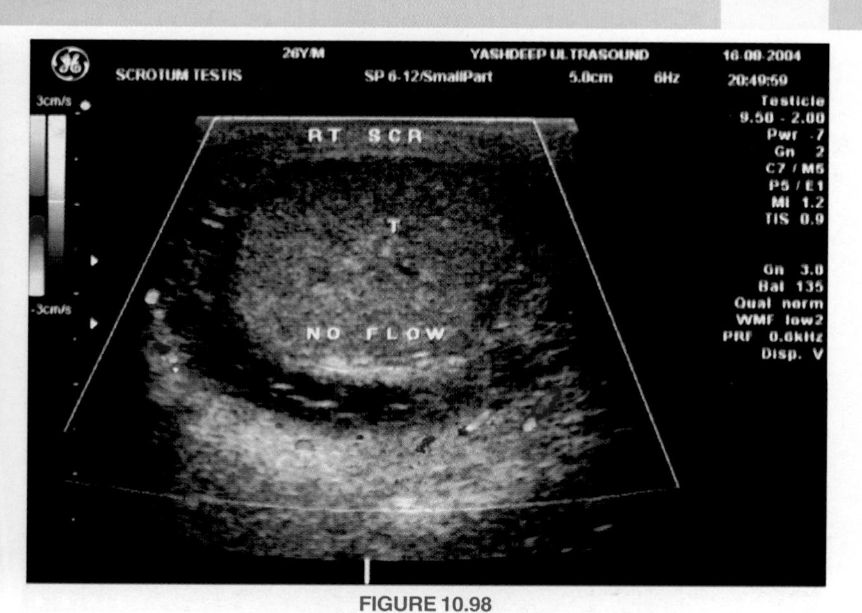

FIGURE 10.98

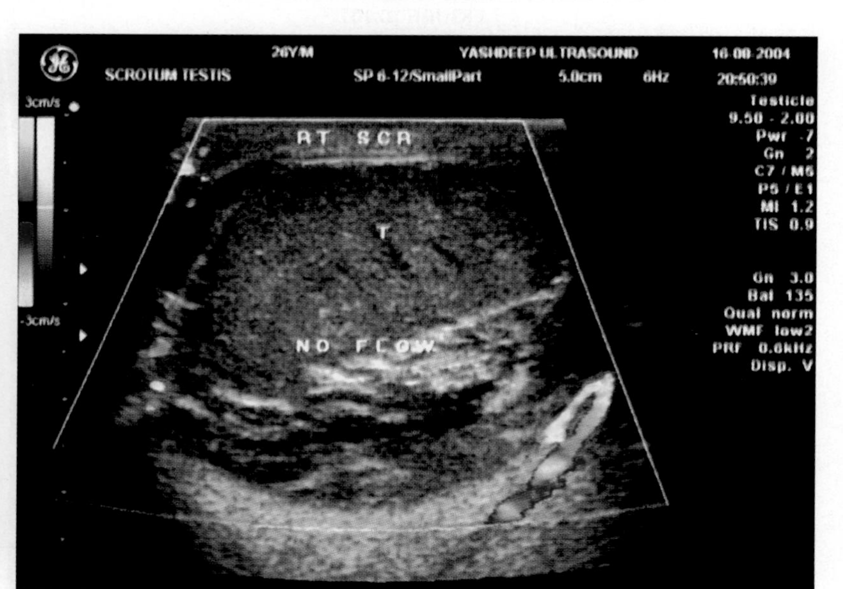

FIGURE 10.99

FIGURES 10.98 and 10.99: On color flow imaging no flow is seen in the testis. Peripheral flow is seen in the edematous scrotal wall.

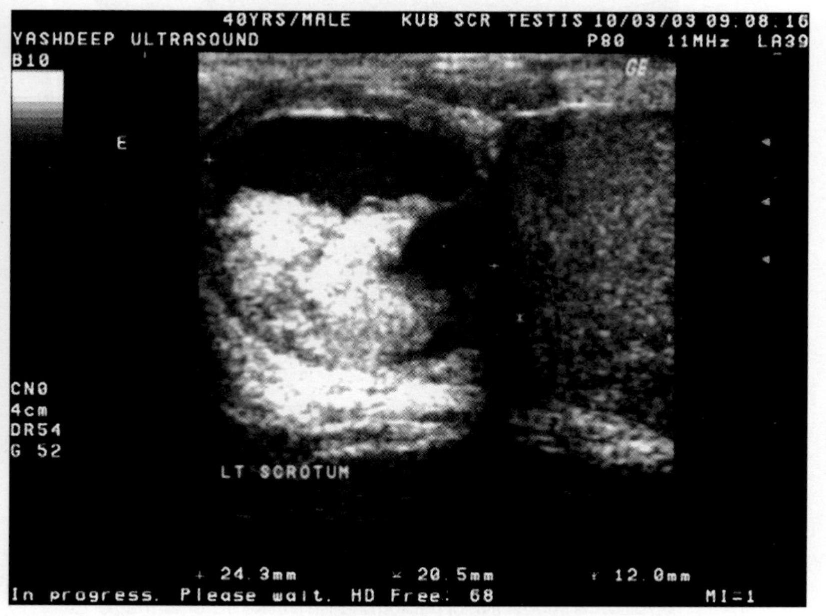

FIGURE 10.100

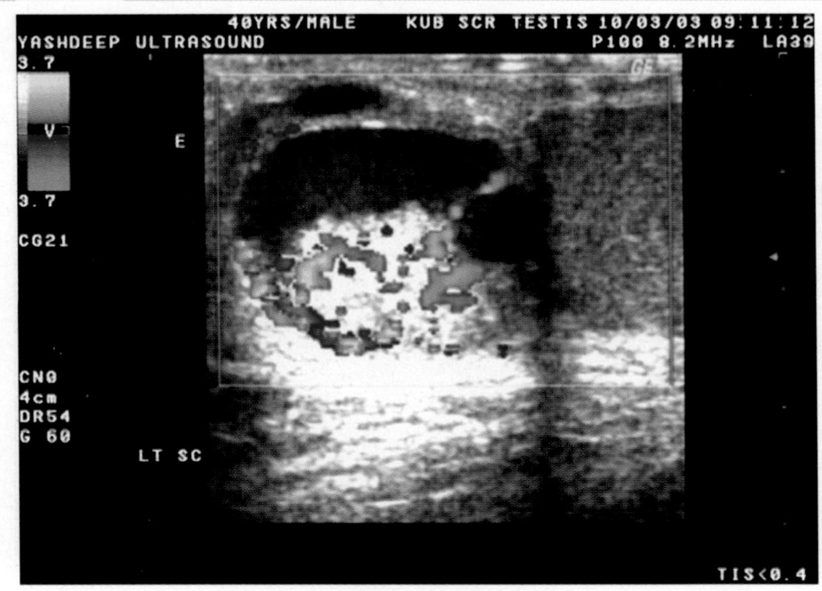

FIGURE 10.101

FIGURES 10.100: and 10.101: Traumatic hemorrhagic epididymitis. HRSG shows traumatic hemorrhagic epididymitis in a patient sustained blunt trauma to the scrotum. The epididymis is enlarged. A big echogenic collection is seen in the epididymis suggestive of fresh blood. On color flow imaging bleeding vessels are seen in the traumatic epididymis.

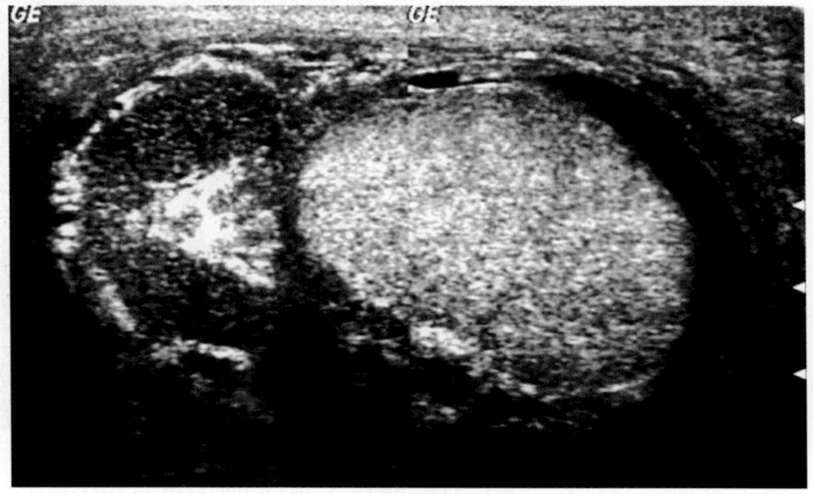

FIGURE 10.102

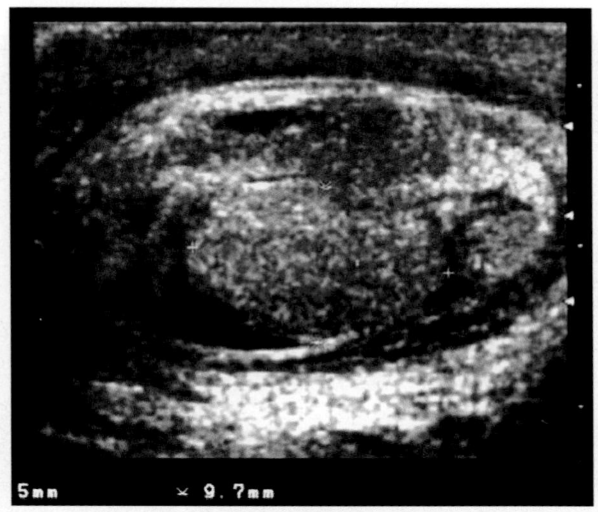

FIGURE 10.103

FIGURES 10.102 and 10.103: Acute epididymitis. A patient presented with acute scrotum. The epididymis is enlarged and irregular in outline. Multiple echogenic specks are seen in epididymis suggestive of hemorrhage in a case of hemorrhagic epididymitis.

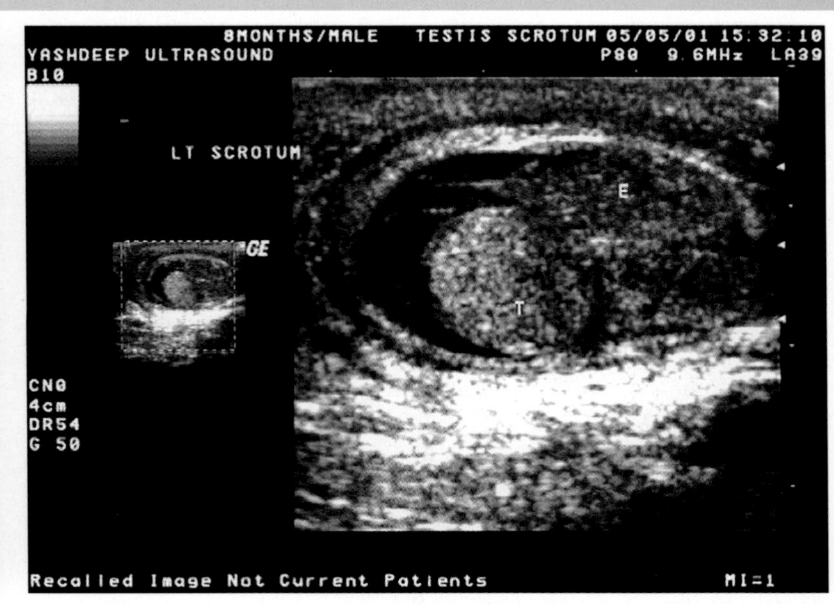

FIGURE 10.104

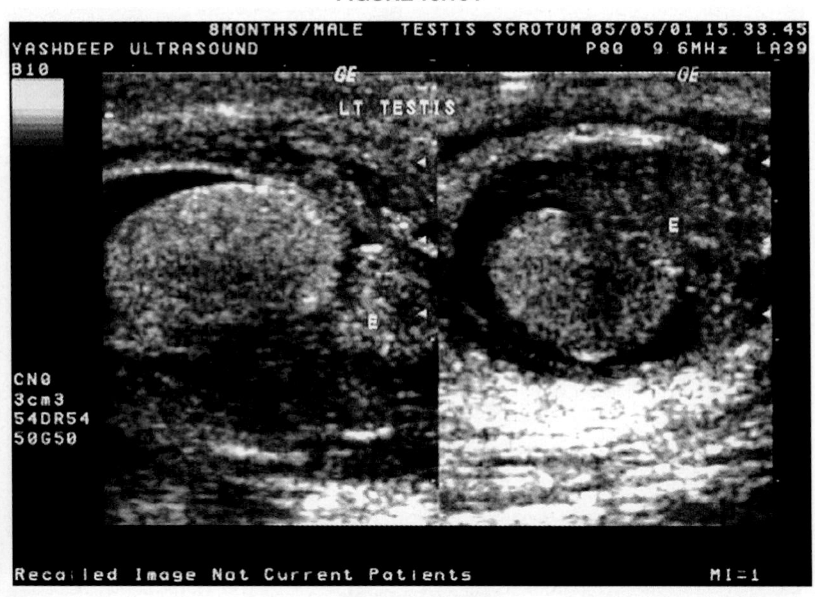

FIGURE 10.105

FIGURES 10.104 and 10.105: Acute epididymitis. HRSG shows irregular hypoechoic enlarged epididymis suggestive of acute epididymitis. However, the testis is seen normally. Reactionary collection is seen in the scrotal sac.

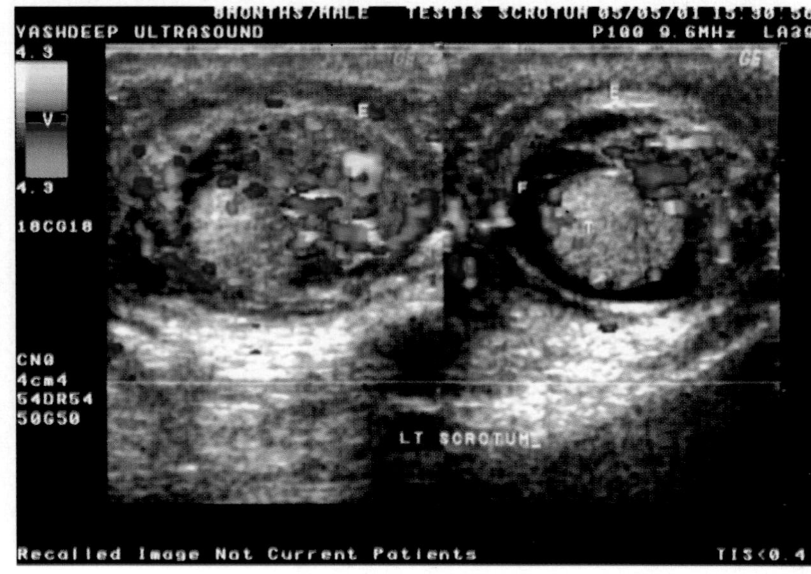

FIGURE 10.106

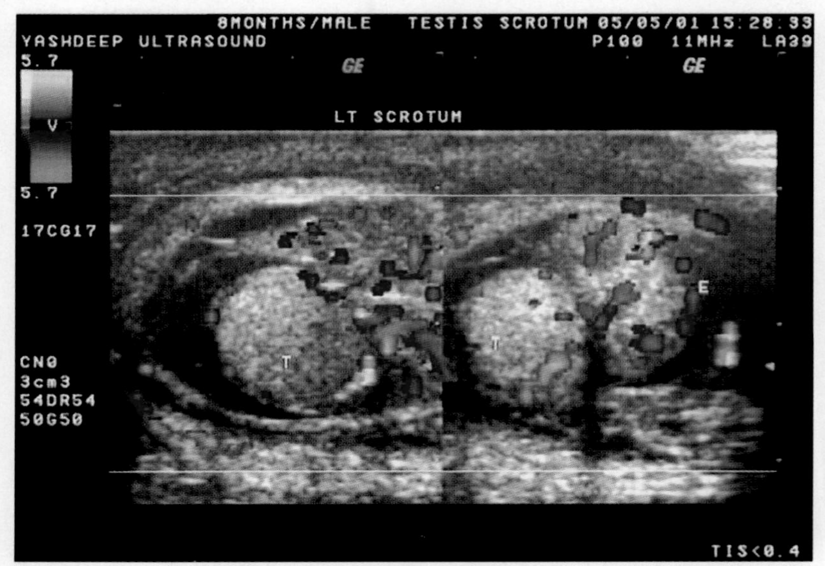

FIGURE 10.107

FIGURES 10.106 and 10.107: Acute epididymitis. Color flow imaging of the same patient shows increased flow in the epididymis. The flow is seen in bizarre fashion. Normal flow is seen in the testis.

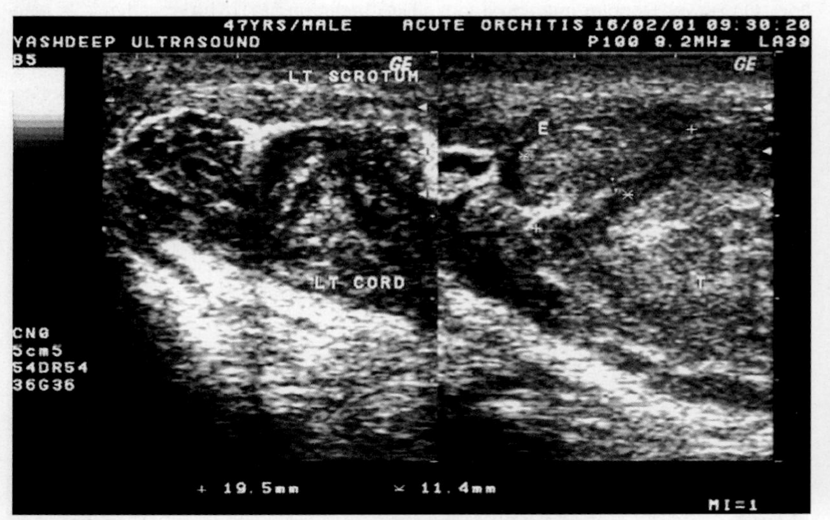

FIGURE 10.108

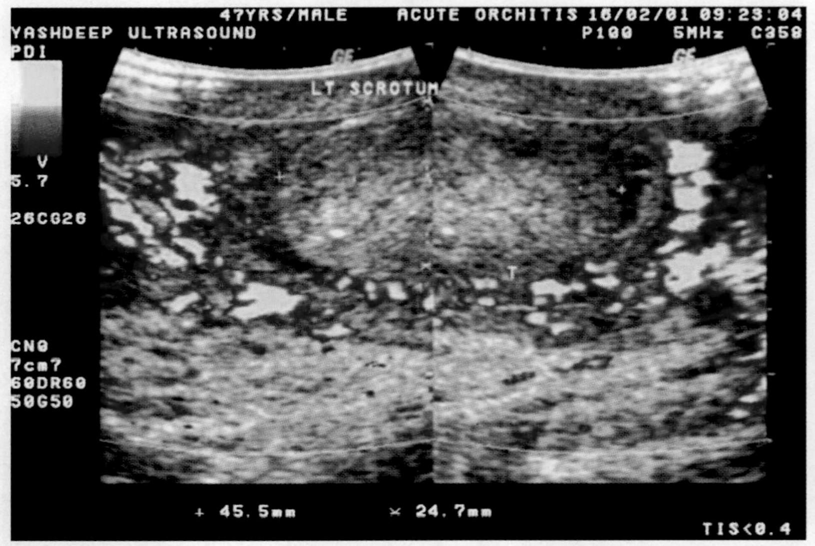

FIGURE 10.109

FIGURES 10.108 and 10.109: Acute epididymo-orchitis. HRSG shows irregular hypoechoic heterogeneous enlarged epididymis. Multiple echogenic specks are seen in it suggestive of hemorrhage. The testis is enlarged and hypoechoic in texture. Color flow imaging of the same patient shows increased flow in the testicular capsule and in the epididymis.

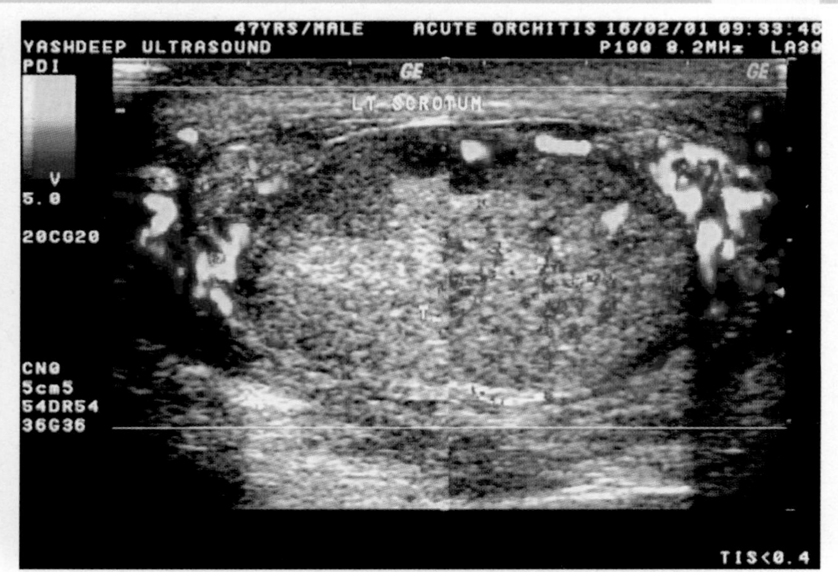

FIGURE 10.110A

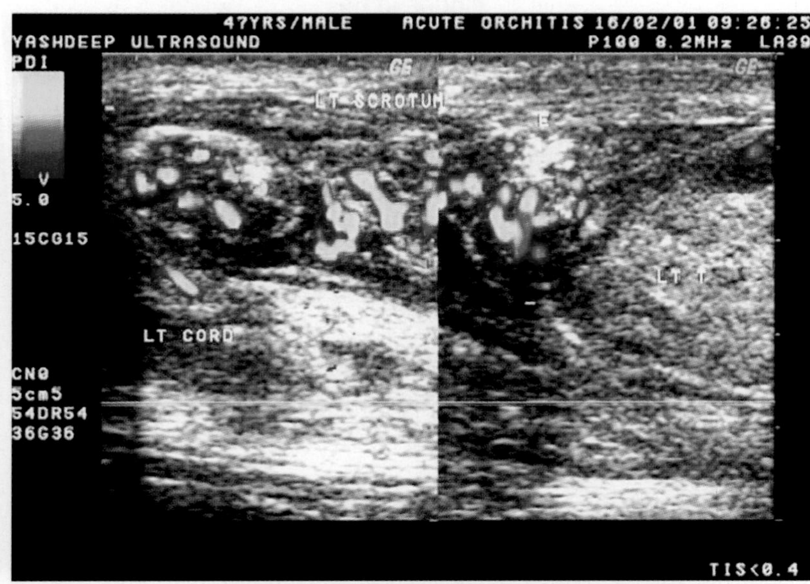

FIGURE 10.110B

FIGURES 10.110A and B: Irregular hypoechoic areas are also seen in the testis in the superior surface suggestive of abscess formation in the testis. The Lt cord is also thickened and multiple dilated vessels are seen along the Lt cord suggestive of cord varices.

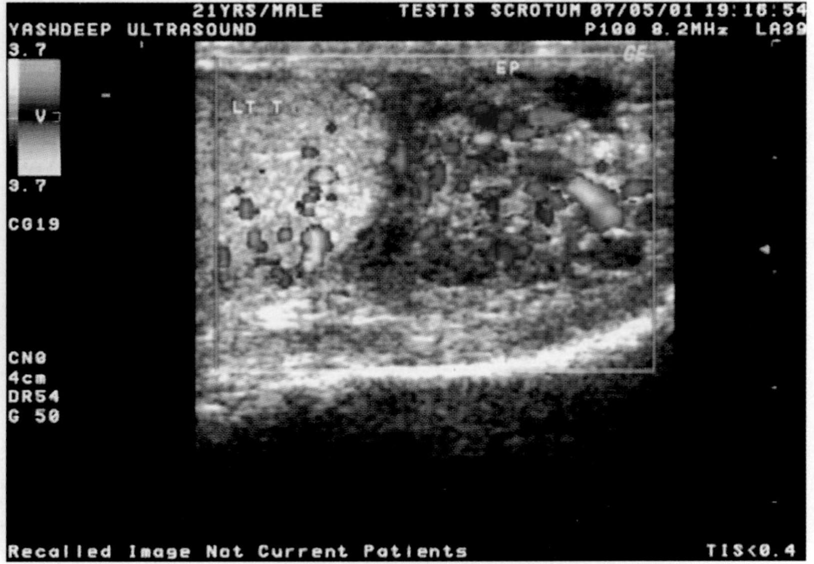

FIGURE 10.111

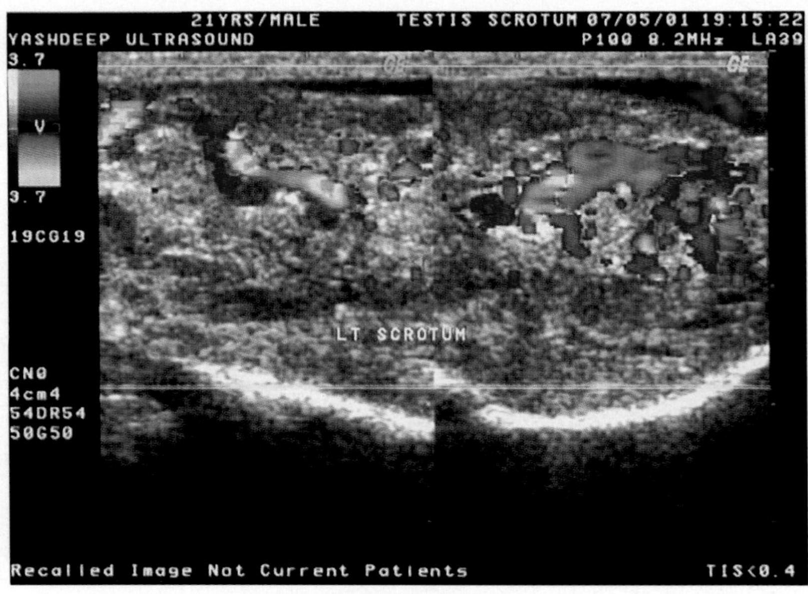

FIGURE 10.112

FIGURES 10.111 and 10.112: Acute epididymo-orchitis. Color Doppler imaging shows enlarged swelling Lt testis and hypoechoic edematous enlarged epididymal head, Color Doppler shows increased flow in testis and epididymis.

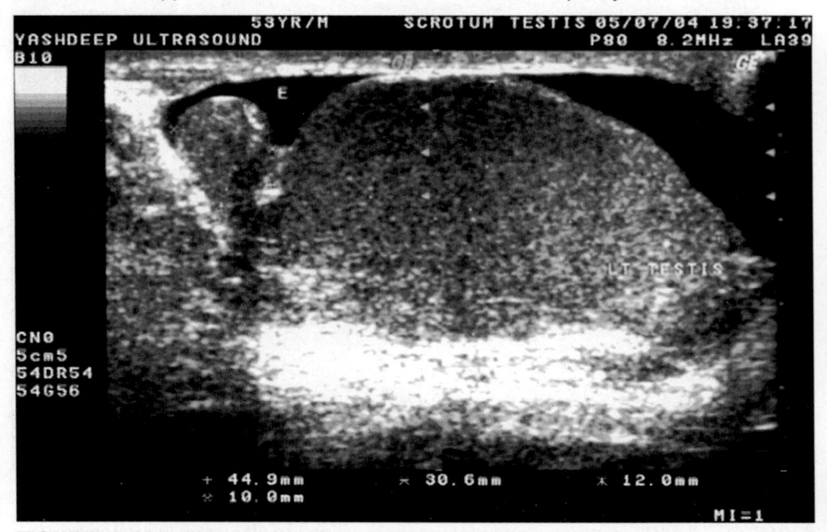

FIGURE 10.113: Acute epididymo-orchitis. A patient presented with acutely painful Lt scrotal swelling. HRSG shows diffusely hypoechoic-enlarged testis. Echo texture is homogeneous. However, the testis is swollen.

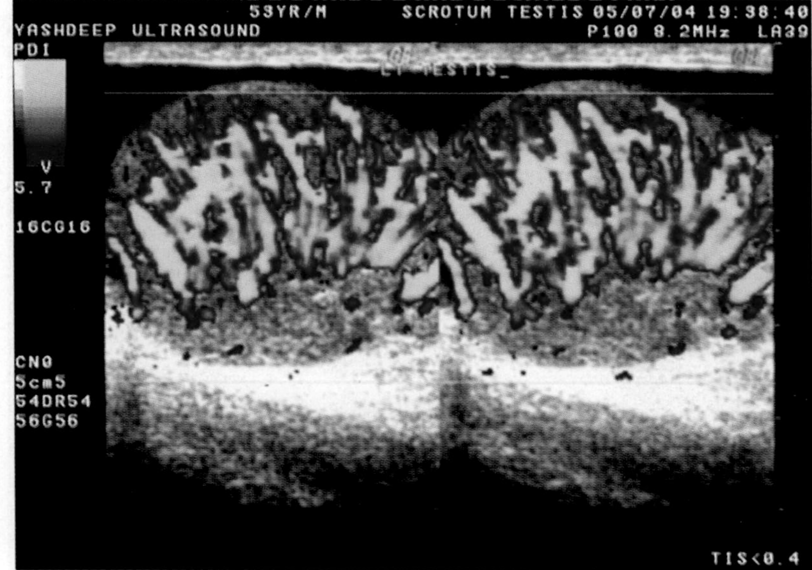

FIGURE 10.114

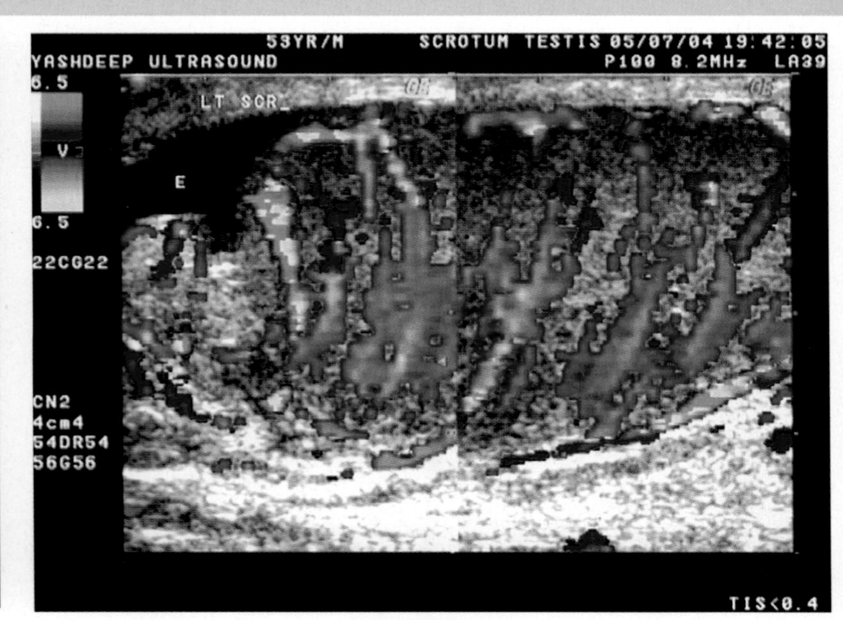

FIGURE 10.115

FIGURES 10.114 and 10.115: On color flow imaging the testis shows marked increased flow. On power Doppler imaging it looks like "testis on fire". The epididymis is also enlarged and hypoechoic. It also shows increased flow. The sonographic signs are suggestive of acute epididymo-orchitis.

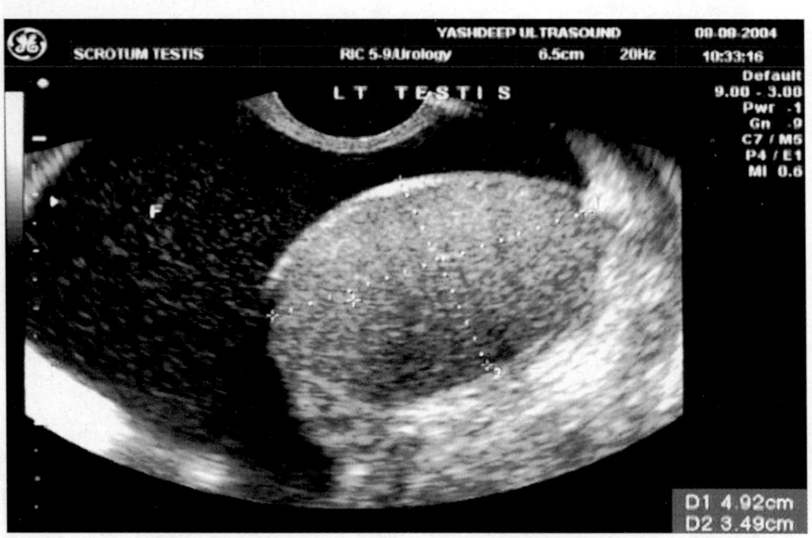

FIGURE 10.116: Acute epididymo-orchitis. A middle aged patient presented with painful Lt scrotal swelling. HRSG shows enlarged hypoechoic testis.

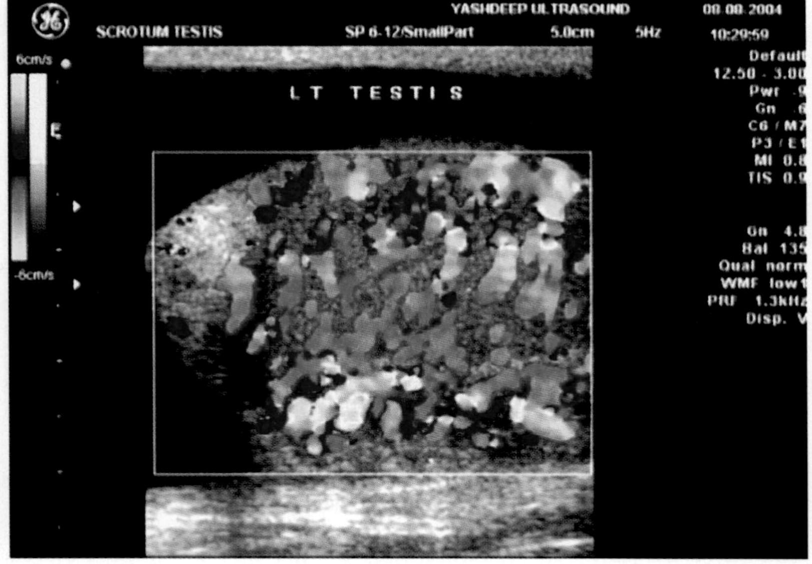

FIGURE 10.117

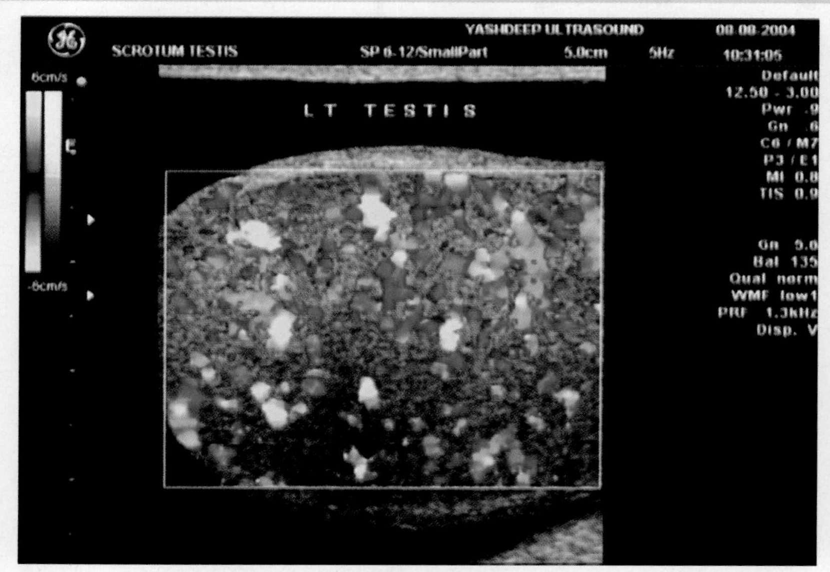

FIGURE 10.118

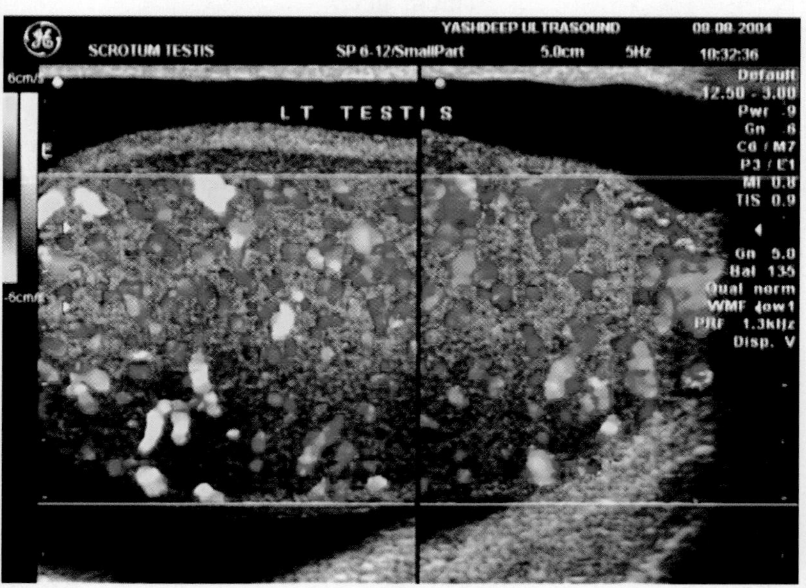

FIGURE 10.119

FIGURES 10.117 to 10.119: On color flow imaging marked increased flow seen on testis. Power Doppler imaging shows multiple vessels suggestive of acute infection.

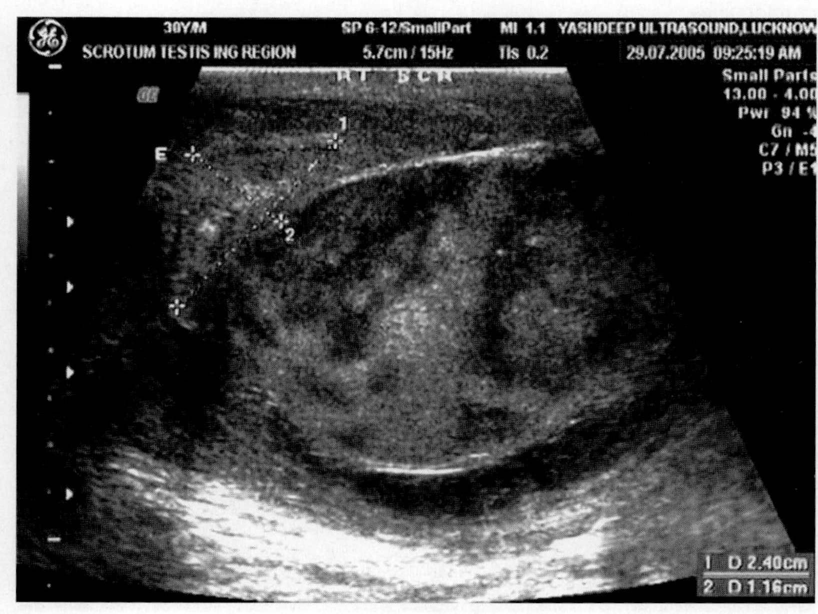

FIGURE 10.120

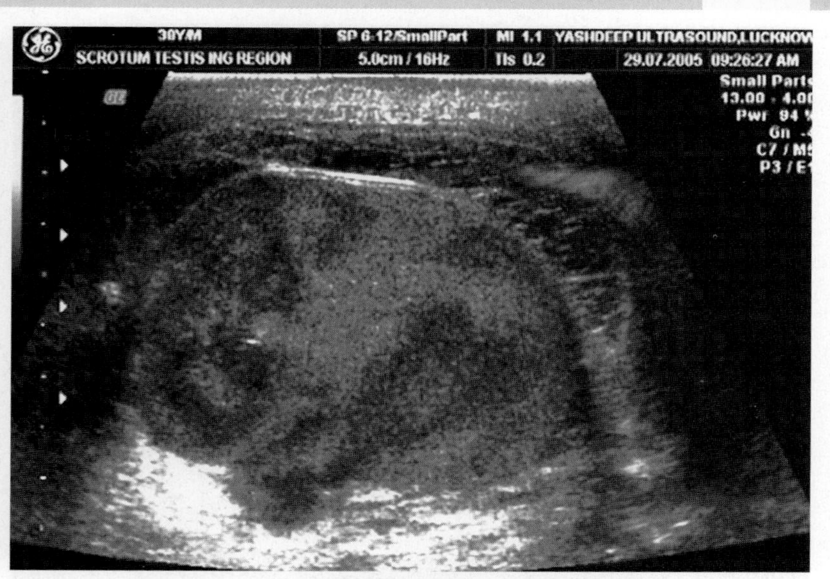

FIGURE 10.121

FIGURES 10.120 and 10.121: Epididymo-orchitis with infected tubercular abscess. A patient presented with Red Hot Rt scrotal swelling with acute pain with high fever. Marked scrotal skin edema was present. HRSG shows diffusely enlarged testis. Multiple irregular hypoechoic areas are seen in the testis suggestive of abscess formation. Low-level echoes are seen in them.

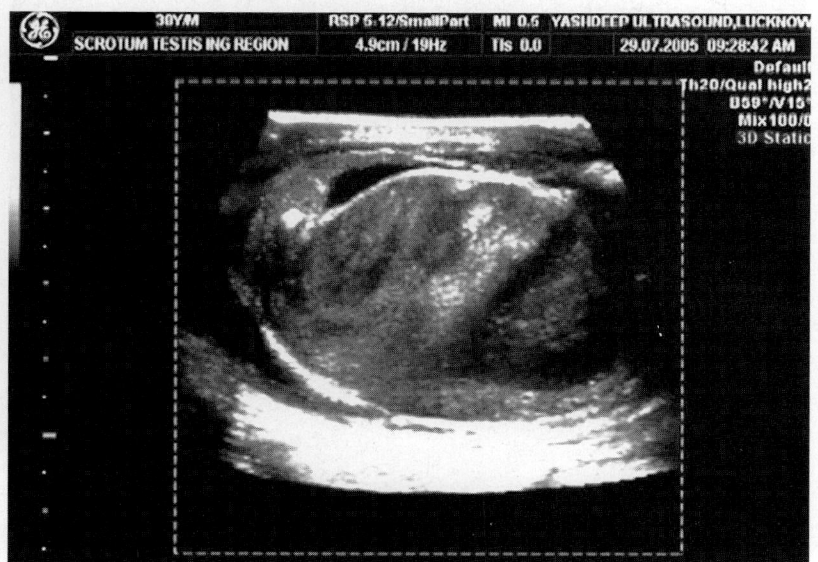

FIGURE 10.122

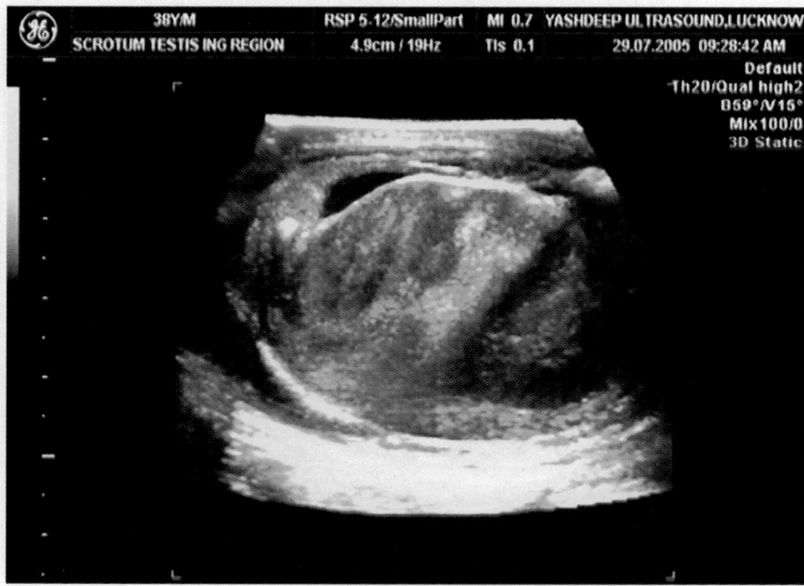

FIGURE 10.123

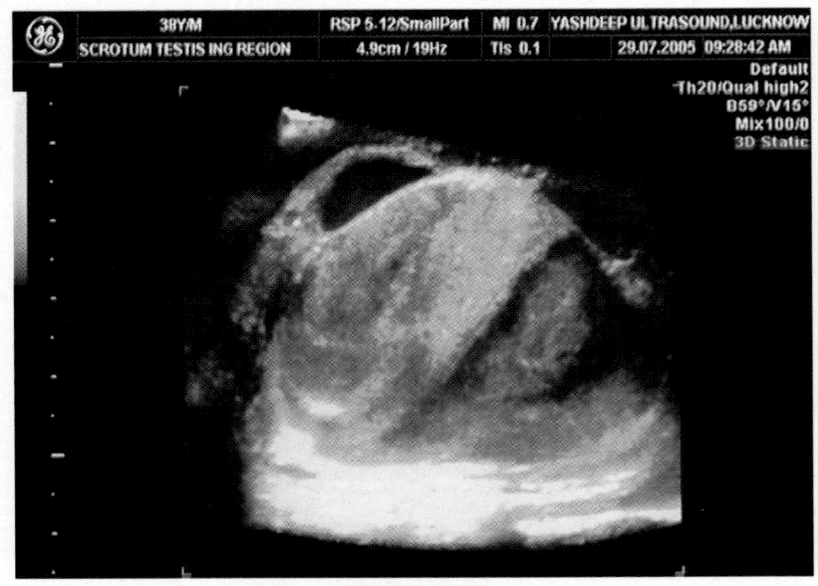

FIGURE 10.124

FIGURES 10.122 to 10.124: 3D surface rendering of the image shows the testis in much better view. The abscess cavity is well marked on 3D imaging.

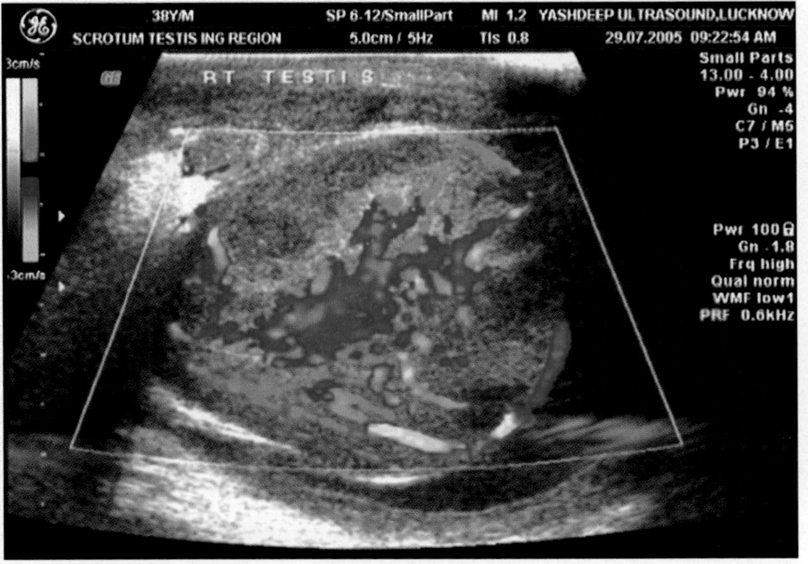

FIGURE 10.125

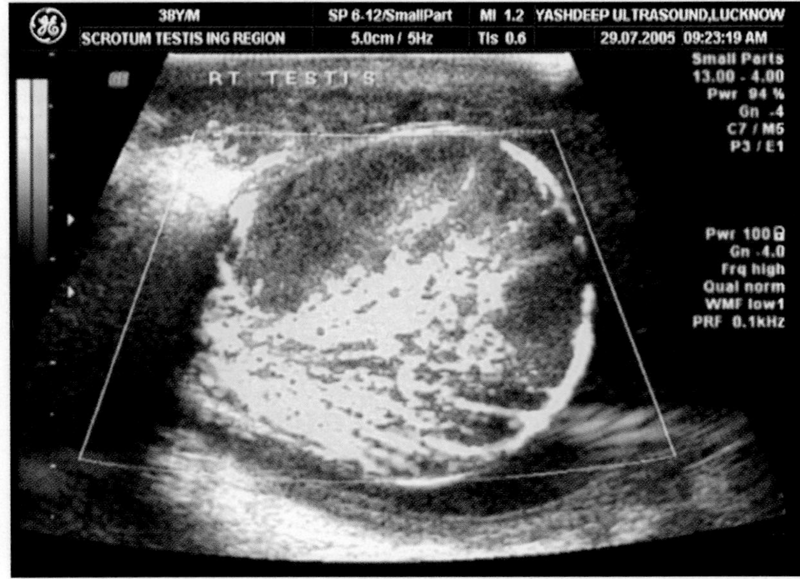

FIGURE 10.126

FIGURES 10.125 and 10.126: On color flow and power Doppler imaging marked increased flow seen in the testis. On surgical exploration it came out to be infected tubercular abscess.

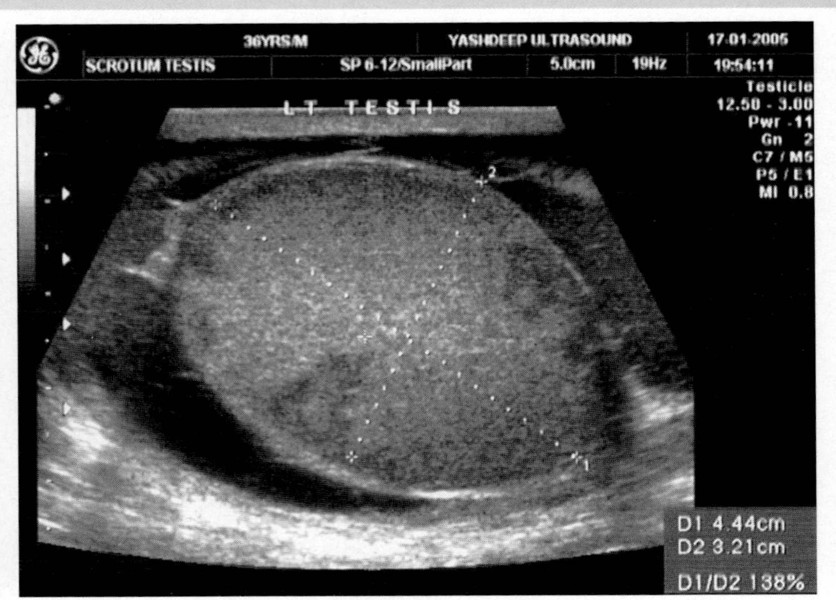

FIGURE 10.127

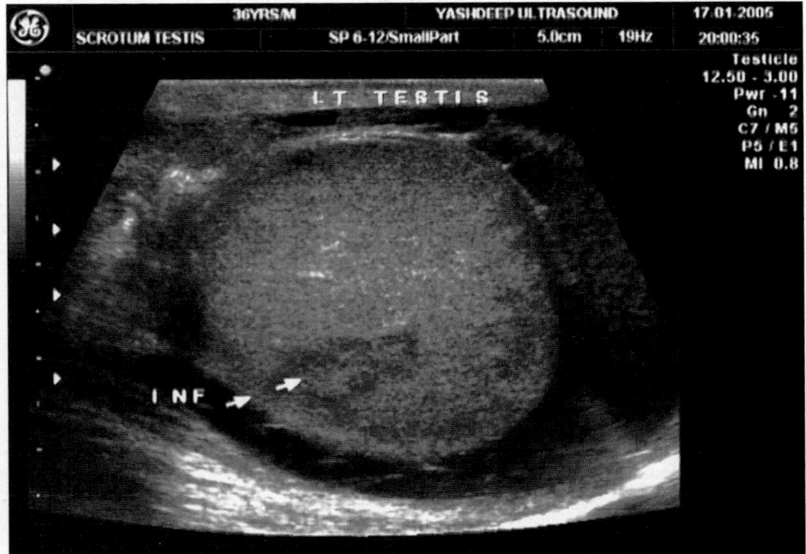

FIGURE 10.128

FIGURES 10.127 and 10.128: **Testicular infarcts.** A patient presented with painful swelling in the Lt scrotum. HRSG shows enlarged swollen Lt testis. Multiple wedge shaped areas are seen in the testis. They are coming from the periphery suggestive of infarctions.

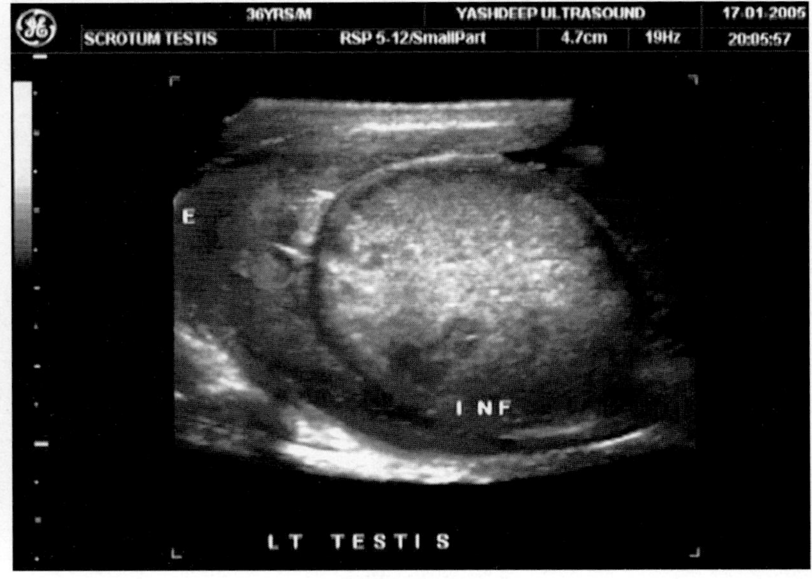

FIGURE 10.129

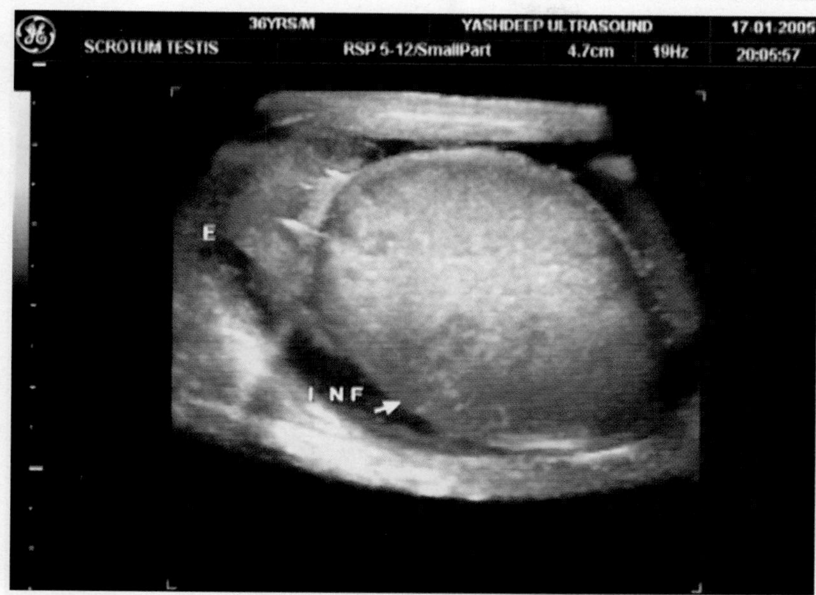

FIGURE 10.130

FIGURES 10.129 and 10.130: 3D surface rendering of the testis shows infarcts in much better view. They are seen as hypoechoic triangular areas.

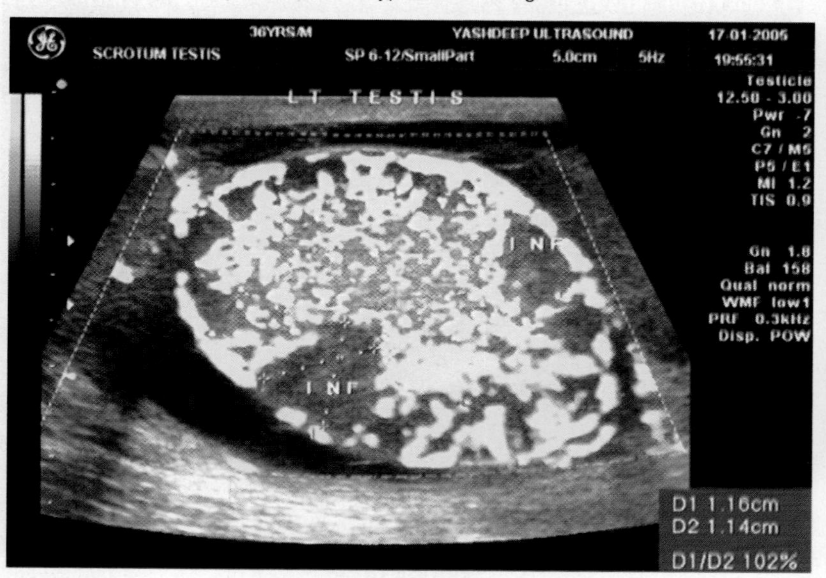

FIGURE 10.131

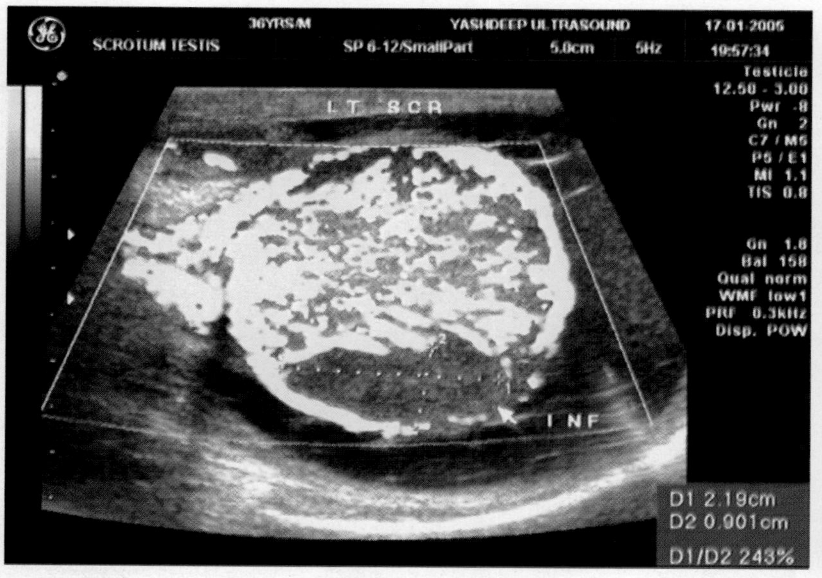

FIGURE 10.132

FIGURES 10.131 and 10.132: On power Doppler imaging the infarcted areas are seen devoid of blood supply. The normal testicular parenchyma shows high flow.

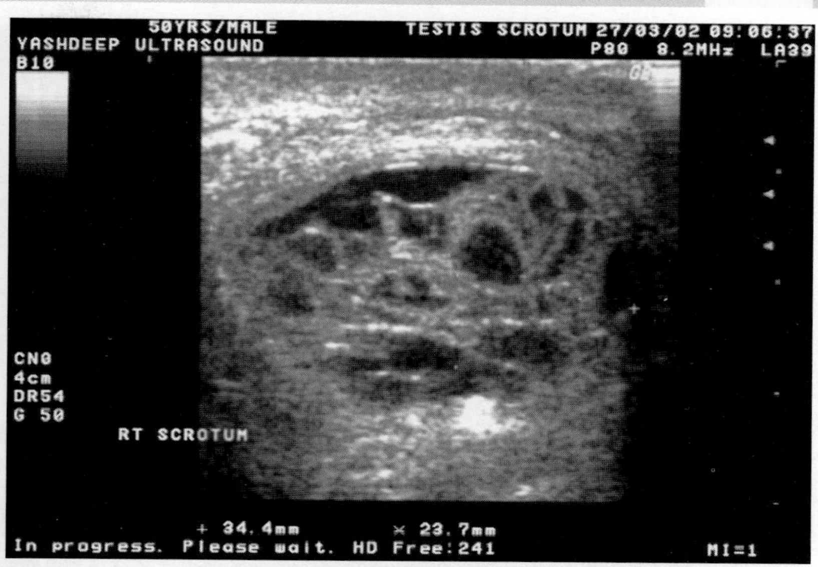

FIGURE 10.133

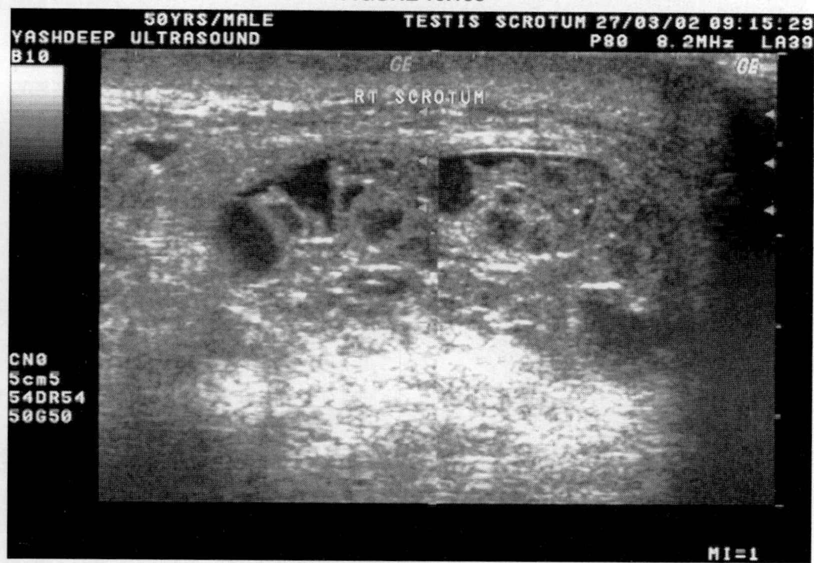

FIGURE 10.134

FIGURES 10.133 and 10.134: Scrotal abscess. A patient presented with painful swelling in the Rt scrotum. HRSG shows markedly edematous scrotal wall. There is evidence of thick low-level echo collection seen in the Rt scrotum. The collection is seen in the extra testicular space. Thick echogenic septa are also seen in it. However, the testis appears to be normal.

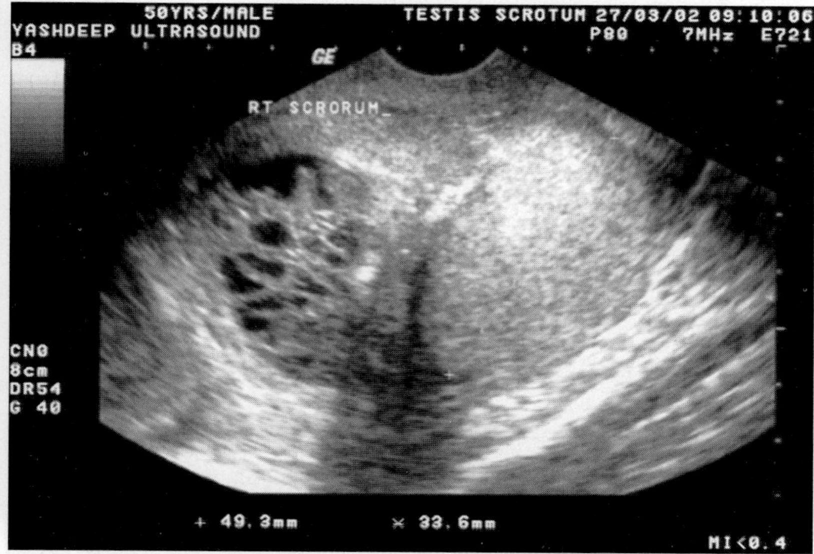

FIGURE 10.135: HRSG shows normal testis pushed to one side by the abscess in the scrotum.

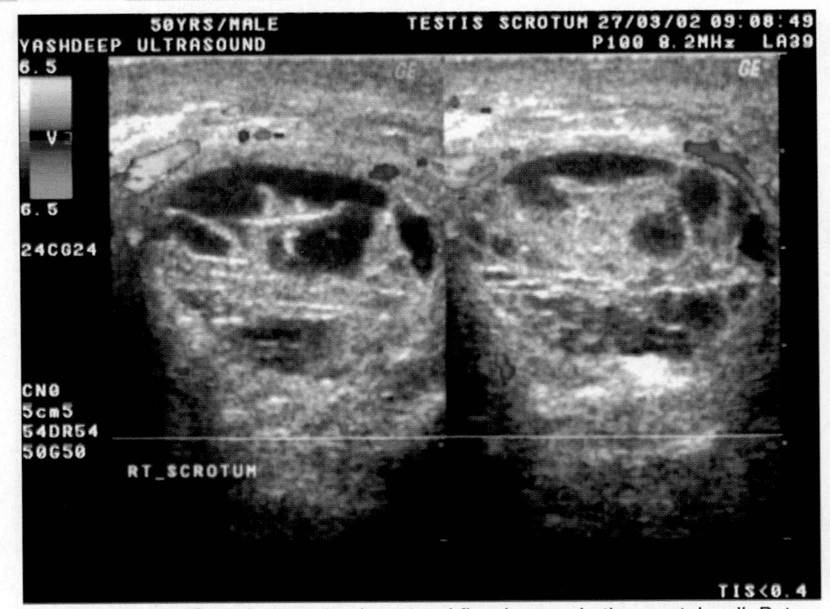

FIGURE 10.136: On color imaging increased flow is seen in the scrotal wall. But no flow is seen in the abscess.

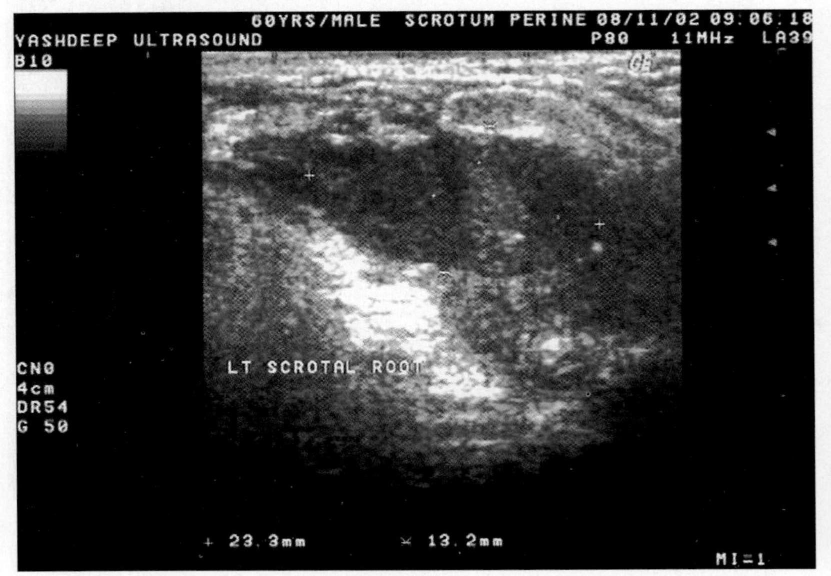

FIGURE 10.137: **Tubercular abscess at the root of the scrotum.** HRSG shows a patient presented with painful swelling at the root of the scrotum. HRSG shows an irregular hypoechoic low-level echo complex mass at the root of the Lt scrotum. Thick echoes are seen in it suggestive of an abscess.

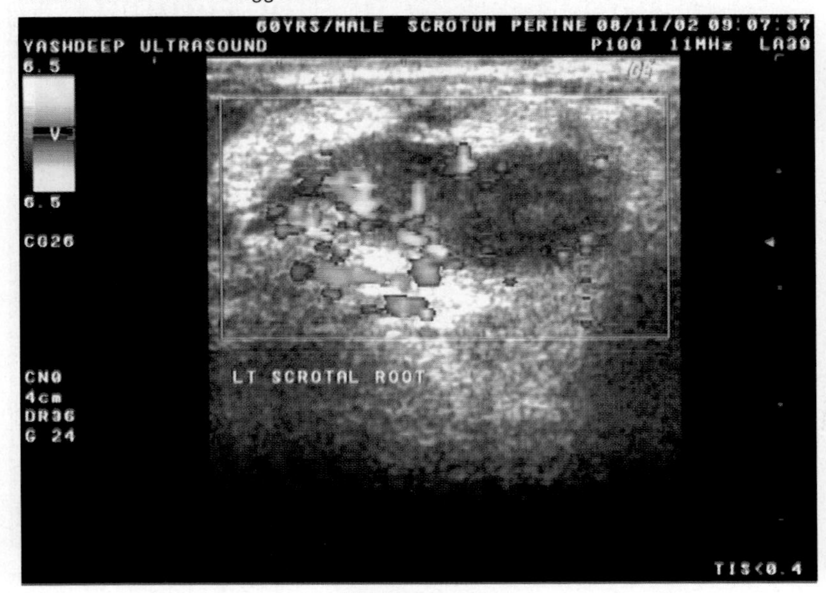

FIGURE 10.138

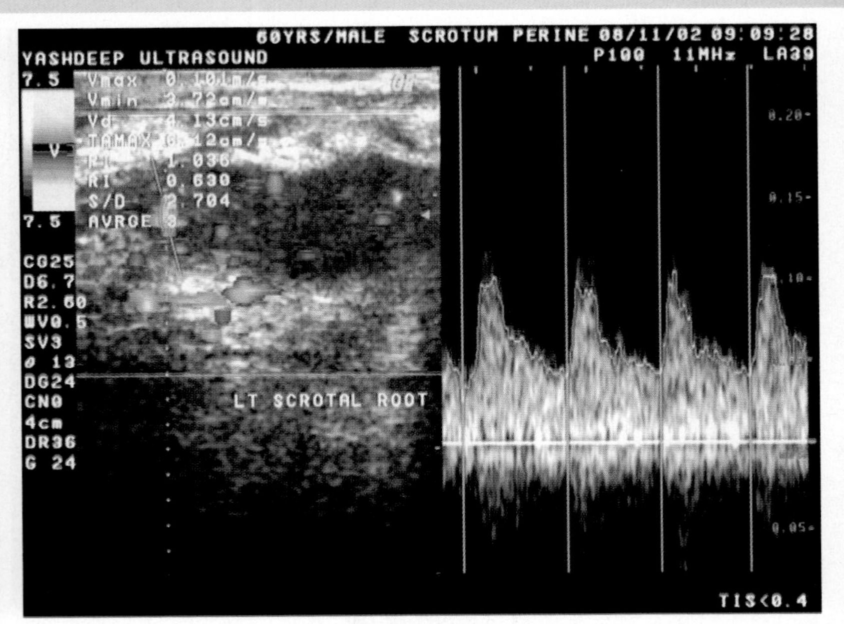

FIGURE 10.139

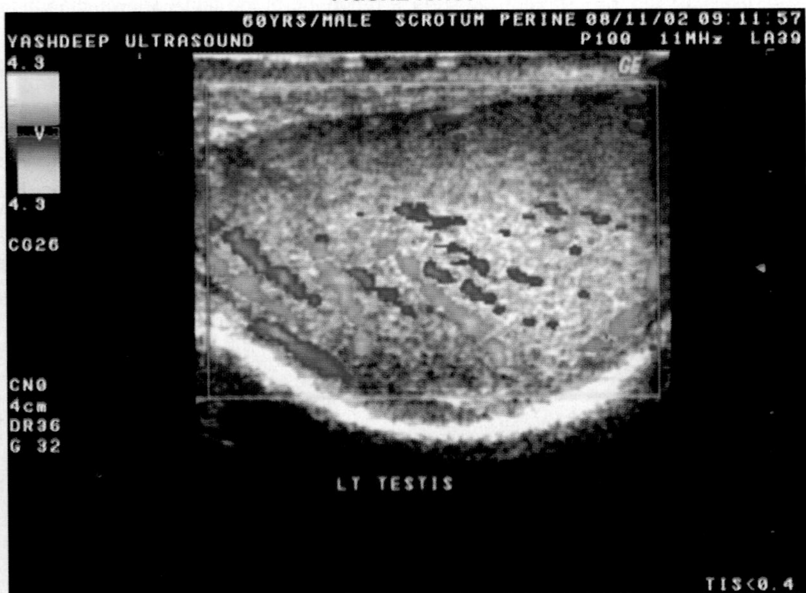

FIGURE 10.140

FIGURES 10.138 to 10.140: On color flow imaging increased flow is seen around the abscess. On spectral Doppler tracing high flow seen in the vessels. However, the testis appears to be normal and it shows normal flow on color flow imaging. The biopsy of the abscess confirmed tubercular abscess.

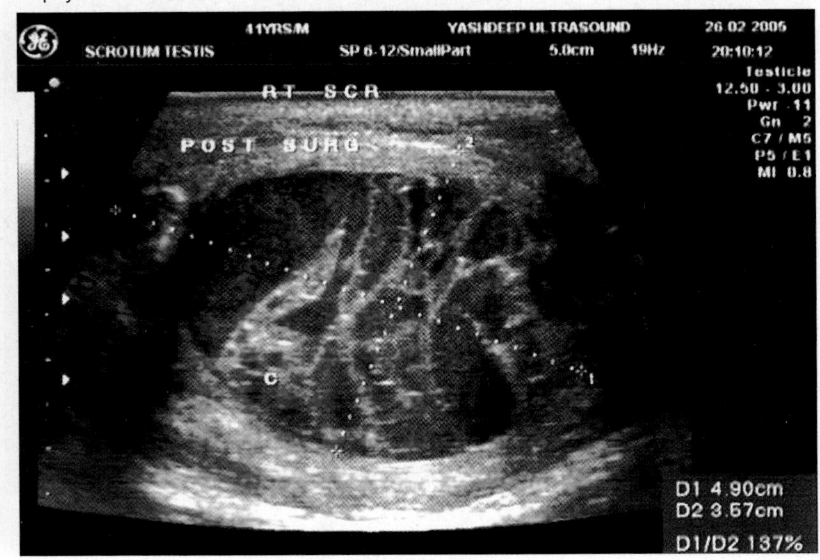

FIGURE 10.141

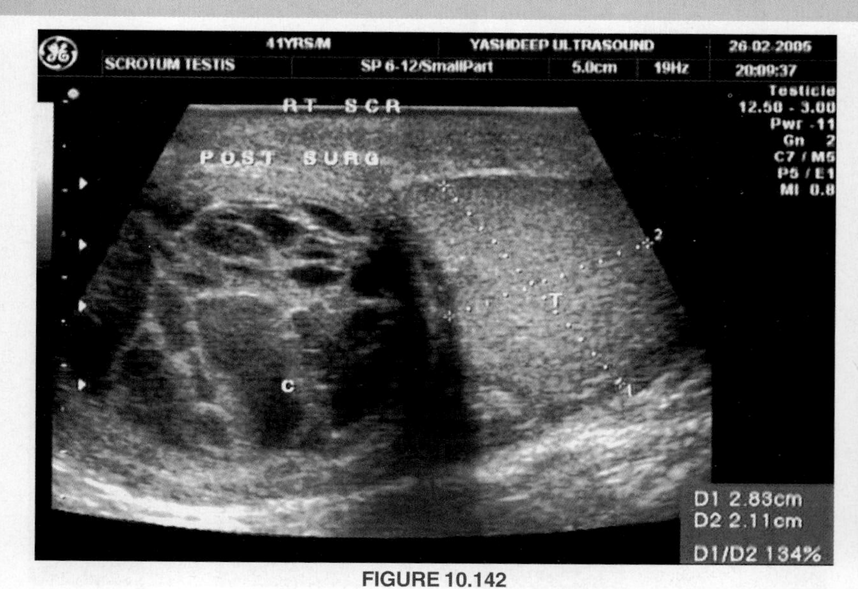

FIGURE 10.142

FIGURES 10.141 and 10.142: Post surgical abscess. A patient developed scrotal swelling with fever after surgery of hydrocele in the Rt scrotum. HRSG shows thick echo collection in the scrotum. The collection is seen in the extra testicular space. Multiple thick echogenic septa are seen in the collection suggestive of abscess formation. The abscess is seen pushing the testis. However, the testis is normal in appearance.

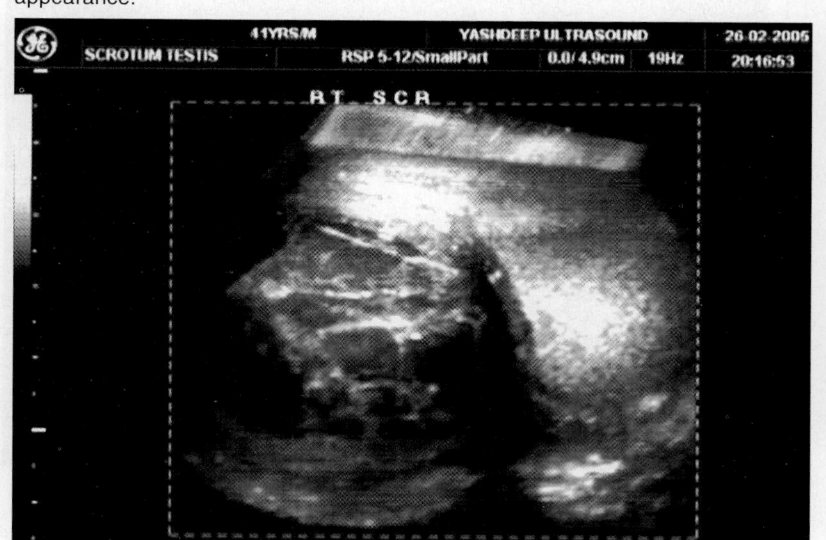

FIGURE 10.143

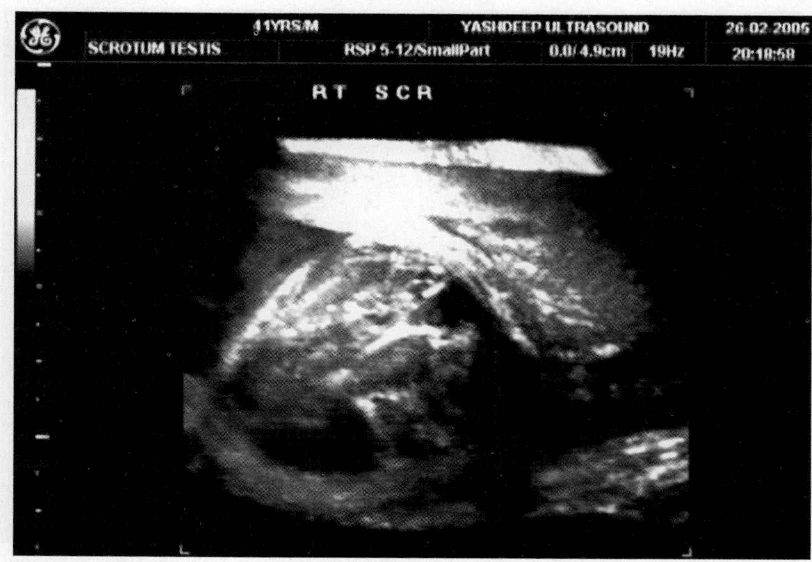

FIGURE 10.144

FIGURES 10.143 and 10.144: 3D surface rendering image of the scrotum shows the global view of abscess. It is seen pushing the testis up and medially abscess in much better view.

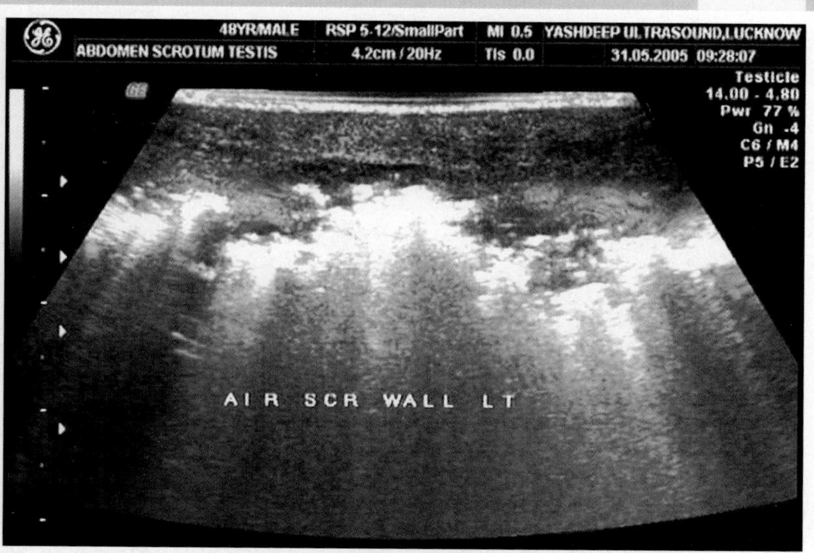

FIGURE 10.145

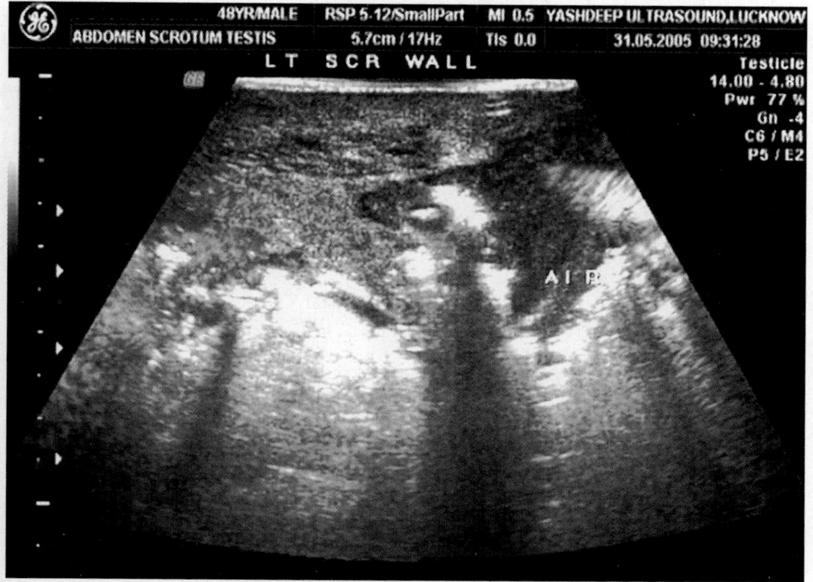

FIGURE 10.146

FIGURES 10.145 and 10.146: Anerobic scrotal wall abscess. A patient presented with acutely painful Lt scrotum. HRSG shows a low attenuating mass in the scrotal wall. Bright air is seen floating in the abscess cavity. Reverberation artifacts are seen due to the floating air.

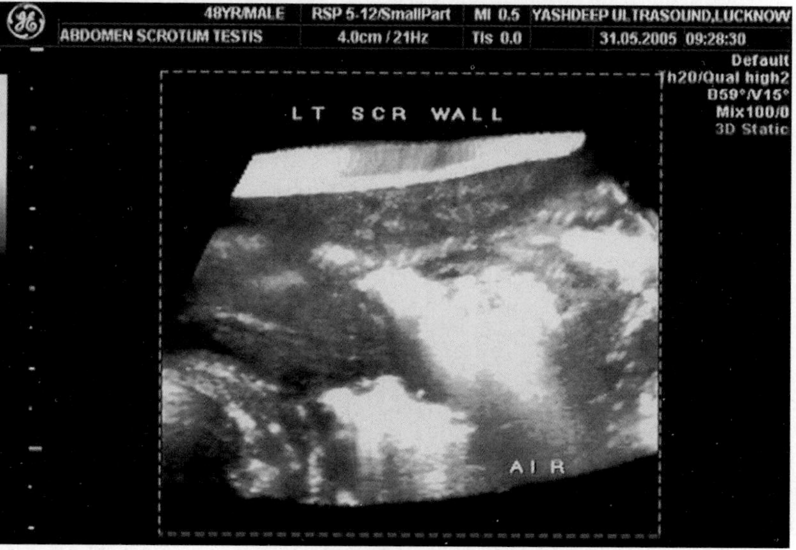

FIGURE 10.147

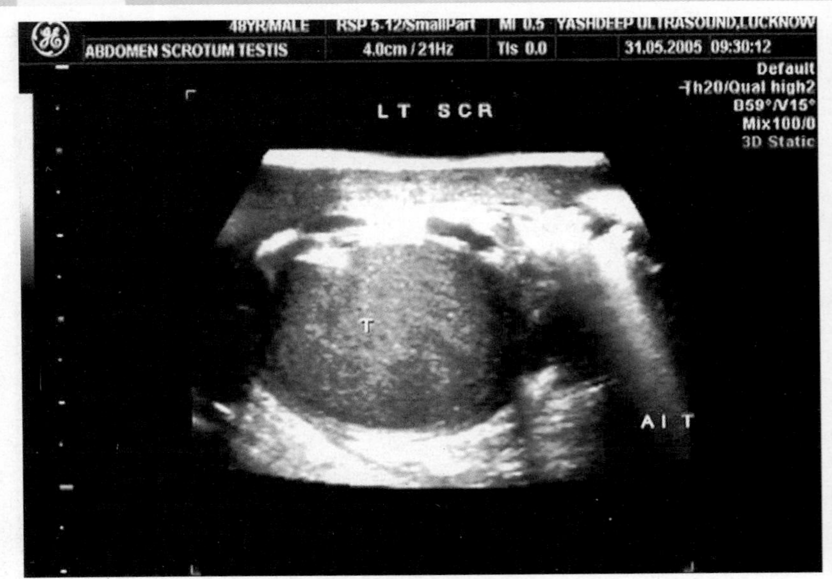

FIGURE 10.148

FIGURES 10.147 and 10.148: 3D imaging of the scrotal wall shows the testis spread from the abscess. It is confined only to the scrotal wall. Air is seen in the scrotal skin.

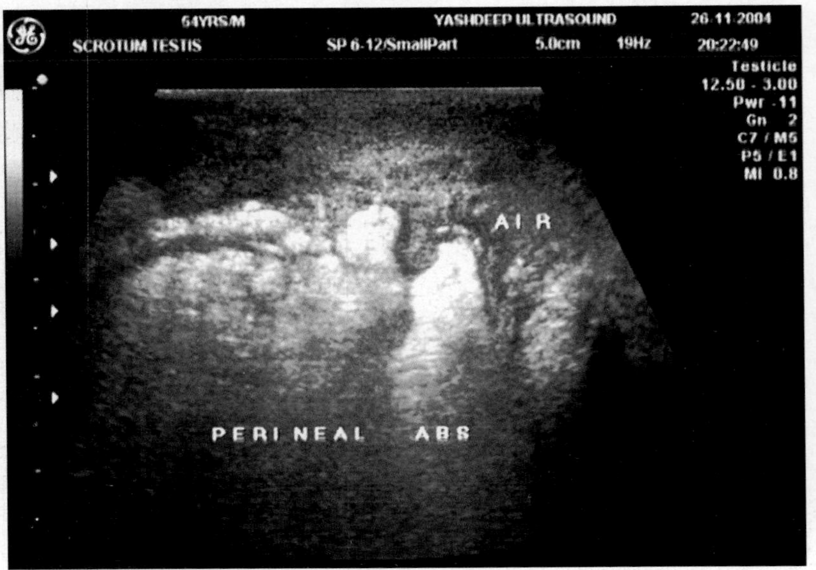

FIGURE 10.149

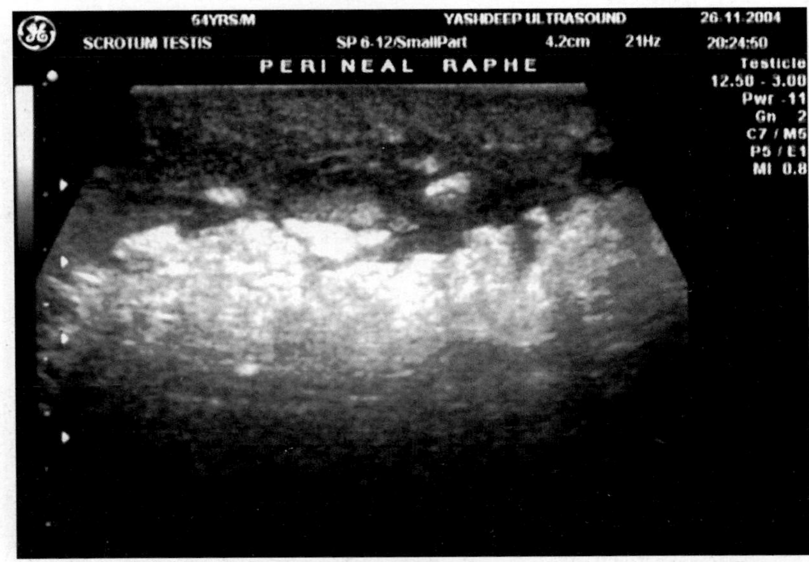

FIGURE 10.150

FIGURES 10.149 and 10.150: Anaerobic perineal abscess. A patient presented with acutely painful Lt scrotum with an abscess formation. HRSG shows an abscess cavity along the perineal raphe. Air is seen floating in the abscess. The abscess is seen.

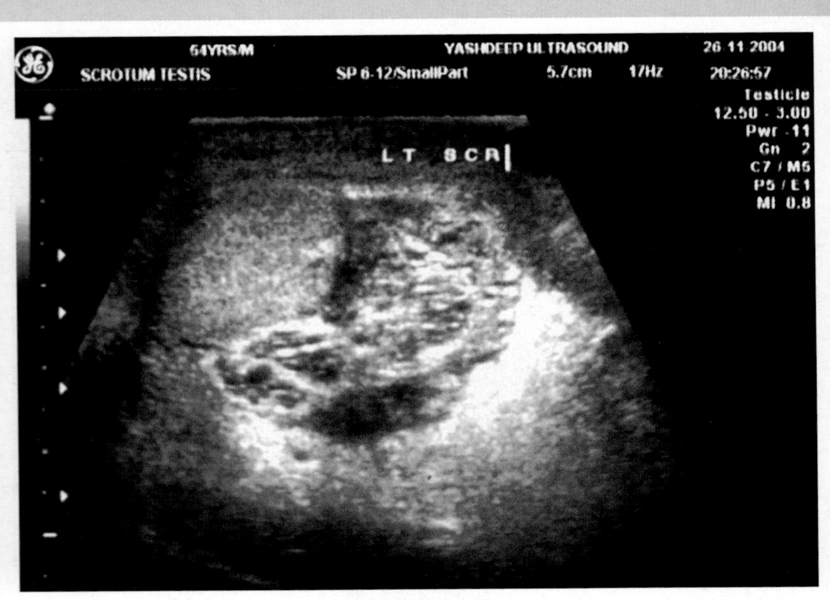

FIGURE 10.151

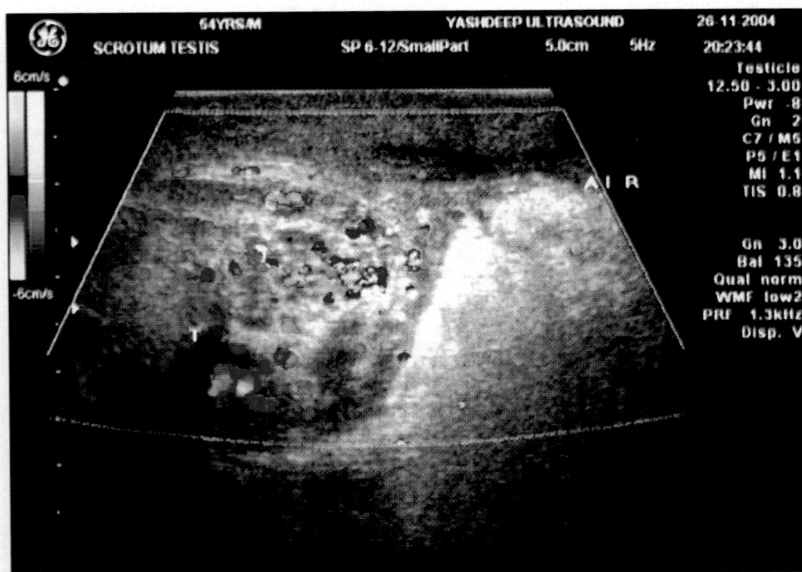

FIGURE 10.152

FIGURES 10.151 and 10.152: The abscess is seen expending in the Lt scrotum. However, the testis appears to be normal. No suggestion of any testis abscess is seen. On color flow imaging increased flow is seen in the scrotal wall.

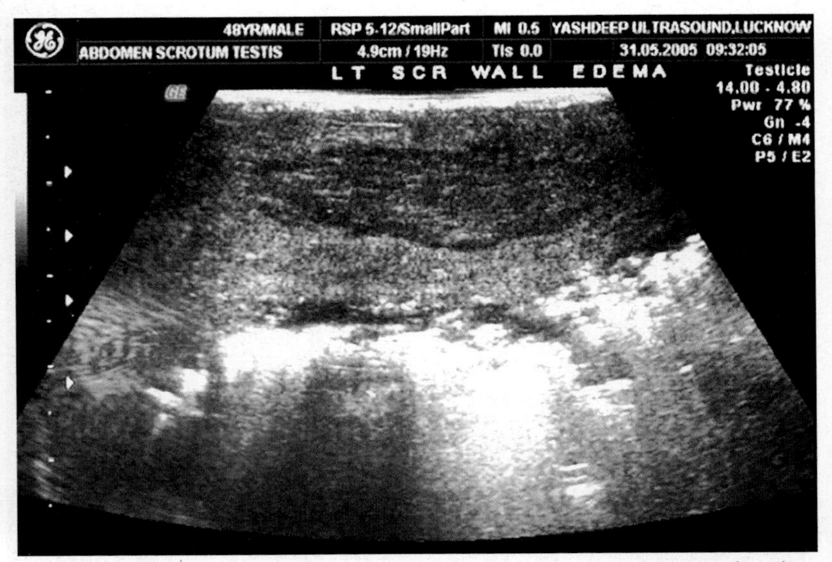

FIGURE 10.153: Impending abscess in the scrotal wall with orchitis. A patient presented with painful swelling in the Lt scrotum. HRSG shows markedly thickened Lt scrotal wall. Low-level echoes are seen in wall of the scrotum suggestive of breaking down of the abscess.

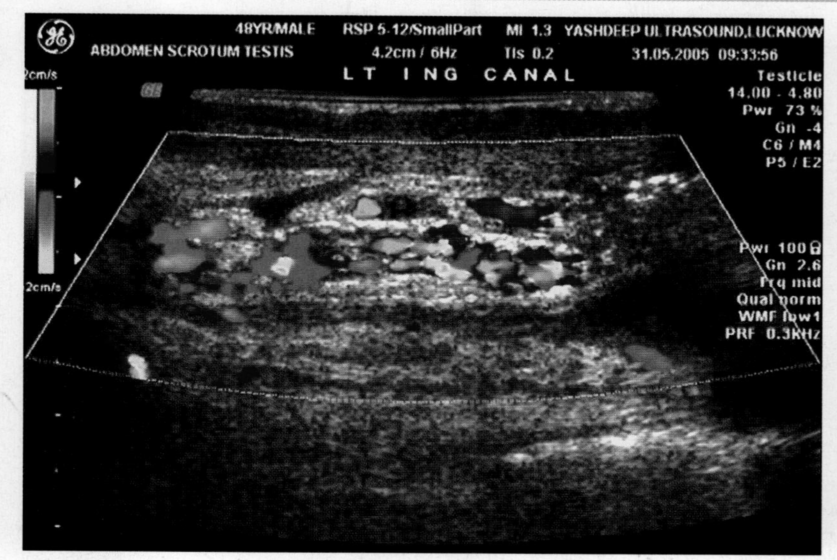

FIGURE 10.154

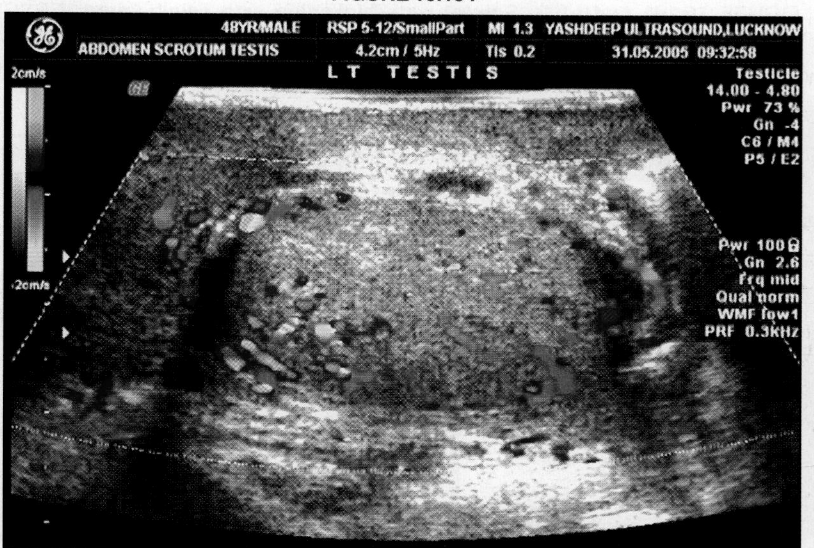

FIGURE 10.155

FIGURES 10.154 and 10.155: On color flow imaging multiple vessels are seen feeding the scrotal wall and inguinal canal. The testis is also enlarged and swollen with hypoechoic texture. Low-level echoes are seen in it suggestive of Orchitis.

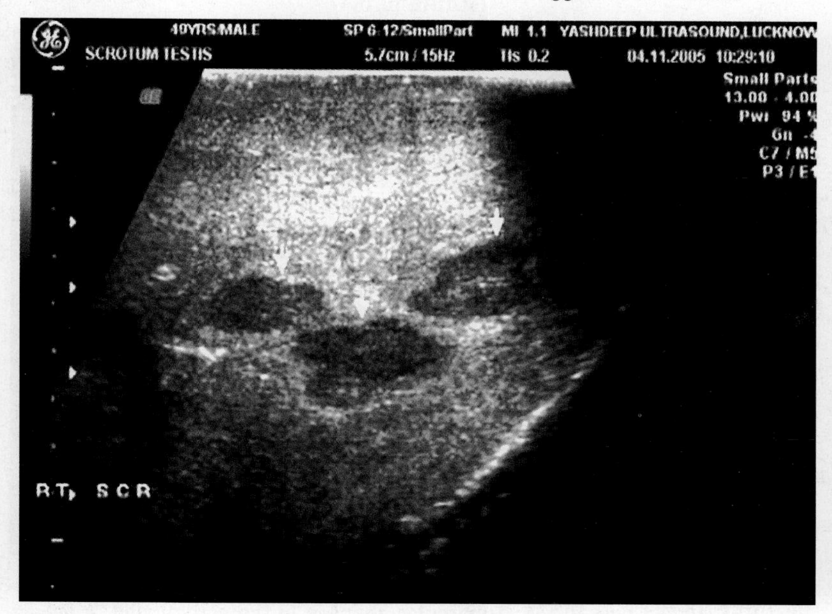

FIGURE 10.156

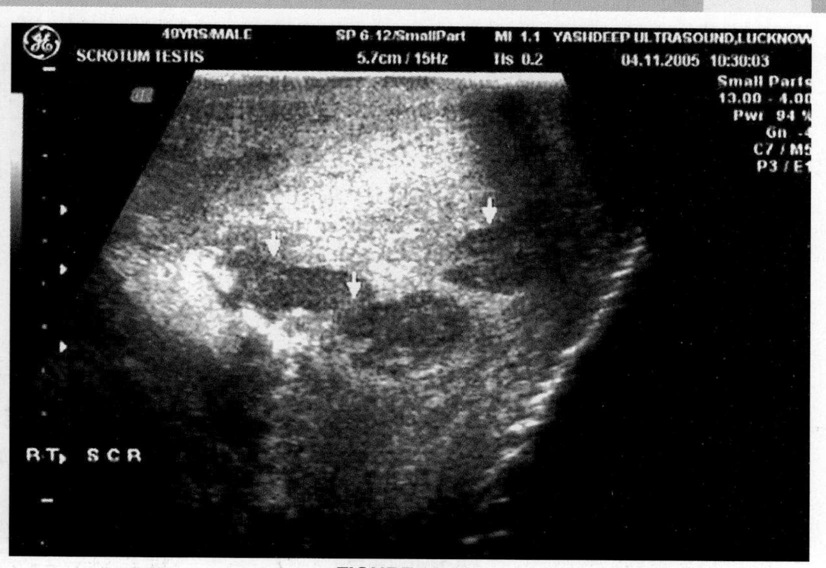

FIGURE 10.157

FIGURES 10.156 and 10.157: Scrotal wall abscess. A patient presented with painful swelling in the Lt scrotum. HRSG shows well defined low-level echo complex masses in the scrotal wall suggestive of abscesses.

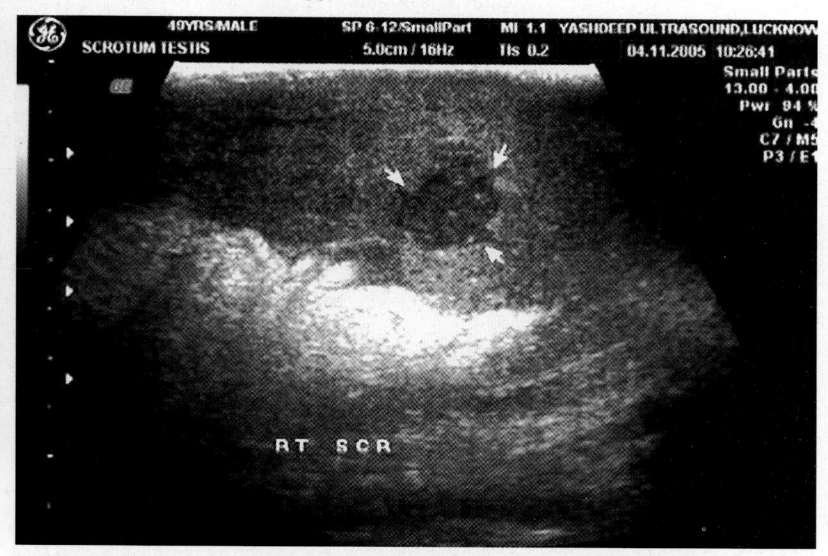

FIGURE 10.158: One abscess shows breaking down of the abscess.

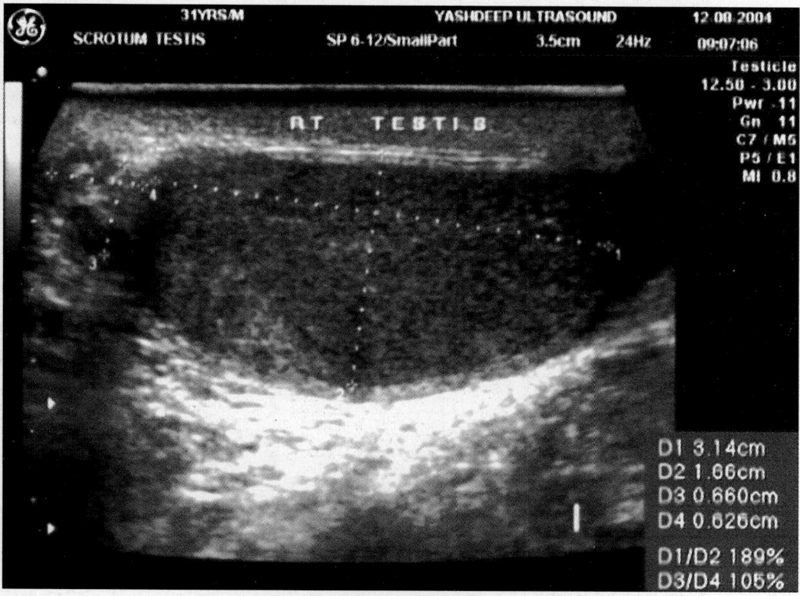

FIGURE 10.159

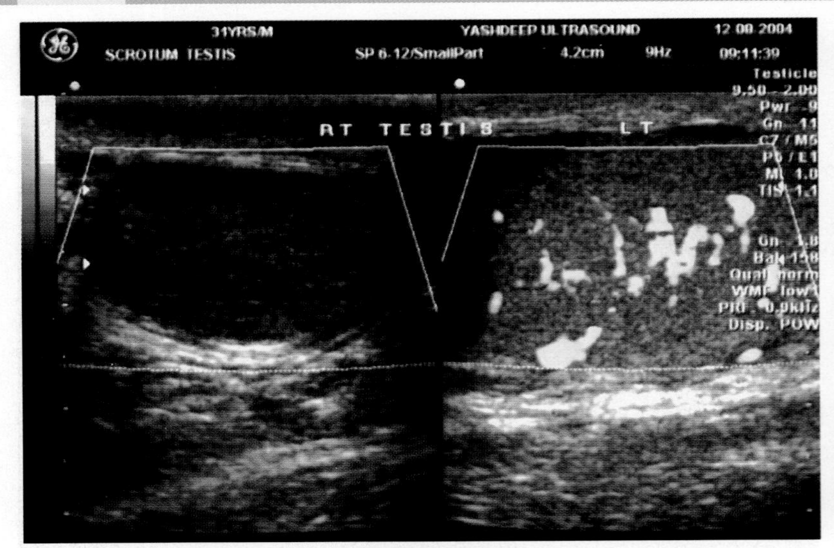

FIGURE 10.160

FIGURE 10.159 and 10.160: Chronic testicular ischemia. A patient presented with chronic pain in the Rt testis. HRSG shows small grossly hypoechoic testis. The testis shows inhomogeneous texture. On color flow imaging no flow in the testis. No edema is seen in the testis. The Lt testis is normal in appearance. Normal flow is seen in the Lt testis suggestive of Rt testicular Ischemia.

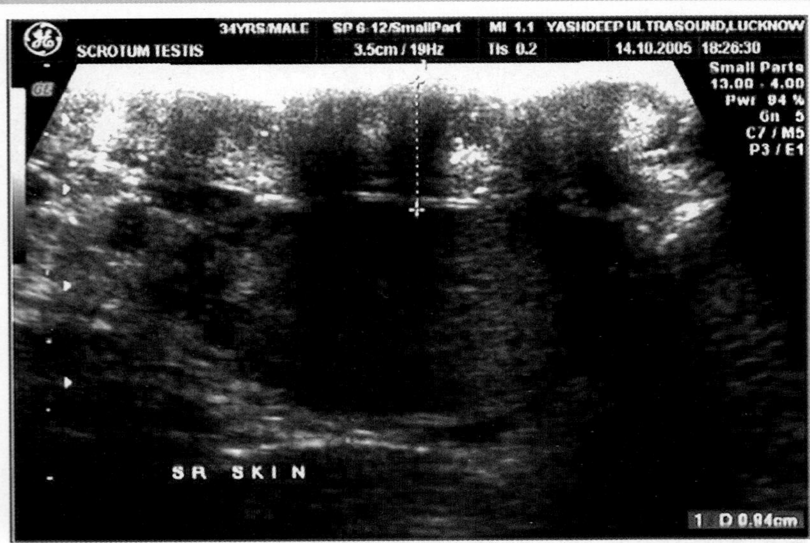

FIGURE 10.163

FIGURE 10.161 to 10.163: Scrotal wall thickening with calcification. HRSG shows A patient presented with markedly thickened scrotal wall. Multiple calcified specks are seen fixed with wall of scrotum. They are accompanied with acoustic shadowing. The testis is also hypoechoic in texture. Findings are suggestive of chronic inflammatory disease.

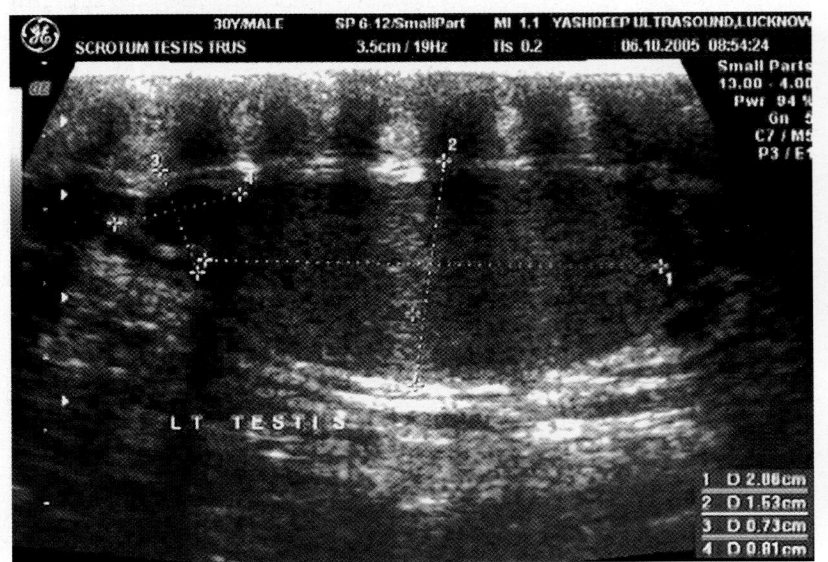

FIGURE 10.161

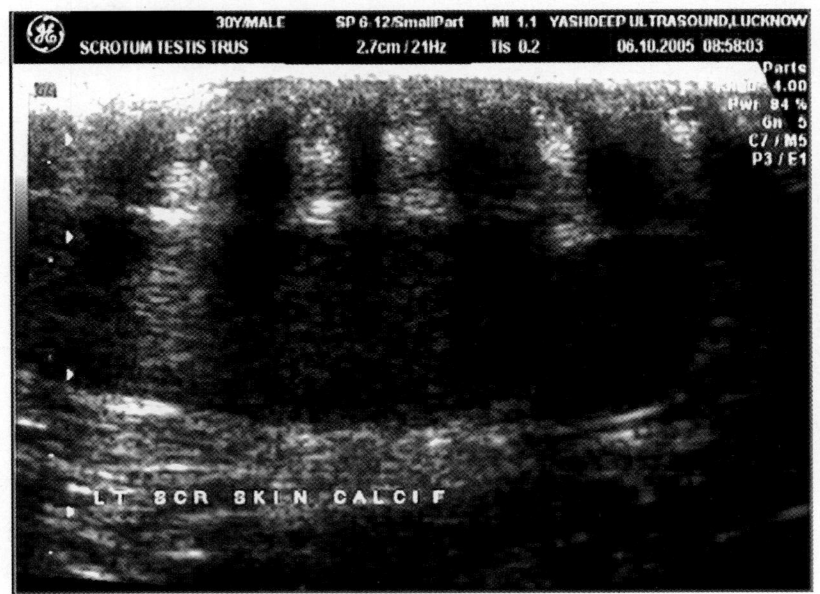

FIGURE 10.162

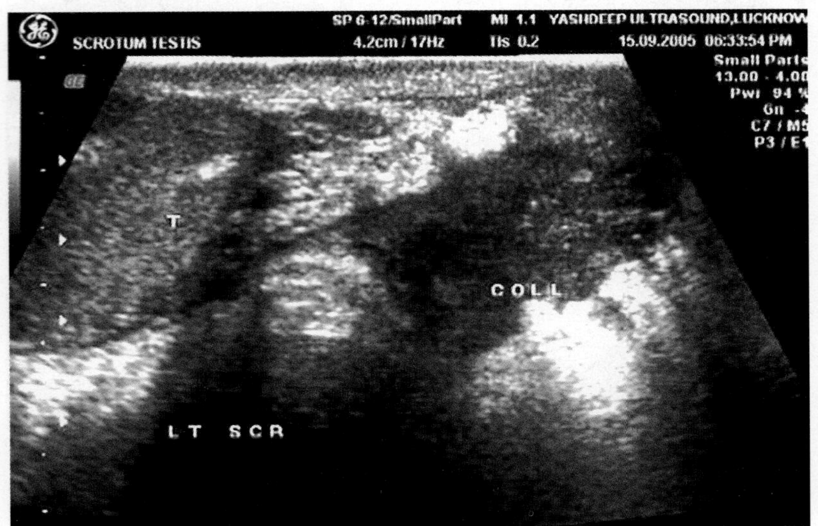

FIGURE 10.164

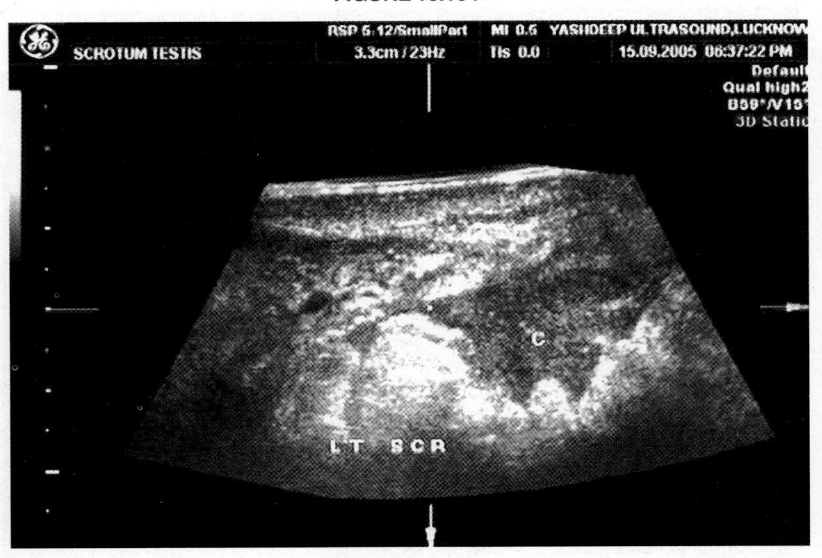

FIGURE 10.165

FIGURES 10.164 and 10.165: Tubercular abscess in the scrotal wall. A patient presented with discharging sinus in the Lt scrotal wall. HRSG shows a thick low-level echo complex in the scrotal wall. It is irregular in outline. 3D imaging of the scrotum shows the abscess in much details.

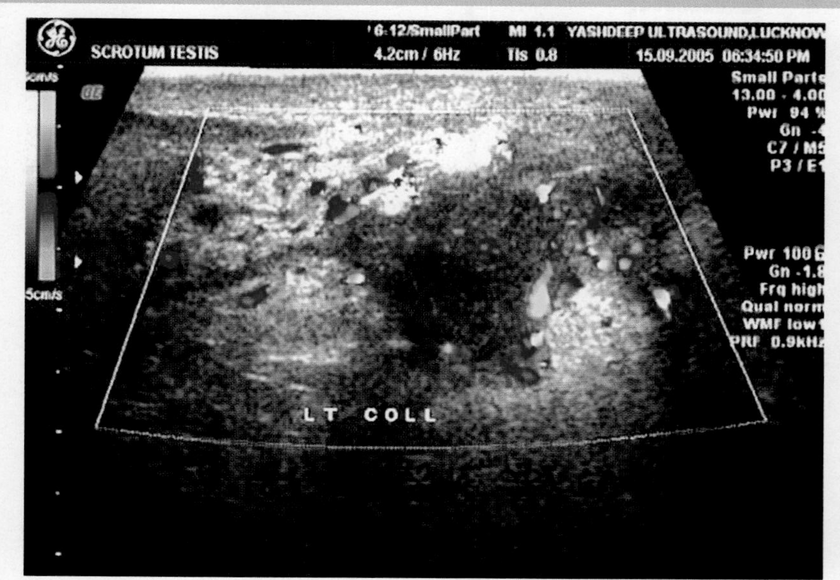

FIGURE 10.166: On color flow imaging moderate flow is seen in the abscess. Biopsy of the abscess turned out to be tubercular.

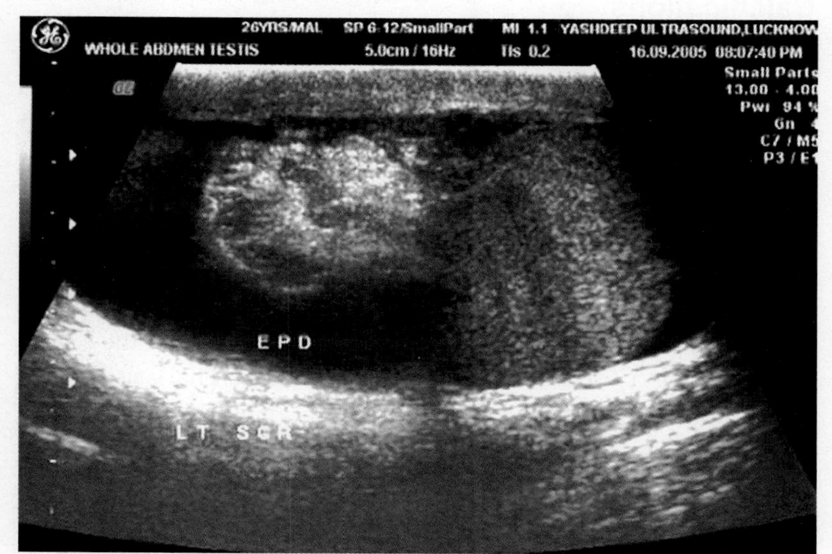

FIGURE 10.167

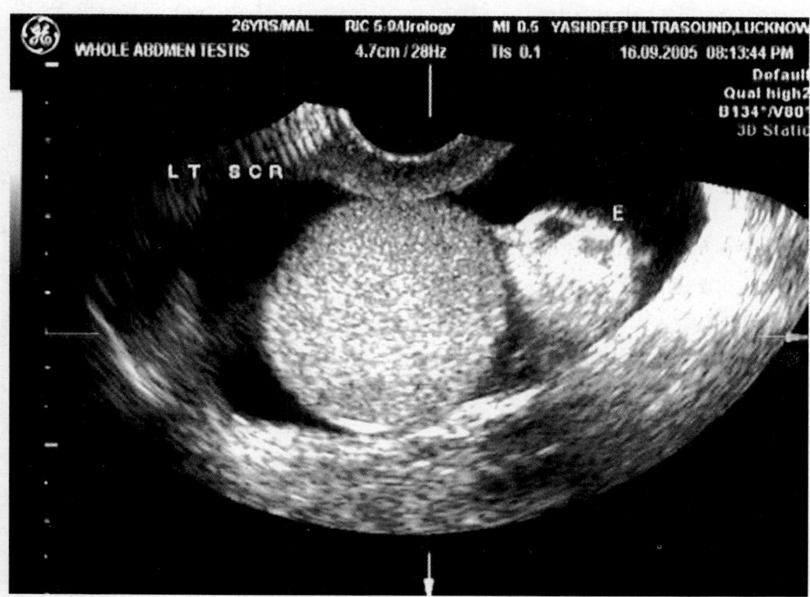

FIGURE 10.168

FIGURES 10.167 and 10.168: Scrotal trauma. A patient sustained painful swelling in the Lt scrotum after trauma. HRSG shows enlarged swollen echogenic epididymis multiple bright echoes are seen over the epididymis suggestive of hemorrhage. Low-level echoes are seen in the scrotum suggestive of reactionary hydrocele.

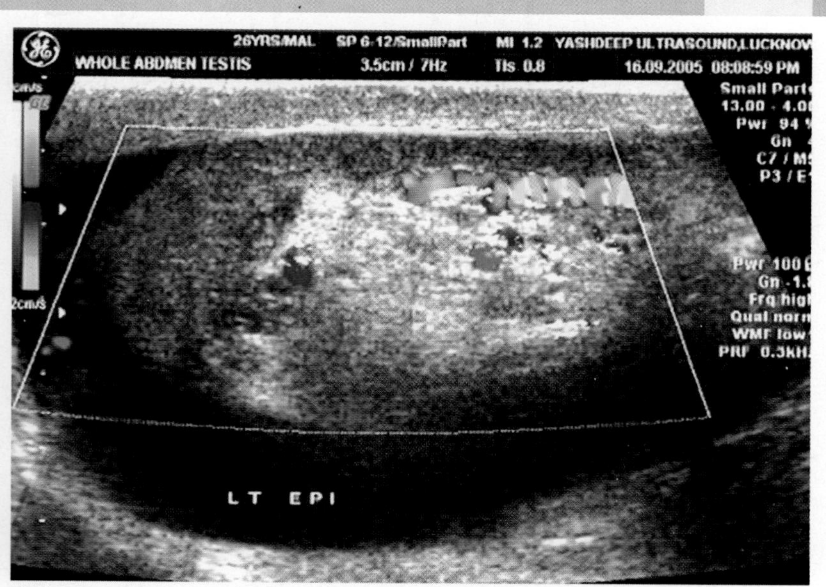

FIGURE 10.169: On color flow imaging: few vessels are seen in the epididymis suggestive of hemorrhagic epididymitis.

Other Fluids

Hematoceles and pyoceles are usually secondary to trauma and inflammation. They present as scrotal swelling. Sonography reveals fluid surrounding the testis. The fluid contains low level echoes, debris levels and internal septations. There may be associated air.

Scrotal lymphoceles are secondary to renal transplantation. They are the result of lymphatic dissection with perirenal lymphocele tracking down to the scrotum. It presents as any other fluid with low-level echoes and septations.

Testicular Tumors

Unless proved otherwise an intra testicular mass must be considered to be malignant. The small number (5% benign tumors include epidermoid cyst, leiomyomas, gonadostromal tumors, lipoma, fibroma, hamartoma and neurofibroma.

The following are the commonly encountered neoplasms of the testis.

1. Seminomatous germ cell tumors
   - Seminoma—40 to 50%
2. Non-seminomatous germ cell tumors
   - Embryonal cell carcinoma—15 to 20%
   - Teratocarcinoma
   - Choriocarcinoma
3. Metastatic tumors—0.6 to 3.6%.

Most tumors are of germ cell origin.

Most patients of testicular tumors present as painless testicular masses. Some of them complain of dragging scrotal pain. There may be associated symptoms of metastatic disease.

The commonly encountered sonographic appearances are as follows:

Focal Alteration in Testicular Echotexture (100% sensitivity)

Nodules as small as 5.0 mm can be resolved with great accuracy due to presently available high-resolution machines. The tumors are generally hypoechoic when under 8 to 10 mm in size. The margins are usually regular, and the tumors are well defined in comparison to the rest of the testicular echo texture. Larger tumors are invariably heterogeneous showing areas of calcifications, necrosis (cystic areas) and dysembryogenetic components.

Seminoma

Seminoma in 40 to 50% are the most common testicular tumor in the adult age group of 40 to 50 years. It is more common in undescended testis. Sonography findings include:

i. Hypoechoic solid mass
ii. Homogeneous echo texture of tumor,
iii. Round or oval shape of tumor with sharp delineation from normal testicular parenchyma, and
iv. Multifocal involvement.

Embryonal Carcinoma

Embryonal cell carcinoma is more aggressive than a typical Seminoma. Pure embryonal cell carcinoma is rare. Commonly it is a component of mixed tumors mainly in association with teratoma or teratocarcinoma. Sonographically, it demonstrates heterogeneous echo texture secondary to necrosis or hemorrhage, but often-cystic components and possibly calcification may occur.

Choriocarcinoma

Choriocarcinoma is a highly malignant tumor prone to hematogeneous metastasis.

Teratoma

Teratoma occurs commonly in infants and children. Sonographically, these lesions are markedly heterogeneous with multiple cystic areas and echogenic foci with or without acoustic shadowing, secondary to bone, fibrosis or hair.

Non-germ Cell Tumor

Non-germ cell tumors are usually benign, but may secrete steroids such as estrogen leading to endocrinological syndromes. These tumors may be malignant in 10% of the cases, and include those of Sertoli cell, Leydig cell or mesenchymal origin.

Metastasis

Metastasis to the testis may occur from renal and prostatic carcinoma, and rarely from bronchogenic, urinary bladder, melanoma and gastrointestinal neoplasms.

Adrenal Rest Tumors

Adrenal rest tumors of the testis may occur in cases of congenital adrenal hyperplasia or Cushing's syndrome leading to increased circulating Adrenocorticotrophic hormone (ACTH).

Color Doppler studies show increased vascularity. However, differentiation between benign tumors and neoplasms cannot be distinguished with certainty of color flow studies. However, the final answer is obtained either by an orchidectomy or a fine-needle aspiration cytology study.

Lesions of the Epididymis

Tumors

Tumors are very rare and are almost benign. They appear as well-outlined and demarcated solid mass usually of increased echogenicity. Adenomatoid tumors are most common epididymal tumors. These are rounded, homogeneous and isoechoic masses with adjacent epididymal tissue. Adenomatoid tumors may arise from the tunica albuginea and may be hyper or hypoechoic.

Cysts

Cysts are the most common scrotal masses and are seen with 20 to 40% are asymptomatic patients. Usually, these are perfectly anechoic, but some contain homogeneous low-level echoes. These may represent spermatocele and are usually seen in patients with previous history of vasectomy and contain spermatozoa.

Calcifications

Calcifications may be seen in the tunica of the testis near the epididymis or free floating within the scrotal sac. These are also called as "scrotal calculi" or scrotal pears and are always benign.

Scrotal Lymph Edema

In scrotal lymph edema, there is marked thickening of the scrotal skin and median raphe. The skin can by up to 4 cm in thickness and shows evidence of tubular channels of varying dimensions. There should be no confusion with varicoceles are lymphatics are within the scrotal wall. Whereas, varicoceles are inside the scrotal sac. Filariasis is one of the common causes of lymph edema in India. Some workers have been successful in isolating live filarial worm from lesion recently.

Tuberculosis

The common ultrasonographic features noted in chronic inflammatory disease as in tuberculosis of the scrotum are as follows:

1. An enlarged deformed epididymis especially in the tail.
2. Heterogeneous echo texture with associated calcification.
3. Extra testicular nodules.
4. Associated hydrocele, which is invariably septate.

Epididymis is the main site of tubercular involvement in the scrotum. There is enlargement of the epididymis followed by abscess formation. Adjacent scrotal wall is thickened with sinus formation and coarse thickening of tunica vaginalis. Involvement of contra lateral epididymis (Probably by lymphatic route) is involvement of testis. Resolution of tubercular epididymitis may leave residual extra testicular masses with large area of calcification.

Though tuberculosis primarily involves the epididymis, there is a definite entity where there is primary involvement of the testis. The testis is then enlarged, hypoechoic and the epididymis is completely normal. An FNAC (fine-needle aspiration cytology) is of great help. Primary testicular tuberculosis is commonly found in patients who are on immunochemotherapy for generalized diseases like leukemia or lymphoma. The color flow study does not show any significant increase in vascularity.

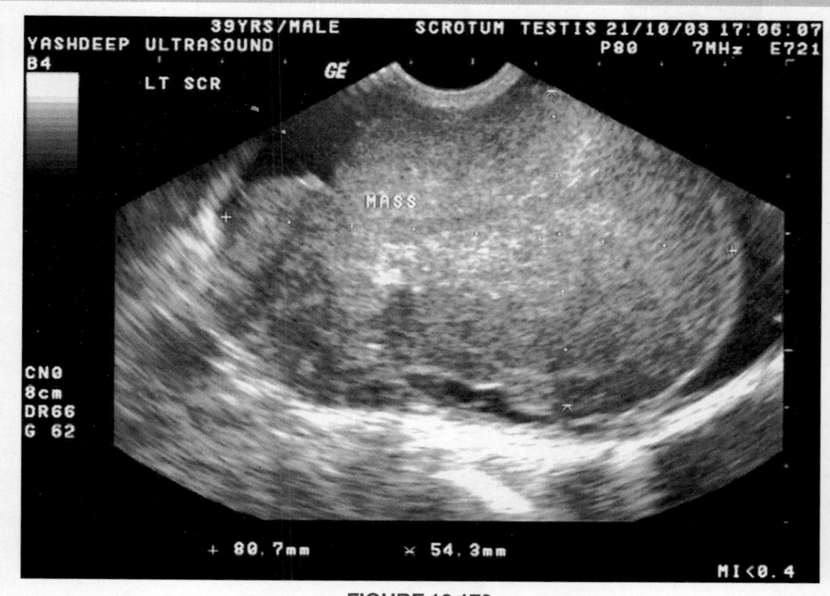

FIGURE 10.170

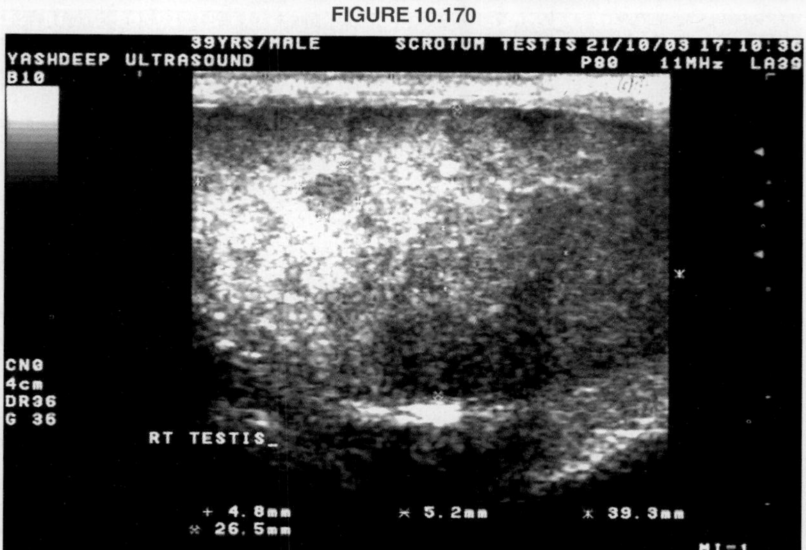

FIGURE 10.171

FIGURES 10.170 and 10.171: Disseminated lymphoma involving both testis. A patient presented with soft tissue swelling in the Lt scrotum. HRSG shows a big heterogeneous mass in the Lt scrotal sac. The mass is seen invading whole of the testis. Normal testis texture is lost. No necrosis is seen in the mass. Rt testis is also shows heterogeneous mass. Normal texture is lost with diffuse enlargement of the testis.

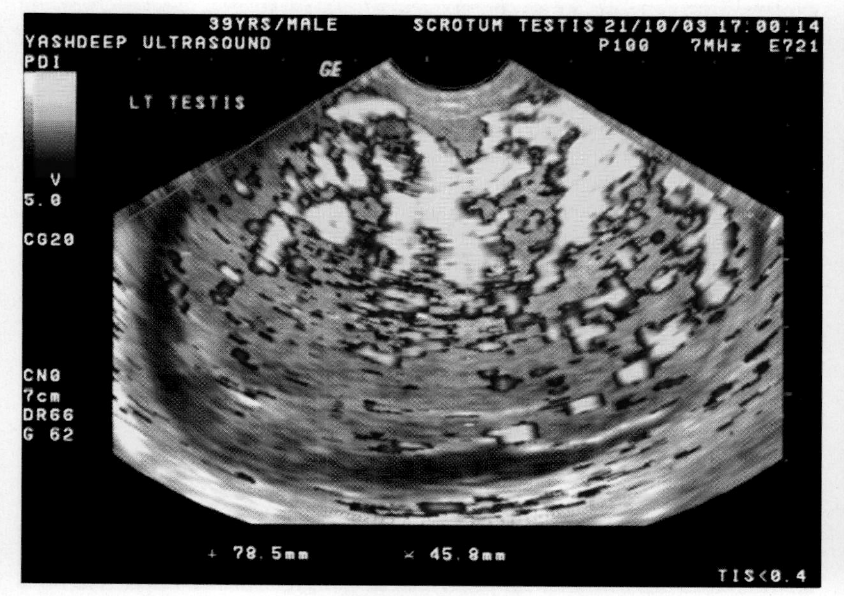

FIGURE 10.172

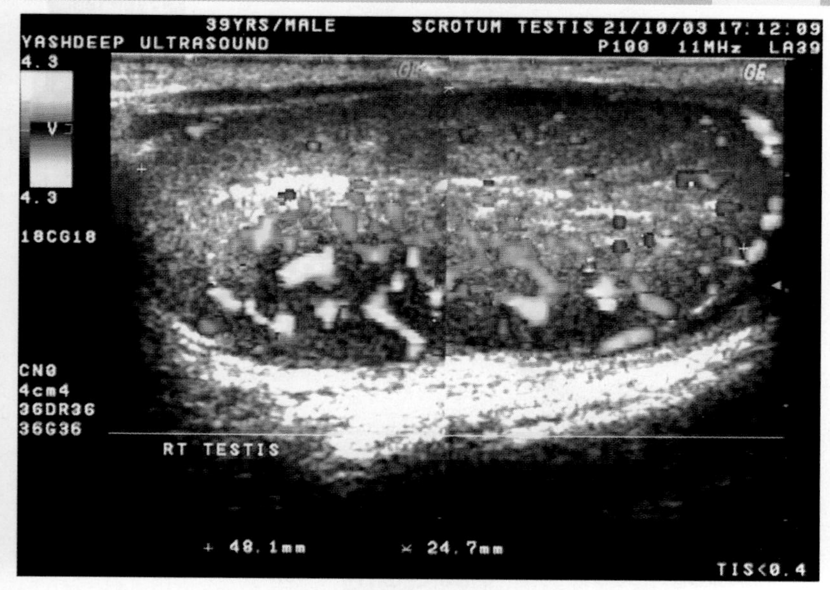

FIGURE 10.173

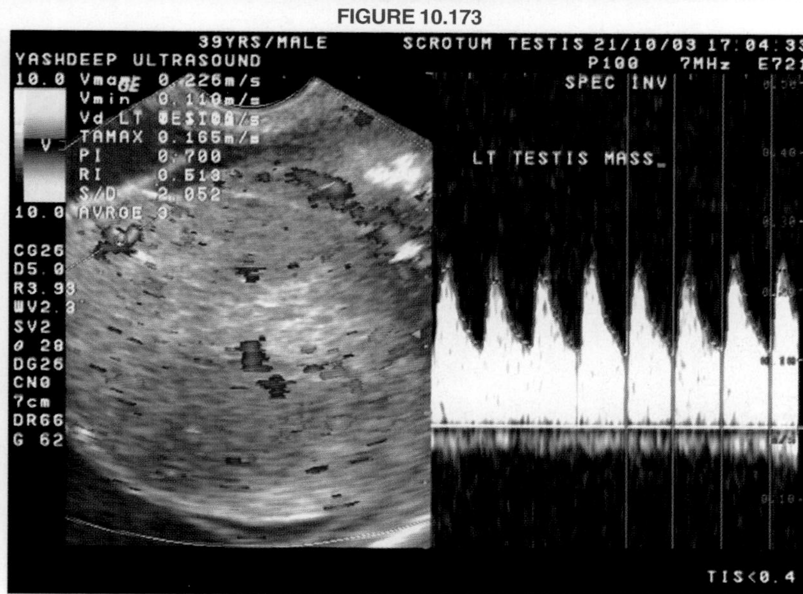

FIGURE 10.174

FIGURES 10.172 to 10.174: On color flow imaging marked increased flow is seen in the testicular mass on both sides. On spectral Doppler tracing high arterial flow seen in the mass.

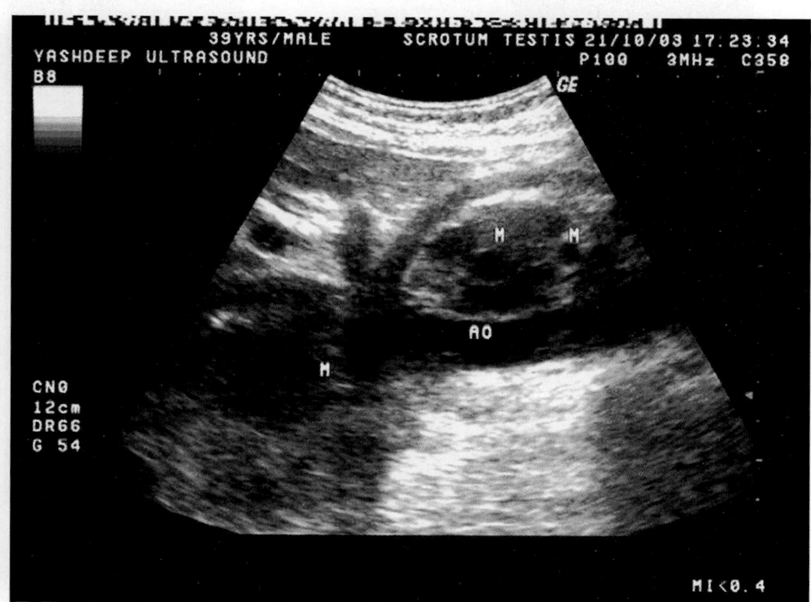

FIGURE 10.175

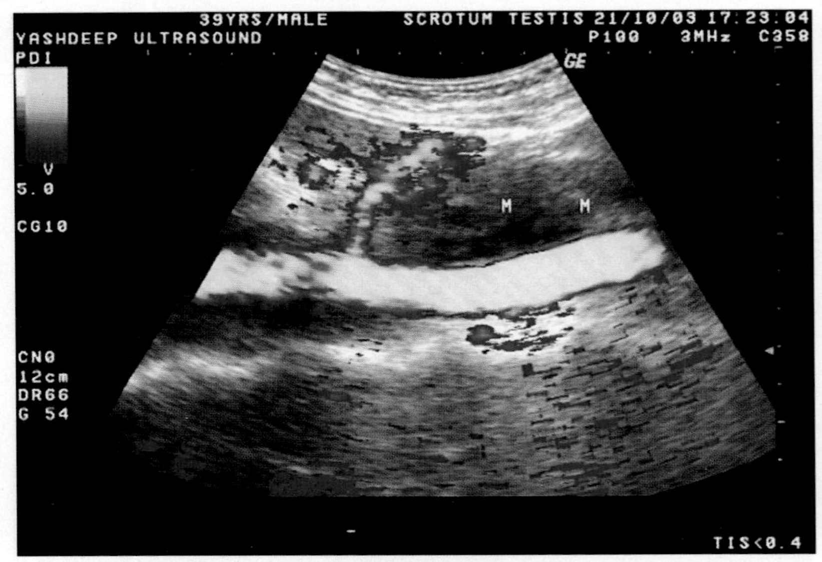

FIGURE 10.176

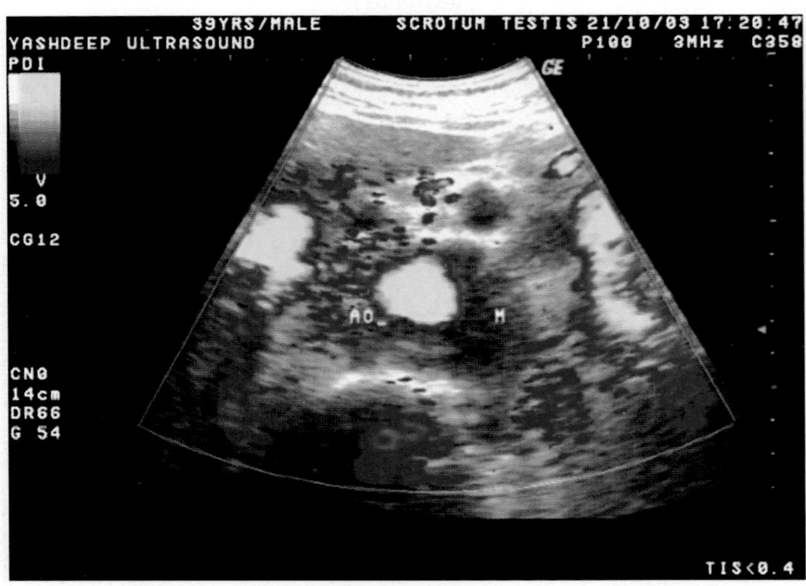

FIGURE 10.177

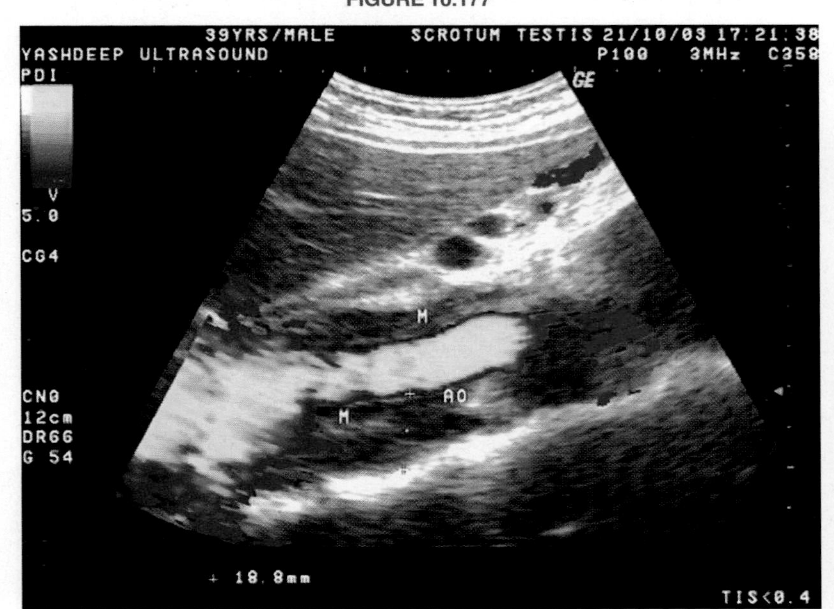

FIGURE 10.178

FIGURES 10.175 to 10.178: Abdominal ultrasound of the same patient shows multiple enlarged nodes in the retroperitoneum. The nodal masses are hypoechoic in texture. Normal lymph node hilum is lost. They are seen lifting the SMA and celiac axis. The lymph nodes masses are also seen anterior and posterior to aorta.

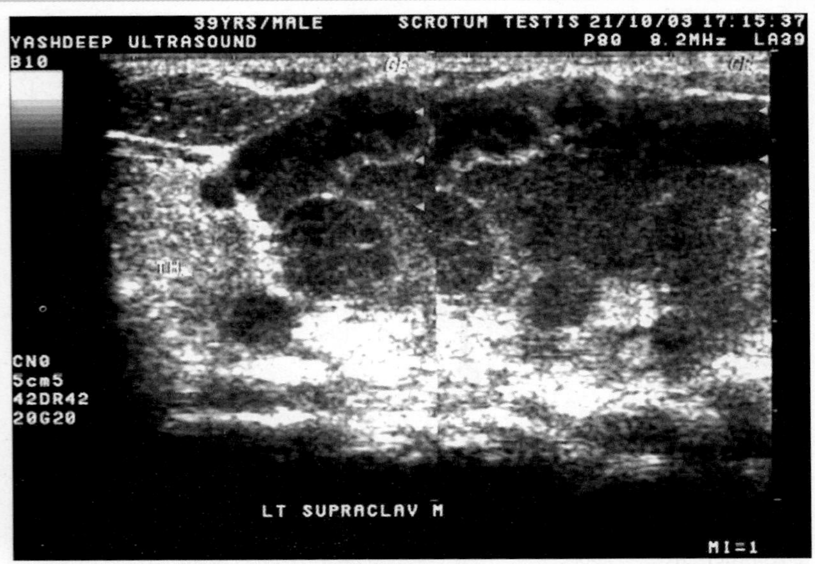

FIGURE 10.179

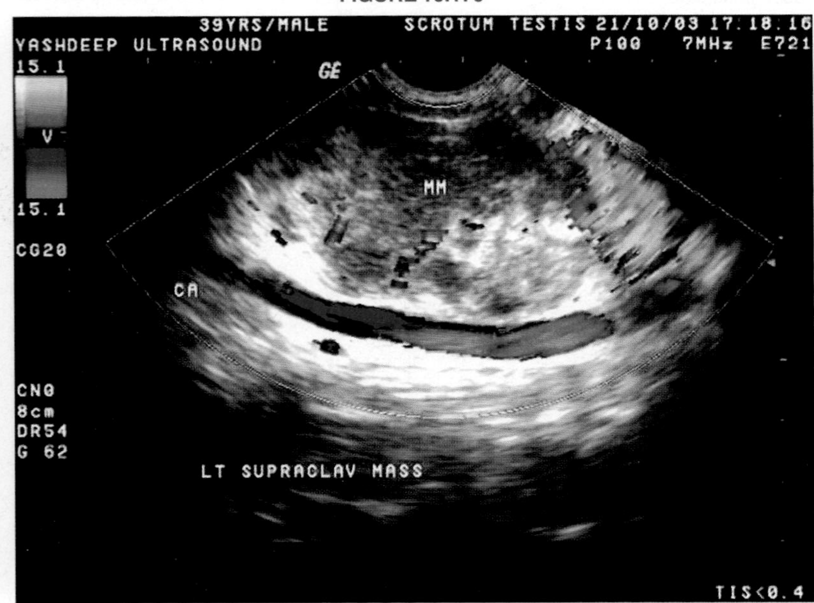

FIGURE 10.180

FIGURES 10.179 and 10.180: HRSG of the neck of the same patient shows multiple nodal masses in the Lt supraclavicular region. Biopsy of the mass confirmed disseminated lymphoma.

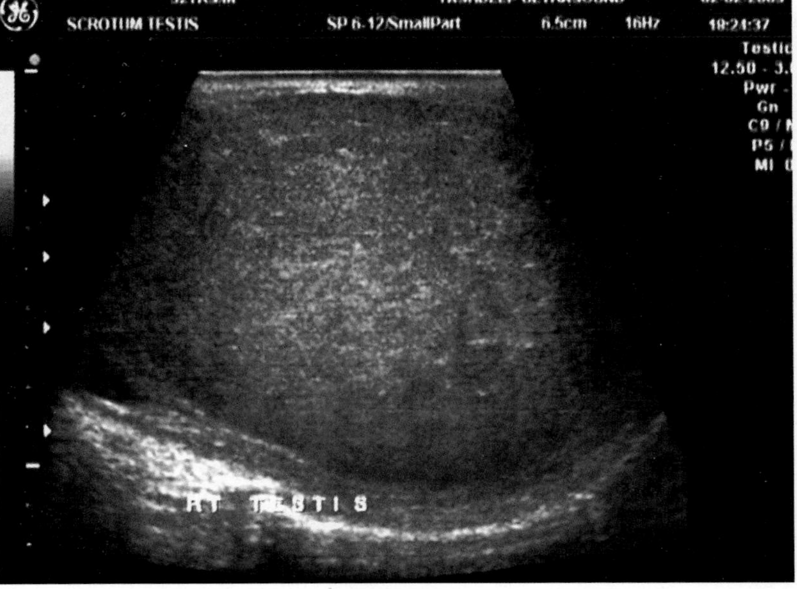

FIGURE 10.181

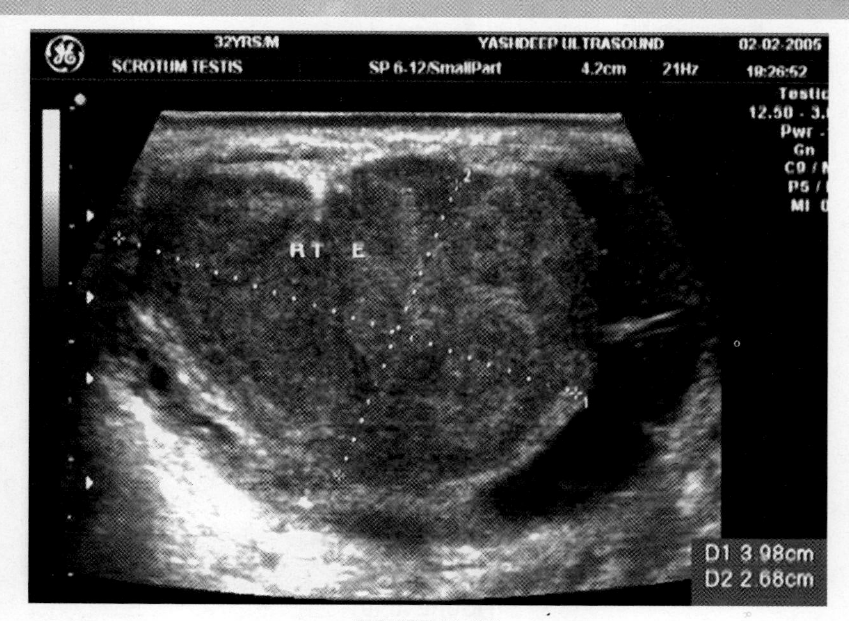

FIGURE 10.182

FIGURES 10.181 and 10.182: Lymphoma testis. A known patient of lymphoma presented diffuse enlargement of Rt scrotal sac. HRSG shows diffuse hypoechoic enlargement of the testis. The epididymis is also enlarged and hypoechoic in texture. Normal testicular echoes are lost. Fine isoechogenic nodules are seen in the mass.

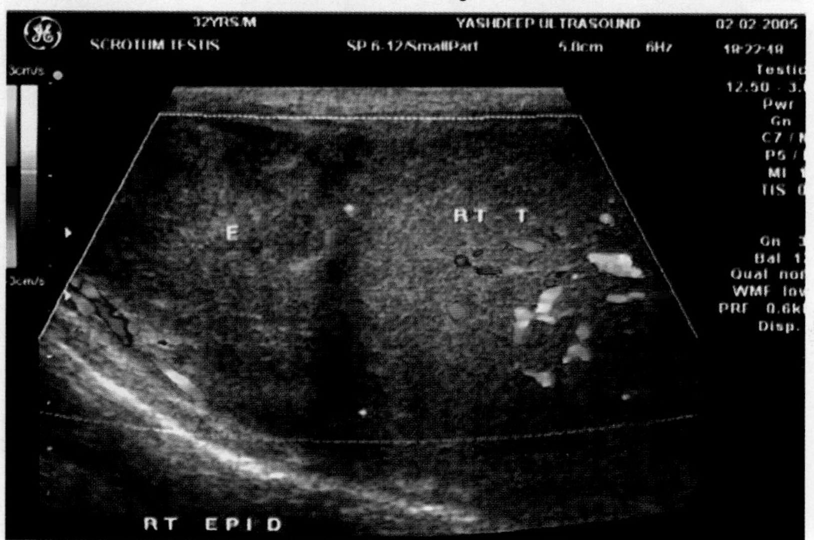

FIGURE 10.183

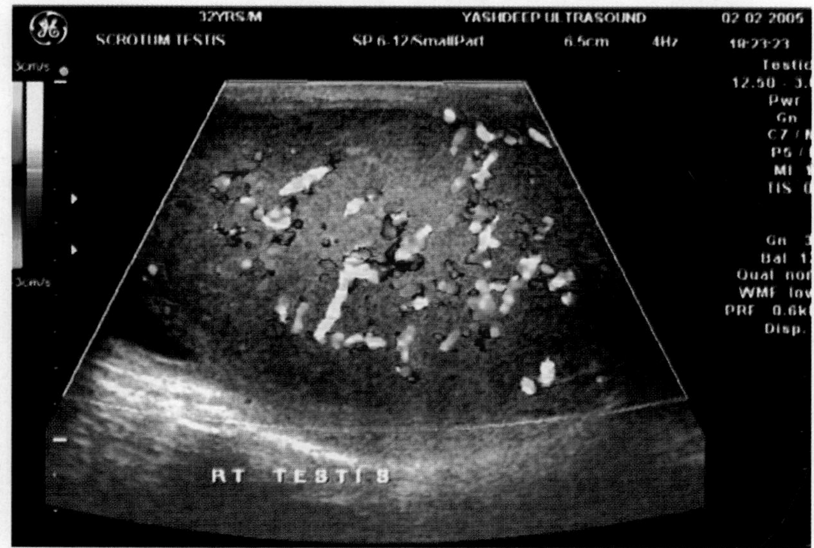

FIGURE 10.184

FIGURES 10.183 and 10.184: On color flow imaging increased flow is seen in the mass. The biopsy of the mass confirmed lymphoma.

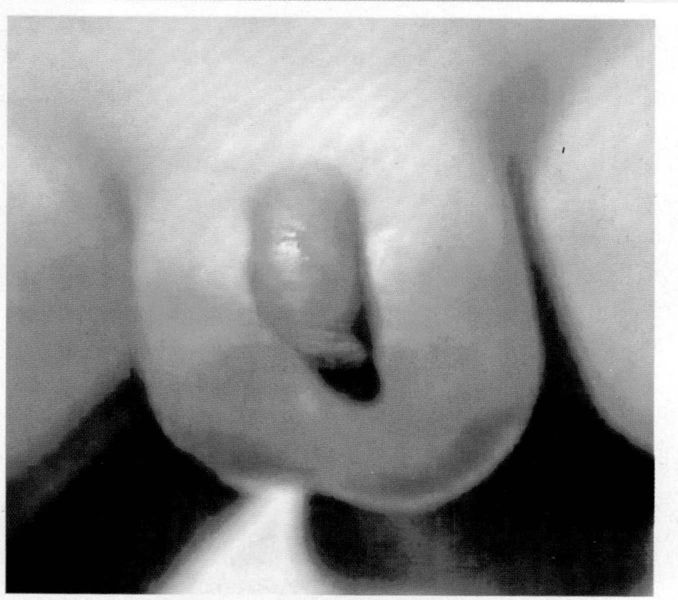

FIGURE 10.185: Acute testicular leukemia. A young child was presented with bilateral painful scrotal swelling.

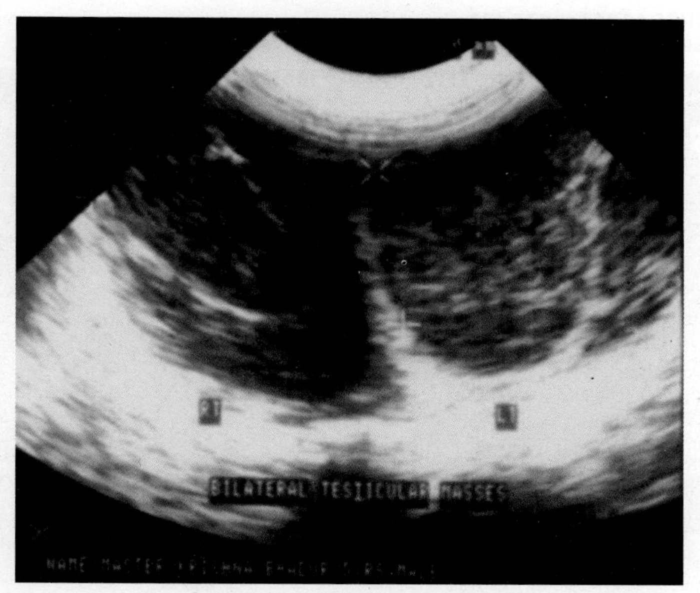

FIGURE 10.186: HRSG shows diffuse enlargement of both testis. They are hypoechoic in texture.

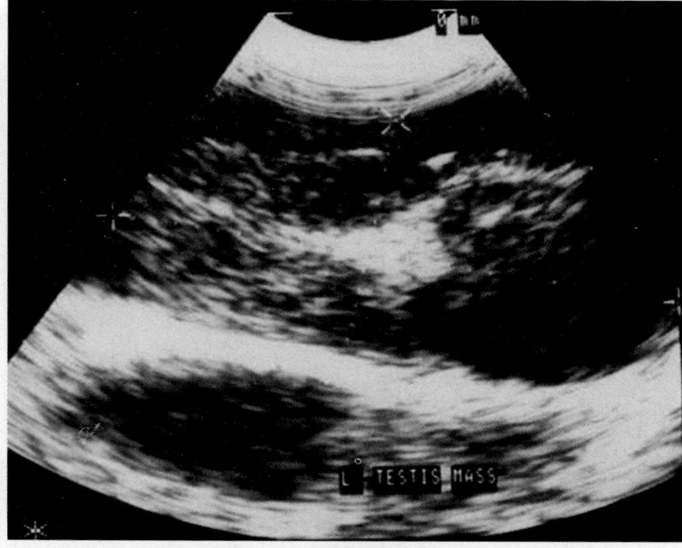

FIGURE 10.187

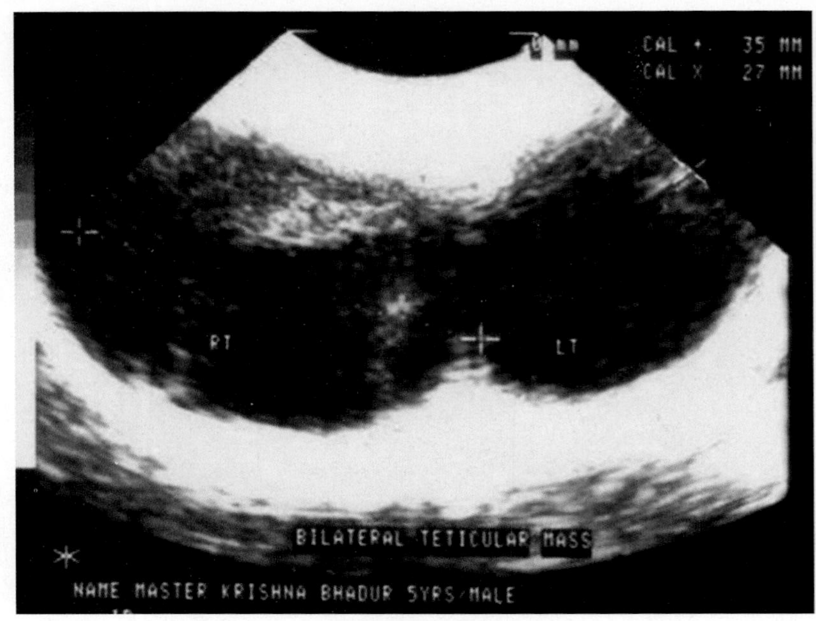

FIGURE 10.188

FIGURES 10.187 and 10.188: Normal testicular texture is lost and they are hypoechoic in texture. Small amount of fluid collection is also seen in the scrotal sac.

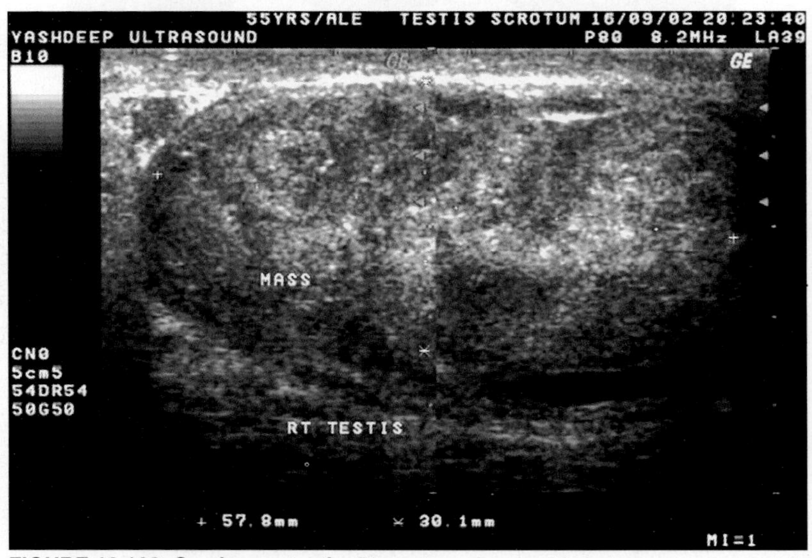

FIGURE 10.189: Seminoma testis. Rt scrotum shows a big heterogeneous mass in the testis. The mass has destroyed normal testicular texture. Multiple hypoechoic areas are seen in the mass. But no cystic degeneration is seen. The mass is seen confined to the tunica albuginea.

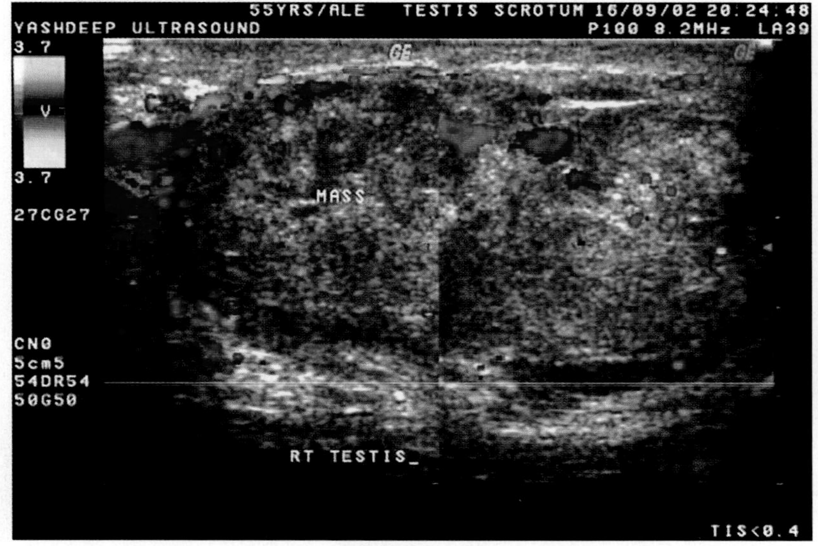

FIGURE 10.190

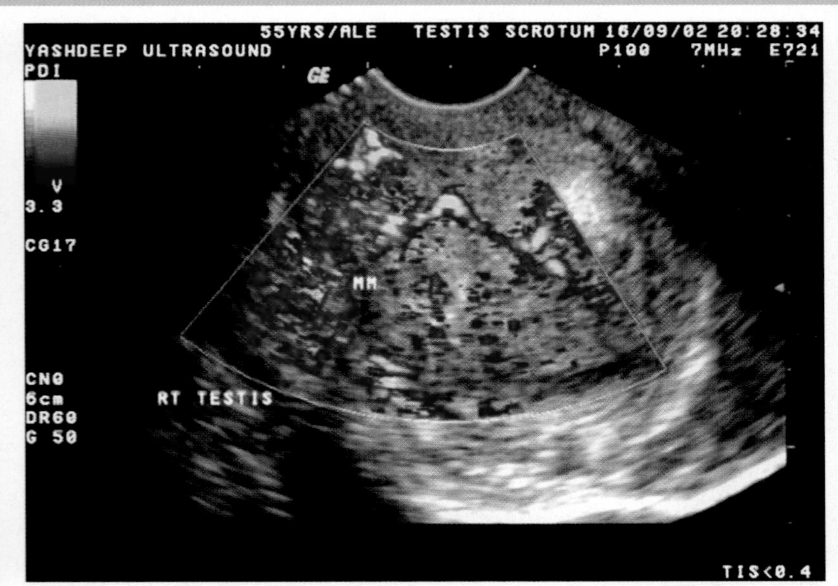

FIGURE 10.191

FIGURES 10.190 and 10.191: Color flow-imaging shows increase flow in the mass. Biopsy of the mass shows Seminoma.

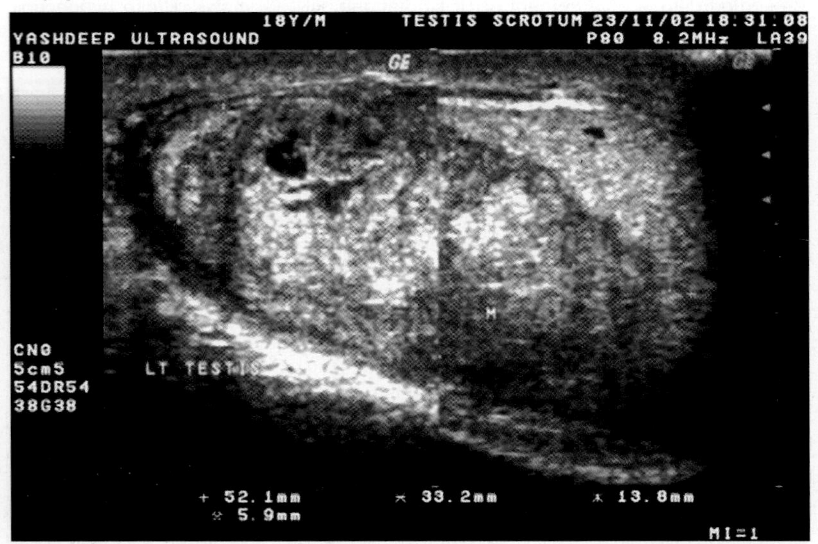

FIGURE 10.192

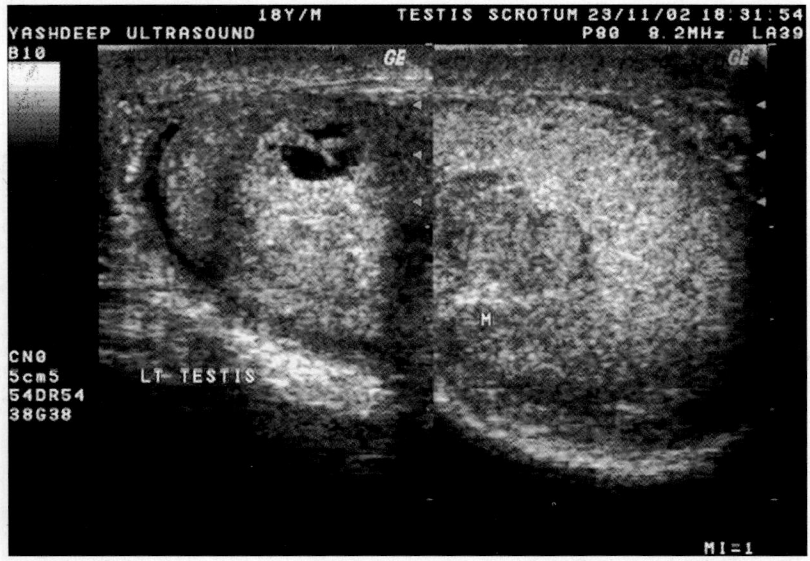

FIGURE 10.193

FIGURES 10.192 and 10.193: A young boy presented with enlarged Lt. scrotal swelling. HRSG shows diffuse enlargement of Lt. testis, echotexture is heterogeneous mass in the testis. Two third of testis is replaced by the mass. Central area of necrosis is also seen in the mass. Biopsy of the mass confirmed seminoma testis.

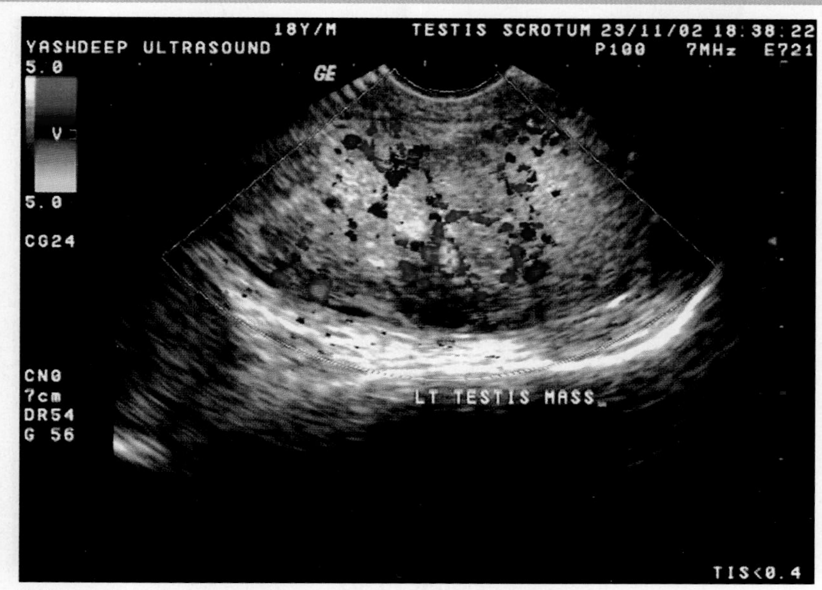

FIGURE 10.194: On color flow imaging increased flow is seen in the mass.

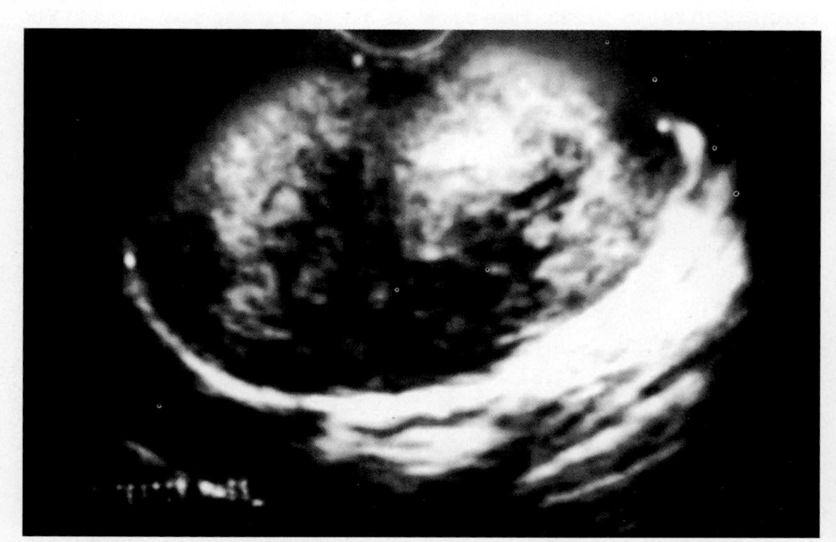

FIGURE 10.195: Seminoma testis with distant metastasis. A patient presented with soft tissue mass in the Rt scrotal sac. HRSG shows a big heterogeneous mass invading the testis. The capsule of the testis is also seen broken.

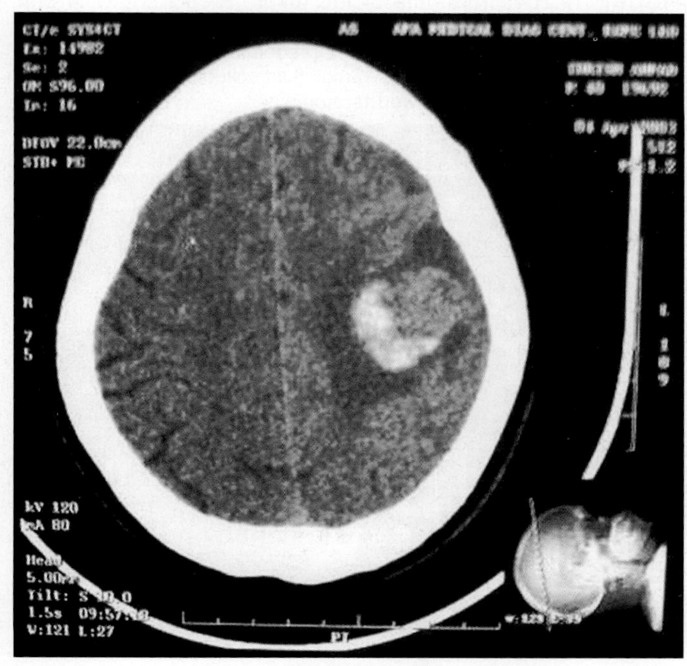

FIGURE 10.196

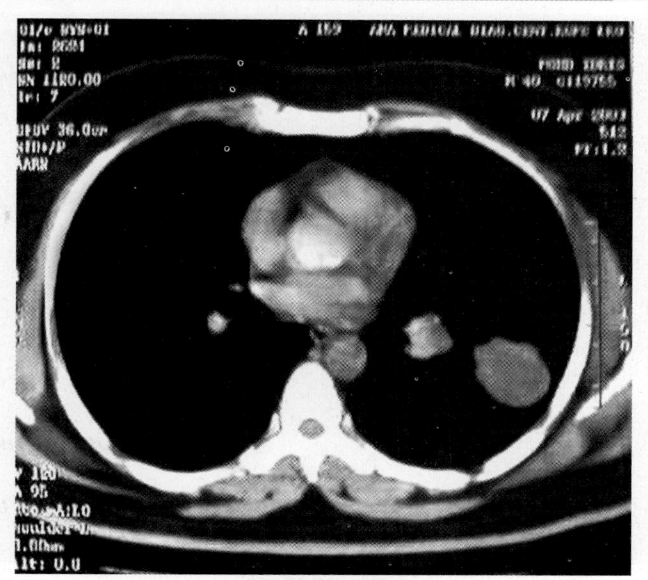

FIGURE 10.197

FIGURES 10.196 and 10.197: The same patient presented with seizure disorder. CT scan of the head shows a big metastatic deposit in the Lt parietal lobe with internal hemorrhage. CT thorax also shows big nodular masses in the Lt lung suggestive of secondaries.

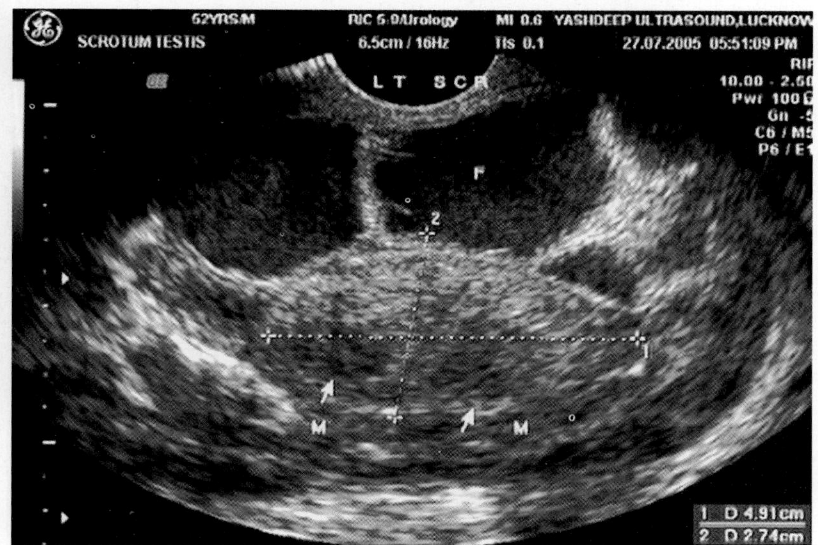

FIGURE 10.198

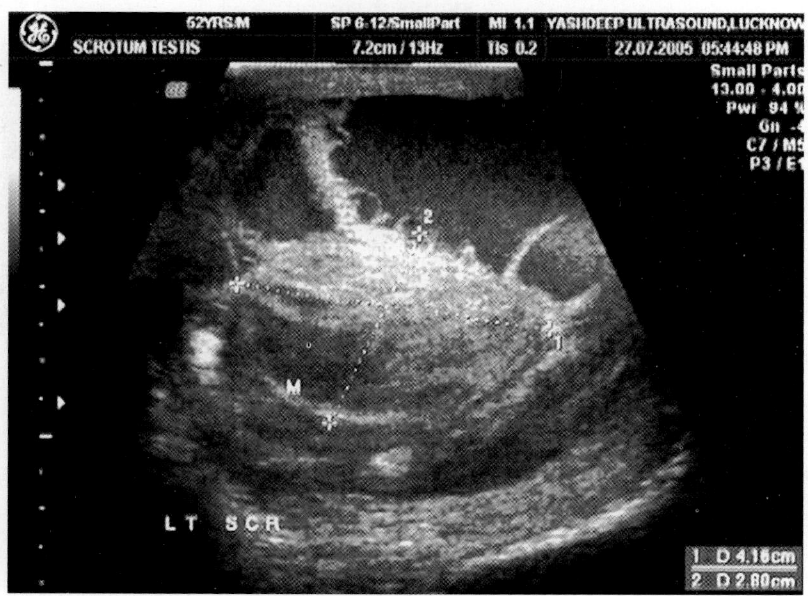

FIGURE 10.199

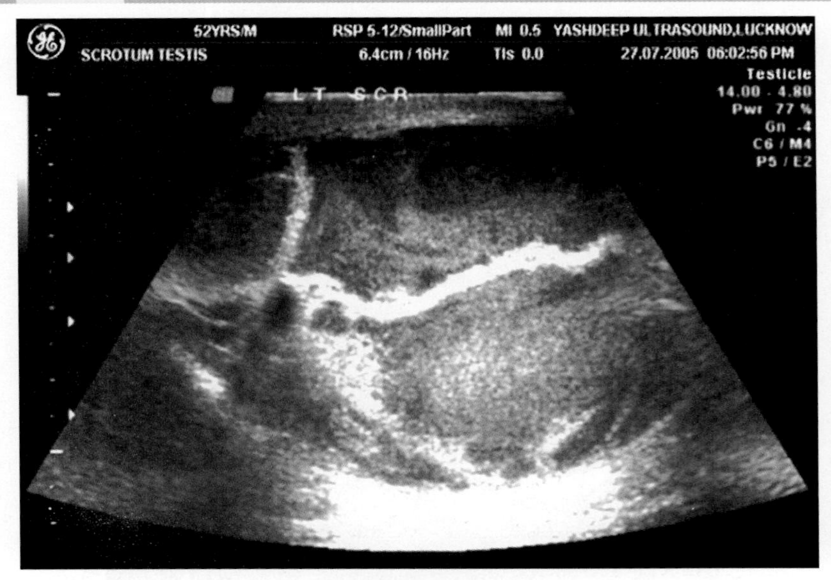

FIGURE 10.200

FIGURES 10.198 to 10.200: Lymphoma testis. A known patient of lymphoma presented with Lt scrotal swelling. HRSG shows low-level echo collection in the Lt scrotum. Echogenic bands are also seen in the scrotal sac. The Lt testis shows irregular hypoechoic foci, which are seen in the testis suggestive of deposit. Thick collection is also seen in the testis.

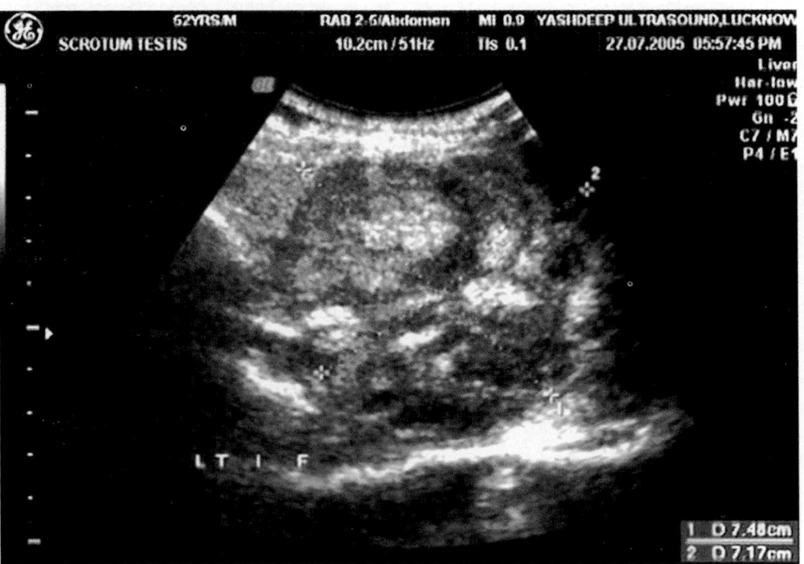

FIGURE 10.201

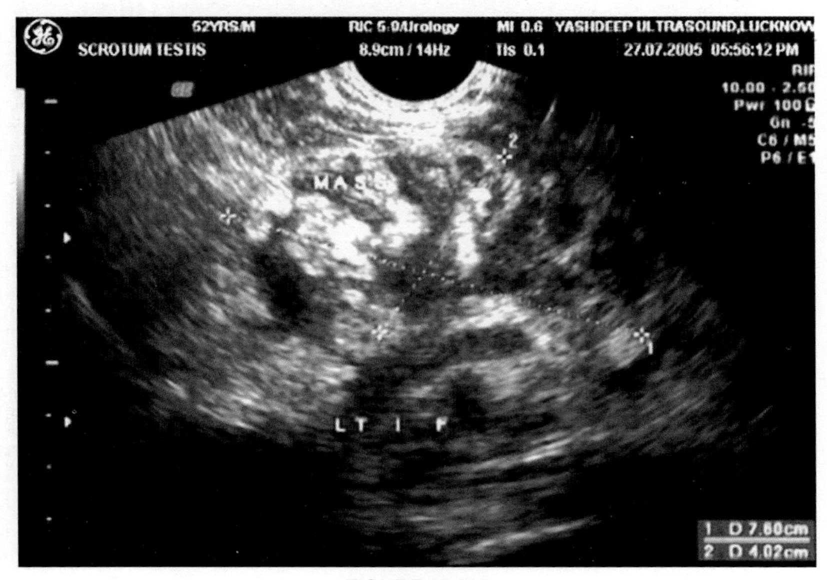

FIGURE 10.202

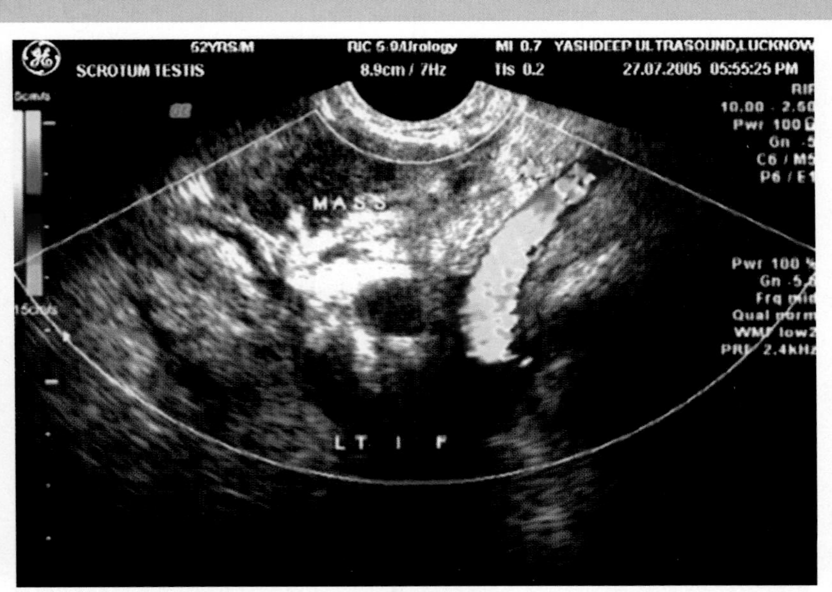

FIGURE 10.203

FIGURES 10.201 to 10.203: Heterogeneous masses are also seen in the Lt iliac fossa. They are seen invading the bowel loops. Nodal masses are also seen in the Lt inguinal region. Biopsy of the mass is confirmed lymphoma.

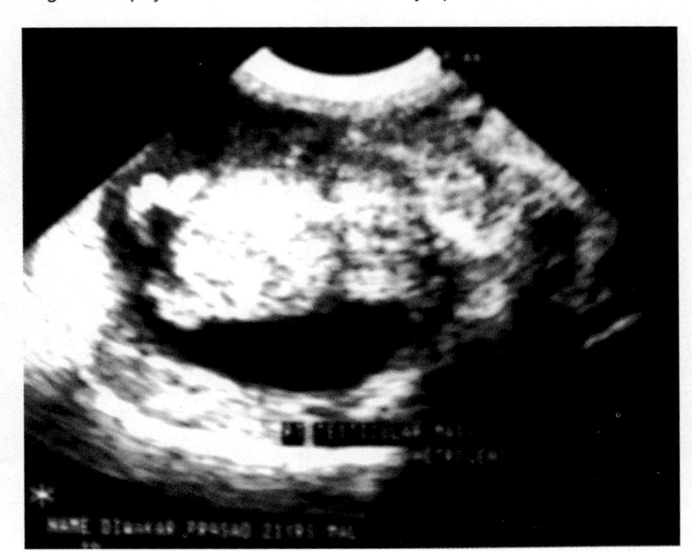

FIGURE 10.204: Teratoma of the testis. HRSG shows a big heterogeneous irregular mass in the Rt testis. The mass has totally destroyed normal scrotal anatomy. It is predominantly solid. Echogenic calcification is seen in the mass. Small fluid collection is also seen in the scrotal sac. Biopsy showed teratoma.

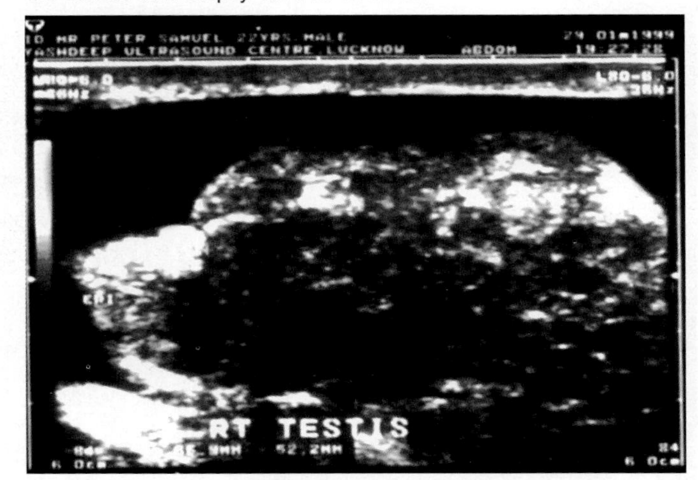

FIGURE 10.205: Testicular teratoma. A big irregular mass is seen in the Rt testis. Multiple calcified specks are also seen in the mass. Echogenic calcified rim is also seen. Normal texture is lost turned out to be a teratoma.

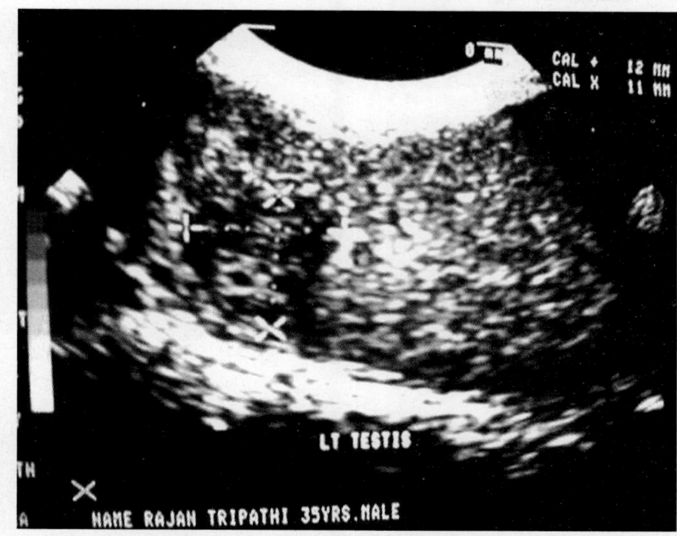

FIGURE 10.206: Occult primary tumor. HRSG shows a small irregular hypoechoic mass in the Lt testis. The mass was not palpable and irregular in outline. HRSG is highly sensitive in picking up occult primary tumors.

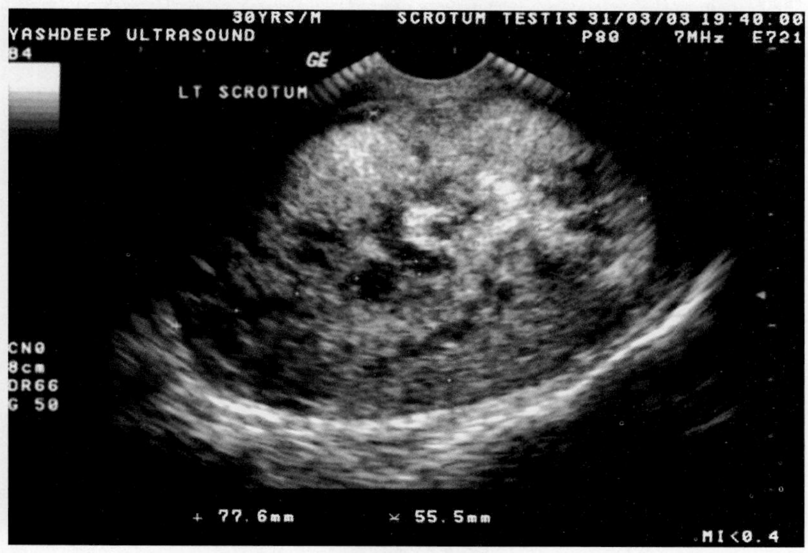

FIGURE 10.207

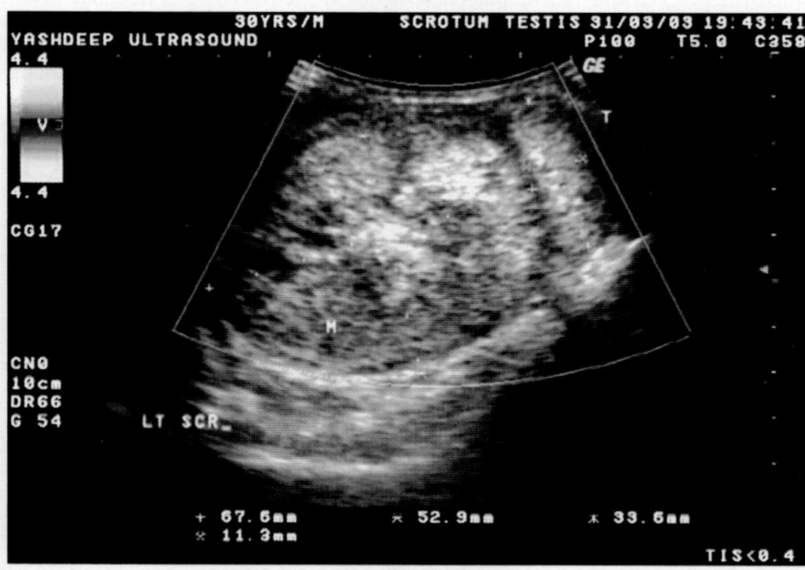

FIGURE 10.208

FIGURES 10.207 and 10.208: Mixed germ cell tumors. HRSG shows irregular heterogeneous mass in the testis. Echogenic shadows are also seen in the mass suggestive of hemorrhage in it. Small hydrocele is also present in the tumor. Mixed germ cell tumors are second most common primary tumors after seminoma and occur in 3rd to 5th decade of the life.

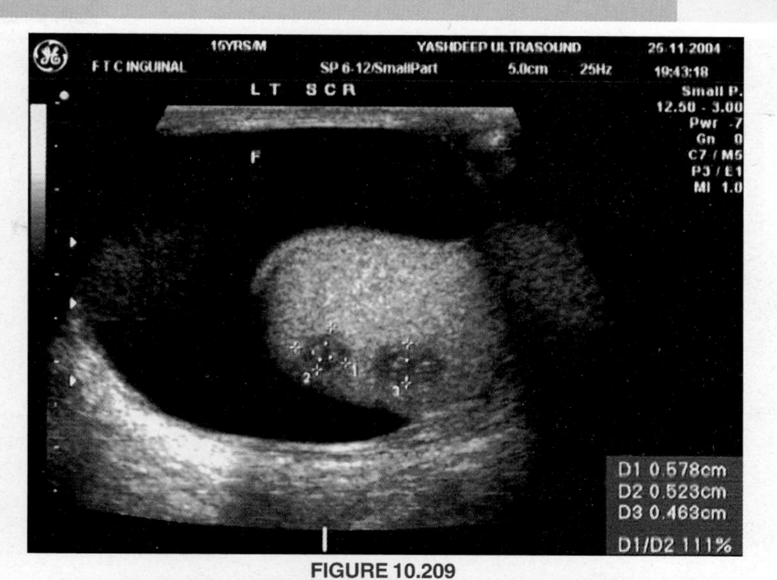

FIGURE 10.209

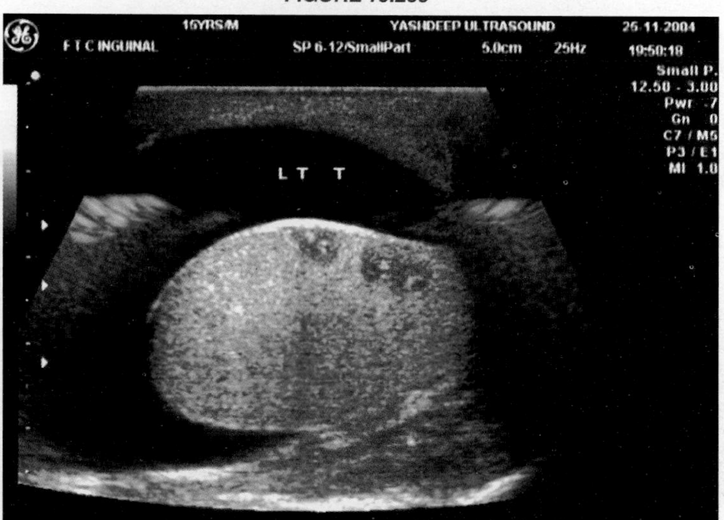

FIGURE 10.210

FIGURES 10.209 and 10.210: Metastatic deposits of leukemia. A known patient of leukemia presented with enlargement of the Lt scrotum with pain. HRSG shows increased amount of fluid in the scrotal sac. The testis shows irregular hypoechoic foci suggestive of deposits.

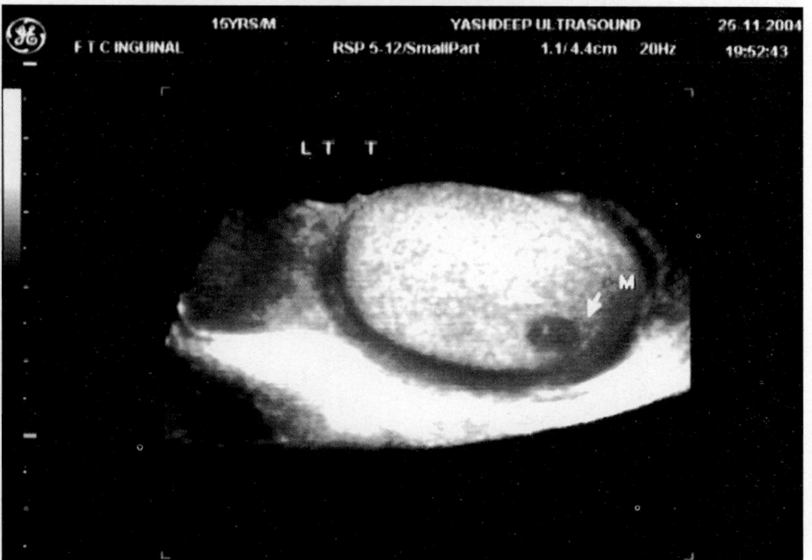

FIGURE 10.211

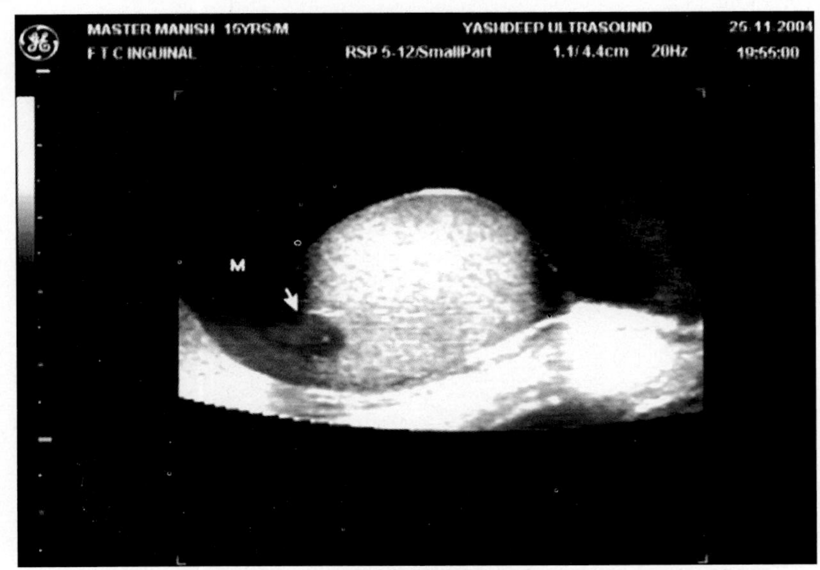

FIGURE 10.212

FIGURES 10.211 and 10.212: 3D surface rendering image of the testis shows the deposits clearly.

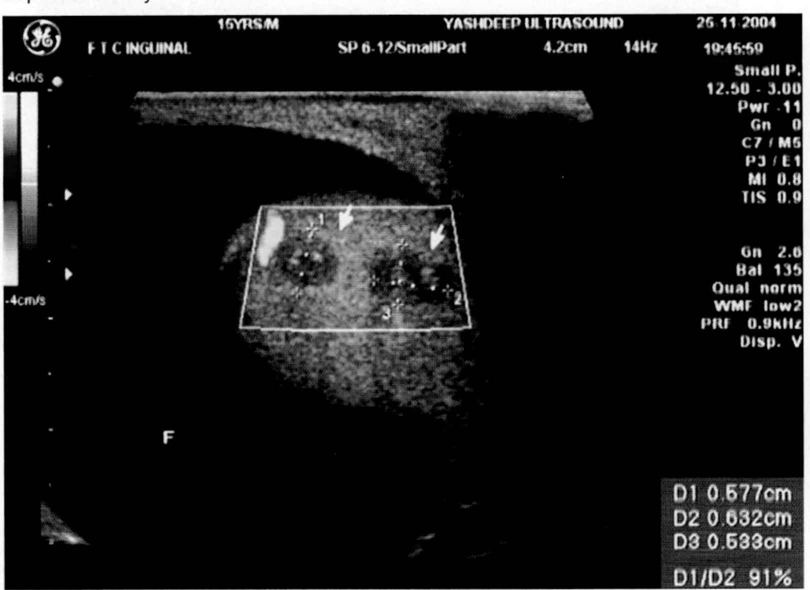

FIGURE 10.213

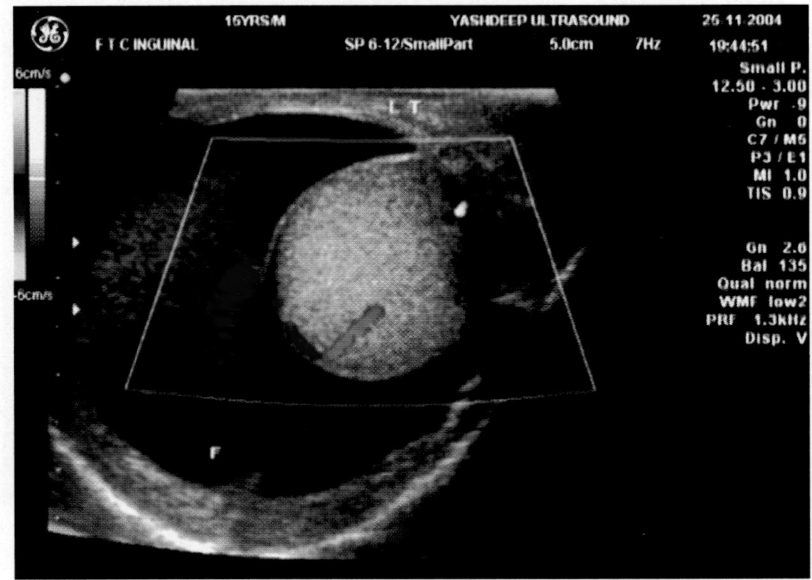

FIGURE 10.214

FIGURES 10.213 and 10.214: On color Doppler imaging shows poor flow in the metastasis deposits.

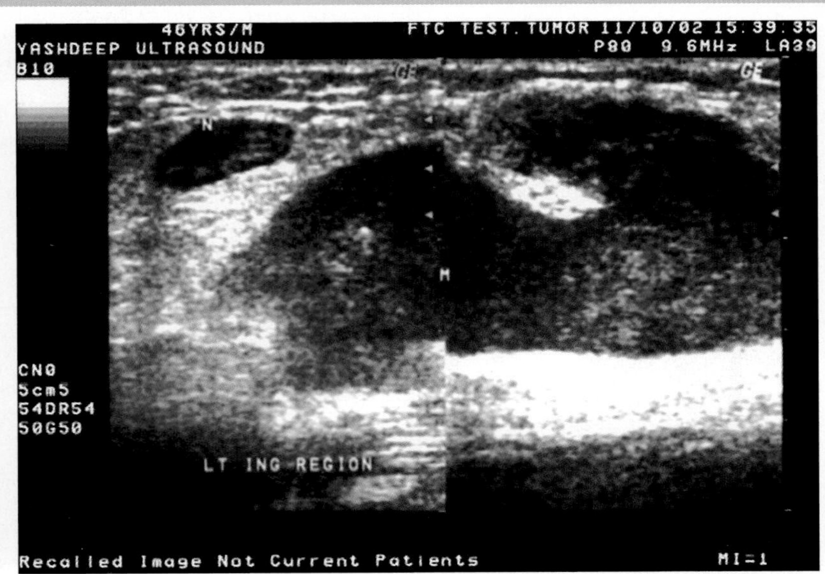

FIGURE 10.215

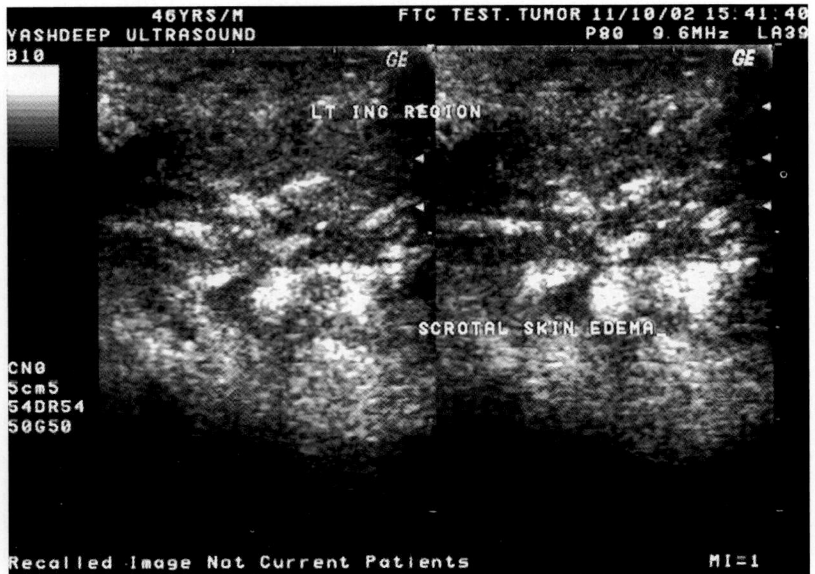

FIGURE 10.216

FIGURES 10.215 and 10.216: Metastatic nodes in operated case of testicular tumor. An operated patient of testicular tumor presented with inguinal masses. HRSG shows big irregular hypoechoic masses in the Lt inguinal region suggestive of enlarged nodes. Normal nodal texture is lost. The scrotal skin also shows marked skin edema with distorted texture. Marked thickening of the skin is seen.

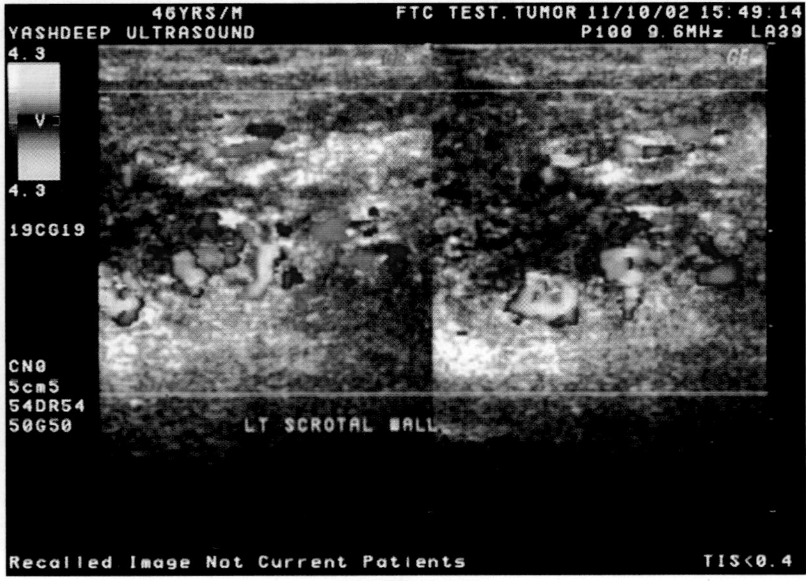

FIGURE 10.217

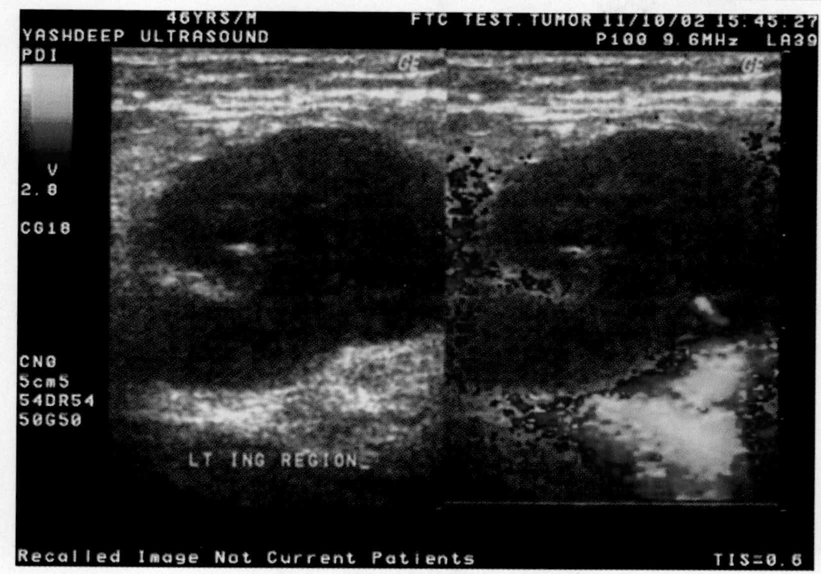

FIGURE 10.218

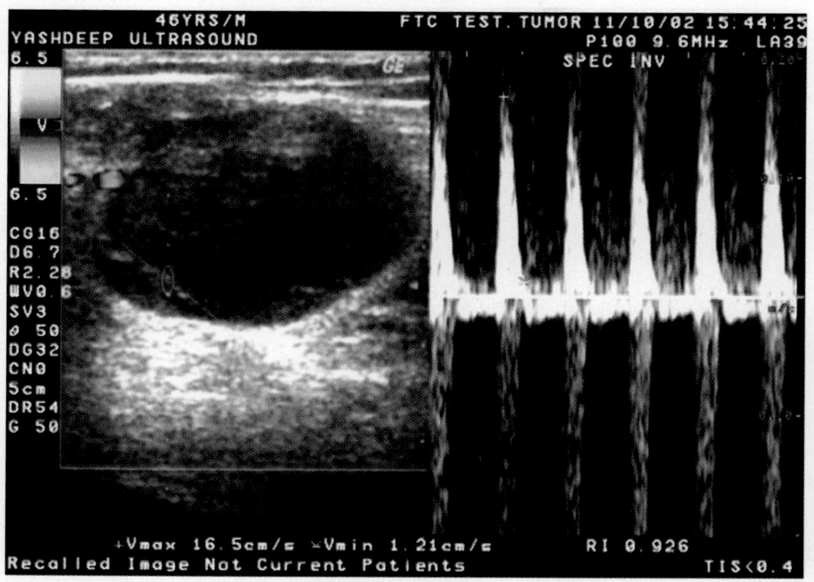

FIGURE 10.219

FIGURES 10.217 to 10.219: On color Doppler imaging increased flow is seen in the edematous scrotal skin suggestive of infiltrations. Enlarged nodes also show high flow. Biopsy of the nodes confirmed metastatic deposits.

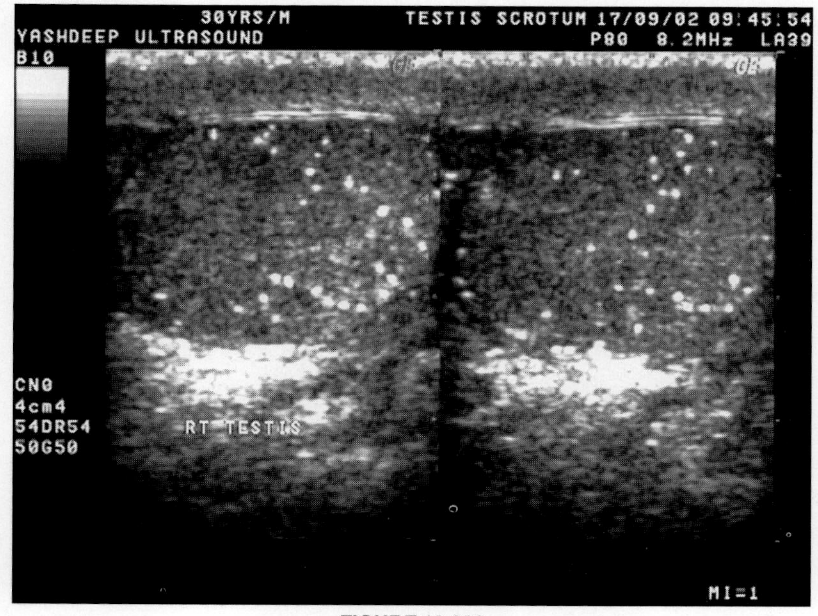

FIGURE 10.220

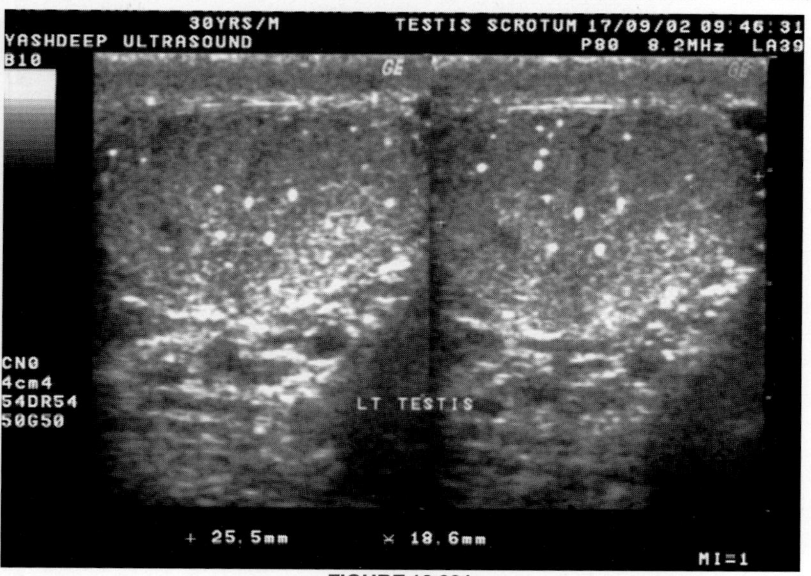

FIGURE 10.221

FIGURES 10.220 and 10.221: Microlithiasis of the testis. Multiple small-calcified specks are seen distributed in both the testis parenchyma suggestive of micro-calcification. They are bright echogenic shadows, which are not accompanied acoustic shadowing. Micro-calcification is known to be precancerous condition of the testis.

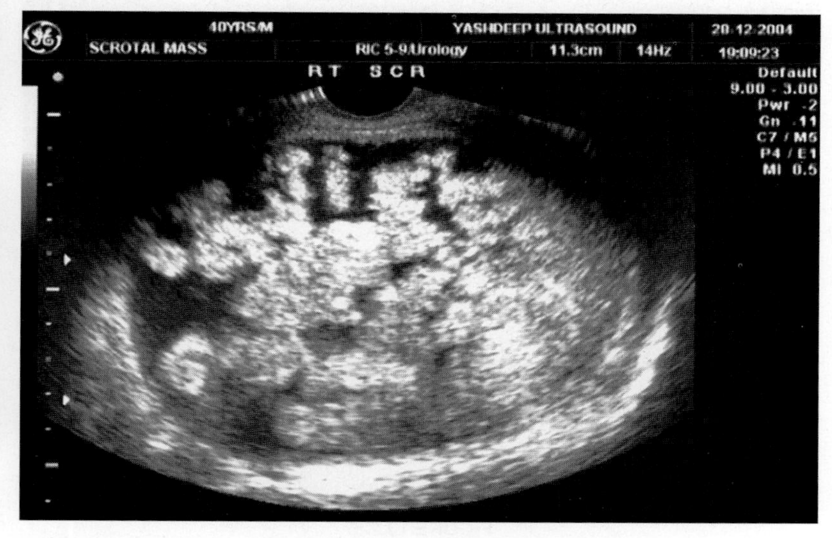

FIGURE 10.222: Huge testicular hematoma presented as mass. A patient with huge Rt scrotal mass. HRSG of the scrotal shows big heterogeneous disorganized mass-filling whole of the scrotal sac. No evidence of any calcification or cystic degeneration is seen in it. Initially the mass was mistaken as testicular tumor. However, the testis was seen lying at the root of the scrotum. It was displaced and small in size.

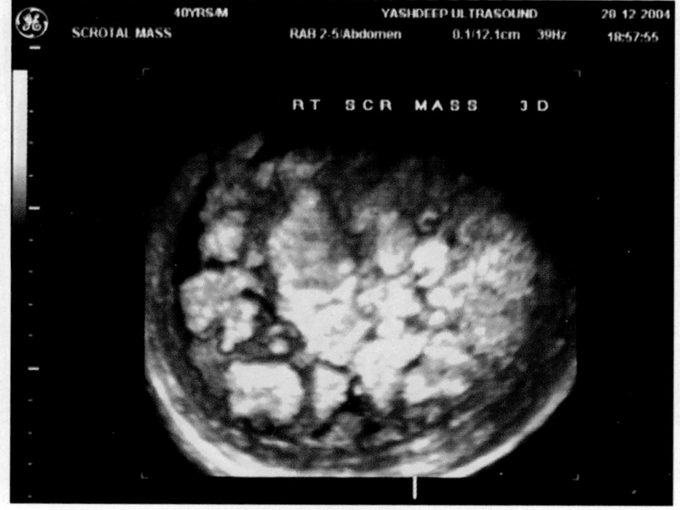

FIGURE 10.223

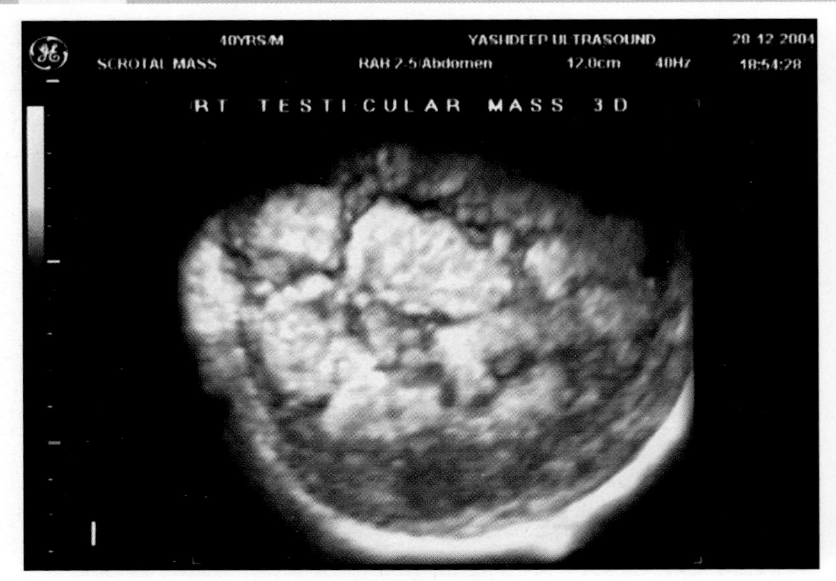

FIGURE 10.224

FIGURES 10.223 and 10.224: 3D surface-rendering image of the mass shows lobulated structure of the mass. The biopsy of the mass confirmed disorganized chronic hematoma.

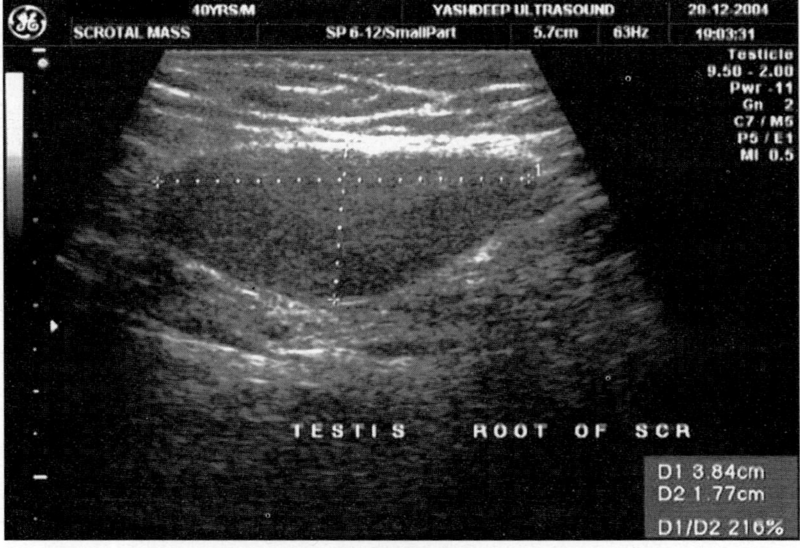

FIGURE 10.225

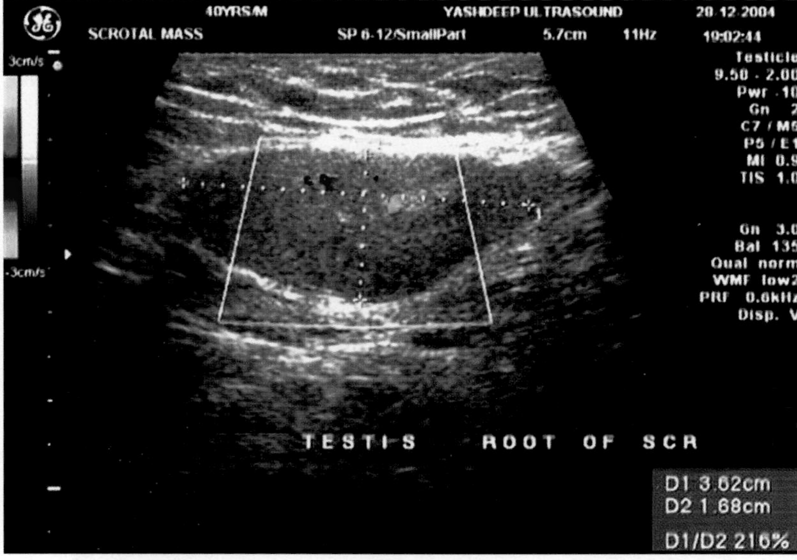

FIGURE 10.226

FIGURES 10.225 and 10.226: The testis was seen lying at root of scrotum separate from hematoma. It was small in size with hypoechoic texture. On color flow imaging low flow seen in the testis.

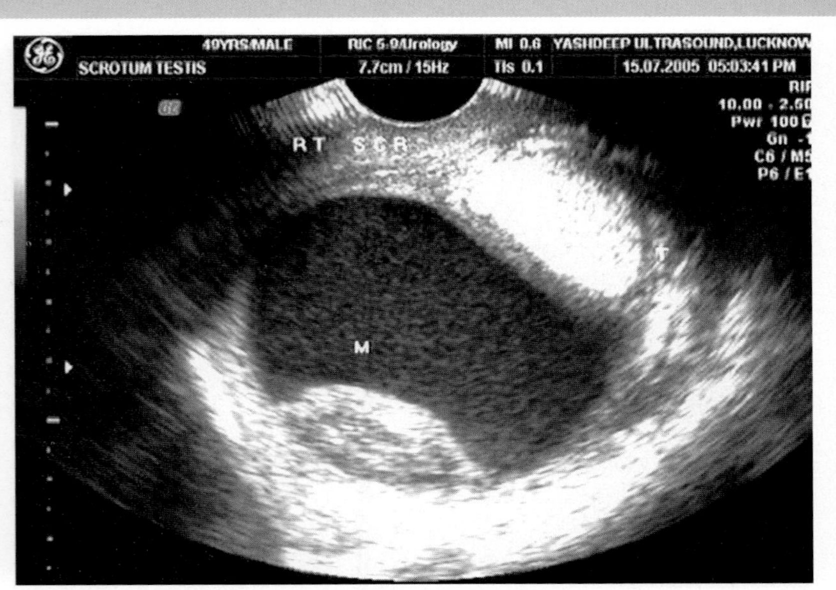

FIGURE 10.227

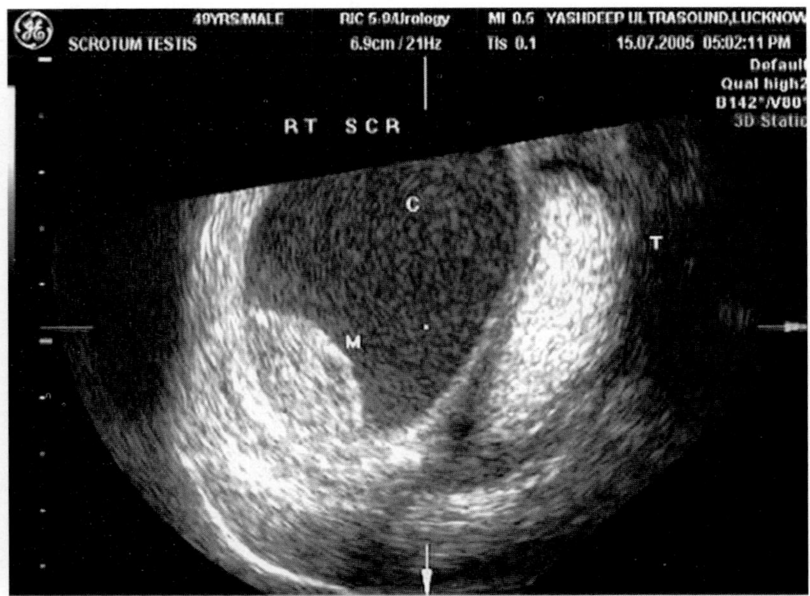

FIGURE 10.228

FIGURES 10.227 and 10.228: Organized hematoma with hydrocele. HRSG shows echogenic-organized hematoma in the Rt scrotum. The mass is seen lying at the posterior pole of scrotum. It is echogenic well organized. Thick echoes are seen in the collection. 3D of the scrotum shows the testis lying outside and displaced.

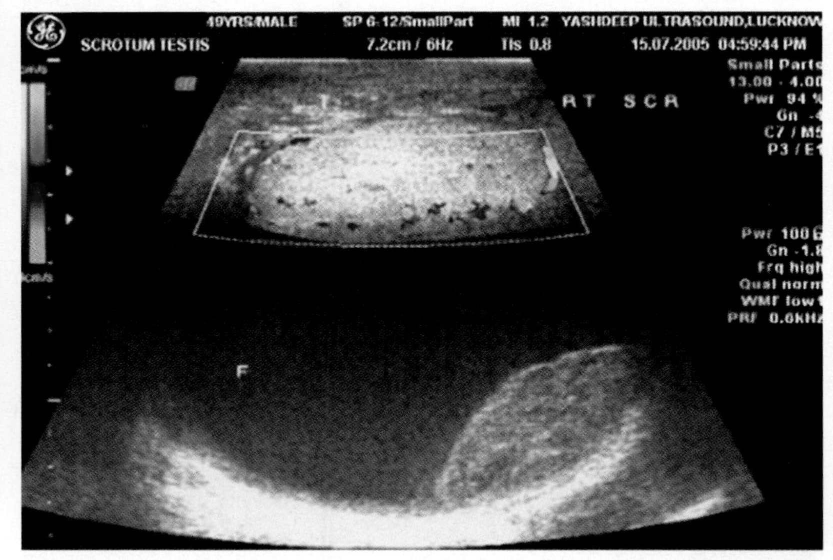

FIGURE 10.229: On color flow imaging normal flow is seen in the testis. However, no flow was seen in the organized hematoma.

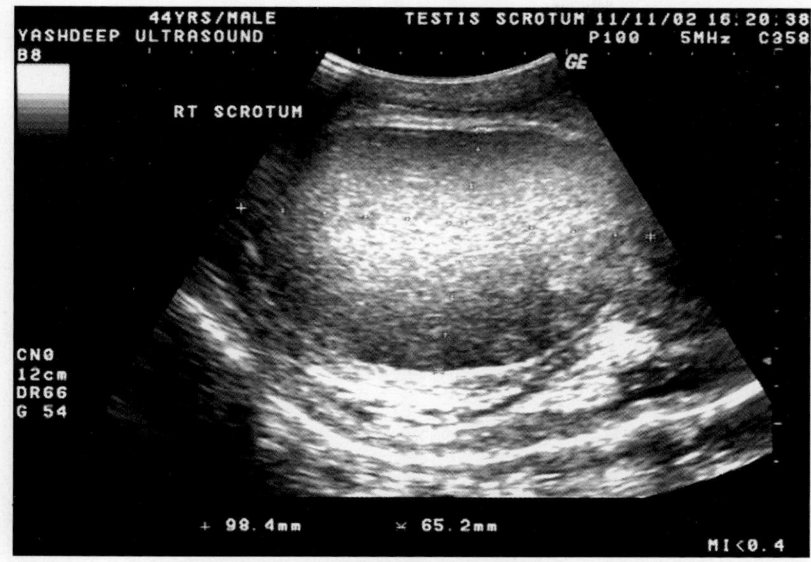

FIGURE 10.230

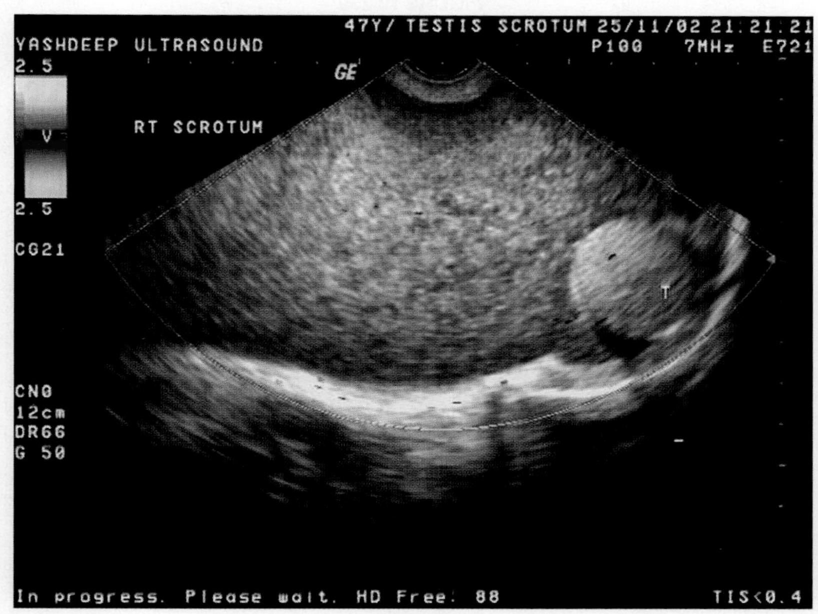

FIGURE 10.231

FIGURES 10.230 and 10.231: Tense hydrocele presented as a mass. A patient presented with tense scrotal swelling. HRSG shows thick collection in the scrotal sac. Homogeneous echoes are seen in it. The testis is seen pushed into one corner of the scrotal sac. It is seen compressed with the fluid pressure.

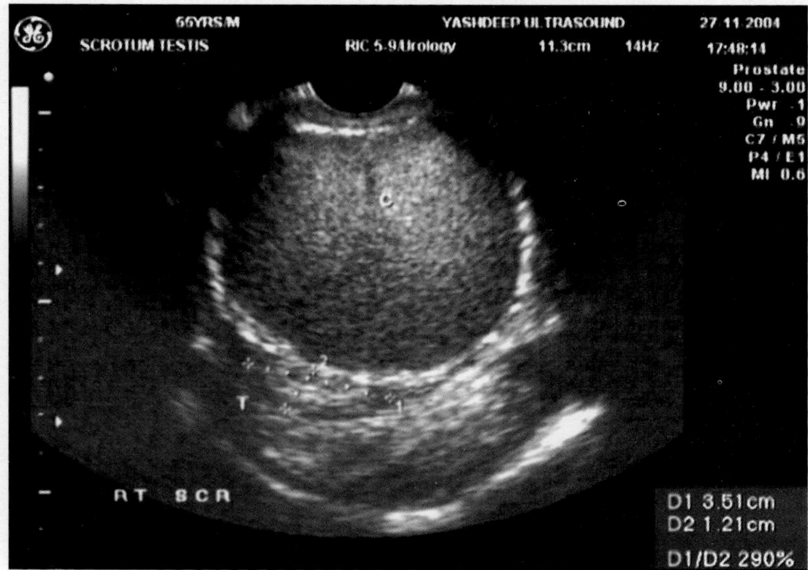

FIGURE 10.232

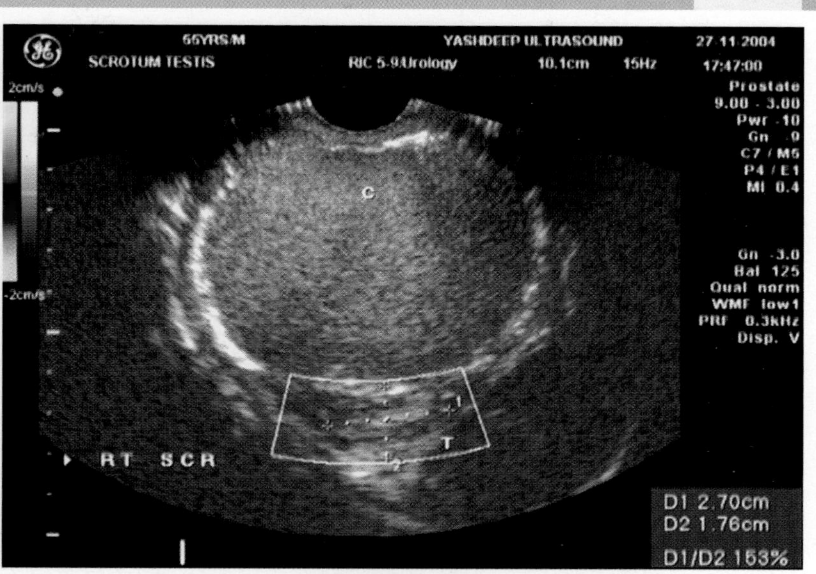

FIGURE 10.233

FIGURES 10.232 and 10.233: Chronic scrotal collection. HRSG shows large amount of fluid collection in the scrotal sac. It is seen in the extra testicular space. The periphery of the collection shows echogenic calcified specks fixed with the peripheral wall. The testis is seen displaced down and compressed by the fluid.

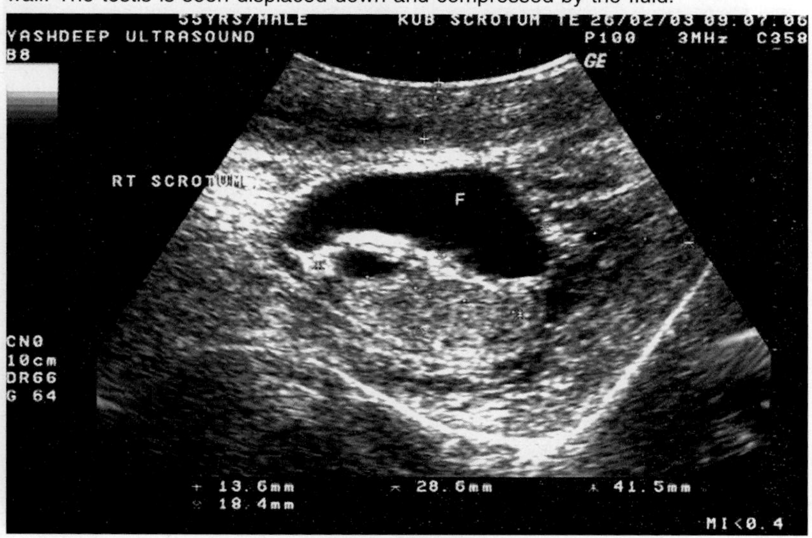

FIGURE 10.234

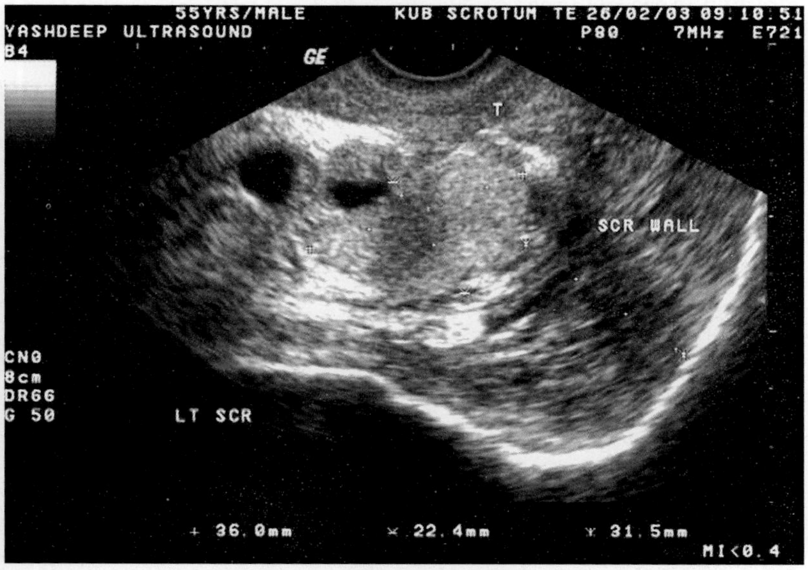

FIGURE 10.235

FIGURES 10.234 and 10.235: Bilateral atrophic testis. HRSG shows bilateral small testis. They are irregular in outline. Cystic necrosis is also seen in them. The scrotal wall shows marked thickening. Therefore the testis was not clinically palpable.

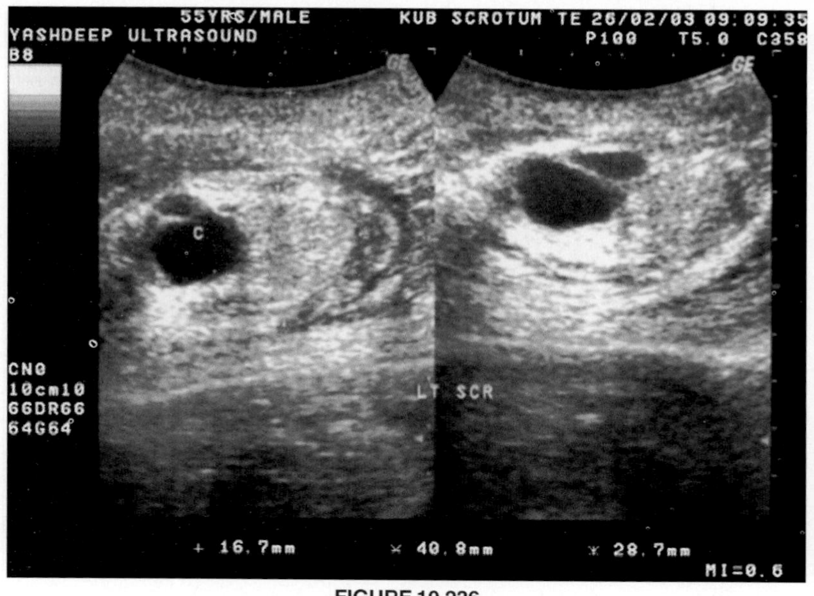

FIGURE 10.236

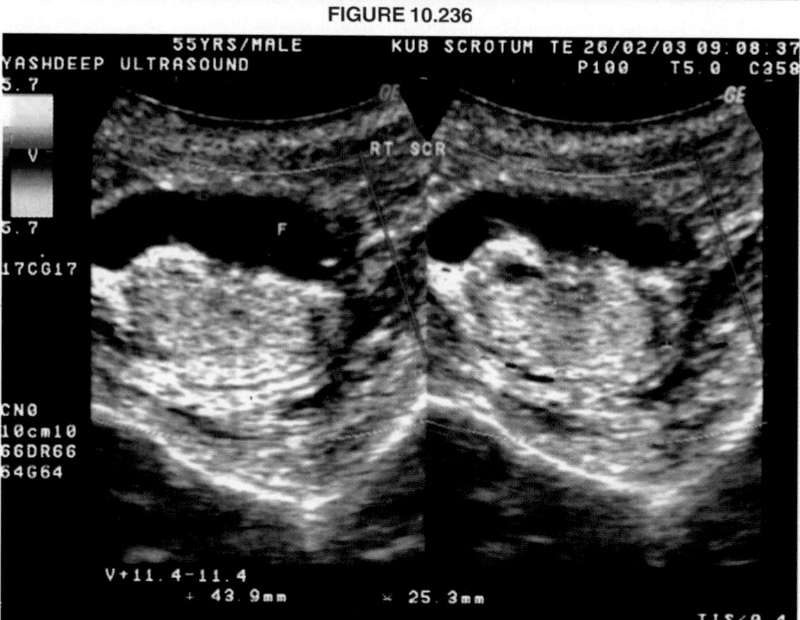

FIGURE 10.237

FIGURES 10.236 and 10.237. There is also evidence of cystic degeneration seen in the Lt testis. On color flow imaging poor flow seen in the testis. Small amount of fluid collection is also seen.

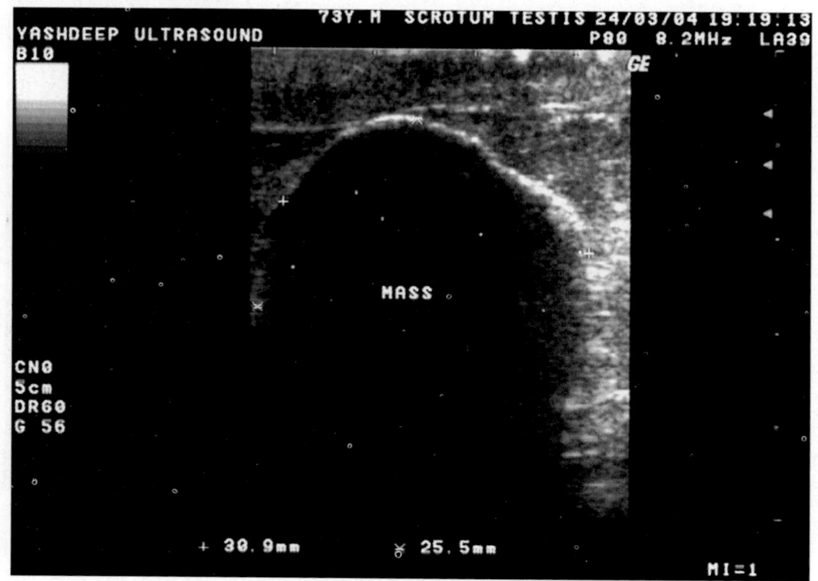

FIGURE 10.238

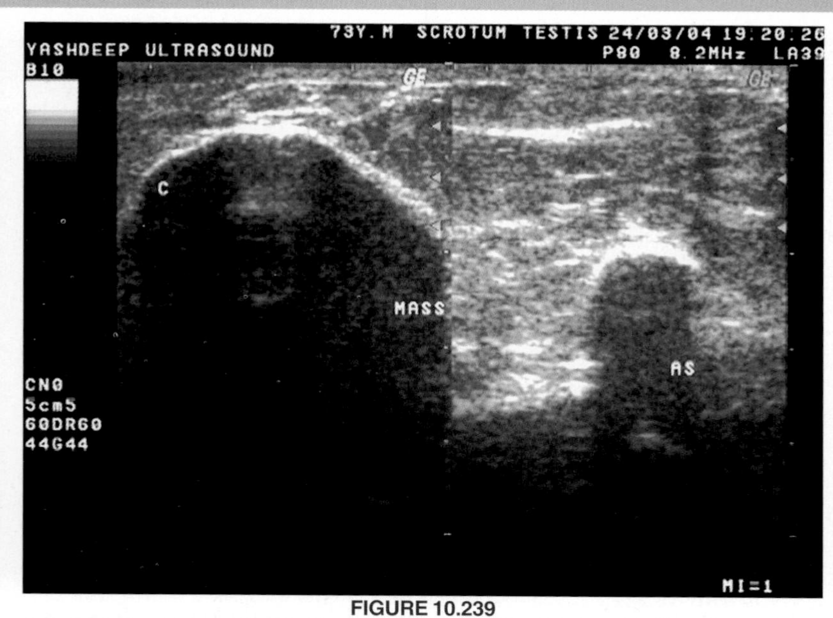

FIGURE 10.239

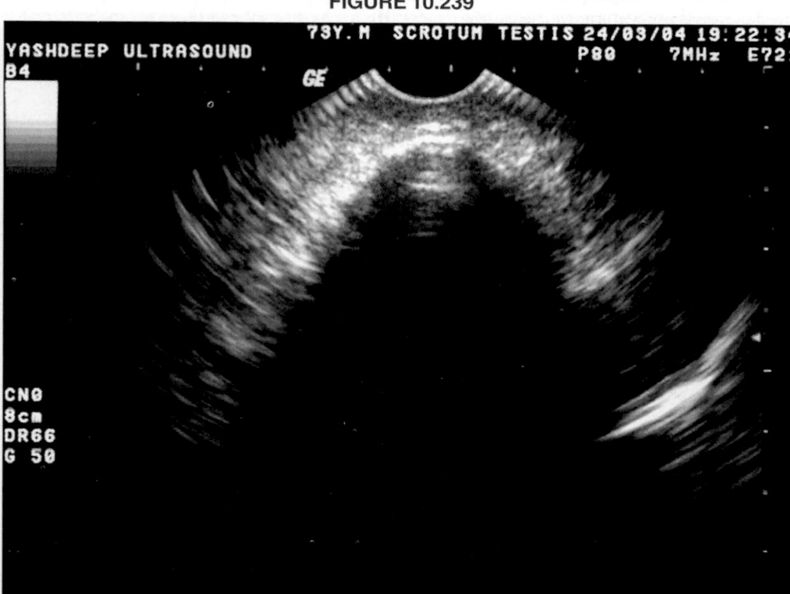

FIGURE 10.240

FIGURES 10.238 to 10.240: Calcified hematoma in the scrotal wall. A patient presented with mass in the Lt scrotal wall. HRSG shows densely calcified mass. The mass shows dense calcification. The calcification makes strong acoustic shadowing, masking the mass. Peripheral ring calcification is seen in the mass. Postoperative findings confirmed calcified hematoma.

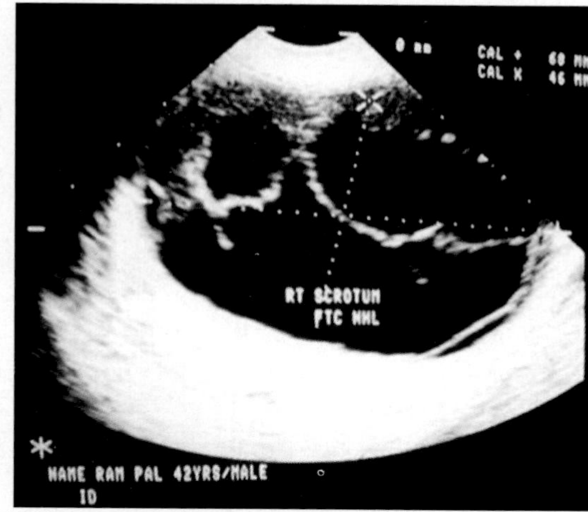

FIGURE 10.241: HRSG shows multiloculated scrotal collection in a case of Non-Hodgkin lymphoma-undergone orchidectomy. The patient presented with tense scrotal swelling. Low-level echo collection with multiple echogenic septa is seen.

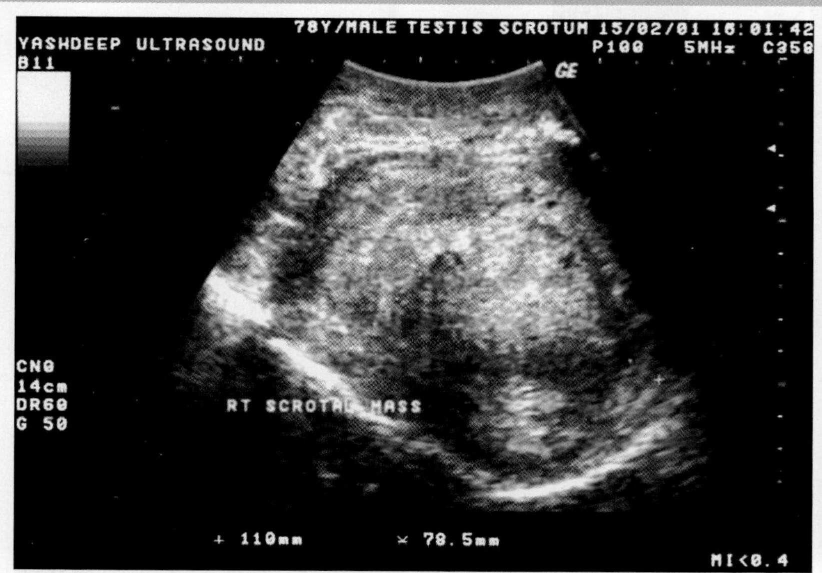

FIGURE 10.242: **Carcinomatous infiltration of scrotal wall.** HRSG shows a marked thickening of the scrotal wall in a case of carcinoma penis. The testis appears to be normal.

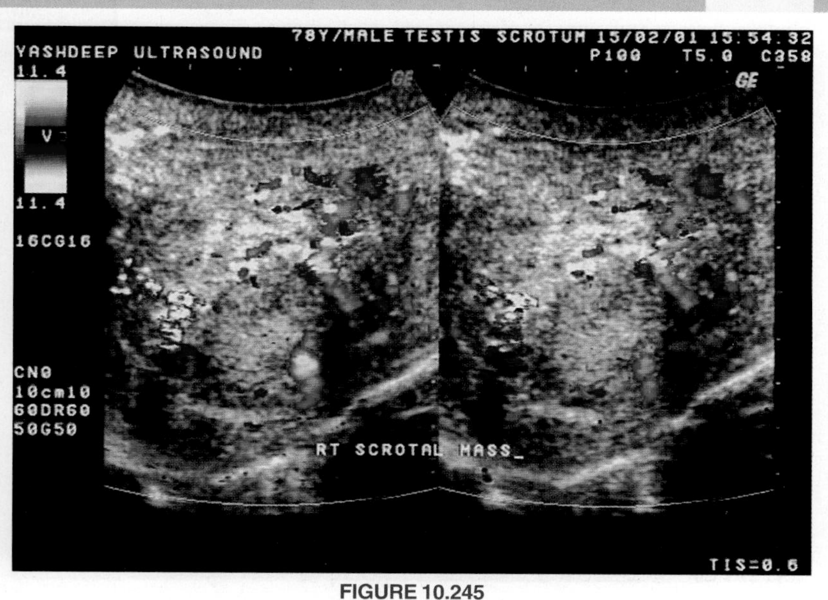

FIGURE 10.245

FIGURES 10.244 and 10.245: Color flow imaging shows normal flow in the testis. However, increased vascularity of scrotal wall is also seen.

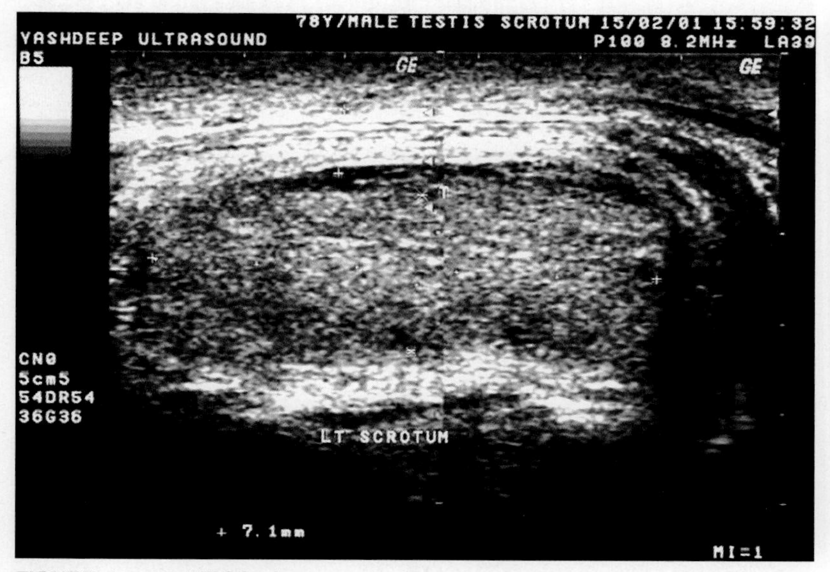

FIGURE 10.243: HRSG shows markedly thickened scrotal wall. However, testis appears to be normal.

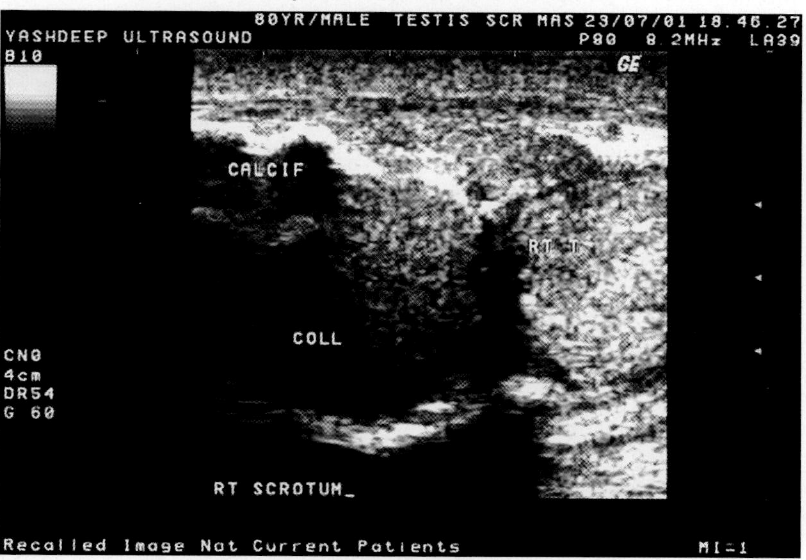

FIGURE 10.246

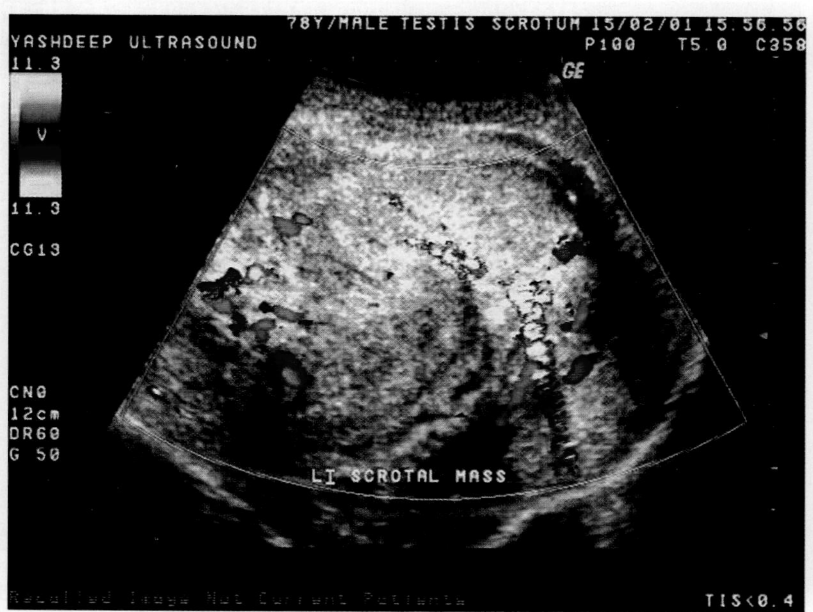

FIGURE 10.244

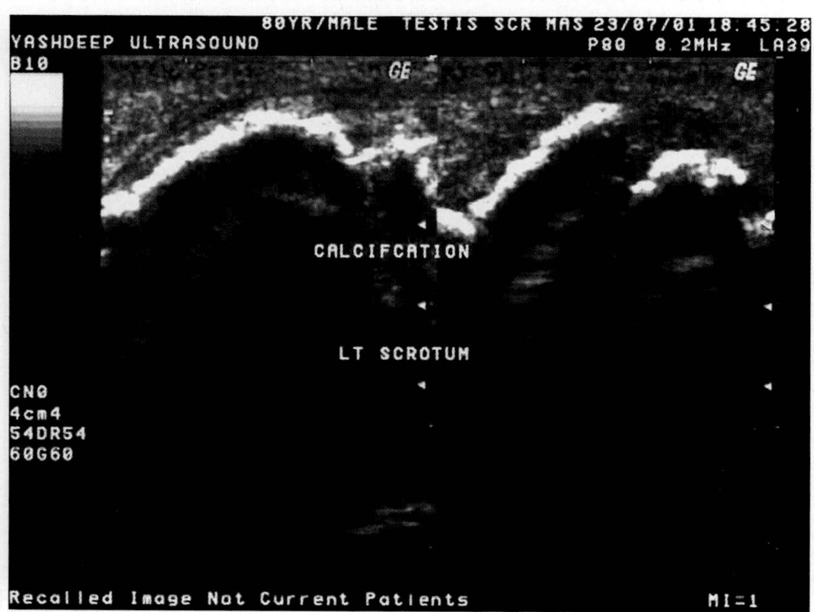

FIGURE 10.247

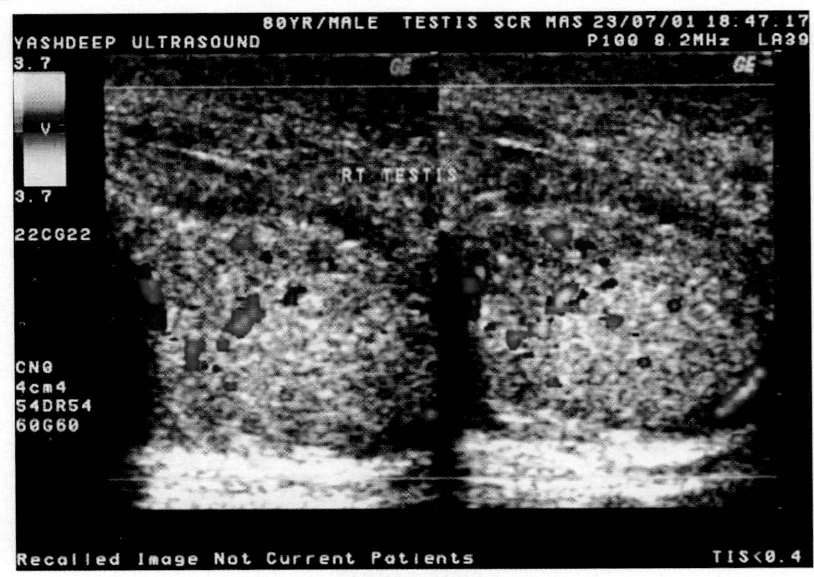

FIGURE 10.248

FIGURES 10.246 to 10.248: Chronic calcified scrotal collection. A patient presented with hard bilateral scrotal swelling. Low-level collection is seen in the scrotal sac. However, dense calcified rim is seen around the collection accompanied with acoustic shadowing.

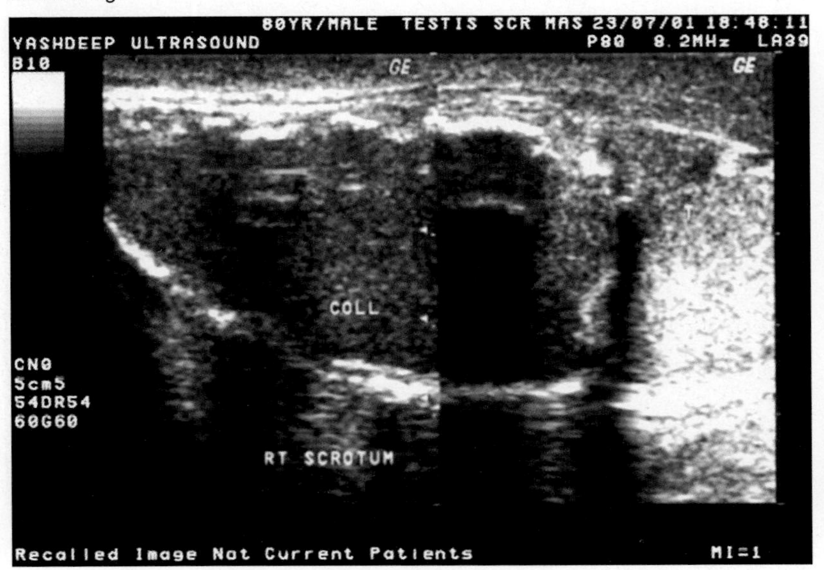

FIGURE 10.249

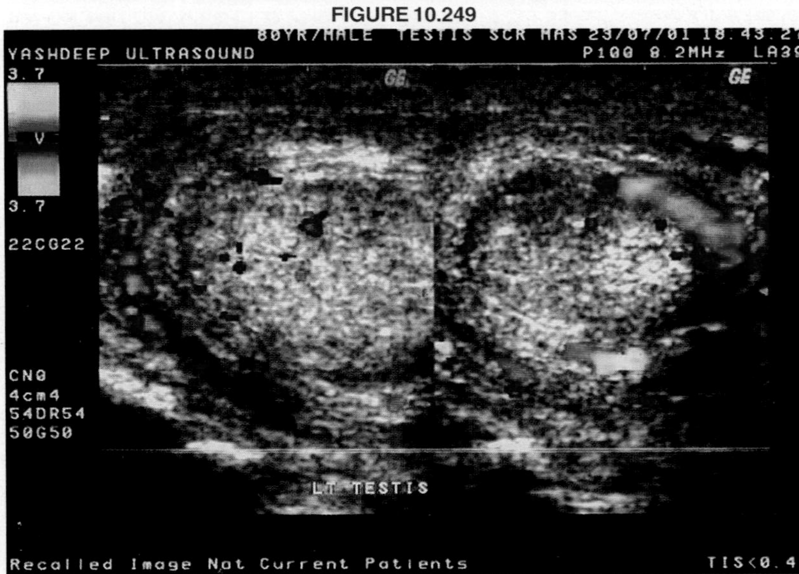

FIGURE 10.250

FIGURES 10.249 and 10.250: Color flow imaging shows normal testicular flow. No evidence of any testicular mass is seen.

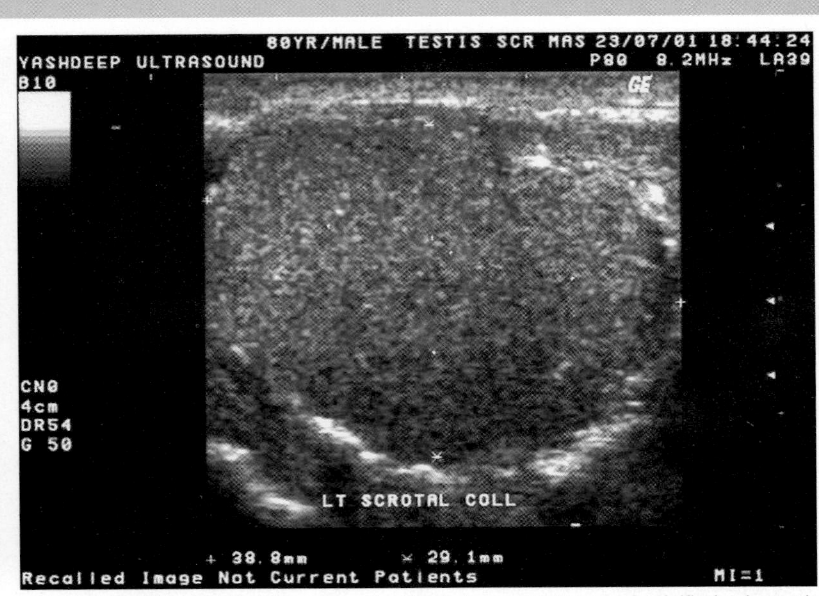

FIGURE 10.251: HRSG shows the collection in the scrotal sac and calcified echogenic rim is seen around the collection. Aspiration showed chronic pyogenic calcification.

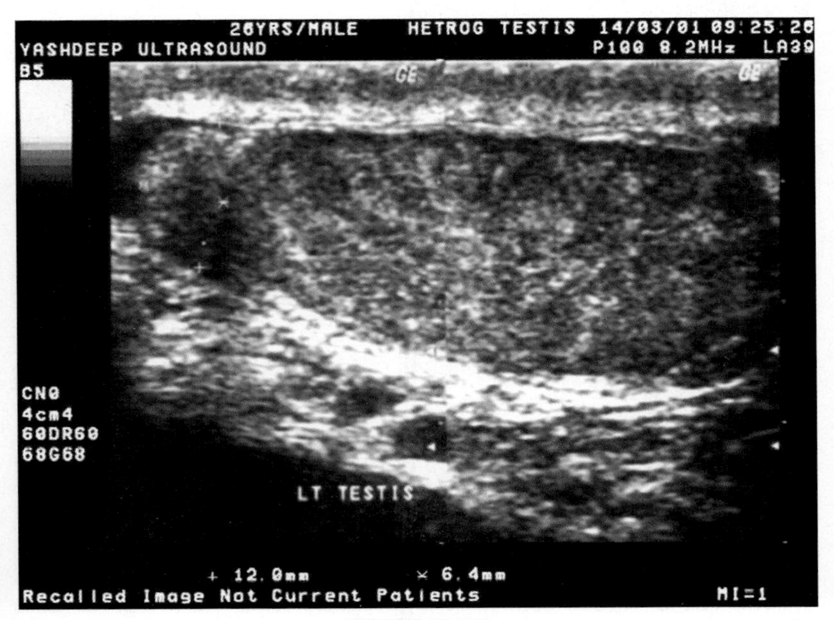

FIGURE 10.252

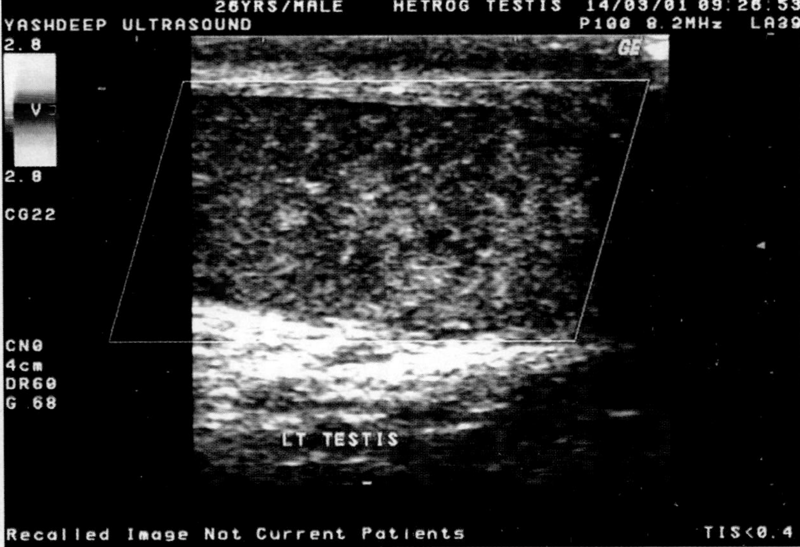

FIGURE 10.253

FIGURES 11.252 and 10.253: Tubercular testis. The testis is irregular in outline and hypoechoic in texture. The normal texture is distorted and multiple hypoechoic areas are seen scattered in the parenchyma. Typical moth eaten appearance is seen. Biopsy of the testis shows tuberculosis of testis.

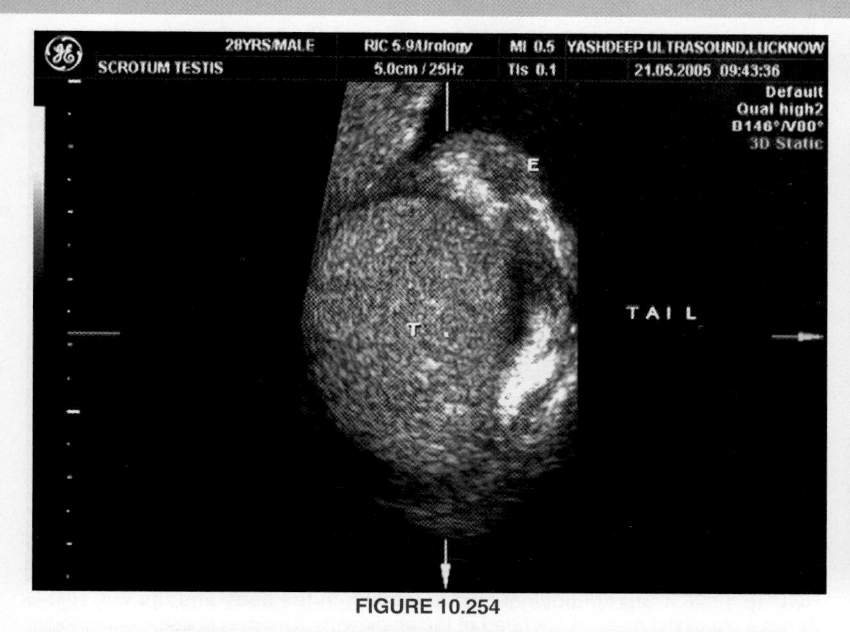

FIGURE 10.254

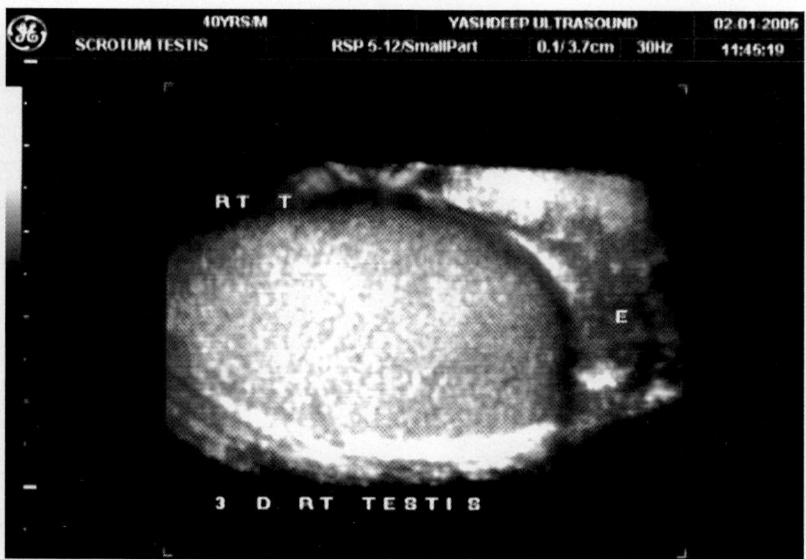

FIGURE 10.255

FIGURES 10.254 and 10.255: 3D surface rendering of epididymis. 3D surface rendering imaging of epididymis show normal head, body and tail of epididymis as homogeneous structure.

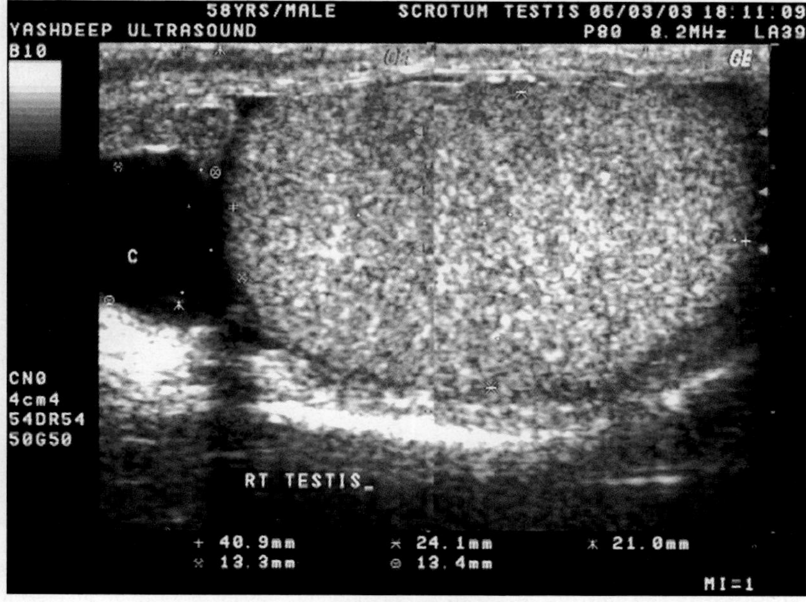

FIGURE 10.256

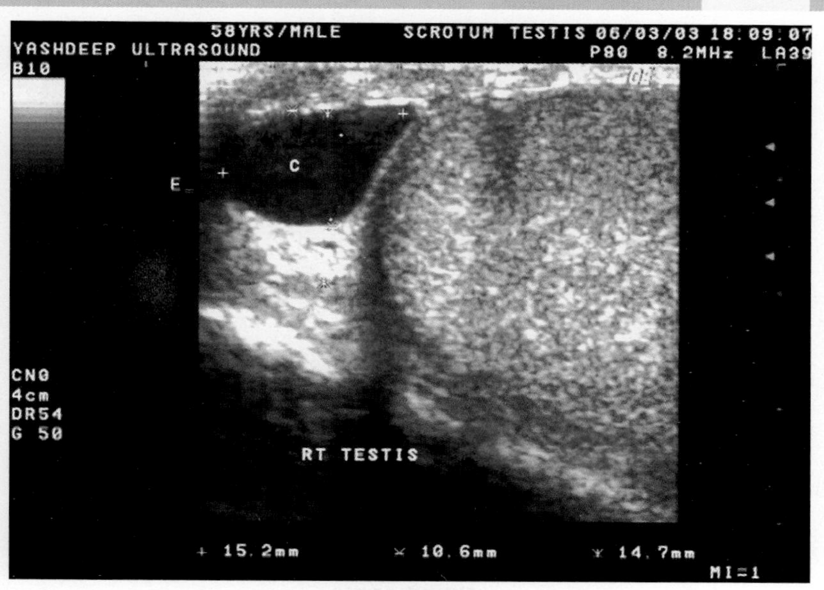

FIGURE 10.257

FIGURES 10.256 and 10.257: Simple epididymal cyst. HRSG shows well defined thin walled cyst in the head of epididymis. No internal echo is seen in the cyst. No calcification or septation is seen. Simple cysts are common and most of the time chance findings on HRSG.

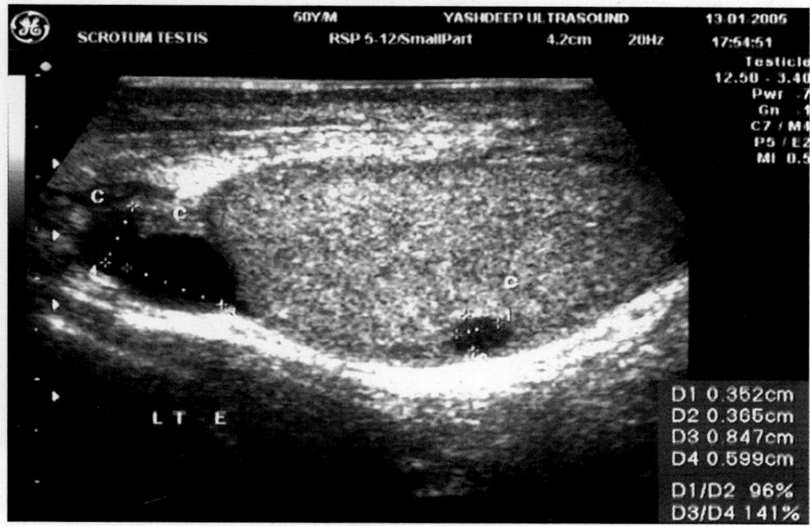

FIGURE 10.258

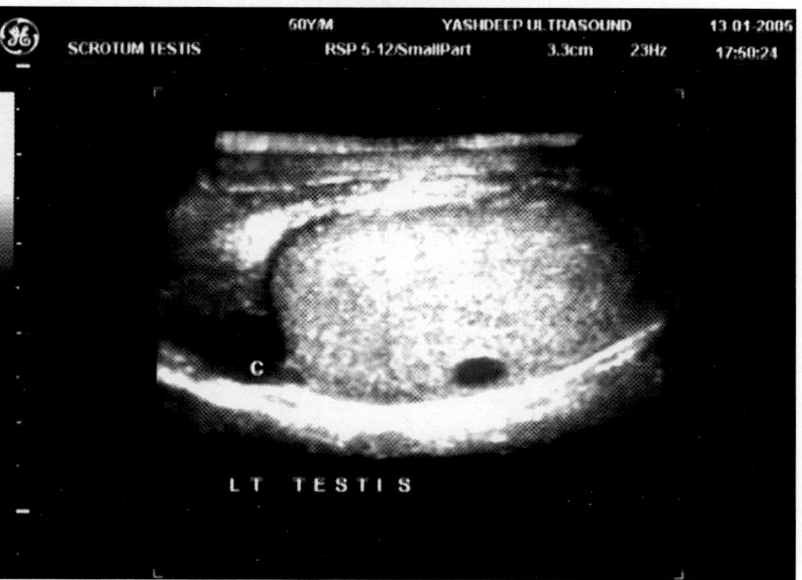

FIGURE 10.259

FIGURES 10.258 and 10.259: Epididymal cyst. HRSG shows well defined cyst in the epididymis. 3D imaging also shows a well-defined cyst in the testis.

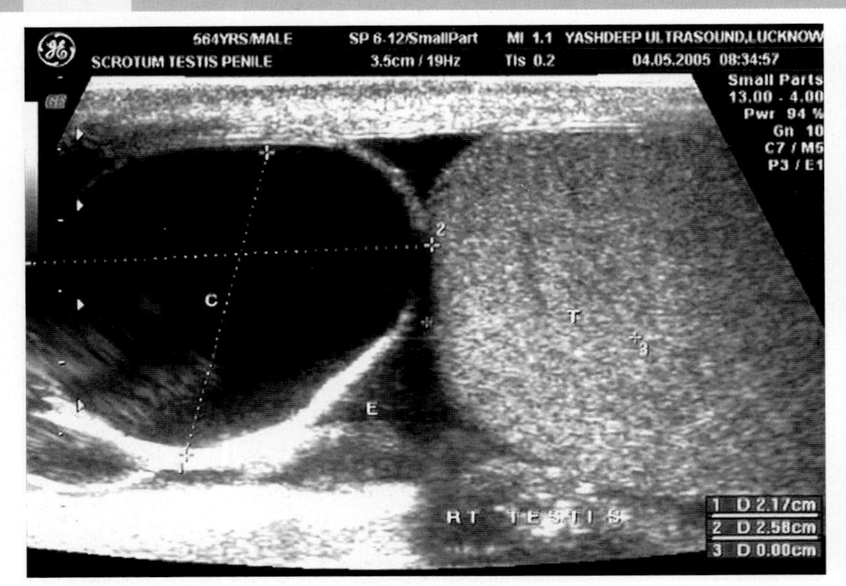

FIGURE 10.260

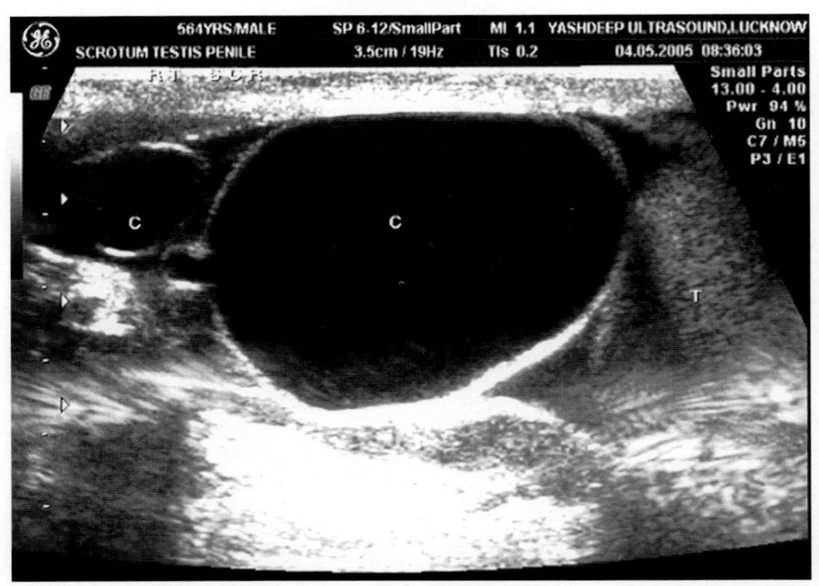

FIGURE 10.261

FIGURES 10.260 and 10.261: Big epididymal cyst. A patient presented with palpable mass in the Rt scrotum. HRSG shows big cyst in the epididymis. No internal echo seen in the cyst. The cyst is seen replacing whole of the epididymis. Small proximal cord cyst is also seen with the big cyst.

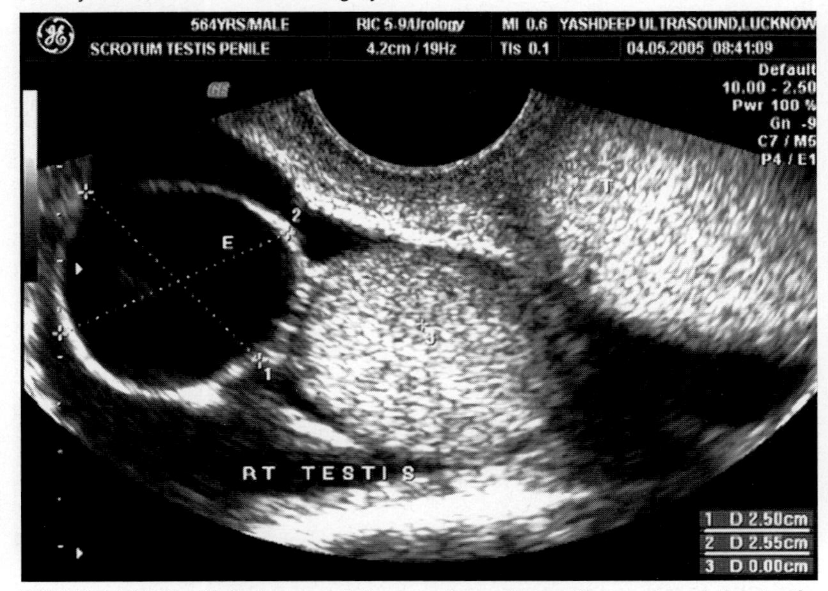

FIGURE 10.262: Global view of the scrotum shows the cyst and the testis. The cyst is seen separate from the testis.

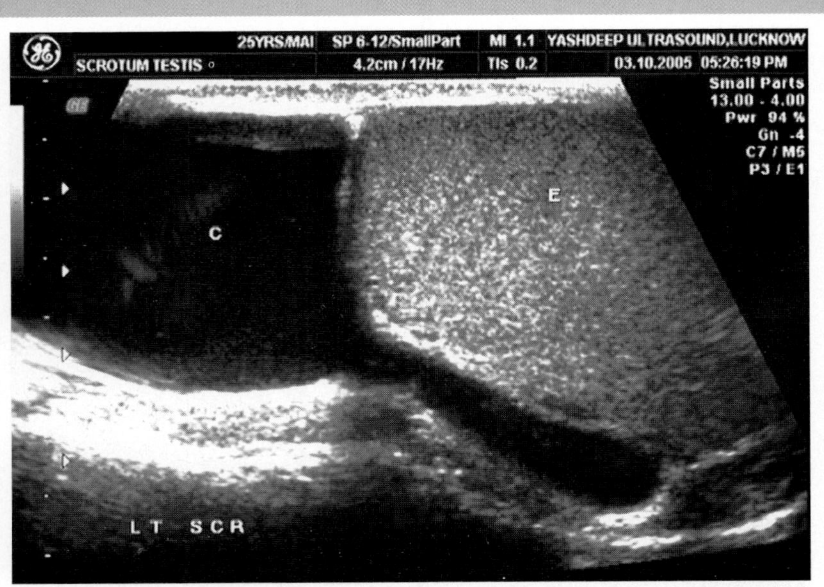

FIGURE 10.263: Big epididymal cyst extending in the body and the tail. HRSG shows a big cystic mass in the head of the epididymis. The cyst is seen extending in the body in the tail of epididymis. It is seen replacing the normal structure.

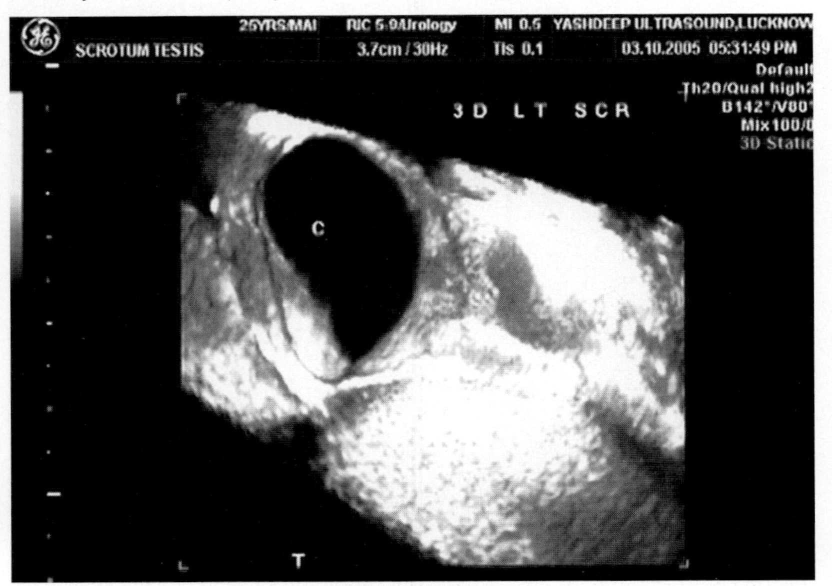

FIGURE 10.264

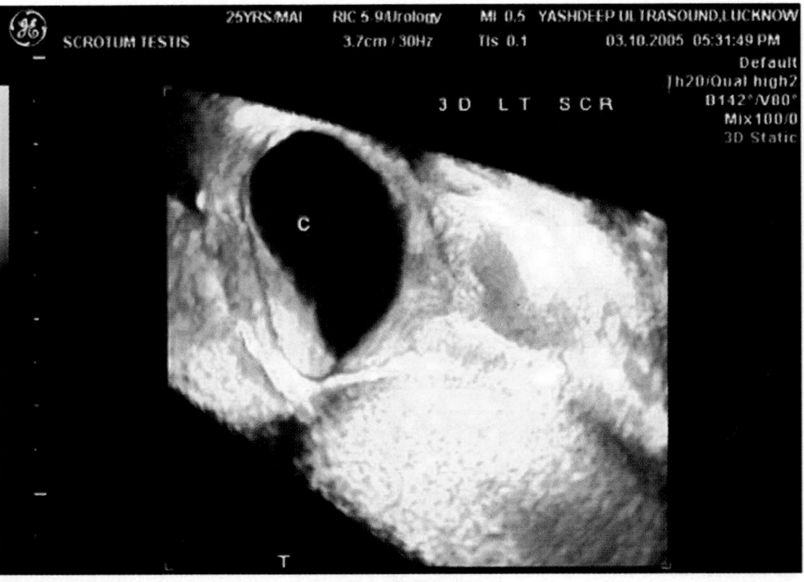

FIGURE 10.265

FIGURES 10.264 and 10.265: 3D Surface rendering imaging of the scrotum shows the cyst in much clinical detail.

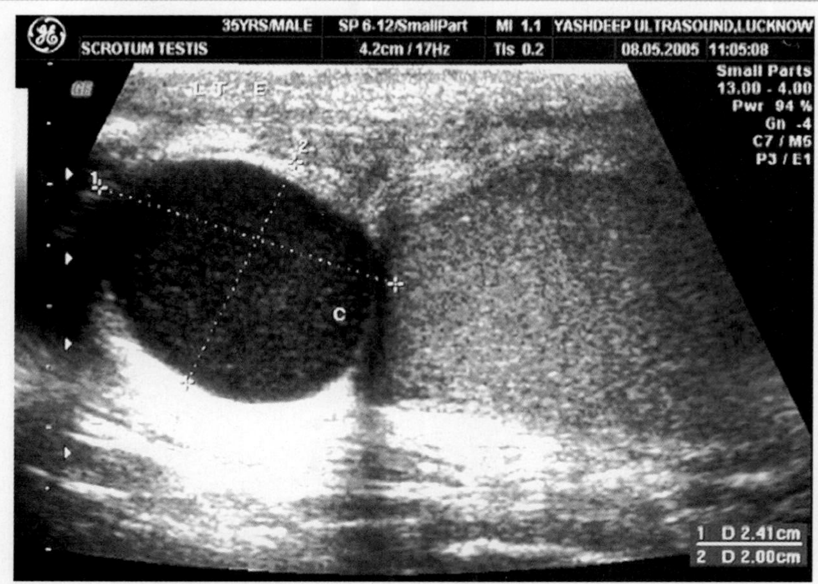

FIGURE 10.266: Spermatocele in the epididymis. A patient of mail infertility presented with oligospermia. HRSG shows a well-defined cyst in the epididymis. The cyst shows homogeneous internal echoes with fine segmentation.

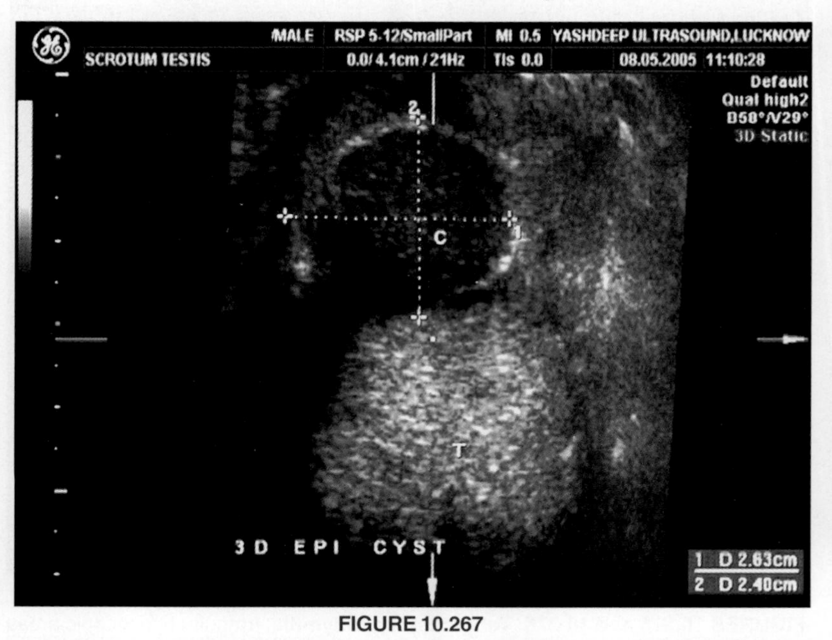

FIGURE 10.267

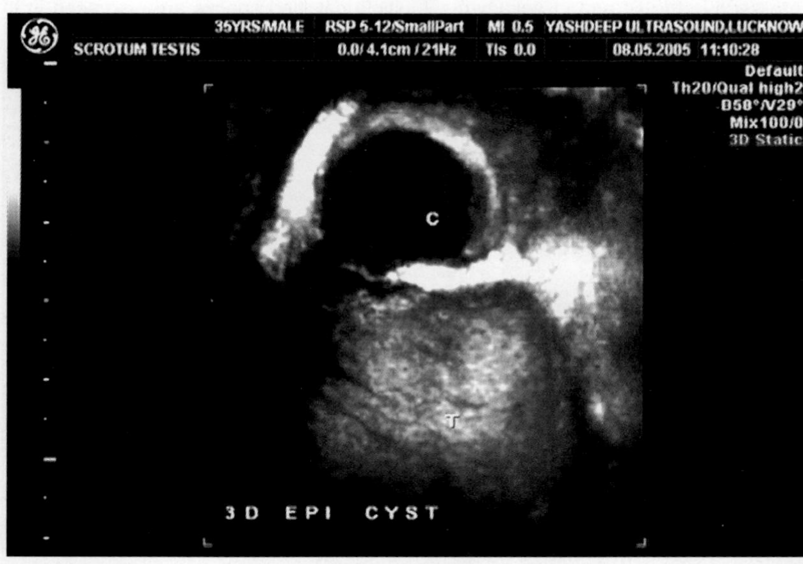

FIGURE 10.268

FIGURES 10.267 and 10.268: 3D imaging of the cyst show the details of the cyst. Aspiration of the cyst confirmed spermatocele.

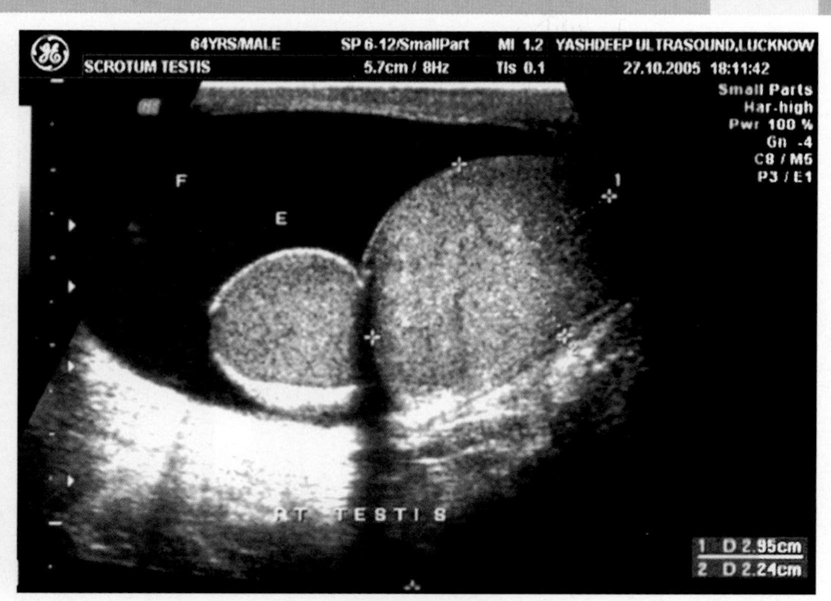

FIGURE 10.269

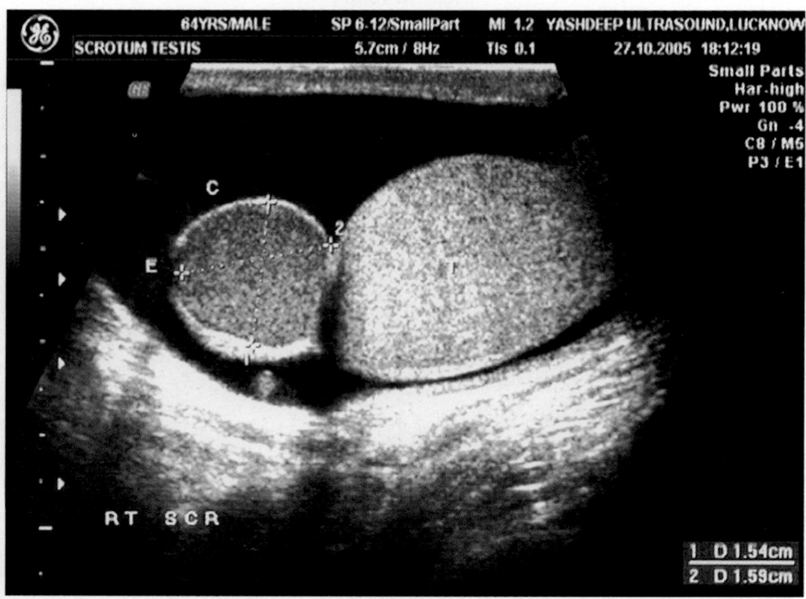

FIGURE 10.270

FIGURES 10.269 and 10.270: Epididymal spermatocele. HRSG shows a thick walled cyst in the Rt scrotum. The cyst shows homogeneous echoes with "ground glass appearance". No septation is seen in the cyst suggestive of spermatocele.

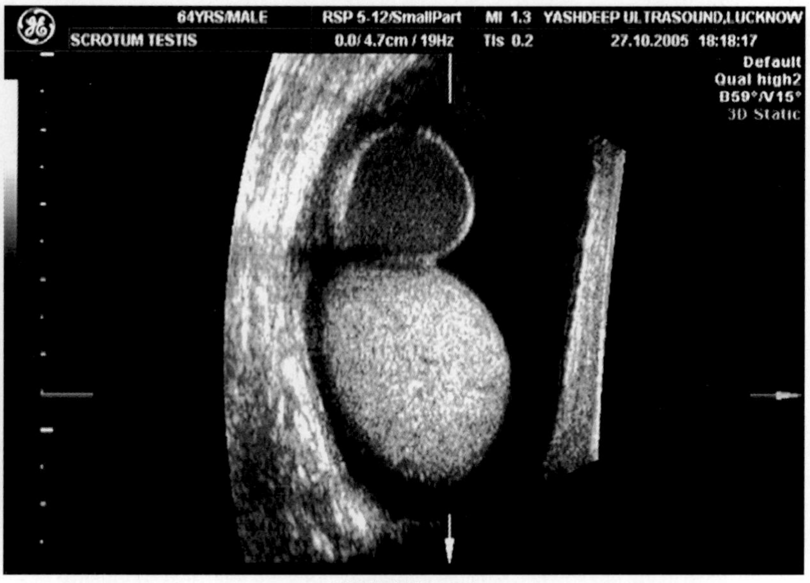

FIGURE 10.271

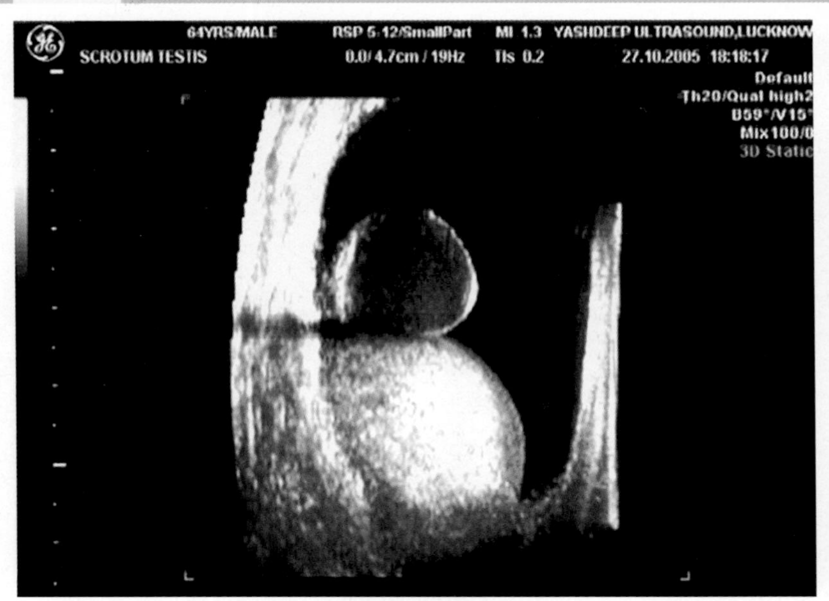

FIGURE 10.272

FIGURES 10.271 and 10.272: 3D surface-rendering image shows the cyst in much better view.

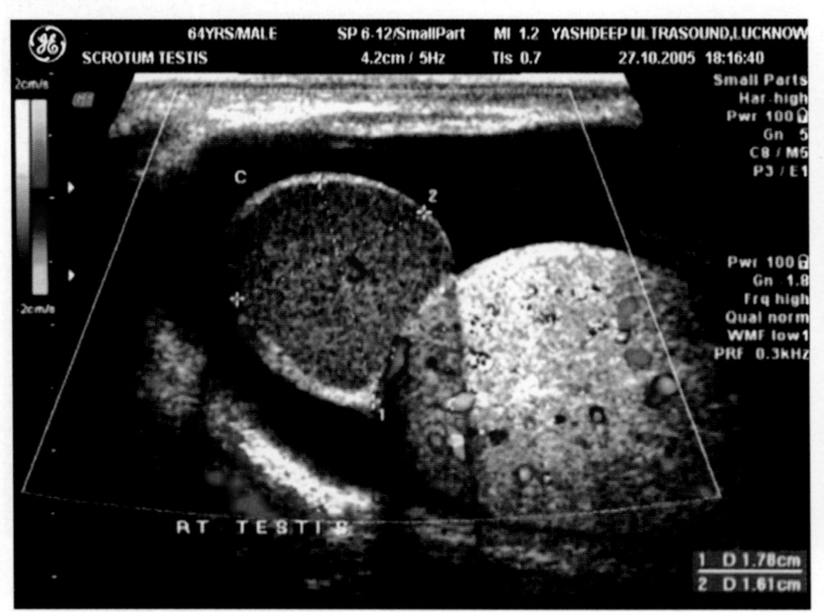

FIGURE 10.273: On color flow imaging no flow seen in the cyst.

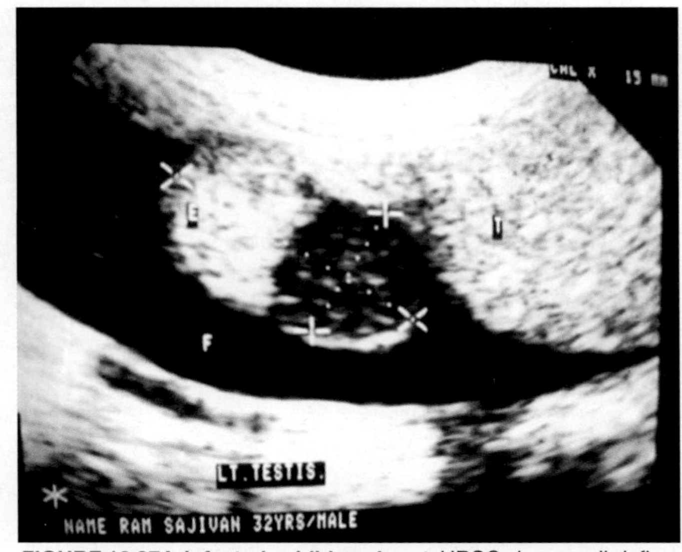

FIGURE 10.274: Infected epididymal cyst. HRSG shows well-defined cyst in the epididymis. Multiple internal echoes are seen in it suggestive of pyogenic collection in the cyst.

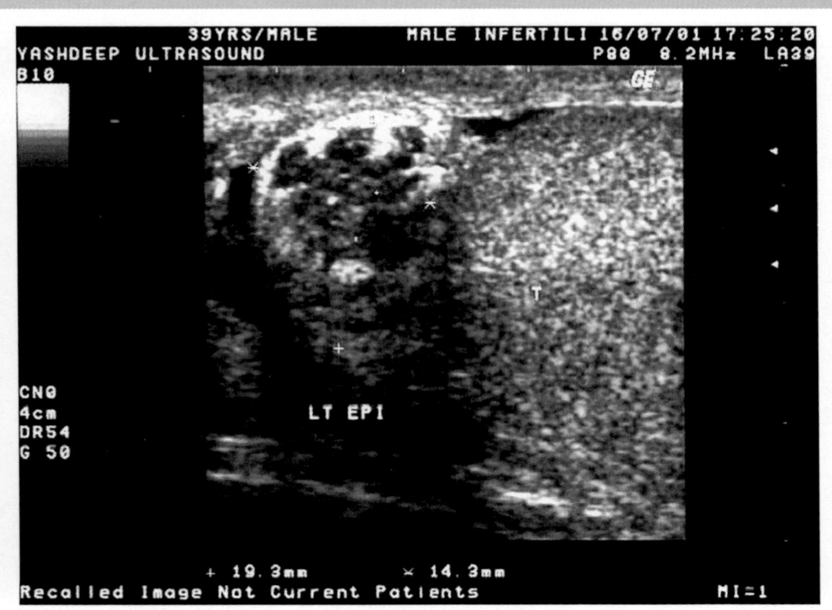

FIGURE 10.275

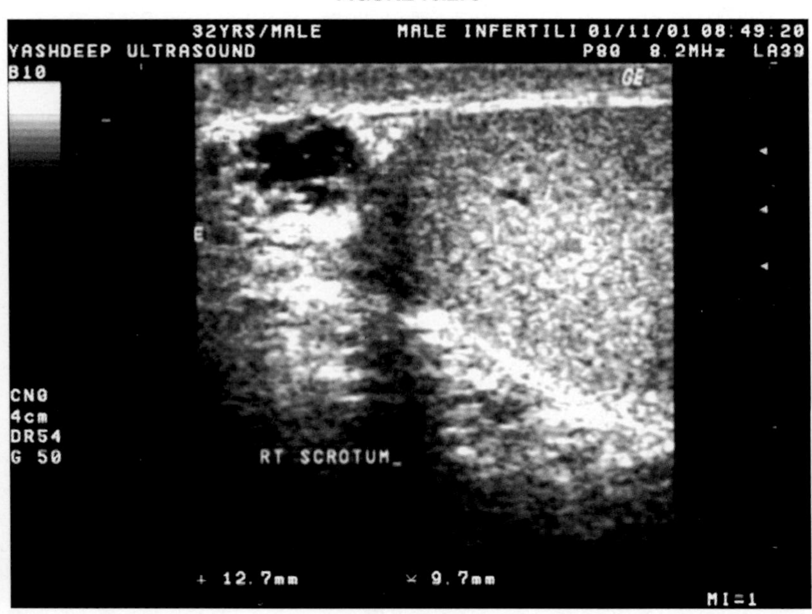

FIGURE 10.276

FIGURES 10.275 and 10.276: Tubercular epididymitis. Irregular hypoechoic areas are seen in the Rt and Lt epididymis. Multiple calcified specks are also seen in it. Biopsy of the mass shows tubercular infiltration.

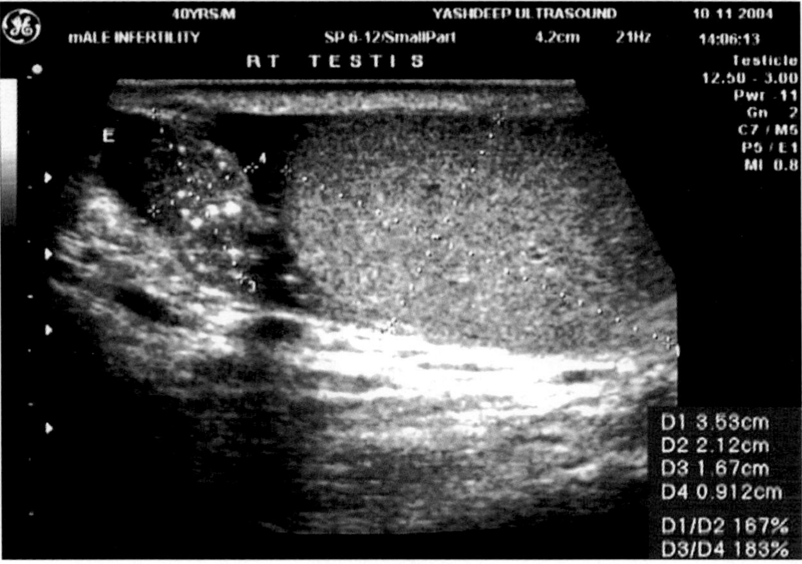

FIGURE 10.277

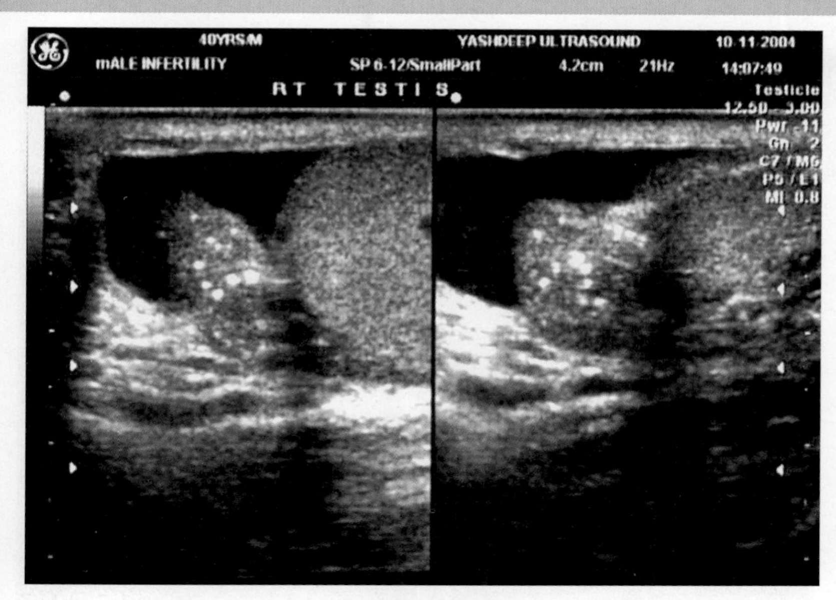

FIGURE 10.278

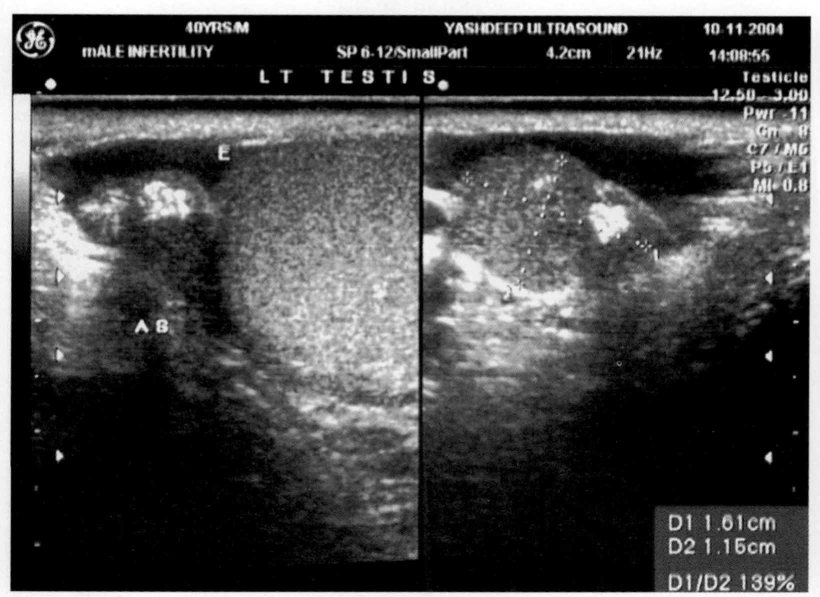

FIGURE 10.279

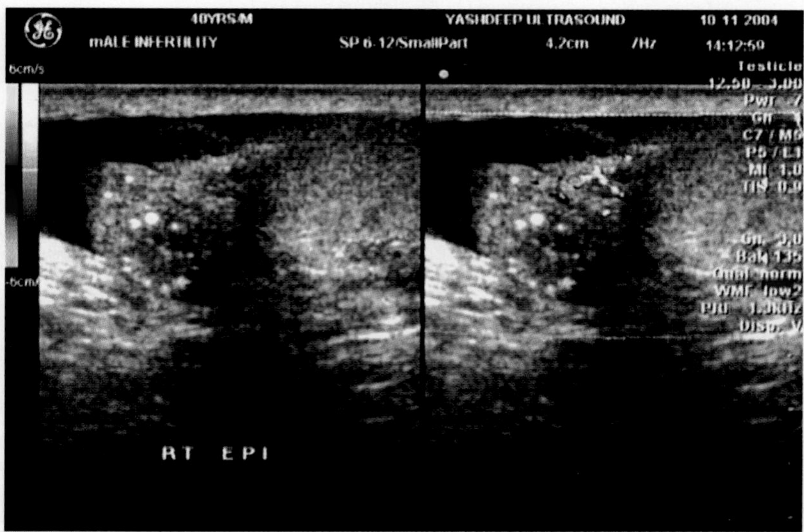

FIGURE 10.280

FIGURES 10.277 to 10.280: Tubercular epididymitis. A patient presented of male infertility presented with oligoasthenospermia. HRSG shows irregular heterogeneous epididymis in both scrotal sac. Multiple calcified specks are seen in the epididymis. Epididymides were irregular in outline. On color flow imaging no flow seen in them. Biopsy of the epididymis confirmed tubercular epididymitis.

Cord Pathology

Varicoceles occur in 2 to 20% of the normal adult population. Varicocele is a potential cause of male infertility due to increased scrotal temperature as a result of venous stasis. Varicocele is more common on the left side due to its direct drainage in the left renal vein. It is bilateral in 10 to 15% of the cases. However, this incidence will surely increase with the advent of color Doppler sonography. The causes of varicoceles are the incompetence of the spermatic vein valves, left renal vein thrombosis or tumoral compression and neoplastic involvement of the retroperitoneum and pelvis.

Ultrasonographic Features

Ultrasonographic diagnosis is usually easy. The spermatic cord is normally examined in supine and erect positions at rest and with Valsalva's maneuver.

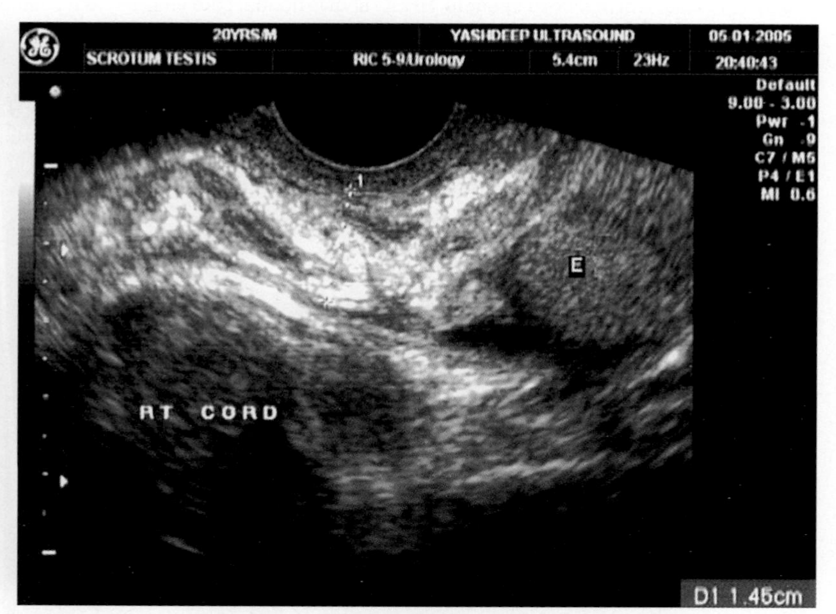

FIGURE 10.281

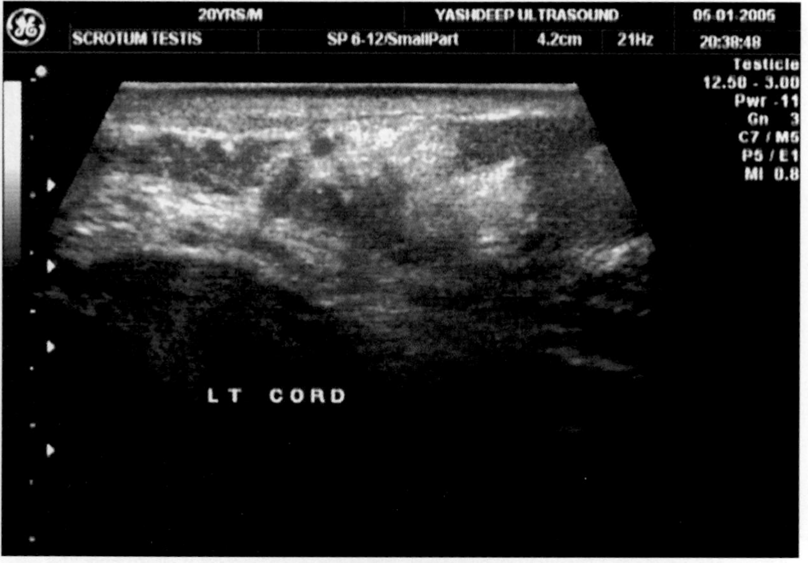

FIGURE 10.282

FIGURES 10.281 and 10.282: Normal cord. HRSG shows the normal cord. It is having mixed echo texture. Hypoechoic tubular shadows are seen in the cord suggestive of vessels. Echogenic fat is also seen.

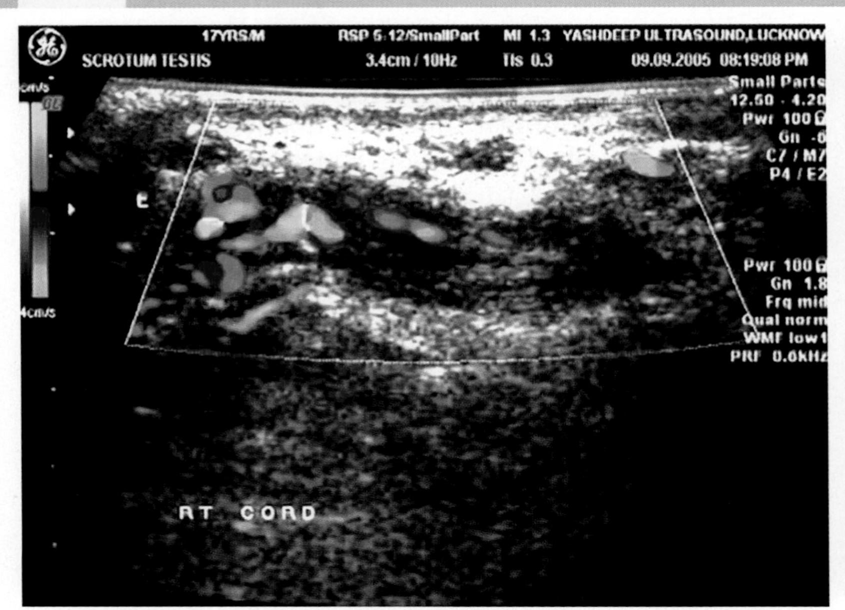

FIGURE 10.283: **Normal cord vessels.** HRSG shows normal cord vessels. Testicular artery is seen. HRSG shows normal cord vessels running to the substance of the cord. Normal pampiniform plexus is seen along the cord.

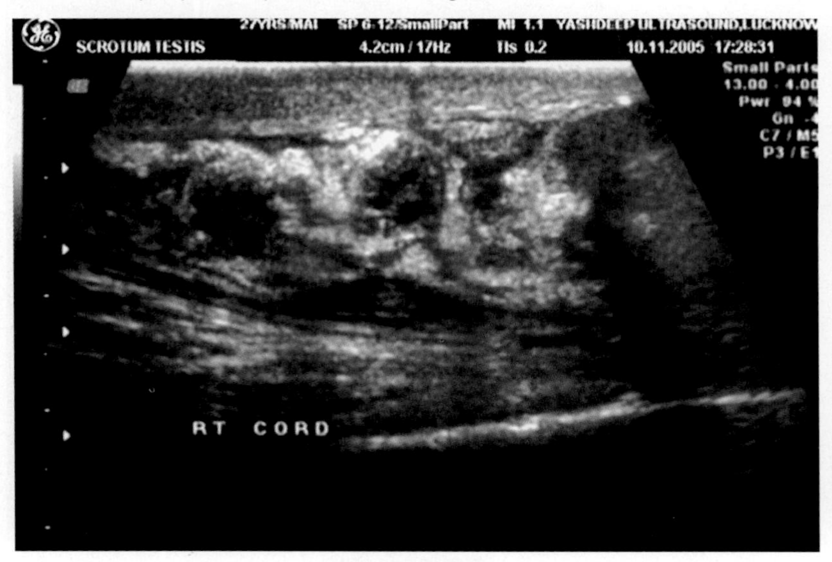

FIGURE 10.284

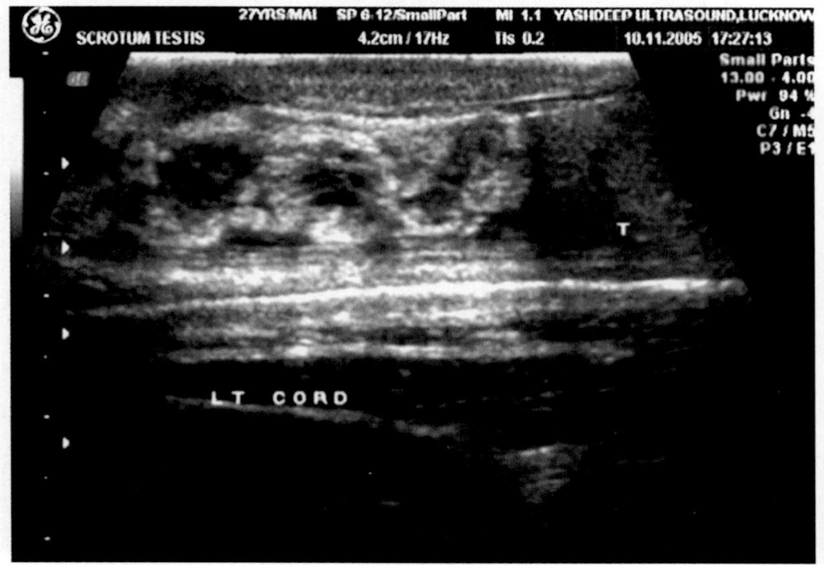

FIGURE 10.285

FIGURES 10.284 and 10.285: Thickened cord. HRSG shows markedly thickened both cords. However, the cord vessels are not dilated. But increased amount of echogenic fat is seen along the cord.

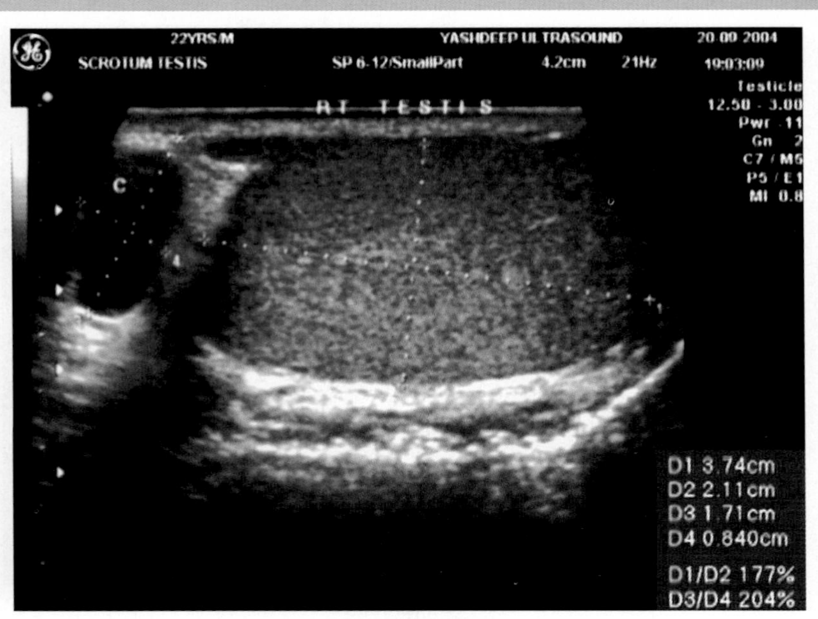

FIGURE 10.286

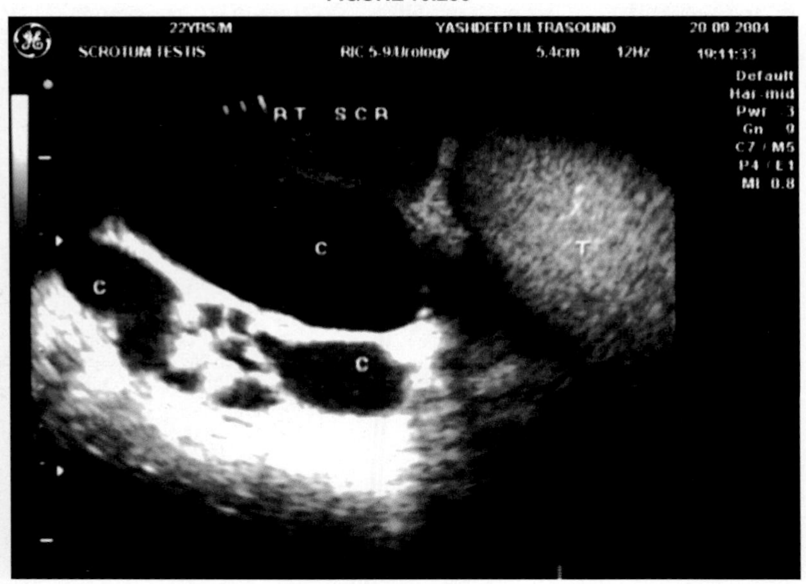

FIGURE 10.287

FIGURES 10.286 and 10.287: Epididymis and cord cyst. HRSG shows multiple cysts along the cord. Normal cord tissue replaced by the cysts. These cord cysts are known as spermatocele. The arrest of the sperm is seen in the cord resulting into oligo or azospermia. Small cyst is also seen in the epididymal head.

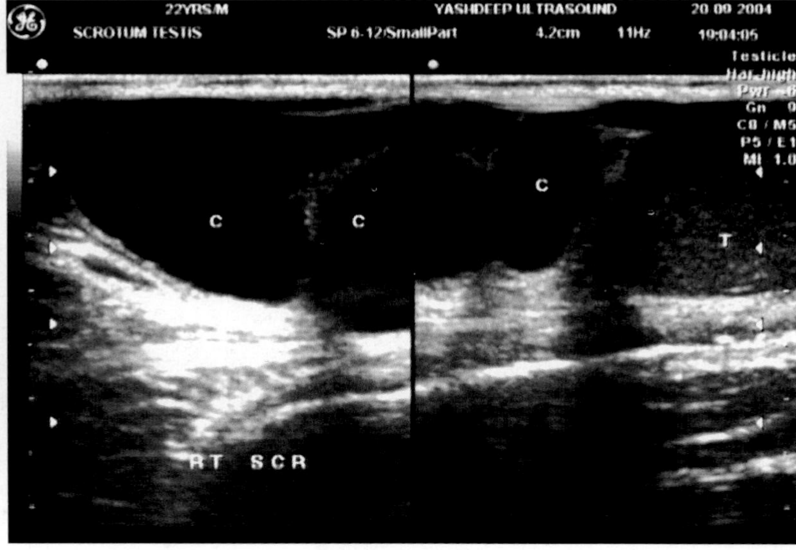

FIGURE 10.288

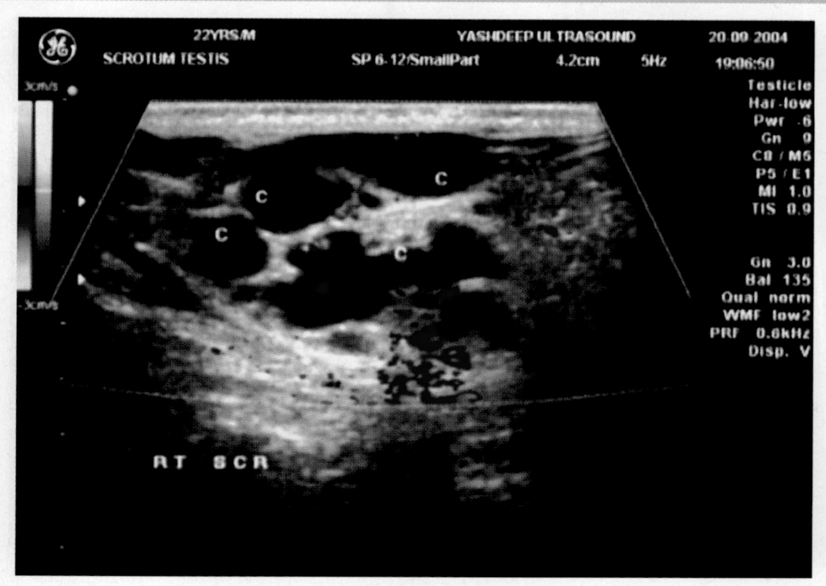

FIGURE 10.289

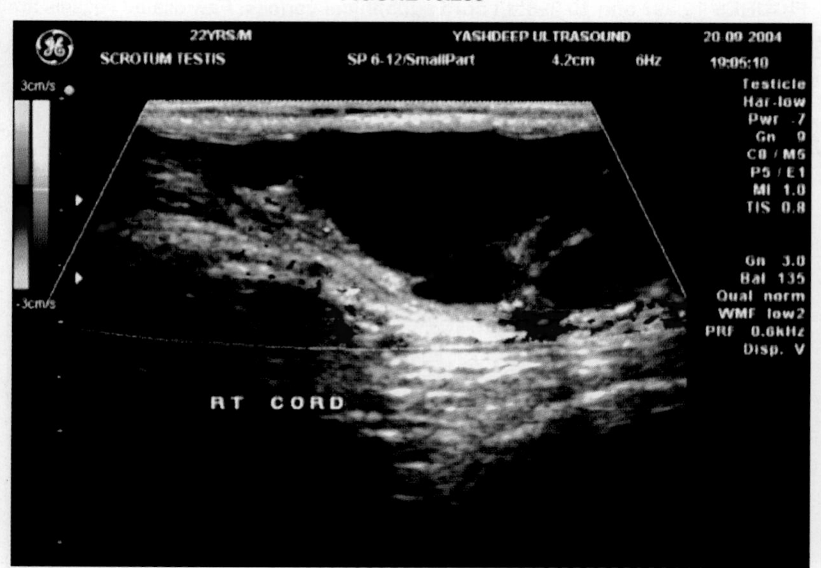

FIGURE 10.290

FIGURES 10.288 to 10.290: Cystic spermatic cord. Whole of the cord replaced by multiple cysts. The cysts are seen non-communicating. On color flow imaging no flow seen in the cyst.

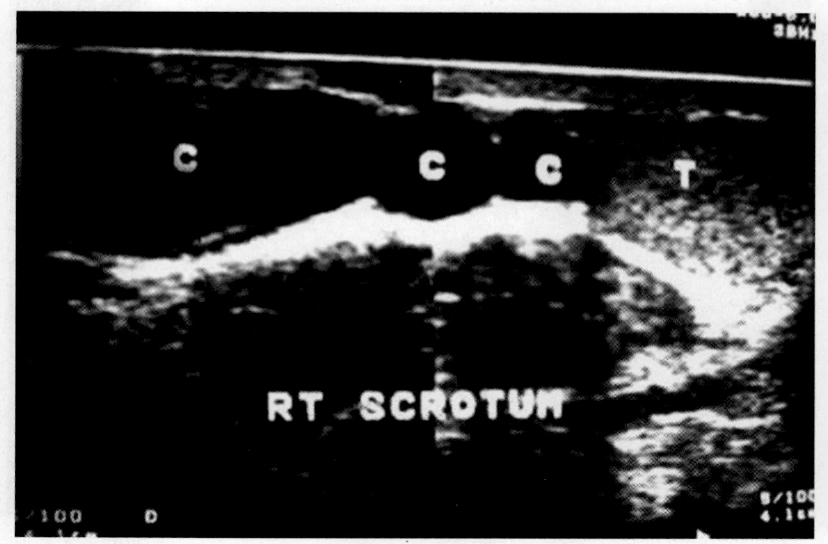

FIGURE 10.291

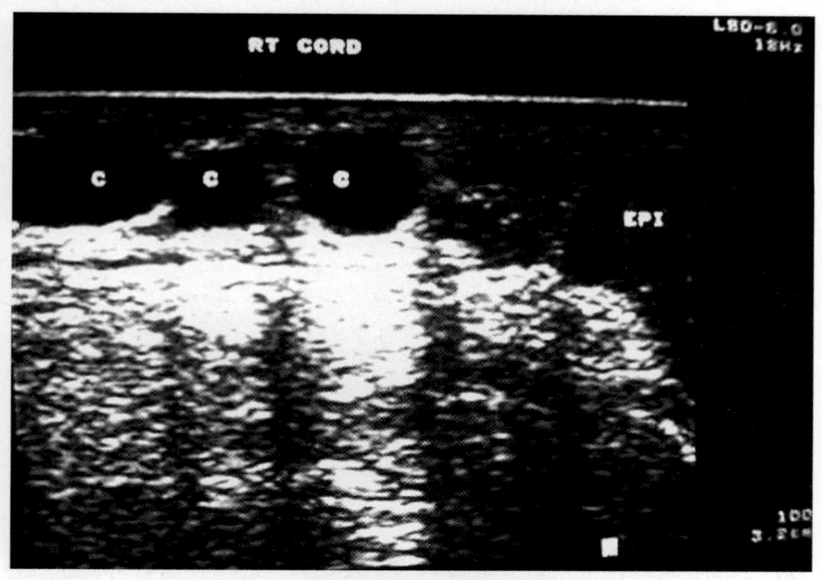

FIGURE 10.292

FIGURES 10.291 and 10.292: Cystic spermatic cord. Multiple well-defined thin walled cysts are seen in Rt cord. They are non-communicating. A big cyst is seen in the distal part of the cord. Post cystic enhancement is also seen in the cyst suggestive of multiple small spermatocele.

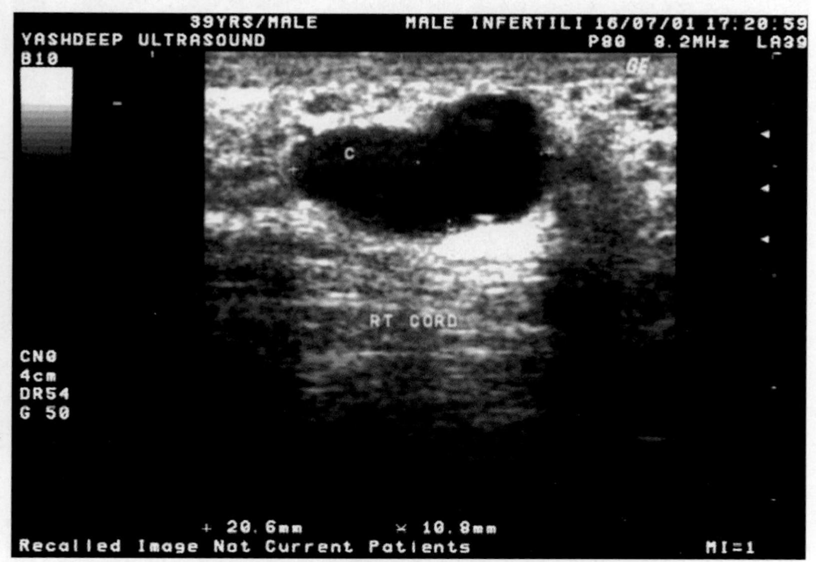

FIGURE 10.293

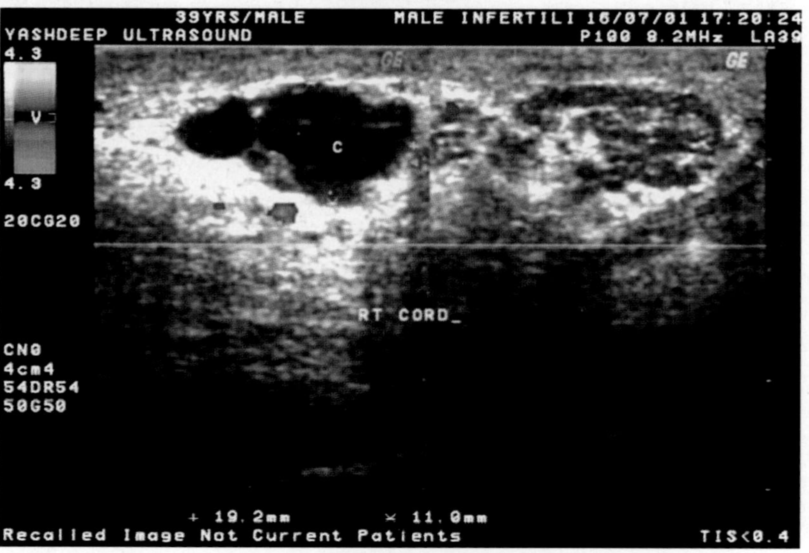

FIGURE 10.294

FIGURES 10.293 and 10.294: Spermatocele in the Rt cord. An irregular well defined thin walled cyst is seen in the proximal part of the cord. Few internal echoes are seen in it. But no septation is seen. FNAC from the cyst shows spermatocele.

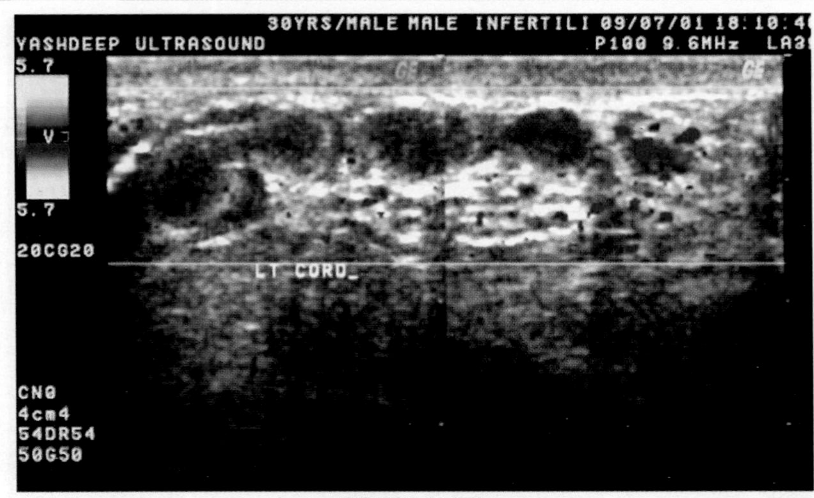

FIGURE 10.295

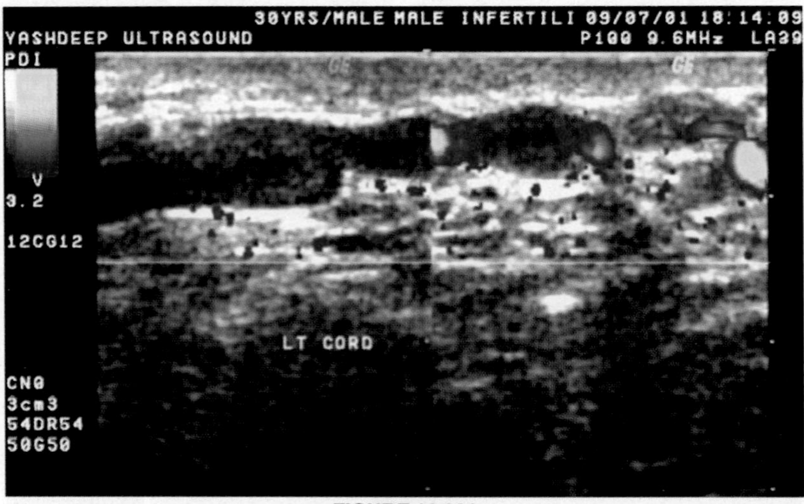

FIGURE 10.296

FIGURES 10.295 and 10.296: Spermatocele in the Lt cord. Multiple cysts are seen along the Lt cord. They are non-communicating and postcystic enhancement is also seen. On color Doppler imaging no flow is seen in the cyst. Thus excluding the dilated vessels. However, power Doppler imaging shows little flow in the wall of the cysts.

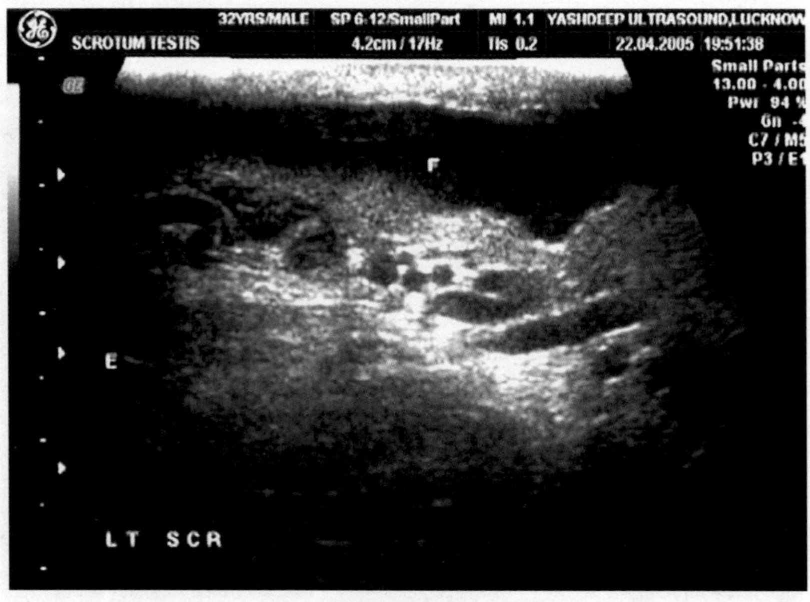

FIGURE 10.297

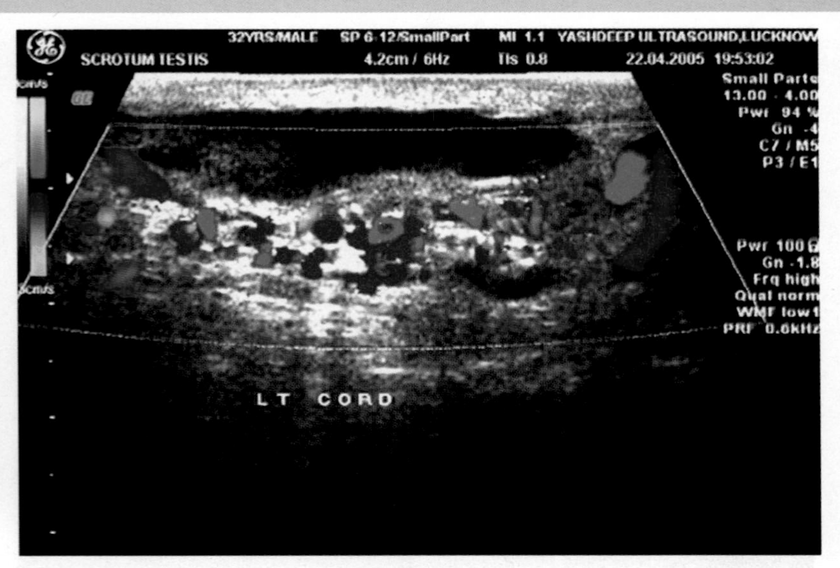

FIGURE 10.298

FIGURES 10.297 and 10.298: Lt cord subclinical varices. Few dilated vessels are seen along the Lt cord. Which were not clinically palpable. Color Doppler imaging shows dilated vessel along the cord confirming the HRSG findings.

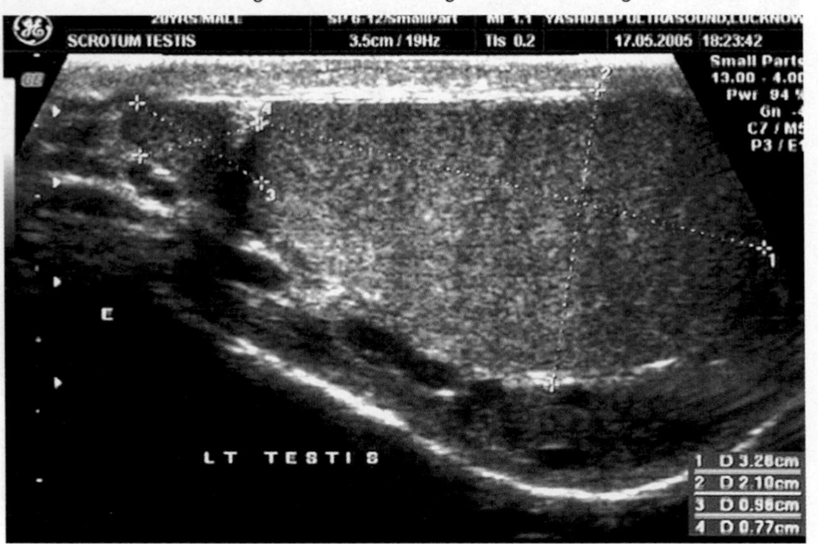

FIGURE 10.299

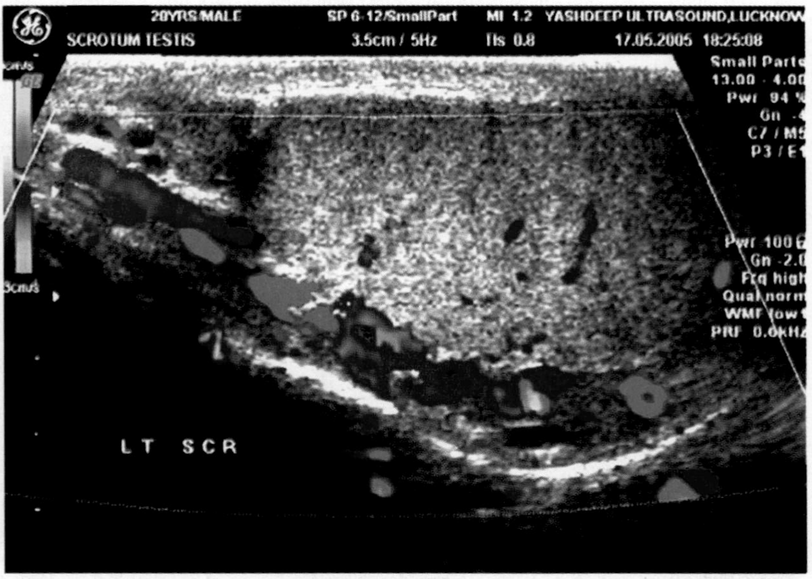

FIGURE 10.300

FIGURES 10.299 and 10.300: Multiple cord varices. HRSG shows multiple dilated vessels along the Lt cord. Fine interlacing is seen and it is seen as a fine net. Color Doppler imaging shows low flow in the dilated vessels suggestive of cord varices.

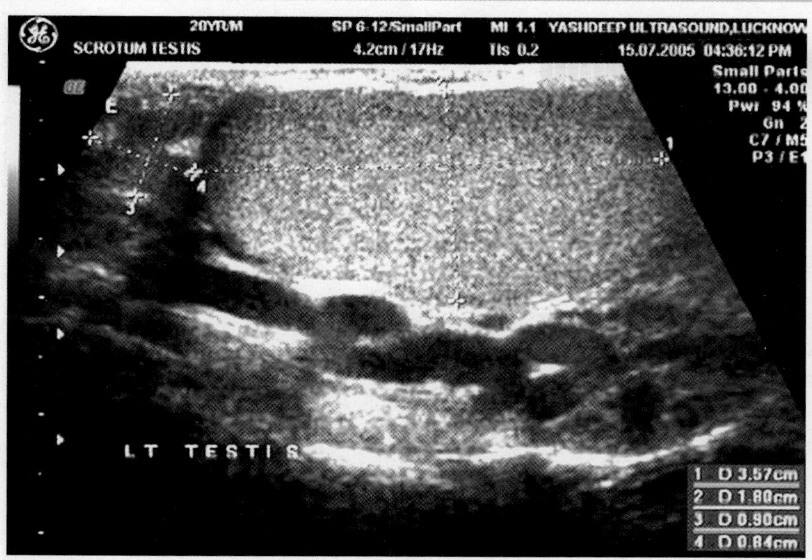

FIGURE 10.301: **Lt scrotal varices.** HRSG shows multiple dilated veins in the Lt scrotum posterior to the testis. They are seen as a bag of worm.

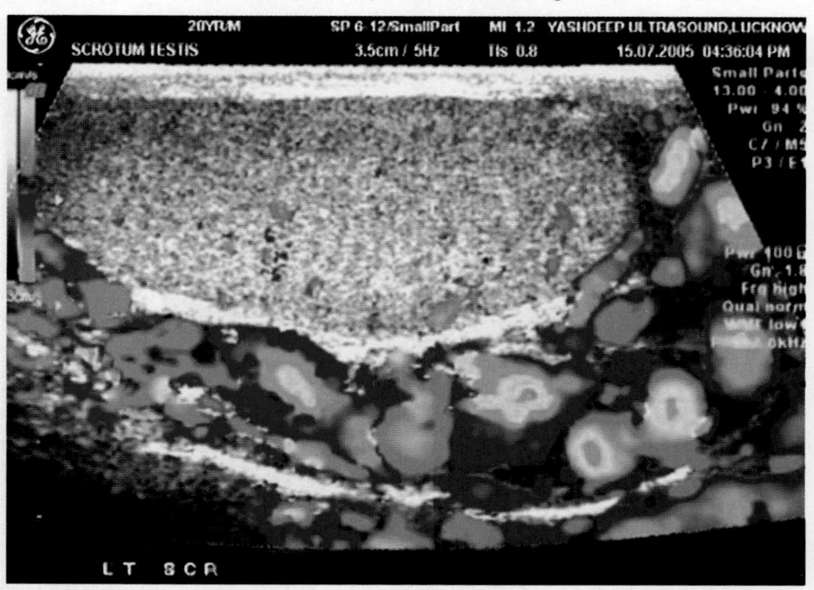

FIGURE 10.302: Color Doppler imaging shows low flow in the varices. Reflux is also seen in them. Marked reflux is seen on Valsalva maneuver. Dilated vessels are also seen along the cord in the same patient suggestive of cord and scrotal varices.

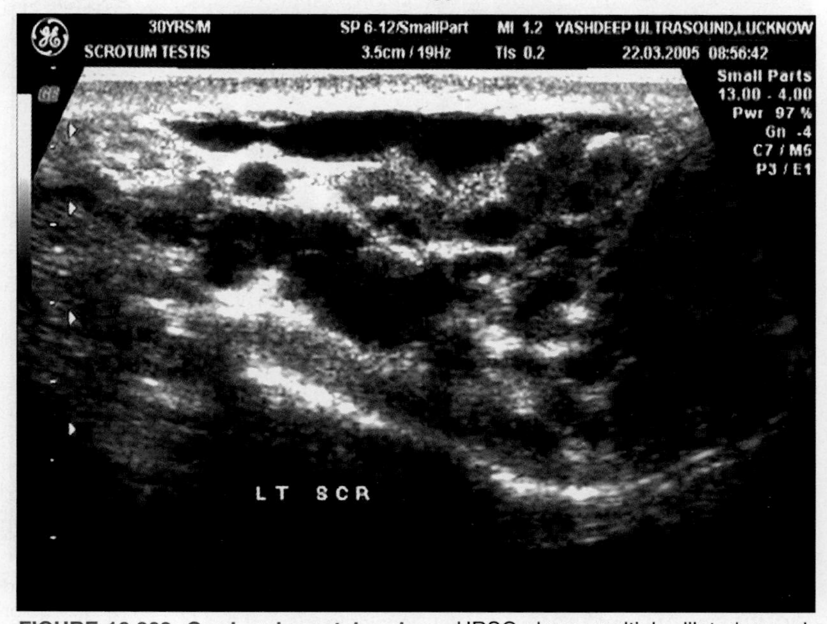

FIGURE 10.303: **Cord and scrotal varices.** HRSG shows multiple dilated vessels along the Lt cord. The dilated vessels look like fine net on gray scale imaging.

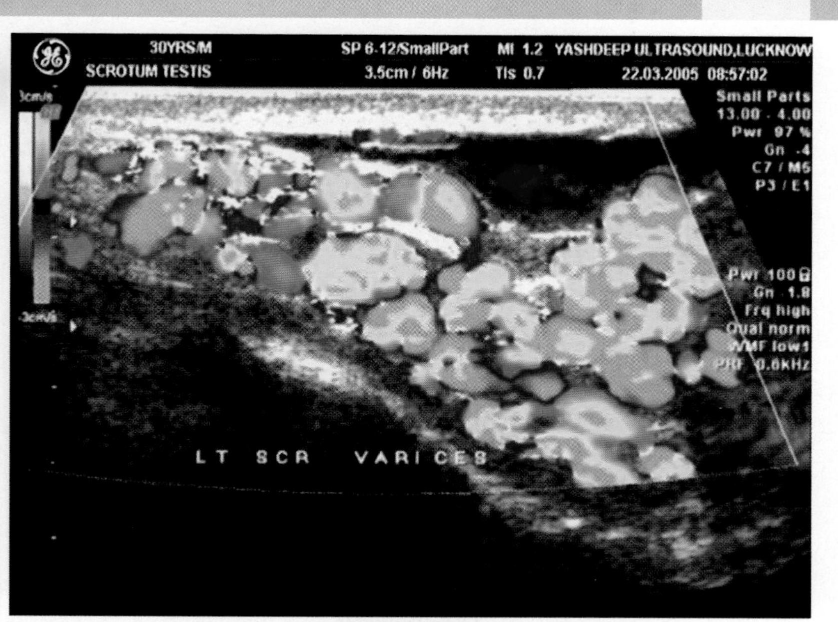

FIGURE 10.304

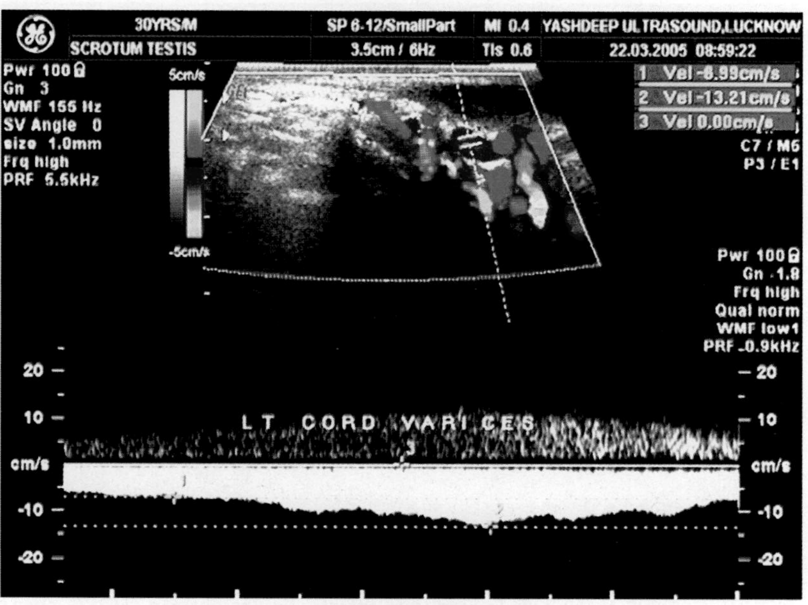

FIGURE 10.305

FIGURES 10.304 and 10.305: On color Doppler imaging the vessels are seen having low venous flow, which is confirmed on spectral Doppler tracing on Valsalva maneuver. High reflux is seen in them.

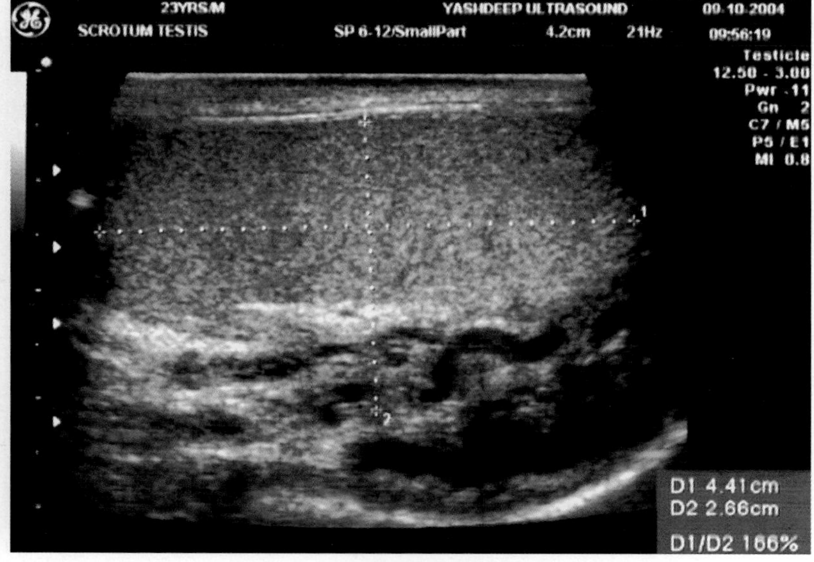

FIGURE 10.306

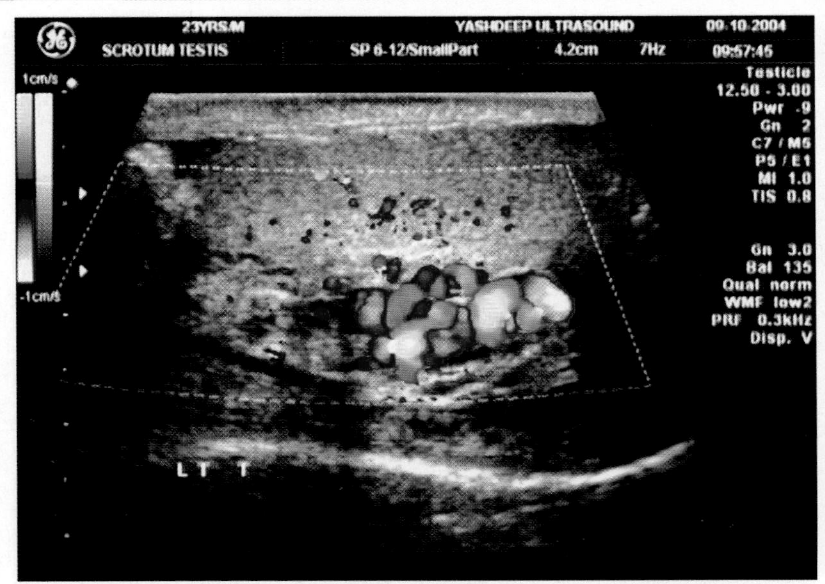

FIGURE 10.307

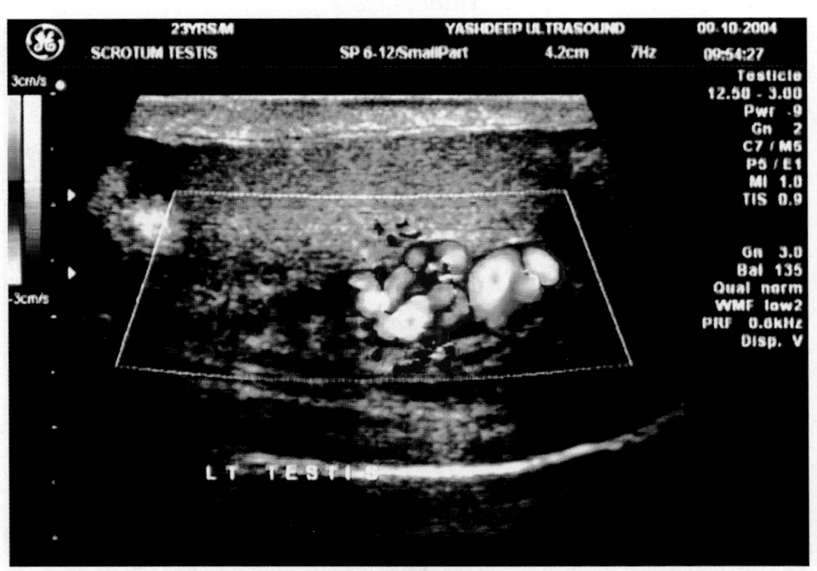

FIGURE 10.308

FIGURES 10.306 to 10.308: Scrotal and intratesticular varices. HRSG shows multiple dilated vessels in the Lt scrotum. There is also evidence of cystic spaces seen in the substance of testis. On color flow imaging they are filled with blood suggestive of scrotal and intratesticular varices.

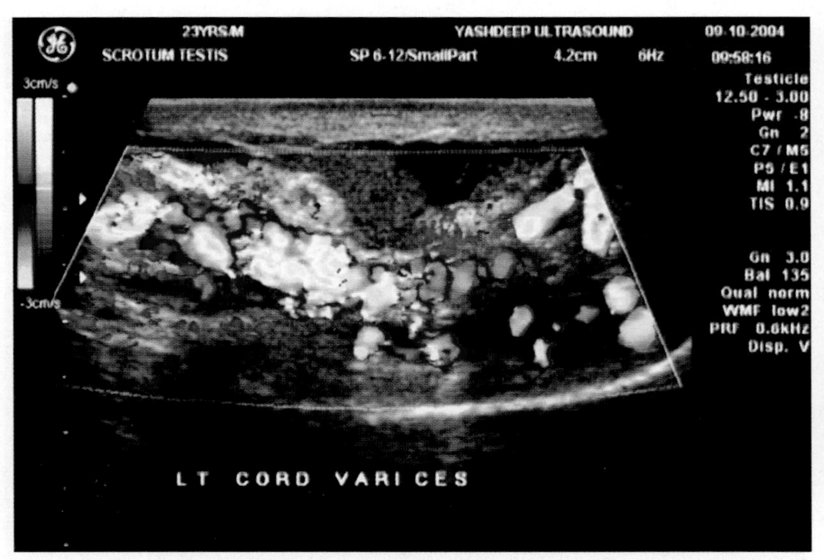

FIGURE 10.309: Lt cord also shows multiple dilated vessels suggestive of grade III cord varices.

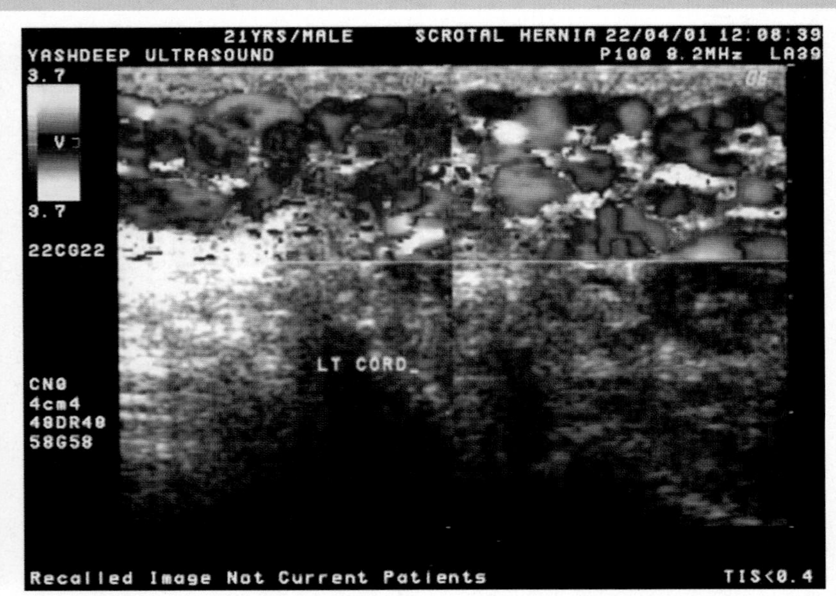

FIGURE 10.310: Arteriovenous malformation. A young man presented with rapidly increasing swelling in the Lt scrotum. HRSG and color flow imaging show multiple dilated vessels in the Lt scrotum and along the Lt cord suggestive of cord and scrotal varices.

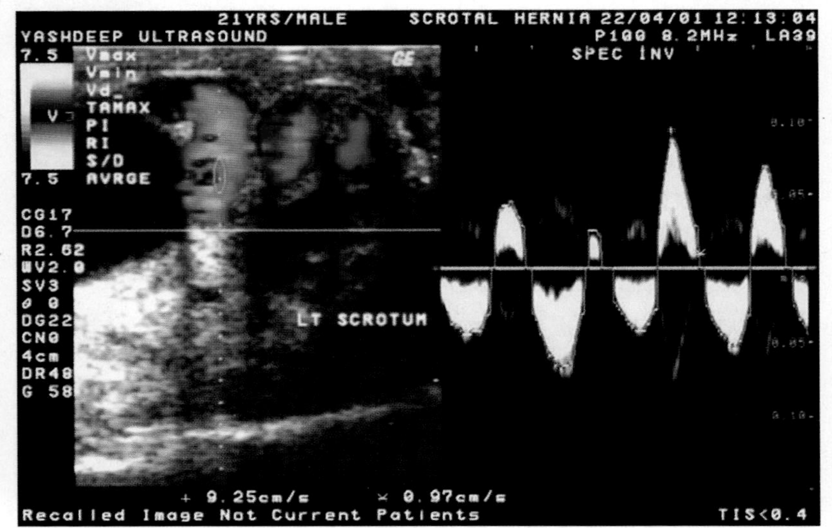

FIGURE 10.311

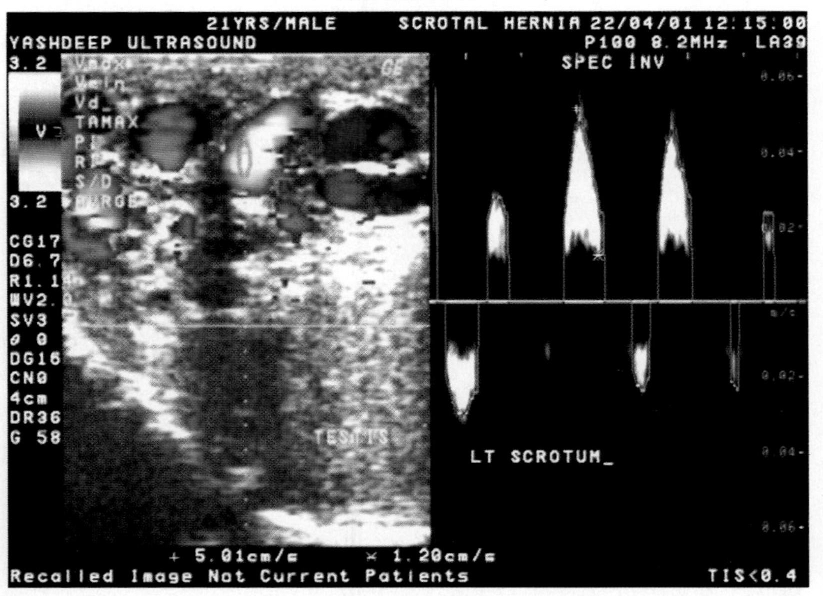

FIGURE 10.312

FIGURES 10.311 and 10.312: Spectral Doppler imaging of the same patient shows arterial flow in the vessels suggestive of arteriovenous malformation.

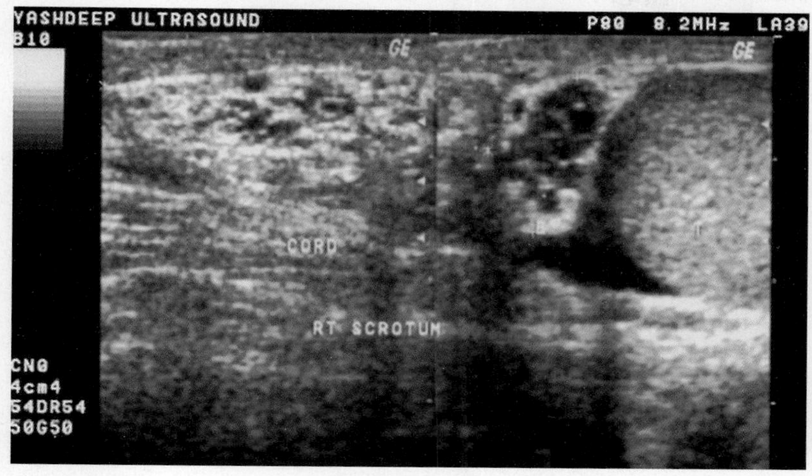

FIGURE 10.313

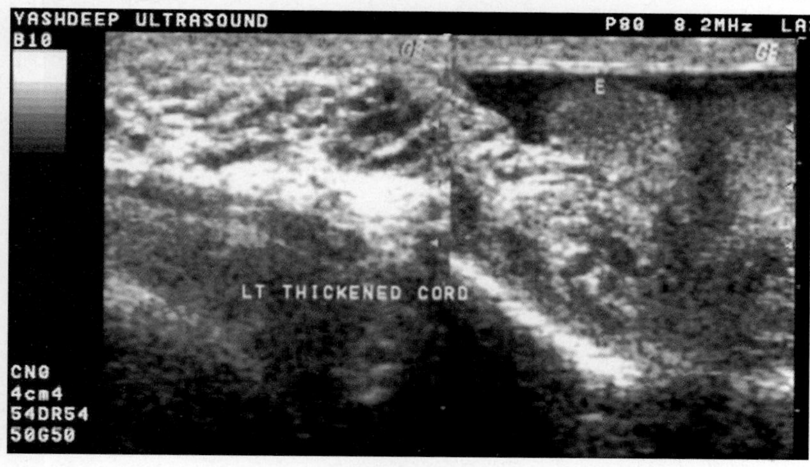

FIGURE 10.314

FIGURES 10.313 and 10.314: Tubercular infiltration of the cord with tuberculosis of epididymis. HRSG shows irregular hypoechoic area in the Lt epididymis. Normal texture is destroyed. The Lt cord is also thickened. Multiple hypoechoic areas are seen in the Lt cord. Biopsy of the epididymis in the cord shows tuberculosis.

Dilated spermatic vessels giving rise to a "bag of worms" appearance is the most common feature. With color Doppler, a varicocele is diagnosed when the vessel has diameter of more than 3 mm. With Valsalva maneuver, the diameter increases significantly, and greater number of dilated vessels are identified. Presence of reflux is seen in varicoceles. Color Doppler studies have been used to grade the venous reflux.

- Grade I — Venous stasis
- Grade II — Intermittent stasis
- Grade III — Continuous stasis
- Ligation of the varicosed vessels often improves the sperm quality, and increased conception rates are reported.

Conclusion

High-resolution sonography has been a boon in the evaluation of scrotal pathologies especially after the advent of color Doppler imaging. It might be in order to say that visualizing the intra scrotal structures and deciphering pathologies of the scrotal contents have greatly facilitated the clinical approach to acute scrotal disorders. As mentioned earlier, the greatest advantage of ultrasonography is to distinguish between intra and extratesticular abnormalities, since most intrascrotal masses should be considered malignant, while extratesticular abnormalities are more commonly benign. An abdominal sonography helps in the evaluation of metastatic disease in the abdomen and also helps in following up of patients after surgery. Though MRI has been of help in assessing the scrotum, ultrasonography will not only remain the primary screening modality but adds good information in making a specific diagnosis in majority of cases.

11

Anterior Abdominal Wall

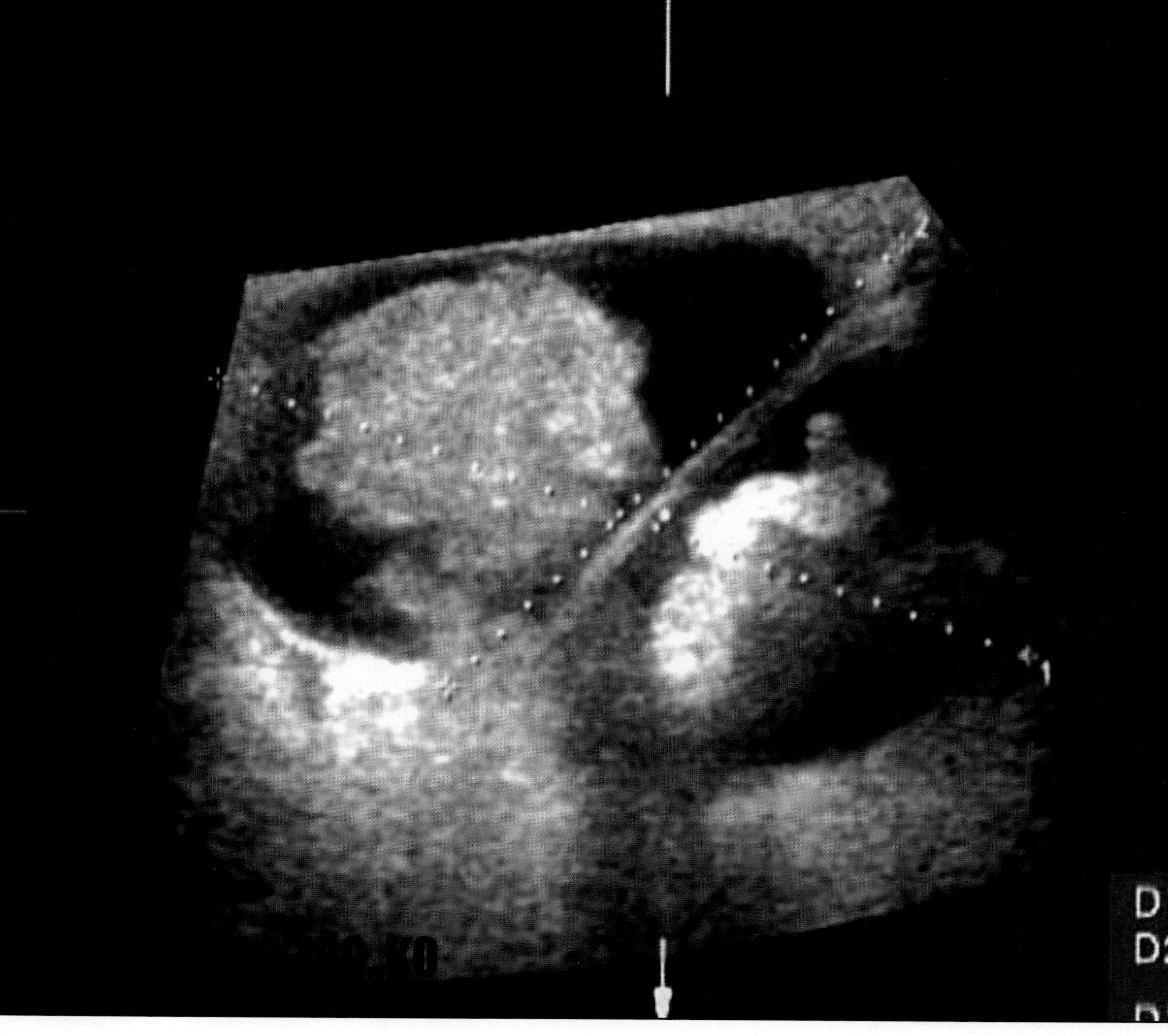

Introduction

Abdominal wall provides the access to ultrasonographic beam to examine internal abdominal organs. A common indication for anterior abdominal wall imaging is to examine a palpable mass in abdominal wall. HRSG can demonstrate about the nature of the mass whether it is solid or cystic. It also tells about the location of the mass whether it is intra-abdominal or in the abdominal wall.

Normal Anatomy

Anterior abdominal wall is a layered structure. The outermost layer is skin, the superficial fascia, subcutaneous fat, muscles, transversalis fascia and extra peritoneal fat. The muscle layer contains paired midline rectus muscles, anteriolateral external oblique, internal oblique and transversus abdominis arise from fifth, sixth and seventh ribs and extend inferiorly and insert in pubic crest. They are wrapped in rectus sheath. The rectus sheath is formed by aponeurosis of internal oblique, external oblique and transversus muscles. The transversalis fascia lies deep to muscle layers. It contains little fat and separates the muscle layers from peritoneum.

Sonographic Appearance of Abdominal Wall

High-resolution sonography can very well demonstrate the normal layer of abdominal wall. A linear transducer of 10 Mhz is choice of transducer to study abdominal wall. The skin stands out clearly as an echogenic line. Subcutaneous fat is seen beneath the skin as hypoechoic layer. Musculofascial layer is more echogenic than subcutaneous fat. High-resolution transducer can show individual muscle bundles. The extraperitoneal fat is seen as an echo poor fat line posterior to the muscle. The rectus abdominis muscles give on lens-shaped appearance on transverse axis. This may give a confusing double image artifact in lower abdomen when imaging is done through rectus abdominis muscle. Therefore, a single gestational sac is seen as two gestational sacs and can be misdiagnosed as twin pregnancy.

Pathological Lesions of Anterior Abdominal wall

Trauma

Blunt injury to abdominal wall may lead to tearing of muscles and pooling of blood in the muscle sheath. It presents as a palpable tender swelling. It is difficult to localize the depth of hematoma on clinical examination. HRSG is good to localize lesion and also status of the lesion. The sono-appearance of hematoma depends upon the age of hematoma. Initially, the hematoma is echogenic with areas of low-level echoes. However, with the passage of time, the blood retracts and liquefaction takes place. Therefore, echo poor mass with echogenic clot retraction is seen on HRSG.

Abscess

Anterior abdominal wall abscess may develop to any reason. However, most common factor is post-surgical abscess formation in abdominal wound. HRSG can demonstrate the different states of abscess. Pus collection is seen in the subcutaneous tissue. The abscess cavity shows thick echo collection. Echogenic debris is also seen on HRSG. Ultrasonographic-guided drainage can be performed to speed up the healing process. At times gas is also seen in the abscess cavity. Communication of abscess cavity to deep layers of abdominal wall can be easily picked up of HRSG. Only removal of stitch can release the pus and helps the wound to heal spontaneously.

Foreign Body

Foreign body entrapped in anterior abdominal wall is very rare and most commonly is seen due to post-surgical complication. Fever, persistent lump and tenderness in local area are most common features complained by the patient. HRSG can very well demonstrate the nature of the mass, its consistency and location. Most common cause is a sponge, which is left in the abdomen after surgery. HRSG can show the layers of sponge. It may be seen as an echogenic mass with granuloma formation. The adherent blood gives a hyper-reflective appearance.

Muscle Tumors

Abdominal wall tumors are uncommon. The most common primary tumors are desmoid tumors of muscles. They take origin from fascia of aponeurosis. They are well-defined hypoechoic masses with homogenous texture. HRSG guided FNAC (fine-needle aspiration cytology) can be done to confirm the diagnosis.

Metastatic Lesions

Metastatic deposits can be found in the abdominal wall. The most common metastatic nodules are seen in metastatic melanoma. However, secondaries from lymphoma, carcinoma of lung, breast, ovary and colon can go the abdominal wall. HRSG shows multiple hypoechoic nodular masses in abdominal wall.

Parasitic Infestations

Worm infestation to muscles and cyst formation can be picked up on HRSG. The most common worm infestation to muscles is cysticercosis. It is caused due to infestation of *Taenia solium*. The worm is about 3 mm in length. It presents as subcutaneous nodule. Typical sonographic findings are seen on HRSG. The scolex is highly reflective nodule, which is surrounded by a cyst. At times two or three cysts are seen with echogenic scolex seen attached to their walls. HRSG is also very good to monitor the treatment response of cysticercosis with medical treatment. The worm dies, the cyst disappears, and only localized thickening of muscle is seen.

Undescended Testis

The testis descends to its normal position in scrotum in the normal development. The descent of testis can arrest at any place from hilum of kidney to inguinal canal. In all undescended testis 80% are palpable and 20% are non palpable. But out of these 20% non-palpable testis, 20% are seen in inguinal canal, and 20% are found in abdomen. HRSG can very well locate the site of the non-palpable testis. Localization of undescended testis is important, as incidence of neoplastic changes is

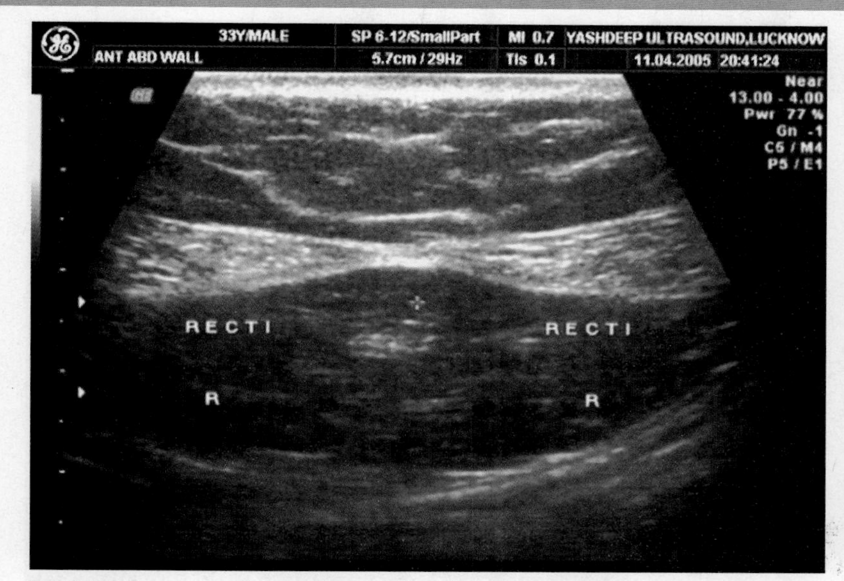

FIGURE 11.1: Normal anatomy. HRSG shows normal anatomy of the anterior abdominal wall. Bright echogenic line is skin. The hypoechoic layer posterior to the skin is subcutaneous fat. The rectus muscle bundles are seen as a lens shaped structure. Extra peritoneal fat is seen as a hypoechoic layer posterior to the muscle bundle.

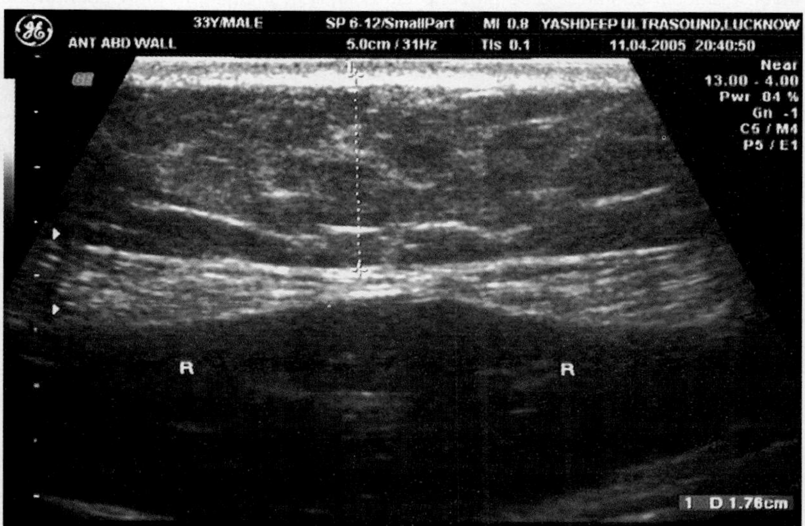

FIGURE 11.2

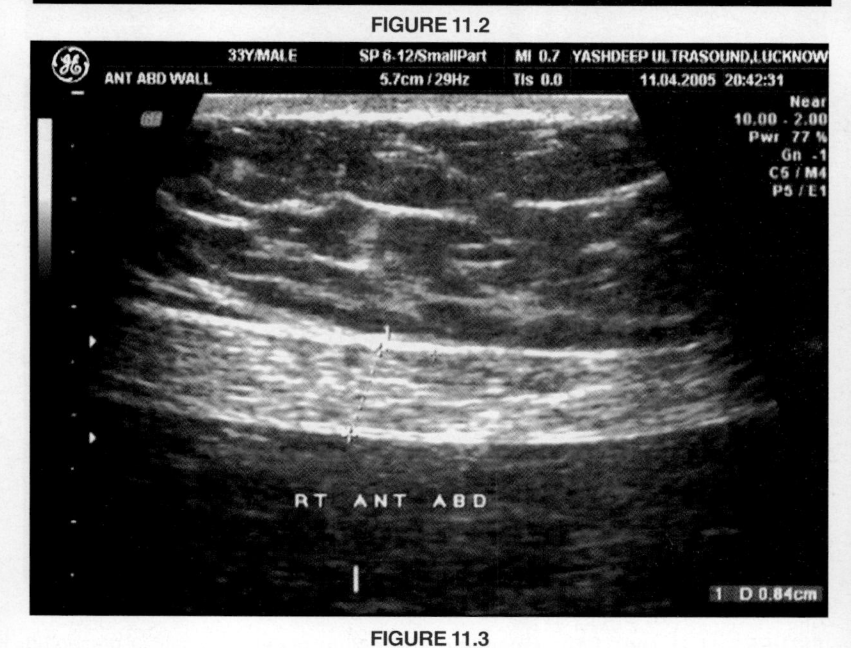

FIGURE 11.3

FIGURES 11.2 and 11.3: Hypertrophy of the subcutaneous fat. HRSG shows extra deposition of fat in the subcutaneous plane presented as a lump.

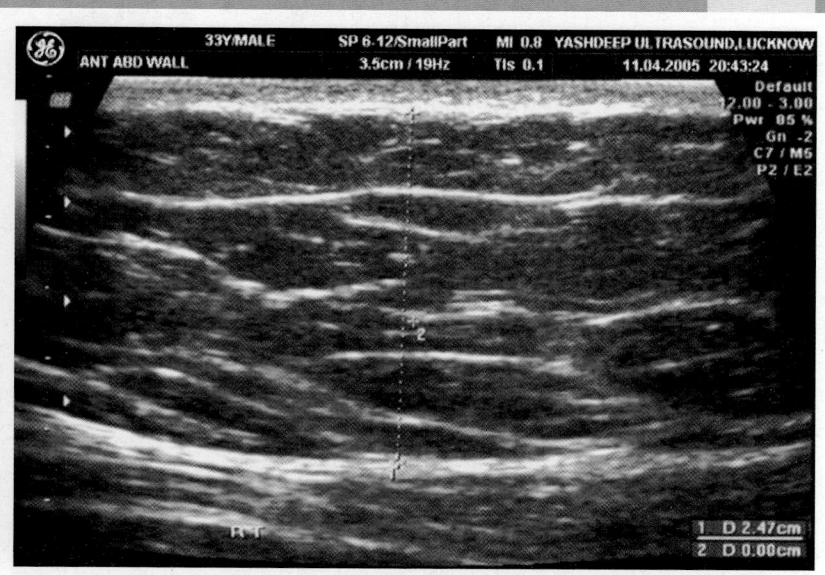

FIGURE 11.4: Thickened subcutaneous fat. A patient presented with lump with epigastric lesion. HRSG shows increased amount of pad of fat in the epigastric region. No evidence of any mass lesion is seen.

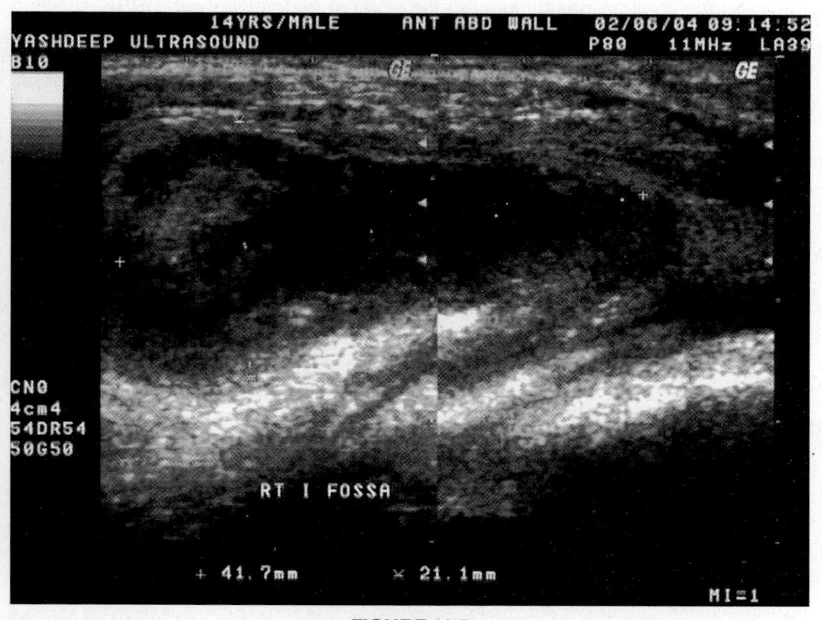

FIGURE 11.5

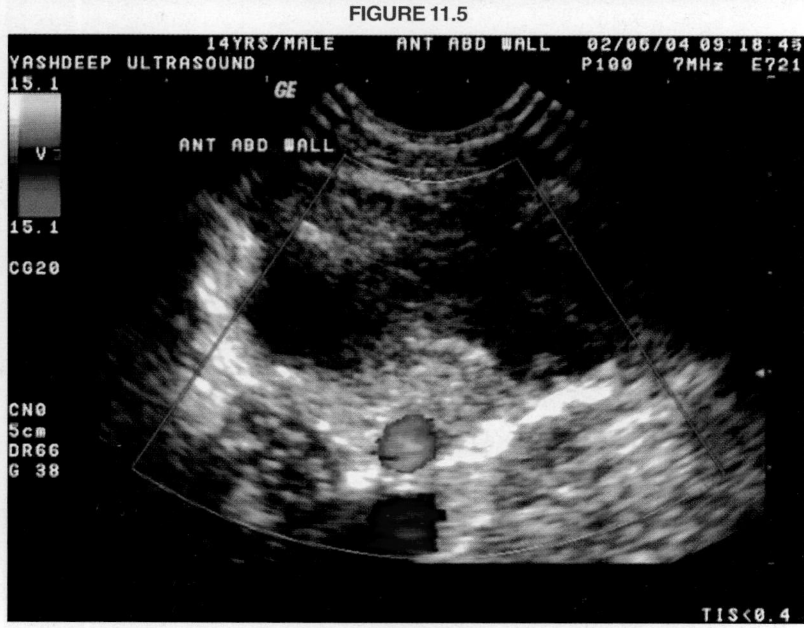

FIGURE 11.6

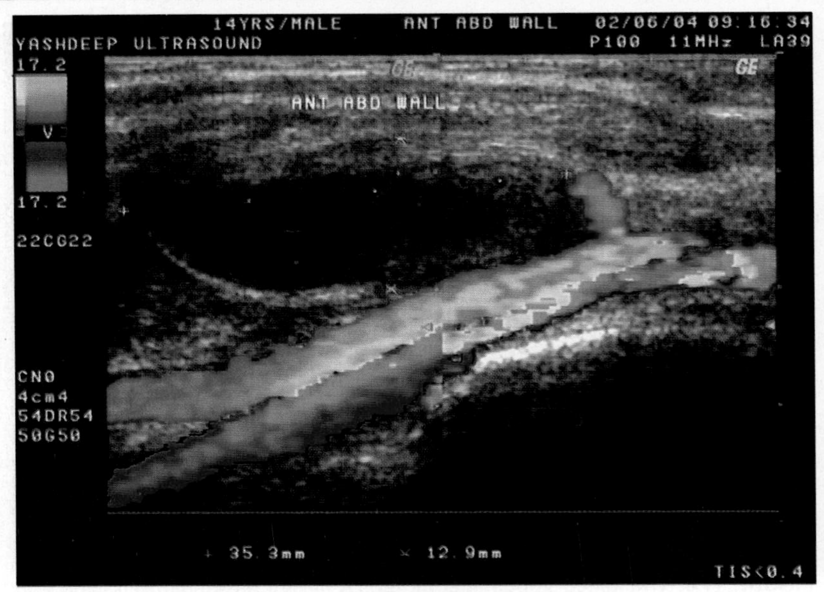

FIGURE 11.7

FIGURES 11.5 to 11.7: Anterior abdominal wall abscess. A young patient presented with high fever with a painful lump in the anterior abdominal wall. HRSG shows a thick low level echo complex mass in the anterior abdominal wall. The mass is seen in the Rt iliac fossa. It is confined to the abdominal wall. Multiple internal echoes are seen in it. On color flow imaging no flow seen in it. Findings are suggestive of big abscess.

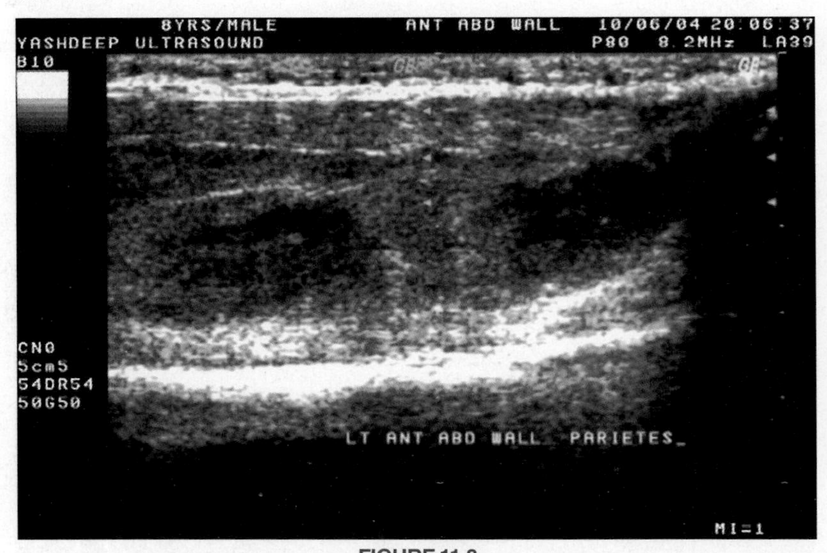

FIGURE 11.8

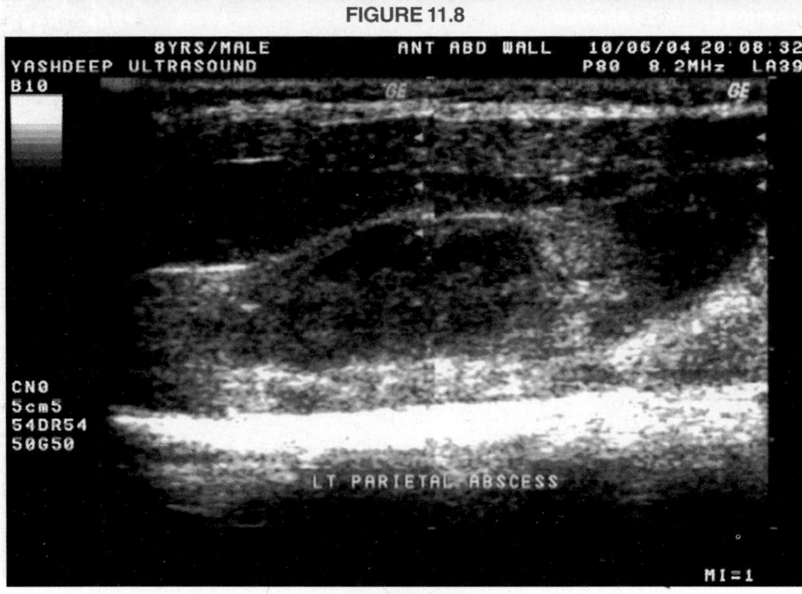

FIGURE 11.9

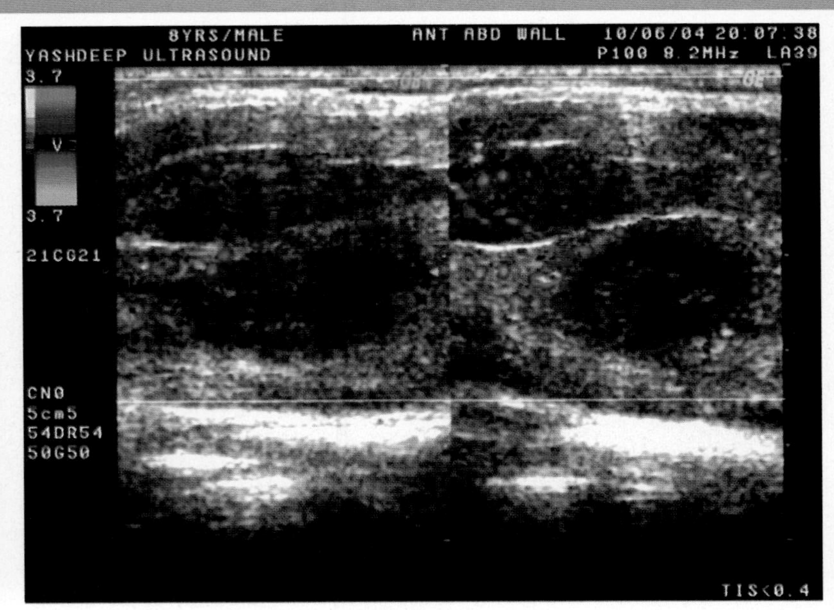

FIGURE 11.10

FIGURES 11.8 to 11.10: Parietal abscess. HRSG shows a thick walled echo complex mass in the Lt anterior abdominal wall. The abscess is seen confined to the Rt parietal layer. Perifocal edema is also present. On color flow imaging no flow seen in the abscess.

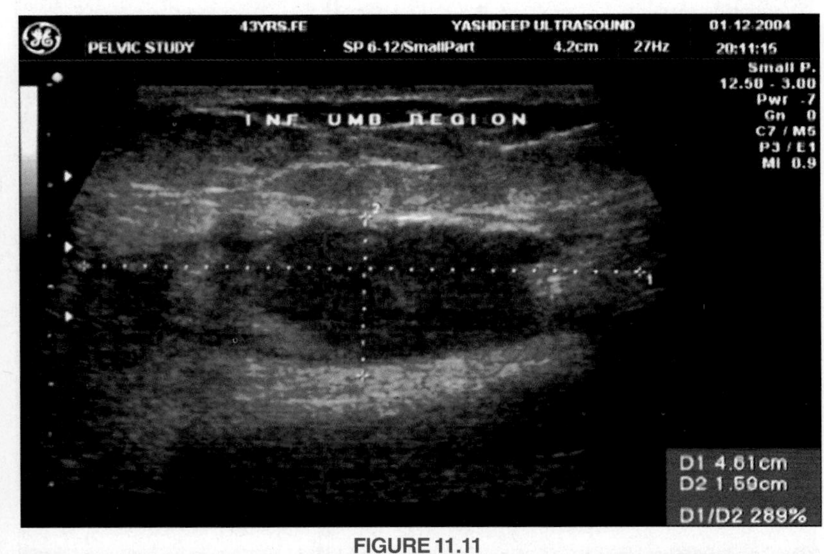

FIGURE 11.11

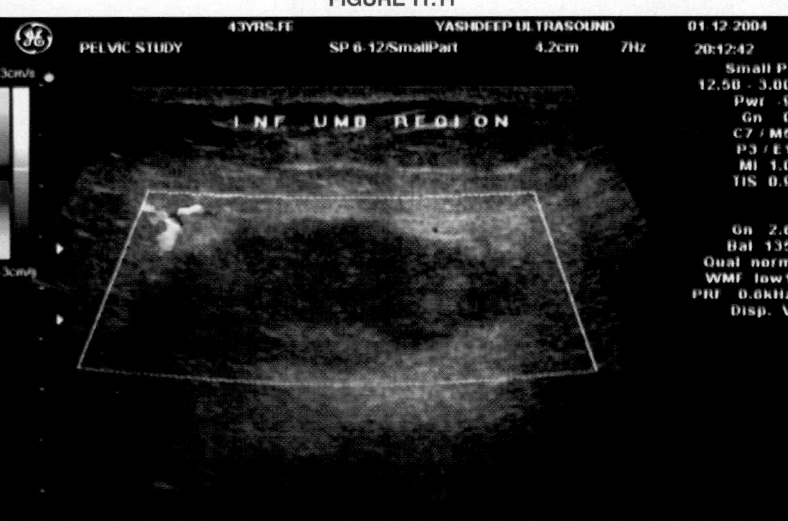

FIGURE 11.12

FIGURES 11.11 and 11.12: Impending abscess in parietes. A patient presented with painful swelling in the infraumbilical region. HRSG shows AN impending abscess in the infraumbilical region. Margins are shaggy. Perifocal edema is present. But no debris is seen.

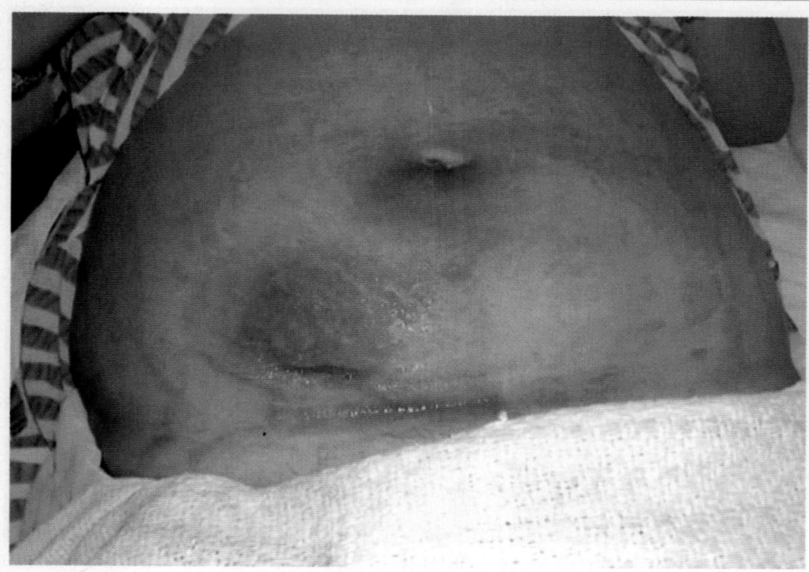

FIGURE 11.13: Anterior abdominal wall abscess after laparoscopic surgery. A patient presented with marked swelling on the anterior abdominal wall in Rt iliac fossa region. It was red and hot with fullness of the anterior abdominal wall.

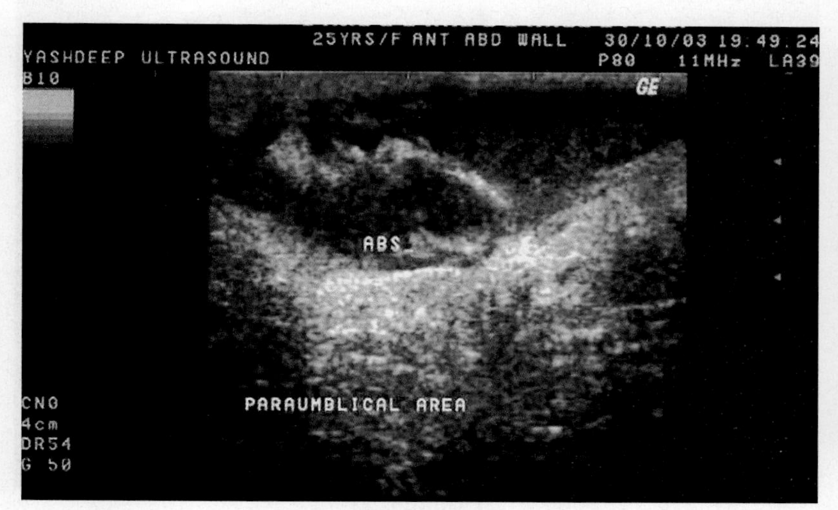

FIGURE 11.14

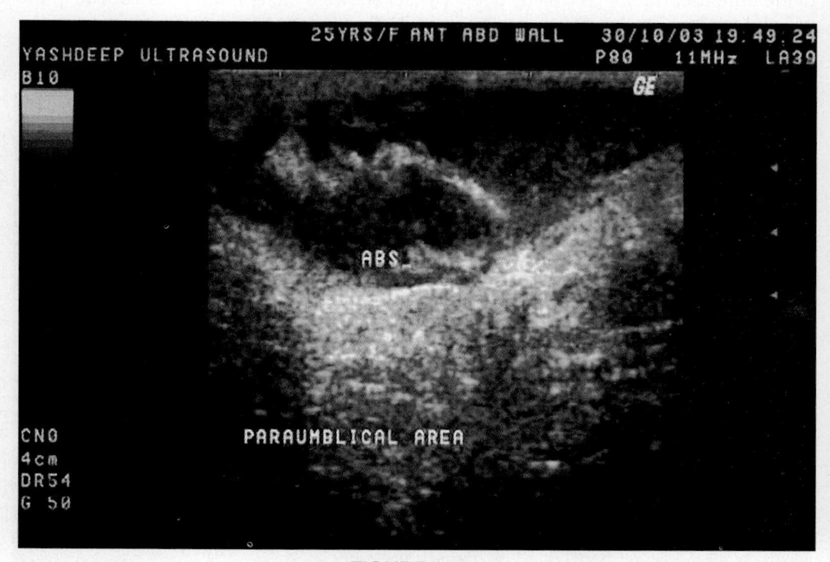

FIGURE 11.15

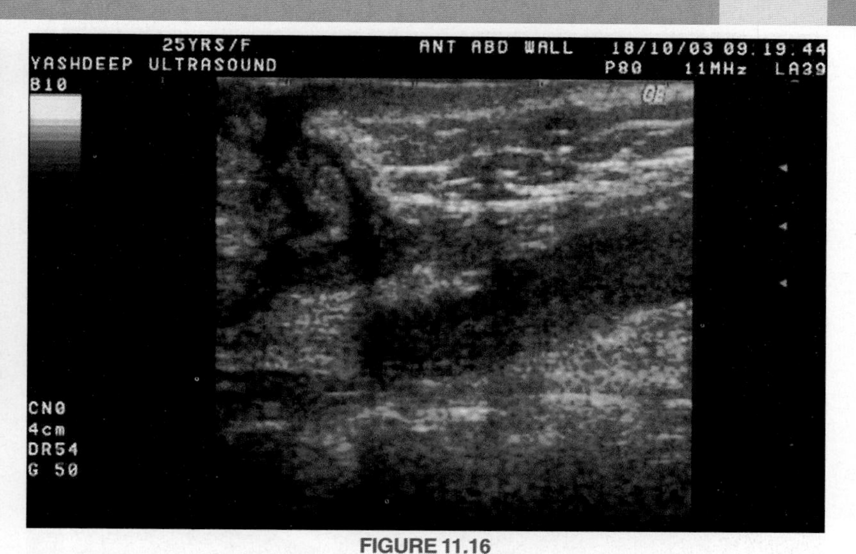

FIGURE 11.16

FIGURES 11.14 to 11.16: HRSG shows thick low-level echo collection in the parietal layers. The collection is seen crossing the mid line and going to the Lt side. Perifocal edema is also presented suggestive of diffuse abscess.

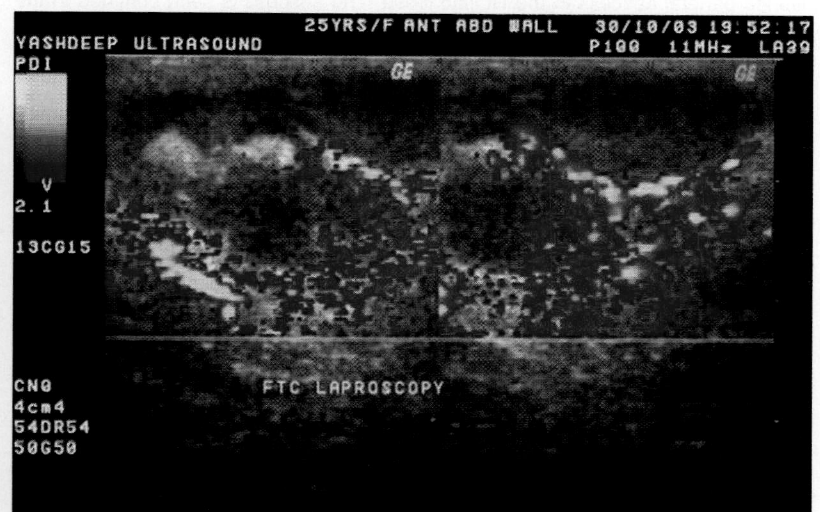

FIGURE 11.17: On color flow imaging peripheral flow seen in the abscess.

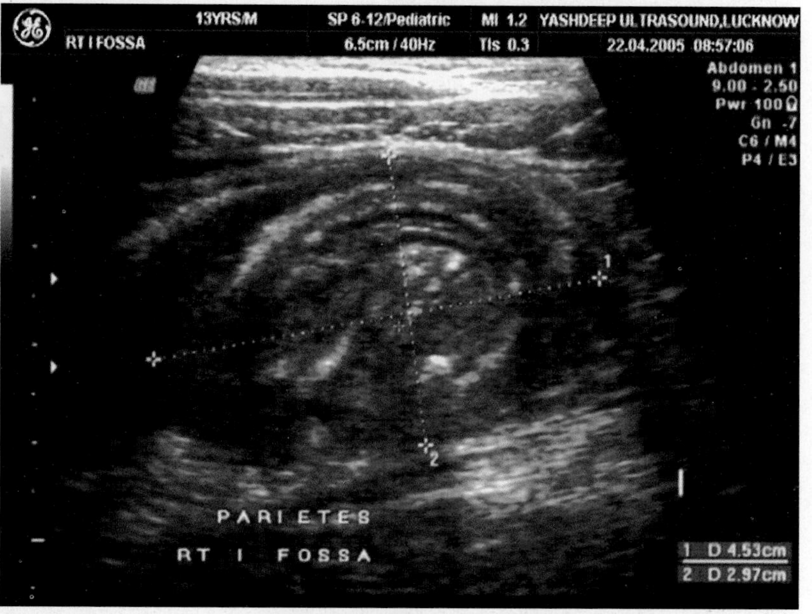

FIGURE 11.18

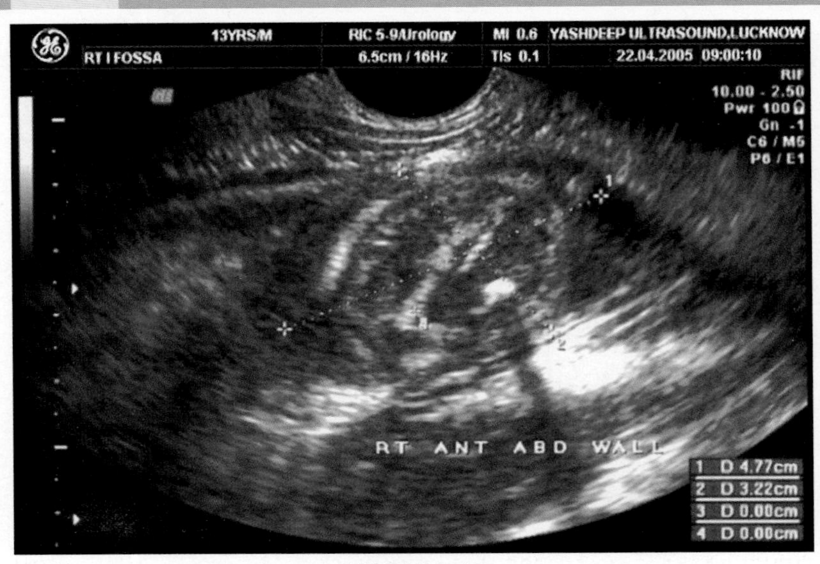

FIGURE 11.19

FIGURES 11.18 and 11.19: Parietal abscess in Rt iliac fossa. HRSG shows well defined parietal abscess in the Rt iliac fossa. The abscess is seen confined to the parietes. Layering sign is positive. Perifocal edema is present. It was mistaken as appendicitis on clinical examination.

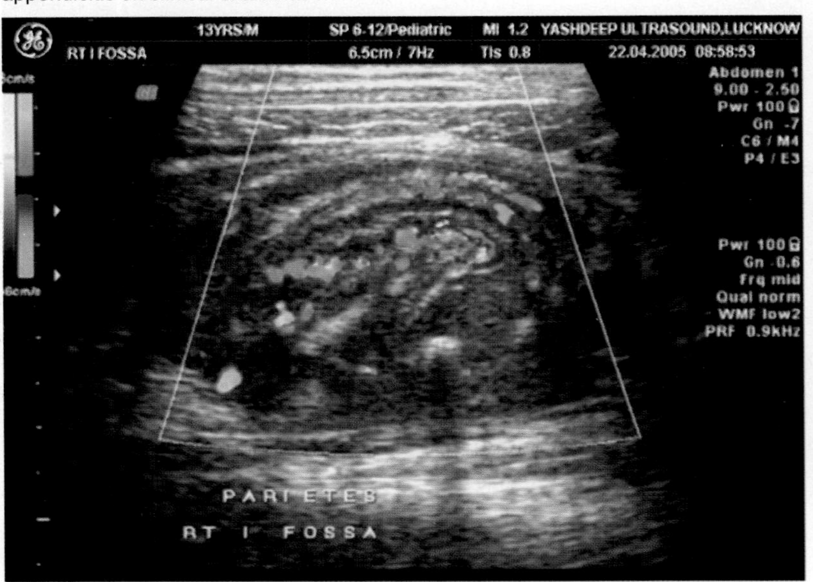

FIGURE 11.20

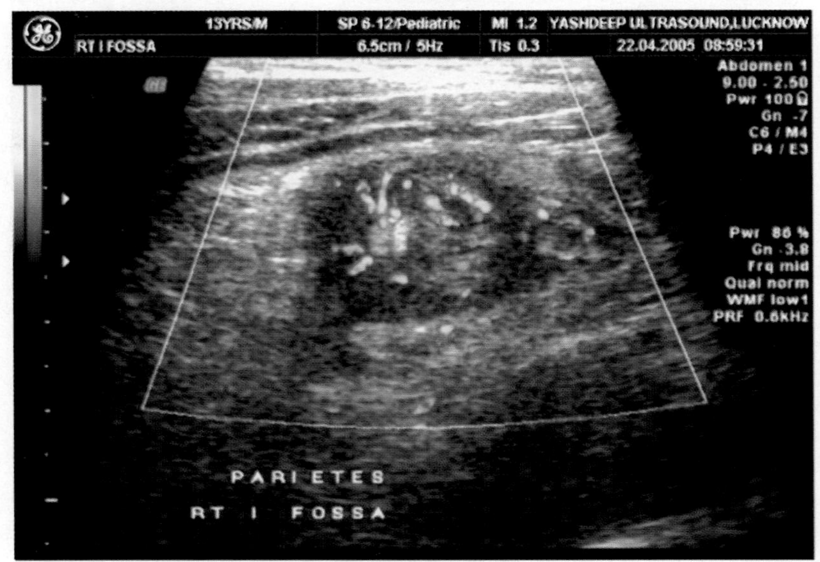

FIGURE 11.21

FIGURES 11.20 and 11.21: On color flow imaging wall hyperemia is seen in the abscess. Multiple vessels are seen feeding the abscess.

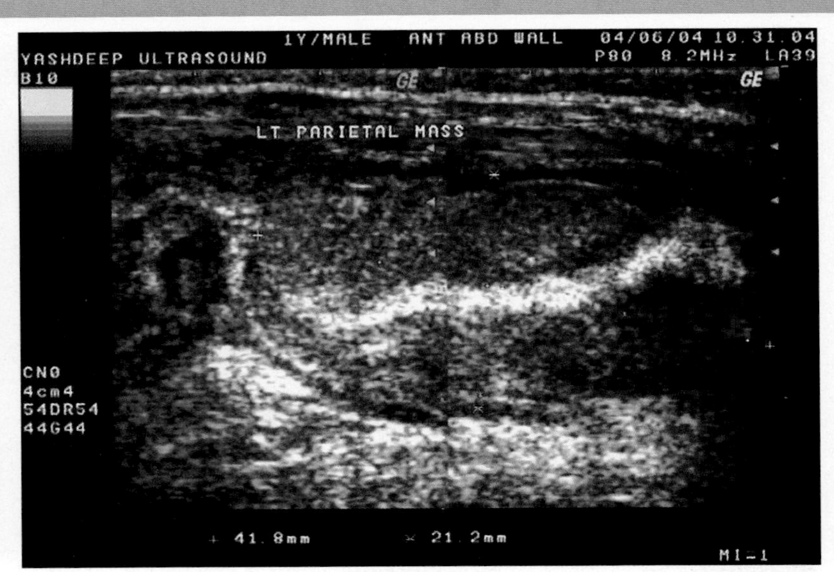

FIGURE 11.22

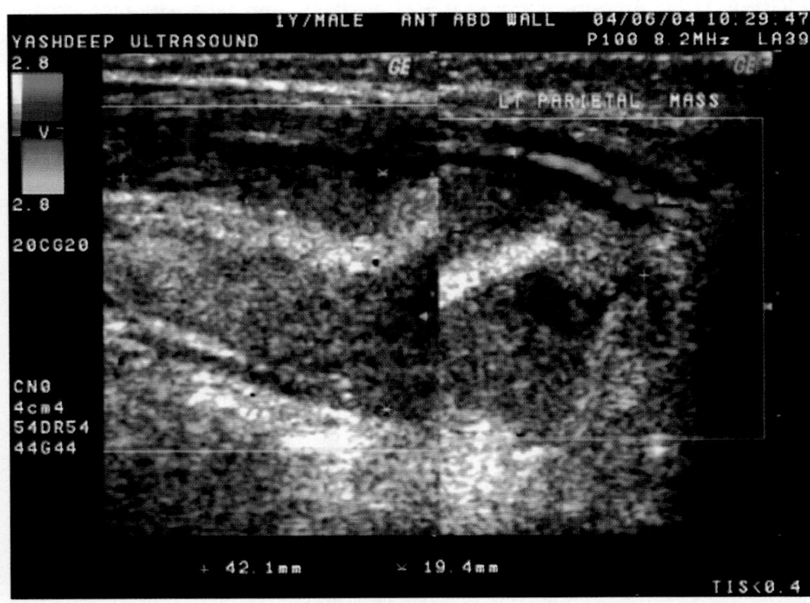

FIGURE 11.23

FIGURES 11.22 and 11.23: Parietal hematoma. A young child sustained blunt injury to the anterior abdominal wall by a fall. HRSG shows a big echogenic collection in the parietal layers. Thick echoes are seen in it. No evidence of any necrosis or cystic degeneration is seen. On color flow imaging no flow is seen in it. Blood came out on aspiration suggestive of fresh hematoma.

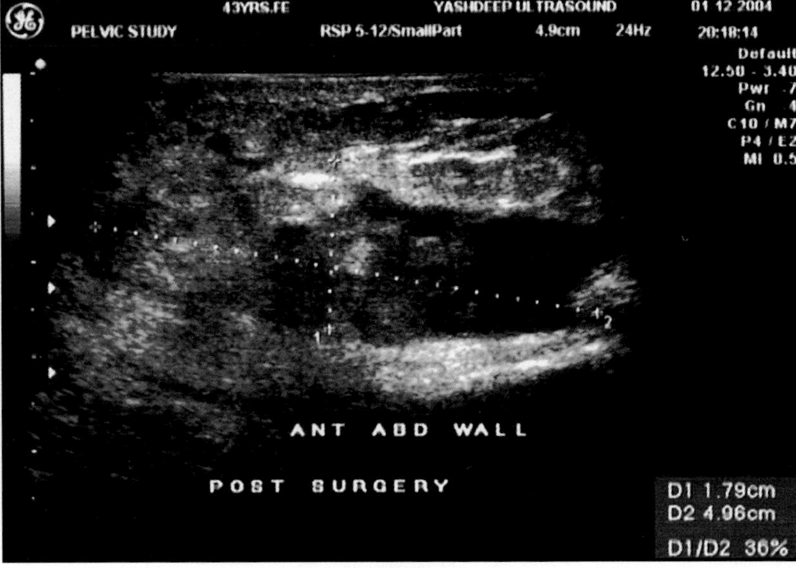

FIGURE 11.24

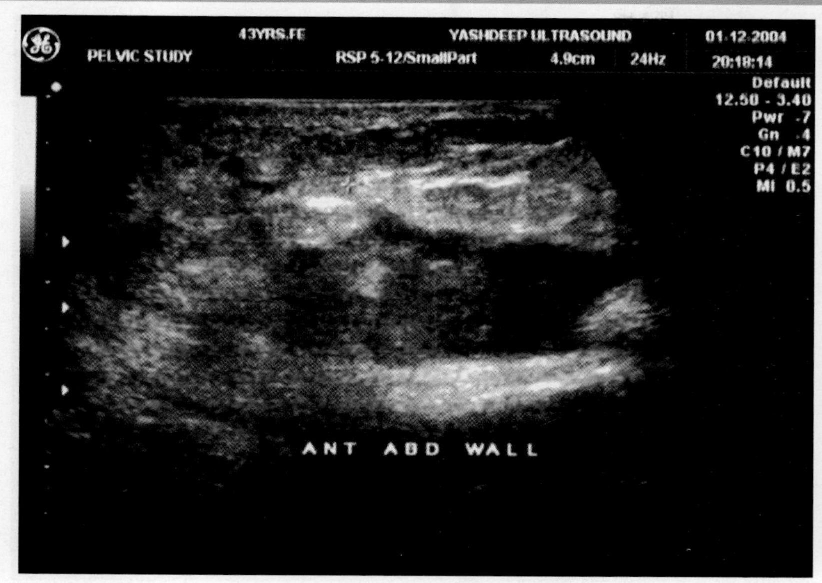

FIGURE 11.25

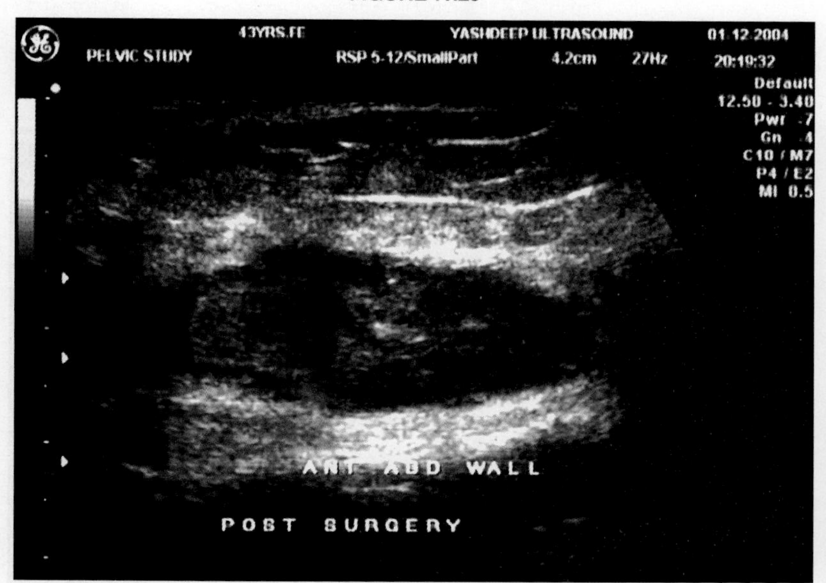

FIGURE 11.26

FIGURES 11.24 to 11.26: Post surgical hematoma. HRSG shows thick collection in the anterior abdominal wall after the surgery. The collection is seen confined in the anterior abdominal wall. Echogenic shadows are seen in them. They are irregular in outline. Aspiration suggests hemorrhagic collection suggestive of hematoma.

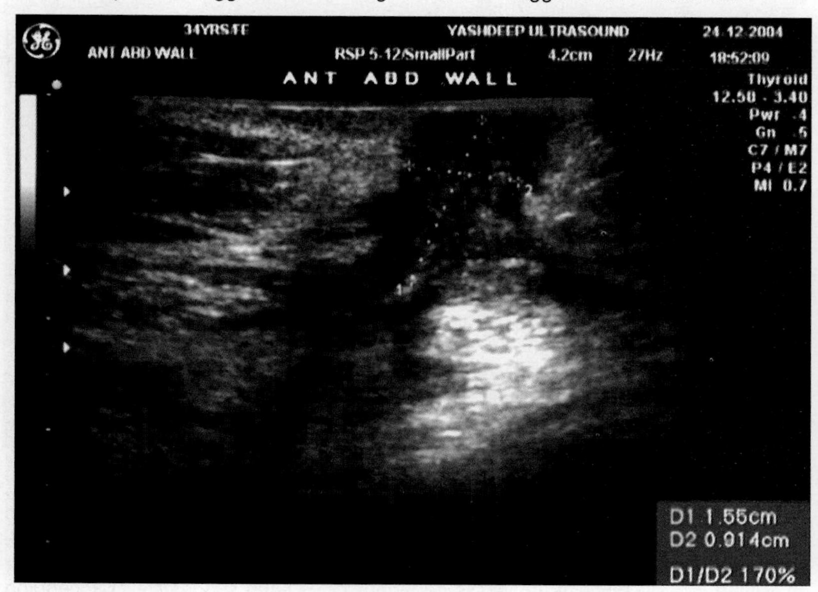

FIGURE 11.27

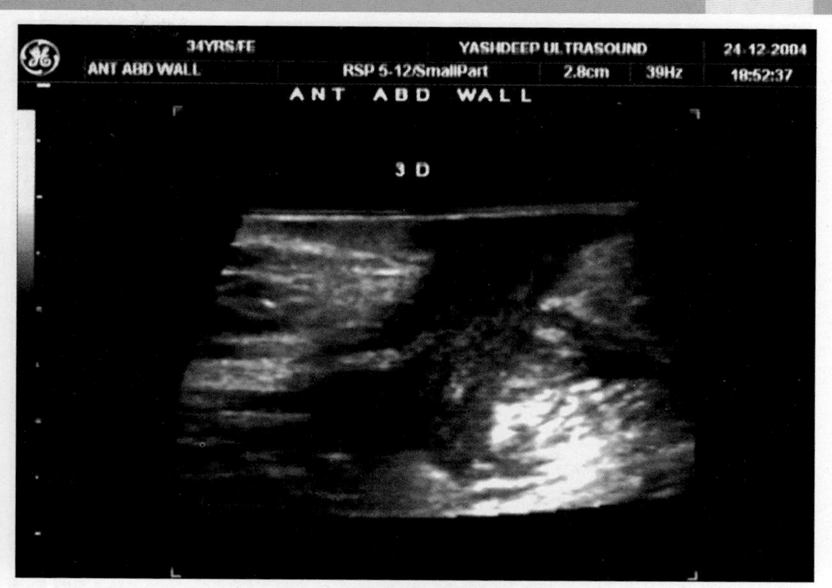

FIGURE 11.28

FIGURES 11.27 and 11.28: Tubercular sinus anterior abdominal wall. HRSG shows an oblique running sinus tract in the anterior abdominal wall, which was seen communicating with small abscess. 3D imaging of the sinus shows the details of the sinus.

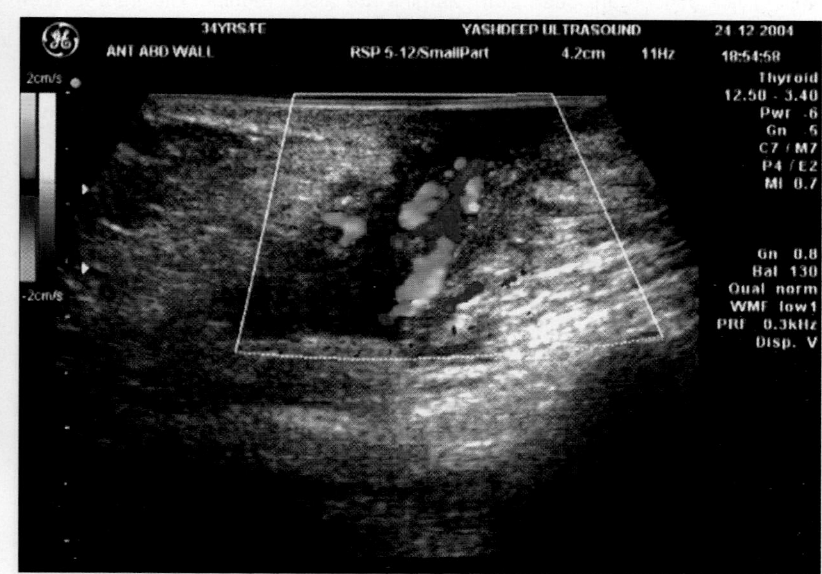

FIGURE 11.29: On color flow imaging two vessels are seen running through the sinus.

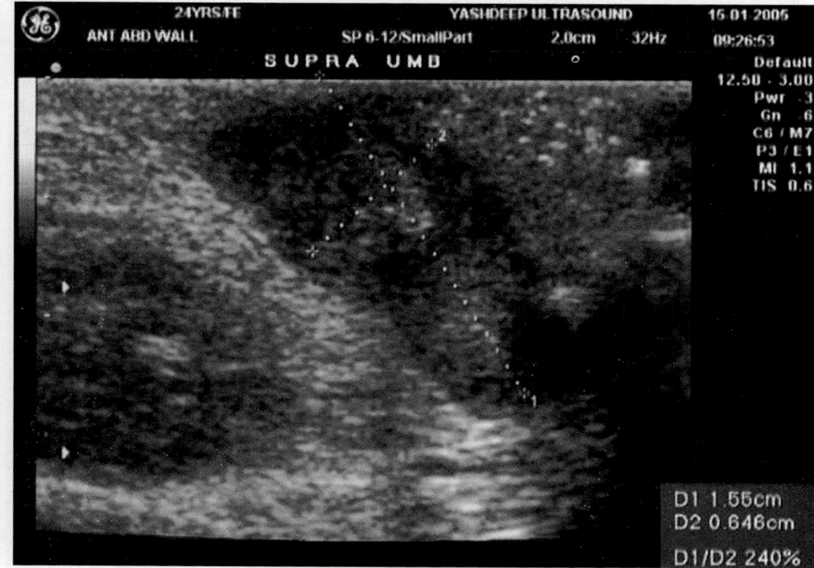

FIGURE 11.30

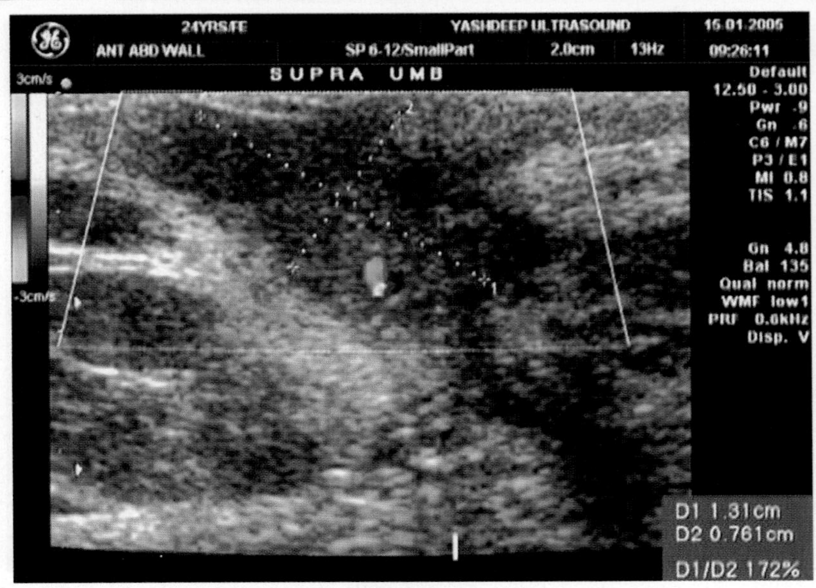

FIGURE 11.31

FIGURES 11.30 and 11.31: Tubercular sinus in the anterior abdominal wall. A patient presented with discharging sinus in the anterior abdominal wall in supra umbilical region. HRSG shows an obliquely running sinus tract in the anterior abdominal wall. Thick echoes are seen in it. Margins are also irregular. The tract is seen communicating with a small abscess cavity. On color flow imaging low flow is seen in it.

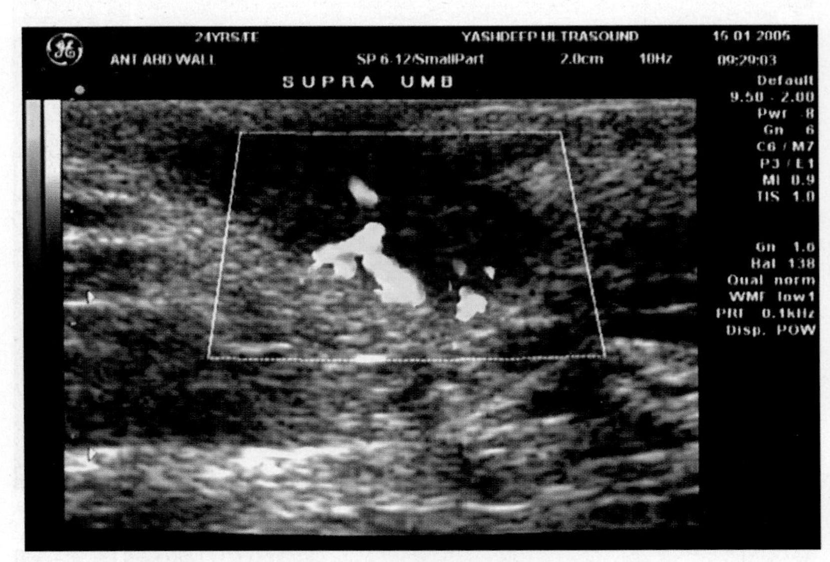

FIGURE 11.32: Biopsy of the mass confirmed tubercular abscess.

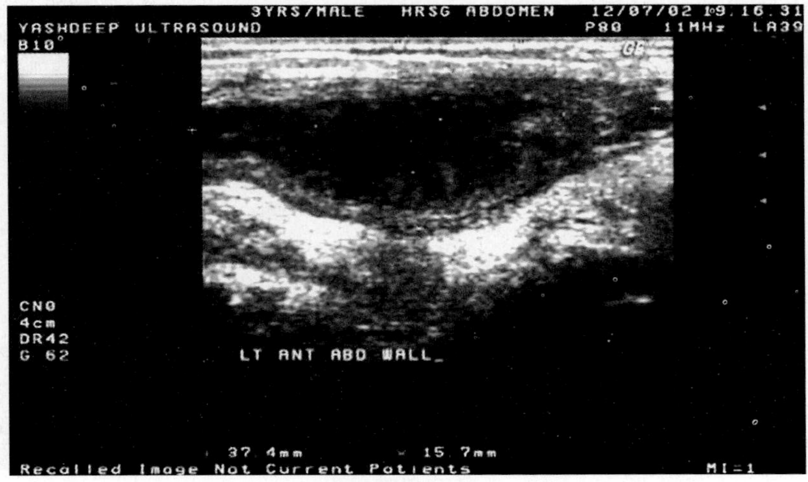

FIGURE 11.33A

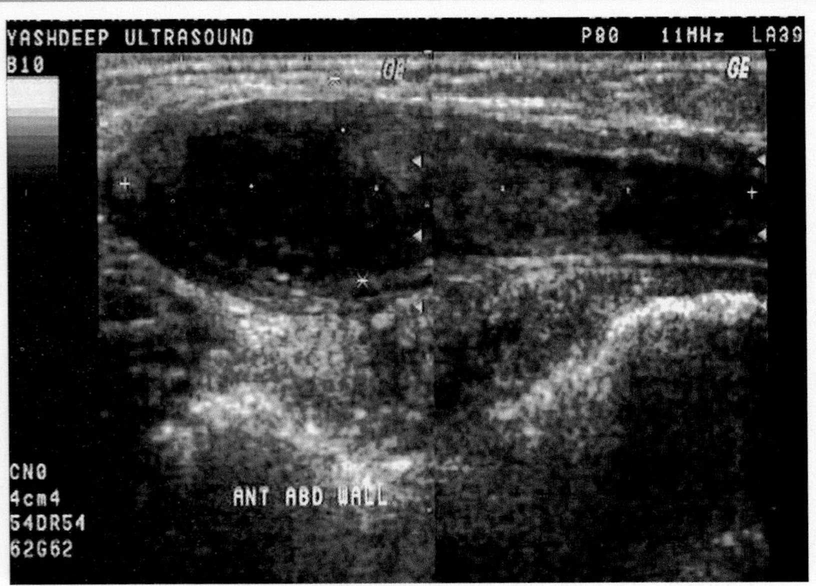

FIGURE 11.33B

FIGURES 11.33A and B: Rectus sheath hematoma. A young child sustained blunt injury abdomen by a fall. HRSG shows thick low-level collection in the rectus sheath. Echogenic shadows are seen floating into it suggestive of rectus sheath hematoma.

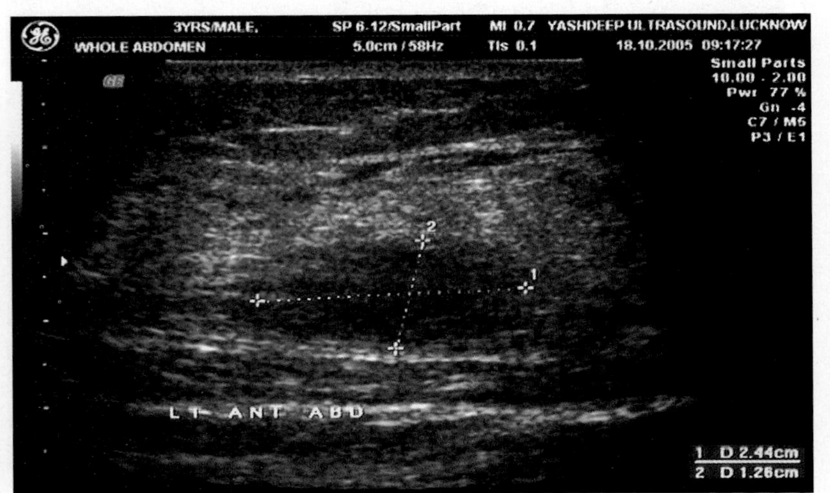

FIGURE 11.34

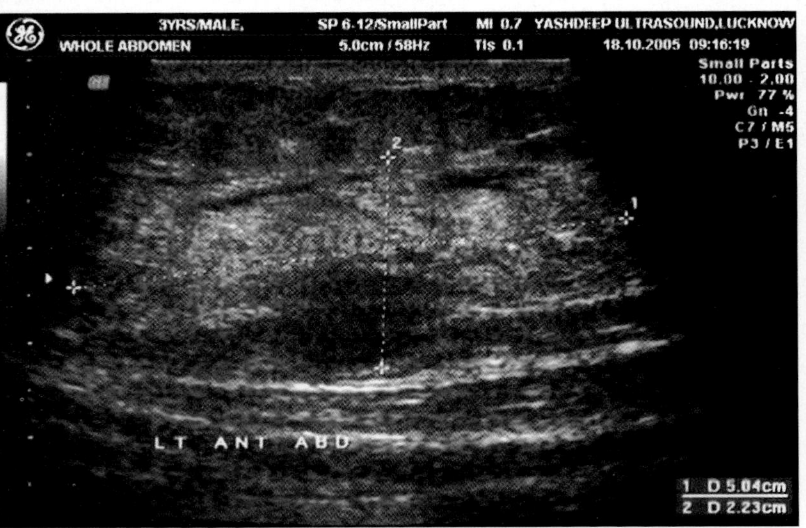

FIGURE 11.35

FIGURES 11.34 and 11.35: Impending abscess of the anterior abdominal wall. A patient presented pain in the anterior abdominal wall. HRSG shows edematous anterior abdominal wall in the Lt para lumbar region. An irregular hypoechoic low-level echo complex mass is seen in the anterior abdominal wall. Margins are irregular in outline. No evidence of any cystic degeneration is seen suggestive of impending abscess.

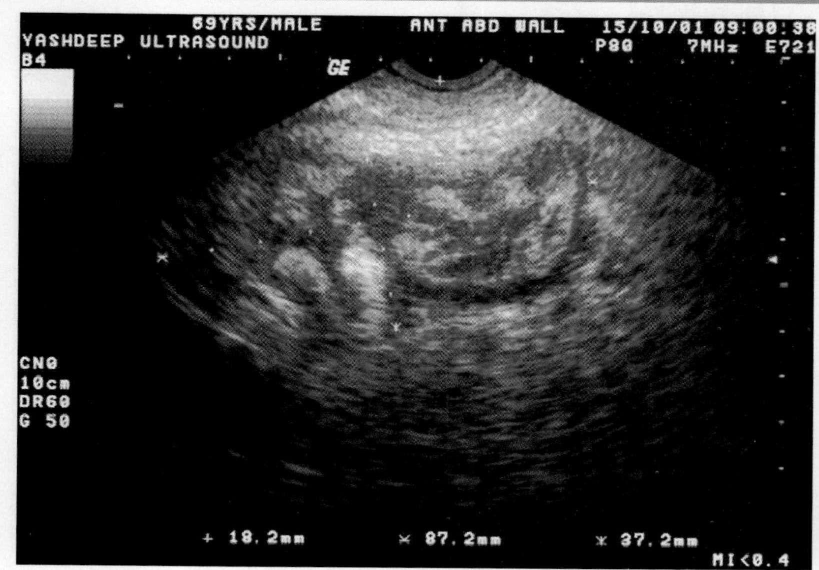

FIGURE 11.36

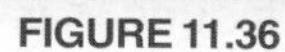

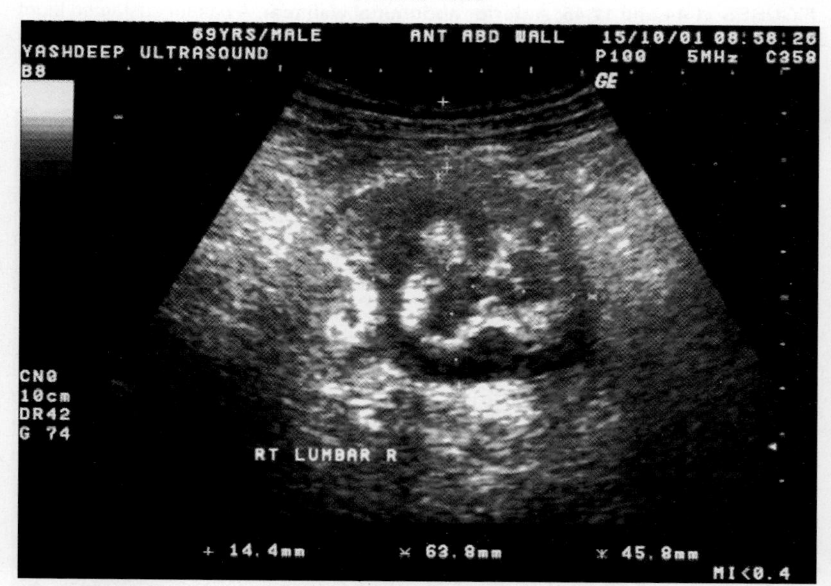

FIGURE 11.37

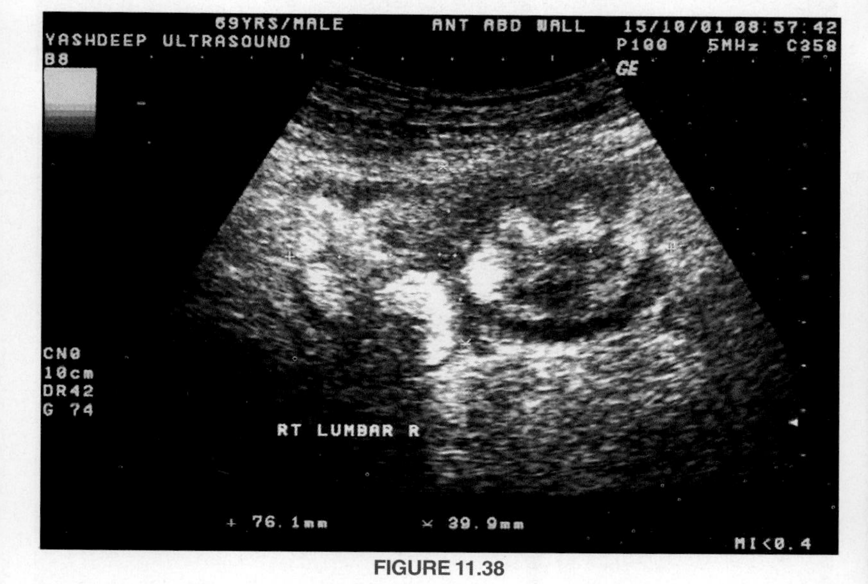

FIGURE 11.38

FIGURES 11.36 to 11.38: Organized hematoma in the Rt anterior abdominal wall. A patient presented a painful swelling in the Rt para lumbar region. HRSG shows a big irregular heterogeneous mass, which was confined into the anterior abdominal wall muscles. Margins are shaggy. Echogenic debris is seen in it. But no evidence of any cystic degeneration is seen. Biopsy of the mass confirmed organized hematoma.

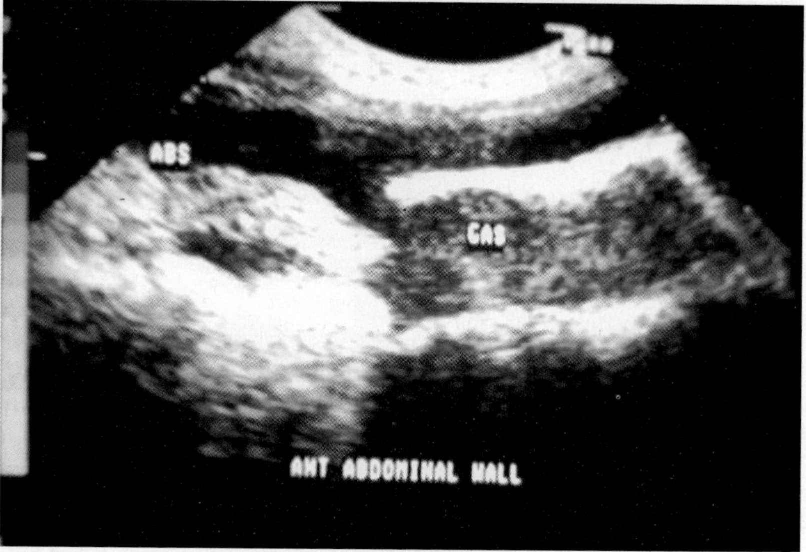

FIGURE 11.39: Gas forming abscess in the anterior abdominal wall. HRSG shows a gas forming abscess in the anterior abdominal wall. The air is seen as an echogenic shadow in the upper part of abscess cavity. Low-level echo collection is seen in the dependent part of the abscess.

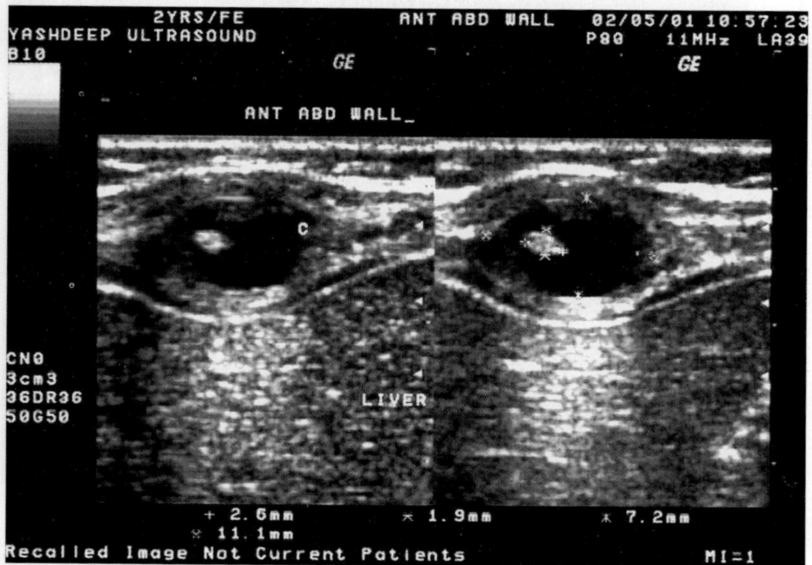

FIGURE 11.40

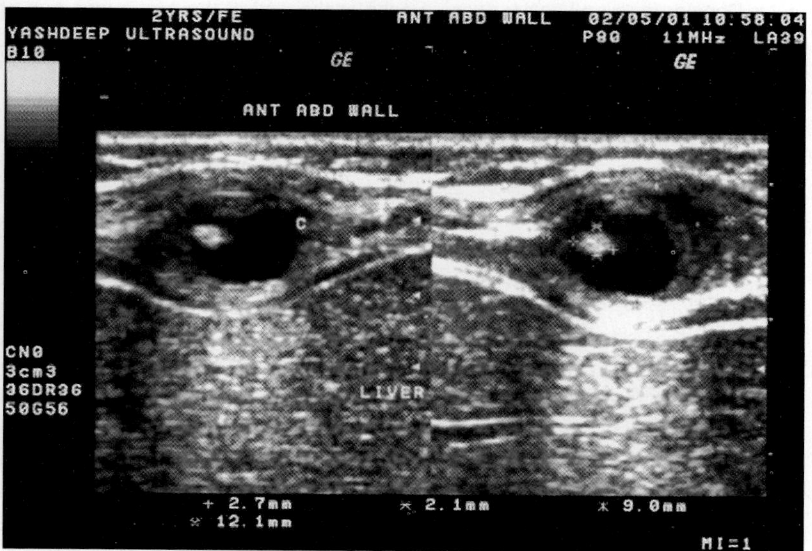

FIGURE 11.41

FIGURES 11.40 and 11.41: Cysticercus cyst in the anterior abdominal wall. HRSG shows a well-defined thin walled cyst in the muscle of the anterior abdominal wall in a young two years old child. An echogenic nidus is seen fixed with the inner wall of the cyst, representing the scolex.

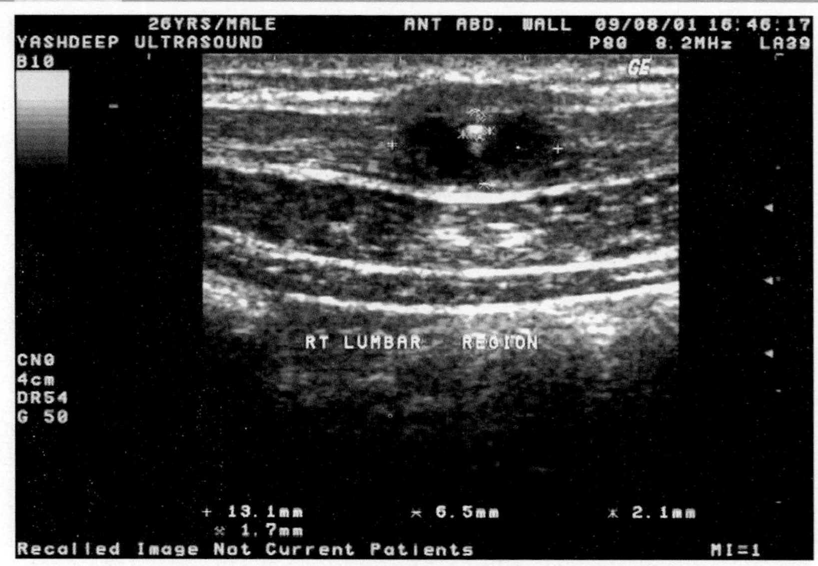

FIGURE 11.42

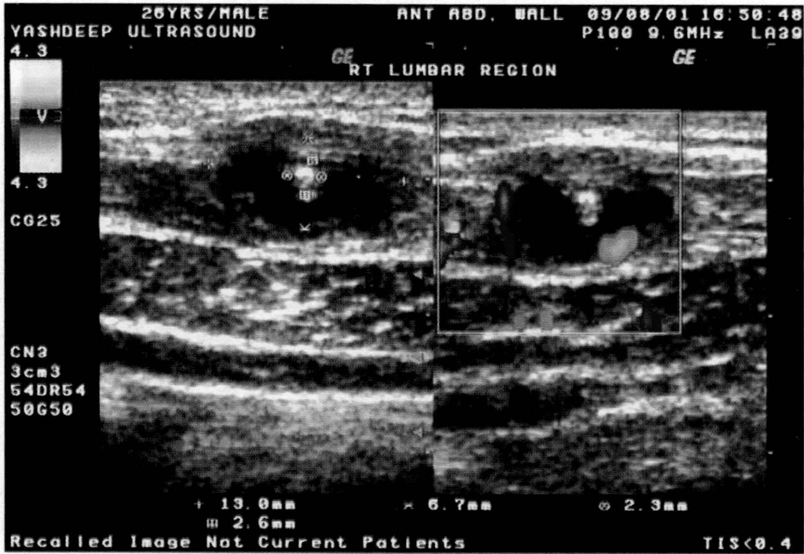

FIGURE 11.43

FIGURES 11.42 and 11.43: Cysticercus cyst in the Rt lumbar region. A well-defined cystic mass is seen in the muscle of the Rt lumbar region in the anterior abdominal wall. An echogenic nidus is seen fixed with the inner wall of the cyst. On color Doppler imaging vascular flow is seen in the wall of the cyst suggestive of typical cysticercus cyst.

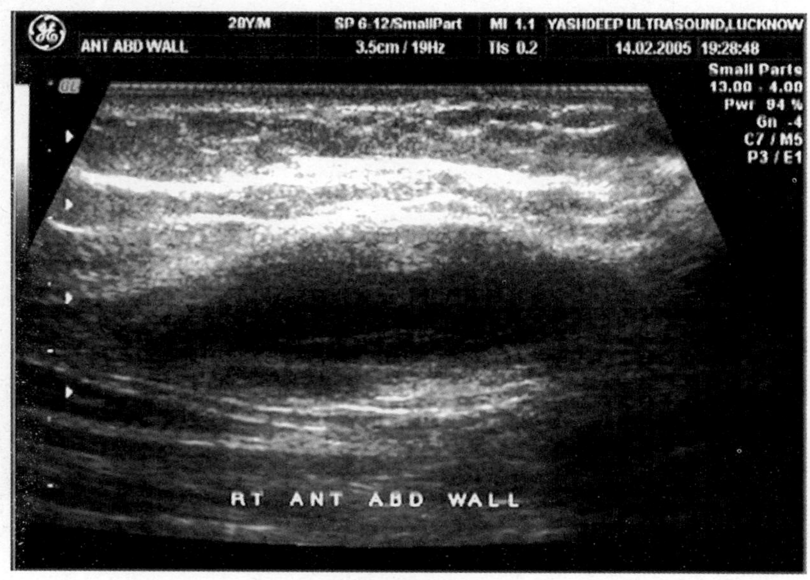

FIGURE 11.44

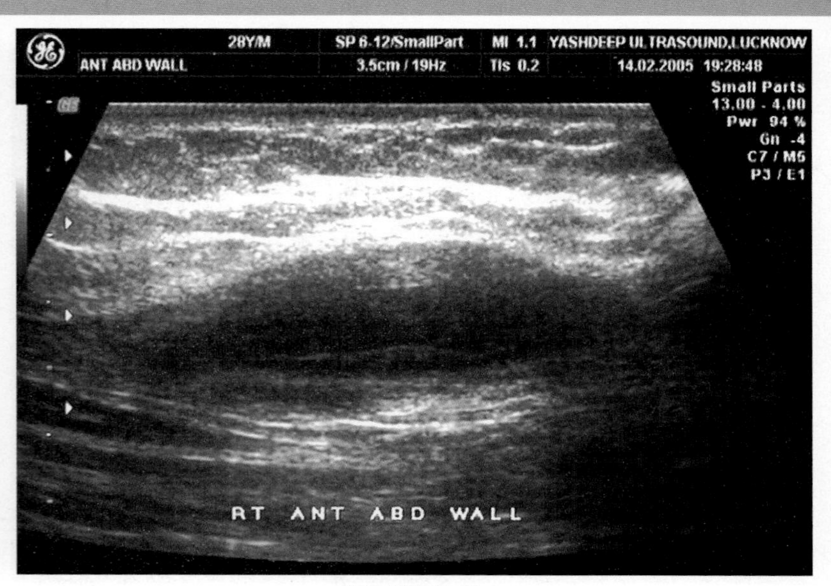

FIGURE 11.45

FIGURES 11.44 and 11.45: Anterior abdominal wall tear. A patient sustained blunt trauma to the anterior abdominal wall in a road traffic accident. HRSG shows a muscle tear in the anterior abdominal wall. The margins are sharp. Fusiform elongated tear is seen in the muscle wall.

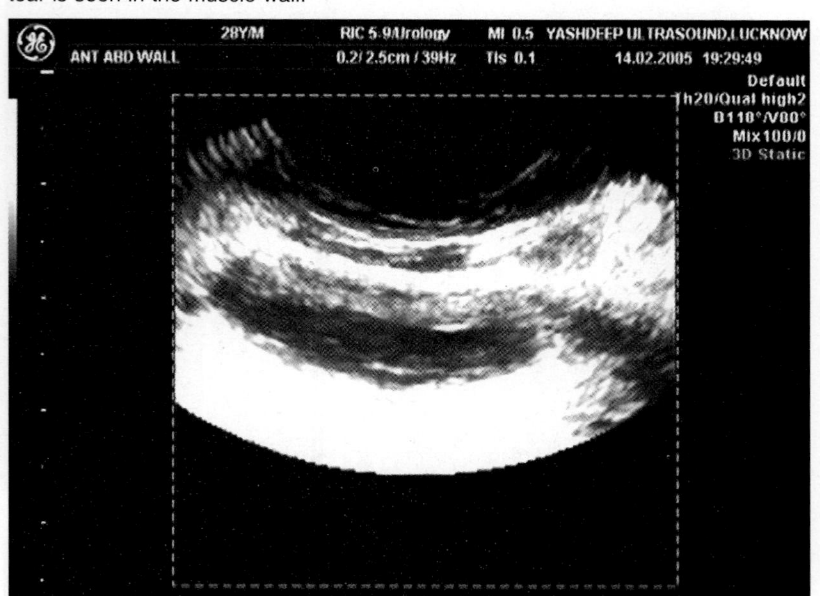

FIGURE 11.46: 3D imaging shows a tear confined to the muscle belly.

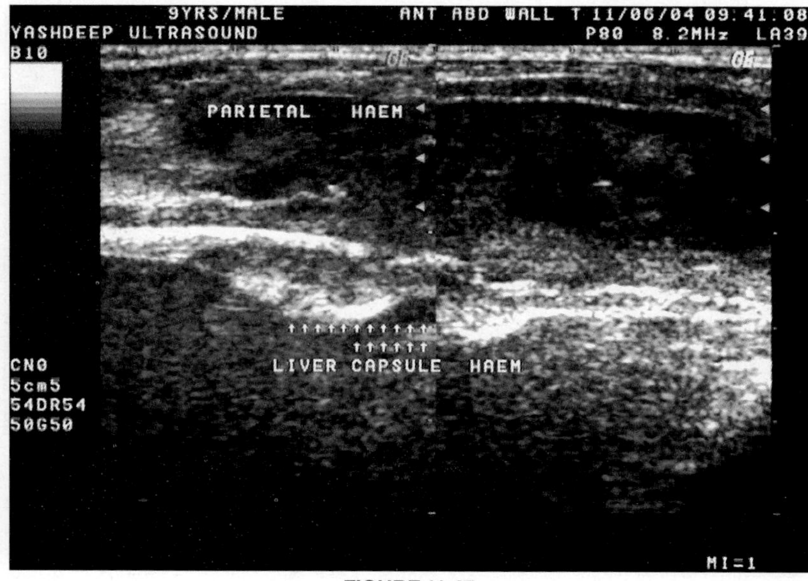

FIGURE 11.47

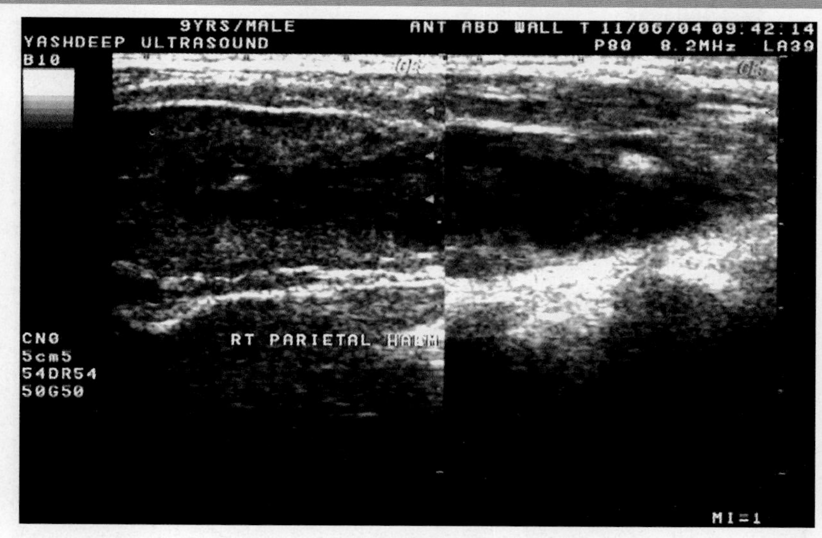

FIGURE 11.48

FIGURES 11.47 and 11.48: Parietal hematoma. A young patient sustained blunt trauma to the abdomen and presented with soft tissue painful swelling over the anterior abdominal wall. HRSG shows a big low level echo collection confined to the parietes suggestive of parietal hematoma. Echogenic shadows are seen floating into it suggestive of hemorrhagic collection.

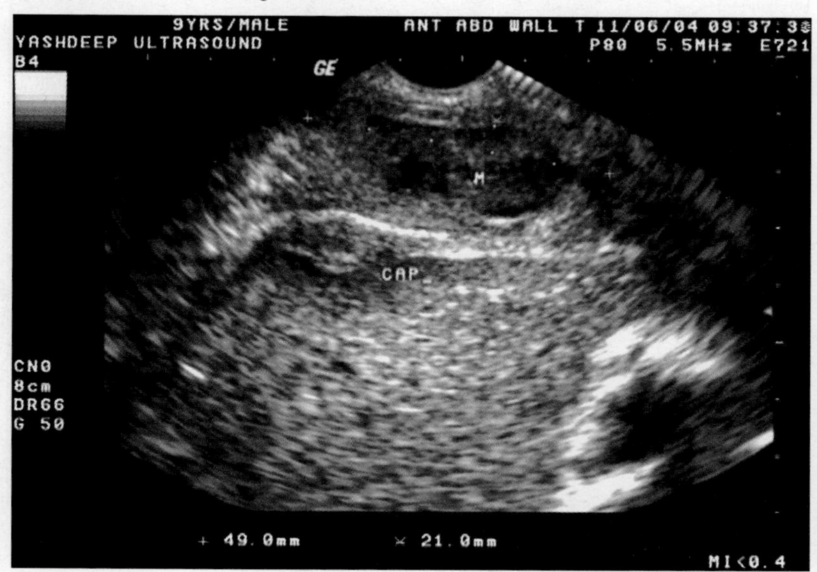

FIGURE 11.49: The hematoma is also seen in the superior capsule of the liver. The capsule is distended. However, the liver is seen normal. HRSG is very good in evaluating anterior abdominal wall trauma and helps in differentiating between intra abdominal hematoma from extra abdominal hematoma.

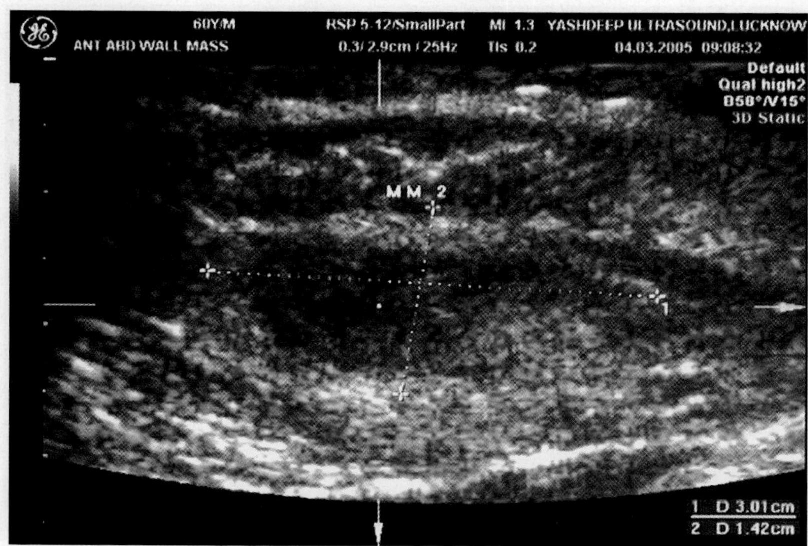

FIGURE 11.50

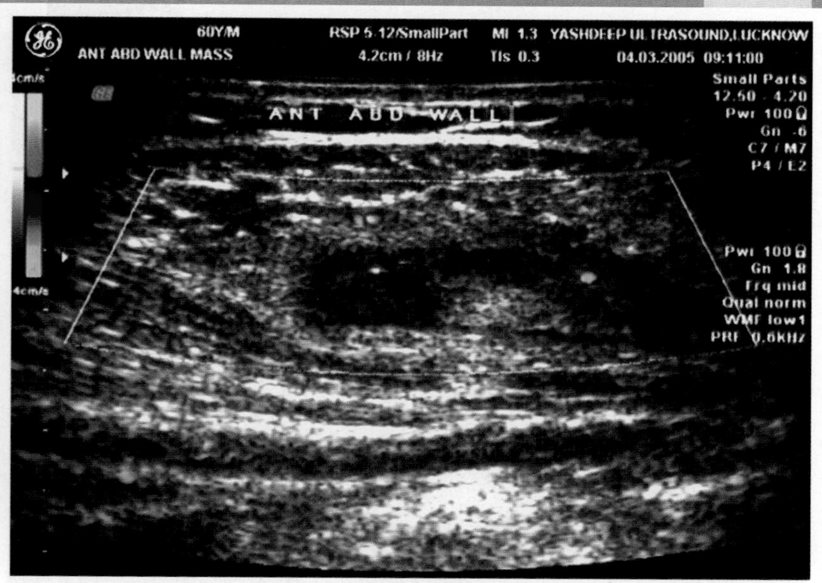

FIGURE 11.51

FIGURES 11.50 and 11.51: Anterior abdominal wall tear with hematoma. HRSG shows a big thick walled tear in the anterior abdominal wall. The tear is confined to the parietes. It is thick walled. Subcutaneous edema is also present. Echogenic collection is seen in the hematoma suggestive of hemorrhagic collection.

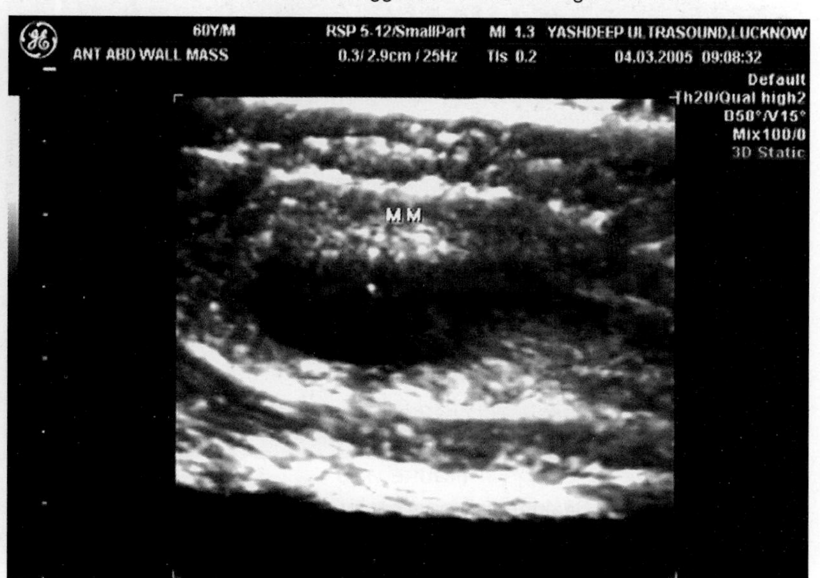

FIGURE 11.52: 3D imaging shows the details of the hemorrhagic tear.

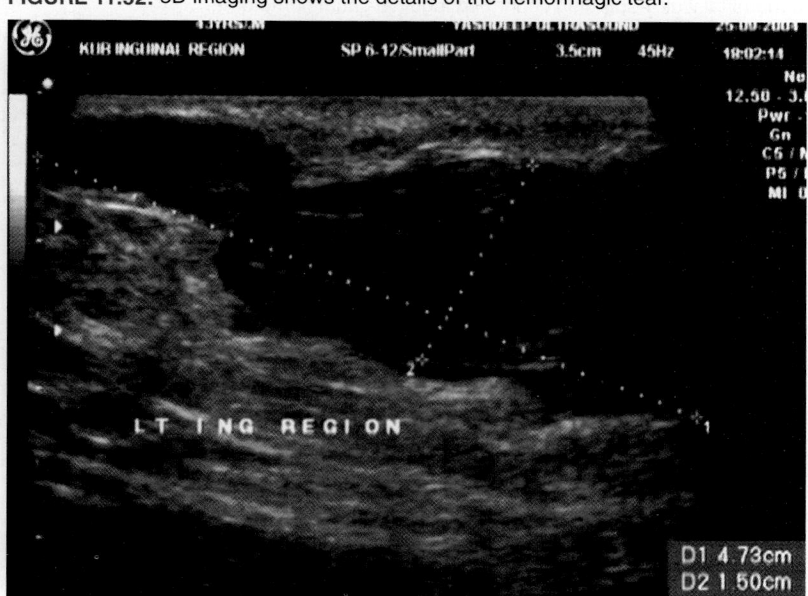

FIGURE 11.53

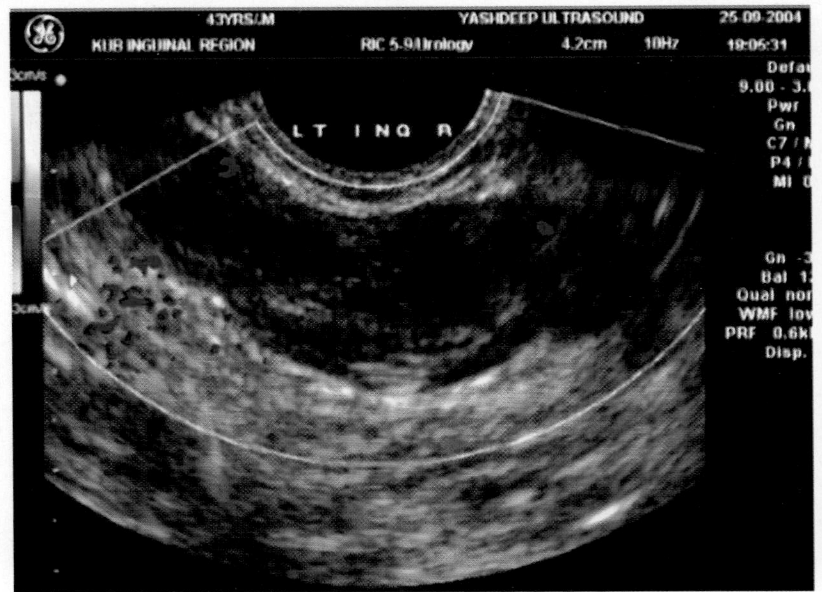

FIGURE 11.54

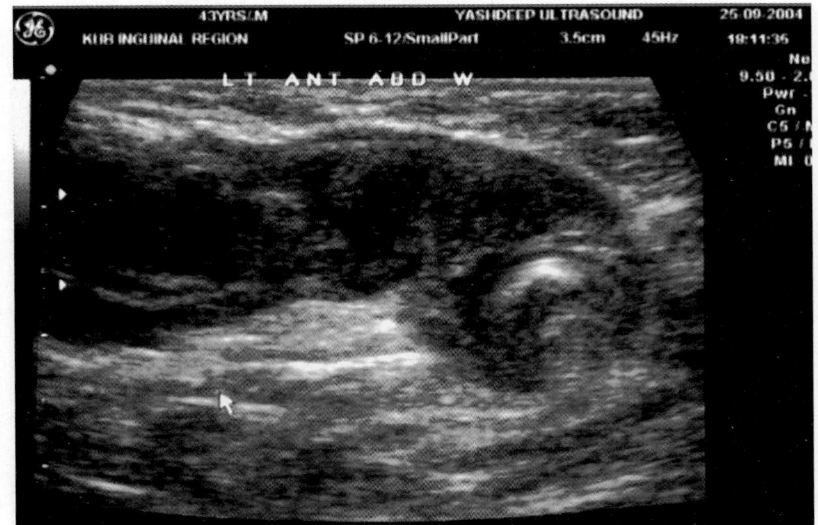

FIGURE 11.55

FIGURES 11.53 to 11.55: Inguinal canal hematoma. A patient presented with blunt injury to abdomen with a painful mass in the Lt inguinal canal. HRSG shows a big low level echo collection in the inguinal region. Thick echoes are seen in it suggestive of inguinal canal hematoma. The hematoma is also seen extending in the Lt anterior abdominal wall. Aspiration confirmed inguinal canal hematoma.

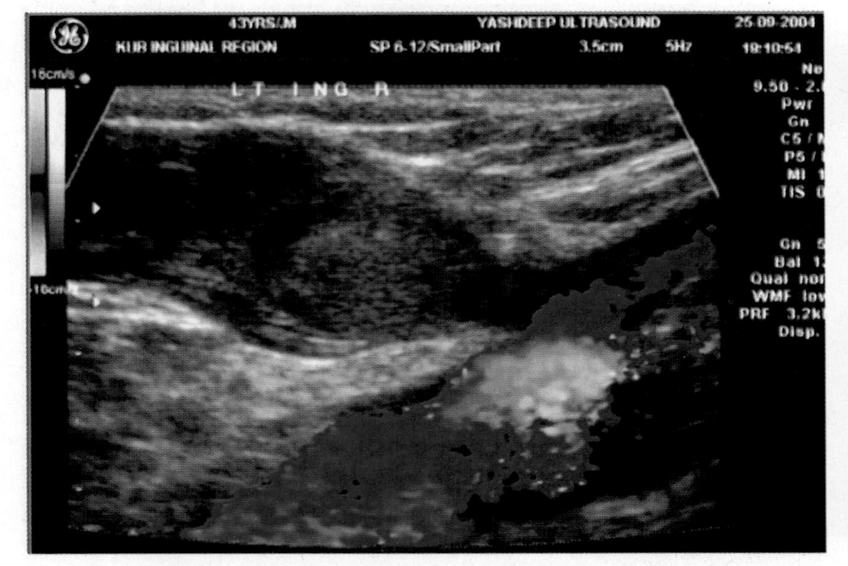

FIGURE 11.56: On color flow imaging no bleeding vessel was seen in the hematoma. The external iliac vein is seen normal.

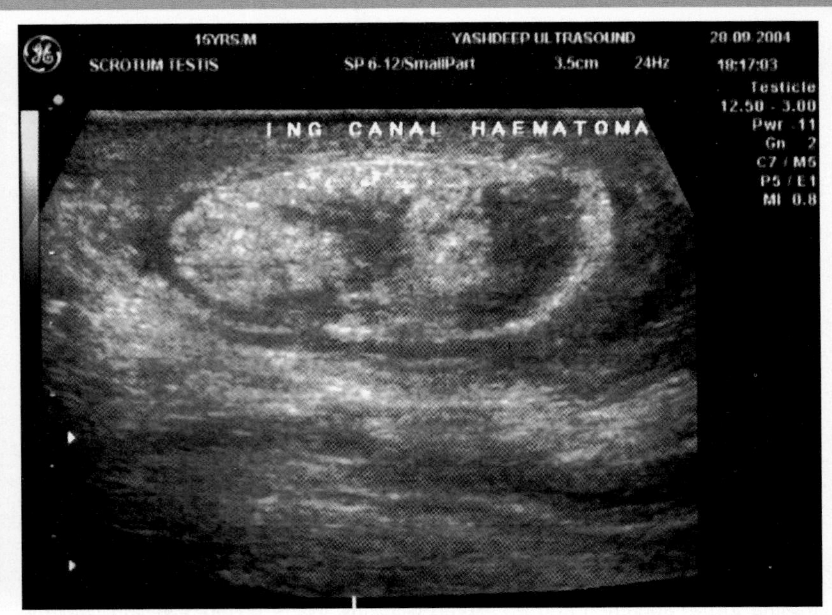

FIGURE 11.57

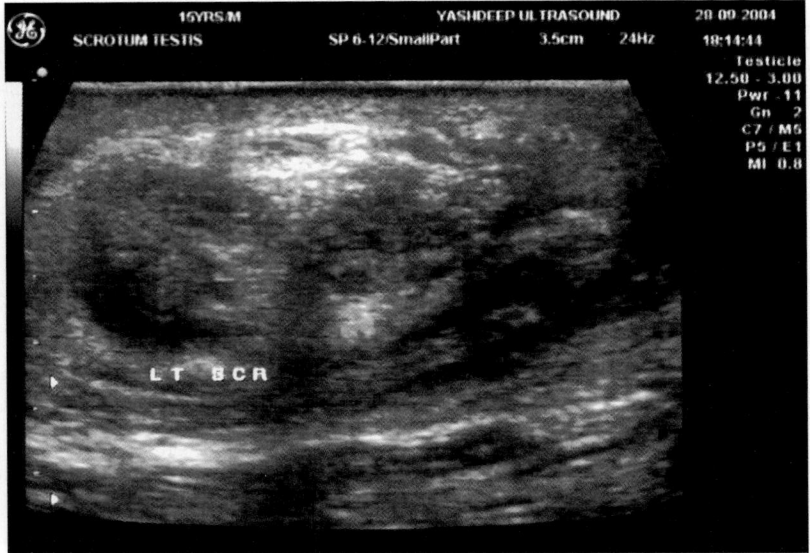

FIGURE 11.58

FIGURES 11.57 and 11.58: Inguinal canal hematoma. A patient presented with painful swelling in the inguinal region after sustaining trauma. HRSG shows echogenic collection in the inguinal region. It shows organization suggestive of fresh blood.

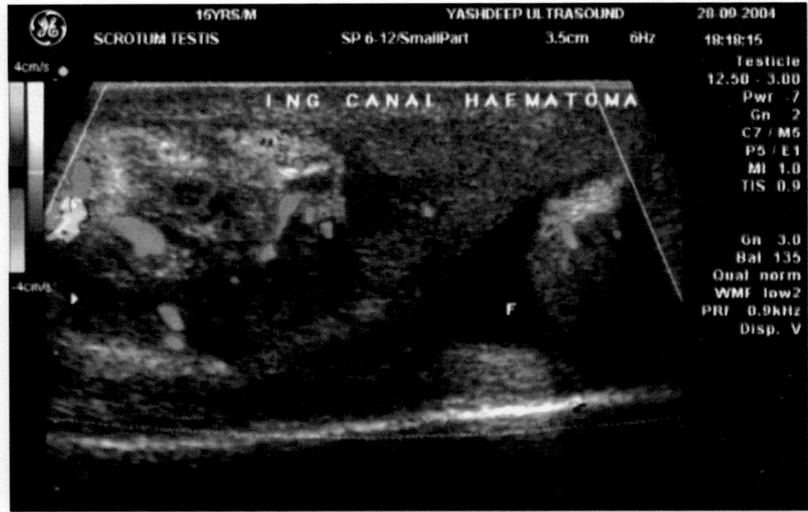

FIGURE 11.59

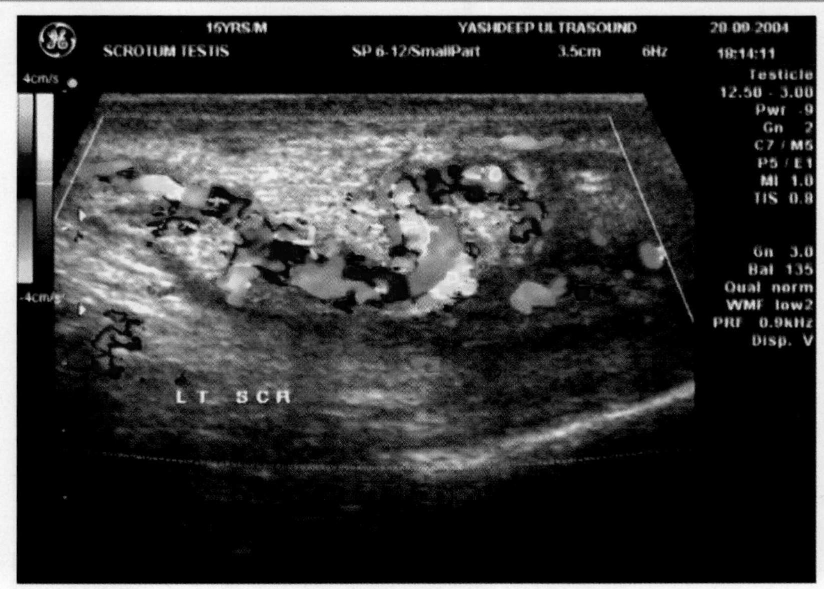

FIGURE 11.60

FIGURES 11.59 and 11.60: On color flow imaging few vessels are seen in the hematoma. The hematoma is seen confined to the inguinal region and it is not seen extending into the Lt scrotal sac.

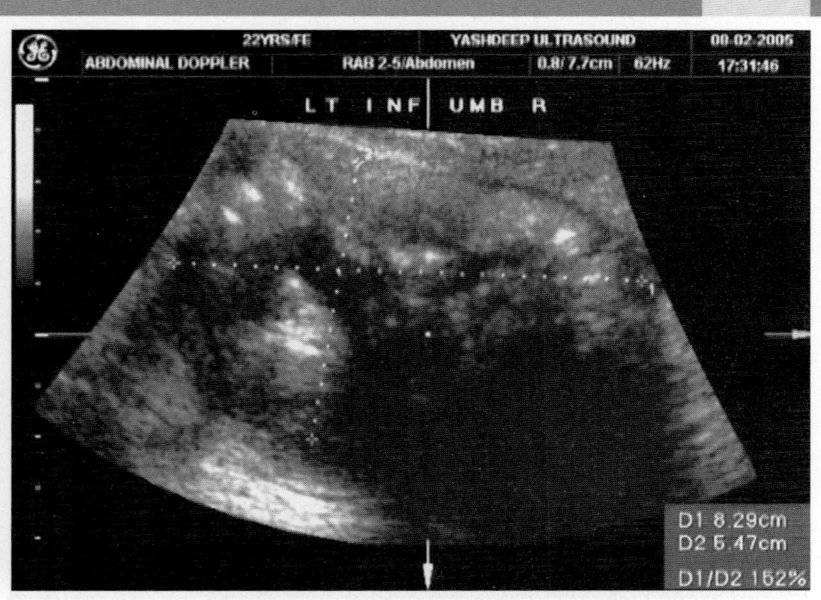

FIGURE 11.62: Foreign body (Sponge) in the anterior abdominal wall. A patient presented with soft tissue mass after the post surgery. HRSG shows a complex echo mass in the Lt infra umbilical region. Echogenic calcification is also seen in it. It casts acoustic shadowing masking the scanning field.

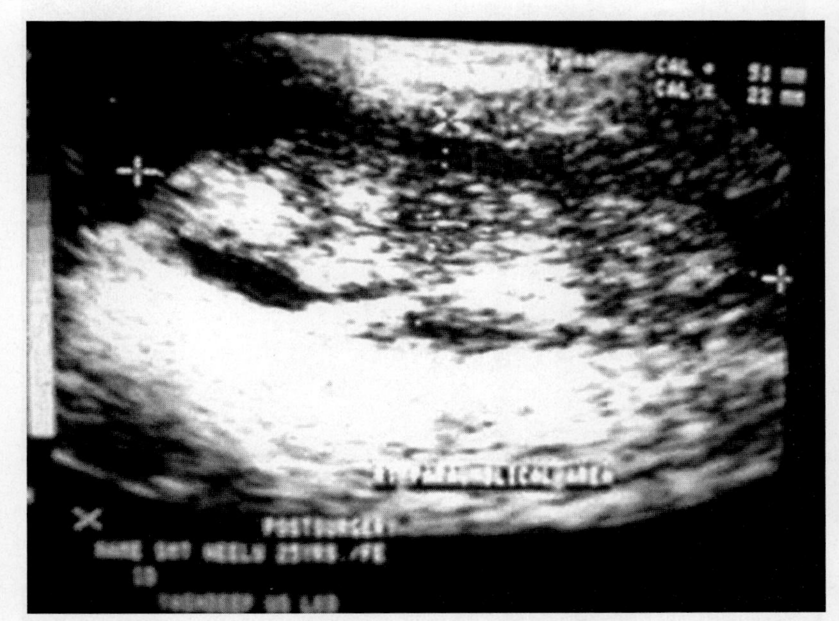

FIGURE 11.61: Foreign body in the anterior abdominal wall—Lt para umbilical region. HRSG shows a well-defined echogenic shadow in the post surgical case presented as a lump. The shadow shows homogeneous echoes with hypoechoic border. Small amount of collection is also seen around it. On exploration it turn out to be a sponge.

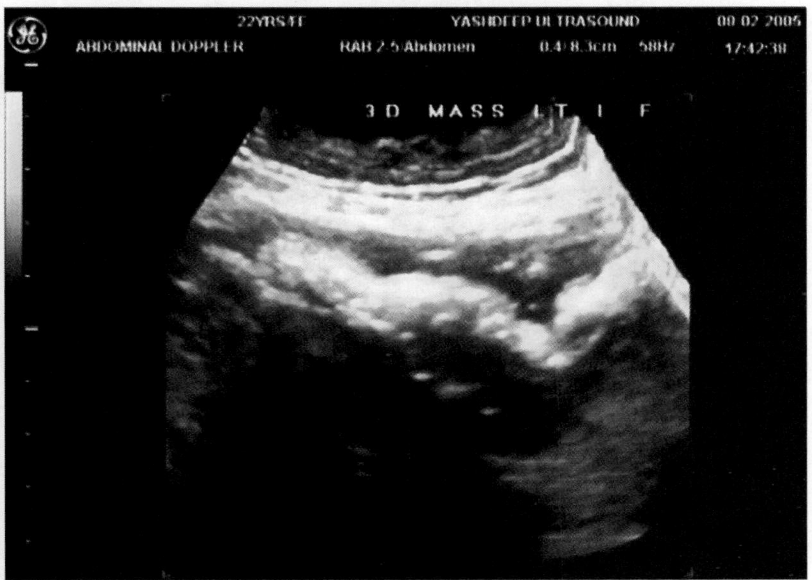

FIGURE 11.63

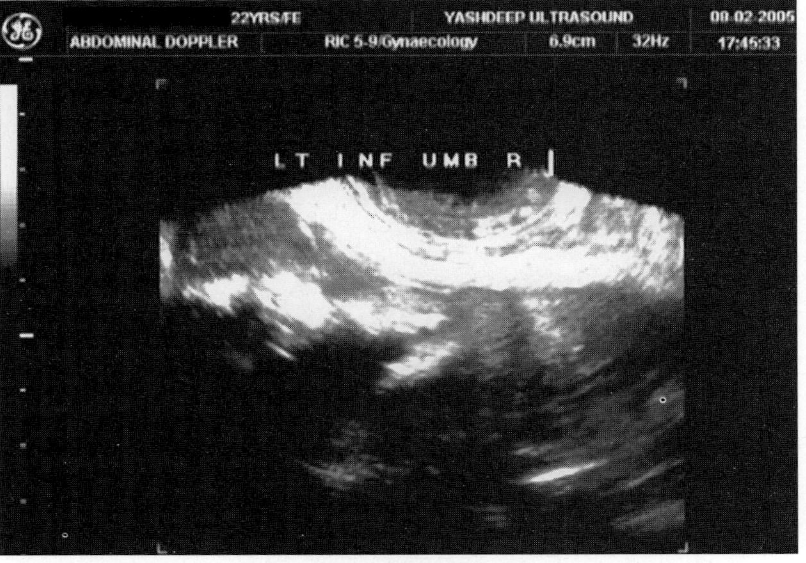

FIGURE 11.64

high in undescended testis. The undescended testis can be easily located in inguinal canal, at root or penis, in suprapubic area and medial side of thigh on HRSG. However, arrest of testis in abdominal cavity is difficult to find out on ultrasonography due to retroperitoneal fat.

Re-canalization of Umbilical Vein

Re-canalization of umbilical vein can be detectable on HRSG. The appearance is irregular and the vein can be traced downward in venous overload.

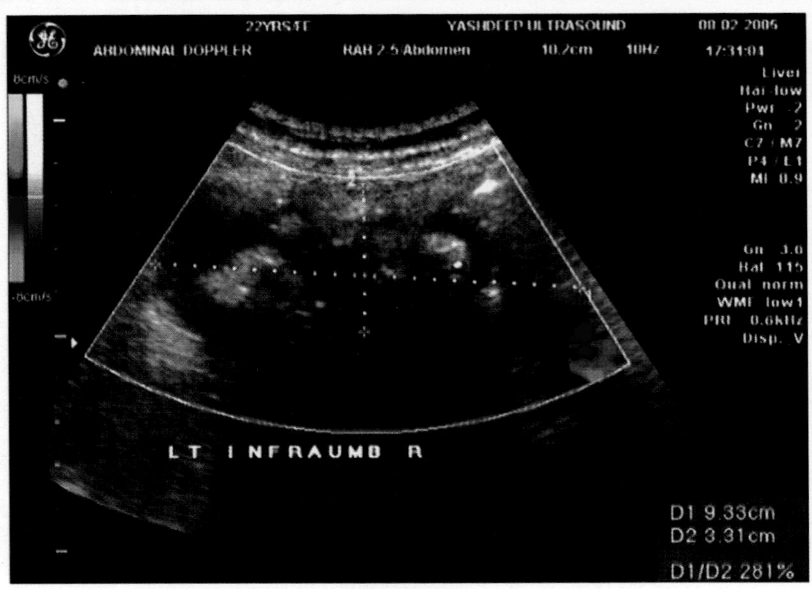

FIGURE 11.65

FIGURES 11.63 to 11.65: 3D imaging of the mass shows a complex echo shadow in the anterior abdominal wall. On color Doppler imaging it shows no flow. Postoperative finding confirmed a sponge in the Lt anterior abdominal wall.

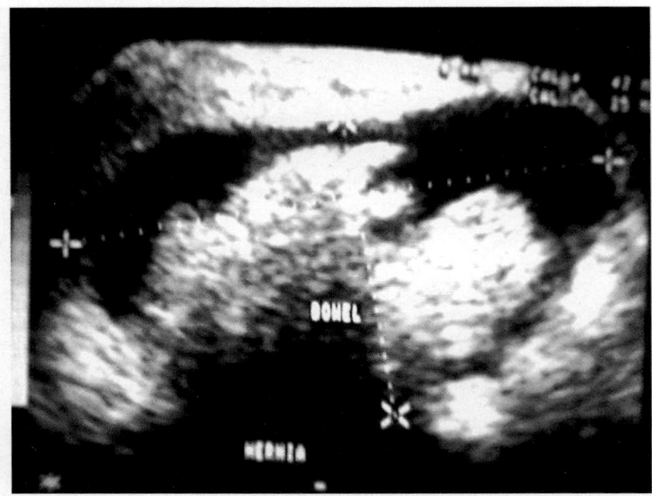

FIGURE 11.66

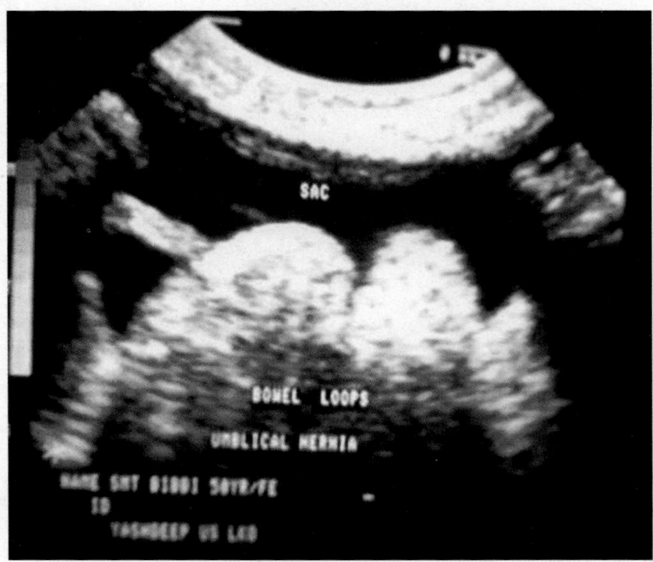

FIGURE 11.67

FIGURES 11.66 and 11.67: Paraumbilical hernia. HRSG shows floating bowel loops in the soft tissue mass in paraumbilical region. Fluid collection is also seen in it suggestive of paraumbilical hernia. Echogenic band is seen in the umbilical hernia obstructing the rèduction of the bowel loops.

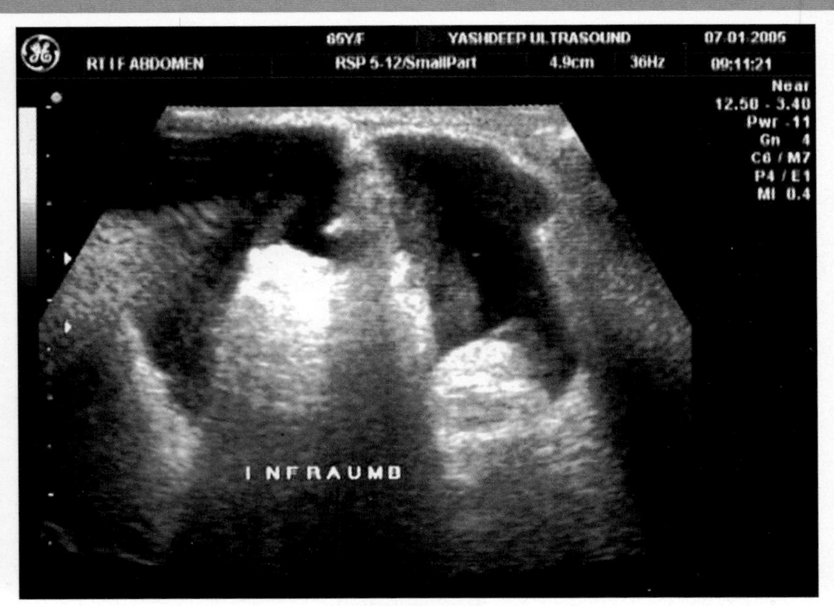

FIGURE 11.68

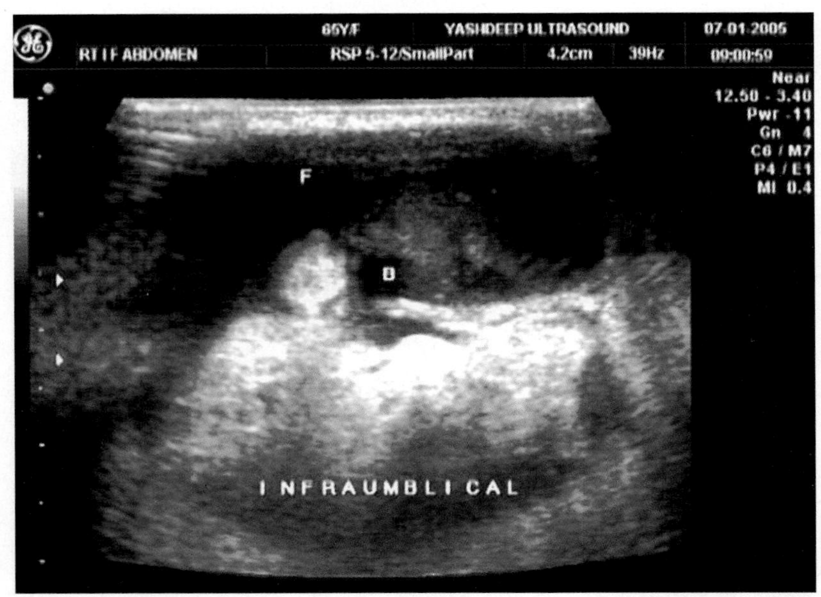

FIGURE 11.69

FIGURES 11.68 and 11.69: Umbilical hernia. HRSG shows herniating bowel loops in infraumbilical region. The bowel loops are seen echogenic floating gas-containing shadows floating in the hernial sac.

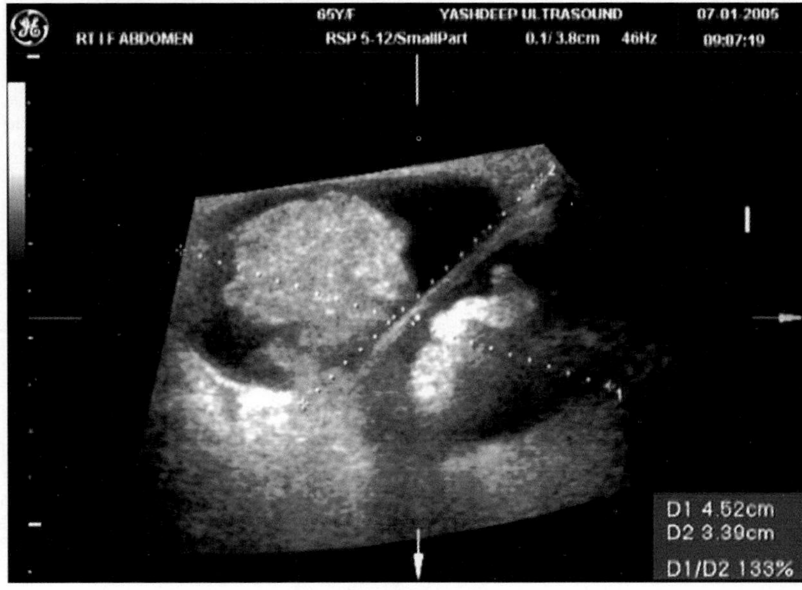

FIGURE 11.70

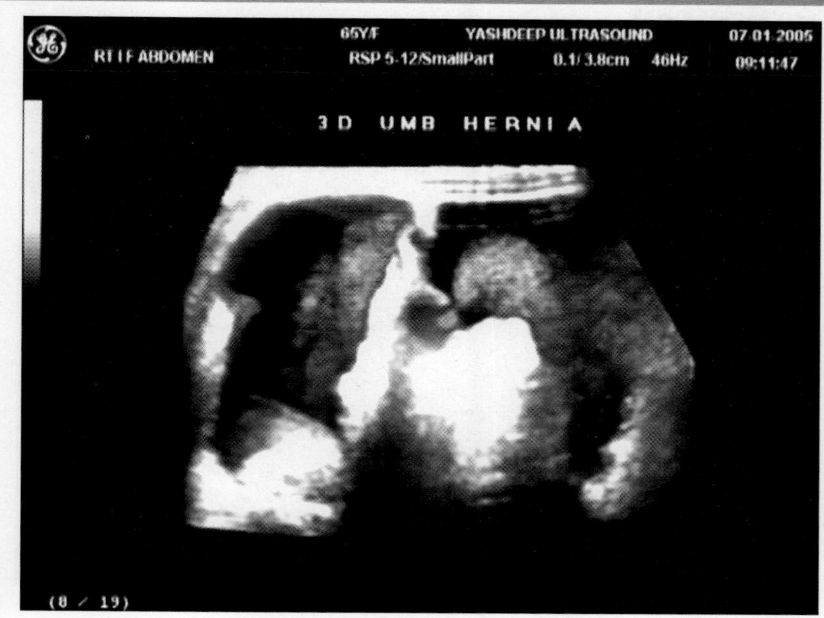

FIGURE 11.71

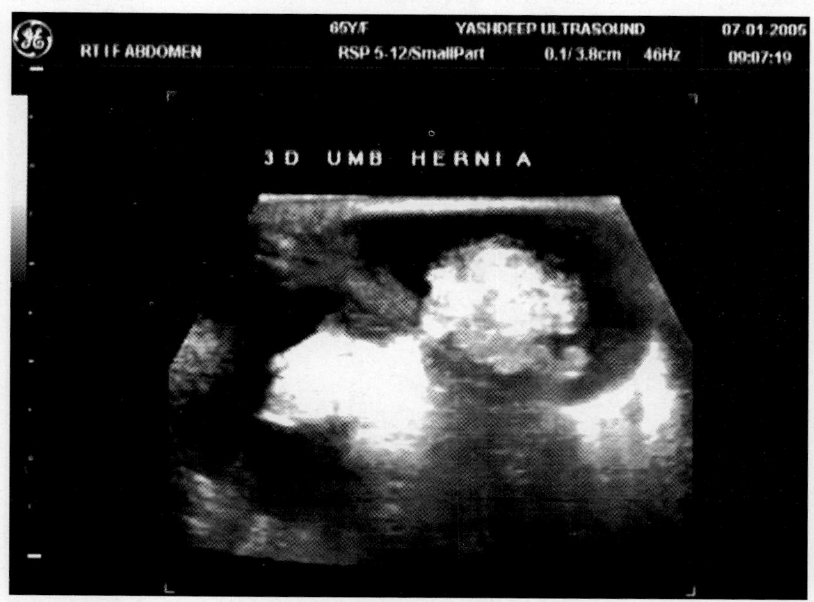

FIGURE 11.72

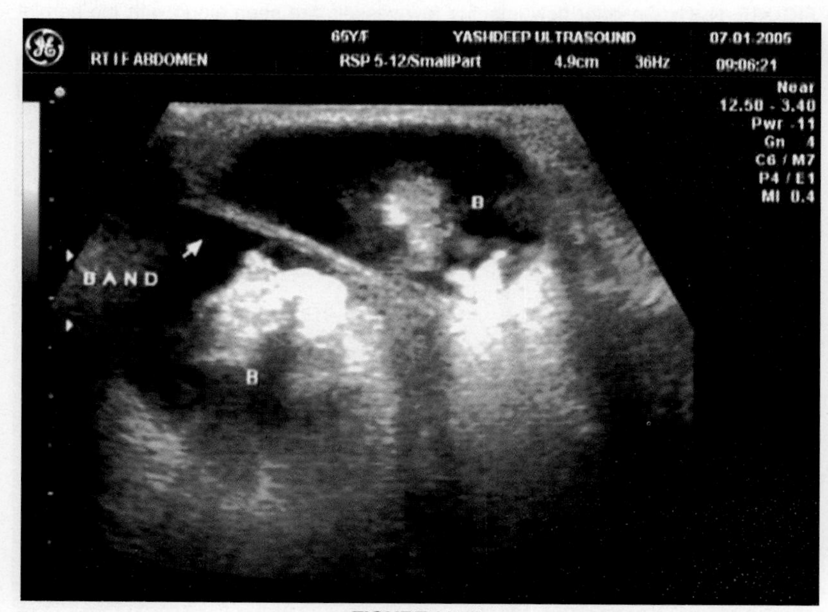

FIGURE 11.73

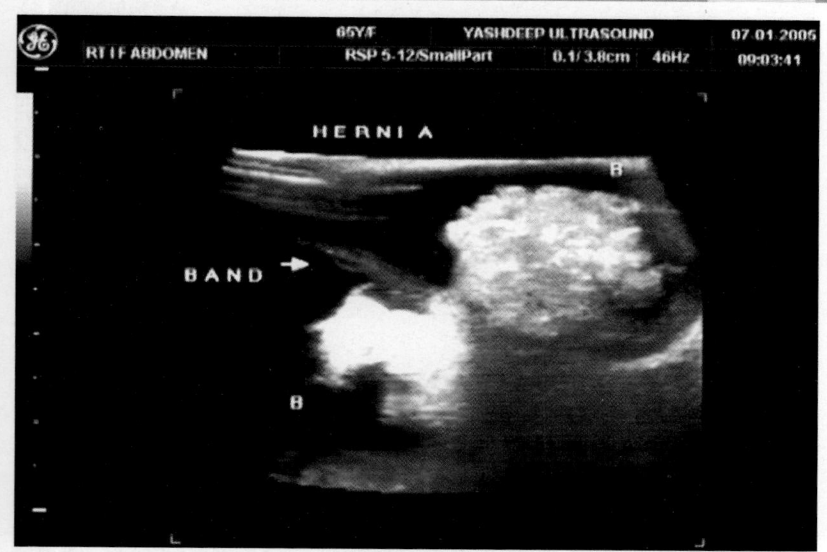

FIGURE 11.74

FIGURES 11.70 to 11.74: 3D imaging with surface rendering shows the details of the hernial sac. An echogenic band is seen dividing the sac into two. The bowel loops are floating in both the sacs. However, no evidence of any knotting is seen. The band resulting into strangulation may entangle the bowel loops. Therefore, color Doppler imaging is important to rule out strangulation.

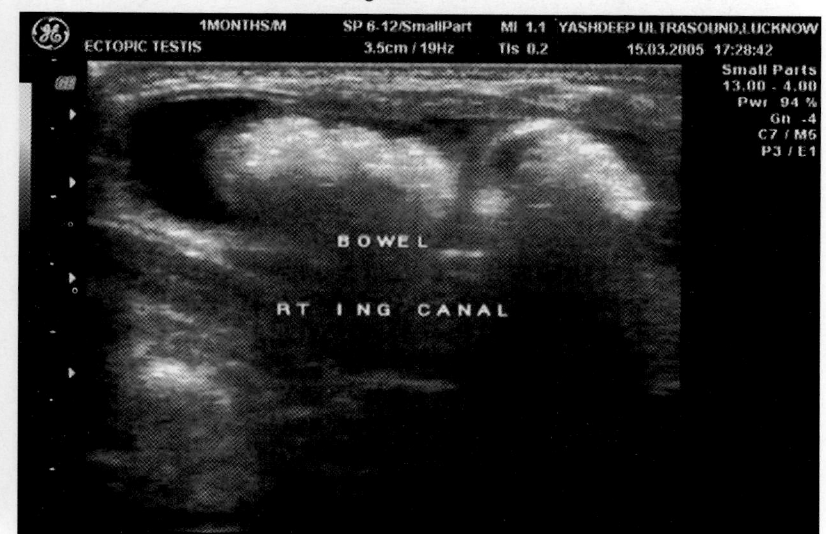

FIGURE 11.75

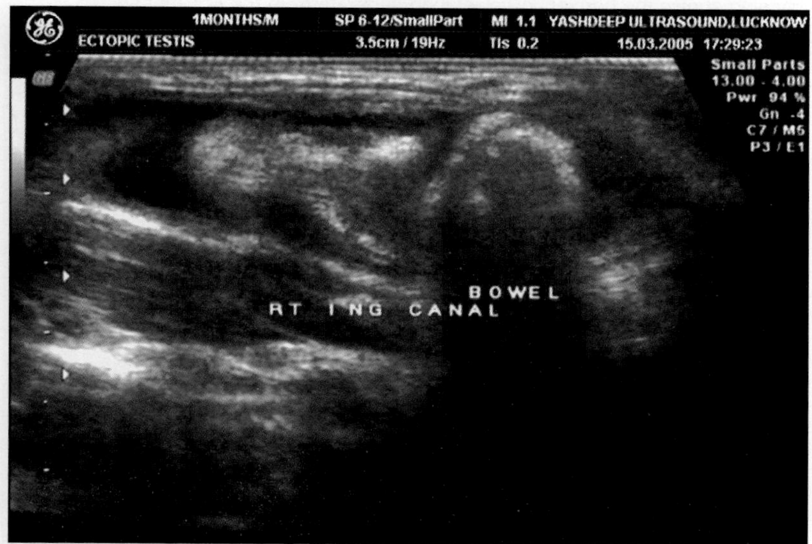

FIGURE 11.76

FIGURES 11.75 and 11.76: Inguinal hernia. A one-month-old child presented with soft tissue swelling in Rt inguinal region. HRSG shows herniating bowel loops in the inguinal region. The bowel loops are seen as echogenic floating shadows in the inguinal region.

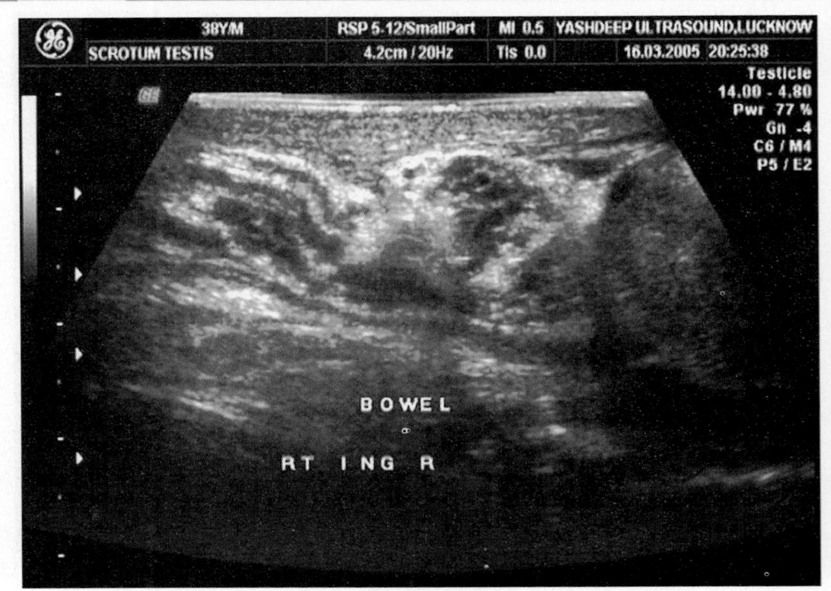

FIGURE 11.77

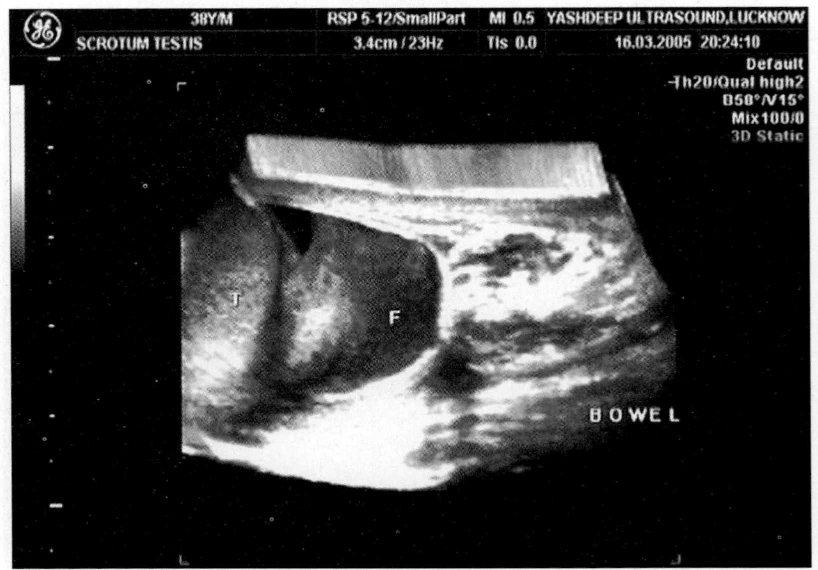

FIGURE 11.78

FIGURES 11.77 and 11.78: Inguinal hernia. HRSG shows an elongated bowel loop herniating in the inguinal canal. However, no herniation is seen in the scrotal sac. 3D imaging shows the herniating bowel loop separate from the scrotal sac.

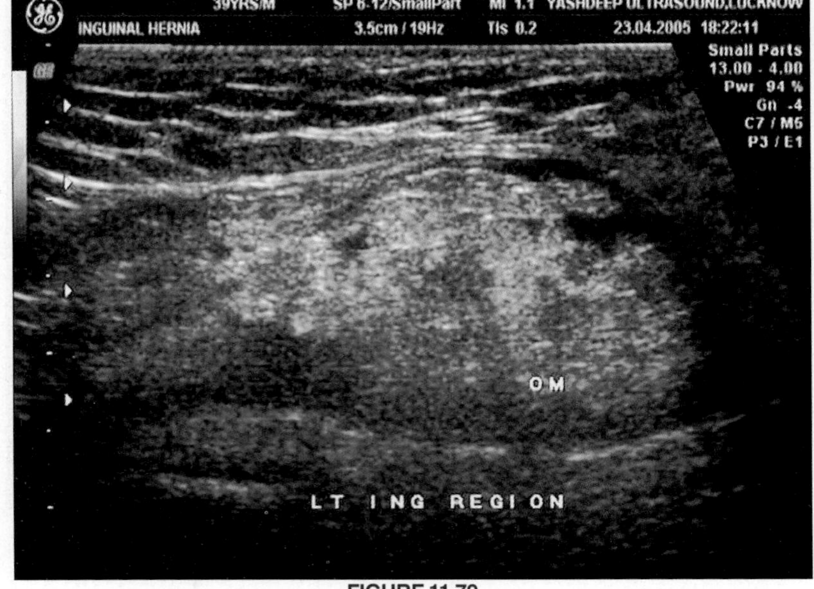

FIGURE 11.79

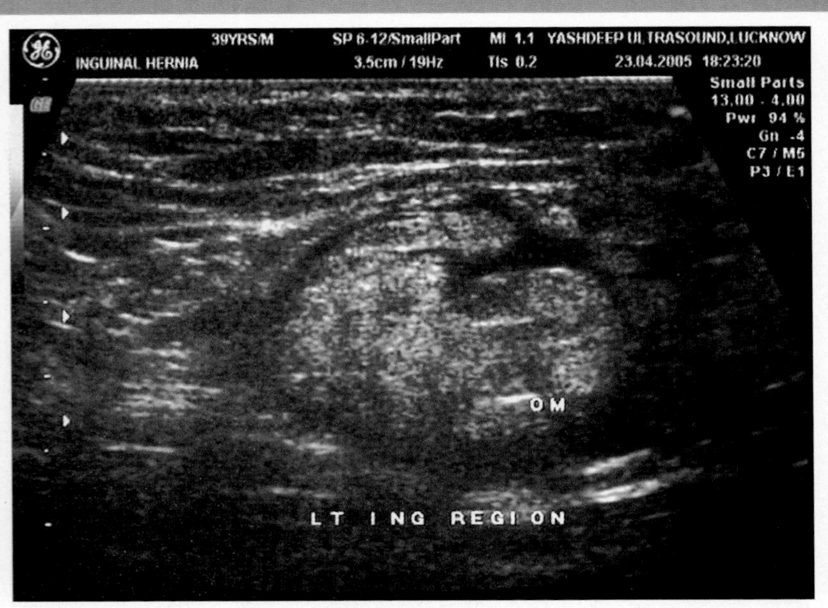

FIGURE 11.80

FIGURES 11.79 and 11.80: Omental hernia. HRSG shows echogenic membranous herniating shadow in the Rt inguinal region suggestive of omental hernia. However, no bowel loop is seen herniating with omentum.

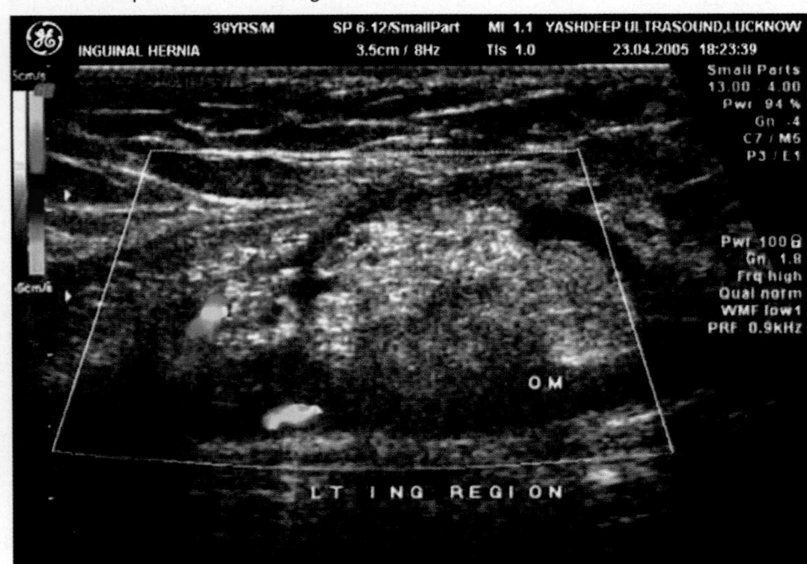

FIGURE 11.81: On color flow imaging few vessels are seen along with the hernial sac.

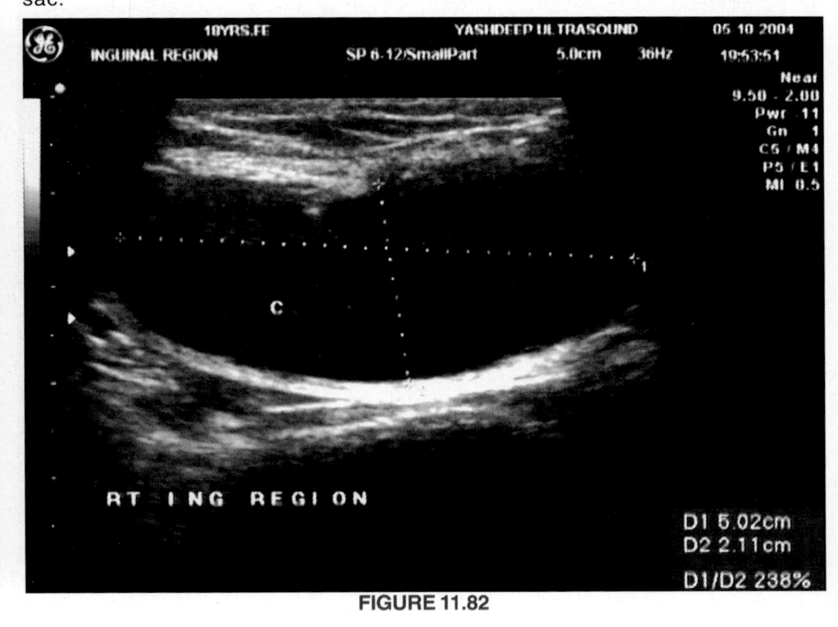

FIGURE 11.82

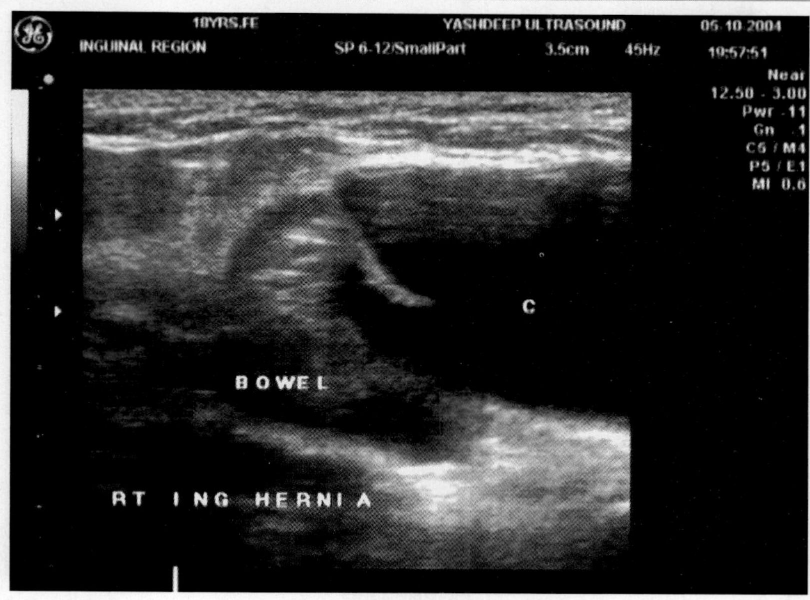

FIGURE 11.83

FIGURES 11.82 and 11.83: Inguinal hernia with fluid collection. HRSG shows a cystic collection in a young girl in the Rt inguinal canal. A bowel loop is also seen coming out with it. However, no evidence of any loculation or septation is seen in it. Inguinal hernias are uncommon in females yet it can be seen in few cases.

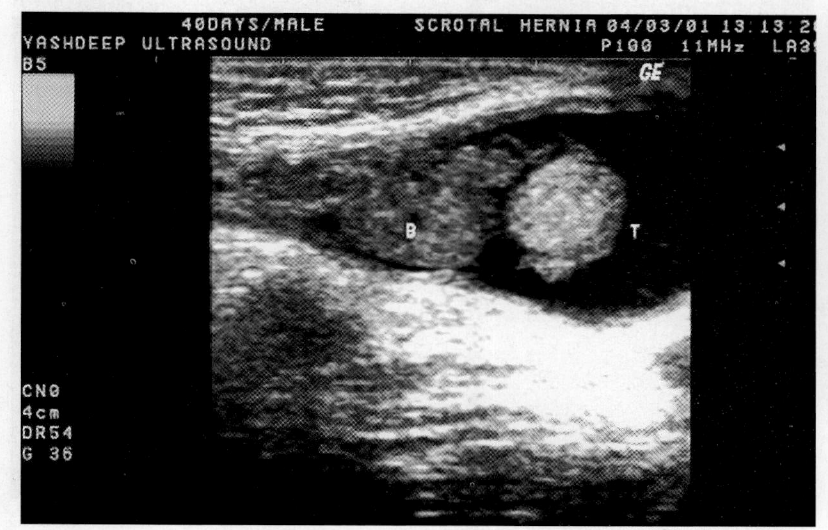

FIGURE 11.84: Inguinal hernia. A bowel loop is seen herniated in the Rt scrotal sac in an infant. The testis is seen normally in the scrotal sac. Fluid is also seen in the scrotal sac and bowel is seen floating in the fluid.

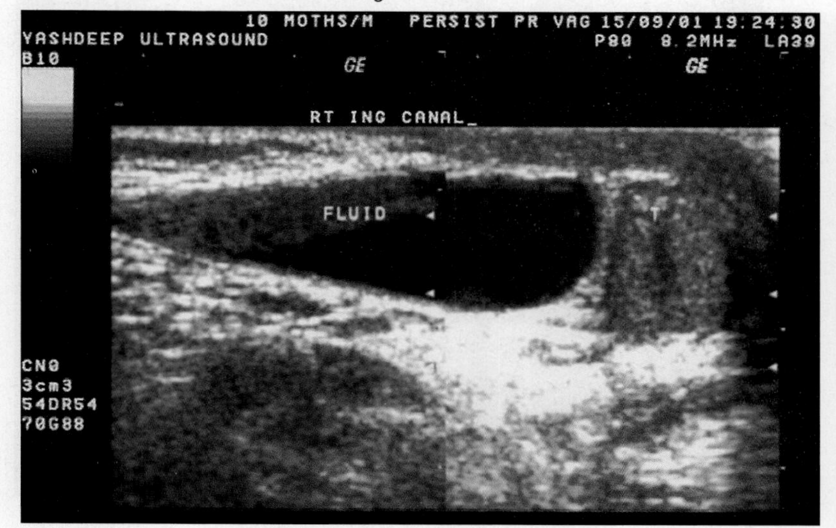

FIGURE 11.85: Fluid in the persistent processes vaginalis. HRSG shows fluid collection in persistent processes vaginalis presented as a Rt sided inguinal swelling. No herniation of bowel loop is seen in it. Testis is seen normally in the scrotal sac.

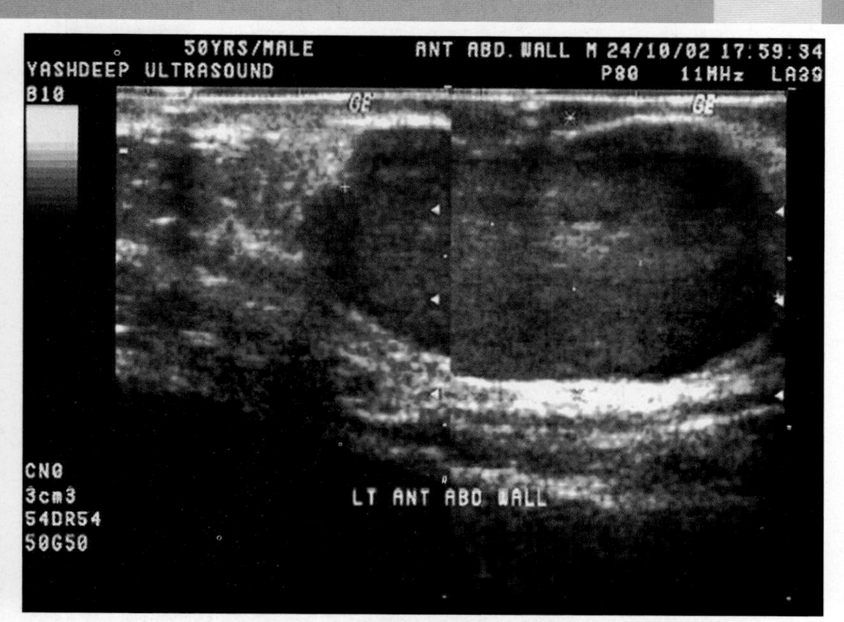

FIGURE 11.86

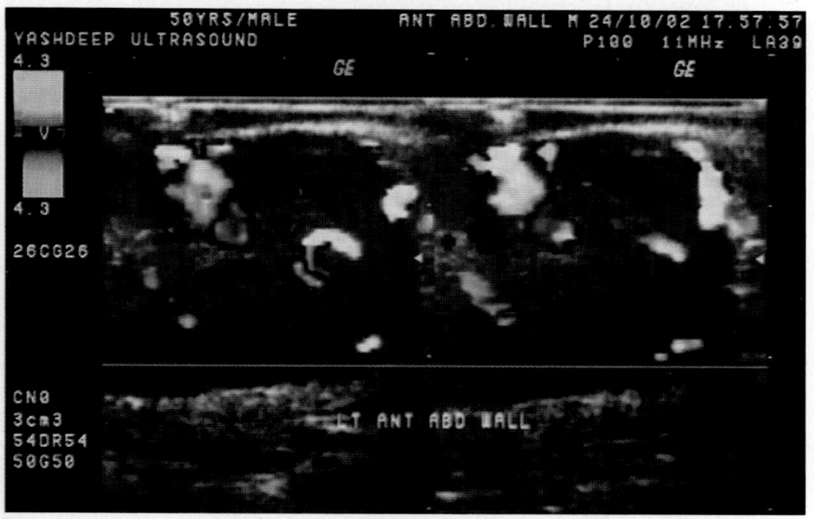

FIGURE 11.87

FIGURES 11.86 and 11.87: Bleeding hematoma in the anterior abdominal wall. A patient presented progressively increasing mass in the anterior abdominal wall after sustaining blunt trauma. HRSG shows a big low level echo collection in the anterior abdominal wall. Homogeneous echoes are seen in it with floating debris. On color flow imaging multiple vessels are seen in the collection. Aspiration of the collection came out to be fresh blood.

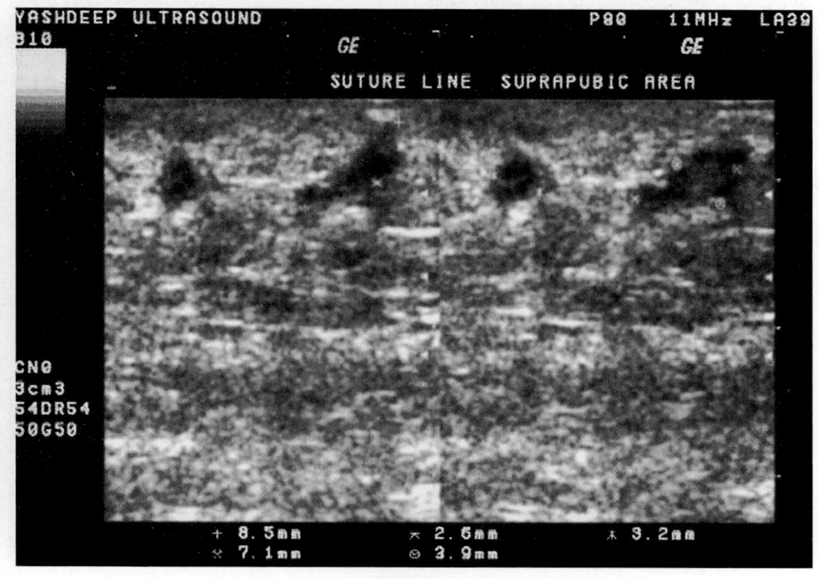

FIGURE 11.88

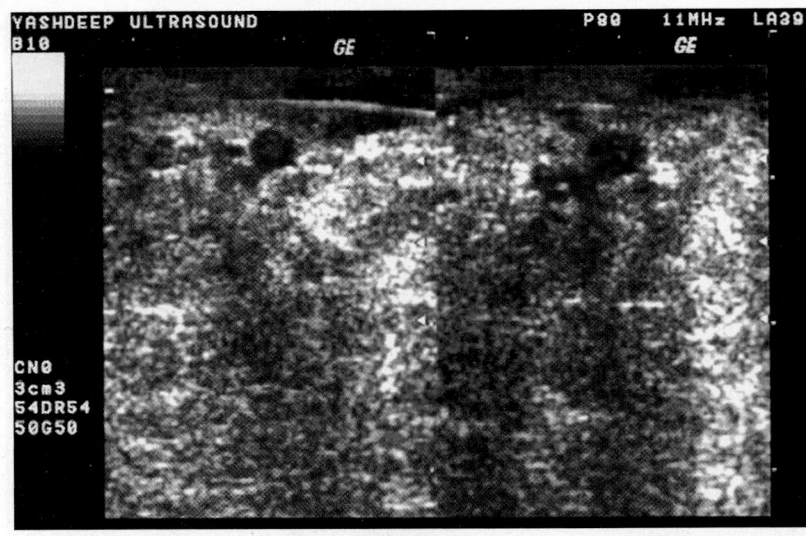

FIGURE 11.89

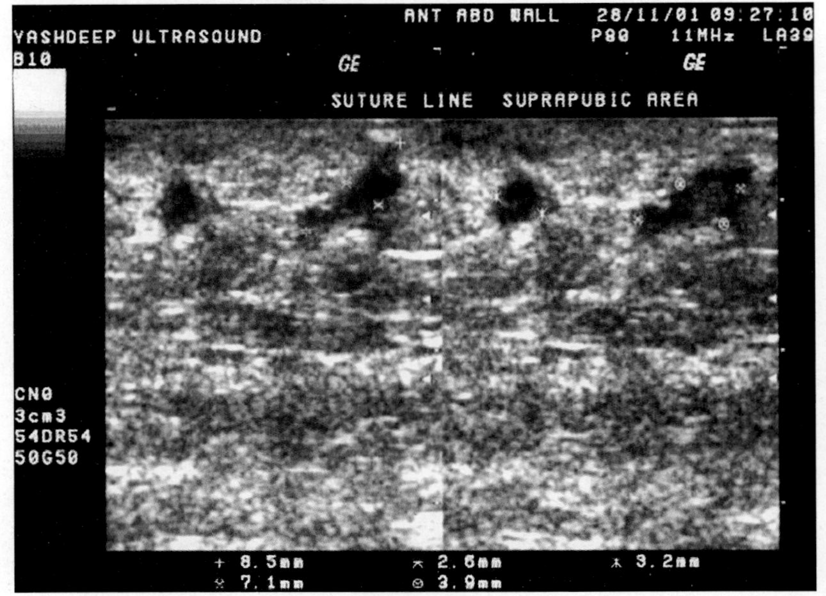

FIGURE 11.90

FIGURES 11.88 to 11.90: Suture line abscess. HRSG shows multiple small hypoechoic pus pockets along the suture line in a post-operated infected suture line. The abscess cavities are seen distributed along the suture line in the subcutaneous planes. Associated soft tissue edema is also present.

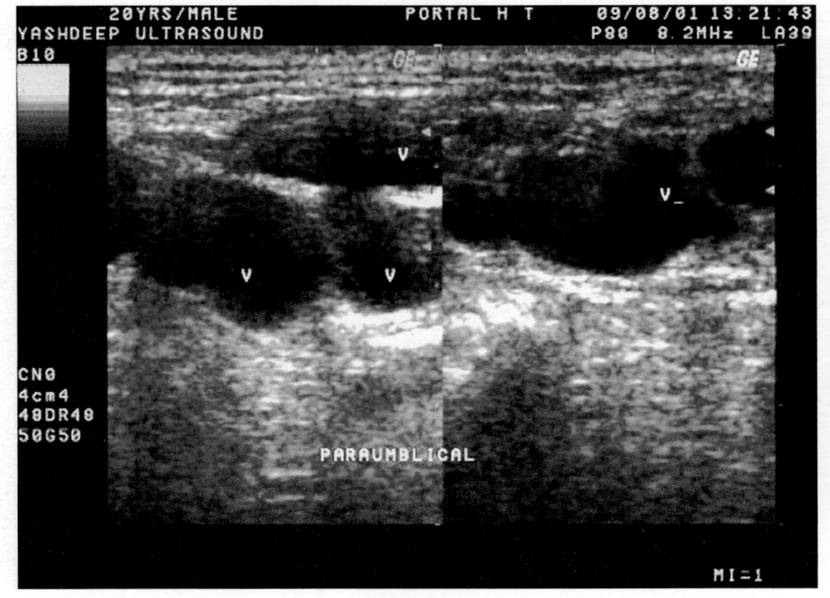

FIGURE 11.91

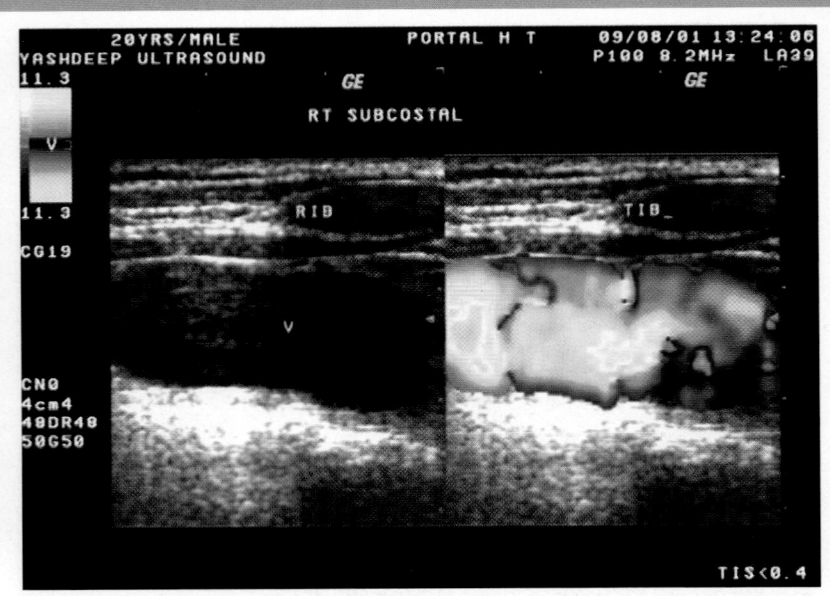

FIGURE 11.92

FIGURES 11.91 and 11.92: Umbilical vein canalization. HRSG shows multiple dilated vessels in the anterior abdominal wall along the paraumbilical region in a known case of portal hypertension. On color Doppler imaging blood flow is seen in it. Multiple dilated varices are also seen along the vein.

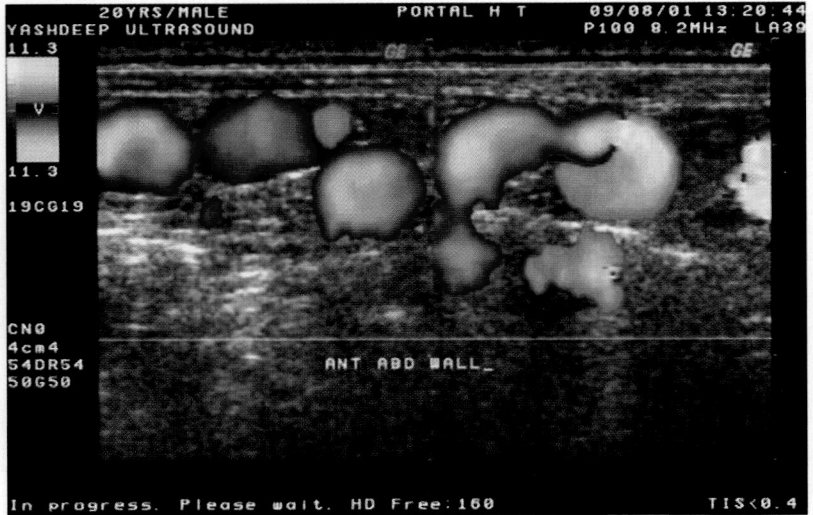

FIGURE 11.93

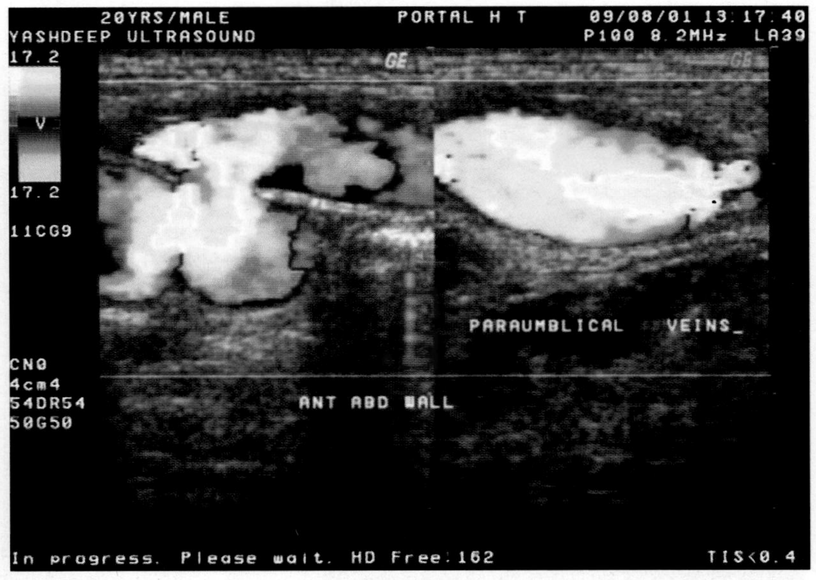

FIGURE 11.94

FIGURES 11.93 and 11.94: On color flow imaging multiple dilated varices are seen along the vein, which are confirmed on color flow imaging.

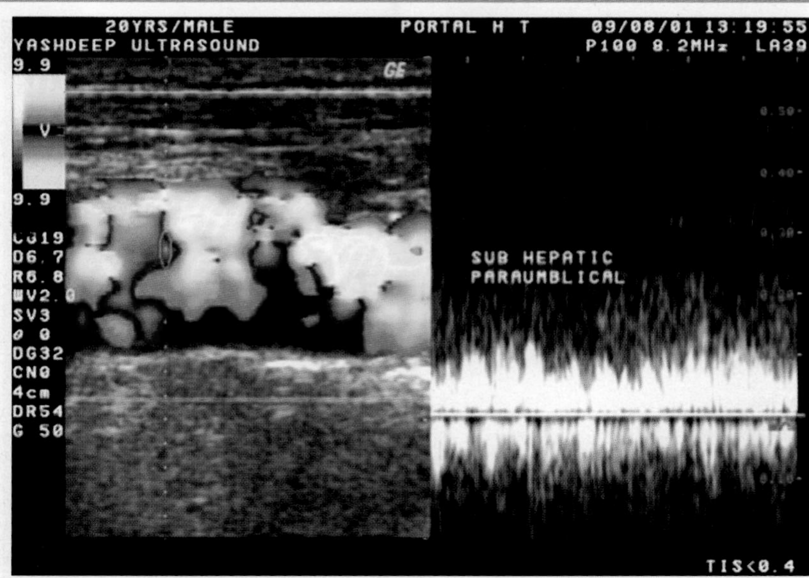

FIGURE 11.95: Spectral Doppler tracing shows high venous flow in it.

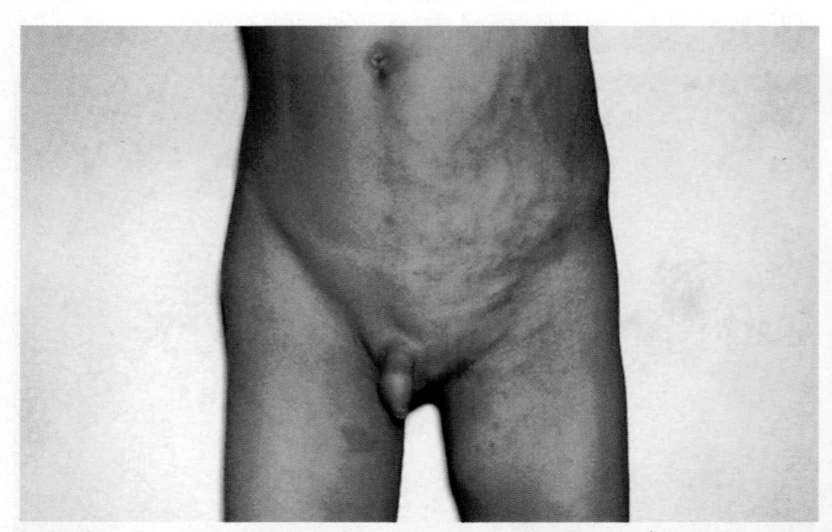

FIGURE 11.96: Local venous malformation anterior abdominal wall. A young child presented with multiple dilated vessels in the Lt anterior abdominal wall, inguinal region and the medial part of upper thigh.

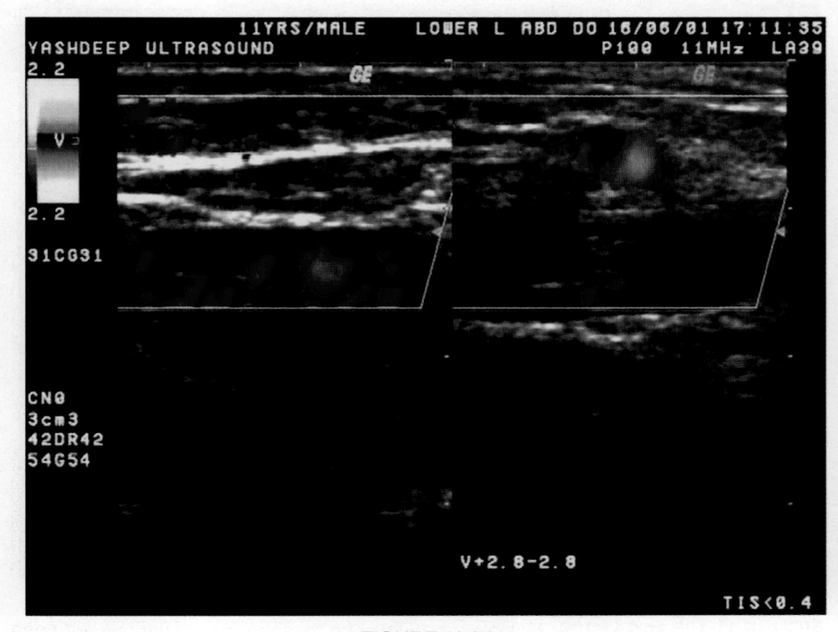

FIGURE 11.97

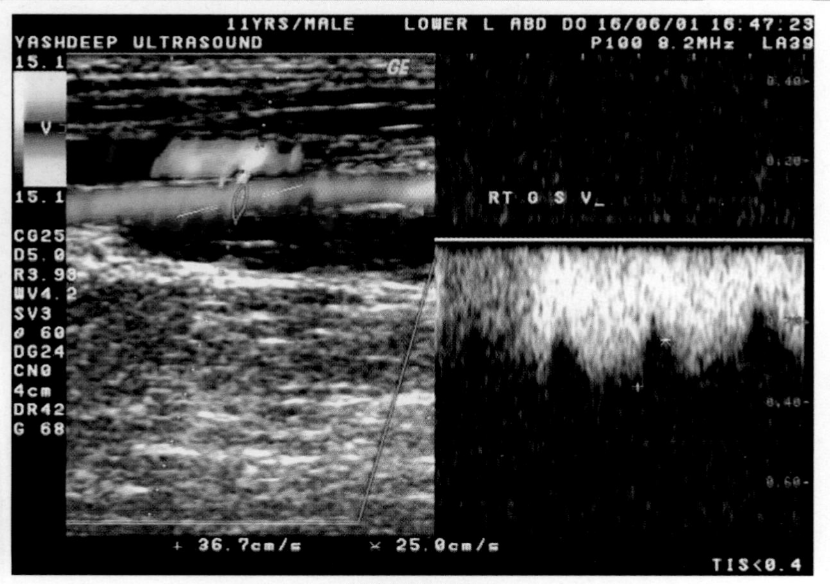

FIGURE 11.98

FIGURES 11.97 and 11.98: Color Doppler imaging shows multiple venous dilatation over the anterior abdominal wall.

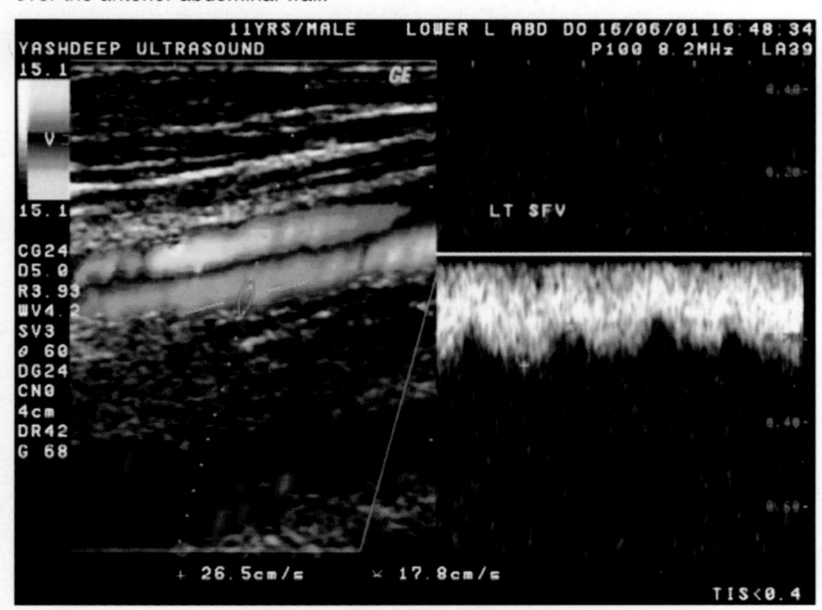

FIGURE 11.99: On spectral Doppler tracing high pulsatile flow is seen in the superficial femoral and common femoral arteries on both sides.

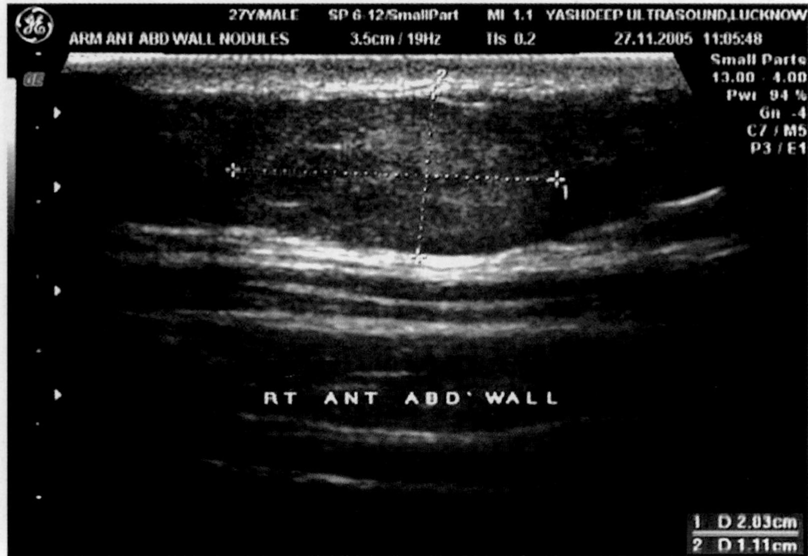

FIGURE 11.100

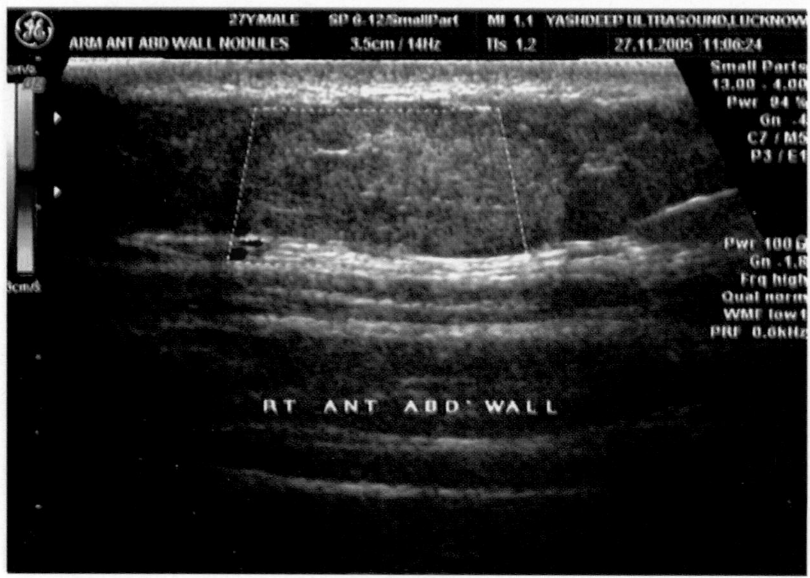

FIGURE 11.101

FIGURES 11.100 and 11.101: Lipoma in anterior abdominal wall. A well-defined echogenic homogeneous mass is seen in the Rt anterior abdominal wall in the Rt hypochondrium. No suggestion of any calcification or cystic degeneration is seen in it. On color flow imaging no flow is seen in it. Sonographic findings are suggestive of lipoma.

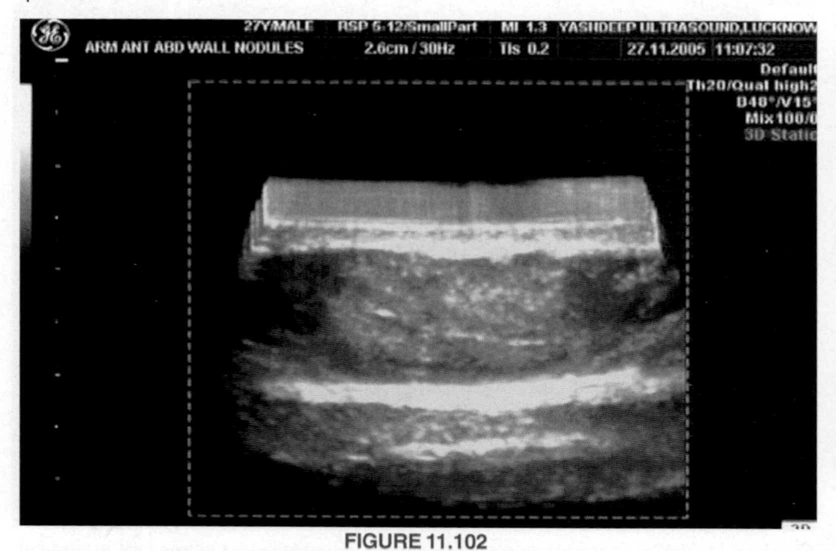

FIGURE 11.102

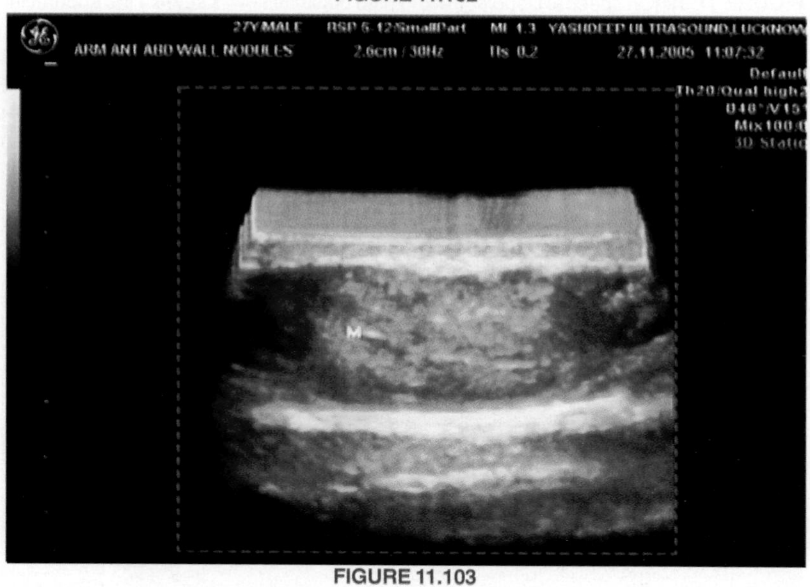

FIGURE 11.103

FIGURES 11.102 and 11.103: 3D imaging of the same patient shows the better details of the lipoma. It is seen confined to the subcutaneous planes. No muscle wall invasion is seen.

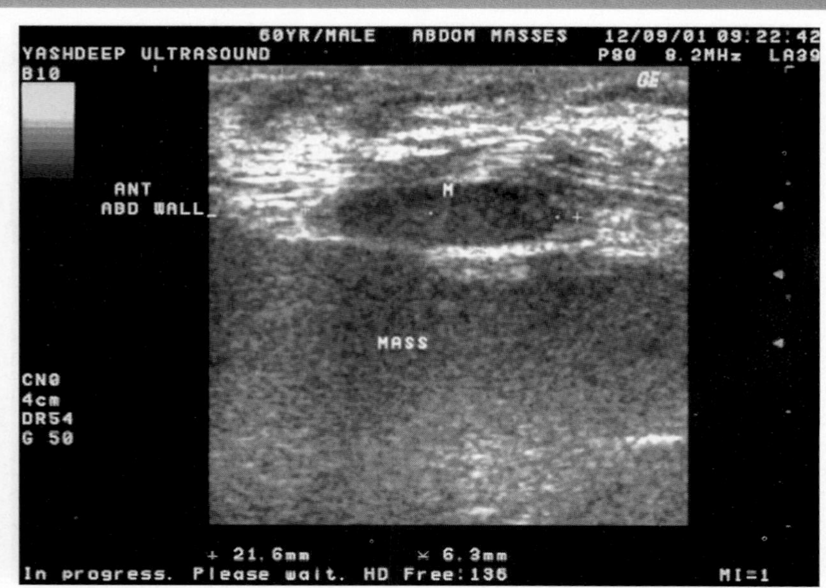

FIGURE 11.104: Hemangiopericytoma infiltrating anterior abdominal wall. A known case of hemangiopericytoma in abdomen presented with a mass in the abdomen and anterior abdominal wall. HRSG shows an irregular hypoechoic mass in the anterior abdominal wall infiltrating it.

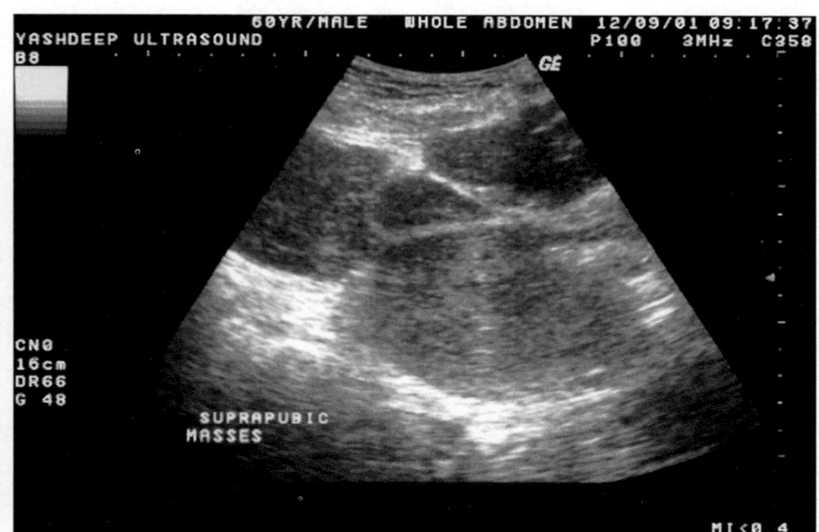

FIGURE 11.105

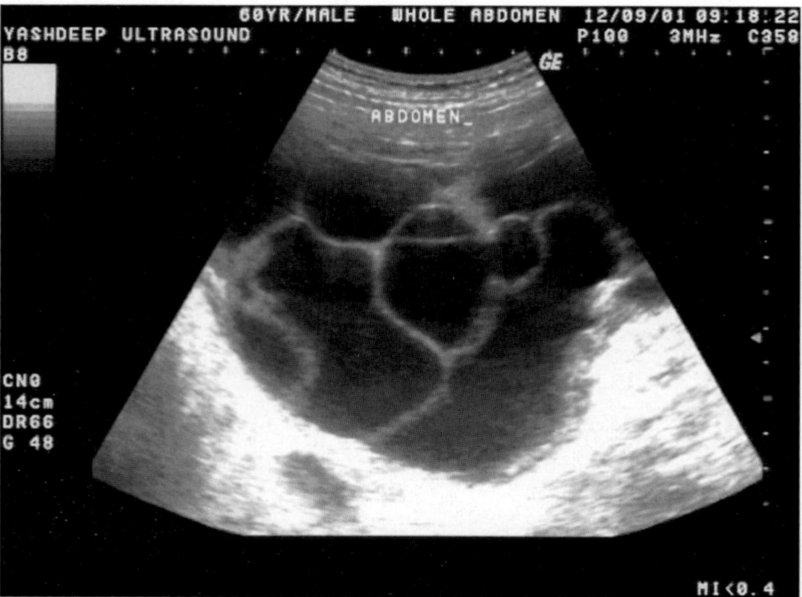

FIGURE 11.106

FIGURE 11.105 and 11.106: Multiple low-level echo complex masses are seen filling the abdominal cavity. Thin echogenic septa are also seen in these masses in the same case.

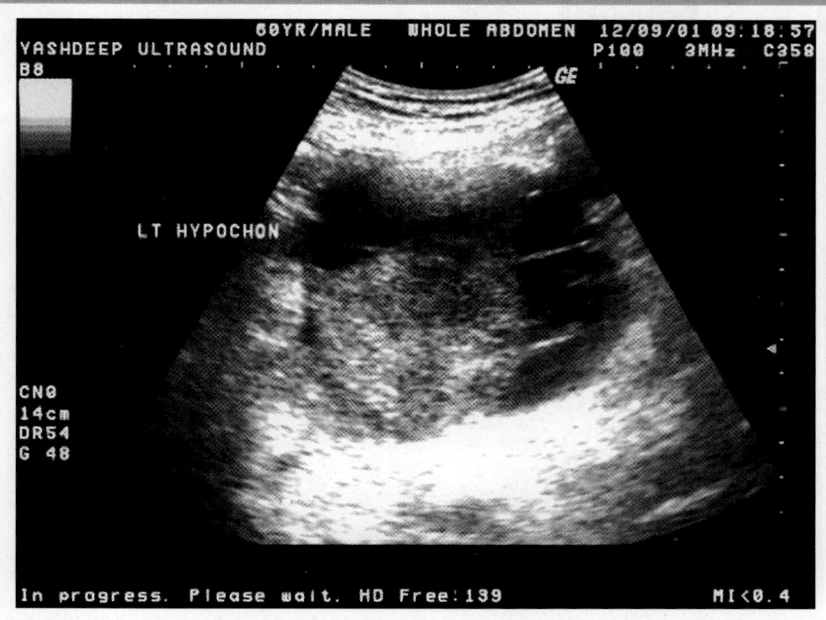

FIGURE 11.107: Soft tissue mass is also seen in the Lt hypochondrium with echogenic debris. Biopsy of the mass shows infiltrating hemangiopericytoma.

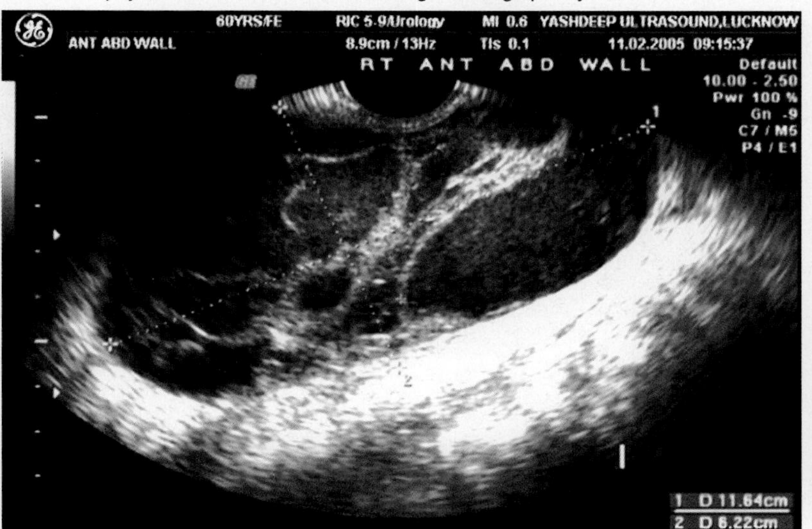

FIGURE 11.108

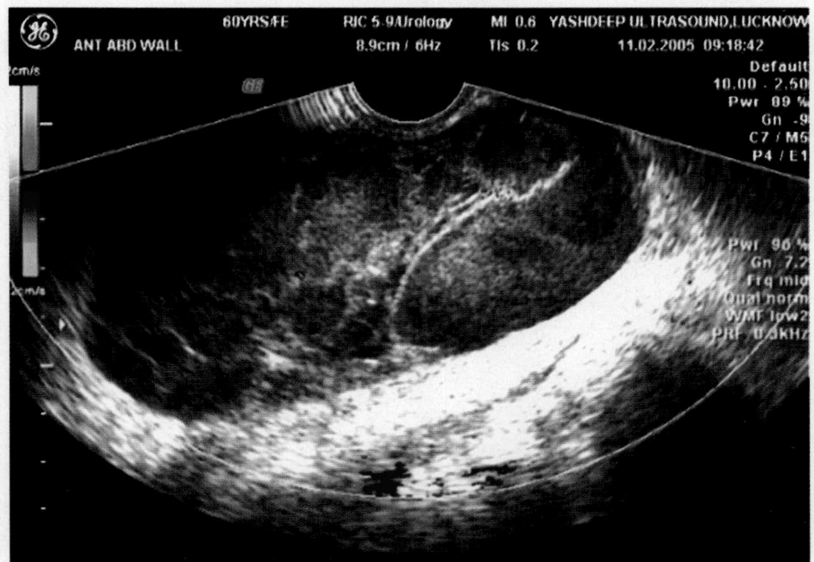

FIGURE 11.109

FIGURES 11.108 and 11.109: FTC operated liposarcoma anterior abdominal wall. An operated patient of liposarcoma presented with cystic mass confined to the anterior abdominal wall on the Rt side. HRSG shows a big loculated collection in the anterior abdominal wall. Thick echoes are seen floating into it. On color flow imaging no flow is seen in the collection. Echogenic septa are also seen in it.

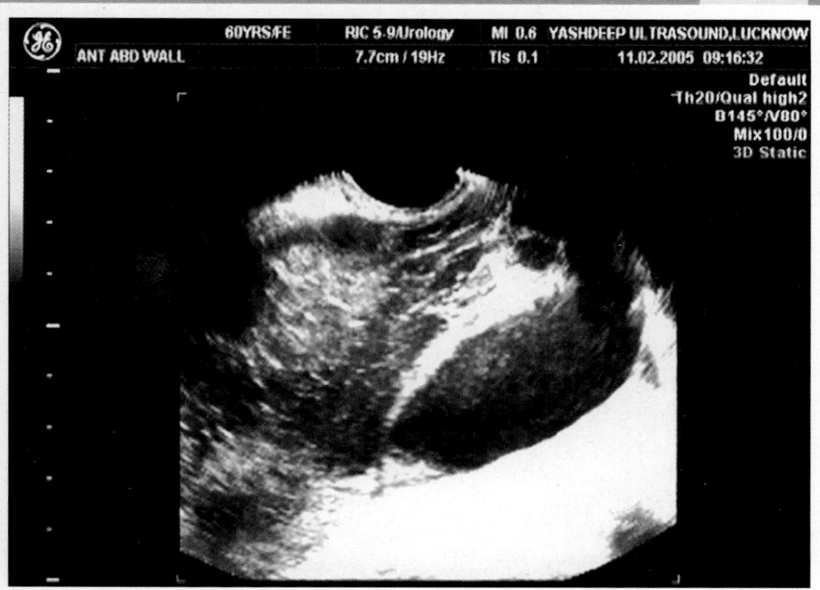

FIGURE 11.110: 3D imaging of the same patient shows the collection confined to the anterior abdominal wall.

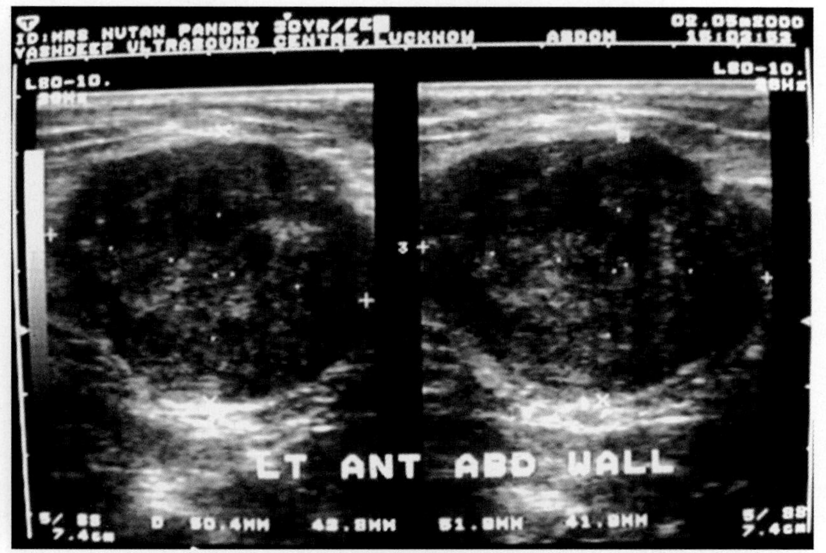

FIGURE 11.111

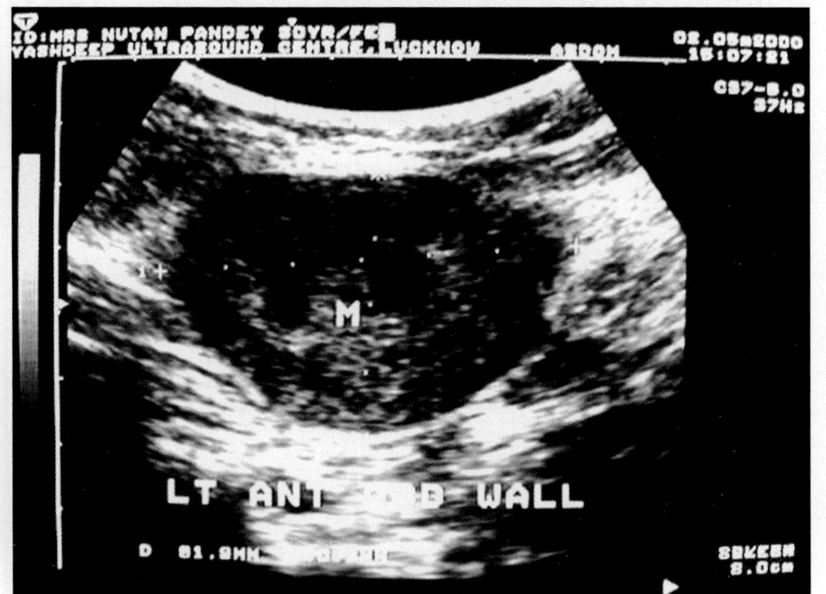

FIGURE 11.112

FIGURES 11.111 and 11.112: Desmoid tumor in anterior abdominal wall. Well-defined homogeneous lobulated soft tissue mass is seen in the muscle of the Lt anterior abdominal wall. No necrosis is seen in it. No cystic degeneration is seen.

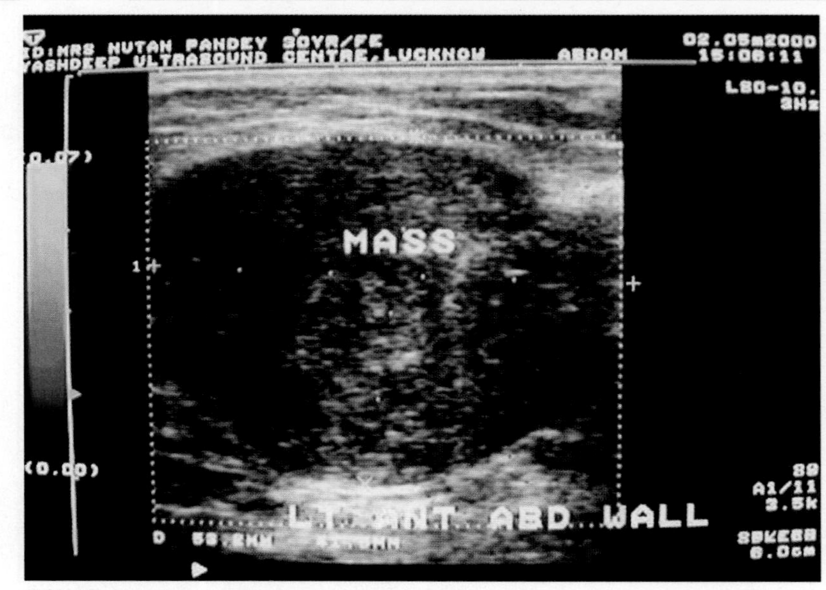

FIGURE 11.113: On color flow imaging poor flow is seen in the mass. Biopsy of the mass shows desmoid tumor.

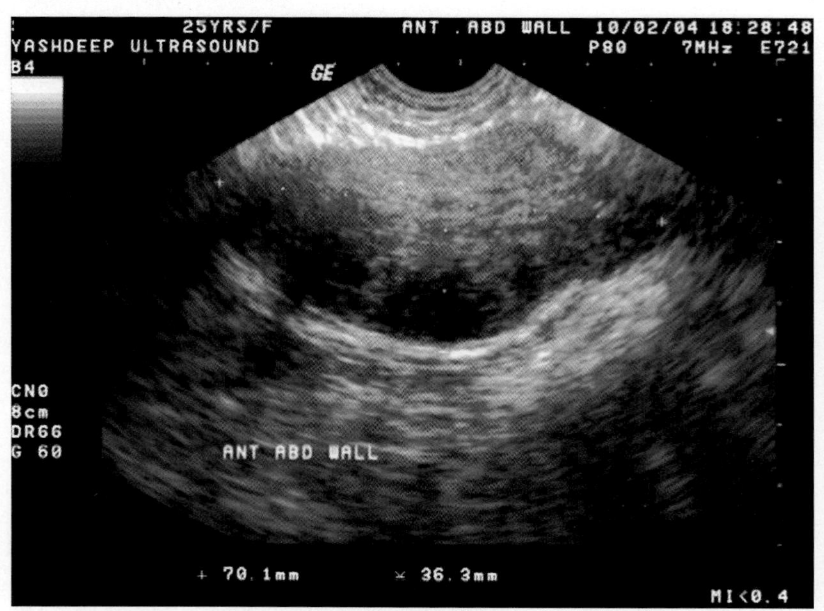

FIGURE 11.114

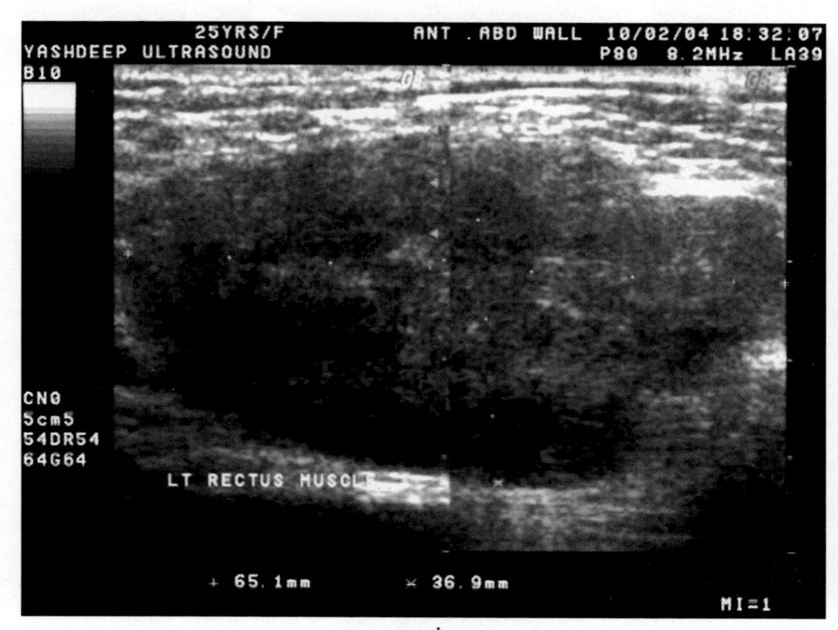

FIGURE 11.115A

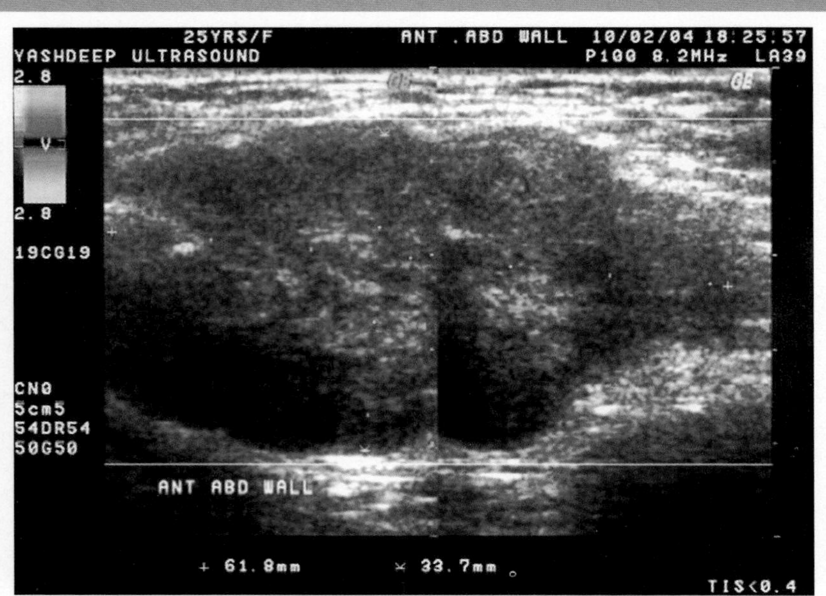

FIGURE 11.115B

FIGURES 11.114 and 11.115: Big desmoid tumor in the anterior abdominal wall. A patient presented with big mass in the anterior abdominal wall. HRSG shows a well-defined lobulated mass in the anterior abdominal wall. The mass is homogeneous in texture. No calcification or necrosis is seen.

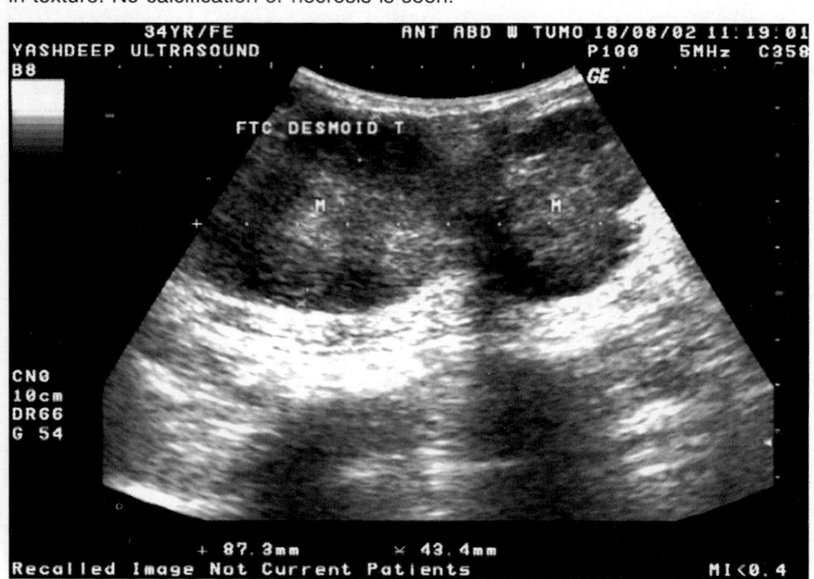

FIGURE 11.116A

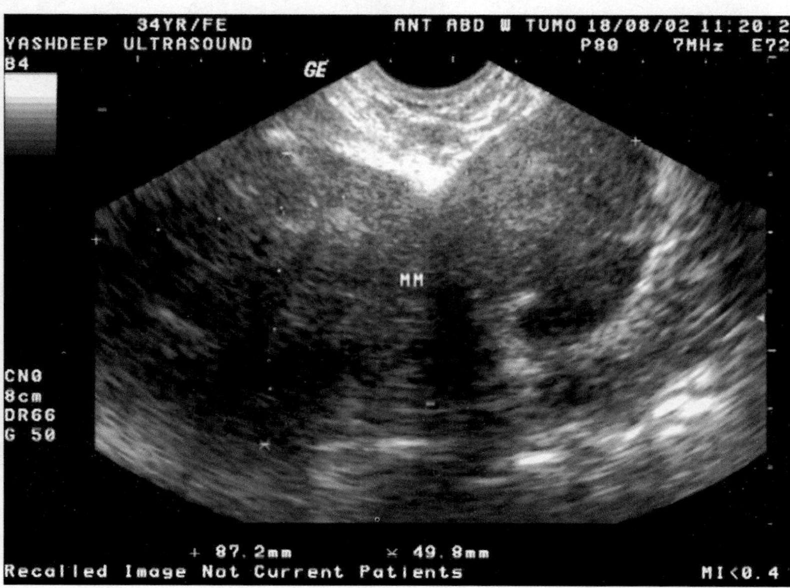

FIGURE 11.116B

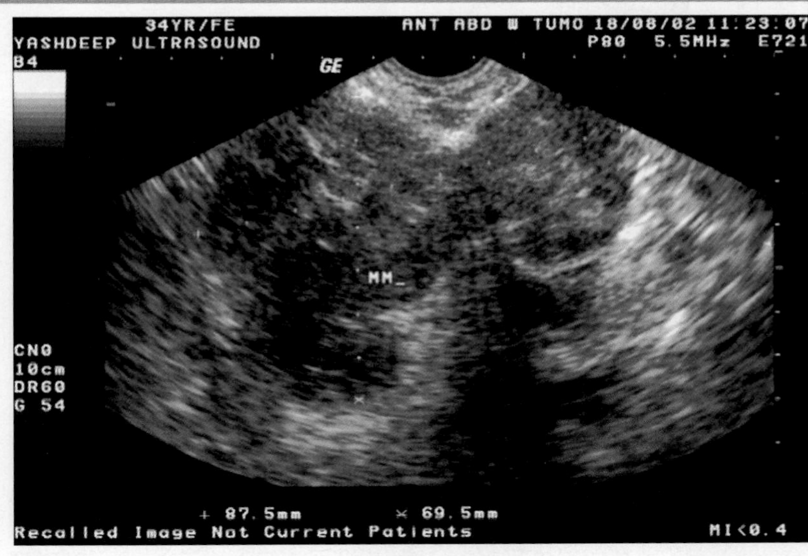

FIGURE 11.116C

FIGURES 11.116A to C: On color flow imaging no flow was seen in the mass. Biopsy of the mass confirmed desmoid tumor.

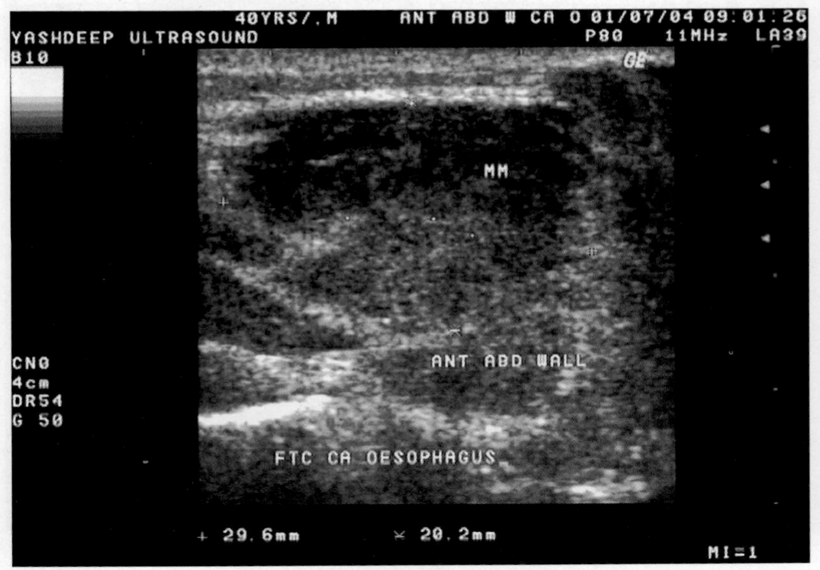

FIGURE 11.117

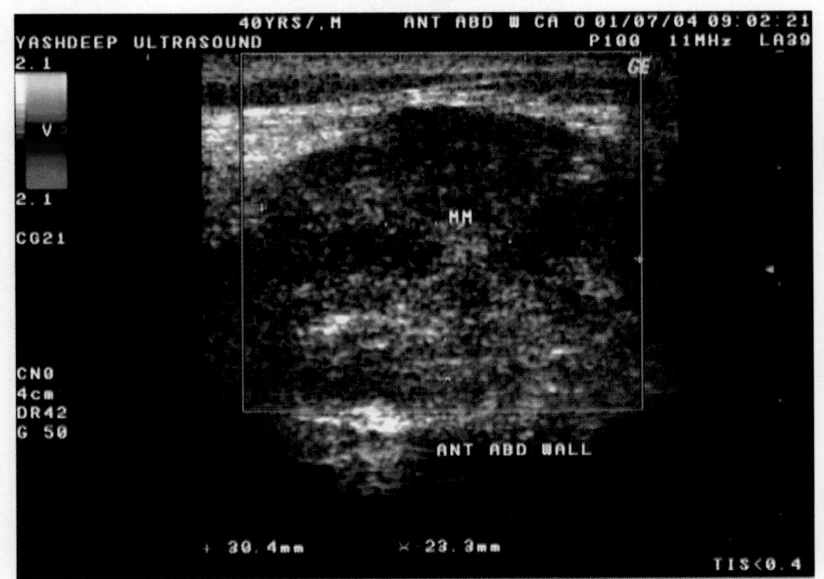

FIGURE 11.118

FIGURES 11.117 and 11.118: Metastatic mass from CA esophagus. A known patient of CA esophagus presented with a soft tissue mass in epigastric region. The mass is irregular in outline with heterogeneous texture. Biopsy of the mass confirmed metastatic deposits. On color flow imaging no flow is seen in the mass.

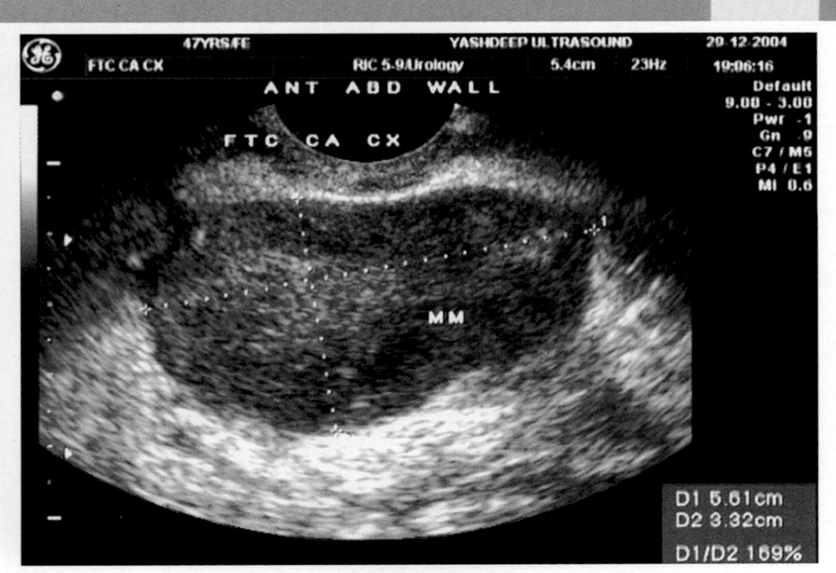

FIGURE 11.119

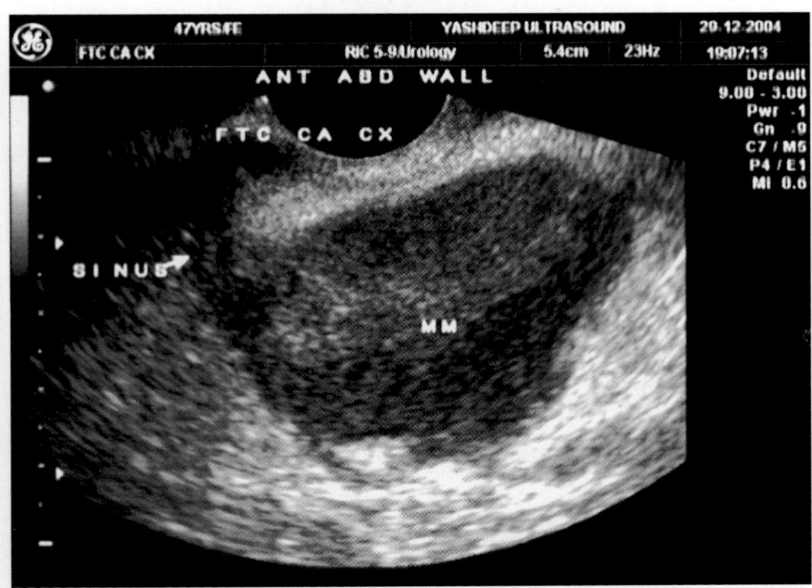

FIGURE 11.120

FIGURES 11.119 and 11.120: Metastatic mass from CA cervix. A known patient of CA cervix presented with soft tissue mass in supra pubic area in the anterior abdominal wall. HRSG shows a well-defined irregular mass in the anterior abdominal wall. The mass was irregular in outline but homogeneous in texture. Biopsy of the mass confirmed metastatic deposits.

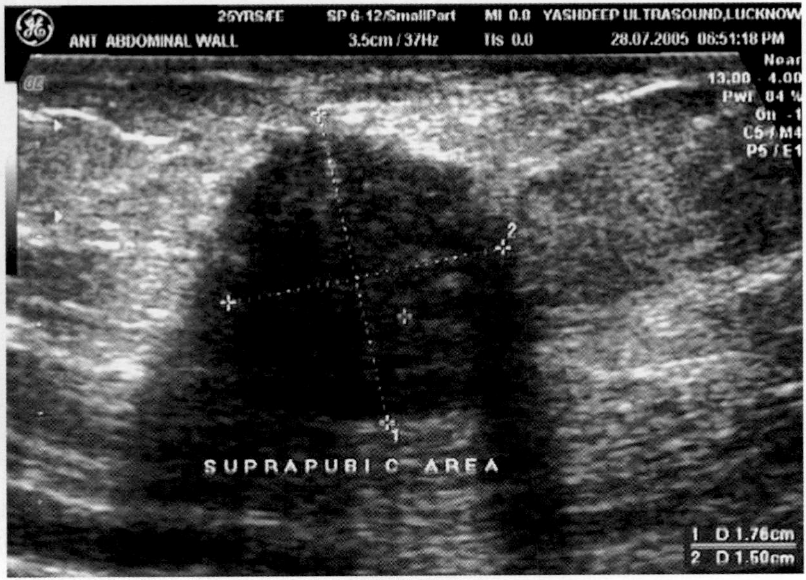

FIGURE 11.121

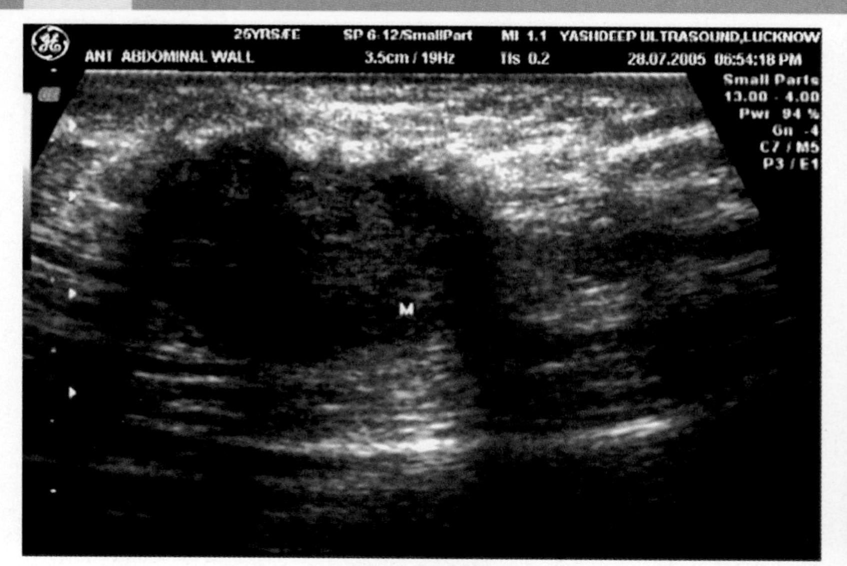

FIGURE 11.122

FIGURES 11.121 and 11.122: Metastatic mass in CA cervix. An operated patient of CA cervix presented with soft tissue mass in the supra pubic area. HRSG shows an irregular heterogeneous mass in the supra pubic area. It was hypoechoic in texture. No necrosis was seen in the mass. No cystic degeneration was seen. Biopsy of the mass confirmed metastatic mass.

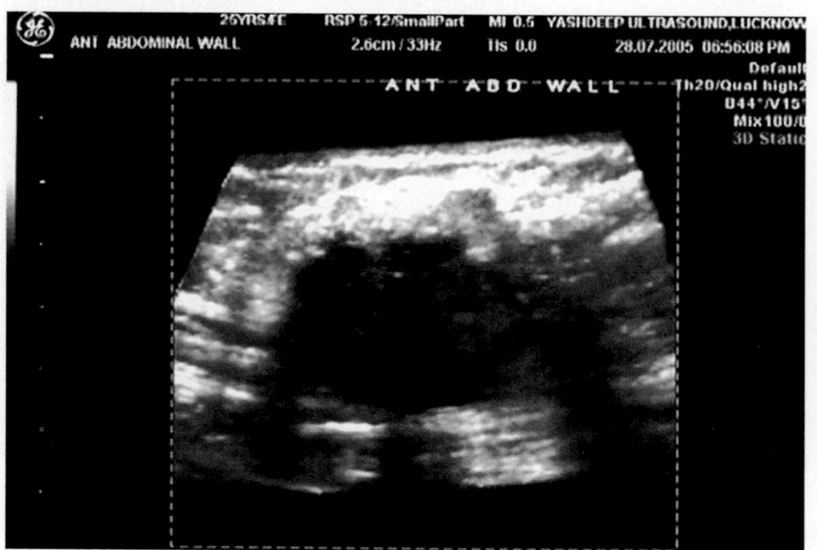

FIGURE 11.123: 3D imaging of the mass shows the mass was confined in the parietes.

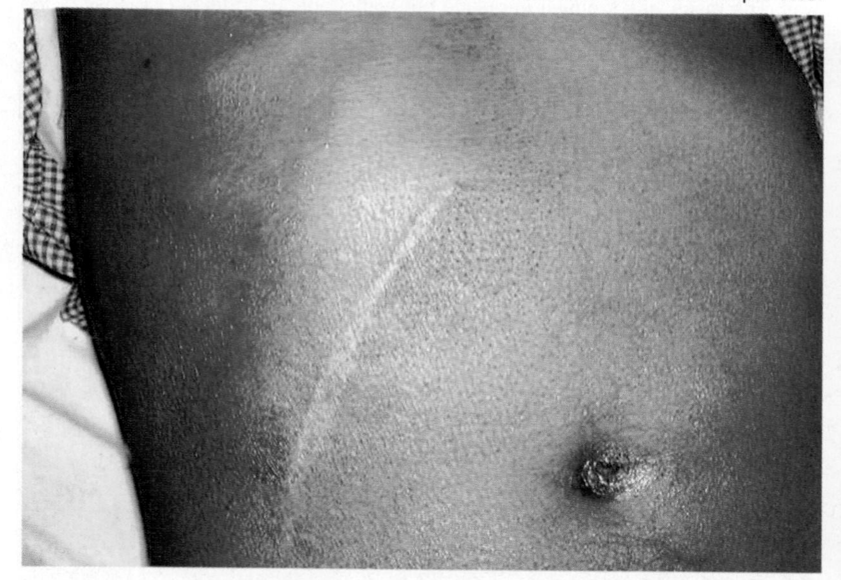

FIGURE 11.124: Scar recurrence of operated carcinoma GB. A young man operated for carcinoma gall bladder presented with soft tissue bulge along the suture line.

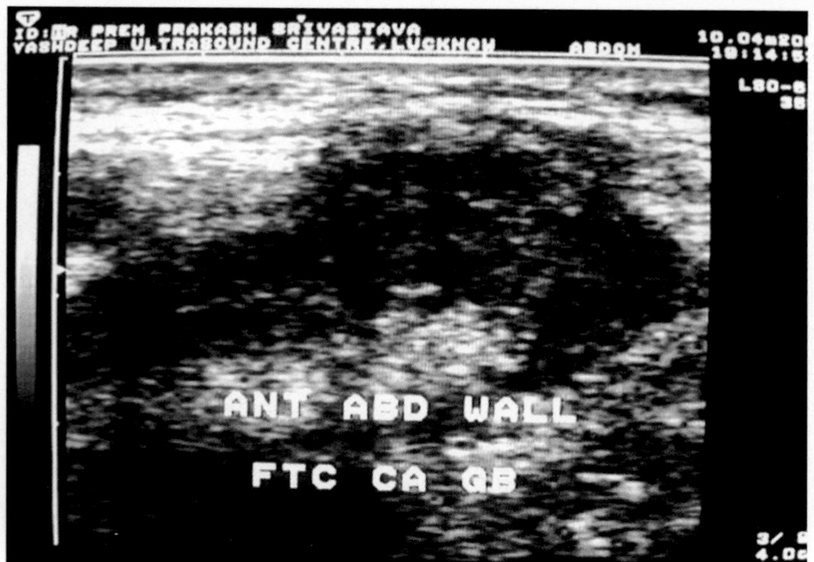

FIGURE 11.125

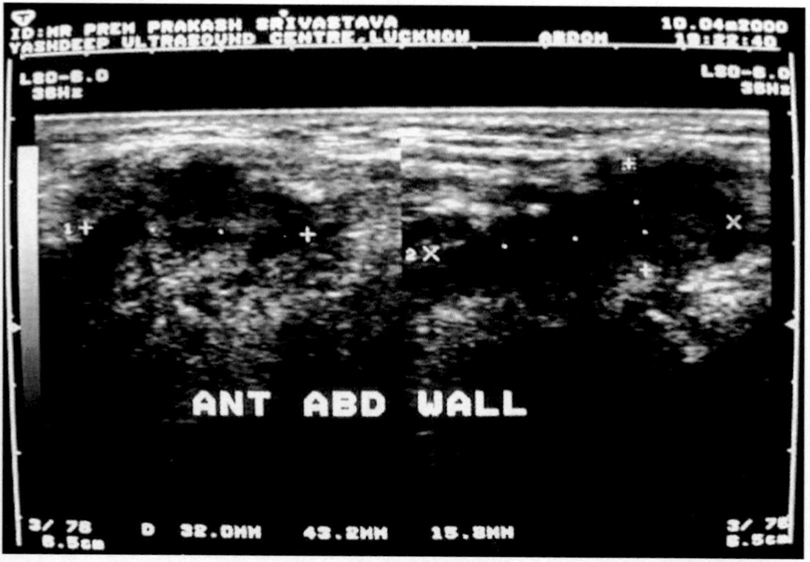

FIGURE 11.126

FIGURES 11.125 and 11.126: HRSG shows infiltrating metastatic deposits in the anterior abdominal wall and along the suture line suggestive of scar recurrence. The mass is also seen infiltrating the parietal layer of the abdominal wall and going posteriorly.

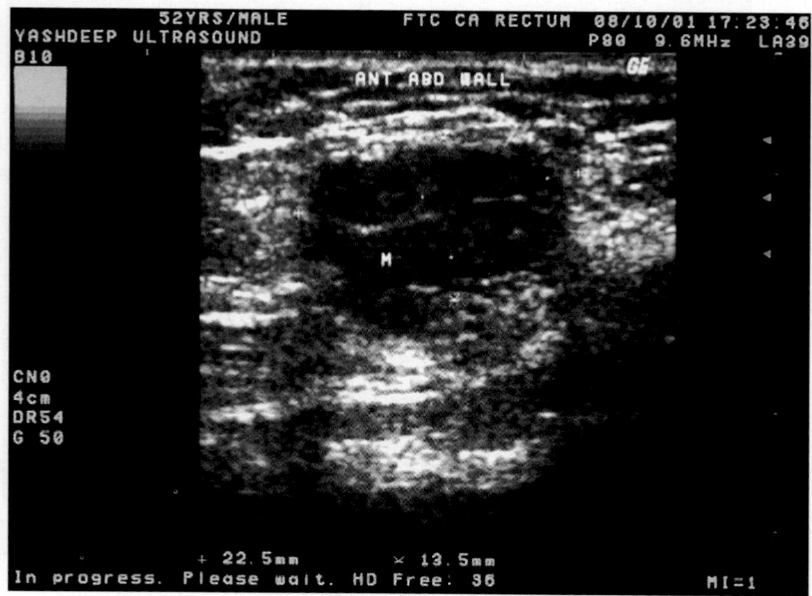

FIGURE 11.127

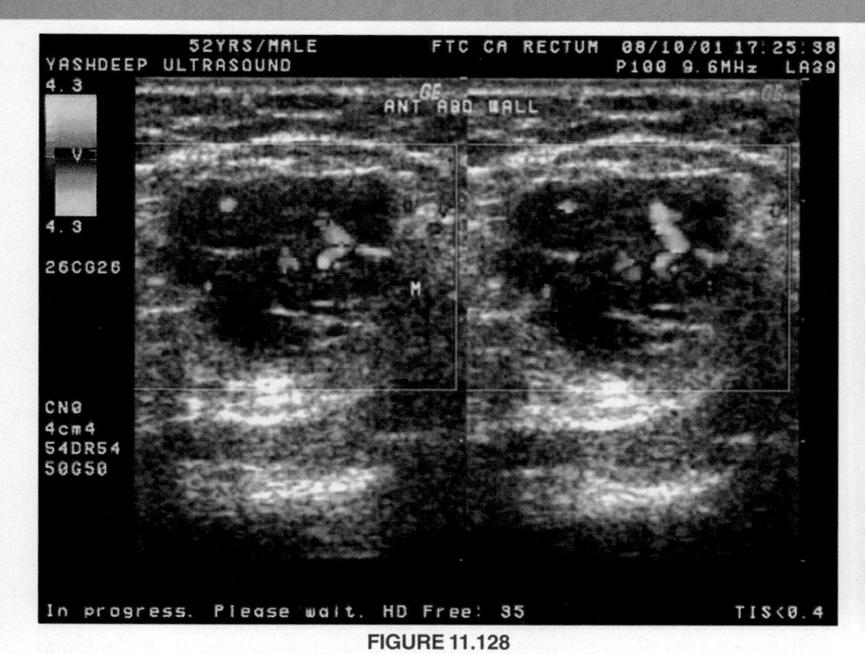

FIGURE 11.128

FIGURES 11.127 and 11.128: Metastatic deposits in carcinoma rectum. Irregular hypoechoic nodular masses are seen infiltrating anterior abdominal wall in a known case of carcinoma rectum. On color flow imaging moderate flow is seen in the mass.

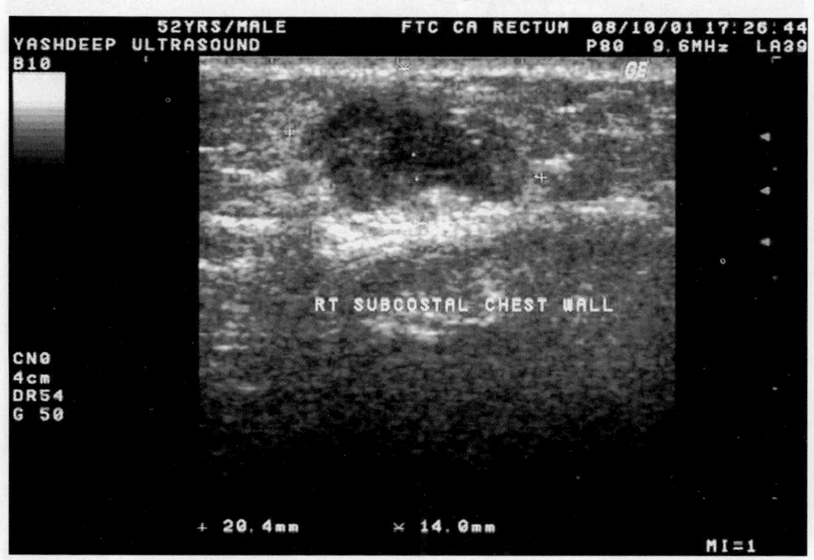

FIGURE 11.129: The metastatic deposit is also seen in Lt subcostal area in the same case.

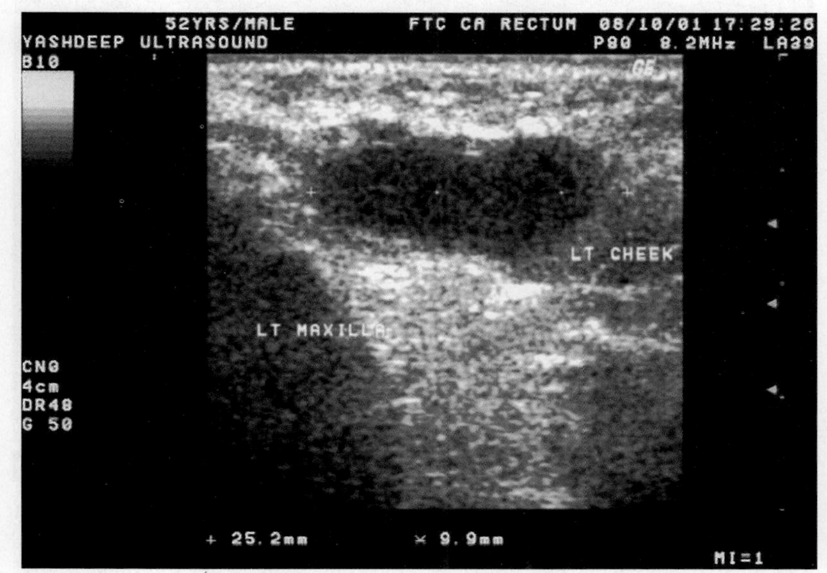

FIGURE 11.130: A hypoechoic irregular mass is also seen over the Lt maxilla suggestive of metastatic deposit in the same case.

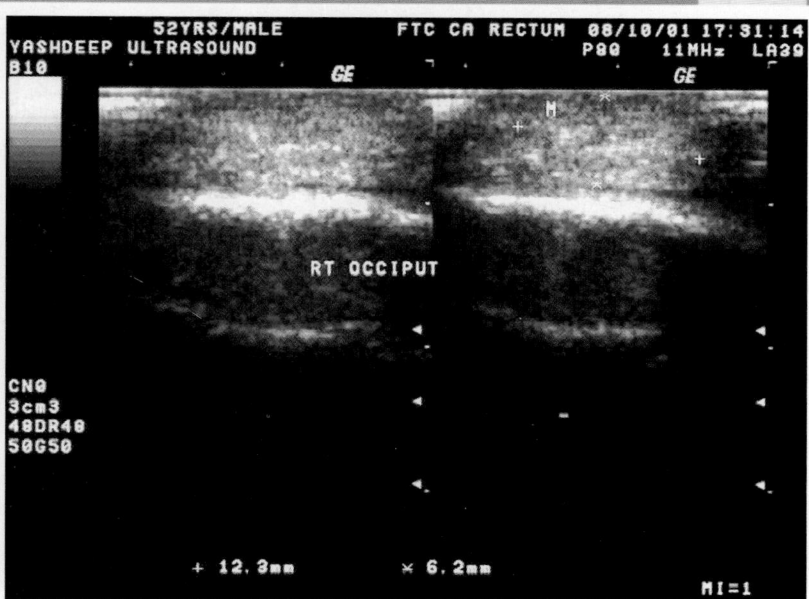

FIGURE 11.131: There is evidence of an echogenic mass seen over the Rt occiput in the same case suggestive of metastasis.

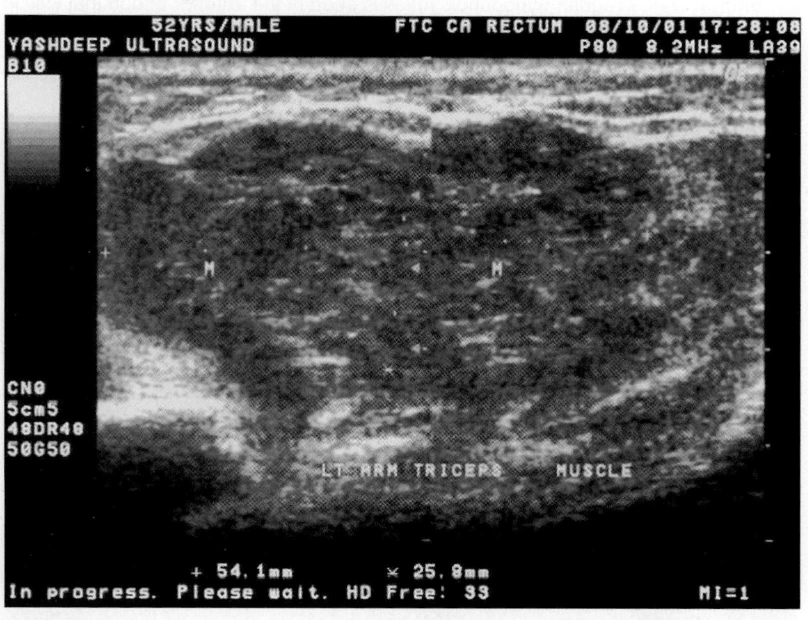

FIGURE 11.132: The muscle infiltration is seen by the metastasis involving the Lt arm triceps muscle. Big irregular mass is seen in the muscle belly distorting the normal muscle texture in the same case of carcinoma rectum.

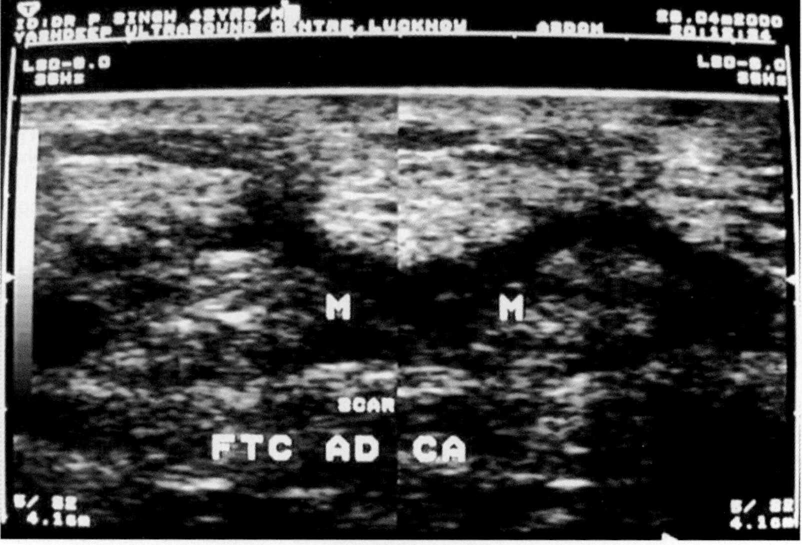

FIGURE 11.133

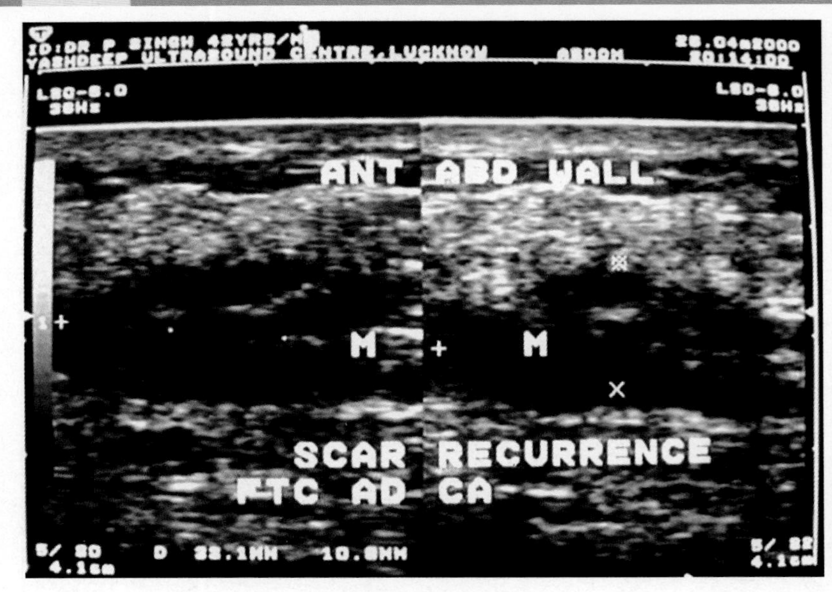

FIGURE 11.134

FIGURES 11.133 and 11.134: Metastasis adenocarcinoma to the anterior abdominal wall. Irregular hypoechoic mass is seen in the anterior abdominal wall in a known case of adenocarcinoma of large bowel. Central area of echogenic focus is seen in the mass. Irregular infiltration of the muscle is seen by the mass.

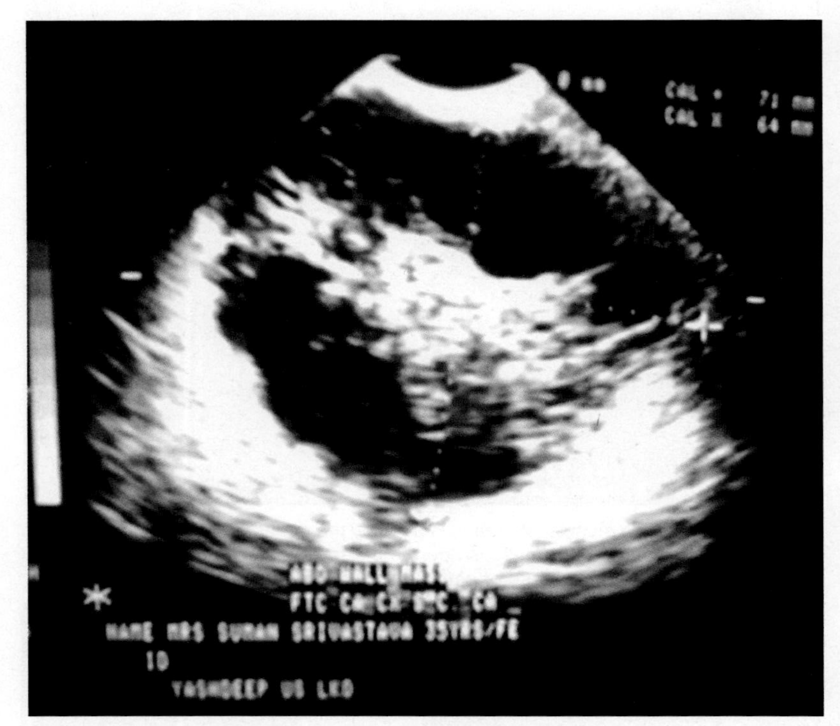

FIGURE 11.135: Metastasis of carcinoma cervix Invading the anterior abdominal wall. A big irregular heterogeneous mass is seen in the supra umbilical area invading the anterior abdominal wall in a known case of carcinoma cervix. The margins are ill-defined. Solid soft tissue component is also seen in the mass. Biopsy of the mass shows metastasis deposits from carcinoma cervix.

Conclusion

Although the pathologies affecting anterior abdominal wall are limited, but it is important to differentiate between abdominal wall and intra-abdominal pathologies as most of the time, intra-abdominal masses mimic as abdominal wall pathologies. HRSG is very good investigation to confirm the localization of mass, its nature and differentiate between intra-abdominal and abdominal mass.

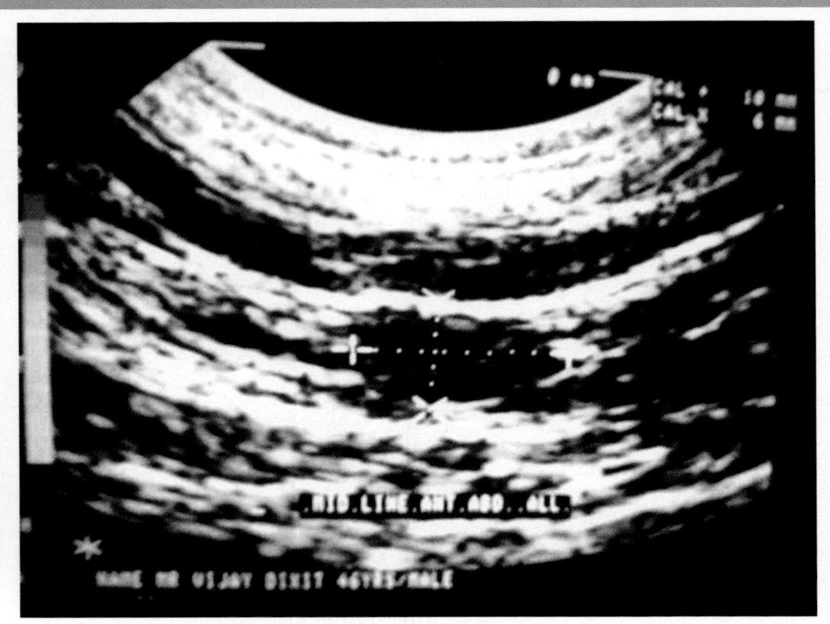

FIGURE 11.136: Metastasis deposits from carcinoma breast. An irregular hypoechoic focus is seen in the anterior abdominal wall in a known case of carcinoma male breast. Biopsy of the mass shows metastatic deposits.

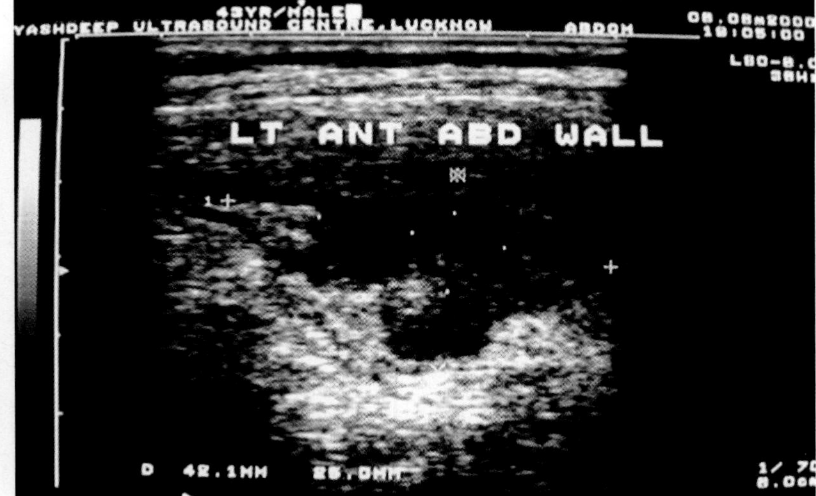

FIGURE 11.137

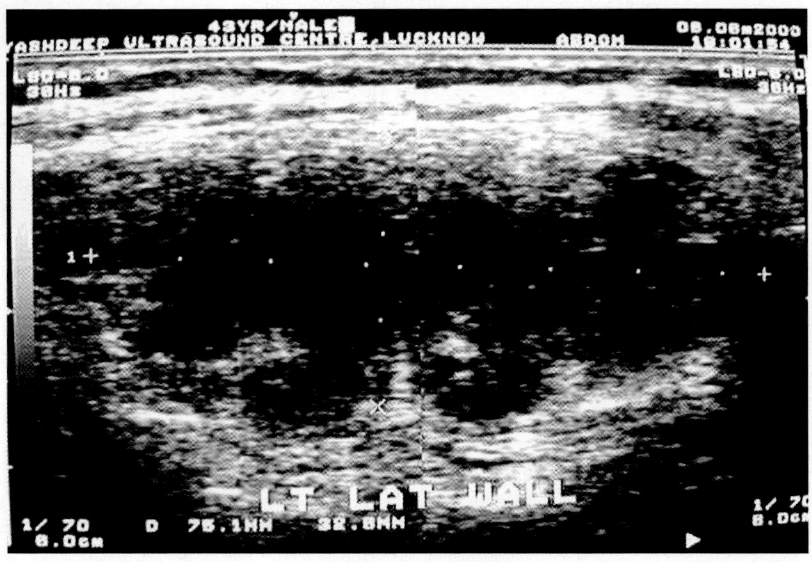

FIGURE 11.138

FIGURES 11.137 and 11.138: Muscle necrosis in the anterior abdominal wall. The anterior abdominal wall muscle shows cystic degeneration of muscle after trauma. Central area of necrosis is seen in it. No loculation or septation is seen. On aspiration serous fluid is seen coming out from the cyst.

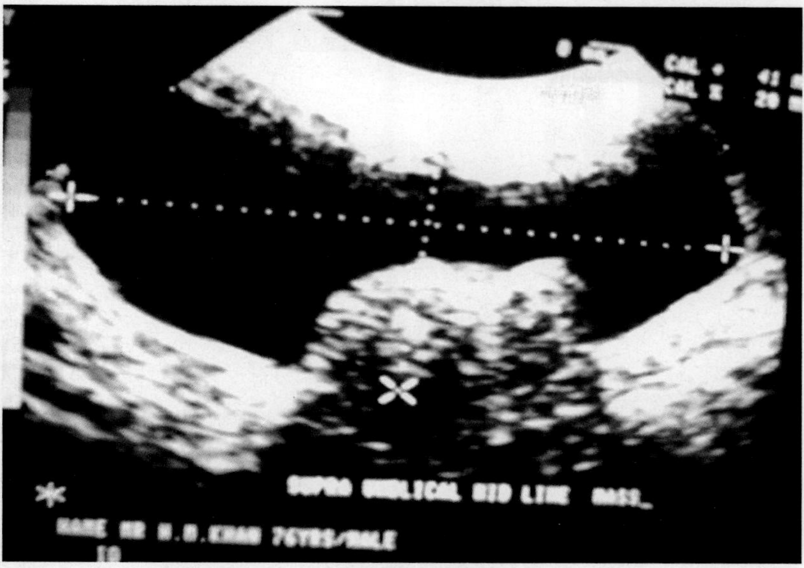

FIGURE 11.139

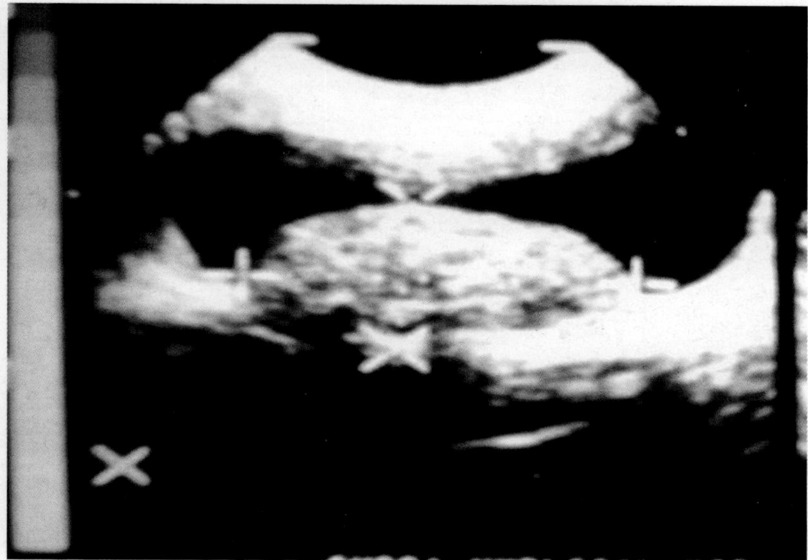

FIGURE 11.140

FIGURES 11.139 and 11.140: Infiltrating adenocarcinoma in the anterior abdominal wall. An echogenic irregular soft tissue mass shadow is seen in the supraumbilical area. The mass is irregular in outline. Fluid is also seen around the mass. Biopsy of the mass shows infiltrating adenocarcinoma of bowel.

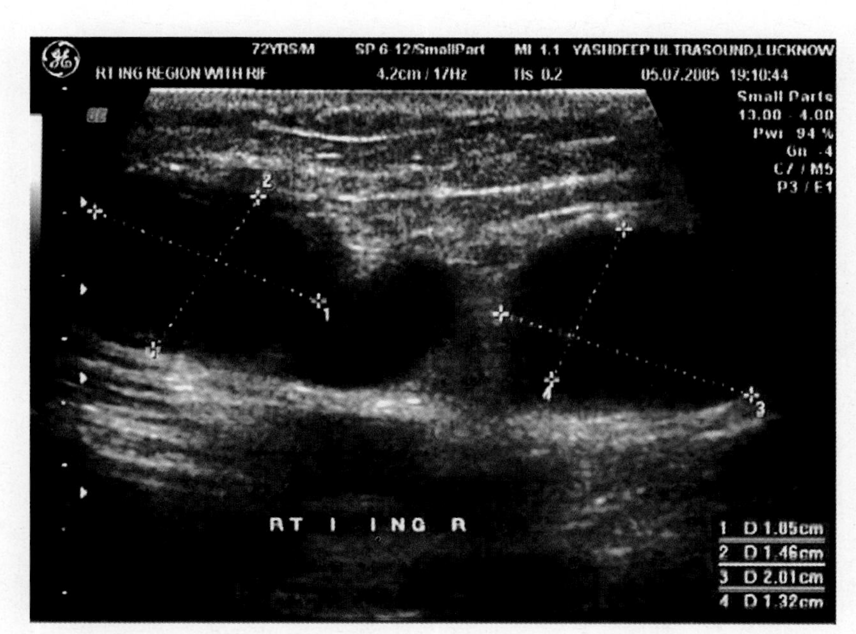

FIGURE 11.141: Malignant nodes in Rt inguinal region. HRSG shows enlarged nodal masses in Rt inguinal region in a case of CA penis. The nodes are hypoechoic in texture. Lymph node hylum is lost. Normal anatomy is distorted.

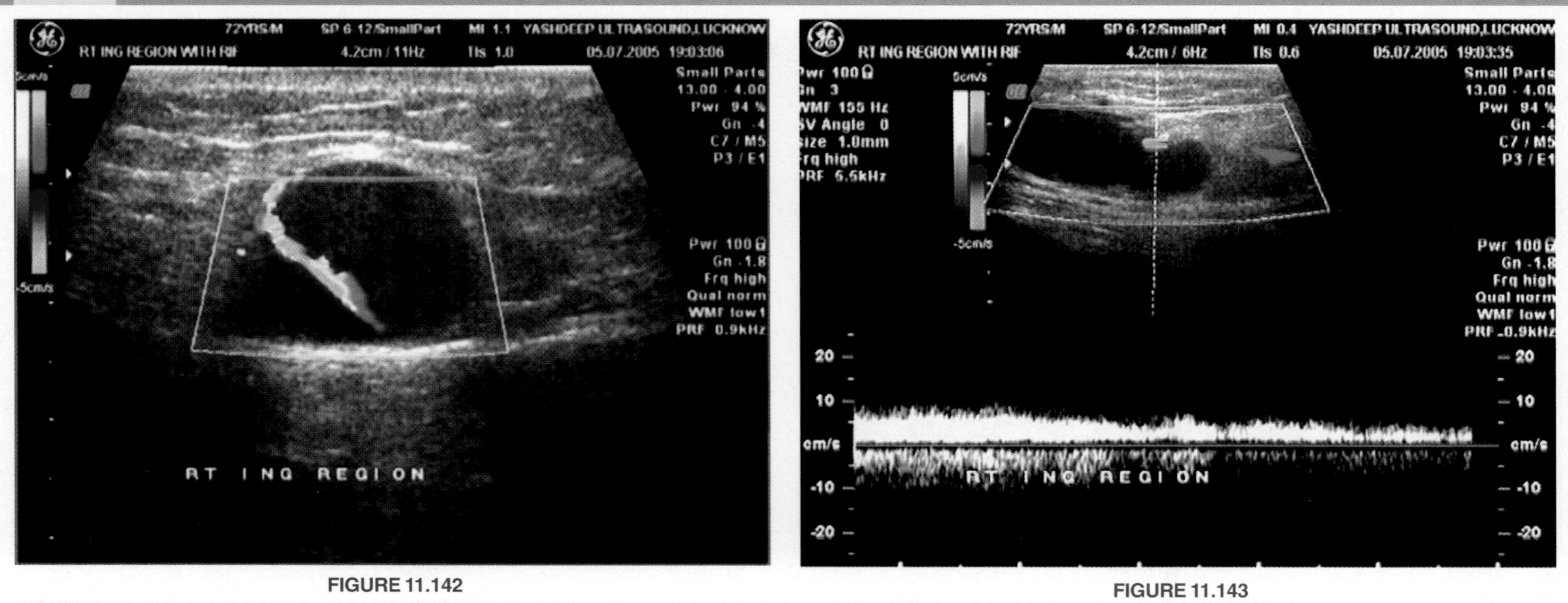

FIGURE 11.142

FIGURE 11.143

FIGURES 11.142 and 11.143: On color flow imaging a big vein is seen feeding the node. On spectral Doppler tracing venous flow was seen in the vein.

12 Penis

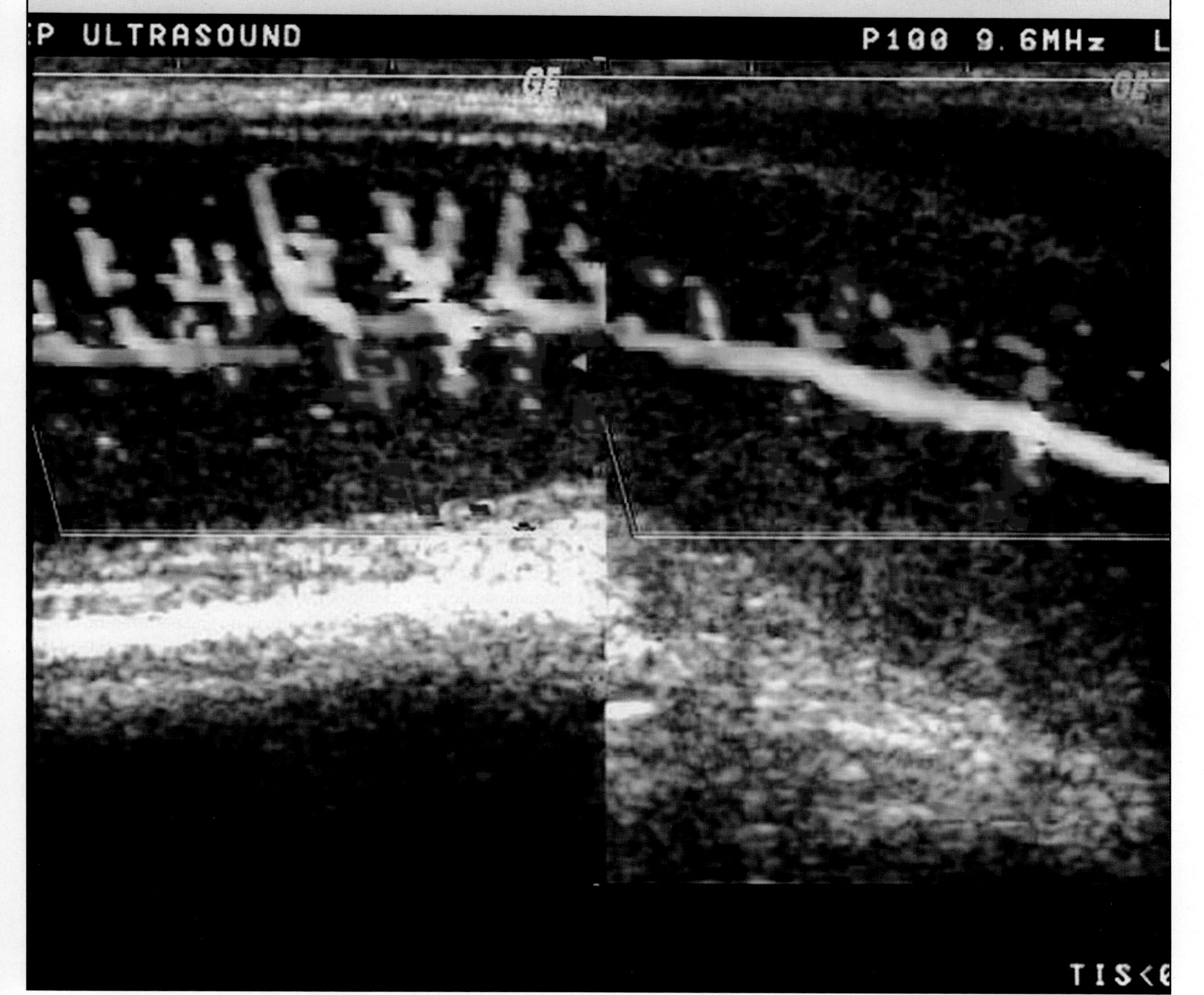

Penis

Introduction

Penis, the male genital organ is an important part of the body to do dual role of providing excretion of urine and also semen, vital for natural human reproduction. In the past, investigation of penis was limited to plain X-ray films, Micturating or retrograde urethrography and cavernography. Recently high resolution sonography of penis and Doppler flow imaging of penile vasculature and blood vessels have made study of penile anatomy, physiological, pathological erectile functional study a practical proposition. Sonourethrography with the saline solution or water as contrast agents, has made HRSG an exciting modality for assessment of penile urethra. Assessment of penile erectile dysfunction with color Doppler study is the most exciting development in field of impotence.

Anatomy

Penis has three major components. There are two paired dorsal columns of corpora cavernosa and one ventral column of corpus spongiosum. They are highly vascular structures and having paired dorsal arteries lying superficially and two deep arteries each running through corpora cavernosa known as cavernosal arteries. These corpora are covered with a much dense connective tissue known as tunica albuginea. The penile urethra runs through the penile shaft through the corpora spongiosa. The venous drainage takes place through the corpora and dorsal vein of penis. It is situated between the two dorsal arteries. HRSG shows normal corpora cavernosa as medium level echo complex bodies running longitudinally. They are best assessed in cross-section. In normal situation low blood flow is seen in the cavernosal arteries. The corpora spongiosum is less echogenic than corpora cavernosa. The urethral lumen can be seen running through it.

The tunica albuginea is thin echogenic layer seen around the corpora and can be easily resolved on HRSG.

Pathological Conditions

Peyronie's Disease

This is an idiopathic condition usually seen in 4th decade onward. There is deposition of fibrous and calcified plaques in corpora cavernosa and tunica albuginea. HRSG shows these plaques. When it is calcified it is accompanied with acoustic shadowing. These plaques can be better assessed and resolved on HRSG than clinical examination and it is an important investigation prior to the surgical removal of the plaque. HRSG can specially tells about the number of plaques, extent of fibrosis and tunica involvement by the disease.

Penile Carcinoma

Carcinoma of penis is epithelial carcinoma. It is almost found in the distal part of penis. The incidence of malignancy is lower in the population where circumcision is common practice. The malignancy is having four states:

State 1: The malignant lesion is confined to fore skin or glans only.

Stage 2: The tumor invades the corpora cavernosa and penile shaft.

Stage 3: The tumor invades the shaft with local spread to lymph nodes.

Stage 4: The tumor invades the shaft and having distant metastasis to para-aortic region.

Almost 50% patients report when tumor spreads to regional inguinal lymph nodes HRSG can very well tell the extent of the primary tumor and its invasion to corporal tissue. Distant metastasis to para-aortic region or to other distant location can be easily assessed on HRSG.

Urethra

The male urethra is having three parts:

1. Prostatic
2. Membranous
3. Penile or spongy urethra.

Posterior urethra is about 3 cm long and seen passing through prostate gland. It is also known as prostatic part of urethra. The membranous urethra is about 1.5 cm in length and narrowest part of urethra. It lies at the level of external urethral sphincter. The penile urethra is the longest part of urethra and measures 15-20 cm in length. It is about 6 mm in diameter. The proximal bulbar and intrabulbar part and the distal part of urethra near the meatus show slight dilatation. HRSG shows all the three part of urethra. Prostatic part of urethra is seen through transabdominal, transperineal approach. However, it is best seen with transrectal approach.

The penile urethra is seen with high frequency transducers having frequency from 7 to 10 MHz. It is seen with direct contact scanning technique. Urethrography is done to evaluate rupture of urethra, urethral fistula and urethral structures. It is done with 8 no. Foley's catheter with balloon inflated in the navicular fossa. The antegrade urethrography is done by asking the patient to void the urine against distally blocked urethral meatus. The retrograde urethrography is done by pushing sterile saline solution through the Foley's catheter. The urethral pathology can very well be evaluated on HRSG. The urethral structure can be identified. Extent of the fibrosis in the surrounding corpus spongiosum can be easily evaluated on HRSG than by conventional X-ray urethrography. HRSG shows narrowing of the lumen with structure of urethra. The fibrosis is seen as an irregular echogenic shadow extending or involving in the spongiosum. It is also helpful in assessing the urethral diverticulam and periurethral abscess formation. Small urethral calculi in the diverticula can be easily picked on HRSG.

Impotence

High-resolution sonography and color flow imaging plays an important role in assessment of pathological causes of impotence. Till recently impotence was thought to be a psychological problem. However, studies show that a majority of the cases were having pathological problems mainly vascular in origin like arterial insufficiency or venous leak. High-resolution sonography plays an important role in assessment in both types of vasculogenic impotence. The duplex color flow imaging is an accurate method for the assessment of arterial flow in the dorsal penile artery, cavernosal artery and venous flow in dorsal penile vein. The arterial flow study can be done before and after injecting vasodilating agents like papaverine injection. The longitudinal examination of corpora cavernosa is done in parasagittal axis. In a flaccid stage, cavernosal artery can be seen in a tortuous course. However, in erect stage it becomes straight. The diameter of cavernosal arteries can also be assessed. 60 mg of papaverine is injected in either of corpora cavernosum. The drug easily disperses from one corpus cavernosum to other. The maximum effect of vasodilating agent is achieved between 5 and 20 minutes after the injection. However, in standard protocol, the study is carried out up to 45 minutes after the injection depending upon the patient response. The velocity measurements are taken at 5 minutes interval for at least 30 minutes after the injection. The normal peak systolic and diastolic measurements should be recorded in both cavernosal arteries at 5 minutes interval.

There are 5 phases of spectral waveform in the cavernosal arteries after the injecting. These 5 phases are seen within 5 to 20 minutes after the injection. However, in a patient having vasculogenic impotence, all the normal 5 phases cannot be achieved due to arterial insufficiency or venous leak.

Arteriogenic Impotence

The normal peak systolic velocity which is achieved after intra cavernosal injection of vasodilating agent is >30 cm/sec. Most of the patients are having moderate to good response on injection have peak systolic velocity >25 cm/sec. However, those patient who are having arterial insufficiency, the peak systolic velocity is <25 cm/sec. This indicates an arterial disease. The peak systolic value between 25 and 30 cm per second should be considered as borderline response. However, value more than 30 cm/sec should be taken as normal. If there is difference of 10 cm/sec between the values of both cavernosal arteries, unilateral arterial disease should be suspected. The diastolic flow in the cavernosal artery may be reversed and should be taken as normal phenomena in latter stages of erection. Collateral vessels to the cavernosal artery may present as a normal variant.

Vengeance Impotence

The vasculogenic impotence may be due to excessive venous leak from the corpora cavernosa. Although the exact cause of this excessive venous leak is not known, however, it is assumed that tunica albuginea stretches beyond the normal limit thus cannot obliterate the emissary veins draining the sinusoids during the erection. Due to the inadequate compression of emissary veins the venous outflow continuous and erection is never fully obtained. A venous leak is suspected when persistent high diastolic flow is seen in the cavernosal arteries and the diastolic flow is >3 cm/sec.

Conclusion

HRSG and color flow imaging are an excellent modality for evaluation of penis and its pathological conditions. The biggest advantage of modality is the study of physiological and pathological erectile function of penis in impotence. It is reliable, cost effective and noninvasive in technique for assessment of vasculogenic impotence.

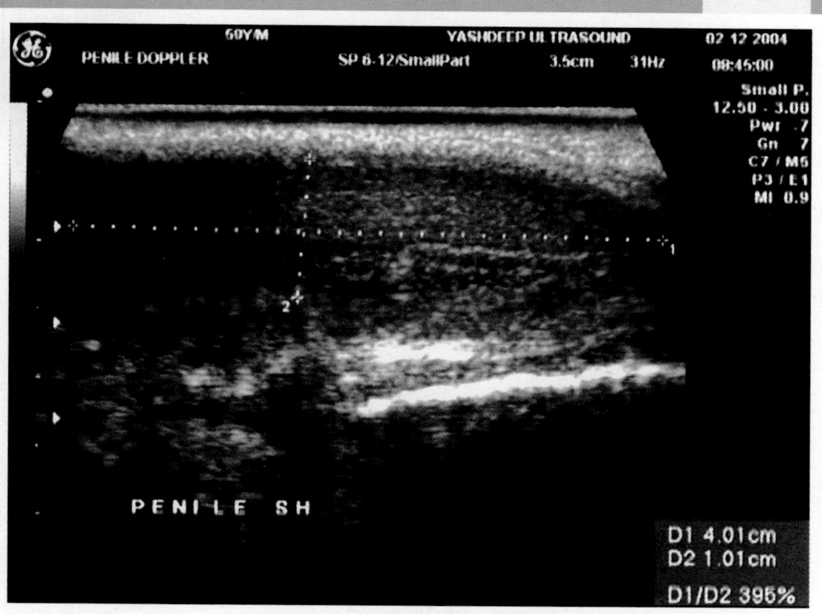

FIGURE 12.1: Normal anatomy (Penile shaft). HRSG shows the normal penile shaft. Homogeneous medial level echoes are seen in the corpora.

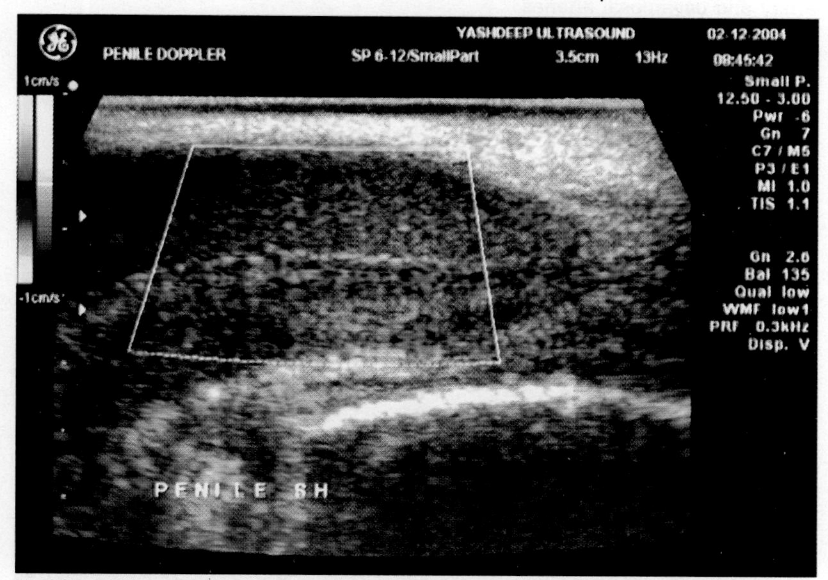

FIGURE 12.2: On color flow imaging very low or no flow is appreciated in penile vessel.

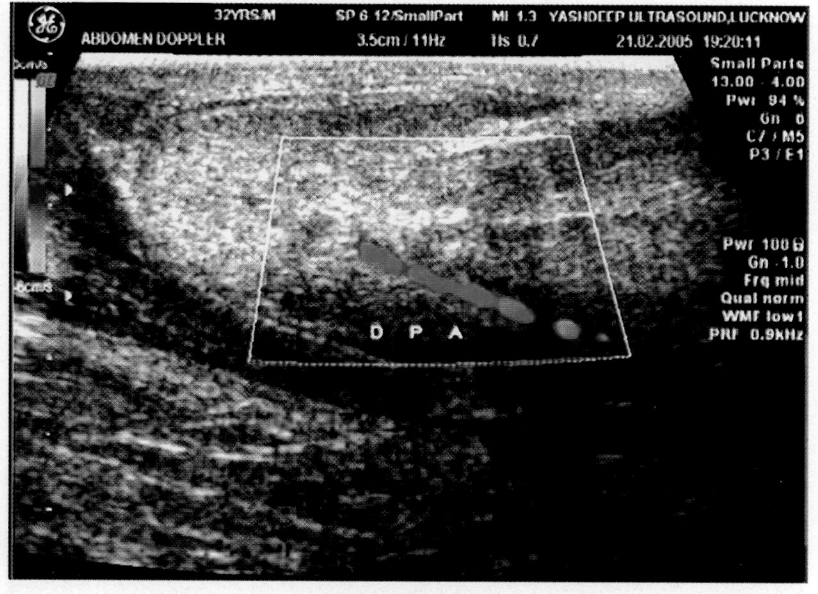

FIGURE 12.3

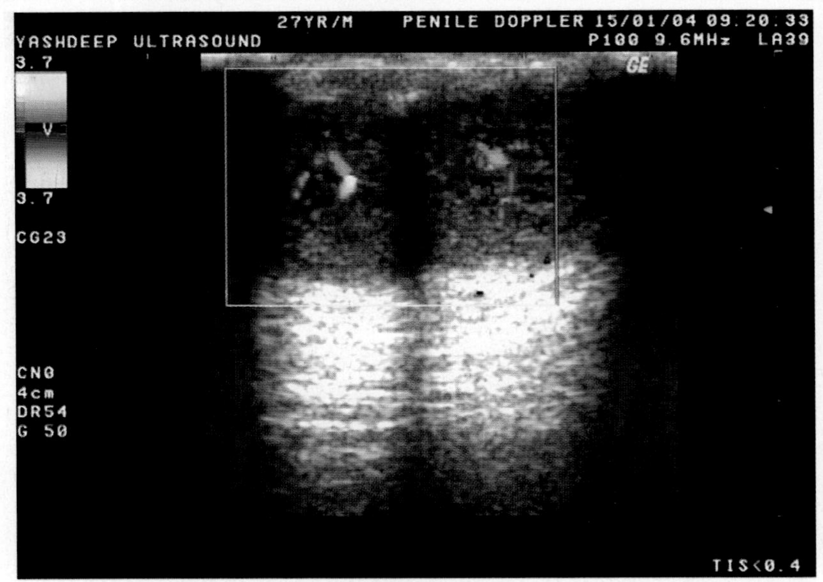

FIGURE 12.4

FIGURES 12.3 and 12.4: Color flow imaging shows low flow in the dorsalis penile artery and cavernosal arteries.

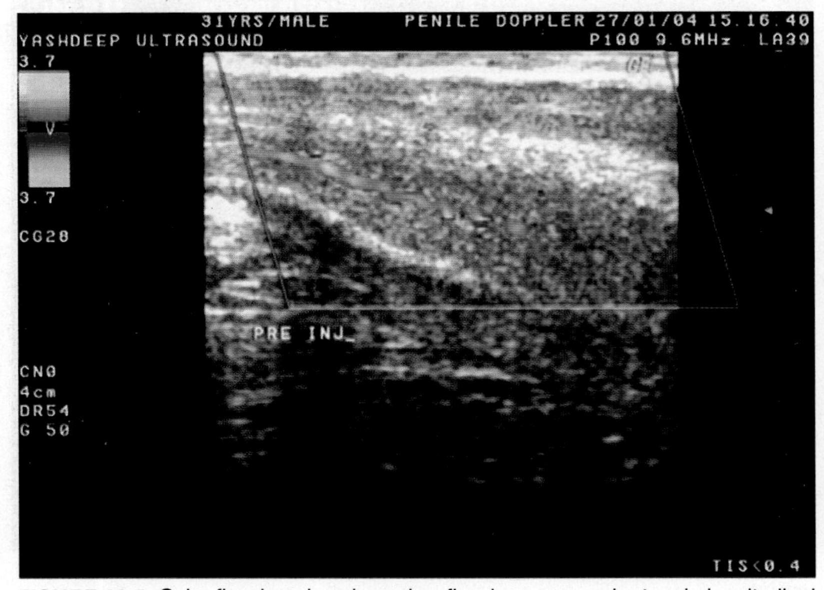

FIGURE 12.5: Color flow imaging shows low flow in cavernosal artery in longitudinal axis.

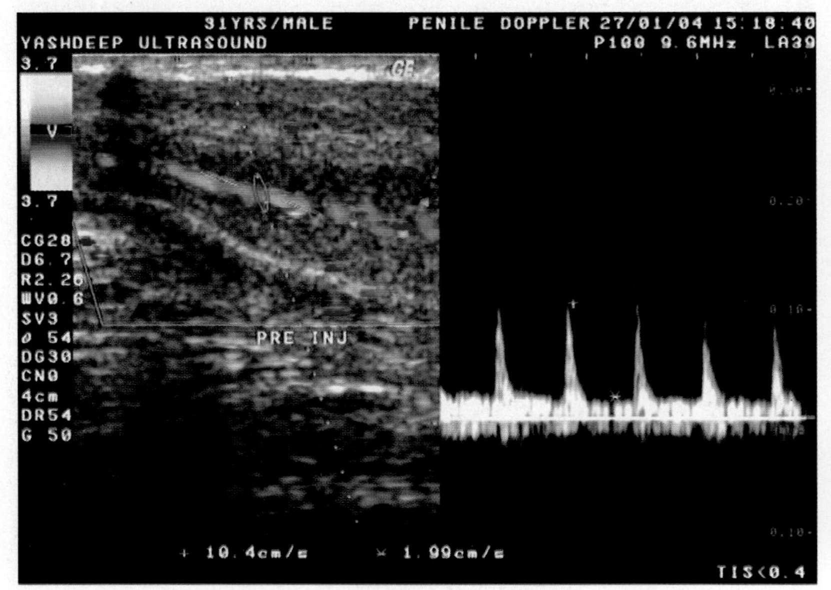

FIGURE 12.6: Spectral Doppler tracing of cavernosal arteries shows the low arterial flow pattern.

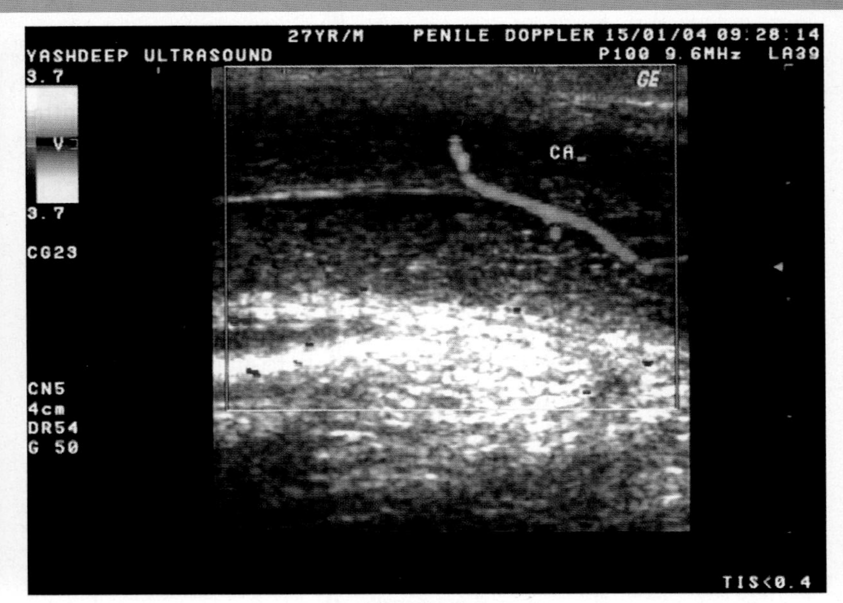

FIGURE 12.7

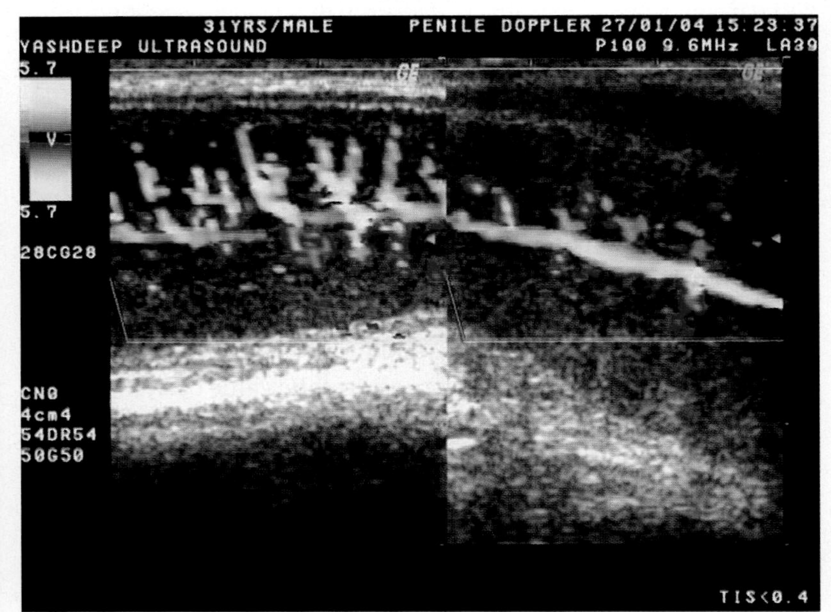

FIGURE 12.8

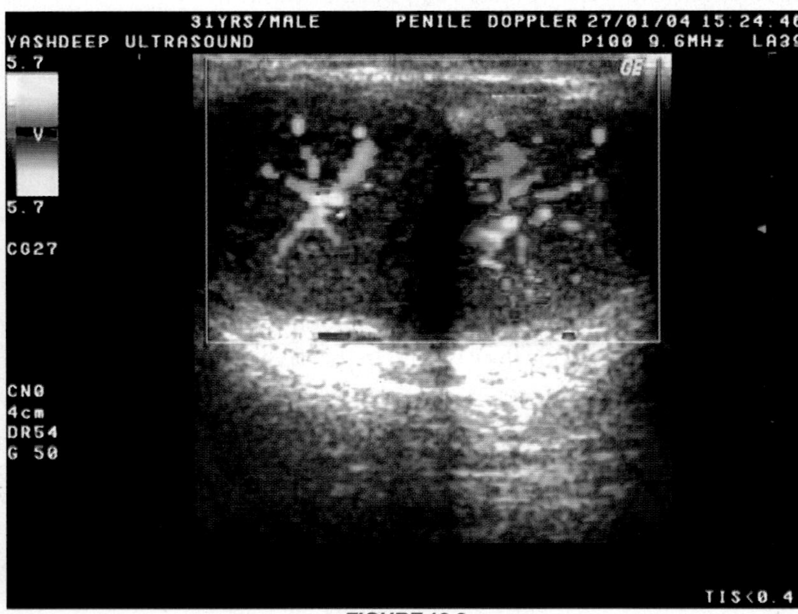

FIGURE 12.9

FIGURES 12.7 to 12.9: Color flow imaging shows multiple small branches coming out from the cavernosal artery and supplying the corpora spongiosa.

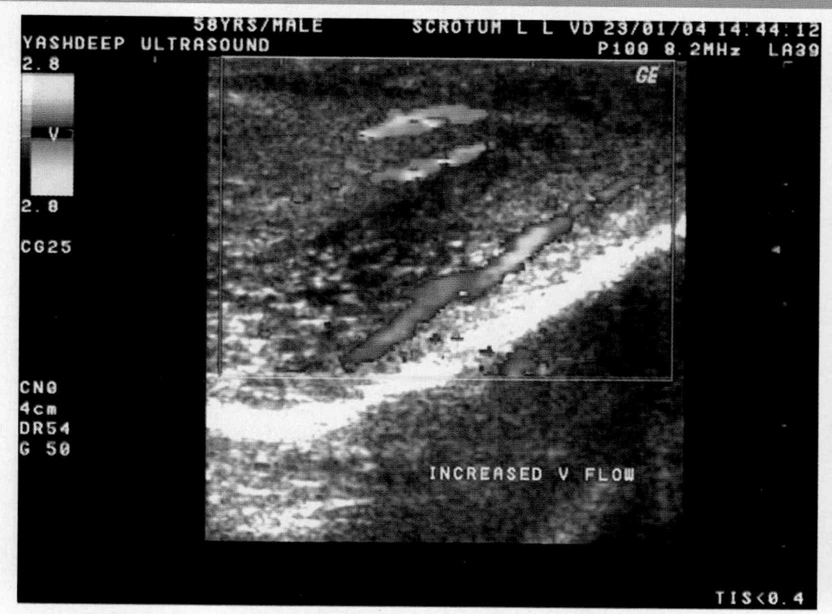

FIGURE 12.10: **The dorsal vein of penis.** HRSG shows dorsal vein of penis running in the substance of penile shaft.

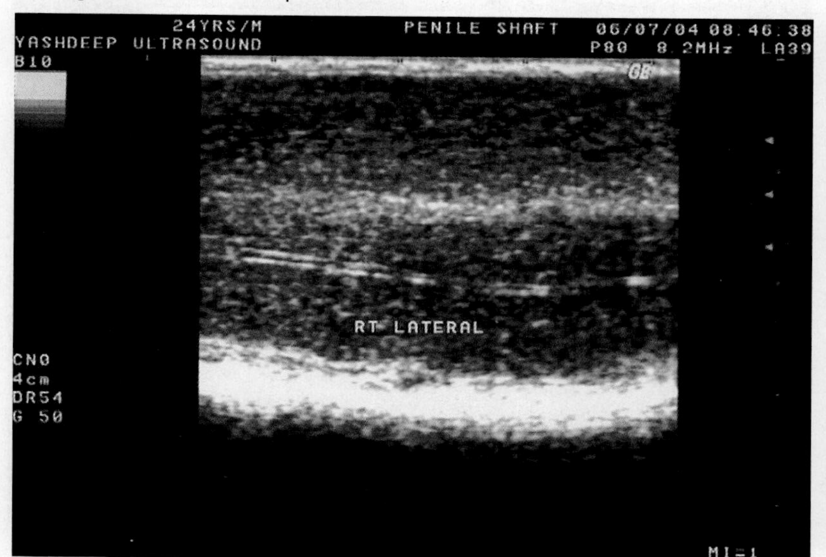

FIGURE 12.11

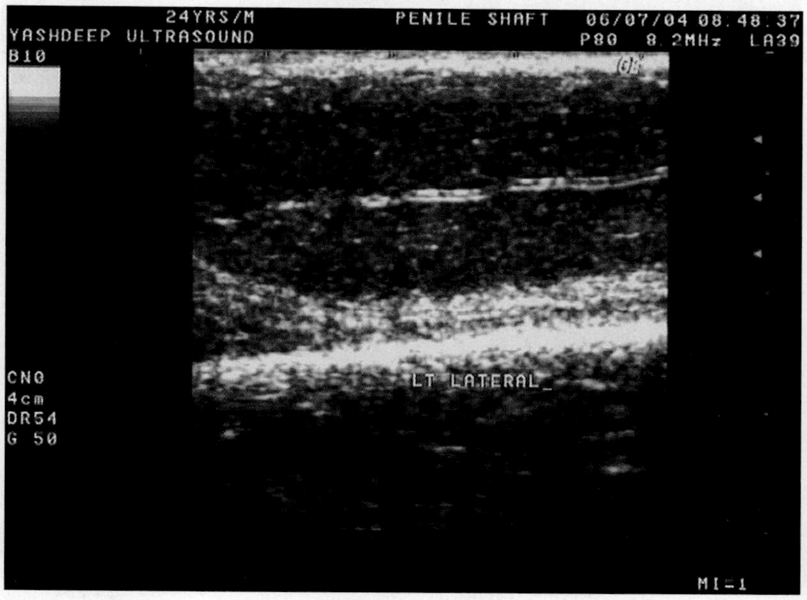

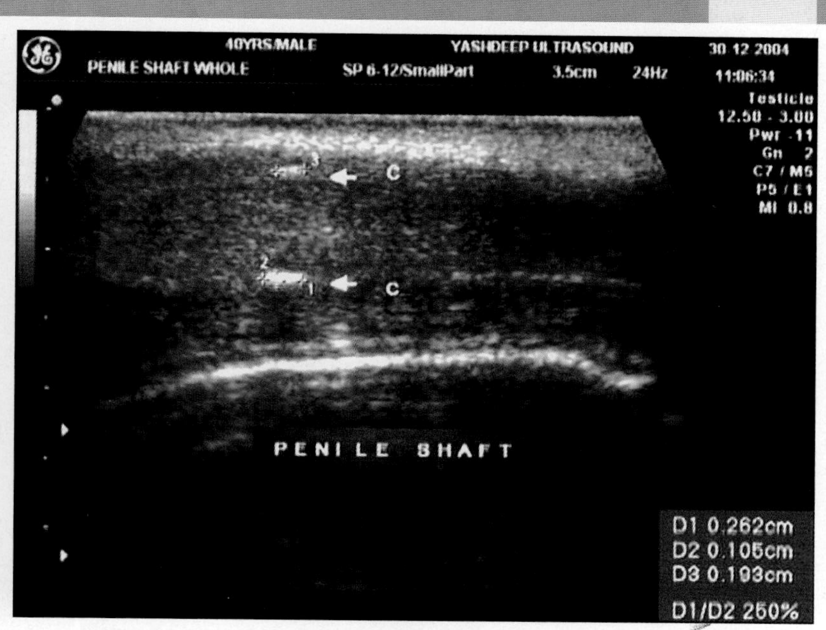

FIGURE 12.13

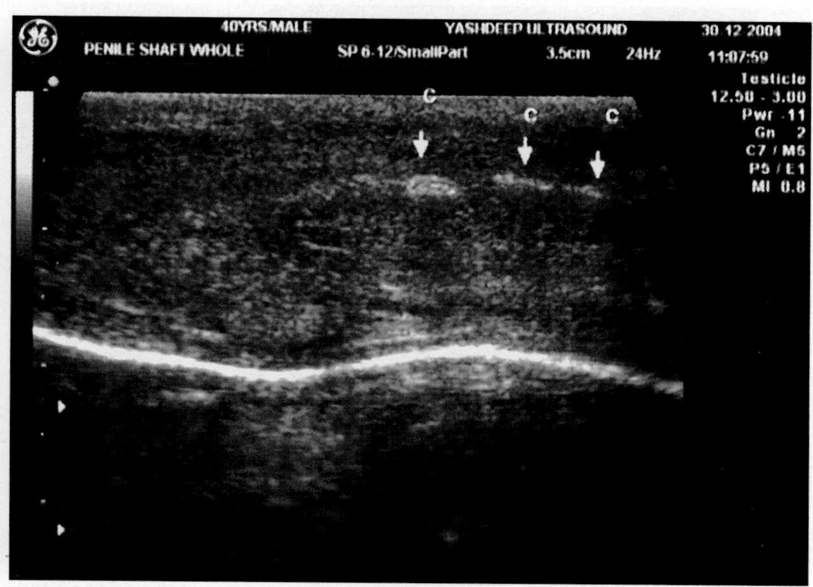

FIGURE 12.14

FIGURES 12.13 and 12.14: Fine calcification in the penile shaft (Peyronie's disease). A patient presented with painful erection with bending. HRSG shows fine calcified echogenic plaque in the corpora spongiosa. HRSG is highly sensitive investigation in picking up early calcification.

FIGURE 12.12

FIGURES 12.11 and 12.12: Normal penile urethra. Urethra is seen as well-defined echogenic tubular shadow running on the ventral side of penis on HRSG.

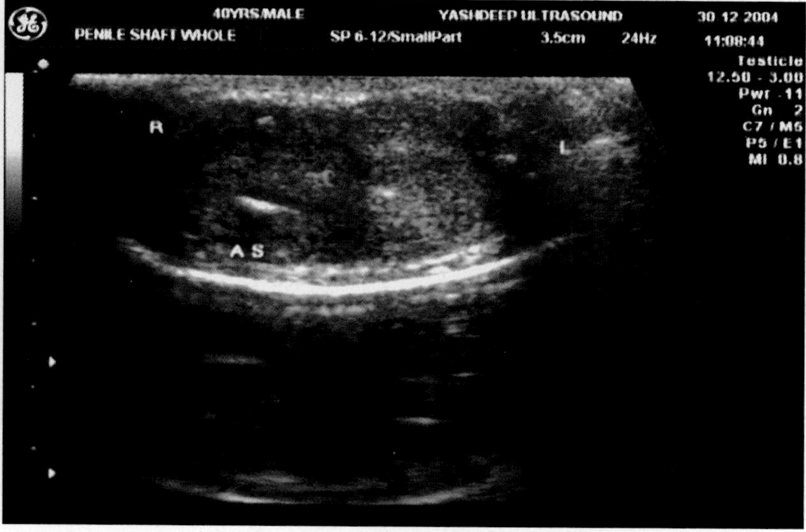

FIGURE 12.15

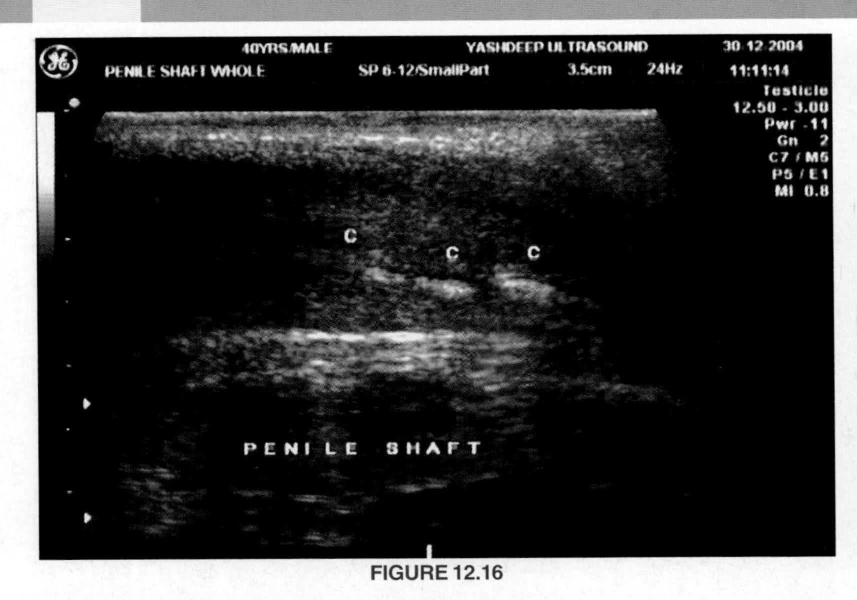

FIGURE 12.16

FIGURES 12.15 and 12.16: Calcified plaques shadow is also seen on the anterior side of the tunica albuginea and also in the substance of corpora cavernosa.

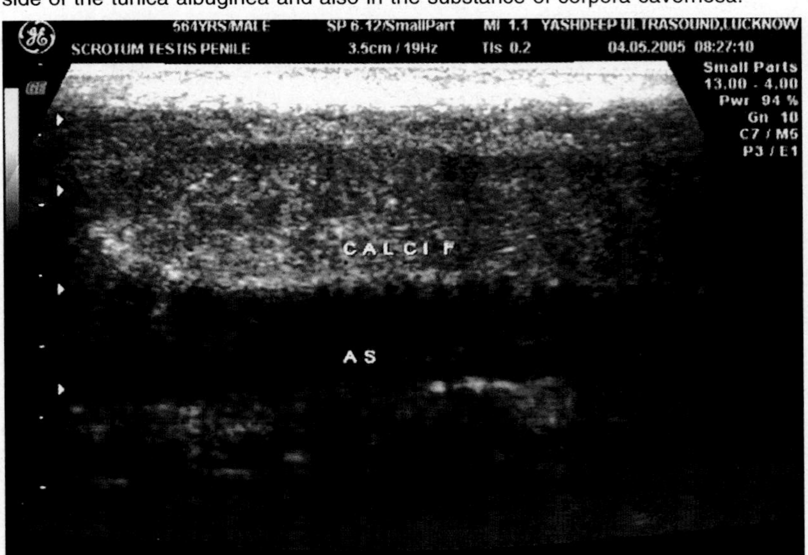

FIGURE 12.17

FIGURE 12.18

FIGURES 12.17 and 12.18: Linear calcification in corpora cavernosa (Peyronie's disease). HRSG shows fine linear calcification in the substance of corpora cavernosa in a patient presented with painful erection and bending. Fine brittle calcified shadows are seen in the penile shaft. They are accompanied with acoustic shadowing.

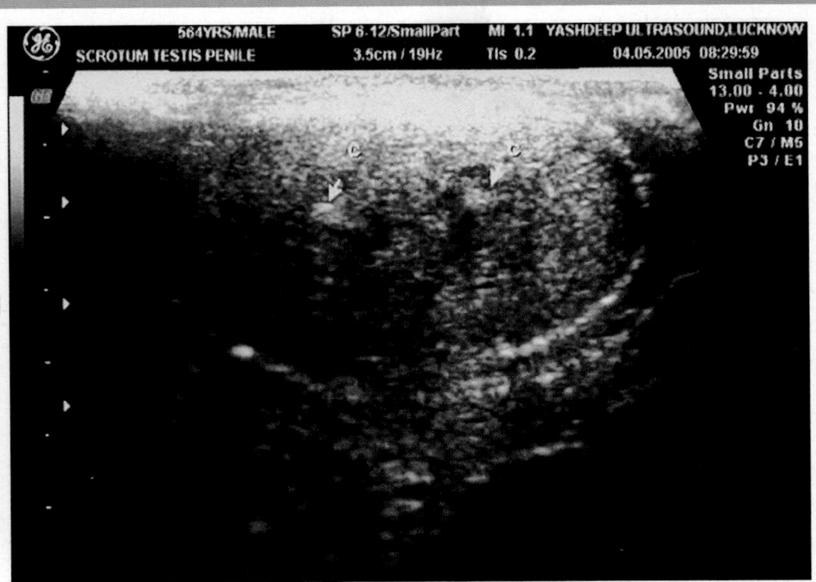

FIGURE 12.19: In cross-section calcified shadows are seen in both corpora cavernosa.

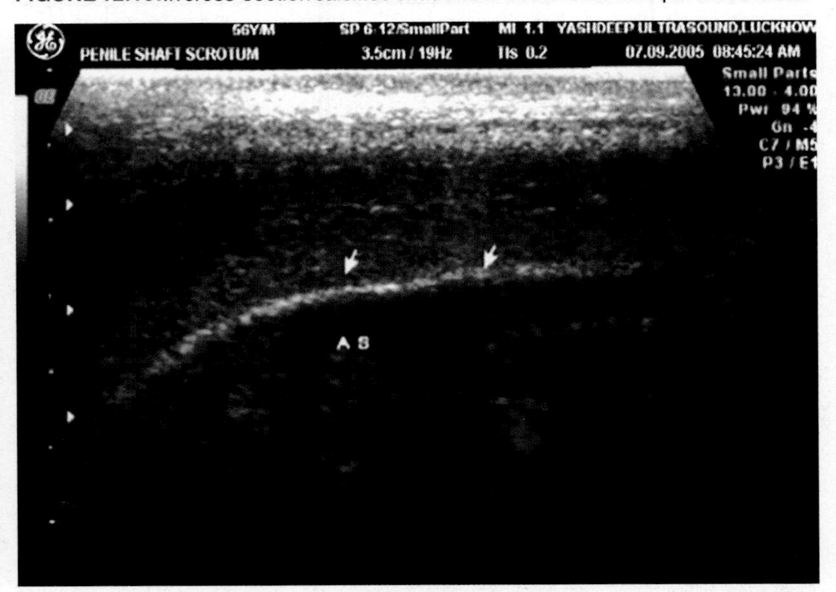

FIGURE 12.20

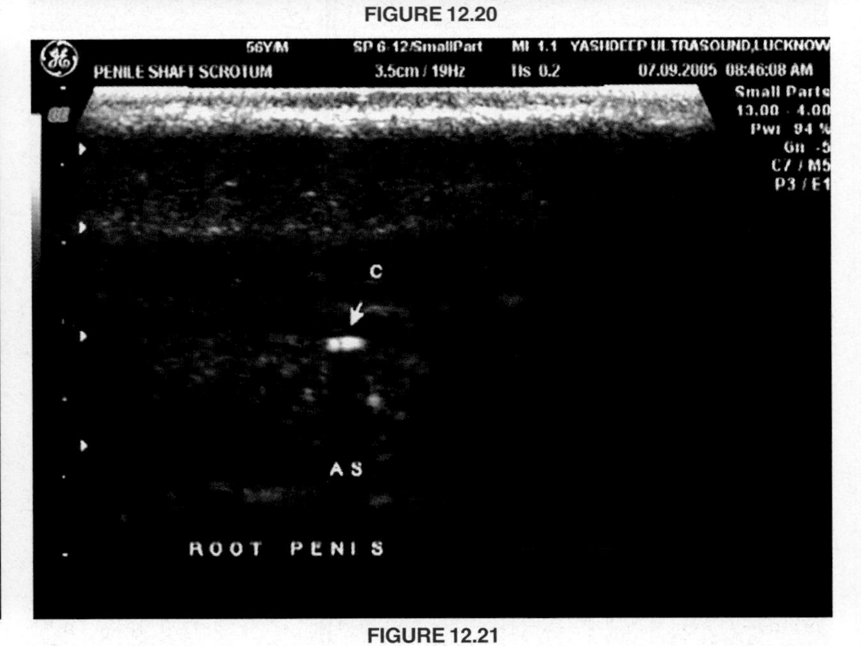

FIGURE 12.21

FIGURES 12.20 and 12.21: Dense echogenic calcified plaques are seen in the mid shaft region in the corpora cavernosa. It is accompanied with dense acoustic shadowing in a case of Peyronie's disease.

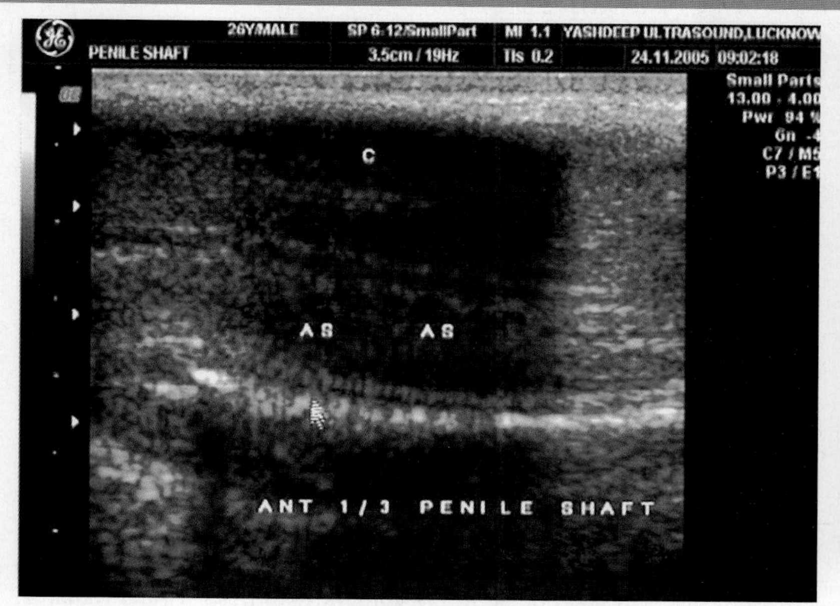

FIGURE 12.22

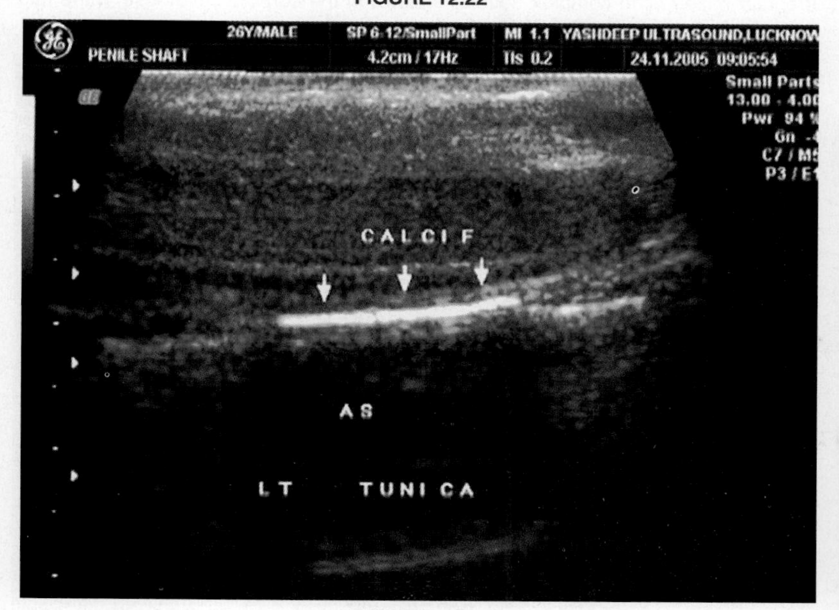

FIGURE 12.23

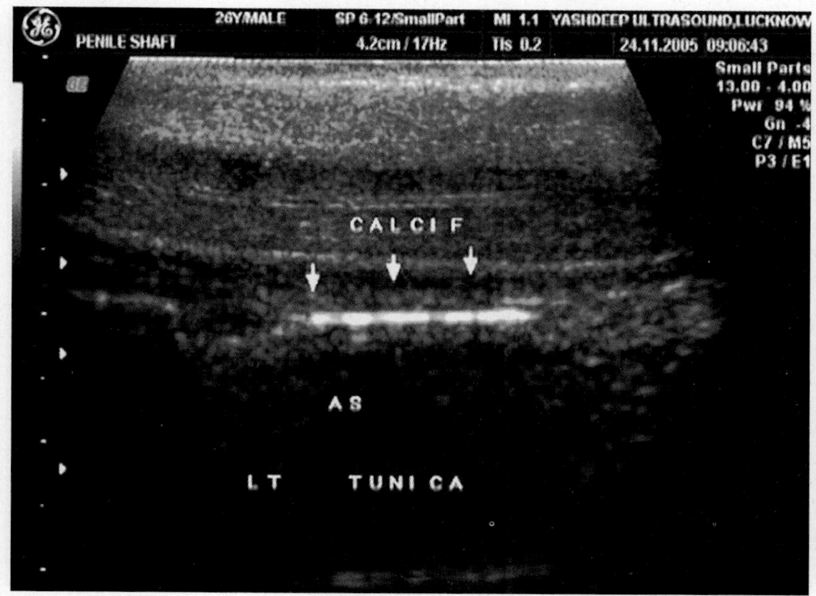

FIGURE 12.24

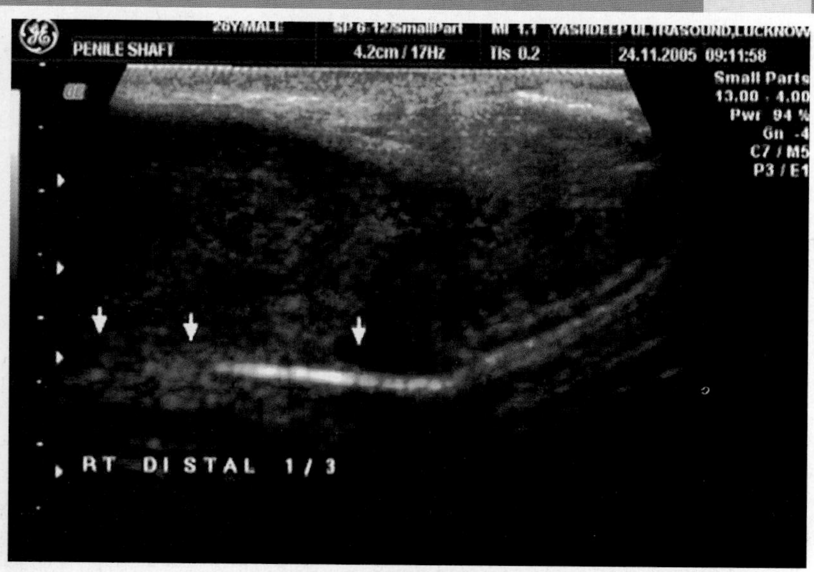

FIGURE 12.25

FIGURES 12.22 to 12.25: Cord like bending calcification in the tunica albuginea. A young patient presented with painful ventral banding of the penis on erection. HRSG shows dense echogenic calcified cord like plaque fixed with on dorsal and ventral side of the tunica albuginea. It is accompanied with dense acoustic shadowing. Highly echogenic calcified plaque is seen on HRSG.

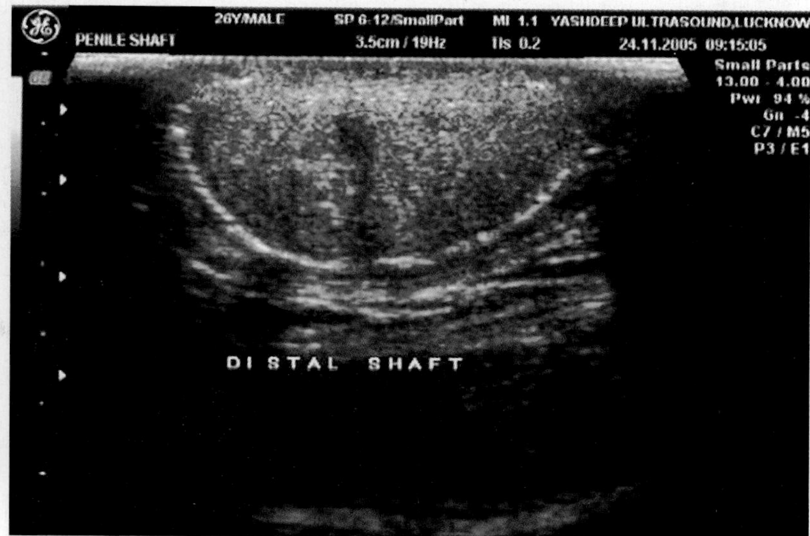

FIGURE 12.26

FIGURE 12.27

FIGURES 12.26 and 12.27: Cross-section imaging of the penile shaft shows the rim like calcification in the tunica albuginea in the same patient.

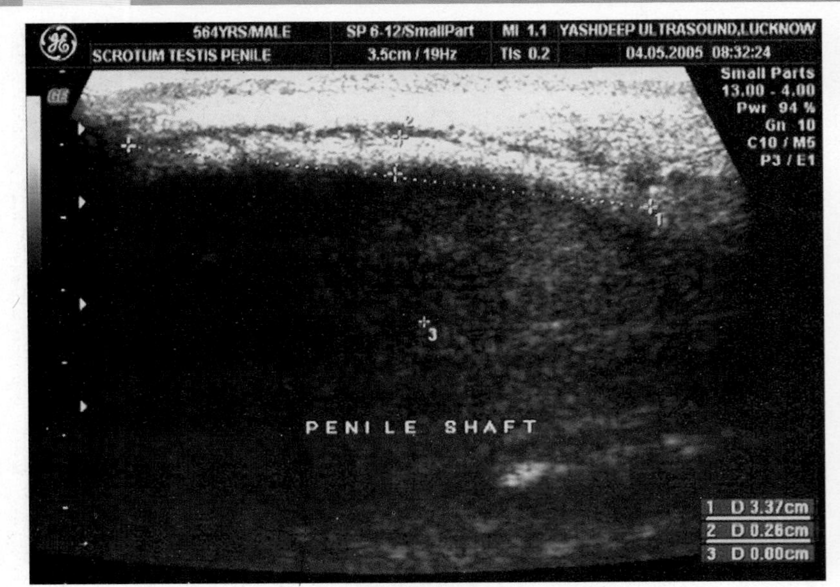

FIGURE 12.28

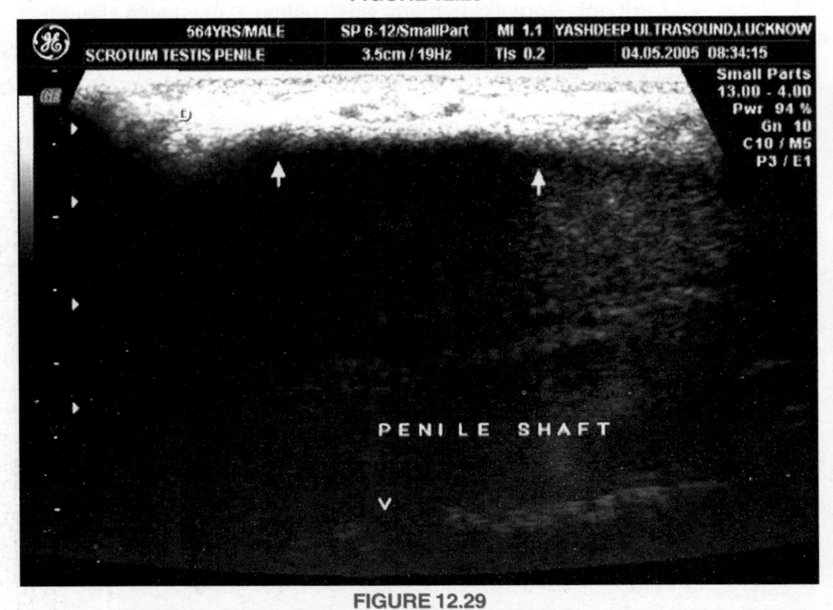

FIGURE 12.29

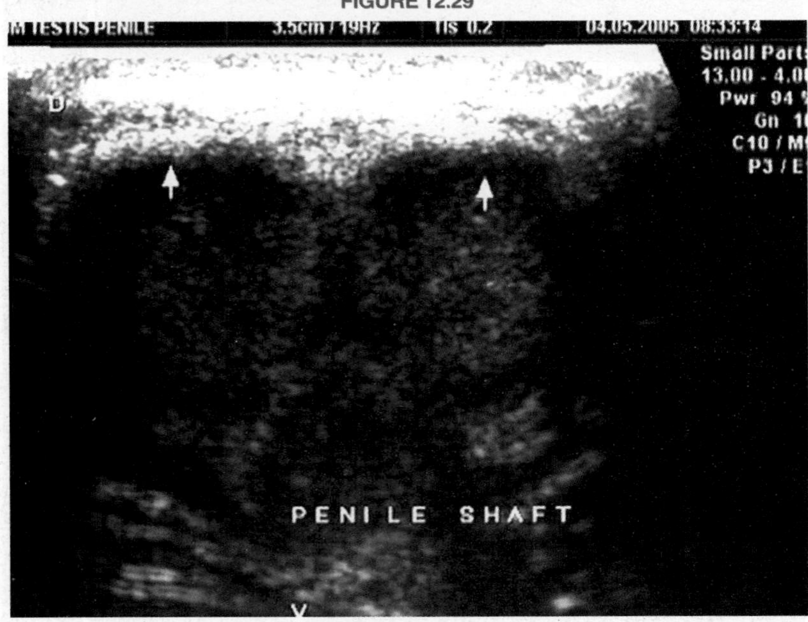

FIGURE 12.30

FIGURES 12.28 to 12.30: Dense thick cord like plaque on the dorsal side of tunica albuginea. HRSG shows a big echogenic thick cord like plaque fixed on dorsal side of the tunica albuginea. It is accompanied with dense acoustic shadowing.

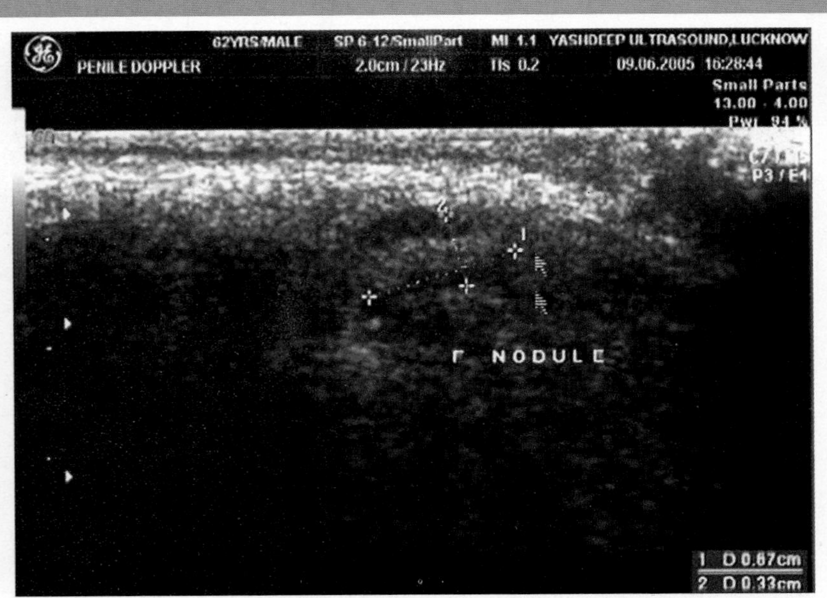

FIGURE 12.31: Fibrotic nodule in the corpora cavernosa. HRSG shows an irregular mixed echo complex nodular shadow in the corpora cavernosa in the mid part of the shaft suggestive of fibrotic nodule.

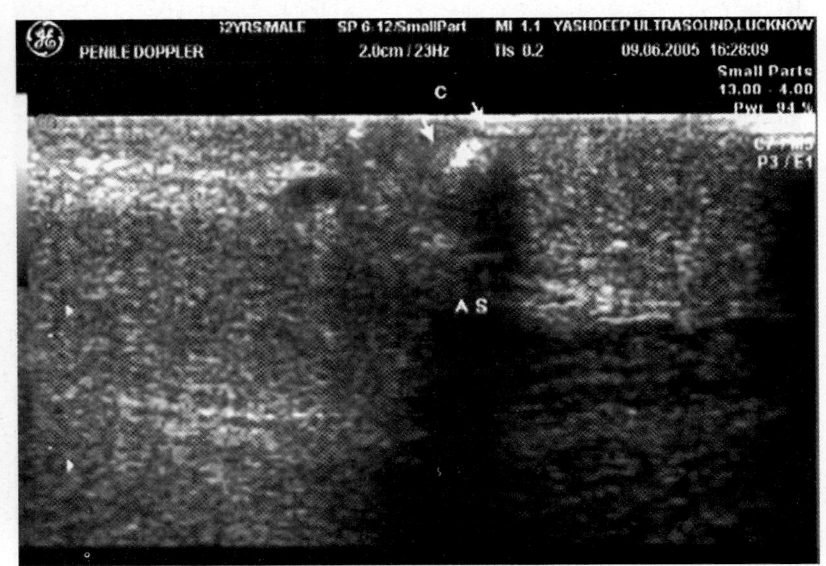

FIGURE 12.32

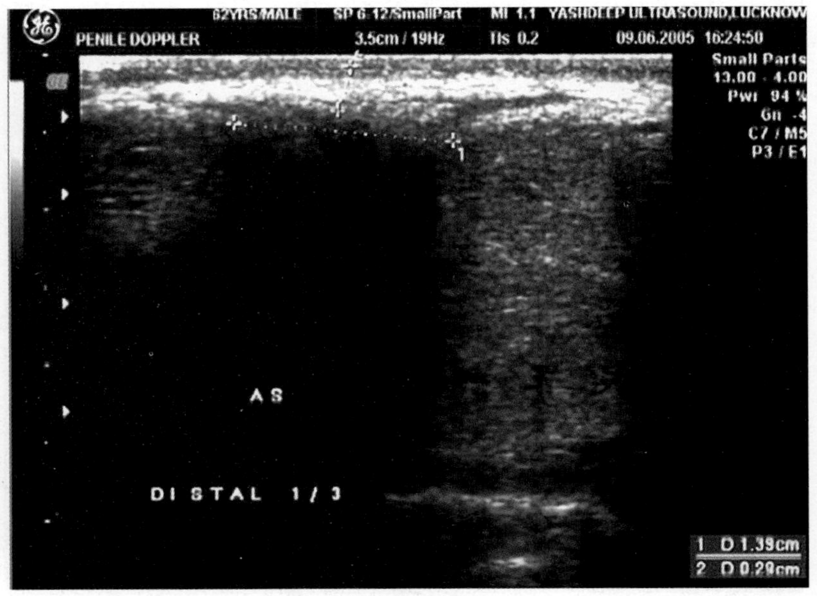

FIGURE 12.33

FIGURES 12.32 and 12.33: Dense echogenic calcified plaque is also seen on the dorsal side of tunica albuginea, which is accompanied with dense acoustic shadowing.

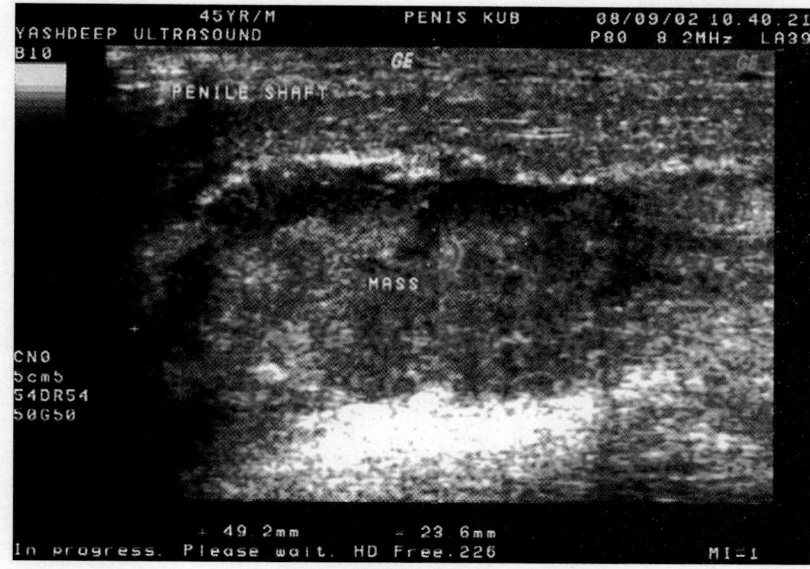

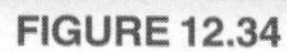

FIGURE 12.34

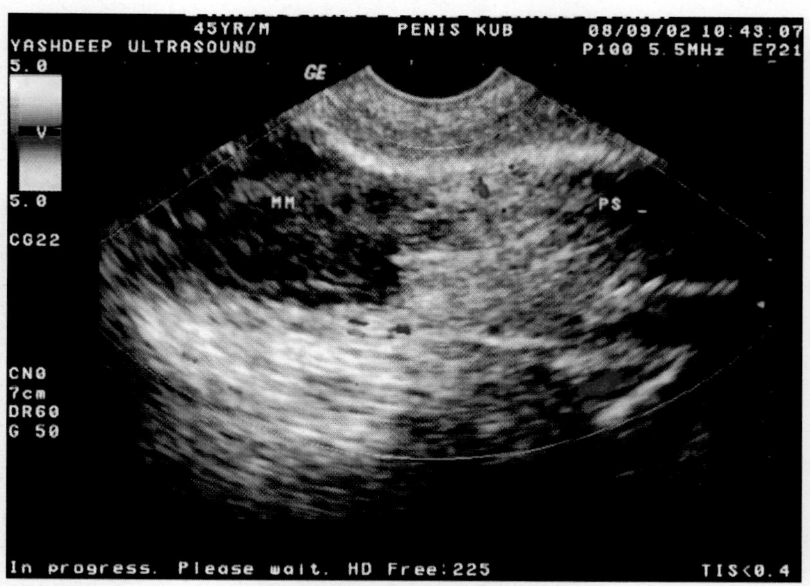

FIGURE 12.35

FIGURES 12.34 and 12.35: Pyogenic abscess in the penis. A middle aged diabetic patient presented with painful soft tissue swelling in the penile shaft. HRSG shows a big low-level echo complex mass in the mid part of the shaft. Thick echoes are seen in it. It is mainly seen on ventral side.

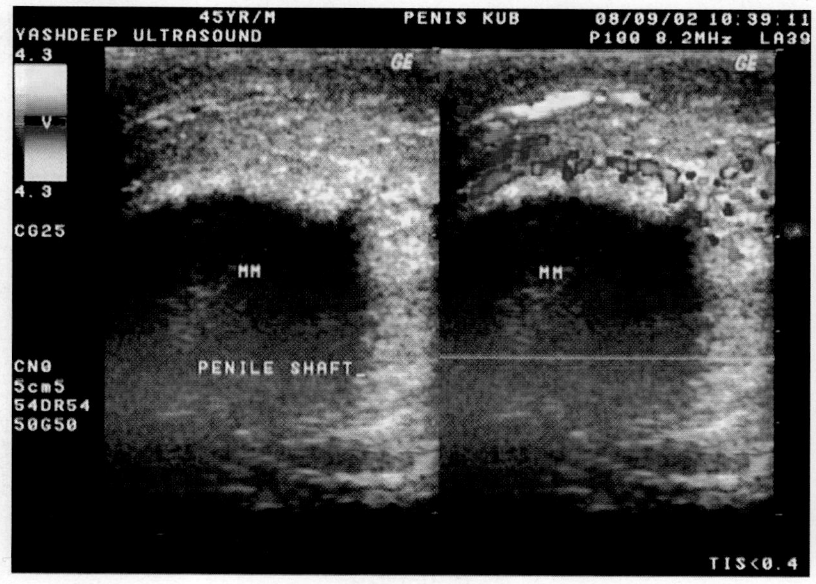

FIGURE 12.36

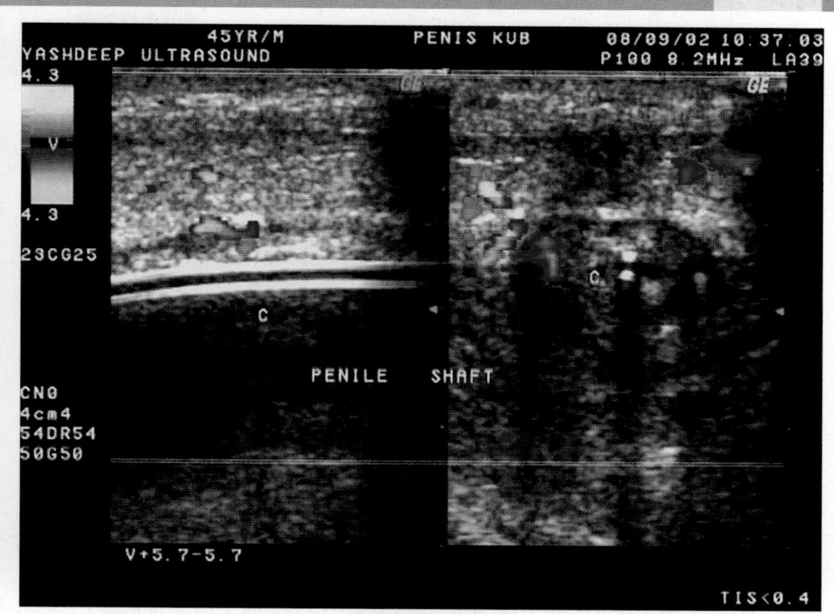

FIGURE 12.37

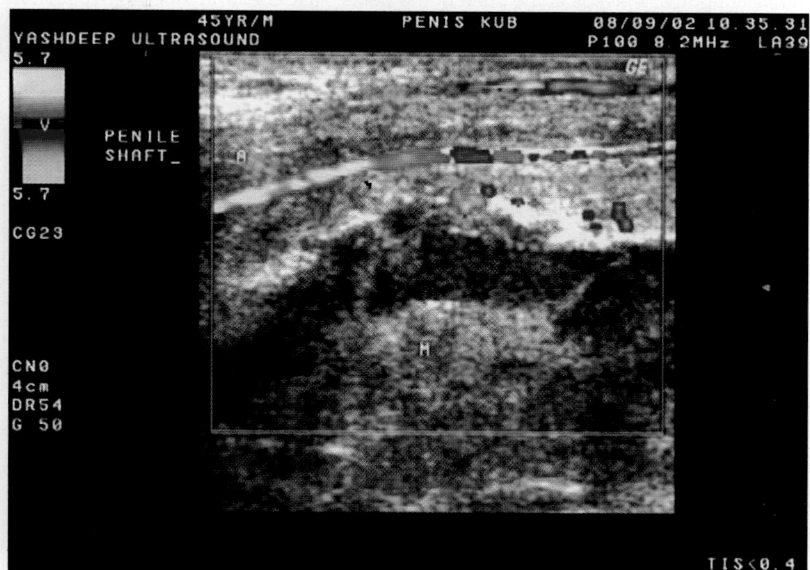

FIGURE 12.38

FIGURES 12.36 to 12.38: On color flow imaging increased flow is seen around the abscess suggestive of hyperemia. However, normal flow is seen in cavernosal artery. But no flow is seen in the abscess. Aspiration of the abscess turned out to be pyogenic collection.

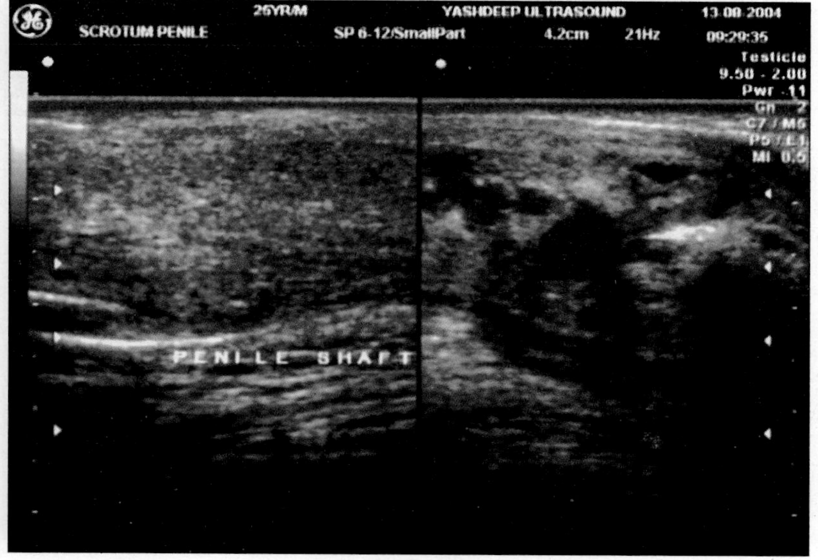

FIGURE 12.39

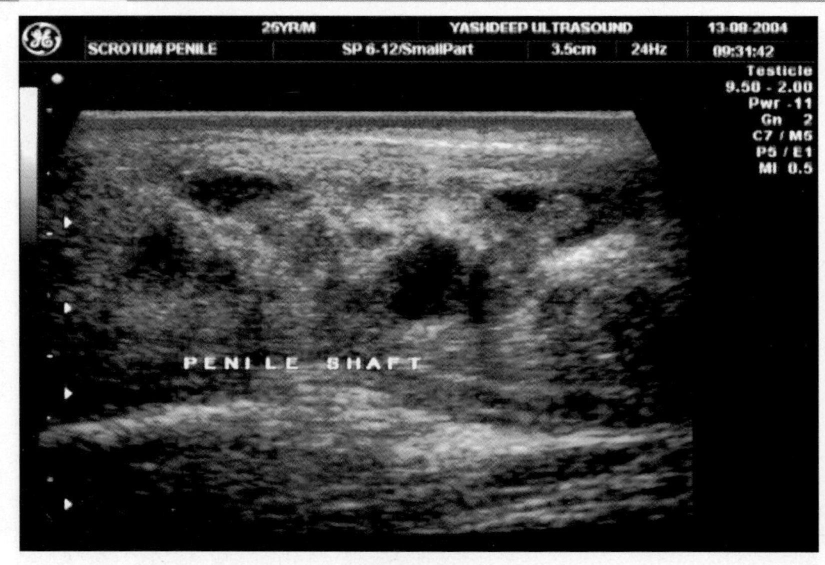

FIGURE 12.40

FIGURES 12.39 and 12.40: Penile tuberculosis. A young patient presented with multiple nodular masses in the shaft of the penis. HRSG shows multiple hypoechoic areas in the proximal and mid part of the shaft. They are discrete in distribution with cystic degeneration. Low-level echoes are seen in them. No calcification is seen. Normal texture is lost and the penile shaft gives mouth eaten appearance.

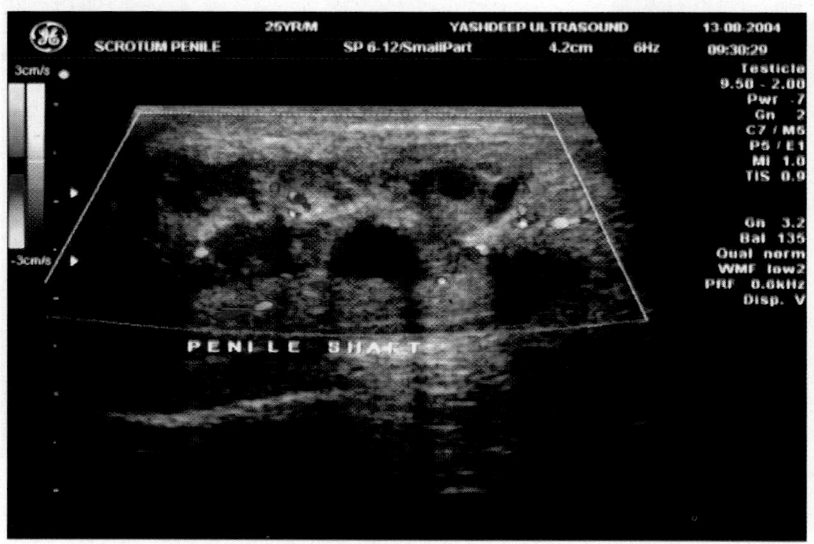

FIGURE 12.41

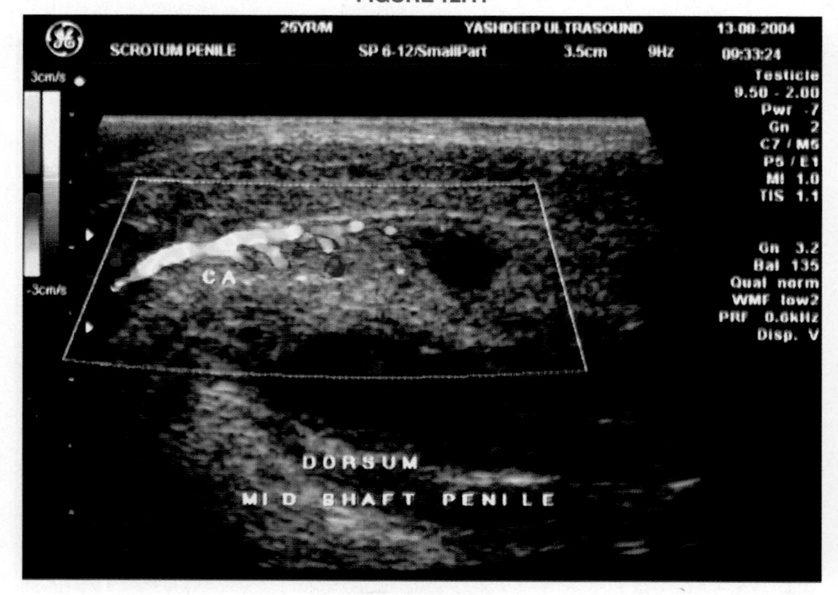

FIGURE 12.42

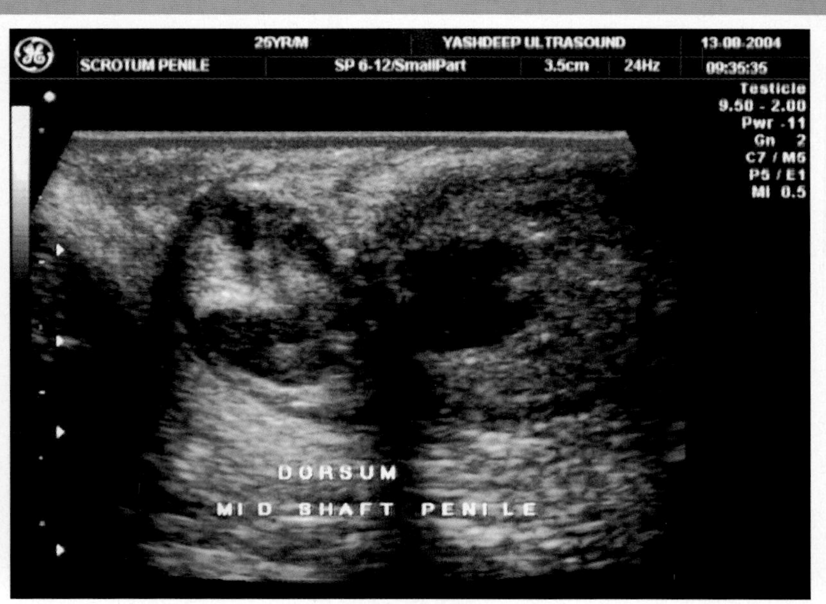

FIGURE 12.43

FIGURES 12.41 to 12.43: On color flow imaging no flow was seen in the abscesses. However, normal flow was seen in the cavernosal arteries. Biopsy of the masses confirmed tuberculosis.

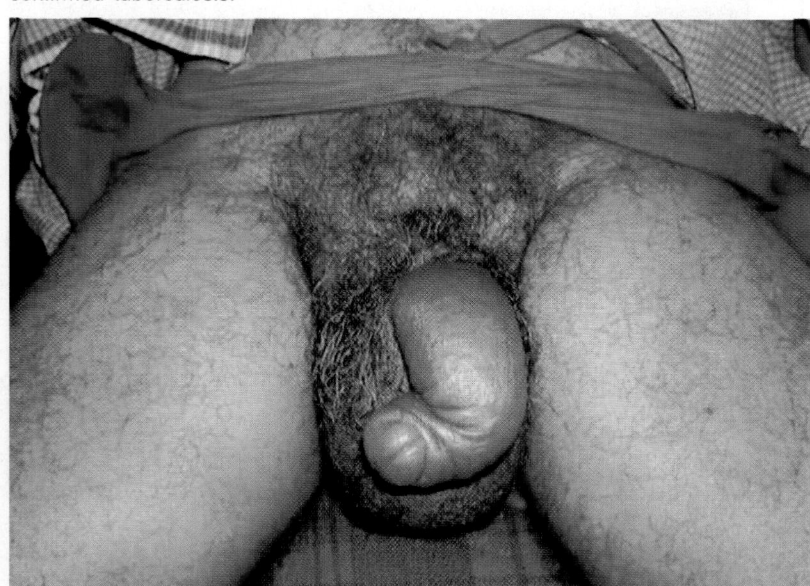

FIGURE 12.44: Penile filariasis a patient presented with marked soft tissue edema over the penile shaft and scrotum with edema over the thigh.

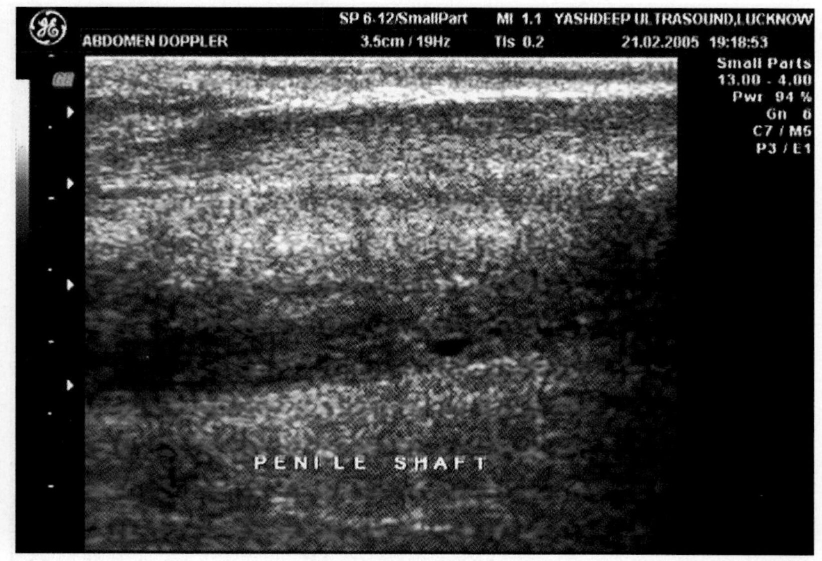

FIGURE 12.45

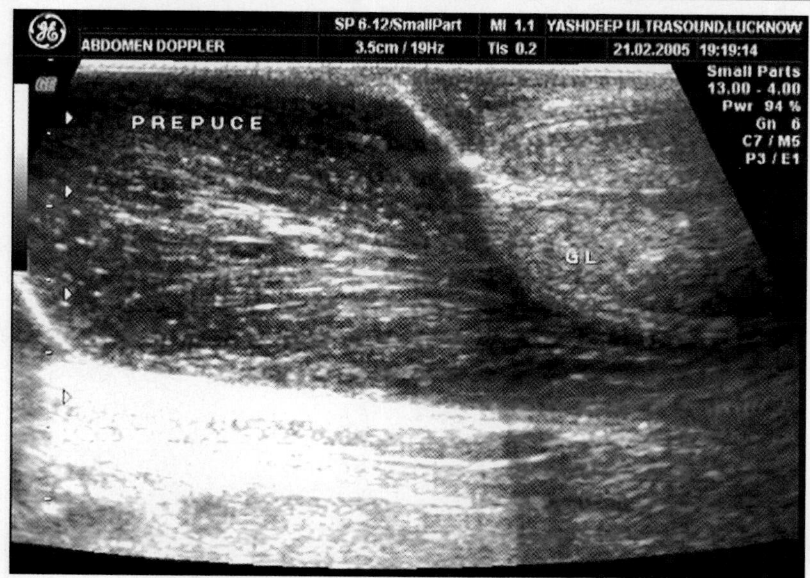

FIGURE 12.46

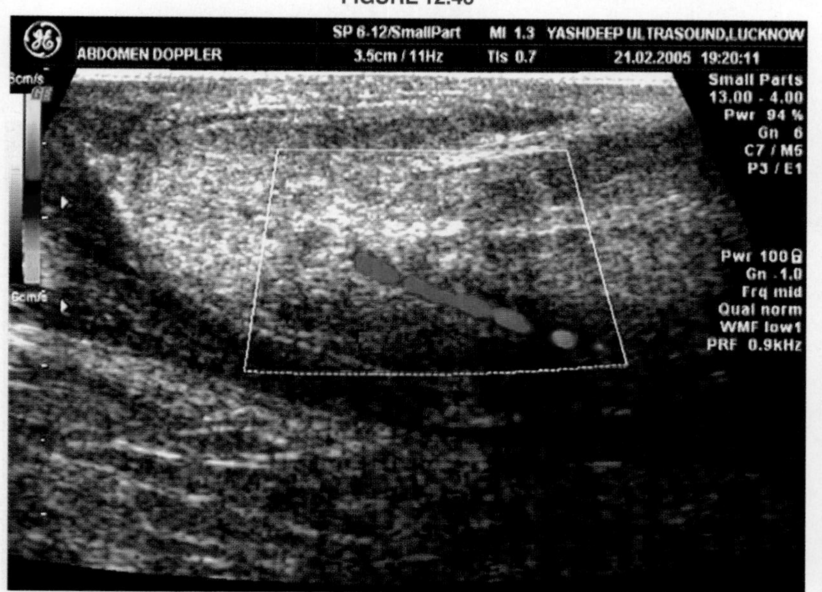

FIGURE 12.47

FIGURES 12.45 to 12.47: HRSG shows marked skin edema over the penile skin. Subcutaneous edema was presented. Marked edema was seen over the prepuce. However, the corpora cavernosa was normal. On color flow imaging normal flow was seen in dorsal penile artery suggestive of filariasis.

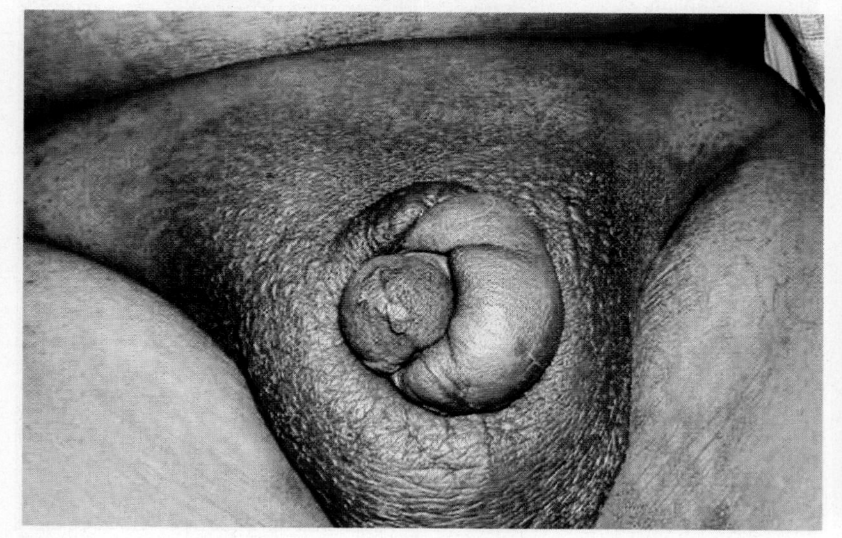

FIGURE 12.48: Filarial penis and scrotum. A young man presented with impotence. The penis is markedly edematous and flaccid in position. It is seen as a coiled collapsed organ. Marked skin edema is present.

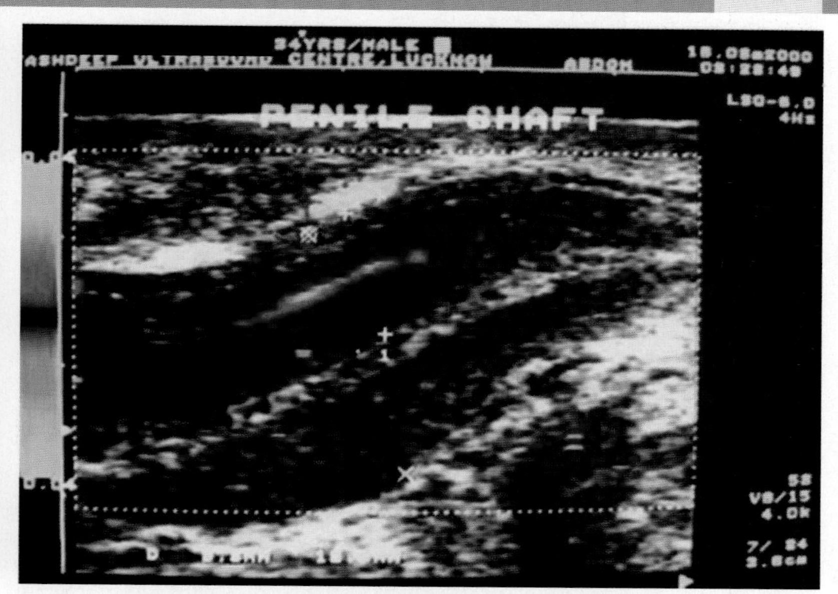

FIGURE 12.49: HRSG and color flow imaging shows hypoechoic corpora cavernosa. Low flow is seen in cavernosal artery after injecting vasodilating agent suggestive of arterial insufficiency and edema in the corpora cavernosa.

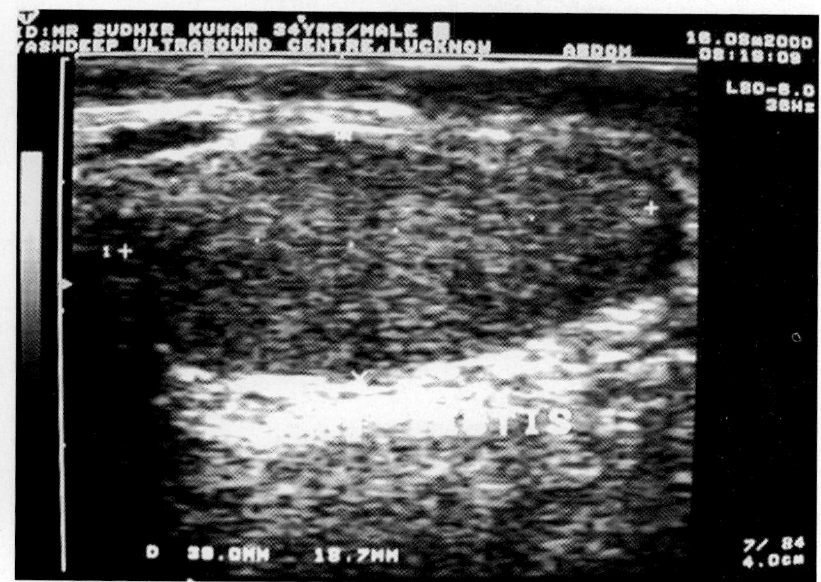

FIGURE 12.50

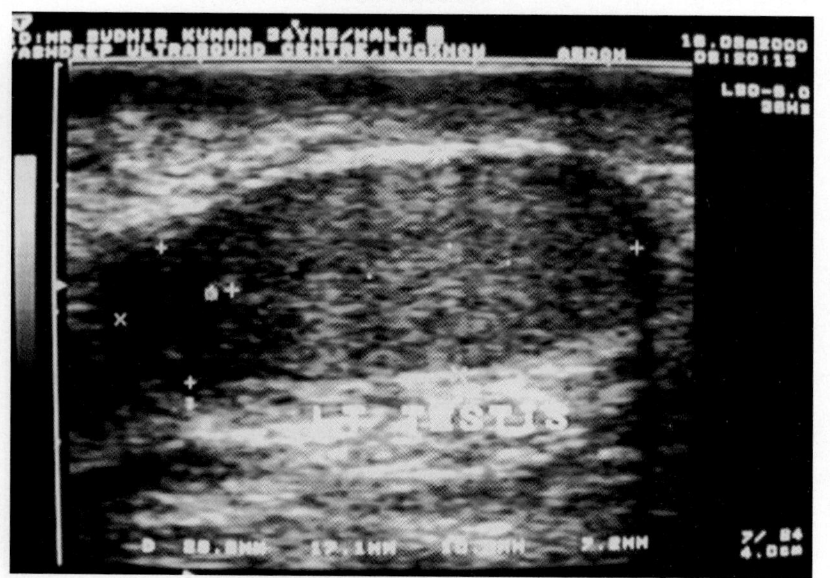

FIGURE 12.51

FIGURES 12.50 and 12.51: Both testis are also small in size and they are hypoechoic in texture. However, echogenicity is homogeneous.

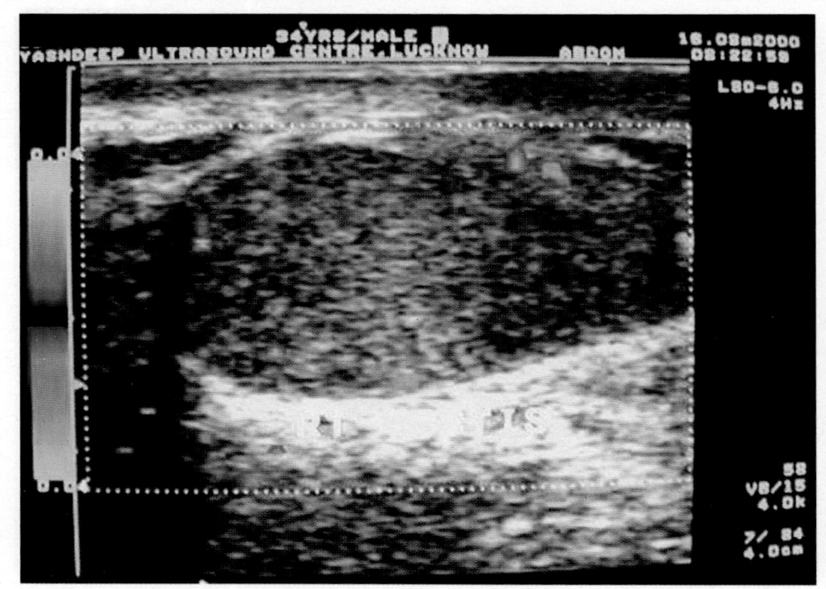

FIGURE 12.52: On color flow imaging poor flow is seen in the testis suggestive of arterial insufficiency.

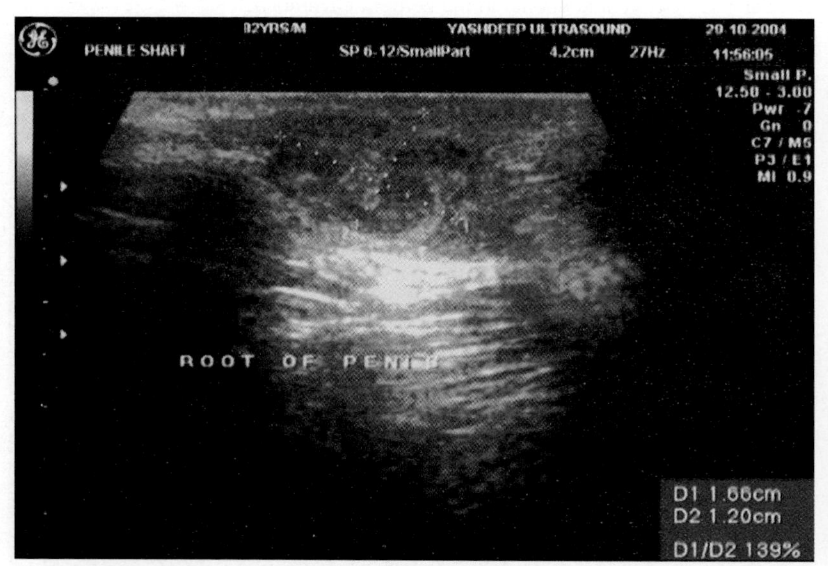

FIGURE 12.53: **Carcinoma penis.** A patient presented with irregular mass at the root of the penis. HRSG showed and irregular heterogeneous mass at the root of the penis. Margins were irregular. It was seen invading the corpora.

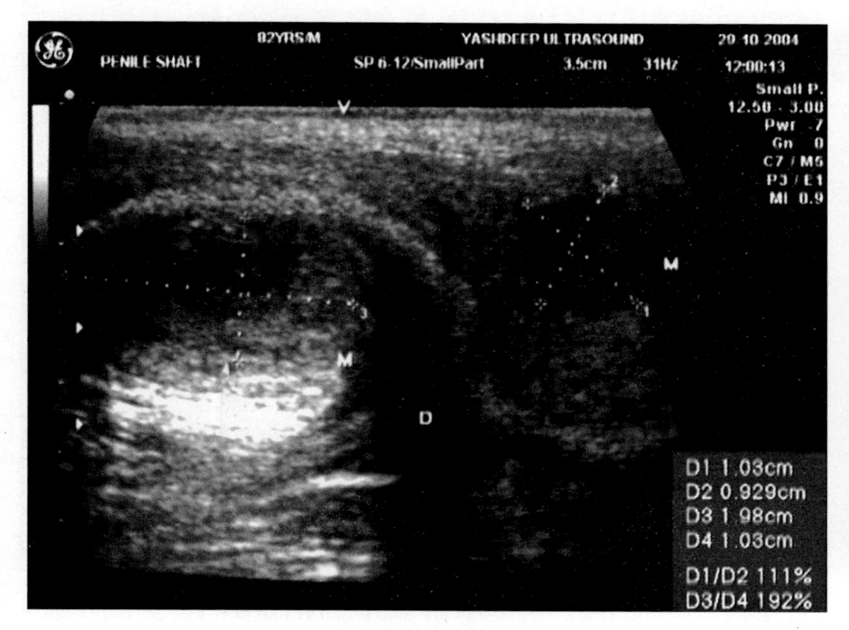

FIGURE 12.54

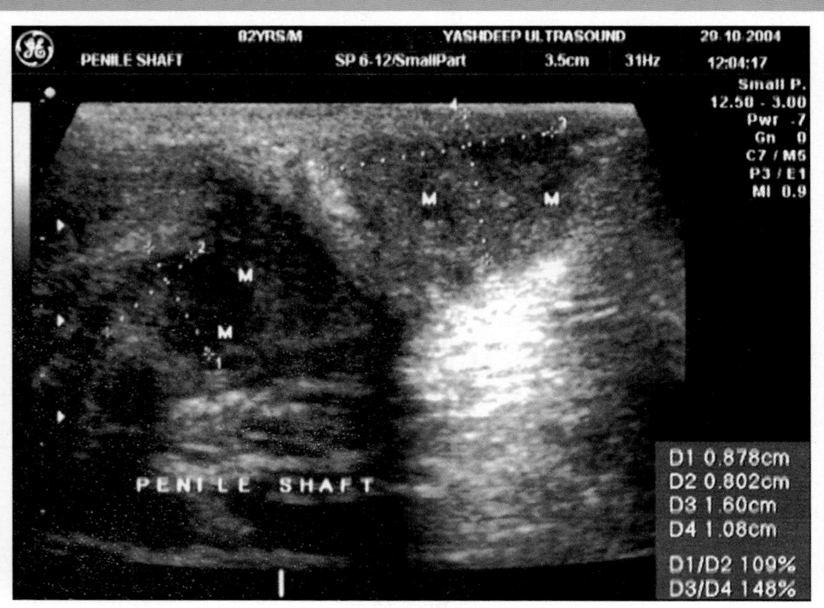

FIGURE 12.55

FIGURES 12.54 and 12.55: The mass was seen invading the root of the penis. There was also evidence of big heterogeneous mass seen invading the ventral side of the penile shaft. The capsule was broken.

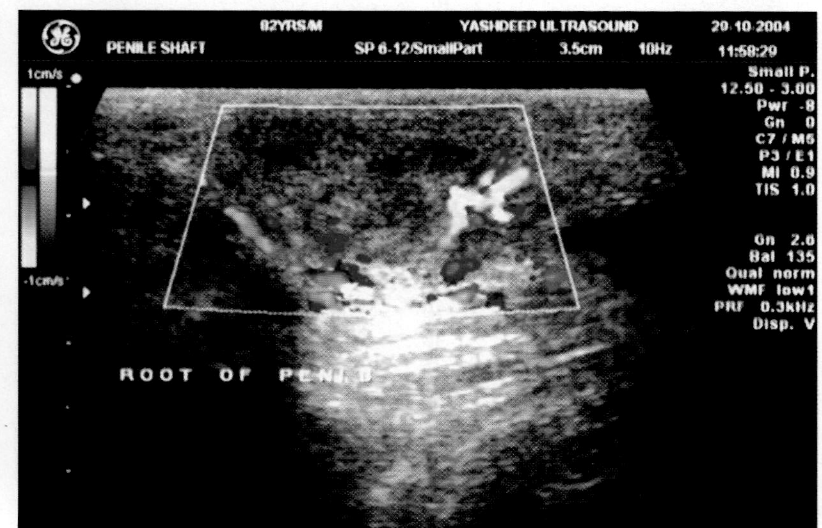

FIGURE 12.56

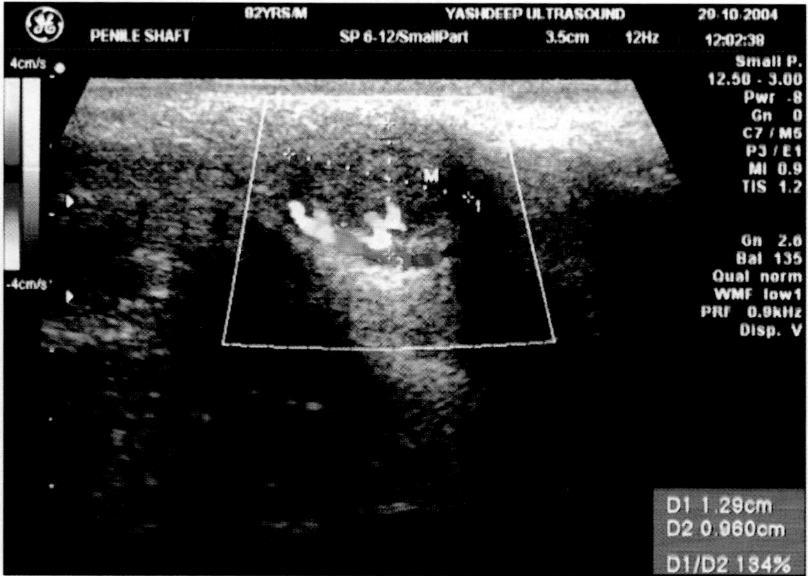

FIGURE 12.57

FIGURES 12.56 and 12.57: On color flow imaging few vessels were seen feeding the mass. Biopsy of the mass confirmed carcinoma penis.

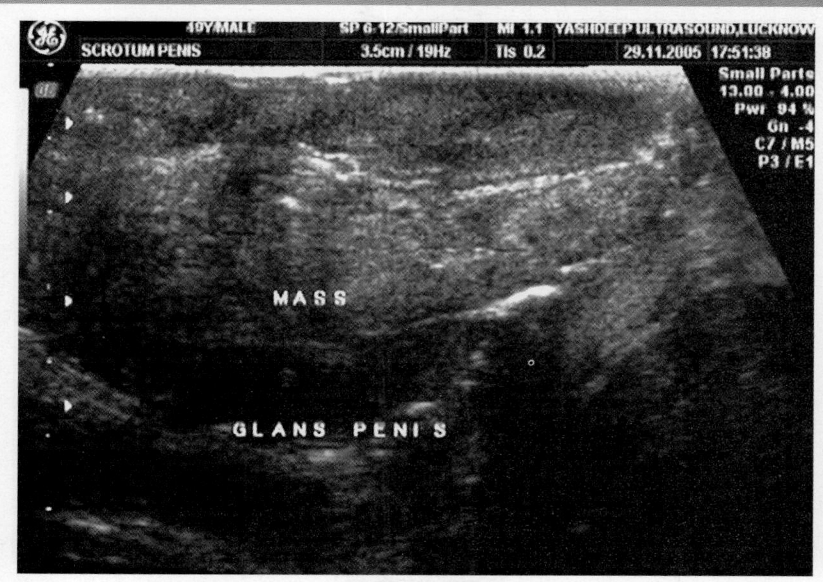

FIGURE 12.58

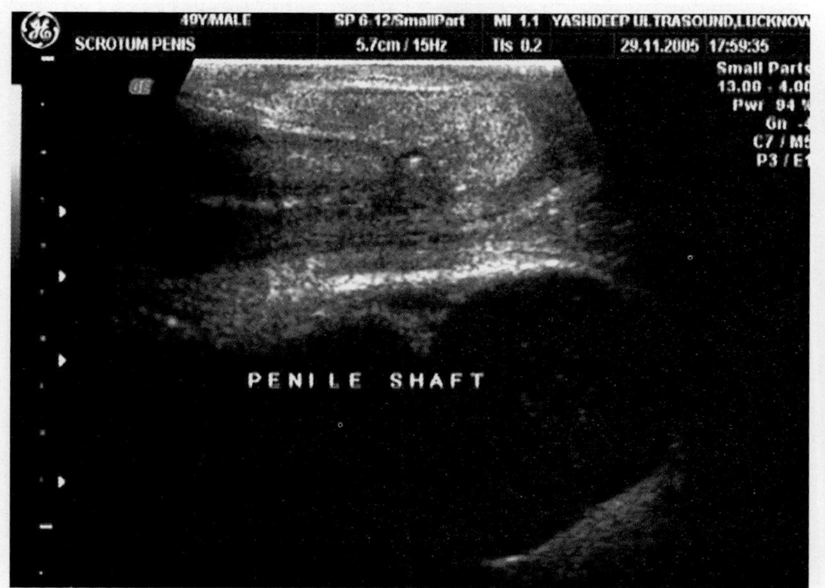

FIGURE 12.59

FIGURES 12.58 and 12.59: Carcinoma of the glans penis. A middle aged patient presented with heterogeneous soft tissue mass over the glans of the penis. Irregular masses were seen over the glans on HRSG. The mass was seen invading the distal penile shaft.

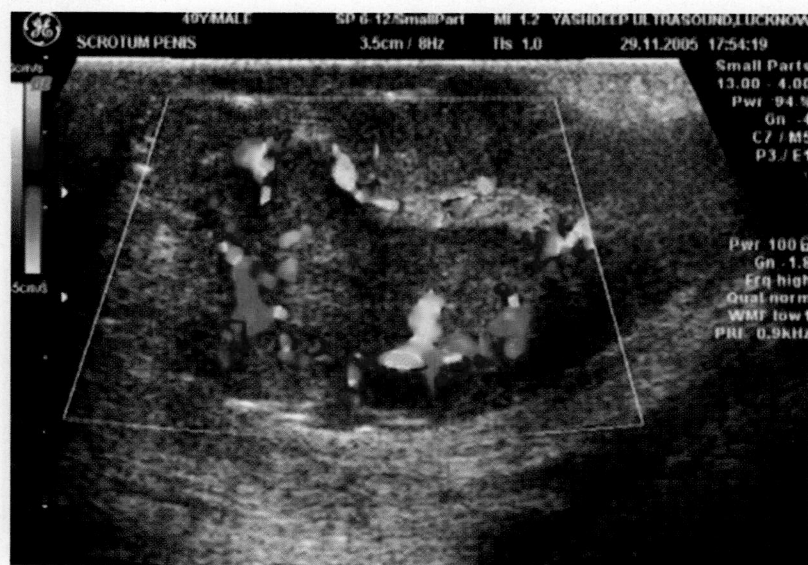

FIGURE 12.60

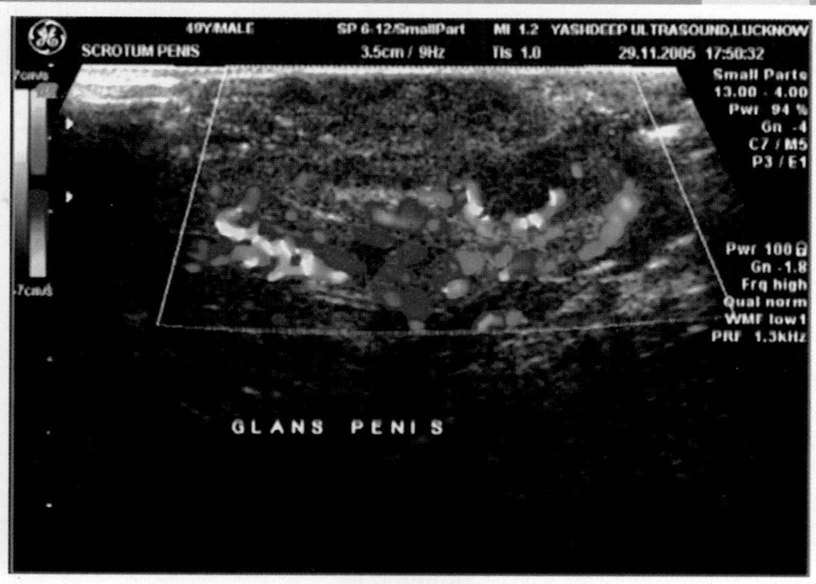

FIGURE 12.61

FIGURES 12.60 and 12.61: On color flow imaging multiple vessels were seen feeding the mass.

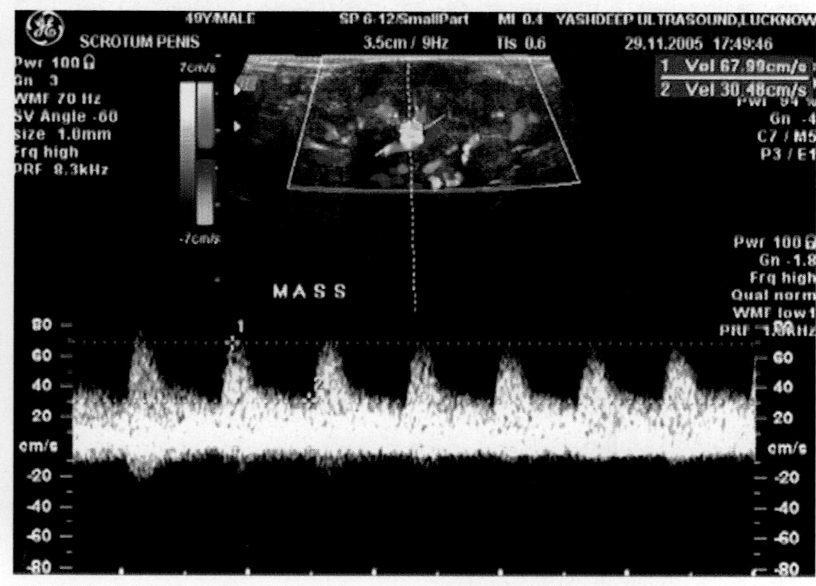

FIGURE 12.62: On spectral Doppler tracing high arterial flow was seen.

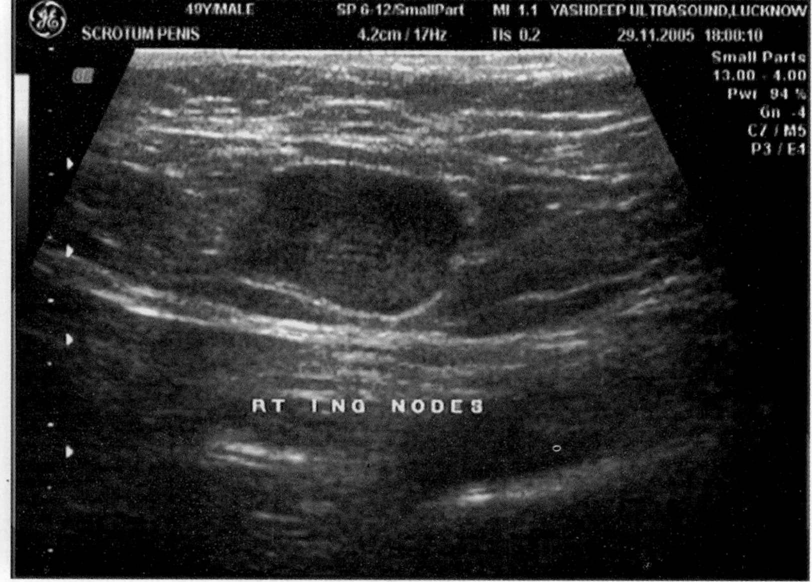

FIGURE 12.63

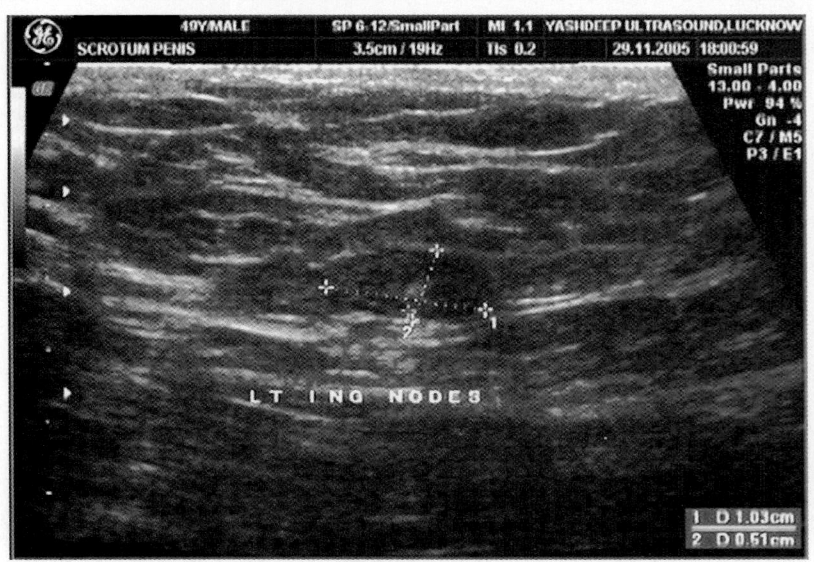

FIGURE 12.64

FIGURES 12.63 and 12.64: Enlarged nodes were also seen in both inguinal region. They were hypoechoic in texture. Lymph node hilum was lost.

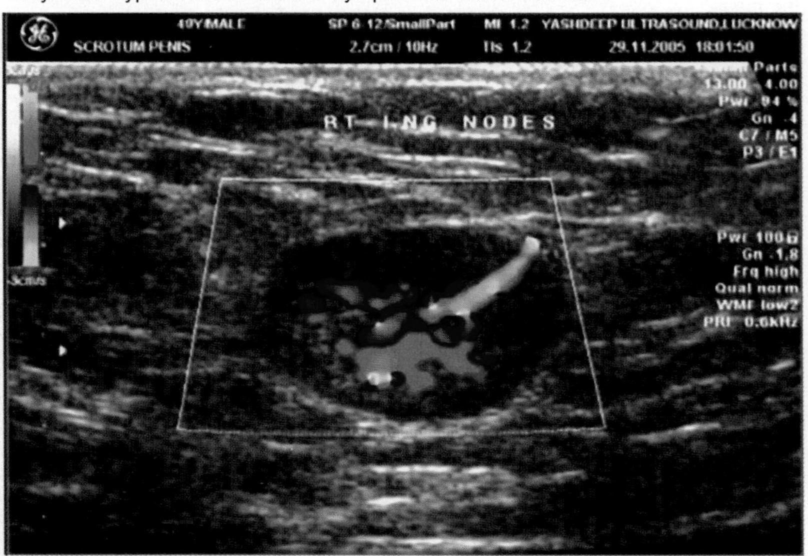

FIGURE 12.65: On color flow imaging high flow was seen in the lymph nodes. Biopsy of the mass confirmed carcinoma penis.

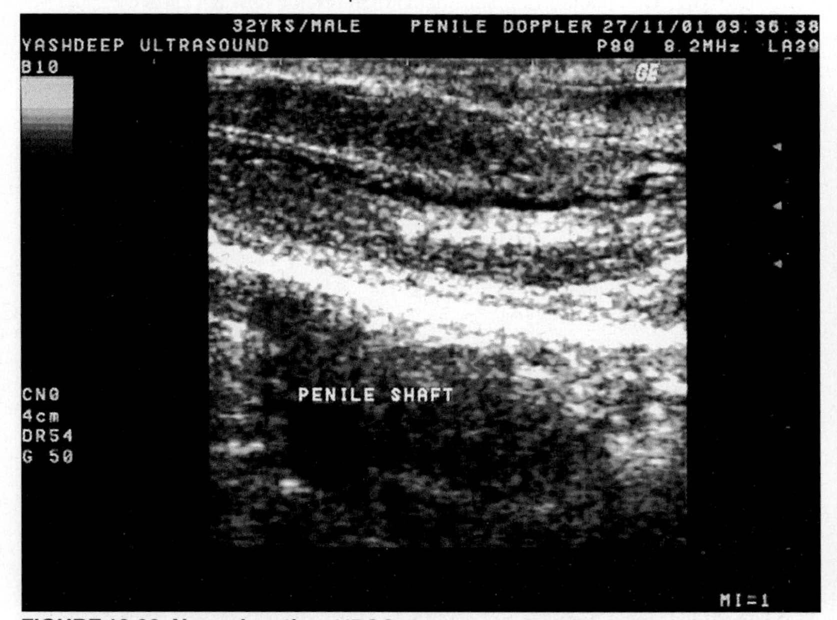

FIGURE 12.66: Normal urethra. HRSG shows normal penile urethra in its middle and distal part. The lumen is anechoic and walls are echogenic due to fibromuscular tissue.

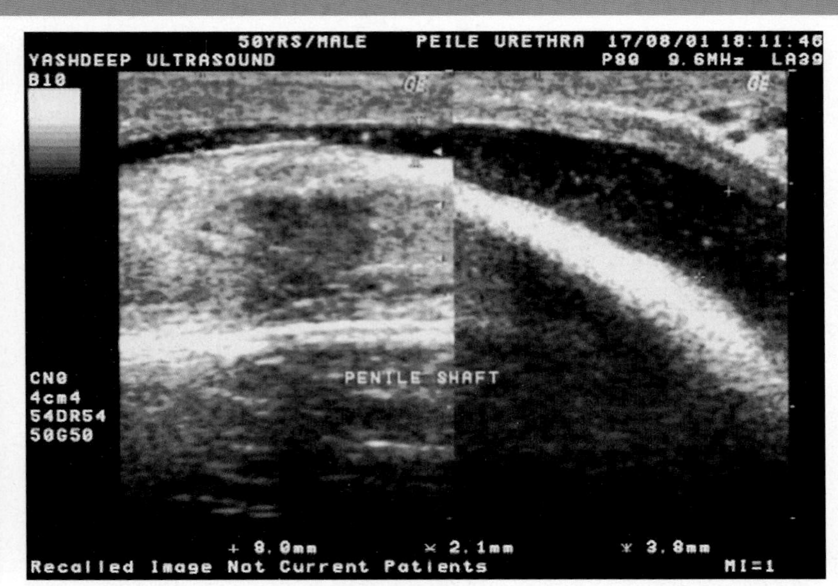

FIGURE 12.67

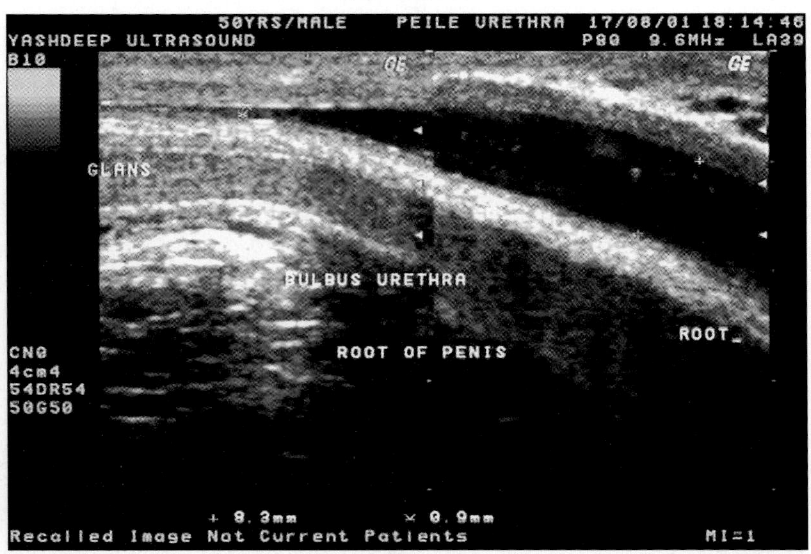

FIGURE 12.68

FIGURES 12.67 and 12.68: Normal sonourethrogram. HRSG shows normal sonourethrogram. It is done by placing no 8 Foley's catheter in the navicular fossa and inflating the bulb thus blocking the mouth. Saline solution is pushed through the catheter and urethra is distended. The urethral cavity is anechoic. No evidence of any septa, fibrous band, or stricture is seen. The urethrogram shows normal anterior, middle and the posterior bulbar part of urethra.

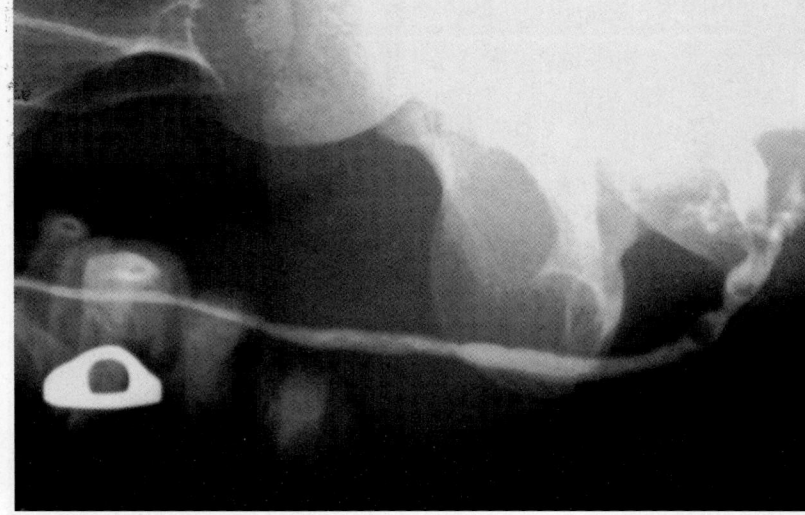

FIGURE 12.69: Urethral structure. X-ray retrograde urethrogram shows a stricture in the bulbar part of urethra with irregular margins.

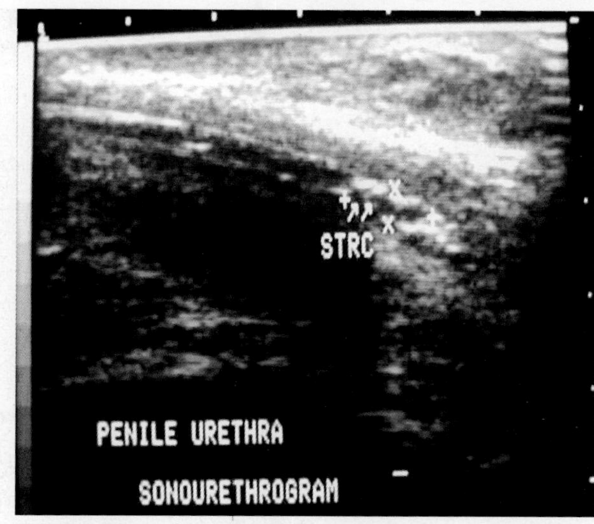

FIGURE 12.70: Sonourethrogram of the same patient shows exact length of the structure with irregular echogenic margins and obliterated lumen. It also shows periurethral fibrosis.

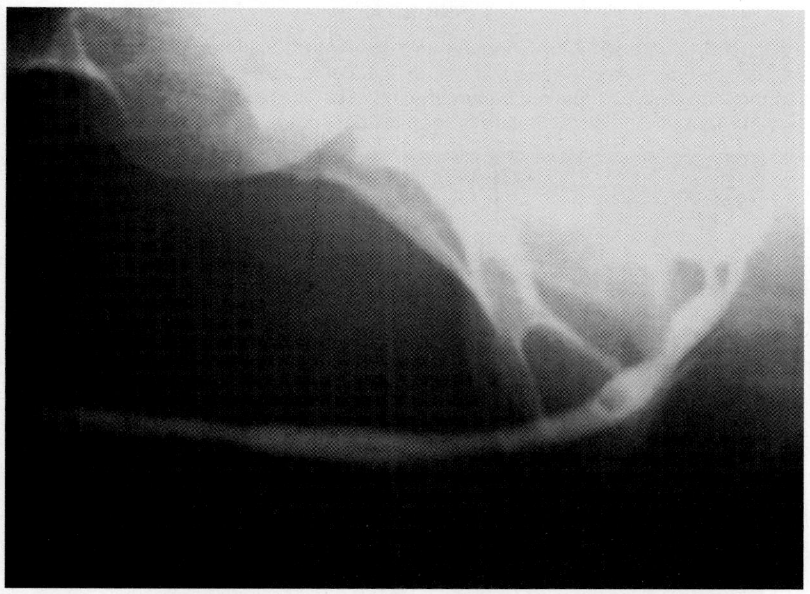

FIGURE 12.71: X-ray urethrogram shows structure urethra in its proximal part.

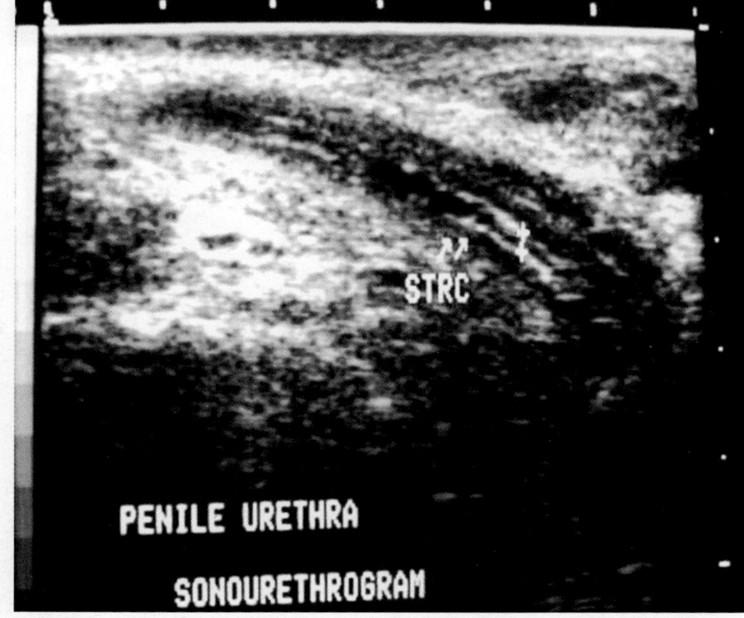

FIGURE 12.72: Sonourethrogram shows the structure urethra better than X-ray urethrogram. The structure is 2 cm in length and it is 2 mm in caliber. Marked periurethral fibrosis is also seen.

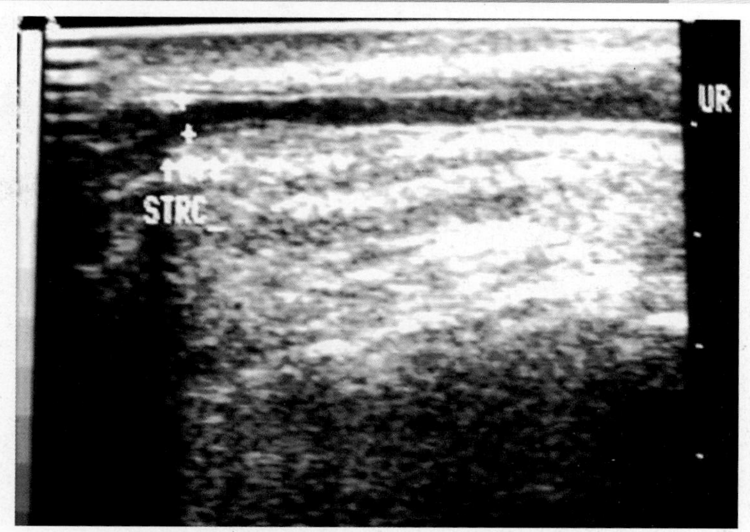

FIGURE 12.73: Sonourethrogram shows a structure urethra in its distal part or anterior part. Hypertrophy mucosal fold is also seen projecting into the lumen part of urethra.

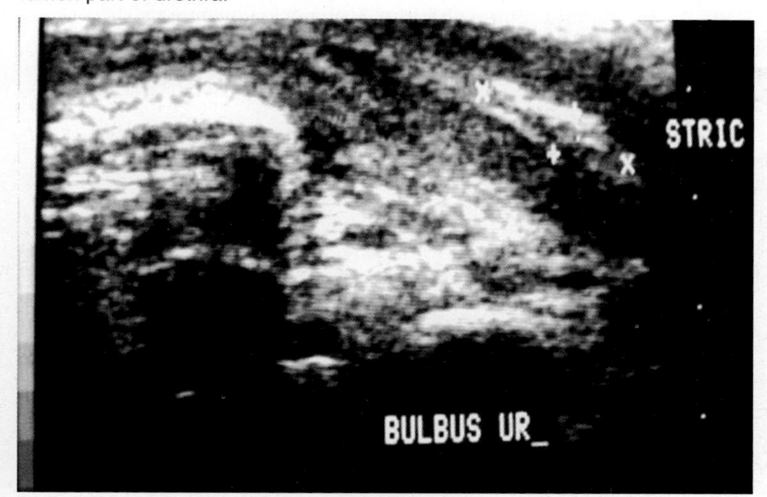

FIGURE 12.74

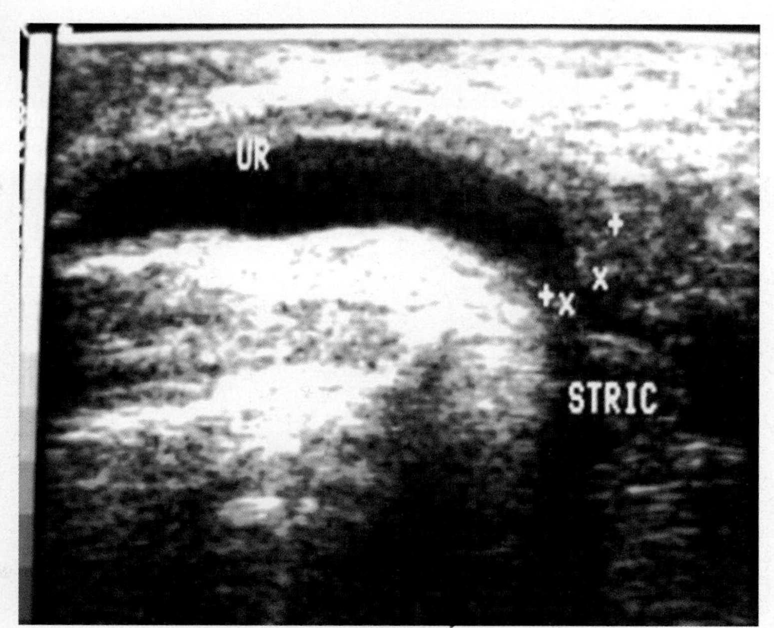

FIGURE 12.75

FIGURES 12.74 and 12.75: Bulbar urethral stricture. Sonourethrogram shows a long stricture in the bulbar part of urethra. The lumen is collapsed. Irregular hypoechoic fibrosis is seen around the urethra compressing the urethra. The proximal part of urethra is distended and high jet of water flow is seen going through the stricture part of urethra.

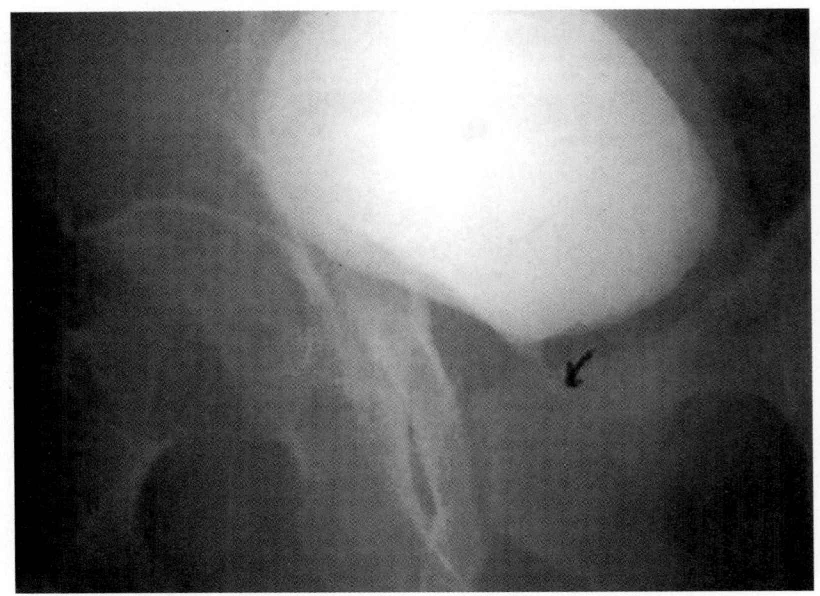

FIGURE 12.76: Bulbar urethral stricture. Micturating urethrogram shows complete stricture of urethra. Beaking of urethral stricture is seen.

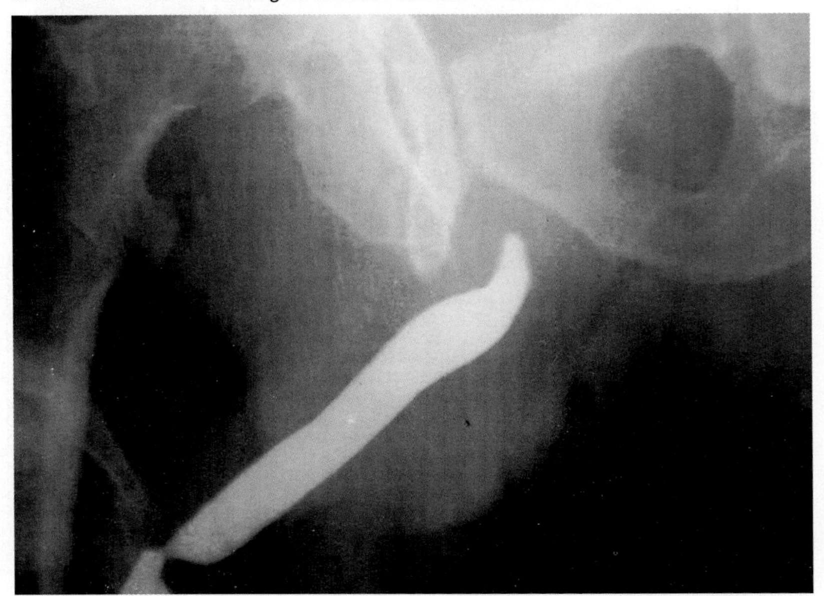

FIGURE 12.77: Retrograde urethrogram shows sudden cut off of the urethral lumen and non-visualization of bulbar urethra in a case of post-traumatic urethral rupture.

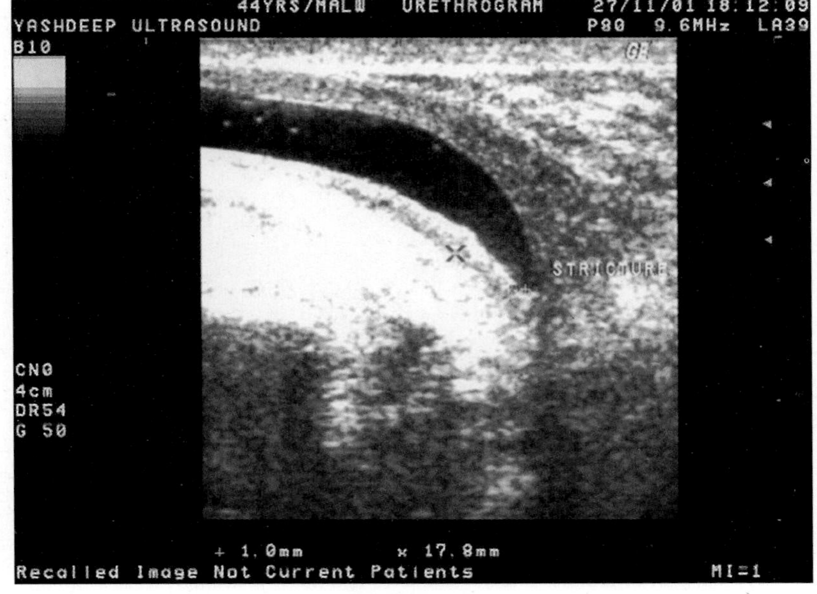

FIGURE 12.78

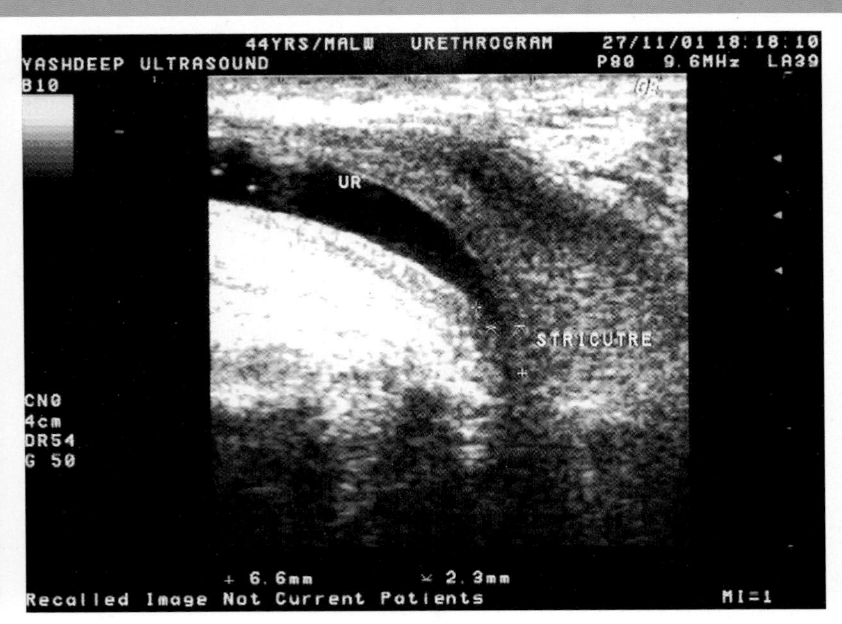

FIGURE 12.79

FIGURES 12.78 and 12.79: Sonourethrogram shows completely obliterated lumen of bulbar urethra. A hypoechoic soft tissue shadow is seen obliterating the lumen of urethra suggestive of marked periurethral fibrosis. The length of urethral structure is around 15 mm. The distal urethra is seen distended.

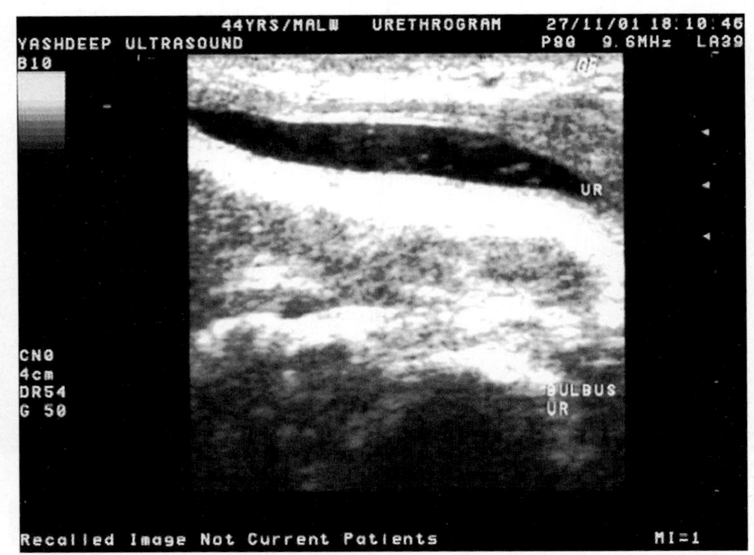

FIGURE 12.80 : The middle and the distal urethra are seen normally with smooth mucosal lining.

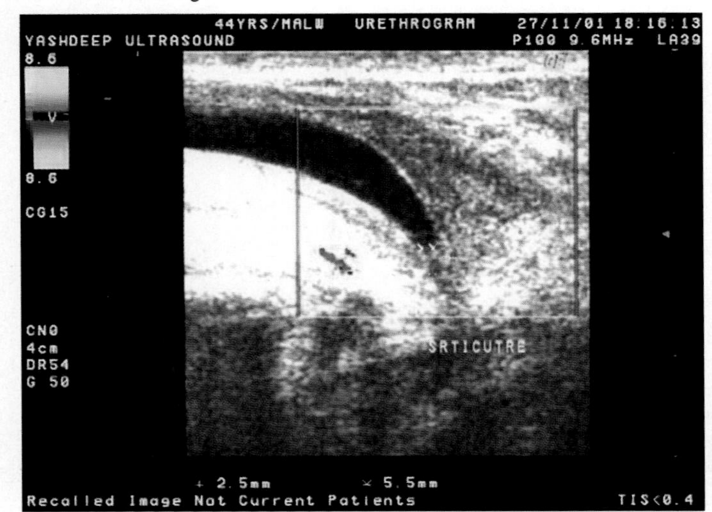

FIGURE 12.81: Color flow imaging through the obliterated urethral structure does not show any flow of the water jet suggestive of completely obliterated lumen by the periurethral fibrosis.

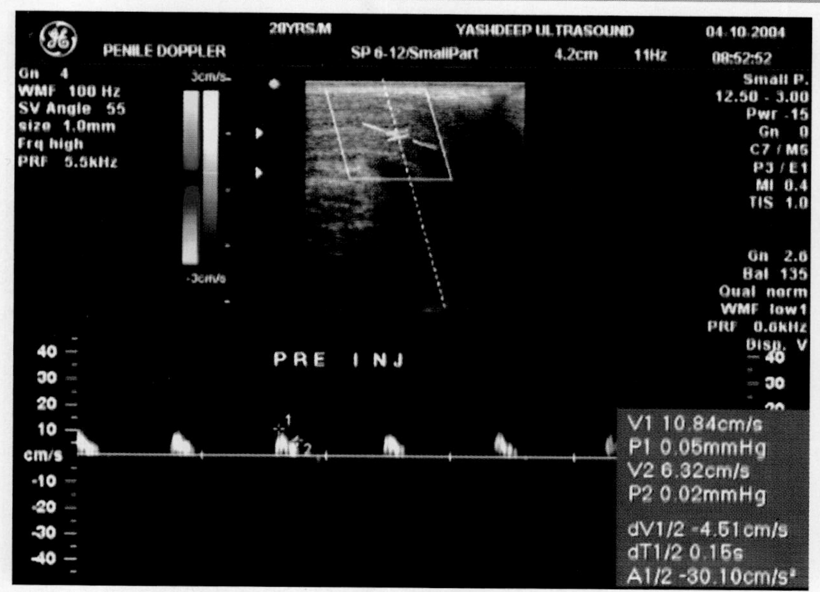

FIGURE 12.82: **Normal cavernosal flow:** Preinjection study shows minimal flow in the cavernosal artery.

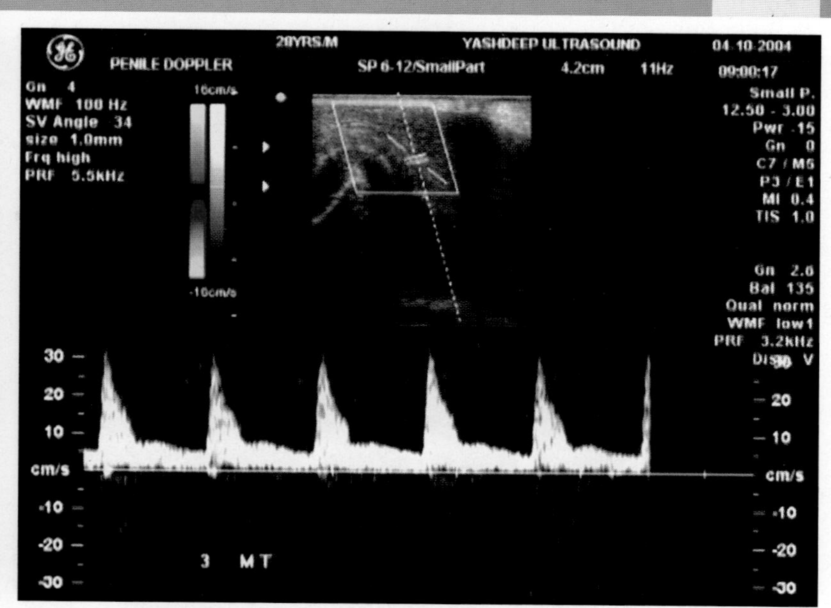

FIGURE 12.83C: Normal sinusoidal flow on color flow imaging. HRSG shows normal filling of the cavernosal sinusoids after injecting vasodilating agents.

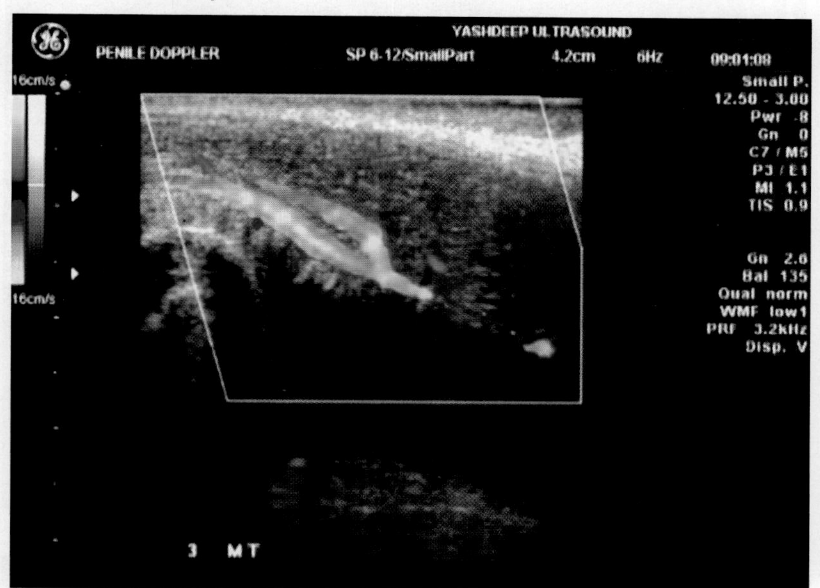

FIGURE 12.83A

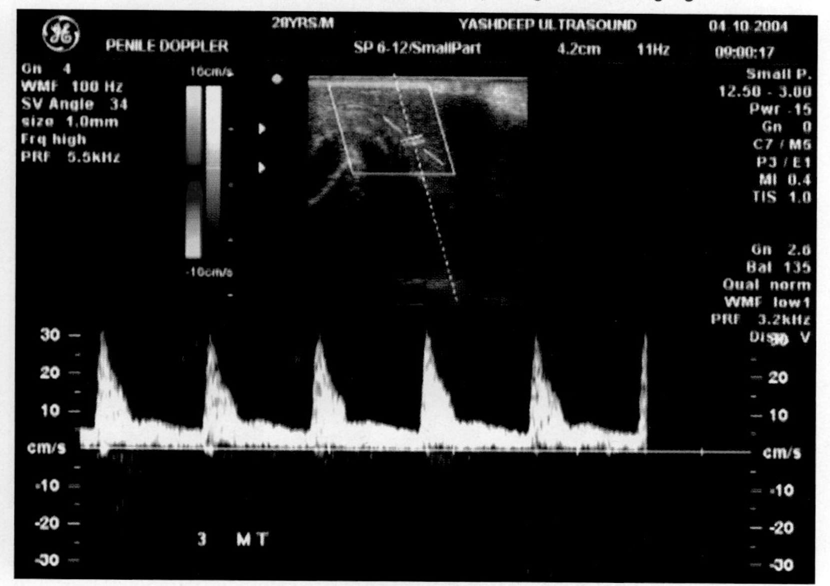

FIGURE 12.84

FIGURE 12.83B

FIGURES 12.83A and B: Cavernosal artery with its branch. HRSG with color flow imaging shows cavernosal artery with its branch. The branches are seen feeding the cavernosal sinuses.

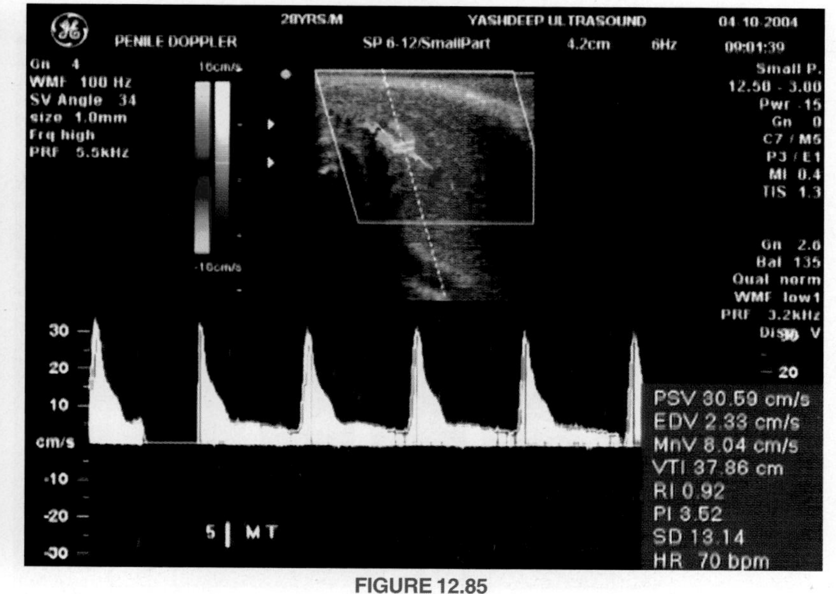

FIGURE 12.85

FIGURES 12.84 and 12.85: Normal penile Doppler sonogram. Normal penile Doppler flow study is seen after injecting vasodilating agent after 3 and 5 minutes. The arterial flow is seen in the cavernosal artery with sharp systolic peak.

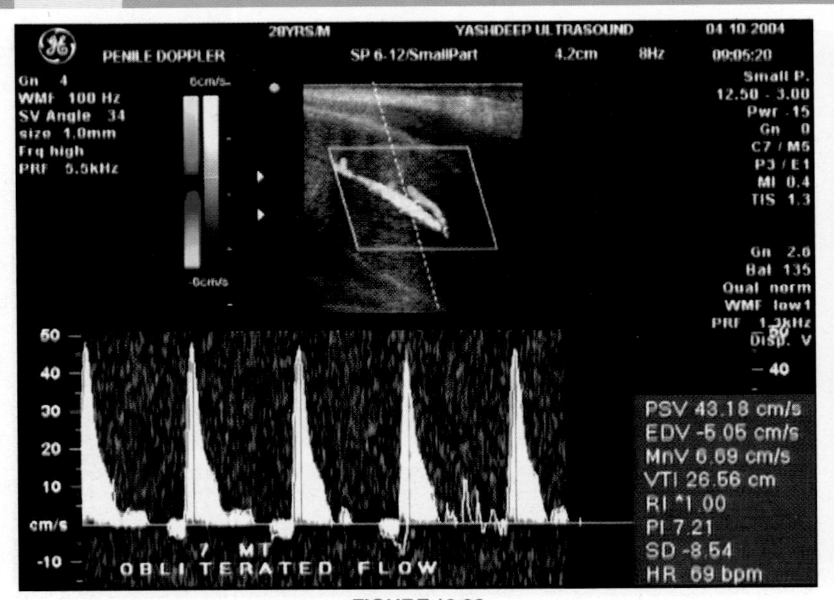

FIGURE 12.86

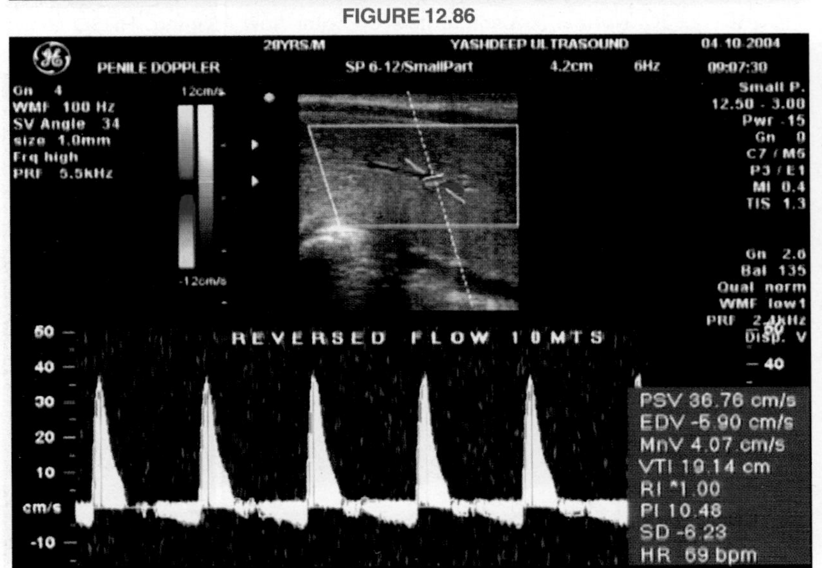

FIGURE 12.87

FIGURES 12.86 and 12.87: The Doppler flow study shows high systolic flow. There is evidence of reversal of the diastolic flow seen due to increased intrapenile pressure. When it exceeds diastolic flow, the reverse diastolic flow is seen and tumescence is achieved.

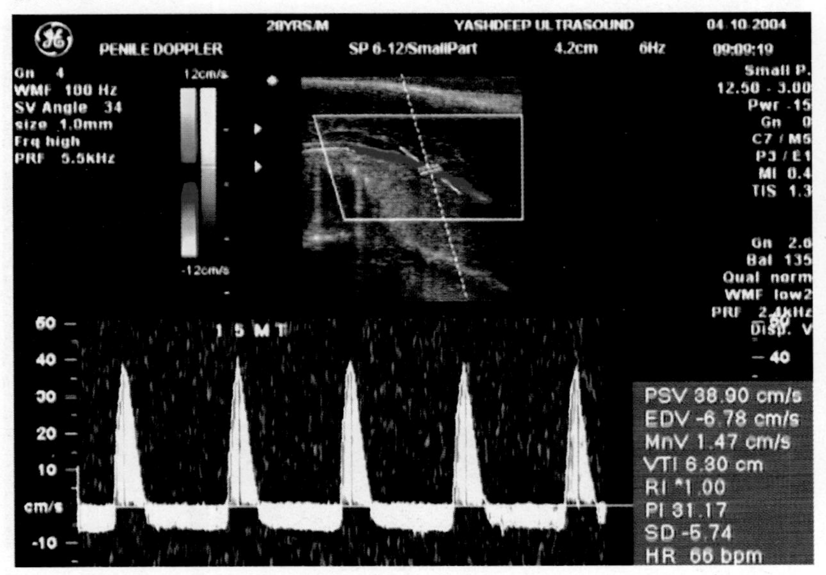

FIGURE 12.88: Doppler spectrum waveform shows high systolic flow with reverse diastolic flow, which also decreases in amount due to the marked increase in the intra-cavernosal pressure resulting in to tumescence.

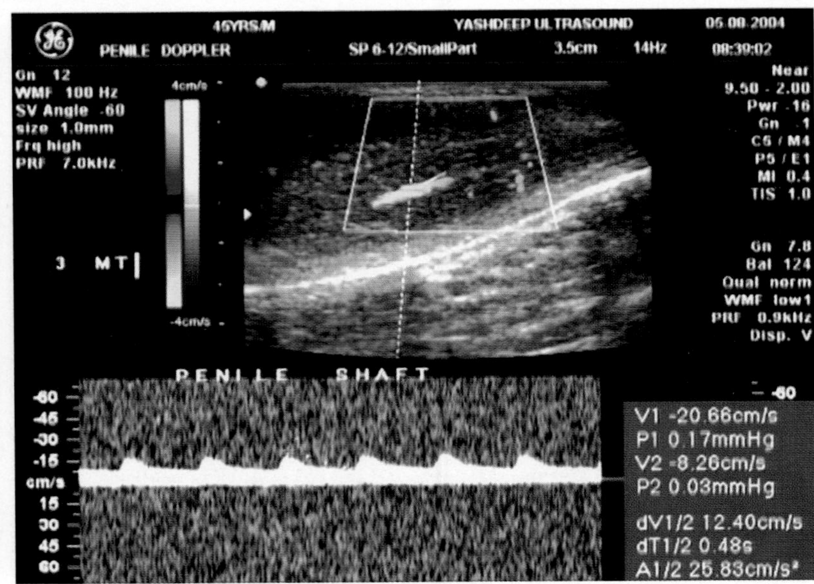

FIGURE 12.89

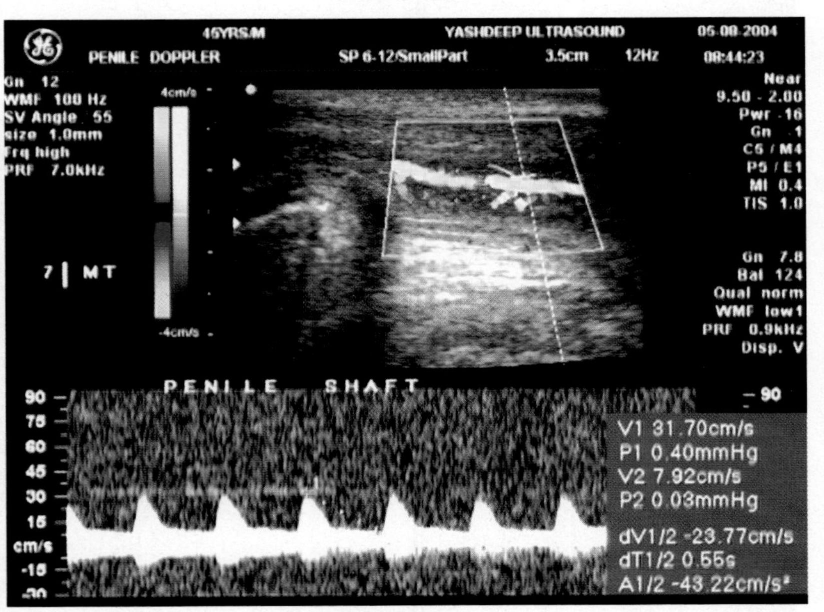

FIGURE 12.90

FIGURE 12.91

FIGURES 12.89 to 12.91: Arterial insufficiency. The Doppler spectral waveform shows increased systolic pressure. However, the systolic peak is less than normal limits and it is 15 cm/sec in its peak state. Partial tumescence is achieved not lasting for long time. Since optimum arterial systolic pressure is not achieved. It suggests of arterial insufficiency. Prolonged diastolic flow is also seen.

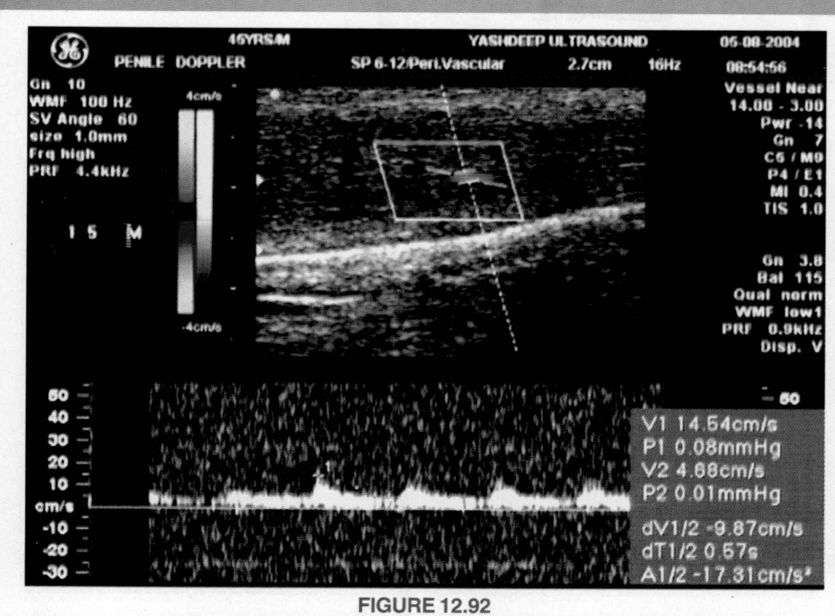

FIGURE 12.92

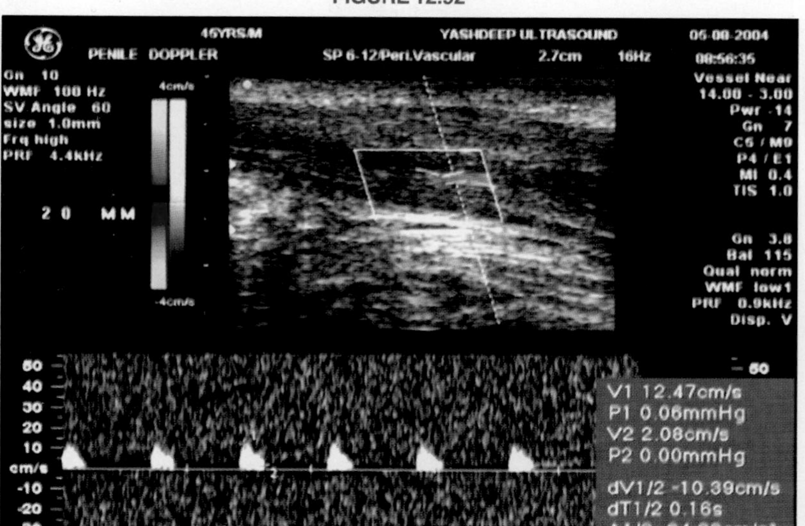

FIGURE 12.93

FIGURES 12.92 and 12.93: After 15 and 20 minutes decreased flow is seen in the cavernosal arteries.

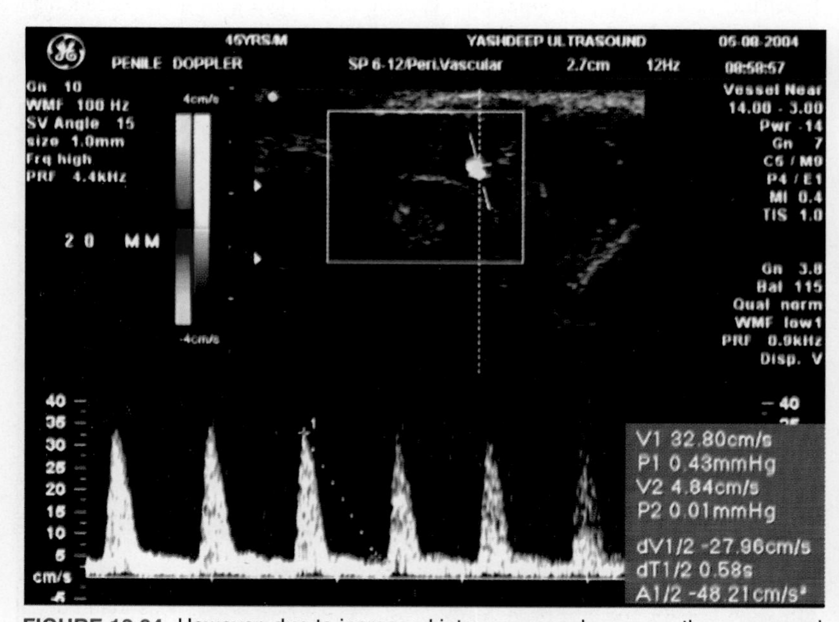

FIGURE 12.94: However, due to increased intracavernosal pressure the cavernosal artery pressure increases to the borderline limit of the normal pressure and partial tumescence was achieved. But it is not sustained for optimum period.

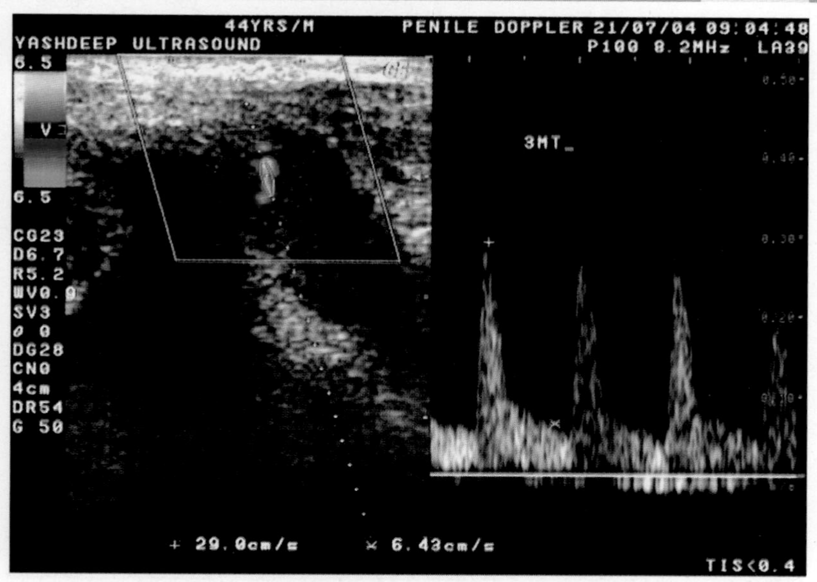

FIGURE 12.95: Venous leak. Persistent high venous flow is seen in a patient after injecting vasodilating agent. The diastolic flow is more than 6 cm/second, which is much more than the normal limits (normal >3 cm/ sec).

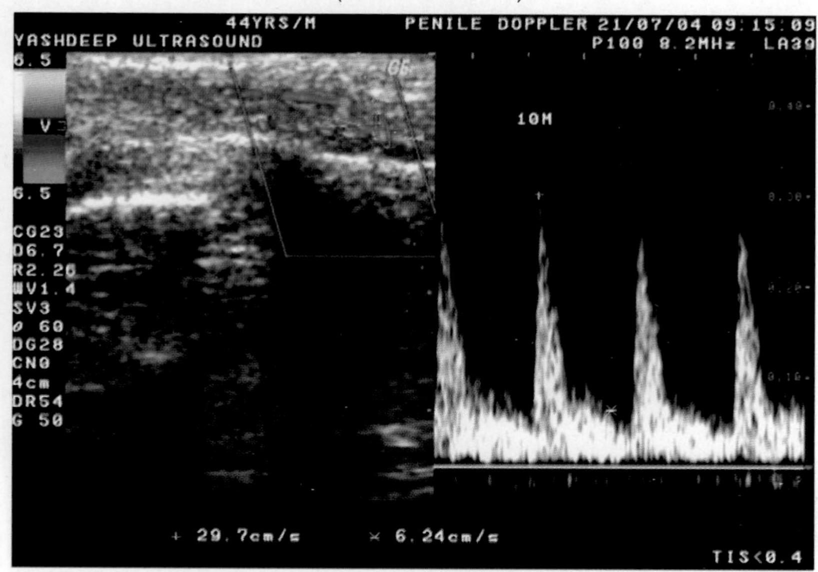

FIGURE 12.96: Venous leak. Persistent high diastolic flow is seen after 10 minutes of injection and it remains more than 6 cm/sec thus indicating venous leak and tumescence is not achieved.

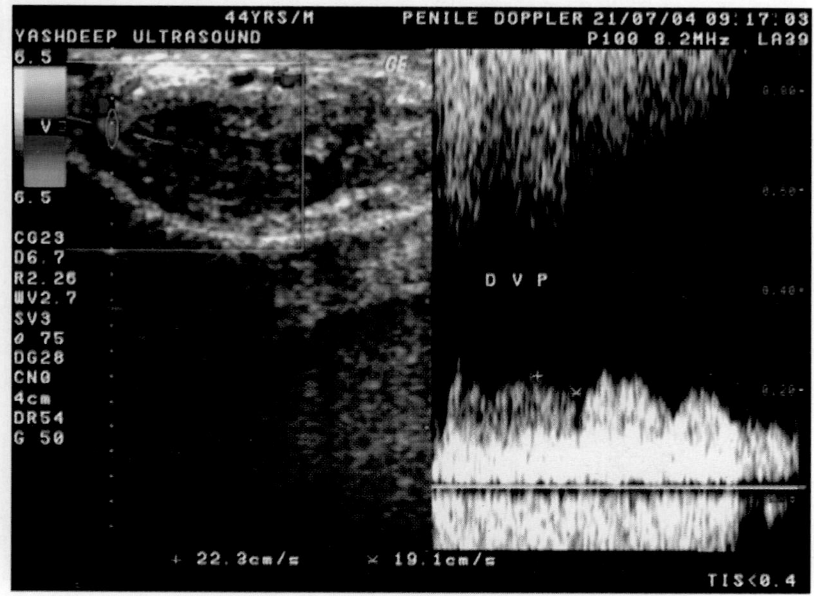

FIGURE 12.97: High flow is seen in the dorsal vein of penis. No obliteration of flow is seen.

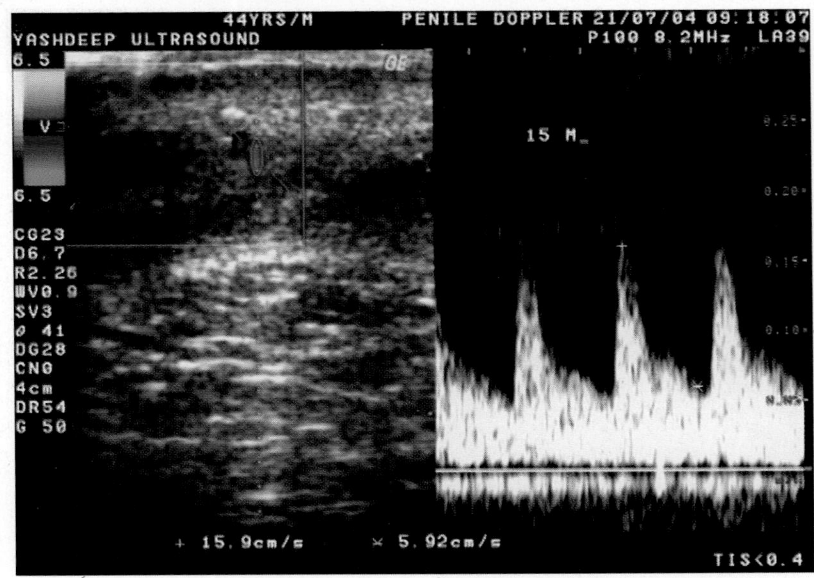

FIGURE 12.98: After 15 minutes of the injection still high diastolic flow is seen in the cavernosal artery and the systolic pressure was much less than the normal limits.:

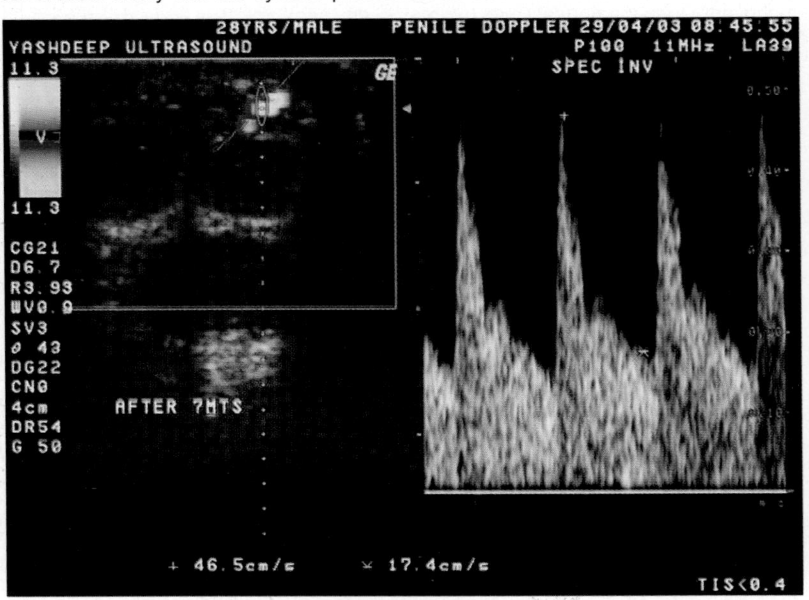

FIGURE 12.99

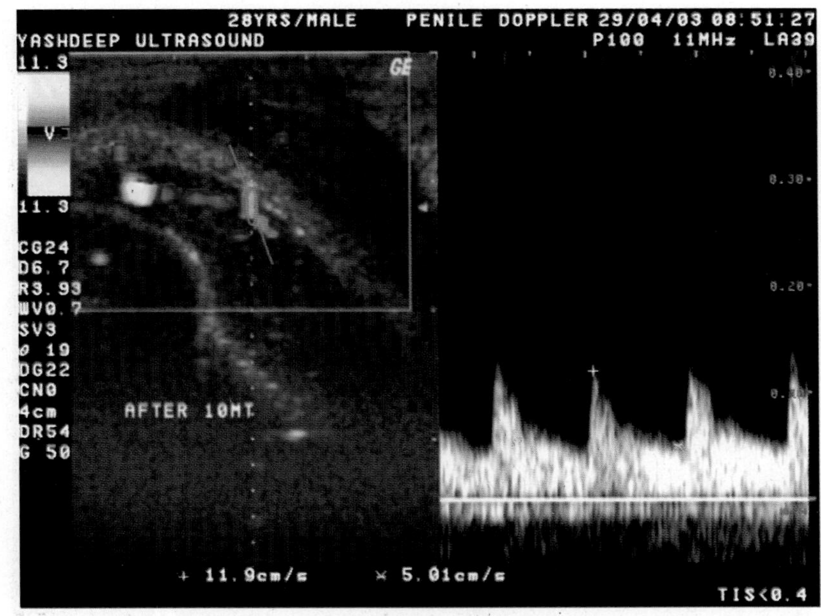

FIGURE 12.100

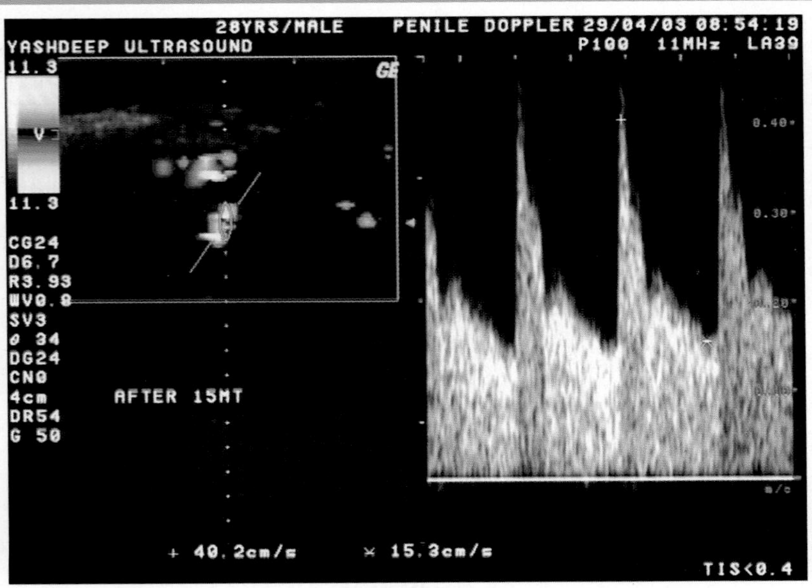

FIGURE 12.101

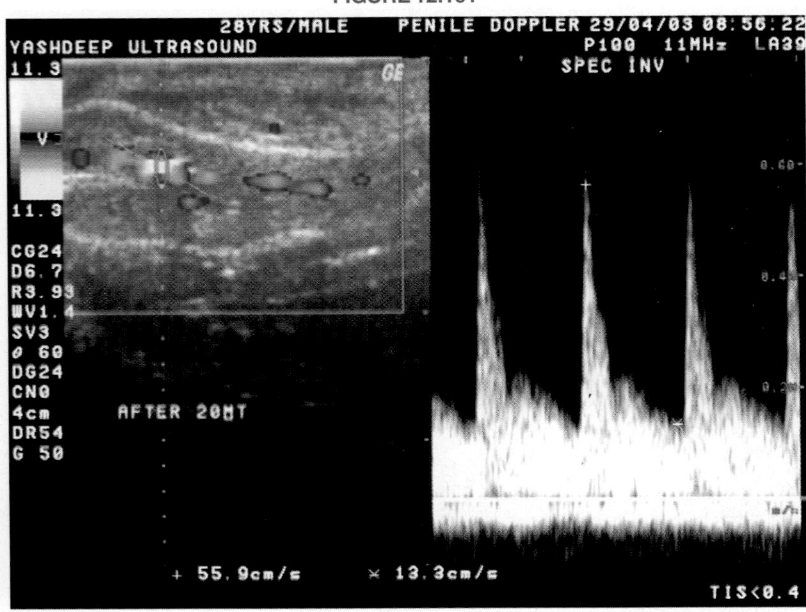

FIGURE 12.102

FIGURES 12.99 to 12.102: Persistent venous leak HRSG showed persistent high diastolic flow in a patient of erectile dysfunction suggestive of venous leak. The persistent high diastolic flow was seen through out the study at 10, 15 and 20 minutes of Doppler tracing suggestive of high venous leak.

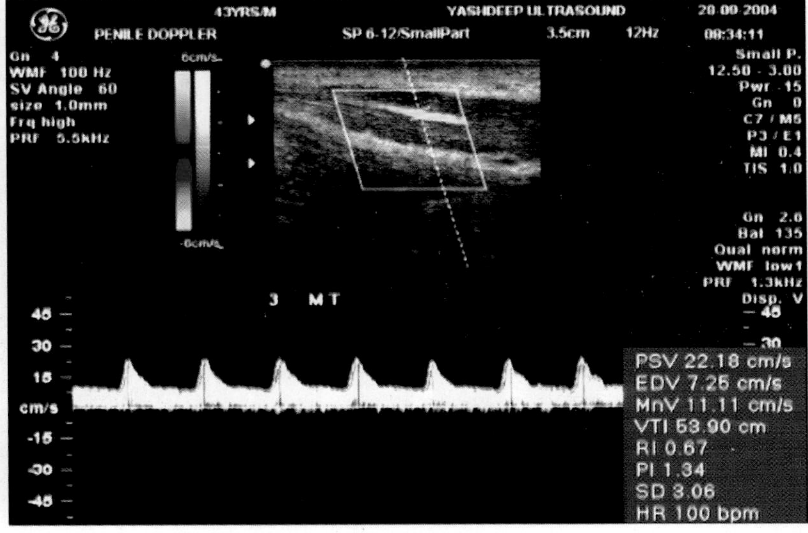

FIGURE 12.103

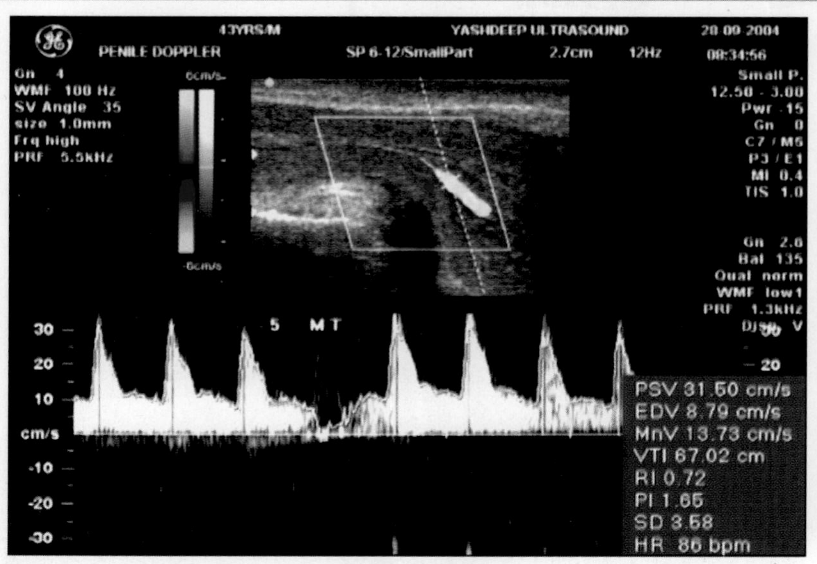

FIGURE 12.104

FIGURES 12.103 and 12.104: Arterial insufficiency with venous leak. A patient of erectile dysfunction showed arterial insufficiency with venous leak. Doppler study shows suboptimal arterial flow after injecting vasodilating drug. The maximum peak arterial velocity was achieved 31.5 cm/sec. However, persistent high diastolic flow was also seen suggestive of venous leak.

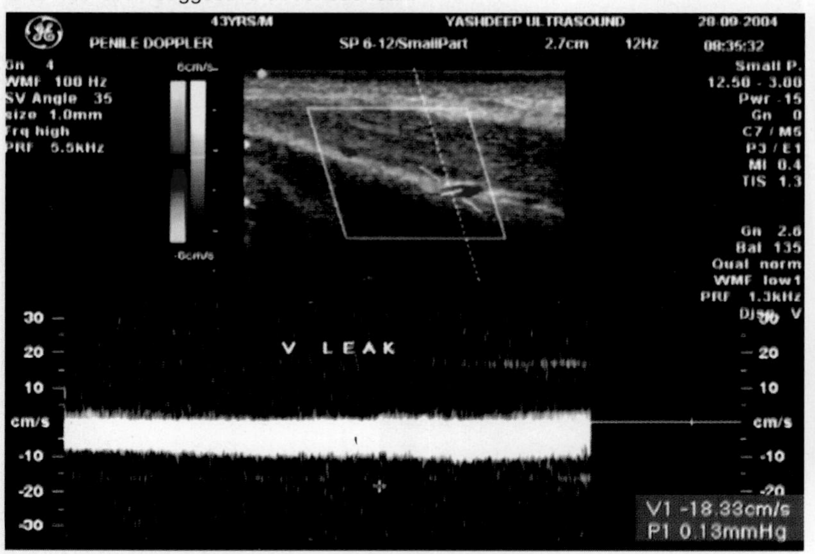

FIGURE 12.105

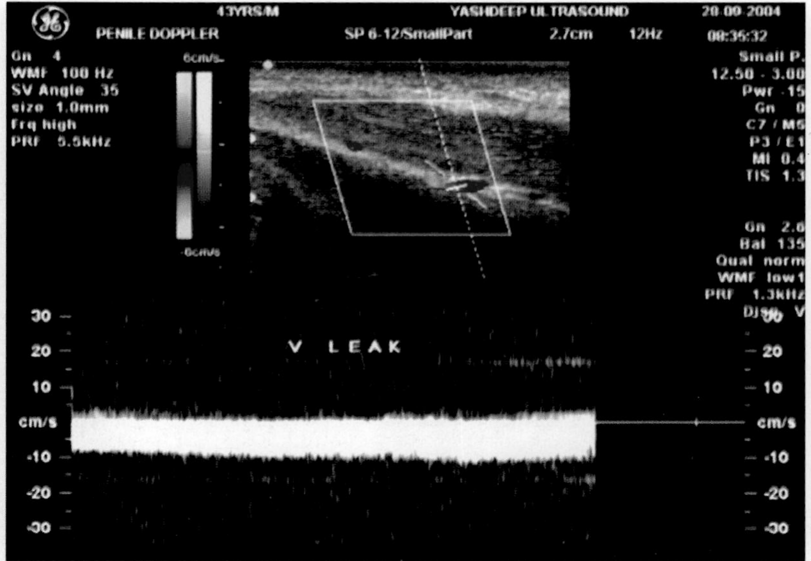

FIGURE 12.106

FIGURES 12.105 and 12.106: There is evidence of high flow seen in the veins of the penis. No obliteration of venous flow seen through out the study.

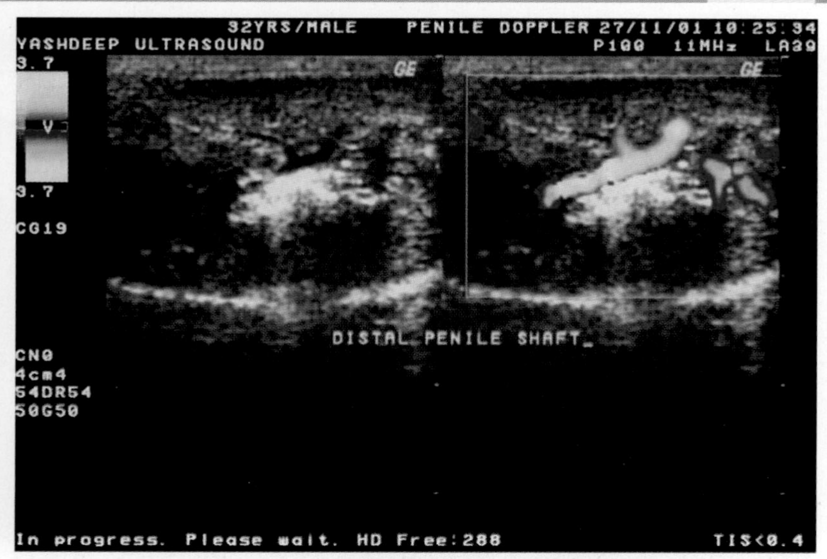

FIGURE 12.107

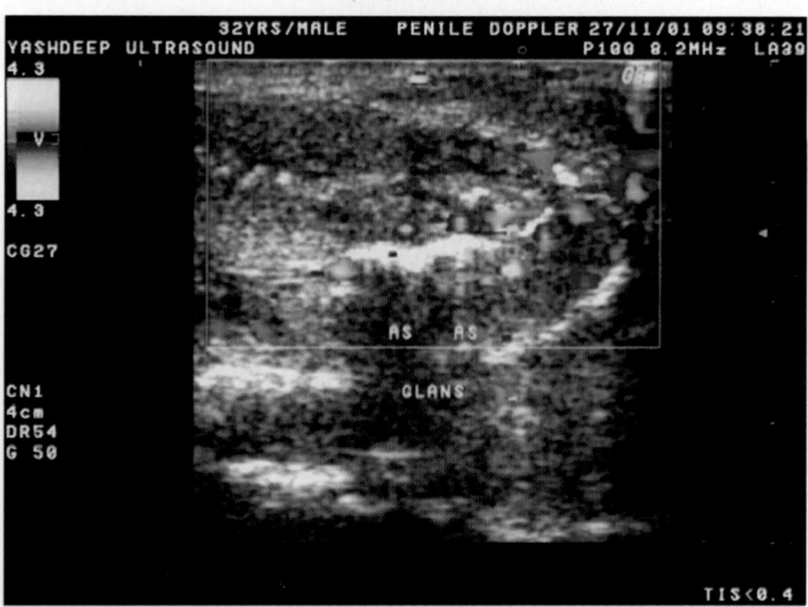

FIGURE 12.108

FIGURES 12.107 and 12.108: Venous malformation. A patient presented with impotence. HRSG and color flow imaging shows multiple dilated vessels in the distal part of the corpora cavernosa. On color flow-imaging pooling of the blood is seen in them with high venous flow suggestive of venous malformation.

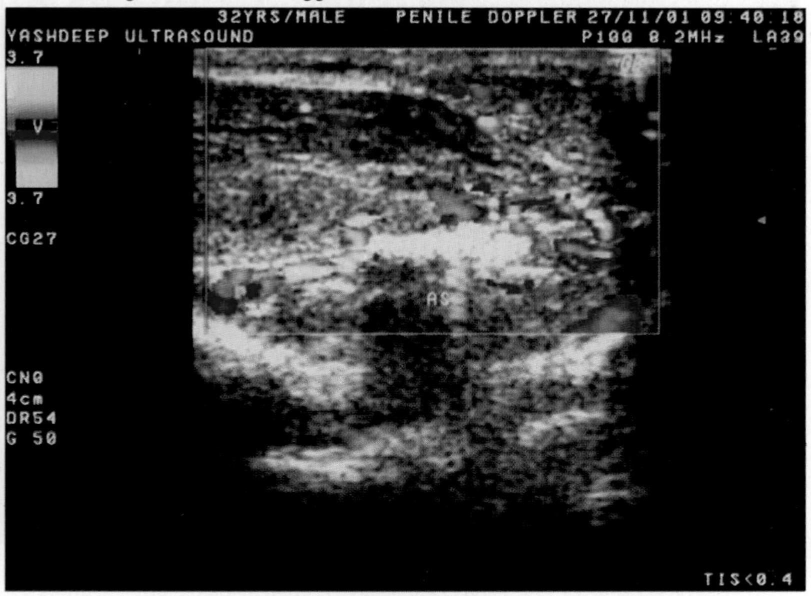

FIGURE 12.109

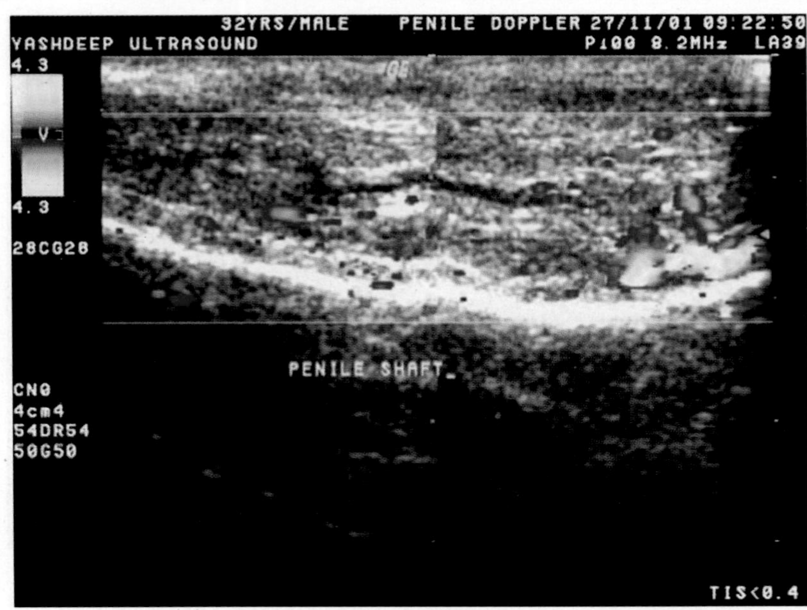

FIGURE 12.110

FIGURES 12.109 and 12.110: Echogenic calcified shadows are also seen in the veins, which are accompanied with acoustic shadowing suggestive phlebolith. Tortuous dilated veins are seen going through the corpora cavernosa.

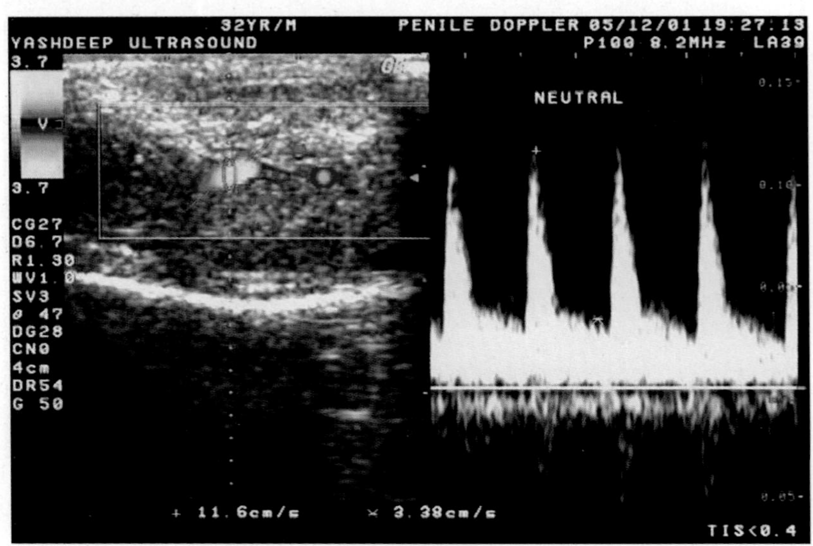

FIGURE 12.111

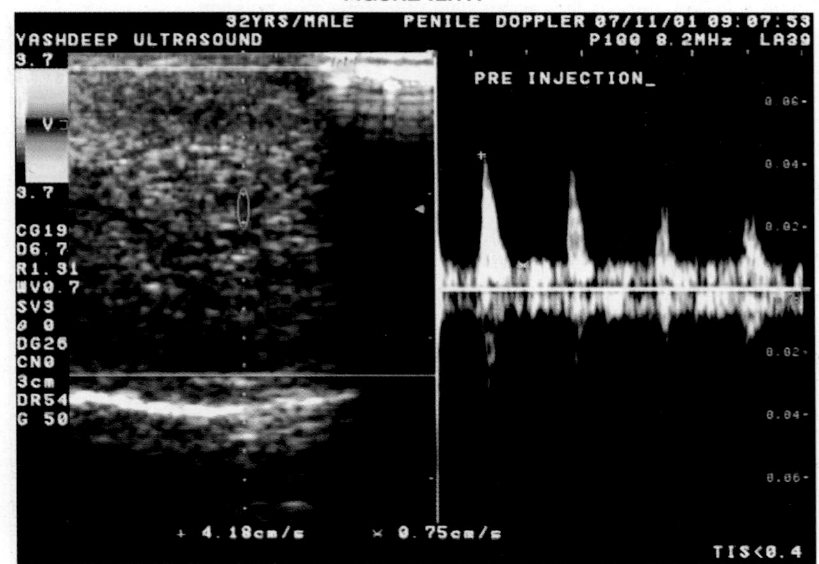

FIGURE 12.112

FIGURES 12.111 and 12.112: High diastolic flow is seen in the cavernosal arteries in neutral position before injecting vasodilating drug. The systolic pressure is low.

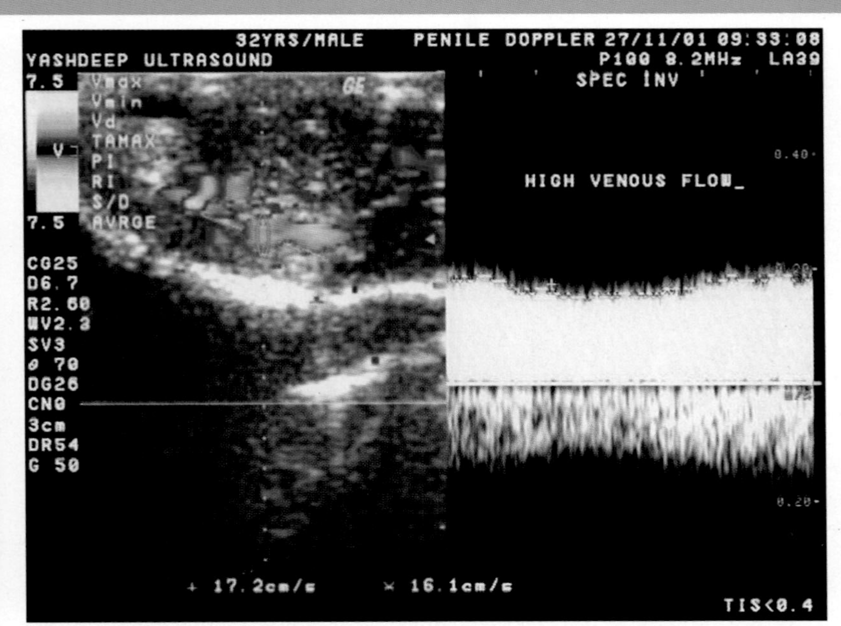

FIGURE 12.113: High venous flow is seen in dilated tortuous vein, typical monophasic venous flow is seen and maximum venous velocity was 17.2 cm/sec after injecting the drug.

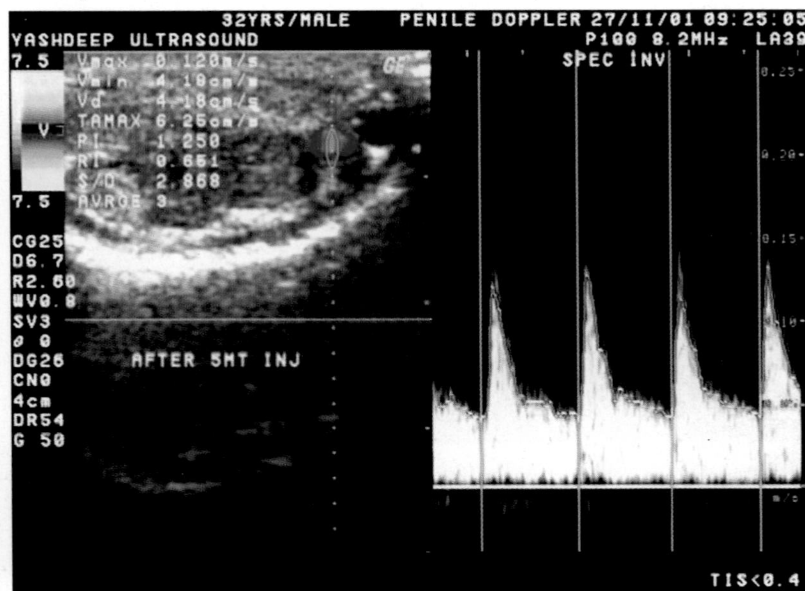

FIGURE 12.114

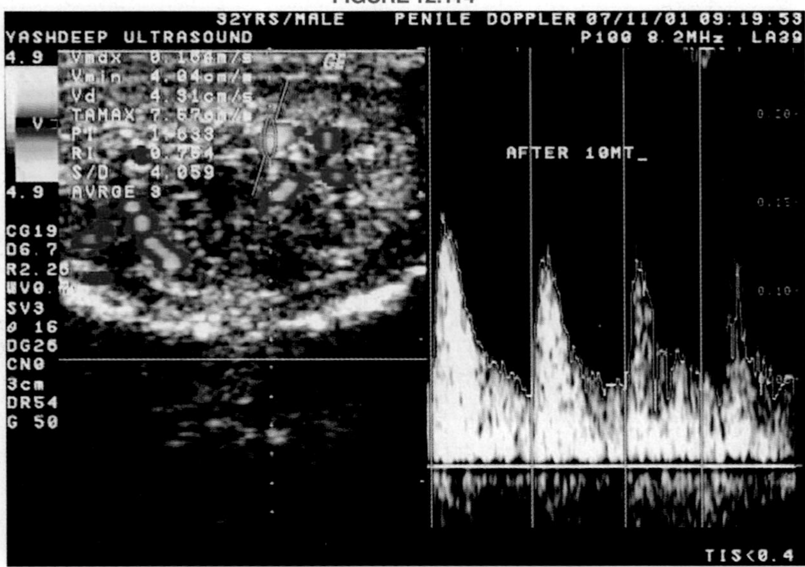

FIGURE 12.115

FIGURES 12.114 and 12.115: Persistent high diastolic flow is seen in the cavernosal arteries on either side. The systolic flow is 12 and 11cm respectively in both cavernosal arteries.

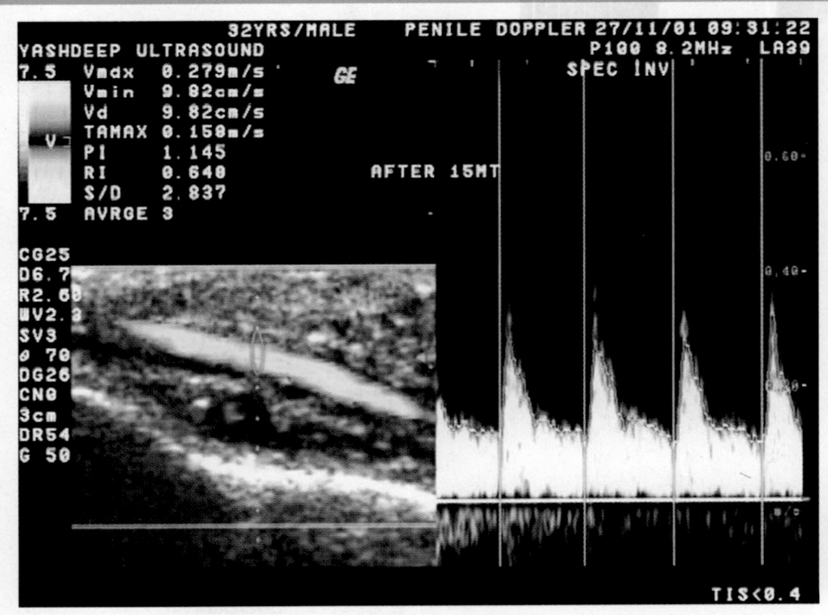

FIGURE 12.116: After 15 minutes persistent high diastolic flow is seen, which is 9 cm/ sec. The systolic pressure is 28 cm.

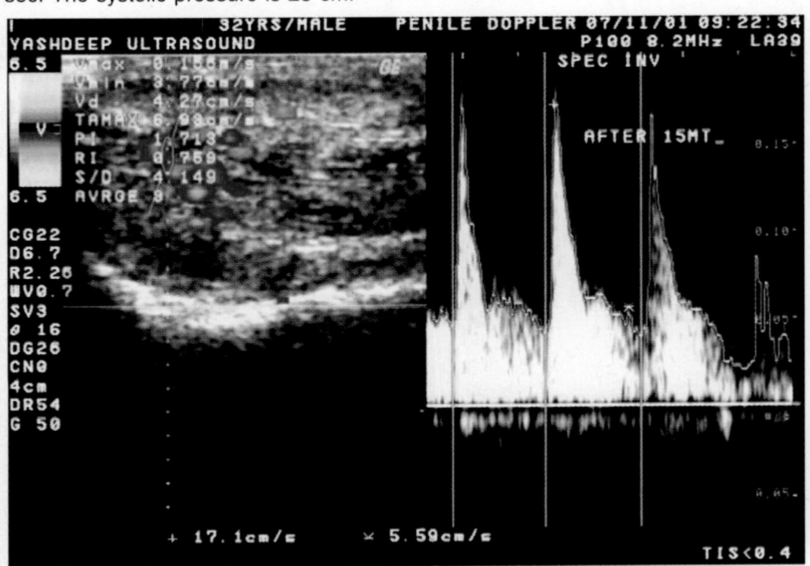

FIGURE 12.117

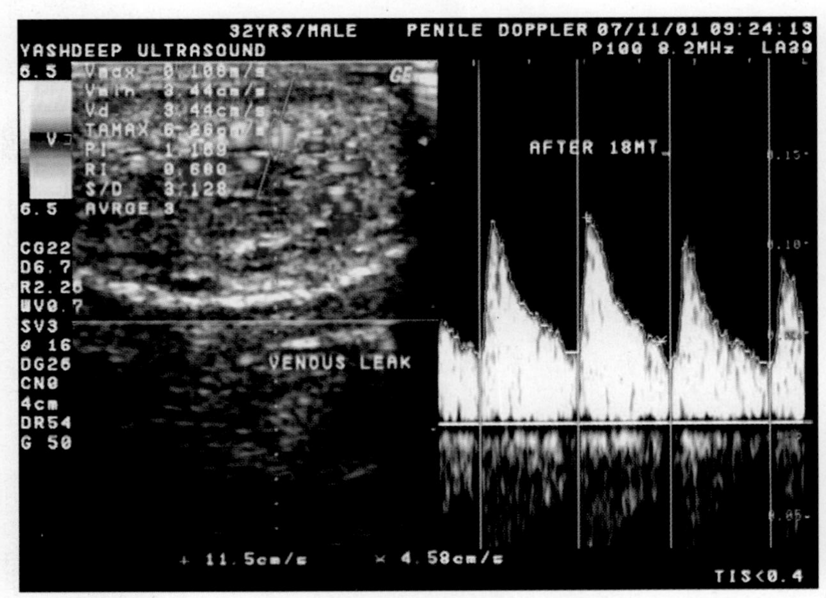

FIGURE 12.118

FIGURES 12.117 and 12.118: Persistent high diastolic flow is seen after 15 and 18 minutes of study.

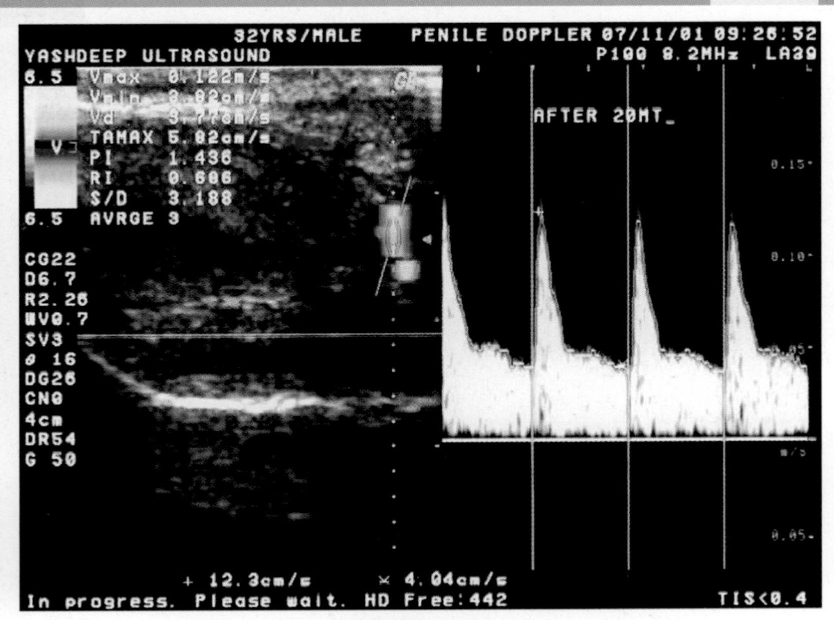

FIGURE 12.119

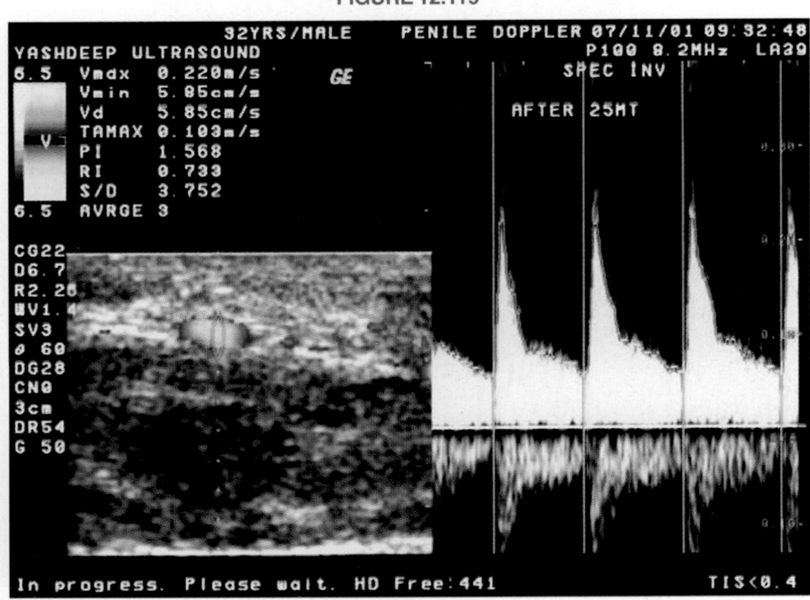

FIGURE 12.120

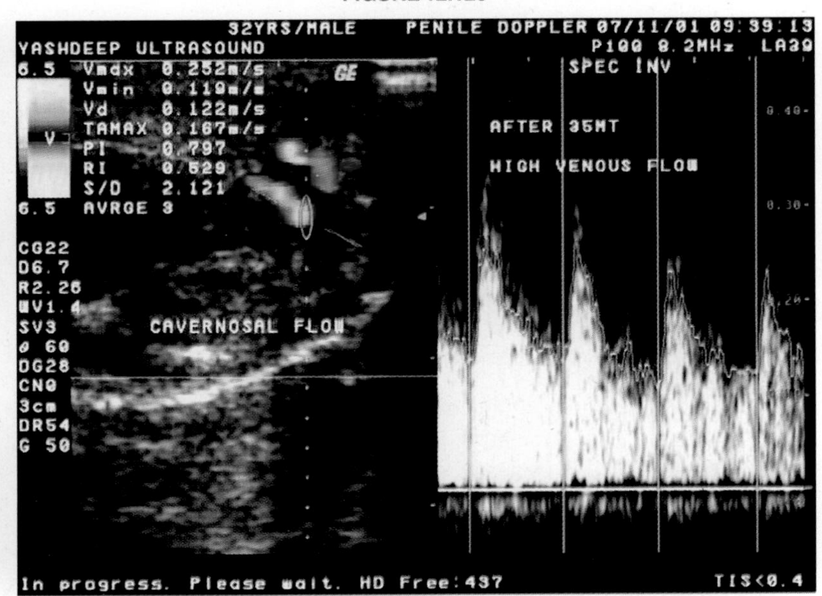

FIGURE 12.121

FIGURES 12.119 to 12.121: Persistent high diastolic flow is seen in the cavernosal arteries recorded at 20, 25 and 35 minutes suggestive of high venous leak. The diastolic flow is 5.85 cm/sec, which is more than the normal.

High Resolution Sonography of Musculoskeletal System

13

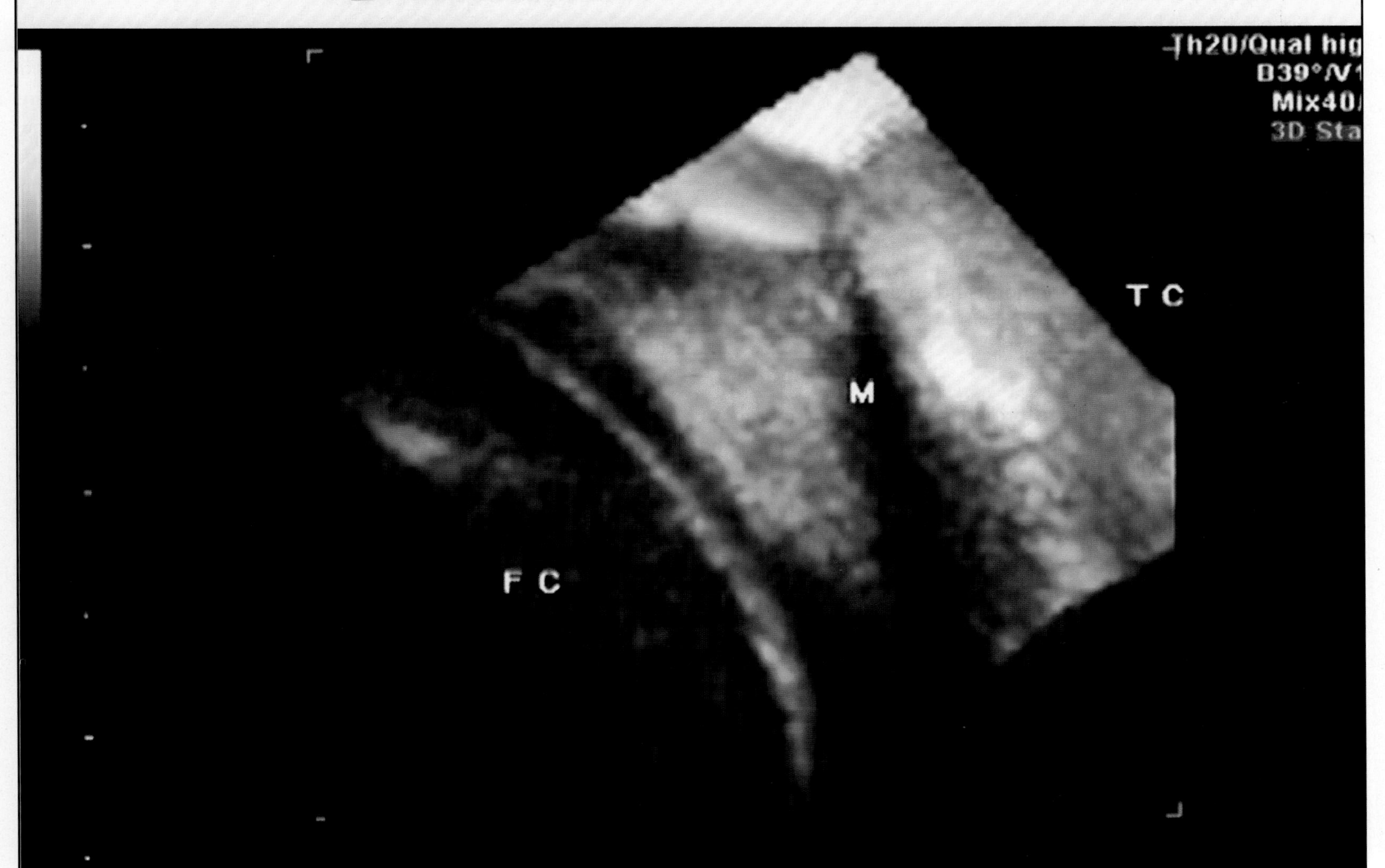

Introduction

High-resolution sonography has become the choice of investigation for evaluation of musculoskeletal system. The multi planner capability, improved tissue characterization and dynamic study of musculoskeletal system have made ultrasound imaging as an important investigative tool for evaluation of musculoskeletal system. With the state of the art equipments and high resolution transducers sonography provides all information available with MRI and more with regards to muscle pathology.

Equipments

High-resolution short focus linear transducers having frequency ranging from 7.5 Mhz to 15 Mhz are the pre requisite for evaluation of musculoskeletal system. Obese patient may require 5 Mhz transducers for better penetration. The equipment with split image facility is must for evaluation of the musculoskeletal system for demonstrating a very long segment or entire muscle.

Indications

Following are the main indication for high-resolution sonography in musculoskeletal system:

1. Evaluation of muscles and other pathologies.
2. Sonography of tendons.
3. Sonography of bursae
4. Sonography of ligaments.
5. Sonography of large synovial joints.
6. Pediatric joint sonography.
7. Sonography of periosteum and bone.
8. Sonography of rheumatoid disease.
9. Evaluation of foreign bodies.
10. Sonography of shoulder joints.
11. Sonography of elbow wrist and hands.
12. Sonography of hip joint, knee joint and ankle joint.
13. Pain syndrome after arthroscopy.

Sonography of Muscles

High-resolution sonography can provide more information about muscles than other imaging modalities like MRI or CT scanning. The availability of examination and low cost make it more practical and affordable technique than MRI. In today era the sport injury makes the largest group, as it constitutes about 40% of the patients who are referred to HRSG evaluation.

Since muscles are dynamic structures therefore, they need dynamic real time study for proper evaluation of muscles and related injury. State of the art equipment with cine loop facility is excellent and very helpful for muscle evaluation. Identification of muscle can be made by its location, origin, insertion and function. All these can be easily done under sonographic vision.

Muscle Pathology

Majority of muscle pathology are traumatic either job related or sports injury. A small group falls into tumor category.

The lesions can be grouped into two:

1. Intramuscular
2. Muscle boundary lesion.

Intramuscular Lesion

The intramuscle lesions are the lesions confined to the muscles. They are as follows:

Muscle rupture: Muscle rupture can be caused by direct trauma or indirect trauma as muscles are crushed against underlying bones. HRSG is good to evaluate compressive muscle rupture. Irregular cavity and shaggy borders characterize USG findings. USG clearly shows disruption of muscle fibers. Complete muscle rupture is rare, partial muscle rupture is more common.

Hematoma: Hematoma formation is hallmark of muscle rupture. Size of hemorrhage is related to extent of injury. USG is very good for evaluation of hemorrhage and resorbption of hemorrhage can be studied on USG. Muscle cysts may be chronic or traumatic. They may be extension of synovial cyst.

Myositis ossificans: Muscle contusion with intramuscular hemorrhage may calcify and then ossify. These lesions are the frequent findings in athletes involved in contact sports. Evaluation of myositis ossificans is easy with sonography. Maturation of these lesions takes about 5-6 months. It is identified as soft tissue mass with disorganized inhomogeneous consistency. It is at time difficult to differentiate it from soft tissue neoplasm.

Myositis: HRSG is very helpful in early detection of myositis when the clinical picture is nonspecific. The muscle becomes hypoechoic due to exudation and muscle edema. Comparison with normal side will differentiate the affected side well. A fulminating. Myositis will result in abscess formation secondary to osteomyelitis.

Tubercular abscess: Tuberculosis of the muscle is not very common. Now it is seen frequently in more resistant type of tuberculosis. Most of the time it is seen as an isolated abscess formation involving the muscle groups in the body. However, it may be associated with systemic disease. Many times it can present as vascular ischemic contracture when seen when involves the muscle group of forearm. HRSG findings are non-specific. But it shows irregular hypoechoic low-level echo complex abscesses with loss of normal muscle texture. Central area of necrosis is seen in the abscess. Perifocal edema may present when superadded infection is seen.

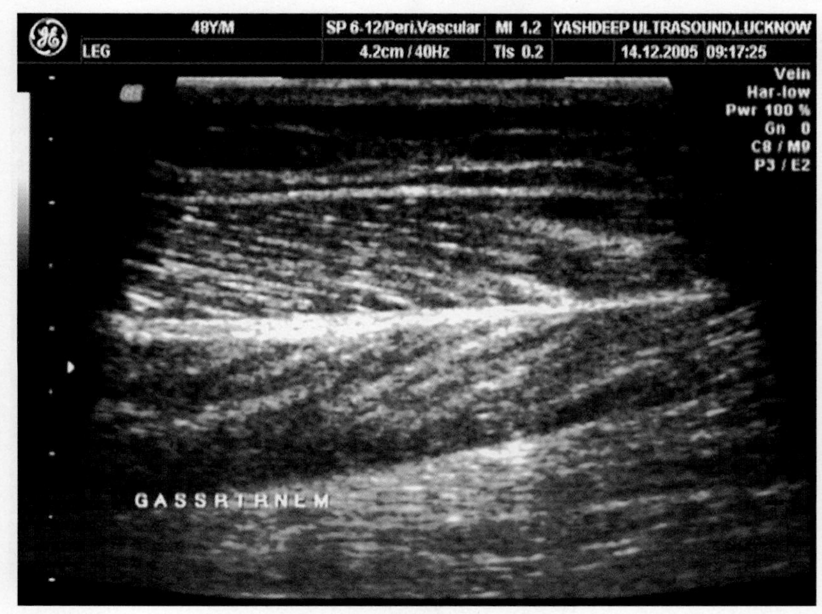

FIGURE 13.1

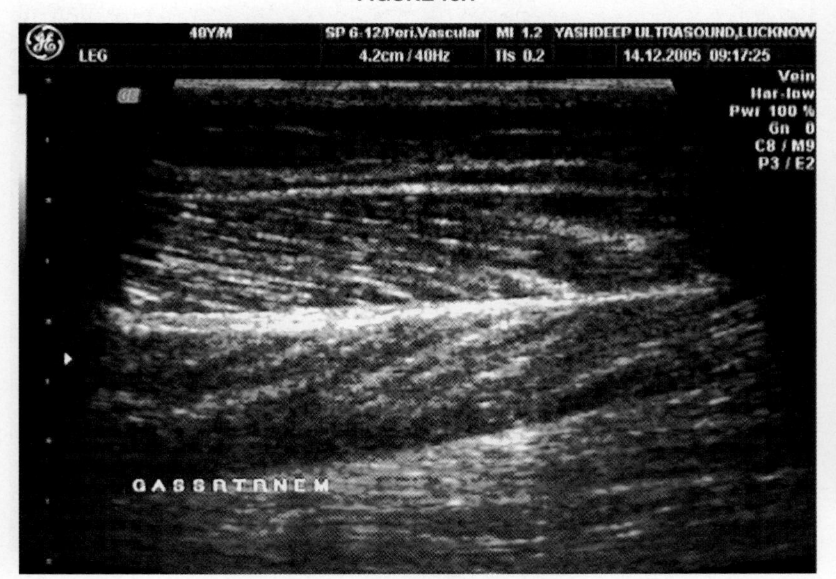

FIGURE 13.2

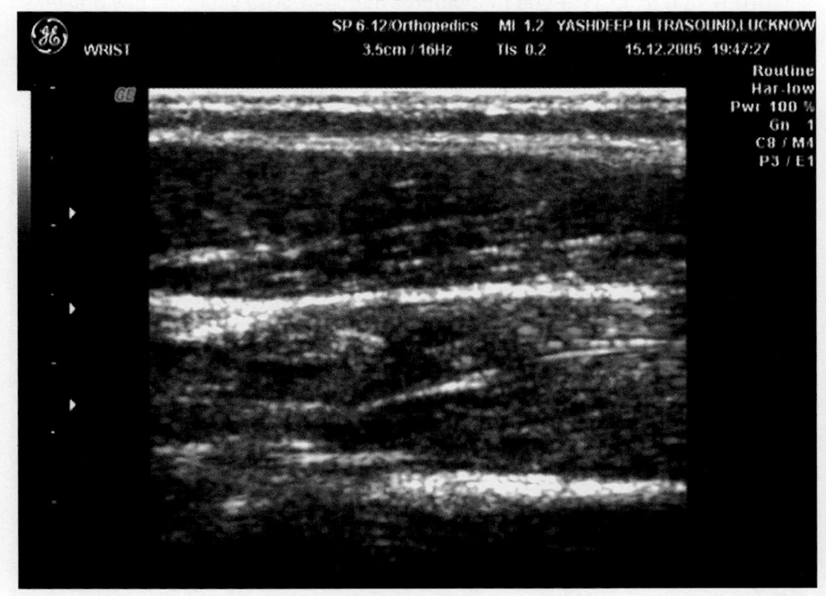

FIGURE 13.3

FIGURES 13.1 to 13.3: Normal anatomy. HRSG shows typical normal bipinnate pattern of skeletal muscle. Echogenic fibrous septa are seen running through the muscle. The muscle fibers are seen arranged in bipinnate fashion.

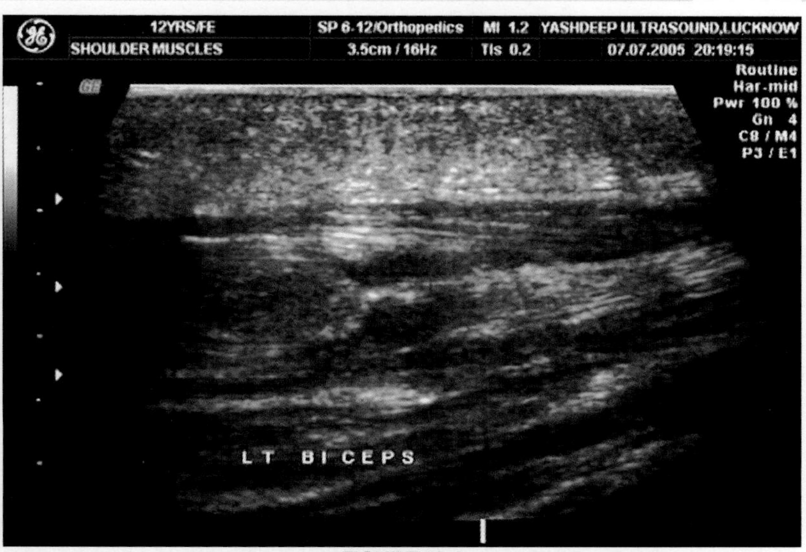

FIGURE 13.4

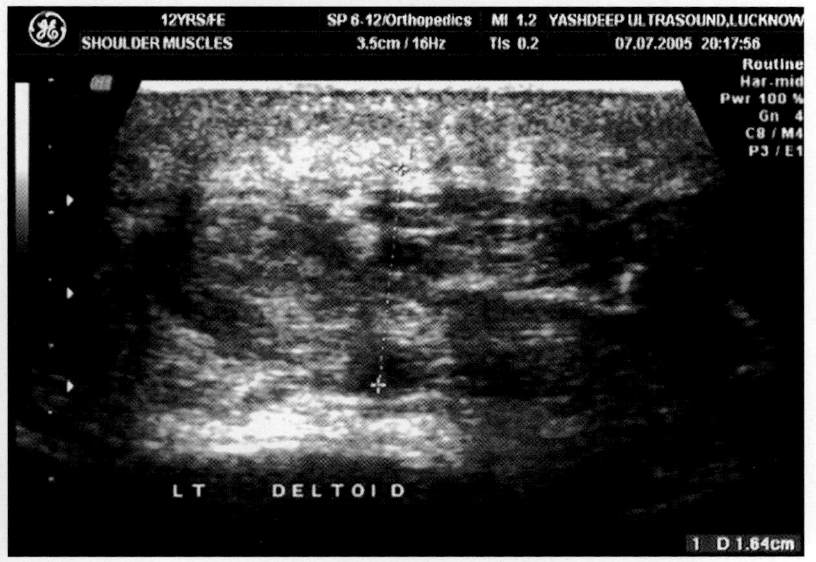

FIGURE 13.5

FIGURES 13.4 and 13.5: Acute myositis. A young girl presented with restricted shoulder movement with acute pain with high fever. HRSG shows marked swollen edematous biceps and deltoid muscles of the Lt shoulder joints. Multiple irregular hypoechoic areas are seen in the belly of the muscles suggestive of micro abscess associated subcutaneous tissue edema.

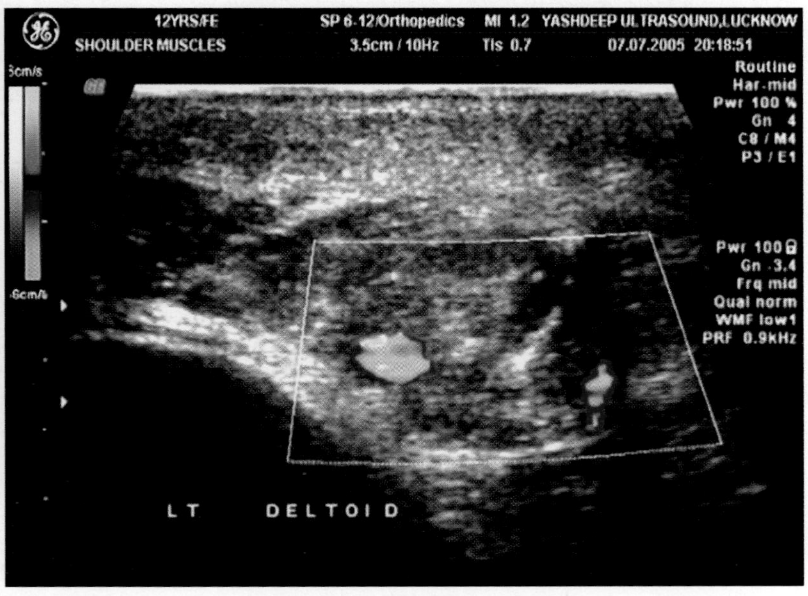

FIGURE 13.6: On color flow imaging multiple vessels are seen in the abscess.

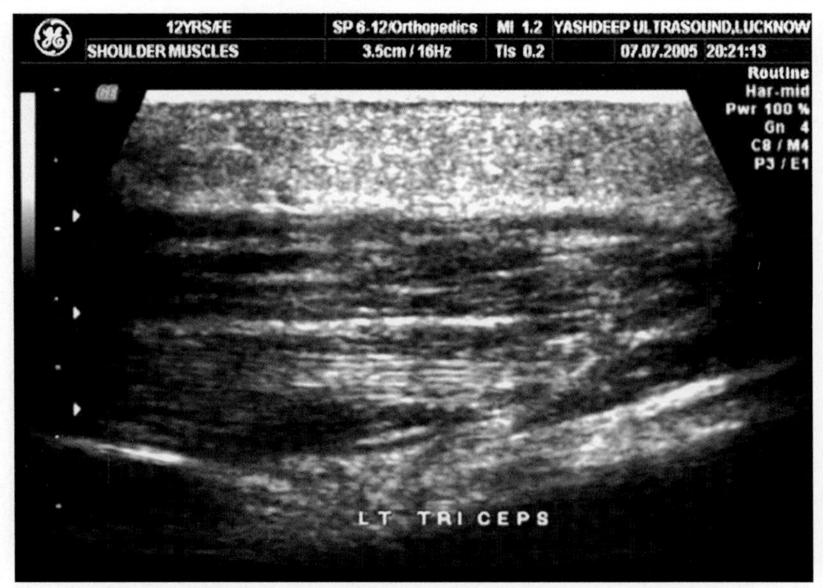

FIGURE 13.7: The triceps muscle of the same patient is also edematous and hypoechoic in texture suggestive of acute myositis.

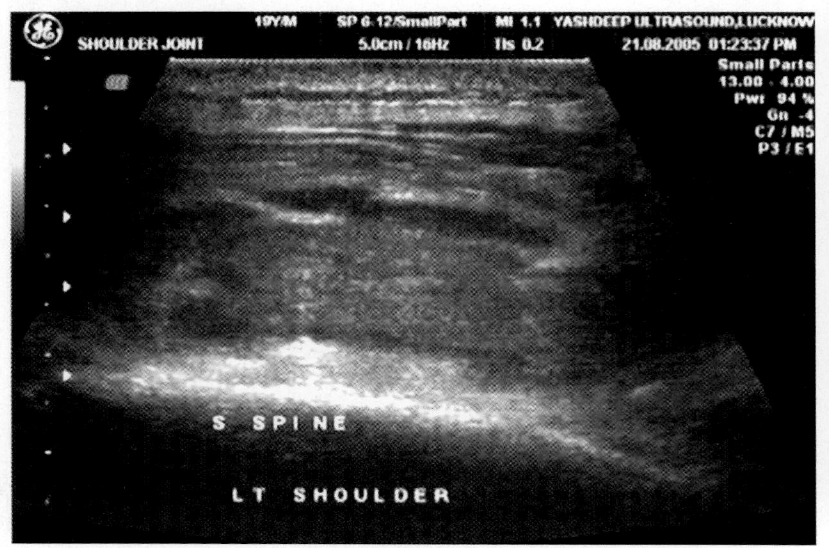

FIGURE 13.8

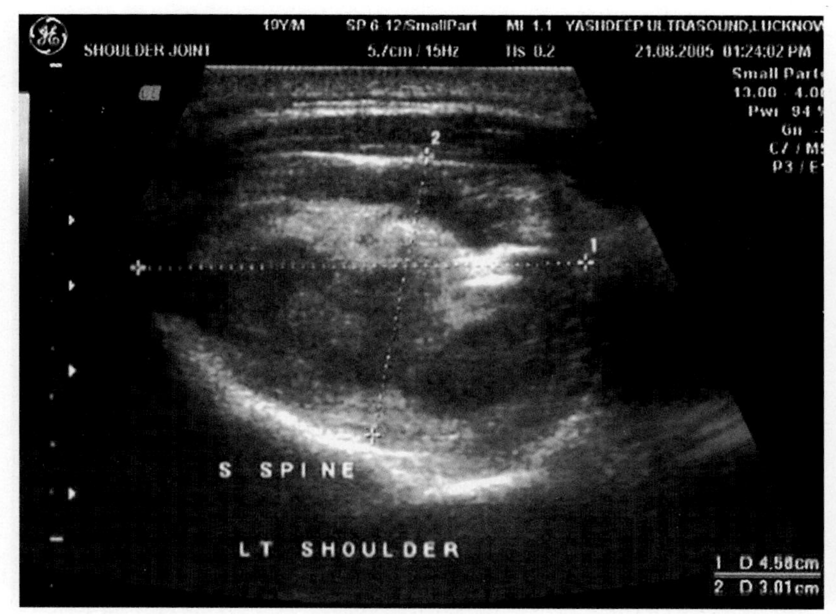

FIGURE 13.9

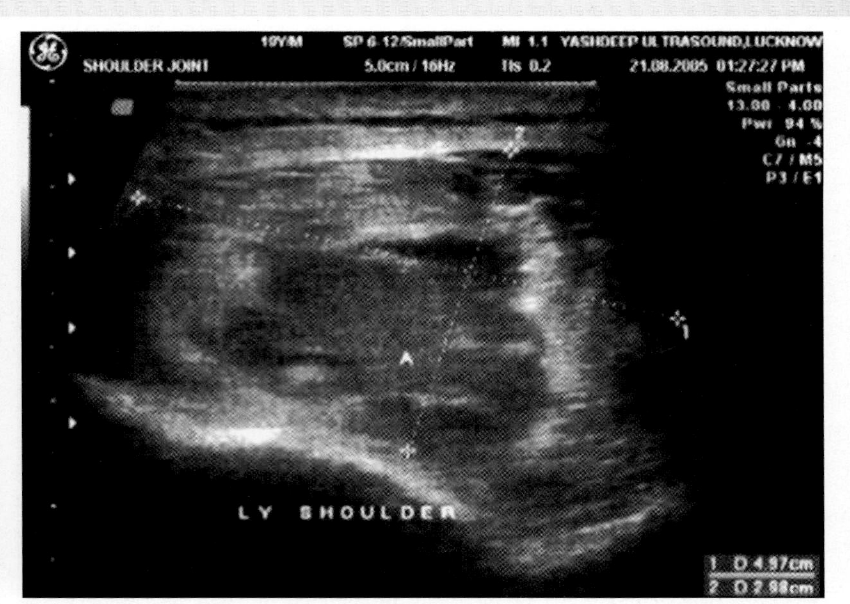

FIGURE 13.10

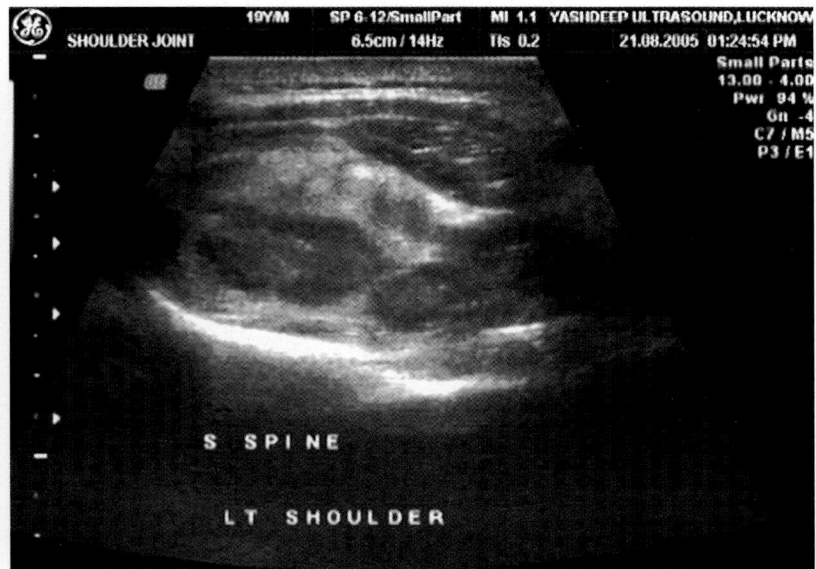

FIGURE 13.11

FIGURES 13.8 to 13.11: Pyogenic muscle abscess. A 19 years old boy presented with acute painful swelling over the Lt shoulder girdle over the scapular region. HRSG shows a big irregular hypoechoic low-level echo complex mass involving the Lt infraspinatous muscles. The muscle was edematous. The thick collection is seen in the belly of muscles suggestive of abscess. Layering sign was positive.

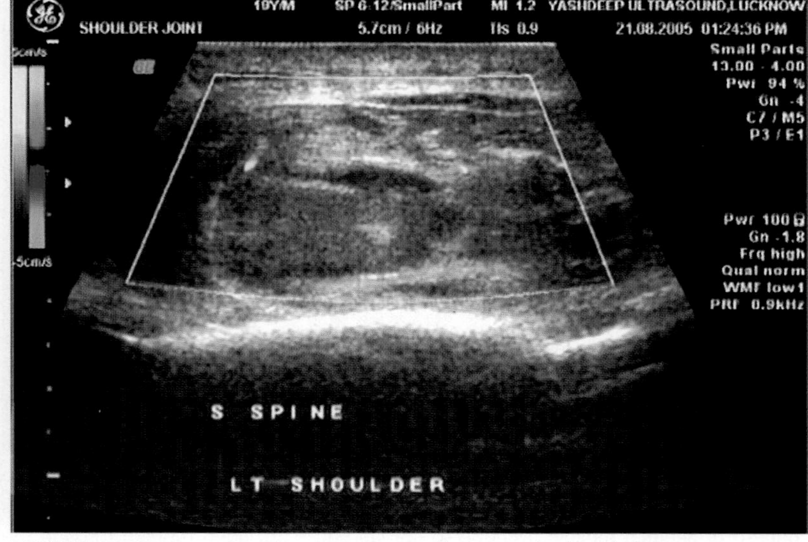

FIGURE 13.12: On color flow imaging no flow seen in the abscess.

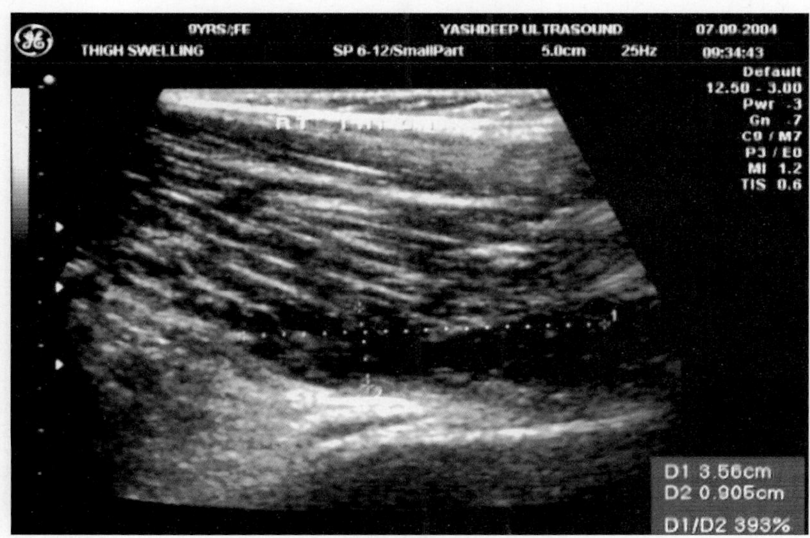

FIGURE 13.13

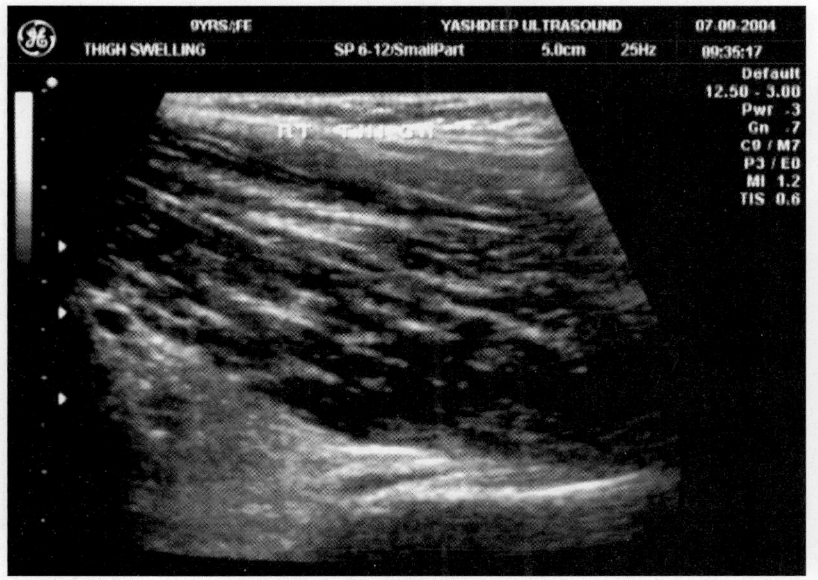

FIGURE 13.14

FIGURES 13.13 and 13.14: Acute myositis. HRSG shows edematous thigh muscle in a 9 years old child. Muscle fibers are sparse and small amount of fluid is seen between them. The muscle is hypoechoic. Hypoechoic pockets are seen in the belly of abscess suggestive of micro abscess.

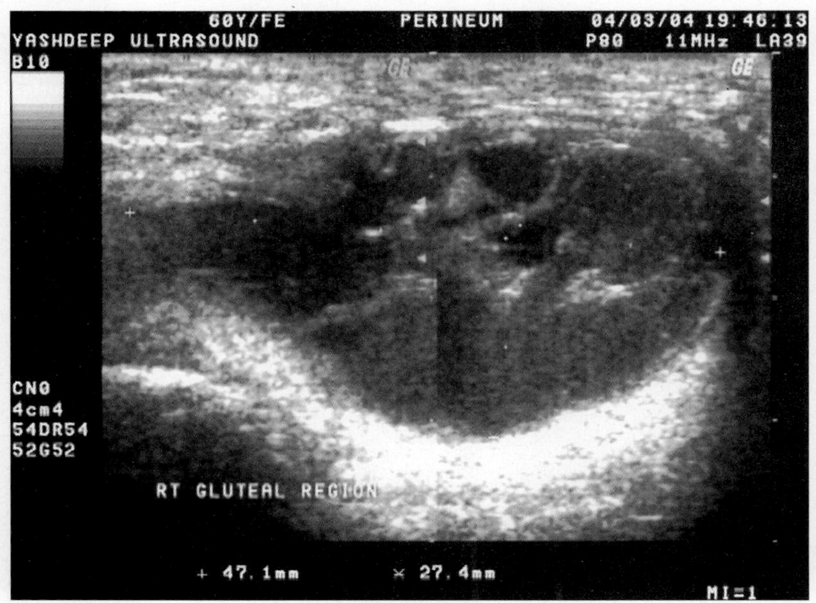

FIGURE 13.15

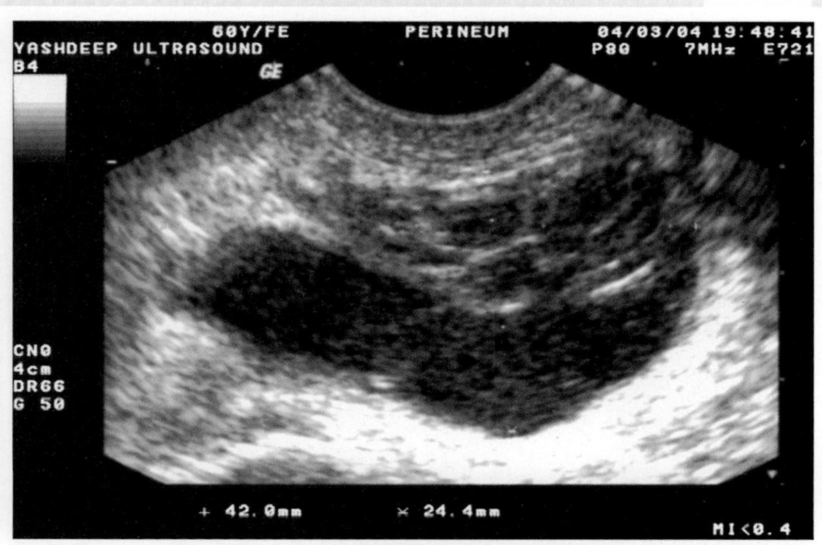

FIGURE 13.16

FIGURES 13.15 and 13.16: Big gluteal abscess. An 60-year-old lady presented with big gluteal swelling with fever. HRSG shows a big mixed echo complex mass in the Rt gluteal muscle. Thick echoes are seen in it with debris suggestive of big abscess.

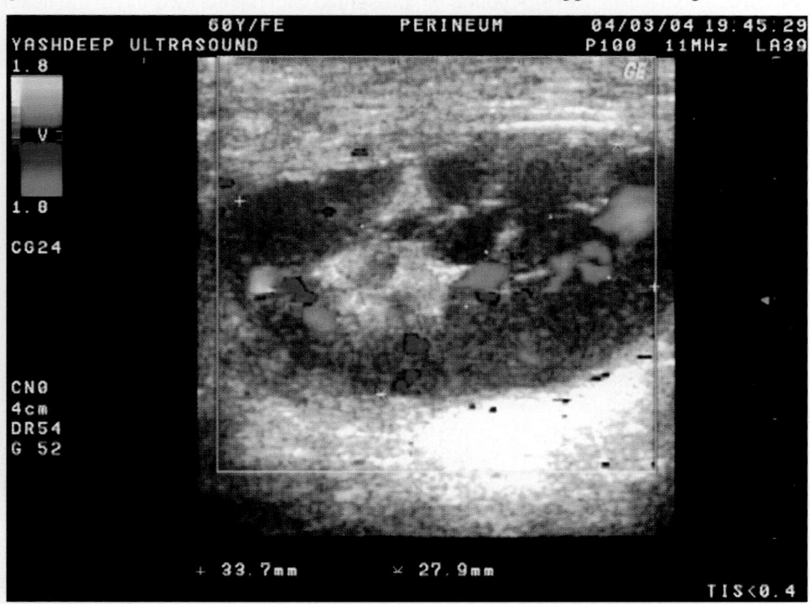

FIGURE 13.17: On color flow imaging few vessels are seen feeding the vessels suggestive of acute inflammation.

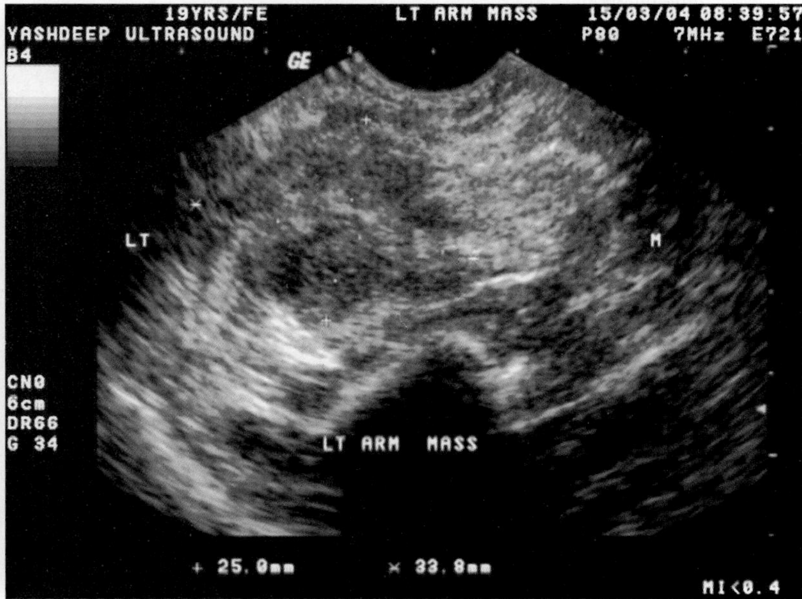

FIGURE 13.18

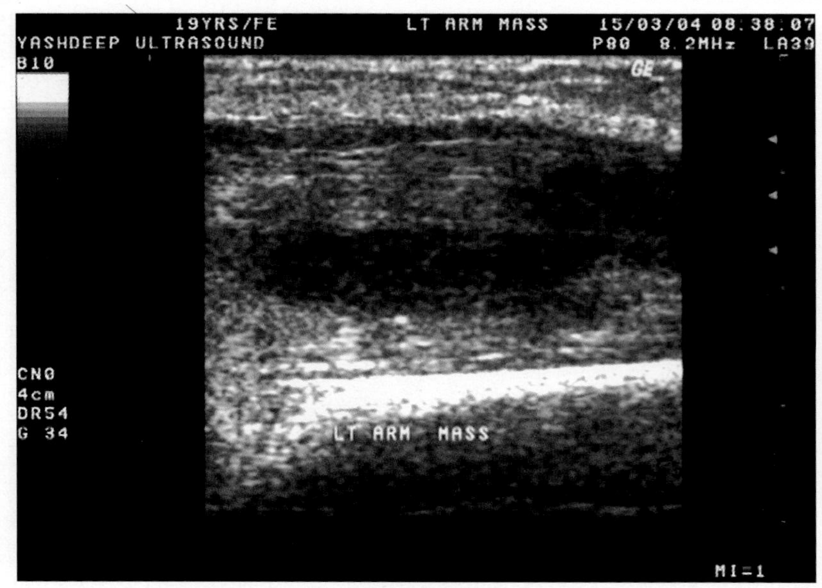

FIGURE 13.19

FIGURES 13.18 and 13.19: A young girl presented with painful swelling in the flexor group of muscle of the Lt arm. HRSG shows thick low-level echo complex mass in the muscle belly. Hypoechoic collection is seen in it suggestive of pyogenic abscess.

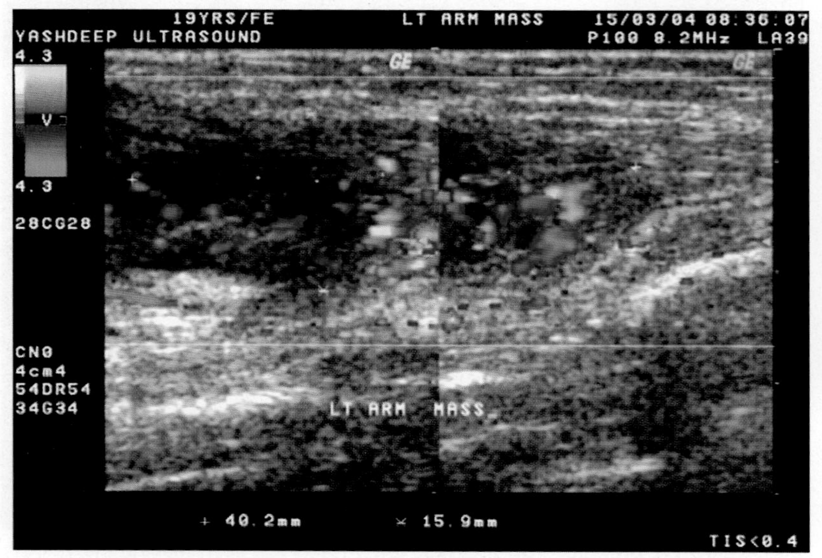

FIGURE 13.20: On color flow imaging high flow seen in the abscess suggestive of pyogenic inflammation.

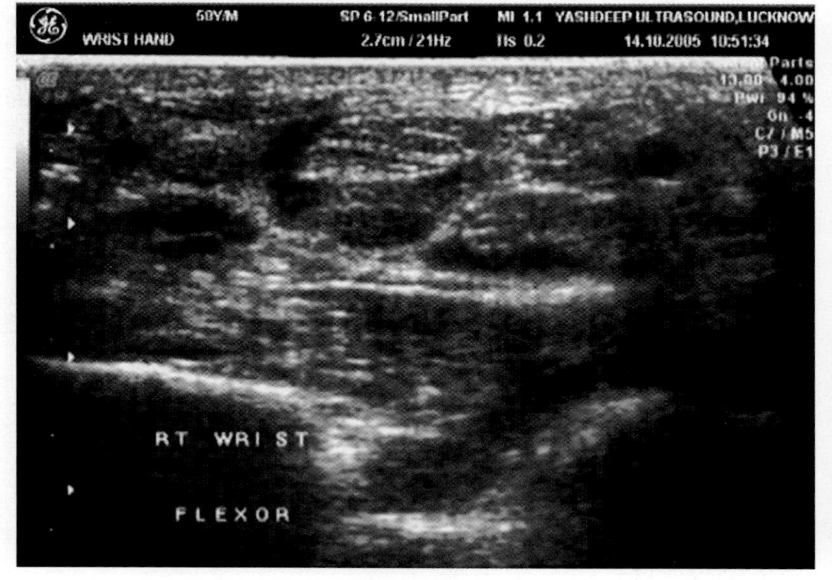

FIGURE 13.21

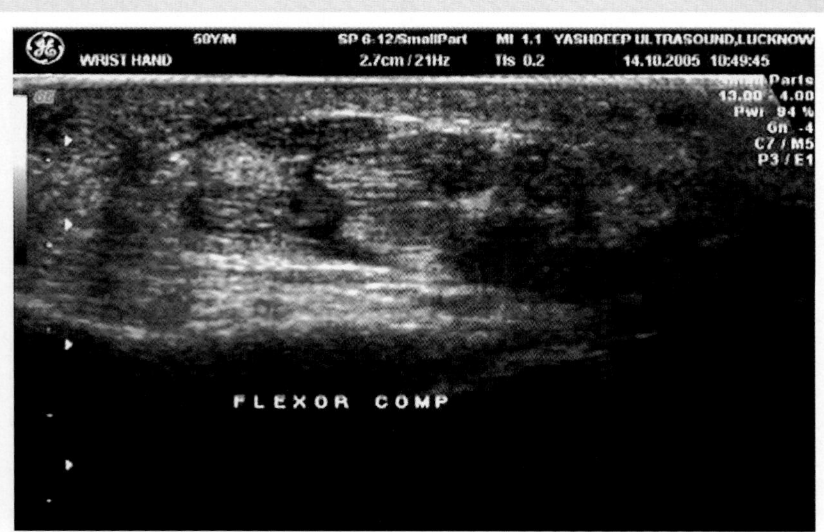

FIGURE 13.22

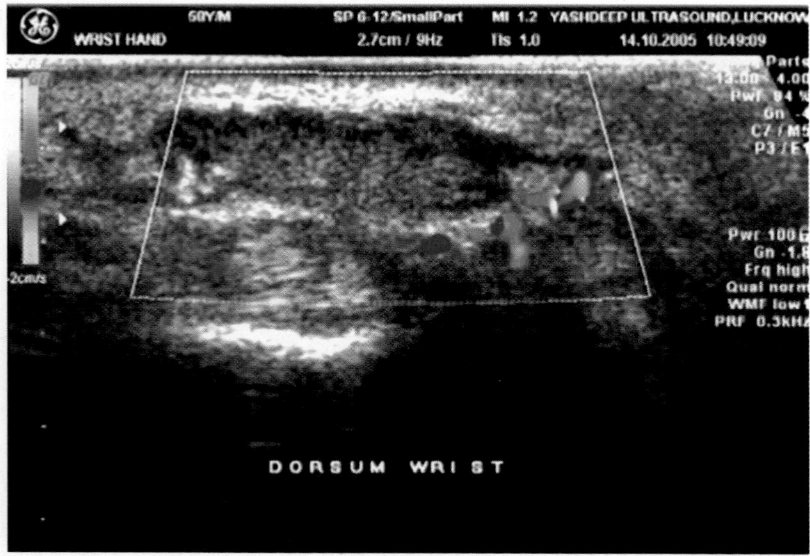

FIGURE 13.23

FIGURES 13.21 to 13.23: Pyogenic abscess in the wrist. A 50-year-old patient presented with acute swelling over the flexor compartment of the Rt wrist with restricted movements. HRSG shows thick collection in flexor compartment of the wrist. The tendons are also edematous and thickened. Low-level echo collection is also seen over the dorsal surface of the wrist suggestive of pyogenic collection. **On color flow imaging** low flow is seen in it.

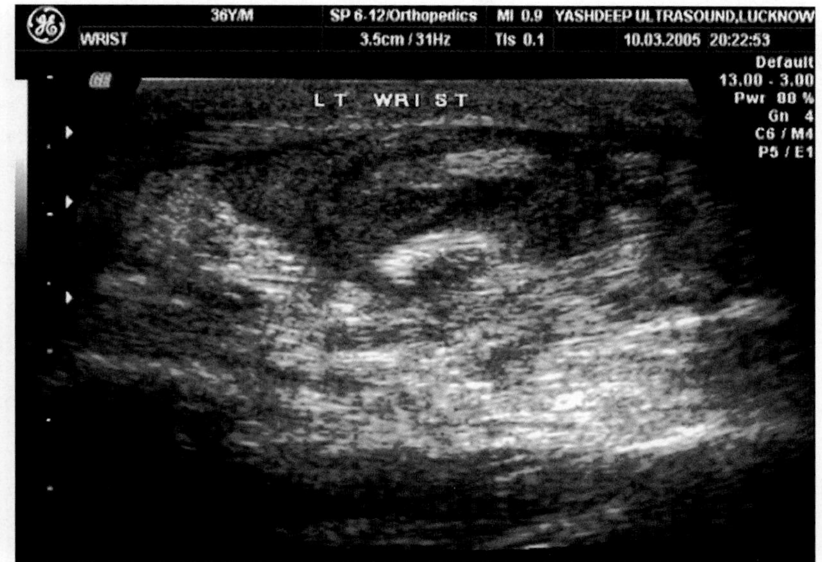

FIGURE 13.24

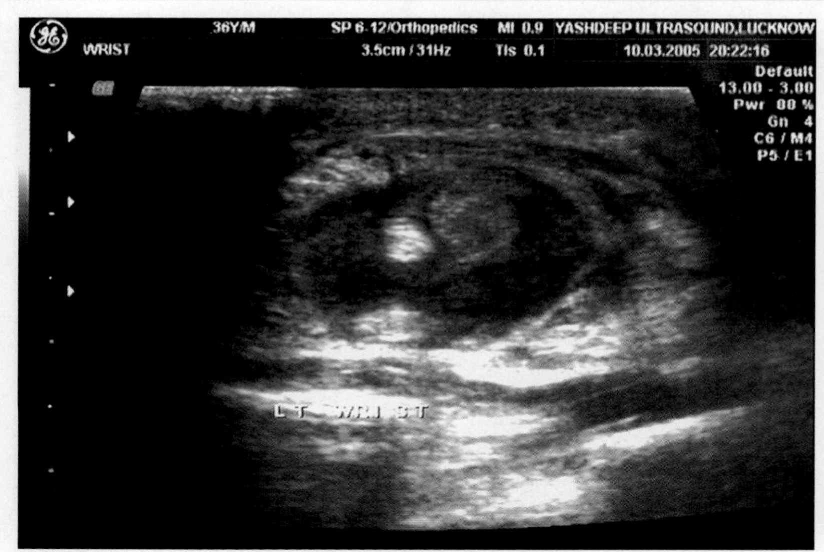

FIGURE 13.25

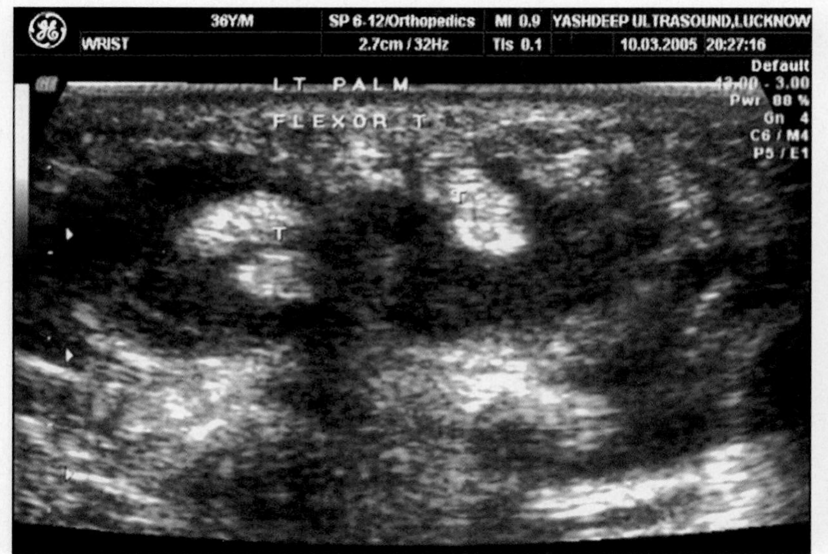

FIGURE 13.26

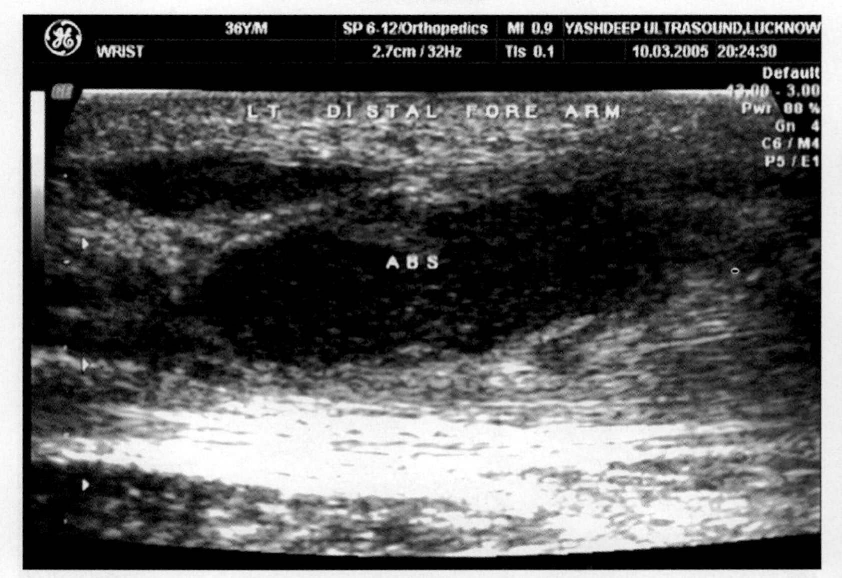

FIGURE 13.27

FIGURES 13.24 to 13.27: Acute pyogenic abscess in the wrist. A man presented with painful red hot swelling over the wrist joint with restricted movements. HRSG shows thick mixed echo complex mass in the flexor compartment of the wrist. Collection is also seen extending into the Lt palmar surface. Tendons are thickened and echogenic. They are also edematous. The collection is also seen seeping through the palmar compartment. It is also seen extending in the belly of distal muscle group. Aspiration confirmed acute pyogenic abscess.

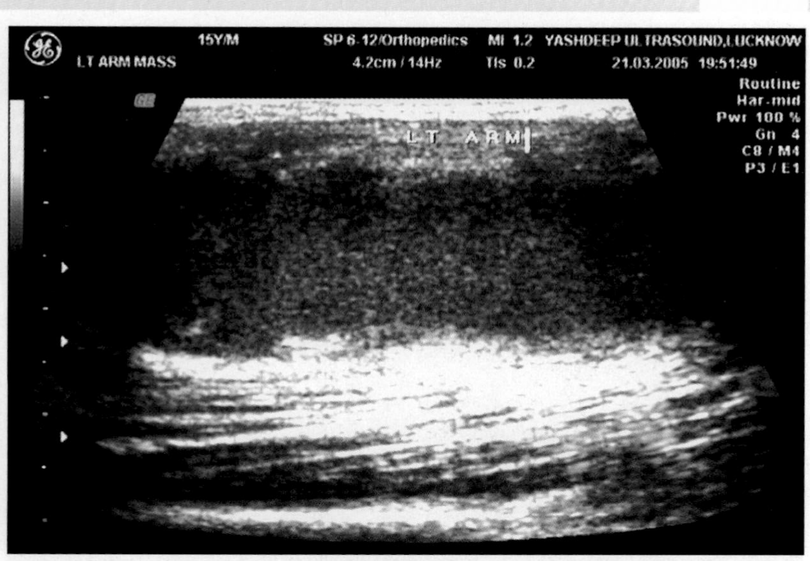

FIGURE 13.28

FIGURE 13.29

FIGURES 13.28 and 13.29: Pyogenic collection in the Lt arm. A young boy presented with painful swelling in the Lt arm. HRSG shows thick collection in the muscle belly in the Lt arm. Homogeneous echoes are seen in the collection. No septation is seen. No focal mass is seen.

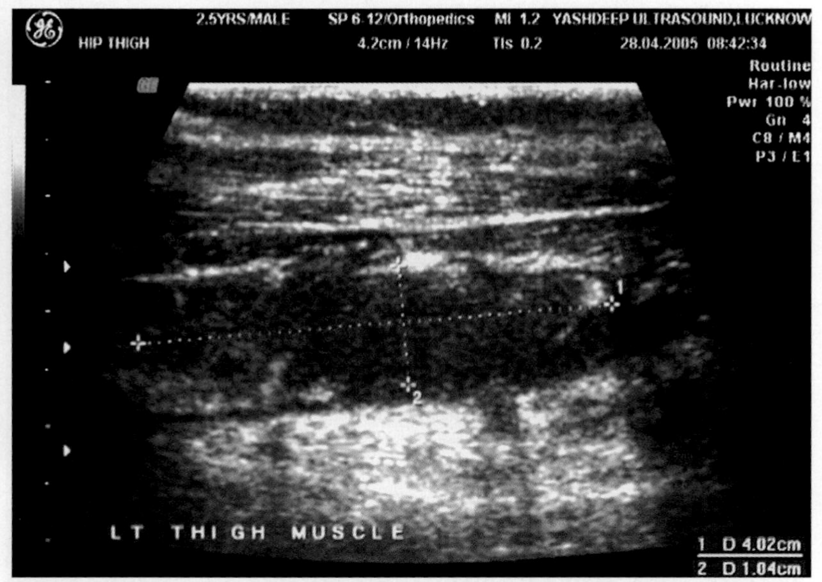

FIGURE 13.30: Acute myositis involving the thigh muscle. A young child presented with acute painful swelling over the Lt thigh. HRSG shows edematous anterior compartment muscles. They are hypoechoic in texture. The posterior belly of the muscles shows breaking down of the muscle tissue with an abscess formation.

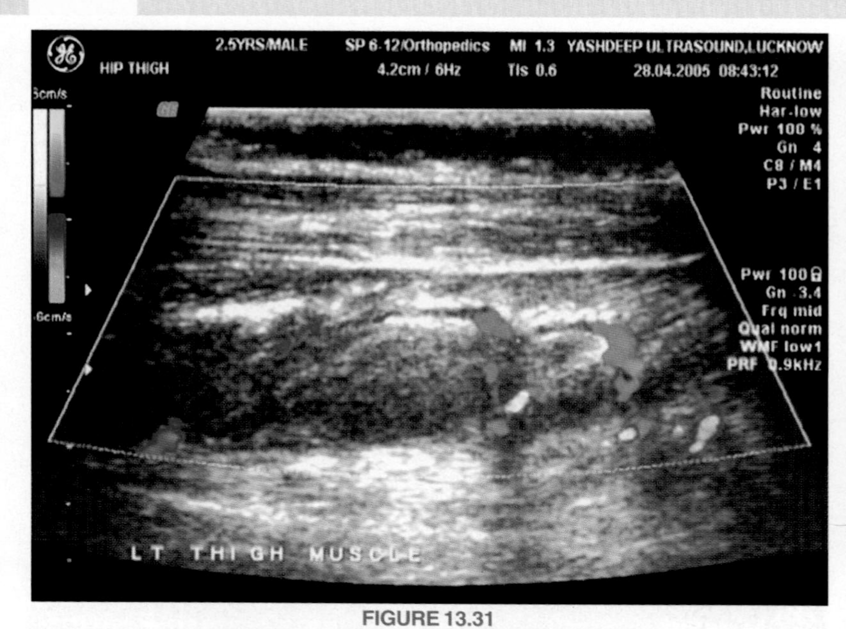

FIGURE 13.31

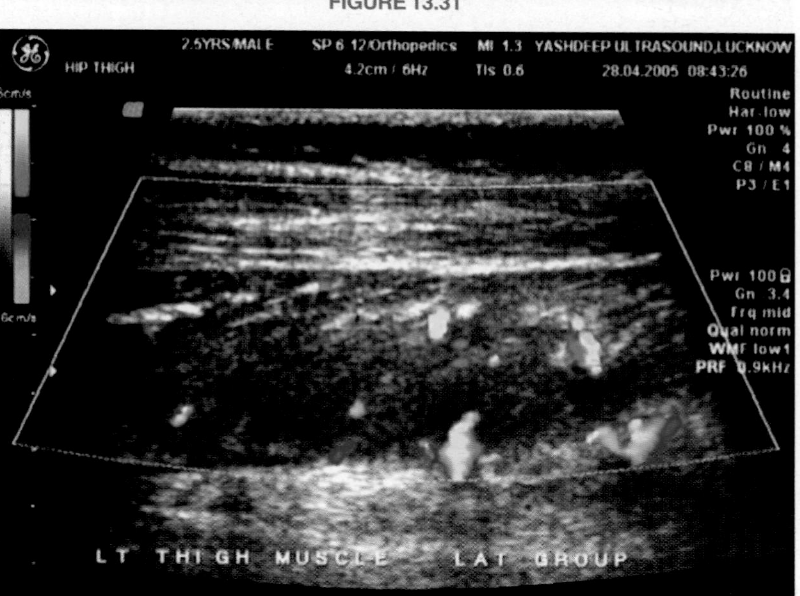

FIGURE 13.32

FIGURES 13.31 and 13.32: On color flow imaging increased flow seen in the muscle belly suggestive of hyperemia.

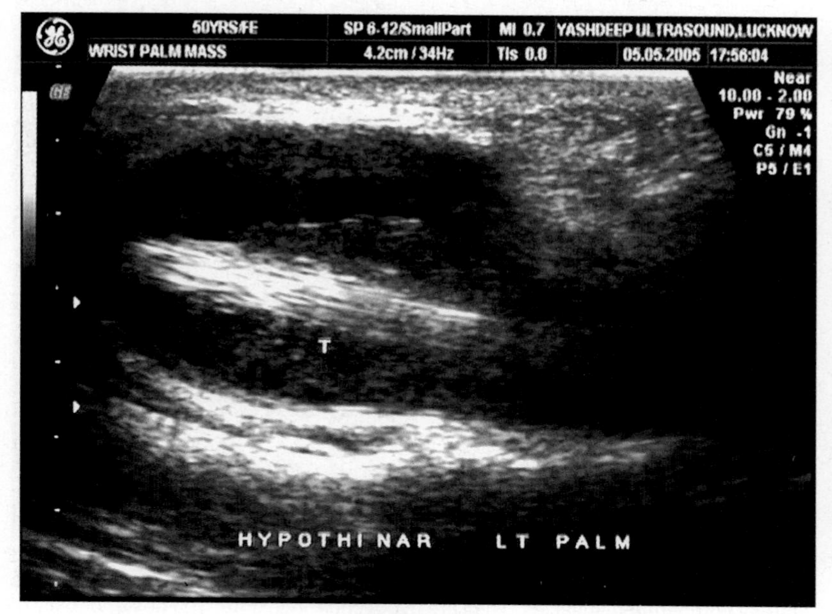

FIGURE 13.33

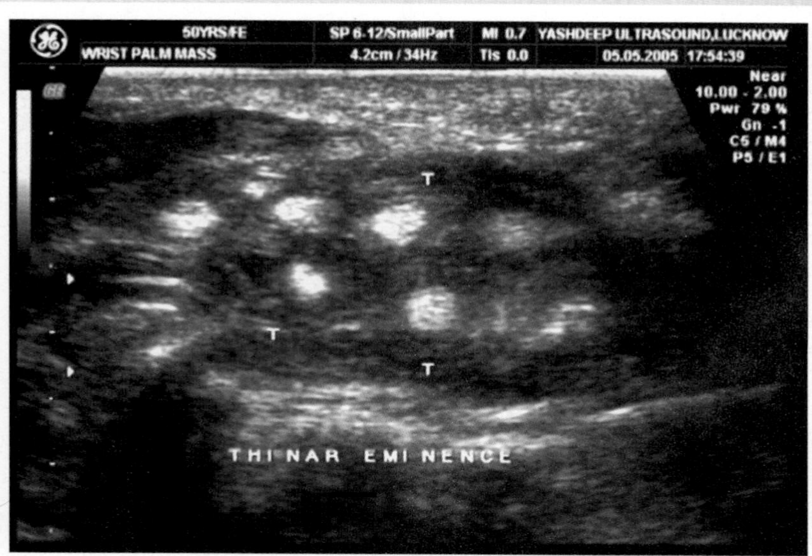

FIGURE 13.34

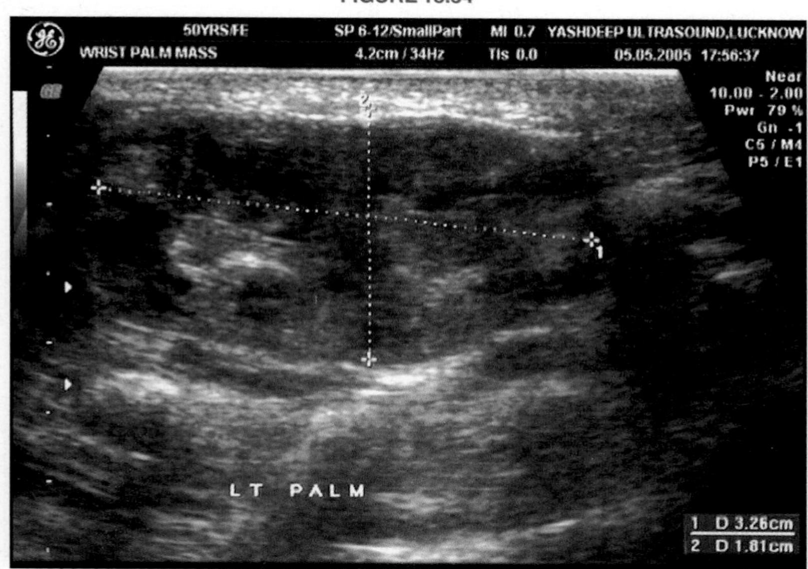

FIGURE 13.35

FIGURES 13.33 to 13.35: Palm abscess. A patient presented with painful swelling over the Lt palm involving thinar and hypothinar eminence. Thick echogenic collection is seen in the palmar space involving both eminences. Fluid collection is also seen in the tendon sheath. Tendon is also thickened. Associated synovial membrane proliferation is also seen. Aspiration confirmed acute fulminating abscess.

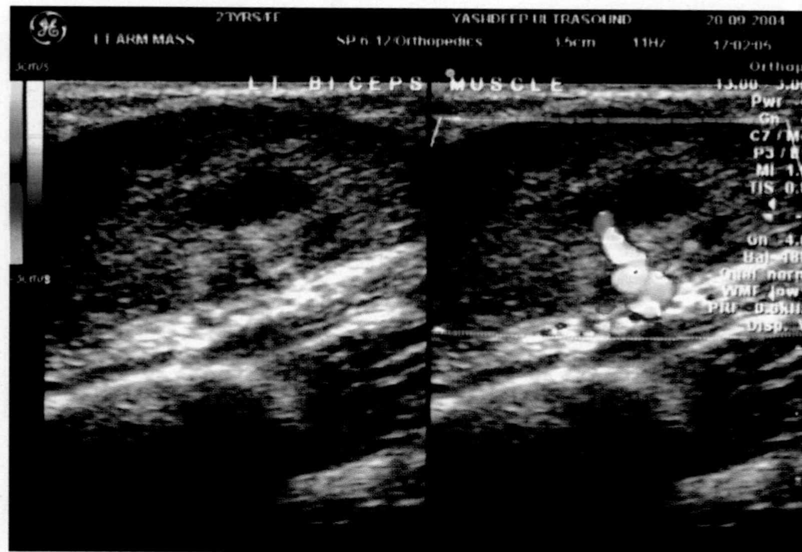

FIGURE 13.36: Biceps muscle abscess HRSG shows thick abscess cavity in the Lt biceps muscle in a patient presented with mass. **On color flow imaging** shows low flow in the mass. Aspiration confirmed abscess.

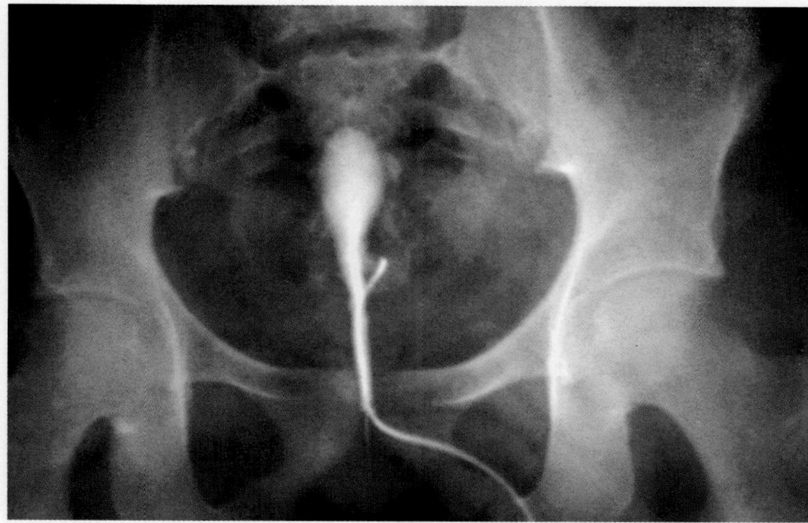

FIGURE 13.37

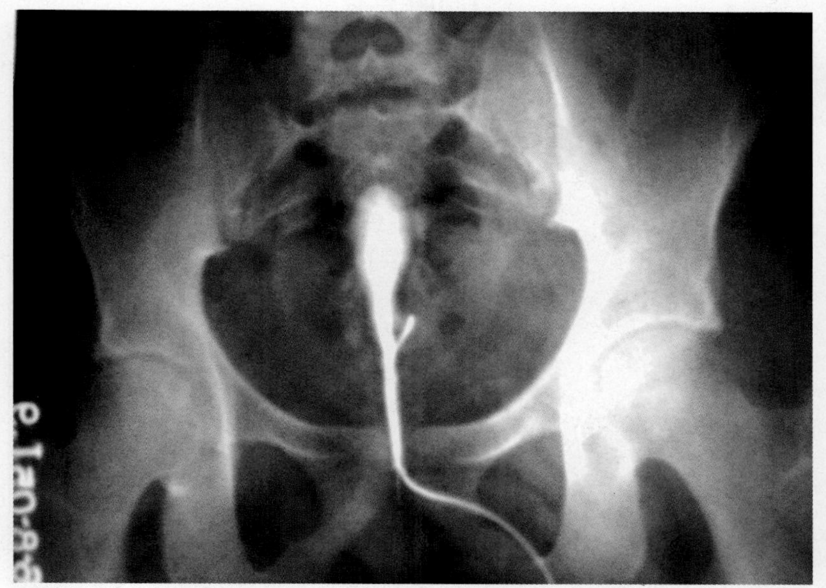

FIGURE 13.38

FIGURES 13.37 and 13.38: Pilonidal sinus with X-ray sinogram. A young boy presented with discharging sinus in the gluteal region just above the anal opening on the Lt side. Sinogram shows a sinus tract with pooling of the dye in the sinus cavity. Branching of the sinus is also seen on X-ray gram.

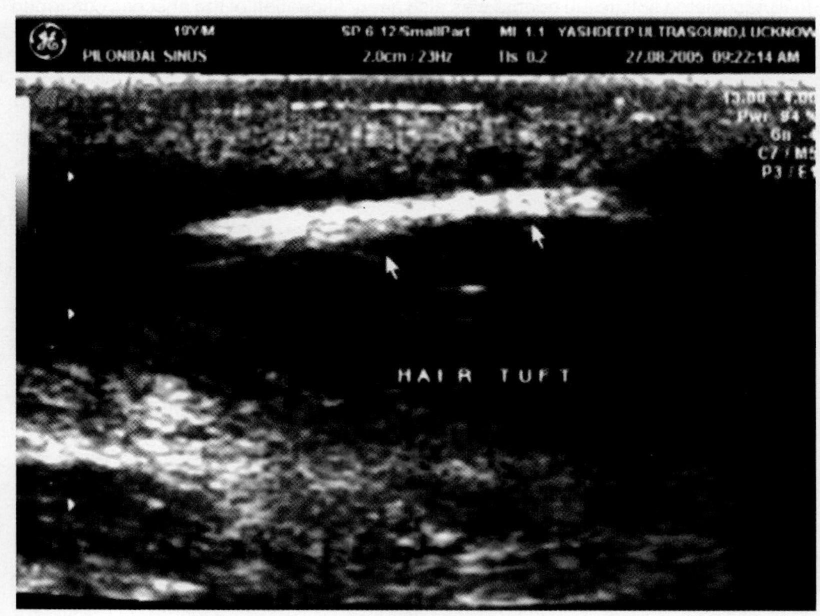

FIGURE 13.39

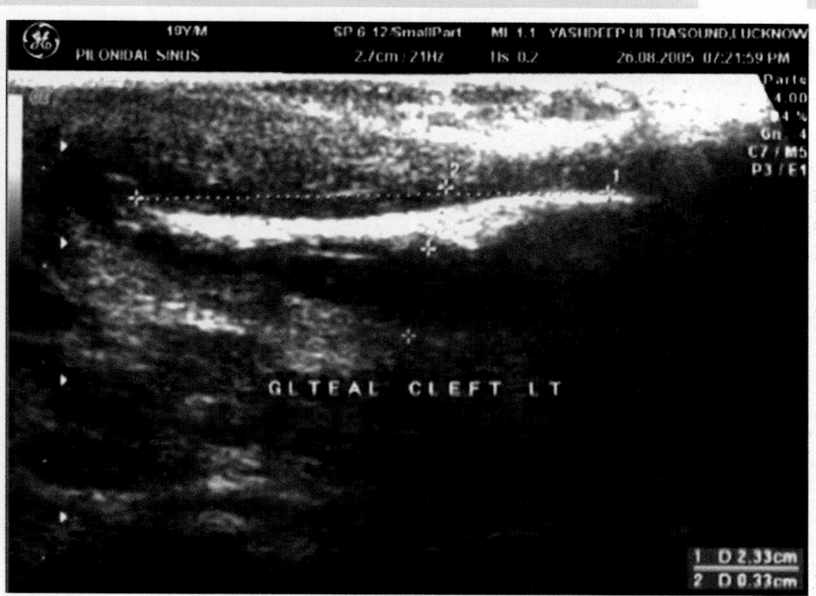

FIGURE 13.40

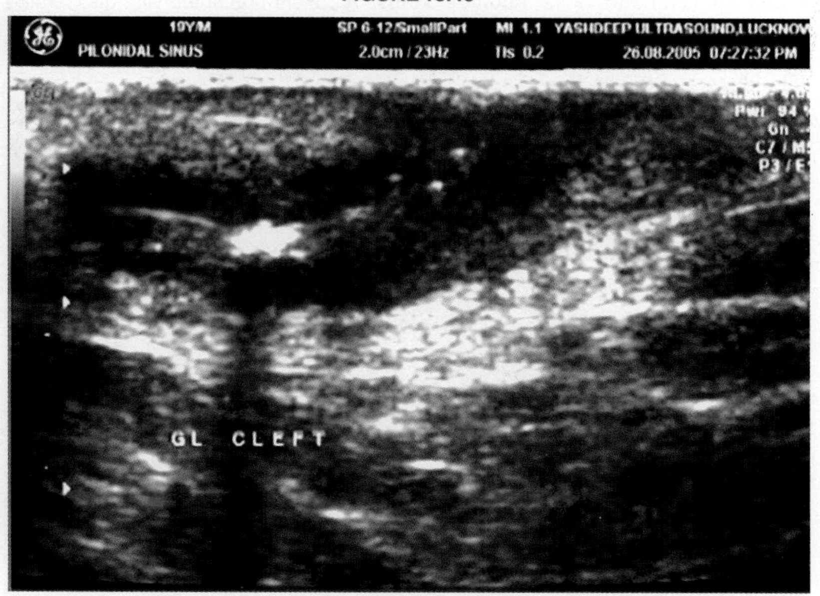

FIGURE 13.41

FIGURES 13.39 to 13.41: Pilonidal sinus in the gluteal cleft. High-resolution sonography of the same patient shows obliquely running sinus in the Lt gluteal cleft. Highly echogenic linear band like shadow is seen embedded in the sinus suggestive of hair tuft. Exploration confirmed the sonographic findings.

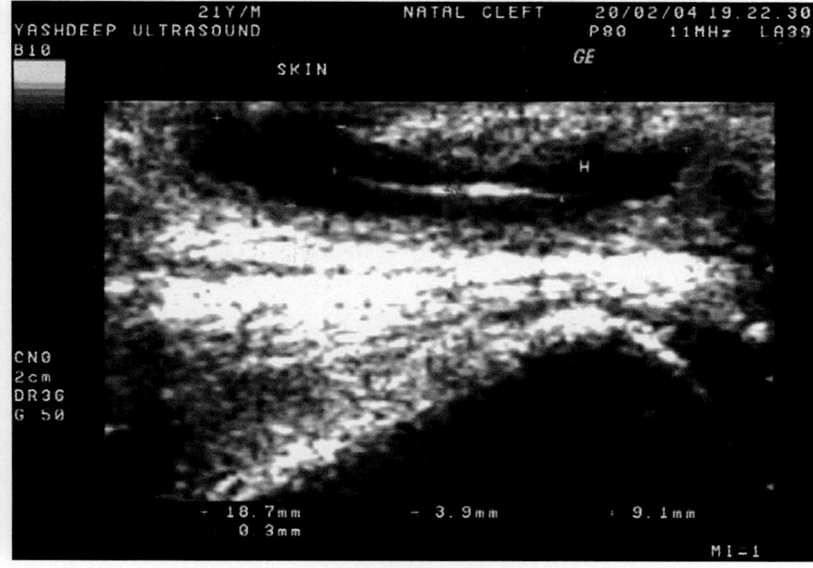

FIGURE 13.42

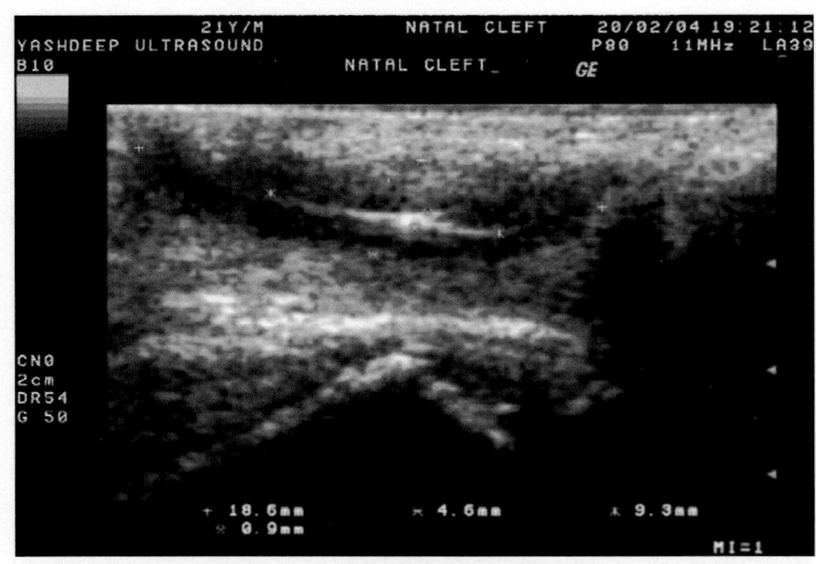

FIGURE 13.43

FIGURES 13.42 and 13.43: Pilonidal sinus. A patient presented with discharging sinus in the gluteal region. HRSG shows obliquely running sinus in the gluteal region. An echogenic linear shadow is seen in it running obliquely in the sinus suggestive of hair tuft.

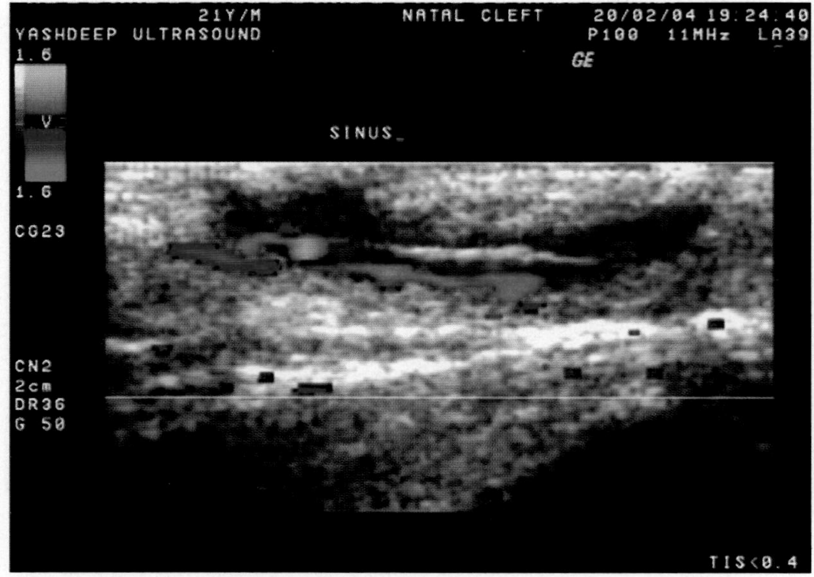

FIGURE 13.44: On color flow Imaging-Low flow is seen in the wall. Exploration confirmed the sonographic findings.

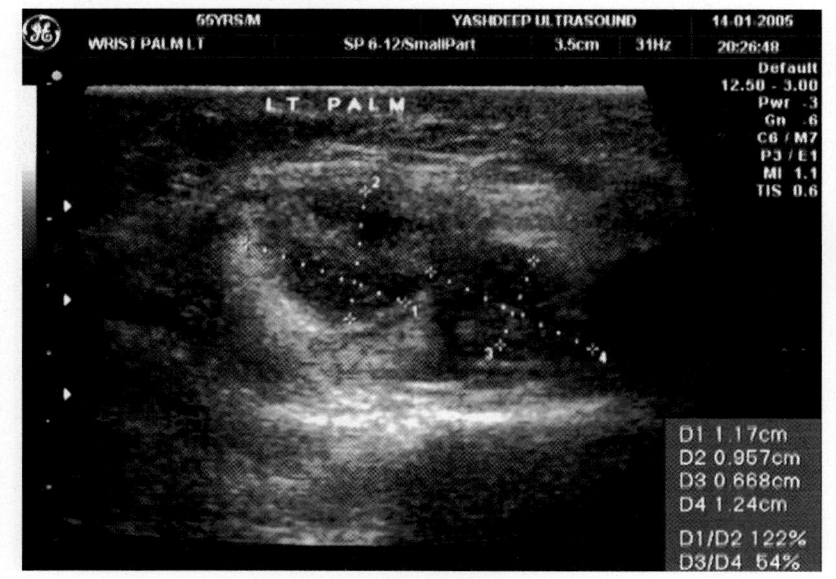

FIGURE 13.45

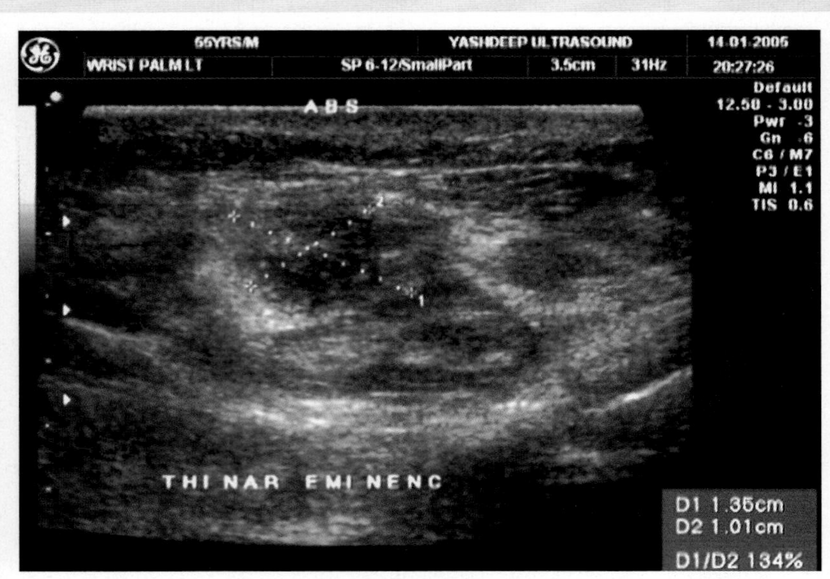

FIGURE 13.46

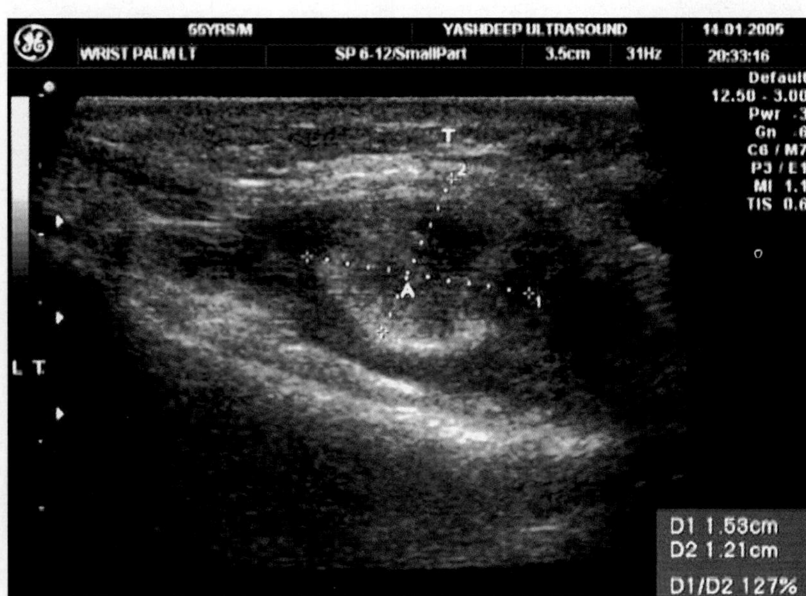

FIGURE 13.47

FIGURES 13.45 to 13.47: Chronic Lt palm abscess. HRSG shows an irregular mixed echo complex mass in the Lt thinar eminence. Multiple internal echoes are seen in it. Associated soft tissue edema was also seen. Aspiration of the abscess confirmed chronic abscess.

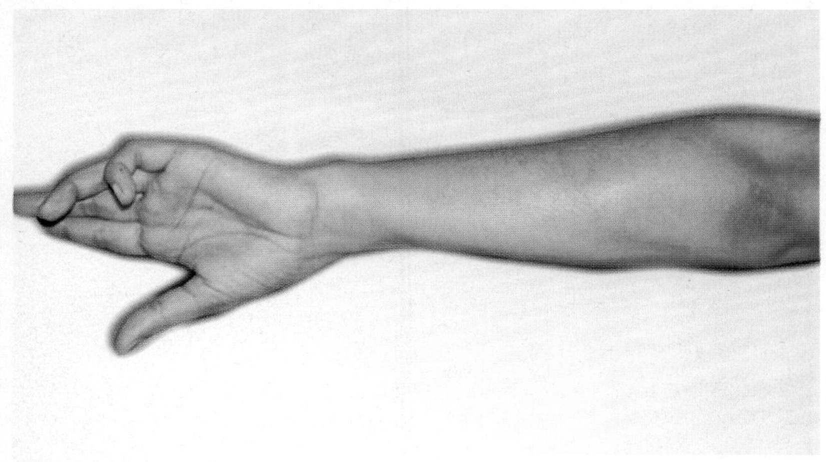

FIGURE 13.48: Tubercular muscle abscess. A young lady presented with a soft tissue mass swelling over the flexor surface of the Lt forearm. The fingers are flexed.

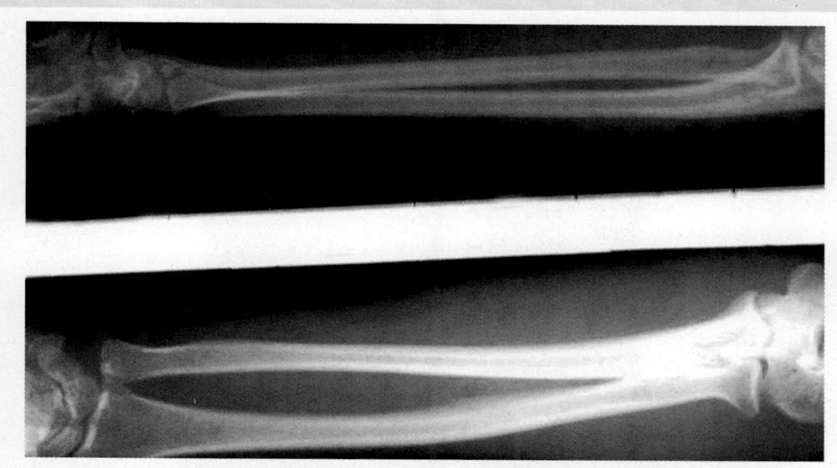

FIGURE 13.49: X-ray of the forearm shows normal bone with soft tissue.

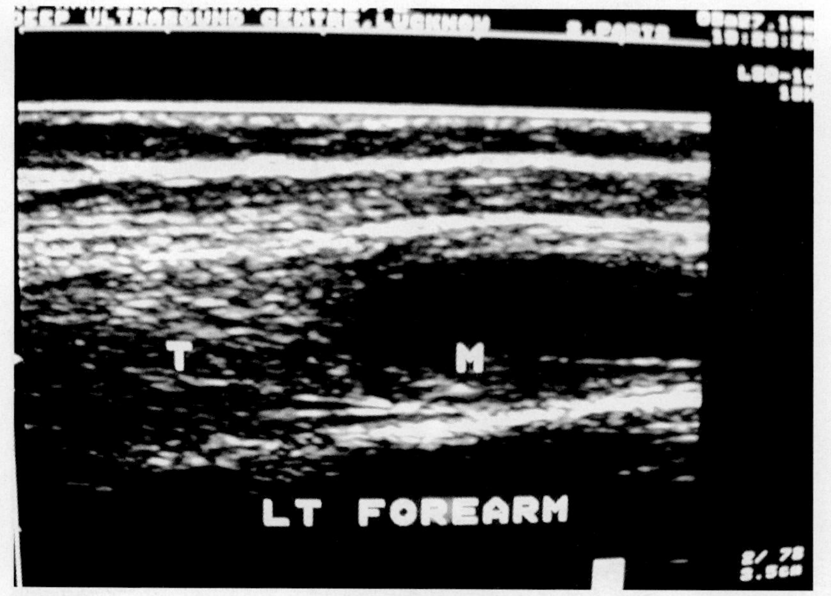

FIGURE 13.50: HRSG shows low-level echo complex mass in the flexor group of the muscles. The normal texture is lost. However, muscle capsule is intact suggestive of an abscess.

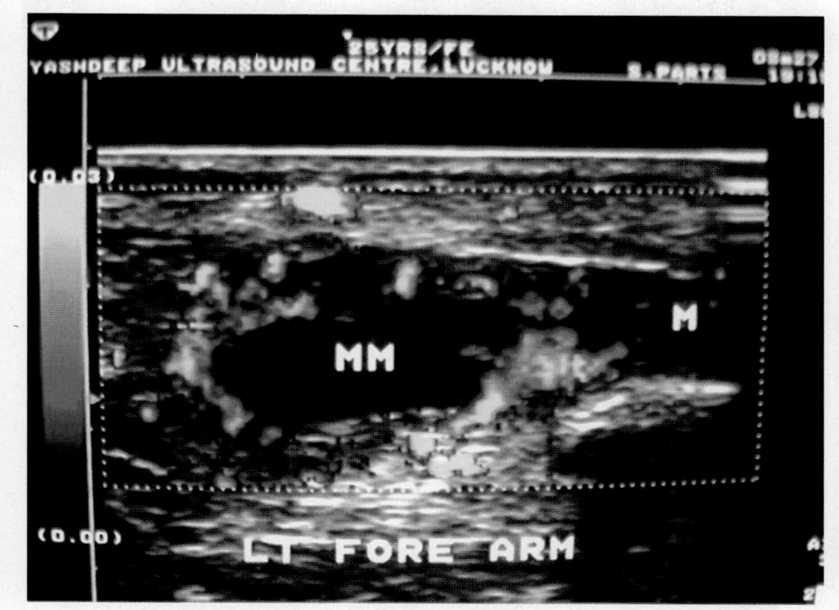

FIGURE 13.51: Power Doppler imaging of the mass shows peripheral flow. The FNAC of the mass was tubercular abscess.

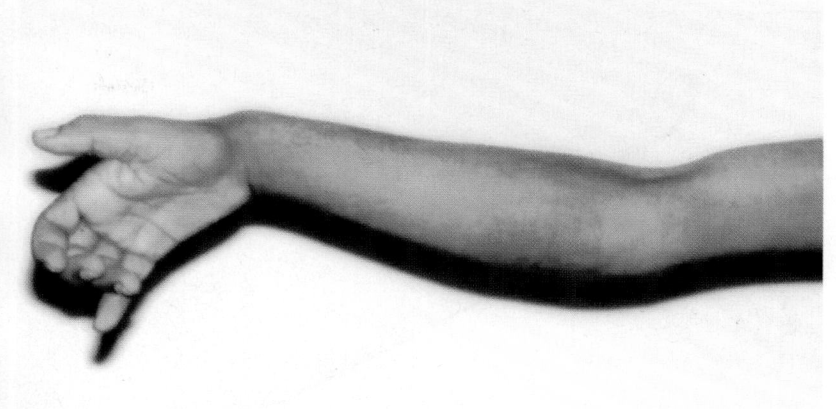

FIGURE 13.52: A young boy presented with flexion deformity of fingers and wrist. Soft tissue swelling is seen over the flexor surface of the arm clinically presented as vascular ischemic contracture (VIC).

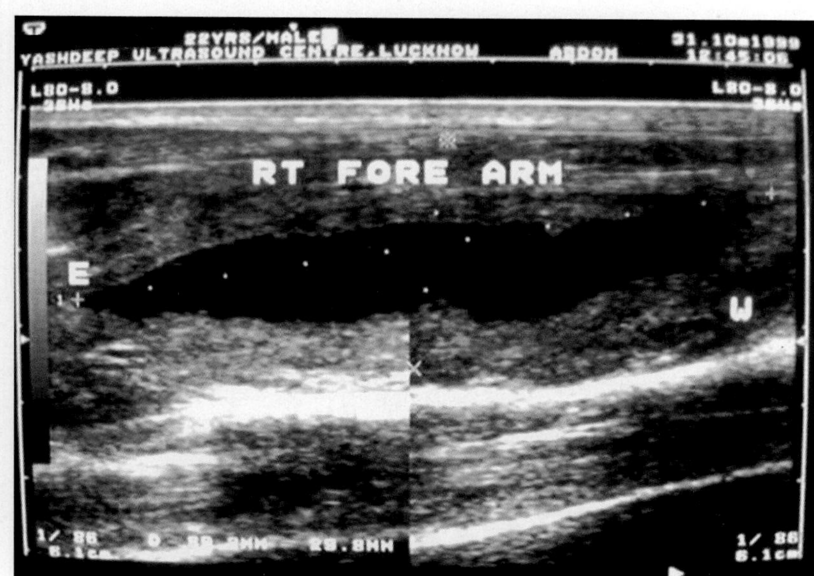

FIGURE 13.53: HRSG shows an abscess in the flexor group of the muscles. Multiple internal echoes are seen in it.

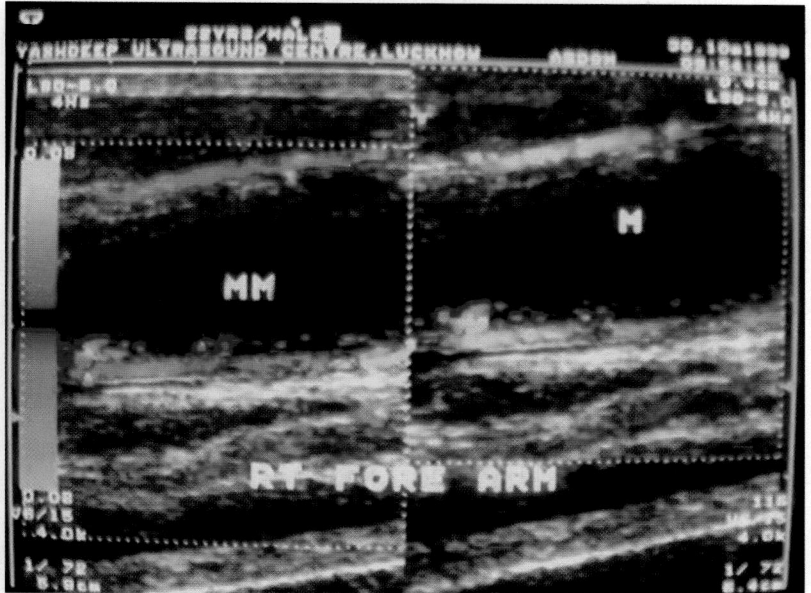

FIGURE 13.54: Color Doppler imaging shows pressure on the vessels. But no vessel wall involvement is seen. FNAC shows tubercular abscess.

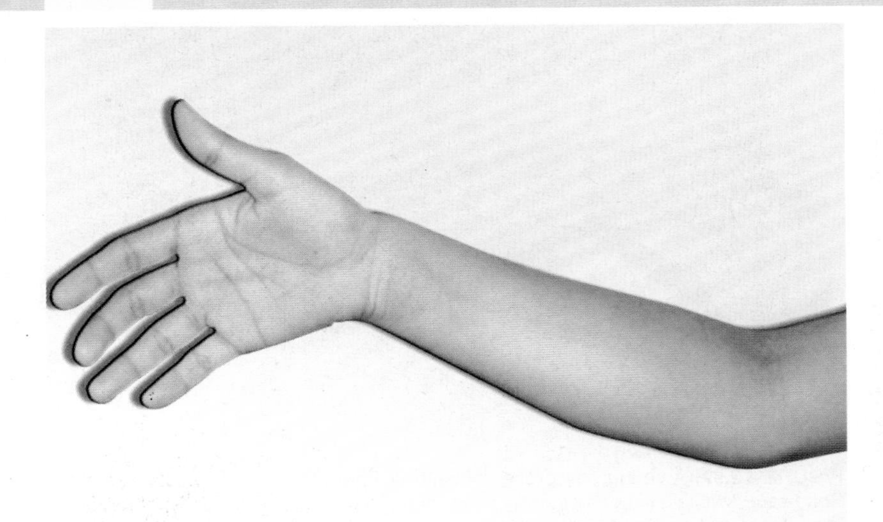

FIGURE 13.55: Forearm abscess presented as VIC. A young girl presented with flexion deformity in the wrist with a palpable lump on the forearm. Clinically it was suspected as vascular ischemic contracture.

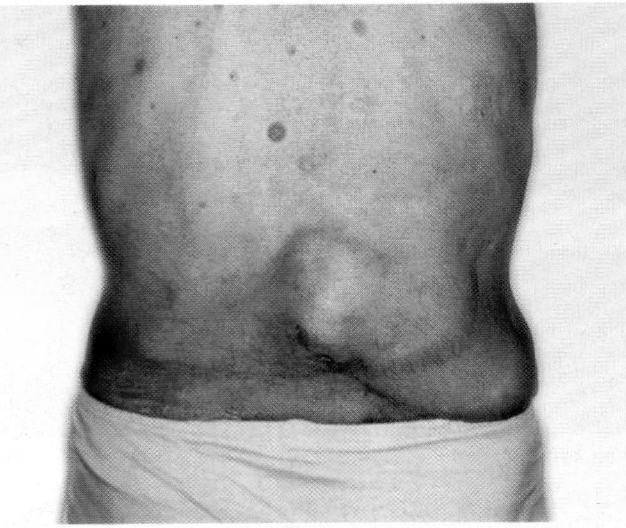

FIGURE 13.58: Paraspinal tubercular abscess. A middle aged man presented with soft tissue swelling over the paraspinal region in the lower back.

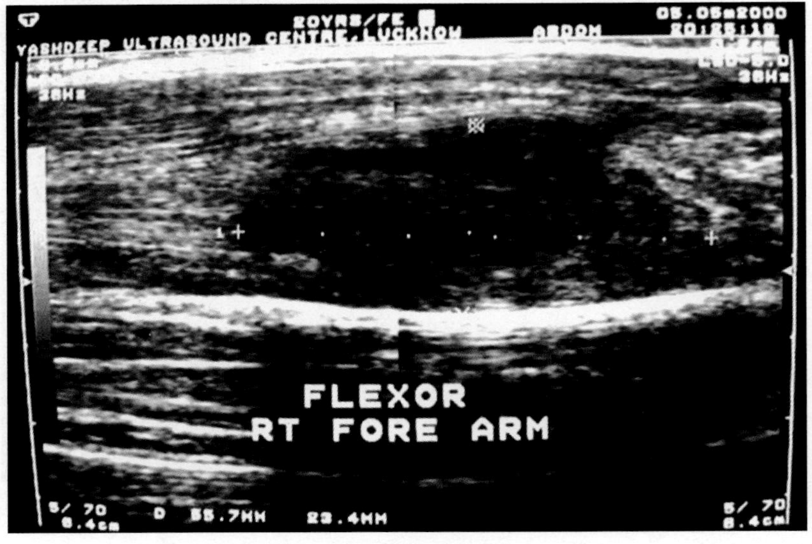

FIGURE 13.56

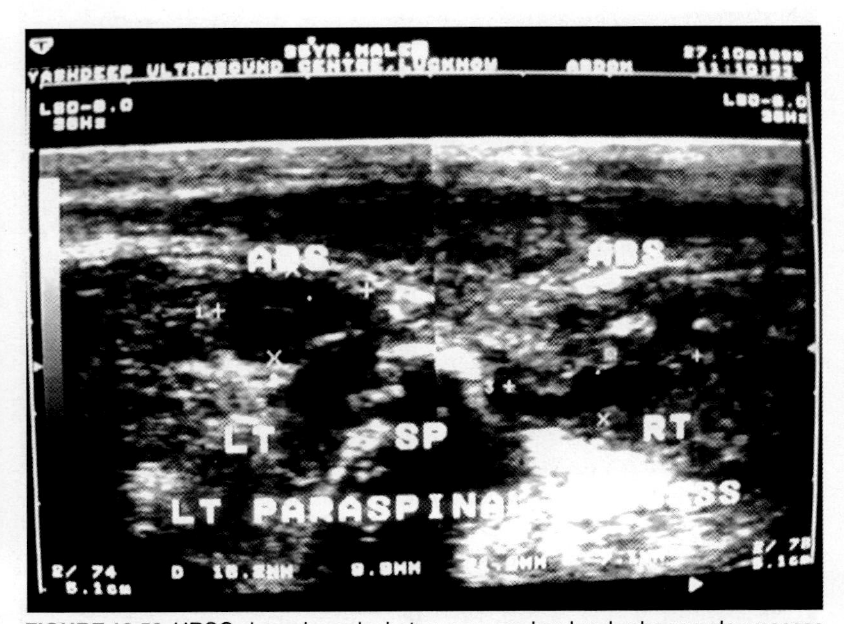

FIGURE 13.59: HRSG shows irregular heterogeneous low-level echo complex masses involving the paraspinal muscles on either side of the spine suggestive of chronic tubercular abscess. There is also evidence of bony destruction of transverse process of vertebra seen on Lt side.

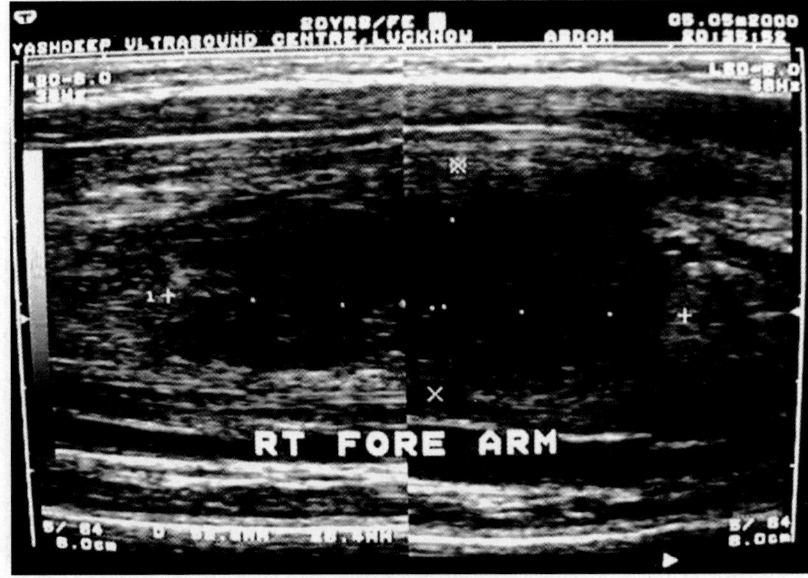

FIGURE 13.57

FIGURES 13.56 and 13.57: HRSG shows low-level echo complex mass in the flexor group of muscle of forearm. Central area of necrosis is seen in it suggestive of an abscess.

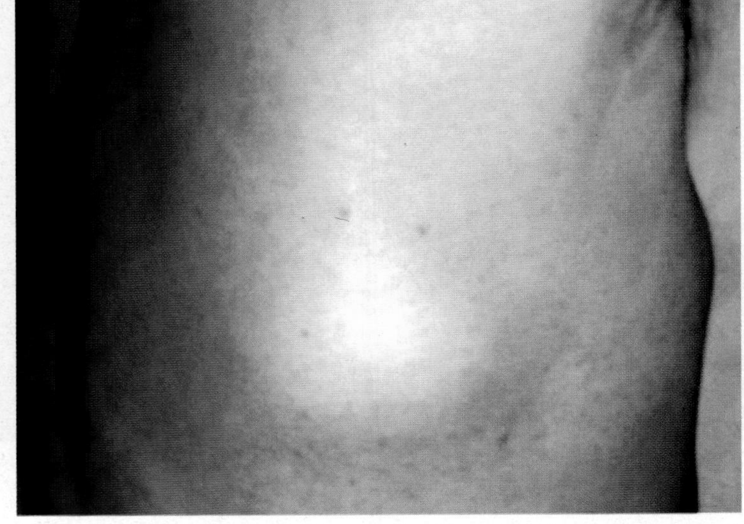

FIGURE 13.60: Tubercular abscess in the back. A middle aged man presented with soft tissue swelling over Rt infrascapular region.

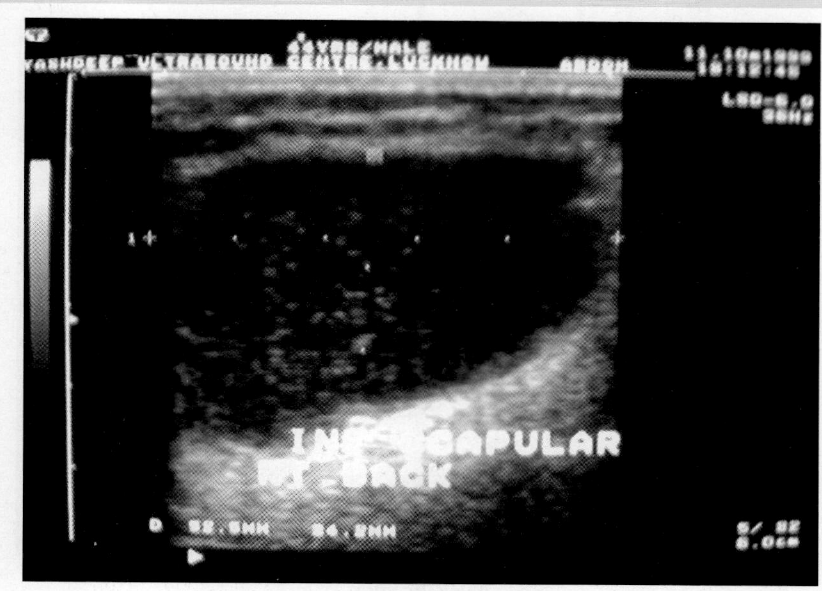

FIGURE 13.61

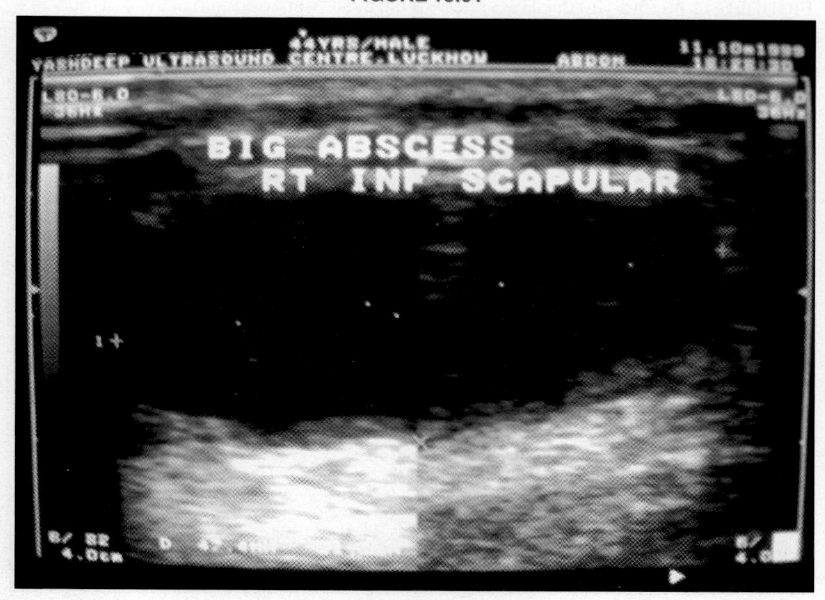

FIGURE 13.62

FIGURES 13.61 and 13.62 HRSG shows big low-level echo collection in the back. The collection is homogeneous. However, debris is seen on the dependent part. Normal muscle texture is lost. Aspiration of the abscess shows tubercular collection.

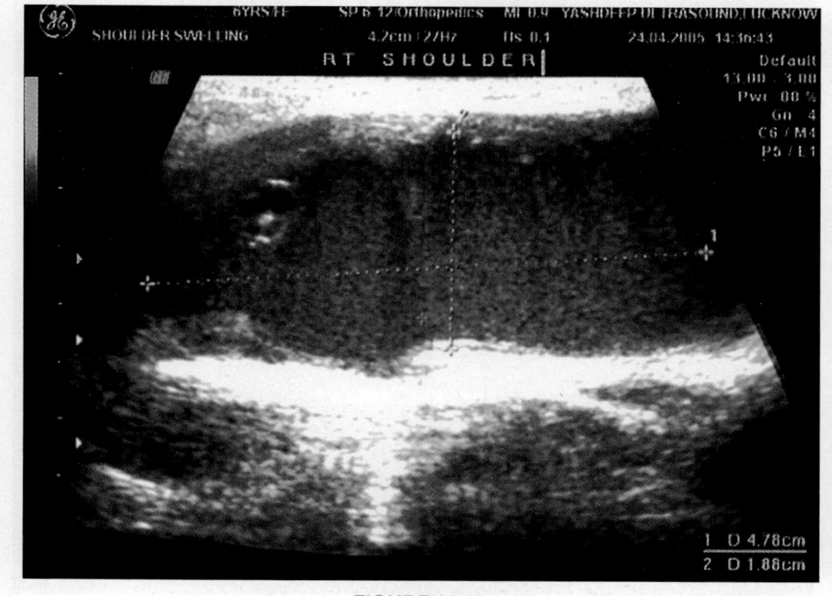

FIGURE 13.63

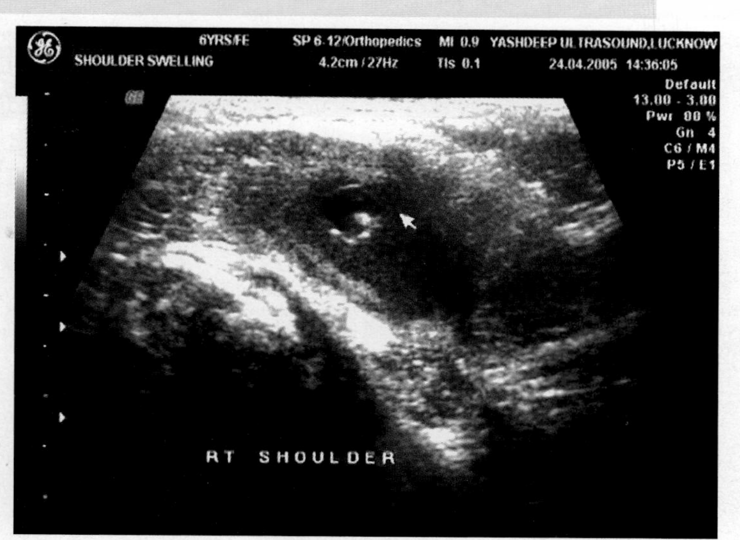

FIGURE 13.64

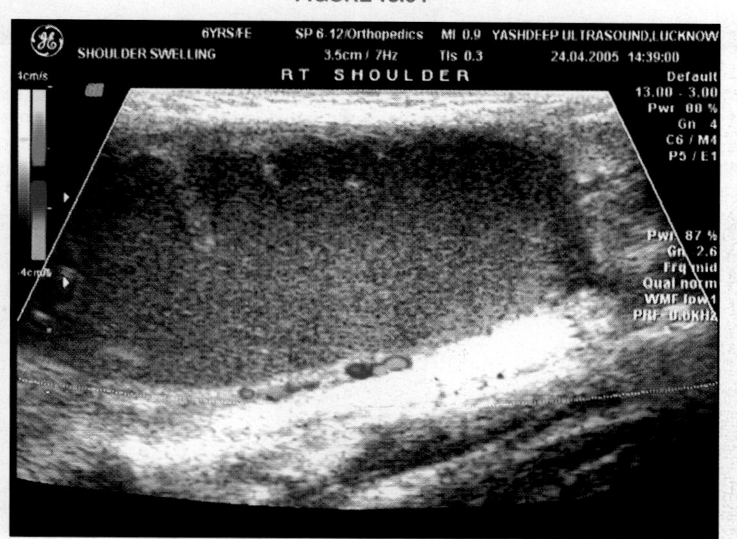

FIGURE 13.65

FIGURES 13.63 to 13.65: Tubercular abscess with cysticercus cyst. A young girl presented with soft tissue swelling over the deltoid muscle in Rt shoulder. A thick collection is seen in the belly of the muscle. Normal muscle texture is lost and it is replaced with tubercular abscess. However, a well define cyst is also seen embedded in the collection. An echogenic nidus is seen fixed with the inner wall of the cyst typical of the cystic cercus cyst. On color flow imaging no flow is seen in the collection. Aspiration of the cyst confirmed tubercular abscess with embedded cysticercus cyst.

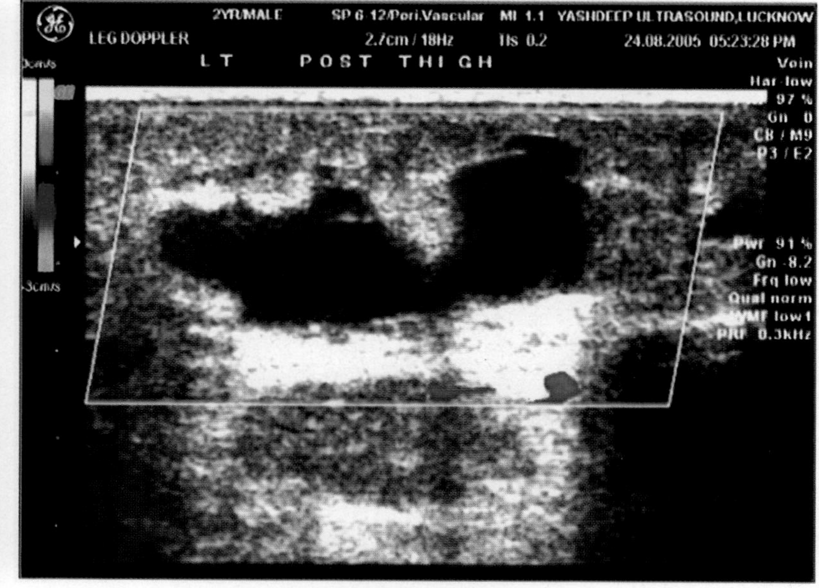

FIGURE 13.66

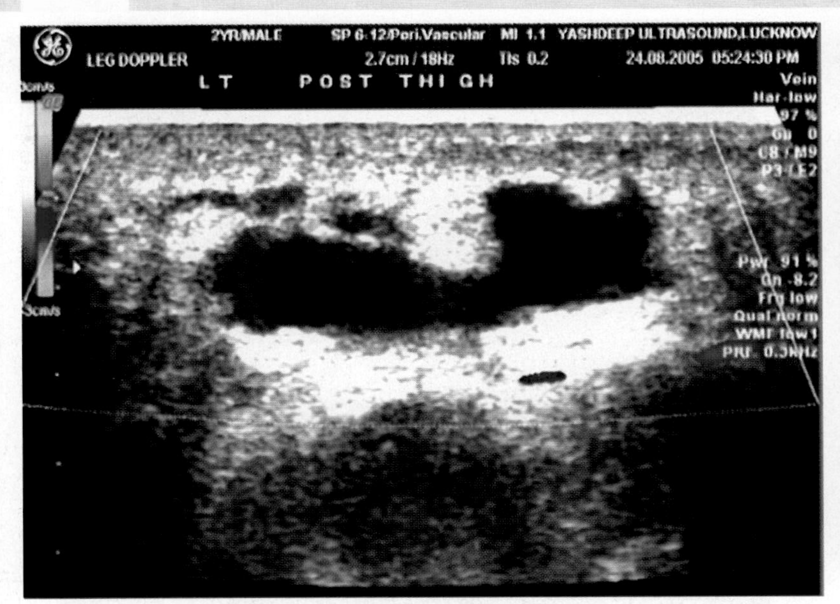

FIGURE 13.67

FIGURE 13.66 and 13.67: Posttraumatic cystic degeneration of the muscles. HRSG shows big irregular hypoechoic cystic mass in the posterior compartment of the thigh. Normal muscle texture is lost and it is replaced with the cystic collection. On color flow imaging no flow was seen in it suggestive of posttraumatic necrosis.

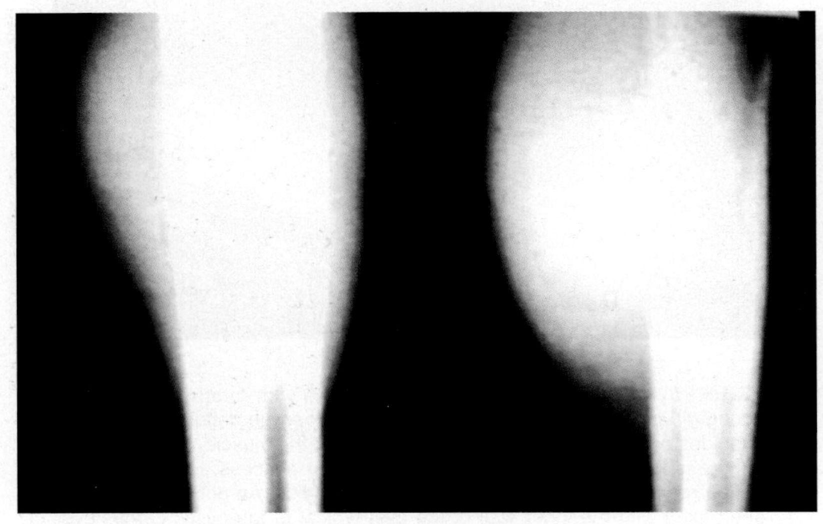

FIGURE 13.68: Cystic degeneration of calf muscles. X-ray of the patient shows soft tissue mass over the Lt calf. The bones are normal.

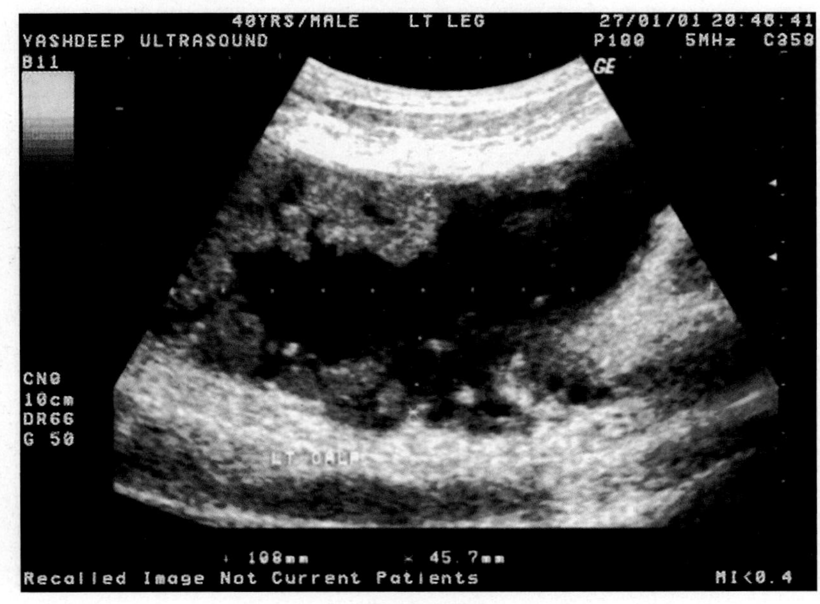

FIGURE 13.69

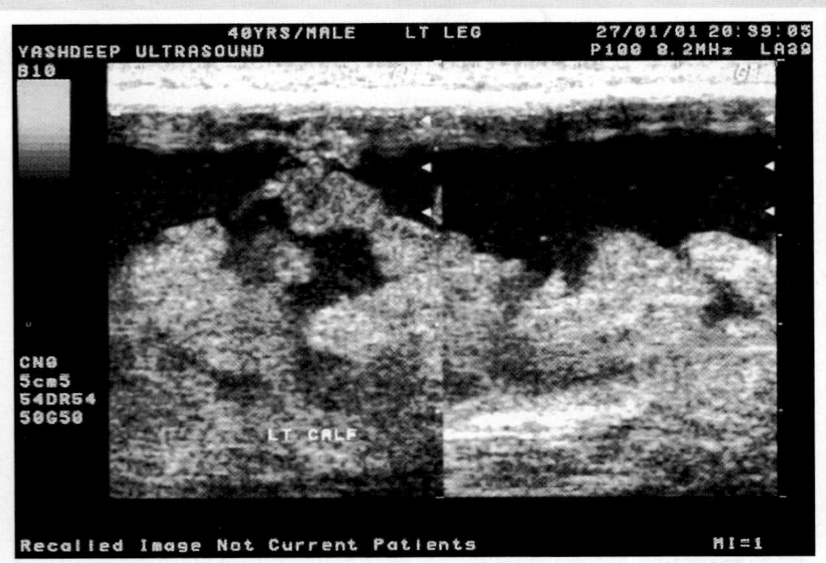

FIGURE 13.70

FIGURES 13.69 and 13.70: HRSG shows a big cystic degeneration of the calf muscles. The degenerated muscle bundles are seen floating in the cyst. Normal muscle texture is lost. Aspiration of the cyst shows chronic degeneration of the muscle.

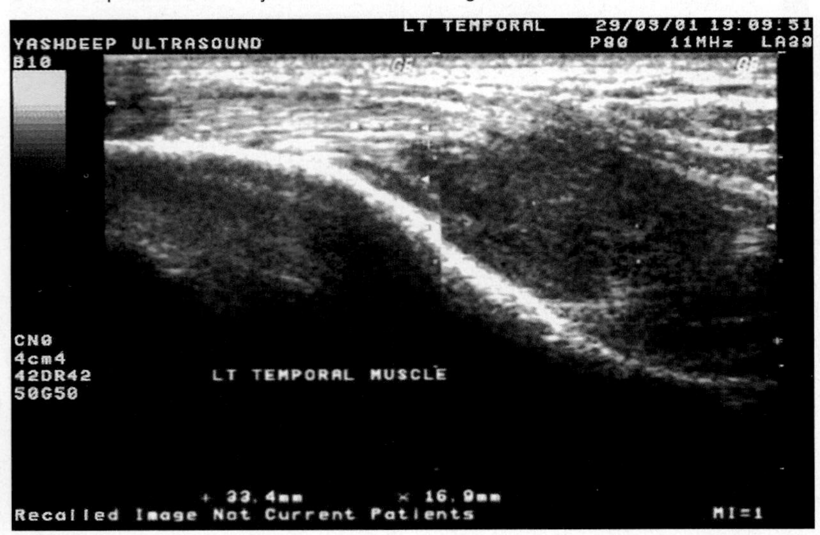

FIGURE 13.71: Pyogenic abscess temporal muscle. HRSG shows inflamed edematous Lt temporal muscle. Muscle edema is present. The muscle fibers are seen separated due to edema. Low-level echo complex shadows are seen in the muscle. Perifocal edema is present. However, muscle capsule is intact.

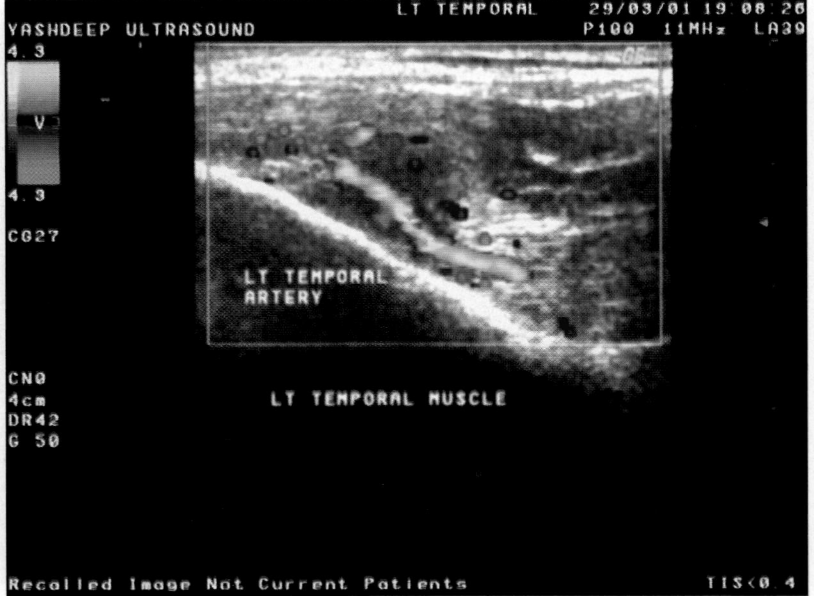

FIGURE 13.72

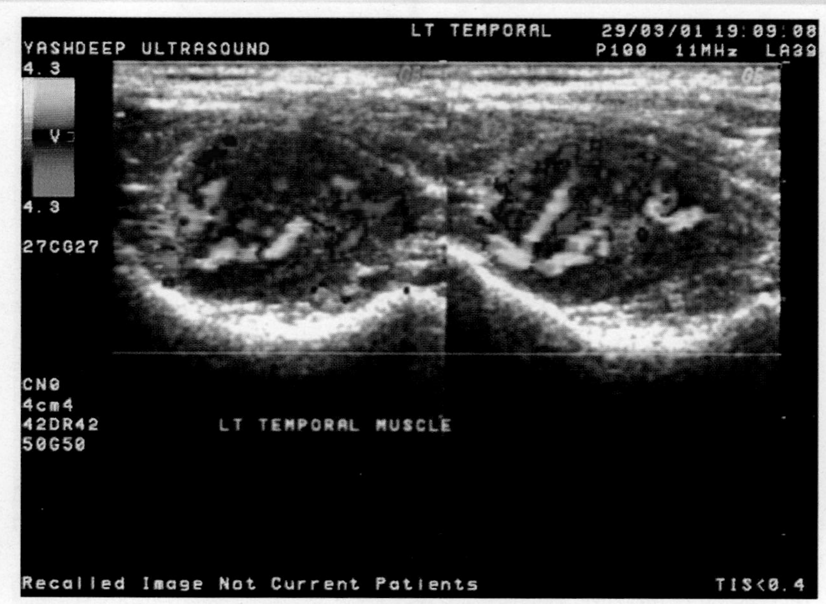

FIGURE 13.73

FIGURES 13.72 and 13.73: Color Doppler imaging shows moderate flow in the abscess suggestive of hyperemia.

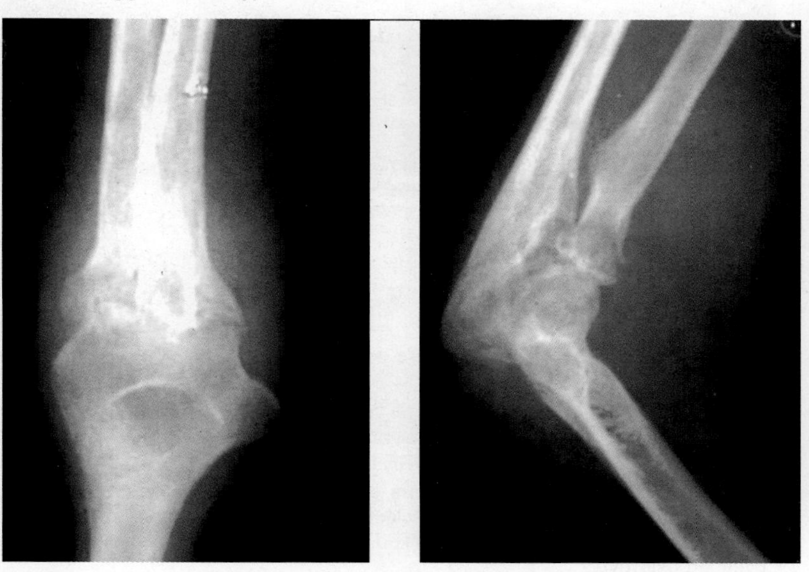

FIGURE 13.74: Cysticercosis elbow. A young boy presented with soft tissue mass over the Lt elbow. X-ray of the patient shows soft tissue swelling. However, bones are normal.

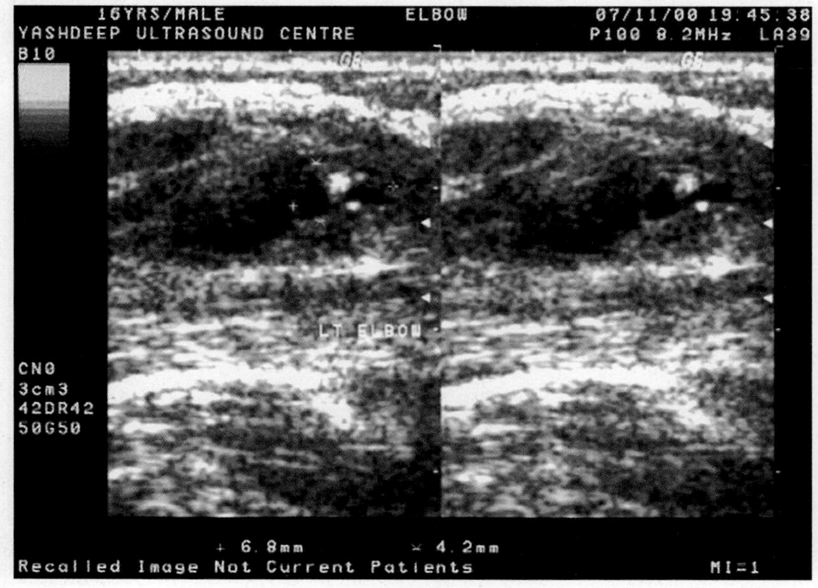

FIGURE 13.75

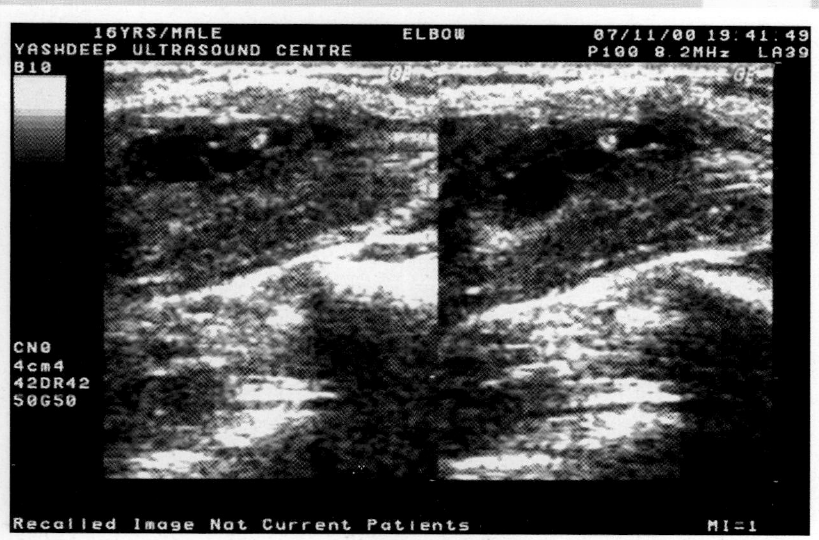

FIGURE 13.76

FIGURES 13.75 and 13.76: HRSG shows markedly thickened muscle belly. Two cysts are seen in the muscle. Echogenic nidus is seen fixed with the inner wall of the cyst suggestive of scloex, typical findings of cysticercosis. Muscle edema is also present.

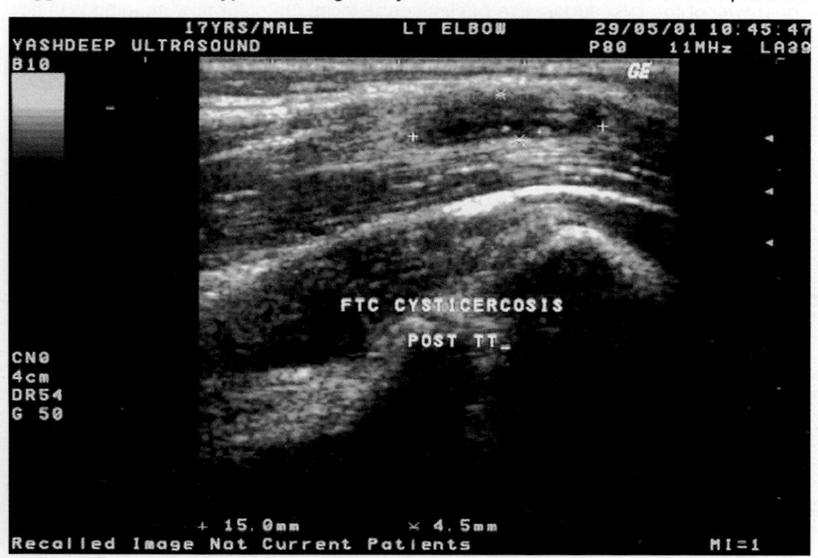

FIGURE 13.77: Post treatment regression of the cyst. HRSG shows localized thickening in the same patient after post albendazole treatment of the disease. The cyst disappeared and collapsed. However, localized thickening of the muscle is seen suggestive of fibrosis in the same patient. HRSG shows collapse of the cyst with focal area of fibrosis in the same patient.

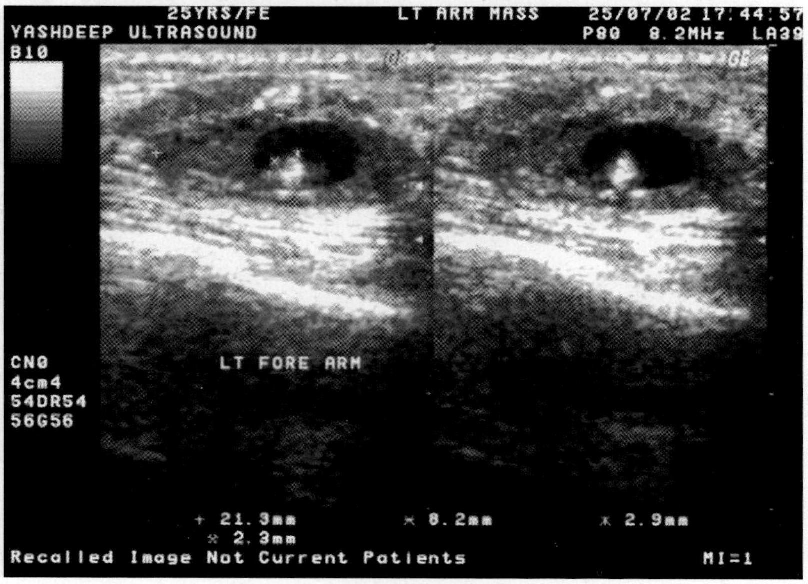

FIGURE 13.78

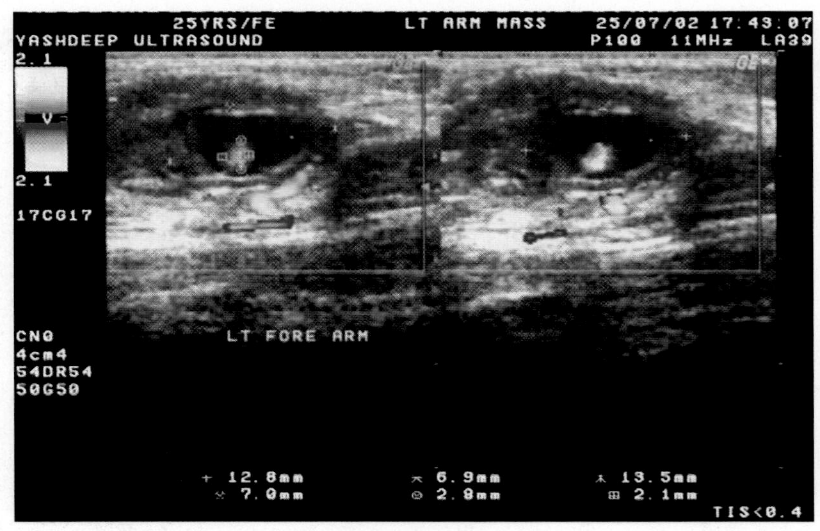

FIGURE 13.79

FIGURES 13.78 and 13.79: Cysticercosis cyst in the forearm muscle. A young lady presented with soft tissue swelling over the Lt forearm. HRSG shows a well-defined cystic mass in the muscle belly. An echogenic nidus is seen fixed with the inner wall of the cyst suggestive of cysticercus cyst. On color flow imaging wall flow is seen in the cyst.

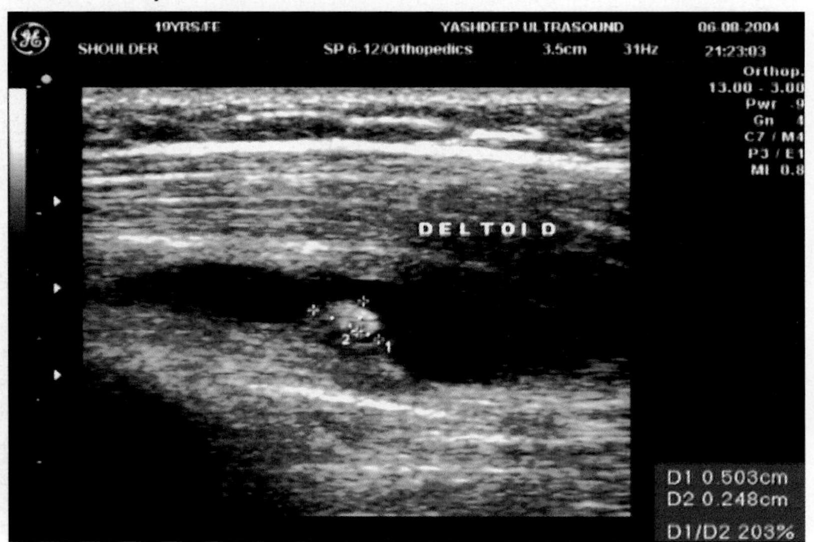

FIGURE 13.80

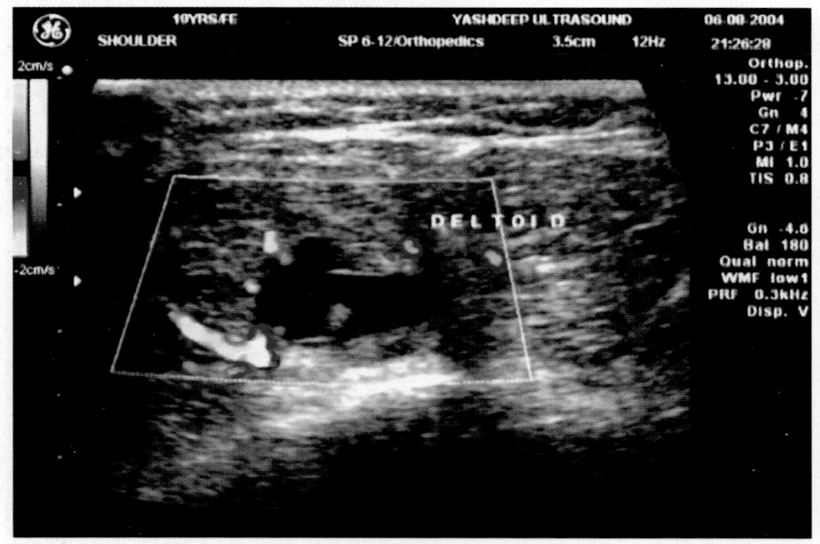

FIGURE 13.81

FIGURES 13.80 and 13.81: Deltoid muscle. HRSG shows a big cystic mass in the deltoid muscle of Rt shoulder. An echogenic nidus is seen fixed with the inner wall of the cyst. Associated muscle edema is also present. **On color flow imaging** few vessels are seen feeding the cyst suggestive of acute inflammation.

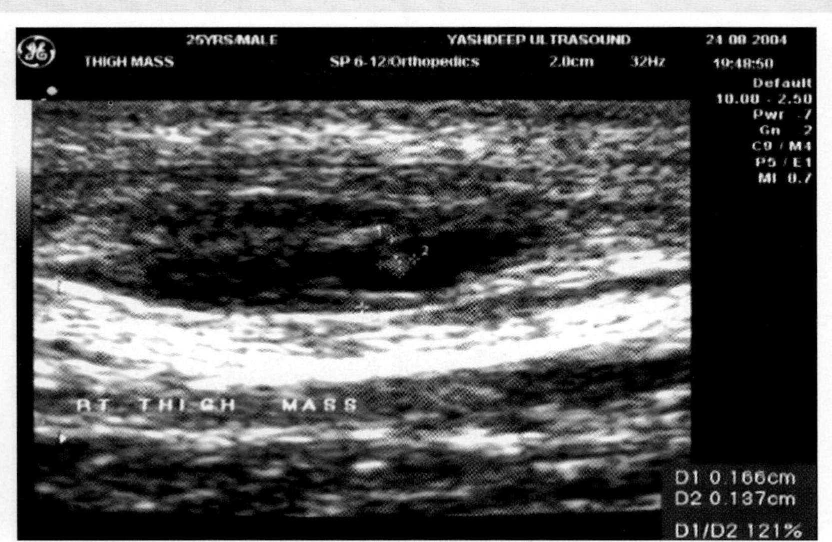

FIGURE 13.82

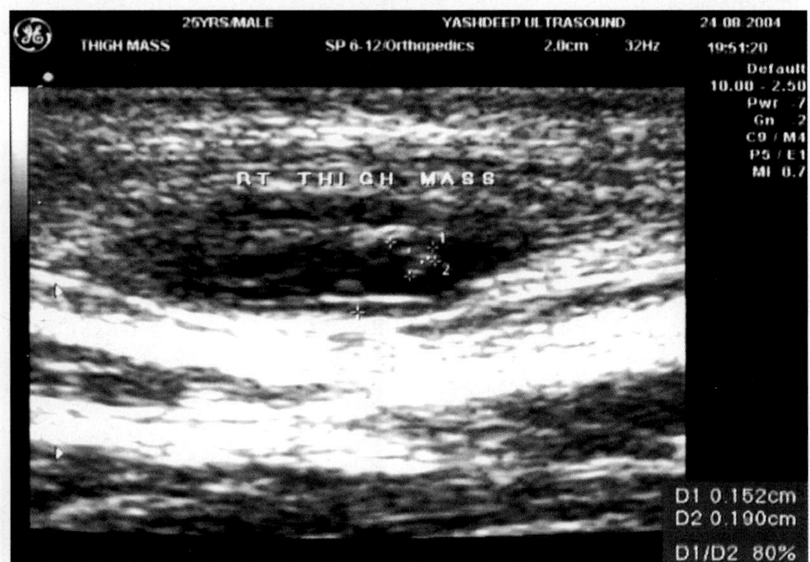

FIGURE 13.83

FIGURES 13.82 and 13.83: Cysticercus cyst in the thigh. A patient presented with nodular mass over the Rt thigh muscle. HRSG shows cystic mass in the belly of the muscle. An echogenic nidus is seen fixed with the inner wall of the cyst suggestive of cysticercosis cyst.

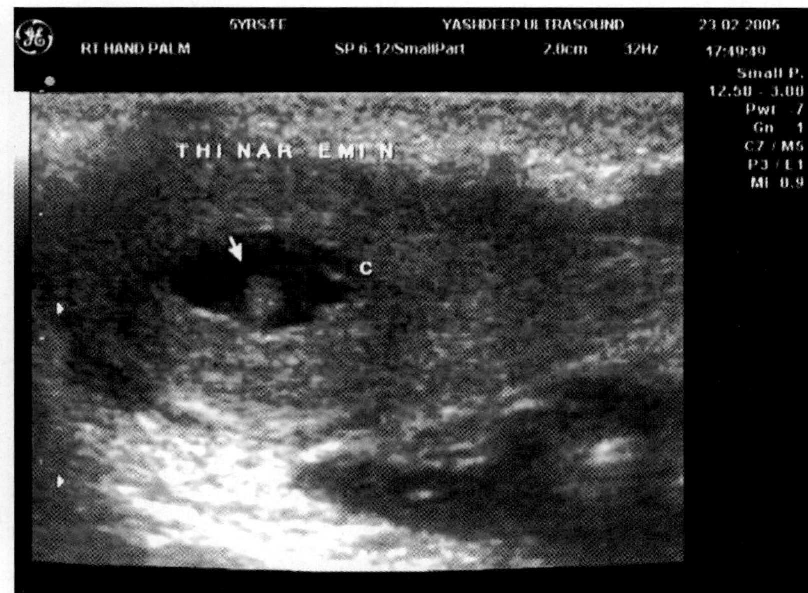

FIGURE 13.84: Cysticercus in the thinar eminence. A patient presented with soft tissue nodular shadow over the Rt thinar eminence in the palm. HRSG shows a well define cystic mass over the thinar eminence.

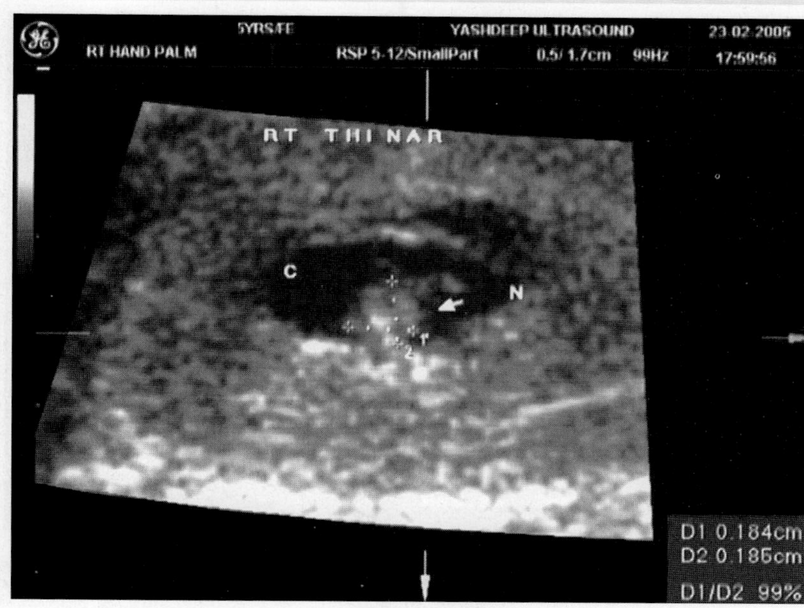

FIGURE 13.85

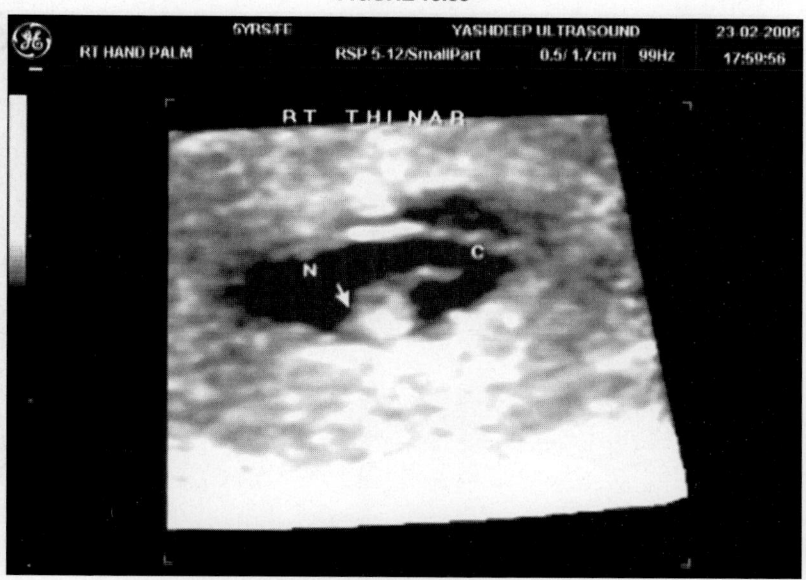

FIGURE 13.86

FIGURES 13.85 and 13.86: 3D imaging of the cyst shows the details of the cyst showing typical feature of cyst. Palmar invasion of the cysticercus cyst is unusual and not seen frequently. I have seen only three cases in last 15 years.

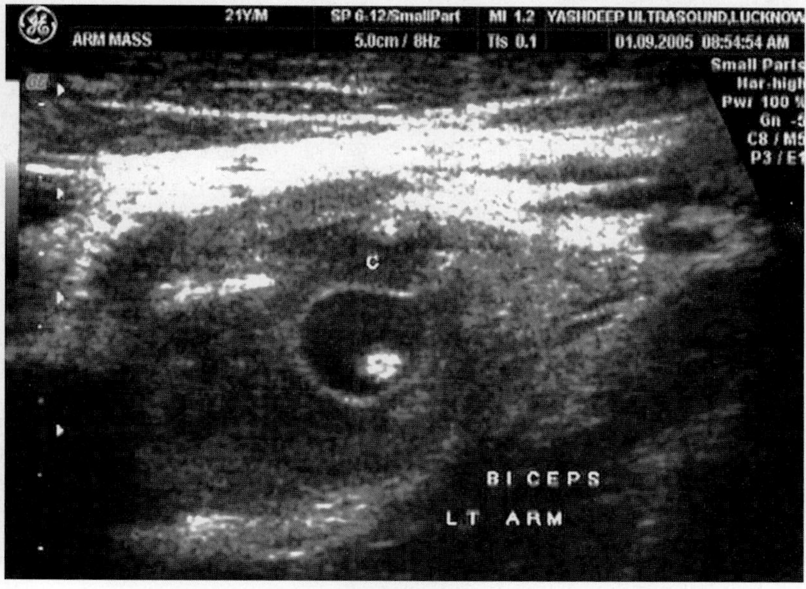

FIGURE 13.87

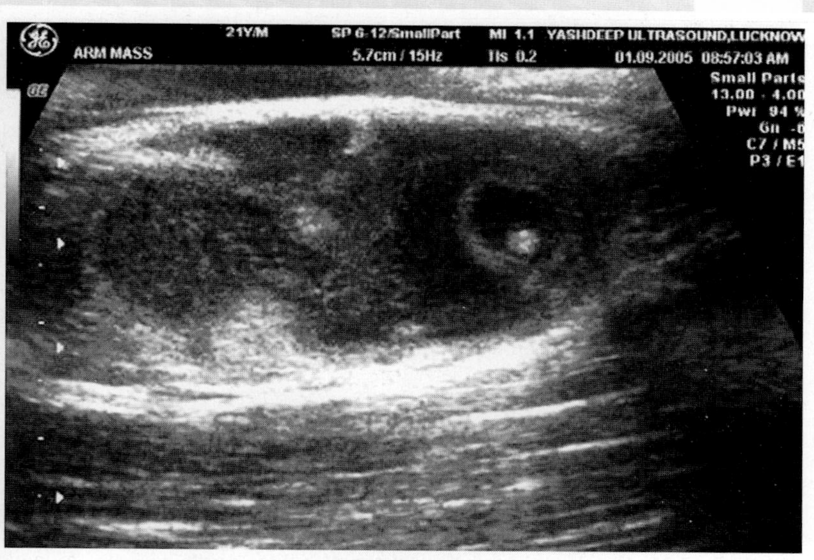

FIGURE 13.88

FIGURES 13.87 and 13.88: Biceps muscle cysticercus cyst. A young man presented with marked soft tissue swelling over the Lt biceps muscle. HRSG shows edematous muscle belly. A well-defined cyst is seen embedded in the muscle. An echogenic nidus is seen fixed with the inner wall of the cyst suggestive of cysticercus cyst.

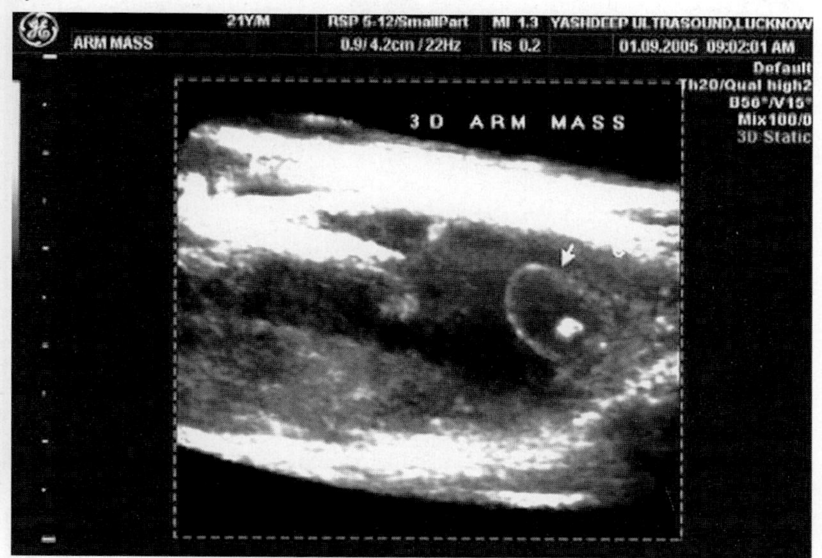

FIGURE 13.89

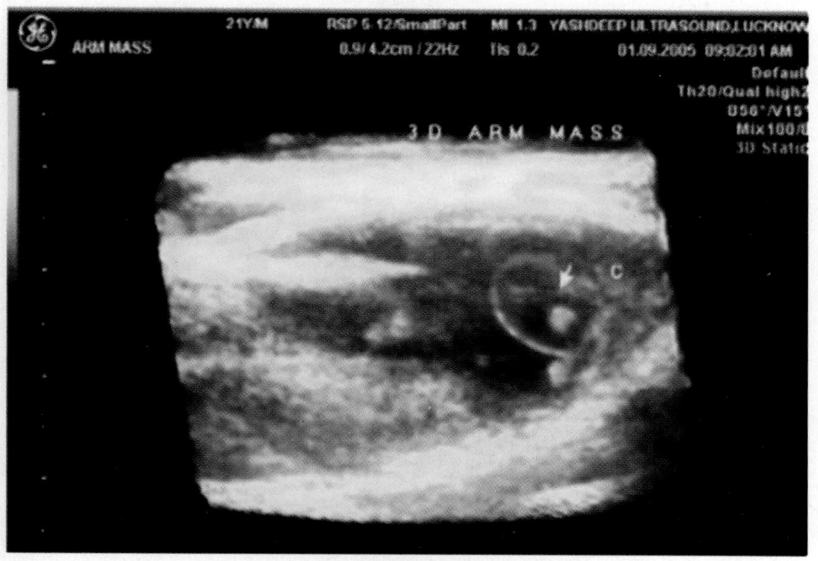

FIGURE 13.90

FIGURES 13.89 and 13.90: 3D surface rendering imaging of the cyst shows the better details of the cyst.

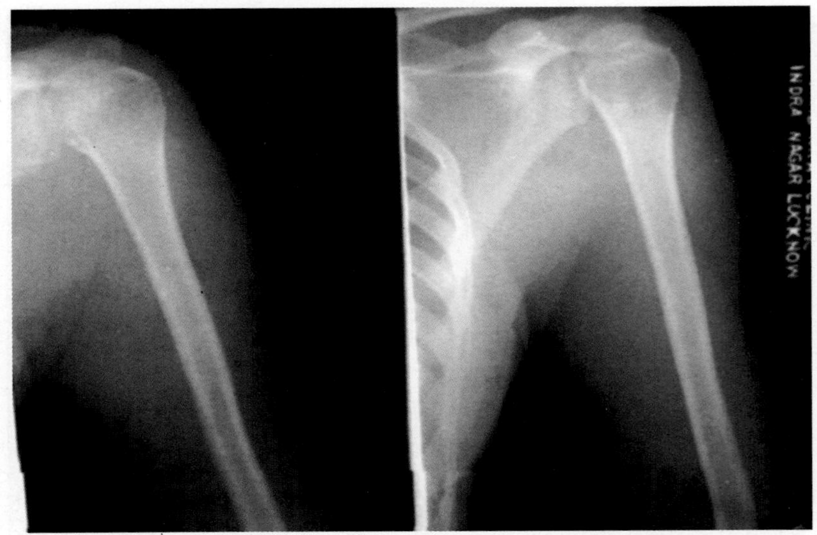

FIGURE 13.91: **Cystic degeneration of the shoulder girdle muscle.** A young girl presented with freely hanging Lt arm. Plane X-ray of the shoulder shows soft tissue mass shadow. However, the bones are absolutely normal.

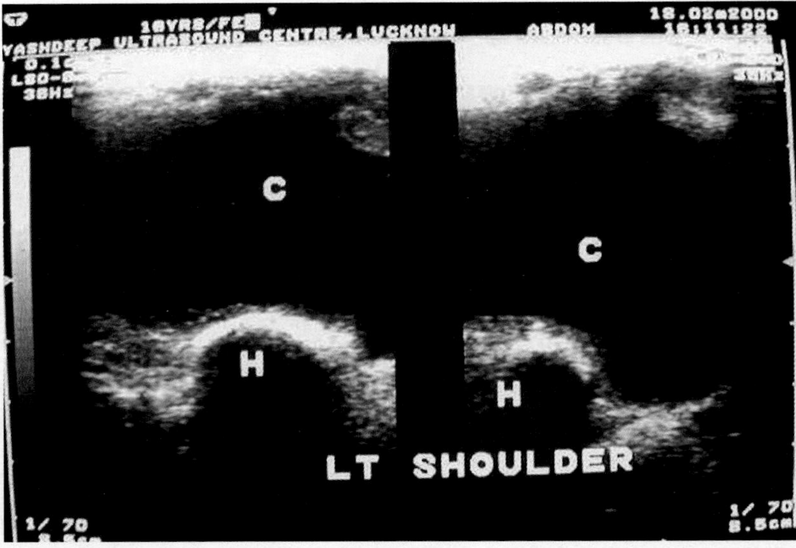

FIGURE 13.92

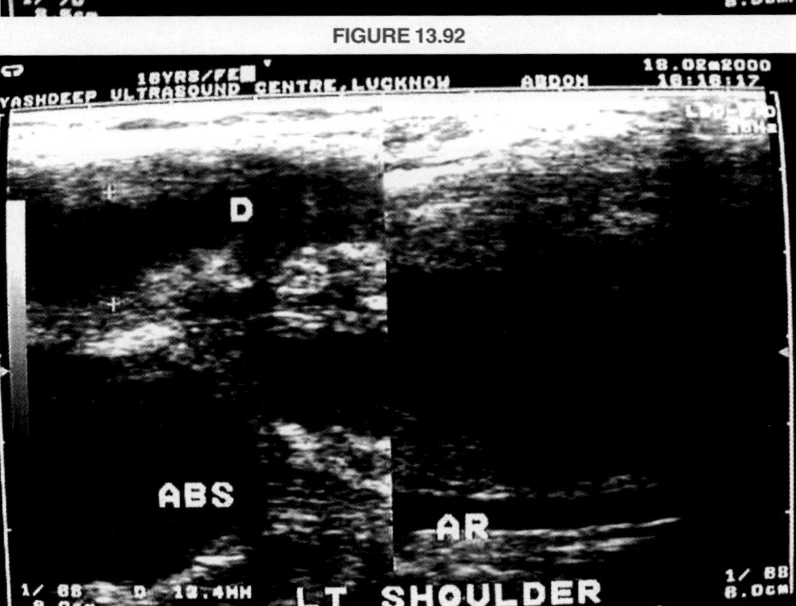

FIGURE 13.93

FIGURES 13.92 and 13.93: HRSG shows cystic degeneration of the muscles involving the Lt shoulder girdle. The deltoid biceps and triceps muscle show cystic degeneration with low-level echo collection in the muscles. Normal muscle anatomy was lost. The collection is also seen anterior to the humerus shaft. The bony shaft was normal.

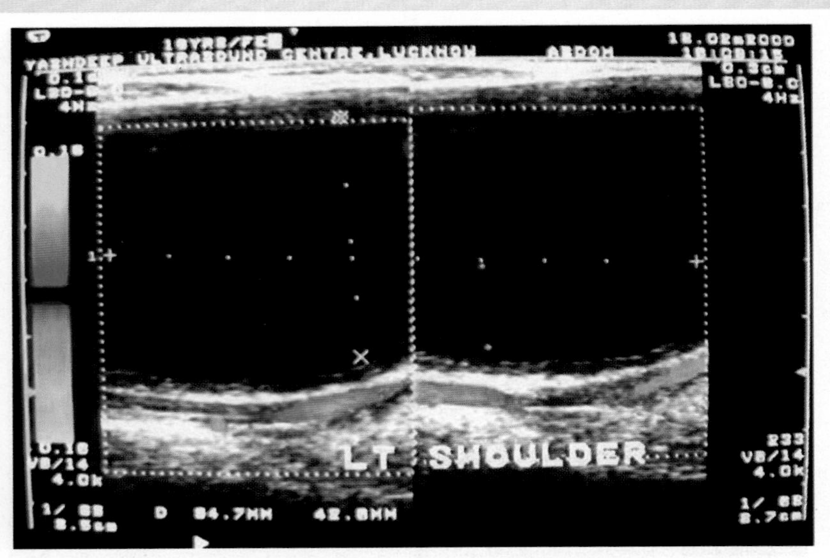

FIGURE 13.94: On color flow imaging the vessels are seen intact.

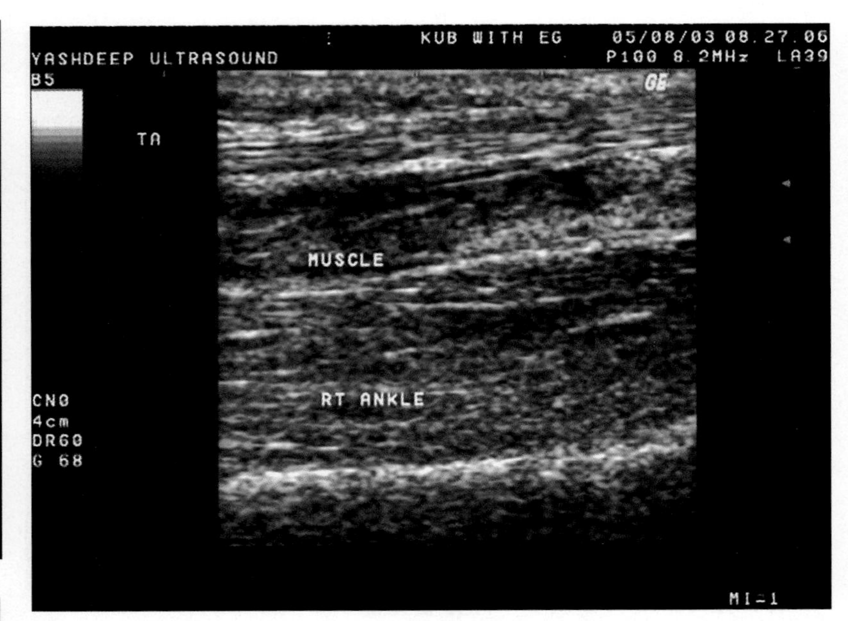

FIGURE 13.95

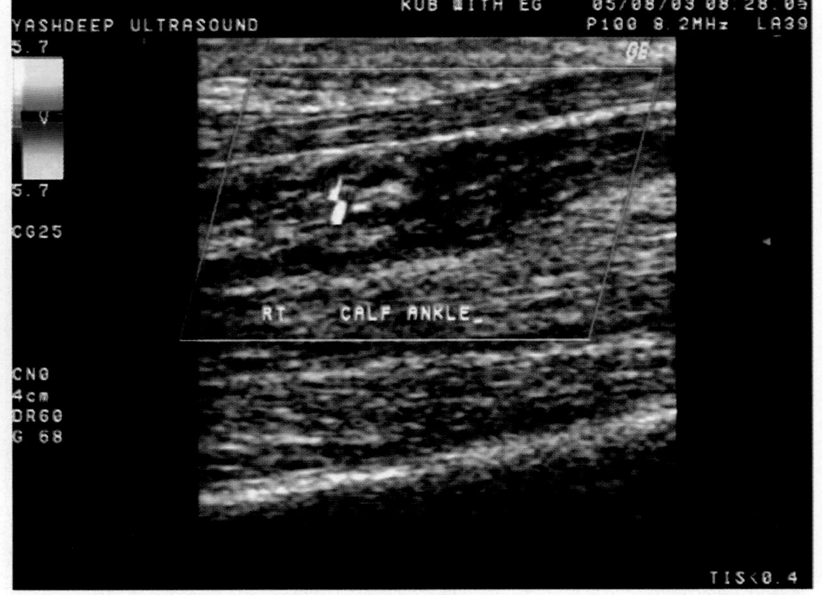

FIGURE 13.96

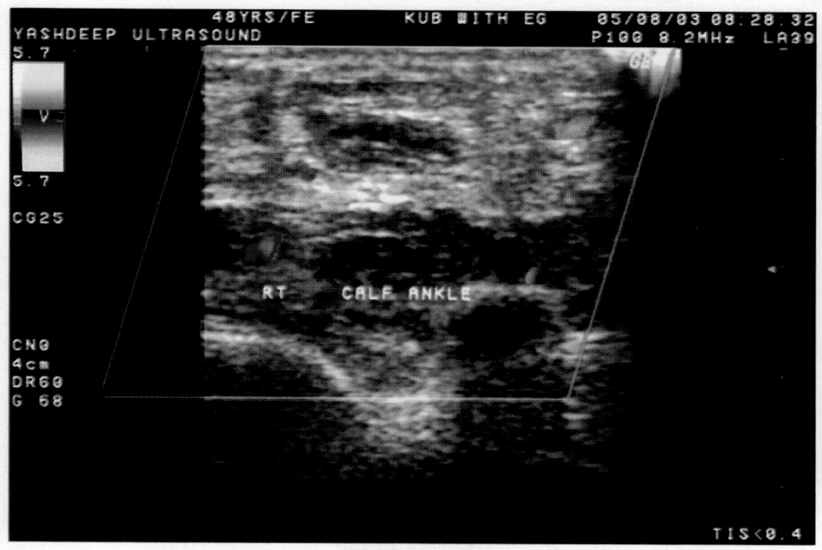

FIGURE 13.97

FIGURES 13.95 to 13.97: Contusion injury to the calf muscle. HRSG shows contusion injury to the calf muscle. Irregular hypoechoic contusion is seen in the muscles. Partial disruption of the muscle fibers is seen. However, no suggestion of any tear is seen. **On color flow imaging** no flow is seen around the contusion.

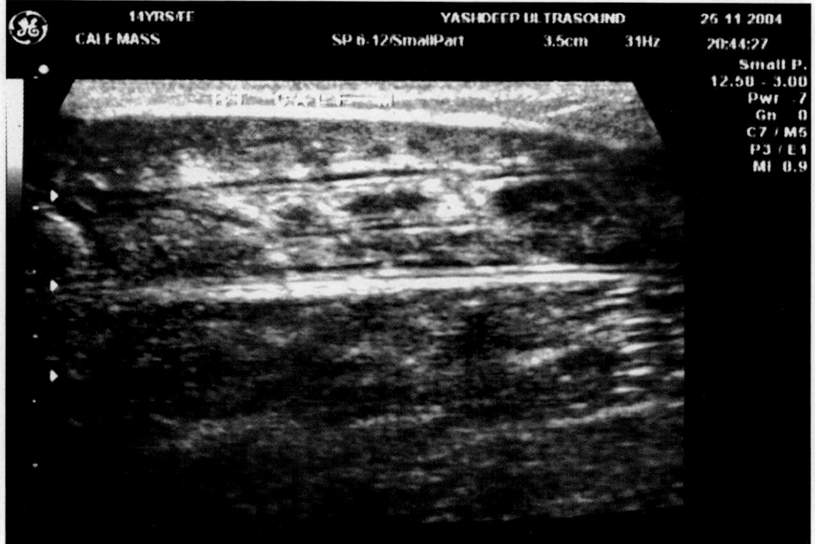

FIGURE 13.98

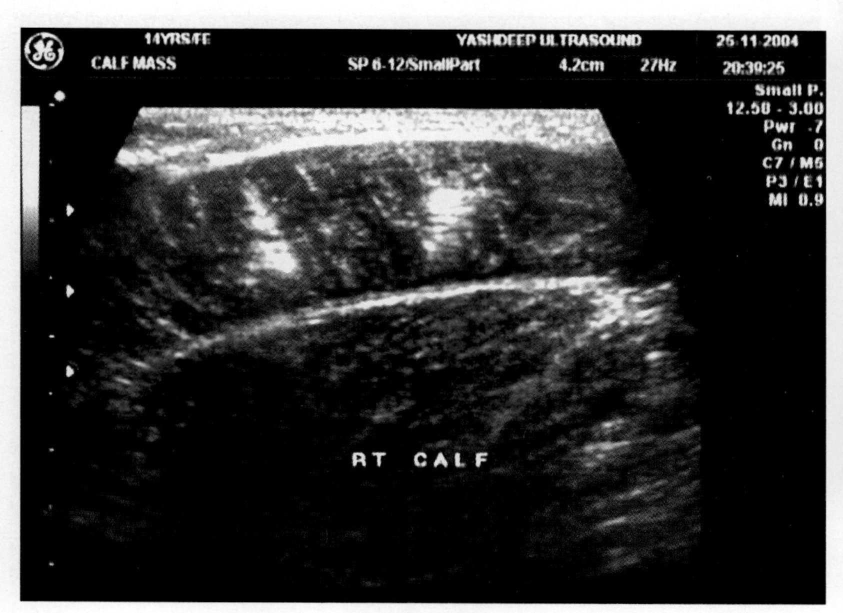

FIGURE 13.99

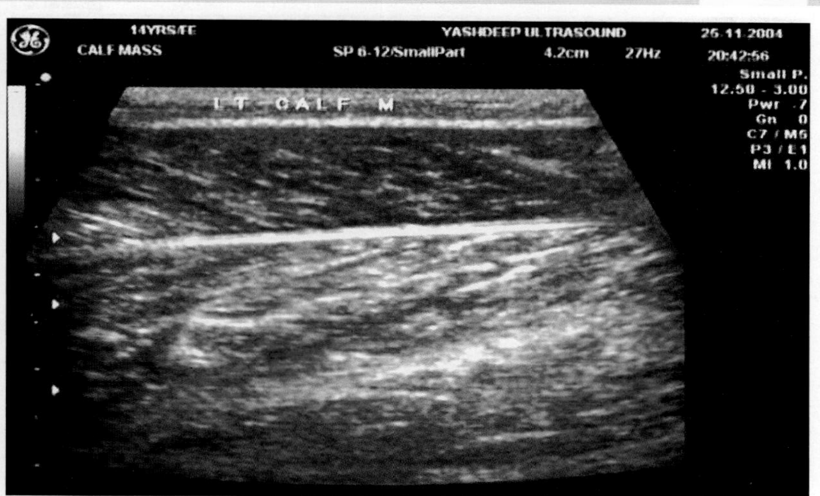

FIGURE 13.100

FIGURES 13.98 to 13.100: Muscle herniation. A patient presented with pain on prolonged walking. HRSG shows the calf muscle at rest and after walking. The resting stage muscle bundles appear to be normal. No evidence of any tear is seen in them. However on prolonged walking, the muscle is seen bulged out. The edema is seen in the muscle bundles. The capsule of the muscle is seen bulging out suggestive of herniation. The Lt calf muscle shows normal texture. No herniation is seen on the Lt side.

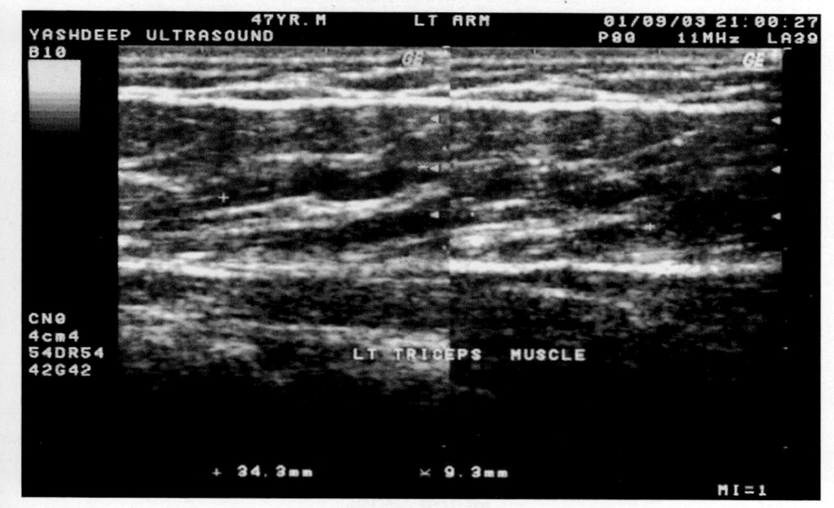

FIGURE 13.101

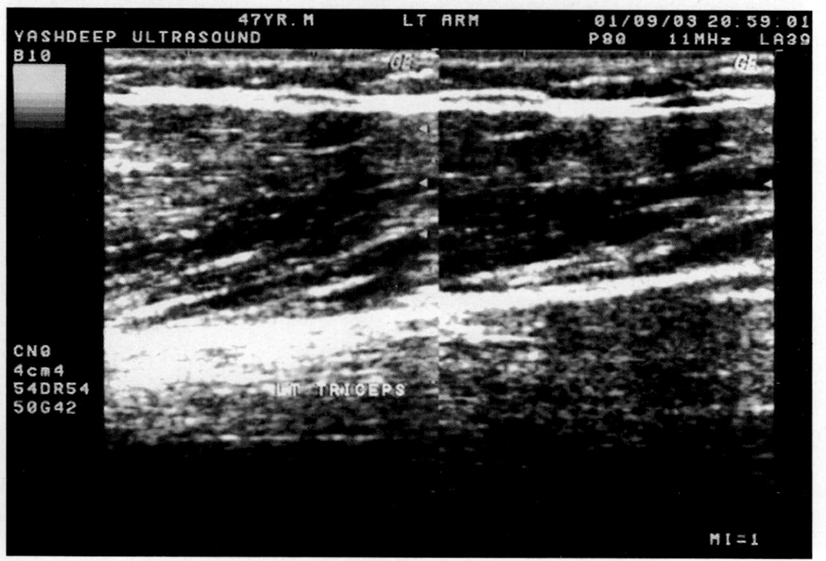

FIGURE 13.102

FIGURES 13.101 and 13.102: Small tear in the muscle. A patient presented with painful movement of the arm with restricted movements. HRSG shows irregular hypoechoic clefts in the triceps muscle. Which are running obliquely in the muscle belly suggestive of tear.

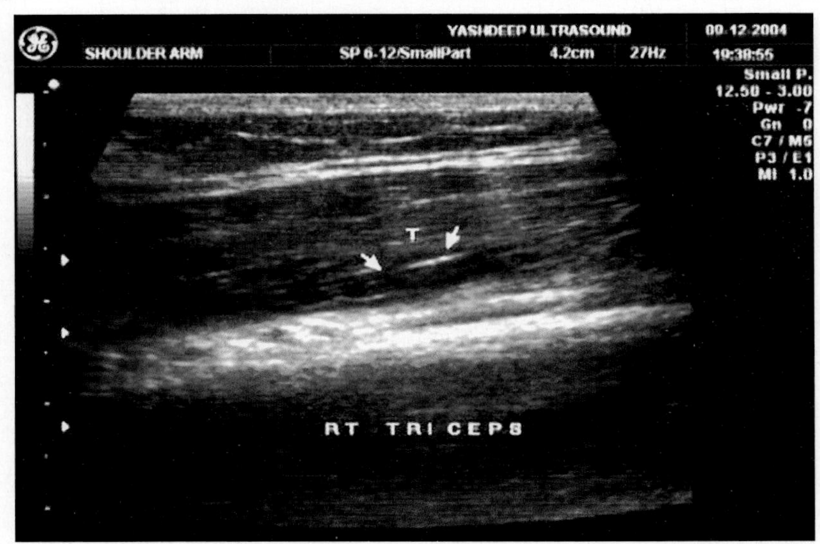

FIGURE 13.103: Triceps muscle tear. HRSG shows a hypoechoic cleft in the triceps muscle of Rt arm in a patient presented with painful movement of the arm. A well-defined cleft is seen running in the belly of the muscle with associated perifocal edema suggestive of a tear.

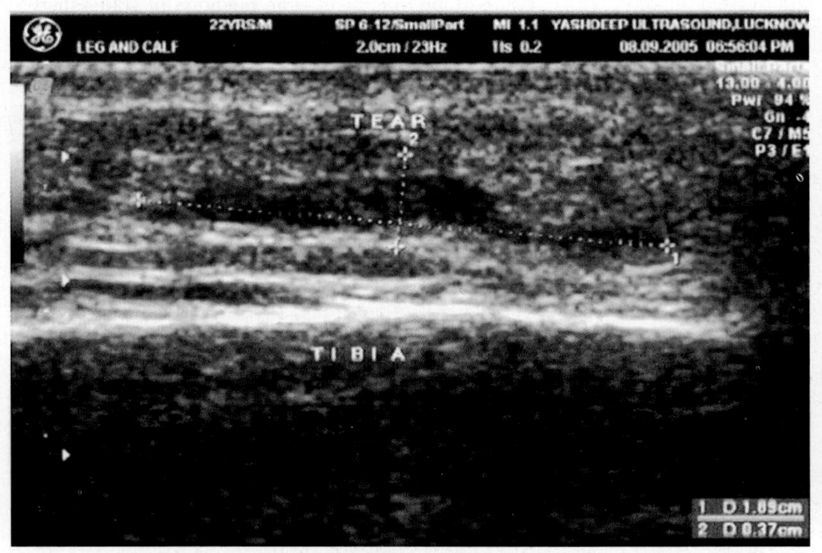

FIGURE 13.104

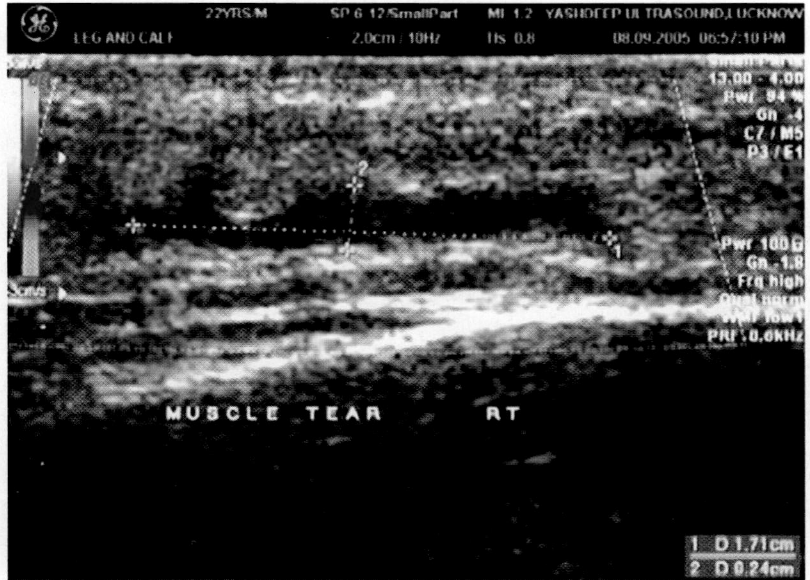

FIGURE 13.105

FIGURES 13.104 and 13.105: Muscle tear with contusion. HRSG shows a hypoechoic cleft in the muscle of the leg. It is seen anterior to the tibia. Perifocal edema is also present. On color flow imaging no flow is seen in the tear.

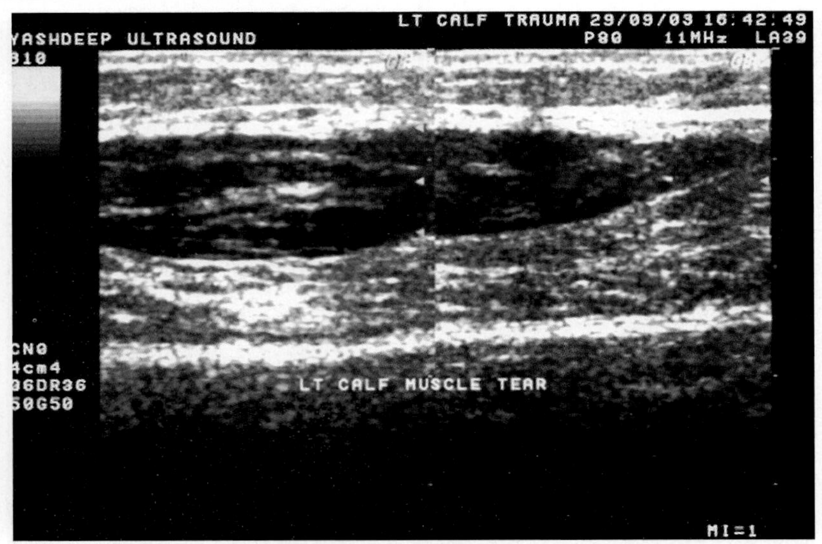

FIGURE 13.106

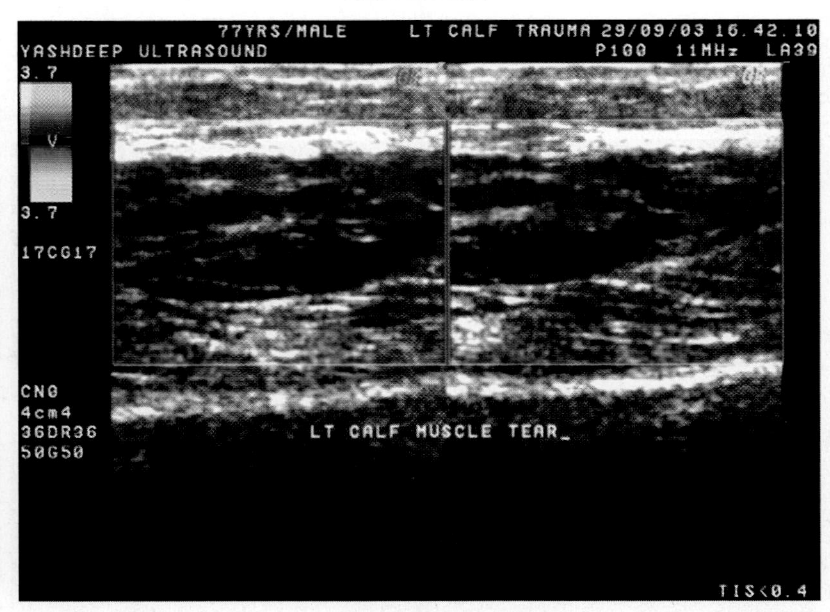

FIGURE 13.107

FIGURES 13.106 and 13.107: Big tear in the calf muscles. HRSG shows a big tear in the calf muscle after sustaining a fall in a trench resulting into shearing calf muscle tear. A longitudinal tear is seen running through the belly of the gastrocnemius muscle. Disruption of the muscle fibers is also seen.

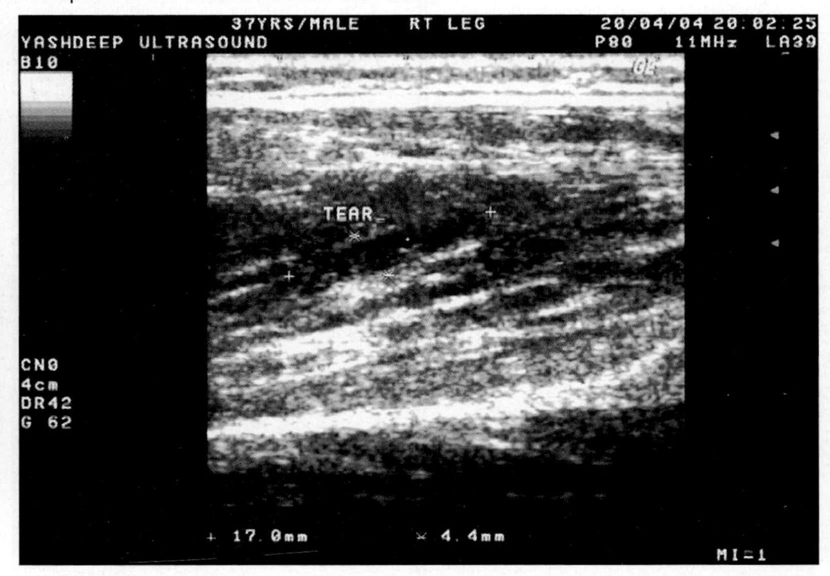

FIGURE 13.108

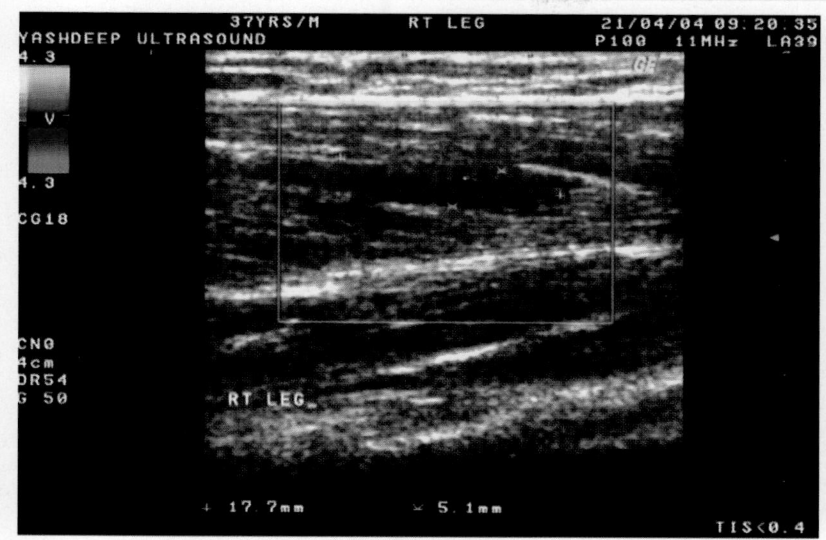

FIGURE 13.109

FIGURES 13.108 and 13.109: A patient presented with pain on long walking. HRSG shows irregular hypoechoic cleft in the muscle belly of the anterior compartment of the leg. The muscle fibers are torned and disrupted. Perifocal edema is present. But no evidence of any calcification is seen. On color flow imaging no flow is seen in the tear.

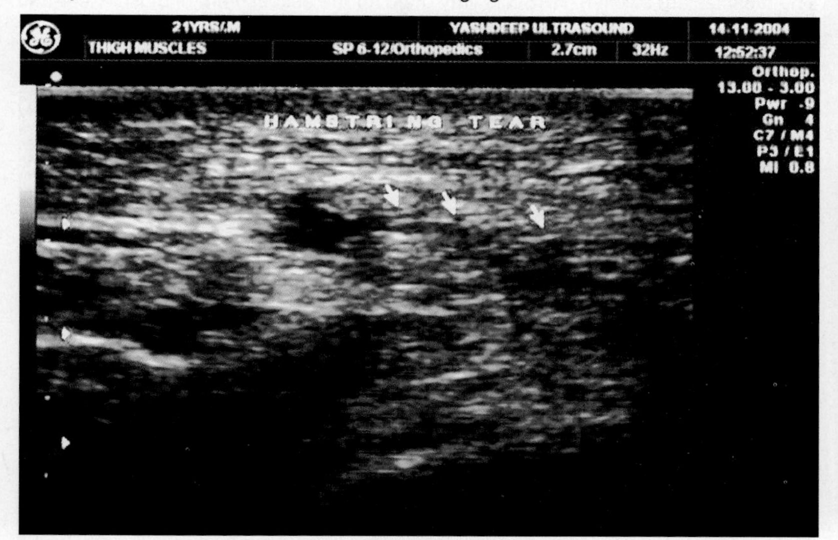

FIGURE 13.110

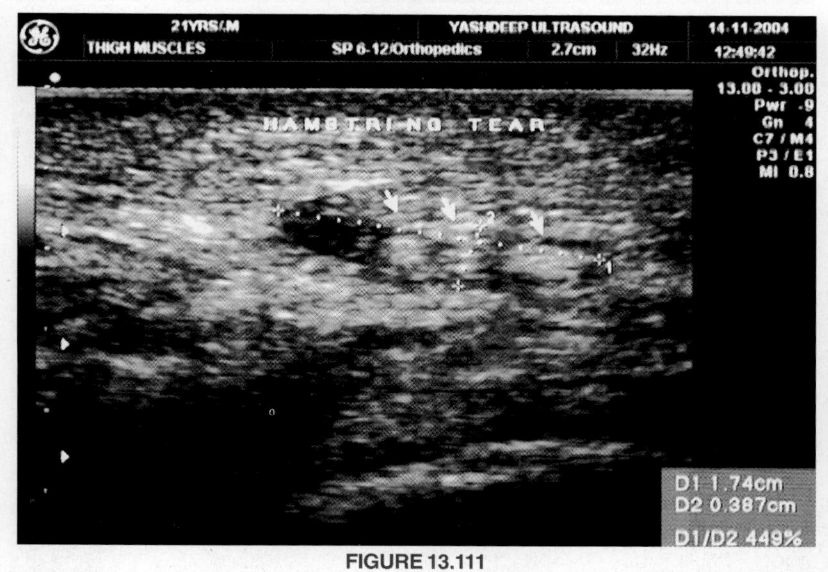

FIGURE 13.111

FIGURES 13.110 and 13.111: Hamstring muscle tear. A patient presented with acute pain by performed exercise near the medial side of the thigh. HRSG shows rupture of the hamstring muscle before its insertion in the ischial tuberosity. Echogenic collection is seen in the rupture suggestive of fresh hemorrhage. The margins are irregular in outline.

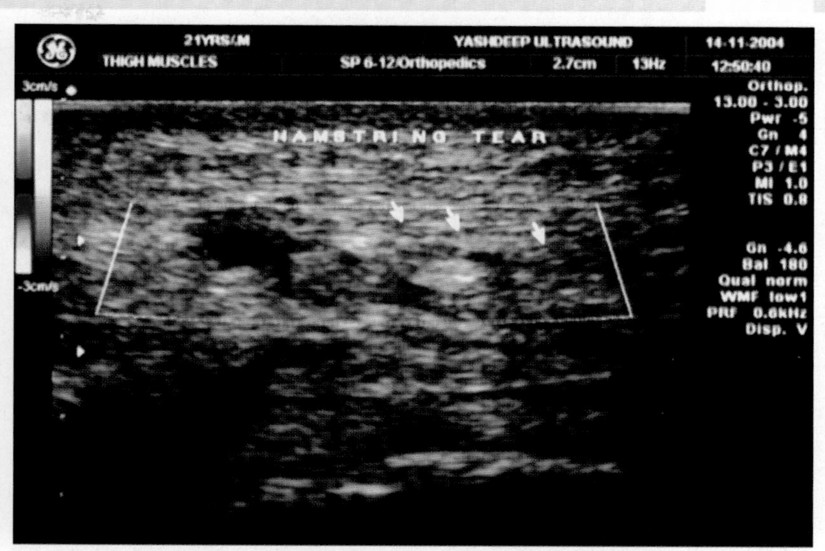

FIGURE 13.112: On color flow imaging no flow is seen in the rupture.

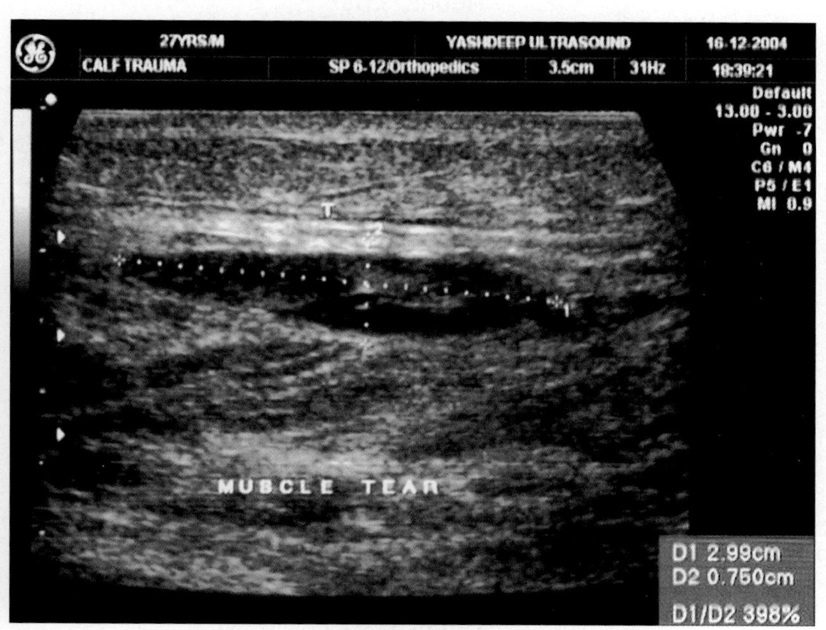

FIGURE 13.113

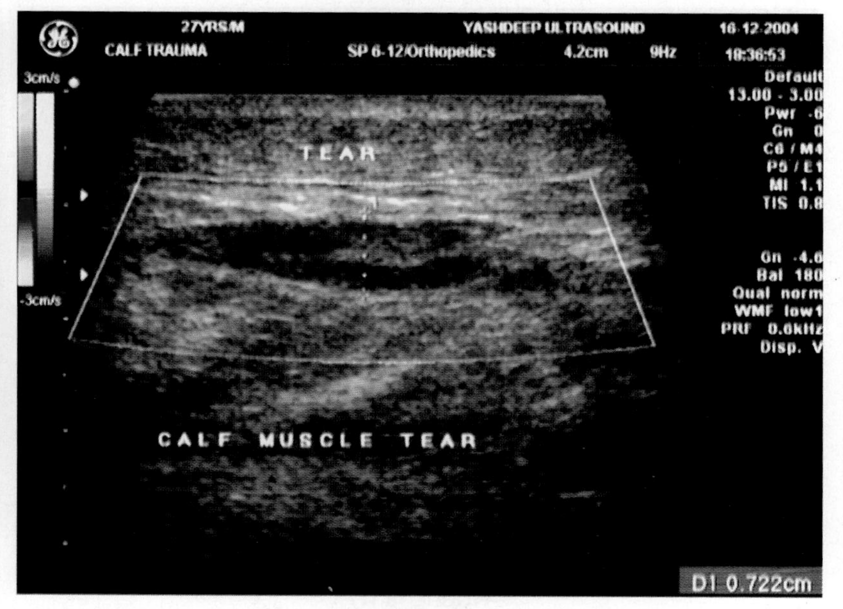

FIGURE 13.114

FIGURES 13.113 and 13.114: Calf muscle tear. HRSG shows a big cleft in the calf muscle of a long distance runner. A fusiform cleft is seen in the gastrocnemius muscle.

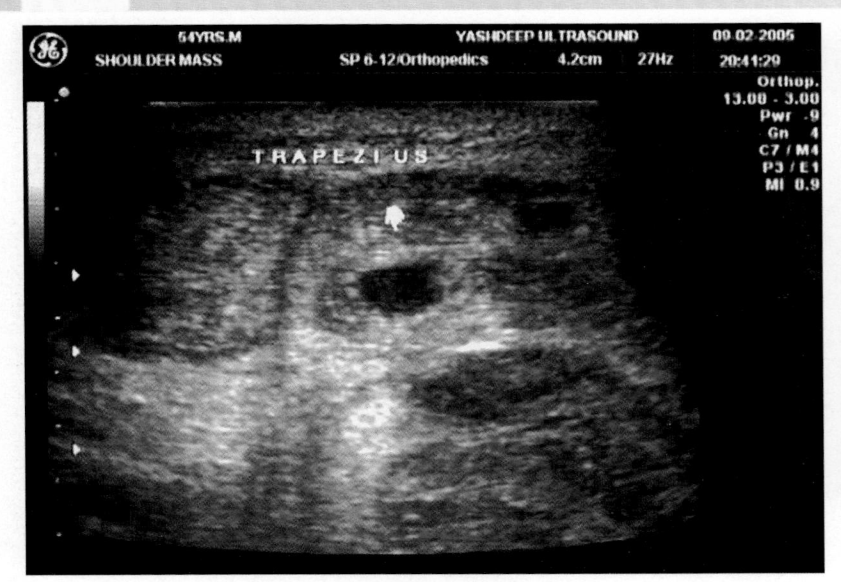

FIGURE 13.115

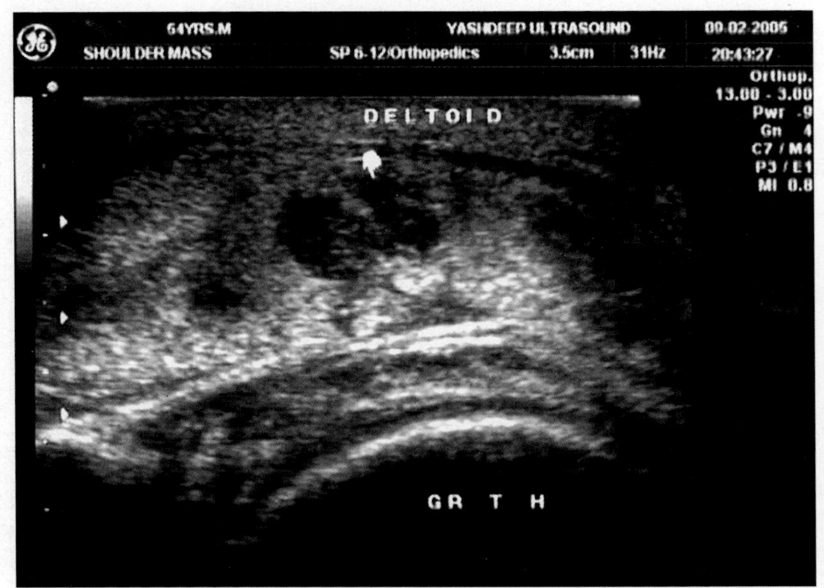

FIGURE 13.116

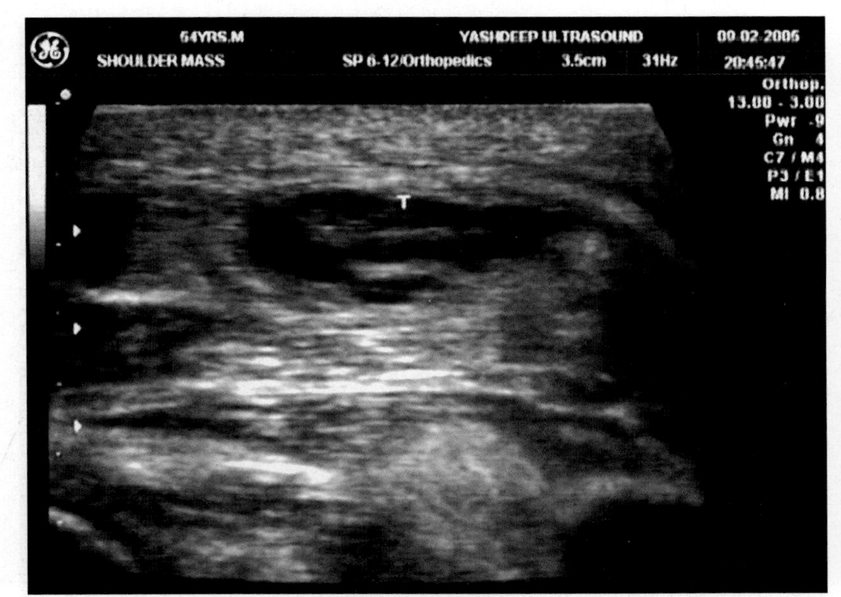

FIGURE 13.117

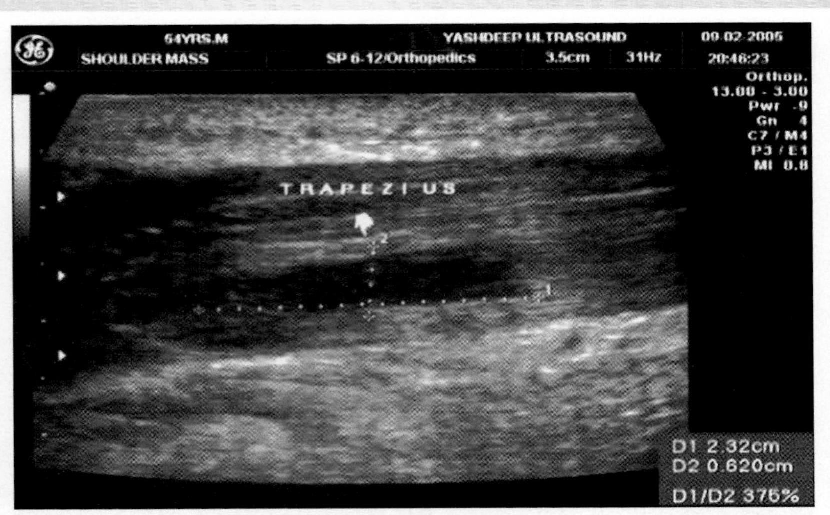

FIGURE 13.118

FIGURES 13.115 to 13.118: Multiple tears in the shoulder girdle muscles. A patient presented with acutely painful shoulder joint with marked swelling. The muscles were red and hot. HRSG shows the shoulder girdle muscles are hypoechoic edematous. They are swollen. Big irregular clefts are seen in trapezius, deltoid, biceps and triceps muscles. The clefts are seen running through the muscle belly. The muscles are swollen. HRSG is very good for the assessment of muscle tear and small tears are very well picked up on HRSG.

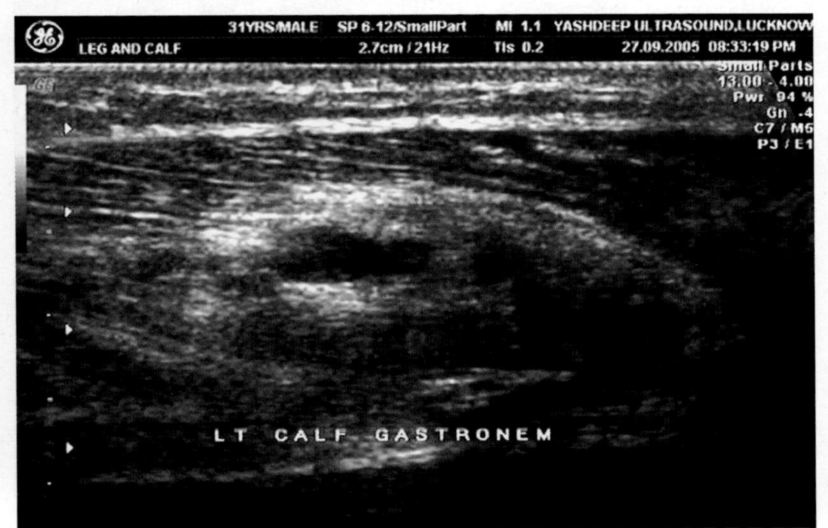

FIGURE 13.119

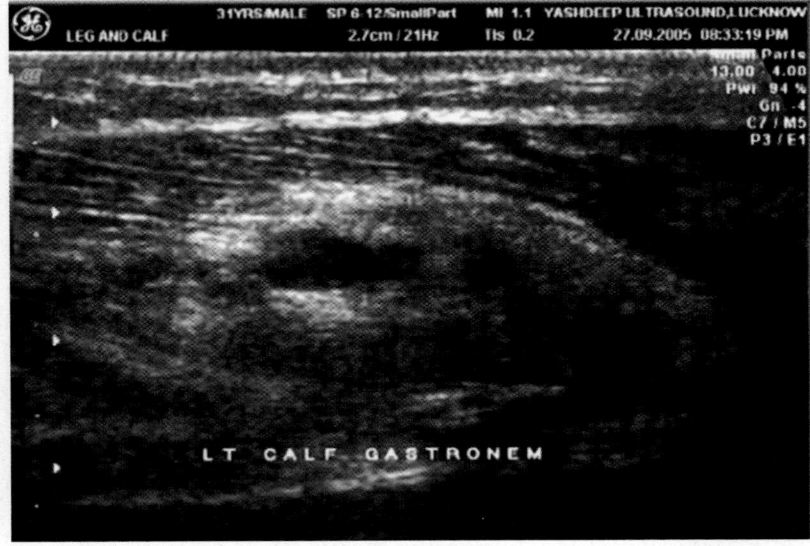

FIGURE 13.120

FIGURES 13.119 and 13.120: Calf muscle tear. A patient presented with blunt injury to the Lt leg and presented with marked painful swelling. HRSG shows calf muscle hematoma. The gastrocnemius muscle is seen torn. A hypoechoic cleft is seen in it suggestive of tear. The muscle is edematous.

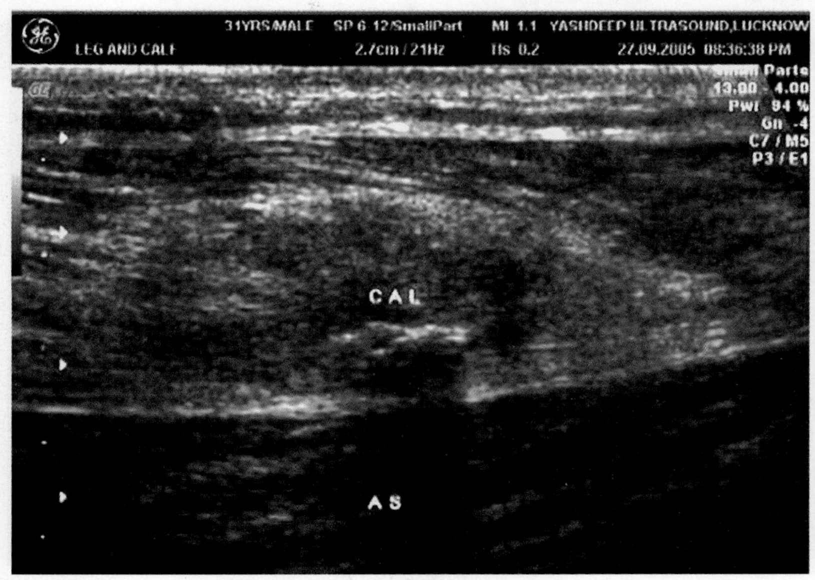

FIGURE 13.121

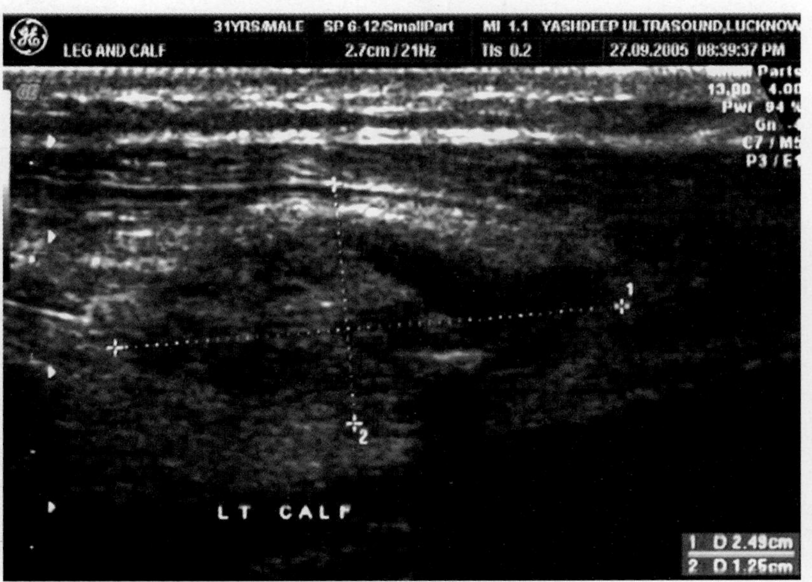

FIGURE 13.122

FIGURES 13.121 and 13.122: Echogenic calcified shadows are also seen in the tear. They are accompanied with acoustic shadowing suggestive of organization of the hemorrhage.

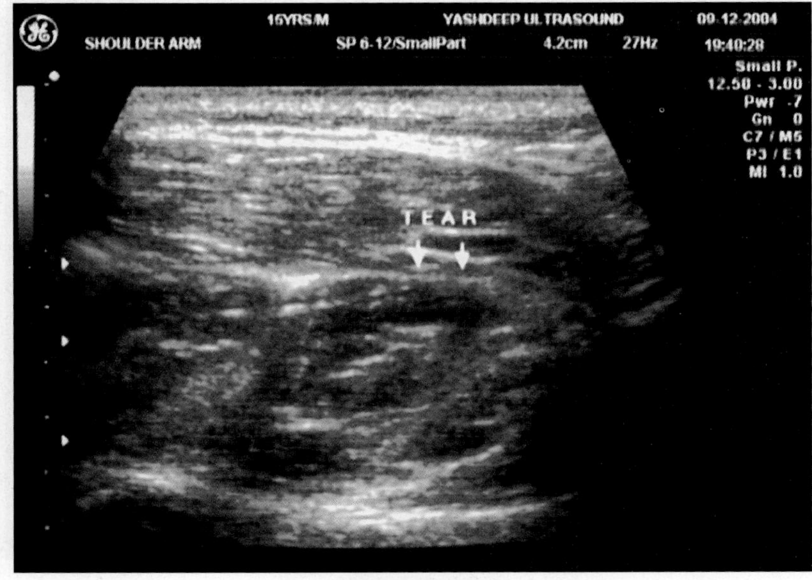

FIGURE 13.123

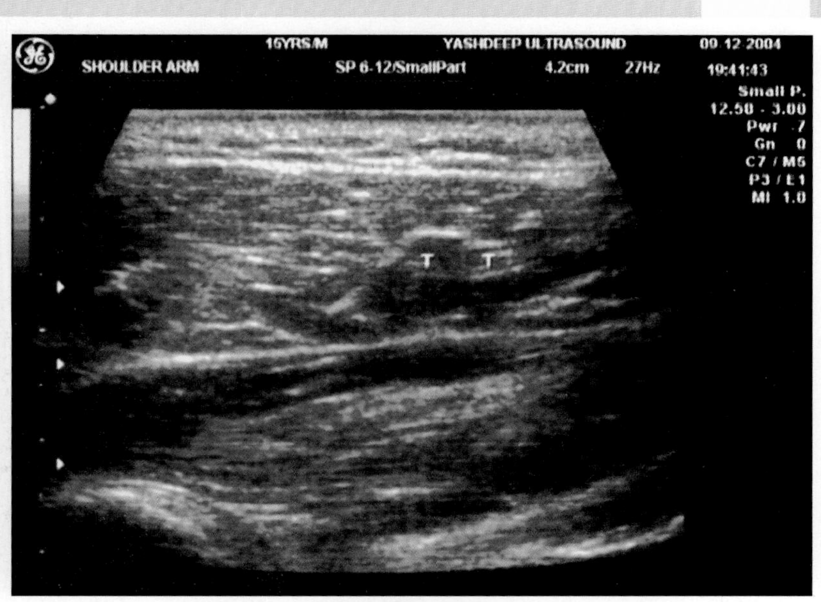

FIGURE 13.124

FIGURES 13.123 and 13.124: Biceps muscle tear. A young boy presented with acute pain with swelling in the arm. HRSG shows swollen edematous biceps muscle. A hypoechoic cleft is seen running through the belly of the muscle. Perifocal edema is presented suggestive of acute recent tear.

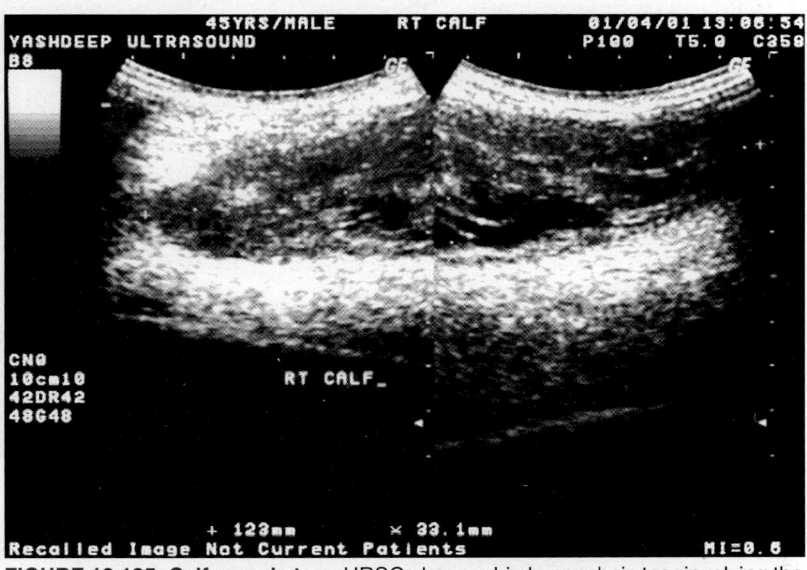

FIGURE 13.125: Calf muscle tear. HRSG shows a big hypoechoic tear involving the Rt calf muscle in blunt injury to the leg. The fibers are torn. Echogenic collection is also seen suggestive of hematoma formation.

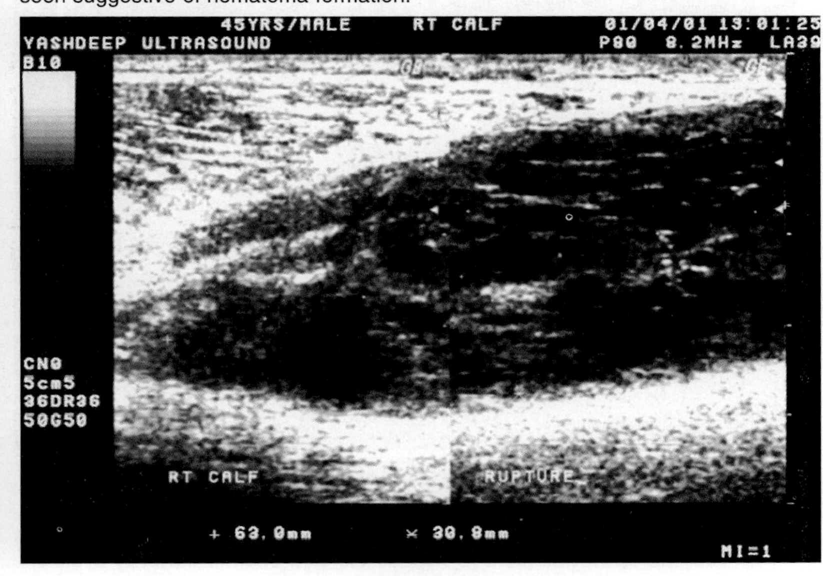

FIGURE 13.126

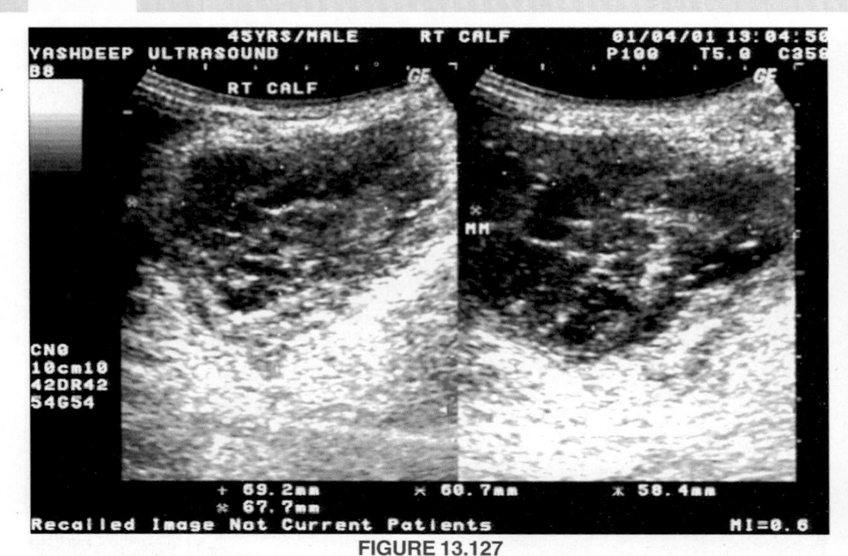

FIGURE 13.127

FIGURES 13.126 and 13.127: HRSG shows big tear in the muscle running through the belly of the muscle.

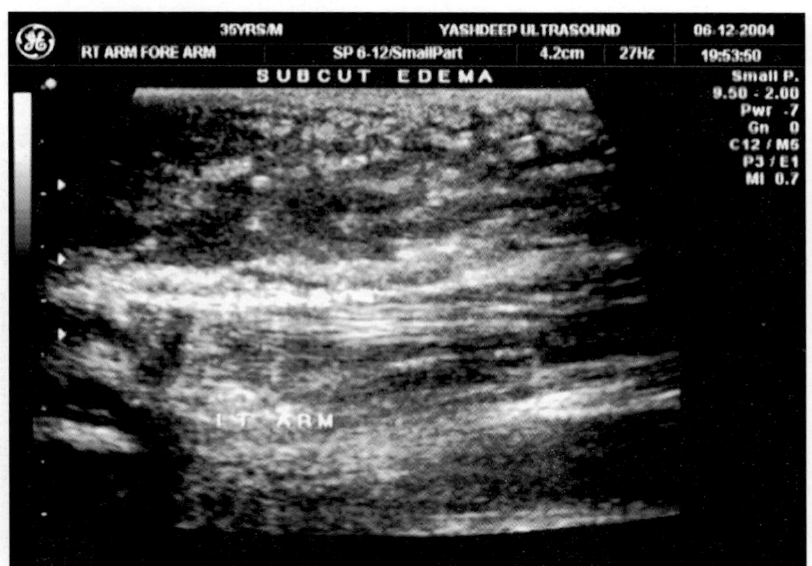

FIGURE 13.128

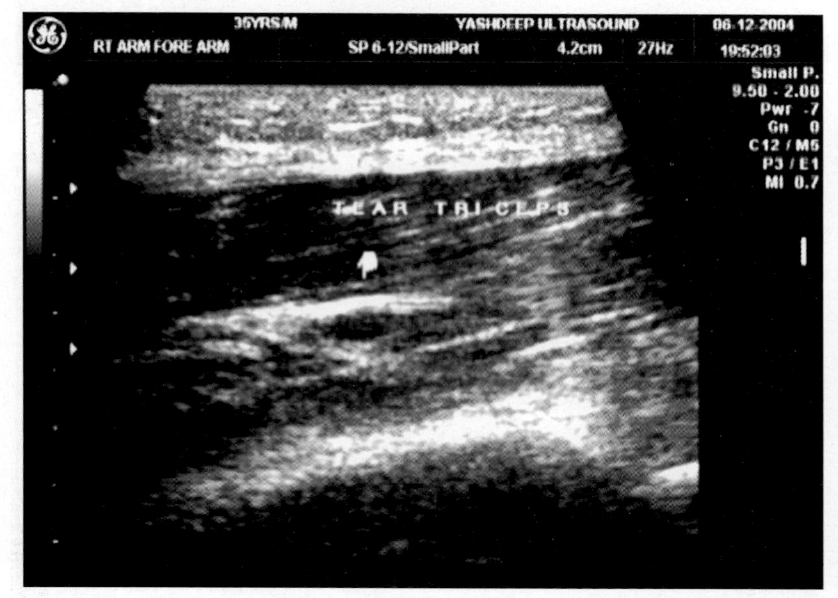

FIGURE 13.129

FIGURES 13.128 and 13.129: Arm muscle tear with subcutaneous contusion. A patient sustained blunt trauma to the Lt elbow and arm. HRSG shows subcutaneous edema. There is also evidence of a hypoechoic cleft seen in the belly of triceps muscle. Muscle was also hypoechoic and edematous.

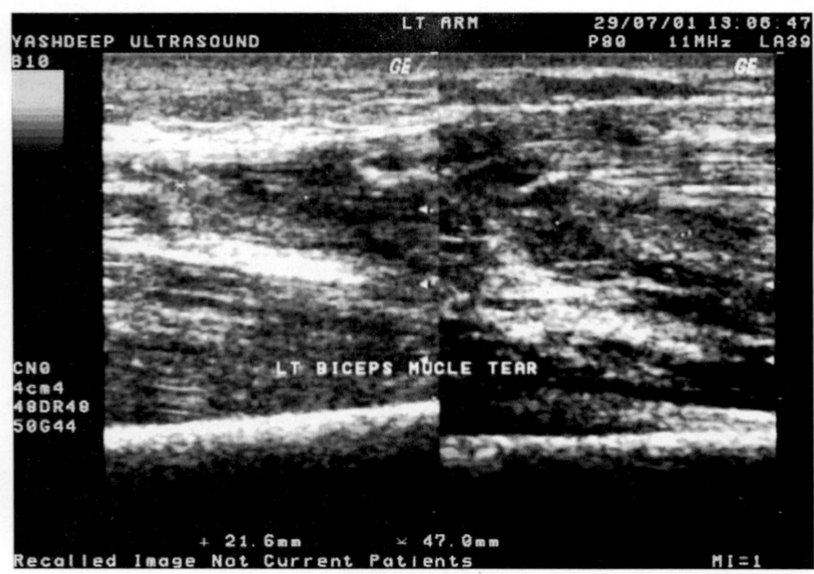

FIGURE 13.130

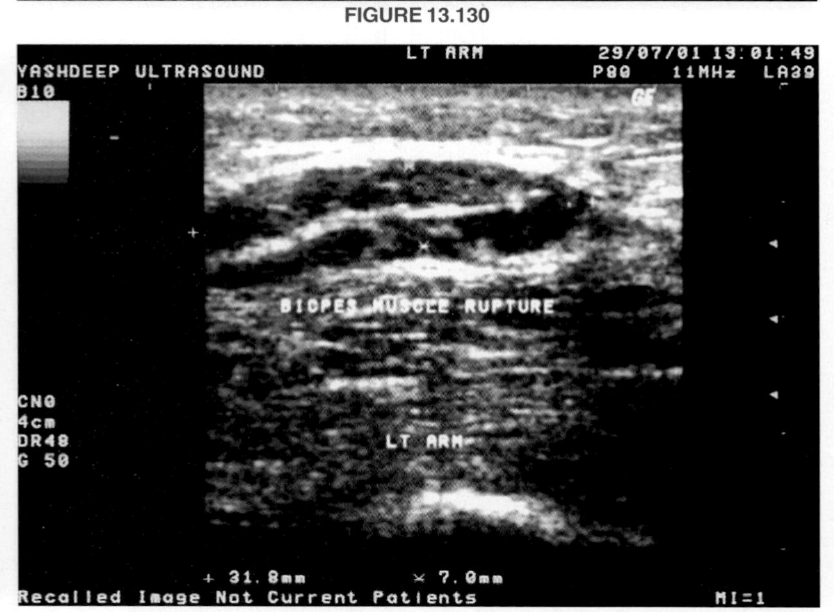

FIGURE 13.131

FIGURES 13.130 and 13.131: Biceps muscle tear. HRSG shows a big tear in the belly of biceps muscle of Rt arm. The muscle capsule is ruptured. Fibers are torn and disrupted suggestive of acute rupture of the muscle.

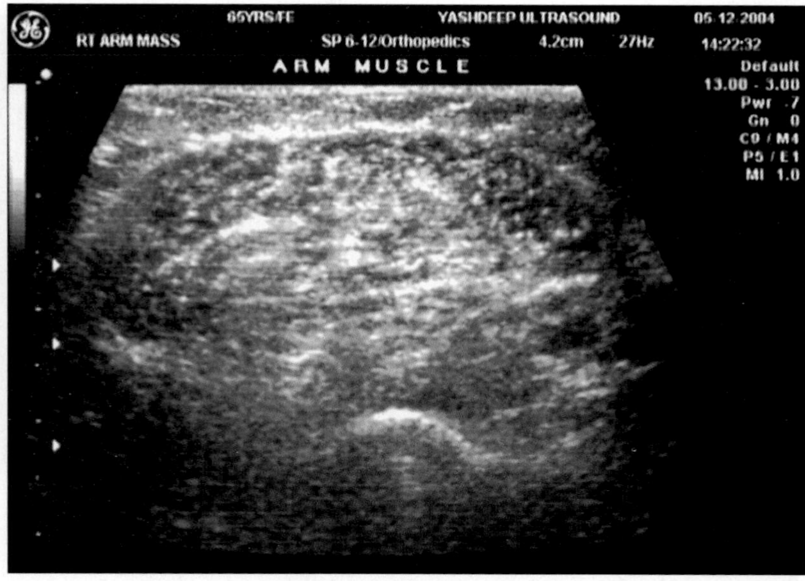

FIGURE 13.132

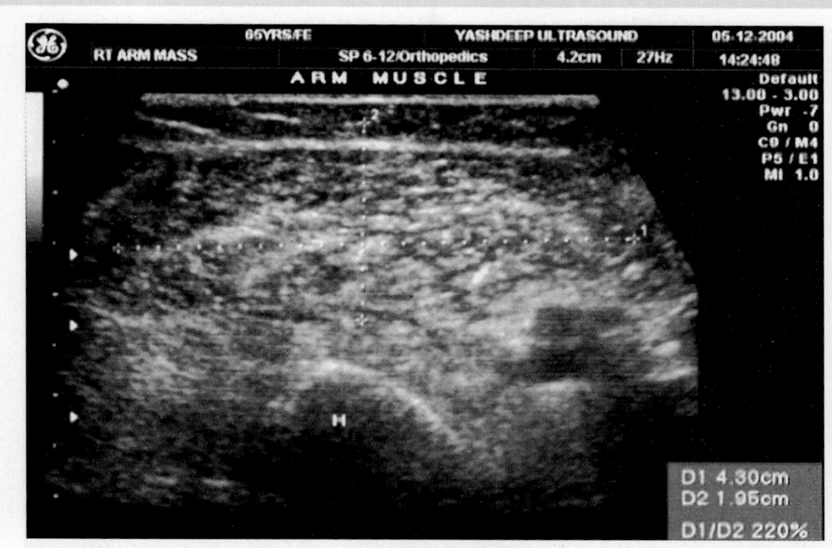

FIGURE 13.133

FIGURES 13.132 and 1.133: Contusion injury to the biceps muscle. An old lady sustained blunt trauma to the Rt arm with painful swelling. HRSG shows edematous biceps muscle of the arm and muscle bundles fibers are sparse with echogenic fluid collection in the muscle belly suggestive of contusion injury.

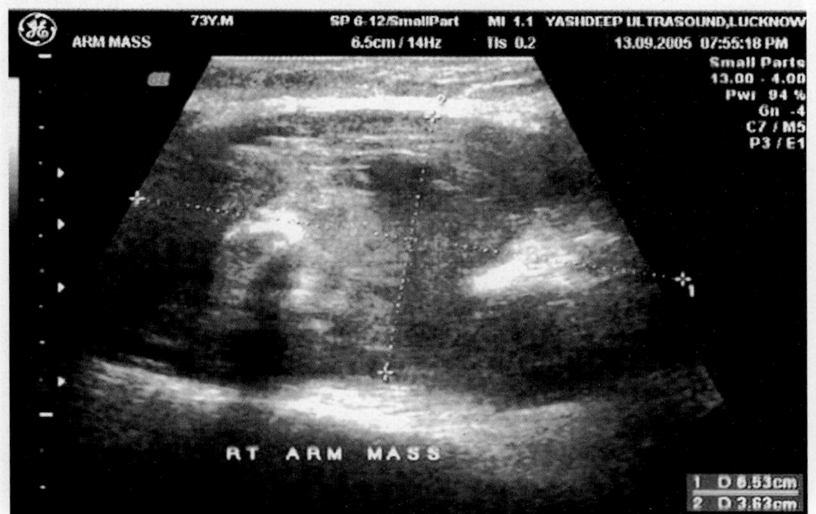

FIGURE 13.134

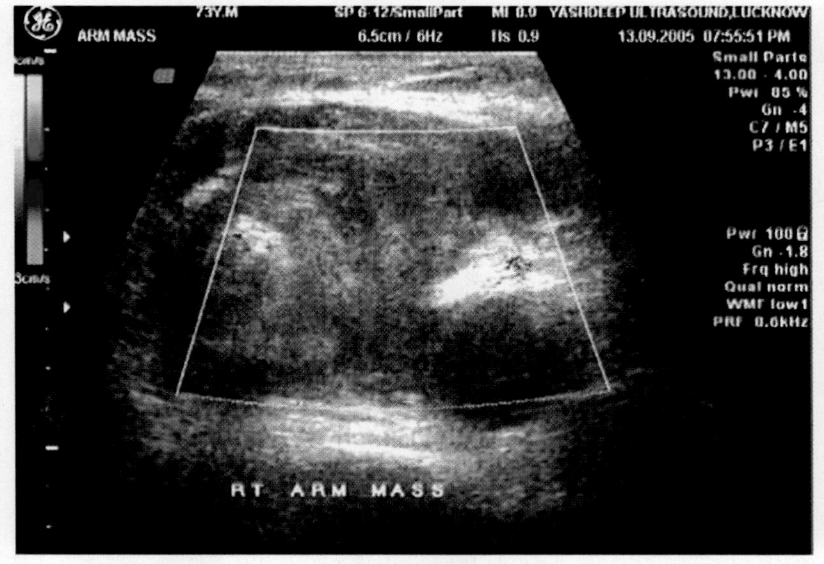

FIGURE 13.135

FIGURES 13.134 and 13.135: Acute rupture of the biceps muscle. A patient presented with marked painful swelling of the Rt arm after sustaining injury. Patient was not able to lift the arm. HRSG shows acute rupture of the biceps muscle. A big echogenic collection is seen in the belly of the biceps muscle suggestive of hemorrhagic collection with hematoma formation.

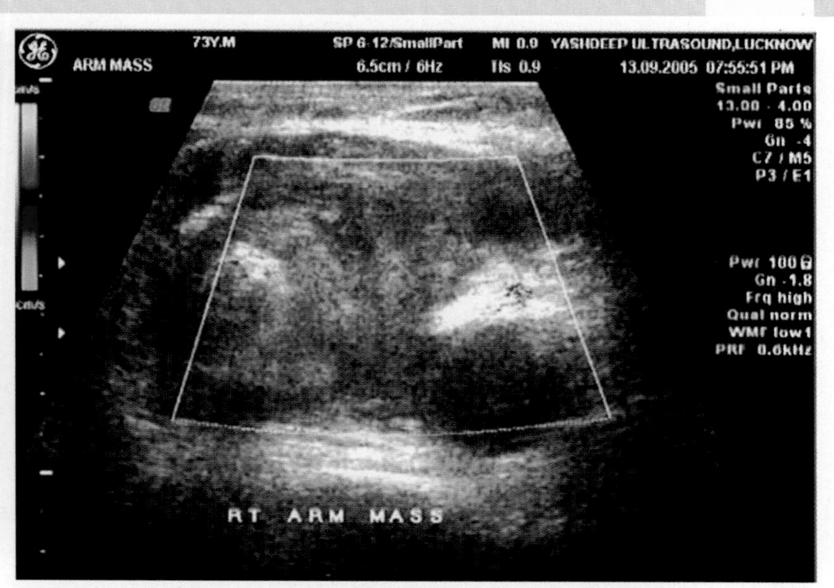

FIGURE 13.136: On color flow imaging no flow seen in the hematoma.

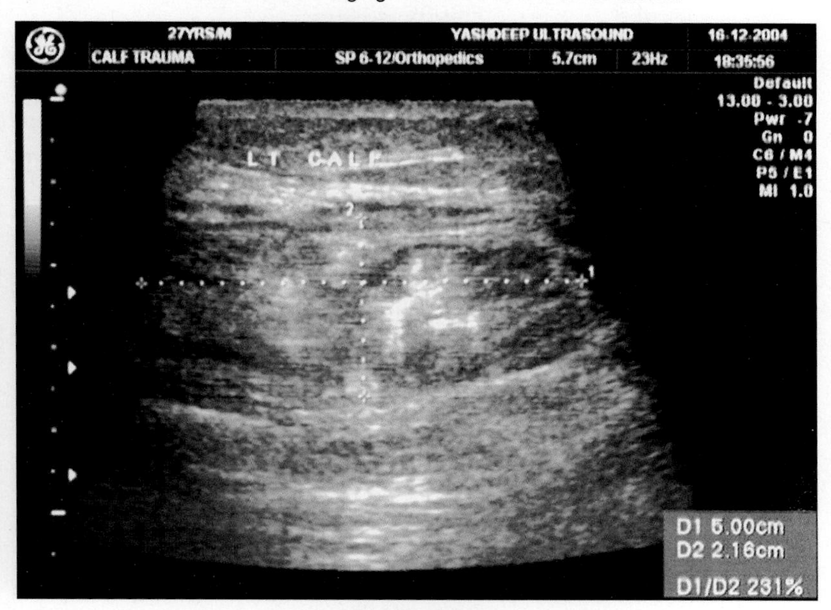

FIGURE 13.137

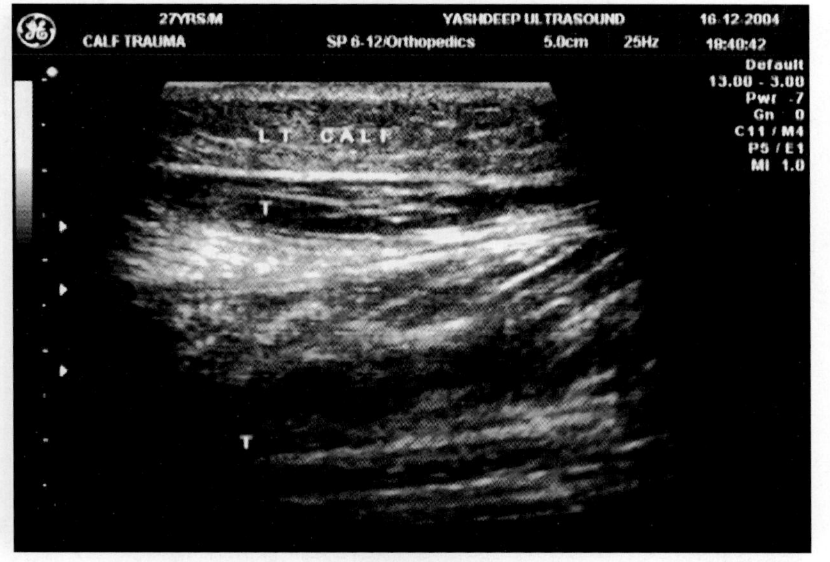

FIGURE 13.138

FIGURES 13.137 and 13.138: Calf muscle hematoma. HRSG shows a big hematoma formation in the Lt calf muscle after sustaining trauma. Echogenic hematoma is seen confined to the muscle belly. There is also evidence of tear seen in the muscle. The tear is running obliquely.

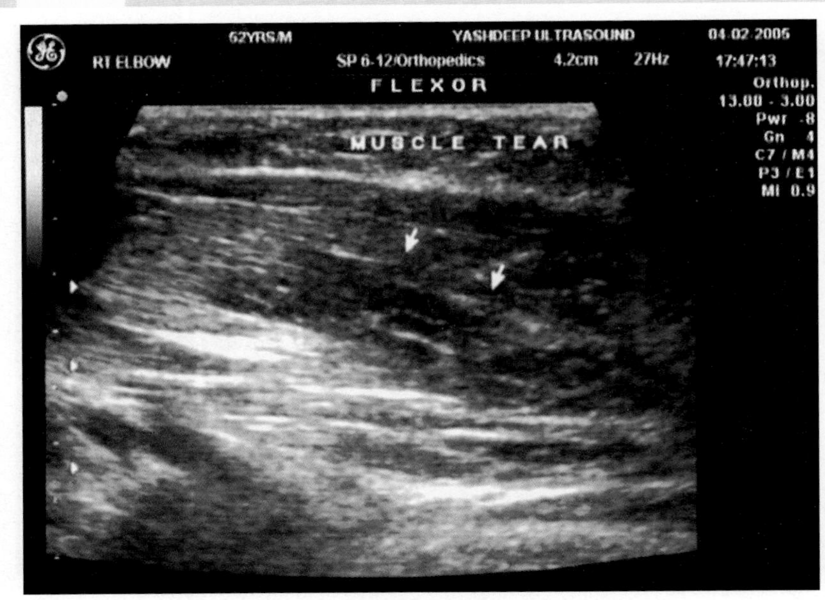

FIGURE 13.139

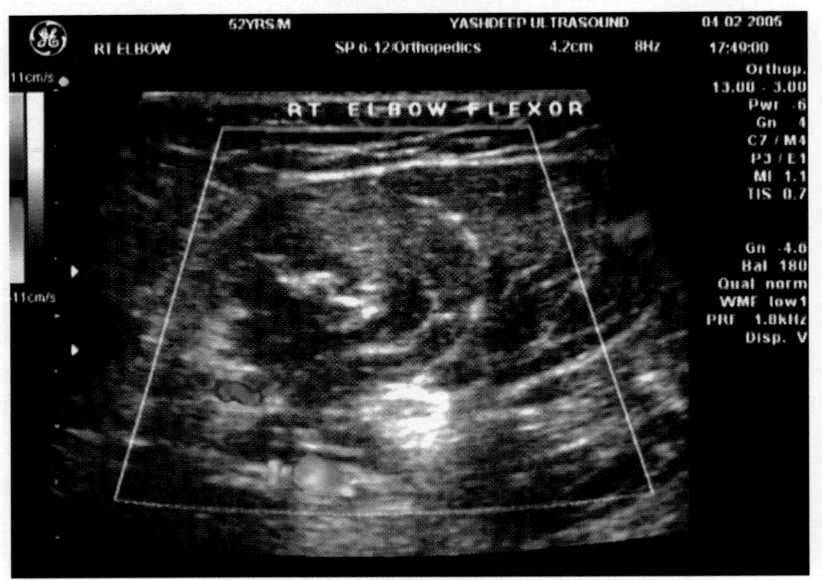

FIGURE 13.140

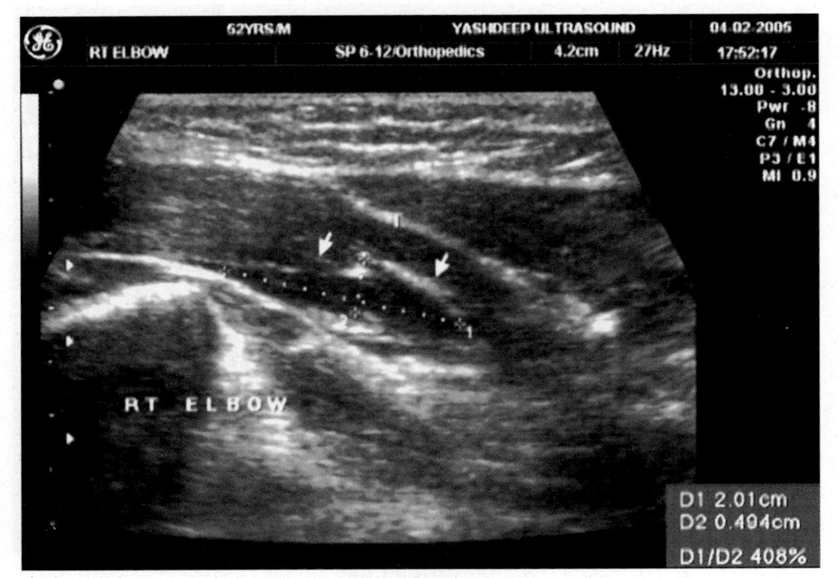

FIGURE 13.141

FIGURES 13.139 to 13.141: Arm muscle hematoma. HRSG shows a big hematoma in the muscle of the flexor group of the elbow. It is seen pressing over the veins. There is also evidence of muscle tear seen in the flexor group of muscle. The muscle belly is edematous. The patient was unable to lift the arm.

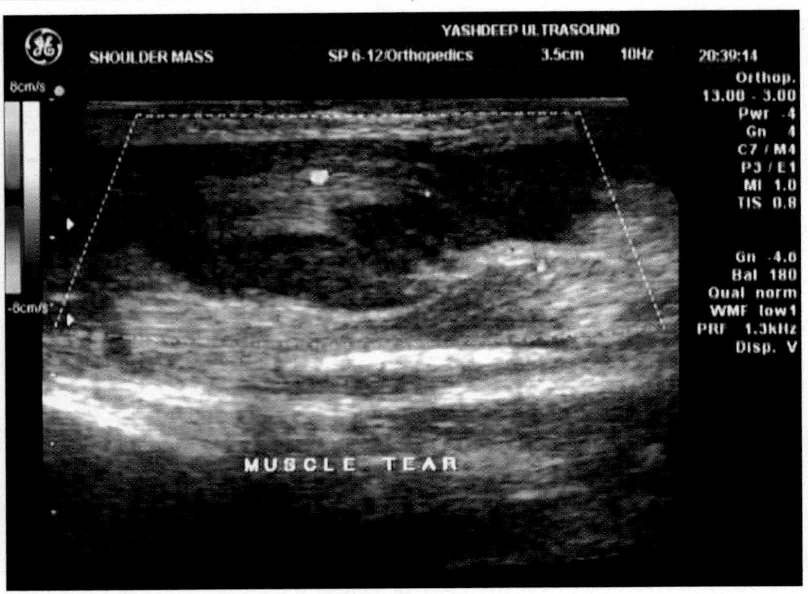

FIGURE 13.142

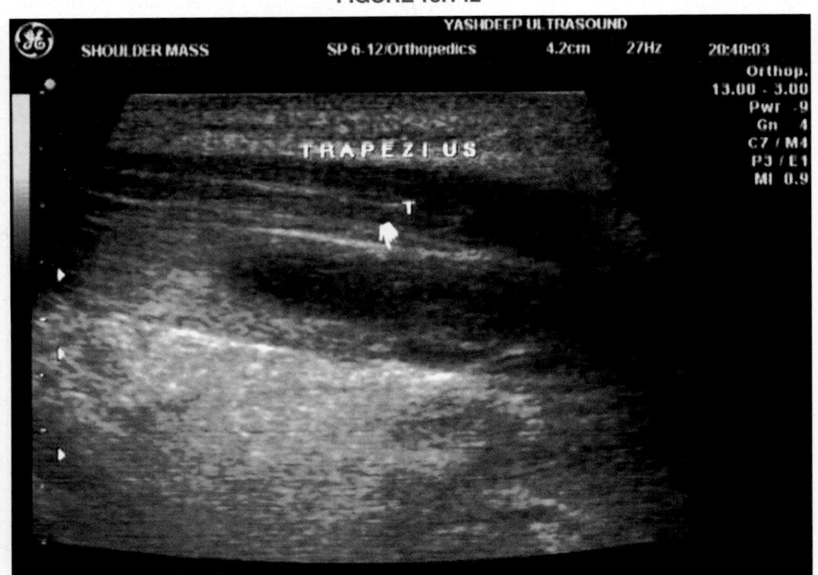

FIGURE 13.143

FIGURES 13.142 and 13.143: Muscle tear presented as shoulder mass. A patient presented with painful swelling in the Lt shoulder with marked restricted movements. HRSG shows a big muscle tear in the deltoid and trapezius muscle. Thick collection is seen in the belly of the muscle suggestive of hemorrhagic collection. The muscles belly is edematous swollen. No evidence of any calcification seen in it. On color flow imaging no flow was seen in the tear.

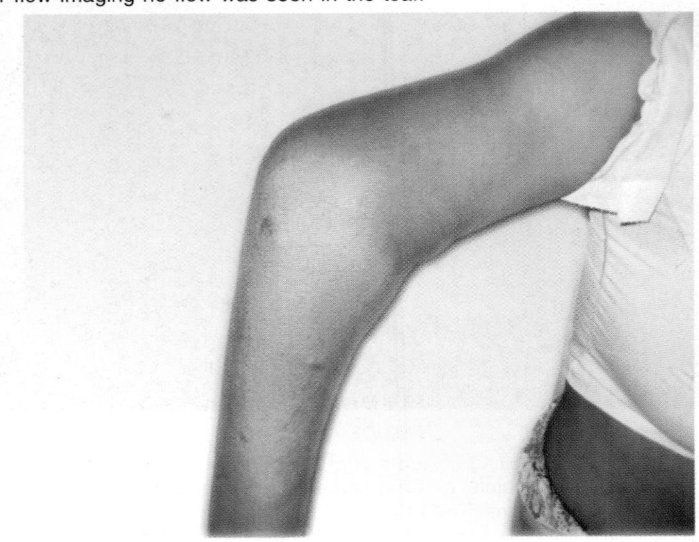

FIGURE 13.144

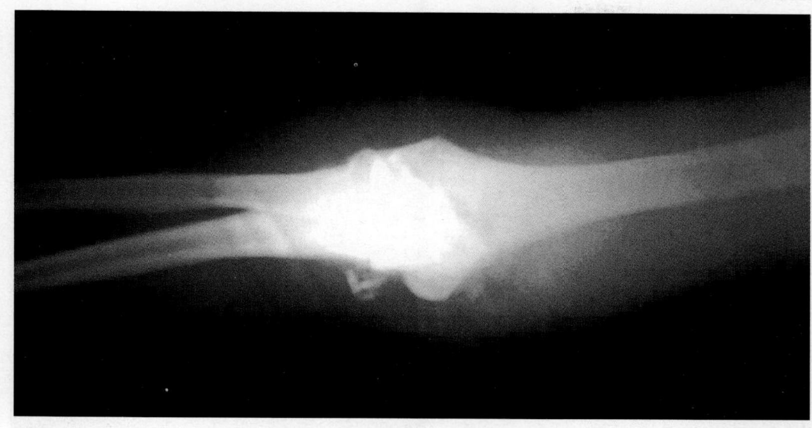

FIGURE 13.145

FIGURES 13.144 and 13.145: Myositis ossificans. A lady presented with hard soft tissue swelling over the Rt elbow after sustained trauma with having plaster cast. X-ray of the elbow showed soft tissue swelling with radio dense calcification seen in the soft tissue.

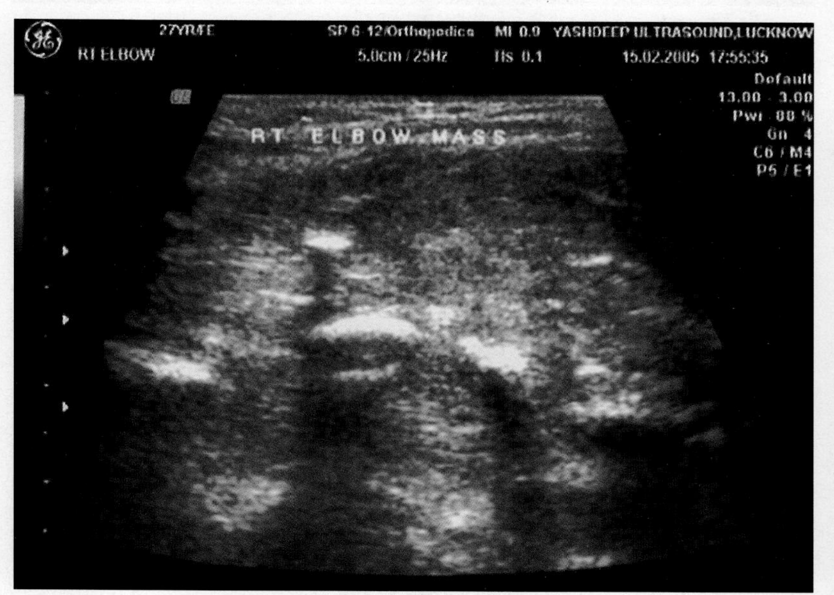

FIGURE 13.146

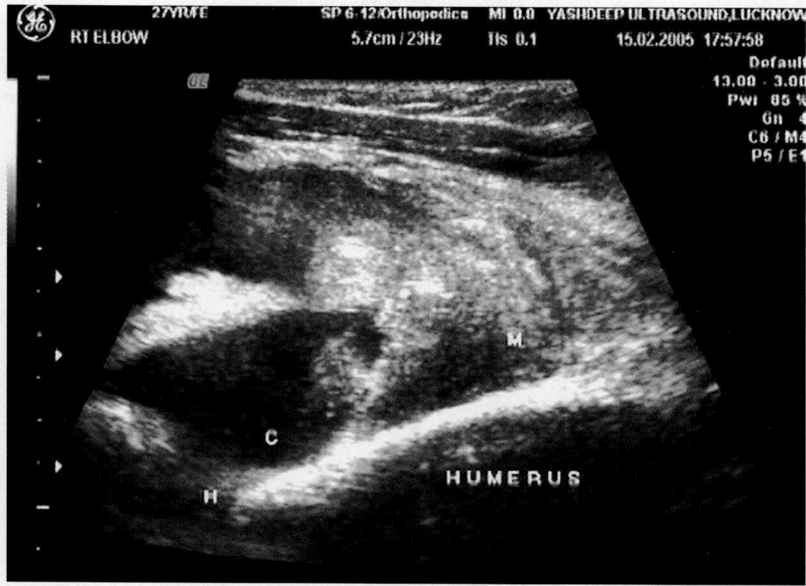

FIGURE 13.147

FIGURES 13.146 and 13.147: HRSG shows a soft tissue mass in the elbow muscle. Multiple echogenic calcified specks are seen in the mass. They are accompanied with acoustic shadowing. Normal muscles texture is lost. The muscles was seen ossified. Small amount of collection is seen in the olecranon bursa.

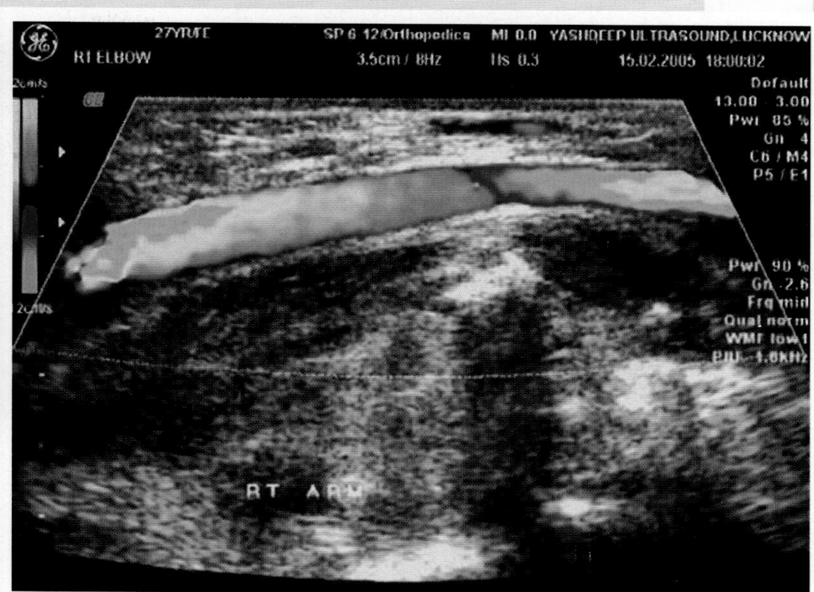

FIGURE 13.148: On color flow imaging. No flow was seen in the mass. The arm vessel was seen separate from the mass. Sonographic findings are suggestive of myositis ossicificans.

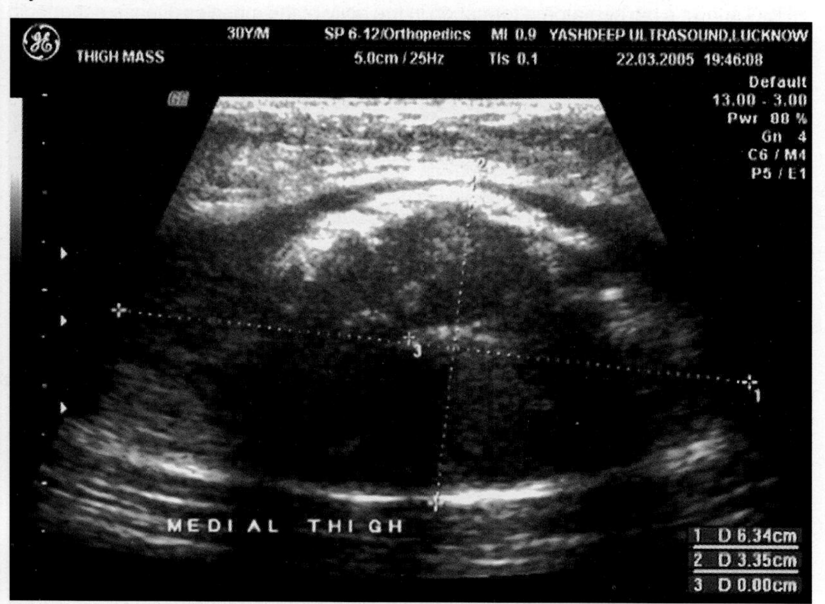

FIGURE 13.149

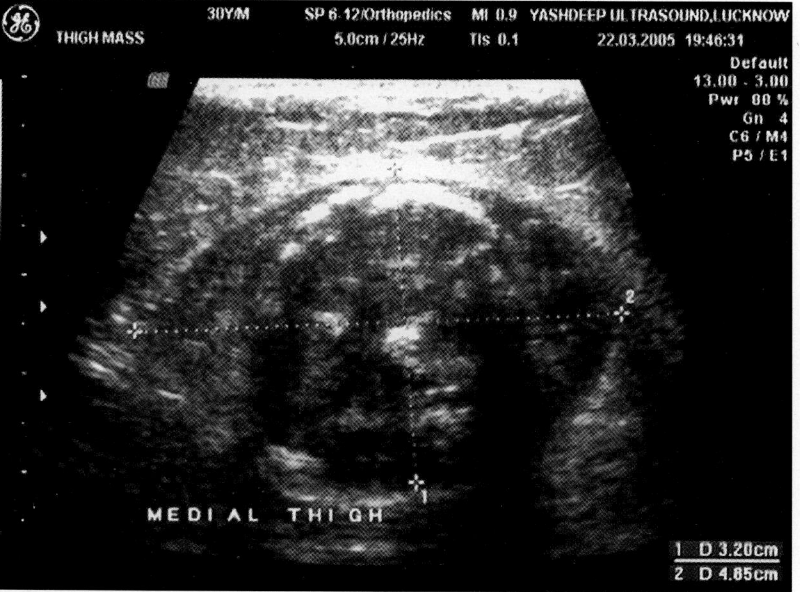

FIGURE 13.150

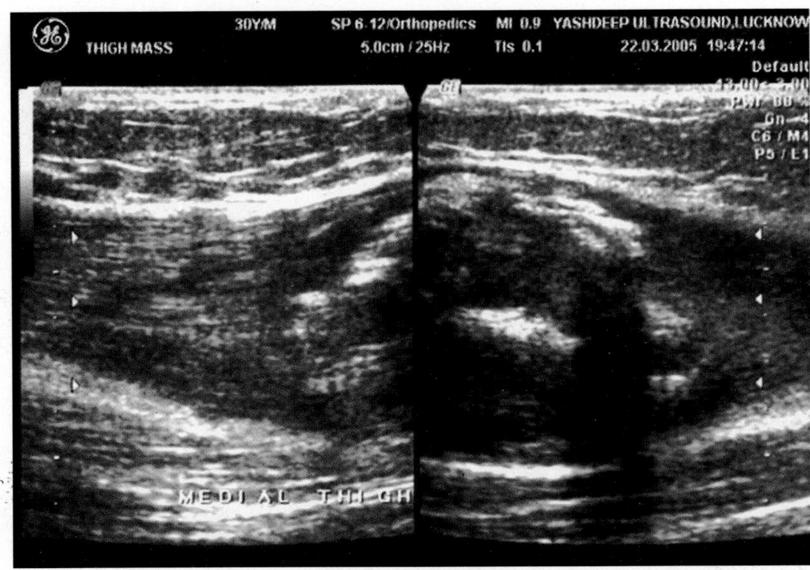

FIGURE 13.151

FIGURES 13.149 to 13.151: Muscle hematoma with calcification. A patient presented with acute rupture of adductor group of muscles after sustaining trauma HRSG shows a big hematoma formation in the belly of the muscles. Dense echogenic rim calcification is seen around the hematoma. Echogenic calcified specks are also seen in the substance of hematoma.

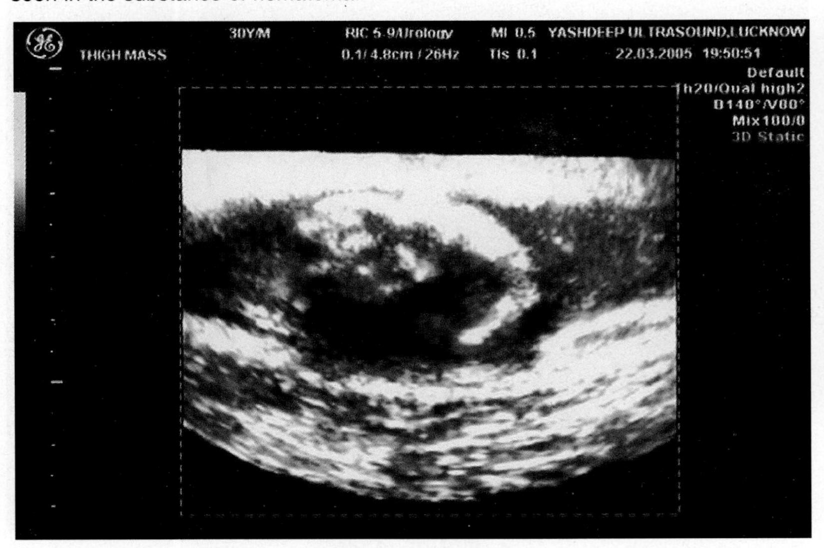

FIGURE 13.152: 3D imaging of the same patient shows the hematoma in much clinical details.

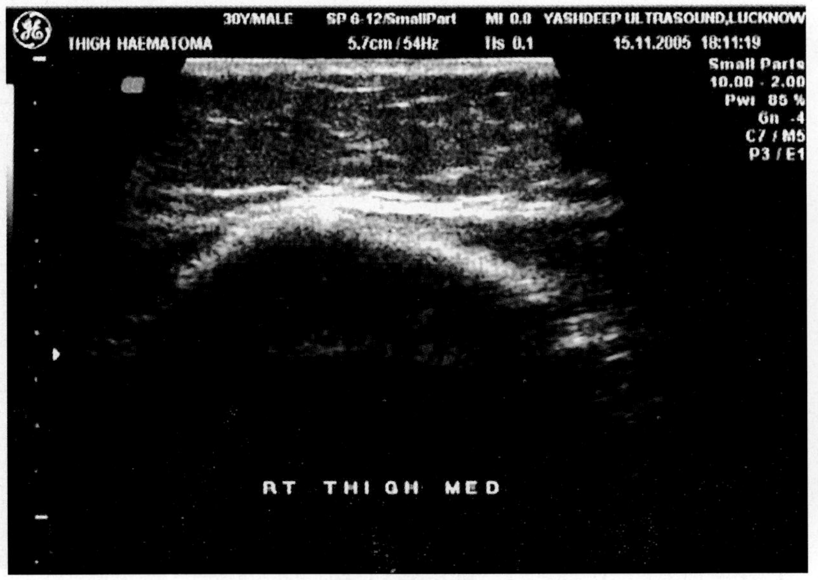

FIGURE 13.153

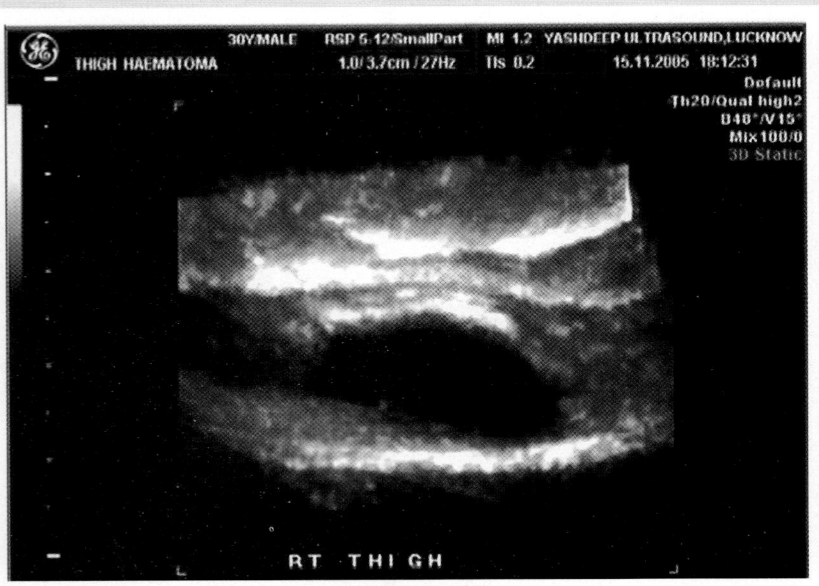

FIGURE 13.154

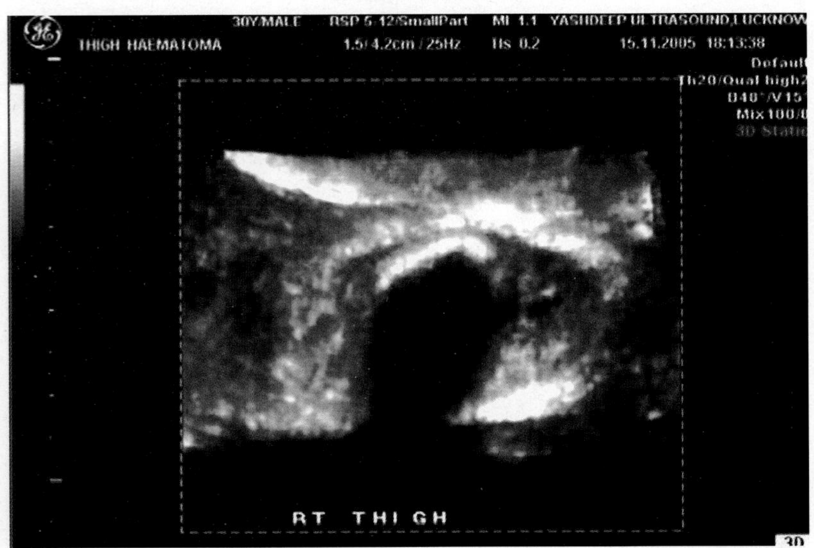

FIGURE 13.155

FIGURES 13.153 to 13.155: Resolving muscle hematoma. Same patient scanned after 8th months of initial injury showed the resolving hematoma. The hematoma remarkably has reduced in size in comparison to the previous study. A thin rim calcification is seen around the hematoma. 3D imaging of the hematoma showed the details of the rim calcification. The extent of hematoma has reduced markedly in comparison to the previous study.

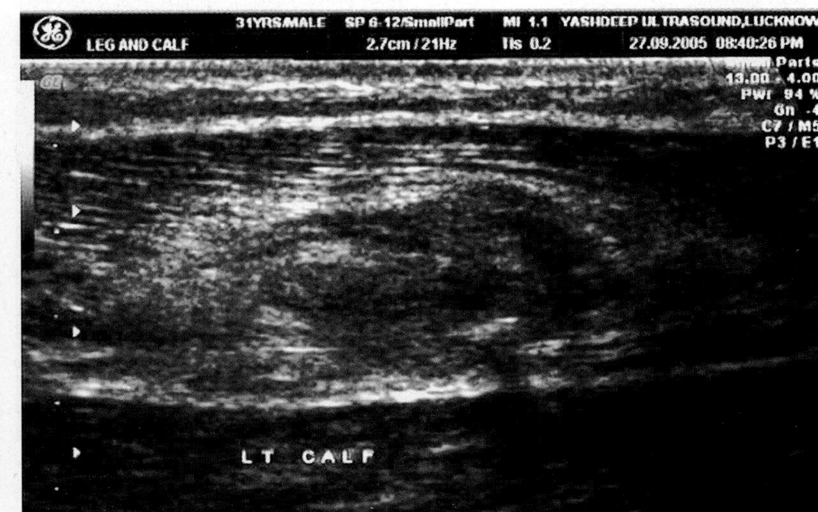

FIGURE 13.156: Calf hematoma with early calcification. HRSG shows an irregular hypoechoic hematoma in the gastrocnemius muscle in the Lt calf. Fine calcification is seen in the hematoma, which is accompanied with acoustic shadowing.

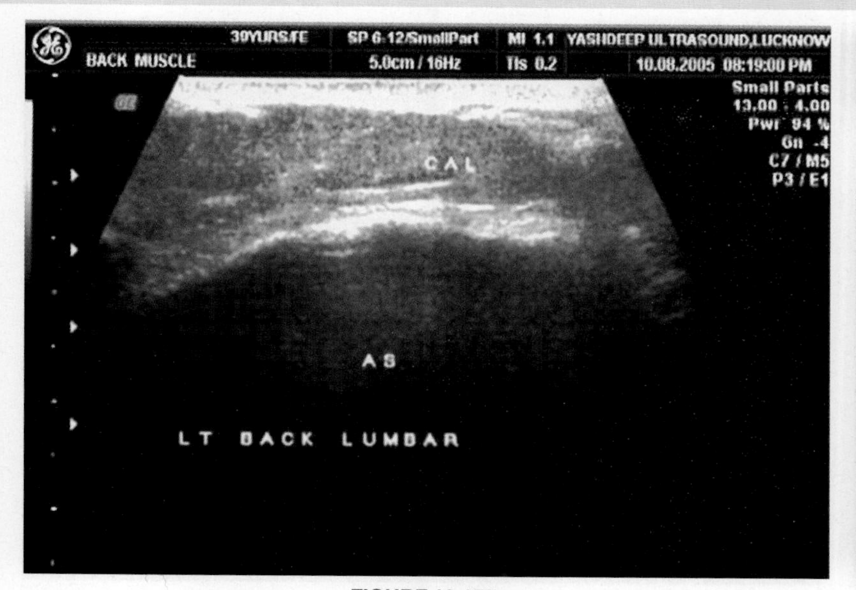

FIGURE 13.157

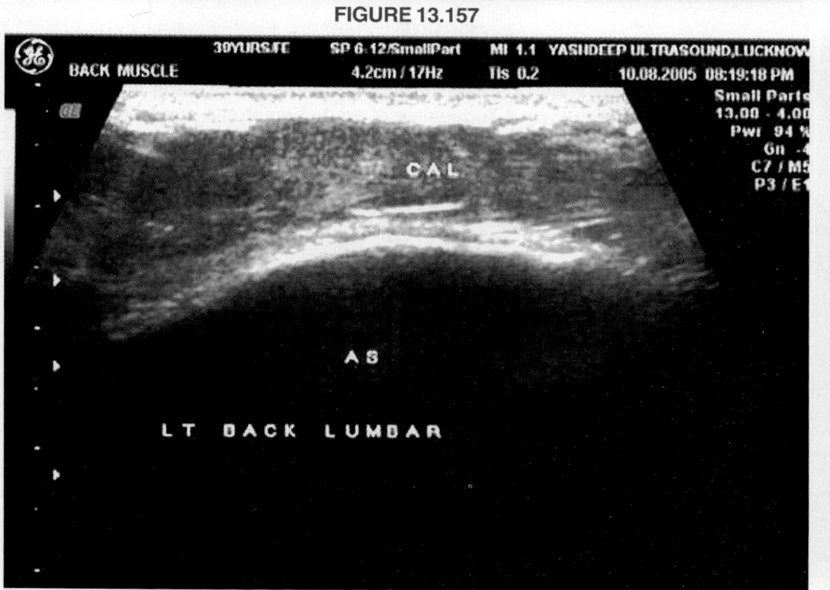

FIGURE 13.158

FIGURES 13.157 and 13.158: Cord like calcification in the lumbar region. A patient presented with cord like lesion in the Lt paralumbar region. HRSG shows a thick echogenic cord like calcified plaque in the muscle. Dense acoustic shadow is seen posterior to it masking the tissues.

FIGURE 13.159: 3D imaging of the same patient shows the calcified plaque in a much better way.

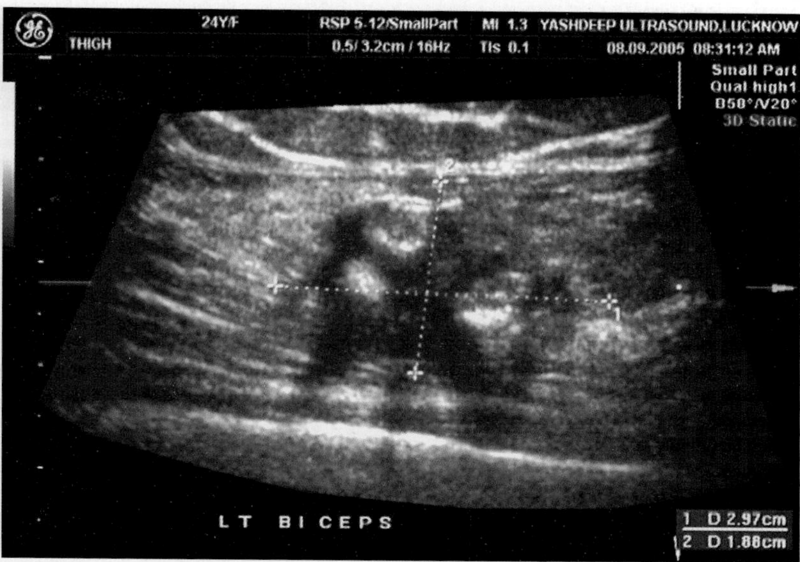

FIGURE 13.160

FIGURE 13.161

FIGURES 13.160 and 13.161: Myositis ossificans in the arm muscle. A young girl presented with painful swelling in the arm muscle. HRSG shows heterogeneous mass in the muscle belly. Dense echogenic calcified specks are seen in it, which are accompanied with acoustic shadowing. 3D imaging shows better details of the calcifications.

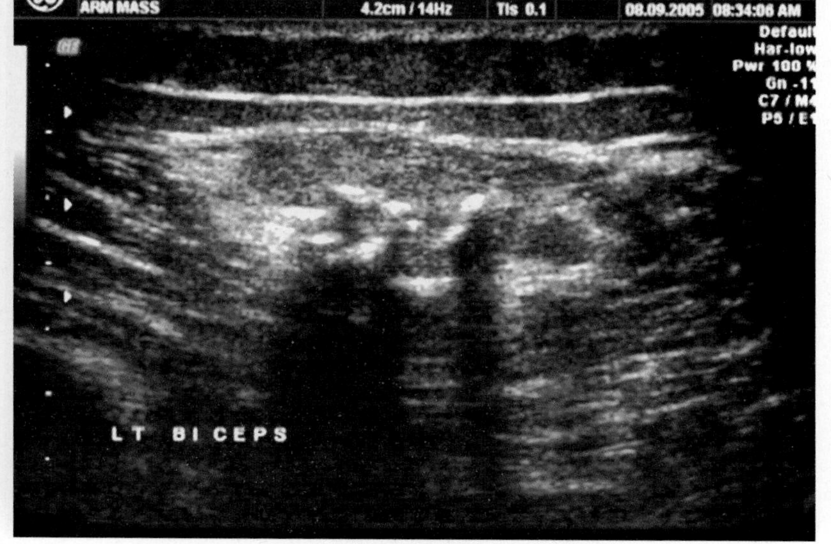

FIGURE 13.162

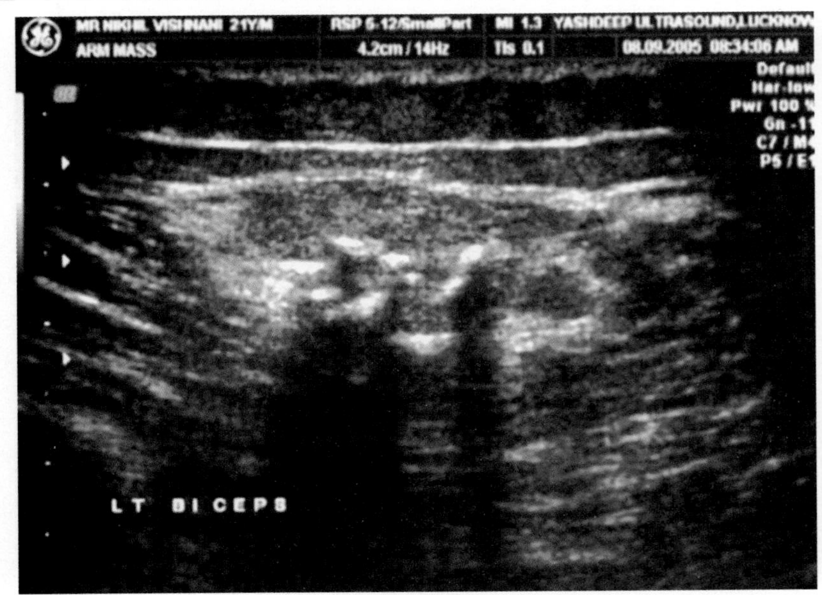

FIGURE 13.163

FIGURES 13.162 and 13.163: Arm muscles calcification. HRSG shows calcified mass in the Lt biceps muscle. The calcified specks are seen casting strong acoustic shadowing suggestive of myositis ossification.

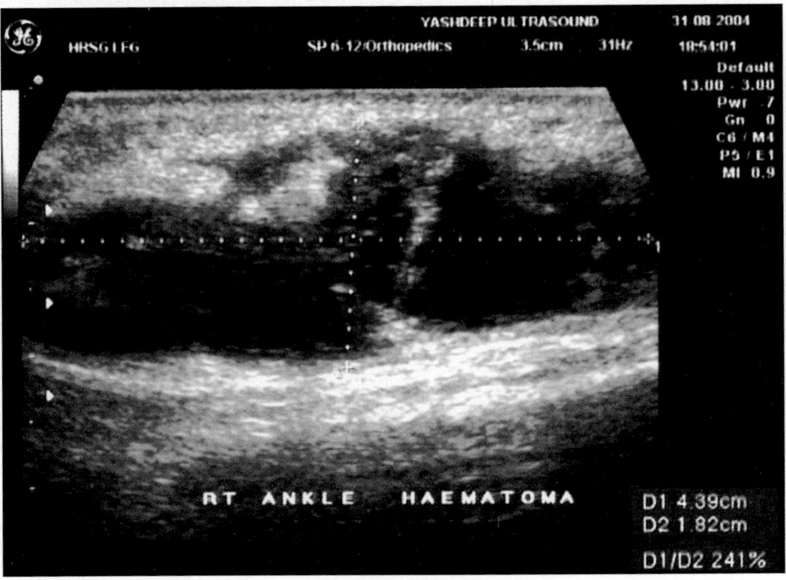

FIGURE 13.164

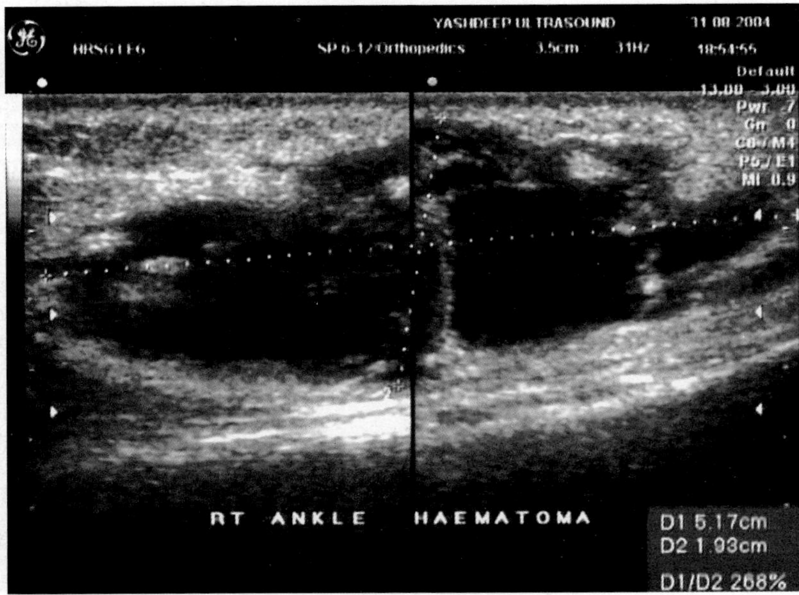

FIGURE 13.165

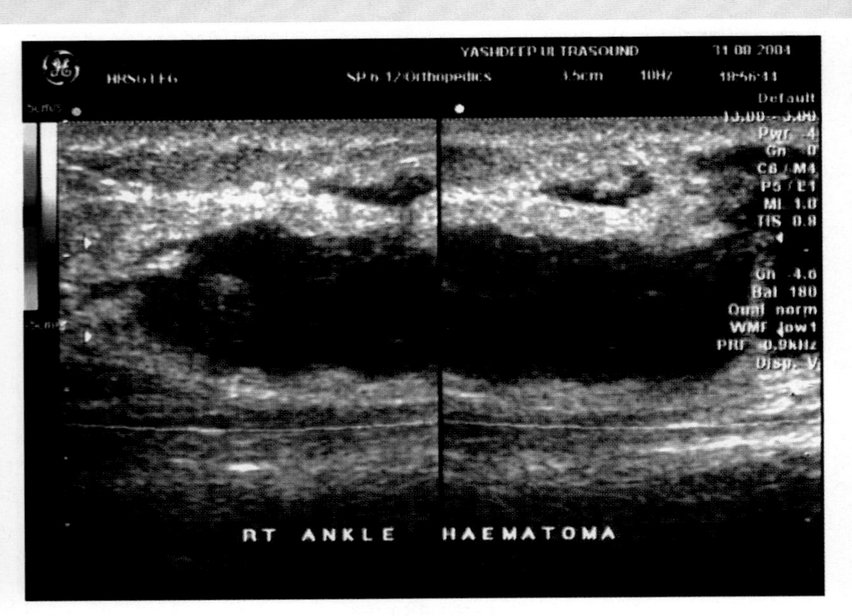

FIGURE 13.166

FIGURE 13.164 to 13.166: Acute rupture with hematoma formation in the ankle muscle. HRSG shows big regular heterogeneous mass in the ankle muscles. Low-level echoes are seen in it suggestive of acute hematoma due to rupture of the muscles. **On color flow imaging** no flow is seen in the ruptured muscle.

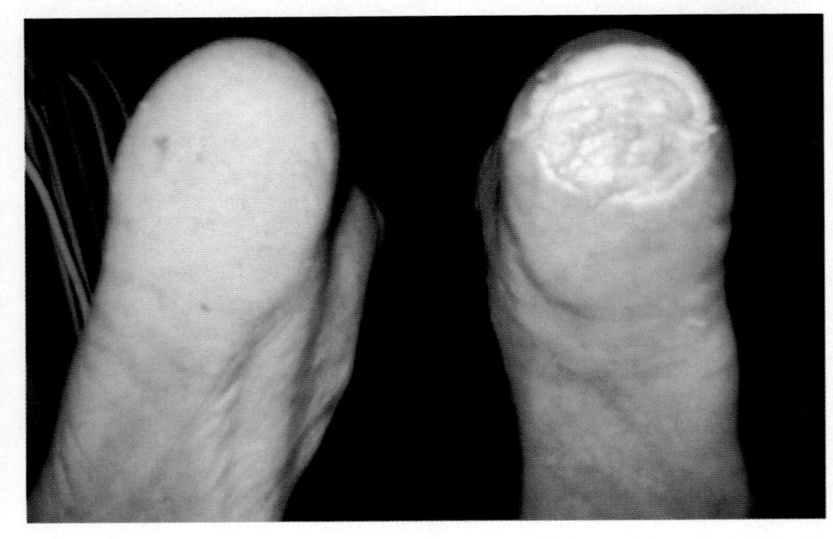

FIGURE 13.167: Heel pad loss with destruction. A drug addict presented with non-healing ulcer over the Rt heel pad.

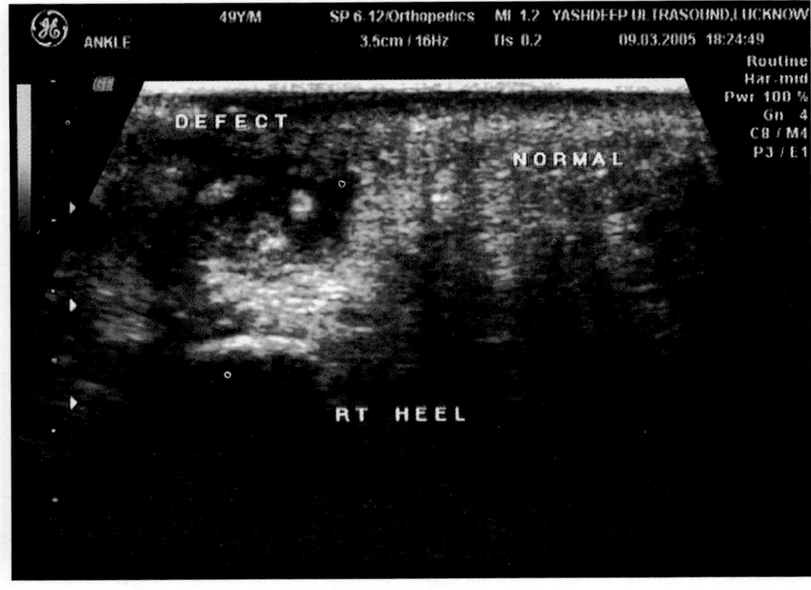

FIGURE 13.168

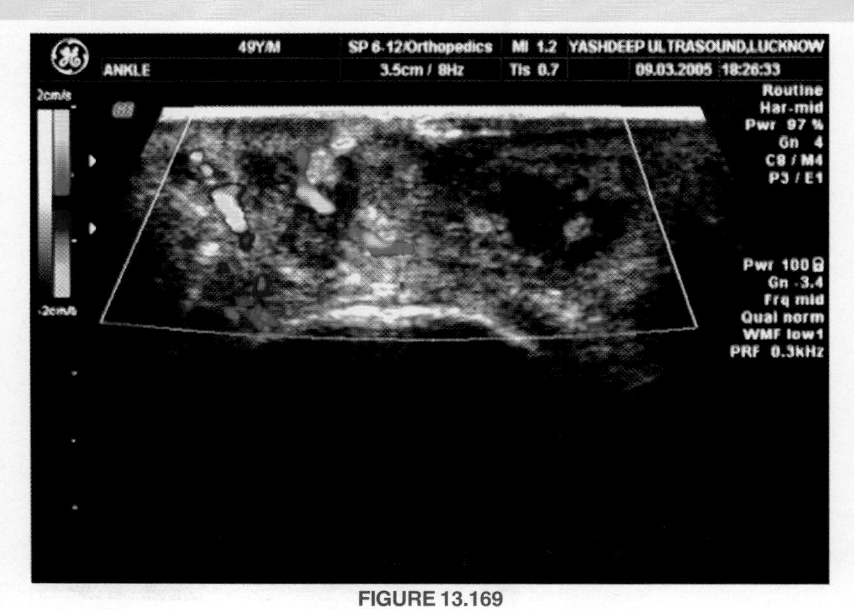

FIGURE 13.169

FIGURES 13.168 and 13.169: HRSG shows a big defect in the heel pad in the fascia. Low-level echoes are seen in to it. **On color flow imaging** few vessels are seen in it in the fascia suggestive of secondary infection.

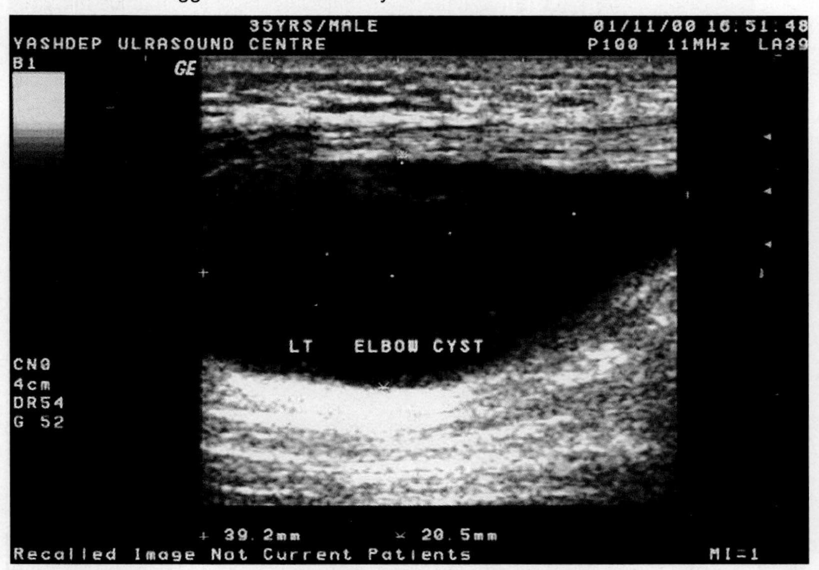

FIGURE 13.170: Lt arm cyst. A big cystic mass is seen in the Lt arm near the elbow. It is unilocular anechoic. The cyst is seen pressing over the arm veins and resulting into venous stasis.

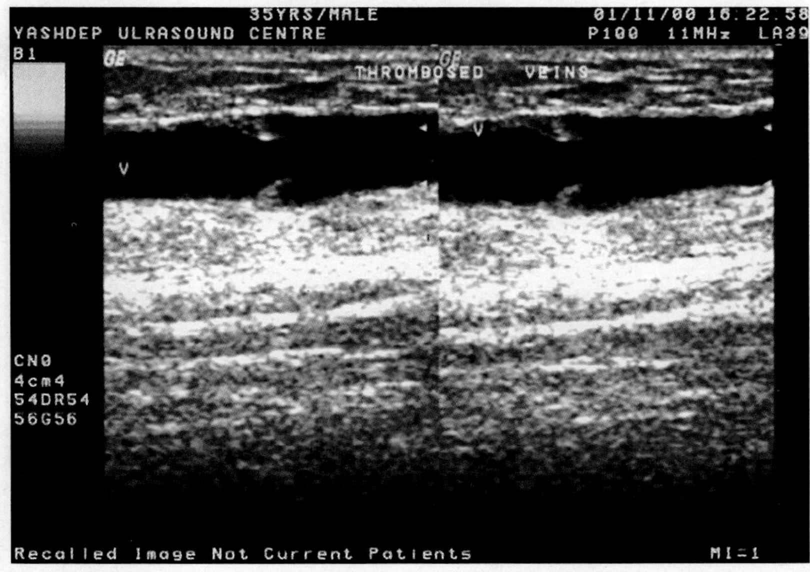

FIGURE 13.171: The lt arm vein is seen thrombosed. The valves are seen rigid with no valvular movement due to long standing pressure exerted by the cyst.

Vascular Masses in the Muscles

Many times vascular malformations or aneurysm can present as muscle pathology. Periarticular aneurysms are common near to the hip and knee joints. However, they may be seen near to any joint. Congenital vascular mass formations are rare. But congenital arterio venous malformations involving the upper limb or lower limb may be seen. HRSG can differentiate very well between the lesions. Lymphangiomas, lymphangitis and other vascular malformations can be very well evaluated on HRSG. The aneurysm may be due to trauma, fungal or mycotic in nature. Fusiform aneurysm is most common in common femoral arteries. Saccular aneurysm is also seen in femoral artery. However, they are more common in popliteal artery. They are usually present as a symptomatic masses in the popliteal fossa and usually non pulsatile in nature. Therefore, it is difficult to differentiate it from baker cysts on clinical examination. These aneurysms may have large thrombus with wall calcification. Blood flow through these aneurysms is very slow.

Post stenotic aneurysm is rare and usually found near to the shoulder joint in subclavian, axillary and brachial arteries. They are usually associated with cervical rib or thoracic outlet syndrome.

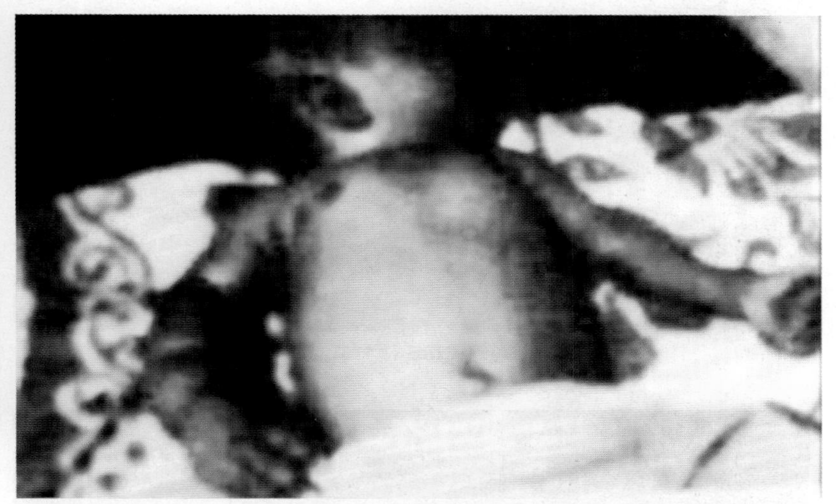

FIGURE 13. 172: Rt upper limb AV malformation. A young child was born with Rt upper limb hypertrophy. The upper limb is more than double the size of Lt upper limb.

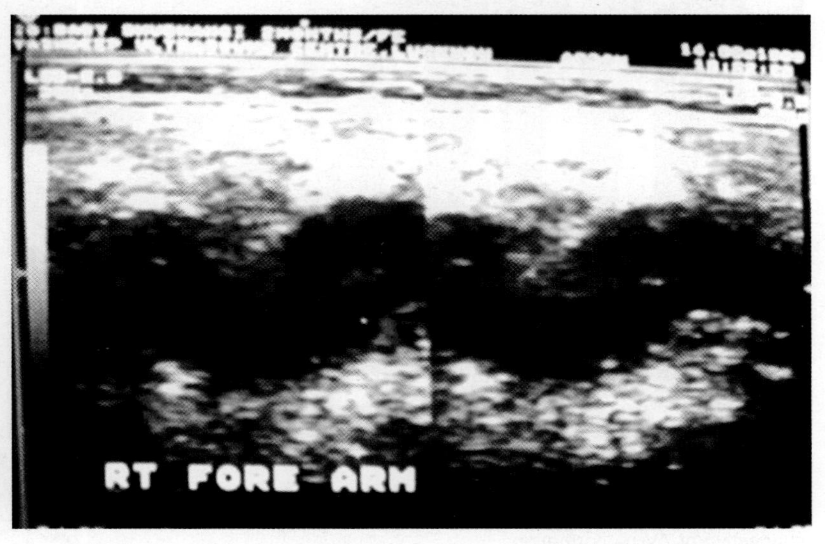

FIGURE 13.173: HRSG shows multiple dilated blood vessels in the Rt upper limb.

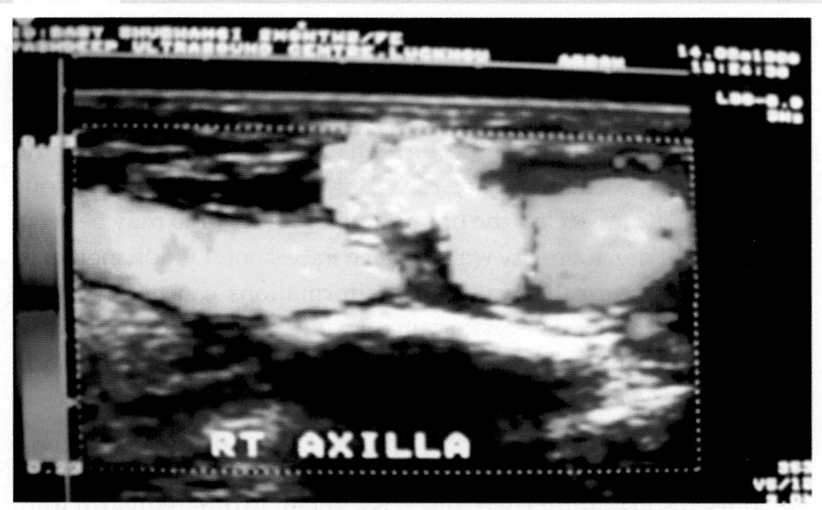

FIGURE 13.174: Color flow imaging shows grossly dilated arteries and veins in the Rt upper limb. They are seen in whole of the limb. High blood flow is seen in them.

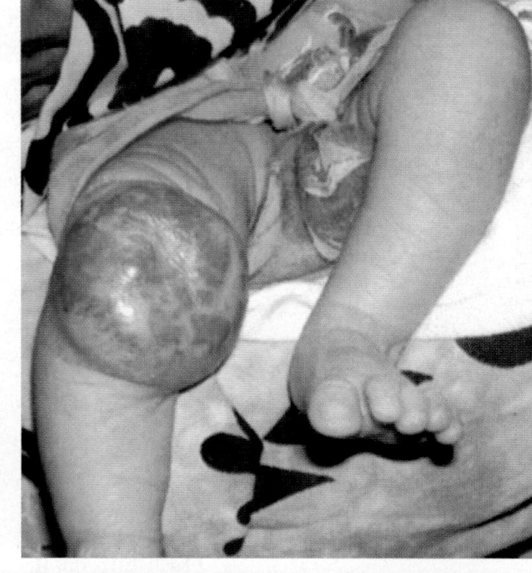

FIGURE 13. 177: AV malformation over the Rt knee. An infant was born with a soft tissue mass over the Rt lower thigh, knee and upper part of leg.

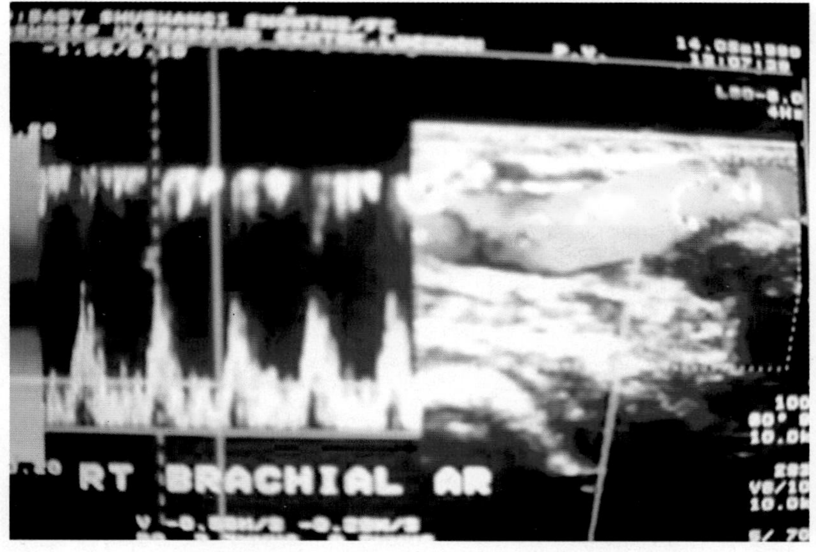

FIGURE 13.175

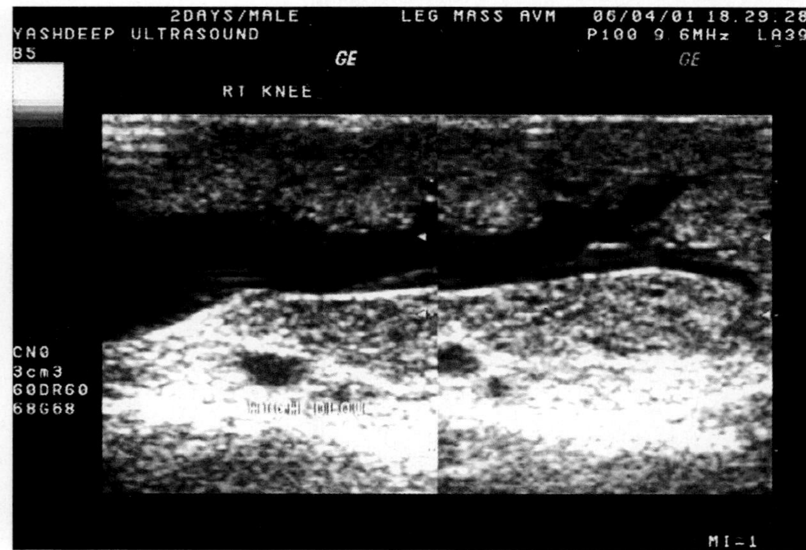

FIGURE 13.178

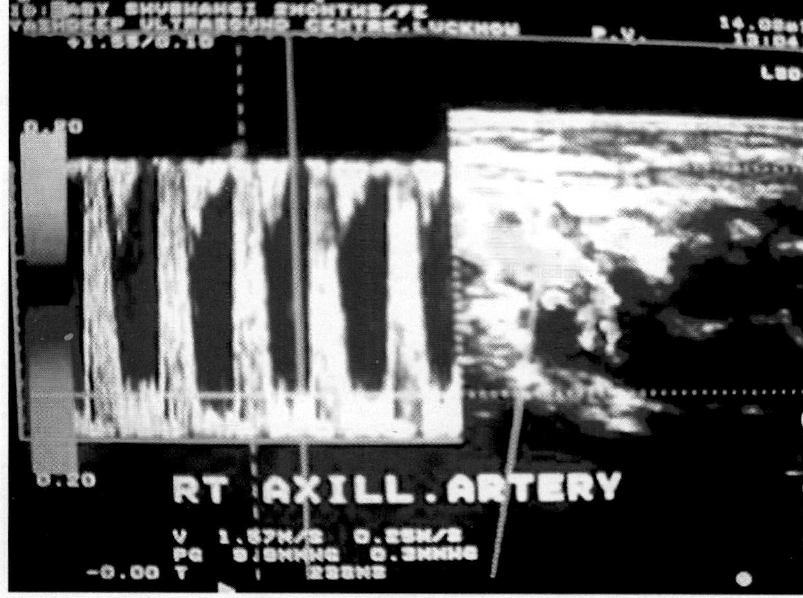

FIGURE 13.176

FIGURES 13.175 and 13.176: Spectral Doppler tracing and color flow imaging show high flow in the vessels. High flow is also seen in the veins with arterial flow pattern suggestive of AV malformation.

FIGURE 13.179

FIGURES 13.178 and 13.179: HRSG shows multiple dilated vessels in the mass. They are seen crisscrossing the mass.

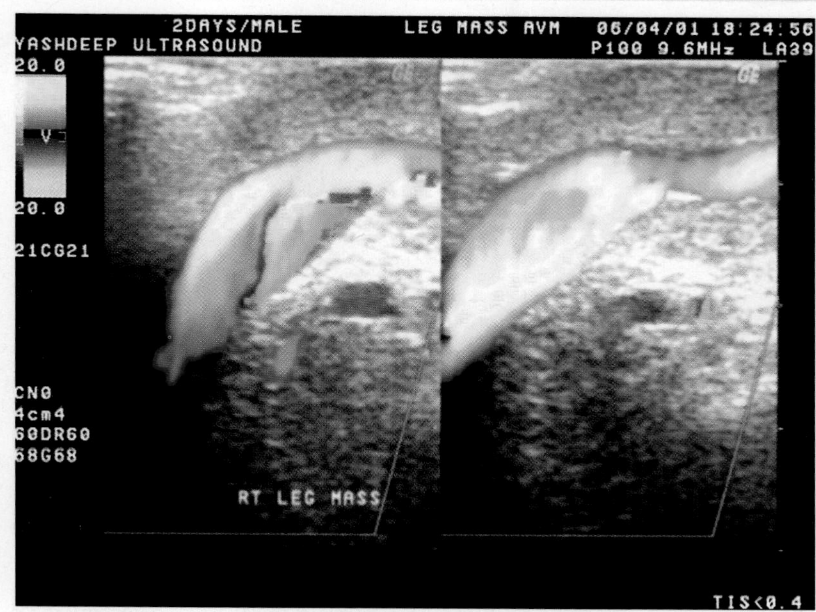

FIGURE 13.180

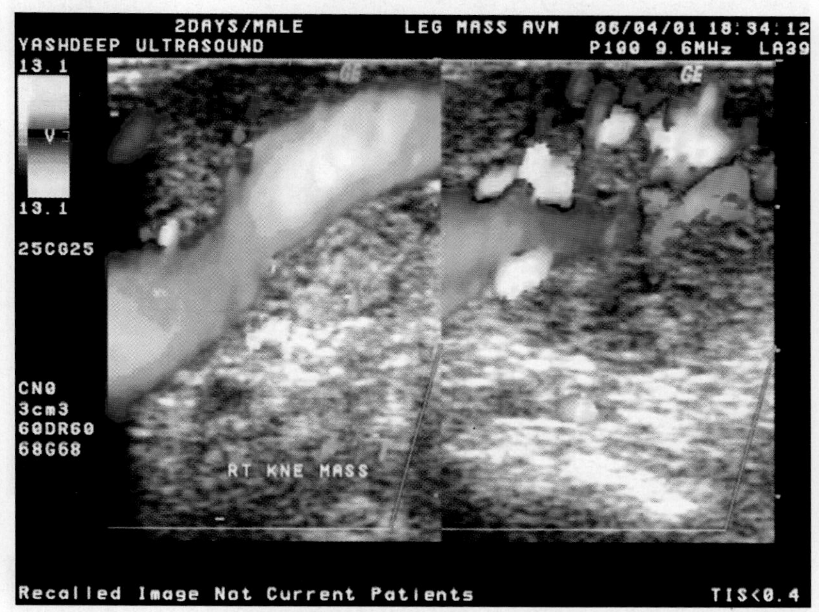

FIGURE 13.181

FIGURES 13.180 and 13.181: Color flow imaging shows high blood flow in the vessels. Venous and arterial flow pattern is seen in the vessels.

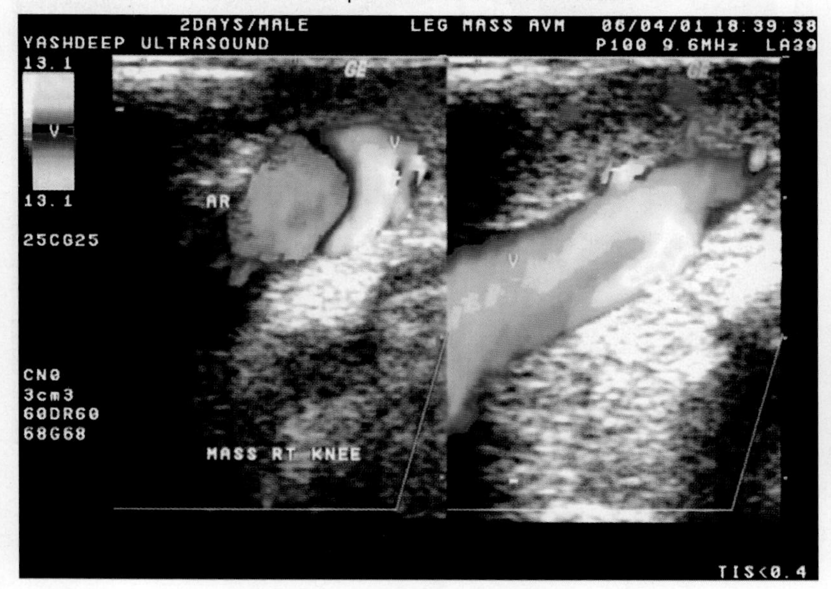

FIGURE 13.182

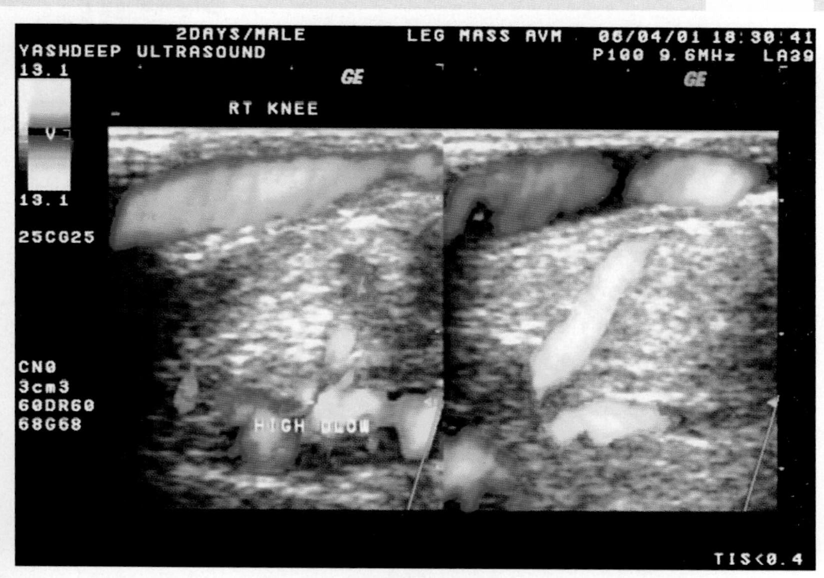

FIGURE 13.183

FIGURES 13.182 and 13.183: Blood flow is seen in the vessels crisscrossing the mass. But no calcification is seen in them.

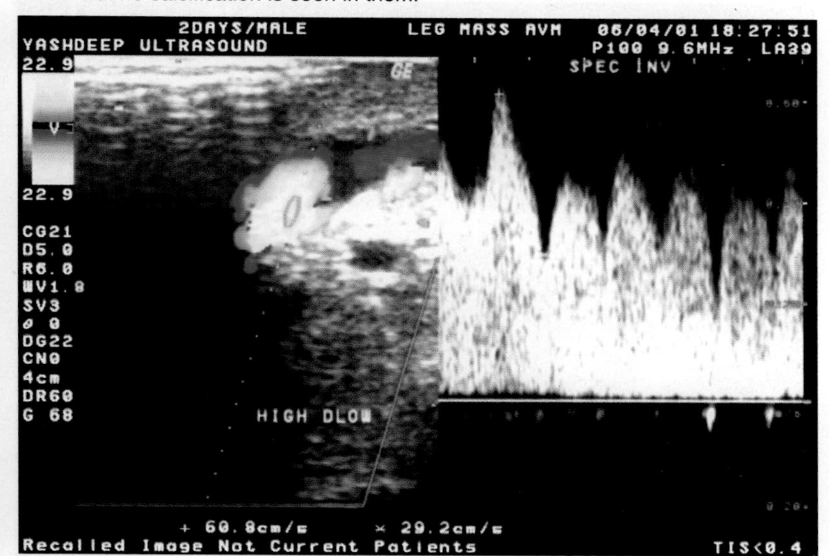

FIGURE 13.184

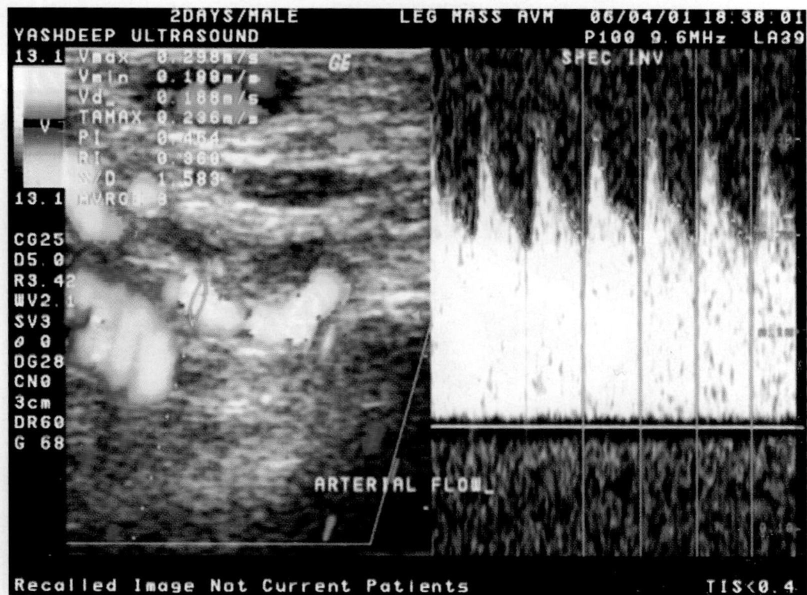

FIGURE 13.185

FIGURES 13.184 and 13.185: Spectral Doppler tracing shows high venous flow with pulsatile waveform in the veins suggestive of arteriovenous malformation. Spectral Doppler tracing shows arterial flow in the vessels.

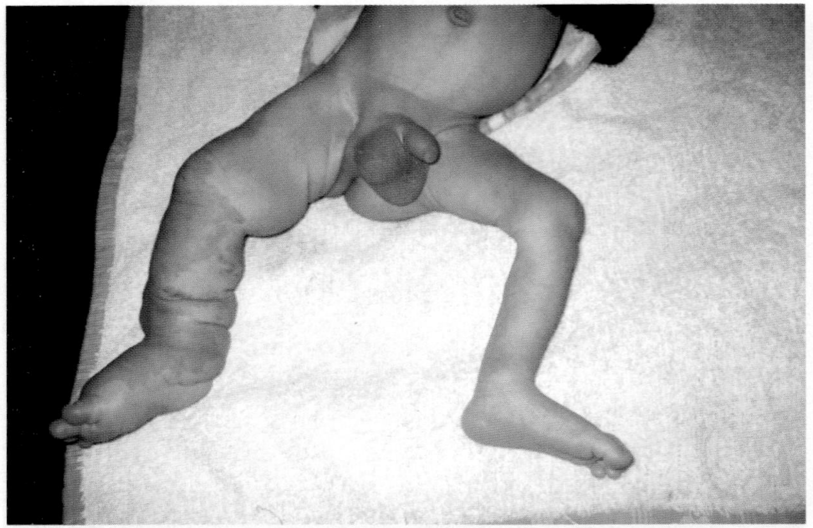

FIGURE 13.186: AV malformation of Rt lower limb. A child was born with marked enlargement of the Rt lower limb. The Rt limb is more than twice the size of the Lt limb.

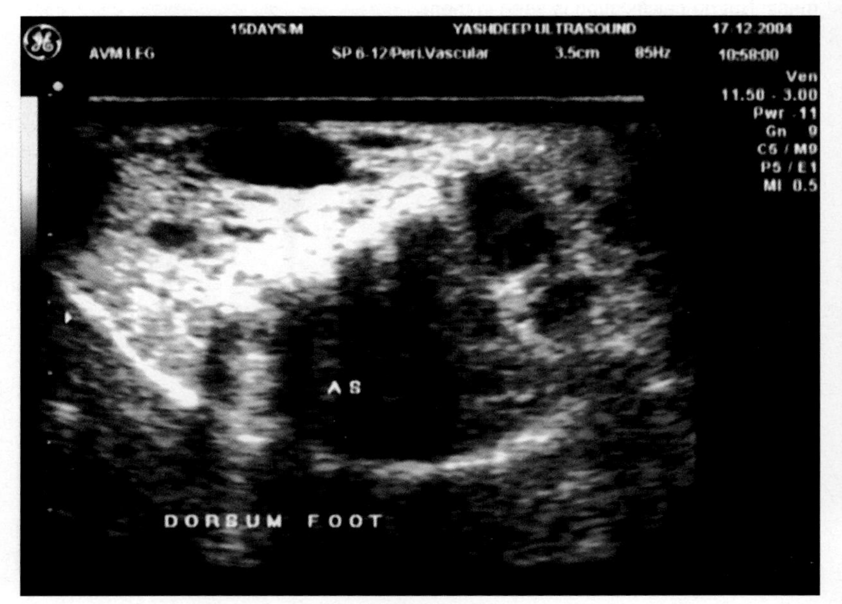

FIGURE 13.187

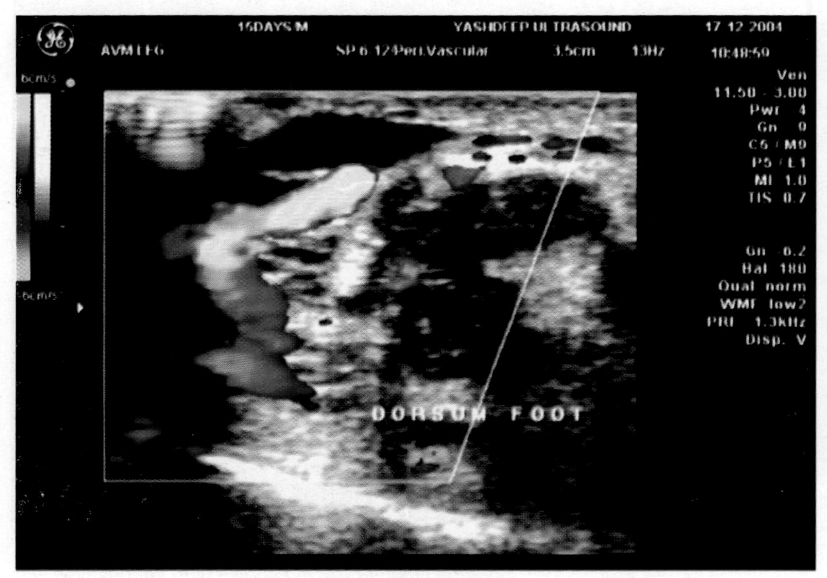

FIGURE 13.188

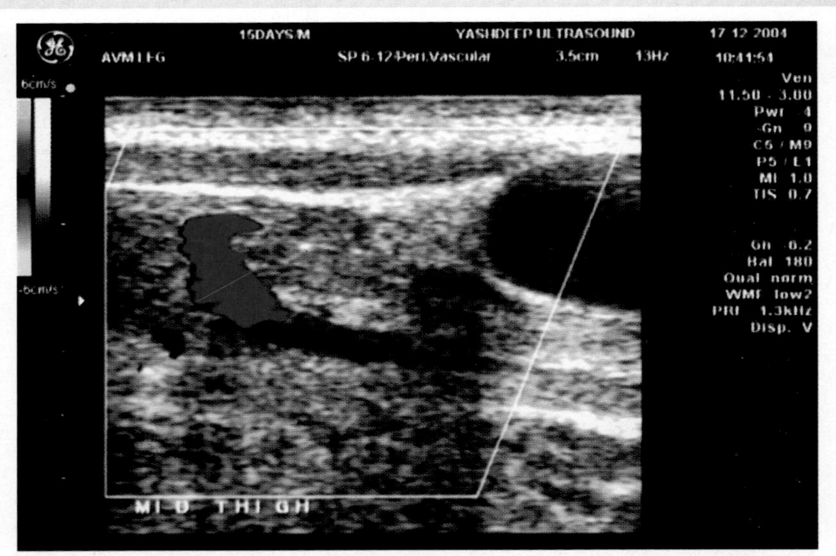

FIGURE 13.189

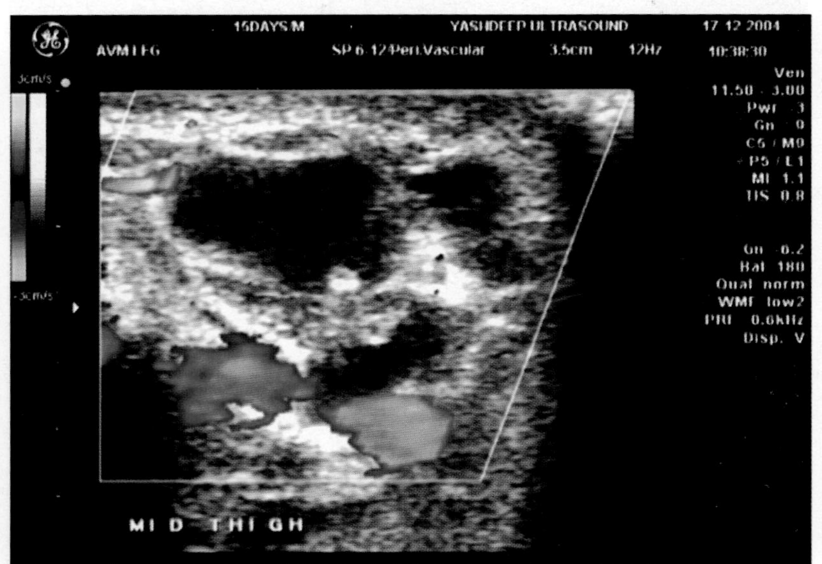

FIGURE 13.190

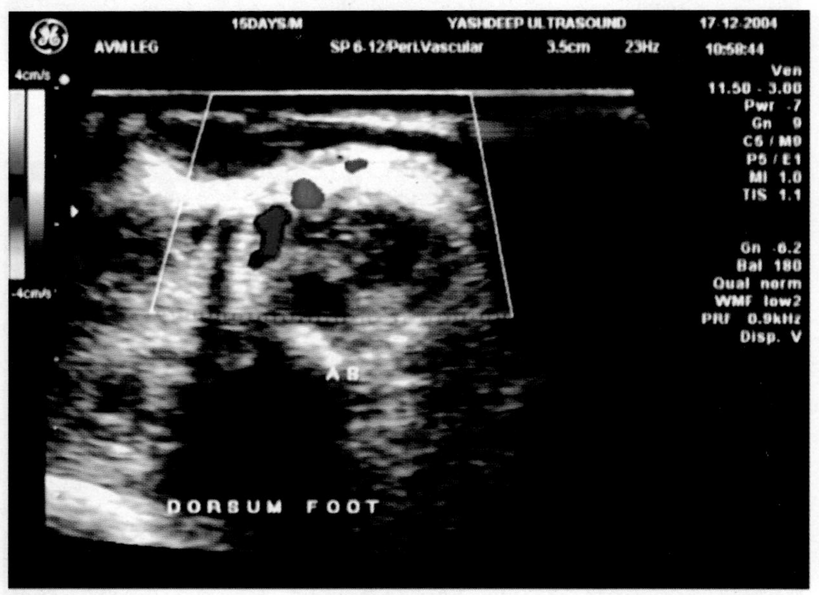

FIGURE 13.191

FIGURES 13.187 to 13.191: HRSG shows multiple dilated vessels in the limb. The dilated vessels are seen in the thigh, leg and also on dorsum of the foot. **On color flow imaging** arterial flow is seen in them. Dilated vessels also show venous flow pattern. Echogenic calcified specks are also seen in the vessels.

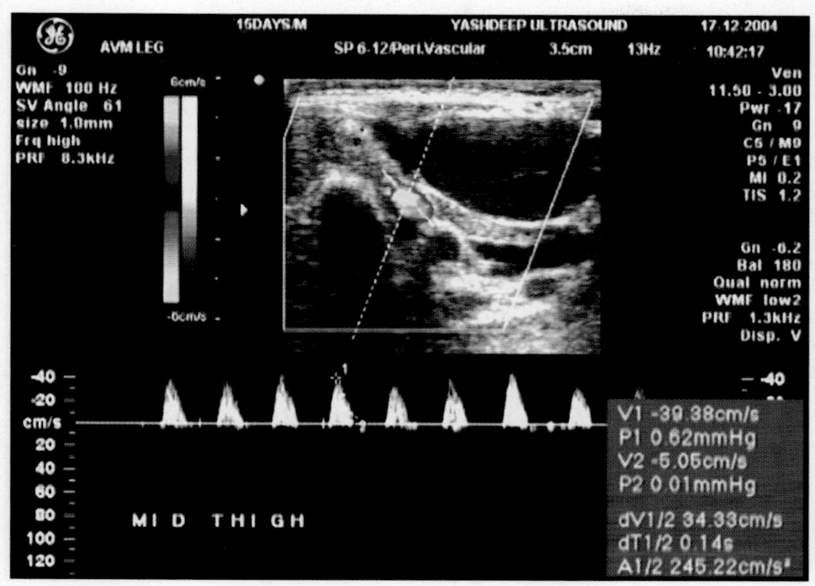

FIGURE 13.192: On spectral Doppler tracing low arterial flow is seen in the vessels. Sonographic findings are consistent with AVM.

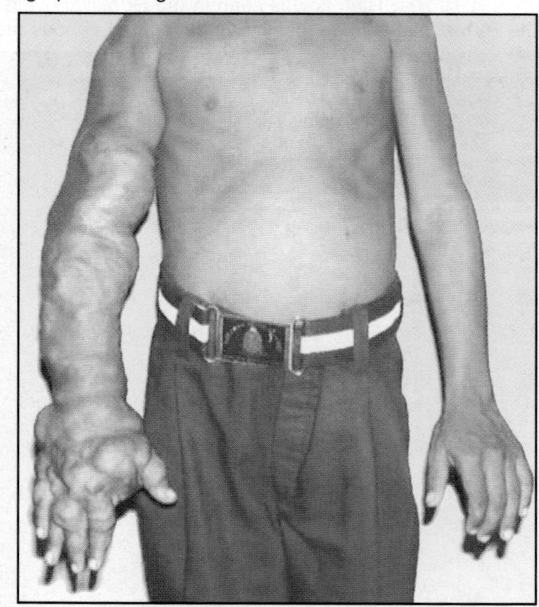

FIGURE 13.193: Cavernous venous malformation involving upper limb. A young boy presented with enlarged hypertrophied Rt upper limb with multiple dilated vessels in the limb.

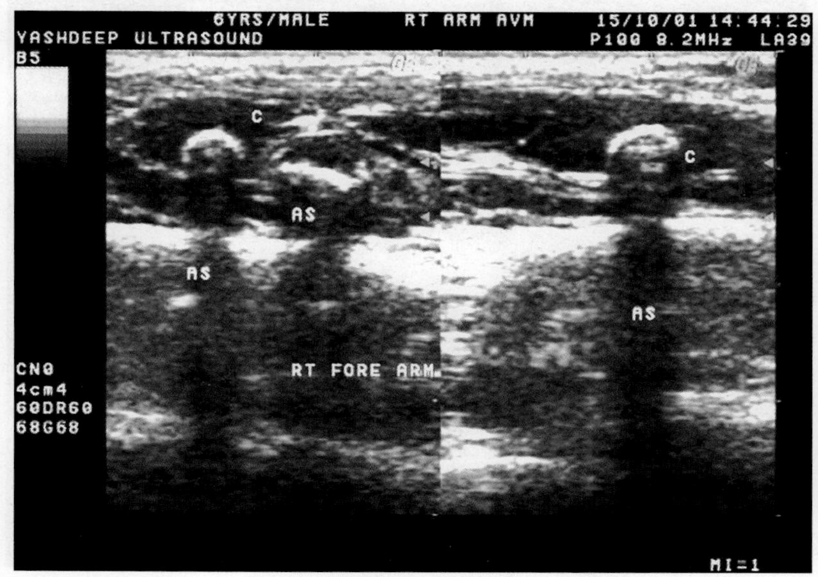

FIGURE 13.194

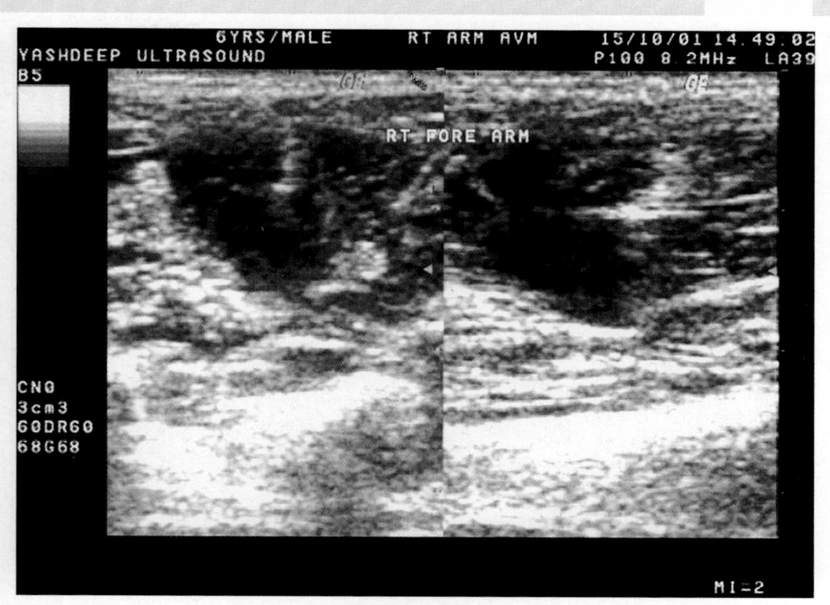

FIGURE 13.195

FIGURES 13.194 and 13.195: HRSG shows multiple dilated vessels in the upper limb. Echogenic calcified shadows are also seen in the vessels suggestive of phlebolith.

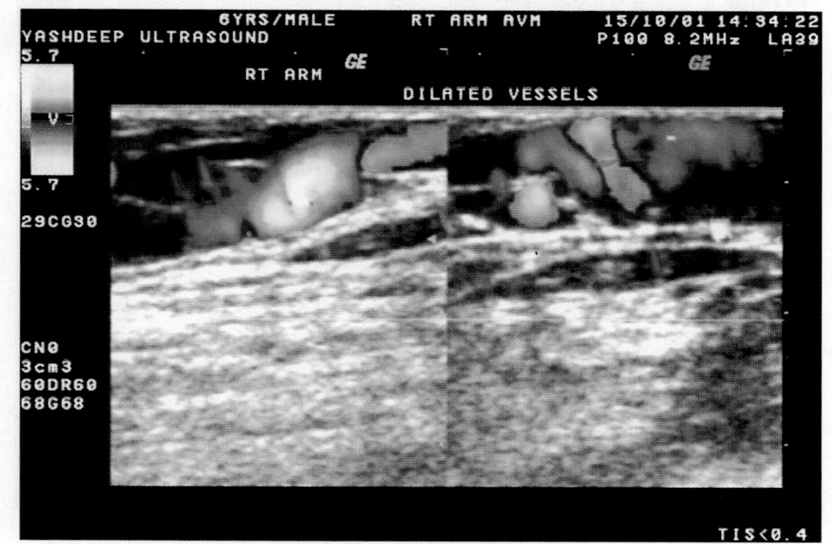

FIGURE 13.196

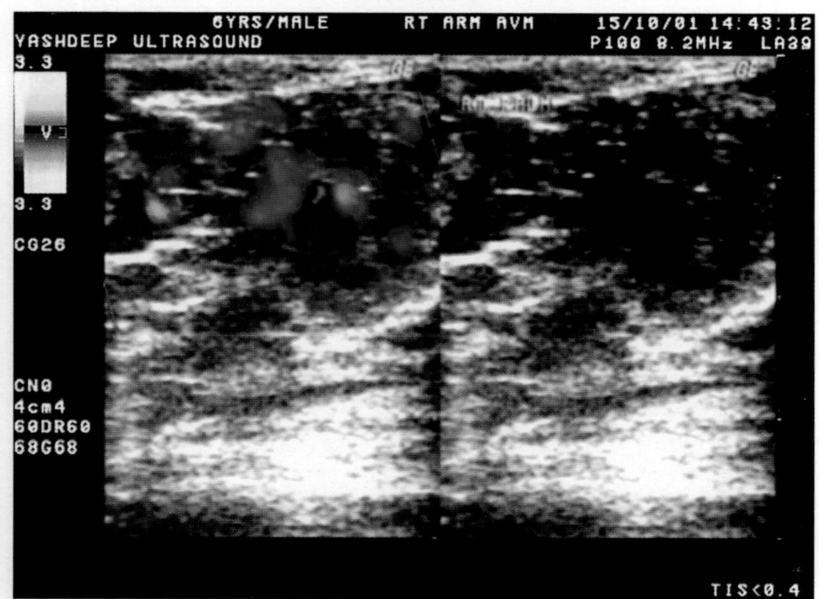

FIGURE 13.197

FIGURES 13.196 and 13.197: On color flow imaging low blood flow is seen in the vessels with venous flow pattern.

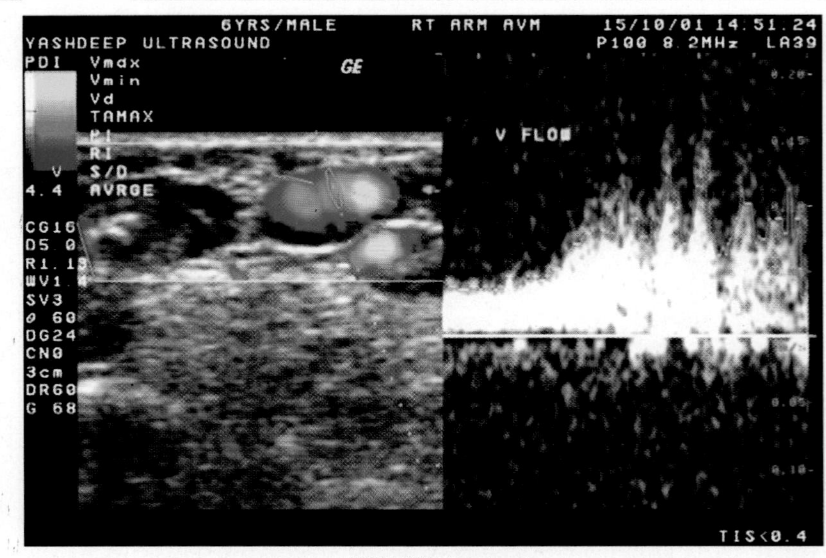

FIGURE 13.198

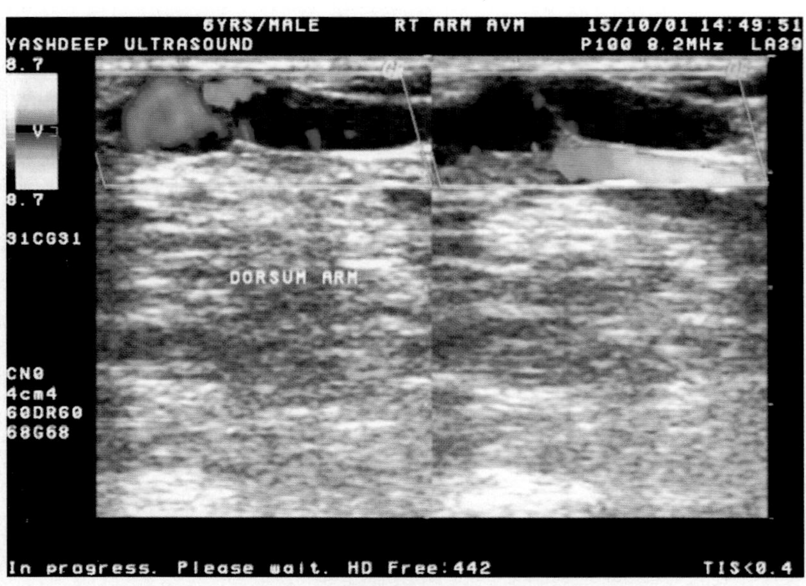

FIGURE 13.199

FIGURES 19.198 and 19.199: Spectral Doppler tracing shows venous flow.

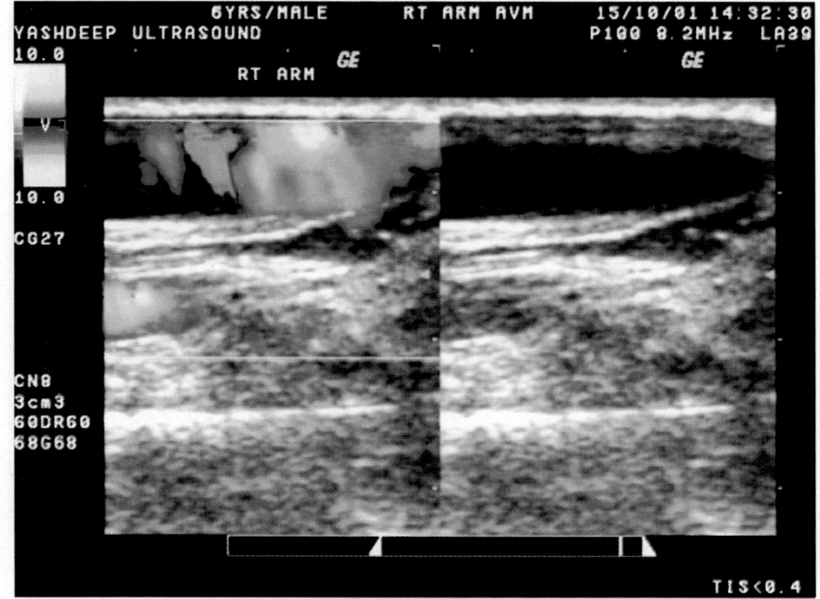

FIGURE 13.200

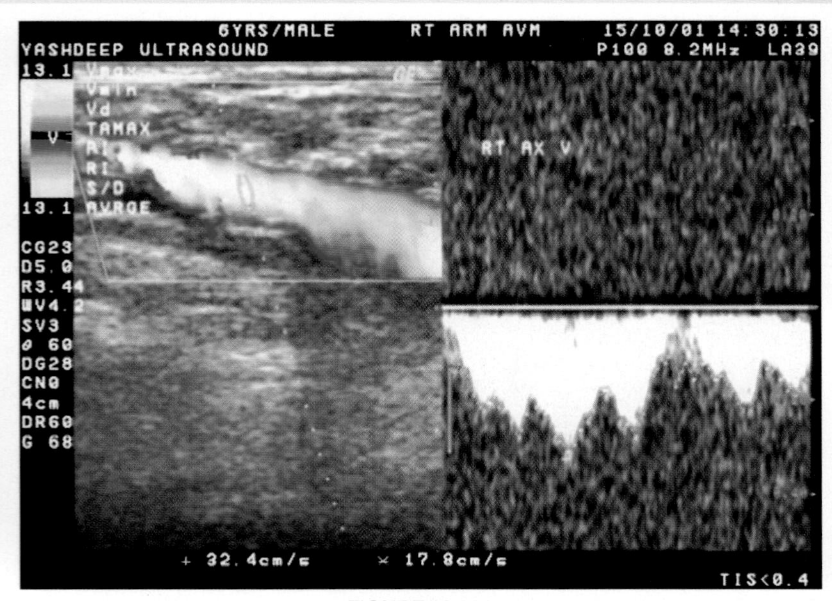

FIGURE 13.201

FIGURES 13.200 and 13.201: High flow is seen in the Rt axillary vein and venous flow is seen in the Rt brachial vein suggestive of cavernous venous malformation.

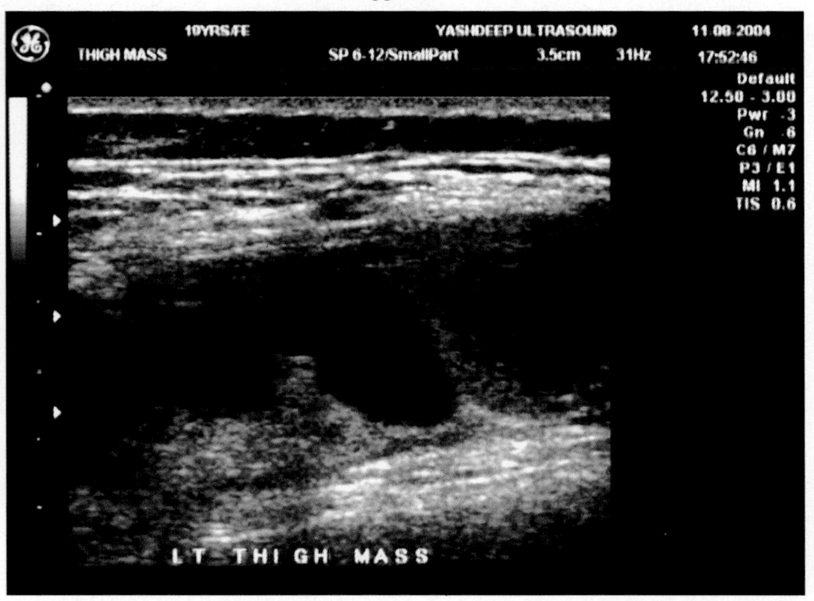

FIGURE 13.202: Localized AVM in thigh. A patient presented with soft tissue mass in the mid thigh region. HRSG shows multiple dilated vessels in the mid thigh region.

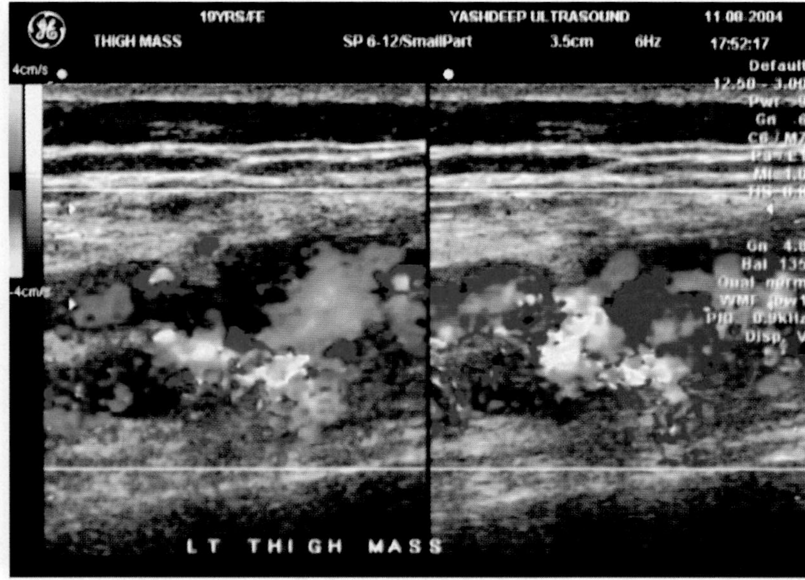

FIGURE 13.203

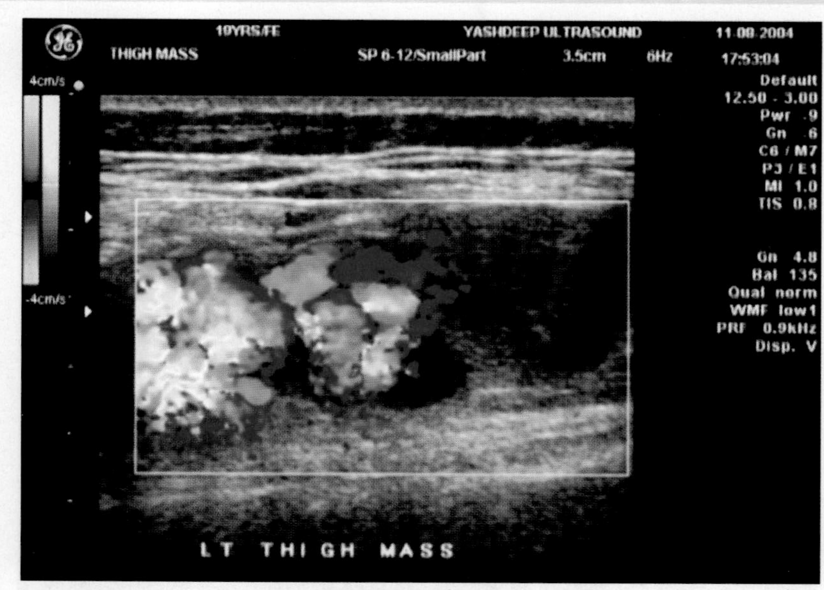

FIGURE 13.204

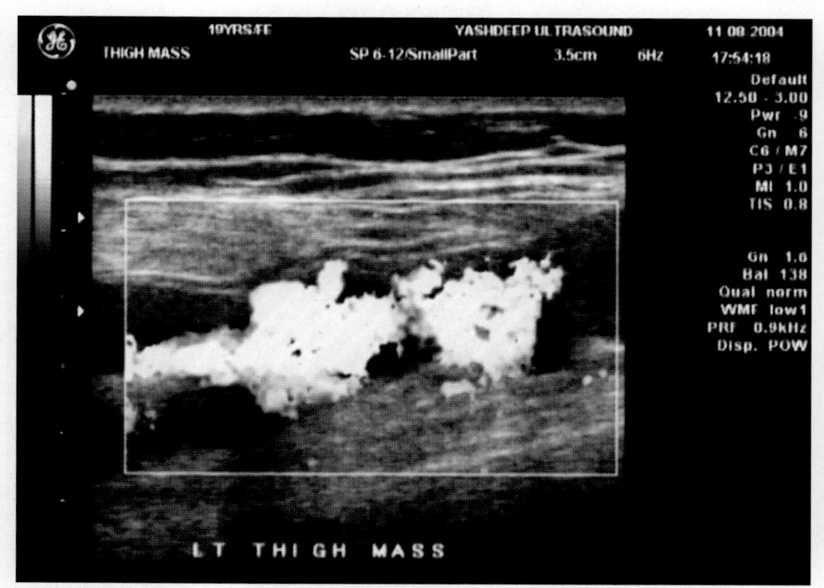

FIGURE 13.205

FIGURES 13.203 to 13.205: On color flow imaging high flow is seen in the vessels suggestive of localized AVM.

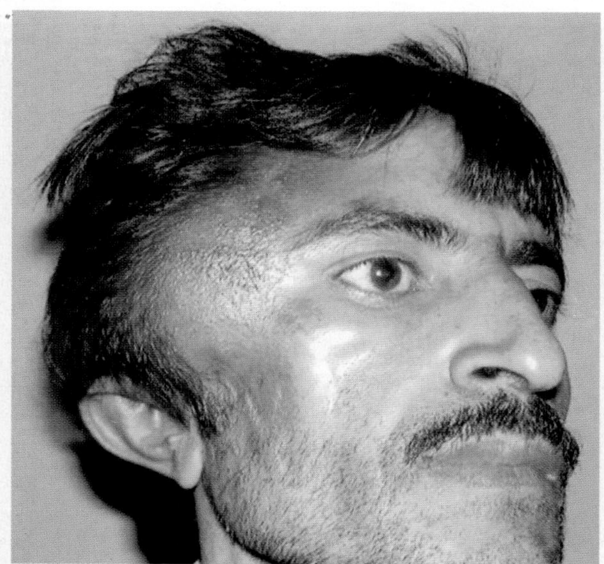

FIGURE 13.206: Temporal bone AV malformation. A middle aged man presented with increasing swelling over the Rt temporal area.

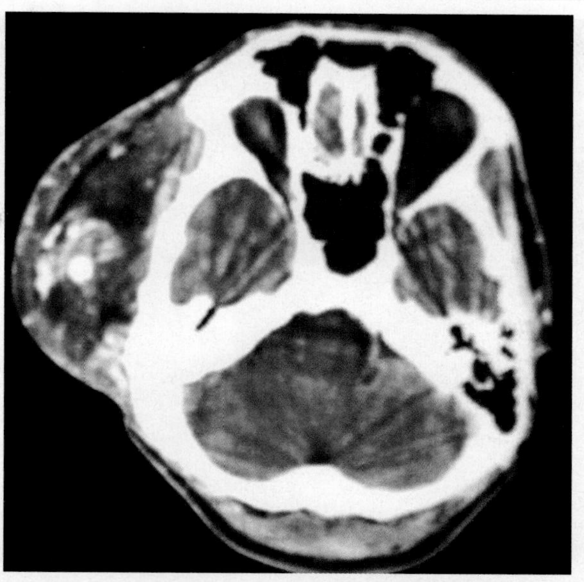

FIGURE 13.207: CT scan of the same patient shows a soft tissue mass over the Rt temporal region. Radio dense shadows are also seen in the mass with a dilated vessel.

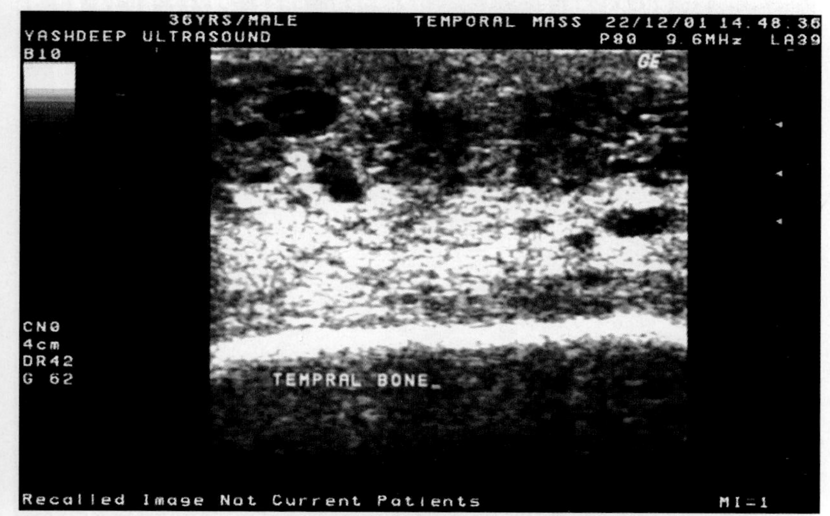

FIGURE 13.208

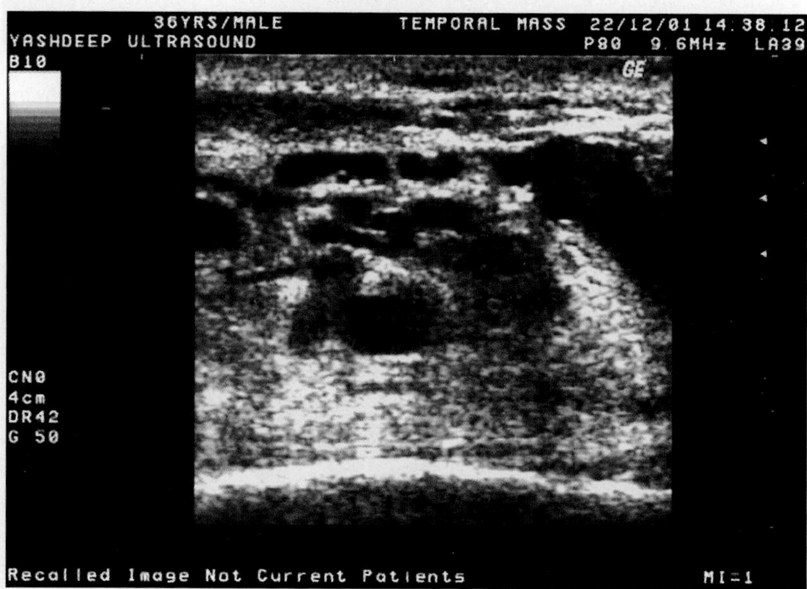

FIGURE 13.209

FIGURES 13.208 and 13.209: HRSG shows a big soft tissue mass over the Rt temporal region. Multiple dilated vessels are seen in the mass. Few echogenic-calcified specks are also seen in it. However, the temporal bone is well visualized. The vessels are not seen invading the temporal bone.

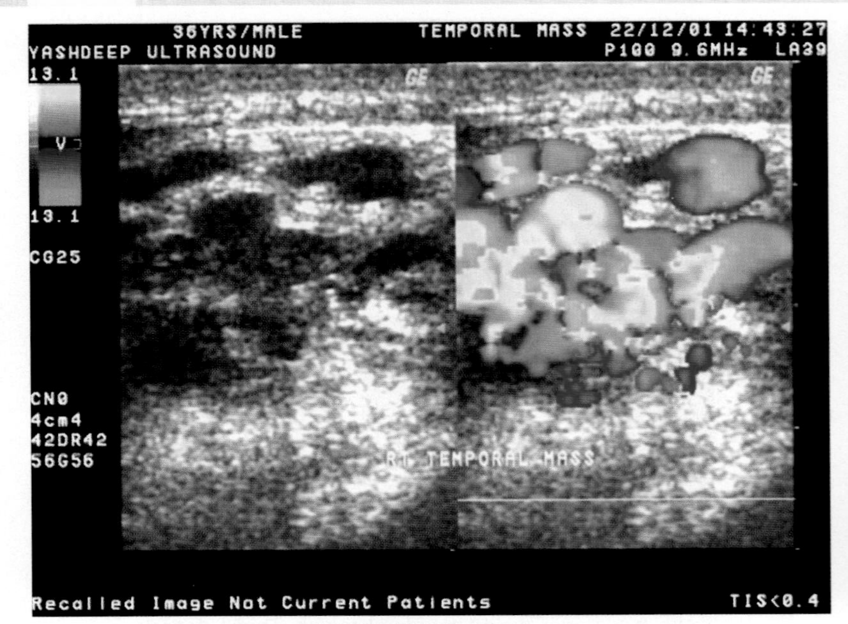

FIGURE 13.210

FIGURE 13.211

FIGURES 13.210 and 13.211: On color Doppler imaging the vessels are showing high blood flow in bizarre fashion.

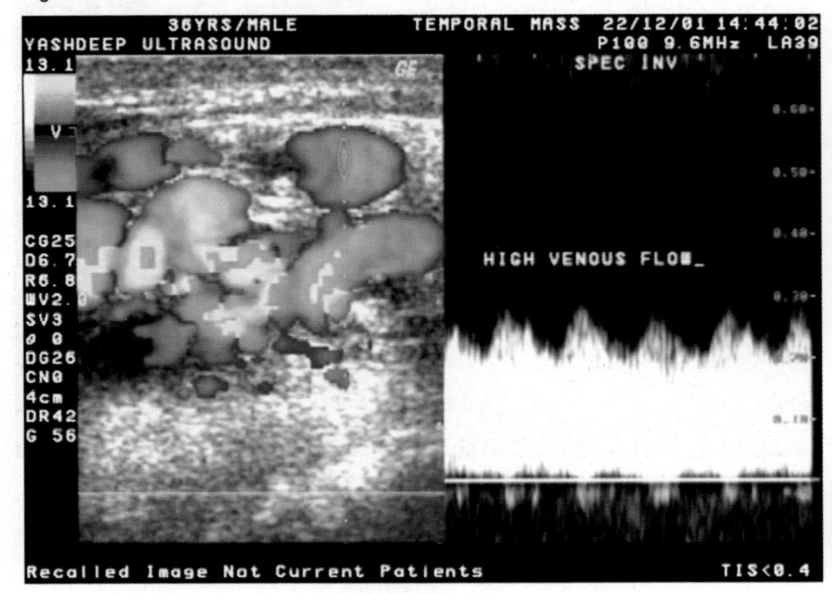

FIGURE 13.212

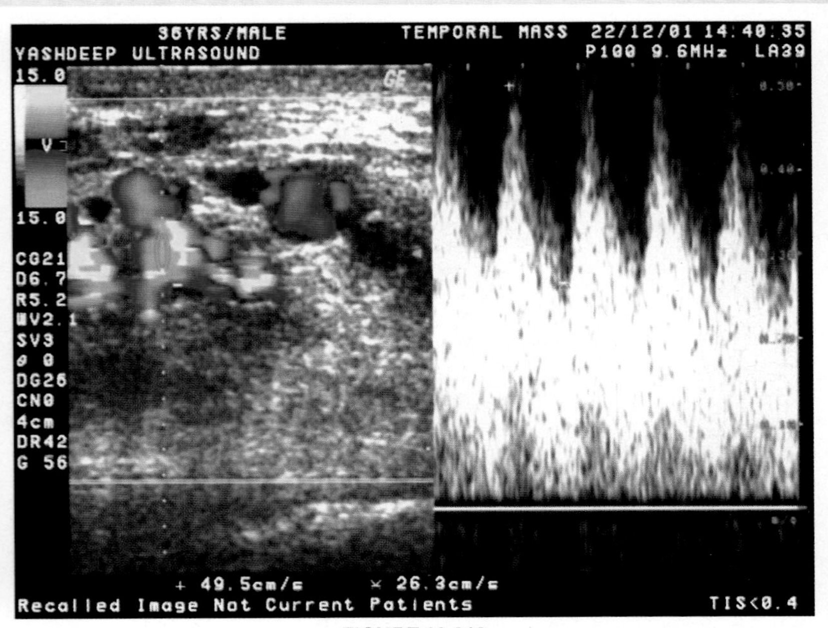

FIGURE 13.213

FIGURES 13.212 and 13.213: Dilated veins are also seen in the vessels and spectral Doppler tracing shows high pulsatile venous flow with arterial flow in the vessels suggestive of arteriovenous malformation.

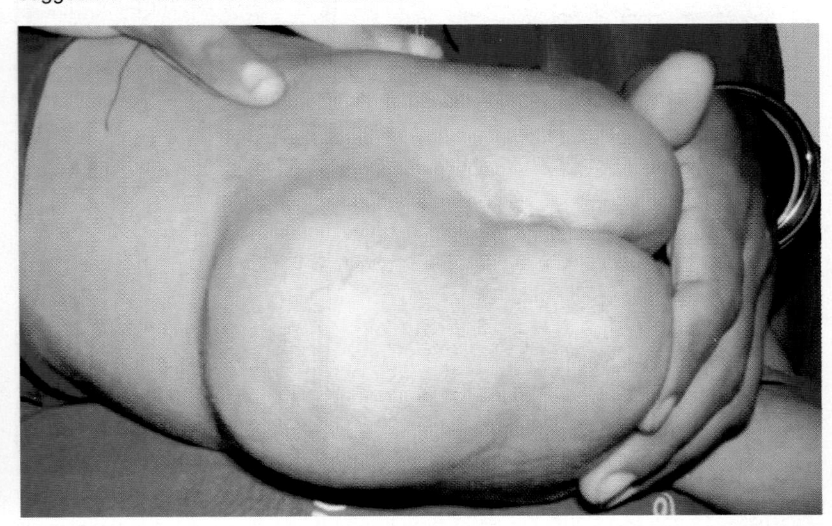

FIGURE 13.214: Vascular mass in the Rt hip. A child was born with a soft tissue mass over the Rt gluteal region.

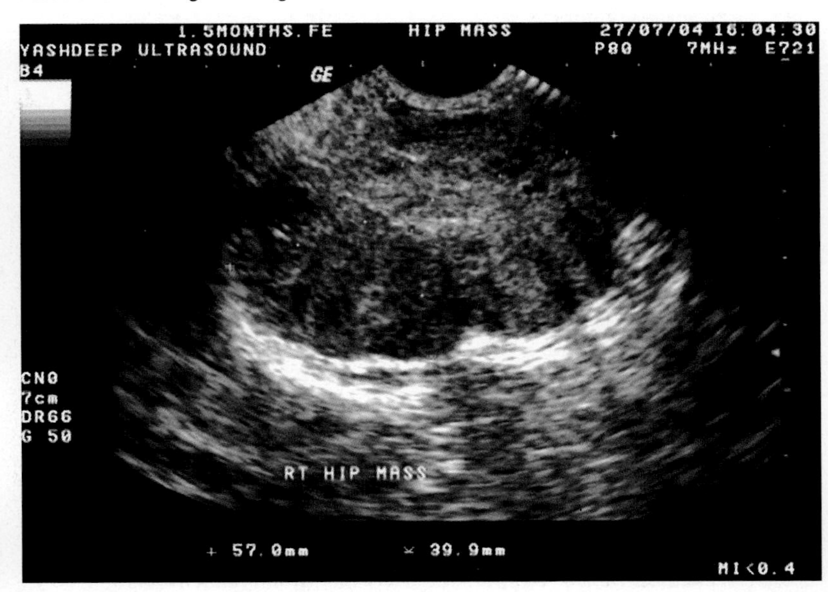

FIGURE 13.215: HRSG shows multiple dilated vessels in the mass. The vessels are running in crisscross fashion.

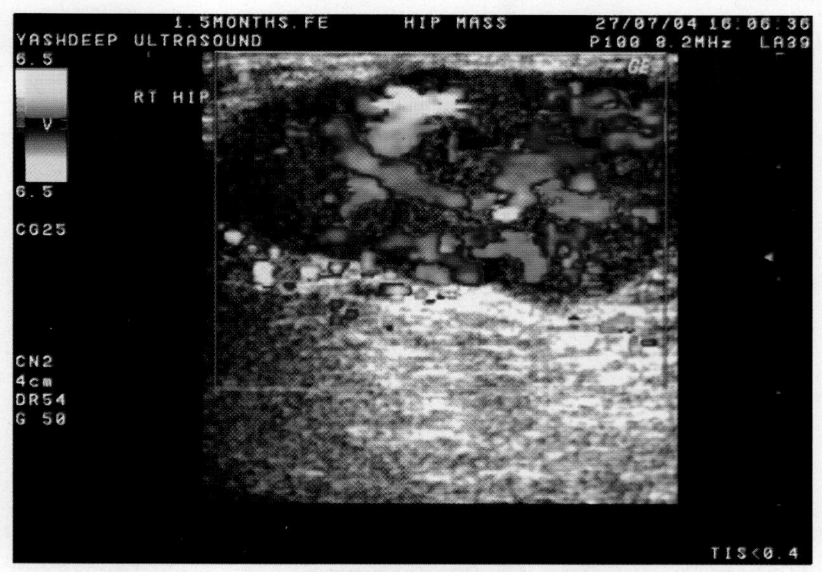

FIGURE 13.216

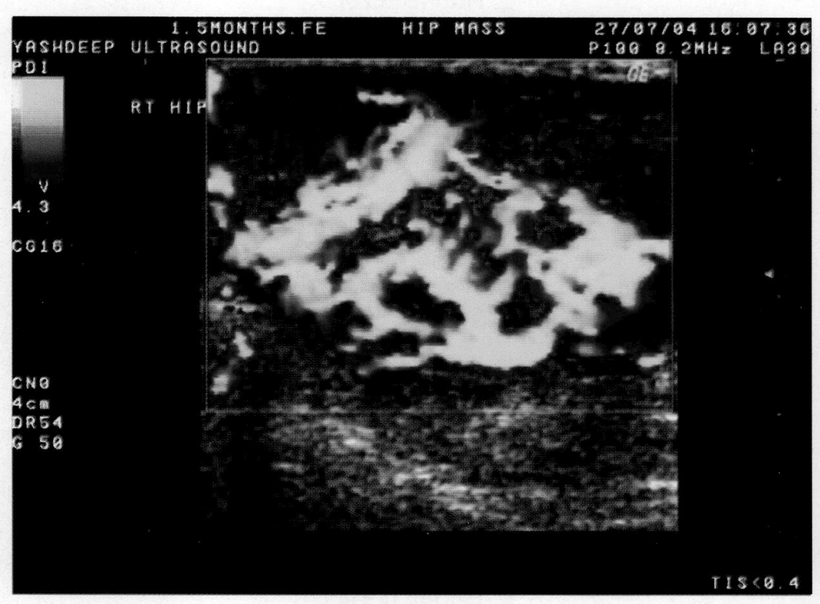

FIGURE 13.217

FIGURES 13.216 and 13.217: Multiple dilated vessels are seen in the mass, which are confirmed on color flow imaging.

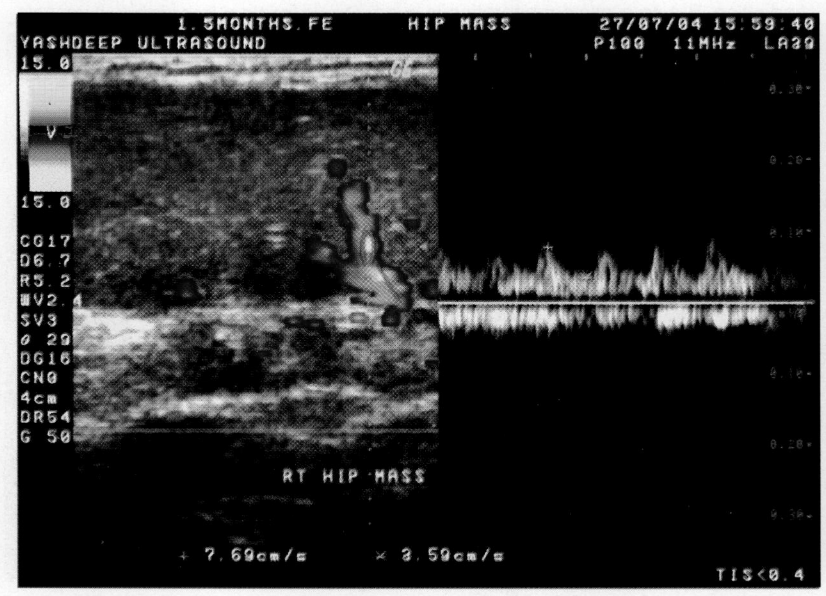

FIGURE 13.218: On spectral Doppler tracing high venous flow seen in it suggestive of AVM.

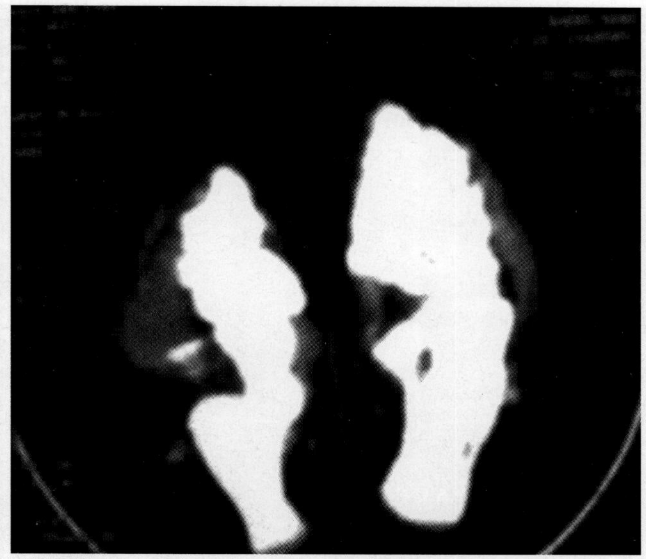

FIGURE 13.219: Local venous malformation over the foot. A young girl presented with soft tissue swelling over the dorsal of the Rt foot. CT scan of the foot shows a soft tissue homogeneous mass on the lateral side of the foot. It shows homogeneous enhancement on contrast examination.

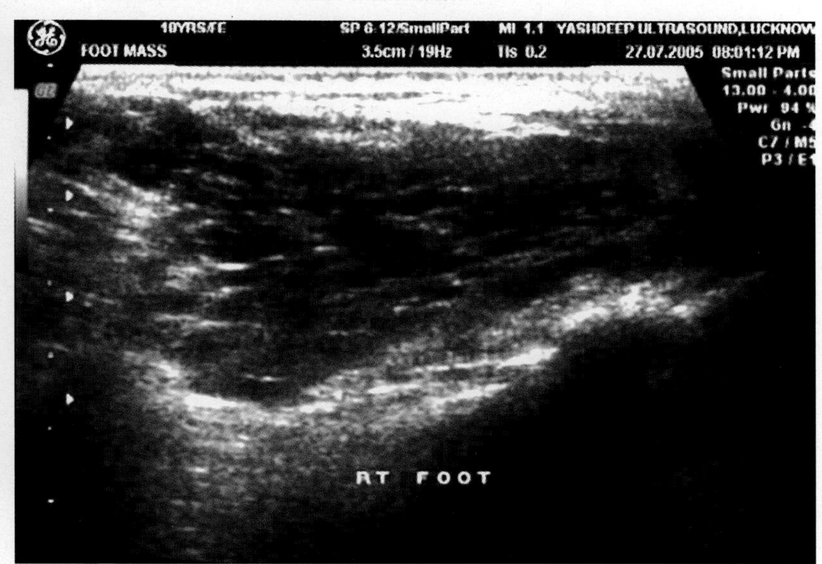

FIGURE 13.220: HRSG shows a soft tissue mass over the foot. Multiple dilated vessels are seen in the mass.

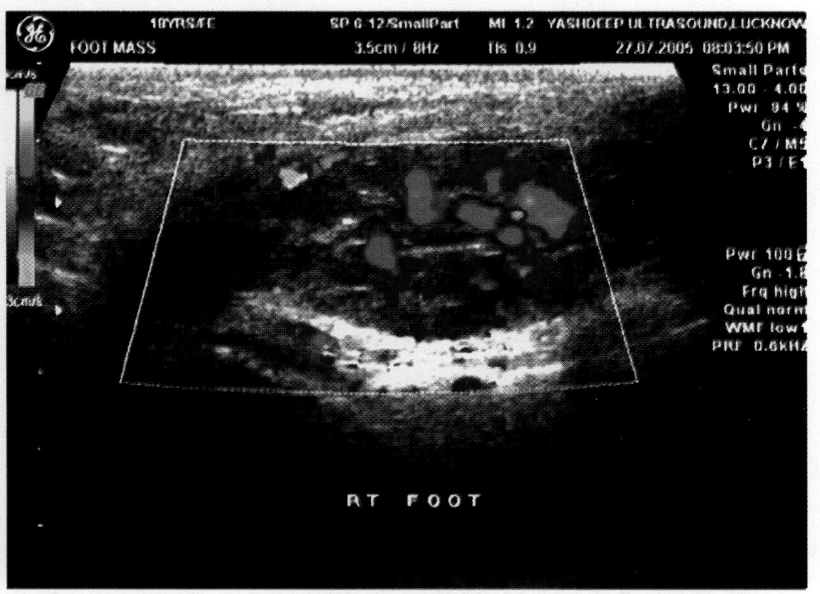

FIGURE 13.221A

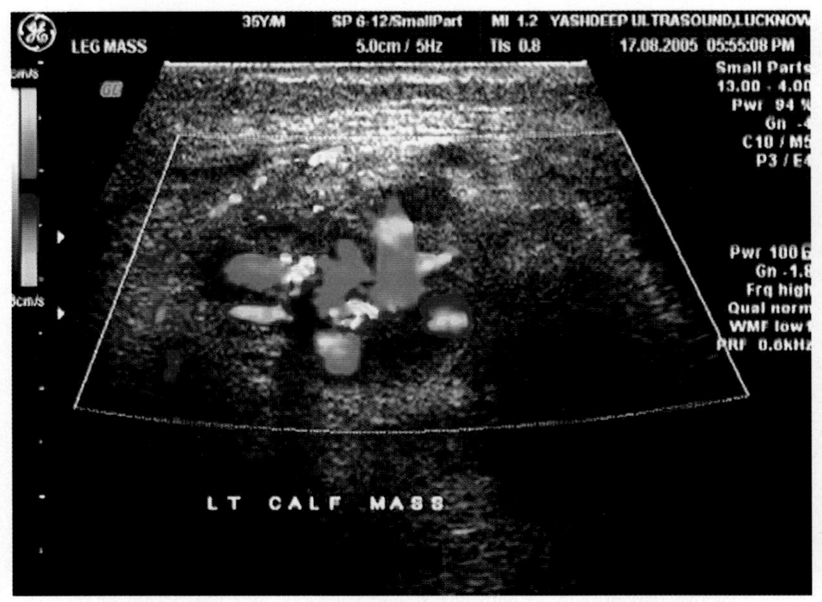

FIGURE 13.221B

FIGURES 13.221A and B: The dilated vessels are conformed. On color flow imaging suggestive of vascular mass.

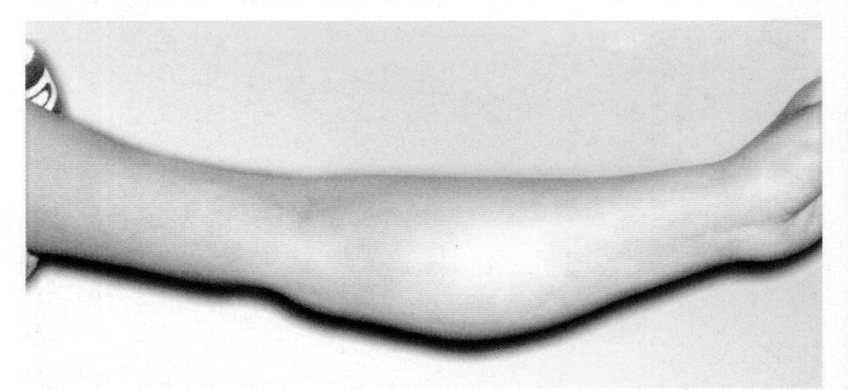

FIGURE 13.222: Angiomatous mass. A young boy presented with progressive swelling over the Lt forearm. It was non pulsatile on clinical examination.

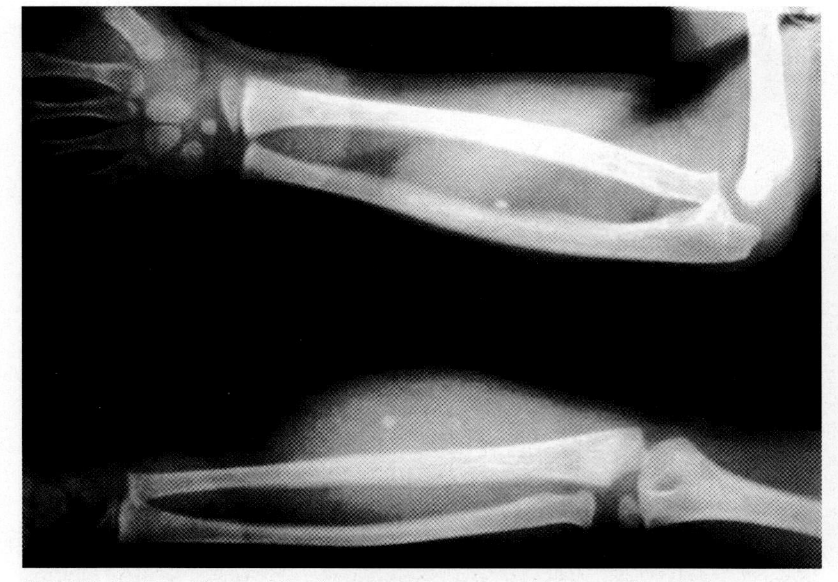

FIGURE 13.223: X-ray of the forearm shows normal bones with soft tissue mass. Calcified radio dense shadows are also seen in the soft tissue mass.

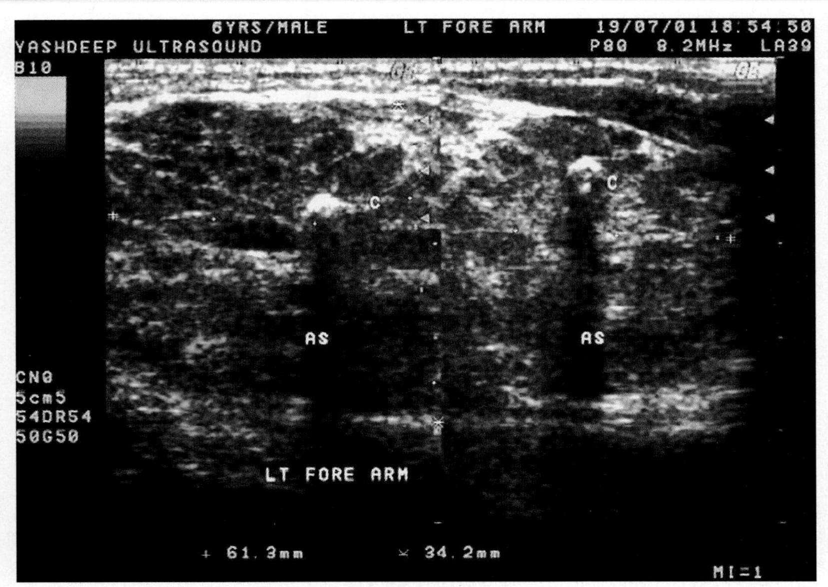

FIGURE 13.224

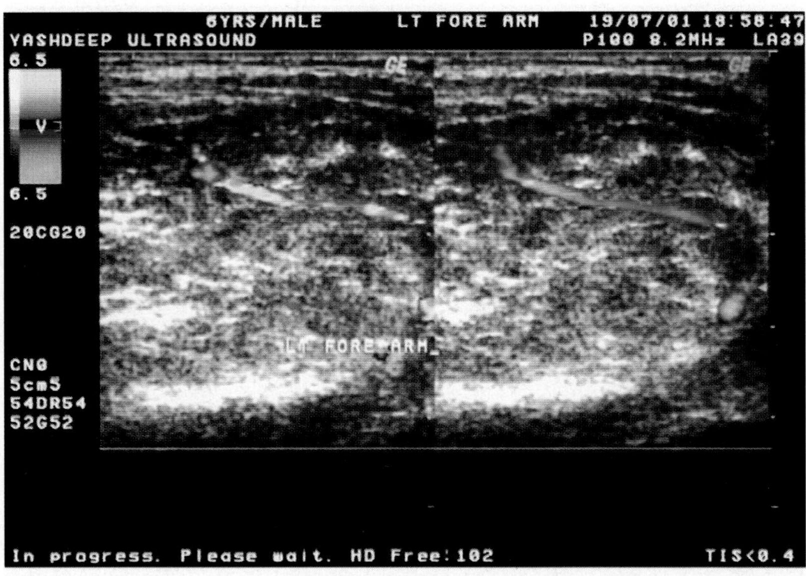

FIGURE 13.225

FIGURES 13.224 and 13.225: HRSG shows a big lobulated mass with heterogeneous texture in the muscles of the forearm. Multiple dilated vessels are seen in it. Echogenic calcified shadows are also seen in the mass. They are accompanied with acoustic shadowing. But no cystic degeneration is seen. On color flow imaging few vessels are seen feeding the mass.

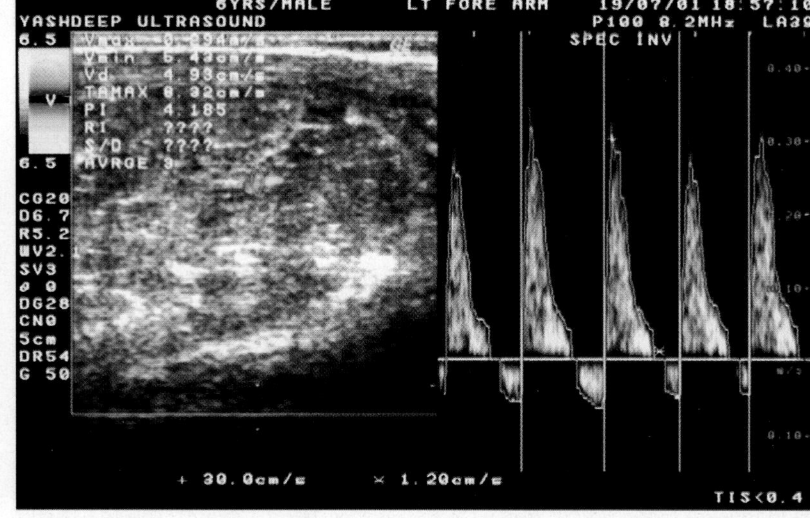

FIGURE 13.226: Spectral Doppler tracing shows arterial flow in the vessel. Biopsy of the mass shows angiomatous mass.

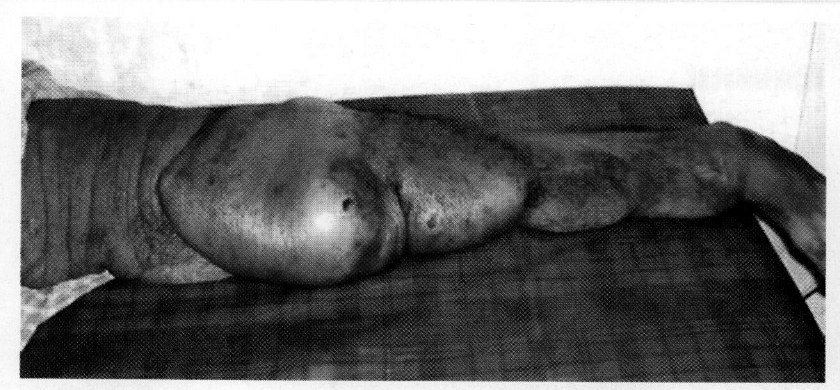

FIGURE 13.227: Lt lower limb gigantism. A patient presented with marked hypertrophy of whole Lt lower limb. It is more than double the size of Rt lower limb. Marked skin edema is present with loss of normal muscle tone. On clinical examination there is also evidence of cystic masses seen over the Lt knee.

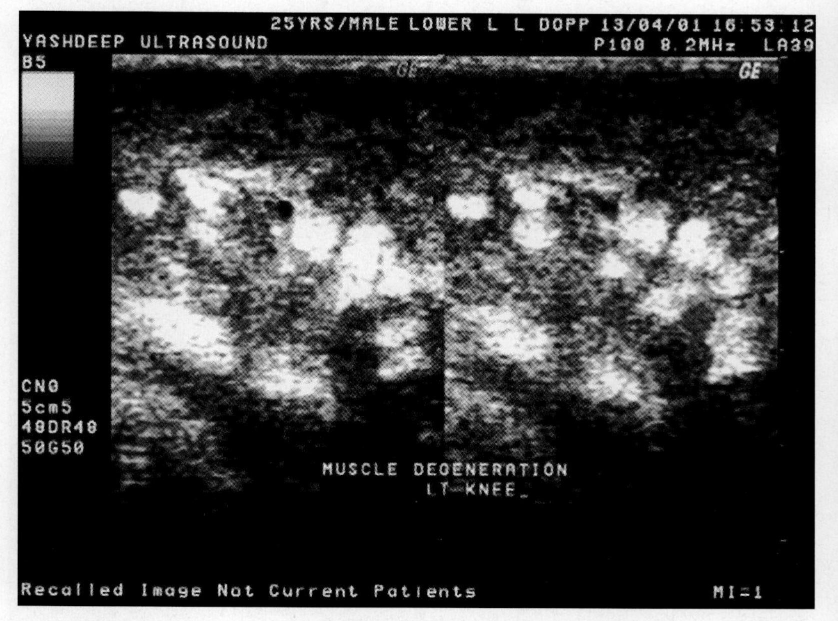

FIGURE 13.228

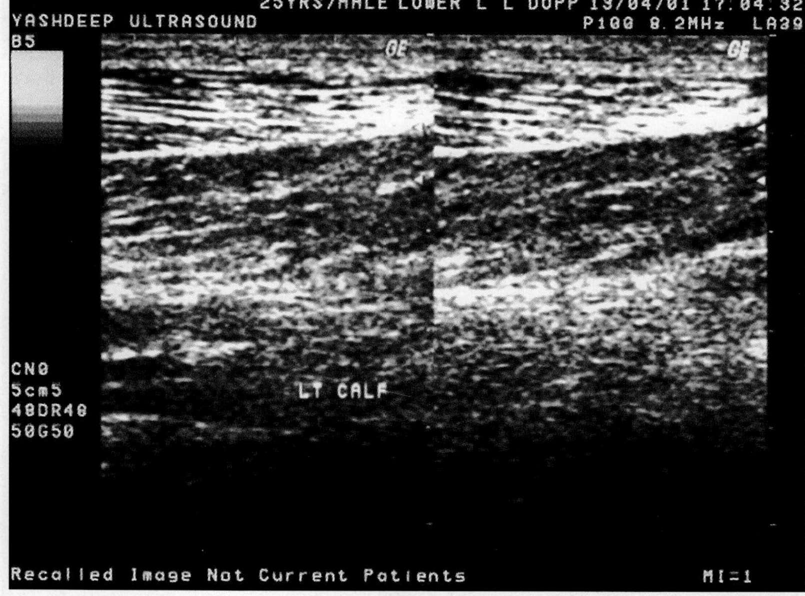

FIGURE 13.229

FIGURES 13.228 and 13.229: HRSG shows gross disorganization of the normal muscle texture involving the thigh muscles. Multiple echogenic globular shadows are seen in the muscles. Normal muscle texture is lost. The muscle capsules are also broken. However, normal residual muscle texture is seen in calf muscles. Marked skin and hypodermis edema is present.

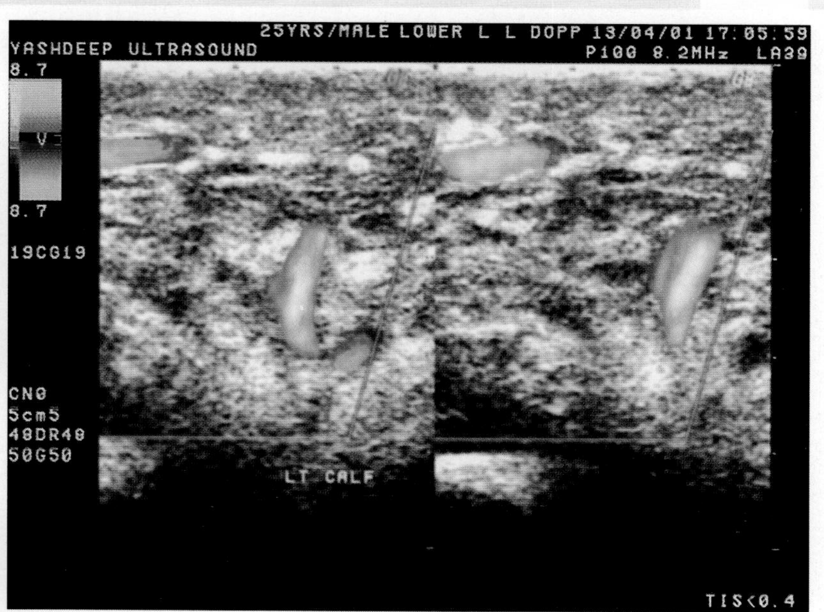

FIGURE 13.230

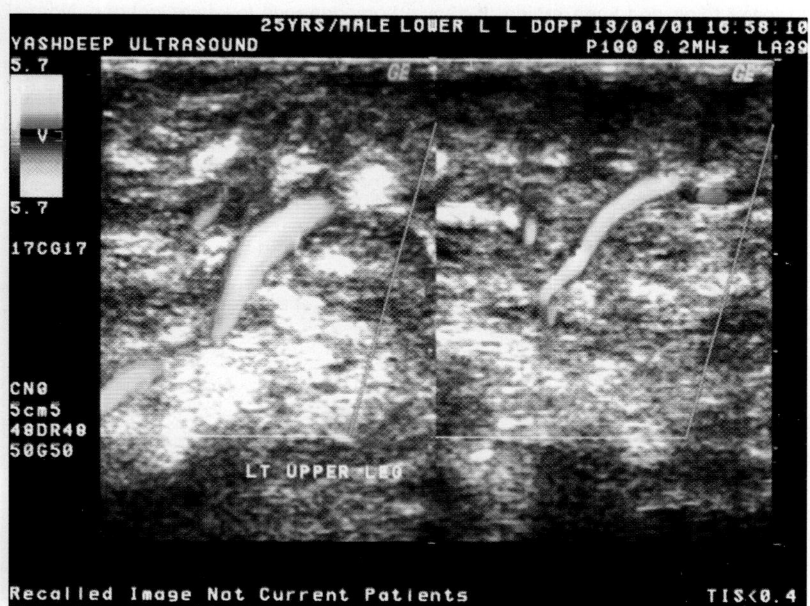

FIGURE 13.231

FIGURES 13.230 and 13.231: On color Doppler imaging few vessels are seen in the muscles with low flow. No vascular abnormality is seen.

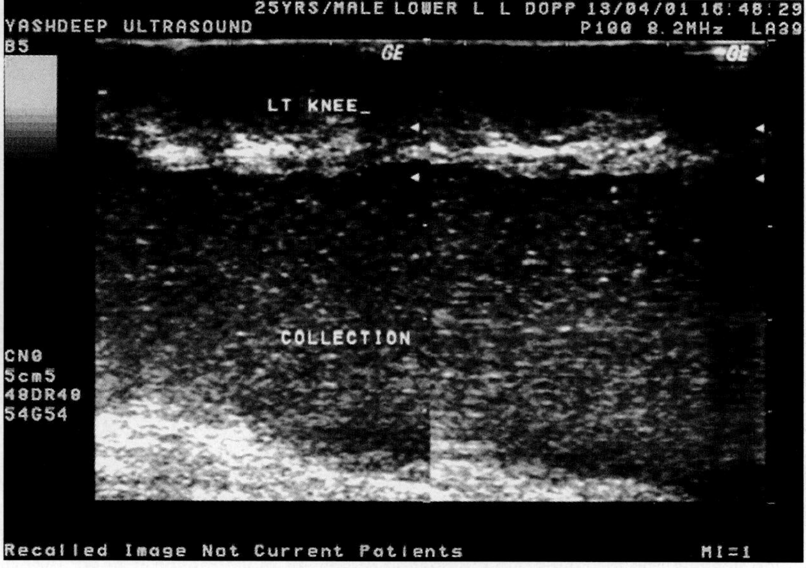

FIGURE 13.232

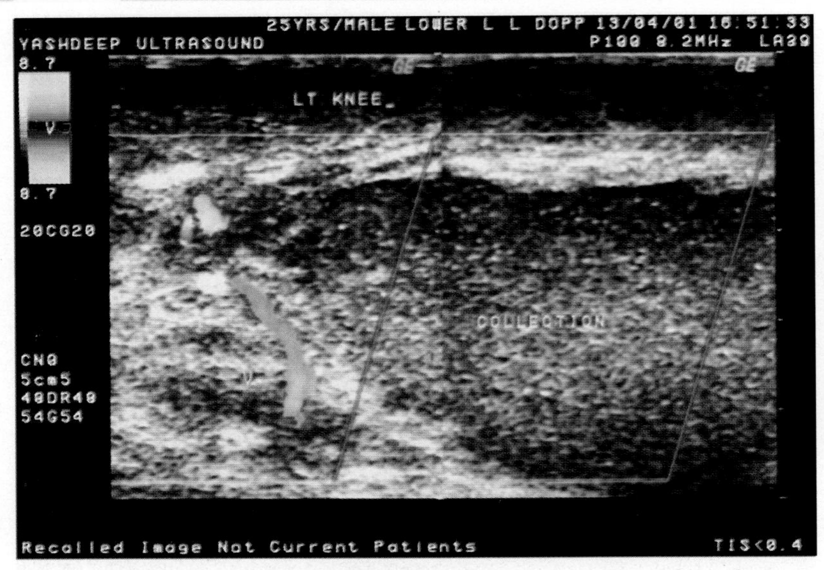

FIGURE 13.233

FIGURES 13.232 and 13.233: HRSG shows gross amount of fluid collection over the Lt lower thigh and knee region. It shows homogeneous low-level echoes. Normal muscles are not seen.

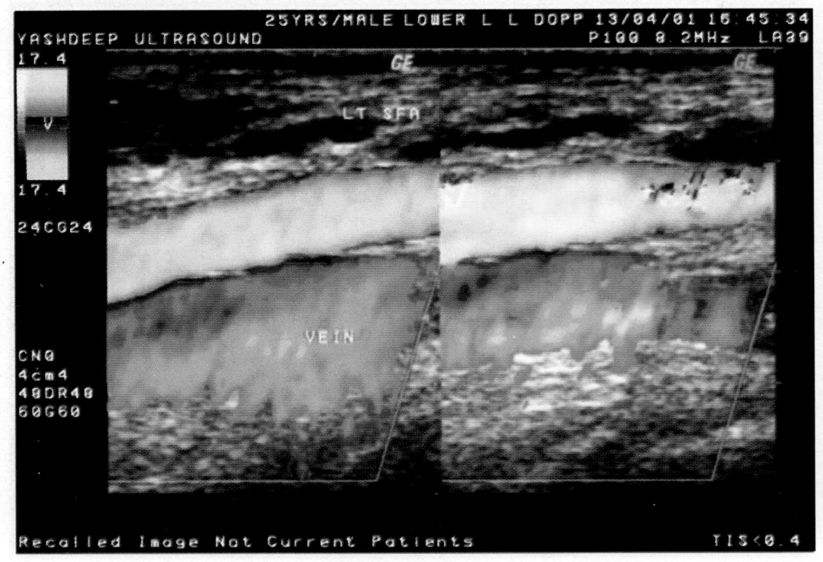

FIGURE 13.234

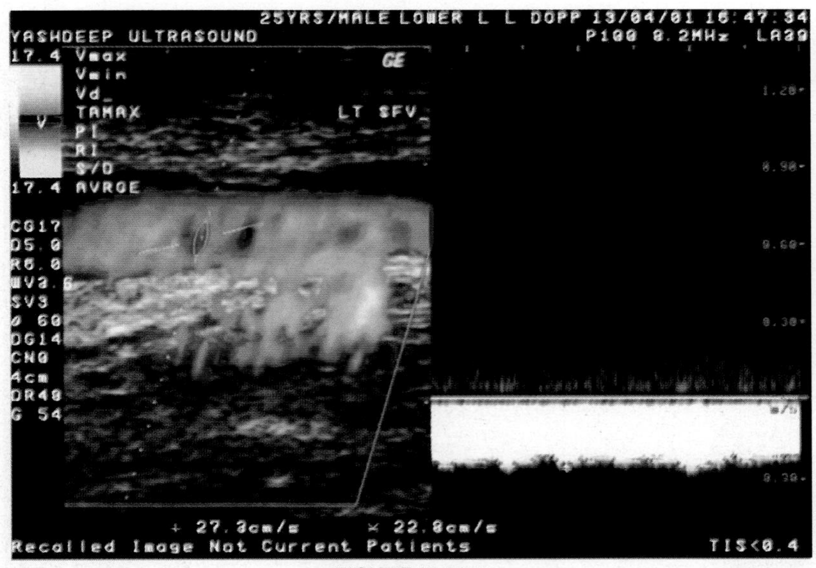

FIGURE 13.235

FIGURES 13.234 and 13.235: On color Doppler imaging normal flow is seen in the thigh vessels and the veins. No suggestion of any AV malformation is seen. Biopsy shows cystic degeneration of the muscle with lymphangitis.

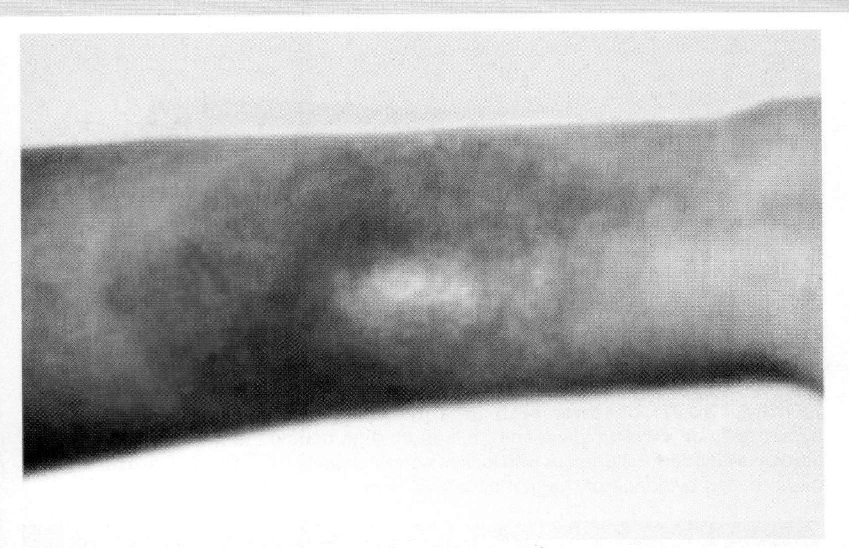

FIGURE 13.236: Localized venous malformation. A young girl presented with soft tissue mass swelling over the forearm.

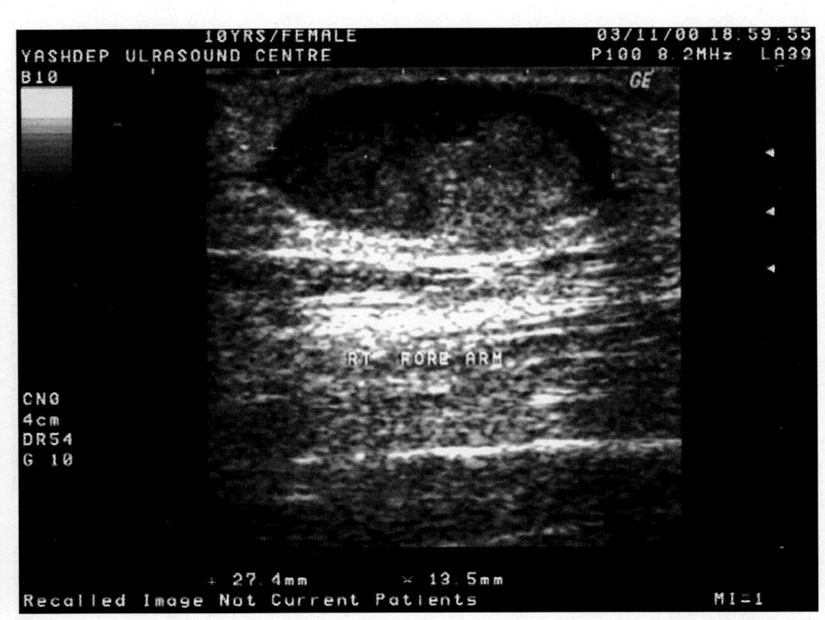

FIGURE 13.237: HRSG shows homogeneous low-level echo complex mass in the Lt forearm on the flexor surface of the arm.

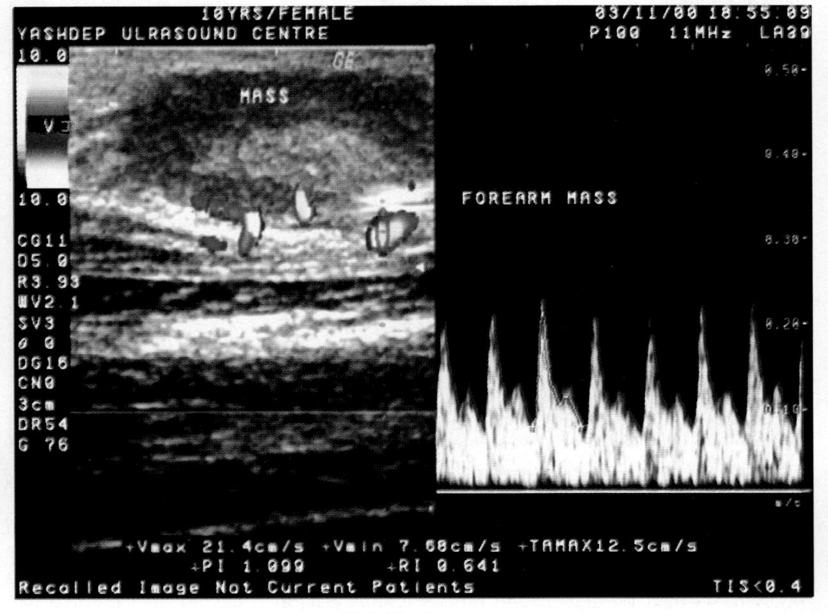

FIGURE 13.238

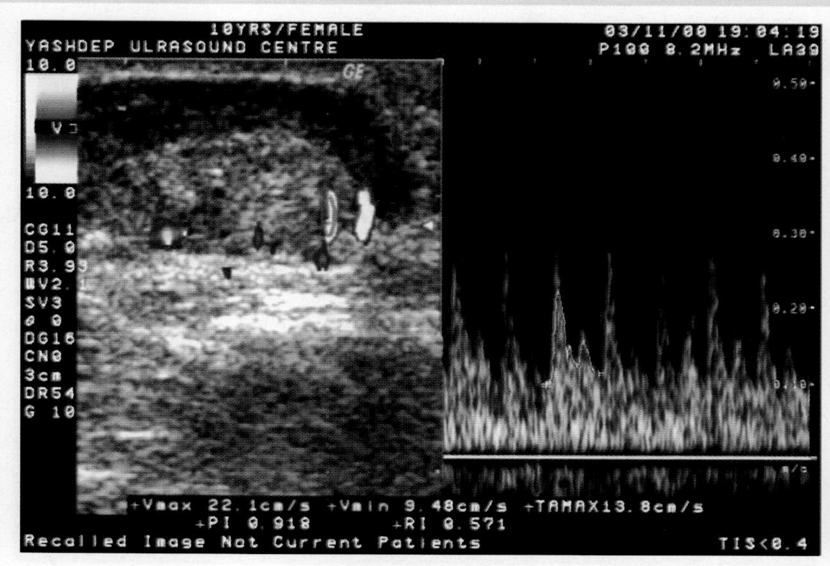

FIGURE 13.239

FIGURES 13.238 and 13.239: On color and spectral flow imaging low flow is seen in the mass with high diastolic pattern. On postoperative findings it turned out to be a localized venous malformation.

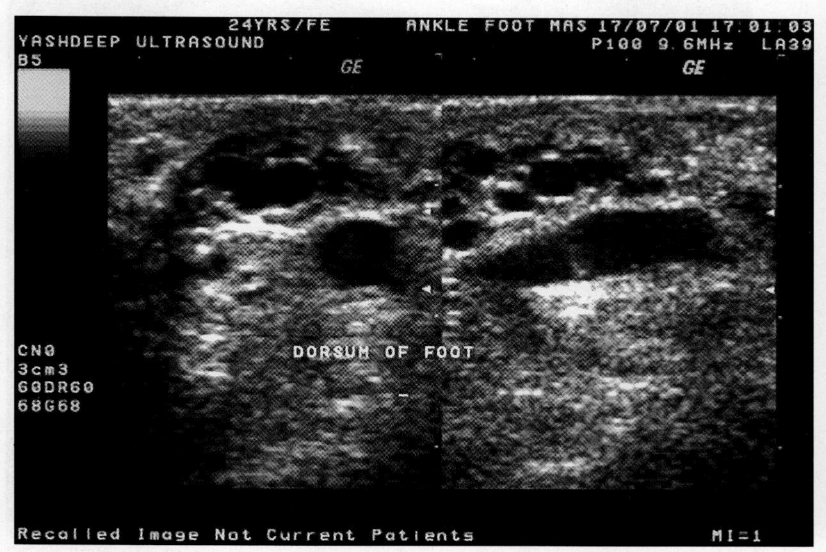

FIGURE 13.240: Hemangioma lower thigh. A young girl presented with a soft tissue mass over Lt thigh lower part lateral surface. HRSG shows multiple dilated blood vessels in the mass.

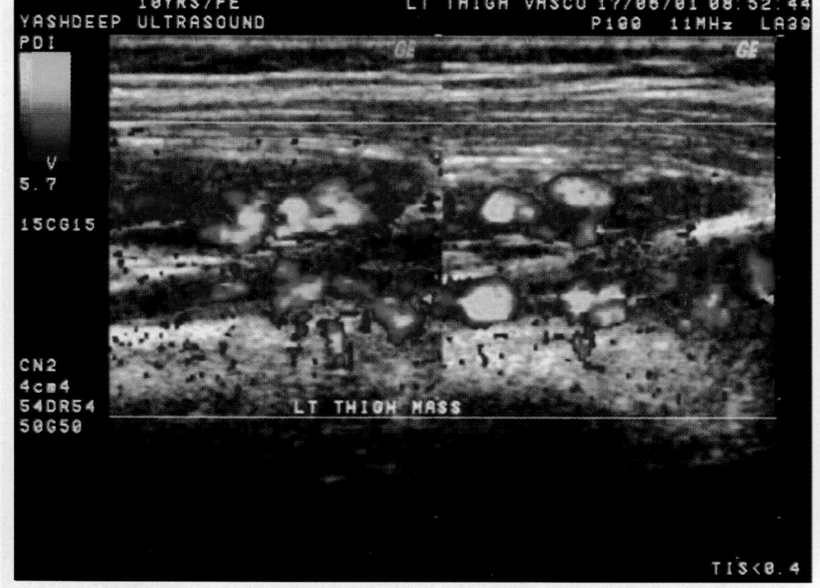

FIGURE 13.241: On color flow imaging low blood flow is seen in the mass. No evidence of any feeding vessel is seen suggestive of hemangioma.

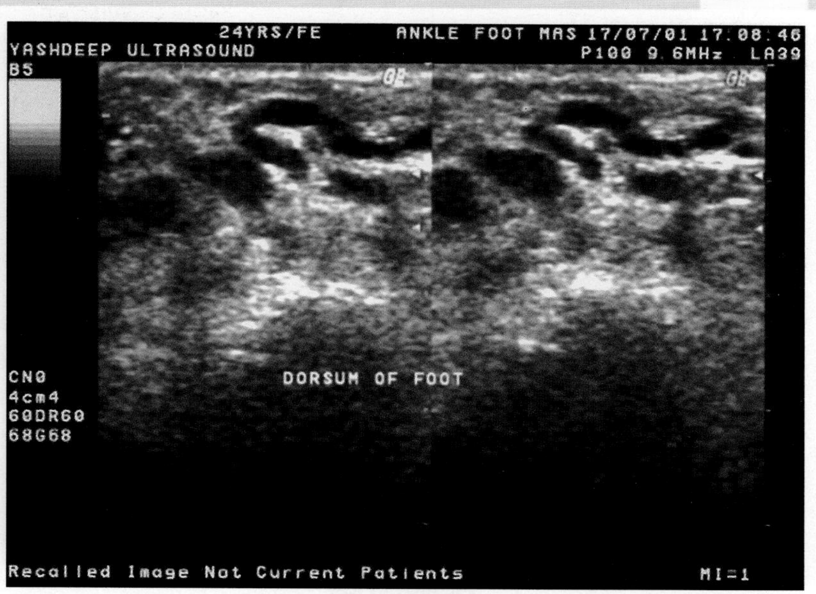

FIGURE 13.242

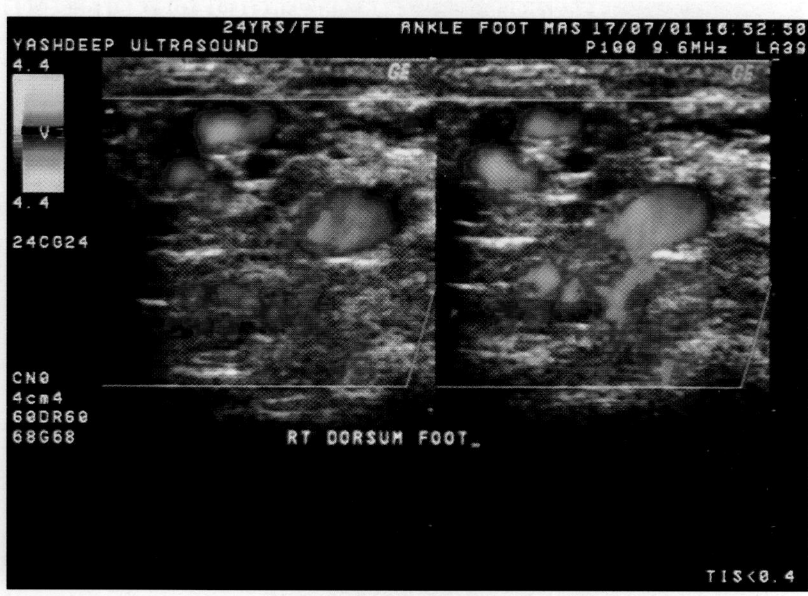

FIGURE 13.243

FIGURES 13.242 and 13.243: AV malformation over the Rt foot. A girl presented with soft tissue mass over the dorsum of Rt foot and lower ankle. Multiple dilated vessels are seen in the mass. **On color flow imaging** high flow is seen in them.

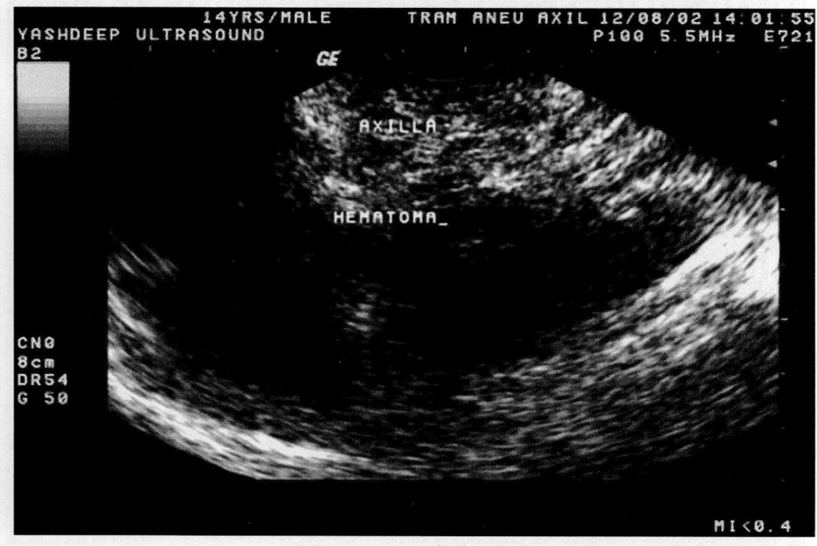

FIGURE 13.244: Traumatic aneurysm Rt axilla. A young child presented progressively increasing soft tissue mass in the Rt axilla after sustaining piercing trauma to the axilla. HRSG shows a big low-level echo complex mass in the axilla.

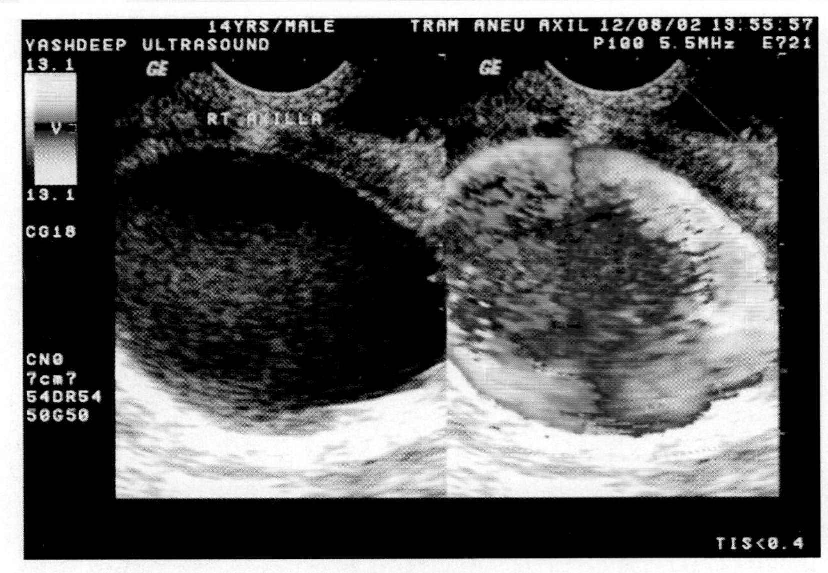

FIGURE 13.245

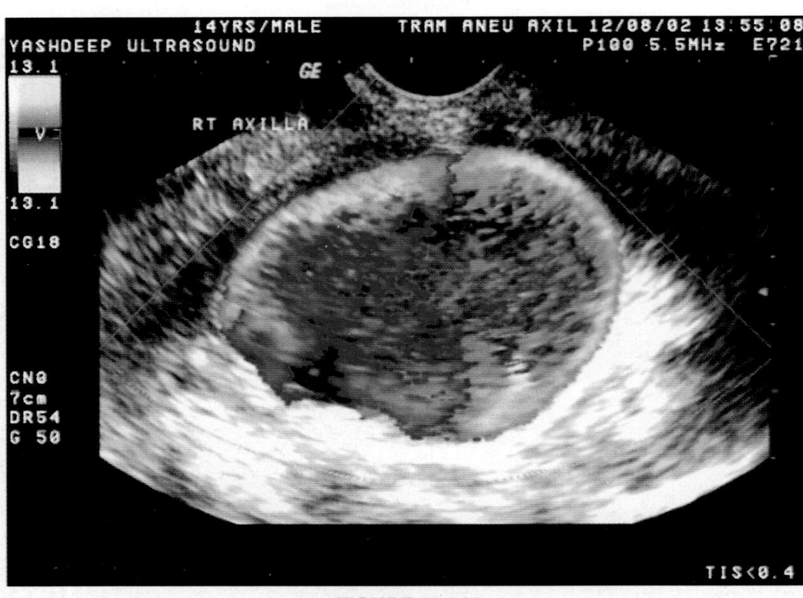

FIGURE 13.246

FIGURES 13.245 and 13.246: On color flow imaging high flow seen in the mass in whirlpool fashion suggestive of aneurysm.

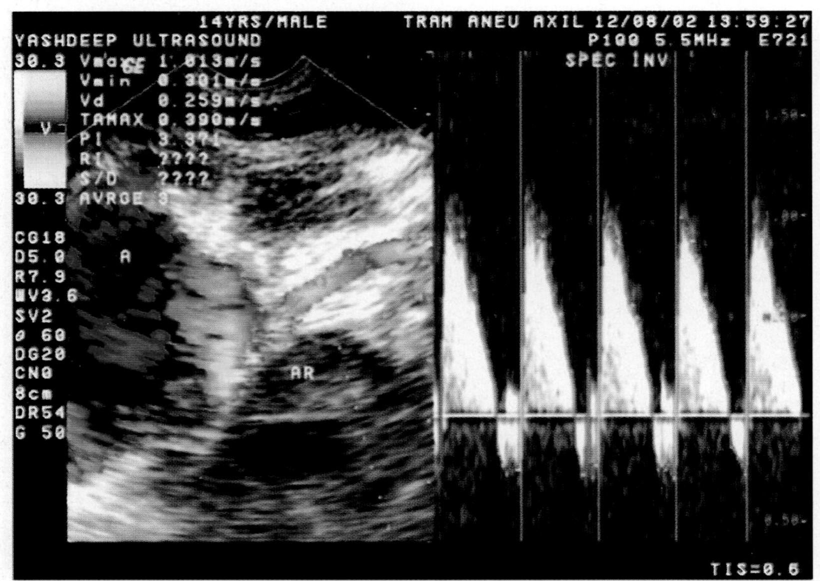

FIGURE 13.247: On spectral Doppler tracing the aneurysm is seen communicating with axillary artery and high flow is seen in the aneurysm. Postoperative findings confirmed traumatic aneurysm of axillary artery.

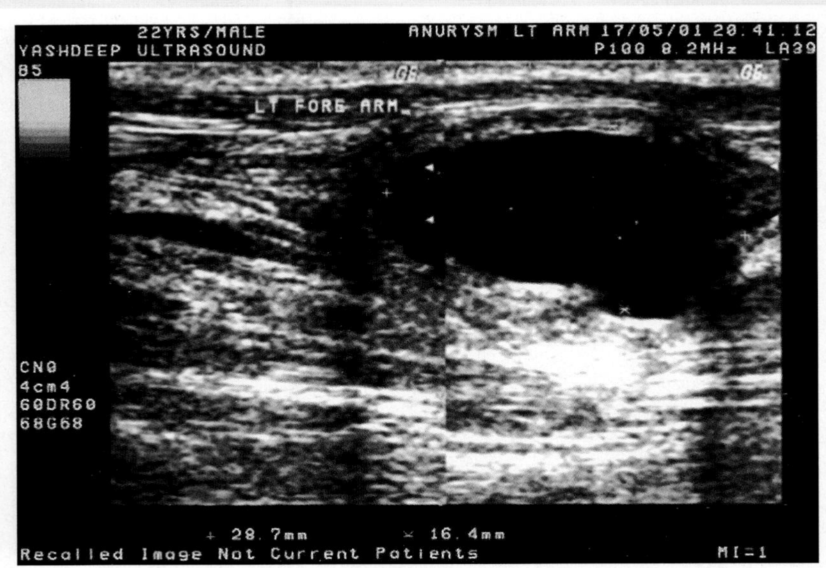

FIGURE 13.248

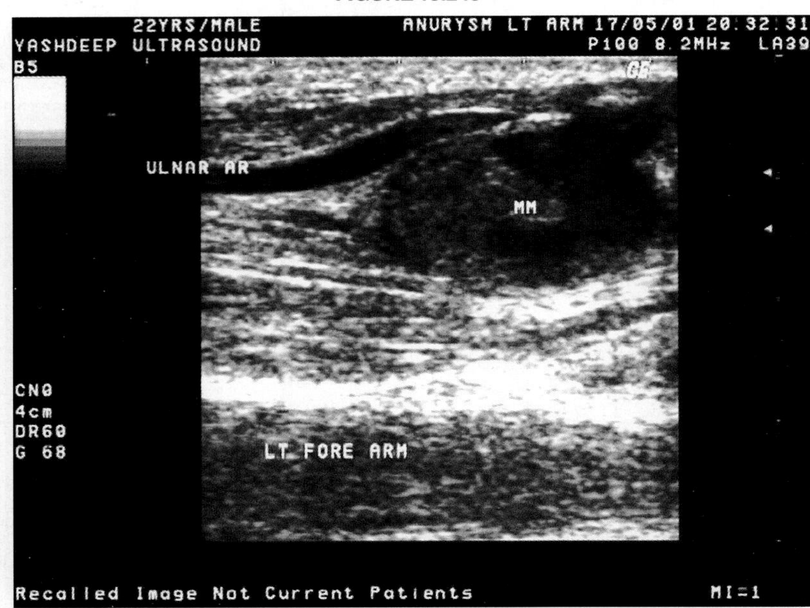

FIGURE 13.249

FIGURES 13.248 and 13.249: Traumatic aneurysm of ulnar artery. A young man presented with soft tissue mass swelling over the distal part of forearm. HRSG shows a cystic mass in the forearm. A feeding vessel is also seen in the mass.

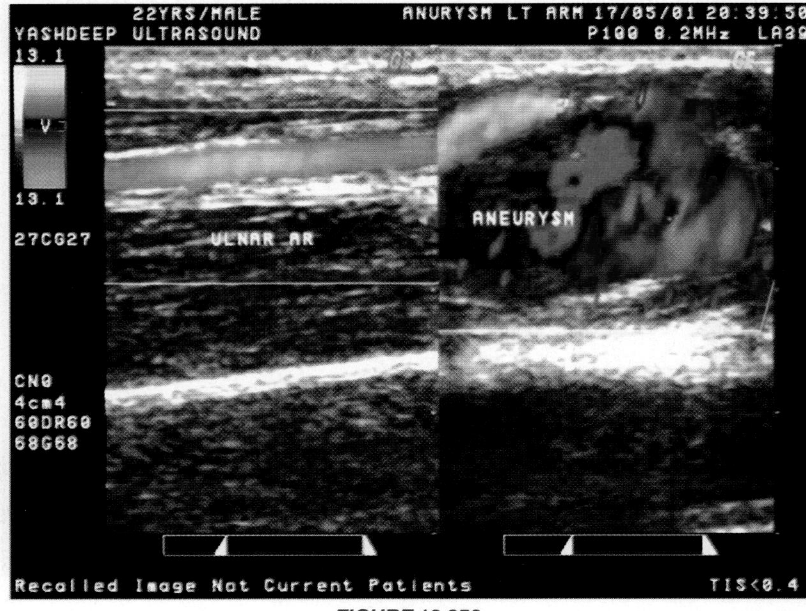

FIGURE 13.250

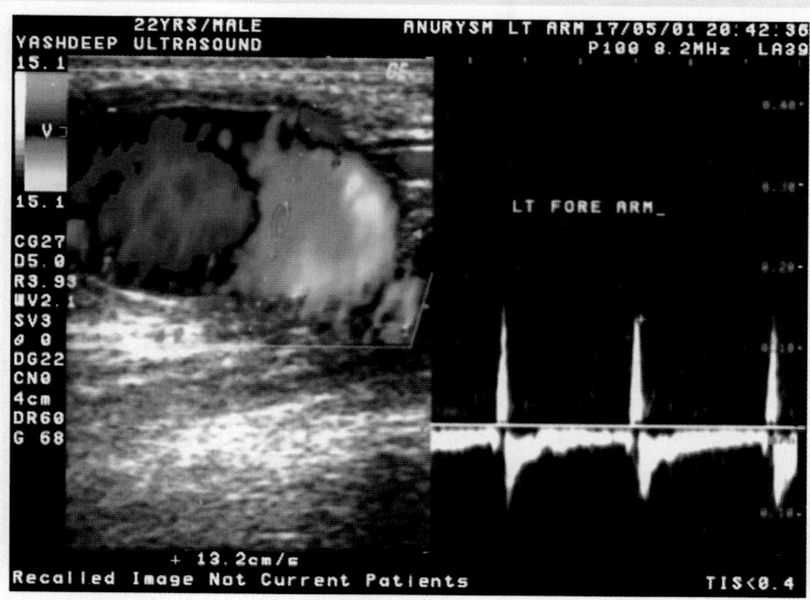

FIGURE 13.251

FIGURES 13.250 and 13.251: On color flow imaging the feeding vessels was the ulnar artery and blood flow is seen in the mass in bizarre fashion suggestive of an aneurysm of the ulnar artery.

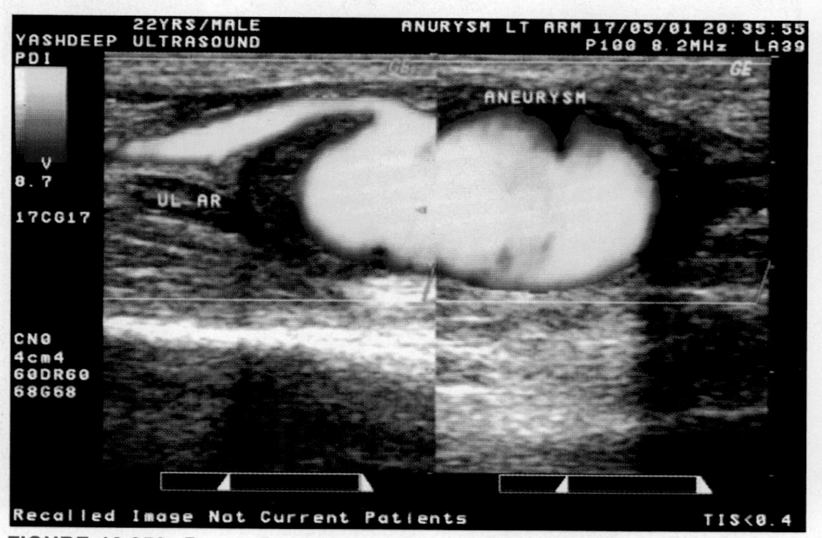

FIGURE 13.252: Power Doppler imaging shows the aneurysm in the feeding vessels very clearly.

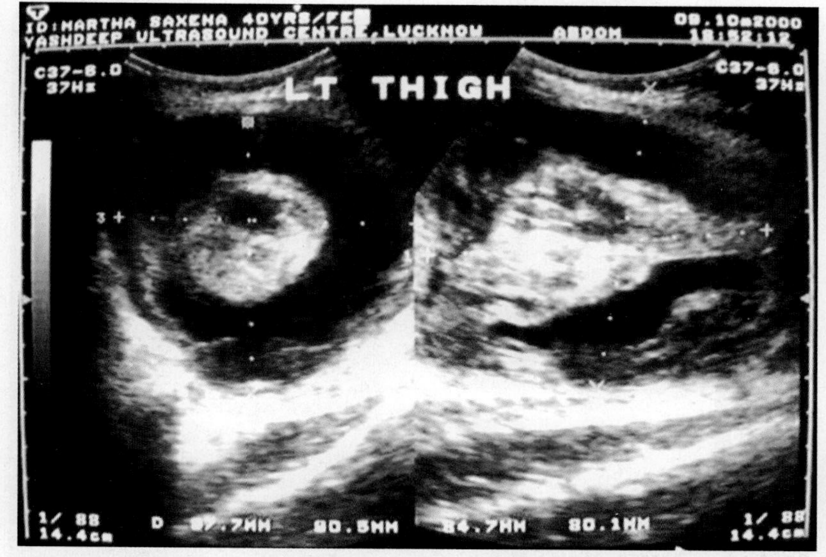

FIGURE 13.253: Aneurysm in the Lt thigh. A patient presented with a semi hard mass in the mid thigh region. HRSG shows a heterogeneous mass in the mid thigh region. It is having a dense echogenic calcified mass in the center. Outer hypoechoic area is seen in it.

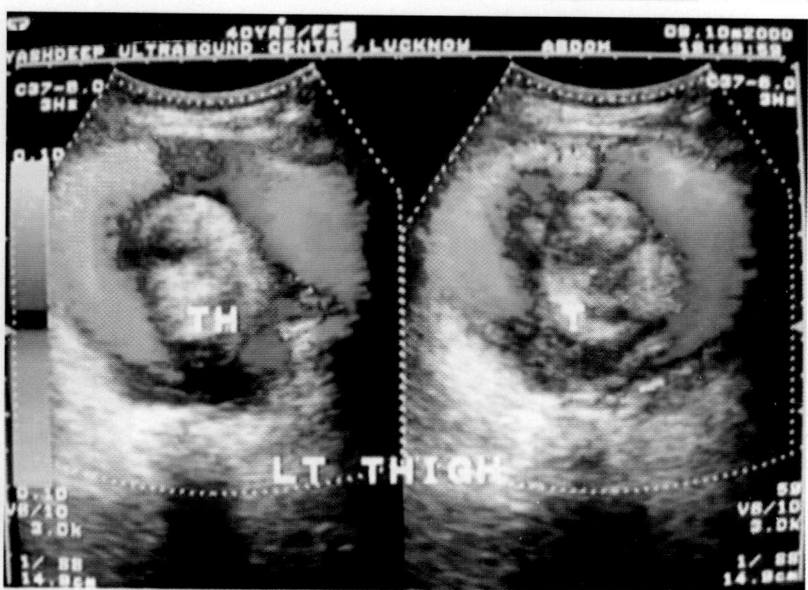

FIGURE 13.254

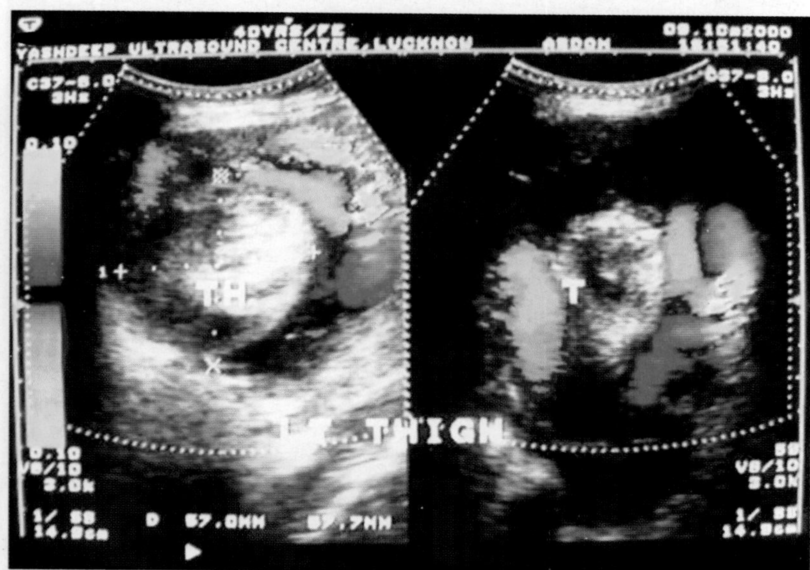

FIGURE 13.255

FIGURES 13.254 and 13.255: On color Doppler imaging the outer hypoechoic region shows blood flow in whirlpool fashion. Echogenic center area does not show any flow suggestive of thrombus. It is the aneurysm of superficial femoral artery. The aneurysm shows central area of thrombosis with echogenic thrombus in the center.

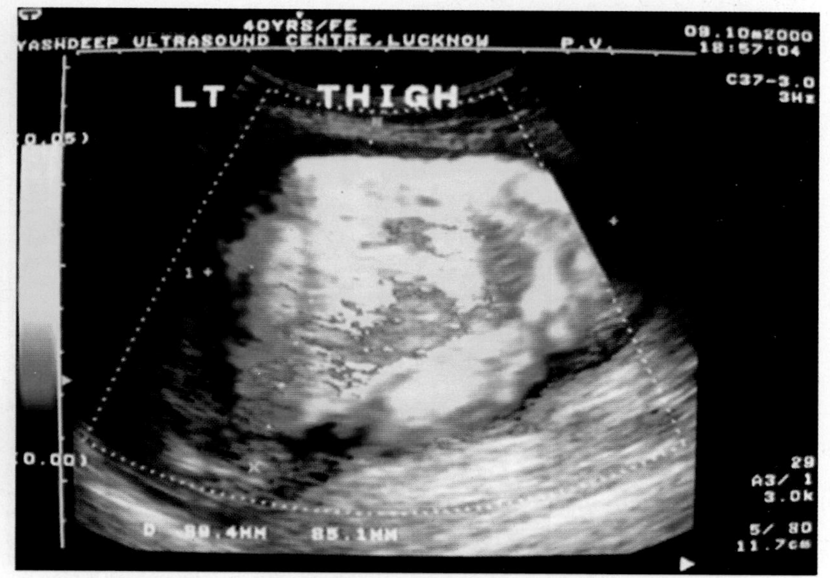

FIGURE 13.256: Power Doppler imaging shows high flow in the aneurysm wall.

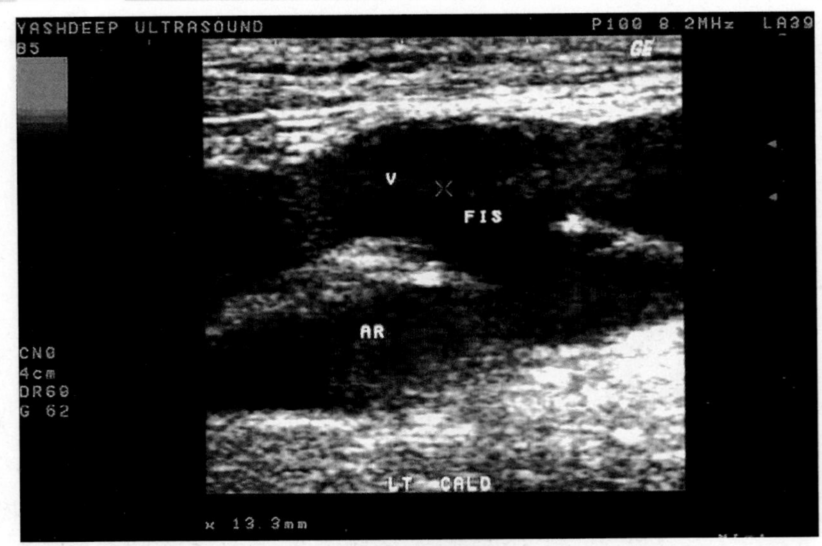

FIGURE 13.257

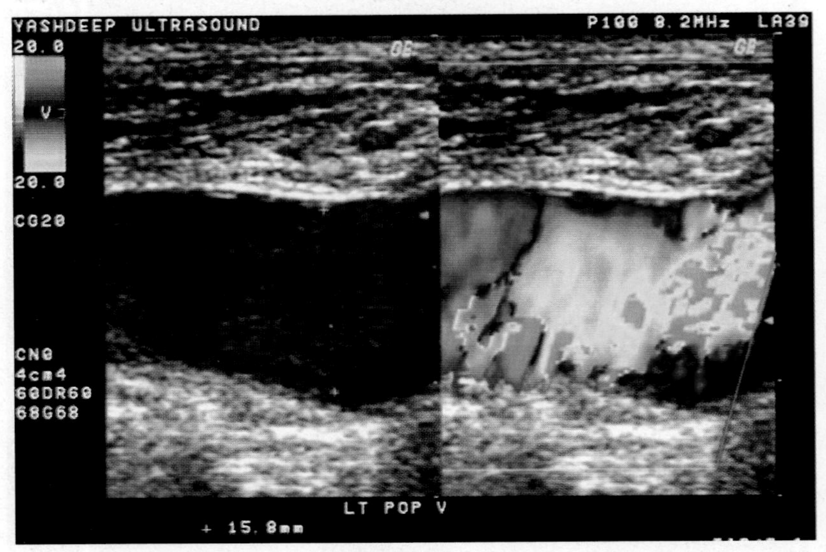

FIGURE 13.258

FIGURES 13.257 and 13.258: Arteriovenous fistula in the thigh. A patient presented with pulsatile swelling in the Lt thigh in its lower part reaching up to the popliteal fossa. HRSG shows well-defined vessels. They are dilated. There is also evidence of communication seen between the two vessels. On color Doppler imaging high flow is seen in the vein with pulsatile flow pattern. The flow is so high that it caused aliasing suggestive of arteriovenous fistula between superficial femoral artery and popliteal vein after trauma. The cause of the fistula was the trauma.

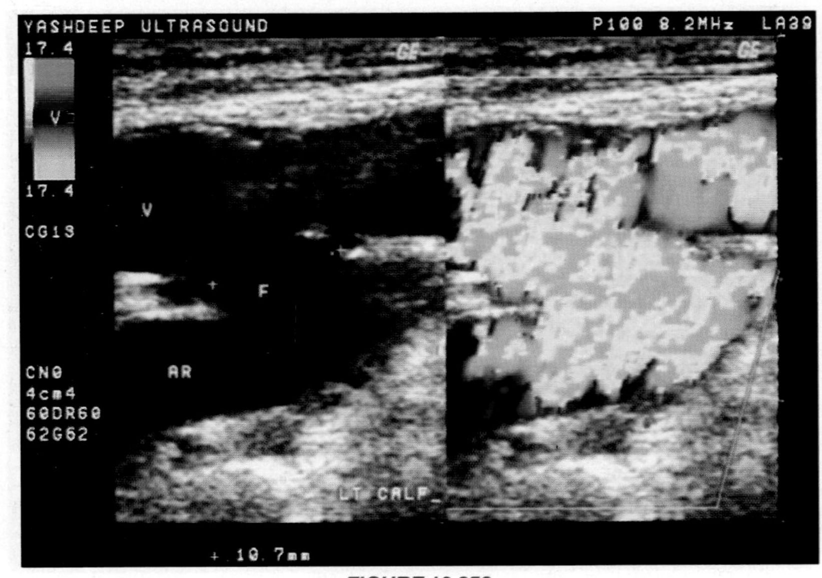

FIGURE 13.259

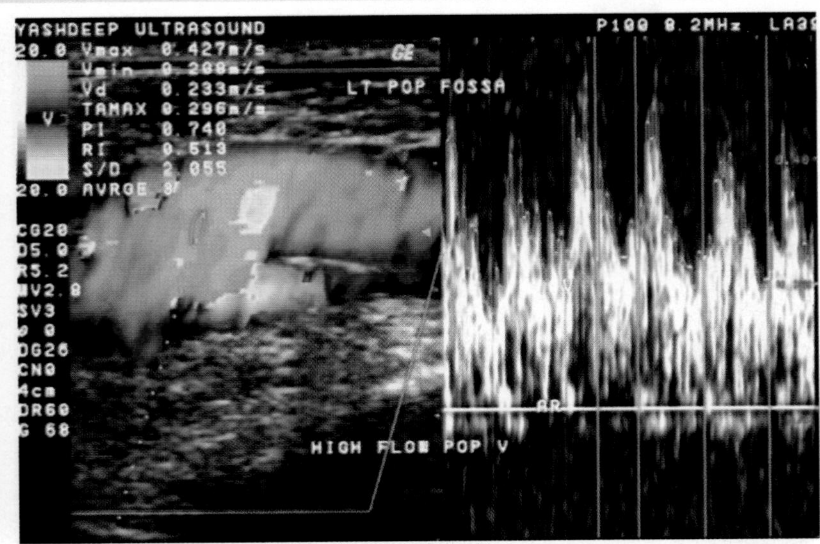

FIGURE 13.260

FIGURES 13.259 and 13.260: On spectral Doppler tracing high flow is seen in the fistula with marked aliasing.

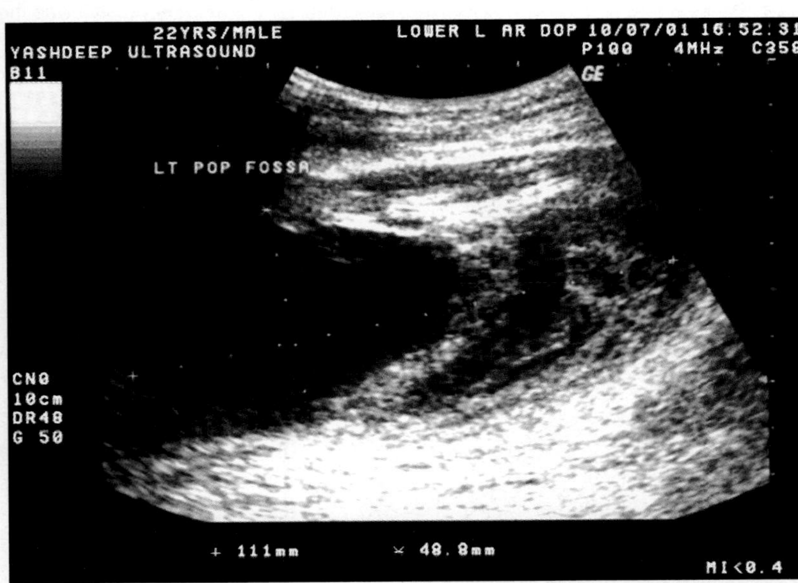

FIGURE 13.261: Popliteal aneurysm. A patient presented with solid mass over the Lt popliteal fossa. HRSG shows a solid cystic mass in the popliteal fossa.

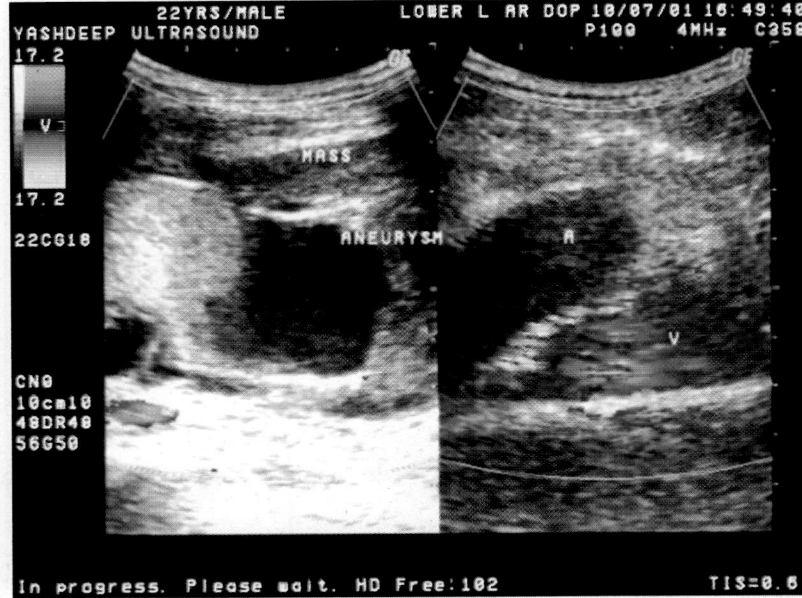

FIGURE 13.262

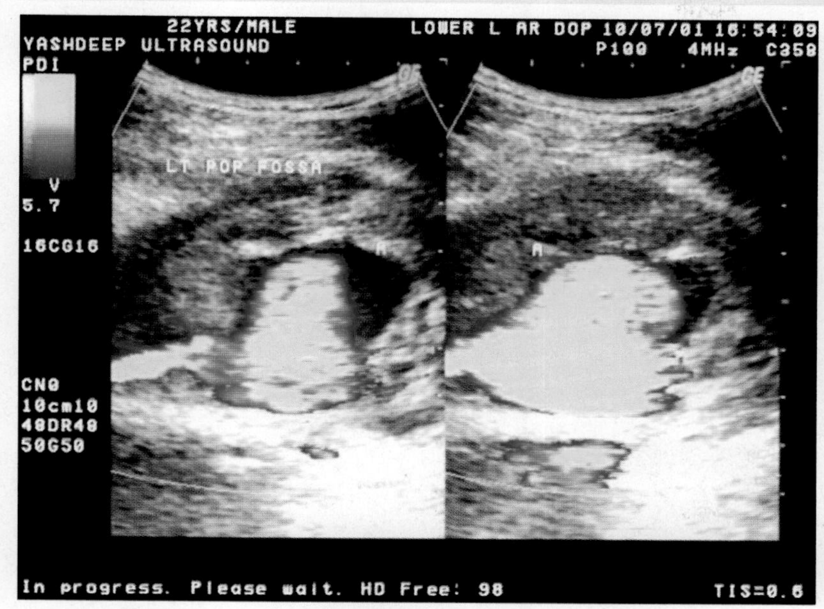

FIGURE 13.263

FIGURES 13.262 and 13.263: On color Doppler imaging the mass shows irregular blood flow. Half of the cavity of the mass is filled with echogenic focus and no flow is seen in it suggestive of thrombosis. On Doppler findings the mass turn out to be a popliteal artery aneurysm. There is also evidence of dissection of the wall seen.

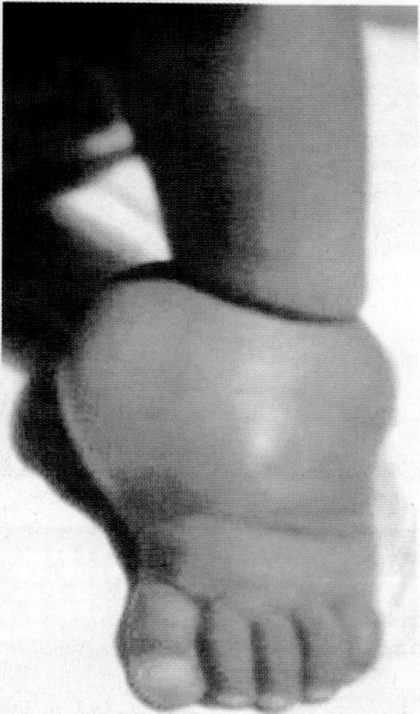

FIGURE 13.264: **Lymphangioma.** A young child presented with soft tissue mass over the leg and dorsum of the foot.

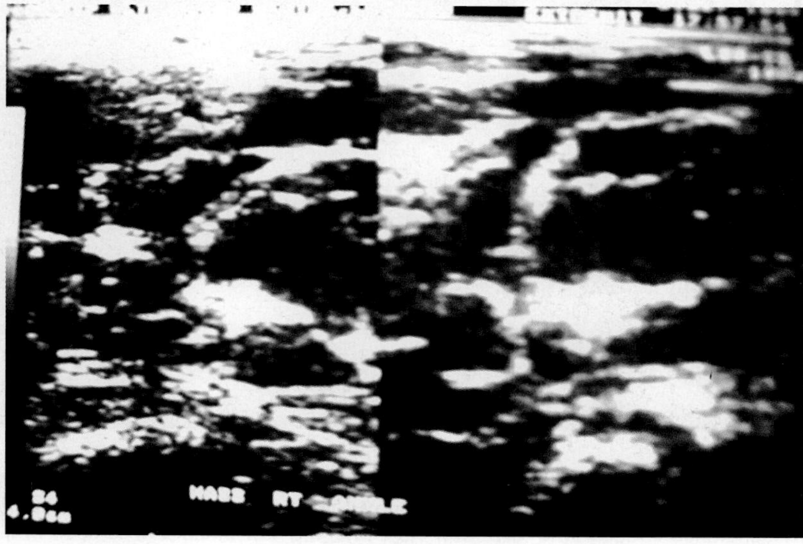

FIGURE 13.265

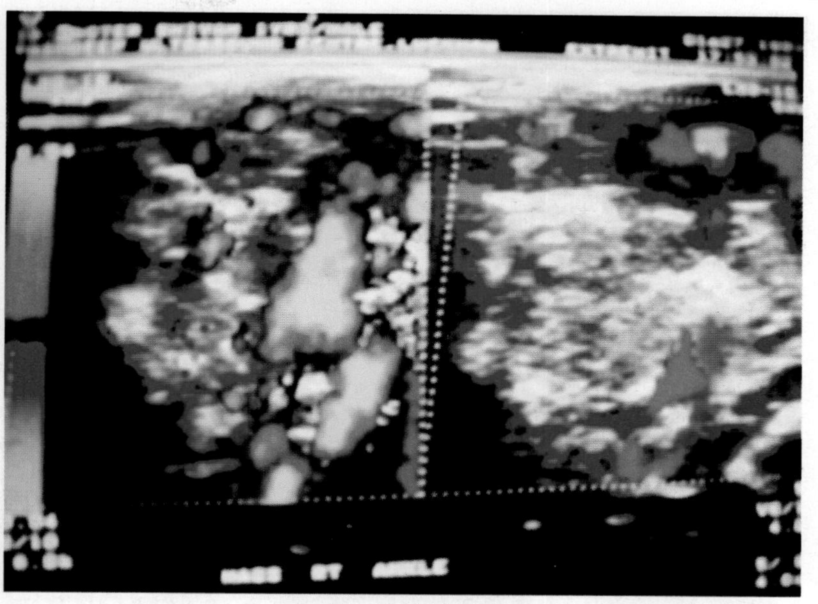

FIGURE 13.266

FIGURES 13.265 and 13.266: HRSG shows multiple dilated vessels in the mass. The mass shows echogenic septa. Few echogenic shadows are also seen in it suggestive of debris. Color Doppler imaging shows little flow in the mass. However, power Doppler imaging shows better flow. Sonographic findings are suggestive of lymphangioma.

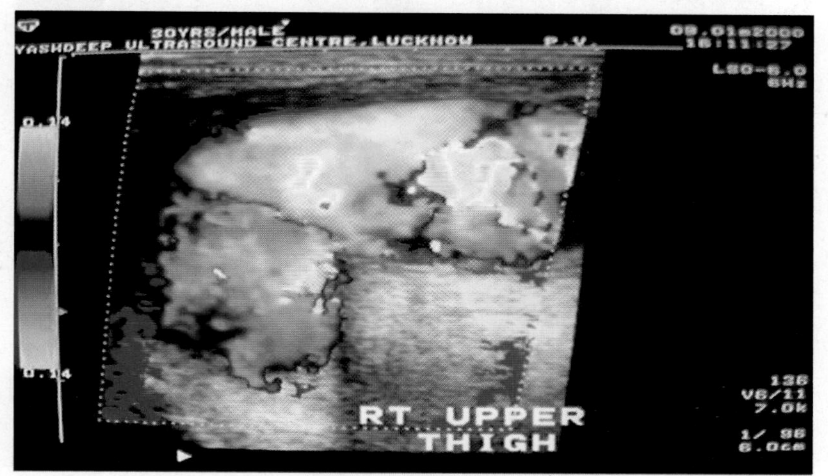

FIGURE 13.267

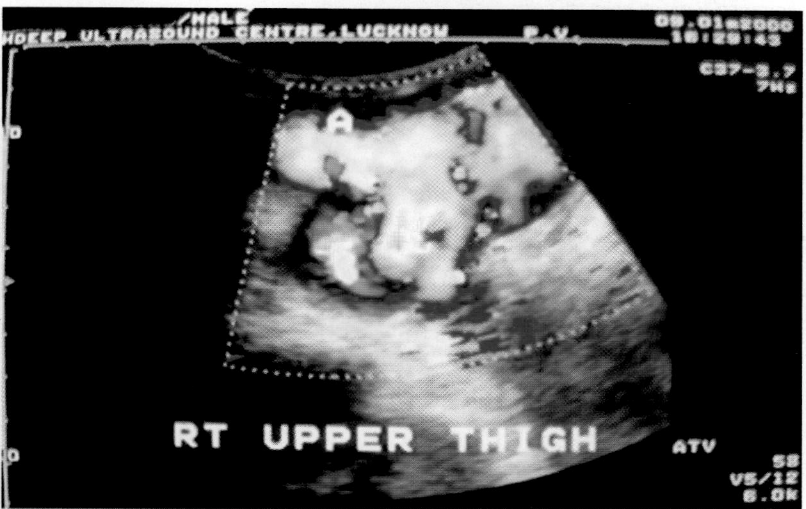

FIGURE 13.268

FIGURES 13.267 and 13.268: **AV fistula with aneurysm in the Rt upper thigh.** A patient presented with a pulsatile mass in Rt upper thigh. Color flow imaging of the mass shows a big aneurysm involving the common femoral artery and also part of proximal superficial femoral artery. There is also evidence of fistular communication seen between common femoral artery and vein.

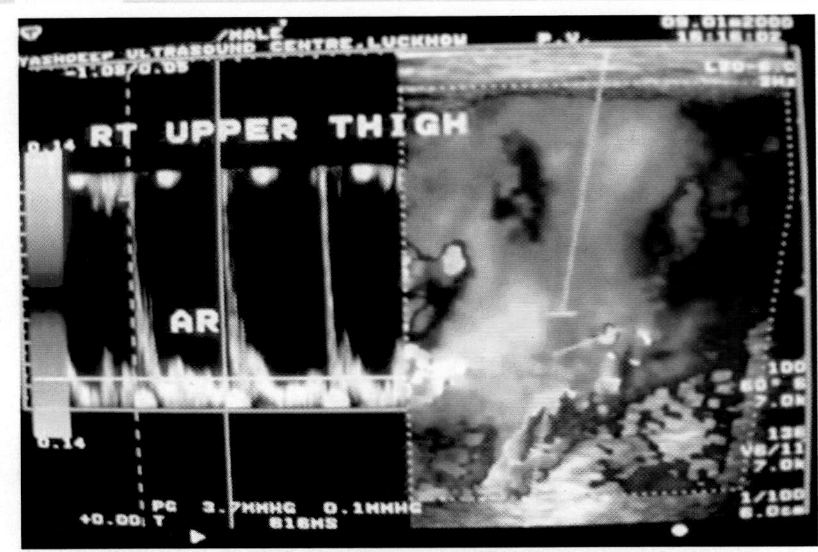

FIGURE 13.269

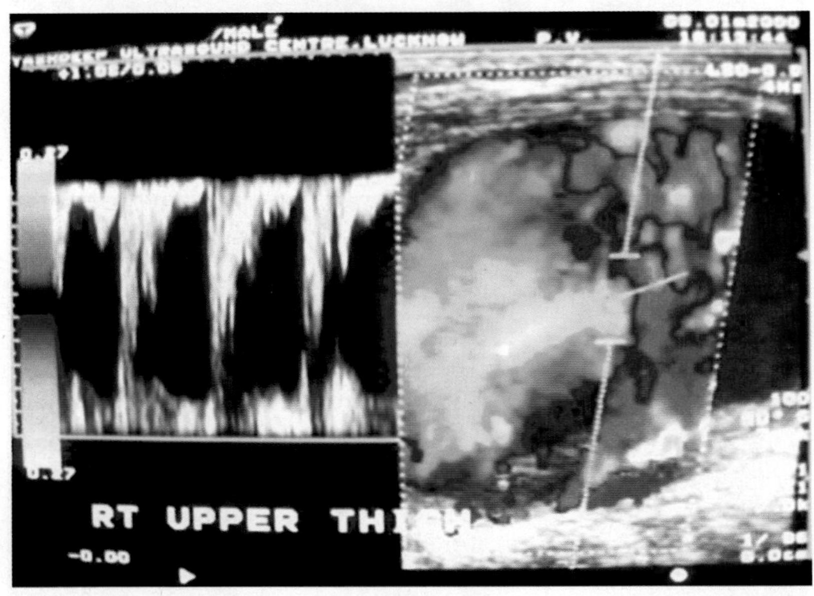

FIGURE 13.270

FIGURES 13.269 and 13.270: On Doppler spectral tracing arterial and venous flow is seen in the fistula with mixing of the blood. Pulsatile is seen in the dilated common femoral vein.

Tumors

Figures 13.271 to 13.336.

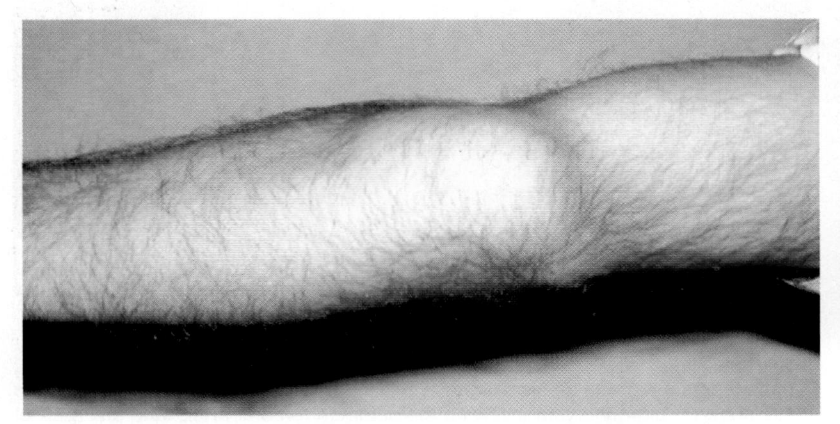

FIGURE 13.271: A middle aged man presented with well-defined mass over the elbow.

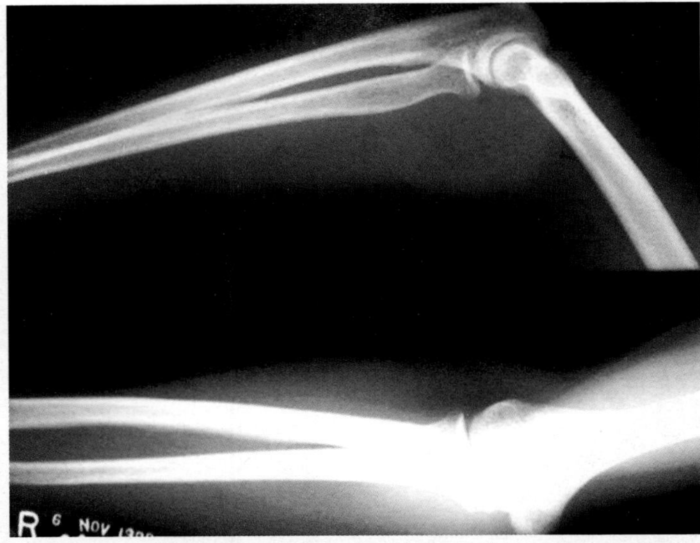

FIGURE 13.272: Plain X-ray of the arm shows soft tissue mass near the elbow. However, bones are normal.

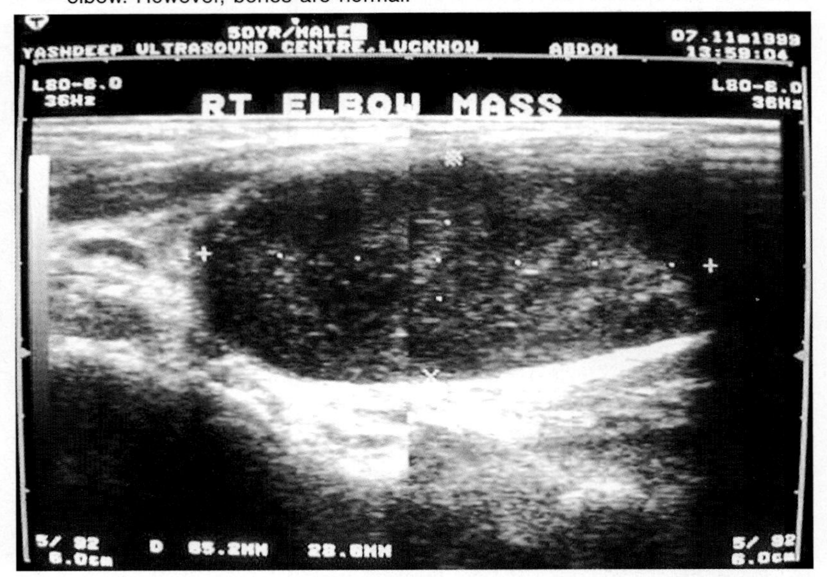

FIGURE 13.273

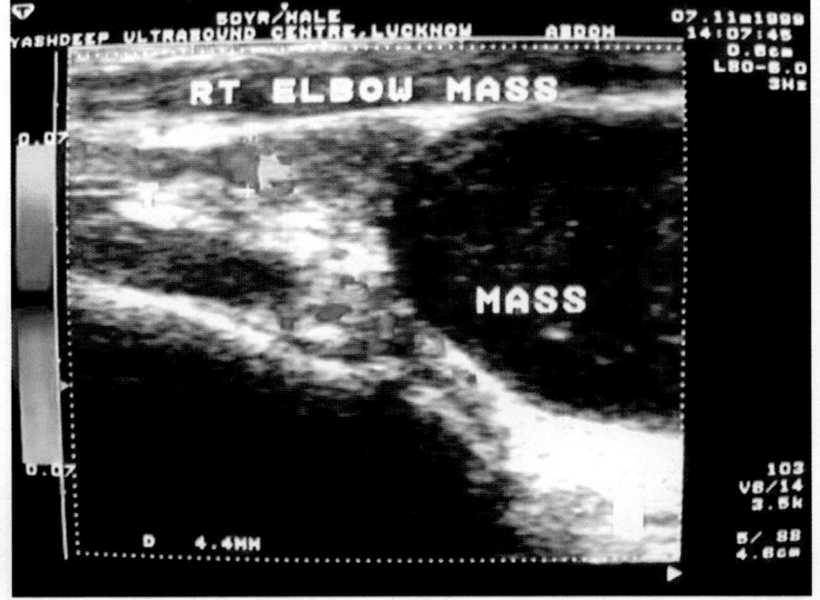

FIGURE 13.274

FIGURES 13.273 and 13.274: HRSG shows well-defined, well and capsulated lobulated mass in the belly of the muscle. It is homogenous and hypoechoic in texture. No cystic degeneration is seen in it. On color flow imaging low blood flow is seen in it. Biopsy shows fibroma.

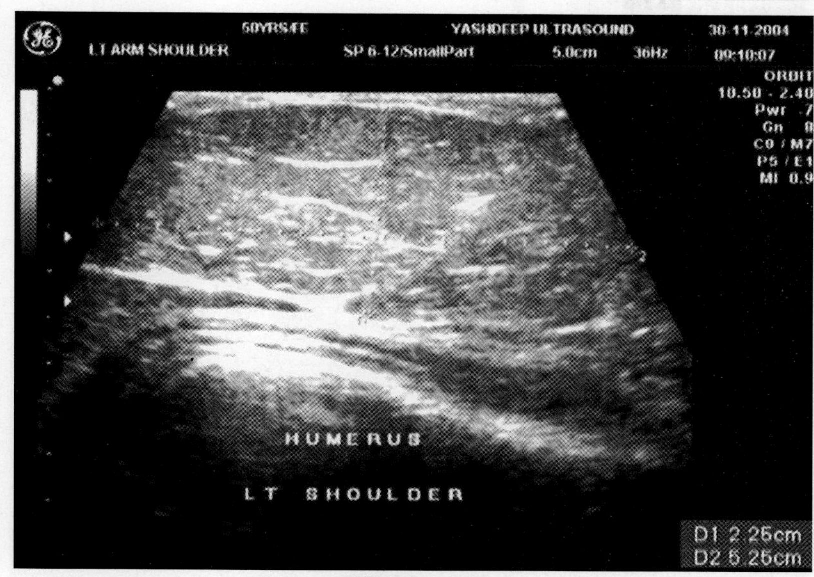

FIGURE 13.275

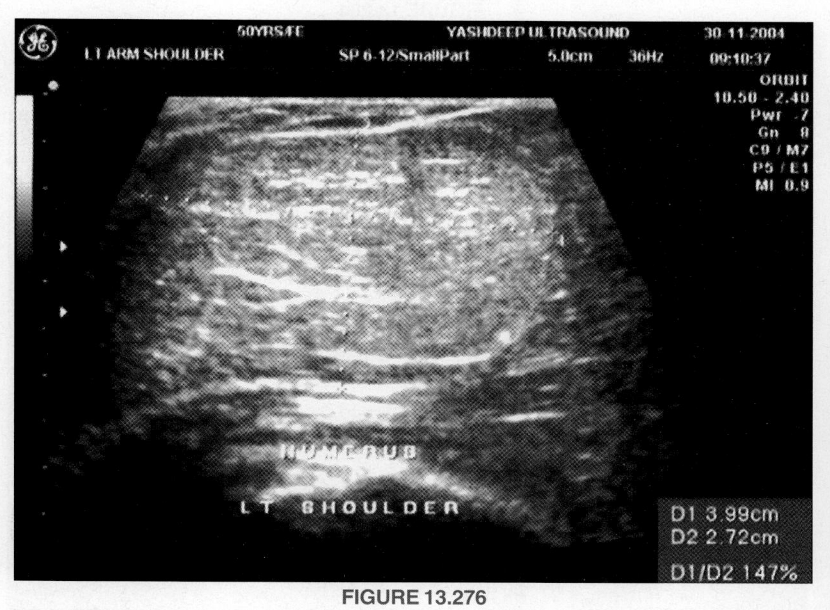

FIGURE 13.276

FIGURES 13.275 and 13.276: Big lipoma over the Lt shoulder. A lady presented with soft tissue mass over the Lt shoulder. HRSG shows a big echogenic soft tissue mass over the shoulder. It is well-defined with sharp margins. It is highly echogenic in texture. No necrosis or calcification seen in it.

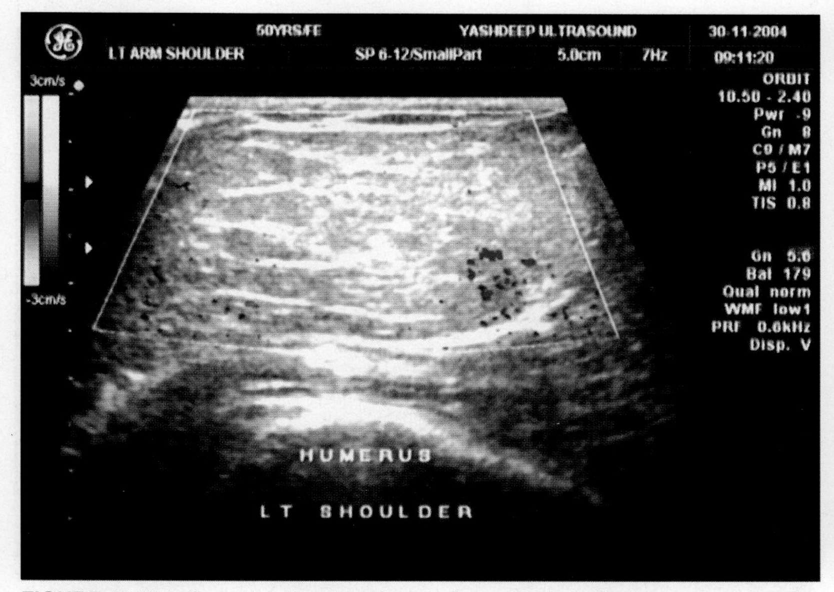

FIGURE 13.277: On color flow imaging no flow seen in it. Findings are suggestive of lipoma.

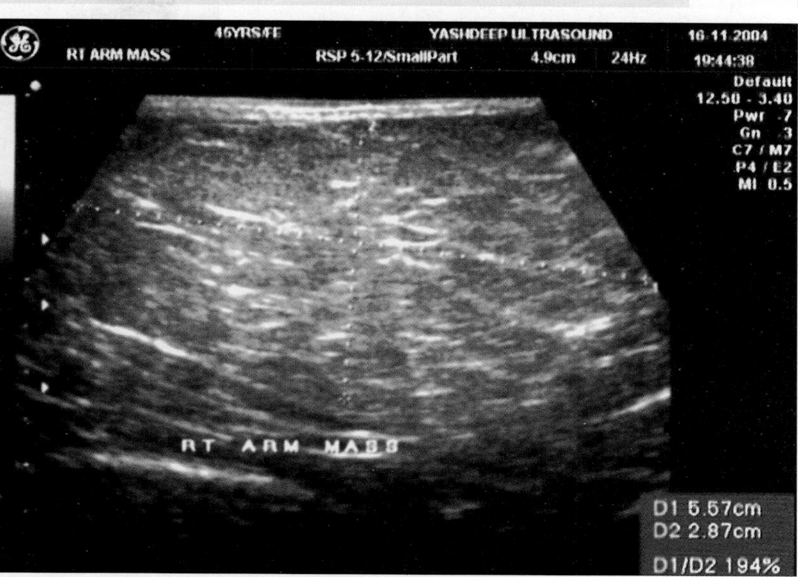

FIGURE 13.278: Big Lipoma over the Rt arm. HRSG shows a big well-defined echogenic mass in the Rt arm. It is homogeneous in texture. It is seen anterior to the humeral shaft suggestive of big lipoma.

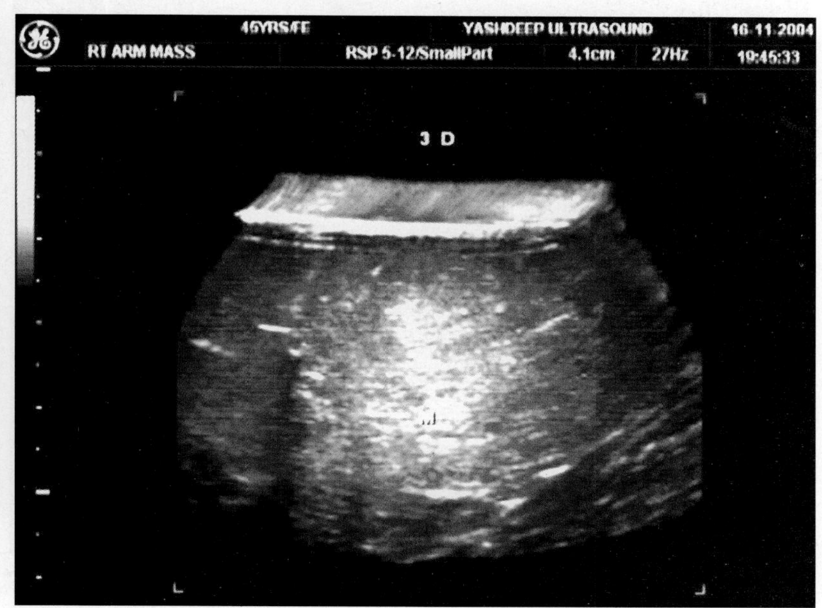

FIGURE 13.279

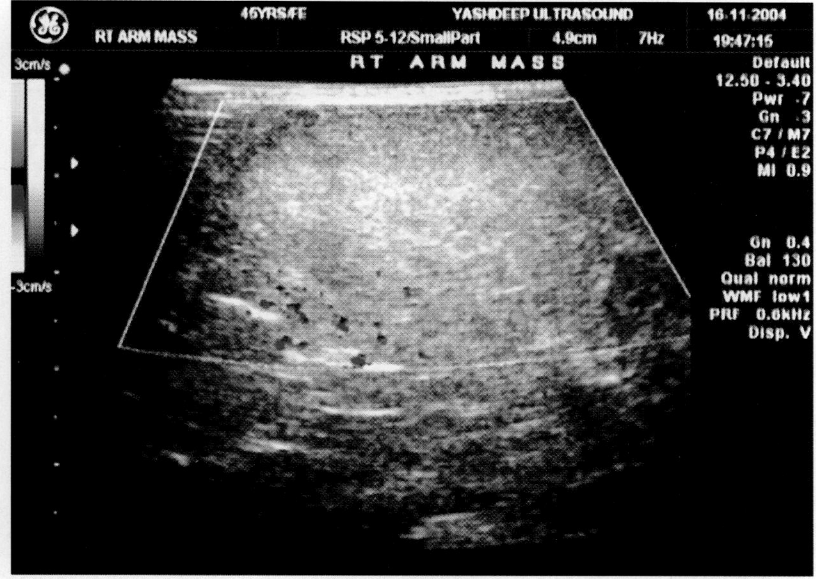

FIGURE 13.280

FIGURES 13.279 and 13.280: 3D imaging of the mass shows that it is confined to the anterior compartment of the muscle. **On color flow imaging** no flow was seen in the mass.

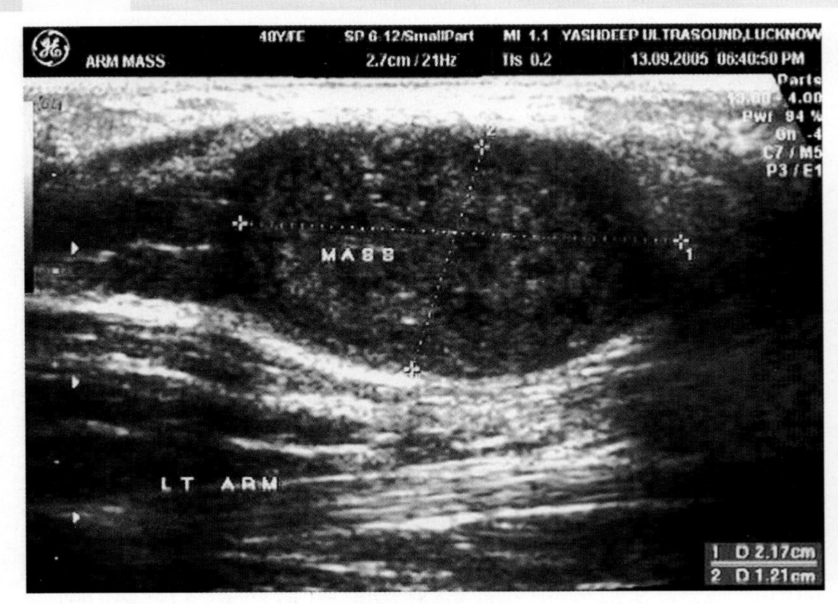

FIGURE 13.281

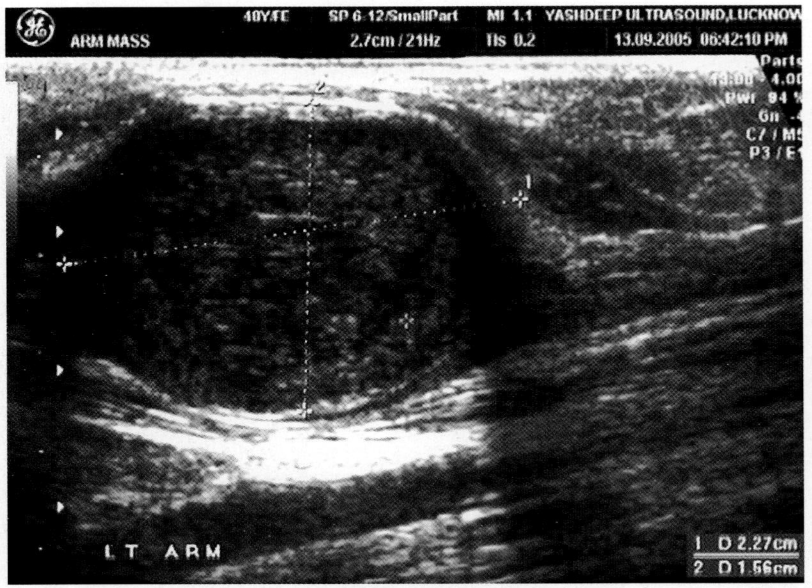

FIGURE 13.282

FIGURES 13.281 and 13.282: Benign muscle tumor. A well-defined homogeneous soft tissue mass seen in the Lt arm. HRSG shows well encapsulated soft tissue mass in the muscle. It is homogeneous in texture. No calcification is seen in it.

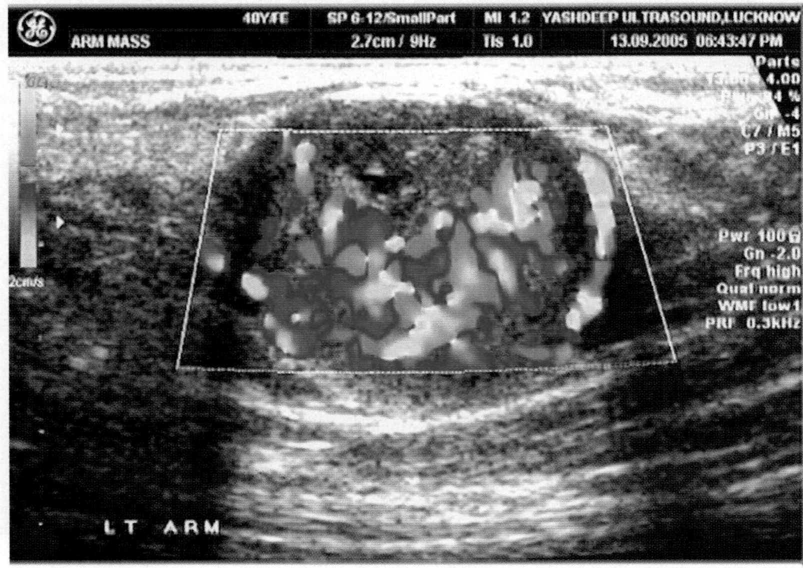

FIGURE 13.283

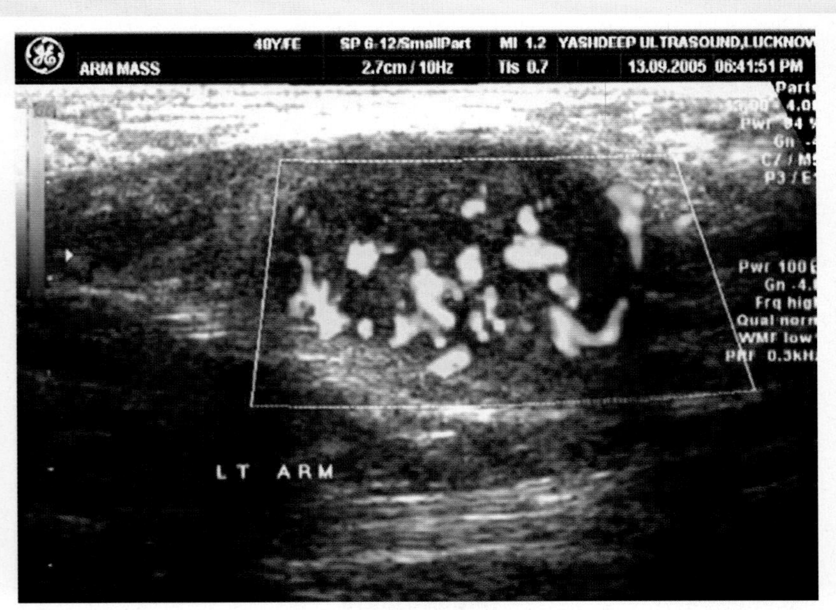

FIGURE 13.284

FIGURES 13.283 and 13.284: On color flow imaging high flow is seen in the mass. Biopsy of the mass confirmed fibroma.

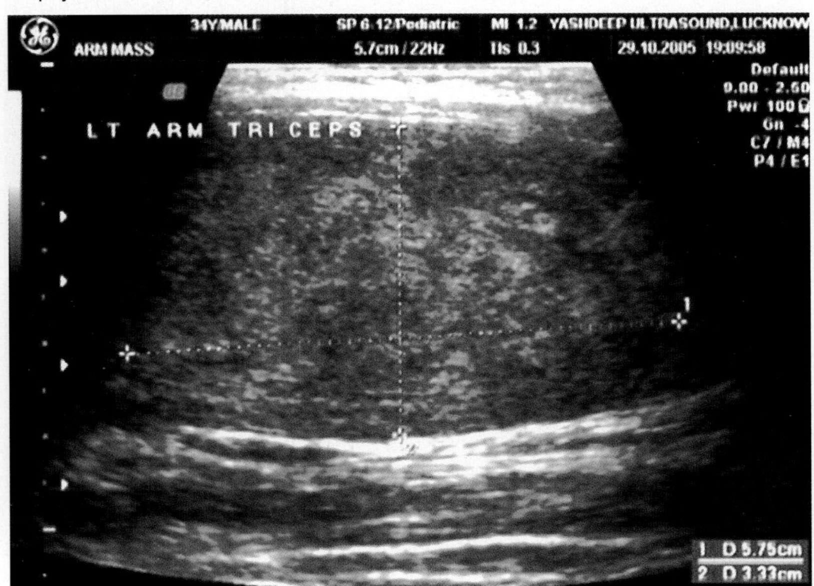

FIGURE 13.285

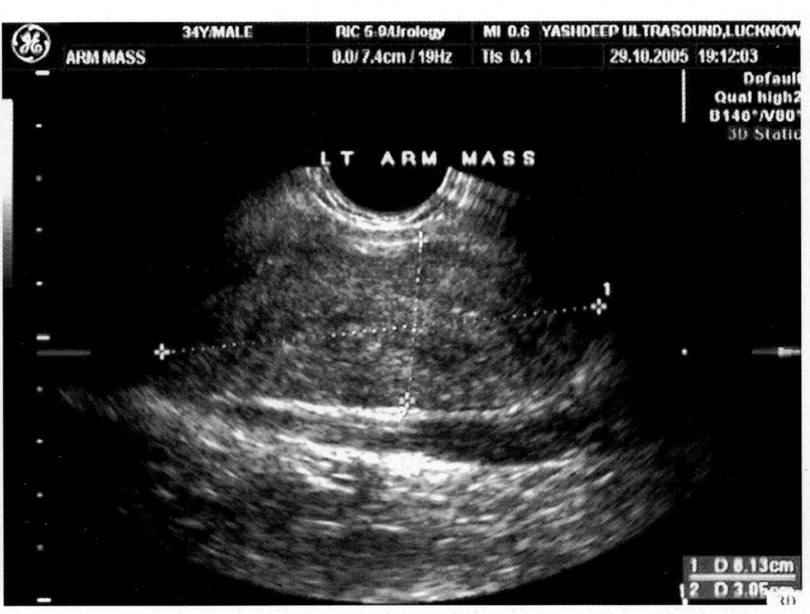

FIGURE 13.286

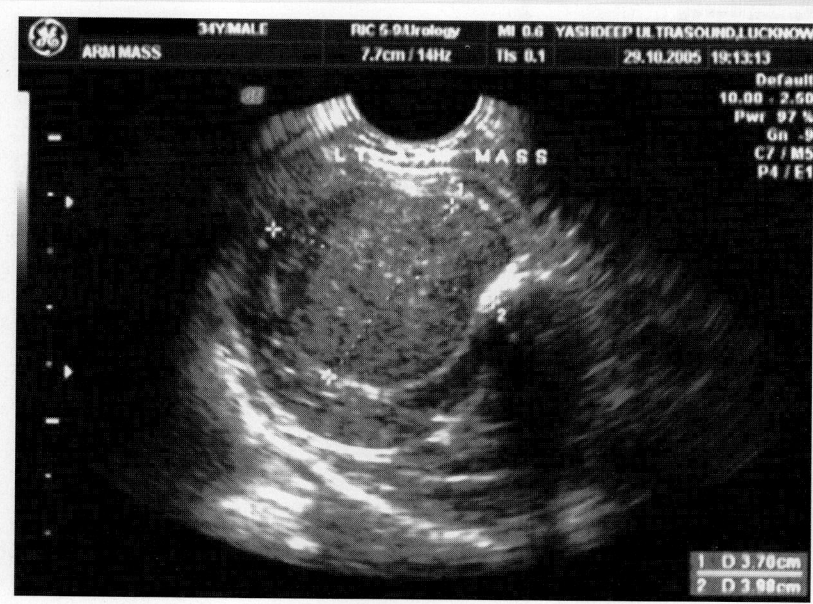

FIGURE 13.287

FIGURES 13.285 to 13.287: Benign tumor in the muscle. A patient presented with soft tissue mass in the triceps muscle in the Lt arm. HRSG shows homogeneous soft tissue mass in the muscle belly.

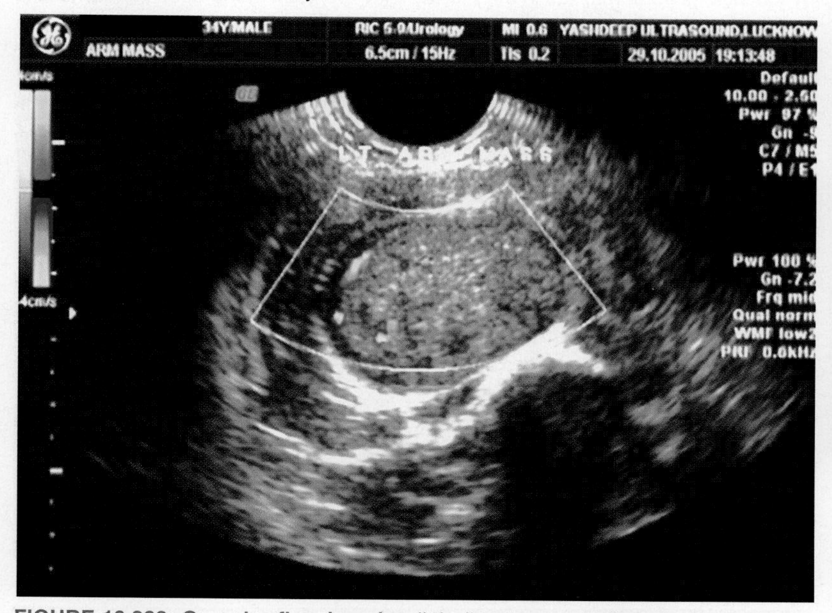

FIGURE 13.288: On color flow imaging little flow is seen in the mass.

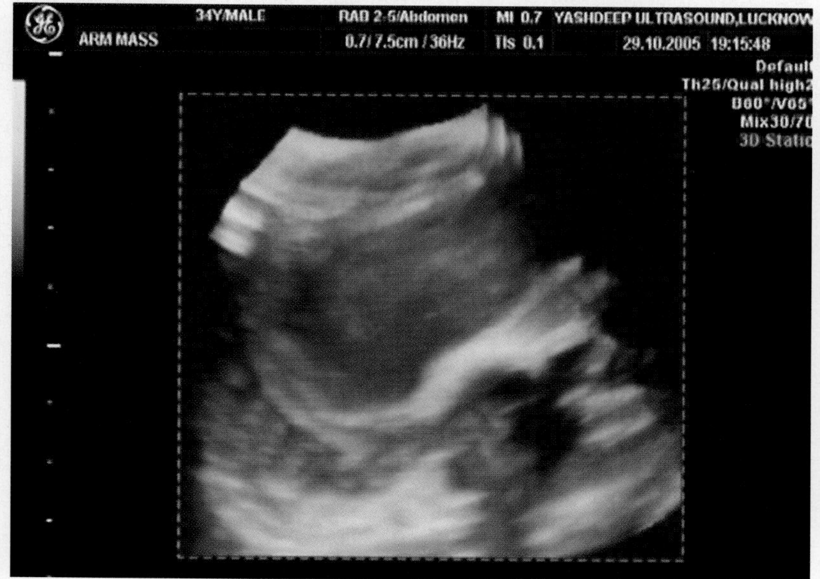

FIGURE 13.289: 3D imaging of the mass shows depth of the mass in the muscle.

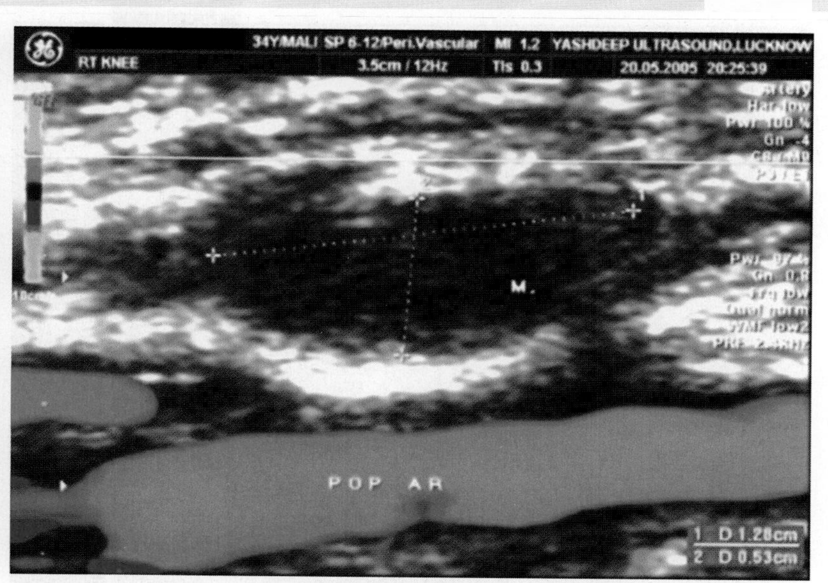

FIGURE 13.290

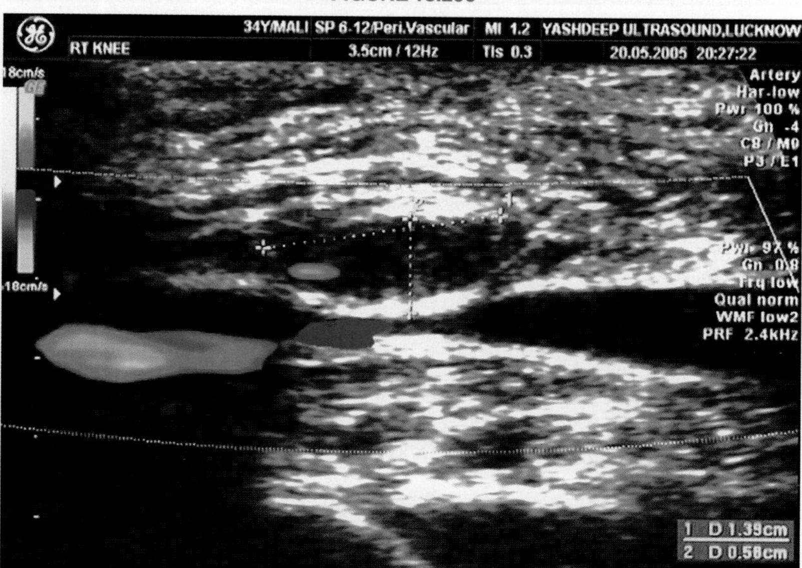

FIGURE 13.291

FIGURES 13.290 and 13.291: Nerve cell tumor. A patient presented with highly sensitive nodular shadow in the popliteal fossa. HRSG shows a well-defined homogeneous fusiform hypoechoic mass in relation to the popliteal nerve. It is seen anterior to the vessels. Capsule is highly echogenic. Biopsy of the mass confirmed popliteal nerve shwannoma.

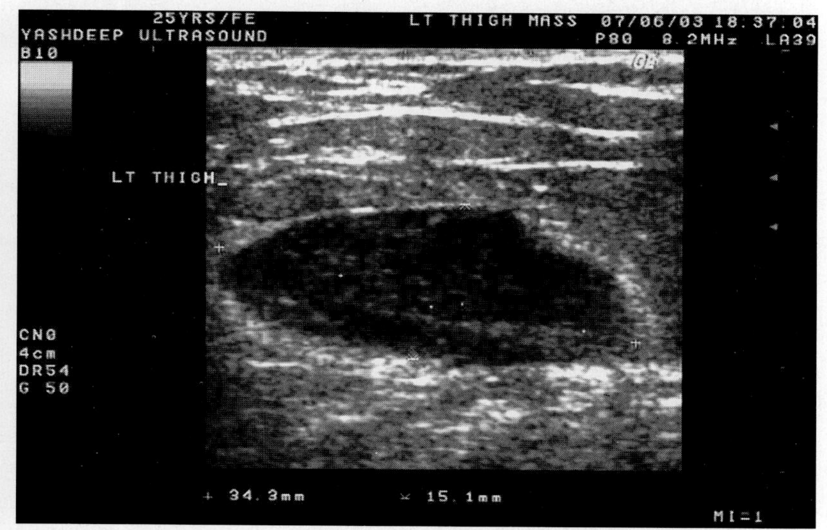

FIGURE 13.292: Nerve cell tumor. Well-defined hypoechoic nodular masses are seen in the mid thigh region in a patient, which was highly painful on clinical examination. HRSG shows homogeneous hypoechoic mass.

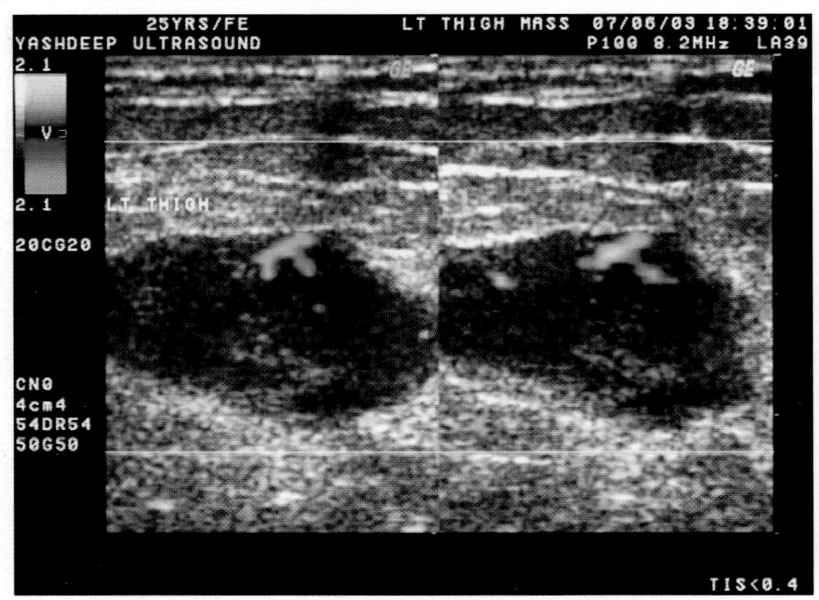

FIGURE 13.293

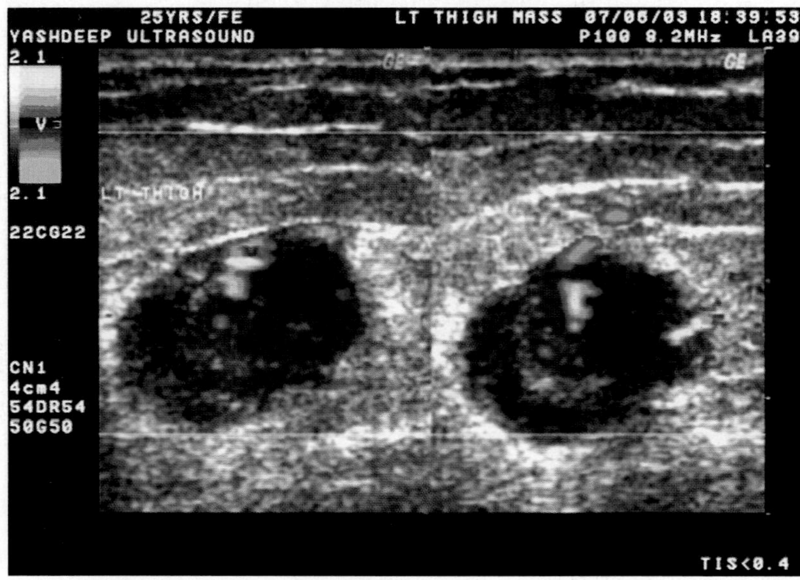

FIGURE 13.294

FIGURES 13.293 and 13.294: On color flow imaging small vessels was seen feeding the mass. Biopsy of the mass confirmed nerve cell tumor.

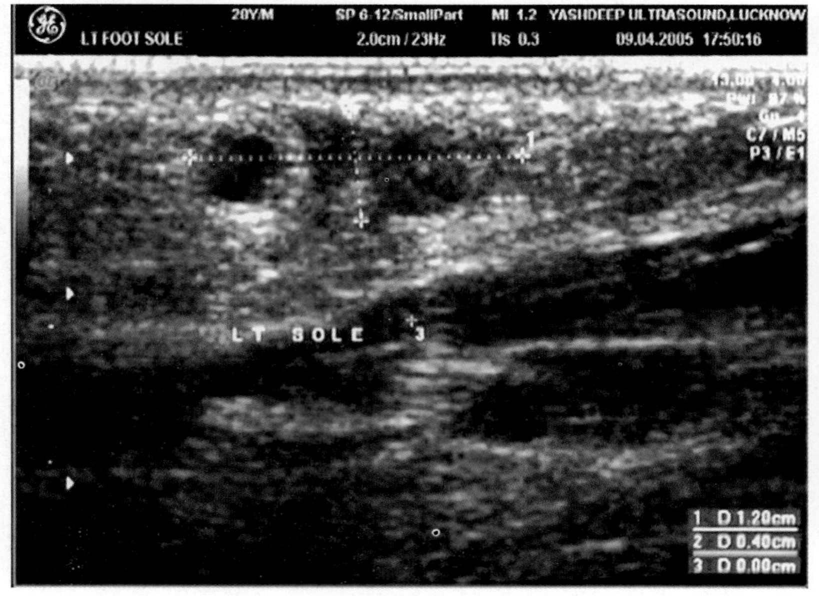

FIGURE 13.295

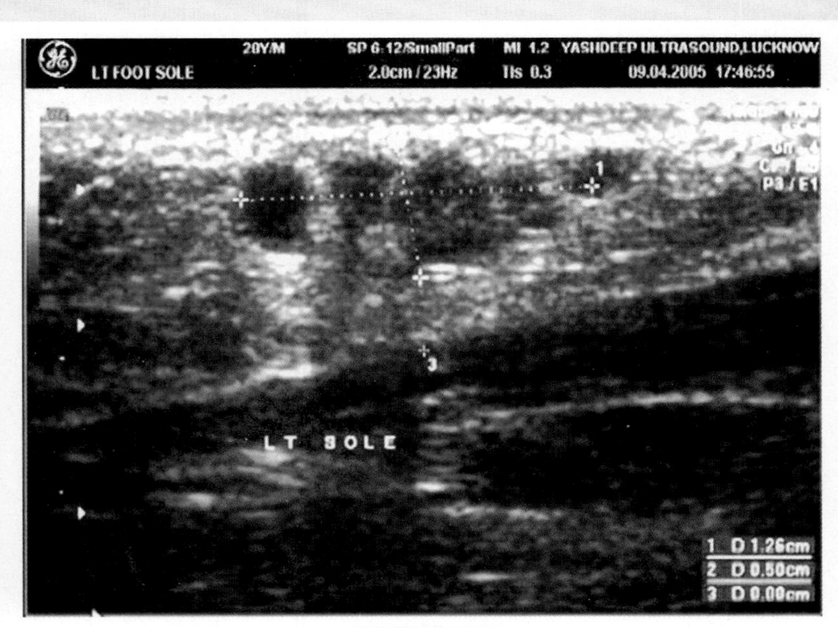

FIGURE 13.296

FIGURES 13.295 and 13.296: Hamartoma in the sole. A patient presented with irregular nodular mass in the sole on the plantar fascia. HRSG shows multiple hypoechoic nodular masses in the fascia. They are seen in the deep plantar fascia. Few nodules show cystic degeneration.

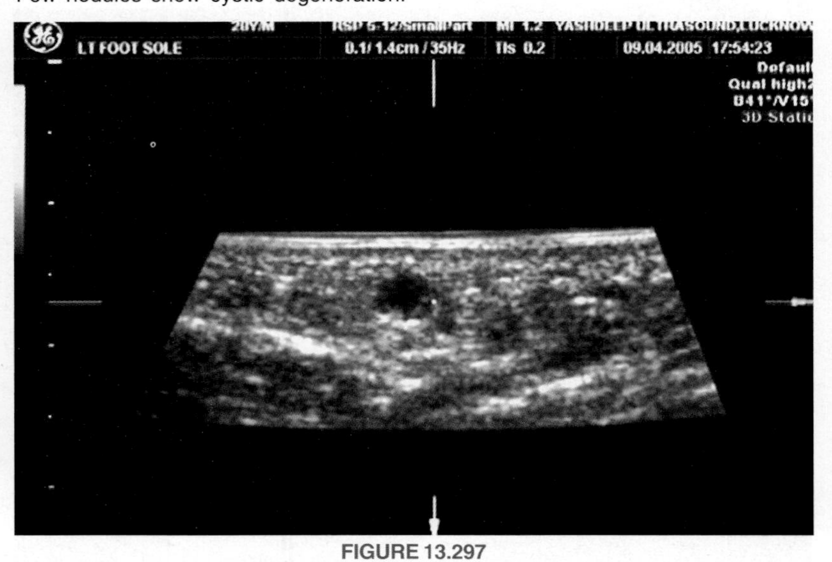

FIGURE 13.297

FIGURE 13.298

FIGURES 13.297 and 13.298: 3D imaging of the nodules shows better clinical details of the nodules.

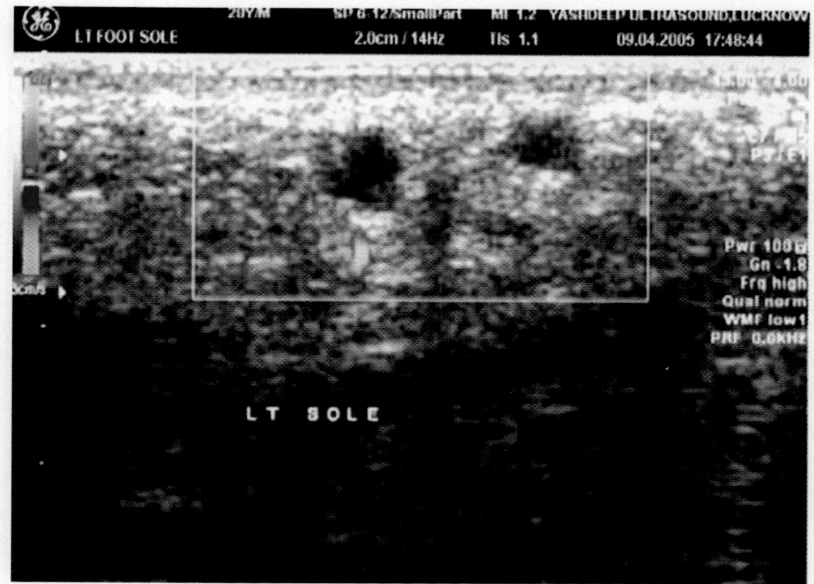

FIGURE 13.299

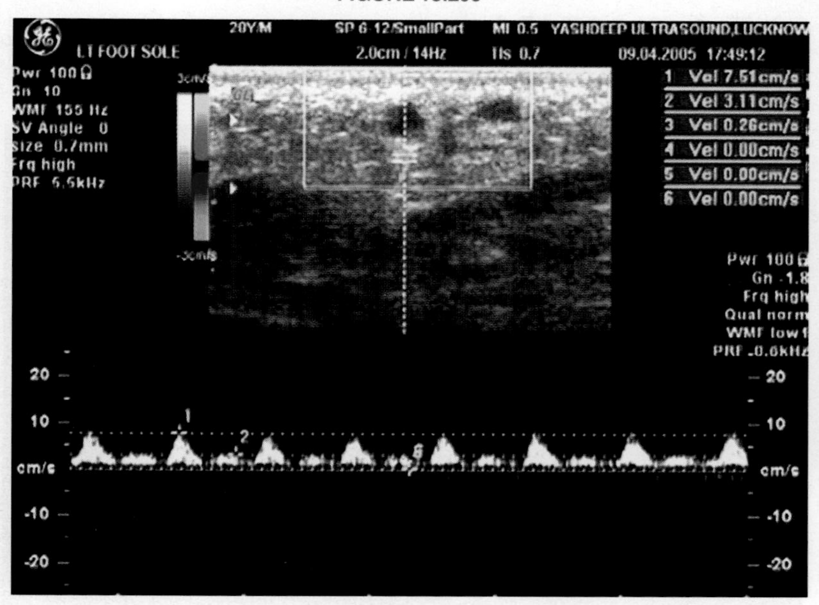

FIGURE 13.300

FIGURES 13.299 and 13.300: On color flow imaging poor flow seen in them. Biopsy of the nodules confirmed hamartoma.

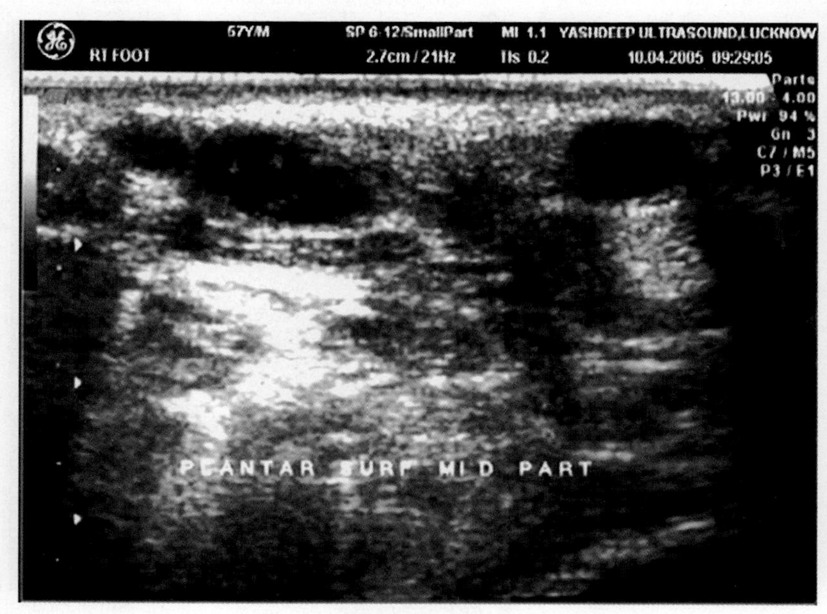

FIGURE 13.301

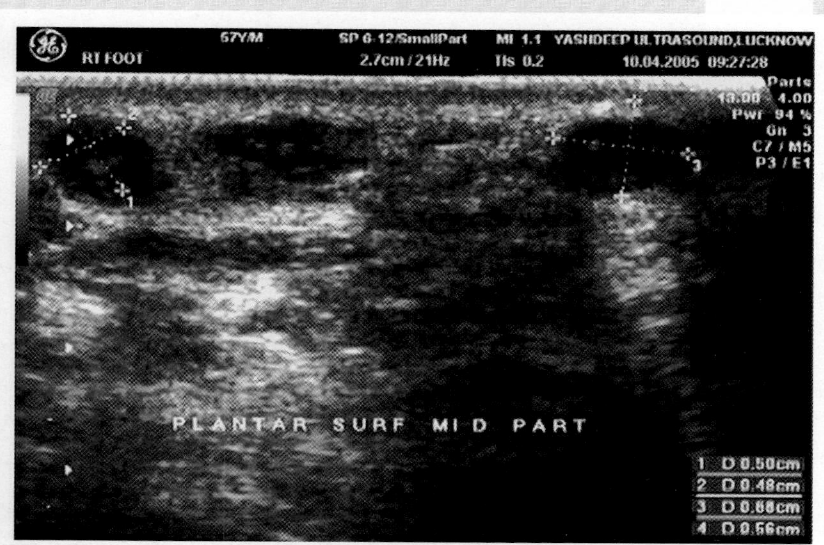

FIGURE 13.302

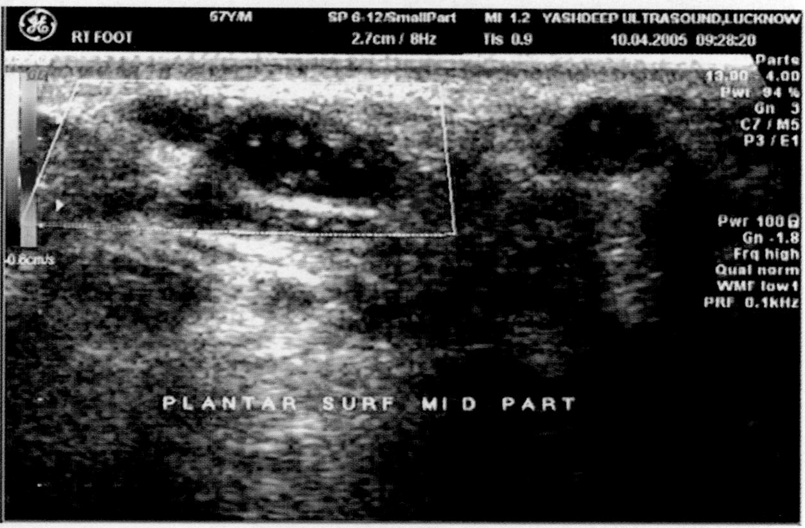

FIGURE 13.303

FIGURES 13.301 to 13.303: Foot nodules (gangliomas). A patient presented with multiple nodules in the mid part of the Rt foot plantar side. HRSG shows well-defined homogeneous hypoechoic nodular shadows in the mid line of the foot. Few nodules show cystic degeneration. **On color flow imaging** no flow is seen in them. Biopsy of the nodules confirmed gangliomas.

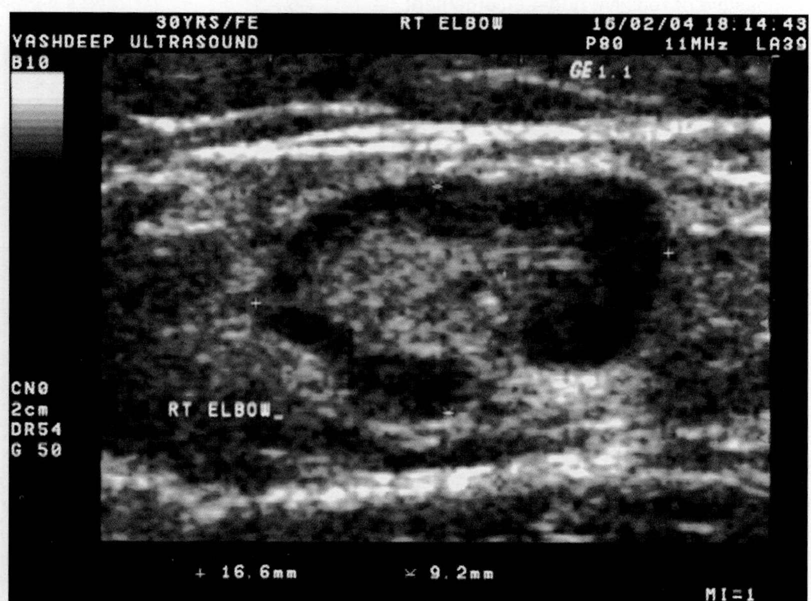

FIGURE 13.304: Enlarged nodes in the elbow. A patient presented wit painful enlarged nodal mass on the medial side of the Rt elbow. HRSG shows well-defined enlarged nodes in the elbow. Lymph node hilum is intact.

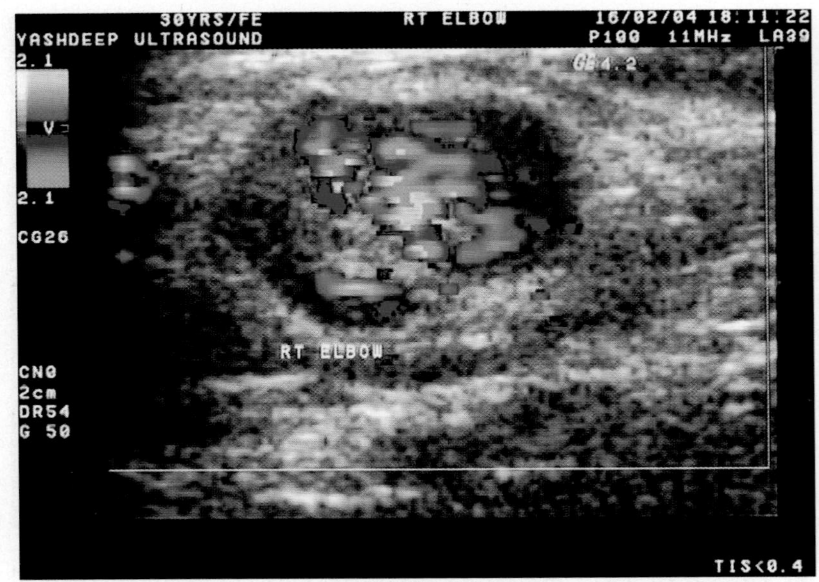

FIGURE 13.305

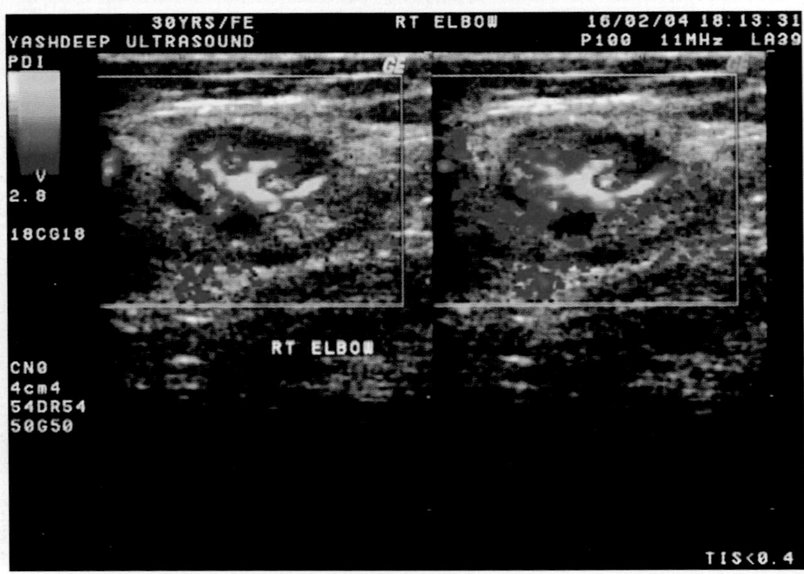

FIGURE 13.306

FIGURES 13.305 and 13.306: On color flow imaging high flow is seen in the node suggestive of reactive nodal enlargement.

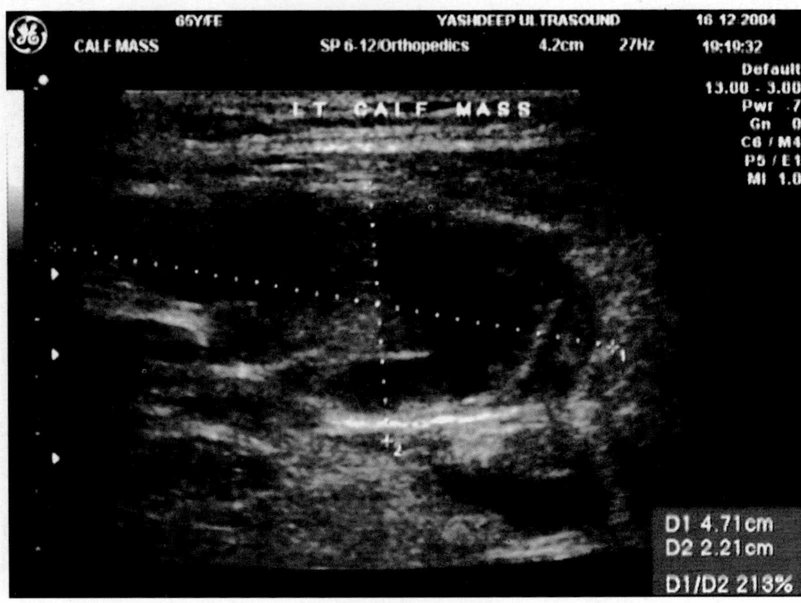

FIGURE 13.307

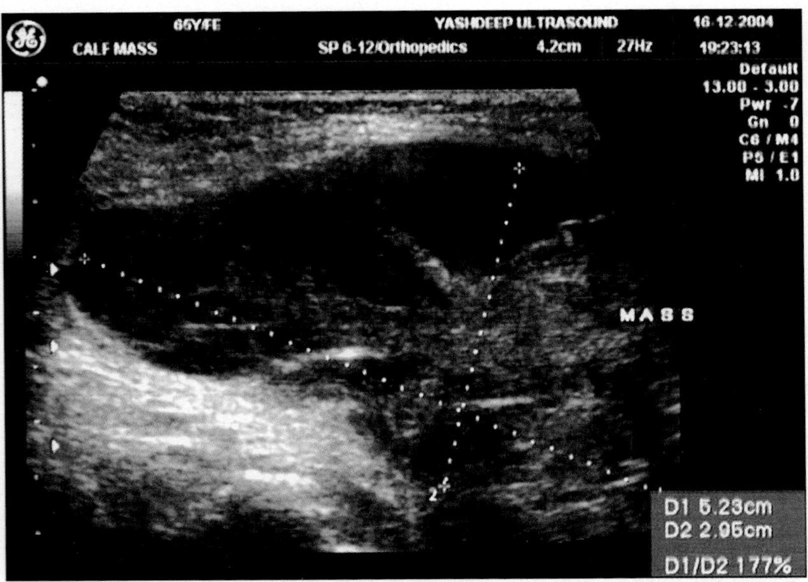

FIGURE 13.308

FIGURES 13.307 and 13.308: Big malignant muscle tumor. A patient presented with big mass in the calf muscle. HRSG shows big heterogeneous mass in the calf muscle. The mass is seen invading the gastrocnemius muscle. The capsule was broken. It was seen invading the fascial planes.

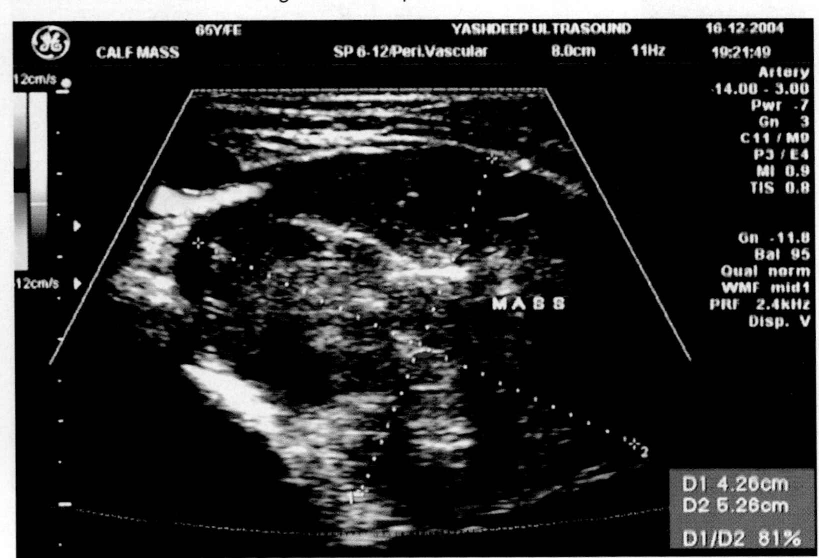

FIGURE 13.309: On color flow imaging poor flow was seen in the mass.

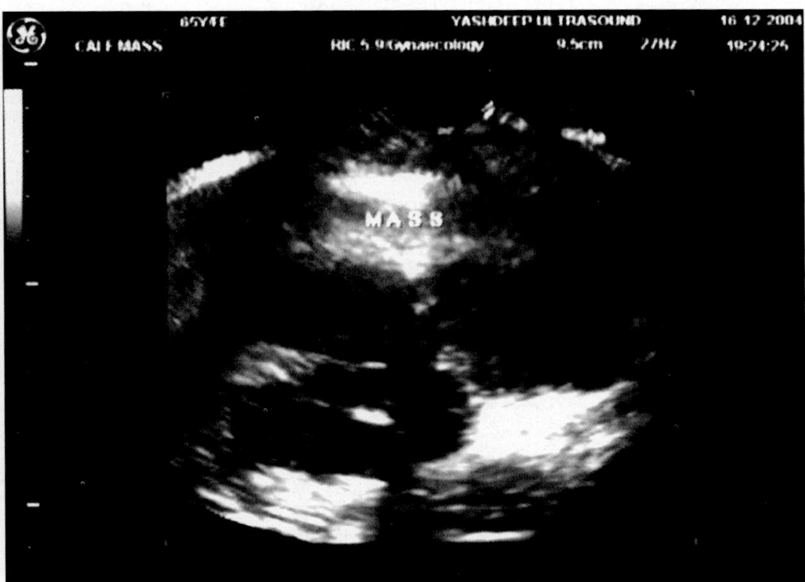

FIGURE 13.310

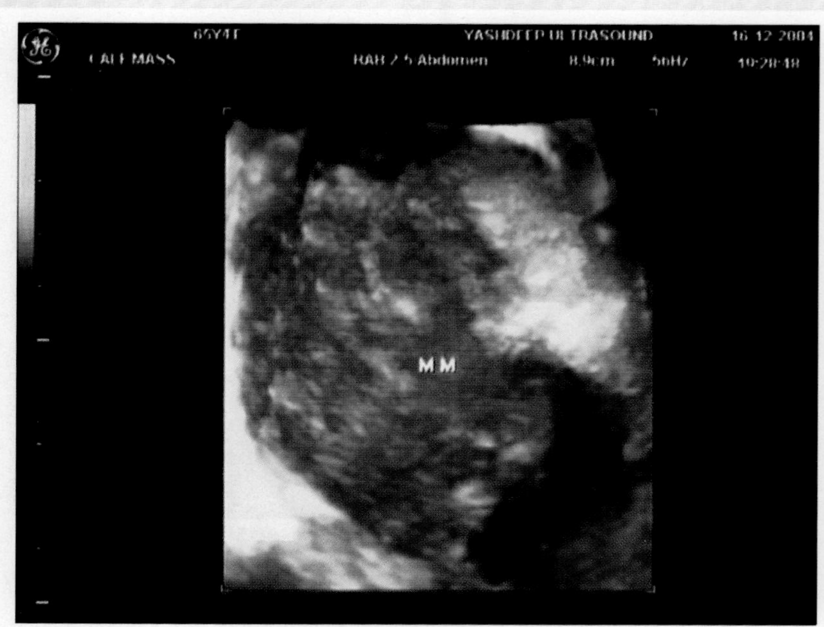

FIGURE 13.311

FIGURES 13.310 and 13.311: 3D surface-rendering image of the tumor shows the invasion of the tumor in the fascial planes. Biopsy of the mass confirmed malignant rhabdomayosarcoma.

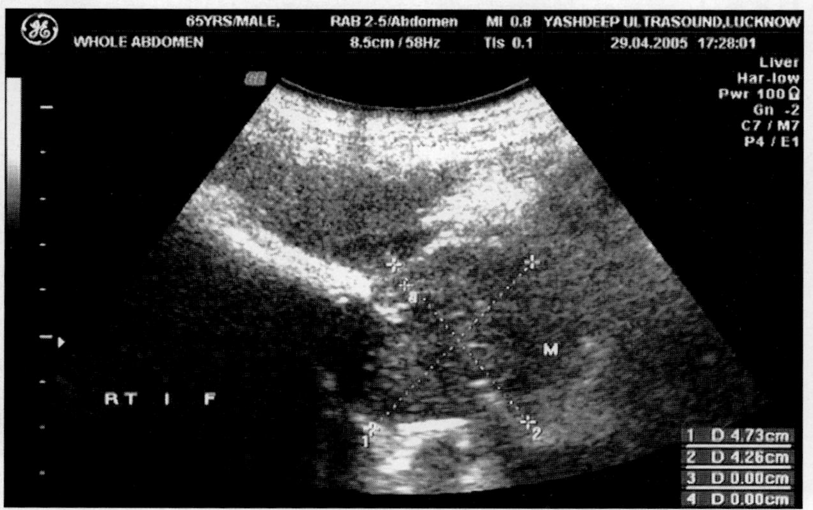

FIGURE 13.312

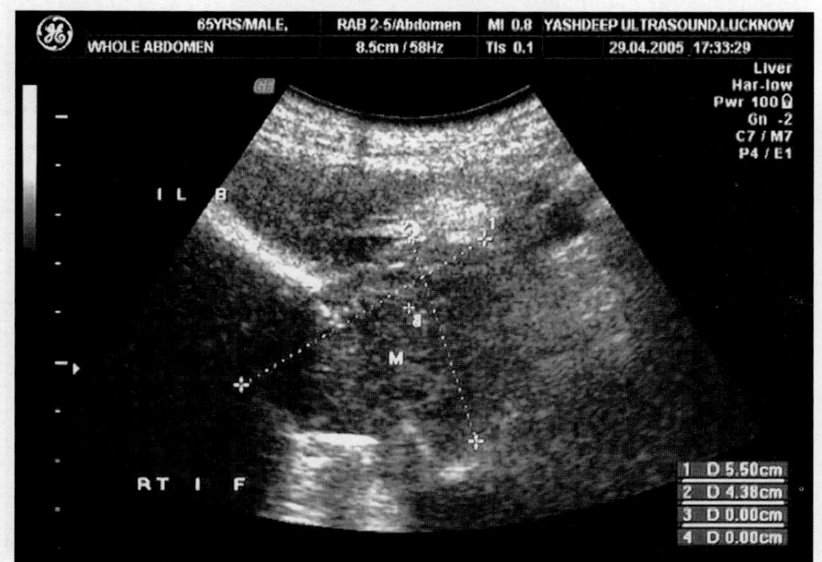

FIGURE 13.313

FIGURES 13.312 and 13.313: Muscle tumor invading the bone. A patient presented with mass in the Rt iliac fossa. The mass was seen invading the iliac bone and also the muscles capsule was broken. It was seen invading in the surrounded tissue.

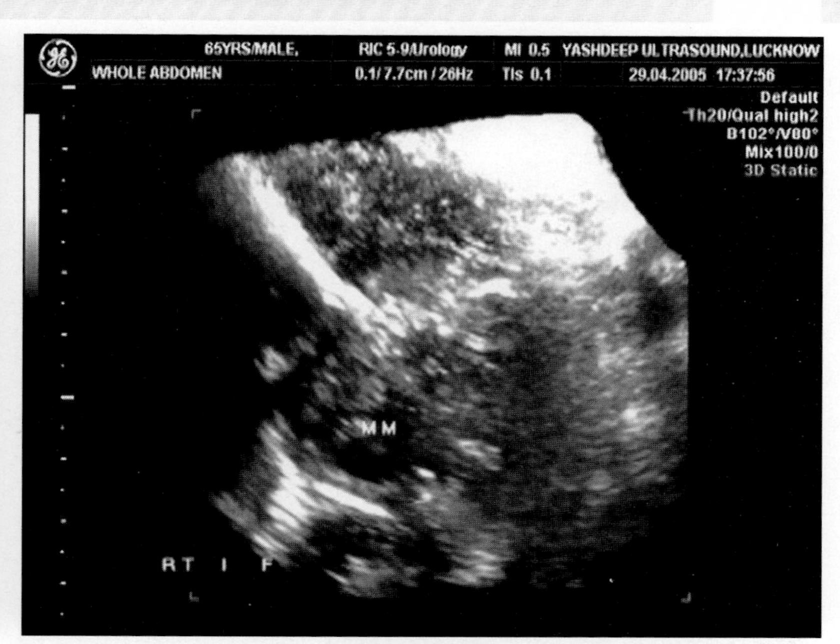

FIGURE 13.314

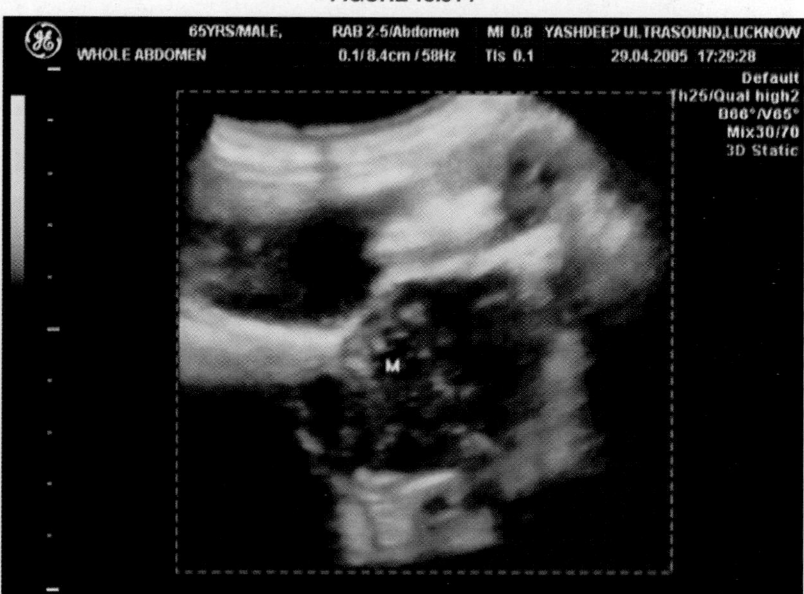

FIGURE 13.315

FIGURES 13.314 and 13.315: 3D imaging of the mass gave the much better details tumor.

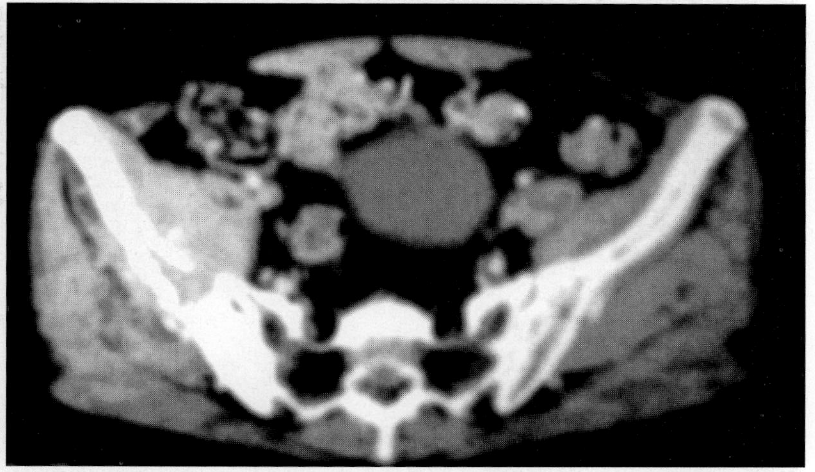

FIGURE 13.316: CT of the same patient showed the details of the tumor and confirmed the ultrasound findings. Biopsy of the mass confirmed malignant sarcoma.

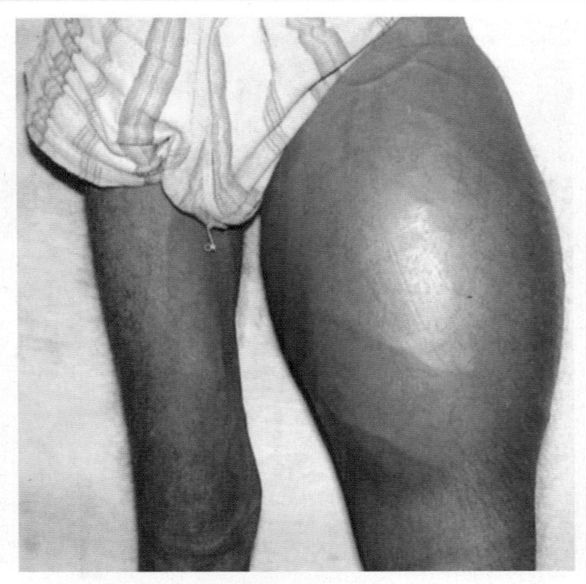

FIGURE 13.317

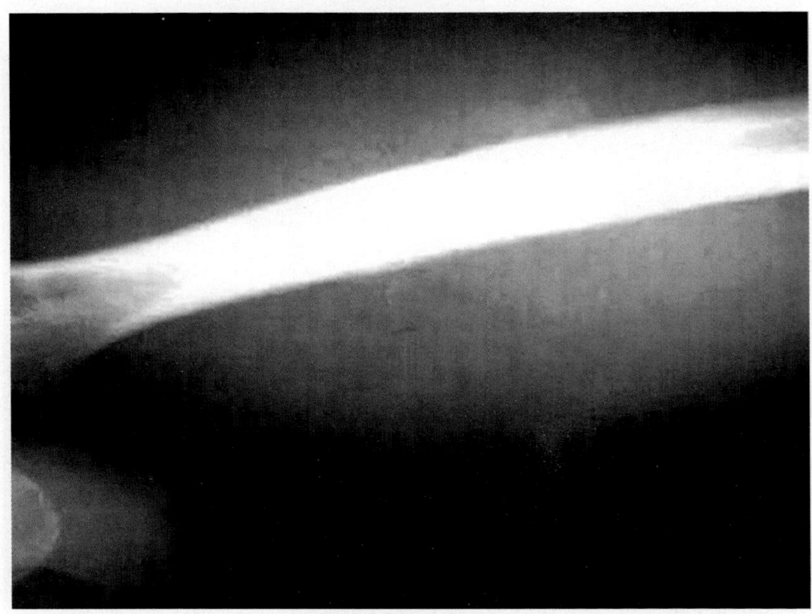

FIGURE 13.318

FIGURES 13.317 and 13.318: Huge thigh muscle tumor. A patient presented with a huge mass in the Lt thigh. X-ray of the thigh shows a huge soft tissue swelling. However, the femur bone appears to be normal. No evidence of any bony erosion seen.

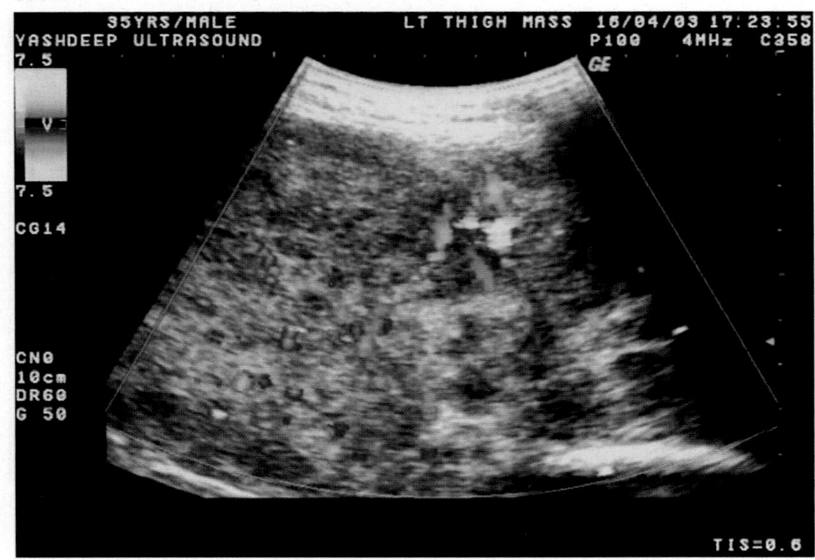

FIGURE 13.319

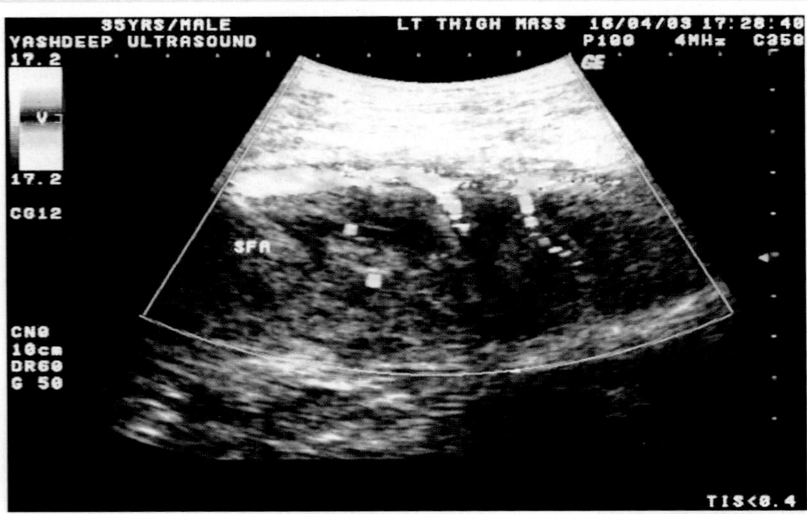

FIGURE 13.320

FIGURES 13.319 and 13.320: HRSG shows a huge mass in the muscle with normal muscle texture lost. The mass was predominantly solid. No cystic necrosis was seen in the mass. **On color flow imaging** superficial femoral artery was seen embedded in the mass and giving the feeding branches to the mass. However, the vessel was seen intact.

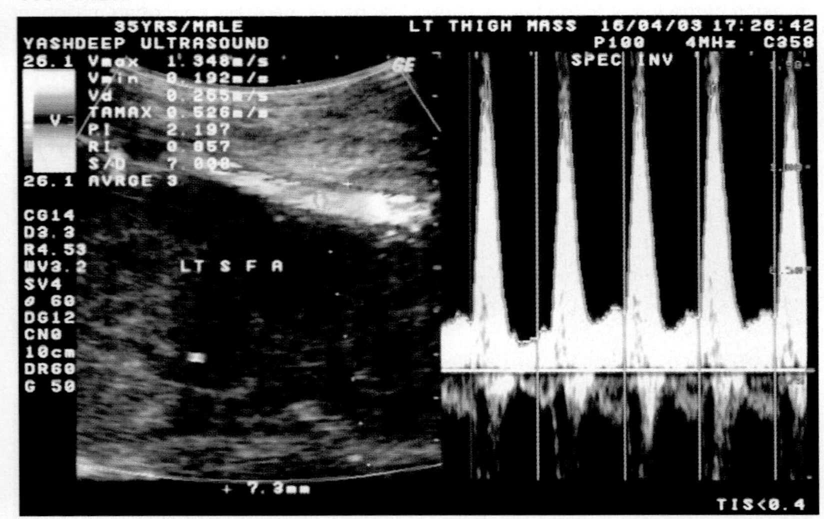

FIGURE 13.321

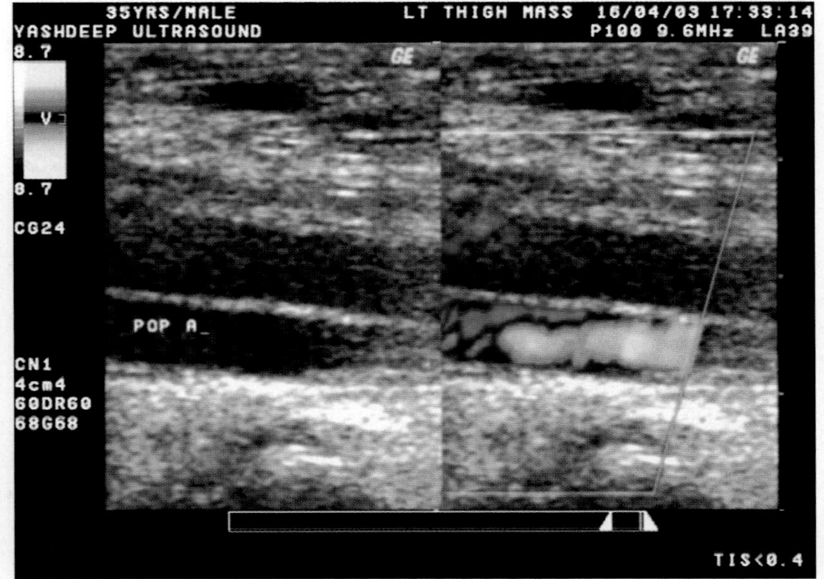

FIGURE 13.322

FIGURES 13.321 and 13.322: On spectral Doppler tracing high resistance flow was seen in the SFA due to pressure on the artery. But no flow was seen in superficial femoral vein. Biopsy of the mass confirmed fibrosarcoma.

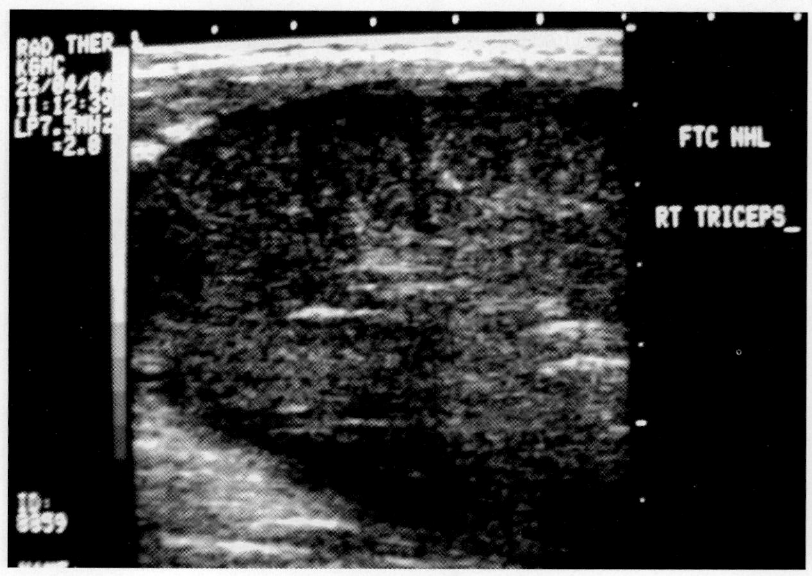

FIGURE 13.323

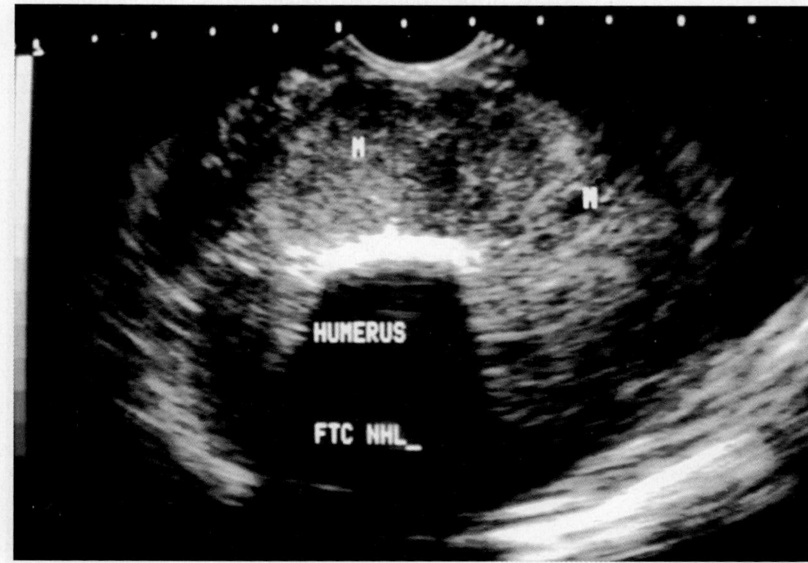

FIGURE 13.324

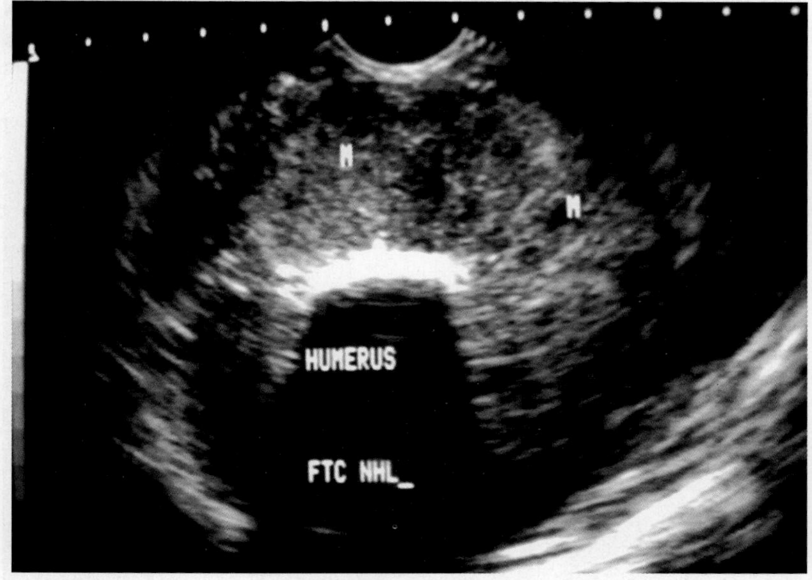

FIGURE 13.325

FIGURES 13.323 to 13.325: Metastasis in the muscle. A known patient of non Hodgkin lymphoma presented with acutely painful swelling in the Lt arm muscle. HRSG shows a big mass in the Lt triceps muscles. The mass was heterogeneous. Hypoechoic areas are seen in it. Biopsy of the mass confirmed metastasis of the NHL.

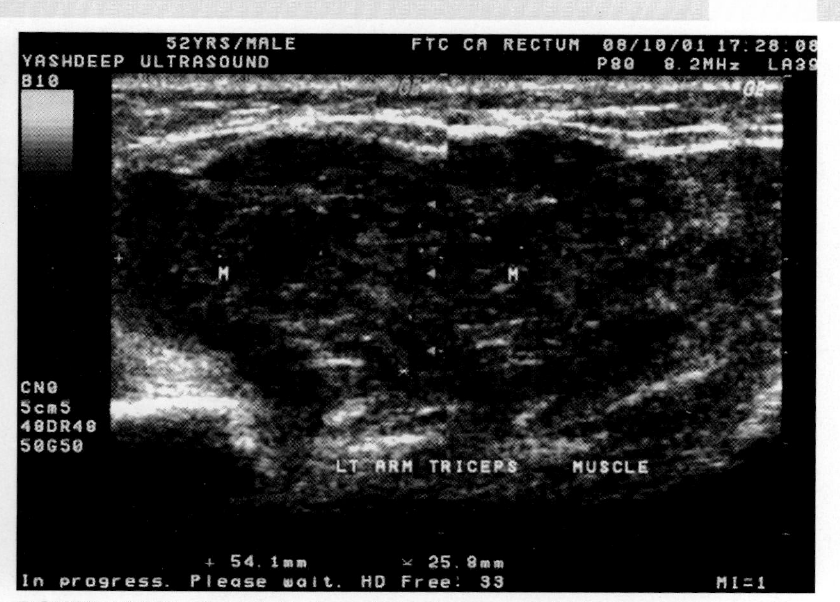

FIGURE 13.326: Metastatic invasion of arm muscle in carcinoma rectum. HRSG shows big irregular heterogeneous mass invading the triceps muscle in a known case of carcinoma rectum. Biopsy of the mass showed metastatic deposits from adenocarcinoma rectum.

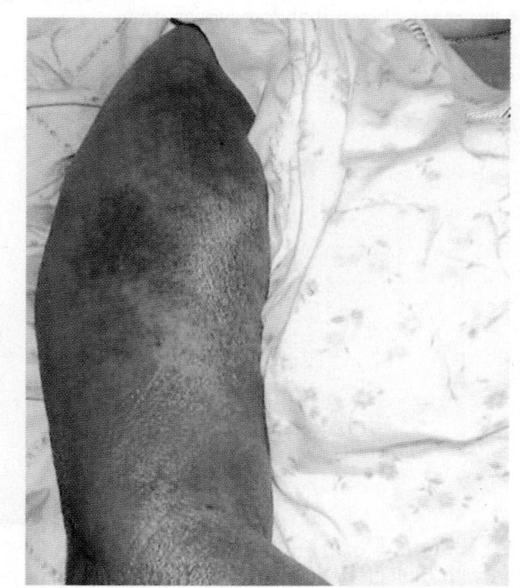

FIGURE 13.327: Malignant tumor of arm. An old lady presented with soft tissue mass over the Rt arm biceps region.

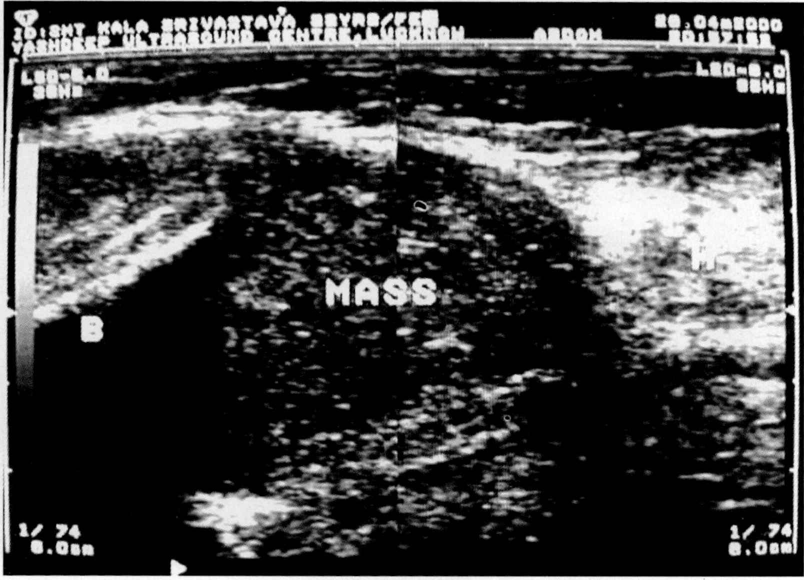

FIGURE 13.328

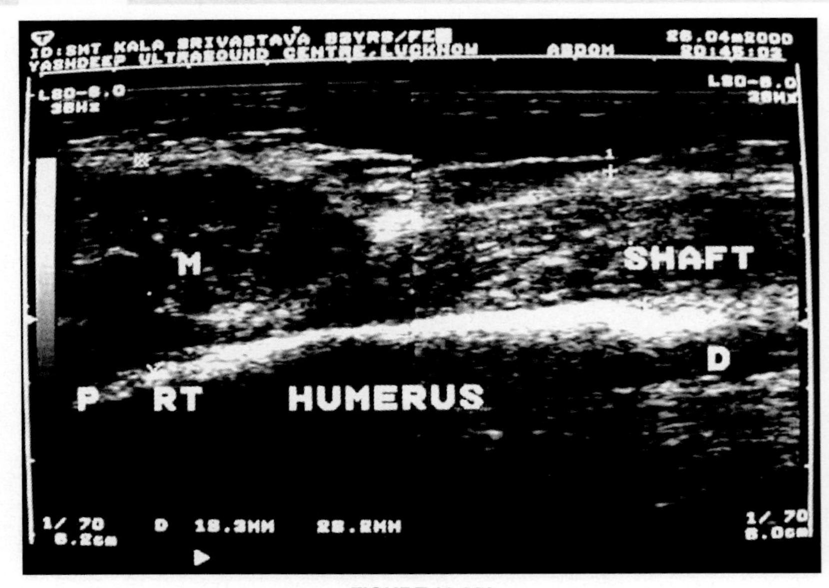

FIGURE 13.329

FIGURES 13.328 and 13.329: HRSG shows big heterogeneous mass involving the muscle of the arm. The capsule is broken. It is heterogeneous in texture. It is also seen invading the muscle plains. However, bone is normal. No cystic degeneration in the mass. Biopsy of the mass shows rhabdomayosarcoma.

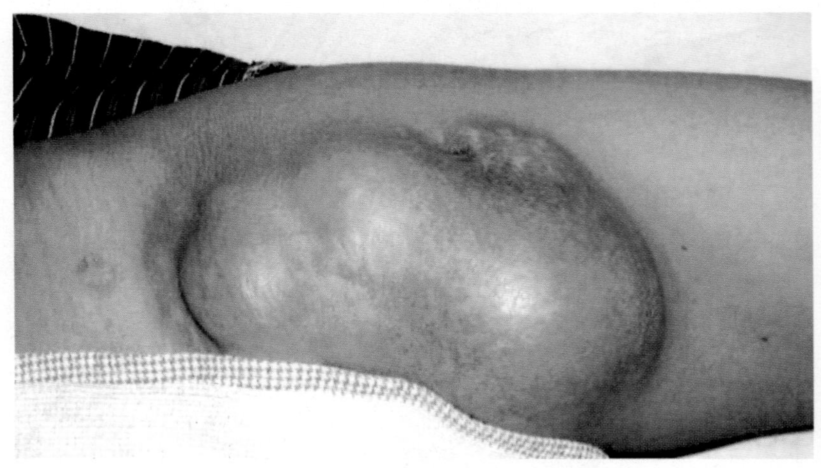

FIGURE 13.330: Lymphoma of medial side of thigh extending into the hip. A patient presented with multiple lobulated masses upper medial part of thigh with restricted hip movements.

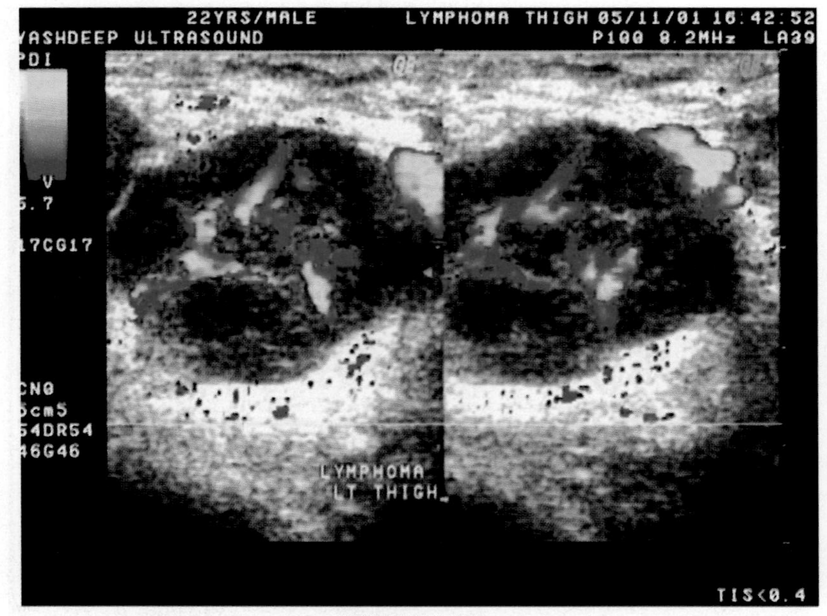

FIGURE 13.331

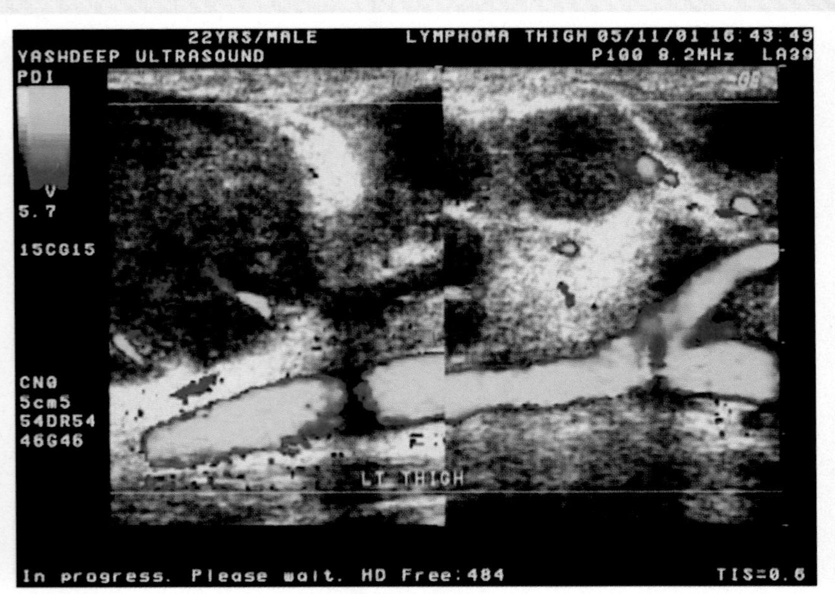

FIGURE 13.332

FIGURES 13.331 and 13.332: HRSG shows big multiple hypoechoic lobulated soft tissue masses. On color Doppler imaging they show moderate flow. Biopsy of the mass shows lymphoma.

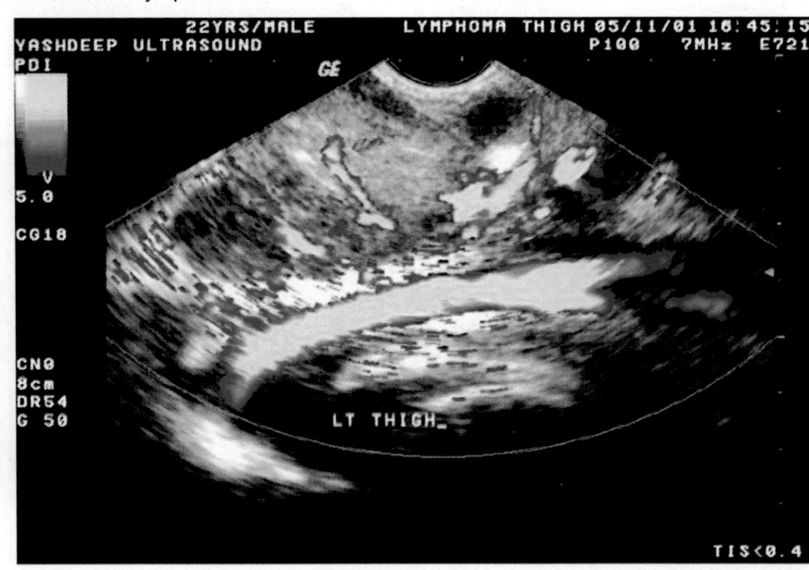

FIGURE 13.333: HRSG shows extension of the masses in the mass of thigh. The masses are seen pressing over the common femoral vessels. Fluid is seen also in the joint capsule.

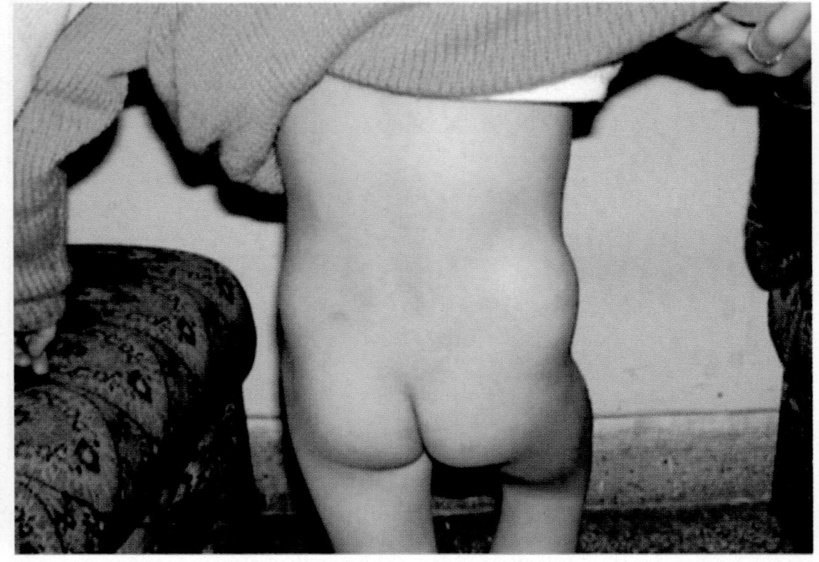

FIGURE 13.334: Rhabdomyosarcoma of Lt lumbar muscle. A young child presented with soft tissue mass over the Lt lumbar region.

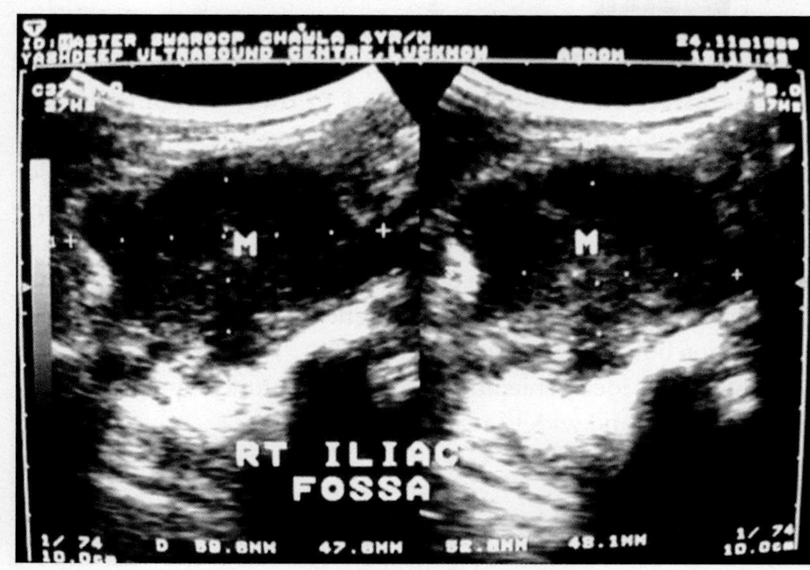

FIGURE 13.335

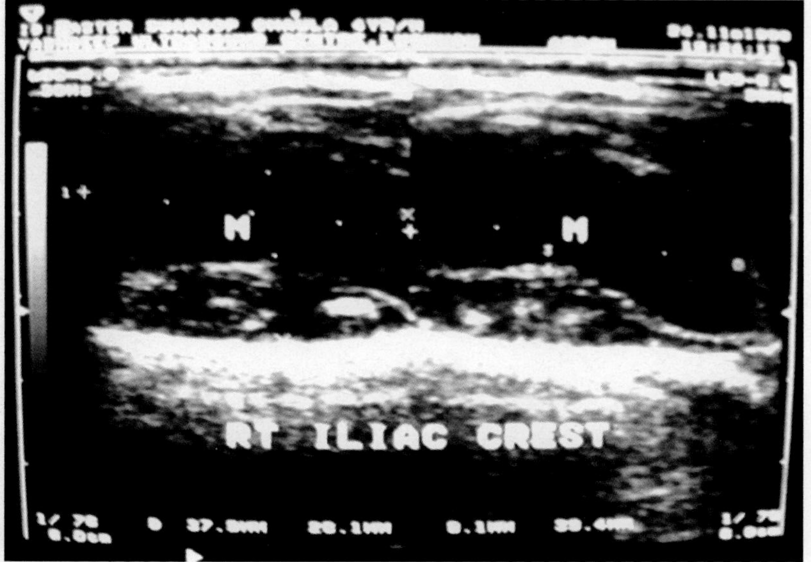

FIGURE 13.336

FIGURES 13.335 and 13.336: HRSG shows a big irregular heterogeneous soft tissue mass involving the Lt lumbar muscles. The mass has destroyed the normal muscle anatomy. Capsule was broken. The mass was seen invading the surrounding tissues. However, no evidence of bony involvement is seen. Biopsy of the mass confirmed invasive rhabdomyosarcoma.

HRSG of Large Synovial Joints

Large synovial joints are difficult to be evaluated clinically and HRSG is frequently requested to evaluate them. Shoulder, hip and knee joints are deep-seated joints and physical examination is difficult in the joints and proper assessment of pathology is not possible. HRSG dramatically improves our skill to evaluate these joints. It is not related to the size of the joint. Even smaller interphalangeal joint can be evaluated on HRSG. The main advantage of HRSG on arthroscopy is to evaluate and examine periarticular soft tissue. Intra-articular disease can be evaluated accurately on HRSG. It can quickly quantify the amount on intra-articular fluid. It also identifies synovial edema, hemarthrosis and loose bodies in the joint. Cartilage and synovial thickness can be evaluated on HRSG. Peripheral meniscal tears can be evaluated with HRSG and it is a valuable tool in diagnosis both peri articular and intra-articular disease.

Sonography of Tendon

Evaluation of tendon requires high contrast and spatial resolution. High frequency transducers 7 to 12 Mhz are mandatory. The following are the indications for tendon evaluation.

1. Tendonitis
2. Tendon rupture
3. Acute tendonitis

Acute tendonitis is characterized with increased fluid collection in the tendon sheath or synovial sheath. It is seen as a halo around the tendon in transverse images. In sub acute cases the tendon gets thickened and become echogenic. Most commonly it is seen in large biceps tendon in swimmers having shallow bicipital groove.

Tendon Rupture

Most commonly it is seen in rotator cuff tear and large head of biceps is most common site. Clinically it is known as shoulder impingement syndrome. Sonographic findings are empty bicipital groove with a fusiform muscle mass located distally. The posterior tibial tendon is also prone for rupture.

Examination of Shoulder

High-resolution sonography is very useful in evaluation chronic shoulder pain and shoulder pathology. Impingement syndrome is the most common cause of chronic shoulder pain and it is caused by the compression of anterior cuff against the anterior acromial edge and coracoacromial ligament. In initial stages edema and petechial hemorrhages are seen which may lead to tendonitis, fibrosis, incomplete and complete tear of rotator cuff may result later on. The estimated rotator cuff tear incidence above 60 years of age is around 30%, which is fairly high. HRSG provides a non invasive, cost effective technique for evaluation of rotator cuff tear. The best clinical indications of rotator cuff disease are a painful shoulder with painful arc, positive impingement test, positive supra spinatous test or limited abduction of shoulder. The examination of the rotator cuff begins by making the patient to sit on a rotating stool. Visualizing biceps tendon is first step and it is taken as the landmark. The biceps tendon is seen within the bicipital groove, which is over laying the anterior surface of humorous with shoulder held in neutral position. After visualizing biceps tendon, the sub scapularis tendon is seen. Few fibers of subscapularis tendon are seen crossing the biceps tendon anteriorly. It is seen as a fan shaped tendon. The supraspinatous tendon is found deep to the deltoid muscle. It is seen more echogenic than the deltoid muscle and curvilinear in shape. It is very homogeneous in appearance and the full tendon is seen in neutral position then abduction and internal rotation of the arm. Any inhomogenicity in the tendon should be taken as pathological changes or tear. The tendon is seen as a triangular structure in coronal planes.

The posterior cuff comprises of infraspinatous tendon and teres minor. It is seen by placing the transducer over the posterior aspect of shoulder at the level of middle of glenohumeral joint. The infraspinatous tendon is triangular beak shape tendon having fine fibril fibers. By moving the transducer 2 to 3 cm below the infraspinatous tendon, the teres minor tendon is seen. The muscle in the tendon has a more rectangular shape and forms the posterioinferior part of the rotator cuff.

Rotator cuff tear and rotator cuff calcification both cause inhomogeneous focal shadows in the cuff. Small punctate calcified specks may be seen in the superior cuff. It may not cause shadowing. Rotator cuff tears are hypoechoic. However, few cases may be hyper echoic due to fibrosis or calcification. Large rotator cuff tear are seen as a big defect in supraspinatous tendon or complete loss of the tendon. The supraspinatus tendon may be replaced with a small amount of fluid. Fluid may be seen in sub deltoid bursa. Large tears when associated with arthritis mainly rheumatoid arthritis are associated with distension of subdeltoid or subacromial bursa. The synovium of the bursa may be irregular thickened and proliferative. Small tear less than 1cm in size are difficult to be seen. HRSG besides rotator cuff tear can tell about other muscle pathologies of the shoulder including deltoid, pectoral, brachial and biceps muscles can be easily evaluated on HRSG. Muscle rupture, hematoma formation, cyst, parasitic infestations, abscesses and degeneration can be easily picked up on HRSG.

The biceps tendon though is not the part of rotator cuff is an important tendon and should be included in rotator cuff study. In impingement syndrome fluid is seen in tendon sheath with edema of the tendon resulting into hypoechoic appearance of the tendon. If fluid is seen in the tendon sheath, tendonitis must be considered. However, the fluid may be due to intra-articular pathology. Rotator cuff tears may lead to fluid accommodation in subacromial, subdeltoid bursa. Degenerative changes can also need to fluid accommodation. Calcifications around the shoulder are also common. HRSG is important in finding the size and location of the calcification. The subscapular tendon calcifications are rare and difficult to be seen on plane X-ray. However, it can be easily picked up on HRSG. HRSG is also important in acute trauma to the shoulder and it can easily pick up greater tuberosity fracture. Biceps tendon dislocation though is uncommon but can be picked up on HRSG.

HRSG is the ideal screening method for rotator cuff pathology as it is non invasive and widely available.

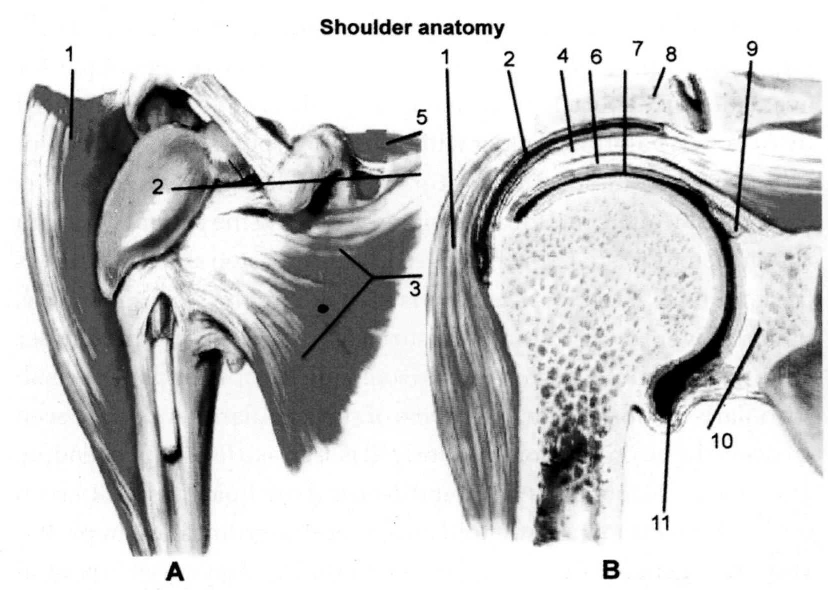

FIGURE 13.337: Diagram shows normal shoulder anatomy and anterior and superior rotator cuff. (A) Anterior view—1. Deltoid muscle, 2. Subdeltoid bursa, 3. Subscapularis muscle, 5. Supraspinatus muscle. (B) Coronal view—1. Deltoid muscle, 2. Subdeltoid bursa 4. Supraspinatus tendon 6. Capsular ligament, 7. Synovial membrane, 8. Acromion, 9. Glenoid labarum, 10. Glenoid fossa and scapula, 11. Axillary recess.

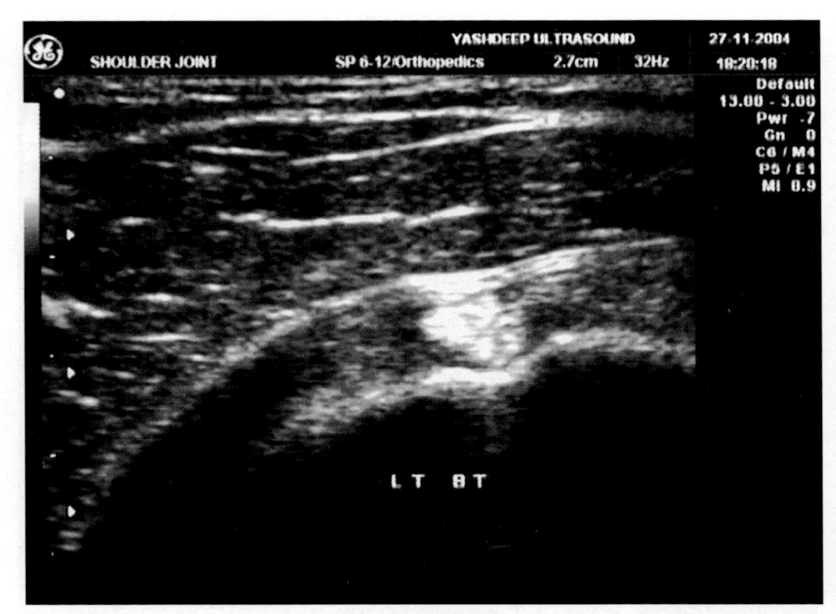

FIGURE 13.338

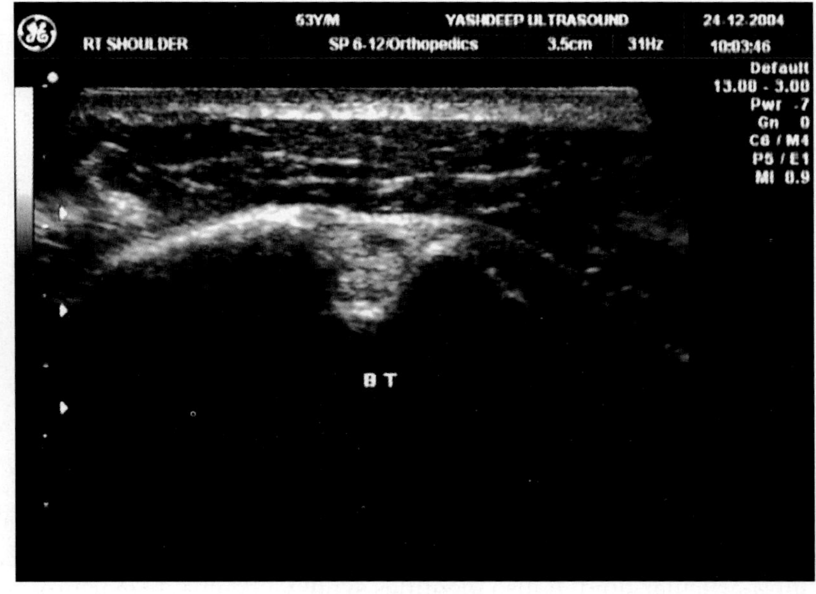

FIGURE 13.339

FIGURES 13.338 and 13.339: Normal biceps tendon. HRSG shows echogenic normal biceps tendon as an echogenic button shadow in transverse axis in the bicipital groove. This is the typical appearance of the long head of the biceps tendon in transverse axis.

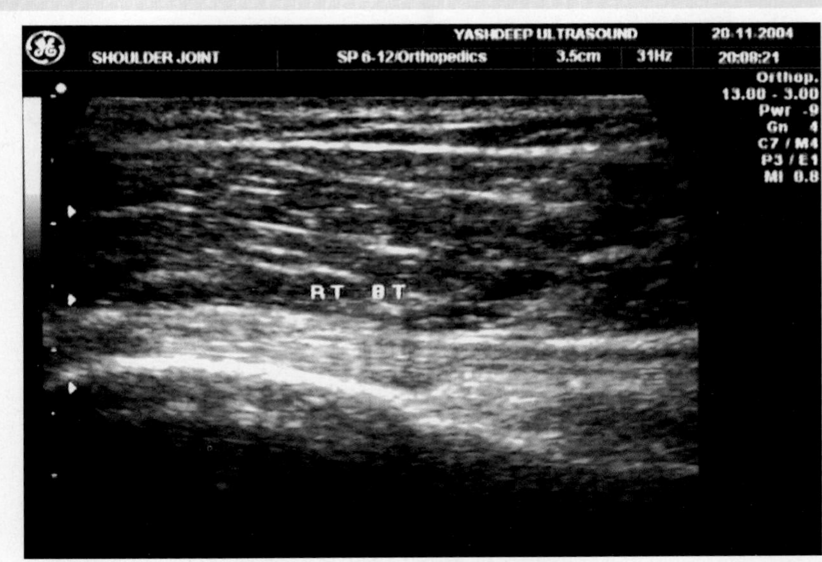

FIGURE 13.340: Normal bicep tendon. HRSG shows normal bicep tendon in the longitudinal axis. The tendon is seen in long axis as a medium level echo band shadow. Fine tendon fibers are seen.

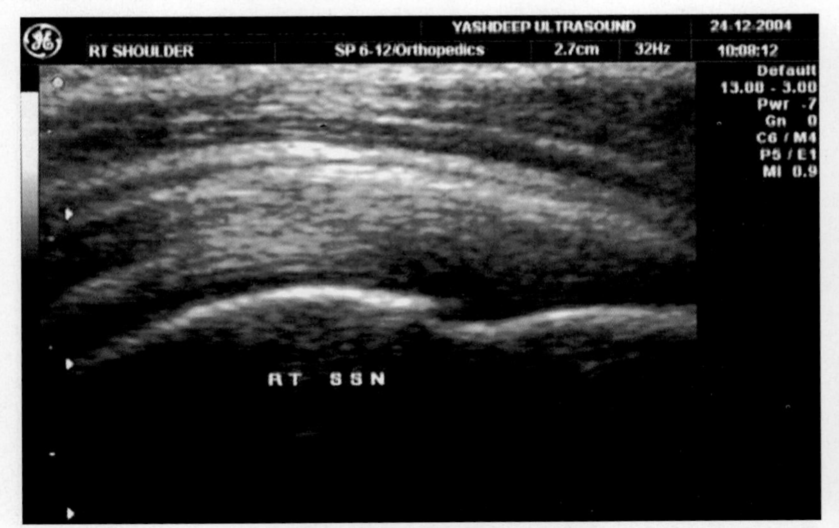

FIGURE 13.341

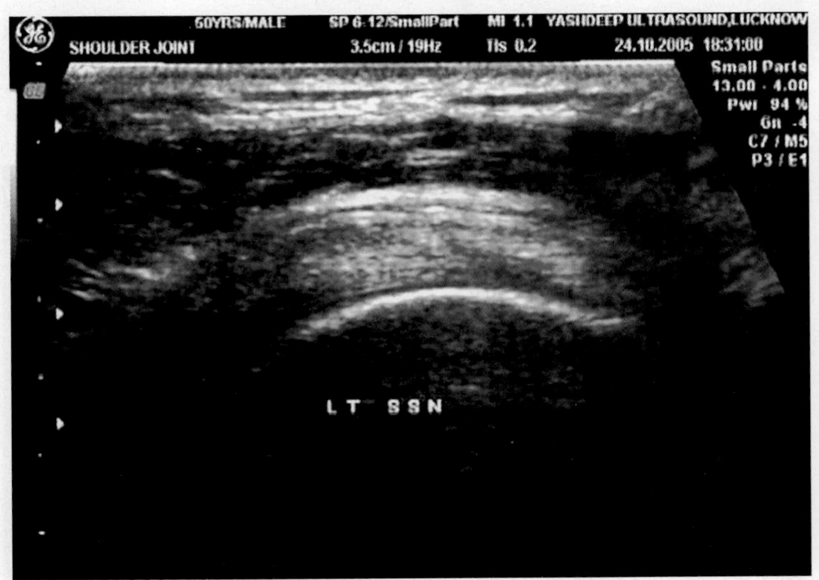

FIGURE 13.342

FIGURES 13.341 and 13.342: Normal supraspinatus tendon. HRSG shows normal supraspinatus tendon. Fine fibers of tendons are seen. The tendon is seen as curvilinear band in coronal plane. Typical "**Parrot Beak Appearance**" is seen in coronal plane. It is seen as a band in transverse plane. The tendon is situated deep to the subacromial bursa. Deltoid muscle is seen above the tendon.

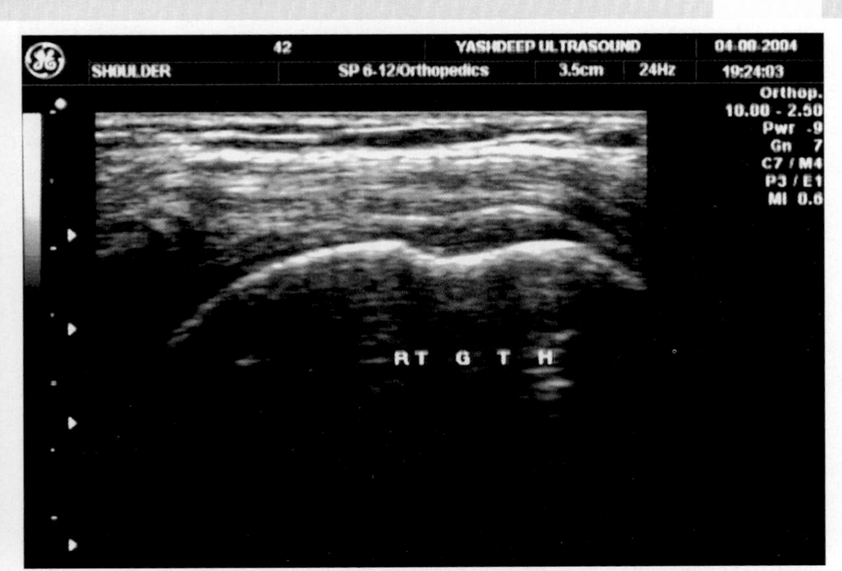

FIGURE 13.343: The tuberosity of humeral head. The articular surface is seen as a wide echogenic convex linear shadow. It shows central area of depression due fovea centralsis. The hyaline cartilage is seen as a hypoechoic band anterior to the humeral head.

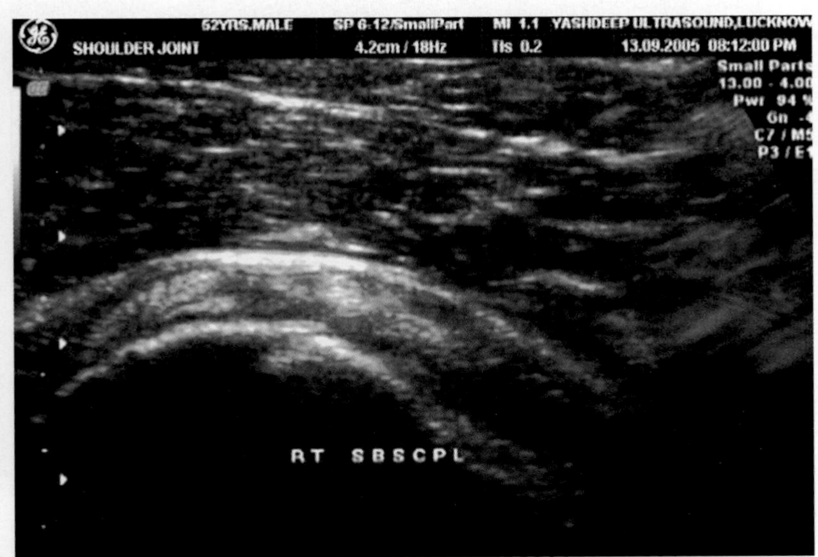

FIGURE 13.344: Normal subscapularis tendon. HRSG shows normal subscapularis tendon. Normal fan shaped appearance of the tendon is seen. The biceps tendon is seen anterior to it.

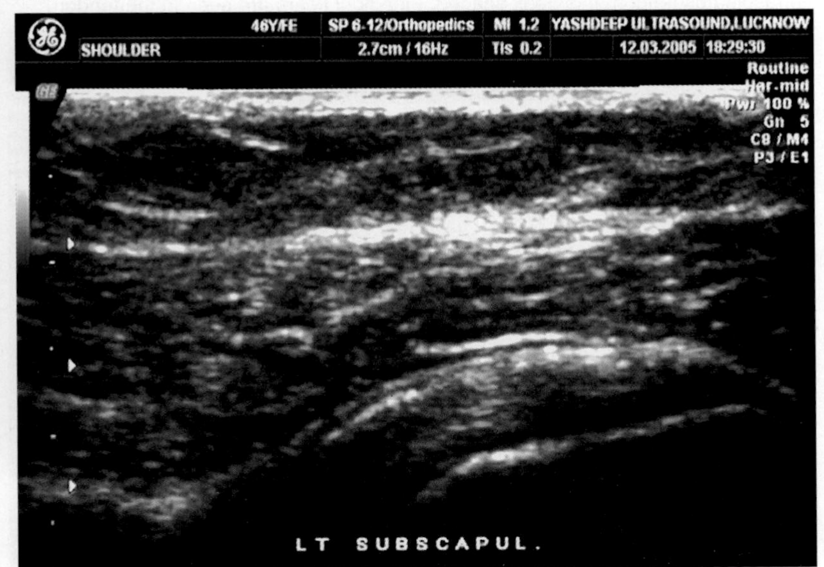

FIGURE 13.345: The subscapularis tendon in abduction. HRSG shows the subscapularis tendon. The tendon is seen in arm abduction as a medium level echo band.

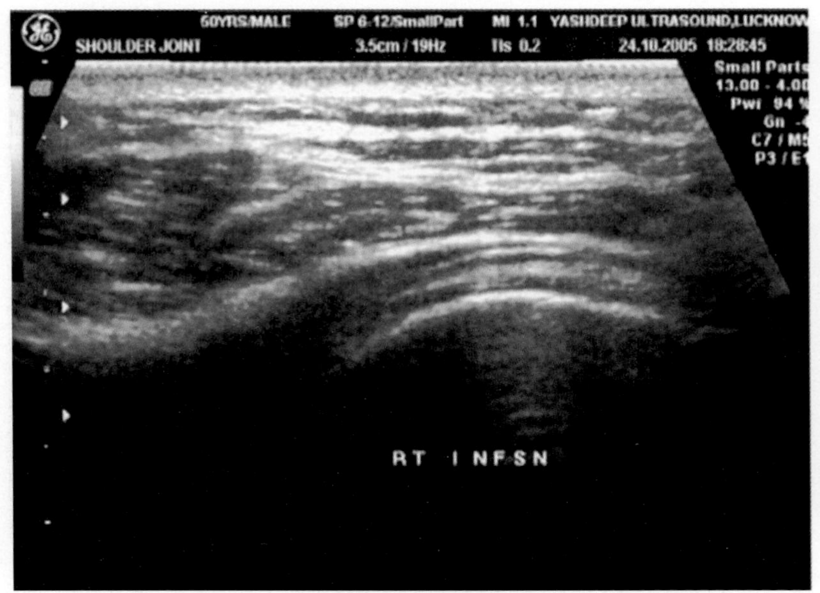

FIGURE 13.346

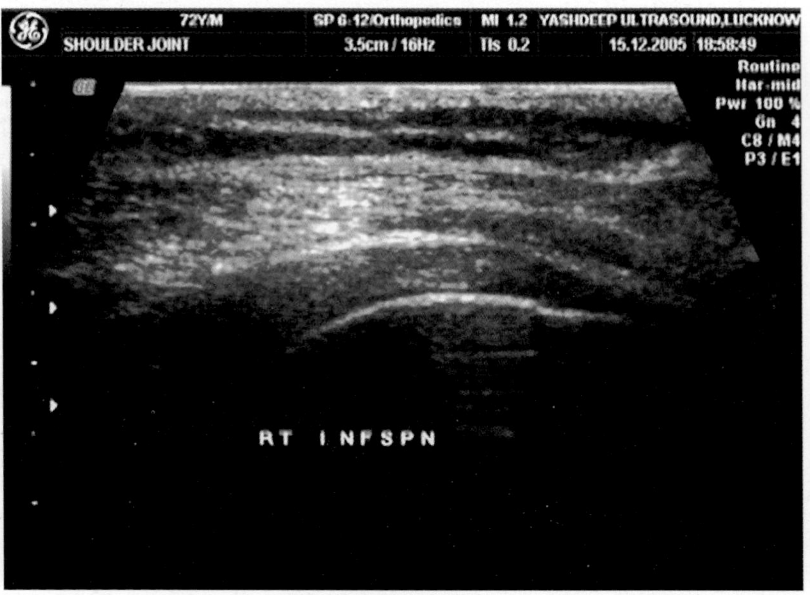

FIGURE 13.347

FIGURES 13.346 and 13.347: Normal infraspinatus tendon. HRSG shows normal infraspinatus tendon. It is beak shaped. Fine fibril fibers are seen in the tendon.

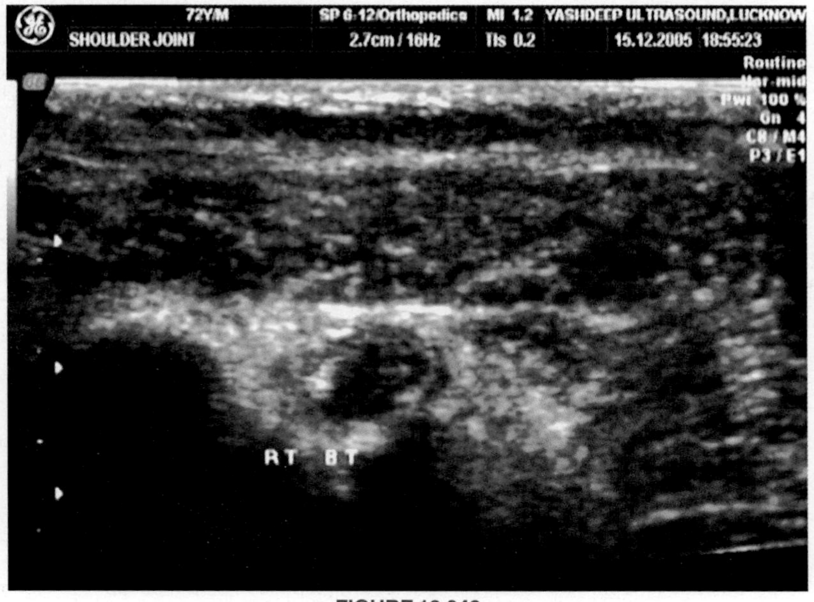

FIGURE 13.348

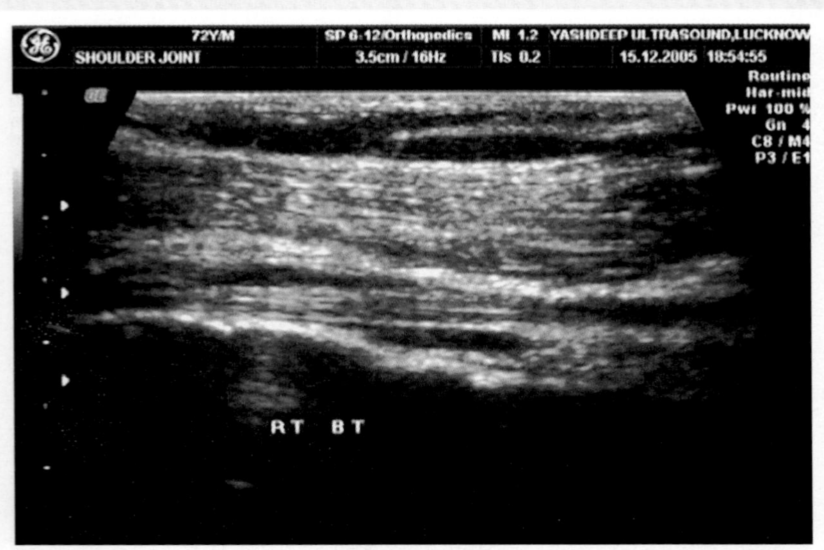

FIGURE 13.349

FIGURES 13.348 and 13.349: Subluxation of biceps tendon head. HRSG shows subluxation of the biceps tendon head. The tendon is seen subluxated from its insertion site. Small amount of fluid is also seen in the bicipital groove. The tendon is also thickened suggestive of tendonitis.

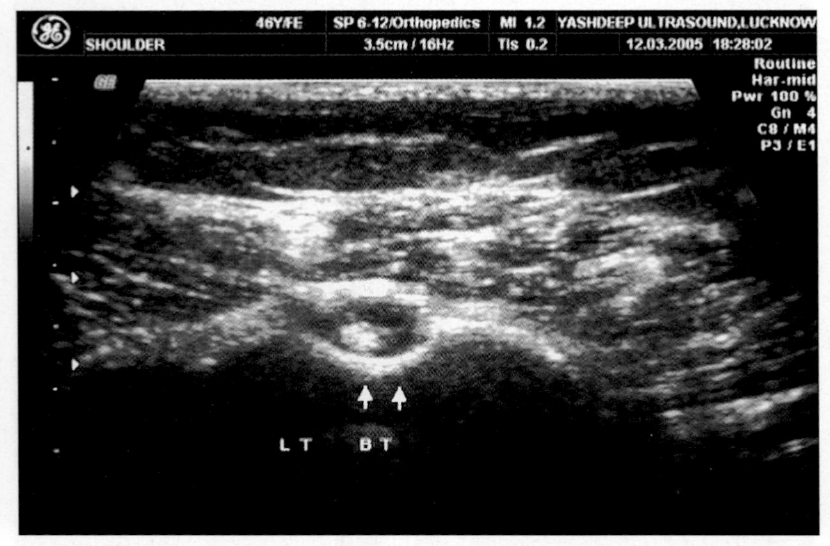

FIGURE 13.350: Biceps tendonitis with adhesive capsulitis. HRSG shows thickening of the biceps tendon. There is evidence of small amount of fluid seen in the bicipital groove. The fluid collection is seen on the medial side of the tendon with fine echoes in a tendon sheath suggestive of adhesive capsulitis.

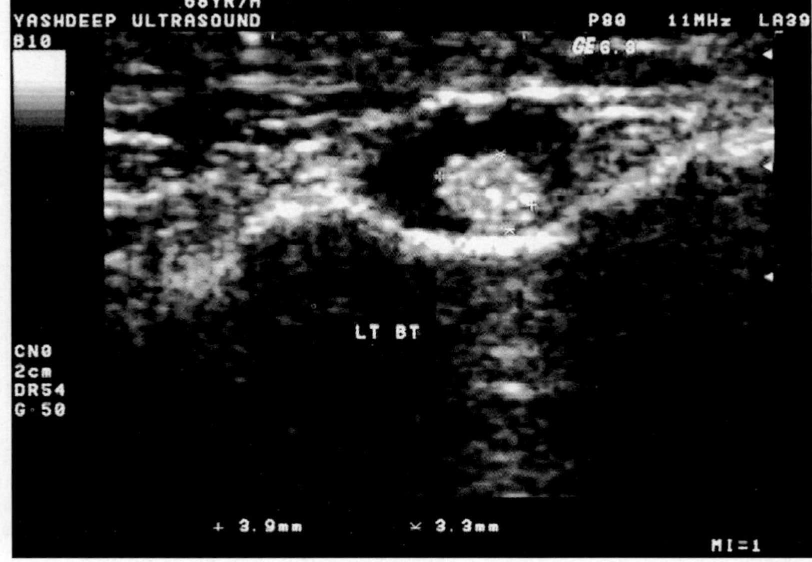

FIGURE 13.351

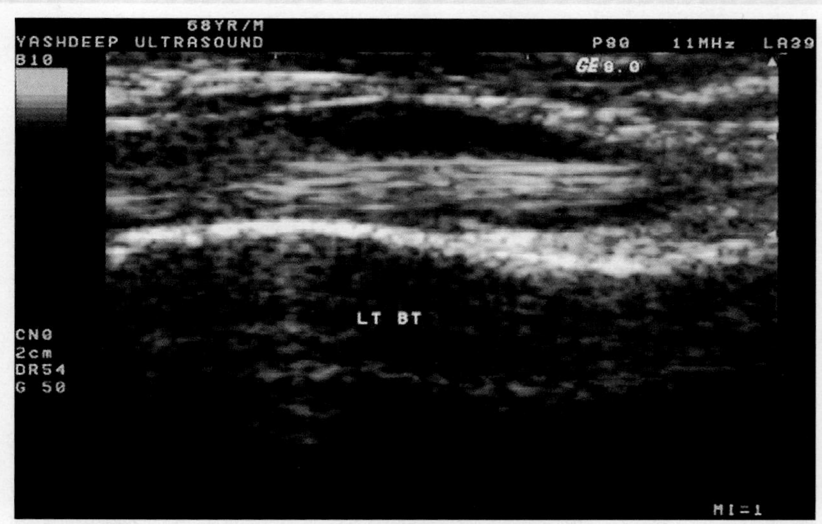

FIGURE 13.352

FIGURES 13.351 and 13.352: Biceps tendonitis. The biceps tendon is well visualized. The tendon is thickened. Moderate amount of fluid collection is seen in the tendon sheath. Fine membranes are also seen in the sheath suggestive of synovial membrane proliferation in a case of biceps tendonitis.

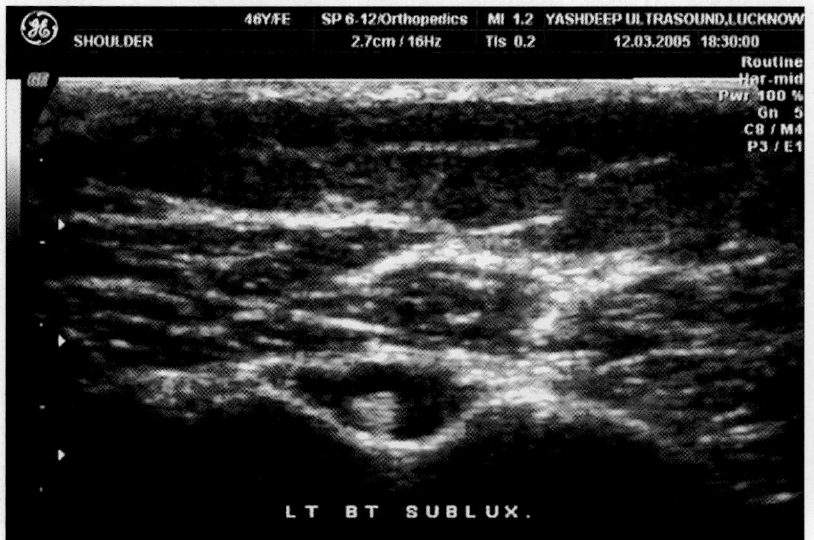

FIGURE 13.353

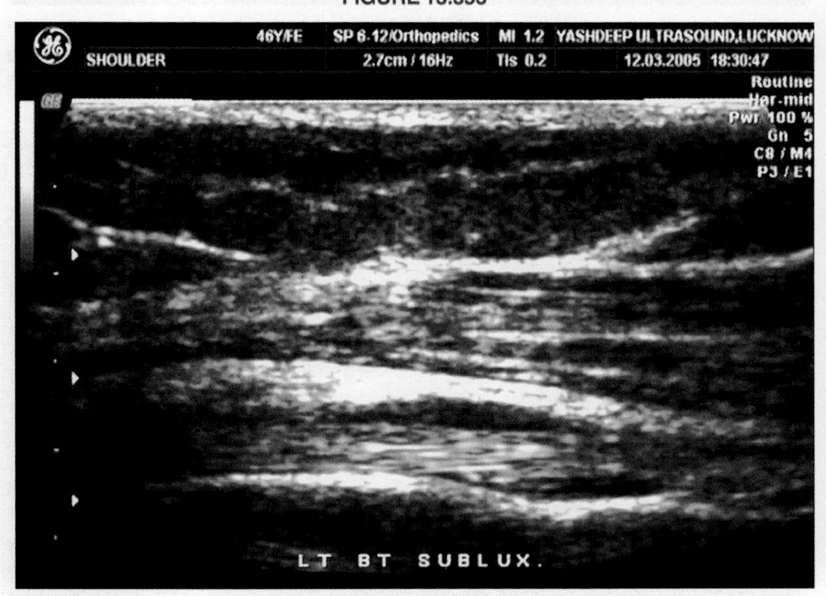

FIGURE 13.354

FIGURES 13.353 and 13.354: Subluxation of the biceps tendon head. HRSG shows displacement of the biceps head near its insertion side in the bicipital groove. Medial displacement is seen. Fluid collection is also seen in the biciptial groove. However, tendon fibers are intact. No evidence of any rupture of the tendon is seen.

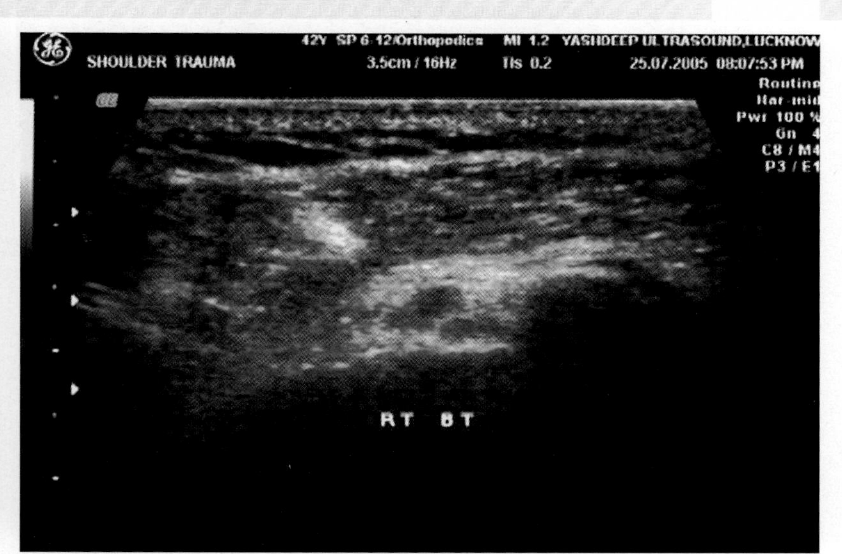

FIGURE 13.355

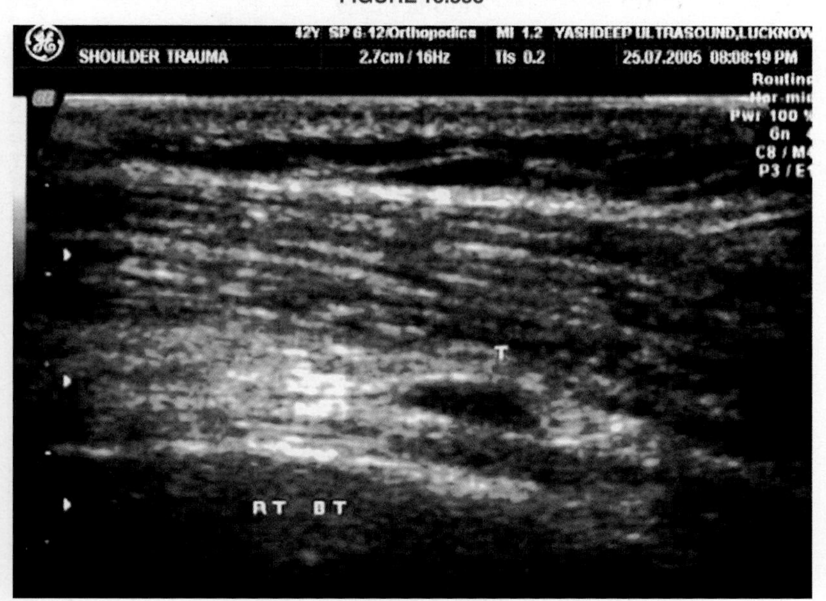

FIGURE 13.356

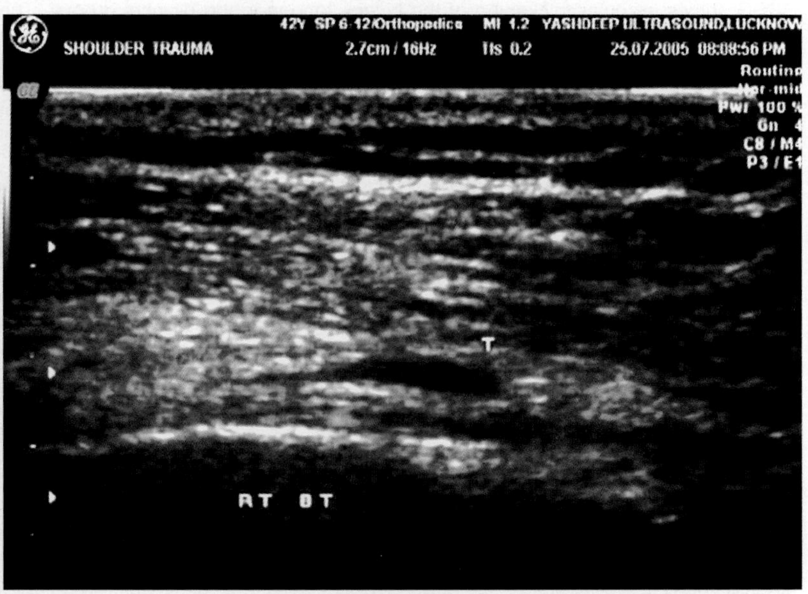

FIGURE 13.357

FIGURES 13.355 to 13.357: Partial tear of the biceps tendon. A patient presented with painful movement in the arm. HRSG shows hypoechoic cleft in the mid part of the tendon. The cleft is seen running in the substance of the tendon suggestive of intra substance tear. The biceps head is also irregular in outline with hazy margins. Displacement of the head is seen from its insertion side.

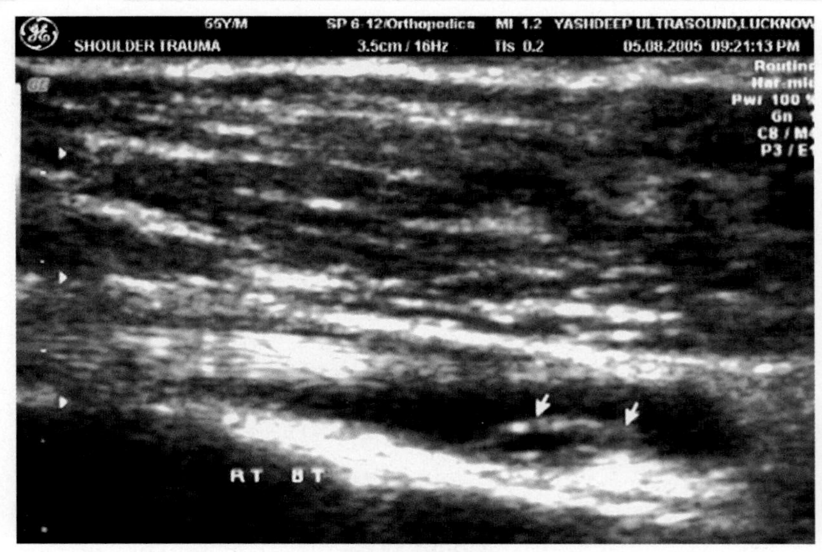

FIGURE 13.358

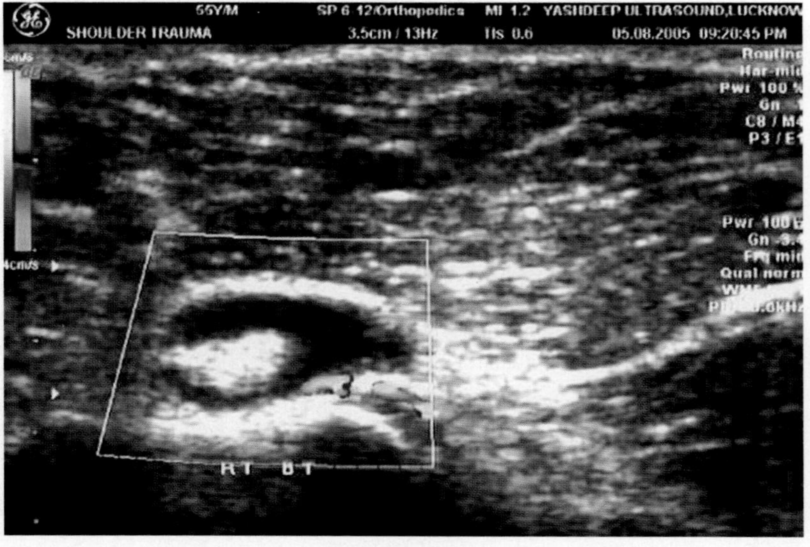

FIGURE 13.359

FIGURES 13.358 and 13.359: Chronic biceps tendon tear with tendonitis. HRSG shows irregular thickening of the biceps tendon in the bicipital groove. The tendon shows thinning in its mid part. Partial rupture of the tendon is seen with synovial membrane proliferation. Fluid collection is also seen in the tendon sheath suggestive of chronic biceps tendonitis with tear. **On color flow imaging** few vessels are seen in the proliferated synovium. Increased vascularity is feature of chronic tear due to synovial membrane proliferation. It is not seen in acute rupture of the tendon.

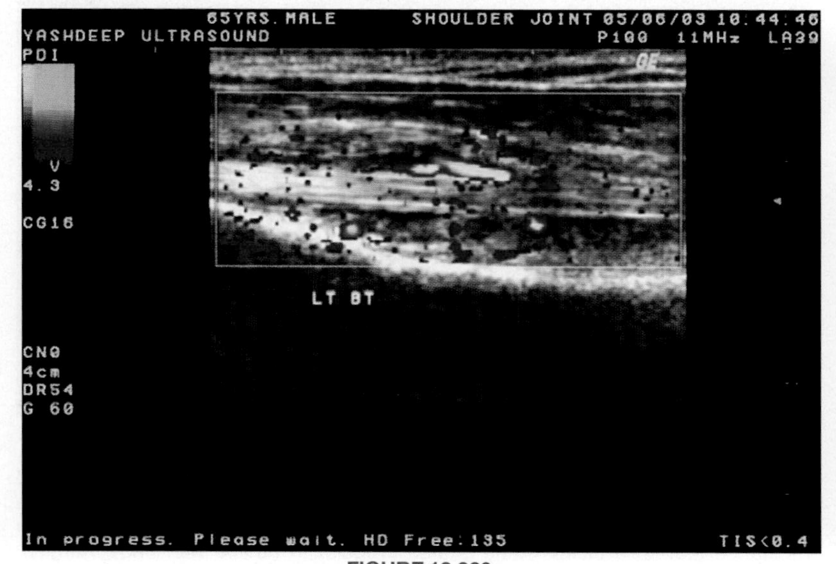

FIGURE 13.360

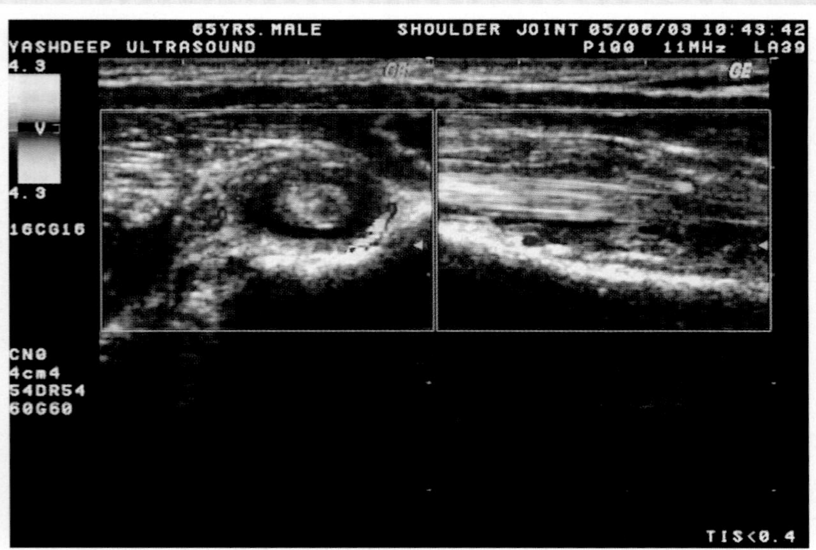

FIGURE 13.361

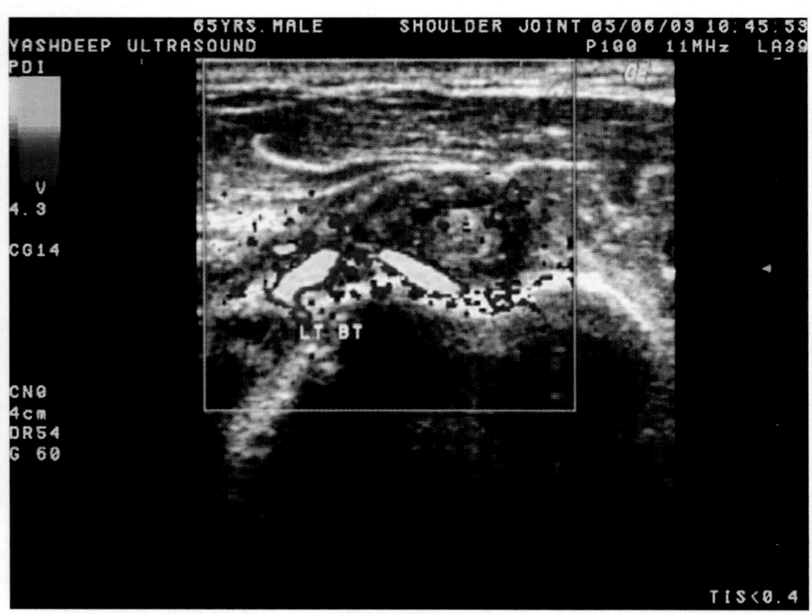

FIGURE 13.362

FIGURES 13.360 to 13.362: Chronic biceps tendonitis. HRSG shows thickened biceps tendon. There is evidence of membranous thickening seen in the bicipital groove. Vessels are also seen in the membranes suggestive of chronicity.

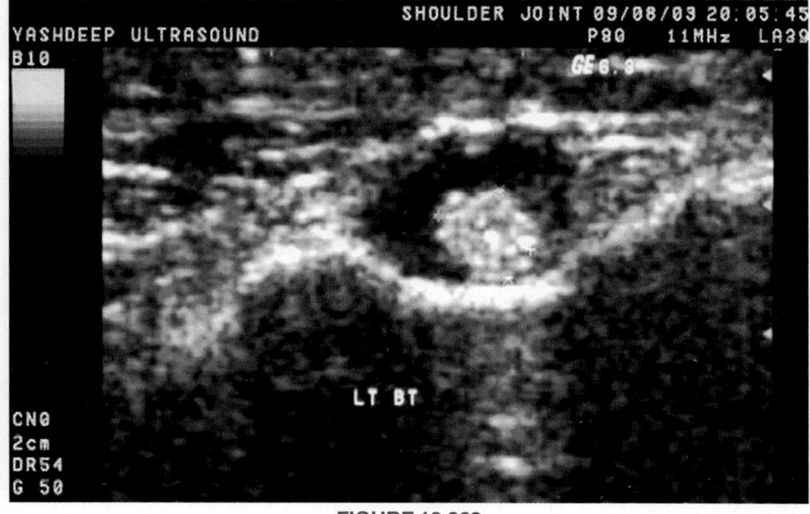

FIGURE 13.363

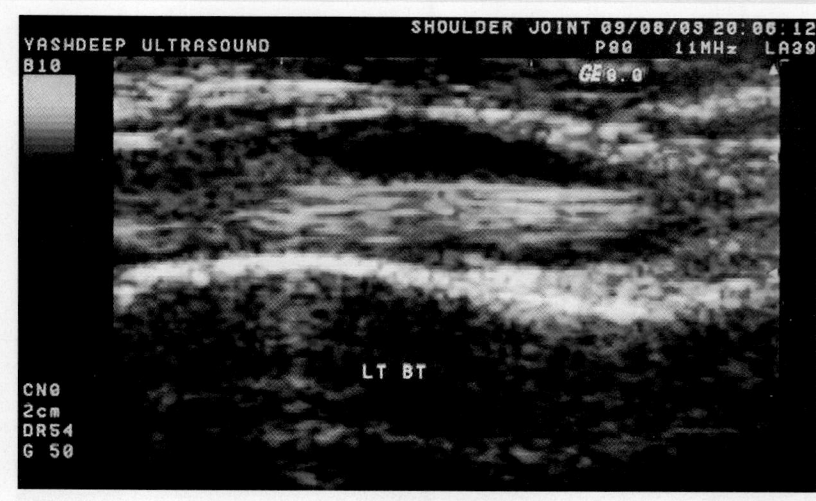

FIGURE 13.364

FIGURES 13.363 and 13.364: Hypertrophy of the biceps tendon. HRSG shows hypertrophy of the biceps tendon in a swimmer. The biceps tendon is thickened. Fluid collection is also seen in the tendon sheath. The patient presented with painful movement in the shoulder joint.

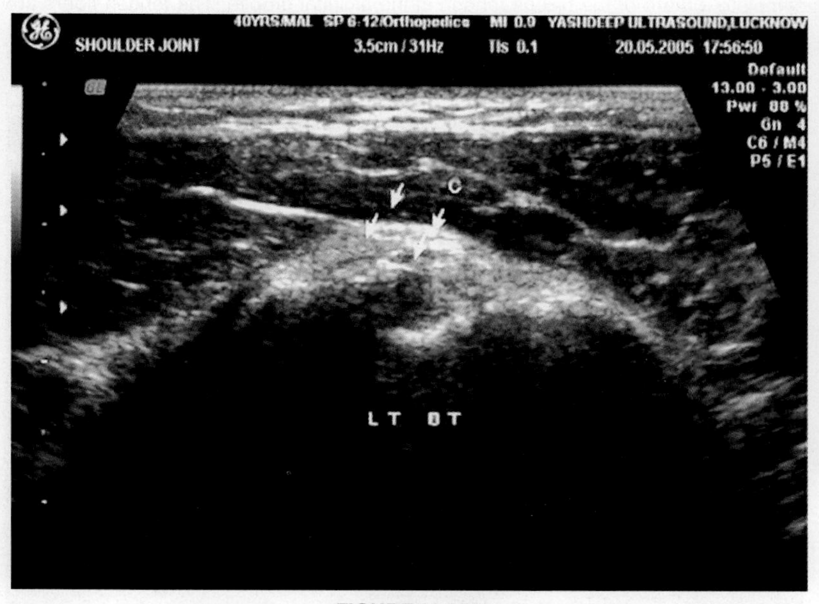

FIGURE 13.365

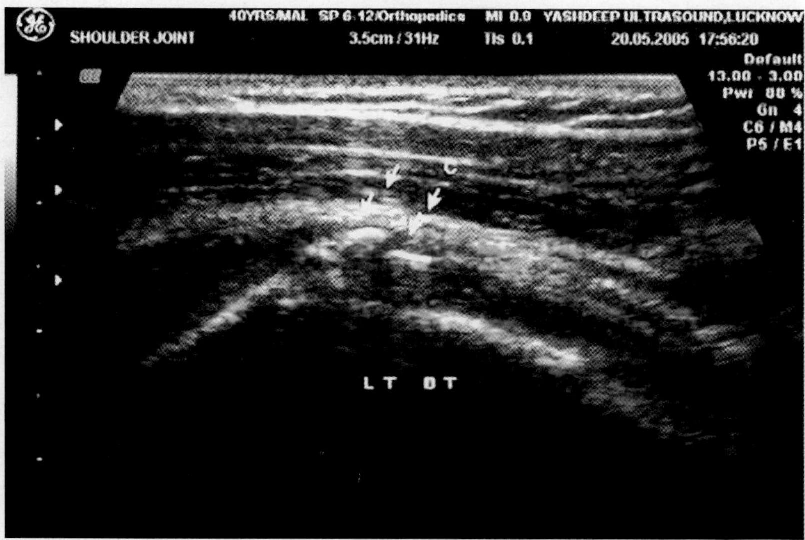

FIGURE 13.366

FIGURES 13.365 and 13.366: Chronic biceps calcific tendonitis with rupture of tendon. A patient presented with chronic pain with limited arm movements. HRSG shows multiple echogenic calcified shadows fixed at the head of biceps tendon. They are accompanied with acoustic shadowing. The tendon is also irregular in outline.

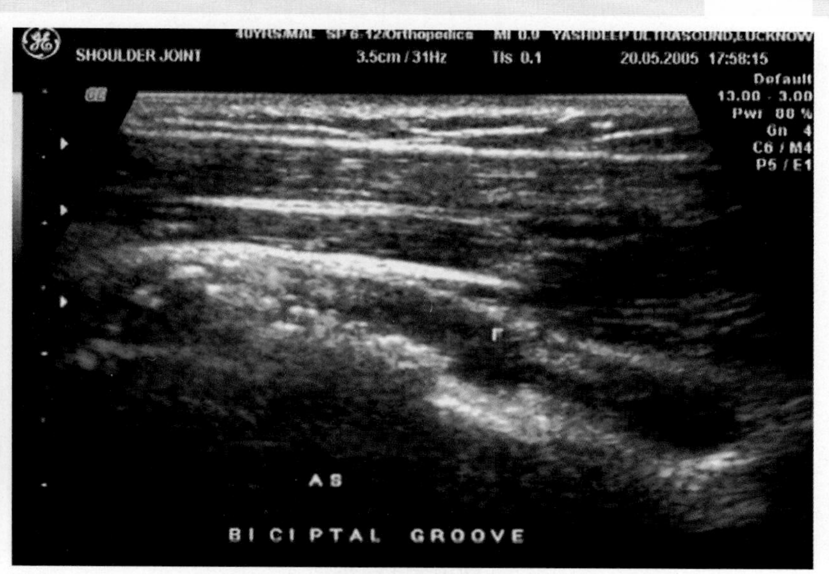

FIGURE 13.367

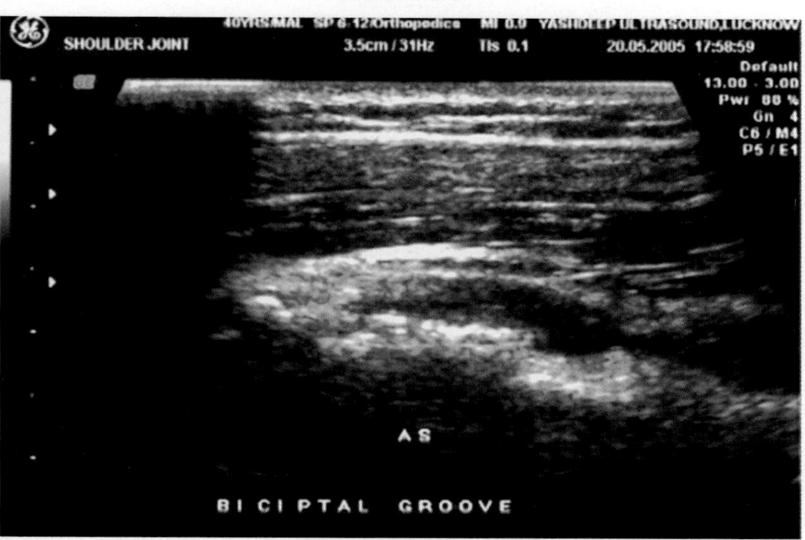

FIGURE 13.368

FIGURES 13.367 and 13.368: Partial rupture of the tendon fibers is also seen in the distal 1/3rd part of the tendon. Echogenic calcified shadows are also seen in the bicipital groove, which are adherent with the membranes.

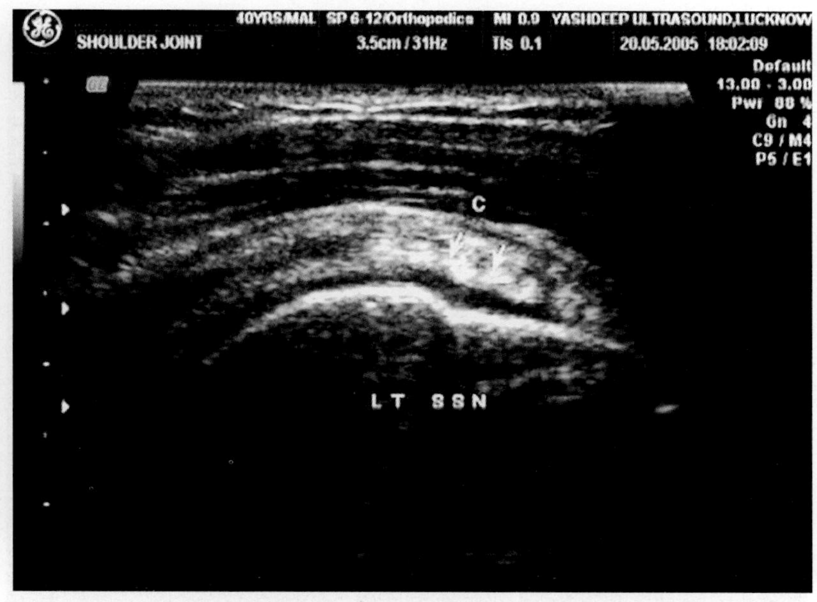

FIGURE 13.369: The supraspinatus tendon also shows multiple calcified specks in the substance of tendon. The thinning of the tendon is also seen with inhomogeneous texture suggestive of calcified tendonitis.

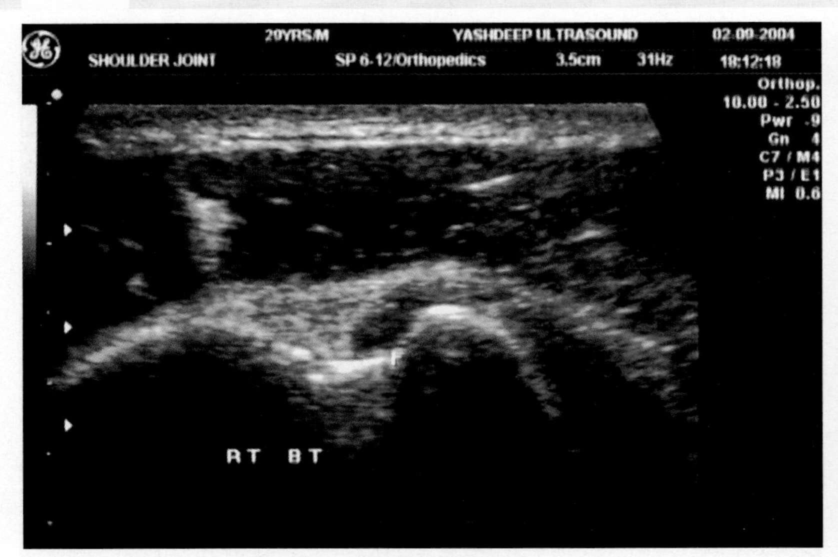

FIGURE 13.370

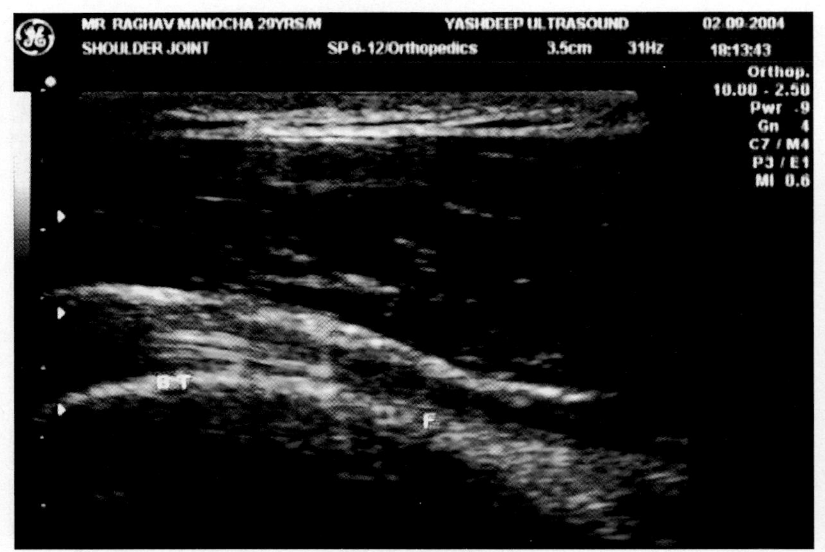

FIGURE 13.371

FIGURES 13.370 and 13.371: Complete biceps tendon tear. HRSG shows complete tear of bicep tendon with retraction of tendon. The bicipital groove is empty and the fluid is seen in the groove. No tendon shadow is seen in bicipital groove.

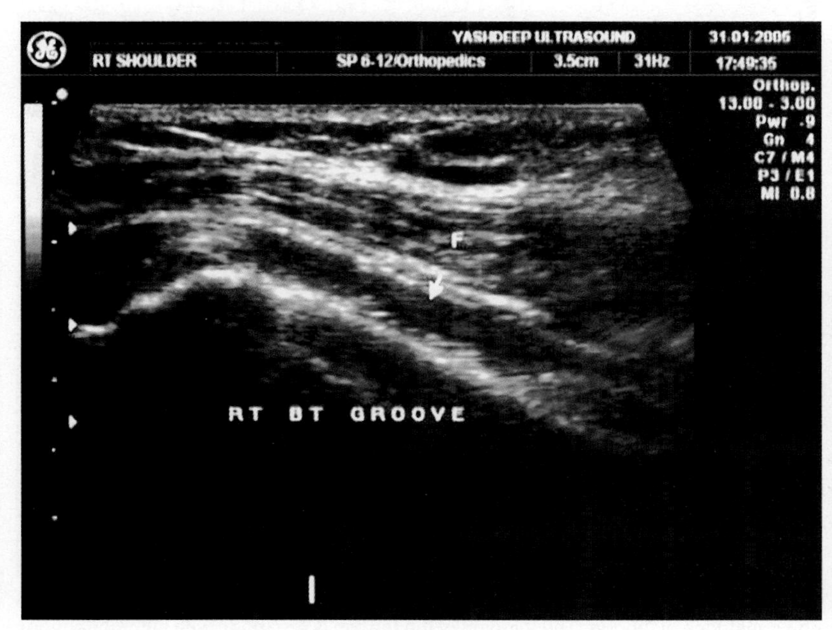

FIGURE 13.372

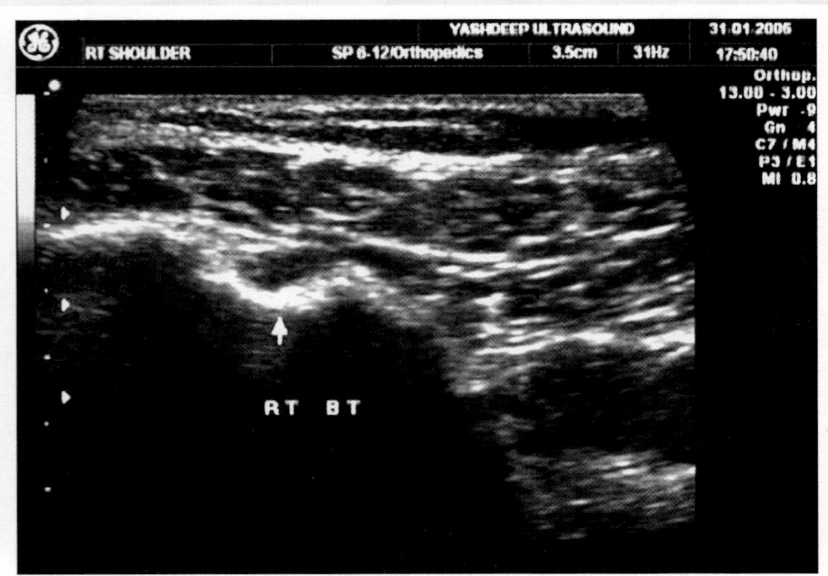

FIGURE 13.373

FIGURES 13.372 and 13.373: Complete biceps tendon rupture. HRSG shows complete rupture of the biceps tendon in the bicipital groove. The tendon fibers are not seen in the bicipital groove. Part of the tendon is seen in the distal end of the tendon. Fluid collection with low-level echo is seen in the bicipital groove suggestive of hemorrhagic collection.

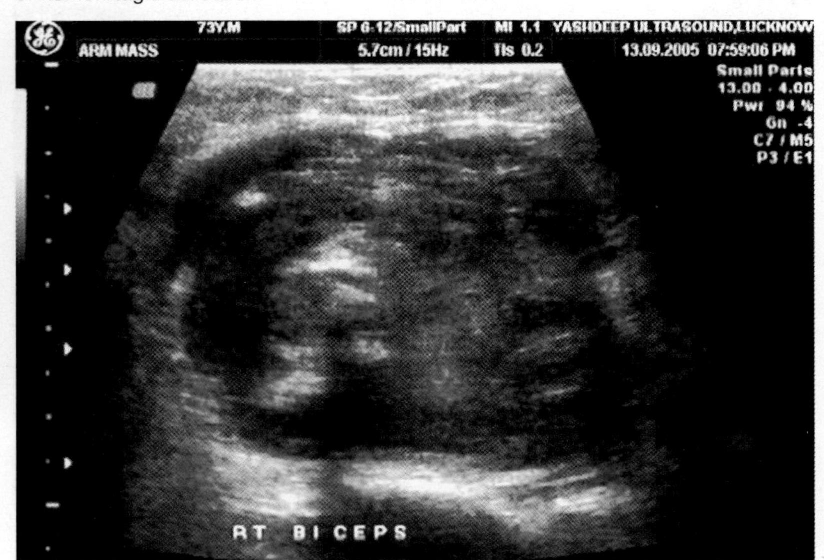

FIGURE 13.374

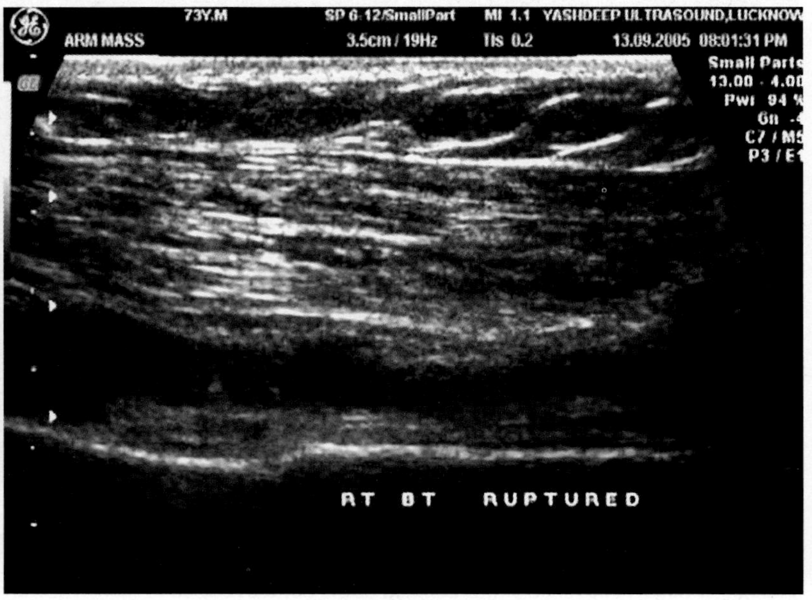

FIGURE 13.375

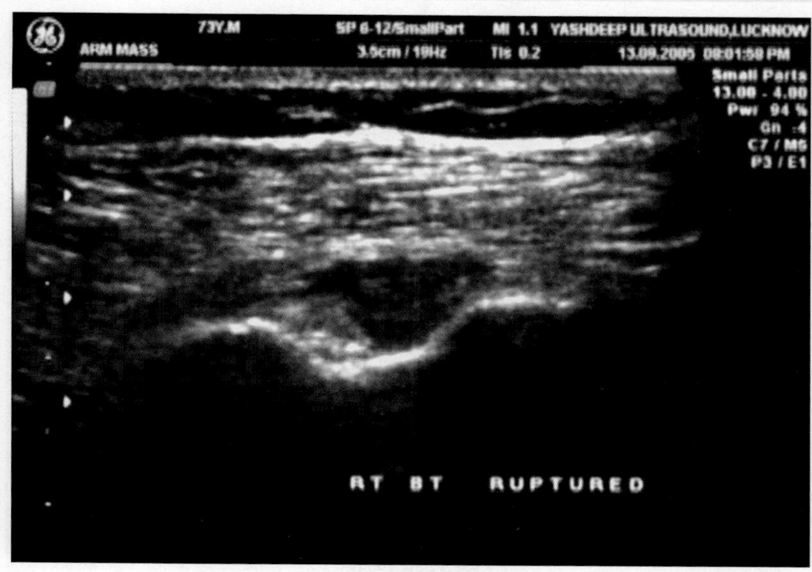

FIGURE 13.376

FIGURES 13.374 to 13.376: Biceps tendon rupture with rupture of the biceps muscles. HRSG shows acute rupture of biceps muscle in an old patient sustained trauma. Big hematoma is seen in the biceps muscle. The biceps tendon is also seen completely rupture from its insertion. The bicipital groove shows hemorrhagic collection. No tendon is seen in the bicipital groove.

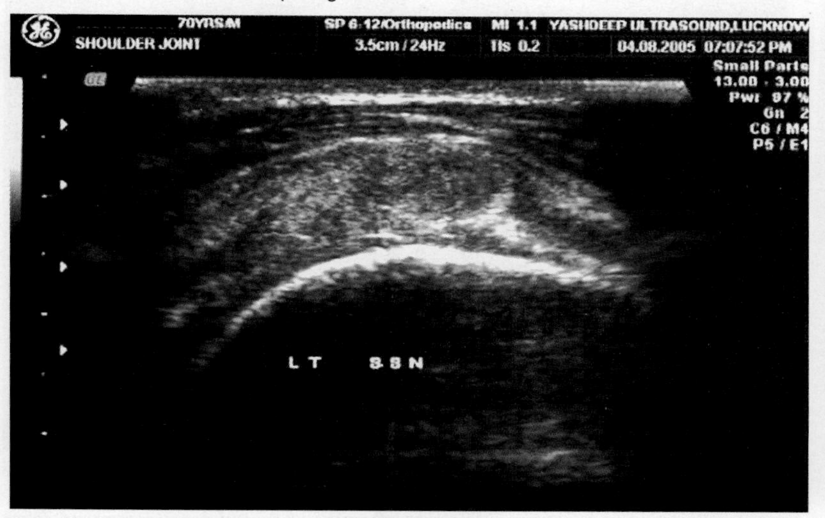

FIGURE 13.377

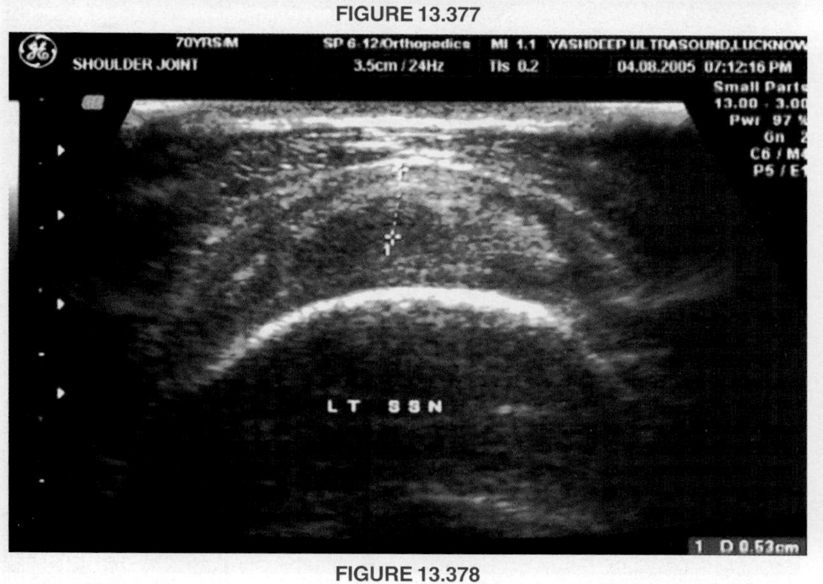

FIGURE 13.378

FIGURES 13.377 and 13.378: Acute supraspinatus tendon tear with hemorrhage. HRSG shows edematous echogenic supraspinatus tendon. The tendon is edematous and echogenic in texture. Normal tendon texture is lost. Hemorrhagic collection is seen in the substance of the tendon suggestive acute injury to the tendon.

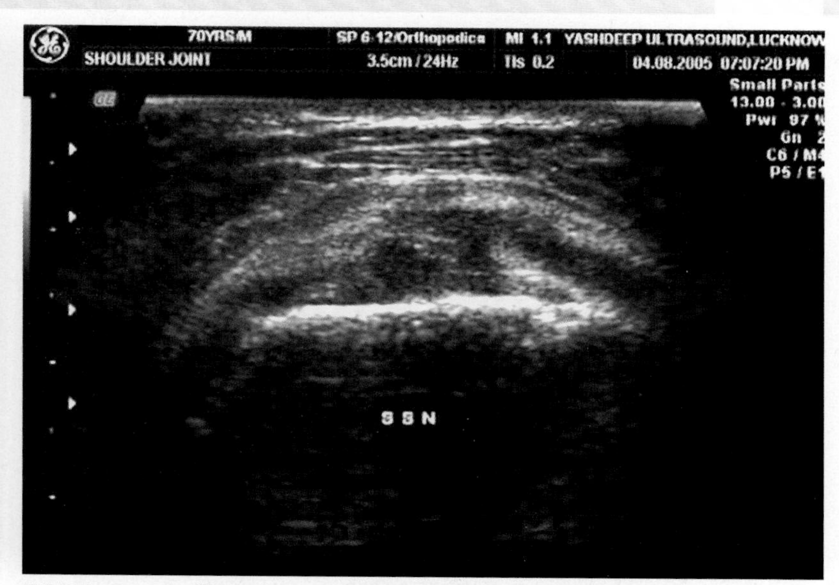

FIGURE 13.379: There is also evidence of irregular hypoechoic tear seen in the substance of tendon in the same patient. The tendon is inhomogeneous and swelled up.

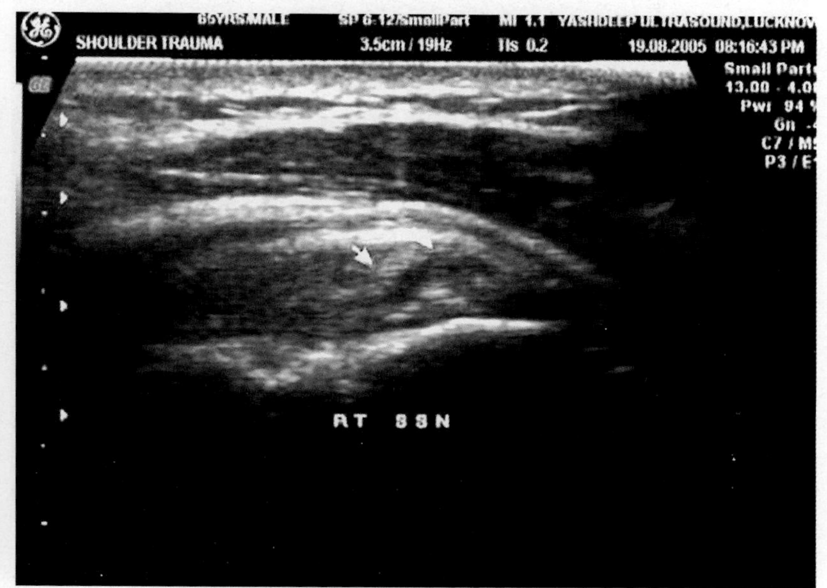

FIGURE 13.380

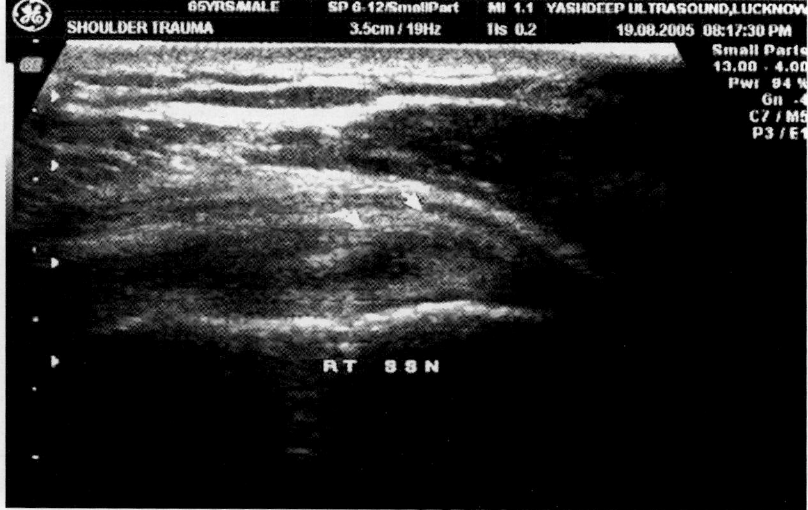

FIGURE 13.381

FIGURES 13.380 and 13.381: Supraspinatus tear. HRSG shows an obliquely running hypoechoic cleft in the substance of tendon in a patient presented with limited shoulder movement with pain. Disruption of the tendon fibers is seen in the tear. However, the tear was seen confined to substance of the tendon.

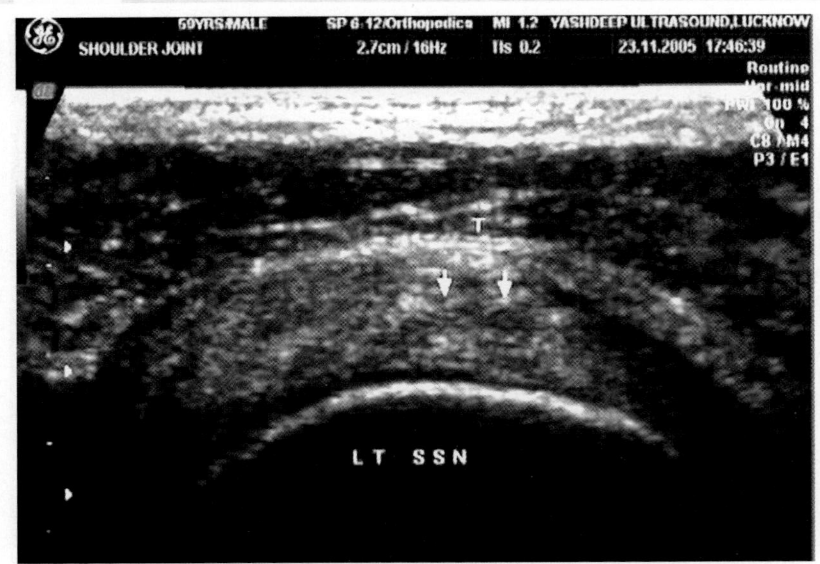

FIGURE 13.382

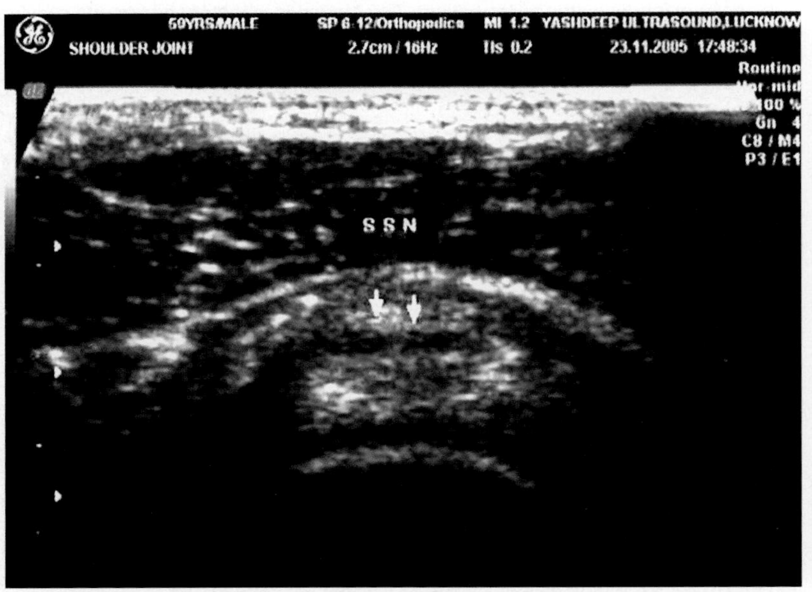

FIGURE 13.383

FIGURES 13.382 and 13.383: Supraspinatus tendon partial tear. HRSG shows partial tear in the supraspinatus tendon. A hypoechoic cleft seen in the supraspinatus tendon. Tendon was inhomogeneous in texture.

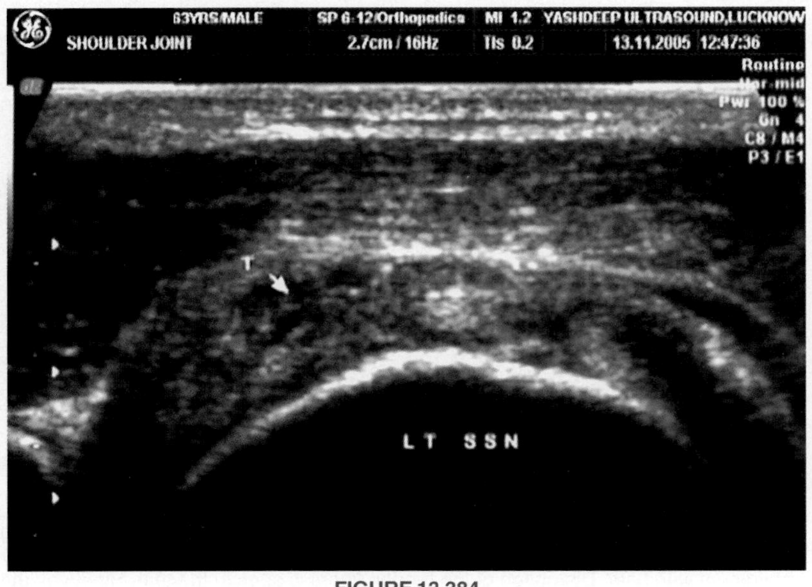

FIGURE 13.384

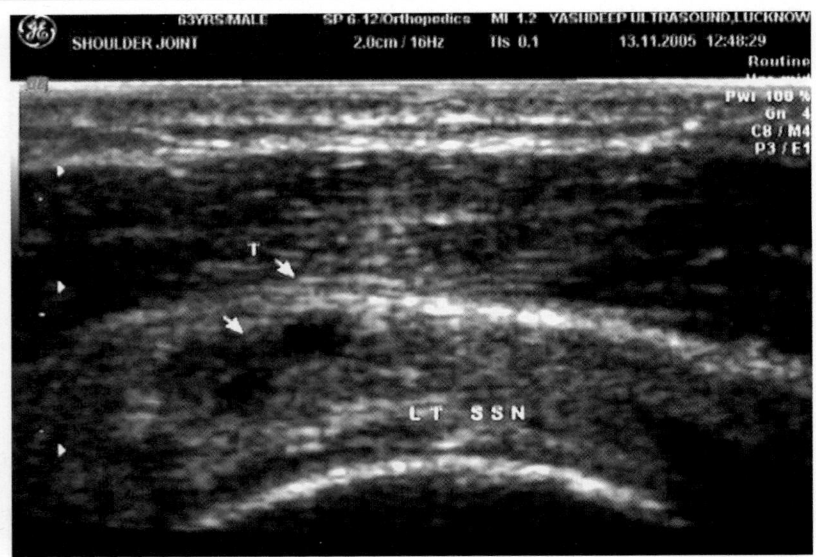

FIGURE 13.385

FIGURES 13.384 and 13.385: Focal supraspinatus tear. HRSG shows focal tear in the supraspinatus tendon in a patient presented with painful shoulder syndrome.

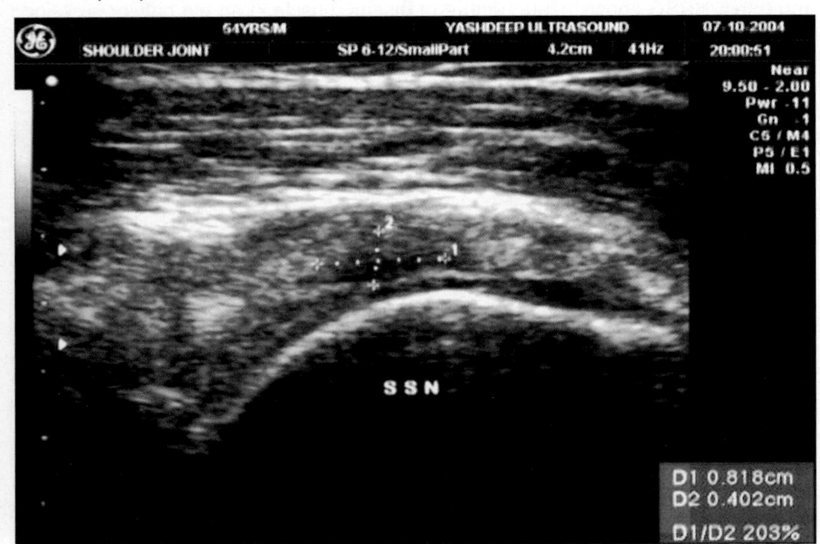

FIGURE 13.386: Focal tear of supraspinatus tendon. HRSG shows small hypoechoic cleft in the supraspinatus tendon in its mid part.

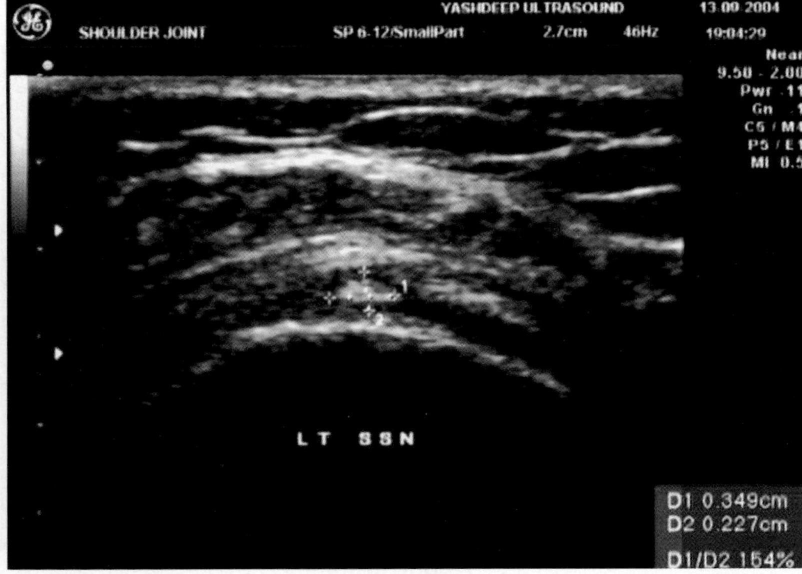

FIGURE 13.387: Echogenic fibers nodule in supraspinatus tendon. A patient presented with painful shoulder movement. HRSG shows an echogenic nodular shadow in the substance of tendon. No calcification seen in it. The tendon is also inhomogeneous in texture suggestive of fibrotic nodule.

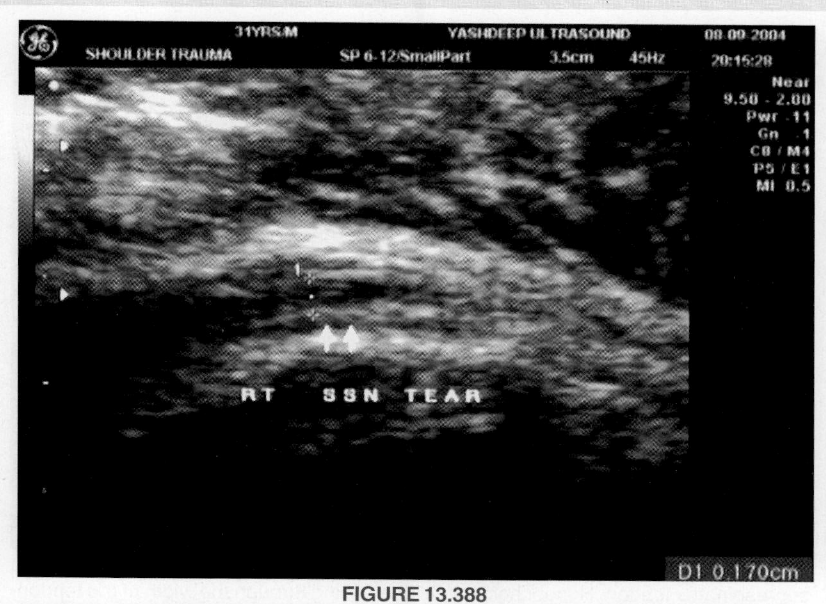

FIGURE 13.388

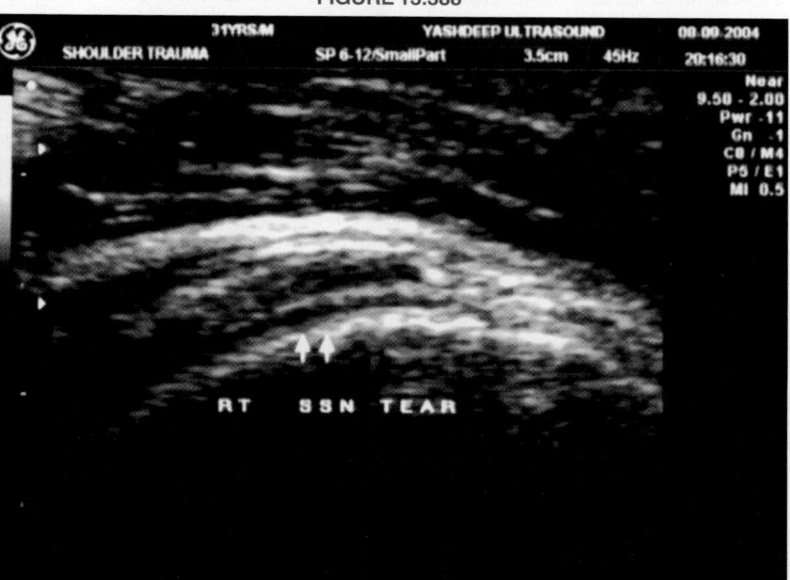

FIGURE 13.389

FIGURES 13.388 and 13.389: Partial tear of supraspinatus tendon. HRSG shows an irregular cleft in the substance of tendon suggestive of partial tear in the tendon.

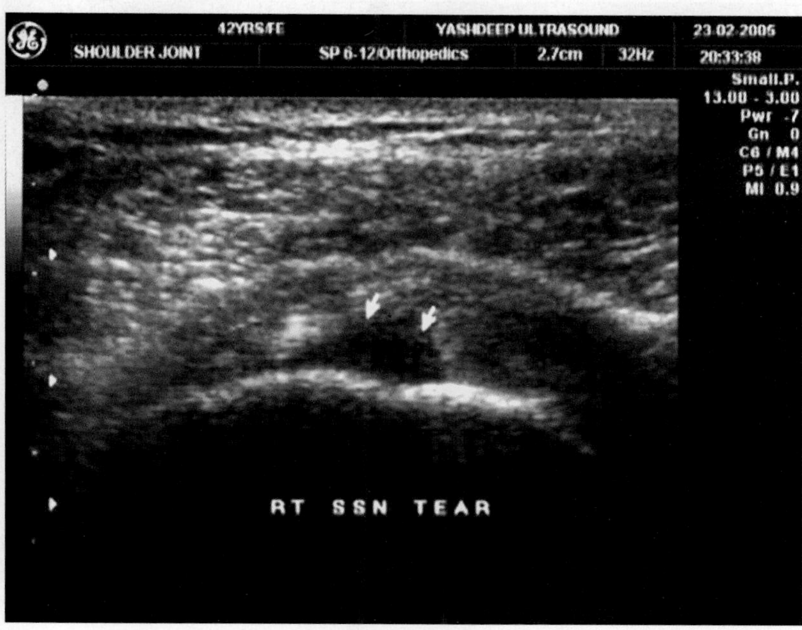

FIGURE 13.390

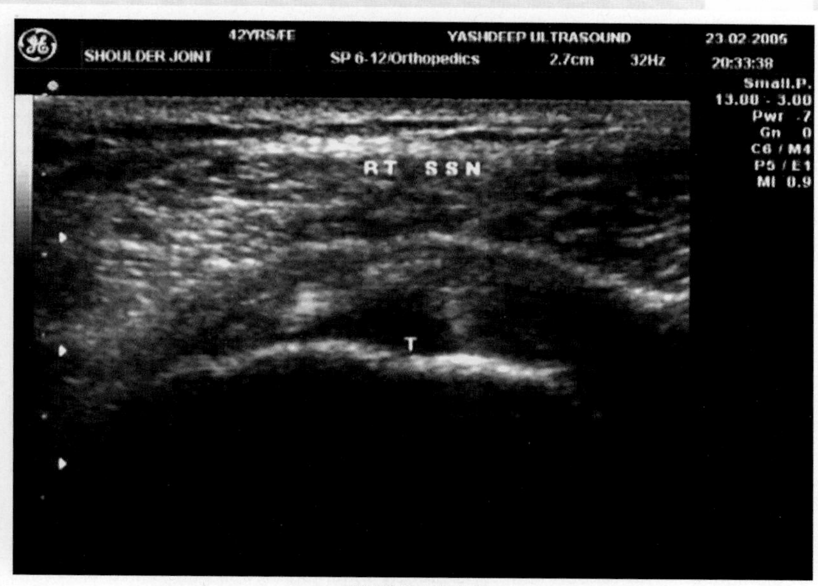

FIGURE 13.391

FIGURES 13.390 and 13.391: Big tear in the supraspinatus tendon. A big irregular hypoechoic defect is seen in the substance of supraspinatus tendon. Normal tendon fibers are lost. Whole of the tendon is swelled up. Small amount of collection is also seen in the cleft suggestive of recent cleft.

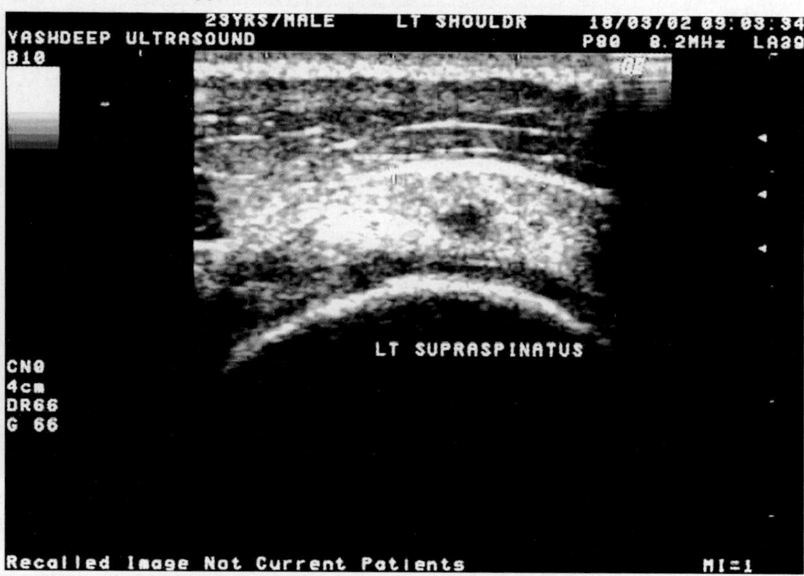

FIGURE 13.392: Focal cleft in Lt supraspinatus tendon. A young badminton player patient presented with painful movements elevation of the shoulder joint. HRSG shows an irregular hypoechoic focus in the substance of the tendon suggestive of focal tear. Margins are irregular in outline. No calcification is seen around it.

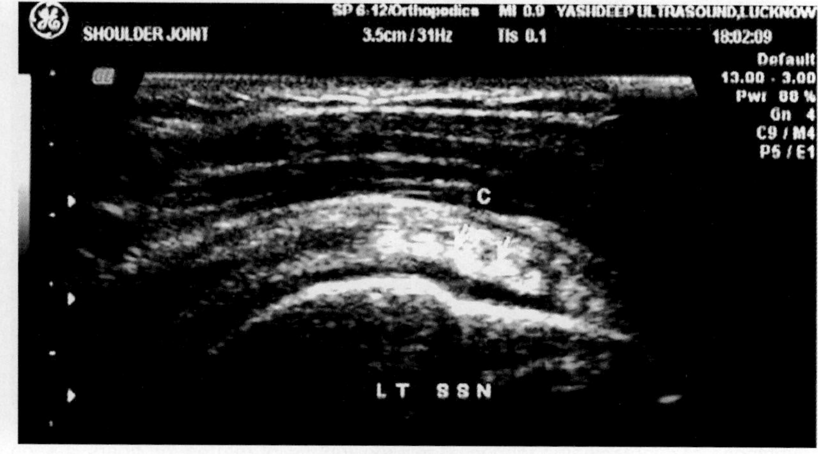

FIGURE 13.393

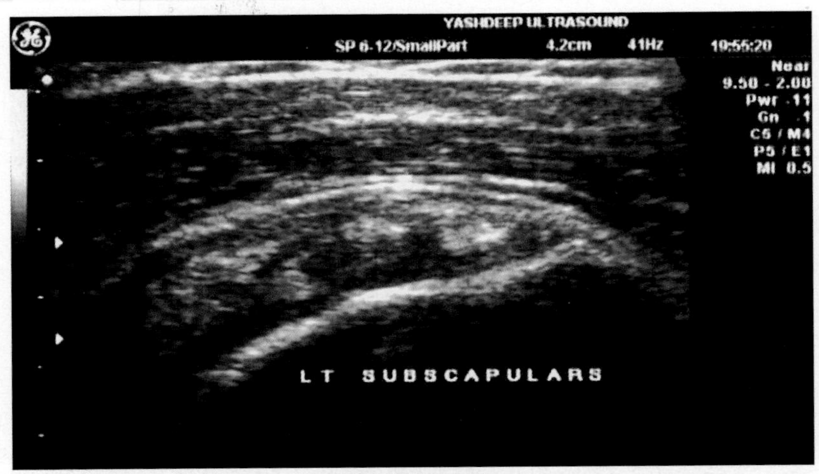

FIGURE 13.394

FIGURES 13.393 and 13.394: Calcific tendonitis. HRSG shows multiple echogenic-calcified shadows in the substance of the supraspinatus tendon. They are discrete in distribution. The tendon texture is grossly inhomogeneous suggestive of calcific tendonitis. The subscapularis tendon in the same patient also shows focal hypoechoic cleft suggestive of tear in the tendon.

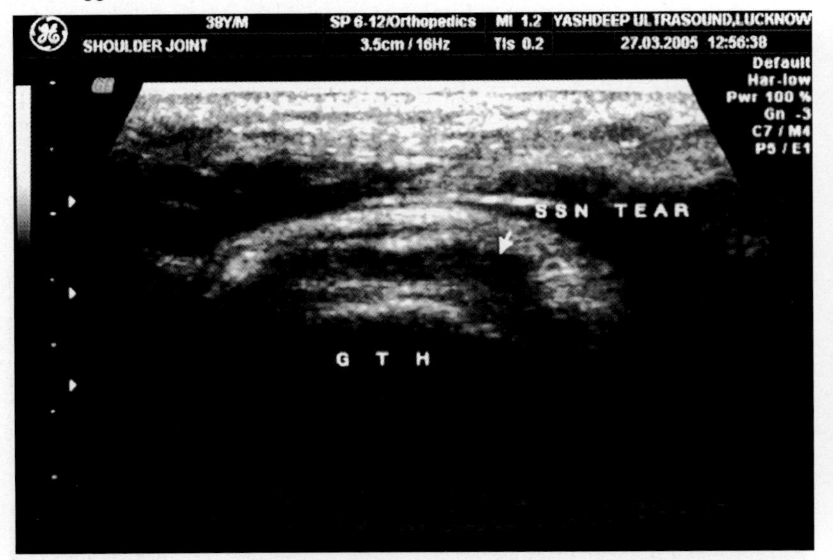

FIGURE 13.395: Big supraspinatus tear. HRSG shows a big hypoechoic cleft in the Rt supraspinatus tendon. Complete loss of the tendon fibers is seen. A big hypoechoic cleft is seen in the tendon suggestive of big tear.

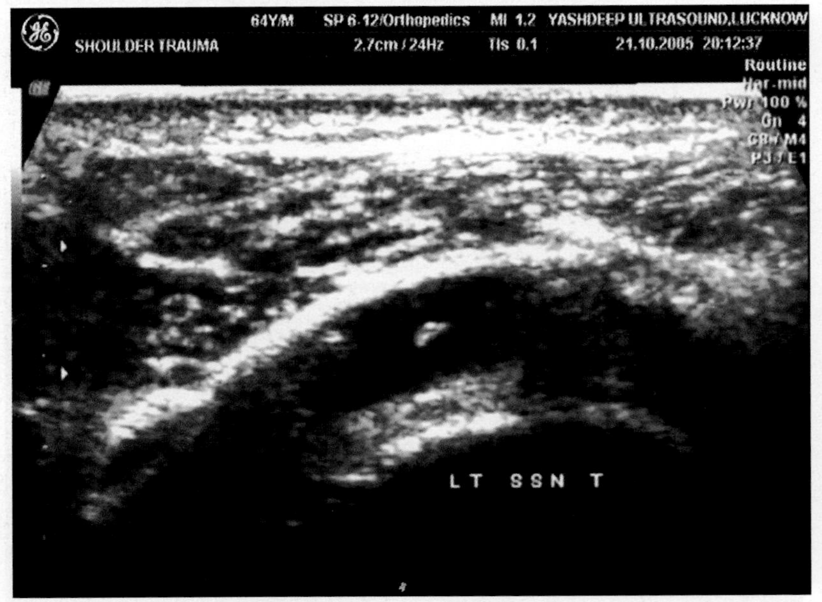

FIGURE 13.396: Big supraspinatus tear. A big irregular hypoechoic tear in the substance of supraspinatus tendon. Complete rupture of the tendon fibers is seen with big hematoma formation.

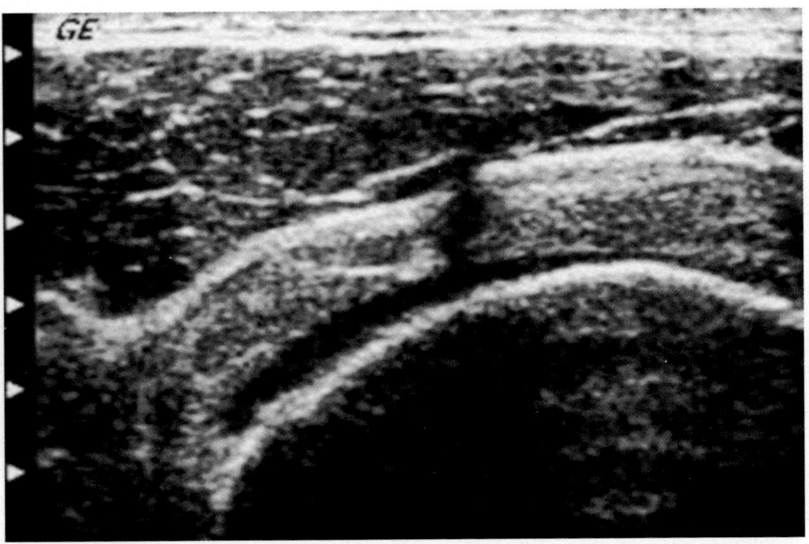

FIGURE 13.397: Complete supraspinatus tear. HRSG shows complete tear in supraspinatus tendon. Hypoechoic tear is seen going through the width of the tendon.

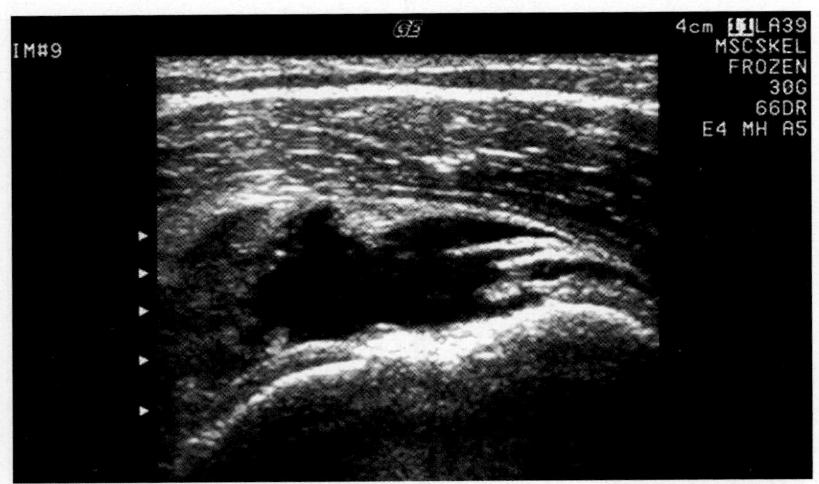

FIGURE 13.398: Acute rupture of tendon. HRSG shows complete acute rupture of supraspinatus tendon. Complete loss of the tendon is seen (*Courtesy*—GE medical system image library).

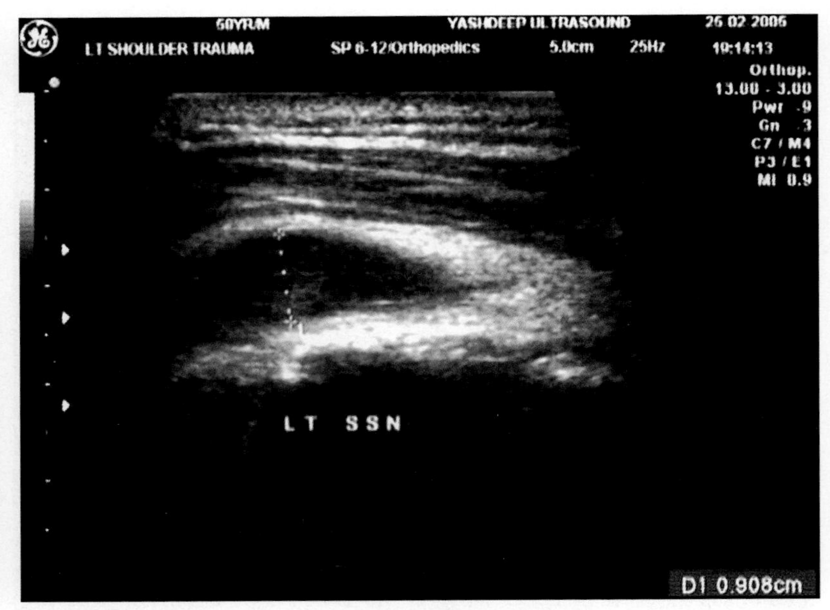

FIGURE 13.399

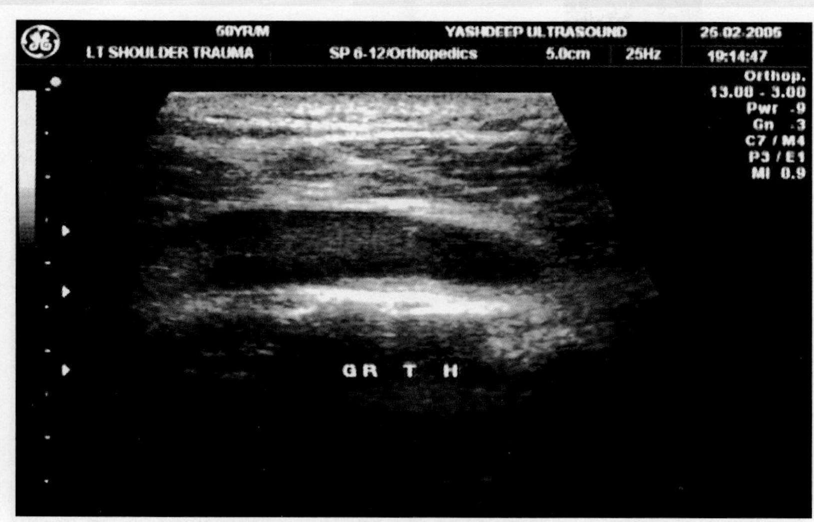

FIGURE 13.400

FIGURES 13.399 and 13.400: Complete acute rupture of supraspinatus tendon. A patient sustained fracture head of Lt humerus. HRSG shows complete rupture of the supraspinatus tendon. Thick echo collection is seen in the substance of tendon suggestive of hematoma formation. The greater tuberosity of the humerus was also flat and normal convex contour is lost.

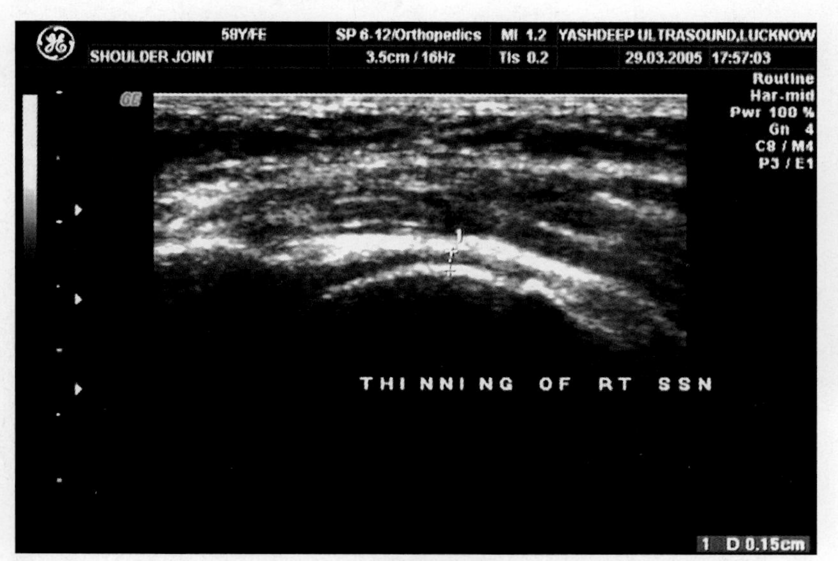

FIGURE 13.401

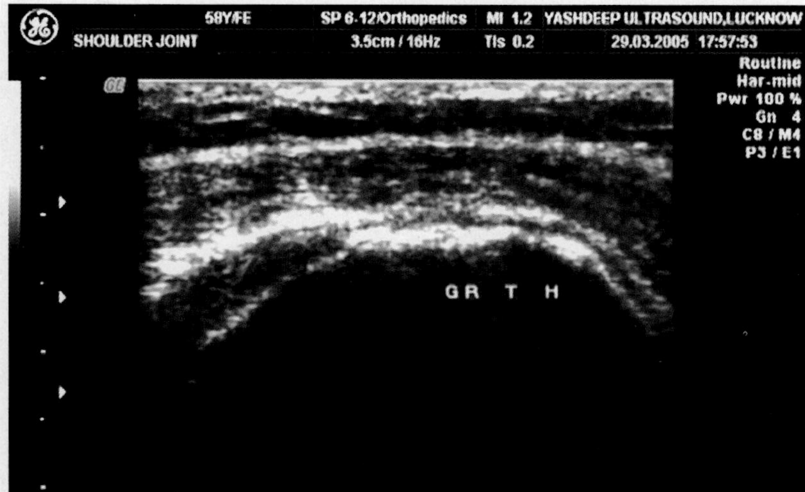

FIGURE 13.402

FIGURES 13.401 and 13.402: Complete loss of rotator cuff. HRSG shows complete loss of supraspinatus tendon in man. Normal tendon fibers are not seen. Marked thinning of the tendon is seen. The articular surface of the greater tuberosity of the humerus shows irregular outline with brittling. Echogenic calcified shadows are seen over the cartilage suggestive of degenerative changes.

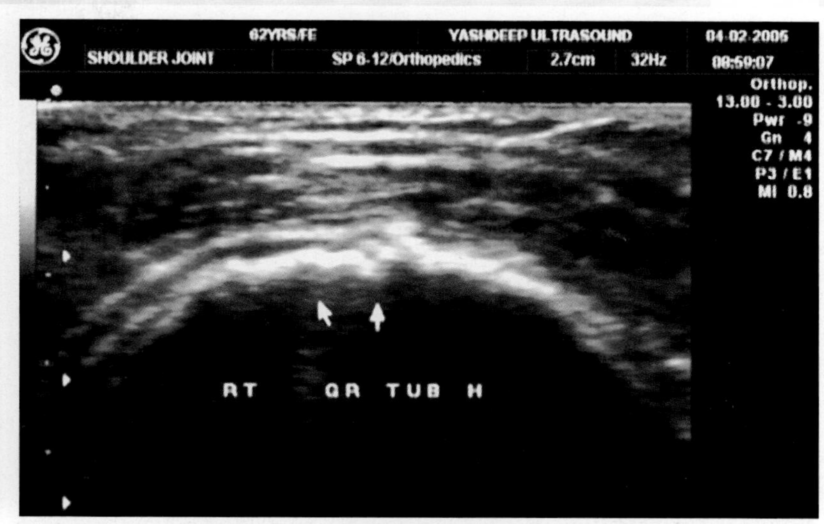

FIGURE 13.403

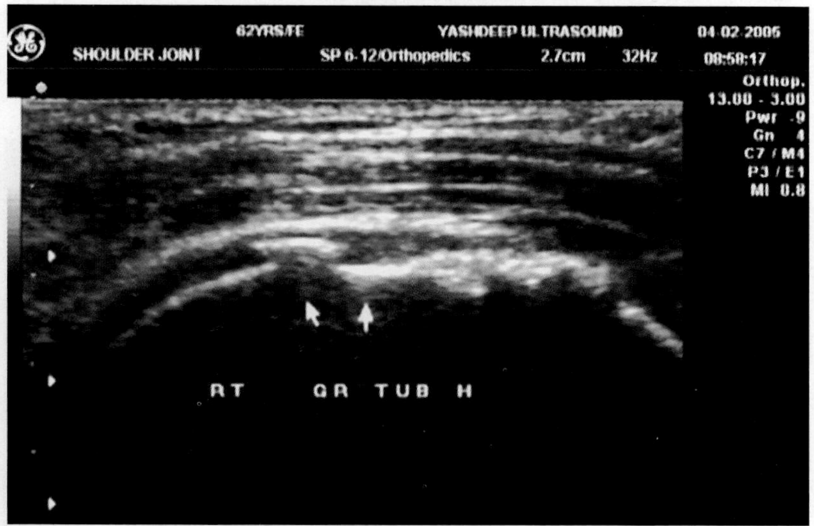

FIGURE 13.404

FIGURES 13.403 and 13.404: Complete rotator cuff tear with fracture of greater tuberosity of humerus. HRSG shows complete loss of the supraspinatus tendon in an old lady. The cartilage is also irregular in outline. There is also evidence of fracture of greater tuberosity of the humerus seen. The fractured segment is seen displaced anteriorly causing pain in the movement.

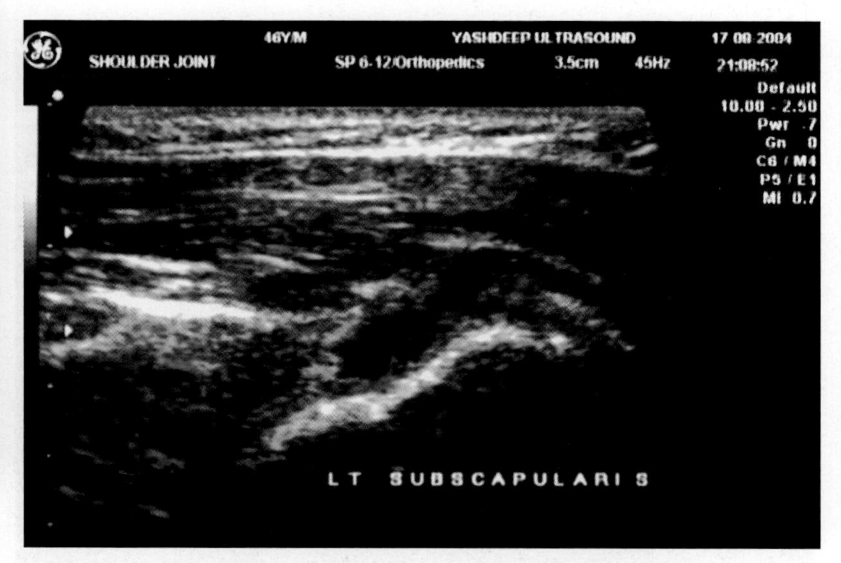

FIGURE 13.405: Lt subscapularis tear (anterior cuff). HRSG shows big irregular hypoechoic cleft in a patient who sustained direct trauma to the anterior cuff by banging the shoulder against a glass door. Big hypoechoic cleft seen in the anterior cleft.

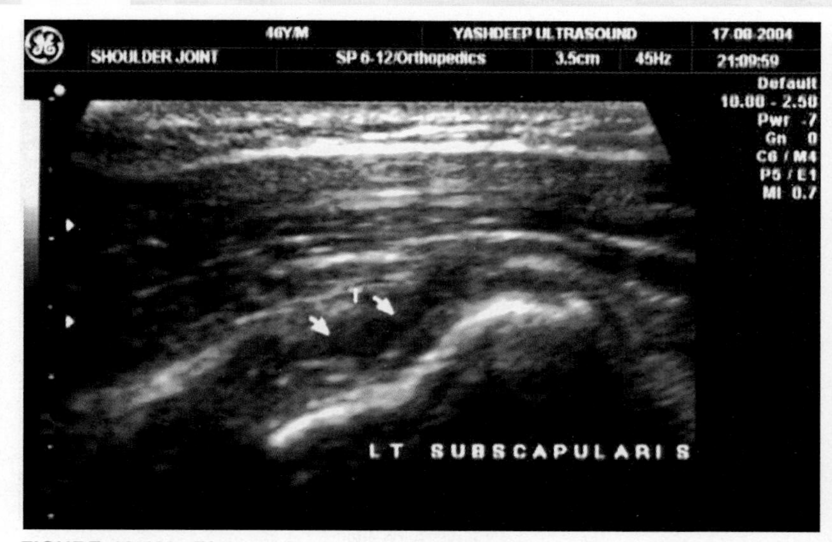

FIGURE 13.406: Big tear in the Lt subscapularis tendon. HRSG shows a big irregular hypoechoic cleft in the Lt subscapularis tendon. The tendon is edematous and swelled up.

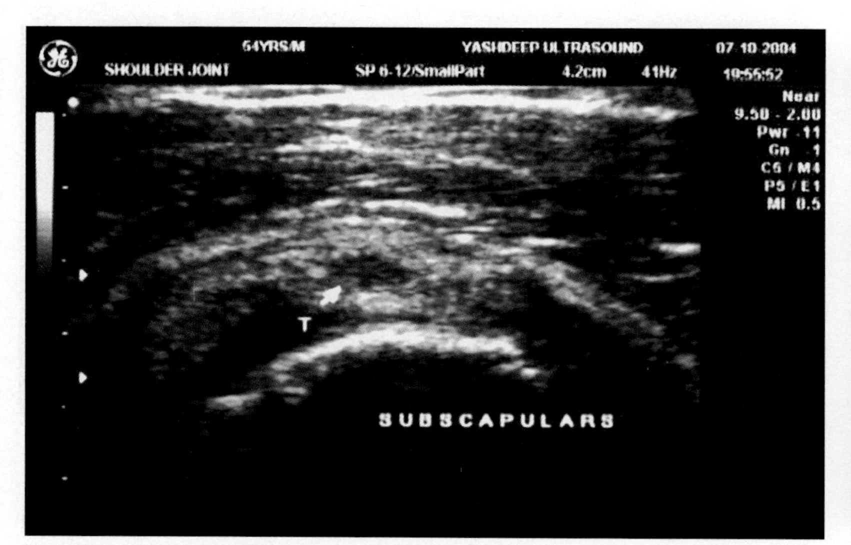

FIGURE 13.407: Focal tear in the Rt supraspinatus tendon. HRSG shows a small focal cleft in the Rt supraspinatus tendon. The tear is seen confined to the substance of the tendon suggestive of intrasubstance tear.

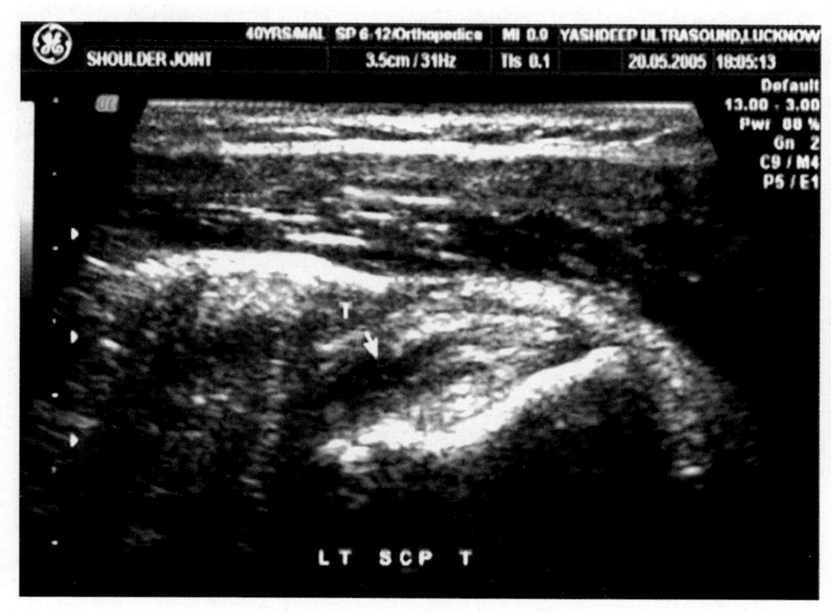

FIGURE 13.408

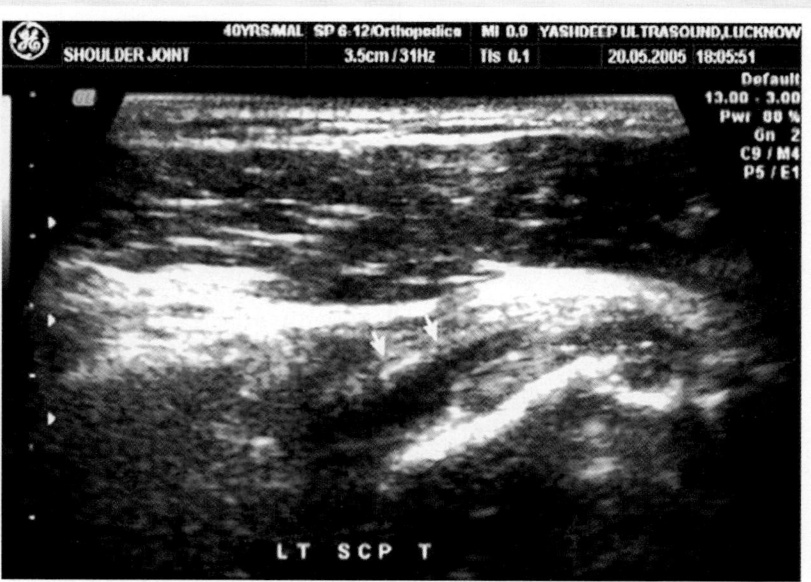

FIGURE 13.409

FIGURES 13.408 and 13.409: Big anterior cuff tear. HRSG shows a big tear in the Lt anterior cuff. Margins are irregular in outline and echogenic. Tear is seen running through substance of the tendon.

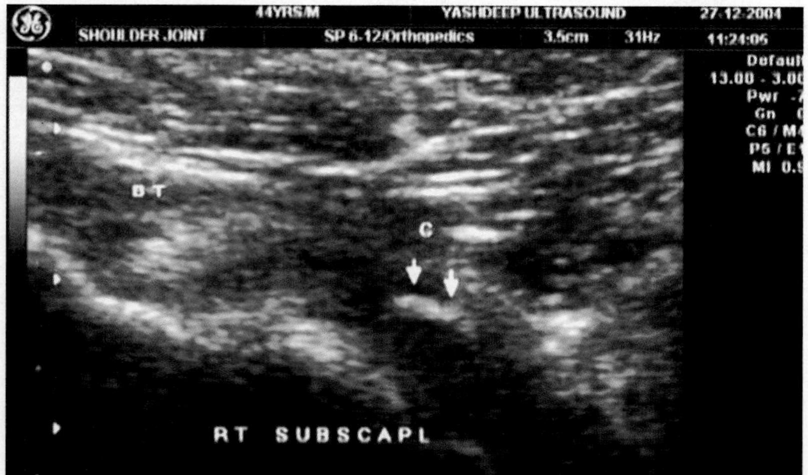

FIGURE 13.410

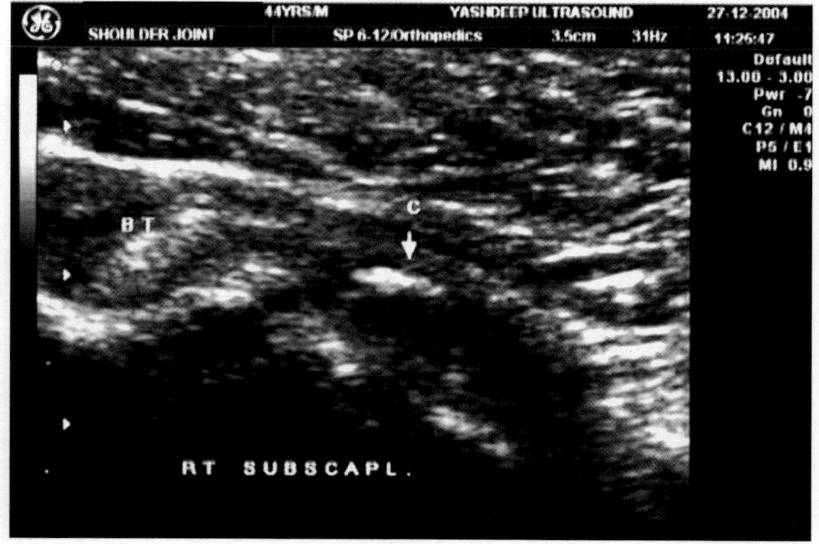

FIGURE 13.411

FIGURES 13.410 and 13.411: Chronic calcific tendonitis. A patient presented with marked painful movement of the Rt shoulder joint mainly on internal rotation of the arm. HRSG shows an echogenic calcified plaque fixed in the substance of the anterior cuff. It is accompanied with acoustic shadowing. The subscapularis tendon is also hypoechoic in texture and irregular in outline suggestive of chronic calcific tendonitis.

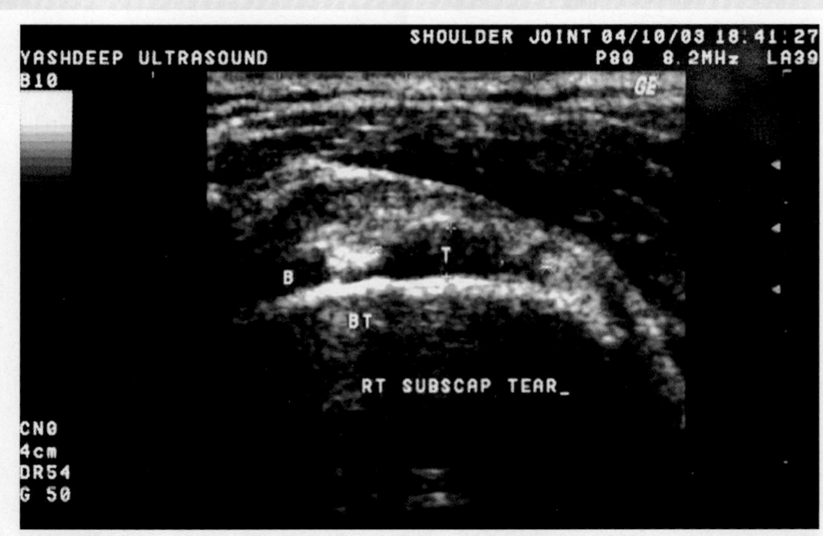

FIGURE 13.412: Acute rupture of the subscapularis tendon. HRSG shows acute rupture of subscapularis tendon in a patient had frequent forward movement with internal rotation of the arm. A big hypoechoic cleft seen in the anterior cuff with loss of tendon fibers.

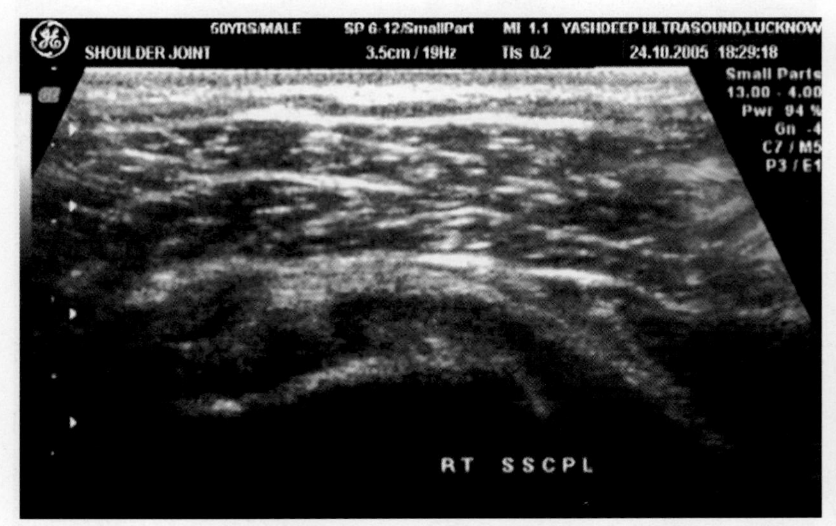

FIGURE 13.413: Acuter tendon trauma. HRSG shows grossly edematous tendon with hypoechoic texture. Normal texture is lost in a patient sustained trauma to the anterior cuff.

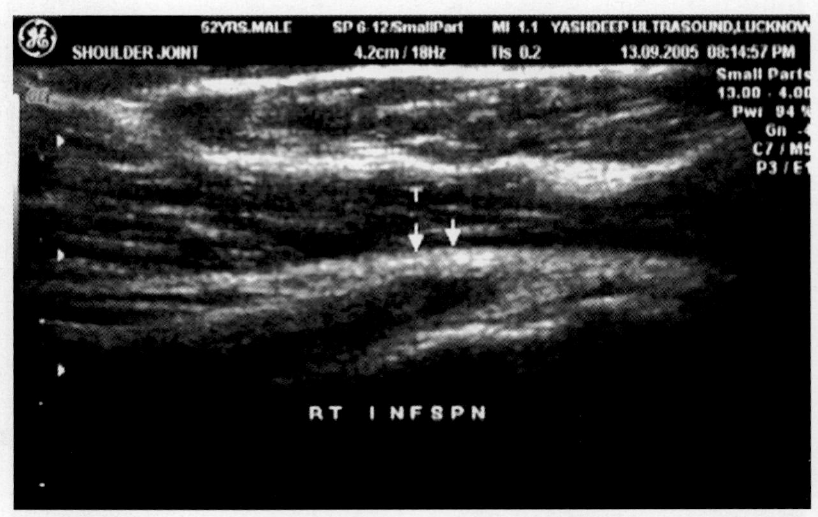

FIGURE 13.414: Posterior cuff tear. HRSG shows a hypoechoic cleft in the infraspinatus tendon in a patient sustained trauma. The cleft is seen running through the substance of the tendon.

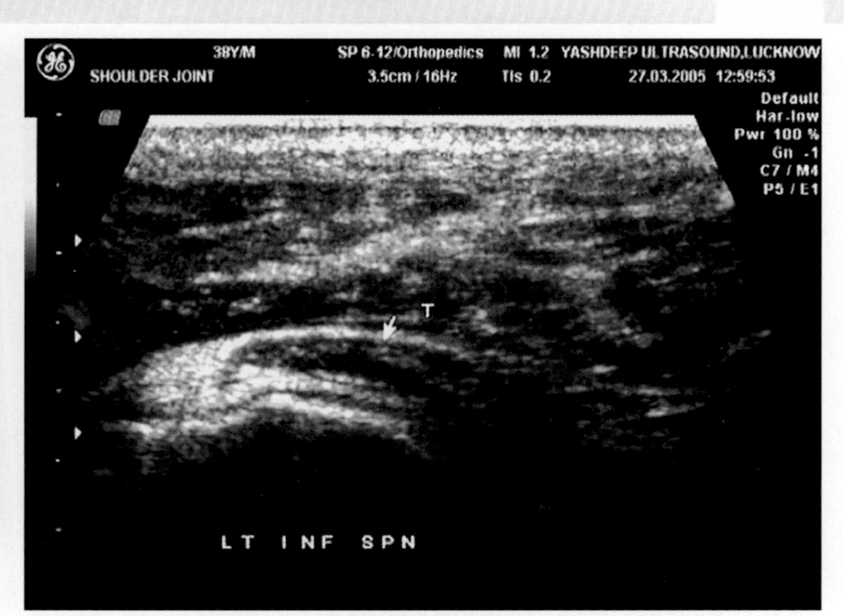

FIGURE 13.415: HRSG shows a big hypoechoic cleft in the substance of the Lt infraspinatus tendon. It is running through to the tendon substance. Tendon is inhomogeneous in texture.

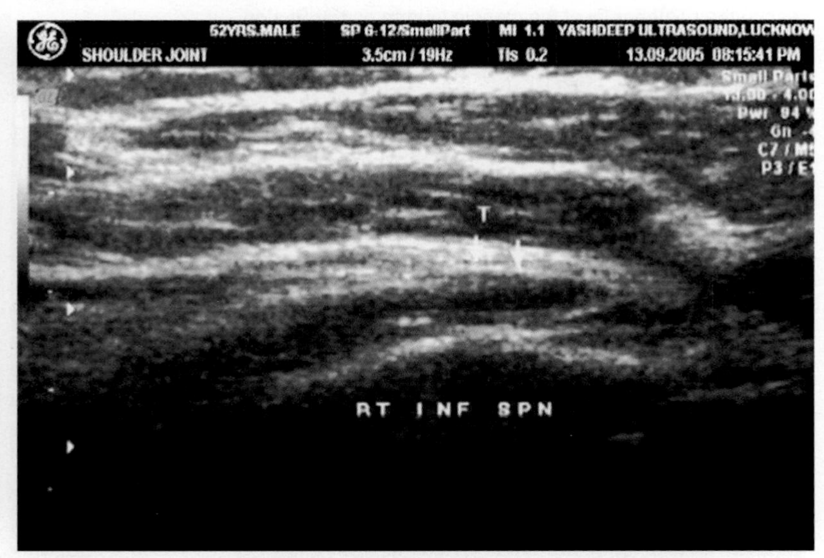

FIGURE 13.416: Rt infraspinatus tear. Well-defined hypoechoic cleft is seen in the substance of Rt infraspinatus tendon near its insertion suggestive of a tear.

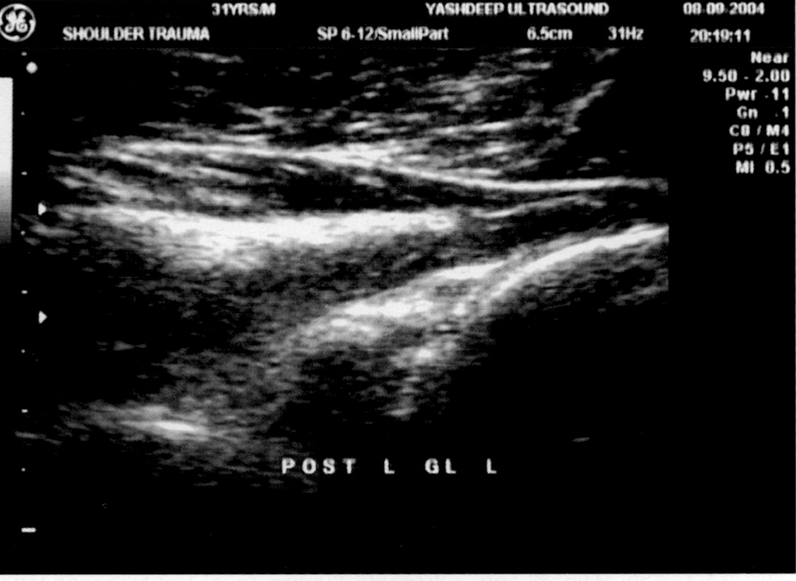

FIGURE 13.417

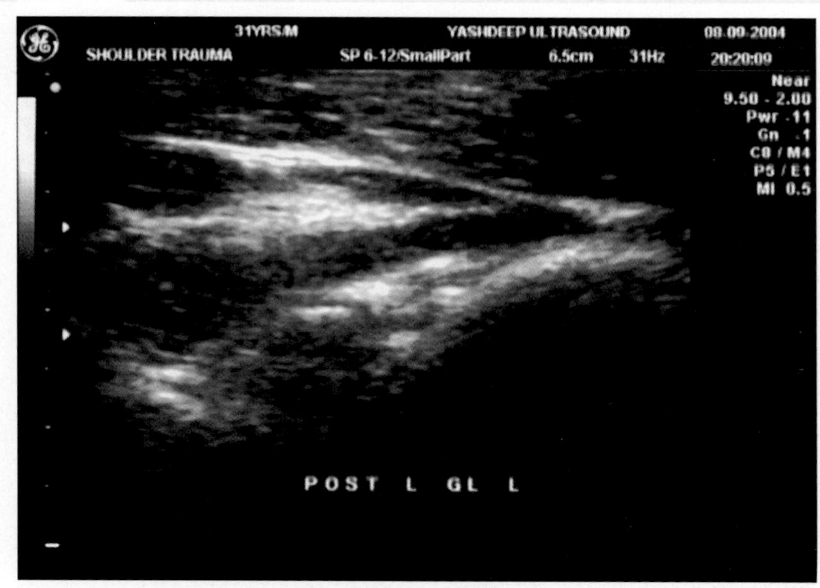

FIGURE 13.418

FIGURES 13.417 and 13.418: Glenoid labrum tear. HRSG shows irregular disrupted posterior glenoid labrum in a patient sustained trauma. There is evidence of discontinuity seen in the posterior lip of the glenoid labrum on the Rt side.

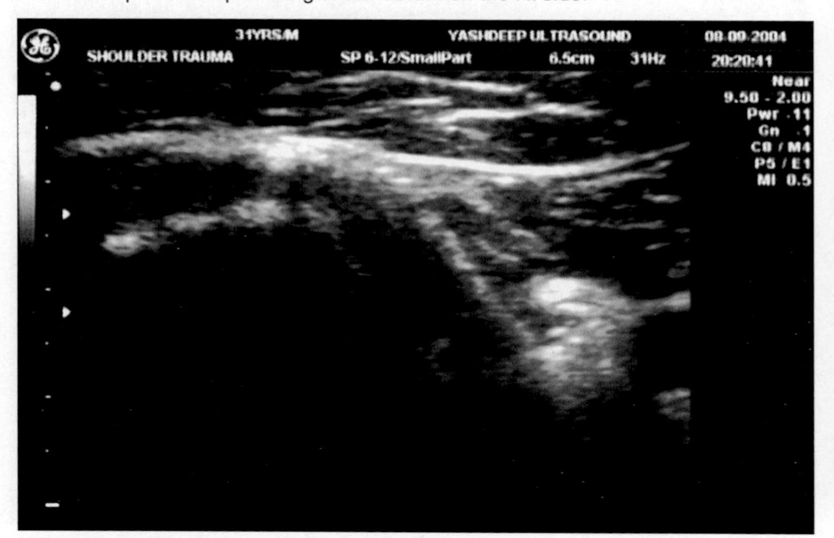

FIGURE 13.419: Lt side of the glenoid labrum shows an smooth echogenic line suggestive of intact glenoid labrum.

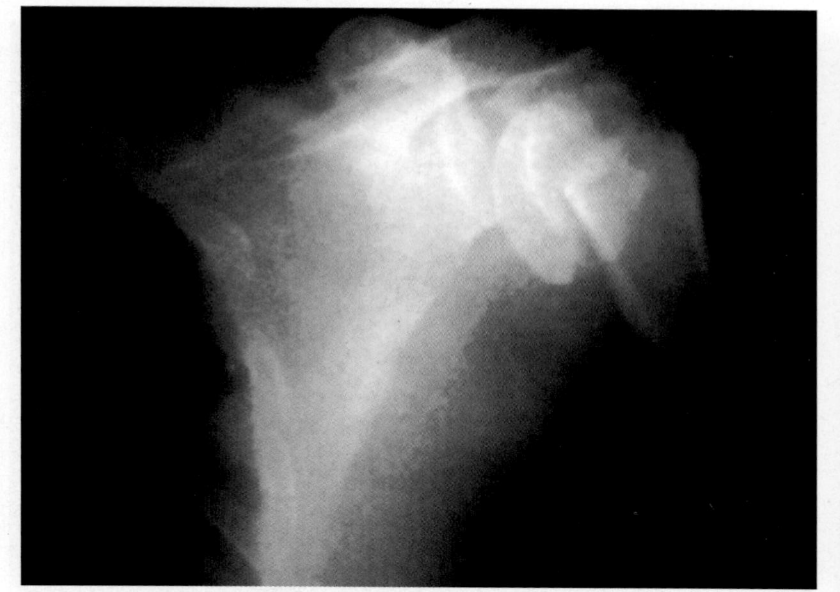

FIGURE 13.420: Hemorrhagic subdeltoid bursitis (post traumatic). X-ray of the patient shows impacted fracture neck of humerus. The head is seen dislocated downward and impacted in the joint cavity.

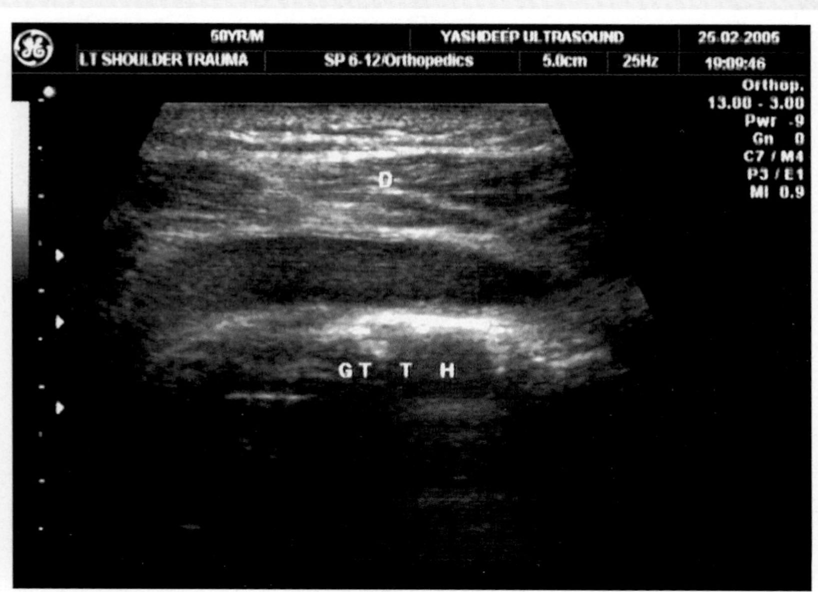

FIGURE 13.421A

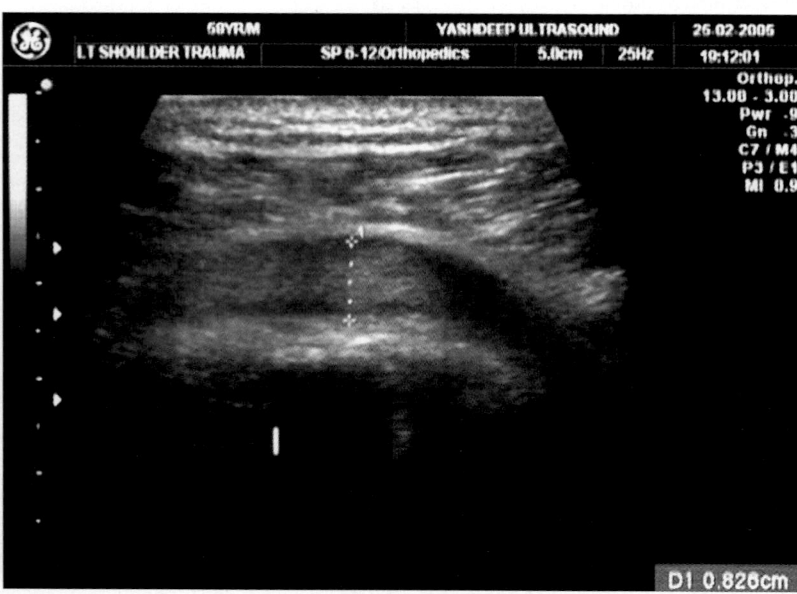

FIGURE 13.421B

FIGURES 13.421A and B: HRSG shows gross amount of fluid collection seen in the deltoid bursa. The supraspinatus tendon is completely ruptured. The greater tuberosity of the humerus is flat and lost its normal convex contour.

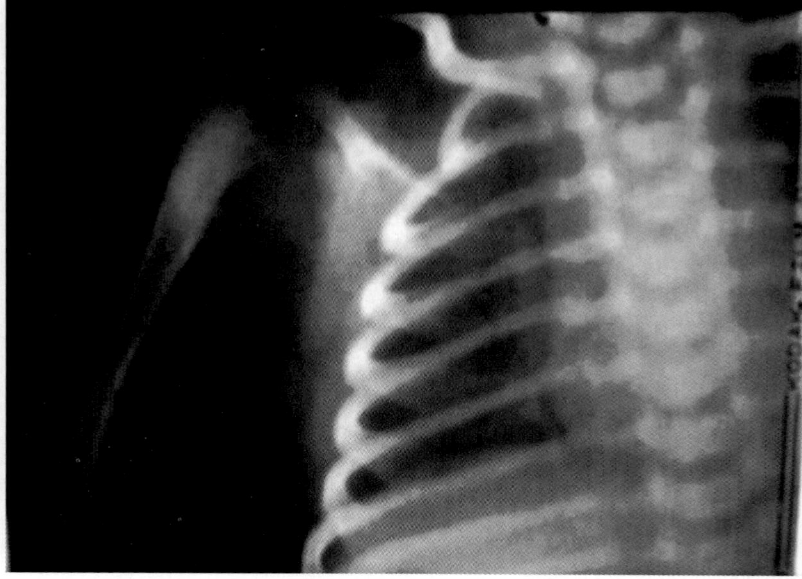

FIGURE 13.422A

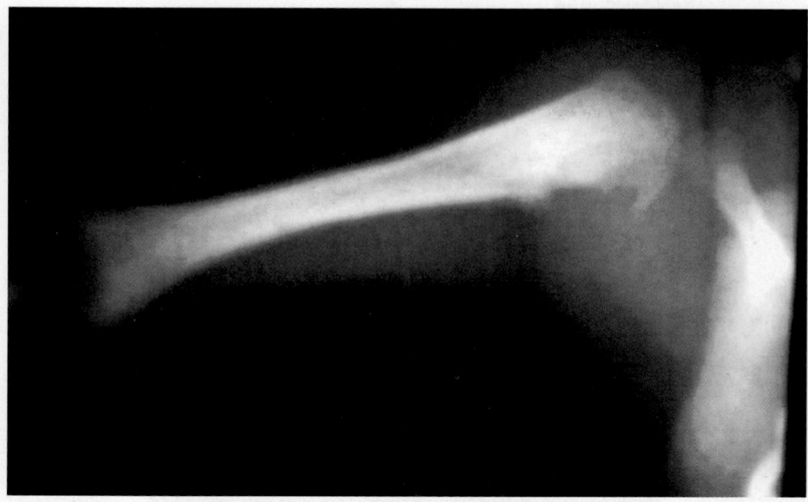

FIGURE 13.422B

FIGURES 13. 422A and B: Septic arthritis of Rt shoulder joint. X-rays of an infant child show irregular radiolucent. The defect is seen in the proximal humeral shaft in the diaphysis.

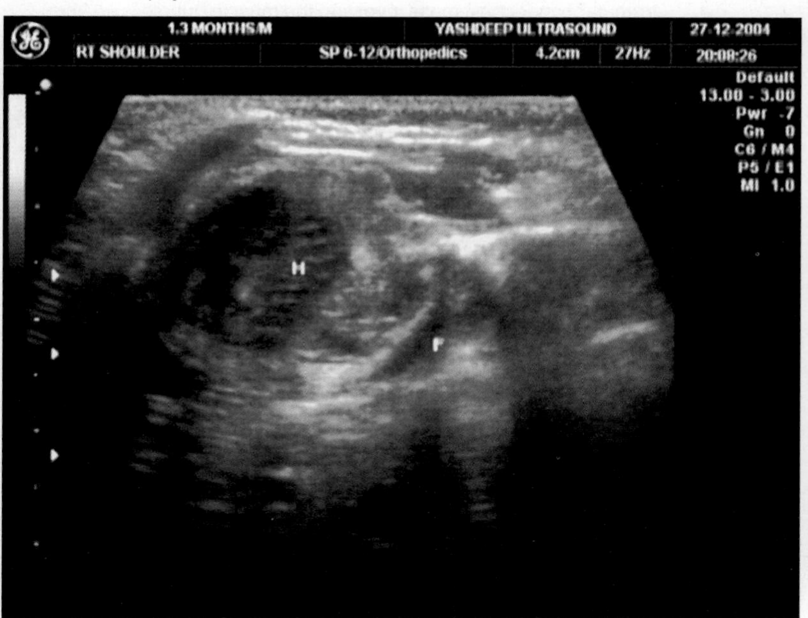

FIGURE 13.423A

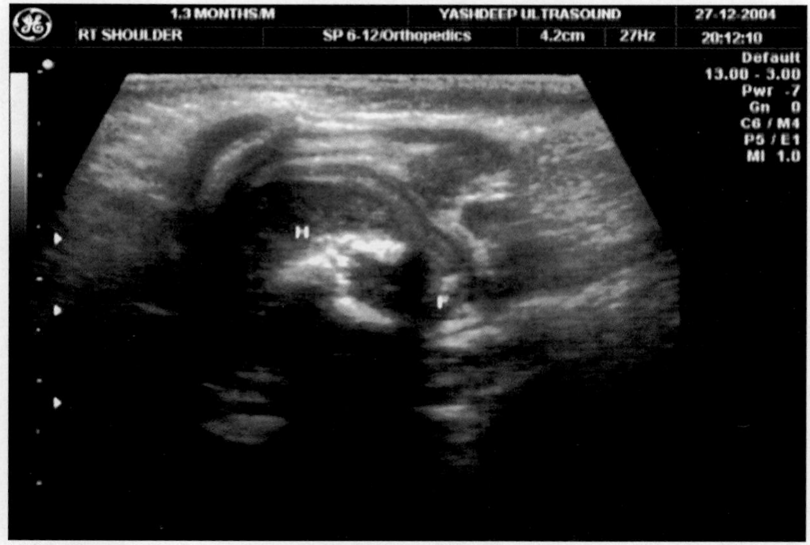

FIGURE 13.423B

FIGURES 13.423A and B: HRSG of the same patient shows irregular humeral head. The size of the head is small. The margins are irregular in outline. The head is De-shaped.

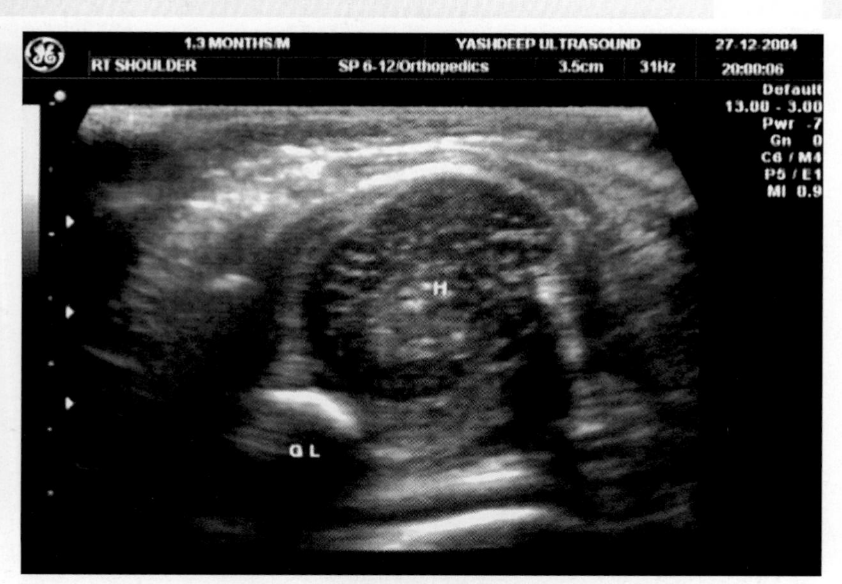

FIGURE 13.424A

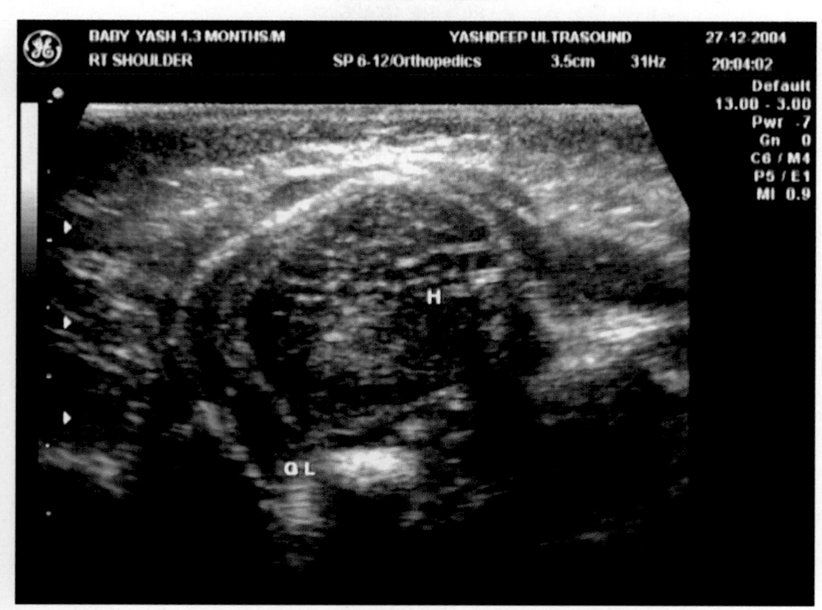

FIGURE 13.424B

FIGURES 13.424A and B: The Lt humeral head of the same patient shows the normal size head with well-defined glenoid cavity. Small amount of the fluid is seen in the joint capsule.

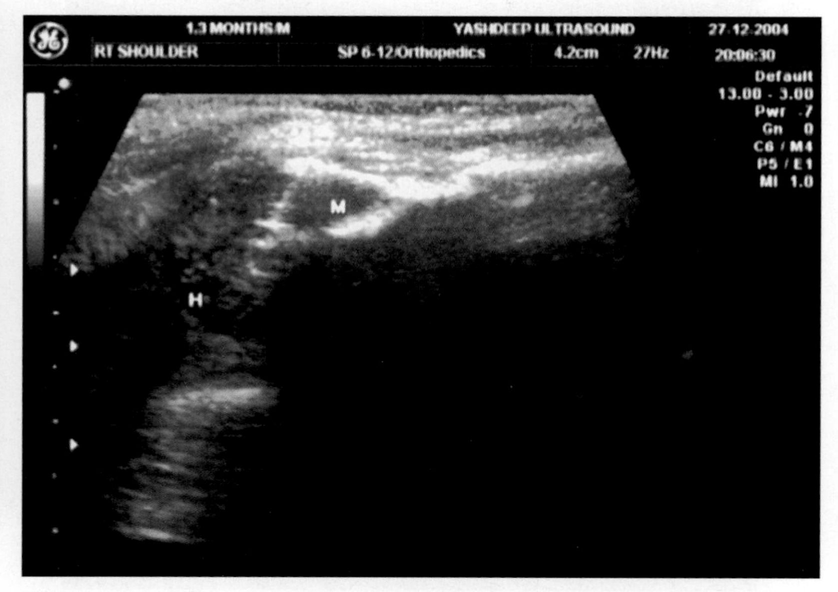

FIGURE 13.425: The proximal humeral shaft is also shows erosion. The shaft bony cortex is irregular in outline with cortical defects seen on HRSG.

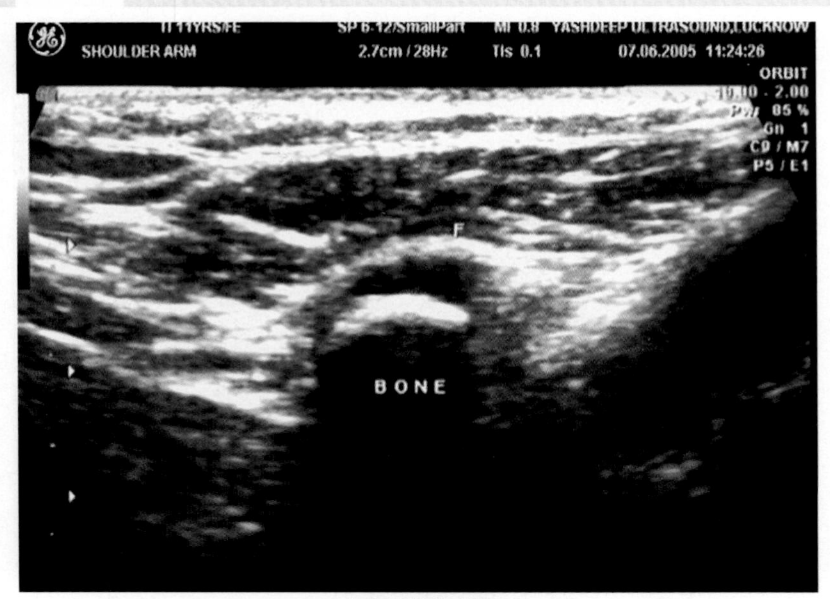

FIGURE 13.426

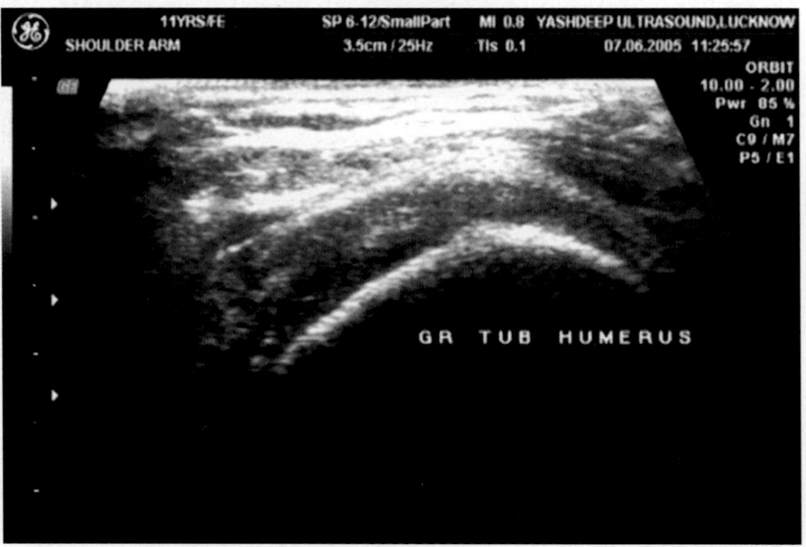

FIGURE 13.427

FIGURES 13.426 and 13.427: A young girl presented with painful shoulder movement. HRSG shows small amount of fluid in the joint capsule suggestive of effusion.

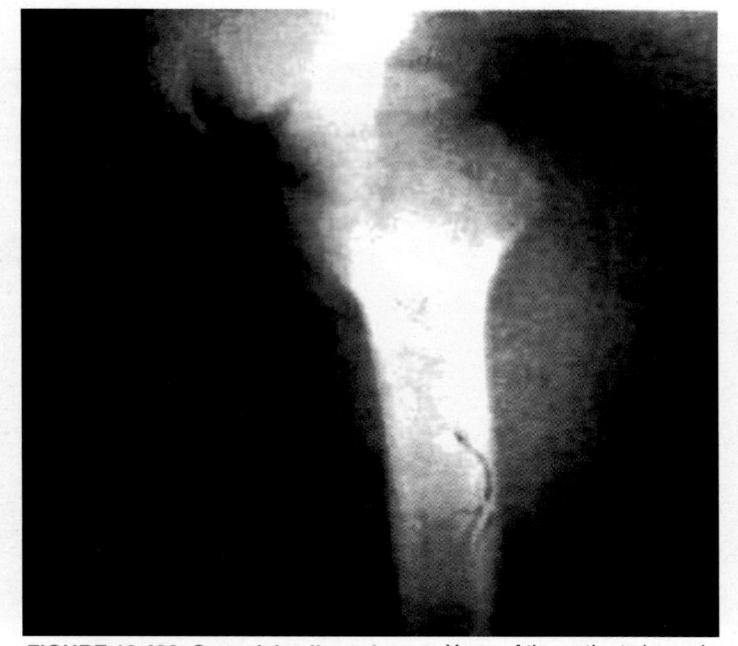

FIGURE 13.428: Synovial cell carcinoma. X-ray of the patient showed a soft tissue mass involving the Rt shoulder.

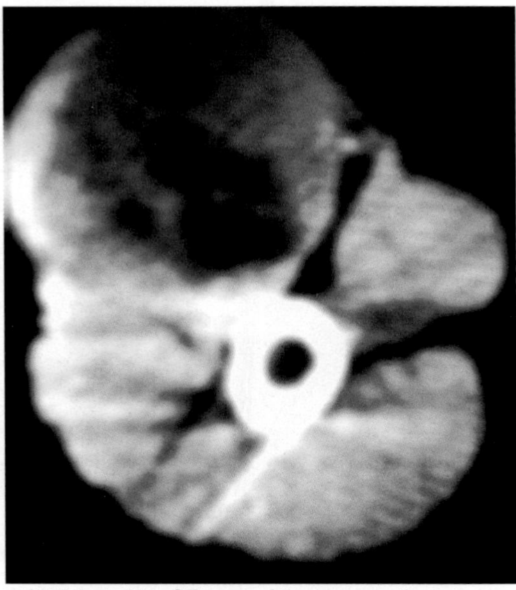

FIGURE 13.429: CT scan of the same patient showed irregular hypoechoic mass involving the muscle bundle of Lt shoulder. Biopsy of the mass showed synovial cell carcinoma.

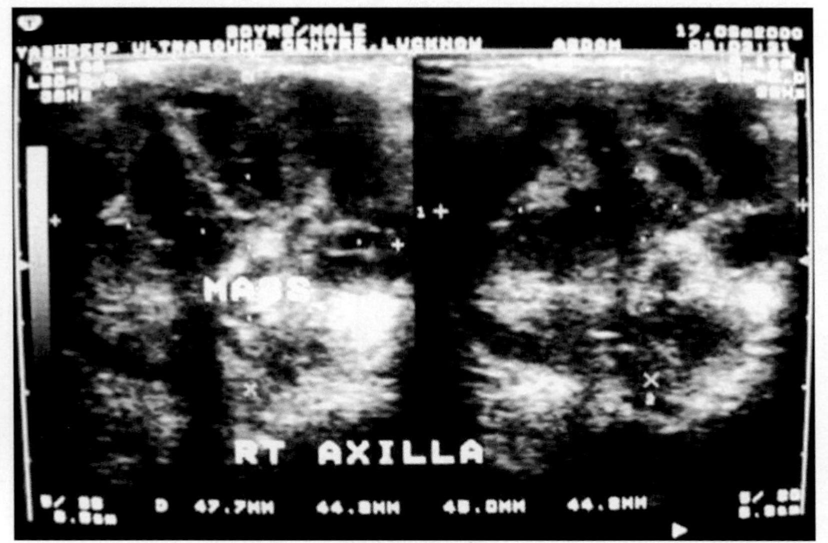

FIGURE 13.430

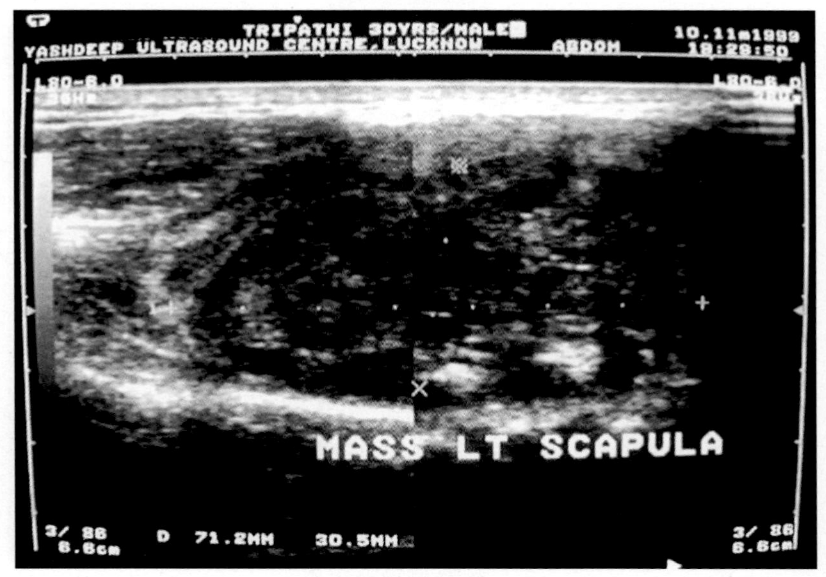

FIGURE 13.431

FIGURES 13.430 and 13.431: HRSG of the same patient shows big irregular heterogeneous soft tissue mass involving the Lt shoulder muscle and axilla.

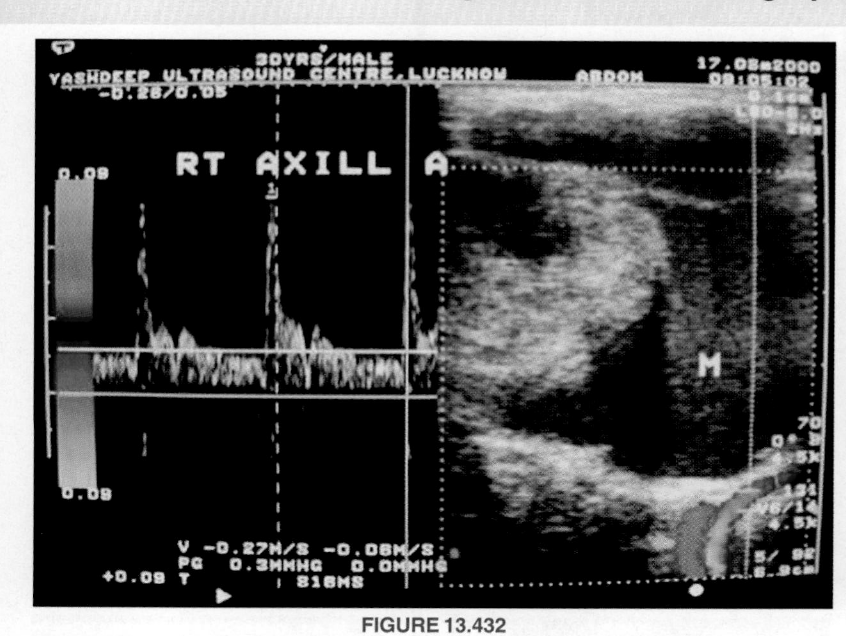

FIGURE 13.432

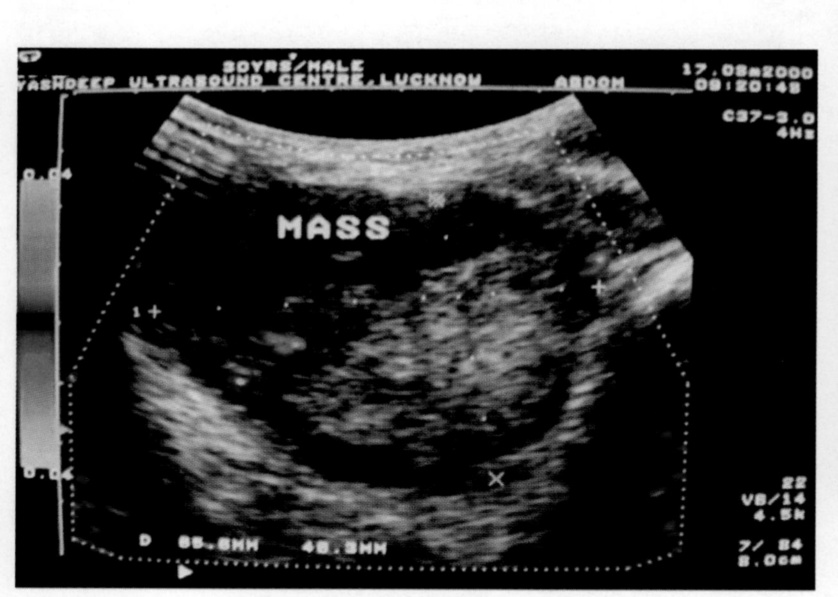

FIGURE 13.433

FIGURES 13.432 and 13.433: HRSG shows normal axillary vessels. The vessels were not invaded by the mass. The request for HRSG was to find out vessel involvement, which was not there.

HRSG of Wrist and Hand

The main clinical indications for ultrasound examination of wrist and hand are to find out the nature of the swelling its origin. Tenosynovitis of the flexor tendon can be very well evaluated on HRSG. It can result in carpal tunnel syndrome. Synovial cyst identification and its origin is possible on ultrasound. These cysts are usually hard and can be mistaken for bony hypertrophy on clinical examination. A 10 MHz transducer is required for evaluation of wrist and hand. The tendons are seen in transverse imaging. Median nerve is also identified anterior to the tendons. Longitudinal scan is performed after the transverse scan. Contralateral normal wrist is always seen for better evaluation of disease pattern. Minimal fluid collection in the tendon sheath can be picked up easily.

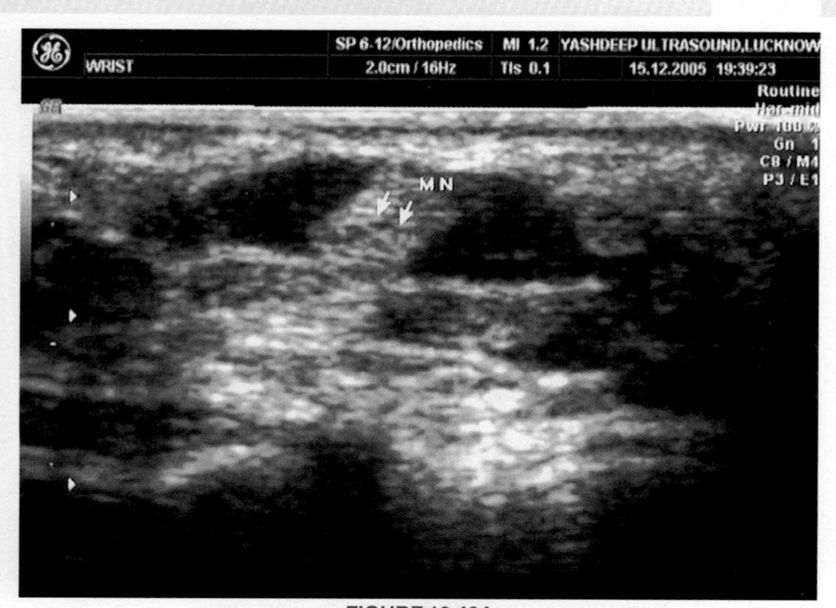

FIGURE 13.434

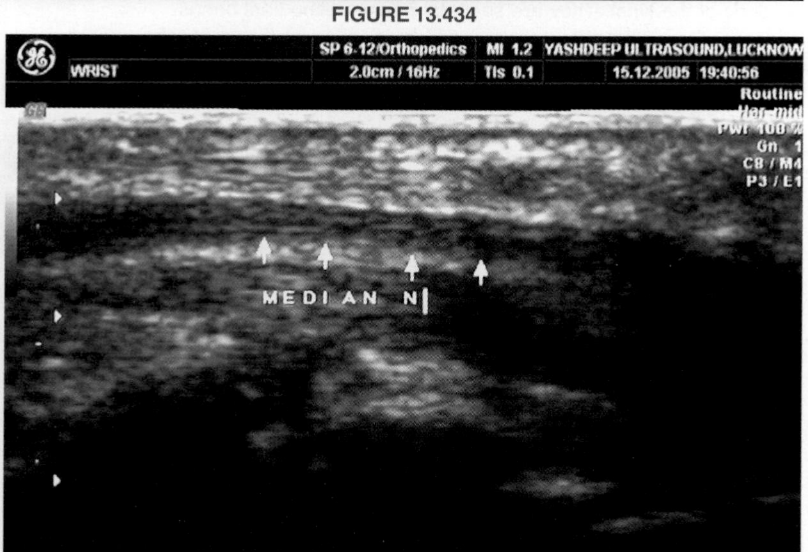

FIGURE 13.435

FIGURES 13.434 and 13.435: Normal median nerve. HRSG shows normal median nerve in longitudinal and transverse axis. It is seen between superficial and deep flexor tendons. It is seen as a hypoechoic band and in transverse axis as an oval shaped shadow.

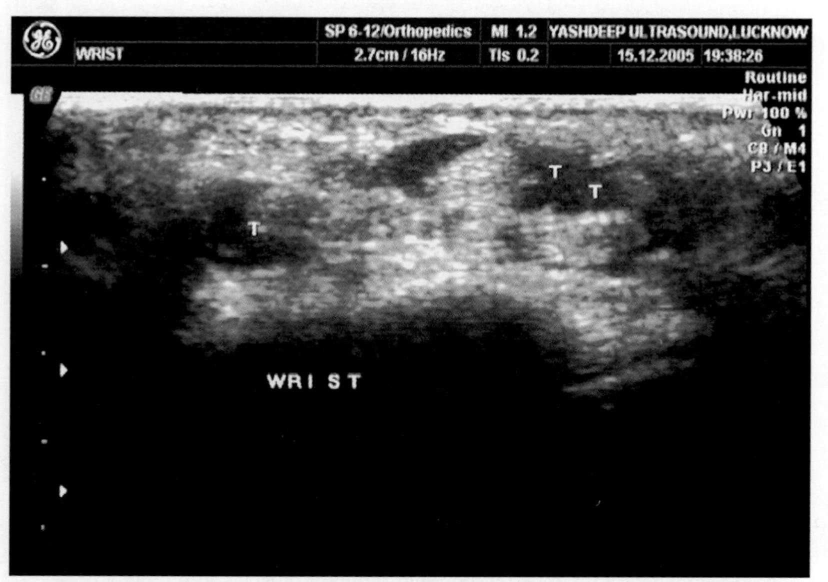

FIGURE 13.436: Normal flexor tendon. HRSG shows hypoechoic oval shaped median nerve anteriorly placed. The superficial flexor tendon is seen posterior to it. The tendon shows bright linear shadows showing the normal fibril texture of the tendon.

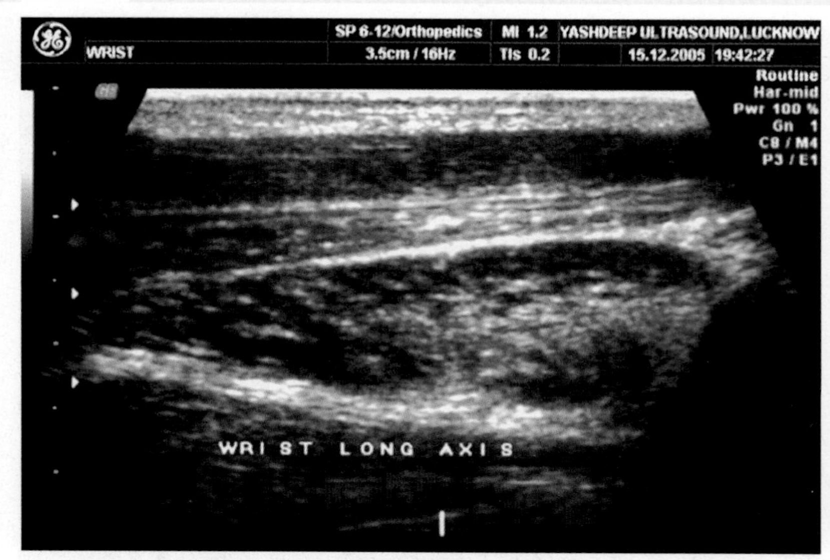

FIGURE 13.437: Normal flexor tendons of the wrist. HRSG shows normal flexor tendons of the wrist. They are seen as medium echo fibril bands.

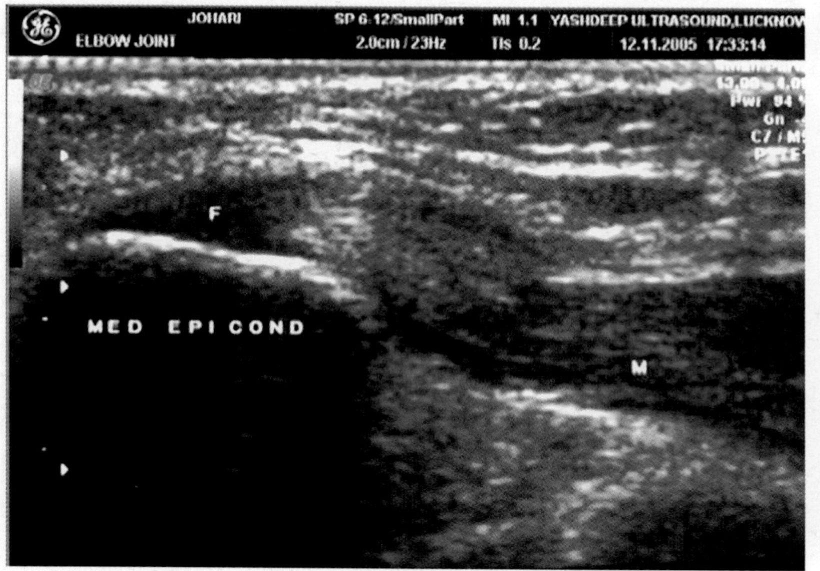

FIGURE 13.438

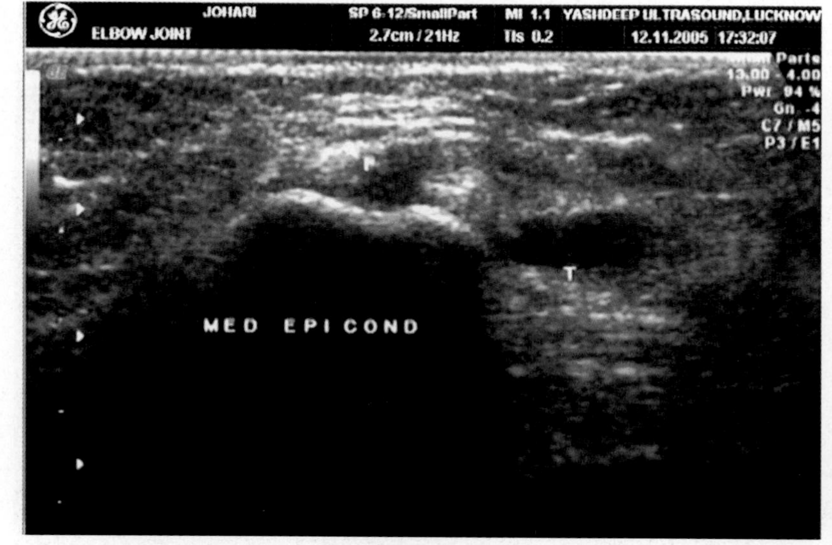

FIGURE 13.439

FIGURES 13.438 and 13.439: Medial epicondylitis (Golfers elbow). A patient presented with acute painful movements of the Rt elbow. HRSG shows thickened edematous extensor tendons on the ulnar group of muscles in a patient presented with painful elbow movement. The proximal ulnar tendons are inflamed. Small amount of fluid is also seen in the joint capsule.

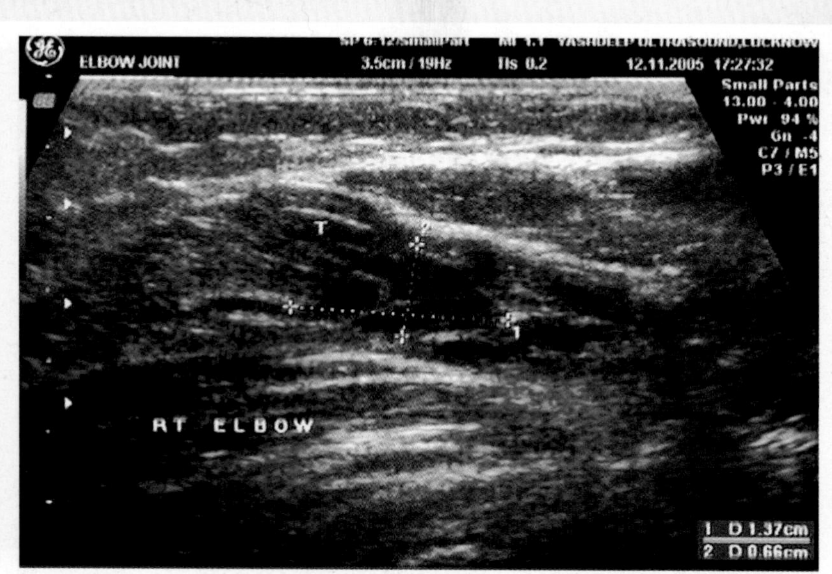

FIGURE 13.440

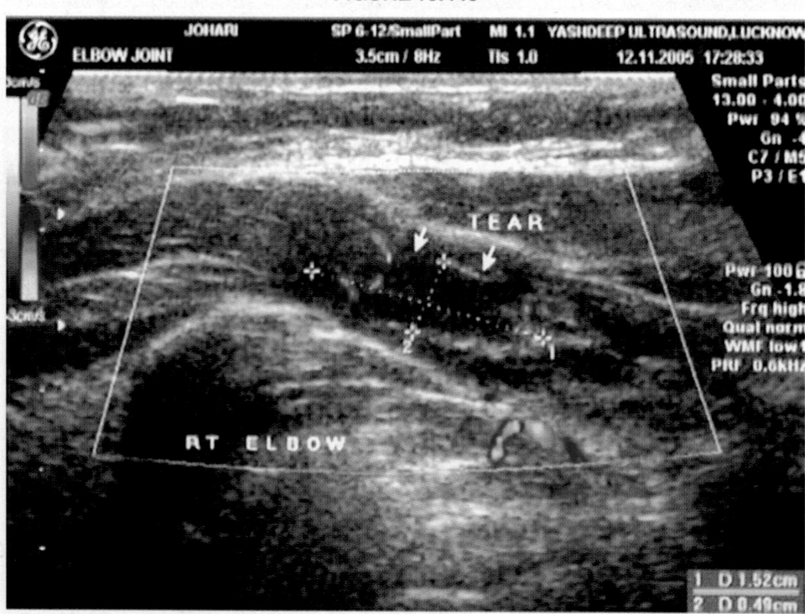

FIGURE 13.441

FIGURES 13.440: and 13.441: There is also evidence of tear seen in the proximal extensor group of tendons. **On color flow imaging** few vessels are seen feeding in the torn muscles suggestive of acute tear.

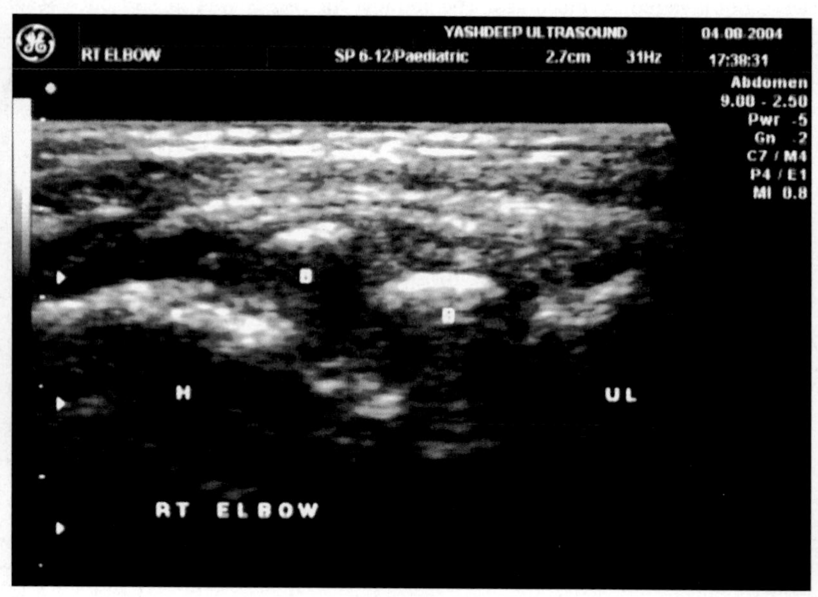

FIGURE 13.442

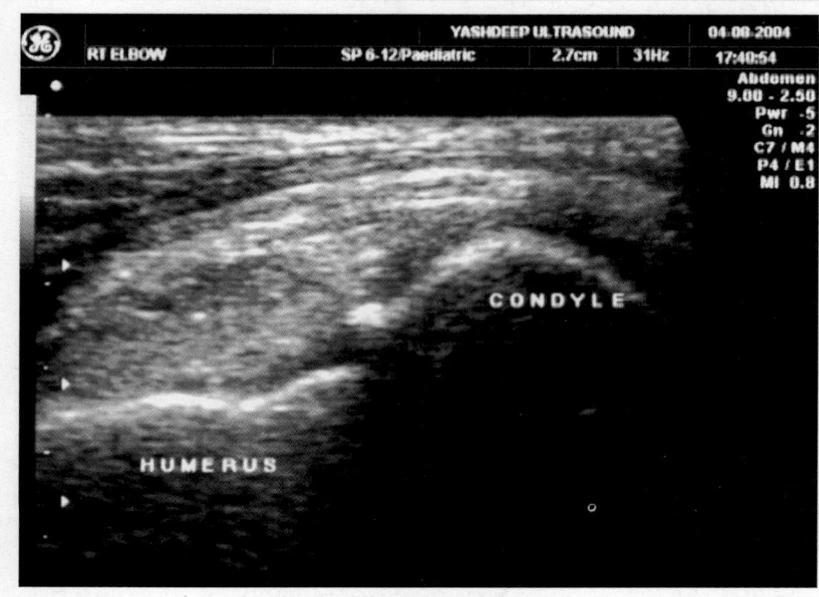

FIGURE 13.443

FIGURES 13.442 and 13.443: Golfer's elbow. HRSG shows small amount of fluid collection in the joint capsule. The proximal extensor group of tendons are edematous and hypoechoic in texture. A small irregular cleft is seen in the tendon suggestive of small tear. The patient presented painful movements.

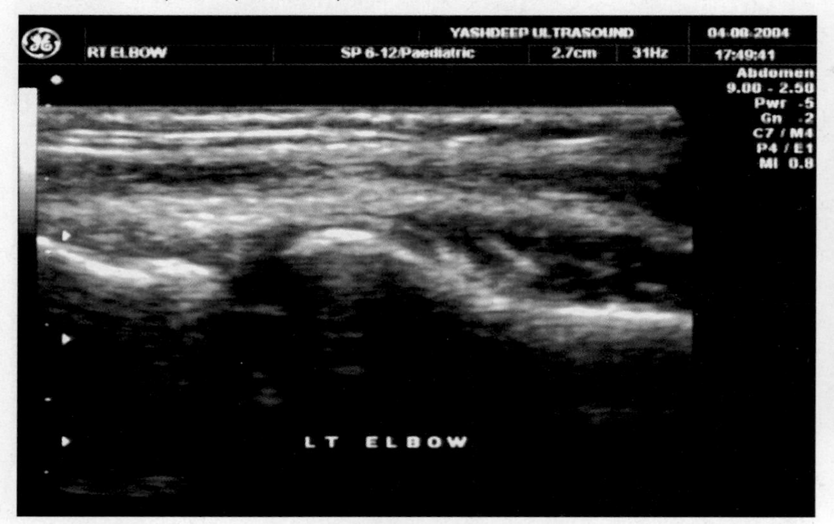

FIGURE 13.444

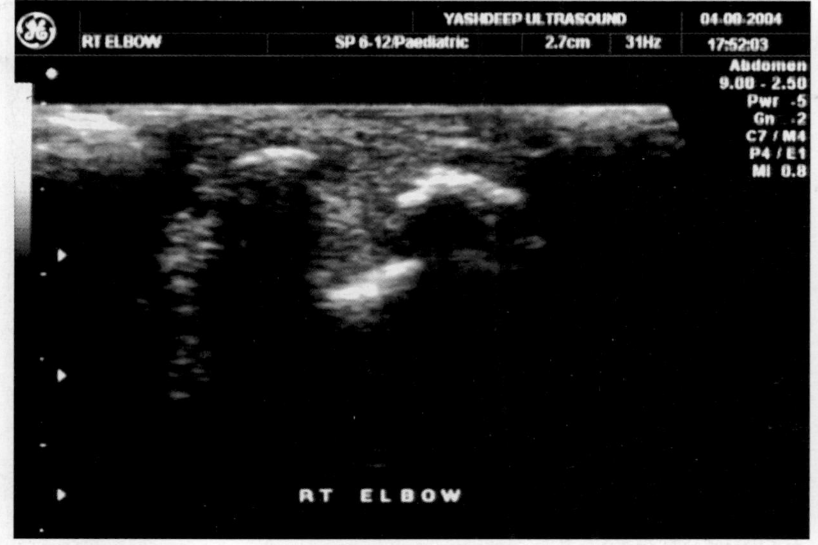

FIGURE 13.445

FIGURES 13.444: and 13.445: The medial epicondyle are seen as bright echogenic shadow. Small amount of fluid is also seen in the joint capsule. The proximal muscles are seen hypoechoic in texture.

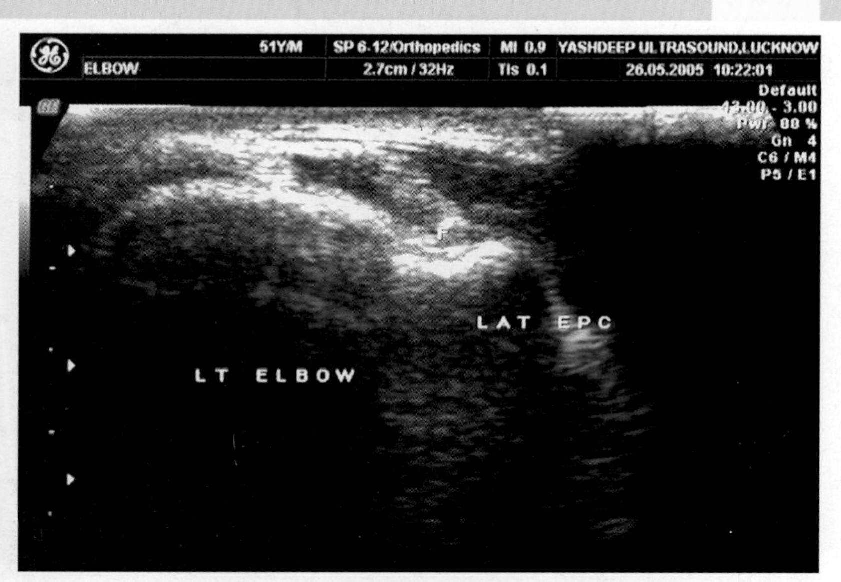

FIGURE 13.446

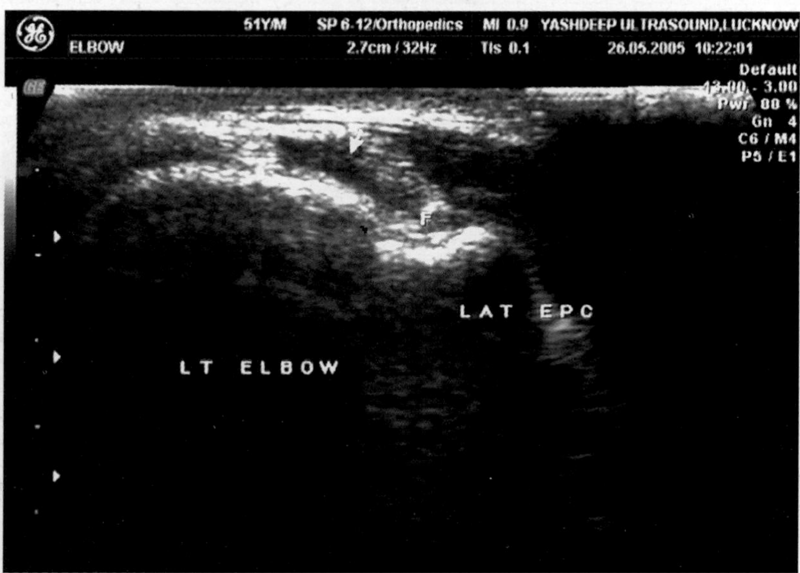

FIGURE 13.447

FIGURES 13.446 and 13.447: Lateral epicondylitis (tennis elbow). A patient presented with marked pain in external rotation of the elbow. HRSG shows thickened edematous radial extensor tendons. Disruption of the tendon fibers is seen suggestive of partial tear.

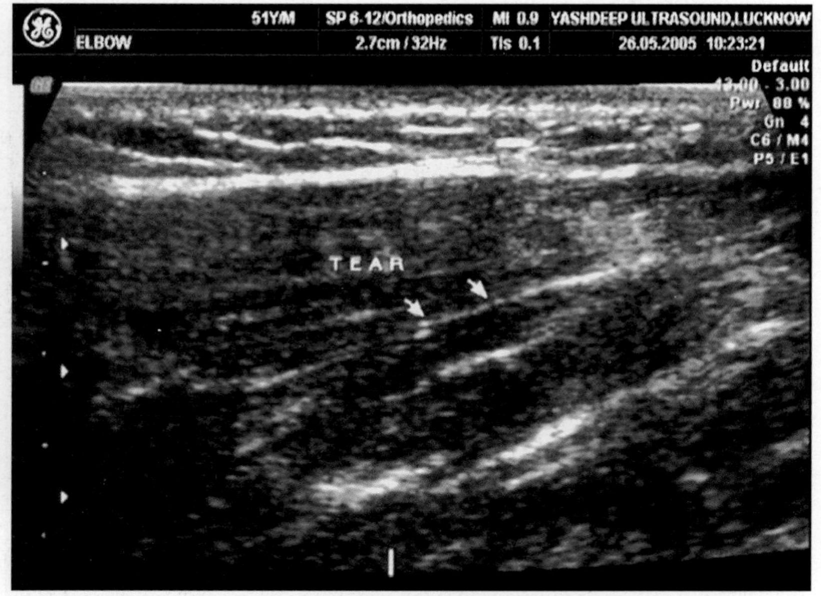

FIGURE 13.448

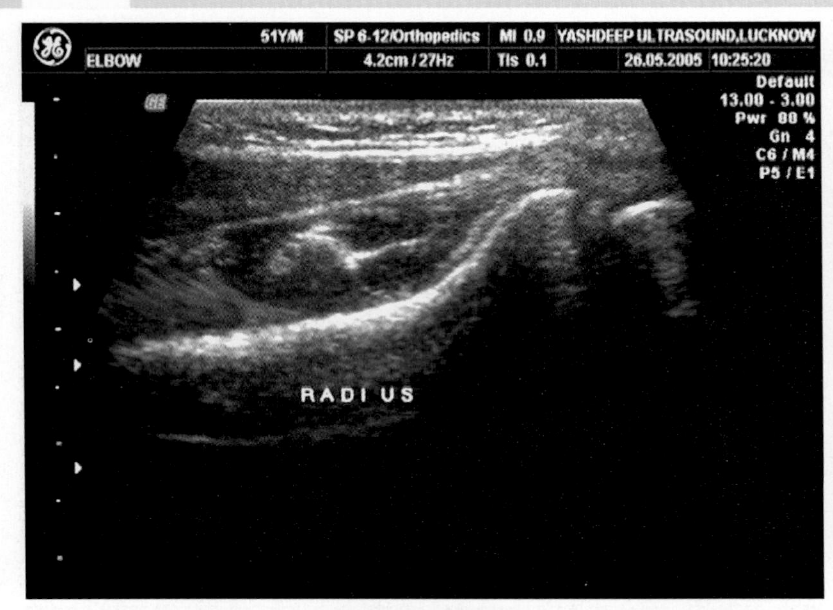

FIGURE 13.449

FIGURES 13.448 and 13.449: HRSG also shows a tear in the radial extensor group of muscle. Hemorrhagic collection is seen in the tear with hematoma formation.

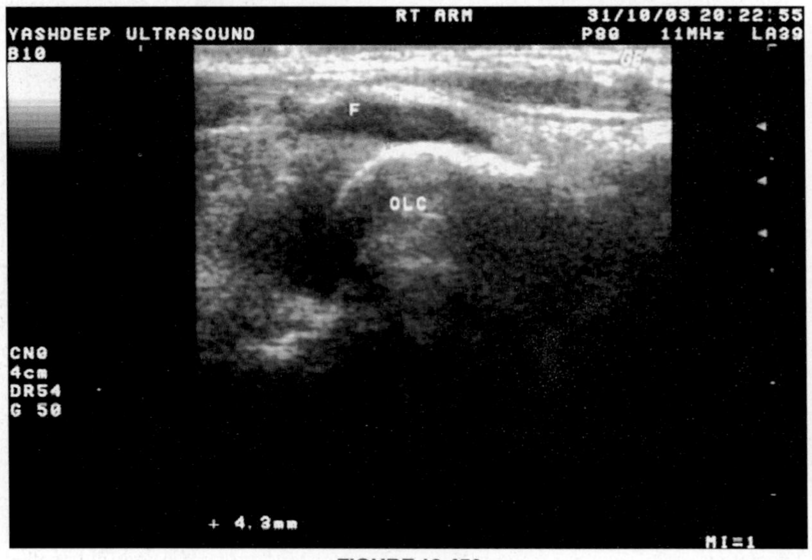

FIGURE 13.450

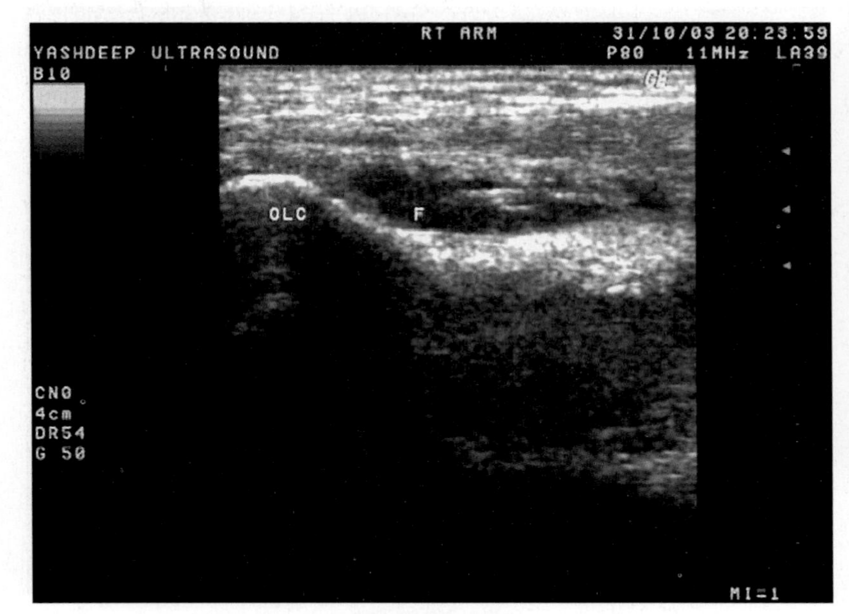

FIGURE 13.451

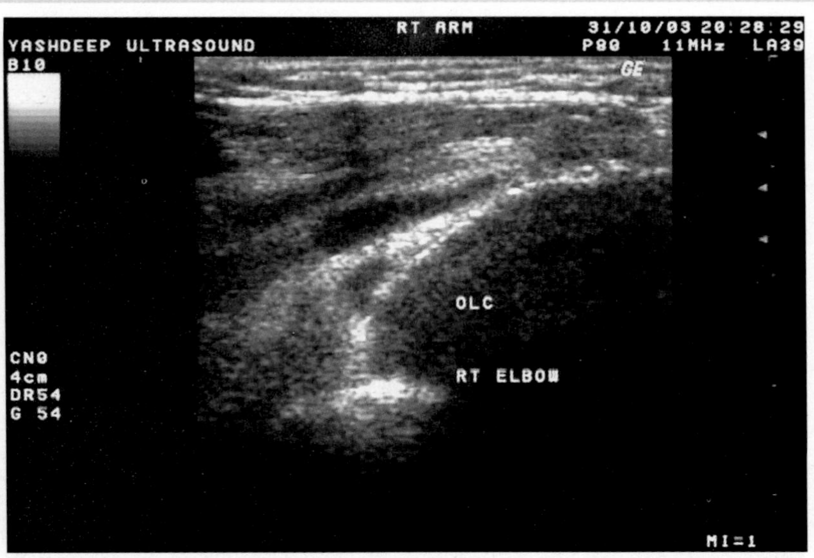

FIGURE 13.452

FIGURES 13.450 to 13.452: Olecranon bursitis. A patient presented with painful elbow syndrome. HRSG shows distended olecranon bursa. Low-level echoes are seen in the bursa. Few echogenic shadows are seen floating in the bursa suggestive of olecranon bursitis.

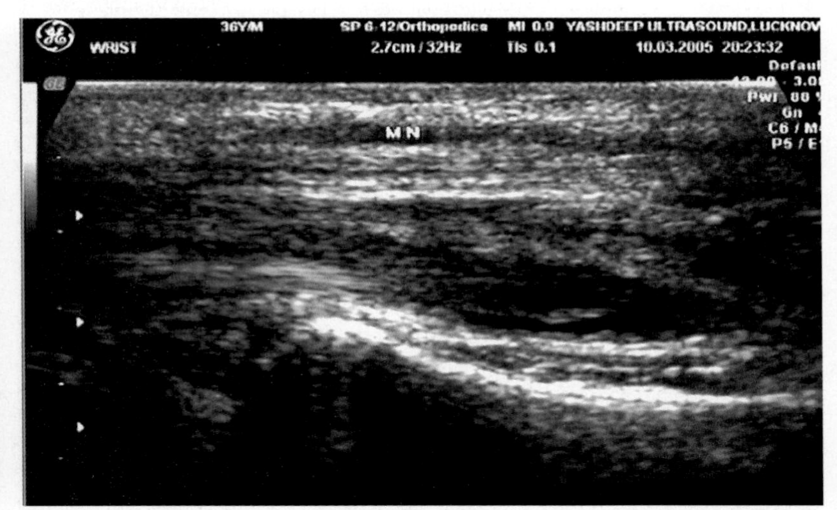

FIGURE 13.453

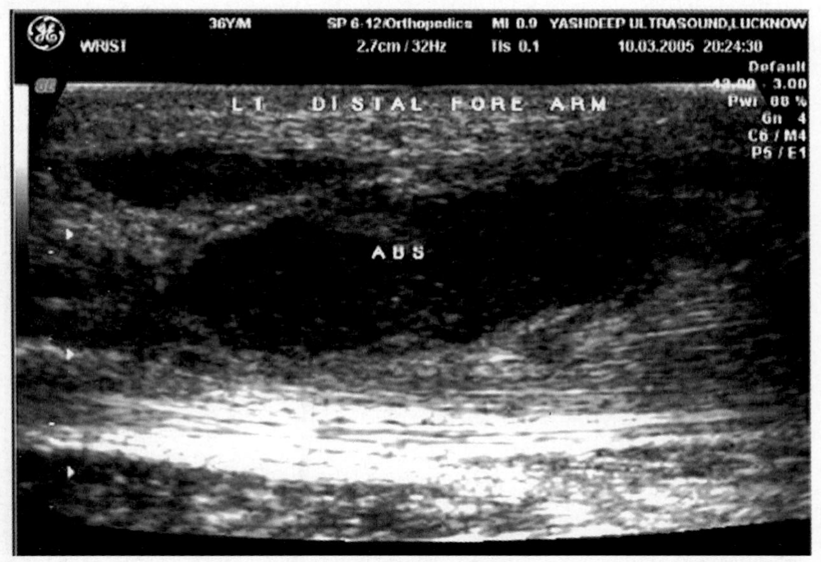

FIGURE 13.454

FIGURES 13.453: and 13.454: Acute pyogenic abscess in the wrist and forearm. A patient presented with acutely painful swelling in the Lt forearm and wrist. HRSG shows well-defined low-level echo complex pocket in the flexor group of muscle in distal forearm. The abscess is seen extending in the flexor compartment of the wrist joint.

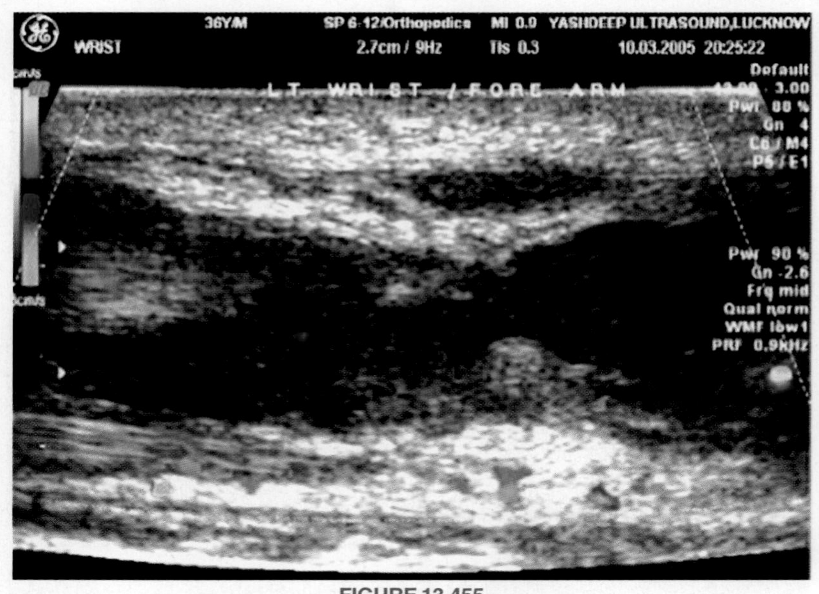

FIGURE 13.455

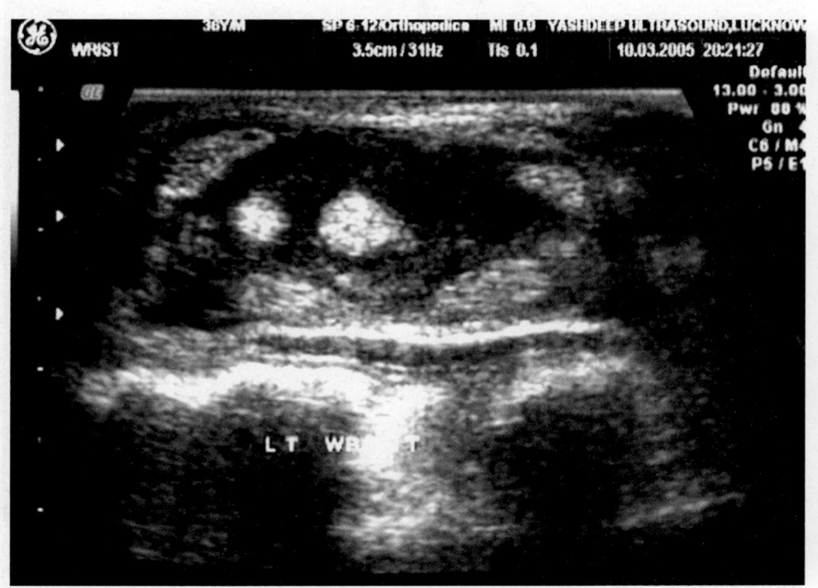

FIGURE 13.456

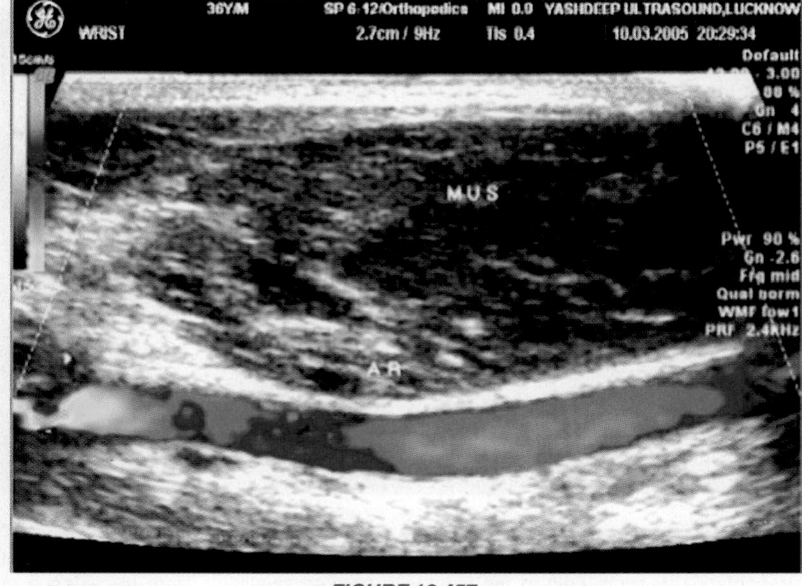

FIGURE 13.457

FIGURES 13.455 to 13.457: The collection is seen extending into the palmar space of the wrist. The median nerve is seen displaced anteriorly by the collection. **On color flow imaging** no flow seen in it.

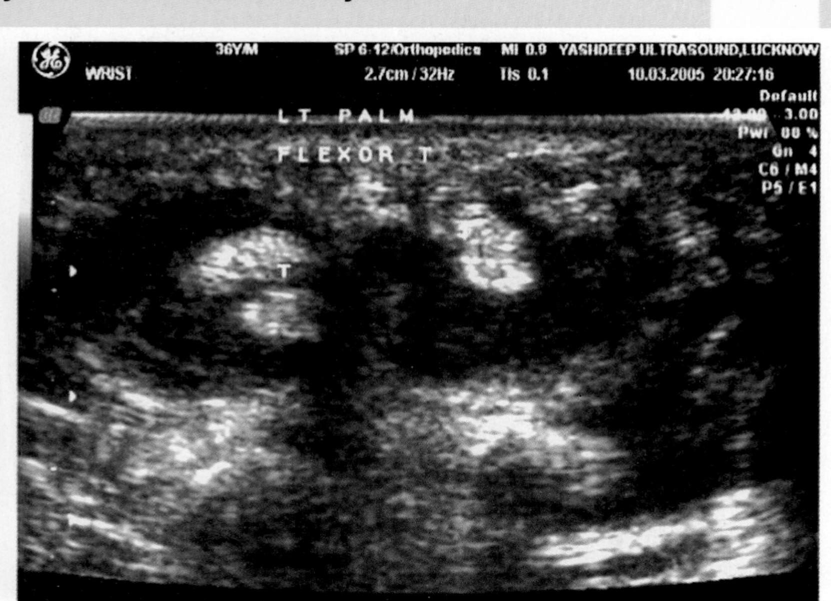

FIGURE 13.458

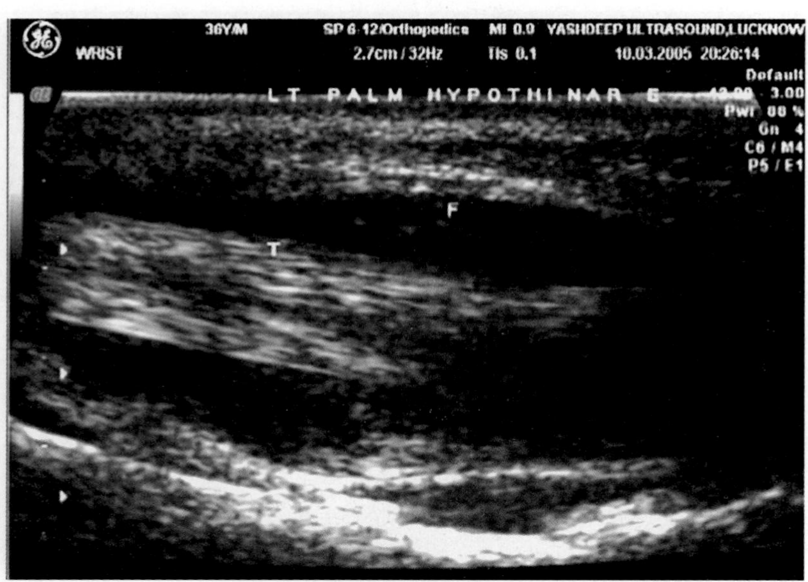

FIGURE 13.459

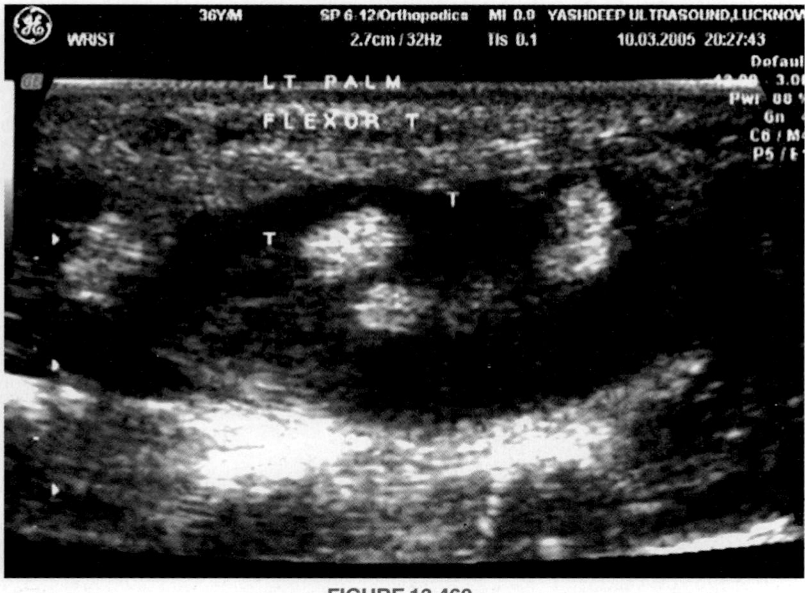

FIGURE 13.460

FIGURES 13.458 to 13.460: Collection is also seen extending in the thinar and hypo thinar eminence of the palm. The tendons are inflamed and edematous. Collection is also seen in the tendon sheath.

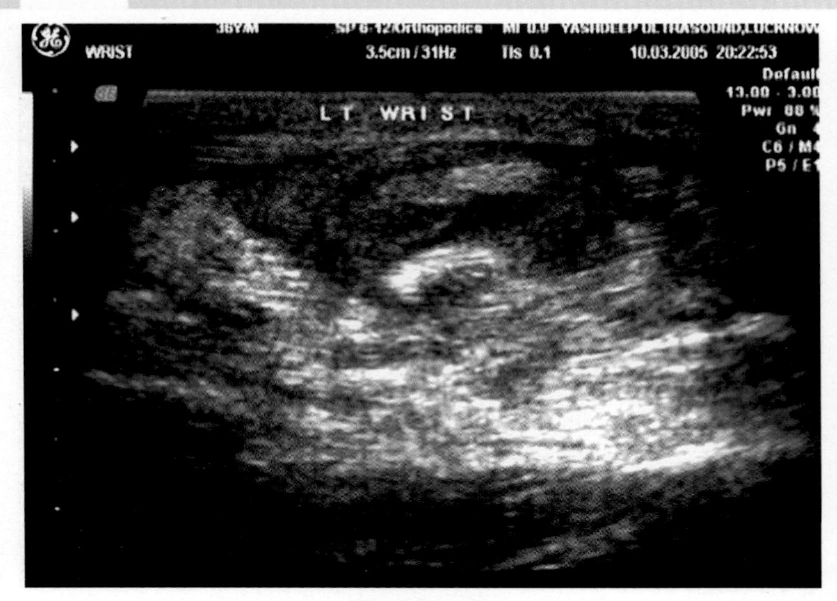

FIGURE 13.461

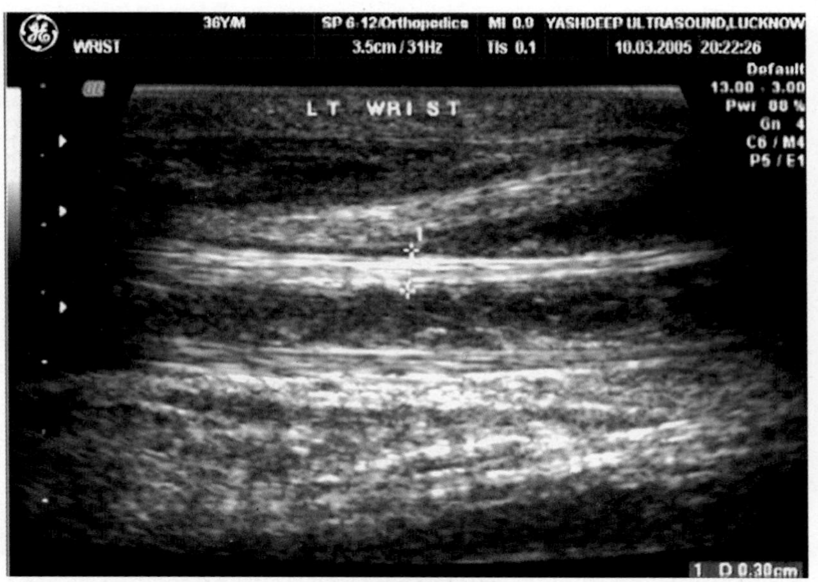

FIGURE 13.462

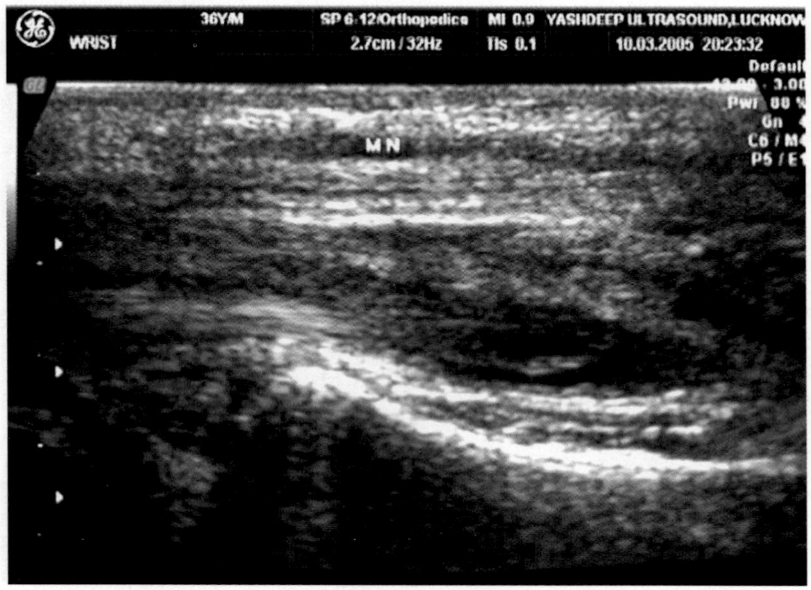

FIGURE 13.463

FIGURES 13.461 to 13.463: A thick walled abscess is also seen in the palm. The tendons are seen in the longitudinal axis. Abscess is seen in the posterior side of the flexor compartment.

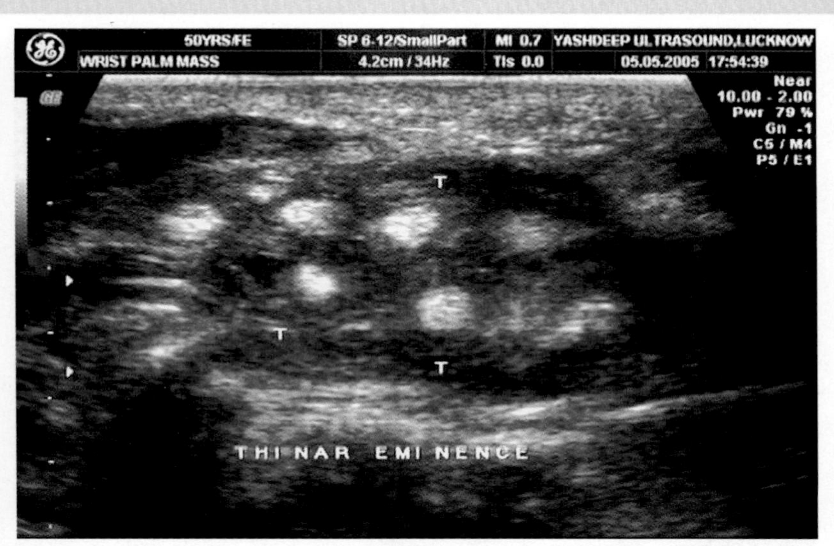

FIGURE 13.464

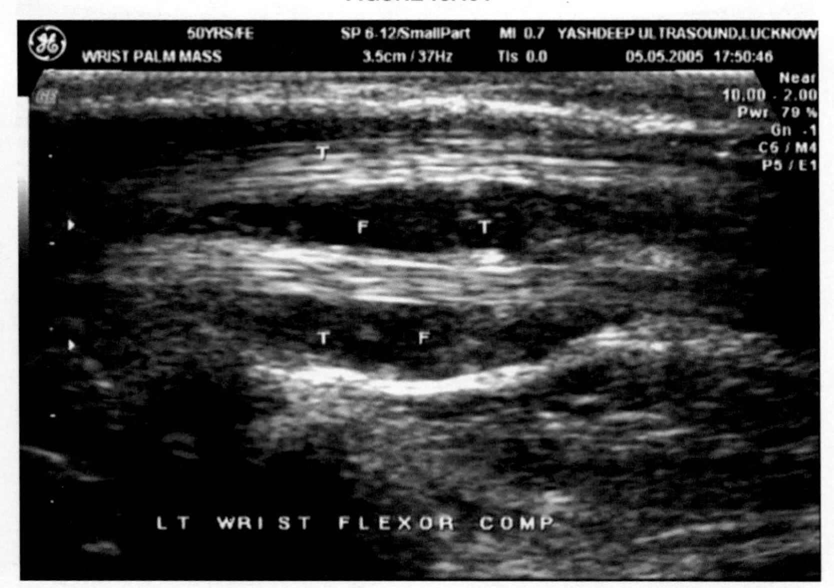

FIGURE 13.465

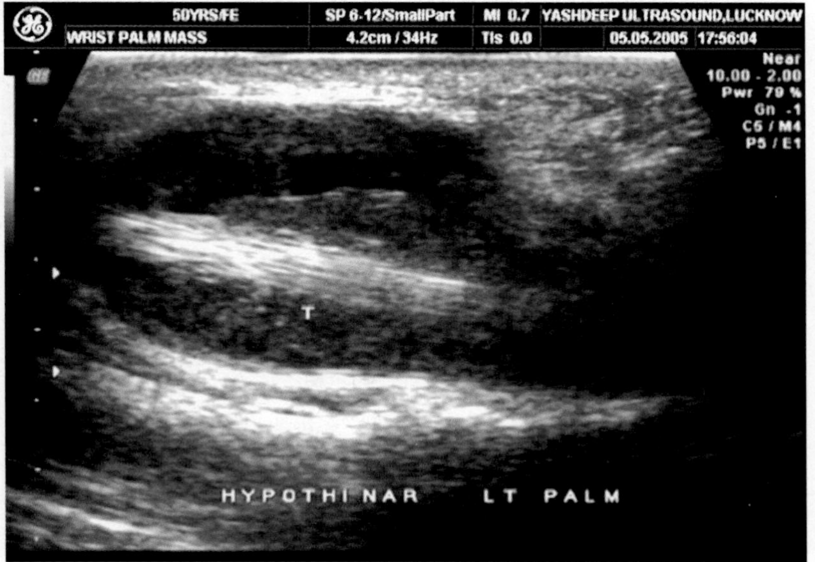

FIGURE 13.466

FIGURES 13.464 to 13.466: Acute on chronic tenosynovitis of wrist. A patient presented with acute pain in the wrist with marked swelling. There is evidence of thick collection seen in the flexor compartment of the wrist. The tendons are inflamed and thickened. They are seen as bright rounded shadows in cross section and echogenic bands in longitudinal section. Thick collection is also seen in the hypo thinar eminence of the wrist. The tendon is seen running through the collection.

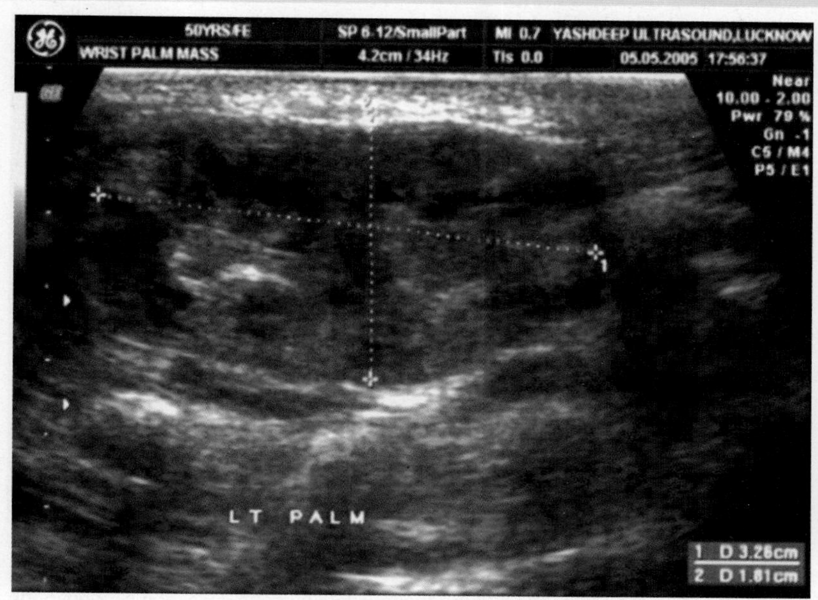

FIGURE 13.467

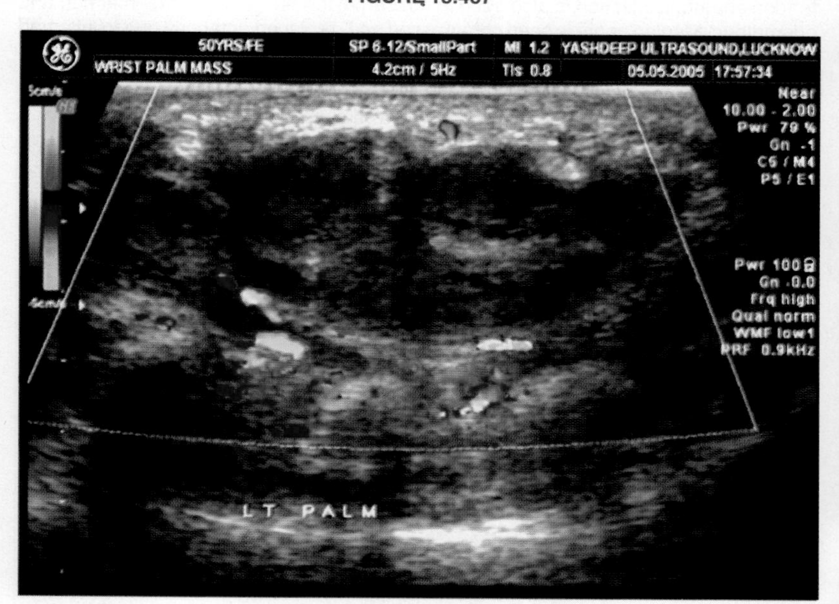

FIGURE 13.468

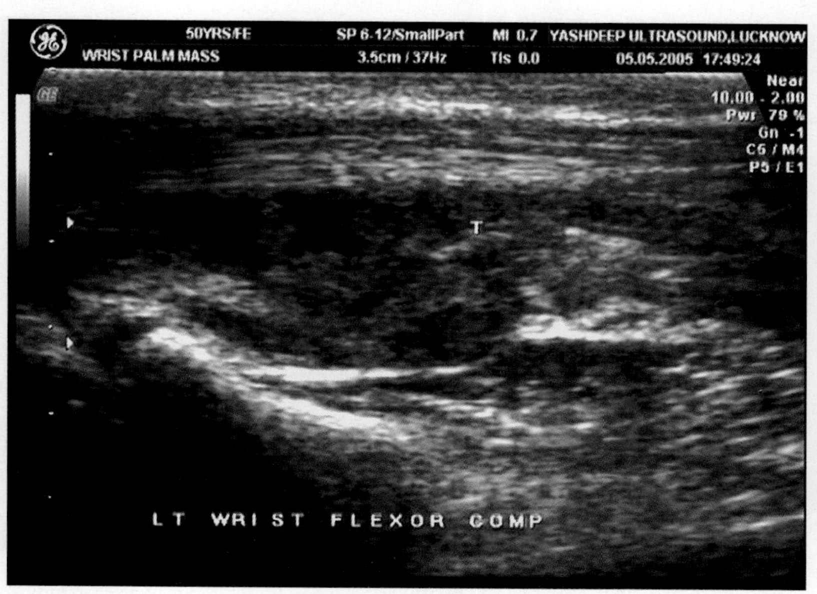

FIGURE 13.469

FIGURES 13.467 to 13.469: On color flow imaging few vessels are seen around the abscess capsule.

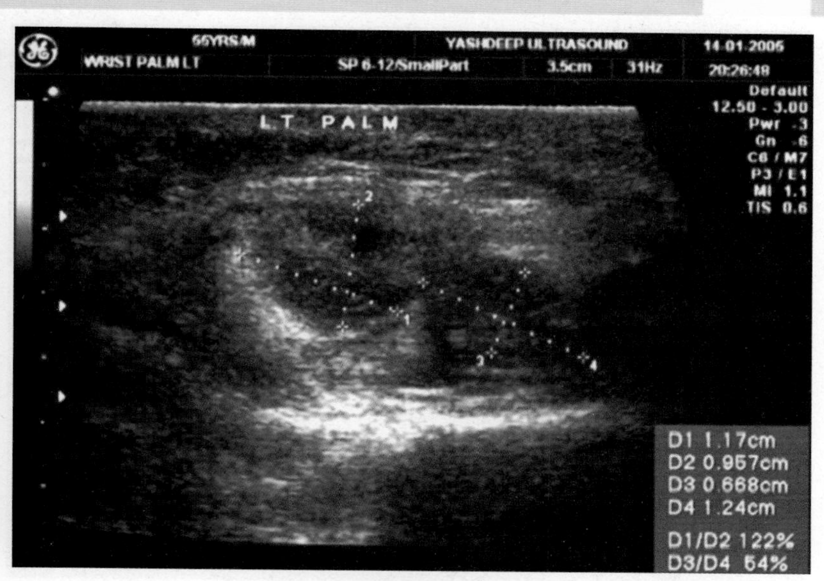

FIGURE 13.470

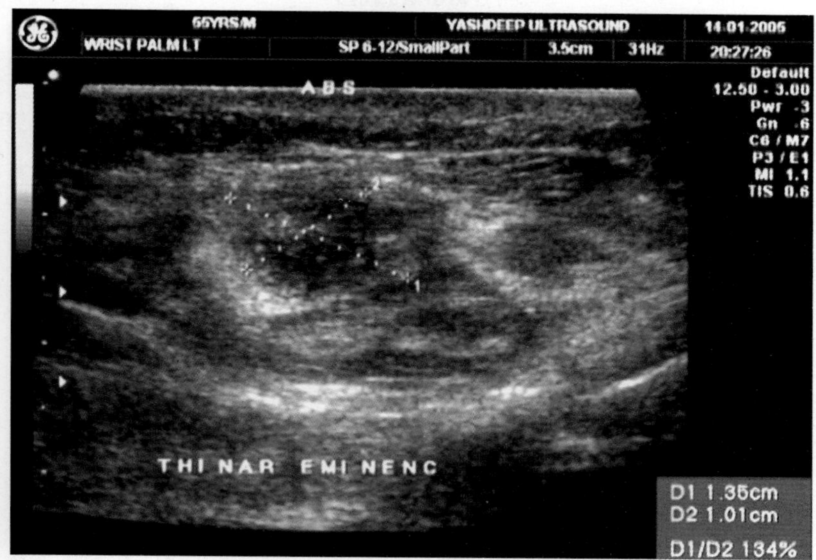

FIGURE 13.471

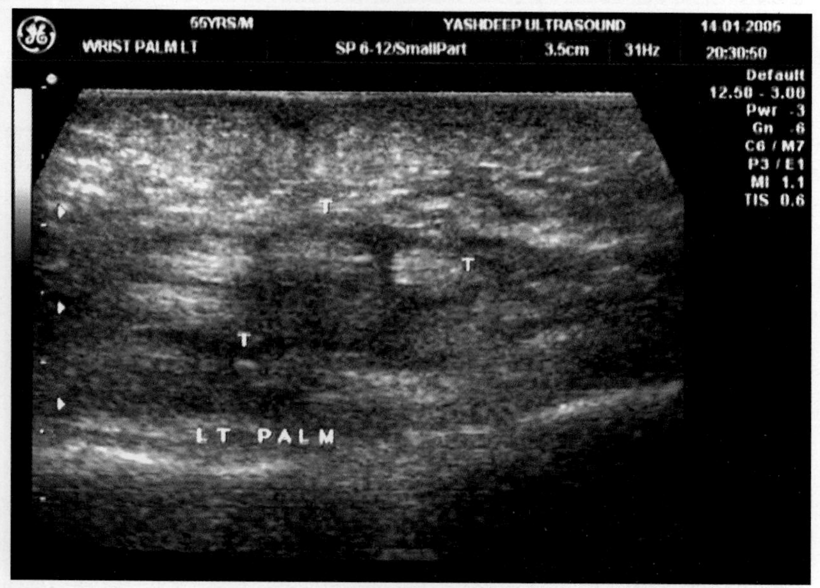

FIGURE 13.472

FIGURES 13.470 to 13.472: Chronic abscess in the palm. HRSG shows an irregular thick walled abscess in the Lt palm. It is seen confined to the thinar eminence. Central area of necrosis is seen in the abscess. But no suggestion of any cystic degeneration is seen. The flexor groups of the tendons are also edematous and hazy in outline. Small abscess is also seen posterior to the tendon.

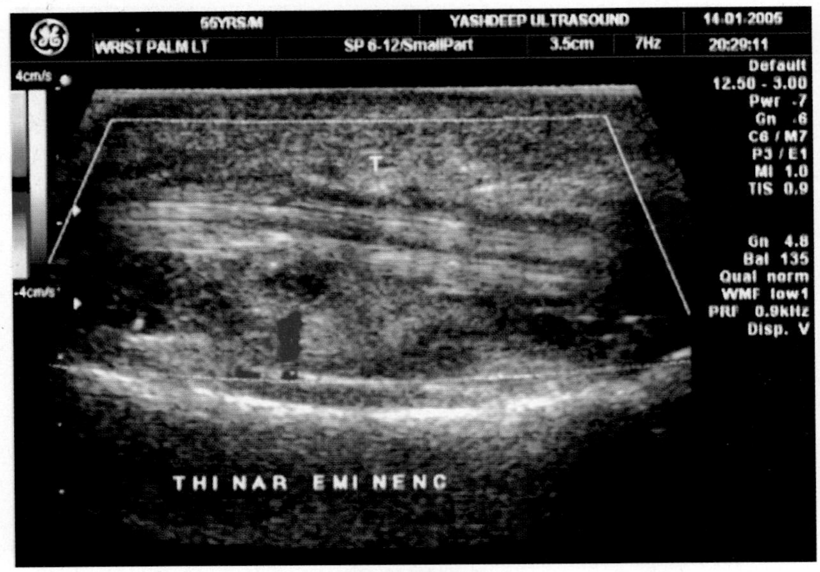

FIGURE 13.473

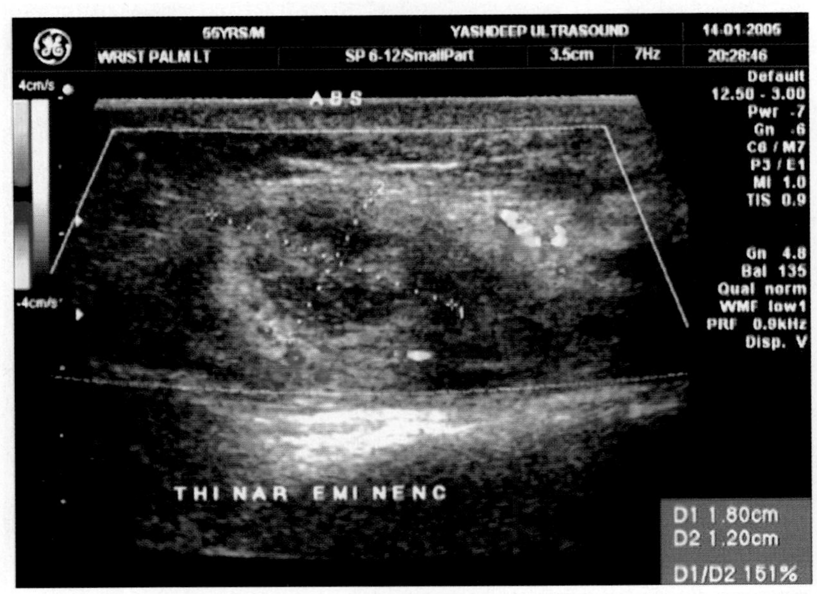

FIGURE 13.474

FIGURES 13.473 and 13.474: On color flow imaging low flow is seen in the abscess. Biopsy of the abscess confirmed chronic tubercular abscess.

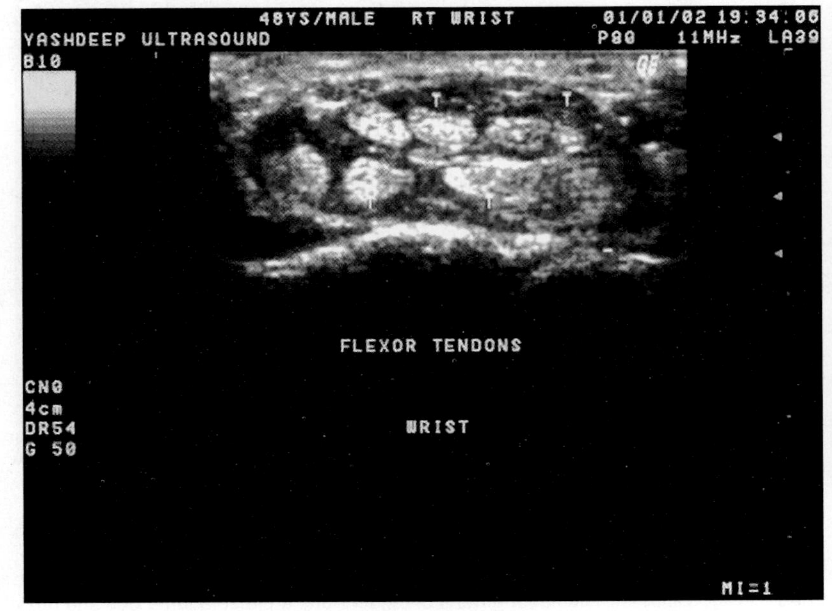

FIGURE 13.475

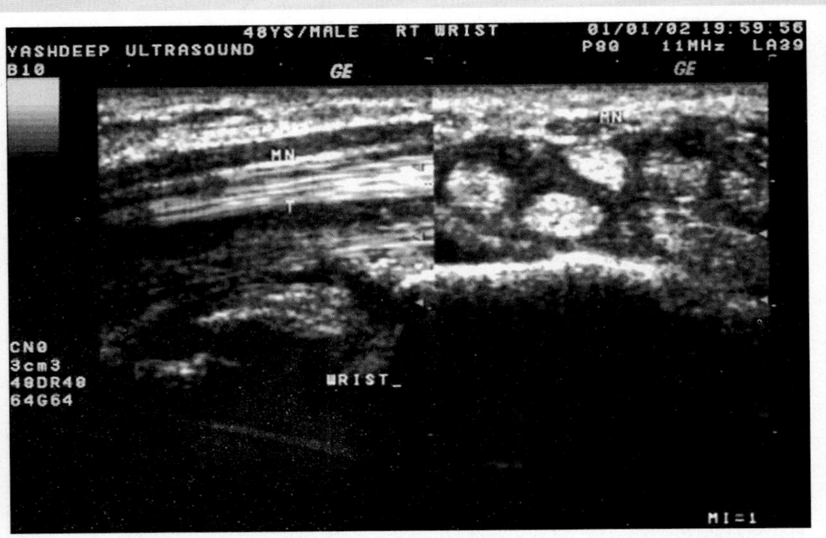

FIGURE 13.476

FIGURES 13.475 and 13.476: Acute tenosynovitis with carpal tunnel syndrome. HRSG shows thickened echogenic tendons. Low-level echo collection is seen in the tendon sheath. Collection is also seen in the carpal tunnel. Tendons are inflamed and edematous. Median nerve is seen compressed and displaced anteriorly due to inflamed edematous tendon.

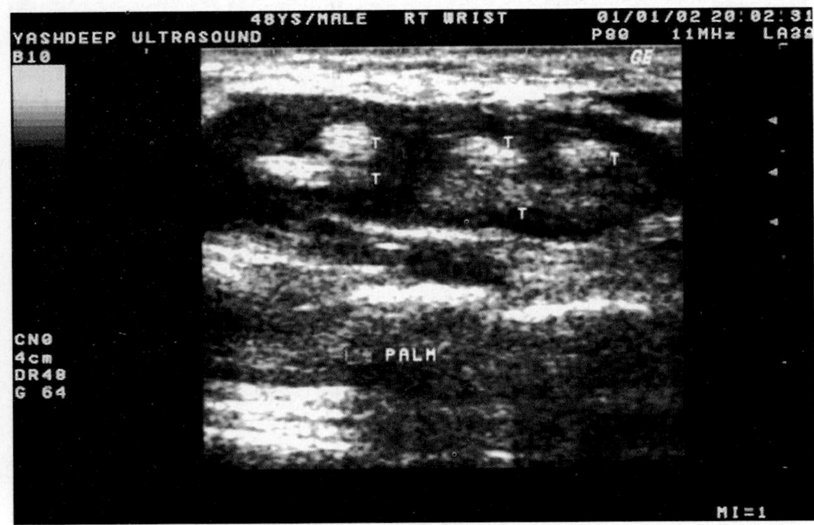

FIGURE 13.477

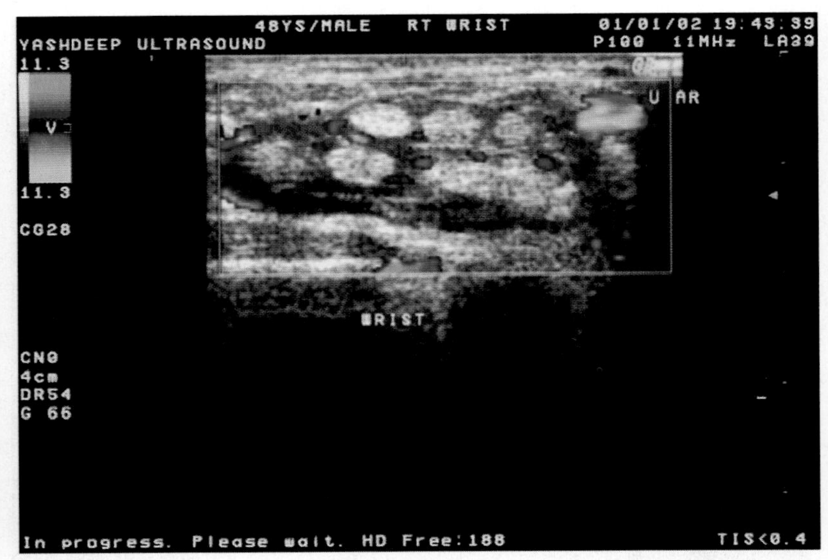

FIGURE 13.478

FIGURES 13.477 and 13.478: Tenosynovitis with carpal tunnel syndrome. HRSG shows inflamed edematous tendons with marked tendon edema. Collection is seen in the tendon sheath. **On color flow imaging** hyperemia is seen around the tendon sheath suggestive of acute inflammation.

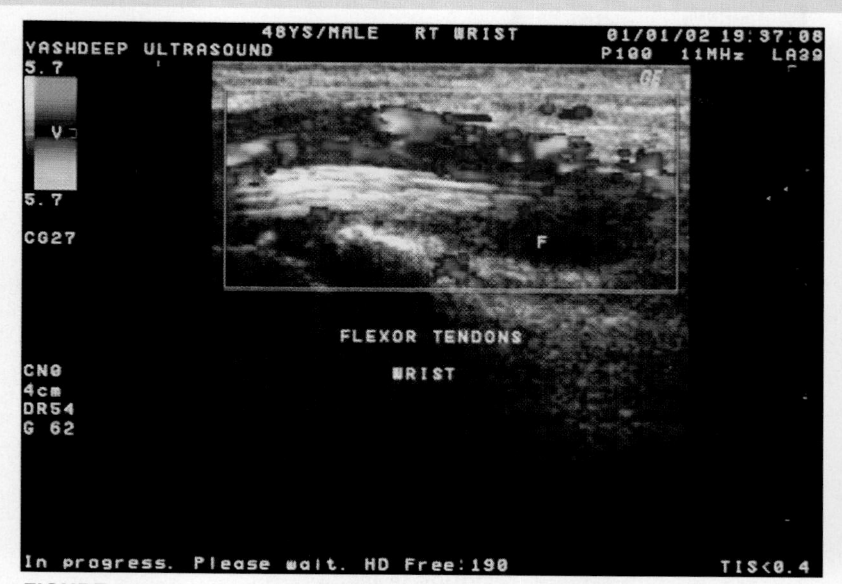

FIGURE 13.479: Acute tenosynovitis with carpal tunnel syndrome. Color Doppler imaging shows increased flow around the tendon sheath suggestive of hyperemia. Tendons are swollen and edematous. The median nerve is seen compressed in between the edematous tendons.

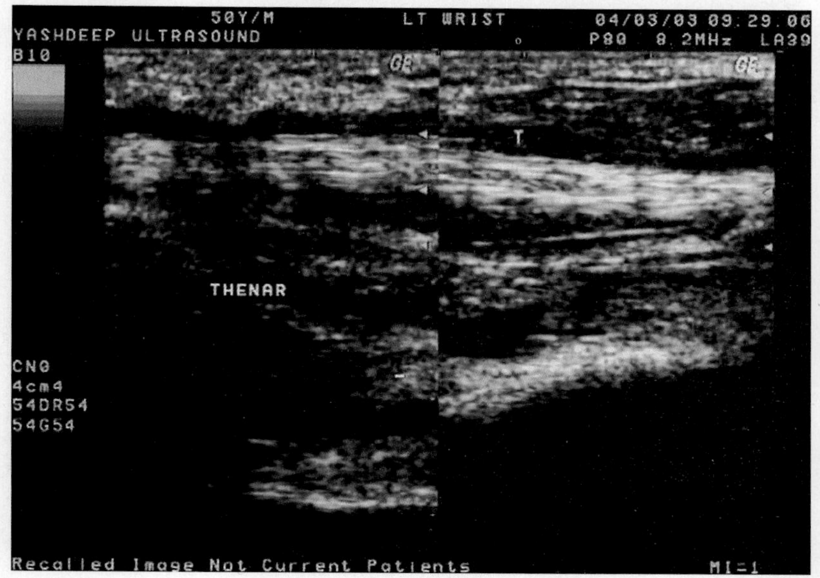

FIGURE 13.480

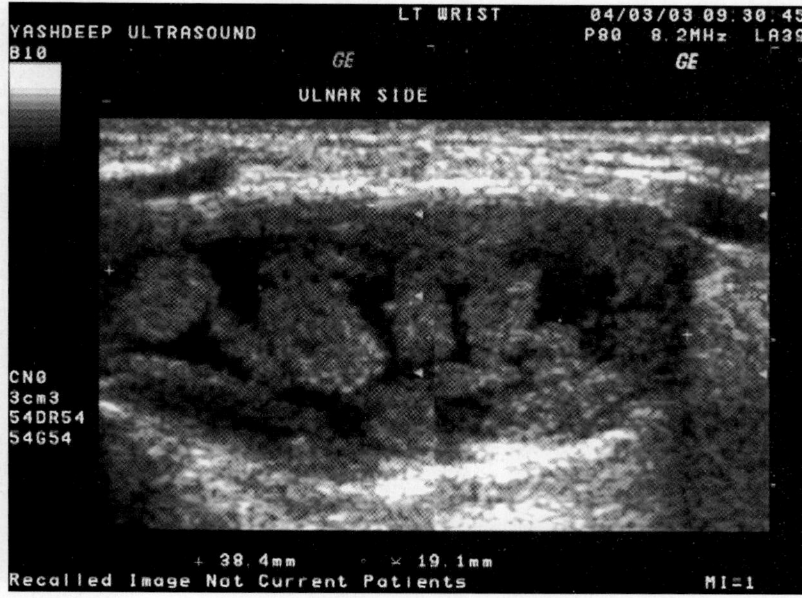

FIGURE 13.481

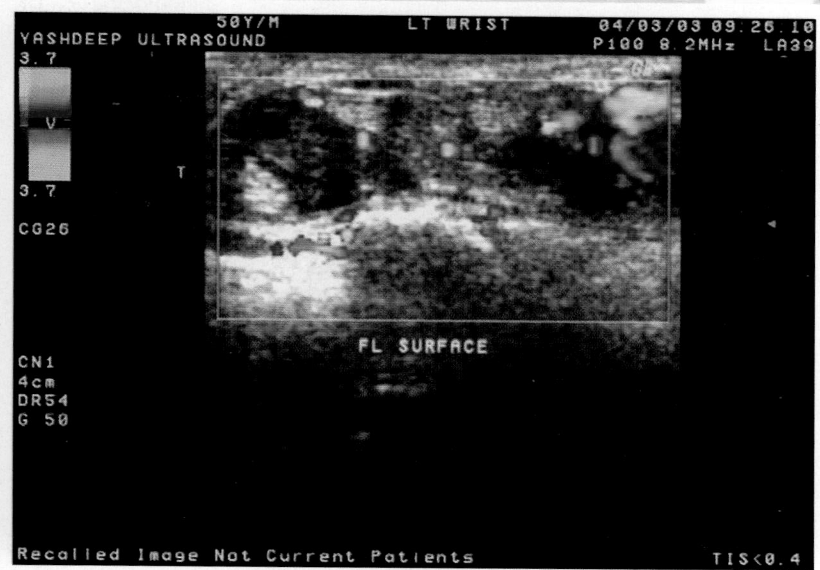

FIGURE 13.482

FIGURES 13.480 to 13.482: The same patient after one year of treatment further evaluated by HRSG. There is evidence of reduction of in the inflammatory exudates seen in comparison to the previous study. However, no complete remission of the disease seen.

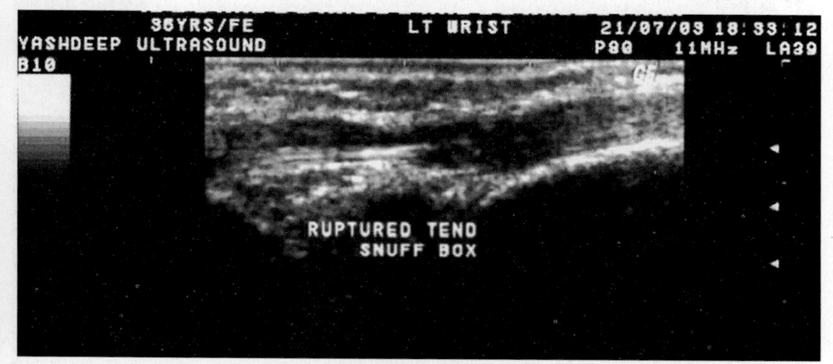

FIGURE 13.483

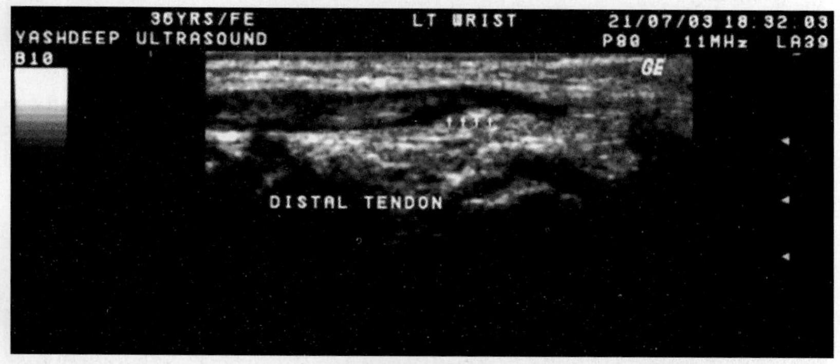

FIGURE 13.484

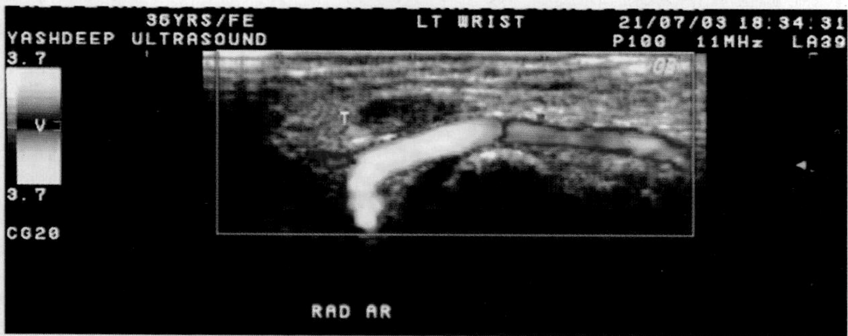

FIGURE 13.485

FIGURES 13.483 to 13.485: Ruptured extensor carpi radialis tendon. HRSG shows rupture of the extensor carpi radialis tendon in snuffbox. The ruptured tendon fibers are seen separate from each other. Small amount of fluid collection is seen in the tendon tunnel.

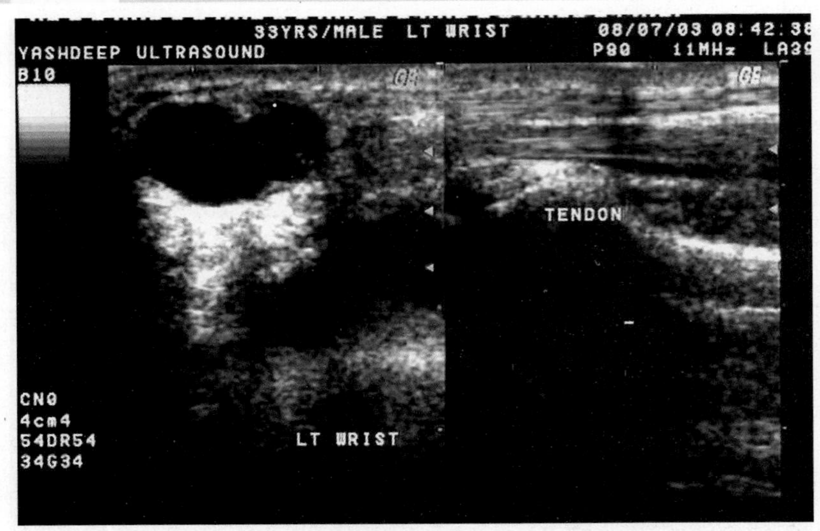

FIGURE 13.486

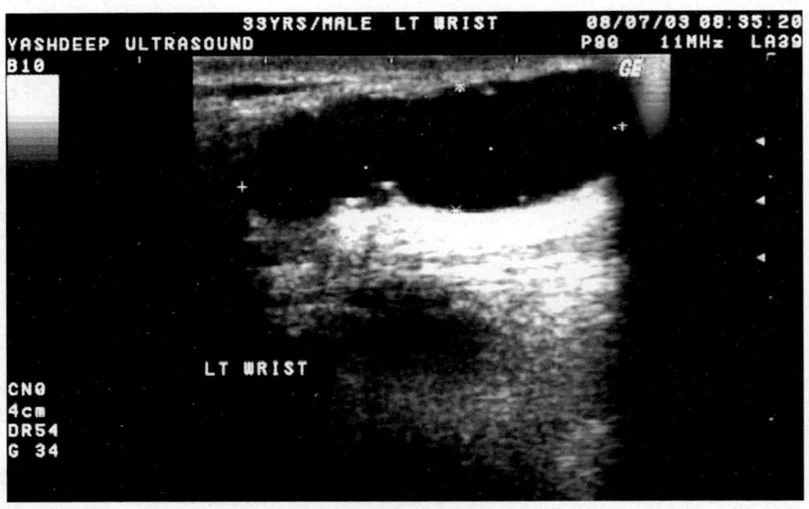

FIGURE 13.487

FIGURES 13.486 and 13.487: Ganglion cyst. HRSG shows a patient presented with soft tissue swelling over the Lt wrist. HRSG shows well-defined cystic mass seen in relation to the tendon. The mass is seen attached with the tendon sheath. No internal echoes seen in the mass. Biopsy of the cyst confirmed ganglion cyst.

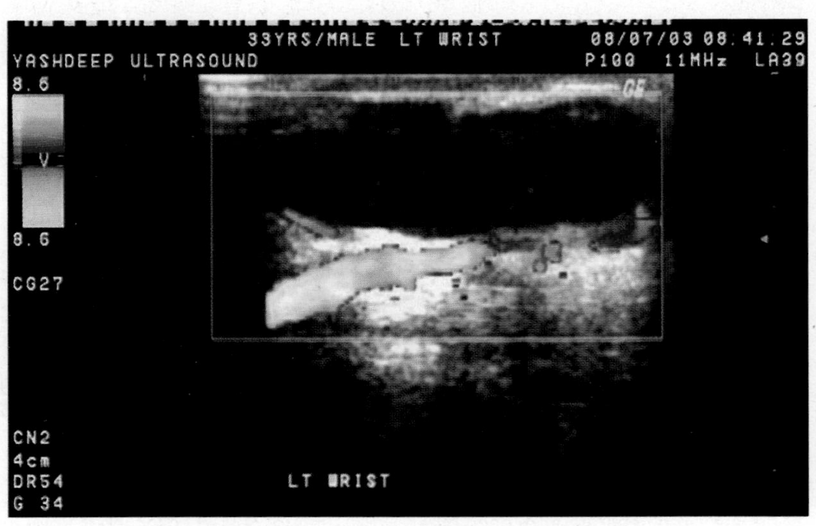

FIGURE 13.488: On color flow imaging no flow seen in the cyst.

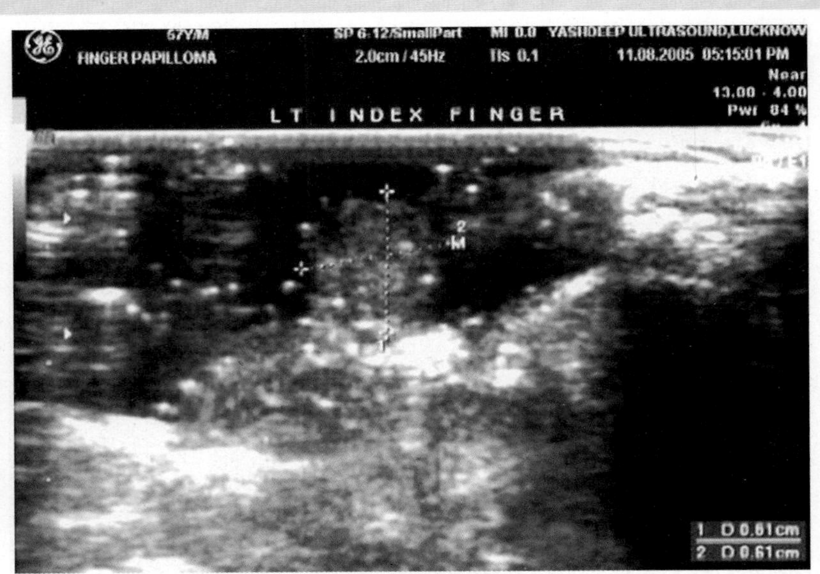

FIGURE 13.489

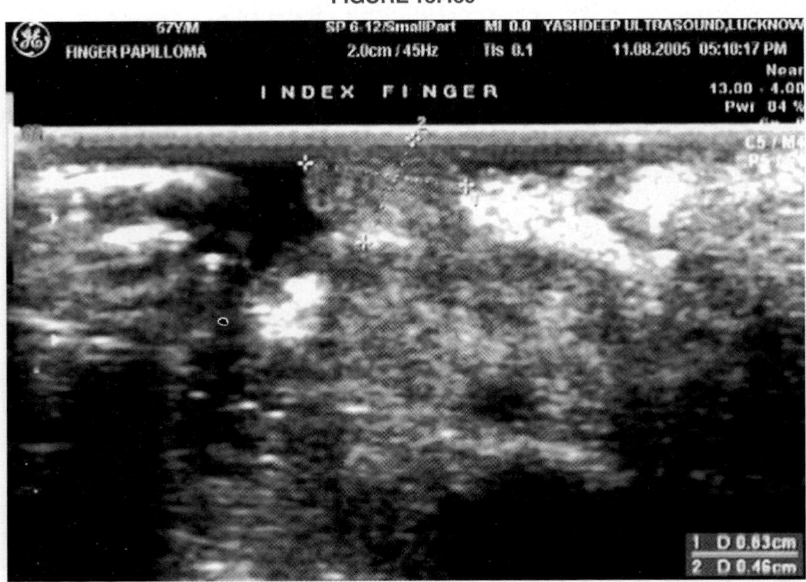

FIGURE 13.490

FIGURES 13.489 and 13.490: Papilloma index finger. HRSG shows a soft tissue mass over the index finger. HRSG shows a well-defined soft tissue mass over the pulp of the index finger. It is homogeneous in texture. However, associated subcutaneous edema is seen.

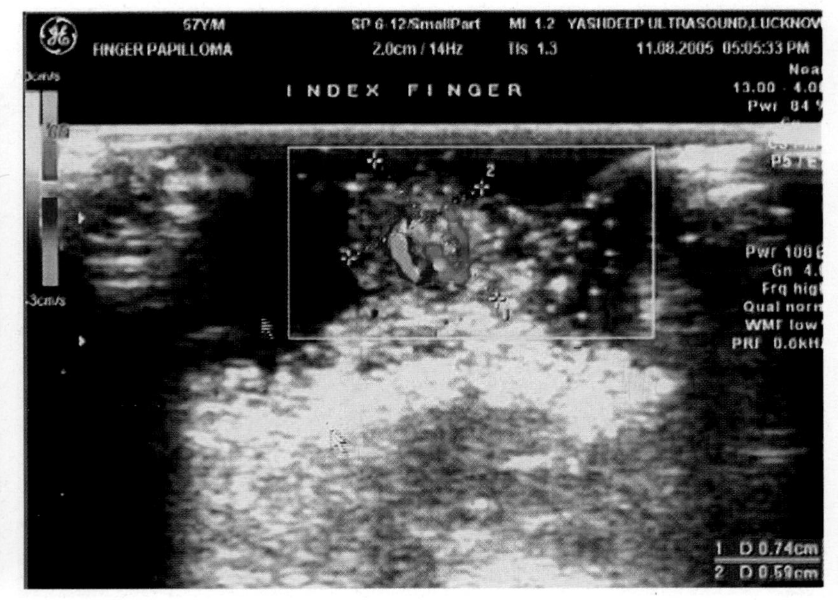

FIGURE 13.491

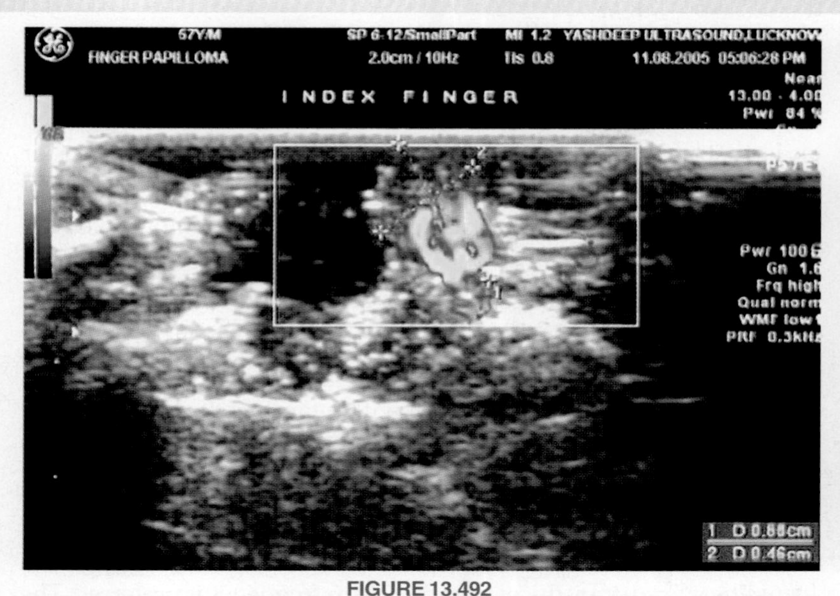

FIGURE 13.492

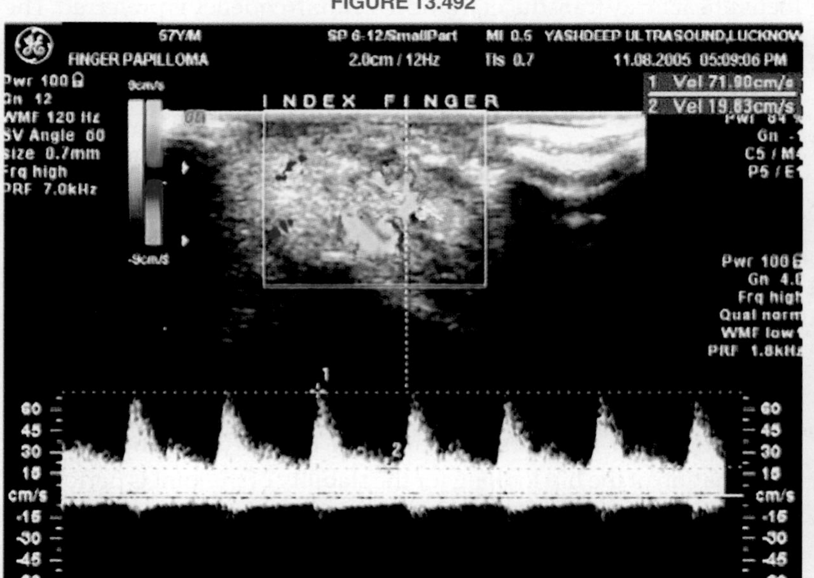

FIGURE 13.493

FIGURES 13.491 to 13.493: On color flow imaging multiple vessels are seen feeding the mass. **On spectral Doppler tracing** high flow seen in it. Biopsy of the mass confirmed papilloma.

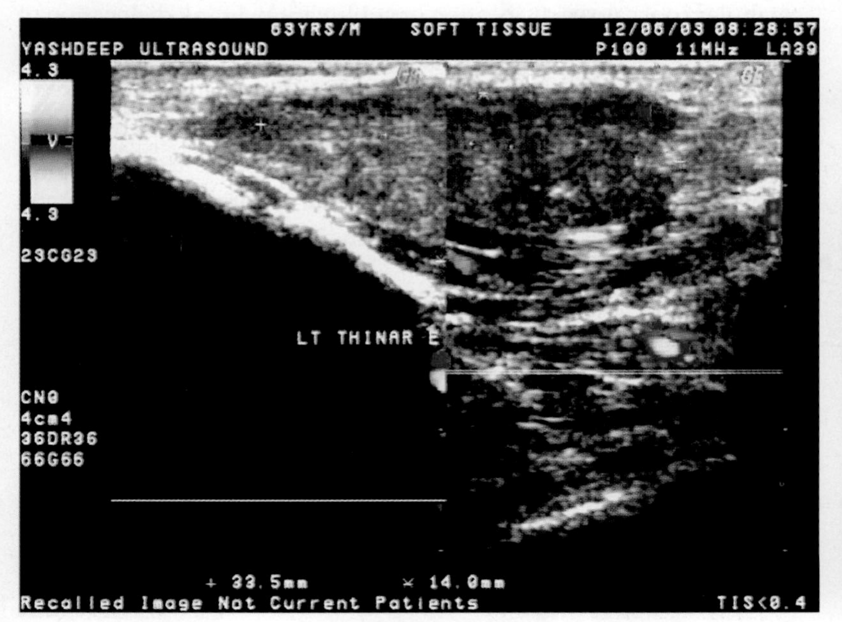

FIGURE 13.494

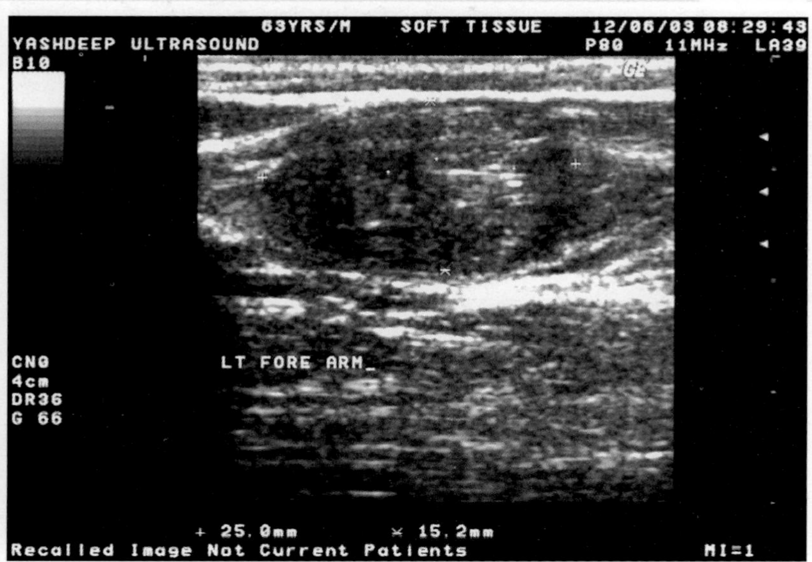

FIGURE 13.495

FIGURE 13.496

FIGURES 13.494 to 13.496: Ganglioma over the forearms. A patient presented with well-defined soft tissue mass over the thinar eminence of Lt palm. HRSG shows well-defined lobulated mass seen in relation to the tendon. It is homogeneous in texture. No calcification is seen in it. No cystic degeneration is seen. **On color flow imaging** high flow seen in it. Biopsy of the mass confirmed ganglioma.

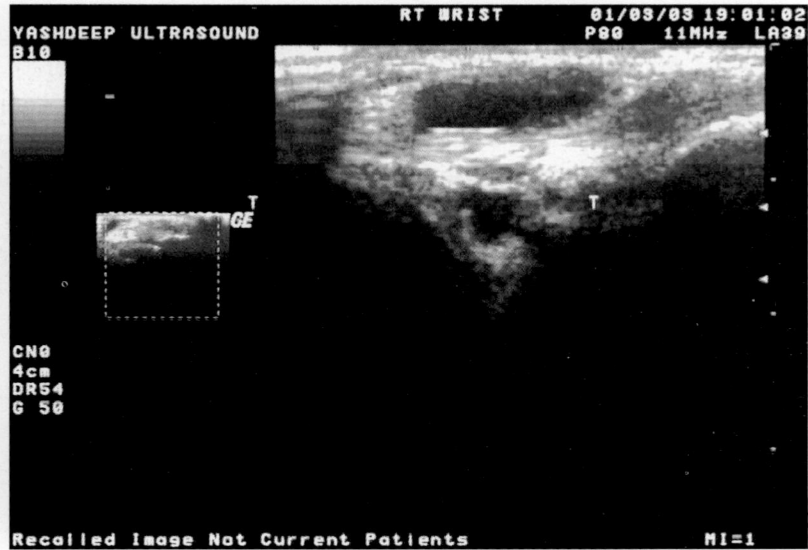

FIGURE 13.497

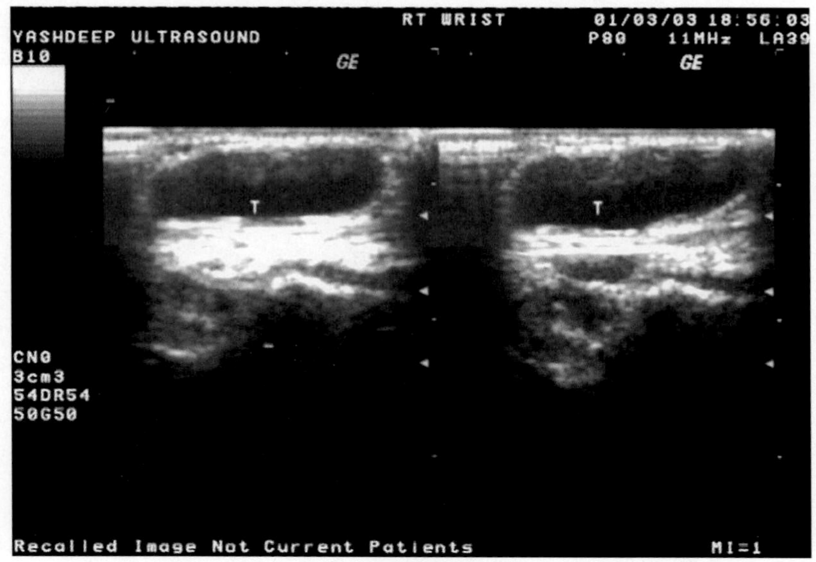

FIGURE 13.498

FIGURE 13.497 and 13.498: Mixoid degeneration of the tendon. A patient presented with soft tissue swelling over the Rt wrist. HRSG shows a well-defined cystic mass in relation to the tendon. Internal echoes are seen in it.

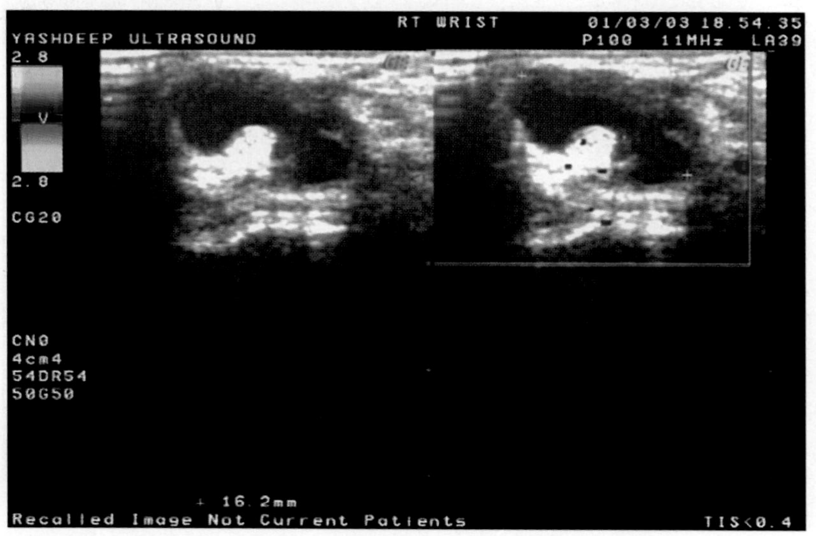

FIGURE 13.499: On color flow imaging no flow seen in it. Biopsy of the cyst confirmed mixoid degeneration.

Sonography of Hip

HRSG is very useful evaluating the hip joints. A number of patients with problems in lumbar spine, pelvis or abdomen complain of pain in the hip. HRSG can differentiate an intra-articular pathology and peri-articular pathology. Joint effusion although non-specific is a clear indication of joint pathology. HRSG is very sensitive in picking up joint effusion and minimal effusion can be picked up on HRSG. Effusions may be present due to infection, inflammation, arthritis, osteonecrosis, osteoarthritis, trauma or tumor or malignant disease. HRSG guided aspiration of fluid can tell the nature of the fluid. It is very good to evaluate postoperative complications and HRSG is very useful in detection and aspiration or hematomas or pyogenic collection after surgery because post surgery, clinical evaluation is not possible.

Congenital Dislocation of Infant Hip (CDH)

Congenital dislocated hip of an infant may be due to mechanical or physiological causes. The mechanical cause may be due to compression and constraint pressure on fetus in utero as in severe oligomnios or breech presentation. The physiological cause of CDH is due to laxity of the hip joint. Ultrasound is an important method to evaluate CDH. It can tell about the following facts:

1. The position of cartilaginous femoral head within the acetabulum
2. Any change in the position of the femoral head within the acetabulum with stress examination.
3. The steepness of acetabular roof with quantification of acetabular angles and assessment of degree of maturity or dysplasia of the acetabulum.

The infant hip is studied from its lateral aspects. The transducer is placed lateral to the femoral head over to the greater trochanter. Short focus linear array transducer from 5 to 7 Mhz frequency is preferred. The hip is examined in coronal and transverse planes with respect to the bony pelvis. In coronal plain femoral head should fit in acetabulum as a ball within a socket. The ossification nucleus is seen as an echogenic focus. It is accompanied with acoustic shadowing. The femoral head is unossified and shows hypoechoic texture with multiple fine bright echoes. The triradiate cartilage of the acetabulum is also seen as a hypoechoic shadow between femoral head and unossified medial wall of acetabulum.

In transverse view the femoral head is normally seen over a wide-open 'V'. The triradiate cartilage separates the two limbs of the "V".

On stress examination of the hip joint laxity, subluxability or dislocation can be assessed with real time study. The dynamic examination of the hip looking for the stability of the joint is performed in transverse plane with the hip flexed. During the procedure pressure is applied to the flexed hip in an attempt to force it posteriorly. It can show posterior subluxation or dislocation of the femur with stress shows a positive maneuver.

The joint mobility is not only important factor in the development of CDH. The acetabular shape plays an important role in CDH. A steep acetabular roof causes dislocation or subluxation. Frank dislocation of the femoral head in which it goes behind the acetabulum along the ileum is rare. The hip is more commonly subluxated at the level of acetabular rim. The femoral head confinement in the acetabulum and coverage by the bony and cartilaginous portions of the acetabular roof are easily assessed on HRSG. The center of the femoral head is normally medial to the lateral bony acetabular rim in infant less than 3 months old. 45% of newborn are having an immature pattern with the femoral head. The bony coverage of the femoral head seen on HRSG can be compared with radiographic acetabular index. All hips with more than 58% of femoral head covered by the bony acetabulum show normal X-rays pictures. All hips with less than 33% coverage show abnormal radiograph. The dysplastic acetabulum associated with CDH is shallow and show a steep acetabular roof. The alpha angle is measured between baseline and a line drawn from the triradiate cartilage to the lateral aspect of bony acetabular roof. The angle is important because it represents acetabular roof ossification, which promotes joint stability. If the acetabulum is less mature and dysplastic, it shows steeper roof and thus the alpha angle is becomes less, which shows dysmaturity. The iliac crest line forms the base line of the angle.

The beta angle is measured between the line drawn from lateral aspect of the labarum to the lateral bony margin of acetabulum and the lateral iliac baseline. The beta angle gives information about the assessment of aversion of the cartilaginous acetabular roof and, hence, superior femoral head displacement. The baseline is drawn along the straight line of lateral aspect of the ileum, which grossly approaches the perpendicular to the radiographic HILGREINER's line. In normal cases, normal alpha angle should be >60 degree and beta angle should be < 77.

HRSG is the choice of investigation for evaluation congenital dislocation of hip and provides required information for the surgical correction or medical treatment for the pathology.

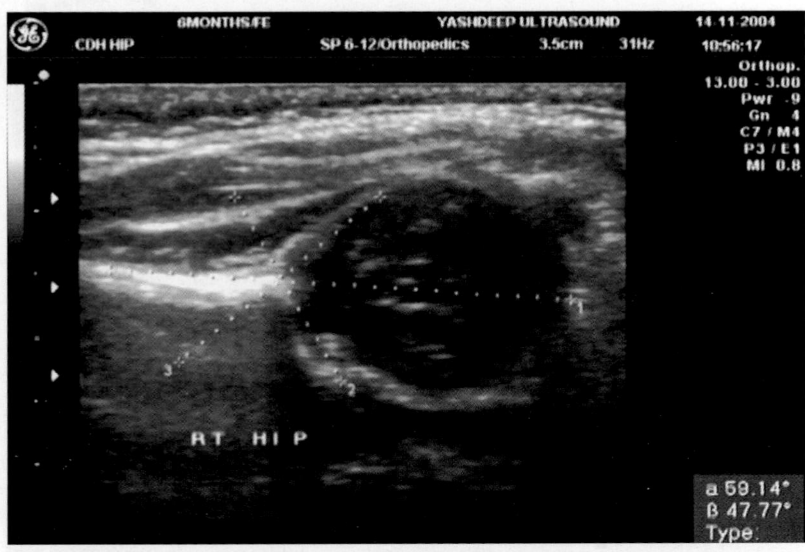

FIGURE 13.502: Normal femoral head in extension. HRSG shows normal femoral head in the acetabulum in extension. The triradiate cartilage is seen intact. The femoral head is seen well placed in the acetabular cavity.

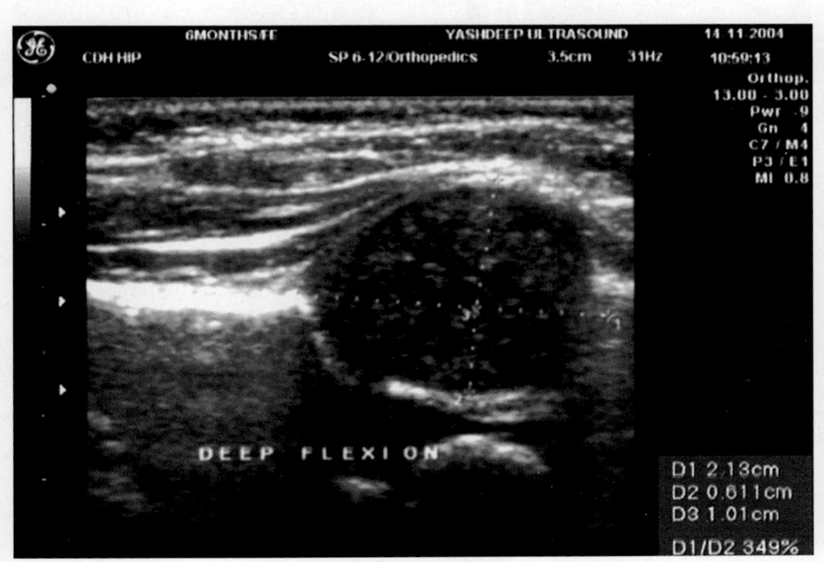

FIGURE 13.500

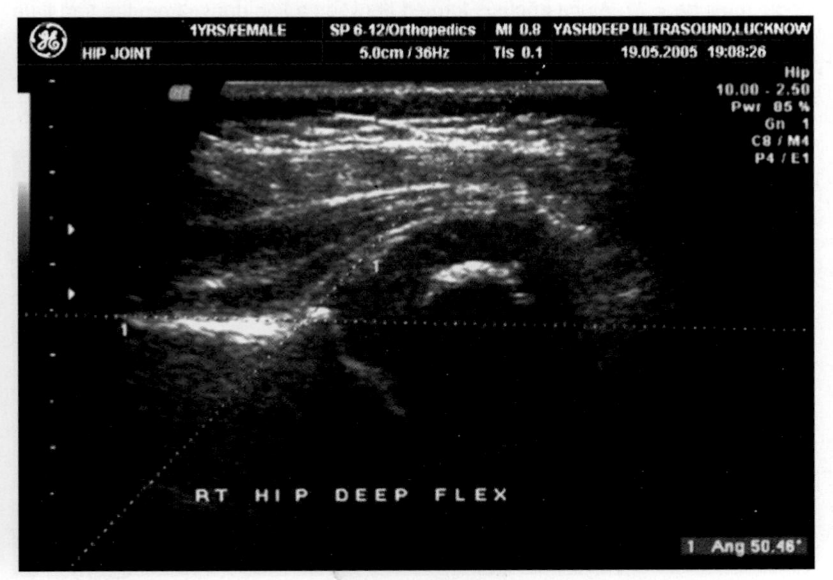

FIGURE 13.503

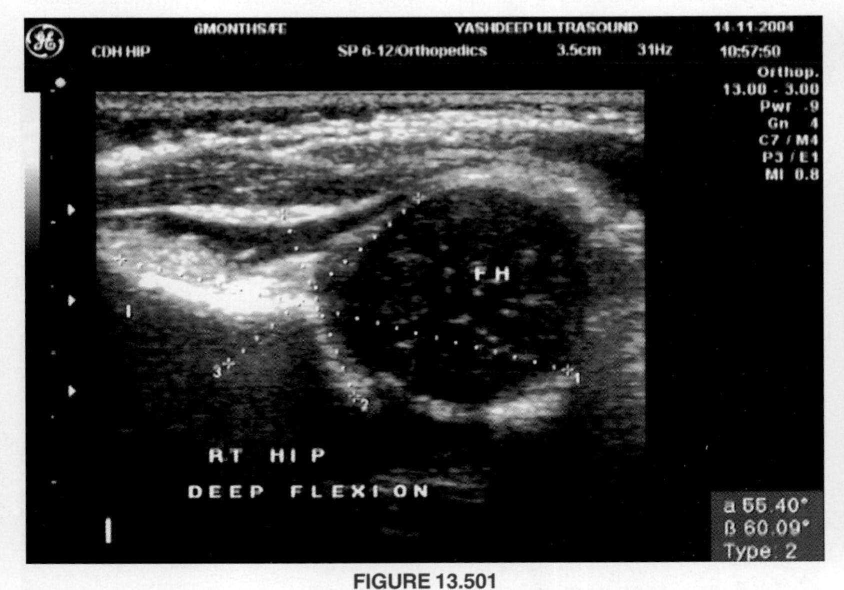

FIGURE 13.501

FIGURES 13.500 and 13.501: Normal infant hip. HRSG shows normal femoral head in acetabulum. The roof of acetabulum is well found. The femoral head is hypoechoic. Iliac bone is seen normal echogenic line.

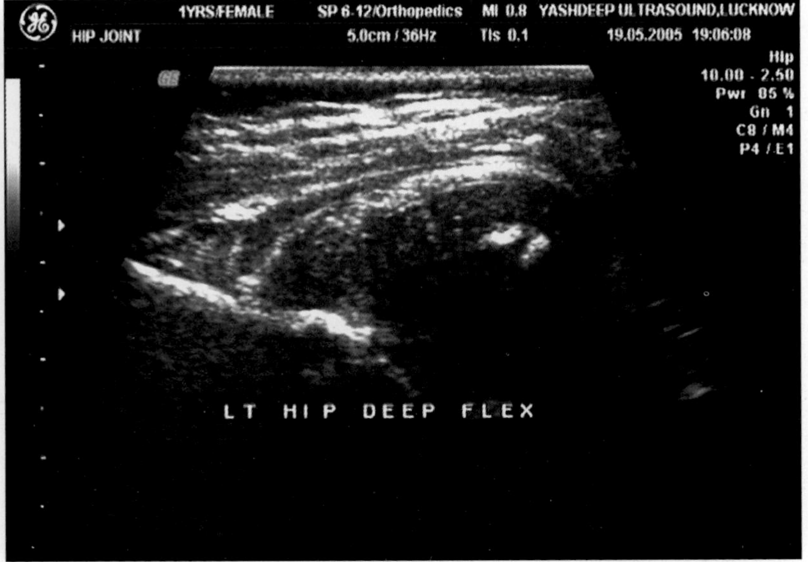

FIGURE 13.504

FIGURES 13.503 and 13.504: Normal femoral head in flexion. HRSG shows normal femoral head movement in flexion. The head is seen in place with upward movement of the femoral head and it is limited with the acetabular roof.

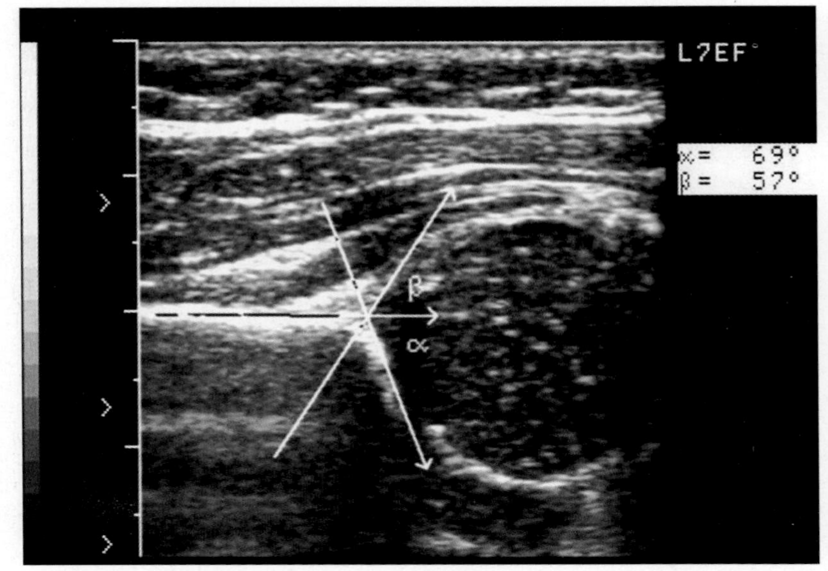

FIGURE 13.505: Normal alpha, beta angle. HRSG shows alpha and beta angle. The angle measurement is important. The normal alpha angle should be > 60. Normal beta angle must be < 77. Alpha angle is formed between a line drawn from the triradiate cartilage to the lateral aspect of the bony acetabular roof. The base line is the line drawn parallel to iliac crest. The beta angle is formed between the line drawn from the lateral aspect of labrum to the lateral bony margin of acetabulum and same lateral iliac base line.

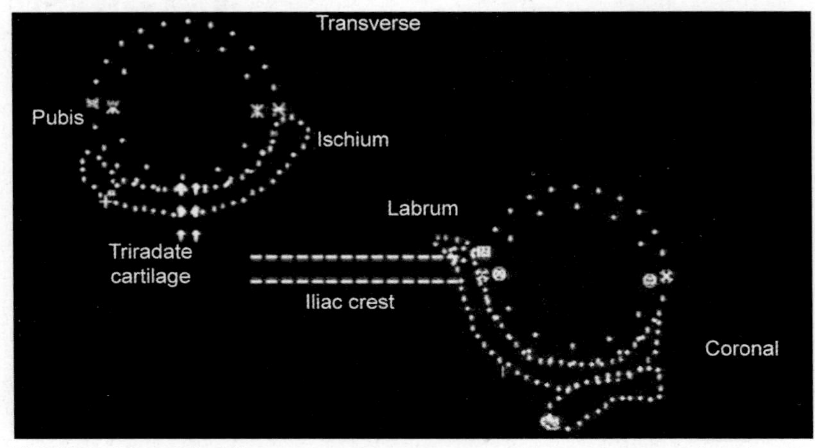

FIGURE 13.506A: Diagrammatic representation of femoral head with acetabulum normal condition. The diagrammatic representation shows normal relation with acetabulum, ischium and iliac crest.

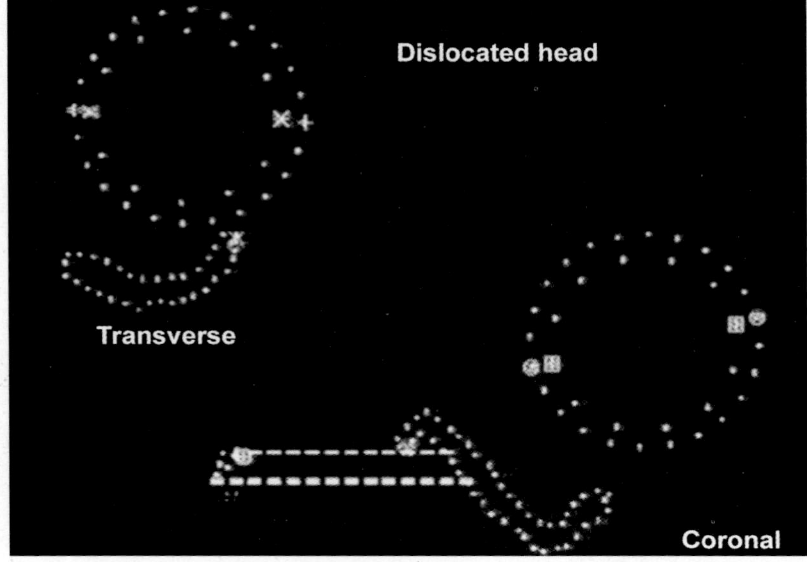

FIGURE 13.506B: Diagrammatic representation of dislocated head. The line diagram shows dislocated head from the acetabulum in transverses and coronal planes.

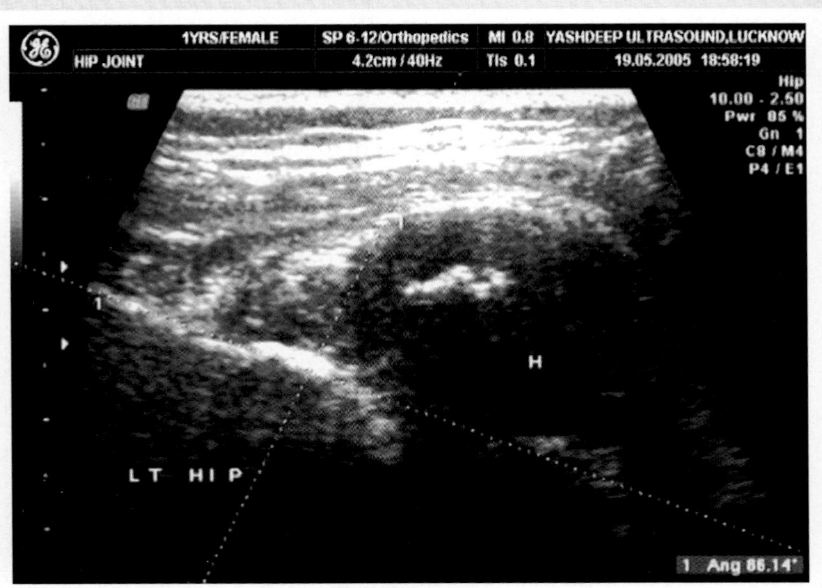

FIGURE 13.507

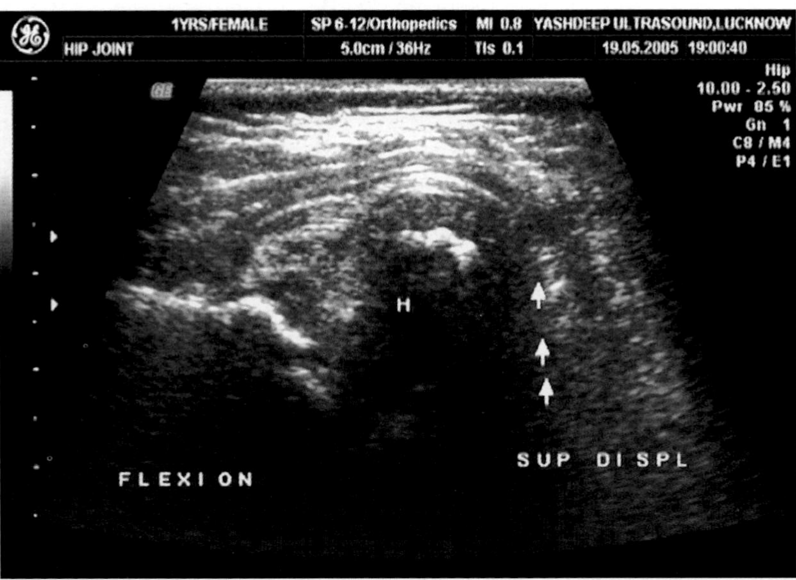

FIGURE 13.508

FIGURES 13.507 and 13.508: Congenital dislocation of hip. HRSG shows dislocation of femoral head in the Lt hip. The head is seen displaced from the acetabular cavity in and seen going up posterio lateraly. On deep flexion it is seen slipping out from the acetabulum.

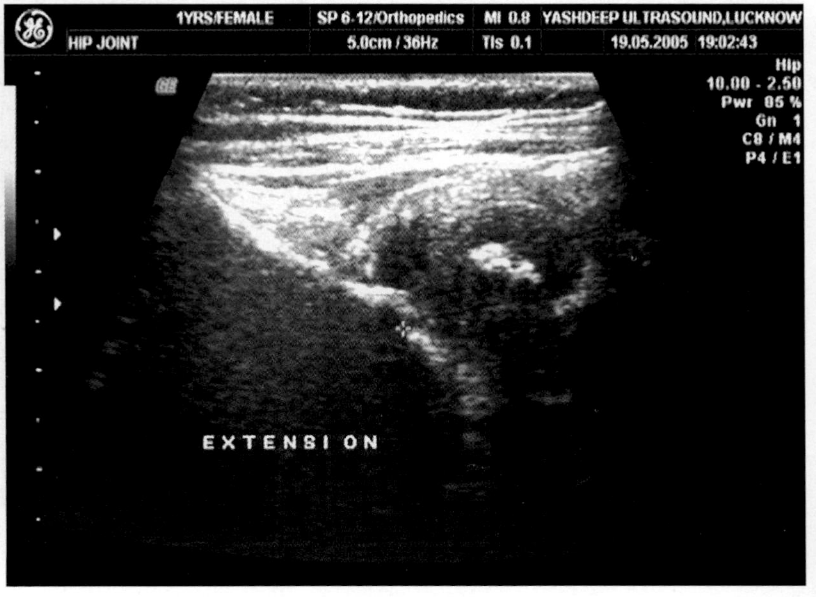

FIGURE 13.509

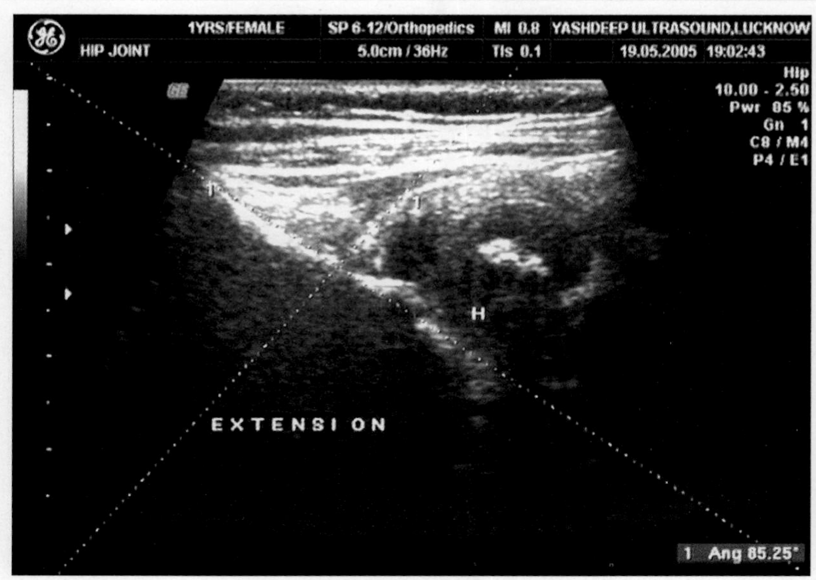

FIGURE 13.510

FIGURES 13.509 and 13.510: On extension the head is seen out side of the acetabulum. The beta angle is more than 85°, which is more than normal limit. (Normal < 77°).

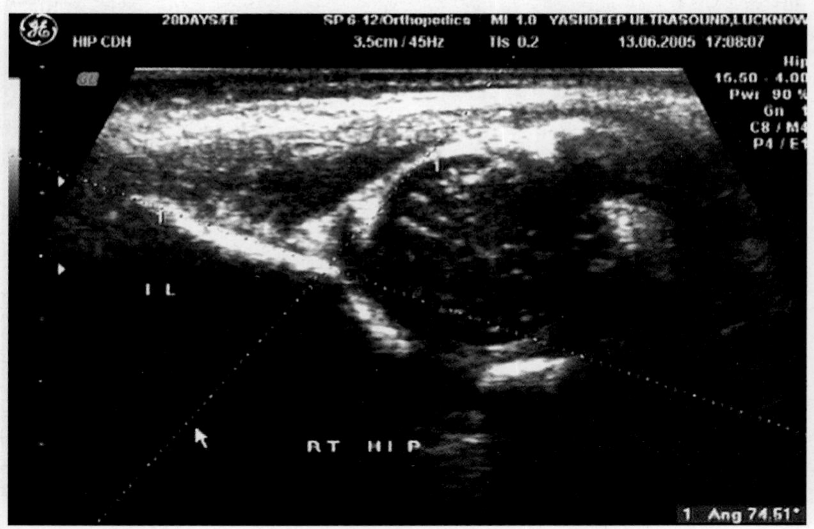

FIGURE 13.511

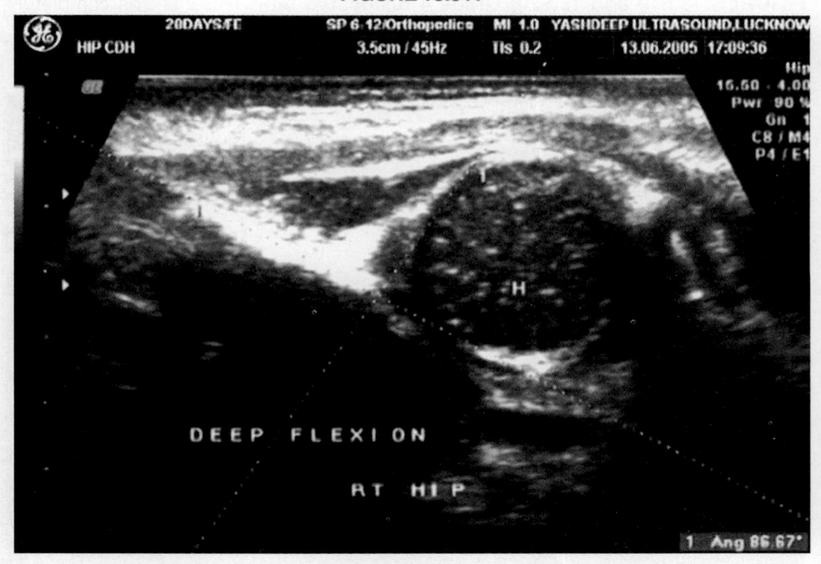

FIGURE 13.512

FIGURES 13.511 and 13.512: Subluxated femoral head. HRSG shows normally placed hip in acetabular cavity in neutral position. The beta angle is 74.5°, which is within normal limits. However, deep flexion of hip the head is seen subluxated in the acetabular cavity. The beta angle is 86.6°, which is more than normal limits.

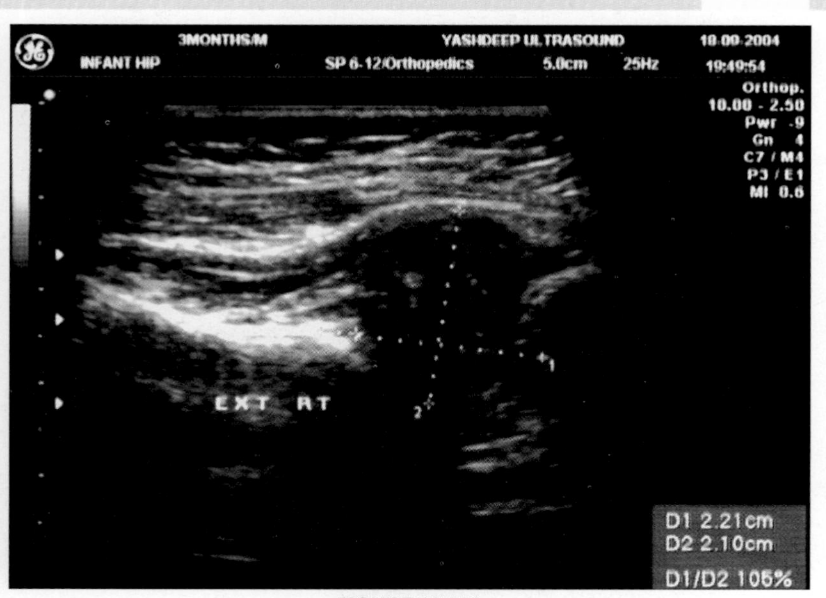

FIGURE 13.513

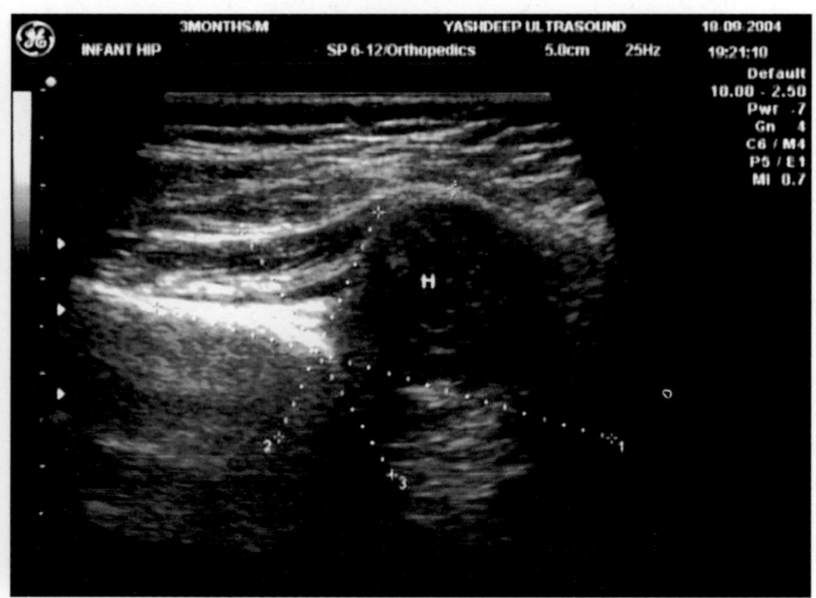

FIGURE 13.514

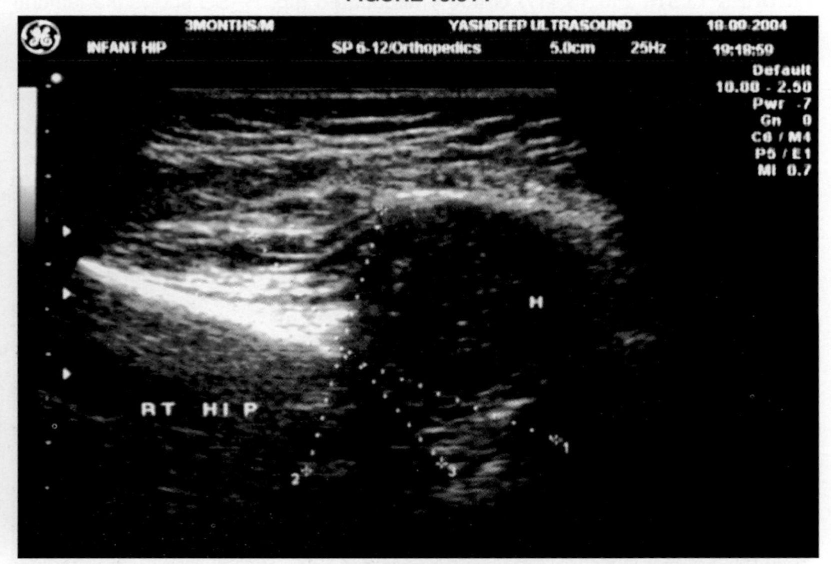

FIGURE 13.515

FIGURES 13.513 to 13.515: Hip dislocation. HRSG shows dislocation of the femoral head in Rt hip joint. More than half of the femoral head is seen lying outside from the acetabular cavity in extension and on deep flexion, the head is seen completely slipping out from the acetabulum.

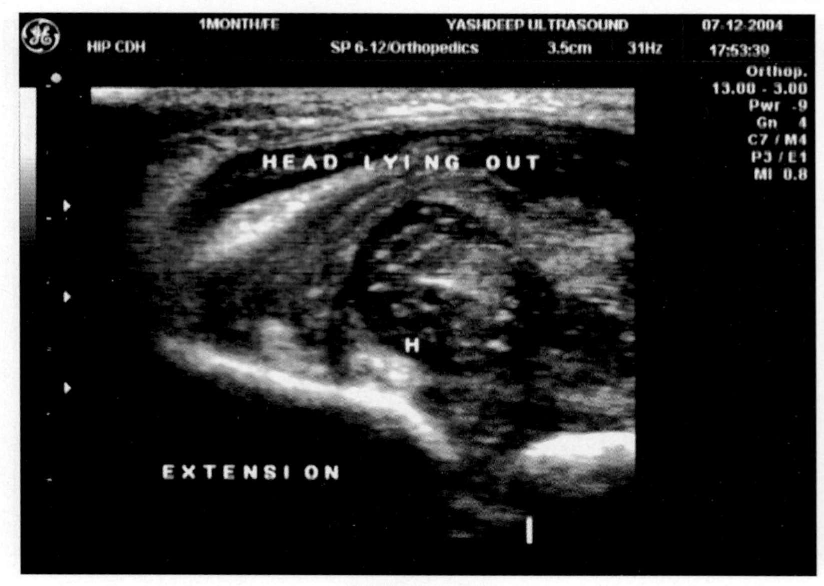

FIGURE 13.516

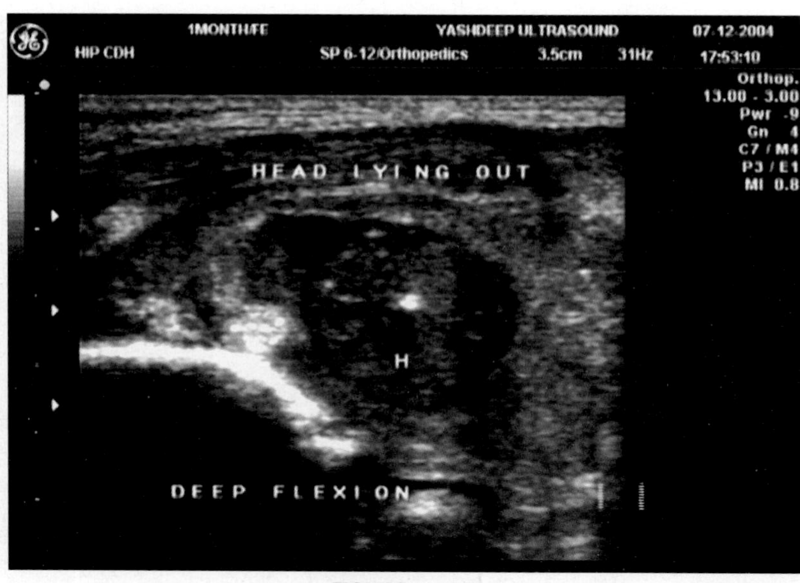

FIGURE 13.517

FIGURES 13.516 and 13.517: Grade 4 dislocated hip. HRSG shows the femoral head lying outside completely from the acetabulum on Rt side in neutral position. It is seen posteriomedially. Normal relations are distorted.

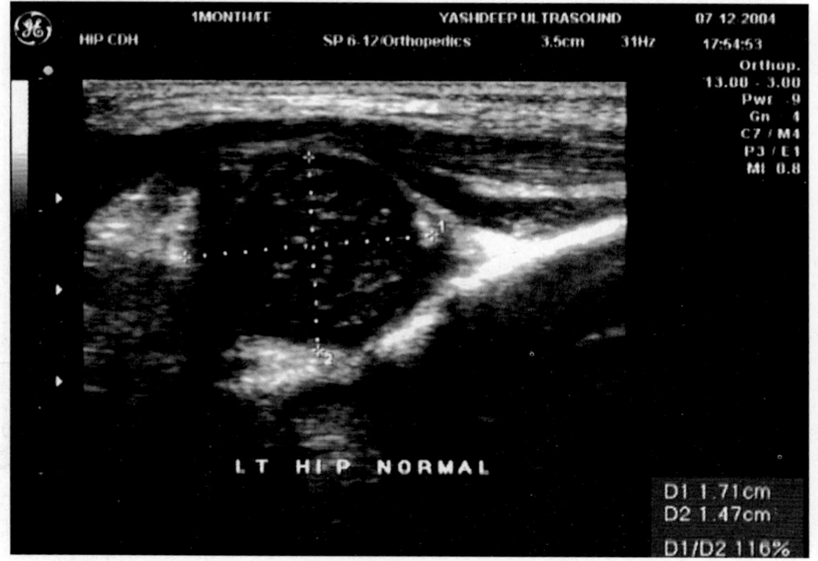

FIGURE 13.518: The Lt hip shows normal placement of the femoral head in the acetabular cavity.

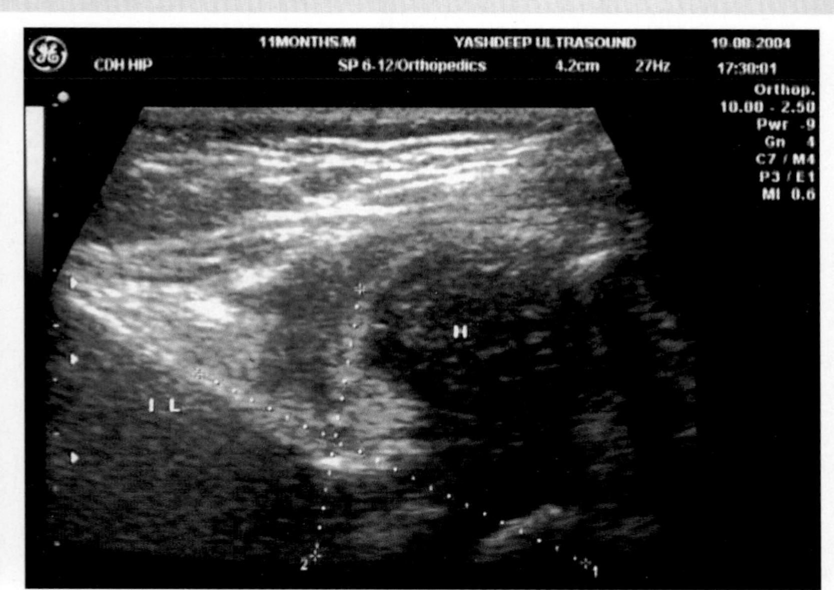

FIGURE 13.519

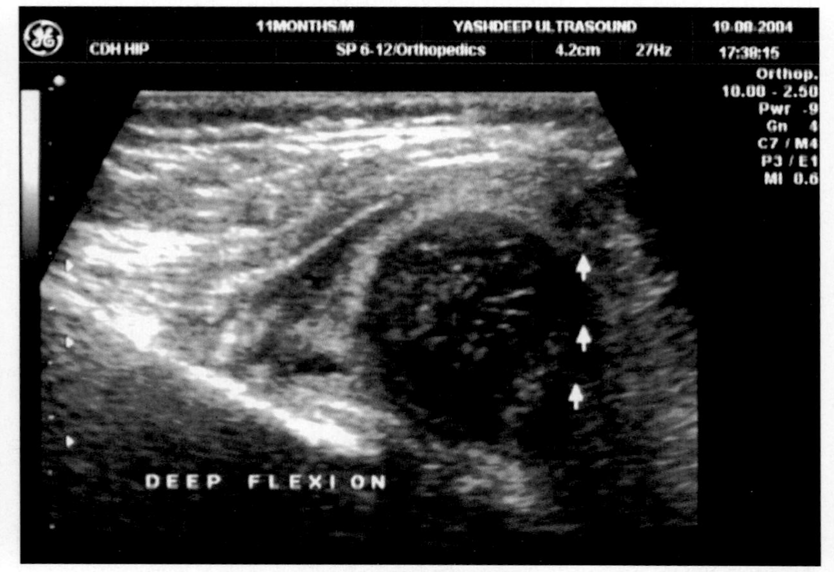

FIGURE 13.520

FIGURES 13.519 and 13.520: Grade 4 dislocated hip. HRSG shows completely missed out femoral head in a child. The head is seen lying outside in neutral position. On deep flexion the head is seen overriding the iliac crest.

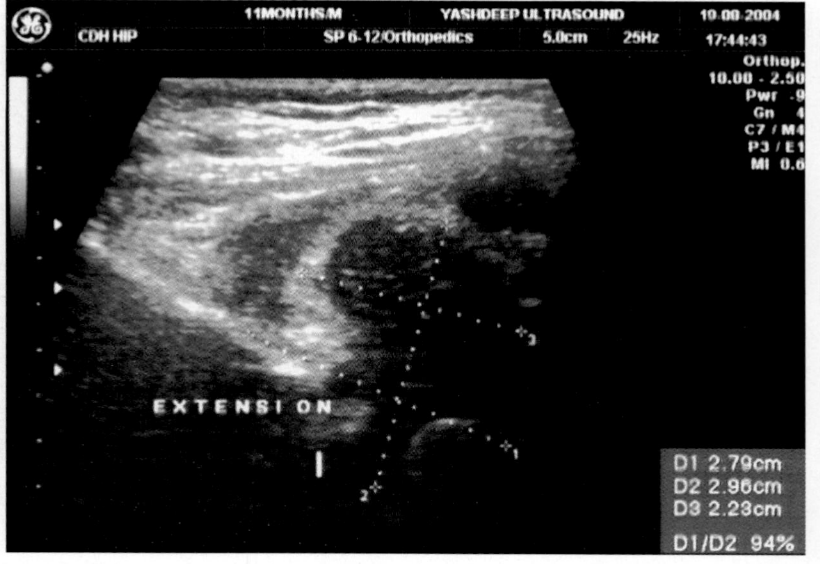

FIGURE 13.521

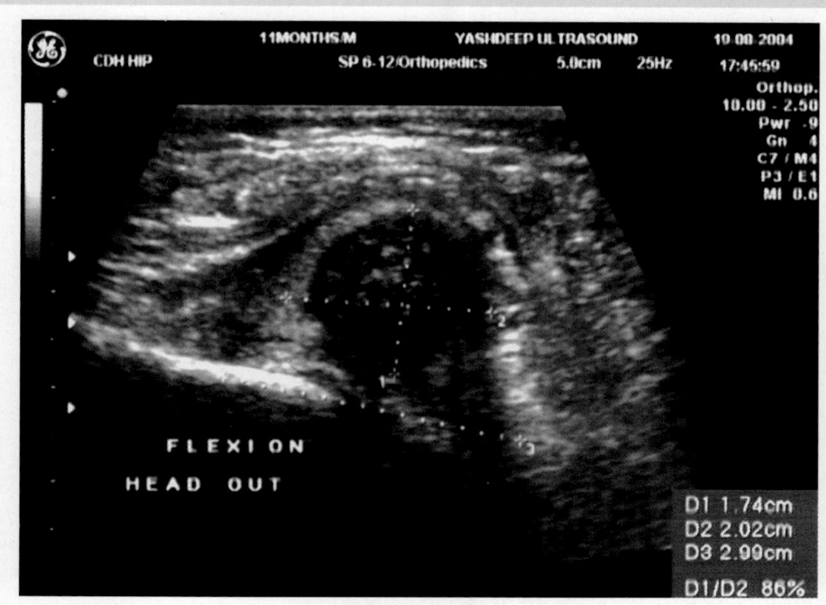

FIGURE 13.522

FIGURES 13.521 and 13.522: Whole of the head is seen lying outside from acetabular cavity in extension as well as in flexion.

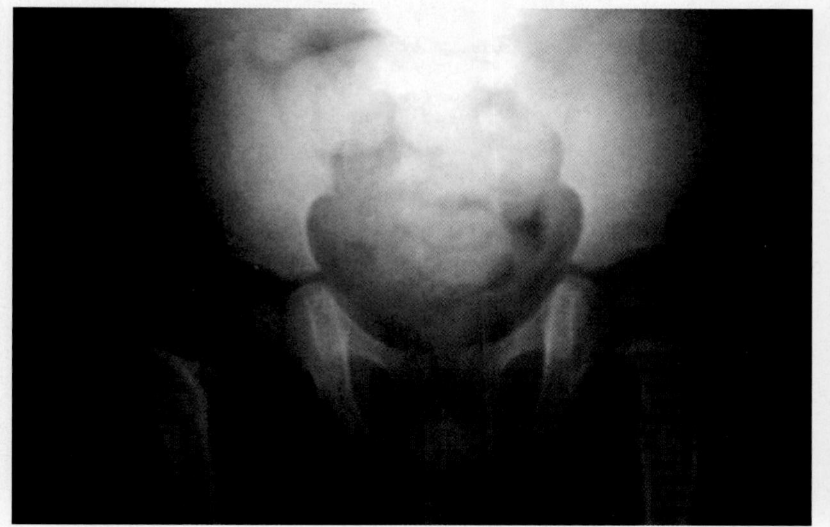

FIGURE 13.523

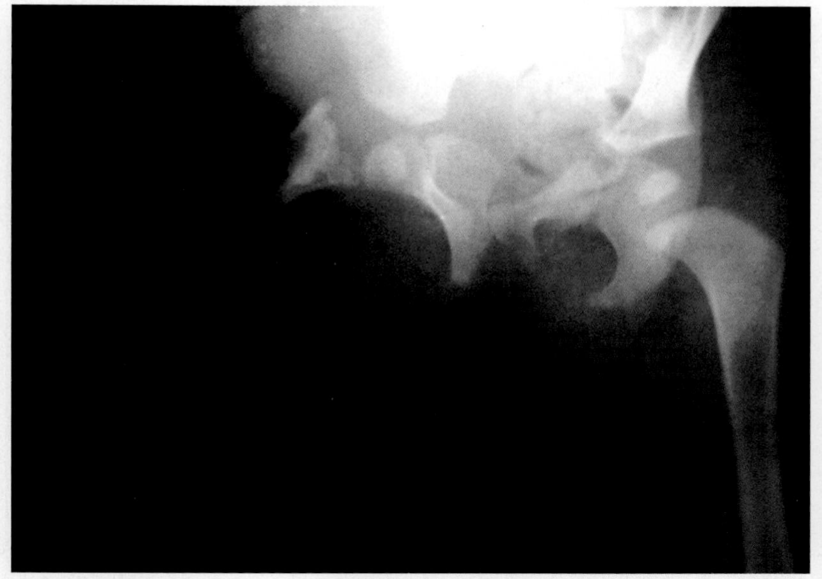

FIGURE 13.524

FIGURES 13.523 and 13.524: Septic arthritis of Rt hip joint. Plain X-ray of hip joint shows irregular radiolucent defect in Rt femoral shaft. The neck of the femur is seen destroyed.

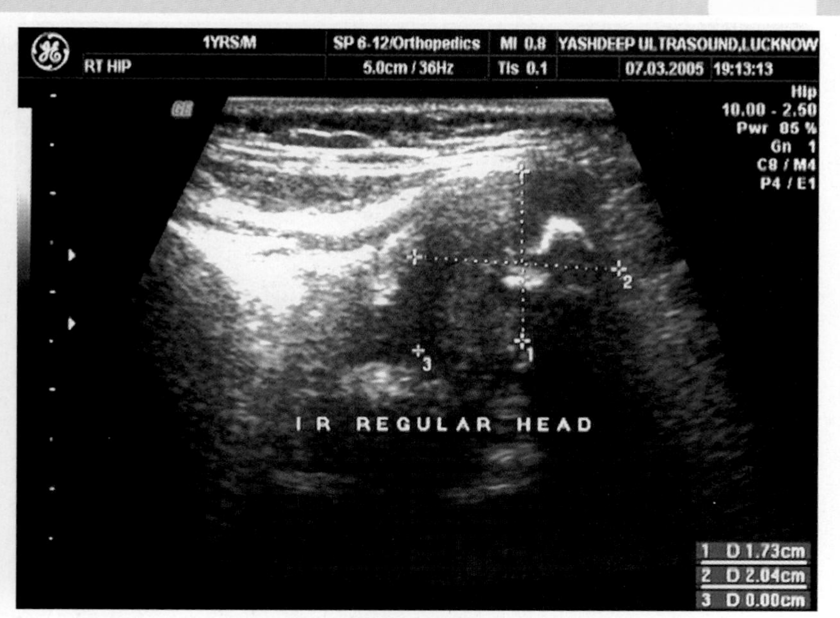

FIGURE 13.525

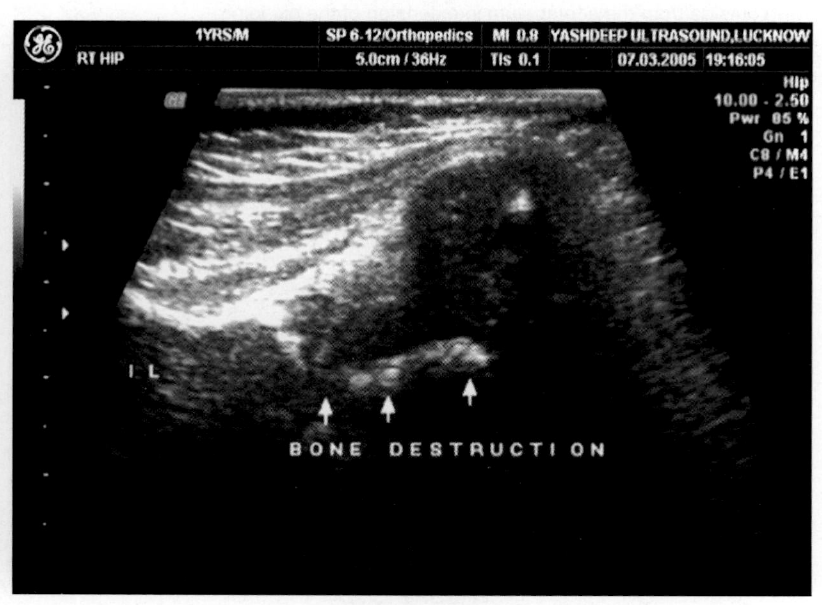

FIGURE 13.526

FIGURES 13.525 and 13.526: HRSG shows bony erosion of the femoral shaft. The femur head is irregular in outline. Normal rounded shape is lost. Echogenic specks are seen in the capsule.

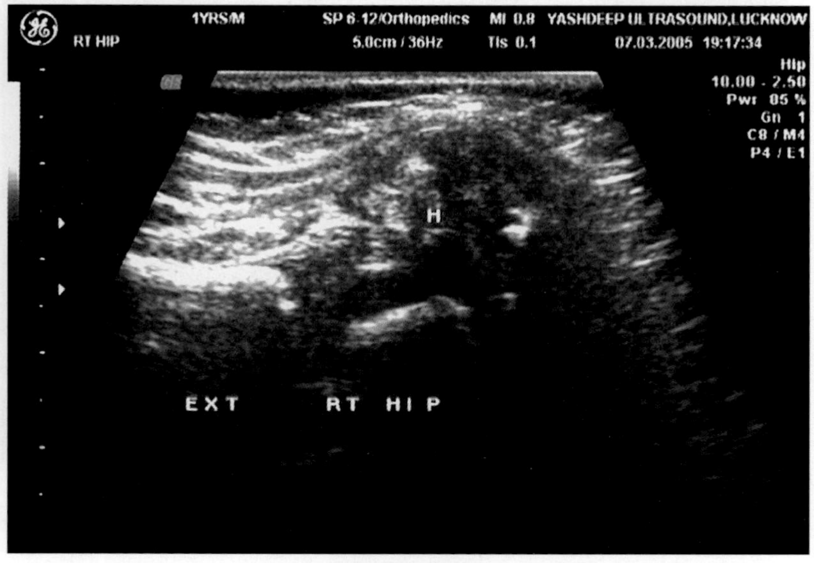

FIGURE 13.527

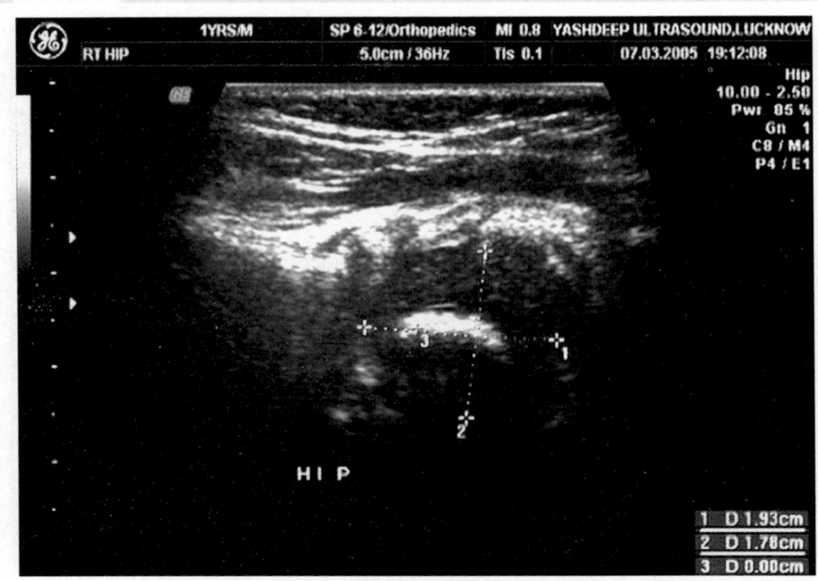

FIGURE 13.528

FIGURES 13.527 and 13.528: The femoral head is also seen dislocated and seen lying outside from the acetabulum in extension of the hip joint.

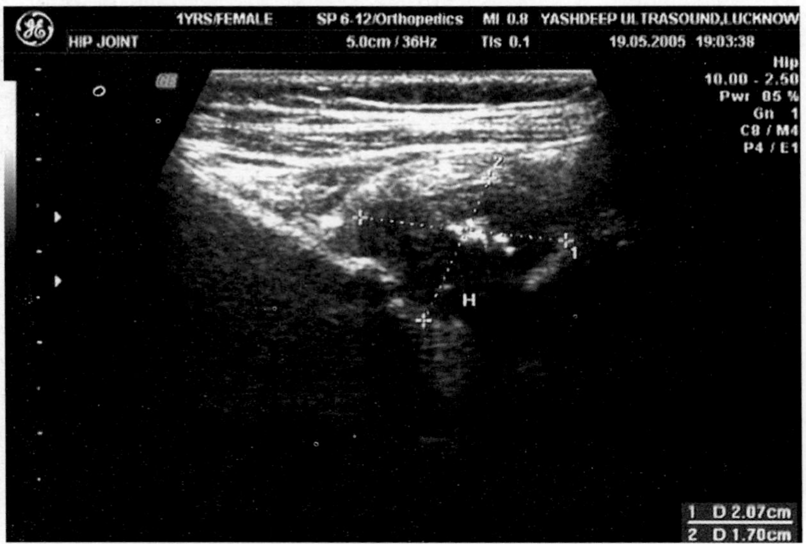

FIGURE 13.529

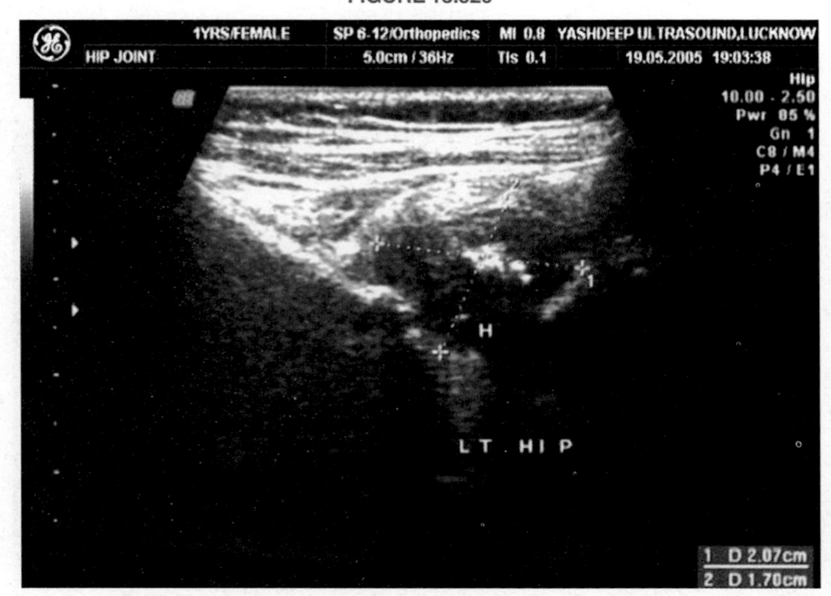

FIGURE 13.530

FIGURES 13.529 and 13.530: The head is also seen displaced out from the acetabular cavity on deep flexion. The acetabulum is also irregular in outline. Margins are echogenic with calcification seen.

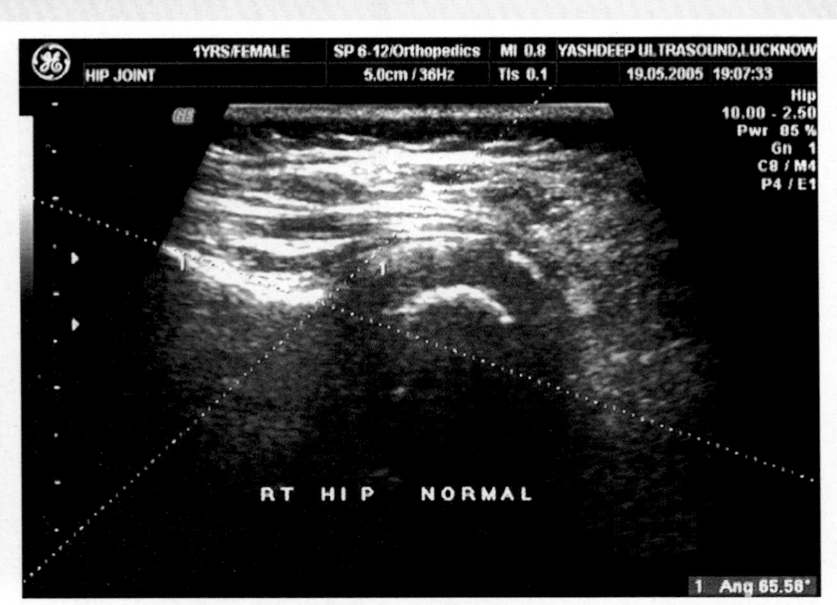

FIGURE 13.531: The Lt hip shows normal femoral head.

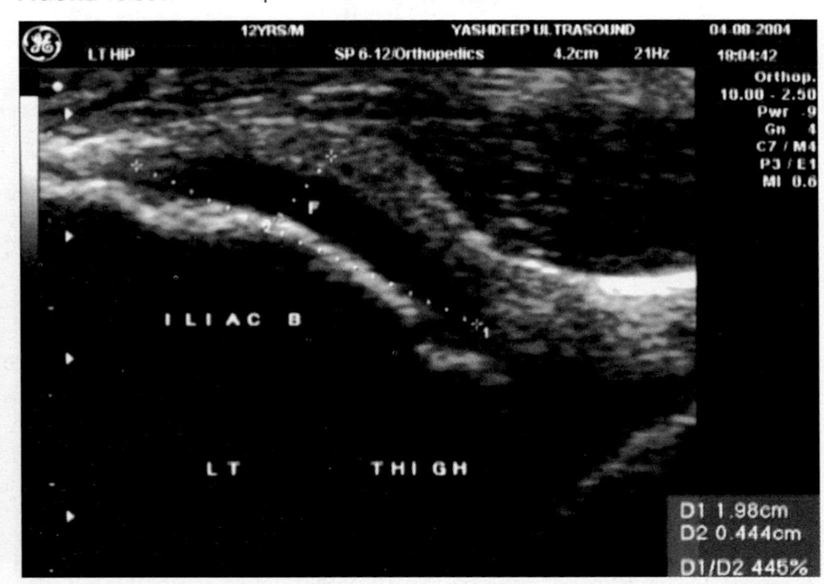

FIGURE 13.532: Septic arthritis with fluid in the capsule. A patient presented with painful hip movement on the Lt side. HRSG shows low-level echo collection in the joint capsule. The collection is also seen going on the medial side of the thigh.

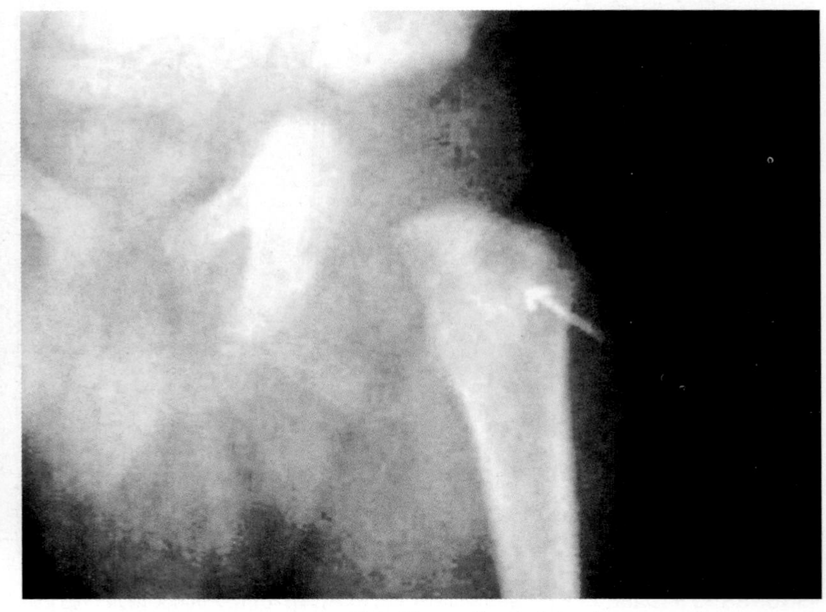

FIGURE 13.533: Septic arthritis. Plain X-ray of Lt hip joint shows a radiolucent area in the upper part of femoral shaft in an infant presented with painful Rt hip.

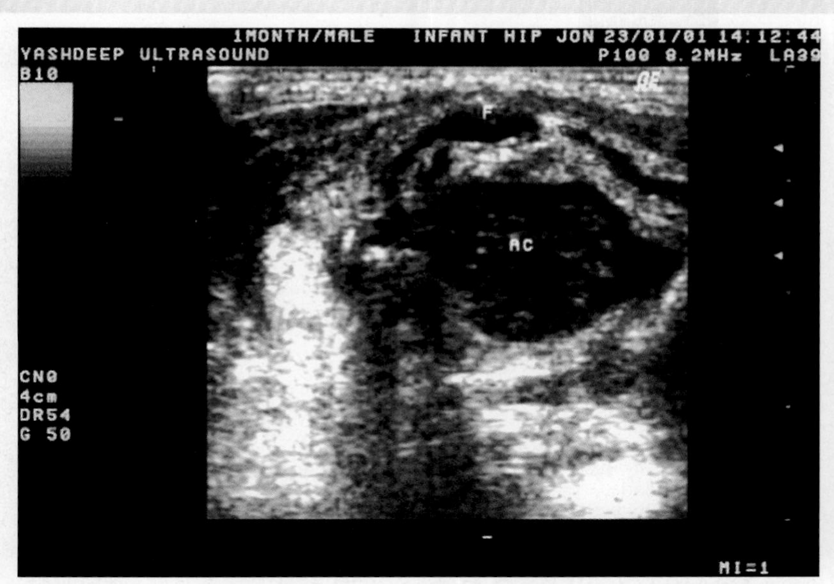

FIGURE 13.534: HRSG shows a hypoechoic low-level echo collection in the acetabular cavity. The femoral head is also irregular in its superior part. Few fine specks of calcification are also seen suggestive of pyogenic collection.

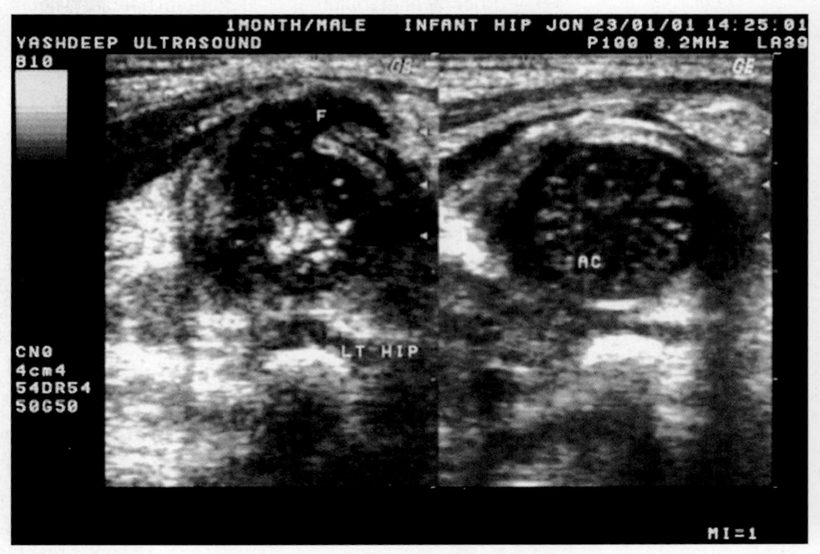

FIGURE 13.535

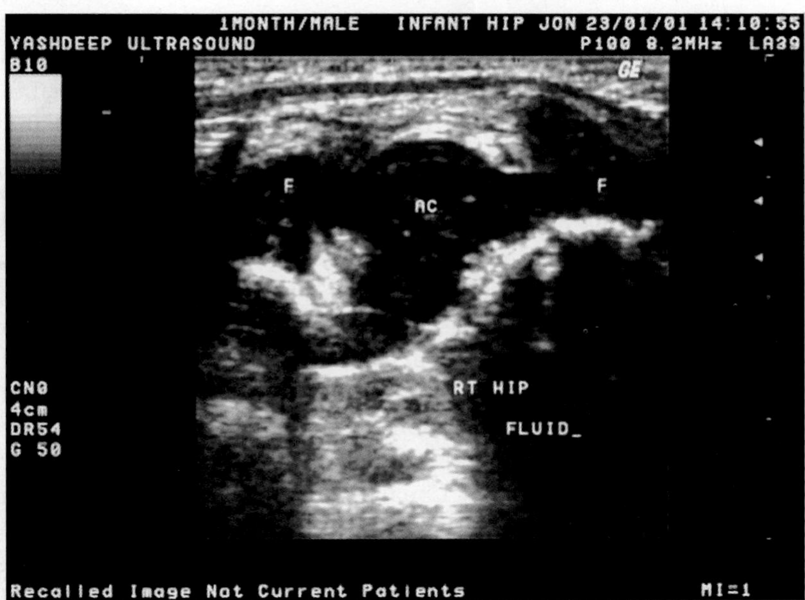

FIGURE 13.536

FIGURES 13.535 and 13.536: Bilateral septic arthritis. HRSG shows irregular hypoechoic echo collection in acetabular cavity and intracapsular part of joint in a patient presented with bilateral swollen joints with painful movements.

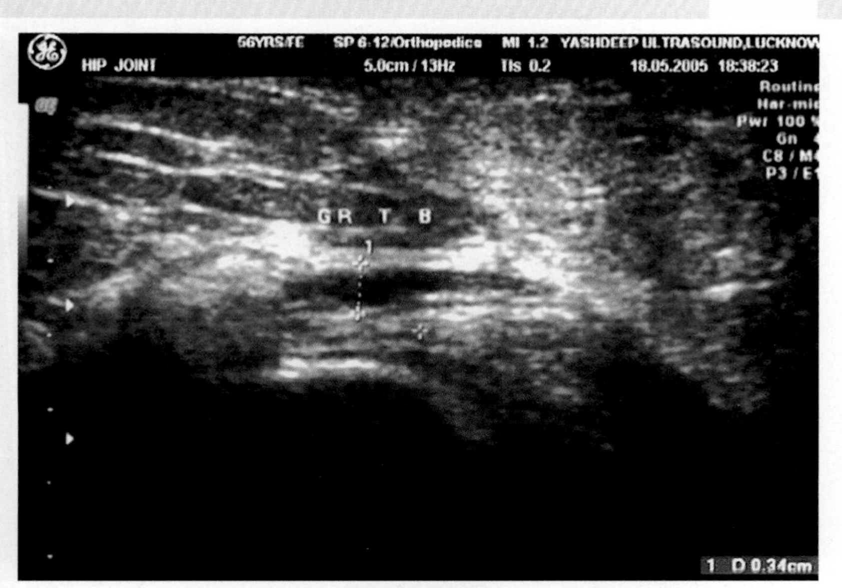

FIGURE 13.537

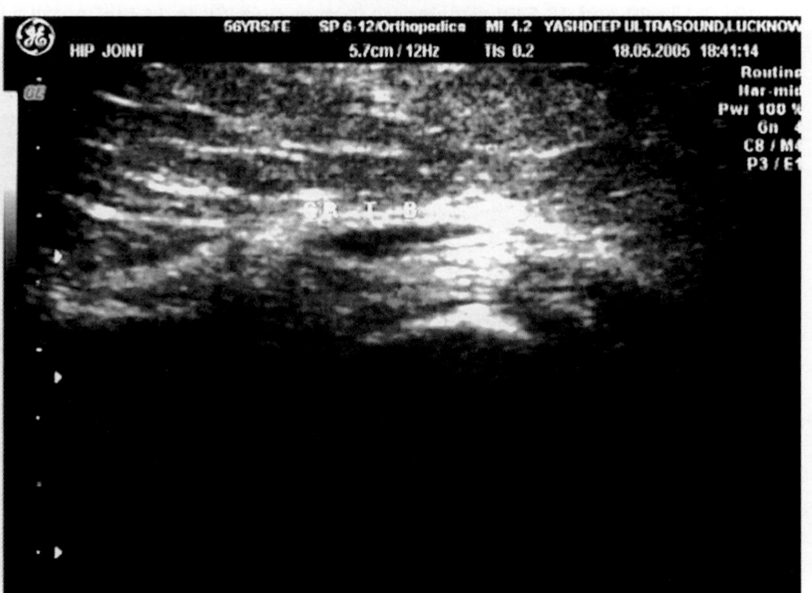

FIGURE 13.538

FIGURES 13.537 and 13.538: Greater trochanter bursitis. A patient presented with painful hip movement. HRSG shows distended greater trochanter bursa. The bursal walls are echogenic. Low-level echo collection is seen in the bursa suggestive of greater trochanter bursitis.

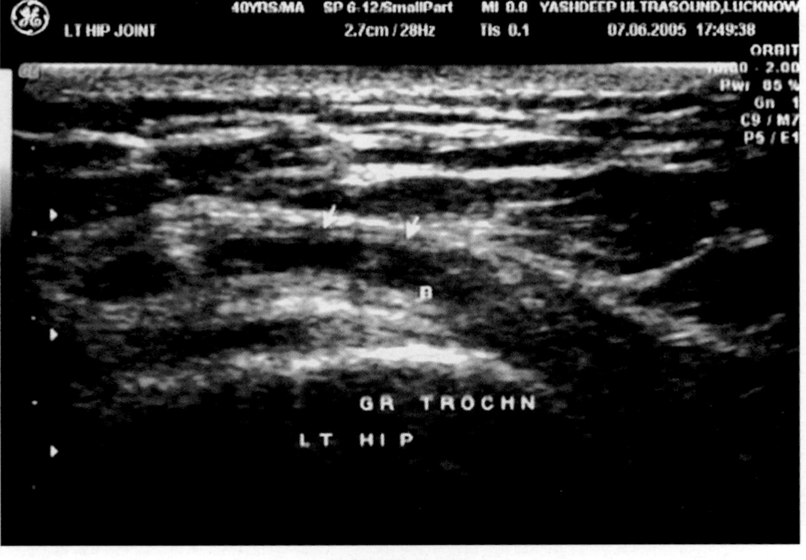

FIGURE 13.539

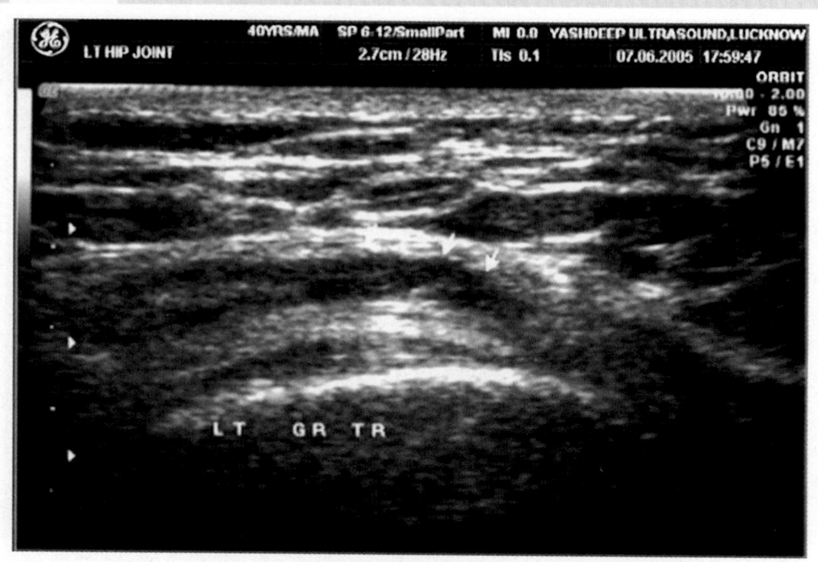

FIGURE 13.540

FIGURES 13.539 and 13.540: Greater trochanter bursitis in Lt hip. A patient presented with painful movement of the Lt hip. HRSG shows distended greater trochanter bursa. Thick echo collection is seen in the bursa. The bursal fluid shows echogenic debris suggestive of bursitis.

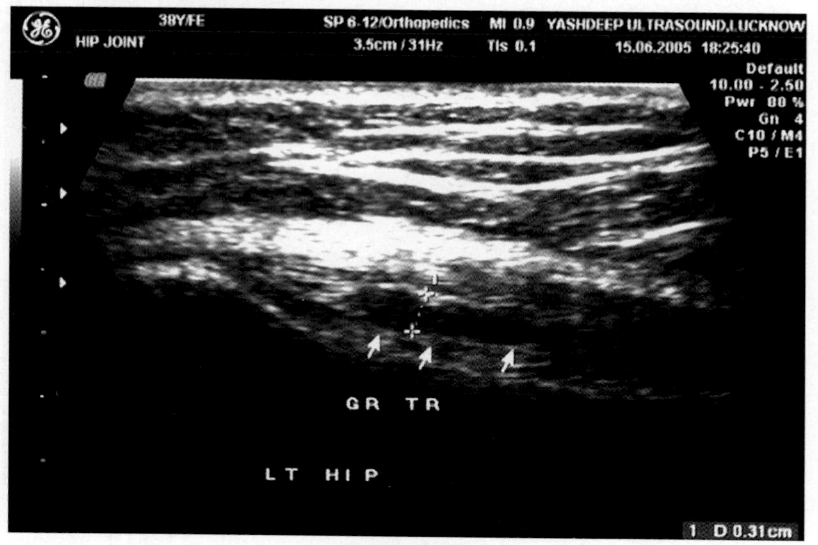

FIGURE 13.541

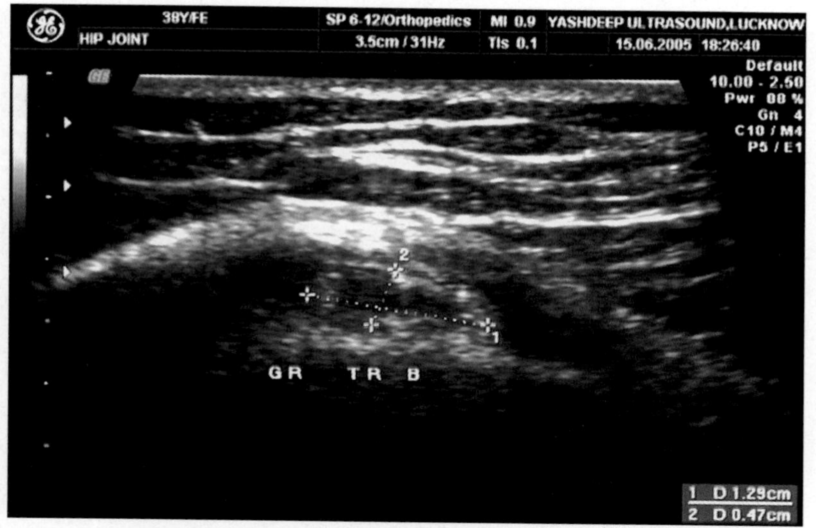

FIGURE 13.542

FIGURES 13.541 and 13.542: Greater trochanter bursitis. HRSG shows echogenic greater trochanter bursa in the Lt hip. Bursal walls are distended. Echogenic shadows are floating in the bursa. Low-level echoes are seen floating into it. Margins are echogenic suggestive of bursitis.

Table 17.1: Graf classification of dislocation of hip types

Type	*Description*	*Alpha Angle (Degrees)*	*Beta Angle (Degrees)*	*Findings*
1	Normal	>60		Should not dislocate in absence Absence of a neuromuscular imbalance with altered biomechanics.
2A	Physiological immature <3 months.	50-59		
2B	Delayed ossification >3 months.	50-59		
2C	Very deficient bony	43-49	<77	At this stage beta angle acetabulum but femoral becomes important. Head still concentric
2D	Femoral head subluxated	43-49	>77	Increased beta angle signifies an elevated, everted labrum and Subluxation.
3	Dislocated	<43	>77	
4	Severe dysplasia/ dislocation	Not measurable; flat, shallow, bonny acetabulum		Labrum inverted; interposed between femoral head and ileum

Sonography of Knee

High Resolution Sonography of the knee is most frequently asked investigation. Trauma to the knee is most frequent and the soft tissue structures of the knee joint like ligaments, tendons, menisci and joint capsules are easily injured by trauma. HRSG is non-invasive technique to evaluate all these structures. Intra-articular and extra-articular disease can be easily evaluated on HRSG and they are most frequent cause of the pain and swelling associated with the knee joint. Most of the time normal radiographic examination is in conclusive. In such situations HRSG plays an important role. It can tell about intra as well as extra articular disease non-invasively. Where as arthrography and arthroscopy are invasive techniques. The biggest advantage of the modality is to get the information about the joint function during real time examination.

Tendonitis is the common cause of knee pain. Most of the time it involves quadriceps, biceps, patellar or tripod tendons. Calcification is also seen in chronic tendonitis. It can tell about the tendon rupture by demonstrating disrupted tendon fibers. Bursitis is also one of the causes of knee pain. It can be traumatic septic hemorrhagic or chemical in origin. Rupture of baker cyst is always associated with pain and swelling of the knee. It may present as deep vein thrombosis.

HRSG is also valuable in evaluating ligaments injuries, rupture, tear or degenerative changes in the ligament. Medial or lateral collateral ligament can be easily evaluated. However, the anterior and posterior cruciate ligaments are difficult to be seen on HRSG. But proper technique and window settings anterior cruciate ligament can be seen in most of the cases. The meniscal tear and rupture can be evaluated very well on HRSG. Different type of meniscal tears, calcification, cyst formation or mixoid degeneration can be picked up on HRSG. The sonographic evaluation of knee starts with evaluation of suprapatellar bursa and quadriceps tendon. Usually a small thin film not more than 2 mm is seen in suprapatellar bursa.

The images of patellar tendon are seen after that. Medial and the collateral ligaments are seen on either side of the knee and both medial and lateral recesses are evaluated. The menisci are seen in lateral and medial decubitus positions with valgus and varous stress technique. It is done by slightly lifting the heel with support and with internal rotation. Thus the menisci are seen. They are evaluated in longitudinal and coronal planes. The popliteal fossa is seen in prompt position and any pathology of popliteal fossa like muscle tear, cyst, baker cyst, vascular malformations, meniscal tear or nerve tumors can be very well evaluated on HRSG.

HRSG is an excellent modality to evaluate degenerative changes involving the intra-articular part of the joint. It can tell about the articular surface erosion, loose body formation and entrapment of the fragments without any invasion.

Sonography of Ligaments

Traumatic injury in knee and ankle are very common in sports. Chronic ligament injury and meniscal tear are very common in athletes. They can be poorly evaluated clinically. HRSG and MRI are both excellent techniques best suited to evaluate ligament injury. They provide multi planner imaging and provide good detail of anatomy. The main ligaments, which can be evaluated on HRSG, are as follows:

Knee Joint

1. Medial collateral ligament.
2. Lateral collateral ligament.
3. Cruciate ligament.

Medial collateral ligament is specialized structure. It is 9 cm long flat and tri-layered structure. It is also known as tibial collateral ligament. It extends from medial femoral condyle to the medial aspect of proximal tibia. HRSG demonstrated medial collateral ligament as trilaminar structure separated by hypo echoic zones.

Ligament Pathology

It can be either intra-articular or extra-articular. Intra-articular tear requires patience and careful technique in knee joint. A 60-degree opening is required to get optimum sonographic window.

Extra-articular Ligaments

HRSG provides reliable means of detecting acute ligament injury. It can detect partial and complete tear with confidence. Complete rupture shows discontinuity of ligament with free ends are separated by hematoma.

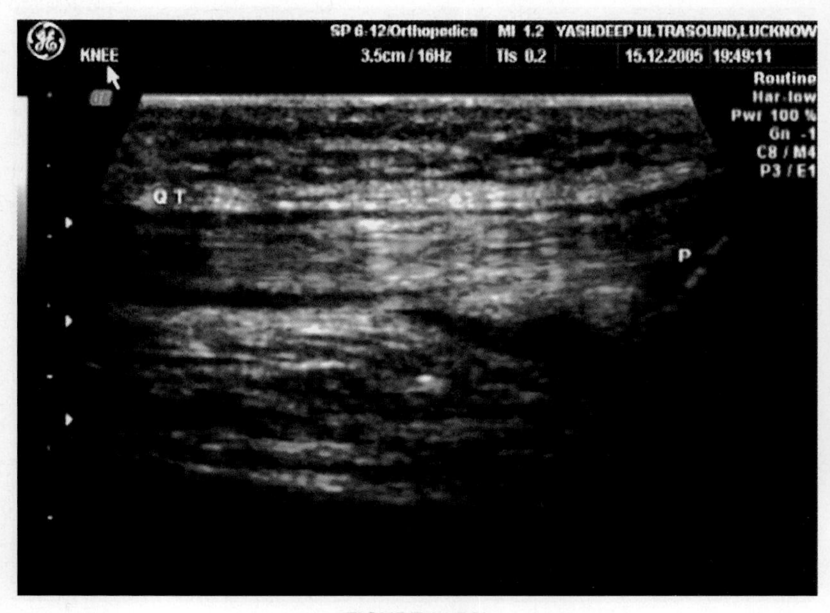

FIGURE 13.543

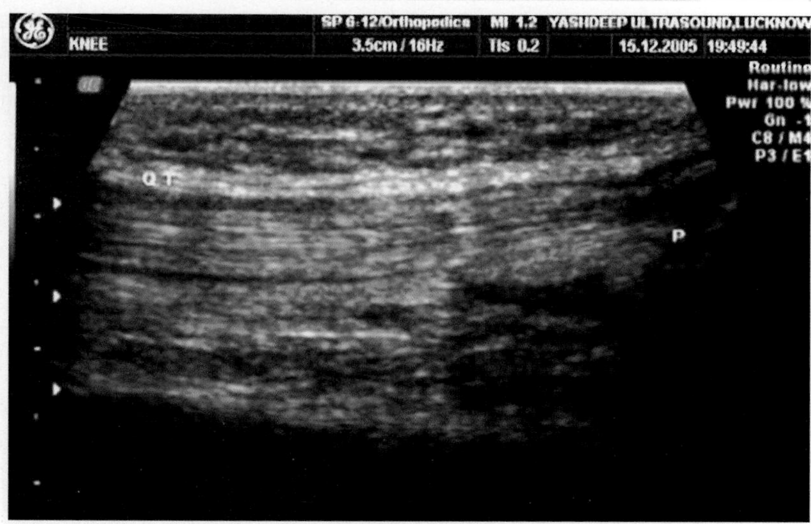

FIGURE 13.544

FIGURES 13.543 and 13.544: Normal quadriceps tendon. HRSG shows normal quadriceps tendon. The tendon is seen as a medium level echo strap with fine fibril texture. The fine tendon sheath is seen thin echogenic membrane covering the tendon. Small amount of fluid is seen in suprapatellar bursa.

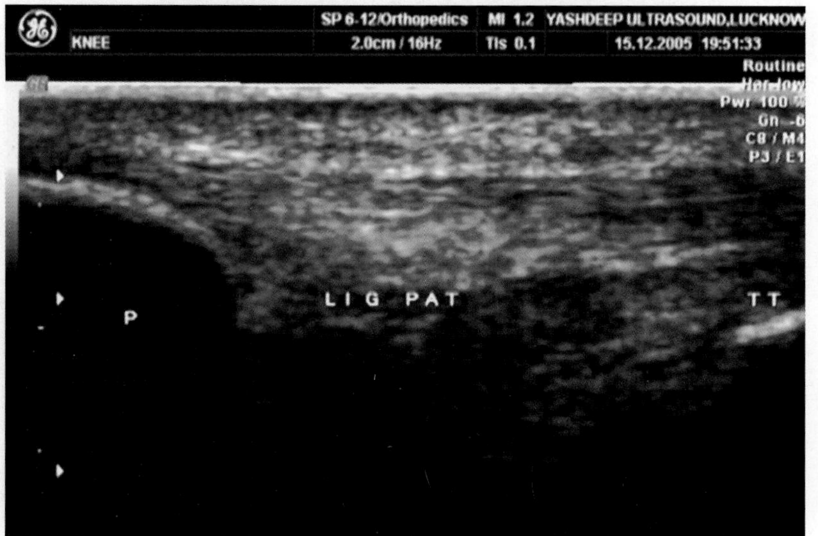

FIGURE 13.545

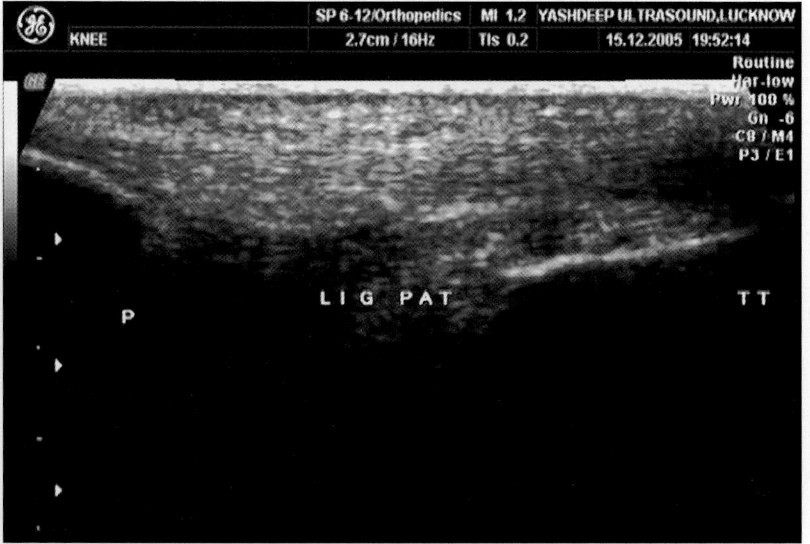

FIGURE 13.546

FIGURES 13.545 and 13.546: Normal patellar tendon. HRSG shows normal patellar tendon. This tendon does not have synovial sheath. The inferior portion of the tendon is attached at tibial tuberosity.

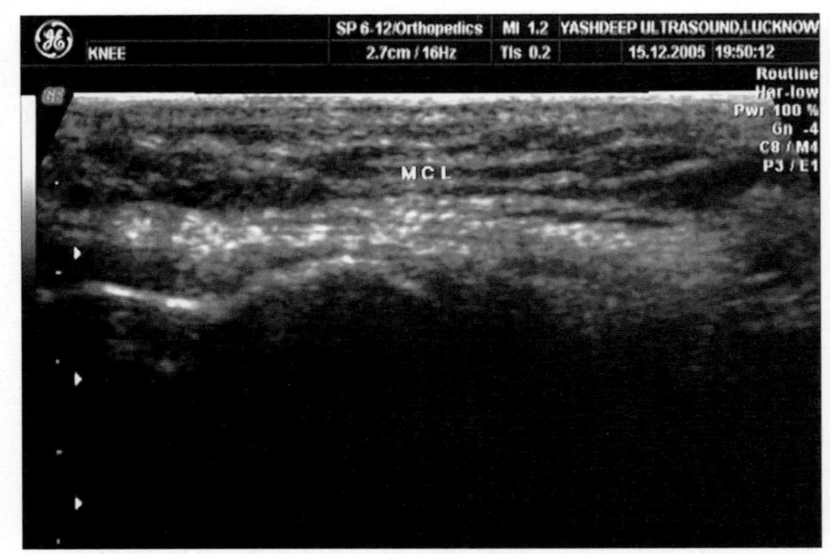

FIGURE 13.547: Normal medial collateral ligament. HRSG shows normal medial collateral ligament as an echogenic band with smooth borders.

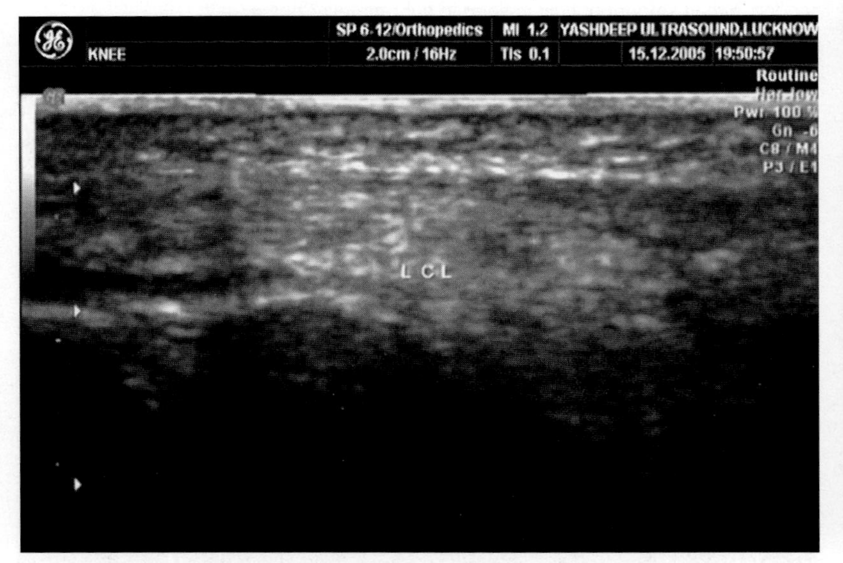

FIGURE 13.548: Normal lateral collateral ligament. HRSG shows normal lateral collateral ligament. It is seen as an echogenic strap with smooth margins.

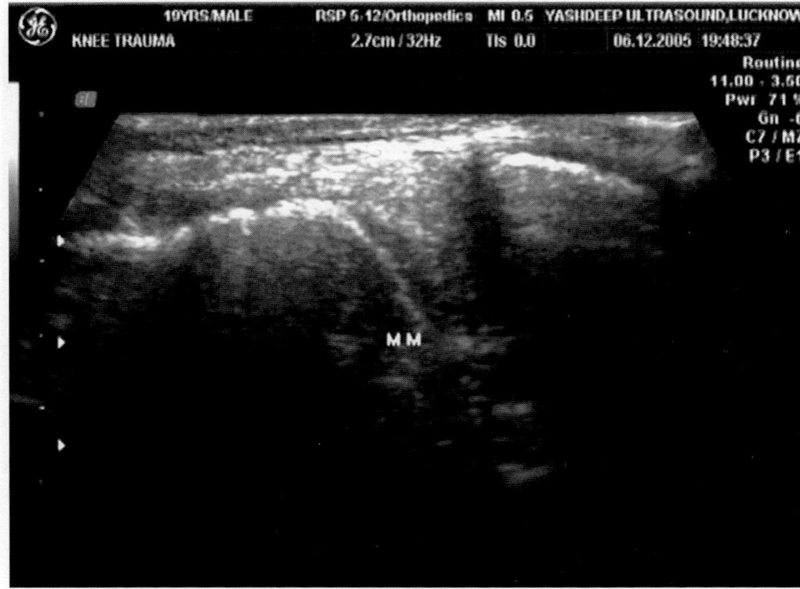

FIGURE 13.549

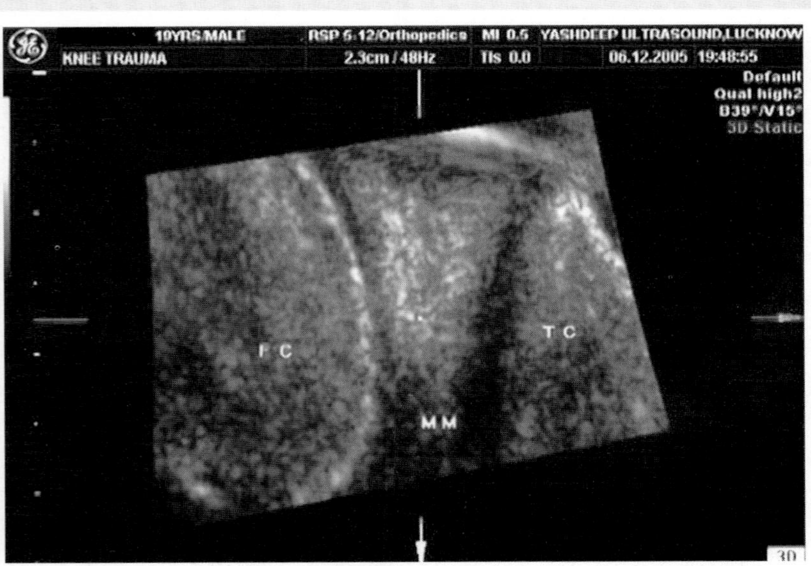

FIGURE 13.550

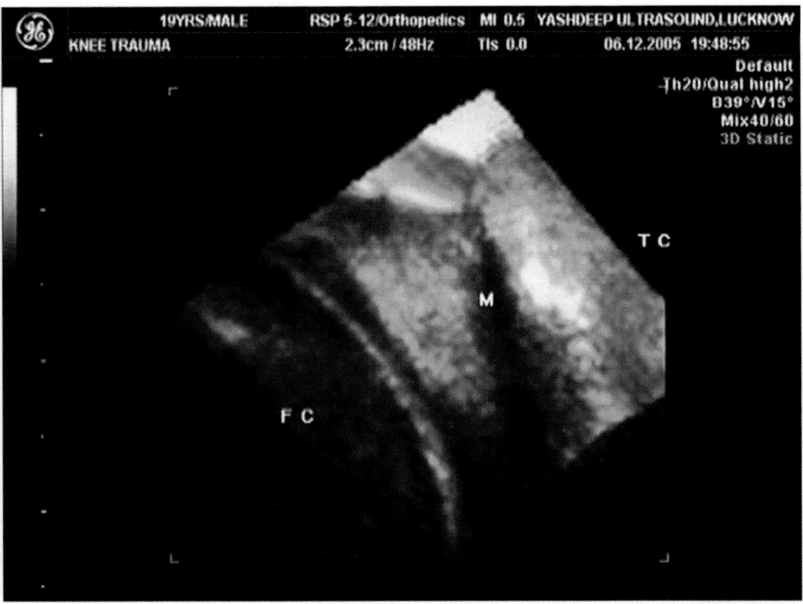

FIGURE 13.551

FIGURES 13.549 to 13.551: Medial meniscus. HRSG shows normal medial meniscus. It is seen echogenic triangular shaped body. It is seen situated between the layers of articular hyaline cartilage.

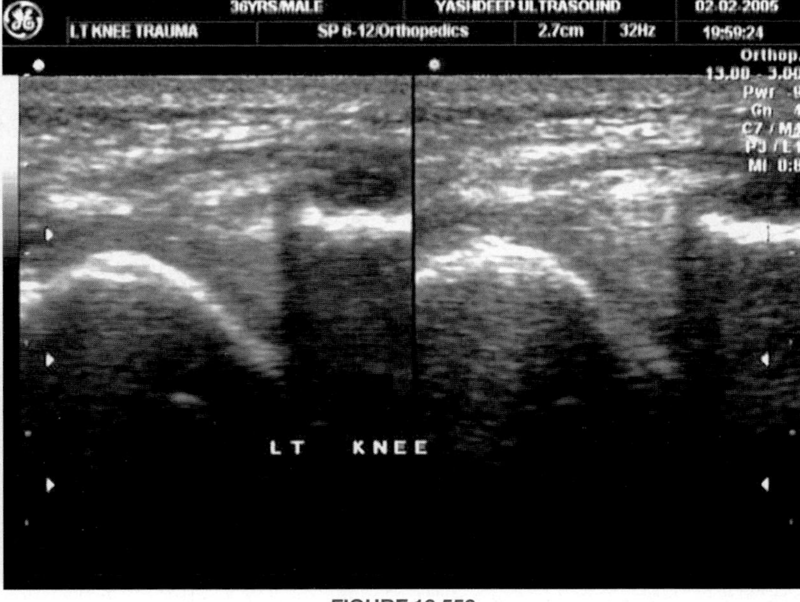

FIGURE 13.552

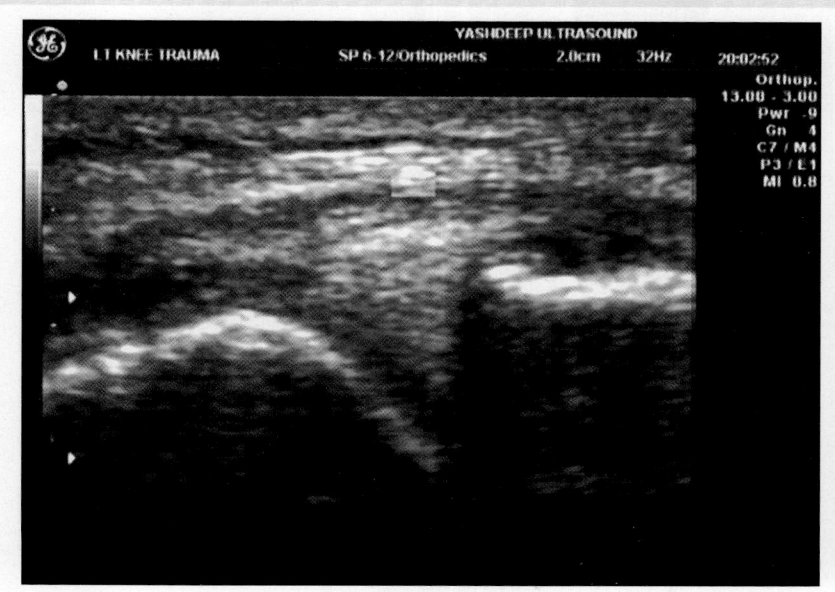

FIGURE 13.553

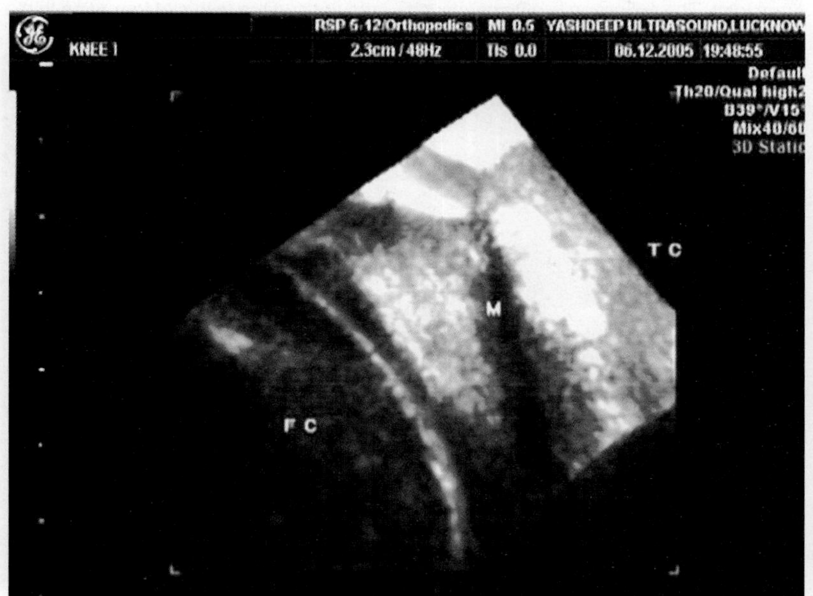

FIGURE 13.554

FIGURES 13.552 to 13.554: Normal lateral meniscus. Normal lateral meniscus is seen in coronal and longitudinal axis. It is seen as a rectangular body in the longitudinal plane and as a triangular shaped body in coronal plane.

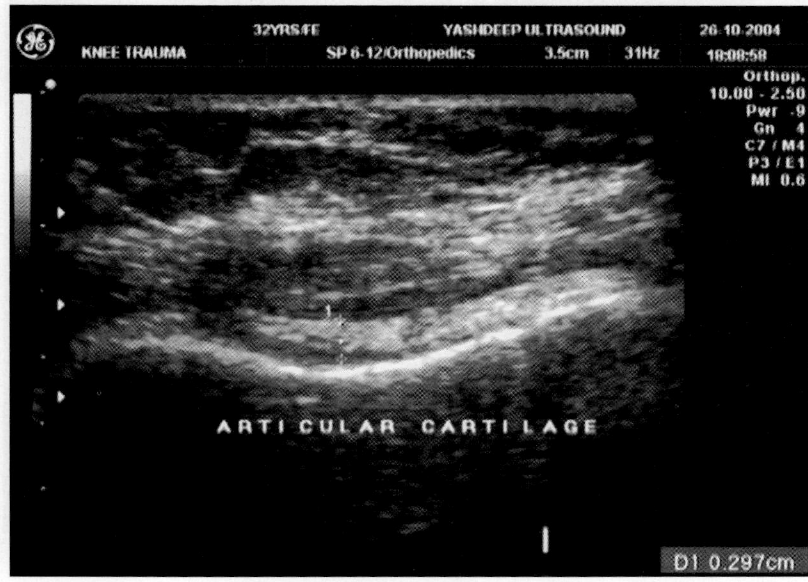

FIGURE 13.555

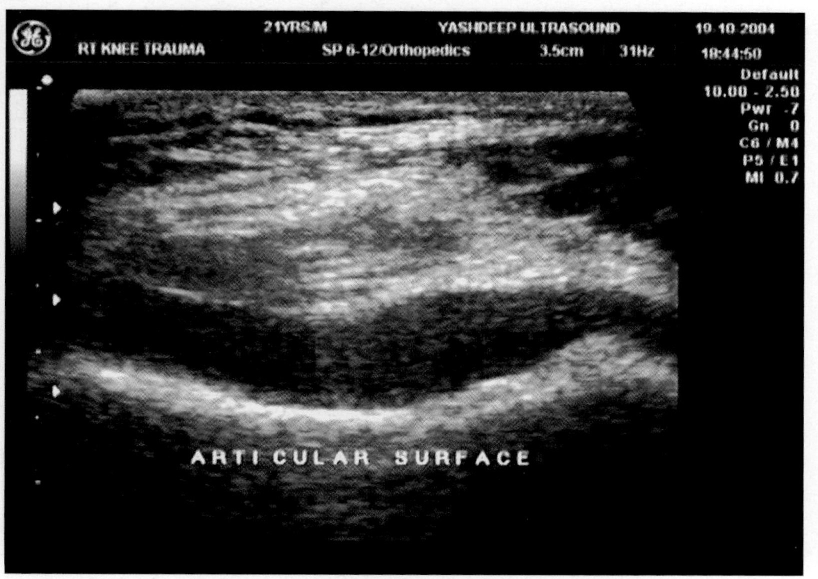

FIGURE 13.556

FIGURES 13.555 and 13.556: Normal articular surface. HRSG shows normal articular surface of condyle. The surface is smooth and hyaline cartilage is seen as a hypoechoic linear shadow.

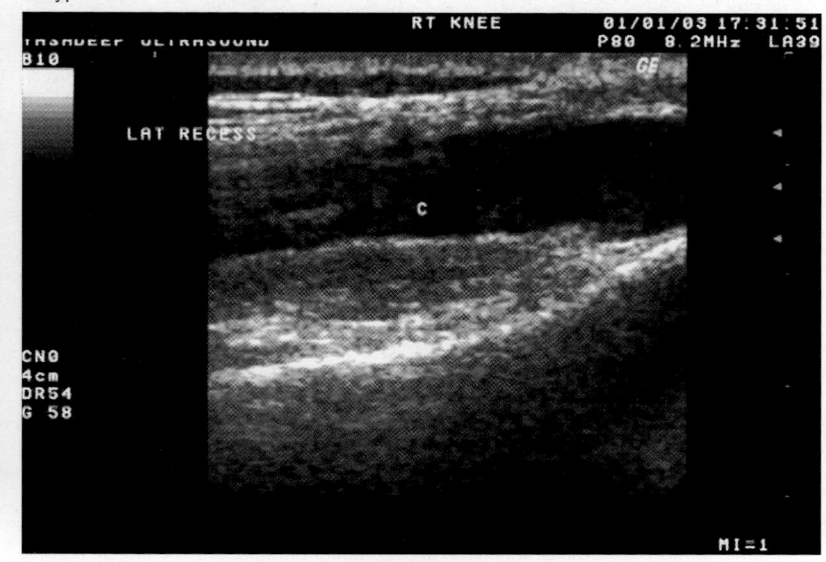

FIGURE 13.557

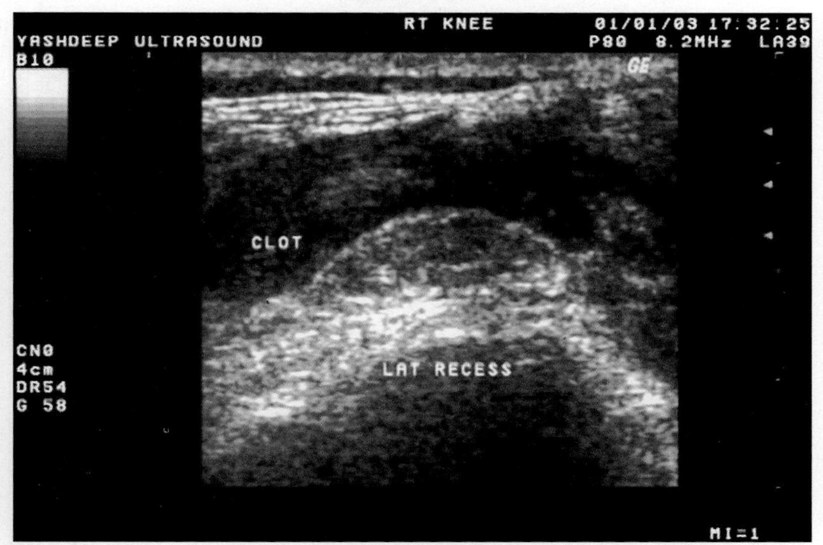

FIGURE 13.558

FIGURES 13.557 and 13.558: Acute hemarthrosis. HRSG shows dense echogenic collection in the intracapsular part of the joint and in the lateral recess suggestive of blood collection in a case of knee trauma suggestive of hemarthrosis. Echogenic clot is seen floating in the recess.

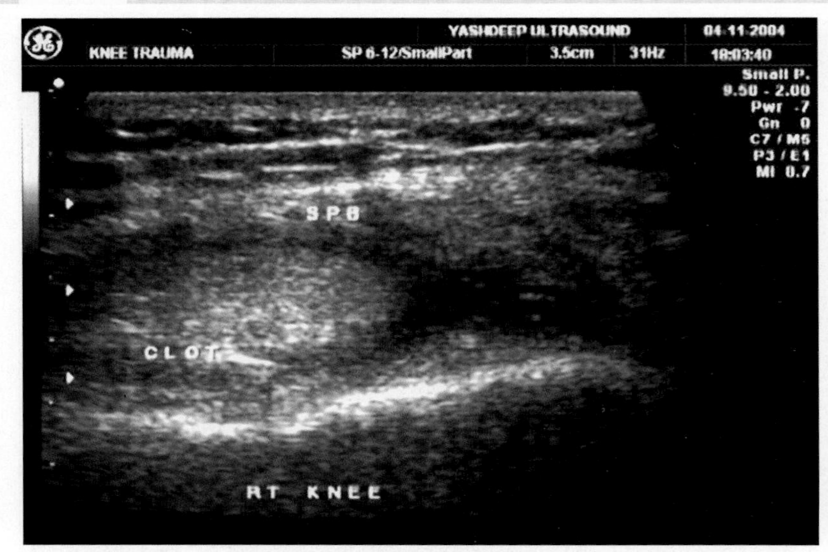

FIGURE 13.559

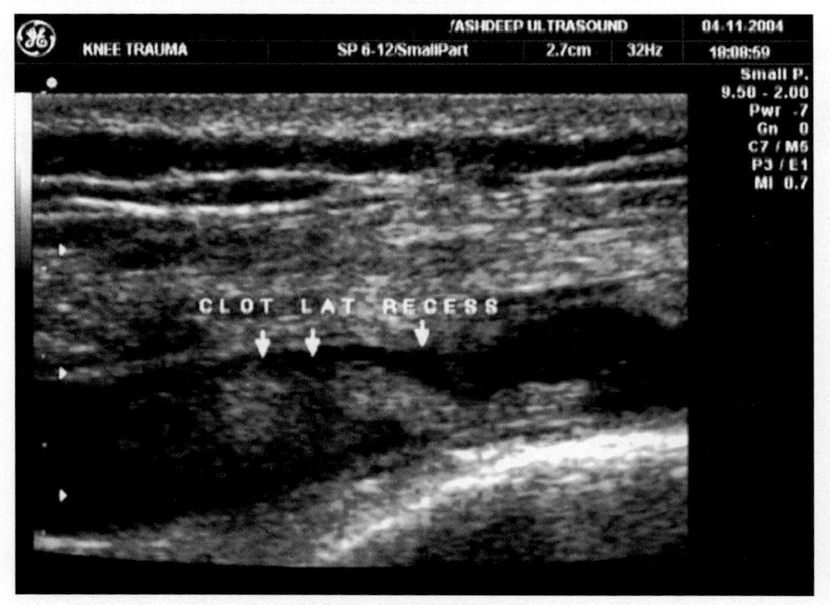

FIGURE 13.560

FIGURES 13.559 and 13.560: Hemorrhagic collection in the knee. HRSG shows echogenic hemorrhagic collection in the suprapatellar bursa and in the lateral recess. Collection is also seen in the lateral recess. Organization of the blood is seen forming the clot.

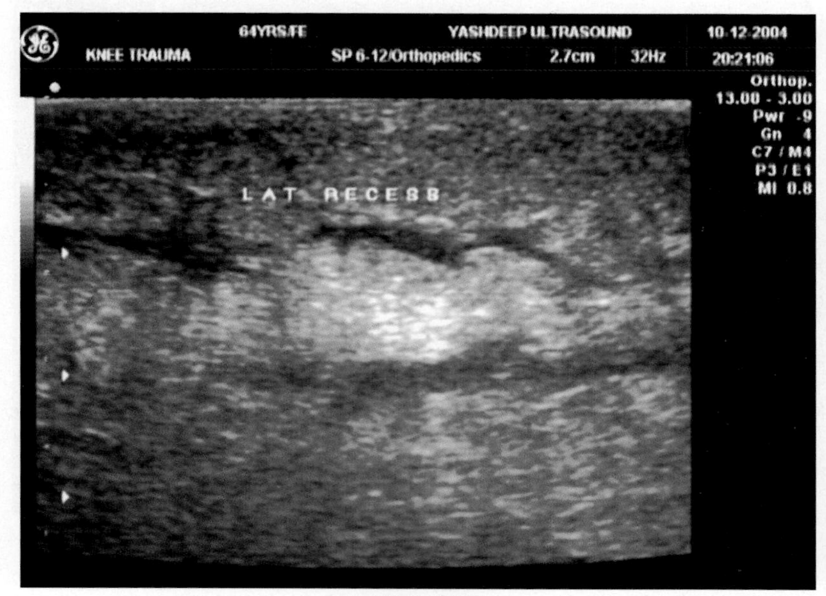

FIGURE 13.561

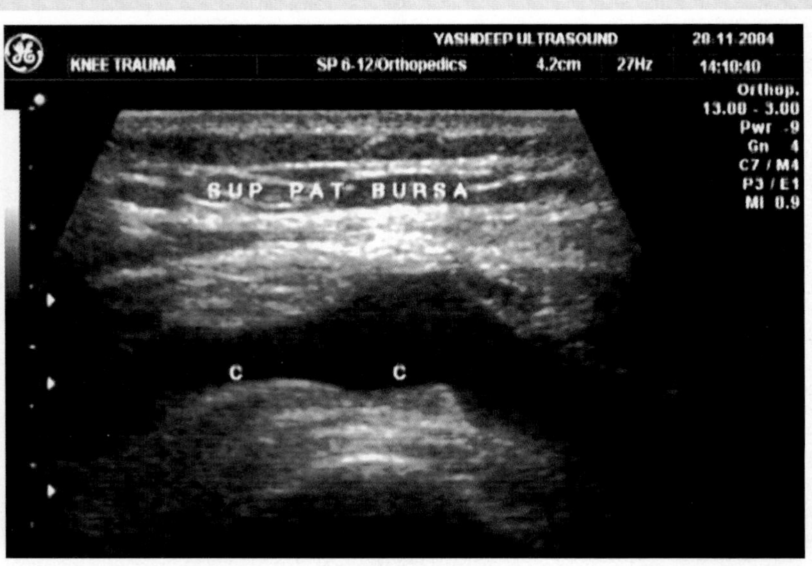

FIGURE 13.562

FIGURES 13.561 and 13.562: A big echogenic clot is seen in the lateral recess of the knee joint. The collection is also seen in the articular surface with clot formation in the suprapatellar bursa.

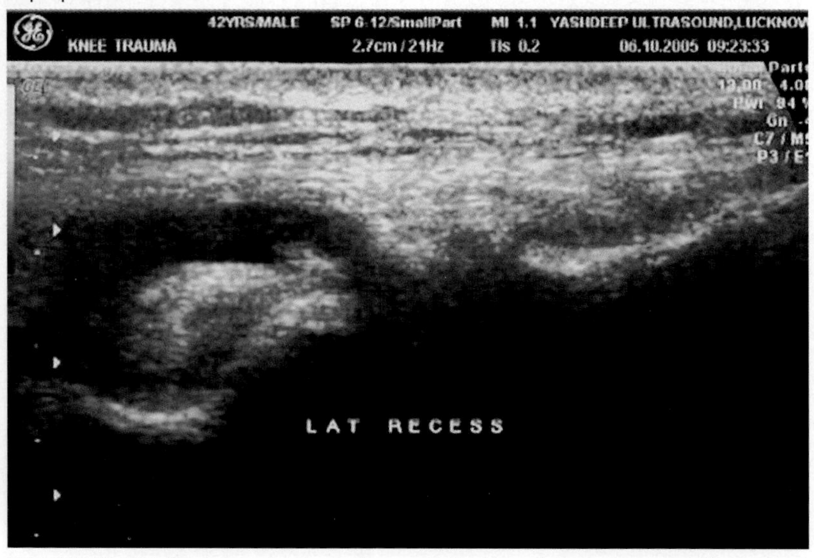

FIGURE 13.563: Big hemorrhagic clot is seen in the lateral recess in a patient sustained trauma to the knee joint. The lateral meniscus is also seen torn.

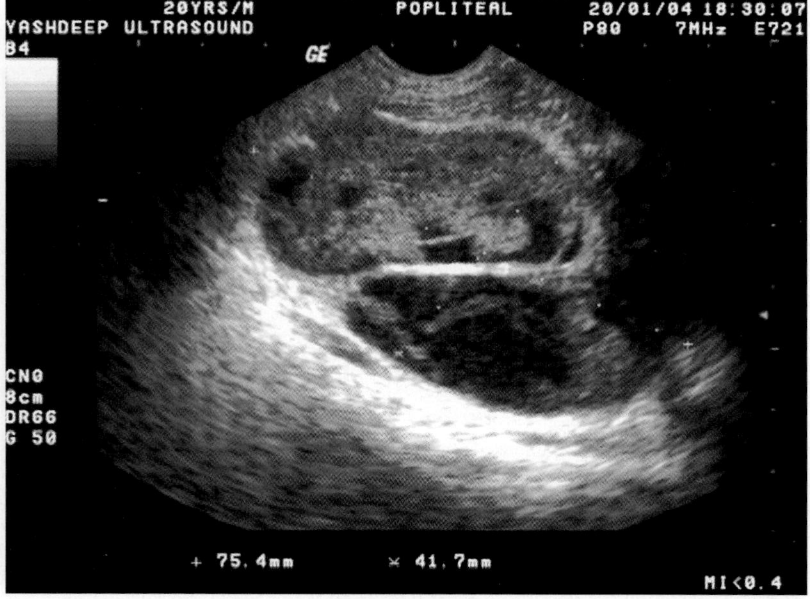

FIGURE 13.564

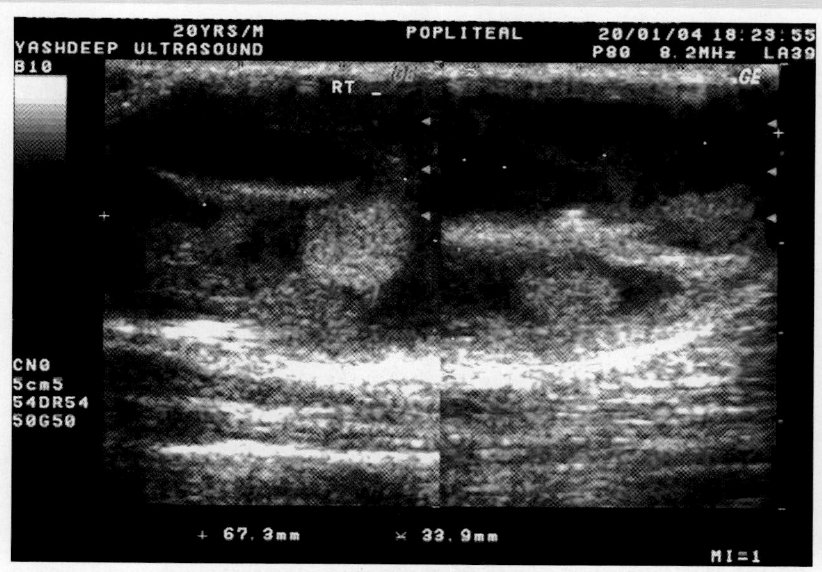

FIGURE 13.565

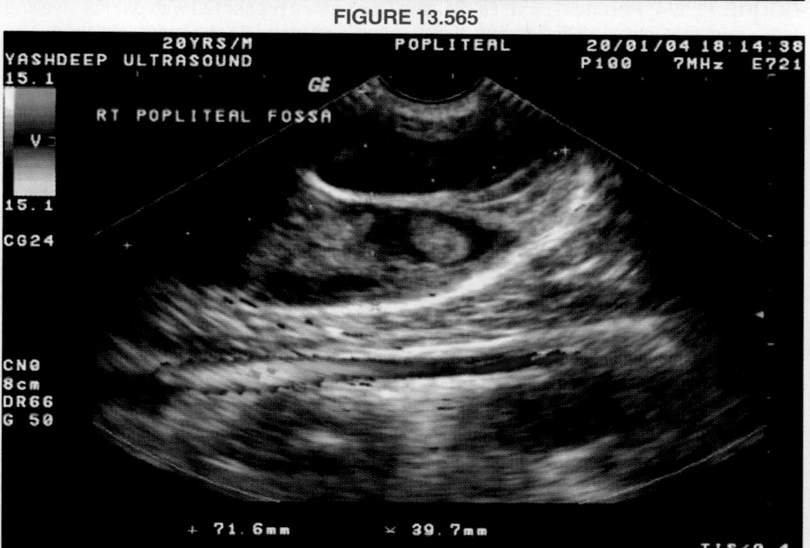

FIGURE 13.566

FIGURES 13.564 to 13.566: Big hematoma in the popliteal fossa. A patient presented with big mass in the popliteal fossa with evidence of arterial sufficiency. HRSG shows a big hematoma formation in the popliteal fossa. The hematoma is heterogeneous in texture. It is seen pressing over the popliteal artery resulting into flow obstruction. The popliteal vein is seen obliterated by the hematoma.

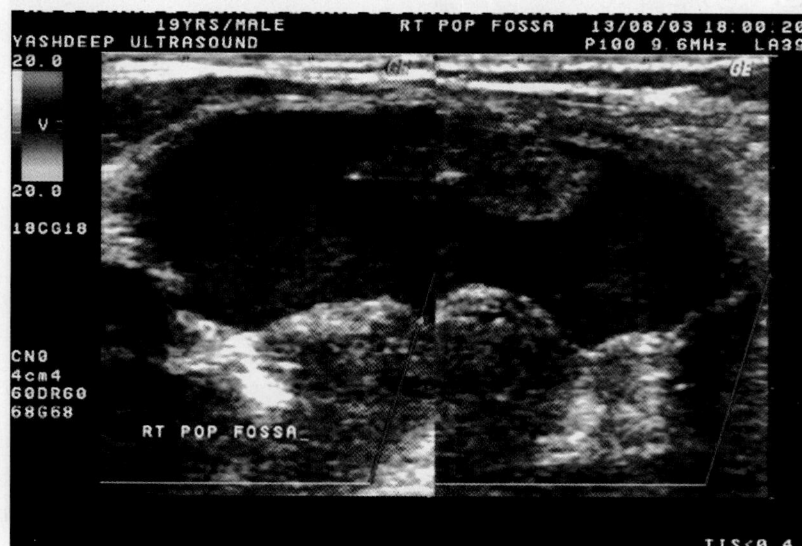

FIGURE 13.567: Big hematoma in the popliteal fossa with rupture of the muscle. HRSG shows a big hematoma in the young boy with evidence of rupture of the muscle. The hematoma is seen in the gastrocnemius muscle. No evidence of any blood flow is seen in the hematoma.

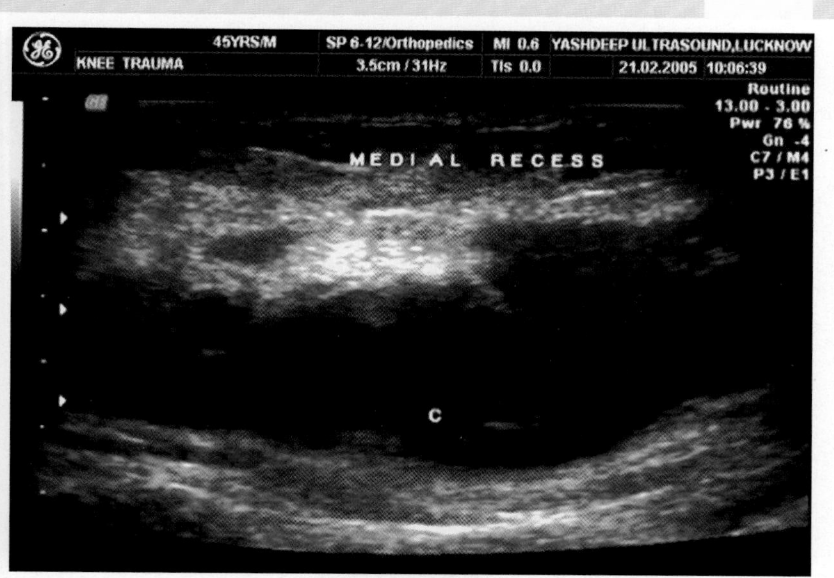

FIGURE 13.568

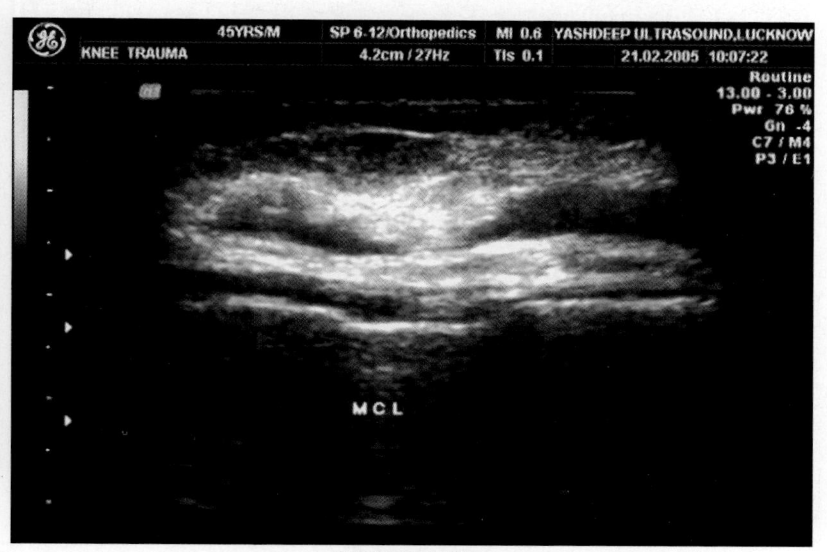

FIGURE 13.569

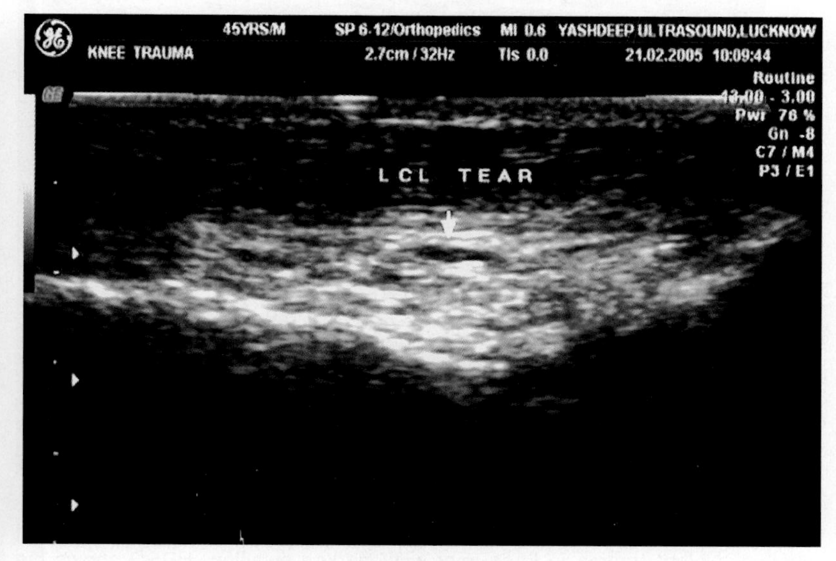

FIGURE 13.570

FIGURES 13.568 to 13.570: Big hemorrhagic collection is seen in the lateral recess of the knee joint. Hemorrhagic clot is also seen in the medial recess. However, the medial collateral ligament is intact. But lateral collateral ligament shows a hypoechoic cleft running through the substance of ligament suggestive of tear.

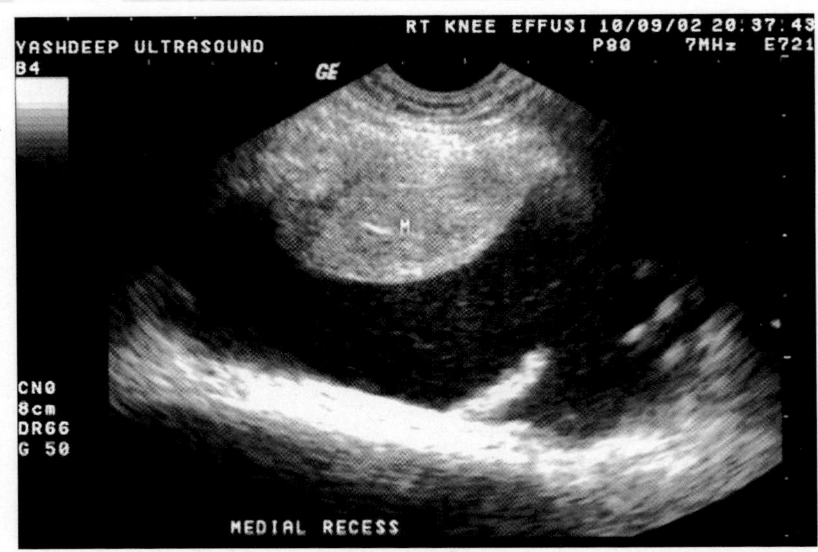

FIGURE 13.571

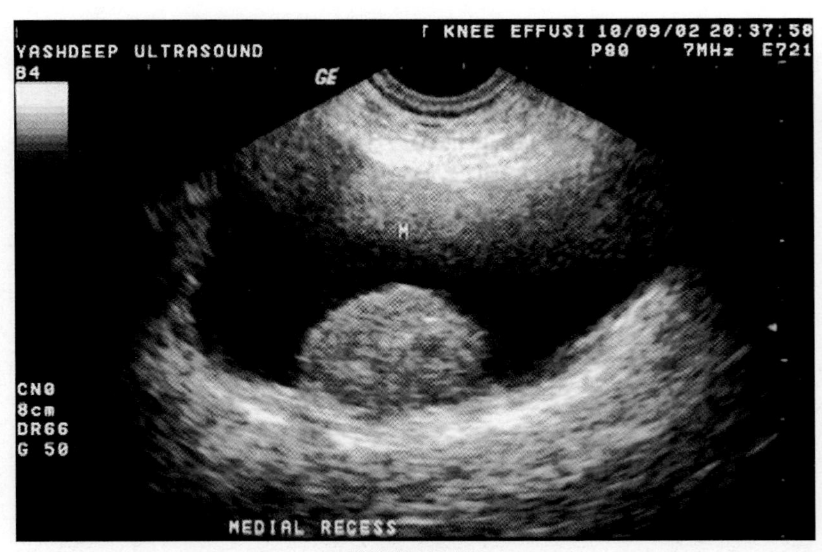

FIGURE 13.572

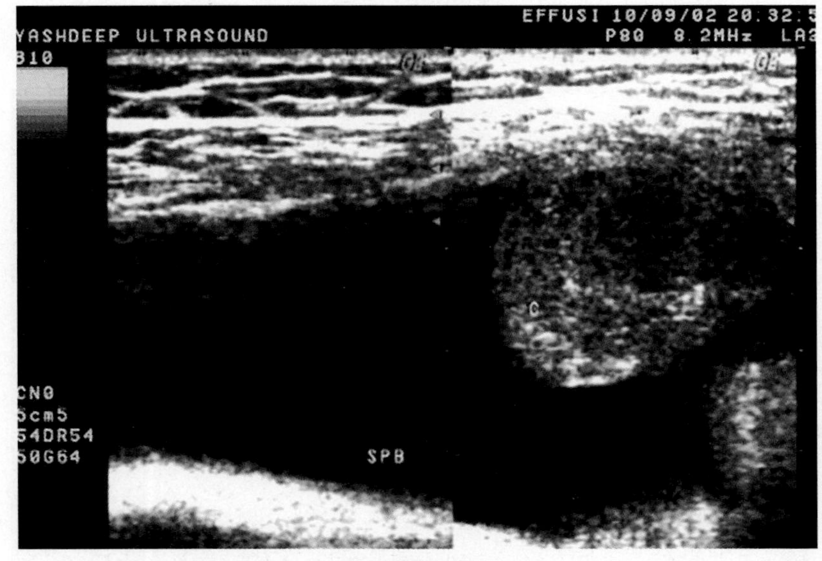

FIGURE 13.573

FIGURES 13.571 to 13.573: Hematoma with synovioma formation. A patient sustained blunt trauma in knee joint with marked swelling. HRSG shows big hemorrhagic collection in the joint capsule. There is evidence of well-defined homogenous rounded masses seen hanging in the medial recess and suprapatellar bursa. They are homogeneous in texture with sharp margins. Biopsy of the masses turns out to be benign synovial tumor suggestive of synovioma.

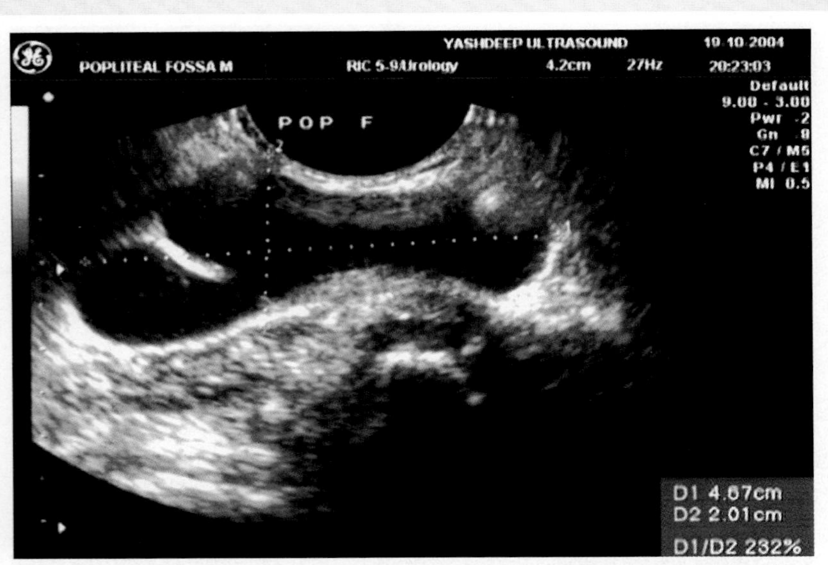

FIGURE 13.574: Popliteal cyst. HRSG shows a patient presented with painful swelling in the popliteal fossa. HRSG shows thick walled cyst in the gastrocnemius bursa. The cyst is seen communicating with the medial recess. Echogenic septa are seen in the cyst. Synovial membrane proliferation is also seen in the cyst suggestive of Bakers cyst.

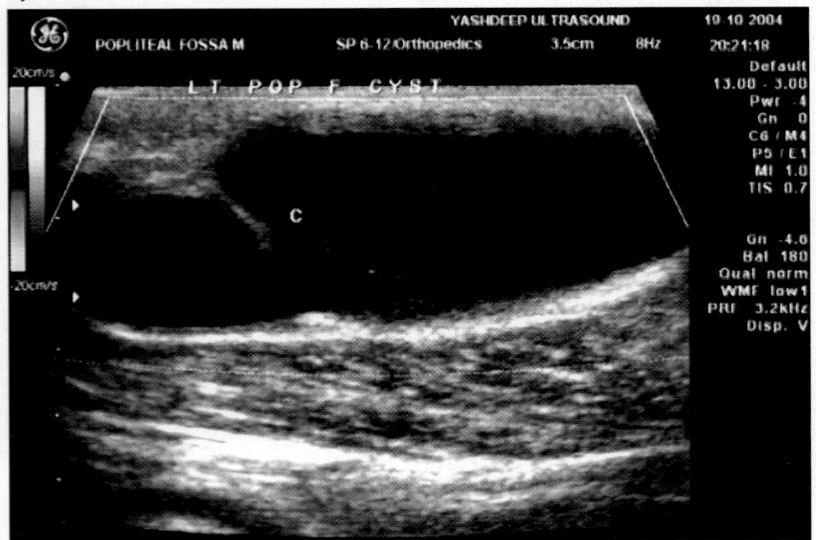

FIGURE 13.575

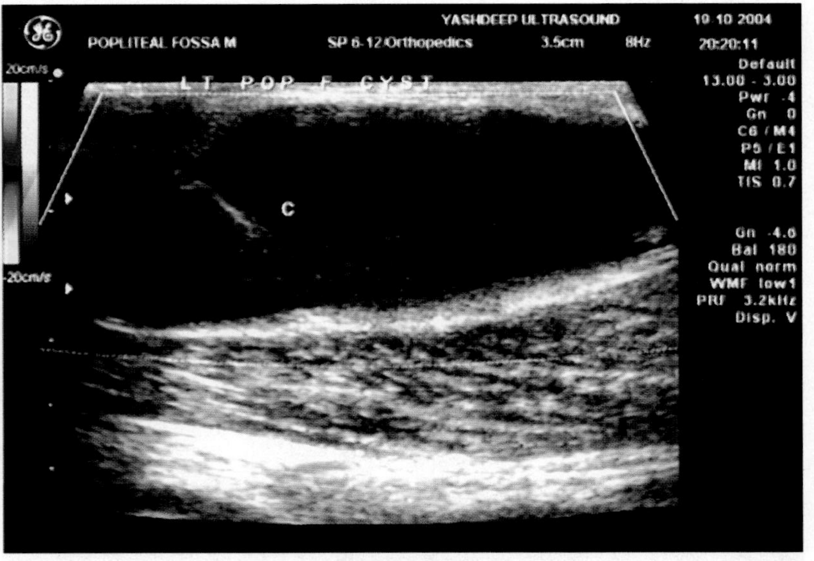

FIGURE 13.576

FIGURES 13.575 and 13.576: Muscles cyst. A patient presented with cystic swelling in the gastrocnemius muscle. HRSG shows a well-defined cyst in the belly of the gastrocnemius muscle in the popliteal fossa. No internal echoes are seen in the cyst. **On color flow imaging** no flow is seen in the cyst.

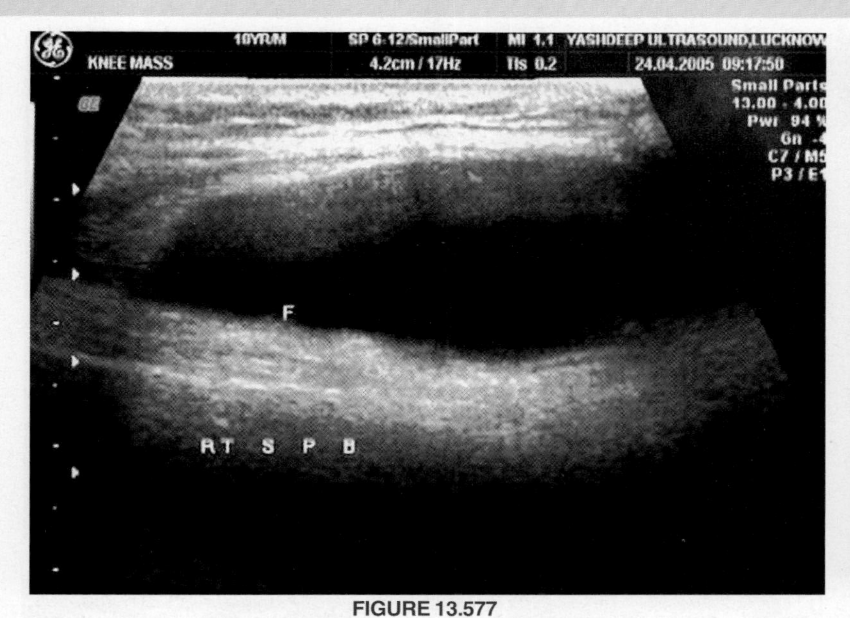

FIGURE 13.577

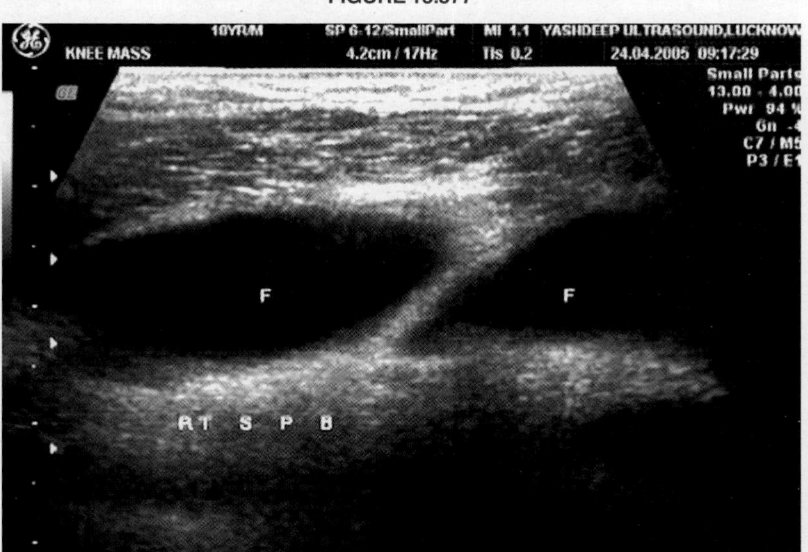

FIGURE 13.578

FIGURES 13.577 and 13.578: Suprapatellar bursal collection. HRSG shows thick low level collection in the suprapatellar bursa. Well-defined echogenic septum is seen running through the bursa.

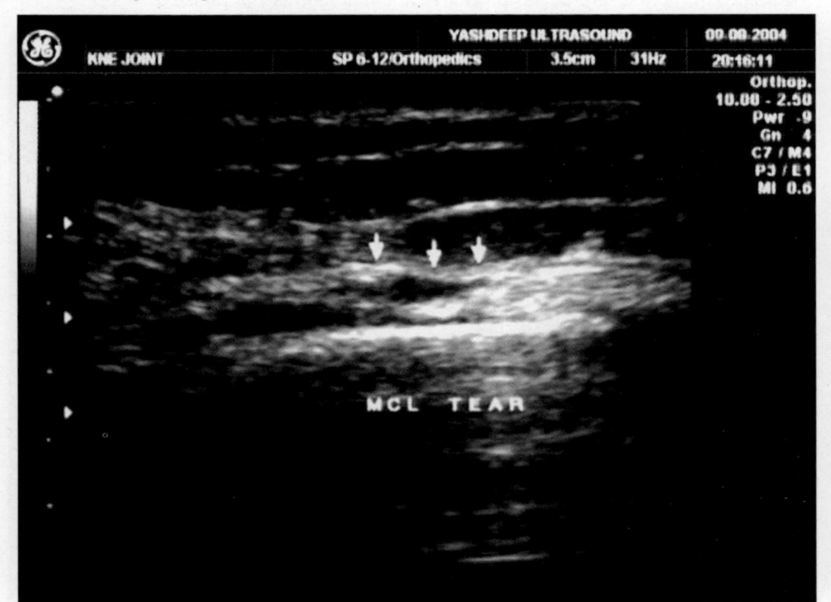

FIGURE 13.579: MCL tear. A hypoechoic cleft seen in the medial collateral ligament. The cleft is seen running through the substance of ligament.

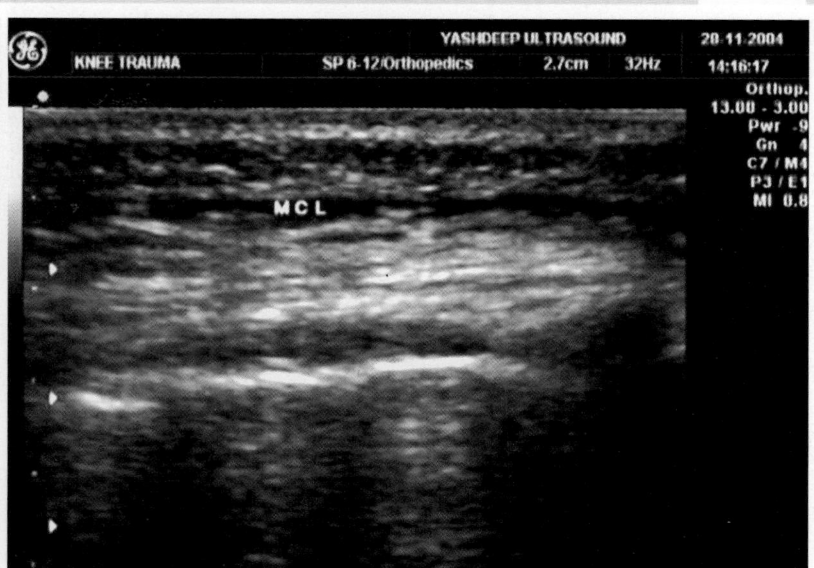

FIGURE 13.580: Partial MCL tear. HRSG shows a thin hypoechoic cleft on the superior border of MCL ligament. Lifting of MCL fibers are seen with small amount of fluid collection suggestive of tear.

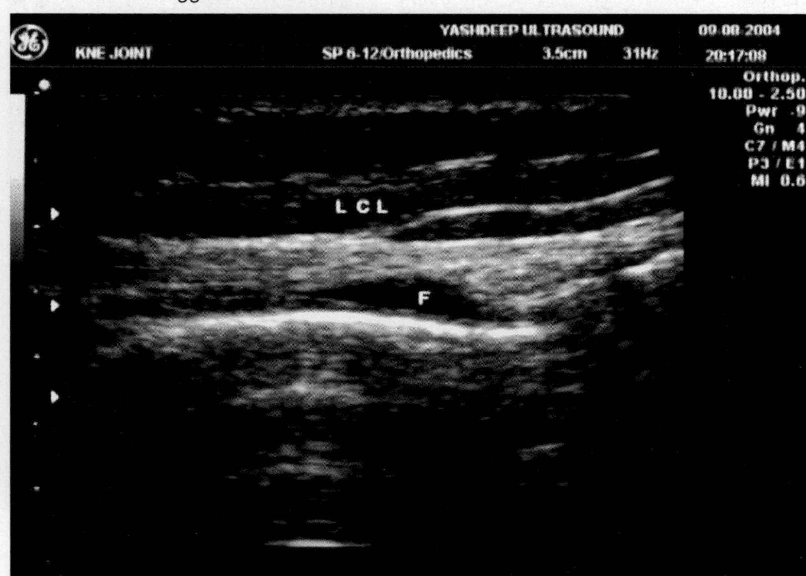

FIGURE 13.581: Lateral collateral ligament tear with thinning. HRSG shows partial tear of the lateral collateral ligament in a sports man sustained injury to the knee joint. Thinning of the lateral collateral ligament is also seen at the site of tear.

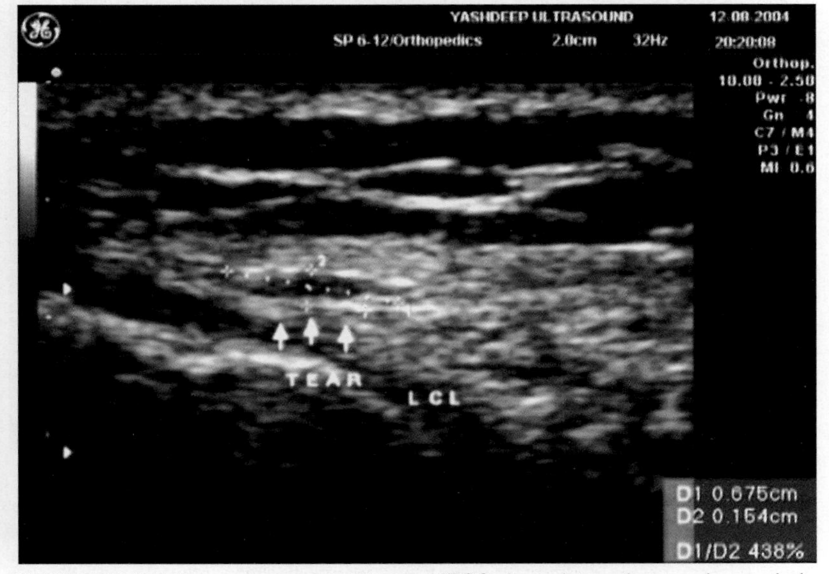

FIGURE 13.582: Intrasubstance LCL tear. HRSG shows intrasubstance hypoechoic cleft in the LCL ligament suggestive of tear. The ligament is also seen thickened.

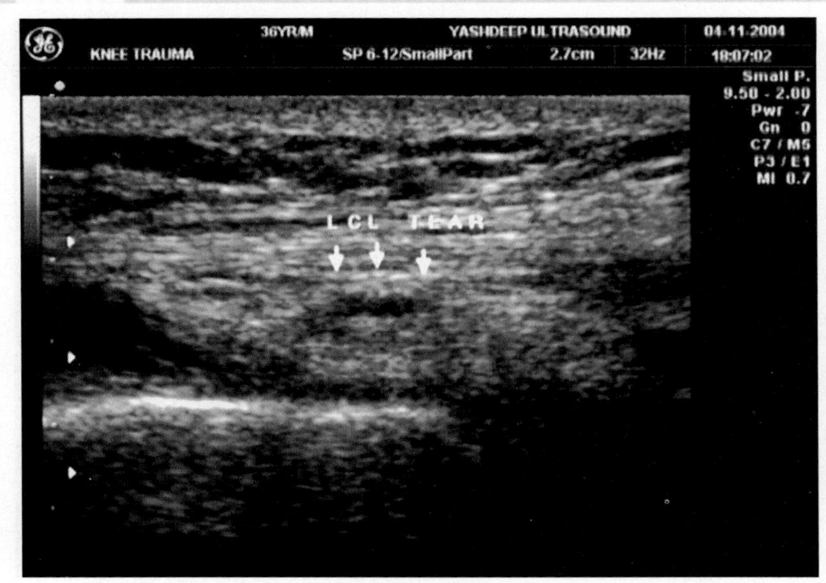

FIGURE 13.583

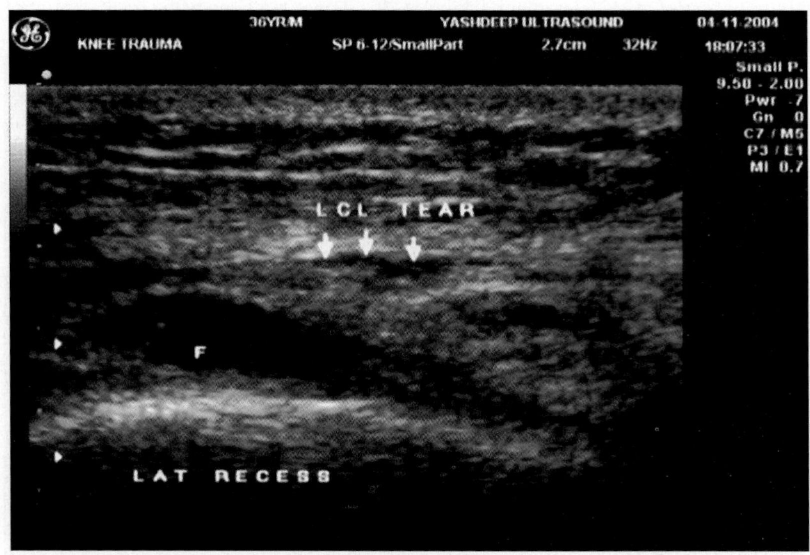

FIGURE 13.584

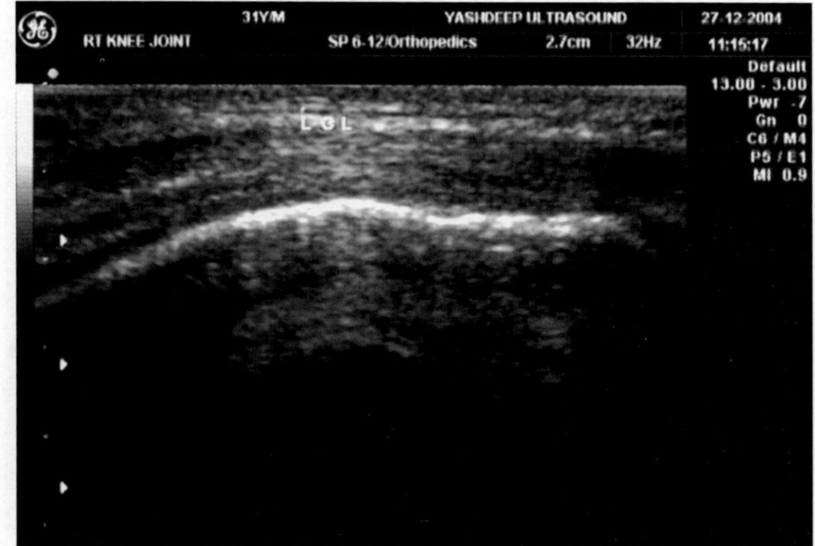

FIGURE 13.585

FIGURES 13.583 to 13.585: Intra substance LCL tear. HRSG shows thickening of the LCL ligament in a patient sustained knee trauma. Hypoechoic cleft is seen in the LCL suggestive of tear. Small amount of fluid collection is also seen in the lateral recess with synovial membrane proliferation. The opposite knee shows the normal LCL.

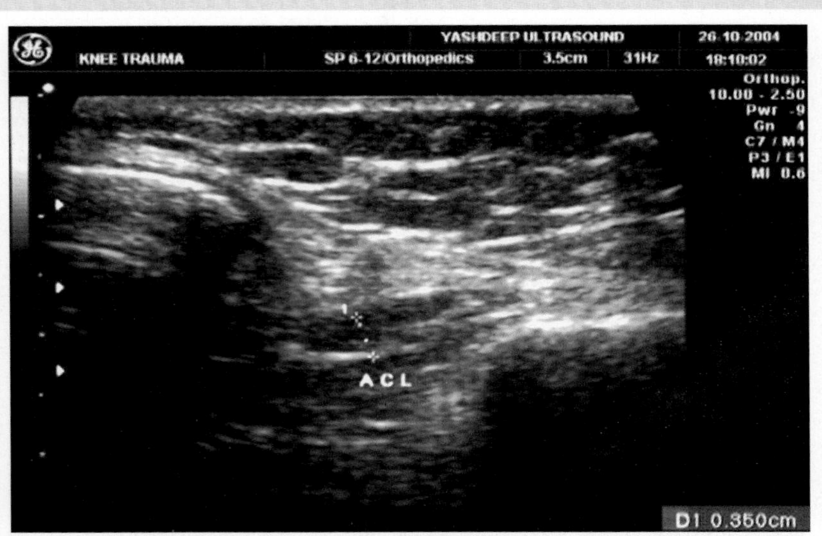

FIGURE 13.586

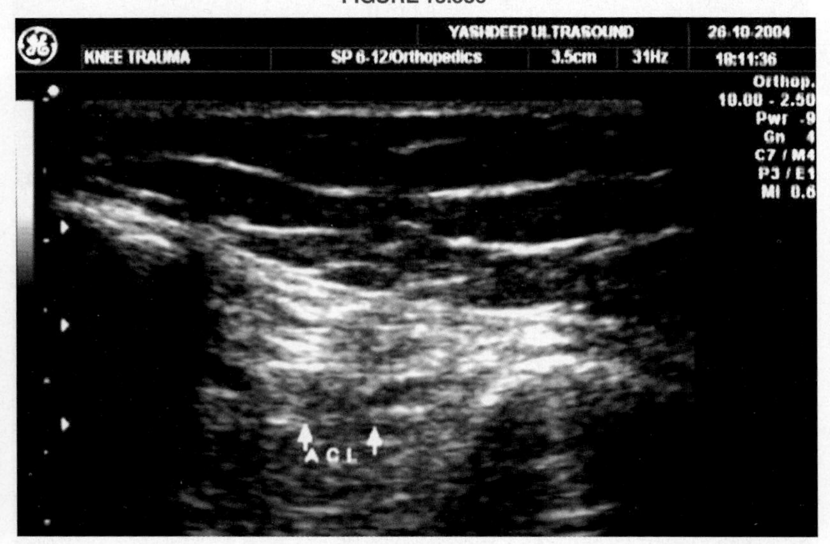

FIGURE 13.587

FIGURES 13.586 and 13.587: Anterior cruciate ligament. A patient presented acutely painful knee joint. HRSG shows thickened hypoechoic anterior cruciate ligament. There is evidence of hypoechoic cleft seen in the ACL. ACL tear is associated with marked pain and patient does not allow proper examination of the knee due to pain. For proper ultrasound evaluation of ACL knee should be flexed to 45° to get proper window. However, due to acute pain patient does not allow to flex the knee. Therefore in most of the cases proper ACL evaluation is not possible.

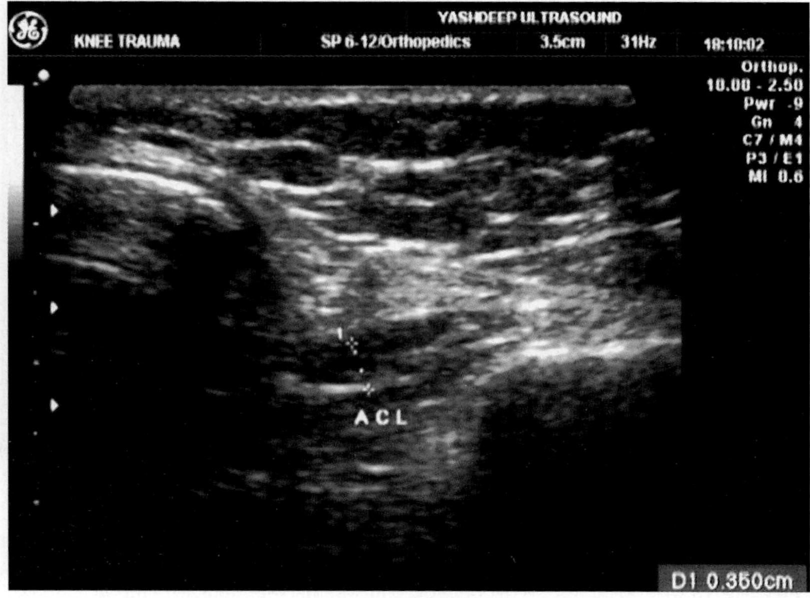

FIGURE 13.588

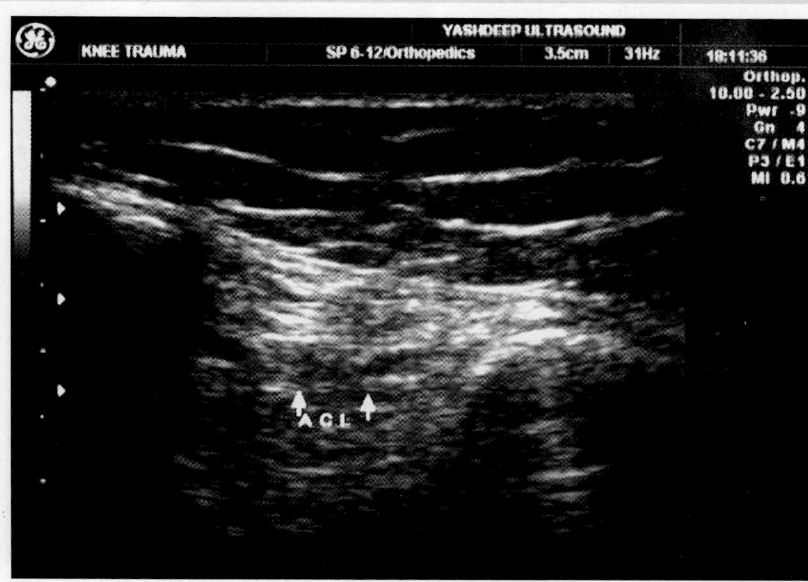

FIGURE 13.589

FIGURES 13.588: and 13.589: Irregular hypoechoic cleft in anterior cruciate ligament in a patient sustained knee trauma.

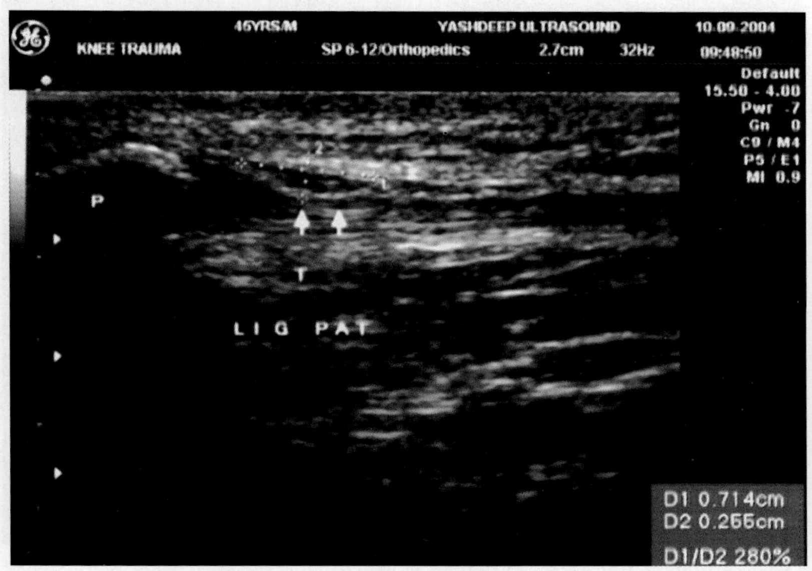

FIGURE 13.590: Pre patellar bursitis. HRSG shows small amount of fluid in pre patellar bursa. The patient presented with pain while climbing the stairs.

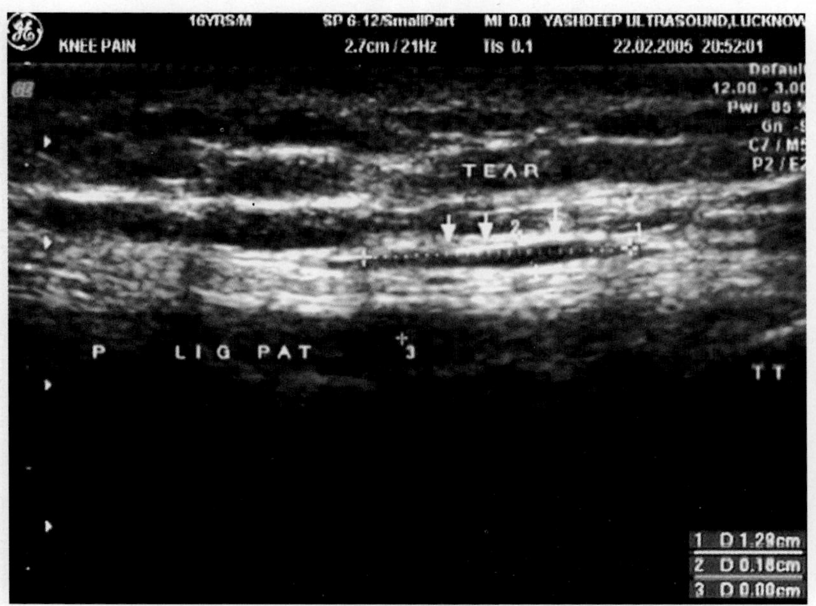

FIGURE 13.591

FIGURE 13.592

FIGURES 13.591 and 13.592: Partial tear of patellar tendon. A high jumper presented with painful movements of the knee while jumping. HRSG shows partial disruption of tear fibers in its mid part. Small amount of fluid is seen in the tear. The tendon is also thickened suggestive of patellar tendonitis with partial tear in the tendon.

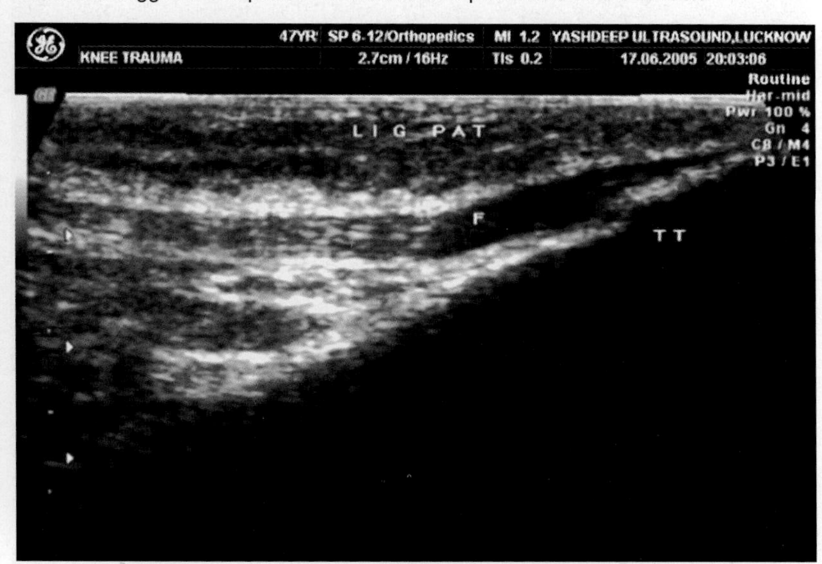

FIGURE 13.593: Ligament patellae tear. HRSG shows disruption of the ligament patellae fibers from its insertion side at tibial tuberosity. Loss of tendon fibers is seen. Fluid collection is seen in the tendon sheath.

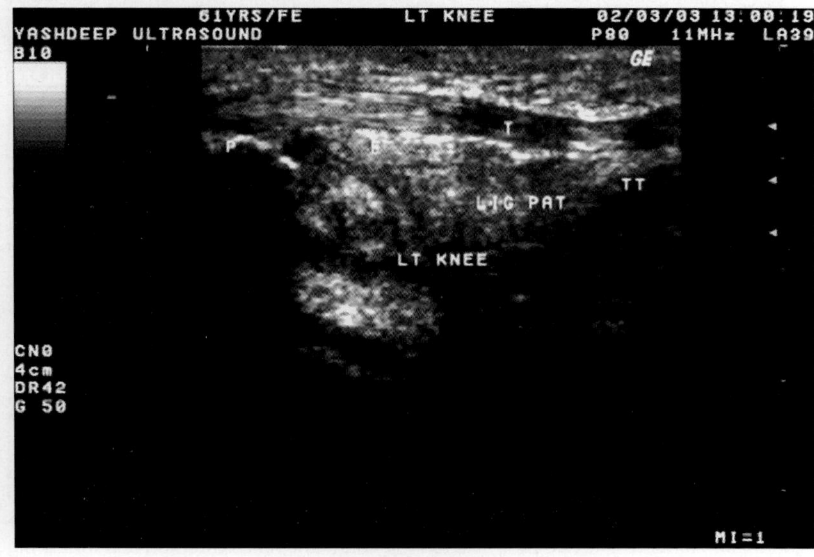

FIGURE 13.594

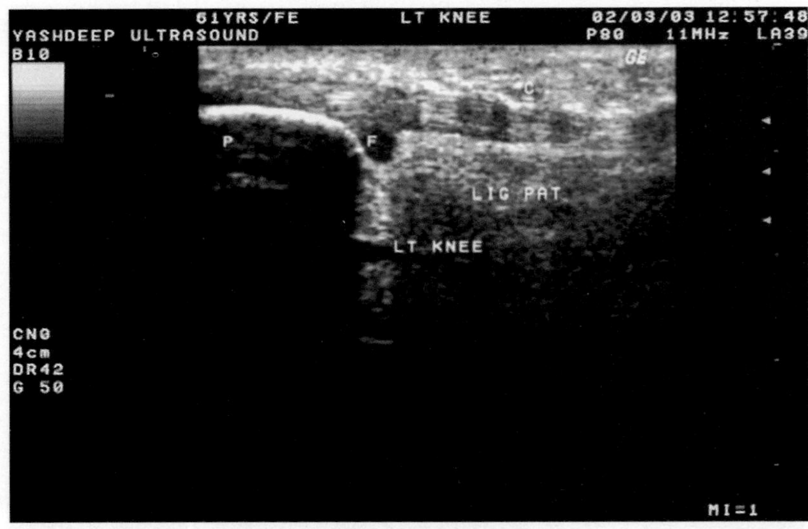

FIGURE 13.595

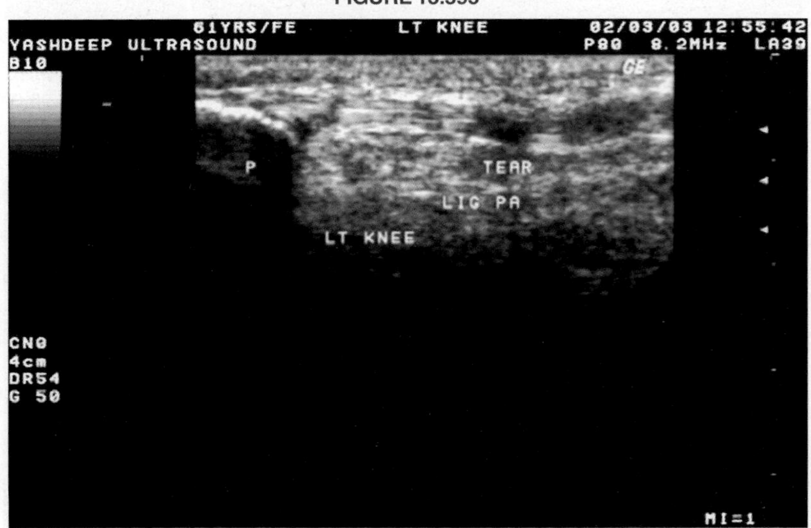

FIGURE 13.596

FIGURES 13.594 to 13.596: Rupture of ligament patellae. An old lady sustained direct trauma to the knee joint over the tibia and presented with marked pain in the knee. HRSG show multiple clefts in the patellar tendon in its mid part and also near the insertion side. Loss of tendon fibers is seen. Fluid collection is seen in the tendon groove. The tendon was inhomogeneous suggestive of tear in ligament patellae.

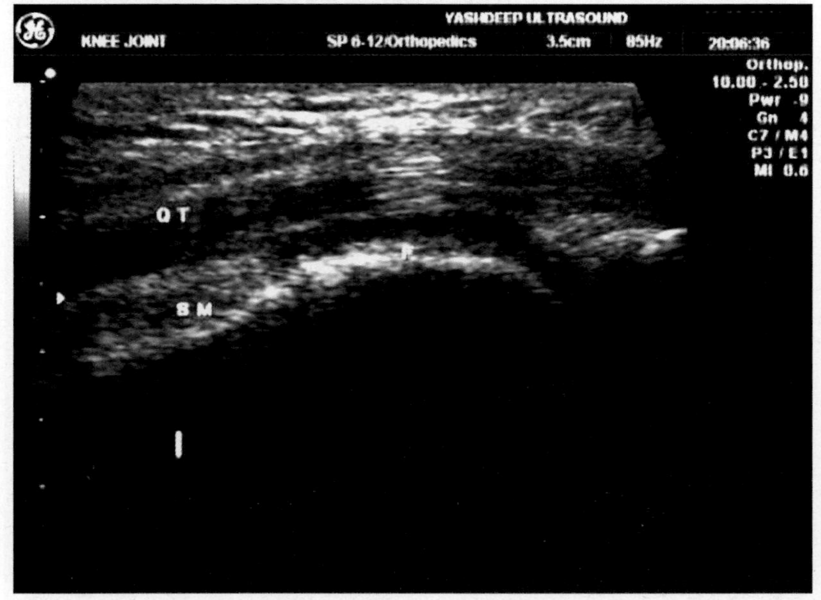

FIGURE 13.597: Quadriceps tendon rupture. HRSG shows rupture of the quadriceps tendon in its mid part. The tendon fibers are seen disrupted and torn. A big cleft is seen in the tendon suggestive of traumatic rupture of the tendon.

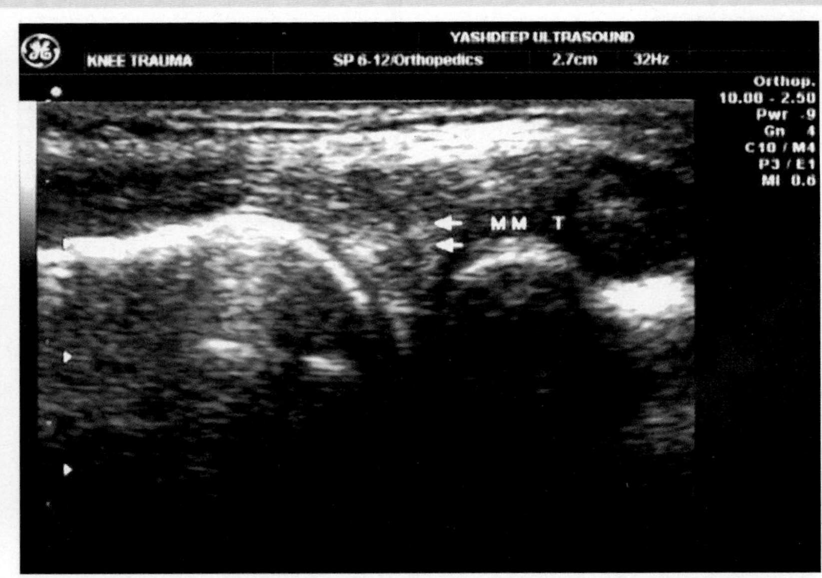

FIGURE 13.598: Meniscal tear. HRSG shows meniscal tear in the same patient in the medial meniscus. A vertical cleft seen running from the apex to the base.

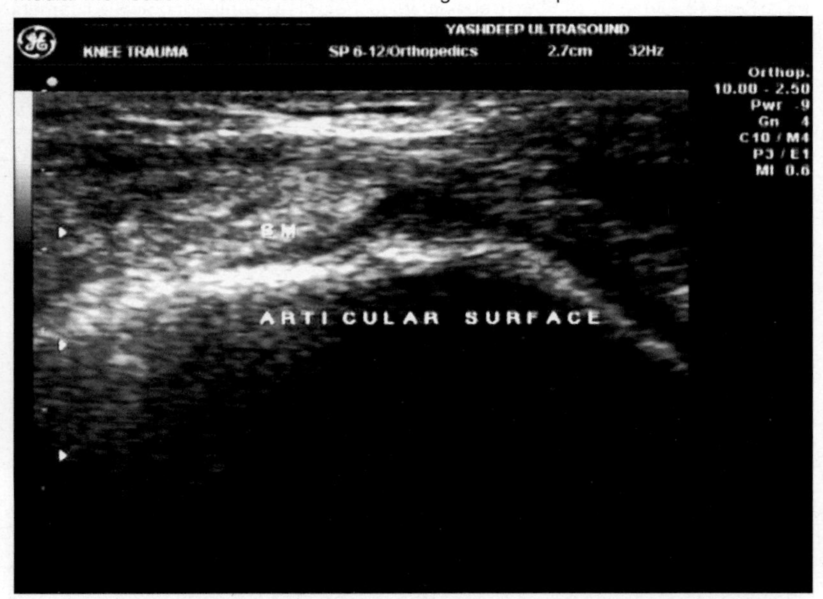

FIGURE 13.599: There is also evidence of hemorrhagic clot seen in the suprapatellar bursa and also in the articular surface of the knee. Echogenic collection is seen into it suggestive of hemarthrosis.

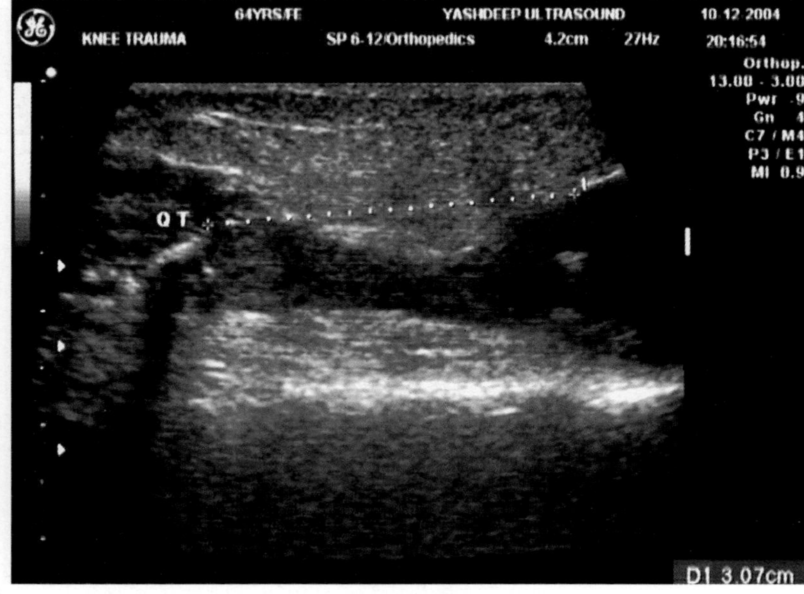

FIGURE 13.600

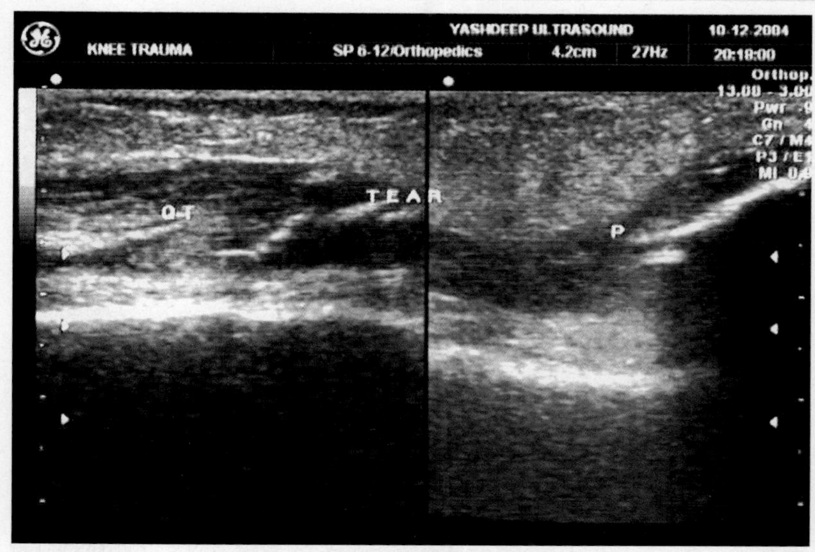

FIGURE 13.601

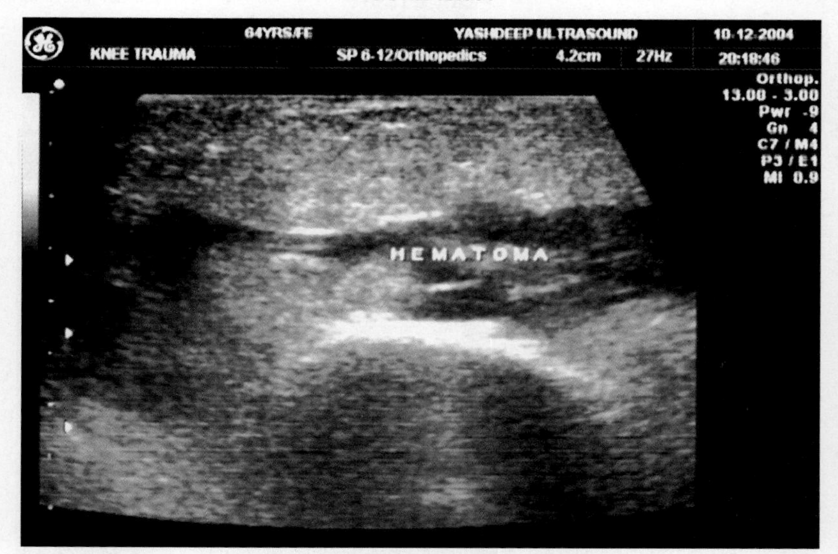

FIGURE 13.602

FIGURES 13.600 to 13.602: Traumatic rupture of quadriceps tendon in the rRt knee. There is big cleft seen in the quadriceps tendon. The tendons fibers are seen torn and there is big defect is seen in the tendon. Echogenic clot is also seen. Part of the tendon is seen attached of the patellar side.

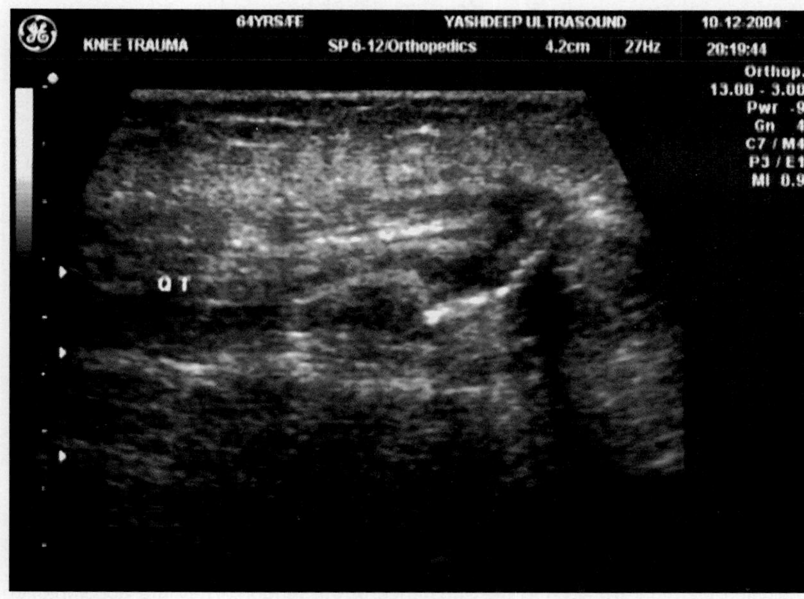

FIGURE 13.603

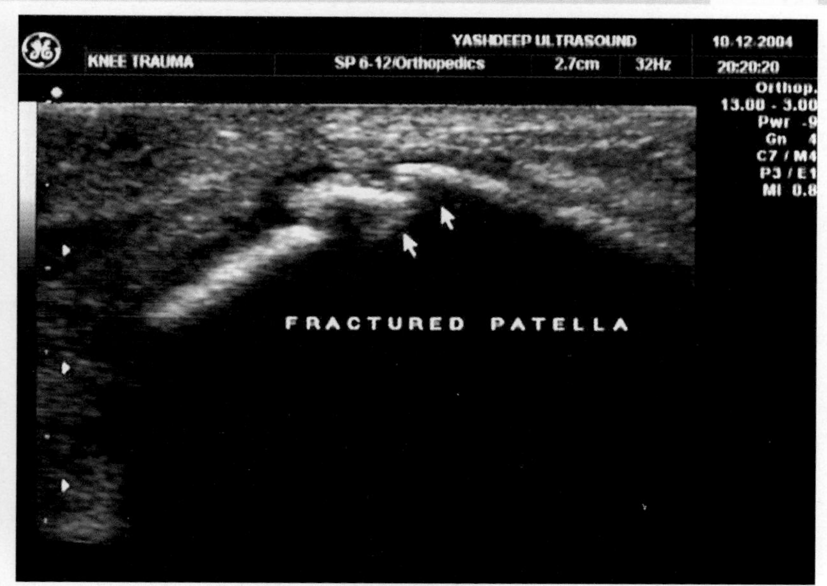

FIGURE 13.604

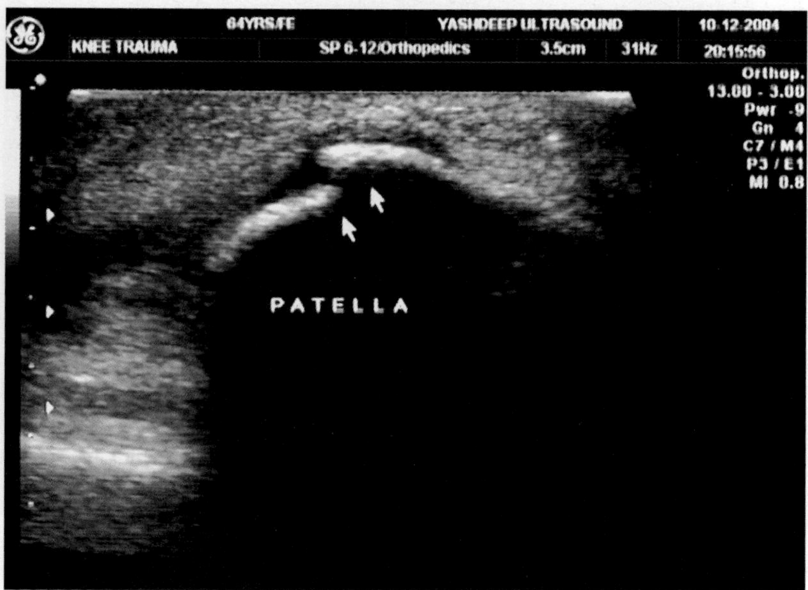

FIGURE 13.605

FIGURES 13.603 to 13.605: There is also evidence of fracture of the patella seen on HRSG. The fracture patellar segments are seen overriding each other.

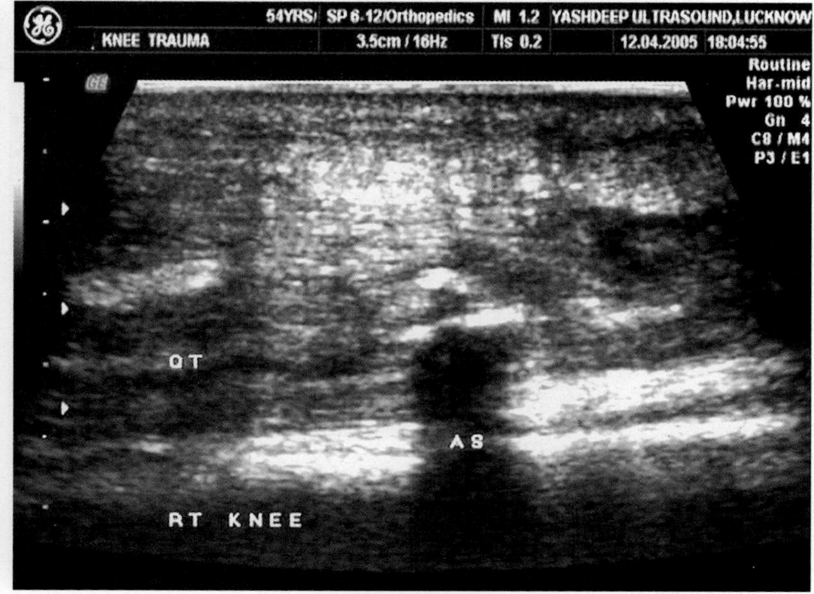

FIGURE 13.606

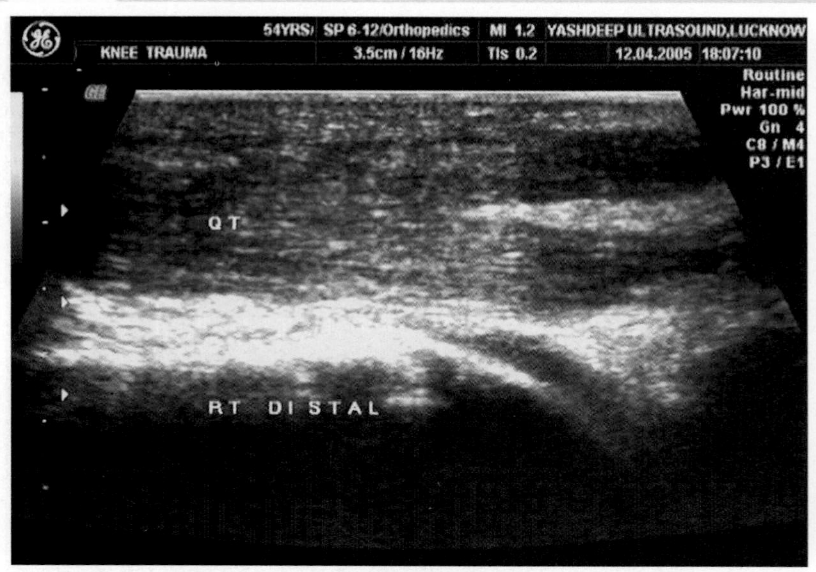

FIGURE 13.607

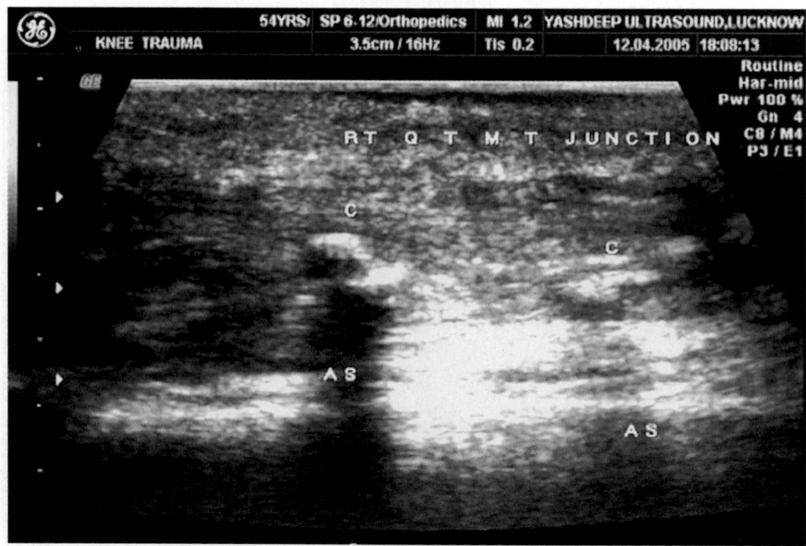

FIGURE 13.608

FIGURES 13.606 to 13.608: Bilateral quadriceps tendon rupture. A patient presented with evidence of bilateral quadriceps tendon rupture and went for surgery for the same. HRSG shows grossly inhomogeneous Rt quadriceps tendon. It is thickened with heterogeneous texture. Dense calcification is seen at musculotendonis junction. Echogenic calcified plaque is seen at the junction. Which is accompanied with acoustic shadowing. There is evidence of hypoechoic cleft is seen in the distal 1/3rd part of the tendon.

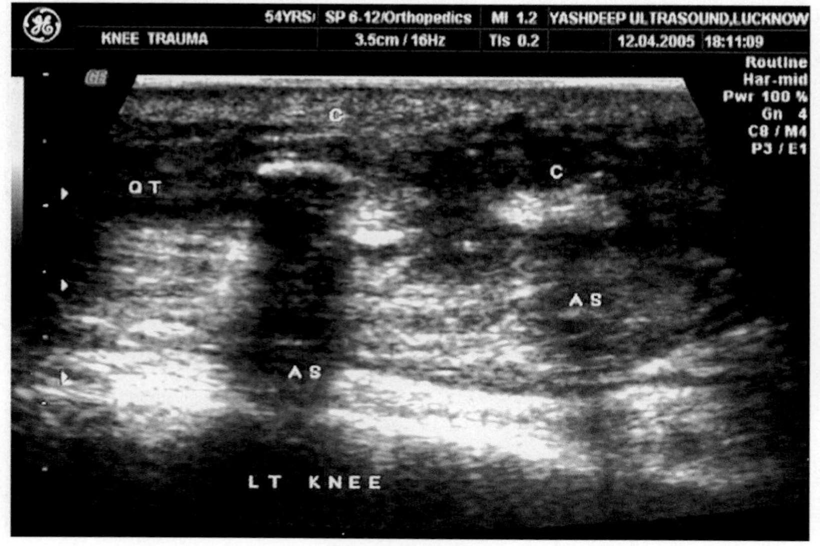

FIGURE 13.609

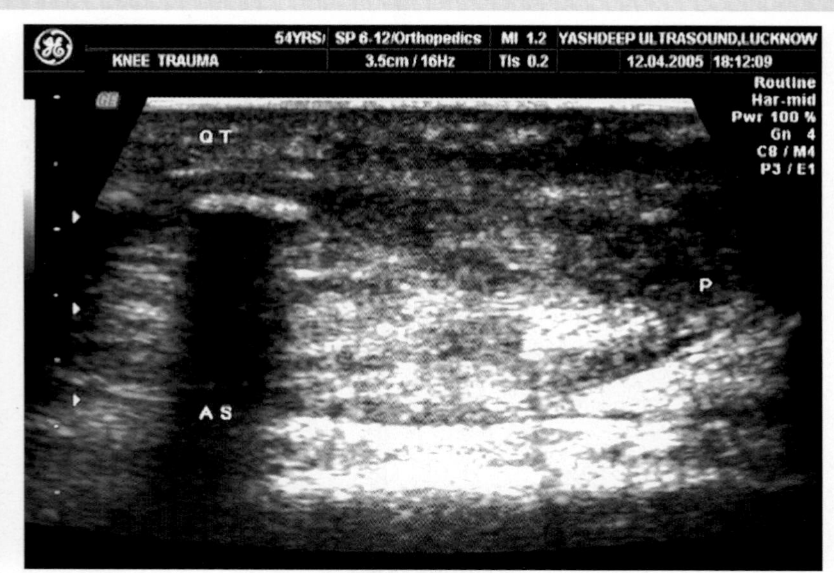

FIGURE 13.610

FIGURES 13.609 and 13.610: The Lt knee also shows heterogeneous edematous Lt tendon. Normal tendon texture is lost. Echogenic calcified specks are seen in the tendon. Hypoechoic cleft is also seen in the tendon. The tendon shows a big tear at musculotendonis junction.

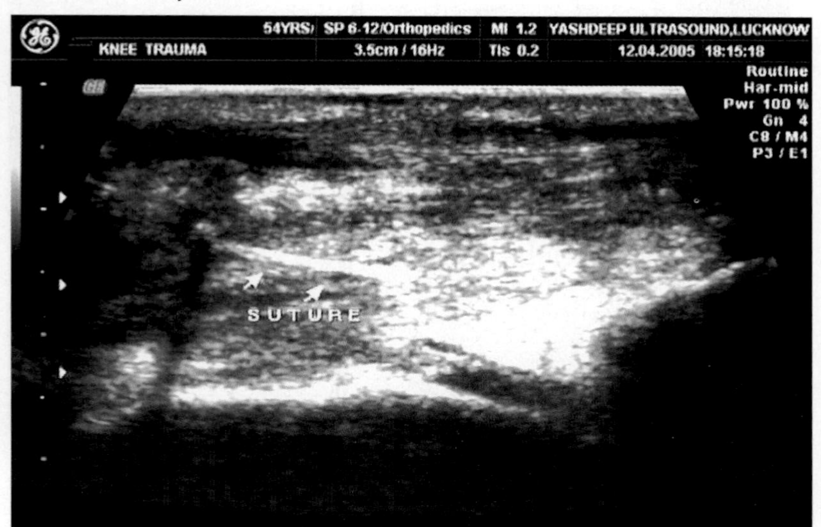

FIGURE 13.611

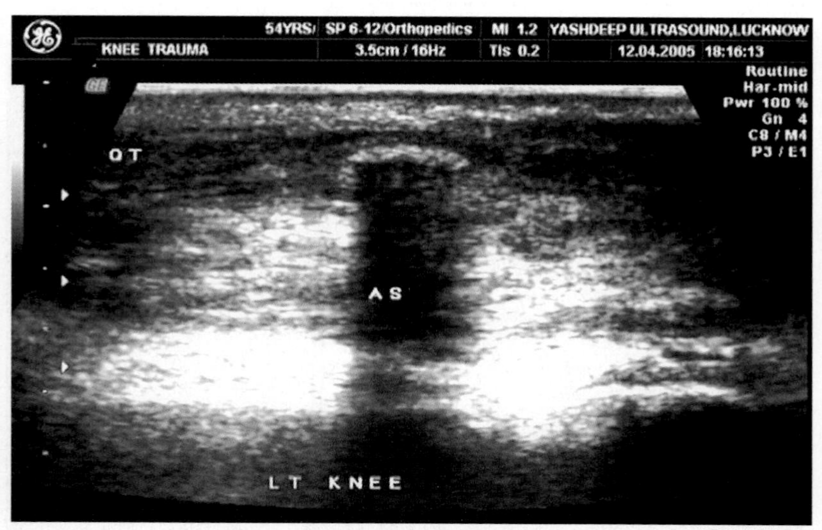

FIGURE 13.612

FIGURES 13.611 and 13.612: The post surgical repair of the tendon. Post surgical repair of the tendon shows echogenic suture thread (shown by the arrows) fastening the tendon segments. Tendon parts are seen closely approximated. However, calcified plaques are seen in the substance of the tendon. They are accompanied with dense acoustic shadowing.

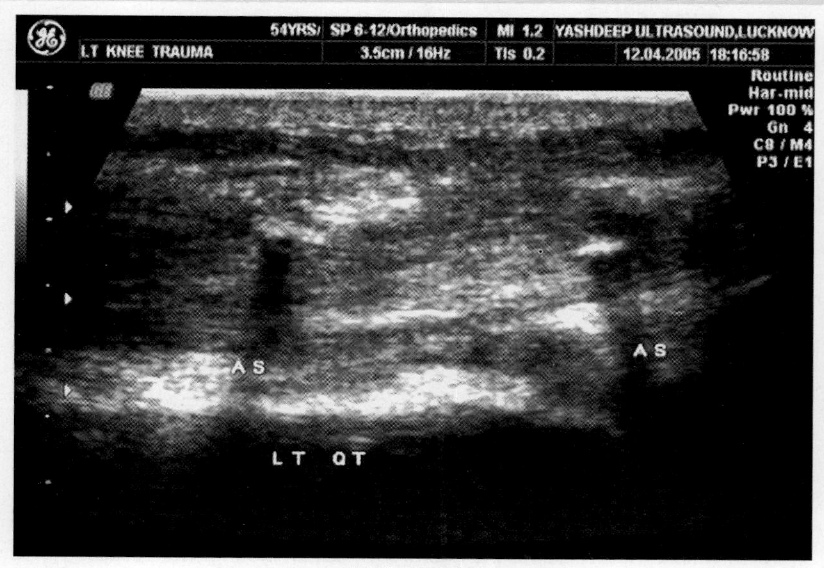

FIGURE 13.613

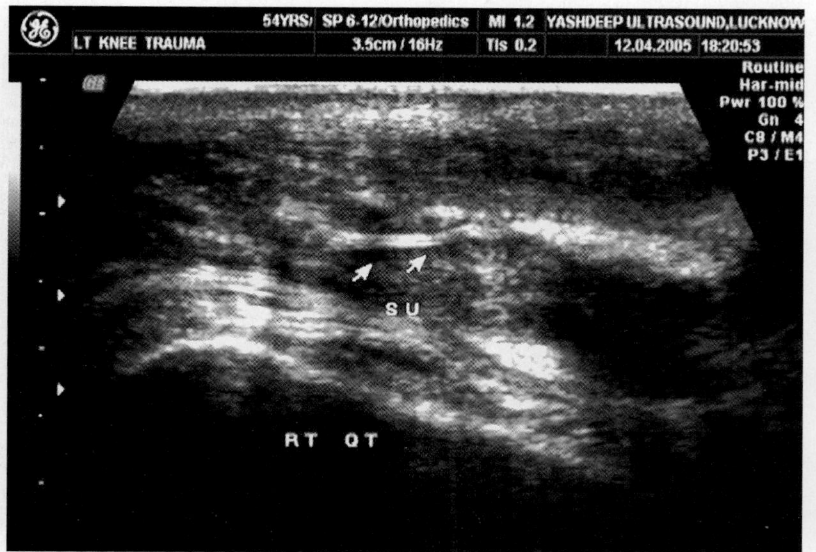

FIGURE 13.614

FIGURES 13.613 and 13.614: Lt tendon also shows bright echogenic suture thread fastening the tendon. However, calcified plaques are also seen in the substance of the tendon. Biopsy of the tendon shows grossly degenerative changes in the tendon.

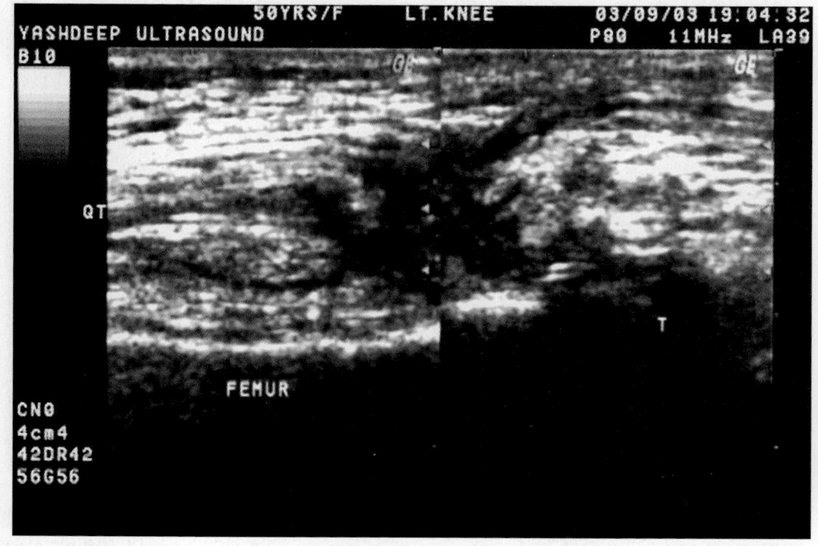

FIGURE 13.615

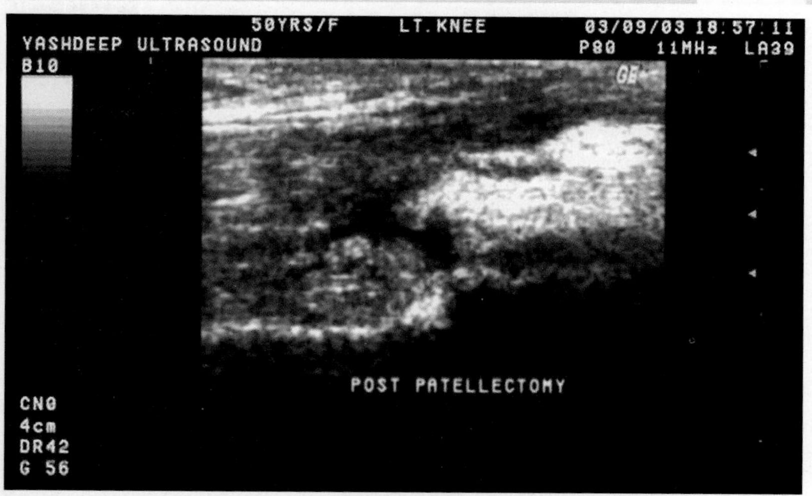

FIGURE 13.616

FIGURES 13.615 and 13.616: Post surgical rupture of the quadriceps tendon. HRSG shows a big hypoechoic cleft in quadriceps tendon in a patient underwent patellectomy. A big gap is seen between the two ends of the quadriceps tendon with loss of tendon fibers. Echogenic suture is also seen torn and snapped from the ends. Hematoma formation is also seen between the ruptured tendon fibers.

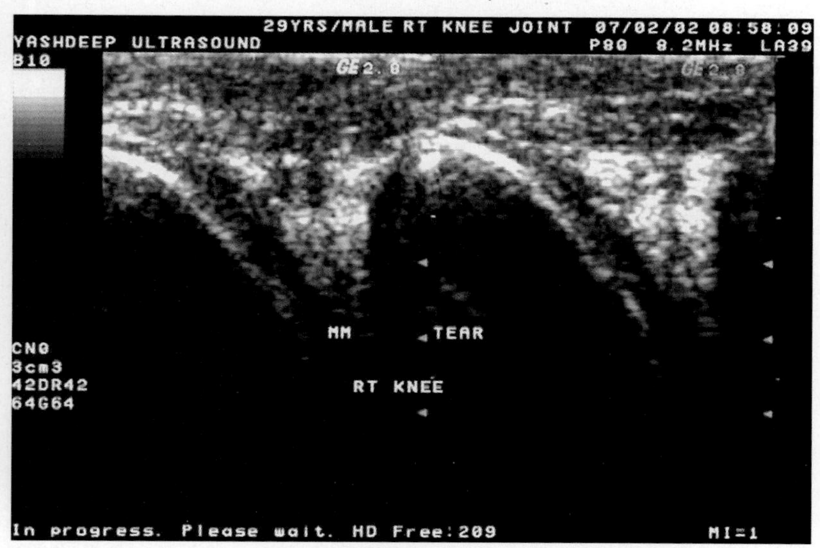

FIGURE 13.617

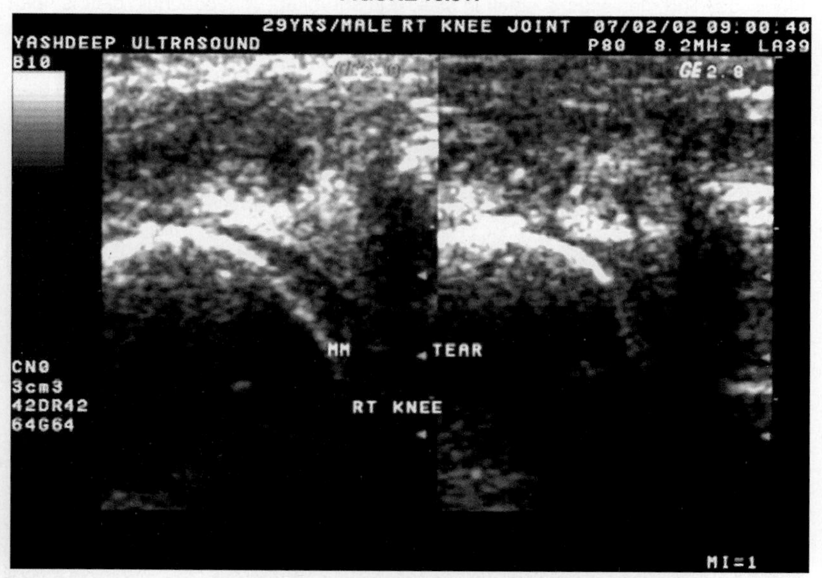

FIGURE 13.618

FIGURES 13.617 and 13.618: Meniscal tear. HRSG shows big irregular hypoechoic cleft running to the medial meniscus suggestive of tear. It is running from the apex to the base of the tear is grown.

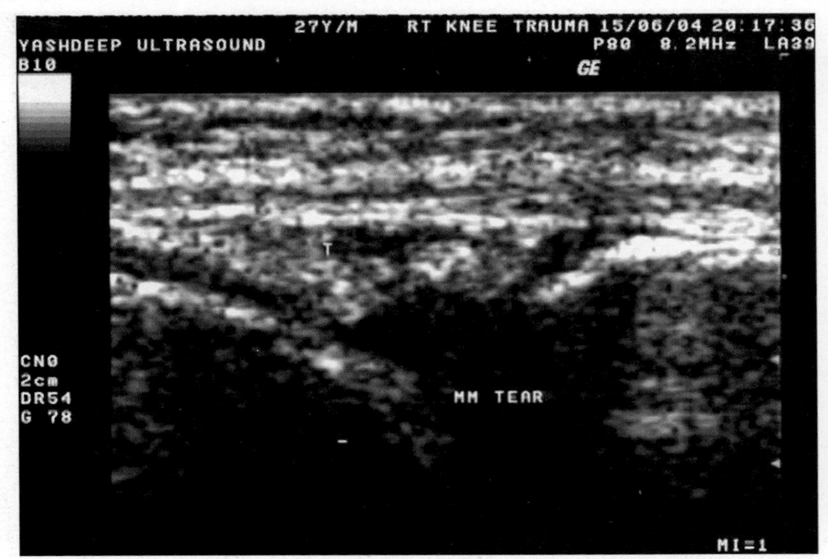

FIGURE 13.619: HRSG shows a big tear at the base of the medial meniscus. The tear is seen running transversely at the base. The apex of the meniscus is also seen torn and normal meniscus is seen lost. Part of the meniscal tissue is lost.

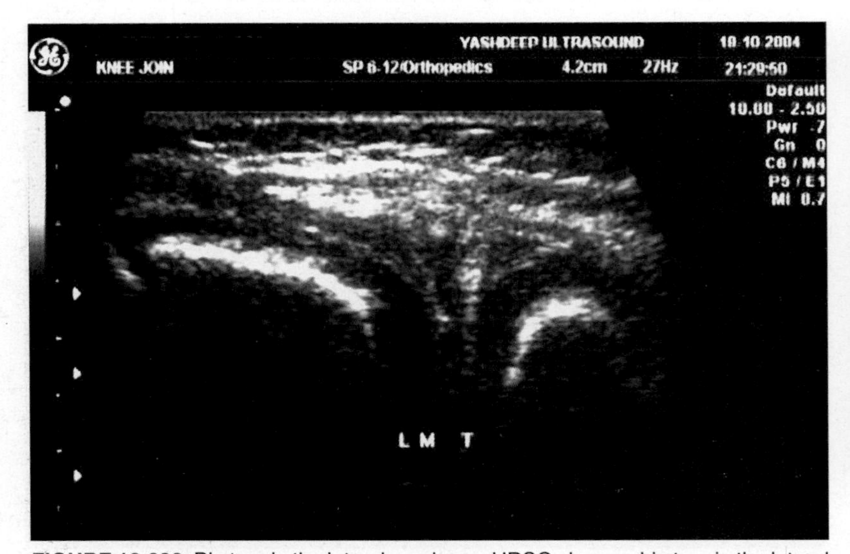

FIGURE 13.620: Big tear in the lateral meniscus. HRSG shows a big tear in the lateral meniscus. The meniscus is small in size. It is shrunken. A hypoechoic cleft is seen running from the apex to the base.

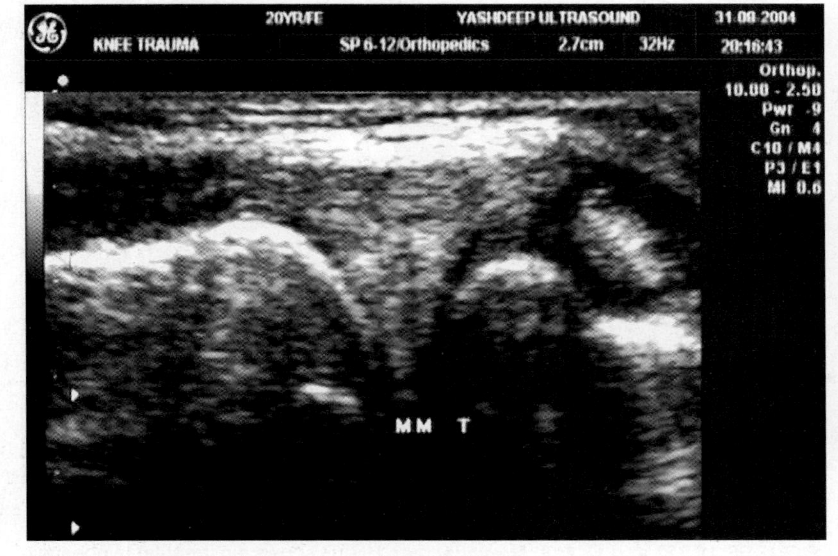

FIGURE 13.621: **Medial meniscus tear.** HRSG shows a hypoechoic cleft running from apex to the base in medial meniscus in a young girl sustained trauma.

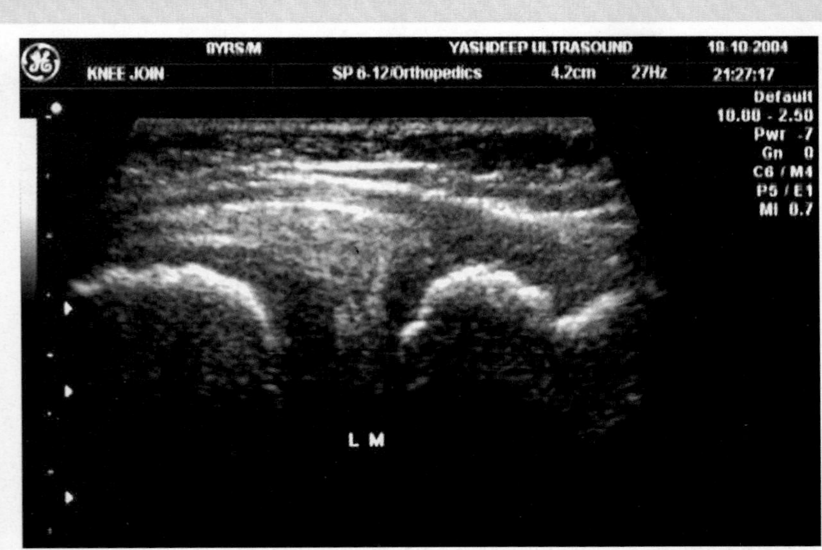

FIGURE 13.622: **Radial tear in the lateral meniscus.** A young child sustained knee trauma. HRSG shows a radial tear in the lateral meniscus. The tear is seen running obliquely in radial fashion.

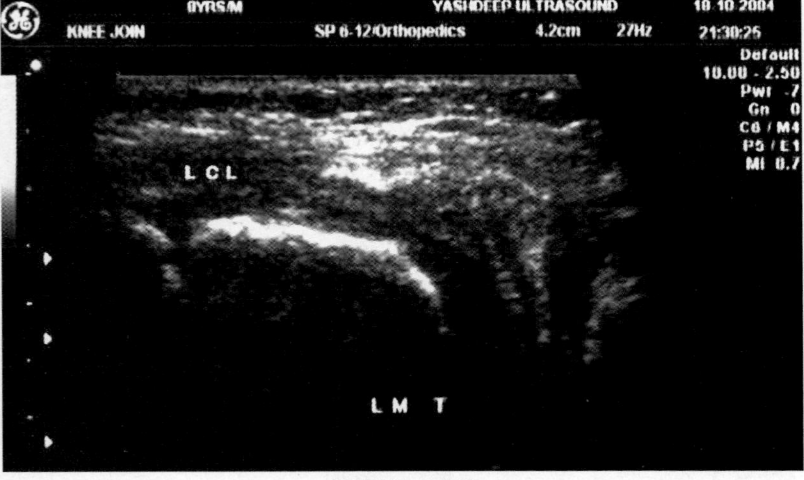

FIGURE 13.623

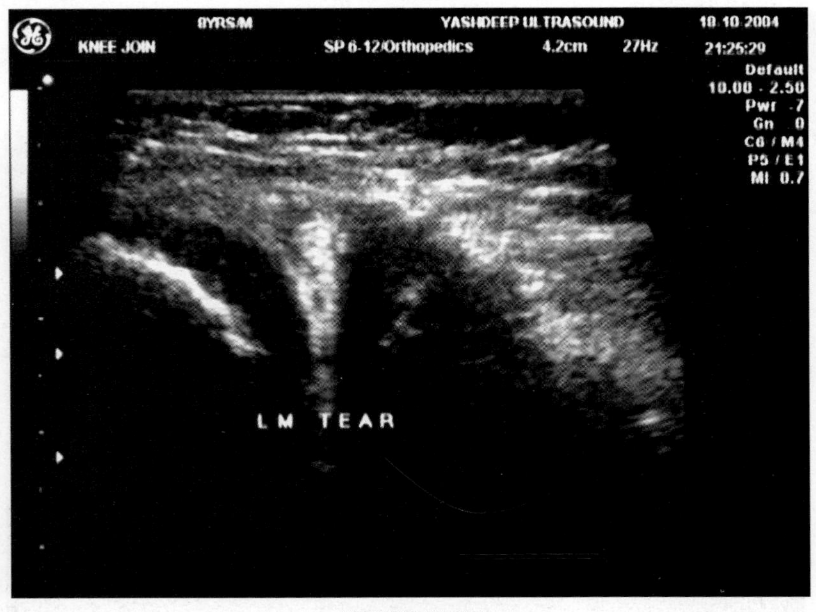

FIGURE 13.624

FIGURES 13.623 and 13.624: **Lateral meniscus tear.** HRSG shows a vertical cleft running in the lateral meniscus from apex to the base. The meniscus is thinned out and lost is normal shape.

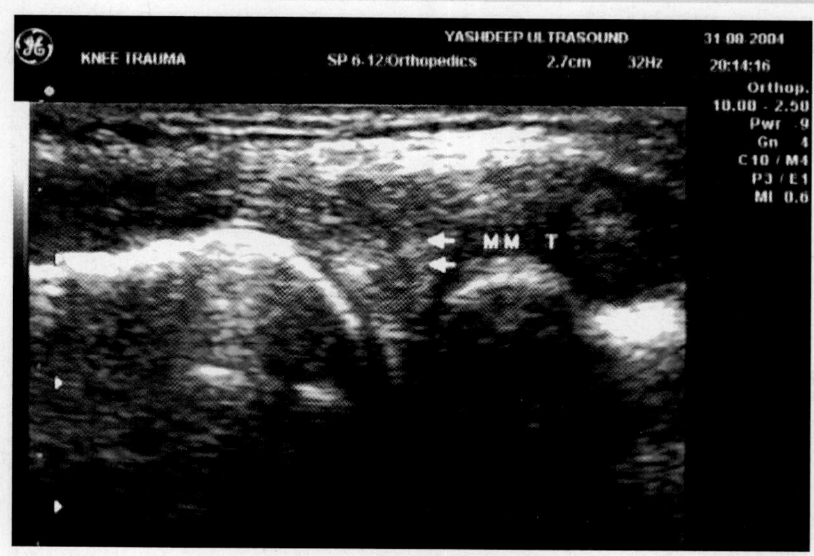

FIGURE 13.625

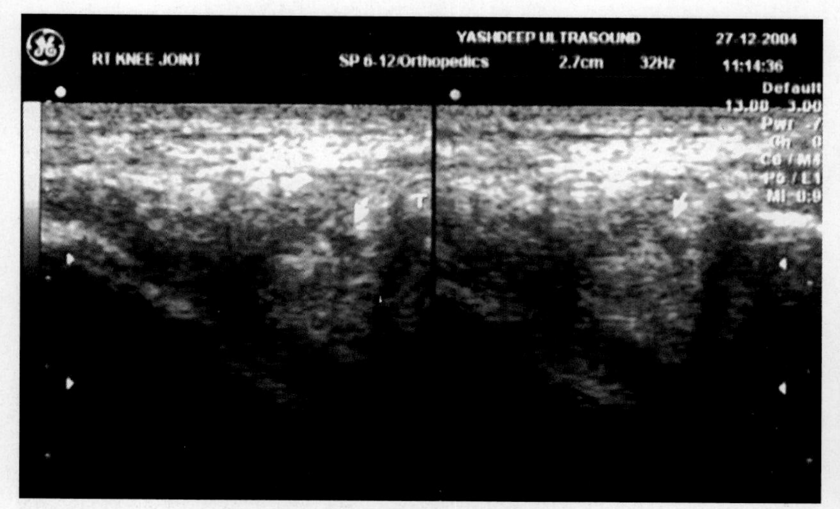

FIGURE 13.626

FIGURES 13.625 and 13.626: Medial meniscus tear. A patient presented with pain in the knee after sustaining trauma. HRSG shows vertical cleft in the knee in the meniscus. It is seen running from apex to the base. Partial tear is seen in meniscus. It is running to the base.

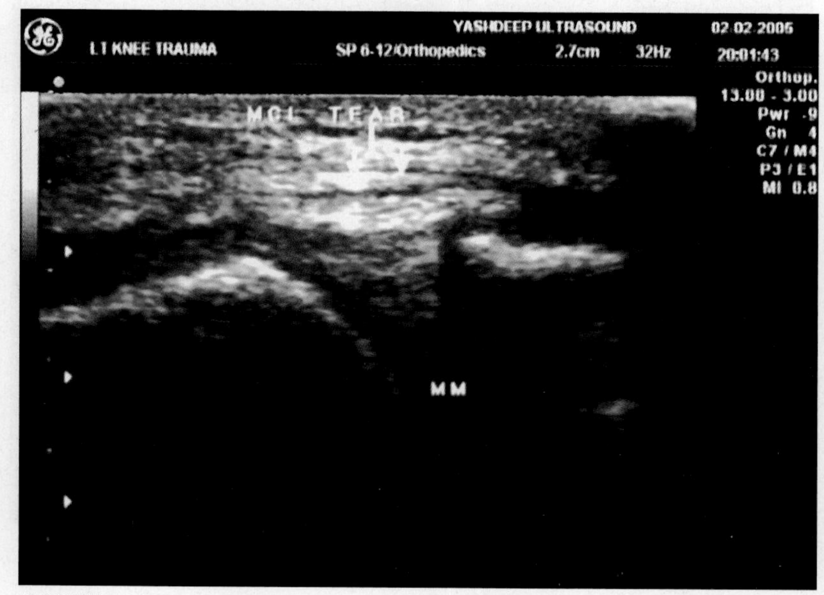

FIGURE 13.627: Ligament tear at meniscoligament junction. HRSG shows a hypoechoic cleft running transversely at meniscoligament junction. It is seen as hypoechoic cleft. The ligament is thickened. Small amount of fluid is also seen in the medial recess.

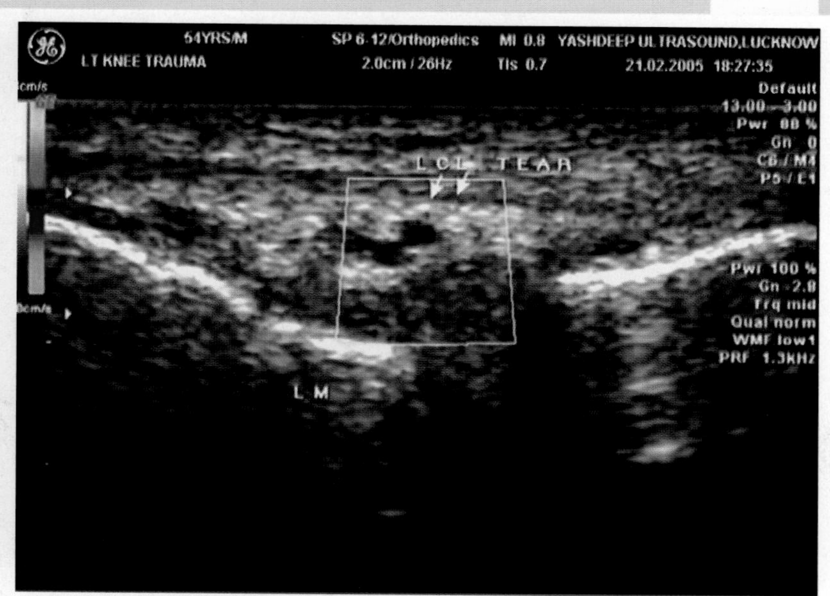

FIGURE 13.628: Small meniscal cyst. HRSG shows a well-defined cyst at the base of the lateral meniscus. Hypoechoic cleft is also seen at the base running obliquely suggestive of tears. **Color Doppler imaging** confirms the tear at times obliquely running vessel may be confused as a tear on gray scale imaging.

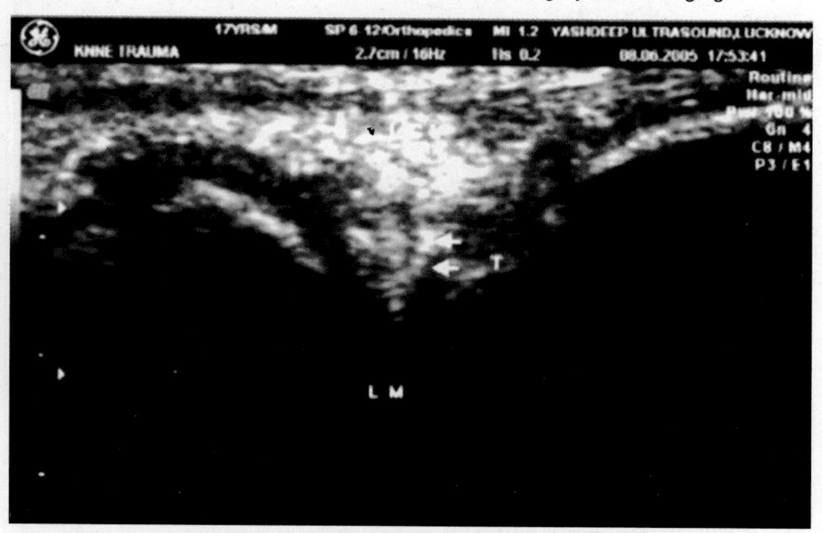

FIGURE 13.629: Chronic meniscal and lateral meniscus tear. The meniscus is irregular in outline and inhomogeneous in texture. The tear margins are also and it is irregular in outline.

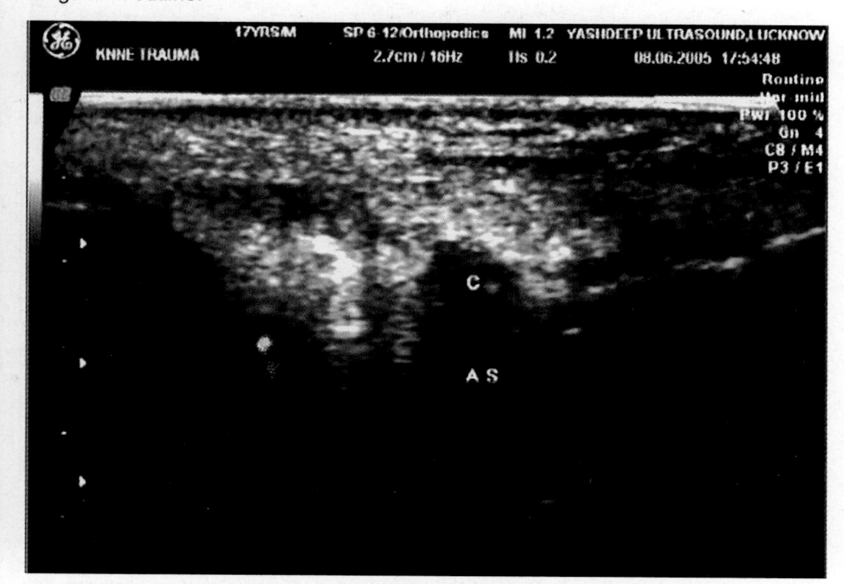

FIGURE 13.630: Chronic meniscal tear with calcification. HRSG shows chronic tear in the medial meniscus. Normal meniscal shape is lost. Echogenic calçified specks are seen at the base of meniscus. They are accompanied with acoustic shadowing.

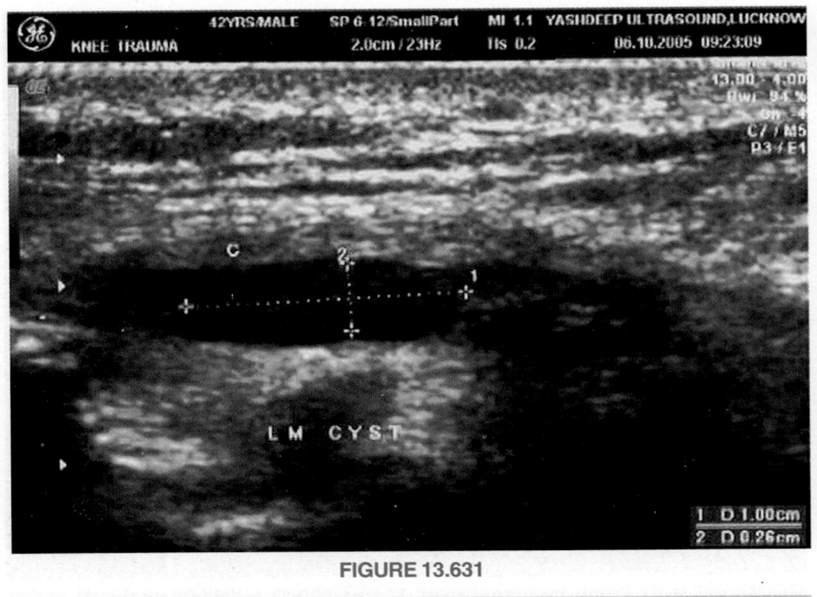

FIGURE 13.631

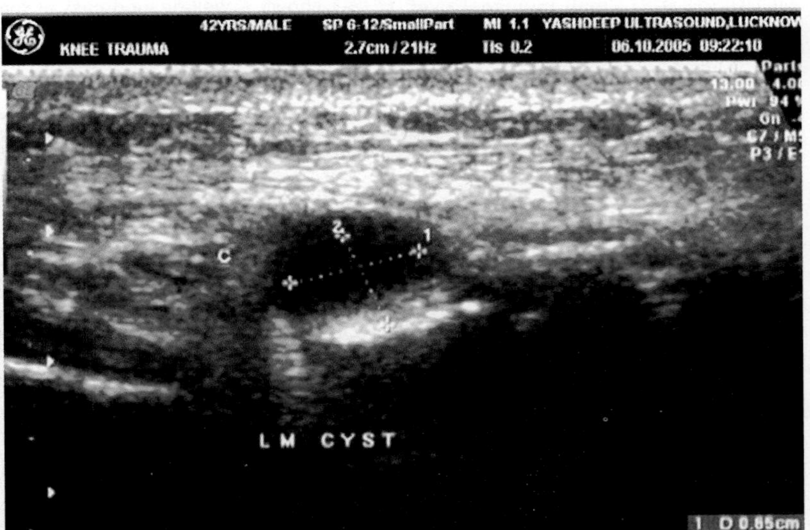

FIGURE 13.632

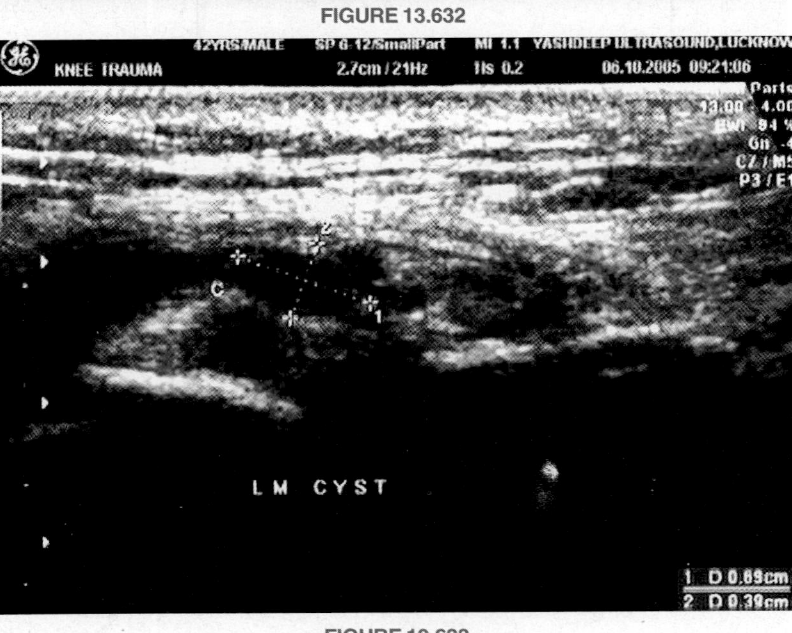

FIGURE 13.633

FIGURES 13.631 to 13.633: Big meniscal cyst. HRSG shows a big cyst at the base of the lateral meniscus. The cyst is covering whole of the base of the meniscus. It is seen going to the inner surface of the meniscal wall. Margins are irregular. The meniscal texture is lost. There is also evidence of multiple hypoechoic areas seen in the meniscus suggestive of degenerative changes in the meniscus.

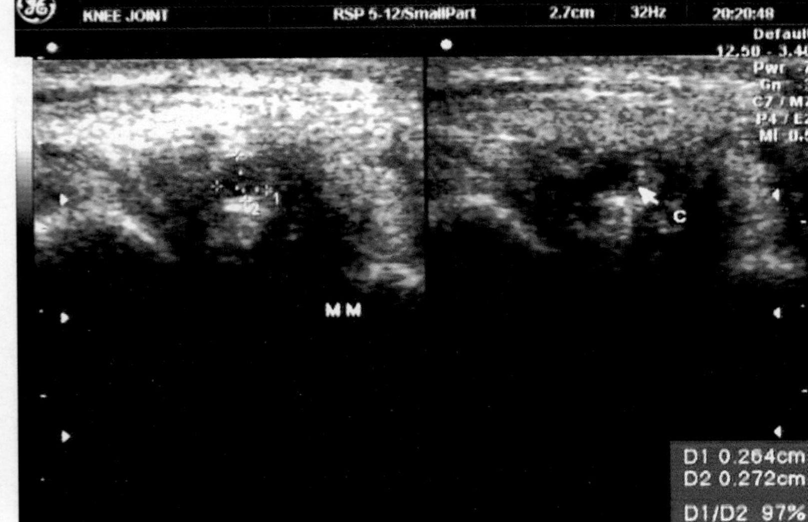

FIGURE 13.634

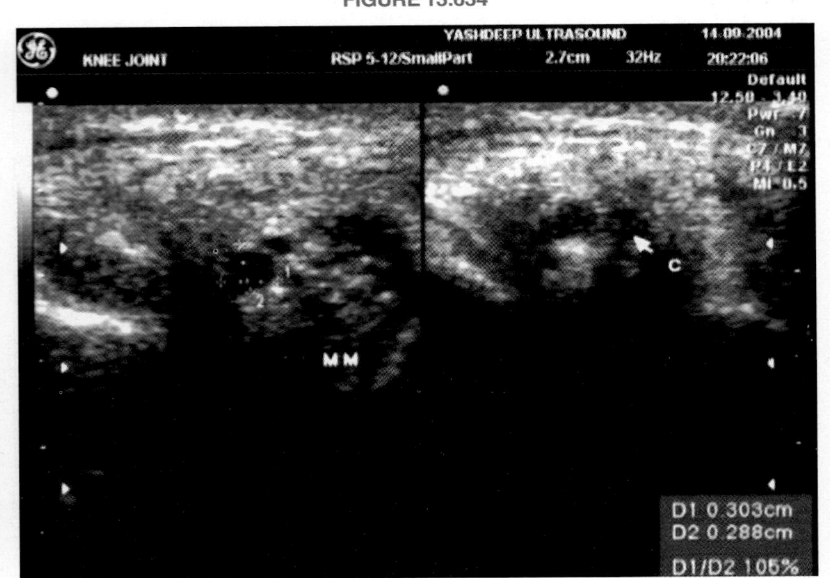

FIGURE 13.635

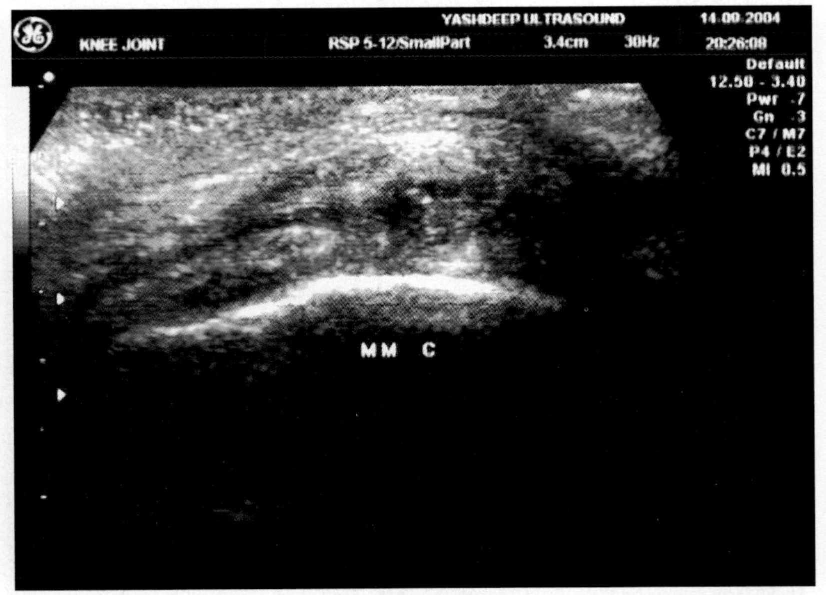

FIGURE 13.636

FIGURES 13.634 to 13.636: Medial meniscal cyst. HRSG shows well-defined cystic mass at the base of the medial meniscus in transverse and longitudinal axis. The cyst wall is echogenic with fine calcification seen in the cyst wall suggestive of chronic meniscal cyst. The meniscus is also thickened and hypoechoic in texture. It is grossly inhomogeneous in texture. Partial tear is also seen in the meniscus at the base.

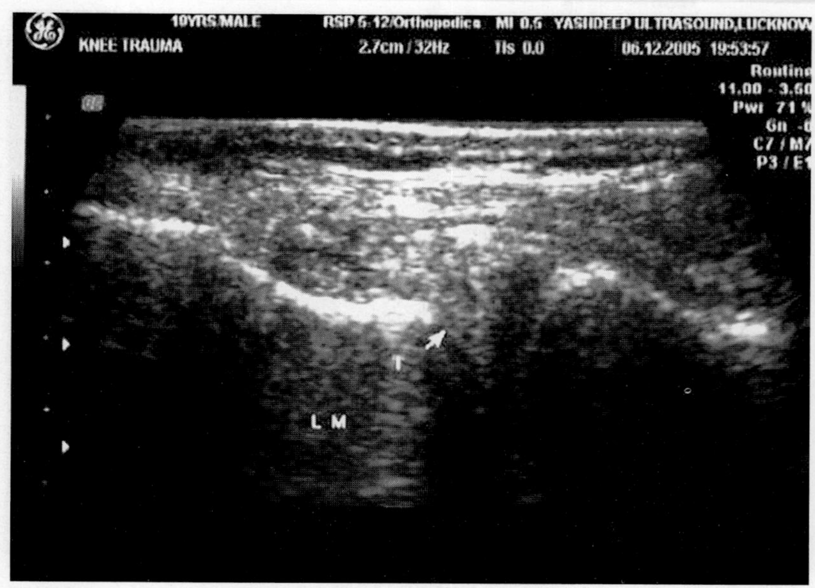

FIGURE 13.637: 2D imaging of the meniscal tear. A young patient sustained trauma to the Lt knee joint. HRSG shows a vertical cleft in the lateral meniscus, which is seen running from the apex to the base. However, the cleft is not well defined.

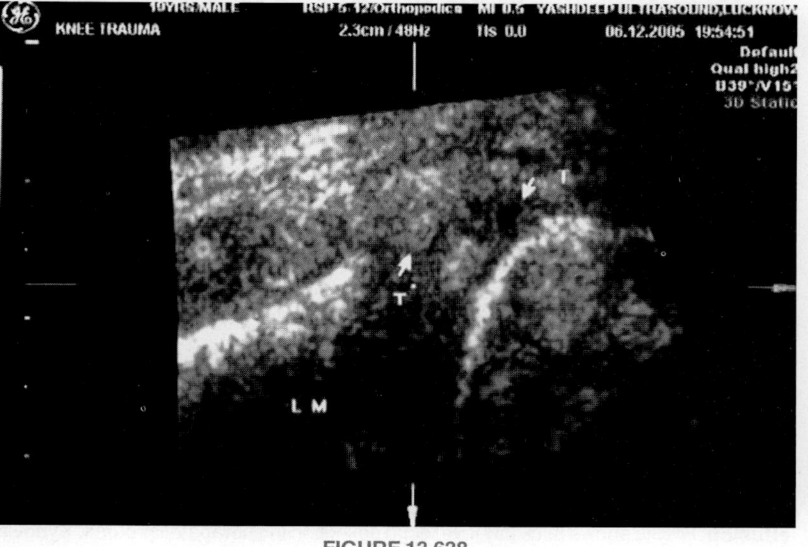

FIGURE 13.638

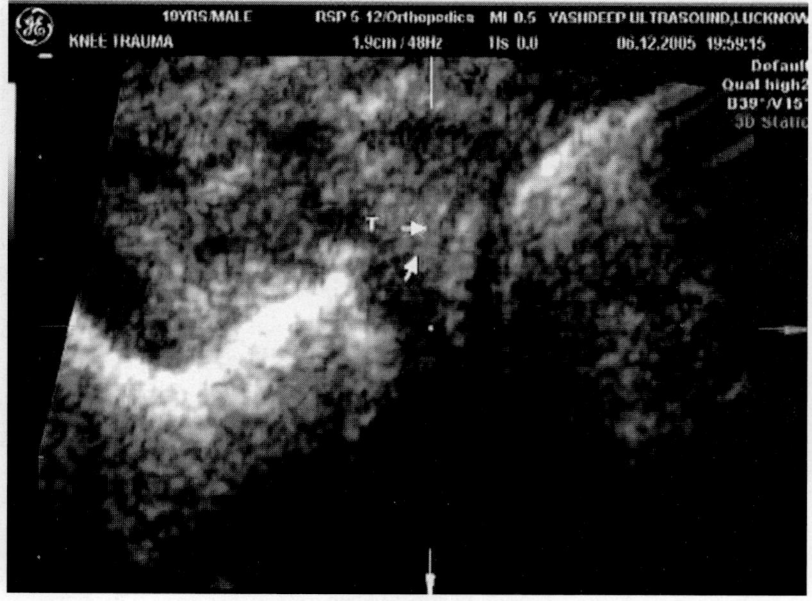

FIGURE 13.639

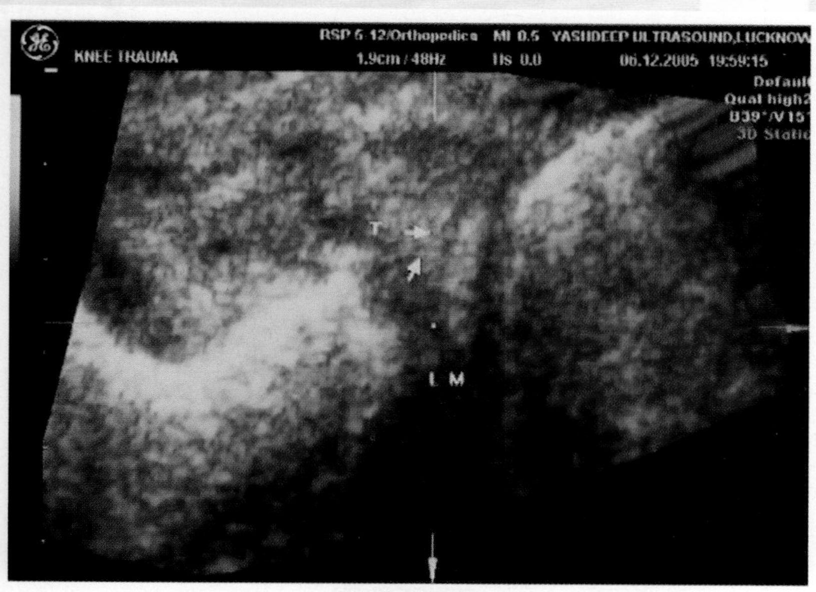

FIGURE 13.640

FIGURES 13.638 to 13.640: 3D surface rendering imaging of the knee clearly shows the tear in much better way. Total extant of the tear is well visualized. The tear is also seen running upto base of the meniscus.

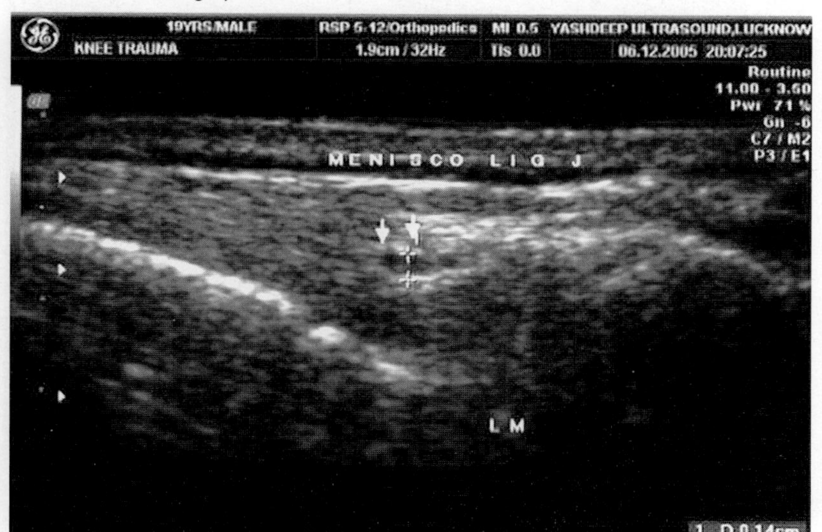

FIGURE 13.641

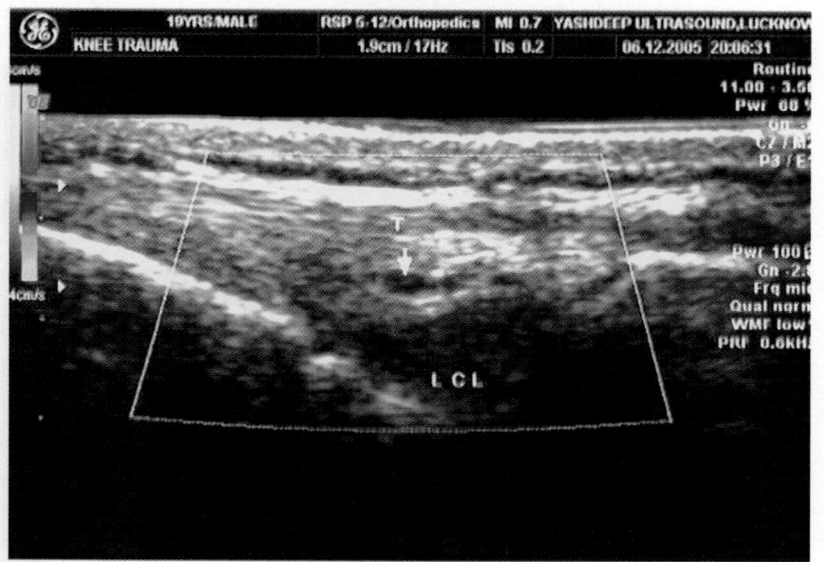

FIGURE 13.642

FIGURES 13.641 and 13.642: Meniscoligament junction tear. HRSG shows a cleft running transversely at meniscoligament junction involving LCL and lateral meniscus. The tear is well defined and color flow imaging rules out the vessels and confirmed the tear.

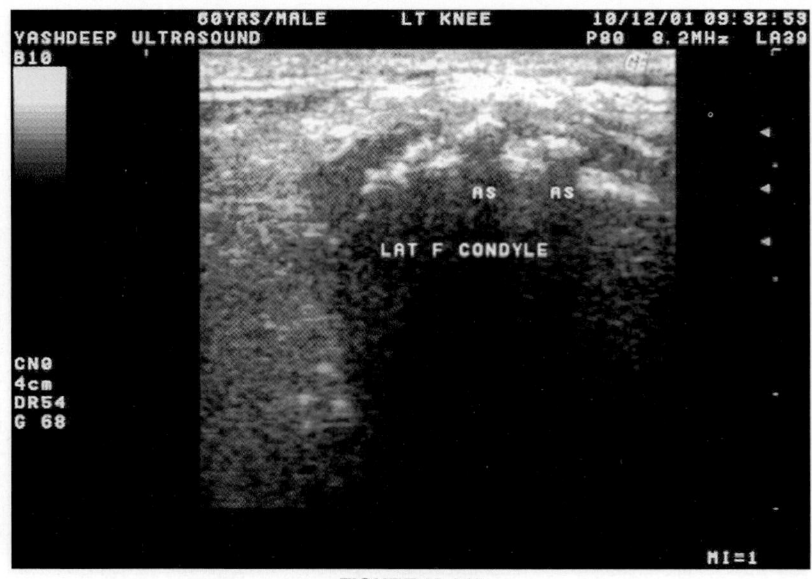

FIGURE 13.643

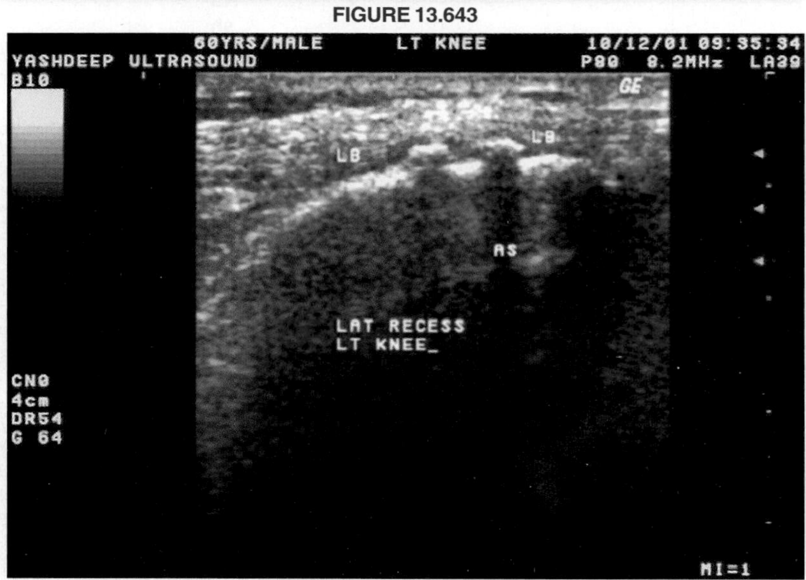

FIGURE 13.644

FIGURES 13.643 and 13.644: Loose bodies in the lateral recess. HRSG shows multiple irregular echogenic shadows in the lateral recess. They are accompanied with acoustic shadowing suggestive of loose bodies. Few bodies are also seen on the lateral articular surface below the lateral femoral condyle.

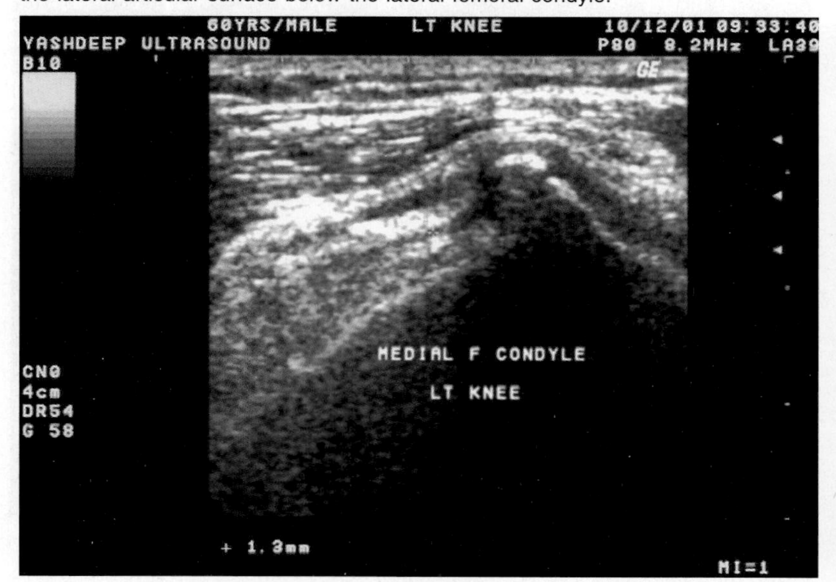

FIGURE 13.645: Irregular femoral articular surface. HRSG shows irregular articular surface of the femoral condyle due to degenerative changes. The hyaline cartilage is thickened.

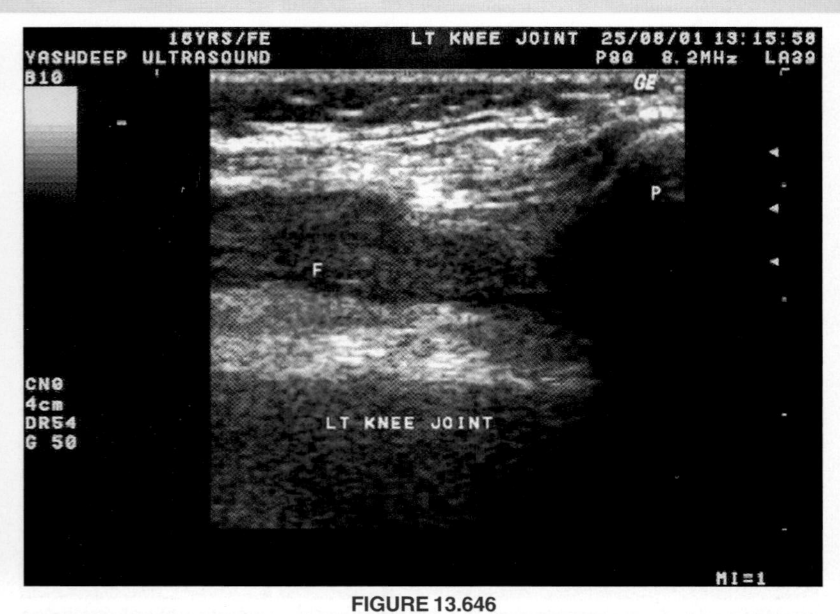

FIGURE 13.646

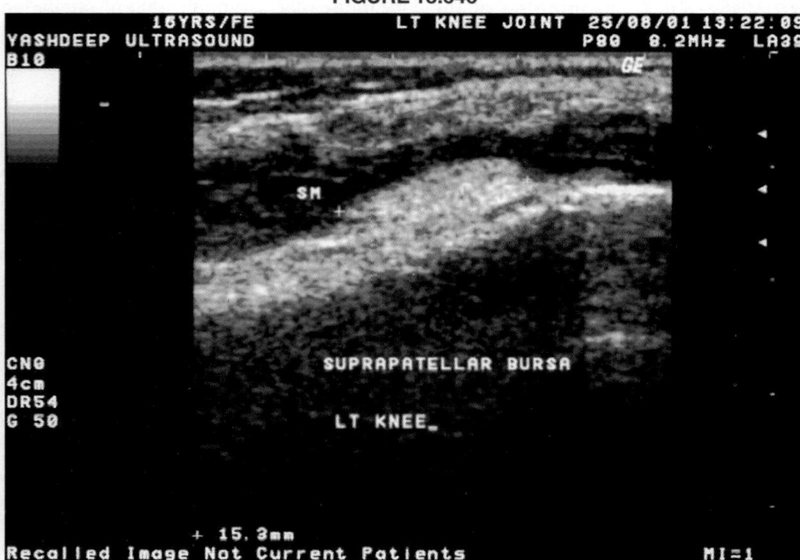

FIGURE 13.647

FIGURES 13.646 and 13.647: Tubercular tenosynovitis of Lt knee joint with effusion. HRSG shows thick low level echo collection in suprapatellar bursa. Synovial membrane proliferation is also seen in a case of tubercular effusion.

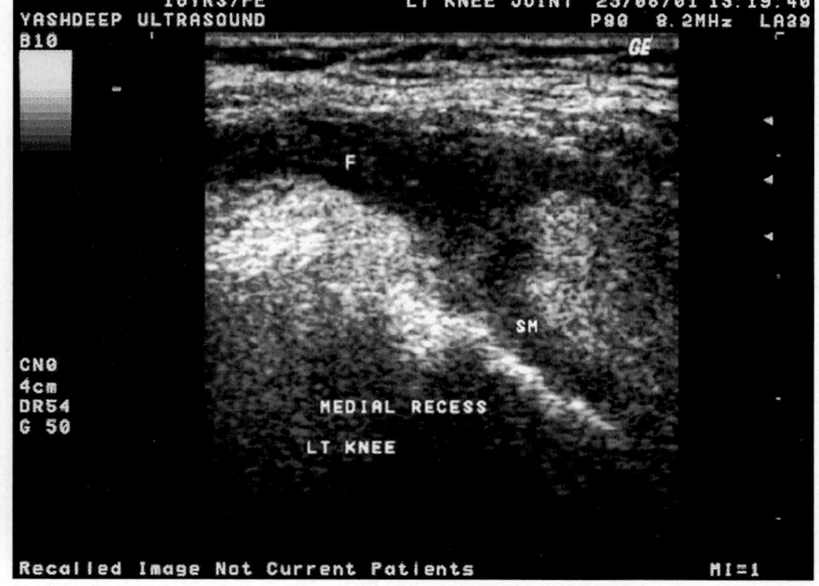

FIGURE 13.648

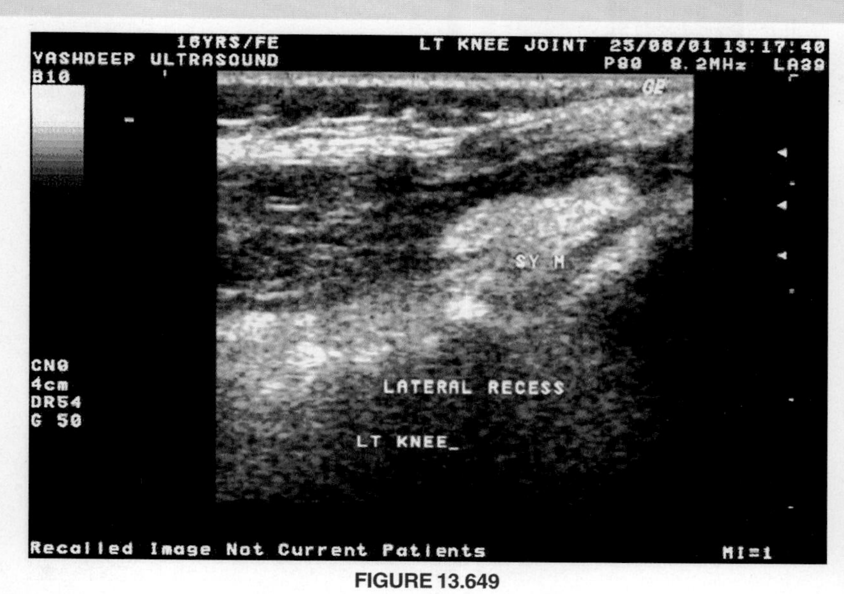

FIGURE 13.649

FIGURES 13.648 and 13.649: Thick echo collection is also seen in lateral and medial recess. Multiple floating shadows are seen in the collection. Echogenic membranous shadow is seen floating in it suggestive of thickened proliferated synovium.

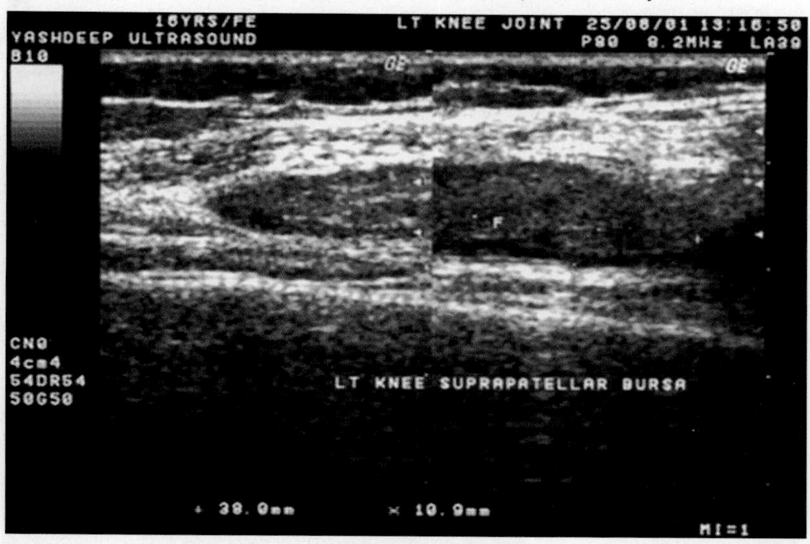

FIGURE 13.650

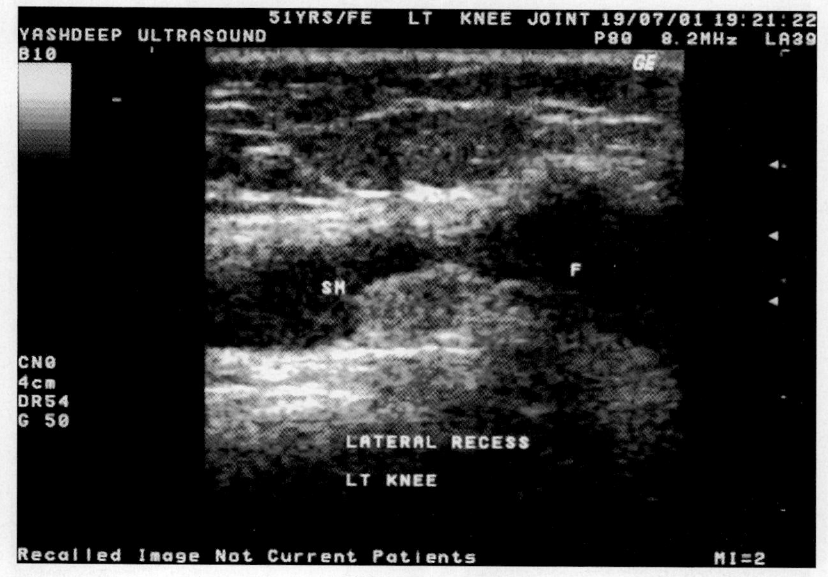

FIGURE 13.651

FIGURES 13.650 and 13.651: HRSG shows thick jelly like collection in the suprapatellar bursa in the lateral recess suggestive of tubercular exudative collection. The synovium is also thickened and proliferative and seen as an echogenic membrane.

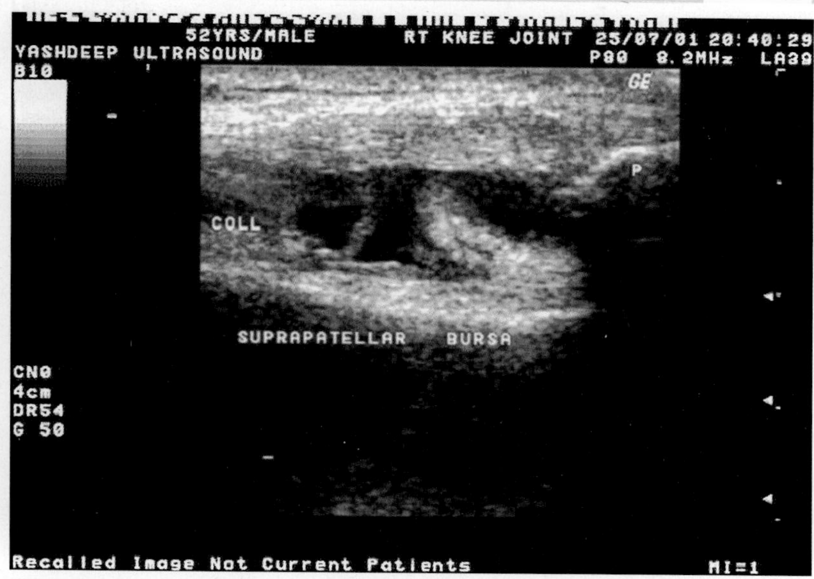

FIGURE 13.652: Pyogenic suprapatellar bursitis. HRSG shows thick collection in the suprapatellar bursa. Low-level echoes are seen floating in it. Echogenic shadows are seen floating in it in a known case of septic arthritis of diabetic patient. Aspiration shows thick purulent collection.

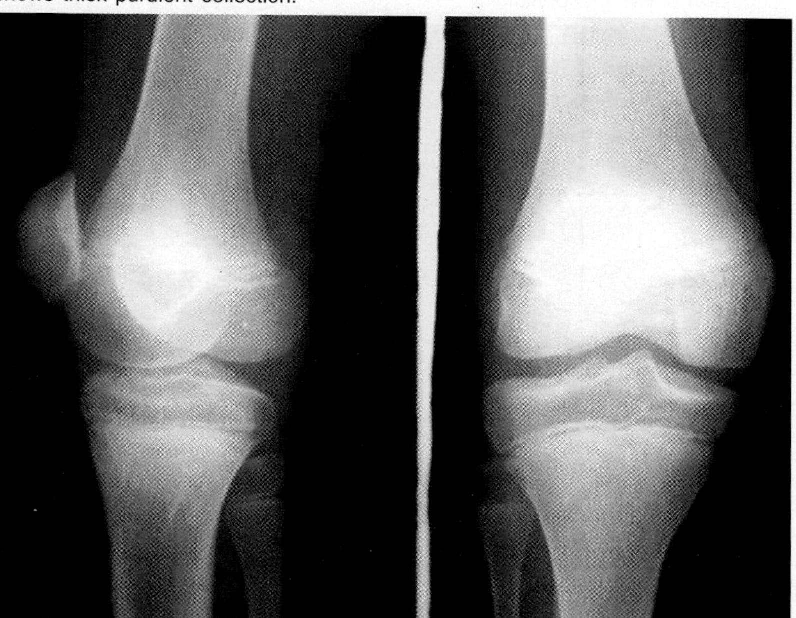

FIGURE 13.653: Rheumatoid arthritis. A young boy presented with bilateral painful nee joint. Plain X-ray of the knee was not showing any bony abnormality.

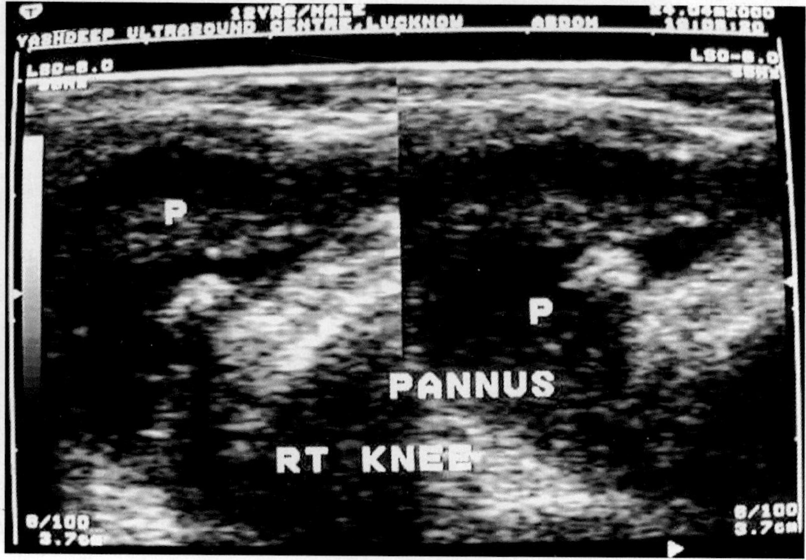

FIGURE 13.654

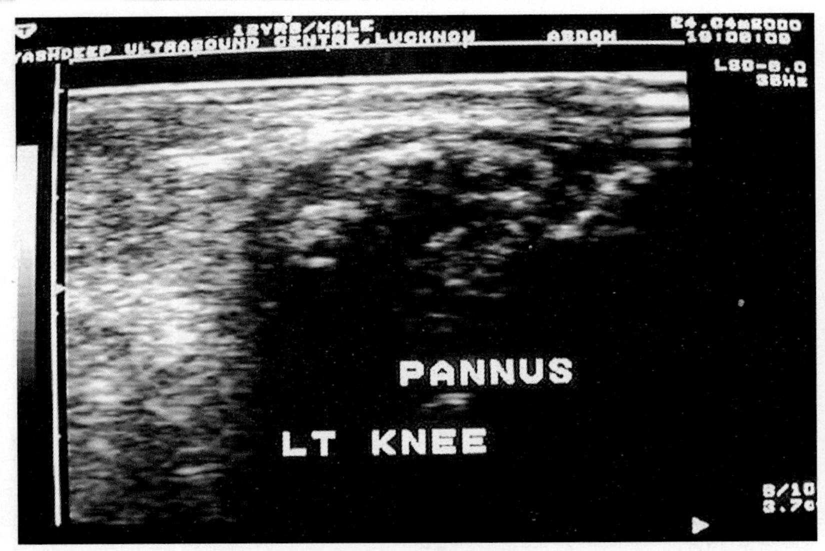

FIGURE 13.655

FIGURES 13.654 and 13.655: HRSG of the knee shows low level echo collection with echogenic proliferative synovial membrane in medial and lateral recess of both knee joint. Echogenic fixed membranous shadow is seen in the Lt knee joint lateral recess suggestive of pannus formation.

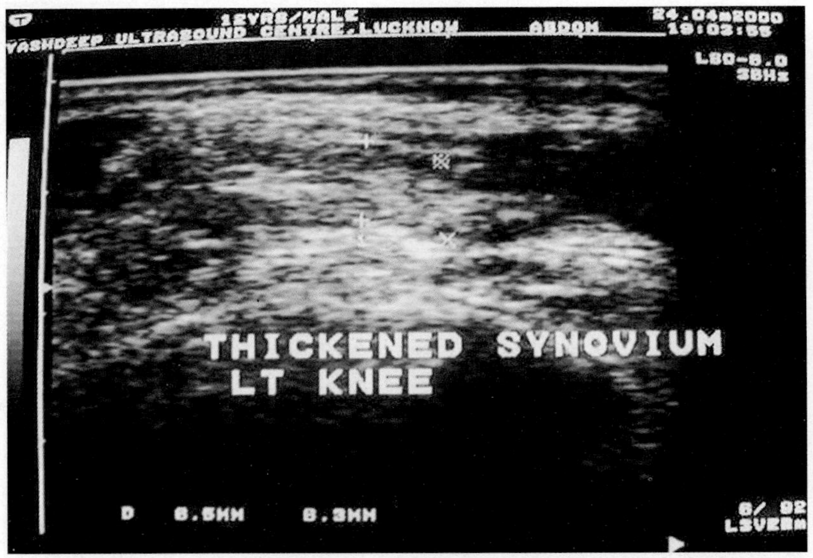

FIGURE 13.656: Markedly thickened synovial membranes is seen in the lateral recess of the same patient. Biopsy of the tissue confirmed rheumatoid arthritis.

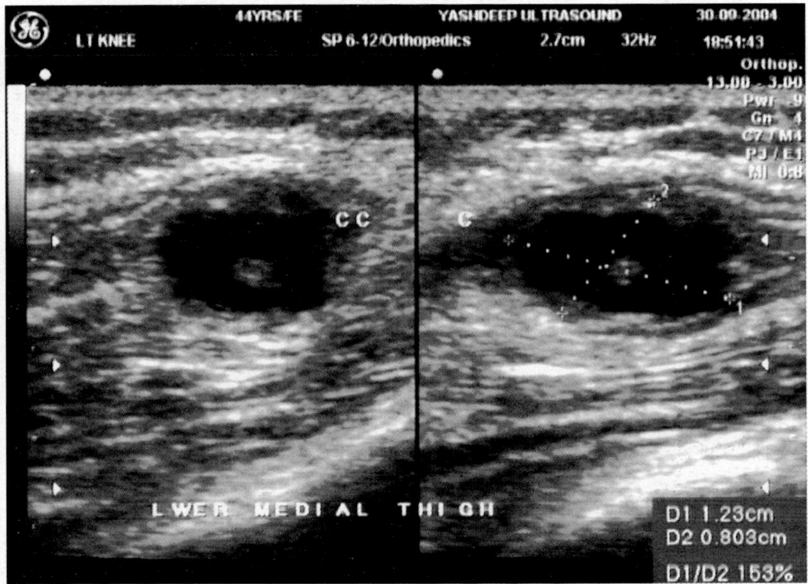

FIGURE 13.657

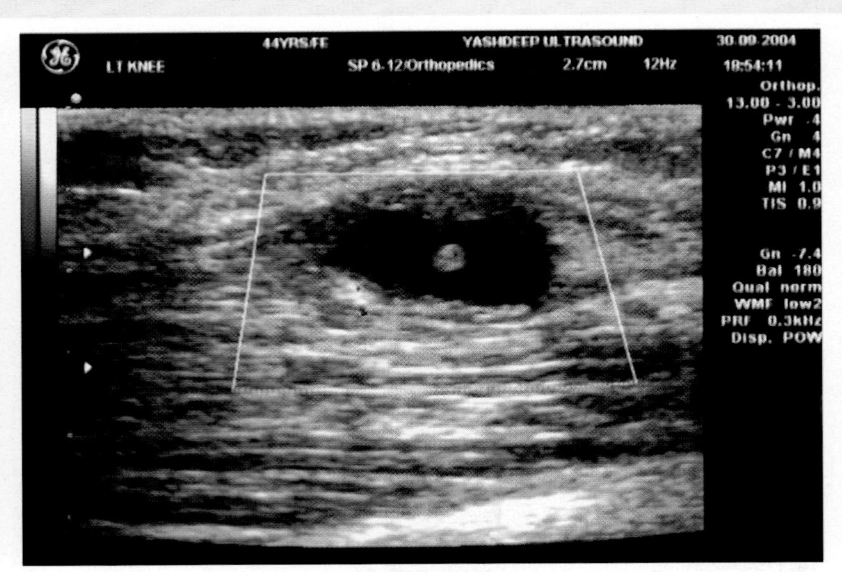

FIGURE 13.658

FIGURES 13.657 and 13.658: Cysticercus cyst. A patient presented with nodular shadow lower medial side of the Lt thigh. HRSG shows well-defined cystic mass in the muscle. A echogenic nidus is seen in the inner wall of the cyst suggestive of cysticercus cyst.

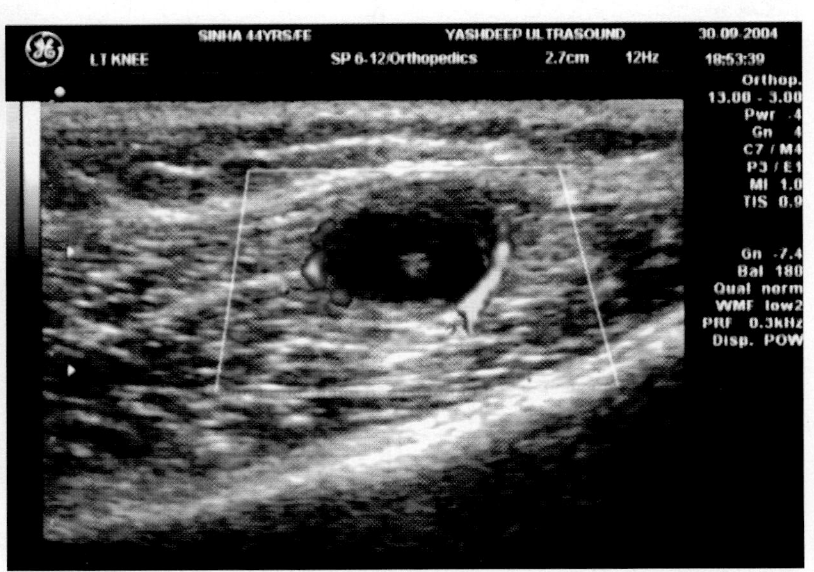

FIGURE 13.659: On color flow imaging peripheral capsular flow is seen in the cyst.

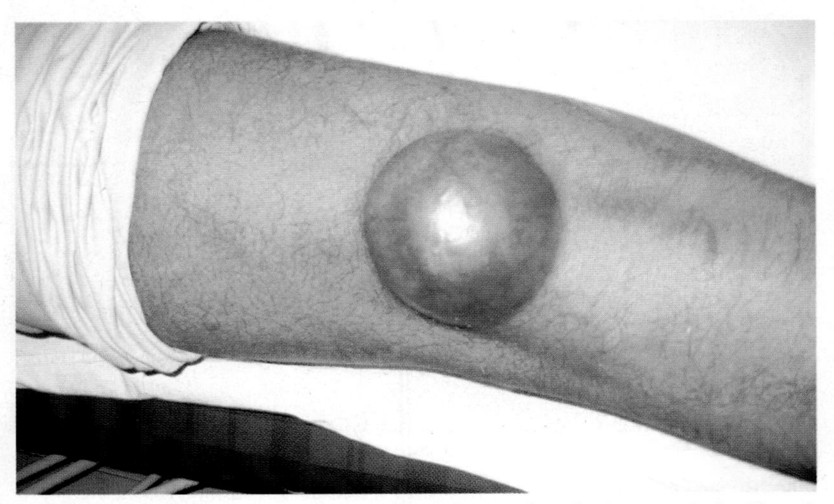

FIGURE 13.660: Lipoma in the popliteal fossa. A patient presented with soft tissue mass with blue discoloration in the popliteal fossa.

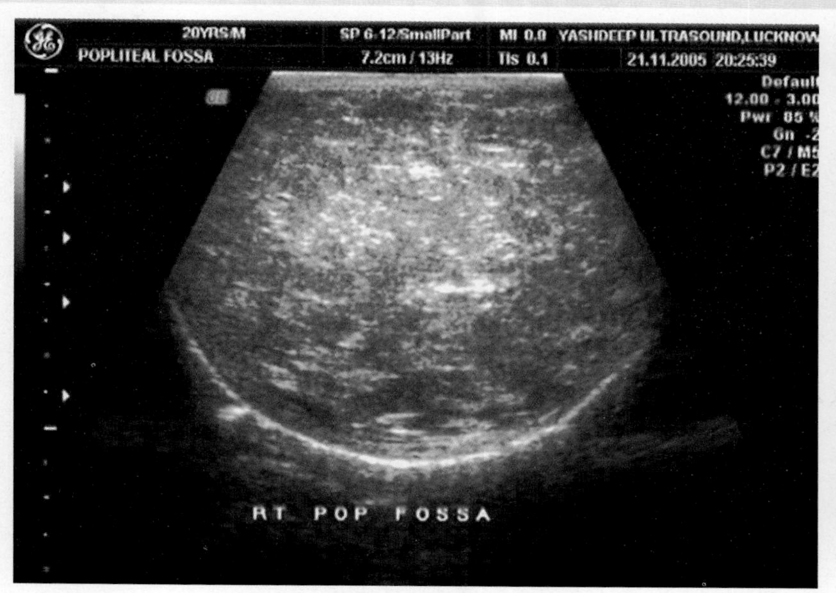

FIGURE 13.661

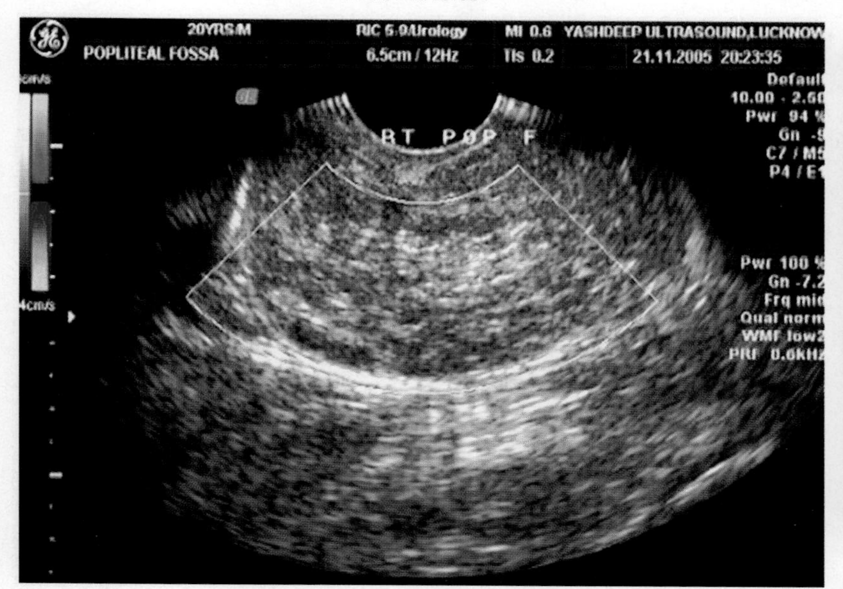

FIGURE 13.662

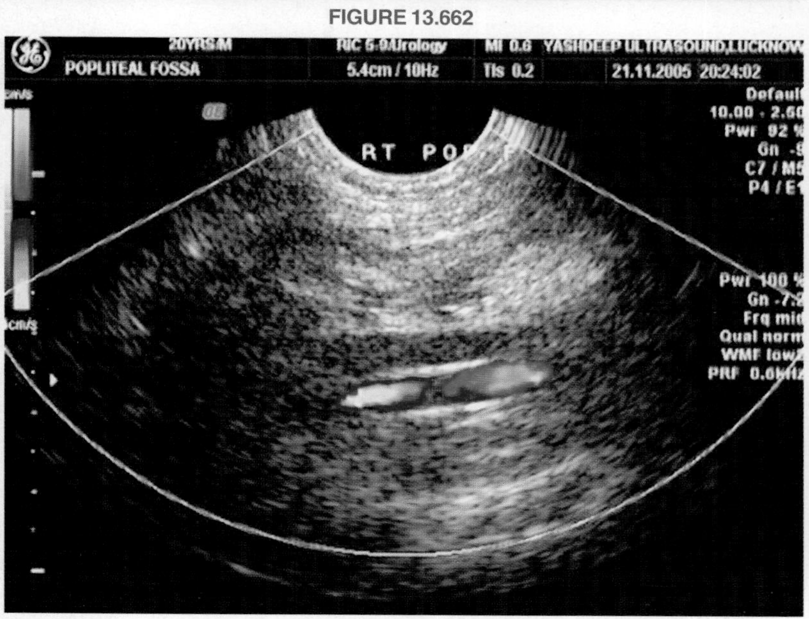

FIGURE 13.663

FIGURES 13.661 to 13.663: HRSG shows a well-defined homogeneous soft tissue mass in the popliteal fossa. Homogeneous echoes are seen in it. No calcification or cystic degeneration is seen in it. **On color flow imaging** no flow is seen in the mass. Biopsy of the mass confirmed lipoma.

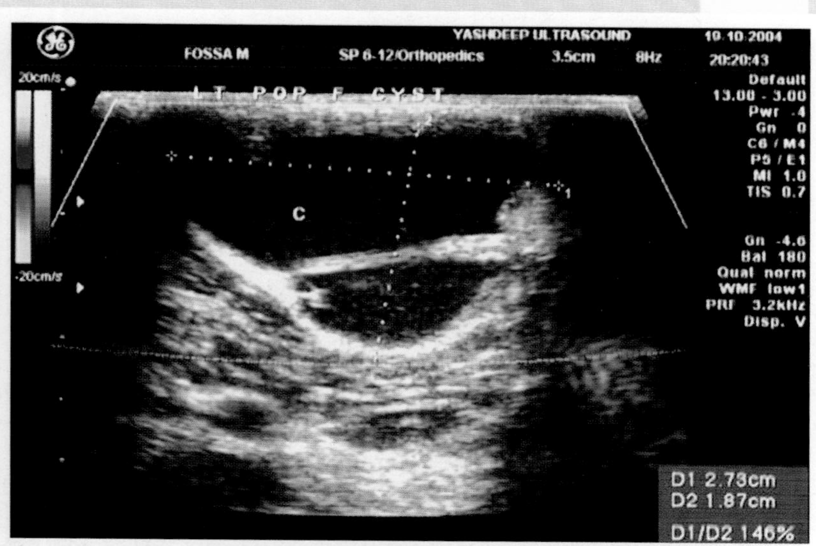

FIGURE 13.664: Popliteal cyst (Bakers cyst). HRSG shows a big cystic mass in the popliteal fossa. The cyst is well defined. It is seen in the gastrocnemius muscle. Echogenic septum is seen in the cyst. **On color flow imaging** no flow seen in the cyst. The aspiration confirmed simple Bakers cyst.

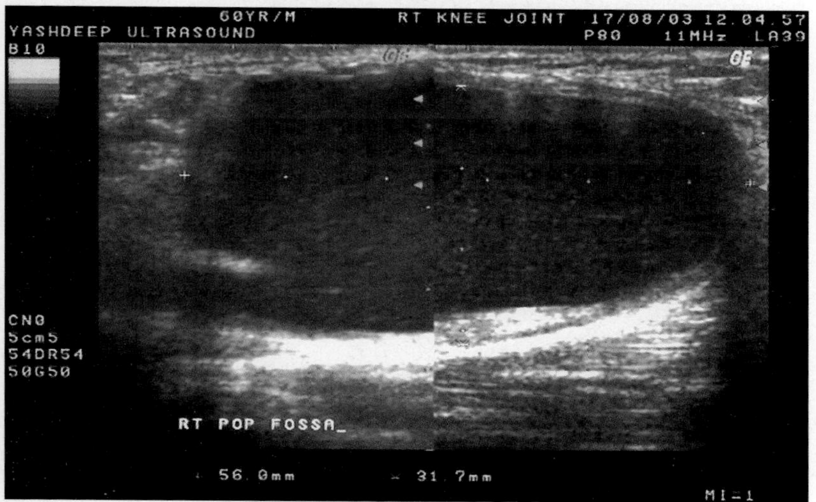

FIGURE 13.665

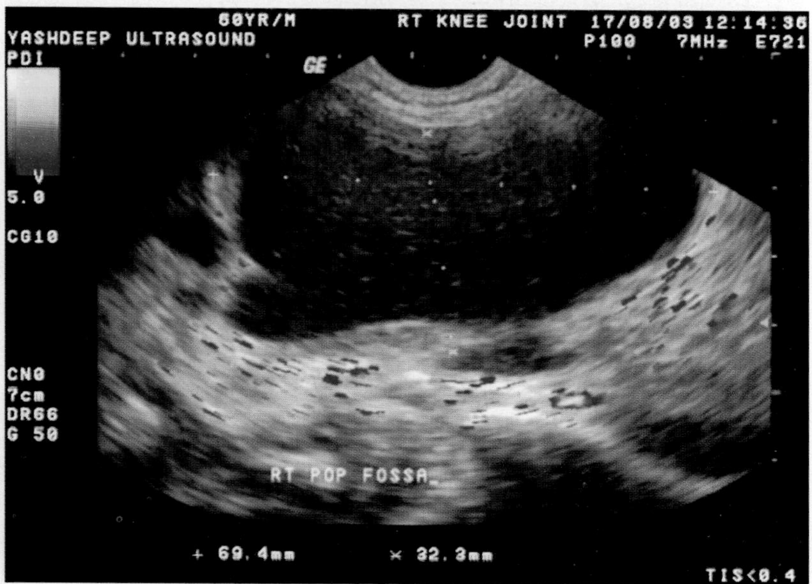

FIGURE 13.666

FIGURES 13.665 and 13.666: Big popliteal cyst. HRSG shows a huge cystic mass in the popliteal fossa. The cyst is seen occupying whole of the popliteal fossa. Internal echoes are seen in the cyst. The cyst is seen over the popliteal vessel. **On color flow imaging** no flow is seen in the cyst. A finger like projection of the cyst is seen communicating to the other side suggestive of huge Baker cyst.

FIGURE 13.667

Examination of Ankle

Acute trauma to the ankle may lead to fractures of malleoli or avulsions. The ligament injury to the opposite side of fracture is common. Most of the time conventional X-ray is inconclusive. Though it can demonstrate soft tissue swelling without fracture. This may give indirect evidence of ligaments injury. HRSG shows direct evidence of ligaments rupture or tear by visualizing the ligament. The anterior fibulo tailor ligament is most frequently injured. Chronic ankle pain in most of the time is due to extra-articular disease. Multiple tendons are presenting adjacent to the joint and may be responsible for the ankle pain.

Retrocalcaneal bursitis or plantar fascia rupture may be the cause of chronic pain. HRSG can demonstrate effusions and loose bodies in anterior synovial recess. Examination of the ankle begins with the evaluation of anterior synovial recess in dorsi-flexed foot position. Joint effusion can be easily picked up on HRSG.

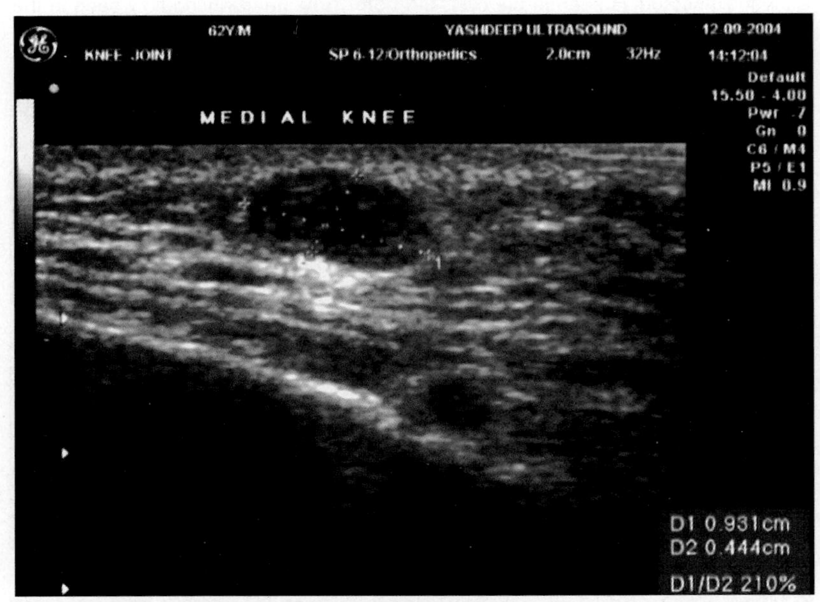

FIGURE 13.668

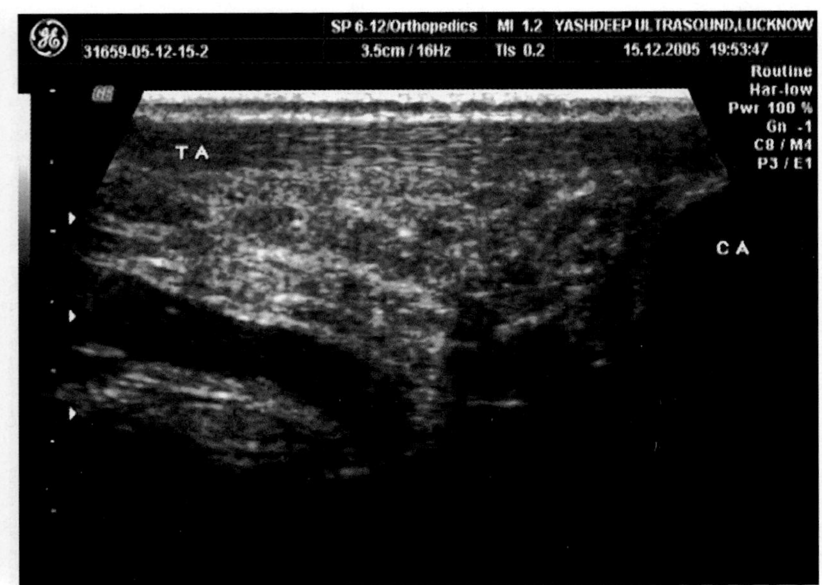

FIGURE 13.670

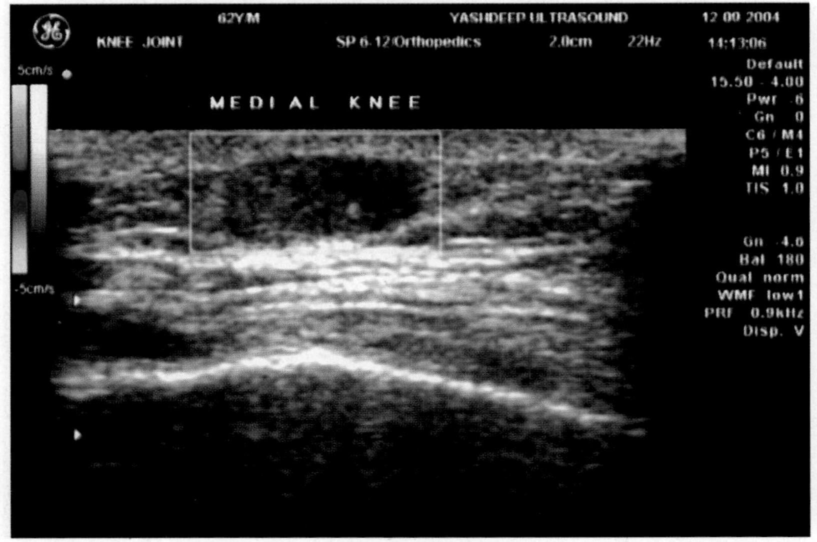

FIGURE 13.669

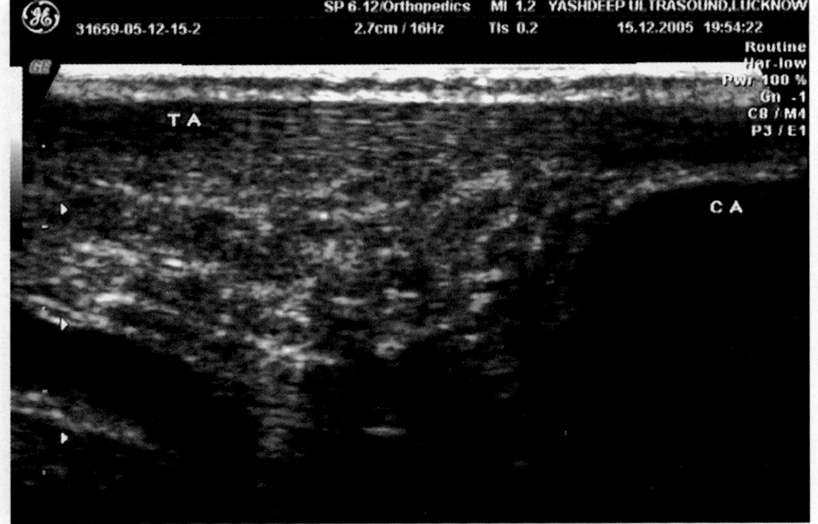

FIGURE 13.671

FIGURES 13.667 to 13.669: Atypical lipoma. A patient presented with well-defined homogeneous soft tissue shadow in the medial side of the Rt knee. HRSG shows a well-defined hypoechoic mass. It shows sharp margins. No calcification is seen in it. No cystic degeneration is seen. **On color flow imaging** no flow seen in it. Biopsy of the mass confirmed atypical lipoma.

FIGURES 13.670 and 13.671: Normal tendon Achilles. HRSG shows normal tendon Achilles. It is seen as homogeneous medium echo strap. Fine tendon fibers are seen in the tendon. The tendon is not having synovial sheath.

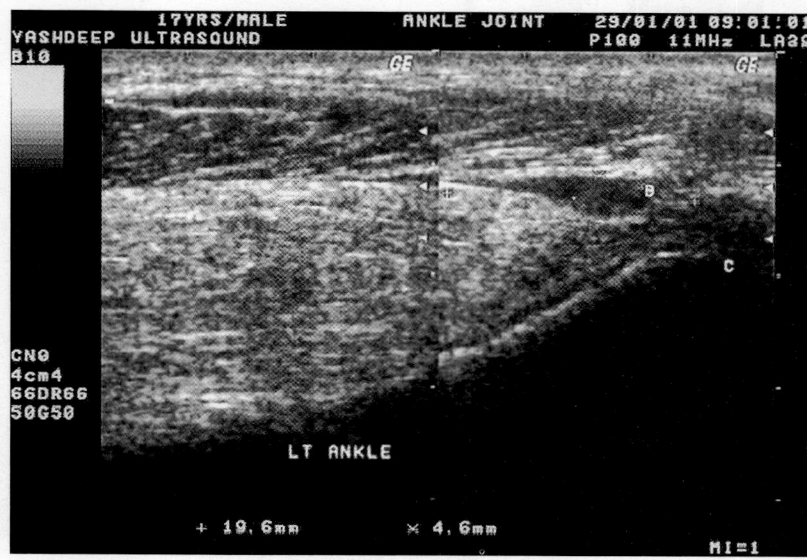

FIGURE 13.672: **Normal tendon Achilles with normal calf muscle.** HRSG shows normal muscle bundle of the calf. The flexor hallusis longus muscle is seen with homogeneous fiber bundle.

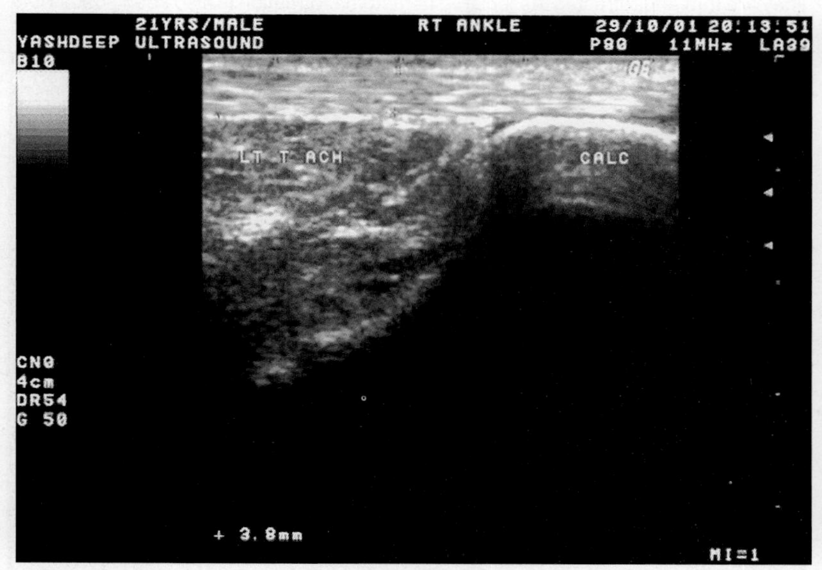

FIGURE 13.673

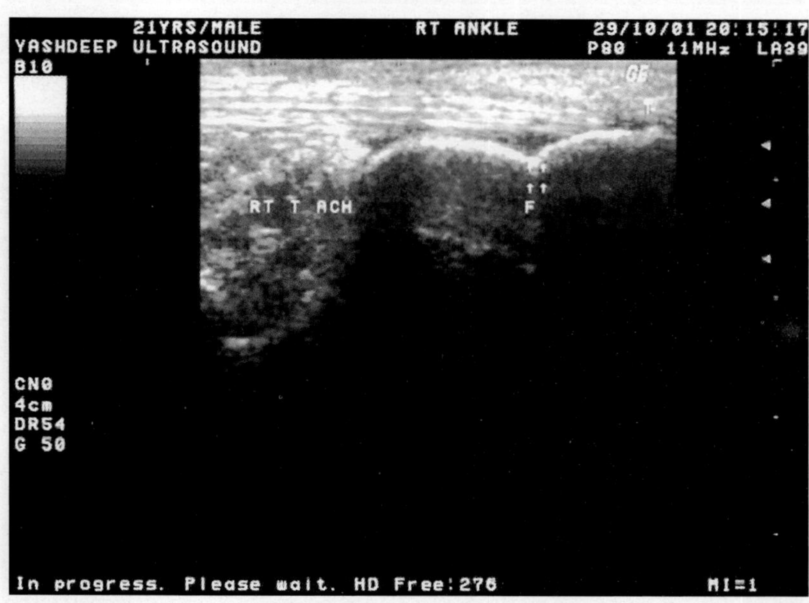

FIGURE 13.674

FIGURES 13.673 and 13.674: **Normal tendon Achilles with subcutaneous tendon bursa.** HRSG shows normal Achilles tendon. Small amount of fluid is seen in the subcutaneous bursa.

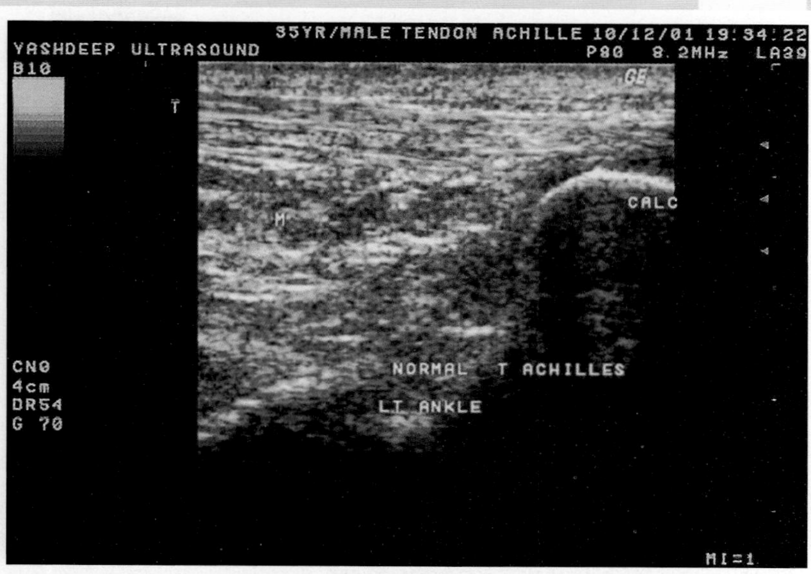

FIGURE 13.675: **Normal tendon insertion over the calcaneum.** HRSG shows insertion of the tendon over the superior surface of the calcanium.

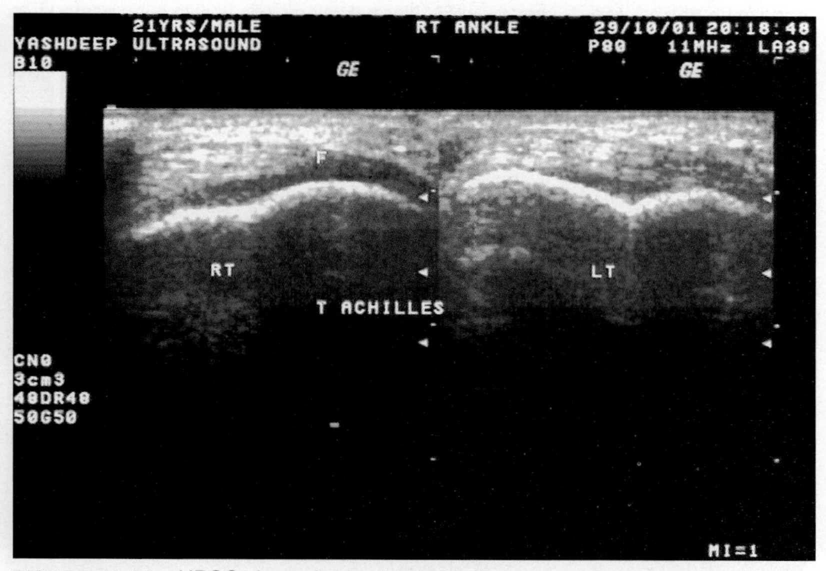

FIGURE 13.676: HRSG shows fluid collection in the retrocalcaneal bursa on Rt side. The fluid pocket is 4 mm in diameter. The Lt bursa is normal.

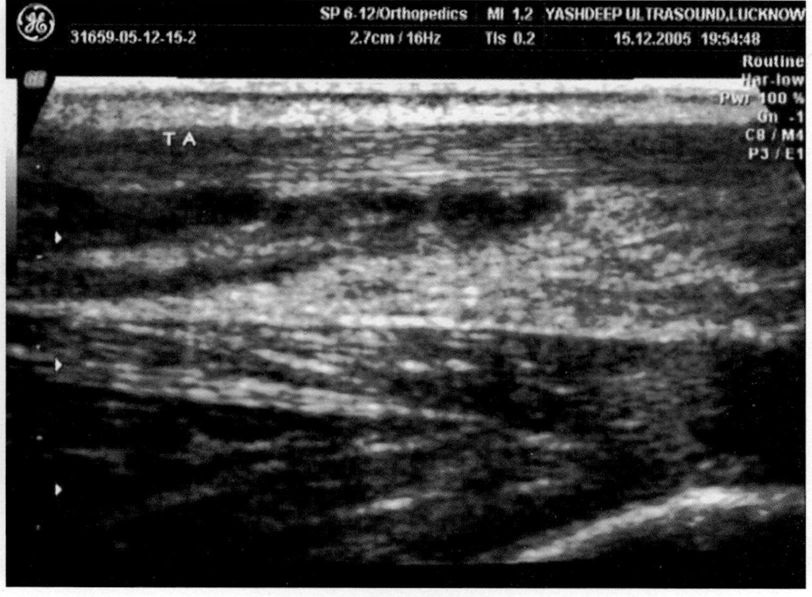

FIGURE 13.677

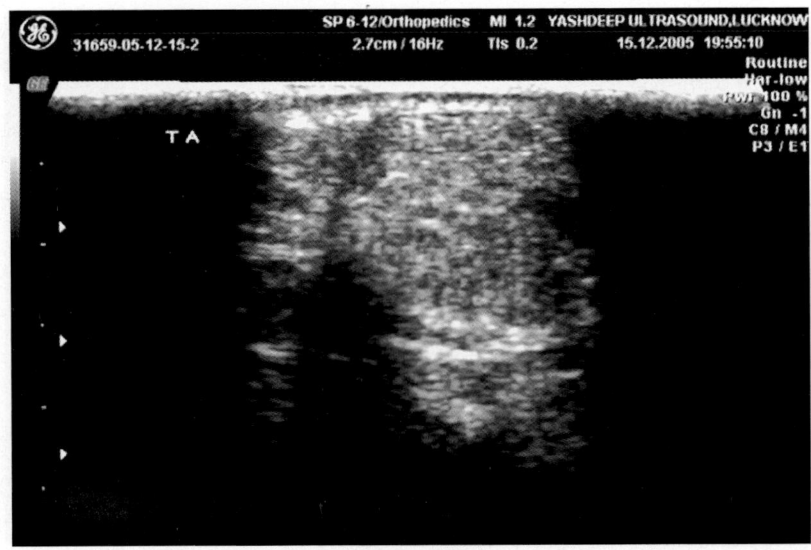

FIGURE 13.678

FIGURES 13.677 and 13.678: Normal tendon in TA and LS. HRSG shows normal tendon Achilles in Transverse axis and longitudinal axis. It looks like as a hypoechoic band in LS and as an oval shadow in TA. Normal calf muscle is also seen posterior to the tendon.

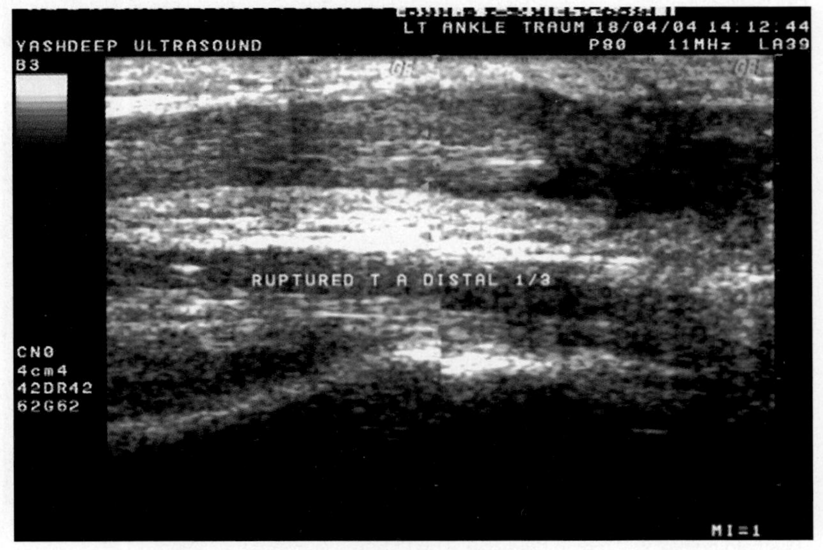

FIGURE 13.679: Partial tear of tendon Achilles. HRSG shows partial tear of the tendon Achilles in its distal 1/3rd part. Disruption of the tendon fibers is seen with hemorrhagic collection.

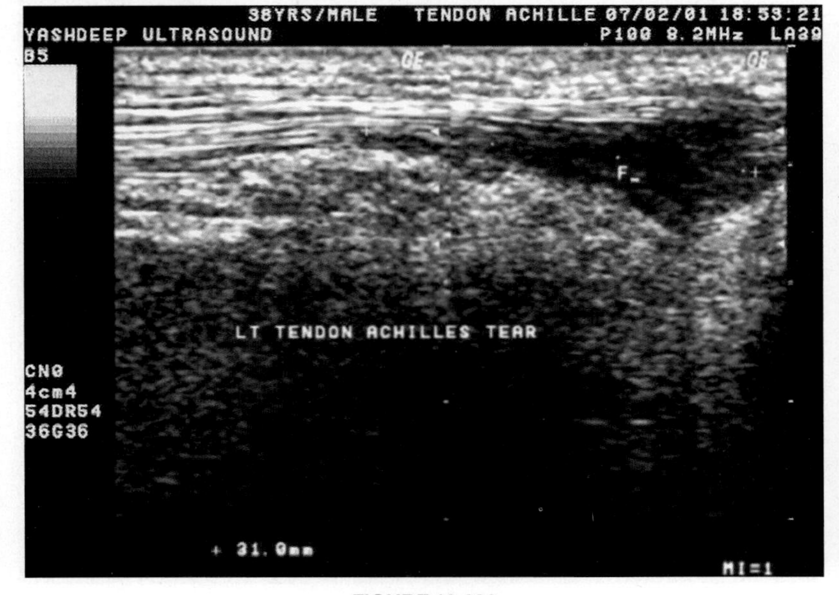

FIGURE 13.680

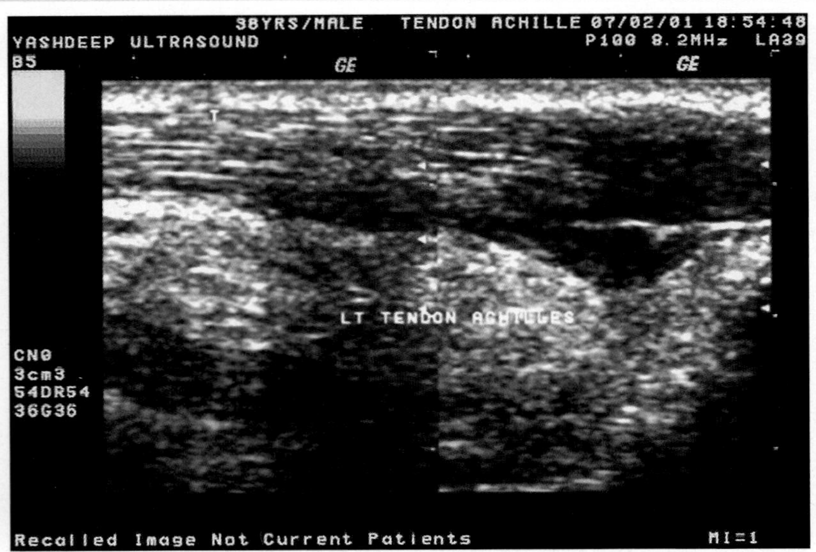

FIGURE 13.681

FIGURES 13.680 and 13.681: Complete rupture of tendon Achilles. HRSG shows complete disruption of tendon from its insertion side. Muscle fibers are disrupted. They are torn and retracted. Associated hematoma formation is also seen posterior to the tendon rupture.

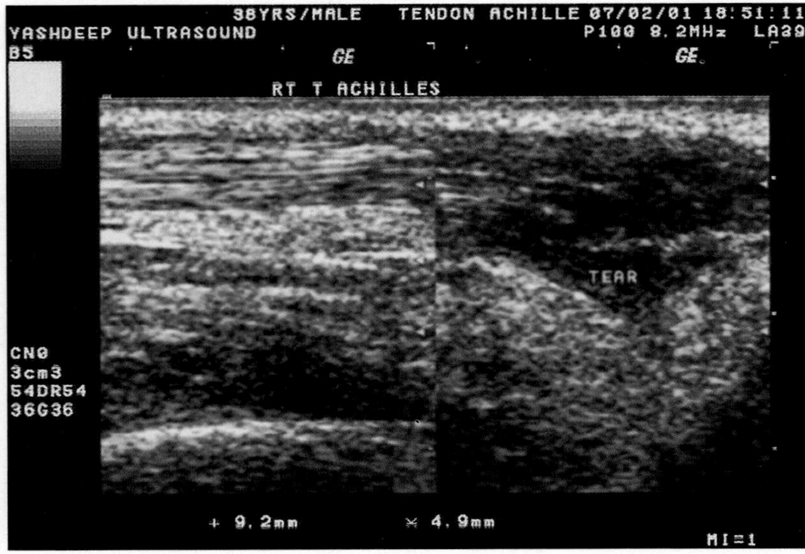

FIGURE 13.682: Acute traumatic rupture of Achilles tendon. HRSG shows a big hypoechoic tear in the distal part of Achilles tendon. There is also evidence of low-level echo collection seen in subcutaneous plane suggestive of hematoma formation. The muscle tear is also seen in the muscle belly.

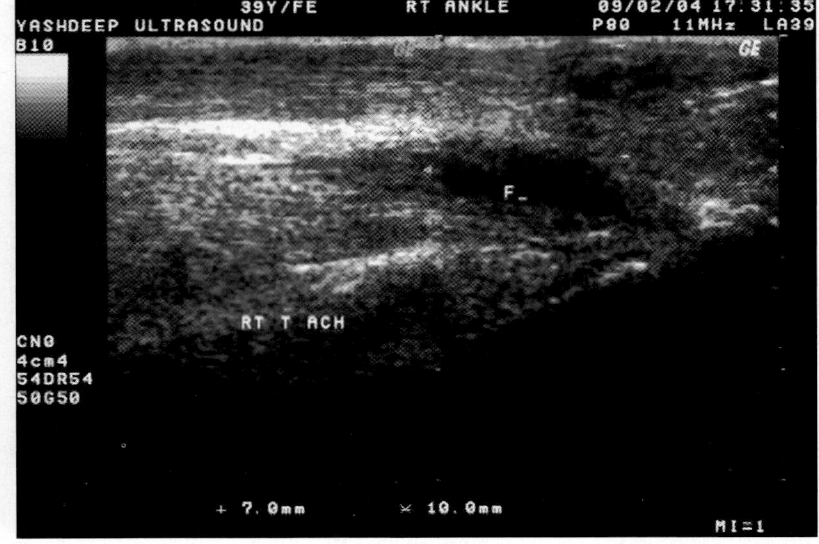

FIGURE 13.683

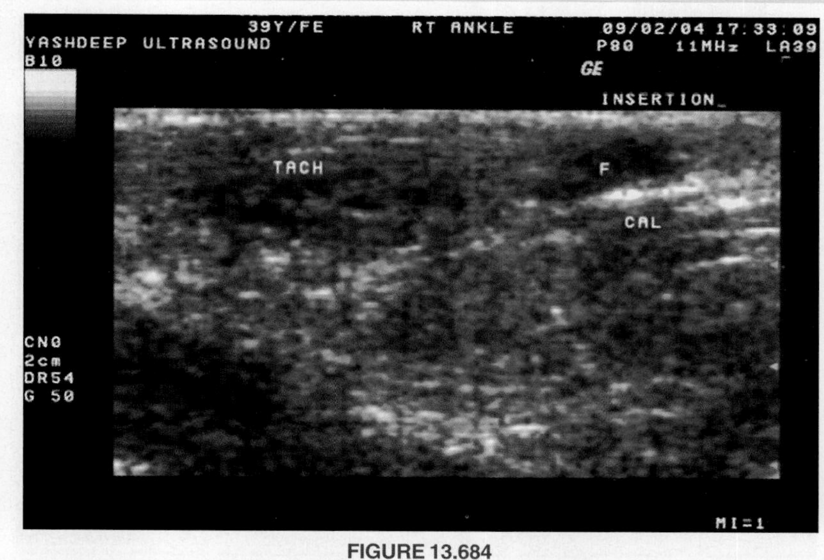

FIGURE 13.684

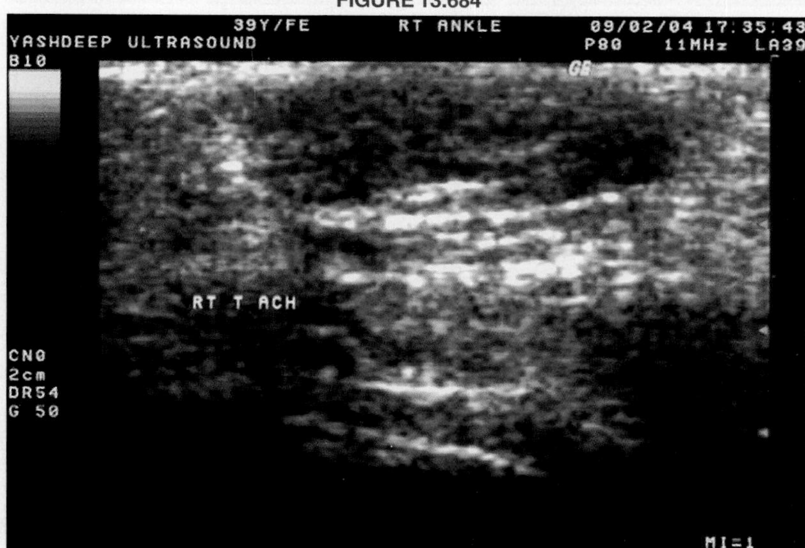

FIGURE 13.685

FIGURES 13.683 to 13.685: Partial rupture of the tendon Achilles. HRSG shows partial rupture of the tendon Achilles from its insertion side. The posterior fibers of the tendon are torned. However, anterior fibers are seen intact. Fluid is also seen in the retrocalcaneal bursa. Hemorrhagic collection is also seen suggestive of hematoma formation.

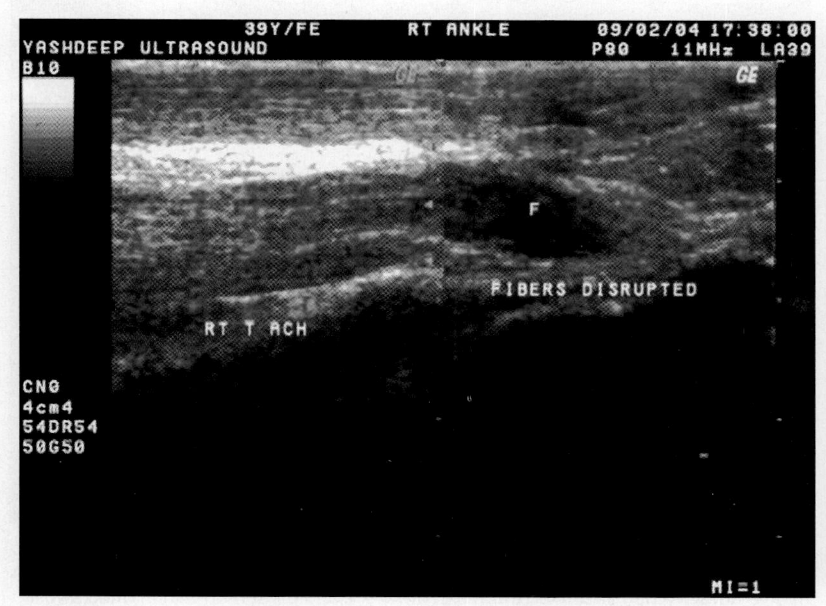

FIGURE 13.686: There is evidence of fluid collection seen in retrocalcaneal bursa. Low-level echoes are seen in the bursal fluid suggestive of hemorrhagic retrocalcaneal bursitis.

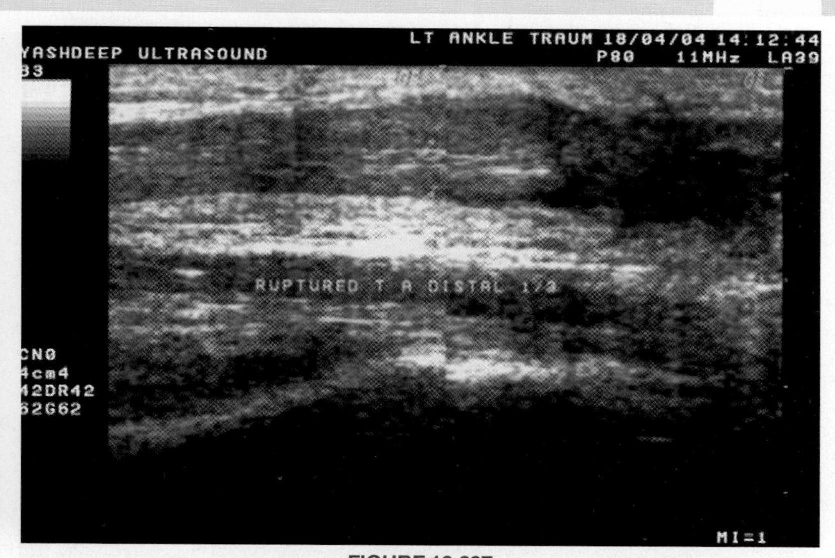

FIGURE 13.687

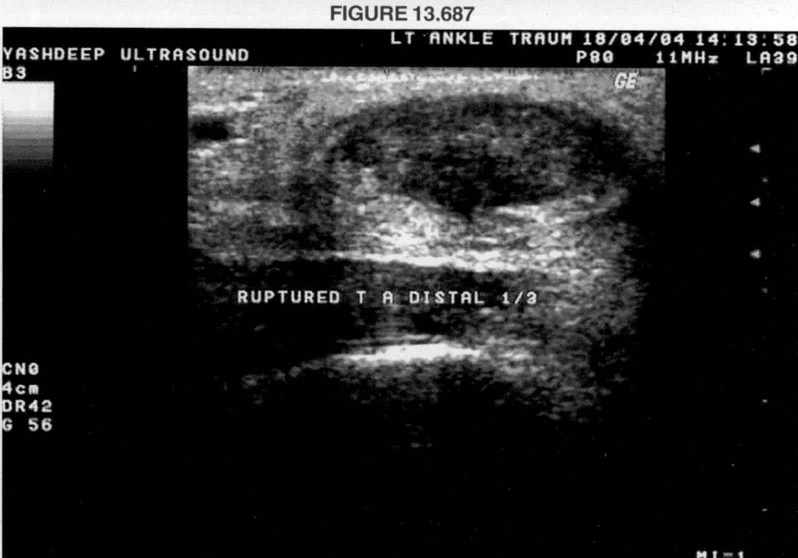

FIGURE 13.688

FIGURES 13.687 and 13.688: Acute rupture of the tendon Achilles. HRSG shows acute rupture of tendon Achilles in its distal 1/3rd part. The tendon is seen avulsed from its insertion site. Normal tendon fibers are lost. Big hemorrhagic collection is seen at the site of insertion. Echogenic shadows are also seen in it suggestive of organization.

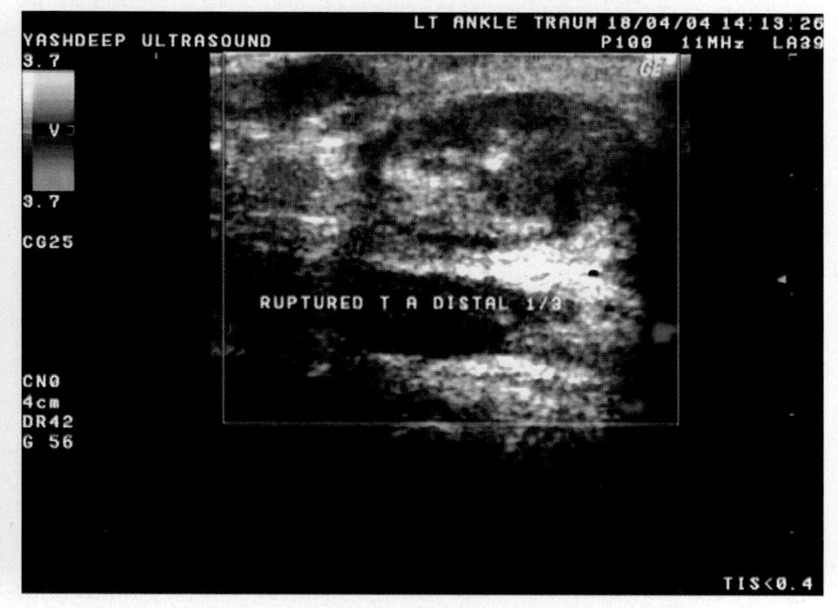

FIGURE 13.689: On color flow imaging no flow is seen in the tendon.

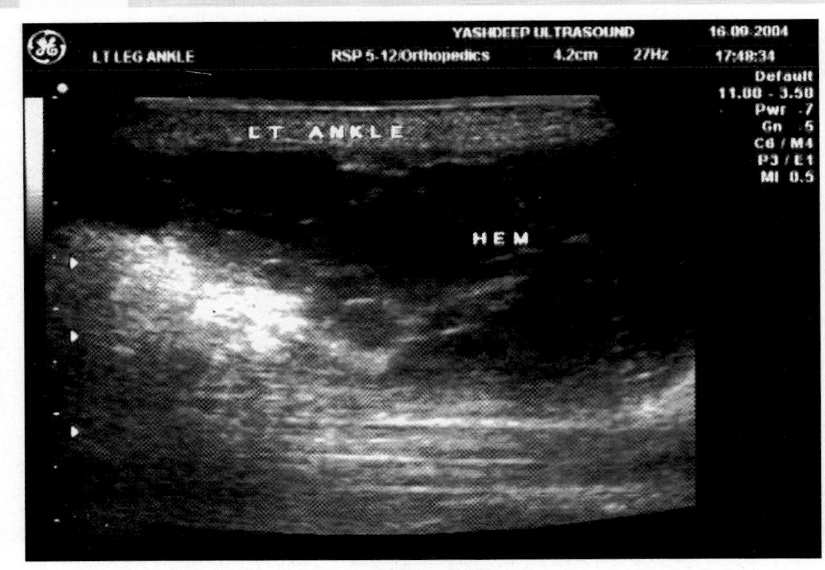

FIGURE 13.690

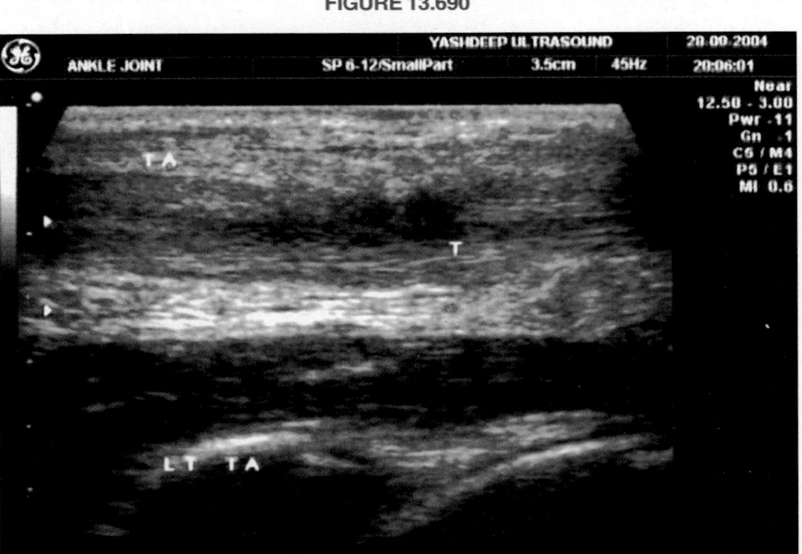

FIGURE 13.691

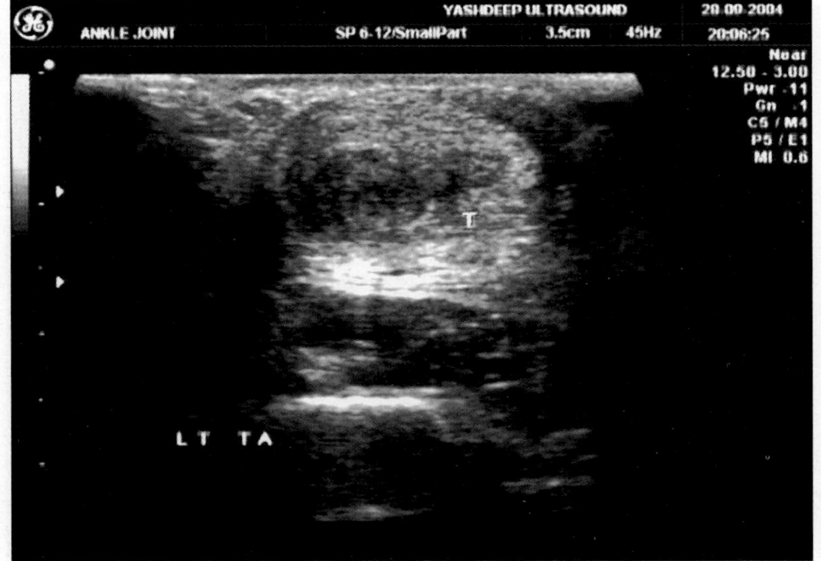

FIGURE 13.692

FIGURES 13.690 to 13.692: Acute ankle trauma with rupture of tendon Achilles. A patient presented with acute rotational injury to the ankle. There is evidence of big hematoma formation seen in the ankle. The tendon Achilles is seen markedly edematous. A big hypoechoic cleft is seen in the tendon suggestive of intrasubstance rupture of the tendon. However, anterior and posterior fibers are intact.

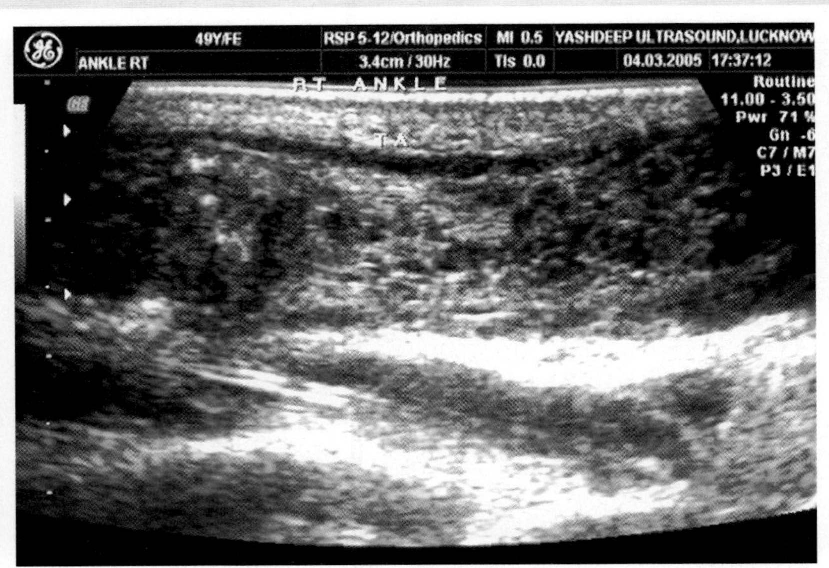

FIGURE 13.693

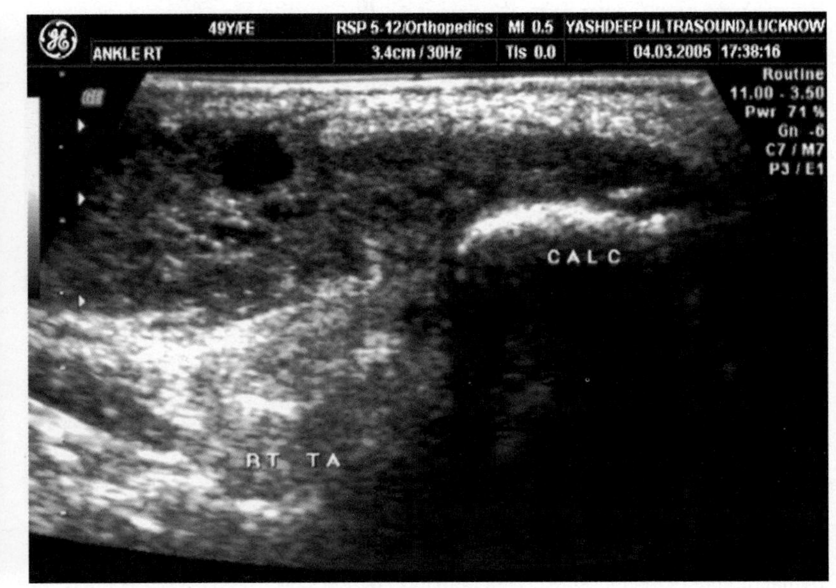

FIGURE 13.694

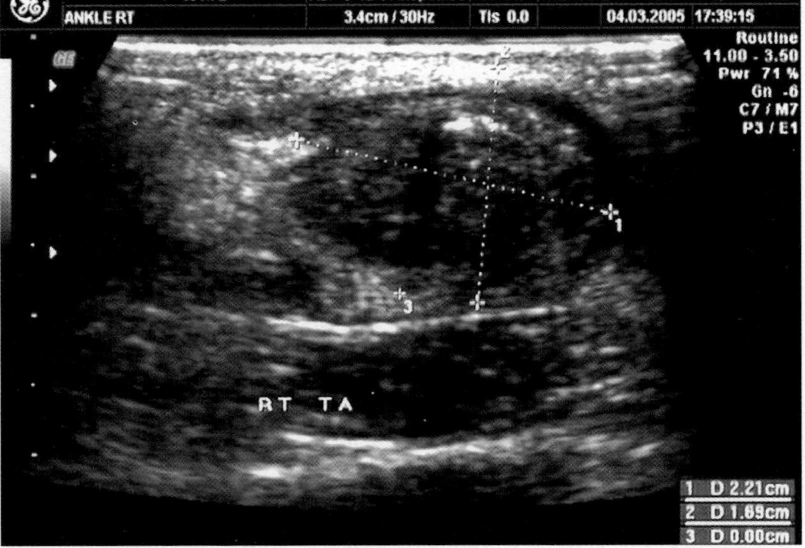

FIGURE 13.695

FIGURES 13.693 to 13.695: Rt ankle trauma with rupture of the tendon. A middle aged woman sustained trauma to the ankle and presented with marked swollen and acute pain in the Rt ankle. HRSG shows grossly edematous hypoechoic swollen tendon Achilles with multiple hypoechoic areas seen in the tendon suggestive of hemorrhagic pockets.

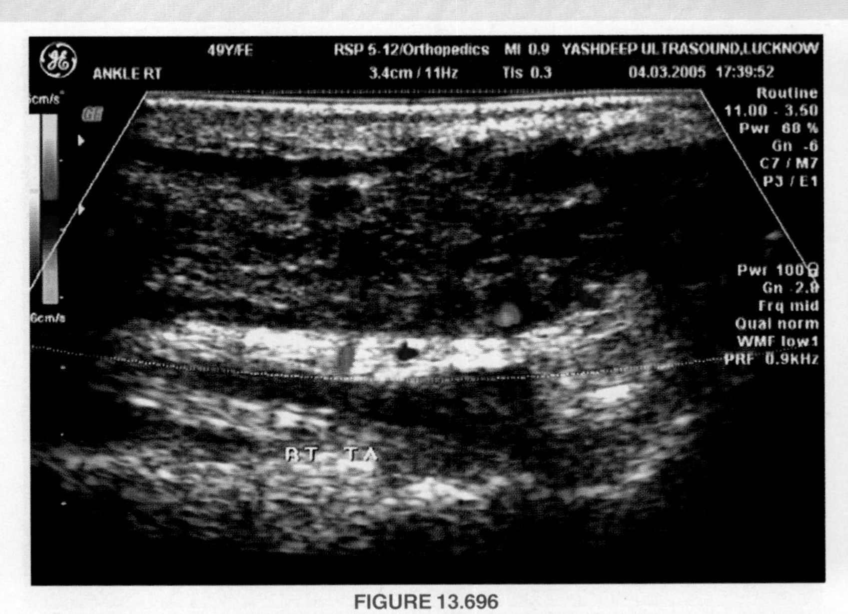

FIGURE 13.696

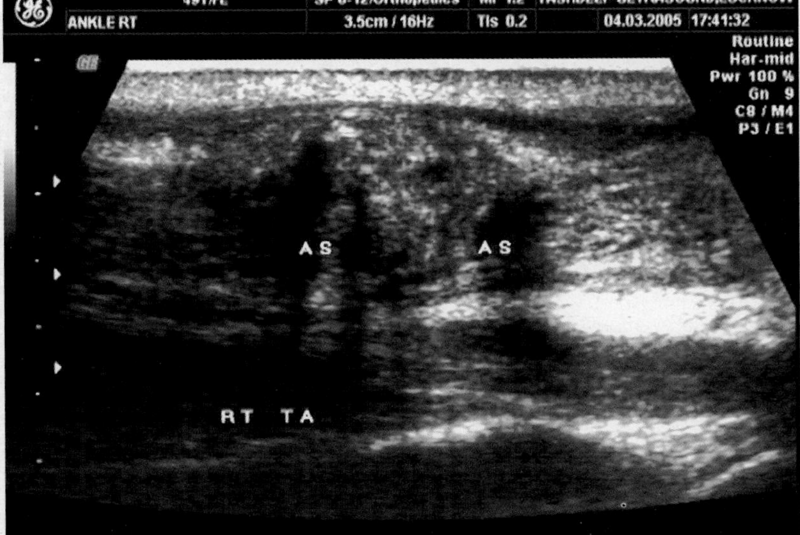

FIGURE 13.697

FIGURES 13.696 and 13.697: The tendon shows intrasubstance tear with big hematoma formation in its distal part. Echogenic calcified specks are also seen in the hematoma. Which are accompanied with acoustic shadowing. **On color flow imaging** no flow is seen in the rupture tendon. However, few vessels are seen around the inflamed subcutaneous tissue.

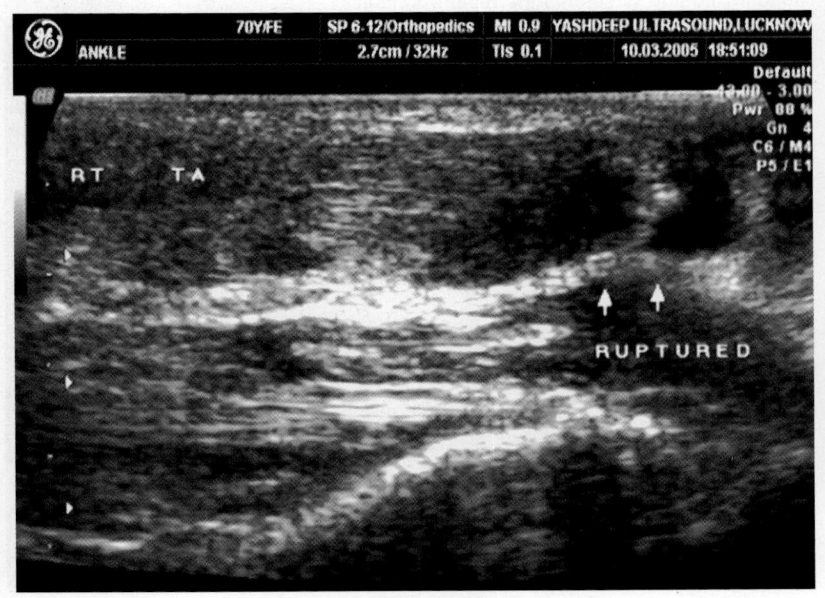

FIGURE 13.698

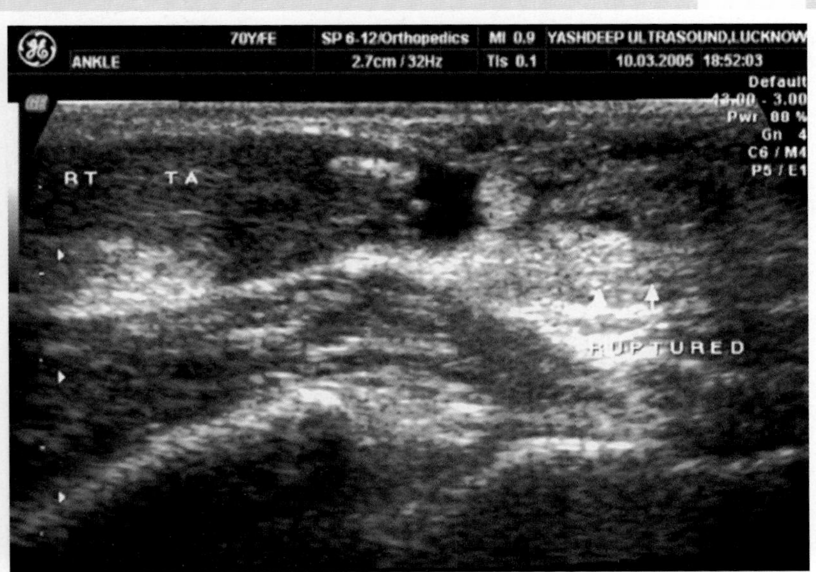

FIGURE 13.699

FIGURES 13.698 and 13.699: Acute rupture of tendon Achilles. HRSG shows edematous hypoechoic swollen Rt tendon Achilles in old lady. Disruptions of the tendon fibers are seen in its distal part.

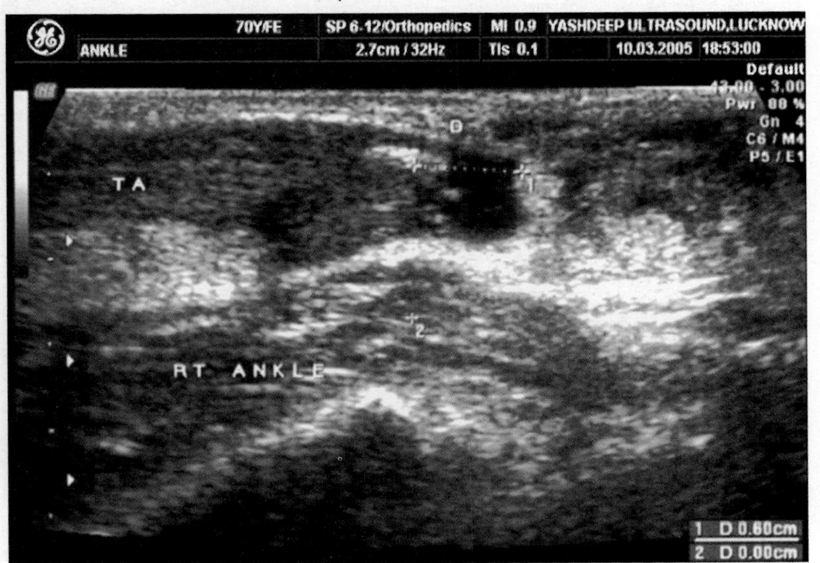

FIGURE 13.700

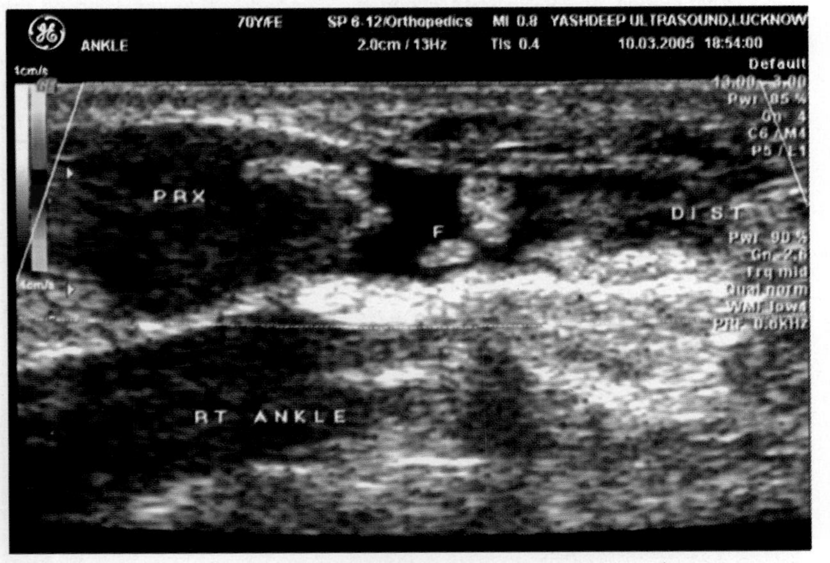

FIGURE 13.701

FIGURES 13.700 and 13.701: There is evidence of loss of tendon fibers seen suggestive of acute rupture with hemorrhagic pockets. Marked soft tissue edema is seen. **On color flow imaging** no flow seen in the tendon.

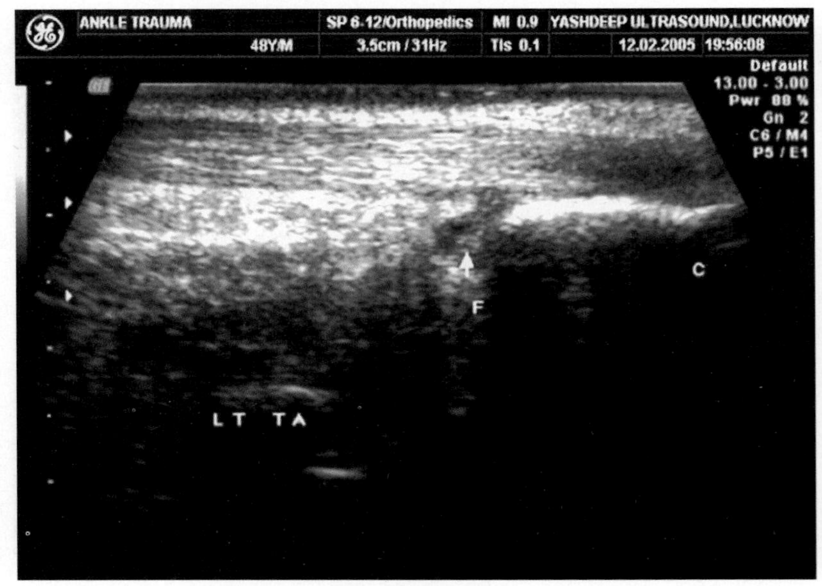

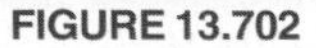
FIGURE 13.702

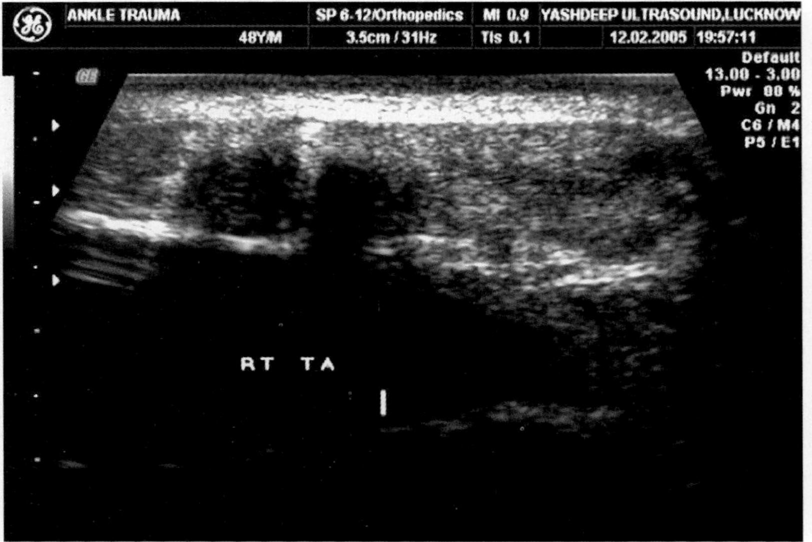

FIGURE 13.703

FIGURES 13.702 and 13.703: Chronic calcific tendonitis with rupture of the tendon. A patient presented with painful ankle movements. HRSG shows edematous hypoechoic tendon Achilles. Dense echogenic calcified specks are seen on the anterior surface of the tendon. They are accompanied with acoustic shadowing. Another small-calcified shadow is seen on the posterior aspect.

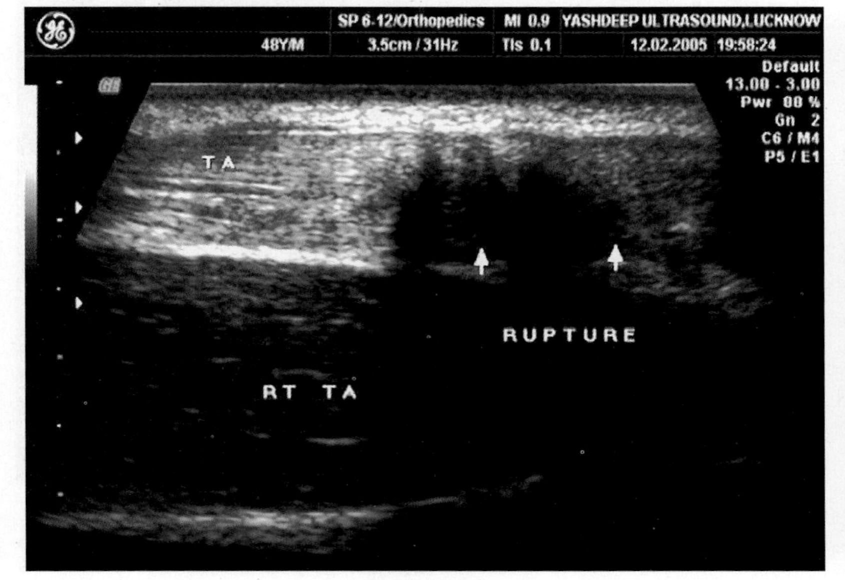

FIGURE 13.704

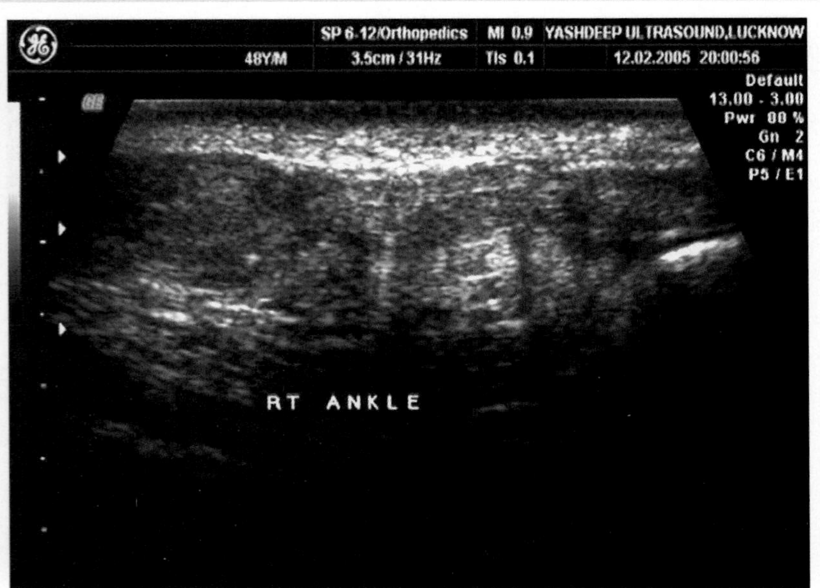

FIGURE 13.705

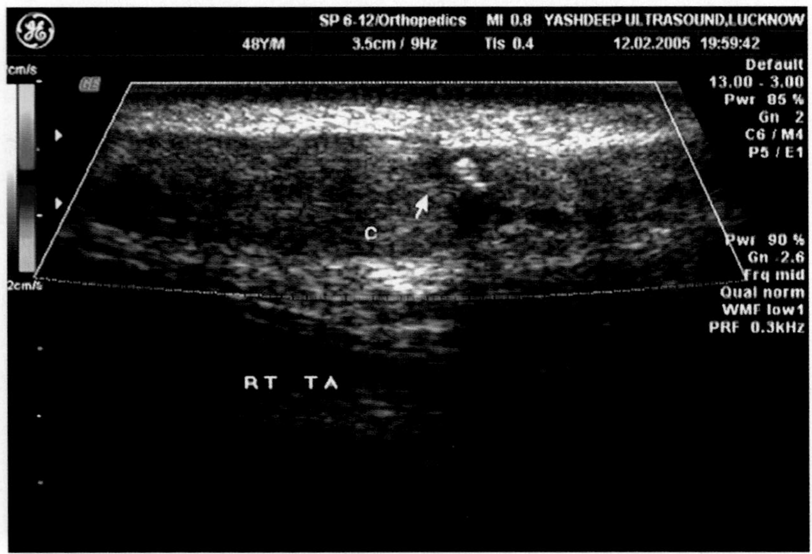

FIGURE 13.706

FIGURES 13.704 to 13.706: There is also evidence of hypoechoic cleft seen in the mid part of the tendon. Loss of tendon fibers are seen on the posterior side suggestive of partial tear of the tendon. A thin sleeve of tendon fibers is seen attached on the superior side of the tendon. The tendon is grossly inhomogeneous. **On color flow imaging** no flow is seen in the rupture tendon.

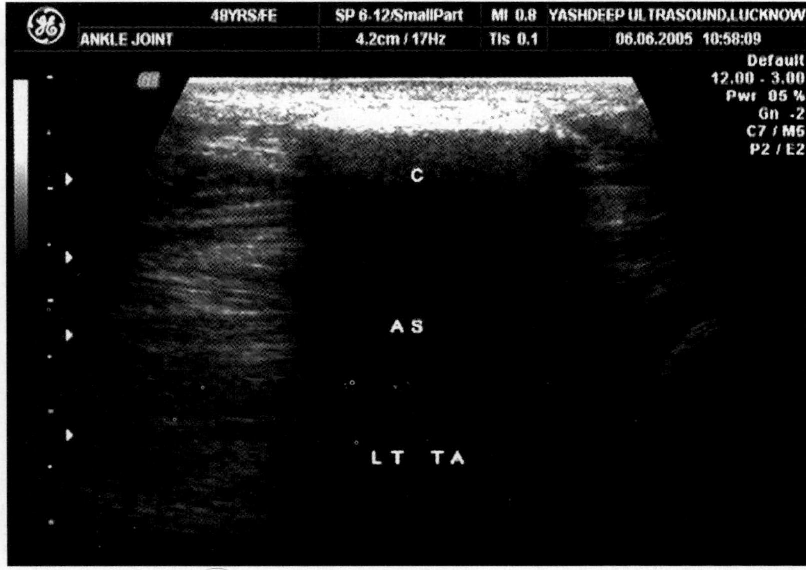

FIGURE 13.707

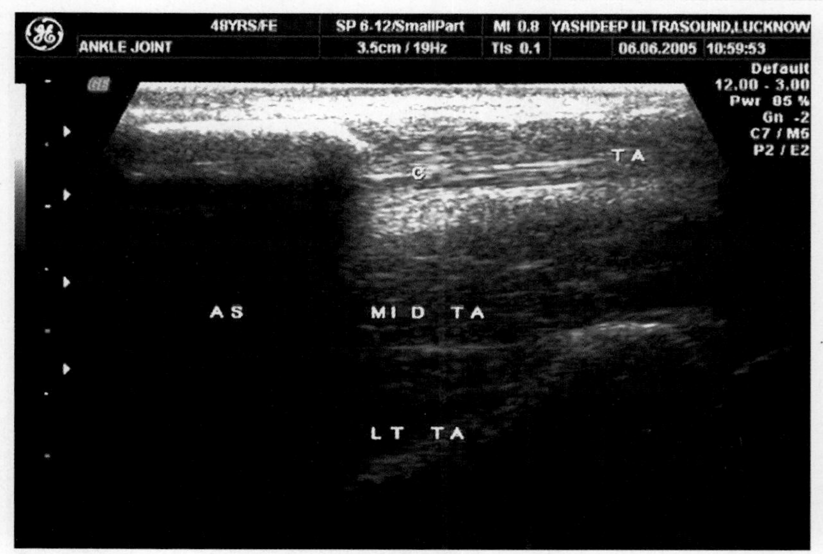

FIGURE 13.708

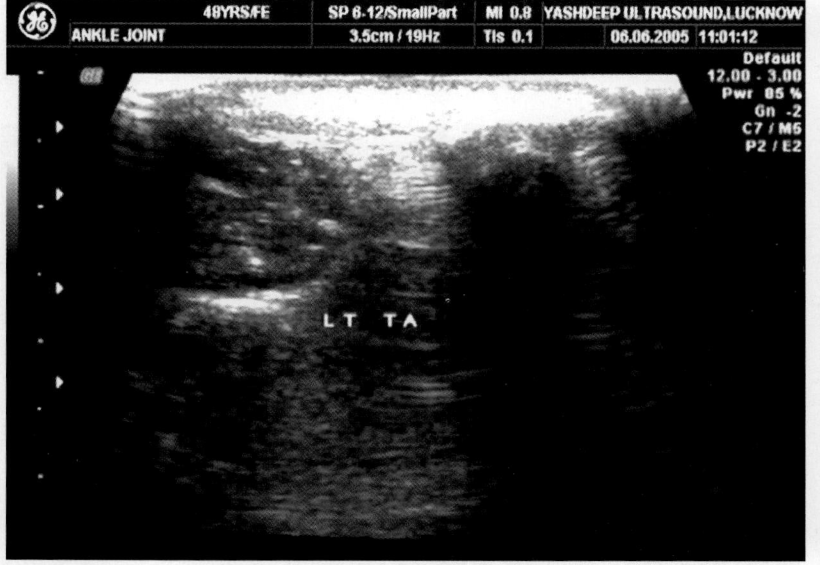

FIGURE 13.709

FIGURES 13.707 to 13.709: Chronic calcific tendonitis. HRSG shows a dense echogenic calcified plaque seen on the anterior side of the distal tendon Achilles. The calcified plaque is seen as a cord like shadow and accompanied with dense acoustic shadowing. The tendon is seen torn in its mid part.

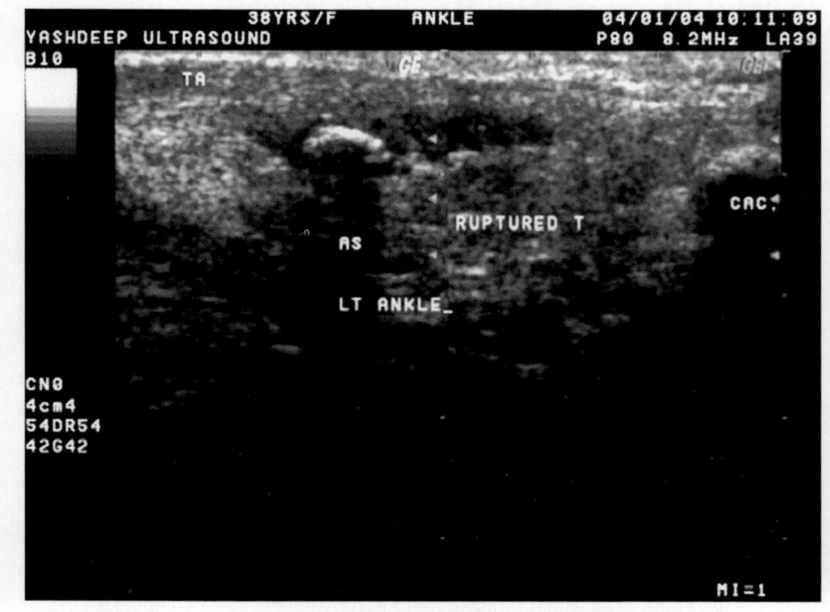

FIGURE 13.710

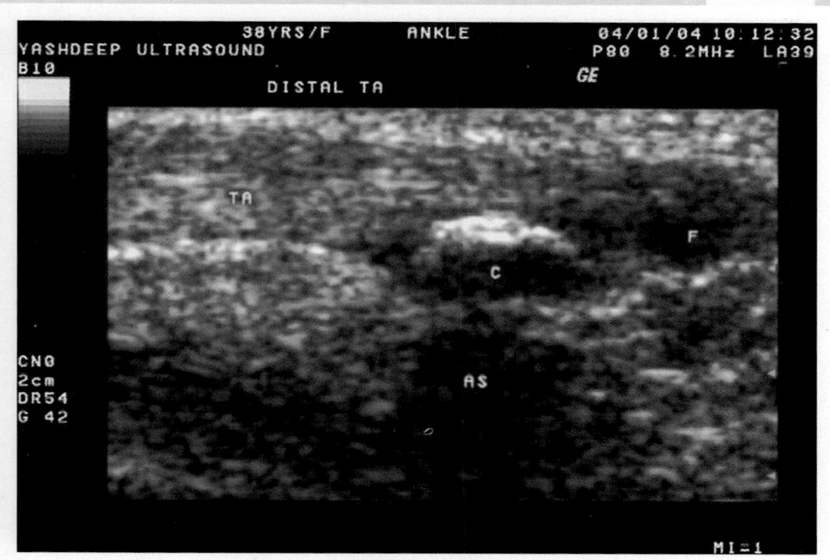

FIGURE 13.711

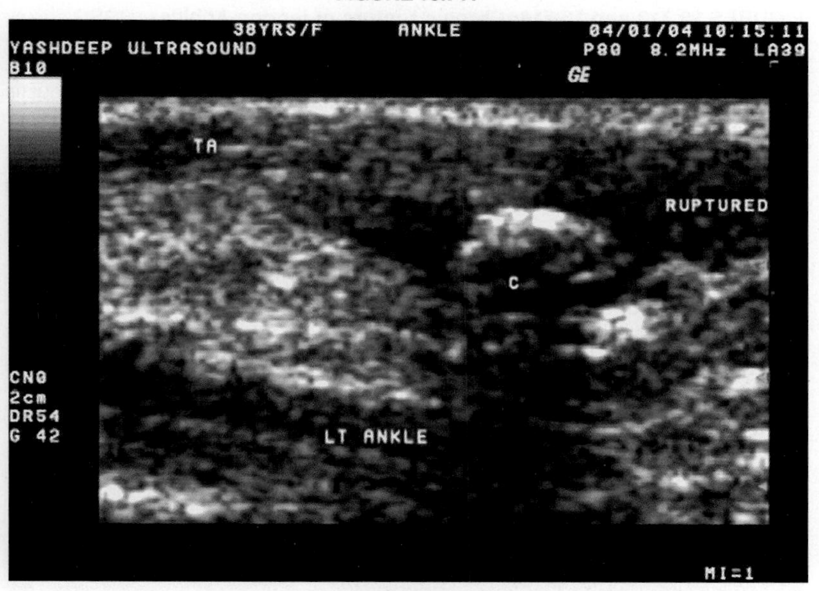

FIGURE 13.712

FIGURES 13.710 to 13.712: Rupture tendon with calcified plaque. HRSG shows rupture of the tendon Achilles in its distal part. Dense echogenic calcified plaque shadow seen in the substance of the tendon. It is accompanied with dense acoustic shadowing. Anterior fibers of the tendon are intact. However, the tendon is seen adhered to the bone by calcified plaque.

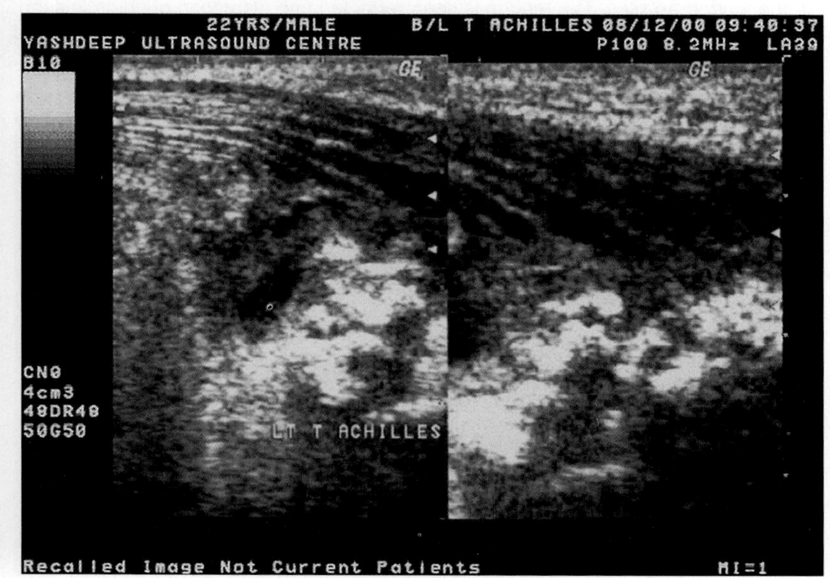

FIGURE 13.713

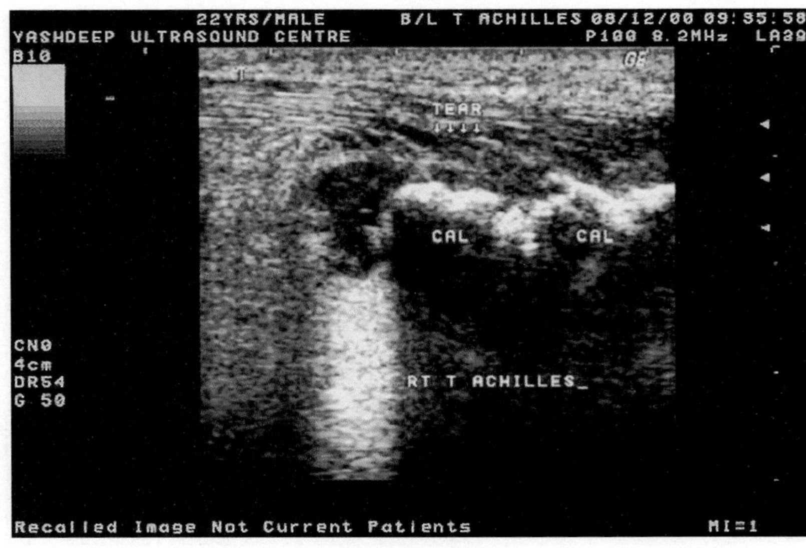

FIGURE 13.714

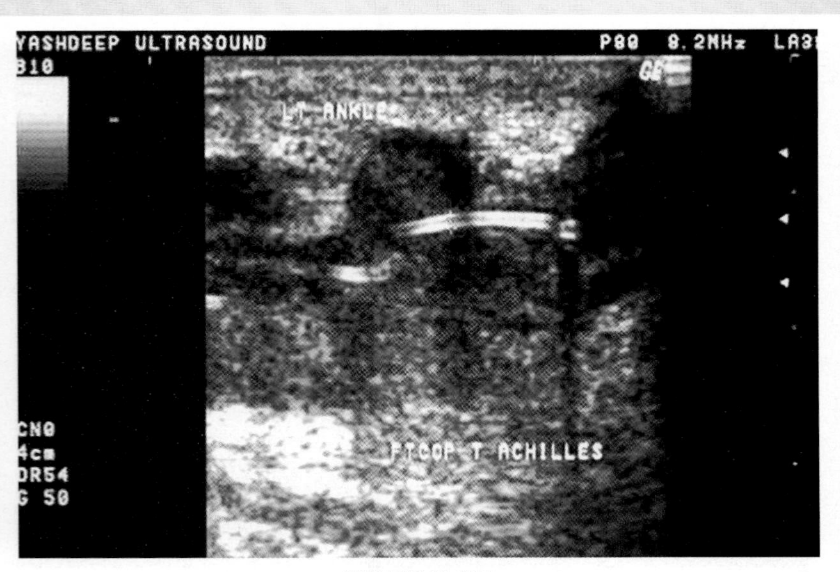

FIGURE 13.717

FIGURES 13.713 and 13.714: Chronic rupture of tendon Achilles. HRSG shows chronic rupture of tendon Achilles from its insertion side. Tendon Achilles fibers are seen dispersed. Multiple calcified echogenic specks are seen between the tendon fibers and calcaneum suggestive of calcification. It is accompanied with acoustic shadowing. The tendon is seen adhered with calcaneum with calcified plaques.

FIGURES 13.716 to and 13.717: Postsurgical rupture of the tendon. A patient underwent tendon repair surgery for the rupture tendon Achilles. However, the tendon got snapped from its suture site. Echogenic suture is seen in the ruptured tendon substance as bright thread line. The ruptured suture ends are seen lying apart.

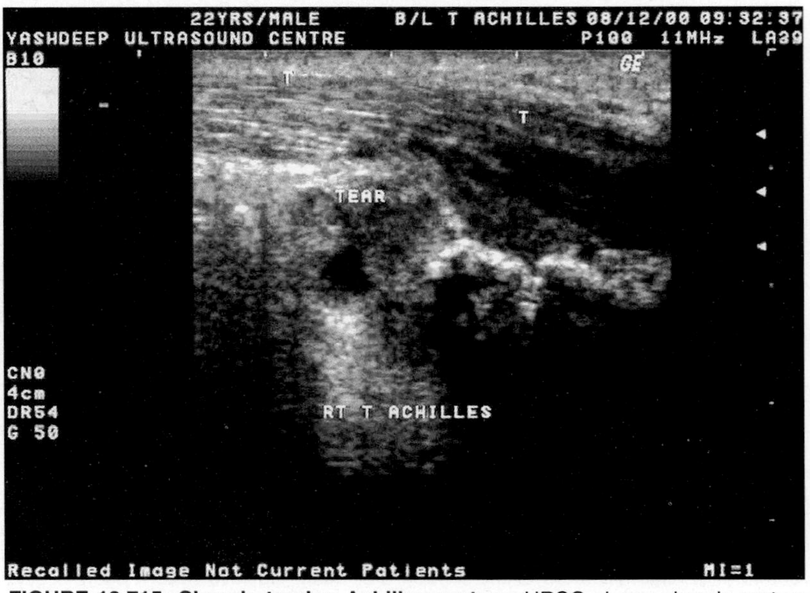

FIGURE 13.715: Chronic tendon Achilles rupture. HRSG shows chronic rupture of the tendon Achilles. The tendon is seen retracted and disruption of fibers is seen. Multiple calcifications are seen posterior to the tendon and it is adhered with calcified plaques.

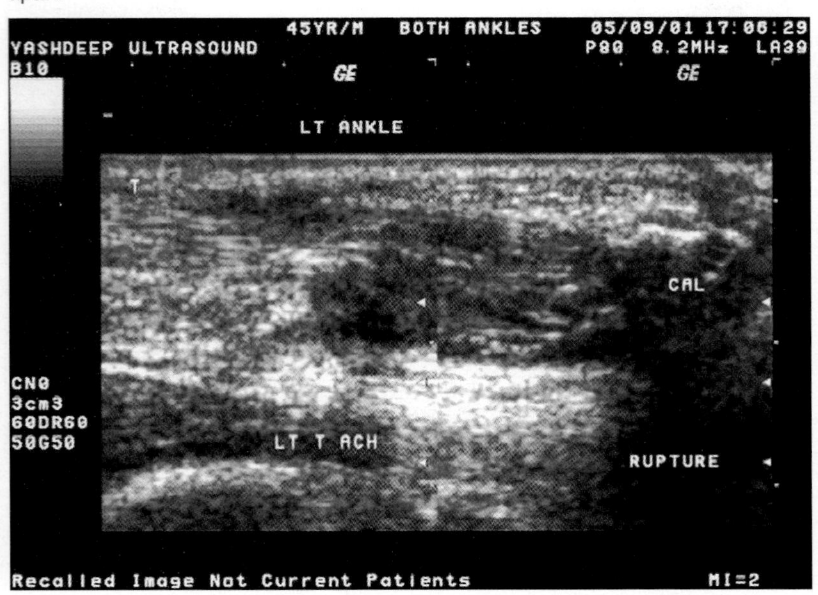

FIGURE 13.718: Chronic rupture of the tendon. HRSG shows chronic rupture of the tendon Achilles in its distal part. Tendon fibers are disrupted and torned. Tendon is edematous. It is retracted from its insertion site. Associated organized hematoma is also seen.

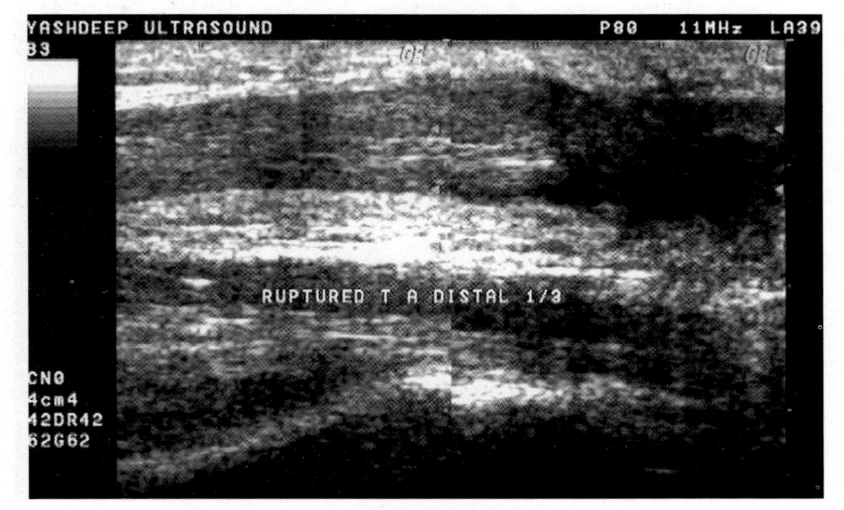

FIGURE 13.716

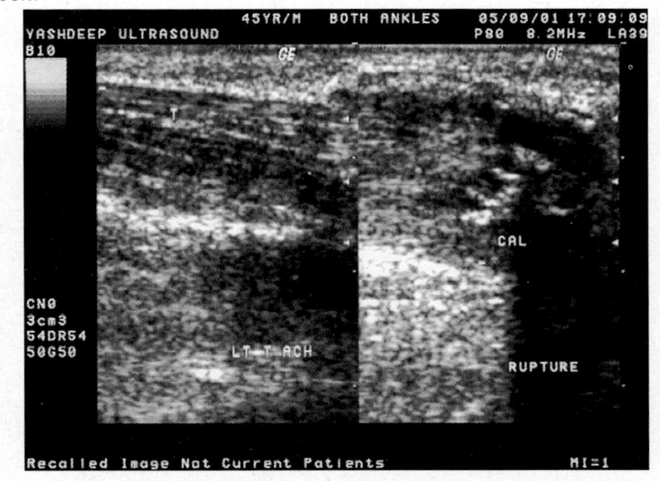

FIGURE 13.719: HRSG shows edematous hypoechoic tendon in the same case with organized hematoma. Multiple calcified specks are seen in the hematoma.

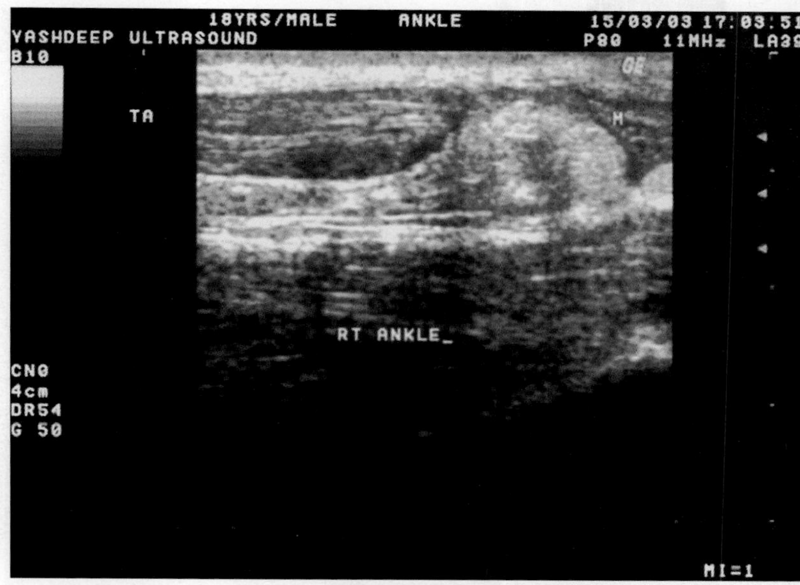

FIGURE 13.720

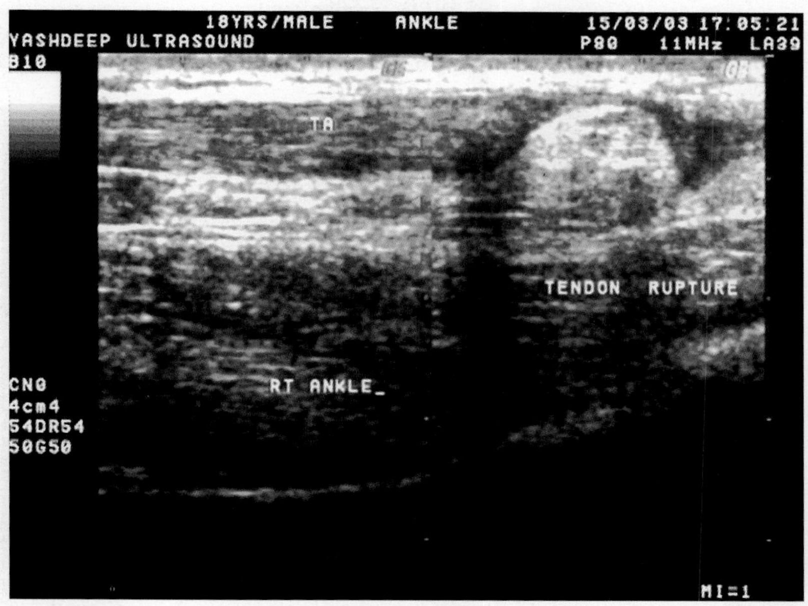

FIGURE 13.721

FIGURES 13.720 and 13.721: Acute tendon rupture. HRSG shows acute rupture of the tendon in its distal part. An echogenic hemorrhagic clot is seen at the rupture site. The tendon fibers are disrupted.

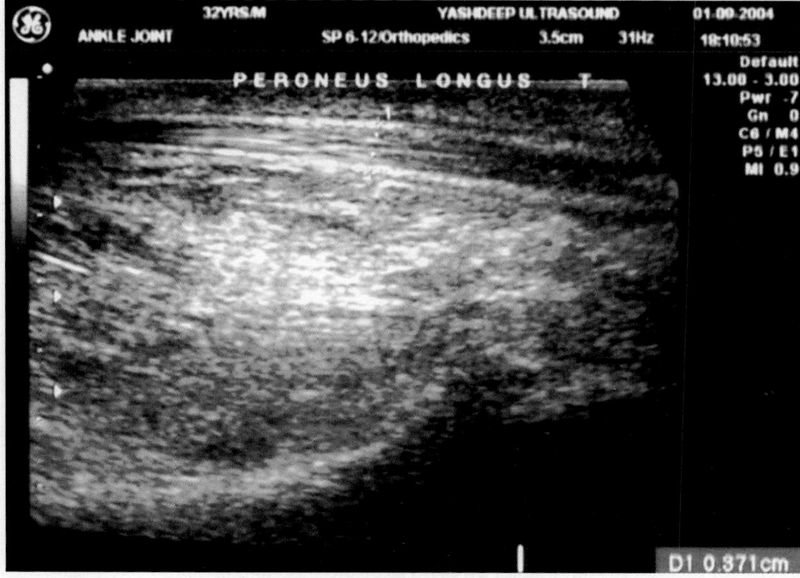

FIGURE 13.722

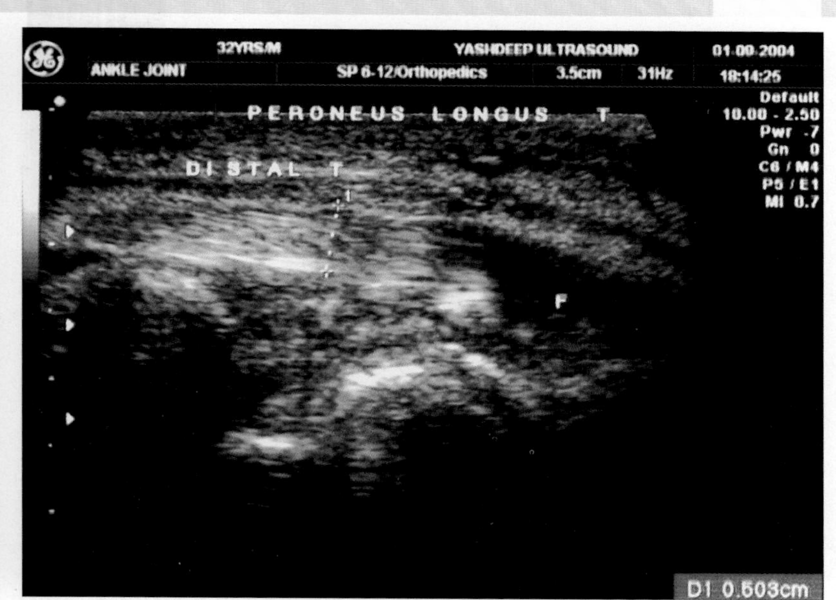

FIGURE 13.723

FIGURES 13.722 and 13.723: Peroneous longus tear. A patient sustained rotational injury to the ankle with marked swelling. HRSG shows edematous peroneous longus tendon. The tendon is seen torned in its distal 1/3rd part. Thinning of the tendon fibers is seen with fluid collection seen in the tendon sheath. Synovial membrane proliferation is also seen.

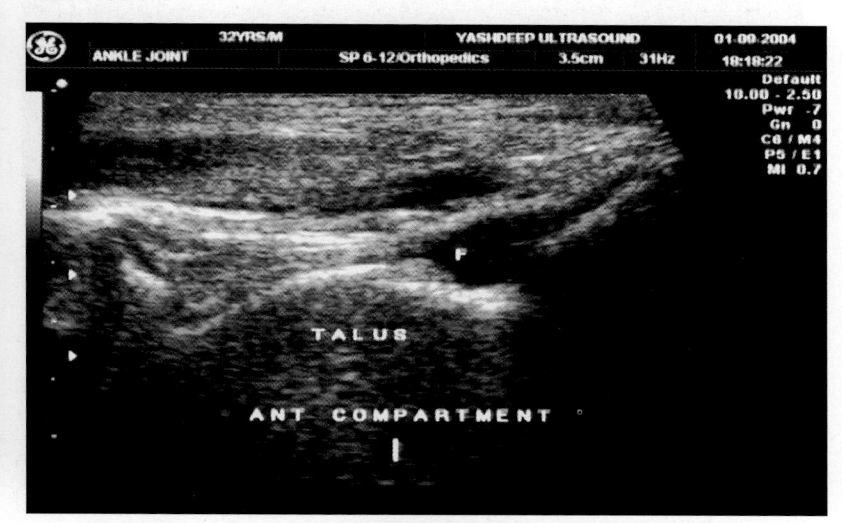

FIGURE 13.724

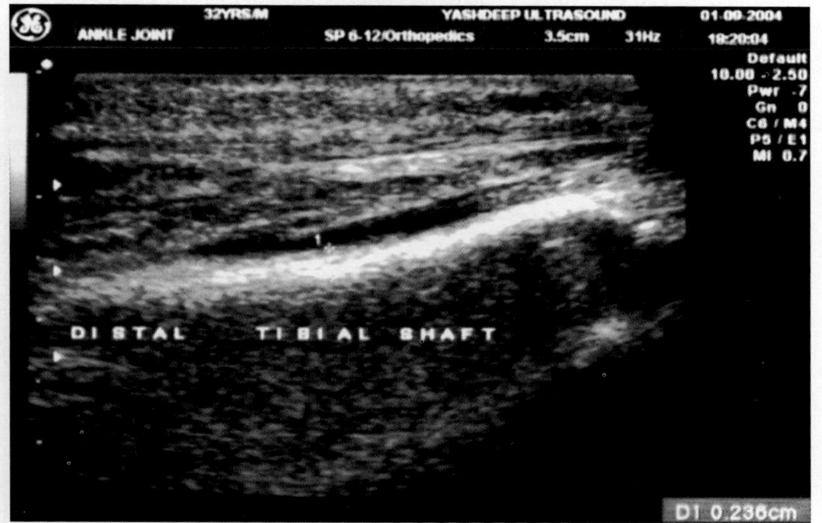

FIGURE 13.725

FIGURES 13.724 and 13.725: Tibialis anterior tendon rupture. HRSG shows partial tear of tibialis anterior tendon rupture at the tibiotailor canal. Fluid is also seen in the canal. Thinning of the tendon fibers is seen. Rupture of the muscle fibers is seen in the anterior compartment of the distal 1/3rd part of the leg.

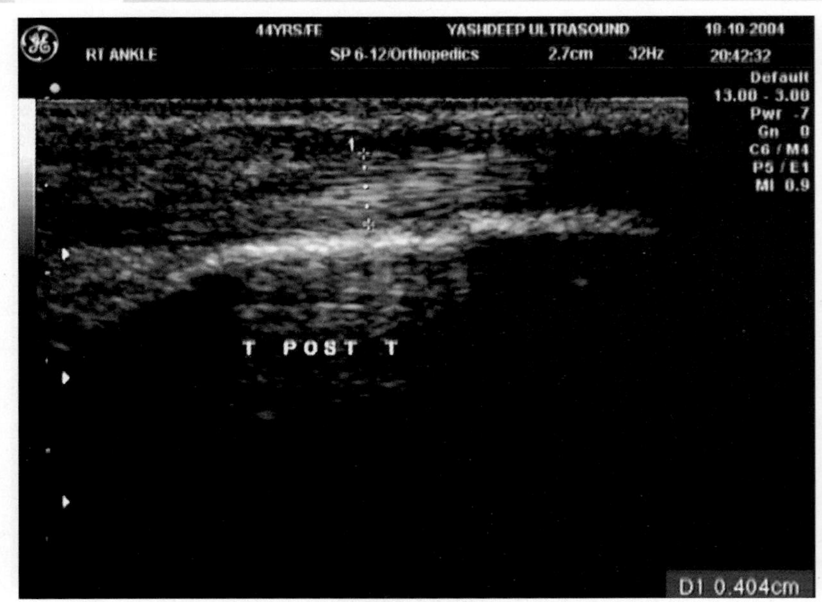

FIGURE 13.726

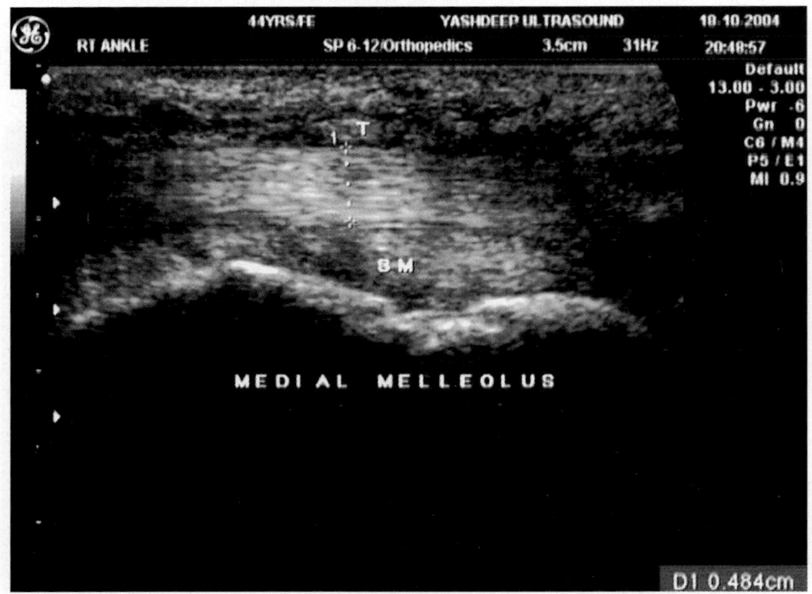

FIGURE 13.727

FIGURES 13.726 and 13.727: Posterior tibial tendonitis. HRSG shows posterior tibial tendon thickening with, tendon sheath edema. Marked synovial membrane thickening is also seen due to synovial proliferation.

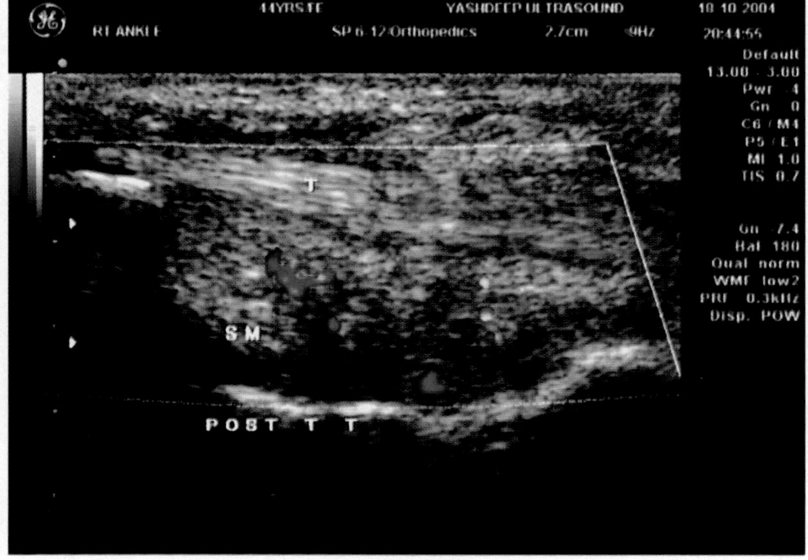

FIGURE 13.728

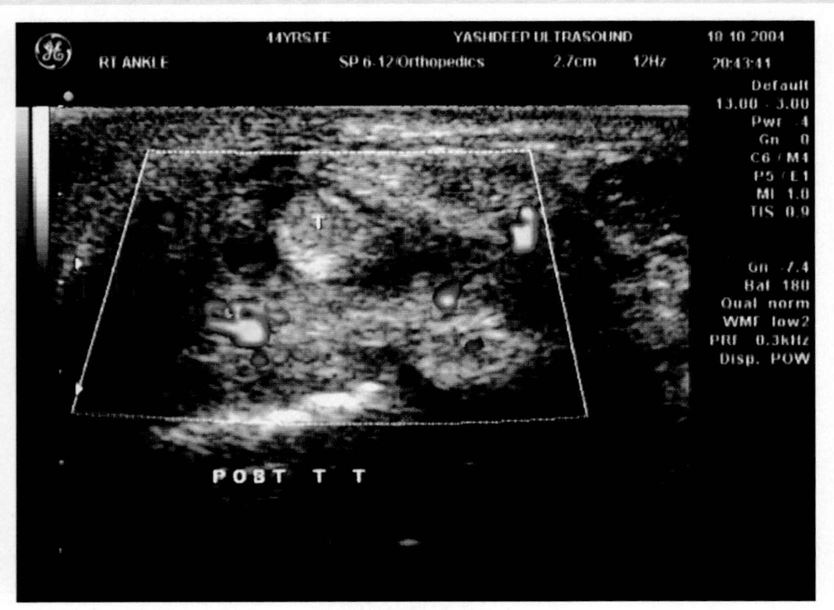

FIGURE 13.729

FIGURES 13.728 and 13.729: On color flow imaging increased flow is seen in the synovium. Multiple vessels are seen in the synovium suggestive of chronic Tenosynovitis.

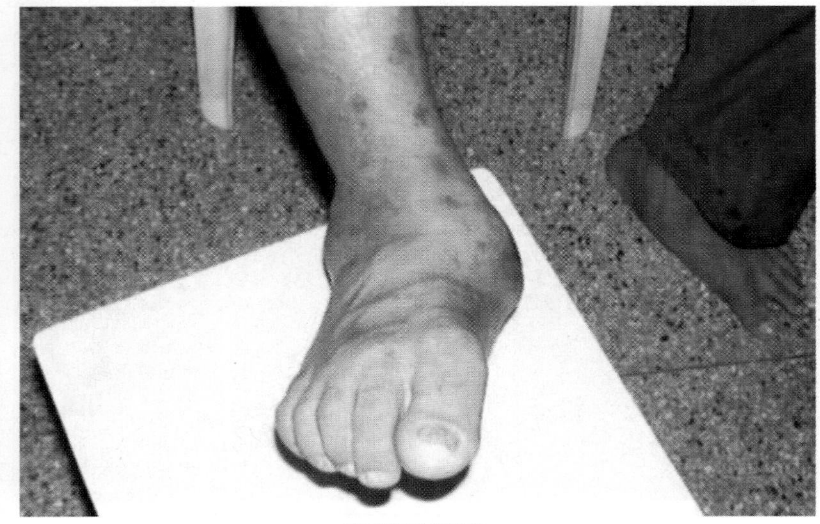

FIGURE 13.730

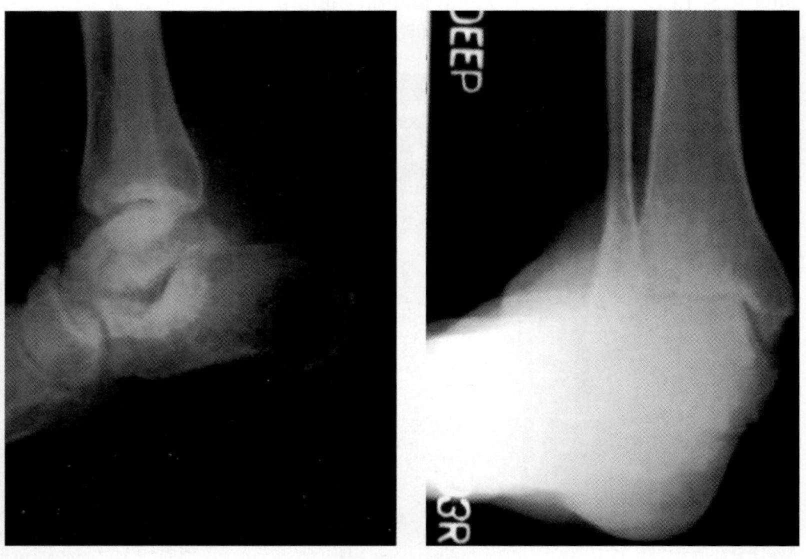

FIGURE 13.731

FIGURES 13.730 and 13.731: Chronic tenosynovitis of posterior tibial tendon. A patient presented with painful swelling over the medial malleolus. The plain X-ray shows normal bones with soft tissue edema.

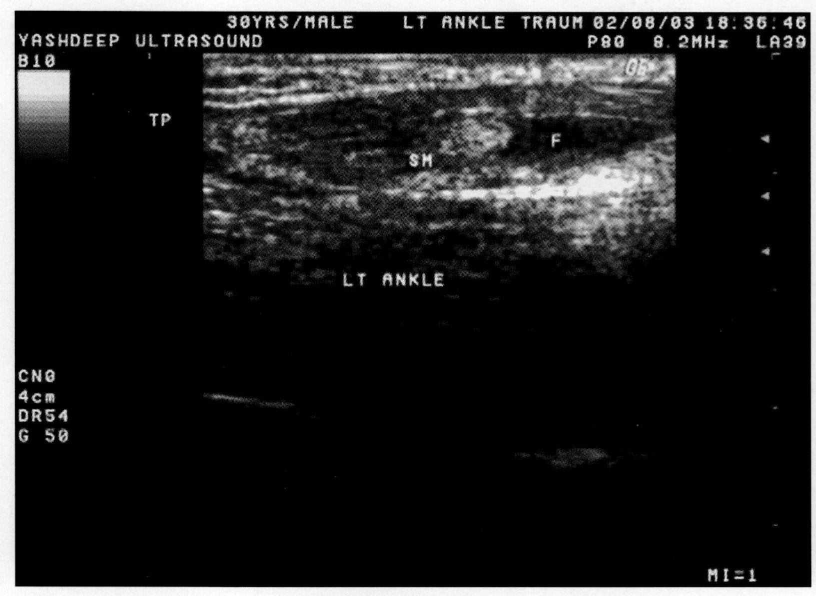

FIGURE 13.732

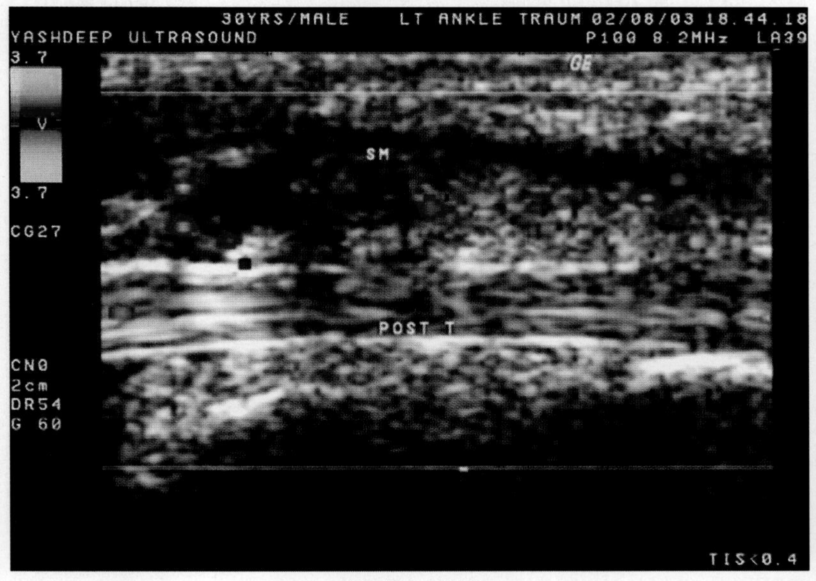

FIGURE 13.733

FIGURES 13.732 and 13.733: HRSG shows marked thickening of the posterior tibial tendon. Tendon sheath shows fluid collection. Synovial membrane proliferation is also suggestive of chronic Tenosynovitis.

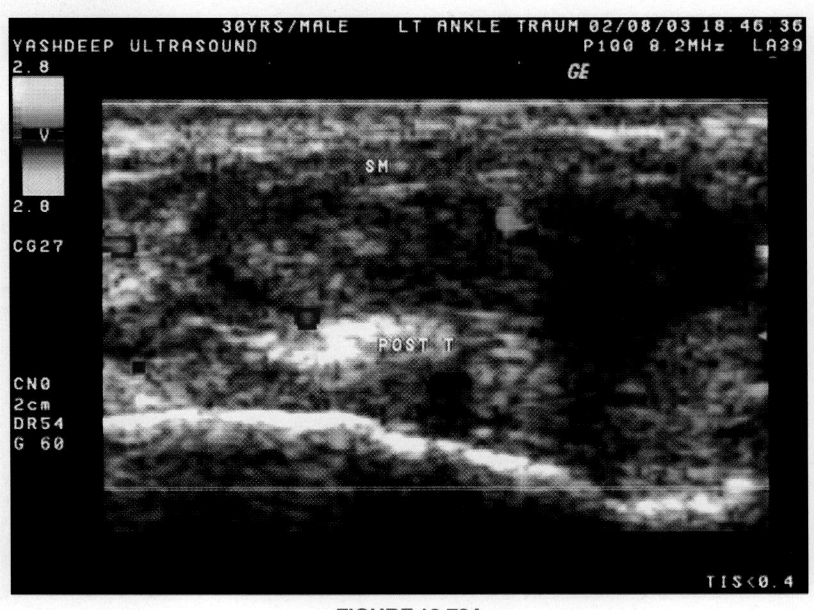

FIGURE 13.734

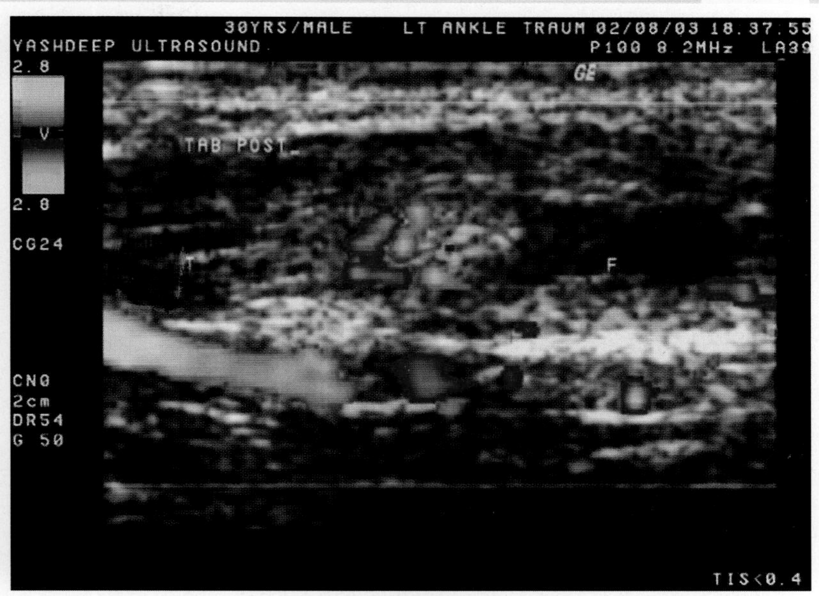

FIGURE 13.735

FIGURES 13.734 and 13.735: On color flow imaging increased flow is seen in the synovial membrane suggestive of chronic Tenosynovitis.

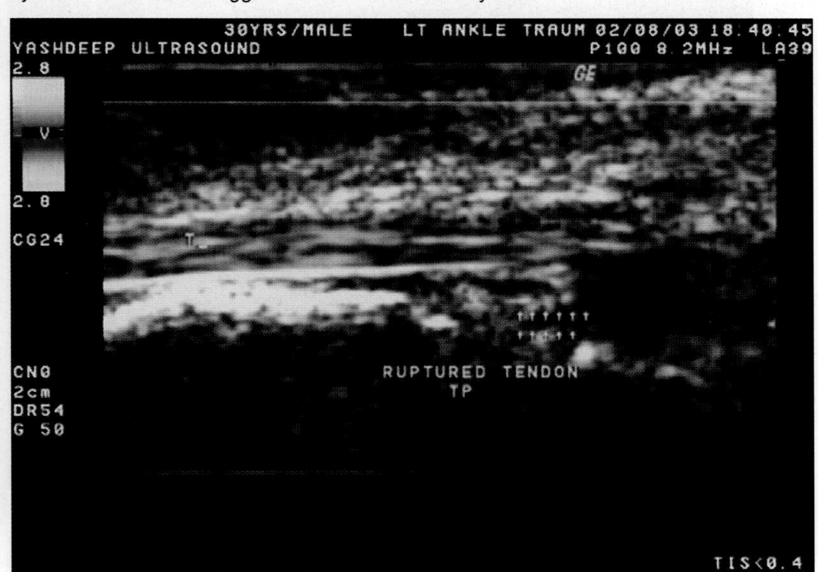

FIGURE 13.736: The distal part of the tendon shows partial tear. The posterior fibers of the tendons are seen torn and disrupted. However, anterior fibers are seen intact. Fluid is also seen in the tendon sheath.

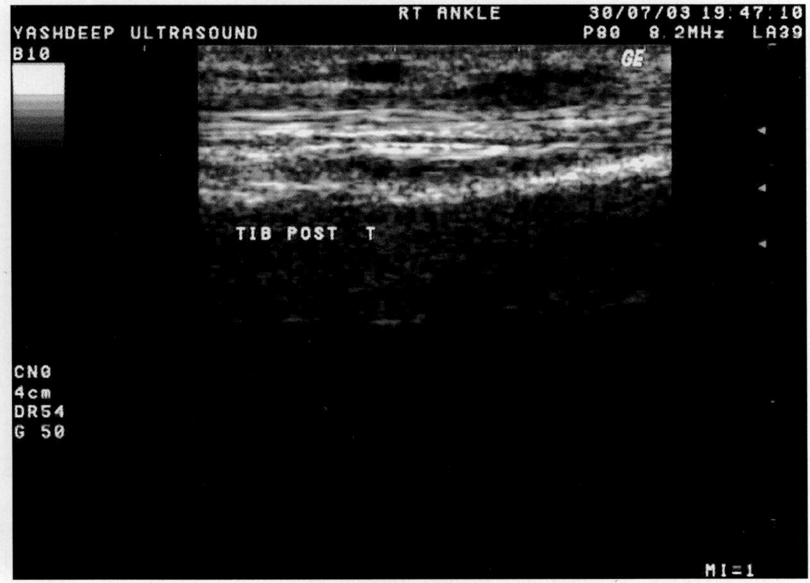

FIGURE 13.737

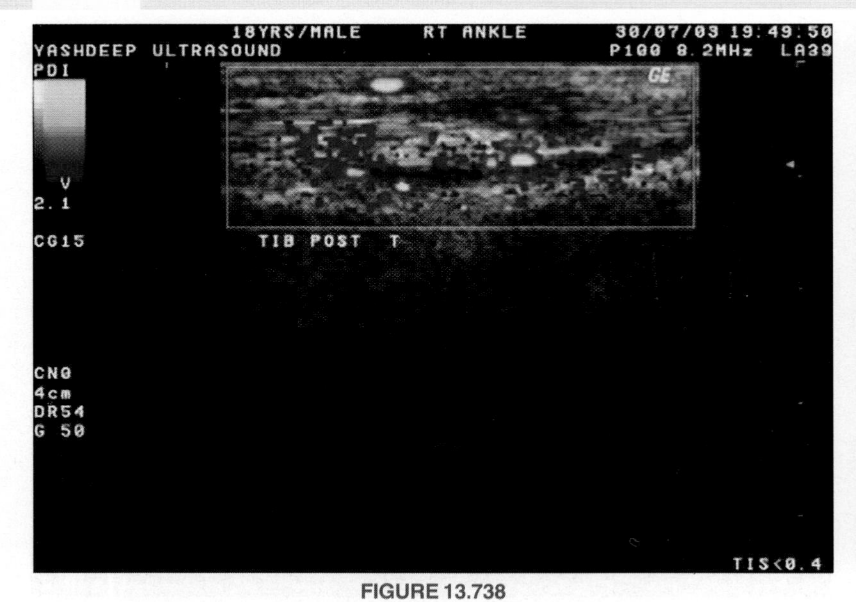

FIGURE 13.738

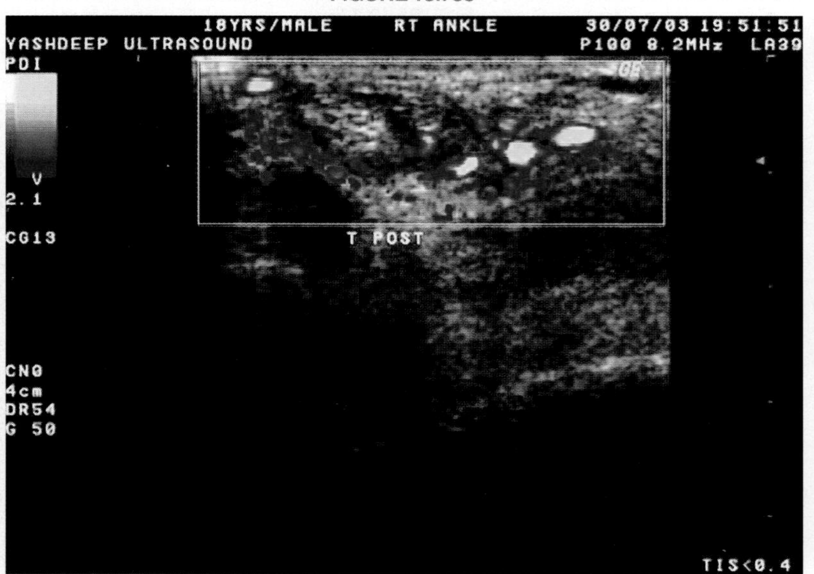

FIGURE 13.739

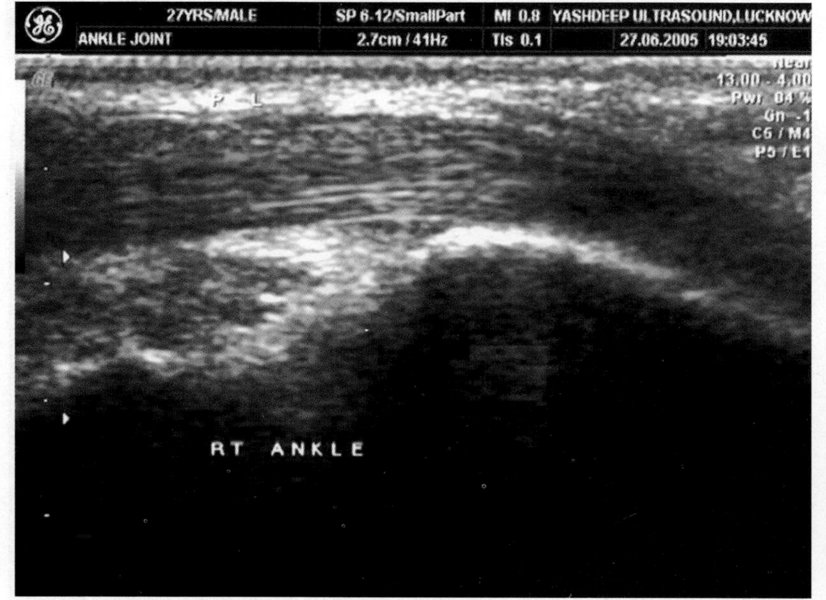

FIGURE 13.740

FIGURES 13.737 to 13.740: Chronic tenosynovitis of posterior tibial tendon. A long distance runner presented with chronic ankle pain. HRSG shows marked thickening of the posterior tibial tendon. Tendon sheath edema is present. Marked synovial membrane proliferation is seen anterior to the medial malleolus.

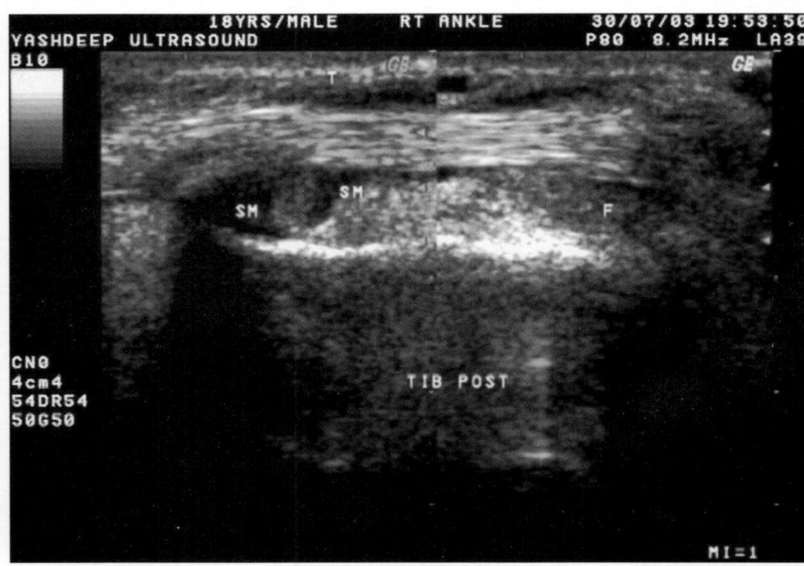

FIGURE 13.741

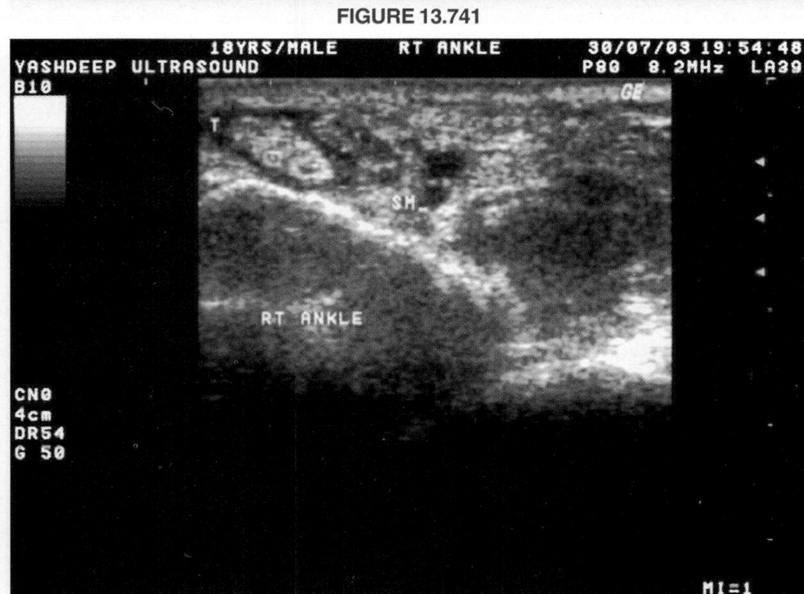

FIGURE 13.742

FIGURES 13.741 and 13.742: On color flow imaging increased flow is seen in the synovium. Flow is also seen in the tendon sheath suggestive of chronic Tenosynovitis.

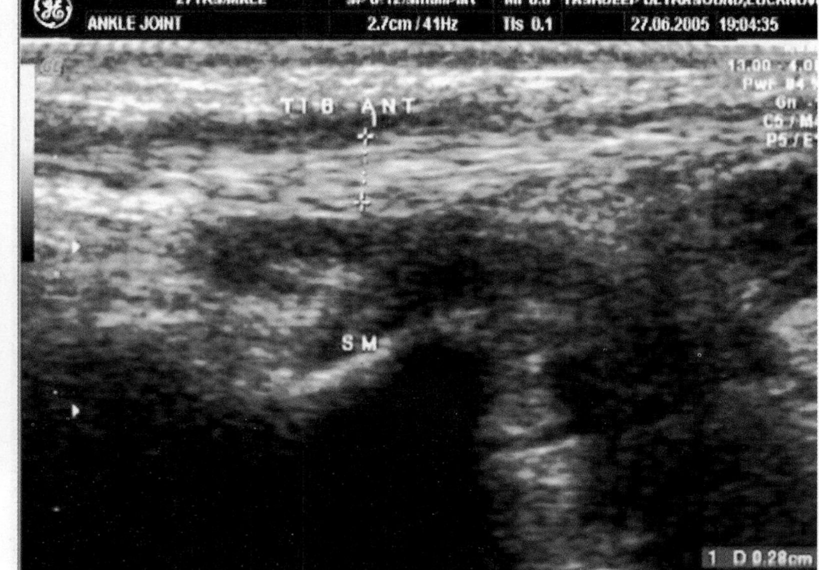

FIGURE 13.743

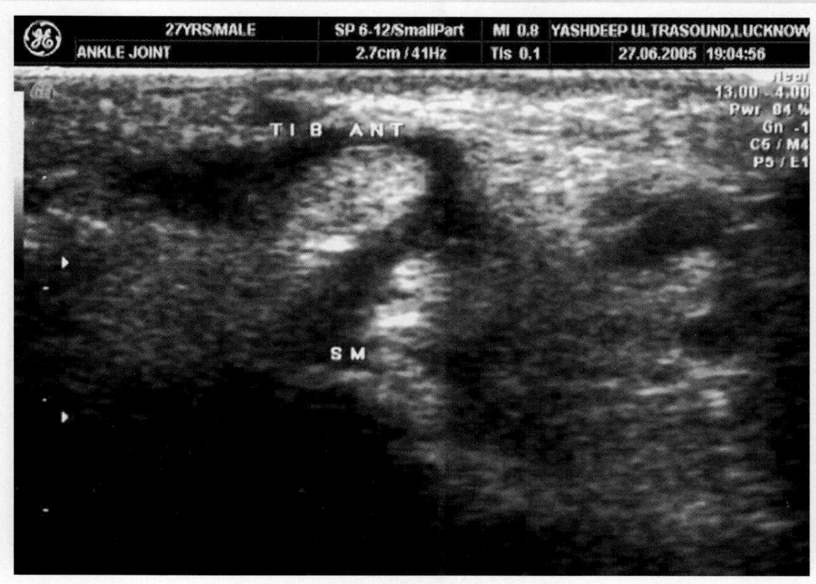

FIGURE 13.744

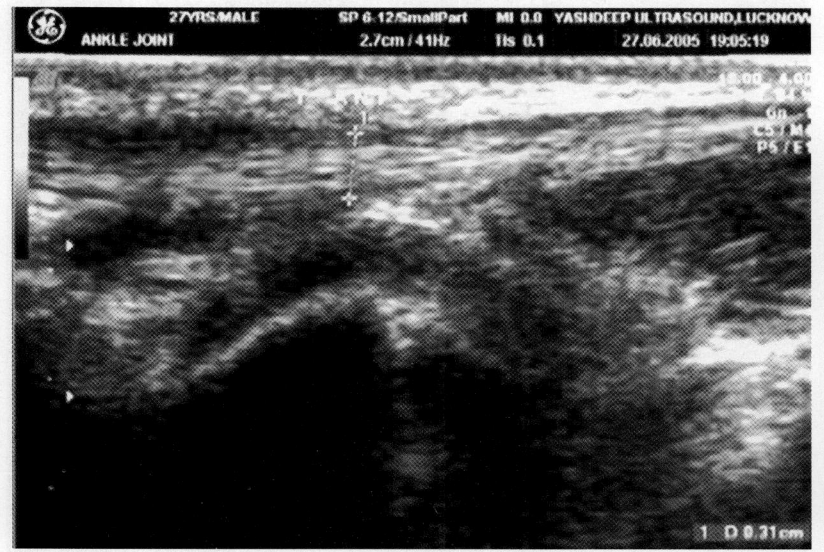

FIGURE 13.745

FIGURES 13.743 to 13.745: Tibialis anterior tenosynovitis. A patient presented with chronic ankle pain. HRSG shows thickened hypoechoic edematous tibialis anterior tendon. The tendon sheath edema is present. Proliferation of the synovial membrane is also seen. But no suggestion of any rupture of the tendon is seen.

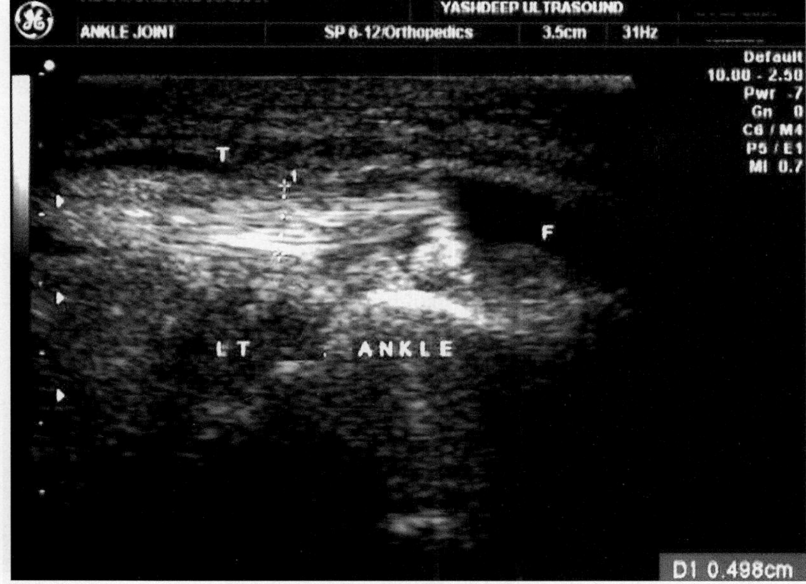

FIGURE 13.746

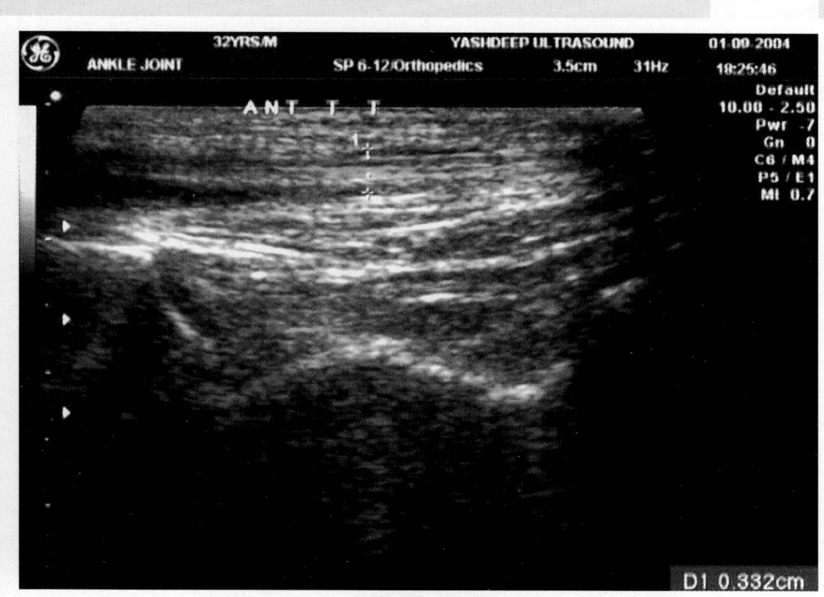

FIGURE 13.747

FIGURES 13.746 and 13.747: Rupture of the anterior tibial tendon. HRSG shows rupture of the anterior tibial tendon in its distal part in a patient sustained trauma to the ankle joint. The fibers are seen torn. Fluid is also seen in the rupture part of the tendon. The rupture tendon is seen retracted. The normal tendon Achilles is seen intact with fine fibril pattern on the other side.

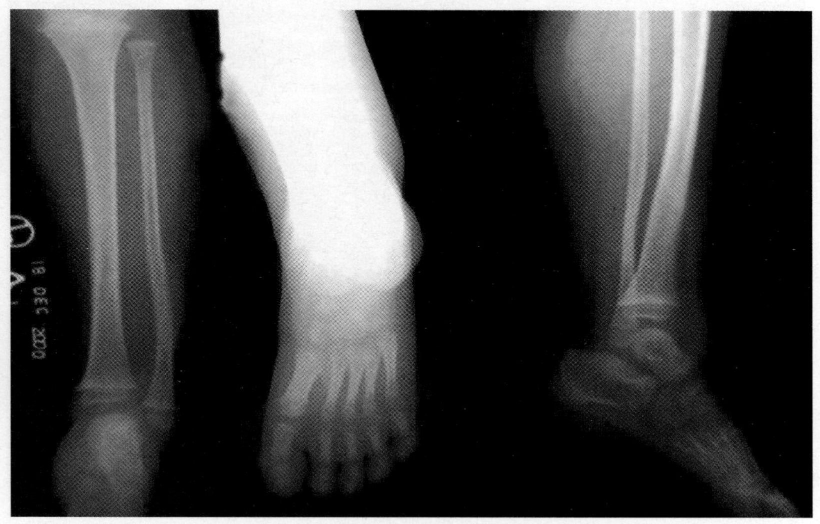

FIGURE 13.748: A young child presented with soft tissue mass over the Lt ankle. Plain X-ray of the ankle shows soft tissue swelling. However, bones were normal.

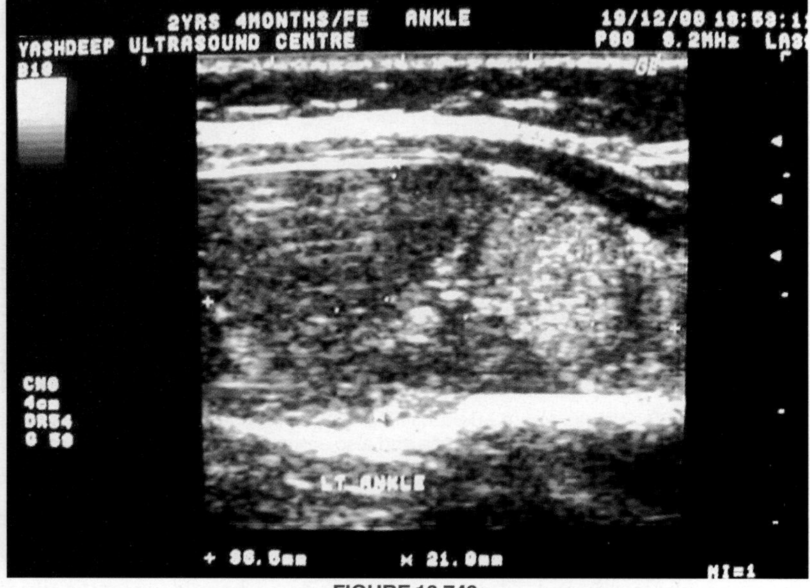

FIGURE 13.749

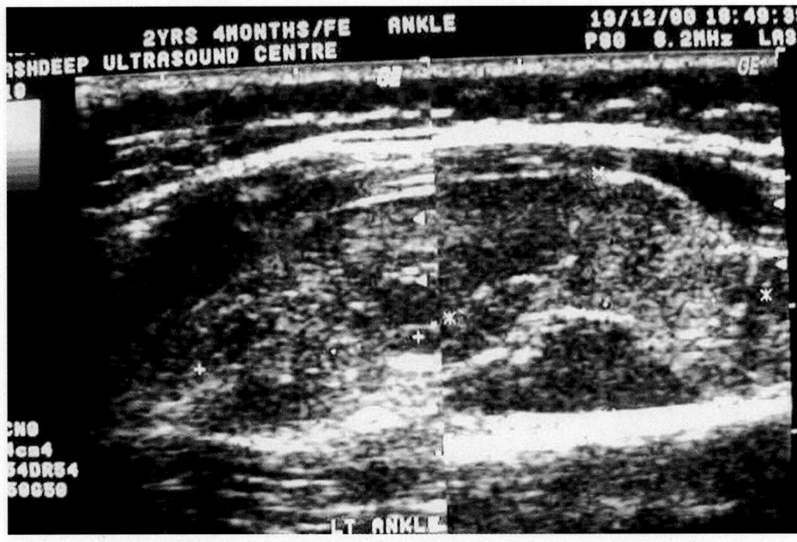

FIGURE 13.750

FIGURES 13.749 and 13.750: HRSG shows a big heterogeneous predominantly solid mass in the lower calf and over the ankle. No cystic degeneration is seen in it. However, few small-calcified specks are seen.

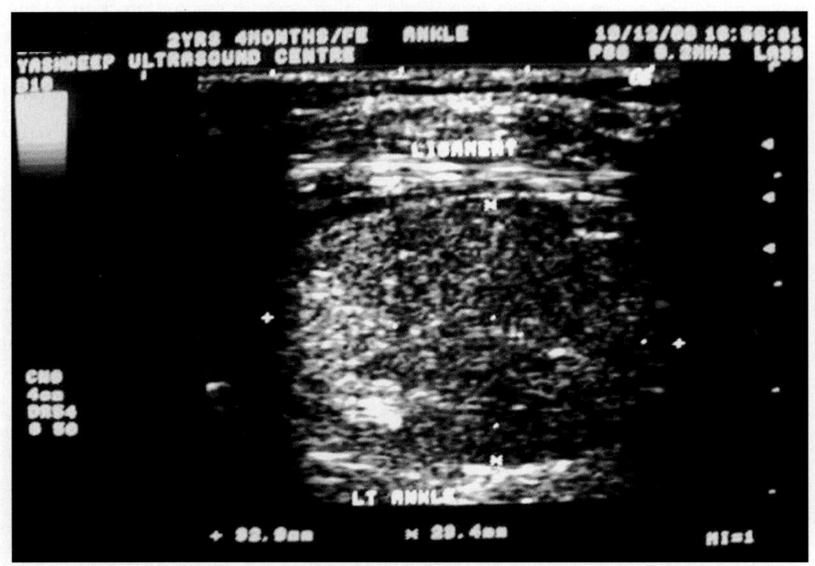

FIGURE 13.751: The mass is homogeneous in texture. Biopsy of the mass shows rhabdomyosarcoma.

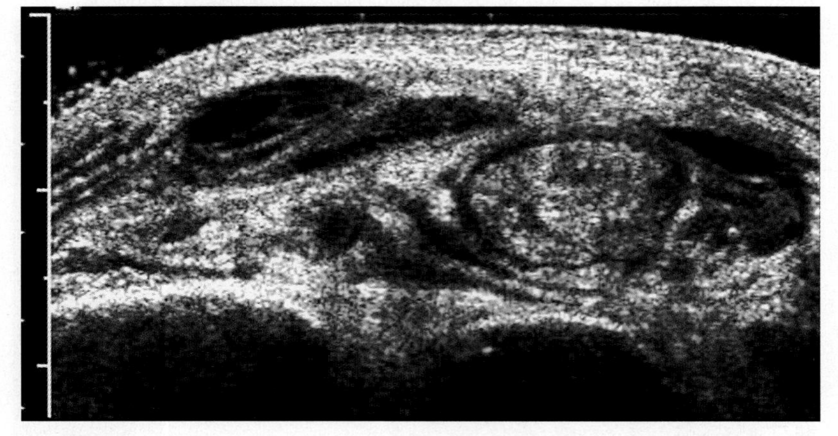

FIGURE 13.752: Diffuse ankle edema. HRSG shows diffuse ankle edema involving the subcutaneous planes, muscle planes and also around the tendon in swollen ankle.

The medial and the lateral ligaments are examined next. The anterior fibulo tailor ligament is most commonly ruptured. Therefore, it should be properly evaluated. The peroneous longus, brevis, anterior tibial

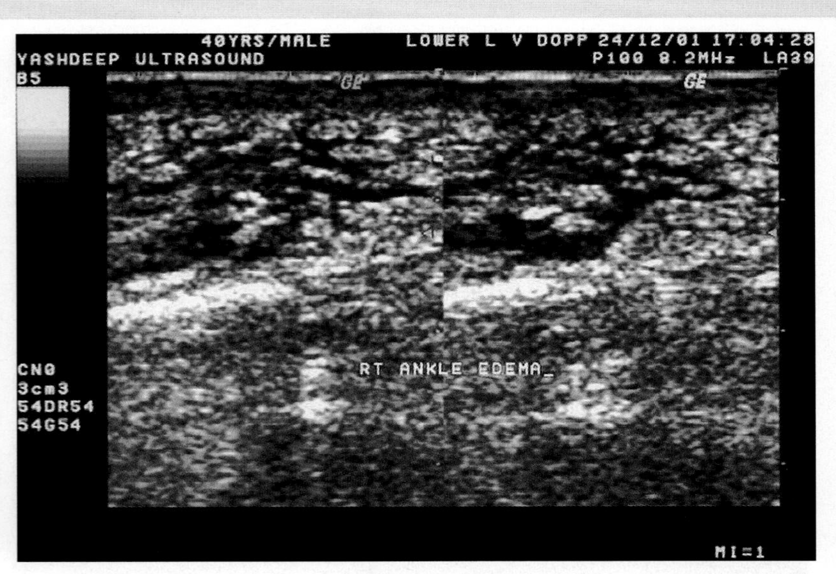

FIGURE 13.753

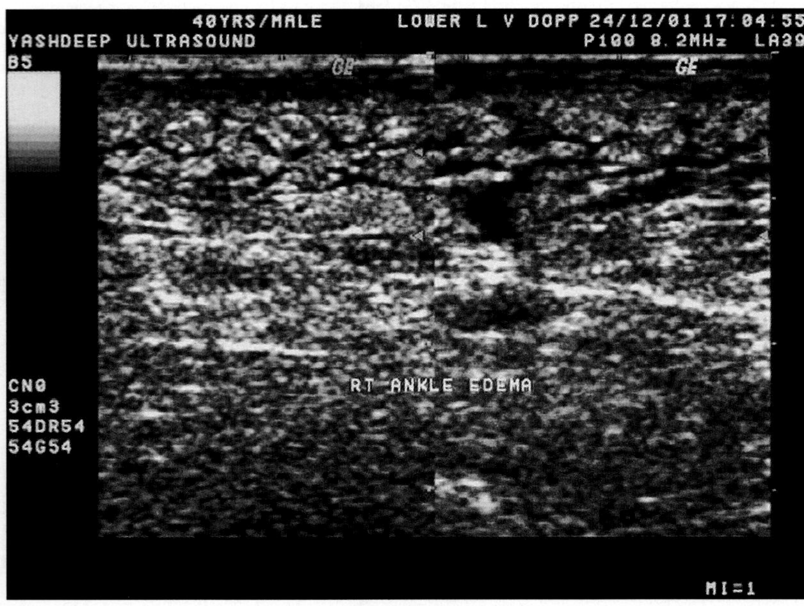

FIGURE 13.754

FIGURES 13.753 and 13.754: Subcutaneous edema. HRSG shows marked subcutaneous edema involving the lower ankle. Dermis and hypodermis edema is present in a case of cellulitis.

and posterior tibial tendons are evaluated in supine position. However, the Achilles tendon, subcutaneous Achilles tendon bursa and retrocalcaneal bursa are evaluated with patient in prompt position.

HRSG of Bursae

Bursae are sac like fluid structures, which are present to facilitate movement of musculoskeletal structures. Bursae are found in the areas where considerable degree of motion is required. HRSG is very sensitive in evaluation of pathological condition of bursae. Following are the main indications:

1. Acute traumatic bursitis.
2. Chronic traumatic bursitis.
3. Hemorrhage bursitis.
4. Chemical bursitis.
5. Rheumatoid and septic bursitis.

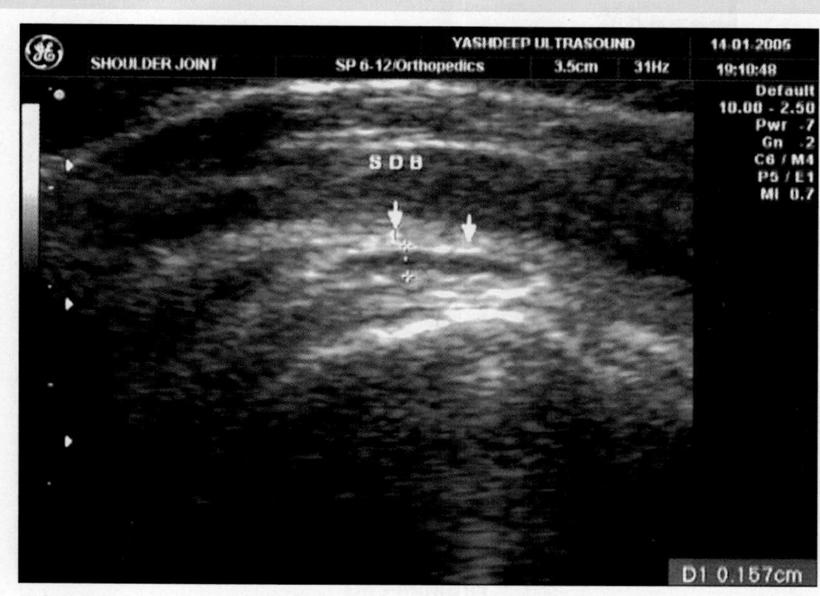

FIGURE 13.755: **Normal subdeltoid bursa.** It is seen as small fluid fill cavity. Bursal ball are normal. Normal size of the bursa should not be more than 3 mm.

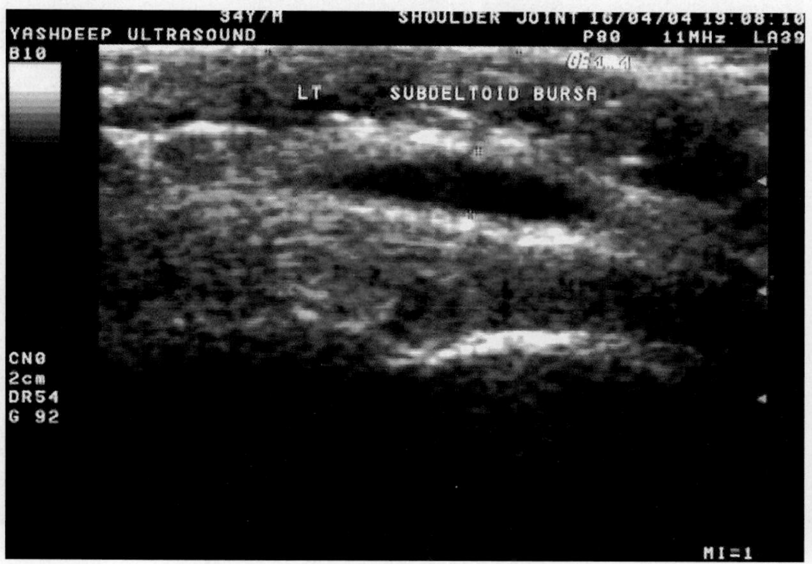

FIGURE 13.756

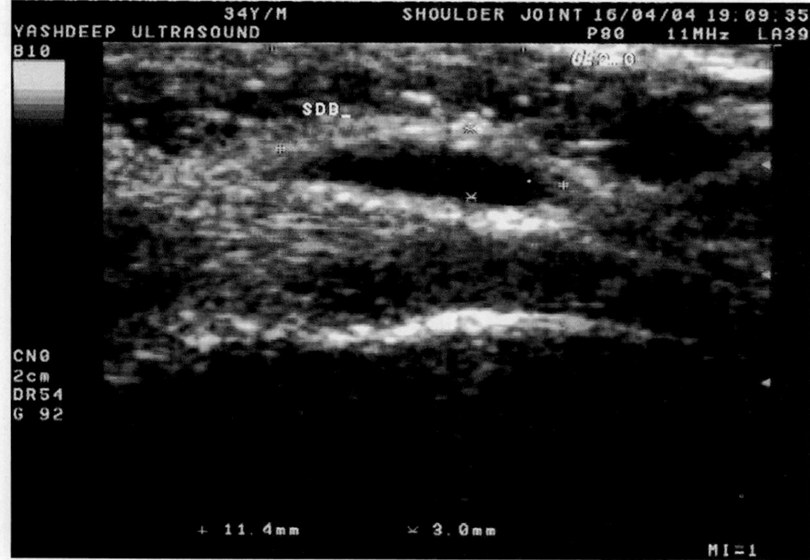

FIGURE 13.757

FIGURES 13.756 and 13.757: **Subdeltoid bursitis.** A patient who was a swimmer presented with painful Rt shoulder movement. HRSG shows the subdeltoid bursa markedly distended. Low-level echo fluid collection is seen in the bursa. Bursal walls are echogenic.

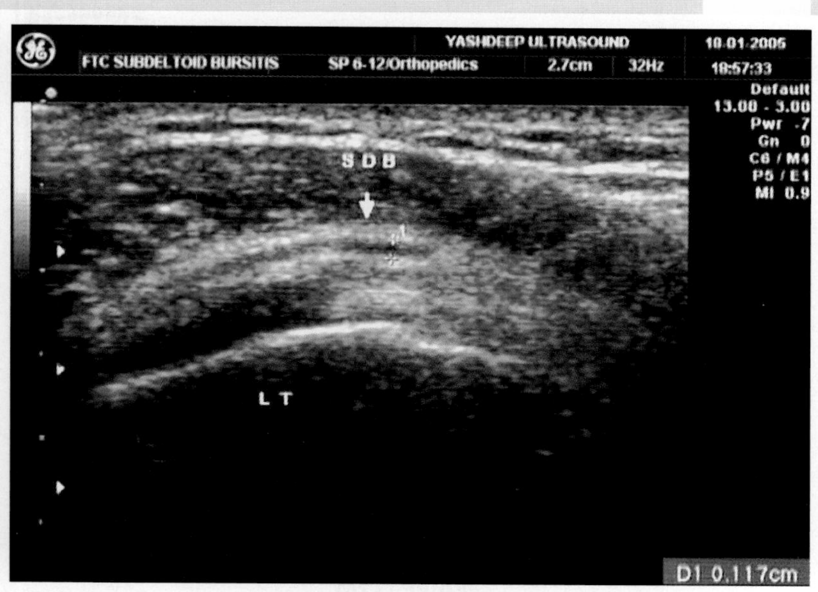

FIGURE 13.758: The Lt shoulder shows the normal bursa in the same patient.

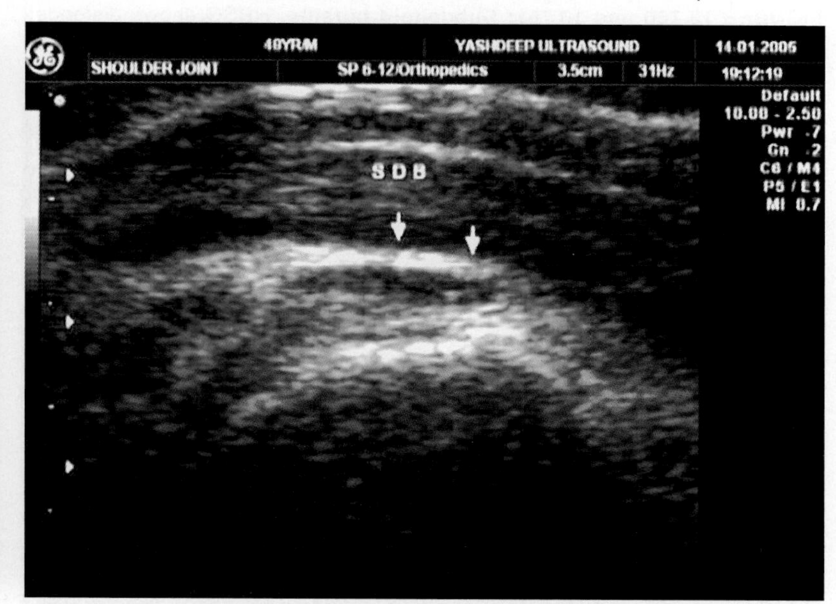

FIGURE 13.759: **Chronic subdeltoid bursitis.** HRSG shows thickened echogenic thickened subdeltoid bursa. Multiple echogenic shadows are seen in the bursal fluid. Bursal walls are echogenic suggestive of chronic subdeltoid bursitis.

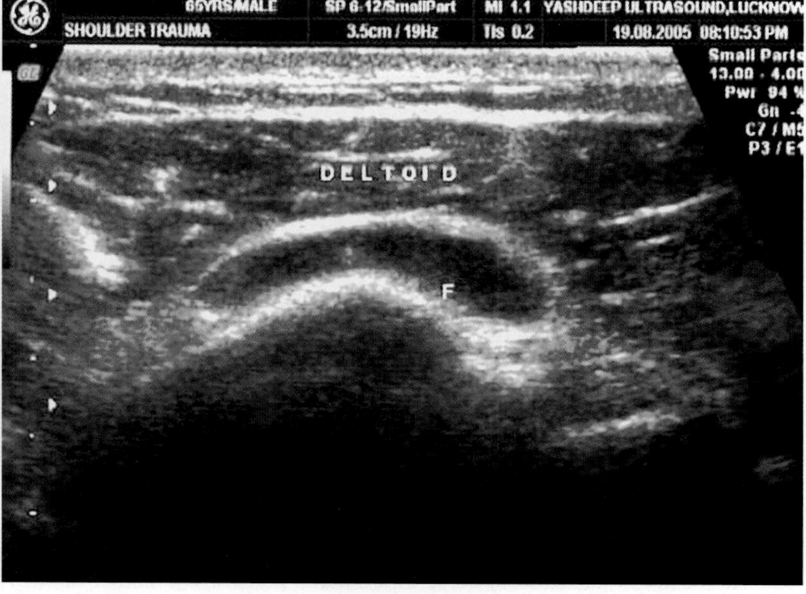

FIGURE 13.760

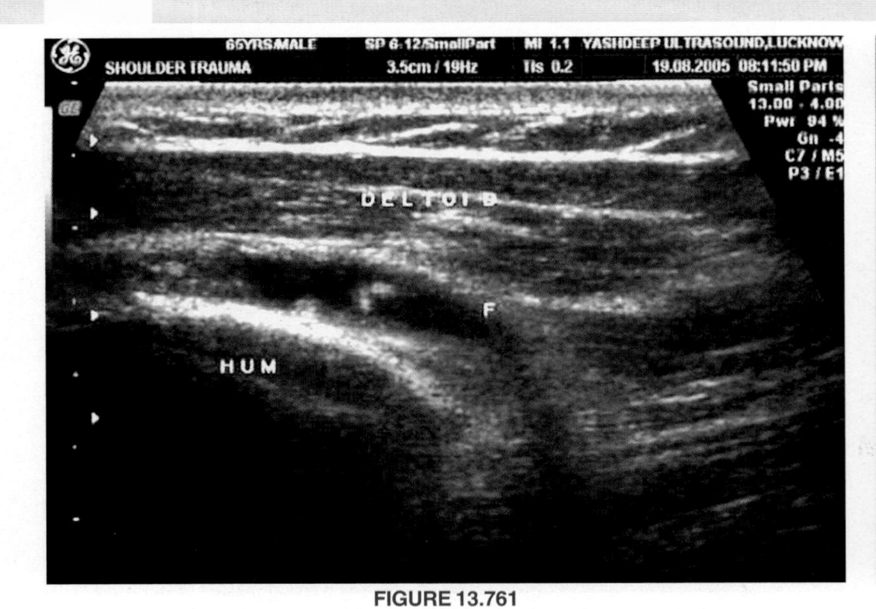

FIGURE 13.761

FIGURES 13.760 and 13.761: Subdeltoid bursitis. HRSG shows distended subdeltoid bursa. Bursal walls are echogenic and thickened. Echogenic floating shadows are seen in the distended bursa suggestive of small hemorrhage.

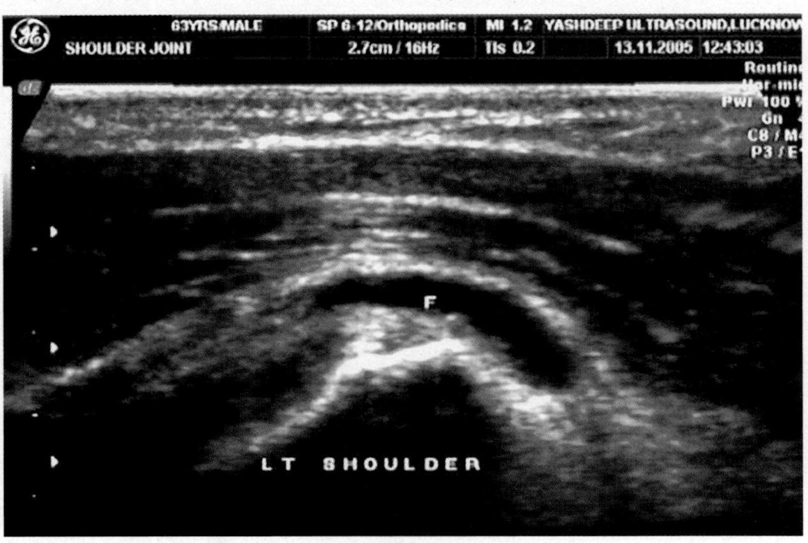

FIGURE 13.762

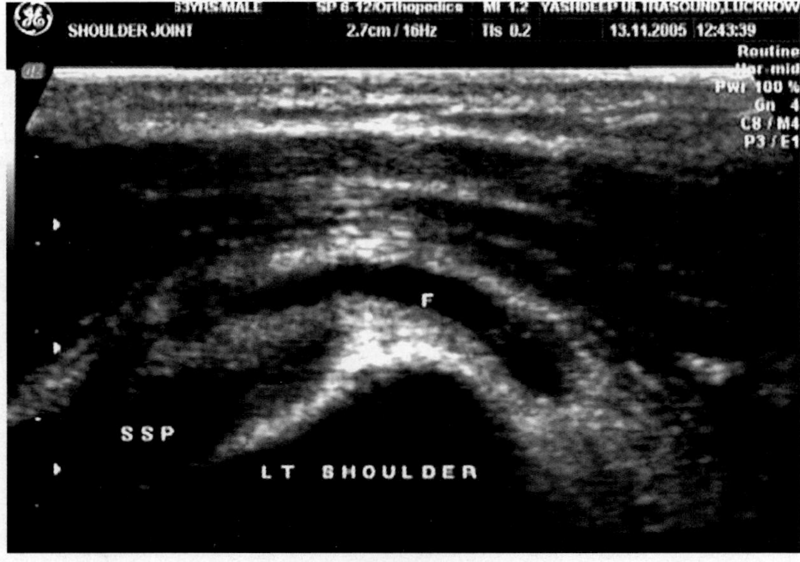

FIGURE 13.763

FIGURES 13.762 and 13.763: Subacromial bursitis. There is also evidence of fluid collection seen in subacromial bursa in the same patient. The bursal walls are echogenic. They are distended.

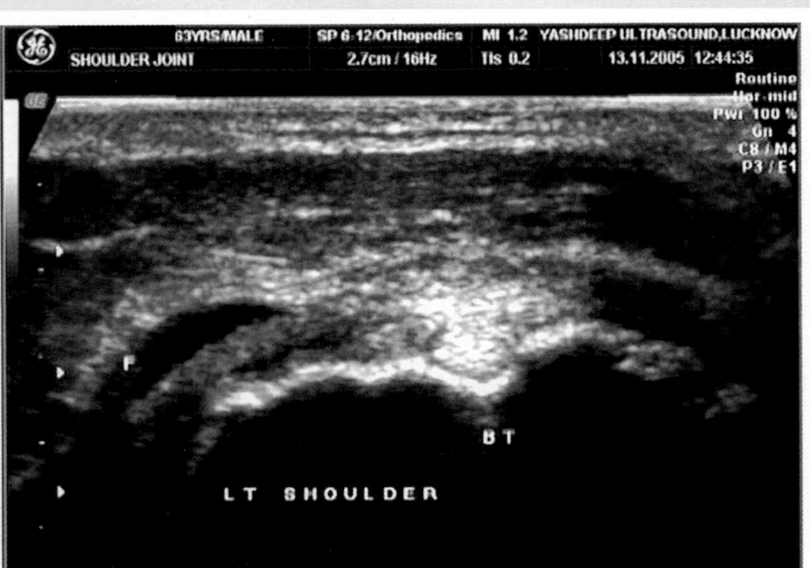

FIGURE 13.764

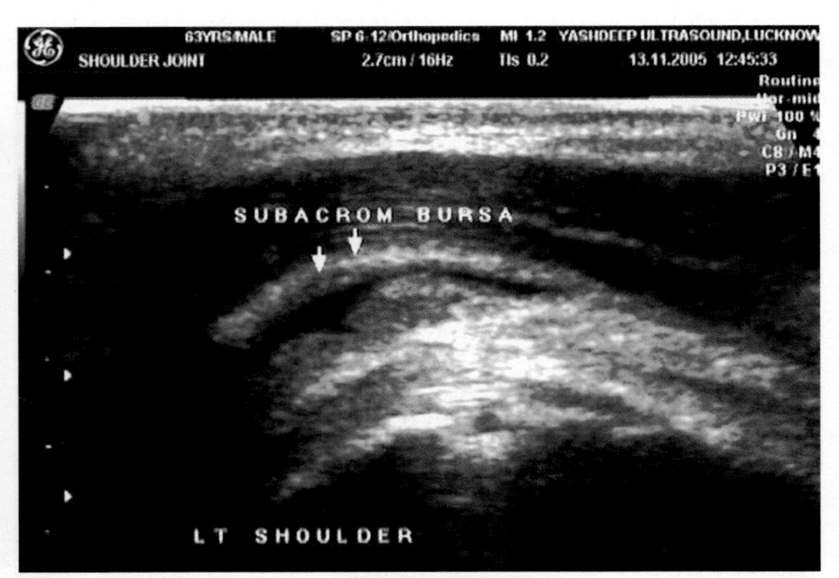

FIGURE 13.765

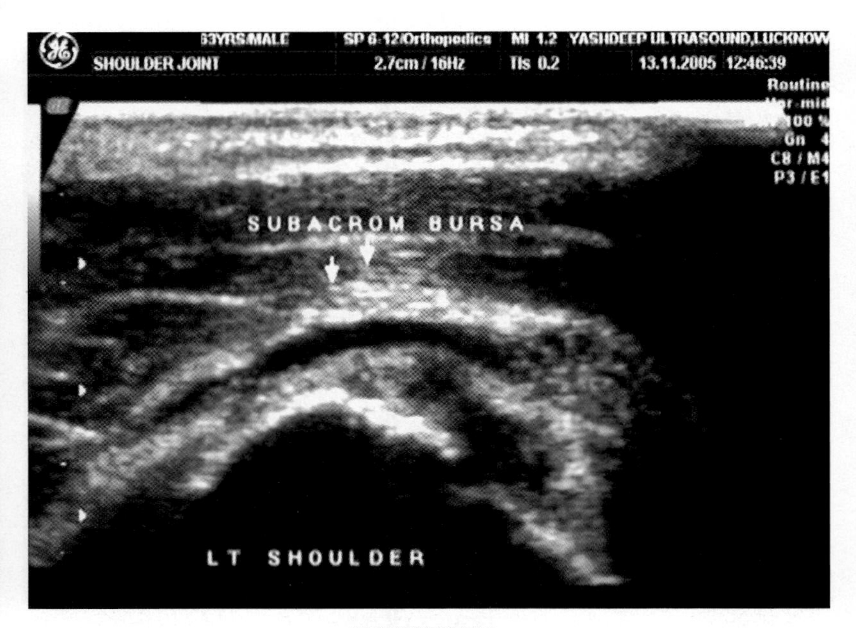

FIGURE 13.766

FIGURES 13.764 to 13.766: Echogenic collection is seen in the subacromial bursa in the same patient. The collection is seen moving with change of posture. Synovial membrane proliferation is also seen in the bursa.

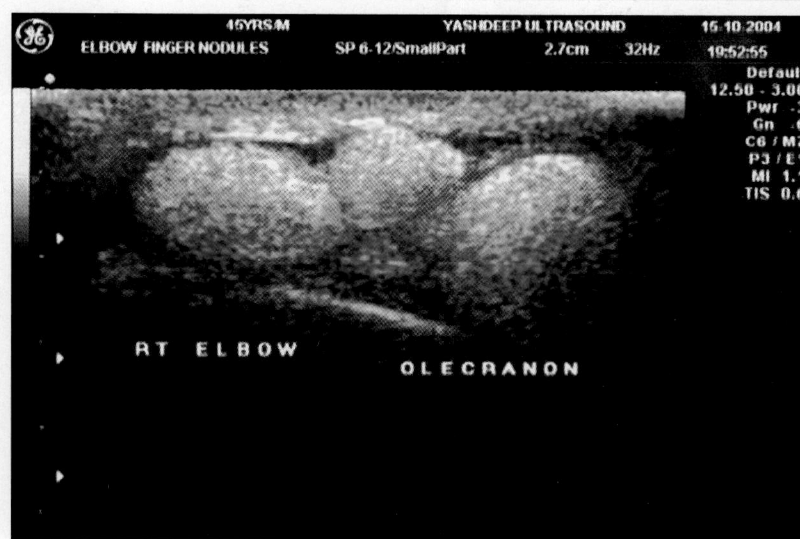

FIGURE 13.767

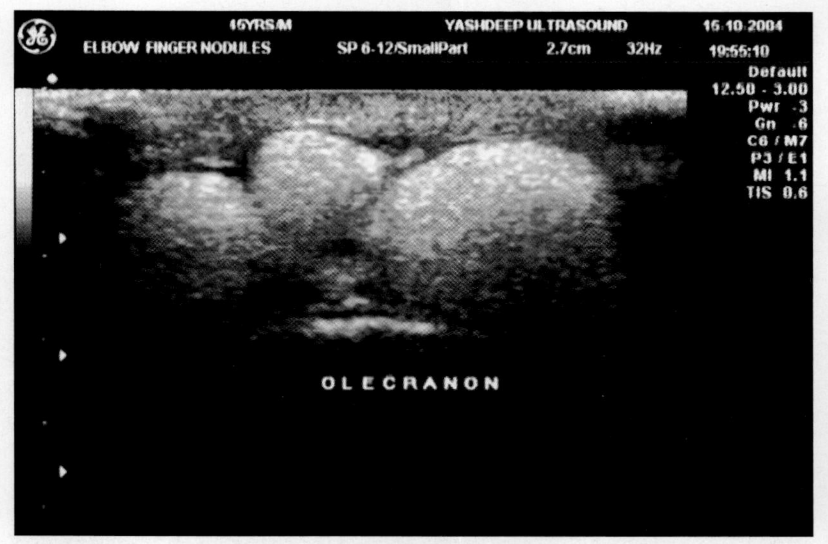

FIGURE 13.768

FIGURES 13.767 and 13.768: Fulminating olecranon bursitis. A patient presented with marked painful swelling in the Rt elbow. HRSG shows markedly distended olecranon bursa. Echogenic floating shadows are seen in the bursal cavity which appear like soft cotton wool like shadows with homogeneous texture. Perifocal edema is present. Aspiration confirmed acute fulminating pyogenic bursitis.

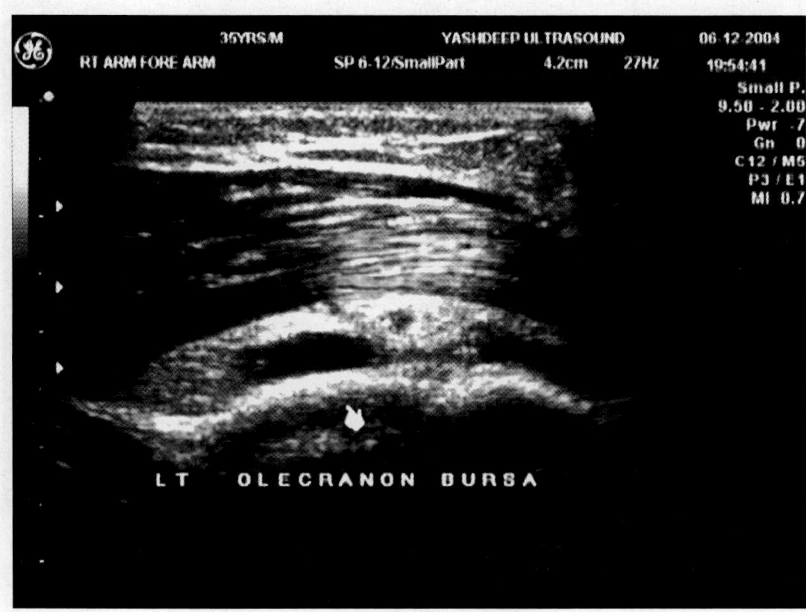

FIGURE 13.769

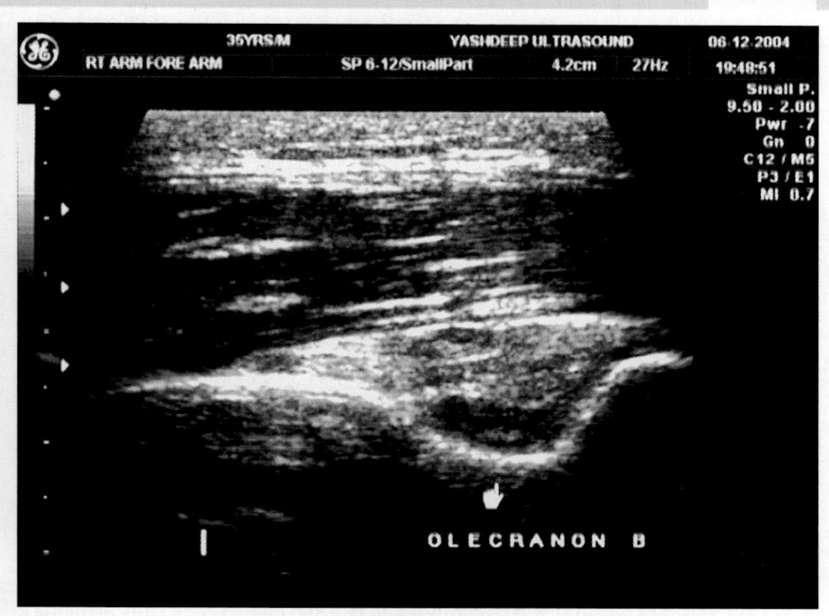

FIGURE 13.770

FIGURES 13.769 and 13.770: Chronic frictional olecranon bursitis. HRSG shows echogenic distended olecranon bursa. Bursal walls are thickened and irregular in outline. Low-level echo collection is seen in bursa. The olecranon cortex is also thickened and echogenic suggestive of frictional bursitis.

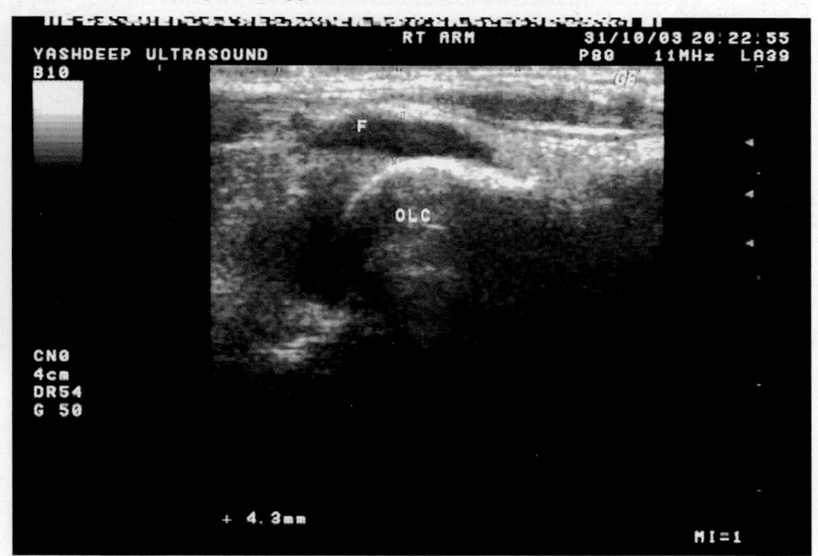

FIGURE 13.771

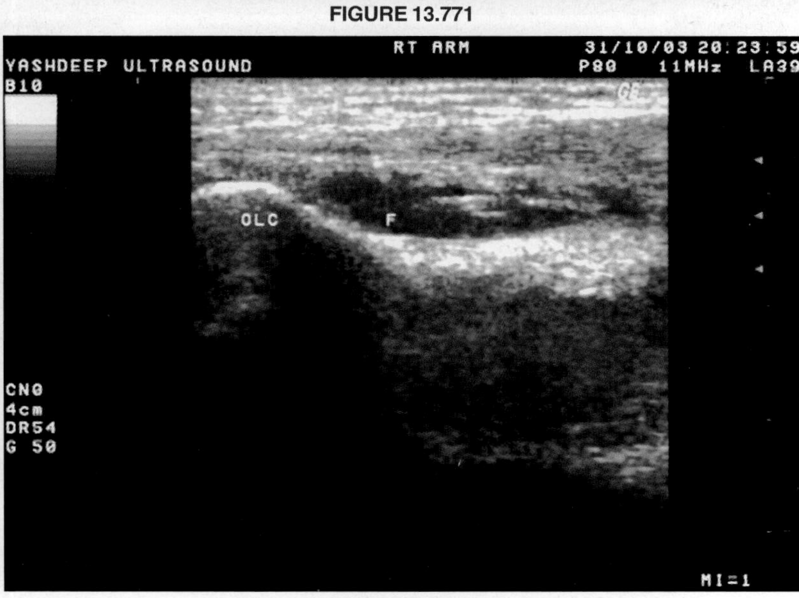

FIGURE 13.772

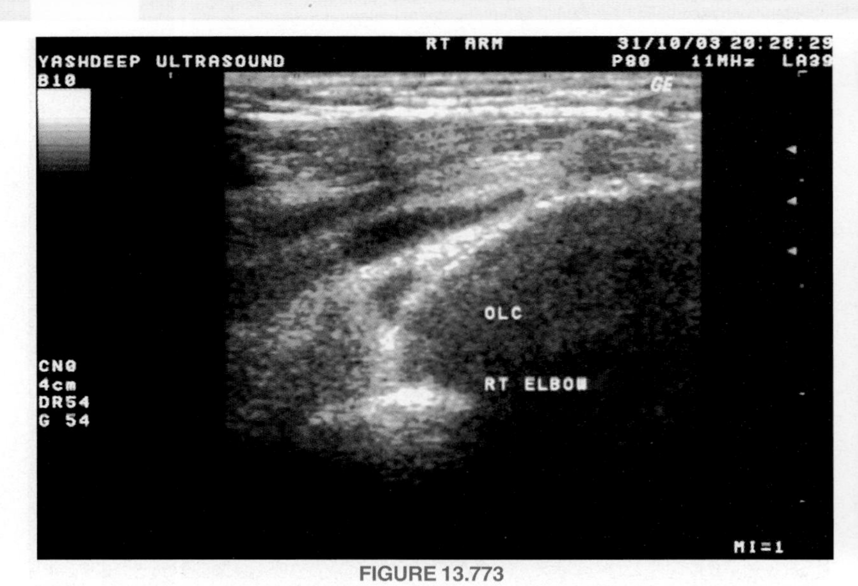

FIGURE 13.773

FIGURES 13.771 and 13.773: Olecranon bursitis. HRSG shows distended olecranon bursa in a patient presented with painful elbow. The olecranon bursa seen distended. Small amount of fluid is seen in the bursa. But no internal echoes are seen in the bursal fluid.

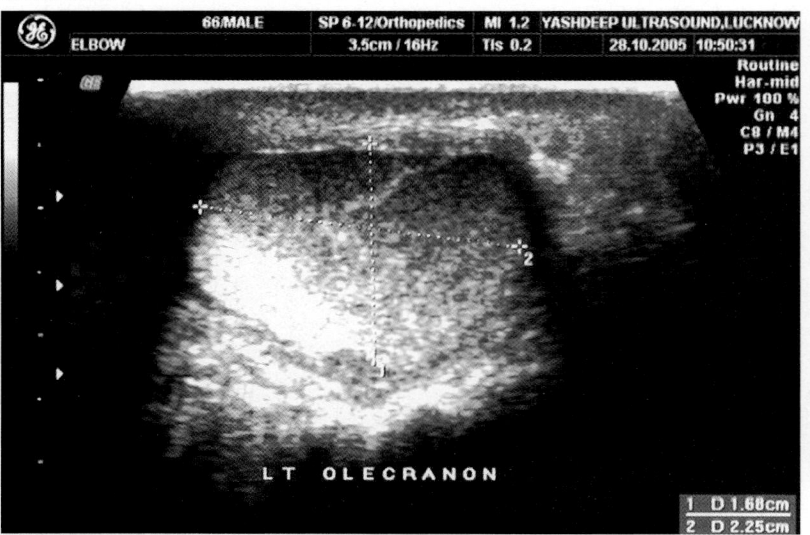

FIGURE 13.774

FIGURE 13.775

FIGURES 13.774 and 13.775: Chronic bursitis with synovial fluid collection. HRSG shows thick low level echo collection in the olecranon bursa in a known patient of olecranon bursitis. Thick collection is seen in the bursa. Echogenic calcified specks are also seen in the bursal fluid, which are fixed on the bursal wall. Synovial membrane proliferation is also seen.

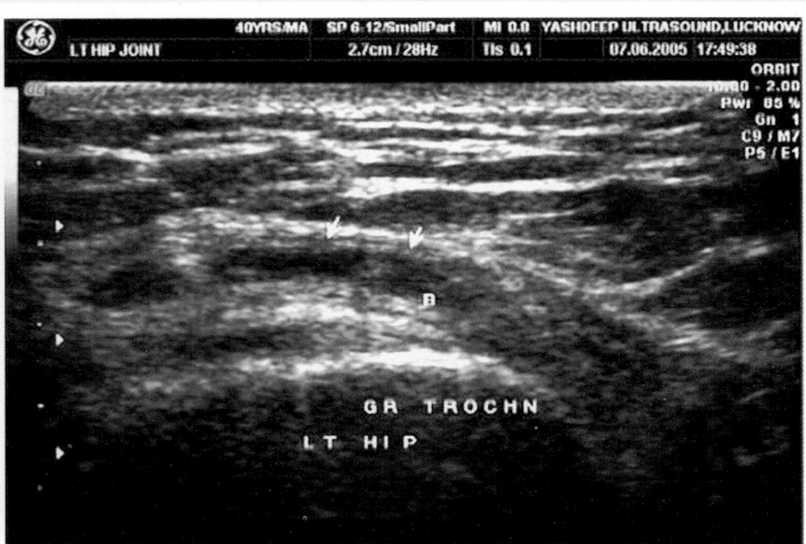

FIGURE 13.776

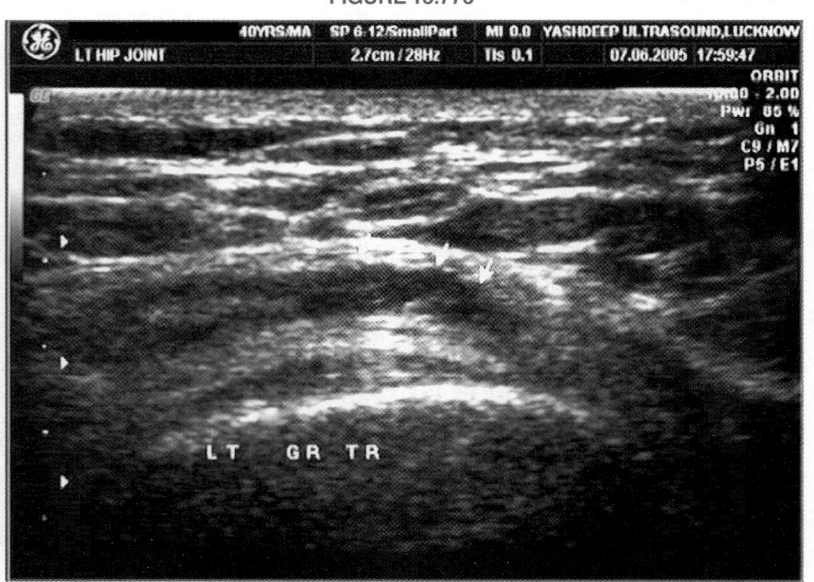

FIGURE 13.777

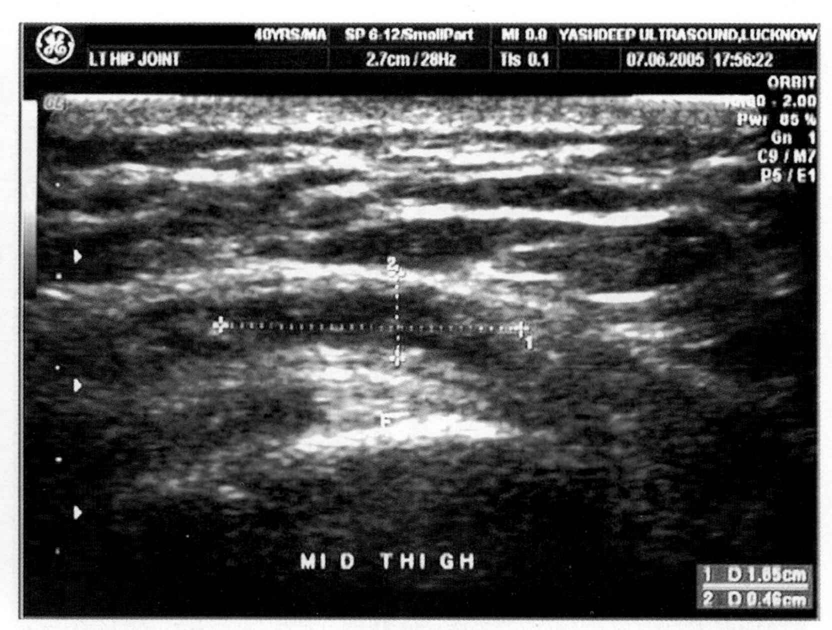

FIGURE 13.778

FIGURES 13.776 to 13.778: Chronic frictional trochanteric bursitis. HRSG shows markedly thickened echogenic wall of trochanteric bursa. The collection is seen in bursa. Multiple internal echoes are seen in it in a case of chronic frictional bursitis.

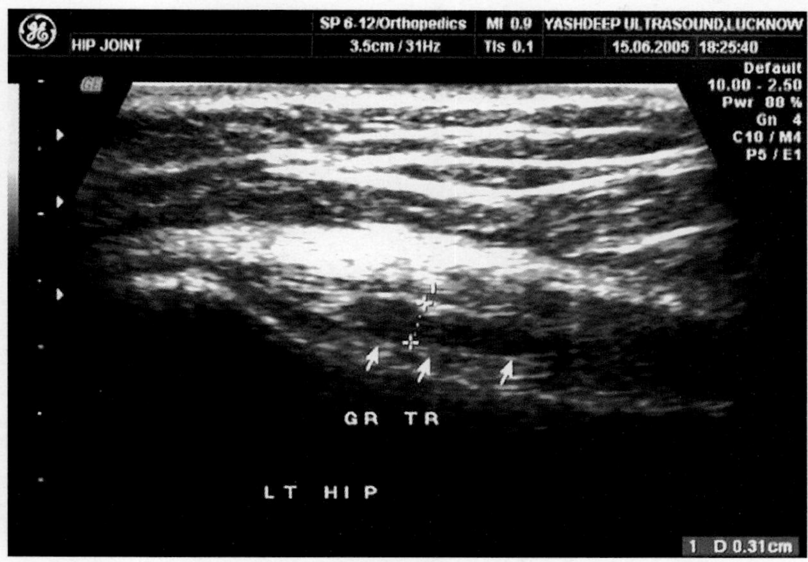

FIGURE 13.779

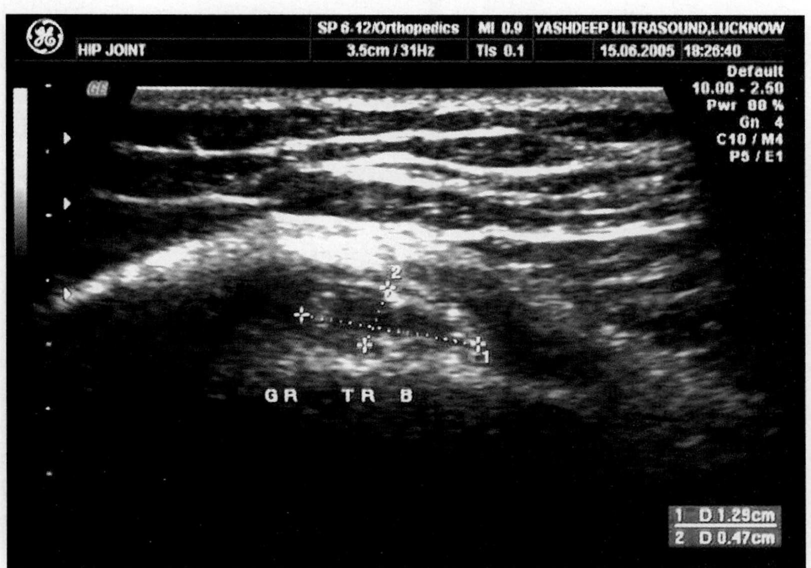

FIGURE 13.780

FIGURES 13.779 and 13.780: Chronic greater trochanteric bursitis. HRSG shows distended greater trochanter bursa. The bursal walls are echogenic and thickened. Echogenic shadows are seen filling the bursal cavity suggestive of chronic bursitis.

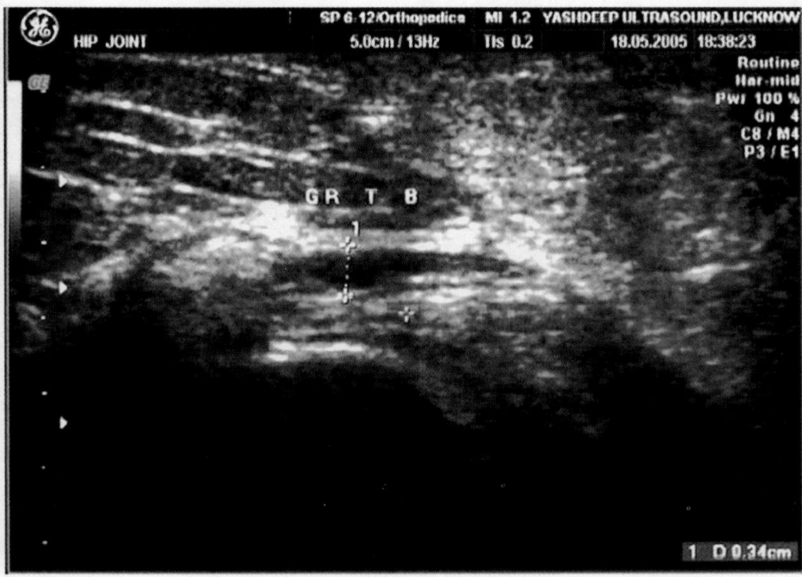

FIGURE 13.781

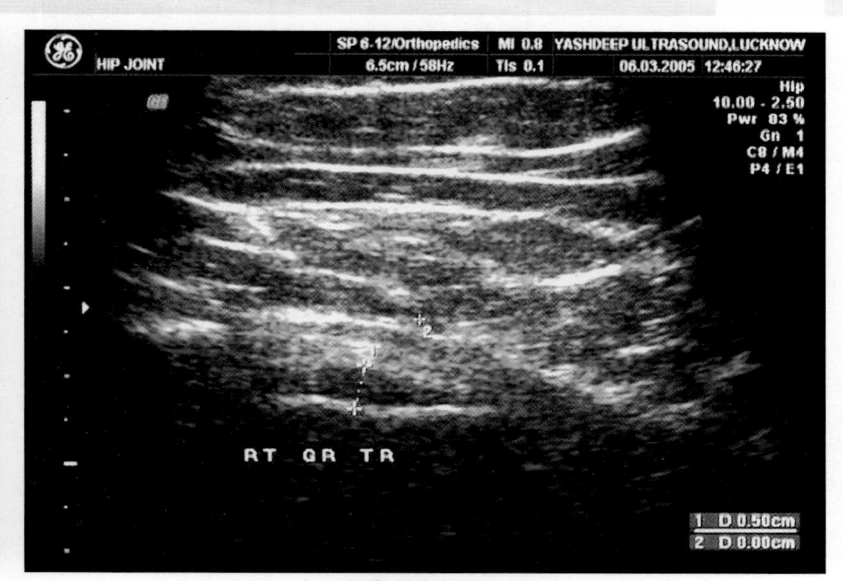

FIGURE 13.782

FIGURES 13.781 and 13.782: Chronic greater trochanteric bursitis. HRSG shows low-level echo collection in the greater trochanter bursa. The bursal walls are echogenic. Margins are irregular and shaggy. Thick collection is seen in the bursal suggestive of chronic bursitis.

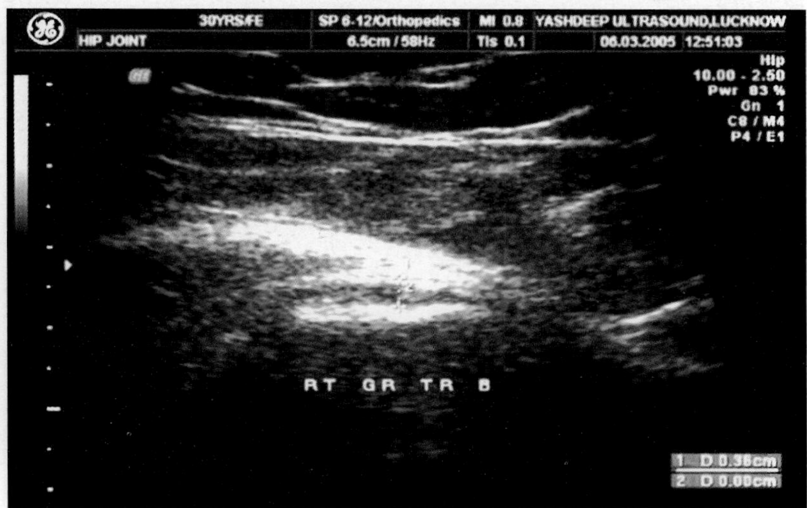

FIGURE 13.783

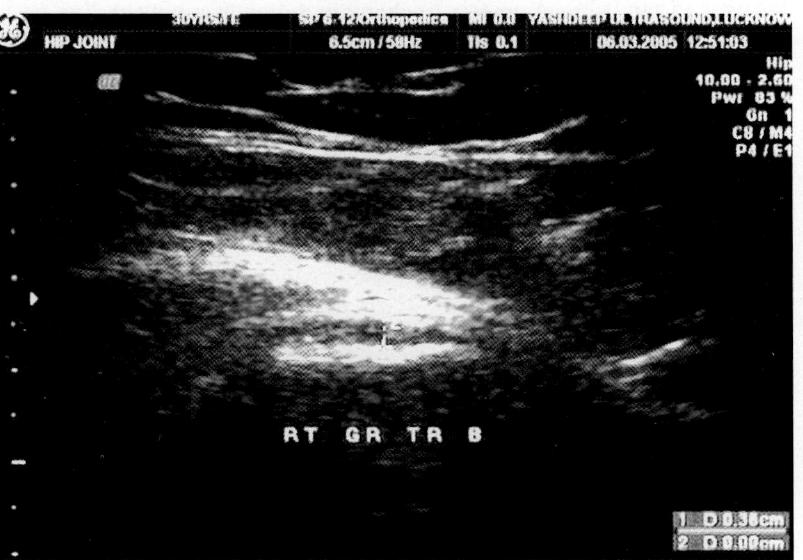

FIGURE 13.784

FIGURES 13.783 and 13.784: Chronic greater trochanter bursitis. HRSG shows irregular thickened Rt greater trochanter bursa. Walls echogenic and thickened. Thick echogenic collection is seen in the bursa. Normal bursal shape is lost suggestive of chronic bursitis.

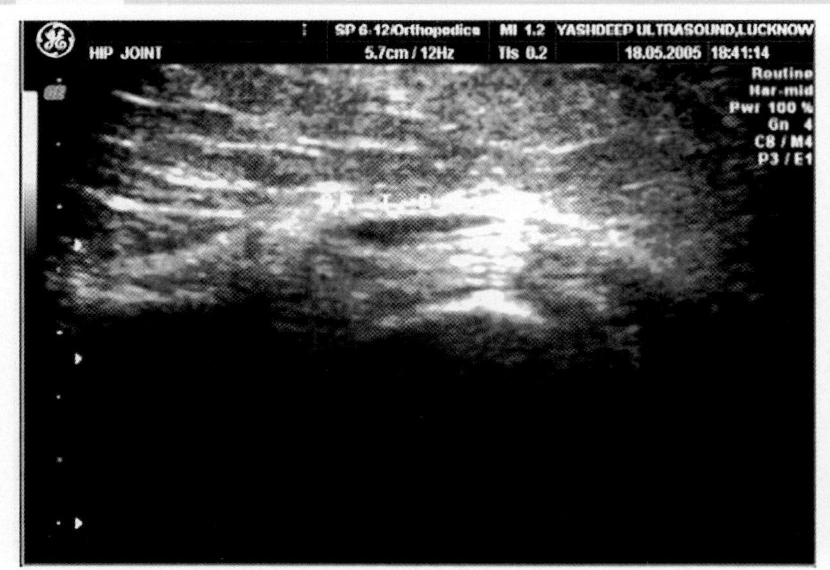

FIGURE 13.785: **Chronic greater trochanter bursitis.** HRSG shows low-level echo collection in the greater trochanteric bursa. The bursal walls are echogenic and irregular in outline. Low-level echoes are seen in it.

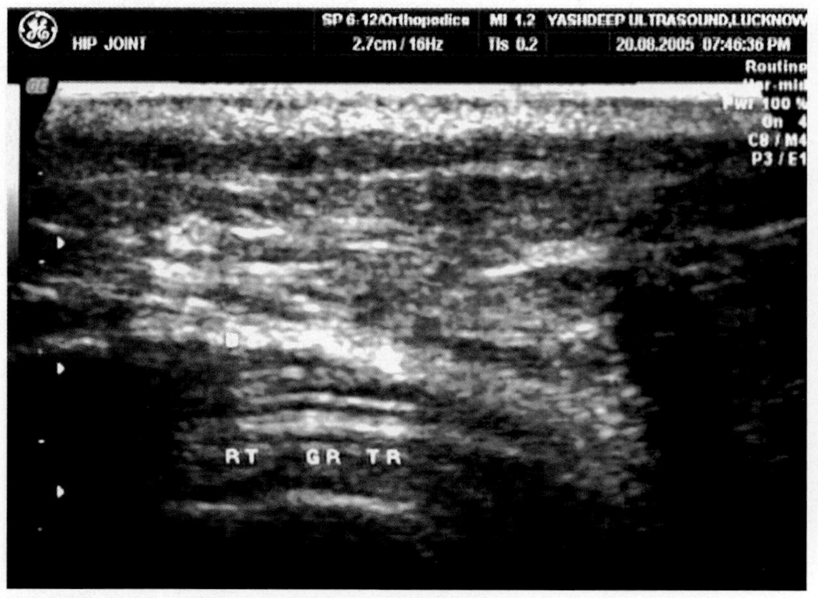

FIGURE 13.786: Normal greater trochanteric bursa in a patient. The bursal walls are normal and smooth in outline. Small amount of fluid is seen in the bursal cavity. No suggestion of any bony irregularity is seen.

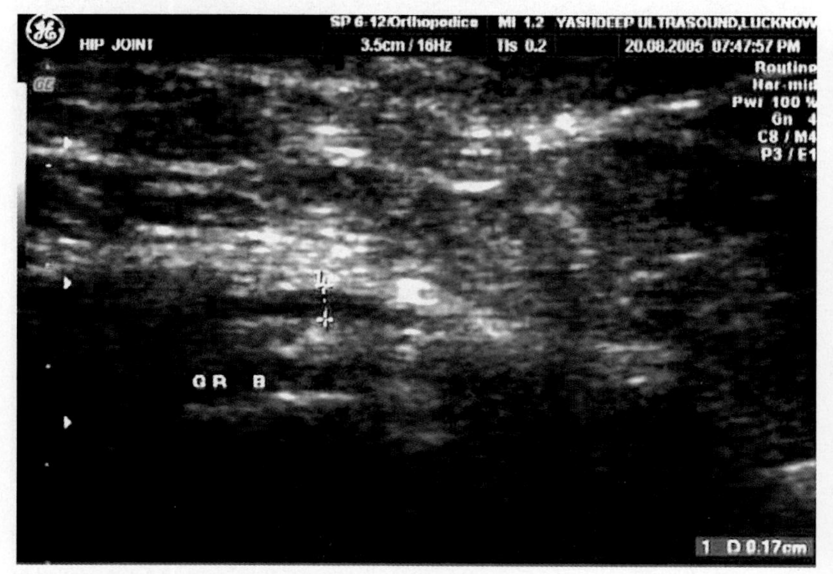

FIGURE 13.787: The contralateral hip shows distended bursa. The bursal walls are thickened and low-level echoes are seen in the bursa.

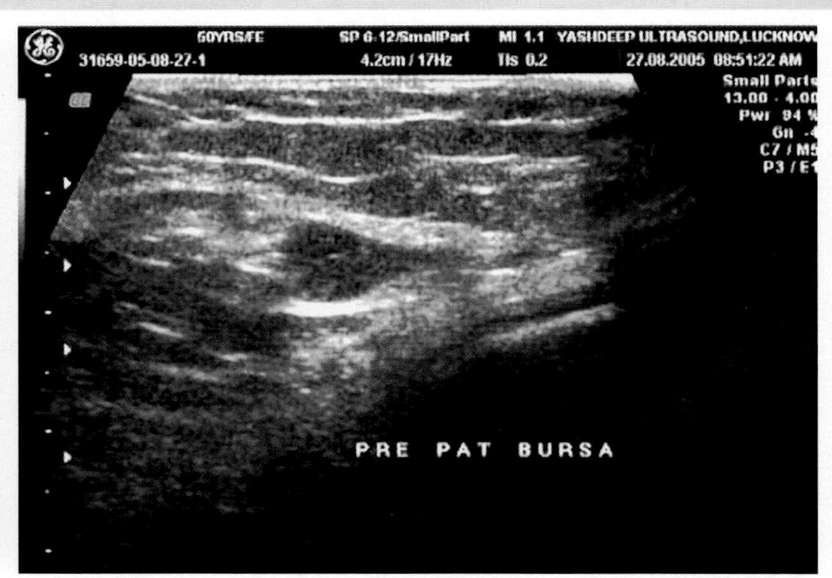

FIGURE 13.788

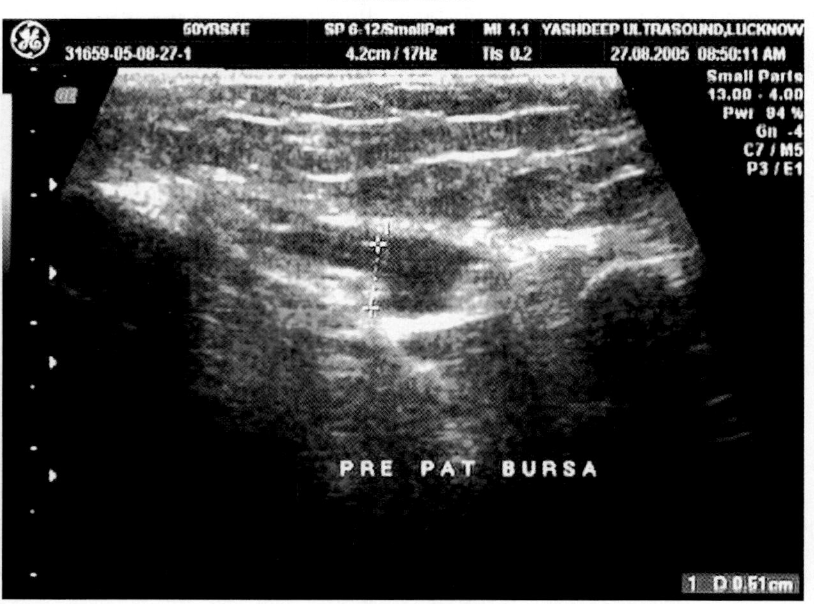

FIGURE 13.789

FIGURES 13.788 and 13.789: Pre patellar bursitis. A patient presented with painful knee movement while climbing the steps. HRSG shows distended pre patellar bursa. Echogenic collection is seen in the bursal cavity. Margins are irregular in outline. Bursal walls are thickened.

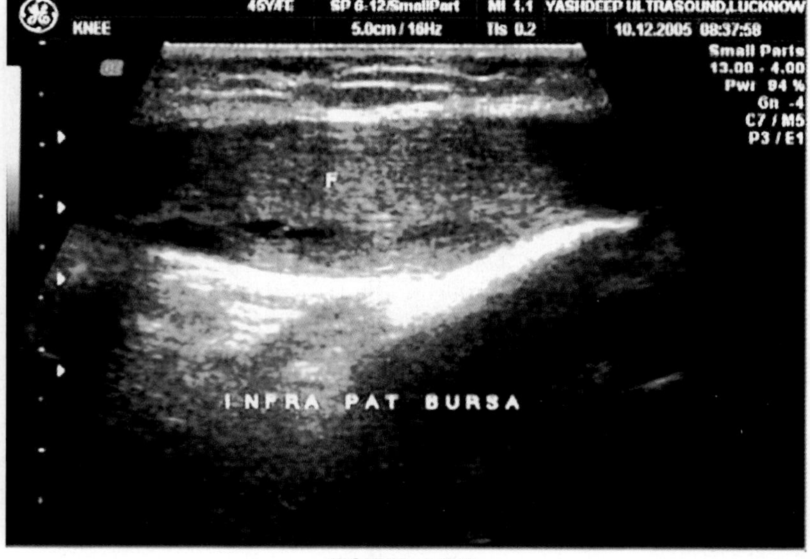

FIGURE 13.790

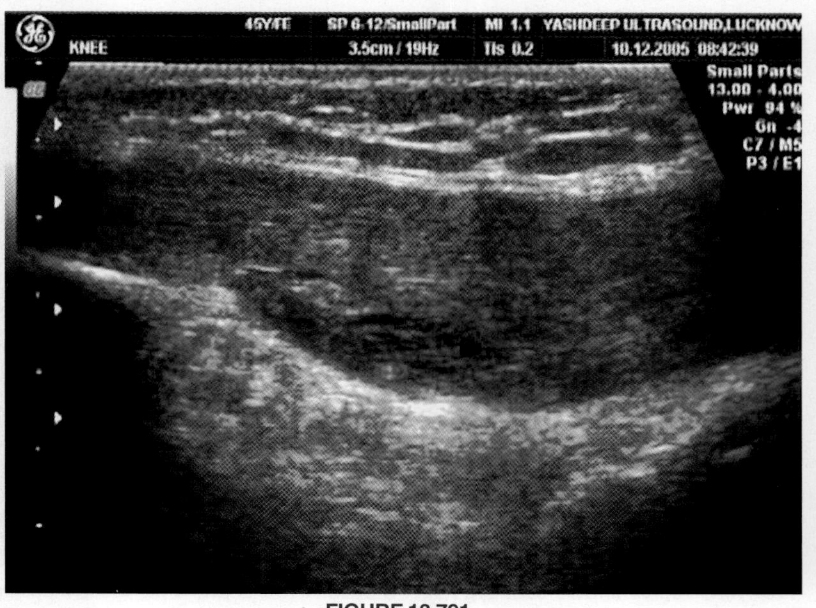

FIGURE 13.791

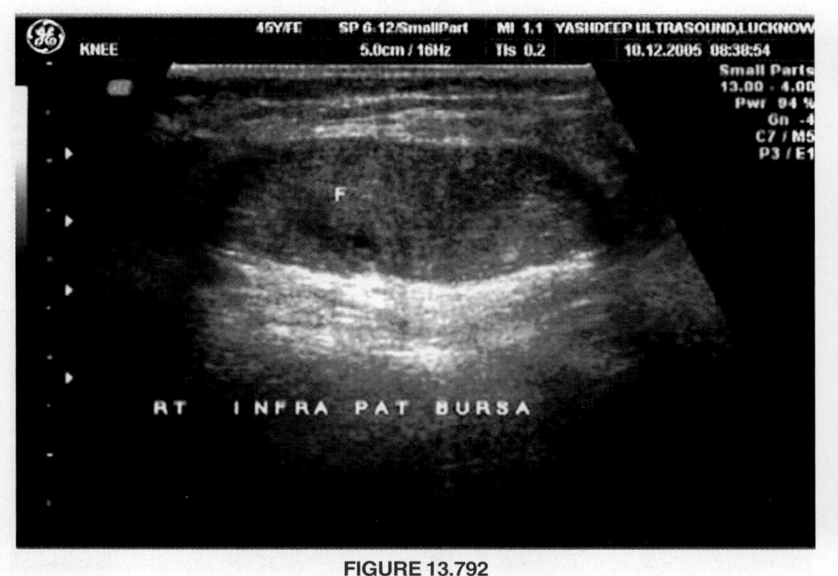

FIGURE 13.792

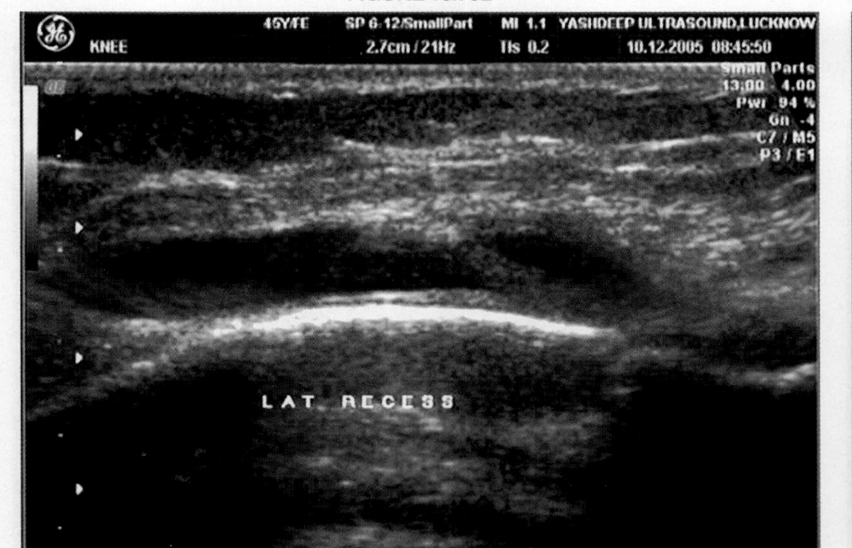

FIGURE 13.793

FIGURES 13.790 to 13.793: Infrapatellar bursitis. A lady presented with marked swelling in the knee. HRSG shows markedly distended infrapatellar bursa. Thick collection is seen in the bursa. The collection is also seen seeping into the lateral recess. Synovial membrane proliferation is also seen. Biopsy confirmed chronic bursitis.

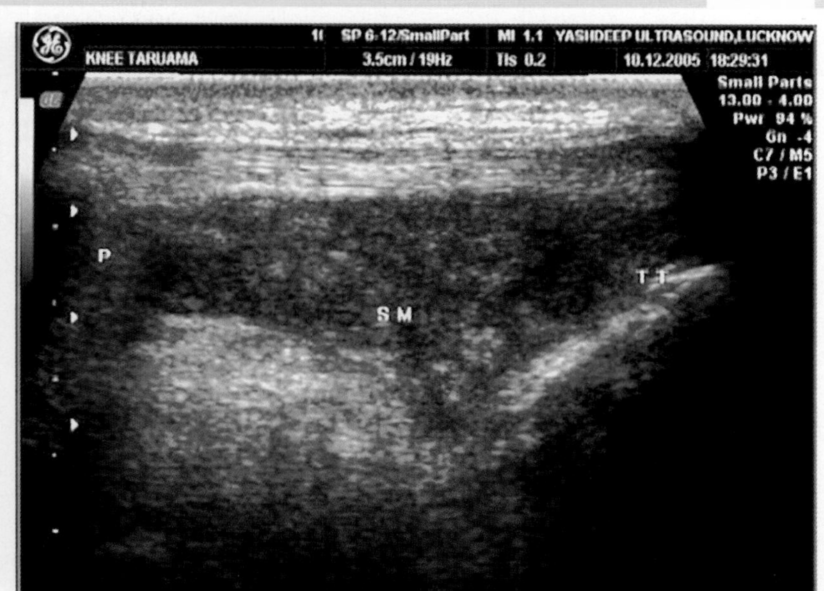

FIGURE 13.794

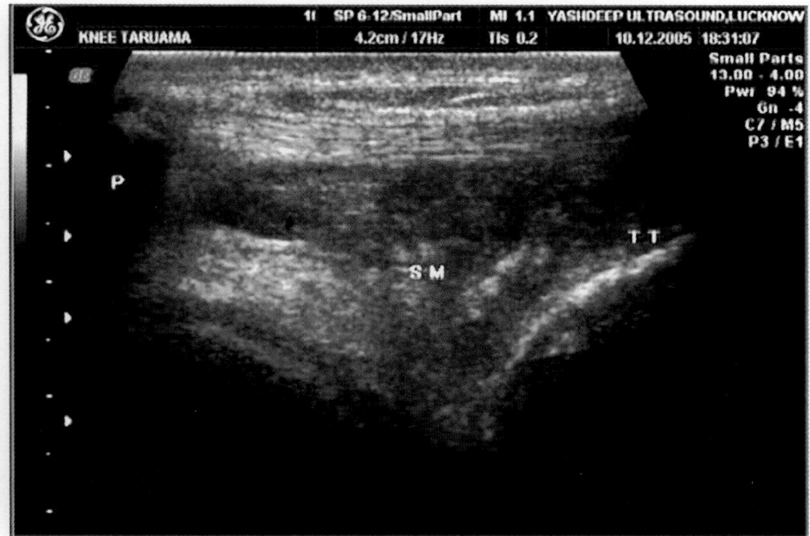

FIGURE 13.795

FIGURE 13.796

FIGURES 13.794 to 13.796: Chronic infrapatellar bursitis with marked synovial membrane proliferation. A patient presented with soft tissue swelling on the anterior and lateral side of the knee. Clinically, it was thought to be a case of meniscal cyst. However, HRSG shows markedly distended infrapatellar bursa. Thick collection is seen in the bursa. Synovial membrane proliferation was also seen. Biopsy of the bursa confirmed chronic inflammatory bursitis.

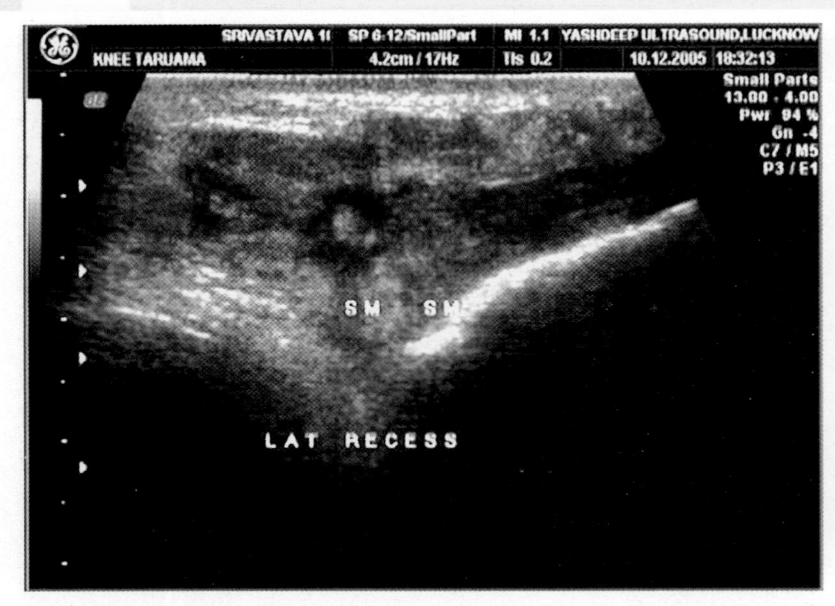

FIGURE 13.797: The collection is also seen in the lateral recess of the joint. Membrane proliferation is seen of the joint.

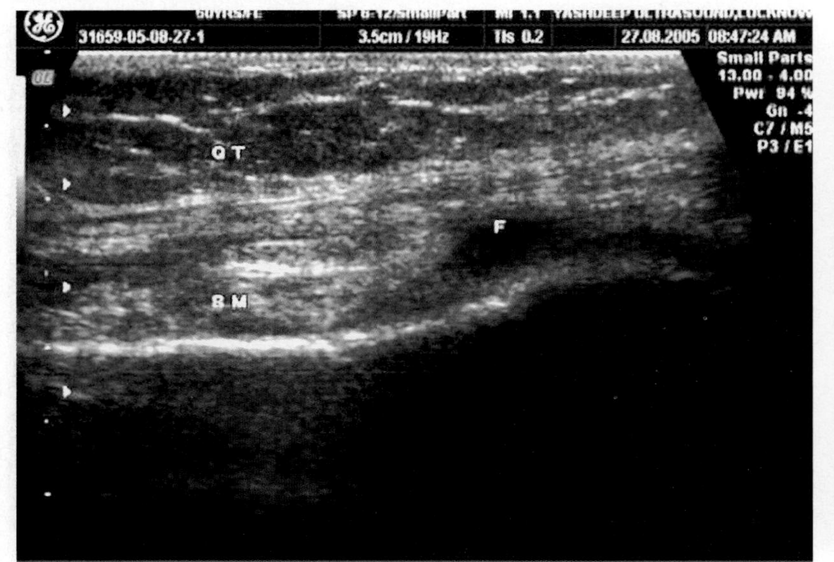

FIGURE 13.798: Suprapatellar bursitis. Echogenic collection is also seen in suprapatellar bursa of the same patient. Marked synovial membrane proliferation is seen in the bursa.

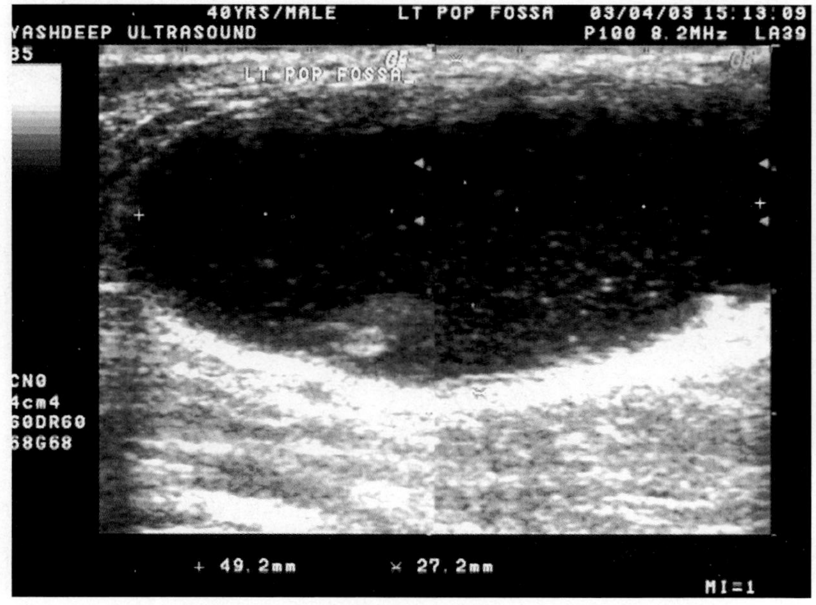

FIGURE 13.799

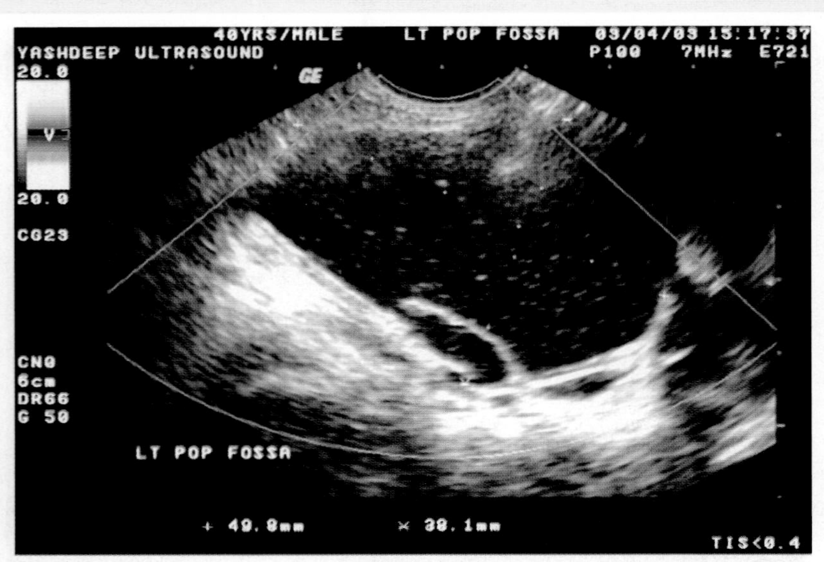

FIGURE 13.800

FIGURES 13.799 and 13.800: Big baker's cyst. HRSG shows a big cystic mass in the Lt popliteal fossa. Multiple internal echoes are seen floating in the cyst. Echogenic debris is also seen in the cyst. A thin septum is also seen in the cyst. Aspiration of the cyst confirmed Baker's cyst.

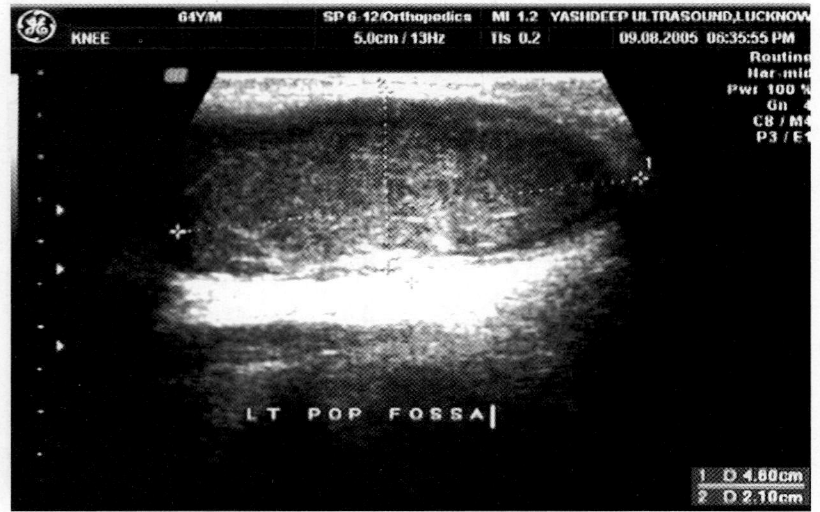

FIGURE 13.801

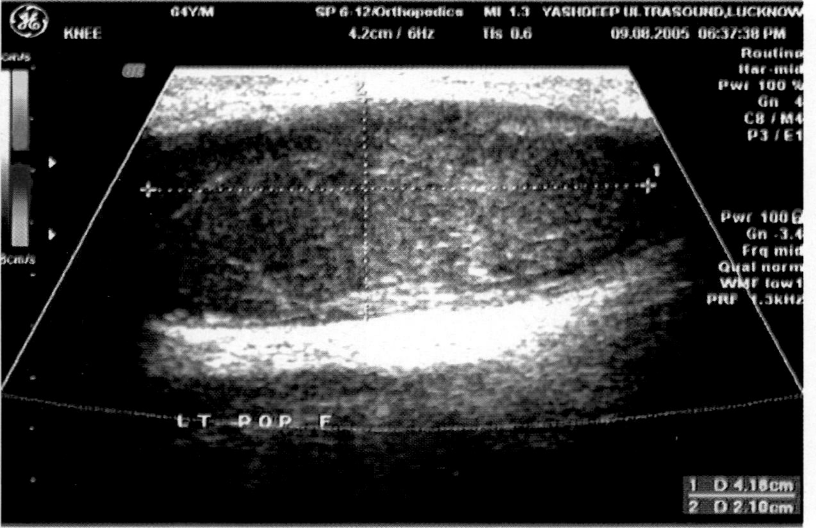

FIGURE 13.802

FIGURES 13.801 and 13.802: Hemorrhagic baker's cyst. A patient presented with acute pain in the knee with marked swelling on the popliteal side. HRSG shows big cystic mass in the Lt popliteal fossa. Thick echogenic collection is seen in the mass. It looks like a ball. On color flow imaging no flow was seen in it. Aspirate of the cyst turned out to be hemorrhagic collection.

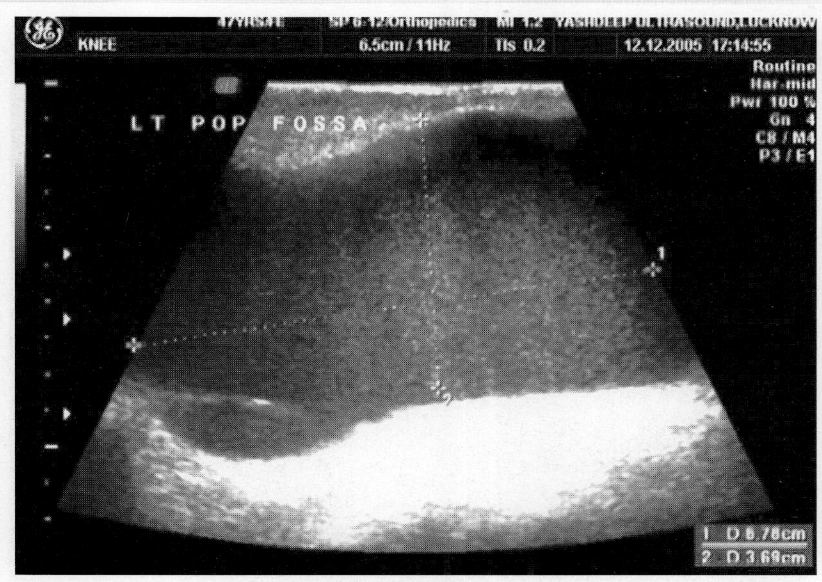

FIGURE 13.803

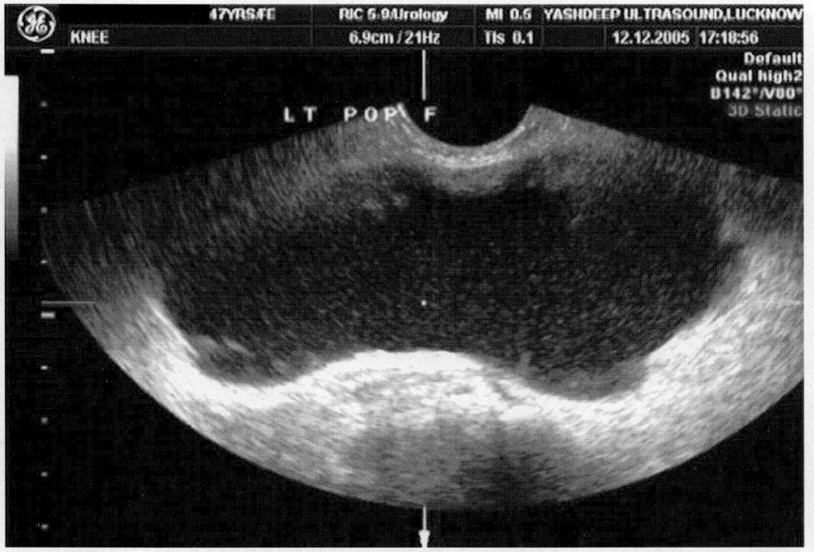

FIGURE 13.804

FIGURES 13.803 and 13.804: Huge baker's cyst in Lt popliteal fossa. HRSG shows a huge baker's cyst in the Lt popliteal fossa. The cyst is seen occupying whole of the popliteal fossa. Internal echoes are seen in the cyst. Pressure is seen on the popliteal vessels. Venous flow obstruction was noted resulting into venous insufficiency.

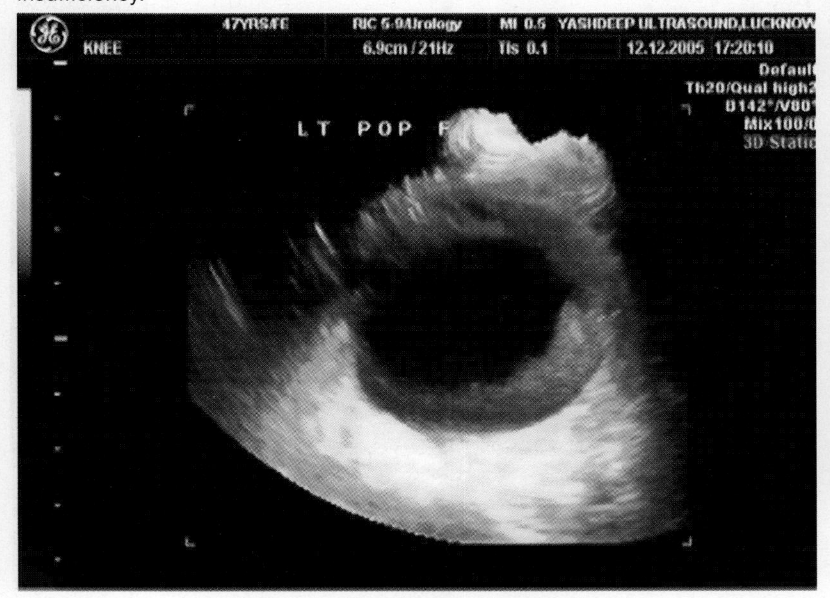

FIGURE 13.805

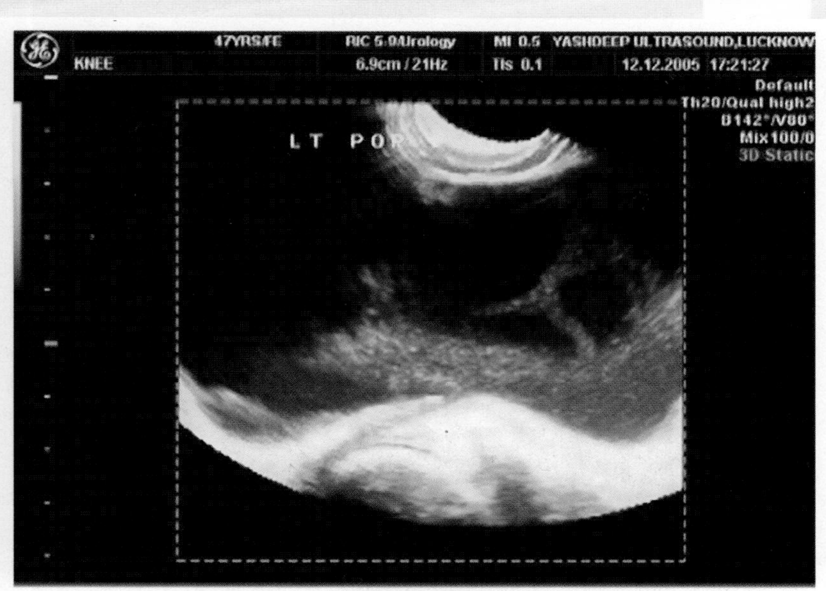

FIGURE 13.806

FIGURES 13.805 and 13.806: 3D and 4D imaging of the cyst shows better details of the cyst with thin septa in the cyst.

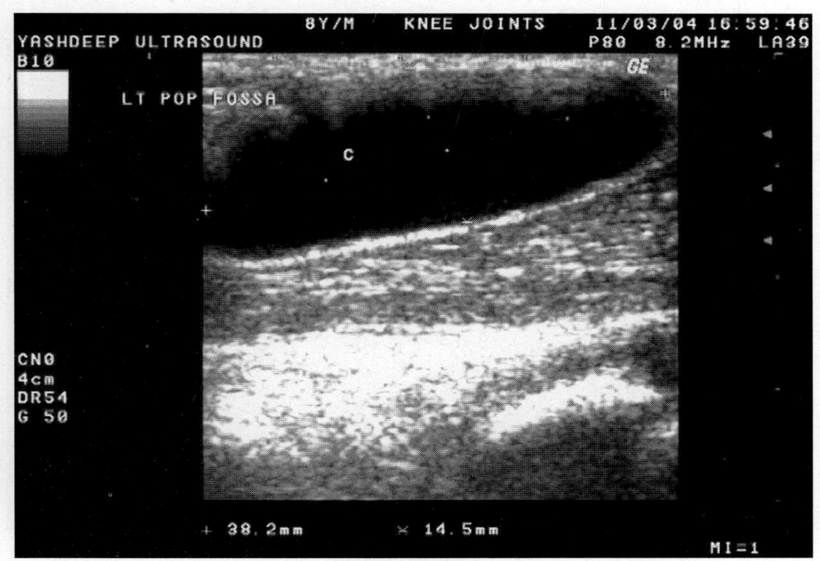

FIGURE 13.807

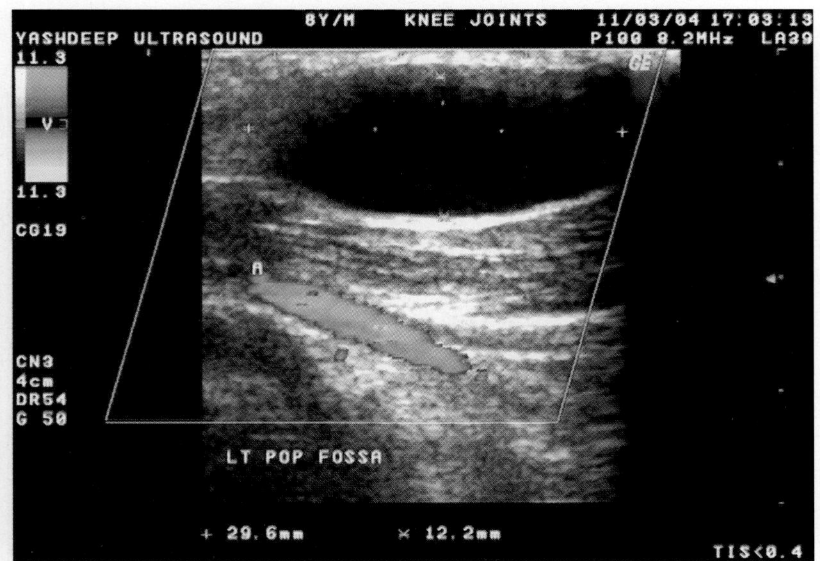

FIGURE 13.808

FIGURES 13.807 and 13.808: Gastrocnemius cyst. HRSG shows well-defined thin walled cyst in the gastrocnemius muscle. No internal echo in the cyst. **On color flow imaging** no flow seen in the cyst wall.

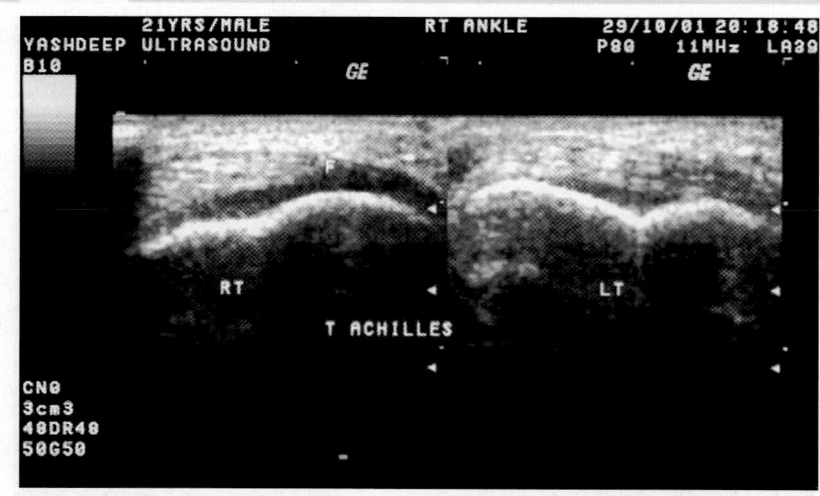

FIGURE 13.809: **Retrocalcaneal bursa.** HRSG shows retrocalcaneal bursa on both sides Rt and Lt. The Lt bursa is normal whereas the Rt bursa shows fluid collection. Normal bursal pocket is 3 mm.

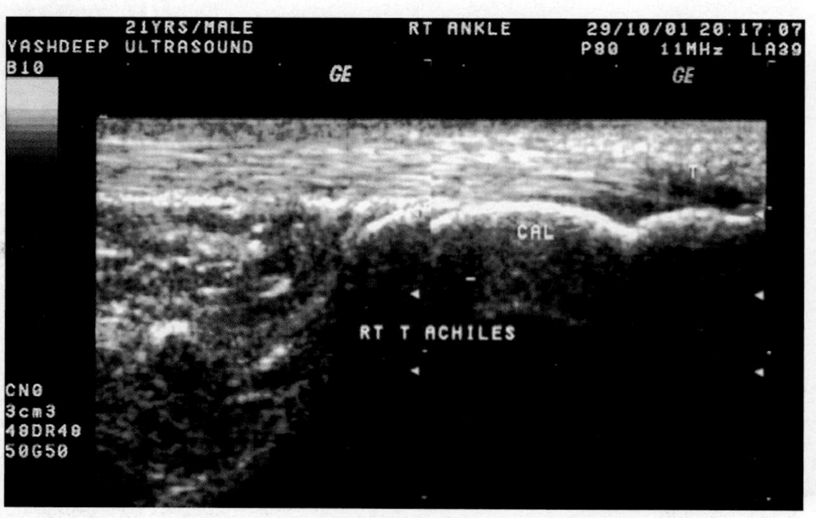

FIGURE 13.810

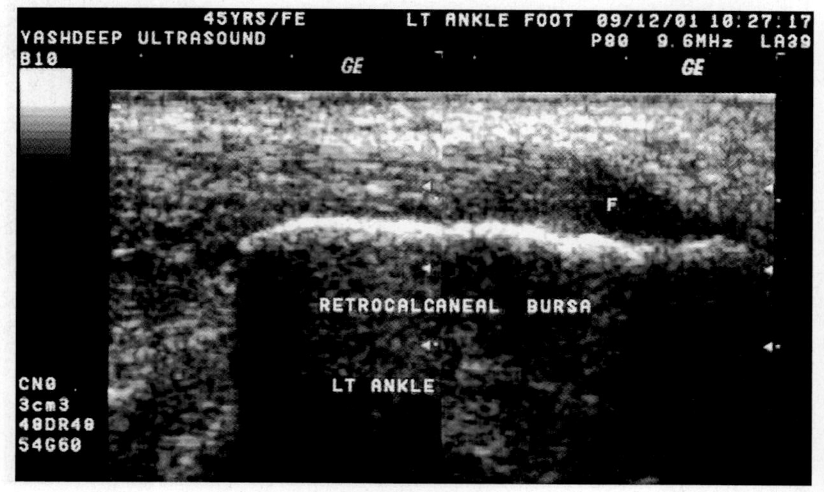

FIGURE 13.811

FIGURES 13.810 and 13.811: Retrocalcaneal bursitis. HRSG shows low-level collection in the retrocalcaneal bursa. The distal fibers of the Achilles tendon are also seen disrupted suggestive of partial tear of tendon Achilles.

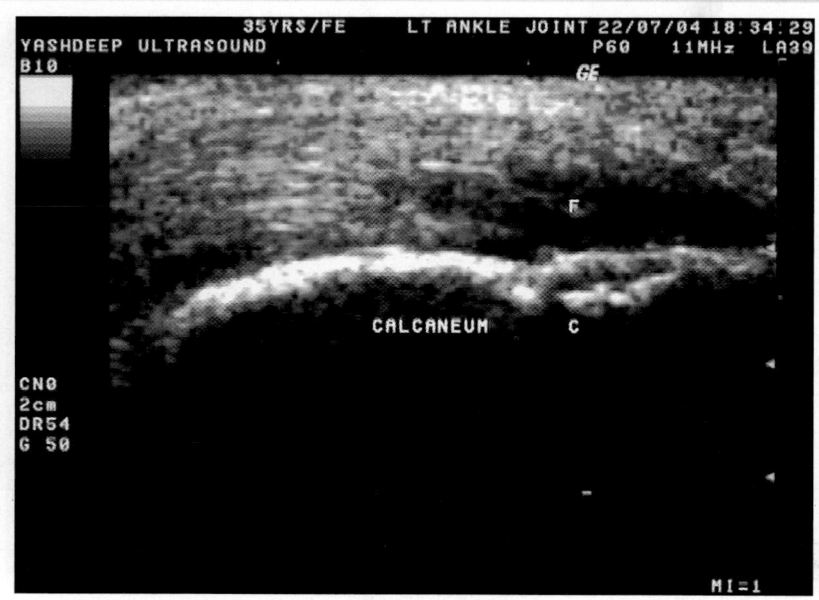

FIGURE 13.812

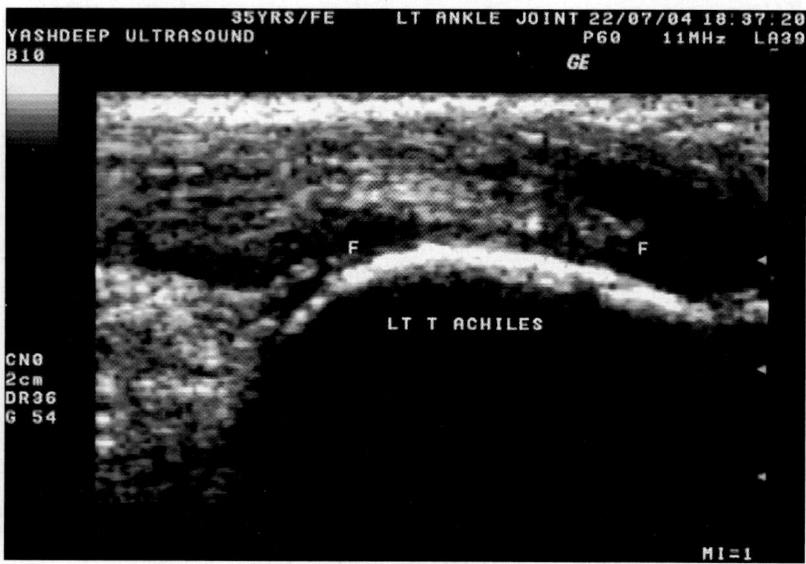

FIGURE 13.813

FIGURES 13.812 and 13.813: Retrocalcaneal bursitis. A patient presented with painful ankle syndrome. HRSG shows fluid collection in the retrocalcaneal bursa. Echogenic shadows are seen floating in the fluid. The distal fibers Achilles are seen torn.

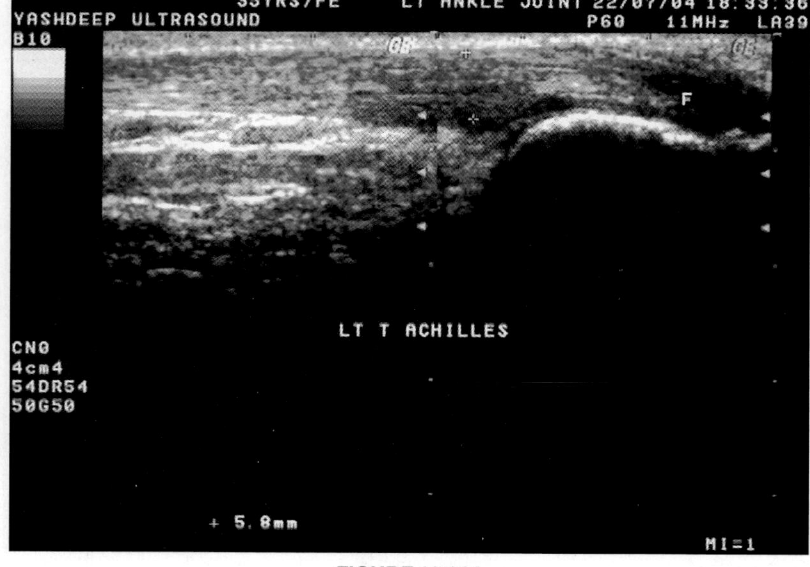

FIGURE 13.814

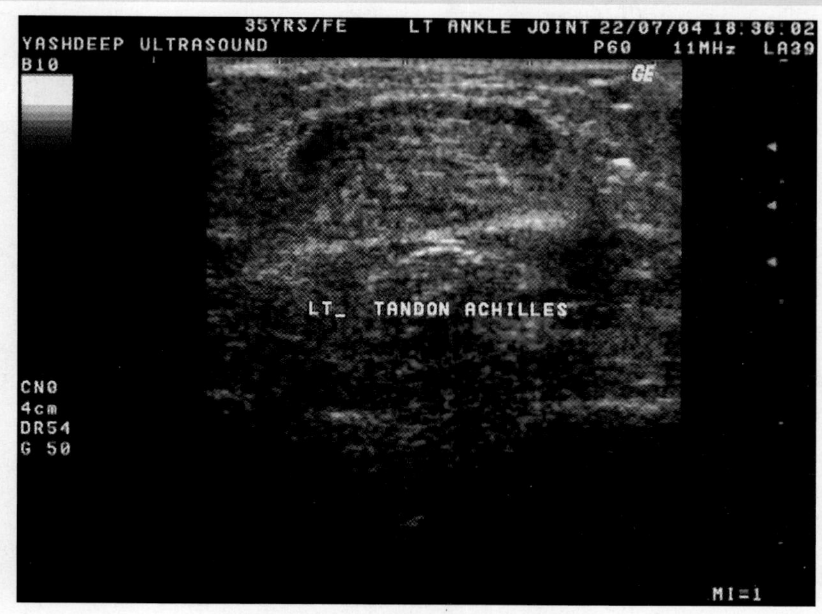

FIGURE 13.815

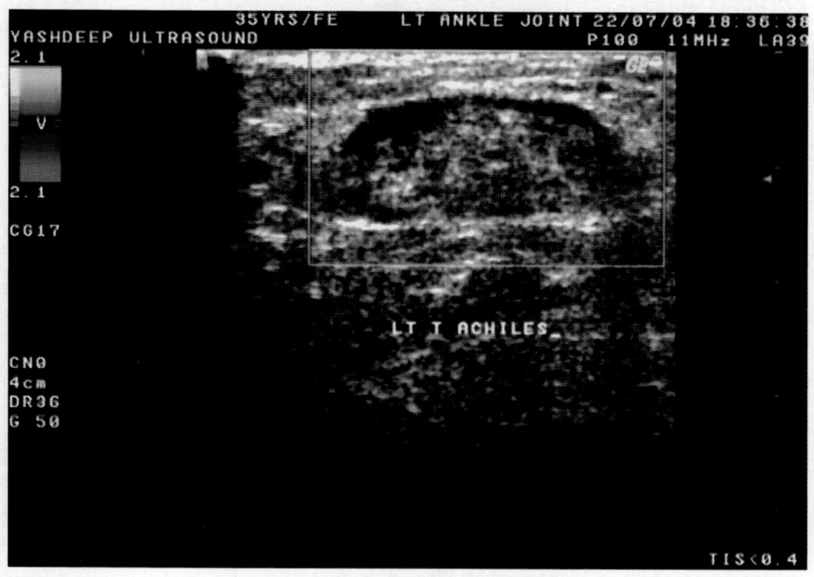

FIGURE 13.816

FIGURES 13.814 to 13.816: The distal 1/3rd of the tendon is seen torn. The tendon is seen retracted. Associated hematoma formation is also seen in the ankle. **On color flow imaging** no flow was seen in the tendon.

Bakers Cyst

Bakers cysts are the enlarged bursa found medially in the popliteal fossa described by Baker. It is the gastrocnemius semimembranous bursa. The location is constant. Fluid collection can take place and develop into a cyst. Hemorrhagic collection may follow after trauma.

HRSG of Periosteum and Bone

Thickening of periosteum is nonspecific finding seen in trauma, fracture, osteotomy, infections, tumors, pagets disease, venous stasis, hypertrophic pulmonary osteoarthopathy and metabolic disorders. HRSG clearly demonstrates periosteal thickening. Usually conventional X-ray is choice of investigation. However in diagnosis of early osteomyelitis and stress fracture, HRSG is superior to the X-ray. Acute osteomyelitis shows elevation of periosteum by a hypoechoic layer of purulent material. It can be easily picked up on HRSG. Results of HRSG are usually positive within 24 hrs following onset of fever in cases of osteomyelitis. Results are positive between 48 to 72 hours in radioisotope bone scanning and radiologically become positive after one week or more.

In chronic osteomyelitis HRSG is not only valuable in making the diagnosis but also helpful in assessment of adjacent soft tissue. In chronic osteomyelitis changes are also seen in adjacent soft tissue with abscess formation. Fluid collection is seen around bony contour in 70% of cases in chronic osteomyelitis. The bone soft tissue interface is highly reflective, which is seen as bright line with acoustic shadowing deep to the interface. The highly reflective cortical bone and tomographic nature of ultrasound imaging make it ideal for evaluation of bony contours in evaluation of all musculoskeletal ultrasound imaging. The identification of grooves, fossae, tuberosities, trochanters and epicondyles guides study. Occult fractures on conventional X-rays can be detected on HRSG. High frequency transducers are very important in these cases. The technique of sonographic palpation is valuable in localizing the region of interest. Often a cortical discontinuity is identified. Associated adjacent subperiosteal hemorrhage will confirm the diagnosis. Sonographic evaluation of bone in children is more informative than conventional X-ray as cartilages can be seen on HRSG.

Ilizarov Leg Lengthening

HRSG is now days used to evaluate new bone formation in patients, who have under gone an ilizarov leg lengthening procedure. This technique corticotomy is performed while leaving the periosteum intact. Fine tensile wires are inserted through the bone and attached to external fixator rim. These two fragments are slowly and gradually stretched apart @ 1mm per day. The new bone is formed along the inter zone and it can be picked up as a thin echogenic line bridging the gap. It is seen much before when they are radiologically seen. HRSG is also used in the callus formation between the fractured segments in the bone. Newly formed callus cannot be seen on X-ray but can be picked up easily on HRSG.

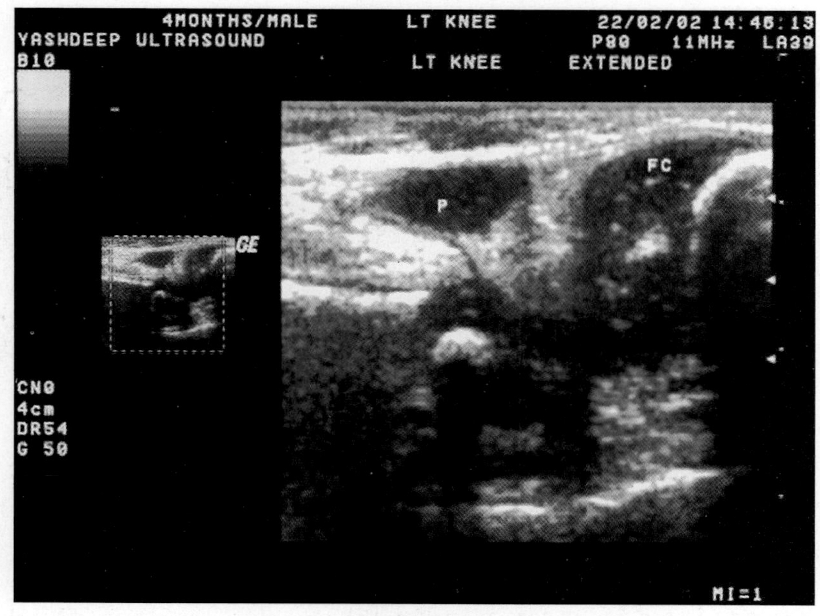

FIGURE 13.817

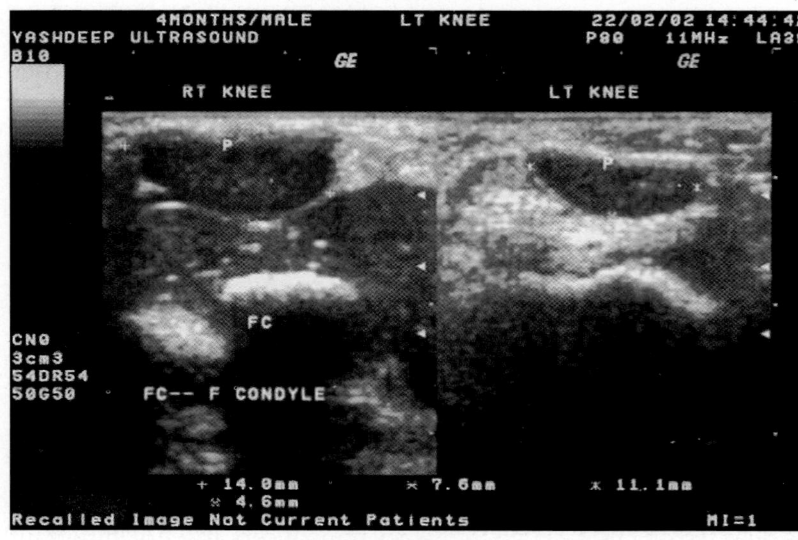

FIGURE 13.818

FIGURES 13.817 and 13.818: Congenital malposition of patella. A newborn child presented with permanently extended left leg. The patella was not palpable in normal position. HRSG shows the patella placed before the femoral condyle. It was seen as well-defined hypoechoic ovoid shadow on HRSG. The right knee shows normal placed patella.

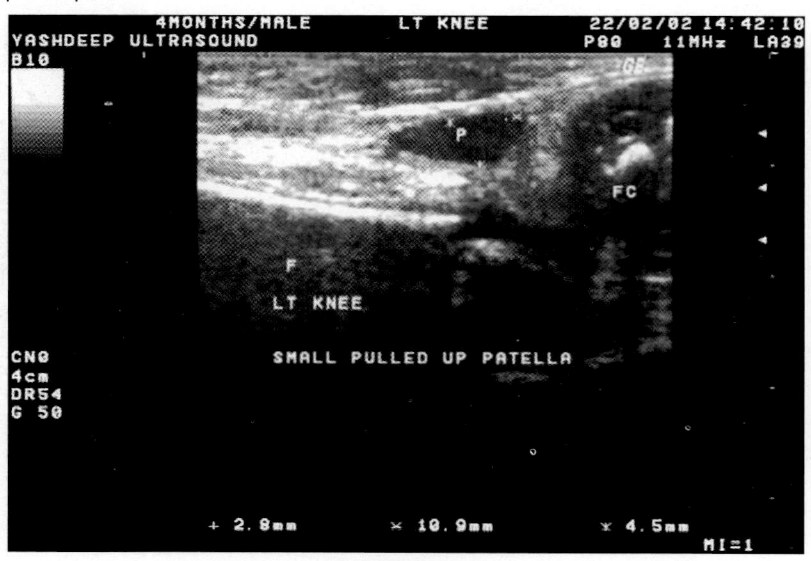

FIGURE 13.819

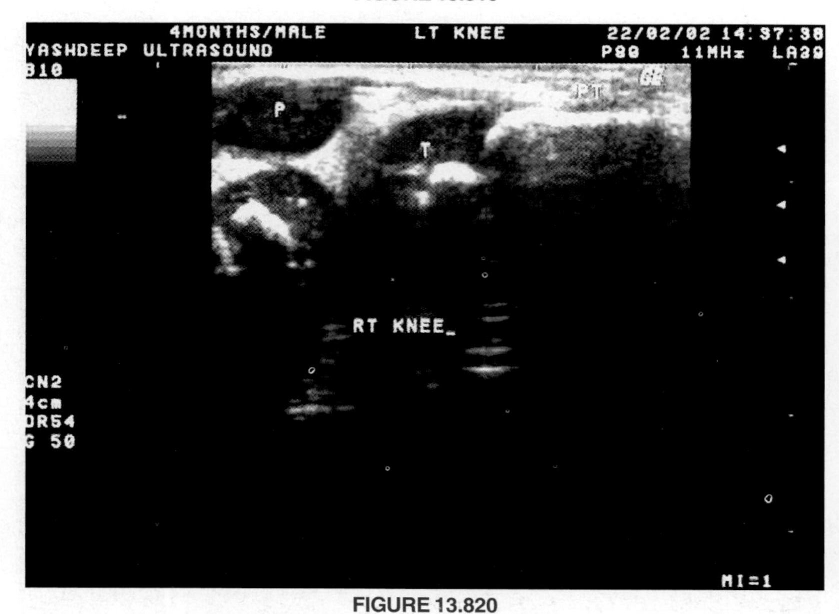

FIGURE 13.820

FIGURES 13.819 and 13.820: The longitudinal view of the knee shows the mal positioned patella on the Lt side and normal patella on the Rt side.

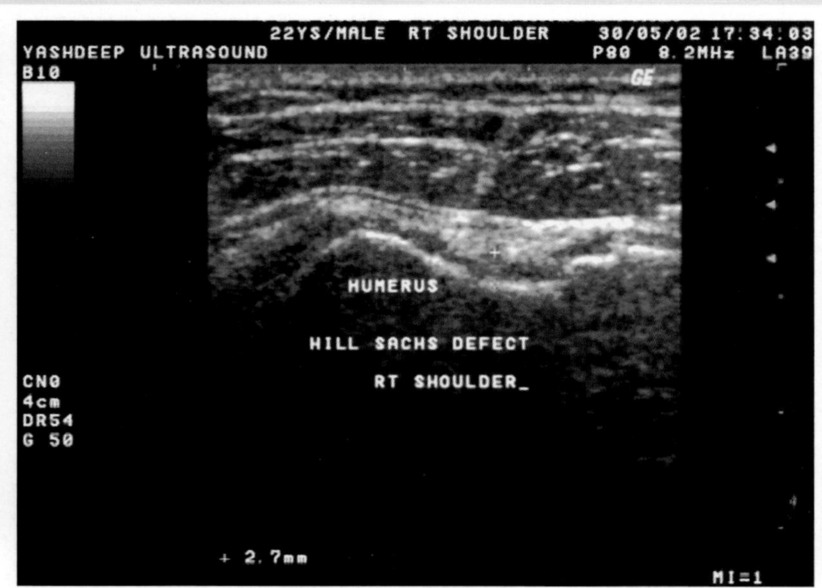

FIGURE 13.821

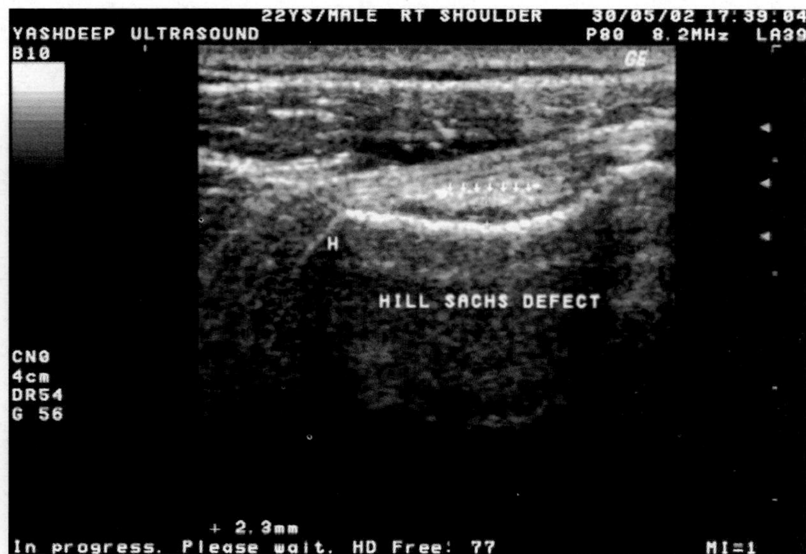

FIGURE 13.822

FIGURES 13.821 and 13.822: Hill Sach's Defect. A patient presented with recurrent dislocation of the Rt shoulder. HRSG shows abnormal shape of greater tuberosity of humerus. A big concavity is seen in the greater tuberosity of humerus instead of normal convex contour. Therefore, it was not properly sitting in the glenoid cavity. The concave shape of humeral head results in the recurrent dislocation of the shoulder joint.

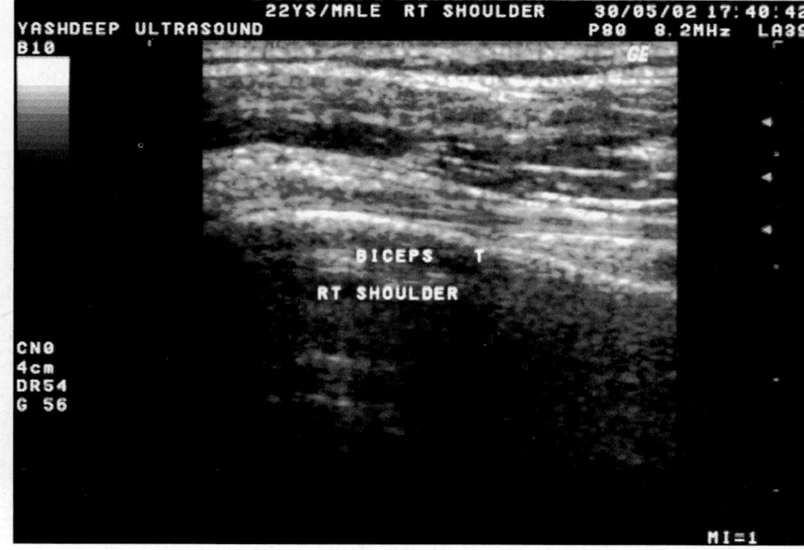

FIGURE 13.823

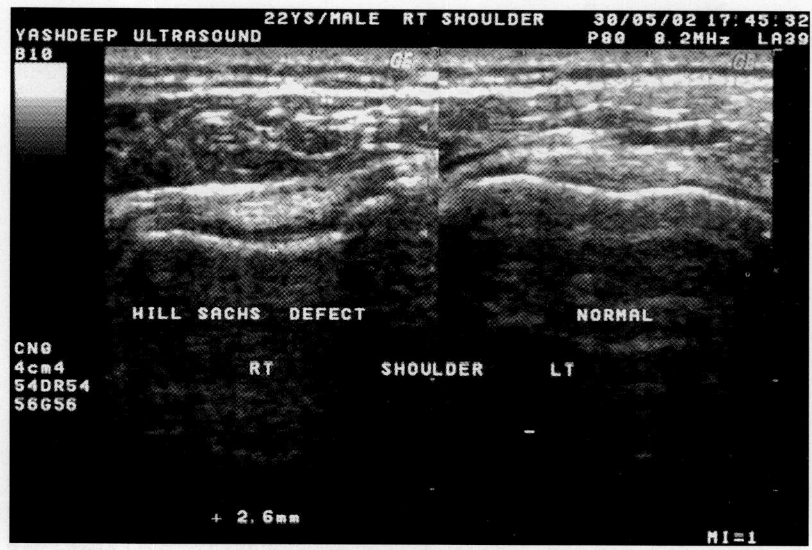

FIGURE 13.824

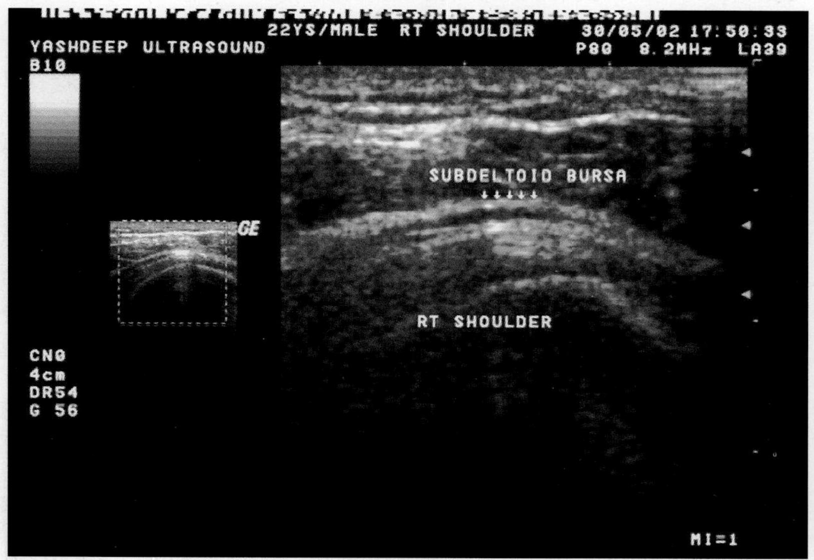

FIGURE 13.825

FIGURES 13.823 to 13.825: HRSG shows the normal contour of greater tuberosity of the Lt side and defective contour of tuberosity on Rt side. The biceps tendon is also seen thickened due to recurrent dislocation resulting into tendonosis. There is also evidence of fluid collection seen in subdeltoid bursa due to recurrent dislocation. The bursal walls are echogenic and irregular in outline resulting into frictional bursitis.

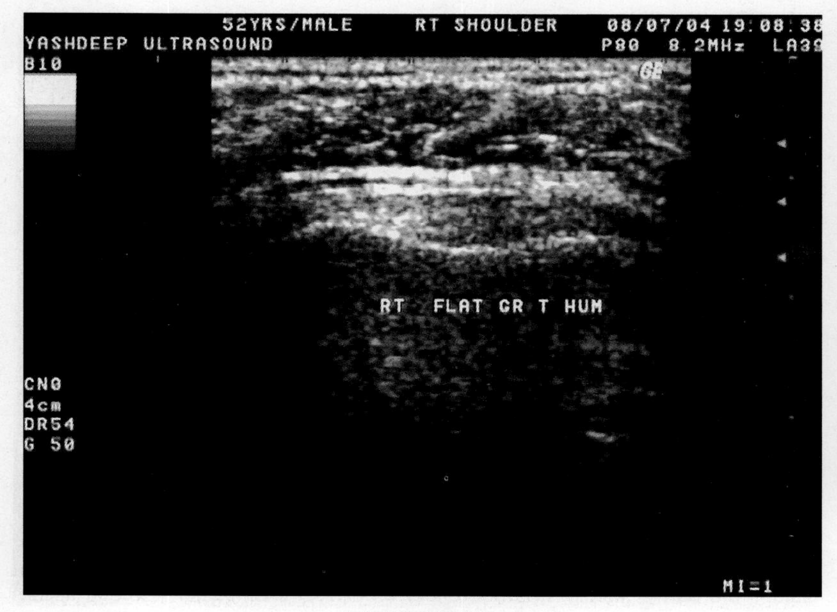

FIGURE 13.826

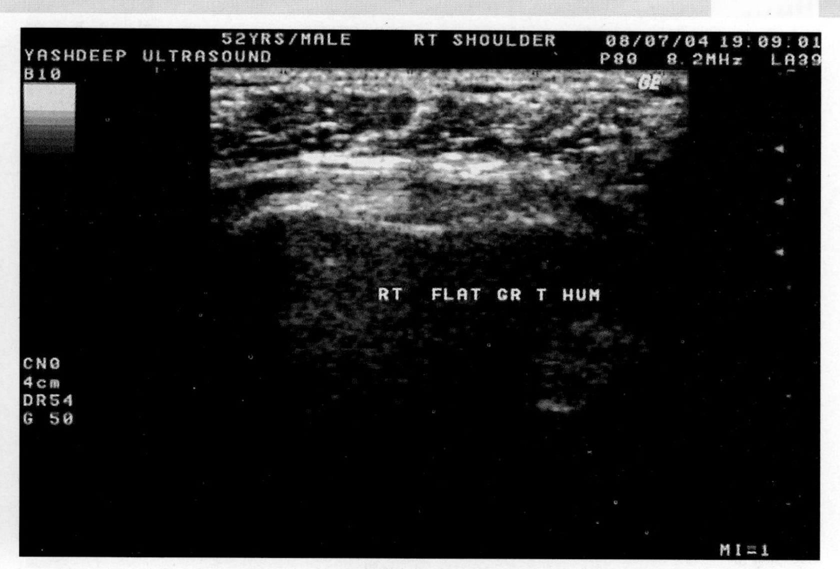

FIGURE 13.827

FIGURES 13.826 and 13.827: Hill Sach's defect. HRSG shows flat articular surface of greater tuberosity of humerus suggestive of Hill Sach's defect. A patient presented with recurrent dislocation of Rt shoulder joint. Normal convex contour of the humeral head is lost. The articular surface shows flattening and concavity suggestive of Hill Sach's defect.

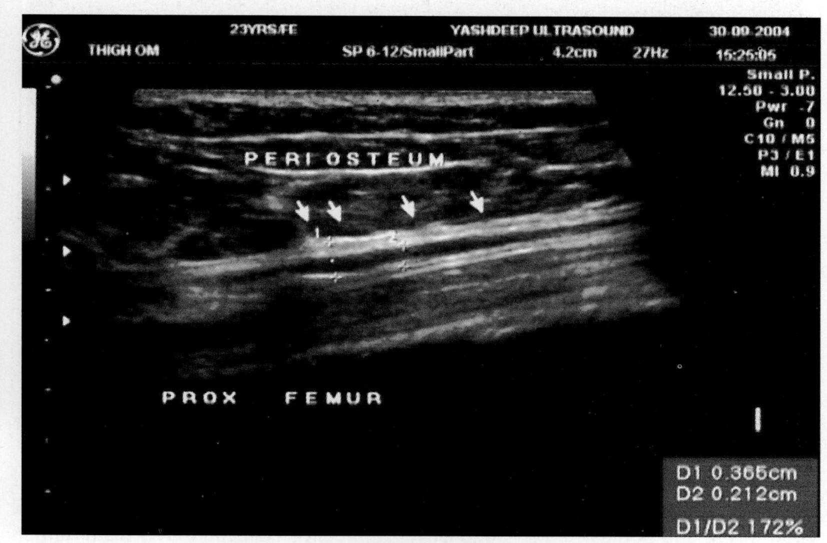

FIGURE 13.828

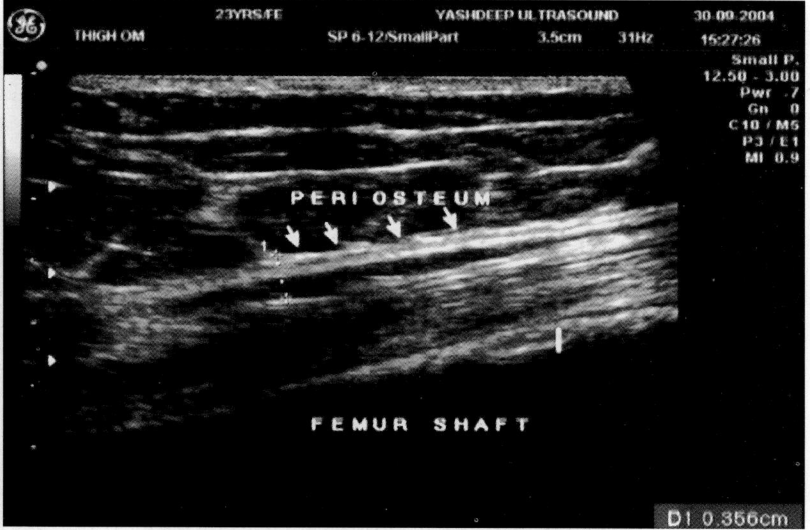

FIGURE 13.829

FIGURES 13.828 and 13.829: Acute osteomyelitis. HRSG shows irregular thickening of femoral cortex. It is echogenic in texture. Lifting of the periosteum with subperiosteal fluid collection is seen.

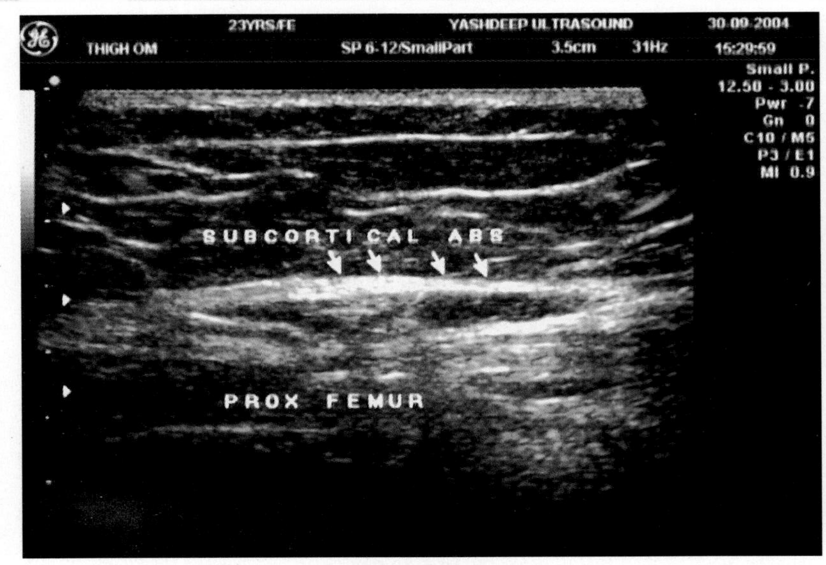

FIGURE 13.830

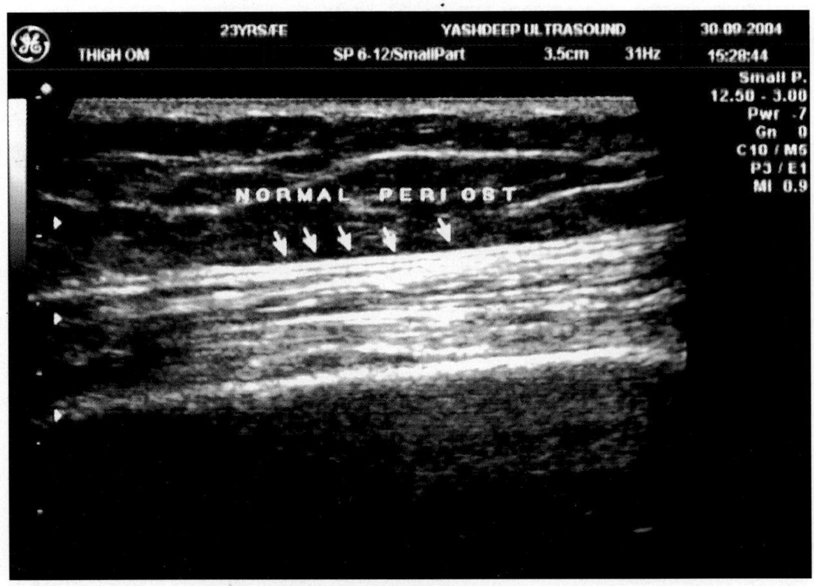

FIGURE 13.831

FIGURES 13.830 and 13.831: There is also evidence of thick echogenic shadow seen in the sub cortical region suggestive of infective collection. The distal femur shows normal bony cortex and bony periosteum.

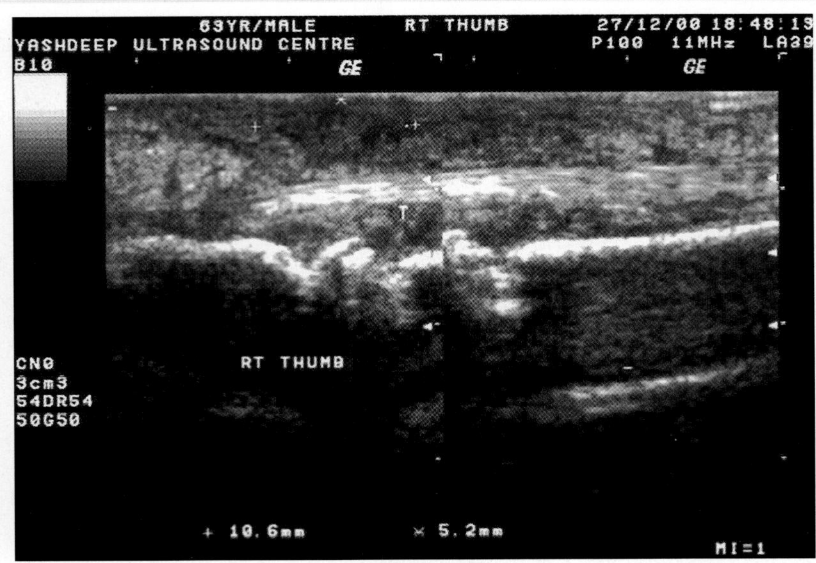

FIGURE 13.833

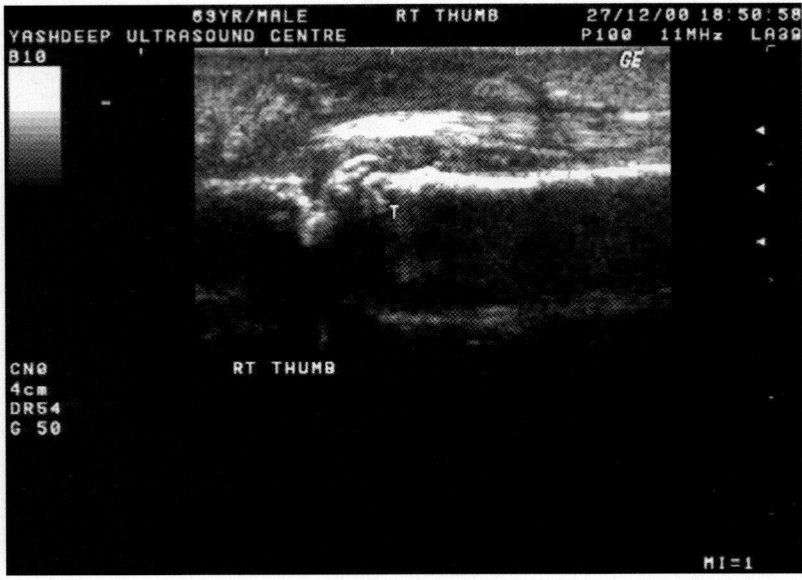

FIGURE 13.834

FIGURES 13.833 and 13.834: HRSG shows bony destruction of head of second phalanx and base of terminal phalanx of the Rt thumb. Associated soft tissue edema is also present. The flexor hallucis longus tendon is also thickened and echogenic suggestive of Tenosynovitis. There is also evidence of small abscess seen in the subcutaneous tissue.

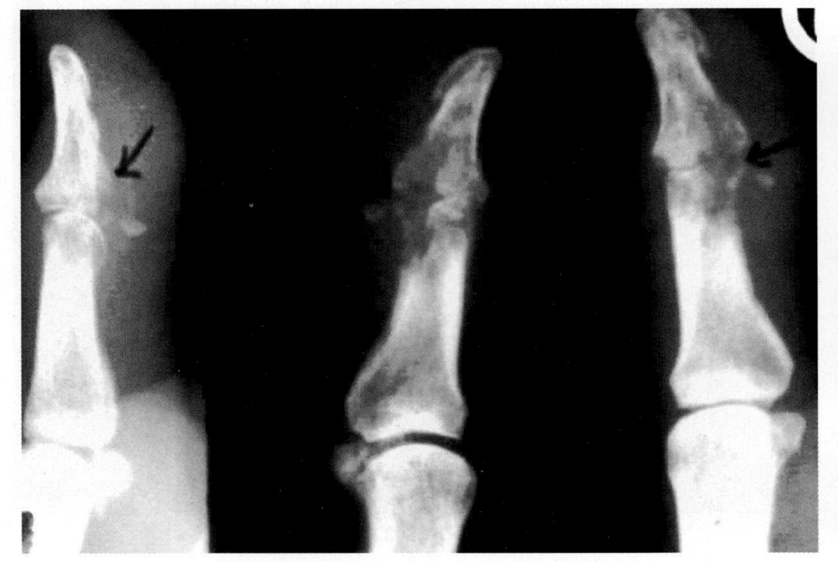

FIGURE 13.832: Chronic osteomyelitis with bony destruction of Rt thumb. Plain X-ray of the patient shows bony destruction of base of terminal phalanx of Rt thumb. Soft tissue swelling is also seen.

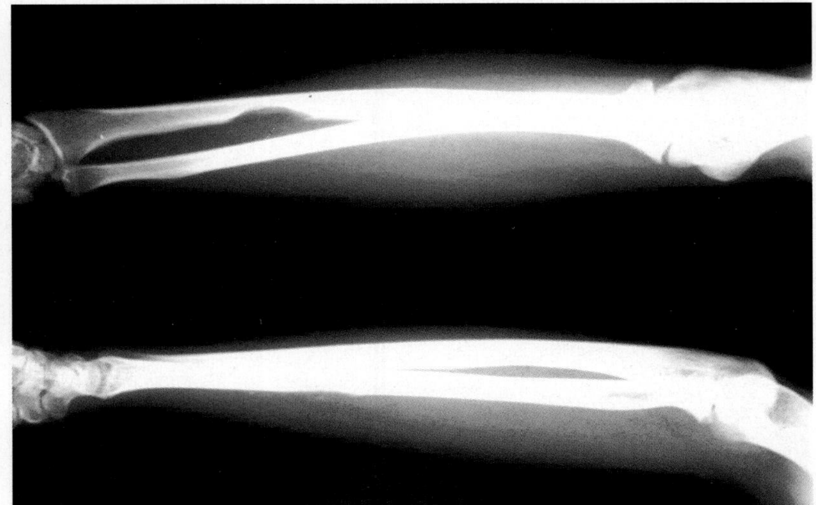

FIGURE 13.835: Plain X-ray of forearm shows a well-defined smooth defect in the distal 1/3rd part of radius. It is not tender on clinical examination. No soft tissue swelling is seen.

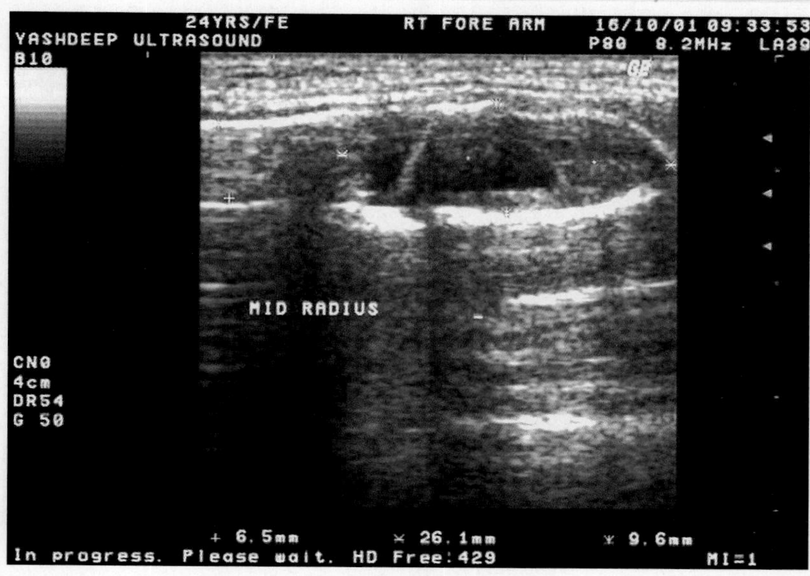

FIGURE 13.836

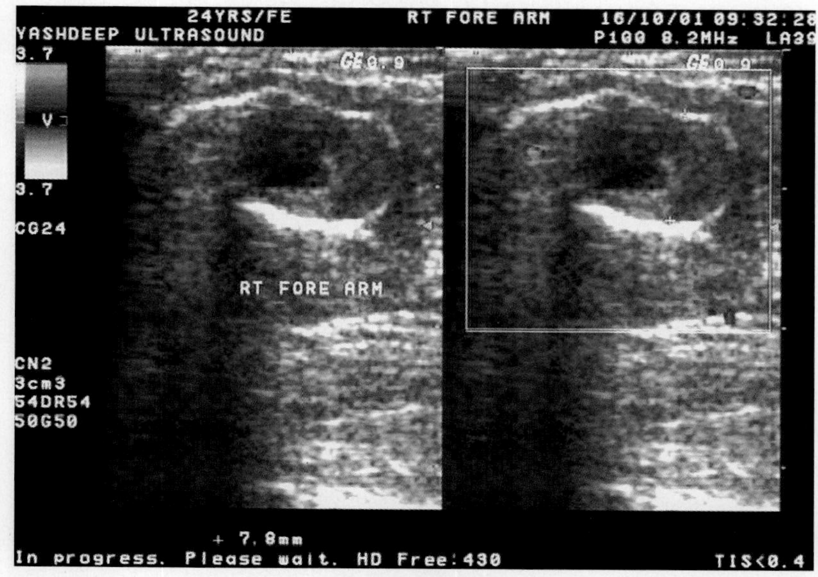

FIGURE 13.837

FIGURES 13.836 and 13.837: HRSG shows a big hypoechoic cystic mass is seen in the bone. The cortex is splayed. Central area of necrosis is seen in the mass. Normal bony medullary texture is lost.

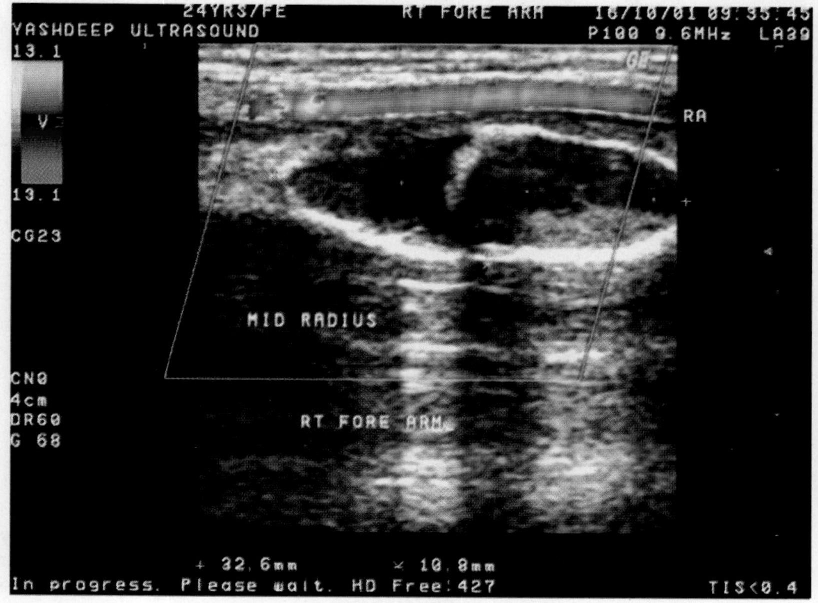

FIGURE 13.838

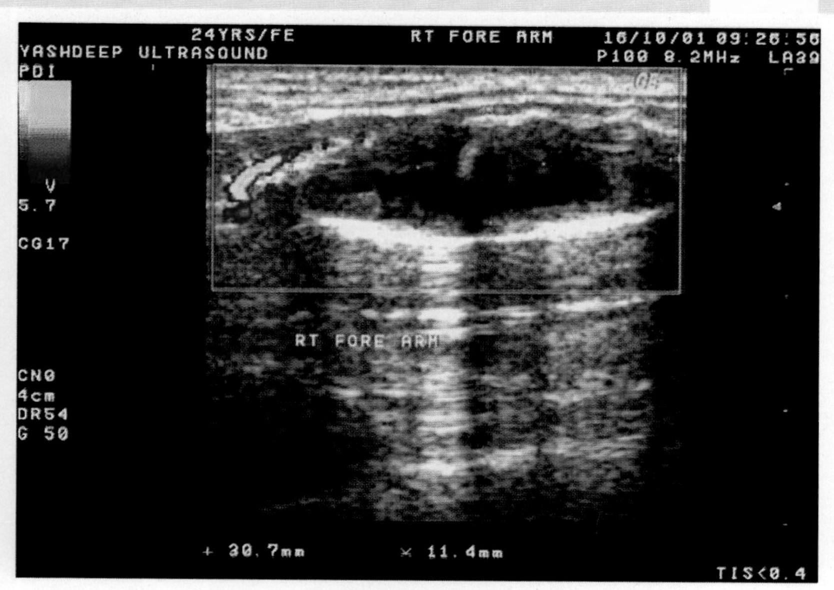

FIGURE 13.839

FIGURES 13.838 and 13.839: On color flow imaging poor flow is seen in the cyst. Biopsy of the cyst shows unilocular bone cyst.

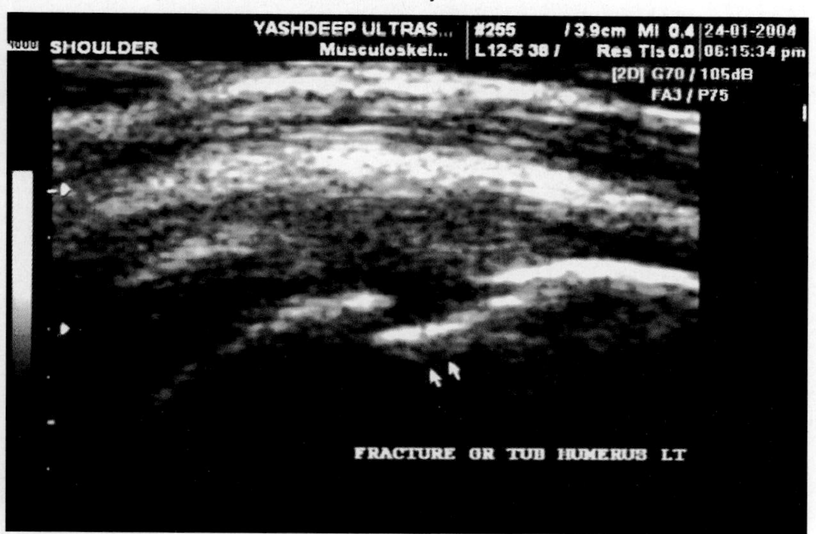

FIGURE 13.840

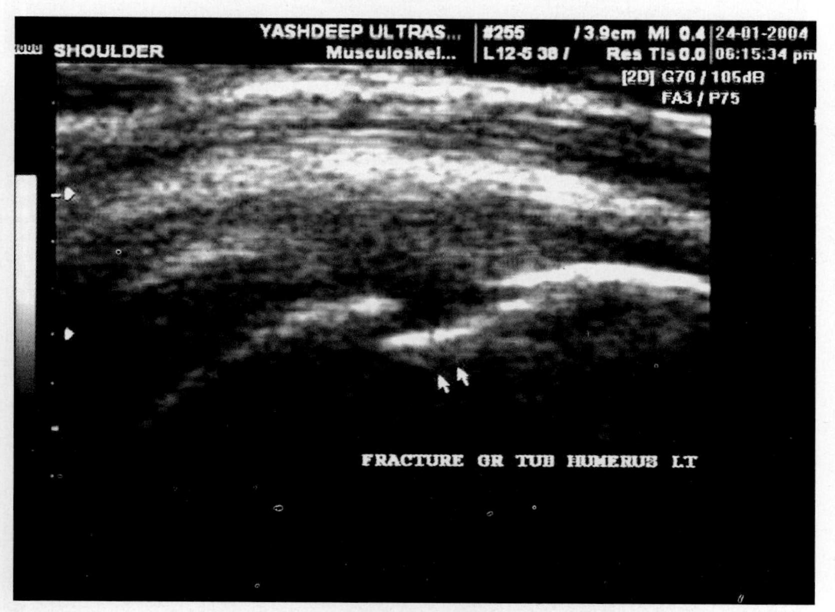

FIGURE 13.841

FIGURES 13.840 and 13.841: Fracture greater tuberosity. HRSG shows fracture of the greater tuberosity of the humerus in a patient sustained blunt trauma. The fracture segment is well visualized on HRSG.

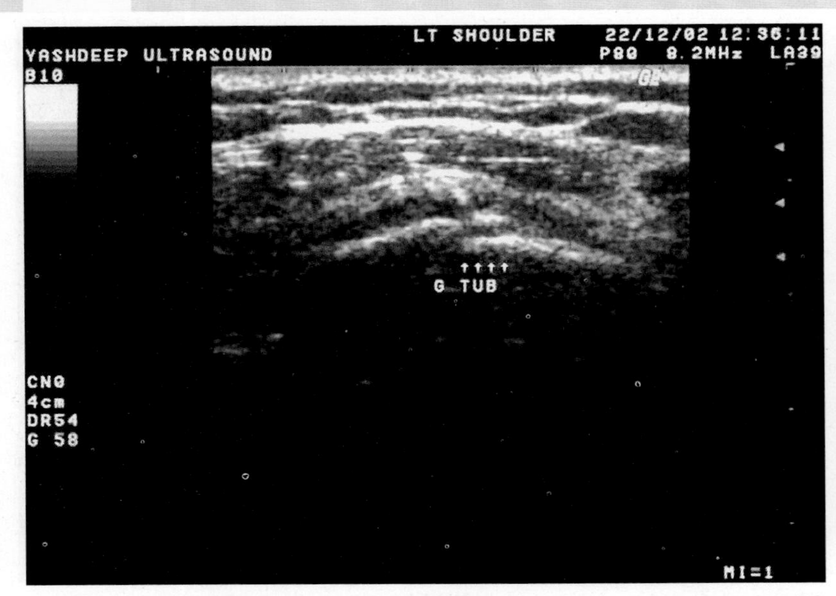

FIGURE 13.842: Subtle fracture of greater tuberosity of humerus. HRSG shows subtle fracture of greater tuberosity of humerus in a patient had painful shoulder movements. The fracture cartilage is seen in close apposition.

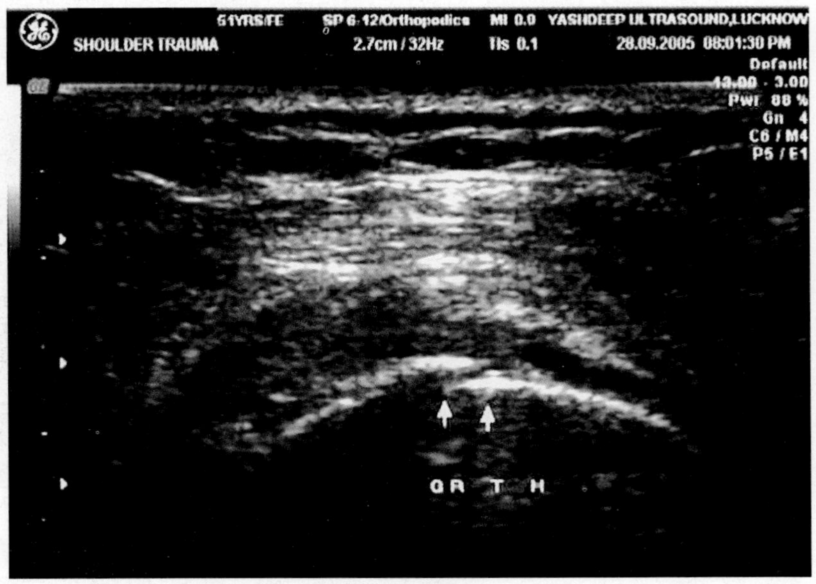

FIGURE 13.843

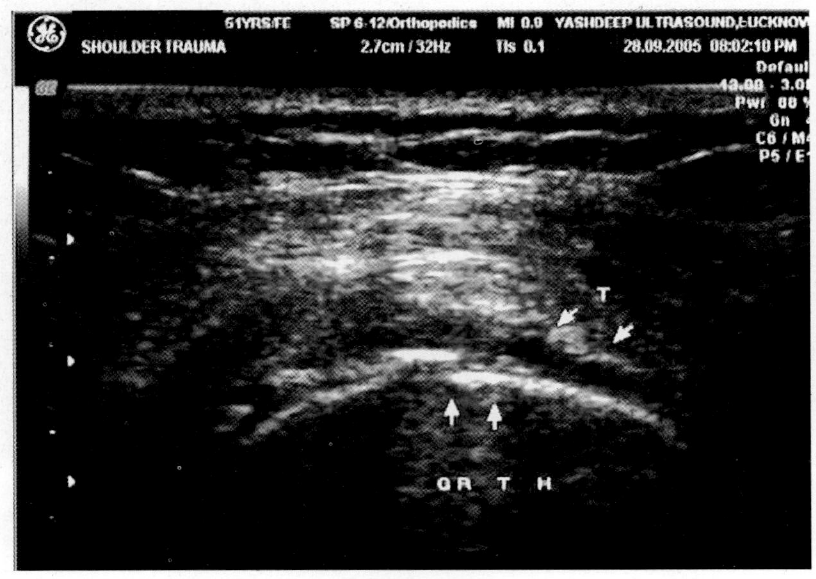

FIGURE 13.844

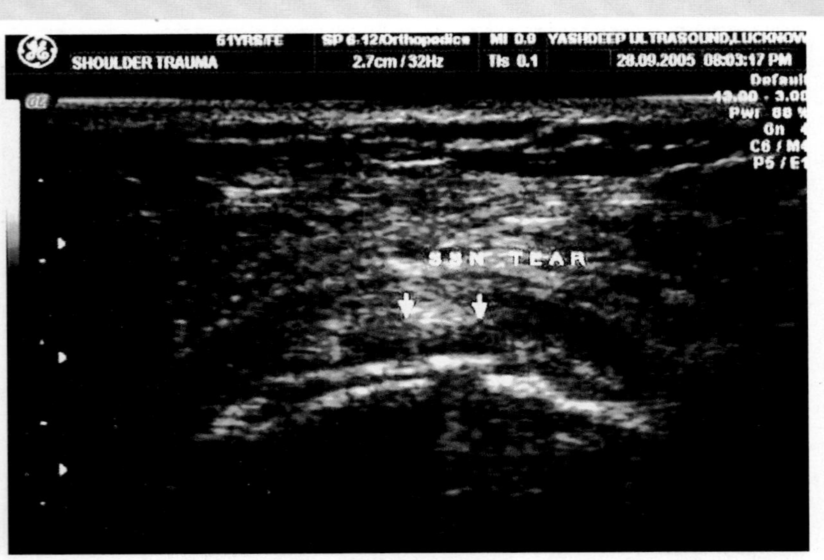

FIGURE 13.845

FIGURES 13.843 to 13.845: Fracture of greater tuberosity of humerus with partial tear of supraspinatus tendon. HRSG shows fracture of greater tuberosity of humerus. Disruption of the bony cartilage is seen. There is also evidence of partial tear of supraspinatus tendon seen.

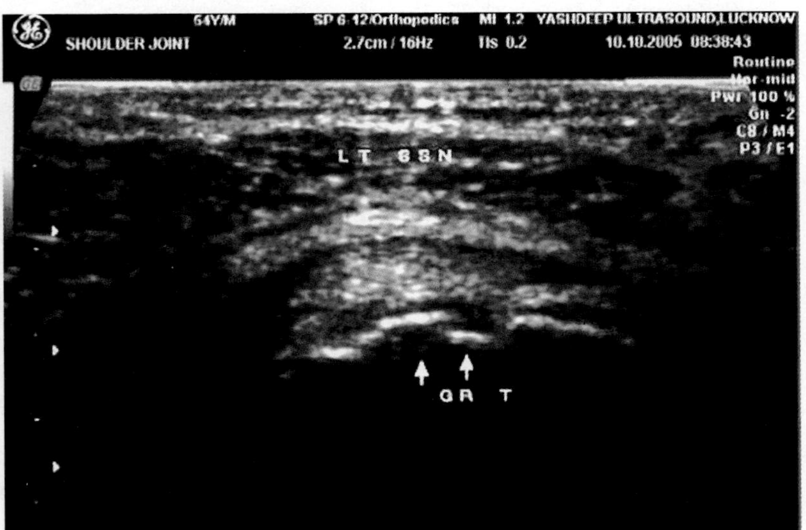

FIGURE 13.846

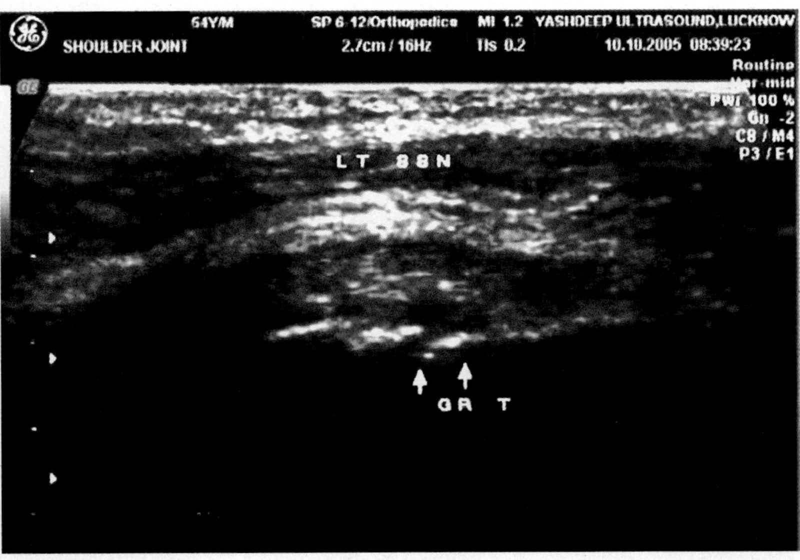

FIGURE 13.847

FIGURES 13.846 and 13.847: Irregular articular surface of greater tuberosity of humerus A patient presented with painful shoulder movements. The articular surface of the greater tuberosity of the humerus is markedly irregular in outline. The cartilage shows erosion with uneven surface. Partial rupture of the cartilage is also seen.

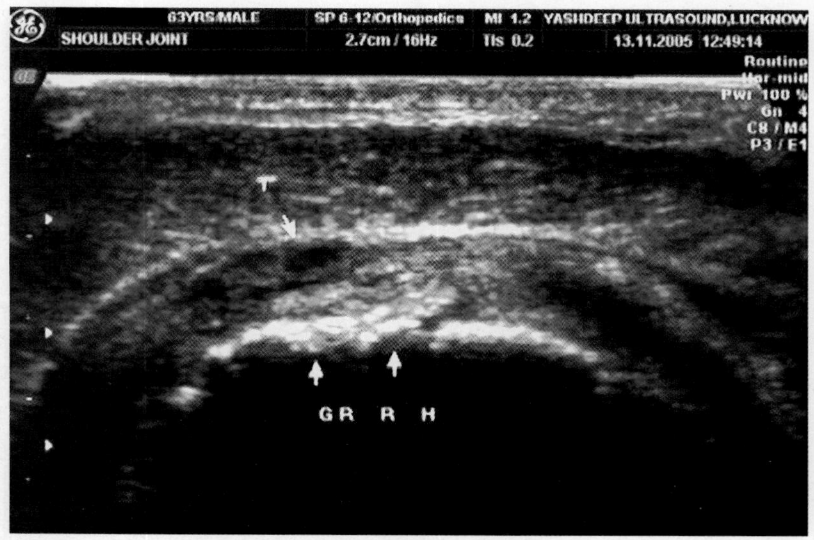

FIGURE 13.848

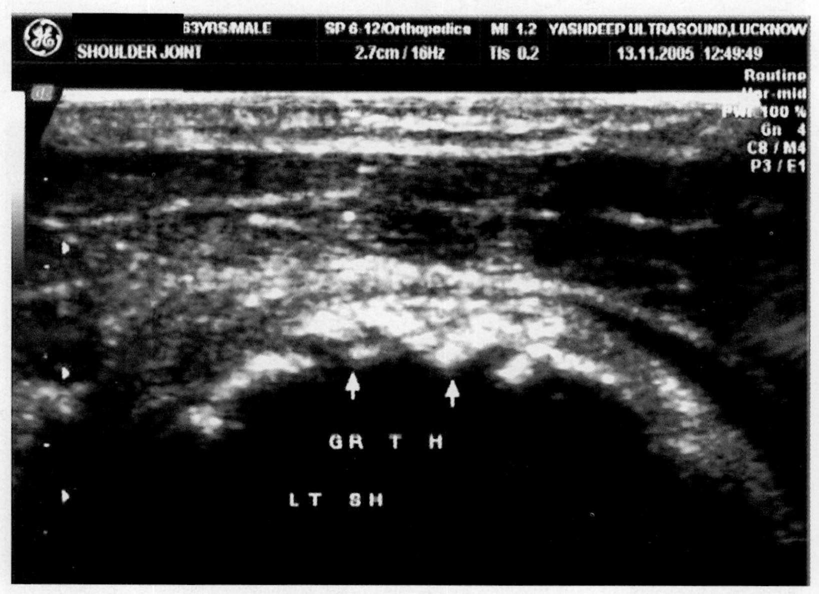

FIGURE 13.849

FIGURES 13.848 and 13.849: Degenerative changes in the greater tuberosity of the humerus. HRSG shows gross irregular surface of the articular cartilage in Lt humeral head. Brittling of the cartilage is seen with marked uneven surface suggestive of degenerative changes.

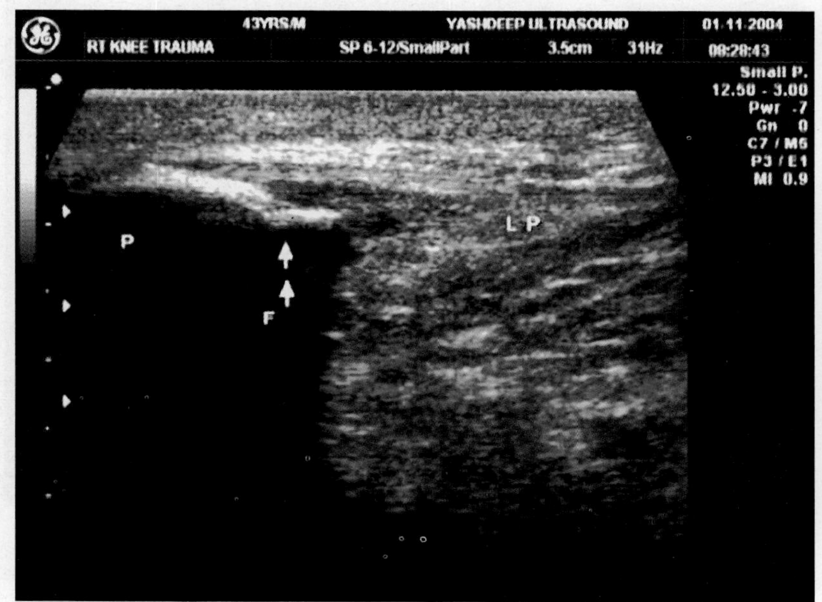

FIGURE 13.850

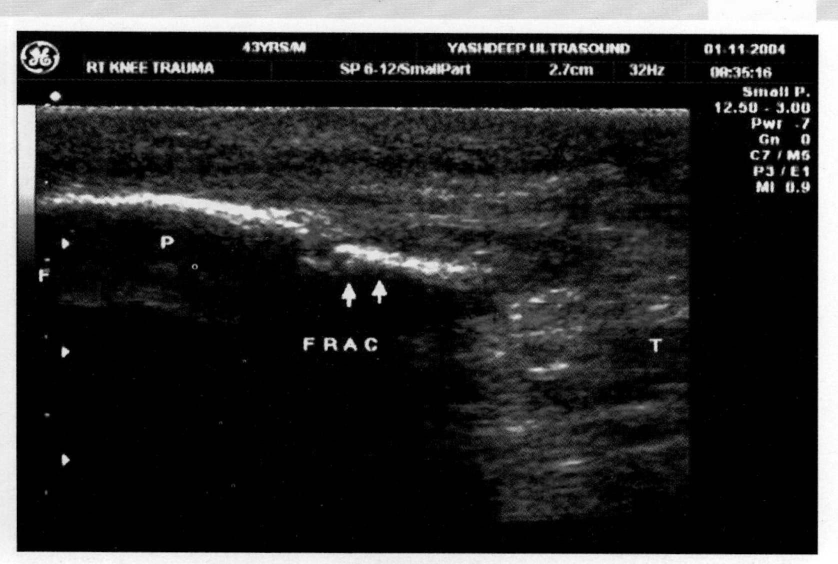

FIGURE 13.851

FIGURES 13.850 and 13.851: Fracture of patella. HRSG shows fracture of the patella in a patient sustained knee trauma. The fracture segment is well visualized on HRSG.

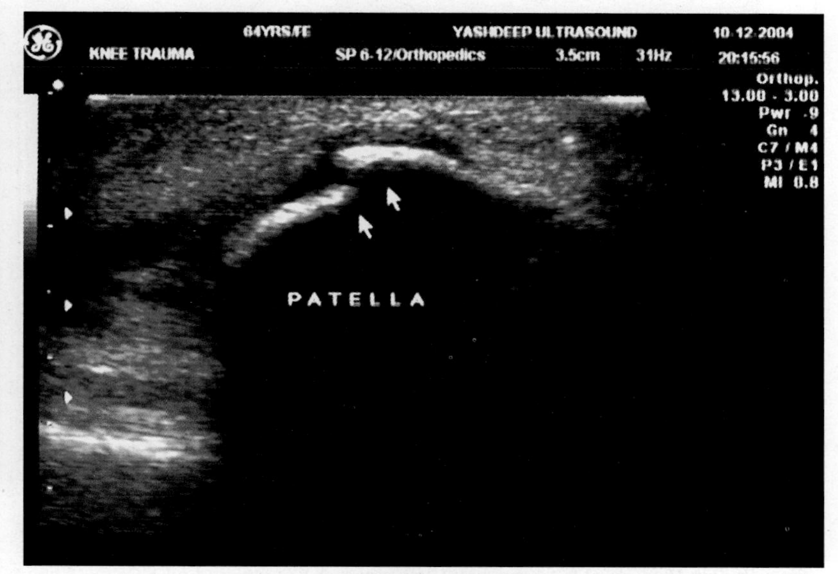

FIGURE 13.852

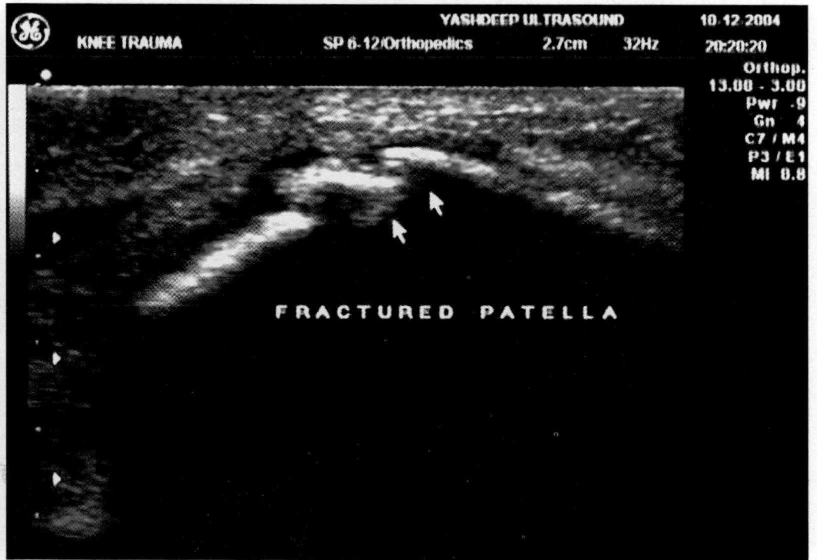

FIGURE 13.853

FIGURES 13.852 and 13.853: Fracture of patella. HRSG shows fracture of patella in a patient sustained direct trauma to the knee joint. The fracture parts are seen in close apposition. Patient also had associated quadriceps tendon rupture.

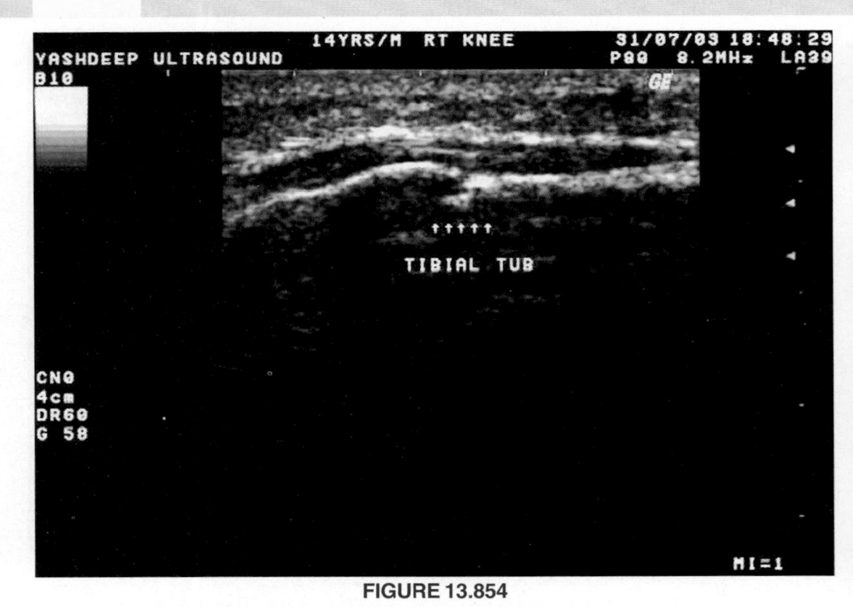

FIGURE 13.854

FIGURE 13.855

FIGURES 13.854 and 13.855: Fracture of tibial tuberosity. A young boy sustained trauma in a knee joint with marked tenderness over tibial tuberosity. HRSG shows avulsion chip fracture of the tuberosity, which is well appreciated on HRSG. Lt tibial shows normal tuberosity.

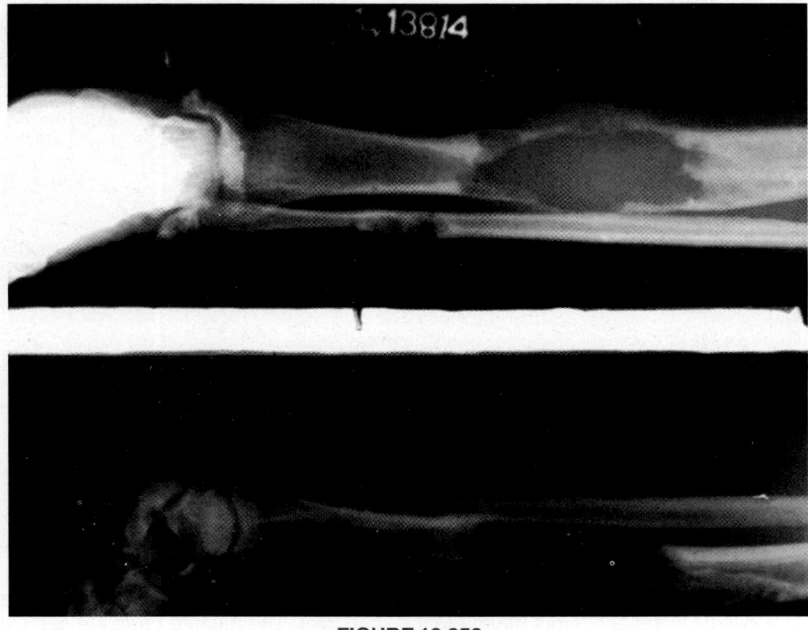

FIGURE 13.856

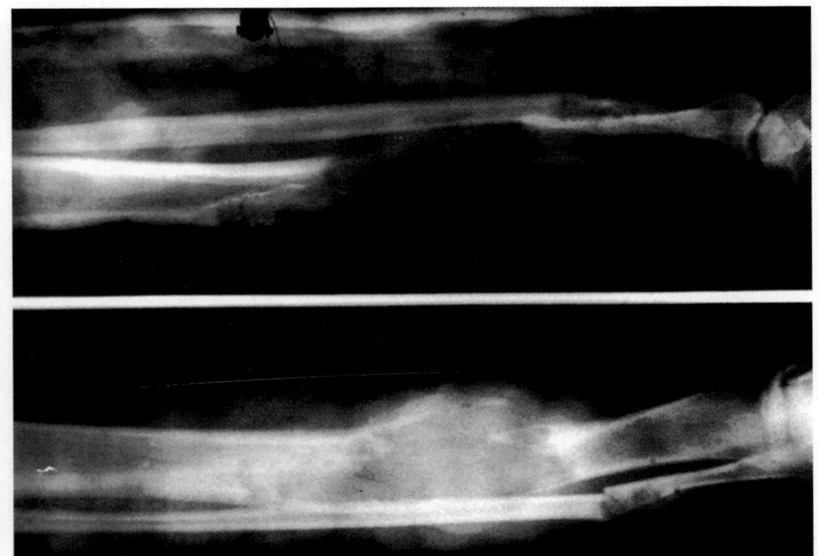

FIGURE 13.857

FIGURES 13.856 and 13.857: Aneurysmal bone cyst. A patient presented with big bony defect in the tibia with marked medullary expansion.

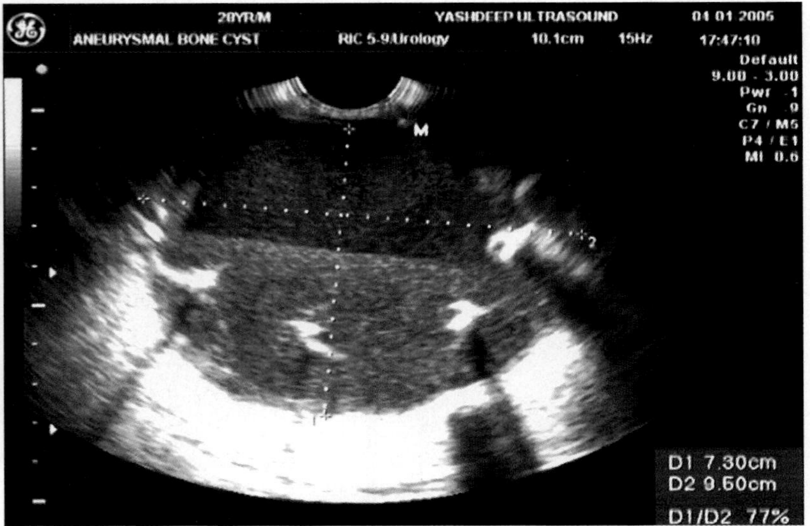

FIGURE 13.858

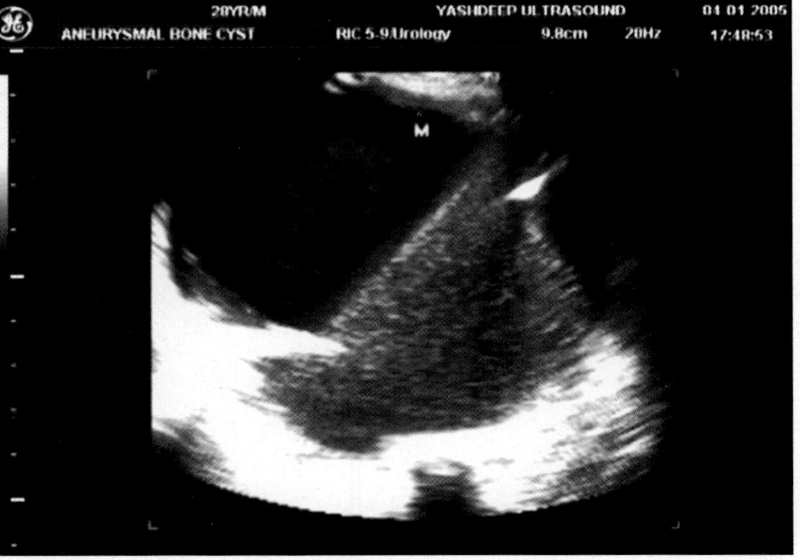

FIGURE 13.859

FIGURES 13.858 and 13.859: HRSG shows a big low level echo complex mass in the tibial shaft. Thick echoes are seen in the mass with debris sedimentation. Echogenic debris is seen floating in the mass.

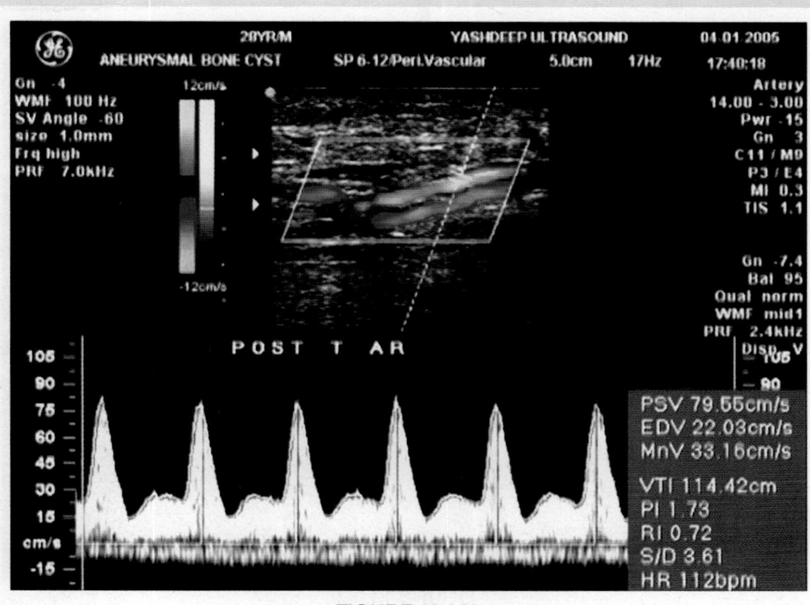

FIGURE 13.860

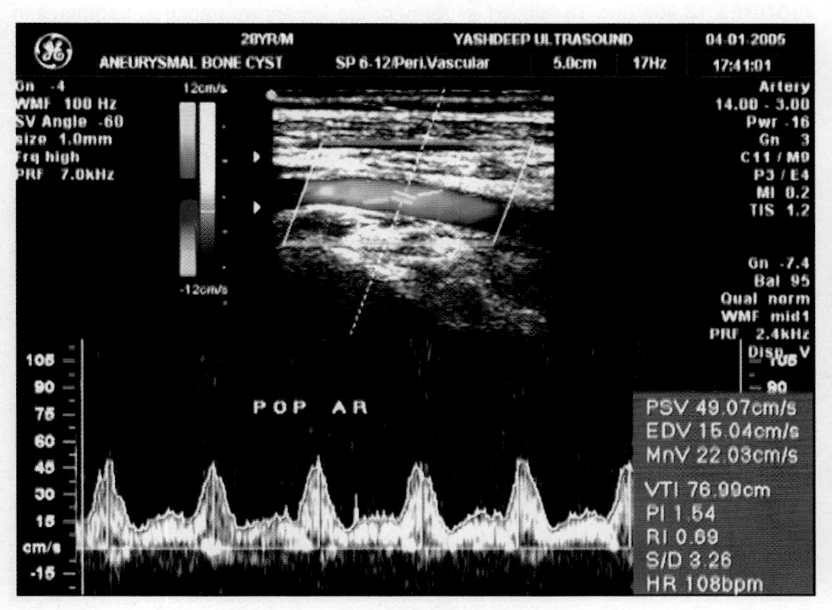

FIGURE 13.861

FIGURES 13.860 and 13.861: On color Doppler imaging the major vessels of the leg are seen intact. They are not invaded by the mass. However, no flow is seen in the cystic mass.

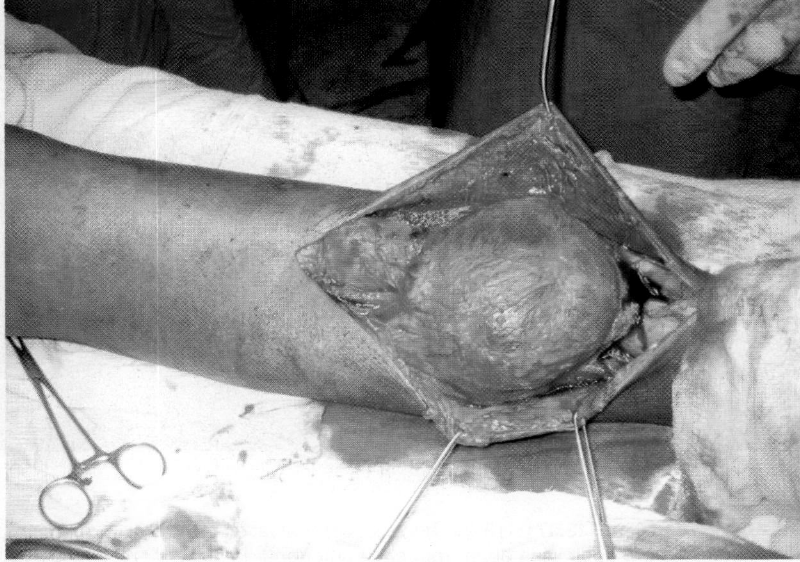

FIGURE 13.862

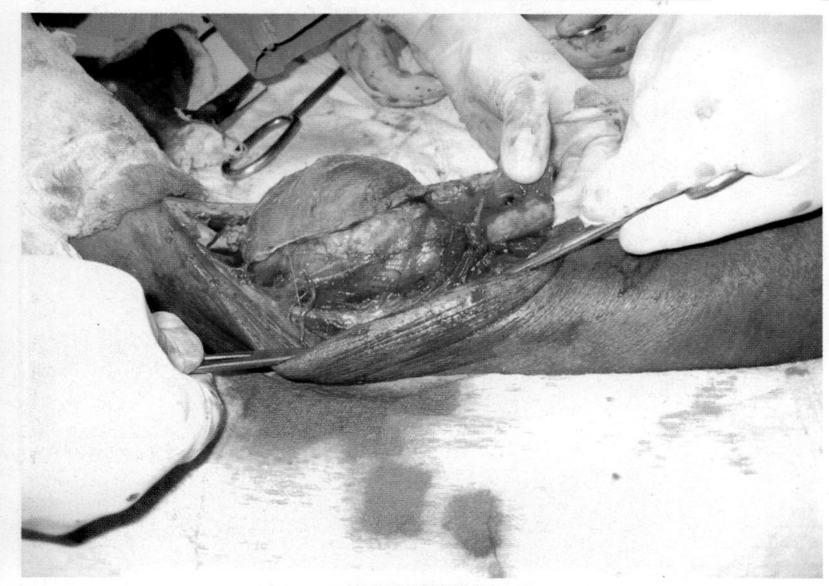

FIGURE 13.863

FIGURES 13.862 and 13.863: Per-operative findings of the mass confirmed aneurysmal bone cyst.

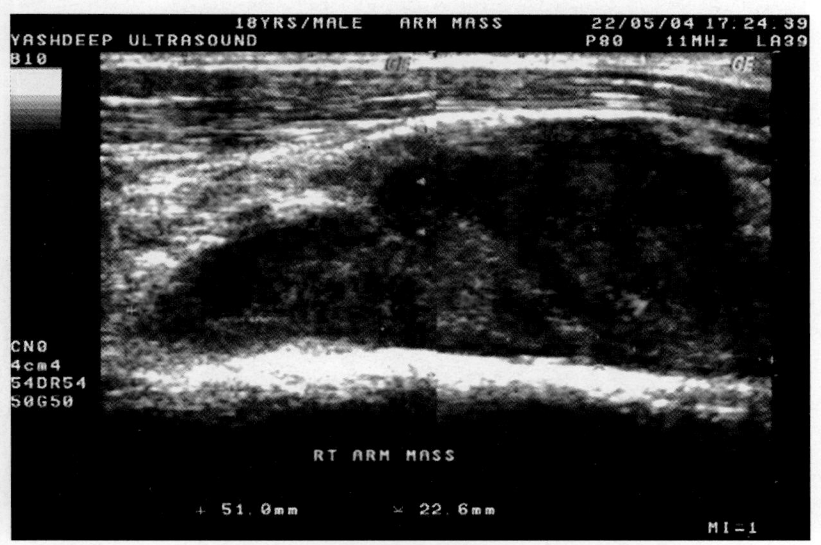

FIGURE 13.864

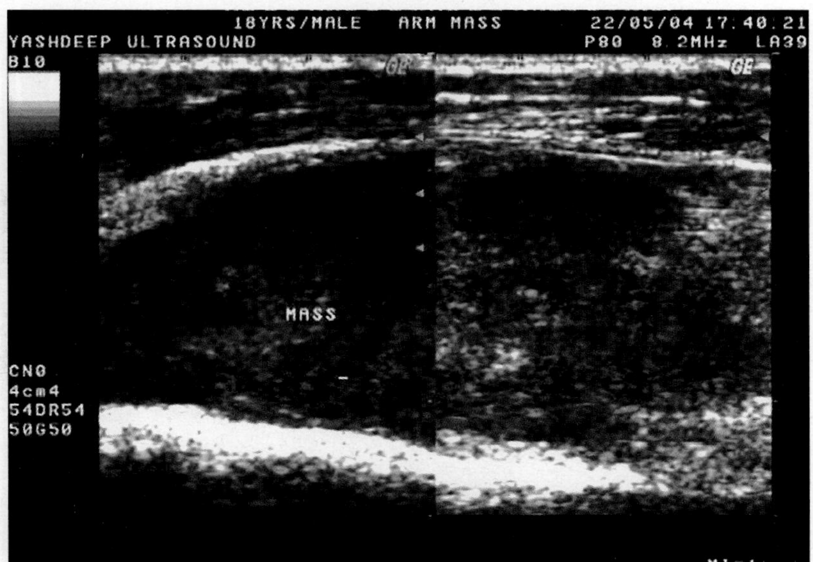

FIGURE 13.865

FIGURES 13.864 and 13.865: Ewing tumors of the arm. A patient presented with soft tissue mass involving the Rt arm. HRSG shows a big predominantly homogeneous expensile mass. The mass is seen invading the cortex of the bone. No cystic degeneration is seen in the mass. No calcification is seen.

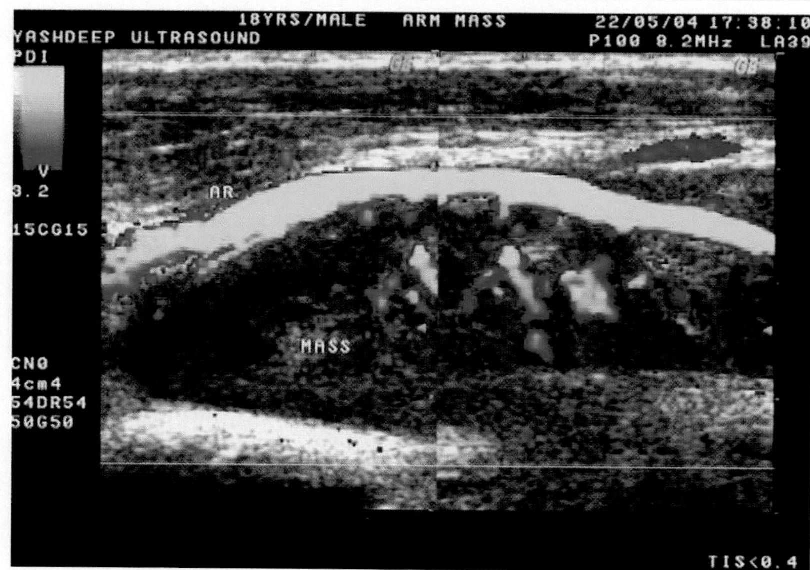

FIGURE 13.866: The mass shows moderate flow however the brachial artery is intact and not invaded by the mass.

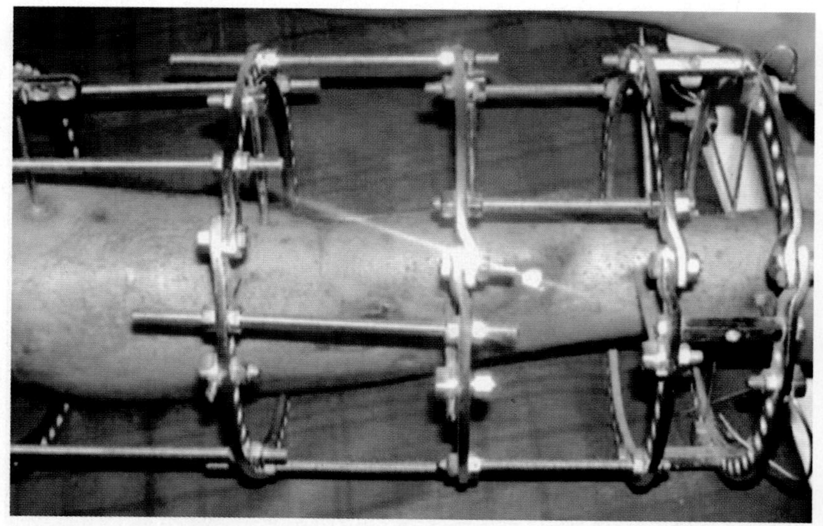

FIGURE 13.867: New bone formation. A patient presented with both bone fracture of lower leg. The fixators are applied to the fractured bones.

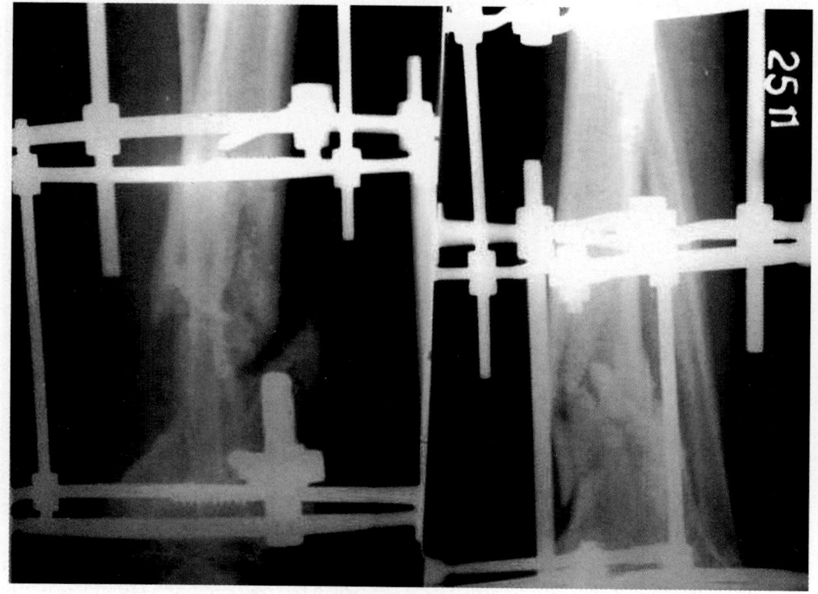

FIGURE 13.868

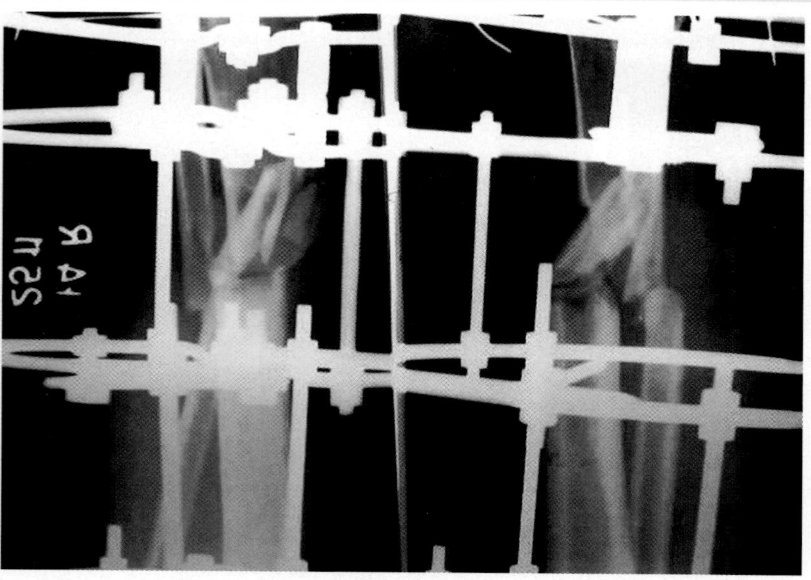

FIGURE 13.869

FIGURES 13.868 and 13.869: Plain X-ray of the leg shows fractured fragments in position. However, new bone formation of callus is not seen radiologically.

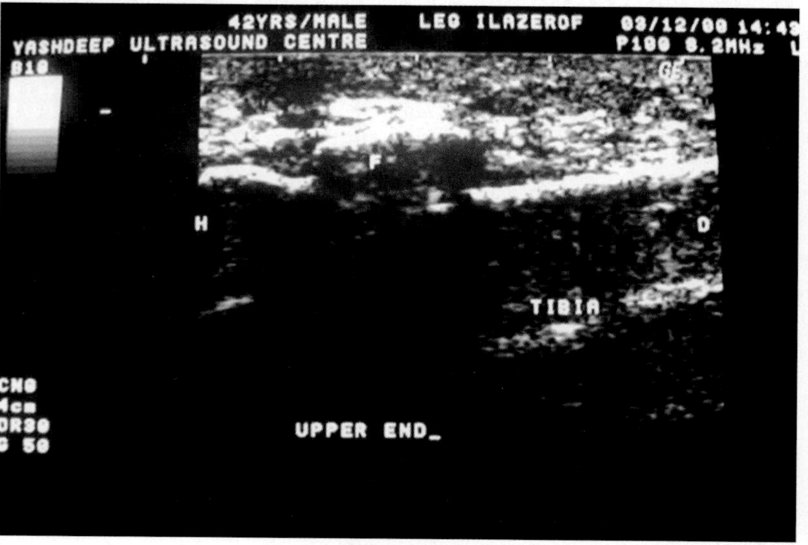

FIGURE 13.870

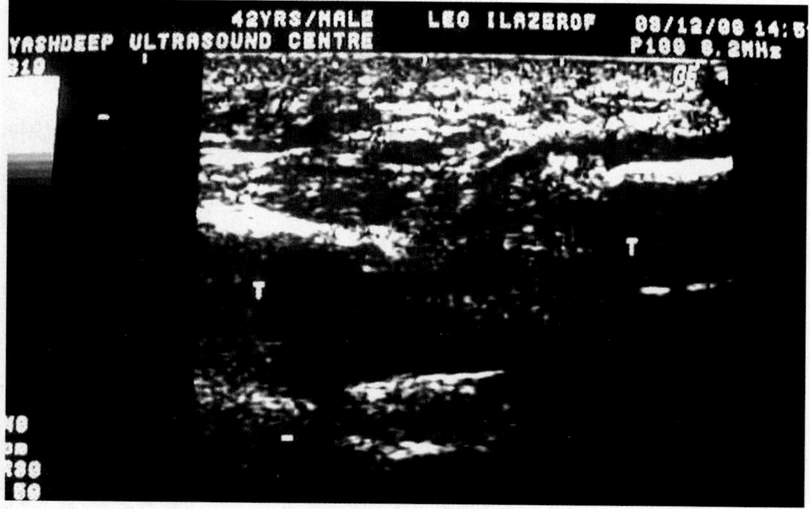

FIGURE 13.871

FIGURES 13.870 and 13.871: HRSG shows proper evaluation of the broken segments. The segments are seen lying apart from each other and no callus formation is seen between the fractured segments.

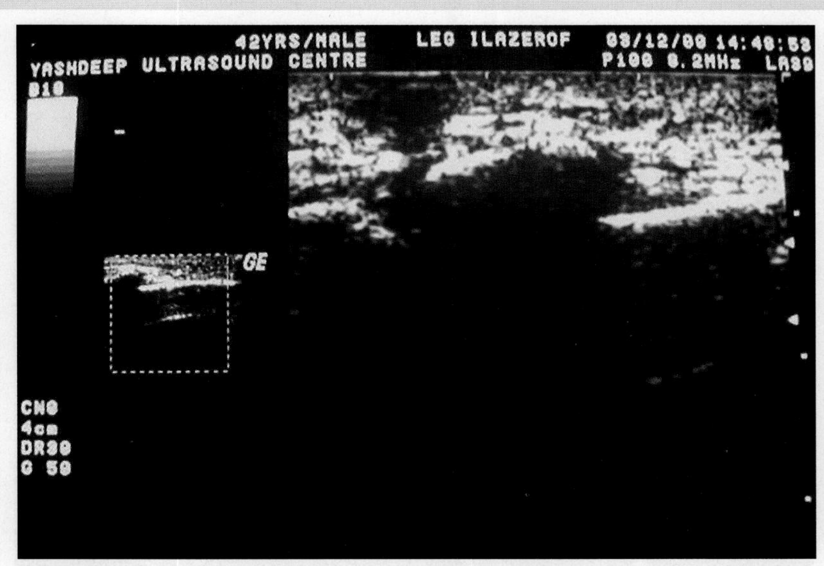

FIGURE 13.872: In the lower part of the tibia the broken segments are seen wide apart. HRSG shows no callus formation between two broken segments. HRSG is a good tool to evaluate new bone and thus helps in evaluating the healing process.

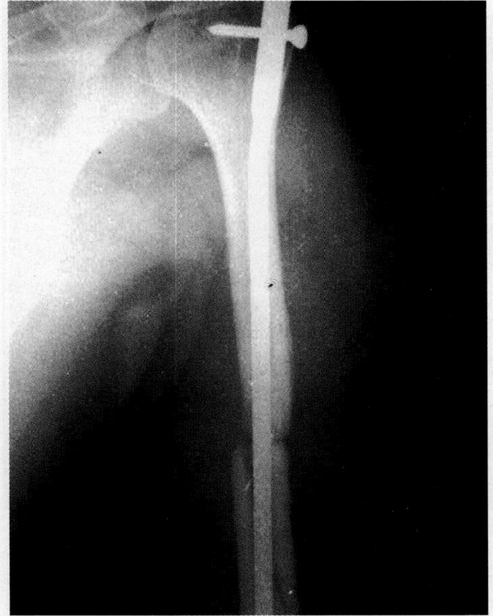

FIGURE 13.873: Plane X-ray of the Lt arm shows broken shaft of humerus. Intramedullary nail is seen. The X-ray shows normal callus formation on the lateral border of the humerus. However, medial border shows irregular defect between the two broken segments.

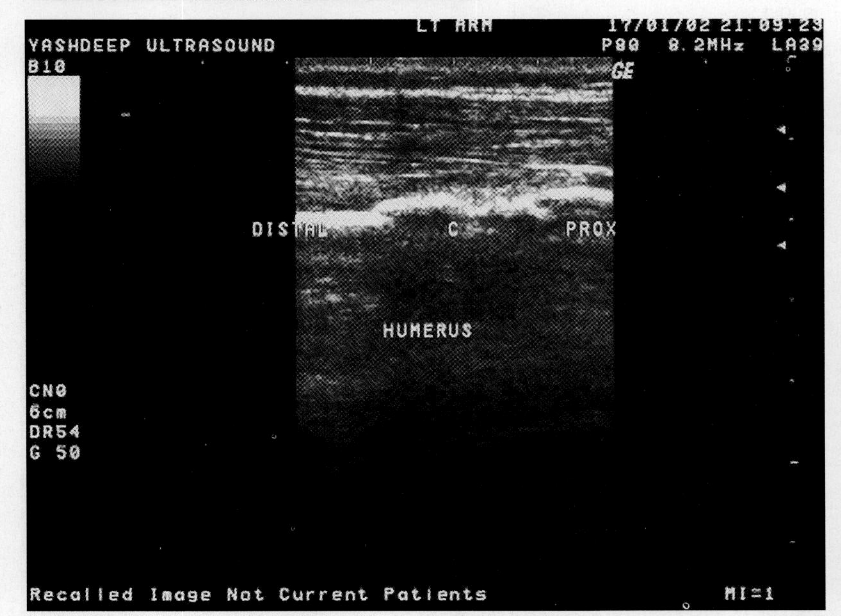

FIGURE 13.874: HRSG of the same patient shows normal callus formation on the lateral border of the humerus. It is seen as an irregular echogenic line joining the two broken segments. No defect is seen between the two broken segments.

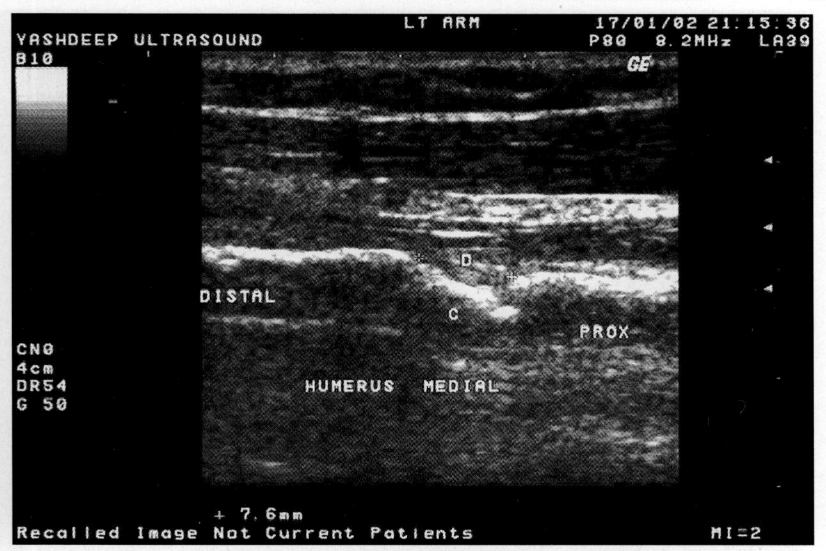

FIGURE 13.875

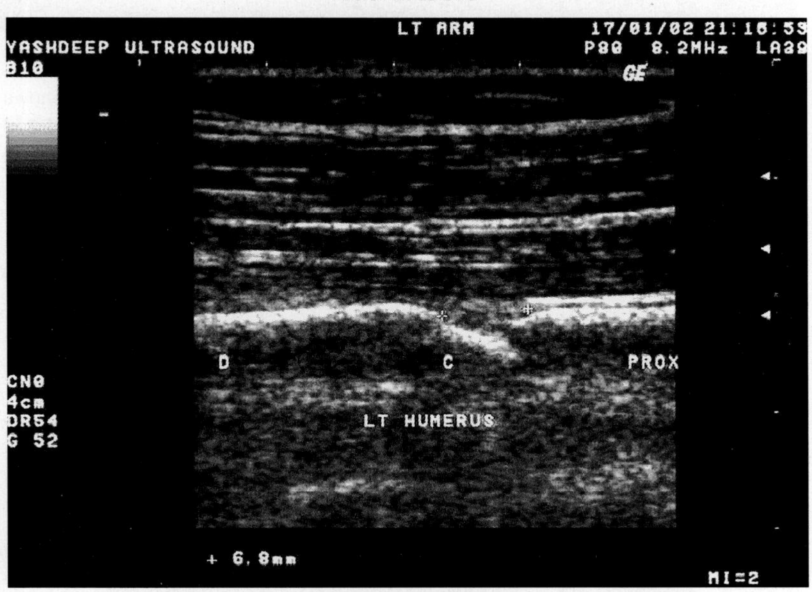

FIGURE 13.876

FIGURES 13.875 and 13.876: HRSG shows deficient callus formation between the two broken segments on the medial border of the Lt humerus. A defect of around 7 mm is seen between the two segments. However, an echogenic callus is seen formed on the inferior surface of the border. But the callus does not yet formed on the upper surface of the border.

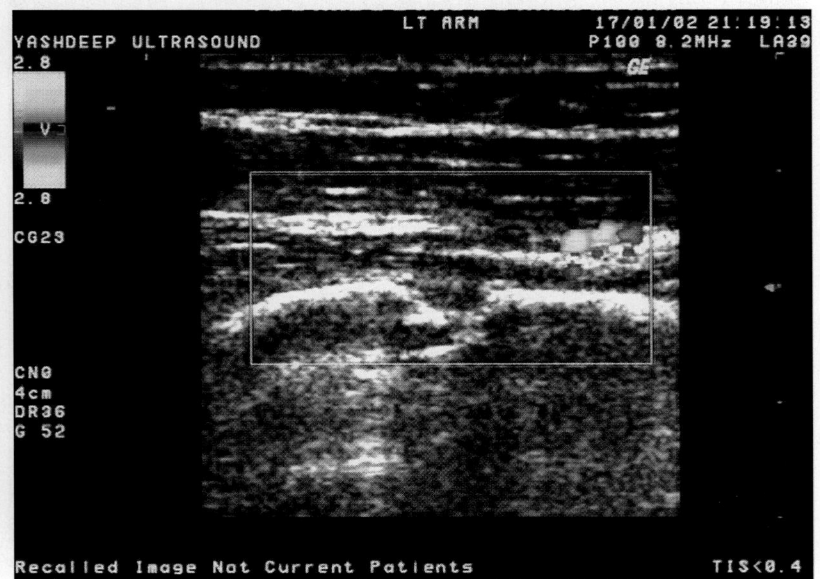

FIGURE 13.877

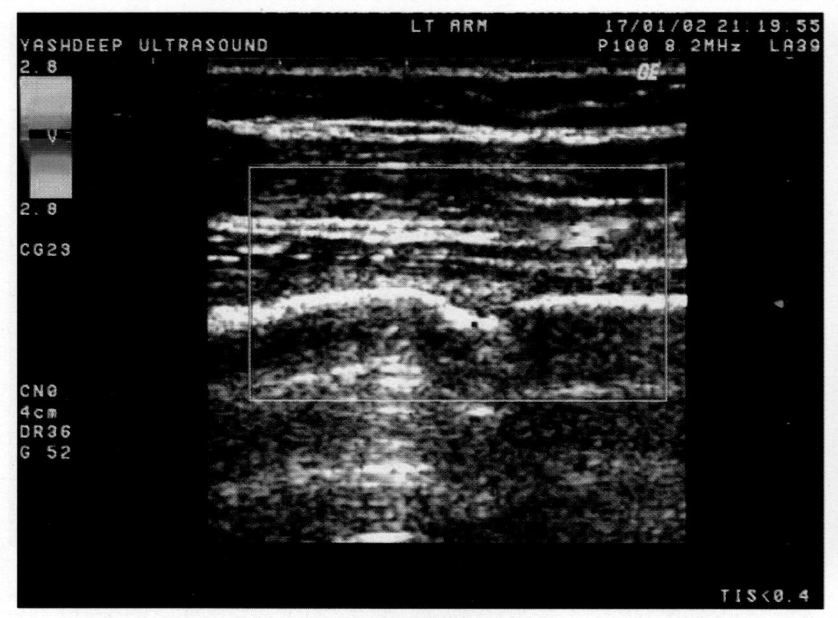

FIGURE 13.878

FIGURES 13.877 and 13.878: Color Doppler imaging of the same patient shows neovascularization at the broken site, which is suggestive of good vascular supply to the broken segments.

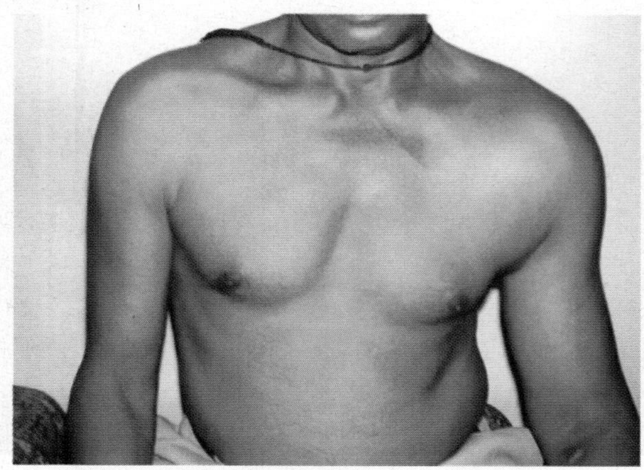

FIGURE 13.879: Neuropathic Lt shoulder joint. A young man presented with limited movement of Lt shoulder joint with drooping of the shoulder. However, the movements were painless. Patient was unable to lift the arm.

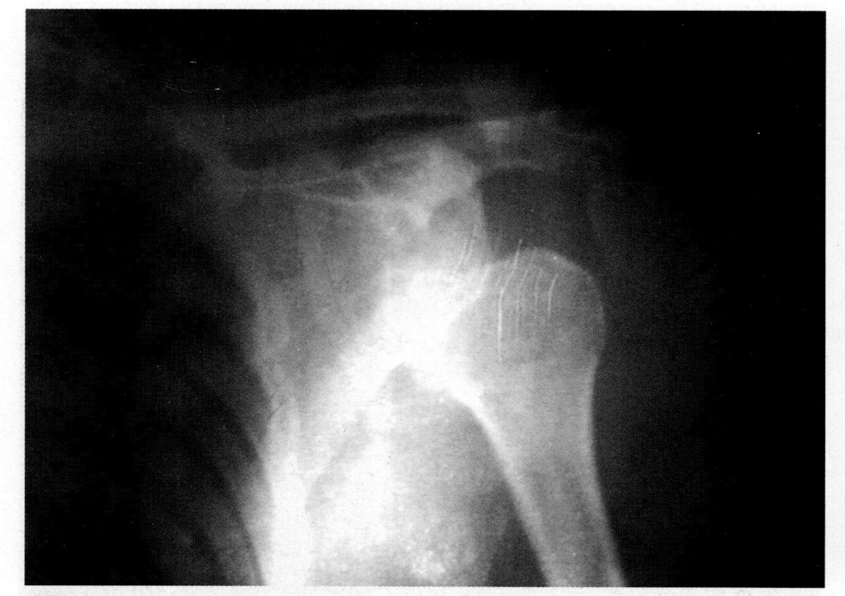

FIGURE 13.880: Plain X-ray of the shoulder shows normal shoulder bones. The humeral head is well-defined. However, soft tissue mass shadow is seen around the shoulder joint.

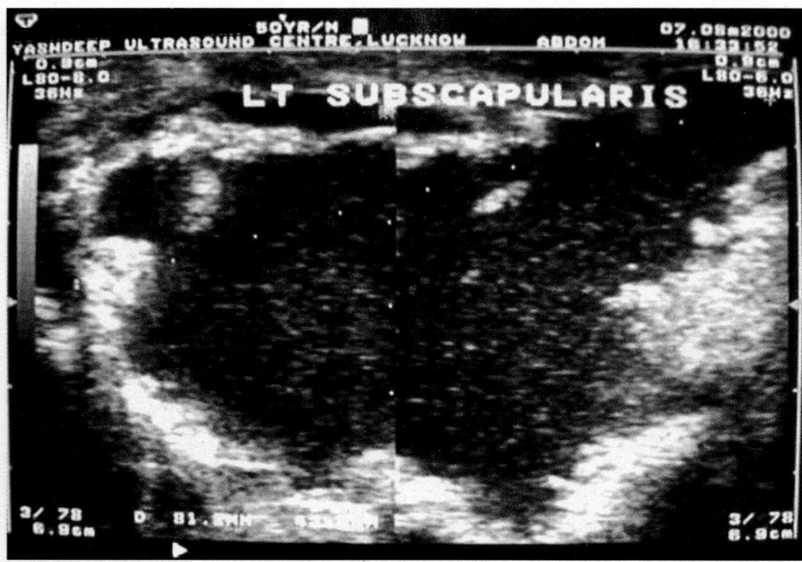

FIGURE 13.881

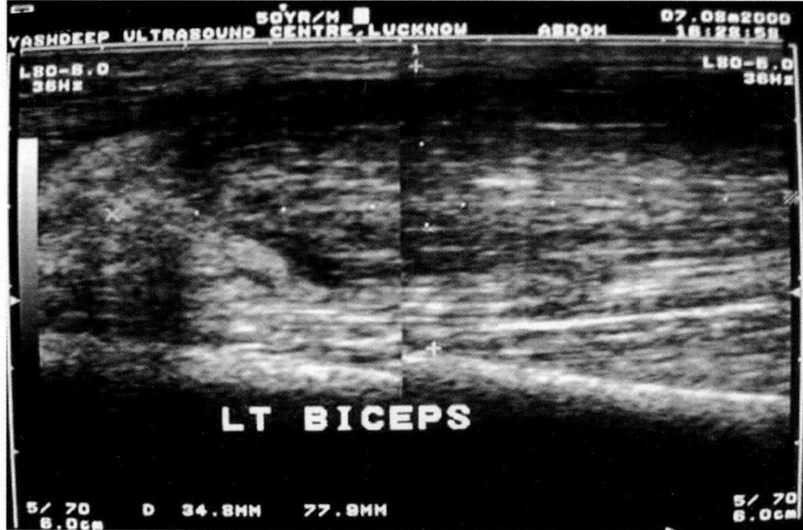

FIGURE 13.882

FIGURES 13.881 and 13.882: HRSG shows cystic degeneration of Lt shoulder joint muscles involving deltoid, biceps, triceps and subscapularis. Normal muscle texture is lost.

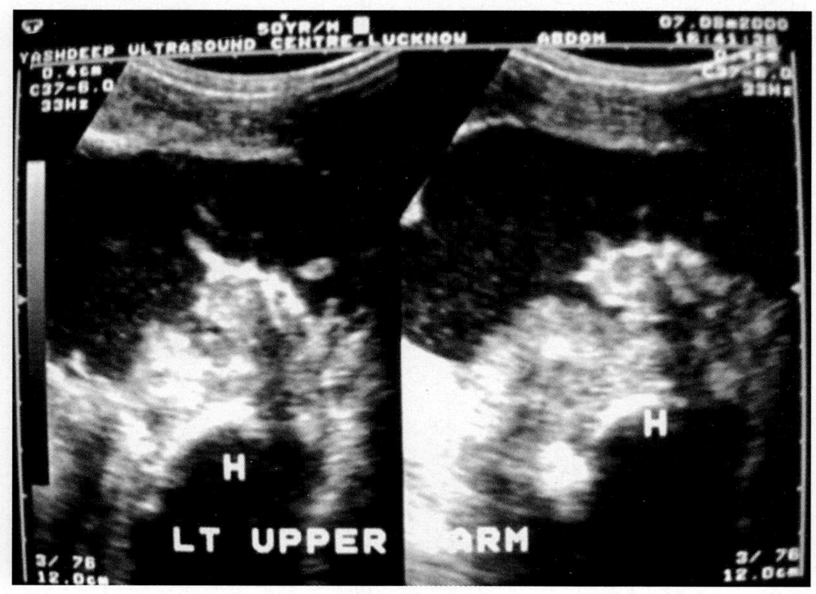

FIGURE 13.883

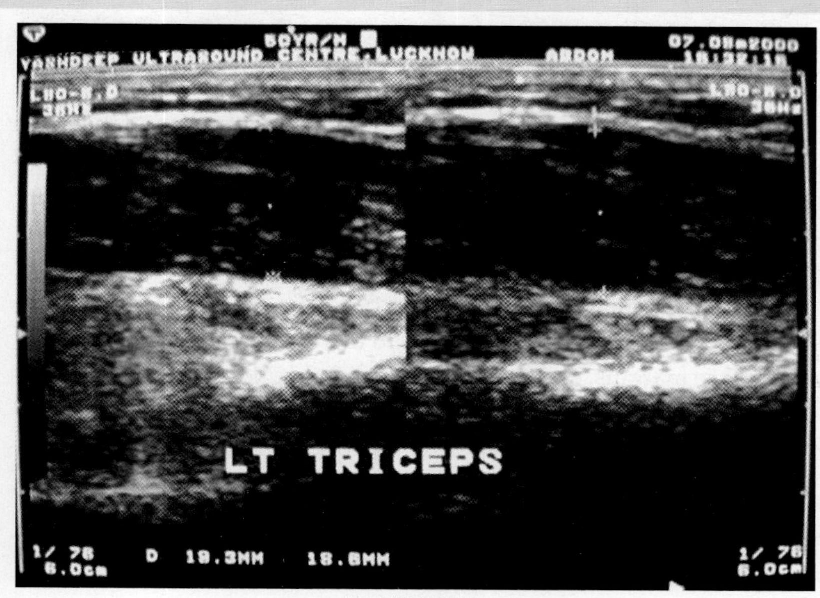

FIGURE 13.884

FIGURES 13.883 and 13.884: Cystic degeneration is seen in Lt deltoid and triceps muscle. Normal muscle texture is lost.

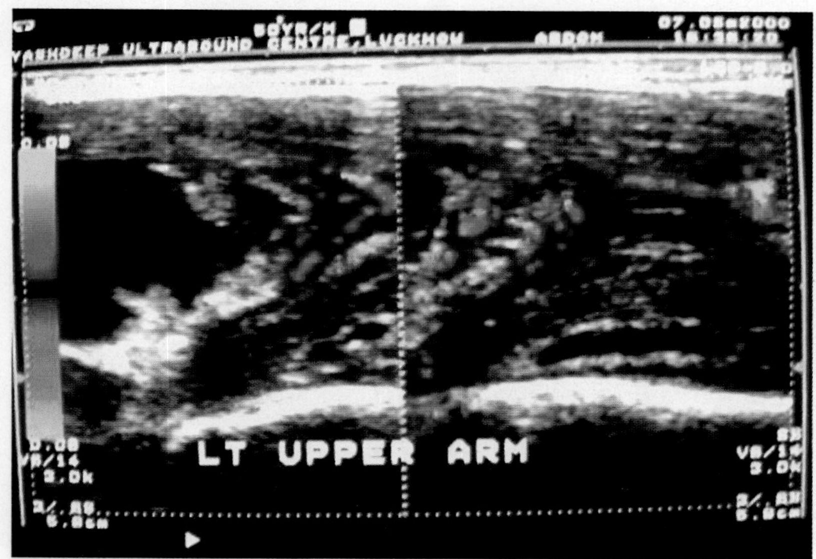

FIGURE 13.885

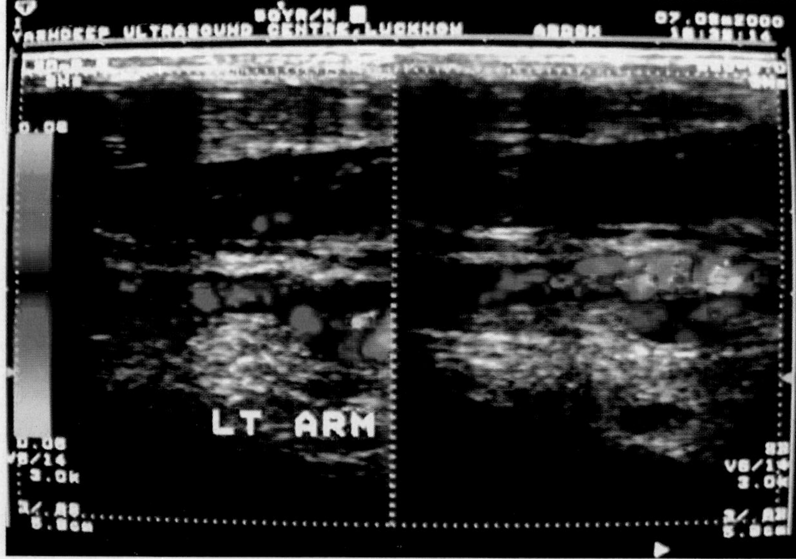

FIGURE 13.886

FIGURES 13.885 and 13.886: Color Doppler imaging of the same patient shows little flow in the shoulder muscle. However axillary vessels are normal.

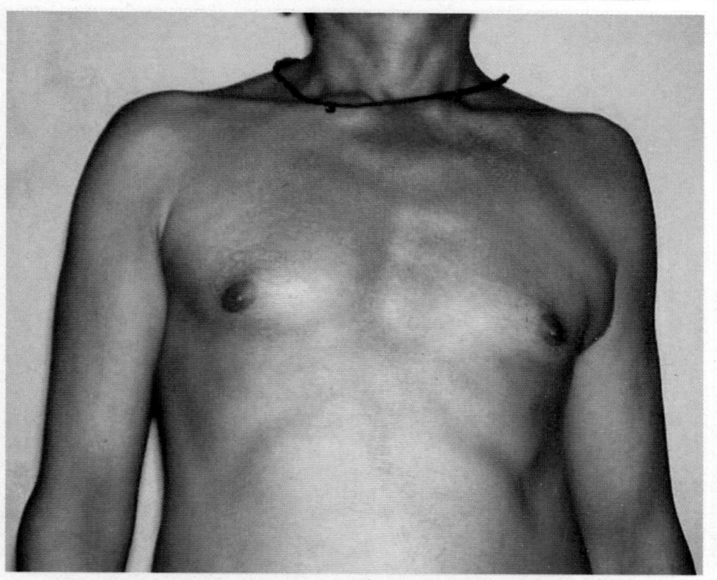

FIGURE 13.887: The same patient presented after 3 months of previous study. The normal shoulder contour is lost. The muscle bulk is reduced in amount. The arm was hanging freely.

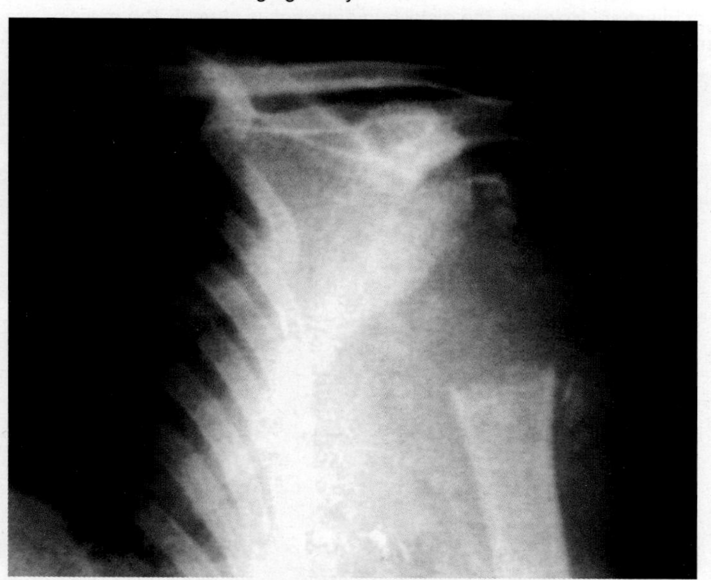

FIGURE 13.888: X-ray of the patient shows complete destruction of the humeral head, surgical neck and upper part of the shaft. The other bones are also destroyed.

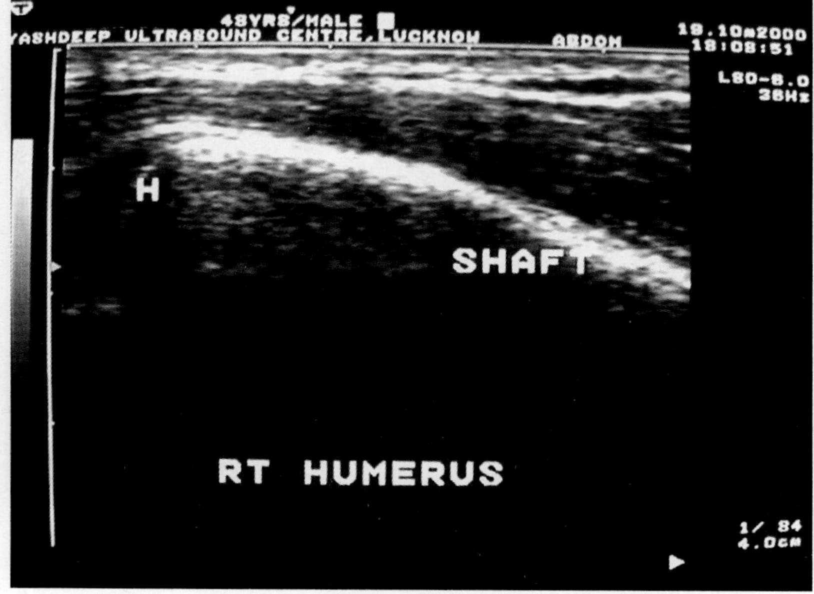

FIGURE 13.889

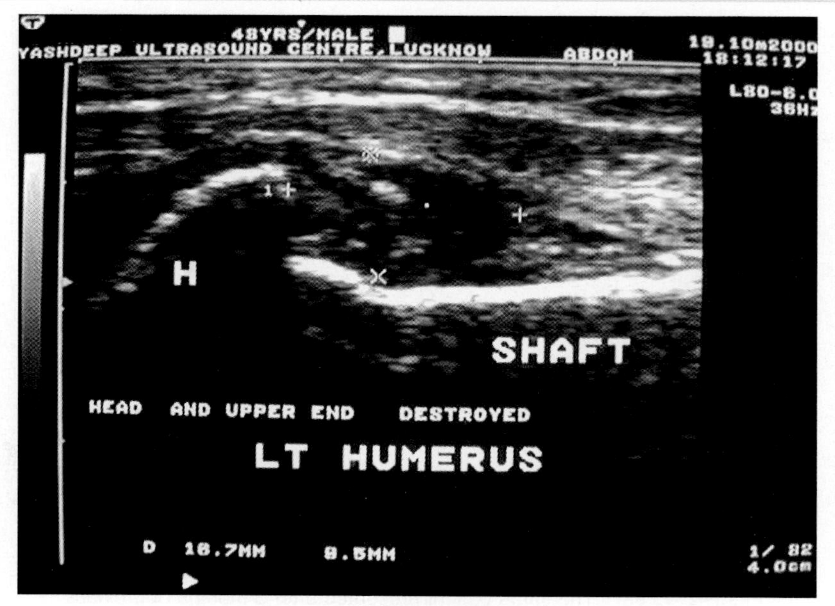

FIGURE 13.890

FIGURES 13.889 and 13.890: HRSG shows destroyed humeral head in the upper part of the shaft. However, the distal humeral shaft is normal. The muscles are small in amount with totally disorganized texture.

HRSG of Nerve and Neural Masses

High-resolution ultrasound is an excellent modality for evaluation of nerve and neural masses. It helps in differentiating them from the other masses. Normal nerve appears to be as linear fibrilar echogenic structure in close relation to the vessels. However, the nerve echotexture merges with the surrounding tissue structures therefore, visualization of the normal nerve depends on surroundings tissue echotexture. Nerves are usually hyperechoic to muscles but hypoechoic to the tendons. Most of the nerve tumors are homogeneous in texture. They are well defined with sharp margins. They are fusiform in shape. No cystic degeneration is seen in them. The neurilemmomas tumors show eccentric growth whereas neurofibromas show central growth. HRSG is very useful in studying nerve pathologies after trauma or inflammation with entrapment syndromes.

FIGURE 13.891

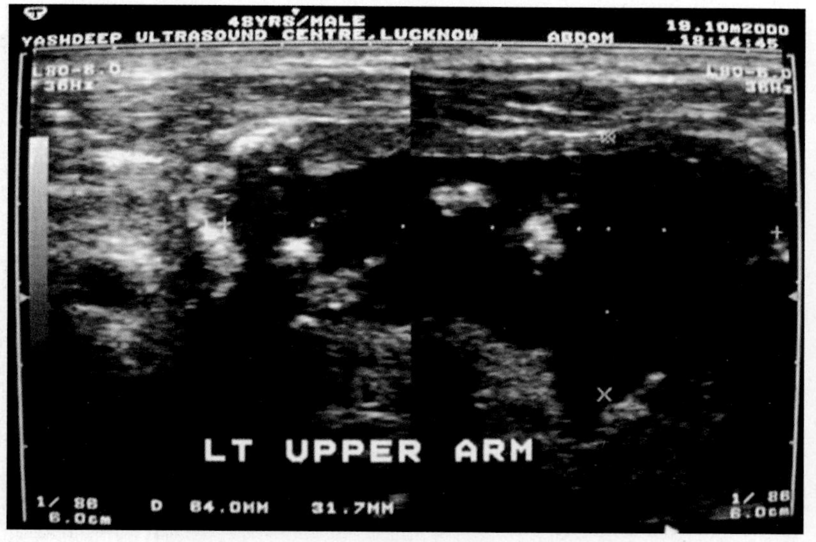

FIGURE 13.892

FIGURES 13.891 and 13.892: The deltoid, bicep and triceps muscles are completely destroyed. Small muscle bundles are seen floating in the cyst. The joint was painless. The findings are suggestive of neuropathic joints.

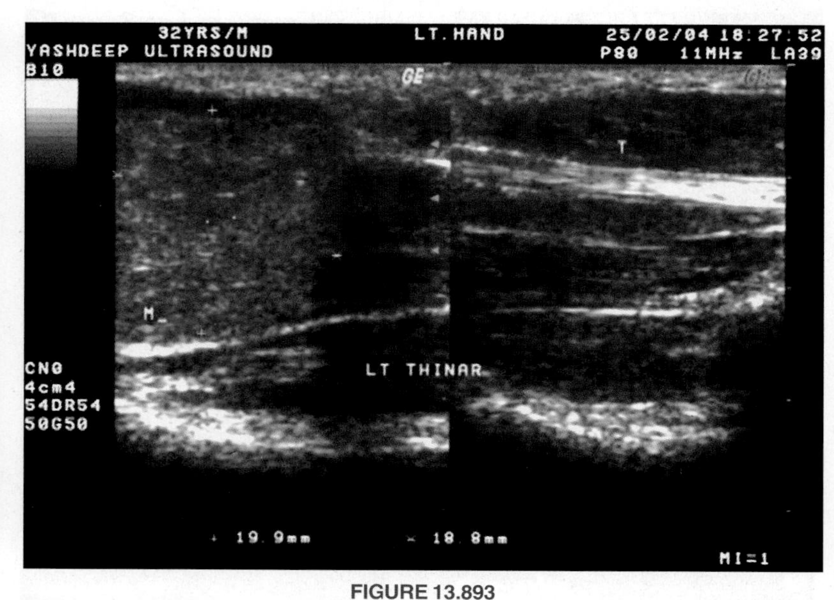

FIGURE 13.893

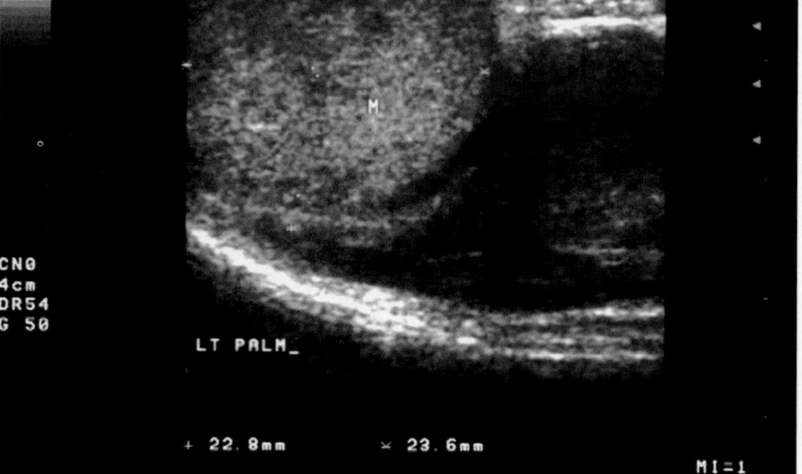

FIGURE 13.894

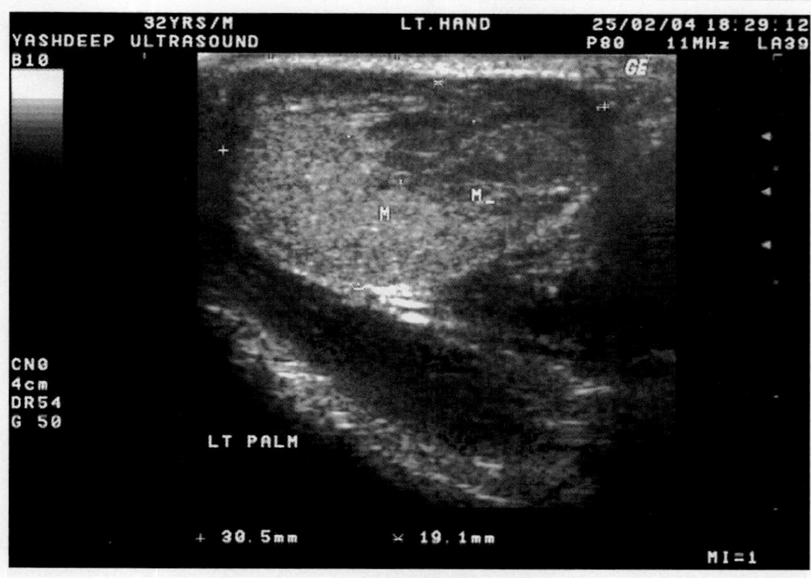

FIGURE 13.895

FIGURES 13.893 to 13.895: A patient presented a soft tissue mass over the Lt palm. HRSG shows well-defined homogeneous rounded mass in relation to the flexor tendons. The mass shows sharp margins. No calcification is seen in it. No cystic degeneration is seen.

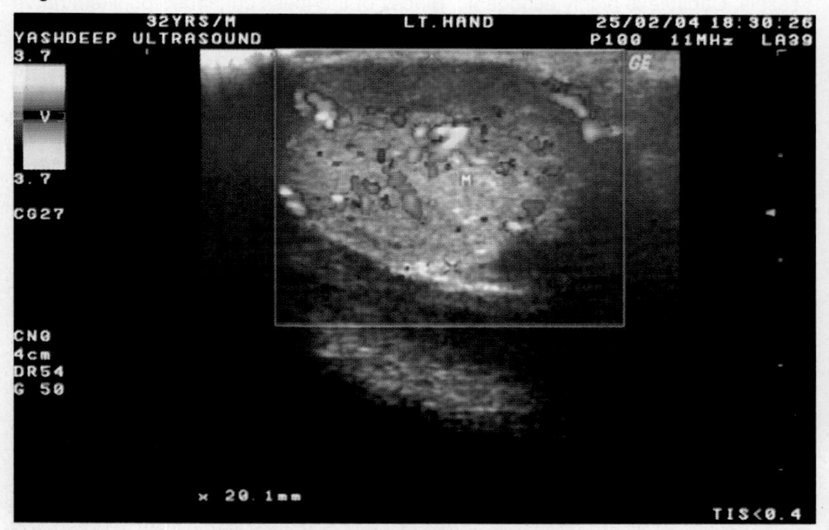

FIGURE 13.896

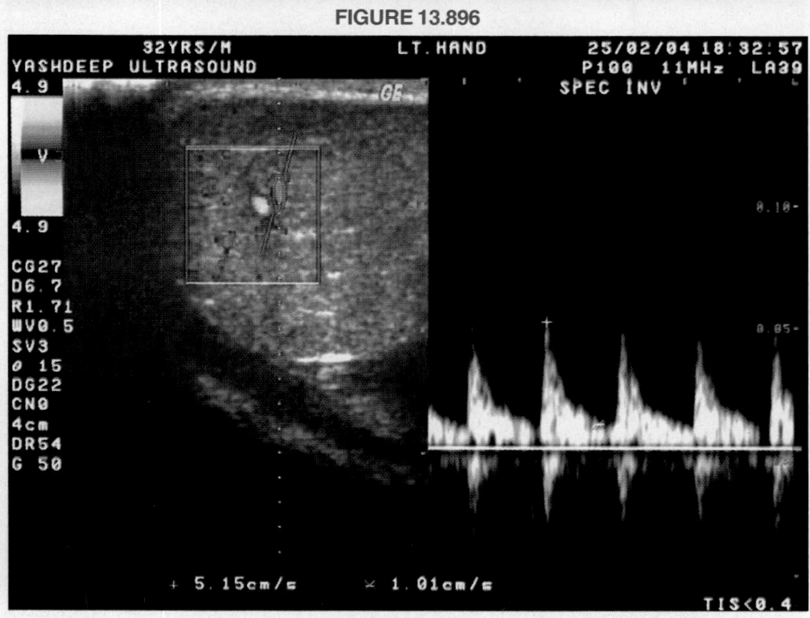

FIGURE 13.897

FIGURES 13.896 and 13.897: **On color flow imaging** increased flow is seen in it. Biopsy of the mass confirmed big ganglioma.

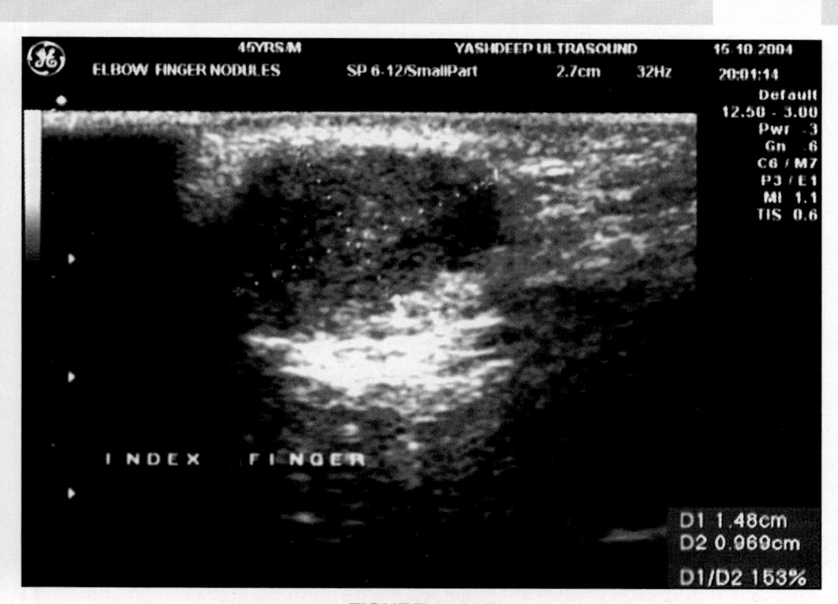

FIGURE 13.898

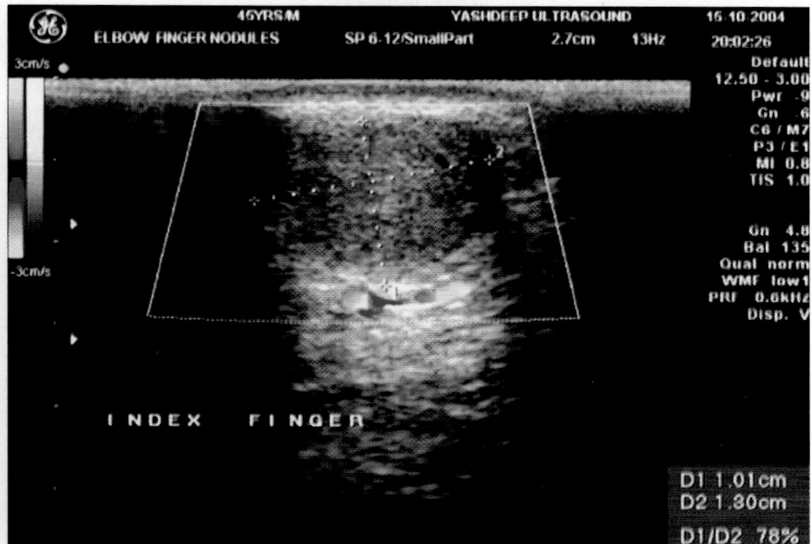

FIGURE 13.899

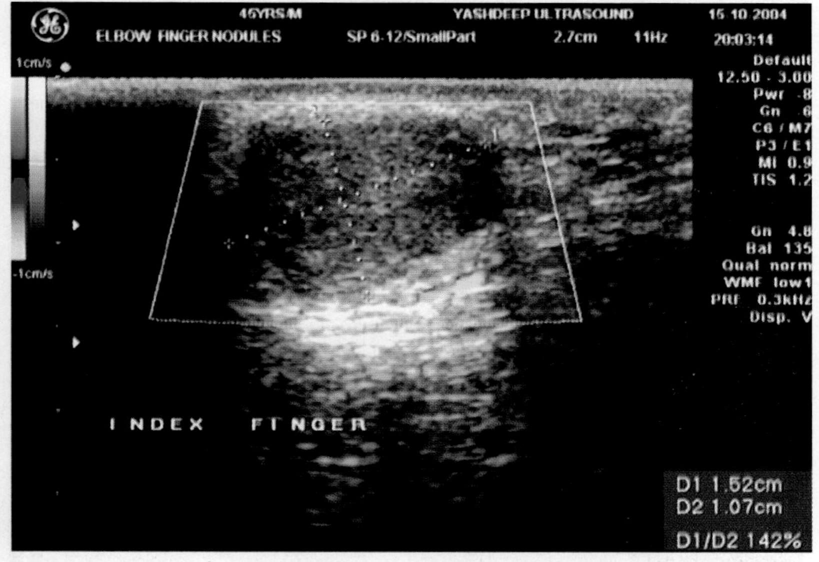

FIGURE 13.900

FIGURES 13.898 to 13.900: A patient presented with painful swelling over the index finger. HRSG shows well-defined homogeneous soft tissue mass overt the index finger. It shows sharp margins. No calcification is seen. No cystic degeneration is seen in it. **On color flow imaging** no flow is seen in the mass. Biopsy of the mass confirmed neuroma.

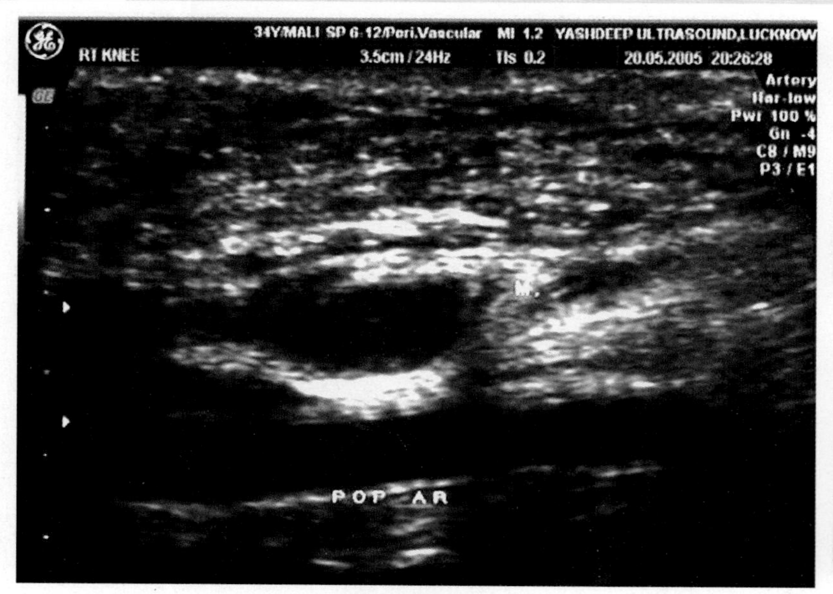

FIGURE 13.901

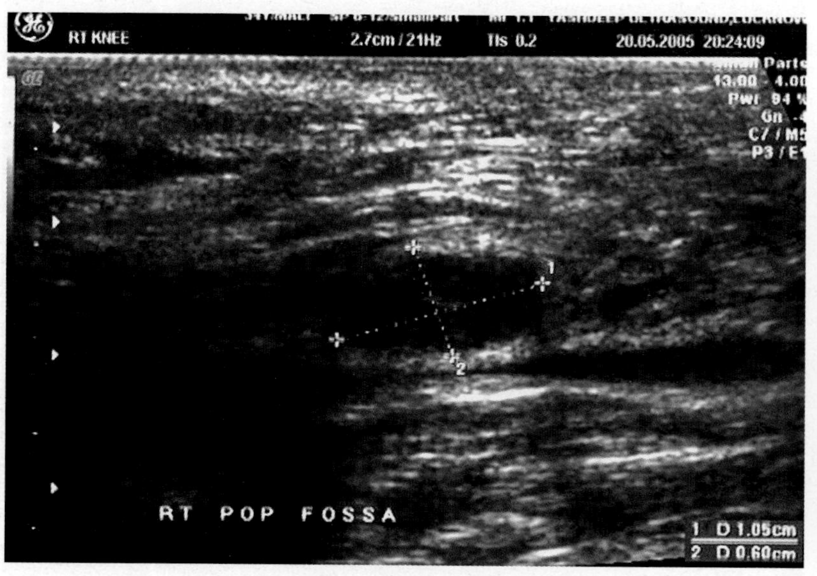

FIGURE 13.902

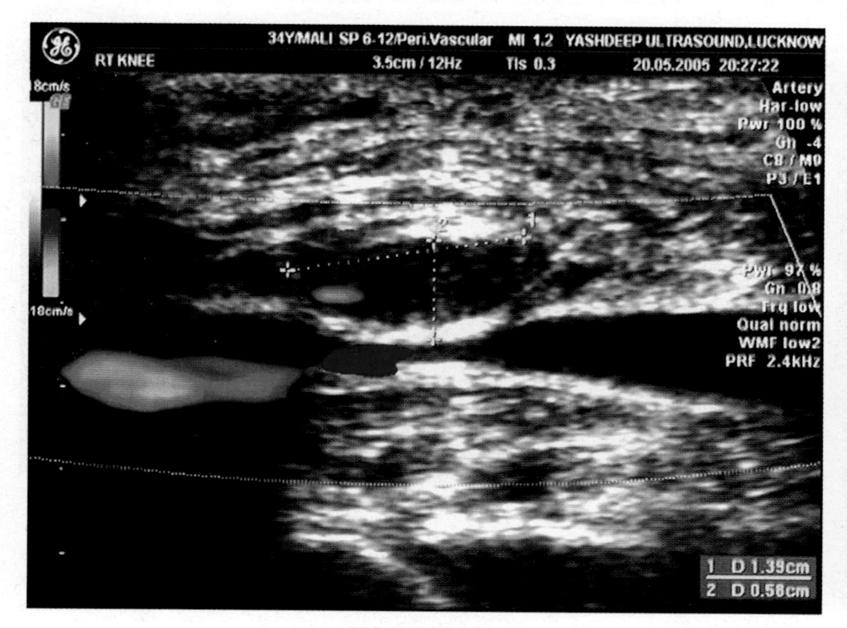

FIGURE 13.903

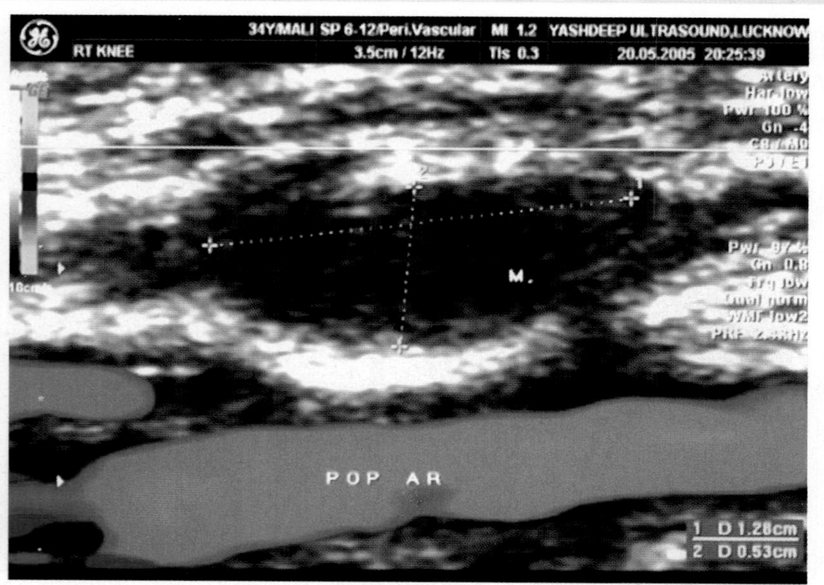

FIGURE 13.904

FIGURES 13.901 to 904: Nerve cell tumor. A patient presented with highly sensitive nodular shadow in the popliteal fossa. HRSG shows a well-defined homogeneous fusiform hypoechoic mass in relation to the popliteal nerve. It is seen anterior to the vessels. Capsule is highly echogenic. Biopsy of the mass confirmed popliteal nerve shwannoma. **On color flow imaging** no flow seen in it.

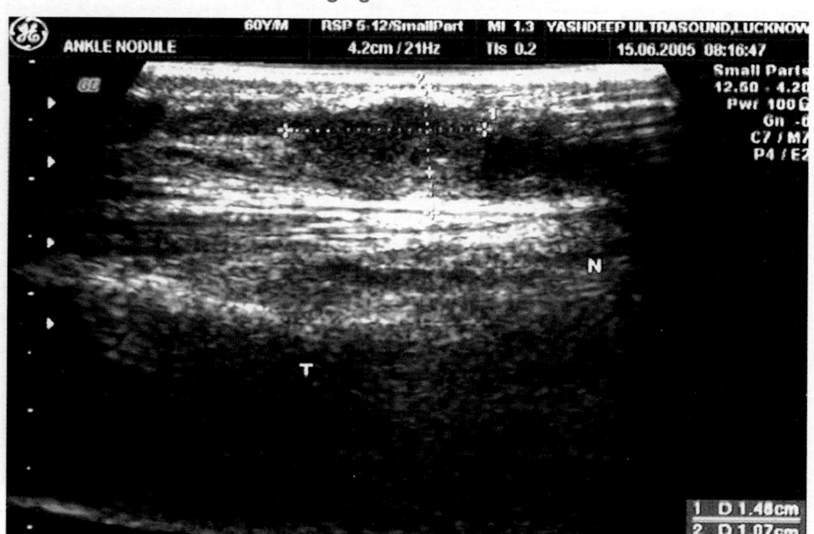

FIGURE 13.905

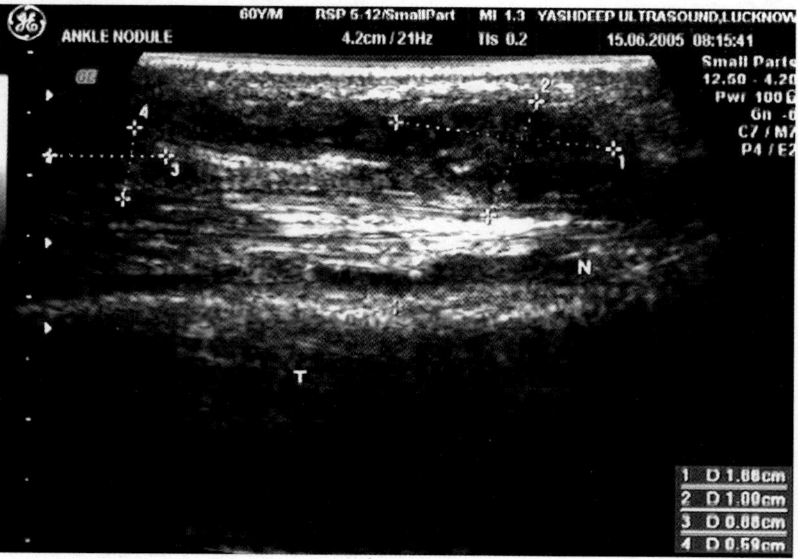

FIGURE 13.906

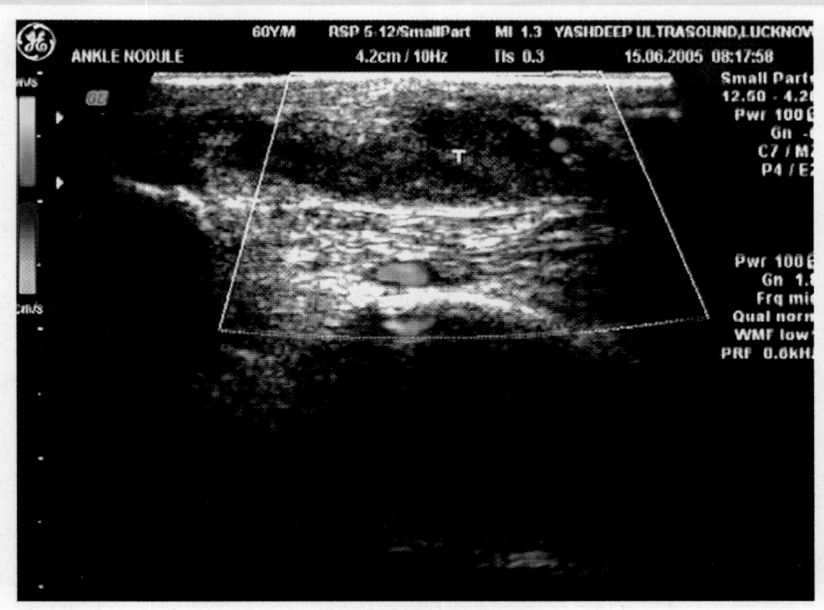

FIGURE 13.907

FIGURES 13.905 to 13.907: Neurofibroma. A patient presented with nodular palpable mass over the anterior compartment of the leg. HRSG shows well-defined homogeneous nodular shadows. They are seen fixed with the tendon sheath. **On color flow imaging** no flow is seen in it. Biopsy confirmed neurofibroma.

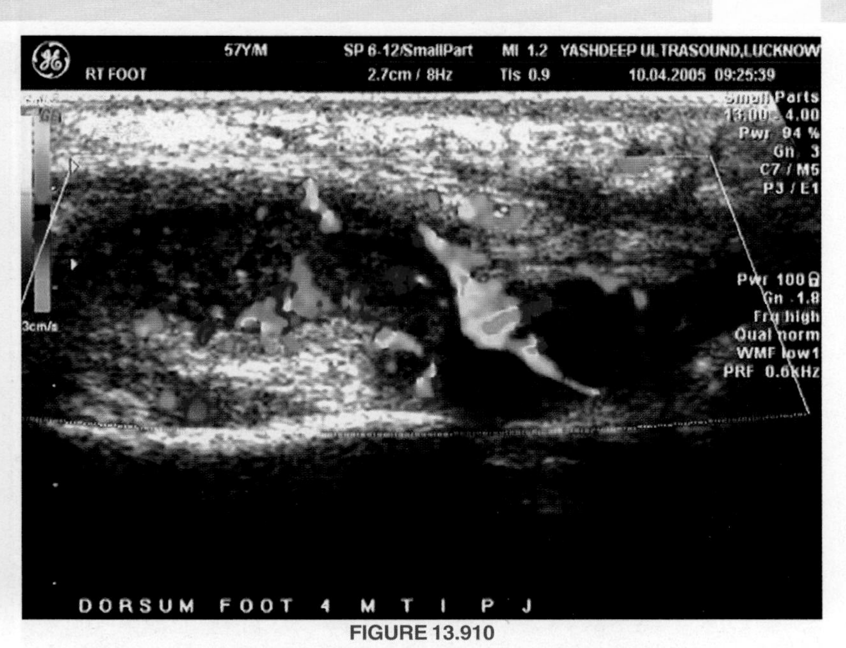

FIGURE 13.910

FIGURES 13.908 to 13.910: Well-defined homogeneous nodular mass is seen over the dorsum of the foot in a patient. The mass shows sharp margins. Perifocal edema is seen around the mass. **On color flow imaging** increased flow seen in it. Biopsy of the mass confirmed infected ganglioma.

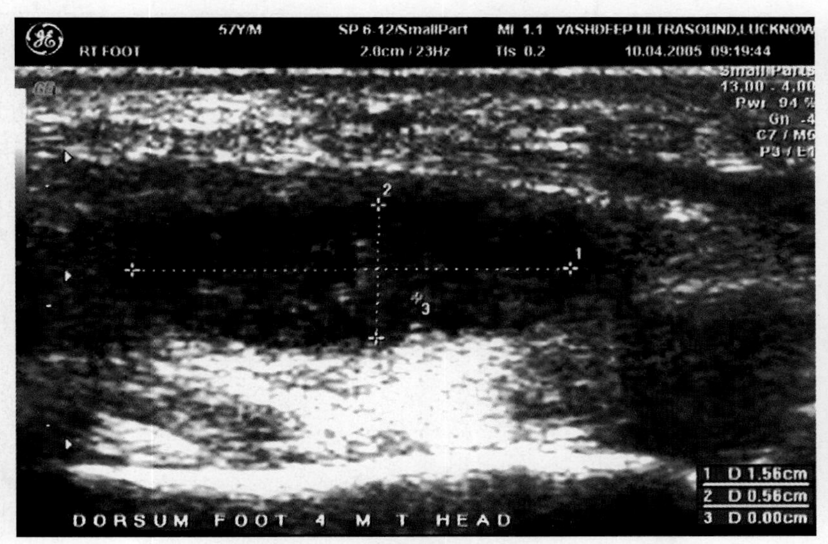

FIGURE 13.908

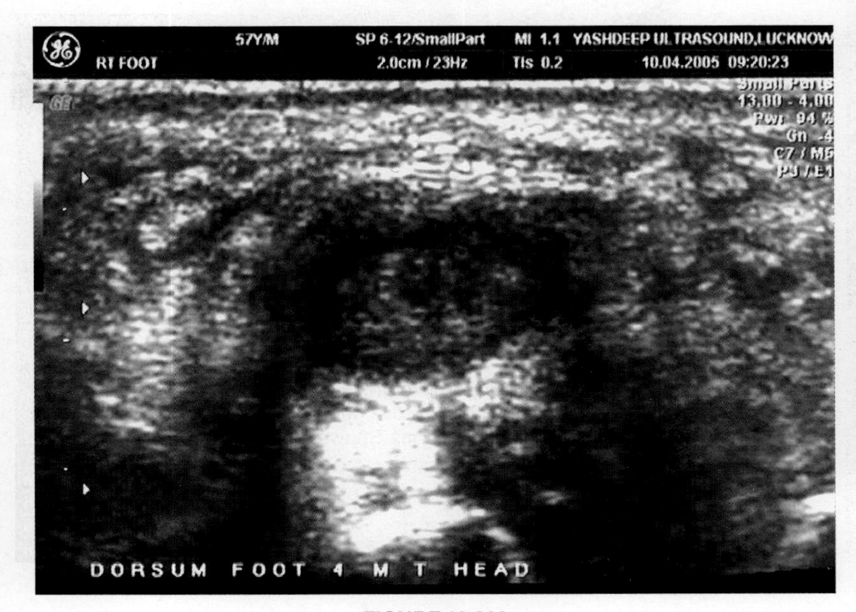

FIGURE 13.909

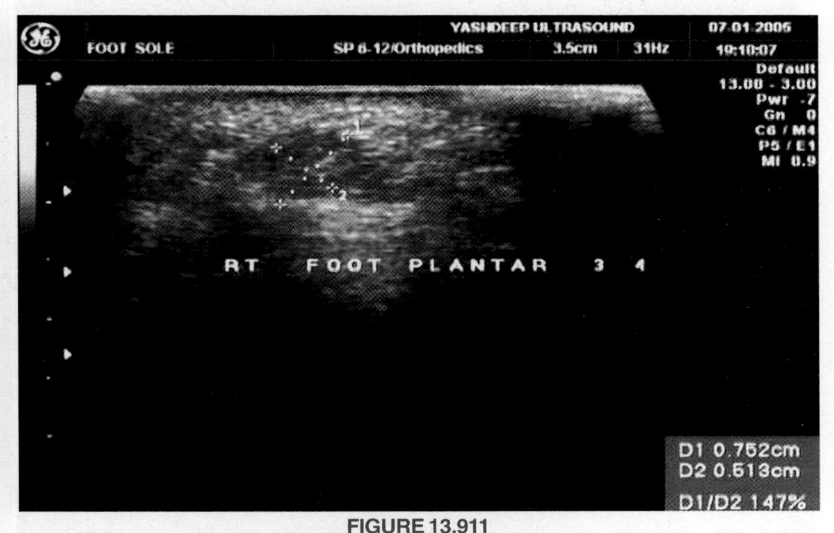

FIGURE 13.911

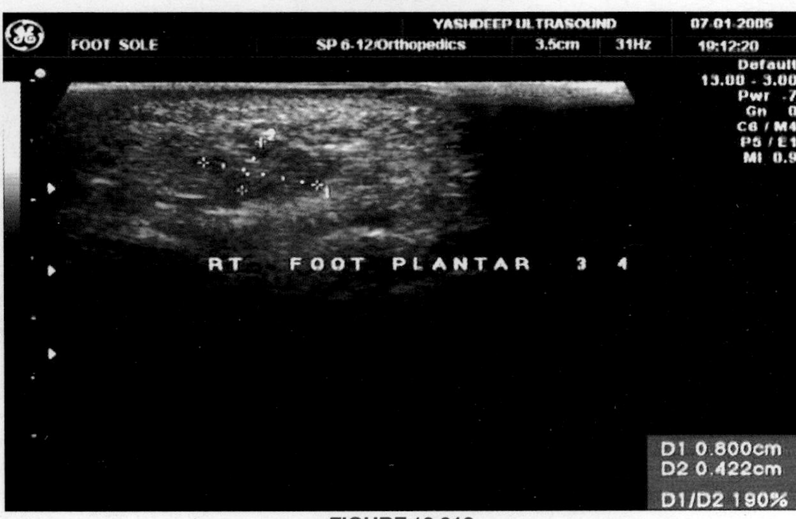

FIGURE 13.912

FIGURES 13.911 and 13.912: Morton's neuroma. A patient presented with marked pain on walking over metatarsal head in the Rt foot. HRSG shows well-defined homogeneous ovoid shadows over the plantar surface of the foot at 3 and 4 metatarsal heads. They are well-defined with sharp margins. No calcification or necrosis is seen in them. The sonographic appearance suggests Morton's neuroma.

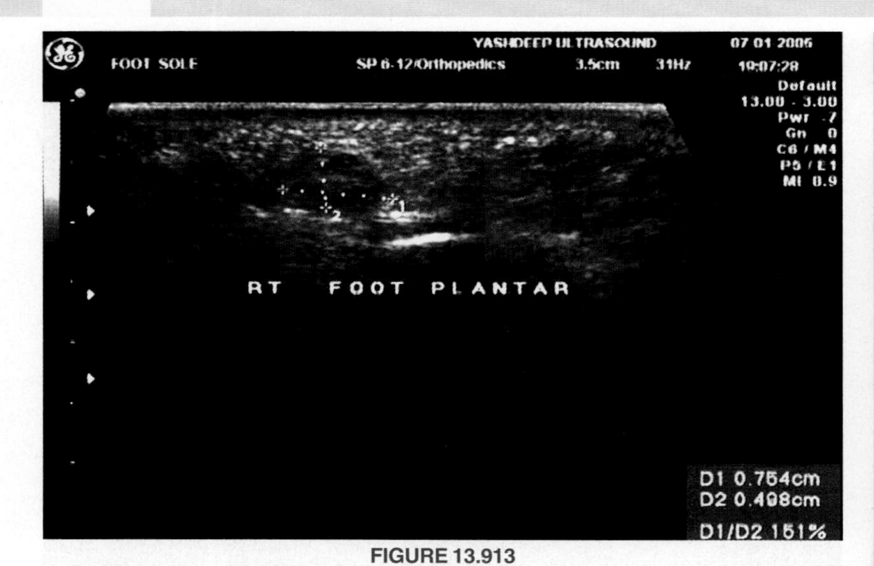

FIGURE 13.913

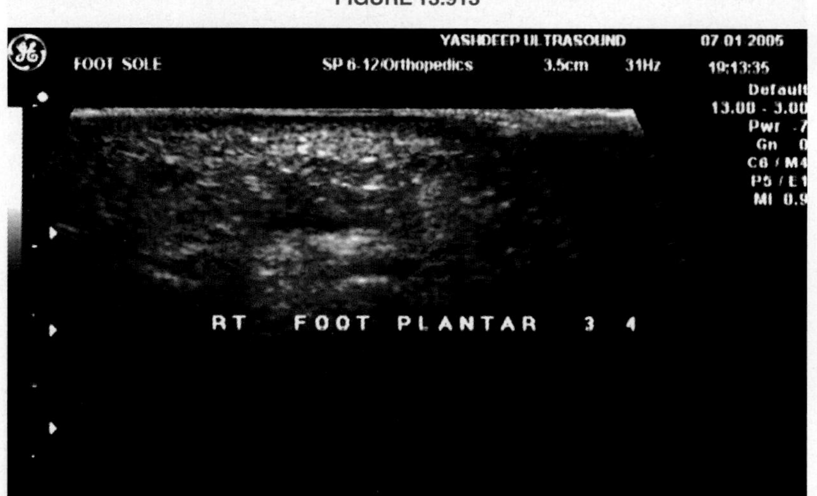

FIGURE 13.914

FIGURES 13.913 and 13.914: Morton's neuroma. HRSG shows typical appearance of Morton's neuroma over 3rd and 4th metatarsal heads.

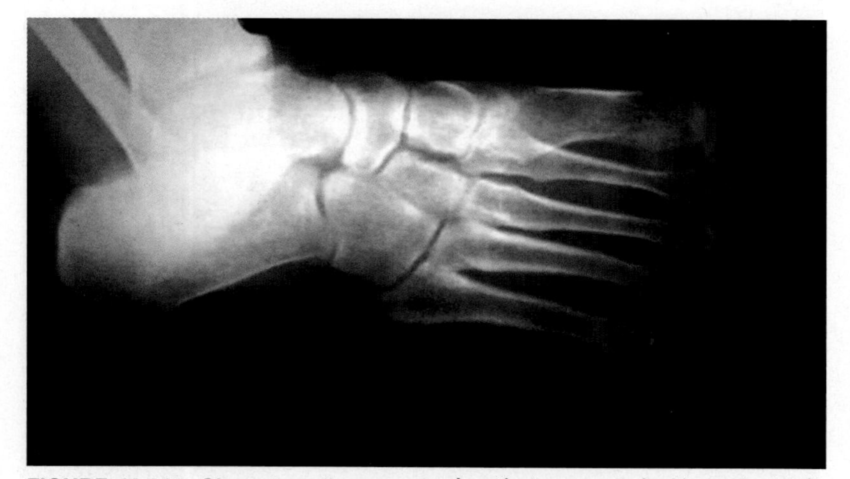

FIGURE 13.915: Shawn's cell neuroma. A patient presented with marked pain while walking. The pain was so intense that he stopped walking by the affected foot and developed disuse atrophy. X-ray of the foot does not show any bony abnormality.

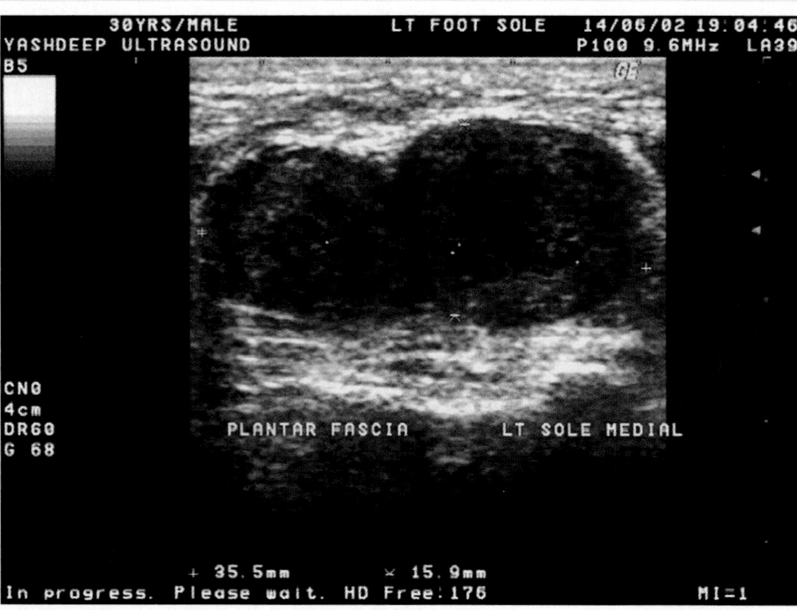

FIGURE 13.916: HRSG shows well-defined homogeneous hypoechoic masses over the medial side of the foot. They showed central area of necrosis.

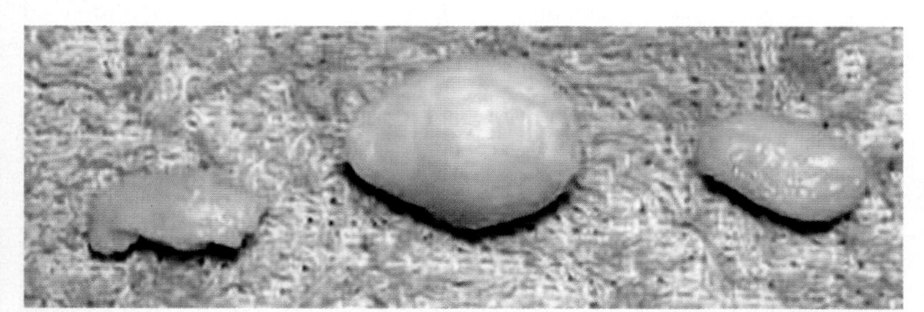

FIGURE 13.917

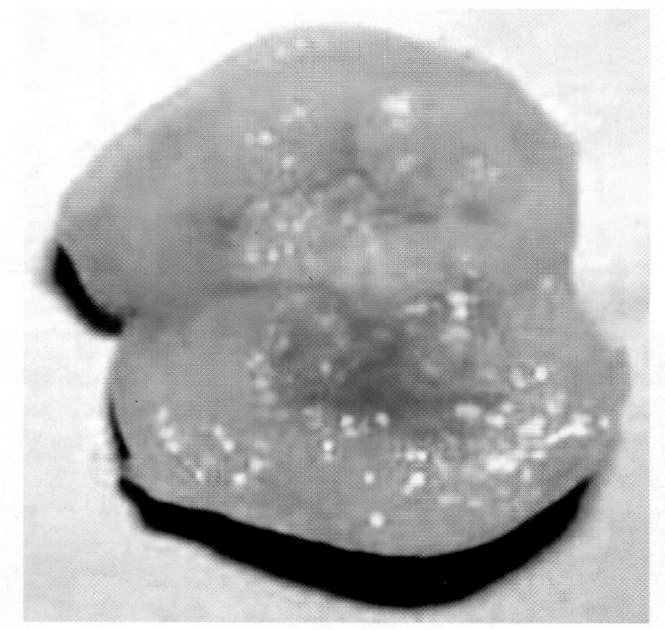

FIGURE 13.918

FIGURES 13.917 and 13.918: Postoperative gross examination of the nodule show well-defined neuromas. The cut section of the neuromas showed central area of haemorrhage. Biopsy of the nodules confirms Shawn's cell neuromas.

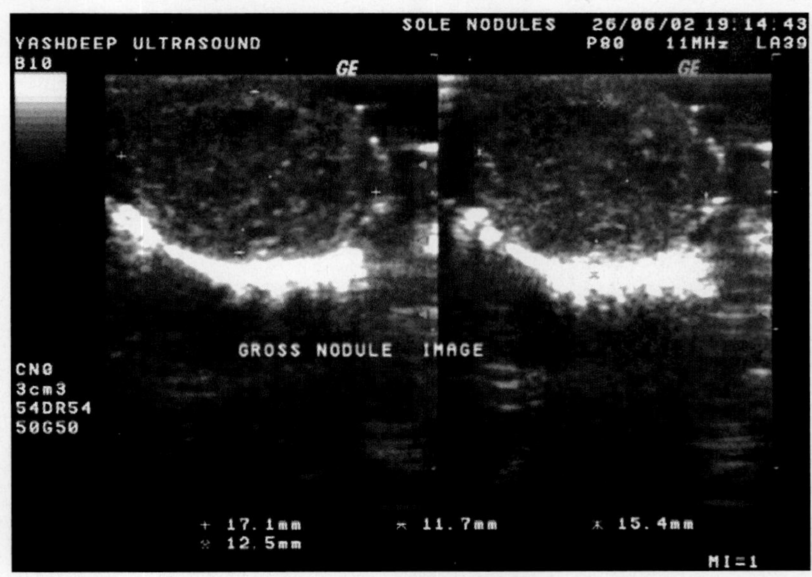

FIGURE 13.919

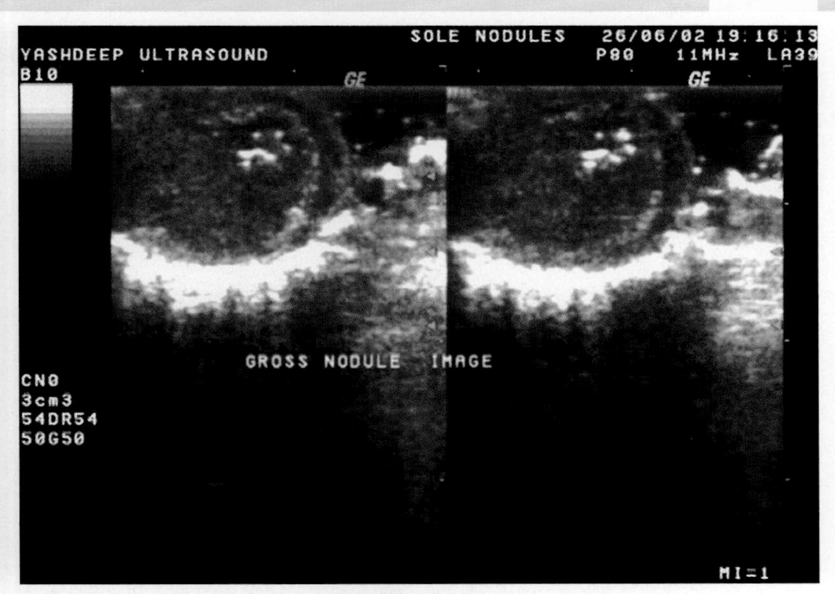

FIGURE 13.920

FIGURES 13.919 and 13.920: HRSG of the gross nodules confirmed the sonographic findings of neuromas.

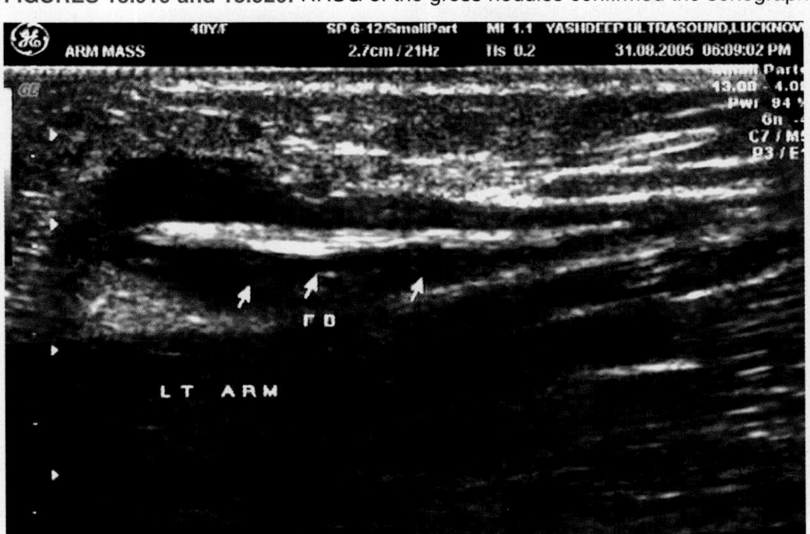

FIGURE 13.921

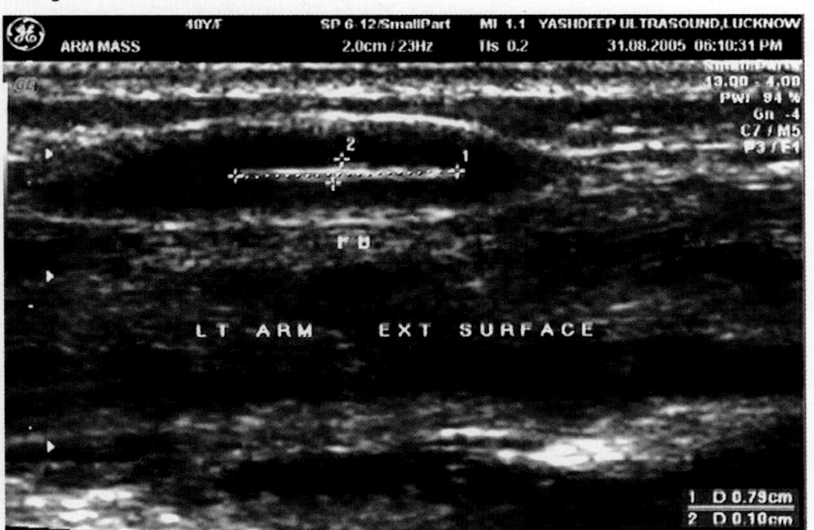

FIGURE 13.922

FIGURES 13.921 and 13.922: Foreign body in the muscles. A patient presented with painful nodules in the muscles of the arms. HRSG shows hyperechoic linear shadows embedded in the arm muscles.

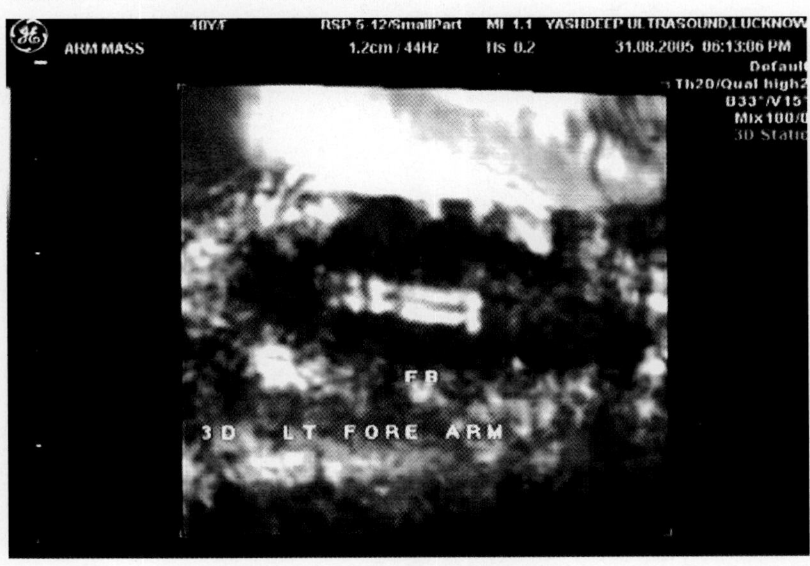

FIGURE 13.923

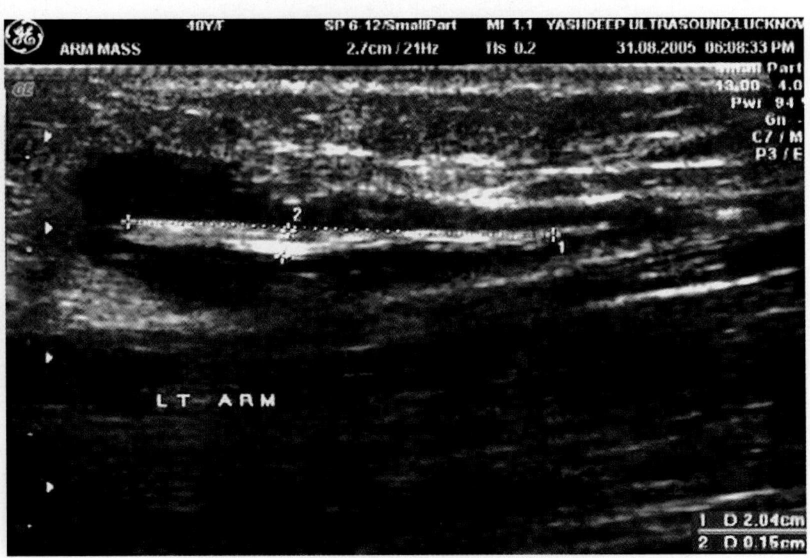

FIGURE 13.924

FIGURES 13.923 and 13.924: The 3D imaging of the arm muscles showed the details of the foreign body. Postoperative findings confirmed embedded thorn in the muscles.

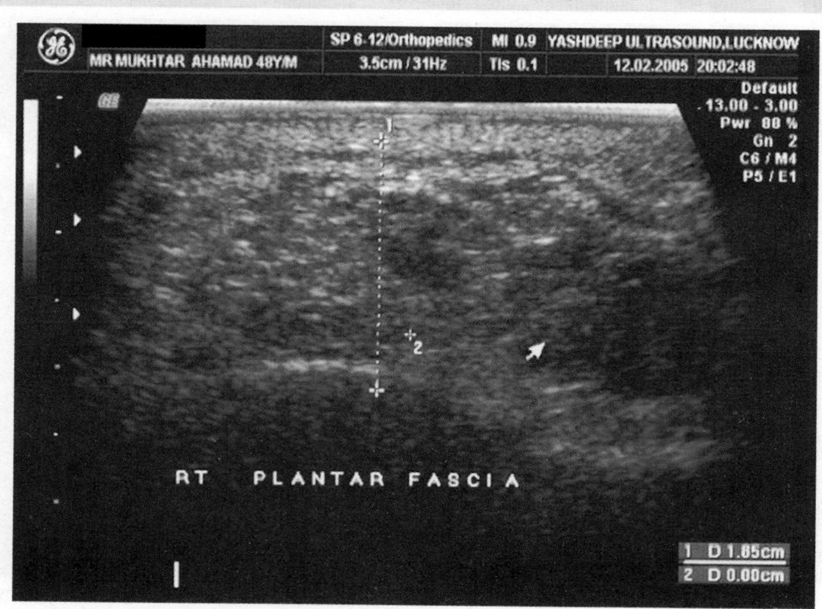

FIGURE 13.925: Plantar fascitis. A patient presented with painful walk. The plantar fascia was markedly thickened. Multiple hypoechoic areas are seen in it suggestive of edema. The fascia was inhomogeneous in texture. Findings are suggestive of plantar fascitis.

Bibliography

1. Eye and Orbit

1. Berlin LA, Zakov ZN: Ultrasonographic evaluation of hemorrhagic choroidal detachments. In Ossoinig KC (Ed): Ophthalmic Echography Dordrecht: Dr W Junk 1987; 313.
2. Bigar F, Bosshard C, Kloti R et al: Combined A and B-scan echography—preoperative evaluation of vitrectomy patients. Mod Probl Ophthalmol 1977; 18:2.
3. Blumenkranz MS, Byrne SF: Standardized echography (ultrasonography) for the detection and characterization of retinal detachment. Ophthalmology 1982; 89: 821.
4. Bronson NR: Techniques of ultrasonic localization and extraction of intraocular and extraocular foreign bodies. Am F Ophthalmol 1965; 60: 596.
5. Byrne BM, van Heuven WA, Lawton AW: Echographic characteristics of benign orbital schwannomas (neurilemmomas). Am F Ophthalmol 1988; 106:195.
6. Byrne SF, Glaser JS: Orbital tissue differentiation with standardized echography. Ophthalmology 1983; 90: 1071.
7. Byrne SF, Hughes JR: Orbital dermoid cysts. In Ossoinig KC (Ed): Ophthalmic Echography Dordrecht: Dr W Junk 1987; 465.
8. Byrne SF: Standardized echography in the differentiation of orbital lesions. Surv Ophthalmol 1984; 29: 226.
9. Byrne SF: Standardized echography of the eye and orbit. In Naidich TB, Quencer R (Eds): Clinical Neurosonography Berlin: Springer-Verlag 1987; 252.
10. Byrne SF: Standardized echography of the eye and orbit. Neurosonography 1986; 28: 618.
11. Clemens S, Kroll P, Rochels R: Ultrasonic findings after treatment of retinal detachment by intravitreal silicon instillation. Am F Ophthalmol 1984; 98: 369.
12. Coleman DJ, Franzen LA: Vitreous surgery—preoperative evaluation and prognostic value of ultrasonic display of vitreous hemorrhage. Arch Ophthalmol 1974; 92: 375.
13. Coleman DJ, Jack RL, Franzen LA: High resolution B-scan ultrasonography of the orbit (II)—Hemangiomas of the orbit. Arch Ophthalmol 1972; 88: 368.
14. Coleman DJ, Lizzi FL, Jack RL: Ultrasonography of the Eye and Orbit Philadelphia: Lea and Febiger 1977; 248.
15. Dallow RL: Ultrasonography in ocular and orbital trauma. Int Ophthalmol Clin 1974; 14(4): 23.
16. Dallow RL: Ultrasonography of the orbit. Int Ophthalmol Clin 1986; 26(3): 51.
17. DiBernardo C, Blodi BA, Byrne SF: Echographic evaluation of retinal tears in patients with spontaneous vitreous hemorrhage. ARVO Abstracts. Invest Ophthalmol Vis Sci 1991; 32(suppl): 878.
18. Fisher YL: Microphthalmus with ocular communicating orbital cyst-ultrasonic diagnosis. Ophthalmology 1978; 85: 1208.
19. Fuller DG, Hutton WL: Presurgical Evaluation of Eyes which Opaque Media New York: Grune and Stratton 1982; 109.
20. Fuller DG, Laqua H, Machemer R: Ultrasonographic diagnosis of massive periretinal proliferation in eyes with opaque Media (triangular retinal detachment). Am F Ophthalmol 1977; 83: 460.
21. Goes F: Ultrasonographic and clinical characteristics of orbital pseudotumors. In Ossoinig KC (Ed): Ophthalmic Echography Dordrecht: Dr W Junk 1987; 499.
22. Green RL, Byrne SF: Diagnostic ophthalmic ultrasound. In Ryan SJ (Ed): Retina St. Louis: CV Mosby 1989; 1: 191.
23. Harr DL, Quencer RM, Abrams GW: Computed tomography and ultrasound in the evaluation of orbital infection and pseudotumor. Radiology 1982; 142: 395.
24. Iijima Y, Asanagi K: A new B scan ultrasonographic technique for observing ciliary body detachment. Am F Ophthalmol 1983; 95: 498.
25. Kerman BM, Coleman DJ: B-scan ultrasonography of retinal detachments. Ann Ophthalmol 1978; 10: 903.
26. Ossoinig KC, Bigar F, Kaefring SL et al: Echographic detection and localization of BB shots in the eye and orbit. Bibl Ophthalmol 1975; 83: 109.
27. Ossoinig KC: Echography of the eye, orbit, and periorbital region. In Arger PH (Ed): Orbit Roentgenology New York. J Wiley and Sons 1977; 224.
28. Ossoinig KC: Standardized Ophthalmic Echography of the Eye, Orbit and Periorbital Region: A comprehensive Slide Set and Study Guide (3rd ed) lowa City: Goodfellow 1985.
29. Ossoinig KC: Seher K: Ultrasonic diagnosis of intraocular foreign bodies. In Gitter KA, Keeney AH, Sarin LK et al (Eds): Ophthalmic Ultrasound St Louis: CV Mosby 1969; 311.
30. Ossoinig KC: Quantitative echography—the basis of tissue differentiation. F Clin Ultrasound 1974;2:33.
31. Ossoinig KC: Standardized echography—basic principles, clinical applications and results. Int Ophthalmol Clin 1979; 19(4): 127.
32. Packer S, Rotman M, Salanitro P: Radiotherapy of choroidal melanoma with iodine-125 Ophthalmology 1980; 87: 582.
33. Poujol J: Intraocular tumours. In de Vleiger M (Ed): Handbook of Clinical Ultrasound New York: John Wiley and Sons 1978; 863.
34. Rootman J, Graeb D: Vascular lesions. In Rootman J (Ed): Diseases of the Orbit Philadelphia: JB Lippincott 1988; 525.
35. Sawada A, Frazier SL, Ossoinig KC: The role of ultrasound in the management of ocular foreign bodies. In White D, Brown RE (Eds): Ultrasound in Medicine New York: Plenum 1977; 3A: 1003.
36. Shalka HW, Callahan MA: Ultrasonically aided percutaneous orbital aspiration. Ophthalmic Surg 1979; 10(11): 41.
37. Shields CL, Shields JA, Pankajkumar S: Retinoblastoma in Older children. Ophthalmology 1991; 98: 395.
38. Shields JA: Diagnosis and Management of Orbital Tumors Philadelphia: WB Saunders 1989.
39. Sklar EL, Quencer RM, Byrne SF et al: Correlative study of the computed tomographic, ultrasonographic, and pathological characteristics of cavernous versus capillary hemangiomas of the orbit. F Clin Neuro Ophthalmol 1986; 6: 14.
40. Spoor TC, Kennerdell JS, Dekker A et al: Orbital fine needle aspiration biopsy with B-scan guidance. Am F Ophthalmol 1980; 89: 274.
41. Taylor D, Moore A: Retinoblastoma. In Pediatric Ophthalmology Boston: Blackwell Scientific 1990; 355.

2. Thyroid

1. Austin CW: Ultrasound evaluation of thyroid and parathyroid disease. Semin Ultrasound 1982; 4: 250.
2. Bartels PC, Boer RO: Subacute thyroiditis (de Quervain) presenting as painless "cold" nodule. J Nucl Med 1987; 28:1488.
3. Blum M, Passalaqua AM, Sackler JP et al: Thyroid echography in subacute thyroiditis. Radiology 1977; 125: 795.
4. Carroll BA: Asymptomatic thyroid nodules—incidental sonographic detection. AJR 1982; 133: 499.
5. Fobbe F, Finke R, Reichenstein R et al: Sonography. Eur J Radiol 1989; 2: 29.
6. Fujii Y, Wakasugi M, Yamada K et al: A study of ultrasonic diagnostic criteria for thyroid nodules: Proceeding of the 57th JSUM Takamastu 1990; 433.
7. Gorman B, Charboneau JW, James EM et al: Medullary thyroid carcinoma—a role of the high-resolution US. Radiology 1987; 162: 147.
8. Graif M, Itzchak Y, Strauss S et al: Parathyroid sonography—diagnostic accuracy related to shape, location and texture of the gland. Br J Radiol 1987; 60: 439.
9. Gritzman N, Grasl MCH, Helmer M et al: Invasion of carotid artery and jugular vein by lymph node metastases—detection with sonography. AJR 1990; 154: 441.
10. Hajeck PC, Salomonowitz E, Turk R et al: Lymph nodes of neck—evaluation with US. Radiology 1986; 158: 739.
11. Karstrup S, Glenthoj A, Torp-Pedersen S et al: Ultrasonically guided fine-needle aspiration of suggested enlarged parathyroid glands. Acta Radiol 1988; 29: 213.
12. Rogers WM: Anomalous development of the thyroid. In Werner SC, Ingbar SH (Eds): The Thyroid New York: Harper and Row 1978; 416.
13. Sakai F, Kiyono K, Sone S et al: Ultrasound evaluation of cervical metastatic lymphadenopathy. J Ultrasound Med 1988; 7: 305.
14. Simeone JF, Mueller PR, Ferrucci JT (Jr) et al: High-resolution real-time sonography of the parathyroid. Radiology 1981; 141: 745.
15. Solbiati L, Ballarati E, Cioffi V et al: Microcalcifications—a clue in the diagnosis of thyroid malignancies: Proceedings of the 76th RSNA, Chicago 1990; 140.
16. Solbiati L, Montali G, Croce F et al: Parathyroid tumors detected by fine-needle aspiration biopsy under ultrasonic guidance. Radiology 1983; 148: 793.
17. Solbiati L, Rizzatto G, Bellotti E et al: High resolution sonography of cervical lymph nodes in head and neck cancer—criteria for differentiation of reactive versus malignant nodes: Proceedings of the 74th RSNA, Chicago 1988; 113.
18. Solbiati L, Volterrani L, Rizzatto G et al: The thyroid gland with low uptake lesions—evaluation by ultrasound. Radiology 1985; 155: 187.
19. Sutton RT, Reading CC, Charboneau JW et al: US-guided biopsy of neck masses in postoperative management of patients with thyroid cancer. Radiology 1988; 168: 769.

3. Neck

1. Badami JP, Athey PA: Sonography in the diagnosis of branchial cysts. AJR 1981; 137: 1245-48.
2. Ballerini G, Mantero M, Sbrocca M: Ultrasonic patterns of parotid masses. JCU 1984; 12: 273-77.
3. Bartlett LJ, Pon M: High-resolution real-time ultrasonography of the submandibular salivary glands. J Ultrasound Med 1984; 3: 433-37.
4. Bruneton JN, Roux P, Caramella E et al: Ear, nose and throat cancer—ultrasound diagnosis of metastasis to cervical lymph nodes. Radiology 1984; 152: 771-73.
5. Friedman AP, Haller JO, Goodman JD et al: Sonographic evaluation of noninflammation neck masses in children. Radiology 1983; 147: 693-97.
6. Gritzmann N: Sonography of the salivary glands. AJR 1989; 153: 161-66.
7. Maragam D, Gooding GAW: Ultrasonic guided aspiration of parotid abscess. Arch Otolaryngol 1981; 107: 546.
8. Reede DL, Whelan MA, Bergeron RT: CT of the soft tissue structures of the neck. Radiol Clin North Am 1984; 1: 239-50.
9. Reede DL, Whelan MA, Bergeron RT: Computed tomography of the infrahyoid neck. Radiology 1982; 145: 397-402.
10. Rubaltelli L, Sponga T, Candiani F et al: Infantile recurrent sialectatic parotitis—the role of sonography and sialography in diagnosis and follow-up. Br J Radiol 1987; 60: 1211-14.
11. Sakai F, Kiyono K, Sone S et al: Ultrasonic evaluation of cervical metastatic lymphadenopathy. J Ultrasound Med 1988; 7: 305-10.
12. Scheible FW, Leopold GR: Diagnostic imaging in head and neck disease—current application of ultrasound. Head Neck Sur 1978; 1: 11.
13. Solbiati L, Rizzatto G, Bellotti E et al: High-resolution sonography of cervical lymph nodes in head and neck cancer—criteria for differentiation of reactive versus malignant nodes: Proceedings 74th RSNA meeting, Chicago 1988; 113.
14. Van den Brekel MWM, Stel HV, Castelijns JA et al: Cervical lymph node metastatis—assessment of radiologic criteria. Radiology 1990; 177: 379-84.

4. Salivary Glands

1. Bartlett LJ, Pon M: High-resolution real-time ultrasonography of the submandibular salivary glands. J Ultrasound Med 1984; 3: 433-37.
2. Bruneton JN, Roux P, Caramella E et al: Ear, nose and throat cancer—ultrasound diagnosis of metastasis to cervical lymph nodes. Radiology 1984; 152: 771-73.
3. Friedman AP, Haller JO, Goodman JD et al: Sonographic evaluation of noninflammation neck masses in children. Radiology 1983; 147: 693-97.
4. Reede DL, Whelan MA, Bergeron RT: CT of the soft tissue structures of the neck. Radiol Clin North Am 1984; 1: 239-50.
5. Rubaltelli L, Sponga T, Candiani F et al: Infantile recurrent sialectatic parotitis—the role of sonography and sialography in diagnosis and follow-up. Br J Radiol 1987; 60: 1211-14.
6. Sakai F, Kiyono K, Sone S et al: Ultrasonic evaluation of cervical metastatic lymphadenopathy. J Ultrasound Med 1988; 7: 305-10.
7. Scheible FW, Leopold GR: Diagnostic imaging in head and neck disease—current application of ultrasound. Head Neck Sur 1978; 1: 11.
8. Van den Brekel MWM, Stel HV, Castelijns JA et al: Cervical lymph node metastatis—assessment of radiologic criteria. Radiology 1990; 177: 379-84.

5. Neurosonography of Neonatal Brain

1. Babcock DS: Sonography of congenital malformations of brain. Clinical Neurosonography Berlin: Springer-Verlag 1987; 62-73.
2. Babcock DS: The normal absent and abnormal corpus callosum—sonographic findings. Radiology 1984; 151(2): 450-53.
3. Beu WE, Mccormick WF: Neurologic Infections in Children Philadelphia: WB Saunders 1981; 3-76.
4. Bowerman RA, Donn SM, Selever TM, Faffee MH: Natural history of periventricular intraventricular haemorrhage and its complications—sonographic observations. AJNR 1984; 5: 527-38.

5. Cardoze JD, Gulstein RB, Filly RA: Exclusion of foetal ventriculomegaly with a single measurement—the width of the lateral ventricle atrium. Radiology 1988; 169: 711-14.
6. Carson SC, Hertzberg BS, Bowie JD et al: Value of sonography in the diagnosis of intracranial haemorrhage and periventricular leucomalacia—a postmortem study of 35 cases. AJNR 1990; 11: 677-83.
7. Chuang S, Harwood-Nash DC: Tumors and cysts. Clinical Neurosonography Berlin: Springer-Verlag 1987; 97-109.
8. Cubbertery DA, Jafte RB, Nexon GW: Sonographic demonstration of galenic arteriovenous malformation in neonate. AJNR 1982; 3: 435-39.
9. De Myer W: Classification of cerebral malformations. Birth Defects 1971; 7: 78-93.
10. Ford LM, Han BL, Steineten J et al: Very low birth weight preterm infants with or without intracranial hemorrhage. Clinical Pediatr 1989; 28: 302-10.
11. Gupta RK, Pant CS: Role of high-resolution ultrasound in newborns. Manual of Practical New Born Care 1992; 144-51.
12. Gupta RK, Pant CS, Sharma A et al: Ultrasound diagnosis of multiple cystic encephalomalacia. Pediatr Radiol 1988; 18: 6-8.
13. Han BK, Babcock DS, Adamsh MC: Bacterial meningitis in infants—sonographic findings. Radiology 1985; 154 (3): 645-50.
14. Harwood-Nash DC, Fitz CE: Neuroradiology in Infants and Children St Louis: CV Mosby 1976.
15. Papile LA, Burstein J, Burstein R et al: Incidence of evolution of subependymal intraventricular haemorrhage—a study of infants with birth weights less than 1500 gm. J Pediatr 1978; 92: 529-34.
16. Reeder JD, Sanders RC: Ventriculitis in the neonate—recognition by sonography. AJNR 1983; 4: 37-41.
17. Rumack CM, Jhonson MC: Perinatal and Infant Brain Imaging Chicago: Year Book Medical Publishers 1984.
18. Shuman WP, Rogers JV, Mack LA: Real-time sonographic sector scanning of the neonatal cranium-technique and normal anatomy. AJNR 1981; 2:349-56.
19. Taylor GA, Sanders RC: Dandy-Walker syndrome recognition by sonography. AJNR 1983; 4: 1203-06.
20. Volpe JJ: Intraventricular haemorrhage in premature infant—current concepts. Ann Neurol 1989; 25: 3-11.

6. Peripheral Chest

1. Bell J, O'Reilly J, Laszlo G, Wilde P, Goddard P: Imaging techniques in the diagnosis of a mediastinal mass. Bristol Med Chir J Oct 1983;176:176-78.
2. Bonhof JA et al: Transoesophageal mediastinal sonography. Ann Radiol (Paris) 1985;28:19-20.
3. Callahan JA, Seward JB, Talik AJ et al: Pericardiocentesis assisted by two-dimensional echocardiography. J Thorac Cardiovasc Surg 1983;85:877-79.
4. Candell-Riera J, Del Castillo H G, Permanyer- Miralda G, Soler-Soler J: Echocardiography features of the interventricular septum in chronic constrictive pericarditis. Circulation 1978; 57(6): 1154-58.
5. Dorne HJ: Differentiation of pulmonary parenchymal consolidation from pleural disease using the sonographic fluid consolidation from pleural disease using the sonographic fluid bronchogram. Radiology 1986;158:41.
6. Engel PJ, Hon H, Fowler NO, Plummer S: Echocardiographic study of right ventricular wall motion in cardiac tamponade. Am J Cardiol 1982; 50: 1018 21.
7. Fataar S: Ultrasound in chest disease: Pleura. Australas Radiol 1988;32:295-301.
8. Ikezoe J, Morimoto S, Arizawa J et al: Ultrasonography of mediastinal teratoma. JCU 1986;14:513-20.
9. Lipscomb DJ, Flower CDR, Hadfield JW: Ultrasound of the pleura: an assessment of its clinical value. Clin Radiol 1981;32: 289-90.
10. Lipscomb DJ, Flower CDR: Ultrasound in the diagnosis and management of pleural disease. Br J Dis Chest 1980;74:353-61.
11. Marks WM, Filly RA, Callen PW: Realtime evaluation of pleural lesions: new observations regarding the probability of obtaining free fluid. Radiology 1982;142:163-64.
12. Martin RP, Bowden R, Filly K, Popp RL: Intrapericardial abnormalities in patients with pericardial effusion. Circulation 1980;61:568-72.
13. Millman A, Meller J, Motro M et al: Pericardial effusion by echocardiography. Ann Intern Med 1977; 86(4): 434-36.
14. Rasmussen OS, Boris P: Ultrasound guided puncture of pleural fluid collection and superficial thoracic masses. Eur J Radiol 1989;9:91-92.
15. Ries T, Umarino G, Nikaidoh H, Kennedy L: Realtime ultrasonography of subcranial bronchogenic cysts in two children. Radiology 1982;145:121-22.
16. Schnittger I, Bowden RE, Abrahams J, Popp RL: Echocardiography: pericardial thickening and constrictive pericarditis. Am J Cardiol 1978; 42: 388-95.
17. Voelkel AG, Pietro DA, Folland ED, Fisher ML, Parisi AF: Echocardiography features of constrictive pericarditis. Circulation 1978;58(5):871-75.
18. Wernecke K, Peters P, Galanski M: Mediastinal tumors, evaluation with suprasternal sonography. Radiology 1986; 159: 405-09

7. Breast Ultrasound

1. Alder DD: Ultrasound of benign breast condidtions. Semin Ultrasound CT, MR 1989; 10: 106-18.
2. Bassett LW, Kimme-Smith C, Southland LK et al: Automated and hand-held breast ultrasound—effect on patient management. Radiology 1987; 165: 103-08.
3. Bohm-Velez M, Mendelson EB: Computed tomography, ultrasound and magnetic resonance imaging in evaluating the breast. Semin-Ultrasound, CT, MR 1989; 10: 171-76.
4. Cole-Beuglet C, Soriano Rz, Krutz AB et al: Sonomammography correlated with pathology in 122 patients. AJR 1983; 140: 369-75.
5. Cole-Beuglet C, Soriano Rz, Krutz AB et al: Ultrasound analysis of 104 primary breast carcinomas classified according to histopathologic type. Radiology 1983; 147: 191-96.
6. Fornage BD, Sneige N, Faroux MJ et al: Sonographic appearance and ultrasound-guided fine-needle aspiration biopsy of breast carcinoma smaller than 1 cm. J Ultrasound Med 1990; 9: 559.
7. Harper AP, Kelly-Fry E, Noe JS et al: Ultrasound in the evalution of solid breast masses. Radiology 1983; 146: 731-36.
8. Hilton SVW, Leopold GR, Olson LK et al: Realtime breast sonography—application in 300 consecutive patients. AJR 1986; 147: 479-86.
9. Jackson VP: Breast neoplasms—duplex sonographic imaging as an adjunct in diagnosis—letter to the editor. Radiology 1989; 170: 578.
10. Jackson VP, Kelly-Fry F, Rothschild et al: Automated breast sonography using a 7.5 MHz PVDF transducer—preliminary clinical evaluation. Radiology 1986; 159: 679-84.
11. Jackson VP, Rothschild PA, Kreipke DL et al: The spectrum of sonographic findings of fibroadenoma of the breast. Invest Radiol 1986; 21: 31-40.
12. Jackson VP: Sonography of malignant breast disease. Semin Ultrasound, CT, MR 1989; 10: 1190-1231.
13. Kopans DB, Mayer JE, Lindors et al: Breast sonography to guide cyst aspiration and wire localisation of occult solid lesions. AJR 1984; 143: 489-512.
14. Logan WW, Hoffman HY: Diabetic fibrous disease. Radiology 1989; 172: 667-70.
15. Mendelson EB, Bohm-Velez M, Bhagwanani DG et al: Sonographic spectrum of fibroadenomas—a guide to clinical management. Paper presented at the Radiological Society of North America Annual Meeting, Chicago, 1988.

16. Rubin E, Miller VE, Berland LL et al: Hand-held real-time breast sonography. AJR 1985; 144: 623-27.
17. Sickles EA, Filly CA, Callen PW: Benign breast lesions—ultrasound detection ad diagnosis. Radiology 1984: 151: 467-70.
18. Sickles EA, Filly CA, Callen PW: Breast cancer detection with sonography and mammography—comparison using state-of-the-art equipment. AJR 1983; 140: 843-45.

8. Gastrointestinal System

1. Abu-Yusef MM, Bleicher JJ, Maher JW et al: High resolution sonography of acute appendicitis. AJR 1987;149:53-58.
2. Bhisitkul DM, Shkolnik A, Donaldson JS, Henricks BD, Feinstein KA, Fernbach SK, Listernick R: Clinical application of ultrasonography in the diagnosis of intussusception. J Pediatr 1992; 121: 182-86.
3. Blumhagen JD, Maclin L, Krauter D, Rosenbaum DM, Weinberger E. Sonographic diagnosis of hypertrophic pyloric stenosis. AJR 1988;150:1367-70.
4. Bowen A, Mazer J, Zarabi M, Fujioka M: Cystic meconium peritonitis: ultrasonographic features. Pediatr Radiol 1984;14:18-22.
5. Bowerman RA, Silver TM, Jaffe MH. Real-time ultrasound diagnosis of intussusception in children. Radiology 1982;143: 527-29.
6. Boychuk RB, Lyons EA, Goodhand TK: Duodenal atresia diagnosed by ultrasound. Radiology 1978;127:500.
7. Carrol BA, Moskowitz PS: Sonographic diagnosis of a neonatal meconium cyst. AJR 1981;137:1262-64.
8. Cohen HL, Haller JO, Mestl Al, Coren C, Schechter S, Eaton DH: Neonatal duodenum: fluid-aided ultrasound examination. Radiology 1987;164:805-809.
9. Cohen HL, Schechter S, Mestel AL, Eaton DH, Haller JO: Ultrasonic "double track" sign in hypertrophic pyloric stenosis. J Ultrasound Med 1987;6:139-43.
10. Dinkel E, Dittrich M, Peters H et al: Real-time ultrasound in Crohn's disease: characteristic features and clinical implications. Pediatr Radiol 1986;16:8-12.
11. Feinstein KA, Myers M, Fernbach SK et al: Peritoneal fluid in children with intussusception: its sonographic detection and relationship to successful reduction. Abdom Imag 1993;18:277.
12. Hayashi K, Futagawa S, Kozaki S, Hirao K, Hombo Z: Ultrasound and CT diagnosis of intramural duodenal haematoma. Pediatr Radiol 1988;18:167-68.
13. Hayden CK Jr, Schwartz MZ, Davis M, Swischuk LE: Combined esophageal and duodenal atresia: sonographic findings. AJR 1983;140:225-26.
14. Hayden CK Jr, Kuchelmeister J, Lipscomb TS: Sonography of acute appendicitis in childhood: perforation versus nonperforation. J Ultrasound Med 1992;11:209.
15. Hayden CK Jr, Boulden TF, Swischuk LE, Lobe TE: Sonographic demonstration of duodenal obstruction with midgut volvulus. AJR 1984;143:9-10.
16. Hernaz-Shulman M, Sells LL, Ambrosino MM, Heller RM, Stein SM, Neblett WW III: Hypertrophic pyloric stenosis in the infant without a palpable olive: accuracy of sonographic diagnosis. Radiology 1994;193:771.
17. Kao SC, Smith WL, Abu-Yousef MM et al: Acute appendicitis in children: sonographic findings. AJR 1989;153:373-79.
18. Kodroff MB, Hartenberg MA, Goldschmidt RA: Ultrasonographic diagnosis of gangrenous bowel in neonatal necrotizing enterocolitis. Pediatr Radiol 1984;14:168-80.
19. Lim JH, Ko YT, Lee DH et al. Sonography of inflammatory bowel disease: findings and value indifferential diagnosis. AJR 1994; 163:343.
20. Malpani A, Ramani SK, Wolverson MK: Role of sonography in trichobezoars. J Ultrasound Med 1988;7:661-63.
21. Naik DR, Bolia A, Boon AW: Demonstration of a lactobezoar by ultrasound. Br J Radiol 1987;60:506-508.
22. Neilson D, Hollman AS: The ultrasonic diagnosis of infantile hypertrophic pyloric stenosis: technique and accuracy. Clini Radiol 1994;49:246-47.
23. Patriquin HB, Garcier J-M, Lafortune M, Yazbeck S, Russo P, Jequier S, Ouimet A, Filiatrault D: Appendicitis in children and young adults: Doppler sonographic-pathologic correlation. AJR 1996;166:629-33.
24. Pracors JP, Tran-Minh VA, Morin de Finfe CH, Deffrenne-Pracros P, Louis D, Basset T: Acute intestinal intussusception in children: contribution of ultrasonography (145 cases). Ann Radiol 1987;30:525-30.
25. Quillin SP, Seigel MJ, Coffin CM: Acute appendicitis in children: value of sonography in detecting perforation. AJR 1992;159:1265-68.
26. Quillin SP, Seigel MJ: Diagnosis of appendiceal abscess in children with acute appendicitis: value of color Doppler sonography. AJR 1995;164:1251-54.
27. Ramachandran P, Sivit CJ, Newman KD, Schwartz MZ: Ultrasonography as an adjunct in the diagnosis of acute appendicitis: a 4-year experience. J Pediatr Surg 1996;31:164-69.
28. Rubin SZ, Martin DJ: Ultrasonography in management of possible appendicitis in childhood. J Pediatr Surg 1990;25:737-40.
29. Sata Y, Pringle KC, Bergman RA et al: Congenital anorectal anomalies: MR imaging. Radiology 1988;168:157-62.
30. Schneider K, Dickerhoff R, Bertele RM: Malignant gastric sarcoma: diagnosis by ultrasound and endoscopy. Pediatr Radiol 1983;16:69-70.
31. Seigel MJ, Carel C, Surratt S: Ultrasonography of acute abdominal pain in children. JAMA 1991;226:1087-89.
32. Shah RS, Kaddu SJ, Kirtane JM: Benign mature teratoma of the large bowel: a case report. J Pediatr Surg 1996;31:701-702.
33. Shimanuki Y, Aihara T, Takano H, Moritani T, Oguma E, , Kuroki H, Shibata A, Nozawa K, Ohkawara K, Hirata A, Imagizumi S: Clockwise whirlpoor sign at color Doppler US: an objective and definite sign of midgut volvulus. Radiology 1996;199:261-64.
34. Steyaert H, Guitard J, Moscovivi J, Juricic M, Vaysse P: Benign lesions that can have a proliferative course. J Pediatr Surg 1996; 31:677-80.
35. Stunden RJ, LeQuesne GW, Little KE: The improved ultrasound diagnosis of hypertrophic pyloric stenosis. Pediatr Radiol 1986;16:200-205.
36. Teele RL, Smith EH: Ultrasound in the diagnosis of idiopathic hypertrophic pyloric stenosis. N Engl J Med 1977;296:1149-50.
37. Verschelden P, Filiatrault D, Garel L, Grignon A, Perreault G, Boisvert J, Dubois J: Intussusception in children: reliability of US in diagnosis—a prospective study. Radiology 1992;184:741-44.
38. Vignault F, Ilitarault D, Brandt ML et al: Acute appendicitis in children: evaluation with US. Radiology 1990;176:501-504.
39. Weinberg B, Rao PS, Shah KD: Ultrasound demonstration of an intramural leiomyoma of the gastric cardia with pathologic correlation. J Clin Ultrasound 1988;16:580-84.
40. Woo SK, Kim JS, Suh SJ. Paik TW, Choi SO: Childhood intussusception: US-guided hydrostatic reduction. Radiology 1992;182:77-80.
41. Worliceck H. Lutz H, Heyder N et al: Ultrasound findings in Crohn's disease and ulcerative colitis: prospective study. J Clin Ultrasound 1987;15:153-58.

9. Prostate and Seminal Vesicles

1. Ajzen SA, Goldenberg SL, Allen GJ et al: Palpable prostatic nodules: Comparison of ultrasound and digital guidance for fine needle aspiration biopsy. Radiology 1989;171:521-23.

2. Andriole GL. Coplen DE, Mikkelsen DJ et al: Sonographic and pathological staging of patients with clinically localized prostate cancer. J Urol 1989;142:1259-61.
3. Ayala AG, Ro JY, Babaian R et al: The prostatic capsule: does is exist? Am J Surg Pathol 1989;13:21-27.
4. Berry SJ, Coffey DS, Walsh PC et al: The development of human benign prostatic hyperplasia with age. J Urol 1984;132:474-79.
5. Brawer MK, Rennels MA, Nagle RB et al: Prostatic intraepithelial neoplasia: A lesion that may be confused with cancer on prostatic ultrasound. J Urol 1989;142:1510-12.
6. Bude R, Bree RL, Adler RS et al: Transrectal ultrasound appearance of granulomatous prostatis. J Ultrasound Med 1990;9: 677-80.
7. Burks DD, Drolshagen LF, Fleischer AC et al: Transrectal sonography of benign and malignant prostate lesions. AJR 1986;146: 1187-91.
8. Carter HB, Hamper UM, Sheth S et al: Evaluation of transrectal ultrasound in the early detection of prostate cancer. J Urol 1989;142:1008-10.
9. Chodak GW: Screening for prostate cancer. Urol Clin North Am 1989;16:657-61.
10. Di Trapani D, Pavone C, Serreta V et al: Chronic prostatitis and prostatodynia: ultrasonographic alterations of the prostate, bladder neck, seminal vesicles and periprostatic venous plexus. Eur Urol 1988;15:230-34.
11. Drago JR, Nesbitt JA, Badalament RA: Use of transrectal ultrasound in detection of prostatic carcinoma: A preliminary report. J Surg Oncol 1989;41:274-77.
12. Dyke CH. Toi A, Sweet JM: Value of random ultrasound guided transrectal prostate biopsy. Radiology 1990;176:345-49.
13. Fuse H, Sumiya H, Ishii H et al: Treatment of hemospermia caused by dilated seminal vesicles by direct drug injection guided by ultrasonography. J Urol 1988;140:991-92.
14. Hamper UM, Sheth S, Walsh PC et al: Bright echogenic foci in early prostatic carcinoma: Sonographic and pathologic correlation. Radiology 1990; 176:339-43.
15. Hendrikx AJ, Doesburg WH, Reintjes AG et al: Determination of prostatic volume by ultrasonography. Urology 1989;33(4):336-39.
16. Hunter PT, Butler SA, Hodge GB et al: Detection of prostatic cancer using transrectal ultrasound and sonographically guided biopsy in 1410 symptomatic patients. J Endourol 1989;3:167-75.
17. Jakobsen H, Torp-Pedersen S, Juul N: Ultrasonic evaluation of age-related human prostatic growth and development of benign prostatic hyperplasia. ScandJ Urol Nephrol 1988;107(suppl):26-31.
18. Lee F, Littrup PJ, Torp-Pedersen S et al: Prostate cancer: Comparison of transrectal ultrasound and digital rectal examination. Radiology 1988;168: 389-94.
19. Lee F, Torp-Pedersen S, Carroll JT et al: Use of transrectal ultrasound and prostate-specific antigen in diagnosis of prostatic intraepithelial neoplasia. Urology 1989;34(suppl):4-8.
20. Lee F, Torp-Pedersen S, Littrup PJ et al: Hypoechoic lesions of the prostate: clinical relevance of tumor size, digital rectal examination, and prostate-specific antigen. Radiology 1989;170:29-32.
21. Lee F, Torp-Pedersen ST, McLeary RD: Diagnosis of prostate cancer by transrectal ultrasound. Urol Clin North Am 1989; 16:663-73.
22. Lee F. Torp-Pedersen ST, Siders DB et al: Transrectal ultrasound in the diagnosis and staging of prostatic carcinoma. Radiology 1989;170:609-15.
23. McNeal JE: The zonal anatomy of the prostate. Prostate 1981;2: 35-39.
24. Muldoon L, Resnick MI: Results of ultrasonography of the prostate. Urol Clin North Am 1989;16:693-702.
25. Palken M, Cobb OE, Warren BH et al: Prostate cancer: Correlation of digital rectal examination, transrectal ultrasound and prostate specific antigen levels with tumor volume in radical prostatectomy specimens. J Urol 1990;143:115-62.
26. Resnick MI: Transrectal ultrasound guided versus digitally directed prostatic biopsy: A comparative study. J Urol 1987;139:754-57.
27. Rifkin MD, McGlynn ET, Choi H: Echogenicity of prostatic cancer correlated with histologic grade and stromal fibrosis: endorectal ultrasound studies. J Urol 1989;170:549-52.
28. Rifkin MD: Endorectal sonography of the prostate: clinical implications. AJR 1987;148:1137-42.
29. Shabsingh R, Lerner S, Fishman IJ et al: The role of transrectal ultrasonography in the diagnosis and management of prostatic and seminal vesicle cysts. J Urol 1989;141:1206-09.
30. Shinohara K, Wheeler TM, Scardino PT: The appearance of prostate cancer on transrectal ultrasonography: Correlation of imaging and pathological examinations. J Urol 1989;142:76-82.
31. Torp-Pedersen S, Lee F, Littrup PJ et al: Transrectal biopsy of the prostate guided with transrectal ultrasound: longitudinal and multiplanner scanning. Radiology 1989;170:23-27.
32. Tzai T, Chang C, Yang CX et al: Transrectal sonography of the prostate and seminal vesicles on patients with hemospermia. J Formosan Med Assoc 1989;88:232-35.

10. Scrotum and Testis

1. Burks DS, Markey BJ et al: Suspected testicular torsion and ischemia—evaluation with color Doppler sonography. Radiology 1990; 175: 815-21.
2. Cattolica EV, Karol JB et al: High testicular salvage rate in torsion of the spermatic cord. J Urol 1982; 128: 589-90.
3. Chin DCP, Holder LE et al: Correlation of radionuclide imaging and diagnostic ultrasound in scrotal diseases. J Nuclear Med 1986; 27: 1774-81.
4. Desgrandchamps F: Testicles non descendus-Des connaisances actuelles. J Urol (Paris) 1990; 96: 407-14.
5. Didionna A, Rizzatto G: Pyocele of the scrotum, sonographic demonstration of fluid level and a pus forming component. J Ult Med 1986; 15: 99-100.
6. Fournier GR, Laing FC et al: Scrotal ultrasonography and the management of testicular trauma. Uro Clin N Am 1989; 16: 377-85.
7. Hill MC, Sander RC: Ultrasonography of benign diseases of the scrotum. In Sanders RC, Hill MC (Eds): Ultrasound Annual. New York: Raven Press 1987; 197-237.
8. Horstman WG, Middleton WD, Melson GL: Scrotal inflammatory disease —color Doppler ultrasound finding. Radiology 1991; 179: 55-59.
9. Horstmann, Middleton WD et al:Color Doppler ultrasound of the scrotum. Radiographics 1991; 11: 1941-57.
10. Jenson MC, Lac KP et al: Color Doppler sonography in testicular torsion. J Clin Ultrasound 1990; 18: 446-48.
11. John Radcliff Hospital Cryptorchidism Study Group. Cryptor-chidism—an apparent substantial increase since 1960. Br Med J 1986; 298: 1401-04.
12. Kim SH, Pollack HM et al: Tuberculous epididymo-orchitis, sonographic findings. J Urol 1993; 150: 81-84.
13. Lerner RM, Mevorach RA et al: Color Doppler ultrasound in the evaluation of acute scrotal disease. Radiology 1990; 176: 355-58.
14. Leung ML, Gooding GAW et al: High-resolution sonograpy of scrotal contents in asymptomatic subjects. AJR 1984; 143: 161-64.
15. Linkowski GD, Avellone A, Gooding GAW: Scrotal calculi—sonographic detection. Radiology 1985; 156: 484.
16. Marks LB, Rutgers GL et al: Testicular seminoma—clinical and pathological features that may predict paraortic lymph node metastasis. J urol 1990; 143: 524-27.

17. Martin B, Tubiana JM: Significance of scrotal calcifications detected by sonography. J Clin Ultrasound 1988; 16: 545-52.
18. Middleton WD, Rumack, Thorne DA, Melson GL: Color Doppler ultrasound of the normal testis. AJR 1989; 152: 293-96.
19. Middleton WD, Siegar BA et al: Acute scrotal disorders—prospective comparison of color Doppler and testicular scintigraphy. Radiographics 1990; 177: 177-97.
20. Middleton WD: Scrotal sonography in 1991. Ultrasound Quaterly 1991; 9: 61-87.
21. Rallis PW, Jenson MC et al: Color Doppler sonography in acute epididymitis and orchitis. J Clin Ultrasound 1990; 18: 383-86.
22. Schustar G: Traumatic rupture of the testicle—a review of the literature. J Urol 1982; 127: 1194-96.
23. Steinfeld AD: Testicular germ cell tumor—review of contemporary evaluation and management. Radiology 1990; 175: 603-06.
24. Steinhardt GF, Broyarsky S et al: Testicular torsion—pitfalls of color Doppler sonography. J Urol 1993; 150: 461-62.
25. Subramanyam BR, Moris SC et al: Diffuse testicular disease, sonographic features and significance.AJR 1985; 145: 1221-24.
26. Tanagho EA: Anatomy and surgical approach to the male genital tract. In Walsh PC, Gittes RF et al (Eds): Campbell's Urology (5th ed). Philadelphia: Saunders, 1986; 65-68.

11. Anterior Abdominal Wall

1. Cervantes J, Sanchez-Cortazar J, Ponte RJ et al: Ultrasound diagnosis of rectus sheath haematoma. American Surgeon 1983; 49: 542.
2. Engel JM, Deitch EA: Sonography of the anterior abdominal wall. AJR 1981; 137: 73.
3. Fried AM, Meeker WR: Incarcerated spingelian hernia—ultrasonic differential diagnosis. AJR 1979; 133: 104.
4. Goldberg BB: Ultrasonic evaluation of superficial masses. JCU 1975; 3: 91.
5. Kaftori JK, Rosenberger A, Pollack S et al: Rectus sheath haematoma—ultrasonographic diagnosis. AJR 1977; 128: 281.
6. Madrazo BL, Klugo RC, Parks JA et al: Ultrasonographic demonstration of the undescended testis. Radiology 1979; 92: 885.
7. Muller N, Cooperberg PL, Rowley A et al: Ultrasonic refraction by the rectus abdominis muscles—the double image artifact. J Ultrasound Med 1984; 3: 515.
8. Rankin RN, Hutton L, Grace DM: Postoperative abdominal wall hematomas have a distinctive appearance on ultrasonography. Can J Surg 1985; 28:84.

12. Penis

1. Aboseif SR, Lue TF: Hemodynamics of penile erection. Urol Clin North Am 1988;15:1-7.
2. Altraffer LF, Jordan JH: Sonographic demonstration of Peyronie's plaques. Urology 1981;17:292-95.
3. Balconi G, Angeli E, Nessi R et al: Ultrasonographic evaluation of Peyronie's disease. Urol Radiol 1988;10:85-88.
4. Benson CB, Vickers MA: Sexual impotence caused by vascular disease: diagnosis with duplex sonography. AJR 1989;153:1149-53.
5. Beulter LE, Gleason DM: Integrating the advances in the diagnosis and treatment of male potency disturbance. J Urol 1981;126:338-42.
6. Bookstein JJ: Cavernosal veno-occlusive insufficiency in male impotence: evaluation of degree and location. Radiology 1987; 164:175-78.
7. Buvat J, Bervat-Hertaut M, Dehaene JL et al: It intravenous injection of papaverine a reliable screening test for vasculogenic impotence? J Urol 1986;135:476-78.
8. Collins WE, McKendry JBR, Silverman M et al: Multidisciplinary survey of erectile dysfunction. Canad Med Assoc J 1982; 128:1393-99.
9. Desai KM, Gingell JC, Skidmore R et al: Application of computerized penile arterial waveform analysis in the diagnosis of arteriogenic impotence: an initial study in potent and impotent men. Br J Urol 1987;60:5:450-56.
10. Fujita T, Shirai M: Mechanism of erection. J Clin Exp Med 1989;148:249.
11. Gelbard M, Sarti D, Kanfman J: Ultrasound imaging of Peyronie's plaques. J Urol 1981;125:44-45.
12. Gluck CD, Bundy Al, Fine C et al: Sonographic urethrogram: a comparison to roentgenographic techniques in 22 patients. J Urol 1988;140:1404-08.
13. Hattery RR, King BF, Lewis RW et al: Vasculogenic impotence: duplex and color Doppler. Radiol Clin North Am 1991.
14. Hricak H, Marotti M, Gilbert TJ et al: Normal penile anatomy and abnormal penile conditions: evaluation with MR imaging. Radiology 1988;169:683-90.
15. Krane RJ, Goldstein I, Saenz de Tejada I: Medical progress: impotence. New Engl J Med 1989;321:1648-59.
16. Krysiewicz S, Mellinger BC: The role of imaging in the diagnostic evaluation of impotence. AJR 1989;153:1133-39.
17. Lue TF, Hricak H, Marich KW et al: Evaluation of arteriogenic impotence with intracorporeal injection of papaverine and the duplex ultrasound scanner. Semin Urol 1985;3:1:43-48.
18. Lue TF, Hricak H, Marich KW et al: Vasculogenic impotence evaluated by high resolution ultrasonography and pulsed Doppler spectrum analysis. Radiology 1985;155:3:777-81.
19. Lue TF, Tanagho EA: Physiology of erection and pharmacological management of impotence. J Urol 1987;137:5:829-36.
20. McAninch JW, Laing FC, Jeffrey RB: Sonourethrography in the evaluation of urethral strictures: a preliminary report. J Urol 1988;139:294-97.
21. Mellinger BC, Vaughan ED Jr, Thompson SL et al: Correlation between intracavernous papaverine injection and Doppler analysis in impotent men. Urology 1987;416-19.
22. Merkle W, Wagner W: Sonography of the distal male urethra-a new diagnostic procedure for urethral stricture: result of a retrospective study. J Urol 1988;140:1409-11.
23. Metz P, Ebbehoj J, Uhrenholdt A et al: Peyronie's disease and erectile failure. J Urol 1983;30:1103-04.
24. Mueller SC, Leu TF: Evaluation of vasculogenic impotence. Urol Clin North Am 1988;15:65-76.
25. Newman HF, Northup JD, Delvin J: Mechanism of human penile erection. Invest Urol 1963;1:350-53.
26. Quam JP, King BF, James EM et al: Duplex and color Doppler ultrasound evaluation of vasculogenic impotence (scientific exhibit). Radiol Soc North Am 1989.
27. Robinson LQ, Woodcock JP, Stephenson TP: Duplex scanning in suspected vasculogenic impotence: a worthwhile exercise? Br J Urol 1989;63:4:432-36.
28. Rollandi GA, Tentarelli T, Vespier M: Computed tomographic findings in Peyronie's disease. Urol Radiol 1985;7:153-56.
29. Scappini P, Piscioli F, Pusiol T, Hofstetter A et al: Penile cancer: aspiration biopsy cytology for staging. Cancer 1986;58:1526-33.
30. Schwartz AN, Wang KY, Mack LA et al: Evaluation of normal erectile function with color flow Doppler sonography. AJR 1989;153:1155-60.
31. Shabsingh R, Fishman IJ, Scott FB: Evaluation of erectile impotence. Urology 1988;32:2:83-90.
32. Shirai M, Ishii N, Mitsukawa S et al: Hemodynamic mechanism of erection in the human penis. Arch Androl 1978;1:345-49.

33. Sufrin G, Huben R: Benign and malignant lesions of the penis. In Gillenwater J, Grayhack J, Howards S et al (Eds): Adult and Pediatric Urology. Chicago: Year Book Medical Publishers Inc; 1987.
34. Tudoriu T, Bourmer H: The hemodynamics of erection at the level of the penis and its local deterioration. J Urol 1983;129:741-45.
35. Vickers M, Benson C, Richie J: High resolution ultrasonography and pulsed wave Doppler for detection of corporovenous incompetence in erectile dysfunction. J Urol 1990;143:1128-30.
36. Virag R, Bouilly P, Frydman D: Is impotence an arterial disorder? A study of arterial risk factors in 440 impotent men. Lancet 1985;1:181-84.
37. Virag R, Frydman D, Legman M et al: Intracavernous injection of papaverine as a diagnostic and therapeutic method in erectile failure Angiology. J Vasc Dis 1984;35:78-87.
38. Wanger G, Uhrenholdt A: Blood flow by clearance in the human corpus cavernosum in the flaccid and erect states. In Zorgniotti AW, Rossi G (Eds): Vasculogenic Impotence: Proceedings of the First International Conference on Corpus Cavernosum Revascularization. Springfield, IL: Charles C. Thomas, Publisher; 1980.

13. High Resolution Sonography of Musculoskeletal System

1. Abiri MM, Kirpekar M, Alblow RC: Osteomyelitis:detection with US. Radiology 1988; 169:795-97.
2. Aisen AM, McCune WJ, MacGuire A et al: Sonographic evaluation of the cartilage of the knee. Radiology 1984; 153:781-84.
3. Berman L, Klenerman L: Ultrasound screening for hip abnormalities:preliminary findings in 1001 neonates. BMJ 1986; 293: 719-22.
4. Boal DK, Schwenkter EP: The infant hip:assessment with real-time US. Radiology 1985; 157:667-72.
5. Bretzke CA, Crass JR, Criag EV, Feinberg SB: Ultrasonography of the rotator cuff:normal and pathologic anatomy. Invest Radiol 1985; 20; 311-15.
6. Burk DL Jr, Karasick D, Kurtz AB et al: Rotator cuff tears: prospective comparison of MR imaging with arthrography, sonography, and surgery. AJR 1989; 153:87-92.
7. Cady EB, Gardener JE, Edwards RH: Ultrasonic tissue characterization of skeletal muscle. Eur J Clin Invest 1983;13:469-73.
8. Chinn D H, Filly RA, Callen PW: Unusual ultrasonic appearance of a solid Schwannoma. JCU 1982; 10:243-45.
9. Clarke NMP, Harcke HT: Real-time ultrasound in the diagnosis of congenital dislocation and dysplasia of the hip. J Bone Joint Surg (Br) 1985;67:406-12.
10. Crass JR: Current concepts in the radiographic evaluation of the rotator cuff. Crit Rev Diagn Imaging 1988;28:23-73.
11. Dahlstrom H, Friberg S: Stability of the hip joint after reduction of late-diagnosed congenital dislocation of the hip. Pediatr Orthop 1987; 7:401-04.
12. Dahlstrom H, Oberg L, Friberg S: Sonography in congenital dislocation of the hip. Acta Orthop 1986; 57:402-06.
13. de Flaviis L, Nessi R, Leonardi M, Ulivi M: Dynamic ultrasonography of capsulo-ligamentous knee joint traumas. JCU 1988; 16:487-92.
14. Fornage BD, Rifkin MC: Ultrasound examination of tendons. Radiol Clin North Am 1988; 26:87-107.
15. Fornage BD, Rifkin MD, Touche DB, Segal PM: Sonography of the patellar tendon:preliminary observations. AJR 1984;143:179-82.
16. Fornage BD, Touche D, Raguet M, Jacob M, Segal PH: Accidents musculaire sportif. Apport original de l'ultrasongraphic. La Nouvelle Presse Medicale 1982; 11:571-75.
17. Fornage BD, Touche DH, Rifkin MD. Small parts real-time sonography: A new 'water-path'. J Ultrasound Med 1984;3:355-57.
18. Fornage BD: Achilles tendon:US examination. Radiology 1986; 159:759-64.
19. Fornage BD: Ultrasonography of muscles and tendons. Examination technique and atlas of normal anatomy of the extremities. New York:Springer-Verlag 1988.
20. Fornage BD: Ultrasonography of muscles and tendons: examination technique and atlas of normal anatomy of the extremities. Springer 1989.
21. Gooding GAW: Tenosynovitis of the wrist. A sonogrpahic demonstration. J Ultrasound Med 1988;7:225-26.
22. Graf R: Classification of hip joint dysplasia by means of sonography. Arch Orthop Trauma Surg 1984;102:248-55.
23. Graf R: Fundamentals of sonographic diagnosis of infant hip dysplasia. Pediatr Orthop 1984; 4:735-40.
24. Harke HT, Soo-Lee M, Sinning L, Clarke NP, Borns PF, MacEwen GD: Ossification of the infant hip:sonographic and radiographic correlation. AJR 1986;147:317-21.
25. Hooper KD, Smazal SF Jr, Ghaed N: CT and ultrasonic evaluation of rectus sheath hematoma: a complication of anticoagulant therapy. Milit Med 1983; 148:447-49.
26. Hugher DG, Wilson DJ: Ultrasound appearances of peripheral nerve tumors. Br J Radiol 1986; 59:1041-43.
27. Ismail AM, Balakrishnan R, Rajakumar MK: Rupture of patellar ligament after steroid infiltration. Report of a case. J Bone Joint Surg 1969; 51B:503-05.
28. Leekam RN, Salsberg BB, Bogoch E, Shankar L: Sonographic diagnosis of partial Achilles tendon rupture and healing. J Ultrasound Med 1986;5:115-16.
29. Lehto M, Alanen A: Healing of a muscle trauma. Correlation of sonographical and histological findings in an experimental study in rats. J Ultrasound Med 1987; 6:425-29.
30. MacLarnon JC, Wilson DJ: Aspects of ultrasound in haemophilia (proceedings of the British Medical Ultrasound society). Br J Radiol 1985;58:692.
31. Mata JM, Donoso L, Olazabal A, Blanch L, Ris J: Whole pelvic osteomyelitis:unusual finding in staphylococcal sepsis. Pediatr Radiol 1987:17:427-28.
32. McMaster PE: Tendon and muscle ruptures. Clinical and experimental studies on the causes and location of subcutaneous ruptures. J Bone Joint Surg 1933; 15:705-22.
33. Middleton WD: Status of rotator cuff sonography. Radiology 1989; 173:307-09.
34. Miller CL, Karasick D, Kurtz AB, Fenlin JM: Limited sensitivity of ultrasound for detection of rotator cuff tears. Skeletal Radiol 1989; 18:179-83.
35. Novick G, Ghelman B, Schneider M: Sonography of neonatal and infant hip. AJR 1983; 141:639-45.
36. Peck RJ, Metrewelli C: Early myositis ossificans:a new echographic sign. Clin Radiol 1988; 39:586-88.
37. Peck RJ: Ultrasound of the painful hip in children. Br J Radiol 1986;59:290-94.
38. Pettersson H, Slone RM, Spanier S, Gillespie T, Fitzsimmons JR, Scott KN: Musculoskeletal tumours:T1 and T2 relaxation times. Radiology 1988; 167:783-85.
39. Terjesen T, Bredland T, Berg V: Ultrasound for hip assessment in the newborn. J Bone Joint Surg (Br) 1989;71:767-73.
40. Wilson DJ, Green DJ, MacLarnon JC: Arthrosonography of the painful hip. Clin Radiol 1984;35:17-19.
41. Wilson DJ. Musculoskeletal ultrasound. Radiology Now 1986; 3:35-37.
42. Wilson DJ: Diagnostic ultrasound in the musculoskeletal system. Current Orthopaedics 1988; 2:41-50.
43. Wilson DJ: Musculoskeletal ultrasound. Radiology Now 1986; 3:35-37.
44. Wilson DJ: Ultrasonic imaging of soft tissues. Clin Radiol 1989;40 341-42.
45. Wyld PJ, Dawson KP, Chisholm RJ: Ultrasound in the assessment of synovial thickening in the hemophilic knee. Aust NZ J Med 1984;14:678-80.
46. Zieger M, Schulz RD: Method and results of ultrasound in hip studies. Ann Radiol (Paris) 1986; 29:383-86.

Index

A

B

C

D

E

F

G

H

I

K

L

M

N

O

P

R

S

T

U

V